A Concordance to Darwin's
ORIGIN OF SPECIES,
FIRST EDITION

A Concordance to Darwin's
ORIGIN OF SPECIES,

FIRST EDITION

Edited by
PAUL H. BARRETT,
DONALD J. WEINSHANK,
and TIMOTHY T. GOTTLEBER

Cornell University Press
ITHACA AND LONDON

Publication of this work has been made possible in
part by a grant from the Andrew W. Mellon Foundation.

First published 1981 by Cornell University Press.
Published in the United Kingdom by Cornell University Press Ltd.,
Ely House, 37 Dover Street, London W1X 4HQ.

International Standard Book Number 0-8014-1319-2
Library of Congress Catalog Card Number 80-66893
Printed in the United States of America
*Librarians: Library of Congress cataloging information appears
on the last page of the book.*

*This work is dedicated to C.R.D., who started it all,
and to Wilma B., Annette W., and Susette D.,
who steadfastly encouraged and supported the project.*

CONTENTS

PREFACE

The ever increasing interest in evolution has led us to produce this concordance. Charles Darwin wrote many books and articles, as well as thousands of pages of letters and manuscripts, on evolution, but without doubt the first edition (1859) of his *On the Origin of Species by Means of Natural Selection* is his best-known scientific contribution, and there is general agreement that this book has had as great an impact on the course of human culture as any scholarly work ever published.*

The ideas Darwin expressed, the various shades of meaning he gave to the words he used—words such as "selection," "creation," "struggle," "survival," and "competition"—have intrigued and puzzled students of evolution ever since *The Origin* first appeared. This concordance, which enables a researcher to find nearly every word printed in the 490 pages of the original text, provides a useful tool not only for scientists who are specialists in evolution, but also for historians and philosophers of science, biologists, linguists, and the general public. Like any concordance, ours enables researchers to locate particular words quickly, and thus helps them make a detailed analysis of Darwin's thoughts at different times and in different contexts. For instance, an analysis of Darwin's philosophical approach to scientific methodology, based on a study of how he uses such terms as "hypothesis," "theory," "fact," "truth," "supposition," "perhaps," "positive," "perfect," "speculate," can reveal much about the mixture of doubt and certainty in Darwin's mind—a subject that has received little attention.

Each word listed in this concordance is arranged alphabetically and printed in capital letters in the center of the page. These indexed words are called keywords, and each of them is surrounded by as much of the original text as possible, with fragments of words regularly occurring at the beginning and the end of lines. In the margin to the left of each line is the number of the page on which the keyword appears in the first edition of *The Origin*. The entries for each keyword are listed according to the order in which they appear in the original text.

*The more familiar title, *The Origin of Species*, was adopted for the sixth edition, a redesigned popular edition published in February 1872.

Words hyphenated in the original are here indexed as two separate words; for example, "fresh-water" is indexed separately under "fresh" and "water."

A few words—conjunctions, articles, prepositions, some pronouns, and several verbs and adverbs—have been omitted from the concordance because their inclusion would have resulted in a much longer and more expensive book. For technical reasons, a few words are suppressed in one form but not in others. Also suppressed are the terms ("A 14," "D," Roman numerals, etc.) which were used as labels for the phylogenetic diagram (pages 116–17 of the first edition). All the words and terms that are not indexed, as well as the number of times each of them is printed in *The Origin of Species*, can be found on the "List of Words Suppressed and Their Frequency."

Linguists and other researchers interested in word-frequency counts or the statistical sampling of word sets and similar projects may contact the editors for information on procedures for such specialized analyses. The programs used to produce this concordance were written at Michigan State University using Control Data Corporation's Fortran Extended Version 4 language and ASCII input/output routines selected from the University of Washington's FASTIO package. The Control Data Corporation Cyber 170 Series, Model 750 computer was used with the Michigan State University SCOPE/HUSTLER operating system. The concordance was printed on a Data Printer Corporation CT4964 ASCII printer.

A facsimile reprint of the first edition of *The Origin of Species*, to which this concordance is indexed, was published by Harvard University Press.

Profound gratitude is due the following faculty and staff of Michigan State University for the support and technical help that made this concordance project possible: John Cantlon, Vice-President for Research Development; Willard G. Warrington, Director of the Undergraduate University Division; Harry Eick, Associate Dean, College of Natural Science; Richard J. Seltin, Chairperson, Department of Natural Science; Julian Kateley, Associate Director, Computer Laboratory; Lewis Greenberg, Assistant Director, Computer Laboratory (Systems); Tom Carroll, Assistant Director, Computer Laboratory (User Information Center); and John Kohmetscher, Operations Manager, Computer Laboratory.

The authors especially acknowledge Rick L. Stevens, Michael Kupfer, and Dianne Kwiatkowski, students of Michigan State University, for their help during the final stages of the production of this concordance. Special appreciation is also due Terry Kidner Kohn and David Kohn for valuable suggestions and encouragement.

We thank all of the undergraduate students who patiently typed the text, proofread and corrected successive printouts, and in general worked meticulously to prepare *The Origin* for production of the concordance. We hope that, in the process, those students came to share our love of Darwin's works.

For needed financial support, credit is due the Michigan State University All-

University Research Initiation Grant Committee, the University College, the College of Natural Science, and the Department of Natural Science, Michigan State University.

PAUL H. BARRETT
DONALD J. WEINSHANK
TIMOTHY T. GOTTLEBER

East Lansing, Michigan

List of Words Suppressed
and Their Frequency

Word Suppressed	Frequency	Word Suppressed	Frequency
A	2462	Could	234
A1	5	D	117
A2	1	Did	27
A5	2	Do	238
A10	4	Does	74
A14	9	Doing	3
Above	40	Done	21
Again	68	Down	27
All	552	Dr	36
Along	22	During	215
Also	100	E	315
Am	68	E14	2
An	506	Each	544
And	4364	Enough	14
Another	183	Etc	53
Any	511	Even	210
Are	1133	Ever	70
Around	9	Every	122
As	1585	F	34
At	680	F8	1
Away	33	F10	1
B	5	F14	8
B14	4	Far	193
Back	14	Few	164
Be	1651	For	1109
Been	931	Four	19
Below	6	From	1117
Both	141	G	51
But	856	Gave	1
By	1322	Get	23
C	35	Getting	6
Came	11	Give	95
Can	514	Given	135
Co	5	Gives	20
Come	81	Giving	17
Comes	26	Go	28
Coming	10	Goes	7

Word Suppressed	Frequency	Word Suppressed	Frequency
Going	7	Next	15
Gone	17	No	321
Got	4	Not	852
H	10	Now	232
Had	244	O14	4
Has	611	Of	7265
Have	1761	Off	26
Having	252	On	1350
He	163	Once	54
Her	33	One	641
Here	114	Ones	8
High	59	Only	265
Him	17	Or	1180
His	90	Other	748
II	1	Others	68
III	1	Our	278
In	3904	Out	71
Inside	2	Outer	13
Into	199	Outside	2
Is	1424	Over	103
It	1054	Own	100
Its	403	P	17
Itself	35	P14	3
IV	1	Part	155
IX	1	Parts	154
J	4	Put	15
Just	33	Puts	1
K	1	Putting	1
Keep	13	Q14	2
Keeps	1	R	3
Kept	42	Re	10
L	124	Rt	1
Left	35	S	335
Less	175	S2	1
Let	58	Said	47
M	16	Same	778
M1	2	Saw	11
M2	1	Say	35
M3	1	Saying	5
M6	1	Says	7
M9	1	Second	18
M10	1	See	271
M14	6	Seeing	12
Many	451	Seen	90
Me	153	She	24
Mm	2	Side	28
More	575	Sir	19
Most	402	So	530
Mr	83	Some	657
Much	229	Sometimes	116
N	438	Soon	32
Near	31	Sq	1

Word Suppressed	Frequency	Word Suppressed	Frequency
St	14	Very	400
Still	109	VI	3
Such	221	VII	1
T	41	VIII	1
Take	35	W	6
Taken	45	W10	1
Takes	10	Was	160
Taking	11	We	1155
Tell	15	Well	152
Than	370	Went	1
That	2081	Were	188
The	10,144	What	148
Their	700	When	412
Them	191	Where	66
Then	145	Which	1228
There	298	While	4
These	494	Whilst	26
They	541	Who	75
Things	8	Whoever	1
This	959	Whom	8
Those	262	Whose	12
Three	91	With	973
Through	100	Within	104
To	3563	Without	68
Together	128	Would	489
Too	25	X	3
Took	7	XI	1
Two	344	XII	1
Under	221	XIII	1
Until	32	XIV	1
Up	92	Y	280
Upper	25	You	3
Us	142	Z10	1
Use	59	Z14	4
Used	36		
V	2		

Total no. of words in text = 153,773

A Concordance to Darwin's
ORIGIN OF SPECIES,
FIRST EDITION

0211 t then began to play with its antennae on the ABDOMEN first of one aphis and then of another: and
0211 felt the antennae, immediately lifted up its ABDOMEN and excreted a limpid drop of sweet juice, w
0239 caste, and they have an enormously developed ABDOMEN which secretes a sort of honey, supplying th
0217 to follow by inheritance the occasional and ABERRANT habit of their mother and in their turn woul
0429 d, will give to us our so called osculant or ABERRANT groups. The more aberrant any form is, the g
0429 called osculant or aberrant groups. The more ABERRANT any form is, the greater must be the number
0429 d utterly lost. And we have some evidence of ABERRANT forms having suffered severely from extincti
0429 siren, for example, would not have been less ABERRANT had each been represented by a dozen species
0429 gation, does not commonly fall to the lot of ABERRANT genera. We can, I think, account for this fa
0429 nk, account for this fact only by looking at ABERRANT forms as failing groups conquered by more su
0486 cies and groups of species, which are called ABERRANT, and which may fancifully be called living f
0186 r the correction of spherical and chromatic ABERRATION, could have been formed by natural selectio
0202 andard under nature. The correction for the ABERRATION of light is said, on high authority, not to
0472 , absolutely perfect: and if some of them be ABHORRENT to our ideas of fitness. We need not marvel
0062 has treated this subject with more spirit and ABILITY than W. Herbert, Dean of Manchester, evident
0224 e to the species, until an ant was formed as ABJECTLY dependent on its slaves as is the Formica ru
0087 umbler pigeons more perish in the egg than are ABLE to get out of it: so that fanciers assist in t
0087 r nature, and if so, natural selection will be ABLE to modify one sex in its functional relations
0094 o that an individual so characterised would be ABLE to obtain its food more quickly, and so have a
0101 hich I have collected, but which I am not here ABLE to give, I am strongly inclined to suspect tha
0106 as soon as it has been much improved, will be ABLE to spread over the open and continuous area, a
0124 f the two groups: and every naturalist will be ABLE to bring some such case before his mind. In th
0136 better for the good swimmers if they had been ABLE to swim still further, whereas it would have b
0136 tter for the bad swimmers if they had not been ABLE to swim at all and had stuck to the wreck. The
0144 that others rarely coexist, without our being ABLE to assign any reason. What can be more singula
0163 i lie under a great disadvantage in not being ABLE to give them. I can only repeat that such case
0195 themselves from these small enemies, would be ABLE to range into new pastures and thus gain a gre
0210 of such gradations: or we ought at least to be ABLE to show that gradations of some kind are possi
0222 nguinea emerge, carrying a pupa: but I was not ABLE to find the desolated nest in the thick heath.
0231 n at the commencement of the first cell. I was ABLE practically to show this fact, by covering the
0245 e degrees of sterility. I hope, however, to be ABLE to show that sterility is not a specially acqu
0257 nalogous facts could be given. No one has been ABLE to point out what kind, or what amount, of dif
0261 ed by systematic affinity, for no one has been ABLE to graft trees together belonging to quite dis
0266 or to the conditions of life. When hybrids are ABLE to breed inter se, they transmit to their offs
0271 estic varieties, and from not wishing or being ABLE to produce recondite and functional difference
0274 the unimportant differences, which Gartner is ABLE to point out, between hybrid and mongrel plant
0298 n would greatly lessen the chance of our being ABLE to trace the stages of transition in any one g
0299 ance, geologists at some future period will be ABLE to prove, that our different breeds of cattle,
0303 o produce many divergent forms, which would be ABLE to spread rapidly and widely throughout the wo
0324 a, the most skilful naturalist would hardly be ABLE to say whether the existing or the pleistocene
0328 the eocene deposits of England and France, is ABLE to draw a close general parallelism between th
0353 , as I believe, is, that mammals have not been ABLE to migrate, whereas some plants, from their va
0353 procued on one side alone, and have not been ABLE to migrate to the other side. Some few familie
0433 : and Milne Edwards has lately insisted, in an ABLE paper, on the high importance of looking to ty
0464 which certainly have existed. We should not be ABLE to recognise a species as the parent of any on
0464 s will be discovered, that naturalists will be ABLE to decide, on the common view, whether or not
0482 , to young and rising naturalists, who will be ABLE to view both sides of the question with impart
0484 ution in natural history. Systematists will be ABLE to pursue their labours as at present: but the
0487 having been of vast duration. But we shall be ABLE to gauge with some security the duration of th
0067 n than any other animal. This subject has been ABLY treated by several authors, and I shall, in my
0356 dispersal. Sir C. Lyell and other authors have ABLY treated this subject. I can give here only the
0427 attention to this distinction, and he has been ABLY followed by Macleay and others. The resemblanc
0024 e wild rock pigeon, yet are certainly highly ABNORMAL in other parts of their structure, we may lo
0024 ally or by chance picked out extraordinarily ABNORMAL species: and further, that these very specie
0027 ing nowhere feral: these species having very ABNORMAL characters in certain respects, as compared
0038 , i think, further understand the frequently ABNORMAL character of our domestic races, and likewis
0039 crop of somewhat unusual size: and the more ABNORMAL or unusual any character was when it first a
0144 ut the two orders of mammalia which are most ABNORMAL in their dermal covering, viz. Cetacea (whal
0144 ers, etc.), that these are likewise the most ABNORMAL in their teeth. I know of no case better ada
0150 lied species. Thus, the bat's wing is a most ABNORMAL structure in the class mammalia: but the rul
0154 eason to doubt. Hence when an organ, however ABNORMAL it may be, has been transmitted in approxima
0255 fertility: and, as we have seen, in certain ABNORMAL cases, even to an excess of fertility, beyon
0259 cture between their parents, exceptional and ABNORMAL individuals sometimes are born, which closel
0154 the course of time cease: and that the most ABNORMALLY developed organs may be made constant, I ca
0139 tition to which the inhabitants of these dark ABODES will probably have been exposed. Acclimatisat
0014 but certainly revert in character to their ABORIGINAL stocks. Hence it has been argued that no de
0014 tate. In many cases we do not know what the ABORIGINAL stock was, and so could not tell whether or
0015 extent, or even wholly, revert to the wild ABORIGINAL stock. Whether or not the experiment would
0018 , that these had descended from a different ABORIGINAL stock from our European cattle: and several
0019 of our several domestic races from several ABORIGINAL stocks, has been carried to an absurd extre
0019 have been produced by the crossing of a few ABORIGINAL species: but by crossing we can get only fo
0023 have descended from at least seven or eight ABORIGINAL stocks: for it is impossible to make the or
0023 characteristic enormous crop? The supposed ABORIGINAL stocks must all have been rock pigeons, tha
0023 of the domestic breeds. Hence the supposed ABORIGINAL stocks must either still exist in the count
0025 her, firstly, that all the several imagined ABORIGINAL stocks were coloured and marked like the ro
0032 een produced by a single variation from the ABORIGINAL stock. We have proofs that this is not so i
0038 ies, do not by a strange chance possess the ABORIGINAL stocks of any useful plants, but that the n
0159 n the feet, characters not possessed by the ABORIGINAL rock pigeon: these then are analogous varia
0164 ds of the horse have descended from several ABORIGINAL species, one of which, the dun, was striped
0254 tic animals have descended from two or more ABORIGINAL species, since commingled by intercrossing.
0254 mingled by intercrossing. On this view, the ABORIGINAL species must either at first have produced
0254 makes me greatly doubt, whether the several ABORIGINAL species would at first have freely bred tog

```
0299  scended from a single stock or from several ABORIGINAL stocks; or, again, whether certain sea shel
0353  d in America and Australia; and some of the ABORIGINAL plants are identically the same at these di
0402  ies are not generally closely allied to the ABORIGINAL inhabitants, but are very distinct species,
0467  whether very many of them are varieties or ABORIGINAL species. There is no obvious reason why the
0016  her competent judges as the descendants of ABORIGINALLY distinct species. If any marked distinctio
0026  othesis so far as to suppose that species, ABORIGINALLY as distinct as carriers, tumblers, pouters
0028  h has attended, are descended from so many ABORIGINALLY distinct species. Ask, as I have asked, a
0043  t doubt that the intercrossing of species, ABORIGINALLY distinct, has played an important part in
0268  hat these dogs have descended from several ABORIGINALLY distinct species. Nevertheless the perfect
0389  idence that the barren island of Ascension ABORIGINALLY possessed under half a dozen flowering pla
0408  ct competition with each other and with the ABORIGINES; and according as the immigrants were capab
0022  e head and tail touch; the oil gland is quite ABORTED. Several other less distinct breeds might ha
0145  en this occurs, the adherent nectary is quite ABORTED; when the colour is absent from only one of
0149  cur to the general subject of rudimentary and ABORTED organs; and I will here only add that their
0450  class of animals. Rudimentary, atrophied, or ABORTED organs. Organs or parts in this strange cond
0451  purposes, many become rudimentary or utterly ABORTED for one, even the more important purpose; an
0452  ain groups. Rudimentary organs may be utterly ABORTED; and this implies, that we find in an animal
0455  y, imperfect, and useless condition, or quite ABORTED, far from presenting a strange difficulty, a
0485  phology, adaptive characters, rudimentary and ABORTED organs, etc., will cease to be metaphorical,
0022  part of the oesophagus, the development and ABORTION of the oil gland; the number of the primary
0145  his difference is often accompanied with the ABORTION of parts of the flower. But, in some Composi
0145  n other parts of the flower had caused their ABORTION; but in some Compositae there is a differenc
0436  y the atrophy and ultimately by the complete ABORTION of certain parts, by the soldering together
0455  lost, and we should have a case of complete ABORTION. The principle, also, of economy, explained
0457  rrence of rudimentary organs and their final ABORTION, present to us no inexplicable difficulties;
0053  y, the species which are most common, that is ABOUND most in individuals, and the species which ar
0070  of some social plants being social, that is, ABOUNDING in individuals, even on the extreme confine
0307  ther and lower stage to the Silurian system, ABOUNDING with new and peculiar species. Traces of Li
0350  h natural selection. Widely ranging species, ABOUNDING in individuals, which have already triumphe
0069  forget that each species, even where it most ABOUNDS, is constantly suffering enormous destructio
0017  cts would have great weight in making us doubt ABOUT the immutability of the many very closely all
0021  wonderful development of the carunculated skin ABOUT the head, and this is accompanied by greatly
0027  of pigeons is in the fifth Aegyptian dynasty, ABOUT 3000 B.C., as was pointed out to me by Profes
0028  geons were much valued by Akber Khan in India, ABOUT the year 1600: never less than 20,000 pigeons
0028  tised before, has improved them astonishingly. ABOUT this same period the Dutch were as eager abou
0028  about this same period the Dutch were as eager ABOUT pigeons as were the old Romans. The paramount
0040  mes been noticed, namely, that we know nothing ABOUT the origin or history of any of our domestic
0147  . the elder Geoffroy and Goethe propounded, at ABOUT the same period, their law of compensation or
0181  ing through the air rather than for flight. If ABOUT a dozen genera of birds had become extinct or
0193  ore serious difficulty; for they occur in only ABOUT a dozen fishes, of which several are widely r
0210  s show. I removed all the ants from a group of ABOUT a dozen aphides on a dockplant, and prevented
0211  mmediately seemed, by its eager way of running ABOUT, to be well aware what a rich flock it had di
0221  r jaws. Another day my attention was struck by ABOUT a score of the slave makers haunting the same
0222  nest of F. flava, and quickly ran away; but in ABOUT a quarter of an hour, shortly after all the l
0222  d the returning file burthened with booty, for ABOUT forty yards, to a very thick clump of heath,
0222  or three individuals of F. fusca were rushing ABOUT in the greatest agitation, and one was perche
0228  ye perfectly true or parts of a sphere, and of ABOUT the diameter of a cell. It was most interesti
0228  sins had acquired the above stated width (i.e. ABOUT the width of an ordinary cell), and were in d
0228  width of an ordinary cell), and were in depth ABOUT one sixth of the diameter of the sphere of wh
0231  o the delicate hexagonal walls, which are only ABOUT one four hundredth of an inch in thickness; t
0231  kness; the plates of the pyramidal basis being ABOUT twice as thick. By this singular manner of bu
0247  not be doubted; for Gartner gives in his table ABOUT a score of cases of plants which he castrated
0283  atom, until reduced in size they can be rolled ABOUT by the waves, and then are more quickly groun
0283  hey are abraded and how seldom they are rolled ABOUT! Moreover, if we follow for a few miles any l
0286  nce from the northern to the southern Downs is ABOUT 22 miles, and the thickness of the several fo
0286  ess of the several formations is on an average ABOUT 1100 feet, as I am informed by Prof. Ramsay.
0299  nary illustration. The Malay Archipelago is of ABOUT the size of Europe from the North Cape to the
0306  ozen years have effected, it seems to me to be ABOUT as rash in us to dogmatize on the succession
0320  tion of new forms has caused the extinction of ABOUT the same number of old forms. The competition
0357  the course of time we know something definite ABOUT the means of distribution, we shall be enable
0360  d perhaps be safer to assume that the seeds of ABOUT 10/100 plants of a flora, after having been d
0361  rth thus completely enclosed by wood in an oak ABOUT 50 years old, three dicotyledonous plants ger
0373  at the structure of a vast mound of detritus, ABOUT 800 feet in height, crossing a valley of the
0378  by Dr. Hooker, that all the flowering plants, ABOUT forty six in number, common to Tierra del Fue
0378  under the ecuator at the level of the sea was ABOUT the same with that now felt there at the heig
0391  rds. Bermuda, on the other hand, which lies at ABOUT the same distance from North America as the G
0418  ts are at work, they do not trouble themselves ABOUT the physiological value of the characters whi
0489  g on the bushes, with various insects flitting ABOUT, and with worms crawling through the damp ear
0283  rine productions, showing how little they are ABRADED and how seldom they are rolled about! Moreov
0032  rists; but the variations are here often more ABRUPT. No one supposes that our choicest production
0296  t on my theory have existed between them, but ABRUPT, though perhaps very slight, changes of form.
0302  arance of whole groups of Allied Species. The ABRUPT manner in which whole groups of species sudde
0304  adding as I thought one more instance of the ABRUPT appearance of a great group of species. But m
0454  ubt whether species under nature ever undergo ABRUPT changes. I believe that disuse has been the m
0174  er a large territory, then becoming somewhat ABRUPTLY rarer and rarer on the confines, and finally
0175  ns, and sometimes it is quite remarkable how ABRUPTLY, as Alph. de Candolle has observed, a common
0303  ss of mammals was always spoken of as having ABRUPTLY come in at the commencement of the tertiary
0311  on of chapters, may represent the apparently ABRUPTLY changed forms of life, entombed in our conse
0314  g all the inhabitants of a country to change ABRUPTLY, or simultaneously, or to an equal degree. T
0316  oup sometimes falsely appear to have come in ABRUPTLY; and I have attempted to give an explanation
0317  at its lower end, not in a sharp point, but ABRUPTLY; it then gradually thickens upwards, sometim
0042  mals vary less than others, yet the rarity or ABSENCE of distinct breeds of the cat, the donkey, p
0135  ed; and I should prefer explaining the entire ABSENCE of the anterior tarsi in Ateuchus, and their
0136  ted on by Mr. Wollaston, of the almost entire ABSENCE of certain large groups of beetles, elsewher
0171  ry of descent with modification. Transitions. ABSENCE or rarity of transitional varieties. Transit
0172  ct and Hybridism in separate chapters. On the ABSENCE or rarity of transitional varieties. As natu
```

Page **(Key Word)**

```
0193 o inheritance from a common ancestor; and its ABSENCE in some of the members to its loss through d
0279 imperfection of the Geological Record. On the  ABSENCE of intermediate varieties at the present day
0279 ntermittence of geological formations. On the   ABSENCE of intermediate varieteies in any one format
0290 ndred feet within the recent period, than the   ABSENCE of any recent deposits sufficiently extensiv
0307 iods. But the difficulty of understanding the   ABSENCE of vast piles of fossiliferous strata, which
0310 in our European formations; the almost entire   ABSENCE, as at present known, of fossiliferous forma
0383 tions. On the inhabitants of oceanic islands.   ABSENCE of the Batrachians and of terrestrial Mammal
0392 nd ultimately into trees. With respect to the   ABSENCE of whole orders on oceanic islands, Bory St.
0393 xplained through glacial agency. This general   ABSENCE of frogs, toads, and newts on so many oceani
0395 presence of endemic bats on islands, with the   ABSENCE of all terrestrial mammals. Besides the abse
0395 sence of all terrestrial mammals. Besides the   ABSENCE of terrestrial mammals in relation to the re
0396 articular classes or sections of classes, the   ABSENCE of whole groups, as of batrachians, and of t
0415 aving one or more ovaria, in the existence or   ABSENCE of albumen, in the imbricate or valvular aes
0465 sidence, more extinction. With respect to the   ABSENCE of fossiliferous formations beneath the lowe
0478 presence of peculiar species of bats, and the   ABSENCE of all other mammals, on oceanic islands, ar
0145 nectary is cuite aborted; when the colour is    ABSENT from only one of the two upper petals, the ne
0294 dant in the neighbouring sea, but are rare or   ABSENT in this particular deposit. It is an excellen
0323 rmations either above or below, are similarly   ABSENT at these distant points of the world. In the
0409 atrachians and terrestrial mammals, should be   ABSENT from oceanic islands, whilst the most isolate
0410 ermediate deposit the forms which are therein   ABSENT, but which occur above and below; so in space
0069 e arctic regions, or snow capped summits, or    ABSOLUTE deserts, the struggle for life is almost exc
0202 m europe. Natural selection will not produce    ABSOLUTE perfection, nor do we always meet, as far as
0206 tural selection will not necessarily produce    ABSOLUTE perfection; nor, as far as we can judge by o
0206 s we can judge by our limited faculties, can    ABSOLUTE perfection be everywhere found. On the theor
0255 uence than so much inorganic dust. From this    ABSOLUTE zero of fertility, the pollen of different s
0398 e verde Archipelagos: but what an entire and   ABSOLUTE difference in their inhabitants! The inhabit
0472 al selection, that more cases of the want of   ABSOLUTE perfection have not been observed. The compl
0003 , and which has flowers with separate sexes    ABSOLUTELY requiring the agency of certain insects to
0032 ring successive generations, of differences   ABSOLUTELY inappreciable by an uneducated eye, differe
0055 ra, with from only one to four species, are   ABSOLUTELY excluded from the tables. These facts are o
0070 elatively to the numbers of its enemies, is   ABSOLUTELY necessary for its preservation. Besides can
0072 arched it for food. Here we see that cattle   ABSOLUTELY determine the existence of the Scotch fir;
0073 set a seed. Many of our orchidaceous plants   ABSOLUTELY require the visits of moths to remove their
0171 f small importance. Organs not in all cases   ABSOLUTELY perfect. The law of Unity of Type and of th
0173 ere they intermingle, they are generally as   ABSOLUTELY distinct from each other in every detail of
0189 sive, slight modifications, my theory would   ABSOLUTELY break down. But I can find out no such case
0195 f cattle and other animals in South America   ABSOLUTELY depends on their power of resisting the att
0199 e variety. This doctrine, if true, would be   ABSOLUTELY fatal to my theory. Yet I fully admit that
0211 , certain instincts cannot be considered as   ABSOLUTELY perfect; but as details on this and other s
0219 ven than his celebrated father. This ant is   ABSOLUTELY dependent on its slaves; without their aid,
0223 s young, and cannot even feed itself: it is   ABSOLUTELY dependent on its numerous slaves. Formica s
0229 eted, and had become perfectly flat: it was   ABSOLUTELY impossible, from the extreme thinness of th
0235 b of the hive bee, as far as we can see, is   ABSOLUTELY perfect in economising wax. Thus, as I beli
0237 ect differing greatly from its parents, yet   ABSOLUTELY sterile; so that it could never have transm
0243 and, the fact that instincts are not always   ABSOLUTELY perfect and are liable to mistakes; that no
0247 which the best botanists rank as varieties,   ABSOLUTELY sterile together; and as he came to the sam
0254 resent state of knowledge, be considered as   ABSOLUTELY universal. Laws governing the Sterility of
0261 pacity; as in hybridisation, is by no means   ABSOLUTELY governed by systematic affinity. Although m
0342 nd of species, preserved in our museums, is   ABSOLUTELY as nothing compared with the incalculable n
0416 he only character, according to Owen, which   ABSOLUTELY distinguishes fishes and reptiles, the infl
0464 e number of specimens in all our museums is   ABSOLUTELY as nothing compared with the countless gene
0472 s in nature be not, as far as we can judge,   ABSOLUTELY perfect; and if some of them be abhorrent t
0453 of rudimentary teeth which are subsequently   ABSORBED, can be of any service to the rapidly growin
0475 ds of resemblance to their parents, in being   ABSORBED into each other by successive crosses, and i
0001 om strong, I have been urged to publish this   ABSTRACT. I have more especially been induced to do t
0002 ome brief extracts from my manuscripts. This   ABSTRACT, which I now publish, must necessarily be im
0004 ns, I shall devote the first chapter of this   ABSTRACT to Variation under Domestication. We shall t
0161 pported by some facts; and I can see no more   ABSTRACT improbability in a tendency to produce any c
0269 he few following cases, which I will briefly   ABSTRACT. The evidence is at least as good as that fr
0356 s subject. I can give here only the briefest   ABSTRACT of the more important facts. Change of clima
0376 ther in a most remarkable manner. This brief   ABSTRACT applies to plants alone: some strictly analo
0481 ws given in this volume under the form of an   ABSTRACT, I by no means expect to convince experience
0019 ral aboriginal stocks, has been carried to an   ABSURD extreme by some authors. They believe that ev
0186 y natural selection, seems, I freely confess,   ABSURD in the highest possible degree. Yet reason te
0062 re, with every fact on distribution, rarity,   ABUNDANCE, extinction, and variation, will be dimly s
0070 em; nor can the birds, though having a super   ABUNDANCE of food at this one season, increase in num
0304 nct tertiary species; from the extraordinary   ABUNDANCE of the individuals of many species all over
0070 f very rare plants being sometimes extremely   ABUNDANT in the few spots where they do occur; and th
0095 le fields of the red clover offer in vain an   ABUNDANT supply of precious nectar to the hive bee. T
0147 e same varieties of the cabbage do not yield   ABUNDANT and nutritious foliage and a copious supply
0267 gain, both with plants and animals, there is   ABUNDANT evidence, that a cross between very distinct
0294 nding sea; or, conversely, that some are now   ABUNDANT in the neighbouring sea, but are rare or abs
0322 e can precisely say why this species is more   ABUNDANT in individuals than that; why this species a
0357 xisting species. It seems to me that we have   ABUNDANT evidence of great oscillations of level in o
0375 istic of the Cordillera. On the mountains of   ABYSSINIA, several European forms and some few repres
0379 getable forms on the mountains of Borneo and   ABYSSINIA. I suspect that this preponderant migration
0439 ze, and the first leaves of the phyllodineous ACACEAS, are pinnate or divided like the ordinary le
0049 definition of these terms has been generally   ACCEPTED, is vainly to beat the air. Many of the case
0254 ich originated with Pallas, has been largely   ACCEPTED by modern naturalists; namely, that most of
0101 n so perfectly enclosed within the body, that   ACCESS from without and the occasional influence of
0145 ulpture; and even the ovary itself, with its   ACCESSORY parts, differs, as has been described by Ca
0190 swimbladder has, also, been worked in as an   ACCESSORY to the auditory organs of certain fish; or,
0165 ppears from what would commonly be called an   ACCIDENT, that I was led solely from the occurence of
0216 d from what we must in our ignorance call an   ACCIDENT. In some cases compulsory habit alone has su
0267 persuade myself that this parallelism is an   ACCIDENT or an illusion. Both series of facts seem to
0364 ing to their feet, which is in itself a rare   ACCIDENT. Even in this case, how small would the chan
```

Page **(Key Word)**

Page **********************************(Key Word)**********************************

```
0094 mind, I can see no reason to doubt that an ACCIDENTAL deviation in the size and form of the body,
0142 rom the few survivors, with care to prevent ACCIDENTAL crosses, and then again get seed from these
0144 se of correlation, I think it can hardly be ACCIDENTAL, that if we pick out the two orders of mamm
0157 unusual nature, the relation can hardly be ACCIDENTAL. The same number of joints in the tarsi is
0179 he intermediate varieties will be liable to ACCIDENTAL extermination; and during the process of fu
0189 power always intently watching each slight ACCIDENTAL alteration in the transparent layers; and c
0209 the natural selection of what may be called ACCIDENTAL variations of instincts; that is of variati
0213 which habit and the selection of so called ACCIDENTAL variations have played in modifying the men
0242 numerous, slight, and as we must call them ACCIDENTAL, variations, which are in any manner profit
0289 ll's Manual, will bring home the truth, how ACCIDENTAL and rare is their preservation, far better
0358 must now say a few words on what are called ACCIDENTAL means, but which more properly might be cal
0364 ese means of transport are sometimes called ACCIDENTAL, but this is not strictly correct: the curr
0364 ly correct: the currents of the sea are not ACCIDENTAL, nor is the direction of prevalent gales of
0384 ossibility of their occasional transport by ACCIDENTAL means; like that of the live fish not rarel
0093 d been effectually fertilised by the bees, ACCIDENTALLY dusted with pollen, having flown from tree
0222 slave making F. sanguinea; and when I had ACCIDENTALLY disturbed both nests, the little ants atta
0036 carefully preserved during famines and other ACCIDENTS, to which savages are so liable, and such c
0326 atal and geographical changes; or on strange ACCIDENTS, but in the long run the dominant forms wil
0405 at climatal and geographical changes and for ACCIDENTS of transport; and consequently for the migr
0131 election; organs of flight and of vision. ACCLIMATISATION. Correlation of growth. Compensation and
0139 k abodes will probably have been exposed. ACCLIMATISATION. Habit is hereditary with plants, as in
0139 , and this leads me to say a few words on ACCLIMATISATION. As it is extremely common for species o
0139 a single parent, if this view be correct, ACCLIMATISATION must be readily effected during long con
0141 circumstances, into play. How much of the ACCLIMATISATION of species to any peculiar climate is du
0142 as tender as ever it was, as proving that ACCLIMATISATION cannot be effected! The case, also, of t
0140 ted to different temperatures, or becoming ACCLIMATISED: thus the pines and rhododendrons, raised
0140 we know that they have subsequently become ACCLIMATISED to their new homes. As I believe that our
0011 s believe that long limbs are almost always ACCOMPANIED by an elongated head. Some instances of co
0021 runculated skin about the head, and this is ACCOMPANIED by greatly elongated eyelids, very large e
0145 ce, the daisy, and this difference is often ACCOMPANIED with the abortion of parts of the flower.
0147 e tuft of feathers on the head is generally ACCOMPANIED by a diminished comb, and a large beard by
0166 genus. The appearance of the stripes is not ACCOMPANIED by any change of form or by any other new
0265 independent of general health, and is often ACCOMPANIED by excess of size or great luxuriance. On
0114 of diversification of structure, with the ACCOMPANYING differences of habit and constitution, det
0116 n and of extinction, will tend to act. The ACCOMPANYING diagram will aid us in understanding this
0312 ological succession of organic beings, better ACCORD with the common view of the immutability of s
0314 seems to me satisfactory. These several facts ACCORD well with my theory. I believe in no fixed la
0331 ee how far these several facts and inferences ACCORD with the theory of descent with modification.
0334 cording to their periods of existence, do not ACCORD in arrangement. The species extreme in charac
0335 affinity, this arrangement would not closely ACCORD with the order in time of their production, a
0396 een developed into trees, etc., seem to me to ACCORD better with the view of occasional means of t
0096 ected so large a body of facts, showing, in ACCORDANCE with the almost universal belief of breeder
0161 s will be governed by natural selection, in ACCORDANCE with the diverse habits of the species, and
0218 r instincts but their structure modified in ACCORDANCE with their parasitic habits: for they do no
0292 ruth of these views, for they are in strict ACCORDANCE with the general principles inculcated by S
0328 france, although he finds in both a curious ACCORDANCE in the numbers of the species belonging to
0329 ons could be arranged in the same order, in ACCORDANCE with the general succession of the form of
0368 will not have been much disturbed, and, in ACCORDANCE with the principles inculcated in this volu
0396 e would have been more equally modified, in ACCORDANCE with the paramount importance of the relati
0404 e found some identical species, showing, in ACCORDANCE with the foregoing view, that at some forme
0446 ecome modified through natural selection in ACCORDANCE with their diverse habits. Then, from the m
0448 age in the same manner with its parents, in ACCORDANCE with their similar habits. Some further exp
0485 ientific and common language will come into ACCORDANCE. In short, we shall have to treat species i
0040 erhaps more in one district than in another, ACCORDING to the state of civilisation of the inhabit
0049 e different geographical ranges; and lastly, ACCORDING to very numerous experiments made during se
0075 of feathers, and all must fall to the ground ACCORDING to definite laws; but how simple is this pr
0089 elegant carriage and beauty to his bantams, ACCORDING to his standard of beauty, I can see no goo
0089 ions, the most melodious or beautiful males, ACCORDING to their standard of beauty, might produce
0091 taking to catch rats; another mice; one cat, ACCORDING to Mr. St. John, bringing home winged game,
0091 shall soon have to return. I may add, that, ACCORDING to Mr. Pierce, there are two varieties of t
0107 perhaps, it comes that the flora of Madeira, ACCORDING to Oswald Heer, resembles the extinct terti
0111 do good and distinct species. Nevertheless, ACCORDING to my view, varieties are species in the pr
0119 ndant: for natural selection will always act ACCORDING to the nature of the places which are often
0120 rked varieties (w10 and z10) or two species, ACCORDING to the amount of change supposed to be repr
0125 form two distinct families, or even orders, ACCORDING to the amount of divergent modification sup
0133 birds which are confined to continents are, ACCORDING to Mr. Gould, brighter coloured than those
0144 e shape of the head of the child. In snakes, ACCORDING to Schlegel, the shape of the body and the
0146 t importance, the seeds being in some cases, ACCORDING to Tausch, orthospermous in the exterior fl
0154 f the wing of the bat, it must have existed, ACCORDING to my theory, for an immense period in near
0158 al selection having more or less completely, ACCORDING to the lapse of time, overmastered the tend
0159 ird may be added, namely, the common turnip. ACCORDING to the ordinary view of each species having
0189 look to much isolated species, round which, ACCORDING to my theory, there has been much extinctio
0205 on, or strength in the battle for life, only ACCORDING to the standard of that country. Hence the
0218 be, and has been, generated. I may add that, ACCORDING to Dr. Gray and to some other observers, th
0221 ther probably in search of aphides or cocci. ACCORDING to Huber, who had ample opportunities for o
0226 two, three, or more perfectly flat surfaces, ACCORDING as the cell adjoins two, three or more othe
0260 al crosses between the same two species, for ACCORDING as the one species or the other is used as
0274 roduced from nearly related species, follows ACCORDING to Gartner the same laws. When two species
0298 d and perfected in some considerable degree. ACCORDING to this view, the chance of discovering in
0301 perhaps extremely slight degree, they would, ACCORDING to the principles followed by many palaeont
0323 been observed by several authors: so it is, ACCORDING to Lyell, with the several European and Nor
0334 rranged by Dr. Falconer in two series, first ACCORDING to their mutual affinities and then accordi
0334 ccording to their mutual affinities and then ACCORDING to their periods of existence, do not accor
0384 owly accustomed to live in fresh water; and, ACCORDING to Valenciennes, there is hardly a single g
0387 of the great southern water lily (probably, ACCORDING to Dr. Hooker, the Nelumbium luteum) in a h
0408 habited by very different forms of life; for ACCORDING to the length of time which has elapsed sin
```

Page **********************************(Key Word)**********************************

Page **************************************(Key Word)**************************************

0408 ed since new inhabitants entered one region: ACCORDING to the nature of the communication which al
0408 enter, either in greater or lesser numbers; ACCORDING or not, as those which entered happened to
0408 with each other and with the aborigines; and ACCORDING as the immigrants were capable of varying m
0416 e nostrils to the mouth, the only character, ACCORDING to Owen, which absolutely distinguishes fis
0425 were seen to come out of the womb of a bear? ACCORDING to all analogy, it would be ranked with bea
0430 most cases is general and not special; thus, ACCORDING to Mr. Waterhouse, of all Rodents, the bizc
0454 hornless breeds of cattle, more especially, ACCORDING to Youatt, in young animals, and the state
0490 that, whilst this planet has gone cycling on ACCORDING to the fixed law of gravity, from so simple
0226 een as perfect as the comb of the hive bee. ACCORDINGLY I wrote to Professor Miller, of Cambridge,
0352 n perhaps in any other organic beings; and, ACCORDINGLY, we find no inexplicable cases of the same
0109 low, intermittent action of natural selection ACCORDS perfectly well with what geology tells us of
0133 egree some of the characters of such species, ACCORDS with our view that species of all kinds are
0290 lands, whence the sediment has been derived, ACCORDS with the belief of vast intervals of time ha
0316 ntain) admit its truth; and the rule strictly ACCORDS with my theory. For as all the species of th
0322 s and whole groups of species become extinct, ACCORDS well with the theory of natural selection. W
0327 the same forms of life throughout the world, ACCORDS well with the principle of new species havin
0338 ly recent times. For this doctrine of Agassiz ACCORDS well with the theory of natural selection. I
0488 has been independently created. To my mind it ACCORDS better with what we know of the laws impress
0003 r to the other, it is equally preposterous to ACCOUNT for the structure of this parasite, with its
0020 intermediate between their parents; and if we ACCOUNT for our several domestic races by this proce
0029 o habit; but he would be a bold man who would ACCOUNT by such agencies for the differences of a dr
0111 a greater degree; but this alone would never ACCOUNT for so habitual and large an amount of diffe
0141 ever, push the foregoing argument too far, on ACCOUNT of the probable origin of some of our domest
0142 of seedling kidney beans ever appear, for an ACCOUNT has been published how much more hardy some
0166 e stated that the most probable hypothesis to ACCOUNT for the reappearance of very ancient charact
0172 profound mathematicians? Fourthly, how can we ACCOUNT for species, when crossed, being sterile and
0198 o them only to show that, if we are unable to ACCOUNT for the characteristic differences of our do
0235 remote in the scale of nature, that we cannot ACCOUNT for their similarity by inheritance from a c
0285 0 to 3000 feet. Prof. Ramsay has published an ACCOUNT of a downthrow in Anglesea of 2300 feet; and
0322 en, we may justly feel surprise why we cannot ACCOUNT for the extinction of this particular specie
0328 emselves differ in a manner very difficult to ACCOUNT for, cons¡dering the proximity of the two ar
0335 g species over the globe, will not attempt to ACCOUNT for the close resemblance of the distinct sp
0339 ica under the same latitude, would attempt to ACCOUNT, on the one hand, by dissimilar physical con
0345 , higher in the scale of nature; and this may ACCOUNT for that vague yet ill defined sentiment, fe
0369 europe; and it may be reasonably asked how I ACCOUNT for the necessary degree of uniformity of th
0372 e continuous shores of the Polar Circle, will ACCOUNT, on the theory of modification, for many clo
0373 s in this island, tell the same story. If one ACCOUNT which has been published can be trusted, we
0391 and we know from Mr. J. M. Jones's admirable ACCOUNT of Bermuda, that very many North American bi
0399 at very closely, as we know from Dr. Hooker's ACCOUNT, to those of America: but on the view that t
0426 ts of life, it assumes high value; but we can ACCOUNT for its presence in so many forms with such
0428 many smaller and feebler groups. Thus we can ACCOUNT for the fact that all organisms, recent and
0429 the lot of aberrant genera. We can, I think, ACCOUNT for this fact only by looking at aberrant fo
0431 the several groups in each class. We may thus ACCOUNT even for the distinctness of whole classes f
0436 mple in form; and then natural selection will ACCOUNT for the infinite diversity in structure and
0196 ay their aquatic origin, may perhaps be thus ACCOUNTED for. A well developed tail having been form
0218 by the males. This instinct may probably be ACCOUNTED for by the fact of the hens laying a large
0240 dimentary in the smaller workers than can be ACCOUNTED for merely by their proportionally lesser s
0249 cially fertilisec hybrids may, I believe, be ACCOUNTED for by close interbreeding having been avoi
0250 ce in their results may, I think, be in part ACCOUNTED for by Herbert's great horticultural skill,
0346 distribution. Present distribution cannot be ACCOUNTED for by differences in physical conditions.
0346 of the inhabitants of various regions can be ACCOUNTED for by their climatal and other physical co
0392 they are elsewhere. Such cases are generally ACCOUNTED for by the physical conditions of the islan
0393 d newts on so many oceanic islands cannot be ACCOUNTED for by their physical conditions; indeed it
0410 t rare, may, as I have attempted to show, be ACCOUNTED for by migration at some former period unde
0456 might even have been anticipated, and can be ACCOUNTED for by the laws of inheritance. Summary. In
0462 . a broken or interrupted range may often be ACCOUNTED for by the extinction of the species in the
0035 merly kept in this country. By comparing the ACCOUNTS given in old pigeon treatises of carriers an
0177 t is the same principle which, as I believe, ACCOUNTS for the common species in each country; as s
0406 fitted for distant transportation, probably ACCOUNTS for a law which has long been observed, and
0453 he scheme of nature? An eminent physiologist ACCOUNTS for the presence of rudimentary organs, by s
0045 y afford materials for natural selection to ACCUMULATE, in the same manner as man can accumulate i
0045 o accumulate, in the same manner as man can ACCUMULATE in any given direction individual differenc
0134 ariability; and natural selection will then ACCUMULATE all profitable variations, however slight,
0300 a distant epoch. Wherever sediment did not ACCUMULATE on the bed of the sea, or where it did not
0300 on the bed of the sea, or where it did not ACCUMULATE at a sufficient rate to protect organic bod
0467 variations given to him by nature, and thus ACCUMULATE them in any desired manner. He thus adapts
0029 to sum up in their minds slight differences ACCUMULATED during many successive generations. May no
0033 ts, there is another means of observing the ACCUMULATED effects of selection namely, by comparing
0037 vated plants, thus slowly and unconsciously ACCUMULATED, explains, as I believe, the well known fa
0084 r will his products be, compared with those ACCUMULATED by nature during whole geological periods.
0085 hrough variation, and the modifications are ACCUMULATED by natural selection for the good of the b
0117 the outer dotted lines) being preserved and ACCUMULATED by natural selection. When a dotted line r
0117 mount of variation is supposed to have been ACCUMULATED to have formed a fairly well marked variet
0143 t variations in any one part occur, and are ACCUMULATED through natural selection, other parts bec
0143 he most obvious case is, that modifications ACCUMULATED solely for the good of the young or larva,
0153 of variability, which has continually been ACCUMULATED by natural selection for the benefit of th
0156 ; for secondary sexual characters have been ACCUMULATED by sexual selection, which is less rigid i
0158 to variations in the same parts having been ACCUMULATED by natural and sexual selection; and thus
0173 ation; and such fossiliferous masses can be ACCUMULATED only where much sediment is deposited on t
0181 useful, each being propagated, until by the ACCUMULATED effects of this process of natural selecti
0205 in its structure could not have been slowly ACCUMULATED by means of natural selection. But we may
0237 cations of structure could have been slowly ACCUMULATED by natural selection. This difficulty, tho
0283 e good to show how slowly the mass has been ACCUMULATED. Let him remember Lyell's profound remark,
0286 e overlying sedimentary deposits might have ACCUMULATED. in thinner masses than elsewhere, the abov
0289 peculiar forms of life, had elsewhere been ACCUMULATED. And if in each separate territory, hardly
0290 hink, safely conclude that sediment must be ACCUMULATED in extremely thick, solid, or extensive ma

Page **************************************(Key Word)**************************************

Page **(Key Word)**

```
0291  ife which then existed: or, sediment may be ACCUMULATED to any thickness and extent over a shallow
0291  ion, has come to the conclusion that it was ACCUMULATED during subsidence. I may add, that the onl
0292  tionary, thick deposits could not have been ACCUMULATED in the shallow parts, which are the most f
0292  k more accurately, the beds which were then ACCUMULATED will have been destroyed by being upraised
0292  ur great deposits rich in fossils have been ACCUMULATED. Nature may almost be said to have guarded
0295  a thick fossiliferous formation can only be ACCUMULATED during a period of subsidence: and to keep
0300  ation would be destroyed, almost as soon as ACCUMULATED, by the incessant coast action, as we now
0307  which on my theory no doubt were somewhere ACCUMULATED before the Silurian epoch, is very great.
0308  rmations would in all probability have been ACCUMULATED from sediment derived from their wear and
0314  al selection, and whether the variations be ACCUMULATED to a greater or lesser amount, thus causin
0315  our formations have been almost necessarily ACCUMULATED at wide and irregularly intermittent inter
0328  contemporaneous formations have often been ACCUMULATED over very wide spaces in the same quarter
0410  been the same, and modifications have been ACCUMULATED by the same power of natural selection. Ch
0446  ve value to each breed, and which have been ACCUMULATED by man's selection, have not generally fir
0465  enough to resist future degradation, can be ACCUMULATED only where much sediment is deposited on t
0469  plex relations of life, would be preserved, ACCUMULATED, and inherited? Why, if man can by patienc
0481  the full effects of many slight variations, ACCUMULATED during an almost infinite number of genera
0001  be made out on this question by patiently ACCUMULATING and reflecting on all sorts of facts which
0004  shall see how great is the power of man in ACCUMULATING his selection successive slight variati
0052  ore, to the action of natural selection in ACCUMULATING (as will hereafter be more fully explained
0209  tural selection preserving and continually ACCUMULATING variations of instinct to any extent that
0216  partly by habit, and by man selecting and ACCUMULATING during successive generations, peculiar me
0243  g conditions of life, in natural selection ACCUMULATING slight modifications of instinct to any ex
0288  e bottom of the sea, where sediment is not ACCUMULATING. I believe we are continually taking a mos
0294  ts, including fossil remains, have gone on ACCUMULATING within the same area during the whole of t
0295  e formation, the deposit must have gone on ACCUMULATING for a very long period, in order to have g
0299  f europe, when most of our formations were ACCUMULATING. The Malay Archipelago is one of the riche
0300  he formations which we suppose to be there ACCUMULATING. I suspect that not many of the strictly l
0471  ation. As natural selection acts solely by ACCUMULATING slight, successive, favourable variations,
0032  nsists in the great effect produced by the ACCUMULATION in one direction, during successive genera
0041  icient, with extreme care, to allow of the ACCUMULATION of a large amount of modification in almos
0061  rganic beings to his own uses, through the ACCUMULATION of slight but useful variations, given to
0086  d modify organic beings at any age, by the ACCUMULATION of profitable variations at that age, and
0095  ction can act only by the preservation and ACCUMULATION of infinitesimally small inherited modific
0119  intervals long enough to have allowed the ACCUMULATION of a considerable amount of divergent vari
0170  ause for each must exist, it is the steady ACCUMULATION, through natural selection, of such differ
0210  selection, except by the slow and gradual ACCUMULATION of numerous, slight, yet profitable, varia
0233  der. As natural selection acts only by the ACCUMULATION of slight modifications of structure or in
0242  cation in structure can be effected by the ACCUMULATION of numerous, slight, and as we must call t
0284  of the time which has elapsed during their ACCUMULATION; yet what time this must have consumed! Go
0284  by the currents of the sea, the process of ACCUMULATION in any one area must be extremely slow. Bu
0284  ces suffered, independently of the rate of ACCUMULATION of the degraded matter, probably offers th
0295  ry, has generally been intermittent in its ACCUMULATION. When we see, as is so often the case, a f
0296  have required an enormous period for their ACCUMULATION; yet no one ignorant of this fact would ha
0296  overlooked, intervals have occurred in its ACCUMULATION. In other cases we have the plainest evide
0300  chipelago, together with a contemporaneous ACCUMULATION of sediment, would exceed the average dura
0303  some cases than the time required for the ACCUMULATION of each formation. These intervals will ha
0315  y, perhaps, be nearly the same: but as the ACCUMULATION of long enduring fossiliferous formations
0329  es have gone on slowly changing during the ACCUMULATION of the several formations and during the l
0336  being closely related, is obvious. As the ACCUMULATION of each formation has often been interrupt
0342  wing to subsidence being necessary for the ACCUMULATION of fossiliferous deposits thick enough to
0459  h analogous with, human reason, but by the ACCUMULATION of innumerable slight variations, each goo
0465  formations have been intermittent in their ACCUMULATION; and their duration, I am inclined to beli
0480  ll slowly changing by the preservation and ACCUMULATION of successive slight favourable variations
0487  made at hazard and at rare intervals. The ACCUMULATION of each great fossiliferous formation will
0290  lations of level. Such thick and extensive ACCUMULATIONS of sediment may be formed in two ways: ei
0303  y series. And now one of the richest known ACCUMULATIONS of fossil mammals belongs to the middle o
0030  n their history. The key is man's power of ACCUMULATIVE selection: nature gives successive variati
0043  e causes of Change I am convinced that the ACCUMULATIVE action of Selection, whether applied metho
0133  ot tell how much of it to attribute to the ACCUMULATIVE action of natural selection, and how much
0002  to the reader reposing some confidence in my ACCURACY. No doubt errors will have crept in, though
0032  to appreciate. Not one man in a thousand has ACCURACY of eye and judgment sufficient to become an
0230  ted elder Huber, but I am convinced of their ACCURACY; and if I had space, I could show that they
0271  ey present analogous facts. Kolreuter, whose ACCURACY has been confirmed by every subsequent obser
0293  roughout Europe been correlated with perfect ACCURACY. With marine animals of all kinds, we may sa
0299  formations which have been examined with any ACCURACY, excepting those of the United States of Ame
0307  y a small portion of the world is known with ACCURACY. M. Barrande has lately added another and lo
0361  onous plants germinated: I am certain of the ACCURACY of this observation. Again, I can show that
0488  e modified: so that we must not overrate the ACCURACY of organic change as a measure of time. Duri
0208  this comparison gives, I think, a remarkably ACCURATE notion of the frame of mind under which an i
0240  not the actual measurements, but a strictly ACCURATE illustration: the difference was the same as
0227  test difficulty, that she can somehow judge ACCURATELY at what distance to stand from her fellow l
0292  ate periods of elevation; or, to speak more ACCURATELY, the beds which were then accumulated will
0384  ce, salt water fish can with care be slowly ACCUSTOMED to live in fresh water; and, according to V
0442  excepting for reproduction. We are so much ACCUSTOMED to see differences in structure between the
0131  olly incorrect expression, but it serves to ACKNOWLEDGE plainly our ignorance of the cause of each
0186  in the principle of natural selection, will ACKNOWLEDGE that every organic being is constantly end
0485  species. Hereafter we shall be compelled to ACKNOWLEDGE that the only distinction between species
0006  extinct species, in the same manner as the ACKNOWLEDGED varieties of any one species are the desce
0040  value of the new sub breed are once fully ACKNOWLEDGED, the principle, as I have called it, of un
0058  t britain. Now, in this same catalogue, 53 ACKNOWLEDGED varieties are recorded, and these range ov
0058  ong range over 14.3 provinces. So that the ACKNOWLEDGED varieties have very nearly the same restri
0112  on having a rather longer beak; and on the ACKNOWLEDGED principle that fanciers do not and will no
0178  or representative species, and likewise of ACKNOWLEDGED varieties), exist in the intermediate zone
0206  y theory be strictly true. It is generally ACKNOWLEDGED that all organic beings have been formed o
0247  h as the Leguminosae, in which there is an ACKNOWLEDGED difficulty in the manipulation) half of th
```

Page **(Key Word)**

Page ***************************************(Key Word)**

```
0415  with this fact; and it has been most fully ACKNOWLEDGED in the writings of almost every author. It
0433  er one species; we use descent in classing ACKNOWLEDGED varieties, however different they may be f
0456  d in ranking together the sexes, ages, and ACKNOWLEDGED varieties of the same species, however dif
0469  acts of creation, and varieties which are ACKNOWLEDGED to have been produced by secondary laws. O
0475  uch points, as do the crossed offspring of ACKNOWLEDGED varieties. On the other hand, these would
0485  is quite possible that forms now generally ACKNOWLEDGED to be merely varieties may hereafter be th
0002  ace prevents my having the satisfaction of ACKNOWLEDGING the generous assistance which I have rece
0033  could give several references to the full ACKNOWLEDGMENT of the importance of the principle in wor
0031  uthorities. Youatt, who was probably better ACQUAINTED with the works of agriculturists than almos
0036  cion existing in the mind of any one at all ACQUAINTED with the subject that the owner of either o
0209  most wonderful instincts with which we are ACQUAINTED, namely, those of the hive bee and of many
0210  r the sole good of another, with which I am ACQUAINTED, is that of aphides voluntarily yielding th
0335  f in the immutability of species. He who is ACQUAINTED with the distribution of existing species o
0003  iting this world have been modified, so as to ACQUIRE that perfection of structure and coadaptatio
0025  he mongrel offspring are very apt suddenly to ACQUIRE these characters; for instance, I crossed so
0078  as necessary, as it seems to be difficult to ACQUIRE. All that we can do, is to keep steadily in
0162  uniform nature occasionally varying so as to ACQUIRE, in some degree, the character of the same p
0386  members: for these latter seem immediately to ACQUIRE, as if in consequence, a very wide range. I
0461  pring of slightly modified forms or varieties ACQUIRE from being crossed increased vigour and fert
0015  endency to reversion, that is, to lose their ACQUIRED characters, whilst kept under unchanged cond
0022  . the period at which the perfect plumage is ACQUIRED varies, as does the state of the down with w
0085  rouse, and in keeping that colour, when once ACQUIRED, true and constant. Nor ought we to think th
0146  eritance; for an ancient progenitor may have ACQUIRED through natural selection some one modificat
0172  itable perfection? Thirdly, can instincts be ACQUIRED and modified through natural selection? What
0182  dicate the natural steps by which birds have ACQUIRED their perfect power of flight; but they serv
0196  under new conditions of life and with newly ACQUIRED habits. To give a few instances to illustrat
0197  character of importance and might have been ACQUIRED through natural selection; as it is, I have
0200  fer that these several bones might have been ACQUIRED through natural selection, subjected formerl
0205  t could not, in its present state, have been ACQUIRED by natural selection, a power which acts sol
0208  ds of time and states of the body. When once ACQUIRED, they often remain constant throughout life.
0209  at the greater number of instincts have been ACQUIRED by habit in one generation, and then transmi
0209  many ants, could not possibly have been thus ACQUIRED. It will be universally admitted that instin
0210  ions by which each complex instinct has been ACQUIRED, for these could be found only in the lineal
0212  in other animals. But fear of man is slowly ACQUIRED, as I have elsewhere shown, by various anima
0216  conclude, that domestic instincts have been ACQUIRED and natural instincts have been lost partly
0224  f raising slaves. When the instinct was once ACQUIRED, if carried out to a much less extent even t
0228  ke its nest, I believe that the hive bee has ACQUIRED, through natural selection, her inimitable a
0228  each other, that by the time the basins had ACQUIRED the above stated width (i.e. about the width
0235  having transmitted by inheritance its newly ACQUIRED economical instinct to new swarms, which in
0236  d must therefore believe that they have been ACQUIRED by independent acts of natural selection. I
0236  umed that all its characters had been slowly ACQUIRED through natural selection; namely, by an ind
0237  it could never have transmitted successively ACQUIRED modifications of structure or instinct to it
0242  nd by inherited tools or weapons, and not by ACQUIRED knowledge and manufactured instruments, a pe
0245  e to them, and therefore could not have been ACQUIRED by the continued preservation of successive
0245  le to show that sterility is not a specially ACQUIRED or endowed quality, but is incidental on oth
0245  endowed quality, but is incidental on other ACQUIRED differences. In treating this subject, two c
0272  ecial endowment, but is incidental on slowly ACQUIRED modifications, more especially in the reprod
0275  relate to characters which have been slowly ACQUIRED by selection. Consequently, sudden reversion
0303  ad been effected, and a few species had thus ACQUIRED a great advantage over other organisms, a co
0444  ant whether most of its characters are fully ACQUIRED a little earlier or later in life. It would
0444  ive modifications, by which each species has ACQUIRED its present structure, may have supervened a
0445  ies, I found that the puppies had not nearly ACQUIRED their full amount of proportional difference
0445  orse, I find that the colts have by no means ACQUIRED their full amount of proportional difference
0446  e desired qualities and structures have been ACQUIRED earlier or later in life, if the full grown
0446  ced tumbler, which when twelve hours old had ACQUIRED its proper proportions, proves that this is
0456  s attempted arrangement, with the grades of ACQUIRED difference marked by the terms varieties, sp
0475  on the view of instincts having been slowly ACQUIRED through natural selection we need not marvel
0481  o be of short duration; and now that we have ACQUIRED some idea of the lapse of time, we are too a
0204  e, there is no logical impossibility in the ACQUIREMENT of any conceivable degree of perfection th
0488  on a new foundation, that of the necessary ACQUIREMENT of each mental power and capacity by grada
0133  e zone of habitation of other species, often ACQUIRING in a very slight degree some of the charact
0071  ouched by the hand of man; but several hundred ACRES of exactly the same nature had been enclosed
0072  ts of view, whence I could examine hundreds of ACRES of the unenclosed heath, and literally I coul
0165  k arabian sire, were much more plainly barred ACROSS the legs than is even the pure quagga. Lastly
0221  t year, however, in the month of July, I came ACROSS a community with an unusually large stock of
0222  ereas they were much terrified when they came ACROSS the pupae, or even the earth from the nest of
0352  le and exceptional. The capacity of migrating ACROSS the sea is more distinctly limited in terrest
0353  heir varied means of dispersal, have migrated ACROSS the vast and broken interspace. The great and
0358  ted for wide dissemination; but for transport ACROSS the sea, the greater or less facilities may b
0359  sank in a few days, they could not be floated ACROSS wide spaces of the sea, whether or not they w
0360  nts belonging to one country might be floated ACROSS 924 miles of sea to another country; and when
0360  ra, after having been dried, could be floated ACROSS a space of sea 900 miles in width; and would
0361  ny kinds are blown by gales to vast distances ACROSS the ocean. We may I think safely assume that
0364  wever, would suffice for occasional transport ACROSS tracts of sea some hundred miles in breadth,
0364  t every year, one or two land birds are blown ACROSS the whole Atlantic Ocean, from North America
0368  lder climate permitted their former migration ACROSS the low intervening tracts, since become too
0385  sure to alight on a pool or rivulet, if blown ACROSS sea to an oceanic island or to any other dist
0387  e flight and go to other waters, or are blown ACROSS the sea; and we have seen that seeds retain t
0391  transported far more easily than land shells, ACROSS three or four hundred miles of open sea. The
0393  ould be great difficulty in their transportal ACROSS the sea, and therefore why they do not exist
0394  for no terrestrial mammal can be transported ACROSS a wide space of sea, but bats can fly across.
0394  across a wide space of sea, but bats can fly ACROSS. Bats have been seen wandering by day far ove
0397  might be floated in chinks of drifted timber ACROSS moderately wide arms of the sea. And I found
0401  . the currents of the sea are rapid and sweep ACROSS the archipelago, and gales of wind are extrao
0008  of variability, whatever they may be, generally ACT: whether during the early or late period of de
0008  tive elements having been affected prior to the ACT of conception. Several reasons make me believe
```

Page ***************************************(Key Word)**

act

Page **(Key Word)**

```
0009 d not be surprised at this system, when it does ACT under confinement, acting not quite regularly,
0010 ted by the treatment of the parent prior to the ACT of conception. These cases anyhow show that va
0010 nected, as some authors have supposed, with the ACT of generation. Seedlings from the same fruit,
0038 rarely cares for what is internal. He can never ACT by selection, excepting on variations which ar
0044 term includes the unknown element of a distinct ACT of creation. The term variety is almost equall
0055 r hand, if we look at each species as a special ACT of creation, there is no apparent reason why m
0074 es it can be shown that widely different checks ACT on the same species in different districts. Wh
0080 nce, discussed too briefly in the last chapter, ACT in regard to variation? Can the principle of s
0080 ply to nature? I think we shall see that it can ACT most effectually. Let it be borne in mind in w
0083 selection, what may not nature effect? Man can ACT only on external and visible characters: natur
0083 far as they may be useful to any being. She can ACT on every internal organ, on every shade of con
0084 y formerly were. Although natural selection can ACT only through and for the good of each being, y
0086 of nature, natural selection will be enabled to ACT on and modify organic beings at any age, by th
0087 o get out of it; so that fanciers assist in the ACT of hatching. Now, if nature had to make the be
0092 he same species would thus get crossed; and the ACT of crossing, we have good reason to believe (a
0095 t lines of inland cliffs. Natural selection can ACT only by the preservation and accumulation of i
0097 eat good, as I believe, of the plant. Bees will ACT like a camel hair pencil, and it is quite suff
0103 nd uniform in character. It will obviously thus ACT far more efficiently with those animals which
0108 new species. That natural selection will always ACT with extreme slowness, I fully admit. Its acti
0108 i do believe that natural selection will always ACT very slowly, often only at long intervals of t
0113 if its natural powers of increase be allowed to ACT, it can succeed in increasing (the country not
0116 tural selection and of extinction, will tend to ACT. The accompanying diagram will aid us in under
0119 d descendant; for natural selection will always ACT according to the nature of the places which ar
0125 in the struggle for existence, it will chiefly ACT on those which already have some advantage; an
0134 show how indirectly the conditions of life must ACT. Again, innumerable instances are known to eve
0146 the manner alone in which natural selection can ACT. For instance, Alph. de Candolle has remarked
0149 al selection, it should never be forgotten, can ACT on each part of each being, solely through and
0150 ne sex, but are not directly connected with the ACT of reproduction. The rule applies to males and
0160 several marks, beyond the influence of the mere ACT of crossing on the laws of inheritance. No dou
0187 which are coated by pigment, and which properly ACT only by excluding lateral pencils of light, ar
0188 light, are convex at their upper ends and must ACT by convergence; and at their lower ends there
0192 a, but which, likewise, very slightly aided the ACT of respiration, have been gradually converted
0194 d why she should not; for natural selection can ACT only by taking advantage of slight successive
0197 ey facilitate, or may be indispensable for this ACT; but as sutures occur in the skulls of young b
0203 al selection will always be very slow, and will ACT, at any one time, only on a very few forms; an
0205 al selection in each well stocked country, must ACT chiefly through the competition of the inhabit
0253 s should have gone on increasing. If we were to ACT thus, and pair brothers and sisters in the cas
0261 ale sexual element of the one will often freely ACT on the female sexual element of the other, but
0262 he union of the male and female elements in the ACT of reproduction, yet that there is a rude degr
0273 nd be super added to that arising from the mere ACT of crossing. The slight degree of variability
0315 on this view, does not mark a new and complete ACT of creation, but only an occasional scene, tak
0437 from the yielding of the separate pieces in the ACT of parturition of mammals, will by no means ex
0447 e of modification, adapted in one descendant to ACT as hands, in another as paddles, in another as
0453 h is formed merely of cellular tissue, can thus ACT? Can we suppose that the formation of rudiment
0469 ariability and a powerful agent always ready to ACT and select, why should we doubt that variation
0471 produce no great or sudden modification; it can ACT only by very short and slow steps. Hence the c
0472 ng produced in such vast numbers for one single ACT, and being then slaughtered by their sterile s
0480 organs. But disuse and selection will generally ACT on each creature, when it has come to maturity
0483 e authors seem no more startled at a miraculous ACT of creation than at an ordinary birth. But do
0483 tissues? Do they believe that at each supposed ACT of creation one individual or many were produc
0014 , and not to its primary cause, which may have ACTED on the ovules or male element; in nearly the
0084 sider as very trifling importance, may thus be ACTED on. When we see leaf eating insects green, an
0127 les. Whether natural selection has really thus ACTED in nature, in modifying and adapting the vari
0127 ils extinction; and how largely extinction has ACTED in the world's history, geology plainly decla
0159 e constitution and tendency to variation, when ACTED on by similar unknown influences. In the vege
0167 ing a comparison, the same laws appear to have ACTED in producing the lesser differences between v
0195 constitutional differences, might assuredly be ACTED on by natural selection. The tail of the gira
0213 ble than natural instincts; but they have been ACTED on by far less rigorous selection, and have b
0216 most cases, probably, habit and selection have ACTED together. We shall, perhaps, best understand
0235 g importance, that they could hardly have been ACTED on by natural selection; cases of instincts a
0266 e beneficial to all living things. We see this ACTED on by farmers and gardeners in their frequent
0280 roportion as this process of extermination has ACTED on an enormous scale, so must the number of i
0319 rse's life, and in what degree, they severally ACTED. If the conditions had gone on, however slowl
0443 riod it is fully displayed. The cause may have ACTED, and I believe generally has acted, even befo
0443 se may have acted, and I believe generally has ACTED, even before the embryo is formed; and the va
0467 o obvious reason why the principles which have ACTED so efficiently under domestication should not
0467 fficiently under domestication should not have ACTED under nature. In the preservation of favoured
0009 affected by unperceived causes as to fail in ACTING, we need not be surprised at this system, whe
0009 s system, when it does act under confinement, ACTING not quite regularly, and producing offspring
0013 it may not be due to the same original cause ACTING on both; but when amongst individuals, appare
0074 case of many species, many different checks, ACTING at different periods of life, and during diff
0082 hange in the conditions of life, by specially ACTING on the reproductive system, causes or increas
0089 ng been chiefly modified by sexual selection, ACTING when the birds have come to the breeding age
0178 the new forms thus produced and the old ones ACTING and reacting on each other. So that, in any o
0241 fore me, I believe that natural selection, by ACTING on the fertile parents, could form a species
0345 life, produced by the laws of variation still ACTING round us, and preserved by Natural Selection.
0467 existence, we see the most powerful and ever ACTING means of selection. The struggle for existenc
0468 e is a strictly limited quantity. Man, though ACTING on external characters alone and often capric
0469 roducts? What limit can be put to this power, ACTING during long ages and rigidly scrutinising the
0480 existence, and will thus have little power of ACTING on an organ during early life; hence the orga
0487 ecies are produced and exterminated by slowly ACTING and still existing causes, and not by miracul
0489 plex a manner, have all been produced by laws ACTING around us. These laws, taken in the largest s
0008 an any other part of the organisation, to the ACTION of any change in the conditions of life. Noth
0010 nd of growth, and of inheritance: for had the ACTION of the conditions been direct, if any of the
0010 variation, we should attribute to the direct ACTION of heat, moisture, light, food, etc., is most
0011 nge may, I think, be attributed to the direct ACTION of the conditions of life, as, in some cases,
```

Page **(Key Word)**

Page **(Key Word)**

```
0015  ect would have to be attributed to the direct  ACTION of the poor soil), that they would to a large
0029  ect may, perhaps, be attributed to the direct   ACTION of the external conditions of life, and some
0043  lieve that the conditions of life, from their   ACTION on the reproductive system, are so far of the
0043  th. Something may be attributed to the direct   ACTION of the conditions of life. Something must be
0043  f change I am convinced that the accumulative   ACTION of Selection, whether applied methodically an
0052  some cases, due merely to the long continued    ACTION of different physical conditions in two diffe
0052  arent to one in which it differs more, to the   ACTION of natural selection in accumulating (as will
0056  ht generally to find the manufactory still in   ACTION, more especially as we have every reason to b
0061  reafter see, is a power incessantly ready for   ACTION, and is as immeasurably superior to man's fee
0068  severe mortality from epidemics with man. The   ACTION of climate seems at first sight to be quite i
0069  d to attribute the whole effect to its direct   ACTION. But this is a very false view: we forget tha
0069  stunted forms, due to the directly injurious    ACTION of climate, than we do in proceeding southwar
0075  ut how simple is this problem compared to the   ACTION and reaction of the innumerable plants and an
0080  ssing, isolation, number of individuals. Slow   ACTION. Extinction caused by Natural Selection. Dive
0080  nts of any small area, and to naturalisation.   ACTION of Natural Selection, through Divergence of C
0090  en called a monstrosity. Illustrations of the   ACTION of Natural Selection. In order to make it cle
0095  e of geology: but we now very seldom hear the   ACTION, for instance, of the coast waves, called a t
0100  nt: for we know of no means, analogous to the   ACTION of insects and of the wind in the case of pla
0108  act with extreme slowness, I fully admit. Its   ACTION depends on there being places in the polity o
0108  er adapted forms having been checked. But the   ACTION of natural selection will probably still ofte
0108  auses are amply sufficient wholly to stop the   ACTION of natural selection. I do not believe so. On
0109  er believe, that this very slow, intermittent   ACTION of natural selection accords perfectly well w
0112  . here, then, we see in man's productions the   ACTION of what may be called the principle of diverg
0129  explained through inheritance and the complex   ACTION of natural selection, entailing extinction an
0133  w much of it to attribute to the accumulative   ACTION of natural selection, and how much to the con
0133  many generations, and how much to the direct   ACTION of the severe climate? for it would appear th
0133  it would appear that climate has some direct   ACTION on the hair of our domestic quadrupeds. Insta
0134  ne me to lay very little weight on the direct   ACTION of the conditions of life. Indirectly, as alr
0136  so many Madeira beetles is mainly due to the   ACTION of natural selection, but combined probably w
0136  n enlarged. This is quite compatible with the   ACTION of natural selection. For when a new insect f
0157  sexual selection, which is less rigid in its   ACTION than ordinary selection, as it does not entai
0157  xual selection will have had a wide scope for   ACTION, and may thus readily have succeeded in givin
0161  e species, and will not be left to the mutual   ACTION of the conditions of life and of a similar in
0197  y be, or it may possibly be due to the mutual   ACTION of putrid matter: but we should be very cauti
0198  so that colour would be thus subjected to the   ACTION of natural selection. But we are far too long
0200  (making some little allowance for the direct   ACTION of physical conditions) may be viewed, either
0206  disuse, being slightly affected by the direct   ACTION of the external conditions of life, and being
0207  and to lay her eggs in other birds' nests. An   ACTION, which we ourselves should require experience
0208  the frame of mind under which an instinctive   ACTION is performed, but not of its origin. How unco
0208  ating a well known song, so in instincts, one   ACTION follows another by a sort of rhythm: if a per
0209  ady finished work. If we suppose any habitual   ACTION to become inherited, and I think it can be sh
0210  stances of an animal apparently performing an   ACTION for the sole good of another, with which I am
0211  ides behaved in this manner, showing that the   ACTION was instinctive, and not the result of experi
0211  ieve that any animal in the world performs an   ACTION for the exclusive good of another of a distin
0211  of such variations, are indispensable for the   ACTION of natural selection, as many instances as po
0214  have taught, the tumbler pigeon to tumble, an   ACTION which, as I have witnessed, is performed by y
0251  e functionally imperfect in their mutual self   ACTION, we must infer that the plants were in an unn
0266  ce occuring in the development, or periodical   ACTION, or mutual relation of the different parts an
0283  d their base. He who most closely studies the   ACTION of the sea on our shores, will, I believe, be
0285  and has been so completely planed down by the   ACTION of the sea, that no trace of these vast dislo
0287  ears: or say three hundred million years. The   ACTION of fresh water on the gently inclined Wealden
0287  s of years as land, and thus have escaped the   ACTION of the sea: when deeply submerged for perhaps
0287  periods, it would, likewise, have escaped the   ACTION of the coast waves. So that in all probabilit
0290  radual rising of the land within the grinding   ACTION of the coast waves. We may, I think, safely c
0290  e masses, in order to withstand the incessant   ACTION of the waves, when first upraised and during
0292  ed and brought within the limits of the coast   ACTION. Thus the geological record will almost neces
0300  s soon as accumulated, by the incessant coast   ACTION, we are now see on the shores of South America
0307  by denudation, or obliterated by metamorphic   ACTION, we ought to find only small remnants of the
0309  er, might have undergone far more metamorphic   ACTION than strata which have always remained nearer
0322  nordinately, and that some check is always in   ACTION, yet seldom perceived by us, the whole econom
0350  number of the former immigrants: and on their   ACTION and reaction, in their mutual struggles for l
0358  nown how far seeds could resist the injurious   ACTION of sea water. To my surprise I found that out
0360  tion and of their resistance to the injurious   ACTION of the salt water. On the other hand he did n
0363  t doubt remain to be discovered, have been in   ACTION year after year, for centuries and tens of th
0364  hen exposed for a great length of time to the   ACTION of sea water: nor could they be long carried
0373  e have some direct evidence of former glacial   ACTION in New Zealand: and the same plants, found on
0373  e trusted, we have direct evidence of glacial   ACTION in the south eastern corner of Australia. Loo
0373  have the clearest evidence of former glacial   ACTION, in huge boulders transported far from their
0374  y at least admit as probable that the glacial   ACTION was simultaneous on the eastern and western s
0408  would be an almost endless amount of organic   ACTION and reaction, and we should find, as we do fi
0412  eader to turn to the diagram illustrating the   ACTION, as formerly explained, of these several prin
0455  as come to maturity and to its full powers of   ACTION, the principle of inheritance at correspondin
0466  growth, by use and disuse, and by the direct   ACTION of the physical conditions of life. There is
0470  xpect, as a general rule, to find it still in   ACTION: and this is the case if varieties be incipie
0481  med, and great valleys excavated, by the slow   ACTION of the coast waves. The mind cannot possibly
0486  the effects of use and disuse, on the direct   ACTION of external conditions, and so forth. The stu
0489  ion: Variability from the indirect and direct   ACTION of the external conditions of life, and from
0207  be easy to show that several distinct mental   ACTIONS are commonly embraced by this term: but ever
0208  f its origin. How unconsciously many habitual   ACTIONS are performed, indeed not rarely in direct o
0213  ep, by shepherd dogs. I cannot see that these   ACTIONS, performed without experience by the young,
0213  leaf of the cabbage, I cannot see that these   ACTIONS differ essentially from true instincts. If w
0213  distant point, we should assuredly call them   ACTIONS instinctive. Domestic instincts, as they may
0214  domestic instincts are sometimes spoken of as   ACTIONS which have become inherited solely from long
0216  ssive generations, peculiar mental habits and   ACTIONS, which at first appeared from what we must i
0056  pression, the manufactory of species has been   ACTIVE, we ought generally to find the manufactory s
0237  period alone when the reproductive system is   ACTIVE, as in the nuptial plumage of many birds, and
```

Page **(Key Word)**

Page **(Key Word)**

```
0439 s; but in the case of larvae, the embryos are ACTIVE, and have been adapted for special lines of l
0440 al during any part of its embryonic career is ACTIVE, and has to provide for itself. The period of
0440 adaptations, the similarity of the larvae or ACTIVE embryos of allied animals is sometimes much o
0440 s. in most cases, however, the larvae, though ACTIVE, still obey more or less closely the law of c
0441 eloped organs of sense, and to reach by their ACTIVE powers of swimming, a proper place on which t
0442 ects, whether adapted to the most diverse and ACTIVE habits, or quite inactive, being fed by their
0443 when the embryo becomes at any period of life ACTIVE and has to provide for itself; of the embryo
0448 ple of inheritance at corresponding ages, the ACTIVE young or larvae might easily be rendered by n
0457 purposes as different as possible. Larvae are ACTIVE embryos, which have become specially modified
0470 for where the manufactory of species has been ACTIVE, we might expect, as a general rule, to find
0169 e manufactory of new specific forms has been ACTIVELY at work; there, on an average, we now find m
0440 and has to provide for itself. The period of ACTIVITY may come on earlier or later in life; but wh
0447 animal, which has come to its full powers of ACTIVITY and has to gain its own living; and the effe
0068 or existence; but in so far as climate chiefly ACTS in reducing food, it brings on the most severe
0069 even when climate, for instance extreme cold, ACTS directly, it will be the least vigorous, or th
0069 st exclusively with the elements. That climate ACTS in main part indirectly by favouring other spe
0090 it clear how, as I believe, natural selection ACTS, I must beg permission to give one or two imag
0104 cts, will be prevented. But isolation probably ACTS more efficiently in checking the immigration o
0109 cted with natural selection. Natural selection ACTS solely through the preservation of variations
0121 stocked country natural selection necessarily ACTS by the selected form having some advantage in
0125 t have been expected: for as natural selection ACTS through one form having some advantage over ot
0159 ner, but to three separate yet closely related ACTS of creation. With pigeons, however, we have an
0172 f transitional varieties. As natural selection ACTS solely by the preservation of profitable modif
0185 e. he who believes in separate and innumerable ACTS of creation will say, that in these cases it h
0194 ttle apparent importance. As natural Selection ACTS by life and death, by the preservation of indi
0201 ing injurious to itself, for natural selection ACTS solely by and for the good of each. No organ w
0203 eral facts, which on the theory of independent ACTS of creation are utterly obscure. We have seen
0205 n acquired by natural selection, a power which ACTS solely by the preservation of profitable varia
0206 le of natural selection. For natural selection ACTS by either now adapting the varying parts of ea
0233 ired than for a cylinder. As natural selection ACTS only by the accumulation of slight modificatio
0236 ve that they have been acquired by independent ACTS of natural selection. I will not here enter on
0269 to alter their general habits of life. Nature ACTS uniformly and slowly during vast periods of ti
0396 explicable relation on the view of independent ACTS of creation. All the foregoing remarks on the
0467 ns to new conditions of life, and then nature ACTS on the organisation, and causes variability. B
0469 only supposed to have been produced by special ACTS of creation, and varieties which are acknowled
0471 n the theory of creation. As natural selection ACTS solely by accumulating slight, successive, fav
0472 en have been anticipated. As natural selection ACTS by competition, it adapts the inhabitants of e
0478 erly inexplicable on the theory of independent ACTS of creation. The existence of closely allied o
0487 d still existing causes, and not by miraculous ACTS of creation and by catastrophes; and as the mo
0034 th europeans. Some of these facts do not show ACTUAL selection, but they show that the breeding of
0035 of this kind could never be recognised unless ACTUAL measurements or careful drawings of the breed
0051 series impresses the mind with the idea of an ACTUAL passage. Hence I look at individual differenc
0058 he occurrence of such links cannot affect the ACTUAL characters of the forms which they connect; a
0147 ess or by disuse, and, on the other hand, the ACTUAL withdrawal of nutriment from one part owing t
0178 ady assigned (namely from what we know of the ACTUAL distribution of closely allied or representat
0210 ructures, we ought to find in nature, not the ACTUAL transitional gradations by which each complex
0240 erence in these workers, by my giving not the ACTUAL measurements, but a strictly accurate illustr
0284 me the maximum thickness, in most cases from ACTUAL measurement, in a few cases from estimate, of
0297 ten, as before explained, that A might be the ACTUAL progenitor of B and C, and yet might not at a
0341 america; and some of these fossils may be the ACTUAL progenitors of living species. It must not be
0360 nner, for he placed the seeds in a box in the ACTUAL sea, so that they were alternately wet and ex
0432 an specify this or that branch, though at the ACTUAL fork the two unite and blend together. We cou
0485 weigh more carefully and to value higher the ACTUAL amount of difference between them. It is quit
0488 ably serves as a fair measure of the lapse of ACTUAL time. A number of species, however, keeping i
0031 ted for breeding. What English breeders have ACTUALLY effected is proved by the enormous prices gi
0047 her, not because the intermediate links have ACTUALLY been found, but because analogy leads the ob
0082 degree of isolation to check immigration, is ACTUALLY necessary to produce new and unoccupied plac
0112 , but like extremes, they both go on (as has ACTUALLY occurred with tumbler pigeons) choosing and
0163 these stripes are sometimes very obscure, or ACTUALLY quite lost, in dark coloured asses. The koul
0191 ulty in believing that natural selection has ACTUALLY converted a swimbladder into a lung, or orga
0191 ient period served for respiration have been ACTUALLY converted into organs of flight. In conside
0195 ge. It is not that the larger quadrupeds are ACTUALLY destroyed (except in some rare cases) by the
0196 hough now become of very slight use; and any ACTUALLY injurious deviations in their structure will
0219 he slaves which determine the migration, and ACTUALLY carry their masters in their jaws. So utterl
0236 ich at first appeared to me insuperable, and ACTUALLY fatal to my whole theory. I allude to the ne
0250 d all the individuals of certain species can ACTUALLY be hybridised much more readily than they ca
0369 species. In illustrating what, as I believe, ACTUALLY took place during the Glacial period, I assu
0374 t was, during a part at least of the period, ACTUALLY simultaneous throughout the world. Without s
0380 aturalised; and if the natives have not been ACTUALLY exterminated, their numbers have been greatl
0397 those of the nearest mainland, without being ACTUALLY the same species. Numerous instances could b
0427 us linnaeus, misled by external appearances, ACTUALLY classed an homopterous insect as a moth. We
0436 n being transformed into another; and we can ACTUALLY see in embryonic crustaceans and in many oth
0438 in the one case and legs in the other, have ACTUALLY been modified into skulls or jaws. Yet so st
0445 he eye, seemed almost to be the case; but on ACTUALLY measuring the old dogs and their six days ol
0466 ently domesticated productions. Man does not ACTUALLY produce variability; he only unintentionally
0257 t any other genus; but Gartner found that N. ACUMINATA, which is not a particularly distinct speci
0185 n profoundly modified. On the other hand, the ACUTEST observer by examining the dead body of the w
0425 posterous; and I might answer by the argumentum AD hominem, and ask what should be done if a perfe
0061 n can certainly produce great results, and can ADAPT organic beings to his own uses, through the c
0086 cotton trees. Natural Selection may modify and ADAPT the larva of an insect to a score of continge
0087 lation to the young. In social animals it will ADAPT the structure of each individual for the bene
0176 obably apply to both; and if we in imagination ADAPT a varying species to a very large area, we sh
0176 species to a very large area, we shall have to ADAPT two varieties to two large areas, and a third
0303 it might require a long succession of ages to ADAPT an organism to some new and peculiar line of
0472 number, with natural selection always ready to ADAPT the slowly varying descendants of each to any
0030 r domesticated races is that we see in them ADAPTATION, not indeed to the animal's or plant's own
```

Page **(Key Word)**

Page **(Key Word)**

```
0038 ous, how it is that our domestic races show ADAPTATION in their structure or in their habits to ma
0097 all, such flowers, there is a very curious ADAPTATION between the structure of the flower and the
0139 ot endure a damp climate. But the degree of ADAPTATION of species to the climates under which they
0140 c beings quite as much as, or more than, by ADAPTATION to particular climates. But whether or not
0140 particular climates. But whether or not the ADAPTATION be generally very close, we have evidence,
0141 orrid zones. Hence I am inclined to look at ADAPTATION to any special climate as a quality readily
0184 in nature. Can a more striking instance of ADAPTATION be given than that of a woodpecker for clim
0197 ought that the green colour was a beautiful ADAPTATION to hide this tree .frequenting bird from its
0197 vulture is generally looked at as a direct ADAPTATION for wallowing in putridity; and so it may b
0197 g mammals have been advanced as a beautiful ADAPTATION for aiding parturition, and no doubt they f
0261 eing evergreen and the other deciduous, and ADAPTATION to widely different climates, does not alwa
0392 er few relations are more striking than the ADAPTATION of hooked seeds for transport by the wool
0406 with the subsequent modification and better ADAPTATION of the colonists to their new homes. Summar
0415 erally been subjected to less change in the ADAPTATION of the species to their conditions of life.
0440 ater in life; but whenever it comes on, the ADAPTATION of the larva to its conditions of life is j
0468 which it comes into competition, or better ADAPTATION in however slight a degree to the surroundi
0060 ise in nature. How have all those exquisite ADAPTATIONS of one part of the organisation to another
0060 , been perfected? We see these beautiful co ADAPTATIONS most plainly in the woodpecker and misslet
0061 gentlest breeze; in short, we see beautiful ADAPTATIONS everywhere and in every part of the organi
0132 e produced the many striking and complex co ADAPTATIONS of structure between one organic being and
0206 them during long past periods of time: the ADAPTATIONS being aided in some cases by use and disus
0206 includes, through the inheritance of former ADAPTATIONS, that of Unity of Type. Chapter VII. Insti
0428 when whales are compared with fishes, being ADAPTATIONS in both classes for swimming through the w
0440 l as in the adult animal. From such special ADAPTATIONS, the similarity of the larvae or active em
0003 ts feet, tail, beak, and tongue, so admirably ADAPTED to catch insects under the bark of trees. In
0077 er beetle, the structure of its legs, so well ADAPTED for diving, allows it to compete with other
0081 ounded by barriers, into which new and better ADAPTED forms could not freely enter, we should then
0082 l the native inhabitants are now so perfectly ADAPTED to each other and to the physical conditions
0084 ctions: that they should be infinitely better ADAPTED to the most complex conditions of life, and
0095 aneously or one after the other, modified and ADAPTED in the most perfect manner to each other, by
0098 r the other towards it, the contrivance seems ADAPTED solely to ensure self fertilisation; and no
0104 ciently in checking the immigration of better ADAPTED organisms, after any physical change, such a
0104 e old inhabitants to struggle for, and become ADAPTED to, through modifications in their structure
0108 y very slow, and on the immigration of better ADAPTED forms having been checked. But the action of
0115 e commonly looked at as specially created and ADAPTED for their own country. It might, also, perha
0115 have belonged to a few groups more especially ADAPTED to certain stations in their new homes. But
0116 o physiologist doubts that a stomach by being ADAPTED to digest vegetable matter alone, or flesh a
0122 into some distinct country, or become quickly ADAPTED to some quite new station, in which child an
0122 t each stage of descent, so as to have become ADAPTED to many related places in the natural econom
0127 on, by assuring to the most vigorous and best ADAPTED males the greatest number of offspring. Sexu
0139 descent. It is notorious that each species is ADAPTED to the climate of its own home: species from
0140 itively know that these animals were strictly ADAPTED to their native climate, but in all ordinary
0142 iduals which are born with constitutions best ADAPTED to their native countries. In treatises on m
0144 rmal in their teeth. I know of no case better ADAPTED to show the importance of the laws of correl
0158 ted by natural and sexual selection, and thus ADAPTED for secondary sexual, and for ordinary speci
0170 led to struggle with each other, and the best ADAPTED to survive. Chapter VI. Difficulties on Theo
0173 the process of modification, each has become ADAPTED to the conditions of life of its own region,
0177 sing three varieties of sheep to be kept, one ADAPTED to an extensive mountainous region: a second
0179 uggle for life, it is clear that each is well ADAPTED in its habits to its place in nature. Look a
0182 ing classes as the Crustacea and Mollusca are ADAPTED to live on the land, and seeing that we have
0184 upply of insects were constant, and if better ADAPTED competitors did not already exist in the cou
0195 at first incredible that this could have been ADAPTED for its present purpose by successive slight
0202 s present purpose, with the poison originally ADAPTED to cause galls subsequently intensified, we
0206 nd inorganic conditions of life; or by having ADAPTED them during long past periods of time: the a
0224 exquisite structure of a comb, so beautifully ADAPTED to its end, without enthusiastic admiration.
0306 in confined, until some of the species became ADAPTED to a cooler climate, and were enabled to dou
0315 though the offspring of one species might be ADAPTED (and no doubt this has occurred in innumerab
0358 works, this or that plant is stated to be ill ADAPTED for wide dissemination; but for transport ac
0385 the sea, and subsequently become modified and ADAPTED to the fresh waters of a distant land. Some
0390 to admit that a sufficient number of the best ADAPTED plants and animals have not been created on
0391 their former homes, and have become mutually ADAPTED to each other; and when settled in their new
0402 elago, many even of the birds, though so well ADAPTED for flying from island to island, are distin
0427 distinct lines of descent, may readily become ADAPTED to similar conditions, and thus assume a clo
0428 s members of distinct classes have often been ADAPTED by successive slight modifications to live u
0430 gical, owing to the phascolomys having become ADAPTED to habits like those of a Rodent. The elder
0438 milar elements, many times repeated, and have ADAPTED them to the most diverse purposes. And as th
0439 larvae, the embryos are active, and have been ADAPTED for special lines of life. A trace of the la
0442 metamorphosis. The larvae of insects, whether ADAPTED to the most diverse and active habits, or qu
0447 may become, by a long course of modification, ADAPTED in one descendant to act as hands, in anothe
0453 ly that most parts and organs are exquisitely ADAPTED for certain purposes, tells us with equal pl
0472 w supposed to have been specially created and ADAPTED for that country, being beaten and supplante
0082 individuals of any of the species, by better ADAPTING them to their altered conditions, would tend
0127 eally thus acted in nature, in modifying and ADAPTING the various forms of life to their several c
0206 on. For natural selection acts by either now ADAPTING the varying parts of each being to its organ
0469 mit to this power, in slowly and beautifully ADAPTING each form to the most complex relations of l
0411 always used in classification. Analogical or ADAPTIVE characters. Affinities, general, complex and
0414 hole life of the being, are ranked as merely ADAPTIVE or analogical characters; but to the conside
0414 ications of these organs to mistake a merely ADAPTIVE for an essential character. So with plants,
0427 on between real affinities and analogical or ADAPTIVE resemblances. Lamarck first called attention
0427 we can clearly understand why analogical or ADAPTIVE character, although of the utmost importance
0430 pials are believed to be real and not merely ADAPTIVE, they are due on my theory to inheritance in
0433 nct group, we summarily reject analogical or ADAPTIVE characters, and yet use these same character
0456 de opposition in value between analogical or ADAPTIVE characters, and characters of true affinity:
0478 viceable than others for classification; why ADAPTIVE characters, though of paramount importance t
0485 p, community of type, paternity, morphology, ADAPTIVE characters, rudimentary and aborted organs,
0467 ccumulate them in any desired manner. He thus ADAPTS animals and plants for his own benefit or ple
```

Page **(Key Word)**

Page **(Key Word)**

```
0472  as natural selection acts by competition, it ADAPTS the inhabitants of each country only in relat
0009  l the choicest productions of the garden. I may ADD, that as some organisms will breed most freely
0015  ions, would be opposed to all experience. I may ADD, that when under nature the conditions of life
0027  ical sub species. In favour of this view, I may ADD, firstly, that C. livia, or the rock pigeon, h
0033  published on the subject; and the result, I may ADD, has been, in a corresponding degree, rapid an
0040  e of civilisation of the inhabitants, slowly to ADD to the characteristic features of the breed, w
0042  ent and formation of new breeds. Pigeons, I may ADD, can be propagated in great numbers and at a v
0046  of a tree. This philosophical naturalist, I may ADD, has also quite recently shown that the muscle
0053  f the varying species. Dr. Hooker permits me to ADD, that after having carefully read my manuscrip
0070  ave each other from utter destruction. I should ADD that the good effects of frequent intercrossin
0091  tercrossing we shall soon have to return. I may ADD, that, according to Mr. Pierce, there are two
0096  he reproduction of their kind. This view, I may ADD, was first suggested by Andrew Knight. We shal
0126  ect in the chapter on Classification, but I may ADD that on this view of extremely few of the more
0145  t subject to peloria; and become regular. I may ADD, as an instance of this, and of a striking cas
0149  entary and aborted organs; and I will here only ADD that their variability seems to be owing to th
0151  ase of hermaphrodite cirripedes; and I may here ADD, that I particularly attended to Mr. Waterhous
0169  nd analogous variation, such modifications will ADD to the beautiful and harmonious diversity of n
0199  ces of man, which are so strongly marked: I may ADD that some little light can apparently be throw
0218  cuckoo could be, and has been, generated. I may ADD that, according to Dr. Gray and to some other
0231  t ultimate economy of wax. It seems at first to ADD to the difficulty of understanding how the cel
0291  hat it was accumulated during subsidence. I may ADD, that the only ancient tertiary formation on t
0325  forms of life in various parts of Europe, they ADD, if struck by this strange sequence, we turn o
0392  s. if so, natural selection would often tend to ADD to the stature of herbaceous plants when growi
0481  the term of a hundred million years; it cannot ADD up and perceive the full effects of many sligh
0017  s. i do not dispute that these capacities have ADDED largely to the value of most of our domestica
0159  distinct species; and to these a third may be ADDED, namely, the common turnip. According to the
0273  variability would often continue and be super ADDED to that arising from the mere act of crossing
0299  est degree. Geological research, though it has ADDED numerous species to existing and extinct gene
0307  is known with accuracy. M. Barrande has lately ADDED another and lower stage to the Silurian syste
0339  t well displayed by them. Other cases could be ADDED, as the relation between the extinct and livi
0413  e creator, it seems to me that nothing is thus ADDED to our knowledge. Such expressions as that fa
0429  iking, that the discovery of Australia has not ADDED a single insect belonging to a new order; and
0429  le kingdom, as I learn from Dr. Hooker, it has ADDED only two or three orders of small size. In th
0486  esting subject for study than one more species ADDED to the infinitude of already recorded species
0200  of other species, as we see in the fang of the ADDER, and in the ovipositor of the ichneumon, by w
0082  as man can certainly produce great results by ADDING up in any given direction mere individual dif
0084  ; rejecting that which is bad, preserving and ADDING up all that is good; silently and insensibly
0225  se their old cocoons to hold honey, sometimes ADDING to them short tubes of wax, and likewise maki
0231  ons always piling up the cut away cement, and ADDING fresh cement, on the summit of the ridge. We
0240  n an intermediate condition. I may digress by ADDING, that if the smaller workers had been the mos
0304  rtiary series. This was a sore trouble to me, ADDING as I thought one more instance of the abrupt
0413  er those common to the dog genus, and then by ADDING a single sentence, a full description is give
0468  oduce within a short period a great result by ADDING up mere individual differences in his domesti
0115  re indigenous, and thus a large proportional ADDITION is made to the genera of these 'States. By co
0225  lls, in which the young are hatched, and, in ADDITION, some large cells of wax for holding honey.
0471  f natura non facit saltum, which every fresh ADDITION to our knowledge tends to make more strictly
0275  , melanism, deficiency of tail or horns, or ADDITIONAL fingers and toes; and do not relate to char
0030  ction: nature gives successive variations; man ADDS them up in certain directions useful to him. I
0227  honey; in the same way as the rude humble bee ADDS cylinders of wax to the circular mouths of her
0338  eaves the embryo almost unaltered, continually ADDS, in the course of successive generations, more
0155  sification. It would be almost superfluous to ADDUCE evidence in support of the above statement, t
0002  ussed in this volume on which facts cannot be ADDUCED, often apparently leading to conclusions dir
0199  ous differences between species. I might have ADDUCED for this same purpose the differences betwee
0248  r varieties, with the evidence from fertility ADDUCED by different hybridisers, or by the same aut
0397  just hatched young occasionally crawl on and ADHERE to the feet of birds roosting on the ground,
0145  upper petals; and that when this occurs, the ADHERENT nectary is quite aborted; when the colour is
0362  quite clean, I can show that earth sometimes ADHERES to them: in one instance I removed twenty tw
0386  ioned that earth occasionally, though rarely, ADHERES in some quantity to the feet and beaks of bi
0363  diterranean; and can we doubt that the earth ADHERING to their feet would sometimes include a few
0385  weed, I have twice seen these little plants ADHERING to its back; and it has happened to me, in r
0386  s (a fresh water shell like a limpet) firmly ADHERING to it; and a water beetle of the same family
0137  n habits, a reduction in their size with the ADHESION of the eyelids and growth of fur over them,
0262  and fundamental difference between the mere ADHESION of grafted stocks, and the union of the male
0192  in their size and the obliteration of their ADHESIVE glands. If all pedunculated cirripedes had b
0144  tube. Hard parts seem to affect the form of ADJOINING soft parts; it is believed by some authors
0147  ed through natural selection and another and ADJOINING part being reduced by this same process or
0147  owing to the excess of growth in another and ADJOINING part. I suspect, also, that some of the cas
0148  necessary compensation the reduction of some ADJOINING part. It seems to be a rule, as remarked by
0168  erhaps it tends to draw nourishment from the ADJOINING parts; and every part of the structure whic
0225  stands in close relation to the presence of ADJOINING cells; and the following view may, perhaps,
0225  r into the composition of the bases of three ADJOINING cells on the opposite side. In the series b
0226  ssarily enter into the construction of three ADJOINING cells. It is obvious that the Melipona save
0226  of building: for the flat walls between the ADJOINING cells are not double, but are of the same t
0226  at the same distance from the centres of the ADJOINING spheres in the other and parallel layer; th
0227  isms have been formed by the intersection of ADJOINING spheres in the same layer, she can prolong
0231  along the plane of intersection between two ADJOINING spheres. I have several specimens showing c
0232  can build up a wall intermediate between two ADJOINING spheres; but, as far as I have seen, they n
0232  ll a large part both of that cell and of the ADJOINING cells has been built. This capacity in bees
0234  t a little: for a wall in common even to two ADJOINING cells, would save some little wax. Hence it
0292  of elevation the area of the land and of the ADJOINING shoal parts of the sea will be increased, a
0402  n this subject: namely, that Madeira and the ADJOINING islet of Porto Santo possess many distinct
0453  r organ is of greater size relatively to the ADJOINING parts in the embryo, than in the adult; so
0226  erfectly flat surfaces, according as the cell ADJOINS two, three or more other cells. When one cel
0186  ye, with all its inimitable contrivances for ADJUSTING the focus to different distances, for admit
0042  e crossing with distinct species) those many ADMIRABLE varieties of the strawberry which have been
0048  haracterized as varieties in Mr. Wollaston's ADMIRABLE work, but which it cannot be doubted would
```

Page **(Key Word)**

Page ***(Key Word)***

```
0115  candolle has well remarked in his great and ADMIRABLE work, that floras gain by naturalisation, p
0242   and nature has, as I believe, effected this ADMIRABLE division of labour in the communities of an
0246  irs and works of those two conscientious and ADMIRABLE observers, Kolreuter and Gartner, who almos
0255  clusions are chiefly drawn up from Gartner's ADMIRABLE work on the hybridisation of plants. I have
0285  asterly memoir on this subject. Yet it is an ADMIRABLE lesson to stand on the North Downs and to l
0287  is admitted by every one. The remark of that ADMIRABLE palaeontologist, the late Edward Forbes, sh
0324  parts of the world, has greatly struck those ADMIRABLE observers, MM. de Verneuil and d'Archiac. A
0328  ve occurred in Europe. Mr. Prestwich, in his ADMIRABLE Memoirs on the eocene deposits of England a
0372  y allied crustaceans (as described in Dana's ADMIRABLE work), of some fish and other marine animal
0375  t in the intermediate torrid regions. In the ADMIRABLE Introduction to the Flora of New Zealand, b
0381   some of the most remarkable are stated with ADMIRABLE clearness by Dr. Hooker in his botanical wo
0391  and bird: and we know from Mr. J. M. Jones's ADMIRABLE account of Bermuda, that very many North Am
0395  e natural history of this archipelago by the ADMIRABLE zeal and researches of Mr. Wallace. I have
0398  ll the plants, as shown by Dr. Hooker in his ADMIRABLE memoir on the Flora of this archipelago. Th
0442  ases, as in that of Aphis, if we look to the ADMIRABLE drawings by Professor Huxley of the develop
0474  ght the principle of gradation throws on the ADMIRABLE architectural powers of the hive bee. Habit
0487  d, we shall surely be enabled to trace in an ADMIRABLE manner the former migrations of the inhabit
0003  r, with its feet, tail, beak, and tongue, so ADMIRABLY adapted to catch insects under the bark of
0406  ong been observed, and which has lately been ADMIRABLY discussed by Alph. de Candolle in regard to
0003   coadaptation which most justly excites our ADMIRATION. Naturalists continually refer to external
0224  ly adapted to its end, without enthusiastic ADMIRATION. We hear from mathematicians that bees have
0112  d principle that fanciers do not and will not ADMIRE a medium standard, but like extremes, they bo
0202  ect organ, the eye. If our reason leads us to ADMIRE with enthusiasm a multitude of inimitable con
0202  ay cause the death of some few members. If we ADMIRE the truly wonderful power of scent by which t
0202  es of many insects find their females, can we ADMIRE the production for this single purpose of tho
0202  sisters? It may be difficult, but we ought to ADMIRE the savage instinctive hatred of the queen be
0203  xorable principle of natural selection. If we ADMIRE the several ingenious contrivances, by which
0051  itting much variation, and the truth of this ADMISSION will often be disputed by other naturalists
0194  f natura non facit saltum. We meet with this ADMISSION in the writings of almost every experienced
0358  d other such facts seem to me opposed to the ADMISSION of such prodigious geographical revolutions
0358  universally volcanic composition favour the ADMISSION that they are the wrecks of sunken continen
0019  eculiar breeds of cattle, sheep, etc., we must ADMIT that many domestic breeds have originated in
0019  omestic dogs of the whole world, which I fully ADMIT have probably descended from several wild spe
0020  everal domestic races by this process, we must ADMIT the former existence of the most extreme form
0029  ediate links in the long lines of descent, yet ADMIT that many of our domestic races have descende
0062  icultural knowledge. Nothing is easier than to ADMIT in words the truth of the universal struggle
0108  will always act with extreme slowness, I fully ADMIT. Its action depends on there being places in
0167  wn parents, but other species of the genus. To ADMIT this view is, as it seems to me, to reject a
0188  t, ought not to hesitate to go further, and to ADMIT that a structure even as perfect as the eye o
0191  mplement to the swimbladder. All physiologists ADMIT that the swimbladder is homologous, or ideall
0198  mestic breeds, which nevertheless we generally ADMIT to have arisen through ordinary generation, w
0199   be absolutely fatal to my theory. Yet I fully ADMIT that many structures are of no direct use to
0239   principle of natural selection, when I do not ADMIT that such wonderful and well established fact
0261  acity is a specially endowed quality, but will ADMIT that it is incidental on differences in the l
0268  and yield perfectly fertile offspring. I fully ADMIT that this is almost invariably the case. But
0282   a revolution in natural science, yet does not ADMIT how imcomprehensibly vast have been the past
0288   taking a most erroneous view, when we tacitly ADMIT to ourselves that sediment is being deposited
0316  strongly opposed to such views as I maintain) ADMIT its truth; and the rule strictly accords with
0320  peat what I published in 1845, namely, that to ADMIT that species generally become rare before the
0320  hen it ceases to exist, is much the same as to ADMIT that sickness in the individual is the foreru
0331  re recent members of the same classes, we must ADMIT that there is some truth in the remark. Let u
0357  several intervening oceanic islands. I freely ADMIT the former existence of many islands, now bur
0374  inct evidence to the contrary, we may at least ADMIT as probable that the glacial action was simul
0389  . I have already stated that I cannot honestly ADMIT Forbes's view on continental extensions, whic
0389  f good Hope or in Australia, we must, I think, ADMIT that something quite independently of any dif
0390  reation of each separate species, will have to ADMIT that a sufficient number of the best adapted
0459  perably great, cannot be considered real if we ADMIT the following propositions, namely, that grad
0465  at the geological record is imperfect all will ADMIT; but that it is imperfect to the degree which
0465  egree which I require, few will be inclined to ADMIT. If we look to long enough intervals of time,
0475  es have been produced by secondary laws. If we ADMIT that the geological record is imperfect in an
0476  t. Looking to geographical distribution, if we ADMIT that there has been during the long course of
0481  he chief cause of our natural unwillingness to ADMIT that one species has given birth to other and
0482   to me a strange conclusion to arrive at. They ADMIT that a multitude of forms, which till lately
0482  l characteristic feature of true species, they ADMIT that these have been produced by variation, b
0482  ich are those produced by secondary laws. They ADMIT variation as a vera causa in one case, they a
0485   manner as these naturalists treat genera, who ADMIT that genera are merely artificial combination
0272  s are more variable than hybrids; but Gartner ADMITS that hybrids from species which have long bee
0272  iking instances of this fact. Gartner further ADMITS that hybrids between very closely allied spec
0389  on equal continental areas: Alph. de Candolle ADMITS this for plants, and Wollaston for insects. I
0390  exterminated many native productions. He who ADMITS the doctrine of the creation of each separate
0468  es in his domestic productions; and every one ADMITS that there are at least individual difference
0013  trange and commoner deviations may be freely ADMITTED to be inheritable. Perhaps the correct way o
0016  s in a state of nature. I think this must be ADMITTED, when we find that there are hardly any dome
0049  aries, or Ireland, be sufficient? It must be ADMITTED that many forms, considered by highly compet
0060  he existence of any well marked varieties be ADMITTED. But the mere existence of individual variab
0156  e only two other remarks. I think it will be ADMITTED, without my entering on details, that second
0156  s are very variable; I think it also will be ADMITTED that species of the same group differ from e
0201  n one which seems to me of any weight. It is ADMITTED that the rattlesnake has a poison fang for i
0209  e been thus acquired. It will be universally ADMITTED that instincts are as important as corporeal
0216  l of all known instincts. It is now commonly ADMITTED that the more immediate and final cause of t
0285  e denudation of the Weald. Though it must be ADMITTED that the denudation of the Weald has been a
0287  tological collections are very imperfect, is ADMITTED by every one. The remark of that admirable p
0341  e descendants. This cannot for an instant be ADMITTED. These huge animals have become wholly extin
0352  n the agency of a miracle. It is universally ADMITTED, that in most cases the area inhabited by a
0357  used by Forbes are to be trusted, it must be ADMITTED that scarcely a single island exists which h
0357  lls standing over them. Whenever it is fully ADMITTED, as I believe it will some day be, that each
```

Page ***(Key Word)***

Page ***(Key Word)***

```
0358  essitated on the view advanced by Forbes and ADMITTED by his many followers. The nature and relati
0374  thern extremity of the continent. If this be ADMITTED, it is difficult to avoid believing that the
0396  e been more complete; and if modification be ADMITTED, all the forms of life would have been more
0416  ogical importance, but which are universally ADMITTED as highly serviceable in the definition of w
0418  si and this doctrine has very generally been ADMITTED as true. The same fact holds good with flowe
0435  pelessness of the attempt has been expressly ADMITTED by Owen in his most interesting work on the
0444  ent. These two principles, if their truth be ADMITTED, will, I believe, explain all the above spec
0460  many graduated steps. There are, it must be  ADMITTED, cases of special difficulty on the theory o
0469  sides such differences, all naturalists have ADMITTED the existence of varieties, which they think
0478  islands to the African mainland. It must be  ADMITTED that these facts receive no explanation on t
0484  ecies, or when analogous views are generally ADMITTED, we can dimly foresee that there will be a c
0051  ut he will succeed in this at the expense of  ADMITTING much variation, and the truth of this admis
0186  usting the focus to different distances, for ADMITTING different amounts of light, and for the cor
0357  best of my judgment we are not authorized in ADMITTING such enormous geographical changes within t
0408  s. if the difficulties be not insuperable in ADMITTING that in the long course of time the individ
0481  tinct species, is that we are always slow in ADMITTING any great change of which we do not see the
0253  n, in opposition to the constantly repeated  ADMONITION of every breeder. And in this case, it is n
0458  e of descent, that I should without hesitation ADOPT this view, even if it were unsupported by oth
0459  al selection may be extended. Effects of its ADOPTION on the study of Natural history. Concluding
0086  the horns of our sheep and cattle when nearly ADULT; so in a state of nature, natural selection w
0086  the laws of correlation, the structure of the ADULT; and probably in the case of those insects wh
0086  r larvae. So, conversely, modifications in the ADULT will probably often affect the structure of t
0127  dify the egg, seed, or young, as easily as the ADULT. Amongst many animals, sexual selection will
0143  fely be concluded, affect the structure of the ADULT; in the same manner as any malconformation af
0143  eriously affects the whole organisation of the ADULT. The several parts of the body which are homo
0338  uture chapter I shall attempt to show that the ADULT differs from its embryo, owing to variations
0338  e generations, more and more difference to the ADULT. Thus the embryo comes to be left as a sort o
0418  f equal importance with those derived from the ADULT, for our classifications of course include al
0418  re important for this purpose than that of the ADULT, which alone plays its full part in the econo
0424  and hermaphrodites of certain cirripedes, when ADULT, and yet no one dreams of separating them. Th
0424  h they may differ from each other and from the ADULT; as he likewise includes the so called altern
0440  is just as perfect and as beautiful as in the  ADULT animal. From such special adaptations, the si
0440  h, or even more, from each other than do their ADULT parents. In most cases, however, the larvae,
0442  rences in structure between the embryo and the ADULT, and likewise a close similarity in the embry
0442  does not at any period differ widely from the  ADULT; thus Owen has remarked in regard to cuttle f
0442  erence in structure between the embryo and the ADULT; of parts in the same individividual embryo, wh
0446  proportions, almost exactly as much as in the  ADULT state. The two principles above given seem to
0448  have seen to be the case with cirripedes. The  ADULT might become fitted for sites or habits, in w
0449  important for classification than that of the  ADULT. For the embryo is the animal in its less mod
0449  of descent, however much the structure of the  ADULT may have been modified and obscured; we have
0453  the adjoining parts in the embryo, than in the ADULT; so that the organ at this early age is less
0453  ntary. Hence, also, a rudimentary organ in the ADULT, is often said to have retained its embryonic
0455  embryo, and their lesser relative size in the  ADULT. But if each step of the process of reduction
0457  mologous parts or organs, though fitted in the ADULT members for purposes as different as possible
0479  so closely alike, and should be so unlike the  ADULT forms. We may cease marvelling at the embryo
0385  e immediately killed by sea water, as are the  ADULTS. I could not even understand how some natural
0447  mble each other much more closely than do the  ADULTS, just as we have seen in the case of pigeons.
0154  case applicable, which most naturalists would  ADVANCE, namely, that specific characters are more v
0194  riations: she can never take a leap, but must  ADVANCE by the shortest and slowest steps. Organs of
0142  ieties have not been produced, has even been   ADVANCED; for it is now as tender as ever it was, as
0147  of the cases of compensation which have been   ADVANCED, and likewise some other facts, may be merge
0197  res in the skulls of young mammals have been   ADVANCED as a beautiful adaptation for aiding parturi
0242  descendants. I am surprised that no one has    ADVANCED this demonstrative case of neuter insects, a
0248  space here to enter on details, the evidence   ADVANCED by our best botanists on the question whethe
0355  e word variety for species) from that lately   ADVANCED in an ingenious paper by Mr. Wallace, in whi
0358  cent period, as are necessitated on the view   ADVANCED by Forbes and admitted by his many followers
0379  reater numbers, and having consequently been   ADVANCED through natural selection and competition to
0385  not be jarred off, though at a somewhat more   ADVANCED age they would voluntarily drop off. These j
0459  lated. That many and grave objections may be   ADVANCED against the theory of descent with modificat
0244  equences of one general law, leading to the    ADVANCING of all organic beings, namely, multiply, v
0069  those which have got least food through the    ADVANCING winter, which will suffer most. When we tra
0201  but they are now rapidly yielding before the   ADVANCING legions of plants and animals introduced fr
0147  er fitted to be wafted further, might get an   ADVANGAGE over those producing seed less fitted for d
0077  ed to the elements of air and water. Yet the  ADVANTAGE of plumed seeds no doubt stands in the clos
0078  ng in number, we should have to give it some  ADVANTAGE over its competitors, or over the animals w
0078  with respect to climate would clearly be an   ADVANTAGE to our plant; but we have reason to believe
0078  country; for we should have to give it some   ADVANTAGE over a different set of competitors or enem
0078  try in our imagination to give any form some  ADVANTAGE over another. Probably in no single instanc
0081  ossibly survive) that individuals having any  ADVANTAGE, however slight, over others, would have th
0082  its of one inhabitant would often give it an  ADVANTAGE over others; and still further modification
0082  kind would often still further increase the   ADVANTAGE. No country can be named in which all the n
0083  at the natives might have been modified with  ADVANTAGE, so as to have better resisted such intrude
0087  ucture of one species, without giving it any  ADVANTAGE, for the good of another species; and thoug
0087  l grown pigeon very short for the bird's own  ADVANTAGE, the process of modification would be very
0089  had, in successive generations, some slight   ADVANTAGE over other males, in their weapons, means o
0093  ess might commence. No naturalist doubts the  ADVANTAGE of what has been called the physiological d
0095  ar to the hive bee. Thus it might be a great  ADVANTAGE to the hive bee to have a slightly longer o
0095  ome rare in any country, it might be a great  ADVANTAGE to the red clover to have a shorter or more
0115  d to be modified, in order to have gained an  ADVANTAGE over the other natives; and we may, I think
0115  ces, would have been profitable to them. The  ADVANTAGE of diversification in the inhabitants of th
0121  sarily acts by the selected form having some  ADVANTAGE in the struggle for life over other forms,
0122  , so that they must originally have had some  ADVANTAGE over most of the other species of the genus
0125  selection acts through one form having some   ADVANTAGE over other forms in the struggle for existe
0125  chiefly act on those which already have some  ADVANTAGE; and the largeness of any group shows that
0125  s have inherited from a common ancestor some  ADVANTAGE in common. Hence, the struggle for the prod
0137  h of fur over them, might in such case be an  ADVANTAGE; and if so, natural selection would constan
```

Page ***(Key Word)***

Page **************************************(Key Word)**************************************

```
0148  h effected by slow steps, would be a decided  ADVANTAGE to each successive individual of the specie
0149  rt of each being, solely through and for its   ADVANTAGE. Rudimentary parts, it has been stated by s
0157  part would, it is highly probable, be taken    ADVANTAGE of by natural and sexual selection, in orde
0169  of the organisation has generally been taken   ADVANTAGE of in giving secondary sexual differences t
0176  m inhabiting larger areas, will have a great    ADVANTAGE over the intermediate variety, which exists
0183  rfection, so as to have given them a decided    ADVANTAGE over other animals in the battle for life.
0186  her in habits or structure, and thus gain an    ADVANTAGE over some other inhabitant of the country,
0190  al selection might easily specialise, it any    ADVANTAGE were thus gained, a part or organ, which ha
0194  orking for the good of each being and taking    ADVANTAGE of analogous variations, has sometimes modi
0194  for natural selection can act only by taking    ADVANTAGE of slight successive variations; she can ne
0195  ange into new pastures and thus gain a great    ADVANTAGE. It is not that the larger quadrupeds are a
0196  f animals having a will, to give one male an    ADVANTAGE in fighting with another or in charming the
0196  nown causes, it may at first have been of no    ADVANTAGE to the species, but may subsequently have b
0196  pecies, but may subsequently have been taken   ADVANTAGE of by the descendants of the species under
0197  of growth, and have been subsequently taken    ADVANTAGE of by the plant undergoing further modifica
0197  from the laws of growth, and has been taken    ADVANTAGE of in the parturition of the higher animals
0200  oughout nature one species incessantly takes   ADVANTAGE of, and profits by, the structure of anothe
0204  isting in greater numbers, will have a great    ADVANTAGE over the less numerous intermediate variety
0205  s to a species, have been subsequently taken   ADVANTAGE of by the still further modified descendant
0211  inct species, yet each species tries to take   ADVANTAGE of the instincts of others, as each takes a
0211  ge of the instincts of others, as each takes   ADVANTAGE of the weaker bodily structure of others. S
0217  , or if the young were made more vigorous by   ADVANTAGE having been taken of the mistaken maternal
0217  ld birds or the fostered young would gain an    ADVANTAGE. And analogy would lead me to believe, that
0218  dy made and stored by another sphex it takes   ADVANTAGE of the prize, and becomes for the occasion
0219  making an occasional habit permanent, if of    ADVANTAGE to the species, and if the insect whose nes
0234  in this case be no doubt that it would be an    ADVANTAGE to our humble bee, if a slight modification
0235  explained by natural selection having taken    ADVANTAGE of numerous, successive, slight modificatio
0243  of other animals, but that each animal takes   ADVANTAGE of the instincts of others; that the canon
0245  lity of hybrids could not possibly be of any    ADVANTAGE to them, and therefore could not have been
0301  o one place, but if possessed of any decided   ADVANTAGE, or when further modified and improved, the
0303  and a few species had thus acquired a great    ADVANTAGE over other organisms, a comparatively short
0314  ll others. Whether such variability be taken   ADVANTAGE of by natural selection, and whether the va
0320  s, is produced and maintained by having some   ADVANTAGE over those with which it comes into competi
0325  ed by new varieties arising, which have some   ADVANTAGE over older forms; and those forms, which ar
0325  ms, which are already dominant, or have some   ADVANTAGE over the other forms in their own country,
0327  inheritance, and to having already had some    ADVANTAGE over their parents or over other species: t
0337  ach new species is formed by having had some   ADVANTAGE in the struggle for life over other and pre
0351  s an independent property, and will be taken   ADVANTAGE of by natural selection, only so far as it
0392  rbaceous plants alone, might readily gain an   ADVANTAGE by growing taller and taller and overtoppin
0402  nication. Undoubtedly if one species has any   ADVANTAGE whatever over another, it will in a very br
0403  european land shells which no doubt had some   ADVANTAGE over the indigenous species. From these con
0468  remote in the scale of nature. The slightest   ADVANTAGE in one being, at any age or during any seas
0468  n the charms of the males: and the slightest   ADVANTAGE will lead to victory. As geology plainly pr
0093  ouri hence we may believe that it would be    ADVANTAGEOUS to a plant to produce stamens alone in one
0094  aration of the sexes of our plant would be    ADVANTAGEOUS on the principle of the division of labour
0099  nal cross with a distinct individual being    ADVANTAGEOUS or indispensable! If several varieties of
0109  the preservation of variations in some way    ADVANTAGEOUS, which consequently endure. But as from th
0127  structure, constitution, and habits, to be    ADVANTAGEOUS to them, I think it would be a most extrao
0145  to attract insects, whose agency is highly    ADVANTAGEOUS in the fertilisation of plants of these tw
0146  d, as it may at first appear' and if it be    ADVANTAGEOUS, natural selection may have come into play
0146  ems impossible that they can be in any way    ADVANTAGEOUS to the plant: yet in the Umbelliferae thes
0201  each part, each will be found on the whole   ADVANTAGEOUS. After the lapse of time, under changing c
0205  the laws of growth, and at first in no way    ADVANTAGEOUS to a species, have been subsequently taken
0212  habits in certain species, which might, if    ADVANTAGEOUS to the species, give rise, through natural
0223  ies which had seized them, if it were more    ADVANTAGEOUS to this species to capture workers than to
0234  ence it would continually be more and more   ADVANTAGEOUS to our humble bee, if she were to make her
0234  d. again, from the same cause, it would be    ADVANTAGEOUS to the Melipona, if she were to make her c
0238  certain members of the community, has been   ADVANTAGEOUS to the community: consequently the fertile
0054  t degree modified, will still inherit those   ADVANTAGES that enabled their parents to become domina
0090  ence, or charms: and have transmitted these  ADVANTAGES to their male offspring. Yet, I would not w
0114  he closest competition with each other, the   ADVANTAGES of diversification of structure, with the a
0118  modified forms, will tend to inherit those   ADVANTAGES which made their common parent (A) more num
0118  will likewise partake of those more general  ADVANTAGES which made the genus to which the parent sp
0119  rge genus, will tend to partake of the same  ADVANTAGES which made their parent successful in life,
0122  ll probably have inherited some of the same  ADVANTAGES: they have also been modified and improved
0179  through natural selection and gain further   ADVANTAGES. Lastly, looking not to any one time, but t
0428  g to the larger genera, tend to inherit the  ADVANTAGES, which made the groups to which they belong
0053  tson, to whom I am much indebted for valuable ADVICE and assistance on this subject, soon convince
0141  ve, both from analogy, and from the incessant ADVICE given in agricultural works, even in the anci
0002  y sketch of 1844, honoured me by thinking it  ADVISABLE to publish, with Mr. Wallace's excellent me
0261  but not in a reversed direction. It will be   ADVISABLE to explain a little more fully by an exampl
0027  iest known record of pigeons is in the fifth  AEGYPTIAN dynasty, about 3000 B.C., as was pointed ou
0418  ch as those for propelling the blood, or for  AERATING it, or those for propagating the race, are f
0184  ich never climbs a tree! Petrels are the most AERIAL and oceanic of birds, yet in the quiet Sounds
0394  rial mammals do not occur on oceanic islands, AERIAL mammals do occur on almost every island. New
0396  trial mammals notwithstanding the presence of AERIAL bats, the singular proportions of certain ord
0409  islands possess their own peculiar species of AERIAL mammals or bats. We can see why there should
0415  ce of albumen, in the imbricate or valvular    AESTIVATION. Any one of these characters singly is fre
0045  tions. These individual differences generally AFFECT what naturalists consider unimportant parts;
0058  orms, and the occurrence of such links cannot  AFFECT the actual characters of the forms which they
0073  ica) the vegetation: this again would largely  AFFECT the insects; and this, as we just have seen i
0081  hange of climate itself, would most seriously  AFFECT many of the others. If the country were open
0086  ure insect. These modifications will no doubt  AFFECT, through the laws of correlation, the structu
0086  odifications in the adult will probably often  AFFECT the structure of the larva; but in all cases
0103  trict to another. The intercrossing will most  AFFECT those animals which unite for each birth, whi
0143  g or larva, will, it may safely be concluded,  AFFECT the structure of the adult; in the same manne
```

Page **************************************(Key Word)**************************************

Page **********************************(Key Word)**********************************

```
0144 f the corolla into a tube. Hard parts seem to AFFECT the form of adjoining soft parts; it is belie
0168 in hard parts and in external parts sometimes AFFECT softer and internal parts. When one part is l
0168 s of structure at an early age will generally AFFECT parts subsequently developed; and there are v
0198 eds; and a mountainous country would probably AFFECT the hind limbs from exercising them more, and
0198 ffected. The shape, also, of the pelvis might AFFECT by pressure the shape of the head of the youn
0286 is source of doubt probably would not greatly AFFECT the estimate as applied to the western extrem
0447 odifying an organ, such influence will mainly AFFECT the mature animal, which has come to its full
0455 at the same age, and consequently will seldom AFFECT or reduce it in the embryo. Thus we can under
0008 and female reproductive elements having been AFFECTED prior to the act of conception. Several reas
0009 aving their reproductive system so seriously AFFECTED by unperceived causes as to fail in acting,
0009 their reproductive system has not been thus AFFECTED; so will some animals and plants withstand d
0010 we see that the treatment of the parent has AFFECTED a bud or offset, and not the ovules or polle
0010 he ovules or pollen, or to both, having been AFFECTED by the treatment of the parent prior to the
0010 ndividuals exposed to certain conditions are AFFECTED in the same way, the change at first appears
0012 rs that white sheep and pigs are differently AFFECTED from coloured individuals by certain vegetab
0081 ns neither useful nor injurious would not be AFFECTED by natural selection, and would be left a fl
0094 a complete separation of the sexes would be AFFECTED. Let us now turn to the nectar feeding insec
0132 e male and female sexual elements seem to be AFFECTED before that union takes place which is to fo
0132 y differ essentially from an ovule, is alone AFFECTED. But why, because the reproductive system is
0198 nt limbs and even the head would probably be AFFECTED. The shape, also, of the pelvis might affect
0206 some cases by use and disuse, being slightly AFFECTED by the direct action of the external conditi
0214 it is known that a cross with a bulldog has AFFECTED for many generations the courage and obstina
0242 e members of a community could possibly have AFFECTED the structure or instincts of the fertile me
0245 sterility various in degree, not universal, AFFECTED by close interbreeding, removed by domestica
0248 t the fertility of pure species is so easily AFFECTED by various circumstances, that for all pract
0256 first crosses and of hybrids, is more easily AFFECTED by unfavourable conditions, than is the fert
0264 to have their reproductive systems seriously AFFECTED. This, in fact, is the great bar to the dome
0265 h, the male element is the most liable to be AFFECTED; but sometimes the female more than the male
0265 r reproductive systems having been specially AFFECTED, though in a lesser degree than when sterili
0265 pendently of the general state of health, is AFFECTED by sterility in a very similar manner. In th
0273 ot had their reproductive systems in any way AFFECTED, and they are not variable; but hybrids them
0273 es have their reproductive systems seriously AFFECTED, and their descendants are highly variable.
0291 evel, and apparently these oscillations have AFFECTED wide spaces. Consequently formations rich in
0328 have reason to believe that large areas are AFFECTED by the same movement, it is probable that st
0328 e, and that large areas have invariably been AFFECTED by the same movements. When two formations h
0336 orms of the inhabitants of the sea have been AFFECTED. On the theory of descent, the full meaning
0337 t doubt that this process of improvement has AFFECTED in a marked and sensible manner the organisa
0373 ntain vegetation, that Siberia was similarly AFFECTED. Along the Himalaya, at points 900 miles apa
0407 d, which I am fully convinced simultaneously AFFECTED the whole world, or at least great meridiona
0443 male and female sexual elements having been AFFECTED by the conditions to which either parent, or
0443 hen the horns of cross bred cattle have been AFFECTED by the shape of the horns of either parent.
0462 climatal and geographical changes which have AFFECTED the earth during modern periods; and such ch
0134 rked, they seem to play an important part in AFFECTING the reproductive system, and in thus induci
0143 t; in the same manner as any malconformation AFFECTING the early embryo, seriously affects the who
0163 one curious and complex case, not indeed as AFFECTING any important character, but from occurring
0435 in some way to the modified form, but often AFFECTING by correlation of growth other parts of the
0443 ly assumed, perhaps from monstrosities often AFFECTING the embryo at a very early period, that sli
0132 ston is convinced that residence near the sea AFFECTS their colours. Moquin Tandon gives a list of
0143 rmation affecting the early embryo, seriously AFFECTS the whole organisation of the adult. The sev
0198 l observers are convinced that a damp climate AFFECTS the growth of the hair, and that with the ha
0277 allied to that sterility which so frequently AFFECTS pure species, when their natural conditions
0484 tance as that the same poison often similarly AFFECTS plants and animals; or that the poison secre
0003 that a naturalist, reflecting on the mutual AFFINITIES of organic beings, on their embryological r
0005 thirteenth, their classification or mutual AFFINITIES, both when mature and in an embryonic condi
0124 only in a slight degree. In this case, its AFFINITIES to the other fourteen new species will be o
0124 en see that the diagram throws light on the AFFINITIES of extinct beings, which, though generally
0128 se principles, I believe, the nature of the AFFINITIES of all organic beings may be explained. It
0129 e have seen illustrated in the diagram. The AFFINITIES of all the beings of the same class have so
0130 which in some small degree connects by its AFFINITIES two large branches of life, and which has a
0138 close similarity in their organisation and AFFINITIES might have been expected; but, as Schiodte
0138 till to see in the cave animals of America, AFFINITIES to the other inhabitants of that continent,
0139 ult to give any rational explanation of the AFFINITIES of the blind cave animals to the other inha
0193 of which several are widely remote in their AFFINITIES. Generally when the same organ appears in s
0305 hat certain much older fishes, of which the AFFINITIES are as yet imperfectly known, are really te
0312 forms of life throughout the world. On the AFFINITIES of extinct species to each other and to liv
0329 esponding stages in the two regions. On the AFFINITIES of extinct species to each other, and to li
0329 living forms. Let us now look to the mutual AFFINITIES of extinct and living species. They all fal
0330 l, as the lepidosiren, is discovered having AFFINITIES directed towards very distinct groups. Yet
0333 , the main facts with respect to the mutual AFFINITIES of the extinct forms of life to each other
0334 two series, first according to their mutual AFFINITIES and then according to their periods of exis
0380 view here given in regard to the range and AFFINITIES of the allied species which live in the nor
0411 of natural selection. Chapter XIII. Mutual AFFINITIES of Organic Beings. Morphology. Embryology.
0411 ication. Analogical or adaptive characters. AFFINITIES, general, complex and radiating. Extinction
0414 fording very clear indications of its true AFFINITIES. We are least likely in the modifications o
0419 s are often plainly influenced by chains of AFFINITIES. Nothing can be easier than to define a num
0422 present in a series, on a flat surface, the AFFINITIES which we discover in nature amongst the bei
0423 nguages, extinct and modern, by the closest AFFINITIES, and would give the filiation and origin of
0427 the very important distinction between real AFFINITIES and analogical or adaptive resemblances. La
0427 der is compared with another, but give true AFFINITIES when the members of the same class or order
0430 r observations on the general nature of the AFFINITIES of distinct orders of plants. On the princi
0431 stand the excessively complex and radiating AFFINITIES by which all the members of the same family
0431 , without the aid of a diagram, the various AFFINITIES which they perceive between the many living
0432 ed together by a long, but broken, chain of AFFINITIES. Extinction has only separated groups: it h
0433 ins that great and universal feature in the AFFINITIES of all organic beings, namely, their subord
0434 by the most complex and radiating lines of AFFINITIES. We shall never, probably, disentangle the
0434 obably, disentangle the inextricable web of AFFINITIES between the members of any one class; but w
```

Page **********************************(Key Word)**********************************

Page ***************************************(Key Word)***

```
0456 complex, radiating, and circuitous lines of AFFINITIES into one grand system: the rules followed a
0478 rinciples we see how it is, that the mutual AFFINITIES of the species and genera within each class
0479 ters are the most valuable of all. The real AFFINITIES of all organic beings are due to inheritanc
0483 sses can be connected together by chains of AFFINITIES, and all can be classified on the same prin
0256 the same conditions. By the term systematic AFFINITY is meant, the resemblance between species in
0257 hem, is largely governed by their systematic AFFINITY. This is clearly shown by hybrids never havi
0257 y, but the correspondence between systematic AFFINITY and the facility of crossing is by no means
0258 n completely independent of their systematic AFFINITY, or of any recognisable difference in their
0259 s is not always governed by their systematic AFFINITY or degree of resemblance to each other. This
0261 fting, the capacity is limited by systematic AFFINITY, for no one has been able to graft trees too
0261 y no means absolutely governed by systematic AFFINITY. Although many distinct genera within the sa
0263 ent, as might have been expected, systematic AFFINITY, by which every kind of resemblance and diss
0265 ncy goes to a certain extent with systematic AFFINITY, for whole groups of animals and plants are
0276 terility does not strictly follow systematic AFFINITY, but is governed by several curious and comp
0277 certain extent, parallel with the systematic AFFINITY of the forms which are subjected to experime
0277 are subjected to experiment; for systematic AFFINITY attempts to express all kinds of resemblance
0335 arranged as well as they could be in serial AFFINITY, this arrangement would not closely accord w
0346 physical conditions. Importance of barriers. AFFINITY of the productions of the same continent. Ce
0349 included in the foregoing statements, is the AFFINITY of the productions of the same continent or
0396 tween the depth of the sea and the degree of AFFINITY of the the mammalian inhabitants of islands
0397 gard to the inhabitants of islands, is their AFFINITY to those of the nearest mainland, without be
0398 sed to have been created here: yet the close AFFINITY of most of these birds to American species i
0398 , and nowhere else, bear so plain a stamp of AFFINITY to those created in America? There is nothin
0399 the commencement of the Glacial period. The AFFINITY, which, though feeble, I am assured by Dr. H
0399 se, and is at present inexplicable: but this AFFINITY is confined to the plants, and will, I do no
0408 in so mysterious a manner linked together by AFFINITY, and are likewise linked to the extinct bein
0416 e highly serviceable in exhibiting the close AFFINITY between Ruminants and Pachyderms. Robert Bro
0416 mportant an aid in determining the degree of AFFINITY of this strange creature to birds and reptil
0420 s which naturalists consider as showing true AFFINITY between any two or more species, are those w
0425 there will certainly be close resemblance or AFFINITY. As descent has universally been used in cla
0428 ke limbs serve as characters exhibiting true AFFINITY between the several members of the whale fam
0430 elonging to one group of animals exhibits an AFFINITY to a quite distinct group, this affinity in
0430 an affinity to a quite distinct group, this AFFINITY in most cases is general and not special; th
0430 ecies more than to another. As the points of AFFINITY of the bizcacha to Marsupials are believed t
0431 related to each other by circuitous lines of AFFINITY of various lengths (as may be seen in the di
0456 adaptive characters, and characters of true AFFINITY; and other such rules: all naturally follow
0485 n interest. The terms used by naturalists of AFFINITY, relationship, community of type, paternity,
0240 ted specimens of these workers, I can AFFIRM that the eyes are far more rucimentary in the
0166 , and was, as we have seen, answered in the AFFIRMATIVE. What now are we to say to these several f
0045 ferences are highly important for us, as they AFFORD materials for natural selection to accumulate
0110 , showing that it is the common species which AFFORD the greatest number of recorded varieties, or
0308 but not one oceanic island is as yet known to AFFORD even a remnant of any palaeozoic or secondary
0422 logical arrangement of the races of man would AFFORD the best classification of the various langua
0440 w of common embryonic resemblance. Cirripedes AFFORD a good instance of this: even the illustrious
0463 y does not every collection of fossil remains AFFORD plain evidence of the gradation and mutation
0470 over, the species of the larger genera, which AFFORD the greater number of varieties or incipient
0004 ugh it be, of variation under domestication, AFFORDED the best and safest clue. I may venture to e
0038 gion inhabited by quite uncivilised man, has AFFORDED us a single plant worth culture. It is not t
0162 intermediate form. But the best evidence is AFFORDED by parts or organs of an important and unifo
0378 e most humid and hottest districts will have AFFORDED an asylum to the tropical natives. The mount
0378 he long line of the Cordillera, seem to have AFFORDED two great lines of invasion: and it is a str
0481 ical record is so perfect that it would have AFFORDED us plain evidence of the mutation of species
0414 food of an animal, I have always regarded as AFFORDING very clear indications of its true affiniti
0248 be shown that neither sterility nor fertility AFFORDS any clear distinction between species and va
0366 l period, which, as we shall immediately see, AFFORDS a simple explanation of these facts. We have
0034 ed by passages in Pliny. The savages in South AFRICA match their draught cattle by colour, as do s
0034 are valued by the negroes of the interior of AFRICA who have not associated with Europeans. Some
0240 same nest of the driver ant (Anomma) of West AFRICA. The reader will perhaps best appreciate the
0306 were enabled to double the southern capes of AFRICA or Australia, and thus reach other and distan
0340 rmerly more closely related in its mammals to AFRICA than it is at the present time. Analogous fac
0347 pare large tracts of land in Australia, South AFRICA, and western South America, between latitudes
0347 n they are to the productions of Australia or AFRICA under nearly the same climate. Analogous fact
0347 ference between the inhabitants of Australia, AFRICA, and South America under the same latitude: f
0348 ng over a hemisphere we come to the shores of AFRICA; and over this vast space we meet with no wel
0349 ands of the Pacific and the eastern shores of AFRICA, on almost exactly opposite meridians of long
0349 y a true ostrich or emeu, like those found in AFRICA and Australia under the same latitude, to the
0357 t recently have been connected with Europe or AFRICA, and Europe likewise with America. Other auth
0375 been discovered in the intertropical parts of AFRICA. On the Himalaya, and on the isolated mountai
0398 cape of Verde Islands are related to those of AFRICA, like those of the Galapagos to America. I be
0399 m america: and the Cape de Verde islands from AFRICA; and that such colonists would be liable to m
0399 of Kerguelen Land, though standing nearer to AFRICA than to America, are related, and that very c
0391 sess one peculiar bird, and many European and AFRICAN birds are almost every year blown there, as
0478 se of the Cape de Verde archipelago and other AFRICAN islands to the African mainland. It must be
0478 archipelago and other African islands to the AFRICAN mainland. It must be admitted that these fac
0001 s which could possibly have any bearing on it. AFTER five years' work I allowed myself to speculat
0003 estices of Creation would, I presume, say that AFTER a certain unknown number of generations, some
0006 long remain obscure, I can entertain no doubt, AFTER the most deliberate study and dispassionate j
0020 ways best to study some special group, I have, AFTER deliberation, taken up domestic pigeons. I ha
0039 f by the value which would now be set on them, AFTER several breeds have once fairly been establis
0053 ng species. Dr. Hooker permits me to add, that AFTER having carefully read my manuscript, and exam
0057 sagacious and most experienced observers, and, AFTER deliberation, they concur in this view. In th
0095 ht slowly become, either simultaneously or one AFTER the other, modified and adapted in the most p
0098 spring towards the pistil, or slowly move one AFTER the other towards it, the contrivance seems a
0101 in the case of flowers, I have as yet failed, AFTER consultation with one of the highest authorit
0104 g the immigration of better adapted organisms, AFTER any physical change, such as of climate or el
0107 he range of each species will thus be checked: AFTER physical changes of any kind, immigration wil
```

Page ***************************************(Key Word)***

Page **(Key Word)**

```
0112 d be noted as forming two sub breeds; finally,   AFTER the lapse of centuries, the sub breeds would
0116 an early and incomplete stage of development.    AFTER the foregoing discussion, which ought to have
0117 pposed all to appear simultaneously, but often   AFTER long intervals of time; nor are they all supp
0117 each had represented ten thousand generations. And  AFTER a thousand generations, species (A) is suppos
0118 rved during the next thousand generations. And   AFTER this interval, variety a1 is supposed in the
0118 for any length of time; some of the varieties,   AFTER each thousand generations, producing only a s
0119 ginary, and might have been inserted anywhere,   AFTER intervals long enough to have allowed the acc
0120 having given off any fresh branches or races.    AFTER ten thousand generations, species (A) is supp
0120 species (I) has produced, by analogous steps,    AFTER ten thousand generations, either two well mar
0121 o be represented between the horizontal lines.   AFTER fourteen thousand generations, six new specie
0137 d professor Silliman thought that it regained,   AFTER living some days in the light, some slight po
0138 kness. By the time that an animal had reached,   AFTER numberless generations, the deepest recesses,
0146 ction some one modification in structure, and,   AFTER thousands of generations, some other and inde
0160 urprising fact that characters should reappear   AFTER having been lost for many, perhaps for hundre
0160 y, for a dozen or even a score of generations.   AFTER twelve generations, the proportion of blood,
0160 cter which has been lost in a breed, reappears   AFTER a great number of generations, the most proba
0160 esis is, not that the offspring suddenly takes   AFTER an ancestor some hundred generations distant,
0173 ese contingencies will concur only rarely, and   AFTER enormously long intervals. Whilst the bed of
0195 f high importance to an early progenitor, and,   AFTER having been slowly perfected at a former peri
0201 each will be found on the whole advantageous.    AFTER the lapse of time, under changing conditions
0211 evented their attendance during several hours.   AFTER this interval, I felt sure that the aphides w
0221 off by the tyrants, who perhaps fancied that,   AFTER all, they had been victorious in their late c
0222 ay; but in about a quarter of an hour, shortly   AFTER all the little yellow ants had crawled away,
0227 er to suppose, but this is no difficulty, that   AFTER hexagonal prisms have been formed by the inte
0231 a multitude of bees all work together; one bee   AFTER working a short time at one cell going to ano
0250 e three first flowers soon ceased to grow, and   AFTER a few days perished entirely, whereas the pod
0264 ever, are differently circumstanced before and   AFTER birth: when born and living in a country wher
0275 I entirely agree with Dr. Prosper Lucas, who,   AFTER arranging an enormous body of facts with resp
0288 on record of a formation conformably covered,   AFTER an enormous interval of time, by another and
0291 gy, and have been surprised to note how author   AFTER author, in treating of this or that great for
0305 merly have had a similarly confined range, and   AFTER having been largely developed in some one sea
0312 on. New species have appeared very slowly, one   AFTER another, both on the land and in the waters.
0316 er or lesser degree. A group does not reappear   AFTER it has once disappeared; or its existence, as
0317 and groups of species gradually disappear, one  AFTER another, first from one spot, the from anothe
0325 able observers, MM. de Verneuil and d'Archiac.  AFTER referring to the parallelism of the palaeozoi
0336 d close of these periods; but we ought to find  AFTER intervals, very long as measured by years, bu
0339 e same areas mean? He would be a bold man, who  AFTER comparing the present climate of Australia an
0340 ffer in nearly the same manner and degree. But  AFTER very long intervals of time and after great g
0340 ree. But after very long intervals of time and  AFTER great geographical changes, permitting much i
0344 nferiors in the struggle for existence. Hence,  AFTER long intervals of time, the productions of th
0348 e innumerable islands as halting places, until  AFTER travelling over a hemisphere we come to the s
0354 ation could be offered of many such cases. But  AFTER some preliminary remarks, I will discuss a fe
0358 se i found that out of 87 kinds, 64 germinated  AFTER an immersion of 28 days; and a few survived a
0359 nger period. So that as 64/87 seeds germinated  AFTER an immersion of 28 days; and as 18/94 plants
0359 ecies as in the foregoing experiment) floated,  AFTER being dried, for above 28 days, as far as we
0360 t the seeds of about 10/100 plants of a flora,  AFTER having been dried, could be floated across a
0362 i know by trial, the germination of seeds; now  AFTER a bird has found and devoured a large supply
0362 some hawks and owls bolt their prey whole, and  AFTER an interval of from twelve to twenty hours, d
0362 let, canary, hemp, clover, and beet germinated  AFTER having been from twelve to twenty one hours i
0362 rent birds of prey; and two seeds of beet grew  AFTER having been thus retained for two days and fo
0362 hing eagles, storks, and pelicans; these birds  AFTER an interval of many hours, either rejected th
0363 ain to be discovered, have been in action year  AFTER year, for centuries and tens of thousands of
0377 on the other hand, the temperate productions,  AFTER migrating nearer to the equator, though they
0387 o of seeds, though they reject many other kinds  AFTER having swallowed them; even small fish swallo
0387 d potamogeton. Herons and other birds, century  AFTER century, have gone on daily devouring fish; t
0390 ready explained, species occasionally arriving  AFTER long intervals in a new and isolated district
0397 one of these shells was the Helix pomatia, and  AFTER it had again hybernated I put it in sea water
0410 r of the world, or to those which have changed  AFTER having migrated into distant quarters, then b
0429 e one; but such richness in species, as I find  AFTER some investigation, does not commonly fall to
0443 imals, cannot positively tell, until some time  AFTER the animal has been born, what its merits or
0445 pigeons of various breeds, within twelve hours  AFTER being hatched; I carefully measured the propo
0475 uccessive intervals; and the amount of change,  AFTER equal intervals of time, is widely different
0475 f their descendants, causes the forms of life,  AFTER long intervals of time, to appear as if they
0484 sence a species. This I feel sure, and I speak  AFTER experience, will be no slight relief. The end
0034 ly tries to get as good dogs as he can, and    AFTERWARDS breeds from his own best dogs, but he has n
0183 habits generally change first and structure    AFTERWARDS; or whether slight modifications of structu
0211 with their antennae; but not one excreted.     AFTERWARDS I allowed an ant to visit them, and it imme
0351 ach other, migrate in a body into a new and    AFTERWARDS isolated country, they will be little liabl
0359 or not they were injured by the salt water.    AFTERWARDS I tried some larger fruits, capsules, etc.,
0359 ut when dried, they floated for 90 days and    AFTERWARDS when planted they germinated: an asparagus
0359 dried it floated for 85 days, and the seeds    AFTERWARDS germinated: the ripe seeds of Heloscladium
0359 n dried they floated for above 90 days, and    AFTERWARDS germinated. Altogether out of the 94 dried
0387 cted in pellets or in excrement, many hours    AFTERWARDS. When I saw the great size of the seeds of
0452 s, can often be detected in the embryo, but    AFTERWARDS wholly disappear. It is also, I believe, a
0062 edge of a desert is said to struggle for life  AGAINST the drought, though more properly it should
0095 the same objections which were at first urged  AGAINST Sir Charles Lyell's noble views on the moder
0100 e valid, but that nature has largely provided  AGAINST it by giving to trees a strong tendency to b
0199 the protest lately made by some naturalists    AGAINST the utilitarian doctrine that every detail o
0202 sp or of the bee as perfect, which, when used  AGAINST many attacking animals, cannot be withdrawn,
0203 ifficulties and objections which may be urged  AGAINST my theory. Many of them are very grave; but
0242 ed this demonstrative case of neuter insects,  AGAINST the well known doctrine of Lamarck. Summary.
0279 chief objections which might be justly urged   AGAINST the views maintained in this volume. Most of
0280 ious and gravest objection which can be urged  AGAINST my theory. The explanation lies, as I believ
0292 ed. Nature may almost be said to have guarded  AGAINST the frequent discovery of her transitional g
0299 of all the many objections which may be urged  AGAINST my views. Hence it will be worth while to su
0308 ei and may be truly urged as a valid argument  AGAINST the views here entertained. To show that it
```

Page **(Key Word)**

Page ***(Key Word)***

```
0365  this, as it seems to me, is no valid argument  AGAINST  what would be effected by occasional means o
0377  ate and could not have presented a firm front  AGAINST  intruders, that a certain number of the more
0400  nts of the islands may be used as an argument  AGAINST  my views; for it may be asked, how has it ha
0459  hat many and grave objections may be advanced  AGAINST  the theory of descent with modification thro
0463  ble of the many objections which may be urged  AGAINST  my theory. Why, again, do whole groups of al
0465  ns and difficulties which may justly be urged  AGAINST  my theory; and I have now briefly recapitula
0139  the cave animals should be very anomalous, as  AGASSIZ  has remarked in regard to the blind fish, th
0302  by several palaeontologists, for instance, by  AGASSIZ, Pictet, and by none more forcibly than by P
0305  owever, that the whole of them did appear, as  AGASSIZ  believes, at the commencement of the chalk f
0310  minent palaeontologists, namely Cuvier, Owen,  AGASSIZ, Barrande, Falconer, E. Forbes, etc., and al
0338  ountries could not have foreseen this result.  AGASSIZ  insists that ancient animals resemble to a c
0338  paratively recent times. For this doctrine of  AGASSIZ  accords well with the theory of natural sele
0366  ht have remained in this same belief, had not  AGASSIZ  and others called vivid attention to the Gla
0418  by those great naturalists, Milne Edwards and  AGASSIZ, that embryonic characters are the most impo
0439  ot be given, than a circumstance mentioned by  AGASSIZ, namely, that having forgotten to ticket the
0449  s of their descendants, our existing species.  AGASSIZ  believes this to be a law of nature; but I a
0013  s to appear in the offspring at a corresponding  AGE, though sometimes earlier. In many cases this
0014  y a peculiarity should appear at any particular  AGE, yet that it does tend to appear in the offspr
0086  blec to act on and modify organic beings at any  AGE, by the accumulation of profitable variations
0086  e accumulation of profitable variations at that  AGE, and by their inheritance at a corresponding a
0086  ge, and by their inheritance at a corresponding  AGE. If it profit a plant to have its seeds more a
0089  acting when the birds have come to the breeding  AGE  or during the breeding season; the modificatio
0127  cal powers of increase of each species, at some  AGE, season, or year, a severe struggle for life,
0168  will be saved. Changes of structure at an early  AGE  will generally affect parts subsequently devel
0173  emains being embedded and preserved to a future  AGE  only in masses of sediment sufficiently thick
0289  r true lacustrine bed is known belonging to the  AGE  of our secondary or palaeozoic formations. But
0290  faunas will probably be preserved to a distant  AGE. A little reflection will explain why along th
0291  which will hardly last to a distant geological  AGE, was certainly deposited during a downward osc
0300  be formed of sufficient thickness to last to an  AGE, as distant in futurity as the secondary forma
0304  one species had been discovered in beds of this  AGE, I concluded that this great group had been su
0306  which must have lived long before the Silurian  AGE, and which probably differed greatly from any
0307  nger than, the whole interval from the Silurian  AGE  to the present day; and that during these vast
0307  nants of the formations next succeeding them in  AGE, and these ought to be very generally in a met
0334  are intermediate in character, intermediate in  AGE. But supposing for an instant, in this and oth
0338  owing to variations supervening at a not early  AGE, and being inherited at a corresponding age. T
0338  rly age, and being inherited at a corresponding  AGE. This process, whilst it leaves the embryo alm
0385  jarred off, though at a somewhat more advanced  AGE  they would voluntarily drop off. These just ha
0411  ained by variations not supervening at an early  AGE, and being inherited at a corresponding age. P
0411  rly age, and being inherited at a corresponding  AGE. Rudimentary Organs; their origin explained. S
0439  reseblance, sometimes lasts till a rather late  AGE: thus birds of the same genus, and of closely
0443  hen an hereditary disease, which appears in old  AGE  alone, has been communicated to the offspring
0444  vidence to render it probable, that at whatever  AGE  any variation first appears in the parent, it
0444  parent, it tends to reappear at a corresponding  AGE  in the offspring. Certain variations can only
0444  ater in life, tend to appear at a corresponding  AGE  in the offspring and parent. I am far from mea
0444  gest sense) which have supervened at an earlier  AGE  in the child than in the parent. These two pri
0446  ed, not at the corresponding, but at an earlier  AGE. Now let us apply these facts and the above tw
0446  of variation having supervened at a rather late  AGE, and having been inherited at a corresponding
0447  e, and having been inherited at a corresponding  AGE, the young of the new species of our supposed
0447  ssive modification supervening at a rather late  AGE, and being inherited at a corresponding late a
0447  ge, and being inherited at a corresponding late  AGE, the fore limbs in the embryos of the several
0447  ced will be inherited at a corresponding mature  AGE. Whereas the young will remain unmodified, or
0448  ly resembling their parents from their earliest  AGE, we can see that this would result from the tw
0448  at the child should be modified at a very early  AGE  in the same manner with its parents, in accord
0450  modification having supervened at a very early  AGE, or by the variations having been inherited at
0453  n in the adult; so that the organ at this early  AGE  is less rudimentary, or even cannot be said to
0455  duce the organ in its reduced state at the same  AGE, and consequently will seldom affect or reduce
0455  were to be inherited, not at the corresponding  AGE, but at an extremely early period of life (as
0468  e, the slightest advantage in one being, at any  AGE  or during any season, over those with which it
0479  e variations not always supervening at an early  AGE, and being inherited at a corresponding not ea
0480  h recuced or rencered rudimentary at this early  AGE. The calf, for instance, has inherited teeth,
0483  formed on the same pattern, and at an embryonic  AGE  the species closely resemble each other. There
0010  lt: my impression is, that with animals such  AGENCIES  have produced very little direct effect, tho
0029  ould be a bold man who would account by such  AGENCIES  for the differences of a dray and race horse
0319  antly being checked by unperceived injurious  AGENCIES; and that these same unperceived agencies ar
0319  us agencies; and that these same unperceived  AGENCIES  are amply sufficient to cause rarity, and fi
0387  aller fresh water animals. Other and unknown  AGENCIES  probably have also played a part. I have sta
0003  with separate sexes absolutely requiring the  AGENCY  of certain insects to bring pollen from one f
0090  attribute all such sexual differences to this  AGENCY: for we see peculiarities arising and becomin
0098  no doubt it is useful for this end: but, the  AGENCY  of insects is often required to cause the sta
0115  in the naturalisation of plants through man's  AGENCY  in foreign lands. It might have been expected
0145  e ray florets serve to attract insects, whose  AGENCY  is highly advantageous in the fertilisation o
0203  ny other plants are fertilised through insect  AGENCY, can we consider as equally perfect the elabo
0252  such alone are fairly treated, for by insect  AGENCY  the several individuals of the same hybrid va
0252  convince himself of the efficiency of insect  AGENCY  by examining the flowers of the more sterile
0320  ated, either locally or wholly, through man's  AGENCY. I may repeat what I published in 1845, namel
0352  n with subsequent migration, and calls in the  AGENCY  of a miracle. It is universally admitted, tha
0380  ed to continental forms, naturalised by man's  AGENCY. I am far from supposing that all difficultie
0381  and not another becomes naturalised by man's  AGENCY  in a foreign land; why one ranges twice or th
0385  resh water shells from the other. But another  AGENCY  is perhaps more effectual: I suspended a duck
0387  of these plants was not very great. The same  AGENCY  may have come into play with the eggs of some
0393  be correct) may be explained through glacial  AGENCY. This general absence of frogs, toads, and ne
0395  turalisation of certain mammals through man's  AGENCY: but we shall soon have much light thrown on
0402  that many species, naturalised through man's  AGENCY, have spread with astonishing rapidity over n
0454  nges. I believe that disuse has been the main  AGENCY; that it has led in successive generations to
0467  this process of selection has been the great  AGENCY  in the production of the most distinct and us
0469  e have under nature variability and a powerful  AGENT  always ready to act and select, why should we
```

Page ***(Key Word)***

```
0361  birds can hardly fail to be highly effective AGENTS in the transportation of seeds. I could give
0007  n cultivated, and which have varied during all AGES under the most different climates and treatmen
0080  cters of trifling importance, its power at all AGES and on both sexes. Sexual Selection. On the ge
0082  ry slight modification, which in the course of AGES chanced to arise, and which in any way favoure
0084   the hand of time has marked the long lapse of AGES, and then so imperfect is our view into long p
0084  mperfect is our view into long past geological AGES, that we only see that the forms of are now di
0089  thus produced being inherited at corresponding AGES or seasons, either by the males alone, or by t
0126  mmary of Chapter. If during the long course of AGES and under varying conditions of life, organic
0127   of qualities being inherited at corresponding AGES, can modify the egg, seed, or young, as easily
0217  ere would be eggs and young birds of different AGES in the same nest. If this were the case, the p
0217  il to be by having eggs and young of different AGES at the same time; then the old birds or the fo
0237  ucture which have become correlated to certain AGES, and to either sex. We have differences correl
0288   of the bottom of the sea not rarely lying for AGES in an unaltered condition. The remains which d
0290  found, though the supply of sediment must for  AGES have been great, from the enormous degradation
0302  cess; and the progenitors must have lived long AGES before their modified descendants. But we cont
0303  ely that it might require a long succession of  AGES to adapt an organism to some new and peculiar
0309  onderant movement have changed in the lapse of  AGES? At a period immeasurably antecedent to the s
0315  troyed, that fanciers, by striving during long  AGES for the same object, might make a new breed ha
0316  group have appeared in the long succession of  AGES, so long must its members have continuously ex
0330  ergone much change in the course of geological  AGES; and it would be difficult to prove the truth
0349  essentially American. We may look back to past  AGES, as shown in the last chapter, and we find Ame
0371  ate productions are concerned, took place long  AGES ago. And as the plants and animals migrated so
0391  ira have been stocked by birds, which for long  AGES have struggled together in their former homes,
0410  r colour. In looking to the long succession of  AGES, as in now looking to distant provinces throug
0410  s of life which have changed during successive  AGES within the same quarter of the world, or to th
0418   for our classifications of course include all  AGES of each species. But it is by no means obvious
0433  ssinc the individuals of both sexes and of all  AGES, although having few characters in common, und
0444  in variations can only appear at corresponding  AGES, for instance, peculiarities in the caterpilla
0448   the principle of inheritance at corresponding  AGES, the active young or larvae might easily be re
0455   the principle of inheritance at corresponding  AGES will reproduce the organ in its reduced state
0456  niversally used in ranking together the sexes,  AGES, and acknowledged varieties of the same specie
0457  modifications being inherited at corresponding  AGES. On this same principle, and bearing in mind,
0464  varieties; but who will pretend that in future  AGES so many fossil links will be discovered, that
0469  t can be put to this power, acting during long  AGES and rigidly scrutinising the whole constitutio
0476   that there has been during the long course of  AGES much migration from one part of the world to a
0480   the principle of inheritance at corresponding  AGES have been inherited from a remote period to th
0481  the amount of variation in the course of long   AGES is a limited quantity; no clear distinction ha
0488  as a mere fragment of time, compared with the   AGES which have elapsed since the first creature, t
0179  rom existing in greater numbers will, in the AGGREGATE, present more variation, and thus be furthe
0417  e or less importance. The value indeed of an AGGREGATE of characters is very evident in natural hi
0417  s universally constant. The importance of an AGGREGATE of characters, even when none are important
0225  pherical and of nearly equal sizes, and are  AGGREGATED into an irregular mass. But the important p
0234  more and more regular, nearer together, and  AGGREGATED into a mass, like the cells of the Melipona
0426  cestor. And we know that such correlated or  AGGREGATED characters have especial value in classific
0220  ly come out, and like their masters are much AGITATED and defend the nest: when the nest is much d
0222  f. fusca were rushing about in the greatest  AGITATION, and one was perched motionless with its ow
0018  s originated there, four or five thousand years AGO? But Mr. Horner's researches have rendered it
0018   of the Nile thirteen or fourteen thousand years AGO, and who will pretend to say how long before t
0032  with drawings made only twenty or thirty years AGO. When a race of plants is once pretty well est
0035  gs of the breeds in question had been made long AGO, which might serve for comparison. In some cas
0048  often called, as geographical races! Many years AGO, when comparing, and seeing others compare, th
0113  r that can be supported in any country has long AGO arrived at its full average. If its natural po
0150  ies, tends to be highly variable. Several years AGO I was much struck with a remark, nearly to the
0258  l crosses between the same two species was long AGO observed by Kolreuter. To give an instance: Mi
0303  geological treatises, published not many years  AGO, the great class of mammals was always spoken
0307  ers, they would almost certainly have been long AGO supplanted and exterminated by their numerous
0329  iffers from living forms. But, as Buckland long AGO remarked, all fossils can be classed either in
0339  he later tertiary periods. Mr. Clift many years AGO showed that the fossil mammals from the Austra
0365   the loftiest mountains of Europe. Even as long AGO as 1747, such facts led Gmelin to conclude tha
0371  productions are concerned, took place long ages AGO. And as the plants and animals migrated southw
0374  silla of Caraccas the illustrious Humboldt long AGO found species belonging to genera characterist
0393  rders on oceanic islands, Bory St. Vincent long AGO remarked that Batrachians (frogs, toads, newts
0399  uth America, and other southern lands were long AGO partially stocked from a nearly intermediate t
0404  more general way. Mr. Gould remarked to me long AGO, that in those genera of birds which range ove
0349   on these same plains of La Plata, we see the AGOUTI and bizcacha, animals having nearly the same
0046  t of variation; and hardly two naturalists can AGREE which forms to rank as species and which as v
0272  ing of varieties. And, on the other hand, they AGREE most closely in very many important respects.
0275  nd naturally produced.' On the whole I entirely AGREE with Dr. Prosper Lucas, who, after arranging
0299  those of the United States of America. I fully AGREE with Mr. Godwin Austen, that the present cond
0346  umpolar land is almost continuous, all authors AGREE that one of the most fundamental divisions in
0428  mbers of the whale family; for these cetaceans AGREE in so many characters, great and small, that
0473  than the generic characters in which they all  AGREE? Why, for instance, should the colour of a fl
0024  is, that the above specified breeds, though AGREEING generally in constitution, habits, voice, co
0101  me family and even of the same genus, though AGREEING closely with each other in almost their whol
0050  in most cases, will bring naturalists to an AGREEMENT how to rank doubtful forms. Yet it must be
0185  al having habits and structure not at all in AGREEMENT. What can be plainer than that the webbed f
0206   . by unity of type is meant that fundamental AGREEMENT in structure, which we see in organic being
0027  stication in Europe and in India; and that it AGREES in habits and in a great number of points of
0030  l and elegant; when we compare the host of AGRICULTURAL, culinary, orchard, and flower garden race
0086   of the many varieties of our culinary and AGRICULTURAL plants; in the caterpillar and cocoon stag
0111  y the short horns (I quote the words of an AGRICULTURAL writer) as if by some murderous pestilence
0141  gy, and from the incessant advice given in AGRICULTURAL works, even in the ancient Encyclopaedias
0031  ple of selection as that which enables the AGRICULTURIST, not only to modify the character of his
0423  used in classing varieties: thus the great AGRICULTURIST Marshall says the horns are very useful f
0031  bably better acquainted with the works of AGRICULTURISTS than almost any other individual, and who
0043  ished, their occasional intercrossing, with the AID of selection, has, no doubt, largely aided in
```

Page ***(Key Word)***

```
0116  will tend to act. The accompanying diagram will  AID  us in understanding this rather perplexing sub
0127  st many animals, sexual selection will give its  AID  to ordinary selection, by assuring to the most
0137  ; and if so, natural selection would constantly  AID  the effects of disuse. It is well known that s
0182  ugh the air, slightly rising and turning by the  AID  of their fluttering fins, might have been modi
0196  a fly flapper, an organ of prehension, or as an  AID  in turning, as with the dog, though the aid mu
0196  an aid in turning, as with the dog, though the   AID  must be slight, for the hare, with hardly any
0197  ay archipelago climbs the loftiest trees by the  AID  of exquisitely constructed hooks clustered aro
0213  ound pointer can no more know that he points to  AID  his master, than the white butterfly knows why
0219  solutely dependent on its slaves; without their  AID, the species would certainly become extinct in
0358  lly unknown. Until I tried, with Mr. Berkeley's  AID, a few experiments, it was not even known how
0392  re by occasional means of transport, and by the  AID, as halting places, of existing and now sunken
0416  been considered by naturalists as important an   AID  in determining the degree of affinity of this
0431  ed of any ancient and noble family, even by the  AID  of a genealogical tree, and almost impossible
0431  , and almost impossible to do this without this  AID, we can understand the extraordinary difficult
0431  sts have experienced in describing, without the  AID  of a diagram, the various affinities which the
0477  es with modification, we can understand, by the  AID  of the Glacial period, the identity of some fe
0479  s to breathe the air dissolved in water, by the  AID  of well developed branchiae. Disuse, aided som
0480  ed by natural selection to browse without their  AID; whereas in the calf, the teeth have been left
0486  h may fancifully be called living fossils, will  AID  us in forming a picture of the ancient forms o
0003  dr. Hooker, who for the last fifteen years has   AIDED  me in every possible way by his large stores
0020  race may be modified by occasional crosses, if   AIDED  by the careful selection of those individual
0041  ings and bred from them, then, there appeared (  AIDED  by some crossing with distinct species) those
0043  h the aid of selection, has, no doubt, largely   AIDED  in the formation of new sub breeds; but the i
0137  ably due to gradual reduction from disuse, but   AIDED  perhaps by natural selection. In South Americ
0137  others have been reduced by natural selection   AIDED  by use and disuse, so in the case of the cave
0190  so as to perform all the work by itself, being   AIDED  during the process of modification by the oth
0192  rous frena, but which, likewise, very slightly   AIDED  the act of respiration, have been gradually c
0205  function, the one having been perfected whilst   AIDED  by the other, must often have largely facilit
0206  ng past periods of time: the adaptations being   AIDED  in some cases by use and disuse, being slight
0224  f. sanguinea, which, as we have seen, is less    AIDED  by its slaves than the same species in Switze
0479  y the aid of well developed branchiae. Disuse,   AIDED  sometimes by natural selection, will often te
0197  e been advanced as a beautiful adaptation for    AIDING  parturition, and no doubt they facilitate, or
0085  se with other coloured flesh. If, with all the   AIDS  of art, these slight differences make a great
0098  in many other cases, far from there being any    AIDS  for self fertilisation, there are special cont
0420  uently discovered. All the foregoing rules and   AIDS  and difficulties in classification are explain
0021  height in a compact flock, and tumbling in the   AIR  head over heels. The runt is a bird of great s
0049  been generally accepted, is vainly to beat the   AIR. Many of the cases of strongly marked varietie
0077  tion seems at first confined to the elements of  AIR  and water. Yet the advantage of plumed seeds n
0180  parachute and allows them to glide through the   AIR  to an astonishing distance from tree to tree.
0181  ks of structure, fitted for gliding through the  AIR, now connect the Galeopithecus with the other
0181  originally constructed for gliding through the   AIR  rather than for flight. If about a dozen gener
0182  at flying fish, which now glide far through the  AIR, slightly rising and turning by the aid of the
0187  and likewise to those coarser vibrations of the  AIR  which produce sound. In looking for the gradat
0190  e fish with gills or branchiae that breathe the  AIR  dissolved in the water, at the same time that
0190  water, at the same time that they breathe free   AIR  in their swimbladders, this latter organ havin
0204  mal which at first could only glide through the  AIR. We have seen that a species may under new con
0204  m bladder has apparently been converted into an  AIR  breathing lung. The same organ having performe
0303  r line of life, for instance to fly through the  AIR; but that when this had been effected, and a f
0360  at they were alternately wet and exposed to the  AIR  like really floating plants. He tried 98 seeds
0385  ir nature, survived on the duck's feet, in damp  AIR, from twelve to twenty hours; and in this leng
0428  nhabit for instance the three elements of land,  AIR, and water, we can perhaps understand how it i
0479  ms. We may cease marvelling at the embryo of an  AIR  breathing mammal or bird having branchial slit
0479  , like those in a fish which has to breathe the  AIR  dissolved in water, by the aid of well develop
0028  pedigree and race. Pigeons were much valued by   AKBER Khan in India, about the year 1600; never les
0011  s of the ear, from the animals not being much    ALARMED by danger, seems probable. There are many la
0013  tance. Every one must have heard of cases of     ALBINISM, prickly skin, hairy bodies, etc., appearing
0275  e, and which have suddenly appeared, such as     ALBINISM, melanism, deficiency of tail or horns, or a
0163  n length and outline. A white ass, but not an    ALBINO, has been described without either spinal or
0415  r more ovaria, in the existence or absence of    ALBUMEN, in the imbricate or valvular aestivation. A
0376  hooker informs me that twenty five species of    ALGAE are common to New Zealand and to Europe, but
0416  of insects are folded, mere colour in certain    ALGAE, mere pubescence on parts of the flower in gr
0185  ate bird, which has all its four toes webbed,    ALIGHT on the surface of the sea. On the other hand,
0385  or seven hundred miles, and would be sure to     ALIGHT on a pool or rivulet, if blown across sea to
0386  n the open ocean; they would not be likely to    ALIGHT on the surface of the sea, so that the dirt w
0186  either living on the dry land or most rarely    ALIGHTING on the water; that there should be long toe
0020  everal generations, hardly two of them will be   ALIKE, and then the extreme difficulty, or rather u
0033  e leaves of the cabbage are, and how extremely   ALIKE the flowers, how unlike the flowers of the he
0033  ike the flowers of the heartsease are, and how   ALIKE the leaves; how much the fruit of the differe
0143  , and which, at an early embryonic period, are   ALIKE, seem liable to vary in an allied manner: we
0155  re variable than those parts which are closely   ALIKE in the several species? I do not see that any
0277  n forms known to be varieties, or sufficiently   ALIKE to considered as varieties, and their mongrel
0349  eir nests similarly constructed, but not quite   ALIKE, with eggs coloured in nearly the same manner
0372  ion. We cannot say that they have been created   ALIKE, in correspondence with the nearly similar ph
0413  g together those living objects which are most   ALIKE, and for separating those which are most unli
0420  g together and separating objects more or less   ALIKE. But I must explain my meaning more fully. I
0424  in the same group, because allied in blood and   ALIKE in some other respects. If it could be proved
0437  erent, are at an early stage of growth exactly   ALIKE. How inexplicable are these facts on the ordi
0439  different purposes, are in the embryo exactly    ALIKE. The embryos, also, of distinct animals withi
0442  purposes, being at this early period of growth   ALIKE; of embryos of different species within the s
0451  udiment and the perfect pistil are essentially   ALIKE in nature. An organ serving for two purposes,
0479  al modification of parts or organs, which were   ALIKE in the early progenitor of each class. On the
0479  rds, reptiles, and fishes should be so closely   ALIKE, and should be so unlike the adult forms. We
0190  me time wholly distinct functions; thus the      ALIMENTARY canal respires, digests, and excretes in th
0064  at the end of the fifth century there would be   ALIVE fifteen million elephants, descended from the
0113  d to feed on new kinds of prey, either dead or   ALIVE; some inhabiting new stations, climbing trees
0130  by some chance has been favoured and is still    ALIVE on its summit, so we occasionally see an anim
```

Page ***(Key Word)***

Page ***(Key Word)***

```
0137  t they were frequently blind: one which I kept ALIVE was certainly in this condition, the cause, a
0432  ch of their descendants, may be supposed to be ALIVE: and the links to be as fine as those between
0477  e identity of some few plants, and the close   ALLIANCE of many others, on the most distant mountain
0477  t different climates; and likewise the close   ALLIANCE of some of the inhabitants of the sea in the
0006  widely and is very numerous, and why another   ALLIED species has a narrow range and is rare? Yet t
0015  plants, and compare them with species closely  ALLIED together, we generally perceive in each domes
0016  e species in nature to which they are nearest   ALLIED. With these exceptions (and with that of the
0016  ost cases in a lesser degree than, do closely   ALLIED species of the same genus in a state of natur
0017  out the immutability of the many very closely   ALLIED and natural species, for instance, of the man
0021  s, others singularly short tails. The barb is   ALLIED to the carrier, but, instead of a very long b
0029  een produced, either from one or from several   ALLIED species. Some little effect may, perhaps, be
0032  is practice, except sometimes amongst closely   ALLIED sub breeds. And when a cross have been made, t
0047  lieve that many of these doubtful and closely   ALLIED forms have permanently retained their charact
0051  he will encounter a greater number of closely  ALLIED forms. But if his observations be widely exte
0051  aturalists. When, moreover, he comes to study   ALLIED forms brought from countries not now continuo
0058  ve, that those species which are very closely  ALLIED to other species, and in so far resemble vari
0058  species, but which he considers as so closely  ALLIED to other species as to be of doubtful value:
0058  ted average range, as have those very closely  ALLIED forms, marked for me by Mr. Watson as doubtfu
0059  species are apt to be closely, but unequally,  ALLIED together, forming little clusters round certa
0059  s round certain species. Species very closely  ALLIED to other species apparently have restricted r
0076  the competition should be most severe between  ALLIED forms, which fill nearly the same place in th
0098  sation, it is well known that if very closely  ALLIED forms or varieties are planted near each othe
0110  gle for Existence that it is the most closely  ALLIED forms, varieties of the same species, and spe
0115  in any land would generally have been closely  ALLIED to the indigenes; for these are commonly look
0123  n species. The new species, moreover, will be  ALLIED to each other in a widely different manner. O
0138  ts of the two continents are not more closely  ALLIED than might have been anticipated from the gen
0139  of the European cave insects are very closely  ALLIED to those of the surrounding country. It would
0143  period, are alike, seem liable to vary in an   ALLIED manner: we see this in the right and left sid
0150  r manner, in comparison with the same part in  ALLIED species, tends to be highly variable. Several
0150  d in comparison with the same part in closely  ALLIED species. Thus, the bat's wing is a most abnor
0155  species, has differed considerably in closely  ALLIED species, that it has also, been variable in t
0158  ence in these same characters between closely  ALLIED species: that secondary sexual and ordinary s
0159  es often assumes some of the characters of an  ALLIED species, or reverts to some of the characters
0162  the character of the same part or organ in an  ALLIED species. I have collected a long list of such
0169  comparison with the same part or organ in the  ALLIED species, must have gone through an extraordin
0173  but it may be urged that when several closely  ALLIED species inhabit the same territory we surely
0173  lly meet at successive intervals with closely  ALLIED or representative species, evidently filling
0173  ropolis inhabited by each. By my theory these  ALLIED species have descended from a common parent:
0175  rply defined. If I am right in believing that  ALLIED or representative species, when inhabiting bo
0176  be eminently liable to the inroads of closely  ALLIED forms existing on both sides of it. But a far
0178  we know of the actual distribution of closely  ALLIED or representative species, and likewise of ac
0180  transitional habits and structures in closely  ALLIED species of the same genus; and of diversified
0203  receding and intermediate gradations. Closely  ALLIED species, now living on a continuous area, mus
0218  ains the origin of a singular instinct in the  ALLIED group of ostriches. For several hen ostriches
0243  instincts: as by that common case of closely  ALLIED, but certainly distinct, species, when inhabi
0249  numbers: and as the parent species, or other  ALLIED hybrids, generally grow in the same garden; t
0256  cal importance and which differ little in the  ALLIED species. Now the fertility of first crosses b
0257  ilies: and on the other hand, by very closely  ALLIED species generally uniting with facility. But
0257  itude of cases could be given of very closely  ALLIED species which will not unite, or only with ex
0261  nct families: and, on the other hand, closely  ALLIED species, and varieties of the same species, c
0263  pecies is placed on the stigma of a distantly  ALLIED species, though the pollen tubes protrude, th
0266  show, is that in two cases, in some respects   ALLIED, sterility is the common result, in the one c
0266  pect that a similar parallelism extends to an  ALLIED yet very different class of facts. It is an o
0272  ther admits that hybrids between very closely  ALLIED species are more variable than those from ver
0273  t when any two species, although most closely  ALLIED to each other, are crossed with a third speci
0277  ounded of two distinct species, seems closely  ALLIED to that sterility which so frequently affects
0281  escended from this species or from some other  ALLIED species, such as C. oenas. So with natural sp
0293  find closely graduated varieties between the   ALLIED species which lived at its commencement and a
0298  cifically different, yet are far more closely  ALLIED to each other than are the species found in m
0302  . on the sudden appearance of whole groups of  ALLIED Species. The abrupt manner in which whole gro
0306  ctions. On the sudden appearance of groups of  ALLIED Species in the lowest known fossiliferous str
0306  wn fossiliferous strata. There is another and  ALLIED difficulty, which is much graver. I allude to
0317  odification and the production of a number of  ALLIED forms must be slow and gradual, one species g
0321  nd thus caused its extermination: and if many  ALLIED forms be developed from the successful intrud
0321  yield their places: and it will generally be   ALLIED forms, which will suffer from some inherited
0322  hich thus yield their places will commonly be  ALLIED, for they will partake of some inferiority in
0324  from containing fossil remains in some degree  ALLIED, and from not including those forms which are
0327  r forms: and as these inferior forms would be  ALLIED in groups by inheritance, whole groups would
0327  e new and victorious forms, will generally be  ALLIED in groups, from inheriting some inferiority i
0331  ll form a small family: b14 and f14 a closely  ALLIED family or sub family; and o14, e14, m14, a th
0336  rately long as measured geologically, closely  ALLIED forms, or, as they have been called by some a
0339  ammals from the Australian caves were closely  ALLIED to the living marsupials of that continent. I
0340  d the southern half was formerly more closely  ALLIED, than it is at present, to the northern half.
0340  c the next succeeding period of time, closely  ALLIED though in some degree modified descendants. I
0341  ther I suppose that the megatherium and other  ALLIED huge monsters have left behind them in South
0341  re are many extinct species which are closely  ALLIED in size and in other characters to the specie
0341  next succeeding formation there be six other  ALLIED or representative genera with the same number
0342  hich must formerly have connected the closely  ALLIED or representative species, found in the sever
0344  in the long run tend to people the world with  ALLIED, but modified, descendants: and these will ge
0345  osely consecutive formations are more closely  ALLIED to each other, than are those of remote forma
0349  ed, replace each other. He hears from closely  ALLIED, yet distinct kinds of birds, notes nearly si
0355  in space and time with a pre existing closely  ALLIED species. And I now know from correspondence,
0355  s of creation do not directly bear on another  ALLIED question, namely whether all the individuals
0368  s, as remarked by Ramond, are more especially  ALLIED to the plants of northern Scandinavia; those
0369  forms, and some few are distinct yet closely   ALLIED or representative species. In illustrating wh
0372  rs as distinct species: and a host of closely  ALLIED or representative forms which are ranked by a
```

Page ***(Key Word)***

Page ***(Key Word)***

```
0372   the theory of modification, for many closely  ALLIED  forms now living in areas completely sundered
0372     the still more striking case of many closely  ALLIED  crustaceans (as described in Cana's admirable
0374      on the present distribution of identical and  ALLIED  species. In America, Dr. Hooker has shown tha
0374     se two points are; and there are many closely  ALLIED  species. On the lofty mountains of equatorial
0379    ustralia, that many more identical plants and   ALLIED  forms have apparently migrated from the north
0380      in regard to the range and affinities of the  ALLIED  species which live in the northern and southe
0382    present distribution both of the same and of   ALLIED  forms of life can be explained. The living wa
0383     ite different classes, an enormous range, but  ALLIED  forms in a remarkable manner throug
0384     lead to this same conclusion. With respect to  ALLIED  fresh water fish occurring at very distant po
0385     resh water shells have a very wide range, and  ALLIED  species, which, on my theory, are descended f
0389      t all the individuals both of the same and of ALLIED  species have descended from a single parent;
0395      in both of the same mammiferous species or of ALLIED  species in a more or less modified condition.
0395     ine banks, and they are inhabited by closely   ALLIED  or identical quadrupeds. No doubt some few an
0400     , though specifically distinct, to be closely  ALLIED  to those of the nearest continent, we sometim
0402     erroneous view of the probability of closely   ALLIED  species invading each other's territory, when
0402     ed in new countries are not generally closely  ALLIED  to the aboriginal inhabitants; but are very d
0402    istinct on each; thus there are three closely   ALLIED  species of mocking thrush, each confined to i
0404     ns, let them be ever so distant; many closely  ALLIED  or representative species occur, there will l
0404     en the two regions. And wherever many closely  ALLIED  species occur, there will be found many forms
0404    t remote points of the world of other species  ALLIED  to it, is shown in another and more general w
0408     ividuals of the same species, and likewise of  ALLIED  species, have proceeded from some one source;
0415     lue, is almost shown by the one fact, that in  ALLIED  groups, in which the same organ, as we have e
0419     the species at both ends, from being plainly   ALLIED  to others, and these to others, and so onward
0419     re especially in very large groups of closely  ALLIED  forms. Temminck insists on the utility or eve
0419     es, at first overlooked, but because numerous  ALLIED  species, with slightly different grades of di
0420    nce in the several branches or groups, though  ALLIED  in the same degree in blood to their common p
0420     will suppose the letters A to L to represent   ALLIED  genera, which lived during the Silurian epoch
0423    ance would keep the forms together which were   ALLIED  in the greatest number of points. In tumbler
0424     tumblers are kept in the same group, because   ALLIED  in blood and alike in some other respects. If
0439     thus birds of the same genus, and of closely  ALLIED  genera, often resemble each other in their fi
0440     similarity of the larvae or active embryos of  ALLIED  animals is sometimes much obscured; and cases
0444     different, are really varieties most closely   ALLIED, and have probably descended from the same wi
0452     t and in other respects. Moreover, in closely  ALLIED  species, the degree to which the same organ h
0456     forms which are considered by naturalists as   ALLIED, together with their modification through nat
0463     s into another district occupied by a closely  ALLIED  species, we have no just right to expect ofte
0463     zones, will be liable to be supplanted by the  ALLIED  forms on either hand; and the latter, from ex
0463     nst my theory. Why, again, do whole groups of  ALLIED  species appear, though certainly they often f
0470     do the species of smaller genera. The closely  ALLIED  species also of the larger genera apparently
0474     h in common, we can understand how it is that  ALLIED  species, when placed under considerably diffe
0476     some degree intermediate between existing and  ALLIED  groups. Recent forms are generally looked at
0476     ife. Lastly, the law of the long endurance of  ALLIED  forms on the same continent, of marsupials in
0476     the recent and the extinct will naturally be   ALLIED  by descent. Looking to geographical distribut
0478     nt acts of creation. The existence of closely  ALLIED  or representative species in any two areas, i
0478     st invariably find that wherever many closely  ALLIED  species inhabit two areas, some identical spe
0478     on to both still exist. Wherever many closely  ALLIED  yet distinct species occur, many doubtful for
0486     uals of the same species, and all the closely  ALLIED  species of most genera, have within a not ver
0171    h habits widely different from those of their  ALLIES. Organs of extreme perfection. Means of trans
0314    ome to differ considerably from their nearest  ALLIES  on the continent of Europe, whereas the marin
0321    n developed from any one species, the nearest  ALLIES  of that species, i.e. the species of the same
0417      been remarked, a species may depart from its  ALLIES  in several characters, both of high physiolog
0426     its most important characteristics, from its   ALLIES, and yet be safely classed with them. This ma
0088    is law of battle descends; I know not; male   ALLIGATORS  have been described as fighting, bellowing,
0418     s which they use in defining a group, or in   ALLOCATING  any particular species. If they find a char
0033    lowed: for hardly any one is so careless as to ALLOW  his worst animals to breed. In regard to plan
0041    re not amply sufficient, with extreme care, to ALLOW  of the accumulation of a large amount of modi
0083    nd short wool to the same climate. He does not  ALLOW  the most vigorous males to struggle for the f
0313    d in the midst of an older formation, and then ALLOW  the pre existing fauna to reappear; but Lyell
0334    rms of life above and below. We must, however, ALLOW  for the entire extinction of some preceding f
0451    thus in plants, the office of the pistil is to ALLOW  the pollen tubes to reach the ovules protecte
0006    in of species and varieties, if he makes due  ALLOWANCE  for our profound ignorance in regard to the
0150    es of error, but I hope that I have made due   ALLOWANCE  for them. It should be understood that the
0200    in every living creature (making some little   ALLOWANCE  for the direct action of physical condition
0210    an do. I have been surprised to find, making   ALLOWANCE  for the instincts of animals having been bu
0286     this will at first appear much too small an   ALLOWANCE; but it is the same as if we were to assume
0287    ntury for the whole length would be an ample   ALLOWANCE. At this rate, on the above data, the denud
0302    and of the United States. We do not make due   ALLOWANCE  for the enormous intervals of time, which h
0406    ave endeavoured to show, that if we make due   ALLOWANCE  for our ignorance of the full effects of al
0334    the successive formations. Subject to these   ALLOWANCES, the fauna of each geological period undoub
0407    from one parent source; if we make the same   ALLOWANCES  as before for our ignorance, and remember t
0001    e any bearing on it. After five years' work I  ALLOWED  myself to speculate on the subject, and drew
0041    tever their quality may be, will generally be   ALLOWED  to breed, and this will effectually prevent
0068    species perished from the other species being  ALLOWED  to grow up freely. The amount of food for ea
0076    up for half a dozen generations, if they were  ALLOWED  to struggle together, like beings in a state
0083    ed by naturalised productions, that they have  ALLOWED  foreigners to take firm possession of the la
0099    age, radish, onion, and some other plants, be  ALLOWED  to seed near each other, a large majority, a
0108    ions of the old inhabitants; and time will be  ALLOWED  for the varieties in each to become well mod
0113    average. If its natural powers of increase be  ALLOWED  to act, it can succeed in increasing (the co
0119    anywhere, after intervals long enough to have  ALLOWED  the accumulation of a considerable amount of
0211    antennae; but not one excreted. Afterwards I   ALLOWED  an ant to visit them, and it immediately see
0229    tance, I put the comb back into the hive, and  ALLOWED  the bees to go on working for a short time,
0252    al individuals of the same hybrid variety are  ALLOWED  to freely cross with each other, and the inj
0356    sibly even continents together, and thus have  ALLOWED  terrestrial productions to pass from one to
0408    ding to the nature of the communication which  ALLOWED  certain forms and not others to enter, eithe
0088    eaving offspring. Sexual selection by always   ALLOWING  the victor to breed might surely give indomi
0216    vidently done for the instinctive purpose of   ALLOWING, as we see in wild ground birds, their mothe
0230    g to make equal spherical hollows, but never   ALLOWING  the spheres to break into each other. Now be
```

Page ***(Key Word)***

Page **(Key Word)**

```
0066  a rapidly fluctuating amount of food, for it  ALLOWS  them rapidly to increase in number. But the r
0077  ture of its legs, so well adapted for diving,  ALLOWS  it to compete with other aquatic insects, to
0180  anse of skin, which serves as a parachute and  ALLOWS  them to glide through the air to an astonishi
0011  hereafter briefly mentioned. I will here only  ALLUDE  to what may be called correlation of growth.
0054  on the side of the larger genera. I will here  ALLUDE  to only two causes of obscurity. Fresh water
0183  s of changed habits it will suffice merely to  ALLUDE  to that of the many British insects which now
0236  ble, and actually fatal to my whole theory. I  ALLUDE  to the neuters or sterile females in insect c
0306  nd allied difficulty, which is much graver. I  ALLUDE  to the manner in which numbers of species of
0014  e, is clearly due to the male element. Having  ALLUDED  to the subject of reversion, I may here refe
0051  ust like the pigeon or poultry fancier before  ALLUDED  to, with the amount of difference in the for
0092  n to believe (as will hereafter be more fully  ALLUDED  to), would produce very vigorous seedlings,
0093  ation of the sexes of plants; presently to be  ALLUDED  to. Some holly trees bear only male flowers,
0106  ws, understand some facts which will be again  ALLUDED  to in our chapter on geographical distributi
0109  n our chapter on Geology; but it must be here  ALLUDED  to from being intimately connected with natu
0134  us. Effects of Use and Disuse. From the facts  ALLUDED  to in the first chapter, I think there can b
0182  that any of the grades of wing structure here  ALLUDED  to, which perhaps may all have resulted from
0198  nd unknown laws of variation; and I have here  ALLUDED  to them only to show that, if we are unable
0264  case is very different. I have more than once  ALLUDED  to a large body of facts, which I have colle
0267  offspring. I believe, indeed, from the facts  ALLUDED  to in our fourth chapter, that a certain amo
0340  si and I have shown in the publications above  ALLUDED  to, that in America the law of distribution
0370  glacial epoch. Believing, from reasons before  ALLUDED  to, that our continents have long remained i
0053  with much brevity, is rather perplexing, and  ALLUSIONS  cannot be avoided to the struggle for exist
0002  tory of the Malay archipelago, has arrived at  ALMOST  exactly the same general conclusions that I h
0005  ; and we shall then see how Natural Selection  ALMOST  inevitably causes much Extinction of the less
0011  rowth. Any change in the embryo or larva will  ALMOST  certainly entail changes in the mature animal
0011  subject. Breeders believe that long limbs are  ALMOST  always accompanied by an elongated head. Some
0012  and thus augmenting, any peculiarity, he will  ALMOST  certainly unconsciously modify other parts of
0013  rs in the child, the mere doctrine of chances  ALMOST  compels us to attribute its reappearance to i
0015  rious breeds, and esculent vegetables, for an  ALMOST  infinite number of generations, would be oppo
0018  f knowledge, I should value more than that of  ALMOST  any one, thinks that all the breeds of poultr
0021  the short faced tumbler has a beak in outline  ALMOST  like that of a finch; and the common tumbler
0027  rought from distant countries, we can make an  ALMOST  perfect series between the extremes of struct
0031  r fully to realise what they have done, it is  ALMOST  necessary to read several of the many treatis
0031  something quite plastic, which they can model  ALMOST  as they please. If I had space I could quote
0031  uainted with the works of agriculturists than  ALMOST  any other individual, and who was himself a v
0031  pedigree; and these have now been exported to  ALMOST  every quarter of the world. The improvement i
0037  final result is concerned, has been followed  ALMOST  unconsciously. It has consisted in always cul
0038  ed man, it should not be overlooked that they  ALMOST  always have to struggle for their own food, a
0041  mutation of a large amount of modification in  ALMOST  any desired direction. But as variations mani
0042  pt up; such breeds as we do sometimes see are  ALMOST  always imported from some other country, ofte
0044  distinct act of creation. The term variety is  ALMOST  equally difficult to define, but here communi
0044  t to define, but here community of descent is  ALMOST  universally implied, though it can rarely be
0046  ity in these main nerves in Coccus, which may  ALMOST  be compared to the irregular branching of the
0050  y attract his attention, varieties of it will  ALMOST  universally be found recorded. These varietie
0051  een his doubtful forms, he will have to trust  ALMOST  entirely to analogy, and his difficulties wil
0058  mr. Watson as doubtful species, but which are  ALMOST  universally ranked by British botanists as do
0065  La Plata, clothing square leagues of surface  ALMOST  to the exclusion of all other plants, have be
0065  ions in their new homes. In a state of nature  ALMOST  every plant produces seed, and amongst animal
0067  so little, and the number of the species will  ALMOST  instantaneously increase to any amount. The f
0069  or absolute deserts, the struggle for life is  ALMOST  exclusively with the elements. That climate a
0075  nd grass feeding quadrupeds. But the struggle  ALMOST  invariably will be most severe between the in
0075  same species, the struggle will generally be  ALMOST  equally severe, and we sometimes see the cont
0081  proportional numbers of its inhabitants would  ALMOST  immediately undergo a change, and some specie
0091  its, and another hunting on marshy ground and  ALMOST  nightly catching woodcocks or snipes. The ten
0095  he preserved being; and as modern geology has  ALMOST  banished such views as the excavation of a gr
0096  ody of facts, showing, in accordance with the  ALMOST  universal belief of breeders, that with anima
0097  close together that self fertilisation seems  ALMOST  inevitable. Many flowers, on the other hand,
0101  s, though agreeing closely with each other in  ALMOST  their whole organisation, yet are not rarely,
0102  the area be large, its several districts will  ALMOST  certainly present different conditions of lif
0107  the natural scale. These anomalous forms may  ALMOST  be called living fossils; they have endured t
0109  ber of specific forms goes on perpetually and  ALMOST  indefinitely increasing, numbers inevitably m
0113  and each variety of grass is annually sowing  ALMOST  countless seeds; and thus, as it may be said,
0129  s, and these round other points, and so on in  ALMOST  endless cycles. On the view that each species
0135  se in their progenitors; for as the tarsi are  ALMOST  always lost in many dung feeding beetles, the
0136  strongly insisted on by Mr. Wollaston, of the  ALMOST  entire absence of certain large groups of bee
0136  umerous, and which groups have habits of life  ALMOST  necessitating frequent flight; these several
0149  hich has to cut all sorts of things may be of  ALMOST  any shape; whilst a tool for some particular
0151  order, and I am fully convinced that the rule  ALMOST  invariably holds good with cirripedes. I shal
0155  in our chapter on Classification. It would be  ALMOST  superfluous to adduce evidence in support of
0160  t we can see to the contrary, transmitted for  ALMOST  any number of generations. When a character w
0167  of God a mere mockery and deception; I would  ALMOST  as soon believe with the old and ignorant cos
0174  period. Geology would lead us to believe that  ALMOST  every continent has been broken up into islan
0175  way insensibly. But when we bear in mind that  ALMOST  every species, even in its metropolis, would
0179  fication through natural selection, they will  ALMOST  certainly be beaten and supplanted by the for
0183  to changed habits; both probably often change  ALMOST  simultaneously. Of cases of changed habits it
0184  (parus major) may be seen climbing branches,  ALMOST  like a creeper. It often, like a shrike, kill
0189  ht alterations, generation will multiply them  ALMOST  infinitely, and natural selection will pick o
0193  , is the same in Orchis and Asclepias, genera  ALMOST  as remote as possible amongst flowering plant
0194  e meet with this admission in the writings of  ALMOST  every experienced naturalist; or, as Milne Ed
0201  , namely, to warn its prey to escape. I would  ALMOST  as soon believe that the cat curls the end of
0203  because the very process of natural selection  ALMOST  implies the continual supplanting and extinct
0205  ated transitions. We are far too ignorant, in  ALMOST  every case, to be enabled to assert that any
0210  canon of Natura non facit saltum applies with  ALMOST  equal force to instincts as to bodily organs.
0216  s under domestication, for the mother hen has  ALMOST  lost by disuse the power of flight. Hence we
0235  d on by natural selection; cases of instincts  ALMOST  identically the same in animals so remote in
0238  d males, but from each other, sometimes to an  ALMOST  incredible degree, and are thus divided into
```

Page **(Key Word)**

Page ***(Key Word)***

```
0246 mirable observers, Kolreuter and Gartner, who ALMOST devoted their lives to this subject, without
0249 , that I cannot doubt the correctness of this ALMOST universal belief amongst breeders. Hybrids ar
0254 als, we must either give up the belief of the ALMOST universal sterility of distinct species of an
0257 been more largely crossed than the species of ALMOST any other genus; but Gartner found that N. ac
0259 of their pure parents; and these hybrids are ALMOST always utterly sterile, even when the other h
0266 ry different class of facts. It is an old and ALMOST universal belief, founded, I think, on a cons
0267 lainly see that great benefit is derived from ALMOST any change in the habits of life. Again, both
0268 fertile offspring. I fully admit that this is ALMOST invariably the case. But if we look to variet
0275 emblances seem chiefly confined to characters ALMOST monstrous in their nature, and which have sud
0285 onsideration of these facts impresses my mind ALMOST in the same manner as does the vain endeavour
0286 le indented length; and we must remember that ALMOST all strata contain harder layers or nodules,
0290 the geological formations of each region are ALMOST invariably intermittent; that is, have not fo
0292 coast action. Thus the geological record will ALMOST necessarily be rendered intermittent. I feel
0292 in fossils have been accumulated. Nature may ALMOST be said to have guarded against the frequent
0297 on, we find that the embedded fossils, though ALMOST universally ranked as specifically different,
0300 h fossiliferous formation would be destroyed, ALMOST as soon as accumulated, by the incessant coas
0305 rouchout the world at this same period. It is ALMOST superfluous to remark that hardly any fossil
0307 n the progenitors of these orders, they would ALMOST certainly have been long ago supplanted and e
0310 pecies appear in our European formations; the ALMOST entire absence, as at present known, of fossi
0315 as whilst subsiding, our formations have been ALMOST necessarily accumulated at wice and irregular
0315 creation, but only an occasional have, taken ALMOST at hazard, in a slowly changing drama. We can
0315 d not be identically the same; for both would ALMOST certainly inherit different characters from t
0316 pigeon, for the newly formed fantail would be ALMOST sure to inherit from its new progenitor some
0320 consequent extinction of less favoured forms ALMOST inevitably follows. It is the same with our d
0321 lian seas; and a few members of the great and ALMOST extinct group of Ganoid fishes still inhabit
0322 oup of species. On the Forms of Life changing ALMOST simultaneously throughout the World. Scarcely
0322 than the fact, that the forms of life change ALMOST simultaneously throughout the world. Thus our
0336 least those inhabiting the sea, have changed ALMOST simultaneously throughout the world, and ther
0338 ge. This process, whilst it leaves the embryo ALMOST unaltered, continually adds, in the course of
0343 me extent. The extinction of old forms is the ALMOST inevitable consequence of the production of n
0346 matal and other physical conditions. Of late, ALMOST every author who has studied the subject has
0346 s conclusion. The case of America alone would ALMOST suffice to prove its truth: for if we exclude
0346 northern parts where the circumpolar land is ALMOST continuous, all authors agree that one of the
0346 ests, marshes, lakes, and great rivers, under ALMOST every temperature. There is hardly a climate
0347 cepting in the northern parts, where the land ALMOST joins, and where, under a slightly different
0347 er the same latitude: for these countries are ALMOST as much isolated from each other as is possib
0349 pacific and the eastern shores of Africa, on ALMOST exactly opposite meridians of longitude. But
0351 ich will have supervened since ancient times, ALMOST any amount of migration is possible. But in m
0354 by their progenitor. If it can be shown to be ALMOST invariably the case, that a region, of which
0357 lly bridged over every ocean, and have united ALMOST every island to some mainland. If indeed the
0357 ow cuite separate, have been continuously, or ALMOST continuously, united with each other, and wit
0358 in the marine faunas on the opposite sides of ALMOST every continent, the close relation of the te
0358 er continuity with continents. Nor does their ALMOST universally volcanic composition favour the a
0358 greater or less facilities may be said to be ALMOST wholly unknown. Until I tried, with Mr. Berke
0363 r many facts could be given showing that soil ALMOST everywhere is charged with seeds. Reflect for
0364 alt water, they could not endure our climate. ALMOST every year, one or two land birds are blown a
0365 had become fully stocked with inhabitants. On ALMOST bare land, with few or no destructive insects
0366 planation of these facts. We have evidence of ALMOST every conceivable kind, organic and inorganic
0368 e species on distant mountain summits, we may ALMOST conclude without other evidence, that a colde
0370 hall see that under the Polar Circle there is ALMOST continuous land from western Europe, through
0370 of the same plants and animals inhabited the ALMOST continuous circumpolar land; and that these p
0371 arts of the Old and New Worlds will have been ALMOST continuously united by land, serving as a bri
0373 es of the world. But we have good evidence in ALMOST every case, that the epoch was included withi
0380 ked with endemic Alpine forms; but these have ALMOST everywhere largely yielded to the more domina
0382 striking passage has speculated, in language ALMOST identical with mine, on the effects of great
0382 p and surviving in the mountian fastnesses of ALMOST every land, which serve as a record, full of
0383 spersal would follow from this capacity as an ALMOST necessary consequence. We can here consider o
0384 continent the species often range widely and ALMOST capriciously: for two river systems will have
0391 bird, and many European and African birds are ALMOST every year blown there, as I am informed by M
0394 n oceanic islands, aerial mammals do occur on ALMOST every island. New Zealand possesses two bats
0397 e instance of one of the cases of difficulty. ALMOST all oceanic islands, even the most isolated a
0398 miles from the shores of South America. Here ALMOST every product of the land and water bears the
0399 alogous facts could be given: indeed it is an ALMOST universal rule that the endemic productions o
0399 fact becomes an anomaly. But this difficulty ALMOST disappears on the view that both New Zealand,
0400 e situated so near each other that they would ALMOST certainly receive immigrants from the same or
0402 l hold their own places and keep separate for ALMOST any length of time. Being familiar with the f
0408 rsified conditions of life, there would be an ALMOST endless amount of organic action and reaction
0415 ps from this cause it has partly arisen, that ALMOST all naturalists lay the greatest stress on re
0415 es not determine its classificatory value, is ALMOST shown by the one fact, that in allied groups,
0415 en most fully acknowledged in the writings of ALMOST every author. It will suffice to quote the hi
0417 both of high physiological importance and of ALMOST universal prevalence, and yet leave us in no
0419 genera, they seem to be, at least at present, ALMOST arbitrary. Several of the best botanists; suc
0426 f beings having different habits, we may feel ALMOST sure, on the theory of descent, that these ch
0427 t importance to the welfare of the being, are ALMOST valueless to the systematist. For animals, be
0428 ng large and their parents dominant, they are ALMOST sure to spread widely, and to seize on more a
0431 , even by the aid of a genealogical tree, and ALMOST impossible to do this without this aid, we ca
0433 ts from the struggle for existence, and which ALMOST inevitably induces extinction and divergence
0434 in homologous organs: the parts may change to ALMOST any extent in form and size, and yet they alw
0436 ws and legs in crustaceans. It is familiar to ALMOST every one, that in a flower the relative posi
0444 of the silk moth; or, again, in the horns of ALMOST full grown cattle. But further than this, var
0445 parents, and this; judging by the eye, seemed ALMOST to be the case; but on actually measuring the
0446 of the other breeds, in all its proportions, ALMOST exactly as much as in the adult state. The tw
0460 fficulty can be mastered. With respect to the ALMOST universal sterility of species when first cro
0460 which forms so remarkable a contrast with the ALMOST universal fertility of varieties when crossed
0466 erations, may continue to be inherited for an ALMOST infinite number of generations. On the other
0468 rally be most severe between them; it will be ALMOST equally severe between the varieties of the s
```

Page ***(Key Word)***

Page ***(Key Word)***

```
0471  and diverging in character, together with the  ALMOST  inevitable contingency of much extinction, ex
0475  s a part in the history of the organic world,   ALMOST  inevitably follows on the principle of natura
0478  parents formerly inhabited both areas; and we   ALMOST  invariably find that wherever many closely al
0481  f that species were immutable productions was   ALMOST  unavoidable as long as the history of the wor
0481  many slight variations, accumulated during an   ALMOST  infinite number of generations. Although I am
0486  study of natural history become! A grand and    ALMOST  untrodden field of inquiry will be opened, on
0487  all causes of organic change is one which is    ALMOST  independent of altered and perhaps suddenly a
0489  rowth with Reproduction: Inheritance which is   ALMOST  implied by reproduction: Variability from the
0002  been cautious in trusting to good authorities   ALONE. I can here give only the general conclusions
0012  hrown on this principle by theoretical writers  ALONE. When a deviation appears not unfrequently, a
0013  tted from one sex to both sexes, or to one sex  ALONE, more commonly but not exclusively to the lik
0013  usively, or in a much greater degree, to males  ALONE. A much more important rule, which I think ma
0019  le, as many sheep, and several goats in Europe  ALONE, and several even within Great Britain. One a
0073  fertilisation of our clovers; but humble bees   ALONE visit the common red clover (Trifolium praten
0078  hey are destroyed by the rigour of the climate  ALONE. Not until we reach the extreme confines of l
0089  esponding ages or seasons, either by the males  ALONE, or by the males and females; but I have not
0093  be advantageous to a plant to produce stamens   ALONE in one flower or on one whole plant, and pist
0093  one flower or on one whole plant, and pistils   ALONE in another flower or on another plant. In pla
0095  on red clover, which is visited by humble bees  ALONE: so that whole fields of the red clover offer
0096  s, i must trust to some general considerations  ALONE. In the first place, I have collected so larg
0097  inishes vigour and fertility; that these facts  ALONE incline me to believe that it is a general la
0111  me character and in a greater degree; but this  ALONE would never account for so habitual and large
0116  ch by being adapted to digest vegetable matter  ALONE, or flesh alone, draws most nutriment from th
0116  ted to digest vegetable matter alone, or flesh  ALONE, draws most nutriment from these substances.
0125  , as now explained, to the formation of genera  ALONE. If, in our diagram, we suppose the amount of
0127  will also give characters useful to the males   ALONE, in their struggles with other males. Whether
0132  pparently differ essentially from an ovule, is  ALONE affected. But why, because the reproductive s
0146  t whole orders, are entirely due to the manner  ALONE in which natural selection can act. For insta
0149  the part has to serve for one special purpose   ALONE. In the same way that a knife which has to cu
0178  es which they tend to connect. From this cause  ALONE the intermediate varieties will be liable to
0190  had performed two functions, for one function  ALONE, and thus wholly change its nature by insensi
0197  ate these latter remarks. If green woodpeckers  ALONE had existed, and we did not know that there w
0215  stand and hunt best. On the other hand, habit   ALONE in some cases has sufficed: no animal is more
0215  , never wish to sit on their eggs. Familiarity  ALONE prevents our seeing how universally and large
0216  ll an accident. In some cases compulsory habit  ALONE has sufficed to produce such inherited mental
0217  ounc would probably have to be fed by the male  ALONE. But the American cuckoo is in this predicame
0221  ith their masters in making the nest; and they  ALONE open and close the doors in the morning and e
0223  exclusive care of the larvae, and the masters   ALONE go on slave making expeditions. In Switzerlan
0223  food for the community. In England the masters  ALONE usually leave the nest to collect building ma
0236  s of eyes, and in instinct. As far as instinct  ALONE is concerned, the prodigious difference in th
0237  not only to one sex, but to that short period   ALONE when the reproductive system is active, as in
0238  rent: in Cryptocerus, the workers of one caste  ALONE carry a wonderful sort of shield on their hea
0239  ividual neuters in the same nest; but in a few  ALONE: and that by the long continued selection of
0242  ure or instincts of the fertile members, which  ALONE leave descendants. I am surprised that no one
0252  raise large beds of the same hybrids; and such  ALONE are fairly treated, for by insect agency th
0274  f several experiments made by Kolreuter. These  ALONE are the unimportant differences, which Gartne
0275  ors on the supposed fact, that mongrel animals  ALONE are born closely like one of their parents: b
0310  cti of this history we possess the last volume  ALONE, relating only to two or three countries. Of
0329  tention either to the living or to the extinct  ALONE, the series is far less perfect than if we co
0335  species are distinct in each stage. This fact   ALONE, from its generality, seems to have shaken Pr
0341  ner case, two or three species of two or three  ALONE of the six older genera will have been the pa
0343  d, when the formations of any one great region  ALONE, as that of Europe, are considered: he may ur
0346  s come to this conclusion. The case of America  ALONE would almost suffice to prove its truth: for
0350  heory, is simply inheritance, that cause which  ALONE, as far as we positively know, produces organ
0353  rity of species have been produced on one side  ALONE, and have not been able to migrate to the oth
0353  each species having been produced in one area  ALONE, and having subsequently migrated from that a
0376  manner. This brief abstract applies to plants  ALONE: some strictly analogous facts could be given
0392  d and having to compete with herbaceous plants  ALONE, might readily gain an advantage by growing t
0417  e of characters, even when none are important,  ALONE explains, I think, that saying of Linnaeus, t
0418  for this purpose than that of the adult, which  ALONE plays its full part in the economy of nature.
0443  n hereditary disease, which appears in old age  ALONE, has been communicated to the offspring from
0449  ved true. It can be proved true in those cases  ALONE in which the ancient state, now supposed to b
0454  purpose. Or an organ might be retained for one  ALONE of its former functions. An organ, when rende
0468  ity. Man, though acting on external characters  ALONE and often capriciously, can produce within a
0053  nd other questions; hereafter to be discussed.  ALPH. de Candolle and others have shown that plants
0115  new homes. But the case is very different; and  ALPH. de Candolle has well remarked in his great an
0146  which natural selection can act. For instance,  ALPH. de Candolle has remarked that winged seeds ar
0175  etimes it is quite remarkable how abruptly, as  ALPH. de Candolle has observed, a common alpine spe
0360  hardly be transported by any other means; as   ALPH. de Candolle has shown that such plants genera
0379  sted on by Hooker in regard to America, and by  ALPH. de Candolle in regard to Australia, that many
0386  ands. This is strikingly shown, as remarked by  ALPH. de Candolle, in large groups of terrestrial p
0387  fine water lily the Nelumbium, and remembered   ALPH. de Candolle's remarks on this plant, I though
0389  ompared with those on equal continental areas:  ALPH. de Candolle admits this for plants, and Wolla
0392  include only herbaceous species; now trees, as  ALPH. de Candolle has shown, generally have, whatev
0402  ng in a large proportion of cases, as shown by  ALPH. de Candolle, to distinct genera. In the Galap
0406  d which has lately been admirably discussed by  ALPH. de Candolle in regard to plants, namely, that
0045  sh waters of the Baltic, or dwarfed plants on   ALPINE summits, or the thicker fur of an animal from
0084  cts green, and bark feeders mottled grey; the   ALPINE ptarmigan white in winter, the red grouse the
0175  , as Alph. de Candolle has observed, a common  ALPINE species disappears. The same fact has been no
0349  lofty peaks of the Cordillera and we find an    ALPINE species of bizcacha: we look to the waters, a
0365  by hundreds of miles of low lands, where the   ALPINE species could not possibly exist, is one of t
0366  e covered with snow and ice, and their former   ALPINE inhabitants would descend to the plains. By t
0367  we can thus also understand the fact that the   ALPINE plants of each mountain range are more especi
0367  generally have been due south and north. The   ALPINE plants, for example, of Scotland, as remarked
0368  tory a manner the present distribution of the   ALPINE and Arctic productions of Europe and America,
0368  een liable to much modification. But with our   ALPINE productions, left isolated from the moment of
```

Page ***(Key Word)***

Page **(Key Word)***

```
0369  probability have become mingled with ancient ALPINE species, which must have existed on the mount
0369  been the case; for if we compare the present ALPINE plants and animals of the several great Europ
0371  tion, for far more modification than with the ALPINE productions, left isolated, within a much mor
0376  from polar towards equatorial latitudes, the ALPINE or mountain floras really become less and les
0380  glacial period they were stocked with endemic ALPINE forms; but these have almost everywhere range
0403  every mountain, in every lake and marsh. For ALPINE species, excepting in so far as the same form
0403  ding lowlands; thus we have in South America, ALPINE humming birds, Alpine rodents, Alpine plants,
0403  have in South America, Alpine humming birds, ALPINE rodents, Alpine plants, etc., all of strictly
0403  merica, Alpine humming birds, Alpine rodents, ALPINE plants, etc., all of strictly American forms,
0406  era themselves ranging widely, such facts, as ALPINE, lacustrine, and marsh productions being rela
0365  same plants living on the snowy regions of the ALPS or Pyrenees, and in the extreme northern parts
0367  e central parts of Europe, as far south as the ALPS and Pyrenees, and even stretching into Spain.
0015  generally perceive in each domestic race, as ALREADY remarked, less uniformity of character than
0042  of new races, at least, in a country which is ALREADY stocked with other races. In this respect en
0054  bitants of the country, the species which are ALREADY dominant will be the most likely to yield of
0057  ge are now manufacturing, many of the species ALREADY manufactured still to a certain extent resem
0063  the plants of the same and other kinds which ALREADY clothe the ground. The missletoe is dependen
0067  which suffer most from germinating in ground ALREADY thickly stocked with other plants. Seedlings
0069  ill increase in numbers, and, as each area is ALREADY fully stocked with inhabitants, the other sp
0077  nds in the closest relation to the land being ALREADY thickly clothed by other plants; so that the
0088  weapons. The males of carnivorous animals are ALREADY well armed; though to them and to others, sp
0094  ight a degree under nature, then as pollen is ALREADY carried regularly from flower to flower, and
0104  nimals which unite for each birth; but I have ALREADY attempted to show that we have reason to bel
0106  e infinitely complex from the large number of ALREADY existing species; and if some of these many
0106  the new forms produced on large areas, which ALREADY have been victorious over many competitors,
0109  increase of all organic beings, each area is ALREADY fully stocked with inhabitants, it follows t
0121  oduced. In each genus, the species, which are ALREADY extremely different in character, will gener
0125  existence, it will chiefly act on those which ALREADY have some advantage; and the largeness of an
0127  dence given in the following chapters. But we ALREADY see how it entails extinction; and how large
0134  ion of the conditions of life. Indirectly, as ALREADY remarked, they seem to play an important par
0162  reversion or from analogous variation) which ALREADY occur in some other members of the same grou
0175  es into competition; and as these species are ALREADY defined objects (however they may have becom
0176  any form existing in lesser numbers would, as ALREADY remarked, run a greater chance of being exte
0178  ese intermediate varieties will, from reasons ALREADY assigned (namely from what we know of the ac
0184  nt, and if better adapted competitors did not ALREADY exist in the country, I can see no difficult
0192  cirripedes had become extinct; and they have ALREADY suffered far more extinction than have sessi
0208  the sixth stage, so that much of its work was ALREADY done for it, far from feeling the benefit of
0208  had left off, and thus tried to complete the ALREADY finished work. If we suppose any habitual ac
0218  on, yet that when this insect finds a burrow ALREADY made and stored by another sphex it takes ad
0227  hat if we could slightly modify the instincts ALREADY possessed by the Melipona, and in themselves
0227  would not be very surprising, seeing that she ALREADY does so to a certain extent, and seeing what
0227  to arrange her cells in level layers, as she ALREADY does her cylindrical cells; and we must furt
0227  several are making their spheres; but she is ALREADY so far enabled to judge of distance, that sh
0249  t this would be injurious to their fertility, ALREADY lessened by their hybrid origin. I am streng
0255  ame rules apply to both kingdoms. It has been ALREADY remarked, that the degree of fertility, both
0279  ore important manner on the presence of other ALREADY defined organic forms, than on climate; and,
0321  ondary period, we must remember what has been ALREADY said on the probable wide intervals of time
0325  over older forms; and those forms, which are ALREADY dominant, or have some advantage over the ot
0326  nt, varying, and far spreading species, which ALREADY have invaded to a certain extent the territo
0326  ations, and that severe competition with many ALREADY existing forms, would be highly favourable,
0327  dominant owing to inheritance, and to having ALREADY had some advantage over their parents or ove
0350  tion of organism to organism being, as I have ALREADY often remarked, the most important of all re
0350  species, abounding in individuals, which have ALREADY triumphed over many competitors in their own
0388  the individuals of the species, however few, ALREADY occupying any pond, yet as the number of kin
0389  o inhabit distant points of the globe. I have ALREADY stated that I cannot honestly admit Forbes's
0390  ight have been expected on my theory, for, as ALREADY explained, species occasionally arriving aft
0439  e appendage, is explained. Embryology. It has ALREADY been casually remarked that certain organs i
0466  son to believe that a modification, which has ALREADY been inherited for many generations, may con
0469  seems to me to be in itself probable. I have ALREADY recapitulated, as fairly as I could, the opp
0473  t, we can understand this fact; for they have ALREADY varied since they branched off from a common
0482  d with much flexibility of mind, and who have ALREADY begun to doubt on the immutability of specie
0486  n one more species added to the infinitude of ALREADY recorded species. Our classifications will c
0072  become feral, and this would certainly greatly ALTER (as indeed I have observed in parts of South
0102  work. But when many men, without intending to ALTER the breed, have a nearly common standard of p
0269  n nearly the same manner; and does not wish to ALTER their general habits of life. Nature acts uni
0329  vered so many fossil links, that he has had to ALTER the whole classification of these two orders;
0435  t of modification there will be no tendency to ALTER the framework of bones or the relative connex
0189  ys intently watching each slight accidental ALTERATION in the transparent layers; and carefully se
0189  parent layers; and carefully selecting each ALTERATION which under varied circumstances, may in an
0189  ing bodies, variation will cause the slight ALTERATIONS, generation will multiply them almost infi
0024  , which is the rock pigeon in a very slightly ALTERED state, has become feral in several places. A
0035  rom the spaniel, and has probably been slowly ALTERED from it. It is known that the English pointe
0082  the species, by better adapting them to their ALTERED conditions, would tend to be preserved; and
0124  retained the form of (F), either unaltered or ALTERED only in a slight degree. In this case, its a
0187  stages of descent, in an unaltered or little ALTERED condition. Amongst existing Vertebrata, we f
0332  slightly modified in relation to its slightly ALTERED conditions of life, and yet retain throughou
0422  might be that some very ancient language had ALTERED little, and had given rise to few new langua
0422  eral races, descended from a common race) had ALTERED much, and had given rise to many new languag
0487  change is one which is almost independent of ALTERED and perhaps suddenly altered physical condit
0487  t independent of altered and perhaps suddenly ALTERED physical conditions, namely, the mutual rela
0034  he has no wish or expectation of permanently ALTERING the breed. Nevertheless I cannot doubt that
0467  l to him at the time, without any thought of ALTERING the breed. It is certain that he can largely
0292  ill less could this have happened during the ALTERNATE periods of elevation; or, to speak more acc
0424  adult; as he likewise includes the so called ALTERNATE generations of Steenstrup, which can only i
0465  on the subsiding bed of the sea. During the ALTERNATE periods of elevation and of stationary leve
0233  n wasp) making hexagonal cells, if she work ALTERNATELY on the inside and outside of two or three
```

Page **(Key Word)***

Page ***(Key Word)***

```
0360  a box in the actual sea, so that they were ALTERNATELY wet and exposed to the air like really flo
0382  entical with mine, on the effects of great ALTERNATIONS of climate on geographical distribution. I
0254  te fertile under domestication. This latter ALTERNATIVE seems to me the most probable, and I am in
0264  then perish at an early period. This latter ALTERNATIVE has not been sufficiently attended to; but
0004  conviction of the high value of such studies, ALTHOUGH they have been very commonly neglected by na
0006  many past geological epochs in its history. ALTHOUGH much remains obscure, and will long remain o
0016  monstrous characteri by which I mean, that, ALTHOUGH differing from each other, and from the othe
0025  re coloured and marked like the rock pigeon, ALTHOUGH no other existing species is thus coloured a
0027  ture with all the domestic breeds. Secondly, ALTHOUGH an English carrier or short faced tumbler di
0042  nal rambling habits, cannot be matched, and, ALTHOUGH so much valued by women and children, we har
0042  from some other country, often from islands. ALTHOUGH I do not doubt that some domestic animals va
0063  , and no prudential restraint from marriage. ALTHOUGH some species may be now increasing, more or
0068  l probability, be less game than at present, ALTHOUGH hundreds of thousands of game animals are no
0084  now different from what they formerly were. ALTHOUGH natural selection can act only through and f
0092  flower to flower, and a cross thus effected, ALTHOUGH nine tenths of the pollen were destroyed, it
0100  parated sexes. When the sexes are separated, ALTHOUGH the male and female flowers may be produced
0105  ll isolated area, such as an oceanic island, ALTHOUGH the total number of the species inhabiting i
0105  l times; and this we are incapable of doing. ALTHOUGH I do not doubt that isolation is of consider
0106  t, have concurred. Finally, I conclude that, ALTHOUGH small isolated areas probably have been in s
0119  er of the descendants will not be increased; ALTHOUGH the amount of divergent modification may hav
0126  vision of the animal and vegetable kingdoms. ALTHOUGH extremely few of the most ancient species ma
0134  species keeping true, or not varying at all, ALTHOUGH living under the most opposite climates. Suc
0169  the characters of their ancient progenitors. ALTHOUGH new and important modifications may not aris
0181  is, also, furnished with an extensor muscle. ALTHOUGH no graduated links of structure, fitted for
0185  and, grebes and coots are eminently aquatic, ALTHOUGH their toes are only bordered by membrane. Wh
0188  eagle might be formed by natural selection. ALTHOUGH in this case he does not know any of the tra
0192  g the ova from being washed out of the sack? ALTHOUGH we must be extremely cautious in concluding
0193  anomalous organ, it should be observed that, ALTHOUGH the general appearance and function of the o
0194  ommon to inheritance from the same ancestor. ALTHOUGH in many cases it is most difficult to conjec
0196  e been transmitted in nearly the same state, ALTHOUGH now become of very slight use; and any actua
0201  ave been produced through natural selection. ALTHOUGH many statements may be found in works on nat
0204  rushes, and petrels with the habits of auks. ALTHOUGH the belief that an organ so perfect as the e
0211  ively excrete for the sole good of the ants. ALTHOUGH I do not believe that any animal in the worl
0218  lately shown good reason for believing that ALTHOUGH the Tachytes nigra generally makes its own b
0220  for information on this and other subjects. ALTHOUGH fully trusting to the statements of Huber an
0222  slaves, as has been described by Mr. Smith. ALTHOUGH so small a species, it is very courageous, a
0251  enera, as Lobelia, Passiflora and Verbascum. ALTHOUGH the plants in these experiments appeared per
0251  experiments appeared perfectly healthy, and ALTHOUGH both the ovules and pollen of the same flowe
0253  assuredly be lost in a very few generations. ALTHOUGH I do not know of any thoroughly well authent
0254  , and I am inclined to believe in its truth, ALTHOUGH it rests on no direct evidence. I believe, f
0261  absolutely governed by systematic affinity. ALTHOUGH many distinct genera within the same family
0262  sed with their own pollen. We thus see, that ALTHOUGH there is a clear and fundamental difference
0263  arious species has been a special endowment; ALTHOUGH in the case of crossing, the difficulty is a
0270  , growing near each other in his garden; and ALTHOUGH these plants have separated sexes, they neve
0273  r further insists that when any two species, ALTHOUGH most closely allied to each other, are cross
0293  they are rare, they may be here passed over. ALTHOUGH each formation has indisputably required a v
0293  onal weight to the following considerations. ALTHOUGH each formation may mark a very long lapse of
0328  tain stages in England with those in France, ALTHOUGH he finds in both a curious accordance in the
0348  no well defined and distinct marine faunas. ALTHOUGH hardly one shell, crab or fish is common to
0362  however, were always killed by this process. ALTHOUGH the beaks and feet of birds are generally qu
0379  perate latitudes of the opposite hemisphere. ALTHOUGH we have reason to believe from geological ev
0387  the Nelumbium luteum) in a heron's stomach; ALTHOUGH I do not know the fact, yet analogy makes me
0388  r egg will have a good chance of succeeding. ALTHOUGH there will always be a struggle for life bet
0390  ar more fully and perfectly than has nature. ALTHOUGH in oceanic islands the number of kinds of in
0399  lso plainly related to South America, which, ALTHOUGH the next nearest continent, is so enormously
0403  some of which live in crevices of stone; and ALTHOUGH large quantities of stone are annually trans
0427  rstand why analogical or adaptive character, ALTHOUGH of the utmost importance to the welfare of t
0433  e individuals of both sexes and of all ages, ALTHOUGH having few characters in common, under one s
0472  prise at the inhabitants of any one country, ALTHOUGH on the ordinary view supposed to have been s
0477  separated by the whole intertropical ocean. ALTHOUGH two areas may present the same physical cond
0481  ng an almost infinite number of generations. ALTHOUGH I am fully convinced of the truth of the vie
0483  marks of nourishment from the mother's womb? ALTHOUGH naturalists very properly demand a full expl
0488  e history of the world, as at present known, ALTHOUGH of a length quite incomprehensible by us, wi
0022  differ to a slight degree from each other. ALTOGETHER at least a score of pigeons might be chosen
0031  he character of his flock, but to change it ALTOGETHER. It is the magician's wand, by means of whi
0234  ies, or on quite distinct causes, and so be ALTOGETHER independent of the quantity of honey which
0284  ...............................  2,240 Feet. making ALTOGETHER.......................72,584 Fee
0359  r above 90 days; and afterwards germinated. ALTOGETHER out of the 94 dried plants, 18 floated for
0387  ewi the plants were of many kinds, and were ALTOGETHER 537 in number; and yet the viscid mud was a
0002  rors will have crept in, though I hope I have ALWAYS been cautious in trusting to good authorities
0011  . breeders believe that long limbs are almost ALWAYS accompanied by an elongated head. Some instan
0020  of the Domestic Pigeon. Believing that it is ALWAYS best to study some special group, I have, aft
0022  e orifice of the nostrils, of the tongue (not ALWAYS in strict correlation with the length of beak
0037  wed almost unconsciously. It has consisted in ALWAYS cultivating the best known variety, sowing it
0038  it should not be overlooked that they almost ALWAYS have to struggle for their own food, at least
0040  have called it, of unconscious selection will ALWAYS tend, perhaps more at one period than at anot
0041  ly to this plant. No doubt the strawberry had ALWAYS varied since it was cultivated, but the sligh
0042  such breeds as we do sometimes see are almost ALWAYS imported from some other country, often from
0047  sumed hybrid nature of the intermediate links ALWAYS remove the difficulty. In very many cases, ho
0057  and, as far as my imperfect results go, they ALWAYS confirm the view. I have also consulted some
0062  troyed by birds and beasts of prey; we do not ALWAYS bear in mind, that though food may be now sup
0066  ecessary to keep the foregoing considerations ALWAYS in mind, never to forget that every single or
0076  me similarity in habits and constitution, and ALWAYS in structure, the struggle will generally be
0088  nce of leaving offspring. Sexual selection by ALWAYS allowing the victor to breed might surely giv
0090  ng, and so be preserved or selected, provided ALWAYS that they retained strength to master their p
0096  s of course obvious that two individuals must ALWAYS unite for each birth; but in the case of herm
```

Page ***(Key Word)***

Page **********************************(Key Word)**********************************

0099	ctly the reverse, for a plant's own pollen is	ALWAYS prepotent over foreign pollen; but to this su
0102	occupied as might be, natural selection will	ALWAYS tend to preserve all the individuals varying
0103	ardly be counterbalanced by natural selection	ALWAYS tending to modify all the individuals in each
0103	districts. On the above principle, nurserymen	ALWAYS prefer getting seed from a large body of plan
0108	duce new species. That natural selection will	ALWAYS act with extreme slowness; I fully admit. Its
0108	ons occur, and variation itself is apparently	ALWAYS a very slow process. The process will often b
0108	and, I do believe that natural selection will	ALWAYS act very slowly, often only at long intervals
0111	species and species of the same genus. As has	ALWAYS been my practice, let us seek light on this h
0114	t varieties of any one species of grass would	ALWAYS have the best chance of succeeding and of inc
0114	individual and individual must be severe, we	ALWAYS find great diversity in its inhabitants. For
0119	diffied descendant; for natural selection will	ALWAYS act according to the nature of the places whi
0135	heir progenitors; for as the tarsi are almost	ALWAYS lost in many dung feeding beetles, they must
0146	xternal structure of the seeds, which are not	ALWAYS correlated with any differences in the flower
0148	s. thus, as I believe, natural selection will	ALWAYS succeed in the long run in reducing and savin
0153	g as selection is rapidly going on, there may	ALWAYS be expected to be much variability in the str
0164	s not considered as purely bred. The spine is	ALWAYS striped; the legs are generally barred; and t
0177	ne. For forms existing in larger numbers will	ALWAYS have a better chance, within any given period
0183	rades of structure in a fossil condition will	ALWAYS be less, from their having existed in lesser
0189	further we must suppose that there is a power	ALWAYS intently watching each slight accidental alte
0196	injurious deviations in their structure will	ALWAYS have been checked by natural selection. Seein
0198	air the horns are correlated. Mountain breeds	ALWAYS differ from lowland breeds; and a mountainous
0202	ll not produce absolute perfection, nor do we	ALWAYS meet, as far as we can judge, with this high
0203	because the process of natural selection will	ALWAYS be very slow, and will act, at any one time,
0218	and wasted eggs. Many bees are parasitic, and	ALWAYS lay their eggs in the nests of bees of other
0224	ection increasing and modifying the instinct,	ALWAYS supposing each modification to be of use to t
0225	tant point to notice, is that these cells are	ALWAYS made at that degree of nearness to each other
0227	so far enabled to judge of distance, that she	ALWAYS describes her spheres so as to intersect larg
0229	bs it has appeared to me that the bees do not	ALWAYS succeed in working at exactly the same rate f
0230	they gnaw into this from the opposite sides,	ALWAYS working circularly as they deepen each cell.
0230	rictly correct; the first commencement having	ALWAYS been a little hood of wax; but I will not her
0231	nner in which the bees build is curious; they	ALWAYS make the first rough wall from ten to twenty
0231	y thin wall is left in the middle; the masons	ALWAYS piling up the cut away cement, and adding fre
0231	have a thin wall steadily growing upward; but	ALWAYS crowned by a gigantic coping. From all the ce
0233	wo or three cells commenced at the same time,	ALWAYS standing at the proper relative distance from
0238	, that I do not doubt that a breed of cattle,	ALWAYS yielding oxen with extraordinarily long horns
0243	e other hand, the fact that instincts are not	ALWAYS absolutely perfect and are liable to mistakes
0247	how that there is any degree of sterility. He	ALWAYS compares the maximum number of seeds produced
0251	o try it during several subsequent years, and	ALWAYS with the same result. This result has, also,
0256	is likewise innately variable; for it is not	ALWAYS the same when the same two species are crosse
0259	rmediate character between their two parents,	ALWAYS closely resemble one of them; and such hybrid
0259	ir pure parents; and these hybrids are almost	ALWAYS utterly sterile, even when the other hybrids
0259	is innately variable. That it is by no means	ALWAYS the same in degree in the first cross and in
0259	a first cross between any two species is not	ALWAYS governed by their systematic affinity or degr
0261	tation to widely different climates, does not	ALWAYS prevent the two grafting together. As in hybr
0267	se be kept under the same conditions of life,	ALWAYS induces weakness and sterility in the progeny
0271	er, and crossed with the Nicotiana glutinosa,	ALWAYS yielded hybrids not so sterile as those which
0276	osses between the same two species. It is not	ALWAYS equal in degree in a first cross and in the h
0280	ological record. In the first place it should	ALWAYS be borne in mind what sort of intermediate fo
0280	m. but this is a wholly false view; we should	ALWAYS look for forms intermediate between each spec
0282	th more ancient species; and so on backwards,	ALWAYS converging to the common ancestor of each gre
0303	any years ago, the great class of mammals was	ALWAYS spoken of as having abruptly come in at the c
0305	ht to suppose that the seas of the world have	ALWAYS been so freely open from south to north as th
0309	ore metamorphic action than strata which have	ALWAYS remained nearer to the surface. The immense a
0310	t have been heated under great pressure, have	ALWAYS seemed to me to require some special explanat
0319	e successful competitor. It is most difficult	ALWAYS to remember that the increase of every living
0322	increase inordinately, and that some check is	ALWAYS in action, yet seldom perceived by us, the wh
0362	of germination. Certain seeds, however, were	ALWAYS killed by this process. Although the beaks an
0367	would seize on the cleared and thawed ground,	ALWAYS ascending higher and higher, as the warmth in
0388	ood chance of succeeding. Although there will	ALWAYS be a struggle for life between the individual
0411	ication. Classification of varieties. Descent	ALWAYS used in classification. Analogical or adaptiv
0413	group, which, from its familiarity, does not	ALWAYS sufficiently strike us, is in my judgment ful
0414	d to the habits and food of an animal, I have	ALWAYS regarded as affording very clear indications
0415	h are important is generally, but by no means	ALWAYS, true. But their importance for classificatio
0417	character, however important that may be, has	ALWAYS failed; for no part of the organisation is un
0418	, aug. St. Hilaire. If certain characters are	ALWAYS found correlated with others, though no appar
0434	ost any extent in form and size, and yet they	ALWAYS remain connected together in the same order.
0437	plex mouth formed of many parts, consequently	ALWAYS have fewer legs; or conversely, those with ma
0443	e this plainly in our own children; we cannot	ALWAYS tell whether the child will be tall or short,
0462	r during very long periods of time there will	ALWAYS be a good chance for wide migration by many m
0469	under nature variability and a powerful agent	ALWAYS ready to act and select, why should we doubt
0472	to increase in number, with natural selection	ALWAYS ready to adapt the slowly varying descendants
0479	on the principle of successive variations not	ALWAYS supervening at an early age, and being inheri
0481	to other and distinct species, is that we are	ALWAYS slow in admitting any great change of which w
0041	ants, are generally far more successful than	AMATEURS in getting new and valuable varieties. The k
0139	s remarked in regard to the blind fish, the	AMBLYOPSIS, and as is the case with the blind Proteus
0001	the distribution of the inhabitants of South	AMERICA and in the geological relations of the prese
0048	how many of those birds and insects in North	AMERICA and Europe, which differ very slightly from
0049	een well asked, will suffice? If that between	AMERICA and Europe is ample, will that between the C
0064	e of slow breeding cattle and horses in South	AMERICA, and latterly in Australia, had not been wel
0065	o the Himalaya, which have been imported from	AMERICA since its discovery. In such cases, and endl
0067	ially in regard to the feral animals of South	AMERICA. Here I will make only a few remarks, just t
0073	(as indeed I have observed in parts of South	AMERICA) the vegetation: this again would largely af
0134	in this state. The loggerheaded duck of South	AMERICA can only flap along the surface of the water
0137	aided perhaps by natural selection. In South	AMERICA, a burrowing rodent, the tuco tuco, or Cteno
0138	resemblance of the other inhabitants of North	AMERICA and Europe. On my view we must suppose that
0138	ht expect still to see in the cave animals of	AMERICA, affinities to the other inhabitants of that

Page **********************************(Key Word)**********************************

Page **(Key Word)**

```
0179  in nature. Look at the Mustela vison of North AMERICA, which has webbed feet and which resembles a
0183  flycatcher (Saurophagus sulphuratus) in South AMERICA, hovering over one spot and then proceeding
0184  thus breaking them like a nuthatch. In North AMERICA the black bear was seen by Hearne swimming f
0184  sects in the chinks of the bark? Yet in North AMERICA there are woodpeckers which feed largely on
0195  xistence of cattle and other animals in South AMERICA absolutely depends on their power of resisti
0210  ut little observed except in Europe and North AMERICA, and for no instinct being known amongst ext
0243  heritance, how it is that the thrush of South AMERICA lines its nest with mud, in the same peculia
0243  is that the male wrens (Troglodytes) of North AMERICA, build cock nests, to roost in, like the mal
0254  of certain indigenous domestic dogs of South AMERICA, all are quite fertile together; and analogy
0289  c. Lyell in the carboniferous strata of North AMERICA. In regard to mammiferous remains, a single
0289  he superimposed formations; so it is in North AMERICA, and in many other parts of the world. The m
0290  the rising coast of the western side of South AMERICA, no extensive formations with recent or tert
0291  tertiary formation on the west coast of South AMERICA, which has been bulky enough to resist such
0294  ewhat earlier in the palaeozoic beds of North AMERICA than in those of Europe; time having apparen
0294  ographical changes occurred in other parts of AMERICA during this space of time. When such beds as
0299  ain sea shells inhabiting the shores of North AMERICA, which are ranked by some conchologists as d
0299  racy, excepting those of the United States of AMERICA. I fully agree with Mr. Godwin Austen, that
0300  action, as we now see on the shores of South AMERICA. During the periods of subsidence there woul
0303  species have been discovered in India, South AMERICA, and in Europe even as far back as the eocen
0308  er immense territories in Russia and in North AMERICA, do not support the view, that the older a f
0308  f the existing continents of Europe and North AMERICA. But we do not know what was the state of th
0309  ome parts of the world, for instance in South AMERICA, of bare metamorphic rocks, which must have
0318  its introduction by the Spaniards into South AMERICA, has run wild over the whole country and has
0319  naturalisation of the domestic horse in South AMERICA, that under more favourable conditions it wo
0323  l chalk itself can be found; namely, in North AMERICA, in equatorial South America, in Tierra del
0323  namely, in North America, in equatorial South AMERICA, in Tierra del Fuego, at the Cape of Good Ho
0323  ormations of Russia, Western Europe and North AMERICA, a similar parallelism in the forms of life
0324  to be compared with those now living in South AMERICA or in Australia, the most skilful naturalist
0324  ted at the present day on the shores of North AMERICA would hereafter be liable to be classed with
0324  ictly modern beds, of Europe, North and South AMERICA, and Australia, from containing fossil remai
0325  ange sequence, we turn our attention to North AMERICA, and there discover a series of analogous ph
0339  living marsupials of that continent. In South AMERICA, a similar relationship is manifest, even to
0339  nt climate of Australia and of parts of South AMERICA under the same latitude, would attempt to ac
0340  pes should have been solely produced in South AMERICA. For we know that Europe in ancient times wa
0340  in the publications above alluded to, that in AMERICA the law of distribution of terrestrial mamma
0340  formerly different from what it now is. North AMERICA formerly partook strongly of the present cha
0341  huge monsters have left behind them in South AMERICA the sloth, armadillo, and anteater, as their
0341  aracters to the species still living in South AMERICA; and some of these fossils may be the actual
0341  parently is the case of the Edentata of South AMERICA, still fewer genera and species will have le
0346  ject has come to this conclusion. The case of AMERICA alone would almost suffice to prove its trut
0347  in australia, South Africa, and western South AMERICA, between latitudes 25 degrees and 35 degrees
0347  again we may compare the productions of South AMERICA south of Lat. 35 degrees with those north of
0347  e inhabitants of Australia, Africa, and South AMERICA under the same latitude: for these countries
0348  stern and western shores of South and Central AMERICA; yet these great faunas are separated only b
0348  isthmus of Panama. Westward of the shores of AMERICA, a wide space of open ocean extends, with no
0348  ree approximate faunas of Eastern and Western AMERICA and the eastern Pacific Islands, yet many fi
0352  ammal common to Europe and Australia or South AMERICA? The conditions of life are nearly the same,
0353  animals and plants have become naturalised in AMERICA and Australia; and some of the aboriginal pl
0357  th europe or Africa, and Europe likewise with AMERICA. Other authors have thus hypothetically brid
0364  ir course, would never bring seeds from North AMERICA to Britain, though they might and do bring s
0364  n across the whole Atlantic Ocean, from North AMERICA to the western shores of Ireland and England
0365  the White Mountains, in the United States of AMERICA, are all the same with those of Labrador, an
0366  t geological period, central Europe and North AMERICA suffered under an Arctic climate. The ruins
0367  od came on a little earlier or later in North AMERICA than in Europe, so will the southern migrati
0368  e alpine and Arctic productions of Europe and AMERICA, that when in other regions we find the same
0369  he lower mountains and on the plains of North AMERICA and Europe; and it may be reasonably asked h
0370  m western Europe, through Siberia, to eastern AMERICA. And to this continuity of the circumpolar l
0371  le identity, between the productions of North AMERICA and Europe, a relationship which is most rem
0371  observers, that the productions of Europe and AMERICA during the later tertiary stages were more c
0372  eastern and western shores of temperate North AMERICA; and the still more striking case of many cl
0372  t inhabitants of the temperate lands of North AMERICA and Europe, are inexplicable on the theory o
0372  compare, for instance, certain parts of South AMERICA with the southern continents of the Old Worl
0373  south eastern corner of Australia. Looking to AMERICA; in the northern half, ice borne fragments o
0373  ntains. In the Cordillera of Equatorial South AMERICA, glaciers once extended far below their pres
0374  ous on the eastern and western sides of North AMERICA, in the Cordillera under the ecuator and und
0374  tribution of identical and allied species. In AMERICA, Dr. Hooker has shown that between forty and
0374  species. On the lofty mountains of ecuatorial AMERICA a host of peculiar species belonging to Euro
0378  del Fuego and to Europe still exist in North AMERICA, which must have late in the line of march.
0379  , strongly insisted on by Hooker in regard to AMERICA, and by Alph. de Candolle in regard to Austr
0382  e means, as I believe, the southern shores of AMERICA, Australia, New Zealand have become slightly
0391  ch lies at about the same distance from North AMERICA as the Galapagos Islands do from South Ameri
0391  merica as the Galapagos Islands do from South AMERICA, and which has a very peculiar soil, does no
0397  en 500 and 600 miles from the shores of South AMERICA. Here almost every product of the land and w
0398  plain a stamp of affinity to those created in AMERICA? There is nothing in the conditions of life,
0398  ose of Africa, like those of the Galapagos to AMERICA. I believe this grand fact can receive no so
0399  ransport or by formerly continuous land, from AMERICA; and the Cape de Verde islands from Africa:
0399  and, though standing nearer to Africa than to AMERICA, are related, and that very closely, as we k
0399  e know from Dr. Hooker's account, to those of AMERICA: but on the view that this island has been m
0399  cted; but it is also plainly related to South AMERICA, which, although the next nearest continent,
0399  ears on the view that both New Zealand, South AMERICA, and other southern lands were long ago part
0403  e surrounding lowlands; thus we have in South AMERICA, Alpine humming birds, Alpine rodents, Alpin
0404  in the blind animals inhabiting the caves of AMERICA and of Europe. Other analogous facts could b
0405  ce, two varieties of the same species inhabit AMERICA and Europe, and the species thus has an imme
0408  er different latitudes; for instance in South AMERICA, the inhabitants of the plains and mountains
0469  many representative forms in Europe and North AMERICA. If then we have under nature variability an
0473  he blind animals inhabiting the dark caves of AMERICA and Europe. In both varieties and species co
```

Page **(Key Word)**

Page ***(Key Word)***

```
0475 y the same instincts: why the thrush of South AMERICA, for instance, lines her nest with mud like
0476 , of marsupials in Australia, of endentata in AMERICA, and other such cases, is intelligible, for
0048 th one with another, and with those from the AMERICAN mainland, I was much struck how entirely vag
0074 ew is this! Every one has heard that when an AMERICAN forest is cut down, a very different vegetat
0138 imals having been separately created for the AMERICAN and European caverns, close similarity in th
0138 and Europe. On my view we must suppose that AMERICAN animals, having ordinary powers of vision, s
0138 inent. And this is the case with some of the AMERICAN cave animals, as I hear from Professor Dana:
0217 ly have to be fed by the male alone. But the AMERICAN cuckoo is in this predicament: for she makes
0217 the same time. It has been asserted that the AMERICAN cuckoo occasionally lays her eggs in other b
0217 of our European cuckoo had the habits of the AMERICAN cuckoo: but that occasionally she laid an eg
0218 l hen ostriches, at least in the case of the AMERICAN species, unite and lay first a few eggs in o
0218 r three days. This instinct, however, of the AMERICAN ostrich has not as yet been perfected: for a
0268 other dogs with foxes, or that certain South AMERICAN indigenous domestic dogs do not readily cros
0290 en examining many hundred miles of the South AMERICAN coasts, which have been upraised several hun
0294 y been required for their migration from the AMERICAN to the European seas. In examining the lates
0323 o Lyell, with the several European and North AMERICAN tertiary deposits. Even if the new fossil sp
0339 there in such numbers, are related to South AMERICAN types. This relationship is even more clearl
0340 ced in Australia: or that Edentata and other AMERICAN types should have been solely produced in So
0346 d old Worlds: yet if we travel over the vast AMERICAN continent, from the central parts of the Uni
0349 gellan are inhabited by one species of Rhea (AMERICAN ostrich), and northward the plains of La Pla
0349 rder of Rodents, but they plainly display an AMERICAN type of structure. We ascerd the lofty peaks
0349 , but the coypu and capybara, rodents of the AMERICAN type. Innumerable other instances could be g
0349 be given. If we look to the islands off the AMERICAN shore, however much they may differ in geolo
0349 may be all peculiar species, are essentially AMERICAN. We may look back to past ages, as shown in
0349 s, as shown in the last chapter, and we find AMERICAN types then prevalent on the American contine
0350 we find American types then prevalent on the AMERICAN continent and in the American seas. We see i
0350 evalent on the American continent and in the AMERICAN seas. We see in these facts some deep organi
0371 glec in the one great region with the native AMERICAN productions, and have had to compete with th
0391 ble account of Bermuda, that very many North AMERICAN birds, during their great annual migrations,
0394 y far over the Atlantic Ocean: and two North AMERICAN species either regularly or occasionally vis
0395 arly 1000 fathoms in depth, and here we find AMERICAN forms, but the species and even the genera a
0398 d water bears the unmistakeable stamp of the AMERICAN continent. There are twenty six land birds,
0398 the close affinity of most of these birds to AMERICAN species in every character, in their habits,
0398 continent, yet feels that he is standing on AMERICAN land. Why should this be so? Why should the
0398 esembles closely the conditions of the South AMERICAN coast: in fact there is a considerable dissi
0403 odents, Alpine plants, etc., all of strictly AMERICAN forms, and it is obvious that a mountain, as
0478 ipelago, of Juan Fernandez, and of the other AMERICAN islands being related in the most striking m
0478 o the plants and animals of the neighbouring AMERICAN mainland: and those of the Cape de Verde arc
0318 termination of whole groups of beings, as of AMMONITES towards the close of the secondary period,
0321 at the close of the palaeozoic period and of AMMONITES at the close of the secondary period, we mu
0004 n the next chapter the Struggle for Existence AMONGST all organic beings throughout the world, whi
0012 of which many remarkable cases could be given AMONGST animals and plants. From the facts collected
0013 same original cause acting on both: but when AMONGST individuals, apparently exposed to the same
0013 rcumstances, appears in the parent, say, once AMONGST several million individuals, and it reappear
0016 t there are hardly any domestic races, either AMONGST animals or plants, which have not been ranke
0032 ly opposed to this practice, except sometimes AMONGST closely allied sub breeds. And when a cross
0039 t differences might, and indeed do now, arise AMONGST pigeons, which are rejected as faults or dev
0046 s. we may instance Rubus, Rosa, and Hieracium AMONGST plants, several genera of insects, and sever
0048 only 112, a difference of 139 doubtful forms! AMONGST animals which unite for each birth, and whic
0060 on. It has been seen in the last chapter that AMONGST organic beings in a state of nature there is
0065 nature almost every plant produces seed, and AMONGST animals there are very few which do not annu
0070 ossibly in part through facility of diffusion AMONGST the crowded animals, been disproportionably
0078 a plant or animal is placed in a new country AMONGST new competitors, though the climate may be e
0088 important for victory, as the sword or spear. AMONGST birds, the contest is often of a more peacef
0104 ross, uniformity of character can be retained AMONGST them, as long as their conditions of life re
0110 hem. We see the same process of extermination AMONGST our domesticated productions, through the se
0127 egg, seed, or young, as easily as the adult. AMONGST many animals, sexual selection will give its
0178 ortions, in which many forms, more especially AMONGST the classes which unite for each birth and w
0179 of their former existence could be found only AMONGST fossil remains, which are preserved, as we s
0181 ying Lemur, which formerly was falsely ranked AMONGST bats. It has an extremely wide flank membran
0187 in an unaltered or little altered condition. AMONGST existing Vertebrata, we find but a small amo
0190 s of some kind. Numerous cases could be given AMONGST the lower animals of the same organ performi
0193 sclepias, genera almost as remote as possible AMONGST flowering plants. In all these cases of two
0210 orth America, and for no instinct being known AMONGST extinct species, how very generally gradatio
0249 e correctness of this almost universal belief AMONGST breeders. Hybrids are seldom raised by exper
0259 h rare exceptions extremely sterile. So again AMONGST hybrids which are usually intermediate in st
0404 ue, though it would be difficult to prove it. AMONGST mammals, we see it strikingly displayed in B
0415 ate Cnestis from Connarus. To give an example AMONGST insects, in one great division of the Hymeno
0417 ves, that this genus should still be retained AMONGST the Malpighiaceae. This case seems to me wel
0419 eir arbitrary value. Instances could be given AMONGST plants and insects, of a group of forms, fir
0422 e, the affinities which we discover in nature AMONGST the beings of the same group. Thus, on the v
0427 both these mammals and fishes, is analogical. AMONGST insects there are innumerable instances: th
0460 ructures have been perfected, more especially AMONGST broken and failing groups of organic beings:
0004 domestication. We shall thus see that a large AMOUNT of hereditary modification is at least possib
0007 w conditions of life to cause any appreciable AMOUNT of variation: and that when the organisation
0011 hances of structure. Nevertheless some slight AMOUNT of change may, I think, be attributed to the
0011 life, as, in some cases, increased size from AMOUNT of food, colour from particular kinds of food
0016 productions. When we attempt to estimate the AMOUNT of structural difference between the domestic
0019 i cannot doubt that there has been an immense AMOUNT of inherited variation. Who can believe that
0028 hese considerations in explaining the immense AMOUNT of variation which pigeons have undergone, wi
0030 only a variety of the wild Dipsacus: and this AMOUNT of change may have suddenly arisen in a seedl
0037 t varieties they could anywhere find. A large AMOUNT of change in our cultivated plants, thus slow
0041 care, to allow of the accumulation of a large AMOUNT of modification in almost any desired directi
0046 c, in which the species present an inordinate AMOUNT of variation: and hardly two naturalists can
0050 there is, as it seems to me, an overwhelming AMOUNT of experimental evidence, showing that they d
0050 hat as varieties: for he knows nothing of the AMOUNT and kind of variation to which the group is s
```

Page ***(Key Word)***

Page **(Key Word)**

```
0051 r poultry fancier before alluded to, with the AMOUNT of difference in the forms which he is contin
0056 e compelled to come to a determination by the AMOUNT of difference between them, judging by analog
0056 n them, judging by analogy whether or not the AMOUNT suffices to raise one or both to the rank of
0056 one or both to the rank of species. Hence the AMOUNT of difference is one very important criterion
0057 n regard to insects, that in large genera the  AMOUNT of difference between the species is often ex
0057 y differ from each other by a less than usual  AMOUNT of difference. Moreover, the species of the l
0057 tween varieties and species; namely, that the  AMOUNT of difference between varieties, when compare
0058 y connect; and except, secondly, by a certain  AMOUNT of difference, for two forms, if differing ve
0059 nking forms have not been discovered; but the  AMOUNT of difference considered necessary to give to
0066 pecies, which depend on a rapidly fluctuating  AMOUNT of food, for it allows them rapidly to increa
0067 s will almost instantaneously increase to any  AMOUNT. The face of Nature may be compared to a yiel
0068 species being allowed to grow up freely. The   AMOUNT of food for each species of course gives the
0082 nothing. Not that, as I believe, any extreme   AMOUNT of variability is necessary; as man can certa
0102 is is an extremely intricate subject. A large  AMOUNT of inheritable and diversified variability is
0102 able variations, will compensate for a lesser  AMOUNT of variability in each individual, and is, I
0102 process of selection, notwithstanding a large  AMOUNT of crossing with inferior animals. Thus it wi
0109 tificial selection, I can see no limit to the  AMOUNT of change, to the beauty and infinite complex
0111 ld never account for so habitual and large an  AMOUNT of difference as that between varieties of th
0114 the truth of the principle, that the greatest  AMOUNT of life can be supported by great diversifica
0117 rked by a small numbered letter, a sufficient  AMOUNT of variation is supposed to have been accumul
0119 ve allowed the accumulation of a considerable  AMOUNT of divergent variation. As all the modified d
0119 scendants will not be increased; although the  AMOUNT of divergent modification may have been incre
0120 d from their common parent. If we suppose the  AMOUNT of change between each horizontal line in our
0120 odification to be more numerous or greater in  AMOUNT, to convert these three forms into well defin
0120 w10 and z10) or two species, according to the  AMOUNT of change supposed to be represented between
0122 iagram be assumed to represent a considerable  AMOUNT of modification, species (A) and all the earl
0123 nt tendency of natural selection, the extreme  AMOUNT of difference in character between species al
0125 era alone. If, in our diagram, we suppose the  AMOUNT of change represented by each successive grou
0125 ct families, or even orders, according to the  AMOUNT of divergent modification supposed to be repr
0139 plants, as in the period of flowering, in the  AMOUNT of rain requisite for seeds to germinate, in
0151 s, pyrgoma, these valves present a marvellous  AMOUNT of diversification: the homologous valves in
0151 ing sometimes wholly unlike in shape; and the  AMOUNT of variation in the individuals of several of
0152 e breeds of the pigeon; see what a prodigious  AMOUNT of difference there is in the beak of the dif
0153 that this part has undergone an extraordinary  AMOUNT of modification, since the period when the sp
0153 than one geological period. An extraordinary   AMOUNT of modification implies an unusually large an
0153 implies an unusually large and long continued  AMOUNT of variability, which has continually been ac
0156 heir organisation; compare, for instance, the  AMOUNT of difference between the males of gallinaceo
0156 l characters are strongly displayed, with the  AMOUNT of difference between their females; and the
0157 ng to the species of the same group a greater  AMOUNT of difference in their sexual characters, tha
0158 of secondary sexual characters, and the great  AMOUNT of difference in these same characters betwee
0169 cies, must have gone through an extraordinary  AMOUNT of modification since the genus arose; and th
0173 thick and extensive to withstand an enormous   AMOUNT of future degradation; and such fossiliferous
0187 ngst existing Vertebrata, we find but a small  AMOUNT of gradation in the structure of the eye, and
0224 he proper shape to hold the greatest possible  AMOUNT of honey, with the least possible consumption
0238 cess has been repeated, until that prodigious  AMOUNT of difference between the fertile and sterile
0240 . the reader will perhaps best appreciate the  AMOUNT of difference in these workers, by my giving
0242 proves that with animals, as with plants, any  AMOUNT of modification in structure can be effected
0242 ercise or habit having come into play. For no  AMOUNT of exercise, or habit, or volition, in the ut
0257 has been able to point out what kind, or what  AMOUNT, of difference in any recognisable character
0267 uced to in our fourth chapter, that a certain  AMOUNT of crossing is indispensable even with hermap
0269 st the evidence of the existence of a certain  AMOUNT of sterility in the few following cases, whic
0272 propagated for several generations an extreme  AMOUNT of variability in their offspring is notoriou
0277 ue to distinct causes; for both depend on the  AMOUNT of difference of some kind between the specie
0281 , whilst its descendants had undergone a vast  AMOUNT of change; and the principle of competition b
0282 t time will not have sufficed for so great an  AMOUNT of organic change, all changes having been ef
0284 h's crust has elsewhere suffered. And what an  AMOUNT of degradation is implied by the sedimentary
0284 any one area must be extremely slow. But the   AMOUNT of denudation which the strata have in many p
0291 on thick enough, when upraised, to resist any  AMOUNT of degradation, may be formed. I am convinced
0293 als of all kinds, we may safely infer a large  AMOUNT of migration during climatal and other change
0294 xcellent lesson to reflect on the ascertained  AMOUNT of migration of the inhabitants of Europe dur
0295 sediment must nearly have counterbalanced the  AMOUNT of subsidence. But this same movement of subs
0295 ancing between the supply of sediment and the  AMOUNT of subsidence is probably a rare contingency;
0296 f such species were to undergo a considerable  AMOUNT of modification during any one geological per
0297 s, but when they meet with a somewhat greater  AMOUNT of difference between any two forms, they ran
0308 europe and of the United States; and, from the AMOUNT of sediment, miles in thickness, of which the
0313 though there are exceptions to this rule. The  AMOUNT of organic change, as Pictet has remarked, do
0314 iations be accumulated to a greater or lesser  AMOUNT, thus causing a greater or lesser amount of m
0314 sser amount, thus causing a greater or lesser  AMOUNT of modification in the varying species, depen
0315 nct. In members of the same class the average  AMOUNT of change, during long and equal periods of t
0315 arly intermittent intervals; consequently the  AMOUNT of organic change exhibited by the fossils em
0326 of spreading into new territories. A certain   AMOUNT of isolation, recurring at long intervals of
0327 tants of each region underwent a considerable  AMOUNT of modification and extinction, and that ther
0328 andinavia; nevertheless he finds a surprising  AMOUNT of difference in the species. If the several
0334 ite new forms by immigration, and for a large  AMOUNT of modification, during the long and blank in
0351 ve supervened since ancient times, almost any  AMOUNT of migration is possible. But in many other c
0356 one on simultaneously changing, and the whole  AMOUNT of modification will not have been due, at ea
0356 ch I have selected as presenting the greatest  AMOUNT of difficulty on the theory of single centres
0370 favourable climate, I attribute the necessary  AMOUNT of uniformity in the sub arctic and northern
0377 ies, this period will have been ample for any  AMOUNT of migration. As the cold came slowly on, all
0377 ants and animals can withstand a considerable  AMOUNT of cold, many might have escaped exterminatio
0389 ch I have selected as presenting the greatest  AMOUNT of difficulty, on the view that all the indiv
0395 cies and even the genera are distinct. As the  AMOUNT of modification in all cases depends to a cer
0400 since their arrival), we find a considerable   AMOUNT of difference in the several islands. This di
0408 ons of life, there would be an almost endless  AMOUNT of organic action and reaction, and we should
0420 alogical in order to be natural: but that the  AMOUNT of difference in the several branches or grou
0421 beings belong to Silurian genera. So that the  AMOUNT or value of the differences between organic b
```

Page **(Key Word)**

Page ++(Key Word)++

```
0425  rgone a certain, and sometimes a considerable AMOUNT of modification, may not this same element of
0435  as to serve as a wing: yet in all this great AMOUNT of modification there will be no tendency to
0438  o the most diverse purposes. And as the whole AMOUNT of modification will have been effected by sl
0445  he puppies had not nearly acquired their full AMOUNT of proportional difference. So, again, I was
0445  he colts have by no means acquired their full AMOUNT of proportional difference. As the evidence a
0464  at there has not been time sufficient for any AMOUNT of organic change; for the lapse of time has
0466  e undergone; but we may safely infer that the AMOUNT has been large, and that modifications can be
0468  sertion is quite incapable of proof, that the AMOUNT of variation under nature is a strictly limit
0470  es; for they differ from each other by a less AMOUNT of difference than do the species of smaller
0474  ched off from a common progenitor, an unusual AMOUNT of variability and modification, and therefor
0475  e slowly and at successive intervals; and the AMOUNT of change, after equal intervals of time, is
0481  to no variation; it cannot be proved that the AMOUNT of variation in the course of long ages is a
0483  ome instances necessarily implies an enormous AMOUNT of modification in the descendants. Throughou
0485  more carefully and to value higher the actual AMOUNT of difference between them. It is quite possi
0488  ermination of of others; it follows, that the AMOUNT of organic change in the fossils of consecuti
0115  ely infer that diversification of structure, AMOUNTING to new generic differences, would have been
0186  different distances, for admitting different AMOUNTS of light, and for the correction of spherica
0049  suffice? If that between America and Europe is AMPLE, will that between the Continent and the Azor
0071  shire, on the estate of a relation where I had AMPLE means of investigation, there was a large and
0096  y, though I have the materials prepared for an AMPLE discussion. All vertebrate animals, all insec
0221  aphides or cocci. According to Huber, who had AMPLE opportunities for observation, in Switzerland
0287  h per century for the whole length would be an AMPLE allowance. At this rate, on the above data, t
0377  in a few centuries, this period will have been AMPLE for any amount of migration. As the cold came
0384  forms, and in such cases there will have been AMPLE time for great geographical changes, and cons
0405  chi so that in such cases there will have been AMPLE time for great climatal and oeographical chan
0116  ng discussion, which ought to have been much AMPLIFIED, we may, I think, assume that the modified
0040  ; not that mere individual differences are not AMPLY sufficient, with extreme care, to allow of th
0108  any will exclaim that these several causes are AMPLY sufficient wholly to stop the action of natur
0319  ; and that these same unperceived agencies are AMPLY sufficient to cause rarity, and finally extin
0453  hate of lime? When a man's fingers have been AMPUTATED, imperfect nails sometimes appear on the st
0247  he found the common red and blue pimpernels (ANAGALLIS arvensis and coerulea), which the best bota
0049  f argument, from geographical distribution. ANALOGICAL variation, hybridism, etc., have been brouo
0051  ing; and he has little general knowledge of ANALOGICAL variation in other groups and in other coun
0407  rom the importance of barriers and from the ANALOGICAL distribution of sub genera, genera, and fam
0411  ies. Descent always used in classification. ANALOGICAL or adaptive characters. Affinities, general
0414  the being, are ranked as merely adaptive or ANALOGICAL characters; but to the consideration of the
0427  ant distinction between real affinities and ANALOGICAL or adaptive resemblances. Lamarck first cal
0427  d between both these mammals and fishes, is ANALOGICAL. Amongst insects there are innumerable ins
0427  veal descent, we can clearly understand why ANALOGICAL or adaptive character, although of the utmo
0427  paradox, that the very same characters are ANALOGICAL when one class or order is compared with an
0428  ape of the body and fin like limbs are only ANALOGICAL when whales are compared with fishes, being
0430  ngly suspected that the resemblance is only ANALOGICAL, owing to the phascolomys having become ada
0433  with a distinct group, we summarily reject ANALOGICAL or adaptive characters, and yet use these s
0456  tance; the wide opposition in value between ANALOGICAL or adaptive characters, and characters of t
0059  rieties. And we can clearly understand these ANALOGIES, if species have once existed as varieties,
0059  es, and have thus originated; whereas, these ANALOGIES are utterly inexplicable if each species ha
0427  d racehorse is hardly more fanciful than the ANALOGIES which have been drawn by some authors betwe
0090  battle, or attractive to the females. We see ANALOGOUS cases under nature, for instance, the tuft
0100  ertilising element; for we know of no means, ANALOGOUS to the action of insects and of the wind in
0112  ic productions. We shall here find something ANALOGOUS. A fancier is struck by a pigeon having a s
0112  on parent. But how, it may be asked, can any ANALOGOUS principle apply in nature? I believe it can
0120  d that a second species (I) has produced, by ANALOGOUS steps, after ten thousand generations, eith
0131  riable. Species of the same genus vary in an ANALOGOUS manner. Reversions to long lost characters.
0140  he has observed similar facts in Ceylon, and ANALOGOUS observations have been made by Mr. H. C. Wa
0146  e founded his main divisions of the order on ANALOGOUS differences. Hence we see that modification
0159  specific purposes. Distinct species present ANALOGOUS variations; and a variety of one species of
0159  y the aboriginal rock pigeon; these then are ANALOGOUS variations in two or more distinct races. T
0159  presume that no one will doubt that all such ANALOGOUS variations are due to the several races of
0159  . in the vegetable kingdom we have a case of ANALOGOUS variation, in the enlarged stems, or roots
0159  this be not so, the case will then be one of ANALOGOUS variation in two so called distinct species
0160  is a case of reversion, and not of a new yet ANALOGOUS variation appearing in the several breeds.
0161  cted that they would occasionally vary in an ANALOGOUS manner; so that a variety of one species wo
0162  our domestic breeds were reversions or only ANALOGOUS variations; but we might have inferred that
0162  ntly existing character and what are new but ANALOGOUS variations, yet we ought, on my theory, som
0162  nd characters (either from reversion or from ANALOGOUS variation) which already occur in some othe
0169  ar influences will naturally tend to present ANALOGOUS variations, and these same species may occa
0169  difications may not arise from reversion and ANALOGOUS variation, such modifications will add to t
0188  r that the eye has been formed by a somewhat ANALOGOUS process. But may not this inference be pres
0193  y been shown that Rays have an organ closely ANALOGOUS to the electric apparatus, and yet do not,
0194  e good of each being and taking advantage of ANALOGOUS variations, has sometimes modified in very
0199  ignorance of the precise cause of the slight ANALOGOUS differences between species. I might have a
0247  came to the same conclusion in several other ANALOGOUS cases; it seems to me that we may well be p
0257  eight other species of Nicotiana. Very many ANALOGOUS facts could be given. No one has been able
0262  run to a certain extent parallel. Something ANALOGOUS occurs in grafting; for Thouin found that t
0271  , i am inclined to suspect that they present ANALOGOUS facts. Kolreuter, whose accuracy has been c
0276  specially endowed with various and somewhat ANALOGOUS degrees of difficulty in being grafted toge
0288  he molluscan genus Chiton offers a partially ANALOGOUS case. With respect to the terrestrial produ
0325  orth America, and there discover a series of ANALOGOUS phenomena, it will appear certain that all
0340  ls to Africa than it is at the present time. ANALOGOUS facts could be given in relation to the dis
0347  lia or Africa under nearly the same climate. ANALOGOUS facts could be given with respect to the in
0375  to the Flora of New Zealand; by Dr. Hooker, ANALOGOUS and striking facts are given in regard to t
0376  tract applies to plants alone: some strictly ANALOGOUS facts could be given on the distribution of
0391  ers of insects in Madeira apparently present ANALOGOUS facts. Oceanic islands are sometimes defici
0395  als are the same on both sides; we meet with ANALOGOUS facts on many islands separated by similar
0399  ll betraying their original birthplace. Many ANALOGOUS facts could be given: indeed it is an almos
0401  archipelago, and in a lesser degree in some ANALOGOUS instances, is that the new species formed i
```

Page ++(Key Word)++

Page **(Key Word)**

```
0404 ng the caves of America and of Europe. Other ANALOGOUS facts could be given. And it will, I believ
0435 r lip, mandibles, and two pairs of maxillae. ANALOGOUS laws govern the construction of the mouths
0444 n embryology. But first let us look at a few ANALOGOUS cases in domestic varieties. Some authors w
0459 perfected, not by means superior to, though ANALOGOUS with, human reason, but by the accumulation
0484 his volume on the origin of species, or when ANALOGOUS views are generally admitted, we can dimly
0047 e links have actually been found, but because ANALOGY leads the observer to suppose either that th
0051 rms, he will have to trust almost entirely to ANALOGY, and his difficulties will rise to a climax.
0056 amount of difference between them, judging by ANALOGY whether or not the amount suffices to raise
0059 the species of large genera present a strong ANALOGY with varieties. And we can clearly understan
0141 has some influence I must believe, both from ANALOGY, and from the incessant advice given in agri
0181 migrate, or old ones become modified, and all ANALOGY would lead us to believe that some at least
0217 e fostered young would gain an advantage. And ANALOGY would lead me to believe, that the young thu
0239 in this case, we may safely conclude from the ANALOGY of ordinary variations, that each successive
0254 america, all are quite fertile together; and ANALOGY makes me greatly doubt, whether the several
0301 e thousands of miles beyond its confines; and ANALOGY leads me to believe that it would be chiefly
0319 species, we might have felt certain from the ANALOGY of all other mammals, even of the slow breed
0387 stomach; although I do not know the fact, yet ANALOGY makes me believe that a heron flying to anot
0425 e out of the womb of a bear? According to all ANALOGY, it would be ranked with bears; but then ass
0452 n animal or plant no trace of an organ, which ANALOGY would lead us to expect to find, and which i
0484 s, and plants from an equal or lesser number. ANALOGY would lead me one step further, namely, to t
0484 s have descended from some one prototype. But ANALOGY may be a deceitful guide. Nevertheless all l
0484 se or oak tree. Therefore I should infer from ANALOGY that probably all the organic beings which h
0013 her or grandmother or other much more remote ANCESTOR; why a peculiarity is often transmitted from
0026 lief that the child ever reverts to some one ANCESTOR, removed by a greater number of generations.
0125 hat its species have inherited from a common ANCESTOR some advantage in common. Hence, the struggl
0160 looc, to use a common expression, of any one ANCESTOR, is only 1 in 2048; and yet, we see, it is g
0160 t that the offspring suddenly takes after an ANCESTOR some hundred generations distant, but that i
0161 never know the exact character of the common ANCESTOR of a group, we could not distinguish these t
0193 te its presence to inheritance from a common ANCESTOR; and its absence in some of the members to i
0194 cture in common to inheritance from the same ANCESTOR. Although in many cases it is most difficult
0282 n backwards, always converging to the common ANCESTOR of each great class. So that the number of i
0426 characters have been inherited from a common ANCESTOR. And we know that such correlated or aggrega
0428 of body and structure of limbs from a common ANCESTOR. So it is with fishes. As members of distinc
0131 , to which the parents and their more remote ANCESTORS have been exposed during several generation
0187 , we ought to look exclusively to its lineal ANCESTORS; but this is scarcely ever possible, and we
0210 for these could be found only in the lineal ANCESTORS of each species, but we ought to find in th
0443 conditions to which either parent, or their ANCESTORS, have been exposed. Nevertheless an effect
0015 onally revert in some of their characters to ANCESTRAL forms, it seems to me not improbable, that
0025 on the well known principle of reversion to ANCESTRAL characters, if all the domestic breeds have
0161 oulc occasionally exhibit reversions to lost ANCESTRAL characters. As, however, we never know the
0190 ssec, we should have to look to very ancient ANCESTRAL forms, long since become extinct. We should
0200 either as having been of special use to some ANCESTRAL form, or as being now of special use to the
0018 domestic animals is, that we find in the most ANCIENT records, more especially on the monuments of
0018 who will pretend to say how long before these ANCIENT periods, savages, like those of Tierra del F
0024 ht species were so thoroughly domesticated in ANCIENT times by half civilized man, as to be quite
0034 le of selection I find distinctly given in an ANCIENT Chinese encyclopaedia. Explicit rules are la
0034 domestic animals was carefully attended to in ANCIENT times, and is now attended to by the lowest
0074 en observed that the trees now growing on the ANCIENT Indian mounds, in the Southern United States
0111 yorkshire, it is historically known that the ANCIENT black cattle were displaced by the long horn
0125 s fact, for the extinct species lived at very ANCIENT epochs when the branching lines of descent h
0125 ve descended from one species of a still more ANCIENT and unknown genus. We have seen that in each
0126 hat on this view of extremely few of the more ANCIENT species having transmitted descendants, and
0126 kingdoms. Although extremely few of the most ANCIENT species may now have living and modified des
0139 rope, I am only surprised that more wrecks of ANCIENT life have not been preserved, owing to the l
0141 vice given in agricultural works, even in the ANCIENT Encyclopaedias of China, to be very cautious
0146 n truth are simply due to inheritance; for an ANCIENT progenitor may have acquired through natural
0164 he above described appearances are all due to ANCIENT crosses with the dun stock. But I am not at
0166 hesis to account for the reappearance of very ANCIENT character, is, that there is a tendency in t
0169 lly revert to some of the characters of their ANCIENT progenitors. Although new and important modi
0190 an has passed, we should have to look to very ANCIENT ancestral forms, long since become extinct.
0191 have descended by ordinary generation from an ANCIENT prototype, of which we know nothing, furnish
0191 s, it is probable that organs which at a very ANCIENT period served for respiration have been actu
0193 e electric organs had been inherited from one ANCIENT progenitor thus provided, we might have expe
0217 her bird's nests. Now let us suppose that the ANCIENT progenitor of our European cuckoo had the ha
0282 their turn been similarly connected with some ANCIENT species; and so on backwards, always converg
0291 n, may be formed. I am convinced that all our ANCIENT formations, which are rich in fossils, have
0291 d during subsidence. I may add, that the only ANCIENT tertiary formation on the west coast of Sout
0296 ous beds 1400 feet thick in Nova Scotia, with ANCIENT root bearing strata, one above the other, at
0301 forms. When such varieties returned to their ANCIENT homes, as they would differ from their forme
0302 ave slowly multiplied before they invaded the ANCIENT archipelagoes of Europe and of the United St
0306 eatly from any known animal. Some of the most ANCIENT Silurian animals, as the Nautilus, Lingula,
0307 silurian epoch, is very great. If these most ANCIENT beds had been wholly worn away by denudation
0312 iving species. On the state of development of ANCIENT forms. On the succession of the same types w
0329 plained on the principle of descent. The more ANCIENT any form is, the more, as a general rule, it
0330 ay from each other by a dozen characters, the ANCIENT members of the same two groups would be dist
0330 ch other. It is a common belief that the more ANCIENT a form is, by so much the more it tends to c
0331 have inherited something in common from their ANCIENT and common progenitor. On the principle of t
0331 s, the more it will generally differ from its ANCIENT progenitor. Hence we can understand the rule
0331 ence we can understand the rule that the most ANCIENT fossils differ most from existing forms. We
0333 hey subsequently diverged. Thus it comes that ANCIENT and extinct genera are often in some slight
0335 ure for corresponding lengths of time: a very ANCIENT form might occasionally last much longer tha
0336 formations, by the physical conditions of the ANCIENT areas having remained nearly the same. Let i
0336 xpect to find. On the state of Development of ANCIENT Forms. There has been much discussion whethe
0336 r recent forms are more highly developed than ANCIENT. I will not here enter on this subject; for
0337 s must, on my theory, be higher than the more ANCIENT; for each new species is formed by having ha
0337 torious forms of life, in comparison with the ANCIENT and beaten forms; but I can see no way of te
```

Page **(Key Word)**

Page **(Key Word)**

```
0338  ve foreseen this result. Agassiz insists that  ANCIENT  animals resemble to a certain extent the emb
0338  sort of picture, preserved by nature, of the   ANCIENT  and less modified condition of each animal.
0340  in South America. For we know that Europe in   ANCIENT  times was peopled by numerous marsupials; an
0344  erstand how it is that all the forms of life,  ANCIENT  and recent, make together one grand system:
0344  ency to divergence of character, why the more  ANCIENT  a form is, the more it generally differs fro
0344  generally differs from those now living. Why   ANCIENT  and extinct forms often tend to fill up gaps
0344  ncing them a little closer together. The more  ANCIENT  a form is, the more often, apparently, it di
0344  ate between groups now distinct; for the more  ANCIENT  a form is, the more nearly it will be relate
0345  ressed. If it should hereafter be proved that  ANCIENT  animals resemble to a certain extent the emb
0351  atal changes which will have supervened since  ANCIENT  times, almost any amount of migration is pos
0369  , in all probability have become mingled with  ANCIENT  Alpine species, which must have existed on t
0373  kim, Dr. Hooker saw maize growing on gigantic  ANCIENT  moraines. South of the equator, we have some
0384  ned: but some fresh water fish belong to very  ANCIENT  forms, and in such cases there will have bee
0394  mmals: many volcanic islands are sufficiently  ANCIENT,  as shown by the stupendous degradation whic
0405  e should bear in mind that some are extremely  ANCIENT,  and must have branched off from a common pa
0412  er one class, for all have descended from one  ANCIENT  but unseen parent, and, consequently, have i
0414  other. It might have been thought (and was in  ANCIENT  times thought) that those parts of the struc
0422  possible one. Yet it might be that some very   ANCIENT  language had altered little, and had given r
0429  cess of modification, how it is that the more  ANCIENT  forms of life often present characters in so
0430  ing the bizcacha, branched off from some very  ANCIENT  Marsupial, which will have had a character i
0430  by inheritance, more of the character of its  ANCIENT  progenitor than have other Rodents: and ther
0431  ationship between the numerous kindred of any  ANCIENT  and noble family, even by the aid of a genea
0431  r vertebrate animals, by the belief that many  ANCIENT  forms of life have been utterly lost, throug
0432  mmediate parents; or those parents from their  ANCIENT  and unknown progenitor. Yet the natural arra
0435  of the several parts. If we suppose that the  ANCIENT  progenitor, the archetype as it may be calle
0449  shows us the structure of their less modified  ANCIENT  progenitors, we can clearly see why ancient
0449  d ancient progenitors, we can clearly see why  ANCIENT  and extinct forms of life should resemble th
0449  proved true in those cases alone in which the  ANCIENT  state, now supposed to be represented in man
0450  mind, that the supposed law of resemblance of  ANCIENT  forms of life to the embryonic stages of rec
0450  earing, in the many descendants from some one  ANCIENT  progenitor, at a very early period in the li
0464  a must somewhere have been deposited at these  ANCIENT  and utterly unknown epochs in the world's hi
0476  s. as the groups which have descended from an  ANCIENT  progenitor have generally diverged in charac
0476  descendants: and thus we can see why the more  ANCIENT  a fossil is, the oftener it stands in some d
0476  at as being, in some vague sense, higher than  ANCIENT  and extinct forms: and they are in so far hi
0486  sils, will aid us in forming a picture of the  ANCIENT  forms of life. Embryology will reveal to us
0487  s of immigration, some light can be thrown on  ANCIENT  geography. The noble science of Geology lose
0007  fferences and Origin. Principle of Selection   ANCIENTLY followed, its Effects. Methodical and Uncon
0017  ions have varied. In the case of most of our   ANCIENTLY domesticated animals and plants, I do not t
0038  e with that given to the plants in countries   ANCIENTLY civilised. In regard to the domestic animal
0162  ft doubtful, what cases are reversions to an   ANCIENTLY existing character and what are new but ana
0466  are still occasionally produced by our most   ANCIENTLY domesticated productions. Man does not actu
0030  d this is known to have been the case with the ANCON sheep. But when we compare the dray horse and
0386  rms me that a Dyticus has been caught with an  ANCYLUS (a fresh water shell like a limpet) firmly a
0373  t 800 feet in height, crossing a valley of the ANDES: and this I now feel convinced was a gigantic
0007  k, some probability in the view propounded by  ANDREW Knight, that this variability may be partly c
0096  this view, I may add, was first suggested by   ANDREW Knight. We shall presently see its importance
0227  sides of the hexagonal prisms will have every  ANGLE identically the same with the best measuremen
0232  fficulty, as when two pieces of comb met at an ANGLE, how often the bees would entirely pull down
0416  hes fishes and reptiles, the inflection of the ANGLE of the jaws of the Marsupials, the manner in
0426  r may be, let it be the mere inflection of the ANGLE of the jaw, the manner in which an insect's w
0224  nceivable how they can make all the necessary  ANGLES and planes, or even perceive when they are co
0225  ed of three rhombs. These rhombs have certain  ANGLES, and the three which form the pyramidal base
0232  seen, they never gnaw away and finish off the  ANGLES of a cell till a large part both of that cell
0235  ch other, than they know what are the several  ANGLES of the hexagonal prisms and of the basal rhom
0285  y has published an account of a downthrow in   ANGLESEA of 2300 feet: and he informs me that he full
0389  dge has 847 plants, and the little island of  ANGLESEA 764, but a few ferns and a few introduced pl
0005  the doctrine of Malthus, applied to the whole  ANIMAL and vegetable kingdoms. As many more individu
0008  of life. Nothing is more easy than to tame an  ANIMAL, and few things more difficult than to get it
0011  e of the effect of use. Not a single domestic  ANIMAL can be named which has not in some country dr
0011  almost certainly entail changes in the mature  ANIMAL. In monstrosities, the correlations between o
0017  savage possibly know, when he first tamed an   ANIMAL, whether it would vary in succeeding generati
0024  ows that it is most difficult to get any wild  ANIMAL to breed freely under domestication: yet on t
0030  we see in them adaptation, not indeed to the   ANIMAL's or plant's own good, but to man's use or fa
0031  the animals. Breeders habitually speak of an   ANIMAL's organisation as something quite plastic, wh
0031  , and who was himself a very good judge of an  ANIMAL, speaks of the principle of selection as that
0036  spring of their domestic animals, yet any one  ANIMAL particularly useful to them, for any special
0041  the most important point of all, is, that the  ANIMAL or plant should be so highly useful to man, o
0045  s on Alpine summits, or the thicker fur of an  ANIMAL from far northwards, would not in some cases
0050  i have been struck with the fact, that if any  ANIMAL or plant in a state of nature be highly usefu
0063  thus applied with manifold force to the whole  ANIMAL and vegetable kingdoms: for in this case ther
0066  reat majority of cases is an early one. If an  ANIMAL can in any way protect its own eggs or young,
0066  that in all cases, the average number of any  ANIMAL or plant depends only indirectly on the numbe
0067  , so incomparably better known than any other  ANIMAL. This subject has been ably treated by severa
0074  quite credible that the presence of a feline  ANIMAL in large numbers in a district might determin
0078  hence, also, we can see that when a plant or   ANIMAL is placed in a new country amongst new compet
0085  o think that the occasional destruction of an  ANIMAL of any particular colour would produce little
0087  vestigation. A structure used only once in an  ANIMAL's whole life, if of high importance to it, mi
0089  lieve, that when the males and females of any  ANIMAL have the same general habits of life, but dif
0100  have not found a single case of a terrestrial  ANIMAL which fertilises itself. We can understand th
0101  to discover a single case of an hermaphrodite  ANIMAL with the organs of reproduction so perfectly
0101  ed to suspect that, both in the vegetable and  ANIMAL kingdoms, an occasional intercross with a dis
0103  t within the same area, varieties of the same  ANIMAL can long remain distinct, from haunting diffe
0113  structure the descendants of our carnivorous  ANIMAL became, the more places they would be enabled
0113  uld be enabled to occupy. What applies to one  ANIMAL will apply throughout all time to all animals
0126  very few classes in each main division of the  ANIMAL and vegetable kingdoms. Although extremely fe
0130  live on its summit, so we occasionally see an  ANIMAL like the Ornithorhynchus or Lepidosiren, whic
```

Page **(Key Word)**

Page **(Key Word)**

```
0137  ammation of the eyes must be injurious to any  ANIMAL, and as eyes are certainly not indispensable
0138  tined for total darkness. By the time that an  ANIMAL had reached, after numberless generations, th
0148  ; for in the struggle for life to which every  ANIMAL is exposed, each individual Proteolepas would
0152  r domestic animals, if any part, or the whole  ANIMAL, be neglected and no selection be applied, th
0164  h commoner in the foal than in the full grown  ANIMAL. Without here entering on further details, I
0167  nds on thousands of generations, and I see an  ANIMAL striped like a zebra, but perhaps otherwise v
0171  ell defined? Secondly, is it possible that an  ANIMAL having, for instance, the structure and habit
0171  have been formed by the modification of some  ANIMAL with wholly different habits? Can we believe
0179  i hold, how, for instance, a land carnivorous  ANIMAL could have been converted into one with aquat
0179  to one with aquatic habits; for how could the  ANIMAL in its transitional state have subsisted? It
0179  rt legs, and form of tail; during summer this  ANIMAL dives for and preys on fish, but during the l
0183  ould be easy for natural selection to fit the  ANIMAL, by some modification of its structure, for i
0185  ly have felt surprise when he has met with an  ANIMAL having habits and structure not at all in agr
0186  odification in the organ be ever useful to an  ANIMAL under changing conditions of life, then the d
0190  ly and in the fish Cobites. In the Hydra, the  ANIMAL may be turned inside out, and the exterior su
0196  veloped tail having been formed in an aquatic  ANIMAL, it might subsequently come to be worked in f
0204  have been formed by natural selection from an  ANIMAL which at first could only glide through the a
0205  ften been retained (as the tail of an aquatic  ANIMAL by its terrestrial descendants), though it ha
0207  to enable us to perform, when performed by an  ANIMAL, more especially by a very young one, without
0210  others. One of the strongest instances of an  ANIMAL apparently performing an action for the sole
0211  the ants. Although I do not believe that any  ANIMAL in the world performs an action for the exclu
0215  d, habit alone in some cases has sufficed; no  ANIMAL is more difficult to tame than the young of t
0215  an the young of the wild rabbit; scarcely any  ANIMAL is tamer than the young of the tame rabbit; b
0236  orking ant or other neuter insect had been an  ANIMAL in the ordinary state, I should have unhesita
0238  lesh and fat to be well marbled together; the  ANIMAL has been slaughtered, but the breeder goes wi
0243  stincts are of the highest importance to each  ANIMAL. Therefore I can see no difficulty, under cha
0243  xclusive good of other animals, but that each  ANIMAL takes advantage of the instincts of others; t
0252  hether any case of a perfectly fertile hybrid  ANIMAL can be considered as thoroughly well authenti
0253  brothers and sisters in the case of any pure  ANIMAL, which from any cause had the least tendency
0265  n tell, till he tries, whether any particular  ANIMAL will breed under confinement or any plant see
0306  hich probably differed greatly from any known  ANIMAL. Some of the most ancient Silurian animals, a
0325  depend on general laws which govern the whole  ANIMAL kingdom. M. Barrande has made forcible remark
0330  osition, for every now and then even a living  ANIMAL, as the lepidosiren, is discovered having aff
0338  e ancient and less modified condition of each  ANIMAL. This view may be true, and yet it may never
0352  a species is continuous; and when a plant or  ANIMAL inhabits two points so distant from each othe
0414  remotely related to the habits and food of an  ANIMAL, I have always regarded as affording very cle
0427  between the dugong, which is a pachydermatous  ANIMAL, and the whale, and between both these mammal
0435  has so pleased the creator to construct each  ANIMAL and plant. The explanation is manifest on the
0438  as we find in the other great classes of the  ANIMAL and vegetable kingdoms. Naturalists frequentl
0439  otten to ticket the embryo of some vertebrate  ANIMAL, he cannot now tell whether it be that of a m
0440  osed. The case, however, is different when an  ANIMAL during any part of its embryonic career is ac
0440  t as perfect and as beautiful as in the adult  ANIMAL. From such special adaptations, the similarit
0441  terpillar. In some cases, however, the mature  ANIMAL is generally considered as lower in the scale
0443  metimes a higher organisation than the mature  ANIMAL, into which it is developed. I believe that a
0443  ot positively tell, until some time after the  ANIMAL has been born, what its merits or form will u
0443  ither parent. For the welfare of a very young  ANIMAL, as long as it remains in its mother's womb,
0446  d earlier or later in life, if the full grown  ANIMAL possesses them. And the cases just given, mor
0447  fer greatly from the fore limbs in the mature  ANIMAL; the limbs in the latter having undergone muc
0447  such influence will mainly affect the mature  ANIMAL, which has come to its full powers of activit
0449  than that of the adult. For the embryo is the  ANIMAL in its less modified state; and in so far it
0449  structure of its progenitor. In two groups of  ANIMAL, however much they may at present differ from
0452  aborted; and this implies, that we find in an  ANIMAL or plant no trace of an organ, which analogy
0457  ructed on the same pattern in each individual  ANIMAL and plant. On the principle of successive sli
0480  we may believe, that the teeth in the mature  ANIMAL were reduced, during successive generations,
0004  probable that a careful study of domesticated  ANIMALS and of cultivated plants would offer the bes
0005  subject of Instinct, or the mental powers of  ANIMALS; thirdly, Hybridism, or the infertility of s
0007  ub variety of our older cultivated plants and  ANIMALS, one of the first points which strikes us, i
0007  flect on the vast diversity of the plants and  ANIMALS which have been cultivated, and which have v
0008  yield new varieties; our oldest domesticated  ANIMALS are still capable of rapid improvement or mo
0008  ases when the male and female unite. How many  ANIMALS there are which will not breed, though livin
0009  laws are which determine the reproduction of  ANIMALS under confinement, I may just mention that c
0009  finement, I may just mention that carnivorous  ANIMALS, even from the tropics, breed in this countr
0009  s. when, on the one hand, we see domesticated  ANIMALS and plants, though often weak and sickly, ye
0009  stem has not been thus affected; so will some  ANIMALS and plants withstand domestication or cultiv
0010  s most difficult: my impression is, that with  ANIMALS such agencies have produced very little dire
0011  n transported from one climate to another. In  ANIMALS it has a more marked effect; for instance, I
0011  he disuse of the muscles of the ear, from the  ANIMALS not being much alarmed by danger, seems prob
0012  many remarkable cases could be given amongst  ANIMALS and plants. From the facts collected by Heus
0012  mperfect teeth; long haired and coarse haired  ANIMALS are apt to have, as is asserted, long or man
0015  hereditary varieties or races of our domestic  ANIMALS and plants, and compare them with species cl
0016  are hardly any domestic races, either amongst  ANIMALS or plants, which have not been ranked by som
0017  assumed that man has chosen for domestication  ANIMALS and plants having an extraordinary inherent
0017  r domestication? I cannot doubt that if other  ANIMALS and plants, equal in number to our domestica
0017  he case of most of our anciently domesticated  ANIMALS and plants, I do not think it is possible to
0018  elieve in the multiple origin of our domestic  ANIMALS is, that we find in the most ancient records
0019  of inherited variation. Who can believe that  ANIMALS closely resembling the Italian greyhound, th
0026  rward one case of the hybrid offspring of two  ANIMALS clearly distinct being themselves perfectly
0029  that all the breeders of the various domestic  ANIMALS and the cultivators of plants, with whom I h
0031  s devoted to this subject, and to inspect the  ANIMALS. Breeders habitually speak of an animal's or
0031  ed is proved by the enormous prices given for  ANIMALS with a good pedigree; and these have now bee
0032  s that deviate from the proper standard. With  ANIMALS this kind of selection is, in fact, also fol
0033  any one is so careless as to allow his worst  ANIMALS to breed. In regard to plants, there is anot
0034  d barbarous periods of English history choice  ANIMALS were often imported, and laws were passed to
0034  esis, it is clear that the colour of domestic  ANIMALS was at that early period attended to. Savage
0034  w sometimes cross their dogs with wild canine  ANIMALS, to improve the breed, and they formerly did
0034  , but they show that the breeding of domestic  ANIMALS was carefully attended to in ancient times,
```

Page **(Key Word)**

Page **(Key Word)**

```
0034  to possess and breed from the best individual  ANIMALS, is more important. Thus, a man who intends
0036  character of the offspring of their domestic   ANIMALS, yet any one animal particularly useful to t
0036  which savages are so liable, and such choice   ANIMALS would thus generally leave more offspring th
0036  s selection going on. We see the value set on  ANIMALS even by the barbarians of Tierra del Fuego,
0038  nciently civilised. In regard to the domestic  ANIMALS kept by uncivilised man, it should not be ov
0040  kes more care than usual in matching his best  ANIMALS and thus improves them, and the improved ind
0042  he last thirty or forty years. In the case of  ANIMALS with separate sexes, facility in preventing
0042  s. although I do not doubt that some domestic  ANIMALS vary less than others, yet the rarity or abs
0043  sum up on the origin of our Domestic Races of  ANIMALS and plants. I believe that the conditions of
0043  , been greatly exaggerated, both in regard to  ANIMALS and to those plants which are propagated by
0048  , a difference of 139 doubtful forms! Amongst  ANIMALS which unite for each birth, and which are hi
0049  by many entomologists. Even Ireland has a few  ANIMALS, now generally regarded as varieties, but wh
0060  rs of increase. Rapid increase of naturalised  ANIMALS and plants. Nature of the checks to increase
0060  mber of individuals. Complex relations of all  ANIMALS and plants throughtout nature. Struggle for
0062  l, but success in leaving progeny. Two canine  ANIMALS in a time of dearth, may be truly said to st
0064  ckoned to be the slowest breeder of all known  ANIMALS, and I have taken some pains to estimate its
0064  f the astonishingly rapid increase of various  ANIMALS in a state of nature, when circumstances hav
0064  re striking is the evidence from our domestic  ANIMALS of many kinds which have run wild in several
0065  , no one supposes that the fertility of these  ANIMALS or plants has been suddenly and temporarily
0065  almost every plant produces seed, and amongst  ANIMALS there are very few which do not annually pai
0065  e may confidently assert, that all plants and  ANIMALS are tending to increase at a geometrical rat
0065  ife. Our familiarity with the larger domestic  ANIMALS tends, I think, to mislead us: we see no gre
0067  ength, more especially in regard to the feral  ANIMALS of South America. Here I will make only a fe
0067  some of the chief points. Eggs or very young   ANIMALS seem generally to suffer most, but this is n
0068  aining food, but the serving as prey to other  ANIMALS, which determines the average numbers of a s
0068  esent, although hundreds of thousands of game  ANIMALS are now annually killed. On the other hand,
0069  plants, nor resist destruction by our native  ANIMALS. When a species, owing to highly favourable
0070  , this seems generally to occur with our game  ANIMALS, often ensue: and here we have a limiting ch
0070  ugh facility of diffusion amongst the crowded  ANIMALS, been disproportionably favoured: and here c
0072  y, which lays its eggs in the navels of these  ANIMALS when first born. The increase of these flies
0073  give one more instance showing how plants and  ANIMALS, most remote in the scale of nature, are bou
0075  nd insect, between insects, snails, and other  ANIMALS with birds and beasts of prey, all striving
0075  on and reaction of the innumerable plants and  ANIMALS which have determined, in the course of cent
0076  arieties of any one of our domestic plants or  ANIMALS have so exactly the same strength, habits, a
0077  prey, and to escape serving as prey to other  ANIMALS. The store of nutriment laid up within the s
0078  e advantage over its competitors, or over the  ANIMALS which preyed on it. On the confines of its g
0078  e reason to believe that only a few plants or  ANIMALS range so far, that they are destroyed by the
0083  les. He does not rigidly destroy all inferior  ANIMALS, but protects during each varying season, as
0087  he parent in relation to the young. In social  ANIMALS it will adapt the structure of each individu
0088  perhaps, severest between males of polygamous  ANIMALS, and these seem oftenest provided with speci
0088  ith special weapons. The males of carnivorous  ANIMALS are already well armed: though to them and t
0090  ming attached to the male sex in our domestic  ANIMALS (as the wattle in male carriers, horn like p
0090  ke the case of a wolf, which preys on various  ANIMALS, securing some by craft, some by strength, a
0090  when they might be compelled to prey on other  ANIMALS. I can see no more reason to doubt this, tha
0091  any change in the proportional numbers of the  ANIMALS on which our wolf preyed, a cub might be bor
0091  ces in the natural tendencies of our domestic  ANIMALS; one cat, for instance, taking to catch rats
0096  introduce a short digression. In the case of  ANIMALS and plants with separate sexes, it is of cou
0096  pared for an ample discussion. All vertebrate  ANIMALS, all insects, and some other large groups of
0096  , all insects, and some other large groups of  ANIMALS, pair for each birth. Modern research has mu
0096  ns us. But still there are many hermaphrodite  ANIMALS which certainly do not habitually pair, and
0096  lmost universal belief of breeders, that with  ANIMALS and plants a cross between different varieti
0100  he subject. Turning for a very brief space to  ANIMALS: on the land there are some hermaphrodites,
0100  y considering the medium in which terrestrial  ANIMALS live, and the nature of the fertilising elem
0100  onal cross could be effected with terrestrial  ANIMALS without the concurrence of two individuals.
0100  he concurrence of two individuals. Of aquatic  ANIMALS, there are many self fertilising hermaphrodi
0101  s a strange anomaly that, in the case of both  ANIMALS and plants, species of the same family and e
0102  n, and all try to get and breed from the best  ANIMALS, much improvement and modification surely bu
0102  ding a large amount of crossing with inferior  ANIMALS. Thus it will be in nature: for within a con
0103  her. The intercrossing will most affect those  ANIMALS which unite for each birth, which wander muc
0103  h do not breed at a very quick rate. Hence in  ANIMALS of this nature, for instance in birds, varie
0103  hich cross only occasionally, and likewise in  ANIMALS which unite for each birth, but which wander
0103  s lessened. Even in the case of slow breeding  ANIMALS, which unite for each birth, we must not ove
0103  usly thus act far more efficiently with those  ANIMALS which unite for each birth; but I have alrea
0104  t occasional intercrosses take place with all  ANIMALS and with all plants. Even if these take plac
0111  uickly new breeds of cattle, sheep, and other  ANIMALS, and varieties of flowers, take the place of
0112  fferences slowly become greater, the inferior  ANIMALS with intermediate characters, being neither
0113  mbers. We can clearly see this in the case of  ANIMALS with simple habits. Take the case of a carni
0113  eizing on places at present occupied by other  ANIMALS: some of them, for instance, being enabled t
0113  animal will apply throughout all time to all  ANIMALS, that is, if they vary, for otherwise natura
0114  e called a simultaneous rotation. Most of the  ANIMALS and plants which live close round any small
0115  s. by considering the nature of the plants or  ANIMALS which have struggled successfully with the i
0116  f any land, the more widely and perfectly the  ANIMALS and plants are diversified for different hab
0116  able of there supporting themselves. A set of  ANIMALS, with their organisation but little diversif
0127  r young, as easily as the adult. Amongst many  ANIMALS, sexual selection will give its aid to ordin
0128  re apt to overlook from familiarity, that all  ANIMALS and all plants throughout all time and space
0132  the effect is extremely small in the case of  ANIMALS, but perhaps rather more in that of plants.
0133  life. Thus, it is well known to furriers that  ANIMALS of the same species have thicker and better
0134  can be little doubt that use in our domestic  ANIMALS strengthens and enlarges certain parts, and
0134  e, for we know not the parent forms: but many  ANIMALS have structures which can be explained by th
0137  nd as eyes are certainly not indispensable to  ANIMALS with subterranean habits, a reduction in the
0137  ects of disuse. It is well known that several  ANIMALS, belonging to the most different classes, wh
0137  uch useless, could be in any way injurious to  ANIMALS living in darkness, I attribute their loss w
0137  ir loss wholly to disuse. In one of the blind  ANIMALS, namely, the cave rat, the eyes are of immen
0138  mate: so that on the common view of the blind  ANIMALS having been separately created for the Ameri
0138  ope. On my view we must suppose that American  ANIMALS, having ordinary powers of vision, slowly mi
0138  cesses of the Kentucky caves, as did European  ANIMALS into the caves of Europe. We have some evide
```

Page **(Key Word)**

Page **(Key Word)**

0138 gradation of habit; for, as Schiodte remarks, ANIMALS not far remote from ordinary forms, prepare
0138 ons, we might expect still to see in the cave ANIMALS of America, affinities to the other inhabita
0138 is is the case with some of the American cave ANIMALS, as I hear from Professor Dana; and some of
0139 planation of the affinities of the blind cave ANIMALS to the other inhabitants of the two continen
0139 om feeling any surprise that some of the cave ANIMALS should be very anomalous, as Agassiz has rem
0140 ur climate, and from the number of plants and ANIMALS brought from warmer countries which here enj
0140 ught from the Azores to England. In regard to ANIMALS, several authentic cases could be given of s
0140 ely; but we do not positively know that these ANIMALS were strictly adapted to their native climat
0140 eir new homes. As I believe that our domestic ANIMALS were originally chosen by uncivilised man be
0140 on and extraordinary capacity in our domestic ANIMALS of not only withstanding the most different
0141 an argument that a large proportion of other ANIMALS, now in a state of nature, could easily be b
0141 f the probable origin of some of our domestic ANIMALS from several wild stocks: the blood, for ins
0141 at and mouse cannot be considered as domestic ANIMALS, but they have been transported by man to ma
0141 lity of constitution, which is common to most ANIMALS. On this view, the capacity of enduring the
0141 t climates by man himself and by his domestic ANIMALS, and such facts as that former species of th
0142 of China, to be very cautious in transposing ANIMALS from one district to another; for it is not
0152 ink we can obtain some light. In our domestic ANIMALS, if any part, or the whole animal, be neglec
0152 pecially concerns us is, that in our domestic ANIMALS those points, which at the present time are
0174 especially with freely crossing and wandering ANIMALS. In looking at species as they are now distr
0179 o show that within the same group carnivorous ANIMALS exist having every intermediate grade betwee
0180 nd preys like other polecats on mice and land ANIMALS. If a different case had been taken, and it
0180 rrels; here we have the finest gradation from ANIMALS with their tails only slightly flattened, an
0182 ight have been modified into perfectly winged ANIMALS. If this had been effected, who would have e
0182 bird for flight, we should bear in mind that ANIMALS displaying early transitional grades of the
0183 ave given them a decided advantage over other ANIMALS in the battle for life. Hence the chance of
0188 earing in mind how small the number of living ANIMALS is in proportion to those which have become
0190 merous cases could be given amongst the lower ANIMALS of the same organ performing at the same tim
0191 cture with the lungs of the higher vertebrate ANIMALS: hence there seems to me to be no great diff
0191 can, indeed, hardly doubt that all vertebrate ANIMALS having true lungs have descended by ordinary
0195 istribution and existence of cattle and other ANIMALS in South America absolutely depends on their
0196 gan of locomotion the tail is in most aquatic ANIMALS, its general presence and use for many purpo
0196 nce and use for many purposes in so many land ANIMALS, which in their lungs or modified swimbladde
0196 e largely modified the external characters of ANIMALS having a will, to give one male an advantage
0197 advantage of in the parturition of the higher ANIMALS. We are profoundly ignorant of the causes pr
0198 differences in the breeds of our domesticated ANIMALS in different countries, more especially in t
0198 ; and again correlation would come into play. ANIMALS kept by savages in different countries often
0200 pper of the seal, are of special use to these ANIMALS. We may safely attribute these structures to
0200 r its progenitors, than they now are to these ANIMALS having such widely diversified habits. There
0202 ng before the advancing legions of plants and ANIMALS introduced from Europe. Natural selection wi
0202 fect, which, when used against many attacking ANIMALS, cannot be withdrawn, owing to the backward
0207 instinct and of the other mental qualities of ANIMALS within the same class. I will not attempt an
0208 ent or reason, often comes into play, even in ANIMALS very low in the scale of nature. Frederick C
0210 o find, making allowance for the instincts of ANIMALS having been but little observed except in Eu
0212 the sight of fear of the same enemy in other ANIMALS. But fear of man is slowly acquired, as I ha
0212 cuired, as I have elsewhere shown, by various ANIMALS inhabiting desert islands; and we may see an
0213 odifying the mental qualities of our domestic ANIMALS. A number of curious and authentic instances
0215 ersally and largely the minds of our domestic ANIMALS have been modified by domestication. It is s
0215 where the savages do not keep these domestic ANIMALS. How rarely, on the other hand, do our civil
0235 s of instincts almost identically the same in ANIMALS so remote in the scale of nature, that we ca
0236 shown that some insects and other articulate ANIMALS in a state of nature occasionally become ste
0242 , is very interesting, as it proves that with ANIMALS, as with plants, any amount of modification
0242 how that the mental qualities of our domestic ANIMALS vary, and that the variations are inherited.
0243 been produced for the exclusive good of other ANIMALS, but that each animal takes advantage of the
0246 state of the male element in both plants and ANIMALS; though the organs themselves are perfect in
0252 ller brought from other flowers. In regard to ANIMALS, much fewer experiments have been carefully
0252 ents can be trusted, that is if the genera of ANIMALS are as distinct from each other, as are the
0252 the genera of plants, we may infer that ANIMALS more widely separated in the scale of nature
0252 however, be borne in mind that, owing to few ANIMALS breeding freely under confinement, few exper
0253 essive generations of the more fertile hybrid ANIMALS, I hardly know of an instance in which two f
0253 thenticated cases of perfectly fertile hybrid ANIMALS, I have some reason to believe that the hybr
0254 aturalists; namely, that most of our domestic ANIMALS have descended from two or more aboriginal s
0254 is view of the origin of many of our domestic ANIMALS, we must either give up the belief of the al
0254 st universal sterility of distinct species of ANIMALS when crossed; or we must look at sterility,
0254 ined facts on the intercrossing of plants and ANIMALS, it may be concluded that some degree of ste
0255 pains to ascertain how far the rules apply to ANIMALS, and considering how scanty our knowledge is
0255 w scanty our knowledge is in regard to hybrid ANIMALS, I have been surprised to find how generally
0264 ts, which I have collected, showing that when ANIMALS and plants are removed from their natural co
0265 act, is the great bar to the domestication of ANIMALS. Between the sterility thus superinduced and
0265 with systematic affinity, for whole groups of ANIMALS and plants are rendered impotent by the same
0267 , and back again. During the convalescence of ANIMALS, we plainly see that great benefit is derive
0267 e habits of life. Again, both with plants and ANIMALS, there is abundant evidence, that a cross be
0269 he most important consideration, new races of ANIMALS and plants are produced under domestication
0274 lieve it to be with varieties of plants. With ANIMALS one variety certainly ofter has this prepote
0274 several remarks are apparently applicable to ANIMALS; but the subject is here excessively complic
0275 me authors on the supposed fact, that mongrel ANIMALS alone are born closely like one of their par
0275 he cases which I have collected of cross bred ANIMALS closely resembling one parent, the resemblan
0275 ing an enormous body of facts with respect to ANIMALS, comes to the conclusion, that the laws of r
0288 ater. I suspect that but few of the very many ANIMALS which live on the beach between high and low
0291 the bottom will be inhabited by extremely few ANIMALS, and the mass when upraised will give a most
0293 correlated with perfect accuracy. With marine ANIMALS of all kinds, we may safely infer a large am
0294 i, within that limit of depth at which marine ANIMALS can flourish; for we know what vast geograph
0298 one other consideration is worth notice: with ANIMALS and plants that can propagate rapidly and ar
0298 cal or confined to some one spot. Most marine ANIMALS have a wide range; and we have seen that wit
0298 rieties; so that with shells and other marine ANIMALS, it is probably those which have had the wid
0300 uspect that not many of the strictly littoral ANIMALS, or of those which lived on naked submarine
0303 owever, is that of the Whale family; as these ANIMALS have huge bones, are marine, and range over

Page **(Key Word)**

Page ***(Key Word)***

```
0305  sed basin, in which any great group of marine  ANIMALS might be multiplied; and here they would rem
0306  own animal. Some of the most ancient Silurian  ANIMALS, as the Nautilus, Lingula, etc., do not diff
0319  is has been the progress of events with those  ANIMALS which have been exterminated, either locally
0324  trict geological sense: for if all the marine  ANIMALS which live at the present day in Europe, and
0329  at palaeontologist, Owen, showing how extinct  ANIMALS fall in between existing groups. Cuvier rank
0330  s that he is every day taught that palaeozoic  ANIMALS, though belonging to the same orders, famili
0337  eviously occupied, we may believe, if all the  ANIMALS and plants of Great Britain were set free in
0337  places now occupied by our native plants and  ANIMALS. Under this point of view, the productions o
0338  een this result. Agassiz insists that ancient  ANIMALS resemble to a certain extent the embryos of
0338  ble to a certain extent the embryos of recent  ANIMALS of the same classes; or that the geological
0338  the present day, it would be vain to look for  ANIMALS having the common embryological character of
0340  ven in relation to the distribution of marine  ANIMALS. On the theory of descent with modification,
0341  cannot for an instant be admitted. These huge  ANIMALS have become wholly extinct, and have left no
0345  if it should hereafter be proved that ancient  ANIMALS resemble to a certain extent the embryos of
0345  o a certain extent the embryos of more recent  ANIMALS of the same class, the fact will be intellig
0349  of La Plata, we see the agouti and bizcacha,  ANIMALS having nearly the same habits as our hares a
0353  rly the same, so that a multitude of European  ANIMALS and plants have become naturalised in Americ
0357  ved as halting places for plants and for many  ANIMALS during their migration. In the coral produci
0365  s. i do not doubt that out of twenty seeds and  ANIMALS transported to an island, even if far less w
0365  acial period. The identity of many plants and  ANIMALS, on mountain summits, separated from each ot
0367  ould likewise be covered by arctic plants and  ANIMALS, and these would be nearly the same with tho
0369  r if we compare the present Alpine plants and  ANIMALS of the several great European mountain range
0370  period, a large number of the same plants and  ANIMALS inhabited the almost continuous circumpolar
0370  s circumpolar land; and that these plants and  ANIMALS, both in the Old and New Worlds, began slowl
0371  ok place long ages ago. And as the plants and  ANIMALS migrated southward, they will have become mi
0372  dmirable work), of some fish and other marine  ANIMALS, in the Mediterranean and in the seas of Jap
0376  d be given on the distribution of terrestrial  ANIMALS. In marine productions, similar cases occur:
0377  what vast spaces some naturalised plants and  ANIMALS have spread within a few centuries, this per
0377  lia. As we know that many tropical plants and  ANIMALS can withstand a considerable amount of cold,
0378  siderable number of plants, a few terrestrial  ANIMALS, and some marine productions, migrated durin
0387  h the eggs of some of the smaller fresh water  ANIMALS. Other and unknown agencies probably have al
0388  bution of fresh water plants and of the lower  ANIMALS, whether retaining the same identical form o
0388  the wide dispersal of their seeds and eggs by  ANIMALS, more especially by fresh water birds, which
0390  on to believe that the naturalised plants and  ANIMALS have nearly or quite exterminated many nativ
0390  ficient number of the best adapted plants and  ANIMALS have not been created on oceanic islands; fo
0393  islands are peculiarly well fitted for these  ANIMALS: for frogs have been introduced into Madeira
0393  lied so as to become a nuisance. But as these  ANIMALS and their spawn are known to be immediately
0393  a terrestrial mammal (excluding domesticated  ANIMALS kept by the natives) inhabiting an island si
0394  appear at a quicker rate than other and lower  ANIMALS. Though terrestrial mammals do not occur on
0398  voice, was manifest. So it is with the other  ANIMALS, and with nearly all the plants, as shown by
0404  orld. We see this same principle in the blind  ANIMALS inhabiting the caves of America and of Europ
0418  al value is set on them. As in most groups of  ANIMALS, important organs, such as those for propell
0418  able in classification: but in some groups of  ANIMALS all these, the most important vital organs,
0418  ost important of any in the classification of  ANIMALS: and this doctrine has very generally been a
0427  n drawn by some authors between very distinct  ANIMALS. On my view of characters being of real impo
0427  are almost valueless to the systematist. For  ANIMALS, belonging to two most distinct lines of des
0430  that, when a member belonging to one group of  ANIMALS exhibits an affinity to a quite distinct gro
0431  instance, of birds from all other vertebrate  ANIMALS, by the belief that many ancient forms of li
0434  n to the homologous bones in widely different  ANIMALS. We see the same great law in the constructi
0436  ee in embryonic crustaceans and in many other  ANIMALS, and in flowers, that organs, which when mat
0439  exactly alike. The embryos, also, of distinct  ANIMALS within the same class are often strikingly s
0439  ure, in which the embryos of widely different  ANIMALS of the same class resemble each other, often
0440  the young blackbird, are of any use to these  ANIMALS, or are related to the conditions to which t
0440  ity of the larvae or active embryos of allied  ANIMALS is sometimes much obscured; and cases could
0442  similarity in the embryos of widely different  ANIMALS within the same class, that we might be led
0442  le in the embryo. And in some whole groups of  ANIMALS and in certain members of other groups, the
0443  breeders of cattle, horses, and various fancy  ANIMALS, cannot positively tell, until some time aft
0444  ei and some direct evidence from our domestic  ANIMALS supports this view. But in other cases it is
0445  ace horses differed as much as the full grown  ANIMALS: and this surprised me greatly, as I think i
0448  ule of development in certain whole groups of  ANIMALS, as with cuttle fish and spiders, and with a
0450  the common parent form of each great class of  ANIMALS. Rudimentary, atrophied, or aborted organs.
0454  ore especially, according to Youatt, in young  ANIMALS, and the state of the whole flower in the ca
0454  me rudimentary, as in the case of the eyes of  ANIMALS inhabiting dark caverns, and of the wings of
0467  te them in any desired manner. He thus adapts  ANIMALS and plants for his own benefit or pleasure.
0468  sical conditions, will turn the balance. With  ANIMALS having separated sexes there will in most ca
0473  vered with skin; or when we look at the blind  ANIMALS inhabiting the dark caves of America and Eur
0474  oves by graduated steps in endowing different  ANIMALS of the same class with their several instinc
0475  mistakes; and at many instincts causing other  ANIMALS to suffer. If species be only well marked an
0477  uld be peculiar. We can clearly see why those  ANIMALS which cannot cross wide spaces of ocean, as
0478  ved. We see this in nearly all the plants and  ANIMALS of the Galapagos archipelago, of Juan Fernan
0478  in the most striking manner to the plants and  ANIMALS of the neighbouring American mainland; and t
0483  ed? Were all the infinitely numerous kinds of  ANIMALS and plants created as eggs or seed, or as fu
0484  the members of the same class. I believe that  ANIMALS have descended from at most only four or fiv
0484  step further, namely, to the belief that all  ANIMALS and plants have descended from some one prot
0484  ame poison often similarly affects plants and  ANIMALS: or that the poison secreted by the gall fly
0490  ceiving, namely, the production of the higher  ANIMALS, directly follows. There is grandeur in this
0191  ists that the branchiae and dorsal scales of  ANNELIDS are homologous with the wings and wingcovers
0201  exclusive good of another species, it would  ANNIHILATE my theory, for such could not have been pro
0239  onderful and well established facts at once  ANNIHILATE my theory. In the simpler case of neuter in
0243  of difficulty, to the best of my judgment,  ANNIHILATE it. On the other hand, the fact that instin
0258  between forms so closely related (as Matthiola  ANNUA and glabra) that many botanists rank them onl
0064  s progeny. Linnaeus has calculated that if an  ANNUAL plant produced only two seeds; and there is n
0257  fruit, and in the cotyledons, can be crossed.  ANNUAL and perennial plants, deciduous and evergreen
0391  many North American birds, during their great  ANNUAL migrations, visit either periodically or occa
0063  be dependent on the moisture. A plant which  ANNUALLY produces a thousand seeds, of which on an av
0065  ngst animals there are very few which do not  ANNUALLY pair. Hence we may confidently assert, that
```

Page ***(Key Word)***

Page **(Key Word)**

```
0065 ng on them, and we forget that thousands are ANNUALLY slaughtered for food, and that in a state of
0065 the only difference between organisms which ANNUALLY produce eggs or seeds by the thousand, and t
0068 undreds of thousands of game animals are now ANNUALLY killed. On the other hand, in some cases, as
0075 ere have gone on during long centuries, each ANNUALLY scattering its seeds by the thousand; what w
0076 of nature, and if the seed or young were not ANNUALLY sorted. As species of the same genus have us
0113 at each species and each variety of grass is ANNUALLY sowing almost countless seeds; and thus, as
0236 the community that a number should have been ANNUALLY born capable of work, but incapable of procr
0363 for a moment on the millions of quails which ANNUALLY cross the Mediterranean; and can we doubt th
0402 nd is well stocked with its own species, for ANNUALLY more eggs are laid there than can possibly b
0403 ; and although large quantities of stone are ANNUALLY transported from Porto Santo to Madeira, yet
0141 n their habits, ought not to be looked at as ANOMALIES, but merely as examples of a very common fl
0155 group, the more subject it is to individual ANOMALIES. On the ordinary view of each species havin
0395 d or identical quadrupeds. No doubt some few ANOMALIES occur in this great archipelago, and there
0107 and in fresh water we find some of the most ANOMALOUS forms now known in the world, as the Ornith
0107 widely separated in the natural scale. These ANOMALOUS forms may almost be called living fossils;
0139 that some of the cave animals should be very ANOMALOUS, as Agassiz has remarked in regard to the b
0184 nally have given rise to new species, having ANOMALOUS habits, and with their structure either sli
0185 e suspected its sub aquatic habits; yet this ANOMALOUS member of the strictly terrestrial thrush f
0193 t species furnished with apparently the same ANOMALOUS organ, it should be observed that, although
0323 d with still living sea shells; but as these ANOMALOUS monsters coexisted with the Mastodon and Ho
0013 tever as the rule, and non inheritance as the ANOMALY. The laws governing inheritance are quite un
0101 ust have struck most naturalists as a strange ANOMALY that, in the case of both animals and plants
0134 fessor Owen has remarked, there is no greater ANOMALY in nature than a bird that cannot fly; yet t
0353 ocal, or confined to one area. What a strange ANOMALY it would be, if, when coming one step lower
0399 rgs, drifted by the prevailing currents, this ANOMALY disappears. New Zealand in its endemic plant
0399 o enormously remote, that the fact becomes an ANOMALY. But this difficulty almost disappears on th
0240 ecimens from the same nest of the driver ant (ANOMMA) of West Africa. The reader will perhaps best
0172 and I will here only state that I believe the ANSWER mainly lies in the record being incomparably
0180 far more difficult, and I could have given no ANSWER. Yet I think such difficulties have very lit
0233 the progenitors of the hive bee? I think the ANSWER is not difficult: it is known that bees are o
0307 rimordial periods, I can give no satisfactory ANSWER. Several of the most eminent geologists, with
0319 urselves why this or that species is rare, we ANSWER that something is unfavourable in its conditi
0343 d of the Silurian system was deposited: I can ANSWER this latter question only hypothetically, by
0353 of the northern and southern hemispheres? The ANSWER, as I believe, is, that mammals have not been
0425 sition is of course preposterous; and I might ANSWER by the argumentum ad hominem, and ask what sh
0437 y of natural selection, we can satisfactorily ANSWER these questions. In the vertebrata, we see a
0464 unknown epochs in the world's history. I can ANSWER these questions and grave objections only on
0483 tion of species. The question is difficult to ANSWER, because the more distinct the forms are whic
0166 r breed of horses, and was, as we have seen, ANSWERED in the affirmative. What now are we to say t
0363 ed erratic boulders on these islands, and he ANSWERED that he had found large fragments of granite
0394 ands? On my view this question can easily be ANSWERED; for no terrestrial mammal can be transporte
0441 increase much in size. In the second stage, ANSWERING to the chrysalis stage of butterflies, they
0465 ory; and I have now briefly recapitulated the ANSWERS and explanations which can be given to them.
0211 ; but not one excreted. Afterwards I allowed an ANT to visit them, and it immediately seemed, by i
0211 sweet juice, which was eagerly devoured by the ANT. Even the quite young aphides behaved in this
0219 observer even than his celebrated father. This ANT is absolutely dependent on its slaves; without
0219 ? if we had not known of any other slave making ANT, it would have been hopeless to have speculate
0219 rst discovered by P. Huber to be a slave making ANT. This species is found in the southern parts o
0224 ification to be of use to the species, until an ANT was formed as abjectly dependent on its slaves
0236 etter exemplified by the hive bee. If a working ANT or other neuter insect had been an animal in t
0237 selected, and so onwards. But with the working ANT we have an insect differing greatly from its p
0240 condition; we should then have had a species of ANT with neuters very nearly in the same condition
0240 rous specimens from the same nest of the driver ANT (Anomma) of West Africa. The reader will perha
0241 been first formed, as in the case of the driver ANT, and then the extreme forms, from being the mo
0354 ges, and at distant points in the arctic and ANTARCTIC regions; and secondly (in the following cha
0363 s from one part to another of the arctic and ANTARCTIC regions, as suggested by Lyell; and during
0381 by Dr. Hooker in his botanical works on the ANTARCTIC regions. These cannot be here discussed. I
0381 commencement of the Glacial period, when the ANTARCTIC lands, now covered with ice, supported a hi
0399 ediate though distant point, namely from the ANTARCTIC islands, when they were clothed with vegeta
0341 m in South America the sloth, armadillo, and ANTEATER, as their degenerate descendants. This canno
0144 ea (whales) and Edentata (armadillos, scaly ANTEATERS, etc.), that these are likewise the most ab
0309 the lapse of ages? At a period immeasurably ANTECEDENT to the Silurian epoch, continents may have
0138 es, such as an increase in the length of the ANTENNAE or palpi, as a compensation for blindness. N
0148 ment attached to the bases of the prehensile ANTENNAE. Now the saving of a large and complex struc
0211 s well as I could, as the ants do with their ANTENNAE; but not one excreted. Afterwards I allowed
0211 d discovered; it then began to play with its ANTENNAE on the abdomen first of one aphis and then o
0211 their; and each aphis, as soon as it felt the ANTENNAE, immediately lifted up its abdomen and excre
0415 n one great division of the Hymenoptera, the ANTENNAE, as Westwood has remarked, are most constant
0416 ation; yet no one probably will say that the ANTENNAE in these two divisions of the same order are
0441 ificent compound eyes, and extremely complex ANTENNAE; but they have a closed and imperfect mouth,
0441 n a well constructed mouth; but they have no ANTENNAE; and their two eyes are now reconverted into
0135 (and I have observed the same fact) that the ANTERIOR tarsi, or feet, of many male dung feeding be
0135 prefer explaining the entire absence of the ANTERIOR tarsi in Ateuchus, and their rudimentary con
0148 pedes consists of the three highly important ANTERIOR segments of the head enormously developed, a
0148 rasitic and protected Proteolepas, the whole ANTERIOR part of the head is reduced to the merest ru
0310 these large areas, the many formations long ANTERIOR to the silurian epoch in a completely metamo
0370 tions of the Old and New Worlds, at a period ANTERIOR to the Glacial epoch. Believing, from reason
0420 d from a species which existed at an unknown ANTERIOR period. Species of three of these genera (A,
0427 in the shape of the body and in the fin like ANTERIOR limbs, between the dugong, which is a pachyd
0436 parts of a certain number of vertebrae. The ANTERIOR and posterior limbs in each member of the ve
0092 re and more pollen, and had larger and larger ANTHERS, would be selected. When our plant, by this
0093 ized pistil, and four stamens with shrivelled ANTHERS, in which not a grain of pollen can be detec
0097 r, yet what a multitude of flowers have their ANTHERS and stigmas fully exposed to the weather! bu
0097 exposure, more especially as the plant's own ANTHERS and pistil generally stand so close together
0097 and it is quite sufficient just to touch the ANTHERS of one flower and then the stigma of another
0098 ollen granules are swept out of the conjoined ANTHERS of each flower, before the stigma of that in
```

Page **(Key Word)**

0099 l has shown, and as I can confirm, either the ANTHERS burst before the stigma is ready for fertili
0249 e (as I know from my own experience) from the ANTHERS of another flower, as from the anthers of th
0249 om the anthers of another flower, as from the ANTHERS of the flower itself which is to be fertilis
0452 of brushing the pollen out of the surrounding ANTHERS. Again, an organ may become rudimentary for
0055 ed and well defined varieties, I was led to ANTICIPATE that the species of the larger genera in ea
0054 species. And this, perhaps, might have been ANTICIPATED; for, as varieties, in order to become in
0054 larger genera. This, again, might have been ANTICIPATED; for the mere fact of many species of the
0100 the United States; and the result was as I ANTICIPATED. On the other hand, Dr. Hooker has recentl
0138 ot more closely allied than might have been ANTICIPATED from the general resemblance of the other
0172 e bee to make cells, which have practically ANTICIPATED the discoveries of profound mathematicians
0242 aith in this principle, I should never have ANTICIPATED that natural selection could have been eff
0456 doctrine of creation, might even have been ANTICIPATED, and can be accounted for by the laws of i
0457 ntrary, their presence might have been even ANTICIPATED. The importance of embryological character
0472 be strange, or perhaps might even have been ANTICIPATED. As natural selection acts by competition,
0055 one having few. To test the truth of this ANTICIPATION I have arranged the plants of twelve count
0089 ay their gorgeous plumage and perform strange ANTICS before the females, which standing by as spec
0376 lance in its crustacea to Great Britain, its ANTIPODE, than to any other part of the world. Sir J.
0020 are very important, as being of considerable ANTIQUITY. I have associated with several eminent fan
0033 importance of the principle in works of high ANTIQUITY. In rude and barbarous periods of English h
0312 most recent beds, though undoubtedly of high ANTIQUITY if measured by years, only one or two speci
0161 ed: for instance, in the common snapdragon (ANTIRRHINUM) a rudiment of a fifth stamen so often app
0452 als of the species. Thus in the snapdragon (ANTIRRHINUM) we generally do not find a rudiment of a
0143 : thus a family of stags once existed with an ANTLER only on one side; and if this had been of any
0207 their origin. Instincts graduated. Aphides and ANTS. Instincts variable. Domestic instincts, their
0207 koo, ostrich, and parasitic bees. Slave making ANTS. Hive bee, its cell making instinct. Difficult
0209 ted, namely, those of the hive bee and of many ANTS, could not possibly have been thus acquired. I
0210 voluntarily yielding their sweet excretion to ANTS: that they do so voluntarily, the following fa
0210 y, the following facts show. I removed all the ANTS from a group of about a dozen aphides on a doc
0211 in the same manner, as well as I could, as the ANTS do with their antennae; but not one excreted.
0211 instinctively excrete for the sole good of the ANTS. Although I do not believe that any animal in
0216 s' nests; the slave making instinct of certain ANTS; and the comb making power of the hive bee: th
0221 (f. fusca) sometimes as many as three of these ANTS clinging to the legs of the slave making F. sa
0222 s, f. flava, with a few of these little yellow ANTS still clinging to the fragments of the nest. T
0222 s, and I have seen it ferociously attack other ANTS. In one instance I found to my surprise an ind
0222 accidentally disturbed both nests, the little ANTS attacked their big neighbours with surprising
0222 f an hour, shortly after all the little yellow ANTS had crawled away, they took heart and carried
0222 y of F. sanguinea, and found a number of these ANTS entering their nest, carrying the dead bodies
0223 nated I will not pretend to conjecture. But as ANTS, which are not slave makers, will as I have se
0223 stored as food might become developed; and the ANTS thus unintentionally reared would then follow
0236 only a single case, that of working or sterile ANTS. How the workers have been rendered sterile is
0236 ulty. The great difficulty lies in the working ANTS differing widely from both the males and the f
0238 ; namely, the fact that the neuters of several ANTS differ, not only from the fertile females and
0239 ttle as they may be called, which our European ANTS guard or imprison. It will indeed be thought t
0239 ow surprisingly the neuters of several British ANTS differ from each other in size and sometimes i
0240 udiments of ocelli, though the male and female ANTS of this genus have well developed ocelli. I ma
0241 mes as big. The jaws, moreover, of the working ANTS of the several sizes differed wonderfully in s
0242 ision of labour is useful to civilised man. As ANTS work by inherited instincts and by inherited t
0242 rable division of labour in the communities of ANTS, by the means of natural selection. But I am b
0244 the young cuckoo ejecting its foster brothers, ANTS making slaves, the larvae of ichneumonidae fee
0460 rs or sterile females in the same community of ANTS; but I have attempted to show how this difficu
0094 or food. I could give many facts, showing how ANXIOUS bees are to save time; for instance, their h
0010 t prior to the act of conception. These cases ANYHOW show that variation is not necessarily connec
0083 nder which they live, that none of them could ANYHOW be improved; for in all countries, the native
0034 make a new strain or sub breed, superior to ANYTHING existing in the country. But, for our purpos
0201 ural selection will never produce in a being ANYTHING injurious to itself, for natural selection a
0208 on be interrupted in a song, or in repeating ANYTHING by rote, he is generally forced to go back t
0282 h sediment, before he can hope to comprehend ANYTHING of the lapse of time, the monuments of which
0299 rwise would have been, yet has done scarcely ANYTHING in breaking down the distinction between spe
0309 normously long periods. If then we may infer ANYTHING from these facts, we may infer that where ou
0351 migration nor isolation in themselves can do ANYTHING. These principles come into play only by bri
0359 d, for above 28 days, as far as we may infer ANYTHING from these scanty facts, we may conclude tha
0037 and preserved the best varieties they could ANYWHERE find. A large amount of change in our cultiv
0119 are imaginary, and might have been inserted ANYWHERE, after intervals long enough to have allowed
0290 rmations with recent or tertiary remains can ANYWHERE be found, though the supply of sediment must
0159 reeds of pigeons, in countries most widely set APART, present sub varieties with reversed feathers
0373 ected. Along the Himalaya, at points 900 miles APART, glaciers have left the marks of their former
0135 not one had even a relic left. In the Onites APELLES the tarsi are so habitually lost, that the i
0022 ence of processes. The size and shape of the APERTURES in the sternum are highly variable; so is t
0207 fferent in their origin. Instincts graduated. APHIDES and ants. Instincts variable. Domestic insti
0210 other, with which I am acquainted, is that of APHIDES voluntarily yielding their sweet excretion t
0210 ed all the ants from a group of about a dozen APHIDES on a dockplant, and prevented their attendan
0211 rs. After this interval, I felt sure that the APHIDES would want to excrete. I watched them for so
0211 rly devoured by the ant. Even the quite young APHIDES behaved in this manner, showing that the act
0211 y viscid, it is probably a convenience to the APHIDES to have it removed; and therefore probably t
0211 o have it removed; and therefore probably the APHIDES do not instinctively excrete for the sole go
0221 they ascended together probably in search of APHIDES or cocci. According to Huber, who had ample
0221 ates, their principal office is to search for APHIDES. This difference in the usual habits of the
0223 es, tend, and milk as it may be called, their APHIDES; and thus both collect food for the communit
0239 , supplying the place of that excreted by the APHIDES, or the domestic cattle as they may be calle
0211 with its antennae on the abdomen first of one APHIS and then of another; and each aphis, as soon
0211 rst of one aphis and then of another; and each APHIS, as soon as it felt the antennae, immediately
0442 elopment; but in some few cases, as in that of APHIS, if we look to the admirable drawings by Prof
0448 members of the great class of insects, as with APHIS. With respect to the final cause of the young
0181 g the hind legs, we perhaps see traces of an APPARATUS originally constructed for gliding through
0188 t natural selection has converted the simple APPARATUS of an optic nerve merely coated with pigmen
0191 s now generally held, a part of the auditory APPARATUS has been worked in as a complement to the s

Page **(Key Word)**

```
0191  h we know nothing, furnished with a floating  APPARATUS or swimbladder. We can thus, as I infer fro
0193  e an organ closely analogous to the electric  APPARATUS, and yet do not, as Matteuchi asserts, disc
0218  or they do not possess the pollen collecting  APPARATUS which would be necessary if they had to sto
0005  h. in the four succeeding chapters, the most  APPARENT and gravest difficulties on the theory will
0014  a wider extension, and that when there is no   APPARENT reason why a peculiarity should appear at an
0020  ther utter hopelessness, of the task becomes  APPARENT. Certainly, a breed intermediate between two
0055  es as a special act of creation, there is no  APPARENT reason why more varieties should occur in a
0146  e umbelliferae these differences are of such  APPARENT importance, the seeds being in some cases, a
0146  manner. So, again, I do not doubt that some   APPARENT correlations, occurring throughout whole ord
0171  of my judgment, the greater number are only   APPARENT, and those that are real are not, I think, f
0194  shortest and slowest steps. Organs of little  APPARENT importance. As natural Selection acts by lif
0313  dentical form never reappears. The strongest  APPARENT exception to this latter rule, is that of th
0316  s continuous. I am aware that there are some  APPARENT exceptions to this rule, but the exceptions
0343  t of Europe, are considered: he may urge the  APPARENT, but often falsely apparent, sudden coming i
0343  he may urge the apparent, but often falsely   APPARENT, sudden coming in of whole groups of species
0365  pecies living at distant points, without the  APPARENT possibility of their having migrated from on
0418  ways found correlated with others, though no  APPARENT bond of connexion can be discovered between
0427  lines of descent. We can also understand the  APPARENT paradox, that the very same characters are a
0487  tants of that continent in relation to their  APPARENT means of immigration, some light can be thro
0002  ume on which facts cannot be adduced, often   APPARENTLY leading to conclusions directly opposite to
0010  d the parents, as Muller has remarked, have   APPARENTLY been exposed to exactly the same conditions
0010  produced very little direct effect, though   APPARENTLY more in the case of plants. Under this poin
0013  ting on both: but when amongst individuals,   APPARENTLY exposed to the same conditions, any very ra
0025  ck bars: some semi domestic breeds and some  APPARENTLY truly wild breeds have, besides the two bla
0059  pecies very closely allied to other species  APPARENTLY have restricted ranges. In all these severa
0089  ar childish to attribute any effect to such  APPARENTLY weak means: I cannot here enter on the deta
0091  case. Certain plants excrete a sweet juice,  APPARENTLY for the sake of eliminating something injur
0106  ent of Australia have formerly yielded, and  APPARENTLY are now yielding, before those of the large
0108  e variations occur, and variation itself is  APPARENTLY always a very slow process. The process wil
0120  english race horse and English pointer have  APPARENTLY both gone on slowly diverging in character
0130  s two large branches of life, and which has  APPARENTLY been saved from fatal competition by having
0132  d, which in its earliest condition does not  APPARENTLY differ essentially from an ovule, is alone
0163  ation and partly under nature. It is a case  APPARENTLY of reversion. The ass not rarely has very d
0193  of two very distinct species furnished with  APPARENTLY the same anomalous organ, it should be obse
0199  arked: I may add that some little light can  APPARENTLY be thrown on the origin of these difference
0204  possible. For instance, a swim bladder has  APPARENTLY been converted into an air breathing lung.
0210  one of the strongest instances of an animal  APPARENTLY performing an action for the sole good of a
0227  ical burrows in wood many insects can make,  APPARENTLY by turning round on a fixed point. We must
0235  ns are known to exist: cases of instinct of  APPARENTLY such trifling importance, that they could h
0247  perimentised on by Gartner were potted, and  APPARENTLY were kept in a chamber in his house. That t
0263  r or lesser difficulty in effecting a union  APPARENTLY depends on several distinct causes. There m
0274  th either parent. These several remarks are  APPARENTLY applicable to animals: but the subject is h
0277  light changes in the conditions of life are  APPARENTLY favourable to the vigour and fertility of a
0279  at the present day, under the circumstances  APPARENTLY most favourable for their presence, namely
0291  ne numerous slow oscillations of level, and  APPARENTLY these oscillations have affected wide space
0294  merica than in those of Europe: time having  APPARENTLY been required for their migration from the
0305  ntly insisted on by palaeontologists of the  APPARENTLY sudden appearance of a whole group of speci
0311  d succession of chapters, may represent the  APPARENTLY abruptly changed forms of life, entombed in
0314  ed unaltered. We can perhaps understand the  APPARENTLY quicker rate of change in terrestrial and i
0318  d the former horse under conditions of life  APPARENTLY so favourable. But how utterly groundless w
0321  ss than its production. With respect to the  APPARENTLY sudden extermination of whole families or o
0326  a. we might therefore expect to find, as we  APPARENTLY do find, a less strict degree of parallel s
0329  he species would not all be the same in the  APPARENTLY corresponding stages in the two regions. On
0329  xample, he dissolves by fine gradations the  APPARENTLY wide difference between the pig and the cam
0341  enera and species decreasing in numbers, as  APPARENTLY is the case of the Edentata of South Americ
0344  the more ancient a form is, the more often,  APPARENTLY, it displays characters in some degree inte
0379  more identical plants and allied forms have  APPARENTLY migrated from the north to the south, than
0383  within the same country, and as the sea is  APPARENTLY a still more impassable barrier, that they
0391  the different orders of insects in Madeira  APPARENTLY present analogous facts. Oceanic islands ar
0391  nt in certain classes, and their places are  APPARENTLY occupied by the other inhabitants: in the G
0405  till less is it meant, that a species which  APPARENTLY has the capacity of crossing barriers and r
0443  nd has to provide for itself: of the embryo  APPARENTLY having sometimes a higher organisation than
0461  d under domestication: and as domestication  APPARENTLY tends to eliminate sterility, we ought not
0470  ly allied species also of the larger genera  APPARENTLY have restricted ranges, and they are cluste
0475  we need not marvel at some instincts being  APPARENTLY not perfect and liable to mistakes, and at
0013  life a peculiarity first appears, it tends to  APPEAR in the offspring at a corresponding age, thou
0014  ed peculiarities in the horns of cattle could  APPEAR only in the offspring when nearly mature; pec
0014  e: peculiarities in the silkworm are known to  APPEAR at the corresponding caterpillar or cocoon st
0014  s no apparent reason why a peculiarity should  APPEAR at any particular age, yet that it does tend
0014  any particular age, yet that it does tend to  APPEAR in the offspring at the same period at which
0037  when a slightly better variety has chanced to  APPEAR, selecting it, and so onwards. But the garden
0041  riations manifestly useful or pleasing to man  APPEAR only occasionally, the chance of their appear
0045  differences, such as are known frequently to  APPEAR in the offspring from the same parents, or wh
0070  e. but even some of these so called epidemics  APPEAR to be due to parasitic worms, which have from
0086  at those variations which under domestication  APPEAR at any particular period of life, tend to rea
0087  al selection. Inasmuch as peculiarities often  APPEAR under domestication in one sex and become her
0089  ently attractive to all his hen birds. It may  APPEAR childish to attribute any effect to such appa
0094  ense and incarnatum) do not on a hasty glance  APPEAR to differ in length: yet the hive bee can eas
0117  ersified nature: they are not supposed all to  APPEAR simultaneously, but often after long interval
0133  ct action of the severe climate? for it would  APPEAR that climate has some direct action on the ha
0142  he constitution of seedling kidney beans ever  APPEAR, for an account has been published how much m
0146  orders, is so far fetched, it may at first   APPEAR: and if it be advantageous, natural selection
0160  these coloured marks are eminently liable to  APPEAR in the crossed offspring of two distinct and
0162  ted with the blue tint, and which it does not  APPEAR probable would all appear together from simpl
0162  d which it does not appear probable would all  APPEAR together from simple variation. More especial
0163  stated by Mr. Blyth and others, occasionally  APPEAR: and I have been informed by Colonel Poole th
```

Page **(Key Word)**

Page ***(Key Word)***

```
0166  horse genus the stripes are either plainer or  APPEAR  more commonly in the young than in the old. C
0167  ns of instituting a comparison, the same laws   APPEAR  to have acted in producing the lesser differe
0176  varieties in the genus Balanus. And it would    APPEAR  from information given me by Mr. Watson, Dr.
0199  ntering on copious details my reasoning would   APPEAR  frivolous. The foregoing remarks lead me to s
0239  itable modification did not probably at first   APPEAR  in all the individual neuters in the same nes
0260  foregoing rules and facts, on the other hand,   APPEAR  to me clearly to indicate that the sterility
0286  of one inch in a century. This will at first    APPEAR  much too small an allowance: but it is the sa
0294  upraised, organic remains will probably first   APPEAR  and disappear at different levels, owing to t
0301  reserved, transitional varieties would merely   APPEAR  as so many distinct species. It is, also, pro
0302  ner in which whole groups of species suddenly   APPEAR  in certain formations, has been urged by seve
0303  in the succeeding formation such species will   APPEAR  as if suddenly created. I may here recall a r
0305  assuming, however, that the whole of them did   APPEAR, as Agassiz believes, at the commencement of
0306  umbers of species of the same group, suddenly   APPEAR  in the lowest known fossiliferous rocks. Most
0308  e nature of the organic remains, which do not   APPEAR  to have inhabited profound depths, in the sev
0310  udden manner in which whole groups of species   APPEAR  in our European formations; the almost entire
0316  that the species of a group sometimes falsely   APPEAR  to have come in abruptly; and I have attempte
0317  es are found, the line will sometimes falsely   APPEAR  to begin at its lower end; not in a sharp poi
0325  over a series of analogous phenomena, it will   APPEAR  certain that all these modifications of speci
0329  the form of life, and the order would falsely   APPEAR  to be strictly parallel; nevertheless the spe
0344  ls of time, the productions of the world will   APPEAR  to have changed simultaneously. We can unders
0394  and on continents it is thought that mammals    APPEAR  and disappear at a quicker rate than other an
0401  fectually separated from each other than they   APPEAR  to be on a map. Nevertheless a good many spec
0415  though here even when all taken together they   APPEAR  insufficient to separate Cnestis from Connaru
0443  ly period, that slight variations necessarily   APPEAR  at an equally early period. But we have littl
0443  even before the formation of the embryo, may    APPEAR  late in life; as when an hereditary disease,
0444  in the offspring. Certain variations can only   APPEAR  at corresponding ages, for instance, peculiar
0444  ve appeared earlier or later in life, tend to   APPEAR  at a corresponding age in the offspring and p
0453  ave been amputated, imperfect nails sometimes   APPEAR  on the stumps: I could as soon believe that t
0459  o them their full force. Nothing at first can   APPEAR  more difficult to believe than that the more
0463  why, again, co whole groups of allied species   APPEAR, though certainly they often falsely appear,
0463  s appear, though certainly they often falsely   APPEAR, to have come in suddenly on the several geol
0465  scovered in a geological formation, they will   APPEAR  as if suddenly created there, and will be sim
0475  rms of life, after long intervals of time, to   APPEAR  as if they had changed simultaneously through
0014  remarks are of course confined to the first    APPEARANCE  of the peculiarity, and not to its primary
0036  wo gentlemen is so great that they have the    APPEARANCE  of being quite different varieties. If ther
0037  iently distinct to be ranked at their first    APPEARANCE  as distinct varieties, and whether or not t
0041  pear only occasionally, the chance of their    APPEARANCE  will be much increased by a large number of
0049  latior. These plants differ considerably in    APPEARANCE: they have a different flavour and emit a d
0102  ividuals, by giving a better chance for the    APPEARANCE  within any given period of profitable varia
0105  selection, by decreasing the chance of the    APPEARANCE  of favourable variations. If we turn to nat
0159  e have another case, namely, the occasional    APPEARANCE  in all the breeds, of slaty blue birds with
0166  ring of the other species of the genus. The    APPEARANCE  of the stripes is not accompanied by any ch
0193  ould be observed that, although the general    APPEARANCE  and function of the organ may be the same,
0220  red masters, so that the contrast in their    APPEARANCE  is very great. When the nest is slightly di
0257  most widely different in habit and general    APPEARANCE, and having strongly marked differences in
0259  e degree in which they resemble in external    APPEARANCE  either parent. And lastly, that the facilit
0268  they may differ from each other in external    APPEARANCE, cross with perfect facility, and yield per
0268  ieties, differing widely from each other in    APPEARANCE, for instance of the pigeon or of the cabba
0279  ieties in any one formation. On the sudden    APPEARANCE  of groups of species. On their sudden appea
0279  rance of groups of species. On their sudden    APPEARANCE  in the lowest known fossiliferous strata. I
0293  iffs are at the present time suffering. The    APPEARANCE  of the surface and the vegetation show that
0302  essed so hardly on my theory. On the sudden    APPEARANCE  of whole groups of Allied Species. The abru
0304  s i thought one more instance of the abrupt    APPEARANCE  of a great group of species. But my work ha
0305  y palaeontologists of the apparently sudden    APPEARANCE  of a whole group of species, is that of the
0306  and range of its productions. On the sudden    APPEARANCE  of groups of Allied Species in the lowest k
0312  organic Beings. On the slow and successive    APPEARANCE  of new species. On their different rates of
0312  cies follow the same general rules in their    APPEARANCE  and disappearance as do single species. On
0312  en; but, as Bronn has remarked, neither the    APPEARANCE  nor disappearance of their many now extinct
0316  ies, follow the same general rules in their    APPEARANCE  and disappearance as do single species, cha
0318  lower process than their production: if the    APPEARANCE  and disappearance of a group of species be
0318  han at its lower end, which marks the first    APPEARANCE  and increase in numbers of the species. In
0320  f other breeds in other countries. Thus the    APPEARANCE  of new forms and the disappearance of old f
0326  viduals, from giving a better chance of the    APPEARANCE  of favourable variations, and that severe c
0335  er such cases, that the record of the first    APPEARANCE  and disappearance of the species was perfec
0342  n has played an important part in the first    APPEARANCE  of new forms in any one area and formation;
0376  sir J. Richardson, also, speaks of the re    APPEARANCE  on the shores of New Zealand, Tasmania, etc
0438  d into skulls or jaws. Yet so strong is the    APPEARANCE  of a modification of this nature having occ
0440  nd sessile, which differ widely in external    APPEARANCE, have larvae in all their several stages ba
0473  on the theory of creation is the occasional    APPEARANCE  of stripes on the shoulder and legs of the
0483  they ignore the whole subject of the first    APPEARANCE  of species in what they consider reverent s
0083  isible characters: nature cares nothing for    APPEARANCES, except in so far as they may be useful to
0164  , was striped; and that the above described    APPEARANCES  are all due to ancient crosses with the du
0427  nstances: thus Linnaeus, misled by external    APPEARANCES, actually classed an homopterous insect as
0014  fspring at the same period at which it first   APPEARED  in the parent. I believe this rule to be of
0039  l or unusual any character was when it first   APPEARED, the more likely it would be to catch his at
0041  st seedlings and bred from them, then, there   APPEARED  (aided by some crossing with distinct specie
0090  rnamental to this bird: indeed, had the tuft   APPEARED  under domestication, it would have been call
0101  to be physically impossible. Cirripedes long   APPEARED  to me to present a case of very great diffic
0137  s certainly in this condition, the cause, as   APPEARED  on dissection, having been inflammation of t
0142  published how much more hardy some seedlings   APPEARED  to be than others. On the whole, I think we
0216  ar mental habits and actions, which at first   APPEARED  from what we must in our ignorance call an a
0229  topping their work. In ordinary combs it has   APPEARED  to me that the bees do not always succeed in
0236  lf to one special difficulty, which at first   APPEARED  to me insuperable, and actually fatal to my
0251  um. Although the plants in these experiments   APPEARED  perfectly healthy, and although both the ovu
0275  ous in their nature, and which have suddenly   APPEARED, such as albinism, melanism, deficiency of t
0294  ll known, for instance, that several species   APPEARED  somewhat earlier in the palaeozoic beds of N
```

Page ***(Key Word)***

Page ***(Key Word)***

```
0302  transitional links between the species which  APPEARED  at the commencement and close of each format
0305  wise be shown that the species of this group  APPEARED  suddenly and simultaneously throughout the w
0312  cent and natural selection. New species have  APPEARED  very slowly, one after another, both on the
0312  d only one or two are new forms, having here  APPEARED  for the first time, either locally, or, as f
0316  hat as long as any species of the group have  APPEARED  in the long succession of ages, so long must
0336  rmediate varieties between the species which  APPEARED  at the commencement and close of these perio
0400  ed, though only in a small degree. This long  APPEARED  to me a great difficulty: but it arises in c
0444  sive modification, or most of them, may have  APPEARED  at an extremely early period. I have stated
0444  s which, for all that we can see, might have  APPEARED  earlier or later in life, tend to appear at
0446  by man's selection, have not generally first  APPEARED  at an early period of life, and have been in
0446   characteristic differences must either have  APPEARED  at an earlier period than usual, or, if not
0447  n earlier period than that at which it first  APPEARED  In either case (as with the short faced tum
0450  earlier period than that at which they first  APPEARED.  It should also be borne in mind, that the s
0453  on believe that these vestiges of nails have  APPEARED,  not from unknown laws of growth; but in ord
0008  ions of the reproductive system; this system  APPEARING  to be far more susceptible than any other p
0013  albinism, prickly skin, hairy bodies, etc.,  APPEARING  in several members of the same family. If s
0013  little importance to us, that peculiarities  APPEARING  in the males of our domestic breeds are oft
0014  ned bull, the greater length of horn, though  APPEARING  late in life, is clearly due to the male el
0027  e blue colour and various marks occasionally  APPEARING  in all the breeds, both when kept pure and
0160  on, and not of a new yet analogous variation  APPEARING  in the several breeds. We may I think confi
0162  his, from the blue colour and marks so often  APPEARING  when distinct breeds of diverse colours are
0228  til they were converted into shallow basins,  APPEARING  to the eye perfectly true or parts of a sph
0237  y natural selection. This difficulty, though  APPEARING  insuperable, is lessened, or, as I believe,
0269  y ought not to expect to find sterility both  APPEARING  and disappearing under nearly the same cond
0293  on on this head. When we see a species first  APPEARING  in the middle of any formation, it would be
0293  her changes; and when we see a species first  APPEARING  in any formation, the probability is that i
0444  tain that the greyhound and bull dog, though  APPEARING  so different, are really varieties most clo
0450  on the principle of slight modifications not  APPEARING,  in the many descendants from some one anci
0459  essor. Nevertheless, this difficulty, though  APPEARING  to our imagination insuperably great, canno
0010  affected in the same way, the change at first  APPEARS  to be directly due to such conditions; but i
0012  ts. From the facts collected by Heusinger, it  APPEARS  that white sheep and pigs are differently af
0013  y theoretical writers alone. When a deviation  APPEARS  not unfrequently, and we see it in the fathe
0013  e extraordinary combination of circumstances,  APPEARS  in the parent, say, once amongst several mil
0013  t whatever period of life a peculiarity first  APPEARS,  it tends to appear in the offspring at a co
0037  e pear, though cultivated in classical times,  APPEARS,  from Pliny's description, to have been a fr
0092  sole object of fertilisation, its destruction  APPEARS  a simple loss to the plant: yet if a little
0161  rhinum) a rudiment of a fifth stamen so often  APPEARS,  that this plant must have an inherited tend
0165  so convinced that not even a stripe of colour  APPEARS  from what would commonly be called an accide
0166  see this tendency strong whenever a dun tint  APPEARS,  a tint which approaches to that of the gene
0193  eir affinities. Generally when the same organ  APPEARS  in several members of the same class, especi
0224  ficulty is not nearly so great as it at first  APPEARS:  all this beautiful work can be shown, I thi
0269  estic varieties less remarkable than at first  APPEARS.  It can, in the first place, be clearly show
0443  in life: as when an hereditary disease, which  APPEARS  in old age alone, has been communicated to t
0444  ble, that at whatever age any variation first  APPEARS  in the parent, it tends to reappear at a cor
0445  t of proportional difference. As the evidence  APPEARS  to me conclusive, that the several domestic
0392  orm an endemic species, having as useless an  APPENDAGE  as any rudimentary organ, for instance, as
0439  descent from true legs, or from some simple  APPENDAGE,  is explained. Embryology. It has already b
0437  nal vertebrae bearing certain processes and  APPENDAGES;  in the articulata, we see the body divided
0437  into a series of segments, bearing external  APPENDAGES;  and in flowering plants, we see a series o
0309  e earliest silurian period. The coloured map  APPENDED  to my volume on Coral Reefs, led me to concl
0029  sorts, for instance a Ribston pippin or Codlin  APPLE,  could ever have proceeded from the seeds of
0063  the ground. The missletoe is dependent on the  APPLE  and a few other trees, but can only in a far
0261  ich is ranked as a distinct genus, than on the  APPLE,  which is a member of the same genus. Even di
0423  stance, not to class two varieties of the pine  APPLE  together, merely because their fruit, though
0029  ecies. Van Mons, in his treatise on pears and  APPLES,  shows how utterly he disbelieves that the se
0023  ve led me to this belief are in some degree  APPLICABLE  in other cases, I will here briefly give th
0024  nt, as it seems to me, of great weight, and  APPLICABLE  in several other cases, is, that the above
0151  e rarely to them. The rule being so plainly  APPLICABLE  in the case of secondary sexual characters,
0154  because an explanation is not in this case  APPLICABLE,  which most naturalists would advance, name
0243  ural history, of natura non facit saltum is  APPLICABLE  to instincts as well as to corporeal struct
0274  arent. These several remarks are apparently  APPLICABLE  to animals; but the subject is here excessi
0147  be maintained that the law is of universal  APPLICATION;  but many good observers, more especially
0151  , as it illustrates the rule in its largest  APPLICATION.  The opercular valves of sessile cirripede
0403  fitted to their new homes, is of the widest  APPLICATION  throughout nature. We see this on every mo
0005  treated of. This is the doctrine of Malthus,  APPLIED  to the whole animal and vegetable kingdoms.
0026  re is some probability in this hypothesis, if  APPLIED  to species closely related together, though
0043  the accumulative action of selection, whether  APPLIED  methodically and more quickly, or unconsciou
0052  son with mere individual differences, is also  APPLIED  arbitrarily, and for mere convenience sake.
0063  itions of life. It is the doctrine of Malthus  APPLIED  with manifold force to the whole animal and
0095  lled a trifling and insignificant cause, when  APPLIED  to the excavation of gigantic valleys or to
0152  hole animal, be neglected and no selection be  APPLIED,  that part (for instance, the comb in the Do
0237  , when it is remembered that selection may be  APPLIED  to the family, as well as to the individual,
0255  pollen of different species of the same genus  APPLIED  to the stigma of some one species, yields a
0286  ably would not greatly affect the estimate as  APPLIED  to the western extremity of the district. If
0457  in the homologous organs, to whatever purpose  APPLIED,  of the different species of a class; or to
0113  places they would be enabled to occupy. What  APPLIES  to one animal will apply throughout all time
0150  hould be understood that the rule by no means  APPLIES  to any part, however unusually developed, un
0150  the other species of the same genus. The rule  APPLIES  very strongly in the case of secondary sexua
0150  secondary sexual characters, used by Hunter,  APPLIES  to characters which are attached to one sex,
0150  nected with the act of reproduction. The rule  APPLIES  to males and females; but as females more ra
0150  fer remarkable secondary sexual character, it  APPLIES  more rarely to them. The rule being so plain
0151  good in this class. I cannot make out that it  APPLIES  to plants, and this would seriously have sha
0155  y remain the same. Something of the same kind  APPLIES  to monstrosities: at least Is. Geoffroy St.
0168  in the second Chapter that the same principle  APPLIES  to the whole individual; for in a district w
0210  covered. The canon of Natura non facit saltum  APPLIES  with almost equal force to instincts as to b
0376  a most remarkable manner. This brief abstract  APPLIES  to plants alone: some strictly analogous fac
```

Page ***(Key Word)***

Page **************************************(Key Word)**************************************

```
0080  we have seen is so potent in the hands of man, APPLY to nature? I think we shall see that it can a
0112  , it may be asked, can any analogous principle APPLY in nature? I believe it can and does apply mo
0112  ole apply in nature? I believe it can and does APPLY most efficiently, from the simple circumstanc
0113  led to occupy. What applies to one animal will APPLY throughout all time to all animals, that is,
0150  he class mammalia; but the rule would not here APPLY, because there is a whole group of bats havin
0150  s a whole group of bats having wings; it would APPLY only if some one species of bat had its wings
0165  isfied with this theory, and should be loth to APPLY it to breeds so distinct as the heavy Belgian
0176  ffer from species, the same rule will probably APPLY to both; and if we in imagination adapt a var
0255  aken much pains to ascertain how far the rules APPLY to animals, and considering how scanty our kn
0255  surprised to find how generally the same rules APPLY to both kingdoms. It has been already remarke
0269  r degree of sterility when crossed; and we may APPLY the same rule to domestic varieties. In the s
0306  same group have descended from one progenitor, APPLY with nearly equal force to the earliest known
0369  day. But the foregoing remarks on distribution APPLY not only to strictly arctic forms, but also t
0446  rresponding, but at an earlier age. Now let us APPLY these facts and the above two principles, whi
0044  her, and in having restricted ranges. Before APPLYING the principles arrived at in the last chapte
0007   to the new conditions of life to cause any APPRECIABLE amount of variation; and that when the org
0112  gence, causing differences, at first barely APPRECIABLE, steadily to increase, and the breeds to d
0134  ht, until they become plainly developed and APPRECIABLE by us. Effects of Use and Disuse. From the
0032  es which I for one have vainly attempted to APPRECIATE. Not one man in a thousand has accuracy of
0240  f west Africa. The reader will perhaps best APPRECIATE the amount of difference in these workers,
0094  f the proboscis, etc., far too slight to be APPRECIATED by us, might profit a bee or other insect,
0417  cters; for this saying seems founded on an APPRECIATION of many trifling points of resemblance, to
0330  orms, the objection is probably valid. But I APPREHEND that in a perfectly natural classification
0415  all their parts, not only in this but, as I APPREHEND, in every natural family, is very unequal,
0423  ause less constant. In classing varieties, I APPREHEND if we had a real pedigree, a genealogical c
0164  from one between brown and black to a close APPROACH to cream colour. I am aware that Colonel Ham
0220  tatements of Huber and Mr. Smith, I tried to APPROACH the subject in a sceptical frame of mind, as
0330  ite distinct, at that period made some small APPROACH to each other. It is a common belief that th
0333  uld in the older formations make some slight APPROACH to each other; so that the older members sho
0417  trange creature to birds and reptiles, as an APPROACH in structure in any one internal and and imp
0221  , and evidently not in search of food; they APPROACHED and were vigorously repulsed by an independ
0166  g whenever a dun tint appears, a tint which APPROACHES to that of the general colouring of the oth
0430  o marsupials; but in the points in which it APPROACHES this order, its relations are general, and
0219  nest and stored food are thus feloniously APPROPRIATED, be not thus exterminated. Slave making in
0348  or fish is common to the above named three APPROXIMATE faunas of Eastern and Western America and
0154  bnormal it may be, has been transmitted in APPROXIMATELY the same condition to many modified desce
0295  eriod of subsidence; and to keep the depth APPROXIMATELY the same, which is necessary in order to
0262  the quince; so do different varieties of the APRICOT and peach on certain varieties of the plum.
0012  eeth; long haired and coarse haired animals are APT to have, as is asserted, long or many horns; p
0025  specified marks, the mongrel offspring are very APT suddenly to acquire these characters; for inst
0059  r of varieties. In large genera the species are APT to be closely, but unequally, allied together,
0084  ng, yet characters and structures, which we are APT to consider as very trifling importance, may t
0128  ruly wonderful fact, the wonder of which we are APT to overlook from familiarity, that all animals
0133  which live exclusively on the sea side are very APT to have fleshy leaves. He who believes in the
0149  by some authors, and I believe with truth, are APT to be highly variable. We shall have to recur
0165  mon mule from the ass and horse is particularly APT to have bars on its legs. I once saw a mule wi
0217  to believe, that the young thus reared would be APT to follow by inheritance the occasional and ab
0217  abit of their mother and in their turn would be APT to lay their eggs in other birds' nests, and t
0402  astonishing rapidity over new countries, we are APT to infer that most species would thus spread;
0406  ny group of organisms is, the more widely it is APT to range. The relations just discussed, namely
0481  ired some idea of the lapse of time, we are too APT to assume, without proof, that the geological
0182  ich; and functionally for no purpose like the APTERYX. Yet the structure of each of these birds is
0385  me, in removing a little duck weed from one AQUARIUM to another, that I have quite unintentionall
0385  of a bird sleeping in a natural pond, in an AQUARIUM, where many ova of fresh water shells were h
0077  d for diving, allows it to compete with other AQUATIC insects, to hunt for its own prey, and to es
0100  ithout the concurrence of two individuals. Of AQUATIC animals, there are many self fertilising her
0179  nimal could have been converted into one with AQUATIC habits; for how could the animal in its tran
0179  having every intermediate grade between truly AQUATIC and strictly terrestrial habits; and as each
0184  rendered, by natural selection, more and more AQUATIC in their structure and habits, with larger a
0185  ater ouzel would never have suspected its sub AQUATIC habits; yet this anomalous member of the str
0185  he other hand, grebes and coots are eminently AQUATIC, although their toes are only bordered by me
0185  oating plants, yet the water hen is nearly as AQUATIC as the coot; and the landrail nearly as terr
0196  nt an organ of locomotion the tail is in most AQUATIC animals, its general presence and use for ma
0196  r lungs or modified swimbladders betray their AQUATIC origin, may perhaps be thus accounted for. A
0196  well developed tail having been formed in an AQUATIC animal, it might subsequently come to be wor
0200  bt were as useful as they now are to the most AQUATIC of existing birds. So we may believe that th
0205  ce has often been retained (as the tail of an AQUATIC animal by its terrestrial descendants), thou
0385  drop off. These just hatched molluscs, though AQUATIC in their nature, survived on the duck's feet
0386  rrestrial plants, which have only a very few AQUATIC members; for these latter seem immediately t
0388  petition will probably be less severe between AQUATIC than between terrestrial species; consequent
0388  han the average for the migration of the same AQUATIC species, we should not forget the probabilit
0471  uld have been created to dive and feed on sub AQUATIC insects; and that a petrel should have been
0035  me to surpass in fleetness and size the parent ARAB stock, so that the latter, by the regulations
0165  ubsequently produced from the mare by a black ARABIAN sire, were much more plainly barred across t
0339  xtinct and living brackish water shells of the ARALO Caspian Sea. Now what does this remarkable La
0052  een that I look at the term species, as one ARBITRARILY given for the sake of convenience to a set
0052  ere individual differences, is also applied ARBITRARILY, and for mere convenience sake. Guided by
0428  llelism of this nature in any one class, by ARBITRARILY raising or sinking the value of the groups
0482  variation as a vera causa in one case, they ARBITRARILY reject it in another, without assigning an
0048  nd, I was much struck how entirely vague and ARBITRARY is the distinction between species and vari
0411  groups. This classification is evidently not ARBITRARY like the grouping of the stars in constella
0419  they seem to be, at least at present, almost ARBITRARY. Several of the best botanists, such as Mr.
0419  and others, have strongly insisted on their ARBITRARY value. Instances could be given amongst pla
0428  shows that this valuation has hitherto been ARBITRARY), could easily extend the parallelism over
0435  we suppose that the ancient progenitor, the ARCHETYPE as it may be called, of all mammals, had it
0325  se admirable observers, MM. de Verneuil and d'ARCHIAC. After referring to the parallelism of the p
```

Page **************************************(Key Word)**************************************

Page ***************************************(Key Word)***

0002 w studying the natural history of the Malay ARCHIPELAGO, has arrived at almost exactly the same ge
0048 from the separate islands of the Galapagos ARCHIPELAGO, both one with another, and with those fro
0164 ; and from Norway in the north to the Malay ARCHIPELAGO in the south. In all parts of the world th
0197 l selection. A trailing bamboo in the Malay ARCHIPELAGO climbs the loftiest trees by the aid of ex
0292 of a continent when first broken up into an ARCHIPELAGO), and consequently during subsidence, thou
0299 under an imaginary illustration. The Malay ARCHIPELAGO is of about the size of Europe from the No
0299 en, that the present condition of the Malay ARCHIPELAGO, with its numerous large islands separated
0299 our formations were accumulating. The Malay ARCHIPELAGO is one of the richest regions of the whole
0300 eve that the terrestrial productions of the ARCHIPELAGO would be preserved in an excessively imper
0300 ecay, no remains could be preserved. In our ARCHIPELAGO, I believe that fossiliferous formations c
0300 of subsidence over the whole or part of the ARCHIPELAGO, together with a contemporaneous accumulat
0301 ; and in these cases the inhabitants of the ARCHIPELAGO would have to migrate, and no closely cons
0301 very many of the marine inhabitants of the ARCHIPELAGO now range thousands of miles beyond its co
0305 at present. Even at this day, if the Malay ARCHIPELAGO were converted into land, the tropical par
0363 and other rocks, which do not occur in the ARCHIPELAGO. Hence we may safely infer that icebergs f
0390 a, or of the endemic birds in the Galapagos ARCHIPELAGO, with the number found on any continent, a
0394 else in the world: Norfolk Island, the Viti ARCHIPELAGO, the Bonin Islands, the Caroline and Maria
0395 s on this head in regard to the great Malay ARCHIPELAGO, which is traversed near Celebes by a spac
0395 oubt some few anomalies occur in this great ARCHIPELAGO, and there is much difficulty in forming a
0395 licht thrown on the natural history of this ARCHIPELAGO by the admirable zeal and researches of Mr
0397 i will give only one, that of the Galapagos ARCHIPELAGO, situated under the equator, between 500 a
0398 n his admirable memoir on the Flora of this ARCHIPELAGO. The naturalist, looking at the inhabitant
0398 posec to have been created in the Galapagos ARCHIPELAGO, and nowhere else, bear so plain a stamp o
0399 the law which causes the inhabitants of an ARCHIPELAGO, though specifically distinct, to be close
0400 sting manner, within the limits of the same ARCHIPELAGO. Thus the several islands of the Galapagos
0400 . thus the several islands of the Galapagos ARCHIPELAGO are tenanted, as I have elsewhere shown, i
0400 look to those inhabitants of the Galapagos ARCHIPELAGO which are found in other parts of the worl
0401 rprising fact in this case of the Galapagos ARCHIPELAGO, and in a lesser degree in some analogous
0401 s of the sea are rapid and sweep across the ARCHIPELAGO, and gales of wind are extraordinarily rar
0401 arts of the world and those confined to the ARCHIPELAGO, are common to the several islands, and we
0402 dolle, to distinct genera. In the Galapagos ARCHIPELAGO, many even of the birds, though so well ad
0403 nhabit the several islands of the Galapagos ARCHIPELAGO, not having universally spread from island
0406 pecies which inhabit the islets of the same ARCHIPELAGO, and especially the striking relation of t
0406 g relation of the inhabitants of each whole ARCHIPELAGO or island to those of the nearest mainland
0409 n clearly see why all the inhabitants of an ARCHIPELAGO, though specifically distinct on the sever
0478 all the plants and animals of the Galapagos ARCHIPELAGO, of Juan Fernandez, and of the other Ameri
0478 an mainland; and those of the Cape de Verde ARCHIPELAGO and other African islands to the African m
0302 multiplied before they invaded the ancient ARCHIPELAGOES of Europe and of the United States. We do
0309 till mainly areas of subsidence, the great ARCHIPELAGOES still areas of oscillations of level, and
0394 e bonin Islands, the Caroline and Marianne ARCHIPELAGOES, and Mauritius, all possess their peculia
0398 s, between the Galapagos and Cape de Verde ARCHIPELAGOS: but what an entire and absolute differenc
0228 through natural selection, her inimitable ARCHITECTURAL powers. But this theory can be tested by
0233 long and graduated succession of modified ARCHITECTURAL instincts, all tending towards the presen
0474 ciple of gradation throws on the admirable ARCHITECTURAL powers of the hive bee. Habit no doubt so
0235 ve bee. Beyond this stage of perfection in ARCHITECTURE, natural selection could not lead; for the
0069 r in descending a mountain. When we reach the ARCTIC regions, or snow capped summits, or absolute
0078 we reach the extreme confines of life, in the ARCTIC regions or on the borders of an utter desert,
0139 the climate of its own home: species from an ARCTIC or even from a temperate region cannot endure
0141 s: the blood, for instance, of a tropical and ARCTIC wolf or wild dog may perhaps be mingled in ou
0304 of many species all over the world, from the ARCTIC regions to the equator, inhabiting various zo
0347 erate forms, as there now is for the strictly ARCTIC productions. We see the same fact in the grea
0354 mountain ranges, and at distant points in the ARCTIC and antarctic regions: and secondly (in the f
0363 sported seeds from one part to another of the ARCTIC and antarctic regions, as suggested by Lyell;
0366 al europe and North America suffered under an ARCTIC climate. The ruins of a house burnt by fire d
0366 as each more southern zone became fitted for ARCTIC beings and ill fitted for their former more t
0366 habitants, the latter would be supplanted and ARCTIC productions would take their places. The inha
0366 reached its maximum, we should have a uniform ARCTIC fauna and flora, covering the central parts o
0367 he united States would likewise be covered by ARCTIC plants and animals, and these would be nearly
0367 the final result. As the warmth returned, the ARCTIC forms would retreat northward, closely follow
0367 w melted from the bases of the mountains, the ARCTIC forms would seize on the cleared and thawed g
0367 when the warmth had fully returned, the same ARCTIC species, which had lately lived in a body tog
0367 tain range are more especially related to the ARCTIC forms living due north or nearly due north of
0368 dor; those of the mountains of Siberia to the ARCTIC regions of that country. These views, grounde
0368 er the present distribution of the Alpine and ARCTIC productions of Europe and America, that when
0368 stribution of the fossil Gnathodon), then the ARCTIC and temperate productions will at a very late
0368 warmer period, since the Glacial period. The ARCTIC forms, during their long southern migration a
0369 erent; for it is not likely that all the same ARCTIC species will have been left on mountain range
0369 eriod, I assumed that at its commencement the ARCTIC productions were as uniform round the polar r
0369 ks on distribution apply not only to strictly ARCTIC forms, but also to many sub arctic and to som
0369 o strictly arctic forms, but also to many sub ARCTIC and to some few northern temperate forms, for
0369 the necessary degree of uniformity of the sub ARCTIC and northern temperate forms round the world,
0369 e glacial period. At the present day, the sub ARCTIC and northern temperate productions of the Old
0370 66 degrees 67 degrees; and that the strictly ARCTIC productions then lived on the broken land sti
0370 the necessary amount of uniformity in the sub ARCTIC and northern temperate productions of the Old
0371 od, on the several mountain ranges and on the ARCTIC lands of the two Worlds. Hence it has come, t
0376 ranges of the intertropical regions, are not ARCTIC, but belong to the northern temperate zones.
0376 r mountain floras really become less and less ARCTIC. Many of the forms living on the mountians of
0377 y the temperate productions, and these by the ARCTIC; but with the latter we are not now concerned
0378 e period when the cold was most intense, when ARCTIC forms had migrated some twenty five degrees o
0379 om geological evidence that the whole body of ARCTIC shells underwent scarcely any modification du
0382 ain summits, in a line gently rising from the ARCTIC lowlands to a great height under the equator.
0394 ed foxes, as so frequently now happens in the ARCTIC regions. Yet it cannot be said that small isl
0069 e, they will increase in numbers, and, as each AREA is already fully stocked with inhabitants, the
0080 d to the diverstiy of inhabitants of any small AREA, and to naturalisation. Action of Natural Sele
0081 nts were in some manner modified: for, had the AREA been open to immigration, these same places wo
0102 us it will be in nature: for within a confined AREA, with some place in its polity not so perfectl

Page ***************************************(Key Word)***

Page ***(Key Word)***

0102 er to fill up the unoccupied place. But if the AREA be large, its several districts will almost ce
0103 to the conditions of each; for in a continuous AREA, the conditions will generally graduate away i
0103 talogue of facts, showing that within the same AREA, varieties of the same animal can long remain
0104 f natural selection. In a confined or isolated AREA, if not very large, the organic and inorganic
0104 ndividuals of a varying species throughout the AREA in the same manner in relation to the same con
0105 ction of new species. If, however, an isolated AREA be very small, either from being surrounded by
0105 these remarks, and look at any small isolated AREA, such as an oceanic island, although the total
0105 ves, for to ascertain whether a small isolated AREA, or a large open area like a continent, has be
0105 whether a small isolated area, or a large open AREA like a continent, has been most favourable for
0105 ole I am inclined to believe that largeness of AREA is of more importance, more especially in the
0105 spreading widely. Throughout a great and open AREA, not only will there be a better chance of fav
0106 be able to spread over the open and continuous AREA, and will thus come into competition with many
0106 evere, on a large than on a small and isolated AREA. Moreover, great areas, though now continuous,
0106 g, before those of the larger Europaeo Asiatic AREA. Thus, also, it is that continental production
0107 esh water basins, taken together, make a small AREA compared with that of the sea or of the land;
0107 present day, from having inhabited a confined AREA, and from having thus been exposed to less sev
0107 or terrestrial productions a large continental AREA, which will probably undergo many oscillations
0107 y to endure long and to spread widely. For the AREA will first have existed as a continent, and th
0108 lands shall be re converted into a continental AREA, there will again be severe competition: the m
0109 powers of increase of all organic beings, each AREA is already fully stocked with inhabitants, it
0114 y natural circumstances. In an extremely small AREA, especially if freely open to immigration, and
0128 ore living beings can be supported on the same AREA the more they diverge in structure, habits, an
0174 be extremely cautious in inferring, because an AREA is now continous, that it has been continuous
0174 pecies as they are now distributed over a wide AREA, we generally find them tolerably numerous ove
0175 entative species, when inhabiting a continuous AREA, are generally so distributed that each has a
0176 nation adapt a varying species to a very large AREA, we shall have to adapt two varieties to two l
0176 er numbers from inhabiting a narrow and lesser AREA; and practically, as far as I can make out, th
0178 in different portions of a strictly continuous AREA, intermediate varieties will, it is probable,
0203 ely allied species, now living on a continuous AREA, must often have been formed when the area was
0203 ous area, must often have been formed when the AREA was not continuous, and when the conditions of
0203 es are formed in two districts of a continuous AREA, an intermediate variety will often be formed,
0279 presence, namely on an extensive and continous AREA with graduated physical conditions. I endeavou
0284 he sea, the process of accumulation in any one AREA must be extremely slow. But the amount of denu
0287 ring oscillations of level, which we know this AREA has undergone, the surface may have existed fo
0291 all geological facts tell us plainly that each AREA has undergone numerous slow oscillations of le
0292 assing notice. During periods of elevation the AREA of the land and of the adjoining shoal parts o
0292 e other hand, during subsidence, the inhabited AREA and number of inhabitants will decrease (excep
0293 became wholly extinct. We forget how small the AREA of Europe is compared with the rest of the wor
0294 s that it only then first immigrated into that AREA. It is well known, for instance, that several
0294 ins, have gone on accumulating within the same AREA during the whole of this period. It is not, fo
0295 modification will have had to live on the same AREA throughout this whole time. But we have seen t
0295 ment of subsidence will often tend to sink the AREA whence the sediment is derived, and thus dimin
0300 other by enormous intervals, during which the AREA would be either stationary or rising; whilst r
0302 rget how large the world is, compared with the AREA over which our geological formations have been
0322 of a new group have taken possession of a new AREA, they will have exterminated in a correspondin
0342 n the first appearance of new forms in any one AREA and formation; that widely ranging species are
0352 s universally admitted, that in most cases the AREA inhabited by a species is continuous; and when
0353 other, are generally local, or confined to one AREA. What a strange anomaly it would be, if, when
0353 ew of each species having been produced in one AREA alone, and having subsequently migrated from t
0353 ne, and having subsequently migrated from that AREA as far as its powers of migration and subsiste
0354 that each species has been produced within one AREA, and has migrated thence as far as it could. I
0354 during some part of their migration) from the AREA inhabited by their progenitor. If it can be sh
0389 ts, only 750 in number, with those on an equal AREA at the Cape of Good Hope or in Australia, we m
0390 r found on any continent, and then compare the AREA of the islands with that of the continent, we
0407 ecies may have ranged continuously over a wide AREA, and then have become extinct in the intermedi
0410 ace, it certainly is the general rule that the AREA inhabited by a single species, or by a group o
0410 r to a certain period of time, or to a certain AREA, are often characterised by trifling character
0412 rsity of the forms of life which, in any small AREA, come into the closest competition; and by loo
0463 me extinct and supplanted form. Even on a wide AREA, which has during a long period remained conti
0478 f high generality that the inhabitants of each AREA are related to the inhabitants of the nearest
0048 the same country, but are common in separated AREAS. How many of those birds and insects in North
0106 on a small and isolated area. Moreover, great AREAS, though now continuous, owing to oscillations
0106 ally, I conclude that, although small isolated AREAS probably have been in some respects highly fa
0106 n will generally have been more rapid on large AREAS; and what is more important, that the new for
0106 mportant, that the new forms produced on large AREAS, which already have been victorious over many
0174 in the form of the land and of climate, marine AREAS now continuous must often have existed within
0174 pecies have been formed on strictly continuous AREAS; though I do not doubt that the formerly brok
0174 ot doubt that the formerly broken condition of AREAS now continous has played an important part in
0176 shall have to adapt two varieties to two large AREAS, and a third variety to a narrow intermediate
0176 exist in larger numbers from inhabiting larger AREAS, will have a great advantage over the interme
0178 anent; and this assuredly we do see. Secondly, AREAS now continuous must often have existed within
0309 onclude that the great oceans are still mainly AREAS of subsidence, the great archipelagoes still
0309 s of subsidence, the great archipelagoes still AREAS of oscillations of level, and the continents
0309 s of oscillations of level, and the continents AREAS of elevation. But have we any right to assume
0309 el, of the force of elevation; but may not the AREAS of preponderant movement have changed in the
0309 ys remained nearer to the surface. The immense AREAS in some parts of the world, for instance in S
0310 may perhaps believe that we see in these large AREAS, the many formations long anterior to the sil
0312 e succession of the same types within the same AREAS. Summary of preceding and present chapters. L
0315 at masses of sediment having been deposited on AREAS whilst subsiding, our formations have been al
0328 world. As we have reason to believe that large AREAS are affected by the same movement, it is prob
0328 s has invariably been the case, and that large AREAS have invariably been affected by the same mov
0328 ount for, considering the proximity of the two AREAS, unless, indeed, it be assumed that an isthmu
0336 ons, by the physical conditions of the ancient AREAS having remained nearly the same. Let it be re
0339 e succession of the same types within the same AREAS, during the later tertiary periods. Mr. Clift
0339 e succession of the same types within the same AREAS mean? He would be a bold man, who after compa
0340 , succession of the same types within the same AREAS, is at once explained; for the inhabitants of

Page ***(Key Word)***

areas

```
0345 of the same types of structure within the same  AREAS during the later geological periods ceases to
0347 r in only a slight degree; for instance, small    AREAS in the Old World could be pointed out hotter
0350 iling throughout space and time, over the same    AREAS of land and water, and independent of their p
0351 ra, and even families are confined to the same    AREAS, as is so commonly and notoriously the case.
0353 but had been produced in two or more distinct     AREAS! Hence it seems to me, as it has to many othe
0371 emarkable, considering the distance of the two    AREAS, and their separation by the Atlantic Ocean.
0372 n, for many closely allied forms now living in    AREAS completely sundered. Thus, I think, we can un
0372 in the Mediterranean and in the seas of Japan,    AREAS now separated by a continent and by nearly a
0372 the nearly similar physical conditions of the     AREAS: for if we compare, for instance, certain par
0380 e more dominant forms, generated in the larger    AREAS and more efficient workshops of the north. In
0380 nd yielded to those produced within the larger    AREAS of the north, just in the same way as the pro
0388 water productions ever can range, over immense    AREAS, and having subsequently become extinct in in
0389 umber compared with those on equal continental   AREAS: Alph. de Candolle admits this for plants, an
0408 of the highest importance, we can see why in two AREAS having nearly the same physical conditions sh
0409 s were probably derived. We can see why in two    AREAS, however distant from each other, there shoul
0409 the present time the differences in different     AREAS. We see this in many facts. The endurance of
0477 by the whole intertropical ocean. Although two    AREAS may present the same physical conditions of l
0477 ost important of all relations, and as the two    AREAS will have received colonists from some third
0477 ortions, the course of modification in the two   AREAS will inevitably be different. On this view of
0478 ly allied or representative species in any two    AREAS, implies, on the theory of descent with modif
0478 that the same parents formerly inhabited both    AREAS; and we almost invariably find that wherever
0478 erever many closely allied species inhabit two   AREAS, some identical species common to both still
0362 nstance I removed twenty two grains of dry        ARGILLACEOUS earth from one foot of a partridge, and in
0046 s are very far from uniform. Authors sometimes    ARGUE in a circle when they state that important or
0193 y, we must own that we are far too ignorant to    ARGUE that no transition of any kind is possible. T
0268 ly ranks them as undoubted species. If we thus    ARGUE in a circle, the fertility of all varieties p
0277 prising, when we remember how liable we are to   ARGUE in a circle with respect to varieties in a st
0364 to maturity! But it would be a great error to    ARGUE that because a well stocked island, like Grea
0014 to their aboriginal stocks. Hence it has been     ARGUED that no deductions can be drawn from domestic
0015 , is not of great importance for our line of      ARGUMENT; for by the experiment itself the conditions
0017 e descended from one or several species. The      ARGUMENT mainly relied on by those who believe in the
0024 s to be quite prolific under confinement. An      ARGUMENT, as it seems to me, of great weight, and app
0049 sideration; for several interesting lines of      ARGUMENT, from geographical distribution, analogical
0141 severer test) under them, may be used as an       ARGUMENT that a large proportion of other animals, no
0141 es. We must not, however, push the foregoing      ARGUMENT too far, on account of the probable origin o
0267 fspring. It may be urged, as a most forcible      ARGUMENT, that there must be some essential distincti
0308 xplicable; and may be truly urged as a valid      ARGUMENT against the views here entertained. To show
0365 ed. But this, as it seems to me, is no valid      ARGUMENT against what would be effected by occasional
0400 inhabitants of the islands may be used as an      ARGUMENT against my views; for it may be asked, how h
0459 ng remarks. As this whole volume is one long      ARGUMENT, it may be convenient to the reader to have
0466 on. Now let us turn to the other side of the      ARGUMENT. Under domestication we see much variability
0002 by fully stating and balancing the facts and      ARGUMENTS on both sides of each question; and this ca
0029 ght differences, yet they ignore all general      ARGUMENTS, and refuse to sum up in their minds slight
0306 owest known fossiliferous rocks. Most of the      ARGUMENTS which have convinced me that all the existi
0310 o do not attach much weight to the facts and      ARGUMENTS of other kinds given in this volume, will u
0357 every island to some mainland. If indeed the     ARGUMENTS used by Forbes are to be trusted, it must b
0458 ven if it were unsupported by other facts or     ARGUMENTS. Chapter XIV. Recapitulation and Conclusion
0469 ns: now let us turn to the special facts and     ARGUMENTS in favour of the theory. On the view that s
0483 ms are which we may consider, by so much the     ARGUMENTS fall away in force. But some arguments of t
0483 h the arguments fall away in force. But some     ARGUMENTS of the greatest weight extend very far. All
0425 rse preposterous; and I might answer by the      ARGUMENTUM ad hominem, and ask what should be done if
0346 ersified conditions; the most humid districts,   ARID deserts, lofty mountains, grassy plains, fores
0039 y slight differences might, and indeed do now,   ARISE amongst pigeons, which are rejected as faults
0060 lps us but little in understanding how species   ARISE in nature. How have all those exquisite adapt
0061 er more than do the species of the same genus,   ARISE? All these results, as we shall more fully se
0082 cation, which in the course of ages chanced to   APISE, and which in any way favoured the individual
0099 dlings are mongrelized? I suspect that it must    ARISE from the pollen of a distinct variety having
0169 though new and important modifications may not   ARISE from reversion and analogous variation, such
0209 nated. As modifications of corporeal structure   ARISE from, and are increased by, use or habit, and
0030 , some variations useful to him have probably    ARISEN suddenly, or by one step; many botanists, for
0030 ; and this amount of change may have suddenly    ARISEN in a seedling. So it has probably been with t
0045 arents, or which may be presumed to have thus    ARISEN, being frequently observed in the individuals
0196 hen a modification of structure has primarily    ARISEN from the above or other unknown causes, it ma
0197 ot climbers, the hooks on the bamboo may have    ARISEN from unknown laws of growth, and have been su
0197 ken egg, we may infer that this structure has    ARISEN from the laws of growth, and has been taken a
0198  which nevertheless we generally admit to have   ARISEN through ordinary generation, we ought not to
0199 h formerly were useful, or which formerly had    ARISEN from correlation of growth, or from other unk
0204 ying to live wherever it can live, how it has    ARISEN that there are upland geese with webbed feet,
0415 world. Perhaps from this cause it has partly     ARISEN, that almost all naturalists lay the greatest
0428 ry, and ternary classifications have probably    ARISEN. As the modified descendants of dominant spec
0400 ong appeared to me a great difficulty: but it    ARISES in chief part from the deeply seated error of
0015 ill determine how far the new characters thus    ARISING shall be preserved. When we look to the numb
0090 nces to this agency: for we see peculiarities    ARISING and becoming attached to the male sex in our
0105 e be a better chance of favourable variations    ARISING from the large number of individuals of the
0168 variable in number and in structure, perhaps    ARISING from such parts not having been closely spec
0273 uld often continue and be super added to that    ARISING from the mere act of crossing. The slight de
0325 tion. New species are formed by new varieties   ARISING, which have some advantage over older forms;
0181 le that the membrane connected fingers and fore ARM of the Galeopithecus might be greatly lengthen
0200 s; we cannot believe that the same bones in the ARM of the monkey, in the fore leg of the horse, i
0434 . we never find, for instance, the bones of the ARM and forearm, or of the thigh and leg, transpos
0339 gigantic pieces of armour like those of the      ARMADILLO, found in several parts of La Plata; and Pr
0341 left behind them in South America the sloth,     ARMADILLO, and anteater, as their degenerate descenda
0144 vering, viz. Cetacea (whales) and Edentata (     ARMADILLOES, scaly anteaters, etc.), that these are li
0088 males of carnivorous animals are already well    ARMED; though to them and to others, special means
0486 e object in view. We possess no pedigrees or     ARMORIAL bearings; and we have to discover and trace
0339 an uneducated eye, in the gigantic pieces of     ARMOUR like those of the armadillo, found in several
```

0022 ree of divergence and relative size of the two ARMS of the furcula. The proportional width of the
0150 fessor Owen, with respect to the length of the ARMS of the ourang outang, that he has come to a ne
0397 hinks of drifted timber across moderately wide ARMS of the sea. And I found that several species d
0401 in sight of each other, are separated by deep ARMS of the sea, in most cases wider than the Briti
0169 rdinary amount of modification since the genus AROSE; and thus we can understand why it should oft
0227 fixed point. We must suppose the Melipona to ARRANGE her cells in level layers, as she already do
0413 judgment fully explained. Naturalists try to ARRANGE the species, genera, and families in each cl
0055 o test the truth of this anticipation I have ARRANGED the plants of twelve countries, and the cole
0226 er, and had made them of equal sizes and had ARRANGED them symmetrically in a double layer, the re
0329 veral formations in the two regions could be ARRANGED in the same order, in accordance with the ge
0334 for instance, mastodons and elephants, when ARRANGED by Dr. Falconer in two series, first accordi
0335 nd extinct races of the domestic pigeon were ARRANGED as well as they could be in serial affinity,
0436 w that they consist of metamorphosed leaves, ARRANGED in a spire. In monstrous plants, we often ge
0260 etween their parents, seems to be a strange ARRANGEMENT. The foregoing rules and facts, on the oth
0334 heir periods of existence, do not accord in ARRANGEMENT. The species extreme in character are not
0335 l as they could be in serial affinity, this ARRANGEMENT would not closely accord with the order in
0420 n my meaning more fully. I believe that the ARRANGEMENT of the groups within each class, in due su
0421 different. Nevertheless their genealogical ARRANGEMENT remains strictly true, not only at the pre
0422 rtain extent their characters. This natural ARRANGEMENT is shown, as far as is possible on paper,
0422 still less possible to have given a natural ARRANGEMENT; and it is notoriously not possible to rep
0422 , the natural system is genealogical in its ARRANGEMENT, like a pedigree; but the degrees of modif
0422 perfect pedigree of mankind, a genealogical ARRANGEMENT of the races of man would afford the best
0422 nging dialects, had to be included; such an ARRANGEMENT would, I think, be the only possible one.
0422 roups; but the proper or even only possible ARRANGEMENT would still be genealogical; and this woul
0432 tural classification, or at least a natural ARRANGEMENT, would be possible. We shall see this by t
0432 ent and unknown progenitor. Yet the natural ARRANGEMENT in the diagram would still hold good; and,
0433 it has been perfected, genealogical in its ARRANGEMENT, with the grades of difference between the
0449 ions were nearly perfect, the only possible ARRANGEMENT, would be genealogical. Descent being on m
0456 system: it is genealogical in its attempted ARRANGEMENT, with the grades of acquired difference ma
0457 cation is intelligible, on the view that an ARRANGEMENT is only so far natural as it is genealogic
0471 ontingency of much extinction, explains the ARRANGEMENT of all the forms of life, in groups subord
0479 scent. The natural system is a genealogical ARRANGEMENT, in which we have to discover the lines of
0252 tried than with plants. If our systematic ARRANGEMENTS can be trusted, that is if the genera of a
0275 ely agree with Dr. Prosper Lucas, who, after ARRANGING an enormous body of facts with respect to a
0413 me authors look at it merely as a scheme for ARRANGING together those living objects which are mos
0150 th of this proposition without giving the long ARRAY of facts which I have collected, and which ca
0362 the English coast destroyed so many on their ARRIVAL. Some hawks and owls bolt their prey whole,
0396 he same specific form or modified since their ARRIVAL, could have reached their present homes. But
0400 how they have come to be modified since their ARRIVAL), we find a considerable amount of differenc
0051 turalists come very near to, but do not quite ARRIVE at the rank of species; or, again, between su
0365 ng there, nearly every seed, which chanced to ARRIVE, would be sure to germinate and survive. Disp
0391 ari; and it is obvious that marine birds could ARRIVE at these islands more easily than land birds.
0482 ted. This seems to me a strange conclusion to ARRIVE at. They admit that a multituce of forms, whi
0002 natural history of the Malay archipelago, has ARRIVED at almost exactly the same general conclusio
0002 only the general conclusions at which I have ARRIVED, with a few facts in illustration; but which
0002 ns directly opposite to those at which I have ARRIVED. A fair result can be obtained only by fully
0044 ricted ranges. Before applying the principles ARRIVED at in the last chapter to organic beings in
0113 can be supported in any country has long ago ARRIVED at its full average. If its natural powers o
0120 only well marked varieties; or they may have ARRIVED at the doubtful category of sub species: but
0136 atural selection. For when a new insect first ARRIVED on the island, the tendency of natural selec
0171 eory of Natural Selection. Long before having ARRIVED at this part of my work, a crowd of difficul
0194 cture by what transitions an organ could have ARRIVED at its present state; yet, considering that
0224 an outline of the conclusions at which I have ARRIVED. He must be a dull man who can examine the e
0248 d, namely, Kolreuter and Gartner, should have ARRIVED at diametrically opposite conclusions in reg
0249 been avoided. Now let us turn to the results ARRIVED at by the third most experienced hybridiser,
0292 by Sir C. Lyell; and E. Forbes independently ARRIVED at a similar conclusion. One remark is here
0407 we are led to this conclusion, which has been ARRIVED at by many naturalists under the designation
0460 instinct, or any whole being, could not have ARRIVED at its present state by many graduated steps
0390 , as already explained, species occasionally ARRIVING after long intervals in a new and isolated d
0037 endid results from such poor materials; but the ART, I cannot doubt, has been simple, and, as far
0061 efforts, as the works of Nature are to those of ART. We will now discuss in a little more detail t
0085 other coloured flesh. If, with all the aids of ART, these slight differences make a great differe
0191 of the neck and the loop like course of the ARTERIES still marking in the embryo their former pos
0440 tebrata the peculiar loop like course of the ARTERIES near the branchial slits are related to simi
0479 ng mammal or bird having branchial slits and ARTERIES running in loops, like those in a fish which
0367 exterminated on all lesser heights) and in the ARTIC regions of both hemispheres. Thus we can unde
0142 ferences to habit. The case of the Jerusalem ARTICHOKE, which is never propagated by seed, and of
0187 by which the eye has been perfected. In the ARTICULATA we can commence a series with an optic nerv
0419 belonging to this, and to no other class of ARTICULATA. Geographical distribution has often been u
0437 ng certain processes and appendages; in the ARTICULATA, we see the body divided into a series of s
0437 ny vertebrae; the unknown progenitor of the ARTICULATA, many segments; and the unknown progenitor
0188 as is possessed by any member of the great ARTICULATE class. He who will go thus far, if he find
0236 it can be shown that some insects and other ARTICULATE animals in a state of nature occasionally b
0436 limbs in each member of the vertebrate and ARTICULATE classes are plainly homologous. We see the
0063 kingdoms; for in this case there can be no ARTIFICIAL increase of food, and no prudential restrai
0109 if feeble man can do much by his powers of ARTIFICIAL selection, I can see no limit to the amount
0183 ow feed on exotic plants, or exclusively on ARTIFICIAL substances. Of diversified habits innumerab
0198 d countries where there has been but little ARTIFICIAL selection. Careful observers are convinced
0249 increases, and goes on increasing. Now, in ARTIFICIAL fertilisation pollen is as often taken by c
0320 isappearance of old forms, both natural and ARTIFICIAL, are bound together. In certain flourishing
0361 the crop of a pigeon, which had floated on ARTIFICIAL salt water for 30 days, to my surprise near
0413 ating those which are most unlike; or as an ARTIFICIAL means for enunciating, as briefly as possib
0423 assing varieties on a natural instead of an ARTIFICIAL system; we are cautioned, for instance, not
0485 at genera, who admit that genera are merely ARTIFICIAL combinations made for convenience. This ma
0195 ion. The tail of the giraffe looks like an ARTIFICIALLY constructed fly flapper; and it seems at f
0237 fferent breeds of cattle in relation to an ARTIFICIALLY imperfect state of the male sex; for oxen

Page **(Key Word)**

0247 of cases of plants which he castrated, and ARTIFICIALLY fertilised with their own pollen, and (exc
0249 , that if even the less fertile hybrids be ARTIFICIALLY fertilised with hybrid pollen of the same
0249 fertility in the successive generations of ARTIFICIALLY fertilised hybrids may, I believe, be acco
0247 he common red and blue pimpernels (Anagallis ARVENSIS and coeruleo), which the best botanists rank
0100 ker tabulated the trees of New Zealand, and Dr. ASA Gray those of the United States, and the resul
0115 e a single instance: in the last edition of Dr. ASA Gray's Manual of the Flora of the Northern Uni
0176 ar from information given me by Mr. Watson, Dr. ASA Gray, and Mr. Wollaston, that generally when v
0365 rador, and nearly all the same, as we hear from ASA Gray, with those on the loftiest mountains of
0372 lds, we find very few identical species (though ASA Gray has lately shown that more plants are ide
0049 hey grow in somewhat different stations; they ASCEND mountains to different heights; they have dif
0349 nly display an American type of structure. We ASCEND the lofty peaks of the Cordillera and we find
0378 turned, these temperate forms would naturally ASCEND the higher mountains, being exterminated on t
0161 nder unknown favorable conditions, gains an ASCENDANCY. For instance, it is probable that in each
0221 tree, twenty five years distant, which they ASCENDED together probably in search of aphides or co
0069 northwards; hence in going northward, or in ASCENDING a mountain, we far oftener meet with stunte
0174 tory proper to each. We see the same fact in ASCENDING mountains, and sometimes it is quite remark
0367 ize on the cleared and thawed ground, always ASCENDING higher and higher, as the warmth increased,
0389 . we have evidence that the barren island of ASCENSION aboriginally possessed under half a dozen f
0105 e may thus greatly deceive ourselves, for to ASCERTAIN whether a small isolated area, or a large o
0222 ith surprising courage. Now I was curious to ASCERTAIN whether F. sanguinea could distinguish the
0251 untains of Chile. I have taken some pains to ASCERTAIN the degree of fertility of some of the comp
0255 sation of plants. I have taken much pains to ASCERTAIN how far the rules apply to animals, and con
0072 close together that all cannot live. When I ASCERTAINED that these young trees had not been sown o
0219 t can be more extraordinary than these well ASCERTAINED facts? If we had not known of any other sl
0254 domestication. Finally, looking to all the ASCERTAINED facts on the intercrossing of plants and a
0289 ns, we may infer that this could nowhere be ASCERTAINED. The frequent and great changes in the min
0294 it is an excellent lesson to reflect on the ASCERTAINED amount of migration of the inhabitants of
0368 grounded as they are on the perfectly well ASCERTAINED occurrence of a former Glacial period, see
0271 these facts; from the great difficulty of ASCERTAINING the infertility of varieties in a state of
0466 tions of life. There is much difficulty in ASCERTAINING how much modification our domestic product
0193 gland at the end, is the same in Orchis and ASCLEPIAS, genera almost as remote as possible amongs
0076 most different climates. In Russia the small ASIATIC cockroach has everywhere driven before it it
0106 yielding, before those of the larger Europaeo ASIATIC area. Thus, also, it is that continental pro
0275 rent varieties, or of distinct species. Laying ASIDE the question of fertility and sterility, in a
0029 ded from so many aboriginally distinct species. ASK, as I have asked, a celebrated raiser of Heref
0166 es on this hybric from the ass and hemionus, to ASK Colonel Poole whether such face stripes ever o
0319 species of all classes, in all countries. If we ASK ourselves why this or that species is rare, we
0342 will rightly reject my whole theory. For he may ASK in vain where are the numberless transitional
0343 en coming in of whole groups of species. He may ASK where are the remains of those infinitely nume
0425 might answer by the argumentum ad hominem, and ASK what should be done if a perfect kangaroo were
0029 aboriginally distinct species. Ask, as I have ASKED, a celebrated raiser of Hereford cattle, whet
0049 t species; but what distance, it has been well ASKED, will suffice? If that between America and Eu
0061 ry part of the organic world. Again, it may be ASKED, how is it that varieties, which I have calle
0096 nts are hermaphrodites. What reason, it may be ASKED, is there for supposing in these cases that t
0112 d from their common parent. But how, it may be ASKED, can any analogous principle apply in nature?
0179 ith peculiar habits and structure. It has been ASKED by the opponents of such views as I hold, how
0180 different case had been taken, and it had been ASKED how an insectivorous quadruped could possibly
0233 r its conditions of life, it may reasonably be ASKED, how a long and graduated succession of modif
0237 ure or instinct to its progeny. It may well be ASKED how is it possible to reconcile this case wit
0260 ween the same two species? Why, it may even be ASKED, has the production of hybrids been permitted
0318 ncreased in numbers at an unparalleled rate, I ASKED myself what could so recently have exterminat
0341 ws of past and present distribution. It may be ASKED in ridicule, whether I suppose that the megat
0369 h america and Europe; and it may be reasonably ASKED how I account for the necessary degree of uni
0394 ll possess their peculiar laws. Why, it may be ASKED, has the supposed creative force produced bat
0400 as an argument against my views; for it may be ASKED, how has it happened in the several islands s
0425 ly included as a single species. But it may be ASKED, what ought we to do, if it could be proved t
0462 ns as fine as our present varieties, it may be ASKED, why do we not see these linking forms all ar
0480 e slight favourable variations. Why, it may be ASKED, have all the most eminent living naturalists
0483 what they consider reverent silence. It may be ASKED how far I extend the doctrine of the modifica
0298 umerous, fine, intermediate, fossil links, by ASKING ourselves whether, for instance, geologists a
0359 afterwards when planted they germinated; an ASPARAGUS plant with ripe berries floated for 23 days
9343 e world may have presented a wholly different ASPECT; and that the older continents, formed of for
0417 d thus laugh at our classification. But when ASPICARPA produced in France, during several years, o
0017 her climates? Has the little variability of the ASS or guinea fowl, or the small power of enduranc
0163 ture. It is a case apparently of reversion. The ASS not rarely has very distinct transverse bars o
0163 ly very variable in length and outline. A white ASS, but not an albino, has been described without
0165 . rollin asserts, that the common mule from the ASS and horse is particularly apt to have bars on
0165 ings, which I have seen, of hybrids between the ASS and zebra, the legs were much more plainly bar
0165 rms me that he knows of a second case) from the ASS and the hemionus; and this hybrid, though the
0165 s and the hemionus: and this hybrid, though the ASS seldom has stripes on its legs and the hemionu
0165 nce of the face stripes on this hybrid from the ASS and hemionus, to ask Colonel Poole whether suc
0166 ke a zebra, or striped on the shoulders like an ASS. In the horse we see this tendency strong when
0167 descended from one or more wild stocks, or the ASS, the hemionus, quagga, and zebra. He who belie
0258 tallion horse being first crossed with a female ASS, and then a male ass with a mare: these two sp
0258 irst crossed with a female ass, and then a male ASS with a mare: these two species may then be sai
0274 those authors are right, who maintain that the ASS has a prepotent power over the horse, so that
0274 t both the mule and the hinny more resemble the ASS than the horse; but that the prepotency runs m
0274 t the prepotency runs more strongly in the male ASS than in the female, so that the mule, which is
0275 at the mule, which is the offspring of the male ASS and mare, is more like an ass, than is the hin
0275 pring of the male ass and mare, is more like an ASS, than is the hinny, which is the offspring of
0275 the hinny, which is the offspring of the female ASS and stallion. Much stress has been laid by som
0015 shadow of evidence in favour of this view: to ASSERT that we could not breed our cart and race hor
0065 o not annually pair. Hence we may confidently ASSERT, that all plants and animals are tending to i
0167 s was independently created, will, I presume, ASSERT that each species has been created with a ten
0205 orant, in almost every case, to be enabled to ASSERT that any part or organ is so unimportant for
0211 ent; but want of space prevents me. I can only ASSERT, that instincts certainly do vary, for instan

Page **(Key Word)**

0240 size; and I fully believe, though I dare not ASSERT so positively, that the workers of intermedia
0012 coarse haired animals are apt to have, as is ASSERTED, long or many horns; pigeons with feathered
0076 ain with the varieties of sheep: it has been ASSERTED that certain mountain varieties will starve
0087 irds, used for breaking the egg. It has been ASSERTED, that of the best short beaked tumbler pigeo
0163 those on the legs of the zebra: it has been ASSERTED that these are plainest in the foal, and fro
0163 i believe this to be true. It has also been ASSERTED that the stripe on each shoulder is sometime
0217 y hatched, all at the same time. It has been ASSERTED that the American cuckoo occasionally lays h
0345 s i believe it to be, and it may at least be ASSERTED that the record cannot be proved to be much
0362 red a large supply of food, it is positively ASSERTED that all the grains do not pass into the giz
0468 on had not come into play. It has often been ASSERTED, but the assertion is quite incapable of pro
0481 w of the mutability of species? It cannot be ASSERTED that organic beings in a state of nature are
0393 e studded. I have taken pains to verify this ASSERTION, and I have found it strictly true. I have,
0468 to play. It has often been asserted, but the ASSERTION is quite incapable of proof, that the amoun
0165 he several species of the horse genus. Rollin ASSERTS, that the common mule from the ass and horse
0193 ctric apparatus, and yet do not, as Matteuchi ASSERTS, discharge any electricity, we must own that
0248 , and in one case for ten generations, yet he ASSERTS positively that their fertility never increa
0251 se hybrids seed freely. For instance, Herbert ASSERTS that a hybrid from Calceolaria integrifolia
0270 ch like the maize has separated sexes, and he ASSERTS that their mutual fertilisation is by so muc
0270 from their own coloured flowers. Moreover, he ASSERTS that when yellow and white varieties of one
0330 e, and a higher authority could not be named, ASSERTS that he is every day taught that palaeozoic
0163 cure, or actually quite lost, in dark coloured ASSES. The koulan of Pallas is said to have been se
0144 ers rarely coexist, without our being able to ASSIGN any reason. What can be more singular than th
0167 n one case out of a hundred can we pretend to ASSIGN any reason why this or that part differs, mor
0261 sap, etc., but in a multitude of cases we can ASSIGN no reason whatever. Great diversity in the si
0293 then lived: but I can by no means pretend to ASSIGN due proportional weight to the following cons
0469 how naturalists differ in the rank which they ASSIGN to the many representative forms in Europe an
0178 mediate varieties will, from reasons already ASSIGNED (namely from what we know of the actual dist
0203 d for an intermediate zone: but from reasons ASSIGNED, the intermediate variety will usually exist
0279 ional links, is a very obvious difficulty. I ASSIGNED reasons why such links do not commonly occur
0297 e gradations. And this from the reasons just ASSIGNED we can seldom hope to effect in any one geol
0482 ey arbitrarily reject it in another, without ASSIGNING any distinction in the two cases. The day w
0087 n are able to get out of it: so that fanciers ASSIST in the act of hatching. Now, if nature had to
0002 satisfaction of acknowledging the generous ASSISTANCE which I have received from very many natura
0048 on, to whom I lie under deep obligation for ASSISTANCE of all kinds, has marked for me 182 British
0053 i am much indebted for valuable advice and ASSISTANCE on this subject, soon convinced me that the
0020 as being of considerable antiquity. I have ASSOCIATED with several eminent fanciers, and have bee
0034 roes of the interior of Africa who have not ASSOCIATED with Europeans. Some of these facts do not
0208 by the will or reason. Habits easily become ASSOCIATED with other habits, and with certain periods
0213 tastes, and likewise of the oddest tricks, ASSOCIATED with certain frames of mind or periods of t
0313 of a similar fact, in an existing crocodile ASSOCIATED with many strange and lost mammals and rept
0398 roportions in which the several classes are ASSOCIATED together, which resembles closely the condi
0390 ed district, and having to compete with new ASSOCIATES, will be eminently liable to modification,
0405 lands in the struggle for life with foreign ASSOCIATES. But on the view of all the species of a ge
0472 lation to the degree of perfection of their ASSOCIATES; so that we need feel no surprise at the in
0488 es and coming into competition with foreign ASSOCIATES, might become modified; so that we must not
0064 atural increase: it will be under the mark to ASSUME that it breeds when thirty years old, and goe
0116 to have been much amplified, we may, I think, ASSUME that the modified descendants of any one spec
0140 native climate, but in all ordinary cases we ASSUME such to be the case: nor do we know that they
0161 tendency in each generation in the plumage to ASSUME this colour. This view is hypothetical, but c
0188 ference be presumptuous? Have we any right to ASSUME that the Creator works by intellectual powers
0286 der to form some crude notion on the subject, ASSUME that the sea would eat into cliffs 500 feet i
0286 llowance; but it is the same as if we were to ASSUME a cliff one yard in height to be eaten back a
0309 areas of elevation. But have we any right to ASSUME that things have thus remained from eternity?
0331 st from existing forms. We must not, however, ASSUME that divergence of character is a necessary c
0360 ments. Therefore it would perhaps be safer to ASSUME that the seeds of about 10/100 plants of a fl
0361 ances across the ocean. We may I think safely ASSUME that under such circumstances their rate of f
0427 ecome adapted to similar conditions, and thus ASSUME a close external resemblance; but such resemb
0472 n varieties enter any zone, they occasionally ASSUME some of the characters of the species proper
0481 idea of the lapse of time, we are too apt to ASSUME, without proof, that the geological record is
0017 nce in favour of this view. It has often been ASSUMED that man has chosen for domestication animal
0024 he multiple origin of our pigeons, it must be ASSUMED that at least seven or eight species were so
0024 s like those of the fantail. Hence it must be ASSUMED not only that half civilized man succeeded i
0047 by intermediate links; nor will the commonly ASSUMED hybrid nature of the intermediate links alwa
0120 one species would vary. In the diagram I have ASSUMED that a second species (I) has produced, by a
0122 may continue to exist. If then our diagram be ASSUMED to represent a considerable amount of modifi
0162 created species, that the one in varying has ASSUMED some of the characters of the other, so as t
0236 ordinary state, I should have unhesitatingly ASSUMED that all its characters had been slowly acqu
0328 imity of the two areas, unless, indeed, it be ASSUMED that an isthmus separated two seas inhabited
0369 ually took place during the Glacial period, I ASSUMED that at its commencement the arctic producti
0443 of descent with modification. It is commonly ASSUMED, perhaps from monstrosities often affecting
0444 best by having a long beak, whether or not it ASSUMED a beak of this particular length, as long as
0009 s mean a single bud or offset, which suddenly ASSUMES a new and sometimes very different character
0159 ariations; and a variety of one species often ASSUMES some of the characters of an allied species,
0166 tain bars and other marks; and when any breed ASSUMES by simple variation a bluish tint, these bar
0426 hose having very different habits of life, it ASSUMES high value; for we can account for its prese
0162 s to find the varying offspring of a species ASSUMING characters (either from reversion or from an
0305 et imperfectly known, are really teleostean. ASSUMING, however, that the whole of them did appear,
0309 nts now stand. Nor should we be justified in ASSUMING that if, for instance, the bed of the Pacifi
0004 oduced perfect as we now see them: but this ASSUMPTION seems to me to be no explanation, for it le
0024 ith the rock pigeon seems to me a very rash ASSUMPTION. Moreover, the several above named domestic
0212 t on the reader's mind. I can only repeat my ASSURANCE, that I do not speak without good evidence.
0137 ranean in its habits than the mole; and I was ASSURED by a Spaniard, who had often caught them, th
0251 he complex crosses of Rhododendrons, and I am ASSURED that many of them are perfectly fertile. Mr.
0253 bred geese must be far more fertile; for I am ASSURED by two eminently capable judges, namely Mr.
0393 found it strictly true. I have, however, been ASSURED that a frog exists on the mountains of the g
0399 iod. The affinity, which, though feeble, I am ASSURED by Dr. Hooker is real, between the flora of

Page **(Key Word)**

```
0449 same or similar embryonic stages, we may feel ASSURED that they have both descended from the same
0486 totypes of each great class. When we can feel ASSURED that all the individuals of the same species
0032 if he wants any of these qualities, he will ASSUREDLY fail. Few would readily believe in the natu
0046 under any other point of view many instances ASSUREDLY can be given. There is one point connected
0073 ins uniform for long periods of time, though ASSUREDLY the merest trifle would often give the vict
0081 places in the economy of nature which would ASSUREDLY be better filled up, if some of the origina
0085 erence in cultivating the several varieties, ASSUREDLY, in a state of nature, where the trees woul
0127 ations useful to any organic being do occur, ASSUREDLY individuals thus characterised will have th
0178 structure in some degree permanent; and this ASSUREDLY we do see. Secondly, areas now continuous m
0179 the species of the same group together, must ASSUREDLY have existed; for the very process of natur
0195 lated with constitutional differences, might ASSUREDLY be acted on by natural selection. The tail
0199 ritance; and consequently, though each being ASSUREDLY is well fitted for its place in nature, man
0213 d driving them to a distant point, we should ASSUREDLY call these actions instinctive. Domestic in
0253 least tendency to sterility, the breed would ASSUREDLY be lost in a very few generations. Although
0268 of all varieties produced under nature will ASSUREDLY have to be granted. If we turn to varieties
0280 tum full of such intermediate links? Geology ASSUREDLY does not reveal any such finely graduated o
0282 ies, must have been inconceivably great. But ASSUREDLY, if this theory be true, such have lived up
0301 fine transitional forms, which on my theory ASSUREDLY have connected all the past and present spe
0319 lowly, becoming less and less favourable, we ASSUREDLY should not have perceived the fact, yet the
0336 uthors, representative species; and these we ASSUREDLY do find. We find, in short, such evidence o
0425 ogy, it would be ranked with bears; but then ASSUREDLY all the other species of the kangaroo famil
0441 ales: and in the latter, the development has ASSUREDLY been retrograde; for the male is a mere sac
0456 rom presenting a strange difficulty, as they ASSUREDLY do on the ordinary doctrine of creation, mi
0127 will give its aid to ordinary selection, by ASSURING to the most vigorous and best adapted males
0194 inct and unknown is very small, I have been ASTONISHED how rarely an organ can be named, towards w
0373 their present level. In central Chile I was ASTONISHED at the structure of a vast mound of detritu
0021 s. the diversity of the breeds is something ASTONISHING. Compare the English carrier and the short
0032 common gooseberry may be quoted. We see an ASTONISHING improvement in many florists' flowers, whe
0180 allows them to glide through the air to an ASTONISHING distance from tree to tree. We cannot doub
0185 uria berardi, in its general habits, in its ASTONISHING power of diving, its manner of swimming, a
0270 uite incredible! but it is the result of an ASTONISHING number of experiments made during many yea
0276 secondary laws, this similarity would be an ASTONISHING fact. But it harmonises perfectly with the
0402 ised through man's agency, have spread with ASTONISHING rapidity over new countries, we are apt to
0472 laughtered by their sterile sisters; at the ASTONISHING waste of pollen by our fir trees; at the i
0028 never practised before, has improved them ASTONISHINGLY. About this same period the Dutch were as
0064 namely, the numerous recorded cases of the ASTONISHINGLY rapid increase of various animals in a st
0021 h it glories in inflating, may well excite ASTONISHMENT and even laughter. The turbit has a very s
0318 late geological period, I was filled with ASTONISHMENT; for seeing that the horse, since its intr
0319 ourable. But how utterly groundless was my ASTONISHMENT! Professor Owen soon perceived that the to
0453 ing on them, every one must be struck with ASTONISHMENT: for the same reasoning power which tells
0378 d and hottest districts will have afforded an ASYLUM to the tropical natives. The mountain ranges
0135 sent, but in a rudimentary condition. In the ATEUCHUS or sacred beetle of the Egyptians, they are
0135 the entire absence of the anterior tarsi in ATEUCHUS, and their rudimentary condition in some oth
0357 forbes insisted that all the islands in the ATLANTIC must recently have been connected with Europ
0359 sical Atlas, the average rate of the several ATLANTIC currents is 33 miles per diem (some currents
0364 or two land birds are blown across the whole ATLANTIC Ocean, from North America to the western sho
0369 worlds are separated from each other by the ATLANTIC Ocean and by the extreme northern part of th
0371 f the two areas, and their separation by the ATLANTIC Ocean. We can further understand the singula
0394 have been seen wandering by day far over the ATLANTIC Ocean; and two North American species either
0359 r power of germination. In Johnston's Physical ATLAS, the average rate of the several Atlantic cur
0132 es are more brightly coloured under a clear ATMOSPHERE, than when living on islands or near the co
0357 ow marked, as I believe, by rings of coral or ATOLLS standing over them. Whenever it is fully admi
0283 d these remaining fixed, have to be worn away, ATOM by atom, until reduced in size they can be rol
0283 remaining fixed, have to be worn away, atom by ATOM, until reduced in size they can be rolled abou
0232 s a painter could have done with his brush, by ATOMS of the coloured wax having been taken from th
0483 riods in the earth's history certain elemental ATOMS have been commanded suddenly to flash into li
0147 g seeds. When the seeds in our fruits become ATROPHIED, the fruit itself gains largely in size and
0416 . again, no one will say that rudimentary or ATROPHIED organs are of high physiological or vital i
0450 of each great class of animals. Rudimentary, ATROPHIED, or aborted organs. Organs or parts in this
0453 th equal plainness that these rudimentary or ATROPHIED organs, are imperfect and useless. In works
0436 o much obscured as to be finally lost, by the ATROPHY and ultimately by the complete abortion of c
0310 record in any degree perfect, and who do not ATTACH much weight to the facts and arguments of oth
0426 xistence, are generally the most constant, we ATTACH especial value to them; but if these same org
0482 fact. Any one whose disposition leads him to ATTACH more weight to unexplained difficulties than
0087 stication in one sex and become hereditarily ATTACHED to that sex, the same fact probably occurs u
0090 or we see peculiarities arising and becoming ATTACHED to the male sex in our domestic animals (as
0148 f the head is reduced to the merest rudiment ATTACHED to the bases of the prehensile antennae. Now
0150 d by Hunter, applies to characters which are ATTACHED to one sex, but are not directly connected w
0153 roduced by man's selection, sometimes become ATTACHED, from causes quite unknown to us, more to on
0391 e can see that their eggs or larvae, perhaps ATTACHED to seaweed or floating timber, or to the fee
0441 swimming, a proper place on which to become ATTACHED and to undergo their final metamorphosis. Wh
0068 even the tiger in India most rarely dares to ATTACK a young elephant protected by its dam. Climat
0215 cat genus, when kept tame, are most eager to ATTACK poultry, sheep and pigs! and this tendency ha
0215 when quite young, require to be taught not to ATTACK poultry, sheep, and pigs! No doubt they occas
0215 d pigs! No doubt they occasionally do make an ATTACK, and are then beaten; and if not cured, they
0222 ry courageous, and I have seen it ferociously ATTACK other ants. In one instance I found to my sur
0222 ntally disturbed both nests, the little ants ATTACKED their big neighbours with surprising courage
0202 ee as perfect, which, when used against many ATTACKING animals, cannot be withdrawn, owing to the
0085 se than yellow plums; whereas another disease ATTACKS yellow fleshed peaches far more than those w
0091 lky, with shorter legs, which more frequently ATTACKS the shepherd's flocks. Let us now take a mor
0195 our of the flesh, which, from determining the ATTACKS of insects or from being correlated with con
0195 utely depends on their power of resisting the ATTACKS of insects: so that individuals which could
0198 , states that in cattle susceptibility to the ATTACKS of flies is correlated with colour, as is th
0401 an in another, and it would be exposed to the ATTACKS of somewhat different enemies. If then it va
0052 ll varieties or incipient species necessarily ATTAIN the rank of species. They may whilst in this
0201 we see that this is the degree of perfection ATTAINED under nature. The endemic productions of New
```

Page **(Key Word)**

Page ***(Key Word)***

```
0016  nces in our domesticated productions. When we  ATTEMPT  to estimate the amount of structural differe
0049  idism, etc., have been brought to bear on the  ATTEMPT  to determine their rank. I will here give on
0136  battling with the winds, or by giving up the   ATTEMPT  and rarely or never flying. As with mariners
0150  nearly similar conclusion. It is hopeless to   ATTEMPT  to convince any one of the truth of this pro
0179  re preserved, as we shall in a future chapter  ATTEMPT  to show, in an extremely imperfect and inter
0207  of animals within the same class. I will not   ATTEMPT  any definition of instinct. It would be easy
0335  of existing species over the globe, will not   ATTEMPT  to account for the close resemblance of the
0338  atural selection. In a future chapter I shall  ATTEMPT  to show that the adult differs from its embr
0339  south America under the same latitude, would   ATTEMPT  to account, on the one hand, by dissimilar p
0435  plants. Nothing can be more hopeless than to   ATTEMPT  to explain this similarity of pattern in mem
0435  rine of final causes. The hopelessness of the  ATTEMPT  has been expressly admitted by Owen in his m
0032  eye, differences which I for one have vainly   ATTEMPTED  to appreciate. Not one man in a thousand ha
0104  ich unite for each birth; but I have already   ATTEMPTED  to show that we have reason to believe that
0242  ons are inherited. Still more briefly I have   ATTEMPTED  to show that instincts vary slightly in a s
0263  anc dissimilarity between organic beings is    ATTEMPTED  to be expressed. The facts by no means seem
0266  itions, is rendered sterile. All that I have   ATTEMPTED  to show, is that in two cases, in some resp
0316  appear to have come in abruptly; and I have    ATTEMPTED  to give an explanation of this fact, which
0336  ations, we ought not to expect to find, as I   ATTEMPTED  to show in the last chapter, in any one or
0341  f the preceding and present Chapters. I have   ATTEMPTED  to show that the geological record is extre
0407  of climatal changes on distribution, as I have ATTEMPTED  to show how important has been the influenc
0410  ceptions, which are not rare, may, as I have   ATTEMPTED  to show, be accounted for by migration at s
0411  n variation and on Natural Selection, I have   ATTEMPTED  to show that it is the widely ranging, the
0412  n increasing indefinitely in size. I further   ATTEMPTED  to show that from the varying descendants o
0412  ooking to certain facts in naturalisation. I   ATTEMPTED  also to show that there is a constant tende
0423  ld be universally preferred; and it has been   ATTEMPTED  by some authors. For we might feel sure, wh
0429  e. in the chapter on geological succession I   ATTEMPTED  to show, on the principle of each group hav
0456  inheritance. Summary. In this chapter I have   ATTEMPTED  to show, that the subordination of group to
0456  he natural system: it is genealogical in its   ATTEMPTED  arrangement, with the grades of acquired di
0460  es in the same community of ants: but I have   ATTEMPTED  to show how this difficulty can be mastered
0462  facilitated migration. As an example, I have   ATTEMPTED  to show how potent has been the influence o
0474  e class with their several instincts. I have   ATTEMPTED  to show how much light the principle of gra
0487  eding organic forms. We must be cautious in    ATTEMPTING  to correlate as strictly contemporaneous tw
0277  ected to experiment: for systematic affinity   ATTEMPTS  to express all kinds of resemblance between
0282  rate formations, and to mark how each author   ATTEMPTS  to give an inadequate idea of the duration o
0041  ry began to vary just when gardeners began to  ATTEND  closely to this plant. No doubt the strawberr
0211  aphides on a dockplant, and prevented their    ATTENDANCE  during several hours. After this interval,
0028  ed that the several breeds to which each has   ATTENDED, are descended from so many aboriginally dis
0033  ers of a century: it has certainly been more   ATTENDED  to of late years, and many treatises have be
0034  of domestic animals was at that early period   ATTENDED  to. Savages now sometimes cross their dogs w
0034  e breeding of domestic animals was carefully   ATTENDED  to in ancient times, and is now attended to
0034  lly attended to in ancient times, and is now   ATTENDED  to by the lowest savages. It would, indeed,
0074  s and nests: and Mr. H. Newman, who has long   ATTENDED  to the habits of humble bees, believes that
0088  more peaceful character. All those who have    ATTENDED  to the subject, believe that there is the se
0089  t attractive partner. Those who have closely   ATTENDED  to birds in confinement well know that they
0151  des: and I may here add, that I particularly   ATTENDED  to Mr. Waterhouse's remark, whilst investiga
0151  remarkably small degree, I have particularly   ATTENDED  to them, and the rule seems to me certainly
0152  ils, etc., these being the points now mainly   ATTENDED  to by English fanciers. Even in the sub bree
0219  n parts of England, and its habits have been   ATTENDED  to by Mr. F. Smith, of the British Museum, t
0264  latter alternative has not been sufficiently   ATTENDED  to: but I believe, from observations communi
0034  would, indeed, have been a strange fact, had   ATTENTION  not been paid to breeding, for the inherita
0039  ed, the more likely it would be to catch his   ATTENTION. But to use such an expression as trying to
0041  , or so much valued by him, that the closest   ATTENTION  should be paid to even the slightest deviat
0041  or structure of each individual. Unless such   ATTENTION  be paid nothing can be effected. I have see
0042  a few being kept by poor people, and little   ATTENTION  paid to their breeding: in peacocks, from n
0050  o man, or from any cause closely attract his   ATTENTION, varieties of it will almost universally be
0050  ere is some variation. But if he confine his   ATTENTION  to one class within one country, he will so
0100  remarks on the sexes of trees simply to call   ATTENTION  to the subject. Turning for a very brief sp
0221  , their slaves in their jaws. Another day my   ATTENTION  was struck by about a score of the slave ma
0273  generations, is a curious fact and deserves   ATTENTION. For it bears on and corroborates the view
0289  he world. The most skilful geologist, if his   ATTENTION  had been exclusively confined to these larg
0292  s extremely imperfect: but if we confine our   ATTENTION  to any one formation, it becomes more diffi
0325  struck by this strange sequence, we turn our   ATTENTION  to North America, and there discover a seri
0329  s, cannot be disputed. For if we confine our   ATTENTION  either to the living or to the extinct alon
0366  ief, had not Agassiz and others called vivid   ATTENTION  to the Glacial Period, which, as we shall i
0427  adaptive resemblances. Lamarck first called   ATTENTION  to this distinction, and he has been ably f
0034  e the breed, and they formerly did so, as is   ATTESTED  by passages in Pliny. The savages in South A
0050  ohly useful to man, or from any cause closely  ATTRACT  his attention, varieties of it will almost u
0089  rivalry between the males of many species to   ATTRACT  by singing the females. The rock thrush of G
0145  sprengel's idea that the ray florets serve to  ATTRACT  insects, whose agency is highly advantageous
0089  g by as spectators, at last choose the most    ATTRACTIVE  partner. Those who have closely attended to
0089  escribed how one pied peacock was eminently    ATTRACTIVE  to all his hen birds. It may appear childis
0090  be either useful to the males in battle, or    ATTRACTIVE  to the females. We see analogous cases unde
0093  ation or natural selection of more and more    ATTRACTIVE  flowers, had been rendered highly attractiv
0093  ttractive flowers, had been rendered highly    ATTRACTIVE  to insects, they would, unintentionally on
0093  on as the plant had been rendered so highly    ATTRACTIVE  to insects that pollen was regularly carrie
0003  this may be true: but it is preposterous to    ATTRIBUTE  to mere external conditions, the structure,
0010  uch, in the case of any variation, we should   ATTRIBUTE  to the direct action of heat, moisture, lig
0013  ere doctrine of chances almost compels us to   ATTRIBUTE  its reappearance to inheritance. Every one
0052  ut I have not much faith in this view; and I   ATTRIBUTE  the passage of a variety, from a state in w
0069  climate being conspicuous, we are tempted to   ATTRIBUTE  the whole effect to its direct action. But
0074  e bees more numerous than elsewhere, which I   ATTRIBUTE  to the number of cats that destroy the mice
0074  lothing an entangled bank, we are tempted to   ATTRIBUTE  their proportional numbers and kinds to wha
0089  all his hen birds. It may appear childish to   ATTRIBUTE  any effect to such apparently weak means: I
0090  eir male offspring. Yet, I would not wish to   ATTRIBUTE  all such sexual differences to this agency:
0132  tionally disturbed in the parents, I chiefly   ATTRIBUTE  the varying or plastic condition of the off
0133  to a being, we cannot tell how much of it to   ATTRIBUTE  to the accumulative action of natural selec
```

Page ***(Key Word)***

attribute

Page **************************************(Key Word)**************************************

```
0137  y injurious to animals living in darkness, I  ATTRIBUTE  their loss wholly to disuse. In one of the
0146  service to the species. We may often falsely  ATTRIBUTE  to correlation of growth, structures which
0155  characters; and these characters in common I  ATTRIBUTE  to inheritance from a common progenitor, fo
0159  een independently created, we should have to  ATTRIBUTE  this similarity in the enlarged stems of th
0193  having very different habits of life, we may  ATTRIBUTE  its presence to inheritance from a common a
0196  nough. In the second place, we may sometimes  ATTRIBUTE  importance to characters which are really o
0200  special use to these animals. We may safely  ATTRIBUTE  these structures to inheritance. But to the
0212  e been most persecuted by man. We may safely  ATTRIBUTE  the greater wildness of our large birds to
0215  ted for tameness; and I presume that we must  ATTRIBUTE  the whole of the inherited change from extr
0319  st surprise at its rarity; for rarity is the  ATTRIBUTE  of a vast number of species of all classes,
0370  migration under a more favourable climate, I  ATTRIBUTE  the necessary amount of uniformity in the s
0384  removed from the water. But I am inclined to  ATTRIBUTE  the dispersal of fresh water fish mainly to
0008  e most frequent cause of variability may be   ATTRIBUTED  to the male and female reproductive element
0008  in their native country! This is generally   ATTRIBUTED  to vitiated instincts; but how many cultiva
0010  rt my view, that variability may be largely   ATTRIBUTED  to the ovules or pollen, or to both, having
0011  me slight amount of change may, I think, be   ATTRIBUTED  to the direct action of the conditions of l
0011  nd I presume that this change may be safely   ATTRIBUTED  to the domestic duck flying much less, and
0015  case, however, some effect would have to be   ATTRIBUTED  to the direct action of the poor soil), tha
0029  pecies. Some little effect may, perhaps, be   ATTRIBUTED  to the direct action of the external condit
0042  t, the donkey, peacock, goose, etc., may be   ATTRIBUTED  in main part to selection not having been b
0043  of correlation of growth. Something may be   ATTRIBUTED  to the direct action of the conditions of l
0043  f the conditions of life. Something must be   ATTRIBUTED  to use and disuse. The final result is thus
0132  ughout nature. Some little influence may be   ATTRIBUTED  to climate, food, etc.: thus, E. Forbes spe
0145  bed by Cassini. These differences have been   ATTRIBUTED  by some authors to pressure, and the shape
0350  the inhabitants of different regions may be   ATTRIBUTED  to modification through natural selection,
0410  he rule are so few, that they may fairly be   ATTRIBUTED  to our not having as yet discovered in an i
0355  om correspondence, that this coincidence he   ATTRIBUTES  to generation with modification. The previo
0286  layers or nodules, which from long resisting  ATTRITION  form a breakwater at the base. Hence, under
0190  also, been worked in as an accessory to the  AUDITORY  organs of certain fish; for I do not kno
0191  ch view is now generally held, a part of the  AUDITORY  apparatus has been worked in as a complement
0185  or never go near the water; and no one except  AUDUBON  has seen the frigate bird, which has all its
0212  , but often from causes wholly unknown to us:  AUDUBON  has given several remarkable cases of differ
0387  tribution must remain quite inexplicable; but  AUDUBON  states that he found the seeds of the great
0397  , but sometimes by species found elsewhere. Dr. AUG. A. Gould has given several interesting cases
0418  e more clearly than by that excellent botanist, AUG. St. Hilaire. If certain characters are always
0111  e lesser difference between varieties become  AUGMENTED  into the greater difference between species
0470  of varieties of the same species, tend to be  AUGMENTED  into the greater differences characteristic
0012  . hence, if man goes on selecting, and thus  AUGMENTING, any peculiarity, he will almost certainly
0220  e nests at various hours during May, June and  AUGUST, both in Surrey and Hampshire, and has never
0220  he slaves, though present in large numbers in  AUGUST, either leave or enter the nest. Hence he con
0185  kes flight, would be mistaken by any one for an  AUK  or grebe; nevertheless, it is essentially a pe
0471  its and structure fitting it for the life of an  AUK  or grebe! and so on in endless other cases. Bu
0186  iving thrushes, and petrels with the habits of  AUKS. Organs of extreme perfection and complication
0204  iving thrushes, and petrels with the habits of  AUKS. Although the belief that an organ so perfect
0250  rtilised! For instance, a bulb of Hippeastrum  AULICUM  produced four flowers; three were fertilised
0299  tes of America. I fully agree with Mr. Godwin  AUSTEN, that the present condition of the Malay Arch
0018  , savages, like those of Tierra del Fuego or  AUSTRALIA, who possess a semi domestic dog, may not h
0038  an, we can understand how it is that neither  AUSTRALIA, the Cape of Good Hope, nor any other regio
0064  and horses in South America, and latterly in  AUSTRALIA, had not been well authenticated, they woul
0100  that he finds that the rule does not hold in  AUSTRALIA; and I have made these few remarks on the s
0106  the productions of the smaller continent of  AUSTRALIA  have formerly yielded, and apparently are n
0215  from countries, such as Tierra del Fuego and  AUSTRALIA, where the savages do not keep these domest
0306  ed to double the southern capes of Africa or  AUSTRALIA, and thus reach other and distant seas. Fro
0306  for five minutes on some one barren point in  AUSTRALIA, and then to discuss the number and range o
0324  with those now living in South America or in  AUSTRALIA, the most skilful naturalist would hardly b
0324  eds, of Europe, North and South America, and  AUSTRALIA, from containing fossil remains in some deg
0339  , who after comparing the present climate of  AUSTRALIA  and of parts of South America under the sam
0340  ould have been chiefly or solely produced in  AUSTRALIA; or that Edentata and other American types
0347  phere, if we compare large tracts of land in  AUSTRALIA, South Africa, and western South America, b
0347  h other, than they are to the productions of  AUSTRALIA  or Africa under nearly the same climate. An
0347  great difference between the inhabitants of  AUSTRALIA, Africa, and South America under the same l
0349  rich or emeu, like those found in Africa and  AUSTRALIA  under the same latitude. On these same plai
0352  ot find a single mammal common to Europe and  AUSTRALIA  or South America? The conditions of life ar
0353  lants have become naturalised in America and  AUSTRALIA; and some of the aboriginal plants are iden
0373  lacial action in the south eastern corner of  AUSTRALIA. Looking to America; in the northern half,
0375  north as Japan. On the southern mountains of  AUSTRALIA, Dr. F. Muller has discovered several Europ
0375  by Dr. Hooker, of European genera, found in  AUSTRALIA, but not in the intermediate torrid regions
0377  cape of Good Hope, and in parts of temperate  AUSTRALIA. As we know that many tropical plants and a
0379  erica, and by Alph. de Candolle in regard to  AUSTRALIA, that many more identical plants and allied
0380  round in La Plata, and in a lesser degree in  AUSTRALIA, and have to a certain extent beaten the na
0380  d during the last thirty or forty years from  AUSTRALIA. Something of the same kind must have occur
0382  s I believe, the southern shores of America,  AUSTRALIA, New Zealand have become slightly tinted by
0389  an equal area at the Cape of Good Hope or in  AUSTRALIA, we must, I think, admit that something qui
0395  y islands separated by similar channels from  AUSTRALIA. The West Indian Islands stand on a deeply
0399  demic plants is much more closely related to  AUSTRALIA, the nearest mainland, than to any other re
0399  een the flora of the south western corner of  AUSTRALIA  and of the Cape of Good Hope, is a far more
0403  us, the south east and south west corners of  AUSTRALIA  have nearly the same physical conditions, a
0429  the fact is striking, that the discovery of  AUSTRALIA  has not added a single insect belonging to
0476  orms on the same continent, of marsupials in  AUSTRALIA, of endentata in America, and other such ca
0116  t may be doubted, for instance, whether the  AUSTRALIAN  marsupials, which are divided into groups d
0116  e with these well pronounced orders. In the  AUSTRALIAN  mammals, we see the process of diversificat
0321  n the secondary formations, survives in the  AUSTRALIAN  seas; and a few members of the great and al
0339  ago showed that the fossil mammals from the  AUSTRALIAN  caves were closely allied to the living mar
0375  ill more striking is the fact that southern  AUSTRALIAN  forms are clearly represented by plants gro
0375  s of the mountains of Borneo. Some of these  AUSTRALIAN  forms, as I hear from Dr. Hooker, extend al
0140  es to England. In regard to animals, several  AUTHENTIC  cases could be given of species within hist
```

Page **************************************(Key Word)**************************************

Page ***(Key Word)**

```
0213  ur comestic animals. A number of curious and AUTHENTIC instances could be given of the inheritance
0064  d latterly in Australia, had not been well AUTHENTICATED, they would have been quite incredible. S
0252  nimal can be considered as thoroughly well AUTHENTICATED. It should, however, be borne in mind tha
0253  hough I do not know of any thoroughly well AUTHENTICATED cases of perfectly fertile hybrid animals
0003  , or of the volition of the plant itself. The AUTHOR of the Vestiges of Creation would, I presume,
0019  e, and several even within Great Britain. One AUTHOR believes that there formerly existed in Great
0050  how closely it has been studied; yet a German AUTHOR makes more than a dozen species out of forms,
0149  curs in lesser numbers, is constant. The same AUTHOR and some botanists have further remarked that
0155  ced in works on natural history, that when an AUTHOR has remarked with surprise that some importan
0248  uced by different hybridisers, or by the same AUTHOR, from experiments made during different years
0282  on separate formations, and to mark how each AUTHOR attempts to give an inadequate idea of the du
0291  geology, and have been surprised to note how AUTHOR after author, in treating of this or that gre
0291  have been surprised to note how author after AUTHOR, in treating of this or that great formation,
0339  of the Cld World. We see the same law in this AUTHOR's restorations of the extinct and gigantic bi
0346  er physical conditions. Cf late, almost every AUTHOR who has studied the subject has come to this
0415  acknowledged in the writings of almost every AUTHOR. It will suffice to quote the highest authori
0002  mperfect. I cannot here give references and AUTHORITIES for my several statements; and I must trus
0002  ve always been cautious in trusting to good AUTHORITIES alone. I can here give only the general co
0031  ssages to this effect from highly competent AUTHORITIES. Youatt, who was probably better acquainte
0035  time of that monarch. Some highly competent AUTHORITIES are convinced that the setter is directly
0050  , and in this country the highest botanical AUTHORITIES and practical men can be quoted to show th
0101  after consultation with one of the highest AUTHORITIES, namely, Professor Huxley, to discover a s
0310  l how rash it is to differ from these great AUTHORITIES, to whom, with others, we owe all our know
0045  of structure, which he could collect on good AUTHORITY, as I have collected, during a course of ye
0202  for the aberration of light is said, on high AUTHORITY, not to be perfect even in that most perfec
0217  n other birds' nests; but I hear on the high AUTHORITY of Dr. Brewer, that this is a mistake. Neve
0310  but I have reason to believe that one great AUTHORITY, Sir Charles Lyell, from further reflexion
0330  to the Invertebrata, Barrande, and a higher AUTHORITY could not be named, asserts that he is ever
0376  example. I may quote a remark by the highest AUTHORITY, Prof. Dana, that it is certainly a wonderf
0415  author. It will suffice to quote the highest AUTHORITY, Robert Brown, who in speaking of certain o
0450  born calves. It has even been stated on good AUTHORITY that rudiments of teeth can be detected in
0357  : but to the best of my judgment we are not AUTHORIZED in admitting such enormous geographical cha
0010  riation is not necessarily connected, as some AUTHORS have supposed, with the act of generation. S
0011  drooping ears; and the view suggested by some AUTHORS, that the drooping is due to the disuse of t
0018  inclined to believe, in opposition to several AUTHORS, that all the races have descended from one
0019  has been carried to an absurd extreme by some AUTHORS. They believe that every race which breeds t
0026  inct being themselves perfectly fertile. Some AUTHORS believe that long continued domestication el
0038  artly explains what has been remarked by some AUTHORS, namely, that the varieties kept by savages
0043  rcumstances, with all organic beings, as some AUTHORS have thought. The effects of variability are
0044  e species, and not generally propagated. Some AUTHORS use the term variation in a technical sense,
0046  of certain insects are very far from uniform. AUTHORS sometimes argue in a circle when they state
0046  t important organs never vary; for these same AUTHORS practically rank that character as important
0050  eties, moreover, will be often ranked by some AUTHORS as species. Look at the common oak, how clos
0067  this subject has been ably treated by several AUTHORS, and I shall, in my future work, discuss som
0131  the cause of each particular variation. Some AUTHORS believe it to be as much the function of the
0143  omologous parts, as has been remarked by some AUTHORS, tend to cohere; this is often seen in monst
0144  adjoining soft parts; it is believed by some AUTHORS that the diversity in the shape of the pelvi
0145  hese differences have been attributed by some AUTHORS to pressure, and the shape of the seeds in t
0149  rudimentary parts, it has been stated by some AUTHORS, and I believe with truth, are apt to be hig
0201  and for the destruction of its prey; but some AUTHORS suppose that at the same time this snake is
0246  which he found two forms, considered by most AUTHORS as distinct species, quite fertile together,
0274  another variety. For instance, I think those AUTHORS are right, who maintain that the ass has a p
0275  d stallion. Much stress has been laid by some AUTHORS on the supposed fact, that mongrel animals a
0318  involved in the most gratuitous mystery. Some AUTHORS have even supposed that as the individual ha
0323  he forms of life has been observed by several AUTHORS: so it is, according to Lyell, with the seve
0336  d forms, or, as they have been called by some AUTHORS, representative species; and these we assure
0346  he circumpolar land is almost continuous, all AUTHORS agree that one of the most fundamental divis
0355  or single hermaphrodite, or whether, as some AUTHORS suppose, from many individuals simultaneousl
0356  l. means of Dispersal. Sir C. Lyell and other AUTHORS have ably treated this subject. I can give h
0357  rica, and Europe likewise with America. Other AUTHORS have thus hypothetically bridged over every
0361  ght would often be 35 miles an hour; and some AUTHORS have given a far higher estimate. I have nev
0413  ystem. But what is meant by this system? Some AUTHORS look at it merely as a scheme for arranging
0423  ed in classifying varieties, as with species. AUTHORS have insisted on the necessity of classing v
0423  preferred; and it has been attempted by some AUTHORS. For we might feel sure, whether there had b
0427  n the analogies which have been drawn by some AUTHORS between very distinct animals. On my view of
0444  w analogous cases in domestic varieties. Some AUTHORS who have written on Dogs, maintain that the
0483  the blindness of preconceived opinion. These AUTHORS seem no more startled at a miraculous act of
0488  thrown on the origin of man and his history. AUTHORS of the highest eminence seem to be fully sat
0240  of neuters in the same species, that I gladly AVAILED myself of Mr. F. Smith's offer of numerous s
0017  ns under domestication, they would vary on an AVERAGE as largely as the parent species of our exis
0055  nt any varieties, invariably present a larger AVERAGE number of varieties than do the species of t
0056  ies, that is of incipient species, beyond the AVERAGE. It is not that all large genera are now var
0056  ny species of a genus have been formed, on an AVERAGE many are still forming; and this holds good.
0057  rieties or incipient species greater than the AVERAGE are now manufacturing, many of the species a
0058  l value: these 63 reputed species range on an AVERAGE over 6.9 of the provinces into which Mr. Wat
0058  arieties have very nearly the same restricted AVERAGE range, as have those very closely allied for
0059  te indefinite. In genera having more than the AVERAGE number of species in any country, the specie
0059  he species of these genera have more than the AVERAGE number of varieties. In large genera the spe
0059  nant species of the larger genera which on an AVERAGE vary most; and varieties, as we shall hereaf
0063  lly produces a thousand seeds, of which on an AVERAGE only one comes to maturity, may be more trul
0066  , a small number may be produced, and yet the AVERAGE stock be fully kept up; but if many eggs or
0066  the full number of a tree, which lived on an AVERAGE for a thousand years, if a single seed were
0066  in a fitting place. So that in all cases, the AVERAGE number of any animal or plant depends only i
0068  s prey to other animals, which determines the AVERAGE numbers of a species. Thus, there seems to b
0068  te plays an important part in determining the AVERAGE numbers of a species, and periodical seasons
0074  potent, but all concurring in determining the AVERAGE number or even the existence of the species.
```

Page ***(Key Word)**

```
0078  ssential manner. If we wished to increase its AVERAGE numbers in its new home, we should have to m
0113  any country has long ago arrived at its full AVERAGE. If its natural powers of increase be allowe
0117  e have seen in the second chapter, that on an AVERAGE more of the species of large genera vary tha
0169  forms has been actively at work, there, on an AVERAGE, we now find most varieties or incipient spe
0177  shown in the second chapter, presenting on an AVERAGE a greater number of well marked varieties th
0247  ossed and by their hybrid offspring, with the AVERAGE number produced by both pure parent species
0286  thickness of the several formations is on an AVERAGE about 1100 feet, as I am informed by Prof. R
0293  y bronn and Woodward, have concluded that the AVERAGE duration of each formation is twice or thric
0293  h formation is twice or thrice as long as the AVERAGE duration of specific forms. But insuperable
0294  e beds, might be tempted to conclude that the AVERAGE duration of life of the embedded fossils had
0300  us accumulation of sediment, would exceed the AVERAGE duration of the same specific forms; and the
0315  ome extinct. In members of the same class the AVERAGE amount of change, during long and equal peri
0342  ormation is, perhaps, short compared with the AVERAGE duration of specific forms: that migration h
0359  ermination. In Johnston's Physical Atlas, the AVERAGE rate of the several Atlantic currents is 33
0360  ng at the rate of 60 miles per diem); on this AVERAGE, the seeds of 14/100 plants belonging to one
0360  ear the sea; and this would have favoured the AVERAGE length of their flotation and of their resis
0388  high; and this will give longer time than the AVERAGE for the migration of the same aquatic specie
0405  ve a wide range, or even that they have on an AVERAGE a wide range; but only that some of the spec
0405  ise to new forms will largely determine their AVERAGE range. For instance, two varieties of the sa
0465  nclined to believe, has been shorter than the AVERAGE duration of specific forms. Successive forma
0057  have endeavoured to test this numerically by AVERAGES, and, as far as my imperfect results go, the
0028  erent breeds can be kept together in the same AVIARY. I have discussed the probable origin of dome
0042  may be kept true, though mingled in the same AVIARY, and this circumstance must have largely favo
0188  startling lengths. It is scarcely possible to AVOID comparing the eye to a telescope. We know tha
0253  the same time from different parents, so as to AVOID the ill effects of close interbreeding. On th
0280  difficult, when looking at any two species, to AVOID picturing to myself, forms directly intermedi
0289  n we follow them in nature, it is difficult to AVOID believing that they are closely consecutive.
0374  inent. If this be admitted, it is difficult to AVOID believing that the temperature of the whole w
0438  e having occurred, that naturalists can hardly AVOID employing language having this plain signific
0053  is rather perplexing, and allusions cannot be AVOIDED to the struggle for existence, divergence of
0249  ounted for by close interbreeding having been AVOIDED. Now let us turn to the results arrived at b
0002  ope in a future work to do this. For I am well AWARE that scarcely a single point is discussed in
0060  s some individual variability; indeed I am not AWARE that this has ever been disputed. It is immat
0095  favourable deviations of structure. I am well AWARE that this doctrine of natural selection, exem
0101  tinct individual is a law of nature. I am well AWARE that there are, on this view, many cases of d
0150  ion that it is a rule of high generality. I am AWARE of several causes of error, but I hope that I
0164  lack to a close approach to cream colour. I am AWARE that Colonel Hamilton Smith, who has written
0211  by its eager way of running about, to be well AWARE what a rich flock it had discovered; it then
0212  lection, to quite new instincts. But I am well AWARE that these general statements, without facts
0233  and so make an isolated hexagon: but I am not AWARE that any such case has been observed; nor wou
0293  isite to change one species into another. I am AWARE that two palaeontologists, whose opinions are
0316  ence, as long as it lasts, is continuous. I am AWARE that there are some apparent exceptions to th
0386  er haunts. I do not believe that botanists are AWARE how charged the mud of ponds is with seeds: I
0441  ganisation: I use this expression, though I am AWARE that it is hardly possible to define clearly
0145  hat in irregular flowers, those nearest to the AXIS are oftenest subject to peloria, and become re
0134  in nearly the same condition as the domestic AYLESBURY duck. As the larger ground feeding birds se
0072  southward and northward in a feral state: and AZARA and Rengger have shown that this is caused by
0307  es and bituminous matter in some of the lowest AZOIC rocks, probably indicates the former existenc
0049  mple, will that between the Continent and the AZORES, or Madeira, or the Canaries, or Ireland, be
0140  n european species of plants brought from the AZORES to England. In regard to animals, several aut
0363  the now temperate regions to another. In the AZORES, from the large number of the species of plan
0393  frogs have been introduced into Madeira, the AZORES, and Mauritius, and have multiplied so as to
0048  a, including the most polymorphic forms, Mr. BABINGTON gives 251 species, whereas Mr. Bentham give
0083  on the same food: he does not exercise a long BACKED or long legged quadruped in any peculiar mann
0202  g animals, cannot be withdrawn, owing to the BACKWARD serratures, and so inevitably causes the dea
0282  nnected with more ancient species: and so on BACKWARDS, always converging to the common ancestor o
0034  id to breeding, for the inheritance of good and BAD qualities is so obvious. At the present time,
0084  on, even the slightest: rejecting that which is BAD, preserving and adding up all that is good: si
0136  ther, whereas it would have been better for the BAD swimmers if they had not been able to swim at
0469  creature, favouring the good and rejecting the BAD? I can see no limit to this power, in slowly
0159  ommonly called, of the Swedish turnip and Ruta BAGA, plants which several botanists rank as variet
0035  ove and modify any breed, in the same way as BAKEWELL, Collins, etc., by this very same process, o
0036  n purely bred from the original stock of Mr. BAKEVELL for upwards of fifty years. There is not a s
0036  any one instance from the pure blood of Mr. BAKEWELL's flock, and yet the difference between the
0127  , must be judged of by the general tenour and BALANCE of evidence given in the following chapters.
0201  r doing an injury to its possessor. If a fair BALANCE be struck between the good and evil caused b
0232  he work of construction seems to be a sort of BALANCE struck between many bees, all instinctively
0291  e of subsidence and supply of sediment nearly BALANCE each other, the sea will remain shallow and
0467  orn than can possibly survive. A grain in the BALANCE will determine which individual shall live a
0468  urrounding physical conditions, will turn the BALANCE. With animals having separated sexes there w
0073  yet in the long run the forces are so nicely BALANCED, that the face of nature remains uniform for
0082  country are struggling together with nicely BALANCED forces, extremely slight modifications in th
0083  ure or constitution may well turn the nicely BALANCED scale in the struggle for life, and so be pr
0147  e same period, their law of compensation or BALANCEMENT of growth; or, as Goethe expressed it, in
0002  lt can be obtained only by fully stating and BALANCING the facts and arguments on both sides of ea
0295  vement continues. In fact, this nearly exact BALANCING between the supply of sediment and the amou
0192  he small frena, serving for respiration. The BALANIDAE or sessile cirripedes, on the other hand, h
0176  te between well marked varieties in the genus BALANUS. And it would appear from information given
0044  ition of shells in the brackish waters of the BALTIC, or dwarfed plants on Alpine summits, or the
0197  use, probably to sexual selection. A trailing BAMBOO in the Malay Archipelago climbs the loftiest
0197  rees which are not climbers, the hooks on the BAMBOO may have arisen from unknown laws of growth,
0009  or variable. Sterility has been said to be the BANE of horticulture; but on this view we owe varia
0095  natural selection, if it be a true principle, BANISH the belief of the continued creation of new o
0095  rved being; and as modern geology has almost BANISHED such views as the excavation of a great vall
0074  at the plants and bushes clothing an entangled BANK, we are tempted to attribute their proportiona
0393  nnot be considered as oceanic, as it lies on a BANK connected with the mainland: moreover, iceberg
```

Page ***(Key Word)***

```
0395  est Indian Islands stand on a deeply submerged BANK, nearly 1000 fathoms in depth, and here we fin
0489   it is interesting to contemplate an entangled BANK, clothed with many plants of many kinds, with
0019  eded from the common wild Indian fowl (Gallus BANKIVA). In regard to ducks and rabbits, the breeds
0359  branches, and that these might be dried on the BANKS, and then by a fresh rise in the stream be wa
0395  ands are situated on moderately deep submarine BANKS, and they are inhabited by closely allied or
0030  ayers which never desire to sit, and with the BANTAM so small and elegant; when we compare the hos
0089  time give elegant carriage and beauty to his BANTAMS, according to his standard of beauty, I can
0025  having it bluish); the tail has a terminal dark BAR, with the bases of the outer feathers external
0025  colour, with the white rump, double black wing BAR, and barred and white edged tail feathers, as
0160  th two black bars on the wings, a white rump, a BAR at the end of the tail, with the outer feather
0265  seriously affected. This, in fact, is the great BAR to the domestication of animals. Between the s
0021  and tails, others singularly short tails. The BARB is allied to the carrier, but, instead of a ve
0023  arrier, the short faced tumbler, the runt, the BARB, pouter, and fantail in the same genus; more e
0024  arrier, or that of the short faced tumbler, or BARB; for reversed feathers like those of the jacob
0025  child of the pure white fantail and pure black BARB was of as beautiful a blue colour, with the wh
0161   it is probable that in each generation of the BARB pigeon, which produces most rarely a blue and
0036  we see the value set on animals even by the BARBARIANS of Tierra del Fuego, by their killing and d
0034  iple in works of high antiquity. In rude and BARBAROUS periods of English history choice animals w
0036  fferent varieties. If there exist savages so BARBAROUS as never to think of the inherited characte
0098  kolreuter has shown to be the case with the BARBERRY; and curiously in this very genus, which see
0025  ormly white fantails with some uniformly black BARBS, and they produced mottled brown and black bi
0445  n the wild stock, in pouters, fantails, runts, BARBS, dragons, carriers, and tumblers. Now some of
0221  usca from another nest, and put them down on a BARE spot near the place of combat; they were eager
0309  f the world, for instance in South America, of BARE metamorphic rocks, which must have been heated
0365  come fully stocked with inhabitants. On almost BARE land, with few or no destructive insects or bi
0051  rst step towards such slight varieties as are BARELY thought worth recording in works on natural h
0112  of divergence, causing differences, at first BARELY appreciable, steadily to increase, and the br
0440  ance, have larvae in all their several stages BARELY distinguishable. The embryo in the course of
0255  gradation can be shown to exist; but only the BAREST outline of the facts can here be given. When
0003  o admirably adapted to catch insects under the BARK of trees. In the case of the misseltoe, which
0084  on. When we see leaf eating insects green, and BARK feeders mottled grey; the alpine ptarmigan whi
0184  s and for seizing insects in the chinks of the BARK? Yet in North America there are woodpeckers wh
0440  e illustrious Cuvier did not perceive that a BARNACLE was, as it certainly is, a crustacean: but a
0151  opercular valves of sessile cirripedes (rock BARNACLES) are, in every sense of the word, very impo
0307  tion of the world is known with accuracy. M. BARRANDE has lately added another and lower stage to
0307  e been detected in the Longmynd beds beneath BARRANDE's so called primordial zone. The presence of
0310  aeontologists, namely Cuvier, Owen, Agassiz, BARRANDE, Falconer, E. Forbes, etc., and all our grea
0313  ule, is that of the so called colonies of M. BARRANDE, which intrude for a period in the midst of
0317  geologists, as Elie de Beaumont, Murchison, BARRANDE, etc., whose general views would naturally l
0325  ws which govern the whole animal kingdom. M. BARRANDE has made forcible remarks to precisely the s
0325  der the most different climates. We must, as BARRANDE has remarked, look to some special law. We s
0328  ns on some of the later tertiary formations. BARRANDE, also, shows that there is a striking genera
0330  nd the camel. In regard to the Invertebrata, BARRANDE, and a higher authority could not be named,
0025  th the white rump, double black wing bar, and BARRED and white edged tail feathers, as any wild ro
0161  , which produces most rarely a blue and black BARRED bird, there has been a tendency in each gener
0163  n the shoulder. The quagga, though so plainly BARRED like a zebra over the body, is without bars o
0164  ine is always striped; the legs are generally BARRED; and the shoulder stripe, which is sometimes
0165  ss and zebra, the legs were much more plainly BARRED than the rest of the body; and in one of them
0165   a black Arabian sire, were much more plainly BARRED across the legs than is even the pure quagga.
0165  ulder stripe, nevertheless, had all four legs BARRED, and had three short shoulder stripes, like t
0473  ds of pigeon have descended from the blue and BARRED rock pigeon! On the ordinary view of each spe
0071  nvestigation, there was a large and extremely BARREN heath, which had never been touched by the ha
0072  ng young firs. Yet the heath was so extremely BARREN, and so extensive that no one would ever have
0262  ther species, when thus grafted were rendered BARREN. On the other hand, certain species of Sorbus
0289  that during the periods which were blank and BARREN in his own country, great piles of sediment,
0295  ologist, that very thick deposits are usually BARREN of organic remains, except near their upper o
0306  turalist to land for five minutes on some one BARREN point in Australia, and then to discuss the n
0386  are occasionally found on the most remote and BARREN islands in the open ocean; they would not be
0389   is not quite fair. We have evidence that the BARREN island of Ascension aboriginally possessed un
0393  situated at a much less distance are equally BARREN. The Falkland Islands, which are inhabited by
0348  a halting place for emigrants; here we have a BARRIER of another kind, and as soon as this is pass
0383  the sea is apparently a still more impassable BARRIER, that they never would have extended to dist
0081  island, or of a country partly surrounded by BARRIERS, into which new and better adapted forms cou
0105  very small, either from being surrounded by BARRIERS, or from having very peculiar physical condi
0346  rences in physical conditions. Importance of BARRIERS. Affinity of the productions of the same con
0347  ch strikes us in our general review is, that BARRIERS of any kind, or obstacles to free migration,
0348  eing separated from each other by impassable BARRIERS, either of land or open sea, they are wholly
0348  s of the Pacific, we encounter no impassable BARRIERS, and we have innumerable islands as halting
0350  f all relations. Thus the high importance of BARRIERS comes into play by checking migration; as do
0353  pace. The great and striking influence which BARRIERS of every kind have had on distribution, is i
0366  ravel southward, unless they were stopped by BARRIERS, in which case they would perish. The mounta
0383  ver systems are separated from each other by BARRIERS of land, it might have been thought that fre
0405  hich apparently has the capacity of crossing BARRIERS and ranging widely, as in the case of certai
0405  idely implies not only the power of crossing BARRIERS, but the more important power of being victo
0407  ions, more especially from the importance of BARRIERS and from the analogical distribution of sub
0408  e can thus understand the high importance of BARRIERS, whether of land or water, which separate ou
0025  lly edged with white; the wings have two black BARS; some semi domestic breeds and some apparently
0025  truly wild breeds have, besides the two black BARS, the wings chequered with black. These several
0159  the breeds, of slaty blue birds with two black BARS on the wings, a white rump, a bar at the end o
0163  he ass not rarely has very distinct transverse BARS on its legs, like those on the legs of the zeb
0163  barred like a zebra over the body, is without BARS on the legs; but Dr. Gray has figured one spec
0163  red one specimen with very distinct zebra like BARS on the hocks. With respect to the horse, I hav
0163  istinct breeds, and of all colours; transverse BARS on the legs are not rare in duns, mouse duns,
0165  the ass and horse is particularly apt to have BARS on its legs. I once saw a mule with its legs s
0166  phical races) of a bluish colour, with certain BARS and other marks; and when any breed assumes by
0166  sumes by simple variation a bluish tint, these BARS and other marks invariably reappear; but witho
```

Page ***(Key Word)***

Page ***(Key Word)***

```
0166  we see a strong tendency for the blue tint and BARS and marks to reappear in the mongrels. I have
0225  is well known, is an hexagonal prism, with the BASAL edges of its six sides bevelled so as to join
0231  nding in position to the planes of the rhombic BASAL plates of future cells. But the rough wall of
0235  eral angles of the hexagonal prisms and of the BASAL rhombic plates. The motive power of the proce
0092  m their sap: this is effected by glands at the BASE of the stipules in some Leguminosae, and at th
0177  hilly tract; and a third to wide plains at the BASE; and that the inhabitants are all trying with
0180  flying squirrels have their limbs and even the BASE of the tail united by a broad expanse of skin,
0225  angles, and the three which form the pyramidal BASE of a single cell on one side of the comb, ente
0229  or i have noticed half completed rhombs at the BASE of a just commenced cell, which were slightly
0230  ey do not make the whole three sided pyramidal BASE of any one cell at the same time, but only the
0283  e of nothing in wearing away rock. At last the BASE of the cliff is undermined, huge fragments fal
0283  ars have elapsed since the waters washed their BASE. He who most closely studies the action of the
0286  g resisting attrition form a breakwater at the BASE. Hence, under ordinary circumstances, I conclu
0378  hat now growing with strange luxuriance at the BASE of the Himalaya, as graphically described by H
0451  ach the ovules protected in the ovarium at its BASE. The pistil consists of a stigma supported on
0488  more important researches. Psychology will be BASED on a new foundation, that of the necessary ac
0025  h); the tail has a terminal dark bar, with the BASES of the outer feathers externally edged with w
0092  et juice or nectar to be excreted by the inner BASES of the petals of a flower. In this case insec
0094  of cutting holes and sucking the nectar at the BASES of certain flowers, which they can, with a ve
0148  reduced to the merest rudiment attached to the BASES of the prehensile antennae. Now the saving of
0160  the outer feathers externally edged near their BASES with white. As all these marks are characteri
0225  the comb, enter into the composition of the BASES of three adjoining cells on the opposite side
0227  hexagonal prisms united together by pyrimidal BASES formed of three rhombs; and the rhombs and th
0283  and, or mud. But how often do we see along the BASES of retreating cliffs rounded boulders, all th
0367  erate regions. And as the snow melted from the BASES of the mountains, the arctic forms would seiz
0368  e moment of the returning warmth, first at the BASES and ultimately on the summits of the mountain
0228  was built upon the festooned edge of a smooth BASIN, instead of on the straight edges of a three
0305  cean would form a large and perfectly enclosed BASIN, in which any great group of marine animals m
0107  nct tertiary flora of Europe. All fresh water BASINS, taken together, make a small area compared w
0228  wider until they were converted into shallow BASINS, appearing to the eye perfectly true or parts
0228  ever several bees had begun to excavate these BASINS near together, they had begun their work at s
0228  istance from each other, that by the time the BASINS had acquired the above stated width (i.e. abo
0228  of which they formed a part, the rims of the BASINS intersected or broke into each other. As soon
0228  wax on the lines of intersection between the BASINS, so that each hexagonal prism was built upon
0228  tantly began on both sides to excavate little BASINS near to each other, in the same way as before
0228  e of wax was so thin, that the bottoms of the BASINS, if they had been excavated to the same depth
0229  ped their excavation in due time: so that the BASINS, as soon as they had been a little deepened,
0229  planes of imaginary intersection between the BASINS on the opposite sides of the ridge of wax. In
0229  ombic plate had been left between the opposed BASINS, but the work, from the unnatural state of th
0229  they circularly gnawed away and deepened the BASINS on both sides, in order to have succeeded in
0229  eeded in thus leaving flat plates between the BASINS, by stopping work along the intermediate plan
0226  a gross imitation of the three sided pyramidal BASIS of the cell of the hive bee. As in the cells
0231  inch in thickness; the plates of the pyramidal BASIS being about twice as thick. By this singular
0450  l in the males of mammals: I presume that the BASTARD wing in birds may be safely considered as a
0150  same part in closely allied species. Thus, the BAT's wing is a most abnormal structure in the cla
0150  ngs; it would apply only if some one species of BAT had its wings developed in some remarkable man
0154  descendants, as in the case of the wing of the BAT, it must have existed, according to my theory,
0171  ng, for instance, the structure and habits of a BAT, could have been formed by the modification of
0180  ould possibly have been converted into a flying BAT, the question would have been far more difficu
0180  ficulty in any particular case like that of the BAT. Look at the family of squirrels; here we have
0181  f flight are concerned, would convert it into a BAT. In bats which have the wing membrane extended
0200  n the fore leg of the horse, in the wing of the BAT, and in the flipper of the seal, are of specia
0200  al bones in the limbs of the monkey, horse, and BAT, which have been inherited from a common proge
0204  life could not graduate into each other; that a BAT, for instance, could not have been formed by n
0434  the paddle of the porpoise, and the wing of the BAT, should all be constructed on the same pattern
0437  eated in the formation of the wing and leg of a BAT, used as they are for such totally different p
0440  the same bones in the hand of a man, wing of a BAT, and fin of a porpoise, are related to similar
0442  obvious reason why, for instance, the wing of a BAT, or the fin of a porpoise, should not have bee
0474  in the most unusual manner, like the wing of a BAT, and yet not be more variable than any other s
0479  es being the same in the hand of a man, wing of BAT, fin of the porpoise, and leg of the horse, th
0479  similarity of pattern in the wing and leg of a BAT, though used for such different purpose, in th
0331  s. yet if we compare the older Reptiles and BATRACHIANS, the older Fish, the older Cephalopods, an
0383  abitants of oceanic islands. Absence of the BATRACHIANS and of terrestrial Mammals. On the relatio
0393  ds, Bory St. Vincent long ago remarked that BATRACHIANS (frogs, toads, newts) have never been foun
0396  classes, the absence of whole groups, as of BATRACHIANS, and of terrestrial mammals notwithstandin
0409  e can see why whole groups of organisms, as BATRACHIANS and terrestrial mammals, should be absent
0431  ms of life which once connected fishes with BATRACHIANS. There has been still less in some other c
0150  here apply, because there is a whole group of BATS having wings; it would apply only if some one
0181  mur, which formerly was falsely ranked amongst BATS. It has an extremely wide flank membrane, stre
0181  are concerned, would convert it into a bat. In BATS which have the wing membrane extended from the
0394  almost every island. New Zealand possesses two BATS found nowhere else in the world: Norfolk Islan
0394  oes, and Mauritius, all possess their peculiar BATS. Why, it may be asked, has the supposed creati
0394  sked, has the supposed creative force produced BATS and no other mammals on remote islands? On my
0394  be transported across a wide space of sea, but BATS can fly across. Bats have been seen wandering
0394  a wide space of sea, but bats can fly across. BATS have been seen wandering by day far over the A
0395  and we can understand the presence of endemic BATS on islands, with the absence of all terrestria
0396  mammals notwithstanding the presence of aerial BATS, the singular proportions of certain orders of
0404  gst mammals, we see it strikingly displayed in BATS, and in a lesser degree in the Felidae and Can
0409  heir own peculiar species of aerial mammals or BATS. We can see why there should be some relation
0477  on the other hand, new and peculiar species of BATS, which can traverse the ocean, should so often
0478  h facts as the presence of peculiar species of BATS, and the absence of all other mammals, on ocea
0030  we compare the game cock, so pertinacious in BATTLE, with other breeds so little quarrelsome, wit
0073  the relations can ever be as simple as this. BATTLE within battle must ever be recurring with var
0073  can ever be as simple as this. Battle within BATTLE must ever be recurring with varying success;
0076  has been victorious over another in the great BATTLE of life. A corollary of the highest importanc
0080  me way to each being in the great and complex BATTLE of life, should sometimes occur in the course
```

Page ***(Key Word)***

Page **(Key Word)***

```
0098  s. how low in the scale of nature this law of BATTLE descends, I know not: male alligators have be
0090  t believe to be either useful to the males in BATTLE, or attractive to the females. We see analogo
0128  ter will be their chance of succeeding in the BATTLE of life. Thus the small differences distingui
0129  ried to overmaster other species in the great BATTLE for life. The limbs divided into great branch
0183  a decided advantage over other animals in the BATTLE for life. Hence the chance of discovering spe
0205  y will produce perfection, or strength in the BATTLE for life, only according to the standard of t
0326  l degree, whenever their inhabitants met, the BATTLE would be prolonged and severe: and some from
0136  er of individuals were saved by successfully BATTLING with the winds, or by giving up the attempt
0164  es be seen in duns, and I have see a trace in a BAY horse. My son made a careful examination and s
0164  ld horses. Colonel Poole has seen both gray and BAY Kattywar horses striped when first foaled. I h
0288  few of the very many animals which live on the BEACH between high and low watermark are preserved.
0001       Introduction When on board H.M.S. BEAGLE, as naturalist, I was much struck with certai
0386  family, a Colymbetes, once flew on board the BEAGLE, when forty five miles distant from the neare
0003  tance, of the woodpecker, with its feet, tail, BEAK, and tongue, so admirably adapted to catch ins
0021  e cape of mouth. The short faced tumbler has a BEAK in outline almost like that of a finch: and th
0021  nt is a bird of great size, with long, massive BEAK and large feet: some of the sub breeds of runt
0021  ed to the carrier, but, instead of a very long BEAK, has a very short and very broad one. The pout
0021  ghter. The turbit has a very short and conical BEAK, with a line of reversed feathers down the bre
0022  lways in strict correlation with the length of BEAK), the size of the crop and of the upper part o
0024  out the whole great family of Columbidae for a BEAK like that of the English carrier, or that of t
0027  h breed, for instance the wattle and length of BEAK of the carrier, the shortness of that of the t
0031  it would take him six years to obtain head and BEAK. In Saxony the importance of the principle of
0087  for opening the cocoon, or the hard tip to the BEAK of nestling birds, used for breaking the egg.
0087  ct of hatching. Now, if nature had to make the BEAK of a full grown pigeon very short for the bird
0112  s struck by a pigeon having a slightly shorter BEAK: another fancier is struck by a pigeon having
0112  r is struck by a pigeon having a rather longer BEAK: and on the acknowledged principle that fancie
0152  rodigious amount of difference there is in the BEAK of the different tumblers, in the beak and wat
0152  in the beak of the different tumblers, in the BEAK and wattle of the different carriers, in the c
0335  xtreme in the important character of length of BEAK originated earlier than short beaked tumblers,
0424  in the important character of having a longer BEAK, yet all are kept together from having the com
0444  which obtained its food best by having a long BEAK, whether or not it assumed a beak of this part
0444  aving a long beak, whether or not it assumed a BEAK of this particular length, as long as it was f
0445  rtions (but will not here give details) of the BEAK, width of mouth, length of nostril and of evel
0445  iffer so extraordinarily in length and form of BEAK, that they would, I cannot doubt, be ranked in
0083  d fitting manner: he feeds a long and a short BEAKED pigeon on the same food: he does not exercise
0087  it has been asserted, that of the best short BEAKED tumbler pigeons more perish in the egg than a
0335  length of beak originated earlier than short BEAKED tumblers, which are at the opposite end of th
0012  n between their outer toes: pigeons with short BEAKS have small feet: and those with long beaks la
0012  ort beaks have small feet: and those with long BEAKS large feet. Hence, if man goes on selecting,
0021  ler, and see the wonderful difference in their BEAKS, entailing corresponding differences in their
0087  e egg, which had the most powerful and hardest BEAKS, for all with weak beaks would inevitably per
0087  powerful and hardest beaks, for all with weak BEAKS would inevitably perish: or, more delicate an
0112  and breeding from birds with longer and longer BEAKS, or with shorter and shorter beaks. Again, we
0112  and longer beaks, or with shorter and shorter BEAKS. Again, we may suppose that at an early perio
0362  re always killed by this process. Although the BEAKS and feet of birds are generally quite clean,
0386  rely, adheres in some quantity to the feet and BEAKS of birds. Wading birds, which frequent the mu
0451  that rudiments of teeth can be detected in the BEAKS of certain embryonic birds. Nothing can be pl
0142  not be effected! The case, also, of the kidney BEAN has been often cited for a similar purpose, an
0077  g plants produced from such seeds (as peas and BEANS), when sown in the midst of long grass, I sus
0142  sow, during a score of generations, his kidney BEANS so early that a very large proportion are des
0142  erences in the constitution of seedling kidney BEANS ever appear, for an account has been publishe
0009  ent, with the exception of the plantigrades or BEAR family; whereas, carnivorous birds, with the r
0019  wild species of sheep peculiar to it! When we BEAR in mind that Britain has now hardly one peculi
0049  riation, hybridism, etc., have been brought to BEAR on the attempt to determine their rank. I will
0062  t least I have found it so, than constantly to BEAR this conclusion in mind. Yet unless it be thor
0062  by birds and beasts of prey: we do not always BEAR in mind, that though food may be now superabun
0084  complex conditions of life, and should plainly BEAR the stamp of far higher workmanship? It may be
0085  effect. It is, however, far more necessary to BEAR in mind that there are many unknown laws of co
0087  ral history, I cannot find one case which will BEAR investigation. A structure used only once in a
0088  fighting all day long: male stag beetles often BEAR wounds from the huge mandibles of other males.
0093  , presently to be alluded to. Some holly trees BEAR only male flowers, which have four stamens pro
0093  n, and a rudimentary pistil: other holly trees BEAR only female flowers: these have a full sized p
0100  nst it by giving to trees a strong tendency to BEAR flowers with separated sexes. When the sexes a
0129  now grown into great branches, yet survive and BEAR all the other branches: so with the species wh
0141  a state of nature, could easily be brought to BEAR widely different climates. We must not, howeve
0175  or depth graduate away insensibly. But when we BEAR in mind that almost every species, even in its
0182  , as the wings of a bird for flight, we should BEAR in mind that animals displaying early transiti
0184  em like a nuthatch. In North America the black BEAR was seen by Hearne swimming for hours with wid
0191  g transitions of organs, it is so important to BEAR in mind the probability of conversion from one
0355  d multiple centres of creation do not directly BEAR on another allied question, namely whether all
0377  into the warmest spots. But the great fact to BEAR in mind is, that all tropical productions will
0389  al; but shall consider some other facts, which BEAR on the truth of the two theories of independen
0398  n the Galapagos Archipelago, and nowhere else, BEAR so plain a stamp of affinity to those created
0405  wide distribution of certain genera, we should BEAR in mind that some are extremely ancient, and m
0407  ly ever been properly experimentised on: if we BEAR in mind how often a species may have ranged co
0417  ertain plants, belonging to the Malpighiaceae, BEAR perfect and degraded flowers; in the latter, a
0425  uced, by a long course of modification, from a BEAR? Ought we to rank this one species with bears,
0425  angaroo were seen to come out of the womb of a BEAR? According to all analogy, it would be ranked
0425  aroo family would have to be classed under the BEAR genus. The whole case is preposterous: for whe
0480  ers of some beetles, should thus so frequently BEAR the plain stamp of inutility! Nature may be sa
0147  accompanied by a diminished comb, and a large BEARD by diminished wattles. With species in a stat
0001  sorts of facts which could possibly have any BEARING on it. After five years' work I allowed myse
0063  orically be said to struggle with other fruit BEARING plants, in order to tempt birds to devour an
0094  very little more trouble, enter by the mouth. BEARING such facts in mind, I can see no reason to d
0147  utritious foliage and a copious supply of oil BEARING seeds. When the seeds in our fruits become a
0188  ersity in the eyes of living crustaceans, and BEARING in mind how small the number of living anima
```

Page **(Key Word)**

Page **(Key Word)***

```
0204  s nearest congeners. Hence we can understand, BEARING in mind that each organic being is trying to
0296  feet thick in Nova Scotia, with ancient root BEARING strata, one above the other, at no less than
0377  an their own. Hence, it seems to me possible, BEARING in mind that the tropical productions were i
0402  ton have communicated to me a remarkable fact BEARING on this subject; namely, that Madeira and th
0408  which formerly inhabited the same continent. BEARING in mind that the mutual relations of organis
0437  ebrata, we see a series of internal vertebrae BEARING certain processes and appendages; in the art
0437  e the body divided into a series of segments, BEARING external appendages; and in flowering plants
0450  s. organs or parts in this strange condition, BEARING the stamp of inutility, are extremely common
0457  rresponding ages. On this same principle, and BEARING in mind, that when organs are reduced in siz
0457  e being has to provide for its own wants, and BEARING in mind how strong is the principle of inher
0483  and in the case of mammals, were they created BEARING the false marks of nourishment from the moth
0486  in view. We possess no pedigrees or armorial BEARINGS; and we have to discover and trace the many
0060  o groups. Chapter III. Struggle For Existence. BEARS on natural selection. The term used in a wide
0060  emarks, to show how the struggle for existence BEARS on Natural Selection. It has been seen in the
0184  country, I can see no difficulty in a race of BEARS being rendered, by natural selection, more an
0233  two just commenced cells, is important, as it BEARS on a fact, which seems at first quite subvers
0273  a curious fact and deserves attention. For it BEARS on and corroborates the view which I have tak
0283  med of worn and rounded pebbles, each of which BEARS the stamp of time, are good to show how slowl
0398  ere almost every product of the land and water BEARS the unmistakeable stamp of the American conti
0425  a bear? Ought we to rank this one species with BEARS, and what should we do with the other species
0425  ording to all analogy, it would be ranked with BEARS: but then assuredly all the other species of
0134  abited several oceanic islands, tenanted by no BEAST of prey, has been caused by disuse. The ostri
0062  r their nestlings, are destroyed by birds and BEASTS of prey; we do not always bear in mind, that
0068  lephant and rhinoceros; none are destroyed by BEASTS of prey: even the tiger in India most rarely
0072  se numbers are probably regulated by hawks or BEASTS of prey) were to increase in Paraguay, the fl
0075  cts, snails, and other animals with birds and BEASTS of prey, all striving to increase, and all fe
0180  wn country, by enabling it to escape birds or BEASTS of prey, or to collect food more quickly, or
0180  on change, let other competing rodents or new BEASTS of prey immigrate, or old ones become modifie
0195  dearth to search for food, or to escape from BEASTS of prey. Organs now of trifling importance ha
0049  erms has been generally accepted, is vainly to BEAT the air. Many of the cases of strongly marked
0075  imate, or are naturally the most fertile, will BEAT the others and so yield more seed, and will co
0177  mmon forms, in the race for life, will tend to BEAT and supplant the less common forms, for these
0379  ial period, the northern forms were enabled to BEAT the less powerful southern forms. Just in the
0471  would not hold them, the more dominant groups BEAT the less dominant. This tendency in the large
0083  land. And as foreigners have thus everywhere BEATEN some of the natives, we may safely conclude t
0110  y given period, and they will consequently be BEATEN in the race for life by the modified descenda
0179  ural selection, they will almost certainly be BEATEN and supplanted by the forms which they connec
0215  occasionally do make an attack, and are then BEATEN: and if not cured, they are destroyed: so tha
0279  e forms which they connect, will generally be BEATEN out and exterminated during the course of fur
0327  nd producing new species. The forms which are BEATEN and which yield their places to the new and v
0337  the eocene fauna or flora would certainly be BEATEN and exterminated: as would a secondary fauna
0337  s of life, in comparison with the ancient and BEATEN forms: but I can see no way of testing this s
0337  not the highest in their own class, may have BEATEN the highest molluscs. From the extraordinary
0345  successive period in the world's history have BEATEN their predecessors in the race for life, and
0380  ee in Australia, and have to a certain extent BEATEN the natives: whereas extremely few southern f
0472  y created and adapted for that country, being BEATEN and supplanted by the naturalised productions
0317  ven up, even by those geologists, as Elie de BEAUMONT, Murchison, Barrande, etc., whose general vi
0025  white fantail and pure black barb was of as BEAUTIFUL a blue colour, with the white rump, double
0030  nt seasons and for different purposes, or so BEAUTIFUL in his eyes, we must, I think, look further
0060  another being, been perfected? We see these BEAUTIFUL co adaptations most plainly in the woodpeck
0061  ted by the gentlest breeze; in short, we see BEAUTIFUL adaptations everywhere and in every part of
0074  the Southern United States, display the same BEAUTIFUL diversity and proportion of kinds as in the
0089  usands of generations, the most melodious or BEAUTIFUL males, according to their standard of beaut
0098  tance, in Lobelia fulgens, there is a really BEAUTIFUL and elaborate contrivance by which every on
0130  vers the surface with its ever branching and BEAUTIFUL ramifications. Chapter V. Laws of Variation
0169  ariation, such modifications will add to the BEAUTIFUL and harmonious diversity of nature. Whateve
0191  falling into the lungs, notwithstanding the BEAUTIFUL contrivance by which the glottis is closed.
0197  uld have thought that the green colour was a BEAUTIFUL adaptation to hide this tree frequenting bi
0197  lls of young mammals have been advanced as a BEAUTIFUL adaptation for aiding parturition, and no d
0224  ly so great as it at first appears: all this BEAUTIFUL work can be shown, I think, to follow from
0440  conditions of life is just as perfect and as BEAUTIFUL as in the adult animal. From such special a
0490  rom so simple a beginning endless forms most BEAUTIFUL and most wonderful have been, and are being
0077  to the hair on the tiger's body. But in the BEAUTIFULLY plumed seed of the dandelion, and in the f
0224  amine the exquisite structure of a comb, so BEAUTIFULLY adapted to its end, without enthusiastic a
0392  by mammals, some of the endemic plants have BEAUTIFULLY hooked seeds; yet few relations are more s
0441  tage of butterflies, they have six pairs of BEAUTIFULLY constructed natatory legs, a pair of magni
0469  n see no limit to this power, in slowly and BEAUTIFULLY adapting each form to the most complex rel
0037  ainly be recognised in the increased size and BEAUTY which we now see in the varieties of the hear
0089  can in a short time give elegant carriage and BEAUTY to his bantams, according to his standard of
0089  to his bantams, according to his standard of BEAUTY, I can see no good reason to doubt that femal
0089  autiful males, according to their standard of BEAUTY, might produce a marked effect. I strongly su
0109  see no limit to the amount of change, to the BEAUTY and infinite complexity of the coadaptations
0199  at very many structures have been created for BEAUTY in the eyes of man, or for mere variety. The
0199  ffects of sexual selection, when displayed in BEAUTY to charm the females, can be called useful on
0349  we look to the waters, and we do not find the BEAVER or musk rat, but the coypu and capybara, rode
0072  er that, as soon as the land was enclosed, it BECAME thickly clothed with vigorously growing young
0073  doubt, that if the whole genus of humble bees BECAME extinct or very rare in England, the heartsea
0086  be in the least degree injurious: for if they BECAME so, they would cause the extinction of the sp
0113  ure the descendants of our carnivorous animal BECAME, the more places they would be enabled to occ
0133  urs for a warm sea: but that this other shell BECAME bright coloured by variation when it ranged i
0135  ere used more, and its wings less, until they BECAME incapable of flight. Kirby has remarked (and
0157  mmon progenitor, or of its early descendants, BECAME variable: variations of this part would, it i
0181  bers or become exterminated, unless they also BECAME modified and improved in structure in a corre
0293  would be equally rash to suppose that it then BECAME wholly extinct. We forget how small the area
0306  ld remain confined, until some of the species BECAME adapted to a cooler climate, and were enabled
0334  fth stage, and are the parents of those which BECAME still more modified at the seventh stage; hen
```

Page **(Key Word)***

Page **(Key Word)**

```
0366   cold came on, and as each more southern zone   BECAME fitted for arctic beings and ill fitted for t
0370   n slowly to migrate southwards as the climate   BECAME less warm, long before the commencement of th
0379   than the southern forms. And thus, when they    BECAME commingled during the Glacial period, the nor
0404   rms, and it is obvious that a mountain, as it    BECAME slowly upheaved, would naturally be colonised
0442   n proper proportion, as soon as any structure   BECAME visible in the embryo. And in some whole grou
0028   eons at some, yet quite insufficient, length;   BECAUSE when I first kept pigeons and watched the se
0047   e form is ranked as a variety of another, not   BECAUSE the intermediate links have actually been fo
0047   ermediate links have actually been found, but   BECAUSE analogy leads the observer to suppose either
0070   y of corn and rape seed, etc., in our fields,   BECAUSE the seeds are in great excess compared with
0116   unequal distances. I have said a large genus,   BECAUSE we have seen in the second chapter, that on
0132   ly from an ovule, is alone affected. But why,   BECAUSE the reproductive system is disturbed, this o
0140   als were originally chosen by uncivilised man   BECAUSE they were useful and bred readily under conf
0140   l and bred readily under confinement, and not   BECAUSE they were subsequently found capable of far
0150   mammalia; but the rule would not here apply,    BECAUSE there is a whole group of bats having wings;
0154   sual circumstance. I have chosen this example   BECAUSE an explanation is not in this case applicabl
0154   ic characters are more variable than generic,   BECAUSE they are taken from parts of less physiologi
0157   ngs is a character of the highest importance,   BECAUSE common to large groups: but in certain gener
0160   i think confidently come to this conclusion,    BECAUSE, as we have seen, these coloured marks are e
0168   special parts or organs being still variable,   BECAUSE they have recently varied and thus come to d
0174   we should be extremely cautious in inferring,   BECAUSE an area is now continous, that it has been c
0177   s of varying and intermediate links: firstly,   BECAUSE new varieties are very slowly formed, for va
0190   n of the swimbladder in fishes is a good one,   BECAUSE it shows us clearly the highly important fac
0203   multitude of intermediate gradations, partly   BECAUSE the process of natural selection will always
0203   ne time, only on a very few forms: and partly   BECAUSE the very process of natural selection almost
0242   the power of natural selection, and likewise   BECAUSE this is by far the most serious special diff
0287   condary period. I have made these few remarks   BECAUSE it is highly important for us to gain some n
0302   of the geological record, and falsely infer,   BECAUSE certain genera or families have not been fou
0364   ! but it would be a great error to argue that   BECAUSE a well stocked island, like Great Britain, h
0390   escendants. But put it by no means follows, that,  BECAUSE in an island nearly all the species of one c
0419   family or family; and this has been done, not   BECAUSE further research has detected important stru
0419   uctural differences, at first overlooked, but   BECAUSE numerous allied species, with slightly diffe
0423   varieties of the pine apple together, merely   BECAUSE their fruit, though the most important part,
0423   are very useful for this purpose with cattle,   BECAUSE they are less variable than the shape or col
0423   th sheep the horns are much less serviceable,   BECAUSE less constant. In classing varieties, I appr
0424   t, these tumblers are kept in the same group,   BECAUSE allied in blood and alike in some other resp
0424   s monsters; he includes varieties, not solely   BECAUSE they closely resemble the parent form, but b
0424   se they closely resemble the parent form, but   BECAUSE they are descended from it. He who believes
0427   classing large and widely distributed genera,   BECAUSE all the species of the same genus, inhabitin
0453   t. would it be thought sufficient to say that   BECAUSE planets revolve in elliptic courses round th
0483   species. The question is difficult to answer,   BECAUSE the more distinct the forms are which we may
0012   h other. The whole organisation seems to have   BECOME plastic, and tends to depart in some small de
0023   ers, seems very improbable: or they must have   BECOME extinct in the wild state. But birds breeding
0024   to their native country: but not one has ever   BECOME wild or feral, though the dovecot pigeon, whi
0024   pigeon in a very slightly altered state, has   BECOME feral in several places. Again, all recent ex
0024   rther, that these very species have since all   BECOME extinct or unknown. So many strange contingen
0026   racter derived from such cross will naturally   BECOME less and less, as in each succeeding generati
0032   as accuracy of eye and judgment sufficient to   BECOME an eminent breeder. If gifted with these qual
0032   l capacity and years of practice requisite to   BECOME even a skilful pigeon fancier. The same princ
0037   ther or not two or more species or races have   BECOME blended together by crossing, may plainly be
0039   med what the descendants of that pigeon would   BECOME through long continued, partly unconscious an
0051   ncy will be to make many species, for he will   BECOME impressed, just like the pigeon or poultry fa
0052   cies. They may whilst in this incipient state   BECOME extinct, or they may endure as varieties for
0053   si and this might have been expected, as they   BECOME exposed to diverse physical conditions, and a
0054   n anticipated: for, as varieties, in order to   BECOME in any degree permanent, necessarily have to
0054   hose advantages that enabled their parents to   BECOME dominant over their compatriots. If the plant
0059   varieties, as we shall hereafter see, tend to   BECOME converted into new and distinct species. The
0059   tinct species. The larger genera thus tend to   BECOME larger: and throughout nature the forms of Li
0059   forms of life which are now dominant tend to   BECOME still more dominant by leaving many modified
0059   us, the forms of life throughout the universe   BECOME divided into groups subordinate to groups. Ch
0061   eties, which I have called incipient species,   BECOME ultimately converted into good and distinct s
0063   ometrical increase, its numbers would quickly   BECOME so inordinately great that no country could s
0064   ould be given of introduced plants which have   BECOME common throughout whole islands in a period o
0066   d, many must be produced, or the species will   BECOME extinct. It would suffice to keep up the full
0069   ctly well endure our climate, but which never   BECOME naturalised, for they cannot compete with our
0072   would decrease, then cattle and horses would   BECOME feral, and this would certainly greatly alter
0074   england, the heartsease and red clover would   BECOME very rare, or wholly disappear. The number of
0081   tely undergo a change, and some species might   BECOME extinct. We may conclude, from what we have s
0087   ten appear under domestication in one sex and   BECOME hereditarily attached to that sex, the same f
0094   e male organs and sometimes the female organs   BECOME more or less impotent: now if we suppose this
0095   surface. Hence, again, if humble bees were to   BECOME rare in any country, it might be a great adva
0095   nderstand how a flower and a bee might slowly   BECOME, either simultaneously or one after the other
0102   conomy of nature, if any one species does not   BECOME modified and improved in a corresponding degr
0104   for the old inhabitants to struggle for, and   BECOME adapted to, through modifications in their st
0106   ng species; and if some of these many species   BECOME modified and improved, others will have to be
0106   that continental productions have everywhere   BECOME so largely naturalised on islands. On a small
0108   will be allowed for the varieties in each to   BECOME well modified and perfected. When, by renewed
0109   so will the less favoured forms decrease and   BECOME rare. Rarity, as geology tells us, is the pre
0109   efinitely increasing, numbers inevitably must   BECOME extinct. That the number of specific forms ha
0110   uarter of the world, some foreign plants have   BECOME naturalised, without causing, as far as we kn
0110   formed through natural selection, others will   BECOME rarer and rarer, and finally extinct. The for
0111   does the lesser difference between varieties   BECOME augmented into the greater difference between
0112   tronger ones by others, the differences would   BECOME greater; and would be noted as forming two su
0112   the lapse of centuries, the sub breeds would   BECOME converted into two well established and disti
0112   nd distinct breeds. As the differences slowly   BECOME greater, the inferior animals with intermedia
0112   ersified the descendants from any one species   BECOME in structure, constitution, and habits, by so
0115   the indigenes of any country, and have there   BECOME naturalised, we can gain some crude idea in w
```

Page **(Key Word)**

became

Page ***(Key Word)***

```
0116 es will succeed by so much the better as they  BECOME  more diversified in structure, and are thus e
0119 tters marking the successive forms which have   BECOME  sufficiently distinct to be recorded as varie
0121 parent species itself, will generally tend to   BECOME  extinct. So it probably will be with many who
0122  a species get into some distinct country, or   BECOME  quickly adapted to some quite new station, to
0122 s (a) and all the earlier varieties will have   BECOME  extinct, having been replaced by eight new sp
0122 nner at each stage of descent, so as to have    BECOME  adapted to many related places in the natural
0123 ed the original species (A) and (I), have all   BECOME, excepting (F), extinct, and have left no des
0126 formerly most extensively developed, have now   BECOME  extinct. Looking still more remotely to the f
0126 er groups, a multitude of smaller groups will   BECOME  utterly extinct, and leave no modified descen
0128 mbers, the more diversified these descendants   BECOME, the better will be their chance of succeedin
0134 itable variations, however slight, until they   BECOME  plainly developed and appreciable by us. Effe
0140 ei nor do we know that they have subsequently   BECOME  acclimatised to their new homes. As I believe
0143 ulated through natural selection, other parts   BECOME  modified. This is a very important subject, m
0145 the axis are oftenest subject to peloria, and  BECOME  regular. I may add, as an instance of this, a
0146 le by the fact that seeds could not gradually   BECOME  winged through natural selection, except in f
0147 l bearing seeds. When the seeds in our fruits   BECOME  atrophied, the fruit itself gains largely in
0148 nt being wasted in developing a structure now   BECOME  useless. Thus, as I believe, natural selectio
0153 cters, produced by man's selection, sometimes  BECOME  attached, from causes quite unknown to us, mo
0154 he species had blue flowers, the colour would  BECOME  a generic character; and its variation would
0166  other new character. We see this tendency to  BECOME  striped most strongly displayed in hybrids fr
0167 on, in this particular manner, so as often to  BECOME  striped like other species of the genus; and
0169 with any extraordinarily developed organ has   BECOME  the parent of many modified descendants, whic
0173  during the process of modification, each has  BECOME  adapted to the conditions of life of its own
0175 lready defined objects (however they may have  BECOME  so), not blending one into another by insensi
0175 neutral territory between them, in which they  BECOME  rather suddenly rarer and rarer; then, as var
0181 or new beasts of prey immigrate, or old ones   BECOME  modified, and all analogy would lead us to be
0181 of the squirrels would decrease in numbers or  BECOME  exterminated, unless they also became modifie
0181 flight. If about a dozen genera of birds had   BECOME  extinct or were unknown, who would have ventu
0185 feet of the upland goose may be said to have   BECOME  rudimentary in function, though not in struct
0188 animals is in proportion to those which have   BECOME  extinct, I can see no very great difficulty (
0190 k to very ancient ancestral forms, long since  BECOME  extinct. We should be extremely cautious in c
0192 ve glands. If all pedunculated cirripedes had  BECOME  extinct, and they have already suffered far m
0196 mitted in nearly the same state, although now  BECOME  of very slight use; and any actually injuriou
0201  modified; or if it be not so, the being will  BECOME  extinct, as myriads have become extinct. Natu
0201 he being will become extinct, as myriads have  BECOME  extinct. Natural selection tends only to make
0205 y its terrestrial descendants), though it has  BECOME  of such small importance that it could not, i
0208 modified by the will or reason. Habits easily  BECOME  associated with other habits, and with certai
0209 ed work. If we suppose any habitual action to  BECOME  inherited, and I think it can be shown that t
0214 sitions are inherited, and how curiously they  BECOME  mingled, is well shown when different breeds
0214 ble natural instincts, which in a like manner  BECOME  curiously blended together, and for a long pe
0214 are sometimes spoken of as actions which have  BECOME  inherited solely from long continued and comp
0215 se breeds of fowls which very rarely or never  BECOME  broody, that is, never wish to sit on their e
0215 ly possible to doubt that the love of man has  BECOME  instinctive in the dog. All wolves, foxes, ja
0216 ut this instinct retained by our chickens has  BECOME  useless under domestication, for the mother h
0216 stand how instincts in a state of nature have  BECOME  modified by selection, by considering a few c
0219 ithout their aid, the species would certainly  BECOME  extinct in a single year. The males and ferti
0223 le that pupae originally stored as food might  BECOME  developed; and the ants thus unintentionally
0229 the rhombic plate had been completed, and had  BECOME  perfectly flat: it was absolutely impossible,
0236 ate animals in a state of nature occasionally  BECOME  sterile; and if such insects had been social,
0237 sorts of differences of structure which have   BECOME  correlated to certain ages, and to either sex
0237 ee no real difficulty in any character having  BECOME  correlated with the sterile condition of cert
0242 and their instincts and structure would have   BECOME  blended. And nature has, as I believe, effect
0254 ite fertile hybrids, or the hybrids must have  BECOME  in subsequent generations quite fertile under
0267 les of the same species which have varied and  BECOME  slightly different, give vigour and fertility
0267 crosses between males and females which have   BECOME  widely or specifically different, produce hyb
0280 two breeds. These two breeds, moreover, have   BECOME  so much modified, that if we had no historica
0288 an unaltered condition. The remains which do   BECOME  embedded, if in sand or gravel, will when the
0294 g species are common in the deposit, but have  BECOME  extinct in the immediately surrounding sea; o
0314 hen many of the inhabitants of a country have  BECOME  modified and improved, we can understand, on
0314 ism to organism, that any form which does not  BECOME  in some degree modified and improved, will be
0315 if we look to wide enough intervals of time,   BECOME  modified; for those which do not change will
0315 modified; for those which do not change will   BECOME  extinct. In members of the same class the ave
0319 ct, yet the fossil horse would certainly have  BECOME  rarer and rarer, and finally extinct; its pla
0320  namely, that to admit that species generally  BECOME  rare before they become extinct, to feel no s
0320 hat species generally become rare before they  BECOME  extinct, to feel no surprise at the rarity of
0322 ch single species and whole groups of species  BECOME  extinct, accords well with the theory of natu
0332 f extinction and divergence of character, has  BECOME  divided into several sub families and familie
0335 between the rock pigeon and the carrier have   BECOME  extinct; and carriers which are extreme in th
0337 se of time a multitude of British forms would  BECOME  thoroughly naturalized there, and would exter
0337 inhabitant of the southern hemisphere having   BECOME  wild in any part of Europe, we may doubt, if
0341 instant be admitted. These huge animals have   BECOME  wholly extinct, and have left no progeny. But
0341 old species and the other whole genera having  BECOME  utterly extinct. In failing orders, with the
0344 y inherited from a common progenitor, tend to  BECOME  extinct together, and to leave no modified of
0344 emble, the common progenitor of groups, since  BECOME  widely divergent. Extinct forms are seldom di
0350 ification and improvement; and thus they will  BECOME  still further victorious, and will produce gr
0351 have migrated over vast spaces, and have not   BECOME  greatly modified. On these views, it is obvio
0353 multitude of European animals and plants have  BECOME  naturalised in America and Australia; and som
0364 a marvellous fact if many plants had not thus  BECOME  widely transported. These means of transport
0364 of distant continents would not by such means  BECOME  mingled in any great degree: but would remain
0365 ould be so well fitted to its new home, as to  BECOME  naturalised. But this, as it seems to me, is
0365 being upheaved and formed, and before it had  BECOME  fully stocked with inhabitants. On almost bar
0366 h case they would perish. The mountains would  BECOME  covered with snow and ice, and their former A
0368 tion across the low intervening tracts, since  BECOME  too warm for their existence. If the climate,
0369 nce; they will, also, in all probability have  BECOME  mingled with ancient Alpine species, which mu
0371 nd animals migrated southward, they will have  BECOME  mingled in the one great region with the nati
0376 titudes, the Alpine or mountain floras really  BECOME  less and less arctic. Many of the forms livin
```

Page ***(Key Word)***

Page **(Key Word)**

```
0380  es; whereas extremely few southern forms have  BECOME  naturalised in any part of Europe, though hid
0382  hores of America, Australia, New Zealand have   BECOME  slightly tinted by the same peculiar forms of
0383  k, in most cases be explained by their having   BECOME  fitted, in a manner highly useful to them, fo
0385  along the shores of the sea, and subsequently   BECOME  modified and adapted to the fresh waters of a
0388  son to believe that such low beings change or   BECOME  modified less quickly than the high; and this
0388  , over immense areas, and having subsequently   BECOME  extinct in intermediate regions. But the wide
0390  half a dozen flowering plants; yet many have    BECOME  naturalised on it, as they have on New Zealan
0390  e seems to depend on the species which do not   BECOME  modified having immigrated with facility and
0391  gled together in their former homes, and have   BECOME  mutually adapted to each other; and when sett
0393  , and Mauritius, and have multiplied so as to   BECOME  a nuisance. But as these animals and their sp
0394  amed on which our smaller quadrupeds have not   BECOME  naturalised and greatly multiplied. It cannot
0402  ; but we should remember that the forms which   BECOME  naturalised in new countries are not generall
0403  to to Madeira, yet this latter island has not   BECOME  colonised by the Porto Santo species: neverth
0406  ll quarters of the world, where they may have   BECOME  slightly modified in relation to their new co
0407  continuously over a wide area, and then have    BECOME  extinct in the intermediate tracts, I think t
0410  means of transport, and by the species having   BECOME  extinct in the intermediate tracts. Both in t
0411  r incipient species, thus produced ultimately   BECOME  converted, as I believe, into new and distinc
0412  ed descendants proceeding from one progenitor   BECOME  broken up into groups subordinate to groups.
0427  o most distinct lines of descent, may readily   BECOME  adapted to similar conditions, and thus assum
0430  y analogical, owing to the phascolomys having   BECOME  adapted to habits like those of a Rodent. The
0435  ht be shortened or widened to any extent, and   BECOME  gradually enveloped in thick membrane, so as
0436  le that the general pattern of an organ might   BECOME  so much obscured as to be finally lost, by th
0437  nd in flowers, that organs, which when mature   BECOME  extremely different, are at an early stage of
0439  n organs in the individual, which when mature   BECOME  widely different and serve for different purp
0441  owers of swimming, a proper place on which to   BECOME  attached and to undergo their final metamorph
0441  rval condition. But in some genera the larvae   BECOME  developed either into hermaphrodites having t
0442  he same individual embryo, which ultimately    BECOME  very unlike and serve for diverse purposes, b
0446  es, and of which the several new species have   BECOME  modified through natural selection in accorda
0447  ich served as legs in the parent species, may   BECOME, by a long course of modification, adapted in
0448  their parents. Such differences might, also,   BECOME  correlated with successive stages of developm
0448  be the case with cirripedes. The adult might   BECOME  fitted for sites or habits, in which organs o
0451  nstances are on record of these organs having   BECOME  well developed in full grown males, and havin
0451  s, but in our domestic cows the two sometimes   BECOME  developed and give milk. In individual plants
0451  ture. An organ serving for two purposes, many   BECOME  rudimentary or utterly aborted for one, even
0452  the surrounding anthers. Again, an organ may   BECOME  rudimentary for its proper purpose, and be us
0452  s proper function of giving buoyancy, but has   BECOME  converted into a nascent breathing organ or l
0454  adual reduction of various organs, until they   BECOME  rudimentary, as in the case of the eyes of an
0454  organ useful under certain conditions, might   BECOME  injurious under others, as with the wings of
0455  n a word, still retained in the spelling, but   BECOME  useless in the pronunciation; but which serve
0456  dification, all the great facts in Morphology   BECOME  intelligible, whether we look to the same pat
0457  the homologous parts, which when matured will   BECOME  widely different from each other in structure
0457  ssible. Larvae are active embryos, which have   BECOME  specially modified in relation to their habit
0467  number, and which shall decrease, or finally   BECOME  extinct. As the individuals of the same speci
0470  abled to increase by so much the more as they   BECOME  more diversified in habits and structure, so
0471  nant forms; so that each large group tends to   BECOME  still larger, and at the same time more diver
0473  arked varieties, of which the characters have   BECOME  in a high degree permanent, we can understand
0479  ll often tend to reduce an organ, when it has   BECOME  useless by changed habits or under changed co
0484  serve a specific name. This latter point will   BECOME  a far more essential consideration than it is
0486  experience, will the study of natural history   BECOME! A grand and almost untrodden field of inquir
0486  tion. The rules for classifying will no doubt   BECOME  simpler when we have a definite object in vie
0488  to competition with foreign associates, might   BECOME  modified; so that we must not overrate the ac
0489  rian system was deposited, they seem to me to   BECOME  ennobled. Judging from the past, we may safel
0489  ny genera, have left no descendants, but have   BECOME  utterly extinct. We can so far take a prophet
0020  ty, or rather utter hopelessness, of the task   BECOMES  apparent. Certainly, a breed intermediate be
0038  nt part which selection by man has played, it   BECOMES  at once obvious, how it is that our domestic
0080  may be truly said that the whole organisation   BECOMES  in some degree plastic. Let it be borne in m
0101  ual species, as far as function is concerned,   BECOMES  very small. From these several consideration
0148  conditions of life a structure before useful   BECOMES  less useful, any diminution, however slight,
0155  of generic value, when it sinks in value and   BECOMES  only of specific value, often becomes variab
0155  lue and becomes only of specific value, often   BECOMES  variable, though its physiological importanc
0173  cies often meet and interlock; and as the one   BECOMES  rarer and rarer, the other becomes more and
0173  as the one becomes rarer and rarer, the other   BECOMES  more and more frequent, till the one replace
0209  n what originally was a habit and an instinct   BECOMES  so close as not to be distinguished. If Moza
0218  er sphex it takes advantage of the prize, and   BECOMES  for the occasion parasitic. In this case, as
0292  onfine our attention to any one formation, it   BECOMES  more difficult to understand, why we do not
0317  great tree from a single stem, till the group   BECOMES  large. On Extinction. We have as yet spoken
0381  il we can say why one species and not another   BECOMES  naturalised by man's agency in a foreign lan
0399  inent, is so enormously remote, that the fact   BECOMES  an anomaly. But this difficulty almost disap
0414  ed with special habits, the more important it   BECOMES  for classification. As an instance: Owen, in
0443  nditions of existence, except when the embryo   BECOMES  at any period of life active and has to prov
0027  ing quite unknown in a wild state, and their   BECOMING  nowhere feral: these species having very abn
0090  agency: for we see peculiarities arising and   BECOMING  attached to the male sex in our domestic ani
0108  ll oftener depend on some of the inhabitants   BECOMING  slowly modified: the mutual relations of man
0113  g trees, frequenting water, and some perhaps   BECOMING  less carnivorous. The more diversified in ha
0115  cted that the plants which have succeeded in   BECOMING  naturalised in any land would generally have
0140  ce, in the case of some few plants, of their   BECOMING, to a certain extent, naturally habituated t
0140  lly habituated to different temperatures, or   BECOMING  acclimatised: thus the pines and rhododendro
0166  ral very distinct species of the horse genus   BECOMING, by simple variation, striped on the legs li
0174  erably numerous over a large territory, then   BECOMING  somewhat abruptly rarer and rarer on the con
0178  rtant degree, on some of the old inhabitants   BECOMING  slowly modified, with the new forms thus pro
0197  he plant undergoing further modification and   BECOMING  a climber. The naked skin on the head of a v
0260  dowed with sterility simply to prevent their   BECOMING  confounded in nature? I think not. For why s
0276  ng grafted together in order to prevent them   BECOMING  inarched in our forests. The sterility of fi
0319  the conditions had gone on, however slowly,   BECOMING  less and less favourable, we assuredly shoul
0392  land by some other means: and the plant then   BECOMING  slightly modified, but still retaining its h
0173  where much sediment is deposited on the shallow  BED  of the sea, whilst it slowly subsides. These c
```

Page **(Key Word)**

bed

Page **(Key Word)**

```
0173 and after enormously long intervals. Whilst the BED of the sea is stationary or is rising, or when
0288 diment is being deposited over nearly the whole BED of the sea, at a rate sufficiently quick to be
0288 her and later formation, without the underlying BED having suffered in the interval any wear and t
0289 eposits; and that not a cave or true lacustrine BED is known belonging to the age of our secondary
0292 ime to decay. On the other hand, as long as the BED of the sea remained stationary, thick deposits
0297 s, and a third, A, to be found in an underlying BED; even if A were strictly intermediate between
0300 ch. Wherever sediment did not accumulate on the BED of the sea, or where it did not accumulate at
0308 ich sediment was not deposited, or again as the BED of an open and unfathomable sea. Looking to th
0309 ustified in assuming that if, for instance, the BED of the Pacific Ocean were now converted into a
0327 t duration occurred during the periods when the BED of the sea was either stationary or rising, an
0343 s which must have existed long before the first BED of the Silurian system was deposited: I can an
0388 a careful gardener, thus takes her seeds from a BED of a particular nature, and drops them in anot
0465 ere much sediment is deposited on the subsiding BED of the sea. During the alternate periods of el
0489 me few beings which lived long before the first BED of the Silurian system was deposited, they see
0032 the best plants, but merely go over their seed BEDS, and pull up the rogues, as they call the plan
0252 ous to nurserymen. Horticulturists raise large BEDS of the same hybrids, and such alone are fairly
0283 h the mind thus impressed, let any one examine BEDS of conglomerate many thousand feet in thicknes
0284 sult: Palaeozoic strata (not including igneous BEDS).....57,154 Feet. Secondary strata............
0284 ions, which are represented in England by thin BEDS, are thousands of feet in thickness on the Con
0285 te, showed at a glance how far the hard, rocky BEDS had once extended into the open ocean. The sam
0288 embedded, if in sand or gravel, will when the BEDS are upraised generally be dissolved by the per
0290 inhabited by a peculiar marine fauna, tertiary BEDS are so scantily developed, that no record of s
0292 f elevation; or, to speak more accurately, the BEDS which were then accumulated will have been des
0294 es appeared somewhat earlier in the palaeozoic BEDS of North America than in those of Europe; time
0294 f america during this space of time. When such BEDS as were deposited in shallow water near the mo
0294 he distant future, a geologist examining these BEDS, might be tempted to conclude that the average
0295 is so often the case, a formation composed of BEDS of different mineralogical composition, we may
0296 has consumed. Many instances could be given of BEDS only a few feet in thickness, representing for
0296 mation. Many cases could be given of the lower BEDS of a formation having been upraised, denuded,
0296 d, submerged, and then re covered by the upper BEDS of the same formation: facts, showing what wid
0296 , messrs. Lyell and Dawson found carboniferous BEDS 1400 feet thick in Nova Scotia, with ancient r
0297 modified descendants from the lower and upper BEDS of a formation, and unless we obtained numerou
0304 specimens are preserved in the oldest tertiary BEDS: from the ease with which even a fragment of a
0304 and as not one species had been discovered in BEDS of this age, I concluded that this great group
0307 ces of life have been detected in the longmynd BEDS beneath Barrande's so called primordial zone.
0307 an epoch, is very great. If these most ancient BEDS had been wholly worn away by denudation, or ob
0312 forms more gradual. In some of the most recent BEDS, though undoubtedly of high antiquity if measu
0313 or in the same degree. In the oldest tertiary BEDS a few living shells may still be found in the
0317 ickness, and ultimately thins out in the upper BEDS, marking the decrease and final extinction of
0323 distant points, the organic remains in certain BEDS present an unmistakeable degree of resemblance
0324 f this be so, it is evident that fossiliferous BEDS deposited at the present day on the shores of
0324 ble to be classed with somewhat older European BEDS. Nevertheless, looking to a remotely future æp
0324 pliocene, the pleistocene and strictly modern BEDS, of Europe, North and South America, and Austr
0338 bryological character of the Vertebrata, until BEDS far beneath the lowest Silurian strata are dis
0094 slight to be appreciated by us, might profit a BEE or other insect, so that an individual so char
0094 glance appear to differ in length; yet the hive BEE can easily suck the nectar out of the incarnat
0095 abundant supply of precious nectar to the hive BEE. Thus it might be a great advantage to the hiv
0095 thus it might be a great advantage to the hive BEE to have a slightly longer or differently const
0095 y divided tube to its corolla, so that the hive BEE could visit its flowers. Thus I can understand
0095 owers. Thus I can understand how a flower and a BEE might slowly become, either simultaneously or
0172 marvellous an instinct as that which leads the BEE to make cells, which have practically anticipa
0202 can we consider the sting of the wasp or of the BEE as perfect, which, when used against many atta
0202 out its viscera? If we look at the sting of the BEE, as having originally existed in a remote prog
0202 mire the savage instinctive hatred of the queen BEE, which urges her instantly to destroy the youn
0207 ch, and parasitic bees. Slave making ants. Hive BEE, its cell making instinct. Difficulties on the
0207 as so wonderful an instinct as that of the hive BEE making its cells could probably have occurred t
0209 ch we are acquainted, namely, those of the hive BEE and of many ants, could not possibly have been
0216 ain ants; and the comb making power of the hive BEE: these two latter instincts have generally, an
0224 ica rufescens. Cell making instinct of the Hive BEE. I will not here enter on minute details on th
0225 end of the series we have the cells of the hive BEE, placed in a double layer: each cell, as is we
0225 the extreme perfection of the cells of the hive BEE and the simplicity of those of the humble bee,
0225 e bee and the simplicity of those of the humble BEE, we have the cells of the Mexican Melipona dom
0225 ediate in structure between the hive and humble BEE, but more nearly related to the latter: for for
0226 e sided pyramidal basis of the cell of the hive BEE. As in the cells of the hive bee, so here, the
0226 ll of the hive bee. As in the cells of the hive BEE, so here, the three plane surfaces in any one
0226 ly have been as perfect as the comb of the hive BEE. Accordingly I wrote to Professor Miller, of C
0227 s which have been made of the cells of the hive BEE. Hence we may safely conclude that if we could
0227 ona, and in themselves not very wonderful, this BEE would make a structure as wonderfully perfect
0227 ture as wonderfully perfect as that of the hive BEE. We must suppose the Melipona to make her cell
0227 ck of honey: in the same way as the rude humble BEE adds cylinders of wax to the circular mouths o
0227 bird to make its nest, I believe that the hive BEE has acquired, through natural selection, her i
0231 that a multitude of bees all work together; one BEE after working a short time at one cell going t
0233 could have profited the progenitors of the hive BEE? I think the answer is not difficult: it is kn
0234 f bees. Of course the success of any species of BEE may be dependent on the number of its parasite
0234 y often does determine, the numbers of a humble BEE which could exist in a country: and let us fur
0234 ubt that it would be an advantage to our humble BEE, if a slight modification of her instinct led
0234 lly be more and more advantageous to our humble BEE, if she were to make her cells more and more r
0235 ould make a comb as perfect as that of the hive BEE. Beyond this stage of perfection in architectu
0235 ection could not lead: for the comb of the hive BEE, as far as we can see, is absolutely perfect i
0235 derful of all known instincts, that of the hive BEE, can be explained by natural selection having
0236 ld have been far better exemplified by the hive BEE. If a working ant or other neuter insect had b
0434 s of a sphinx moth, the curious folded one of a BEE or bug, and the great jaws of a beetle? yet al
0472 fitness. We need not marvel at the sting of the BEE causing the bee's own death; at drones being p
0472 not marvel at the sting of the bee causing the BEE's own death; at drones being produced in such
0472 r trees; at the instinctive hatred of the queen BEE for her own fertile daughters; at ichneumonida
0474 the admirable architectural powers of the hive BEE. Habit no doubt sometimes comes into play in m
```

Page **(Key Word)**

bees

Page ***(Key Word)***

```
0073  m. i have, also, reason to believe that humble  BEES  are indispensable to the fertilisation of the
0073  of the heartsease (Viola tricolor), for other    BEES  do not visit this flower. From experiments whi
0073  i have tried, I have found that the visits of    BEES, if not indispensable, are at least highly ben
0073  o the fertilisation of our clovers; but humble   BEES  alone visit the common red clover (Trifolium p
0073  mmon red clover (Trifolium pratense), as other   BEES  cannot reach the nectar. Hence I have very lit
0073  ittle doubt, that if the whole genus of humble   BEES  became extinct or very rare in England, the he
0074  are, or wholly disappear. The number of humble   BEES  in any district depends in a great degree on t
0074  who has long attended to the habits of humble    BEES, believes that more than two thirds of them ar
0074  d small towns I have found the nests of humble   BEES  more numerous than elsewhere, which I attribut
0074  ugh the intervention first of mice and then of   BEES, the frequency of certain flowers in that dist
0093  nd boisterous, and therefore not favourable to   BEES, nevertheless every female flower which I exam
0093  xamined had been effectually fertilised by the   BEES, accidentally dusted with pollen, having flown
0094  . i could give many facts, showing how anxious   BEES  are to save time; for instance, their habit of
0095  common red clover, which is visited by humble   BEES  alone; so that whole fields of the red clover
0095  hat the fertility of clover greatly depends on   BEES  visiting and moving parts of the corolla, so a
0095  the stigmatic surface. Hence, again, if humble   BEES  were to become rare in any country, it might h
0097  tructure of the flower and the manner in which   BEES  suck the nectar; for, in doing this, they eith
0097  another flower. So necessary are the visits of   BEES  to papilionaceous flowers, that I have found,
0097  e prevented. Now, it is scarcely possible that   BEES  should fly from flower to flower, and not carr
0097  to the great good, as I believe, of the plant.  BEES  will act like a camel hair pencil, and it is q
0098  ertilisation; but it must not be supposed that   BEES  would thus produce a multituce of hybrids betw
0098  lobelia growing close by, which is visited by   BEES, seeds freely. In very many other cases, thoug
0207  nstincts of the cuckoo, ostrich, and parasitic  BEES. Slave making ants. Hive bee, its cell making
0218  no less than twenty lost and wasted eggs. Many  BEES  are parasitic, and always lay their eggs in th
0218  tic, and always lay their eggs in the nests of  BEES  of other kinds. This case is more remarkable t
0218  remarkable than that of the cuckoo; for these   BEES  have not only their instincts but their struct
0224  c admiration. We hear from mathematicians that  BEES  have practically solved a recondite problem, a
0224  hough this is perfectly effected by a crowd of  BEES  working in a dark hive. Grant whatever instinc
0225  k. at one end of a short series we have humble  BEES, which use their old cocoons to hold honey, so
0225  en completed; but this is never permitted, the  BEES  building perfectly flat walls of wax between t
0228  en them a long thick, square strip of wax: the  BEES  instantly began to excavate minute circular pi
0228  resting to me to observe that wherever several  BEES  had begun to excavate these basins near togeth
0228  into each other. As soon as this occurred, the  BEES  ceased to excavate, and began to build up flat
0228  nife edged ridge, coloured with vermilion. The  BEES  instantly began on both sides to excavate litt
0228  n into each other from the opposite sides. The  BEES, however, did not suffer this to happen, and t
0229  of things, had not been neatly performed. The   BEES  must have worked at very nearly the same rate
0229  do not see that there is any difficulty in the  BEES, whilst at work on the two sides of a strip of
0229  ordinary combs it has appeared to me that the   BEES  do not always succeed in working at exactly th
0229  concave on one side, where I suppose that the   BEES  had excavated too quickly, and convex on the o
0229  kly, and convex on the opposite side, where the BEES  had worked less quickly. In one well marked in
0229  t the comb back into the hive, and allowed the  BEES  to go on working for a short time, and again e
0230  g away the convex side; and I suspect that the  BEES  in such cases stand in the opposed cells and p
0230  vermilion wax, we can clearly see that if the   BEES  were to build for themselves a thin wall of wa
0230  wing the spheres to break into each other. Now  BEES, as may be clearly seen by examining the edge
0230  it would be a great error to suppose that the   BEES  cannot build up a rough wall of wax in the pro
0231  ed away on both sides. The manner in which the  BEES  build is curious; they always make the first r
0231  ng thus crowned by a strong coping of wax, the  BEES  can cluster and crawl over the comb without in
0231  ng how the cells are made, that a multitude of  BEES  all work together; one bee after working a sho
0232  the colour was most delicately diffused by the  BEES, as delicately as a painter could have done wi
0232  ms to be a sort of balance struck between many  BEES, all instinctively standing at the same relati
0232  pieces of comb met at an angle, how often the   BEES  would entirely pull down and rebuild in differ
0232  a shape which they had at first rejected. When  BEES  have a place on which they can stand in their
0232  lt over one face of the slip, in this case the  BEES  can lay the foundations of one wall of a new h
0232  he other completed cells. It suffices that the  BEES  should be enabled to stand at their proper rel
0232  joining cells has been built. This capacity in  BEES  of laying down under certain circumstances a r
0233  the answer is not difficult: it is known that   BEES  are often hard pressed to get sufficient necta
0233  pounds of dry sugar are consumed by a hive of   BEES  for the secretion of each pound of wax; so tha
0233  d nectar must be collected and consumed by the  BEES  in a hive for the secretion of the wax necessa
0234  he construction of their combs. Moreover, many  BEES  have to remain idle for many days during the p
0234  y is indispensable to support a large stock of  BEES  during the winter; and the security of the hiv
0234  is known mainly to depend on a large number of  BEES  being supported. Hence the saving of wax by la
0234  important element of success in any family of   BEES. Of course the success of any species of bee m
0234  independent of the quantity of honey which the  BEES  could collect. But let us suppose that this la
0235  slow degrees, more and more perfectly, led the  BEES  to sweep equal spheres at a given distance fro
0235  the wax along the planes of intersection. The   BEES, of course, no more knowing that they swept th
0362  oat, wheat, millet, canary, hemp, clover, and   BEET  germinated after having been from twelve to tw
0362  s of different birds of prey; and two seeds of  BEET  grew after having been thus retained for two d
0061  r feathers of a bird; in the structure of the   BEETLE  which dives through the water; in the plumed
0077  n the flattened and fringed legs of the water   BEETLE,  the relation seems at first confined to the
0077  d and fall on unoccupied ground. In the water   BEETLE,  the structure of its legs, so well adapted f
0085  smooth skinned fruits suffer far more from a    BEETLE,  a curculio, than those with down: that purpl
0135  imentary condition. In the Ateuchus or sacred   BEETLE  of the Egyptians, they are totally deficient.
0136  nds of successive generations each individual   BEETLE  which flew least, either from its wings havin
0386  a limpet) firmly adhering to it; and a water    BEETLE  of the same family, a Colymbetes, once flew o
0434  one of a bee or bug, and the great jaws of a    BEETLE? yet all these organs, serving for such diffe
0088  ve been seen fighting all day long; male stag   BEETLES  often bear wounds from the huge mandibles of
0135  ior tarsi, or feet, of many male dung feeding   BEETLES  are very often broken off; he examined seven
0135  i are almost always lost in many dung feeding   BEETLES,  they must be lost early in life, and theref
0135  n has discovered the remarkable fact that 200   BEETLES,  out of the 550 species inhabiting Madeira,
0135  n this condition! Several facts, namely, that   BEETLES  in many parts of the world are very frequent
0135  frequently blown to sea and perish; that the   BEETLES  in Madeira, as observed by Mr. Wollaston, li
0136  e sun shines; that the proportion of wingless   BEETLES  is larger on the exposed Dezertas than in Ma
0136  ost entire absence of certain large groups of   BEETLES,  elsewhere excessively numerous, and which q
0136  hat the wingless condition of so many Madeira   BEETLES  is mainly due to the action of natural selec
0136  own out to sea; and, on the other hand, those  BEETLES  which most readily took to flight will often
0157  cter generally common to very large groups of   BEETLES, but in the Engidae, as Westwood has remarke
```

Page ***(Key Word)***

65   beetles

Page **(Key Word)**

```
0392  nes under the soldered elytra of many insular  BEETLES. Again, islands often possess trees or bushe
0404  e compare the distribution of butterflies and  BEETLES. So it is with most fresh water productions,
0439  eptile. The vermiform larvae of moths, flies,  BEETLES, etc., resemble each other much more closely
0451  n quite unmistakeable: for instance there are  BEETLES of the same genus (and even of the same spec
0454   injurious under others, as with the wings of  BEETLES living on small and exposed islands: and in
0480  wings under the soldered wing covers of some  BEETLES, should thus so frequently bear the plain st
0018  ars ago, and who will pretend to say how long  BEFORE these ancient periods, savages, like those of
0028   the breeds, which method was never practised  BEFORE, has improved them astonishingly. About this
0044   each other, and in having restricted ranges.  BEFORE applying the principles arrived at in the las
0049  they are rightly called species or varieties,  BEFORE any definition of these terms has been genera
0051  ssed, just like the pigeon or poultry fancier  BEFORE alluded to, with the amount of difference in
0060  organism the most important of all relations.  BEFORE entering on the subject of this chapter, I mu
0076  small Asiatic cockroach has everywhere driven  BEFORE it its great congener. One species of charloc
0089  r gorgeous plumage and perform strange antics  BEFORE the females, which standing by as spectators,
0098   out of the conjoined anthers of each flower,  BEFORE the stigma of that individual flower is ready
0099  nd as I can confirm, either the anthers burst  BEFORE the stigma is ready for fertilisation, or the
0099  ady for fertilisation, or the stigma is ready  BEFORE the pollen of that flower is ready, so that t
0106  rly yielded, and apparently are now yielding,  BEFORE those of the larger Europaeo Asiatic area. Th
0124  turalist will be able to bring some such case  BEFORE his mind. In the diagram, each horizontal lin
0132  nd female sexual elements seem to be affected  BEFORE that union takes place which is to form a new
0148  uncer changed conditions of life a structure  BEFORE useful becomes less useful, any diminution, h
0163  ected a long list of such cases: but here, as  BEFORE, I lie under a great disadvantage in not bein
0171  aced by the theory of Natural Selection. Long  BEFORE having arrived at this part of my work, a cro
0195  ct as driving away flies: yet we should pause  BEFORE being too positive even in this case, for we
0201  th another: but they are now rapidly yielding  BEFORE the advancing legions of plants and animals i
0228  basins near to each other, in the same way as  BEFORE: but the ridge of wax was so thin, that the b
0241  orkers of the several sizes. With these facts  BEFORE me, I believe that natural selection, by acti
0264  brids, however, are differently circumstanced  BEFORE and after birth: when born and living in a co
0264  and constitution of its mother, and therefore  BEFORE birth, as long as it is nourished within its
0282  ing down old rocks and making fresh sediment,  BEFORE he can hope to comprehend anything of the lap
0292  shallow and to embed and preserve the remains  BEFORE they had time to decay. On the other hand, as
0293   so again when we find a species disappearing  BEFORE the uppermost layers have been deposited, it
0295  en nearly far greater, that is extending from  BEFORE the glacial epoch to the present day. In orde
0297  ate varieties. Nor should it be forgotten, as  BEFORE explained, that A might be the actual progeni
0302  and the progenitors must have lived long ages  BEFORE their modified descendants. But we continuall
0302  eath a certain stage, that they did not exist  BEFORE that stage. We continually forget how large t
0302  have long existed and have slowly multiplied  BEFORE they invaded the ancient archipelagoes of Eur
0304  e of whales in the upper greensand, some time  BEFORE the close of the secondary period. I may give
0306  me one crustacean, which must have lived long  BEFORE the Silurian age, and which probably differed
0307  if my theory be true, it is indisputable that  BEFORE the lowest Silurian stratum was deposited, lo
0307  my theory no doubt were somewhere accumulated  BEFORE the Silurian epoch, is very great. If these m
0318  e to the present day: some having disappeared  BEFORE the close of the palaeozoic period. No fixed
0318  ance of a group of species be represented, as  BEFORE, by a vertical line of varying thickness, the
0320  t to admit that species generally become rare  BEFORE they become extinct, to feel no surprise at t
0326  f time, would probably be also favourable, as  BEFORE explained. One quarter of the world may have
0332  xtinct and recent, descended from A, make, as  BEFORE remarked, one order: and this order, from the
0332   less distinct from each other than they were  BEFORE the discovery of the fossils. If, for instanc
0343  merous organisms which must have existed long  BEFORE the first bed of the Silurian system was depo
0343   ever since the Silurian epoch: but that long  BEFORE that period, the world may have presented a w
0356  ing many individuals during many generations.  BEFORE discussing the three classes of facts, which
0365  an island was being upheaved and formed, and  BEFORE it had become fully stocked with inhabitants.
0369  ies, which must have existed on the mountains  BEFORE the commencement of the Glacial epoch, and wh
0370  elieve that during the newer Pliocene period,  BEFORE the Glacial epoch, and whilst the majority of
0370  to the Glacial epoch. Believing, from reasons  BEFORE alluded to, that our continents have long rem
0370  thwards as the climate became less warm, long  BEFORE the commencement of the Glacial period. We no
0380  rrec on the intertropical mountains: no doubt  BEFORE the Glacial period they were stocked with end
0380   on the land; and the intertropical mountains  BEFORE the Glacial period must have been completely
0381  rn hemisphere, to a former and warmer period,  BEFORE the commencement of the Glacial period, when
0381  y peculiar and isolated flora. I suspect that  BEFORE this flora was exterminated by the Glacial ep
0388  means of dispersal explain this fact. I have  BEFORE mentioned that earth occasionally, though rar
0399  ands, when they were clothed with vegetation,  BEFORE the commencement of the Glacial period. The a
0406  roductions being related (with the exceptions  BEFORE specified) to those on the surrounding low la
0407  ent source: if we make the same allowances as  BEFORE for our ignorance, and remember that some for
0442  the cephalopodic character is manifested long  BEFORE the parts of the embryo are completed: and aq
0443  cted, and I believe generally has acted, even  BEFORE the embryo is formed; and the variation may b
0443  fect thus caused at a very early period, even  BEFORE the formation of the embryo, may appear late
0488  scendants of some few beings which lived long  BEFORE the first bed of the Silurian system was depo
0489   lineal descendants of those which lived long  BEFORE the Silurian epoch, we may feel certain that
0090  w, as I believe, natural selection acts, I must BEG permission to give one or two imaginary illust
0041  that it was most fortunate that the strawberry  BEGAN to vary just when gardeners began to attend c
0041  e strawberry began to vary just when gardeners  BEGAN to attend closely to this plant. No doubt the
0073  s in ever increasing circles of complexity. We  BEGAN this series by insectivorous birds, and we ha
0211  e what a rich flock it had discovered: it then  BEGAN to play with its antennae on the abdomen firs
0228  thick, square strip of wax: the bees instantly  BEGAN to excavate minute circular pits in it: and a
0228  his occurred, the bees ceased to excavate, and  BEGAN to build up flat walls of wax on the lines of
0228  e, coloured with vermilion. The bees instantly  BEGAN on both sides to excavate little basins near
0370  s and animals, both in the Old and New Worlds,  BEGAN slowly to migrate southwards as the climate b
0231  o pile up a broad ridge of cement, and then to  BEGAN cutting it away equally on both sides near th
0317  und, the line will sometimes falsely appear to  BEGIN at its lower end, not in a sharp point, but a
0490  o the fixed law of gravity, from so simple a  BEGINNING endless forms most beautiful and most wonde
0083  s in his power, all his productions. He often  BEGINS his selection by some half monstrous form! or
0248  ay where perfect fertility ends and sterility  BEGINS. I think no better evidence of this can be re
0007  ation: and that when the organisation has once  BEGUN to vary, it generally continues to vary for m
0185  rane between the toes shows that structure has  BEGUN to change. He who believes in separate and in
0228  o me to observe that wherever several bees had  BEGUN to excavate these basins near together, they
0228   excavate these basins near together, they had  BEGUN their work at such a distance from each other
```

Page **(Key Word)**

begun

Page **************************************(Key Word)**************************************

```
0233  tive distance from the parts of the cells just BEGUN, sweeping spheres or cylinders, and building
0482  much flexibility of mind, and who have already BEGUN to doubt on the *mmutability of species, may
0220  very few in number, I thought that they might BEHAVE differently when more numerous; but Mr. Smith
0211  ured by the ant. Even the quite young aphides BEHAVED in this manner, showing that the action was
0341  rium and other allied huge monsters have left BEHIND them in South America the sloth, armadillo, a
0361  frequently enclosed in their interstices and BEHIND them, so perfectly that not a particle could
0062  will be dimly seen or quite misunderstood. We BEHOLD the face of nature bright with gladness, we o
0128  e to group, in the manner which we everywhere BEHOLD, namely, varieties of the same species most c
0221  r, and it was a most interesting spectacle to BEHOLD the masters carefully carrying, as Huber has
0287  logical museums, and what a paltry display we BEHOLD! On the poorness of our Palaeontological coll
0005  ng struggle for existence, it follows that any BEING, if it vary however slightly in any manner pr
0005  transitions, or in understanding how a simple BEING or a simple organ can be changed and perfecte
0005  changed and perfected into a highly developed BEING or elaborately constructed organ; secondly, t
0008  enerations. No case is on record of a variable BEING ceasing to be variable under cultivation. Our
0011  f the muscles of the ear, from the animals not BEING much alarmed by danger, seems probable. There
0016  ters of generic value; all such valuations as BEING at present empirical. Moreover, on the view o
0020  geons, and some of them are very important, as BEING of considerable antiquity. I have associated
0026  brid offspring of two animals clearly distinct BEING themselves perfectly fertile. Some authors be
0027  ly under domestication; these supposed species BEING quite unknown in a wild state, and their beco
0027  t pure and when crossed; the mongrel offspring BEING perfectly fertile; from these several reasons
0029  eride the idea of species in a state of nature BEING lineal descendants of other species? Selectio
0036  n is so great that they have the appearance of BEING quite different varieties. If there exist sav
0038  domestic races, and likewise their differences BEING so great in external characters and relativel
0040  hey will hardly have a distinct name, and from BEING only slightly valued, their history will be d
0041  uch increased by a large number of individuals BEING kept; and hence this comes to be of the highe
0042  y in pairing them; in donkeys, from only a few BEING kept by poor people, and little attention pai
0042  paid to their breeding; in peacocks, from not BEING very easily reared and a large stock not kept
0042  red and a large stock not kept; in geese, from BEING valuable only for two purposes, food and feat
0044  ies of the larger genera resemble varieties in BEING very closely, but unequally, related to each
0045  or which may be presumed to have thus arisen, BEING frequently observed in the individuals of the
0050  rdly wish for better evidence of the two forms BEING specifically distinct. On the other hand, the
0051  systematist, as of high importance for us, as BEING the first step towards such slight varieties
0054  o equal masses, all those in the larger genera BEING placed on one side, and all those in the smal
0060  onditions of life, and of one distinct organic BEING to another being, been perfected? We see thes
0060  , and of one distinct organic being to another BEING, been perfected? We see these beautiful co ad
0062  etaphorical sense, including dependence of one BEING on another, and including (which is more impo
0063  ich all organic beings tend to increase. Every BEING, which during its natural lifetime produces s
0064  is no exception to the rule that every organic BEING naturally increases at so high a rate, that i
0066  ind, never to forget that every single organic BEING around us may be said to be striving to the u
0067  nwards by incessant blows, sometimes one wedge BEING struck, and then another with greater force.
0068  ) nine species perished from the other species BEING allowed to grow up freely. The amount of food
0069  inally disappearing; and the change of climate BEING conspicuous, we are tempted to attribute the
0069  the cause lies quite as much in other species BEING favoured, as in this one being hurt. So it is
0069  n other species being favoured, as in this one BEING hurt. So it is when we travel northward, but
0070  ts in nature, such as that of very rare plants BEING sometimes extremely abundant in the few spots
0070  they do occur; and that of some social plants BEING social, that is, abounding in individuals, ev
0073  le would often give the victory to one organic BEING over another. Nevertheless so profound is our
0073  l when we hear of the extinction of an organic BEING; and as we do not see the cause, we invoke ca
0074  bly come into play; some one check or some few BEING generally the most potent, but all concurring
0075  ld Indian ruins! The dependency of one organic BEING on another, as of a parasite on its prey, lie
0077  s, namely, that the structure of every organic BEING is related, in the most essential yet often h
0077  ubt stands in the closest relation to the land BEING already thickly clothed by other plants; so t
0078  is to keep steadily in mind that each organic BEING is striving to increase at a geometrical rati
0080  at other variations useful in some way to each BEING in the great and complex battle of life, shou
0083  except in so far as they may be useful to any BEING. She can act on every internal organ, on ever
0083  for his own good: Nature only for that of the BEING which she tends. Every selected character is
0083  d character is fully exercised by her; and the BEING is placed under well suited conditions of lif
0084  ity offers, at the improvement of each organic BEING in relation to its organic and inorganic cond
0084  can act only through and for the good of each BEING, yet characters and structures, which we are
0085  rsons are warned not to keep white pigeons, as BEING the most liable to destruction. Hence I can s
0086  lated by natural selection for the good of the BEING, will cause other modifications, often of the
0086  wind, I can see no greater difficulty in this BEING effected through natural selection, than in t
0087  might be selected, the thickness of the shell BEING known to vary like every other structure. Sex
0089  eeding season; the modifications thus produced BEING inherited at corresponding ages or seasons, e
0095  odifications, each profitable to the preserved BEING; and as modern geology has almost banished su
0097  be of the meaning of the law) that no organic BEING self fertilises itself for an eternity of gen
0098  lly cross. In many other cases, far from there BEING any aids for self fertilisation, there are sp
0099  an occasional cross with a distinct individual BEING advantageous or indispensable! If several var
0099  d that this is part of the general law of good BEING derived from the intercrossing of distinct in
0100  ; and this will give a better chance of pollen BEING occasionally carried from tree to tree. That
0100  ial plants, on the view of an occasional cross BEING indispensable, by considering the medium in w
0105  r, an isolated area be very small, either from BEING surrounded by barriers, or from having very p
0108  ss, I fully admit. Its action depends on there BEING places in the polity of nature, which can be
0108  ual relations of many of the other inhabitants BEING thus disturbed. Nothing can be effected, unle
0109  n geology; but it must be here alluded to from BEING intimately connected with natural selection.
0109  si for as new forms are continually and slowly BEING produced, unless we believe that the number o
0112  inferior animals with intermediate characters, BEING neither very swift nor very strong, will have
0113  by other animals: some of them, for instance, BEING enabled to feed on new kinds of prey, either
0116  ards. No physiologist doubts that a stomach by BEING adapted to digest vegetable matter alone, or
0116  let us see how this principle of great benefit BEING derived from divergence of character, combine
0117  ere the importance of the principle of benefit BEING derived from divergence of character comes in
0117  ations (represented by the outer dotted lines) BEING preserved and accumulated by natural selectio
0118  parents varied. Moreover, these two varieties, BEING only slightly modified forms, will tend to in
0122  s so generally the case in nature; species (A) BEING more nearly related to B, C, and D, than to t
0127  no variation ever had occurred useful to each BEING's own welfare, in the same way as so many var
0127  o man. But if variations useful to any organic BEING do occur, assuredly individuals thus characte
```

Page **************************************(Key Word)**************************************

```
0127 hus characterised will have the best chance of BEING preserved in the struggle for life; and from
0127 tural selection, on the principle of qualities BEING inherited at corresponding ages, can modify t
0132 in the conditions of life; and to this system BEING functionally disturbed in the parents, I chie
0132 that union takes place which is to form a new BEING. In the case of sporting plants, the bud, whi
0132 erence of climate, food, etc., produces on any BEING is extremely doubtful. My impression is, that
0132 o adaptations of structure between one organic BEING and another, which we see everywhere througho
0133 when a variation is of the slightest use to a BEING, we cannot tell how much of it to attribute t
0133 . instances could be given of the same variety BEING produced under conditions of life as differen
0133 and, on the other hand, of different varieties BEING produced from the same species under the same
0136 have had the best chance of surviving from not BEING blown out to sea; and, on the other hand, tho
0140 ithstanding the most different climates but of BEING perfectly fertile (a far severer test) under
0144 y, and that others rarely coexist, without our BEING able to assign any reason. What can be more s
0146 ces are of such apparent importance, the seeds BEING in some cases, according to Tausch, orthosper
0146 unknown laws of correlated growth, and without BEING, as far as we can see, of the slightest servi
0147 etween the effects, on the one hand, of a part BEING largely developed through natural selection, a
0147 tural selection and another and adjoining part BEING reduced by this same process or by disuse, an
0148 chance of supporting itself, by less nutriment BEING wasted in developing a structure now become u
0149 ver he forgotten, can act on each part of each BEING, solely through and for its advantage. Rudime
0151 cter, it applies more rarely to them. The rule BEING so plainly applicable in the case of secondar
0151 the homologous valves in the different species BEING sometimes wholly unlike in shape; and the amo
0152 carriage and tail of our fantails, etc., these BEING the points now mainly attended to by English
0155 ation can be given. But on the view of species BEING only strongly marked and fixed varieties, we
0158 all principles closely connected together. All BEING mainly due to the species of the same group h
0158 o parts which have recently and largely varied BEING more likely still to go on varying than parts
0158 nd to further variability, to sexual selection BEING less rigid than ordinary selection, and to va
0161 ability in a tendency to produce any character BEING inherited for an endless number of generation
0161 s, than in quite useless or rudimentary organs BEING, as we all know them to be, thus inherited. I
0161 characters another species; this other species BEING on my view only a well marked and permanent v
0163 efore, I lie under a great disadvantage in not BEING able to give them. I can only repeat that suc
0168 higher in the scale. Rudimentary organs, from BEING useless, will be disregarded by natural selec
0168 ks we have referred to special parts or organs BEING still variable, because they have recently va
0171 o this day I can never reflect on them without BEING staggered; but, to the best of my judgment, t
0171 all nature in confusion instead of the species BEING, as we see them, well defined? Secondly, is i
0172 how can we account for species, when crossed, BEING sterile and producing sterile offspring, wher
0172 i believe the answer mainly lies in the record BEING incomparably less perfect than is generally s
0172 rally supposed: the imperfection of the record BEING chiefly due to organic beings not inhabiting
0173 ofound depths of the sea, and to their remains BEING embedded and preserved to a future age only i
0173 or is rising, or when very little sediment is BEING deposited; there will be blanks in our geolog
0175 s prey for others; in short, that each organic BEING is either directly or indirectly related in t
0176 , as already remarked, run a greater chance of BEING exterminated than one existing in large numbe
0181 and fuller flank membranes, each modification BEING useful, each being propagated, until by the a
0181 embranes, each modification being useful, each BEING propagated, until by the accumulated effects
0182 ight exclusively, as far as we know, to escape BEING devoured by other fish? When we see any struc
0184 ry, I can see no difficulty in a race of bears BEING rendered, by natural selection, more and more
0185 s wings under water. He who believes that each BEING has been created as we now see it, must occas
0186 se cases it has pleased the Creator to cause a BEING of one type to take the place of one of anoth
0186 selection, will acknowledge that every organic BEING is constantly endeavouring to increase in num
0186 ng to increase in numbers; and that if any one BEING vary ever so little, either in habits or stru
0186 e to one very imperfect and simple, each grade BEING useful to its possessor, can be shown to exis
0190 aving a ductus pneumaticus for its supply, and BEING divided by highly vascular partitions. In the
0190 ected so as to perform all the work by itself, BEING aided during the process of modification by t
0192 existed as organs for preventing the ova from BEING washed out of the sack? Although we must be e
0194 atural selection, working for the good of each BEING and taking advantage of analogous variations,
0195 regard to the whole economy of any one organic BEING, to say what slight modifications would be of
0195 rom determining the attacks of insects or from BEING correlated with constitutional differences, m
0195 driving away flies; yet we should pause before BEING too positive even in this case, for we know t
0199 at the chief part of the organisation of every BEING is simply due to inheritance; and consequentl
0199 to inheritance; and consequently, though each BEING assuredly is well fitted for its place in nat
0200 n of special use to some ancestral form, or as BEING now of special use to the descendants of this
0201 ses. Natural selection will never produce in a BEING anything injurious to itself, for natural sel
0201 , it will be modified; or if it be not so, the BEING will become extinct, as myriads have become e
0201 ural selection tends only to make each organic BEING as perfect as, or slightly more perfect than,
0206 understand, bearing in mind that each organic BEING is trying to live wherever it can live, how i
0206 either now adapting the varying parts of each BEING to its organic and inorganic conditions of li
0206 ing long past periods of time: the adaptations BEING aided in some cases by use and disuse, being
0206 s being aided in some cases by use and disuse, BEING slightly affected by the direct action of the
0206 action of the external conditions of life, and BEING in all cases subjected to the several laws of
0210 europe and North America, and for no instinct BEING known amongst extinct species, how very gener
0213 ger delight by each breed, and without the end BEING known, for the young pointer can no more know
0221 untries, probably depends merely on the slaves BEING captured in greater numbers in Switzerland th
0226 ith three other cells, which, from the spheres BEING nearly of the same size, is very frequently a
0231 f wax has in every case to be finished off, by BEING largely gnawed away on both sides. The manner
0231 both those just commenced and those completed, BEING thus crowned by a strong coping of wax, the b
0231 n thickness; the plates of the pyramidal basis BEING about twice as thick. By this singular manner
0233 ould any good be derived from a single hexagon BEING built, as in its construction more materials
0234 own mainly to depend on a large number of bees BEING supported. Hence the saving of wax by largely
0236 h the males and fertile females, and yet, from BEING sterile, they cannot propagate their kind. Th
0236 on, I can see no very great difficulty in this BEING effected by natural selection. But I must pas
0236 tructure, as in the shape of the thorax and in BEING destitute of wings and sometimes of eyes, and
0237 cht profitable modification of structure, this BEING inherited by its offspring, which again varie
0238 to each other, but are perfectly well defined; BEING as distinct from each other, as are any two s
0241 e driver ant, and then the extreme forms, from BEING the most useful to the community, having been
0242 ould be effected with them only by the workers BEING sterile; for had they been fertile, they woul
0246 red over, owing to the sterility in both cases BEING looked on as a special endowment, beyond the
0246 t devoted their lives to this subject, without BEING deeply impressed with the high generality of
0247 t, must be secluded in order to prevent pollen BEING brought to it by insects from other plants. N
```

0254 ndelible characteristic, but as one capable of BEING removed by domestication. Finally, looking to
0258 an the case, for instance, of a stallion horse BEING first crossed with a female ass, and then a m
0259 tions in excess. That their fertility, besides BEING eminently susceptible to favourable and unfav
0260 the species which are crossed. The differences BEING of so peculiar and limited a nature, that, in
0261 e fully by an example what I mean by sterility BEING incidental on other differences, and not a sp
0261 great diversity in the size of two plants, one BEING woody and the other herbaceous, one being eve
0261 one being woody and the other herbaceous, one BEING evergreen and the other deciduous, and adapta
0262 fferent individuals of the same two species in BEING grafted together. As in reciprocal crosses, t
0271 ct domestic varieties, and from not wishing or BEING able to produce recondite and functional diff
0273 ely, that it is due to the reproductive system BEING eminently sensitive to any change in the cond
0273 itive to any change in the conditions of life, BEING thus often rendered either impotent or at lea
0276 nd somewhat analogous degrees of difficulty in BEING grafted together in order to prevent them bec
0277 stem and their whole organisation disturbed by BEING compounded of two distinct species, seems clo
0277 y of the hybrids produced, and the capacity of BEING grafted together, though this latter capacity
0279 distinctness of specific forms, and their not BEING blended together by innumerable transitional
0283 icker rate than many other deposits; yet, from BEING formed of worn and rounded pebbles, each of w
0288 we tacitly admit to ourselves that sediment is BEING deposited over nearly the whole bed of the se
0289 foregoing; namely, from the several formations BEING separated from each other by wide intervals o
0292 e then accumulated will have been destroyed by BEING upraised and brought within the limits of the
0298 s again would greatly lessen the chance of our BEING able to trace the stages of transition in any
0298 erhaps, best perceive the improbability of our BEING enabled to connect species by numerous, fine,
0311 n which the history is supposed to be written, BEING more or less different in the interrupted suc
0314 many complex contingencies, on the variability BEING of a beneficial nature, on the power of inter
0316 es are certainly exceptional; the general rule BEING a gradual increase in number, till the group
0317 no rise first to two or three varieties, these BEING slowly converted into species, which in their
0319 arer and rarer, and finally extinct; its place BEING seized on by some more successful competitor.
0319 to remember that the increase of every living BEING is constantly being checked by unperceived in
0319 e increase of every living being is constantly BEING checked by unperceived injurious agencies; an
0321 he sufferers may often long be preserved, from BEING fitted to some peculiar line of life, or from
0326 e process of diffusion may often be very slow, BEING dependent on climatal and geographical change
0327 ely and varying; the new species thus produced BEING themselves dominant owing to inheritance, and
0330 ted to any extinct species or group of species BEING considered as intermediate between living spe
0331 too simple, too few species and too few species BEING given, but this is unimportant for us. The ho
0331 pends solely on the descendants from a species BEING thus enabled to seize on many and different p
0331 ome Silurian forms, that a species might go on BEING slightly modified in relation to its slightly
0336 formations, though ranked as distinct species, BEING closely related, is obvious. As the accumulat
0338 variations supervening at a not early age, and BEING inherited at a corresponding age. This proces
0342 a single formation; that, owing to subsidence BEING necessary for the accumulation of fossilifero
0348 other, under corresponding climates; but from BEING separated from each other by impassable barri
0350 for life; the relation of organism to organism BEING, as I have already often remarked, the most i
0359 as in the foregoing experiment) floated, after BEING dried, for above 28 days, as far as we may in
0361 from the roots of drifted trees, these stones BEING a valuable royal tax. I find on examination,
0361 ds, when floating on the sea, sometimes escape BEING immediately devoured; and seeds of many kinds
0365 lapse of geological time, whilst an island was BEING upheaved and formed, and before it had become
0376 the southern hemisphere are of doubtful value, BEING ranked by some naturalists as specifically di
0378 s would naturally ascend the higher mountains, BEING exterminated on the lowlands; those which had
0379 untains, and in the southern hemisphere. These BEING surrounded by strangers will have had to comp
0397 nity to those of the nearest mainland, without BEING actually the same species. Numerous instances
0402 d keep separate for almost any length of time. BEING familiar with the fact that many species, nat
0405 sing barriers, but the more important power of BEING victorious in distant lands in the struggle f
0406 ther with the seeds and eggs of many low forms BEING very minute and better fitted for distant tra
0406 , as alpine, lacustrine, and marsh productions BEING related (with the exceptions before specified
0407 e change most slowly, enormous periods of time BEING thus granted for their migration, I do not th
0409 overning the succession of forms in past times BEING nearly the same with those governing at the p
0411 ariations not supervening at an early age, and BEING inherited at a corresponding age. Rudimentary
0414 habits of life, and the general place of each BEING in the economy of nature, would be of very hi
0414 ntimately connected with the whole life of the BEING, are ranked as merely adaptive or analogical
0414 ing of the dugong, says, The generative organs BEING those which are most remotely related to the
0415 r important they may be for the welfare of the BEING in relation to the outer world. Perhaps from
0415 aturalist can have worked at any group without BEING struck with this fact; and it has been most f
0417 f trifling characters, mainly depends on their BEING correlated with several other characters of m
0419 in common; yet the species at both ends, from BEING plainly allied to others, and these to others
0420 o their common progenitor, may differ greatly, BEING due to the different degrees of modification
0420 undergone; and this is expressed by the forms BEING ranked under different genera, families, sect
0427 ery distinct animals. On my view of characters BEING of real importance for classification, only i
0427 of the utmost importance to the welfare of the BEING, are almost valueless to the systematist. For
0428 alogical when whales are compared with fishes, BEING adaptations in both classes for swimming thro
0433 ral System. On this idea of the natural system BEING, in so far as it has been perfected, genealog
0435 inary view of the independent creation of each BEING, we can only say that so it is; that it has s
0435 essive slight modifications, each modification BEING profitable in some way to the modified form,
0436 ndibles, and two pair of maxillae, these parts BEING perhaps very simple in form; and then natural
0436 irect evidence of the possibility of one organ BEING transformed into another; and we can actually
0441 fine clearly what is meant by the organisation BEING higher or lower. But no one probably will dis
0442 diverse and active habits, or quite inactive, BEING fed by their parents or placed in the midst o
0442 me very unlike and serve for diverse purposes, BEING at this early period of growth alike; of embr
0442 each other; of the structure of the embryo not BEING closely related to its conditions of existenc
0445 s of various breeds, within twelve hours after BEING hatched; I carefully measured the proportions
0447 fication supervening at a rather late age, and BEING inherited at a corresponding age, the fo
0449 le arrangement, would be genealogical. Descent BEING on my view the hidden bond of connexion which
0450 ch, though perhaps caused at the earliest, and BEING inherited at a corresponding not early period
0455 an organ, and this will generally be when the BEING has come to maturity and to its full powers o
0457 upervening at a very early period of life, and BEING inherited at a corresponding period, we can u
0457 f life, through the principle of modifications BEING inherited at corresponding ages. On this same
0457 l generally be at that period of life when the BEING has to provide for its own wants, and bearing
0460 aying that any organ, or instinct, or any whole BEING, could not have arrived at its present state
0461 of life, we need not feel surprise at hybrids BEING in some degree sterile, for their constitutio

Page **(Key Word)**

```
0461 ns can hardly fail to have been disturbed from  BEING compounded of two distinct organisations. Thi
0461 ghtly modified forms or varieties acquire from  BEING crossed increased vigour and fertility. So th
0465 remains from consecutive formations invariably  BEING much more closely related to each other, than
0466 ms to be mainly due to the reproductive system  BEING eminently susceptible to changes in the condi
0468 cale of nature. The slightest advantage in one  BEING, at any age or during any season, over those
0472 pecially created and adapted for that country,  BEING beaten and supplanted by the naturalised prod
0472 the bee causing the bee's own death; at drones  BEING produced in such vast numbers for one single
0472 d in such vast numbers for one single act, and  BEING then slaughtered by their sterile sisters; at
0475 selection we need not marvel at some instincts  BEING apparently not perfect and liable to mistakes
0475 and kinds of resemblance to their parents, in  BEING absorbed into each other by successive crosse
0475 e fact of the fossil remains of each formation  BEING in some degree intermediate in character betw
0476 roups, follows from the living and the extinct  BEING the offspring of common parents. As the group
0476 roups. Recent forms are generally looked at as  BEING, in some vague sense, higher than ancient and
0477 we need feel no surprise at their inhabitants  BEING widely different, if they have been for a lon
0478 n fernandez, and of the other American islands  BEING related in the most striking manner to the pl
0478 racters, though of paramount importance to the  BEING, are of hardly any importance in classificati
0479 rudimentary parts, though of no service to the  BEING, are often of high classificatory value; and
0479 ital importance may be. The framework of bones  BEING the same in the hand of a man, wing of bat, f
0479 ns not always supervening at an early age, and  BEING inherited at a corresponding not early period
0480 o the present day. On the view of each organic  BEING and each separate organ having been specially
0485 fication. When we no longer look at an organic  BEING as a savage looks at a ship, as at something
0486 merous workmen; when we thus view each organic  BEING, how far more interesting, I speak from exper
0487 f organism to organism, the improvement of one  BEING entailing the improvement or the exterminatio
0489 ction works solely by and for the good of each  BEING, all corporeal and mental endowments will ten
0489 nd us. These laws, taken in the largest sense,  BEING Growth with Reproduction; Inheritance which i
0490 eautiful and most wonderful have been, and are  BEING, evolved.

0003 eflecting on the mutual affinities of organic  BEINGS, on their embryological relations, their geog
0003 ith its relations to several distinct organic  BEINGS, by the effects of external conditions, or of
0004 aves the case of the coadaptations of organic  BEINGS to each other and to their physical condition
0004 he struggle for Existence amongst all organic  BEINGS throughout the world, which inevitably follow
0005 consider the geological succession of organic  BEINGS throughout time; in the eleventh and twelfth,
0006 in regard to the mutual relations of all the  BEINGS which live around us. Who can explain why one
0007 s of food. It seems pretty clear that organic  BEINGS must be exposed during several generations to
0043 cy, under all circumstances, with all organic  BEINGS, as some authors have thought. The effects of
0044 les arrived at in the last chapter to organic  BEINGS in a state of nature, we must briefly discuss
0053 circumstance) with different sets of organic  BEINGS. But my tables further show that, in any limi
0060 seen in the last chapter that amongst organic  BEINGS in a state of nature there is some individual
0061 infinitely complex relations to other organic  BEINGS and to external nature, will tend to the pres
0061 produce great results, and can adapt organic  BEINGS to his own uses, through the accumulation of
0062 ly and philosophically shown that all organic  BEINGS are exposed to severe competition. In regard
0063 llows from the high rate at which all organic  BEINGS tend to increase. Every being, which during i
0071 are the checks and relations between organic  BEINGS, which have to struggle together in the same
0075 parasite on its prey, lies generally between  BEINGS remote in the scale of nature. This is often
0076 they were allowed to struggle together, like  BEINGS in a state of nature, and if the seed or youn
0077 n hidden manner, to that of all other organic  BEINGS, with which it comes into competition for foo
0078 orance on the mutual relations of all organic  BEINGS; a conviction as necessary, as it seems to be
0080 parent'. Explains the grouping of all organic  BEINGS. How will the struggle for existence, discuss
0080 tting are the mutual relations of all organic  BEINGS to each other and to their physical condition
0086 will be enabled to act on and modify organic  BEINGS at any age, by the accumulation of profitable
0096 lief of the continued creation of new organic  BEINGS, or of any great and sudden modification in t
0101 ly then, we may conclude that in many organic  BEINGS, a cross between two individuals is an obviou
0102 rant an indefinite period; for as all organic  BEINGS are striving, it may be said, to seize on eac
0104 ervals, will be great. If there exist organic  BEINGS which never intercross, uniformity of charact
0109 xity of the coadaptations between all organic  BEINGS, one with another and with their physical con
0109 geometrical powers of increase of all organic  BEINGS, each area is already fully stocked with inha
0116 abled to encroach on places occupied by other  BEINGS. Now let us see how this principle of great b
0119 unoccupied or not perfectly occupied by other  BEINGS; and this will depend on infinitely complex r
0124 ram throws light on the affinities of extinct  BEINGS; which, though generally belonging to the sam
0126 re, we can predict that the groups of organic  BEINGS which are now large and triumphant, and which
0126 and under varying conditions of life, organic  BEINGS vary at all in the several parts of their org
0127 te complexity of the relations of all organic  BEINGS to each other and to their conditions of exis
0128 s to divergence of character; for more living  BEINGS can be supported on the same area the more th
0128 , the nature of the affinities of all organic  BEINGS may be explained. It is a truly wonderful fac
0129 eat fact in the classification of all organic  BEINGS; but, to the best of my judgment, it is expla
0129 ted in the diagram. The affinities of all the  BEINGS of the same class have sometimes been represe
0131 ariations, so common and multiform in organic  BEINGS under domestication, and in a lesser degree i
0140 ir ranges by the competition of other organic  BEINGS quite as much as, or more than, by adaptation
0149 the very general opinion of naturalists, that  BEINGS low in the scale of nature are more variable
0168 is probably from this same cause that organic  BEINGS low in the scale of nature are more variable
0170 ations of structure, by which the innumerable  BEINGS on the face of this earth are enabled to stru
0172 on of the record being chiefly due to organic  BEINGS not inhabiting profound depths of the sea, an
0175 in the most important manner to other organic  BEINGS, we must see that the range of the inhabitant
0179 ord. On the origin and transitions of organic  BEINGS with peculiar habits and structure. It has be
0194 arly the same manner two parts in two organic  BEINGS, which owe but little of their structure in c
0194 all the parts and organs of many independent  BEINGS, each supposed to have been separately create
0206 it is generally acknowledged that all organic  BEINGS have been formed on two great laws, Unity of
0206 reement in structure, which we see in organic  BEINGS of the same class, and which is quite indepen
0244 aw, leading to the advancement of all organic  BEINGS, namely, multiply, vary, let the strongest li
0263 resemblance and dissimilarity between organic  BEINGS is attempted to be expressed. The facts by no
0264 rly period; more especially as all very young  BEINGS seem eminently sensitive to injurious or unna
0265 or less sterile hybrids. Lastly, when organic  BEINGS are placed during several generations under c
0265 t has observed. Thus we see that when organic  BEINGS are placed under new and unnatural conditions
0267 in the conditions of life benefit all organic  BEINGS, and on the other hand, that slight crosses,
0267 of a particular nature, often render organic  BEINGS in some degree sterile; and that greater cros
0277 le to the vigour and fertility of all organic  BEINGS. It is not surprising that the degree of diff
0300 richest regions of the whole world in organic  BEINGS; yet if all the species were to be collected
```

Page **(Key Word)**

beings

Page ***************************************(Key Word)***************************************

```
0306  us to dogmatize on the succession of organic BEINGS throughout the world, as it would be for a na
0312  er x. On the Geological Succession of Organic BEINGS. On the slow and successive appearance of new
0312  ating to the geological succession of organic BEINGS, better accord with the common view of the im
0314  , by the more complex relations of the higher BEINGS to their organic and inorganic conditions of
0318  however, the extermination of whole groups of BEINGS, as of ammonites towards the close of the sec
0325  treat of the present distribution of organic  BEINGS, and find how slight is the relation between
0342  th care; that only certain classes of organic BEINGS have been largely preserved in a fossil state
0346  d. In considering the distribution of organic BEINGS over the.face of the globe, the first great f
0349  k by the manner in which successive groups of  BEINGS, specifically distinct, yet clearly related,
0352  al mammals, than perhaps in any other organic  BEINGS: and, accordingly, we find no inexplicable ca
0355  ls simultaneously created. With those organic  BEINGS which never intercross (if such exist), the s
0366  h more southern zone became fitted for arctic  BEINGS and ill fitted for their former more temperat
0382  a great height under the equator. The various  BEINGS thus left stranded may be compared with savag
0383  dissimilarity of the surrounding terrestrial  BEINGS, compared with those of Britain. But this pow
0388  that we have reason to believe that such low   BEINGS change or become modified less quickly than t
0408  inity, and are likewise linked to the extinct  BEINGS which formerly inhabited the same continent.
0408  we should find, as we do find, some groups of  BEINGS greatly, and some only slightly modified, som
0409  ation to the means of migration, one group of  BEINGS, even within the same class, should have all
0411  n. chapter XIII. Mutual Affinities of Organic  BEINGS. Morphology. Embryology. Rudimentary Organs.
0411  ary. From the first dawn of life, all organic  BEINGS are found to resemble each other in descendin
0413  only known cause of the similarity of organic  BEINGS, is the bond, hidden as it is by various degr
0416  same important organ within the same group of  BEINGS. Again, no one will say that rudimentary or a
0421  enus F; just as some few still living organic  BEINGS belong to Silurian genera. So that the amount
0421  t or value of the differences between organic  BEINGS all related to each other in the same degree
0422  ities which we discover in nature amongst the  BEINGS of the same group. Thus, on the view which I
0426  g, occur together throughout a large group of  BEINGS having different habits, we may feel almost s
0433  rsal feature in the affinities of all organic  BEINGS, namely, their subordination in group under g
0448  be said to be retrograde. As all the organic   BEINGS, extinct and recent, which have ever lived on
0456  relationship, by which all living and extinct  BEINGS are united by complex, radiating, and circuit
0456  ertainly known cause of similarity in organic  BEINGS, we shall understand what is meant by the nat
0457  able species, genera, and families of organic  BEINGS, with which this world is peopled, have all d
0460  amongst broken and failing groups of organic  BEINGS; but we see so many strange gradations in nat
0461  that the vigour and fertility of all organic   BEINGS are increased by slight changes in their cond
0462  forms all around us? Why are not all organic   BEINGS blended together in an inextricable chaos? Wi
0464  has been geologically explored. Only certain   BEINGS of certain classes can be preserved in a foss
0467  lity; he only unintentionally exposes organic  BEINGS to new conditions of life, and then nature ac
0467  io of increase which is common to all organic  BEINGS. This high rate of increase is proved by calc
0468  he struggle will often be very severe between  BEINGS most remote in the scale of nature. The sligh
0468  changes, we might have expected that organic   BEINGS would have varied under nature, in the same w
0469  we doubt that variations in any way useful to  BEINGS, under their excessively complex relations of
0471  the grand fact of the grouping of all organic  BEINGS seems to me utterly inexplicable on the theor
0476  cent. The grand fact that all extinct organic  BEINGS belong to the same system with recent beings,
0476  beings belong to the same system with recent  BEINGS, falling either into the same or into interme
0476  conquered the older and less improved organic  BEINGS in the struggle for life. Lastly, the law of
0476  a parallelism in the distribution of organic  BEINGS throughout space, and in their geological suc
0476  ession throughout time; for in both cases the  BEINGS have been connected by the bond of ordinary g
0478  have seen, that all past and present organic   BEINGS constitute one grand natural system, with gra
0479  le of all. The real affinities of all organic  BEINGS are due to inheritance or community of descen
0481  f species? It cannot be asserted that organic  BEINGS in a state of nature are subject to no variat
0484  er from analogy that probably all the organic  BEINGS which have ever lived on this earth have desc
0488  and death of the individual. When I view all  BEINGS not as special creations, but as the lineal d
0488  ns, but as the lineal descendants of some few  BEINGS which lived long before the first bed of the
0489  futurity; for the manner in which all organic  BEINGS are grouped, shews that the greater number of
0164  areful examination and sketch for me of a dun  BELGIAN cart horse with a double stripe on each shou
0165  o apply it to breeds so distinct as the heavy  BELGIAN cart horse, Welch ponies, cobs, the lanky Ka
0304  ch he had himself extracted from the chalk of  BELGIUM. And, as if to make the case as striking as
0012  itance: like produces like is his fundamental  BELIEF: doubts have been thrown on this principle by
0023  eral of the reasons which have led me to this  BELIEF are in some degree applicable in other cases,
0026  ons, for we know of no fact countenancing the  BELIEF that the child ever reverts to some one ances
0052  called an incipient species; but whether this  BELIEF be justifiable must be judged of by the gener
0079  uggle, we may console ourselves with the full  BELIEF, that the war of nature is not incessant, tha
0095  ection, if it be a true principle, banish the  BELIEF of the continued creation of new organic bein
0096  wing, in accordance with the almost universal  BELIEF of breeders, that with animals and plants a c
0097  at very long intervals, indispensable. On the  BELIEF that this is a law of nature, we can, I think
0151  ants, and this would seriously have shaken my  BELIEF in its truth, had not the great variability i
0193  ch other. Nor does geology at all lead to the  BELIEF that formerly most fishes had electric organs
0204  petrels with the habits of auks. Although the  BELIEF that an organ so perfect as the eye could hav
0249  oubt the correctness of this almost universal  BELIEF amongst breeders. Hybrids are seldom raised b
0254  domestic animals, we must either give up the  BELIEF of the almost universal sterility of distinct
0266  s of facts. It is an old and almost universal  BELIEF, founded, I think, on a considerable body of
0290  e sediment has been derived, accords with the  BELIEF of vast intervals of time having elapsed betw
0302  ofessor Sedgwick, as a fatal objection to the  BELIEF in the transmutation of species. If numerous
0304  ondary formation, seemed fully to justify the  BELIEF that this great and distinct order had been s
0320  heory of natural selection is grounded on the  BELIEF that each new variety, and ultimately each ne
0330  small approach to each other. It is a common  BELIEF that the more ancient a form is, by so much t
0335  s to have shaken Professor Pictet in his firm  BELIEF in the immutability of species. He who is acq
0354  grave a nature, that we ought to give up the  BELIEF, rendered probable by general considerations,
0354  nd various occasional means of transport, the  BELIEF that this has been the universal law, seems t
0358  ic islands likewise seem to me opposed to the  BELIEF of their former continuity with continents. N
0366  ints; and we might have remained in this same  BELIEF, had not Agassiz and others called vivid atte
0376  can be thrown on the foregoing facts, on the  BELIEF, supported as it is by a large body of geolog
0389  legitimately followed out, would lead to the  BELIEF that within the recent period all existing is
0431  rds from all other vertebrate animals, by the  BELIEF that many ancient forms of life have been utt
0459  mstances in its favour. Causes of the general  BELIEF in the immutability of species. How far the t
0481  a special endowment and sign of creation. The  BELIEF that species were immutable productions was a
0482  nent naturalists have of late published their  BELIEF that a multitude of reputed species in each g
```

Page ***************************************(Key Word)***************************************

Page **************************************(Key Word)***

```
0484 ould lead me one step further, namely, to the  BELIEF that all animals and plants have descended fr
0006 they determine the present welfare, and, as I  BELIEVE, the future success and modification of ever
0008 he act of conception. Several reasons make me  BELIEVE in this; but the chief one is the remarkable
0011 laire's great work on this subject. Breeders  BELIEVE that long limbs are almost always accompanie
0014 editary diseases and some other facts make me  BELIEVE that the rule has a wider extension, and tha
0014 d at which it first appeared in the parent. I  BELIEVE this rule to be of the highest importance in
0017 ing different quarters of the world. I do not  BELIEVE, as we shall presently see, that all our dog
0017 s, the argument mainly relied on by those who  BELIEVE in the multiple origin of our domestic anima
0018 european cattle; and several competent judges  BELIEVE that these latter have had more than one wil
0018 cannot give here, I am doubtfully inclined to  BELIEVE, in opposition to several authors, that all
0019 ed to an absurd extreme by some authors. They  BELIEVE that every race which breeds true, let the d
0019 mmense amount of inherited variation. Who can  BELIEVE that animals closely resembling the Italian
0020 mely different races or species, I can hardly  BELIEVE. Sir J. Sebright expressly experimentised fo
0022 m as well defined species. Moreover, I do not  BELIEVE that any ornithologist would place the Engli
0026 ng themselves perfectly fertile. Some authors  BELIEVE that long continued domestication eliminates
0030 or by one step; many botanists, for instance,  BELIEVE that the fuller's teazle, with its hooks, wh
0032 es, he will assuredly fail. Few would readily  BELIEVE in the natural capacity and years of practic
0035 ed has been less improved. There is reason to  BELIEVE that King Charles's spaniel has been unconsc
0037 and unconsciously accumulated, explains, as I  BELIEVE, the well known fact, that in a vast number
0043 f our Domestic Races of animals and plants. I  BELIEVE that the conditions of life, from their acti
0043 t importance as causing variability. I do not  BELIEVE that variability is an inherent and necessar
0043 mportance of the crossing of varieties has, I  BELIEVE, been greatly exaggerated, both in regard to
0047 ost important for us. We have every reason to  BELIEVE that many of these doubtful and closely alli
0052 cture in certain definite directions. Hence I  BELIEVE a well marked variety may be justly called a
0056 n, more especially as we have every reason to  BELIEVE the process of manufacturing new species to
0058 t to be reversed. But there is also reason to  BELIEVE, that those species which are very closely a
0067 , from some observations which I have made, I  BELIEVE that it is the seedlings which suffer most f
0068 iodical seasons of extreme cold or drought, I  BELIEVE to be the most effective of all checks. I es
0070 ame species for its preservation, explains, I  BELIEVE, some singular facts in nature, such as that
0070 nes of their range. For in such cases, we may  BELIEVE, that a plant could exist only where the con
0073 us to fertilise them. I have, also, reason to  BELIEVE that humble bees are indispensable to the fe
0078 advantage to our plant; but we have reason to  BELIEVE that only a few plants or animals range so f
0082 or the work of improvement. We have reason to  BELIEVE, as stated in the first chapter, that a chan
0082 ural selection can do nothing. Not that, as I  BELIEVE, any extreme amount of variability is necess
0082 parably longer time at her disposal. Nor do I  BELIEVE that any great physical change, as of climat
0084 the black grouse that of peaty earth, we must  BELIEVE that these tints are of service to these bir
0089 . all those who have attended to the subject,  BELIEVE that there is the severest rivalry between t
0089 re to enter on this subject. Thus it is, as I  BELIEVE, that when the males and females of any anim
0090 cocks of certain fowls, etc.) which we cannot  BELIEVE to be either useful to the males in battle,
0090 election. In order to make it clear how, as I  BELIEVE, natural selection acts, I must beg permissi
0092 d the act of crossing, we have good reason to  BELIEVE (as will hereafter be more fully alluded to)
0093 hysiological division of labour; hence we may  BELIEVE that it would be advantageous to a plant to
0096 vious. Nevertheless I am strongly inclined to  BELIEVE that with all hermaphrodites two individuals
0097 rtility; that these facts alone incline me to  BELIEVE that it is a general law of nature (utterly
0097 rom one to the other, to the great good, as I  BELIEVE, of the plant. Bees will act like a camel ha
0100 stinct individuals only in a limited sense. I  BELIEVE this objection to be valid, but that nature
0102 diversified variability is favourable, but I  BELIEVE mere individual differences suffice for the
0102 of variability in each individual, and is, I  BELIEVE, an extremely important element of success.
0103 e confined to separated countries; and this I  BELIEVE to be the case. In hermaphrodite organisms w
0104 eady attempted to show that we have reason to  BELIEVE that occasional intercrosses take place with
0105 of new species, on the whole I am inclined to  BELIEVE that largeness of area is of more importance
0108 top the action of natural selection. I do not  BELIEVE so. On the other hand, I do believe that nat
0108 i do not believe so. On the other hand, I do  BELIEVE that natural selection will always act very
0108 f the same region at the same time. I further  BELIEVE, that this very slow, intermittent action of
0109 tinually and slowly being produced, unless we  BELIEVE that the number of specific forms goes on pe
0111 h importance on my theory, and explains, as I  BELIEVE, several important facts. In the first place
0112 an any analogous principle apply in nature? I  BELIEVE it can and does apply most efficiently, from
0120 4 and m14, all descended from (A). Thus, as I  BELIEVE, species are multiplied and genera are forme
0123 en as distinct sub families. Thus it is, as I  BELIEVE, that two or more genera are produced by des
0128 mediate forms of life. On these principles, I  BELIEVE, the nature of the affinities of all organic
0129 sometimes been represented by a great tree. I  BELIEVE this simile largely speaks the truth. The gr
0130 des many a feebler branch, so by generation I  BELIEVE it has been with the great Tree of Life, whi
0131 se of each particular variation. Some authors  BELIEVE it to be as much the function of the reprodu
0131 r cultivation, than under nature, leads me to  BELIEVE that deviations of structure are in some way
0134 seldom take flight except to escape danger, I  BELIEVE that the nearly wingless condition of severa
0135 re is not sufficient evidence to induce us to  BELIEVE that mutilations are ever inherited; and I s
0136 hti these several considerations have made me  BELIEVE that the wingless condition of so many Madei
0139 it very hot and very cold countries; and as I  BELIEVE that all the species of the same genus have
0140 ich here enjoy good health. We have reason to  BELIEVE that species in a state of nature are limite
0140 become acclimatised to their new homes. As I  BELIEVE that our domestic animals were originally ch
0141 hat habit or custom has some influence I must  BELIEVE, both from analogy, and from the incessant a
0144 versity in the shape of their kidneys. Others  BELIEVE that the shape of the pelvis in the human mo
0147 ny good observers, more especially botanists,  BELIEVE in its truth. I will not, however, here give
0148 ng a structure now become useless. Thus, as I  BELIEVE, natural selection will always succeed in th
0149 ts, it has been stated by some authors, and I  BELIEVE with truth, are apt to be highly variable. W
0154 an those commonly used for classing genera. I  BELIEVE this explanation is partly, yet only indirec
0163 foal, and from inquiries which I have made, I  BELIEVE this to be true. It has also been asserted t
0167 mockery and deception; I would almost as soon  BELIEVE with the old and ignorant cosmogonists, that
0171 e animal with wholly different habits? Can we  BELIEVE that natural selection could produce, on the
0172 cal record; and I will here only state that I  BELIEVE the answer mainly lies in the record being i
0174 uring a long period. Geology would lead us to  BELIEVE that almost every continent has been broken
0174 is way of escaping from the difficulty; for I  BELIEVE that many perfectly defined species have bee
0176 but a far more important consideration, as I  BELIEVE, is that, during the process of further modi
0177 mproved. It is the same principle which, as I  BELIEVE, accounts for the common species in each cou
0177 nted, intermediate hill variety. To sum up, I  BELIEVE that species come to be tolerably well defin
0180 t food more quickly, or as there is reason to  BELIEVE, by lessening the danger from occasional fal
```

Page **************************************(Key Word)***

Page **(Key Word)**

```
0181  me modified, and all analogy would lead us to  BELIEVE  that some at least of the squirrels would de
0189  of individuals of many kinds; and may we not   BELIEVE  that a living optical instrument might thus
0193  e can generally be detected. I am inclined to  BELIEVE  that in nearly the same way as two men have
0198  in high regions would, we have some reason to  BELIEVE, increase the size of the chest; and again c
0198  er different climates; and there is reason to  BELIEVE  that constitution and colour are correlated.
0199  produced for the good of its possessor. They   BELIEVE  that very many structures have been created
0199  of life of each species. Thus, we can hardly   BELIEVE  that the webbed feet of the upland goose or
0200  are of special use to these birds; we cannot   BELIEVE  that the same bones in the arm of the monkey
0200  the most aquatic of existing birds. So we may  BELIEVE  that the progenitor of the seal had not a fl
0200  ng or grasping; and we may further venture to  BELIEVE  that the several bones in the limbs of the m
0201  rn its prey to escape. I would almost as soon  BELIEVE  that the cat curls the end of its tail when
0205  of natural selection. But we may confidently  BELIEVE  that many modifications, wholly due to the l
0205  ed descendants of this species. We may, also,  BELIEVE  that a part formerly of high importance has
0209  tent that may be profitable. It is thus, as I  BELIEVE, that all the most complex and wonderful ins
0209  o not doubt it has been with instincts. But I  BELIEVE  that the effects of habit are of quite subor
0211  the sole good of the ants. Although I do not  BELIEVE  that any animal in the world performs an act
0214  that have never seen a pigeon tumble. We may   BELIEVE  that some one pigeon showed a slight tendenc
0217  in an advantage. And analogy would lead me to  BELIEVE, that the young thus reared would be apt to
0217  ung. By a continued process of this nature, I  BELIEVE  that the strange instinct of our cuckoo coul
0227  those which guide a bird to make its nest, I   BELIEVE  that the hive bee has acquired, through natu
0235  lutely perfect in economising wax. Thus, as I  BELIEVE, the most wonderful of all known instincts,
0236  ance from a common parent, and must therefore  BELIEVE  that they have been acquired by independent
0237  appearing insuperable, is lessened, or, as I  BELIEVE, disappears, when it is remembered that sele
0238  x could ever have propagated its kind. Thus I  BELIEVE  it has been with social insects: a slight mo
0238  e members having the same modification. And I  BELIEVE  that this process has been repeated, until t
0239  have been rendered by natural selection, as I  BELIEVE  to be quite possible, different from the fer
0240  their proportionally lesser size; and I fully  BELIEVE, though I dare not assert so positively, tha
0241  several sizes. With these facts before me, I  BELIEVE  that natural selection, by acting on the fer
0241  ermediate structure were produced. Thus, as I  BELIEVE, the wonderful fact of two distinctly define
0242  uld have become blended. And nature has, as I  BELIEVE, effected this admirable division of labour
0247  nd cowslip, which we have such good reason to  BELIEVE  to be varieties, and only once or twice succ
0248  in the first few generations. Nevertheless I  BELIEVE  that in all these experiments the fertility
0249  ons of artificially fertilised hybrids may, I  BELIEVE, be accounted for by close interbreeding hav
0253  fertile hybrid animals, I have some reason to  BELIEVE  that the hybrids from Cervulus vaginalis and
0254  to me the most probable, and I am inclined to  BELIEVE  in its truth, although it rests on no direct
0254  h, although it rests on no direct evidence. I  BELIEVE, for instance, that our does have descended
0254  fertile hybrids. So again there is reason to  BELIEVE  that our European and the humped Indian catt
0263  differences in their vegetative systems; so I  BELIEVE  that the still more complex laws governing t
0264  has not been sufficiently attended to; but I  BELIEVE, from observations communicated to me by Mr.
0264  rst crosses. I was at first very unwilling to  BELIEVE  in this view; as hybrids, when once born, ar
0265  extremely liable to vary, which is due, as I  BELIEVE, to their reproductive systems having been s
0267  ives vigour and fertility to the offspring. I  BELIEVE, indeed, from the facts alluded to in our fo
0269  in the second place, some eminent naturalists  BELIEVE  that a long course of domestication tends to
0269  nce is at least as good as that from which we  BELIEVE  in the sterility of a multitude of species.
0270  as the plants have separated sexes. No one, I  BELIEVE, has suspected that these varieties of maize
0274  pressing its likeness on the hybrid; and so I  BELIEVE  it to be with varieties of plants. With anim
0280  against my theory. The explanation lies, as I  BELIEVE, in the extreme imperfection of the geologic
0283  with sand or pebbles; for there is reason to  BELIEVE  that pure water can effect little of nothing
0283  the action of the sea on our shores, will, I  BELIEVE, be most deeply impressed with the slowness
0286  fallen fragments. On the other hand, I do not  BELIEVE  that any line of coast, ten or twenty miles
0288  he sea, where sediment is not accumulating. I  BELIEVE  we are continually taking a most erroneous v
0300  ory of the world! But we have every reason to  BELIEVE  that the terrestrial productions of the arch
0300  ins could be preserved. In our archipelago, I  BELIEVE  that fossiliferous formations could be forme
0301  beyond its confines! and analogy leads me to  BELIEVE  that it would be chiefly these far ranging s
0305  stage further back; and some palaeontologists  BELIEVE  that certain much older fishes, of which the
0310  some special explanation; and we may perhaps  BELIEVE  that we see in these large areas, the many f
0310  immutability of species. But I have reason to  BELIEVE  that one great authority, Sir Charles Lyell,
0313  erved in Switzerland. There is some reason to  BELIEVE  that organisms, considered high in the scale
0313  from the face of the earth, we have reason to  BELIEVE  that the same identical form never reappears
0314  e several facts accord well with my theory. I  BELIEVE  in no fixed law of development; causing all
0315  troyed, and in nature we have every reason to  BELIEVE  that the parent form will generally be suppl
0317  ion. On the contrary, we have every reason to  BELIEVE, from the study of the tertiary formations,
0318  any single genus endures. There is reason to  BELIEVE  that the complete extinction of the species
0320  eriods, so that looking to later times we may  BELIEVE  that the production of new forms has caused
0321  the most liable to extermination. Thus, as I  BELIEVE, a number of new species descended from one
0324  so, again, several highly competent observers  BELIEVE  that the existing productions of the United
0328  ther parts of the world. As we have reason to  BELIEVE  that large areas are affected by the same mo
0335  the species was perfect, we have no reason to  BELIEVE  that forms successively produced necessarily
0337  ch must have been previously occupied, we may  BELIEVE, if all the animals and plants of Great Brit
0345  en the geological record be as imperfect as I  BELIEVE  it to be; and it may at least be asserted th
0351  as is so commonly and notoriously the case. I  BELIEVE, as was remarked in the last chapter, in no
0351  many other cases, in which we have reason to  BELIEVE  that the species of a genus have been produc
0353  rn and southern hemispheres? The answer, as I  BELIEVE, is, that mammals have not been able to migr
0356  for each birth, or which often intercross, I  BELIEVE  that during the slow process of modification
0357  eans such sunken islands are now marked, as I  BELIEVE, by rings of coral or atolls standing over t
0357  ver them. Whenever it is fully admitted, as I  BELIEVE  it will some day be, that each species has p
0357  he former extension of the land. But I do not  BELIEVE  that it will ever be proved that within the
0368  sent (as some geologists in the United States  BELIEVE  to have been the case, chiefly from the dist
0369  sentative species. In illustrating what, as I  BELIEVE, actually took place during the Glacial peri
0370  pletely separated by wider spaces of ocean. I  BELIEVE  the above difficulty may be surmounted by lo
0370  of an opposite nature. We have good reason to  BELIEVE  that during the newer Pliocene period, befor
0371  ement of the Glacial period. We now see, as I  BELIEVE, their descendants, mostly in a modified con
0378  e pyrenees. At this period of extreme cold, I  BELIEVE  that the climate under the equator at the le
0378  s graphically described by Hooker. Thus, as I  BELIEVE, a considerable number of plants, a few ther
0379  posite hemisphere. Although we have reason to  BELIEVE  from geological evidence that the whole body
0380  iod must have been completely isolated; and I  BELIEVE  that the productions of these islands on the
```

Page **(Key Word)**

```
0381 as kerguelen Land, New Zealand, and Fuegia, I BELIEVE that towards the close of the Glacial period
0382 ial period, by these means, as I BELIEVE, the southern shores of America, Australia,
0382 ns of climate on geographical distribution. I BELIEVE that the world has recently felt one of his
0384 nsider only a few cases. In regard to fish, I BELIEVE that the same species never occur in the fre
0386 to their natural fresh water haunts. I do not BELIEVE that botanists are aware how charged the mud
0387 i do not know the fact, yet analogy makes me BELIEVE that a heron flying to another pond and gett
0388 e scale of nature, and that we have reason to BELIEVE that such low beings change or become modifi
0388 identical form or in some degree modified, I BELIEVE mainly depends on the wide dispersal of thei
0390 an be named. In St. Helena there is reason to BELIEVE that the naturalised plants and animals have
0392 a little doubtful. Facility of immigration, I BELIEVE, has been at least as important as the natur
0398 ca, like those of the Galapagos to America. I BELIEVE this grand fact can receive no sort of expla
0404 nalogous facts could be given. And it will, I BELIEVE, be universally found to be true, that where
0405 points of the world, we ought to find, and I BELIEVE as a general rule we do find, that some at l
0406 ew conditions. There is, also, some reason to BELIEVE from geological evidence that organisms low
0411 us produced ultimately become converted, as I BELIEVE, into new and distinct species; and these, o
0413 ing more is meant by the Natural System; they BELIEVE that it reveals the plan of the Creator; but
0413 our classification, than mere resemblance. I BELIEVE that something more is included; and that or
0415 e. but their importance for classification, I BELIEVE, depends on their greater constancy througho
0420 . but I must explain my meaning more fully. I BELIEVE that the arrangement of the groups within ea
0425 e, and has taken a longer time to complete? I BELIEVE it has thus been unconsciously used; and onl
0433 ifferent they may be from their parenti and I BELIEVE this element of descent is the hidden bond o
0436 ns in the same individual. Most physiologists BELIEVE that the bones of the skull are homologous w
0437 ttle modified forms; therefore we may readily BELIEVE that the unknown progenitor of the vertebrat
0440 a frog under water. We have no more reason to BELIEVE in such a relation, than we have to believe
0440 o believe in such a relation, than we have to BELIEVE that the same bones in the hand of a man, wi
0443 mature animal, into which it is developed. I BELIEVE that all these bones can be explained, as fo
0443 ly displayed. The cause may have acted, and I BELIEVE generally has acted, even before the embryo
0444 inciples, if their truth be admitted, will, I BELIEVE, explain all the above specified leading fac
0452 ut afterwards wholly disappear. It is also, I BELIEVE, a universal rule, that a rudimentary part o
0453 metimes appear on the stumps: I could as soon BELIEVE that these vestiges of nails have appeared,
0454 s under nature ever undergo abrupt changes. I BELIEVE that disuse has been the main agency; that i
0455 rly period of life (as we have good reason to BELIEVE to be possible) the rudimentary part would t
0459 nothing at first can appear more difficult to BELIEVE than that the more complex organs and instin
0462 have been effected. Yet, as we have reason to BELIEVE that some species have retained the same spe
0463 the intermediate zone. For we have reason to BELIEVE that only a few species are undergoing chang
0464 rd is far more imperfect than most geologists BELIEVE. It cannot be objected that there has not be
0465 ulation; and their duration, I am inclined to BELIEVE, has been shorter than the average duration
0466 ns of life remain the same, we have reason to BELIEVE that a modification, which has already been
0473 rids! How simply is this fact explained if we BELIEVE that these species have descended from a str
0480 nitor having well developed teeth; and we may BELIEVE, that the teeth in the mature animal were re
0482 question with impartiality. Whoever is led to BELIEVE that species are mutable will do good servic
0483 than at an ordinary birth. But do they really BELIEVE that at innumerable periods in the earth's h
0483 uddenly to flash into living tissues? Do they BELIEVE that at each supposed act of creation one in
0483 xplanation of every difficulty from those who BELIEVE in the mutability of species, on their own s
0484 embraces all the members of the same class. I BELIEVE that animals have descended from at most onl
0035 tury, and in this case the change has, I BELIEVED, been chiefly effected by crosses with the f
0066 he fulmar petrel lays but one egg, yet it is BELIEVED to be the most numerous bird in the world. O
0143 imbs, varying together, for the lower jaw is BELIEVED to be homologous with the limbs. These tende
0144 fect the form of adjoining soft parts; it is BELIEVED by some authors that the diversity in the sh
0160 1 in 2048; and yet, we see, it is generally BELIEVED that a tendency to reversion is retained by
0246 of varieties, that is of the forms known or BELIEVED to have descended from common parents; and o
0375 pe of Good Hope a very few European species, BELIEVED not to have been introduced by man, and on t
0423 t the classification of varieties, which are BELIEVED or known to have descended from one species.
0430 f affinity of the bizcacha to Marsupials are BELIEVED to be real and not merely adaptive, they are
0485 varieties is, that the latter are known, or BELIEVED, to be connected at the present day by inter
0019 everal even within Great Britain. One author BELIEVES that there formerly existec in Great Britain
0074 long attended to the habits of humble bees, BELIEVES that more than two thirds of them are thus d
0132 further north or from greater depths. Gould BELIEVES that birds of the same species are more brig
0133 e are very apt to have fleshy leaves. He who BELIEVES in the creation of each species, will have t
0164 lton Smith, who has written on this subject, BELIEVES that the several breeds of the horse have de
0167 ass, the hemionus, quagga, and zebra. He who BELIEVES that each equine species was independently c
0185 feet and using its wings under water. He who BELIEVES that each being has been created as we now s
0185 s that structure has begun to change. He who BELIEVES in separate and innumerable acts of creation
0186 ating the fact in dignified language. He who BELIEVES in the struggle for existence and in the pri
0247 ly so sterile, when intercrossed, as Gartner BELIEVES. It is certain, on the one hand, that the st
0252 ty in each successive generation, as Gartner BELIEVES to be the case, the fact would have been not
0262 the same two species in crossing; so Sagaret BELIEVES this to be the case with different individua
0285 f 2300 feet; and he informs me that he fully BELIEVES there is one in Merionethshire of 12,000 fee
0305 hat the whole of them did appear, as Agassiz BELIEVES, at the commencement of the chalk formation,
0424 t because they are descended from it. He who BELIEVES that the cowslip is descended from the primr
0449 r descendants, our existing species. Agassiz BELIEVES this to be a law of nature: but I am bound t
0020 ormec. On the Breeds of the Domestic Pigeon. BELIEVING that it is always best to study some specia
0028 hey bred, I felt fully as much difficulty in BELIEVING that they could ever have descended from a
0175 still more sharply defined. If I am right in BELIEVING that allied or representative species, when
0181 i see any insuperable difficulty in further BELIEVING it possible that the membrane connected fin
0186 g conditions of life, then the difficulty of BELIEVING that a perfect and complex eye could be for
0188 han in the case of many other structures) in BELIEVING that natural selection has converted the si
0191 ere seems to me to be no great difficulty in BELIEVING that natural selection has actually convert
0218 nd M. Fabre has lately shown good reason for BELIEVING that although the Tachytes nigra generally
0289 low them in nature, it is difficult to avoid BELIEVING that they are closely consecutive. But we k
0327 ct worth making. I have given my reasons for BELIEVING that all our greater fossiliferous formatio
0351 odification, there is not much difficulty in BELIEVING that they may have migrated from the same r
0370 , at a period anterior to the Glacial epoch. BELIEVING, from reasons before alluded to, that our c
0374 f be admitted, it is difficult to avoid BELIEVING that the temperature of the whole world was
0407 rmediate tracts, I think the difficulties in BELIEVING that all the individuals of the same specie
0088 alligators have been described as fighting, BELLOWING, and whirling round, like Indians in a war
```

Page **(Key Word)**

0041 of parts of Yorkshire, that as they generally BELONG to poor people, and are mostly in small lots,
0054 o the size of the genera to which the species BELONG. Again, plants low in the scale of organisati
0058 whereas, the species to which these varieties BELONG range over 14.3 provinces. So that the acknow
0114 other most closely, shall, as a general rule, BELONG to what we call different genera and orders.
0115 naturalised plants are enumerated, and these BELONG to 162 genera. We thus see that these natural
0306 f all the species of the orders to which they BELONG, for they do not present characters in any de
0323 one species is identically the same, but they BELONG to the same families, genera, and sections of
0338 st known mammals, reptiles, and fish strictly BELONG to their own proper classes, though some of l
0354 of its inhabitants are closely related to, or BELONG to the same genera with the species of a seco
0376 he intertropical regions, are not arctic, but BELONG to the northern temperate zones. As Mr. H. C.
0384 esent be explained: but some fresh water fish BELONG to very ancient forms, and in such cases ther
0421 just as some few still living organic beings BELONG to Silurian genera. So that the amount or val
0428 vantages, which made the groups to which they BELONG large and their parents dominant, they are al
0433 ate and define the groups to which such types BELONG. Finally, we have seen that natural selection
0476 he grand fact that all extinct organic beings BELONG to the same system with recent beings, fallin
0114 upported twenty species of plants, and these BELONGED to eighteen genera and to eight orders, whic
0115 expected that naturalised plants would have BELONGED to a few groups more especially adapted to c
0118 h made the genus to which the parent species BELONGED, a large genus in its own country. And these
0319 , though so like that of the existing horse, BELONGED to an extinct species. Had this horse been s
0392 growing on an island, to whatever order they BELONGED, and thus convert them first into bushes and
0006 at species are not immutable: but that those BELONGING to what are called the same genera are line
0017 number to our domesticated productions, and BELONGING to equally diverse classes and countries, w
0025 erfectly developed. Moreover, when two birds BELONGING to two distinct breeds are crossed, neither
0100 onally carried from tree to tree. That trees BELONGING to all Orders have their sexes more often s
0114 can raise most food by a rotation of plants BELONGING to the most different orders: nature follow
0117 ommon, widely diffused, and varying species, BELONGING to a genus large in its own country. The li
0119 s from a common and widely diffused species, BELONGING to a large genus, will tend to partake of t
0124 s of extinct beings, which, though generally BELONGING to the same orders, or families, or genera,
0128 widely diffused, and widely ranging species, BELONGING to the larger genera, which vary most; and
0137 suse. It is well known that several animals, BELONGING to the most different classes, which inhabi
0193 resence of luminous organs in a few insects, BELONGING to different families and orders, offers a
0261 no one has been able to graft trees together BELONGING to quite distinct families: and, on the oth
0289 ree. For instance, not a land shell is known BELONGING to either of these vast periods, with one e
0289 t not a cave or true lacustrine bed is known BELONGING to the age of our secondary or palaeozoic f
0302 ansmutation of species. If numerous species, BELONGING to the same genera or families, have really
0321 a new genus, comes to supplant an old genus, BELONGING to the same family. But it must often have
0321 must often have happened that a new species BELONGING to some one group will have seized on the p
0321 ve seized on the place occupied by a species BELONGING to a distinct group, and thus caused its ex
0321 riority in common. But whether it be species BELONGING to the same or to a distinct class, which y
0328 ous accordance in the numbers of the species BELONGING to the same genera, yet the species themsel
0330 y day taught that palaeozoic animals, though BELONGING to the same orders, families, or genera wit
0330 a between living genera, even between genera BELONGING to distinct families. The most common case,
0349 the same habits as our hares and rabbits and BELONGING to the same order of Rodents, but they plai
0360 on this average, the seeds of 14/100 plants BELONGING to one country might be floated across 924
0374 quatorial America a host of peculiar species BELONGING to European genera occur. On the highest mo
0374 illustrious Humboldt long ago found species BELONGING to genera characteristic of the Cordillera.
0381 existence of several quite distinct species, BELONGING to genera exclusively confined to the south
0383 rse. Not only have many fresh water species, BELONGING to quite different classes, an enormous ran
0392 again, islands often possess trees or bushes BELONGING to orders which elsewhere include only herb
0394 n time for the production of endemic species BELONGING to other classes: and on continents it is t
0402 inhabitants, but are very distinct species. BELONGING in a large proportion of cases, as shown by
0410 s of maximum development. Groups of species, BELONGING either to a certain period of time, or to a
0410 some organisms differ little, whilst others BELONGING to a different class, or to a different ord
0411 sed and common, that is the dominant species BELONGING to the larger genera, which vary most. The
0417 e, too slight to be defined. Certain plants, BELONGING to the Malpighiaceae, bear perfect and degr
0419 onwards, can be recognised as unequivocally BELONGING to this, and to no other class of Articulat
0427 t valueless to the systematist. For animals, BELONGING to two most distinct lines of descent, may
0428 he modified descendants of dominant species, BELONGING to the larger genera, tend to inherit the a
0429 y of Australia has not added a single insect BELONGING to a new order: and that in the vegetable k
0430 waterhouse has remarked that, when a member BELONGING to one group of animals exhibits an affinit
0449 can at once be recognised by their larvae as BELONGING to the great class of crustaceans. As the e
0470 fined and distinct objects. Dominant species BELONGING to the larger groups tend to give birth to
0489 ill be the common and widely spread species, BELONGING to the larger and dominant groups, which wi
0303 richest known accumulations of fossil mammals BELONGS to the middle of the secondary series; and o
0374 was at the same time lower along certain broad BELTS of longitude. On this view of the whole world
0374 whole world, or at least of broad longitudinal BELTS, having been simultaneously colder from pole
0407 the whole world, or at least great meridional BELTS. As showing how diversified are the means of
0230 cases stand in the opposed cells and push and BEND the ductile and warm wax (which as I have trie
0124 agram, this is indicated by the broken lines, BENEATH the capital letters, converging in sub branc
0187 class we should probably have to descend far BENEATH the lowest known fossiliferous stratum to di
0188 arent tissue, with a nerve sensitive to light BENEATH, and then suppose every part of this layer t
0222 dependent community of F. flava under a stone BENEATH a nest of the slave making F. sanguinea: and
0302 ertain genera or families have not been found BENEATH a certain stage, that they did not exist bef
0307 life have been detected in the longmynd beds BENEATH Barrande's so called primordial zone. The pr
0310 at present known, of fossiliferous formations BENEATH the Silurian strata, are all undoubtedly of
0331 sive geological formations, and all the forms BENEATH the uppermost line may be considered as exti
0332 s, for instance, above No. VI., but none from BENEATH this line, then only the two families on the
0338 l character of the Vertebrata, until beds far BENEATH the lowest Silurian strata are discovered, a
0357 former existence of many islands, now buried BENEATH the sea, which may have served as halting pl
0386 blespoons of mud from three different points, BENEATH water, on the edge of a little pond; this mu
0463 ges? Why do we not find great piles of strata BENEATH the Silurian system, stored with the remains
0465 ct to the absence of fossiliferous formations BENEATH the lowest Silurian strata. I can only recur
0073 , if not indispensable, are at least highly BENEFICIAL to the fertilisation of our clovers; but hu
0170 atural selection, of such differences, when BENEFICIAL to the individual, that gives rise to all t
0266 light changes in the conditions of life are BENEFICIAL to all living things. We see this acted on
0314 ontingencies, on the variability being of a BENEFICIAL nature, on the power of intercrossing, on t

Page **(Key Word)**

Page **(Key Word)**

```
0087 dapt the structure of each individual for the    BENEFIT of the community: if each in consequence pro
0116 s. now let us see how this principle of great     BENEFIT being derived from divergence of character,
0117 . and here the importance of the principle of     BENEFIT being derived from divergence of character c
0153 been accumulated by natural selection for the     BENEFIT of the species. But as the variability of th
0208 was already done for it, far from feeling the     BENEFIT of this, it was much embarrassed, and, in or
0267 escence of animals, we plainly see that great     BENEFIT is derived from almost any change in the hab
0267 and, slight changes in the conditions of life     BENEFIT all organic beings, and on the other hand, t
0437 ped pieces of bone? As Owen has remarked, the     BENEFIT derived from the yielding of the separate pi
0467 he thus adapts animals and plants for his own     BENEFIT or pleasure. He may do this methodically, or
0091 light innate change of habit or of structure     BENEFITED an individual wolf, it would have the best
0048 mr. Babington gives 251 species, whereas Mr.     BENTHAM gives only 112, a difference of 139 doubtful
0419 y. several of the best botanists, such as Mr.     BENTHAM and others, have strongly insisted on their
0185 et sounds of Tierra del Fuego, the Puffinuria    BERARDI, in its general habits, in its astonishing p
0359 most wholly unknown. Until I tried, with Mr.     BERKELEY's aid, a few experiments, it was not even kn
0391 at these islands more easily than land birds.     BERMUDA, on the other hand, which lies at about the
0391 w from Mr. J. M. Jones's admirable account of     BERMUDA, that very many North American birds, during
0391 e. V. Harcourt. So that these two islands of      BERMUDA and Madeira have been stocked by birds, whic
0394 pecies either regularly or occasionally visit     BERMUDA, at the distance of 600 miles from the mainl
0359 they germinated; an asparagus plant with ripe     BERRIES floated for 23 days, when dried it floated f
0023 breeding or willingly perching on trees. But      BESIDES C. Livia, with its geographical sub species,
0025 s and some apparently truly wild breeds have,     BESIDES the two black bars, the wings chequered with
0259 n conditions in excess. That their fertility,     BESIDES being eminently susceptible to favourable an
0271 ties of Verbascum present no other difference     BESIDES the mere colour of the flower; and one varie
0395 with the absence of all terrestrial mammals.     BESIDES the absence of terrestrial mammals in relati
0468 ual differences in species under nature. But,     BESIDES such differences, all naturalists have admit
0288 the ocean, the bright blue tint of the water     BESPEAKS its purity. The many cases on record of a fo
0004 imals and of cultivated plants would offer the    BEST chance of making out this obscure problem. Nor
0004 of variation under domestication, afforded the    BEST and safest clue. I may venture to express my c
0012 , in two large volumes, is the fullest and the    BEST on this subject. No breeder doubts how strong
0020 e domestic Pigeon. Believing that it is always    BEST to study some special group, I have, after del
0031 each time marked and classed, so that the very    BEST may ultimately be selected for breeding. What
0032 ally due to crossing different breeds: all the    BEST breeders are strongly opposed to this practice
0032 ablished, the seed raisers do not pick out the    BEST plants, but merely go over their seed beds, an
0034 every one trying to possess and breed from the    BEST individual animals, is more important. Thus, a
0034 as he can, and afterwards breeds from his own     BEST dogs, but he has no wish or expectation of per
0036 nt, through the occasional preservation of the    BEST individuals, whether or not sufficiently disti
0037 ly. It has consisted in always cultivating the    BEST known variety, sowing its seeds, and, when a s
0037 rs of the classical period, who cultivated the    BEST pear they could procure, never thought what sp
0037 heir having naturally chosen and preserved the    BEST varieties they could anywhere find. A large am
0040 or takes more care than usual in matching his     BEST animals and thus improves them, and the improv
0041 seedlings from them, and again picked out the     BEST seedlings and bred from them, then, there appe
0050 s. yet it must be confessed, that it is in the    BEST known countries that we find the greatest numb
0075 ed seed be resown, some of the varieties which    BEST suit the soil or climate, or are naturally the
0081 e, however slight, over others, would have the    BEST chance of surviving and of procreating their k
0081 ee in the species called polymorphic. We shall    BEST understand the probable course of natural sele
0087 ing the egg. It has been asserted, that of the    BEST short beaked tumbler pigeons more perish in th
0088 ally, the most vigorous males, those which are    BEST fitted for their places in nature, will leave
0088 improve his breed by careful selection of the    BEST cocks. How low in the scale of nature this law
0090 he swiftest and slimmest wolves would have the    BEST chance of surviving, and so be preserved or se
0091 which results from each man trying to keep the    BEST dogs without any thought to modifying the bree
0091 enefited an individual wolf, it would have the    BEST chance of surviving and of leaving offspring.
0091 the continued preservation of the individuals    BEST fitted for the two sites, two varieties might
0092 s seedlings, which consequently would have the    BEST chance of flourishing and surviving. Some of t
0102 fection, and all try to get and breed from the    BEST animals, much improvement and modification sur
0110 are most numerous in individuals will have the    BEST chance of producing within any given period fa
0114 any one species of grass would always have the    BEST chance of succeeding and of increasing in numb
0121 modified descendants: for these will have the     BEST chance of filling new and widely different pla
0127 y individuals thus characterized will have the    BEST chance of being preserved in the struggle for
0127 election, by assuring to the most vigorous and    BEST adapted males the greatest number of offspring
0129 ssification of all organic beings: but, to the    BEST of my judgment, it is explained through inheri
0136 oned or from indolent habit, will have had the    BEST chance of surviving from not being blown out t
0142 individuals which are born with constitutions    BEST adapted to their native countries. In treatise
0162 o as to produce the intermediate form. But the    BEST evidence is afforded by parts or organs of an
0170 e enabled to struggle with each other, and the    BEST adapted to survive. Chapter VI. Difficulties o
0171 t on them without being staggered; but, to the    BEST of my judgment, the greater number are only ap
0180 act that the structure of each squirrel is the    BEST that it is possible to conceive under all natu
0182 e by a struggle: but. it is not necessarily the    BEST possible under all possible conditions. It mus
0198 slightly different constitutions would succeed    BEST under different climates; and there is reason
0214 , and that the long continued selection of the    BEST individuals in successive generations made tum
0215 rove the breed, dogs which will stand and hunt    BEST. On the other hand, habit alone in some cases
0216 ection have acted together. We shall, perhaps,    BEST understand how instincts in a state of nature
0219 e, but with plenty of the food which they like    BEST, and with their larvae and pupae to stimulate
0227 have every angle identically the same with the    BEST measurements which have been made of the cells
0235 oney in the secretion of wax, having succeeded    BEST, and having transmitted by inheritance its new
0235 swarms, which in their turn will have had the     BEST chance of succeeding in the struggle for exist
0240 nomma) of West Africa. The reader will perhaps    BEST appreciate the amount of difference in these w
0243 y: but none of the cases of difficulty, to the    BEST of my judgment, annihilate it. On the other ha
0247 s (anagallis arvensis and coerulea), which the    BEST botanists rank as varieties, absolutely steril
0248 enter on details, the evidence advanced by our    BEST botanists on the question whether certain doub
0268 d cowslip, which are considered by many of our    BEST botanists as varieties, are said by Gartner no
0284 on of the degraded matter, probably offers the    BEST evidence of the lapse of time. I remember havi
0298 fected by palaeontologists. We shall, perhaps,    BEST perceive the improbability of our being enable
0302 ow poor a record of the mutations of life, the    BEST preserved geological section presented, had no
0326 species, should be those which would have the     BEST chance of spreading still further, and of givi
0333 psi and this by the concurrent evidence of our    BEST palaeontologists seems frequently to be the ca
0350 their own widely extended homes will have the     BEST chance of seizing on new places, when they spr
```

Page **(Key Word)**

Page ***(Key Word)***

0357 nts, and removes many a difficulty: but to the BEST of my judgment we are not authorized in admitt
0390 have to admit that a sufficient number of the BEST adapted plants and animals have not been creat
0401 ganisms: a plant, for instance, would find the BEST fitted ground more perfectly occupied by disti
0419 t at present, almost arbitrary. Several of the BEST botanists, such as Mr. Bentham and others, hav
0420 families, sections, or orders. The reader will BEST understand what is meant, if he will take the
0422 rangement of the races of man would afford the BEST classification of the various languages now sp
0425 les and guides which have been followed by our BEST systematists. We have no written pedigrees; we
0444 or instance, to a bird which obtained its food BEST by having a long beak, whether or not it assum
0449 e been connected by the finest gradations, the BEST, or indeed, if our collections were nearly per
0196 which in their lungs or modified swimbladders BETRAY their aquatic origin, may perhaps be thus acc
0399 fication; the principle of inheritance still BETRAYING their original birthplace. Many analogous f
0426 characters, let them be ever so unimportant, BETRAYS the hidden bond of community of descent. Let
0005 times varying conditions of life, will have a BETTER chance of surviving, and thus be naturally se
0031 mpetent authorities. Youatt, who was probably BETTER acquainted with the works of agriculturists t
0037 riety, sowing its seeds, and, when a slightly BETTER variety has chanced to appear, selecting it,
0038 nstitutions or structure, would often succeed BETTER in the one country than in the other, and thu
0041 dual plants with slightly larger, earlier, or BETTER fruit, and raised seedlings from them, and ag
0050 ith much difficulty. We could hardly wish for BETTER evidence of the two forms being specifically
0061 spring. The offspring, also, will thus have a BETTER chance of surviving, for, of the many individ
0064 s, descended from the first pair. But we have BETTER evidence on this subject than mere theoretica
0067 d, even in regard to mankind, so incomparably BETTER known than any other animal. This subject has
0081 ly surrounded by barriers, into which new and BETTER adapted forms could not freely enter, we shou
0081 he economy of nature which would assuredly be BETTER filled up, if some of the original inhabitant
0082 red the individuals of any of the species, by BETTER adapting them to their altered conditions, wo
0082 favourable to natural selection, by giving a BETTER chance of profitable variations occurring; an
0083 e been modified with advantage, so as to have BETTER resisted such intruders. As man can produce a
0084 s productions; that they should be infinitely BETTER adapted to the most complex conditions of lif
0094 o obtain its food more quickly, and so have a BETTER chance of living and leaving descendants. Its
0100 d from flower to flower' and this will give a BETTER chance of pollen being occasionally carried f
0102 k. a large number of individuals, by giving a BETTER chance for the appearance within any given pe
0102 direction, though in different degrees, so as BETTER to fill up the unoccupied place. But if the a
0104 ued self fertilisation, that they will have a BETTER chance of surviving and propagating their kin
0104 re efficiently in checking the immigration of BETTER adapted organisms, after any physical change,
0105 great and open area, not only will there be a BETTER chance of favourable variations arising from
0108 places in the polity of nature, which can be BETTER occupied by some of the inhabitants of the co
0108 enerally very slow, and on the immigration of BETTER adapted forms having been checked. But the ac
0112 titution, and habits, by so much will they be BETTER enabled to seize on many and widely diversifi
0116 f any one species will succeed by so much the BETTER as they become more diversified in structure,
0117 thousand generations; but it would have been BETTER if each had represented ten thousand generati
0128 ore diversified these descendants become, the BETTER will be their chance of succeeding in the bat
0133 animals of the same species have thicker and BETTER fur the more severe the climate is under whic
0136 shipwrecked near a coast, it would have been BETTER for the good swimmers if they had been able t
0136 wim still further, whereas it would have been BETTER for the bad swimmers if they had not been abl
0142 ieties are said to withstand certain climates BETTER than others: this is very strikingly shown in
0144 st abnormal in their teeth. I know of no case BETTER adapted to show the importance of the laws of
0147 al plants producing seeds which were a little BETTER fitted to be wafted further, might get an adv
0148 sed, each individual Proteolepas would have a BETTER chance of supporting itself, by less nutrimen
0149 whilst a tool for some particular object had BETTER be of some particular shape. Natural selectio
0177 existing in larger numbers will always have a BETTER chance, within any given period, of presentin
0178 e in the natural polity of the country can be BETTER filled by some modification of some one or mo
0184 f the supply of insects were constant, and if BETTER adapted competitors did not already exist in
0189 the million; and each to be preserved till a BETTER be produced, and then the old ones to be dest
0195 pose by successive slight modifications, each BETTER and better, for so trifling an object as driv
0195 cessive slight modifications, each better and BETTER, for so trifling an object as driving away fl
0219 mica (Polyerges) rufescens by Pierre Huber, a BETTER observer even than his celebrated father. Thi
0236 and the perfect females, would have been far BETTER exemplified by the hive bee. If a working ant
0248 rtility ends and sterility begins. I think no BETTER evidence of this can be required than that th
0289 ccidental and rare is their preservation, far BETTER than pages of detail. Nor is their rarity sur
0312 the geological succession of organic beings, BETTER accord with the common view of the immutabili
0326 e that a number of individuals, from giving a BETTER chance of the appearance of favourable variat
0360 m. Martens tried similar ones, but in a much BETTER manner, for he placed the seeds in a box in t
0388 the waters of a foreign country, would have a BETTER chance of seizing on a place, than in the cas
0396 eloped into trees, etc., seem to me to accord BETTER with the view of occasional means of transpor
0403 lonists having been subsequently modified and BETTER fitted to their new homes, is of the widest a
0406 consequently the lower forms will have had a BETTER chance of ranging widely and of still retaini
0406 eggs of many low forms being very minute and BETTER fitted for distant transportation, probably a
0406 together with the subsequent modification and BETTER adaptation of the colonists to their new home
0425 n this view are as good as, or even sometimes BETTER than, other parts of the organisation. We car
0439 he same class are often strikingly similar: a BETTER proof of this cannot be given, than a circums
0468 hose with which it comes into competition, or BETTER adaptation in however slight a degree to the
0487 igrated from some one birthplace; and when we BETTER know the many means of migration, then, by th
0488 independently created. To my mind it accords BETTER with what we know of the laws impressed on ma
0007 estic Varieties. Difficulty of distinguishing BETWEEN Varieties and Species. Origin of Domestic Va
0010 logists that there is no essential difference BETWEEN a bud and an ovule in their earliest stages
0011 re animal. In monstrosities, the correlations BETWEEN quite distinct parts are very curious; and m
0012 horns; pigeons with feathered feet have skin BETWEEN their outer toes; pigeons with short beaks h
0016 ct species. If any marked distinction existed BETWEEN domestic races and species, this source of d
0016 estimate the amount of structural difference BETWEEN the domestic races of the same species, we a
0019 an get only forms in some degree intermediate BETWEEN their parents; and if we account for our sev
0020 a race could be obtained nearly intermediate BETWEEN two extremely different races or species, I
0020 nd failed. The offspring from the first cross BETWEEN two pure breeds is tolerably and sometimes (
0020 mes apparent. Certainly, a breed intermediate BETWEEN two very distinct breeds could not be got wi
0022 cutellae on the toes, the development of skin BETWEEN the toes, are all points of structure which
0023 ld be shown him. Great as the differences are BETWEEN the breeds of pigeons, I am fully convinced
0026 ritance. Lastly, the hybrids or mongrels from BETWEEN all the domestic breeds of pigeons are perfe
0027 untries, we can make an almost perfect series BETWEEN the extremes of structure. Thirdly, those ch

Page ***(Key Word)***

Page **(Key Word)**

0029	y are strongly impressed with the differences BETWEEN the several races; and though they well know
0036	mr. Bakewell's flock, and yet the difference BETWEEN the sheep possessed by these two gentlemen i
0048	tirely vague and arbitrary is the distinction BETWEEN species and varieties. On the islets of the
0049	es peculiar to Great Britain. A wide distance BETWEEN the homes of two doubtful forms leads many n
0049	it has been well asked, will suffice? If that BETWEEN America and Europe is ample, will that betwe
0049	etween America and Europe is ample, will that BETWEEN the Continent and the Azores, or Madeira, or
0051	an hardly hope to find the intermediate links BETWEEN his doubtful forms, he will have to trust al
0051	ear line of demarcation has as yet been drawn BETWEEN species and sub species, that is, the forms
0051	ite arrive at the rank of species; or, again, BETWEEN sub species and well marked varieties, or be
0051	een sub species and well marked varieties, or BETWEEN lesser varieties and individual differences.
0056	nd this holds good. There are other relations BETWEEN the species of large genera and their record
0056	which intermediate links have not been found BETWEEN doubtful forms, naturalists are compelled to
0056	o a determination by the amount of difference BETWEEN them, judging by analogy whether or not the
0057	that in large genera the amount of difference BETWEEN the species is often exceedingly small. I ha
0057	ere is one most important point of difference BETWEEN varieties and species; namely, that the amou
0057	pecies; namely, that the amount of difference BETWEEN varieties, when compared with each other or
0057	their parent species, is much less than that BETWEEN the species of the same genus. But when we c
0058	be explained, and how the lesser differences BETWEEN varieties will tend to increase into the gre
0058	tend to increase into the greater differences BETWEEN species. There is one other point which seem
0060	ughtout nature. Struggle for life most severe BETWEEN individuals and varieties of the same specie
0060	d varieties of the same species; often severe BETWEEN species of the same genus. The relation of o
0065	omehow to be disposed of. The only difference BETWEEN organisms which annually produce eggs or see
0068	g food, it brings on the most severe struggle BETWEEN the individuals, whether of the same or of d
0070	avoured: and here comes in a sort of struggle BETWEEN the parasite and its prey. On the other hand
0071	x and unexpected are the checks and relations BETWEEN organic beings, which have to struggle toget
0072	he old planted clumps. But on looking closely BETWEEN the stems of the heath, I found a multitude
0074	e surrounding virgin forests. What a struggle BETWEEN the several kinds of trees must here have go
0075	cattering its seeds by the thousand; what war BETWEEN insect and insect, between insects, snails,
0075	thousand; what war between insect and insect, BETWEEN insects, snails, and other animals with bird
0075	as of a parasite on its prey, lies generally BETWEEN beings remote in the scale of nature. This i
0075	truggle almost invariably will be most severe BETWEEN the individuals of the same species, for the
0076	e, the struggle will generally be more severe BETWEEN species of the same genus, when they come in
0076	y come into competition with each other, than BETWEEN species of distinct genera. We see this in t
0076	see why the competition should be most severe BETWEEN allied forms, which fill nearly the same pla
0078	ly cold or dry, yet there will be competition BETWEEN some few species, or between the individuals
0078	l be competition between some few species, or BETWEEN the individuals of the same species, for the
0080	selection. On the generality of intercrosses BETWEEN individuals of the same species. Circumstanc
0085	in looking at many small points of difference BETWEEN species, which, as far as our ignorance perm
0088	n a struggle for existence, but on a struggle BETWEEN the males for possession of the females; the
0088	of other males. The war is, perhaps, severest BETWEEN males of polygamous animals, and these seem
0089	t, believe that there is the severest rivalry BETWEEN the males of many species to attract by sing
0096	reeders, that with animals and plants a cross BETWEEN different varieties, or between individuals
0096	lants a cross between different varieties, or BETWEEN individuals of the same variety but of anoth
0097	h flowers, there is a very curious adaptation BETWEEN the structure of the flower and the manner i
0098	ees would thus produce a multitude of hybrids BETWEEN distinct species; for if you bring on the sa
0101	rcross with other individuals, the difference BETWEEN hermaphrodites and unisexual species, as far
0101	conclude that in many organic beings, a cross BETWEEN two individuals is an obvious necessity for
0103	ver intercrossing took place would be chiefly BETWEEN the individuals of the same new variety. A l
0107	the land; and, consequently, the competition BETWEEN fresh water productions will have been less
0109	and infinite complexity of the coadaptations BETWEEN all organic beings, one with another and wit
0111	pecies. How, then, does the lesser difference BETWEEN varieties become augmented into the greater
0111	become augmented into the greater difference BETWEEN species? That this does habitually happen, w
0111	ual and large an amount of difference as that BETWEEN varieties of the same species and species of
0114	ly open to immigration, and where the contest BETWEEN individual and individual must be severe, we
0117	of record in a systematic work. The intervals BETWEEN the horizontal lines in the diagram, may rep
0120	on parent. If we suppose the amount of change BETWEEN each horizontal line in our diagram to be ex
0120	, we get eight species, marked by the letters BETWEEN a14 and m14, all descended from (A). Thus, a
0120	e amount of change supposed to be represented BETWEEN the horizontal lines. After fourteen thousan
0121	the competition will generally be most severe BETWEEN those forms which are most nearly related to
0121	d structure. Hence all the intermediate forms BETWEEN the earlier and later states, that is betwee
0121	between the earlier and later states, that is BETWEEN the less and more improved state of a specie
0123	the extreme amount of difference in character BETWEEN species a14 and z14 will be much greater tha
0123	es a14 and z14 will be much greater than that BETWEEN the most different of the original eleven sp
0124	ure. Having descended from a form which stood BETWEEN the two parent species (A) and (I), now supp
0124	l be in some degree intermediate in character BETWEEN the two groups descended from these species.
0124	ecies (F14) will not be directly intermediate BETWEEN them, but rather between types of the two gr
0124	irectly intermediate between them, but rather BETWEEN types of the two groups; and every naturalis
0124	en, in some degree, intermediate in character BETWEEN existing groups; and we can understand this
0125	new and modified descendants, will mainly lie BETWEEN the larger groups, which are all trying to i
0128	ll they come to equal the greater differences BETWEEN species of the same genus, or even of distin
0132	iking and complex co adaptations of structure BETWEEN one organic being and another, which we see
0144	. what can be more singular than the relation BETWEEN blue eyes and deafness in cats, and the tort
0144	h the female sex; the feathered feet and skin BETWEEN the outer toes in pigeons, and the presence
0144	our of their plumage; or, again, the relation BETWEEN the hair and teeth in the naked Turkish dog,
0144	atural selection, than that of the difference BETWEEN the outer and inner flowers in some Composit
0147	s, for I see hardly any way of distinguishing BETWEEN the effects, on the one hand, of a part bein
0152	truly said to be a constant struggle going on BETWEEN, on the one hand, the tendency to reversion
0153	am convinced, is the case. That the struggle BETWEEN natural selection on the one hand, and the t
0156	mpare, for instance, the amount of difference BETWEEN the males of gallinaceous birds, in which se
0156	ngly displayed, with the amount of difference BETWEEN their females; and the truth of this proposi
0157	e fact, that the secondary sexual differences BETWEEN the two sexes of the same species are genera
0158	amount of difference in these same characters BETWEEN closely allied species; that secondary sexua
0162	e, also, could be given of forms intermediate BETWEEN two other forms, which themselves must be do
0164	a large range of colour is included, from one BETWEEN brown and black to a close approach to cream
0165	oured drawings, which I have seen, of hybrids BETWEEN the ass and zebra, the legs were much more p
0166	riped most strongly displayed in hybrids from BETWEEN several of the most distinct species. Now ob

Page **(Key Word)**

Page **(Key Word)**

```
0167 ave acted in producing the lesser differences BETWEEN varieties of the same species, and the great
0167 the same species, and the greater differences BETWEEN species of the same genus. The external cond
0173 nal parent and all the transitional varieties BETWEEN its past and present states. Hence we ought
0174 lly disappearing. Hence the neutral territory  BETWEEN two representative species is generally narr
0175 with a comparatively narrow neutral territory  BETWEEN them, in which they become rather suddenly r
0176 he rule in the case of varieties intermediate  BETWEEN well marked varieties in the genus Balanus.
0176 n, that generally when varieties intermediate  BETWEEN two other forms occur, they are much rarer n
0178 species. In this case, intermediate varieties  BETWEEN the several representative species and their
0179 animals exist having every intermediate grade  BETWEEN truly aquatic and strictly terrestrial habit
0183 ore, we may conclude that transitional grades  BETWEEN structures fitted for very different habits
0185 the frigate bird, the deeply scooped membrane  BETWEEN the toes shows that structure has begun to c
0199 ise cause of the slight analogous differences   BETWEEN species. I might have adduced for this same
0199 adduced for this same purpose the differences  BETWEEN the races of man, which are so strongly mark
0201 to its possessor. If a fair balance be struck  BETWEEN the good and evil caused by each part, each
0208 out life. Several other points of resemblance  BETWEEN instincts and habits could be pointed out. A
0209 s does sometimes happen, then the resemblance  BETWEEN what originally was a habit and an instinct
0225 ing cells on the opposite side. In the series  BETWEEN the extreme perfection of the cells of the h
0225 melipona itself is intermediate in structure   BETWEEN the hive and humble bee, but more nearly rel
0226 the bees building perfectly flat walls of wax  BETWEEN the spheres which thus tend to intersect. He
0226 y this manner of building; 'for the flat walls BETWEEN the adjoining cells are not double, but are
0226 rallel layer; then, if planes of intersection  BETWEEN the several spheres in both layers be formed
0228 r. tegetmeier, I separated two combs, and put  BETWEEN them a long thick, square strip of wax: the
0228 lat walls of wax on the lines of intersection  BETWEEN the basins, so that each hexagonal prism was
0229 ly along the planes of imaginary intersection  BETWEEN the basins on the opposite sides of the ridg
0229 rge portions of a rhombic plate had been left  BETWEEN the opposed basins, but the work, from the u
0229 to have succeeded in thus leaving flat plates  BETWEEN the basins, by stopping work along the inter
0231 ion, that is, along the plane of intersection  BETWEEN two adjoining spheres. I have several specim
0232 truction seems to be a sort of balance struck  BETWEEN many bees, all instinctively standing at the
0232 leaving ungnawed, the planes of intersection   BETWEEN these spheres. It was really curious to note
0232 pheres, they can build up a wall intermediate  BETWEEN two adjoining spheres; but, as far as I have
0232 ircumstances a rough wall in its proper place  BETWEEN two just commenced cells, is important, as i
0236 ed, the prodigious difference in this respect  BETWEEN the workers and the perfect females, would h
0238 d, until that prodigious amount of difference  BETWEEN the fertile and sterile females of the same
0240 d gradations in important points of structure  BETWEEN the different castes of neuters in the same
0245 est die. Chapter VIII. Hybridism. Distinction  BETWEEN the sterility of first crosses and of hybrid
0245 of first crosses and of hybrids. Parallelism   BETWEEN the effects of changed conditions of life an
0246 t seems to make a broad and clear distinction  BETWEEN varieties and species. First, for the steril
0248 y nor fertility affords any clear distinction  BETWEEN species and varieties; but that the evidence
0249 lf which is to be fertilised; so that a cross  BETWEEN two flowers, though probably on the same pla
0250 ter and Gartner that some degree of sterility  BETWEEN distinct species is a universal law of natur
0250 commonly perfect, fertility in a first cross   BETWEEN two distinct species. This case of the Crinu
0252 t he raises stocks for grafting from a hybrid  BETWEEN Rhod. Ponticum and Catawbiense, and that thi
0252 ave no right to expect that the first crosses  BETWEEN them and the canary, or that their hybrids,
0256 e generally very sterile; but the parallelism  BETWEEN the difficulty of making a first cross, and
0256 systematic affinity is meant, the resemblance  BETWEEN species in structure and in constitution, mo
0257 d species. Now the fertility of first crosses  BETWEEN species, and of the hybrids produced from th
0257 rly shown by hybrids never having been raised  BETWEEN species ranked by systematists in distinct f
0257 uniting with facility. But the correspondence  BETWEEN systematic affinity and the facility of cros
0257 t persevering efforts have failed to produce   BETWEEN extremely close species a single hybrid. Eve
0258 n be crossed with ease. By a reciprocal cross  BETWEEN two species, I mean the case, for instance,
0258 ifference in the result of reciprocal crosses  BETWEEN the same two species was long ago observed b
0258 n in a lesser degree. He has observed it even  BETWEEN forms so closely related (as Matthiola annua
0259 aving, as is usual, an intermediate character  BETWEEN their two parents, always closely resemble o
0259 s which are usually intermediate in structure  BETWEEN their parents; exceptional and abnormal indi
0259 ly, that the facility of making a first cross  BETWEEN any two species is not always governed by th
0260 ement is clearly proved by reciprocal crosses  BETWEEN the same two species, for according as the o
0260 ifference in the result of a reciprocal cross  BETWEEN the same two species? Why, it may even be as
0260 ly related to the facility of the first union  BETWEEN their parents, seems to be a strange arrange
0261 limited a nature, that, in reciprocal crosses  BETWEEN two species the male sexual element of the o
0262 h there is a clear and fundamental difference  BETWEEN the mere adhesion of grafted stocks, and the
0263 h every kind of resemblance and dissimilarity  BETWEEN organic beings is attempted to be expressed.
0265 he great bar to the domestication of animals.  BETWEEN the sterility thus superinduced and that of
0267 als, there is abundant evidence, that a cross  BETWEEN very distinct individuals of the same specie
0267 inct individuals of the same species, that is  BETWEEN members of different strains or sub breeds,
0267 breeding continued during several generations  BETWEEN the nearest relations, especially if these b
0267 er hand, that slight crosses, that is crosses  BETWEEN the males and females of the same species wh
0267 le; and that greater crosses, that is crosses  BETWEEN males and females which have become widely o
0268 that there must be some essential distinction  BETWEEN species and varieties, and that there must b
0269 learly shown that mere external dissimilarity  BETWEEN two species does not determine their greater
0271 species, more seed is produced by the crosses  BETWEEN the same coloured flowers, than between thos
0271 osses between the same coloured flowers, than  BETWEEN those which are differently coloured. Yet th
0272 curence; or to form a fundamental distinction  BETWEEN varieties and species. The general fertility
0272 wish was to draw a marked line of distinction  BETWEEN species and varieties, could find very few a
0272 it seems to me, quite unimportant differences  BETWEEN the so called hybrid offspring of species, a
0272 his fact. Gartner further admits that hybrids  BETWEEN very closely allied species are more variabl
0274 ferences, which Gartner is able to point out,  BETWEEN hybrid and mongrel plants. On the other hand
0276 e view that there is no essential distinction  BETWEEN species and varieties. Summary of Chapter. F
0276 varieties. Summary of Chapter. First crosses   BETWEEN forms sufficiently distinct to be ranked as
0276 times widely different, in reciprocal crosses  BETWEEN the same two species. It is not always equal
0276 n our forests. The sterility of first crosses  BETWEEN pure species, which have their reproductive
0277 pend on the amount of difference of some kind  BETWEEN the species which are crossed. Nor is it sur
0277 attempts to express all kinds of resemblance   BETWEEN all species. First crosses between forms kno
0277 esemblance between all species. First crosses  BETWEEN forms known to be varieties, or sufficiently
0278 rtility, there is a close general resemblance  BETWEEN hybrids and mongrels. Finally, then, the fac
0279 iew, that there is no fundamental distinction  BETWEEN species and varieties. Chapter IX. On the Im
0280 turing to myself, forms directly intermediate  BETWEEN them. But this is a wholly false view; we sh
```

Page **(Key Word)**

Page **(Key Word)***

```
0280  we should always look for forms intermediate  BETWEEN  each species and a common but unknown progen
0280  ted, we should have an extremely close series  BETWEEN  both and the rock pigeon: but we should have
0280  hould have no varieties directly intermediate  BETWEEN  the fantail and pouter: none, for instance,
0281  that links ever existed directly intermediate  BETWEEN  them, but between each and an unknown common
0281  isted directly intermediate between them, but  BETWEEN  each and an unknown common parent. The commo
0281  e direct intermediate links will have existed  BETWEEN  them. But such a case would imply that one f
0281  t of change; and the principle of competition  BETWEEN  organism and organism, between child and par
0281  of competition between organism and organism,  BETWEEN  child and parent, will render this a very ra
0281  genus, by differences not greater than we see  BETWEEN  the varieties of the same species at the pre
0282  umber of intermediate and transitional links,  BETWEEN  all living and extinct species, must have be
0284  feet in thickness on the Continent. Moreover,  BETWEEN  each successive formation, we have, in the o
0288  the very many animals which live on the beach  BETWEEN  high and low watermark are preserved. For in
0289  sia, what wide gaps there are in that country  BETWEEN  the superimposed formations: so it is in Nor
0289  ormed of the length of time which has elapsed  BETWEEN  the consecutive formations, we may infer tha
0290  lief of vast intervals of time having elapsed  BETWEEN  each formation. But we can, I think, see why
0293  not therein find closely graduated varieties  BETWEEN  the allied species which lived at its commen
0293  hould not include a graduated series of links  BETWEEN  the species which then lived: but I can by n
0295  sent day. In order to get a perfect gradation  BETWEEN  two forms in the upper and lower parts of th
0295  ntinues. In fact, this nearly exact balancing  BETWEEN  the supply of sediment and the amount of sub
0296  adations which must on my theory have existed  BETWEEN  them, but abrupt, though perhaps very slight
0297  with a somewhat greater amount of difference  BETWEEN  any two forms, they rank both as species, un
0297  ing bed; even if A were strictly intermediate  BETWEEN  B and C, it would simply be ranked as a thir
0297  t at all necessarily be strictly intermediate  BETWEEN  them in all points of structure. So that we
0298  ne country all the early stages of transition  BETWEEN  any two forms, is small, for the successive
0299  nd extinct genera, and has made the intervals  BETWEEN  some few groups less wide than they otherwis
0299  ely anything in breaking down the distinction  BETWEEN  species, by connecting them together by nume
0301  eservation of all the transitional gradations  BETWEEN  any two or more species. If such gradations
0302  ot discovering innumerable transitional links  BETWEEN  the species which appeared at the commenceme
0303  ntervals of time, which have probably elapsed  BETWEEN  our consecutive formations, longer perhaps.i
0304  er had been suddenly produced in the interval  BETWEEN  the latest secondary and earliest tertiary f
0306  present characters in any degree intermediate  BETWEEN  them. If, moreover, they had been the progen
0308  what was the state of things in the intervals  BETWEEN  the successive formations: whether Europe an
0310  ations infinitely numerous transitional links  BETWEEN  the many species which now exist or have exi
0312  s: and every year tends to fill up the blanks  BETWEEN  them, and to make the percentage system of l
0313  cession of our geological formations: so that  BETWEEN  each two consecutive formations, the forms o
0320  rmerly explained and illustrated by examples,  BETWEEN  the forms which are most like each other in
0322  y said on the probable wide intervals of time  BETWEEN  our consecutive formations: and in these int
0325  c beings, and find how slight is the relation  BETWEEN  the physical conditions of various countries
0328  , is able to draw a close general parallelism  BETWEEN  the successive stages in the two countries:
0329  mations and during the long intervals of time  BETWEEN  them: in this case, the several formations i
0329  e classed either in still existing groups, or  BETWEEN  them. That the extinct forms of life help to
0329  ms of life help to fill up the wide intervals  BETWEEN  existing genera, families, and orders, canno
0329  st, Owen, showing how extinct animals fall in  BETWEEN  existing groups. Cuvier ranked the Ruminants
0330  ine gradations the apparently wide difference  BETWEEN  the pig and the camel. In regard to the Inve
0330  p of species being considered as intermediate  BETWEEN  living species or groups. If by this term it
0330  s directly intermediate in all its characters  BETWEEN  two living forms, the objection is probably
0330  ation many fossil species would have to stand  BETWEEN  living species, and some extinct genera betw
0330  tween living species, and some extinct genera  BETWEEN  living genera, even between genera belonging
0330  me extinct genera between living genera, even  BETWEEN  genera belonging to distinct families. The m
0333  some slight degree intermediate in character  BETWEEN  their modified descendants, or between their
0333  racter between their modified descendants, or  BETWEEN  their collateral relations. In nature the ca
0334  ory will be intermediate in general character  BETWEEN  that which preceded and that which succeeded
0334  y fail to be nearly intermediate in character  BETWEEN  the forms of life above and below. We must,
0334  fication, during the long and blank intervals  BETWEEN  the successive formations. Subject to these
0334  iod undoubtedly is intermediate in character,  BETWEEN  the preceding and succeeding faunas. I need
0334  palaeontologists as intermediate in character  BETWEEN  those of the overlying carboniferous, and un
0334  te, as unequal intervals of time have elapsed  BETWEEN  consecutive formations. It is no real object
0334  s a whole is nearly intermediate in character  BETWEEN  the preceding and succeeding faunas, that ce
0335  ent rock pigeon now lives: and many varieties  BETWEEN  the rock pigeon and the carrier have become
0336  , and as long blank intervals have intervened  BETWEEN  successive formations, we ought not to expec
0336  two formations all the intermediate varieties  BETWEEN  the species which appeared at the commenceme
0339  wonderful relationship in the same continent  BETWEEN  the dead and the living. Professor Owen has
0339  . other cases could be added, as the relation  BETWEEN  the extinct and living land shells of Madeir
0339  xtinct and living land shells of Madeira; and  BETWEEN  the extinct and living brackish water shells
0342  tion, enormous intervals of time have elapsed  BETWEEN  the successive formations; that there has pr
0342  enormous intervals of time which have elapsed  BETWEEN  our consecutive formations: he may overlook
0344  and extinct forms often tend to fill up gaps  BETWEEN  existing forms, sometimes blending two group
0344  splays characters in some degree intermediate  BETWEEN  groups now distinct; for the more ancient a
0345  xtinct forms are seldom directly intermediate  BETWEEN  existing forms; but are intermediate only by
0346  ivisions in geographical distribution is that  BETWEEN  the New and Old Worlds: yet if we travel ove
0347  lia, South Africa, and western South America,  BETWEEN  latitudes 25 degrees and 35 degrees, we shal
0347  close and important manner to the differences  BETWEEN  the productions of various regions. We see t
0347  we see the same fact in the great difference  BETWEEN  the inhabitants of Australia, Africa, and So
0358  egree of relation (as we shall hereafter see)  BETWEEN  the distribution of mammals and the depth of
0371  the relationship, with very little identity,  BETWEEN  the productions of North America and Europe,
0374  pecies. In America, Dr. Hooker has shown that  BETWEEN  forty and fifty of the flowering plants of T
0388  ough there will always be a struggle for life  BETWEEN  the individuals of the species; however few,
0388  the competition will probably be less severe  BETWEEN  aquatic than between terrestrial species: co
0388  probably be less severe between aquatic than  BETWEEN  terrestrial species; consequently an intrude
0395  to a certain extent independent of distance,  BETWEEN  the depth of the sea separating an island fr
0396  nels, we can understand the frequent relation  BETWEEN  the depth of the sea and the degree of affin
0397  agos Archipelago, situated under the equator,  BETWEEN  500 and 600 miles from the shores of South A
0398  in climate, height, and size of the islands,  BETWEEN  the Galapagos and Cape de Verde Archipelagos
0399  h feeble, I am assured by Dr. Hooker is real,  BETWEEN  the flora of the south western corner of Aus
0400  e, or from each other. But this dissimilarity  BETWEEN  the endemic inhabitants of the islands may b
0404  here has been intercommunication or migration  BETWEEN  the two regions. And wherever many closely a
```

Page **(Key Word)***

Page **(Key Word)**

```
0404  in the process of modification. This relation BETWEEN the power and extent of migration of a speci
0409  we can see why there should be some relation BETWEEN the presence of mammals, in a more or less m
0409  modified condition, and the depth of the sea BETWEEN an island and the mainland. We can clearly s
0411  ion separates and defines groups. Morphology, BETWEEN members of the same class, between parts of
0411  orphology, between members of the same class, BETWEEN parts of the same individual. Embryology, la
0416  serviceable in exhibiting the close affinity BETWEEN Ruminants and Pachyderms. Robert Brown has s
0418  apparent bond of connexion can be discovered BETWEEN them, especial value is set on them. As in m
0420  naturalists consider as showing true affinity BETWEEN any two or more species, are those which hav
0421  o that the amount or value of the differences BETWEEN organic beings all related to each other in
0421  or f originally was intermediate in character BETWEEN A and I, and the several genera descended fr
0427  n these views, the very important distinction BETWEEN real affinities and analogical or adaptive r
0427  the body and in the fin like anterior limbs, BETWEEN the dugong, which is a pachydermatous animal
0427  s a pachydermatous animal, and the whale, and BETWEEN both these mammals and fishes, is analogical
0427  alogies which have been drawn by some authors BETWEEN very distinct animals. On my view of charact
0428  serve as characters exhibiting true affinity BETWEEN the several members of the whale family; for
0428  rical parallelism has sometimes been observed BETWEEN the sub groups in distinct classes. A natura
0429  characters in some slight degree intermediate BETWEEN existing groups. A few old and intermediate
0431  t is difficult to show the blood relationship BETWEEN the numerous kindred of any ancient and nobl
0431  m, the various affinities which they perceive BETWEEN the many living and extinct members of the s
0431  t part in defining and widening the intervals BETWEEN the several groups in each class. We may thu
0432  ould blend together by steps as fine as those BETWEEN the finest existing varieties; nevertheless
0432  modified descendants. Every intermediate link BETWEEN these eleven genera and there primordial par
0432  e alive; and the links to be as fine as those BETWEEN the finest varieties. In this case it would
0432  general idea of the value of the differences BETWEEN them. This is what we should be driven to, i
0433  ts arrangement, with the grades of difference BETWEEN the descendants from a common parent, expres
0434  isentangle the inextricable web of affinities BETWEEN the members of any one class; but when we ha
0442  ch accustomed to see differences in structure BETWEEN the embryo and the adult, and likewise a clo
0442  al, but not universal difference in structure BETWEEN the embryo and the adult; of parts in the sa
0445  y, as I think it probable that the difference BETWEEN these two breeds has been wholly caused by s
0456  f no importance; the wide opposition in value BETWEEN analogical or adaptive characters, and chara
0461  changes in the conditions of life and crosses BETWEEN greatly modified forms, lessen fertility; an
0461  changes in the conditions of life and crosses BETWEEN less modified forms, increase fertility. Tur
0463  cases) to discover directly connecting links BETWEEN them, but only between each and some extinct
0463  ectly connecting links between them, but only BETWEEN each and some extinct and supplanted form. E
0463  ination of an infinitude of connecting links, BETWEEN the living and extinct inhabitants of the wo
0463  s of the world, and at each successive period BETWEEN the extinct and still older species, why is
0464  wise possessed many of the intermediate links BETWEEN their past or parent and present states; and
0464  s are varieties? As long as most of the links BETWEEN any two species are unknown, if any one link
0466  know all the possible transitional gradations BETWEEN the simplest and the most perfect organs; it
0468  r, the struggle will generally be most severe BETWEEN them; it will be almost equally severe betwe
0468  etween them; it will be almost equally severe BETWEEN the varieties of the same species, and next
0468  ies of the same species, and next in severity BETWEEN the species of the same genus. But the strug
0468  s. but the struggle will often be very severe BETWEEN beings most remote in the scale of nature. T
0468  sexes there will in most cases be a struggle BETWEEN males for possession of the females. The mos
0469  works. No one can draw any clear distinction BETWEEN individual differences and slight varieties;
0469  dividual differences and slight varieties; or BETWEEN more plainly marked varieties and sub specie
0469  t is that no line of demarcation can be drawn BETWEEN species, commonly supposed to have been prod
0475  eing in some degree intermediate in character BETWEEN the fossils in the formations above and belo
0476  oftener it stands in some degree intermediate BETWEEN existing and allied groups. Recent forms are
0478  oup, and with extinct groups often falling in BETWEEN recent groups, is intelligible on the theory
0481  clear distinction has been, or can be, drawn BETWEEN species and well marked varieties. It cannot
0483  sometimes tend to fill up very wide intervals BETWEEN existing orders. Organs in a rudimentary con
0485  at present; for differences, however slight, BETWEEN any two forms, if not blended by intermediat
0485  lled to acknowledge that the only distinction BETWEEN species and well marked varieties is, that t
0485  present existence of intermediate gradations BETWEEN any two forms, we shall be led to weigh more
0485  value higher the actual amount of difference BETWEEN them. It is quite possible that forms now ge
0487  nce of circumstances; and the blank intervals BETWEEN the successive stages as having been of vast
0225  prism, with the basal edges of its six sides BEVELLED so as to join on to a pyramid, formed of thr
0056  r of varieties, that is of incipient species, BEYOND the average. It is not that all large genera
0160  ce of the slaty blue, with the several marks, BEYOND the influence of the mere act of crossing on
0232  gon, in its strictly proper place, projecting BEYOND the other completed cells. It suffices that t
0235  ke a comb as perfect as that of the hive bee, BEYOND this stage of perfection in architecture, nat
0246  cases being looked on as a special endowment, BEYOND the province of our reasoning powers. The fer
0255  normal cases, even to an excess of fertility, BEYOND that which the plant's own pollen will produc
0301  the archipelago now range thousands of miles BEYOND its confines; and analogy leads me to believe
0306  r ignorance of the geology of other countries BEYOND the confines of Europe and the United States;
0485  avage looks at a ship, as at something wholly BEYOND his comprehension; when we regard every produ
0222  rbed both nests, the little ants attacked their BIG neighbours with surprising courage. Now I was
0241  orkmen had heads four instead of three times as BIG as those of the smaller men, and jaws nearly f
0241  the smaller men, and jaws nearly five times as BIG. The jaws, moreover, of the working ants of th
0027  . birch informs me that pigeons are given in a BILL of fare in the previous dynasty. In the time o
0027  ointed out to me by Professor Lepsius; but Mr. BIRCH informs me that pigeons are given in a bill o
0004  a certain unknown number of generations, some BIRD had given birth to a woodpecker, and some plan
0021  skulls. The carrier, more especially the male BIRD, is also remarkable from the wonderful develop
0021  ling in the air head over heels. The runt is a BIRD of great size, with long, massive beak and lar
0039  artly methodical selection. Perhaps the parent BIRD of all fantails had only fourteen tailfeathers
0061  s to the hairs of a quadruped or feathers of a BIRD; in the structure of the beetle which dives th
0066  ag, yet it is believed to be the most numerous BIRD in the world. One fly deposits hundreds of egg
0087  beak of a full grown pigeon very short for the BIRD's own advantage, the process of modification w
0090  hardly be either useful or ornamental to this BIRD; indeed, had the tuft appeared under domestica
0134  , there is no greater anomaly in nature than a BIRD that cannot fly; yet there are several in this
0153  we do not expect to fail so far as to breed a BIRD as coarse as a common tumbler from a good shor
0161  h produces most rarely a blue and black barred BIRD, there has been a tendency in each generation
0182  ed for any particular habit, as the wings of a BIRD for flight, we should bear in mind that animal
0185  and no one except Audubon has seen the frigate BIRD, which has all its four toes webbed, alight on
0185  ction, though not in structure. In the frigate BIRD, the deeply scooped membrane between the toes
```

Page **(Key Word)**

Page **(Key Word)**

```
0197 tiful adaptation to hide this tree frequenting BIRD from its enemies; and consequently that it was
0200 bed feet of the upland goose or of the frigate BIRD are of special use to these birds; we cannot b
0200 genitor of the upland goose and of the frigate BIRD, webbed feet no doubt were as useful as they n
0217  known occasionally to lay their eggs in other BIRD's nests. Now let us suppose that the ancient p
0217 t that occasionally she laid an egg in another BIRD's nest.If the old bird profited by this occasi
0217  laid an egg in another bird's nest.If the old BIRD profited by this occasional habit, or if the y
0217 n of the mistaken maternal instinct of another BIRD, than by their own mother's care, encumbered a
0227 hardly more wonderful than those which guide a  BIRD to make its nest, I believe that the hive bee
0243  a habit wholly unlike that of any other known  BIRD. Finally, it may not be a logical deduction, b
0252 ve been fairly tried: for instance, the canary  BIRD has been crossed with nine other finches, but
0361 ious seeds passing through the intestines of a  BIRD; but hard seeds of fruit will pass uninjured t
0362 y trial, the germination of seeds; now after a  BIRD has found and devoured a large supply of food,
0362 ss into the gizzard for 12 or even 18 hours. A  BIRD in this interval might easily be blown to the
0363 rried brushwood, bones, and the nest of a land  BIRD, I can hardly doubt that they must occasionall
0385 duck's feet, which might represent those of a   BIRD sleeping in a natural pond, in an aquarium, wh
0387 digested; or the seeds might be dropped by the  BIRD whilst feeding its young, in the same way as f
0390 hus in the Galapagos Islands nearly every land  BIRD, but only two out of the eleven marine birds,
0391 culiar soil, does not possess one endemic land  BIRD; and we know from Mr. J. M. Jones's admirable
0391  island. Madeira does not possess one peculiar  BIRD, and many European and African birds are almos
0439 annot now tell whether it be that of a mammal,  BIRD, or reptile. The vermiform larvae of moths, fl
0440 d in the womb of its mother, in the egg of the  BIRD which is hatched in a nest, and in the spawn o
0444 life. It would not signify, for instance, to a  BIRD which obtained its food best by having a long
0471 cable on this theory. How strange it is that a  BIRD, under the form of woodpecker, should have bee
0479 ng at the embryo of an air breathing mammal or  BIRD having branchial slits and arteries running in
0003  has seeds that must be transported by certain  BIRDS, and which has flowers with separate sexes ab
0009 ntigrades or bear family; whereas, carnivorous  BIRDS, with the rarest exceptions, hardly ever lay
0021 xpanded, and are carried so erect that in good  BIRDS the head and tail touch; the oil gland is qui
0022  the state of the down with which the nestling  BIRDS are clothed when hatched. The shape and size
0022 hologist, and he were told that they were wild  BIRDS, would certainly, I think, be ranked by him a
0023 ust have become extinct in the wild state. But  BIRDS breeding on precipices, and good fliers, are
0025 e domestic breeds, taking thoroughly well bred  BIRDS, all the above marks, even to the white edgin
0025 concur perfectly developed. Moreover, when two  BIRDS belonging to two distinct breeds are crossed,
0025 rbs, and they produced mottled brown and black  BIRDS; these I again crossed together, and one gran
0028 rchs of Iran and Turan sent him some very rare  BIRDS; and, continues the courtly historian, His Ma
0028 y species of finches, or other large groups of  BIRDS, in nature. One circumstance has struck me mu
0042 numbers and at a very quick rate, and inferior  BIRDS may be freely rejected, as when killed they s
0048 e common in separated areas. How many of those  BIRDS and insects in North America and Europe, whic
0048 when comparing, and seeing others compare, the  BIRDS from the separate islands of the Galapagos Ar
0062 of food; we do not see, or we forget, that the  BIRDS which are idly singing round us mostly live o
0062 eir eggs, or their nestlings, are destroyed by  BIRDS and beasts of prey; we do not always bear in
0063 ach other. As the missletoe is disseminated by  BIRDS, its existence depends on birds; and it may m
0063 isseminated by birds, its existence depends on  BIRDS; and it may metaphorically be said to struggl
0063  other fruit bearing plants, in order to tempt  BIRDS to devour and thus disseminate its seeds rath
0068 winter of 1854 55 destroyed four fifths of the  BIRDS in my own grounds; and this is a tremendous d
0070 re in great excess compared with the number of  BIRDS which feed on them; nor can the birds, though
0070 umber of birds which feed on them; nor can the  BIRDS, though having a super abundance of food at t
0071 have been still greater, for six insectivorous  BIRDS were very common in the plantations, which we
0071 quented by two or three distinct insectivorous  BIRDS. Here we see how potent has been the effect o
0072 habitually checked by some means, probably by   BIRDS. Hence, if certain insectivorous birds (whose
0072 ably by birds. Hence, if certain insectivorous  BIRDS (whose numbers are probably regulated by hawk
0073  have seen in Staffordshire, the insectivorous  BIRDS, and so onwards in ever increasing circles of
0073 plexity. We began this series by insectivorous  BIRDS, and we have ended with them. Not that in nat
0075 etween insects, snails, and other animals with  BIRDS and beasts of prey, all striving to increase,
0084 lieve that these tints are of service to these  BIRDS and insects in preserving them from danger. G
0084 numbers; they are known to suffer largely from  BIRDS of prey; and hawks are guided by their eyesig
0087 ocoon, or the hard tip to the beak of nestling  BIRDS, used for breaking the egg. It has been asser
0087 ously the most rigorous selection of the young  BIRDS within the egg, which had the most powerful a
0088 nt for victory, as the sword or spear. Amongst  BIRDS, the contest is often of a more peaceful char
0089 inging the females. The rock thrush of Guiana,  BIRDS of Paradise, and some others, congregate; and
0089 ve partner. Those who have closely attended to  BIRDS in confinement well know that they often take
0089 eacock was eminently attractive to all his hen  BIRDS. It may appear childish to attribute any effe
0089 i can see no good reason to doubt that female   BIRDS, by selecting, during thousands of generation
0089 with respect to the plumage of male and female  BIRDS, in comparison with the plumage of the young,
0089 modified by sexual selection, acting when the   BIRDS have come to the breeding age or during the b
0103 nce in animals of this nature, for instance in  BIRDS, varieties will generally be confined to sepa
0112 th tumbler pigeons) choosing and breeding from  BIRDS with longer and longer beaks, or with shorter
0132 h or from greater depths. Gould believes that   BIRDS of the same species are more brightly coloure
0133 an those confined to cold and deeper seas. The  BIRDS which are confined to continents are, accordi
0134 c aylesbury duck. As the larger ground feeding  BIRDS seldom take flight except to escape danger, I
0134  that the nearly wingless condition of several  BIRDS, which now inhabit or have lately inhabited s
0144 at the diversity in the shape of the pelvis in  BIRDS causes the remarkable diversity in the shape
0144 the presence of more or less down on the young  BIRDS when first hatched, with the future colour of
0151 s than do other species of distinct genera. As  BIRDS within the same country vary in a remarkably
0156 f difference between the males of gallinaceous  BIRDS, in which secondary sexual characters are str
0159 al appearance in all the breeds, of slaty blue  BIRDS with two black bars on the wings, a white rum
0180 l in its own country, by enabling it to escape  BIRDS or beasts of prey, or to collect food more qu
0181 er than for flight. If about a dozen genera of  BIRDS had become extinct or were unknown, who would
0182 who would have ventured to have surmised that   BIRDS might have existed which used their wings sol
0182 he apteryx. Yet the structure of each of these  BIRDS is good for it, under the conditions of life
0182 om disuse, indicate the natural steps by which  BIRDS have acquired their perfect power of flight!
0182 ve on the land, and seeing that we have flying  BIRDS and mammals, flying insects of the most diver
0184 creeper. It often, like a shrike, kills small   BIRDS by blows on the head; and I have many times s
0184 ee! Petrels are the most aerial and oceanic of  BIRDS, yet in the quiet Sounds of Tierra del Fuego,
0186 urprise that there should be geese and frigate  BIRDS with webbed feet, either living on the dry la
0197 l; but as sutures occur in the skulls of young  BIRDS and reptiles, which have only to escape from
0200 f the frigate bird are of special use to these  BIRDS: we cannot believe that the same bones in the
```

Page **(Key Word)**

Page **(Key Word)**

```
0200 s they now are to the most aquatic of existing BIRDS. So we may believe that the progenitor of the
0207 cuckoo to migrate and to lay her eggs in other BIRDS' nests. An action, which we ourselves should
0212 in its total loss. So it is with the nests of BIRDS, which vary partly in dependence on the situa
0212 nstinctive quality, as may be seen in nestling BIRDS, though it is strenghthened by experience, an
0212 land, in the greater wildness of all our large BIRDS than of our small birds: for the large birds
0212 dness of all our large birds than of our small BIRDS: for the large birds have been most persecute
0212 e birds than of our small birds: for the large BIRDS have been most persecuted by man. We may safe
0212 ly attribute the greater wildness of our large BIRDS to this cause; for in uninhabited islands lar
0212 o this cause; for in uninhabited islands large BIRDS are not more fearful than small: and the magp
0214 ch, as I have witnessed, is performed by young BIRDS, that have never seen a pigeon tumble. We may
0216 purpose of allowing, as we see in wild ground BIRDS, their mother to fly away. But this instinct
0216 hich leads the cuckoo to lay her eggs in other BIRDS' nests: the slave making instinct of certain
0217 unincubated, or there would be eggs and young BIRDS of different ages in the same nest. If this w
0217 can cuckoo occasionally lays her eggs in other BIRDS' nests; but I hear on the high authority of D
0217 ess, I could give several instances of various BIRDS which have been known occasionally to lay the
0217 different ages at the same time: then the old BIRDS or the fostered young would gain an advantage
0217 r turn would be apt to lay their eggs in other BIRDS' nests, and thus be successful in rearing the
0218 for her own offspring. The occasional habit of BIRDS laying their eggs in other birds' nests eithe
0218 onal habit of birds laying their eggs in other BIRDS' nests either of the same or of a distinct sp
0237 m is active, as in the nuptial plumage of many BIRDS, and in the hooked jaws of the male salmon. W
0253 but from different hatches; and from these two BIRDS he raised no less than eight hybrids (grandch
0264 d great experience in hybridising gallinaceous BIRDS, that the early death of the embryo is a very
0314 inent of Europe, whereas the marine shells and BIRDS have remained unaltered. We can perhaps under
0339 hor's restorations of the extinct and gigantic BIRDS of New Zealand. We see it also in the birds o
0339 ic birds of New Zealand. We see it also in the BIRDS of the caves of Brazil. Mr. Woodward has show
0349 ars from closely allied, yet distinct kinds of BIRDS, notes nearly similar, and sees their nests s
0361 ation. Again, I can show that the carcasses of BIRDS, when floating on the sea, sometimes escape b
0361 d seeds of many kinds in the crops of floating BIRDS long retain their vitality: peas and vetches,
0361 , to my surprise nearly all germinated. Living BIRDS can hardly fail to be highly effective agents
0361 i could give many facts showing how frequently BIRDS of many kinds are blown by gales to vast dist
0361 kinds of seeds, out of the excrement of small BIRDS, and these seemed perfect, and some of them,
0361 following fact is more important: the crops of BIRDS do not secrete gastric juice, and do not in t
0362 les, and hawks are known to look out for tired BIRDS, and the contents of their torn crops might t
0362 twenty one hours in the stomachs of different BIRDS of prey; and two seeds of beet grew after hav
0362 water plants: fish are frequently devoured by BIRDS, and thus the seeds might be transported from
0362 to fishing eagles, storks, and pelicans; these BIRDS after an interval of many hours, either rejec
0362 y this process. Although the beaks and feet of BIRDS are generally quite clean, I can show that ea
0364 be long carried in the crops or intestines of BIRDS. These means, however, would suffice for occa
0364 ur climate. Almost every year, one or two land BIRDS are blown across the whole Atlantic Ocean, fr
0365 re land, with few or no destructive insects or BIRDS living there, nearly every seed, which chance
0385 their ova are not likely to be transported by BIRDS, and they are immediately killed by sea water
0386 eres in some quantity to the feet and beaks of BIRDS. Wading birds, which frequent the muddy edges
0386 uantity to the feet and beaks of birds. Wading BIRDS, which frequent the muddy edges of ponds, if
0386 , would be the most likely to have muddy feet. BIRDS of this order I can show are the greatest wan
0387 would be an inexplicable circumstance if water BIRDS did not transport the seeds of fresh water pl
0387 w water lily and Potamogeton. Herons and other BIRDS, century after century, have gone on daily de
0388 ges by animals, more especially by fresh water BIRDS, which have large powers of flight, and natur
0390 emic land shells in Madeira, or of the endemic BIRDS in the Galapagos Archipelago, with the number
0390 nd bird, but only two out of the eleven marine BIRDS, are peculiar; and it is obvious that marine
0391 s, are peculiar; and it is obvious that marine BIRDS could arrive at these islands more easily tha
0391 arrive at these islands more easily than land BIRDS. Bermuda, on the other hand, which lies at ab
0391 ount of Bermuda, that very many North American BIRDS, during their great annual migrations, visit
0391 e peculiar bird, and many European and African BIRDS are almost every year blown there, as I am in
0391 ds of Bermuda and Madeira have been stocked by BIRDS, which for long ages have struggled together
0391 d or floating timber, or to the feet of wading BIRDS, might be transported far more easily than la
0391 reptiles, and in New Zealand gigantic wingless BIRDS, take the place of mammals. In the plants of
0397 ccasionally crawl on and adhere to the feet of BIRDS roosting on the ground, and thus get transpor
0398 american continent. There are twenty six land BIRDS, and twenty five of these are ranked by Mr. G
0398 here; yet the close affinity of most of these BIRDS to American species in every character, in th
0402 in the Galapagos Archipelago, many even of the BIRDS, though so well adapted for flying from islan
0403 nhabited by a vast number of distinct mammals, BIRDS, and plants. The principle which determines t
0403 thus we have in South America, Alpine humming BIRDS, Alpine rodents, Alpine plants, etc., all of
0404 marked to me long ago, that in those genera of BIRDS which range over the world, many of the speci
0405 y, as in the case of certain powerfully winged BIRDS, will necessarily range widely; for we should
0417 degree of affinity of this strange creature to BIRDS and reptiles, as an approach in structure in
0419 to define a number of characters common to all BIRDS; but in the case of crustaceans, such definit
0419 ecessity of this practice in certain groups of BIRDS; and it has been followed by several entomolo
0431 hole classes from each other, for instance, of BIRDS from all other vertebrate animals, by the bel
0431 y lost, through which the early progenitors of BIRDS were formerly connected with the early progen
0437 explain the same construction in the skulls of BIRDS. Why should similar bones have been created i
0439 , sometimes lasts till a rather late age: thus BIRDS of the same genus, and of closely allied gene
0445 ons, carriers, and tumblers. Now some of these BIRDS, when mature, differ so extraordinarily in le
0445 een natural productions. But when the nestling BIRDS of these several breeds were placed in a row,
0445 were incomparably less than in the full grown BIRDS. Some characteristic points of difference, fo
0446 s in a state of nature. Let us take a genus of BIRDS, descended on my theory from some one parent
0450 of mammals: I presume that the bastard wing in BIRDS may be safely considered as a digit in a rudi
0451 be detected in the beaks of certain embryonic BIRDS. Nothing can be plainer than that wings are f
0454 s inhabiting dark caverns, and of the wings of BIRDS inhabiting oceanic islands, which have seldom
0479 we can clearly see why the embryos of mammals, BIRDS, reptiles, and fishes should be so closely al
0489 , clothed with many plants of many kinds, with BIRDS singing on the bushes, with various insects f
0004 own number of generations, some bird had given BIRTH to a woodpecker, and some plant to the missel
0048 ul forms! Amongst animals which unite for each BIRTH, and which are highly locomotive, doubtful fo
0096 hat two individuals must always unite for each BIRTH; but in the case of hermaphrodites this is fa
0096 e other large groups of animals, pair for each BIRTH. Modern research has much diminished the numb
0101 o individuals is an obvious necessity for each BIRTH; in many others it occurs perhaps only at lon
0103 most affect those animals which unite for each BIRTH, which wander much, and which do not breed at
```

Page **(Key Word)**

Page **(Key Word)**

```
0103  , and likewise in animals which unite for each BIRTH, but which wander little and which can increa
0103  of slow breeding animals, which unite for each BIRTH, we must not overrate the effects of intercro
0104  iently with those animals which unite for each BIRTH: but I have already attempted to show that we
0178  ially amongst the classes which unite for each BIRTH and wander much, may have separately been ren
0264  are differently circumstanced before and after BIRTH: when born and living in a country where thei
0264  nstitution of its mother, and therefore before BIRTH, as long as it is nourished within its mother
0356  all organisms which habitually unite for each BIRTH, or which often intercross: I believe that du
0470  es belonging to the larger groups tend to give BIRTH to new and dominant forms: so that each large
0481  illingness to admit that one species has given BIRTH to other and distinct species, is that we are
0483  miraculous act of creation than at an ordinary BIRTH. But do they really believe that at innumerab
0488  o secondary causes, like those determining the BIRTH and death of the individual. When I view all
0326  be prolonged and severe: and some from the other might be victorious BIRTHPLACE.
0354  each species having migrated from a single BIRTHPLACE; then, considering our ignorance with respe
0357  at each species has proceeded from a single BIRTHPLACE; and when in the course of time we know som
0389  therefore have all proceeded from a common BIRTHPLACE, notwithstanding that in the course of time
0399  inheritance still betraying their original BIRTHPLACE. Many analogous facts could be given: indee
0487  one parent, and have migrated from some one BIRTHPLACE; and when we better know the many means of
0229  des of the ridge of wax. In parts, only little BITS, in other parts, large portions of a rhombic p
0307  one. The presence of phosphatic nodules and BITUMINOUS matter in some of the lowest azoic rocks, p
0349  me plains of La Plata, we see the agouti and BIZCACHA, animals having nearly the same habits as ou
0349  cordillera and we find an alpine species of BIZCACHA: we look to the waters, and we do not find t
0430  rding to Mr. Waterhouse, of all Rodents, the BIZCACHA is most nearly related to Marsupials: but in
0430  to another. As the points of affinity of the BIZCACHA to Marsupials are believed to be real and no
0430  ppose either that all Rodents, including the BIZCACHA, branched off from some very ancient Marsupi
0430  ions. On either view we may suppose that the BIZCACHA has retained, by inheritance, more of the ch
0025  xternally edged with white: the wings have two BLACK bars; some semi domestic breeds and some appa
0025  rently truly wild breeds have, besides the two BLACK bars, the wings chequered with black. These's
0025  s the two black bars, the wings chequered with BLACK. These several marks do not occur together in
0025  e uniformly white fantails with some uniformly BLACK barbs, and they produced mottled brown and bl
0025  ack barbs, and they produced mottled brown and BLACK birds; these I again crossed together, and on
0025  grandchild of the pure white fantail and pure BLACK barb was of as beautiful a blue colour, with
0025  ful a blue colour, with the white rump, double BLACK wing bar, and barred and white edged tail fea
0084  the red grouse the colour of heather, and the BLACK grouse that of peaty earth, we must believe t
0085  destroy every lamb with the faintest trace of BLACK. In plants the down of the fruit and the colo
0111  ire, it is historically known that the ancient BLACK cattle were displaced by the long horns, and
0159  n all the breeds, of slaty blue birds with two BLACK bars on the wings, a white rump, a bar at the
0161  pigeon, which produces most rarely a blue and BLACK barred bird, there has been a tendency in eac
0164  colour is included, from one between brown and BLACK to a close approach to cream colour. I am awa
0165  pring subsequently produced from the mare by a BLACK Arabian sire, were much more plainly barred a
0184  ing them like a nuthatch. In North America the BLACK bear was seen by Hearne swimming for hours wi
0197  sted, and we did not know that there were many BLACK and pied kinds, I dare say that we should hav
0220  d in the nests of F. sanguinea. The slaves are BLACK and not above half the size of their red mast
0440  e whelp of a lion, or the spots on the young BLACKBIRD, are of any use to these animals, or are re
0204  n are at least possible. For instance, a swim BLADDER has apparently been converted into an air br
0452  r a distinct object: in certain fish the swim BLADDER seems to be rudimentary for its proper funct
0284  he opinion of most geologists, enormously long BLANK periods. So that the lofty pile of sedimentar
0289  e suspected that during the periods which were BLANK and barren in his own country, great piles of
0292  during such periods there will generally be a BLANK in the geological record. On the other hand,
0327  posited during periods of subsidence: and that BLANK intervals of vast duration occurred during th
0327  reserve organic remains. During these long and BLANK intervals I suppose that the inhabitants of e
0329  ation in one region often corresponding with a BLANK interval in the other, and if in both regions
0334  ge amount of modification, during the long and BLANK intervals between the successive formations.
0336  mation has often been interrupted, and as long BLANK intervals have intervened between successive
0465  ions are separated from each other by enormous BLANK intervals of time: for fossiliferous formatio
0465  ion and of stationary level the record will be BLANK. During these latter periods there will proba
0487  unusual concurrence of circumstances, and the BLANK intervals between the successive stages as ha
0173  le sediment is being deposited, there will be BLANKS in our geological history. The crust of the e
0312  y stages: and every year tends to fill up the BLANKS between them, and to make the percentage syst
0051  and individual differences. These differences BLEND into each other in an insensible series: and
0091  wly be formed. These varieties would cross and BLEND where they met: but to this subject of interc
0356  ve been submerged, and the two faunas will now BLEND or may formerly have blended: where the sea n
0432  distinguished from other groups, as all would BLEND together by steps as fine as those between th
0432  h, though at the actual fork the two unite and BLEND together. We could not, as I have said, defin
0037  not two or more species or races have become BLENDED together by crossing, may plainly be recogni
0214  ncts, which in a like manner become curiously BLENDED together, and for a long period exhibit trac
0242  eir instincts and structure would have become BLENDED. And nature has, as I believe, effected this
0266  rent structures and constitutions having been BLENDED into one. For it is scarcely possible that t
0279  ctness of specific forms, and their not being BLENDED together by innumerable transitional links,
0355  of improved varieties, which will never have BLENDED with other individuals or varieties, but wil
0356  wo faunas will now blend or may formerly have BLENDED: where the sea now extends, land may at a fo
0462  all around us? Why are not all organic beings BLENDED together in an inextricable chaos? With resp
0485  however slight, between any two forms, if not BLENDED by intermediate gradations, are looked at by
0015  , so that free intercrossing might check, by BLENDING together, any slight deviations of structure
0175  jects (however they may have become so), not BLENDING one into another by insensible gradations, t
0255  lity, in order to prevent their crossing and BLENDING together in utter confusion. The following r
0260  e it would be equally important to keep from BLENDING together? Why should the degree of sterili
0276  es of sterility to prevent them crossing and BLENDING in nature, than to think that trees have be
0344  ll up gaps between existing forms, sometimes BLENDING two groups previously classed as distinct in
0019  greyhound, the bloodhound, the bull dog, or BLENHEIM spaniel, etc., so unlike all wild Canidae, e
0137  d often caught them, that they were frequently BLIND: one which I kept alive was certainly in this
0137  habit the caves of Styria and of Kentucky, are BLIND. In some of the crabs the foot stalk for the
0137  ute their loss wholly to disuse. In one of the BLIND animals, namely, the cave rat, the eyes are o
0138  lar climate: so that on the common view of the BLIND animals having been separately created for th
0139  rational explanation of the affinities of the BLIND cave animals to the other inhabitants of the
0139  lous, as Agassiz has remarked in regard to the BLIND fish, the Amblyopsis, and as is the case with
0139  h, the Amblyopsis, and as is the case with the BLIND Proteus with reference to the reptiles of Eur
```

Page **(Key Word)**

```
0404  whole world. We see this same principle in the BLIND animals inhabiting the caves of America and o
0473  the burrowing tucutucu, which is occasionally BLIND, and then at certain moles, which are habitua
0473  nd then at certain moles, which are habitually BLIND and have their eyes covered with skin; or whe
0473  eyes covered with skin; or when we look at the BLIND animals inhabiting the dark caves of America
0138  the antennae or palpi, as a compensation for BLINDNESS. Notwithstanding such modifications, we mig
0483  ll be given as a curious illustration of the BLINDNESS of preconceived opinion. These authors seem
0026  g generation there will be less of the foreign BLOOD; but when there has been no cross with a dist
0036  has deviated in any one instance from the pure BLOOD of Mr. Bakewell's flock, and yet the differen
0141  domestic animals from several wild stocks: the BLOOD, for instance, of a tropical and arctic wolf
0160  s. after twelve generations, the proportion of BLOOD, to use a common expression, of any one ances
0160  ained by this very small proportion of foreign BLOOD. In a breed which has not been crossed, but i
0184  ndulatory flight, told me plainly of its close BLOOD relationship to our common species; yet it is
0341  wer genera and species will have left modified BLOOD descendants. Summary of the preceding and pre
0410  d the more nearly any two forms are related in BLOOD, the nearer they will generally stand to each
0418  rtant organs, such as those for propelling the BLOOD, or for aerating it, or those for propagating
0420  or groups, though allied in the same degree in BLOOD to their common progenitor, may differ greatl
0420  single species, are represented as related in BLOOD or descent to the same degree; they may metap
0421  ll related to each other in the same degree in BLOOD, has come to be widely different. Nevertheles
0424  are kept in the same group, because allied in BLOOD and alike in some other respects. If it could
0427  not reveal, will rather tend to conceal their BLOOD relationship to their proper lines of descent
0431  y predecessors. As it is difficult to show the BLOOD relationship between the numerous kindred of
0017  ance, it could be shown that the greyhound, BLOODHOUND, terrier, spaniel, and bull dog, which we a
0019  osely resembling the Italian greyhound, the BLOODHOUND, the bull dog, or Blenheim spaniel, etc., s
0020  st extreme forms, as the Italian greyhound, BLOODHOUND, bull dog, etc., in the wild state. Moreove
0029  s of a dray and race horse, a greyhound and BLOODHOUND, a carrier and tumbler pigeon. One of the m
0135  in many parts of the world are very frequently BLOWN to sea and perish; that the beetles in Madeir
0136  ad the best chance of surviving from not being BLOWN out to sea; and, on the other hand, those bee
0136  readily took to flight will oftenest have been BLOWN to sea and thus have been destroyed. The inse
0360  sea to another country; and when stranded, if BLOWN to a favourable spot by an inland gale, they
0361  showing how frequently birds of many kinds are BLOWN by gales to vast distances across the ocean.
0362  hours. A bird in this interval might easily be BLOWN to the distance of 500 miles, and hawks are k
0364  . almost every year, one or two land birds are BLOWN across the whole Atlantic Ocean, from North A
0385  uld be sure to alight on a pool or rivulet, if BLOWN across sea to an oceanic island or to any oth
0387  hen take flight and go to other waters, or are BLOWN across the sea; and we have seen that seeds r
0391  ropean and African birds are almost every year BLOWN there, as I am informed by Mr. F. V. Harcourt
0402  ose the mocking thrush of Chatham Island to be BLOWN to Charles Island, which has its own mocking
0067  close together and driven inwards by incessant BLOWS, sometimes one wedge being struck, and then a
0184  it often, like a shrike, kills small birds by BLOWS on the head; and I have many times seen and h
0012  orrelation are quite whimsical: thus cats with BLUE eyes are invariably deaf; colour and constitut
0025  e consideration. The rock pigeon is of a slaty BLUE, and has a white rump (the Indian sub species,
0025  stinct breeds are crossed, neither of which is BLUE or has any of the above specified marks, the m
0025  tail and pure black barb was of as beautiful a BLUE colour, with the white rump, double black wing
0027  in most other respects to the rock pigeon; the BLUE colour and various marks occasionally appearin
0144  can be more singular than the relation between BLUE eyes and deafness in cats, and the tortoise sh
0154  if some species in a large genus of plants had BLUE flowers and some had red, the colour would by
0154  r, and no one would be surprised at one of the BLUE species varying into red, or conversely; but i
0154  red, or conversely; but if all the species had BLUE flowers, the colour would become a generic cha
0159  asional appearance in all the breeds, of slaty BLUE birds with two black bars on the wings, a whit
0160  of life to cause the reappearance of the slaty BLUE, with the several marks, beyond the influence
0161  the barb pigeon, which produces most rarely a BLUE and black barred bird, there has been a tenden
0162  of the markings, which are correlated with the BLUE tint, and which it does not appear probable wo
0162  pecially we might have inferred this, from the BLUE colour and marks so often appearing when disti
0166  are crossed, we see a strong tendency for the BLUE tint and bars and marks to reappear in the mon
0247  a fertile seed; as he found the common red and BLUE pimpernels (Anagallis arvensis and coerulea),
0268  most naturalists as species. For instance, the BLUE and red pimpernel, the primrose and cowslip, w
0288  usly large proportion of the ocean, the bright BLUE tint of the water bespeaks its purity. The man
0473  estic breeds of pigeon have descended from the BLUE and barred rock pigeon! On the ordinary view o
0162  iations; but we might have inferred that the BLUENESS was a case of reversion, from the number of
0025  ecies, C. intermedia of Strickland, having it BLUISH); the tail has a terminal dark bar, with the
0166  three sub species or geographical races) of a BLUISH colour, with certain bars and other marks; an
0166  when any breed assumes by simple variation a BLUISH tint, these bars and other marks invariably r
0486  ur, the experience, the reason, and even the BLUNDERS of numerous workmen; when we thus view each
0018  ld think, from facts communicated to me by Mr. BLYTH, on the habits, voice, and constitution, etc.
0018  races have descended from one wild stock. Mr. BLYTH, whose opinion, from his large and varied sto
0163  der stripe; but traces of it, as stated by Mr. BLYTH and others, occasionally appear: and I have b
0253  ed by two eminently capable judges, namely Mr. BLYTH and Capt. Hutton, that whole flocks of these
0254  theri but from facts communicated to me by Mr. BLYTH, I think they must be considered as distinct
0098  the mane to the lion, the shoulder pad to the BOAR, and the hooked jaw to the male salmon; for th
0001                   Introduction When on BOARD H.M.S. Beagle, as naturalist, I was much stru
0386  of the same family, a Colymbetes; once flew on BOARD the Beagle, when forty five miles distant fro
0013  ard of cases of albinism, prickly skin, hairy BODIES, etc., appearing in several members of the sa
0180  as remarked, with the posterior part of their BODIES rather wide and with the skin on their flanks
0188  he find on finishing this treatise that large BODIES of facts, otherwise inexplicable, can be expl
0189  then the old ones to be destroyed. In living BODIES, variation will cause the slight alterations,
0201  by which its eggs are deposited in the living BODIES of other insects. If it could be proved that
0221  their small opponents, and carried their dead BODIES as food to their nest, twenty nine yards dist
0222  e ants entering their nest, carrying the dead BODIES of F. fusca (showing that it was not a migrat
0240  ermediate condition. So that we here have two BODIES of sterile workers in the same nest, differin
0244  rvae of ichneumonidae feeding within the live BODIES of caterpillars, not as specially endowed or
0300  ulate at a sufficient rate to protect organic BODIES from decay, no remains could be preserved. In
0362  he stomachs of dead fish, and then gave their BODIES to fishing eagles, storks, and pelicans; thes
0472  ers: at ichneumonidae feeding within the live BODIES of caterpillars; and at other such cases. The
0209  own causes which produce slight deviations of BODILY structure. No complex instinct can possibly b
0210  es with almost equal force to instincts as to BODILY organs. Changes of instinct may sometimes be
0211  others, as each takes advantage of the weaker BODILY structure of others. So again, in some few ca
0015  conditions, and whilst kept in a considerable BODY, so that free intercrossing might check, by bl
```

Page **(Key Word)**

```
0021 ery broad one. The pouter has a much elongated BODY, wings, and legs; and its enormously developed
0022 ngth of wing and tail to each other and to the BODY; the relative length of leg and of the feet; t
0035 selection; and by careful training, the whole BODY of English racehorses have come to surpass in
0077 rasite which clings to the hair on the tiger's BODY. But in the beautifully plumed seed of the dan
0094 cidental deviation in the size and form of the BODY, or in the curvature and length of the probosc
0096 n the first place, I have collected so large a BODY of facts, showing, in accordance with the almo
0101 reproduction so perfectly enclosed within the BODY, that access from without and the occasional i
0103 one spot, and might there maintain itself in a BODY, so that whatever intercrossing took place wou
0103 erymen always prefer getting seed from a large BODY of plants of the same variety, as the chance o
0115 of labour in the organs of the same individual BODY, a subject so well elucidated by Milne Edwards
0135 cessive generations the size and weight of its BODY, its legs were used more, and its wings less,
0143 isation of the adult. The several parts of the BODY which are homologous, and which, at an early e
0143 we see this in the right and left sides of the BODY varying in the same manner; in the front and h
0144 nakes, according to Schlegel, the shape of the BODY and the manner of swallowing determine the pos
0163 though so plainly barred like a zebra over the BODY, is without bars on the legs: but Dr. Gray has
0165 much more plainly barred than the rest of the BODY; and in one of them there was a double shoulde
0185 nd, the acutest observer by examining the dead BODY of the water ouzel would never have suspected
0192 es have no branchiae, the whole surface of the BODY and sack, including the small frena, serving f
0208 with certain periods of time and states of the BODY. When once acquired, they often remain constan
0248 ose interbreeding. I have collected so large a BODY of facts, showing that close interbreeding les
0264 rent. I have more than once alluded to a large BODY of facts, which I have collected, showing that
0266 al belief, founded, I think, on a considerable BODY of evidence, that slight changes in the condit
0275 rosper Lucas, who, after arranging an enormous BODY of facts with respect to animals, comes to the
0351 rect competition with each other, migrate in a BODY into a new and afterwards isolated country, th
0367 me arctic species, which had lately lived in a BODY together on the lowlands of the Old and New Wo
0368 cially to be noticed, they will have kept in a BODY together; consequently their mutual relations
0376 , on the belief, supported as it is by a large BODY of geological evidence, that the whole world,
0379 elieve from geological evidence that the whole BODY of arctic shells underwent scarcely any modifi
0390 ified having immigrated with facility and in a BODY, so that their mutual relations have not been
0423 less variable than the shape or colour of the BODY, etc.; whereas with sheep the horns are much l
0427 d others. The resemblance, in the shape of the BODY and in the fin like anterior limbs, between th
0428 mpared one with another: thus the shape of the BODY and fin like limbs are only analogical when wh
0428 imming through the water; but the shape of the BODY and fin like limbs serve as characters exhibit
0428 hat they have inherited their general shape of BODY and structure of limbs from a common ancestor.
0437 and appendages; in the articulata, we see the BODY divided into a series of segments, bearing ext
0488 me. A number of species, however, keeping in a BODY might remain for a long period unchanged, whil
0328 lelism in the successive Silurian deposits of BOHEMIA and Scandinavia; nevertheless he finds a sur
0093 been carried. The weather had been cold and BOISTEROUS, and therefore not favourable to bees, neve
0029 e, and some little to habit; but he would be a BOLD man who would account by such agencies for the
0339 ypes within the same areas mean? He would be a BOLD man, who after comparing the present climate o
0014 facts the above statement has so often and so BOLDLY been made. There would be great difficulty in
0362 so many on their arrival. Some hawks and owls BOLT their prey whole, and after an interval of fro
0144 the most important viscera. The nature of the BOND of correlation is very frequently quite obscur
0267 connected together by some common but unknown BOND, which is essentially related to the principle
0350 seas. We see in these facts some deep organic BOND, prevailing throughout space and time, over th
0350 curiosity, who is not led to inquire what this BOND is. This bond, on my theory, is simply inherit
0350 is not led to inquire what this bond is. This BOND, on my theory, is simply inheritance, that cau
0410 hin each class have been connected by the same BOND of ordinary generation; and the more nearly an
0413 se of the similarity of organic beings, is the BOND, hidden as it is by various degrees of modific
0418 und correlated with others, though no apparent BOND of connexion can be discovered between them, e
0420 gical; that community of descent is the hidden BOND which naturalists have been unconsciously seek
0426 hem be ever so unimportant, betrays the hidden BOND of community of descent. Let two forms have no
0433 believe this element of descent is the hidden BOND of connexion which naturalists have sought und
0449 ealogical. Descent being on my view the hidden BOND of connexion which naturalists have been seeki
0476 th cases the beings have been connected by the BOND of ordinary generation, and the means of modif
0303 range over the world, the fact of not a single BONE of a whale having been discovered in any secon
0437 rous and such extraordinarily shaped pieces of BONE? As Owen has remarked, the benefit derived fro
0011 instance, I find in the domestic duck that the BONES of the wing weigh less and the bones of the l
0011 that the bores of the wing weigh less and the BONES of the leg more, in proportion to the whole s
0011 ortion to the whole skeleton, than do the same BONES in the wild duck; and I presume that this cha
0022 of the several breeds, the development of the BONES of the face in length and breadth and curvatu
0200 o these birds: we cannot believe that the same BONES in the arm of the monkey, in the fore leg of
0200 ay further venture to believe that the several BONES in the limbs of the monkey, horse, and bat, w
0200 its. Therefore we may infer that these several BONES might have been acquired through natural sele
0288 anism wholly soft can be preserved. Shells and BONES will decay and disappear when left on the bot
0289 when we remember how large a proportion of the BONES of tertiary mammals have been discovered eith
0303 f the Whale family; as these animals have huge BONES, are marine, and range over the world, the fa
0339 rly seen in the wonderful collection of fossil BONES made by MM. Lund and Clausen in the caves of
0363 h and stones, and have even carried brushwood, BONES, and the nest of a land bird, I can hardly do
0416 ws of young ruminants, and certain rudimentary BONES of the leg, are highly serviceable in exhibit
0434 the same pattern, and should include the same BONES, in the same relative positions? Geoffroy St.
0434 e same order. We never find, for instance, the BONES of the arm and forearm, or of the thigh and l
0434 the same names can be given to the homologous BONES in widely different animals. We see the same
0435 e original pattern, or to transpose parts. The BONES of a limb might be shortened or widened to an
0435 as a fin? or a webbed foot might have all its BONES, or certain bones, lengthened to any extent,
0435 bbed foot might have all its bones, or certain BONES, lengthened to any extent, and the membrane c
0435 will be no tendency to alter the framework of BONES or the relative connexion of the several part
0436 ndividual. Most physiologists believe that the BONES of the skull are homologous with, that is cor
0437 ion in the skulls of birds. Why should similar BONES have been created in the formation of the win
0440 elation, than we have to believe that the same BONES in the hand of a man, wing of a bat, and fin
0452 ell shown in the drawings given by Owen of the BONES of the leg of the horse, ox, and rhinoceros.
0479 heir vital importance may be. The framework of BONES being the same in the hand of a man, wing of
0394 rld: Norfolk Island, the Viti Archipelago, the BONIN Islands, the Caroline and Marianne Archipelag
0222 ae. I traced the returning file burthened with BOOTY, for about forty yards, to a very thick clump
0185 nently aquatic, although their toes are only BORDERED by membrane. What seems plainer than that th
0078 ines of life, in the arctic regions or on the BORDERS of an utter desert, will competition cease.
```

Page **(Key Word)**

```
0081  f the others. If the country were open on its BORDERS, new forms would certainly immigrate, and th
0251  ous growth and rapid progress to maturity, and MORE good seed, which vegetated freely. In a letter
0202  riginally existed in a remote progenitor as a BORING and serrated instrument, like that in so many
0005  . as many more individuals of each species are BORN than can possibly survive; and as, consequentl
0061  ividuals of any species which are periodically BORN, but a small number can survive. I have called
0072  eggs in the navels of these animals when first BORN. The increase of these flies, numerous as they
0081  bt (remembering that many more individuals are BORN than can possibly survive) that individuals ha
0091  imals on which our wolf preyed, a cub might be BORN with an innate tendency to pursue certain kind
0142  y tend to preserve those individuals which are BORN with constitutions best adapted to their nativ
0152  to perfection, and frequently individuals are BORN which depart widely from the standard. There m
0203  troy the young queens her daughters as soon as BORN, or to perish herself in the combat: for undou
0212  isposition of individuals of the same species, BORN in a state of nature, is extremely diversified
0236  munity that a number should have been annually BORN capable of work, but incapable of procreation,
0237  election; namely, by an individual having been BORN with some slight profitable modification of st
0259  ptional and abnormal individuals sometimes are BORN, which closely resemble one of their pure pare
0264  to believe in this view: as hybrids, when once BORN, are generally healthy and long lived, as we s
0264  tly circumstanced before and after birth: when BORN and living in a country where their two parent
0275  supposed fact, that mongrel animals alone are BORN closely like one of their parents; but it can
0443  ell, until some time after the animal has been BORN, what its merits or form will ultimately turn
0467  ned in the third chapter. More individuals are BORN than can possibly survive. A grain in the bala
0080  ee that it can act most effectually. Let it be BORNE in mind in what an endless number of strange
0080  tion becomes in some degree plastic. Let it be BORNE in mind how infinitely complex and close fitt
0193  urious contrivance of a mass of pollen grains, BORNE on a foot stalk with a sticky gland at the en
0252  hly well authenticated. It should, however, be BORNE in mind that, owing to few animals breeding f
0280  record. In the first place it should always be BORNE in mind what sort of intermediate forms must,
0363  t these islands had been partly stocked by ice BORNE seeds, during the glacial epoch. At my reques
0373  looking to America: in the northern half, ice BORNE fragments of rock have been observed on the e
0450  t which they first appeared. It should also be BORNE in mind, that the supposed law of resemblance
0456  ring this view of classification. it should be BORNE in mind that the element of descent has been
0375  ts crowing on the summits of the mountains of BORNEO. Some of these Australian forms, as I hear fr
0379  southern vegetable forms on the mountains of BORNEO and Abyssinia. I suspect that this preponderan
0035  panish pointer certainly came from Spain, Mr. BORROW has not seen, as I am informed by him, any na
0393  he absence of whole orders on oceanic islands, BORY St. Vincent long ago remarked that Batrachians
0451  wo rudimentary teats in the udders of the genus BOS, but in our domestic cows the two sometimes be
0304  published, when a skilful palaeontologist, M. BOSQUET, sent me a drawing of a perfect specimen of
0050  s varieties, and in this country the highest BOTANICAL authorities and practical men can be quoted
0053  ciently well marked to have been recorded in BOTANICAL works. Hence it is the most flourishing, or
0358  n. i shall here confine myself to plants. In BOTANICAL works, this or that plant is stated to be i
0381  ith admirable clearness by Dr. Hooker in his BOTANICAL works on the antarctic regions. These canno
0408  r, which separate our several zoological and BOTANICAL provinces. We can thus understand the local
0048  sing number of forms have been ranked by one BOTANIST as good species; and by another as mere vari
0418  by none more clearly than by that excellent BOTANIST, Aug. St. Hilaire. If certain characters are
0030  obably arisen suddenly, or by one step: many BOTANISTS, for instance, believe that the fuller's te
0048  of the United States, drawn up by different BOTANISTS, and see what a surprising number of forms
0048  varieties, but which have all been ranked by BOTANISTS as species; and in making this list he has
0048  which nevertheless have been ranked by some BOTANISTS as species; and he has entirely omitted sev
0058  ich are almost universally ranked by British BOTANISTS as good and true species. Finally, then, va
0085  nd the colour of the flesh are considered by BOTANISTS as characters of the most trifling importan
0147  on; but many good observers, more especially BOTANISTS, believe in its truth. I will not, however,
0149  mbers, is constant. The same author and some BOTANISTS have further remarked that multiple parts a
0159  h turnip and Ruta baga, plants which several BOTANISTS rank as varieties produced by cultivation f
0247  allis arvensis and coerulea), which the best BOTANISTS rank as varieties, absolutely sterile toget
0248  n details, the evidence advanced by our best BOTANISTS on the question whether certain doubtful fo
0258  ed (as Matthiola annua and glabra) that many BOTANISTS rank them only as varieties. It is also a r
0268  ip, which are considered by many of our best BOTANISTS as varieties, are said by Gartner not to be
0386  al fresh water haunts. I do not believe that BOTANISTS are aware how charged the mud of ponds is w
0419  s been followed by several entomologists and BOTANISTS. Finally, with respect to the comparative v
0419  esent, almost arbitrary. Several of the best BOTANISTS, such as Mr. Bentham and others, have stron
0192  ovigerous frena, the eggs lying loose at the BOTTOM of the sack, in the well enclosed shell; but
0288  nes will decay and disappear when left on the BOTTOM of the sea, where sediment is not accumulatin
0288  tear, seem explicable only on the view of the BOTTOM of the sea not rarely lying for ages in an un
0291  arches of E. Forbes, we may conclude that the BOTTOM will be inhabited by extremely few animals, a
0291  ed to any thickness and extent over a shallow BOTTOM, if it continue slowly to subside. In this la
0296  ls. Hence, when the same species occur at the BOTTOM, middle, and top of a formation, the probabil
0228  e: but the ridge of wax was so thin, that the BOTTOMS of the basins, if they had been excavated to
0229  had been a little deepened, came to have flat BOTTOMS; and these flat bottoms, formed by thin litt
0229  ed, came to have flat bottoms: and these flat BOTTOMS, formed by thin little plates of the vermili
0283  along the bases of retreating cliffs rounded BOULDERS, all thickly clothed by marine productions,
0363  g to inquire whether he had observed erratic BOULDERS on these islands, and he answered that he ha
0366  cored flanks, polished surfaces, and perched BOULDERS, of the icy streams with which their valleys
0366  t a large part of the United States, erratic BOULDERS, and rocks scored by drifted icebergs and co
0373  nt, as far south as lat. 46 degrees; erratic BOULDERS have, also, been noticed on the Rocky Mounta
0373  t evidence of former glacial action, in huge BOULDERS transported far from their parent source. We
0393  ainland; moreover, icebergs formerly brought BOULDERS to its western shores, and they may have for
0073  imals, most remote in the scale of nature, are BOUND together by a web of complex relations. I sha
0081  r in which the inhabitants of each country are BOUND together, that any change in the numerical pr
0234  e bounding surface of each cell would serve to BOUND other cells, and much wax would be saved. Aga
0242  s, by the means of natural selection. But I am BOUND to confess that, with all my faith in this p
0320  of old forms, both natural and artificial, are BOUND together. In certain flourishing groups, the
0449  believes this to be a law of nature: but I am BOUND to confess that I only hope to see the law he
0234  lipona: for in this case a large part of the BOUNDING surface of each cell would serve to bound ot
0360  uch better manner, for he placed the seeds in a BOX in the actual sea, so that they were alternate
0437  creation! Why should the brain be enclosed in a BOX composed of such numerous and such extraordina
0046  al genera of insects, and several genera of BRACHIOPOD shells. In most polymorphic genera some of
0046  other countries, and likewise, judging from BRACHIOPOD shells, at former periods of time. These fa
0044  that the dwarfed condition of shells in the BRACKISH waters of the Baltic, or dwarfed plants on A
```

Page ***************************************(Key Word)**

```
0339  madeira; and between the extinct and living BRACKISH water shells of the Aralo Caspian Sea. Now w
0437  the ordinary view of creation! Why should the BRAIN be enclosed in a box composed of such numerou
0484  whether or not some fifty species of British BRAMBLES are true species will cease. Systematists wi
0063  issletoes, growing close together on the same BRANCH, may more truly be said to struggle with each
0129  of growth all the growing twigs have tried to BRANCH out on all sides, and to overtop and kill the
0129  the first growth of the tree, many a limb and BRANCH has decayed and dropped off; and these lost b
0130  e. as we here and there see a thin straggling BRANCH springing from a fork low down in a tree, and
0130  growth to fresh buds, and these, if vigorous, BRANCH out and overtop on all sides many a feebler b
0130  h out and overtop on all sides many a feebler BRANCH, so by generation I believe it has been with
0184  heard it hammering the seeds of the yew on a BRANCH, and thus breaking them like a nuthatch. In N
0356  hall, however, presently have to discuss this BRANCH of the subject in some detail. Changes of lev
0421  s from I; so will it be with each subordinate BRANCH of descendants, at each successive period. If
0432  l parent, and every intermediate link in each BRANCH and sub branch of their descendants, may be s
0432  very intermediate link in each branch and sub BRANCH of their descendants, may be supposed to be a
0432  common. In a tree we can specify this or that BRANCH, though at the actual fork the two unite and
0436  bscured. There is another and equally curious BRANCH of the present subject; namely, the compariso
0123  will be nearly related from having recently BRANCHED off from a10; b14 and f14, from having diver
0153  ification, since the period when the species BRANCHED off from the common progenitor of the genus.
0156  od, since that period when the species first BRANCHED off from their common progenitor, and subseq
0168  since the several species of the same genus BRANCHED off from a common parent, are more variable
0338  in regard to subordinate groups, which have BRANCHED off from each other within comparatively rec
0405  at some are extremely ancient, and must have BRANCHED off from a common parent at a remote epoch:
0430  er that all Rodents, including the bizcacha, BRANCHED off from some very ancient Marsupial, which
0430  supials; or that both Rodents and Marsupials BRANCHED off from a common progenitor, and that both
0473  act; for they have already varied since they BRANCHED off from a common progenitor in certain char
0474  art has undergone, since the several species BRANCHED off from a common progenitor, an unusual amo
0093  gmas of twenty flowers, taken from different BRANCHES, under the microscope, and on all, without e
0119  nted in the diagram by the several divergent BRANCHES proceeding from (A). The modified offspring
0119  ring from the later and more highly improved BRANCHES in the lines of descent, will, it is probabl
0119  nd so destroy, the earlier and less improved BRANCHES: this is represented in the diagram by some
0119  resented in the diagram by some of the lower BRANCHES not reaching to the upper horizontal lines.
0120  s, without either having given off any fresh BRANCHES or races. After ten thousand generations, sp
0124  neath the capital letters, converging in sub BRANCHES downwards towards a single point; this point
0129  o overtop and kill the surrounding twigs and BRANCHES, in the same manner as species and groups of
0129  attle for life. The limbs divided into great BRANCHES, and these into lesser and lesser branches,
0129  t branches; and these into lesser and lesser BRANCHES, were themselves once, when the tree was sma
0129  of the former and present buds by ramifying BRANCHES may well represent the classification of all
0129  ush, only two or three, now grown into great BRANCHES, yet survive and bear all the other branches
0129  branches, yet survive and bear all the other BRANCHES; so with the species which lived during long
0129  has decayed and dropped off; and these lost BRANCHES of various sizes may represent those whole o
0130  cegree connects by its affinities two large BRANCHES of life, and which has apparently been saved
0130  f life, which fills with its dead and broken BRANCHES the crust of the earth, and covers the surfa
0184  titmouse (Parus major) may be seen climbing BRANCHES, almost like a creeper. It often, like a shr
0197  ucted hooks clustered around the ends of the BRANCHES, and this contrivance, no doubt, is of the h
0359  to me that floods might wash down plants or BRANCHES, and that these might be dried on the banks,
0359  nto the sea. Hence I was led to dry stems in BRANCHES of 94 plants with ripe fruit, and to place t
0360  hand he did not previously dry the plants or BRANCHES with the fruit; and this, as we have seen, w
0420  that the amount of difference in the several BRANCHES or groups, though allied in the same degree.
0190  e one instance, there are fish with gills or BRANCHIAE that breathe the air dissolved in the water
0191  ttis is closed. In the higher Vertebrata the BRANCHIAE have wholly disappeared, the slits on the s
0191  it is conceivable that the now utterly lost BRANCHIAE might have been gradually worked in by natu
0191  iew entertained by some naturalists that the BRANCHIAE and dorsal scales of Annelids are homologou
0192  ed within the sack. These cirripedes have no BRANCHIAE, the whole surface of the body and sack, in
0192  l enclosed shell; but they have large folded BRANCHIAE. Now I think no one will dispute that the o
0192  one family are strictly homologous with the BRANCHIAE of the other family; indeed, they graduate
0192  radually converted by natural selection into BRANCHIAE, simply through an increase in their size a
0192  pedes, who would ever have imagined that the BRANCHIAE in this latter family had originally existe
0479  olved in water, by the aid of well developed BRANCHIAE. Disuse, aided sometimes by natural selecti
0440  ar loop like course of the arteries near the BRANCHIAL slits are related to similar conditions, in
0479  yo uf an air breathing mammal or bird having BRANCHIAL slits and arteries running in loops, like t
0045  ecies. I should never have expected that the BRANCHING of the main nerves close to the great centr
0046  hich may almost be compared to the irregular BRANCHING of the stem of a tree. This philosophical n
0125  pecies lived at very ancient epochs when the BRANCHING lines of descent had diverged less. I see n
0126  r and more highly perfected sub groups, from BRANCHING out and seizing on many new places in the p
0130  earth, and covers the surface with its ever BRANCHING and beautiful ramifications. Chapter V. Law
0156  and come to differ within the period of the BRANCHING off of the species from a common progenitor
0187  low stage, numerous gradations of structure, BRANCHING off in two fundamentally different lines, c
0301  species of the same group into one long and BRANCHING chain of life. We ought only to look for a
0317  low steps other species, and so on, like the BRANCHING of a great tree from a single stem, till th
0422  agram, but in much too simple a manner. If a BRANCHING diagram had not been used, and only the nam
0133  a coasts, as every collector knows, are often BRASSY or lurid. Plants which live exclusively on th
0339  made by MM. Lund and Clausen in the caves of BRAZIL. I was so much impressed with these facts tha
0339  . we see it also in the birds of the caves of BRAZIL. Mr. Woodward has shown that the same law hol
0341  and have left no progeny. But in the caves of BRAZIL, there are many extinct species which are clo
0374  ean genera occur. On the highest mountains of BRAZIL, some few European genera were found by Gardn
0383  when first collecting in the fresh waters of BRAZIL, feeling much surprise at the similarity of t
0022  opment of the bones of the face in length and BREADTH and curvature differs enormously. The shape,
0022  differs enormously. The shape, as well as the BREADTH and length of the ramus of the lower jaw, va
0022  ber of the ribs, together with their relative BREADTH and the presence of processes. The size and
0364  rt across tracts of sea some hundred miles in BREADTH, or from island to island, or from a contine
0059  o be explained, the larger genera also tend to BREAK up into smaller genera. And thus, the forms o
0189  ight modifications, my theory would absolutely BREAK down. But I can find out no such case. No dou
0230  cal hollows, but never allowing the spheres to BREAK into each other. Now bees, as may be clearly
0286  f a lofty cliff would be more rapid from the BREAKAGE of the fallen fragments. On the other hand,
0387  nd yet the viscid mud was all contained in a BREAKFAST cup! Considering these facts, I think it wo
0087  tip to the beak of nestling birds, used for BREAKING the egg. It has been asserted, that of the b
```

Page ***************************************(Key Word)**

Page ***(Key Word)***

```
0184  g the seeds of the yew on a branch, and thus BREAKING them like a nuthatch. In North America the b
0299  have been, yet has done scarcely anything in BREAKING down the distinction between species, by con
0119  stinct to be recorded as varieties. But these BREAKS are imaginary, and might have been inserted a
0286   which from long resisting attrition form a BREAKWATER at the base. Hence, under ordinary circumst
0021  ak, with a line of reversed feathers down the BREAST; and it has the habit of continually expandin
0090  nature, for instance, the tuft of hair on the BREAST of the turkey cock, which can hardly be eithe
0190  , there are fish with gills or branchiae that BREATHE the air dissolved in the water, at the same
0190  lved in the water, at the same time that they BREATHE free air in their swimbladders, this latter
0479  o ir loops, like those in a fish which has to BREATHE the air dissolved in water, by the aid of we
0484  e primordial form, into which life was first BREATHED. When the views entertained in this volume o
0490  h its several powers, having been originally BREATHED into a few forms or into one; and that, whil
0182  ble. Seeing that a few members of such water BREATHING classes as the Crustacea and Mollusca are a
0198  head of the young in the womb. The laborious BREATHING necessary in high regions would, we have so
0204  er has apparently been converted into an air BREATHING lung. The same organ having performed simul
0452  ncy, but has become converted into a nascent BREATHING organ or lung. Other similar instances coul
0479  may cease marvelling at the embryo of an air BREATHING mammal or bird having branchial slits and a
0025  of the domestic breeds, taking thoroughly well BRED birds, all the above marks, even to the white
0028  the several kinds, knowing well how true they BRED, I felt as much difficulty in believing
0036  rgess, as Mr. Youatt remarks, have been purely BRED from the original stock of Mr. Bakewell for up
0041  m, and again picked out the best seedlings and BRED from them, then, there appeared (aided by some
0140  y uncivilised man because they were useful and BRED readily under confinement, and not because the
0164  se without stripes is not considered as purely BRED. The spine is always striped; the legs are gen
0166  call the breeds of pigeons, some of which have BRED true for centuries; species; and how exactly p
0253  enerally ranked in distinct genera, have often BRED in this country with either pure parent, and i
0253  e parent, and in one single instance they have BRED inter se. This was effected by Mr. Eyton, who
0253  from one nest. In India, however, these cross BRED geese must be far more fertile; for I am assur
0254  aboriginal species would at first have freely BRED together and have produced quite fertile hybri
0275  g to the cases which I have collected of cross BRED animals closely resembling one parent, the res
0443  e parent. Or again, as when the horns of cross BRED cattle have been affected by the shape of the
0008  nd few things more difficult than to get it to BREED freely under confinement, even in the many ca
0008  ite. How many animals there are which will not BREED, though living long under not very close conf
0009  at carnivorous animals, even from the tropics, BREED in this country pretty freely under confineme
0009  garden. I may add, that as some organisms will BREED most freely under the most unnatural conditio
0015  vour of this view: to assert that we could not BREED our cart and race horses, long and short horn
0017  n from a state of nature, and could be made to BREED for an equal number of generations under dome
0020  ss, of the task becomes apparent. Certainly, a BREED intermediate between two very distinct breeds
0020  , taken up domestic pigeons. I have kept every BREED which I could purchase or obtain, and have be
0024  it is most difficult to get any wild animal to BREED freely under domestication; yet on the hypoth
0025  coloured and marked, so that in each separate BREED there might be a tendency to revert to the ve
0026  colours and markings. Or, secondly, that each BREED, even the purest, has within a dozen or, at m
0026  moved by a greater number of generations. In a BREED which has been crossed only once with some di
0026  has been crossed only once with some distinct BREED, the tendency to reversion to any character d
0026  t when there has been no cross with a distinct BREED, and there is a tendency in both parents to r
0027  seven or eight supposed species of pigeons to BREED freely under domestication: these supposed sp
0027  haracters which are mainly distinctive of each BREED, for instance the wattle and length of beak o
0027  r of tail feathers in the fantail, are in each BREED eminently variable; and the explanation of th
0029  er, who was not fully convinced that each main BREED was descended from a distinct species. Van Mo
0030  land or mountain pasture, with the wool of one BREED good for one purpose, and that of another bre
0030  reed good for one purpose, and that of another BREED for another purpose; when we compare the many
0033  s so careless as to allow his worst animals to BREED. In regard to plants, there is another means
0034  dogs with wild canine animals, to improve the BREED, and they formerly did so, as is attested by
0034  ct object in view, to make a new strain or sub BREED, superior to anything existing in the country
0034  h results from every one trying to possess and BREED from the best individual animals, is more imp
0034  ish or expectation of permanently altering any BREED. Nevertheless I cannot doubt that this proces
0035  during centuries, would improve and modify any BREED, in the same way as Bakewell, Collins, etc.,
0035  or but little changed individuals of the same BREED may be found in less civilised districts, whe
0035  e found in less civilised districts, where the BREED has been less improved. There is reason to be
0039  anciers, as it is not one of the points of the BREED. Nor let it be thought that some great deviat
0039  ations from the standard of perfection of each BREED. The common goose has not given rise to any m
0039  varieties; hence the Thoulouse and the common BREED, which differ only in colour, that most fleet
0040  of any of our domestic breeds. But, in fact, a BREED, like a dialect of a language, can hardly be
0040  on, the spreading and knowledge of any new sub BREED will be a slow process. As soon as the points
0040  as soon as the points of value of the new sub BREED are once fully acknowledged, the principle, a
0040  aps more at one period than at another, as the BREED rises or falls in fashion, perhaps more in on
0040  y to adapt to the characteristic features of the BREED, whatever they may be. But the chance will be
0041  under favourable conditions of life, so as to BREED freely in that country. When the individuals
0041  r quality may be, will generally be allowed to BREED, and this will effectually prevent selection.
0042  ts of open plains rarely possess more than one BREED of the same species. Pigeons can be mated for
0042  en and children, we hardly ever see a distinct BREED kept up; such breeds as we do sometimes see a
0065  that nearly all the young have been enabled to BREED. In such cases the geometrical ratio of incre
0088  ual selection by always allowing the victor to BREED might surely give indomitable courage, length
0088  ighter, who knows well that he can improve his BREED by careful selection of the best cocks. How l
0091  best dogs without any thought to modifying the BREED. Even without any change in the proportional
0102  when many men, without intending to alter the BREED, have a nearly common standard of perfection,
0102  standard of perfection, and all try to get and BREED from the best animals, much improvement and m
0103  ach birth, which wander much, and which do not BREED at a very quick rate. Hence in animals of thi
0143  ; and if this had been of any great use to the BREED it might probably have been rendered permanen
0152  ce, the comb in the Dorking fowl) or the whole BREED will cease to have a nearly uniform character
0152  cease to have a nearly uniform character. The BREED will then be said to have degenerated. In rud
0152  faced tumbler, it is notoriously difficult to BREED them nearly to perfection, and frequently ind
0153  and, the power of steady selection to keep the BREED true. In the long run selection gains the day
0153  day, and we do not expect to fail so far as to BREED a bird as coarse as a common tumbler from a g
0160  erhaps for hundreds of generations. But when a BREED has been crossed only once by some other bree
0160  breed has been crossed only once by some other BREED, the offspring occasionally show a tendency t
0160  tendency to revert in character to the foreign BREED for many generations, some say, for a dozen o
0160  s very small proportion of foreign blood. In a BREED which has not been crossed, but in which both
```

Page ***(Key Word)***

```
0160 ons. When a character which has been lost in a BREED, reappears after a great number of generation
0164 . in the north west part of India the Kattywar BREED of horses is so generally striped, that, as I
0164 as i hear from Colonel Poole, who examined the BREED for the Indian Government, a horse without st
0166 s ever occur in the eminently striped Kattywar BREED of horses, and was, as we have seen, answered
0166 ith certain bars and other marks; and when any BREED assumes by simple variation a bluish tint, th
0177 nd consequently the improved mountain or plain BREED will soon take the place of the less improved
0177  soon take the place of the less improved hill BREED; and thus the two breeds, which originally ex
0213 dividual, performed with eager delight by each BREED, and without the end being known, for the you
0215 s to procure, without intending to improve the BREED, dogs which will stand and hunt best. On the
0238 owers of selection, that I do not doubt that a BREED of cattle, always yielding oxen with extraord
0253 cause had the least tendency to sterility, the BREED would assuredly be lost in a very few generat
0265 l he tries, whether any particular animal will BREED under confinement or any plant seed freely un
0266 e conditions of life. When hybrids are able to BREED inter se, they transmit to their offspring fr
0315 ong ages for the same object, might make a new BREED hardly distinguishable from our present fanta
0316 le that a fantail, identical with the existing BREED, could be raised from any other species of pi
0356 differ slightly from the horses of every other BREED; but they do not owe their difference and sup
0424  common habit of tumbling; but the short faced BREED has nearly or quite lost this habit; neverthe
0446 cteristic differences which give value to each BREED, and which have been accumulated by man's sel
0467  the time, without any thought of altering the BREED. It is certain that he can largely influence
0467 at he can largely influence the character of a BREED by selecting, in each successive generation,
0012  the fullest and the best on this subject. No BREEDER doubts how strong is the tendency to inherit
0029 less of the laws of inheritance than does the BREEDER, and knowing no more than he does of the int
0031 hen had given it existence. That most skilful BREEDER, Sir John Sebright, used to say, with respec
0032 and judgment sufficient to become an eminent BREEDER. If gifted with these qualities, and he stud
0064 s. the elephant is reckoned to be the slowest BREEDER of all known animals, and I have taken some
0102 xterminated. In man's methodical selection, a BREEDER selects for some definite object, and free i
0238 her; the animal has been slaughtered, but the BREEDER goes with confidence to the same family. I h
0253 o the constantly repeated admonition of every BREEDER. And in this case, it is not at all surprisi
0011 oy st. Hilaire's great work on this subject. BREEDERS believe that long limbs are almost always ac
0028 nce has struck me much; namely, that all the BREEDERS of the various domestic animals and the cult
0030 l. it is certain that several of our eminent BREEDERS have, even within a single lifetime, modifie
0031 to this subject, and to inspect the animals. BREEDERS habitually speak of an animal's organisation
0031 e pleases. Lord Somerville, speaking of what BREEDERS have done for sheep, says: It would seem as
0031 ately be selected for breeding. What English BREEDERS have actually effected is proved by the enor
0032 e to crossing different breeds; all the best BREEDERS are strongly opposed to this practice, excep
0034 is so obvious. At the present time, eminent BREEDERS try by methodical selection, with a distinct
0036 s unconsciously followed, in so far that the BREEDERS could never have expected or even have wishe
0065 ich produce extremely few, is, that the slow BREEDERS would require a few more years to people, un
0096 cordance with the almost universal belief of BREEDERS, that with animals and plants a cross betwee
0112 ontinued selection of swifter horses by some BREEDERS, and of stronger ones by others, the differe
0238 ntly expects to get nearly the same variety; BREEDERS of cattle wish the flesh and fat to be well
0249 ness of this almost universal belief amongst BREEDERS. Hybrids are seldom raised by experimentalis
0443 ints the other way; for it is notorious that BREEDERS of cattle, horses, and various fancy animals
0445 pies differed from each other: I was told by BREEDERS that they differed just as much as their par
0009 nd plants, though often weak and sickly, yet BREEDING quite freely under confinement; and when, on
0023 ust all have been rock pigeons, that is, not BREEDING or willingly perching on trees. But besides
0023  become extinct in the wild state. But birds BREEDING on precipices, and good fliers, are unlikely
0031 the very best may ultimately be selected for BREEDING. What English breeders have actually effecte
0032 n separating some very distinct variety, and BREEDING from it, the principle would be so obvious a
0034 how actual selection, but they show that the BREEDING of domestic animals was carefully attended t
0034 strange fact, had attention not been paid to BREEDING; for the inheritance of good and bad qualiti
0042 r people, and little attention paid to their BREEDING; in peacocks, from not being very easily rea
0064 d by the progeny of a single pair. Even slow BREEDING man has doubled in twenty five years, and at
0064 it breeds when thirty years old, and goes on BREEDING till ninety years old, bringing forth three
0064 e statements of the rate of increase of slow BREEDING cattle and horses in South America, and latt
0089 tion, acting when the birds have come to the BREEDING age or during the breeding season; the modif
0089  have come to the breeding age or during the BREEDING season; the modifications thus produced bein
0103 s is thus lessened. Even in the case of slow BREEDING animals, which unite for each birth, we must
0103 inct, from haunting different stations, from BREEDING at slightly different seasons, or from varie
0112  occurred with tumbler pigeons) choosing and BREEDING from birds with longer and longer beaks, or
0252 be borne in mind that, owing to few animals BREEDING freely under confinement, few experiments ha
0314 n the power of intercrossing, on the rate of BREEDING, on the slowly changing physical conditions
0319 alogy of all other mammals, even of the slow BREEDING elephant, and from the history of the natura
0446 select their horses, dogs, and pigeons, for BREEDING, when they are nearly grown up: they are ind
0013 rities appearing in the males of our domestic BREEDS are often transmitted either exclusively, or
0015 d short horned cattle, and poultry of various BREEDS, and esculent vegetables, for an almost infin
0018 the monuments of Egypt, much diversity in the BREEDS; and that some of the breeds closely resemble
0018 diversity in the breeds; and that some of the BREEDS closely resemble, perhaps are identical with,
0018 case, what does it show, but that some of our BREEDS originated there, four or five thousand years
0018 n that of almost any one, thinks that all the BREEDS of poultry have proceeded from the common wil
0019 bankiva). In regard to ducks and rabbits, the BREEDS of which differ considerably from each other
0019 e authors. They believe that every race which BREEDS true, Let the distinctive characters be ever
0019 of these kingdoms possesses several peculiar BREEDS of cattle, sheep, etc., we must admit that ma
0019 sheep, etc., we must admit that many domestic BREEDS have originated in Europe; for whence could t
0020 fspring from the first cross between two pure BREEDS is tolerably and sometimes (as I have found w
0020  breed intermediate between two very distinct BREEDS could not be got without extreme care and lon
0020 ermanent race having been thus formed. On the BREEDS of the Domestic Pigeon. Believing that it is
0021 the London Pigeon Clubs. The diversity of the BREEDS is something astonishing. Compare the English
0021 massive beak and large feet: some of the sub BREEDS of runts have very long necks, others very lo
0021 ss, utter a very different coo from the other BREEDS. The fantail has thirty or even forty tail fe
0022 is quite aborted. Several other less distinct BREEDS might have been specified. In the skeletons o
0022 en specified. In the skeletons of the several BREEDS, the development of the bones of the face in
0022 of flight differs remarkably: as does in some BREEDS the voice and disposition. Lastly, in certain
0022 the voice and disposition. Lastly, in certain BREEDS, the males and females have come to differ to
0023 me genus; more especially as in each of these BREEDS several truly inherited sub breeds, or specie
0023 h of these breeds several truly inherited sub BREEDS, or species as he might have called them, cou
```

Page **(Key Word)**

```
0023  him. Great as the differences are between the  BREEDS  of pigeons, I am fully convinced that the com
0023  i will here briefly give them. If the several   BREEDS  are not varieties, and have not proceeded fro
0023  it is impossible to make the present domestic   BREEDS  by the crossing of any lesser number: how, fo
0023  e, could a pouter be produced by crossing two   BREEDS  unless one of the parent stocks possessed the
0023  ave not any of the characters of the domestic   BREEDS. Hence the supposed aboriginal stocks must ei
0023  , which has the same habits with the domestic   BREEDS, has not been exterminated even on several of
0024  oreover, the several above named domesticated   BREEDS have been transported to all parts of the wor
0024  ral other cases, is, that the above specified   BREEDS, though agreeing generally in constitution, h
0025  wings have two black bars; some semi domestic   BREEDS and some apparently truly wild breeds have, b
0025  omestic breeds and some apparently truly wild   BREEDS have, besides the two black bars, the wings c
0025  ole family. Now, in every one of the domestic   BREEDS, taking thoroughly well bred birds, all the a
0025  ver, when two birds belonging to two distinct   BREEDS are crossed, neither of which is blue or has
0025  to ancestral characters, if all the domestic   BREEDS have descended from the rock pigeon. But if w
0026  ids or mongrels from between all the domestic   BREEDS of pigeons are perfectly fertile. I can state
0026  rvations, purposely made on the most distinct   BREEDS. Now, it is difficult, perhaps impossible, to
0027  rious marks occasionally appearing in all the   BREEDS, both when kept pure and when crossed: the mo
0027  er. I can feel no doubt that all our domestic   BREEDS have descended from the Columba livia with it
0027  of points of structure with all the domestic   BREEDS. Secondly, although an English carrier or sho
0027  rock pigeon, yet by comparing the several sub   BREEDS of these breeds, more especially those brough
0027  by comparing the several sub breeds of these   BREEDS, more especially those brought from distant c
0028  ourtly historian, His Majesty by crossing the   BREEDS, which method was never practised before, has
0028  . we shall then, also, see how it is that the   BREEDS so often have a somewhat monstrous character.
0028  e circumstance for the production of distinct   BREEDS, that male and female pigeons can be easily m
0028  be easily mated for life; and thus different   BREEDS can be kept together in the same aviary. I ha
0028  e read, are firmly convinced that the several   BREEDS to which each has attended, are descended fro
0030  e horse, the dromedary and camel, the various   BREEDS of sheep fitted either for cultivated land or
0030  for another purpose; when we compare the many   BREEDS of dogs, each good for man in very different
0030  e cock, so pertinacious in battle, with other   BREEDS so little quarrelsome, with everlasting layer
0030  e variability. We cannot suppose that all the   BREEDS were suddenly produced as perfect and as usef
0030  nse he may be said to make for himself useful   BREEDS. The great power of this principle of selecti
0031  gle lifetime, modified to a large extent some   BREEDS of cattle and sheep. In order fully to realis
0031  no means generally due to crossing different   BREEDS; all the best breeders are strongly opposed t
0032  , except sometimes amongst closely allied sub   BREEDS. And when a cross has been made, the closest
0034  ogs. Livingstone shows how much good domestic   BREEDS are valued by the negroes of the interior of
0034  to get as good dogs as he can, and afterwards   BREEDS from his own dogs, but he has no wish or
0035  ctual measurements or careful drawings of the   BREEDS in question had been made long ago, which mig
0036  treatises of carriers and tumblers with these   BREEDS as now existing in Britain, India, and Persia
0038  ll hereafter be more fully explained, two sub   BREEDS might be formed. This, perhaps, partly explai
0039  il, or like individuals of other and distinct   BREEDS, in which as many as seventeen tail feathers
0039  which would now be set on them, after several   BREEDS have once fairly been established. Many sligh
0040  the origin or history of any of our domestic   BREEDS. But, in fact, a breed, like a dialect of a l
0040  ve had a definite origin. A man preserves and   BREEDS from an individual with some slight deviation
0042  favoured the improvement and formation of new   BREEDS. Pigeons, I may add, can be propagated in gre
0042  ardly ever see a distinct breed kept up: such   BREEDS as we do sometimes see are almost always impo
0042  others, yet the rarity or absence of distinct   BREEDS of the cat, the donkey, peacock, goose, etc.,
0042  e having been felt in the display of distinct   BREEDS. To sum up on the origin of our Domestic Race
0043  uctions. When in any country several domestic   BREEDS have once been established, their occasional
0043  bt, largely aided in the formation of new sub   BREEDS; but the importance of the crossing of variet
0064  : it will be under the mark to assume that it   BREEDS when thirty years old, and goes on breeding t
0111  tances could be given showing how quickly new   BREEDS of cattle, sheep, and other animals, and vari
0112  reater, and would be noted as forming two sub   BREEDS; finally, after the lapse of centuries, the s
0112  inally, after the lapse of centuries, the sub   BREEDS would become converted into two well establis
0112  verted into two well established and distinct   BREEDS. As the differences slowly become greater, th
0112  ly appreciable, steadily to increase, and the   BREEDS to diverge in character both from each other
0141  ld dog may perhaps be mingled in our domestic   BREEDS. The rat and mouse cannot be considered as do
0142  an should have succeeded in selecting so many   BREEDS and sub breeds with constitutions specially f
0142  succeeded in selecting so many breeds and sub   BREEDS with constitutions specially fitted for their
0152  so eminently liable to variation. Look at the   BREEDS of the pigeon; see what a prodigious amount o
0152  ended to by English fanciers. Even in the sub   BREEDS, as in the short faced tumbler, it is notorio
0159  king to our domestic races. The most distinct   BREEDS of pigeons, in countries most widely set apar
0159  namely, the occasional appearance in all the   BREEDS, of slaty blue birds with two black bars on t
0160  analogous variation appearing in the several   BREEDS. We may I think confidently come to this conc
0160  ring of two distinct and differently coloured   BREEDS; and in this case there is nothing in the ext
0162  old, whether these characters in our domestic   BREEDS were reversions or only analogous variations:
0162  ur and marks so often appearing when distinct   BREEDS of diverse colours are crossed. Hence, though
0163  spinal stripe in horses of the most distinct   BREEDS, and of all colours; transverse bars on the l
0164  shoulder stripes in horses of very different   BREEDS, in various countries from Britain to Eastern
0164  en on this subject, believes that the several   BREEDS of the horse have descended from several abor
0165  his theory, and should be loth to apply it to   BREEDS so distinct as the heavy Belgian cart horse,
0166  species. Now observe the case of the several   BREEDS of pigeons: they are descended from a pigeon
0166  form or character. When the oldest and truest   BREEDS of various colours are crossed, we see a stro
0166  mmonly in the young than in the old. Call the   BREEDS of pigeons, some of which have bred true for
0177  he mountains or on the plains improving their   BREEDS more quickly than the small holders on the in
0177  he less improved hill breed; and thus the two   BREEDS, which originally existed in greater numbers,
0198  this by reflecting on the differences in the   BREEDS of our domesticated animals in different coun
0198  h the hair the horns are correlated. Mountain   BREEDS always differ from lowland breeds: and a moun
0198  d. mountain breeds always differ from lowland   BREEDS: and a mountainous country would probably aff
0198  he characteristic differences of our domestic   BREEDS, which nevertheless we generally admit to hav
0213  t us look to the familiar case of the several   BREEDS of dogs: it cannot be doubted that young poin
0214  become mingled, is well shown when different   BREEDS of dogs are crossed. Thus it is known that a
0215  remarkable instance of this is seen in those   BREEDS of fowls which very rarely or never become br
0237  slight differences in the horns of different   BREEDS of cattle in relation to an artificially impe
0237  ct state of the male sex: for oxen of certain   BREEDS have longer horns than in other breeds, in co
0237  ertain breeds have longer horns than in other   BREEDS, in comparison with the horns of the bulls or
0237  the horns of the bulls or cows of these same   BREEDS. Hence I can see no real difficulty in any ch
0252  finches, but as not one of these nine species   BREEDS freely in confinement, we have no right to ex
```

Page **(Key Word)**

Page **(Key Word)**

```
0267 s between members of different strains or sub BREEDS, gives vigour and fertility to the offspring.
0280 ged, the characteristic features of these two BREEDS. These two breeds, moreover, have become so m
0280 istic features of these two breeds. These two BREEDS, moreover, have become so much modified, that
0299 iod will be able to prove, that our different BREEDS of cattle, sheep, horses, and dogs have desce
0320 ort horn cattle, and takes the place of other BREEDS in other countries. Thus the appearance of ne
0445 robable that the difference between these two BREEDS has been wholly caused by selection under dom
0445 s to me conclusive, that the several domestic BREEDS of Pigeon have descended from one wild specie
0445 species, I compared young pigeons of various BREEDS, within twelve hours after being hatched; I c
0445 but when the nestling birds of these several BREEDS were placed in a row, though most of them cou
0446 ounc of the wild rock pigeon and of the other BREEDS, in all its proportions, almost exactly as mu
0454 oductions, as the stump of a tail in tailless BREEDS, the vestige of an ear in earless breeds, the
0454 less breeds, the vestige of an ear in earless BREEDS, the reappearance of minute dangling horns in
0454 pearance of minute dangling horns in hornless BREEDS of cattle, more especially, according to Youa
0467 tion of the most distinct and useful domestic BREEDS. That many of the breeds produced by man have
0467 and useful domestic breeds. That many of the BREEDS produced by man have to a large extent the ch
0473 r, in the same manner as the several domestic BREEDS of pigeon have descended from the blue and ba
0061 e plumed seed which is wafted by the gentlest BREEZE; in short, we see beautiful adaptations every
0203 that a few granules may be wafted by a chance BREEZE on to the ovules? Summary of Chapter. We have
0214 w there are house tumblers, as I hear from Mr. BRENT, which cannot fly eighteen inches high withou
0362 rn crops might thus readily get scattered. Mr. BRENT informs me that a friend of his had to give u
0367 igher, as the warmth increased, whilst their BRETHERN were pursuing their northern journey. Hence,
0379 till plainly related by inheritance to their BRETHREN of the northern or southern hemispheres, now
0053 , treated as it necessarily here is with much BREVITY, is rather perplexing, and allusions cannot
0096 ut I must here treat the subject with extreme BREVITY, though I have the materials prepared for an
0127 preservation, I have called, for the sake of BREVITY, Natural Selection. Natural selection, on th
0272 shall here discuss this subject with extreme BREVITY. The most important distinction is, that in
0217 ests: but I hear on the high authority of Dr. BREWER, that this is a mistake. Nevertheless, I coul
0371 ost continuously united by land, serving as a BRIDGE, since rendered impassable by cold, for the i
0357 erica. Other authors have thus hypothetically BRIDGED over every ocean, and have united almost eve
0002 ish, with Mr. Wallace's excellent memoir, some BRIEF extracts from my manuscripts. This Abstract,
0006 condition. In the last chapter I shall give a BRIEF recapitulation of the whole work, and a few c
0100 l attention to the subject. Turning for a very BRIEF space to animals: on the land there are some
0376 o each other in a most remarkable manner. This BRIEF abstract applies to plants alone: some strict
0402 ntage whatever over another, it will in a very BRIEF time wholly or in part supplant it; but if bo
0356 eated this subject. I can give here only the BRIEFEST abstract of the more important facts. Change
0004 y, be compelled to treat this subject far too BRIEFLY, as it can be treated properly only by givin
0011 hich can be dimly seen, and will be hereafter BRIEFLY mentioned. I will here only allude to what m
0023 degree applicable in other cases, I will here BRIEFLY give them. If the several breeds are not var
0029 dants of other species? Selection. Let us now BRIEFLY consider the steps by which domestic races h
0044 organic beings in a state of nature, we must BRIEFLY discuss whether these latter are subject to
0080 ill the struggle for existence, discussed too BRIEFLY in the last chapter, act in regard to variat
0151 f the more remarkable cases: I will here only BRIEFLY give one, as it illustrates the rule in its
0188 ous substance. With these facts, here far too BRIEFLY and imperfectly given, which show that there
0212 in a state of nature will be strengthened by BRIEFLY considering a few cases under domestication.
0242 trine of Lamarck. Summary. I have endeavoured BRIEFLY in this chapter to show that the mental qual
0242 that the variations are inherited. Still more BRIEFLY I have attempted to show that instincts vary
0269 lity in the few following cases, which I will BRIEFLY abstract. The evidence is at least as good a
0278 ybrids and mongrels. Finally, then, the facts BRIEFLY given in this chapter do not seem to me oppo
0413 or as an artificial means for enunciating, as BRIEFLY as possible, general propositions, that is,
0459 ader to have the leading facts and inferences BRIEFLY recapitulated. That many and grave objection
0465 ly be urged against my theory; and I have now BRIEFLY recapitulated the answers and explanations w
0062 e misunderstood. We behold the face of nature BRIGHT with gladness, we often see superabundance of
0133 at this shell, for instance, was created with BRIGHT colours for a warm sea; but that this other s
0133 a warm sea; but that this other shell became BRIGHT coloured by variation when it ranged into war
0288 enormously large proportion of the ocean, the BRIGHT blue tint of the water bespeaks its purity. T
0133 d to tropical and shallow seas are generally BRIGHTER coloured than those confined to cold and dee
0133 d to continents are, according to Mr. Gould, BRIGHTER coloured than those of islands. The insect s
0132 , and when living in shallow water, are more BRIGHTER coloured than those of the same species furt
0132 eves that birds of the same species are more BRIGHTLY coloured under a clear atmosphere, than when
0003 ely requiring the agency of certain insects to BRING pollen from one flower to the other, it is ec
0026 . now, it is difficult, perhaps impossible, to BRING forward one case of the hybrid offspring of t
0050 ties. Close investigation, in most cases, will BRING naturalists to an agreement how to rank doubt
0097 push the flowers own pollen on the stigma, or BRING pollen from another flower. So necessary are
0098 f hybrids between distinct species; for if you BRING on the same brush a plant's own pollen and po
0103 sses in retarding natural selection; for I can BRING a considerable catalogue of facts, showing th
0124 o groups; and every naturalist will be able to BRING some such case before his mind. In the diagra
0289 shed in the Supplement to Lyell's Manual, will BRING home the truth, how accidental and rare is th
0364 . the currents, from their course, would never BRING seeds from North America to Britain, though t
0364 h america to Britain, though they might and do BRING seeds from the West Indies to our western sho
0064 and goes on breeding till ninety years old, BRINGING forth three pair of young in this interval;
0091 er mice; one cat, according to Mr. St. John, BRINGING home winged game, another hares or rabbits,
0220 s, on the other hand, may be constantly seen BRINGING in materials for the nest, and food of all k
0223 slaves and masters work together, making and BRINGING materials for the nest: both, but chiefly th
0344 as distinct into one: but more commonly only BRINGING them a little closer together. The more anci
0351 ing. These principles come into play only by BRINGING organisms into new relations with each other
0068 as climate chiefly acts in reducing food, it BRINGS on the most severe struggle between the indiv
0019 n europe alone, and several even within Great BRITAIN. One author believes that there formerly exi
0019 believes that there formerly existed in Great BRITAIN eleven wild species of sheep peculiar to it!
0019 eep peculiar to it! When we bear in mind that BRITAIN has now hardly one peculiar mammal, and Fran
0036 tumblers with these breeds as now existing in BRITAIN, India, and Persia, we can, I think, clearly
0048 disputed. Compare the several floras of Great BRITAIN, of France or of the United States, drawn up
0049 it as an undoubted species peculiar to Great BRITAIN. A wide distance between the homes of two do
0058 inces into which Mr. Watson has divided Great BRITAIN. Now, in this same catalogue, 53 acknowledge
0164 y different breeds, in various countries from BRITAIN to Eastern China; and from Norway in the nor
0284 of each formation in different parts of Great BRITAIN; and this is the result: Palaeozoic strata (
0284 o that the lofty pile of sedimentary rocks in BRITAIN, gives but an inadequate idea of the time wh
```

Page **(Key Word)**

Page **(Key Word)**

```
0299  the North Cape to the Mediterranean, and from BRITAIN to Russia: and therefore equals all the geol
0337  lieve, if all the animals and plants of Great BRITAIN were set free in New Zealand, that in the co
0337  uctions of New Zealand were set free in Great BRITAIN, whether any considerable number would be en
0337  this point of view, the productions of Great BRITAIN may be said to be higher than those of New Z
0352  ll feel any difficulty in such cases as Great BRITAIN having been formerly united to Europe, and c
0364  would never bring seeds from North America to BRITAIN, though they might and do bring seeds from t
0364  hat because a well stocked island, like Great BRITAIN, has not, as far as is known (and it would b
0365  an island, even if far less well stocked than BRITAIN, scarcely more than one would be so well fit
0373  f the cold period, from the western whores of BRITAIN to the Oural range, and southward to the Pyr
0376  closer resemblance in its crustacea to Great BRITAIN, its antipode, than to any other part of the
0383  no terrestrial beings, compared with those of BRITAIN. But this power in fresh water productions o
0395  ne, the relation generally holds good. We see BRITAIN separated by a shallow channel from Europe,
0024  n exterminated even on several of the smaller BRITISH islets, or on the shores of the Mediterrania
0048  ssistance of all kinds, has marked for me 182 BRITISH plants, which are generally considered as va
0049  most experienced ornithologists consider our BRITISH red grouse as only a strongly marked race of
0058  s, but which are almost universally ranked by BRITISH botanists as good and true species. Finally,
0060  e, the two or three hundred doubtful forms of BRITISH plants are entitled to hold, if the existenc
0183  suffice merely to allude to that of the many BRITISH insects which now feed on exotic plants, or
0219  have been attended to by Mr. F. Smith, of the BRITISH Museum, to whom I am much indebted for infor
0224  ed out to a much less extent even than in our BRITISH F. sanguinea, which, as we have seen, is Les
0239  shown how surprisingly the neuters of several BRITISH ants differ from each other in size and some
0243  mud, in the same peculiar manner as does our BRITISH thrush: how it is that the male wrens (Trogl
0284  t is, very nearly thirteen and three quarters BRITISH miles. Some of these formations, which are r
0337  nd, that in the course of time a multitude of BRITISH forms would become thoroughly naturalized th
0401  arms of the sea, in most cases wider than the BRITISH Channel, and there is no reason to suppose t
0475  or instance, lines her nest with mud like our BRITISH species. On the view of instincts having bee
0484  disputes whether or not some fifty species of BRITISH brambles are true species will cease. System
0021  of a very long beak, has a very short and very BROAD one. The pouter has a much elongated body, wi
0180  imbs and even the base of the tail united by a BROAD expanse of skin, which serves as a parachute
0231  y work, by supposing masons first to pile up a BROAD ridge of cement, and then to begin cutting it
0246  e sterility of species: for it seems to make a BROAD and clear distinction between varieties and s
0374  ature was at the same time lower along certain BROAD belts of longitude. On this view of the whole
0374  n this view of the whole world, or at least of BROAD longitudinal belts, having been simultaneousl
0418  f subordinated value. This principle has been BROADLY confessed by some naturalists to be the true
0228  a part, the rims of the basins intersected or BROKE into each other. As soon as this occurred, th
0087  bly perish: or, more delicate and more easily BROKEN shells might be selected, the thickness of th
0106  level, will often have recently existed in a BROKEN condition, so that the good effects of isolat
0107  consequently will exist for long periods in a BROKEN condition, will be the most favourable for th
0119  sed. In our diagram the line of succession is BROKEN at regular intervals by small numbered letter
0124  nus. In our diagram, this is indicated by the BROKEN lines, beneath the capital letters, convergin
0126  rlier and less improved sub groups. Small and BROKEN groups and sub groups will finally tend to di
0126  now large and triumphant, and which are least BROKEN up, that is, which as yet have suffered least
0130  t tree of Life, which fills with its dead and BROKEN branches the crust of the earth, and covers t
0135  many male dung feeding beetles are very often BROKEN off: he examined seventeen specimens in his o
0174  believe that almost every continent has been BROKEN up into islands even during the later tertiar
0174  reas: though I do not doubt that the formerly BROKEN condition of areas now continous has played a
0178  on parent, must formerly have existed in each BROKEN portion of the land, but these links will hav
0197  nd reptiles, which have only to escape from a BROKEN egg, we may infer that this structure has ari
0225  ch other, that they would have intersected or BROKEN into each other, if the spheres had been comp
0228  depth as in the former experiment, would have BROKEN into each other from the opposite sides. The
0287  ies are known and named from single and often BROKEN specimens, or from a few specimens collected
0292  tions on the shores of a continent when first BROKEN up into an archipelago), and consequently dur
0312  st gradual. The secondary formations are more BROKEN: but, as Bronn has remarked, neither the appe
0333  of the geological record, and that in a very BROKEN condition, we have no right to expect, except
0344  reappear: for the link of generation has been BROKEN. We can understand how the spreading of the d
0353  dispersal, have migrated across the vast and BROKEN interspace. The great and striking influence
0370  strictly arctic productions then lived on the BROKEN land still nearer to the pole. Now if we look
0412  endants proceeding from one progenitor become BROKEN up into groups subordinate to groups. In the
0421  m each other. The forms descended from A, now BROKEN up into two or three families, constitute a d
0421  tinct order from those descended from I, also BROKEN up into two families. Nor can the existing sp
0431  mmon parent of a whole family of species, now BROKEN up by extinction into distinct groups and sub
0432  forms are still tied together by a long, but BROKEN, chain of affinities. Extinction has only sep
0460  have been perfected, more especially amongst BROKEN and failing groups of organic beings: but we
0462  od chance for wide migration by many means. A BROKEN or interrupted range may often be accounted f
0475  he chain of ordinary generation has once been BROKEN. The gradual diffusion of dominant forms, wit
0489  succession by generation has never once been BROKEN, and that no cataclysm has desolated the whol
0293  opinions are worthy of much deference, namely BRONN and Woodward, have concluded that the average
0312  secondary formations are more broken: but, as BRONN has remarked, neither the appearance nor disa
0215  ds of fowls which very rarely or never become BROODY, that is, never wish to sit on their eggs. Fa
0244  ncts as the young cuckoo ejecting its foster BROTHERS, ants making slaves, the larvae of ichneumon
0253  cts of close interbreeding. On the contrary, BROTHERS and sisters have usually been crossed in eac
0253  increasing. If we were to act thus, and pair BROTHERS and sisters in the case of any pure animal,
0027  breeds of these breeds, more especially those BROUGHT from distant countries, we can make an almos
0042  ted in main part to selection not having been BROUGHT into play: in cats, from the difficulty in p
0049  logical variation, hybridism, etc., have been BROUGHT to bear on the attempt to determine their ra
0051  hen, moreover, he comes to study allied forms BROUGHT from countries not now continuous, in which
0140  te, and from the number of plants and animals BROUGHT from warmer countries which here enjoy good
0140  r. h. C. Watson on European species of plants BROUGHT from the Azores to England. In regard to ani
0141  ls, now in a state of nature, could easily be BROUGHT to bear widely different climates. We must n
0141  of a very common flexibility of constitution, BROUGHT, under peculiar circumstances, into play. Ho
0215  been found incurable in dogs which have been BROUGHT home as puppies from countries, such as Tier
0247  be secluded in order to prevent pollen being BROUGHT to it by insects from other plants. Nearly a
0252  e will find on their stigmas plenty of pollen BROUGHT from other flowers. In regard to animals, mu
0290  re continually worn away, as soon as they are BROUGHT up by the slow and gradual rising of the lan
0292  ill have been destroyed by being upraised and BROUGHT within the limits of the coast action. Thus
0323  , mylodon, Macrauchenia, and Toxodon had been BROUGHT to Europe from La Plata, without any informa
```

Page **(Key Word)**

0352 om parents specifically distinct. We are thus BROUGHT to the question which has been largely discu
0363 nd it is at least possible that they may have BROUGHT thither the seeds of northern plants. Consid
0393 ith the mainland; moreover, icebergs formerly BROUGHT boulders to its western shores, and they may
0399 this island has been mainly stocked by seeds BROUGHT with earth and stones on icebergs, drifted b
0401 seed, for instance, of one plant having been BROUGHT to one island, and that of another plant to
0424 state of nature, every naturalist has in fact BROUGHT descent into his classification; for he incl
0427 e. geographical distribution may sometimes be BROUGHT usefully into play in classing large and wid
0025 iformly black barbs, and they produced mottled BROWN and black birds; these I again crossed togeth
0164 range of colour is included, from one between BROWN and black to a close approach to cream colour
0415 suffice to quote the highest authority, Robert BROWN, who in speaking of certain organs in the Pro
0416 inity between Ruminants and Pachyderms. Robert BROWN has strongly insisted on the fact that the ru
0480 te having been fitted by natural selection to BROWSE without their aid; whereas in the calf, the t
0067 the case would be the same with turf closely BROWSED by quadrupeds, be let to grow, the more vigo
0072 and little trees, which had been perpetually BROWSED down by the cattle. In one square yard, at a
0097 r and then the stigma of another with the same BRUSH to ensure fertilisation; but it must not be s
0098 distinct species; for if you bring on the same BRUSH a plant's own pollen and pollen from another
0232 licately as a painter could have done with his BRUSH, by atoms of the coloured wax having been tak
0452 s as in other compositae, for the purpose of BRUSHING the pollen out of the surrounding anthers. A
0363 with earth and stones, and have even carried BRUSHWOOD, bones, and the nest of a land bird, I can
0088 to strike in the spurred leg, as well as the BRUTAL cockfighter, who knows well that he can impro
0329 rule, it differs from living forms. But, as BUCKLAND long ago remarked, all fossils can be classe
0036 the two flocks of Leicester sheep kept by Mr. BUCKLEY and Mr. Burgess, as Mr. Youatt remarks, have
0010 case of plants. Under this point of view, Mr. BUCKMAN's recent experiments on plants seem extremel
0009 ng plants; by this term gardeners mean a single BUD or offset, which suddenly assumes a new and so
0010 that the treatment of the parent has affected a BUD or offset, and not the ovules or pollen. But i
0010 that there is no essential difference between a BUD and an ovule in their earliest stages of forma
0132 new being. In the case of sporting plants, the BUD, which in its earliest condition does not appa
0261 as the capacity of one plant to be grafted or BUDDED on another is so entirely unimportant for its
0129 imile largely speaks the truth. The green and BUDDING twigs may represent existing species; and th
0129 ere themselves once, when the tree was small, BUDDING twigs; and this connexion of the former and
0010 acter from that of the rest of the plant. Such BUDS can be propagated by grafting, etc., and somet
0043 which are temporarily propagated by cuttings, BUDS, etc., the importance of the crossing both of
0129 ; and this connexion of the former and present BUDS by ramifying branches may well represent the c
0130 on by having inhabited a protected station. As BUDS give rise by growth to fresh buds, and these,
0130 station. As buds give rise by growth to fresh BUDS, and these, if vigorous, branch out and overto
0434 sphinx moth, the curious folded one of a bee or BUG, and the great jaws of a beetle? yet all these
0223 those of the F. rufescens. The latter does not BUILD its own nest, does not determine its own migr
0228 red, the bees ceased to excavate, and began to BUILD up flat walls of wax on the lines of intersec
0230 x, we can clearly see that if the bees were to BUILD for themselves a thin wall of wax, they could
0230 a great error to suppose that the bees cannot BUILD a rough wall of wax in the proper position
0231 ay on both sides. The manner in which the bees BUILD is curious: they always make the first rough
0232 then, by striking imaginary spheres, they can BUILD up a wall intermediate between two adjoining
0235 ance from each other in a double layer, and to BUILD up and excavate the wax along the planes of i
0243 the male wrens (Troglodytes) of North America, BUILD cock nests, to roost in, like the males of ou
0148 dividual not to have its nutriment wasted in BUILDING up an useless structure. I can thus only und
0223 ters alone usually leave the nest to collect BUILDING materials and food for themselves, their sla
0226 leted; but this is never permitted, the bees BUILDING perfectly flat walls of wax between the sphe
0226 hat the Melipona saves wax by this manner of BUILDING; for the flat walls between the adjoining ce
0231 t twice as thick. By this singular manner of BUILDING, strength is continually given to the comb,
0232 all trying to sweep equal spheres, and then BUILDING up, or leaving ungnawed, the planes of inter
0233 st begun, sweeping spheres or cylinders, and BUILDING up intermediate planes. It is even conceivab
0240 e same as if we were to see a set of workmen BUILDING a house of whom many were five feet four inc
0228 n the basins, so that each hexagonal prism was BUILT upon the festooned edge of a smooth basin, in
0232 b growing downwards so that the comb has to be BUILT over one face of the slip, in this case the b
0232 that cell and of the adjoining cells has been BUILT. This capacity in bees of laying down under c
0233 ny good be derived from a single hexagon being BUILT, as in its construction more materials would
0250 n they can be self fertilised? For instance, a BULB of H¹ppeastrum aulicum produced four flowers;
0091 e form, which pursues deer, and the other more BULKY, with shorter legs, which more frequently att
0112 rred swifter horses; another stronger and more BULKY horses. The early differences would be very s
0291 he west coast of South America, which has been BULKY enough to resist such degradation as it has a
0014 pring from a short horned cow by a long horned BULL, the greater length of horn, though appearing
0017 e greyhound, bloodhound, terrier, spaniel, and BULL dog, which we all know propagate their kind so
0019 ing the Italian greyhound, the bloodhound, the BULL dog, or Blenheim spaniel, etc., so unlike all
0020 e forms, as the Italian greyhound, bloodhound, BULL dog, etc., in the wild state. Moreover, the po
0444 itten on Dogs, maintain that the greyhound and BULL cog, though appearing so different, are really
0214 crossed. Thus it is known that a cross with a BULLDOG has affected for many generations the courag
0237 er breeds, in comparison with the horns of the BULLS or cows of these same breeds. Hence I can see
0238 formed by carefully watching which individual BULLS and cows, when matched, produced oxen with th
0359 well known what a difference there is in the BUOYANCY of green and seasoned timber; and it occurre
0452 udimentary for its proper function of giving BUOYANCY, but has become converted into a nascent bre
0036 f Leicester sheep kept by Mr. Buckley and Mr. BURGESS, as Mr. Youatt remarks, have as purely bre
0339 iking manner that most of the fossil mammals, BURIED there in such numbers, are related to South A
0343 l be in a metamorphosed condition, or may lie BURIED under the ocean. Passing from these difficult
0357 mit the former existence of many islands, now BURIED beneath the sea, which may have served as hal
0366 under an Arctic climate. The ruins of a house BURNT by fire do not tell their tale more plainly,
0218 gh the Tachytes nigra generally makes its own BURROW and stores it with paralysed prey for its own
0218 to feed on, yet that when this insect finds a BURROW already made and stored by another sphex it t
0137 to the wreck. The eyes of moles and of some BURROWING rodents are rudimentary in size, and in som
0137 ps by natural selection. In South America, a BURROWING rodent, the tuco tuco, or Ctenomys, is even
0473 in the domestic duck; or when we look at the BURROWING tucutucu, which is occasionally blind, and
0227 extent, and seeing what perfectly cylindrical BURROWS in wood many insects can make, apparently by
0099 hown, and as I can confirm, either the anthers BURST before the stigma is ready for fertilisation,
0222 numerous pupae. I traced the returning file BURTHENED with booty, for about forty yards, to a ver
0363 er that icebergs formerly landed their rocky BURTHENS on the shores of these mid ocean islands, an
0129 wics which flourished when the tree was a mere BUSH, only two or three, now grown into great branc
0074 ent districts. When we look at the plants and BUSHES clothing an entangled bank, we are tempted to

Page ***(Key Word)***

```
0392  eetles. Again, islands often possess trees or BUSHES belonging to orders which elsewhere include o
0392  ey belonged, and thus convert them first into BUSHES and ultimately into trees. With respect to th
0489  ants of many kinds, with birds singing on the BUSHES, with various insects flitting about, and wit
0135  tor of the ostrich had habits like those of a BUSTARD, and that as natural selection increased in
0404  e see it, if we compare the distribution of BUTTERFLIES and beetles. So it is with most fresh wate
0441  stage, answering to the chrysalis stage of BUTTERFLIES, they have six pairs of beautifully constr
0213  he points to aid his master, than the white BUTTERFLY knows why she lays her eggs on the leaf of
0441  r. but no one probably will dispute that the BUTTERFLY is higher than the caterpillar. In some cas
0270  rieties as specifically distinct. Girou de BUZAREINGUES crossed three varieties of gourd, which li
0015  ions, the several races, for instance, of the CABBAGE, in very poor soil (in which case, however,
0033  arieties. See how different the leaves of the CABBAGE are, and how extremely alike the flowers, ho
0099  or indispensable! If several varieties of the CABBAGE, radish, onion, and some other plants, be al
0099  re not perfectly true. Yet the pistil of each CABBAGE flower is surrounded not only by its own six
0147  to fatten readily. The same varieties of the CABBAGE do not yield abundant and nutritious foliage
0213  nows why she lays her eggs on the leaf of the CABBAGE, I cannot see that these actions differ esse
0268  earance, for instance of the pigeon or of the CABBAGE, is a remarkable fact; more especially when
0099  ongrels: for instance, I raised 233 seedling CABBAGES from some plants of different varieties grow
0397  ctly recovered. As this species has a thick CALCAREOUS operculum, I removed it, and when it had fo
0251  manner the species of Pelargonium, Fuchsia, CALCEOLARIA, Petunia, Rhododendron, etc., have been cr
0251  nstance, Herbert asserts that a hybrid from CALCEOLARIA integrifolia and plantaginea, species most
0064  standing room for his progeny. Linnaeus has CALCULATED that if an annual plant procuced only two s
0467  gs. This high rate of increase is proved by CALCULATION, by the effects of a succession of peculia
0064  ence on this subject than mere theoretical CALCULATIONS, namely, the numerous recorded cases of th
0453  f any service to the rapidly growing embryonic CALF by the excretion of precious phosphate of lime
0480  or rendered rudimentary at this early age. The CALF, for instance, has inherited teeth, which neve
0480  on to browse without their aid; whereas in the CALF, the teeth have been left untouched by selecti
0480  is that parts, like the teeth in the embryonic CALF or like the shrivelled wings under the soldere
0032  eir seed beds, and pull up the rogues, as they CALL the plants that deviate from the proper standa
0051  ly be enabled to make up his own mind which to CALL varieties and which species; but he will succe
0057  ut when we come to discuss the principle, as I CALL it, of Divergence of Character, we shall see h
0074  heir proportional numbers and kinds to what we CALL chance. But how false a view is this! Every on
0081  s and the rejection of injurious variations, I CALL Natural Selection. Variations neither useful n
0088  and this leads me to say a few words on what I CALL Sexual Selection. This depends, not on a strug
0100  se few remarks on the sexes of trees simply to CALL attention to the subject. Turning for a very b
0111  ll defined differences. Mere chance, as we may CALL it, might cause one variety to differ in some
0114  y, shall, as a general rule, belong to what we CALL different genera and orders. The same principl
0166  ar more commonly in the young than in the old. CALL the breeds of pigeons, some of which have bred
0213  o them to a distant point, we should assuredly CALL these actions instinctive. Domestic instincts,
0216  st appeared from what we must in our ignorance CALL an accident. In some cases compulsory habit al
0242  cumulation of numerous, slight, and as we must CALL them accidental, variations, which are in any
0232  minants and pachyderms. Yet he who objected to CALL the extinct genera, which thus linked the livi
0001  es, that mystery of mysteries, as it has been CALLED by one of our greatest philosophers. On my re
0005  proved forms of life, and induces what I have CALLED Divergence of Character. In the next chapter
0006  mutable: but that those belonging to what are CALLED the same genera are lineal descendants of som
0011  ioned. I will here only allude to what may be CALLED correlation of growth. Any change in the embr
0023  rited sub breeds, or species as he might have CALLED them, could be shown him. Great as the differ
0034  ur purpose, a kind of Selection, which may be CALLED Unconscious, and which results from every one
0040  fully acknowledged, the principle, as I have CALLED it, of unconscious selection will always tend
0044  t can rarely be proved. We have also what are CALLED monstrosities; but they graduate into varieti
0045  in this case I presume that the form would be CALLED a variety. Again, we have many slight differe
0045  we have many slight differences which may be CALLED individual differences, such as are known fre
0045  catalogue of facts, that parts which must be CALLED important, whether viewed under a physiologic
0046  fer to those genera which have sometimes been CALLED protean or polymorphic, in which the species
0048  y another as varieties; or, as they are often CALLED, as geographical races! Many years ago, when
0049  cies. But to discuss whether they are rightly CALLED species or varieties, before any definition o
0052  i believe a well marked variety may be justly CALLED an incipient species: but whether this belief
0053  t is the most flourishing, or, as they may be CALLED, the dominant species; those which range wide
0060  r us whether a multitude of doubtful forms be CALLED species or sub species or varieties; what ran
0061  asked, how is it that varieties, which I have CALLED incipient species, become ultimately converte
0061  groups of species, which constitute what are CALLED distinct genera, and which differ from each o
0061  born, but a small number can survive. I have CALLED this principle, by which each slight variatio
0070  struggle for life. But even some of these so CALLED epidemics appear to be due to parasitic worms
0081  ing element, as perhaps we see in the species CALLED polymorphic. We shall best understand the pro
0090  eared under domestication, it would have been CALLED a monstrosity. Illustrations of the action of
0093  uralist doubts the advantage of what has been CALLED the physiological division of labour; hence w
0095  the action, for instance, of the coast waves, CALLED a trifling and insignificant cause, when appl
0107  al scale. These anomalous forms may almost be CALLED living fossils; they have endured to the pres
0111  n the process of formation, or are, as I have CALLED them, incipient species. How, then, does the
0112  n man's productions the action of what may be CALLED the principle of divergence, causing differen
0114  different orders: nature follows what may be CALLED a simultaneous rotation. Most of the animals
0127  rised. This principle of preservation, I have CALLED, for the sake of brevity, Natural Selection.
0154  find the generative variability, as it may be CALLED, still present in a high degree. For in this
0155  fer from the species of some other genus, are CALLED generic characters; and these characters in c
0156  , in exactly the same manner: and as these so CALLED generic characters have been inherited from a
0156  fer from other species of the same genus, are CALLED specific characters; and as these specific ch
0159  , in the enlarged stems, or roots as commonly CALLED, of the Swedish turnip and Ruta baga, plants
0159  then be one of analogous variation in two so CALLED distinct species; and to these a third may be
0165  of colour appears from what would commonly be CALLED an accident, that I was led solely from the o
0180  e skin on their flanks rather full, to the so CALLED flying squirrels; and flying squirrels have t
0181  is process of natural selection, a perfect so CALLED flying squirrel was produced. Now look at the
0192  ted cirripedes have two minute folds of skin, CALLED by me the ovigerous frena, which serve, throu
0199  played in beauty to charm the females, can be CALLED useful only in rather a forced sense. But by
0209  fects of the natural selection of what may be CALLED accidental variations of instincts; that is o
0213  ive parts which habit and the selection of so CALLED accidental variations have played in modifyin
0213  stinctive. Domestic instincts, as they may be CALLED, are certainly far less fixed or invariable t
0214  coming in a straight line to his master when CALLED. Domestic instincts are sometimes spoken of a
```

Page ***(Key Word)***

Page **(Key Word)**

```
0223  iefly the slaves, tend, and milk as it may be  CALLED,  their aphides; and thus both collect food fo
0239  phides, or the domestic cattle as they may be  CALLED,  which our European ants guard or imprison. I
0272  quite unimportant differences between the so  CALLED  hybrid offspring of species, and the so calle
0272  alled hybrid offspring of species, and the so  CALLED  mongrel offspring of varieties. And, on the o
0299  ieties, are really varieties or are, as it is  CALLED,  specifically distinct. This could be effecte
0307  ed in the longmynd beds beneath Barrande's so  CALLED  primordial zone. The presence of phosphatic n
0313  eption to this latter rule, is that of the so  CALLED  colonies of M. Barrande, which intrude for a
0336  , closely allied forms, or, as they have been  CALLED  by some authors, representative species; and
0358  atter. I must now say a few words on what are  CALLED  accidental means, but which more properly mig
0358  ental means, but which more properly might be  CALLED  occasional means of distribution. I shall her
0364  orted. These means of transport are sometimes  CALLED  accidental, but this is not strictly correct:
0366  this same belief, had not Agassiz and others  CALLED  vivid attention to the Glacial Period, which,
0413  enera, and families in each class, on what is  CALLED  the Natural System. But what is meant by this
0421  o the same degree; they may metaphorically be  CALLED  cousins to the same millionth degree; yet the
0422  expressed by ranking them under different so  CALLED  genera, sub families, families, sections, ord
0424  rom the adult; as he likewise includes the so  CALLED  alternate generations of Steenstrup, which ca
0427  gical or adaptive resemblances. Lamarck first  CALLED  attention to this distinction, and he has bee
0429  s but little modified, will give to us our so  CALLED  osculant or aberrant groups. The more aberran
0435  ncient progenitor, the archetype as it may be  CALLED,  of all mammals, had its limbs constructed on
0441  g the ordinary structure, or into what I have  CALLED  complemental males: and in the latter, the de
0442  ain in spiders, there is nothing worthy to be  CALLED  a metamorphosis. The larvae of insects, wheth
0472  laws which have governed the production of so  CALLED  specific forms. In both cases physical condit
0486  s and will then truly give what may be  CALLED  the plan of creation. The rules for classifyi
0486  res. Species and groups of species, which are  CALLED  aberrant, and which may fancifully be called
0486  called aberrant, and which may fancifully be  CALLED  living fossils, will aid us in forming a pict
0352  nary generation with subsequent migration, and  CALLS  in the agency of a miracle. It is universally
0450  ugh the gums, in the upper jaws of our unborn  CALVES.  It has even been stated on good authority th
0226  accordingly I wrote to Professor Miller, of  CAMBRIDGE,  and this geometer has kindly read over the
0389  erence in number. Even the uniform county of  CAMBRIDGE  has 847 plants; and the little island of An
0017  mth by the rein deer, or of cold by the common  CAMEL,  prevented their domestication? I cannot doub
0030  e dray horse and race horse, the dromedary and  CAMEL,  the various breeds of sheep fitted either fo
0097  i believe, of the plant. Bees will act like a  CAMEL  hair pencil, and it is quite sufficient just
0330  rently wide difference between the pig and the  CAMEL.  In regard to the Invertebrata, Barrande, and
0241  as Mr. Lubbock made drawings for me with the  CAMERA  lucida of the jaws which I had dissected from
0190  wholly distinct functions; thus the alimentary  CANAL  respires, digests, and excretes in the larva
0049  continent and the Azores, or Madeira, or the  CANARIES,  or Ireland, be sufficient? It must be admit
0252  nts have been fairly tried: for instance, the  CANARY  bird has been crossed with nine other finches
0252  t that the first crosses between them and the  CANARY,  or that their hybrids, should be perfectly f
0362  nation. Some seeds of the oat, wheat, millet,  CANARY,  hemp, clover, and beet germinated after havi
0053  estions, hereafter to be discussed. Alph. de  CANDOLLE  and others have shown that plants which have
0062  serves, at much greater length. The elder De  CANDOLLE  and Lyell have largely and philosophically s
0115  but the case is very different; and Alph. de  CANDOLLE  has well remarked in his great and admirable
0146  us in the central flowers, that the elder De  CANDOLLE  founded his main divisions of the order on a
0146  al selection can act. For instance, Alph. de  CANDOLLE  has remarked that winged seeds are never fou
0175  s quite remarkable how abruptly, as Alph. de  CANDOLLE  has observed, a common alpine species disapp
0360  transported by any other means; and Alph. de  CANDOLLE  has shown that such plants generally have re
0379  hooker in regard to America, and by Alph. de  CANDOLLE  in regard to Australia, that many more ident
0386  is strikingly shown, as remarked by Alph. de  CANDOLLE,  in large groups of terrestrial plants, whic
0387  lily the Nelumbium, and remembered Alph. de  CANDOLLE's  remarks on this plant, I thought that its
0389  h those on equal continental areas: Alph. de  CANDOLLE  admits this for plants, and Wollaston for in
0392  y herbaceous species; now trees, as Alph. de  CANDOLLE  has shown, generally have, whatever the caus
0402  ge proportion of cases, as shown by Alph. de  CANDOLLE,  to distinct genera. In the Galapagos Archip
0406  lately been admirably discussed by Alph. de  CANDOLLE  in regard to plants, namely, that the lower
0430  habits like those of a Rodent. The elder De  CANDOLLE  has made nearly similar observations on the
0019  or blenheim spaniel, etc., so unlike all wild  CANIDAE,  ever existed freely in a state of nature? I
0404  ts, and in a lesser degree in the Felidae and  CANIDAE.  We see it, if we compare the distribution o
0034  ages now sometimes cross their dogs with wild  CANINE  animals, to improve the breed, and they forme
0062  dividual, but success in leaving progeny. Two  CANINE  animals in a time of dearth, may be truly sai
0002  now publish, must necessarily be imperfect. I  CANNOT  here give references and authorities for my s
0002  nt is discussed in this volume on which facts  CANNOT  be adduced, often apparently leading to concl
0002  ents on both sides of each question; and this  CANNOT  possibly be here done. I much regret that wan
0002  sts, some of them personally unknown to me. I  CANNOT,  however, let this opportunity pass without e
0008  mbryo causes monstrosities; and monstrosities  CANNOT  be separated by any clear line of distinction
0008  rmine whether or not the plant sets a seed. I  CANNOT  here enter on the copious details which I hav
0013  ly, and we see it in the father and child, we  CANNOT  tell whether it may not be due to the same or
0017  ommon camel, prevented their domestication? I  CANNOT  doubt that if other animals and plants, equal
0018  with respect to horses, from reasons which I  CANNOT  give here, I am doubtfully inclined to believ
0019  obably descended from several wild species, I  CANNOT  doubt that there has been an immense amount o
0030  at the fuller's teazle, with its hooks, which  CANNOT  be rivalled by any mechanical contrivance, is
0030  nk, look further than to mere variability. We  CANNOT  suppose that all the breeds were suddenly and
0033  e some differences: but, as a general rule, I  CANNOT  doubt that the continued selection of slight
0034  ermanently altering the breed. Nevertheless I  CANNOT  doubt that this process, continued during cen
0037  ults from such poor materials: but the art, I  CANNOT  doubt, has been simple, and, as far as the fi
0037  known fact, that in a vast number of cases we  CANNOT  recognise, and therefore do not know, the wil
0042  , cats, from their nocturnal rambling habits,  CANNOT  be matched, and, although so much valued by w
0048  of this doubtful nature are far from uncommon  CANNOT  be disputed. Compare the several floras of Gr
0048  mr. Wollaston's admirable work, but which it  CANNOT  be doubted would be ranked as distinct specie
0053  brevity, is rather perplexing, and allusions  CANNOT  be avoided to the struggle for existence, div
0058  same general characters as species, for they  CANNOT  be distinguished from species, except, firstl
0058  nking forms, and the occurrence of such links  CANNOT  affect the actual characters of the forms whi
0064  easing, more or less rapidly, in numbers, all  CANNOT  do so, for the world would not hold them. The
0069  but which never become naturalised, for they  CANNOT  compete with our native plants, nor resist de
0071  up in multitudes, so close together that all  CANNOT  live. When I ascertained that these young tre
0073  ed clover (Trifolium pratense), as other bees  CANNOT  reach the nectar. Hence I have very little do
0076  ve out other mountain varieties, so that they  CANNOT  be kept together. The same result has followe
0087  y the selected change. What natural selection  CANNOT  do, is to modify the structure of one species
```

Page **(Key Word)**

Page **(Key Word)***

```
0087  t may be found in works of natural history, I  CANNOT  find one case which will bear investigation.
0089  e any effect to such apparently weak means: I  CANNOT  here enter on the details necessary to suppor
0090  in the cocks of certain fowls, etc.) which we  CANNOT  believe to be either useful to the males in b
0114  most to increase its numbers. Consequently, I  CANNOT  doubt that in the course of many thousands of
0127  parts of their organisation, and I think this  CANNOT  be disputed; if there be, owing to the high g
0127  severe struggle for life, and this certainly  CANNOT  be disputed; then, considering the infinite c
0128  . the several subordinate groups in any class.  CANNOT  be ranked in a single file, but seem rather t
0131  chapter, but a long catalogue of facts which  CANNOT  be here given would be necessary to show the
0132  t least, safely conclude that such influences  CANNOT  have produced the many striking and complex c
0133  iation is of the slightest use to a being, we  CANNOT  tell how much of it to attribute to the accum
0134  no greater anomaly in nature than a bird that  CANNOT  fly; yet there are several in this state. The
0134  inents and is exposed to danger from which it  CANNOT  escape by flight, but by kicking it can defen
0135  hey must be lost early in life, and therefore  CANNOT  be much used by these insects. In some cases
0135  eira, are so far deficient in wings that they  CANNOT  fly; and that of the twenty nine endemic gene
0139  rom an arctic or even from a temperate region  CANNOT  endure a tropical climate, or conversely. So
0139  r conversely. So again, many succulent plants  CANNOT  endure a damp climate. But the degree of adap
0141  led in our domestic breeds. The rat and mouse  CANNOT  be considered as domestic animals, but they h
0142  of these varieties are of recent origin, they  CANNOT  owe their constitutional differences to habit
0142  ever it was, as proving that acclimatisation  CANNOT  be effected! The case, also, of the kidney be
0142  gs, with the same precautions, the experiment  CANNOT  be said to have been even tried. Nor let it b
0150  ay of facts which I have collected, and which  CANNOT  possibly be here introduced. I can only state
0151  to me certainly to hold good in this class. I  CANNOT  make out that it applies to plants, and this
0152  uch cases natural selection either has not or  CANNOT  come into full play, and thus the organisatio
0180  an astonishing distance from tree to tree. We  CANNOT  doubt that each structure is of use to each k
0200  te bird are of special use to these birds; we  CANNOT  believe that the same bones in the arm of the
0200  the complex laws of growth. Natural selection  CANNOT  possibly produce any modification in any one
0201  in works on natural history to this effect, I  CANNOT  find even one which seems to me of any weight
0202  ch, when used against many attacking animals,  CANNOT  be withdrawn, owing to the backward serrature
0211  o again, in some few cases, certain instincts  CANNOT  be considered as absolutely perfect: but as d
0213  miliar case of the several breeds of dogs: it  CANNOT  be doubted that young pointers (I have myself
0213  of at, a flock of sheep, by shepherd dogs. I  CANNOT  see that these actions, performed without exp
0213  e lays her eggs on the leaf of the cabbage, I  CANNOT  see that these actions differ essentially fro
0214  use tumblers, as I hear from Mr. Brent, which  CANNOT  fly eighteen inches high without going head o
0223  not collect food for itself or its young, and  CANNOT  even feed itself: it is absolutely dependent
0230  uld be a great error to suppose that the bees  CANNOT  build up a rough wall of wax in the proper po
0235  eory of natural selection, cases, in which we  CANNOT  see how an instinct could possibly have origi
0235  als so remote in the scale of nature, that we  CANNOT  account for their similarity by inheritance f
0236  le females, and yet, from being sterile, they  CANNOT  propagate their kind. The subject well deserv
0247  e often injurious to the fertility of a plant  CANNOT  be doubted: for Gartner gives in his table ab
0249  vidual or variety increases fertility, that I  CANNOT  doubt the correctness of this almost universa
0254  , is an extremely general result: but that it  CANNOT, under our present state of knowledge, be con
0262  rafting: the common gooseberry, for instance,  CANNOT  be grafted on the currant, whereas the curran
0264  these facts, any more than why certain trees  CANNOT  be grafted on others. Lastly, an embryo may b
0266  ishes. It must, however, be confessed that we  CANNOT  understand, excepting on vague hypotheses, se
0267  which are generally sterile in some degree. I  CANNOT  persuade myself that this parallelism is an a
0286  e to have denuded the Weald. This, of course,  CANNOT  be done; but we may, in order to form some cr
0287  nfinite number of generations, which the mind  CANNOT  grasp, must have succeeded each other in the
0292  g forms. From the foregoing considerations it  CANNOT  be doubted that the geological record, viewed
0305  t would certainly be highly remarkable; but I  CANNOT  see that it would be an insuperable difficult
0306  o the earliest known species. For instance, I  CANNOT  doubt that all the Silurian trilobites have d
0306  o not differ much from living species; and it  CANNOT  on my theory be supposed, that these old spec
0322  till then, we may justly feel surprise why we  CANNOT  account for the extinction of this particular
0325  extinction, and the introduction of new ones,  CANNOT  be owing to mere changes in marine currents o
0329  etween existing genera, families, and orders,  CANNOT  be disputed. For if we confine our attention
0341  teater, as their degenerate descendants. This  CANNOT  for an instant be admitted. These huge animal
0345  d it may at least be asserted that the record  CANNOT  be proved to be much more perfect, the main o
0346  ographical Distribution. Present distribution  CANNOT  be accounted for by differences in physical c
0346  a climate or condition in the Old World which  CANNOT  be paralleled in the New, at least as closely
0353  le. Undoubtedly many cases occur, in which we  CANNOT  explain how the same species could have passe
0372  re inexplicable on the theory of creation. We  CANNOT  say that they have been created alike, in cor
0381  forms, and others have remained unaltered. We  CANNOT  hope to explain such facts, until we can say
0381  tanical works on the antarctic regions. These  CANNOT  be here discussed. I will only say that as fa
0381  ome of these species are so distinct, that we  CANNOT  suppose that there has been time since the co
0384  he world, no doubt there are many cases which  CANNOT  at present be explained: but some fresh water
0389  ts of the globe. I have already stated that I  CANNOT  honestly admit Forbes's view on continental e
0393  , toads, and newts on so many oceanic islands  CANNOT  be accounted for by their physical conditions
0393  come nearest to an exception: but this group  CANNOT  be considered as oceanic, as it lies on a ban
0394  tly now happens in the arctic regions. Yet it  CANNOT  be said that small islands will not support s
0394  become naturalised and greatly multiplied. It  CANNOT  be said, on the ordinary view of creation, th
0400  ost important for its inhabitants: whereas it  CANNOT, I think, be disputed that the nature of the
0400  ide for the moment the endemic species, which  CANNOT  be here fairly included, as we are considerin
0428  so many characters, great and small, that we  CANNOT  doubt that they have inherited their general
0439  en strikingly similar: a better proof of this  CANNOT  be given, than a circumstance mentioned by Ag
0439  cket the embryo of some vertebrate animal, we  CANNOT  now tell whether it be that of a mammal, bird
0440  relation to their conditions of existence. We  CANNOT, for instance, suppose that in the embryos of
0441  t they have a closed and imperfect mouth, and  CANNOT  feed: their function at this stage is, to sea
0443  of cattle, horses, and various fancy animals,  CANNOT  positively tell, until some time after the an
0443  . we see this plainly in our own children: we  CANNOT  always tell whether the child will be tall or
0445  n length and form of beak, that they would, I  CANNOT  doubt, be ranked in distinct genera, had they
0452  compositae, the male florets, which of course  CANNOT  be fecundated, have a pistil, which is in a r
0453  t this early age is less rudimentary, or even  CANNOT  be said to be in any degree rudimentary. Henc
0455  ess, may well be variable, for its variations  CANNOT  be checked by natural selection. At whatever
0459  pearing to our imagination insuperably great,  CANNOT  be considered real if we admit the following
0459  or instinct. The truth of these propositions  CANNOT, I think, be disputed. It is, no doubt, extre
0460  n intercrossed and of their mongrel offspring  CANNOT  be considered as universal; nor is their very
0462  f the species in the intermediate regions. It  CANNOT  be denied that we are as yet very ignorant of
```

Page **(Key Word)***

```
0464  re imperfect than most geologists believe. It CANNOT be objected that there has not been time suff
0466   the simplest and the most perfect organs: it CANNOT be pretended that we know all the varied mean
0471  ore divergent in character. But as all groups CANNOT thus succeed in increasing in size, for the w
0477  r. we can clearly see why those animals which CANNOT cross wide spaces of ocean, as frogs and terr
0480  ed this view of the mutability of species? It CANNOT be asserted that organic beings in a state of
0481  ate of nature are subject to no variation; it CANNOT be proved that the amount of variation in the
0481  between species and well marked varieties. It CANNOT be maintained that species when intercrossed
0481   the slow action of the coast waves. The mind CANNOT possibly grasp the full meaning of the term o
0481  ng of the term of a hundred million years; it CANNOT add up and perceive the full effects of many
0483  cies closely resemble each other. Therefore I CANNOT doubt that the theory of descent with modific
0194  uth of this remark is indeed shown by that old CANON in natural history of Natura non facit saltum
0206  learly understand the full meaning of that old CANON in natural history, Natura non facit saltum.
0206  natural history, Natura non facit saltum. This CANON if we look only to the present inhabitants of
0210  most complex instincts, can be discovered. The CANON of Natura non facit saltum applies with almos
0243  advantage of the instincts of others; that the CANON in natural history, of natura non facit saltu
0460  gradations in nature, as is proclaimed by the CANON, Natura non facit saltum, that we ought to be
0471  t only by very short and slow steps. Hence the CANON of Natura non facit saltum, which every fresh
0006  tudy and dispassionate judgment of which I am CAPABLE, that the view which most naturalists entert
0008  es: our oldest domesticated animals are still CAPABLE of rapid improvement or modification. It has
0027  c. livia, or the rock pigeon, has been found CAPABLE of domestication in Europe and in India: and
0105  n the production of species, which will prove CAPABLE of enduring for a long period, and of spread
0116  e, so will a greater number of individuals be CAPABLE of there supporting themselves. A set of ani
0140  and not because they were subsequently found CAPABLE of far extended transportation, I think the
0141  r species of the elephant and rhinoceros were CAPABLE of enduring a glacial climate, whereas the l
0183  g fish, it does not seem probable that fishes CAPABLE of true flight would have been developed und
0236   that a number should have been annually born CAPABLE of work, but incapable of procreation, I can
0245  could hardly have kept distinct had they been CAPABLE of crossing freely. The importance of the fa
0253  re fertile; for I am assured by two eminently CAPABLE judges, namely Mr. Blyth and Capt. Hutton, t
0254  ot as an indelible characteristic; but as one CAPABLE of being removed by domestication. Finally,
0338  his view may be true, and yet it may never be CAPABLE of full proof. Seeing, for instance, that th
0360  his seeds floated for 42 days, and were then CAPABLE of germination. But I do not doubt that plan
0362  made in the Zoological Gardens, include seeds CAPABLE of germination. Some seeds of the oat, wheat
0408  rigines: and according as the immigrants were CAPABLE of varying more or less rapidly, there would
0484  constant and distinct from other forms, to be CAPABLE of definition: and if definable, whether the
0490  d death, the most exalted object which we are CAPABLE of conceiving, namely, the production of the
0017  verse climates. I do not dispute that these CAPACITIES have added largely to the value of most of
0032  il. Few would readily believe in the natural CAPACITY and years of practice requisite to become ev
0140  tation, I think the common and extraordinary CAPACITY in our domestic animals of not only withstan
0141  is common to most animals. On this view, the CAPACITY of enduring the most different climates by m
0232  of the adjoining cells has been built. This CAPACITY in bees of laying down under certain circums
0258  re highly important, for they prove that the CAPACITY in any two species to cross is often complet
0258  ther hand, these cases clearly show that the CAPACITY for crossing is connected with constitutiona
0261  and not a specially endowed quality. As the CAPACITY of one plant to be grafted or budded on anot
0261  i presume that no one will suppose that this CAPACITY is a specially endowed quality, but will adm
0261  . as in hybridisation, so with grafting, the CAPACITY is limited by systematic affinity, for no on
0261  t invariably, be grafted with ease. But this CAPACITY, as in hybridisation, is by no means absolut
0276  in the same manner as in grafting trees, the CAPACITY of one species or variety to take on another
0277  e fertility of the hybrids produced, and the CAPACITY of being grafted together, though this latte
0277  f being grafted together, though this latter CAPACITY evidently depends on widely different circum
0352  as something remarkable and exceptional. The CAPACITY of migrating across the sea is more distinct
0383  ity to wide dispersal would follow from this CAPACITY as an almost necessary consequence. We can h
0405  ant, that a species which apparently has the CAPACITY of crossing barriers and ranging widely, as
0488  cessary acquirement of each mental power and CAPACITY by gradation. Light will be thrown on the or
0038  derstand how it is that neither Australia, the CAPE of Good Hope, nor any other region inhabited b
0065  ge in India, as I hear from Dr. Falconer, from CAPE Comorin to the Himalaya, which have been impor
0110  no region is as yet fully stocked, for at the CAPE of Good Hope, where more species of plants are
0299  is of about the size of Europe from the North CAPE to the Mediterranean, and from Britain to Russ
0323  ial South America, in Tierra del Fuego, at the CAPE of Good Hope, and in the peninsula of India. F
0375  w representatives of the peculiar flora of the CAPE of Good Hope occur. At the Cape of Good Hope a
0375  r flora of the Cape of Good Hope occur. At the CAPE of Good Hope a very few European species, beli
0377  see at the present day crowded together at the CAPE of Good Hope, and in parts of temperate Austra
0389  in number, with those on an equal area at the CAPE of Good Hope or in Australia, we must, I think
0398  size of the islands, between the Galapagos and CAPE de Verde Archipelagos: but what an entire and
0398  e in their inhabitants! The inhabitants of the CAPE ce Verde Islands are related to those of Afric
0399  ormerly continuous land, from America: and the CAPE de Verde islands from Africa: and that such co
0399  e south western corner of Australia and of the CAPE of Good Hope, is a far more remarkable case, a
0478  ighbouring American mainland; and those of the CAPE de Verde archipelago and other African islands
0250  , namely, that every ovule in a pod of Crinum CAPENSE fertilised by C. revolutum produced a plant,
0306  imate, and were enabled to double the southern CAPES of Africa or Australia, and thus reach other
0121  nd species. The other nine species (marked by CAPITAL letters) of our original genus, may for a lo
0124  is indicated by the broken lines, beneath the CAPITAL letters, converging in sub branches downward
0069  in. When we reach the Arctic regions, or snow CAPPED summits, or absolute deserts, the struggle fo
0384   the species often range widely and almost CAPRICIOUSLY; for two river systems will have some fish
0468  ing on external characters alone and often CAPRICIOUSLY, can produce within a short period a great
0256  incividuals raised from seed out of the same CAPSULE and exposed to exactly the same conditions.
0259  other hybrids raised from seed from the same CAPSULE have a considerable degree of fertility. The
0359  sake I chiefly tried small seeds, without the CAPSULE or fruit; and as all of these sank in a few
0359  ater. Afterwards I tried some larger fruits, CAPSULES, etc., and some of these floated for a long
0253  eminently capable judges, namely Mr. Blyth and CAPT. Hutton, that whole flocks of these crossed ge
0352  s was first produced within a single region CAPTIVATES the mind. He who rejects it, rejects the ve
0222  little and furious F.flava, which they rarely CAPTURE, and it was evident that they did at once di
0224  it were more advantageous to this species to CAPTURE workers than to procreate them, the habit of
0221  probably depends mainly on the slaves being CAPTURED in greater numbers in Switzerland than in En
0219  les, though most energetic and courageous in CAPTURING slaves, do no other work. They are incapabl
0349  nd the beaver or musk rat, but the coypu and CAPYBARA, rodents of the American type. Innumerable o
0374  ntervening hot countries. So on the Silla of CARACCAS the illustrious Humboldt long ago found spec
```

Page **(Key Word)**

0148 ses more or less completely its own shell or CARAPACE. This is the case with the male Ibla, and in
0148 rdinary manner with the Proteolepas: for the CARAPACE in all other cirripedes consists of the thre
0289 xception discovered by Sir C. Lyell in the CARBONIFEROUS strata of North America. In regard to mam
0296 rved: thus, Messrs. Lyell and Dawson found CARBONIFEROUS beds 1400 feet thick in Nova Scotia, with
0334 n character between those of the overlying CARBONIFEROUS, and underlying Silurian system. But each
0361 this observation. Again, I can show that the CARCASSES of birds, when floating on the sea, sometim
0020 stinct breeds could not be got without extreme CARE and long continued selection; nor can I find a
0027 have been watched, and tended with the utmost CARE, and loved by many people. They have been dome
0040 e slight deviation of structure, or takes more CARE than usual in matching his best animals and th
0041 erences are not amply sufficient, with extreme CARE, to allow of the accumulation of a large amoun
0142 then collect seed from the few survivors, with CARE to prevent accidental crosses; and then again
0217 ct of another bird, than by their own mother's CARE, encumbered as she can hardly fail to be by ha
0218 koo has not utterly lost all maternal love and CARE for her own offspring. The occasional habit of
0223 england the slaves seem to have the exclusive CARE of the larvae, and the masters alone go on sla
0288 ogically explored, and no part with sufficient CARE, as the important discoveries made every year
0341 the globe has been geologically explored with CARE; that only certain classes of organic beings h
0356 cescent from any single pair, but to continued CARE in selecting and training many individuals dur
0384 in the second place, salt water fish can with CARE be slowly accustomed to live in fresh water; a
0426 tter than, other parts of the organisation. We CARE not how trifling a character may be, let it be
0440 en an animal during any part of its embryonic CAREER is active, and has to provide for itself. The
0004 observations it seemed to me probable that a CAREFUL study of domesticated animals and of cultiva
0020 dified by occasional crosses, if aided by the CAREFUL selection of those individual mongrels, whic
0035 r be recognised unless actual measurements or CAREFUL drawings of the breeds in question had been
0035 er. By a similar process of selection, and by CAREFUL training, the whole body of English racehors
0050 iments made during several years by that most CAREFUL observer Gartner, they can be crossed only w
0088 o knows well that he can improve his breed by CAREFUL selection of the best cocks. How low in the
0090 an improve the fleetness of his greyhounds by CAREFUL and methodical selection, or by that unconsc
0164 we see a trace in a bay horse. My son made a CAREFUL examination and sketch for me of a dun Belgi
0198 ere has been but little artificial selection. CAREFUL observers are convinced that a damp climate
0249 r complicated experiments are in progress, so CAREFUL an observer as Gartner would have castrated
0276 ees; and is often so slight that the two most CAREFUL experimentalists who have ever lived, have c
0388 often distant piece of water. Nature, like a CAREFUL gardener, thus takes her seeds from a bed of
0445 selection under domestication: but having had CAREFUL measurements made of the dam and of a three
0012 plex and diversified. It is well worth while CAREFULLY to study the several treatises published on
0034 ow that the breeding of domestic animals was CAREFULLY attended to in ancient times, and is now at
0036 l to them, for any special purpose, would be CAREFULLY preserved during famines and other accident
0053 hooker permits me to add, that after having CAREFULLY read my manuscript, and examined the tables
0149 rejected each little deviation of form less CAREFULLY than when the part has to serve for one spe
0189 al alteration in the transparent layers; and CAREFULLY selecting each alteration which under varie
0221 interesting spectacle to behold the masters CAREFULLY carrying, as Huber has described, their sla
0225 the cells of the Mexican Melipona domestica, CAREFULLY described and figured by Pierre Huber. The
0238 narily long horns, could be slowly formed by CAREFULLY watching which individual bulls and cows, w
0239 w few neuter insects out-of Europe have been CAREFULLY examined. Mr. F. Smith has shown how surpri
0240 orkers have their ocelli rudimentary. Having CAREFULLY dissected several specimens of these worker
0247 and in many other cases, Gartner is obliged CAREFULLY to count the seeds, in order to show that t
0248 gh gartner was enabled to rear some hybrids, CAREFULLY guarding them from a cross with either pure
0249 e same garden, the visits of insects must be CAREFULLY prevented during the flowering season: henc
0252 to animals, much fewer experiments have been CAREFULLY tried than with plants. If our systematic a
0302 er which our geological formations have been CAREFULLY examined: we forget that groups of species
0393 mmals offer another and similar case. I have CAREFULLY searched the oldest voyages, but have not f
0445 , within twelve hours after being hatched; I CAREFULLY measured the proportions (but will not here
0485 any two forms, we shall be led to weigh more CAREFULLY and to value higher the actual amount of di
0033 act, also followed; for hardly any one is so CARELESS as to allow his worst animals to breed. In r
0038 as is externally visible; and indeed he rarely CARES for what is internal. He can never act by sel
0083 nly on external and visible characters: nature CARES nothing for appearances, except in so far as
0071 e species of plants (not counting grasses and CARICES) flourished in the plantations, which could
0413 all mammals, by another those common to all CARNIVORA, by another those common to the dog genus,
0009 under confinement, I may just mention that CARNIVOROUS animals, even from the tropics, breed in t
0009 f the planticrades or bear family: whereas, CARNIVOROUS birds, with the rarest exceptions, hardly
0088 provided with special weapons. The males of CARNIVOROUS animals are already well armed; though to
0113 mals with simple habits. Take the case of a CARNIVOROUS quadruped, of which the number that can be
0113 nting water, and some perhaps becoming less CARNIVOROUS. The more diversified in habits and struct
0113 habits and structure the descendants of our CARNIVOROUS animal became, the more places they would
0116 r. waterhouse and others have remarked, our CARNIVOROUS, ruminant, and rodent mammals, could succe
0179 views as I hold, how, for instance, a land CARNIVOROUS animal could have been converted into one
0179 be easy to show that within the same group CARNIVOROUS animals exist having every intermediate gr
0394 the Viti Archipelago, the Bonin Islands, the CAROLINE and Marianne Archipelagoes, and Mauritius, a
0089 but if man can in a short time give elegant CARRIAGE and beauty to his bantams, according to his
0152 and wattle of the different carriers, in the CARRIAGE and tail of our fantails, etc., these being
0019 aces from several aboriginal stocks, has been CARRIED to an absurd extreme by some authors. They b
0021 and these feathers are kept expanded, and are CARRIED so erect that in good birds the head and tai
0024 , and, therefore, some of them must have been CARRIED back again into their native country; but no
0035 ollins, etc., by this very same process, only CARRIED on more methodically, did greatly modify, ev
0092 oss to the plant; yet if a little pollen were CARRIED, at first occasionally and then habitually,
0093 ale tree, the pollen could not thus have been CARRIED. The weather had been cold and boisterous, a
0093 tractive to insects that pollen was regularly CARRIED from flower to flower, another process might
0094 egree under nature, then as pollen is already CARRIED regularly from flower to flower, and as a mo
0099 t may be objected that pollen could seldom be CARRIED from tree to tree, and at most only from flo
0100 ree, we can see that pollen must be regularly CARRIED from flower to flower! and this will give a
0100 a better chance of pollen being occasionally CARRIED from tree to tree. That trees belonging to a
0221 ruthlessly killed their small opponents, and CARRIED their dead bodies as food to their nest, twe
0221 lace of combat: they were eagerly seized, and CARRIED off by the tyrants, who perhaps fancied that
0222 ow ants had crawled away, they took heart and CARRIED off the pupae. One evening I visited another
0224 aves. When the instinct was once acquired, if CARRIED out to a much less extent even than in our B
0269 is difference in the process of selection, as CARRIED on by man and nature, we need not be surpris
0305 xisting species. Lately, Professor Pictet has CARRIED their existence one sub stage further back:

Page **(Key Word)**

carried

Page ***(Key Word)***

```
0363  s loaded with earth and stones, and have even  CARRIED  brushwood, bones, and the rest of a land bir
0364  e action of sea water: nor could they be long   CARRIED  in the crops or intestines of birds. These m
0448  om the young, during a course of modification   CARRIED  on for many generations, having to provide f
0021  is something astonishing. Compare the English   CARRIER  and the short faced tumbler, and see the won
0021  orresponding differences in their skulls. The   CARRIER, more especially the male bird, is also rema
0021  ularly short tails. The barb is allied to the   CARRIER, but, instead of a very long beak, has a ver
0023  hat any ornithologist would place the English   CARRIER, the short faced tumbler, the runt, the barb
0024  olumbidae for a beak like that of the English   CARRIER, or that of the short faced tumbler, or barb
0027  omestic breeds. Secondly, although an English   CARRIER or short faced tumbler differs immensely in
0027  instance the wattle and length of beak of the   CARRIER, the shortness of that of the tumbler, and t
0029  and race horse, a greyhound and bloodhound, a   CARRIER and tumbler pigeon. One of the most remarkab
0335  any varieties between the rock pigeon and the   CARRIER have become extinct; and carriers which are
0362  me that a friend of his had to give up flying   CARRIER pigeons from France to England, as the hawks
0026  se that species, aboriginally as distinct as   CARRIERS, tumblers, pouters, and fantails now are, sh
0035  he accounts given in old pigeon treatises of   CARRIERS and tumblers with these breeds as now existi
0090  our domestic animals (as the wattle in male   CARRIERS, horn like protuberances in the cocks of cer
0152  ers, in the beak and wattle of the different   CARRIERS, in the carriage and tail of our fantails, e
0153  rally to the male sex, as with the wattle of   CARRIERS and the enlarged crop of pouters. Now let us
0335  eon and the carrier have become extinct; and  CARRIERS which are extreme in the important character
0445  in pouters, fantails, runts, barbs, dragons,  CARRIERS, and tumblers. Now some of these birds, when
0035  odwood Races, are favoured in the weights they  CARRY. Lord Spencer and others have shown how the c
0093  oulc, unintentionally on their part, regularly  CARRY  pollen from flower to flower; and that they c
0097  bees should fly from flower to flower, and not  CARRY  pollen from one to the other, to the great co
0219  es which determine the migration, and actually  CARRY  their masters in their jaws. So utterly helpl
0223  be formed, and when they migrate, the masters  CARRY  the slaves. Both in Switzerland. and England t
0223  ich are not slave makers, will as I have seen,  CARRY  off pupae of other species, if scattered near
0238  in cryptocerus, the workers of one caste alone  CARRY  a wonderful sort of shield on their heads, th
0364  ved that scarcely any means of transport would  CARRY  seeds for very great distances: for seeds do
0380  hough hides, wool, and other objects likely to  CARRY  seeds have been largely imported into Europe
0220  ves work energetically with their masters in  CARRYING  them away to a place of safety. Hence, it is
0221  ng spectacle to behold the masters carefully  CARRYING, as Huber has described, their slaves in the
0222  a number of these ants entering their nest,  CARRYING  the dead bodies of F. fusca (showing that it
0222  the last incividual of F. sanguinea emerge,  CARRYING  a pupa: but I was not able to find the desol
0015  is view: to assert that we could not breed our  CART  and race horses, long and short horned cattle,
0164  examination and sketch for me of a dun Belgian  CART  horse with a double stripe on each shoulder an
0165  it to breeds so distinct as the heavy Belgian  CART  horse, Welch ponies, cobs, the lanky Kattywar
0445  rence. So, again, I was told that the foals of  CART  and race horses differed as much as the full g
0445  d of a three days old colt of a race and heavy  CART  horse, I find that the colts have by no means
0021  able from the wonderful development of the  CARUNCULATED  skin about the head, and this is accompani
0003  catch insects under the bark of trees. In the  CASE  of the misseltoe, which draws its nourishment
0004  to me to be no explanation; for it leaves the  CASE  of the coadaptations of organic beings to each
0008  lly continues to vary for many generations. No  CASE  is on record of a variable being ceasing to be
0010  t far from rare under cultivation; and in this  CASE  we see that the treatment of the parent has af
0010  in the same manner. To judge how much, in the  CASE  of any variation, we should attribute to the d
0010  e direct effect, though apparently more in the  CASE  of plants. Under this point of view, Mr. Buckm
0015  e, of the cabbage, in very poor soil (in which  CASE, however, some effect would have to be attribu
0015  r, any slight deviations of structure, in such  CASE, I grant that we could deduce nothing from dom
0017  scended from any one wild species: but, in the  CASE  of some other domestic races, there is presump
0017  g domesticated productions have varied. In the  CASE  of most of our anciently domesticated animals
0018  and generally true than seems to me to be the  CASE, what does it show, but that some of our breed
0019  parent stocks? So it is in India. Even in the  CASE  of the domestic dogs of the whole world, which
0020  g continued selection; nor can I find a single  CASE  on record of a permanent race having been thus
0026  cult, perhaps impossible, to bring forward one  CASE  of the hybrid offspring of two animals clearly
0030  rnspit dog; and this is known to have been the  CASE  with the ancon sheep. But when we compare the
0033  oints: this is hardly ever, perhaps never, the  CASE. The Laws of correlation of growth, the import
0035  y changed within the last century, and in this  CASE  the change has, it is believed, been chiefly a
0036  spring than the inferior ones; so that in this  CASE  there would be a kind of unconscious selection
0042  during the last thirty or forty years. In the  CASE  of animals with separate sexes, facility in pr
0045  for at least some few generations? and in this  CASE  I presume that the form would be called a vari
0051  ht from countries not now continuous, in which  CASE  he can hardly hope to find the intermediate li
0052  very long periods, as has been shown to be the  CASE  by Mr. Wollaston with the varieties of certain
0055  r side, and it has invariably proved to be the  CASE  that a larger proportion of the species on the
0056  es to be a slow one. And this certainly is the  CASE, if varieties be looked at as incipient specie
0057  n do the species of the smaller genera. Or the  CASE  may be put in another way, and it may be said,
0063  than can possibly survive, there must in every  CASE  be a struggle for existence, either one indivi
0063  ole animal and vegetable kingdoms: for in this  CASE  there can be no artificial increase of food, a
0067  o suffer most, but this is not invariably the  CASE. With plants there is a vast destruction of se
0067  cts. If turf which has long been mown, and the  CASE  would be the same with turf closely browsed by
0070  other such plants in a garden! I have in this  CASE  lost every single seed. This view of the neces
0074  cy of certain flowers in that district! In the  CASE  of many species, many different checks, acting
0075  mote in the scale of nature. This is often the  CASE  with those which may strictly be said to strug
0075  uggle with each other for existence, as in the  CASE  of locusts and grass feeding quadrupeds. But t
0075  d, and are exposed to the same dangers. In the  CASE  of varieties of the same species, the struggle
0076  the economy of nature: but probably in no one  CASE  could we precisely say why one species has bee
0078  or colder, damper or drier districts. In this  CASE  we can clearly see that if we wished in imagin
0081  able course of natural selection by taking the  CASE  of a country undergoing some physical change,
0081  ree or mammal has been shown to be. But in the  CASE  of an island, or of a country partly surrounde
0082  oulc have been seized on by intruders. In such  CASE, every slight modification, which in the cours
0082  or increases variability; and in the foregoing  CASE  the conditions of life are supposed to have un
0086  he structure of the adult: and probably in the  CASE  of those insects which live only for a few hou
0087  in works of natural history, I cannot find one  CASE  which will bear investigation. A structure use
0087  of life in the two sexes, as is sometimes the  CASE  with insects. And this leads me to say a few w
0090  r two imaginary illustrations. Let us take the  CASE  of a wolf, which preys on various animals, sec
0091  pherd's flocks. Let us now take a more complex  CASE. Certain plants excrete a sweet juice, apparen
0092  inner bases of the petals of a flower. In this  CASE  insects in seeking the nectar would get dusted
0092  favoured or selected. We might have taken the  CASE  of insects visiting flowers for the sake of co
```

Page ***(Key Word)***

Page **(Key Word)***

```
0093  . i will give only one, not as a very striking  CASE, but as likewise illustrating one step in the
0093  arch of nectar. But to return to our imaginary   CASE: as soon as the plant had been rendered so hig
0094  to the nectar feeding insects in our imaginary   CASE: we may suppose the plant of which we have bee
0096  must here introduce a short digression. In the   CASE of animals and plants with separate sexes, it
0096  s must always unite for each birth; but in the   CASE of hermaphrodites this is far from obvious. Ne
0098  ring forward, as Kolreuter has shown to be the   CASE with the barberry; and curiously in this very
0099  species. When distinct species are crossed the   CASE is directly the reverse, for a plant's own pol
0099  ct we shall return in a future chapter. In the   CASE of a gigantic tree covered with innumerable fl
0100  separated than other plants, I find to be the   CASE in this country; and at my request Dr. Hooker
0100  ese all pair. As yet I have not found a single   CASE of a terrestrial animal which fertilises itsel
0100  o the action of insects and of the wind in the   CASE of plants, by which an occasional cross could
0100  means for an occasional cross. And, as in the   CASE of flowers, I have as yet failed, after consul
0101  namely, Professor Huxley, to discover a single   CASE of an hermaphrodite animal with the organs of
0101  e. cirripedes long appeared to me to present a   CASE of very great difficulty under this point of v
0101  naturalists as a strange anomaly that, in the   CASE of both animals and plants, species of the sam
0102  e species on the confines of each. And in this   CASE the effects of intercrossing can hardly be cou
0103  arated countries; and this I believe to be the   CASE. In hermaphrodite organisms which cross only o
0103  other varieties is thus lessened. Even in the   CASE of slow breeding animals, which unite for each
0113  ase in numbers. We can clearly see this in the   CASE of animals with simple habits. Take the case o
0113  e case of animals with simple habits. Take the   CASE of a carnivorous quadruped, of which the numbe
0115  o certain stations in their new homes. But the   CASE is very different; and Alph. de Candolle has w
0116  her in unequal degrees, as is so generally the   CASE in nature, and is represented in the diagram b
0120  increased in the successive generations. This   CASE would be represented in the diagram, if all th
0122  her in unequal degrees, as is so generally the   CASE in nature; species (A) being more nearly relat
0124  ed or altered only in a slight degree. In this   CASE, its affinities to the other fourteen new spec
0124  ery naturalist will be able to bring some such   CASE before his mind. In the diagram, each horizont
0132  kes place which is to form a new being. In the   CASE of sporting plants, the bud, which in its earl
0132  is, that the effect is extremely small in the   CASE of animals, but perhaps rather more in that of
0137  ids and growth of fur over them, might in such   CASE be an advantage; and if so, natural selection
0137  l selection aided by use and disuse, so in the   CASE of the cave rat natural selection seems to hav
0138  odte and others have remarked, this is not the   CASE, and the cave insects of the two continents ar
0138  nts of the European continent. And this is the   CASE with some of the American cave animals, as I h
0139  the blind fish, the Amblyopsis; and as is the   CASE with the blind Proteus with reference to the r
0140  generally very close, we have evidence, in the   CASE of some few plants, of their becoming, to a ce
0140  in all ordinary cases we assume such to be the   CASE; nor do we know that they have subsequently be
0142  their constitutional differences to habit. The   CASE of the Jerusalem artichoke, which is never pro
0142  g that acclimatisation cannot be effected! The   CASE, also, of the kidney bean has been often cited
0143  most imperfectly understood. The most obvious   CASE is, that modifications accumulated solely for
0144  y comes into play? With respect to this latter   CASE of correlation, I think it can hardly be accid
0144  the most abnormal in their teeth. I know of no   CASE better adapted to show the importance of the l
0145  compositae countenances this idea; but, in the   CASE of the corolla of the Umbelliferae, it is by n
0145  add, as an instance of this, and of a striking   CASE of correlation, that I have recently observed
0148  pletely its own shell or carapace. This is the   CASE with the male Ibla, and in a truly extraordina
0149  ich are higher. I presume that lowness in this   CASE means that the several parts of the organisati
0150  e genus. The rule applies very strongly in the   CASE of secondary sexual characters, when displayed
0151  m. the rule being so plainly applicable in the   CASE of secondary sexual characters, may be due to
0151  dary sexual characters is clearly shown in the   CASE of hermaphrodite cirripedes; and I may here ad
0152  to that species; nevertheless the part in this   CASE is eminently liable to variation. Why should t
0152  rphic groups, we see a nearly parallel natural   CASE; for in such cases natural selection either ha
0153  rly constant. And this, I am convinced, is the   CASE. That the struggle between natural selection o
0154  dition to many modified descendants, as in the   CASE of the wing of the bat, it must have existed,
0154  d, still present in a high degree. For in this   CASE the variability will seldom as yet have been f
0154  example because an explanation is not in this   CASE applicable, which most naturalists would advan
0155  hich have thus come to differ. Or to state the   CASE in another manner: the points in which all the
0159  influences. In the vegetable kingdom we have a   CASE of analogous variation, in the enlarged stems,
0159  n from a common parent: if this be not so, the   CASE will then be one of analogous variation in two
0159  eation. With pigeons, however, we have another   CASE, namely, the occasional appearance in all the
0160  presume that no one will doubt that this is a   CASE of reversion, and not of a new yet analogous v
0160  t and differently coloured breeds; and in this   CASE there is nothing in the external conditions of
0162  we might have inferred that the blueness was a   CASE of reversion, from the number of the markings,
0162  of the same group. And this undoubtedly is the   CASE in nature. A considerable part of the difficul
0163  i will, however, give one curious and complex   CASE, not indeed as affecting any important charact
0163  domestication and partly under nature. It is a   CASE apparently of reversion. The ass not rarely ha
0165  a. Lastly, and this is another most remarkable   CASE, a hybrid has been figured by Dr. Gray (and he
0165  y (and he informs me that he knows of a second   CASE) from the ass and the hemionus; and this hybri
0166  of the most distinct species. Now observe the   CASE of the several breeds of pigeons: they are des
0166  ries, species; and how exactly parallel is the   CASE with that of the horse genus! For myself, I ven
0167  the laws of variation is profound. Not in one   CASE out of a hundred can we pretend to assign any
0169  ocess, requiring a long lapse of time, in this   CASE, natural selection may readily have succeeded
0173  many transitional forms. Let us take a simple   CASE: in travelling from north to south over a cont
0176  met with striking instances of the rule in the   CASE of varieties intermediate between well marked
0176  sting in large numbers; and in this particular   CASE the intermediate form would be eminently liabl
0177  their stocks by selection; the chances in this   CASE will be strongly in favour of the great holder
0178  nct to rank as representative species. In this   CASE, intermediate varieties between the several re
0180  ecats on mice and land animals. If a different   CASE had been taken, and it had been asked how an i
0180  ent to lessen the difficulty in any particular   CASE like that of the bat. Look at the family of sq
0181  ch had been formed by the same steps as in the   CASE of the less perfectly gliding squirrels; and t
0183  having existed in lesser numbers, than in the   CASE of species with fully developed structures. I
0183  e individuals of the same species. When either   CASE occurs, it would be easy for natural selection
0184  le, insects in the water. Even in so extreme a   CASE as this, if the supply of insects were constan
0186  ariations be inherited, which is certainly the   CASE; and if any variation or modification in the o
0187  rcely ever possible, and we are forced in each   CASE to look to species of the same group, that is
0188  no very great difficulty (not more than in the   CASE of many other structures) in believing that na
0188  formed by natural selection, although in this   CASE he does not know any of the transitional grade
0189  olutely break down. But I can find out no such   CASE. No doubt many organs exist of which we do not
0189  e members of a large class, for in this latter   CASE the organ must have been first formed at an ex
```

Page **(Key Word)***

Page ***(Key Word)**

```
0192  either the males or fertile females; but this  CASE  will be treated of in the next chapter. The el
0192  r. the electric organs of fishes offer another  CASE  of special difficulty; it is impossible to con
0193  fferent families and orders, offers a parallel  CASE  of difficulty. Other cases could be given: for
0195  a very different kind, on this head, as in the  CASE  of an organ as perfect and complex as the eye.
0195  d pause before being too positive even in this  CASE,  for we know that the distribution and existen
0204  ore than enough to stagger any one; yet in the  CASE  of any organ, if we know of a long series of g
0205  ions. We are far too ignorant, in almost every  CASE,  to be enabled to assert that any part or orga
0210  , yet profitable, variations. Hence, as in the  CASE  of corporeal structures, we ought to find in n
0210  under different circumstances, etc.; in which  CASE  either one or the other instinct might be pres
0210  n be shown to occur in nature. Again as in the  CASE  of corporeal structure, and conformably with m
0213  riods of time. But let us look to the familiar  CASE  of the several breeds of dogs: it cannot be do
0217  ferent ages in the same nest. If this were the  CASE,  the process of laying and hatching might be i
0218  es. For several hen ostriches, at least in the  CASE  of the American species, unite and lay first a
0218  s laying a large number of eggs; but as in the  CASE  of the cuckoo, at intervals of two or three da
0218  eggs in the nests of bees of other kinds. This  CASE  is more remarkable than that of the cuckoo; fo
0218  nd becomes for the occasion parasitic. In this  CASE,  as with the supposed case of the cuckoo, I ca
0218  parasitic. In this case, as with the supposed  CASE  of the cuckoo, I can see no difficulty in natu
0226  e size, is very frequently and necessarily the  CASE,  the three flat surfaces are united into a pyr
0226  forms a part of two cells. Reflecting on this  CASE,  it occurred to me that if the Melipona had ma
0228  aight edges of a three sided pyramid as in the  CASE  of ordinary cells. I then put into the hive, i
0230  reme growing margin, or the two plates, as the  CASE  may be; and they never complete the upper edge
0231  cells. But the rough wall of wax has in every  CASE  to be finished off, by being largely gnawed aw
0232  to be built over one face of the slip, in this  CASE  the bees can lay the foundations of one wall o
0233  great difficulty in a single insect (as in the  CASE  of a queen wasp) making hexagonal cells, if sh
0233  ated hexagon: but I am not aware that any such  CASE  has been observed; nor would any good be deriv
0234  y required a store of honey: there can in this  CASE  be no doubt that it would be an advantage to o
0234  s, like the cells of the Melipona; for in this  CASE  a large part of the bounding surface of each c
0236  eat length, but I will here take only a single  CASE,  that of working or sterile ants. How the work
0237  be asked how is it possible to reconcile this  CASE  with the theory of natural selection? First, l
0239  s at once annihilate my theory. In the simpler  CASE  of neuter insects all of one caste or of the s
0239  nt from the fertile males and females, in this  CASE,  we may safely conclude from the analogy of or
0240  ve well developed ocelli. I may give one other  CASE: so confidently did I expect to find gradation
0241  ted series having been first formed, as in the  CASE  of the driver ant, and then the extreme forms,
0242  een efficient in so high a degree, had not the  CASE  of these neuter insects convinced me of the fa
0242  of the fact. I have, therefore, discussed this  CASE,  at some little but wholly insufficient length
0242  fficulty, which my theory has encountered. The  CASE,  also, is very interesting, as it proves that
0242  ed that no one has advanced this demonstrative  CASE  of neuter insects, against the well known doct
0243  acts in regard to instincts; as by that common  CASE  of closely allied, but certainly distinct, spe
0245  riters. On the theory of natural selection the  CASE  is especially important, inasmuch as the steri
0246  as far as the microscope reveals. In the first  CASE  the two sexual elements which go to form the e
0246  to form the embryo are perfect; in the second  CASE  they are either not at all developed, or are i
0248  ther pure parent, for six or seven; and in one  CASE  for ten generations, yet he asserts positively
0248  eased. I do not doubt that this is usually the  CASE,  and that the fertility often suddenly decreas
0250  ant, which (he says) I never saw to occur in a  CASE  of its natural fecundation. So that we here ha
0250  first cross between two distinct species. This  CASE  of the Crinum leads me to refer to a most sing
0251  also, been confirmed by other observers in the  CASE  of Hippeastrum with its sub genera, and in the
0251  of hippeastrum with its sub genera, and in the  CASE  of some other genera, as Lobelia, Passiflora a
0252  sive generation, as Gartner believes to be the  CASE,  the fact would have been notorious to nursery
0252  nature can be more easily crossed than in the  CASE  of plants; but the hybrids themselves are, I t
0252  re, I think, more sterile. I doubt whether any  CASE  of a perfectly fertile hybrid animal can be co
0253  eated admonition of every breeder. And in this  CASE,  it is not at all surprising that the inherent
0253  act thus; and pair brothers and sisters in the  CASE  of any pure animal, which from any cause had t
0258  ciprocal cross between two species, I mean the  CASE,  for instance, of a stallion horse being first
0262  n crossing: so Sagaret believes this to be the  CASE  with different individuals of the same two spe
0262  in an imperfect condition, is a very different  CASE  from the difficulty of uniting two pure specie
0262  inded by this latter fact of the extraordinary  CASE  of Hippeastrum, Lobelia, etc., which seeded mu
0263  has been a special endowment; although in the  CASE  of crossing, the difficulty is as important fo
0263  nce and stability of specific forms, as in the  CASE  of grafting it is unimportant for their welfar
0263  le element reaching the ovule, as would be the  CASE  with a plant having a pistil too long for the
0264  ryo to be developed, as seems to have been the  CASE  with some of Thuret's experiments on Fuci. No
0264  rally healthy and long lived, as we see in the  CASE  of the common mule. Hybrids, however, are diff
0264  sexual elements are imperfectly developed, the  CASE  is very different. I have more than once allud
0265  sterility in a very similar manner. In the one  CASE,  the conditions of life have been disturbed, t
0266  ree as to be inappreciable by us; in the other  CASE,  or that of hybrids, the external conditions h
0266  ed, sterility is the common result; in the one  CASE  from the conditions of life having been distur
0266  ns of life having been disturbed, in the other  CASE  from the organisation having been disturbed by
0268  fully admit that this is almost invariably the  CASE.  But if we look to varieties produced under na
0270  roduced only five grains. Manipulation in this  CASE  could not have been injurious, as the plants h
0270  st of infertility, as varieties. The following  CASE  is far more remarkable, and seems at first qui
0281  or instance, a horse from a tapir; and in this  CASE  direct intermediate links will have existed be
0281  nks will have existed between them. But such a  CASE  would imply that one form had remained for a v
0285  ea of eternity. I am tempted to give one other  CASE,  the well known one of the denudation of the W
0288  scan genus Chiton offers a partially analogous  CASE.  With respect to the terrestrial productions w
0291  ither, in profound depths of the sea, in which  CASE,  judging from the researches of E. Forbes, we
0291  it continue slowly to subside. In this latter  CASE,  as long as the rate of subsidence and supply
0295  accumulation. When we see, as is so often the  CASE,  a formation composed of beds of different min
0298  ve been collected from many places; and in the  CASE  of fossil species this could rarely be effecte
0303  or back as the eocene stage. The most striking  CASE,  however, is that of the Whale family: as thes
0304  m the chalk of Belgium. And, as if to make the  CASE  as striking as possible, this sessile cirriped
0305  of our many tertiary and existing species. The  CASE  most frequently insisted on by palaeontologist
0308  extremity of denudation and metamorphism. The  CASE  at present must remain inexplicable: and may b
0312  ble to resist the evidence on this head in the  CASE  of the several tertiary stages: and every year
0313  ; but Lyell's explanation, namely,that it is a  CASE  of temporary migration from a distinct geograp
0328  to conclude that this has invariably been the  CASE,  and that large areas have invariably been aff
0329  e long intervals of time between them; in this  CASE,  the several formations in the two regions cou
0330  elonging to distinct families. The most common  CASE,  especially with respect to very distinct grou
```

Page ***(Key Word)**

Page ***(Key Word)***

0331 e it is quite possible, as we have seen in the CASE of some Silurian forms, that a species might g
0333 from each other by a dozen characters, in this CASE the genera, at the early period marked VI., wo
0333 ween their collateral relations. In nature the CASE will be far more complicated than is represent
0333 st palaeontologists seems frequently to be the CASE. Thus, on the theory of descent with modificat
0335 where subsequently produced, especially in the CASE of terrestrial productions inhabiting separate
0341 ny. Or, which would probably be a far commoner CASE, two or three species of two or three alone of
0341 es decreasing in numbers, as apparently is the CASE of the Edentata of South America, still fewer
0346 d the subject has come to this conclusion. The CASE of America alone would almost suffice to prove
0346 ecies generally require: for it is a most rare CASE to find a group of organisms confined to any s
0350 ces organisms quite like, or, as we see in the CASE of varieties nearly like each other. The dissi
0351 e areas, as is so commonly and notoriously the CASE. I believe, as was remarked in the last chapte
0351 ave descended from the same progenitor. In the CASE of those species, which have undergone during
0354 if it can be shown to be almost invariably the CASE, that a region, of which most of its inhabitan
0364 ich is in itself a rare accident. Even in this CASE, how small would the chance be of a seed falli
0366 unless they were stopped by barriers, in which CASE they would perish. The mountains would become
0368 in the United States believe to have been the CASE, chiefly from the distribution of the fossil G
0368 ltimately on the summits of the mountains, the CASE will have been somewhat different: for it is n
0369 to modification: and this we find has been the CASE: for if we compare the present Alpine plants a
0372 ate North America; and the still more striking CASE of many closely allied crustaceans (as describ
0373 rld. But we have good evidence in almost every CASE, that the epoch was included within the latest
0379 hern migration and re migration northward, the CASE may have been wholly different with those inter
0381 escent with modification, a far more remakable CASE of difficulty. For some of these species are s
0383 ld have extended to distant countries. But the CASE is exactly the reverse. Not only have many fre
0386 nts, but will here give only the most striking CASE: I took in February three tablespoons of mud f
0388 tter chance of seizing on a place, than in the CASE of terrestrial colonists. We should, also, rem
0392 portal by the wool and fur of quadrupeds. This CASE presents no difficulty on my view, for a hooke
0393 to explain. Mammals offer another and similar CASE. I have carefully searched the oldest voyages,
0399 he cape of Good Hope, is a far more remarkable CASE, and is at present inexplicable: but this affi
0401 g the same. The really surprising fact in this CASE of the Galapagos Archipelago, and in a lesser
0405 rossing barriers and ranging widely, as in the CASE of certain powerfully winged birds, will neces
0407 are insuperable: though they often are in this CASE, and in that of the individuals of the same sp
0411 nother on vegetable matter, and so on; but the CASE is widely different in nature: for it is notor
0417 ll be retained amongst the Malpighiaceae. This CASE seems to me well to illustrate the spirit with
0419 of characters common to all birds: but in the CASE of crustaceans, such definition has hitherto b
0421 letely lost traces of their parentage, in this CASE, their places in a natural classification will
0422 ate this view of classification, by taking the CASE of languages. If we possessed a perfect pedigr
0425 to be classed under the bear genus. The whole CASE is preposterous: for where there has been clos
0430 ies, but the general order of Rodents. In this CASE, however, it may be strongly suspected that th
0432 as those between the finest varieties. In this CASE it would be quite impossible to give any defin
0438 rdial organs of any kind, vertebrae in the one CASE and legs in the other, have actually been modi
0439 closely than do the mature insects; but in the CASE of larvae, the embryos are active, and have be
0440 the conditions to which they are exposed. The CASE, however, is different when an animal during a
0440 but a glance at the larva shows this to be the CASE in an unmistakeable manner. So again the two m
0444 m far from meaning that this is invariably the CASE: and I could give a good many cases of variati
0445 s, judging by the eye, seemed almost to be the CASE: but on actually measuring the old dogs and th
0446 g at a corresponding not early period. But the CASE of the short faced tumbler, which when twelve
0447 han do the adults, just as we have seen in the CASE of pigeons. We may extend this view to whole f
0447 han that at which it first appeared. In either CASE (as with the short faced tumbler) the young or
0448 habits of life with their parents: for in this CASE, it would be indispensable for the existence o
0448 in the second stage, as we have seen to be the CASE with cirripedes. The adult might become fitted
0448 he senses, etc., would be useless: and in this CASE the final metamorphosis would be said to be re
0451 are merely not developed: this seems to be the CASE with the mammae of male mammals; for many inst
0454 gans, until they become rudimentary, as in the CASE of the eyes of animals inhabiting dark caverns
0454 ving on small and exposed islands; and in this CASE natural selection would continue slowly to red
0455 d tend to be wholly lost, and we should have a CASE of complete abortion. The principle, also, of
0461 the sterility of hybrids is a very different CASE from that of first crosses, for their reproduc
0470 e, to find it still in action; and this is the CASE if varieties be incipient species. Moreover, t
0474 inherited for a very long period: for in this CASE it will have been rendered constant by long co
0474 tainly is not indispensable, as we see, in the CASE of neuter insects, which leave no progeny to i
0482 s, they admit variation as a vera causa in one CASE, they arbitrarily reject it in another, withou
0483 as eggs or seed, or as full grown? and in the CASE of mammals, were they created bearing the fals
0485 as with the primrose and cowslip: and in this CASE scientific and common language will come into
0002 ts in illustration, but which, I hope, in most CASES will suffice. No one can feel more sensible t
0004 appointed: in this and in all other perplexing CASES I have invariably found that our knowledge, i
0008 eed freely under confinement, even in the many CASES when the male and female unite. How many anim
0008 and yet rarely or never seed! In some few such CASES it has been found out that very trifling chan
0010 e parent prior to the act of conception. These CASES anyhow show that variation is not necessarily
0010 e directly due to such conditions: but in some CASES it can be shown that quite opposite condition
0011 action of the conditions of life, as, in some CASES, increased size from amount of food, colour f
0012 iarities go together, of which many remarkable CASES could be given amongst animals and plants. Fr
0013 e to inheritance. Every one must have heard of CASES of albinism, prickly skin, hairy bodies, etc.
0013 ponding age, though sometimes earlier. In many CASES this could not be otherwise: thus the inherit
0014 uld not possibly live in a wild state. In many CASES we do not know what the aboriginal stock was,
0016 each other in the same manner as, only in most CASES in a lesser degree than, do closely allied sp
0023 belief are in some degree applicable in other CASES, I will here briefly give them. If the severa
0024 great weight, and applicable in several other CASES, is, that the above specified breeds, though
0026 nite number of generations. These two distinct CASES are often confounded in treatises on inherita
0030 useful as we now see them; indeed, in several CASES, we know that this has not been their history
0032 s far more indispensable even than in ordinary CASES. If selection consisted merely in separating
0032 ck. We have proofs that this is not so in some CASES, in which exact records have been kept; thus,
0035 ago, which might serve for comparison. In some CASES, however, unchanged or but little changed ind
0037 the well known fact, that in a vast number of CASES we cannot recognise, and therefore do not kno
0039 o make a fantail, is, I have no doubt, in most CASES, utterly incorrect. The man who first selecte
0043 t is thus rendered infinitely complex. In some CASES, I do not doubt that the intercrossing of spe
0043 and the frequent sterility of hybrids: but the CASES of plants not propagated by seed are of littl
0045 animal from far northwards, would not in some CASES be inherited for at least some few generation

Page ***(Key Word)***

Page **************************************(Key Word)**************************************

```
0045  ralist would be surprised at the number of the  CASES of variability, even in important parts of st
0047  the species, and the other as the variety. But   CASES of great difficulty, which I will not here en
0047  nks always remove the difficulty. In very many   CASES, however, one form is ranked as a variety of
0047  nly guide to follow. We must, however, in many   CASES, decide by a majority of naturalists, for few
0049  cepted, is vainly to beat the air. Many of the   CASES of strongly marked varieties or doubtful spec
0050  ked as varieties. Close investigation, in most   CASES, will bring naturalists to an agreement how t
0051  ge of his observations, he will meet with more   CASES of difficulty; for he will encounter a greate
0052  ce to another and higher stage may be, in some   CASES, due merely to the long continued action of d
0056  pecies and well marked varieties; and in those   CASES in which intermediate links have not been fou
0061  into good and distinct species, which in most   CASES obviously differ from each other far more tha
0064  al calculations, namely, the numerous recorded   CASES of the astonishingly rapid increase of variou
0064  e been quite incredible. So it is with plants:   CASES could be given of introduced plants which hav
0065  rted from America since its discovery. In such   CASES, and endless instances could be given, no one
0065  the young have been enabled to breed. In such   CASES the geometrical ratio of increase, the result
0066  life; and this period in the great majority of   CASES is an early one. If an animal can in any way
0066  o germinate in a fitting place. So that in all   CASES, the average number of any animal or plant de
0068  ow annually killed. On the other hand, in some   CASES, as with the elephant and rhinoceros, none ar
0070  asite and its prey. On the other hand, in many   CASES, a large stock of individuals of the same spe
0070  e extreme confines of their range. For in such   CASES we may believe, that a plant could exist onl
0071  ding, probably come into play in some of these   CASES; but on this intricate subject I will not her
0071  ntricate subject I will not here enlarge. Many   CASES are on record showing how complex and unexpec
0074  or even the existence of the species. In some   CASES it can be shown that widely different checks
0076  harlock will supplant another, and so in other   CASES. We can dimly see why the competition should
0086  affect the structure of the larva; but in all   CASES natural selection will ensure that modificati
0088  n nature, will leave most progeny. But in many   CASES, victory will depend not on general vigour, b
0090  or attractive to the females. We see analogous   CASES under nature, for instance, the tuft of hair
0096  may be asked, is there for supposing in these   CASES that two individuals ever concur in reproduct
0098  largely do they naturally cross. In many other   CASES, far from there being any aids for self ferti
0098  ited by bees, seeds freely. In very many other   CASES, though there be no special mechanical contri
0099  rpose of self fertilisation, should in so many   CASES be mutually useless to each other! How simply
0101  well aware that there are, on this view, many   CASES of difficulty, some of which I am trying to i
0111  es, as is shown by the hopeless doubts in many   CASES how to rank them, yet certainly differ from e
0119  eaching to the upper horizontal lines. In some   CASES I do not doubt that the process of modificati
0132  hough not elsewhere fleshy. Several other such   CASES could be given. The fact of varieties of one
0135  cannot be much used by these insects. In some   CASES we might easily put down to disuse modificati
0137  o rodents are rudimentary in size, and in some   CASES are quite covered up by skin and fur. This st
0140  gland. In regard to animals, several authentic   CASES could be given of species within historical t
0140  d to their native climate, but in all ordinary   CASES we assume such to be the case; nor do we know
0143  ude that habit, use, and disuse, have, in some   CASES, played a considerable part in the modificati
0146  h apparent importance, the seeds being in some   CASES, according to Tausch, orthospermous in the ex
0147  oining part. I suspect, also, that some of the   CASES of compensation which have been advanced, and
0151  uture work, give a list of the more remarkable   CASES; I will here only briefly give one, as it ill
0152  ee a nearly parallel natural case; for in such   CASES natural selection either has not or cannot co
0154  than any other structure. It is only in those   CASES in which the modification has been comparativ
0157  nd on my list; and as the differences in these   CASES are of a very unusual nature, the relation ca
0162  of a group, we could not distinguish these two   CASES: if, for instance, we did not know that the r
0162  ature it must generally be left doubtful, what   CASES are reversions to an anciently existing chara
0162  species. I have collected a long list of such CASES; but here, as before, I lie under a great dis
0163  able to give them. I can only repeat that such   CASES certainly do occur, and seem to me very remar
0163  s. with respect to the horse, I have collected   CASES in England of the spinal stripe in horses of
0164  her details, I may state that I have collected   CASES of leg and shoulder stripes in horses of very
0169  ow process, and natural selection will in such   CASES not as yet have had time to overcome the tend
0171  ns of extreme perfection. Means of transition.   CASES of difficulty. Natura non facit saltum. Organ
0171  organs of small importance. Organs not in all   CASES absolutely perfect. The Law of Unity of Type
0180  avy disadvantage, for out of the many striking   CASES which I have collected, I can give only one o
0180  me that nothing less than a long list of such   CASES is sufficient to lessen the difficulty in any
0183  robably often change almost simultaneously. Of   CASES of changed habits it will suffice merely to a
0185  terrestrial as the quail or partridge. In such   CASES, and many others could be given, habits have
0185  rable acts of creation will say, that in these   CASES it has pleased the Creator to cause a being o
0190  transitional gradations of some kind. Numerous   CASES could be given amongst the lower animals of t
0190  l then digest and the stomach respire. In such   CASES natural selection might easily specialise, if
0190  ivided by highly vascular partitions. In these   CASES, one of the two organs might with ease be mod
0192  ansitional gradations, yet, undoubtedly, grave   CASES of difficulty occur, some of which will be di
0193  s, offers a parallel case of difficulty. Other   CASES could be given; for instance in plants, the v
0193  ossible amongst flowering plants. In all these   CASES of two very distinct species furnished with a
0194  tance from the same ancestor. Although in many   CASES it is most difficult to conjecture by what tr
0195  ds are actually destroyed (except in some rare   CASES) by the flies; but they are incessantly haras
0195  w of trifling importance have probably in some   CASES been of high importance to an early progenito
0201  not space here to enter on this and other such   CASES. Natural selection will never produce in a be
0204  f perfection through natural selection. In the   CASES in which we know of no intermediate or transi
0205  ighly injurious to another species, but in all   CASES at the same time useful to the owner. Natural
0206  s of time: the adaptations being aided in some   CASES by use and disuse, being slightly affected by
0206  external conditions of life, and being in all   CASES subjected to the several laws of growth. Henc
0211  ily structure of others. So again, in some few   CASES, certain instincts cannot be considered as ab
0212  wn to us: Audubon has given several remarkable   CASES of differences in nests of the same species i
0212  can be shown by a multitude of facts. Several   CASES also, could be given, of occasional and stran
0212  l be strengthened by briefly considering a few   CASES under domestication. We shall thus also be en
0215  t best. On the other hand, habit alone in some   CASES has sufficed; no animal is more difficult to
0216  ust in our ignorance call an accident. In some   CASES compulsory habit alone has sufficed to produc
0216  roduce such inherited mental changes; in other   CASES compulsory habit has done nothing, and all ha
0216  th methodically and unconsciously; but in most   CASES, probably, habit and selection have acted tog
0216  me modified by selection, by considering a few   CASES. I will select only three, out of the several
0230  nvex side; and I suspect that the bees in such   CASES stand in the opposed cells and push and bend
0232  hese spheres. It was really curious to note in   CASES of difficulty, as when two pieces of comb met
0235  be opposed to the theory of natural selection,   CASES, in which we cannot see how an instinct could
0235  ow an instinct could possibly have originated;   CASES, in which no intermediate gradations are know
```

Page **************************************(Key Word)**************************************

Page **(Key Word)**

```
0235  no intermediate gradations are known to exist;  CASES of instinct of apparently such trifling impor
0235  ardly have been acted on by natural selection;  CASES of instincts almost identically the same in a
0236  ection. I will not here enter on these several  CASES, but will confine myself to one special diffi
0243  o any extent, in any useful direction. In some  CASES habit or use and disuse have probably come in
0243  in any great degree my theory; but none of the  CASES of difficulty, to the best of my judgment, an
0246  e of the sterility, which is common to the two  CASES, has to be considered. The distinction has pr
0246  n slurred over, owing to the sterility in both  CASES being looked on as a special endowment, beyon
0246  iversal; but then he cuts the knot, for in ten  CASES in which he found two forms, considered by mo
0247  sputes the entire fertility of Kolreuter's ten  CASES. But in these and in many other cases, Gartne
0247  er's ten cases. But in these and in many other  CASES, Gartner is obliged carefully to count the se
0247  or gartner gives in his table about a score of  CASES of plants which he castrated, and artificiall
0247  ised with their own pollen; and (excluding all  CASES such as the Leguminosae, in which there is an
0247  the same conclusion in several other analogous  CASES; it seems to me that we may well be permitted
0253  not know of any thoroughly well authenticated  CASES of perfectly fertile hybrid animals, I have s
0255  ity; and, as we have seen, in certain abnormal  CASES, even to an excess of fertility, beyond that
0255  t, a single fertile seed: but in some of these  CASES a first trace of fertility may be detected, b
0256  ogether, is by no means strict. There are many  CASES, in which two pure species can be united with
0256  , for instance in Dianthus, these two opposite  CASES occur. The fertility, both of first crosses a
0257  crossing is by no means strict. A multitude of  CASES could be given of very closely allied species
0258  he facility of making reciprocal crosses. Such  CASES are highly important, for they prove that the
0258  r whole organisation. On the other hand, these  CASES clearly show that the capacity for crossing i
0258  utterly failed. Several other equally striking  CASES could be given. Thuret has observed the same
0261  ture of their sap, etc., but in a multitude of  CASES we can assign no reason whatever. Great diver
0261  me family have been grafted together, in other  CASES species of the same genus will not take on ea
0262  ductive organs perfect: yet these two distinct  CASES run to a certain extent parallel. Something a
0263  productive systems. These differences, in both  CASES, follow to a certain extent, as might have be
0263  ity of first crosses and of hybrids. These two  CASES are fundamentally different, for, as just rem
0265  , there are many points of similarity. In both  CASES, the sterility is independent of general healt
0265  by excess of size or great luxuriance. In both  CASES, the sterility occurs in various degrees; in
0266  that I have attempted to show, is that in two  CASES, in some respects allied, sterility is the co
0269  rtain amount of sterility in the few following  CASES, which I will briefly abstract. The evidence
0270  rived from hostile witnesses, who in all other  CASES consider fertility and sterility as safe crit
0273  in their offspring is notorious: but some few  CASES both of hybrids and mongrels long retaining u
0273  n natural varieties), and this implies in most  CASES that there has been recent variability; and t
0275  ith hybrids than with mongrels. Looking to the  CASES which I have collected of cross bred animals
0277  ms to depend on several circumstances; in some  CASES largely on the early death of the embryo. The
0281  they differ from each other. Hence in all such  CASES, we should be unable to recognise the parent
0281  will render this a very rare event; for in all  CASES the new and improved forms of life will tend
0283  the process of degradation. The tides in most  CASES reach the cliffs only for a short time twice
0284  ay has given them the maximum thickness, in most  CASES from actual measurement, in a few cases from
0284  n most cases from actual measurement, in a few  CASES from estimate, of each formation in different
0285  in merionethshire of 12,000 feet: yet in these  CASES there is nothing on the surface to show such
0288  int of the water bespeaks its purity. The many  CASES on record of a formation conformably covered,
0293  ved at its commencement and at its close. Some  CASES are on record of the same species presenting
0296  ime represented by the thinner formation. Many  CASES could be given of the lower beds of a formati
0296  ls have occurred in its accumulation. In other  CASES we have the plainest evidence in great fossil
0301  vene during such lengthy periods; and in these  CASES the inhabitants of the archipelago would have
0303  consecutive formations, longer perhaps in some  CASES than the time required for the accumulation o
0316  ue would have been fatal to my views. But such  CASES are certainly exceptional: the general rule b
0318  nd increase in numbers of the species. In some  CASES, however, the extermination of whole groups o
0319  rarity, and finally extinction. We see in many  CASES in the more recent tertiary formations, that
0323  hat the same species are met with; for in some  CASES not one species is identically the same, but
0328  n, extinction, and immigration. I suspect that  CASES of this nature have occurred in Europe. Mr. P
0333  e have no right to expect, except in very rare  CASES, to fill up wide intervals in the natural sys
0335  pposing for an instant, in this and other such  CASES, that the record of the first appearance and
0339  luscs, it is not well displayed by them. Other  CASES could be added, as the relation between the e
0351  nt of migration is possible. But in many other  CASES, in which we have reason to believe that the
0352  rth's surface. Undoubtedly there are very many  CASES of extreme difficulty, in understanding how t
0352  acle. It is universally admitted, that in most  CASES the area inhabited by a species is continuous
0352  ngs; and, accordingly, we find no inexplicable  CASES of the same mammal inhabiting distant points
0352  no geologist will feel any difficulty in such  CASES as Great Britain having been formerly united
0353  mitted, is the most probable. Undoubtedly many  CASES occur, in which we cannot explain how the sam
0354  elessly tedious to discuss all the exceptional  CASES of the same species, now living at distant an
0354  any explanation could be offered of many such  CASES. But after some preliminary remarks, I will d
0355  heritance to the inhabitants of the continent.  CASES of this nature are common; and are, as we sha
0356  d from a single parent. But in the majority of  CASES, namely, with all organisms which habitually
0365  ot possibly exist, is one of the most striking  CASES known of the same species living at distant p
0372  nearly a hemisphere of equatorial ocean. These  CASES of relationship, without identity, of the inh
0376  strial animals. In marine productions, similar  CASES occur; as an example, I may quote a remark by
0383  y, though so unexpected, can, I think, in most  CASES be explained by their having become fitted, i
0383  y consequence. We can here consider only a few  CASES. In regard to fish, I believe that the same s
0384  t points of the world, no doubt there are many  CASES which cannot at present be explained: but som
0384  fish belong to very ancient forms, and in such  CASES there will have been ample time for great geo
0392  y different from what they are elsewhere. Such  CASES are generally accounted for by the physical c
0395  much difficulty in forming a judgment in some  CASES owing to the probable naturalisation of certa
0395  distinct. As the amount of modification in all  CASES depends to a certain degree on the lapse of t
0397  will here give a single instance of one of the  CASES of difficulty. Almost all oceanic islands, ev
0397  r. aug. A. Gould has given several interesting  CASES in regard to the land shells of the islands o
0401  are separated by deep arms of the sea, in most  CASES wider than the British Channel, and there is
0402  ct species, belonging in a large proportion of  CASES, as shown by Alph. de Candolle, to distinct g
0405  mmon parent at a remote epoch; so that in such  CASES there will have been ample time for great cli
0410  ge less than the higher; but there are in both  CASES marked exceptions to the rule. On my theory t
0410  having migrated into distant quarters, in both  CASES the forms within each class have been connect
0410  stand to each other in time and space; in both  CASES the laws of variation have been the same, and
0415  y natural family, is very unequal; and in some  CASES seems to be entirely lost. Again in another w
0425  nd genera under higher groups, though in these  CASES the modification has been greater in degree,
```

Page **(Key Word)**

cases

Page ***(Key Word)***

```
0430  a quite distinct group, this affinity in most CASES is general and not special; thus, according t
0438  as metamorphosed leaves; but it would in these CASES probably be more correct, as Professor Huxley
0440  allied animals is sometimes much obscured; and CASES could be given of the larvae of two species,
0440  ach other than do their adult parents. In most CASES, however, the larvae, though active, still ob
0441  terfly is higher than the caterpillar. In some CASES, however, the mature animal is generally cons
0442  orm like stage of development; but in some few CASES, as in that of Aphis, if we look to the admir
0444  estic animals supports this view. But in other CASES it is quite possible that each successive mod
0444  ariably the case; and I could give a good many CASES of variations (taking the word in the largest
0444  logy. But first let us look at a few analogous CASES in domestic varieties. Some authors who have
0446   the full grown animal possesses them. And the CASES just given, more especially that of pigeons,
0447  , by the effects of use and disuse. In certain CASES the successive steps of variation might super
0448  spect to the final cause of the young in these CASES not undergoing any metamorphosis, or closely
0449  er proved true. It can be proved true in those CASES alone in which the ancient state, now suppose
0450  ents of the pelvis and hind limbs. Some of the CASES of rudimentary organs are extremely curious;
0451  ble of flight, and not rarely lying under wing CASES, firmly soldered together! The meaning of rud
0454  udimentary organs is simple. We have plenty of CASES of rudimentary organs in our comestic product
0454   in monsters. But I doubt whether any of these CASES throw light on the origin of rudimentary orga
0460  aduated steps. There are, it must be admitted, CASES of special difficulty on the theory of natura
0462   we have no right to expect (excepting in rare CASES) to discover directly connecting links betwee
0468  mals having separated sexes there will in most CASES be a struggle between males for possession of
0471  of an auk or grebe! and so on in endless other CASES. But on the view of each species constantly t
0472  live bodies of caterpillars; and at other such CASES. The wonder indeed is, on the theory of natur
0472   on the theory of natural selection, that more CASES of the want of absolute perfection have not b
0472  roduction of so called specific forms. In both CASES physical conditions seem to have produced but
0476  ralia, of endentata in America, and other such CASES, is intelligible, for within a confined count
0476  ogical succession throughout time; for in both CASES the beings have been connected by the bond of
0477  rinciple of former migration, combined in most CASES with modification, we can understand, by the
0482  , without assigning any distinction in the two CASES. The day will come when this will be given as
0339  and living brackish water shells of the Aralo CASPIAN Sea. Now what does this remarkable law of th
0145  sory parts, differs, as has been described by CASSINI. These differences have been attributed by s
0045  at all the individuals of the same species are CAST in the very same mould. These individual diffe
0238  different: in Cryptocerus, the workers of one CASTE alone carry a wonderful sort of shield on the
0239   the Mexican Myrmecocystus, the workers of one CASTE never leave the nest: they are fed by the wor
0239  e nest: they are fed by the workers of another CASTE, and they have an enormously developed abdome
0239   the simpler case of neuter insects all of one CASTE or of the same kind, which have been rendered
0238  , and are thus divided into two or even three CASTES. The castes, moreover, do not generally gradu
0238  us divided into two or even three castes. The CASTES, moreover, do not generally graduate into eac
0240  ant points of structure between the different CASTES of neuters in the same species, that I gladly
0241  , that though the workers can be grouped into CASTES of different sizes, yet they graduate insensi
0241   the wonderful fact of two distinctly defined CASTES of sterile workers existing in the same nest,
0460  hese the existence of two or three defined CASTES of workers or sterile females in the same com
0247  introduced: a plant to be hybridised must be CASTRATED, and, what is often more important, must be
0247  le about a score of cases of plants which he CASTRATED, and artificially fertilised with their own
0249  so careful an observer as Gartner would have CASTRATED his hybrids, and this would have insured in
0439  s explained. Embryology. It has already been CASUALLY remarked that certain organs in the individu
0042  the rarity or absence of distinct breeds of the CAT, the donkey, peacock, goose, etc., may be attr
0091  natural tendencies of our domestic animals: one CAT, for instance, taking to catch rats, another m
0091  stance, taking to catch rats, another mice; one CAT, according to Mr. St. John, bringing home wing
0201  escape. I would almost as soon believe that the CAT curls the end of its tail when preparing to sp
0215   all wolves, foxes, jackals, and species of the CAT genus, when kept tame, are most eager to attac
0215  lost, wholly by habit, that fear of the dog and CAT which no doubt was originally instinctive in t
0439  he spotted feathers in the thrush group. In the CAT tribe, most of the species are striped or spot
0489  tion has never once been broken, and that no CATACLYSM has desolated the whole world. Hence we may
0073  ; and as we do not see the cause, we invoke CATACLYSMS to desolate the world, or invent laws on th
0044  o treat this subject at all properly, a long CATALOGUE of dry facts should be given, but these I s
0045  nimportant parts; but I could show by a long CATALOGUE of facts, that parts which must be called i
0058   has marked for me in the well sifted London CATALOGUE of plants (4th edition) 63 plants which are
0058  has divided Great Britain. Now, in this same CATALOGUE, 53 acknowledged varieties are recorded, an
0103  al selection; for I can bring a considerable CATALOGUE of facts, showing that within the same area
0131  ve remarked in the first chapter, but a long CATALOGUE of facts which cannot be here given would b
0162  er species of the same genus. A considerable CATALOGUE, also, could be given of forms intermediate
0004  can be treated properly only by giving long CATALOGUES of facts. We shall, however, be enabled to
0424  orchidean forms (Monochanthus, Myanthus, and CATASETUM), which had previously been ranked as three
0317  g been swept away at successive periods by CATASTROPHES, is very generally given up, even by those
0487   not by miraculous acts of creation and by CATASTROPHES; and as the most important of all causes o
0252  ng from a hybrid between Rhod. Ponticum and CATAWBIENSE, and that this hybrid seeds as freely as i
0003  ail, beak, and tongue, so admirably adapted to CATCH insects under the bark of trees. In the case
0039  first appeared, the more likely it would be to CATCH his attention. But to use such an expression
0039  t deviation of structure would be necessary to CATCH the fancier's eye: he perceives extremely sma
0083  least by some modification prominent enough to CATCH his eye, or to be plainly useful to him. Unde
0091  stic animals: one cat, for instance, taking to CATCH rats, another mice; one cat, according to Mr.
0091  catching woodcocks or snipes. The tendency to CATCH rats rather than mice is shown to be inherite
0132  ant. Nevertheless, we can here and there dimly CATCH a faint ray of light, and we may feel sure th
0091  hunting on marshy ground and almost nightly CATCHING woodcocks or snipes. The tendency to catch r
0184  mming for hours with widely open mouth, thus CATCHING, like a whale, insects in the water. Even in
0120  es; or they may have arrived at the doubtful CATEGORY of sub species; but we have only to suppose
0014  rm are known to appear at the corresponding CATERPILLAR or cocoon stage. But hereditary diseases a
0086  ur culinary and agricultural plants: in the CATERPILLAR and cocoon stages of the varieties of the
0208  of thought: so P. Huber found it was with a CATERPILLAR, which makes a very complicated hammock; f
0208   very complicated hammock; for if he took a CATERPILLAR which had completed its hammock up to, say
0208  k completed up only to the third stage, the CATERPILLAR simply re performed the fourth, fifth, and
0208  sixth stages of construction. If however, a CATERPILLAR were taken out of a hammock made up, for i
0441  spute that the butterfly is higher than the CATERPILLAR. In some cases, however, the mature animal
0444  ng ages, for instance, peculiarities in the CATERPILLAR, cocoon, or imago states of the silk moth:
0244  umonidae feeding within the live bodies of CATERPILLARS, not as specially endowed or created insti
0472  umonidae feeding within the live bodies of CATERPILLARS; and at other such cases. The wonder indee
```

Page ***(Key Word)***

```
0012 ances of correlation are quite whimsical: thus CATS with blue eyes are invariably deaf; colour and
0042 killed they serve for food. On the other hand, CATS, from their nocturnal rambling habits, cannot
0042 election not having been brought into play: in CATS, from the difficulty in pairing them: in donke
0074 ependent, as every one knows, on the number of CATS; and Mr. Newman says, near villages and small
0074 elsewhere, which I attribute to the number of CATS that destroy the mice. Hence it is quite credi
0144 the relation between blue eyes and deafness in CATS, and the tortoise shell colour with the female
0216 have lost all fear, but fear only of dogs and CATS, for if the hen gives the danger chuckle, they
0091 are two varieties of the wolf inhabiting the CATSKILL Mountains in the United States, one with a l
0014 s the inherited peculiarities in the horns of CATTLE could appear only in the offspring when nearl
0015 r cart and race horses, long and short horned CATTLE, and poultry of various breeds, and esculent
0018 and constitution, etc., of the humped Indian CATTLE, that these had descended from a different ab
0018 different aboriginal stock from our European CATTLE; and several competent judges believe that th
0019 e existed at least a score of species of wild CATTLE, as many sheep, and several goats in Europe a
0019 kingdoms possesses several peculiar breeds of CATTLE, sheep, etc., we must admit that many domesti
0029 i have asked, a celebrated raiser of Hereford CATTLE, whether his cattle might not have descended
0029 brated raiser of Hereford cattle, whether his CATTLE might not have descended from long horns, and
0031 me, modified to a large extent some breeds of CATTLE and sheep. In order fully to realise what the
0034 e savages in South Africa match their draught CATTLE by colour, as do some of the Esquimaux their
0035 n lifetimes, the forms and qualities of their CATTLE. Slow and insensible changes of this kind cou
0035 y. Lord Spencer and others have shown how the CATTLE of England have increased in weight and in ea
0064 ents of the rate of increase of slow breeding CATTLE and horses in South America, and latterly in
0071 tion that the land had been enclosed, so that CATTLE could not enter. But how important an element
0072 hich had been perpetually browsed down by the CATTLE. In one square yard, at a point some hundred
0072 ive that no one would ever have imagined that CATTLE would have so closely and effectually searche
0072 tually searched it for food. Here we see that CATTLE absolutely determine the existence of the Sco
0072 the world insects determine the existence of CATTLE. Perhaps Paraguay offers the most curious ins
0072 st curious instance of this: for here neither CATTLE nor horses nor dogs have ever run wild, thoug
0072 e in Paraguay, the flies would decrease, then CATTLE and horses would become feral, and this would
0086 their chickens: in the horns of our sheep and CATTLE when nearly adult; so in a state of nature, n
0111 ld be given showing how quickly new breeds of CATTLE, sheep, and other animals, and varieties of f
0111 is historically known that the ancient black CATTLE were displaced by the long horns, and that th
0195 e know that the distribution and existence of CATTLE and other animals in South America absolutely
0198 elated. A good observer, also, states that in CATTLE susceptibility to the attacks of flies is cor
0237 fferences in the horns of different breeds of CATTLE in relation to an artificially imperfect stat
0238 s to get nearly the same variety: breeders of CATTLE wish the flesh and fat to be well marbled tog
0238 election, that I do not doubt a breed of CATTLE, always yielding oxen with extraordinarily lo
0239 that excreted by the aphides, or the domestic CATTLE as they may be called, which our European ant
0254 lieve that our European and the humped Indian CATTLE are quite fertile together: but from facts co
0299 e able to prove, that our different breeds of CATTLE, sheep, horses, and dogs have descended from
0320 transported far and near, like our short horn CATTLE, and takes the place of other breeds in other
0423 e horns are very useful for this purpose with CATTLE, because they are less variable than the shap
0443 her way: for it is notorious that breeders of CATTLE, horses, and various fancy animals, cannot po
0443 nt. Or again, as when the horns of cross bred CATTLE have been affected by the shape of the horns
0444 or, again, in the horns of almost full grown CATTLE. But further than this, variations which, for
0454 f minute dangling horns in hornless breeds of CATTLE, more especially, according to Youatt, in you
0022 a highly remarkable manner. The number of the CAUDAL and sacral vertebrae vary: as does the number
0022 oil gland: the number of the primary wing and CAUDAL feathers: the relative length of wing and tai
0137 nd i was assured by a Spaniard, who had often CAUGHT them, that they were frequently blind: one wh
0386 Lyell also informs me that a Dyticus has been CAUGHT with an Ancylus (a fresh water shell like a l
0454 s, and the state of the whole flower in the CAULIFLOWER. We often see rudiments of various parts i
0159 d stems of these three plants, not to the vera CAUSA of community of descent, and a consequent ten
0352 the mind. He who rejects it, rejects the vera CAUSA of ordinary generation with subsequent migrat
0482 secondary laws. They admit variation as a vera CAUSA in one case, they arbitrarily reject it in an
0003 h as climate, food, etc., as the only possible CAUSE of variation. In one very limited sense, as w
0007 l generations to the new conditions of life to CAUSE any appreciable amount of variation: and that
0008 gly inclined to suspect that the most frequent CAUSE of variability may be attributed to the male
0009 ut on this view we owe variability to the same CAUSE which produces sterility; and variability is
0013 whether it may not be due to the same original CAUSE acting on both; but when amongst individuals,
0014 nce of the peculiarity, and not to its primary CAUSE, which may have acted on the ovules or male e
0050 of nature be highly useful to man, or from any CAUSE closely attract his attention, varieties of i
0055 close relation to the size of the genera. The CAUSE of lowly organised plants ranging widely will
0061 ny variation, however slight and from whatever CAUSE proceeding, if it be in any degree profitable
0069 creasing in numbers, we may feel sure that the CAUSE lies quite as much in other species being fav
0070 e due to parasitic worms, which have from some CAUSE, possibly in part through facility of diffusi
0073 of an organic being: and as we do not see the CAUSE, we invoke cataclysms to desolate the world,
0086 ural selection for the good of the being, will CAUSE other modifications, often of the most expect
0086 e injurious: for if they became so, they would CAUSE the extinction of the species. Natural select
0095 ast waves, called a trifling and insignificant CAUSE, when applied to the excavation of gigantic v
0098 ut, the agency of insects is often required to CAUSE the stamens to spring forward, as Kolreuter h
0111 erences. Mere chance, as we may call it, might CAUSE one variety to differ in some character from
0131 es to acknowledge plainly our ignorance of the CAUSE of each particular variation. Some authors be
0132 , and we may feel sure that there must be some CAUSE for each deviation of structure, however slig
0137 ept alive was certainly in this condition, the CAUSE, as appeared on dissection, having been infla
0156 truth of this proposition will be granted. The CAUSE of the original variability of secondary sexu
0157 princ to the less favoured males. Whatever the CAUSE may be of the variability of secondary sexual
0160 nothing in the external conditions of life to CAUSE the reappearance of the slaty blue, with the
0167 eal for an unreal, or at least for an unknown, CAUSE. It makes the works of God a mere mockery and
0168 tural selection. It is probably from this same CAUSE that organic beings low in the scale of natur
0170 d harmonious diversity of nature. Whatever the CAUSE may be of each slight difference in the offsp
0170 nce in the offspring from their parents, and a CAUSE for each must exist, it is the steady accumul
0175 elements of distribution, these facts ought to CAUSE surprise, as climate and height or depth grad
0178 arieties which they tend to connect. From this CAUSE alone the intermediate varieties will be liab
0186 t in these cases it has pleased the Creator to CAUSE a being of one type to take the place of one
0186 nt it may be from its own place. Hence it will CAUSE him no surprise that there should be geese an
0189 be destroyed. In living bodies, variation will CAUSE the slight alterations, generation will multi
0195 ich the importance does not seem sufficient to CAUSE the preservation of successively varying indi
```

```
0197   that the colour is due to some quite distinct CAUSE, probably to sexual selection. A trailing bam
0199   oo much stress on our ignorance of the precise CAUSE of the slight analogous differences between s
0199   m correlation of growth, or from other unknown CAUSE, may reappear from the law of reversion, thou
0202   purpose, with the poison originally adapted to CAUSE galls subsequently intensified, we can perhap
0202   t is that the use of the sting should so often CAUSE the insect's own death: for if on the whole t
0202   quirements of natural selection, though it may CAUSE the death of some few members. If we admire t
0212   he greater wildness of our large birds to this CAUSE: for in uninhabited islands large birds are n
0216   nly admitted that the more immediate and final CAUSE of the cuckoo's instinct is, that she lays he
0234   much wax would be saved. Again, from the same  CAUSE, it would be advantageous to the Melipona, if
0246   loped. This distinction is important, when the CAUSE of the sterility, which is common to the two
0247   nt species in a state of nature. But a serious CAUSE of error seems to me to be here introduced: a
0248   ertility has been diminished by an independent CAUSE, namely, from close interbreeding. I have col
0253   in the case of any pure animal, which from any CAUSE had the least tendency to sterility, the bree
0264   e early death of the embryo is a very frequent CAUSE of sterility in first crosses. I was at first
0273   orroborates the view which I have taken on the CAUSE of ordinary variability: namely, that it is d
0279   further modification and improvement. The main CAUSE, however, of innumerable intermediate links n
0289   mainly results from another and more important CAUSE than any of the foregoing: namely, from the s
0319   e unperceived agencies are amply sufficient to CAUSE rarity, and finally extinction. We see in man
0321   diffied descendants of a species will generally CAUSE the extermination of the parent species; and
0325   climate, or other physical conditions, as the  CAUSE of these great mutations in the forms of life
0327   here to prevail. As they prevailed, they would CAUSE the extinction of other and inferior forms: a
0350   ond, on my theory, is simply inheritance, that CAUSE which alone, as far as we positively know, pr
0392   ndolle has shown, generally have, whatever the  CAUSE may be, confined ranges. Hence trees would be
0413   nd that propinquity of descent, the only known CAUSE of the similarity of organic beings, is the b
0415   relation to the outer world. Perhaps from this CAUSE it has partly arisen, that almost all natural
0443   but at what period it is fully displayed. The  CAUSE may have acted, and I believe generally has a
0448   ects, as with Aphis. With respect to the final CAUSE of the young in these cases not undergoing an
0455   ly often come into play; and this will tend to CAUSE the entire obliteration of a rudimentary orga
0456   s element of descent; the only certainly known CAUSE of similarity in organic beings, we shall und
0481   if they had undergone mutation. But the chief  CAUSE of our natural unwillingness to admit that on
0072   and Azara and Rengger have shown that this is  CAUSED by the greater number in Paraguay of a certa
0076   nited States of one species of swallow having  CAUSED the decrease of another species. The recent i
0076   of the missel thrush in parts of Scotland has  CAUSED the decrease of the song thrush. How frequent
0080   umber of individuals. Slow action. Extinction  CAUSED by Natural Selection. Divergence of Character
0089   r ornament, such differences have been mainly  CAUSED by sexual selection: that is, individual male
0134   lands, tenanted by no beast of prey, has been  CAUSED by disuse. The ostrich indeed inhabits contin
0145   nt from certain other parts of the flower had  CAUSED their abortion; but in some Compositae there
0201   r balance be struck between the good and evil  CAUSED by each part, each will be found on the whole
0320   believe that the production of new forms has  CAUSED the extinction of about the same number of ol
0321   ecies belonging to a distinct group, and thus CAUSED its extermination; and if many allied forms b
0360   fruit; and this, as we have seen, would have  CAUSED some of them to have floated much longer. The
0384   ecent period in the level of the land, having CAUSED rivers to flow into each other. Instances, al
0389   of any difference in physical conditions has  CAUSED so great a difference in number. Even the uni
0443   at what period of life any variation has been CAUSED, but at what period it is fully displayed. Th
0443   ave been exposed. Nevertheless an effect thus CAUSED at a very early period, even before the forma
0445   ence between these two breeds has been wholly CAUSED by selection under domestication; but having
0450   ly period in the life of each, though perhaps CAUSED at the earliest, and being inherited at a cor
0005   n see how Natural Selection almost inevitably CAUSES much Extinction of the less improved forms of
0007   ion. Variation Under Domestication Chapter I. CAUSES of Variability. Effects of Habit. Correlation
0008   has been disputed at what period of life the CAUSES of variability, whatever they may be, general
0008   s show that unnatural treatment of the embryo CAUSES monstrosities; and monstrosities cannot be se
0009   e system so seriously affected by unperceived CAUSES as to fail in acting, we need not be surprise
0043   r endurance is only temporary. Over all these CAUSES of Change I am convinced that the accumulativ
0054   ional number of dominant species. But so many CAUSES tend to obscure this result, that I am surpri
0054   larger genera. I will here allude to only two CAUSES of obscurity. Fresh water and salt loving pla
0082   specially acting on the reproductive system, CAUSES or increases variability; and in the foregoin
0108   rossing. Many will exclaim that these several CAUSES are amply sufficient wholly to stop the actio
0144   diversity in the shape of the pelvis in birds CAUSES the remarkable diversity in the shape of thei
0150   ule of high generality. I am aware of several CAUSES of error, but I hope that I have made due all
0153   's selection, sometimes become attached, from CAUSES quite unknown to us, more to one sex than to
0160   aracter, and that this tendency, from unknown CAUSES, sometimes prevails. And we have just seen th
0196   nd which have originated from quite secondary CAUSES, independently of natural selection. We shoul
0196   marily arisen from the above or other unknown CAUSES, it may at first have been of no advantage to
0197   er animals. We are profoundly ignorant of the CAUSES producing slight and unimportant variations;
0202   to the backward serratures, and so inevitably CAUSES the death of the insect by tearing out its vi
0209   is of variations produced by the same unknown CAUSES which produce slight deviations of bodily str
0212   ture of the country inhabited, but often from CAUSES wholly unknown to us: Audubon has given sever
0234   asites or other enemies, or on quite distinct CAUSES, and so be altogether independent of the quan
0245   dowment, but incidental on other differences. CAUSES of the sterility of first crosses and of hybr
0251   hese facts show on what slight and mysterious CAUSES the lesser or greater fertility of species wh
0263   grafting it is unimportant for their welfare. CAUSES of the Sterility of first Crosses and of Hybr
0263   may now look a little closer at the probable  CAUSES of the sterility of first crosses and of hybr
0263   union apparently depends on several distinct  CAUSES. There must sometimes be a physical impossibi
0277   generally correspond, though due to distinct  CAUSES: for both depend on the amount of difference
0325   g to mere changes in marine currents or other CAUSES more or less local and temporary, but depend
0328   same period, we should find in both, from the CAUSES explained in the foregoing paragraphs, the sa
0342   ies have at first often been local. All these CAUSES taken conjointly, must have tended to make th
0399   t doubt, be some day explained. The law which CAUSES the inhabitants of an archipelago, though spe
0435   class, by utility or by the doctrine of final CAUSES. The hopelessness of the attempt has been exp
0447   sive steps of variation might supervene, from CAUSES of which we are wholly ignorant, at a very ea
0459   eral and special circumstances in its favour. CAUSES of the general belief in the immutability of
0464   and varieties are often at first local, both  CAUSES rendering the discovery of intermediate links
0467   and then nature acts on the organisation, and CAUSES variability. But man can and does select the
0475   h the slow modification of their descendants, CAUSES the forms of life, after long intervals of ti
0486   odden field of inquiry will be opened, on the CAUSES and laws of variation, on correlation of grow
0487   erminated by slowly acting and still existing CAUSES, and not by miraculous acts of creation and b
```

Page ***(Key Word)***

```
0487  atastrophes: and as the most important of all  CAUSES of organic change is one which is almost inde
0488  f the world should have been due to secondary  CAUSES, like those determining the birth and death o
0043  stem, are so far of the highest importance as  CAUSING variability. I do not believe that variabili
0110  reign plants have become naturalised, without  CAUSING, as far as we know, the extinction of any na
0112  at may be called the principle of divergence,  CAUSING differences, at first barely appreciable, st
0127  h other and to their conditions of existence,  CAUSING an infinite diversity in structure, constitu
0148  is rendered superfluous, without by any means  CAUSING some other part to be largely developed in a
0201  ed, as Paley has remarked, for the purpose of  CAUSING pain or for doing an injury to its possessor
0255  the pollen of one of the pure parent species  CAUSING the flower of the hybrid to wither earlier t
0264  reach the female element, but be incapable of  CAUSING an embryo to be developed, as seems to have
0314  ry. I believe in no fixed law of development,  CAUSING all the inhabitants of a country to change a
0314  cumulated to a greater or lesser amount, thus  CAUSING a greater or lesser amount of modification i
0472  s. we need not marvel at the sting of the bee  CAUSING the bee's own death: at drones being produce
0475  and liable to mistakes, and at many instincts  CAUSING other animals to suffer. If species be only
0029  same parents, may they not learn a lesson of  CAUTION, when they deride the idea of species in a s
0423  ural instead of an artificial system; we are  CAUTIONED, for instance, not to class two varieties o
0002  e crept in, though I hope I have always been  CAUTIOUS in trusting to good authorities alone. I can
0141  ancient Encyclopaedias of China, to be very  CAUTIOUS in transposing animals from one district to
0174  d. In the first place we should be extremely  CAUTIOUS in inferring, because an area is now contino
0190  since become extinct. We should be extremely  CAUTIOUS in concluding that an organ could not have b
0192  t of the sack? Although we must be extremely  CAUTIOUS in concluding that any organ could not possi
0197  tion of putrid matter: but we should be very  CAUTIOUS in drawing any such inference, when we see t
0204  ng it. We have have seen in this chapter how  CAUTIOUS we should be in concluding that the most dif
0204  te or transitional states, we should be very  CAUTIOUS in concluding that none could have existed,
0460  facit saltum, that we ought to be extremely  CAUTIOUS in saying that any organ or instinct, or any
0487  ing and succeeding organic forms. We must be  CAUTIOUS in attempting to correlate as strictly conte
0340  in a similar manner we know from Falconer and  CAUTLEY's discoveries, that Northern India was forme
0137  suse. In one of the blind animals, namely, the  CAVE rat, the eyes are of immense size: and Profess
0137  aided by use and disuse, so in the case of the  CAVE rat natural selection seems to have struggled
0138  s have remarked, this is not the case, and the  CAVE insects of the two continents are not more clo
0138  fications, we might expect still to see in the  CAVE animals of America, affinities to the other in
0138  and this is the case with some of the American  CAVE animals, as I hear from Professor Dana: and so
0139  from Professor Dana: and some of the European  CAVE insects are very closely allied to those of th
0139  nal explanation of the affinities of the blind  CAVE animals to the other inhabitants of the two co
0139  far from feeling any surprise that some of the  CAVE animals should be very anomalous, as Agassiz h
0289  aves or in lacustrine deposits: and that not a  CAVE or true lacustrine bed is known belonging to t
0138  ions of life more similar than deep limestone  CAVERNS under a nearly similar climate: so that on t
0138  arately created for the American and European  CAVERNS, close similarity in their organisation and
0454  e case of the eyes of animals inhabiting dark  CAVERNS, and of the wings of birds inhabiting oceani
0137  the most different classes, which inhabit the  CAVES of Styria and of Kentucky, are blind. In some
0138  whereas with all the other inhabitants of the  CAVES, disuse by itself seems to have done its work
0138  the deeper and deeper recesses of the Kentucky  CAVES, as did European animals into the caves of Eu
0138  ntucky caves, as did European animals into the  CAVES of Europe. We have some evidence of this grad
0139  eation. That several of the inhabitants of the  CAVES of the Old and New Worlds should be closely r
0289  ertiary mammals have been discovered either in  CAVES or in lacustrine deposits: and that not a cav
0339  ed that the fossil mammals from the Australian  CAVES were closely allied to the living marsupials
0339  ssil bones made by MM. Lund and Clausen in the  CAVES of Brazil. I was so much impressed with these
0339  ew zealand. We see it also in the birds of the  CAVES of Brazil. Mr. Woodward has shown that the sa
0341  extinct, and have left no progeny. But in the  CAVES of Brazil, there are many extinct species whi
0404  principle in the blind animals inhabiting the  CAVES of America and of Europe. Other analogous fac
0473  look at the blind animals inhabiting the dark  CAVES of America and Europe. In both varieties and
0078  e borders of an utter desert, will competition  CEASE. The land may be extremely cold or dry, yet t
0152  b in the Dorking fowl) or the whole breed will  CEASE to have a nearly uniform character. The breed
0154  on the other hand, will in the course of time  CEASE: and that the most abnormally developed organ
0326  umphant course, or even their existence, would  CEASE. We know not at all precisely what are all th
0466  en it has once come into play, does not wholly  CEASE: for new varieties are still occasionally pro
0472  d or ill occupied place in nature, these facts  CEASE to be strange, or perhaps might even have bee
0479  nd should be so unlike the adult forms. We may  CEASE marvelling at the embryo of an air breathing
0484  cies of British brambles are true species will  CEASE. Systematists will have only to decide (not t
0485  rs, rudimentary and aborted organs, etc., will  CEASE to be metaphorical, and will have a plain sig
0228  ach other. As soon as this occurred, the bees  CEASED to excavate, and began to build up flat walls
0250  t the ovaries of the three first flowers soon  CEASED to grow, and after a few days perished entire
0374  ach point. The cold may have come on, or have  CEASED, earlier at one point of the globe than at an
0320  a species, and yet to marvel greatly when it  CEASES to exist, is much the same as to admit that s
0345  ame areas during the later geological periods  CEASES to be mysterious, and is simply explained by
0008  ons. No case is on record of a variable being  CEASING to be variable under cultivation. Our oldest
0395  at malay Archipelago, which is traversed near  CELEBES by a space of deep ocean: and this space sep
0029  y distinct species. Ask, as I have asked, a  CELEBRATED raiser of Hereford cattle, whether his catt
0219  erre Huber, a better observer even than his  CELEBRATED father. This ant is absolutely dependent on
0230  ements differ from those made by the justly  CELEBRATED elder Huber, but I am convinced of their ac
0207  rasitic bees. Slave making ants. Hive bee, its  CELL making instinct. Difficulties on the theory of
0224  ent on its slaves as is the Formica rufescens.  CELL making instinct of the Hive Bee. I will not he
0225  waterhouse, who has shown that the form of the  CELL stands in close relation to the presence of ad
0225  f the hive bee, placed in a double layer: each  CELL, as is well known, is an hexagonal prism, with
0225  hree which form the pyramidal base of a single  CELL on one side of the comb, enter into the compos
0226  heres which thus tend to intersect. Hence each  CELL consists of an outer spherical portion and of
0226  more perfectly flat surfaces, according as the  CELL adjoins two, three or more other cells. When o
0226  joins two, three or more other cells. When one  CELL comes into contact with three other cells, whi
0226  tion of the three sided pyramidal basis of the  CELL of the hive bee. As in the cells of the hive b
0226  , so here, the three plane surfaces in any one  CELL necessarily enter into the construction of thr
0228  ts of a sphere, and of about the diameter of a  CELL. It was most interesting to me to observe that
0228  ted width (i.e. about the width of an ordinary  CELL), and were in depth about one sixth of the dia
0229  mpleted rhombs at the base of a just commenced  CELL, which were slightly concave on one side, wher
0229  rking for a short time, and again examined the  CELL, and I found that the rhombic plate had been c
0230  always working circularly as they deepen each  CELL. They do not make the whole three sided pyrami
0230  he whole three sided pyramidal base of any one  CELL at the same time, but only the one rhombic pla
```

Page ***(Key Word)***

```
0230  theory. Huber's statement that the very first  CELL is excavated out of a little parallel sided wa
0231  than the excessively thin finished wall of the  CELL, which will ultimately be left. We shall under
0231  her: one bee after working a short time at one   CELL going to another, so that, as Huber has stated
0231  als work even at the commencement of the first   CELL. I was able practically to show this fact, by
0232  g the edges of the hexagonal walls of a single   CELL, or the extreme margin of the circumferential
0232  ll down and rebuild in different ways the same   CELL, sometimes recurring to a shape which they had
0232  never gnaw away and finish off the angles of a   CELL till a large part both of that cell and of the
0232  ngles of a cell till a large part both of that   CELL and of the adjoining cells has been built. Thi
0233  t, by fixing on a point at which to commence a   CELL, and then moving outside, first to one point,
0234  e a large part of the bounding surface of each   CELL would serve to bound other cells, and much wax
0172  n instinct as that which leads the bee to make   CELLS, which have practically anticipated the disco
0207  an instinct as that of the hive bee making its   CELLS will probably have occurred to many readers,
0219  o work, fed and saved the survivors: made some   CELLS and tended the larvae, and put all to rights.
0224  olved a recondite problem, and have made their   CELLS of the proper shape to hold the greatest poss
0224  measures, would find it very difficult to make   CELLS of wax of the true form, though this is perfe
0225  in close relation to the presence of adjoining   CELLS; and the following view may, perhaps, be cons
0225  ise making separate and very irregular rounded   CELLS of wax. At the other end of the series we hav
0225  ax. At the other end of the series we have the   CELLS of the hive bee, placed in a double layer: ea
0225  he composition of the bases of three adjoining   CELLS on the opposite side. In the series between t
0225  e series between the extreme perfection of the   CELLS of the hive bee and the simplicity of those o
0225  licity of those of the humble bee, we have the   CELLS of the Mexican Melipona domestica, carefully
0225  rms a nearly regular waxen comb of cylindrical   CELLS, in which the young are hatched, and, in addi
0225  oung are hatched, and, in addition, some large   CELLS of wax for holding honey. These latter cells
0225  e cells of wax for holding honey. These latter   CELLS are nearly spherical and of nearly equal size
0225  t the important point to notice, is that these   CELLS are always made at that degree of nearness to
0226  g as the cell adjoins two, three or more other   CELLS. When one cell comes into contact with three
0226  n one cell comes into contact with three other   CELLS, which, from the spheres being nearly of the
0226  l basis of the cell of the hive bee. As in the   CELLS of the hive bee, so here, the three plane sur
0226  enter into the construction of three adjoining   CELLS. It is obvious that the Melipona saves wax by
0226  ding; for the flat walls between the adjoining   CELLS are not double, but are of the same thickness
0226  and yet each flat portion forms a part of two   CELLS. Reflecting on this case, it occurred to me t
0227  best measurements which have been made of the   CELLS of the hive bee. Hence we may safely conclude
0227  bee. We must suppose the Melipona to make her   CELLS truly spherical, and of equal sizes: and this
0227  t. we must suppose the Melipona to arrange her   CELLS in level layers, as she already does her cyli
0227  el layers, as she already does her cylindrical   CELLS; and we must further suppose, and this is the
0228  three sided pyramid as in the case of ordinary   CELLS. I then put into the hive, instead of a thick
0230  at the bees in such cases stand in the opposed   CELLS and push and bend the ductile and warm wax (w
0230  lves a thin wall of wax, they could make their   CELLS of the proper shape, by standing at the prope
0230  rt excavation plays in the construction of the   CELLS; but it would be a great error to suppose tha
0231  e planes of the rhombic basal plates of future   CELLS. But the rough wall of wax has in every case
0231  ays crowned by a gigantic coping. From all the   CELLS, both those just commenced and those complete
0231  add to the difficulty of understanding how the   CELLS are made, that a multitude of bees all work t
0232  aced, and worked into the growing edges of the   CELLS all round. The work of construction seems to
0232  r place, projecting beyond the other completed   CELLS. It suffices that the bees should be enabled
0232  other and from the walls of the last completed   CELLS, and then, by striking imaginary spheres, the
0232  ge part both of that cell and of the adjoining   CELLS has been built. This capacity in bees of layi
0233  in its proper place between two just commenced   CELLS, is important, as it bears on a fact, which s
0233  sive of the foregoing theory; namely, that the   CELLS on the extreme margin of wasp combs are somet
0233  in the case of a queen wasp) making hexagonal   CELLS, if she work alternately on the inside and ou
0233  tely on the inside and outside of two or three   CELLS commenced at the same time, always standing a
0233  proper relative distance from the parts of the   CELLS just begun, sweeping spheres or cylinders, an
0234  tion of her instinct led her to make her waxen   CELLS near together, so as to intersect a little: f
0234  le: for a wall in common even to two adjoining   CELLS, would save some little wax. Hence it would c
0234  ous to our humble bee, if she were to make her   CELLS more and more regular, nearer together, and a
0234  together, and aggregated into a mass, like the   CELLS of the Melipona: for in this case a large par
0234  urface of each cell would serve to bound other   CELLS, and much wax would be saved. Again, from the
0234  geous to the Melipona, if she were to make her   CELLS closer together, and more regular in every wa
0453  male flowers, and which is formed merely of     CELLULAR tissue, can thus act? Can we suppose that th
0484  composition, their germinal vesicles, their    CELLULAR structure, and their laws of growth and repr
0231  sing masons first to pile up a broad ridge of   CEMENT, and then to begin cutting it away equally on
0231  dle: the masons always piling up the cut away   CEMENT, and adding fresh cement, on the summit of th
0231  ling up the cut away cement, and adding fresh   CEMENT, on the summit of the ridge. We shall thus ha
0068  ous destruction, when we remember that ten per  CENT. is an extraordinarily severe mortality from e
0045  anching of the main nerves close to the great   CENTRAL ganglion of an insect would have been variab
0145  every one knows the difference in the ray and  CENTRAL florets of, for instance, the daisy, and thi
0145  fference in the flow of nutriment towards the   CENTRAL and external flowers: we know, at least, tha
0145  bserved in some garden pelargoniums, that the   CENTRAL flower of the truss often loses the patches
0145  spect to the difference in the corolla of the  CENTRAL and exterior flowers of a head or umbel, I d
0146  the exterior flowers and coelospermous in the  CENTRAL flowers, that the elder De Candolle founded
0233  ts, at the proper relative distances from the  CENTRAL point and from each other, strike the planes
0346  el over the vast American continent, from the  CENTRAL parts of the United States to its extreme so
0348  f the eastern and western shores of South and  CENTRAL America: yet these great faunas are separate
0366  that within a very recent geological period,   CENTRAL Europe and North America suffered under an A
0366  uniform arctic fauna and flora, covering the   CENTRAL parts of Europe, as far south as the Alps an
0371  dants, mostly in a modified condition, in the  CENTRAL parts of Europe and the United States. On th
0373  ce extended far below their present level. In  CENTRAL Chile I was astonished at the structure of a
0226  ntres placed in two parallel layers: with the  CENTRE of each sphere at the distance of radius x so
0309  a which had subsided some miles nearer to the  CENTRE of the earth, and which had been pressed on b
0381  migrated in radiating lines from some common   CENTRE: and I am inclined to look in the southern, a
0226  mber of equal spheres be described with their  CENTRES placed in two parallel layers: with the cent
0226  1.41421 (or at some lesser distance) from the  CENTRES of the six surrounding spheres in the same l
0226  same layer: and at the same distance from the  CENTRES of the adjoining spheres in the other and pa
0346  ity of the productions of the same continent.  CENTRES of creation. Means of dispersal, by changes
0355  . the previous remarks on single and multiple  CENTRES of creation do not directly bear on another
0356  amount of difficulty on the theory of single   CENTRES of creation, I must say a few words on the m
0407  y naturalists under the designation of single  CENTRES of creation, by some general considerations,
```

Page **(Key Word)**

```
0035  ot doubt that this process, continued during CENTURIES, would improve and modify any breed, in the
0037  flower and kitchen gardens. If it has taken CENTURIES or thousands of years to improve or modify
0075  of trees must here have gone on during long CENTURIES, each annually scattering its seeds by the
0075  mals which have determined, in the course of CENTURIES, the proportional numbers and kinds of tree
0112  two sub breeds; finally, after the lapse of CENTURIES, the sub breeds would become converted into
0166  of pigeons, some of which have bred true for CENTURIES, species; and how exactly parallel is the c
0363  ed, have been in action year after year, for CENTURIES and tens of thousands of years; it would I
0364  to prove this), received within the last few CENTURIES, through occasional means of transport, imm
0377  plants and animals have spread within a few CENTURIES, this period will have been ample for any a
0380  ted into Europe during the last two or three CENTURIES from La Plata, and during the last thirty o
0033  ce for scarcely more than three quarters of a CENTURY; it has certainly been more attended to of l
0035  nter has been greatly changed within the last CENTURY, and in this case the change has, it is beli
0064  erval; if this be so, at the end of the fifth CENTURY there would be alive fifteen million elephan
0286  0 feet in height at the rate of one inch in a CENTURY. This will at first appear much too small an
0287  feet in height, a denudation of one inch per CENTURY for the whole length would be an ample allow
0387  lily and Potamogeton. Herons and other birds, CENTURY after century, have gone on daily devouring
0387  ogeton. Herons and other birds, century after CENTURY, have gone on daily devouring fish; they the
0442  uttle fish; there is no metamorphosis; the CEPHALOPODIC character is manifested long before the pa
0331  and Batrachians, the older Fish, the older CEPHALOPODS, and the Eocene mammals, with the more rec
0001  beagle, as naturalist, I was much struck with CERTAIN facts in the distribution of the inhabitants
0003  e misseltoe, which draws its nourishment from CERTAIN trees, which has seeds that must be transpor
0003  , which has seeds that must be transported by CERTAIN birds, and which has flowers with separate s
0003  rate sexes absolutely requiring the agency of CERTAIN insects to bring pollen from one flower to t
0003  f creation would, I presume, say that after a CERTAIN unknown number of generations, some bird had
0010  all or nearly all the individuals exposed to CERTAIN conditions are affected in the same way, the
0012  erently affected from coloured individuals by CERTAIN vegetable poisons. Hairless dogs have imperf
0013  etimes not so; why the child often reverts in CERTAIN characters to its grandfather or grandmother
0022  breeds the voice and disposition. Lastly, in CERTAIN breeds, the males and females have come to d
0027  se species having very abnormal characters in CERTAIN respects, as compared with all other Columbi
0027  r or short faced tumbler differs immensely in CERTAIN characters from the rock pigeon, yet by comp
0030  es successive variations; man adds them up in CERTAIN directions useful to him. In this sense he m
0030  ciple of selection is not hypothetical. It is CERTAIN that several of our eminent breeders have, e
0034  xportation: the destruction of horses under a CERTAIN size was ordered, and this may be compared t
0038  struggle for their own food, at least during CERTAIN seasons. And in two countries very different
0046  ently shown that the muscles in the larvae of CERTAIN insects are very far from uniform. Authors s
0052  fully explained) differences of structure in CERTAIN definite directions. Hence I believe a well
0052  e case by Mr. Wollaston with the varieties of CERTAIN fossil land shells in Madeira. If a variety
0053  erent consideration from wide range, and to a CERTAIN extent from commonness), often give rise to
0057  f the species already manufactured still to a CERTAIN extent resemble varieties, for they differ f
0057  re generally clustered like satellites around CERTAIN other species. And what are varieties but gr
0057  ly related to each other, and clustered round CERTAIN forms, that is, round their parent species?
0058  hich they connect; and except, secondly, by a CERTAIN amount of difference, for two forms, if diff
0059  llied together, forming little clusters round CERTAIN species. Species very closely allied to othe
0072  caused by the greater number in Paraguay of a CERTAIN fly, which lays its eggs in the navels of th
0072  d by some means, probably by birds. Hence, if CERTAIN insectivorous birds (whose numbers are proba
0074  st of mice and then of bees, the frequency of CERTAIN flowers in that district! In the case of man
0076  varieties of sheep: it has been asserted that CERTAIN mountain varieties will starve out other mou
0085  own; that purple plums suffer far more from a CERTAIN disease than yellow plums; whereas another d
0087  on; for instance, the great jaws possessed by CERTAIN insects, and used exclusively for opening th
0090  iers, horn like protuberances in the cocks of CERTAIN fowls, etc.) which we cannot believe to be e
0091  ght be born with an innate tendency to pursue CERTAIN kinds of prey. Nor can this be thought very
0091  flocks. Let us now take a more complex case. CERTAIN plants excrete a sweet juice, apparently for
0094  ued selection, to be a common plant; and that CERTAIN insects depended in main part on its nectar
0094  holes and sucking the nectar at the bases of CERTAIN flowers, which they can, with a very little
0106  ood effects of isolation will generally, to a CERTAIN extent, have concurred. Finally, I conclude
0107  epidosiren, which, like fossils, connect to a CERTAIN extent orders now widely separated in the na
0115  ed to a few groups more especially adapted to CERTAIN stations in their new homes. But the case is
0134  our domestic animals strengthens and enlarges CERTAIN parts, and disuse diminishes them; and that
0136  r. wollaston, of the almost entire absence of CERTAIN large groups of beetles, elsewhere excessive
0140  e of some few plants, of their becoming, to a CERTAIN extent, naturally habituated to different te
0142  treatises on many kinds of cultivated plants, CERTAIN varieties are said to withstand certain clim
0142  ants, certain varieties are said to withstand CERTAIN climates better than others: this is very st
0142  rees published in the United States, in which CERTAIN varieties are habitually recommended for the
0144  ffroy St. Hilaire has forcibly remarked, that CERTAIN malconformations very frequently, and that o
0145  of the ray petals by drawing nourishment from CERTAIN other parts of the flower had caused their a
0147  the other side. I think this holds true to a CERTAIN extent with our domestic productions: if nou
0157  tance, because common to large groups; but in CERTAIN genera the neuration differs in the differen
0166  geographical races) of a bluish colour, with CERTAIN bars and other marks; and when any breed ass
0187  ach a moderately high stage of perfection. In CERTAIN crustaceans, for instance, there is a double
0190  in as an accessory to the auditory organs of CERTAIN fish; or, for I do not know which view is no
0198  ir own subsistence, and would be exposed to a CERTAIN extent to natural selection, and individuals
0198  colour, as is the liability to be poisoned by CERTAIN plants; so that colour would be thus subject
0208  become associated with other habits, and with CERTAIN periods of time and states of the body. When
0211  cture of others. So again, in some few cases, CERTAIN instincts cannot be considered as absolutely
0212  be given, of occasional and strange habits in CERTAIN species, which might, if advantageous to the
0213  ikewise of the oddest tricks, associated with CERTAIN frames of mind or periods of time. But let u
0216  er birds' nests: the slave making instinct of CERTAIN ants; and the comb making power of the hive
0225  id, formed of three rhombs. These rhombs have CERTAIN angles; and the three which form the pyramid
0227  prising, seeing that she already does so to a CERTAIN extent, and seeing what perfectly cylindrica
0232  t. this capacity in bees of laying down under CERTAIN circumstances a rough wall in its proper pla
0237  of structure which have become correlated to CERTAIN ages; and to either sex. We have differences
0237  imperfect state of the male sex: for oxen of CERTAIN breeds have longer horns than in other breed
0237  come correlated with the sterile condition of CERTAIN members of insect communities: the difficult
0238  nct, correlated with the sterile condition of CERTAIN members of the community, has been advantage
0248  when intercrossed, as Gartner believes. It is CERTAIN, on the one hand, that the sterility of vari
0248  by our best botanists on the question whether CERTAIN doubtful forms should be ranked as species o
```

Page **(Key Word)***

Page **************************************(Key Word)**************************************

```
0250  ly, that there are individual plants, as with  CERTAIN  species of Lobelia, and with all the species
0250  , for it fertilised distinct species. So that  CERTAIN  individual plants and all the individuals of
0250  individual plants and all the individuals of   CERTAIN  species can actually be hybridised much more
0254  ld stocks; yet, with perhaps the exception of  CERTAIN  indigenous domestic dogs of South America, a
0255  complete fertility; and, as we have seen, in   CERTAIN  abnormal cases, even to an excess of fertili
0258  given. Thuret has observed the same fact with  CERTAIN  sea weeds or Fuci. Gartner, moreover, found
0259  not at all necessarily go together. There are  CERTAIN  hybrids which instead of having, as is usual
0259  perfect fertility, or even to fertility under  CERTAIN  conditions in excess. That their fertility,
0262  fferent varieties of the apricot and peach on  CERTAIN  varieties of the plum. As Gartner found that
0262  erfect; yet these two distinct cases run to a  CERTAIN  extent parallel. Something analogous occurs
0262  fted were rendered barren. On the other hand,  CERTAIN  species of Sorbus, when grafted on other spe
0263  these differences, in both cases, follow to a  CERTAIN  extent, as might have been expected, systema
0264  an be given of these facts, any more than why  CERTAIN  trees cannot be grafted on others. Lastly, a
0265  han the male. In both, the tendency goes to a  CERTAIN  extent with systematic affinity, for whole g
0265  of conditions with unimpaired fertility; and  CERTAIN  species in a group will produce unusually fe
0267  acts alluded to in our fourth chapter, that a  CERTAIN  amount of crossing is indispensable even wit
0268  re easily than other dogs with foxes, or that  CERTAIN  South American indigenous domestic dogs do n
0269  to resist the evidence of the existence of a  CERTAIN  amount of sterility in the few following case
0271  other. From observations which I have made on  CERTAIN  varieties of hollyhock, I am inclined to sus
0277  different circumstances, should all run, to a  CERTAIN  extent, parallel with the systematic affinit
0299  several aboriginal stocks: or, again, whether  CERTAIN  sea shells inhabiting the shores of North Am
0302  ch whole groups of species suddenly appear in  CERTAIN  formations, has been urged by several palaeo
0302  geological record, and falsely infer, because  CERTAIN  genera or families have not been found benea
0302  era or families have not been found beneath a  CERTAIN  stage, that they did not exist before that s
0305  back; and some palaeontologists believe that  CERTAIN  much older fishes, of which the affinities a
0319  xisting as a rare species, we might have felt  CERTAIN  from the analogy of all other mammals, even
0320  atural and artificial, are bound together. In  CERTAIN  flourishing groups, the number of new specif
0323  these distant points, the organic remains in  CERTAIN  beds present an unmistakeable degree of rese
0324  related to those which lived in Europe during  CERTAIN  later tertiary stages, than to those which n
0325  series of analogous phenomena, it will appear  CERTAIN  that all these modifications of species, the
0326  ding species, which already have invaded to a  CERTAIN  extent the territories of other species, sho
0326  he power of spreading into new territories. A  CERTAIN  amount of isolation, recurring at long inter
0328  es in the two countries; but when he compares  CERTAIN  stages in England with those in France, alth
0329  ification of these two orders; and has placed  CERTAIN  pachyderms in the same sub order with rumina
0334  een the preceding and succeeding faunas, that  CERTAIN  genera offer exceptions to the rule. For ins
0338  iz insists that ancient animals resemble to a  CERTAIN  extent the embryos of recent animals of the
0342  en geologically explored with care; that only  CERTAIN  classes of organic beings have been largely
0345  be proved that ancient animals resemble to a  CERTAIN  extent the embryos of more recent animals of
0351  om an enormously remote geological period, so  CERTAIN  species have migrated over vast spaces, and
0358  and even seas to their present inhabitants, a  CERTAIN  degree of relation (as we shall hereafter se
0361  three dicotyledonous plants germinated: I am  CERTAIN  of the accuracy of this observation. Again,
0362  se seeds retained their power of germination.  CERTAIN  seeds, however, were always killed by this p
0372  f the areas; for if we compare, for instance,  CERTAIN  parts of South America with the southern con
0374  temperature was at the same time lower along  CERTAIN  broad belts of longitude. On this view of th
0377  tropical productions will have suffered to a  CERTAIN  extent. On the other hand, the temperate pro
0377  onditions, will have suffered less. And it is  CERTAIN  that many temperate plants, if protected fro
0377  sented a firm front against intruders, that a  CERTAIN  number of the more vigorous and dominant tem
0380  n a lesser degree in Australia, and have to a  CERTAIN  extent beaten the natives; whereas extremely
0380  nes and means of migration, or the reason why  CERTAIN  speciesand not other have migrated: why cert
0381  rtain speciesand not other have migrated; why  CERTAIN  species have been modified and have given ri
0391  s. oceanic islands are sometimes deficient in  CERTAIN  classes, and their places are apparently occ
0392  habitants of remote islands. For instance, in  CERTAIN  islands not tenanted by mammals, some of the
0395  om continents, there is also a relation, to a  CERTAIN  extent independent of distance, between the
0395  cases owing to the probable naturalisation of  CERTAIN  mammals through man's agency; but we shall s
0396  unt of modification in all cases depends to a  CERTAIN  degree on the lapse of time, and as during c
0396  e of aerial bats, the singular proportions of  CERTAIN  orders of plants, herbaceous forms having be
0402  to the several islands, and we may infer from  CERTAIN  facts that these have probably spread from s
0405  arriers and ranging widely, as in the case of  CERTAIN  powerfully winged birds, will necessarily ra
0405  cies. In considering the wide distribution of  CERTAIN  genera, we should bear in mind that some are
0408  the nature of the communication which allowed  CERTAIN  forms and not others to enter, either in gre
0410  ent. Groups of species, belonging either to a  CERTAIN  period of time, or to a certain area, are of
0410  g either to a certain period of time, or to a  CERTAIN  area, are often characterised by trifling ch
0412  to the closest competition; and by looking to  CERTAIN  facts in naturalisation. I attempted also to
0415  t authority, Robert Brown, who in speaking of  CERTAIN  organs in the Proteaceae, says their generic
0416  eth in the upper jaws of young ruminants, and  CERTAIN  rudimentary bones of the leg, are highly ser
0416  e wings of insects are folded, mere colour in  CERTAIN  Algae, mere pubescence on parts of the flowe
0417  nts of resemblance, too slight to be defined.  CERTAIN  plants, belonging to the Malpighiaceae, bear
0418  that excellent botanist, Aug. St. Hilaire. If  CERTAIN  characters are always found correlated with
0419  utility or even necessity of this practice in  CERTAIN  groups of birds; and it has been followed by
0422  rom these two genera will have inherited to a  CERTAIN  extent their characters. This natural arrang
0424  in common of the males and hermaphrodites of  CERTAIN  cirripedes, when adult, and yet no one dream
0425  in classing varieties which have undergone a  CERTAIN, and sometimes a considerable amount of modi
0433  ing so perfect a collection: nevertheless, in  CERTAIN  classes, we are tending in this direction; a
0433  lassification. We can understand why we value  CERTAIN  resemblances far more than others; why we ar
0435  or a webbed foot might have all its bones, or  CERTAIN  bones, lengthened to any extent, and the mem
0436  hy and ultimately by the complete abortion of  CERTAIN  parts, by the soldering together of other pa
0436  ct gigantic sea lizards, and in the mouths of  CERTAIN  suctorial crustaceans, the general pattern s
0436  general pattern seems to have been thus to a  CERTAIN  extent obscured. There is another and equall
0436  tive connexion with, the elemental parts of a  CERTAIN  number of vertebrae. The anterior and poster
0437  we see a series of internal vertebrae bearing  CERTAIN  processes and appendages; in the articulata,
0438  urse of modification, should have seized on a  CERTAIN  number of the primordially similar elements,
0438  der at discovering in such parts or organs, a  CERTAIN  degree of fundamental resemblance, retained
0439  y. It has already been casually remarked that  CERTAIN  organs in the individual, which when mature
0441  as lower in the scale than the larva, as with  CERTAIN  parasitic crustaceans. To refer once again t
0442  o. and in some whole groups of animals and in  CERTAIN  members of other groups, the embryo does not
0444  pear at a corresponding age in the offspring.  CERTAIN  variations can only appear at corresponding
```

Page **************************************(Key Word)**************************************

```
0447  degree, by the effects of use and disuse. In CERTAIN cases the successive steps of variation migh
0448  seen that this is the rule of development in CERTAIN whole groups of animals, as with cuttle fish
0451  ents of teeth can be detected in the beaks of CERTAIN embryonic birds. Nothing can be plainer than
0452  urpose, and be used for a distinct object: in CERTAIN fish the swim bladder seems to be rudimentar
0452  the state of the wings of the female moths in CERTAIN groups. Rudimentary organs may be utterly ab
0453  parts and organs are exquisitely adapted for CERTAIN purposes, tells us with equal plainness that
0454  power of flying. Again, an organ useful under CERTAIN conditions, might become injurious under oth
0464  geologically explored. Only organic beings of CERTAIN classes can be preserved in a fossil conditi
0467  hout any thought of altering the breed. It is CERTAIN that he can largely influence the character
0470  f varieties or incipient species, retain to a CERTAIN degree the character of varieties: for they
0473  ucu, which is occasionally blind, and then at CERTAIN moles, which are habitually blind and have t
0473  they branched off from a common progenitor in CERTAIN characters, by which they have come to be sp
0478  ass are so complex and circuitous. We see why CERTAIN characters are far more serviceable than oth
0482  ned difficulties than to the explanation of a CERTAIN number of facts will certainly reject my the
0483  at innumerable periods in the earth's history CERTAIN elemental atoms have been commanded suddenly
0489  d long before the Silurian epoch, we may feel CERTAIN that the ordinary succession by generation h
0011  ny change in the embryo or larva will almost CERTAINLY entail changes in the mature animal. In mon
0012  augmenting, any peculiarity, he will almost CERTAINLY unconsciously modify other parts of the str
0014  stic varieties, when run wild, gradually but CERTAINLY revert in character to their aboriginal sto
0015  its new home. Nevertheless, as our varieties CERTAINLY do occasionally revert in some of their cha
0020  hopelessness, of the task becomes apparent. CERTAINLY, a breed intermediate between two very dist
0022  e were told that they were wild birds, would CERTAINLY, I think, be ranked by him as well defined
0024  tructure, with the wild rock pigeon, yet are CERTAINLY highly abnormal in other parts of their str
0033  ore than three quarters of a century; it has CERTAINLY been more attended to of late years, and ma
0035  tually, that, though the old Spanish pointer CERTAINLY came from Spain, Mr. Borrow has not seen, a
0051  and his difficulties will rise to a climax. CERTAINLY no clear line of demarcation has as yet bee
0056  uring new species to be a slow one. And this CERTAINLY is the case, if varieties be looked at as i
0061  tion. We have seen that man by selection can CERTAINLY produce great results, and can adapt organi
0072  nd horses would become feral, and this would CERTAINLY greatly alter (as indeed I have observed in
0081  ry were open on its borders, new forms would CERTAINLY immigrate, and this also would seriously di
0082  ount of variability is necessary; as man can CERTAINLY produce great results by adding up in any g
0083  isted such intruders. As man can produce and CERTAINLY has produced a great result by his methodic
0092  ctar would get dusted with pollen, and would CERTAINLY often transport the pollen from one flower
0096  l there are many hermaphrodite animals which CERTAINLY do not habitually pair, and a vast majority
0102  be large, its several districts will almost CERTAINLY present different conditions of life; and t
0111  s doubts in many cases how to rank them, yet CERTAINLY differ from each other far less than do goo
0127  r year, a severe struggle for life, and this CERTAINLY cannot be disputed; then, considering the i
0137  frequently blind; one which I kept alive was CERTAINLY in this condition, the cause, as appeared o
0137  be injurious to any animal, and as eyes are CERTAINLY not indispensable to animals with subterran
0151  y attended to them, and the rule seems to me CERTAINLY to hold good in this class. I cannot make o
0157  l the species of the same genus as having as CERTAINLY descended from the same progenitor, as have
0163  give them. I can only repeat that such cases CERTAINLY do occur, and seem to me very remarkable. I
0163  is sometimes double. The shoulder stripe is CERTAINLY very variable in length and outline. A whit
0179  through natural selection, they will almost CERTAINLY be beaten and supplanted by the forms which
0186  y, and the variations be inherited, which is CERTAINLY the case; and if any variation or modificat
0210  tions of some kind are possible; and this we CERTAINLY can do. I have been surprised to find, maki
0211  events me. I can only assert, that instincts CERTAINLY do vary, for instance, the migratory instin
0212  ited States. Fear of any particular enemy is CERTAINLY an instinctive quality, as may be seen in n
0213  time that they are taken out; retrieving is CERTAINLY in some degree inherited by retrievers; and
0213  mestic instincts, as they may be called, are CERTAINLY far less fixed or invariable than natural i
0219  slaves; without their aid, the species would CERTAINLY become extinct in a single year. The males
0243  s by that common case of closely allied, but CERTAINLY distinct, species, when inhabiting distant
0245  he confusion of all organic forms. This view CERTAINLY seems at first probable, for species within
0253  either pure parent species exists, they must CERTAINLY be highly fertile. A doctrine which origina
0273  her parent form; but this, if it be true, is CERTAINLY only a difference in degree. Gartner furthe
0274  arieties of plants. With animals one variety CERTAINLY often has this prepotent power over another
0291  hardly last to a distant geological age, was CERTAINLY deposited during a downward oscillation of
0304  ted during the secondary periods, they would CERTAINLY have been preserved and discovered: and as
0305  ement of the chalk formation, the fact would CERTAINLY be highly remarkable; but I cannot see that
0307  ogenitors of these orders, they would almost CERTAINLY have been long ago supplanted and extermina
0315  identically the same: for both would almost CERTAINLY inherit different characters from their dis
0316  e been fatal to my views. But such cases are CERTAINLY exceptional; the general rule being a gradu
0319  rceived the fact, yet the fossil horse would CERTAINLY have become rarer and rarer, and finally ex
0337  her quarter, the eocene fauna or flora would CERTAINLY be beaten and exterminated: as would a seco
0353  eographical and climatal changes, which have CERTAINLY occurred within recent geological times, mu
0376  he highest authority, Prof. Dana, that it is CERTAINLY a wonderful fact that New Zealand should ha
0376  stinct, by others as varieties; but some are CERTAINLY identical, and many, though closely related
0400  ed so near each other that they would almost CERTAINLY receive immigrants from the same original s
0407  ate and of the level of the land, which have CERTAINLY occurred within the recent period, and of o
0410  which occur above and below: so in space, it CERTAINLY is the general rule that the area inhabited
0425  has been close descent in common, there will CERTAINLY be close resemblance or affinity. As descen
0432  ived throughout all time and space. We shall CERTAINLY never succeed in making so perfect a collec
0440  did not perceive that a barnacle was, as it CERTAINLY is, a crustacean; but a glance at the larva
0456  the use of this element of descent, the only CERTAINLY known cause of similarity in organic beings
0463  hole groups of allied species appear, though CERTAINLY they often falsely appear, to have come in
0463  itors of the Silurian groups of fossils? For CERTAINLY on my theory such strata must somewhere hav
0464  tless generations of countless species which CERTAINLY have existed. We should not be able to reco
0474  mes into play in modifying instincts; but it CERTAINLY is not indispensable, as we see, in the cas
0482  xplanation of a certain number of facts will CERTAINLY reject my theory. A few naturalists, endowe
0253  some reason to believe that the hybrids from CERVULUS vaginalis and Reevesii, and from Phasianus c
0144  most abnormal in their dermal covering, viz. CETACEA (whales) and Edentata (armadilloes, scaly an
0428  veral members of the whale family; for these CETACEANS agree in so many characters, great and smal
0140  orms me that he has observed similar facts in CEYLON, and analogous observations have been made by
0375  of the peninsula of India, on the heights of CEYLON, and on the volcanic cones of Java, many plan
0280  s not reveal any such finely graduated organic CHAIN; and this, perhaps, is the most obvious and o
0281  nless at the same time we had a nearly perfect CHAIN of the intermediate links. It is just possibl
```

Page **************************************(Key Word)**************************************

```
0301   of the same group into one long and branching CHAIN of life. We ought only to look for a few link
0426   hese extreme forms are connected together by a CHAIN of intermediate groups, we may at once infer
0432   are still tied together by a long, but broken, CHAIN of affinities. Extinction has only separated
0475   pecies nor groups of species reappear when the CHAIN of ordinary generation has once been broken.
0476   xplained by their intermediate position in the CHAIN of descent. The grand fact that all extinct o
0348   ind different productions; though as mountain CHAINS, deserts, etc., are not as impassable, or lik
0419   assifications are often plainly influenced by CHAINS of affinities. Nothing can be easier than to
0483   of whole classes can be connected together by CHAINS of affinities, and all can be classified on t
0286   mited a period as since the latter part of the CHALK formation. The distance from the northern to
0286   ars. I doubt whether any rock, even as soft as CHALK, would yield at this rate excepting on the mo
0288   n that the genus Chthamalus existed during the CHALK period. The molluscan genus Chiton offers a p
0304   ipede, which he had himself extracted from the CHALK of Belgium. And, as if to make the case as st
0305   that of the teleostean fishes, low down in the CHALK period. This group includes the large majorit
0305   s agassiz believes, at the commencement of the CHALK formation, the fact would certainly be highly
0322   eously throughout the world. Thus our European CHALK formation can be recognised in many distant p
0322   climates, where not a fragment of the mineral CHALK itself can be found; namely, in North America
0323   takeable degree of resemblance to those of the CHALK. It is not that the same species are met with
0323   reover other forms, which are not found in the CHALK of Europe, but which occur in the formations
0335   organic remains from the several stages of the CHALK formation, though the species are distinct in
0031   for sheep, says: It would seem as if they had CHALKED out upon a wall a form perfect in itself, an
0247   er were potted, and apparently were kept in a CHAMBER in his house. That these processes are often
0004   and of cultivated plants would offer the best CHANCE of making out this obscure problem. Nor have
0005   arying conditions of life, will have a better CHANCE of surviving, and thus be naturally selected.
0024   eral species, but that he intentionally or by CHANCE picked out extraordinarily abnormal species;
0038   ries, so rich in species, do not by a strange CHANCE possess the aboriginal stocks of any useful p
0040   s of the breed, whatever they may be. But the CHANCE will be infinitely small of any record having
0041   pleasing to man appear only occasionally, the CHANCE of their appearance will be much increased by
0061   the offspring, also, will thus have a better CHANCE of surviving, for, of the many individuals of
0074   roportional numbers and kinds to what we call CHANCE. But how false a view is this! Every one has
0081   ever slight, over others, would have the best CHANCE of surviving and of procreating their kind? O
0082   able to natural selection, by giving a better CHANCE of profitable variations occurring; and unles
0088   nless stag or spurless cock would have a poor CHANCE of leaving offspring. Sexual selection by alw
0090   ftest and slimmest wolves would have the best CHANCE of surviving, and so be preserved or selected
0091   ed an individual wolf, it would have the best CHANCE of surviving and of leaving offspring. Some o
0092   lincs, which consequently would have the best CHANCE of flourishing and surviving. Some of these s
0094   n its food more quickly, and so have a better CHANCE of living and leaving descendants. Its descen
0100   flower to flower; and this will give a better CHANCE of pollen being occasionally carried from tre
0101   view; but I have been enabled, by a fortunate CHANCE, elsewhere to prove that two individuals, tho
0102   rge number of individuals, by giving a better CHANCE for the appearance within any given period of
0103   ge body of plants of the same variety, as the CHANCE of intercrossing with other varieties is thus
0104   f fertilisation, that they will have a better CHANCE of surviving and propagating their kind; and
0105   through natural selection, by decreasing the CHANCE of the appearance of favourable variations. I
0105   nd open area, not only will there be a better CHANCE of favourable variations arising from the lar
0109   s or in the number of its enemies, run a good CHANCE of utter extinction. But we may go further th
0110   st numerous in individuals will have the best CHANCE of producing within any given period favourab
0111   sent slight and ill defined differences. Mere CHANCE, as we may call it, might cause one variety t
0114   e species of grass would always have the best CHANCE of succeeding and of increasing in numbers, a
0121   ied descendants; for these will have the best CHANCE of filling new and widely different places in
0125   roup, reduce its numbers, and thus lessen the CHANCE of further variation and improvement. Within
0127   viduals thus characterised will have the best CHANCE of being preserved in the struggle for life;
0128   descendants become, the better will be their CHANCE of succeeding in the battle of life. Thus the
0130   a fork low down in a tree, and which by some CHANCE has been favoured and is still alive on its s
0131   n those in a state of nature. had been due to CHANCE. This, of course, is a wholly incorrect expre
0136   r from indolent habit, will have had the best CHANCE of surviving from not being blown out to sea;
0148   ch individual Proteolepas would have a better CHANCE of supporting itself, by less nutriment being
0176   ers would, as already remarked, run a greater CHANCE of being exterminated than one existing in la
0177   g in larger numbers will always have a better CHANCE, within any given period, of presenting furth
0177   on can do nothing until favourable variations CHANCE to occur, and until a place in the natural po
0183   her animals in the battle for life. Hence the CHANCE of discovering species with transitional grad
0187   see what gradations are possible, and for the CHANCE of some gradations having been transmitted fr
0203   order that a few granules may be wafted by a CHANCE breeze on to the ovules? Summary of Chapter.
0235   s, which in their turn will have had the best CHANCE of succeeding in the struggle for existence.
0249   ial fertilisation pollen is as often taken by CHANCE (as I know from my own experience) from the a
0298   siderable degree. According to this view, the CHANCE of discovering in a formation in any one coun
0298   cies; and this again would greatly lessen the CHANCE of our being able to trace the stages of tran
0326   es, should be those which would have the best CHANCE of spreading still further, and of giving ris
0326   a number of individuals, from giving a better CHANCE of the appearance of favourable variations, a
0338   rata are discovered, a discovery of which the CHANCE is very small. On the Succession of the same
0350   own widely extended homes will have the best CHANCE of seizing on new places, when they spread in
0364   ident. Even in this case, how small would the CHANCE be of a seed falling on favourable soil, and
0388   ed; and a single seed or egg will have a good CHANCE of succeeding. Although there will always be
0388   ers of a foreign country, would have a better CHANCE of seizing on a place, than in the case of te
0392   an herbaceous plant, though it would have no CHANCE of successfully competing in stature with a f
0406   uently the lower forms will have had a better CHANCE of ranging widely and of still retaining the
0462   g periods of time there will always be a good CHANCE for wide migration by many means. A broken or
0037   eeds, and, when a slightly better variety has CHANCED to appear, selecting it, and so onwards. But
0082   ght modification, which in the course of ages CHANCED to arise, and which in any way favoured the
0221   erland than in England. One day I fortunately CHANCED to witness a migration from one nest to anot
0296   r even have been suspected, had not the trees CHANCED to have been preserved: thus, Messrs. Lyell
0365   birds living there, nearly every seed, which CHANCED to arrive, would be sure to germinate and su
0013   reappears in the child, the mere doctrine of CHANCES almost compels us to attribute its reappeara
0177   ill to improve their stocks by selection; the CHANCES in this case will be strongly in favour of t
0008   art of the organisation, to the action of any CHANGE in the conditions of life. Nothing is more ea
0010   conditions are affected in the same way, the CHANGE at first appears to be directly due to such c
0011   structure. Nevertheless some slight amount of CHANGE may, I think, be attributed to the direct act
0011   nes in the wild duck; and I presume that this CHANGE may be safely attributed to the domestic duck
```

Page **************************************(Key Word)**************************************

```
0011 what may be called correlation of growth. Any CHANGE in the embryo or larva will almost certainly
0015 t when under nature the conditions of life do CHANGE, variations and reversions of character proba
0030 iety of the wild Dipsacus: and this amount of CHANGE may have suddenly arisen in a seedling. So it
0031 to modify the character of his flock, but to CHANGE it altogether. It is the magician's wand, by
0035 within the last century, and in this case the CHANGE has, it is believed, been chiefly effected by
0035 fox hound: but what concerns us is, that the CHANGE has been effected unconsciously and gradually
0037 s they could anywhere find. A large amount of CHANGE in our cultivated plants, thus slowly and unc
0043 e is only temporary. Over all these causes of CHANGE I am convinced that the accumulative action o
0069 and rarer, and finally disappearing: and the CHANGE of climate being conspicuous, we are tempted
0069 be in the least degree favoured by any slight CHANGE of climate, they will increase in numbers, an
0071 s previously and planted with Scotch fir. The CHANGE in the native vegetation of the planted part
0078 on the confines of its geographical range, a CHANGE of constitution with respect to climate would
0081 he case of a country undergoing some physical CHANGE, for instance, of climate. The proportional n
0081 nhabitants would almost immediately undergo a CHANGE, and some species might become extinct. We ma
0081 of each country are bound together, that any CHANGE in the numerical proportions of some of the i
0081 some of the inhabitants, independently of the CHANGE of climate itself, would most seriously affec
0082 lieve, as stated in the first chapter, that a CHANGE in the conditions of life, by specially actin
0082 ions of life are supposed to have undergone a CHANGE, and this would manifestly be favourable to n
0082 sal. Nor do I believe that any great physical CHANGE, as of climate, or any unusual degree of isol
0087 f each in consequence profits by the selected CHANGE. What natural selection cannot do, is to modi
0090 etest prey, a deer for instance, had from any CHANGE in the country increased in numbers, or that
0091 ught to modifying the breed. Even without any CHANGE in the proportional numbers of the animals on
0091 wn to be inherited. Now, if any slight innate CHANGE of habit or of structure benefited an individ
0104 proper type: but if their conditions of life CHANGE and they undergo modification,uniformity of c
0104 better adapted organisms, after any physical CHANGE, such as of climate or elevation of the land,
0109 election, I can see no limit to the amount of CHANGE, to the beauty and infinite complexity of the
0113 in increasing (the country not undergoing any CHANGE in its conditions) only by its varying descen
0120 ir common parent. If we suppose the amount of CHANGE between each horizontal line in our diagram t
0120 0) or two species, according to the amount of CHANGE supposed to be represented between the horizo
0125 if, in our diagram, we suppose the amount of CHANGE represented by each successive group of diver
0152 hich at the present time are undergoing rapid CHANGE by continued selection, are also eminently li
0166 ance of the stripes is not accompanied by any CHANGE of form or by any other new character. We see
0166 ks invariably reappear: but without any other CHANGE of form or character. When the oldest and tru
0180 al conditions. Let the climate and vegetation CHANGE, let other competing rodents or new beasts of
0183 d immaterial for us, whether habits generally CHANGE first and structure afterwards: or whether sl
0183 e lead to changed habits: both probably often CHANGE almost simultaneously. Of cases of changed ha
0185 , habits have changed without a corresponding CHANGE of structure. The webbed feet of the upland o
0185 en the toes shows that structure has begun to CHANGE. He who believes in separate and innumerable
0190 ions, for one function alone, and thus wholly CHANGE its nature by insensible steps. Two distinct
0204 at a species may under new conditions of life CHANGE its habits, or have diversified habits, with
0215 we must attribute the whole of the inherited CHANGE from extreme wildness to extreme tameness, si
0267 that great benefit is derived from almost any CHANGE in the habits of life. Again, both with plant
0273 ctive system being eminently sensitive to any CHANGE in the conditions of life, being thus often r
0281 ts descendants had undergone a vast amount of CHANGE: and the principle of competition between org
0282 ve sufficed for so great an amount of organic CHANGE, all changes having been effected very slowly
0293 s short compared with the period requisite to CHANGE one species into another. I am aware that two
0294 t changes of level, the inordinately great CHANGE of climate, on the prodigious lapse of time,
0295 of deposition has been much interrupted, as a CHANGE in the currents of the sea and a supply of se
0297 this view we do find the kind of evidence of CHANGE which on my theory we ought to find. Moreover
0312 e of new species. On their different rates of CHANGE. Species once lost do not reappear. Groups of
0313 greatly. The productions of the land seem to CHANGE at a quicker rate than those of the sea, of w
0313 isms, considered high in the scale of nature, CHANGE more quickly than those that are low: though
0313 xceptions to this rule. The amount of organic CHANGE, as Pictet has remarked, does not strictly co
0313 species will be found to have undergone some CHANGE. When a species has once disappeared from the
0314 , causing all the inhabitants of a country to CHANGE abruptly, or simultaneously, or to an equal d
0314 than others: or, if changing, that it should CHANGE less. We see the same fact in geographical di
0314 aps understand the apparently quicker rate of CHANGE in terrestrial and in more highly organised p
0315 time, become modified: for those which do not CHANGE will become extinct. In members of the same c
0315 mbers of the same class the average amount of CHANGE, during long and equal periods of time, may,
0315 intervals: consequently the amount of organic CHANGE exhibited by the fossils embedded in consecut
0322 triking than the fact, that the forms of life CHANGE almost simultaneously throughout the world. T
0323 he productions of the land and of fresh water CHANGE at distant points in the same parallel manner
0330 ted to those groups which have undergone much CHANGE in the course of geological ages: and it woul
0343 ecies of different classes do not necessarily CHANGE together, or at the same rate, or in the same
0356 riefest abstract of the more important facts. CHANGE of climate must have had a powerful influence
0382 has recently felt in its great cycles of CHANGE: and that on this view, combined with modific
0384 is having occurred during floods, without any CHANGE of level. We have evidence in the loess of th
0388 e have reason to believe that such low beings CHANGE or become modified less quickly than the high
0406 the scale within each great class, generally CHANGE at a slower rate than the higher forms: and c
0407 norance, and remember that some forms of life CHANGE most slowly, enormous periods of time being t
0410 ace the lower members of each class generally CHANGE less than the higher: but there are in both c
0415 rgans having generally been subjected to less CHANGE in the adaptation of the species to their con
0434 connexion in homologous organs: the parts may CHANGE to almost any extent in form and size, and ye
0454 til it rendered harmless and rudimentary. Any CHANGE in function, which can be effected by insensi
0463 hich the climate and other conditions of life CHANGE insensibly in going from a district occupied
0463 elieve that only a few species are undergoing CHANGE at any one period: and all changes are slowly
0464 een time sufficient for any amount of organic CHANGE: for the lapse of time has been so great as t
0474 characters which have been inherited without CHANGE for an enormous period. It is inexplicable on
0475 nd at successive intervals: and the amount of CHANGE, after equal intervals of time, is widely dif
0481 hat we are always slow in admitting any great CHANGE of which we do not see the intermediate steps
0487 s the most important of all causes of organic CHANGE is one which is almost independent of altered
0488 thers: it follows, that the amount of organic CHANGE in the fossils of consecutive formations prob
0488 we must not overrate the accuracy of organic CHANGE as a measure of time. During early periods of
0488 were probably fewer and simpler, the rate of CHANGE was probably slower: and at the first dawn of
0488 f the simplest structure existed, the rate of CHANGE may have been slow in an extreme degree. The
0005 g how a simple being or a simple organ can be CHANGED and perfected into a highly developed being
```

Page **(Key Word)**

```
0015   experiment itself the conditions of life are   CHANGED. If it could be shown that our domestic vari
0035   some cases, however, unchanged or but little   CHANGED individuals of the same breed may be found i
0035   own that the English pointer has been greatly   CHANGED within the last century, and in this case th
0071   ional numbers of the heath plants were wholly   CHANGED, but twelve species of plants (not counting
0078   the conditions of its life will generally be   CHANGED in an essential manner. If we wished to incr
0108   itants of the renewed continent will again be   CHANGED; and again there will be a fair field for na
0109   r at which the inhabitants of this world have   CHANGED. Slow though the process of selection may be
0148   e in every part of the organisation. If under   CHANGED conditions of life a structure before useful
0183   two or three instances of diversified and of   CHANGED habits in the individuals of the same specie
0183   y some modification of its structure, for its   CHANGED habits, or exclusively for one of its severa
0183   her slight modifications of structure lead to   CHANGED habits; both probably often change almost si
0183   ten change almost simultaneously. Of cases of   CHANGED habits it will suffice merely to allude to t
0185   , and many others could be given, habits have   CHANGED without a corresponding change of structure.
0209   , under its present conditions of life. Under   CHANGED conditions of life, it is at least possible
0245   f hybrids. Parallelism between the effects of   CHANGED conditions of life and crossing. Fertility o
0309   y not the areas of preponderant movement have   CHANGED in the lapse of ages? At a period immeasurab
0311   apters, may represent the apparently abruptly   CHANGED forms of life, entombed in our consecutive,
0313   cies of different genera and classes have not   CHANGED at the same rate, or in the same degree. In
0313   ilurian Molluscs and all the Crustaceans have   CHANGED greatly. The productions of the land seem to
0313   ive formations, the forms of life have seldom   CHANGED in exactly the same degree. Yet if we compar
0323   l manner. We may doubt whether they have thus   CHANGED: if the Megatherium, Mylodon, Macrauchenia,
0324   marine forms of life are spoken of as having   CHANGED simultaneously throughout the world, it must
0336   life, at least those inhabiting the sea, have   CHANGED almost simultaneously throughout the world,
0344   productions of the world will appear to have   CHANGED simultaneously. We can understand how it is
0366   filled. So greatly has the climate of Europe   CHANGED, that in Northern Italy, gigantic moraines,
0410   ether we look to the forms of life which have   CHANGED during successive ages within the same quart
0410   quarter of the world, or to those which have   CHANGED after having migrated into distant quarters,
0454   selection; so that an organ rendered, during   CHANGED habits of life, useless or injurious for one
0465   eology plainly declares that all species have   CHANGED; and they have changed in the manner which m
0465   that all species have changed; and they have   CHANGED in the manner which my theory requires, for
0465   anner which my theory requires, for they have   CHANGED slowly and in a graduated manner. We clearly
0468   e way as they generally have varied under the   CHANGED conditions of domestication. And if there be
0475   g intervals of time, to appear as if they had   CHANGED simultaneously throughout the world. The fac
0479   educe an organ, when it has become useless by   CHANGED habits or under changed conditions of life;
0479   has become useless by changed habits or under   CHANGED conditions of life: and we can clearly under
0480   ave thoroughly convinced me that species have   CHANGED, and are still slowly changing by the preser
0008   ases it has been found out that very trifling   CHANGES, such as a little more or less water at some
0011   hat quite opposite conditions produce similar   CHANGES of structure. Nevertheless some slight amoun
0011   embryo or larva will almost certainly entail   CHANGES in the mature animal. In monstrosities, the
0035   ualities of their cattle. Slow and insensible   CHANGES of this kind could never be recognised unles
0040   eserved of such slow, varying, and insensible   CHANGES. I must now say a few words on the circumsta
0045   the same species; I should have expected that   CHANGES of this nature could have been effected only
0084   ditions of life. We see nothing of those slow   CHANGES in progress, until the hand of time has mark
0086   is merely the correlated result of successive   CHANGES in the structure of their larvae. So, conver
0095   sir Charles Lyell's noble views on the modern   CHANGES of the earth, as illustrative of geology; bu
0107   species will thus be checked: after physical   CHANGES of any kind, immigration will be prevented,
0108   of such places will often depend on physical   CHANGES, which are generally very slow, and on the i
0131   productive system is eminently susceptible to   CHANGES in the conditions of life; and to this syste
0138   ural selection will often have effected other   CHANGES, such as an increase in the length of the an
0168   t detriment to the individual, will be saved.   CHANGES of structure at an early age will generally
0174   ieties existing in the intermediate zones. By   CHANGES in the form of the land and of climate, mari
0178   ants. And such new places will depend on slow   CHANGES of climate, or on the occasional immigration
0199   ften have entailed on other parts diversified   CHANGES of no direct use. So again characters which
0210   equal force to instincts as to bodily organs.   CHANGES of instinct may sometimes be facilitated by
0216   has sufficed to produce such inherited mental   CHANGES; in other cases compulsory habit has done no
0265   pecies in a group will sometimes resist great   CHANGES of conditions with unimpaired fertility; and
0266   a considerable body of evidence, that slight   CHANGES in the conditions of life are beneficial to
0267   hence it seems that, on the one hand, slight   CHANGES in the conditions of life benefit all organi
0267   the offspring. But we have seen that greater   CHANGES, or changes of a particular nature, often re
0267   ng. But we have seen that greater changes, or   CHANGES of a particular nature, often render organic
0277   fertility of their offspring! and that slight   CHANGES in the conditions of life are apparently fav
0282   for so great an amount of organic change, all   CHANGES having been effected very slowly through nat
0290   owhere be ascertained. The frequent and great   CHANGES in the mineralogical composition of consecut
0290   secutive formations, generally implying great   CHANGES in the geography of the surrounding lands, w
0293   amount of migration during climatal and other   CHANGES; and when we see a species first appearing i
0294   period; and likewise to reflect on the great   CHANGES of level, on the inordinately great change o
0294   flourish; for we know what vast geographical   CHANGES occurred in other parts of America during th
0294   the migration of species and to geographical   CHANGES. And in the distant future, a geologist exam
0296   will generally have been due to geographical   CHANGES requiring much time. Nor will the closest in
0296   they grew, of many long intervals of time and   CHANGES of level during the process of deposition, w
0296   them, but abrupt, though perhaps very slight,   CHANGES of form. It is all important to remember tha
0298   n any two forms, is small, for the successive   CHANGES are supposed to have been local or confined
0301   cillations of level, and that slight climatal   CHANGES would intervene during such lengthy periods:
0312   ingle species. On Extinction. On simultaneous   CHANGES in the forms of life throughout the world. O
0312   vations of Philippi in Sicily, the successive   CHANGES in the marine inhabitants of that island hav
0325   oduction of new ones, cannot be owing to mere   CHANGES in marine currents or other causes more or l
0325   ffect. It is, indeed, quite futile to look to   CHANGES of currents, climate, or other physical cond
0326   being dependent on climatal and geographical   CHANGES, or on strange accidents; but in the long ru
0340   ntervals of time and after great geographical   CHANGES, permitting much inter migration, the feeble
0346   . centres of creation. Means of dispersal, by   CHANGES of climate and of the level of the land, and
0351   for during the vast geographical and climatal   CHANGES which will have supervened since ancient tim
0353   the other. But the geographical and climatal   CHANGES, which have certainly occurred within recent
0354   h respect to former climatal and geographical   CHANGES and various occasional means of transport, t
0356   ss this branch of the subject in some detail.   CHANGES of level in the land must also have been hig
0357   rized in admitting such enormous geographical   CHANGES within the period of existing species. It se
0357   level in our continents: but not of such vast   CHANGES in their position and extension, as to have
```

Page **(Key Word)**

Page **(Key Word)**

```
0366  tantially as follows. But we shall follow the  CHANGES  more readily, by supposing a new glacial per
0370  may be surmounted by looking to still earlier   CHANGES  of climate of an opposite nature. We have go
0384  ispersal of fresh water fish mainly to slight   CHANGES  within the recent period in the level of the
0384  nce in the loess of the Rhine of considerable   CHANGES  of level in the land within a very recent ge
0384  l have been ample time for great geographical   CHANGES, and consequently time and means for much mi
0396  in degree on the lapse of time, and as during   CHANGES  of level it is obvious that islands separate
0405  mple time for great climatal and geographical   CHANGES  and for accidents of transport; and conseque
0407  our ignorance of the full effects of all the    CHANGES  of climate and of the level of the land, whi
0407  ithin the recent period, and of other similar   CHANGES  which may have occurred within the same peri
0407  rave. As exemplifying the effects of climatal   CHANGES  on distribution, I have attempted to show ho
0435  of growth other parts of the organisation. In   CHANGES  of this nature, there will be little or no t
0454  ther species under nature ever undergo abrupt   CHANGES. I believe that disuse has been the main age
0461  of all organic beings are increased by slight   CHANGES  in their conditions of life, and that the of
0461  ility. So that, on the one hand, considerable   CHANGES  in the conditions of life and crosses betwee
0461  ssen fertility; and on the other hand, lesser   CHANGES  in the conditions of life and crosses betwee
0462  tent of the various climatal and geographical   CHANGES  which have affected the earth during modern
0462  ted the earth during modern periods; and such   CHANGES  will obviously have greatly facilitated migr
0463  uncergoing change at any one period; and all    CHANGES  are slowly effected. I have also shown that
0466  ductive system being eminently susceptible to   CHANGES  in the conditions of life; so that this syst
0468  s that each land has undergone great physical   CHANGES, we might have expected that organic beings
0476  er, owing to former climatal and geographical   CHANGES  and to the many occasional and unknown means
0487  throws, and will continue to throw, on former   CHANGES  of climate and of the level of the land, we
0106  and will thus play an important part in the     CHANGING  history of the organic world. We can, perhap
0175  y no means exclusively depends on insensibly    CHANGING  physical conditions, but in large part on th
0181  can see no difficulty, more especially under    CHANGING  conditions of life, in the continued preserv
0186  the organ be ever useful to an animal under     CHANGING  conditions of life, then the difficulty of b
0189  e every part of this layer to be continually    CHANGING  slowly in density, so as to separate into la
0189  , and with the surfaces of each layer slowly    CHANGING  in form. Further we must suppose that there
0201  advantageous. After the lapse of time, under    CHANGING  conditions of life, if any part comes to be
0204  ity each good for its possessor, then, under     CHANGING  conditions of life, there is no logical impo
0243  al. Therefore I can see no difficulty, under     CHANGING  conditions of life, in natural selection acc
0310  the world imperfectly kept, and written in a    CHANGING  dialect; of this history we possess the last
0311  d there a few lines. Each word of the slowly     CHANGING  language, in which the history is supposed t
0314  sing, on the rate of breeding, on the slowly     CHANGING  physical conditions of the country, and more
0314  entical form much longer than others; or, if    CHANGING, that it should change less. We see the same
0315  l scene, taken almost at hazard, in a slowly     CHANGING  drama. We can clearly understand why a speci
0316  ance and disappearance as do single species,    CHANGING  more or less quickly, and in a greater or le
0322  es or group of species. On the Forms of Life    CHANGING  almost simultaneously throughout the World.
0324  logical sense. The fact of the forms of life    CHANGING  simultaneously, in the above large sense, at
0329  both regions the species have gone on slowly    CHANGING  during the accumulation of the several forma
0356  individuals will have gone on simultaneously    CHANGING, and the whole amount of modification will n
0406  tions just discussed, namely, low and slowly    CHANGING  organisms ranging more widely than the high,
0422  t languages; and all intermediate and slowly    CHANGING  dialects, had to be included; such an arreng
0469  e fail in selecting variations useful, under    CHANGING  conditions of life, to her living products?
0480  t species have changed, and are still slowly    CHANGING  by the preservation and accumulation of succ
0395  s good. We see Britain separated by a shallow   CHANNEL  from Europe, and the mammals are the same on
0401  the sea, in most cases wider than the British   CHANNEL, and there is no reason to suppose that they
0395  s facts on many islands separated by similar    CHANNELS  from Australia. The West Indian Islands stan
0396  is obvious that islands separated by shallow    CHANNELS  are more likely to have been continuously un
0396  he mainland than islands separated by deeper    CHANNELS, we can understand the frequent relation bet
0177  not at any one period present an inextricable   CHAOS  of varying and intermediate links: firstly, b
0462  nic beings blended together in an inextricable  CHAOS? With respect to existing forms, we should re
008   healthy, and the happy survive and multiply.    CHAPTER  IV. Natural Selection. Natural Selection, it
0004  hese considerations, I shall devote the first   CHAPTER  of this Abstract to Variation under Domestic
0004  are most favourable to variation. In the next   CHAPTER  the Struggle for Existence amongst all organ
0005  will be treated at some length in the fourth    CHAPTER; and we shall then see how Natural Selection
0005  e called Divergence of Character. In the next   CHAPTER  I shall discuss the complex and little known
0005  fection of the Geological Record. In the next   CHAPTER  I shall consider the geological succession o
0005  re and in an embryonic condition. In the last   CHAPTER  I shall give a brief recapitulation of the w
0007  f modification. Variation Under Domestication   CHAPTER  I. Causes of Variability. Effects of Habit.
0044  efficiently, is by far the predominant Power.   CHAPTER  II. Variation Under Nature. Variability. Ind
0044  pplying the principles arrived at in the last   CHAPTER  to organic beings in a state of nature, we m
0055  lants ranging widely will be discussed in our   CHAPTER  on geographical distribution. From looking a
0060  me divided into groups subordinate to groups.   CHAPTER  III. Struggle For Existence. Bears on natura
0060  tions. Before entering on the subject of this   CHAPTER, I must make a few preliminary remarks, to s
0060  tural Selection. It has been seen in the last   CHAPTER  that amongst organic beings in a state of na
0061  sults, as we shall more fully see in the next   CHAPTER, follow inevitably from the struggle for lif
0080  existence, discussed too briefly in the last   CHAPTER, act in regard to variation? Can the princip
0082  ave reason to believe, as stated in the first   CHAPTER, that a change in the conditions of life, by
0099  t to this subject we shall return in a future   CHAPTER. In the case of a gigantic tree covered with
0105  be found to be small, as we shall see in our    CHAPTER  on geographical distribution: yet of these s
0106  e facts which will be again alluded to in our   CHAPTER  on geographical distribution; for instance,
0109  s subject will be more fully discussed in our   CHAPTER  on Geology; but it must be here alluded to f
0110  nce of this, in the facts given in the second   CHAPTER, showing that it is the common species which
0110  aturally suffer most. And we have seen in the   CHAPTER  on the Struggle for Existence that it is the
0116  rge genus, because we have seen in the second   CHAPTER, that on an average more of the species of l
0124  xtinct remains. We shall, when we come to our   CHAPTER  on Geology, have to refer again to this subj
0126  i shall have to return to this subject in the   CHAPTER  on Classification, but I may add that on thi
0126  nd classes, as at the present day. Summary of   CHAPTER. If during the long course of ages and under
0131  s ever branching and beautiful ramifications.   CHAPTER  V. Laws of Variation. Effects of external co
0131  ral generations. I have remarked in the first   CHAPTER, but a long catalogue of facts which cannot
0134  isuse. From the facts alluded to in the first   CHAPTER, I think there can be little doubt that use
0155  owever, have to return to this subject in our   CHAPTER  on Classification. It would be almost superf
0168  o differ; but we have also seen in the second   CHAPTER  that the same principle applies to the whole
0171  each other, and the best adapted to survive.   CHAPTER  VI. Difficulties on Theory. Difficulties on
0172  re convenient to discuss this question in the   CHAPTER  on the Imperfection of the geological record
```

Page **(Key Word)**

Page ***************************************(Key Word)***

```
0177  ecies in each country, as shown in the second CHAPTER, presenting on an average a greater number o
0179  which are preserved, as we shall in a future CHAPTER attempt to show, in an extremely imperfect a
0192  but this case will be treated of in the next CHAPTER. The electric organs of fishes offer another
0195  ns would be of importance or not. In a former CHAPTER I have given instances of most trifling char
0203  a chance breeze on to the ovules? Summary of CHAPTER. We have in this chapter discussed some of t
0203  e ovules? Summary of Chapter. We have in this CHAPTER discussed some of the difficulties and objec
0204  d exterminating it. We have have seen in this CHAPTER how cautious we should be in concluding that
0207  of former adaptations, that of Unity of Type. CHAPTER VII. Instinct. Instincts comparable with hab
0242  . summary. I have endeavoured briefly in this CHAPTER to show that the mental qualities of our dom
0243  i do not pretend that the facts given in this CHAPTER strengthen in any great degree my theory; bu
0245  , let the strongest live and the weakest die. CHAPTER VIII. Hybridism. Distinction between the ste
0267  deed, from the facts alluded to in our fourth CHAPTER, that a certain amount of crossing is indisp
0276  ion between species and varieties. Summary of CHAPTER. First crosses between forms sufficiently di
0278  inally, then, the facts briefly given in this CHAPTER do not seem to me opposed to, but even rathe
0279  al distinction between species and varieties. CHAPTER IX. On the Imperfection of the Geological Re
0279  west known fossiliferous strata. In the sixth CHAPTER I enumerated the chief objections which migh
0298  bject I shall have to return in the following CHAPTER. One other consideration is worth notice: wi
0310  . of this volume, only here and there a short CHAPTER has been preserved; and of each page, only h
0312  ed are greatly diminished, or even disappear. CHAPTER X. On the Geological Succession of Organic B
0314  conditions of life, as explained in a former CHAPTER. When many of the inhabitants of a country h
0316  to the present day. We have seen in the last CHAPTER that the species of a group sometimes falsel
0331  e reader to turn to the diagram in the fourth CHAPTER. We may suppose that the numbered letters re
0336  t to find, as I attempted to show in the last CHAPTER, in any one or two formations all the interm
0338  the theory of natural selection. In a future CHAPTER I shall attempt to show that the adult diffe
0346  round us, and preserved by Natural Selection. CHAPTER XI. Geographical Distribution. Present distr
0349  look back to past ages, as shown in the last CHAPTER, and we find American types then prevalent o
0351  case. I believe, as was remarked in the last CHAPTER, in no law of necessary development. As the
0351  sical conditions. As we have seen in the last CHAPTER that some forms have retained nearly the sam
0352  first produced: for, as explained in the last CHAPTER, it is incredible that individuals identical
0354  rctic regions; and secondly (in the following CHAPTER), the wide distribution of fresh water produ
0383  rmer inhabitants of the surrounding lowlands. CHAPTER XII. Geographical Distribution, continued. D
0411  lated by the same power of natural selection. CHAPTER XIII. Mutual Affinities of Organic Beings. M
0420  ble of referring to the diagram in the fourth CHAPTER. We will suppose the letters A to L to repre
0429  nly two or three orders of small size. In the CHAPTER on geological succession I attempted to show
0431  ss. Extinction, as we have seen in the fourth CHAPTER, has played an important part in defining an
0444  mely early period. I have stated in the first CHAPTER, that there is some evidence to render it pr
0455  iple, also, of economy, explained in a former CHAPTER, by which the materials forming any part or
0456  by the laws of inheritance. Summary. In this CHAPTER I have attempted to show, that the subordina
0457  s of facts which have been considered in this CHAPTER, seem to me to proclaim so plainly, that the
0459  were unsupported by other facts or arguments. CHAPTER XIV. Recapitulation and Conclusion. Recapitu
0460  n of the facts given at the end of the eighth CHAPTER, which seem to me conclusively to show that
0465  ly recur to the hypothesis given in the ninth CHAPTER. That the geological record is imperfect all
0467  of naturalisation, as explained in the third CHAPTER. More individuals are born than can possibly
0005  orrelation of growth. In the four succeeding CHAPTERS, the most apparent and gravest difficulties
0127  d balance of evidence given in the following CHAPTERS. But we already see how it entails extinctio
0172  iscussed, Instinct and Hybridism in separate CHAPTERS. On the absence or rarity of transitional va
0207  nct might have been worked into the previous CHAPTERS; but I have thought that it would be more co
0311  s different in the interrupted succession of CHAPTERS, may represent the apparently abruptly chang
0312  same areas. Summary of preceding and present CHAPTERS. Let us now see whether the several facts an
0341  ndants. Summary of the preceding and present CHAPTERS. I have attempted to show that the geologica
0383  odification. Summary of the last and present CHAPTERS. As lakes and river systems are separated fr
0406  their new homes. Summary of last and present CHAPTERS. In these chapters I have endeavoured to sho
0406  mmary of last and present Chapters. In these CHAPTERS I have endeavoured to show, that if we make
0411  e different habits. In our second and fourth CHAPTERS, on Variation and on Natural Selection, I ha
0005  and induces what I have called Divergence of CHARACTER. In the next chapter I shall discuss the co
0007  f habit. Correlation of Growth. Inheritance. CHARACTER of Domestic Varieties. Difficulty of distin
0009  y assumes a new and sometimes very different CHARACTER from that of the rest of the plant. Such bu
0013  ould be, to look at the inheritance of every CHARACTER whatever as the rule, and non inheritance a
0014  run wild, gradually but certainly revert in CHARACTER to their aboriginal stocks. Hence it has he
0015  life do change, variations and reversions of CHARACTER probably do occur; but natural selection, a
0015  ace, as already remarked, less uniformity of CHARACTER than in true species. Domestic races of the
0016  ecies, also, often have a somewhat monstrous CHARACTER; by which I mean, that, although differing
0020  dividual mongrels, which present any desired CHARACTER; but that a race could be obtained nearly i
0026  inct breed, the tendency to reversion to any CHARACTER derived from such cross will naturally beco
0026  is a tendency in both parents to revert to a CHARACTER, which has been lost during some former gen
0028  he breeds so often have a somewhat monstrous CHARACTER. It is also a most favourable circumstance
0031  es the agriculturist, not only to modify the CHARACTER of his flock, but to change it altogether.
0036  barbarous as never to think of the inherited CHARACTER of the offspring of their domestic animals,
0038  e varieties kept by savages have more of the CHARACTER of species than the varieties kept in civil
0038  , further understand the frequently abnormal CHARACTER of our domestic races, and likewise their d
0039  l size; and the more abnormal or unusual any CHARACTER was when it first appeared, the more likely
0046  for these same authors practically rank that CHARACTER as important (as some few naturalists have
0047  hich possess in some considerable degree the CHARACTER of species, but which are so closely simila
0049  t judges as varieties, have so perfectly the CHARACTER of species that they are ranked by other hi
0053  to the struggle for existence, divergence of CHARACTER, and other questions, hereafter to be discu
0057  he principle, as I call it, of Divergence of CHARACTER, we shall see how this may be explained, an
0080  n caused by Natural Selection. Divergence of CHARACTER, related to the diverstiy of inhabitants of
0080  of Natural Selection, through Divergence of CHARACTER and Extinction, on the descendants from a c
0083  of the being which she tends. Every selected CHARACTER is fully exercised by her; and the being is
0083  e country; he seldom exercises each selected CHARACTER in some peculiar and fitting manner; he fee
0084  nature's productions should be far truer in CHARACTER than man's productions; that they should be
0088  rds, the contest is often of a more peaceful CHARACTER. All those who have attended to the subject
0103  or of the same variety, true and uniform in CHARACTER. It will obviously thus act far more effici
0104  beings which never intercross, uniformity of CHARACTER can be retained amongst them, as long as th
0104  and they undergo modification,uniformity of CHARACTER can be given to their modified offspring, s
0111  by some murderous pestilence. Divergence of CHARACTER. The principle, which I have designated by
```

Page ***************************************(Key Word)***

Page **(Key Word)**

```
0111  y marked ones, though having somewhat of the  CHARACTER  of species, as is shown by the hopeless dou
0111  t, might cause one variety to differ in some   CHARACTER  from its parents, and the offspring of this
0111  n to differ from its parent in the very same   CHARACTER  and in a greater degree; but this alone wou
0112  ly to increase, and the breeds to diverge in   CHARACTER  both from each other and from their common
0116  eat benefit being derived from divergence of   CHARACTER, combined with the principles of natural se
0117  of benefit being derived from divergence of    CHARACTER  comes in; for this will generally lead to t
0118  go on increasing in number and diverging in    CHARACTER. In the diagram the process is represented
0119  ultiplying in number as well as diverging in   CHARACTER: this is represented in the diagram by the
0120  apparently both gone on slowly diverging in    CHARACTER  from their original stocks, without either
0120  f10, and m10, which, from having diverged in   CHARACTER  during the successive generations, will hav
0121  es, which are already extremely different in   CHARACTER, will generally tend to produce the greates
0123  lection, the extreme amount of difference in   CHARACTER  between species a14 and z14 will be much or
0124  s worth while to reflect for a moment on the   CHARACTER  of the new species f14, which is supposed n
0124  ich is supposed not to have diverged much in   CHARACTER, but to have retained the form of (F), eith
0124  n, it will be in some degree intermediate in   CHARACTER  between the two groups descended from these
0124  s these two groups have gone on diverging in   CHARACTER  from the type of their parents, the new spe
0124  t are often, in some degree, intermediate in   CHARACTER  between existing groups; and we can underst
0125  wo genera, both from continued divergence of   CHARACTER  and from inheritance from a different paren
0128  ural selection, also, leads to divergence of   CHARACTER: for more living beings can be supported on
0128  s just been remarked, leads to divergence of   CHARACTER  and to much extinction of the less improved
0129  tion, entailing extinction and divergence of   CHARACTER, as we have seen illustrated in the diagram
0150  ore rarely offer remarkable secondary sexual   CHARACTER, it applies more rarely to them. The rule b
0152  le breed will cease to have a nearly uniform   CHARACTER. The breed will then be said to have degene
0154  had red, the colour would by only a specific   CHARACTER, and no one would be surprised at one of th
0154  e flowers, the colour would become a generic   CHARACTER, and its variation would be a more unusual
0155  e of the species. And this fact shows that a   CHARACTER, which is generally of generic value, when
0157  the same number of joints in the tarsi is a   CHARACTER  generally common to very large groups of be
0157  a, the manner of neuration of the wings is a   CHARACTER  of the highest importance, because common t
0160  ng occasionally show a tendency to revert in   CHARACTER  to the foreign breed for many generations,
0160  ed, but in which both parents have lost some   CHARACTER  which their progenitor possessed, the tende
0160  hether strong or weak, to reproduce the lost   CHARACTER  might be, as was formerly remarked, for all
0160  for almost any number of generations. When a   CHARACTER  which has been lost in a breed, reappears a
0161  n there has been a tendency to reproduce the   CHARACTER  in question, which at last, under unknown f
0161  t improbability in a tendency to produce any   CHARACTER  being inherited for an endless number of ge
0161  acters. As, however, we never know the exact   CHARACTER  of the common ancestor of a group, we could
0162  ases are reversions to an anciently existing   CHARACTER  and what are new but analogous variations,
0162  arying so as to acquire, in some degree, the   CHARACTER  of the same part or organ in an allied spec
0163  case, not indeed as affecting any important   CHARACTER, but from occurring in several species of t
0166  ed by any change of form or by any other new   CHARACTER. We see this tendency to become striped mos
0166  eari but without any other change of form or   CHARACTER. When the oldest and truest breeds of vario
0166  account for the reappearance of very ancient   CHARACTER, is, that there is a tendency in the young
0166  ccessive generation to produce the long lost   CHARACTER, and that this tendency, from unknown cause
0169  may readily have succeeded in giving a fixed   CHARACTER  to the organ, in however extraordinary a a
0197  its enemies; and consequently that it was a   CHARACTER  of importance and might have been acquired
0237  s. hence I can see no real difficulty in any   CHARACTER  having become correlated with the sterile c
0239  neuters ultimately came to have the desired   CHARACTER. On this view we ought occasionally to find
0257  at amount, of difference in any recognisable   CHARACTER  is sufficient to prevent two species crossi
0259  tead of having, as is usual, an intermediate   CHARACTER  between their two parents, always closely r
0273  ds and mongrels long retaining uniformity of   CHARACTER  could be given. The variability, however, i
0275  nsequently, sudden reversions to the perfect   CHARACTER  of either parent would be more likely to oc
0275  ften suddenly produced and semi monstrous in   CHARACTER, than with hybrids, which are descended fro
0331  e of the continued tendency to divergence of   CHARACTER, which was formerly illustrated by this dia
0331  must not, however, assume that divergence of   CHARACTER  is a necessary contingency; it depends sole
0332  nued effects of extinction and divergence of   CHARACTER, has become divided into several sub famili
0332  of three families together, intermediate in   CHARACTER, would be justified, as they are intermedia
0333  y stage of descent they have not diverged in   CHARACTER  from the common progenitor of the order, ne
0333  often in some slight degree intermediate in   CHARACTER  between their modified descendants, or betw
0334  th's history will be intermediate in general   CHARACTER  between that which preceded and that which
0334  uld hardly fail to be nearly intermediate in   CHARACTER  between the forms of life above and below.
0334  ogical period undoubtedly is intermediate in   CHARACTER, between the preceding and succeeding fauna
0334  nised by palaeontologists as intermediate in   CHARACTER  between those of the overlying carboniferou
0334  period as a whole is nearly intermediate in   CHARACTER  between the preceding and succeeding faunas
0334  ccord in arrangement. The species extreme in   CHARACTER  are not the oldest, or the most recent; nor
0334  enti nor are those which are intermediate in   CHARACTER, intermediate in age. But supposing for an
0335  carriers which are extreme in the important   CHARACTER  of length of beak originated earlier than s
0335  formation are in some degree intermediate in   CHARACTER, is the fact, insisted on by all palaeontol
0338  for animals having the common embryological   CHARACTER  of the Vertebrata, until beds far beneath t
0340  ica formerly partook strongly of the present   CHARACTER  of the southern half of the continent; and
0344  from the continued tendency to divergence of   CHARACTER, why the more ancient a form is, the more i
0345  n intermediate formation are intermediate in   CHARACTER. The inhabitants of each successive period
0351  hat some forms have retained nearly the same   CHARACTER  from an enormously remote geological period
0363  mr. H. C. Watson) from the somewhat northern   CHARACTER  of the flora in comparison with the latitud
0398  of these birds to American species in every   CHARACTER, in their habits, gestures, and tones of vo
0401  owever, might spread and yet retain the same   CHARACTER  throughout the group, just as we see on con
0403  . the principle which determines the general   CHARACTER  of the fauna and flora of oceanic islands,
0406  ely and of still retaining the same specific   CHARACTER. This fact, together with the seeds and egg
0412  ch are increasing in number and diverging in   CHARACTER, to supplant and exterminate the less diver
0414  o mistake a merely adaptive for an essential   CHARACTER. So with plants, how remarkable it is that
0416  age from the nostrils to the mouth, the only   CHARACTER, according to Owen, which absolutely distin
0416  instead of hair, this external and trifling   CHARACTER  would, I think, have been considered by nat
0417  that a classification founded on any single   CHARACTER, however important that may be, has always
0418  ating any particular species. If they find a   CHARACTER  nearly uniform, and common to a great numbe
0419  site ends of the series, which have hardly a   CHARACTER  in common; yet the species at both ends, fr
0421  sition: for F originally was intermediate in   CHARACTER  between A and I, and the several genera des
0424  ties differ from the others in the important   CHARACTER  of having a longer beak, yet all are kept t
0426  the organisation. We care not how trifling a   CHARACTER  may be, let it be the mere inflection of th
```

Page **(Key Word)**

Page ***(Key Word)***

```
0426  of descent. Let two forms have not a single  CHARACTER in common, yet if these extreme forms are c
0427  learly understand why analogical or adaptive  CHARACTER, although of the utmost importance to the w
0429  each group having generally diverged much in  CHARACTER during the long continued process of modifi
0430  ery ancient Marsupial, which will have had a  CHARACTER in some degree intermediate with respect to
0430  ha has retained, by inheritance, more of the  CHARACTER of its ancient progenitor than have other R
0430  rsupials, from having partially retained the  CHARACTER of their common progenitor, or of an early
0430  the multiplication and gradual divergence in  CHARACTER of the species descended from a common pare
0433  vitably induces extinction and divergence of  CHARACTER in the many descendants from one dominant p
0442  there is no metamorphosis; the cephalopodic  CHARACTER is manifested long before the parts of the
0456  ontingencies of extinction and divergence of  CHARACTER. In considering this view of classification
0467  is certain that he can largely influence the  CHARACTER of a breed by selecting, in each successive
0467  s produced by man have to a large extent the  CHARACTER of natural species, is shown by the inextri
0470  ient species, retain to a certain degree the  CHARACTER of varieties: for they differ from each oth
0471  rger, and at the same time more divergent in  CHARACTER. But as all groups cannot thus succeed in i
0471  to go on increasing in size and diverging in  CHARACTER, together with the almost inevitable contin
0475  rmation being in some degree intermediate in  CHARACTER between the fossils in the formations above
0476  ncient progenitor have generally diverged in  CHARACTER, the progenitor with its early descendants
0476  ly descendants will often be intermediate in  CHARACTER in comparison with its later descendants; a
0478  ontingencies of extinction and divergence of  CHARACTER. On these same principles we see how it is,
0490  o natural Selection, entailing Divergence of  CHARACTER and the Extinction of less improved forms.
0094  or other insect, so that an individual so  CHARACTERISED would be able to obtain its food more qui
0127  being do occur, assuredly individuals thus  CHARACTERISED will have the best chance of being preser
0127  y will tend to produce offspring similarly  CHARACTERISED. This principle of preservation, I have c
0323  ons of genera, and sometimes are similarly  CHARACTERISED in such trifling points as mere superfici
0410  d of time, or to a certain area, are often  CHARACTERISED by trifling characters in common, as of s
0023  ss one of the parent stocks possessed the  CHARACTERISTIC enormous crop? The supposed aboriginal st
0040  of the inhabitants, slowly to add to the  CHARACTERISTIC features of the breed, whatever they may
0160  bases with white. As all these marks are  CHARACTERISTIC of the parent rock pigeon, I presume that
0198  that, if we are unable to account for the  CHARACTERISTIC differences of our domestic breeds, which
0254  st look at sterility, not as an indelible  CHARACTERISTIC, but as one capable of being removed by d
0280  panded with a crop somewhat enlarged, the  CHARACTERISTIC features of these two breeds. These two b
0316  herit from its new progenitor some slight  CHARACTERISTIC differences. Groups of species, that is,
0348  nces are very inferior in degree to those  CHARACTERISTIC of distinct continents. Turning to the se
0375  ong ago found species belonging to genera  CHARACTERISTIC of the Cordillera. On the mountains of Ab
0437  n of the same part or organ is the common  CHARACTERISTIC (as Owen has observed) of all low or litt
0445  y less than in the full grown birds. Some  CHARACTERISTIC points of difference, for instance, that
0446  ly that of pigeons, seem to show that the  CHARACTERISTIC differences which give value to each bree
0446  s is not the universal rule; for here the  CHARACTERISTIC differences must either have appeared at
0470  of modification, the slight differences,  CHARACTERISTIC of varieties of the same species, tend to
0470  be augmented into the greater differences  CHARACTERISTIC of species of the same genus. New and imp
0482  nd which consequently have every external  CHARACTERISTIC feature of true species, they admit that
0332  throughout a vast period the same general  CHARACTERISTICS. This is represented in the diagram by t
0426  depart, in several of its most important  CHARACTERISTICS, from its allies; and yet be safely clas
0048  ira group there are many insects which are  CHARACTERIZED as varieties in Mr. Wollaston's admirable
0013  so; why the child often reverts in certain  CHARACTERS to its grandfather or grandmother or other
0015  nly do occasionally revert in some of their  CHARACTERS to ancestral forms, it seems to me not impr
0015  reversion, that is, to lose their acquired  CHARACTERS, whilst kept under unchanged conditions, an
0015  e explained, will determine how far the new  CHARACTERS thus arising shall be preserved. When we lo
0016  stic races do not differ from each other in  CHARACTERS of generic value. I think it could be shown
0016  ists differ most widely in determining what  CHARACTERS are of generic value: all such valuations b
0019  race which breeds true, let the distinctive  CHARACTERS be ever so slight, has had its wild prototy
0023  ns are known: and these have not any of the  CHARACTERS of the domestic breeds. Hence the supposed
0023  sidering their size, habits, and remarkable  CHARACTERS, seems very improbable; or they must have b
0025  ring are very apt suddenly to acquire these  CHARACTERS; for instance, I crossed some uniformly whi
0025  l known principle of reversion to ancestral  CHARACTERS, if all the domestic breeds have descended
0027  e feral: these species having very abnormal  CHARACTERS in certain respects, as compared with all o
0027  faced tumbler differs immensely in certain  CHARACTERS from the rock pigeon, yet by comparing the
0027  n the extremes of structure. Thirdly, those  CHARACTERS which are mainly distinctive of each breed,
0033  differing from each other chiefly in these  CHARACTERS. It may be objected that the principle of s
0038  heir differences being so great in external  CHARACTERS and relatively so slight in internal parts
0040  iffer only in colour, that most fleeting of  CHARACTERS, have lately been exhibited as distinct at
0045  pleased at finding variability in important  CHARACTERS, and that there are not many men who will l
0046  some of the species have fixed and definite  CHARACTERS. Genera which are polymorphic in one countr
0047  llied forms have permanently retained their  CHARACTERS in their own country for a long time: for a
0047  orms together by others having intermediate  CHARACTERS, he treats the one as a variety of the othe
0058  ally, then, varieties have the same general  CHARACTERS as species, for they cannot be distinguishe
0058  ence of such links cannot affect the actual  CHARACTERS of the forms which they connect: and except
0080  compared with man's selection, its power on  CHARACTERS of trifling importance, its power at all ag
0083  t? man can act only on external and visible  CHARACTERS: nature cares nothing for appearances, exce
0084  through and for the good of each being, yet  CHARACTERS and structures, which we are apt to conside
0085  of the flesh are considered by botanists as  CHARACTERS of the most trifling importance: yet we hea
0112  ter, the inferior animals with intermediate  CHARACTERS, being neither very swift nor very strong,
0127  offspring. Sexual selection will also give  CHARACTERS useful to the males alone, in their struggl
0131  nusual manner are highly variable: specific  CHARACTERS more variable than generic: secondary sexua
0131  ore variable than generic: secondary sexual  CHARACTERS variable. Species of the same genus vary in
0131  n analogous manner. Reversions to long lost  CHARACTERS. Summary. I have hitherto sometimes spoken
0133  quiring in a very slight degree some of the  CHARACTERS of such species, accords with our view that
0150  ry strongly in the case of secondary sexual  CHARACTERS, when displayed in any unusual manner. The
0150  unusual manner. The term, secondary sexual  CHARACTERS, used by Hunter, applies to characters whic
0150  xual characters, used by Hunter, applies to  CHARACTERS which are attached to one sex, but are not
0151  applicable in the case of secondary sexual  CHARACTERS, may be due to the great variability of the
0151  ay be due to the great variability of these  CHARACTERS, whether or not displayed in any unusual ma
0151  ur rule is not confined to secondary sexual  CHARACTERS is clearly shown in the case of hermaphrodi
0151  arieties differ more from each other in the  CHARACTERS of these important valves than do other spe
0153  further deserves notice that these variable  CHARACTERS, produced by man's selection, sometimes bec
0154  be extended. It is notorious that specific  CHARACTERS are more variable than generic. To explain
```

Page **(Key Word)***************************************

```
0154  alists would advance, namely, that specific  CHARACTERS are more variable than generic, because the
0155  pport of the above statement, that specific  CHARACTERS are more variable than generic; but I have
0155  ies of some other genus, are called generic  CHARACTERS; and these characters in common I attribute
0155  s, are called generic characters; and these  CHARACTERS in common I attribute to inheritance from a
0156  same manner: and as these so called generic  CHARACTERS have been inherited from a remote period, s
0156  cies of the same genus, are called specific  CHARACTERS; and as these specific characters have vari
0156  specific characters; and as these specific  CHARACTERS have varied and come to differ within the p
0156  entering on details, that secondary sexual  CHARACTERS are very variable; I think it also will be
0156  other more widely in their secondary sexual  CHARACTERS, than in other parts of their organisation;
0156  llinaceous birds, in which secondary sexual  CHARACTERS are strongly displayed, with the amount of
0156  he original variability of secondary sexual  CHARACTERS is not manifest; but we can see why these c
0156  s is not manifest; but we can see why these  CHARACTERS should not have been rendered as constant a
0156  s of the organisation: for secondary sexual  CHARACTERS have been accumulated by sexual selection,
0157  y be of the variability of secondary sexual  CHARACTERS, as they are highly variable, sexual select
0157  reater amount of difference in their sexual  CHARACTERS, than in other parts of their structure. It
0158  de that the greater variability of specific  CHARACTERS, or those which distinguish species from sp
0158  guish species from species, than of generic  CHARACTERS, or those which the species possess in comm
0158  t the great variability of secondary sexual  CHARACTERS, and the great amount of difference in thes
0158  he great amount of difference in these same  CHARACTERS between closely allied species; that second
0159  ty of one species often assumes some of the  CHARACTERS of an allied species, or reverts to some of
0159  n allied species, or reverts to some of the  CHARACTERS of an early progenitor. These propositions
0159  thers on the head and feathers on the feet,  CHARACTERS not possessed by the aboriginal rock pigeon
0160  no doubt it is a very surprising fact that  CHARACTERS should reappear after having been lost for
0161  f one species would resemble in some of its  CHARACTERS another species; this other species being o
0161  ly a well marked and permanent variety. But  CHARACTERS thus gained would probably be of an importa
0161  t nature, for the presence of all important  CHARACTERS will be governed by natural selection, in a
0161  onally exhibit reversions to lost ancestral  CHARACTERS. As, however, we never know the exact chara
0162  wned, we could not have told, whether these  CHARACTERS in our domestic breeds were reversions or o
0162  the varying offspring of a species assuming  CHARACTERS (either from reversion or from analogous va
0162  the one in varying has assumed some of the  CHARACTERS of the other, so as to produce the intermed
0168  , and hence probably are variable. Specific  CHARACTERS, that is, the characters which have come to
0168  variable. Specific characters, that is, the  CHARACTERS which have come to differ since the several
0168  mmon parent, are more variable than generic  CHARACTERS, or those which have long been inherited, a
0169  ties or incipient species. Secondary sexual  CHARACTERS are highly variable, and such characters di
0169  al characters are highly variable, and such  CHARACTERS differ much in the species of the same grou
0169  cies may occasionally revert to some of the  CHARACTERS of their ancient progenitors. Although new
0195  ter I have given instances of most trifling  CHARACTERS, such as the down on fruit and the colour o
0196  e, we may sometimes attribute importance to  CHARACTERS which are really of very little importance,
0196  direct influence on the organisation; that  CHARACTERS reappear from the law of reversion; that co
0196  ll often have largely modified the external  CHARACTERS of animals having a will, to give one male
0199  ersified changes of no direct use. So again  CHARACTERS which formerly were useful, or which former
0208  nctive. But I could show that none of these  CHARACTERS of instinct are universal. A little dose, a
0236  ld have unhesitatingly assumed that all its  CHARACTERS had been slowly acquired through natural se
0271  s species: from man selecting only external  CHARACTERS in the production of the most distinct dome
0274  owing to the existence of secondary sexual  CHARACTERS; but more especially owing to prepotency in
0275  , the resemblances seem chiefly confined to  CHARACTERS almost monstrous in their nature, and which
0275  onal fingers and toes: and do not relate to  CHARACTERS which have been slowly acquired by selectio
0306  which they belong, for they do not present  CHARACTERS in any degree intermediate between them. If
0315  th would almost certainly inherit different  CHARACTERS from their distinct progenitors. For instan
0330  ct form is directly intermediate in all its  CHARACTERS between two living forms, the objection is
0330  the present day from each other by a dozen  CHARACTERS, the ancient members of the same two groups
0330  istinguished by a somewhat lesser number of  CHARACTERS, so that the two groups, though formerly qu
0330  the more it tends to connect by some of its  CHARACTERS groups now widely separated from each other
0333  milies to differ from each other by a dozen  CHARACTERS, in this case the genera, at the early peri
0333  ked VI., would differ by a lesser number of  CHARACTERS; for at this early stage of descent they ha
0333  iffer less from each other in some of their  CHARACTERS than do the existing members of the same gr
0341  ich are closely allied in size and in other  CHARACTERS to the species still living in South Americ
0344  is, the more often, apparently, it displays  CHARACTERS in some degree intermediate between groups
0410  n area, are often characterised by trifling  CHARACTERS in common, as of sculpture or colour. In lo
0411  d in classification. Analogical or adaptive  CHARACTERS. Affinities, general, complex and radiating
0412  ture, there is a constant tendency in their  CHARACTERS to diverge. This conclusion was supported b
0413  tions, that is, by one sentence to give the  CHARACTERS common, for instance, to all mammals, by an
0413  in a more or less concealed form, that the  CHARACTERS do not make the genus; but that the genus g
0413  ake the genus, but that the genus gives the  CHARACTERS; seem to imply that something more is inclu
0414  are ranked as merely adaptive or analogical  CHARACTERS; but to the consideration of these resembla
0415  e or valvular aestivation. Any one of these  CHARACTERS singly is frequently of more than generic i
0416  asses. Numerous instances could be given of  CHARACTERS derived from parts which must be considered
0417  importance, for classification, of trifling  CHARACTERS, mainly depends on their being correlated w
0417  n their being correlated with several other  CHARACTERS of more or less importance. The value indee
0417  rtance. The value indeed of an aggregate of  CHARACTERS is very evident in natural history. Hence,
0417  ecies may depart from its allies in several  CHARACTERS, both of high physiological importance and
0417  constant. The importance of an aggregate of  CHARACTERS, even when none are important, alone explai
0417  i think, that saying of Linnaeus, that the  CHARACTERS do not give the genus, but the genus gives
0417  not give the genus, but the genus gives the  CHARACTERS; for this saying seems founded on an apprec
0417  ieu has remarked, the greater number of the  CHARACTERS proper to the species; to the genus, to the
0418  selves about the physiological value of the  CHARACTERS which they use in defining a group, or in a
0418  lent botanist, Aug. St. Hilaire. If certain  CHARACTERS are always found correlated with others, th
0418  important vital organs, are found to offer  CHARACTERS of quite subordinate value. We can see why
0418  of quite subordinate value. We can see why  CHARACTERS derived from the embryo should be of equal
0418  , milne Edwards and Agassiz, that embryonic  CHARACTERS are the most important of any in the classi
0418  the two main divisions have been founded on  CHARACTERS derived from the embryo, on the number and
0419  ussion on embryology, we shall see why such  CHARACTERS are so valuable, on the view of classificat
0419  ng can be easier than to define a number of  CHARACTERS common to all birds; but in the case of cru
0420  nded on descent with modification; that the  CHARACTERS which naturalists consider as showing true
0422  ll have inherited to a certain extent their  CHARACTERS. This natural arrangement is shown, as far
0424  might differ in colour and other important  CHARACTERS from negroes. With species in a state of na
```

0424 hese sometimes differ in the most important CHARACTERS, is known to every naturalist: scarcely a s
0425 nces of any kind. Therefore we choose those CHARACTERS which, as far as we can judge, are the leas
0426 ingle points of structure, but when several CHARACTERS, let them be ever so trifling, occur togeth
0426 sure, on the theory of descent, that these CHARACTERS have been inherited from a common ancestor.
0426 we know that such correlated or aggregated CHARACTERS have especial value in classification. We c
0426 ten done, as long as a sufficient number of CHARACTERS, let them be ever so unimportant, betrays t
0426 ter, I think, clearly see why embryological CHARACTERS are of such high classificatory importance.
0427 etween very distinct animals. On my view of CHARACTERS being of real importance for classification
0427 nd the apparent paradox, that the very same CHARACTERS are analogical when one class or order is c
0428 ape of the body and fin like limbs serve as CHARACTERS exhibiting true affinity between the severa
0428 amily; for these cetaceans agree in so many CHARACTERS, great and small, that we cannot doubt that
0429 he more ancient forms of life often present CHARACTERS in some slight degree intermediate between
0431 with their retention by inheritance of some CHARACTERS in common, we can understand the excessivel
0431 b groups, will have transmitted some of its CHARACTERS, modified in various ways and degrees, to a
0432 t types, or forms, representing most of the CHARACTERS of each group, whether large or small, and
0433 sexes and of all ages, although having few CHARACTERS in common, under one species: we use descen
0433 we summarily reject analogical or adaptive CHARACTERS, and yet use these same characters within t
0433 adaptive characters, and yet use these same CHARACTERS within the limits of the same group. We can
0439 for instance, of a crab retaining numerous CHARACTERS, which they would probably have retained th
0443 st be quite unimportant whether most of its CHARACTERS are fully acquired a little earlier or late
0456 n their classifications; the value set upon CHARACTERS, if constant and prevalent, whether of high
0456 ion in value between analogical or adaptive CHARACTERS, and characters of true affinity; and other
0456 ween analogical or adaptive characters, and CHARACTERS of true affinity: and other such rules: all
0457 nticipated. The importance of embryological CHARACTERS and of rudimentary organs in classification
0468 ed quantity. Man, though acting on external CHARACTERS alone and often capriciously, can produce w
0472 zone, they occasionally assume some of the CHARACTERS of the species proper to that zone. In both
0473 rieties and species reversions to long lost CHARACTERS occur. How inexplicable on the theory of cr
0473 ependently created, why should the specific CHARACTERS, or those by which the species of the same
0473 ch other, be more variable than the generic CHARACTERS in which they all agree? Why, for instance,
0473 re only well marked varieties, of which the CHARACTERS have become in a high degree permanent, we
0473 hed off from a common progenitor in certain CHARACTERS, by which they have come to be specifically
0474 t from each other; and therefore these same CHARACTERS would be more likely still to be variable t
0474 ikely still to be variable than the generic CHARACTERS which have been inherited without change fo
0478 complex and circuitous. We see why certain CHARACTERS are far more serviceable than others for cl
0478 han others for classification; why adaptive CHARACTERS, though of paramount importance to the bein
0479 ardly any importance in classification; why CHARACTERS derived from rudimentary parts, though of n
0479 classificatory value: and why embryological CHARACTERS are the most valuable of all. The real affi
0479 the lines of descent by the most permanent CHARACTERS, however slight their vital importance may
0485 ty of type, paternity, morphology, adaptive CHARACTERS, rudimentary and aborted organs, etc., will
0486 s of descent in our natural genealogies, by CHARACTERS of any kind which have long been inherited.
0283 nd the waves eat into them only when they are CHARGED with sand or pebbles; for there is reason to
0289 in his own country, great piles of sediment, CHARGED with new and peculiar forms of life, had els
0363 given showing that soil almost everywhere is CHARGED with seeds. Reflect for a moment on the mill
0386 i do not believe that botanists are aware how CHARGED the mud of ponds is with seeds: I have tried
0463 pecies, why is not every geological formation CHARGED with such links? Why does not every collecti
0002 with a request that I would forward it to Sir CHARLES Lyell, who sent it to the Linnean Society, a
0035 mproved. There is reason to believe that King CHARLES's spaniel has been unconsciously modified to
0095 ections which were at first urged against Sir CHARLES Lyell's noble views on the modern changes of
0282 rehend the lapse of time. He who can read Sir CHARLES Lyell's grand work on the Principles of Geol
0310 ason to believe that one great authority, Sir CHARLES Lyell, from further reflexion entertains gra
0385 nic island or to any other distant point. Sir CHARLES Lyell also informs me that a Dyticus has bee
0402 cking thrush of Chatham Island to be blown to CHARLES Island, which has its own mocking thrush: wh
0402 ishing itself there? We may safely infer that CHARLES Island is well stocked with its own species,
0402 may infer that the mocking thrush peculiar to CHARLES Island is at least as well fitted for its ho
0076 before it its great congener. One species of CHARLOCK will supplant another, and so in other cases
0199 sexual selection, when displayed in beauty to CHARM the females, can be called useful only in rat
0196 an advantage in fighting with another or in CHARMING the females. Moreover when a modification of
0090 males, in their weapons, means of defence, or CHARMS; and have transmitted these advantages to the
0468 pecial weapons or means of defence, or on the CHARMS of the males: and the slightest advantage wil
0184 n fruit, and others with elongated wings which CHASE insects on the wing: and on the plains of La
0402 and. Now let us suppose the mocking thrush of CHATHAM Island to be blown to Charles Island, which
0402 ed for its home as is the species peculiar to CHATHAM Island. Sir C. Lyell and Mr. Wollaston have
0015 derable body, so that free intercrossing might CHECK, by blending together, any slight deviations
0066 eration or at recurrent intervals. Lighten any CHECK, mitigate the destruction ever so little, and
0070 mals, often ensue: and here we have a limiting CHECK independent of the struggle for life. But eve
0074 ns or years, probably come into play: some one CHECK or some few being generally the most potent,
0082 climate, or any unusual degree of isolation to CHECK immigration, is actually necessary to produce
0149 refore to natural selection having no power to CHECK deviations in their structure. Thus rudimenta
0322 tends to increase inordinately, and that some CHECK is always in action, yet seldom perceived by
0065 the geometrical tendency to increase must be CHECKED by destruction at some period of life. Our f
0070 y to the supply of seed, as their numbers are CHECKED during winter: but any one who has tried, kn
0072 ies, numerous as they are, must be habitually CHECKED by some means: probably by birds. Hence, if
0075 lants which first clothed the ground and thus CHECKED the growth of the trees! Throw up a handful
0107 nes of the range of each species will thus be CHECKED: after physical changes of any kind, immigra
0108 migration of better adapted forms having been CHECKED. But the action of natural selection will pr
0168 hat their modifications have not been closely CHECKED by natural selection. It is probably from th
0196 ions in their structure will always have been CHECKED by natural selection. Seeing how important a
0319 d what the unfavourable conditions were which CHECKED its increase, whether some one or several co
0319 ase of every living being is constantly being CHECKED by unperceived injurious agencies: and that
0455 ell be variable, for its variations cannot be CHECKED by natural selection. At whatever period of
0104 isolation probably acts more efficiently in CHECKING the immigration of better adapted organisms,
0105 ture and constitution. Lastly, isolation, by CHECKING immigration and consequently competition, wi
0350 ch importance of barriers comes into play by CHECKING migration; as does time for the slow process
0403 ion has probably played an important part in CHECKING the commingling of species under the same co
0060 naturalised animals and plants. Nature of the CHECKS to increase. Competition universal. Effects o
0067 ck, and then another with greater force. What CHECKS the natural tendency of each species to incre

```
0067 rther increased. We know not exactly what the CHECKS are in even one single instance. Nor will thi
0067 shall, in my future work, discuss some of the CHECKS at considerable length, more especially in re
0068 ht, I believe to be the most effective of all CHECKS. I estimated that the winter of 1854 55 destr
0071 rd showing how complex and unexpected are the CHECKS and relations between organic beings, which h
0074 ! in the case of many species, many different CHECKS, acting at different periods of life, and dur
0074 e cases it can be shown that widely different CHECKS act on the same species in different district
0485 ons made for convenience. This may not be a CHEERING prospect; but we shall have at least be free
0484 living things have much in common, in their CHEMICAL composition, their germinal vesicles, their
0025 have, besides the two black bars, the wings CHEQUERED with black. These several marks do not occu
0198 me reason to believe, increase the size of the CHEST; and again correlation would come into play.
0163 n duns, mouse duns, and in one instance in a CHESTNUT: a faint shoulder stripe may sometimes be se
0165 ripe. In Lord Moreton's famous hybrid from a CHESTNUT mare and male quagga, the hybrid, and even t
0086 ltry, and in the colour of the down of their CHICKENS: in the horns of our sheep and cattle when n
0215 heritance our dogs. On the other hand, young CHICKENS have lost, wholly by habit, that fear of the
0216 s, though reared under a hen. It is not that CHICKENS have lost all fear, but fear only of dogs an
0216 fly away. But this instinct retained by our CHICKENS has become useless under domestication, for
0008 veral reasons make me believe in this; but the CHIEF one is the remarkable effect which confinemen
0067 ust to recall to the reader's mind some of the CHIEF points. Eggs or very young animals seem gener
0077 in the midst of long grass, I suspect that the CHIEF use of the nutriment in the seed is to favour
0199 r the most important consideration is that the CHIEF part of the organisation of every being is si
0255 sterility of first crosses and of hybrids. Our CHIEF object will be to see whether or not the rule
0279 strata. In the sixth chapter I enumerated the CHIEF objections which might be justly urged agains
0345 ished or disappear. On the other hand, all the CHIEF laws of palaeontology plainly proclaim, as it
0400 red to me a great difficulty: but it arises in CHIEF part from the deeply seated error of consider
0465 other in time. Such is the sum of the several CHIEF objections and difficulties which may justly
0480 l not understand. I have now recapitulated the CHIEF facts and considerations which have thoroughl
0481 ecies, if they had undergone mutation. But the CHIEF cause of our natural unwillingness to admit t
0033 will produce races differing from each other CHIEFLY in these characters. It may be objected that
0035 his case the change has, it is believed, been CHIEFLY effected by crosses with the fox hound: but
0067 t of the 357 no less than 295 were destroyed, CHIEFLY by slugs and insects. If turf which has long
0068 grouse, and hares on any large estate depends CHIEFLY on the destruction of vermin. If not one hea
0068 uggle for existence: but in so far as climate CHIEFLY acts in reducing food, it brings on the most
0089 explained on the view of plumage having been CHIEFLY modified by sexual selection, acting when th
0103 at whatever intercrossing took place would be CHIEFLY between the individuals of the same new vari
0125 forms in the struggle for existence, it will CHIEFLY act on those which already have some advanta
0132 eing functionally disturbed in the parents, I CHIEFLY attribute the varying or plastic condition o
0172 upposed; the imperfection of the record being CHIEFLY due to organic beings not inhabiting profoun
0199 be thrown on the origin of these differences, CHIEFLY through sexual selection of a particular kin
0205 ection in each well stocked country, must act CHIEFLY through the competition of the inhabitants o
0223 nd bringing materials for the nest: both, but CHIEFLY the slaves, tend, and milk as it may be call
0255 sion. The following rules and conclusions are CHIEFLY drawn up from Gartner's admirable work on th
0260 cidental or dependent on unknown differences, CHIEFLY in the reproductive systems, of the species
0263 osses, are incidental on unknown differences, CHIEFLY in their reproductive systems. These differe
0275 resembling one parent, the resemblances seem CHIEFLY confined to characters almost monstrous in t
0301 analogy leads me to believe that it would be CHIEFLY these far ranging species which would oftene
0306 s. from these and similar considerations, but CHIEFLY from our ignorance of the geology of other c
0340 mmutable law that marsupials should have been CHIEFLY or solely produced in Australia; or that Ede
0359 immersion of 137 days. For convenience sake I CHIEFLY tried small seeds, without the capsule or fr
0368 united States believe to have been the case, CHIEFLY from the distribution of the fossil Gnathodo
0379 rd towards their former homes: but the forms, CHIEFLY northern, which had crossed the equator, wou
0403 ecies, excepting in so far as the same forms, CHIEFLY of plants, have spread widely throughout the
0013 unfrequently, and we see it in the father and CHILD, we cannot tell whether it may not be due to
0013 l million individuals, and it reappears in the CHILD, the mere doctrine of chances almost compels
0013 etimes inherited and sometimes not so; why the CHILD often reverts in certain characters to its gr
0026 w of no fact countenancing the belief that the CHILD ever reverts to some one ancestor, removed by
0122 ly adapted to some quite new station, in which CHILD and parent do not come into competition, we h
0131 slight deviations of structure, as to make the CHILD like its parents. But the much greater variab
0144 ences by pressure the shape of the head of the CHILD. In snakes, according to Schlegel, the shape
0275 onclusion, that the laws of resemblance of the CHILD to its parents are the same, whether the two
0281 etition between organism and organism, between CHILD and parent, will render this a very rare even
0443 wn children: we cannot always tell whether the CHILD will be tall or short, or what its precise fe
0444 which have supervened at an earlier age in the CHILD than in the parent. These two principles, if
0448 ble for the existence of the species, that the CHILD should be modified at a very early age in the
0089 tractive to all his hen birds. It may appear CHILDISH to attribute any effect to such apparently w
0042 d, and, although so much valued by women and CHILDREN, we hardly ever see a distinct breed kept up
0443 ely turn out. We see this plainly in our own CHILDREN: we cannot always tell whether the child wil
0251 d been a natural species from the mountains of CHILE. I have taken some pains to ascertain the deg
0373 nded far below their present level. In central CHILE I was astonished at the structure of a vast m
0141 l works, even in the ancient Encyclopaedias of CHINA, to be very cautious in transposing animals f
0164 , in various countries from Britain to Eastern CHINA: and from Norway in the north to the Malay Ar
0034 lection I find distinctly given in an ancient CHINESE encyclopaedia. Explicit rules are laid down
0253 ctly fertile. The hybrids from the common and CHINESE geese (A. cygnoides), species which are so d
0184 climbing trees and for seizing insects in the CHINKS of the bark? Yet in North America there are w
0397 r the mouth of the shell, might be floated in CHINKS of drifted timber across moderately wide arms
0288 during the chalk period. The molluscan genus CHITON offers a partially analogous case. With respe
0034 rude and barbarous periods of English history CHOICE animals were often imported, and laws were pa
0036 nts, to which savages are so liable, and such CHOICE animals would thus generally leave more offsp
0009 ty; and variability is the source of all the CHOICEST productions of the garden. I may add, that a
0032 often more abrupt. No one supposes that our CHOICEST productions have been produced by a single v
0067 dug and cleared, and where there could be no CHOKING from other plants, I marked all the seedling
0089 les, which standing by as spectators, at last CHOOSE the most attractive partner. Those who have c
0421 s, at each successive period. If, however, we CHOOSE to suppose that any of the descendants of A o
0425 ent by resemblances of any kind. Therefore we CHOOSE those characters which, as far as we can judg
0112 has actually occurred with tumbler pigeons) CHOOSING and breeding from birds with longer and long
0360 d 98 seeds, mostly different from mine: but he CHOSE many large fruits and likewise seeds from pla
0017 view. It has often been assumed that man has CHOSEN for domestication animals and plants having a
```

Page **(Key Word)**

```
0022 together at least a score of pigeons might be CHOSEN, which if shown to an ornithologist, and he w
0037 some small degree, to their having naturally CHOSEN and preserved the best varieties they could a
0121 polity of nature: hence in the diagram I have CHOSEN the extreme species (A), and the nearly extre
0140 eve that our domestic animals were originally CHOSEN by uncivilised man because they were useful a
0154 would be a more unusual circumstance. I have CHOSEN this example because an explanation is not in
0212 h vary partly in dependence on the situations CHOSEN and on the nature and temperature of the coun
0256 of the individuals which happen to have been CHOSEN for the experiment. So it is with hybrids, fo
0186 ght, and for the correction of spherical and CHROMATIC aberration, could have been formed by natur
0441 size. In the second stage, answering to the CHRYSALIS stage of butterflies, they have six pairs o
0288 . for instance, the several species of the CHTHAMALINAE (a sub family of sessile cirripedes) coat
0288 rmation: yet it is now known that the genus CHTHAMALUS existed during the chalk period. The mollus
0304 g as possible, this sessile cirripede was a CHTHAMALUS, a very commom, large, and ubiquitous genus
0216 ogs and cats, for if the hen gives the danger CHUCKLE, they will run (more especially young turkey
0046 ar from uniform. Authors sometimes argue in a CIRCLE when they state that important organs never v
0268 m as undoubted species. If we thus argue in a CIRCLE, the fertility of all varieties produced unde
0277 n we remember how liable we are to argue in a CIRCLE with respect to varieties in a state of natur
0370 ne period lived further north under the polar CIRCLE, in latitude 66 degrees 67 degrees; and that
0370 at a globe, we shall see that under the Polar CIRCLE there is almost continuous land from western
0371 w and Old Worlds, migrated south of the Polar CIRCLE, they must have been completely cut off from
0372 form along the continuous shores of the Polar CIRCLE, will account, on the theory of modification,
0073 rous birds, and so onwards in ever increasing CIRCLES of complexity. We began this series by insec
0124 urteen new species will be of a curious and CIRCUITOUS nature. Having descended from a form which
0332 diate, not directly, but only by a long and CIRCUITOUS course through many widely different forms.
0345 msi but are intermediate only by a long and CIRCUITOUS course through many extinct and very differ
0431 ll consequently be related to each other by CIRCUITOUS lines of affinity of various lengths (as ma
0456 eings are united by complex, radiating, and CIRCUITOUS lines of affinities into one grand system:
0478 genera within each class are so complex and CIRCUITOUS. We see why certain characters are far more
0227 rude humble bee adds cylinders of wax to the CIRCULAR mouths of her old cocoons. By such modificat
0228 the bees instantly began to excavate minute CIRCULAR pits in it; and as they deepened these littl
0229 ides of the ridge of vermilion wax, as they CIRCULARLY gnawed away and deepened the basins on both
0230 his from the opposite sides, always working CIRCULARLY as they deepen each cell. They do not make
0230 edge of a growing comb, do make a rough, CIRCUMFERENTIAL wall or rim all round the comb; and they
0231 y that they can do this. Even in the rude CIRCUMFERENTIAL rim or wall of wax round a growing comb,
0232 single cell, or the extreme margin of the CIRCUMFERENTIAL rim of a growing comb, with an extremely
0346 if we exclude the northern parts where the CIRCUMPOLAR land is almost continuous, all authors agr
0367 same with those of Europe; for the present CIRCUMPOLAR inhabitants, which we suppose to have ever
0370 tern America. And to this continuity of the CIRCUMPOLAR land, and to the consequent freedom for in
0370 and animals inhabited the almost continuous CIRCUMPOLAR land; and that these plants and animals, b
0028 us character. It is also a most favourable CIRCUMSTANCE for the production of distinct breeds, tha
0028 ther large groups of birds, in nature. One CIRCUMSTANCE has struck me much; namely, that all the b
0042 hough mingled in the same aviary, and this CIRCUMSTANCE must have largely favoured the improvement
0053 all hereafter see, is a far more important CIRCUMSTANCE) with different sets of organic beings. Bu
0112 es apply most efficiently, from the simple CIRCUMSTANCE that the more diversified the descendants
0154 and its variation would be a more unusual CIRCUMSTANCE. I have chosen this example because an exp
0234 llect. But let us suppose that this latter CIRCUMSTANCE determined, as it probably often does dete
0387 facts, I think it would be an inexplicable CIRCUMSTANCE if water birds did not transport the seeds
0439 tter proof of this cannot be given, than a CIRCUMSTANCE mentioned by Agassiz, namely, that having
0484 duction. We see this even in so trifling a CIRCUMSTANCE as that the same poison often similarly af
0038 ons. And in two countries very differently CIRCUMSTANCED, individuals of the same species, having
0104 inhabited the surrounding and differently CIRCUMSTANCED districts, will be prevented. But isolati
0264 on mule. Hybrids, however, are differently CIRCUMSTANCED before and after birth: when born and liv
0326 ions have been for a long period favourably CIRCUMSTANCED in an equal degree, whenever their inhabi
0004 shall, however, be enabled to discuss what CIRCUMSTANCES are most favourable to variation. In the
0007 nknown Origin of our Domestic Productions. CIRCUMSTANCES favourable to Man's power of Selection. W
0013 , due to some extraordinary combination of CIRCUMSTANCES, appears in the parent, say, once amongst
0040 changes. I must now say a few words on the CIRCUMSTANCES, favourable, or the reverse, to man's pow
0043 erent and necessary contingency, under all CIRCUMSTANCES, with all organic beings, as some authors
0055 genus have been formed through variation, CIRCUMSTANCES have been favourable for variation; and h
0055 iation; and hence we might expect that the CIRCUMSTANCES would generally be still favourable to va
0064 various animals in a state of nature, when CIRCUMSTANCES have been favourable to them during two o
0070 when a species, owing to highly favourable CIRCUMSTANCES, increases inordinately in numbers in a s
0080 s between individuals of the same species. CIRCUMSTANCES favourable and unfavourable to Natural Se
0090 hardest pressed for food. I can under such CIRCUMSTANCES see no reason to doubt that the swiftest
0101 n self fertilisation go on for perpetuity. CIRCUMSTANCES favourable to Natural Selection. This is
0107 to less severe competition. To sum up the CIRCUMSTANCES favourable and unfavourable to natural se
0114 n of structure, is seen under many natural CIRCUMSTANCES. In an extremely small area, especially i
0118 large genus in its own country. And these CIRCUMSTANCES we know to be favourable to the productio
0141 y of constitution, brought, under peculiar CIRCUMSTANCES, into play. How much of the acclimatisati
0189 lecting each alteration which under varied CIRCUMSTANCES, may in any way, or in any degree, tend t
0210 f the year, or when placed under different CIRCUMSTANCES, etc.; in which case either one or the ot
0232 acity in bees of laying down under certain CIRCUMSTANCES, a rough wall in its proper place between
0248 e species is so easily affected by various CIRCUMSTANCES, that for all practical purposes it is mo
0255 l now consider a little more in detail the CIRCUMSTANCES and rules governing the sterility of firs
0256 ame two species are crossed under the same CIRCUMSTANCES, but depends in part upon the constitutio
0277 ystems perfect, seems to depend on several CIRCUMSTANCES: in some cases largely on the early death
0277 city evidently depends on widely different CIRCUMSTANCES, should all run, to a certain extent; par
0279 mmonly occur at the present day, under the CIRCUMSTANCES apparently most favourable for their pres
0286 akwater at the base. Hence, under ordinary CIRCUMSTANCES, I conclude that for a cliff 500 feet in
0292 and new stations will often be formed; all CIRCUMSTANCES most favourable, as previously explained,
0304 a valve can be recognised; from all these CIRCUMSTANCES, I inferred that had sessile cirripedes e
0361 may I think safely assume that under such CIRCUMSTANCES their rate of flight would often be 35 mi
0428 modifications to live under nearly similar CIRCUMSTANCES. Mr. Waterhouse has remarked that, when a
0429 by some unusual coincidence of favourable CIRCUMSTANCES. Mr. Waterhouse has remarked that, when a
0459 recapitulation of the general and special CIRCUMSTANCES in its favour. Causes of the general beli
0487 ving depended on an unusual concurrence of CIRCUMSTANCES, and the blank intervals between the succ
0148 nstances could be given: namely, that when a CIRRIPEDE is parasitic within another and is thus pro
```

Page **(Key Word)**

```
0304 perfect specimen of an unmistakeable sessile CIRRIPEDE, which he had himself extracted from the ch
0304 e case as striking as possible, this sessile CIRRIPEDE was a Chthamalus, a very commom, large, and
0101 l can be shown to be physically impossible. CIRRIPEDES long appeared to me to present a case of ve
0148 with which I was much struck when examining CIRRIPEDES, and of which many other instances could be
0148 proteolepas: for the carapace in all other CIRRIPEDES consists of the three highly important ante
0151 clearly shown in the case of hermaphrodite CIRRIPEDES; and I may here add, that I particularly at
0151 the rule almost invariably holds good with CIRRIPEDES. I shall, in my future work, give a list of
0151 pplication. The opercular valves of sessile CIRRIPEDES (rock barnacles) are, in every sense of the
0191 i will give one more instance. Pedunculated CIRRIPEDES have two minute folds of skin, called by me
0192 til they are hatched within the sack. These CIRRIPEDES have no branchiae, the whole surface of the
0192 g for respiration. The Balanidae or sessile CIRRIPEDES, on the other hand, have no ovigerous frena
0192 their adhesive glands. If all pedunculated CIRRIPEDES had become extinct, and they have already s
0288 f the Chthamalinae (a sub family of certain CIRRIPEDES) coat the rocks all over the world in infin
0304 ch struck me. In a memoir on Fossil Sessile CIRRIPEDES, I have stated that, from the number of exi
0304 circumstances, I inferred that had sessile CIRRIPEDES existed during the secondary periods, they
0305 . hence we now positively know that sessile CIRRIPEDES existed during the secondary period; and th
0305 sted during the secondary period; and these CIRRIPEDES might have been the progenitors of our many
0424 of the males and hermaphrodites of certain CIRRIPEDES, when adult, and yet no one dreams of separ
0440 ly the law of common embryonic resemblance. CIRRIPEDES afford a good instance of this: even the il
0440 manner. So again the two main divisions of CIRRIPEDES, the pedunculated and sessile, which differ
0441 rasitic crustaceans. To refer once again to CIRRIPEDES: the larvae in the first stage have three p
0441 eye spot. In this last and completed state, CIRRIPEDES may be considered as either more highly or
0448 stage, as we have seen to be the case with CIRRIPEDES. The adult might become fitted for sites or
0449 obscured; we have seen, for instance, that CIRRIPEDES can at once be recognised by their larvae a
0142 ing, also, of the kidney bean has been often CITED for a similar purpose, and with much greater
0040 than in another, according to the state of CIVILISATION of the inhabitants, slowly to add to the c
0422 ing and subsequent isolation and states of CIVILISATION of the several races, descended from a com
0035 duals of the same breed may be found in less CIVILISED districts, where the breed has been less im
0038 t given to the plants in countries anciently CIVILISED. In regard to the domestic animals kept by
0038 racter of species than the varieties kept in CIVILISED countries. On the view here given of the al
0040 bly first receive a provincial name. In semi CIVILISED countries, with little free communication,
0215 imals. How rarely, on the other hand, do our CIVILISED dogs, even when quite young, require to be
0242 ple that the division of labour is useful to CIVILISED man. As ants work by inherited instincts an
0215 ree of selection, has probably concurred in CIVILISING by inheritance our dogs. On the other hand,
0018 n some degree probable that man sufficiently CIVILIZED to have manufactured pottery existed in the
0024 oughly domesticated in ancient times by half CIVILIZED man, as to be quite prolific under confinem
0024 hence it must be assumed not only that half CIVILIZED man succeeded in thoroughly domesticating s
0198 erent countries, more especially in the less CIVILIZED countries where there has been but little a
0133 h of this difference may be due to the warmest CLAD individuals having been favoured and preserved
0050 iation. But if he confine his attention to one CLASS within one country, he will soon make up his
0126 l the descendants of the same species making a CLASS, we can understand how it is that there exist
0128 classes. The several subordinate groups in any CLASS cannot be ranked in a single file, but seem r
0129 . the affinities of all the beings of the same CLASS have sometimes been represented by a great tr
0150 bat's wing is a most abnormal structure in the CLASS mammalia; but the rule would not here apply,
0151 ule seems to me certainly to hold good in this CLASS. I cannot make out that it applies to plants,
0187 can learn nothing on this head. In this great CLASS we should probably have to descend far beneat
0188 ossessed by any member of the great Articulate CLASS. He who will go thus far, if he find on finis
0189 an organ common to all the members of a Large CLASS, for in this latter case the organ must have
0189 eriod, since which all the many members of the CLASS have been developed; and in order to discover
0193 e organ appears in several members of the same CLASS, especially if in members having very differe
0206 re, which we see in organic beings of the same CLASS, and which is quite independent of their habi
0207 er mental qualities of animals within the same CLASS. I will not attempt any definition of instinc
0266 lelism extends to an allied yet very different CLASS of facts. It is an old and almost universal b
0282 onverging to the common ancestor of each great CLASS. So that the number of intermediate and trans
0303 tises, published not many years ago, the great CLASS of mammals was always spoken of as having abr
0315 ge will become extinct. In members of the same CLASS the average amount of change, during long and
0321 species belonging to the same or to a distinct CLASS, which yield their places to other species wh
0337 ns, for instance, not the highest in their own CLASS, may have beaten the highest molluscs. From t
0345 the embryos of more recent animals of the same CLASS, the fact will be intelligible. The successio
0372 formerly supposed), but we find in every great CLASS many forms, which some naturalists rank as ge
0390 use in an island nearly all the species of one CLASS are peculiar, those of another class, or of a
0390 es of one class are peculiar, those of another CLASS, or of another section of the same class, are
0390 other class, or of another section of the same CLASS, are peculiar; and this difference seems to d
0406 t organisms low in the scale within each great CLASS; generally change at a slower rate than the h
0409 ion, one group of beings, even within the same CLASS, should have all its species endemic, and ano
0410 little, whilst others belonging to a different CLASS, or to a different order, or even only to a d
0410 both time and space the lower members of each CLASS generally change less than the higher; but th
0410 quarters, in both cases the forms within each CLASS have been connected by the same bond of ordin
0411 roups. Morphology, between members of the same CLASS, between parts of the same individual. Embryo
0412 all the genera on this line form together one CLASS, for all have descended from one ancient but
0413 ies, families, and orders, all united into one CLASS. Thus, the grand fact in natural history of t
0413 ange the species, genera, and families in each CLASS, on what is called the Natural System. But wh
0417 e species, to the genus, to the family, to the CLASS, disappear, and thus laugh at our classificat
0419 equivocally belonging to this, and to no other CLASS of Articulata. Geographical distribution has
0420 that the arrangement of the groups within each CLASS, in due subordination and relation to the oth
0423 system; we are cautioned, for instance, not to CLASS two varieties of the pine apple together, mer
0426 of descent; and we put them all into the same CLASS. As we find organs of high physiological impo
0427 e very same characters are analogical when one CLASS or order is compared with another, but give t
0428 e true affinities when the members of the same CLASS or order are compared one with another: thus
0428 uck by a parallelism of this nature in any one CLASS, by arbitrarily raising or sinking the value
0431 and extinct members of the same great natural CLASS. Extinction, as we have seen in the fourth ch
0431 e intervals between the several groups in each CLASS. We may thus account even for the distinctnes
0432 to succeed in collecting all the forms in any CLASS which have lived throughout all time and spac
0433 at system; and how the several members of each CLASS are connected together by the most complex an
0434 h of affinities between the members of any one CLASS; but when we have a distinct object in view,
```

Page **(Key Word)**

```
0434  ogy. We have seen that the members of the same CLASS, independently of their habits of life, resem
0434  rts and organs in the different species of the CLASS are homologous. The whole subject is included
0435  s similarity of pattern in members of the same CLASS, by utility or by the doctrine of final cause
0435  construction of the limbs throughout the whole CLASS. So with the mouths of insects, we have only
0436  not of the same part in different members of a CLASS, but of the different parts or organs in the
0438  strong principle of inheritance. In the great  CLASS of molluscs, though we can homologise the par
0438  in molluscs, even in the lowest members of the CLASS, we do not find nearly so much indefinite rep
0439  yos, also, of distinct animals within the same CLASS are often strikingly similar: a better proof
0439  mbryos of widely different animals of the same CLASS resemble each other, often have no direct rel
0442  os of widely different animals within the same CLASS, that we might be led to look at these facts
0442  f embryos of different species within the same CLASS, generally, but not universally, resembling e
0448  d spiders, and with a few members of the great CLASS of insects, as with Aphis. With respect to th
0449  ised by their larvae as belonging to the great CLASS of crustaceans. As the embryonic state of eac
0450  cured, of the common parent form of each great CLASS of animals. Rudimentary, atrophied, or aborte
0452  ies of the same part in different members of a CLASS, nothing is more common, or more necessary, t
0457  purpose applied, of the different species of a CLASS; or to the homologous parts constructed on th
0457  and the resemblance in different species of a  CLASS of the homologous parts or organs, though fit
0458  opied, have all descended, each within its own CLASS or group, from common parents, and have all b
0461  ed by another parallel, but directly opposite,  CLASS of facts; namely, that the vigour and fertili
0474  teps in endowing different animals of the same CLASS with their several instincts. I have attempte
0477  hes, most of the inhabitants within each great CLASS are plainly related: for they will generally
0478  finities of the species and genera within each CLASS are so complex and circuitous. We see why cer
0479  ich were alike in the early progenitor of each CLASS. On the principle of successive variations no
0484  ification embraces all the members of the same CLASS. I believe that animals have descended from a
0486  gree obscured, or the prototypes of each great CLASS. When we can feel assured that all the indivi
0031  onths, and the sheep are each time marked and  CLASSED, so that their very best may ultimately be sel
0171  ory. These difficulties and objections may be  CLASSED under the following heads: Firstly, why if s
0324  north America would hereafter be liable to be  CLASSED with somewhat older European beds. Neverthel
0329  uckland long ago remarked, all fossils can be  CLASSED either in still existing groups, or between
0344  rms, sometimes blending two groups previously  CLASSED as distinct into one; but more commonly only
0411  er in descending degrees, so that they can be  CLASSED in groups under groups. This classification
0424  descended from the Negro, I think he would be  CLASSED under the Negro group, however much he might
0425  ecies of the kangaroo family would have to be  CLASSED under the bear genus. The whole case is prep
0426  teristics, from its allies, and yet be safely  CLASSED with them. This may be safely done, and is o
0427  eus, misled by external appearances, actually  CLASSED an homopterous insect as a moth. We see some
0449  hich have ever lived on this earth have to be  CLASSED together, and as all have been connected by
0464  iate variety be discovered, it will simply be  CLASSED as another and distinct species. Only a smal
0465  if suddenly created there, and will be simply  CLASSED as new species. Most formations have been in
0017  productions, and belonging to equally diverse  CLASSES and countries, were taken from a state of na
0097  re, we can, I think, understand several large  CLASSES of facts, such as the following, which on an
0126  stand how it is that there exist but very few  CLASSES in each main division of the animal and vege
0126  species of many genera, families, orders, and CLASSES, as at the present day. Summary of Chapter.
0128  , forming sub families, families, orders, sub  CLASSES, and classes. The several subordinate groups
0128  families, families, orders, sub classes, and  CLASSES. The several subordinate groups in any class
0137  eral animals, belonging to the most different  CLASSES, which inhabit the caves of Styria and of Ke
0178  which many forms, more especially amongst the  CLASSES which unite for each birth and wander much,
0182  ng that a few members of such water breathing  CLASSES as the Crustacea and Mollusca are adapted to
0245  ed differences. In treating this subject, two  CLASSES of facts, to a large extent fundamentally di
0256  e sterility of the hybrids thus produced, two  CLASSES of facts which are generally confounded toge
0313  te formation. Species of different genera and  CLASSES have not changed at the same rate, or in the
0319  attribute of a vast number of species of all  CLASSES, in all countries. If we ask ourselves why t
0331  als, with the more recent members of the same CLASSES, we must admit that there is some truth in t
0338  ent the embryos of recent animals of the same CLASSES; or that the geological succession of extinc
0338  and fish strictly belong to their own proper  CLASSES, though some of these old forms are in a sli
0342  gically explored with care; that only certain CLASSES of organic beings have been largely preserve
0343  ly and successively; how species of different CLASSES do not necessarily change together, or at th
0354  ks, I will discuss a few of the most striking CLASSES of facts; namely, the existence of the same
0356  many generations. Before discussing the three CLASSES of facts, which I have selected as presentin
0383  h water species, belonging to quite different CLASSES, an enormous range, but allied species preva
0388  islands. We now come to the last of the three CLASSES of facts, which I have selected as presentin
0391  ic islands are sometimes deficient in certain CLASSES, and their places are apparently occupied by
0394  duction of endemic species belonging to other CLASSES; and on continents it is thought that mammal
0396  , the richness in endemic forms in particular CLASSES or sections of classes, the absence of whole
0396  ic forms in particular classes or sections of CLASSES, the absence of whole groups, as of batrachi
0398  e, or in the proportions in which the several CLASSES are associated together, which resembles clo
0422  sub families, families, sections, orders, and CLASSES. It may be worth while to illustrate this vi
0428  mpared with fishes, being adaptations in both CLASSES for swimming through the water: but the shap
0428  so it is with fishes. As members of distinct  CLASSES have often been adapted by successive slight
0428  n observed between the sub groups in distinct  CLASSES. A naturalist, struck by a parallelism of th
0428  g or sinking the value of the groups in other CLASSES (and all our experience shows that this valu
0429  d under a few great orders, under still fewer CLASSES, and all in one great natural system. As sho
0431  us account even for the distinctness of whole CLASSES from each other, for instance, of birds from
0431  the early progenitors of the other vertebrate CLASSES. There has been less entire extinction of th
0431  ians. There has been still less in some other CLASSES, as in that of the Crustacea, for here the m
0433  erfect a collection: nevertheless, in certain CLASSES, we are tending in this direction; and Milne
0436  each member of the vertebrate and articulate  CLASSES are plainly homologous. We see the same law
0438  f any one part, as we find in the other great CLASSES of the animal and vegetable kingdoms. Natura
0447  ay extend this view to whole families or even CLASSES. The fore limbs, for instance, which served
0456  eties, species, genera, families, orders, and CLASSES. On this same view of descent with modificat
0457  l as it is genealogical. Finally, the several CLASSES of facts which have been considered in this
0464  ally explored. Only organic beings of certain CLASSES can be preserved in a fossil condition, at l
0471  subordinate to groups, all within a few great CLASSES, which we now see everywhere around us, and
0483  ght extend very far. All the members of whole CLASSES can be connected together by chains of affin
0483  fication in the descendants. Throughout whole CLASSES various structures are formed on the same pa
0034  cit rules are laid down by some of the Roman  CLASSICAL writers. From passages in Genesis, it is cl
0037  garden stock. The pear, though cultivated in  CLASSICAL times, appears, from Pliny's description, t
```

Page **(Key Word)**

classical

Page **(Key Word)**

```
0037  it, and so onwards. But the gardeners of the CLASSICAL period, who cultivated the best pear they c
0414  which is partially revealed to us by our CLASSIFCICATIONS. Let us now consider the rules followed
0005  hroughout space: in the thirteenth, their CLASSIFICATION or mutual affinities, both when mature an
0126  return to this subject in the chapter on CLASSIFICATION, but I may add that on this view of extre
0129  no explanation of this great fact in the CLASSIFICATION of all organic beings: but, to the best o
0129  ramifying branches may well represent the CLASSIFICATION of all extinct and living species in grou
0155  return to this subject in our chapter on CLASSIFICATION. It would be almost superfluous to adduce
0270  ranked by Sagaret, who mainly founds his CLASSIFICATION by the test of infertility, as varieties.
0329  links, that he has had to alter the whole CLASSIFICATION of these two orders; and has placed certa
0330  t i apprehend that in a perfectly natural CLASSIFICATION many fossil species would have to stand b
0411  rphology. Embryology. Rudimentary Organs. CLASSIFICATION, groups subordinate to groups. Natural sy
0411  natural system. Rules and difficulties in CLASSIFICATION, explained on the theory of descent with
0411  the theory of descent with modification. CLASSIFICATION of varieties. Descent always used in clas
0411  tion of varieties. Descent always used in CLASSIFICATION. Analogical or adaptive characters. Affin
0411  n be classed in groups under groups. This CLASSIFICATION is evidently not arbitrary like the group
0413  ly that something more is included in our CLASSIFICATION, than mere resemblance. I believe that so
0414  let us now consider the rules followed in CLASSIFICATION, and the difficulties which are encounter
0414  es which are encountered on the view that CLASSIFICATION either gives some unknown plan of creatio
0414  ture, would be of very high importance in CLASSIFICATION. Nothing can be more false. No one regard
0414  habits, the more important it becomes for CLASSIFICATION. As an instance: Owen, in speaking of the
0415  ns always, true. But their importance for CLASSIFICATION, I believe, depends on their greater cons
0416  erences are of quite subordinate value in CLASSIFICATION; yet no one probably will say that the an
0416  e given of the varying importance for the CLASSIFICATION of the same important organ within the sa
0416  this condition are often of high value in CLASSIFICATION. No one will dispute that the rudimentary
0416  rets are of the highest importance in the CLASSIFICATION of the Grasses. Numerous instances could
0417  and important organ. The importance, for CLASSIFICATION, of trifling characters, mainly depends o
0417  d. hence, also, it has been found, that a CLASSIFICATION founded on any single character, however
0417  e class, disappear, and thus laugh at our CLASSIFICATION. But when Aspicarpa produced in France, d
0418  y are considered as highly serviceable in CLASSIFICATION; but in some groups of animals all these,
0418  ters are the most important of any in the CLASSIFICATION of animals: and this doctrine has very ge
0419  haracters are so valuable, on the view of CLASSIFICATION tacitly including the idea of descent. Ou
0419  d, though perhaps not quite logically, in CLASSIFICATION, more especially in very large groups of
0420  egoing rules and aids and difficulties in CLASSIFICATION are explained, if I do not greatly deceiv
0420  a common parent, and, in so far, all true CLASSIFICATION is genealogical; that community of descen
0421  , in this case, their places in a natural CLASSIFICATION will have been more or less completely lo
0422  be worth while to illustrate this view of CLASSIFICATION, by taking the case of languages. If we p
0422  of the races of man would afford the best CLASSIFICATION of the various languages now spoken throu
0423  mation of this view, let us glance at the CLASSIFICATION of varieties, which are believed or known
0423  if we had a real pedigree, a genealogical CLASSIFICATION would be universally preferred; and it ha
0424  list has in fact brought descent into his CLASSIFICATION; for he includes in his lowest grade, or
0426  regated characters have especial value in CLASSIFICATION. We can understand why a species or a gro
0426  r much, we at once value them less in our CLASSIFICATION. We shall hereafter, I think, clearly see
0427  f characters being of real importance for CLASSIFICATION, only in so far as they reveal descent, w
0432  xisting varieties, nevertheless a natural CLASSIFICATION, or at least a natural arrangement, would
0433  s which we are compelled to follow in our CLASSIFICATION. We can understand why we value certain r
0449  of the embryo is even more important for CLASSIFICATION than that of the adult. For the embryo is
0455  n understand, on the genealogical view of CLASSIFICATION, how it is that systematists have found r
0456  of character. In considering this view of CLASSIFICATION, it should be borne in mind that the elem
0457  l characters and of rudimentary organs in CLASSIFICATION is intelligible, on the view that an arra
0478  are far more serviceable than others for CLASSIFICATION; why adaptive characters, though of param
0479  he being, are of hardly any importance in CLASSIFICATION; why characters derived from rudimentary
0417  l to illustrate the spirit with which our CLASSIFICATIONS are sometimes necessarily founded. Pract
0418  ith those derived from the adult, for our CLASSIFICATIONS of course include all ages of each speci
0419  acitly including the idea of descent. Our CLASSIFICATIONS are often plainly influenced by chains o
0428  ptenary, quinary, quaternary, and ternary CLASSIFICATIONS have probably arisen. As the modified de
0456  lties encountered by naturalists in their CLASSIFICATIONS; the value set upon characters, if const
0486  finitude of already recorded species. Our CLASSIFICATIONS will come to be, as far as they can be s
0045  , whether viewed under a physiological or CLASSIFICATORY point of view, sometimes vary in the indi
0415  cal importance. No doubt this view of the CLASSIFICATORY importance of organs which are important
0415  rtance of an organ does not determine its CLASSIFICATORY value, is almost shown by the one fact, t
0415  nearly the same physiological value, its CLASSIFICATORY value is widely different. No naturalist
0426  embryological characters are of such high CLASSIFICATORY importance. Geographical distribution may
0479  o service to the being, are often of high CLASSIFICATORY value: and why embryological characters a
0483  her by chains of affinities, and all can be CLASSIFIED on the same principle, in groups sub ordina
0414  ount importance! We must not, therefore, in CLASSIFYING, trust to resemblances in parts of the org
0423  tion. Nearly the same rules are followed in CLASSIFYING varieties, as with species. Authors have i
0486  called the plan of creation. The rules for CLASSIFYING will no doubt become simpler when we have
0154  ical importance than those commonly used for CLASSING genera. I believe this explanation is partly
0423  s. authors have insisted on the necessity of CLASSING varieties on a natural instead of an artific
0423  art is found to be most constant, is used in CLASSING varieties: thus the great agriculturist Mars
0423  less serviceable, because less constant. In CLASSING varieties, I apprehend if we had a real pedi
0425  ity. As descent has universally been used in CLASSING together the individuals of the same species
0425  remely different: and as it has been used in CLASSING varieties which have undergone a certain, an
0427  y sometimes be brought usefully into play in CLASSING large and widely distributed genera, because
0433  nder group. We use the element of descent in CLASSING the individuals of both sexes and of all age
0433  common, under one species: we use descent in CLASSING acknowledged varieties, however different th
0339  llection of fossil bones made by MM. Lund and CLAUSEN in the caves of Brazil. I was so much impres
0077  lons of the tiger: and in that of the legs and CLAWS of the parasite which clings to the hair on t
0197  , when we see that the skin on the head of the CLEAN feeding male turkey is likewise naked. The su
0362  he beaks and feet of birds are generally quite CLEAN, I can show that earth sometimes adheres to t
0004  therefore, of the highest importance to gain a CLEAR insight into the means of modification and co
0007  connected with excess of food. It seems pretty CLEAR that organic beings must be exposed during se
0008  ; and monstrosities cannot be separated by any CLEAR line of distinction from mere variations. But
0034  sical writers. From passages in Genesis, it is CLEAR that the colour of domestic animals was at th
0051  fficulties will rise to a climax. Certainly no CLEAR line of demarcation has as yet been drawn bet
0090  tion of Natural Selection. In order to make it CLEAR how, as I believe, natural selection acts, I
```

Page **(Key Word)**

Page ***(Key Word)***

```
0132  ame species are more brightly coloured under a CLEAR atmosphere, than when living on islands or ne
0157  sexes of the same species. This relation has a CLEAR meaning on my view of the subject: I look at
0179  d as each exists by a struggle for life, it is  CLEAR that each is well adapted in its habits to it
0220  g them away to a place of safety. Hence, it is  CLEAR, that the slaves feel quite at home. During t
0246  y of species: for it seems to make a broad and  CLEAR distinction between varieties and species. Fi
0248  at neither sterility nor fertility affords any  CLEAR distinction between species and varieties; bu
0262  pollen. We thus see, that although there is a   CLEAR and fundamental difference between the mere a
0304  pplement to Lyell's Manual, published in 1858,  CLEAR evidence of the existence of whales in the up
0309  e existed where oceans are now spread out; and  CLEAR and open oceans may have existed where our co
0316  e group have descended from one species, it is  CLEAR that as long as any species of the group have
0414  imal, I have always regarded as affording very  CLEAR indications of its true affinities. We are le
0469  ecord in systematic works. No one can draw any  CLEAR distinction between individual differences an
0481  course of long ages is a limited quantity; no  CLEAR distinction has been, or can be, drawn betwee
0016  al parent species. This point, if it could be  CLEARED up, would be interesting; if, for instance,
0067  ground three feet long and two wide, dug and   CLEARED, and where there could be no choking from ot
0367  ountains, the arctic forms would seize on the  CLEARED and thawed ground, always ascending higher a
0373  s to the southernmost extremity, we have the   CLEAREST evidence of former glacial action, in huge b
0014  th of horn, though appearing late in life, is  CLEARLY due to the male element. Having alluded to t
0026  e case of the hybrid offspring of two animals  CLEARLY distinct being themselves perfectly fertile.
0036  britain, India, and Persia, we can, I think,  CLEARLY trace the stages through which they have ins
0056  looked at as incipient species; for my tables  CLEARLY show as a general rule that, wherever many s
0059  t a strong analogy with varieties. And we can  CLEARLY understand these analogies, if species have
0069  indirectly by favouring other species, we may  CLEARLY see in the prodigious number of plants in ou
0078  amper or drier districts. In this case we can  CLEARLY see that if we wished in imagination to give
0078  of constitution with respect to climate would  CLEARLY be an advantage to our plant; but we have re
0113  so be enabled to increase in numbers. We can  CLEARLY see this in the case of animals with simple
0151  ot confined to secondary sexual characters is  CLEARLY shown in the case of hermaphrodite cirripede
0190  in fishes is a good one, because it shows us   CLEARLY the highly important fact that an organ orig
0194  e? on the theory of natural selection, we can  CLEARLY understand why she should not: for natural s
0206  nd. On the theory of natural selection we can  CLEARLY understand the full meaning of that old cano
0209  eritance to succeeding generations. It can be  CLEARLY shown that the most wonderful instincts with
0230  eriment of the ridge of vermilion wax, we can  CLEARLY see that if the bees were to build for thems
0230  to break into each other. Now bees, as may be  CLEARLY seen by examining the edge of a growing comb
0231  ing spheres. I have several specimens showing  CLEARLY that they can do this. Even in the rude circ
0246  ctive organs functionally impotent, as may be  CLEARLY seen in the state of the male element in bot
0257  overned by their systematic affinity. This is  CLEARLY shown by hybrids never having been raised be
0258  organisation. On the other hand, these cases  CLEARLY show that the capacity for crossing is conne
0260  lance to each other. This latter statement is  CLEARLY proved by reciprocal crosses between the sam
0260  es and facts, on the other hand, appear to me  CLEARLY to indicate that the sterility both of first
0269  first appears. It can, in the first place, be  CLEARLY shown that mere external dissimilarity betwe
0315  at hazard, in a slowly changing drama. We can  CLEARLY understand why a species when once lost shou
0325  k to some special law. We shall see this more  CLEARLY when we treat of the present distribution of
0326  ew and dominant species; but we can, I think,  CLEARLY see that a number of individuals, from givin
0339  merican types. This relationship is even more  CLEARLY seen in the wonderful collection of fossil b
0345  many extinct and very different forms. We can  CLEARLY see why the organic remains of closely conse
0345  closely linked together by generation: we can  CLEARLY see why the remains of an intermediate forma
0349  groups of beings, specifically distinct, yet  CLEARLY related, replace each other. He hears from c
0355  n, my theory will be strengthened: for we can  CLEARLY understand, on the principle of modification
0375  s the fact that southern Australian forms are  CLEARLY represented by plants growing on the summits
0409  ea between an island and the mainland. We can  CLEARLY see why all the inhabitants of an archipelag
0418  uralists to be the true one; and by none more  CLEARLY than by that excellent botanist, Aug. St. Hi
0426  classification. We shall hereafter, I think,  CLEARLY see why embryological characters are of such
0427  only in so far as they reveal descent, we can  CLEARLY understand why analogical or adaptive charac
0433  s within the limits of the same group. We can  CLEARLY see how it is that all living and extinct fo
0441  am aware that it is hardly possible to define  CLEARLY what is meant by the organisation being high
0449  eir less modified ancient progenitors, we can  CLEARLY see why ancient and extinct forms of life sh
0465  changed slowly and in a graduated manner. We  CLEARLY see this in the fossil remains from consecut
0477  f these, that many should be peculiar. We can  CLEARLY see why those animals which cannot cross wid
0479  orresponding not early period of life, we can  CLEARLY see why the embryos of mammals, birds, repti
0480  under changed conditions of life; and we can  CLEARLY understand on this view the meaning of rudim
0366  anto of Europe, as explained with remarkable  CLEARNESS by Edward Forbes, is substantially as follo
0381  he most remarkable are stated with admirable  CLEARNESS by Dr. Hooker in his botanical works on the
0283  in wearing away rock. At last the base of the  CLIFF is undermined, huge fragments fall down, and
0283  if we follow for a few miles any line of rocky  CLIFF, which is undergoing degradation, we find tha
0286  at which the sea commonly wears away a line of  CLIFF of any given height, we could measure the tim
0286  ; but it is the same as if we were to assume a  CLIFF one yard in height to be eaten back along a w
0286  ts; though no doubt the degradation of a lofty  CLIFF would be more rapid from the breakage of the
0286  ordinary circumstances, I conclude that for a  CLIFF 500 feet in height, a denudation of one inch
0095  the formation of the longest lines of inland  CLIFFS. Natural selection can act only by the preser
0283  egradation. The tides in most cases reach the  CLIFFS only for a short time twice a day, and the wa
0283  often do we see along the bases of retreating  CLIFFS rounded boulders, all thickly clothed by mari
0283  short length or round a promontory, that the  CLIFFS are at the present time suffering. The appear
0285  waves and pared all round into perpendicular  CLIFFS of one or two thousand feet in height; for th
0286  e subject, assume that the sea would eat into  CLIFFS 500 feet in height at the rate of one inch in
0481  yell first insisted that long lines of inland  CLIFFS had been formed, and great valleys excavated,
0339  areas, during the later tertiary periods. Mr.  CLIFT many years ago showed that the fossil mammals
0293  ely infer a large amount of migration during  CLIMATAL and other changes; and when we see a species
0301  ed by oscillations of level, and that slight  CLIMATAL changes would intervene during such lengthy
0326  n may often be very slow, being dependent on  CLIMATAL and geographical changes, or on strange acci
0346  arious regions can be accounted for by their  CLIMATAL and other physical conditions. Of late, almo
0351  region; for during the vast geographical and  CLIMATAL changes which will have supervened since and
0353  point to the other. But the geographical and  CLIMATAL changes, which have certainly occurred withi
0354  idering our ignorance with respect to former  CLIMATAL and geographical changes and various occasio
0369  lso, have been exposed to somewhat different  CLIMATAL influences. Their mutual relations will thus
0405  es there will have been ample time for great  CLIMATAL and geographical changes and for accidents o
0407  remely grave. As exemplifying the effects of  CLIMATAL changes on distribution, I have attempted to
```

Page ***(Key Word)***

Page ★★★(Key Word)★★

```
0462  y ignorant of the full extent of the various CLIMATAL and geographical changes which have affected
0476  art of the world to another, owing to former  CLIMATAL and geographical changes and to the many occ
0003  inually refer to external conditions, such as CLIMATE, food, etc., as the only possible cause of v
0011  light, and perhaps the thickness of fur from  CLIMATE. Habit also has a decided influence, as in t
0011  owering with plants when transported from one CLIMATE to another. In animals it has a more marked
0060  o increase. Competition universal. Effects of CLIMATE. Protection from the number of individuals.
0068  attack a young elephant protected by its dam. CLIMATE plays an important part in determining the a
0068  tality from epidemics with man. The action of CLIMATE seems at first sight to be quite independent
0068  the struggle for existence; but in so far as  CLIMATE chiefly acts in reducing food, it brings on
0068  h subsist on the same kind of food. Even when  CLIMATE, for instance extreme cold, acts directly, i
0069  , and finally disappearing; and the change of  CLIMATE being conspicuous, we are tempted to attribu
0069  least degree favoured by any slight change of  CLIMATE, they will increase in numbers, and, as each
0069  orms, due to the directly injurious action of  CLIMATE, than we do in proceeding southwards or in d
0069  is almost exclusively with the elements. That CLIMATE acts in main part indirectly by favouring ot
0069  r gardens which can perfectly well endure our  CLIMATE, but which never become naturalised, for the
0075  of the varieties which best suit the soil or  CLIMATE, or are naturally the most fertile, will bea
0078  nge, a change of constitution with respect to CLIMATE would clearly be an advantage to our plant;
0078  that they are destroyed by the rigour of the  CLIMATE alone. Not until we reach the extreme confin
0078  w country amongst new competitors, though the CLIMATE may be exactly the same as in its former hom
0081  rooing some physical change, for instance, of CLIMATE. The proportional numbers of its inhabitants
0081  e inhabitants, independently of the change of CLIMATE itself, would most seriously affect many of
0082  believe that any great physical change, as of CLIMATE, or any unusual degree of isolation to check
0083  es sheep with long and short wool to the same CLIMATE. He does not allow the most vigorous males t
0085  be quite unimportant, we must not forget that CLIMATE, food, etc., probably produce some slight an
0104  anisms, after any physical change, such as of CLIMATE or elevation of the land, etc.; and thus new
0132  slight. How much direct effect difference of  CLIMATE, food, etc., produces on any being is extrem
0132  e. some little influence may be attributed to CLIMATE, food, etc.: thus, E. Forbes speaks confiden
0133  ve thicker and better fur the more severe the CLIMATE is under which they have lived: but who can
0133  d how much to the direct action of the severe CLIMATE? for it would appear that climate has some d
0133  the severe climate? for it would appear that  CLIMATE has some direct action on the hair of our do
0138  deep limestone caverns under a nearly similar CLIMATE: so that on the common view of the blind ani
0139  notorious that each species is adapted to the CLIMATE of its own home: species from an arctic or e
0139  m a temperate region cannot endure a tropical CLIMATE, or conversely. So again, many succulent pla
0139  n, many succulent plants cannot endure a damp CLIMATE. But the degree of adaptation of species to
0140  ther or not an imported plant will endure our CLIMATE, and from the number of plants and animals b
0140  animals were strictly adapted to their native CLIMATE, but in all ordinary cases we assume such to
0141  any other rodent, living free under the cold  CLIMATE of Faroe in the north and of the Falklands i
0141  inclined to look at adaptation to any special CLIMATE as a quality readily grafted on an innate wi
0141  rhinoceros were capable of enduring a glacial CLIMATE, whereas the living species are now all trop
0141  he acclimatisation of species to any peculiar CLIMATE is due to mere habit, and how much to the na
0167  me genus. The external conditions of life, as CLIMATE and food, etc., seem to have induced some sl
0174  es. By changes in the form of the land and of CLIMATE, marine areas now continuous must often have
0175  the sea with the dredge. To those who look at CLIMATE and the physical conditions of life as the a
0175  tion, these facts ought to cause surprise, as CLIMATE and height or depth graduate away insensibly
0178  uch new places will depend on slow changes of CLIMATE, or on the occasional immigration of new inh
0180  onceive under all natural conditions. Let the CLIMATE and vegetation change, let other competing r
0196  of natural selection. We should remember that CLIMATE, food, etc., probably have some.little direc
0198  . careful observers are convinced that a damp CLIMATE affects the growth of the hair, and that wit
0267  hanges of seed, tubers, etc, from one soil or CLIMATE to another, and back again. During the conva
0279  other already defined organic forms, than on CLIMATE; and, therefore, that the really governing c
0294  of level, on the inordinately great change of CLIMATE, on the prodigious lapse of time, all includ
0306  ome of the species became adapted to a cooler CLIMATE, and were enabled to double the southern cap
0325  quite futile to look to changes of currents, CLIMATE, or other physical conditions, as the cause
0336  ions. Consider the prodigious vicissitudes of CLIMATE during the pleistocene period, which include
0337  nd preceding forms. If under a nearly similar CLIMATE, the eocene inhabitants of one quarter of th
0339  e a bold man, who after comparing the present CLIMATE of Australia and of parts of South America u
0346  f creation. Means of dispersal, by changes of CLIMATE and of the level of the land, and by occasio
0346  r almost every temperature. There is hardly a CLIMATE or condition in the Old World which cannot b
0347  consequently inhabit a considerably different CLIMATE, and they will be found incomparably more cl
0347  of Australia or Africa under nearly the same CLIMATE. Analogous facts could be given with respect
0347  joins; and where, under a slightly different CLIMATE, there might have been free migration for th
0356  stract of the more important facts. Change of CLIMATE must have had a powerful influence on migrat
0356  ful influence on migration: a region when its CLIMATE was different may have been a high road for
0364  sion in salt water, they could not endure our CLIMATE. Almost every year, one or two land birds ar
0366  pe and North America suffered under an Arctic CLIMATE. The ruins of a house burnt by fire do not t
0366  alleys were lately filled. So greatly has the CLIMATE of Europe changed, that in Northern Italy, g
0366  d period. The former influence of the glacial CLIMATE on the distribution of the inhabitants of Eu
0368  onclude without other evidence, that a colder CLIMATE permitted their former migration across the
0368  e become too warm for their existence. If the CLIMATE, since the Glacial period, has ever been in
0368  rd, will have been exposed to nearly the same CLIMATE, and, as is especially to be noticed, they w
0370  ounted by looking to still earlier changes of CLIMATE of an opposite nature. We have good reason t
0370  world were specifically the same as now, the CLIMATE was warmer than at the present day. Hence we
0370  ppose that the organisms now living under the CLIMATE of latitude 60 degrees, during the Pliocene
0370  om for intermigration under a more favourable CLIMATE, I attribute the necessary amount of uniform
0370  ds, began slowly to migrate southwards as the CLIMATE became less warm, long before the commenceme
0373  , and on the shores of the Pacific, where the CLIMATE is now so different, as far south as lat. 46
0377  s of competitors, can withstand a much warmer CLIMATE than their own. Hence, it seems to me possib
0378  y favoured by high land, and perhaps by a dry CLIMATE: for Dr. Falconer informs me that it is the
0378  tructive to perennial plants from a temperate CLIMATE. On the other hand, the the most humid and h
0378  is period of extreme cold, I believe that the CLIMATE under the equator at the level of the sea wa
0382  mine, on the effects of great alternations of CLIMATE on geographical distribution. I believe that
0398  cal nature of the islands, in their height or CLIMATE, or in the proportions in which the several
0398  blance in the volcanic nature of the soil, in CLIMATE, height, and size of the islands, between th
0400  the same geological nature, the same height, CLIMATE, etc., that many of the immigrants should ha
0407  nce of the full effects of all the changes of CLIMATE and of the level of the land, which have cer
0463  period remained continuous, and of which the CLIMATE and other conditions of life change insensib
```

Page ★★★(Key Word)★★

Page **************************************(Key Word)**************************************

0487 will continue to throw, on former changes of CLIMATE and of the level of the land, we shall surel
0007 ied during all ages under the most different CLIMATES and treatment, I think we are driven to conc
0017 y to vary, and likewise to withstand diverse CLIMATES. I do not dispute that these capacities have
0017 nerations, and whether it would endure other CLIMATES? Has the little variability of the ass or gu
0076 of another species under the most different CLIMATES. In Russia the small Asiatic cockroach has a
0083 tions of life. Man keeps the natives of many CLIMATES in the same country; he seldom exercises eac
0134 all, although living under the most opposite CLIMATES. Such considerations as these incline me to
0139 t the degree of adaptation of species to the CLIMATES under which they live is often overrated. We
0140 s, or more than, by adaptation to particular CLIMATES. But whether or not the adaptation be genera
0140 of not only withstanding the most different CLIMATES but of being perfectly fertile (a far severe
0141 d easily be brought to bear widely different CLIMATES. We must not, however, push the foregoing ar
0141 the capacity of enduring the most different CLIMATES by man himself and by his domestic animals,
0142 tain varieties are said to withstand certain CLIMATES better than others: this is very strikingly
0198 titutions would succeed best under different CLIMATES; and there is reason to believe that constit
0257 stations and fitted for extremely different CLIMATES, can often be crossed with ease. By a recipr
0261 eciduous, and adaptation to widely different CLIMATES, does not always prevent the two grafting to
0322 parts of the world, under the most different CLIMATES, where not a fragment of the mineral chalk i
0325 rouchout the world, under the most different CLIMATES. We must, as Barrande has remarked, look to
0336 orld, and therefore under the most different CLIMATES and conditions. Consider the prodigious vici
0348 not far from each other, under corresponding CLIMATES; but from being separated from each other by
0477 distant mountains, under the most different CLIMATES; and likewise the close alliance of some of
0051 analogy, and his difficulties will rise to a CLIMAX. Certainly no clear line of demarcation has a
0238 nsects. But we have not as yet touched on the CLIMAX of the difficulty; namely, the fact that the
0241 fferent structure; or lastly, and this is our CLIMAX of difficulty, one set of workers of one size
0197 ndergoing further modification and becoming a CLIMBER. The naked skin on the head of a vulture is
0197 ly similar hooks on many trees which are not CLIMBERS, the hooks on the bamboo may have arisen fro
0113 dead or alive; some inhabiting new stations, CLIMBING trees, frequenting water, and some perhaps b
0184 he larger titmouse (Parus major) may be seen CLIMBING branches, almost like a creeper. It often, l
0184 ation be given than that of a woodpecker for CLIMBING trees and for seizing insects in the chinks
0184 n species; yet it is a woodpecker which never CLIMBS a tree! Petrels are the most aerial and ocean
0197 n. a trailing bamboo in the Malay Archipelago CLIMBS the loftiest trees by the aid of exquisitely
0221 ca) sometimes as many as three of these ants CLINGING to the legs of the slave making F. sanguinea
0222 with a few of these little yellow ants still CLINGING to the fragments of the nest. This species i
0060 e less plainly in the humblest parasite which CLINGS to the hairs of a quadruped or feathers of a
0077 t of the legs and claws of the parasite which CLINGS to the hair on the tiger's body. But in the b
0008 l not breed, though living long under not very CLOSE confinement in their native country! This is
0045 expected that the branching of the main nerves CLOSE to the great central ganglion of an insect wo
0050 and consequently must be ranked as varieties. CLOSE investigation, in most cases, will bring natu
0055 igher in the scale; and here again there is no CLOSE relation to the size of the genera. The cause
0063 die. But several seedling missletoes, growing CLOSE together on the same branch, may more truly b
0067 surface, with ten thousand sharp wedges packed CLOSE together and driven inwards by incessant blow
0071 frequent intercrossing; and the ill effects of CLOSE interbreeding, probably come into play in som
0071 wn firs are now springing up in multitudes, so CLOSE together that all cannot live. When I ascerta
0075 o keep up a mixed stock of even such extremely CLOSE varieties as the variously coloured sweet pea
0080 it be borne in mind how infinitely complex and CLOSE fitting are the mutual relations of all organ
0096 to the offspring; and on the other hand, that CLOSE interbreeding diminishes vigour and fertility
0097 nt's own anthers and pistil generally stand so CLOSE together that self fertilisation seems almost
0098 and whilst another species of Lobelia growing CLOSE by, which is visited by bees, seeds freely. I
0099 c surface of the same flower, though placed so CLOSE together, as if for the very purpose of self
0114 ion. Most of the animals and plants which live CLOSE round any small piece of ground, could live o
0138 created for the American and European caverns, CLOSE similarity in their organisation and affiniti
0140 hether or not the adaptation be generally very CLOSE, we have evidence, in the case of some few pl
0164 ncluded, from one between brown and black to a CLOSE approach to cream colour. I am aware that Col
0174 ate conditions of life, why do we not now find CLOSE linking intermediate varieties? This difficul
0177 lly existed in greater numbers, will come into CLOSE contact with each other, without the interpos
0184 and undulatory flight, told me plainly of its CLOSE blood relationship to our common species; yet
0209 ginally was a habit and an instinct becomes so CLOSE as not to be distinguished. If Mozart, instea
0215 e tameness, simply to habit and long continued CLOSE confinement. Natural instincts are lost under
0221 rs in making the nest, and they alone open and CLOSE the doors in the morning and evening; and as
0222 thick heath. The nest, however, must have been CLOSE at hand, for two or three individuals of F. f
0225 has shown that the form of the cell stands in CLOSE relation to the presence of adjoining cells;
0245 various in degree, not universal, affected by CLOSE interbreeding, removed by domestication. Laws
0248 minished by an independent cause, namely, from CLOSE interbreeding. I have collected so large a bo
0249 llected so large a body of facts, showing that CLOSE interbreeding lessens fertility, and, on the
0249 ed hybrids may, I believe, be accounted for by CLOSE interbreeding having been avoided. Now let us
0252 ith each other, and the injurious influence of CLOSE interbreeding is thus prevented. Any one may
0253 ent parents, so as to avoid the ill effects of CLOSE interbreeding. On the contrary, brothers and
0257 forts have failed to produce between extremely CLOSE species a single hybrid. Even within the limi
0267 dispensable even with hermaphrodites; and that CLOSE interbreeding continued during several genera
0275 other respects there seems to be a general and CLOSE similarity in the offspring of crossed specie
0278 ther respects, excluding fertility, there is a CLOSE general resemblance between hybrids and mongr
0280 have ever existed, we should have an extremely CLOSE series between both and the rock pigeon; but
0282 ave been the past periods of time, may at once CLOSE this volume. Not that it suffices to study th
0285 the northern and southern escarpments meet and CLOSE, one can safely picture to oneself the great
0290 tent; that is, have not followed each other in CLOSE sequence. Scarcely any fact struck me more wh
0293 ies which lived at its commencement and its CLOSE. Some cases are on record of the same species
0297 s they are enabled to connect them together by CLOSE intermediate gradations. And this from the re
0301 ch other; and these links, let them be ever so CLOSE, if found in different stages of the same for
0302 species which appeared at the commencement and CLOSE of each formation, pressed so hardly on my th
0304 s in the upper greensand, some time before the CLOSE of the secondary period. I may give another i
0318 resent day: some having disappeared before the CLOSE of the palaeozoic period. No fixed law seems
0318 groups of beings, as of ammonites towards the CLOSE of the secondary period, has been wonderfully
0321 le families or orders, as of Trilobites at the CLOSE of the palaeozoic period and of Ammonites at
0321 the palaeozoic period and of Ammonites at the CLOSE of the secondary period, we must remember wha
0328 osits of England and France, is able to draw a CLOSE general parallelism between the successive st
0335 the globe, will not attempt to account for the CLOSE resemblance of the distinct species in closel

Page **************************************(Key Word)**************************************

Page *********************************(Key Word)*********************************

```
0336  species which appeared at the commencement and  CLOSE  of these periods: but we ought to find after
0347  obstacles to free migration, are related in a    CLOSE  and important manner to the differences betwe
0358  opposite sides of almost every continent, the    CLOSE  relation of the tertiary inhabitants of sever
0381  ealand, and Fuegia, I believe that towards the   CLOSE  of the Glacial period, icebergs, as suggested
0394  y parts of the world on very small islands, if   CLOSE  to a continent: and hardly an island can be n
0398  s, supposed to have been created here: yet the   CLOSE  affinity of most of these birds to American s
0406  ough these stations are so different, the very   CLOSE  relation of the distinct species which inhabi
0416  leg, are highly serviceable in exhibiting the    CLOSE  affinity between Ruminants and Pachyderms. Ro
0425  case is preposterous: for where there has been   CLOSE  descent in common, there will certainly be cl
0425  ose descent in common, there will certainly be   CLOSE  resemblance or affinity. As descent has unive
0427  apted to similar conditions, and thus assume a   CLOSE  external resemblance: but such resemblances w
0442  tween the embryo and the adult, and likewise a   CLOSE  similarity in the embryos of widely different
0477  riod, the identity of some few plants, and the   CLOSE  alliance of many others, on the most distant
0477  the most different climates: and likewise the    CLOSE  alliance of some of the inhabitants of the se
0191  beautiful contrivance by which the glottis is    CLOSED  In the higher Vertebrata the branchiae have
0441  d extremely complex antennae: but they have a    CLOSED  and imperfect mouth, and cannot feed: their f
0015  als and plants, and compare them with species    CLOSELY  allied together, we generally perceive in ea
0016  nly in most cases in a lesser degree than, do    CLOSELY  allied species of the same genus in a state
0017  doubt about the immutability of the many very    CLOSELY  allied and natural species, for instance, of
0018  ty in the breeds: and that some of the breeds    CLOSELY  resemble, perhaps are identical with, those
0019  rited variation. Who can believe that animals    CLOSELY  resembling the Italian greyhound, the bloodh
0026  ity in this hypothesis, if applied to species    CLOSELY  related together, though it is unsupported b
0032  ed to this practice, except sometimes amongst    CLOSELY  allied sub breeds. And when a cross has been
0041  n to vary just when gardeners began to attend    CLOSELY  to this plant. No doubt the strawberry had a
0044  arger genera resemble varieties in being very    CLOSELY, but unequally, related to each other, and i
0047  ee the character of species, but which are so    CLOSELY  similar to some other forms, or are so close
0047  losely similar to some other forms, or are so    CLOSELY  linked to them by intermediate gradations, t
0047  on to believe that many of these doubtful and    CLOSELY  allied forms have permanently retained their
0047  m as a variety of another, even when they are    CLOSELY  connected by intermediate links: nor will th
0050  re be highly useful to man, or from any cause    CLOSELY  attract his attention, varieties of it will
0050  thors as species. Look at the common oak, how    CLOSELY  it has been studied: yet a German author mak
0051  ty: for he will encounter a greater number of    CLOSELY  allied forms. But if his observations be wid
0052  e sake of convenience to a set of individuals    CLOSELY  resembling each other, and that it does not
0055  cies of the smaller genera: for wherever many    CLOSELY  related species (i.e. species of the same ge
0058  to believe, that those species which are very    CLOSELY  allied to other species, and in so far resem
0058  nked as species, but which he considers as so    CLOSELY  allied to other species as to be of doubtful
0058  restricted average range, as have those very    CLOSELY  allied forms, marked for me by Mr. Watson as
0059  es. In large genera the species are apt to be    CLOSELY, but unequally, allied together, forming lit
0059  clusters round certain species. Species very    CLOSELY  allied to other species apparently have rest
0067  own, and the case would be the same with turf    CLOSELY  browsed by quadrupeds, be let to grow, the m
0072  except the old planted clumps. But on looking    CLOSELY  between the stems of the heath, I found a mu
0072  ever have imagined that cattle would have so    CLOSELY  and effectually searched it for food. Here w
0089  e the most attractive partner. Those who have    CLOSELY  attended to birds in confinement well know t
0097  her hand, have their organs of fructification    CLOSELY  enclosed, as in the great papilionaceous or
0098  fertilisation, it is well known that if very    CLOSELY  allied forms or varieties are planted near e
0101  y and even of the same genus, though agreeing    CLOSELY  with each other in almost their whole organi
0110  he struggle for Existence that it is the most    CLOSELY  allied forms, varieties of the same species,
0114  nhabitants, which thus jostle each other most    CLOSELY, shall, as a general rule, belong to what we
0115  raised in any land would generally have been    CLOSELY  allied to the indigenes: for these are commo
0122  one (F), of the two species which were least    CLOSELY  related to the other nine original species,
0128  d, namely, varieties of the same species most    CLOSELY  related together, species of the same genus
0128  ated together, species of the same genus less    CLOSELY  and unequally related together, forming sect
0128  genera, species of distinct genera much less    CLOSELY  related, and genera related in different deg
0138  ve insects of the two continents are not more    CLOSELY  allied than might have been anticipated from
0139  nd some of the European cave insects are very    CLOSELY  allied to those of the surrounding country.
0139  the caves of the Old and New Worlds should be    CLOSELY  related, we might expect from the well known
0150  developed in comparison with the same part in    CLOSELY  allied species. Thus, the bat's wing is a mo
0155  oups of species, has differed considerably in    CLOSELY  allied species, that it has also, been varia
0155  , be more variable than those parts which are    CLOSELY  alike in the several species? I do not see t
0158  f difference in these same characters between    CLOSELY  allied species: that secondary sexual and or
0158  parts of the organisation, are all principles    CLOSELY  connected together. All being mainly due to
0159  y in a like manner, but to three separate yet    CLOSELY  related acts of creation. With pigeons, howe
0168  rhaps arising from such parts not having been    CLOSELY  specialised to any particular function, so t
0168  on, so that their modifications have not been    CLOSELY  checked by natural selection. It is probably
0173  remote. But it may be urged that when several    CLOSELY  allied species inhabit the same territory we
0173  e generally meet at successive intervals with    CLOSELY  allied or representative species, evidently
0176  m would be eminently liable to the inroads of    CLOSELY  allied forms existing on both sides of it. B
0178  om what we know of the actual distribution of    CLOSELY  allied or representative species, and likewi
0179  mberless intermediate varieties, linking most    CLOSELY  all the species of the same group together,
0180  nces of transitional habits and structures in    CLOSELY  allied species of the same genus: and of div
0193  thers have remarked, their intimate structure    CLOSELY  resembles that of common muscle: and as it h
0193  has lately been shown that Rays have an organ    CLOSELY  analogous to the electric apparatus, and yet
0203  ion of preceding and intermediate gradations.    CLOSELY  allied species, now living on a continuous a
0243  egard to instincts: as by that common case of    CLOSELY  allied, but certainly distinct, species, whe
0257  inct families: and on the other hand, by very    CLOSELY  allied species generally uniting with facili
0257  . a multitude of cases could be given of very    CLOSELY  allied species which will not unite, or only
0258  ree. He has observed it even between forms so    CLOSELY  related (as Matthiola annua and glabra) that
0259  e character between their two parents, always    CLOSELY  resemble one of them: and such hybrids, thou
0259  bnormal individuals sometimes are born, which    CLOSELY  resemble one of their pure parents: and thes
0261  te distinct families: and, on the other hand,    CLOSELY  allied species, and varieties of the same sp
0266  which occasionally and exceptionally resemble    CLOSELY  either pure parent. Nor do I pretend that th
0268  are, which, though resembling each other most    CLOSELY, are utterly sterile when intercrossed. Seve
0272  ties. And, on the other hand, they agree most    CLOSELY  in very many important respects. I shall her
0272  tner further admits that hybrids between very    CLOSELY  allied species are more variable than those
0273  ists that when any two species, although most    CLOSELY  allied to each other, are crossed with a thi
0274  ciprocal cross, generally resemble each other    CLOSELY: and so it is with mongrels from a reciproca
```

Page *********************************(Key Word)*********************************

closely

Page **************************************(Key Word)**************************************

```
0275  sed fact, that mongrel animals alone are born  CLOSELY  like one of their parents; but it can be sho
0275  which I have collected of cross bred animals    CLOSELY  resembling one parent, the resemblances seem
0277  ing compounded of two distinct species, seems   CLOSELY  allied to that sterility which so frequently
0281  t form of any two or more species, even if we   CLOSELY  compared the structure of the parent with th
0283  nce the waters washed their base. He who most   CLOSELY  studies the action of the sea on our shores,
0289  is difficult to avoid believing that they are   CLOSELY  consecutive. But we know, for instance, from
0293  ult to understand, why we do not therein find   CLOSELY  graduated varieties between the allied speci
0297  ies, unless at the same time it could be most   CLOSELY  connected with either one or both forms by i
0298  d as specifically different, yet are far more   CLOSELY  allied to each other than are the species fo
0301  the archipelago would have to migrate, and no   CLOSELY  consecutive record of their modifications co
0301  ought only to look for a few links, some more   CLOSELY, some more distantly related to each other;
0313  me degree. Yet if we compare any but the most   CLOSELY  related formations, all the species will be
0324  istocene inhabitants of Europe resembled most   CLOSELY  those of the southern hemisphere. So, again,
0324  ing productions of the United States are more   CLOSELY  related to those which lived in Europe durin
0331  p14, will form a small family; b14 and f14 a   CLOSELY  allied family or sub family; and o14, e14, m
0332  disinterred, these three families would be so   CLOSELY  linked together that they probably would hav
0335  n serial affinity, this arrangement would not  CLOSELY  accord with the order in time of their produ
0335  osite end of the series in this same respect.   CLOSELY  connected with the statement, that the organ
0335   from two consecutive formations are far more   CLOSELY  related to each other, than are the fossils
0335  close resemblance of the distinct species in   CLOSELY  consecutive formations, by the physical cond
0336  ll meaning of the fact of fossil remains from   CLOSELY  consecutive formations, though ranked as dis
0336  ons, though ranked as distinct species, being  CLOSELY  related, is obvious. As the accumulation of
0336  nly moderately long as measured geologically,  CLOSELY  allied forms, or, as they have been called b
0339  fossil mammals from the Australian caves were  CLOSELY  allied to the living marsupials of that cont
0340  nent; and the southern half was formerly more  CLOSELY  allied, than it is at present, to the northe
0340  veries, that Northern India was formerly more  CLOSELY  related in its mammals to Africa than it is
0340  r, during the next succeeding period of time,  CLOSELY  allied though in some degree modified descen
0341  zil, there are many extinct species which are  CLOSELY  allied in size and in other characters to th
0342  links which must formerly have connected the   CLOSELY  allied or representative species, found in t
0345  we can clearly see why the organic remains of  CLOSELY  consecutive formations are more closely alli
0345  ns of closely consecutive formations are more  CLOSELY  allied to each other, than are those of remo
0345  of remote formations; for the forms are more   CLOSELY  linked together by generation: we can clearl
0346  cannot be paralleled in the New, at least as   CLOSELY  as the same species generally require: for i
0347  ate, and they will be found incomparably more  CLOSELY  related to each other, than they are to the
0349  ly related, replace each other. He hears from  CLOSELY  allied, yet distinct kinds of birds, notes n
0353  or those genera in which the species are most  CLOSELY  related to each other, are generally local,
0354  region, of which most of its inhabitants are   CLOSELY  related to, or belong to the same genera wit
0355  nt both in space and time with a pre existing  CLOSELY  allied species. And I now know from correspo
0367  ed, the arctic forms would retreat northward,  CLOSELY  followed up in their retreat by the producti
0369  doubtful forms, and some few are distinct yet  CLOSELY  allied or representative species. In illustr
0371  ca during the later tertiary stages were more  CLOSELY  related to each other than they are at the p
0372  and others as distinct species; and a host of  CLOSELY  allied or representative forms which are ran
0372  ount, on the theory of modification, for many  CLOSELY  allied forms now living in areas completely
0372  ica; and the still more striking case of many  CLOSELY  allied crustaceans (as described in Dana's a
0372  continents of the Old World, we see countries  CLOSELY  corresponding in all their physical conditio
0374  e as these two points are; and there are many  CLOSELY  allied species. On the lofty mountains of eq
0376  ome are certainly identical, and many, though  CLOSELY  related to northern forms, must be ranked as
0395  p submarine banks, and they are inhabited by   CLOSELY  allied or identical quadrupeds. No doubt som
0398  sses are associated together, which resembles  CLOSELY  the conditions of the South American coast:
0399  a than to America, are related, and that very  CLOSELY, as we know from Dr. Hooker's account, to th
0399  ew zealand in its endemic plants is much more  CLOSELY  related to Australia, the nearest mainland,
0400  hipelago, though specifically distinct, to be  CLOSELY  allied to those of the nearest continent, we
0400  shown, in a quite marvellous manner, by very   CLOSELY  related species; so that the inhabitants of
0402  hink, an erroneous view of the probability of  CLOSELY  allied species invading each other's territo
0402  aturalised in new countries are not generally  CLOSELY  allied to the aboriginal inhabitants, but ar
0402  d, are distinct on each; thus there are three  CLOSELY  allied species of mocking thrush, each confi
0404  wo regions, let them be ever so distant, many  CLOSELY  allied or representative species occur, ther
0404  on between the two regions. And wherever many  CLOSELY  allied species occur, there will be found ma
0409  lly distinct on the several islets, should be  CLOSELY  related to each other, and likewise be relat
0409  each other, and likewise be related; but less  CLOSELY, to those of the nearest continent or other
0419  tion, more especially in very large groups of  CLOSELY  allied forms. Temminck insists on the utilit
0424  e includes varieties, not solely because they  CLOSELY  resemble the parent form, but because they a
0439  beetles, etc., resemble each other much more   CLOSELY  than do the mature insects: but in the case
0439  ate age: thus birds of the same genus, and of  CLOSELY  allied genera, often resemble each other in
0440  arvae, though active, still obey more or less  CLOSELY  the law of common embryonic resemblance. Cir
0442  heri of the structure of the embryo not being  CLOSELY  related to its conditions of existence, exce
0444  aring so different, are really varieties most  CLOSELY  allied, and have probably descended from the
0447  ifestly tend to resemble each other much more  CLOSELY  than do the adults, just as we have seen in
0447  parent species will still resemble each other  CLOSELY, for they will not have been modified. But i
0447  hort faced tumbler; the young or embryo would  CLOSELY  resemble the mature parent form. We have see
0448  se cases not undergoing any metamorphosis, or  CLOSELY  resembling their parents from their earliest
0449  lar parents, and are therefore in that degree  CLOSELY  related. Thus, community in embryonic struct
0451  the same species) resembling each other most  CLOSELY  in all respects, one of which will have full
0452  velopment and in other respects. Moreover, in  CLOSELY  allied species, the degree to which the same
0463  e species into another district occupied by a  CLOSELY  allied species, we have no just right to exp
0464  re species if we were to examine them ever so  CLOSELY, unless we likewise possessed many of the in
0465  ecutive formations invariably being much more  CLOSELY  related to each other, than are the fossils
0470  ce than do the species of smaller genera. The  CLOSELY  allied species also of the larger genera app
0478  ndependent acts of creation. The existence of  CLOSELY  allied or representative species in any two
0478  we almost invariably find that wherever many  CLOSELY  allied species inhabit two areas, some ident
0478  ies common to both still exist. Wherever many  CLOSELY  allied yet distinct species occur, many doub
0479  als, birds, reptiles, and fishes should be so  CLOSELY  alike, and should be so unlike the adult for
0483  pattern, and at an embryonic age the species   CLOSELY  resemble each other. Therefore I cannot doub
0486  individuals of the same species, and all the  CLOSELY  allied species of most genera, have within a
0423  same with varieties as with species, namely,  CLOSENESS  of descent with various degrees of modifica
0234  o the Melipona, if she were to make her cells  CLOSER  together, and more regular in every way than
```

Page **************************************(Key Word)**************************************

closer

Page **************************************(Key Word)**************************************

```
0263  sses and of Hybrids. We may now look a little  CLOSER  at the probable causes of the sterility of fi
0344  but more commonly only bringing them a little  CLOSER  together. The more ancient a form is, the mor
0376  wonderful fact that New Zealand should have a  CLOSER  resemblance in its crustacea to Great Britain
0400  stly distinct, are related in an incomparably  CLOSER  degree to each other than to the inhabitants
0032  b breeds. And when a cross has been made, the  CLOSEST selection is far more indispensable even tha
0041  ul to man, or so much valued by him, that the  CLOSEST attention should be paid to even the slighte
0077  antage of plumed seeds no doubt stands in the  CLOSEST relation to the land being already thickly c
0110  and finally extinct. The forms which stand in  CLOSEST competition with those undergoing modificati
0114  ut, it is seen, that where they come into the  CLOSEST competition with each other, the advantages
0296  cal changes recuiring much time. Nor will the  CLOSEST inspection of a formation give any idea of t
0412  life which, in any small area, come into the  CLOSEST competition, and by looking to certain facts
0423  her all languages, extinct and modern, by the  CLOSEST affinities, and would give the filiation and
0468  he same species come in all respects into the  CLOSEST competition with each other, the struggle wi
0063  nts of the same and other kinds which already  CLOTHE the ground. The missletoe is dependent on the
0022  of the down with which the nestling birds are  CLOTHED when hatched. The shape and size of the eggs
0072  n as the land was enclosed, it became thickly  CLOTHED with vigorously growing young firs. Yet the
0075  seedlings, or on the other plants which first  CLOTHED the ground and thus checked the growth of th
0077  st relation to the land being already thickly  CLOTHED by other plants: so that the seeds may be wi
0283  treating cliffs rounded boulders, all thickly  CLOTHED by marine productions, showing how little th
0366  antic moraines, left by old glaciers, are now  CLOTHED by the vine and maize. Throughout a large pa
0378  at large spaces of the tropical lowlands were  CLOTHED with a mingled tropical and temperate vegeta
0399  ly from the antarctic islands, when they were  CLOTHED with vegetation, before the commencement of
0452  : but the style remains well developed and is  CLOTHED with hairs as in other compositae, for the p
0489  interesting to contemplate an entangled bank,  CLOTHED with many plants of many kinds, with birds s
0065  t numerous over the wide plains of La Plata,  CLOTHING square leagues of surface almost to the excl
0074  ricts. When we look at the plants and bushes  CLOTHING an entangled bank, we are tempted to attribu
0203  ect the elaboration by our fir trees of dense  CLOUDS of pollen, in order that a few granules may b
0073  s: but humble bees alone visit the common red  CLOVER (Trifolium pratense), as other bees cannot re
0074  very rare in England, the heartsease and red  CLOVER would become very rare, or wholly disappear.
0094  n easily suck the nectar out of the incarnate  CLOVER, but not out of the common red clover, which
0095  carnate clover, but not out of the common red  CLOVER, which is visited by humble bees alone: so th
0095  e bees alone: so that whole fields of the red  CLOVER offer in vain an abundant supply of precious
0095  ave found by experiment that the fertility of  CLOVER greatly depends on bees visiting and moving p
0095  try, it might be a great advantage to the red  CLOVER to have a shorter or more deeply divided tube
0362  eeds of the oat, wheat, millet, canary, hemp,  CLOVER, and beet germinated after having been from t
0073  highly beneficial to the fertilisation of our  CLOVERS: but humble bees alone visit the common red
0094  the corollas of the common red and incarnate  CLOVERS (Trifolium pratense and incarnatum) do not o
0021  een permitted to join two of the London Pigeon  CLUBS. The diversity of the breeds is something ast
0004  er domestication, afforded the best and safest  CLUE. I may venture to express my conviction of the
0455  ess in the pronunciation, but which serve as a  CLUE in seeking for its derivation. On the view of
0222  booty, for about forty yards, to a very thick  CLUMP of heath, whence I saw the last individual of
0071  . here there are extensive heaths, with a few  CLUMPS of old Scotch firs on the distant hilltops: w
0072  e a single Scotch fir, except the old planted  CLUMPS. But on looking closely between the stems of
0072  ome hundred yards distant from one of the old  CLUMPS, I counted thirty two little trees: and one o
0385  d just hatched shells crawled on the feet, and  CLUNG to them so firmly that when taken out of the
0231  owned by a strong coping of wax, the bees can  CLUSTER and crawl over the comb without injuring the
0057  rked, little groups of species are generally  CLUSTERED like satellites around certain other specie
0057  forms, unequally related to each other, and  CLUSTERED round certain forms, that is, round their p
0129  nked in a single file, but seem rather to be  CLUSTERED round points, and these round other points,
0197  by the aid of exquisitely constructed hooks  CLUSTERED around the ends of the branches, and this c
0470  arently have restricted ranges, and they are  CLUSTERED in little groups round other species, in wh
0059  t unequally, allied together, forming little  CLUSTERS round certain species. Species very closely
0415  together they appear insufficient to separate  CNESTIS from Connarus. To give an example amongst in
0003  o acquire that perfection of structure and  COADAPTATION which most justly excites our admiration.
0004  insight into the means of modification and  COADAPTATION. At the commencement of my observations it
0004  explanation, for it leaves the case of the  COADAPTATION of organic beings to each other and to th
0109  the beauty and infinite complexity of the  COADAPTATIONS between all organic beings, one with anot
0012  ss dogs have imperfect teeth: long haired and  COARSE haired animals are apt to have, as is asserte
0153  t expect to fail so far as to breed a bird as  COARSE as a common tumbler from a good short faced s
0187  red sensitive to light, and likewise to those  COARSER vibrations of the air which produce sound. I
0095  y seldom hear the action, for instance, of the  COAST waves, called a trifling and insignificant ca
0132  phere, than when living on islands or near the  COAST. So with insects, Wollaston is convinced that
0136  er flying. As with mariners shipwrecked near a  COAST, it would have been better for the good swimm
0282  nd us. It is good to wander along lines of sea  COAST, when formed of moderately hard rocks, and ma
0286  height to be eaten back along a whole line of  COAST at the rate on one yard in nearly every twent
0286  other hand, I do not believe that any line of  COAST, ten or twenty miles in length, ever suffers
0287  ould, likewise, have escaped the action of the  COAST waves. So that in all probability a far longe
0290  short 'geological period. Along the whole west  COAST, which is inhabited by a peculiar marine faun
0290  e reflection will explain why along the rising  COAST of the western side of South America, no exte
0290  en great, from the enormous degradation of the  COAST rocks and from muddy streams entering the sea
0290  of the land within the grinding action of the  COAST waves. We may, I think, safely conclude that
0291  he only ancient tertiary formation on the west  COAST of South America, which has been bulky enough
0292  upraised and brought within the limits of the  COAST action. Thus the geological record will almos
0300  lmost as soon as accumulated, by the incessant  COAST action, as we now see on the shores of South
0362  france to England, as the hawks on the English  COAST destroyed so many on their arrival. Some hawk
0366  ders, and rocks scored by drifted icebergs and  COAST ice, plainly reveal a former cold period. The
0398  s closely the conditions of the South American  COAST: in fact there is a considerable dissimilarit
0481  t valleys excavated, by the slow action of the  COAST waves. The mind cannot possibly grasp the ful
0133  f islands. The insect species confined to sea  COASTS, as every collector knows, are often brassy o
0283  impressed with the slowness with which rocky  COASTS are worn away. The observations on this head
0286  ld at this rate excepting on the most exposed  COASTS: though no doubt the degradation of a lofty c
0290  ning many hundred miles of the South American  COASTS, which have been upraised several hundred fee
0288  hamalinae (a sub family of sessile cirripedes)  COAT the rocks all over the world in infinite numbe
0187  commence a series with an optic nerve merely  COATED with pigment, and without any other mechanism
0187  r crustaceans the transparent cones which are  COATED by pigment, and which properly act only by ex
0188  the simple apparatus of an optic nerve merely  COATED with pigment and invested by transparent memb
```

Page **************************************(Key Word)**************************************

0190 in the larva of the dragonfly and in the fish COBITES. In the Hydra, the animal may be turned insi
0165 as the heavy Belgian cart horse, Welch ponies, COBS, the lanky Kattywar race, etc., inhabiting the
0221 nded together probably in search of aphides or COCCI. According to Huber, who had ample opportunit
0046 degree of variability in these main nerves in COCCUS, which may almost be compared to the irregula
0030 very different ways; when we compare the game COCK, so pertinacious in battle, with other breeds
0088 d to the male sex. A hornless stag or spurless COCK would have a poor chance of leaving offspring.
0090 , the tuft of hair on the breast of the turkey COCK, which can hardly be either useful or ornament
0243 le wrens (Troglodytes) of North America, build COCK nests, to roost in, like the males of our dist
0088 e in the spurred leg, as well as the brutal COCKFIGHTER, who knows well that he can improve his br
0076 ferent climates. In Russia the small Asiatic COCKROACH has everywhere driven before it its great c
0088 ove his breed by careful selection of the best COCKS. How low in the scale of nature this law of b
0090 male carriers, horn like protuberances in the COCKS of certain fowls, etc.) which we cannot belie
0014 to appear at the corresponding caterpillar or COCOON stage. But hereditary diseases and some other
0086 d agricultural plants; in the caterpillar and COCOON stages of the varieties of the silkworm; in t
0087 insects, and used exclusively for opening the COCOON, or the hard tip to the beak of nestling bird
0444 r instance, peculiarities in the caterpillar, COCOON, or imago states of the silk moth; or, again,
0225 ries we have humble bees, which use their old COCOONS to hold honey, sometimes adding to them shor
0227 ders of wax to the circular mouths of her old COCOONS. By such modifications of instincts in thems
0029 veral sorts, for instance a Ribston pippin or CODLIN apple, could ever have proceeded from the see
0146 orthospermous in the exterior flowers and COELOSPERMOUS in the central flowers, that the elder De
0247 and blue pimpernels (Anagallis arvensis and COERULEA), which the best botanists rank as varieties
0091 ight be formed which would either supplant or COEXIST with the parent form of wolf. Or, again, the
0144 tions very frequently, and that others rarely COEXIST, without our being able to assign any reason
0323 n, no one would have suspected that they had COEXISTED with still living sea shells; but as these
0323 sea shells; but as these anomalous monsters COEXISTED with the Mastodon and Horse, it might at le
0143 as has been remarked by some authors, tend to COHERE; this is often seen in monstrous plants; and
0168 in the same way, and homologous parts tend to COHERE. Modifications in hard parts and in external
0355 d I now know from correspondence, that this COINCIDENCE he attributes to generation with modificat
0429 ith a few members preserved by some unusual COINCIDENCE of favourable circumstances. Mr. Waterhous
0355 that every species has come into existence COINCIDENT both in space and time with a pre existing
0253 s vaginalis and Reevesii, and from Phasianus COLCHICUS with P. torquatus and with P. versicolor ar
0017 of endurance of warmth by the rein deer, or of COLD by the common camel, prevented their domestica
0068 f a species, and periodical seasons of extreme COLD or drought, I believe to be the most effective
0069 food. Even when climate, for instance extreme COLD, acts directly, it will be the least vigorous,
0077 perfectly well withstand a little more heat or COLD, dampness or dryness, for elsewhere it ranges
0081 l competition cease. The land may be extremely COLD or dry, yet there will be competition between
0093 t thus have been carried. The weather had been COLD and boisterous, and therefore not favourable t
0133 rally brighter coloured than those confined to COLD and deeper seas. The birds which are confined
0139 of the same genus to inhabit very hot and very COLD countries, and as I believe that all the speci
0140 s different constitutional powers of resisting COLD. Mr. Thwaites informs me that he has observed
0141 e than any other rodent, living free under the COLD climate of Faroe in the north and of the Falkl
0366 cebergs and coast ice, plainly reveal a former COLD period. The former influence of the glacial cl
0366 d then pass away, as formerly occurred. As the COLD came on, and as each more southern zone became
0366 ld descend to the plains. By the time that the COLD had reached its maximum, we should have a unif
0367 ly cue north of them: for the migration as the COLD came on, and the re migration on the returning
0371 ving as a bridge, since rendered impassable by COLD, for the inter migration of their inhabitants.
0373 in europe we have the plainest evidence of the COLD period, from the western whores of Britain to
0374 time, as measured by years, at each point. The COLD may have come on, or have ceased, earlier at o
0377 been ample for any amount of migration. As the COLD came slowly on, all the tropical plants and ot
0377 animals can withstand a considerable amount of COLD, many might have escaped extermination during
0378 lowlands of the tropics at the period when the COLD was most intense, when arctic forms had migrat
0378 oot of the Pyrenees. At this period of extreme COLD, I believe that the climate under the equator
0477 er the most diverse conditions, under heat and COLD, on mountain and lowland, on deserts and marsh
0078 r elsewhere it ranges into slightly hotter or COLDER, damper or drier districts. In this case we c
0368 lmost conclude without other evidence, that a COLDER climate permitted their former migration acro
0374 ongitudinal belts, having been simultaneously COLDER from pole to pole, much light can be thrown o
0377 during the Glacial period simultaneously much COLDER than at present. The Glacial period, as measu
0369 nt of the Glacial epoch, and which during its COLDEST period will have been temporarily driven dow
0378 f six or seven thousand feet. During this the COLDEST period, I suppose that large spaces of the t
0136 d feeders, and which, as the flower feeding COLEOPTERA and lepidoptera, must habitually use their
0055 ed the plants of twelve countries, and the COLEOPTEROUS insects of two districts, into two nearly
0314 tion; for instance, in the land shells and COLEOPTEROUS insects of Madeira having come to differ c
0121 nct. So it probably will be with many whole COLLATERAL lines of descent, which will be conquered b
0187 o species of the same group, that is to the COLLATERAL descendants from the same original parent f
0210 f each species, but we ought to find in the COLLATERAL lines of descent some evidence of such grad
0333 heir modified descendants, or between their COLLATERAL relations. In nature the case will be far m
0045 important parts of structure, which he could COLLECT on good authority, as I have collected, duri
0142 e proportion are destroyed by frost, and then COLLECT seed from the few survivors, with care to pr
0180 o it to escape birds or beasts of prey, or to COLLECT food more quickly, or as there is reason to
0223 es not determine its own migrations, does not COLLECT food for itself or its young, and cannot eve
0223 t may be called, their aphides; and thus both COLLECT food for the community. In England the maste
0223 d the masters alone usually leave the nest to COLLECT building materials and food for themselves,
0234 of the quantity of honey which the bees could COLLECT. But let us suppose that this latter circums
0008 re enter on the copious details which I have COLLECTED on this curious subject; but to show how si
0012 n amongst animals and plants. From the facts COLLECTED by Heusinger, it appears that white sheep a
0045 e could collect on good authority, as I have COLLECTED, during a course of years. It should be rem
0096 iderations alone. In the first place, I have COLLECTED so large a body of facts, showing, in accor
0101 and from the many special facts which I have COLLECTED, but which I am not here able to give; I am
0140 he pines and rhododendrons, raised from seed COLLECTED by Dr. Hooker from trees growing at differe
0150 giving the long array of facts which I have COLLECTED, and which cannot possibly be here introduc
0162 e part or organ in an allied species. I have COLLECTED a long list of such cases; but here, as bef
0163 the hocks. With respect to the horse, I have COLLECTED cases in England of the spinal stripe in ho
0164 on further details, I may state that I have COLLECTED cases of leg and shoulder stripes in horses
0180 out of the many striking cases which I have COLLECTED, I can give only one or two instances of tr
0233 prodigious quantity of fluid nectar must be COLLECTED and consumed by the bees in a hive for the
0248 se, namely, from close interbreeding. I have COLLECTED so large a body of facts, showing that clos

```
0264  luded to a large body of facts, which I have COLLECTED, showing that when animals and plants are r
0275  mongrels. Looking to the cases which I have COLLECTED of cross bred animals closely resembling on
0287  en broken specimens, or from a few specimens COLLECTED on some one spot. Only a small portion of t
0298  same species, until many specimens have been COLLECTED from many places; and in the case of fossil
0300  ic beings; yet if all the species were to be COLLECTED which ever lived there, how imperfectly wou
0375  tervening hot lowlands. A list of the genera COLLECTED on the loftier peaks of Java raises a pictu
0092  of insects visiting flowers for the sake of COLLECTING pollen instead of nectar; and as pollen is
0218  habits; for they do not possess the pollen COLLECTING apparatus which would be necessary if they
0224  orkers than to procreate them, the habit of COLLECTING pupae originally for food might by natural
0383  hout the world. I well remember, when first COLLECTING in the fresh waters of Brazil, feeling much
0432  be driven to, if we were ever to succeed in COLLECTING all the forms in any class which have lived
0175  he examined seventeen specimens in his own COLLECTION, and not one had even a relic left. In the
0339  is even more clearly seen in the wonderful COLLECTION of fossil bones made by MM. Lund and Clause
0375  loftier peaks of Java raises a picture of a COLLECTION made on a hill in Europe! Still more striki
0433  tainly never succeed in making so perfect a COLLECTION: nevertheless, in certain classes, we are t
0463  charged with such links? Why does not every COLLECTION of fossil remains afford plain evidence of
0487  d at as a well filled museum, but as a poor COLLECTION made at hazard and at rare intervals. The a
0173  the earth is a vast museum; but the natural COLLECTIONS have been made only at intervals of time i
0279  on. On the poorness of our palaeontological COLLECTIONS. On the intermittence of geological format
0287  ld! On the poorness of our Palaeontological COLLECTIONS. That our palaeontological collections are
0287  ical collections. That our palaeontological COLLECTIONS are very imperfect, is admitted by every o
0449  est gradations, the best, or indeed, if our COLLECTIONS were nearly perfect, the only possible arr
0133  ect species confined to sea coasts, as every COLLECTOR knows, are often brassy or lurid. Plants wh
0035  odify any breed, in the same way as Bakewell, COLLINS, etc., by this very same process, only carri
0163  asionally appear: and I have been informed by COLONEL Poole that the foals of this species are gen
0164  is so generally striped, that, as I hear from COLONEL Poole, who examined the breed for the Indian
0164  ose approach to cream colour. I am aware that COLONEL Hamilton Smith, who has written on this subj
0166  this hybrid from the ass and hemionus, to ask COLONEL Poole whether such face stripes ever occur i
0313  o this latter rule, is that of the so called COLONIES of M. Barrande, which intrude for a period i
0383  lands to those of the nearest mainland. On COLONISATION from the nearest source with subsequent mo
0406  species, but are explicable on the view of COLONISATION from the nearest and readiest source, toge
0403  deira, yet this latter island has not become COLONISED by the Porto Santo species: nevertheless bo
0403  species: nevertheless both islands have been COLONISED by some European land shells which no doubt
0404  t became slowly upheaved, would naturally be COLONISED from the surrounding lowlands. So it is wit
0355  receive from it in the course of time a few COLONISTS, and their descendants, though modified, wo
0365  remote from the mainland, would not receive COLONISTS by similar means. I do not doubt that out o
0388  on a place, than in the case of terrestrial COLONISTS. We should, also, remember that some, perha
0398  galapagos Islands would be likely to receive COLONISTS, whether by occasional means of transport o
0399  de Verde islands from Africa: and that such COLONISTS would be liable to modification; the princi
0403  ted to the inhabitants of that region whence COLONISTS could most readily have been derived, the c
0403  ts could most readily have been derived, the COLONISTS having been subsequently modified and bette
0406  nt modification and better adaptation of the COLONISTS to their new homes. Summary of last and pre
0477  escendants of the same progenitors and early COLONISTS. On this same principle of former migration
0477  ons, and as the two areas will have received COLONISTS from some third source or from each other,
0011  me cases, increased size from amount of food, COLOUR from particular kinds of food and from light,
0012  thus cats with blue eyes are invariably deaf: COLOUR and constitutional peculiarities go together,
0025  nd pure black barb was of as beautiful a blue COLOUR, with the white rump, double black wing bar,
0027  t other respects to the rock pigeon; the blue COLOUR and various marks occasionally appearing in a
0033  fferent kinds of gooseberries differ in size, COLOUR, shape, and hairiness, and yet the flowers pr
0034  rom passages in Genesis, it is clear that the COLOUR of domestic animals was at that early period
0034  in south Africa match their draught cattle by COLOUR, as do some of the Esquimaux their teams of d
0039  se and the common breed, which differ only in COLOUR, that most fleeting of characters, have latel
0084  ptarmigan white in winter, the red grouse the COLOUR of heather, and the black grouse that of peat
0085  might be most effective in giving the proper COLOUR to each kind of grouse, and in keeping that c
0085  r to each kind of grouse, and in keeping that COLOUR, when once acquired, true and constant. Nor o
0085  al destruction of an animal of any particular COLOUR would produce little effect: we should rememb
0086  silkworm; in the eggs of poultry, and in the COLOUR of the down of their chickens; in the horns o
0089  eral habits of life, but differ in structure, COLOUR, or ornament, such differences have been main
0144  and deafness in cats, and the tortoise shell COLOUR with the female sex; the feathered feet and s
0144  ung birds when first hatched, with the future COLOUR of their plumage; or, again, the relation bet
0145  f the truss often loses the patches of darker COLOUR in the two upper petals; and that when this o
0145  e adherent nectary was quite aborted; when the COLOUR is absent from only one of the two upper peta
0154  plants had blue flowers and some had red, the COLOUR would by only a specific character, and no on
0154  but if all the species had blue flowers, the COLOUR would become a generic character, and its var
0161  each generation in the plumage to assume this COLOUR. This view is hypothetical, but could be supp
0162  ly we might have inferred this, from the blue COLOUR and marks so often appearing when distinct br
0164  mouse duns; by the term dun a large range of COLOUR is included, from one between brown and black
0164  brown and black to a close approach to cream COLOUR. I am aware that Colonel Hamilton Smith, who
0165  i was so convinced that not even a stripe of COLOUR appears from what would commonly be called an
0166  ub species or geographical races) of a bluish COLOUR, with certain bars and other marks: and when
0195  characters, such as the down on fruit and the COLOUR of the flesh, which, from determining the att
0197  ay that we should have thought that the green COLOUR was a beautiful adaptation to hide this tree
0197  selection: as it is, I have no doubt that the COLOUR is due to some quite distinct cause, probably
0198  re is reason to believe that constitution and COLOUR are correlated. A good observer, also, states
0198  ty to the attacks of flies is correlated with COLOUR, as is the liability to be poisoned by certai
0198  ity to be poisoned by certain plants: so that COLOUR would be thus subjected to the action of natu
0232  ermilion wax; and I invariably found that the COLOUR was most delicately diffused by the bees, as
0239  ffer from each other in size and sometimes in COLOUR; and that the extreme forms can sometimes be
0271  present no other difference besides the mere COLOUR of the flower; and one variety can sometimes
0410  ling characters in common, as of sculpture or COLOUR. In looking to the long succession of ages, a
0416  n which the wings of insects are folded, mere COLOUR in certain Algae, mere pubescence on parts of
0423  ause they are less variable than the shape or COLOUR of the body, etc.; whereas with sheep the hor
0424  negro group, however much he might differ in COLOUR and other important characters from negroes.
0473  they all agree? Why, for instance, should the COLOUR of a flower be more likely to vary in any one
```

Page **************************************(Key Word)**************************************

```
0012 sheep and pigs are differently affected from COLOURED individuals by certain vegetable poisons. Ha
0025 the several imagined aboriginal stocks were COLOURED and marked like the rock pigeon, although no
0025 , although no other existing species is thus COLOURED and marked, so that in each separate breed t
0075 h extremely close varieties as the variously COLOURED sweet peas, they must be each year harvested
0085 eshed peaches far more than those with other COLOURED flesh. If, with all the aids of art, these s
0132 n living in shallow water, are more brightly COLOURED than those of the same species further north
0132 birds of the same species are more brightly COLOURED under a clear atmosphere, than when living o
0133 ical and shallow seas are generally brighter COLOURED than those confined to cold and deeper seas.
0133 inents are, according to Mr. Gould, brighter COLOURED than those of islands. The insect species co
0133 sea; but that this other shell became bright COLOURED by variation when it ranged into warmer or s
0160 conclusion, because, as we have seen, these COLOURED marks are eminently liable to appear in the
0160 ed offspring of two distinct and differently COLOURED breeds; and in this case there is nothing in
0163 ery obscure, or actually quite lost, in dark COLOURED asses. The koulan of Pallas is said to have
0165 as given a figure of a similar mule. In four COLOURED drawings, which I have seen, of hybrids betw
0228 f wax, a thin and narrow, knife edged ridge, COLOURED with vermilion. The bees instantly began on
0232 ld have done with his brush, by atoms of the COLOURED wax having been taken from the spot on which
0270 tercrossed produce less seed, than do either COLOURED varieties when fertilised with pollen from t
0270 s when fertilised with pollen from their own COLOURED flowers. Moreover, he asserts that when yell
0271 is produced by the crosses between the same COLOURED flowers, than between those which are differ
0271 rs, than between those which are differently COLOURED. Yet these varieties of Verbascum present no
0309 vel, since the earliest silurian period. The COLOURED map appended to my volume on Coral Reefs, le
0349 constructed, but not quite alike, with eggs COLOURED in nearly the same manner. The plains near t
0473 been created independently, have differently COLOURED flowers, than if all the species of the genu
0473 f all the species of the genus have the same COLOURED flowers? If species are only well marked var
0024 ng generally in constitution, habits, voice, COLOURING, and in most parts of their structure, with
0025 highest degree. Some facts in regard to the COLOURING of pigeons well deserve consideration. The
0166 tint which approaches to that of the general COLOURING of the other species of the genus. The appe
0184 ential part of its organisation, even in its COLOURING, in the harsh tone of its voice, and undula
0025 ight be a tendency to revert to the very same COLOURS and markings. Or, secondly, that each breed,
0132 ced that residence near the sea affects their COLOURS. Moquin Tandon gives a list of plants which
0133 shell, for instance, was created with bright COLOURS for a warm sea; but that this other shell be
0162 ten appearing when distinct breeds of diverse COLOURS are crossed. Hence, though under nature it m
0163 orses of the most distinct breeds, and of all COLOURS; transverse bars on the legs are not rare in
0166 when the oldest and truest breeds of various COLOURS are crossed, we see a strong tendency for th
0445 ements made of the dam and of a three days old COLT of a race and heavy cart horse, I find that th
0445 f a race and heavy cart horse, I find that the COLTS have by no means acquired their full amount o
0023 that all have descended from the rock pigeon (COLUMBA livia), including under this term several ge
0027 l our domestic breeds have descended from the COLUMBA livia with its geographical sub species. In
0024 n vain throughout the whole great family of COLUMBIDAE for a beak like that of the English carrier
0027 ertain respects, as compared with all other COLUMBIDAE, though so like in most other respects to t
0386 t; and a water beetle of the same family, a COLYMBETES, once flew on board the Beagle, when forty
0147 head is generally accompanied by a diminished COMB, and a large beard by diminished wattles. With
0152 ction be applied, that part (for instance, the COMB in the Dorking fowl) or the whole breed will c
0216 slave making instinct of certain ants; and the COMB making power of the hive bee: these two latter
0224 n who can examine the exquisite structure of a COMB, so beautifully adapted to its end, without en
0225 midal base of a single cell on one side of the, COMB, enter into the composition of the bases of th
0225 to the latter: it forms a nearly regular waxen COMB of cylindrical cells, in which the young are h
0226 ure would probably have been as perfect as the COMB of the hive bee. Accordingly I wrote to Profes
0229 uickly. In one well marked instance, I put the COMB back into the hive, and allowed the bees to go
0230 learly seen by examining the edge of a growing COMB, do make a rough, circumferential wall or rim
0230 ugh, circumferential wall or rim all round the COMB; and they gnaw into this from the opposite sid
0231 umferential rim or wall of wax round a growing COMB, flexures may sometimes be observed, correspon
0231 f wax, the bees can cluster and crawl over the COMB without injuring the delicate hexagonal walls,
0231 building, strength is continually given to the COMB, with the utmost ultimate economy of wax. It s
0232 margin of the circumferential rim of a growing COMB, with an extremely thin layer of melted vermil
0232 in cases of difficulty, as when two pieces of COMB met at an angle, how often the bees would enti
0232 of wood, placed directly under the middle of a COMB growing downwards so that the comb has to be b
0232 middle of a comb growing downwards so that the COMB has to be built over one face of the slip, in
0235 plane surfaces; and the Melipona would make a COMB as perfect as that of the hive bee. Beyond thi
0235 ure, natural selection could not lead; for the COMB of the hive bee, as far as we can see, is abso
0203 as soon as born, or to perish herself in the COMBAT; for undoubtedly this is for the good of the
0221 ut them down on a bare spot near the place of COMBAT; they were eagerly seized, and carried off by
0221 r all, they had been victorious in their late COMBAT. At the same time I laid on the same place a
0013 y rare deviation, due to some extraordinary COMBINATION of circumstances, appears in the parent, s
0485 ho admit that genera are merely artificial COMBINATIONS made for convenience. This may not be a c
0329 ne, the series is far less perfect than if we COMBINE both into one general system. With respect t
0116 being derived from divergence of character, COMBINED with the principles of natural selection and
0131 ects of external conditions. Use and disuse, COMBINED with natural selection; organs of flight and
0136 due to the action of natural selection, but COMBINED probably with disuse. For during thousands o
0141 te constitutions, and how much to both means COMBINED, is a very obscure question. That habit or c
0143 ts of use and disuse have often been largely COMBINED with, and sometimes overmastered by, the nat
0382 eat cycles of change; and that on this view, COMBINED with modification through natural selection,
0477 on this same principle of former migration, COMBINED in most cases with modification, we can unde
0280 the fantail and pouter; none, for instance, COMBINING a tail somewhat expanded with a crop somewh
0074 the number of field mice, which destroy their COMBS and nests; and Mr. H. Newman, who has long at
0228 the example of Mr. Tegetmeier, I separated two COMBS, and put between them a long thick, square st
0229 ess, and then stopping their work. In ordinary COMBS it has appeared to me that the bees do not al
0233 , that the cells on the extreme margin of wasp COMBS are sometimes strictly hexagonal; but I have
0234 he wax necessary for the construction of their COMBS. Moreover, many bees have to remain idle for
0250 ral skill, and by his having hothouses at his COMMAND. Of his many important statements I will her
0483 's history certain elemental atoms have been COMMANDED suddenly to flash into living tissues? Do t
0093 from flower to flower, another process might COMMENCE. No naturalist doubts the advantage of what
0187 has been perfected. In the Articulata we can COMMENCE a series with an optic nerve merely coated w
0233 sect might, by fixing on a point at which to COMMENCE a cell, and then moving outside, first to on
0229 half completed rhombs at the base of a just COMMENCED cell, which were slightly concave on one si
0230 hombic plates, until the hexagonal walls are COMMENCED. Some of these statements differ from those
```

Page **************************************(Key Word)**************************************

```
0231  coping. From all the cells, both those just COMMENCED and those completed, being thus crowned by
0232  gh wall in its proper place between two just COMMENCED cells, is important, as it bears on a fact,
0233  the inside and outside of two or three cells COMMENCED at the same time, always standing at the pr
0004  s of modification and coadaptation. At the   COMMENCEMENT of my observations it seemed to me probabl
0123  her, but from having diverged at the first   COMMENCEMENT of the process of modification, will be wi
0230  s I have seen, strictly correct: the first   COMMENCEMENT having always been a little hood of wax: b
0231  d, a score of individuals work even at the   COMMENCEMENT of the first cell. I was able practically
0293  ween the allied species which lived at its   COMMENCEMENT and at its close. Some cases are on record
0302  between the species which appeared at the   COMMENCEMENT and close of each formation, pressed so ha
0303  poken of as having abruptly come in at the   COMMENCEMENT of the tertiary series. And now one of the
0303  red in the new red sandstone at nearly the   COMMENCEMENT of this great series. Cuvier used to urge
0304  t group had been suddenly developed at the   COMMENCEMENT of the tertiary series. This was a sore tr
0305  em did appear, as Agassiz believes, at the   COMMENCEMENT of the chalk formation, the fact would cer
0336  between the species which appeared at the   COMMENCEMENT and close of these periods: but we ought t
0369  t have existed on the mountains before the   COMMENCEMENT of the Glacial epoch, and which during its
0369  the Glacial period, I assumed that at its   COMMENCEMENT the arctic productions were as uniform rou
0369  rn temperate forms round the world, at the   COMMENCEMENT of the Glacial period. At the present day,
0370  climate became less warm, long before the   COMMENCEMENT of the Glacial period. We now see, as I be
0381  suppose that there has been time since the   COMMENCEMENT of the Glacial period for their migration,
0381  to a former and warmer period, before the   COMMENCEMENT of the Glacial period, when the antarctic
0382  and now sunken islands, and perhaps at the   COMMENCEMENT of the Glacial period, by icebergs. By the
0399  y were clothed with vegetation, before the   COMMENCEMENT of the Glacial period. The affinity, which
0050  s or mere varieties. When a young naturalist COMMENCES the study of a group of organisms quite unk
0254  from two or more aboriginal species, since   COMMINGLED by intercrossing. On this view, the aborigi
0379  southern forms. And thus, when they became   COMMINGLED during the Glacial period, the northern for
0403  ly played an important part in checking the   COMMINGLING of species under the same conditions of li
0304  is sessile cirripede was a Chthamalus, a very COMMON, large, and ubiquitous genus, of which not on
0017  of warmth by the rein deer, or of cold by the COMMON camel, prevented their domestication? I canno
0018  the breeds of poultry have proceeded from the COMMON wild Indian fowl (Gallus bankiva). In regard
0019  t doubt that they all have descended from the COMMON wild duck and rabbit. The doctrine of the ori
0021  outline almost like that of a finch; and the COMMON tumbler has the singular and strictly inherit
0023  eds of pigeons, I am fully convinced that the COMMON opinion of naturalists is correct, namely, th
0023  ers, are unlikely to be exterminated, and the COMMON rock pigeon, which has the same habits with t
0028  ng that they could ever have descended from a COMMON parent, as any naturalist could in coming to
0032  instance, the steadily increasing size of the COMMON gooseberry may be quoted. We see an astonishi
0039  the standard of perfection of each breed. The COMMON goose has not given rise to any marked variet
0039  marked varieties; hence the Thoulouse and the COMMON breed, which differ only in colour, that most
0044  ful species. Wide ranging, much diffused, and COMMON species vary most. Species of the larger gene
0047  e as a variety of the other, ranking the most COMMON, but sometimes the one first described, as th
0048  ely be found within the same country, but are COMMON in separated areas. How many of those birds a
0050  ntal evidence, showing that they descend from COMMON parents, and consequently must be ranked as v
0050  anked by some authors as species. Look at the COMMON oak, how closely it has been studied; yet a G
0053  y limited country, the species which are most COMMON, that is abound most in individuals; and the
0054  er side, a somewhat larger number of the very COMMON and much diffused or dominant species will be
0064  given of introduced plants which have become COMMON throughout whole islands in a period of less
0071  reater, for six insectivorous birds were very COMMON in the plantations, which were not to be seen
0073  our clovers; but humble bees alone visit the COMMON red clover (Trifolium pratense), as other bee
0080  ter and Extinction, on the descendants from a COMMON parent. Explains the grouping of all organic
0092  guminosae, and at the back of the leaf of the COMMON laurel. This juice, though small in quantity,
0094  nc the nectar by continued selection, to be a COMMON plant; and that certain insects depended in m
0094  f structure. The tubes of the corollas of the COMMON red and incarnate clovers (Trifolium pratense
0094  t of the incarnate clover, but not out of the COMMON red clover, which is visited by humble bees a
0102  t intending to alter the breed, have a nearly COMMON standard of perfection, and all try to get an
0110  in the second chapter, showing that it is the COMMON species which afford the greatest number of r
0112  character both from each other and from their COMMON parent. But how, it may be asked, can any ana
0117  species with restricted ranges. Let (A) be a COMMON, widely diffused, and varying species, belong
0118  to inherit those advantages which made their COMMON parent (A) more numerous than most of the oth
0118  each other, and more considerably from their COMMON parent (A). We may continue the process by si
0118  or modified descendants, proceeding from the COMMON parent (A), will generally go on increasing i
0119  ation. As all the modified descendants from a COMMON and widely diffused species, belonging to a l
0120  aps unequally, from each other and from their COMMON parent. If we suppose the amount of change be
0122  es (A) and (I), were also supposed to be very COMMON and widely diffused species, so that they mus
0125  shows that its species have inherited from a COMMON ancestor some advantage in common. Hence, the
0125  ited from a common ancestor some advantage in COMMON. Hence, the struggle for the production of ne
0128  distinct genera. We have seen that it is the COMMON, the widely diffused, and widely ranging spec
0131  rto sometimes spoken as if the variations, so COMMON and multiform in organic beings under domesti
0138  nder a nearly similar climate; so that on the COMMON view of the blind animals having been separat
0139  words on acclimatisation. As it is extremely COMMON for species of the same genus to inhabit very
0140  e of far extended transportation, I think the COMMON and extraordinary capacity in our domestic an
0141  te wide flexibility of constitution, which is COMMON to most animals. On this view, the capacity o
0141  s anomalies, but merely as examples of a very COMMON flexibility of constitution, brought, under p
0143  seen in monstrous plants; and nothing is more COMMON than the union of homologous parts in normal
0146  o correlation of growth, structures which are COMMON to whole groups of species, and which in trut
0153  fail so far as to breed a bird as coarse as a COMMON tumbler from a good short faced strain. But a
0153  period when the species branched off from the COMMON progenitor of the genus. This period will sel
0155  d generic characters; and these characters in COMMON I attribute to inheritance from a common prog
0155  s in common I attribute to inheritance from a COMMON progenitor, for it can rarely have happened t
0156  hen the species first branched off from their COMMON progenitor, and subsequently have not varied
0156  od of the branching off of the species from a COMMON progenitor, it is probable that they should s
0157  joints in the tarsi is a character generally COMMON to very large groups of beetles; but in the E
0157  character of the highest importance, because COMMON to large groups; but in certain genera the ne
0157  uently, whatever part of the structure of the COMMON progenitor, or of its early descendants, beca
0158  acters, or those which the species possess in COMMON; that the frequent extreme variability of any
0158  extraordinarily it may be developed, if it be COMMON to a whole group of species; that the great v
0158  ies of the same group having descended from a COMMON progenitor, from whom they have inherited muc
0158  enitor, from whom they have inherited much in COMMON, to parts which have recently and largely var
```

Page ***(Key Word)***

```
0159 l races of the pigeon having inherited from a COMMON parent the same constitution and tendency to
0159 k as varieties produced by cultivation from a COMMON parent: if this be not so, the case will then
0159 nd to these a third may be added, namely, the COMMON turnip. According to the ordinary view of eac
0160 enerations, the proportion of blood, to use a COMMON expression, of any one ancestor, is only 1 in
0161 ce a rudiment inherited: for instance, in the COMMON snapdragon (Antirrhinum) a rudiment of a fift
0161 posed, on my theory, to have descended from a COMMON parent, it might be expected that they would
0161 ver, we never know the exact character of the COMMON ancestor of a group, we could not distinguish
0164 is sometimes double and sometimes treble, is COMMON: the side of the face, moreover, is sometimes
0165 of the horse genus. Rollin asserts, that the COMMON mule from the ass and horse is particularly a
0167 s otherwise very differently constructed, the COMMON parent of our domestic horse, whether or not
0168 species of the same genus branched off from a COMMON parent, are more variable than generic charac
0169 nheriting nearly the same constitution from a COMMON parent and exposed to similar influences will
0173 ry these allied species have descended from a COMMON parent; and during the process of modificatio
0175 bruptly, as Alph. de Candolle has observed, a COMMON alpine species disappears. The same fact has
0177 hich exist in lesser numbers. Hence, the more COMMON forms, in the race for life, will tend to bea
0177 life, will tend to beat and supplant the less COMMON forms, for these will be more slowly modified
0177 inciple which, as I believe, accounts for the COMMON species in each country, as shown in the seco
0178 the several representative species and their COMMON parent, must formerly have existed in each br
0184 lainly of its close blood relationship to our COMMON species; yet it is a woodpecker which never c
0189 extinction. Or again, if we look to an organ COMMON to all the members of a large class, for in t
0193 intimate structure closely resembles that of COMMON muscle; and as it has lately been shown that
0193 attribute its presence to inheritance from a COMMON ancestor; and its absence in some of the memb
0194 s, which owe but little of their structure in COMMON to inheritance from the same ancestor. Althou
0200 se, and bat, which have been inherited from a COMMON progenitor, were formerly of more special use
0234 r, so as to intersect a little: for a wall in COMMON even to two adjoining cells, would save some
0236 nt for their similarity by inheritance from a COMMON parent, and must therefore believe that they
0243 ther facts in regard to instincts; as by that COMMON case of closely allied, but certainly distinc
0246 nt, when the cause of the sterility, which is COMMON to the two cases, has to be considered. The d
0246 orms known or believed to have descended from COMMON parents, when intercrossed, and likewise the
0247 eded in getting fertile seed; as he found the COMMON red and blue pimpernels (Anagallis arvensis a
0253 r are perfectly fertile. The hybrids from the COMMON and Chinese geese (A. cygnoides), species whi
0258 ity in making reciprocal crosses is extremely COMMON in a lesser degree. He has observed it even b
0262 om equal, so it sometimes is in grafting; the COMMON gooseberry, for instance, cannot be grafted o
0264 and long lived, as we see in the case of the COMMON mule. Hybrids, however, are differently circu
0266 es, in some respects allied, sterility is the COMMON result, in the one case from the conditions o
0267 f facts seem to be connected together by some COMMON but unknown bond, which is essentially relate
0271 the remarkable fact, that one variety of the COMMON tobacco is more fertile, when crossed with a
0280 forms intermediate between each species and a COMMON but unknown progenitor; and the progenitor wi
0281 between them, but between each and an unknown COMMON parent. The common parent will have had in it
0281 etween each and an unknown common parent. The COMMON parent will have had in its whole organisatio
0282 and so on backwards, always converging to the COMMON ancestor of each great class. So that the num
0294 ted, that some few still existing species are COMMON in the deposit, but have become extinct in th
0312 ion of organic beings, better accord with the COMMON view of the immutability of species, or with
0321 ill suffer from some inherited inferiority in COMMON. But whether it be species belonging to the s
0322 for they will partake of some inferiority in COMMON. Thus, as it seems to me, the manner in which
0323 its. Even if the new fossil species which are COMMON to the Old and New Worlds be kept wholly dist
0327 n groups, from inheriting some inferiority in COMMON; and therefore as new and improved groups spr
0330 nera belonging to distinct families. The most COMMON case, especially with respect to very distinc
0330 de some small approach to each other. It is a COMMON belief that the more ancient a form is, by so
0331 der: for all will have inherited something in COMMON from their ancient and common progenitor. On
0331 ed something in common from their ancient and COMMON progenitor. On the principle of the continued
0333 they have not diverged in character from the COMMON progenitor of the order, nearly so much as th
0338 would be vain to look for animals having the COMMON embryological character of the Vertebrata, un
0344 oups, from their inferiority inherited from a COMMON progenitor, tend to become extinct together,
0344 be related to, and consequently resemble, the COMMON progenitor of groups, since become widely div
0348 stinct, with hardly a fish, shell, or crab in COMMON, than those of the eastern and western shores
0348 s. although hardly one shell, crab or fish is COMMON to the above named three approximate faunas o
0348 ic into the Indian Ocean, and many shells are COMMON to the eastern islands of the Pacific and the
0352 te points, why do we not find a single mammal COMMON to Europe and Australia or South America? The
0354 which on my theory have all descended from a COMMON progenitor, can have migrated (undergoing mod
0355 ts of the continent. Cases of this nature are COMMON, and are, as we shall hereafter more fully se
0363 rom the large number of the species of plants COMMON to Europe, in comparison with the plants of o
0371 he pliocene period, as soon as the species in COMMON, which inhabited the New and Old Worlds, migr
0374 inconsiderable part of its scanty flora, are COMMON to Europe, enormously remote as these two poi
0376 orms me that twenty five species of Algae are COMMON to New Zealand and to Europe, but have not be
0378 flowering plants, about forty six in number, COMMON to Tierra del Fuego and to Europe still exist
0381 e or thrice as far, and is twice or thrice as COMMON, as another species within their own homes. I
0381 es have migrated in radiating lines from some COMMON centre; and I am inclined to look in the sout
0384 for two river systems will have some fish in COMMON and some different. A few facts seem to favou
0385 es, which, on my theory, are descended from a COMMON parent and must have proceeded from a single
0389 rent; and therefore have all proceeded from a COMMON birthplace, notwithstanding that in the cours
0401 ld and those confined to the archipelago, are COMMON to the several islands, and we may infer from
0405 have been ranked as distinct species, and the COMMON range would have been greatly reduced. Still
0405 ly ancient, and must have branched off from a COMMON parent at a remote epoch; so that in such cas
0409 and another group should have all its species COMMON to other quarters of the world. We can see wh
0410 often characterised by trifling characters in COMMON, as of sculpture or colour. In looking to the
0411 is the widely ranging, the much diffused and COMMON, that is the dominant species belonging to th
0412 nd, consequently, have inherited something in COMMON. But the three genera on the left hand have,
0412 ft hand have, on this same principle, much in COMMON, and form a sub family, distinct from that in
0412 nera on the right hand, which diverged from a COMMON parent at the fifth stage of descent. These f
0412 e five genera have also much, though less, in COMMON; and they form a family distinct from that in
0413 at is, by one sentence to give the characters COMMON, for instance, to all mammals, by another tho
0413 or instance, to all mammals, by another those COMMON to all carnivora, by another those common to
0413 ose common to all carnivora, by another those COMMON to the dog genus, and then by adding a single
0418 if they find a character nearly uniform, and COMMON to a great number of forms, and not common to
0418 nd common to a great number of forms, and not COMMON to others, they use it as one of high value;
```

Page ***(Key Word)***

Page **************************************(Key Word)**************************************

```
0418  others, they use it as one of high value; if COMMON to some lesser number, they use it as of subo
0419  easier than to define a number of characters COMMON to all birds; but in the case of crustaceans,
0419  the series, which have hardly a character in COMMON; yet the species at both ends, from being pla
0420  s, are those which have been inherited from a COMMON parent, and, in so far, all true classificati
0420  h allied in the same degree in blood to their COMMON progenitor, may differ greatly, being due to
0421  dants from A will have inherited something in COMMON from their common parent, as will all the des
0421  have inherited something in common from their COMMON parent, as will all the descendants from I; s
0422  sation of the several races, descended from a COMMON race) had altered much, and had given rise to
0423  nearly identical; no one puts the swedish and COMMON turnips together, though the esculent and thi
0424  ak, yet all are kept together from having the COMMON habit of tumbling; but the short faced breed
0424  : scarcely a single fact can be predicated in COMMON of the males and hermaphrodites of certain ci
0425  us; for where there has been close descent in COMMON, there will certainly be close resemblance or
0426  ferent habits, only by its inheritance from a COMMON parent. We may err in this respect in regard
0426  t these characters have been inherited from a COMMON ancestor. And we know that such correlated or
0426  let two forms have not a single character in  COMMON, yet if these extreme forms are connected tog
0427  c varieties, as in the thickened stems of the COMMON and swedish turnip. The resemblance of the gr
0428  l shape of body and structure of limbs from a COMMON ancestor. So it is with fishes. As members of
0430  , these are due on my theory to inheritance in COMMON. Therefore we must suppose either that all Ro
0430  th rodents and Marsupials branched off from a COMMON progenitor, and that both groups have since u
0430  ing partially retained the character of their COMMON progenitor, or of an early member of the grou
0431  in character of the species descended from a  COMMON parent, together with their retention by inhe
0431  etention by inheritance of some characters in COMMON, we can understand the excessively complex an
0431  higher group are connected together. For the  COMMON parent of a whole family of species, now brok
0432  ed from A, or from I, would have something in  COMMON. In a tree we can specify this or that branch
0433  f all ages, although having few characters in COMMON, under one species; we use descent in classin
0433  of difference between the descendants from a  COMMON parent, expressed by the terms genera, famili
0436  f insects, we have only to suppose that their COMMON progenitor had an upper lip, mandibles, and t
0437  e repetition of the same part or organ is the COMMON characteristic (as Owen has observed) of all
0438  phosed, not one from the other, but from some COMMON element. Naturalists, however, use such langu
0440  e, still obey more or less closely the law of COMMON embryonic resemblance. Cirripedes afford a go
0450  o as a picture, more or less obscured, of the COMMON parent form of each great class of animals. R
0450  bearing the stamp of inutility, are extremely COMMON throughout nature. For instance, rudimentary
0452  different members of a class, nothing is more COMMON, or more necessary, than the use and discover
0456  ules; all naturally follow on the view of the COMMON parentage of those forms which are considered
0458  ded, each within its own class or group, from COMMON parents, and have all been modified in the co
0461  r even higher group, must have descended from COMMON parents; and therefore, in however distant an
0464  at naturalists will be able to decide, on the COMMON view, whether or not these doubtful forms are
0467  e high geometrical ratio of increase which is COMMON to all organic beings. This high rate of incr
0473  already varied since they branched off from a COMMON progenitor in certain characters, by which th
0474  since the several species branched off from a COMMON progenitor, an unusual amount of variability
0474  able than any other structure, if the part be COMMON to many subordinate forms, that is, if it has
0474  ies of the same genus having descended from a COMMON parent, and having inherited much in common,
0474  a common parent, and having inherited much in COMMON, we can understand how it is that allied spec
0476  living and the extinct being the offspring of COMMON parents. As the groups which have descended f
0478  ies inhabit two areas, some identical species COMMON to both still exist. Wherever many closely al
0484  , nevertheless all living things have much in COMMON, in their chemical composition, their germina
0485  and cowslip: and in this case scientific and COMMON language will come into accordance. In short,
0489  to futurity so to foretel that it will be the COMMON and widely spread species, belonging to the l
0013  ucture are truly inherited, less strange and  COMMONER deviations may be freely admitted to be inhe
0110  for life by the modified descendants of the   COMMONER species. From these several considerations I
0164  english race horse the spinal stripe is much  COMMONER in the foal than in the full grown animal. W
0341  o progeny. Or, which would probably be a far  COMMONER case, two or three species of two or three a
0117  , also, seen that the species, which are the  COMMONEST and the most widely diffused, vary more tha
0325  lants which are dominant, that is, which are  COMMONEST in their own homes, and are most widely dif
0004  f such studies, although they have been very  COMMONLY neglected by naturalists. From these conside
0013  sex to both sexes, or to one sex alone, more  COMMONLY but not exclusively to the like sex. It is a
0047  onnected by intermediate links; nor will the  COMMONLY assumed hybrid nature of the intermediate li
0115  osely allied to the indigenes: for these are  COMMONLY looked at as specially created and adapted f
0154  of less physiological importance than those  COMMONLY used for classing genera. I believe this exp
0159  ariation, in the enlarged stems, or roots as  COMMONLY called, of the Swedish turnip and Ruta baga,
0165  n a stripe of colour appears from what would  COMMONLY be called an accident, that I was led solely
0166  he stripes are either plainer or appear more  COMMONLY in the young than in the old. Call the breed
0207  how that several distinct mental actions are  COMMONLY embraced by this term; but every one underst
0216  wonderful of all known instincts. It is now  COMMONLY admitted that the more immediate and final c
0250  that we here have perfect, or even more than  COMMONLY perfect, fertility in a first cross between
0271  . he experimentised on five forms, which are  COMMONLY reputed to be varieties, and which he tested
0279  ty. I assigned reasons why such links do not  COMMONLY occur at the present day, under the circumst
0286  if, then, we knew the rate at which the sea  COMMONLY wears away a line of cliff of any given heig
0322  the forms which thus yield their places will  COMMONLY be allied, for they will partake of some inf
0344  ously classed as distinct into one; but more  COMMONLY only bringing them a little closer together.
0351  ies are confined to the same areas; as is so  COMMONLY and notoriously the case. I believe, as was
0411  different in nature; for it is notorious how  COMMONLY members of even the same sub group have diff
0429  as i find after some investigation, does not  COMMONLY fall to the lot of aberrant genera. We can,
0443  the view of descent with modification. It is  COMMONLY assumed, perhaps from monstrosities often af
0469  of demarcation can be drawn between species;  COMMONLY supposed to have been produced by special ac
0053  om wide range, and to a certain extent from  COMMONNESS), often give rise to varieties sufficiently
0018  orm no opinion. I should think, from facts  COMMUNICATED to me by Mr. Blyth, on the habits, voice,
0254  are quite fertile together; but from facts  COMMUNICATED to me by Mr. Blyth, I think they must be c
0264  ended to; but I believe, from observations  COMMUNICATED to me by Mr. Hewitt, who has had great exp
0378  nvasion: and it is a striking fact, lately  COMMUNICATED to me by Dr. Hooker, that all the flowerin
0402  sland. Sir C. Lyell and Mr. Wollaston have  COMMUNICATED to me a remarkable fact bearing on this su
0443  , which appears in old age alone, has been  COMMUNICATED to the offspring from the reproductive ele
0040  semi civilised countries, with little free  COMMUNICATION, the spreading and knowledge of any new s
0408  one region; according to the nature of the  COMMUNICATION which allowed certain forms and not other
0220  species are found only in their own proper  COMMUNITIES, and have never been observed in the nests
0236  to the neuters or sterile females in insect  COMMUNITIES: for these neuters often differ widely in
```

Page **************************************(Key Word)**************************************

Page **(Key Word)**

```
0237 rile condition of certain members of insect COMMUNITIES: the difficulty lies in understanding how
0242 ed this admirable division of labour in the COMMUNITIES of ants, by the means of natural selection
0044 almost equally difficult to define, but here COMMUNITY of descent is almost universally implied, t
0087 re of each individual for the benefit of the COMMUNITY; if each in consequence profits by the sele
0159 these three plants, not to the vera causa of COMMUNITY of descent, and a consequent tendency to va
0202 whole the power of stinging be useful to the COMMUNITY, it will fulfil all the requirements of nat
0202 of drones, which are utterly useless to the COMMUNITY for any other end, or which are ultimately
0203 for undoubtedly this is for the good of the COMMUNITY: and maternal love or maternal hatred, thou
0221 wever, in the month of July, I came across a COMMUNITY with an unusually large stock of slaves, an
0221 d were vigorously repulsed by an independent COMMUNITY of the slave species (F. fusca) sometimes a
0222 stance I found to my surprise, an independent COMMUNITY of F. flava under a stone beneath a nest of
0222 off the pupae. One evening I visited another COMMUNITY of F. sanguinea, and found a number of thes
0223 aphides; and thus both collect food for the COMMUNITY. In England the masters alone usually leave
0234 country; and let us further suppose that the COMMUNITY lived throughout the winter, and consequent
0236 en social, and it had been profitable to the COMMUNITY that a number should have been annually bor
0238 sterile condition of certain members of the COMMUNITY, has been advantageous to the community: co
0238 the community, has been advantageous to the COMMUNITY: consequently the fertile males and females
0238 ly the fertile males and females of the same COMMUNITY flourished, and transmitted to their fertil
0240 ller workers had been the most useful to the COMMUNITY, and those males and females had been conti
0241 eme forms; from being the most useful to the COMMUNITY, having been produced in greater and greate
0241 l their production may have been to a social COMMUNITY of insects, on the same principle that the
0242 olition, in the utterly sterile members of a COMMUNITY could possibly have affected the structure
0420 ll true classification is genealogical; that COMMUNITY of descent is the hidden bond which natural
0425 ve no written pedigrees; we have to make out COMMUNITY of descent by resemblances of any kind. The
0426 r so unimportant, betrays the hidden bond of COMMUNITY of descent. Let two forms have, not a single
0426 ermediate groups, we may at once infer their COMMUNITY of descent; and we put them all into the sa
0449 refore in that degree closely related. Thus, COMMUNITY in embryonic structure reveals community of
0449 us, community in embryonic structure reveals COMMUNITY of descent. It will reveal this community o
0449 ls community of descent. It will reveal this COMMUNITY of descent; however much the structure of t
0460 es of workers or sterile females in the same COMMUNITY of ants; but I have attempted to show how t
0479 all organic beings are due to inheritance or COMMUNITY of descent. The natural system is a genealo
0485 ed by naturalists of affinity, relationship, COMMUNITY of type, paternity, morphology, adaptive ch
0065 india, as I hear from Dr. Falconer, from Cape COMORIN to the Himalaya, which have been imported fr
0021 erited habit of flying at a great height in a COMPACT flock, and tumbling in the air head over hee
0038 ed selection up to a standard of perfection COMPARABLE with that given to the plants in countries
0207 y of Type. Chapter VII. Instinct. Instincts COMPARABLE with habits, but different in their origin.
0419 and botanists. Finally, with respect to the COMPARATIVE value of the various groups of species, su
0154 e cases in which the modification has been COMPARATIVELY recent and extraordinarily great that we
0175 ributed that each has a wide range, with a COMPARATIVELY narrow neutral territory between them, in
0177 xtensive mountainous region; a second to a COMPARATIVELY narrow, hilly tract; and a third to wide
0303 a great advantage over other organisms, a COMPARATIVELY short time would be necessary to produce
0338 h have branched off from each other within COMPARATIVELY recent times. For this doctrine of Agassi
0351 ecies of a genus have been produced within COMPARATIVELY recent times, there is great difficulty o
0015 races of our domestic animals and plants, and COMPARE them with species closely allied together, w
0021 rsity of the breeds is something astonishing. COMPARE the English carrier and the short faced tumb
0030 en the case with the ancon sheep. But when we COMPARE the dray horse and race horse, the dromedary
0030 of another breed for another purpose: when we COMPARE the many breeds of dogs, each good for man i
0030 good for man in very different ways: when we COMPARE the game cock, so pertinacious in battle, wi
0030 with the bantam so small and elegant: when we COMPARE the host of agricultural, culinary, orchard,
0045 ly examine internal and important organs, and COMPARE them in many specimens of the same species.
0048 ure are far from uncommon cannot be disputed. COMPARE the several floras of Great Britain, of Fran
0048 years ago, when comparing, and seeing others COMPARE, the birds from the separate islands of the
0151 y in plants made it particularly difficult to COMPARE their relative degrees of variability. When
0156 s, than in other parts of their organisation; COMPARE, for instance, the amount of difference betw
0173 t, till the one replaces the other. But if we COMPARE these species where they intermingle, they a
0188 llectual powers like those of man? If we must COMPARE the eye to an optical instrument, we ought i
0248 same species. It is also most instructive to COMPARE, but I have not space here to enter on detai
0313 changed in exactly the same degree. Yet if we COMPARE any but the most closely related formations,
0330 ected towards very distinct groups. Yet if we COMPARE the older Reptiles and Batrachians, the olde
0335 roductions inhabiting separated districts. To COMPARE small things with great: if the principal li
0347 roductions! In the southern hemisphere, if we COMPARE large tracts of land in Australia, South Afr
0347 oras more utterly dissimilar. Or again we may COMPARE the productions of South America south of la
0369 and this we find has been the case: for if we COMPARE the present Alpine plants and animals of the
0371 e two Worlds. Hence it has come, that when we COMPARE the now living productions of the temperate
0372 r physical conditions of the areas: for if we COMPARE, for instance, certain parts of South Americ
0389 nd, extending over 780 miles of latitude, and COMPARE its flowering plants, only 750 in number, wi
0390 in the world) is often extremely large. If we COMPARE, for instance, the number of the endemic lan
0390 h the number found on any continent, and then COMPARE the area of the islands with that of the con
0404 in the Felidae and Canidae. We see it, if we COMPARE the distribution of butterflies and beetles.
0016 n extreme degree in some one part, both when COMPARED one with another, and more especially when c
0016 d one with another, and more especially when COMPARED with all the species in nature to which they
0027 abnormal characters in certain respects, as COMPARED with all other Columbidae, though so like in
0032 ers, when the flowers of the present day are COMPARED with drawings made only twenty or thirty yea
0034 a certain size was ordered, and this may be COMPARED to the roguing of plants by nurserymen. The
0035 e increased in weight and in early maturity, COMPARED with the stock formerly kept in this country
0037 pelargonium, dahlia, and other plants, when COMPARED with the older varieties or with their paren
0046 e main nerves in Coccus, which may almost be COMPARED to the irregular branching of the stem of a
0057 amount of difference between varieties, when COMPARED with each other or with their parent species
0067 ase to any amount. The face of Nature may be COMPARED to a yielding surface, with ten thousand sha
0070 ields, because the seeds are in great excess COMPARED with the number of birds which feed on them:
0075 efinite laws: but how simple is this problem COMPARED to the action and reaction of the innumerabl
0080 ural Selection. Natural Selection, its power COMPARED with man's selection, its power on character
0084 consequently how poor will his products be, COMPARED with those accumulated by nature during whol
0107 er basins, taken together, make a small area COMPARED with that of the sea or of the land: and, co
0153 an extraordinary manner in any one species, COMPARED with the other species of the same genus, we
0201 f new Zealand, for instance, are perfect one COMPARED with another: but they are now rapidly yield
```

Page **(Key Word)**

Page **(Key Word)**

0208	and several of the older metaphysicians have	COMPARED instinct with habit. This comparison gives,
0239	ls taken out of the same nest: I have myself	COMPARED perfect gradations of this kind. It often ha
0245	ffspring not universal. Hybrids and mongrels	COMPARED independently of their fertility. Summary. T
0272	orms which are crossed. Hybrids and Mongrels	COMPARED, independently of their fertility. Independe
0272	crossed and of varieties when crossed may be	COMPARED in several other respects. Gartner, whose st
0281	any two or more species, even if we closely	COMPARED the structure of the parent with that of its
0293	y long lapse of years, each perhaps is short	COMPARED with the period requisite to change one spec
0293	t. we forget how small the area of Europe is	COMPARED with the rest of the world; nor have the sev
0302	e continually forget how large the world is,	COMPARED with the area over which our geological form
0314	ial and in more highly organised productions	COMPARED with marine and lower productions, by the mo
0324	cluding the whole glacial epoch), were to be	COMPARED with those now living in South America or in
0342	ved in our museums, is absolutely as nothing	COMPARED with the incalculable number of generations
0342	uration of each formation is, perhaps, short	COMPARED with the average duration of specific forms:
0382	the various beings thus left stranded may be	COMPARED with savage races of man, driven up and surv
0383	arity of the surrounding terrestrial beings,	COMPARED with those of Britain. But this power in fre
0398	y pond, yet as the number of kinds is small,	COMPARED with those on the land, the competition will
0389	ch inhabit oceanic islands are few in number	COMPARED with those on equal continental areas: Alph.
0427	rs are analogical when one class or order is	COMPARED with another, but give true affinities when
0428	n the members of the same class or order are	COMPARED one with another: thus the shape of the body
0428	ke limbs are only analogical when whales are	COMPARED with fishes, being adaptations in both class
0445	geon have descended from one wild species, I	COMPARED young pigeons of various breeds, within twel
0455	ogical importance. Rudimentary organs may be	COMPARED with the letters in a word, still retained i
0464	in all our museums is absolutely as nothing	COMPARED with the countless generations of countless
0488	er be recognised as a mere fragment of time,	COMPARED with the ages which have elapsed since the f
0247	there is any degree of sterility. He always	COMPARES the maximum number of seeds produced by two
0328	ive stages in the two countries: but when he	COMPARES certain stages in England with those in Fran
0027	tain characters from the rock pigeon, yet by	COMPARING the several sub breeds of these breeds, mor
0033	accumulated effects of selection namely, by	COMPARING the diversity of flowers in the different v
0035	the stock formerly kept in this country. By	COMPARING the accounts given in old pigeon treatises
0048	as geographical races! Many years ago, when	COMPARING, and seeing others compare, the birds from
0188	ng lengths. It is scarcely possible to avoid	COMPARING the eye to a telescope. We know that this i
0339	reas mean? He would be a bold man, who after	COMPARING the present climate of Australia and of par
0433	f trifling physiological importance: why, in	COMPARING one group with a distinct group, we summari
0436	e plainly homological. We see the same law in	COMPARING the wonderfully complex jaws and legs in cr
0487	ants of the whole world. Even at present, by	COMPARING the differences of the inhabitants of the s
0010	ct effects of the conditions of life are in	COMPARISON with the laws of reproduction, and of growt
0011	ntries where they are habitually milked, in	COMPARISON with the state of these organs in other cou
0033	r part is valued, in the kitchen garden, in	COMPARISON with the flowers of the same varieties; and
0033	ruit of the same species in the orchard, in	COMPARISON with the leaves and flowers of the same set
0035	d been made long ago, which might serve for	COMPARISON. In some cases, however, unchanged or but l
0052	ctuating forms. The term variety, again, in	COMPARISON with mere individual differences, is also a
0089	to the plumage of male and female birds, in	COMPARISON with the plumage of the young, can be expla
0105	of new organic forms, we ought to make the	COMPARISON within equal times; and this we are incapab
0134	der free nature, we can have no standard of	COMPARISON, by which to judge of the effects of long c
0150	es in an extraordinary degree or manner, in	COMPARISON with the same part in allied species, tends
0150	eloped, unless it be unusually developed in	COMPARISON with the same part in closely allied specie
0150	ings developed in some remarkable manner in	COMPARISON with the other species of the same genus. T
0158	in a species in an extraordinary manner in	COMPARISON with the same part in its congeners; but in
0167	whenever we have the means of instituting a	COMPARISON, the same laws appear to have acted in prod
0169	nary size or in an extraordinary manner, in	COMPARISON with the same part or organ in the allied s
0174	presentative species is generally narrow in	COMPARISON with the territory proper to each. We see t
0208	ans have compared instinct with habit. This	COMPARISON gives, I think, a remarkably accurate notio
0237	have longer horns than in other breeds, in	COMPARISON with the horns of the bulls or cows of thes
0251	eater fertility of species when crossed, in	COMPARISON with the same species when self fertilised,
0273	s are highly variable. But to return to our	COMPARISON of mongrels and hybrids: Gartner states tha
0281	een possible to have determined from a mere	COMPARISON of their structure with that of the rock pi
0285	ion of the Weald has been a mere trifle, in	COMPARISON with that which has removed masses of our p
0337	ore recent and victorious forms of life, in	COMPARISON with the ancient and beaten forms: but I ca
0363	the species of plants common to Europe, in	COMPARISON with the plants of other oceanic islands ne
0363	somewhat northern character of the flora in	COMPARISON with the latitude, I suspected that these f
0389	ants are included in these numbers, and the	COMPARISON in some other respects is not quite fair. W
0436	branch of the present subject; namely, the	COMPARISON not of the same part in different members o
0476	will often be intermediate in character by a	COMPARISON with its later descendants; and thus we can
0487	curity the duration of these intervals by a	COMPARISON of the preceding and succeeding organic for
0136	l reduced, but even enlarged. This is quite	COMPATIBLE with the action of natural selection. For w
0054	their parents to become dominant over their	COMPATRIOTS. If the plants inhabiting a country and de
0004	te of nature: but I shall, unfortunately, be	COMPELLED to treat this subject far too briefly, as i
0056	ound between doubtful forms, naturalists are	COMPELLED to come to a determination by the amount of
0090	other period of the year, when they might be	COMPELLED to prey on other animals. I can see no more
0297	eir relationship, and should consequently be	COMPELLED to rank them all as distinct species. It is
0433	c., we can understand the rules which we are	COMPELLED to follow in our classification. We can und
0485	o the rank of species. Hereafter we shall be	COMPELLED to acknowledge that the only distinction be
0013	he child, the mere doctrine of chances almost	COMPELS us to attribute its reappearance to inherita
0102	given period of profitable variations, will	COMPENSATE for a lesser amount of variability in each
0131	n. acclimatisation. Correlation of growth.	COMPENSATION and economy of growth. False correlations.
0138	the length of the antennae or palpi, as a	COMPENSATION for blindness. Notwithstanding such modifi
0147	ed, at about the same period, their law of	COMPENSATION or balancement of growth: or, as Goethe ex
0147	i suspect, also, that some of the cases of	COMPENSATION which have been advanced, and likewise som
0148	ny organ, without requiring as a necessary	COMPENSATION the reduction of some adjoining part. It s
0069	ich never become naturalised, for they cannot	COMPETE with our native plants, nor resist destructi
0077	egs, so well adapted for diving, allows it to	COMPETE with other aquatic insects, to hunt for its
0116	nisation but little diversified, could hardly	COMPETE with a set more perfectly diversified in str
0116	inant, and rodent mammals, could successfully	COMPETE with these well pronounced orders. In the Au
0371	native American productions, and have had to	COMPETE with them; and in the other great region, wi
0379	eing surrounded by strangers will have had to	COMPETE with many new forms of life; and it is proba
0390	in a new and isolated district, and having to	COMPETE with new associates, will be eminently liabl

Page **(Key Word)**

Page ***(Key Word)**

```
0392  , when established on an island and having to COMPETE with herbaceous plants alone, might readily
0400  the other inhabitants, with which each has to COMPETE, is at least as important, and generally a f
0401  n the different islands, for it would have to COMPETE with different sets of organisms: a plant, f
0016  r plants, which have not been ranked by some COMPETENT judges as mere varieties, and by other comp
0016  etent judges as mere varieties, and by other COMPETENT judges as the descendants of aboriginally d
0018  stock from our European cattle; and several COMPETENT judges believe that these latter have had m
0031  numerous passages to this effect from highly COMPETENT authorities. Youatt, who was probably bette
0035  since the time of that monarch. Some highly COMPETENT authorities are convinced that the setter i
0047  not been ranked as species by at least some COMPETENT judges. That varieties of this doubtful nat
0049  mitted that many forms, considered by highly COMPETENT judges as varieties, have so perfectly the
0049  species that they are ranked by other highly COMPETENT judges as good and true species. But to dis
0307  he dawn of life on this planet. Other highly COMPETENT judges, as Lyell and the late E. Forbes, di
0324  uthern hemisphere. So, again, several highly COMPETENT observers believe that the existing product
0175  immensely in numbers, were it not for other COMPETING species; that nearly all either prey on or
0180  the climate and vegetation change, let other COMPETING rodents or new beasts of prey immigrate, or
0392  ough it would have no chance of successfully COMPETING in stature with a fully developed tree, whe
0053  physical conditions, and as they come into COMPETITION (which, as we shall hereafter see, is a fa
0060  d plants. Nature of the checks to increase. COMPETITION universal. Effects of climate. Protection
0062  at all organic beings are exposed to severe COMPETITION. In regard to plants, no one has treated t
0076  cies of the same genus, when they come into COMPETITION with each other, than between species of d
0076  so in other cases. We can dimly see why the COMPETITION should be most severe between allied forms
0077  er organic beings, with which it comes into COMPETITION for food or residence, or from which it ha
0078  or on the borders of an utter desert, will COMPETITION cease. The land may be extremely cold or d
0078  be extremely cold or dry, yet there will be COMPETITION between some few species, or between the i
0105  n, by checking immigration and consequently COMPETITION, will give time for any new variety to be
0106  nd continuous area, and will thus come into COMPETITION with many others. Hence more new places wi
0106  nce more new places will be formed, and the COMPETITION to fill them will be more severe, on a lar
0107  sea or of the land; and, consequently, the COMPETITION between fresh water productions will have
0107  rom having thus been exposed to less severe COMPETITION. To sum up the circumstances favourable an
0107  ds, will have been subjected to very severe COMPETITION. When converted by subsidence into large s
0108  ontinental area, there will again be severe COMPETITION: the most favoured or improved varieties w
0110  y extinct. The forms which stand in closest COMPETITION with those undergoing modification and imp
0110  nd habits, generally come into the severest COMPETITION with each other. Consequently, each new va
0114  seen, that where they come into the closest COMPETITION with each other, the advantages of diversi
0121  arent. For it should be remembered that the COMPETITION will generally be most severe between thos
0122  in which child and parent do not come into COMPETITION, both may continue to exist. If then our d
0130  which has apparently been saved from fatal COMPETITION by having inhabited a protected station. A
0139  ot been preserved, owing to the less severe COMPETITION to which the inhabitants of these dark abo
0140  f nature are limited in their ranges by the COMPETITION of other organic beings quite as much as,
0172  ess favoured forms with which it comes into COMPETITION. Thus extinction and natural selection wil
0175  hich it is destroyed, or with it comes into COMPETITION; and as these species are already defined
0205  ocked country, must act chiefly through the COMPETITION of the inhabitants one with another, and c
0206  iduals, and more diversified forms, and the COMPETITION will have been severer, and thus the stand
0281  vast amount of change; and the principle of COMPETITION between organism and organism, between chi
0314  s with which the varying species comes into COMPETITION. Hence it is by no means surprising that o
0314  ved, we can understand, on the principle of COMPETITION, and on that of the many all important rel
0320  vantage over those with which it comes into COMPETITION; and the consequent extinction of less fav
0320  of about the same number of old forms. The COMPETITION will generally be most severe, as formerly
0321  ted station, where they have escaped severe COMPETITION. For instance, a single species of Trigoni
0326  e of favourable variations, and that severe COMPETITION with many already existing forms, would be
0337  s of one quarter of the world were put into COMPETITION with the existing inhabitants of the same
0351  a number of species, which stand in direct COMPETITION with each other, migrate in a body into a
0379  been advanced through natural selection and COMPETITION to a higher stage of perfection or dominat
0388  small, compared with those on the land, the COMPETITION will probably be less severe between aquat
0408  red happened to come in more or less direct COMPETITION with each other and with the aborigines: a
0412  h, in any small area, come into the closest COMPETITION, and by looking to certain facts in natura
0468  ecies come in all respects into the closest COMPETITION with each other, the struggle will general
0468  season, over those with which it comes into COMPETITION, or better adaptation in however slight a
0472  n anticipated. As natural selection acts by COMPETITION, it adapts the inhabitants of each country
0488  igrating into new countries and coming into COMPETITION with foreign associates, might become modi
0088  the result is not death to the unsuccessful COMPETITOR, but few or no offspring. Sexual selection
0319  ace being seized on by some more successful COMPETITOR. It is most difficult always to remember th
0069  me period of its life, from enemies or from COMPETITORS for the same place and food; and if these
0069  ame place and food; and if these enemies or COMPETITORS be in the least degree favoured by any sli
0069  r of species of all kinds, and therefore of COMPETITORS, decreases northwards; hence in going nort
0078  uld have to give it some advantage over its COMPETITORS, or over the animals which preyed on it. O
0078  imal is placed in a new country amongst new COMPETITORS, though the climate may be exactly the sam
0078  e it some advantage over a different set of COMPETITORS or enemies. It is good thus to try in our
0102  improved in a corresponding degree with its COMPETITORS, it will soon be exterminated. In man's me
0106  hich already have been victorious over many COMPETITORS, will be those that will spread most widel
0184  nsects were constant, and if better adapted COMPETITORS did not already exist in the country, I ca
0350  als, which have already triumphed over many COMPETITORS in their own widely extended homes will ha
0377  te plants, if protected from the inroads of COMPETITORS, can withstand a much warmer climate than
0429  failing groups conquered by more successful COMPETITORS, with a few members preserved by some unus
0191  auditory apparatus has been worked in as a COMPLEMENT to the swimbladder. All physiologists admit
0441  nary structure, or into what I have called COMPLEMENTAL males: and in the latter, the development
0001  s it will take me two or three more years to COMPLETE it, and as my health is far from strong, I h
0094  gularly from flower to flower, and as a more COMPLETE separation of the sexes of our plant would b
0094  nually favoured or selected, until at last a COMPLETE separation of the sexes would be affected. L
0208  s, it was much embarrassed, and, in order to COMPLETE its hammock, seemed forced to start from the
0208  ge, where it had left off, and thus tried to COMPLETE the already finished work. If we suppose any
0214  ing in each successive generation would soon COMPLETE the work; and unconscious selection is still
0230  o plates, as the case may be; and they never COMPLETE the upper edges of the rhombic plates, until
0255  n the number of seeds produced, up to nearly COMPLETE or even quite complete fertility; and, as we
0255  roduced, up to nearly complete or even quite COMPLETE fertility; and, as we have seen, in certain
0315  ation, on this view, does not mark a new and COMPLETE act of creation, but only an occasional scen
```

Page ***(Key Word)**

Page **(Key Word)**

```
0318  endures. There is reason to believe that the  COMPLETE  extinction of the species of a group is gene
0396  the migration would probably have been more     COMPLETE; and if modification be admitted, all the fo
0425  er in degree, and has taken a longer time to    COMPLETE? I believe it has thus been unconsciously us
0436  y lost, by the atrophy and ultimately by the    COMPLETE abortion of certain parts, by the soldering
0453  ted for the sake of symmetry, or in order to    COMPLETE the scheme of nature; but this seems to me n
0453  he planets, for the sake of symmetry, and to    COMPLETE the scheme of nature? An eminent physiologis
0455  be wholly lost, and we should have a case of    COMPLETE abortion. The principle, also, of economy, e
0208  mock; for if he took a caterpillar which had    COMPLETED its hammock up to, say, the sixth stage of
0208  e of construction, and put it into a hammock    COMPLETED up only to the third stage, the caterpillar
0225  ken into each other, if the spheres had been    COMPLETED; but this is never permitted, the bees buil
0229  the opposite sides: for I have noticed half    COMPLETED rhombs at the base of a just commenced cell
0229  and I found that the rhombic plate had been    COMPLETED, and had become perfectly flat: it was abso
0231  e cells, both those just commenced and those    COMPLETED, being thus crowned by a strong coping of w
0232  ly proper place, projecting beyond the other    COMPLETED cells. It suffices that the bees should be
0232  om each other and from the walls of the last    COMPLETED cells, and then, by striking imaginary sphe
0441  ergo their final metamorphosis. When this is    COMPLETED they are fixed for life: their legs are now
0441  , and very simple eye spot. In this last and    COMPLETED state, cirripedes may be considered as eith
0442  sted long before the parts of the embryo are    COMPLETED; and again in spiders, there is nothing wor
0098  epotent effect, that it will invariably and     COMPLETELY destroy, as has been shown by Gartner, any
0143  do not doubt, may be mastered more or less     COMPLETELY by natural selection: thus a family of stag
0148  nd is thus protected, it loses more or less     COMPLETELY its own shell or carapace. This is the case
0158  d, to natural selection having more or less     COMPLETELY, according to the lapse of time, overmaster
0258  pacity in any two species to cross is often     COMPLETELY independent of their systematic affinity, o
0259  e degree of fertility. These facts show how     COMPLETELY fertility in the hybrid is independent of i
0285  racked, the surface of the land has been so     COMPLETELY planed down by the action of the sea, that
0310  ns long anterior to the silurian epoch in a     COMPLETELY metamorphosed condition. The several diffic
0361  ort: out of one small portion of earth thus     COMPLETELY enclosed by wood in an oak about 50 years o
0370  at present, they must have been still more     COMPLETELY separated by wider spaces of ocean. I belie
0371  th of the Polar Circle, they must have been     COMPLETELY cut off from each other. This separation, a
0372  ny closely allied forms now living in areas     COMPLETELY sundered. Thus, I think, we can understand
0380  ns before the Glacial period must have been     COMPLETELY isolated; and I believe that the production
0384  y period must have parted river systems and     COMPLETELY prevented their inosculation, seems to lead
0421  en so much modified as to have more or less     COMPLETELY lost traces of their parentage, in this cas
0421  classification will have been more or less     COMPLETELY lost, as sometimes seems to have occurred w
0477  ferent, if they have been for a long period     COMPLETELY separated from each other; for as the relat
0005  in any manner profitable to itself, under the   COMPLEX and sometimes varying conditions of life, wi
0005  cter. In the next chapter I shall discuss the   COMPLEX and little known laws of variation and of co
0012  or dimly seen laws of variation is infinitely   COMPLEX and diversified. It is well worth while care
0043  the final result is thus rendered infinitely   COMPLEX. In some cases, I do not doubt that the inte
0060  e. protection from the number of individuals.   COMPLEX relations of all animals and plants throught
0061  individual of any species, in its infinitely   COMPLEX relations to other organic beings and to ext
0071  enlarge. Many cases are on record showing how   COMPLEX and unexpected are the checks and relations
0073  ale of nature, are bound together by a web of   COMPLEX relations. I shall hereafter have occasion t
0080  astic. Let it be borne in mind how infinitely   COMPLEX and close fitting are the mutual relations o
0080  ul in some way to each being in the great and   COMPLEX battle of life, should sometimes occur in th
0081  e, from what we have seen of the intimate and   COMPLEX manner in which the inhabitants of each coun
0084  ould be infinitely better adapted to the most   COMPLEX conditions of life, and should plainly bear
0091  the shepherd's flocks. Let us now take a more   COMPLEX case. Certain plants excrete a sweet juice,
0106  ed, but the conditions of life are infinitely   COMPLEX from the large number of already existing sp
0119  er beings; and this will depend on infinitely   COMPLEX relations. But as a general rule, the more d
0129  , it is explained through inheritance and the   COMPLEX action of natural selection, entailing extin
0132  es cannot have produced the many striking and   COMPLEX co adaptations of structure between one orga
0148  nsile antennae. Now the saving of a large and   COMPLEX structure, when rendered superfluous by the
0163  rkable. I will, however, give one curious and   COMPLEX case, not indeed as affecting any important
0186  hat if numerous gradations from a perfect and   COMPLEX eye to one very imperfect and simple, each g
0186  he difficulty of believing that a perfect and   COMPLEX eye could be formed by natural selection, th
0189  of man? If it could be demonstrated that any   COMPLEX organ existed, which could not possibly have
0195  ad, as in the case of an organ as perfect and   COMPLEX as the eye. In the first place, we are much
0200  m, either directly, or indirectly through the   COMPLEX laws of growth. Natural selection cannot pos
0209  . it is thus, as I believe, that all the most   COMPLEX and wonderful instincts have originated. As
0209  uce slight deviations of bodily structure. No   COMPLEX instinct can possibly be produced through na
0210  actual transitional gradations by which each   COMPLEX instinct has been acquired, for these could
0210  ery generally gradations, leading to the most   COMPLEX instincts, can be discovered. The canon of N
0251  ertain the degree of fertility of some of the   COMPLEX crosses of Rhododendrons, and I am assured t
0260  osses often differ in fertility. Now do these   COMPLEX and singular rules indicate that species hav
0263  ecies. And as we must look at the curious and   COMPLEX laws governing the facility with which trees
0263  ive systems, so I believe that the still more   COMPLEX laws governing the facility of first crosses
0276  inity, but is governed by several curious and   COMPLEX laws. It is generally different, and sometim
0314  ation in the varying species, depends on many   COMPLEX contingencies, on the variability being of a
0314  ith marine and lower productions, by the more   COMPLEX relations of the higher beings to their orga
0322  ning for a moment that we understand the many   COMPLEX contingencies, on which the existence of eac
0331  with modification. As the subject is somewhat   COMPLEX, I must request the reader to turn to the di
0343  tion is necessarily slow, and depends on many   COMPLEX contingencies. The dominant species of the l
0351  ly so far as it profits the individual in its   COMPLEX struggle for life, so the degree of modifica
0411  or adaptive characters. Affinities, general,   COMPLEX and radiating. Extinction separates and defi
0431  in common, we can understand the excessively   COMPLEX and radiating affinities by which all the me
0433  each class are connected together by the most   COMPLEX and radiating lines of affinities. We shall
0436  see the same law in comparing the wonderfully   COMPLEX jaws and legs in crustaceans. It is familiar
0437  should one crustacean, which has an extremely   COMPLEX mouth formed of many parts, consequently alw
0441  r of magnificent compound eyes, and extremely   COMPLEX antennae; but they have a closed and imperfe
0456  h all living and extinct beings are united by   COMPLEX, radiating, and circuitous lines of affiniti
0459  more difficult to believe than that the more   COMPLEX organs and instincts should have been perfec
0466  parent form. Variability is governed by many   COMPLEX laws, by correlation of growth, by use and d
0469  way useful to beings, under their excessively   COMPLEX relations of life, would be preserved, accum
0469  nd beautifully adapting each form to the most   COMPLEX relations of life. The theory of natural sel
0472  solute perfection have not been observed. The   COMPLEX and little known laws governing variation ar
```

Page **(Key Word)**

Page **************************************(Key Word)**************************************

```
0475  heir crossed offspring should follow the same COMPLEX laws in their degrees and kinds of resemblan
0478  e species and genera within each class are so COMPLEX and circuitous. We see why certain character
0485   has had a history; when we contemplate every COMPLEX structure and instinct as the summing up of
0499  each other, and dependent on each other in so COMPLEX a manner, have all been produced by laws act
0073  nd so onwards in ever increasing circles of COMPLEXITY. We began this series by insectivorous bird
0109  mount of change, to the beauty and infinite COMPLEXITY of the coadaptations between all organic be
0127  be disputed; then, considering the infinite COMPLEXITY of the relations of all organic beings to e
0204  f we know of a long series of gradations in COMPLEXITY each good for its possessor, then, under ch
0208   was with a caterpillar, which makes a very COMPLICATED hammock; for if he took a caterpillar whic
0249   would be thus effected. Moreover, whenever COMPLICATED experiments are in progress, so careful an
0251  deserve some notice. It is notorious in how COMPLICATED a manner the species of Pelargonium, Fuchs
0274  nimals; but the subject is here excessively COMPLICATED, partly owing to the existence of secondar
0333  ations. In nature the case will be far more COMPLICATED than is represented in the diagram; for th
0186   of auks. Organs of extreme perfection and COMPLICATION. To suppose that the eye, with all its ini
0295  we see, as is so often the case, a formation COMPOSED of beds of different mineralogical compositi
0308  es in thickness, of which the formations are COMPOSED, we may infer that from first to last large
0437  n! why should the brain be enclosed in a box COMPOSED of such numerous and such extraordinarily sh
0145  ape of the seeds in the ray florets in some COMPOSITAE countenances this idea; but, in the case of
0145  ower had caused their abortion; but in some COMPOSITAE there is a difference in the seeds of the o
0452  f a stigma supported on the style; but some COMPOSITAE, the male florets, which of course cannot b
0452  loped and is clothed with hairs as in other COMPOSITAE, for the purpose of brushing the pollen out
0225  ell on one side of the comb, enter into the COMPOSITION of the bases of three adjoining cells on t
0290  uent and great changes in the mineralogical COMPOSITION of consecutive formations, generally imply
0295  composed of beds of different mineralogical COMPOSITION, we may reasonably suspect that the proces
0358  nor does their almost universally volcanic COMPOSITION favour the admission that they are the wre
0484  ings have much in common, in their chemical COMPOSITION, their germinal vesicles, their cellular s
0144  between the outer and inner flowers in some COMPOSITOUS and Umbelliferous plants. Every one knows
0145  ortion of parts of the flower. But, in some COMPOSITOUS plants, the seeds also differ in shape and
0250  s subsequently fertilised by the pollen of a COMPOUND hybrid descended from three other and distin
0441  tructed natatory legs, a pair of magnificent COMPOUND eyes, and extremely complex antennae; but th
0258  d from reciprocal crosses, though of course COMPOUNDED of the very same two species, the one speci
0266  y possible that two organisations should be COMPOUNDED into one, without some disturbance occuring
0266  ring from generation to generation the same COMPOUNDED organisation, and hence we need not be surp
0266   disturbed by two organisations having been COMPOUNDED into one. It may seem fanciful, but I suspe
0277  their whole organisation disturbed by being COMPOUNDED of two distinct species, seems closely alli
0461  rdly fail to have been disturbed from being COMPOUNDED of two distinct organisations. This paralle
0282  ogist, the facts leading the mind feebly to COMPREHEND the lapse of time. He who can read Sir Char
0282  aking fresh sediment, before he can hope to COMPREHEND anything of the lapse of time, the monument
0485   a ship, as at something wholly beyond his COMPREHENSION; when we regard every production of natur
0214  me inherited solely from long continued and COMPULSORY habit, but this, I think, is not true. No o
0214  ical selection and the inherited effects of COMPULSORY training in each successive generation woul
0216  r ignorance call an accident. In some cases COMPULSORY habit alone has sufficed to produce such in
0216  ch inherited mental changes; in other cases COMPULSORY habit has done nothing, and all has been th
0229  of a just commenced cell, which were slightly CONCAVE on one side, where I suppose that the bees h
0216  especially young turkeys) from under her, and CONCEAL themselves in the surrounding grass or thick
0427  mblances will not reveal, will rather tend to CONCEAL their blood relationship to their proper lin
0135  eira, as observed by Mr. Wollaston, lie much CONCEALED, until the wind lulls and the sun shines; t
0413  d which we often meet with in a more or less CONCEALED form, that the characters do not make the g
0003  sidering the Origin of Species, it is quite CONCEIVABLE that a naturalist, reflecting on the mutua
0182  es, and formerly had flying reptiles, it is CONCEIVABLE that flying fish, which now glide far thro
0191  the embryo their former position. But it is CONCEIVABLE that the now utterly lost branchiae might
0204  cal impossibility in the acquirement of any CONCEIVABLE degree of perfection through natural selec
0233  building up intermediate planes. It is even CONCEIVABLE that an insect might, by fixing on a point
0366  ese facts. We have evidence of almost every CONCEIVABLE kind, organic and inorganic, that within a
0436   the mouths of insects. Nevertheless, it is CONCEIVABLE that the general pattern of an organ might
0448  dered by natural selection different to any CONCEIVABLE extent from their parents. Such difference
0180  squirrel is the best that it is possible to CONCEIVE under all natural conditions. Let the climat
0192  e of special difficulty; it is impossible to CONCEIVE by what steps these wondrous organs have bee
0133  nditions of life as different as can well be CONCEIVED; and, on the other hand, of different varie
0490  most exalted object which we are capable of CONCEIVING, namely, the production of the higher anima
0000  lopment of the embryo, or at the instant of CONCEPTION. Geoffroy St. Hilaire's experiments show th
0008  ts having been affected prior to the act of CONCEPTION. Several reasons make me believe in this; b
0010  treatment of the parent prior to the act of CONCEPTION. These cases anyhow show that variation is
0086  tingencies, wholly different from those which CONCERN the mature insect. These modifications will
0037  n simple, and, as far as the final result is CONCERNED, has been followed almost unconsciously. It
0101  and unisexual species, as far as function is CONCERNED, becomes very small. From these several con
0181  and this, as far as the organs of flight are CONCERNED, would convert it into a bat. In bats which
0207  than I have with that of life itself. We are CONCERNED only with the diversities of instinct and o
0236  and in instinct. As far as instinct alone is CONCERNED, the prodigious difference in this respect
0371  as far as the more temperate productions are CONCERNED, took place long ages ago. And as the plant
0377  e arctic; but with the latter we are not now CONCERNED. The tropical plants probably suffered much
0381  gs, as suggested by Lyell, have been largely CONCERNED in their dispersal. But the existence of se
0414  hat the less any part of the organisation is CONCERNED with special habits, the more important it
0035  cted by crosses with the fox hound; but what CONCERNS us is, that the change has been effected unc
0096  ly unite for reproduction, which is all that CONCERNS us. But still there are many hermaphrodite a
0152  ing condition. But what here more especially CONCERNS us is, that in our domestic animals those po
0187  nerve comes to be sensitive to light, hardly CONCERNS us more than how life itself first originate
0297  es of the same formation. Some experienced CONCHOLOGISTS are now sinking many of the very fine spe
0299  of north America, which are ranked by some CONCHOLOGISTS as distinct species from their European r
0299  eir European representatives, and by other CONCHOLOGISTS as only varieties, are really varieties o
0007  ates and treatment, I think we are driven to CONCLUDE that this greater variability is simply due
0014  fficulty in proving its truth: we may safely CONCLUDE that very many of the most strongly marked d
0081  nd some species might become extinct. We may CONCLUDE, from what we have seen of the intimate and
0083  re beaten some of the natives, we may safely CONCLUDE that the natives might have been modified wi
0101  trying to investigate. Finally then, we may CONCLUDE that in many organic beings, a cross between
0106  a certain extent, have concurred. Finally, I CONCLUDE that, although small isolated areas probably
```

Page **************************************(Key Word)**************************************

```
0107   extreme intricacy of the subject permits. I CONCLUDE, looking to the future, that for terrestrial
0132   in that of plants. We may, at least, safely CONCLUDE that such influences cannot have produced th
0142   be than others. On the whole, I think we may CONCLUDE that habit, use, and disuse, have, in some c
0153   the other species of the same genus, we may CONCLUDE that this part has undergone an extraordinar
0158   possession of the females. Finally, then, I CONCLUDE that the greater variability of specific cha
0176   st these facts and inferences, and therefore CONCLUDE that varieties linking two other varieties t
0182   rough natural selection. Furthermore, we may CONCLUDE that transitional grades between structures
0216   by disuse the power of flight. Hence we may CONCLUDE, that domestic instincts have been acquired
0227   e cells of the hive bee. Hence we may safely CONCLUDE that if we could slightly modify the instinc
0239   les and females, in this case, we may safely CONCLUDE from the analogy of ordinary variations, tha
0286   base. Hence, under ordinary circumstances, I CONCLUDE that for a cliff 500 feet in height, a denud
0290   of the coast waves. We may, I think, safely CONCLUDE that sediment must be accumulated in extreme
0291   ing from the researches of E. Forbes, we may CONCLUDE that the bottom will be inhabited by extreme
0294   st examining these beds, might be tempted to CONCLUDE that the average duration of life of the emb
0309   e oscillations of level, which we may fairly CONCLUDE must have intervened during these enormously
0309   ended to my volume on Coral Reefs, led me to CONCLUDE that the great oceans are still mainly areas
0328   rldi but we are far from having any right to CONCLUDE that this has invariably been the case, and
0341   with the same number of species, then we may CONCLUDE that only one species of each of the six old
0359   fer anything from these scanty facts, we may CONCLUDE that the seeds of 14/100 plants of any count
0365   s long ago as 1747, such facts led Gmelin to CONCLUDE that the same species must have been indepen
0368   s on distant mountain summits, we may almost CONCLUDE without other evidence, that a colder climat
0444   long as it was fed by its parents. Hence, I CONCLUDE, that it is quite possible, that each of the
0455   he view of descent with modification, we may CONCLUDE that the existence of organs in a rudimentar
0143   f the young or larva, will, it may safely be CONCLUDED, affect the structure of the adult; in the
0254   tercrossing of plants and animals, it may be CONCLUDED that some degree of sterility, both in firs
0293   h deference, namely Bronn and Woodward, have CONCLUDED that the average duration of each formation
0304   s had been discovered in beds of this age, I CONCLUDED that this great group had been suddenly dev
0355   ingenious paper by Mr. Wallace, in which he CONCLUDES, that every species has come into existence
0006   recapitulation of the whole work, and a few CONCLUDING remarks. No one ought to feel surprise at m
0190   extinct. We should be extremely cautious in CONCLUDING that an organ could not have been formed by
0192   ? although we must be extremely cautious in CONCLUDING that any organ could not possibly have been
0204   n this chapter how cautious we should be in CONCLUDING that the most different habits of life coul
0204   ional states, we should be very cautious in CONCLUDING that none could have existed, for the homol
0459   s adoption on the study of Natural history. CONCLUDING remarks. As this whole volume is one long a
0003   on, and other such facts, might come to the CONCLUSION that each species had not been independentl
0003   s, from other species. Nevertheless, such a CONCLUSION, even if well founded, would be unsatisfact
0017   hink it is possible to come to any definite CONCLUSION, whether they have descended from one or se
0028   any naturalist could in coming to a similar CONCLUSION in regard to the many species of finches, o
0062   e found it so, than constantly to bear this CONCLUSION in mind. Yet unless it be thoroughly engrai
0150   utang, that he has come to a nearly similar CONCLUSION. It is hopeless to attempt to convince any
0160   ds. We may I think confidently come to this CONCLUSION, because, as we have seen, these coloured m
0247   terile together; and as he came to the same CONCLUSION in several other analogous cases; it seems
0250   d rev. W. Herbert. He is as emphatic in his CONCLUSION that some hybrids are perfectly fertile, as
0274   s, the hybrids do not differ much. But this CONCLUSION, as far as I can make out, is founded on a
0275   facts with respect to animals, comes to the CONCLUSION, that the laws of resemblance of the child
0291   is or that great formation, has come to the CONCLUSION that it was accumulated during subsidence.
0292   . forbes independently arrived at a similar CONCLUSION. One remark is here worth a passing notice.
0293   seems to me, prevent us coming to any just CONCLUSION on this head. When we see a species first a
0307   lyell and the late E. Forbes, dispute this CONCLUSION. We should not forget that only a small por
0317   ral views would naturally lead them to this CONCLUSION. On the contrary, we have every reason to b
0346   ho has studied the subject has come to this CONCLUSION. The case of America alone would almost suf
0384   ir inosculation, seems to lead to this same CONCLUSION. With respect to allied fresh water fish oc
0407   are not insuperable. And we are led to this CONCLUSION, which has been arrived at by many naturali
0412   ndency in their characters to diverge. This CONCLUSION was supported by looking at the great diver
0459   arguments. Chapter XIV. Recapitulation and CONCLUSION. Recapitulation of the difficulties on the
0460   ercrossed species. We see the truth of this CONCLUSION in the vast difference in the result, when
0472   effect; for it is difficult to resist this CONCLUSION when we look, for instance, at the logger h
0482   ndently created. This seems to me a strange CONCLUSION to arrive at. They admit that a multitude o
0001   ese I enlarged in 1844 into a sketch of the CONCLUSIONS, which then seemed to me probable: from th
0002   arrived at almost exactly the same general CONCLUSIONS that I have on the origin of species. Last
0002   ies alone. I can here give only the general CONCLUSIONS at which I have arrived, with a few facts
0002   all the facts, with references, on which my CONCLUSIONS have been grounded; and I hope in a future
0002   not be adduced, often apparently leading to CONCLUSIONS directly opposite to those at which I have
0224   ect, but will merely give an outline of the CONCLUSIONS at which I have arrived. He must be a dull
0248   ould have arrived at diametrically opposite CONCLUSIONS in regard to the very same species. It is
0255   in utter confusion. The following rules and CONCLUSIONS are chiefly drawn up from Gartner's admira
0276   lived; have come to diametrically opposite CONCLUSIONS in ranking forms by this test. The sterili
0445   l difference. As the evidence appears to me CONCLUSIVE, that the several domestic breeds of Pigeon
0460   nd of the eighth chapter, which seem to me CONCLUSIVELY to show that this sterility is no more a s
0025   edging of the outer tail feathers, sometimes CONCUR perfectly developed. Moreover, when two birds
0057   nced observers, and, after deliberation, they CONCUR in this view. In this respect, therefore, the
0096   dividuals, either occasionally or habitually, CONCUR for the reproduction of their kind. This view
0096   sing in these cases that two individuals ever CONCUR in reproduction? As it is impossible here to
0173   it slowly subsides. These contingencies will CONCUR only rarely, and after enormously long interv
0106   on will generally, to a certain extent, have CONCURRED. Finally, I conclude that, although small i
0215   with some degree of selection, has probably CONCURRED in civilising by inheritance our dogs. On t
0100   fected with terrestrial animals without the CONCURRENCE of two individuals. Of aquatic animals, th
0487   recognised as having depended on an unusual CONCURRENCE of circumstances; and the blank intervals
0333   members of the same groups; and this by the CONCURRENT evidence of our best palaeontologists seems
0074   ew being generally the most potent; but all CONCURRING in determining the average number or even t
0118   o the ten thousandth generation, and under a CONDENSED and simplified form up to the fourteen thou
0120   of generations (as shown in the diagram in a CONDENSED and simplified manner), we get eight specie
0005   nities, both when mature and in an embryonic CONDITION. In the last chapter I shall give a brief r
0009   pollen utterly worthless, in the same exact CONDITION as in the most sterile hybrids. When, on th
0044   inherited; but who can say that the dwarfed CONDITION of shells in the brackish waters of the Bal
0106   will often have recently existed in a broken CONDITION, so that the good effects of isolation will
```

Page ***************************************(Key Word)***************************************

0107 ntly will exist for long periods in a broken CONDITION, will be the most favourable for the produc
0118 gle variety, but in a more and more modified CONDITION, some producing two or three varieties, and
0132 , i chiefly attribute the varying or plastic CONDITION of the offspring. The male and female sexua
0132 rting plants, the bud, which in its earliest CONDITION does not apparently differ essentially from
0134 water, and has its wings in nearly the same CONDITION as the domestic Aylesbury duck. As the larg
0134 e danger, I believe that the nearly wingless CONDITION of several birds, which now inhabit or have
0135 enera they are present, but in a rudimentary CONDITION. In the Ateuchus or sacred beetle of the Eg
0135 ior tarsi in Ateuchus, and their rudimentary CONDITION in some other genera, by the long continued
0135 three genera have all their species in this CONDITION! Several facts, namely, that beetles in man
0136 tions have made me believe that the wingless CONDITION of so many Madeira beetles is mainly due to
0137 one which I kept alive was certainly in this CONDITION, the cause, as appeared on dissection, havi
0152 s the organisation is left in a fluctuatuing CONDITION. But what here more especially concerns us
0154 s been transmitted in approximately the same CONDITION to many modified descendants, as in the cas
0154 ding to revert to a former and less modified CONDITION. The principle included in these remarks ma
0174 there, and may be embedded there in a fossil CONDITION. But in the intermediate region, having int
0174 t times in a far less continuous and uniform CONDITION than at present. But I will pass over this
0174 ough I do not doubt that the formerly broken CONDITION of areas now continous has played an import
0183 transitional grades of structure in a fossil CONDITION will always be less, from their having exis
0187 f descent, in an unaltered or little altered CONDITION. Amongst existing Vertebrata, we find but a
0237 er having become correlated with the sterile CONDITION of certain members of insect communities: t
0238 re, or instinct, correlated with the sterile CONDITION of certain members of the community, has be
0240 have their ocelli in an exactly intermediate CONDITION. So that we here have two bodies of sterile
0240 ected by some few members in an intermediate CONDITION. I may digress by adding, that if the small
0240 until all the workers had come to be in this CONDITION; we should then have had a species of ant w
0240 of ant with neuters very nearly in the same CONDITION with those of Myrmica. For the workers of M
0246 se their organs of reproduction in a perfect CONDITION, yet when intercrossed they produce either
0262 ve their reproductive organs in an imperfect CONDITION, is a very different case from the difficul
0288 ea not rarely lying for ages in an unaltered CONDITION. The remains which do become embedded, if i
0299 ree with Mr. Godwin Austen, that the present CONDITION of the Malay Archipelago, with its numerous
0308 uoht to be very generally in a metamorphosed CONDITION. But the descriptions which we now possess
0310 silurian epoch in a completely metamorphosed CONDITION. The several difficulties here discussed, n
0333 geological record, and that in a very broken CONDITION, we have no right to expect, except in very
0338 by nature, of the ancient and less modified CONDITION of each animal. This view may be true, and
0343 own to us, may now all be in a metamorphosed CONDITION, or may lie buried under the ocean. Passing
0346 ry temperature. There is hardly a climate or CONDITION in the Old World which cannot be paralleled
0371 eve, their descendants, mostly in a modified CONDITION, in the central parts of Europe and the Uni
0395 of allied species in a more or less modified CONDITION. Mr. Windsor Earl has made some striking ob
0409 sence of mammals, in a more or less modified CONDITION, and the depth of the sea between an island
0416 importance; yet, undoubtedly, organs in this CONDITION are often of high value in classification.
0441 lowly organised then they were in the larval CONDITION. But in some genera the larvae become devel
0450 rted organs. Organs or parts in this strange CONDITION, bearing the stamp of inutility, are extrem
0453 is often said to have retained its embryonic CONDITION. I have now given the leading facts with re
0455 ans in a rudimentary, imperfect, and useless CONDITION, or quite aborted, far from presenting a st
0461 es the organs on both sides are in a perfect CONDITION. As we continually see that organisms of al
0464 certain classes can be preserved in a fossil CONDITION, at least in any great number. Widely rangi
0473 ings incapable of flight, in nearly the same CONDITION as in the domestic duck; or when we look at
0483 een existing orders. Organs in a rudimentary CONDITION plainly show that an early progenitor had t
0003 . naturalists continually refer to external CONDITIONS, such as climate, food, etc., as the only p
0003 preposterous to attribute to mere external CONDITIONS, the structure, for instance, of the woodpe
0003 organic beings, by the effects of external CONDITIONS, or of habit, or of the volition of the pla
0004 beings to each other and to their physical CONDITIONS of life, untouched and unexplained. It is,
0005 if, under the complex and sometimes varying CONDITIONS of life, will have a better chance of survi
0007 mestic productions having been raised under CONDITIONS of life not so uniform as, and somewhat dif
0007 posed during several generations to the new CONDITIONS of life to cause any appreciable amount of
0008 isation, to the action of any change in the CONDITIONS of life. Nothing is more easy than to tame
0009 breed most freely under the most unnatural CONDITIONS (for instance, the rabbit and ferret kept i
0010 apparently been exposed to exactly the same CONDITIONS of life; and this shows how unimportant the
0010 s how unimportant the direct effects of the CONDITIONS of life are in comparison with the laws of
0010 d of inheritance; for had the action of the CONDITIONS been direct, if any of the young had varied
0010 arly all the individuals exposed to certain CONDITIONS are affected in the same way, the change at
0010 at first appears to be directly due to such CONDITIONS; but in some cases it can be shown that qui
0010 e cases it can be shown that quite opposite CONDITIONS produce similar changes of structure. Never
0011 , be attributed to the direct action of the CONDITIONS of life, as, in some cases, increased size
0013 individuals, apparently exposed to the same CONDITIONS, any very rare deviation, due to some extra
0015 argument; for by the experiment itself the CONDITIONS of life are changed. If it could be shown t
0015 red characters, whilst kept under unchanged CONDITIONS, and whilst kept in a considerable body, so
0015 ence. I may add, that when under nature the CONDITIONS of life do change, variations and reversion
0029 ibuted to the direct action of the external CONDITIONS of life, and some little to habit; but he w
0041 e species should be placed under favourable CONDITIONS of life, so as to breed freely in that coun
0043 s of animals and plants. I believe that the CONDITIONS of life, from their action on the reproduct
0043 y be attributed to the direct action of the CONDITIONS of life. Something must be attributed to us
0044 a modification directly due to the physical CONDITIONS of life, and variations in this sense are s
0046 s kind of variability is independent of the CONDITIONS of life. I am inclined to suspect that we s
0052 long continued action of different physical CONDITIONS in two different regions; but I have not mu
0053 as they become exposed to diverse physical CONDITIONS, and as they come into competition (which,
0054 re is something in the organic or inorganic CONDITIONS of that country favourable to the genus: an
0060 he organisation to another part, and to the CONDITIONS of life, and of one distinct organic being
0063 s of distinct species, or with the physical CONDITIONS of life. It is the doctrine of Malthus appl
0065 degree. The obvious explanation is that the CONDITIONS of life have been very favourable, and that
0066 few more years to people, under favourable CONDITIONS, a whole district, let it be ever so large.
0070 ve, that a plant could exist only where the CONDITIONS of its life were so favourable that many co
0078 tly the same as in its former home, yet the CONDITIONS of its life will generally be changed in an
0080 beings to each other and to their physical CONDITIONS of life. Can it, then, be thought improbabl
0082 s, by better adapting them to their altered CONDITIONS, would tend to be preserved; and natural se
0082 in the first chapter, that a change in the CONDITIONS of life, by specially acting on the reprodu
0082 variability; and in the foregoing case the CONDITIONS of life are supposed to have undergone a ch

Page ***************************************(Key Word)***************************************

conditions

0082 y adapted to each other and to the physical CONDITIONS under which they live, that none of them co
0083 ; and the being is placed under well suited CONDITIONS of life. Man keeps the natives of many clim
0084 finitely better adapted to the most complex CONDITIONS of life, and should plainly bear the stamp
0084 ng in relation to its organic and inorganic CONDITIONS of life. We see nothing of those slow chang
0094 n plants under culture and placed under new CONDITIONS of life, sometimes the male organs and some
0102 cts will almost certainly present different CONDITIONS of life; and then if natural selection be m
0103 district in exactly the same manner to the CONDITIONS of each; for in a continuous area, the cond
0103 ions of each; for in a continuous area, the CONDITIONS will generally graduate away insensibly fro
0104 be retained amongst them, as long as their CONDITIONS of life remain the same, only through the p
0104 h depart from the proper type; but if their CONDITIONS of life change and they undergo modificatio
0104 f not very large, the organic and inorganic CONDITIONS of life will generally be in a great degree
0104 in the same manner in relation to the same CONDITIONS. Intercrosses, also, with the individuals o
0105 iers, or from having very peculiar physical CONDITIONS, the total number of the individuals suppor
0106 f the same species there supported, but the CONDITIONS of life are infinitely complex from the lar
0109 s, one with another and with their physical CONDITIONS of life, which may be effected in the long
0113 he country not undergoing any change in its CONDITIONS) only by its varying descendants seizing on
0114 exposed for many years to exactly the same CONDITIONS, supported twenty species of plants, and th
0117 enerally continue to be exposed to the same CONDITIONS which made their parents variable, and the
0126 a the long course of ages and under varying CONDITIONS of life, organic beings vary at all in the
0127 l organic beings to each other and to their CONDITIONS of existence, causing an infinite diversity
0127 the various forms of life to their several CONDITIONS and stations, must be judged of by the gene
0131 r v. Laws of Variation. Effects of external CONDITIONS. Use and disuse, combined with natural sele
0131 re are in some way due to the nature of the CONDITIONS of life, to which the parents and their mor
0131 is eminently susceptible to changes in the CONDITIONS of life; and to this system being functiona
0133 n of natural selection, and how much to the CONDITIONS of life. Thus, it is well known to furriers
0133 en of the same variety being produced under CONDITIONS of life as different as can well be conceiv
0133 oduced from the same species under the same CONDITIONS. Such facts show how indirectly the conditi
0134 ditions. Such facts show how indirectly the CONDITIONS of life must act. Again, innumerable instan
0134 y little weight on the direct action of the CONDITIONS of life. Indirectly, as already remarked, t
0138 e done its work. It is difficult to imagine CONDITIONS of life more similar than deep limestone ca
0148 part of the organisation. If under changed CONDITIONS of life a structure before useful becomes l
0160 this case there is nothing in the external CONDITIONS of life to cause the reappearance of the sl
0161 ion, which at last, under unknown favorable CONDITIONS, gains an ascendancy. For instance, it is p
0161 ill not be left to the mutual action of the CONDITIONS of life and of a similar inherited constitu
0167 een species of the same genus. The external CONDITIONS of life, as climate and food, etc., seem to
0171 erfect. The law of Unity of Type and of the CONDITIONS of Existence embraced by the theory of Natu
0173 odification, each has become adapted to the CONDITIONS of life of its own region, and has supplant
0174 he intermediate region, having intermediate CONDITIONS of life, why do we not now find close linki
0175 those who look at climate and the physical CONDITIONS of life as the all important elements of di
0175 ely depends on insensibly changing physical CONDITIONS, but in large part on the presence of other
0180 t is possible to conceive under all natural CONDITIONS. Let the climate and vegetation change, let
0181 difficulty, more especially under changing CONDITIONS of life, in the continued preservation of i
0182 ch of these birds is good for it, under the CONDITIONS of life to which it is exposed, for each ha
0182 sarily the best possible under all possible CONDITIONS. It must not be inferred from these remarks
0186 be ever useful to an animal under changing CONDITIONS of life, then the difficulty of believing t
0196 by the descendants of the species under new CONDITIONS of life and with newly acquired habits. To
0199 no direct use to their possessors. Physical CONDITIONS probably have had some little effect on str
0200 allowance for the direct action of physical CONDITIONS) may be viewed, either as having been of sp
0201 us. After the lapse of time, under changing CONDITIONS of life, if any part comes to be injurious,
0203 n the area was not continuous, and when the CONDITIONS of life did not insensibly graduate away fr
0204 . we have seen that a species may under new CONDITIONS of life change its habits, or have diversif
0204 ood for its possessor, then, under changing CONDITIONS of life, there is no logical impossibility
0206 d on two great laws, Unity of Type, and the CONDITIONS of Existence. By unity of type is meant tha
0206 ined by unity of descent. The expression of CONDITIONS of existence, so often insisted on by the i
0206 of each being to its organic and inorganic CONDITIONS of life; or by having adapted them during l
0206 fected by the direct action of the external CONDITIONS of life, and being in all cases subjected t
0206 s of growth. Hence, in fact, the Law of the CONDITIONS of Existence is the higher law; as it inclu
0209 welfare of each species, under its present CONDITIONS of life. Under changed conditions of life,
0209 s present conditions of life. Under changed CONDITIONS of life, it is at least possible that sligh
0213 comparably shorter period, under less fixed CONDITIONS of life. How stongly these domestic instinc
0233 each profitable to the individual under its CONDITIONS of life, it may reasonably be asked, how a
0243 ore I can see no difficulty, under changing CONDITIONS of life, in natural selection accumulating
0243 rld and living under considerably different CONDITIONS of life, yet often retaining nearly the sam
0245 parallelism between the effects of changed CONDITIONS of life and crossing. Fertility of varietie
0256 ds, is more easily affected by unfavourable CONDITIONS, than is the fertility of pure species. But
0256 ame capsule and exposed to exactly the same CONDITIONS. By the term systematic affinity is meant,
0259 rtility, or even to fertility under certain CONDITIONS in excess. That their fertility, besides be
0259 susceptible to favourable and unfavourable CONDITIONS, is innately variable. That it is by no mea
0264 e, they are generally placed under suitable CONDITIONS of life. But a hybrid partakes of only half
0264 roduced by the mother, it may be exposed to CONDITIONS in some degree unsuitable, and consequently
0264 inently sensitive to injurious or unnatural CONDITIONS of life. In regard to the sterility of hybr
0264 s and plants are removed from their natural CONDITIONS, they are extremely liable to have their re
0265 are rendered impotent by the same unnatural CONDITIONS; and whole groups of species tend to produc
0265 roup will sometimes resist great changes of CONDITIONS with unimpaired fertility; and certain spec
0265 are placed during several generations under CONDITIONS not natural to them, they are extremely lia
0265 c beings are placed under new and unnatural CONDITIONS, and when hybrids are produced by the unnat
0265 a very similar manner. In the one case, the CONDITIONS of life have been disturbed, though often i
0266 ther case, or that of hybrids, the external CONDITIONS have remained the same, but the organisatio
0266 parts and organs one to another, or to the CONDITIONS of life. When hybrids are able to breed int
0266 hy an organism, when placed under unnatural CONDITIONS, is rendered sterile. All that I have attem
0266 the common result, in the one case from the CONDITIONS of life having been disturbed, in the other
0266 ody of evidence, that slight changes in the CONDITIONS of life are beneficial to all living things
0267 especially if these be kept under the same CONDITIONS of life, always induces weakness and steril
0267 hat, on the one hand, slight changes in the CONDITIONS of life benefit all organic beings, and
0269 ring and disappearing under nearly the same CONDITIONS of life. Lastly, and this seems to me by fa
0273 ng eminently sensitive to any change in the CONDITIONS of life, being thus often rendered either i

```
0276   susceptible of favourable and unfavourable CONDITIONS. The degree of sterility does not strictly
0277   ly affects pure species, when their natural CONDITIONS of life have been disturbed. This view is s
0277   r offspring; and that slight changes in the CONDITIONS of life are apparently favourable to the vi
0279   and continous area with graduated physical CONDITIONS. I endeavoured to show, that the life of ea
0279   ; and, therefore, that the really governing CONDITIONS of life do not graduate away quite insensib
0314   f breeding, on the slowly changing physical CONDITIONS of the country, and more especially on the
0314   igher beings to their organic and inorganic CONDITIONS of life, as explained in a former chapter.
0315   hould never reappear, even if the very same CONDITIONS of life, organic and inorganic, should recu
0318   ly have exterminated the former horse under CONDITIONS of life apparently so favourable. But how u
0319   nswer that something is unfavourable in its CONDITIONS of life; but what that something is, we can
0319   n south America, that under more favourable CONDITIONS it would in a very few years have stocked t
0319   e could not have told what the unfavourable CONDITIONS were which checked its increase, whether so
0319   n what degree, they severally acted. If the CONDITIONS had gone on, however slowly, becoming less
0325   ges of currents, climate, or other physical CONDITIONS, as the cause of these great mutations in t
0325   slight is the relation between the physical CONDITIONS of various countries, and the nature of the
0326   know not at all precisely what are all the CONDITIONS most favourable for the multiplication of n
0332   odified in relation to its slightly altered CONDITIONS of life, and yet retain throughout a vast p
0336   ely consecutive formations, by the physical CONDITIONS of the ancient areas having remained nearly
0336   efore under the most different climates and CONDITIONS. Consider the prodigious vicissitudes of cl
0339   nt, on the one hand, by dissimilar physical CONDITIONS for the dissimilarity of the inhabitants of
0340   s, and, on the other hand, by similarity of CONDITIONS, for the uniformity of the same types in ea
0346   be accounted for by differences in physical CONDITIONS. Importance of barriers. Affinity of the br
0346   ed for by their climatal and other physical CONDITIONS. Of late, almost every author who has studi
0346   rn point, we meet with the most diversified CONDITIONS; the most humid districts, arid deserts, lo
0346   rganisms confined to any small spot, having CONDITIONS peculiar in only a slight degree; for insta
0347   ra. Notwithstanding this parallelism in the CONDITIONS of the Old and New Worlds, how widely diffe
0347   l find parts extremely similar in all their CONDITIONS, yet it would not be possible to point out
0350   nd water, and independent of their physical CONDITIONS. The naturalist must feel little curiosity,
0350   the direct influence of different physical CONDITIONS. The degree of dissimilarity will depend on
0350   their new homes they will be exposed to new CONDITIONS, and will frequently undergo further modifi
0351   lesser degree with the surrounding physical CONDITIONS. As we have seen in the last chapter that s
0352   europe and Australia or South America? The CONDITIONS of life are nearly the same, so that a mult
0353   tion and subsistence under past and present CONDITIONS permitted, is the most probable. Undoubtedl
0372   espondence with the nearly similar physical CONDITIONS of the areas: for if we compare, for instan
0372   closely corresponding in all their physical CONDITIONS, but with their inhabitants utterly dissimi
0377   ey will have been placed under somewhat new CONDITIONS, will have suffered less. And it is certain
0389   independently of any difference in physical CONDITIONS has caused so great a difference in number.
0392   are generally accounted for by the physical CONDITIONS of the islands; but this explanation seems
0392   at least as important as the nature of the CONDITIONS. Many remarkable little facts could be give
0393   s cannot be accounted for by their physical CONDITIONS; indeed it seems that islands are peculiarl
0398   created in America? There is nothing in the CONDITIONS of life, in the geological nature of the,is
0398   iated together, which resembles closely the CONDITIONS of the South American coast: in fact there
0400   ly seated error of considering the physical CONDITIONS of a country as the most important for its
0401   t would undoubtedly be exposed to different CONDITIONS of life in the different islands, for it wo
0403   g the commingling of species under the same CONDITIONS of life. Thus, the south east and south wes
0403   of Australia have nearly the same physical CONDITIONS, and are united by continuous land, yet the
0404   some former period under different physical CONDITIONS, and the existence at remote points of the
0405   sion, and should place itself under diverse CONDITIONS favourable for the conversion of its offspr
0406   slightly modified in relation to their new CONDITIONS. There is, also, some reason to believe fro
0408   y two areas having nearly the same physical CONDITIONS should often be inhabited by very different
0408   nt regions, independently of their physical CONDITIONS, infinitely diversified conditions of life,
0408   physical conditions, infinitely diversified CONDITIONS of life, there would be an almost endless a
0410   ation at some former period under different CONDITIONS or by occasional means of transport, and by
0415   e in the adaptation of the species to their CONDITIONS of life. That the mere physiological import
0425   ly to have been modified in relation to the CONDITIONS of life to which each species has been rece
0426   rve to preserve life under the most diverse CONDITIONS of existence, are generally the most consta
0427   cent, may readily become adapted to similar CONDITIONS, and thus assume a close external resemblan
0439   her, often have no direct relation to their CONDITIONS of existence. We cannot, for instance, supp
0440   the branchial slits are related to similar CONDITIONS, in the young mammal which is nourished in
0440   d fir of a porpoise, are related to similar CONDITIONS of life. No one will suppose that the strip
0440   use to these animals, or are related to the CONDITIONS to which they are exposed. The case, howeve
0440   omes on, the adaptation of the larva to its CONDITIONS of life is just as perfect and as beautiful
0442   the embryo not being closely related to its CONDITIONS of existence, except when the embryo become
0443   sexual elements having been affected by the CONDITIONS to which either parent, or their ancestors,
0454   lying. Again, an organ useful under certain CONDITIONS, might become injurious under others, as wi
0461   een disturbed by slightly different and new CONDITIONS of life, we need not feel surprise at hybri
0461   gs are increased by slight changes in their CONDITIONS of life, and that the offspring of slightly
0461   n the one hand, considerable changes in the CONDITIONS of life and crosses between greatly modifie
0461   nd on the other hand, lesser changes in the CONDITIONS of life and crosses between less modified f
0463   tinuous, and of which the climate and other CONDITIONS of life change insensibly in going from a d
0466   ing eminently susceptible to changes in the CONDITIONS of life: so that this system, when not rend
0466   e, and by the direct action of the physical CONDITIONS of life. There is much difficulty in ascert
0466   inherited for long periods. As long as the CONDITIONS of life remain the same, we have reason to
0467   intentionally exposes organic beings to new CONDITIONS of life, and then nature acts on the organi
0468   slight a degree to the surrounding physical CONDITIONS, will turn the balance. With animals having
0468   have most successfully struggled with their CONDITIONS of life, will generally leave most progeny.
0468   hey generally have varied under the changed CONDITIONS of domestication. And if there be any varia
0469   selecting variations useful, under changing CONDITIONS of life, to her living products? What limit
0472   lled specific forms. In both cases physical CONDITIONS seem to have produced but little direct eff
0474   s, when placed under considerably different CONDITIONS of life, yet should follow nearly the same
0477   the same continent, under the most diverse CONDITIONS, under heat and cold, on mountain and lowla
0477   ugh two areas may present the same physical CONDITIONS of life, we need feel no surprise at their
0479   useless by changed habits or under changed CONDITIONS of life: and we can clearly understand on t
0486   nd disuse, on the direct action of external CONDITIONS, and so forth. The study of domestic produc
0487   tered and perhaps suddenly altered physical CONDITIONS, namely, the mutual relation of organism to
0489   indirect and direct action of the external CONDITIONS of life, and from use and disuse; a Ratio o
```

Page ***(Key Word)***

```
0066  whole district, let it be ever so large. The CONDOR lays a couple of eggs and the ostrich a score
0066  rich a score, and yet in the same country the CONDOR may be the more numerous of the two: the Fulm
0187  swelling. In other crustaceans the transparent CONES which are coated by pigment, and which proper
0375  on the heights of Ceylon, and on the volcanic CONES of Java, many plants occur, either identicall
0186  formed by natural selection, seems, I freely CONFESS, absurd in the highest possible degree. Yet
0242  means of natural selection. But I am bound to CONFESS, that, with all my faith in this principle,
0449  this to be a law of nature; but I am bound to CONFESS that I only hope to see the law hereafter pr
0046  rtant (as some few naturalists have honestly CONFESSED) which does not vary; and, under this point
0050  t how to rank doubtful forms. Yet it must be CONFESSED, that it is in the best known countries tha
0266  ble, rarely diminishes. It must, however, be CONFESSED that we cannot understand, excepting on vac
0418  nated value. This principle has been broadly CONFESSED by some naturalists to be the true one; and
0466  ections relate to questions on which we are CONFESSEDLY ignorant; nor do we know how ignorant we a
0002  nd I must trust to the reader reposing some CONFIDENCE in my accuracy. No doubt errors will have c
0238  been slaughtered, but the breeder goes with CONFIDENCE to the same family. I have such faith in th
0239  ndeed be thought that I have an overweening CONFIDENCE in the principle of natural selection, when
0292  arily be rendered intermittent. I feel much CONFIDENCE in the truth of these views, for they are i
0482  influenced by this volume; but I look with CONFIDENCE to the future, to young and rising naturali
0489  he whole world. Hence we may look with some CONFIDENCE to a secure future of equally inappreciable
0065  ew which do not annually pair. Hence we may CONFIDENTLY assert, that all plants and animals are te
0132  climate, food, etc.: thus, F. Forbes speaks CONFIDENTLY that shells at their southern limit, and w
0160  aring in the several breeds. We may I think CONFIDENTLY come to this conclusion, because, as we ha
0167  at of the horse genus! For myself, I venture CONFIDENTLY to look back thousands on thousands of gen
0205  d by means of natural selection. But we may CONFIDENTLY believe that many modifications, wholly du
0237  culturist sows seeds of the same stock, and CONFIDENTLY expects to get nearly the same variety: br
0240  loped ocelli. I may give one other case: so CONFIDENTLY did I expect to find gradations in importa
0241  different structure of their jaws. I speak CONFIDENTLY on this latter point, as Mr. Lubbock made
0050  generally there is some variation. But if he CONFINE his attention to one class within one countr
0236  t here enter on these several cases, but will CONFINE myself to one special difficulty, which at f
0292  as a whole, is extremely imperfect; but if we CONFINE our attention to any one formation, it becom
0329  es, and orders, cannot be disputed. For if we CONFINE our attention either to the living or to the
0358  ccasional means of distribution. I shall here CONFINE myself to plants. In botanical works, this o
0389  uctions. In the following remarks I shall not CONFINE myself to the mere question of dispersal; bu
0014  s of embryology. These remarks are of course CONFINED to the first appearance of the peculiarity,
0045  uals of the same species inhabiting the same CONFINED locality. No one supposes that all the indiv
0077  he water beetle, the relation seems at first CONFINED to the elements of air and water. Yet the ad
0088  neral vigour; but on having special weapons, CONFINED to the male sex. A hornless stag or spurless
0102  als. Thus it will be in nature: for within a CONFINED area, with some place in its polity not so p
0103  stance in birds, varieties will generally be CONFINED to separated countries: and this I believe t
0104  nt in the process of natural selection. In a CONFINED or isolated area, if not very large, the org
0107  to the present day, from having inhabited a CONFINED area, and from having thus been exposed to l
0119  ubt that the process of modification will be CONFINED to a single line of descent, and the number
0133  ieties. Thus the species of shells which are CONFINED to tropical and shallow seas are generally b
0133  s are generally brighter coloured than those CONFINED to cold and deeper seas. The birds which are
0133  to cold and deeper seas. The birds which are CONFINED to continents are, according to Mr. Gould, b
0133  ed than those of islands. The insect species CONFINED to sea coasts, as every collector knows, are
0151  an be little doubt. But that our rule is not CONFINED to secondary sexual characters is clearly sh
0258  utional differences imperceptible by us, and CONFINED to the reproductive system. This difference
0275  ng one parent, the resemblances seem chiefly CONFINED to characters almost monstrous in their natu
0289  ogist, if his attention had been exclusively CONFINED to these large territories, would never have
0298  e changes are supposed to have been local or CONFINED to some one spot. Most marine animals have a
0301  rieties would at first generally be local or CONFINED to one place, but if possessed of any decide
0305  europe. Some few families of fish now have a CONFINED range; the teleostean fish might formerly ha
0305  ean fish might formerly have had a similarly CONFINED range, and after having been largely develop
0306  ht be multiplied; and here they would remain CONFINED, until some of the species became adapted to
0346  most rare case to find a group of organisms CONFINED to any small spot, having conditions peculia
0351  genera, whole genera, and even families are CONFINED to the same areas, as is so commonly and not
0353  ill greater number of sections of genera are CONFINED to a single region; and it has been observed
0353  lated to each other, are generally local, or CONFINED to one area. What a strange anomaly it would
0381  nct species, belonging to genera exclusively CONFINED to the south, at these and other distant poi
0384  es, there is hardly a single group of fishes CONFINED exclusively to fresh water, so that we may i
0391  lls, whereas not one species of sea shell is CONFINED to its shores: now, though we do not know ho
0392  , generally have, whatever the cause may be, CONFINED ranges. Hence trees would be little likely t
0399  t present inexplicable: but this affinity is CONFINED to the plants, and will, I do not doubt, be
0401  found in other parts of the world and those CONFINED to the archipelago, are common to the severa
0402  osely allied species of mocking thrush, each CONFINED to its own island. Now let us suppose the mo
0476  er such cases, is intelligible, for within a CONFINED country, the recent and the extinct will nat
0008  he chief one is the remarkable effect which CONFINEMENT or cultivation has on the functions of the
0008  ficult than to get it to breed freely under CONFINEMENT, even in the many cases when the male and
0008  ed, though living long under not very close CONFINEMENT in their native country! This is generally
0009  determine the reproduction of animals under CONFINEMENT, I may just mention that carnivorous anima
0009  , breed in this country pretty freely under CONFINEMENT, with the exception of the plantigrades or
0009  and sickly, yet breeding quite freely under CONFINEMENT; and when, on the other hand, we see indiv
0009  ised at this system, when it does act under CONFINEMENT, acting not quite regularly, and producing
0024  ivilized man, as to be quite prolific under CONFINEMENT. An argument, as it seems to me, of great
0089  those who have closely attended to birds in CONFINEMENT well know that they often take individual
0140  use they were useful and bred readily under CONFINEMENT, and not because they were subsequently fo
0215  s, simply to habit and long continued close CONFINEMENT. Natural instincts are lost under domestic
0252  owing to few animals breeding freely under CONFINEMENT, few experiments have been fairly tried: f
0252  one of these nine species breeds freely in CONFINEMENT, we have no right to expect that the first
0265  ther any particular animal will breed under CONFINEMENT or any plant seed freely under culture; no
0070  bounding in individuals, even on the extreme CONFINES of their range. For in such cases, we may be
0078  over the animals which preyed on it. On the CONFINES of its geographical range, a change of const
0078  limate alone. Not until we reach the extreme CONFINES of life, in the arctic regions or on the bor
0102  other individuals of the same species on the CONFINES of each. And in this case the effects of int
0107  species on each island: intercrossing on the CONFINES of the range of each species will thus be ch
0174  ing somewhat abruptly rarer and rarer on the CONFINES, and finally disappearing. Hence the neutral
```

Page ***(Key Word)***

```
0175  arply defined. Moreover, each species on the CONFINES of its range, where it exists in lessened nu
0301  lago now range thousands of miles beyond its CONFINES; and analogy leads me to believe that it wou
0306  of the geology of other countries beyond the CONFINES of Europe and the United States; and from th
0057  s far as my imperfect results go, they always CONFIRM the view. I have also consulted some sagacio
0099  et, as C. C. Sprengel has shown, and as I can CONFIRM, either the anthers burst before the stigma
0223  ch are the facts, though they did not need CONFIRMATION by me, in regard to the wonderful instinct
0423  he filiation and origin of each tongue. In CONFIRMATION of this view, let us glance at the classif
0251  the same result. This result has, also, been CONFIRMED by other observers in the case of Hippeastr
0271  us facts. Kolreuter, whose accuracy has been CONFIRMED by every subsequent observer, has proved th
0338  oved. Yet I fully expect to see it hereafter CONFIRMED, at least in regard to subordinate groups,
0230  if I had space, I could show that they are CONFORMABLE with my theory. Huber's statement that the
0317  umber of the species of a group is strictly CONFORMABLE with my theory; as the species of the same
0210  as in the case of corporeal structure, and CONFORMABLY with my theory, the instinct of each speci
0288  ty. The many cases on record of a formation CONFORMABLY covered, after an enormous interval of tim
0026  rations. These two distinct cases are often CONFOUNDED in treatises on inheritance. Lastly, the hy
0174  ties? This difficulty for a long time quite CONFOUNDED me. But I think it can be in large part exp
0245  undamentally different, have generally been CONFOUNDED together; namely, the sterility of two spec
0256  d, two classes of facts which are generally CONFOUNDED together, is by no means strict. There are
0260  sterility simply to prevent their becoming CONFOUNDED in nature? I think not. For why should the
0171  transitional forms? Why is not all nature in CONFUSION instead of the species being, as we see the
0245  uality of sterility, in order to prevent the CONFUSION of all organic forms. This view certainly s
0255  heir crossing and blending together in utter CONFUSION. The following rules and conclusions are ch
0076  ch has everywhere driven before it its great CONGENER. One species of charlock will supplant anoth
0158  nner in comparison with the same part in its CONGENERS; and the not great degree of variability in
0204  some habits very unlike those of its nearest CONGENERS. Hence we can understand, bearing in mind t
0283  hus impressed, let any one examine beds of CONGLOMERATE many thousand feet in thickness, which, th
0089  guiana, birds of Paradise, and some others, CONGREGATE; and successive males display their gorgeou
0021  ven laughter. The turbit has a very short and CONICAL beak, with a line of reversed feathers down
0047  here a wide door for the entry of doubt and CONJECTURE is opened. Hence, in determining whether a
0194  hough in many cases it is most difficult to CONJECTURE by what transitions an organ could have arr
0223  sanguinea originated I will not pretend to CONJECTURE. But as ants, which are not slave makers, w
0460  t is, no doubt, extremely difficult even to CONJECTURE by what gradations many structures have bee
0462  others. We are often wholly unable even to CONJECTURE how this could have been effected. Yet, as
0482  o not pretend that they can define, or even CONJECTURE, which are the created forms of life, and w
0098  umerous pollen granules are swept out of the CONJOINED anthers of each flower, before the stigma o
0342  st often been local. All these causes taken CONJOINTLY, must have tended to make the geological re
0415  in another work he says, the genera of the CONNARACEAE differ in having one or more ovaria, in th
0415  appear insufficient to separate Cnestis from CONNARUS. To give an example amongst insects, in one
0058  the actual characters of the forms which they CONNECT: and except, secondly, by a certain amount o
0107  hynchus and Lepidosiren, which, like fossils, CONNECT to a certain extent orders now widely separa
0176  h rarer numerically than the forms which they CONNECT. Now, if we may trust these facts and infere
0176  d in lesser numbers than the forms which they CONNECT, then, I think, we can understand why interm
0178  numbers than the varieties which they tend to CONNECT. From this cause alone the intermediate vari
0179  beaten and supplanted by the forms which they CONNECT; for these from existing in greater numbers
0181  ture, fitted for gliding through the air, now CONNECT the Galeopithecus with the other Lemuridae,
0279  g in lesser numbers than the forms which they CONNECT, will generally be beaten out and exterminat
0297  k both as species, unless they are enabled to CONNECT them together by close intermediate gradatio
0298  ive the improbability of our being enabled to CONNECT species by numerous, fine, intermediate, fos
0330  nt a form is, by so much the more it tends to CONNECT by some of its characters groups now widely
0423  d this would be strictly natural, as it would CONNECT together all languages, extinct and modern,
0007  knight, that this variability may be partly CONNECTED with excess of food. It seems pretty clear
0010  nyhow show that variation is not necessarily CONNECTED, as some authors have supposed, with the ac
0046  s assuredly can be given. There is one point CONNECTED with individual differences, which seems to
0047  riety of another, even when they are closely CONNECTED by intermediate links; nor will the commonl
0054  and are much diffused, but this seems to be CONNECTED with the nature of the stations inhabited b
0109  ust be here alluded to from being intimately CONNECTED with natural selection. Natural selection a
0123  is is a very important consideration), which CONNECTED the original species (A) and (I), have all
0145  . possibly, these several differences may be CONNECTED with some difference in the flow of nutrime
0149  low organisation; the foregoing remark seems CONNECTED with the very general opinion of naturalist
0150  re attached to one sex; but are not directly CONNECTED with the act of reproduction. The rule appl
0158  the organisation, are all principles closely CONNECTED together. All being mainly due to the growt
0181  ther believing it possible that the membrane CONNECTED fingers and fore arm of the Galeopithecus m
0240  in size, but in their organs of vision, yet CONNECTED by some few members in an intermediate cond
0258  early show that the capacity for crossing is CONNECTED with constitutional differences imperceptib
0267  an illusion. Both series of facts seem to be CONNECTED together by some common but unknown bond, w
0281  tural selection all living species have been CONNECTED with the parent species of each genus, by d
0282  y extinct, have in their turn been similarly CONNECTED with more ancient species; and so on backwa
0297  ss at the same time it could be most closely CONNECTED with either one or both forms by intermedia
0298  ens for examination, two forms can seldom be CONNECTED by intermediate varieties and thus proved t
0301  nal forms, which on my theory assuredly have CONNECTED all the past and present species of the sam
0317  ion of new and improved forms are intimately CONNECTED together. The old notion of all the inhabit
0327  end to correspond. There is one other remark CONNECTED with this subject worth making. I have give
0335  of the series in this same respect. Closely CONNECTED with the statement, that the organic remain
0342  transitional links which must formerly have CONNECTED the closely allied or representative specie
0344  make together one grand system; for all are CONNECTED by generation. We can understand, from the
0356  ow extends, land may at a former period have CONNECTED islands or possibly even continents togethe
0357  ands in the Atlantic must recently have been CONNECTED with Europe or Africa, and Europe likewise
0393  considered as oceanic, as it lies on a bank CONNECTED with the mainland; moreover, icebergs forme
0396  all our oceanic islands having been formerly CONNECTED by continuous land with the nearest contine
0410  cases the forms within each class have been CONNECTED by the same bond of ordinary generation; an
0414  ce. These resemblances, though so intimately CONNECTED with the whole life of the being, are ranke
0426  er in common, yet if these extreme forms are CONNECTED together by a chain of intermediate groups,
0431  mbers of the same family or higher group are CONNECTED together. For the common parent of a whole
0431  the early progenitors of birds were formerly CONNECTED with the early progenitors of the other ver
0431  e extinction of the forms of life which once CONNECTED fishes with batrachians. There has been sti
0433  nd how the several members of each class are CONNECTED together by the most complex and radiating
```

Page **(Key Word)**

```
0434  in form and size, and yet they always remain CONNECTED together in the same order. We never find,
0449  to be classed together, and as all have been CONNECTED by the finest gradations, the best, or inde
0476  time; for in both cases the beings have been CONNECTED by the bond of ordinary generation, and the
0483  far. All the members of whole classes can be CONNECTED together by chains of affinities, and all c
0485  hat the latter are known, or believed, to be CONNECTED at the present day by intermediate gradatio
0485  adations, whereas species were formerly thus CONNECTED. Hence, without quite rejecting the conside
0282  fossil remains of such infinitely numerous CONNECTING Links, it may be objected, that time will n
0299  ng down the distinction between species, by CONNECTING them together by numerous, fine, intermedia
0342  why we do not find interminable varieties, CONNECTING together all the extinct and existing forms
0429  form is, the greater must be the number of CONNECTING forms which on my theory have been extermin
0435  lenghtened to any extent, and the membrane CONNECTING them increased to any extent, so as to serv
0462  cepting in rare cases) to discover directly CONNECTING links between them, but only between each a
0463  ne of the extermination of an infinitude of CONNECTING links, between the living and extinct inhab
0130  s or Lepidosiren, which in some small degree CONNECTS by its affinities two large branches of life
0204  n lesser numbers than the two forms which it CONNECTS; consequently the two latter, during the cou
0129  the tree was small, budding twigs; and this CONNEXION of the former and present buds by ramifying
0156  for a very long period remained constant. In CONNEXION with the present subject, I will make only
0418  ated with others, though no apparent bond of CONNEXION can be discovered between them, especial va
0433  his element of descent is the hidden bond of CONNEXION which naturalists have sought under the ter
0434  strongly on the high importance of relative CONNEXION in homologous organs: the parts may change
0435  alter the framework of bones or the relative CONNEXION of the several parts. If we suppose that th
0436  that is correspond in number and in relative CONNEXION with, the elemental parts of a certain numb
0449  descent being on my view the hidden bond of CONNEXION which naturalists have been seeking under t
0031  table and are studied, like a picture by a CONNOISSEUR; this is done three times at intervals of
0125  crease in number. One large group will slowly CONQUER another large group, reduce its numbers, and
0188  the transitional grades. His reason ought to CONQUER his imagination: though I have felt the diff
0083  all countries, the natives have been so far CONQUERED by naturalised productions, that they have
0121  e collateral lines of descent, which will be CONQUERED by later and improved lines of descent. If,
0429  looking at aberrant forms as failing groups CONQUERED by more successful competitors, with a few
0476  er as the later and more improved forms have CONQUERED the older and less improved organic beings
0246  the several memoirs and works of those two CONSCIENTIOUS and admirable observers, Kolreuter and Ga
0482  ecies are mutable will do good service by CONSCIENTIOUSLY expressing his conviction; for only thus
0198  tant variations; and we are immediately made CONSCIOUS of this by reflecting on the differences in
0208  ndeed not rarely in direct opposition to our CONSCIOUS will! yet they may be modified by the will
0289  lt to avoid believing that they are closely CONSECUTIVE. But we know, for instance, from Sir R. Mu
0289  ength of time which has elapsed between the CONSECUTIVE formations, we may infer that this could n
0290  changes in the mineralogical composition of CONSECUTIVE formations, generally implying great chang
0297  er wider intervals, namely, to distinct but CONSECUTIVE stages of the same great formation, we fin
0301  elago would have to migrate, and no closely CONSECUTIVE record of their modifications could be pre
0303  me, which have probably elapsed between our CONSECUTIVE formations, longer perhaps in some cases t
0311  ptly changed forms of life, entombed in our CONSECUTIVE, but widely separated, formations. On this
0313  ogical formations; so that between each two CONSECUTIVE formations, the forms of life have seldom
0315  change exhibited by the fossils embedded in CONSECUTIVE formations is not equal. Each formation, o
0322  probable wide intervals of time between our CONSECUTIVE formations; and in these intervals there m
0334  qual intervals of time have elapsed between CONSECUTIVE formations. It is no real objection to the
0335  all palaeontologists, that fossils from two CONSECUTIVE formations are far more closely related to
0335  emblance of the distinct species in closely CONSECUTIVE formations, by the physical conditions of
0336  of the fact of fossil remains from closely CONSECUTIVE formations, though ranked as distinct spec
0342  vals of time which have elapsed between our CONSECUTIVE formations; he may overlook how important
0345  arly see why the organic remains of closely CONSECUTIVE formations are more closely allied to each
0465  clearly see this in the fossil remains from CONSECUTIVE formations invariably being much more clos
0488  amount of organic change in the fossils of CONSECUTIVE formations probably serves as a fair measu
0087  or the benefit of the community; if each in CONSEQUENCE profits by the selected change. What natur
0343  ction of old forms is the almost inevitable CONSEQUENCE of the production of new forms. We can und
0383  w from this capacity as an almost necessary CONSEQUENCE. We can here consider only a few cases. In
0386  atter seem immediately to acquire, as if in CONSEQUENCE, a very wide range. I think favourable mea
0490  as to lead to a Struggle for Life, and as a CONSEQUENCE to Natural Selection, entailing Divergence
0244  endowed or created instincts, but as small CONSEQUENCES of one general law, leading to the advance
0086  al selection will ensure that modifications CONSEQUENT on other modifications at a different perio
0159  e vera causa of community of descent, and a CONSEQUENT tendency to vary in a like manner, but to t
0320  th which it comes into competition; and the CONSEQUENT extinction of less favoured forms almost in
0370  tinuity of the circumpolar land, and to the CONSEQUENT freedom for intermigration under a more fav
0005  re born than can possibly survive; and as, CONSEQUENTLY, there is a frequently recurring struggle
0046  ce or disservice to the species, and which CONSEQUENTLY have not been seized on and rendered defin
0050  that they descend from common parents, and CONSEQUENTLY must be ranked as varieties. Close investi
0054  that country favourable to the genus; and, CONSEQUENTLY, we might have expected to have found in t
0065  e been very favourable, and that there has CONSEQUENTLY been less destruction of the old and young
0073  england, is never visited by insects, and CONSEQUENTLY, from its peculiar structure, never can se
0075  he others and so yield more seed, and will CONSEQUENTLY in a few years quite supplant the other va
0084  nd efforts of man! how short his time! and CONSEQUENTLY how poor will his products be, compared wi
0092  uld produce very vigorous seedlings, which CONSEQUENTLY would have the best chance of flourishing
0105  ly, isolation, by checking immigration and CONSEQUENTLY competition, will give time for any new va
0107  with that of the sea or of the land; and, CONSEQUENTLY, the competition between fresh water produ
0107  ergo many oscillations of level, and which CONSEQUENTLY will exist for long periods in a broken co
0109  variations in some way advantageous, which CONSEQUENTLY endure. But as from the high geometrical p
0110  ved within any given period, and they will CONSEQUENTLY be beaten in the race for life by the modi
0110  the severest competition with each other. CONSEQUENTLY, each new variety or species, during the p
0113  riving its utmost to increase its numbers. CONSEQUENTLY, I cannot doubt that in the course of many
0118  cy to variability is in itself hereditary, CONSEQUENTLY they will tend to vary, and generally to v
0126  ct, and leave no modified descendants; and CONSEQUENTLY that of the species living at any one peri
0142  is never propagated by seed; and of which CONSEQUENTLY new varieties have not been produced, has
0157  e the two sexes of any one of the species. CONSEQUENTLY, whatever part of the structure of the com
0176  termediate zone. The intermediate variety, CONSEQUENTLY, will exist in lesser numbers from inhabit
0177  the intermediate narrow, hilly tract; and CONSEQUENTLY the improved mountain or plain breed will
0179  e parent forms and the intermediate links. CONSEQUENTLY evidence of their former existence could b
0197  ree frequenting bird from its enemies; and CONSEQUENTLY that it was a character of importance and
```

Page **(Key Word)**

Page **(Key Word)**

```
0199 ry being is simply due to inheritance; and CONSEQUENTLY, though each being assuredly is well fitte
0204 bers than the two forms which it connects; CONSEQUENTLY the two latter, during the course of furth
0205 n of the inhabitants one with another, and CONSEQUENTLY will produce perfection, or strength in th
0234 community lived throughout the winter, and CONSEQUENTLY required a store of honey: there can in th
0238 y, has been advantageous to the community: CONSEQUENTLY the fertile males and females of the same
0264 conditions in some degree unsuitable, and CONSEQUENTLY be liable to perish at an early period; mo
0268 t to be quite fertile when crossed, and he CONSEQUENTLY ranks them as undoubted species. If we thu
0275 ch have been slowly acquired by selection. CONSEQUENTLY, sudden reversions to the perfect characte
0291 se oscillations have affected wide spaces. CONSEQUENTLY formations rich in fossils and sufficientl
0292 first broken up into an archipelago), and CONSEQUENTLY during subsidence, though there will be mu
0297 t recognise their relationship, and should CONSEQUENTLY be compelled to rank them all as distinct
0307 y their numerous and improved descendants. CONSEQUENTLY, if my theory be true, it is indisputable
0315 de and irregularly intermittent intervals; CONSEQUENTLY the amount of organic change exhibited by
0344 the more nearly it will be related to, and CONSEQUENTLY resemble, the common progenitor of groups,
0347 rees with those north of 25 degrees, which CONSEQUENTLY inhabit a considerably different climate,
0352 having been formerly united to Europe, and CONSEQUENTLY possessing the same quadrupeds. But if the
0368 d, they will have kept in a body together; CONSEQUENTLY their mutual relations will not have been
0369 l thus have been in some degree disturbed; CONSEQUENTLY they will have been liable to modification
0371 great region, with those of the Old World. CONSEQUENTLY we have here everything favourable for muc
0379 r own homes in greater numbers, and having CONSEQUENTLY been advanced through natural selection an
0384 e time for great geographical changes, and CONSEQUENTLY time and means for much migration. In the
0387 esh water plants to vast distances, and if CONSEQUENTLY the range of these plants was not very gre
0388 aquatic than between terrestrial species: CONSEQUENTLY an intruder from the waters of a foreign c
0391 o their proper places and habits, and will CONSEQUENTLY have been little liable to modification. M
0405 hanges and for accidents of transport; and CONSEQUENTLY for the migration of some of the species i
0406 t a slower rate than the higher forms; and CONSEQUENTLY the lower forms will have had a better cha
0412 to produce other new and dominant species. CONSEQUENTLY the groups which are now large, and which
0412 d from one ancient but unseen parent, and, CONSEQUENTLY, have inherited something in common. But t
0428 tend to go on increasing in size; and they CONSEQUENTLY supplant many smaller and feebler groups.
0431 rees, to all; and the several species will CONSEQUENTLY be related to each other by circuitous lin
0437 remely complex mouth formed of many parts, CONSEQUENTLY always have fewer legs; or conversely, tho
0437 ly liable to vary in number and structure; CONSEQUENTLY it is quite probable that natural selectio
0448 different from those of their parent, and CONSEQUENTLY to be constructed in a slightly different
0455 in its reduced state at the same age, and CONSEQUENTLY will seldom affect or reduce it in the emb
0462 en possible during a very long period; and CONSEQUENTLY the difficulty of the wide diffusion of sp
0482 by the majority of naturalists, and which CONSEQUENTLY have every external characteristic feature
0005 ological Record. In the next chapter I shall CONSIDER the geological succession of organic beings
0029 other species? Selection. Let us now briefly CONSIDER the steps by which domestic races have been
0045 ifferences generally affect what naturalists CONSIDER unimportant parts; but I could show by a lon
0049 sts. Several most experienced ornithologists CONSIDER our British red grouse as only a strongly ma
0050 h perplexed to determine what differences to CONSIDER as specific, and what as varieties; for he k
0054 nest produce well marked varieties, or, as I CONSIDER them, incipient species. And this, perhaps,
0084 aracters and structures, which we are apt to CONSIDER as very trifling importance, may thus be act
0202 other contrivances are less perfect. Can we CONSIDER the sting of the wasp or of the bee as perfe
0203 are fertilised through insect agency, can we CONSIDER as equally perfect the elaboration by our fi
0254 of first Crosses and of Hybrids. We will now CONSIDER a little more in detail the circumstances an
0270 om hostile witnesses, who in all other cases CONSIDER fertility and sterility as safe criterions o
0270 ile: so that even Gartner did not venture to CONSIDER the two varieties as specifically distinct.
0336 the most different climates and conditions. CONSIDER the prodigious vicissitudes of climate durin
0353 e of many species. So that we are reduced to CONSIDER whether the exceptions to continuity of rang
0354 ect, we shall be enabled at the same time to CONSIDER a point equally important for us, namely, wh
0383 an almost necessary consequence. We can here CONSIDER only a few cases. In regard to fish, I belie
0389 to the mere question of dispersal; but shall CONSIDER some other facts, which bear on the truth of
0414 ed to us by our classificications. Let us now CONSIDER the rules followed in classification, and th
0420 ation; that the characters which naturalists CONSIDER as showing true affinity between any two or
0459 ction of any organ or instinct, which we may CONSIDER, either do now exist or could have existed,
0483 the first appearance of species in what they CONSIDER reverent silence. It may be asked how far I
0483 the more distinct the forms are which we may CONSIDER, by so much the arguments fall away in force
0012 ructure, both those of slight and those of CONSIDERABLE physiological importance is endless. Dr. P
0015 unchanged conditions, and whilst kept in a CONSIDERABLE body, so that free intercrossing might che
0020 me of them are very important, as being of CONSIDERABLE antiquity. I have associated with several
0044 . by a monstrosity I presume is meant some CONSIDERABLE deviation of structure in one part, either
0047 plained. Those forms which possess in some CONSIDERABLE degree the character of species, but which
0067 future work, discuss some of the checks at CONSIDERABLE length, more especially in regard to the f
0103 rding natural selection; for I can bring a CONSIDERABLE catalogue of facts, showing that within th
0105 though I do not doubt that isolation is of CONSIDERABLE importance in the production of new specie
0119 ough to have allowed the accumulation of a CONSIDERABLE amount of divergent variation. As all the
0122 then our diagram be assumed to represent a CONSIDERABLE amount of modification, species (A) and al
0143 and disuse, have, in some cases, played a CONSIDERABLE part in the modification of the constituti
0162 this undoubtedly is the case in nature. A CONSIDERABLE part of the difficulty in recognising a va
0162 of the other species of the same genus. A CONSIDERABLE catalogue, also, could be given of forms i
0259 sed from seed from the same capsule have a CONSIDERABLE degree of fertility. These facts show how
0266 t universal belief, founded, I think, on a CONSIDERABLE body of evidence, that slight changes in t
0291 ward oscillation of level, and thus gained CONSIDERABLE thickness. All geological facts tell us pl
0296 so that if such species were to undergo a CONSIDERABLE amount of modification during any one geol
0298 y have been modified and perfected in some CONSIDERABLE degree. According to this view, the chance
0327 the inhabitants of each region underwent a CONSIDERABLE amount of modification and extinction, and
0337 ere set free in Great Britain, whether any CONSIDERABLE number would be enabled to seize on places
0377 ropical plants and animals can withstand a CONSIDERABLE amount of cold, many might have escaped ex
0378 described by Hooker. Thus, as I believe, a CONSIDERABLE number of plants, a few terrestrial animal
0384 have evidence in the loess of the Rhine of CONSIDERABLE changes of level in the land within a very
0398 e south American coast: in fact there is a CONSIDERABLE dissimilarity in all these respects. On th
0398 se respects. On the other hand, there is a CONSIDERABLE degree of resemblance in the volcanic natu
0400 e modified since their arrival), we find a CONSIDERABLE amount of difference in the several island
0425 have undergone a certain, and sometimes a CONSIDERABLE amount of modification, may not this same
0461 r and fertility. So that, on the one hand, CONSIDERABLE changes in the conditions of life and cros
```

Page **(Key Word)**

Page ***(Key Word)***

```
0484  we can dimly foresee that there will be a  CONSIDERABLE revolution in natural history. Systematist
0010  young of the same litter, sometimes differ  CONSIDERABLY from each other, though both the young and
0019  ks and rabbits, the breeds of which differ  CONSIDERABLY from each other in structure, I do not dou
0049  ula veris and elatior. These plants differ  CONSIDERABLY in appearance; they have a different flavo
0118  nd s2, differing from each other, and more  CONSIDERABLY from their common parent (A). We may conti
0123  rom (I) will, owing to inheritance, differ  CONSIDERABLY from the eight descendants from (A); the t
0155  hout large groups of species, has differed  CONSIDERABLY in closely allied species, that it has als
0194  nd with their structure either slightly or  CONSIDERABLY modified from that of their proper type. A
0243  istant parts of the world and living under  CONSIDERABLY different conditions of life, yet often re
0281  some points of structure may have differed  CONSIDERABLY from both, even perhaps more than they dif
0314  s insects of Madeira having come to differ  CONSIDERABLY from their nearest allies on the continent
0347  f 25 degrees, which consequently inhabit a  CONSIDERABLY different climate, and they will be found
0464  o other and distant regions until they are  CONSIDERABLY modified and improved; and when they do sp
0474  is that allied species, when placed under  CONSIDERABLY different conditions of life, yet should f
0025  d to the colouring of pigeons well deserve  CONSIDERATION. The rock pigeon is of a slaty blue, and
0049  varieties or doubtful species well deserve  CONSIDERATION; for several interesting lines of argumen
0053  their own country (and this is a different  CONSIDERATION from wide range, and to a certain extent
0123  pecies, also (and this is a very important  CONSIDERATION), which connected the original species (A
0176  both sides of it. But a far more important  CONSIDERATION, as I believe, is that, during the proces
0199  orced sense. But by far the most important  CONSIDERATION is that the chief part of the organisatio
0269  this seems to me by far the most important  CONSIDERATION, new races of animals and plants are prod
0285  side having been smoothly swept away. The  CONSIDERATION of these facts impresses my mind almost i
0298  return in the following chapter. One other  CONSIDERATION is worth notice: with animals and plants
0414  ptive or analogical characters; but to the  CONSIDERATION of these resemblances we shall have to re
0484  ter point will become a far more essential  CONSIDERATION than it is at present; for differences, h
0485  nected. Hence, without quite rejecting the  CONSIDERATION of the present existence of intermediate
0004  only neglected by naturalists. From these  CONSIDERATIONS, I shall devote the first chapter of this
0018  , state that, from geographical and other  CONSIDERATIONS, I think it highly probable that our dome
0028  romans. The paramount importance of these  CONSIDERATIONS in explaining the immense amount of varia
0053  e convenience sake. Guided by theoretical  CONSIDERATIONS, I thought that some interesting results
0066  t is most necessary to keep the foregoing  CONSIDERATIONS always in mind, never to forget that ever
0096  on details, I must trust to some general  CONSIDERATIONS alone. In the first place, I have collect
0101  d, becomes very small. From these several  CONSIDERATIONS and from the many special facts which I n
0110  the commoner species. From these several  CONSIDERATIONS I think it inevitably follows, that as ne
0134  ng under the most opposite climates. Such  CONSIDERATIONS as these incline me to lay very little we
0136  essitating frequent flight; these several  CONSIDERATIONS have made me believe that the wingless co
0268  tterly sterile when intercrossed. Several  CONSIDERATIONS, however, render the fertility of domesti
0271  e reproductive system; from these several  CONSIDERATIONS and facts, I do not think that the very g
0292  onal or linking forms. From the foregoing  CONSIDERATIONS it cannot be doubted that the geological
0293  due proportional weight to the following  CONSIDERATIONS. Although each formation may mark a very
0306  and distant seas. From these and similar  CONSIDERATIONS, but chiefly from our ignorance of the ge
0354  the belief, rendered probable by general  CONSIDERATIONS, that each species has been produced with
0403  e over the indigenous species. From these  CONSIDERATIONS I think we need not greatly marvel at the
0407  ngle centres of creation, by some general  CONSIDERATIONS, more especially from the importance of b
0480  ave now recapitulated the chief facts and  CONSIDERATIONS which have thoroughly convinced me that s
0036  ects of a course of selection, which may be  CONSIDERED as unconsciously followed, in so far that t
0048  me 182 British plants, which are generally  CONSIDERED as varieties, but which have all been ranke
0049  cient? It must be admitted that many forms,  CONSIDERED by highly competent judges as varieties, ha
0050  cies out of forms, which are very generally  CONSIDERED as varieties, and in this country the highe
0059  en discovered; but the amount of difference  CONSIDERED necessary to give to two forms the rank of
0085  f the fruit and the colour of the flesh are  CONSIDERED by botanists as characters of the most trif
0100  e, and that flowers on the same tree can be  CONSIDERED as distinct individuals only in a limited s
0141  omestic breeds. The rat and mouse cannot be  CONSIDERED as domestic animals, but they have been tra
0159  sixteen tail feathers in the pouter, may be  CONSIDERED as a variation representing the normal stru
0162  i and this shows, unless all these forms be  CONSIDERED as independently created species, that the
0164  government, a horse without stripes is not  CONSIDERED as purely bred. The spine is always striped
0187  superable by our imagination, can hardly be  CONSIDERED real. How a nerve comes to be sensitive to
0211  some few cases, certain instincts cannot be  CONSIDERED as absolutely perfect; but as details on th
0225  ls; and the following view may, perhaps, be  CONSIDERED only as a modification of his theory. Let u
0246  which is common to the two cases, has to be  CONSIDERED. The distinction has probably been slurred
0246  r in ten cases in which he found two forms,  CONSIDERED by most authors as distinct species, quite
0252  of a perfectly fertile hybrid animal can be  CONSIDERED as thoroughly well authenticated. It should
0254  ed to me by Mr. Blyth, I think they must be  CONSIDERED as distinct species. On this view of the or
0254  t, under our present state of knowledge, be  CONSIDERED as absolutely universal. Laws governing the
0259  rids, we see that when forms, which must be  CONSIDERED as good and distinct species, are united, t
0268  pernel, the primrose and cowslip, which are  CONSIDERED by many of our best botanists as varieties,
0277  n to be varieties, or sufficiently alike to  CONSIDERED as varieties, and their mongrel offspring,
0313  e is some reason to believe that organisms,  CONSIDERED high in the scale of nature, change more qu
0330  y extinct species or group of species being  CONSIDERED as intermediate between living species or g
0331  the forms beneath the uppermost line may be  CONSIDERED as extinct. The three existing genera, a14,
0343  great region alone, as that of Europe, are  CONSIDERED; he may urge the apparent, but often falsel
0393  t to an exception; but this group cannot be  CONSIDERED as oceanic, as it lies on a bank connected
0416  characters derived from parts which must be  CONSIDERED of very trifling physiological importance,
0416  rifling character would, I think, have been  CONSIDERED by naturalists as important an aid in deter
0418  he race, are found nearly uniform, they are  CONSIDERED as highly serviceable in classification; bu
0424  rup, which can only in a technical sense be  CONSIDERED as the same individual. He includes monster
0441  es, however, the mature animal is generally  CONSIDERED as lower in the scale than the larva, as wi
0441  last and completed state, cirripedes may be  CONSIDERED as either more highly or more lowly organis
0450  hat the bastard wing in birds may be safely  CONSIDERED as a digit in a rudimentary state: in very
0456  e common parentage of those forms which are  CONSIDERED by naturalists as allied, together with the
0457  he several classes of facts which have been  CONSIDERED in this chapter, seem to me to proclaim so
0459  ur imagination insuperably great, cannot be  CONSIDERED real if we admit the following propositions
0460  ed and of their mongrel offspring cannot be  CONSIDERED as universal; nor is their very general fer
0003  of knowledge and his excellent judgment. In  CONSIDERING the Origin of Species, it is quite conceiv
0023  yet unknown to ornithologists! and this,  CONSIDERING their size, habits, and remarkable charact
0100  an occasional cross being indispensable, by  CONSIDERING the medium in which terrestrial animals li
```

Page ***(Key Word)***

Page **(Key Word)**

0115 n is made to the genera of these States. By CONSIDERING the nature of the plants or animals which
0127 nd this certainly cannot be disputed; then, CONSIDERING the infinite complexity of the relations o
0191 ctually converted into organs of flight. In CONSIDERING transitions of organs, it is so important
0194 uld have arrived at its present state; yet, CONSIDERING that the proportion of living and known fo
0212 e of nature will be strengthened by briefly CONSIDERING a few cases under domestication. We shall
0216 ature have become modified by selection; by CONSIDERING a few cases. I will select only three, out
0229 ermediate planes or planes of intersection. CONSIDERING how flexible thin wax is, I do not see tha
0239 structure; and this we do find, even often, CONSIDERING how few neuter insects out of Europe have
0255 ain how far the rules apply to animals, and CONSIDERING how scanty our knowledge is in regard to h
0259 external resemblance to either pure parent. CONSIDERING the several rules now given, which govern
0284 this estimate may be quite erroneous; yet, CONSIDERING over what wide spaces very fine sediment i
0328 in a manner very difficult to account for, CONSIDERING the proximity of the two areas, unless, in
0346 cial period co extensive with the world. In CONSIDERING the distribution of organic beings over th
0354 ng migrated from a single birthplace; then, CONSIDERING our ignorance with respect to former clima
0363 ought thither the seeds of northern plants. CONSIDERING that the several above means of transport,
0371 e, a relationship which is most remarkable, CONSIDERING the distance of the two areas, and their s
0387 d mud was all contained in a breakfast cup! CONSIDERING these facts, I think it would be an inexpl
0387 fish are known sometimes to be dropped. In CONSIDERING these several means of distribution, it sh
0400 chief part from the deeply seated error of CONSIDERING the physical conditions of a country as th
0400 h cannot be here fairly included, as we are CONSIDERING how they have come to be modified since th
0405 rieties and ultimately into new species. In CONSIDERING the wide distribution of certain genera, w
0456 extinction and divergence of character. In CONSIDERING this view of classification, it should be
0058 are therein ranked as species, but which he CONSIDERS as so closely allied to other species as to
0220 st, either leave or enter the nest. Hence he CONSIDERS them as strictly household slaves. The most
0436 cture, are intelligible on the view that they CONSIST of metamorphosed leaves, arranged in a spire
0032 le even than in ordinary cases. If selection CONSISTED merely in separating some very distinct var
0037 s been followed almost unconsciously. It has CONSISTED in always cultivating the best known variet
0358 ssiliferous or other such rocks, instead of CONSISTING of mere piles of volcanic matter. I must no
0032 ardly to be worth notice; but its importance CONSISTS in the great effect produced by the accumula
0148 as: for the carapace in all other cirripedes CONSISTS of the three highly important anterior segme
0226 hich thus tend to intersect. Hence each cell CONSISTS of an outer spherical portion and of two, th
0451 ected in the ovarium at its base. The pistil CONSISTS of a stigma supported on the style; but some
0079 ion. When we reflect on this struggle, we may CONSOLE ourselves with the full belief, that the war
0069 sappearing; and the change of climate being CONSPICUOUS, we are tempted to attribute the whole eff
0475 hole groups of species, which has played so CONSPICUOUS a part in the history of the organic world
0415 ication, I believe, depends on their greater CONSTANCY throughout large groups of species; and thi
0415 throughout large groups of species; and this CONSTANCY depends on such organs having generally bee
0085 ng that colour, when once acquired, true and CONSTANT. Nor ought we to think that the occasional d
0121 e for life over other forms, there will be a CONSTANT tendency in the improved descendants of any
0149 organ, when it occurs in lesser numbers, is CONSTANT. The same author and some botanists have fur
0152 he standard. There may be truly said to be a CONSTANT struggle going on between, on the one hand,
0153 ave remained for a much longer period nearly CONSTANT. And this, I am convinced, is the case. That
0154 most abnormally developed organs may be made CONSTANT, I can see no reason to doubt. Hence when an
0155 rtant organ or part, which is generally very CONSTANT throughout large groups of species, has diff
0156 n which have for a very long period remained CONSTANT. In connexion with the present subject, I wi
0156 characters should not have been rendered as CONSTANT and uniform as other parts of the organisati
0180 ame genus; and of diversified habits, either CONSTANT or occasional, in the same species. And it s
0184 case as this, if the supply of insects were CONSTANT, and if better adapted competitors did not a
0208 body. When once acquired, they often remain CONSTANT throughout life. Several other points of res
0412 ossible in the economy of nature, there is a CONSTANT tendency in their characters to diverge. Thi
0412 on. I attempted also to show that there is a CONSTANT tendency in the forms which are increasing i
0415 antennae, as Westwood has remarked, are most CONSTANT in structure; in another division they diffe
0417 r no part of the organisation is universally CONSTANT. The importance of an aggregate of character
0423 o similar. Whatever part is found to be most CONSTANT, is used in classing varieties: thus the gre
0423 orns are much less serviceable, because less CONSTANT. In classing varieties, I apprehend if we ha
0426 ditions of existence, are generally the most CONSTANT; we attach especial value to them; but if th
0456 fications; the value set upon characters, if CONSTANT and prevalent, whether of high vital importa
0470 es in the economy of nature, there will be a CONSTANT tendency in natural selection to preserve th
0474 for in this case it will have been rendered CONSTANT by long continued natural selection. Glancin
0484 ll be easy) whether any form be sufficiently CONSTANT and distinct from other forms, to be capable
0062 ifficult, at least I have found it so, than CONSTANTLY to bear this conclusion in mind. Yet unless
0062 stly live on insects or seeds, and are thus CONSTANTLY destroying life; or we forget how largely t
0069 ach species, even where it most abounds, is CONSTANTLY suffering enormous destruction at some peri
0126 ny new places in the polity of Nature, will CONSTANTLY tend to supplant and destroy the earlier an
0137 vantage; and if so, natural selection would CONSTANTLY aid the effects of disuse. It is well known
0179 ; but the very process of natural selection CONSTANTLY tends, as has been so often remarked, to ex
0186 ill acknowledge that every organic being is CONSTANTLY endeavouring to increase in numbers; and th
0220 ves. The masters, on the other hand, may be CONSTANTLY seen bringing in materials for the nest, an
0253 successive generation, in opposition to the CONSTANTLY repeated admonition of every breeder. And i
0319 that the increase of every living being is CONSTANTLY being checked by unperceived injurious agen
0467 favoured individuals and races, during the CONSTANTLY recurrent Struggle for Existence, we see th
0472 ther cases. But on the view of each species CONSTANTLY trying to increase in number, with natural
0411 bitrary like the grouping of the stars in CONSTELLATIONS. The existence of groups would have been
0061 cies? How do those groups of species, which CONSTITUTE what are called distinct genera, and which
0123 ferent from the other five species, and may CONSTITUTE a sub genus or even a distinct genus. The s
0421 , now broken up into two or three families, CONSTITUTE a distinct order from those descended from
0478 n, that all past and present organic beings CONSTITUTE one grand natural system, with group subord
0341 lder genera has left modified descendants. CONSTITUTING the six new genera. The other seven specie
0012 o note the endless points in structure and CONSTITUTION in which the varieties and sub varieties d
0018 me by Mr. Blyth, on the habits, voice, and CONSTITUTION, etc., of the humped Indian cattle, that t
0024 ified breeds, though agreeing generally in CONSTITUTION, habits, voice, colouring, and in most par
0076 so exactly the same strength, habits, and CONSTITUTION, that the original proportions of a mixed
0076 invariably, some similarity in habits and CONSTITUTION, and always in structure, the struggle wil
0078 nes of its geographical range, a change of CONSTITUTION with respect to climate would clearly be a
0083 , the slightest difference of structure or CONSTITUTION may well turn the nicely balanced scale in
0105 rough modifications in their structure and CONSTITUTION. Lastly, isolation, by checking immigratio

Page **(Key Word)**

Page **(Key Word)**

```
0110  ch, from having nearly the same structure,  CONSTITUTION, and habits, generally come into the sever
0112  from any one species become in structure,   CONSTITUTION, and habits, by so much will they be bette
0114   the accompanying differences of habit and  CONSTITUTION, determine that the inhabitants, which thu
0121  st nearly related to each other in habits,   CONSTITUTION, and structure. Hence all the intermediate
0127  ausing an infinite diversity in structure,   CONSTITUTION, and habits, to be advantageous to them, I
0128  ore they diverge in structure, habits, and   CONSTITUTION, of which we see proof by looking at the i
0141  y crafted on an innate wide flexibility of   CONSTITUTION, which is common to most animals. On this
0141  s examples of a very common flexibility of   CONSTITUTION, brought, under peculiar circumstances, in
0142  it be supposed that no differences in the    CONSTITUTION of seedling kidney beans ever appear, for
0143  nsiderable part in the modification of the   CONSTITUTION, and of the structure of various organs; b
0159  ng inherited from a common parent the same   CONSTITUTION and tendency to variation, when acted on b
0161  ditions of life and of a similar inherited   CONSTITUTION. It might further be expected that the spe
0169  eloped. Species inheriting nearly the same   CONSTITUTION from a common parent and exposed to simila
0198  mates; and there is reason to believe that   CONSTITUTION and colour are correlated. A good observer
0256  ircumstances, but depends in part upon the   CONSTITUTION of the individuals which happen to have be
0256  blance between species in structure and in   CONSTITUTION, more especially in the structure of parts
0264  id partakes of only half of the nature and   CONSTITUTION of its mother, and therefore before birth,
0469  ng ages and rigidly scrutinising the whole   CONSTITUTION, structure, and habits of each creature, f
0012  blue eyes are invariably deaf; colour and    CONSTITUTIONAL peculiarities go together, of which many
0083  n every internal organ, on every shade of    CONSTITUTIONAL difference, on the whole machinery of lif
0140  ound in this country to possess different    CONSTITUTIONAL powers of resisting cold. Mr. Thwaites in
0142  e of recent origin, they cannot owe their    CONSTITUTIONAL differences to habit. The case of the Jer
0167  slight modifications. Habit in producing     CONSTITUTIONAL differences, and use in strengthening, an
0195  of insects or from being correlated with     CONSTITUTIONAL differences, might assuredly be acted on
0248  ree as is the evidence derived from other    CONSTITUTIONAL and structural differences. In regard to
0258  e capacity for crossing is connected with    CONSTITUTIONAL differences imperceptible by us, and conf
0269  nces in the reproductive system, or other    CONSTITUTIONAL differences correlated with the reproduct
0460  ed together; but that it is incidental on    CONSTITUTIONAL differences in the reproductive systems o
0038  he same species, having slightly different   CONSTITUTIONS or structure, would often succeed better
0141  ction of varieties having different innate   CONSTITUTIONS, and how much to both means combined, is
0142  lecting so many breeds and sub breeds with   CONSTITUTIONS specially fitted for their own districts:
0142  erve those individuals which are born with   CONSTITUTIONS best adapted to their native countries. I
0198  n, and individuals with slightly different   CONSTITUTIONS would succeed best under different climat
0266  disturbed by two different structures and   CONSTITUTIONS having been blended into one. For it is s
0379  ifications in their structure, habits, and   CONSTITUTIONS will have profited them. Thus many of the
0460  er that it is not likely that either their   CONSTITUTIONS or their reproductive systems should have
0461  rendered in some degree sterile from their   CONSTITUTIONS having been disturbed by slightly differe
0461  ds being more or less degree sterile, for   CONSTITUTIONS can hardly fail to have been disturbed fr
0435  it is; that it has so pleased the creator to  CONSTRUCT each animal and plant. The explanation is m
0005  nto a highly developed being or elaborately  CONSTRUCTED organ; secondly, the subject of Instinct,
0095  ee to have a slightly longer or differently  CONSTRUCTED proboscis. On the other hand, I have found
0138  ght to darkness. Next follow those that are  CONSTRUCTED for twilight; and, last of all, those dest
0167  bra, but perhaps otherwise very differently  CONSTRUCTED, the common parent of our domestic horse,
0181  rhaps see traces of an apparatus originally  CONSTRUCTED for gliding through the air rather than fo
0190  hly important fact that an organ originally  CONSTRUCTED for one purpose, namely flotation, may be
0192  r insects, which are often very differently  CONSTRUCTED from either the males or fertile females;
0195  l of the giraffe looks like an artificially  CONSTRUCTED fly flapper; and it seems at first incredi
0197  he loftiest trees by the aid of exquisitely  CONSTRUCTED hooks clustered around the ends of the bra
0349  rly similar, and sees their nests similarly  CONSTRUCTED, but not quite alike, with eggs coloured l
0434  ise, and the wing of the bat, should all be  CONSTRUCTED on the same pattern, and should include th
0435  ay be called, of all mammals, had its limbs  CONSTRUCTED on the existing general pattern, for whate
0437  for such widely different purposes, be all  CONSTRUCTED on the same pattern? On the theory of natu
0441  erflies, they have six pairs of beautifully  CONSTRUCTED natatory legs, a pair of magnificent compo
0441  prehensile organs; they again obtain a well  CONSTRUCTED mouth; but they have no antennae, and thei
0448  ose of their parent, and consequently to be  CONSTRUCTED in a slightly different manner, then, on t
0457  cies of a class; or to the homologous parts  CONSTRUCTED on the same pattern in each individual and
0489  arth, and to reflect that these elaborately  CONSTRUCTED forms, so different from each other, and d
0208  its hammock up to, say, the sixth stage of   CONSTRUCTION, and put it into a hammock completed up on
0208  med the fourth, fifth, and sixth stages of  CONSTRUCTION. If however, a caterpillar were taken out
0224  sible consumption of precious wax in their   CONSTRUCTION. It has been remarked that a skilful workm
0226  in any one cell necessarily enter into the  CONSTRUCTION of three adjoining cells. It is obvious th
0230  w important a part excavation plays in the  CONSTRUCTION of the cells; but it would be a great erro
0232  edges of the cells all round. The work of   CONSTRUCTION seems to be a sort of balance struck betwe
0233  om a single hexagon being built, as in its  CONSTRUCTION more materials would be required than for
0233  ending towards the present perfect plan of  CONSTRUCTION, could have profited the progenitors of th
0234  the secretion of the wax necessary for the  CONSTRUCTION of their combs. Moreover, many bees have t
0434  animals. We see the same great law in the  CONSTRUCTION of the mouths of insects: what can be more
0435  irs of maxillae. Analogous laws govern the  CONSTRUCTION of the mouths and limbs of crustaceans. So
0435  the plain signification of the homologous  CONSTRUCTION of the limbs throughout the whole class. S
0437  mammals, will by no means explain the same  CONSTRUCTION in the skulls of birds. Why should similar
0101  se of flowers, I have as yet failed, after  CONSULTATION with one of the highest authorities, namel
0057  o, they always confirm the view. I have also  CONSULTED some sagacious and most experienced observe
0233  om twelve to fifteen pounds of dry sugar are  CONSUMED by a hive of bees for the secretion of each
0233  antity of fluid nectar must be collected and  CONSUMED by the bees in a hive for the secretion of t
0284  r accumulation; yet what time this must have  CONSUMED! Good observers have estimated that sediment
0296  ny idea of the time which its deposition has  CONSUMED. Many instances could be given of beds only
0224  le amount of honey, with the least possible  CONSUMPTION of precious wax in their construction. It
0177  sted in greater numbers, will come into close  CONTACT with each other, without the interposition o
0226  or more other cells. When one cell comes into  CONTACT with three other cells, which, from the sphe
0286  ; and we must remember that almost all strata  CONTAIN harder layers or nodules, which from long re
0387  37 in number; and yet the viscid mud was all  CONTAINED in a breakfast cup! Considering these facts
0324  orth and South America, and Australia, from  CONTAINING fossil remains in some degree allied, and f
0387  d probably reject from its stomach a pellet  CONTAINING the seeds of the Nelumbium undigested; or t
0485  ure as one which has had a history; when we  CONTEMPLATE every complex structure and instinct as th
0489  ss towards perfection. It is interesting to  CONTEMPLATE an entangled bank, clothed with many plant
0300  part of the archipelago, together with a    CONTEMPORANEOUS accumulation of sediment, would exceed t
```

Page **(Key Word)**

Page **(Key Word)***

```
0328  me movement, it is probable that strictly CONTEMPORANEOUS formations have often been accumulated o
0328  rated two seas inhabited by distinct, but CONTEMPORANEOUS, faunas. Lyell has made similar observat
0374  endured for long at each, and that it was CONTEMPORANEOUS in a geological sense, it seems to me pr
0487  us in attempting to correlate as strictly CONTEMPORANEOUS two formations, which include few identi
0362  e known to look out for tired birds, and the CONTENTS of their torn crops might thus readily get s
0075  most equally severe, and we sometimes see the CONTEST soon decided: for instance, if several varie
0088  ry, as the sword or spear. Amongst birds, the CONTEST is often of a more peaceful character. All t
0114  if freely open to immigration, and where the CONTEST between individual and individual must be se
0001  the present to the past inhabitants of that CONTINENT. These facts seemed to me to throw some lig
0049  a and Europe is ample, will that between the CONTINENT and the Azores, or Madeira, or the Canaries
0085  their prey, so much so, that on parts of the CONTINENT persons are warned not to keep white pigeon
0105  l isolated area, or a large open area like a CONTINENT, has been most favourable for the productio
0106  nstance, that the productions of the smaller CONTINENT of Australia have formerly yielded, and app
0107  y, for the area will first have existed as a CONTINENT, and the inhabitants, at this period numero
0108  rs of the various inhabitants of the renewed CONTINENT will again be changed: and again there will
0138  affinities to the other inhabitants of that CONTINENT, and in those of Europe, to the inhabitants
0138  f europe, to the inhabitants of the European CONTINENT. And this is the case with some of the Amer
0173  se: in travelling from north to south over a CONTINENT, we generally meet at successive intervals
0174  y would lead us to believe that almost every CONTINENT has been broken up into islands even during
0284  s, are thousands of feet in thickness on the CONTINENT. Moreover, between each successive formatio
0292  excepting the productions on the shores of a CONTINENT when first broken up into an archipelago),
0309  the Pacific Ocean were now converted into a CONTINENT, we should there find formations older than
0314  onsiderably from their nearest allies on the CONTINENT of Europe, whereas the marine shells and bi
0319  d in a very few years have stocked the whole CONTINENT. But we could not have told what the unfavo
0339  sely allied to the living marsupials of that CONTINENT. In South America, a similar relationship i
0339  , on this wonderful relationship in the same CONTINENT between the dead and the living. Professor
0340  resent character of the southern half of the CONTINENT; and the southern half was formerly more cl
0340  ified descendants. If the inhabitants of one CONTINENT formerly differed greatly from those of ano
0340  merly differed greatly from those of another CONTINENT, so will their modified descendants still d
0346  ers. Affinity of the productions of the same CONTINENT. Centres of creation. Means of dispersal, b
0346  lds: yet if we travel over the vast American CONTINENT, from the central parts of the United State
0347  ated from each other as is possible. On each CONTINENT, also, we see the same fact: for on the opp
0349  the affinity of the productions of the same CONTINENT or sea, though the species themselves are d
0349  is a law of the widest generality, and every CONTINENT offers innumerable instances. Nevertheless
0350  merican types then prevalent on the American CONTINENT and in the American seas. We see in these f
0355  e distance of a few hundreds of miles from a CONTINENT, would probably receive from it in the cour
0355  ted by inheritance to the inhabitants of the CONTINENT. Cases of this nature are common, and are,
0357  s which has not recently been united to some CONTINENT. This view cuts the Gordian knot of the dis
0358  faunas on the opposite sides of almost every CONTINENT, the close relation of the tertiary inhabit
0364  breadth, or from island to island, or from a CONTINENT to a neighbouring island, but not from one
0364  eighbouring island, but not from one distant CONTINENT to another. The floras of distant continent
0365  ansport, immigrants from Europe or any other CONTINENT, that a poorly stocked island, though stand
0372  the seas of Japan, areas now separated by a CONTINENT and by nearly a hemisphere of equatorial oc
0373  glacier. Further south on both sides of the CONTINENT, from lat. 41 degrees to the southernmost e
0374  both sides of the southern extremity of the CONTINENT. If this is admitted, it is difficult to av
0384  aters of distant continents. But on the same CONTINENT the species often range widely and almost c
0389  nds have been nearly or quite joined to some CONTINENT. This view would remove many difficulties,
0390  os archipelago, with the number found on any CONTINENT, and then compare the area of the islands w
0390  are the area of the islands with that of the CONTINENT, we shall see that this is true. This fact
0393  ng an island situated above 300 miles from a CONTINENT or great continental island: and many islan
0394  e world on very small islands, if close to a CONTINENT; and hardly an island can be named on which
0396  ants of islands with those of a neighbouring CONTINENT, an inexplicable relation on the view of in
0396  onnected by continuous land with the nearest CONTINENT: for on this latter view the migration woul
0398  ears the unmistakeable stamp of the American CONTINENT. There are twenty six land birds, and twent
0398  ific, distant several hundred miles from the CONTINENT, yet feels that he is standing on American
0399  islands are related to those of the nearest CONTINENT, or of other near islands. The exceptions a
0399  th america, which, although the next nearest CONTINENT, is so enormously remote, that the fact bec
0400  to be closely allied to those of the nearest CONTINENT, we sometimes see displayed on a small scal
0403  ces, as in the several districts of the same CONTINENT, pre occupation has probably played an impo
0408  nct beings which formerly inhabited the same CONTINENT. Bearing in mind that the mutual relations
0409  d, but less closely, to those of the nearest CONTINENT or other source whence immigrants were prob
0476  e long endurance of allied forms on the same CONTINENT, of marsupials in Australia, of endentata i
0477  ck every traveller, namely, that on the same CONTINENT, under the most diverse conditions, under h
0477  ten be found on islands far distant from any CONTINENT. Such facts as the presence of peculiar spe
0487  itants of the sea on the opposite sides of a CONTINENT, and the nature of the various inhabitants
0487  he nature of the various inhabitants of that CONTINENT in relation to their apparent means of immi
0106  ropaeo Asiatic area. Thus, also, it is that CONTINENTAL productions have everywhere become so larg
0107  e, that for terrestrial productions a large CONTINENTAL area, which will probably undergo many osc
0108  n, the islands shall be re converted into a CONTINENTAL area, there will again be severe competiti
0308  d secondary periods, neither continents nor CONTINENTAL islands existed where our oceans now exten
0380  l islands have everywhere lately yielded to CONTINENTAL forms, naturalised by man's agency. I am f
0389  at i cannot honestly admit Forbes's view on CONTINENTAL extensions; which, if legitimately followe
0389  few in number compared with those on equal CONTINENTAL areas: Alph. de Candolle admits this for p
0393  d above 300 miles from a continent or great CONTINENTAL island: and many islands situated at a muc
0133  eeper seas. The birds which are confined to CONTINENTS are, according to Mr. Gould, brighter colou
0134  used by disuse. The ostrich indeed inhabits CONTINENTS and is exposed to danger from which it cann
0138  t the case, and the cave insects of the two CONTINENTS are not more closely allied than might have
0139  animals to the other inhabitants of the two CONTINENTS on the ordinary view of their independent c
0308  curred in the neighbourhood of the existing CONTINENTS of Europe and North America. But we do not
0308  e palaeozoic and secondary periods, neither CONTINENTS nor continental islands existed where our o
0309  y record: and on the other hand, that where CONTINENTS now exist, large tracts of land have existe
0309  ill areas of oscillations of level, and the CONTINENTS areas of elevation. But have we any right t
0309  hings have thus remained from eternity? Our CONTINENTS seem to to have been formed by a prepondera
0309  asurably antecedent to the silurian epoch, CONTINENTS may have existed where oceans are now sprea
0309  and open oceans may have existed where our CONTINENTS now stand. Nor should we be justified in as
0326  ith the terrestrial inhabitants of distinct CONTINENTS than with the marine inhabitants of the con
```

Page **(Key Word)**

Page **(Key Word)**

```
0339  ssimilarity of the inhabitants of these two CONTINENTS, and, on the other hand, by similarity of c
0343  period extended, and where our oscillating CONTINENTS now stand they have stood ever since the Si
0343  wholly different aspect: and that the older CONTINENTS, formed of formations older than any known
0348  ve endured so long as the oceans separating CONTINENTS, the differences are very inferior in degre
0348  degree to those characteristic of distinct CONTINENTS. Turning to the sea, we find the same law.
0356  iod have connected islands or possibly even CONTINENTS together, and thus have allowed terrestrial
0357  dence of great oscillations of level in our CONTINENTS: but not of such vast charges in their posi
0357  ver be proved that within the recent period CONTINENTS which are now quite separate, have been con
0358  the belief of their former continuity with CONTINENTS. Nor does their almost universally volcanic
0358  dmission that they are the wrecks of sunken CONTINENTS: if they had originally existed as mountain
0364  continent to another. The floras of distant CONTINENTS would not by such means become mingled in a
0370  g, from reasons before alluded to, that our CONTINENTS have long remained in nearly the same relat
0372  in parts of South America with the southern CONTINENTS of the Old World, we see countries closely
0384  never occur in the fresh waters of distant CONTINENTS. But on the same continent the species ofte
0386  ater and even marsh species have, both over CONTINENTS and to the most remote oceanic islands. Thi
0394  species belonging to other classes; and on CONTINENTS it is thought that mammals appear and disap
0394  cies have enormous ranges, and are found on CONTINENTS and on far distant islands. Hence we have o
0395  relation to the remoteness of islands from CONTINENTS, there is also a relation, to a certain ext
0401  ter throughout the group, just as we see on CONTINENTS some species spreading widely and remaining
0024  become extinct or unknown. So many strange CONTINGENCIES seem to me improbable in the highest degr
0086  adapt the larva of an insect to a score of CONTINGENCIES, wholly different from those which concer
0173  the sea, whilst it slowly subsides. These CONTINGENCIES will concur only rarely, and after enormo
0300  tion of the same specific forms: and these CONTINGENCIES are indispensable for the preservation of
0314  e varying species, depends on many complex CONTINGENCIES, on the variability being of a beneficial
0319  its increase, whether some one or several CONTINGENCIES, and at what period of the horse's life,
0322  moment that we understand the many complex CONTINGENCIES, on which the existence of each species d
0343  essarily slow, and depends on many complex CONTINGENCIES. The dominant species of the larger domin
0448  t this would result from the two following CONTINGENCIES: firstly, from the young, during a course
0456  cation through natural selection, with its CONTINGENCIES of extinction and divergence of character
0478  n the theory of natural selection with its CONTINGENCIES of extinction and divergence of character
0043  at variability is an inherent and necessary CONTINGENCY, under all circumstances, with all organic
0295  the amount of subsidence is probably a rare CONTINGENCY: for it has been observed by more than one
0331  that divergence of character is a necessary CONTINGENCY: it depends solely on the descendants from
0471  racter, together with the almost inevitable CONTINGENCY of much extinction, explains the arrangeme
0442  e led to look at these facts as necessarily CONTINGENT in some manner on growth. But there is no o
0174  autious in inferring, because an area is now CONTINUOUS, that it has been continuous during a long
0174  the formerly broken condition of areas now CONTINUOUS has played an important part in the formati
0279  r their presence, namely on an extensive and CONTINUOUS area with graduated physical conditions. I
0203  cess of natural selection almost implies the CONTINUAL supplanting and extinction of preceding and
0003  justly excites our admiration. Naturalists CONTINUALLY refer to external conditions, such as clim
0021  rs down the breast: and it has the habit of CONTINUALLY expanding slightly the upper part of the o
0051  ount of difference in the forms which he is CONTINUALLY studying; and he has little general knowle
0094  tendency more and more increased, would be CONTINUALLY favoured or selected, until at last a comp
0109  go further than this: for as new forms are CONTINUALLY and slowly being produced, unless we belie
0113  to go on varying, and those varieties were CONTINUALLY selected which differed from each other in
0142  reason to doubt that natural selection will CONTINUALLY tend to preserve those individuals which a
0147  rinciple, namely, that natural selection is CONTINUALLY trying to economise in every part of the o
0153  continued amount of variability, which has CONTINUALLY been accumulated by natural selection for
0189  then suppose every part of this layer to be CONTINUALLY changing slowly in density, so as to separ
0209  ficulty in natural selection preserving and CONTINUALLY accumulating variations of instinct to any
0231  is singular manner of building, strength is CONTINUALLY given to the comb, with the utmost ultimat
0234  would save some little wax. Hence it would CONTINUALLY be more and more advantageous to our humbl
0240  unity, and those males and females had been CONTINUALLY selected, which produced more and more of
0280  ural selection, through which new varieties CONTINUALLY take the places of and exterminate their p
0288  iment is not accumulating. I believe we are CONTINUALLY taking a most erroneous view, when we taci
0290  the littoral and sub littoral deposits are CONTINUALLY worn away, as soon as they are brought up
0302  s before their modified descendants. But we CONTINUALLY over rate the perfection of the geological
0302  at they did not exist before that stage. We CONTINUALLY forget how large the world is, compared wi
0338  ilst it leaves the embryo almost unaltered, CONTINUALLY adds, in the course of successive generati
0461  oth sides are in a perfect condition. As we CONTINUALLY see that organisms of all kinds are render
0117  1 and m1. These two varieties will generally CONTINUE to be exposed to the same conditions which m
0118  derably from their common parent (A). We may CONTINUE the process by similar steps for any length
0121  of our original genus, may for a long period CONTINUE transmitting unaltered descendants: and this
0122  arent do not come into competition, both may CONTINUE to exist. If then our diagram be assumed to
0126  red least extinction, will for a long period CONTINUE to increase. But which groups will ultimatel
0182  sitional grades of the structure will seldom CONTINUE to exist to the present day, for they will h
0273  ght expect that such variability would often CONTINUE and be super added to that arising from the
0291  ness and extent over a shallow bottom, if it CONTINUE slowly to subside. In this latter case, as l
0454  dsi and in this case natural selection would CONTINUE slowly to reduce the organ, until it rendere
0466  ady been inherited for many generations, may CONTINUE to be inherited for an almost infinite numbe
0487  the light which geology now throws, and will CONTINUE to throw, on former changes of climate and o
0020  uld not be got without extreme care and long CONTINUED selection: nor can I find a single case on
0026  ctly fertile. Some authors believe that long CONTINUED domestication eliminates this strong tenden
0029  e explanation, I think, is simple: from long CONTINUED study they are strongly impressed with the
0033  , as a general rule, I cannot doubt that the CONTINUED selection of slight variations, either in t
0035  vertheless I cannot doubt that this process, CONTINUED during centuries, would improve and modify
0038  the native plants have not been improved by CONTINUED selection up to a standard of perfection co
0039  nts of that pigeon would become through long CONTINUED, partly unconscious and partly methodical s
0052  ay be, in some cases, due merely to the long CONTINUED action of different physical conditions in
0091  forced to hunt different prey: and from the CONTINUED preservation of the individuals best fitted
0093  cted. When our plant, by this process of the CONTINUED preservation or natural selection of more a
0094  we have been slowly increasing the nectar by CONTINUED selection, to be a common plant: and that c
0095  he most perfect manner to each other, by the CONTINUED preservation of individuals presenting mutu
0095  e a true principle, banish the belief of the CONTINUED creation of new organic beings, or of any g
0104  r and fertility over the offspring from long CONTINUED self fertilisation, that they will have a b
0112  very slight: in the course of time, from the CONTINUED selection of swifter horses by some breeder
```

Page **(Key Word)**

Page **(Key Word)**

```
0125  ); and as these latter two genera, both from  CONTINUED  divergence of character and from inheritanc
0126  he future, we may predict that, owing to the  CONTINUED  and steady increase of the larger groups, a
0134  on, by which to judge of the effects of long  CONTINUED  use or disuse, for we know not the parent f
0135  condition in some other genera, by the long  CONTINUED  effects of disuse in their progenitors; for
0139  isation must be readily effected during long  CONTINUED  descent. It is notorious that each species
0150  rious laws of growth, to the effects of long  CONTINUED  disuse, and to the tendency to reversion. A
0152  present time are undergoing rapid change by  CONTINUED  selection, are also eminently liable to var
0153  fication implies an unusually large and long  CONTINUED  amount of variability, which has continuall
0153  ped part or organ has been so great and long  CONTINUED  within a period not excessively remote, we
0154  ty will seldom as yet have been fixed by the  CONTINUED  selection of the individuals varying in the
0154  n the required manner and degree, and by the  CONTINUED  rejection of those tending to revert to a f
0169  ee than other parts; for variation is a long  CONTINUED  and slow process, and natural selection wil
0181  ly under changing conditions of life, in the  CONTINUED  preservation of individuals with fuller and
0188  is instrument has been perfected by the long  CONTINUED  efforts of the highest human intellects; an
0214  which have become inherited solely from long  CONTINUED  and compulsory habit, but this, I think, is
0214  ncy to this strange habit, and that the long  CONTINUED  selection of the best individuals in succes
0215  o extreme tameness, simply to habit and long  CONTINUED  close confinement. Natural instincts are lo
0217  s be successful in rearing their young. By a  CONTINUED  process of this nature, I believe that the
0239  st, but in a few alone; and that by the long  CONTINUED  selection of the fertile parents which prod
0245  herefore could not have been acquired by the  CONTINUED  preservation of successive profitable degre
0251  led the experiment during five years, and he  CONTINUED  to try it curing several subsequent years,
0267  hermaphrodites; and that close interbreeding  CONTINUED  during several generations between the near
0331  d common progenitor. On the principle of the  CONTINUED  tendency to divergence of character, which
0332  emarked, one order; and this order, from the  CONTINUED  effects of extinction and divergence of cha
0344  d by generation. We can understand, from the  CONTINUED  tendency to divergence of character, why th
0356  rity to descent from any single pair, but to  CONTINUED  care in selecting and training many individ
0383  nds. Chapter XII. Geographical Distribution,  CONTINUED. Distribution of fresh water productions. O
0429  y diverged much in character during the long  CONTINUED  process of modification, how it is that the
0438  obable that natural selection, during a long  CONTINUED  course of modification, should have seized
0447  r paddles, or wings. Whatever influence long  CONTINUED  exercise or use on the one hand, and disuse
0470  ring of any one species. Hence during a long  CONTINUED  course of modification, the slight differen
0474  it will have been rendered constant by long  CONTINUED  natural selection. Glancing at instincts, m
0474  ve no progeny to inherit the effects of long  CONTINUED  habit. On the view of all the species of th
0007  isation has once begun to vary, it generally  CONTINUES  to vary for many generations. No case is on
0028  nd turan sent him some very rare birds; and,  CONTINUES  the courtly historian, His Majesty by cross
0295  nish the supply whilst the downward movement  CONTINUES. In fact, this nearly exact balancing betwe
0120  rger differences distinguishing species. By  CONTINUING  the same process for a greater number of ge
0155  ight surely expect to find them still often  CONTINUING  to vary in those parts of their structure w
0354  duced to consider whether the exceptions to  CONTINUITY  of range are so numerous and of so grave a
0358  to me opposed to the belief of their former  CONTINUITY  with continents. Nor does their almost univ
0370  gh siberia, to eastern America. And to this  CONTINUITY  of the circumpolar land, and to the conseq
0051  allied forms brought from countries not now  CONTINUOUS, in which case he can hardly hope to find t
0103  manner to the conditions of each; for in a  CONTINUOUS  area, the conditions will generally graduat
0106  d, will be able to spread over the open and  CONTINUOUS  area, and will thus come into competition w
0106  ted area. Moreover, great areas, though now  CONTINUOUS, owing to oscillations of level, will often
0174  an area is now continous, that it has been  CONTINUOUS  during a long period. Geology would lead us
0174  f the land and of climate, marine areas now  CONTINUOUS  must often have existed within recent times
0174  e existed within recent times in a far less  CONTINUOUS  and uniform condition than at present. But
0174  efined species have been formed on strictly  CONTINUOUS  areas; though I do not doubt that the forme
0175  r representative species, when inhabiting a  CONTINUOUS  area, are generally so distributed that eac
0178  is assuredly we do see. Secondly, areas now  CONTINUOUS  must often have existed within the recent p
0178  formed in different portions of a strictly  CONTINUOUS  area, intermediate varieties will, it is pr
0203  ns. Closely allied species, now living on a  CONTINUOUS  area, must often have been formed when the
0203  ften have been formed when the area was not  CONTINUOUS, and when the conditions of life did not in
0203  varieties are formed in two districts of a  CONTINUOUS  area, an intermediate variety will often be
0316  ; or its existence, as long as it lasts, is  CONTINUOUS. I am aware that there are some apparent ex
0326  nts than with the marine inhabitants of the  CONTINUOUS, all authors agree that one of the most fun
0346  parts where the circumpolar land is almost  CONTINUOUS, all authors agree that one of the most fun
0348  act; for on the opposite sides of lofty and  CONTINUOUS  mountain ranges, and of great deserts, and
0352  st cases the area inhabited by a species is  CONTINUOUS; and when a plant or animal inhabits two po
0353  pted or rendered discontinuous the formerly  CONTINUOUS  range of many species. So that we are reduc
0370  that under the Polar Circle there is almost  CONTINUOUS  land from western Europe, through Siberia,
0370  ame plants and animals inhabited the almost  CONTINUOUS  circumpolar land; and that these plants and
0372  arlier period, was nearly uniform along the  CONTINUOUS  shores of the Polar Circle, will account, o
0384  difference of the fish on opposite sides of  CONTINUOUS  mountain ranges, which from an early period
0396  c islands having been formerly connected by  CONTINUOUS  land with the nearest continent; for on thi
0399  ccasional means of transport or by formerly  CONTINUOUS  land, from America; and the Cape de Verde i
0403  same physical conditions, and are united by  CONTINUOUS  land, yet they are inhabited by a vast numb
0409  nce of each species and group of species is  CONTINUOUS  in time; for the exceptions to the rule are
0410  ingle species, or by a group of species, is  CONTINUOUS; and the exceptions, which are not rare, ma
0463  ea, which has during a long period remained  CONTINUOUS, and of which the climate and other conditi
0316  ion of ages, so long must its members have  CONTINUOUSLY  existed, in order to have generated either
0316  the genus Lingula, for instance, must have  CONTINUOUSLY  existed by an unbroken succession of gener
0342  ti that each single formation has not been  CONTINUOUSLY  deposited; that the duration of each forma
0357  ts which are now quite separate, have been  CONTINUOUSLY, or almost continuously, united with each
0357  eparate, have been continuously, or almost  CONTINUOUSLY, united with each other, and with the many
0371  e old and New Worlds will have been almost  CONTINUOUSLY  united by land, serving as a bridge, since
0388  of many species having formerly ranged as  CONTINUOUSLY  as fresh water productions ever can range,
0396  llow channels are more likely to have been  CONTINUOUSLY  united within a recent period to the mainl
0401  e that they have at any former period been  CONTINUOUSLY  united. The currents of the sea are rapid
0407  n mind how often a species may have ranged  CONTINUOUSLY  over a wide area, and then have become ext
0026  his tendency, for all that we can see to the  CONTRARY, may be transmitted undiminished for an inde
0160  rly remarked, for all that we can see to the  CONTRARY, transmitted for almost any number of genera
0253  e ill effects of close interbreeding. On the  CONTRARY, brothers and sisters have usually been cros
0317  turally lead them to this conclusion. On the  CONTRARY, we have every reason to believe, from the s
0374  world. Without some distinct evidence to the  CONTRARY, we may at least admit as probable that the
```

Page **(Key Word)**

Page **************************************(Key Word)***************************************

0457 t to us no inexplicable difficulties: on the CONTRARY, their presence might have been even anticip
0100 is remarkable fact, which offers so strong a CONTRAST with terrestrial plants, on the view of an o
0220 f the size of their red masters, so that the CONTRAST in their appearance is very great. When the
0223 of making slaves. Let it be observed what a CONTRAST the instinctive habits of F. sanguinea prese
0273 e first cross or in the first generation, in CONTRAST with their extreme variability in the succee
0460 n first crossed, which forms so remarkable a CONTRAST with the almost universal fertility of varie
0030 which cannot be rivalled by any mechanical CONTRIVANCE, is only a variety of the wild Dipsacus; a
0098 ly move one after the other towards it, the CONTRIVANCE seems adapted solely to ensure self fertil
0098 s very genus, which seems to have a special CONTRIVANCE for self fertilisation, it is well known t
0098 , there is a really beautiful and elaborate CONTRIVANCE by which every one of the infinitely numer
0098 ases, though there be no special mechanical CONTRIVANCE to prevent the stigma of a flower receivin
0191 to the lungs, notwithstanding the beautiful CONTRIVANCE by which the glottis is closed. In the hig
0193 n; for instance in plants, the very curious CONTRIVANCE of a mass of pollen grains, borne on a foo
0197 d around the ends of the branches, and this CONTRIVANCE, no doubt, is of the highest service to th
0098 for self fertilisation, there are special CONTRIVANCES, as I could show from the writings of C. C
0186 pose that the eye, with all its inimitable CONTRIVANCES for adjusting the focus to different dista
0202 with enthusiasm a multitude of inimitable CONTRIVANCES in nature, this same reason tells us, thou
0202 easily err on both sides, that some other CONTRIVANCES are less perfect. Can we consider the stin
0203 ection. If we admire the several ingenious CONTRIVANCES, by which the flowers of the orchis and of
0472 er land. Nor ought we to marvel if all the CONTRIVANCES in nature be not, as far as we can judge,
0486 ure and instinct as the summing up of many CONTRIVANCES, each useful to the possessor, nearly in t
0267 ate to another, and back again. During the CONVALESCENCE of animals, we plainly see that great ben
0042 can be mated for life, and this is a great CONVENIENCE to the fancier, for thus many races may be
0052 s, as one arbitrarily given for the sake of CONVENIENCE to a set of individuals closely resembling
0052 , is also applied arbitrarily, and for mere CONVENIENCE sake. Guided by theoretical considerations
0063 nses, which pass into each other, I use for CONVENIENCE sake the general term of struggle for exis
0211 etion is extremely viscid, it is probably a CONVENIENCE to the aphides to have it removed; and the
0359 few survived an immersion of 137 days. For CONVENIENCE sake I chiefly tried small seeds, without
0172 he crust of the earth? It will be much more CONVENIENT to discuss this question in the chapter on
0207 s; but I have thought that it would be more CONVENIENT to treat the subject separately, especially
0459 hole volume is one long argument, it may be CONVENIENT to the reader to have the leading facts and
0188 convex at their upper ends and must act by CONVERGENCE; and at their lower ends there seems to be
0124 broken lines, beneath the capital letters, CONVERGING in sub branches downwards towards a single
0282 ncient species; and so on backwards, always CONVERGING to the common ancestor of each great class.
0028 cultivators of plants, with whom I have ever CONVERSED, or whose treatises I have read, are firmly
0019 but few distinct from those of Germany as CONVERSELY, and so with Hungary, Spain, etc., but that
0086 anges in the structure of their larvae. So, CONVERSELY, modifications in the adult will probably o
0139 region cannot endure a tropical climate, or CONVERSELY. So again, many succulent plants cannot end
0140 range from warmer to cooler latitudes, and CONVERSELY; but we do not positively know that these a
0148 y developed in a corresponding degree. And, CONVERSELY, that natural selection may perfectly well
0154 ne of the blue species varying into red, or CONVERSELY; but if all the species had blue flowers, t
0294 nct in the immediately surrounding sea; or, CONVERSELY, that some are now abundant in the neighbou
0424 cowslip is descended from the primrose, or CONVERSELY, ranks them together as a single species, a
0437 ts, consequently always have fewer legs; or CONVERSELY, from with many legs have simpler mouths?
0191 mportant to bear in mind the probability of CONVERSION from one function to another, that I will g
0405 under diverse conditions favourable for the CONVERSION of its offspring, firstly into new varietie
0120 to be more numerous or greater in amount, to CONVERT these three forms into well defined species:
0181 as the organs of flight are concerned, would CONVERT it into a bat. In bats which have the wing m
0392 nd, to whatever order they belonged, and thus CONVERT them first into bushes and ultimately into t
0059 s, as we shall hereafter see, tend to become CONVERTED into new and distinct species. The larger g
0061 called incipient species, become ultimately CONVERTED into good and distinct species, which in mo
0107 n subjected to very severe competition. When CONVERTED by subsidence into large separate islands,
0108 y renewed elevation, the islands shall be re CONVERTED into a continental area, there will again b
0112 se of centuries, the sub breeds would become CONVERTED into two well established and distinct bree
0176 wo varieties are supposed on my theory to be CONVERTED and perfected into two distinct species, th
0179 e, a land carnivorous animal could have been CONVERTED into one with aquatic habits; for how could
0180 ctivorous quadruped could possibly have been CONVERTED into a flying bat, the question would have
0188 res) in believing that natural selection has CONVERTED the simple apparatus of an optic nerve mere
0190 ed for one purpose, namely flotation, may be CONVERTED into one for a wholly different purpose, na
0191 elieving that natural selection has actually CONVERTED a swimbladder into a lung, or organ used ex
0191 od served for respiration have been actually CONVERTED into organs of flight. In considering trans
0192 the act of respiration, have been gradually CONVERTED by natural selection into branchiae, simply
0204 instance, a swim bladder has apparently been CONVERTED into an air breathing lung. The same organ
0228 ey make them wider and wider until they were CONVERTED into shallow basins, appearing to the eye p
0305 n at this day, if the Malay Archipelago were CONVERTED into land, the tropical parts of the Indian
0309 tance, the bed of the Pacific Ocean were now CONVERTED into a continent, we should there find form
0317 o two or three varieties, these being slowly CONVERTED into species, which in their turn produce b
0411 ent species, thus produced ultimately become CONVERTED, as I believe, into new and distinct specie
0441 they are fixed for life: their legs are now CONVERTED into prehensile organs; they again obtain a
0447 er late period of life, and having thus been CONVERTED into hands, or paddles, or wings. Whatever
0452 function of giving buoyancy, but has become CONVERTED into a nascent breathing organ or lung. Oth
0187 ly by excluding lateral pencils of light, are CONVEX at their upper ends and must act by convergen
0229 that the bees had excavated too quickly, and CONVEX on the opposed side, where the bees had worke
0230 could have effected this by gnawing away the CONVEX side: and I suspect that the bees in such cas
0004 nd safest clue. I may venture to express my CONVICTION of the high value of such studies, although
0078 e mutual relations of all organic beings; a CONVICTION as necessary, as it seems to be difficult t
0150 bly be here introduced. I can only state my CONVICTION that it is a rule of high generality. I am
0249 ir hybrid origin. I am strengthened in this CONVICTION by a remarkable statement repeatedly made b
0482 d service by conscientiously expressing his CONVICTION; for only thus can the load of prejudice by
0485 re merely artificial combinations made for CONVENIENCE. This may not be a cheering prospect: but
0078 e know what to do, so as to succeed. It will CONVINCE us of our ignorance on the mutual relations
0150 lar conclusion. It is hopeless to attempt to CONVINCE any one of the truth of this proposition wit
0252 eding is thus prevented. Any one may readily CONVINCE himself of the efficiency of insect agency b
0481 form of an abstract, I by no means expect to CONVINCE experienced naturalists whose minds are stoc
0006 ependently created, is erroneous. I am fully CONVINCED that species are not immutable: but that th
0006 scendants of that species. Furthermore, I am CONVINCED that Natural Selection has been the main bu

Page **************************************(Key Word)***************************************

Page ***(Key Word)**

```
0023  re between the breeds of pigeons, I am fully CONVINCED that the common opinion of naturalists is c
0028  , or whose treatises I have read, are firmly CONVINCED that the several breeds to which each has a
0029  r duck, or rabbit fancier, who was not fully CONVINCED that each main breed was descended from a d
0035  narch. Some highly competent authorities are CONVINCED that the setter is directly derived from th
0043  porary. Over all these causes of Change I am CONVINCED that the accumulative action of Selection,
0045  in the individuals of the same species. I am CONVINCED that the most experienced naturalist would
0053  advice and assistance on this subject, soon  CONVINCED me that there were many difficulties, as di
0062  it be thoroughly engrained in the mind, I am  CONVINCED that the whole economy of nature, with ever
0104  hese take place only at long intervals, I am  CONVINCED that the young thus produced will gain so m
0132  ear the coast. So with insects, Wollaston is  CONVINCED that residence near the sea affects their c
0151  lst investigating this Order, and I am fully  CONVINCED that the rule almost invariably holds good
0153  onger period nearly constant. And this, I am  CONVINCED, is the case. That the struggle between nat
0165  ce. With respect to this last fact, I was so  CONVINCED that not even a stripe of colour appears fr
0198  artificial selection. Careful observers are  CONVINCED that a damp climate affects the growth of t
0230  the justly celebrated elder Huber, but I am  CONVINCED of their accuracy; and if I had space, I co
0242  ee, had not the case of these neuter insects  CONVINCED me of the fact. I have, therefore, discusse
0249  ion by their own individual pollen; and I am  CONVINCED that this would be injurious to their ferti
0291  y amount of degradation, may be formed. I am  CONVINCED that all our ancient formations, which are
0306  rous rocks. Most of the arguments which have  CONVINCED me that all the existing species of the sam
0307  ts, with Sir R. Murchison at their head, are  CONVINCED that we see in the organic remains of the l
0372  immediate subject, the Glacial period. I am  CONVINCED that Forbes's view may be largely extended.
0373  g a valley of the Andes; and this I now feel  CONVINCED was a gigantic moraine, left far below any
0407  the modern Glacial period, which I am fully  CONVINCED simultaneously affected the whole world, or
0480  cts and considerations which have thoroughly  CONVINCED me that species have changed, and are still
0481  e number of generations. Although I am fully  CONVINCED of the truth of the views given in this vol
0021  as their names express, utter a very different COO from the other breeds. The fantail has thirty
0237  ired end. Thus, a well flavoured vegetable is COOKED, and the individual is destroyed; but the hor
0140  g largely extended their range from warmer to COOLER latitudes, and conversely; but we do not posi
0306  until some of the species became adapted to a COOLER climate, and were enabled to double the south
0374  whole world was at this period simultaneously COOLER. But it would suffice for my purpose, if the
0185  yet the water hen is nearly as aquatic as the COOT; and the landrail nearly as terrestrial as the
0185  face of the sea. On the other hand, grebes and COOTS are eminently aquatic, although their toes ar
0231  wing upward; but always crowned by a gigantic COPING of wax, the bees can cluster and crawl over t
0231  ose completed; being thus crowned by a strong COPING of wax, the bees can cluster and crawl over t
0008  plant sets a seed. I cannot here enter on the COPIOUS details which I have collected on this curio
0147  t yield abundant and nutritious foliage and a COPIOUS supply of oil bearing seeds. When the seeds
0199  particular kind, but without here entering on COPIOUS details my reasoning would appear frivolous.
0309  iod. The coloured map appended to my volume on CORAL Reefs, led me to conclude that the great ocea
0357  or many animals during their migration. In the CORAL producing oceans such sunken islands are now
0357  ands are now marked, as I believe, by rings of CORAL or atolls standing over them. Whenever it is
0360  t of the widest oceans; and the natives of the CORAL islands in the Pacific, procure stones for th
0349  structure. We ascend the lofty peaks of the CORDILLERA and we find an alpine species of bizcacha;
0373  been noticed on the Rocky Mountains. In the CORDILLERA of Equatorial South America, glaciers once
0374  and western sides of North America, in the CORDILLERA under the equator and the warmer temp
0375  s belonging to genera characteristic of the CORDILLERA. On the mountains of Abyssinia, several Eur
0378  t of the Himalaya, and the long line of the CORDILLERA, seem to have afforded two great lines of i
0070  eservation. Thus we can easily raise plenty of CORN and rape seed, etc., in our fields, because th
0186  n the water; that there should be long toed CORNCRAKES living in meadows instead of in swamps; tha
0187  crustaceans, for instance; there is a double CORNEA, the inner one divided into facets, within ea
0373  idence of glacial action in the south eastern CORNER of Australia. Looking to America; in the nort
0399  rcel, between the flora of the south western CORNER of Australia and of the Cape of Good Hope, is
0181  mely wide flank membrane, stretching from the CORNERS of the jaw to the tail, and including the li
0403  of life. Thus, the south east and south west CORNERS of Australia have nearly the same physical c
0095  ends on bees visiting and moving parts of the COROLLA, so as to push the pollen on to the stigmati
0095  a shorter or more deeply divided tube to its COROLLA, so that the hive bee could visit its flower
0144  structures, as the union of the petals of the COROLLA into a tube. Hard parts seem to affect the f
0145  untenances this idea; but, in the case of the COROLLA of the Umbelliferae, it is by no means, as D
0145  d inner florets without any difference in the COROLLA. Possibly, these several differences may be
0145  rtened. With respect to the difference in the COROLLA of the central and exterior flowers of a hea
0077  over another in the great battle of life. A COROLLARY of the highest importance may be deduced fr
0094  ght deviation of structure. The tubes of the COROLLAS of the common red and incarnate clovers (Tri
0209  admitted that instincts are as important as CORPOREAL structure for the welfare of each species,
0209  stincts have originated. As modifications of CORPOREAL structure arise from, and are increased by,
0210  itable, variations. Hence, as in the case of CORPOREAL structures, we ought to find in nature, not
0210  to occur in nature. Again as in the case of CORPOREAL structure, and conformably with my theory,
0243  tum is applicable to instincts as well as to CORPOREAL structure, and is plainly explicable on the
0474  , they offer no greater difficulty than does CORPOREAL structure on the thoery of the natural sele
0489  olely by and for the good of each being, all CORPOREAL and mental endowments will tend to progress
0013  reely admitted to be inheritable. Perhaps the CORRECT way of viewing the whole subject, would be,
0016  could be shown that this statement is hardly CORRECT; but naturalists differ most widely in deter
0023  ced that the common opinion of naturalists is CORRECT, namely, that all have descended from the ro
0051  er groups and in other countries, by which to CORRECT his first impressions. As he extends the ran
0139  scended from a single parent, if this view be CORRECT, acclimatisation must be readily effected du
0206  ent inhabitants of the world, is not strictly CORRECT, but if we include all those of past times,
0226  information, and tells me that it is strictly CORRECT: If a number of equal spheres be described a
0230  wax, is not, as far as I have seen, strictly CORRECT: the first commencement having always been a
0364  s called accidental, but this is not strictly CORRECT: the currents of the sea are not accidental,
0393  ct that this exception (if the information be CORRECT) may be explained through glacial agency. Th
0438  but it would in these cases probably be more CORRECT, as Professor Huxley has remarked, to speak
0471  to our knowledge tends to make more strictly CORRECT, is on this theory simply intelligible. We c
0186  ing different amounts of light, and for the CORRECTION of spherical and chromatic aberration, coul
0202  , with this high standard under nature. The CORRECTION for the aberration of light is said, on his
0224  s and planes, or even perceive when they are CORRECTLY made. But the difficulty is not nearly so g
0324  d in the older underlying deposits, would be CORRECTLY ranked as simultaneous in a geological sens
0249  ncreases fertility, that I cannot doubt the CORRECTNESS of this almost universal belief amongst br
0487  forms. We must be cautious in attempting to CORRELATE as strictly contemporaneous two formations,
```

Page ***(Key Word)***************************************

Page **(Key Word)***

```
0086  large part of their structure is merely the  CORRELATED  result of successive changes in the structu
0146  tructure of the seeds, which are not always  CORRELATED  with any differences in the flowers, it see
0146  value, may be wholly due to unknown laws of  CORRELATED  growth, and without being, as far as we can
0146  se habits, would naturally be thought to be  CORRELATED  in some necessary manner. So, again, I do n
0162   from the number of the markings, which are  CORRELATED  with the blue tint, and which it does not a
0195  mining the attacks of insects or from being  CORRELATED  with constitutional differences, might assu
0198  hair, and that with the hair the horns are  CORRELATED.  Mountain breeds always differ from lowland
0198  to believe that constitution and colour are  CORRELATED.  A good observer, also, states that in catt
0198  e susceptibility to the attacks of flies is  CORRELATED  with colour, as is the liability to be pois
0237  differences of structure which have become  CORRELATED  to certain ages, and to either sex. We have
0237  ges, and to either sex. We have differences  CORRELATED  not only to one sex, but to that short peri
0237  l difficulty in any character having become  CORRELATED  with the sterile condition of certain membe
0237  e difficulty lies in understanding how such  CORRELATED  modifications of structure could have been
0238  ght modification of structure, or instinct,  CORRELATED  with the sterile condition of certain membe
0269  system, or other constitutional differences  CORRELATED  with the reproductive system. He supplies h
0293  f the same formation throughout Europe been  CORRELATED  with perfect accuracy. With marine animals
0323  and the several formations could be easily  CORRELATED.  These observations, however, relate to the
0417  g characters, mainly depends on their being  CORRELATED  with several other characters of more or le
0418  ire. If certain characters are always found  CORRELATED  with others, though no apparent bond of con
0426  om a common ancestor. And we know that such  CORRELATED  or aggregated characters have especial valu
0448  rents. Such differences might, also, become  CORRELATED  with successive stages of development: so t
0005  x and little known laws of variation and of  CORRELATION  of growth. In the four succeeding chapters
0007  1. causes of Variability. Effects of Habit.  CORRELATION  of Growth. Inheritance. Character of Domes
0011  will here only allude to what may be called  CORRELATION  of growth. Any change in the embryo or lar
0011  ied by an elongated head. Some instances of  CORRELATION  are quite whimsical: thus cats with blue e
0012  ucture, owing to the mysterious laws of the  CORRELATION  of growth. The result of the various, quit
0022  strils, of the tongue (not always in strict  CORRELATION  with the length of beak), the size of the
0033  ever, perhaps never, the case. The laws of  CORRELATION  of growth, the importance of which should
0043  ny unknown laws, more especially by that of  CORRELATION  of growth. Something may be attributed to
0085  in mind that there are many unknown laws of  CORRELATION  of growth, which, when one part of the org
0086  s will no doubt affect, through the laws of  CORRELATION,  the structure of the adult: and probably
0131  s of flight and of vision. Acclimatisation.  CORRELATION  of growth. Compensation and economy of gro
0143  he natural selection of innate differences.  CORRELATION  of Growth. I mean by this expression that
0144  mportant viscera. The nature of the bond of  CORRELATION  is very frequently quite obscure. M. Is. G
0144  o play? With respect to this latter case of  CORRELATION,  I think it can hardly be accidental, that
0144  apted to show the importance of the laws of  CORRELATION  in modifying important structures, indepen
0145  instance of this, and of a striking case of  CORRELATION,  that I have recently observed in some car
0146  species. We may often falsely attribute to  CORRELATION  of growth, structures which are common to
0196  rs reappear from the law of reversion: that  CORRELATION  of growth will have had a most important i
0198  , increase the size of the chest: and again  CORRELATION  would come into play. Animals kept by sava
0199  uite independently of any good thus gained.  CORRELATION  of growth has no doubt played a most impor
0199  e useful, or which formerly had arisen from  CORRELATION  of growth, or from other unknown cause, ma
0200  the several laws of inheritance, reversion,  CORRELATION  of growth, etc. Hence every detail of stru
0269  ectly, or more probably indirectly, through  CORRELATION,  modify the reproductive system in the sev
0409  distant from each other, there should be a  CORRELATION,  in the presence of identical species, of
0435  o the modified form, but often affecting by  CORRELATION  of growth other parts of the organisation.
0466  bility is governed by many complex laws, by  CORRELATION  of growth, by use and disuse, and by the d
0473  a and Europe. In both varieties and species  CORRELATION  of growth seems to have played a most impo
0486  ed, on the causes and laws of variation, on  CORRELATION  of growth, on the effects of use and disus
0011  n the mature animal. In monstrosities, the  CORRELATIONS  between quite distinct parts are very curi
0131  compensation and economy of growth. False  CORRELATIONS.  Multiple, rudimentary, and lowly organise
0146  , again, I do not doubt that some apparent  CORRELATIONS,  occurring throughout whole orders, are en
0168  y developed: and there are very many other  CORRELATIONS  of growth, the nature of which we are utte
0277  of their hybrid offspring should generally  CORRESPOND,  though due to distinct causes; for both de
0313  , as Pictet has remarked, does not strictly  CORRESPOND  with the succession of our geological forma
0327  forms in both ways will everywhere tend to  CORRESPOND.  There is one other remark connected with t
0328  of life; but the species would not exactly  CORRESPOND:  for there will have been a little more tim
0436  s of the skull are homologous with, that is  CORRESPOND  in number and in relative connexion with, t
0257  generally uniting with facility. But the  CORRESPONDENCE  between systematic affinity and the facil
0355  osely allied species. And I now know from  CORRESPONDENCE,  that this coincidence he attributes to g
0372  say that they have been created alike, in  CORRESPONDENCE  with the nearly similar physical conditio
0013  , it tends to appear in the offspring at a  CORRESPONDING  age, though sometimes earlier. In many ca
0014  in the silkworm are known to appear at the  CORRESPONDING  caterpillar or cocoon stage. But heredita
0021  erful difference in their beaks, entailing  CORRESPONDING  differences in their skulls. The carrier,
0033  and the result, I may add, has been, in a  CORRESPONDING  degree, rapid and important. But it is ve
0086  at that age, and by their inheritance at a  CORRESPONDING  age. If it profit a plant to have its see
0089  fications thus produced being inherited at  CORRESPONDING  ages or seasons, either by the males alon
0102  does not become modified and improved in a  CORRESPONDING  degree with its competitors, it will soon
0106  oved, others will have to be improved in a  CORRESPONDING  degree or they will be exterminated. Each
0127  principle of qualities being inherited at  CORRESPONDING  ages, can modify the egg, seed, or young,
0148  me other part to be largely developed in a  CORRESPONDING  degree. And, conversely, that natural sel
0181  me modified and improved in structure in a  CORRESPONDING  manner. Therefore, I can see no difficult
0185  ld be given, habits have changed without a  CORRESPONDING  change of structure. The webbed feet of t
0231  comb, flexures may sometimes be observed,  CORRESPONDING  in position to the planes of the rhombic
0329  t periods, a formation in one region often  CORRESPONDING  with a blank interval in the other, and i
0329  ould not all be the same in the apparently  CORRESPONDING  stages in the two regions. On the Affinit
0335  ccessively produced necessarily endure for  CORRESPONDING  lengths of time: a very ancient form migh
0338  a not early age, and being inherited at a  CORRESPONDING  age. This process, whilst it leaves the e
0348  allel lines not far from each other, under  CORRESPONDING  climates: but from being separated from e
0372  of the Old World, we see countries closely  CORRESPONDING  in all their physical conditions, but wit
0411  at an early age, and being inherited at a  CORRESPONDING  age. Rudimentary Organs; their origin exp
0444  s in the parent, it tends to reappear at a  CORRESPONDING  age in the offspring. Certain variations
0444  ing. Certain variations can only appear at  CORRESPONDING  ages, for instance, peculiarities in the
0444  lier or later in life, tend to appear at a  CORRESPONDING  age in the offspring and parent. I am far
0446  have been inherited by the offspring at a  CORRESPONDING  not early period. But the case of the sho
0446  ences must have been inherited, not at the  CORRESPONDING,  but at an earlier age. Now let us apply
```

Page **(Key Word)***

Page ***(Key Word)***

```
0446  r late age, and having been inherited at a  CORRESPONDING  age, the young of the new species of our
0447  rather late age, and being inherited at a  CORRESPONDING  late age, the fore limbs in the embryos o
0447  fects thus produced will be inherited at a  CORRESPONDING  mature age. Whereas the young will remain
0448  , then, on the principle of inheritance at  CORRESPONDING  ages, the active young or larvae might ea
0450  at the earliest, and being inherited at a  CORRESPONDING  not early period. Embryology rises greatl
0455  of action, the principle of inheritance at  CORRESPONDING  ages will reproduce the organ in its redu
0455  reduction were to be inherited, not at the  CORRESPONDING  age, but at an extremely early period of
0457  y period of life, and being inherited at a  CORRESPONDING  period, we can understand the great leadi
0457  nciple of modifications being inherited at  CORRESPONDING  ages. On this same principle, and bearing
0479  at an early age, and being inherited at a  CORRESPONDING  not early period of life, we can clearly
0480  se, and on the principle of inheritance at  CORRESPONDING  ages have been inherited from a remote pe
0322  ew area, they will have exterminated in a  CORRESPONDINGLY  rapid manner many of the old inhabitants
0243  but is otherwise inexplicable, all tend to  CORROBORATE  the theory of natural selection. This theo
0273  nd deserves attention. For it bears on and  CORROBORATES  the view which I have taken on the cause o
0167  as soon believe with the old and ignorant  COSMOGONISTS,  that fossil shells had never lived, but h
0086  fected through natural selection, than in the  COTTON  planter increasing and improving by selection
0086  ving by selection the down in the pods on his  COTTON  trees. Natural selection may modify and adapt
0257  ven in the pollen, in the fruit, and in the  COTYLEDONS,  can be crossed. Annual and perennial plant
0419  ber and position of the embryonic leaves or  COTYLEDONS,  and on the mode of development of the plum
0247  y other cases, Gartner is obliged carefully to  COUNT  the seeds, in order to show that there is any
0039  as many as seventeen tail feathers have been  COUNTED.  Perhaps the first pouter pigeon did not inf
0072  d yards distant from one of the old clumps, I  COUNTED  thirty two little trees; and one of them, ju
0145  eeds in the ray florets in some Compositae  COUNTENANCES  this idea; but, in the case of the corolla
0026  twenty generations, for we know of no fact  COUNTENANCING  the belief that the child ever reverts to
0102  he effects of intercrossing can hardly be  COUNTERBALANCED  by natural selection always tending to m
0295  , the supply of sediment must nearly have  COUNTERBALANCED  the amount of subsidence. But this same
0071  y changed, but twelve species of plants (not  COUNTING  grasses and carices) flourished in the plant
0386  p in my study for six months, pulling up and  COUNTING  each plant as it grew; the plants were of ma
0084  ome period of their lives, would increase in  COUNTLESS  numbers: they are known to suffer largely f
0113  h variety of grass is annually sowing almost  COUNTLESS  seeds; and thus, as it may be said, is stri
0172  existed, why do we not find them embedded in  COUNTLESS  numbers in the crust of the earth? It will
0464  s is absolutely as nothing compared with the  COUNTLESS  generations of countless species which cert
0464  g compared with the countless generations of  COUNTLESS  species which certainly have existed. We sh
0011  velopment of the udders in cows and goats in  COUNTRIES  where they are habitually milked, in compar
0011  'son with the state of these organs in other  COUNTRIES,  is another instance of the effect of use.
0017  and belonging to equally diverse classes and  COUNTRIES,  were taken from a state of nature, and cou
0019  uld they have been derived, as these several  COUNTRIES  do not possess a number of peculiar species
0023  iginal stocks must either still exist in the  COUNTRIES  where they were originally domesticated, an
0027  , more especially those brought from distant  COUNTRIES,  we can make an almost perfect series betwe
0038  le plant worth culture. It is not that these  COUNTRIES,  so rich in species, do not by a strange ch
0038  comparable with that given to the plants in  COUNTRIES  anciently civilised. In regard to the domes
0038  at least during certain seasons. And in two  COUNTRIES  very differently circumstanced, individuals
0038  species than the varieties kept in civilised  COUNTRIES.  On the view here given of the all importan
0040  receive a provincial name. In semi civilised  COUNTRIES,  with little free communication, the spread
0046  th some few exceptions, polymorphic in other  COUNTRIES,  and likewise, judging from Brachiopod shel
0050  t be confessed, that it is in the best known  COUNTRIES  that we find the greatest number of forms o
0051  gical variation in other groups and in other  COUNTRIES,  by which to correct his first impressions.
0051  he comes to study allied forms brought from  COUNTRIES  not now continuous, in which case he can ha
0055  ipation I have arranged the plants of twelve  COUNTRIES,  and the coleopterous insects of two distri
0083  of them could anyhow be improved: for in all  COUNTRIES,  the natives have been so far conquered by
0103  ties will generally be confined to separated  COUNTRIES;  and this I believe to be the case. In herm
0128  y which now makes them dominant in their own  COUNTRIES.  Natural selection, as has just been remark
0139  same genus to inhabit very hot and very cold  COUNTRIES,  and as I believe that all the species of t
0140  er of plants and animals brought from warmer  COUNTRIES  which here enjoy good health. We have reaso
0142  h constitutions best adapted to their native  COUNTRIES.  In treatises on many kinds of cultivated p
0159  ces. The most distinct breeds of pigeons, in  COUNTRIES  most widely set apart, present sub varietie
0164  horses of very different breeds, in various  COUNTRIES  from Britain to Eastern China; and from Nor
0198  eds of our domesticated animals in different  COUNTRIES,  more especially in the less civilized coun
0198  tries, more especially in the less civilized  COUNTRIES  where there has been but little artificial
0198  o play. Animals kept by savages in different  COUNTRIES  often have to struggle for their own subsis
0215  which have been brought home as puppies from  COUNTRIES,  such as Tierra del Fuego and Australia, wh
0221  habits of the masters and slaves in the two  COUNTRIES,  probably depends merely on the slaves bein
0284  implied by the sedimentary deposits of many  COUNTRIES! Professor Ramsay has given me the maximum
0306  y from our ignorance of the geology of other  COUNTRIES  beyond the confines of Europe and the Unite
0310  volume alone, relating only to two or three  COUNTRIES. Of this volume, only here and there a shor
0319  ast number of species of all classes, in all  COUNTRIES. If we ask ourselves why this or that speci
0320  and takes the place of other breeds in other  COUNTRIES. Thus the appearance of new forms and the d
0325  n between the physical conditions of various  COUNTRIES, and the nature of their inhabitants. This
0326  ing still further, and of giving rise in new  COUNTRIES to new varieties and species. The process o
0328  ism between the successive stages in the two  COUNTRIES; but when he compares certain stages in Eng
0338  rom an examination of the species of the two  COUNTRIES could not have foreseen this result. Agassi
0347  h america under the same latitude: for these  COUNTRIES are almost as much isolated from each other
0350  ing on new places, when they spread into new  COUNTRIES. In their new homes they will be exposed to
0372  southern continents of the Old World, we see  COUNTRIES closely corresponding in all their physical
0374  ich do not exist in the wide intervening hot  COUNTRIES. So on the Silla of Caraccas the illustriou
0383  at they never would have extended to distant  COUNTRIES. But the case is exactly the reverse. Not o
0402  ve spread with astonishing rapidity over new  COUNTRIES, we are apt to infer that most species woul
0402  at the forms which become naturalised in new  COUNTRIES are not generally closely allied to the abo
0488  eral of these species, by migrating into new  COUNTRIES and coming into competition with foreign as
0008  er not very close confinement in their native  COUNTRY! This is generally attributed to vitiated in
0009  animals, even from the tropics, breed in this  COUNTRY pretty freely under confinement, with the ex
0011  tic animal can be named which has not in some  COUNTRY drooping ears; and the view suggested by som
0024  ave been carried back again into their native  COUNTRY; but not one has ever become wild or feral,
0034  b breed, superior to anything existing in the  COUNTRY. But, for our purpose, a kind of Selection,
0035  compared with the stock formerly kept in this  COUNTRY. By comparing the accounts given in old pige
0038  ucture, would often succeed better in the one  COUNTRY than in the other, and thus by a process of
```

Page ***(Key Word)***

Page **(Key Word)**

```
0041  rge number of individuals of a species in any  COUNTRY  requires that the species should be placed u
0041  itions of life, so as to breed freely in that  COUNTRY. When the individuals of any species are sca
0042  in the formation of new races, at least, in a  COUNTRY  which is already stocked with other races. I
0042  ee are almost always imported from some other  COUNTRY, often from islands. Although I do not doubt
0043  igin of our domestic productions. When in any  COUNTRY  several domestic breeds have once been estab
0044  ary most. Species of the larger genera in any  COUNTRY  vary more than the species of the smaller ge
0046  aracters. Genera which are polymorphic in one  COUNTRY  seem to be, with some few exceptions, polymo
0047  nently retained their characters in their own  COUNTRY  for a long time; for as long, as far as we k
0048  variety, can rarely be found within the same   COUNTRY, but are common in separated areas. How many
0050  enerally considered as varieties, and in this  COUNTRY  the highest botanical authorities and practi
0050  confine his attention to one class within one  COUNTRY, he will soon make up his mind how to rank m
0053  t my tables further show that, in any limited  COUNTRY, the species which are most common, that is
0053  ich are most widely diffused within their own  COUNTRY  (and this is a different consideration from
0054  the world, are the most diffused in their own  COUNTRY, and are the most numerous in individuals, w
0054  to struggle with the other inhabitants of the  COUNTRY, the species which are already dominant will
0054  their compatriots. If the plants inhabiting a  COUNTRY  and described in any Flora be divided into t
0054  many species of the same genus inhabiting any  COUNTRY, shows that there is something in the organi
0054  n the organic or inorganic conditions of that  COUNTRY  favourable to the genus; and, consequently,
0055  that the species of the larger genera in each  COUNTRY  would oftener present varieties, than the sp
0059  ore than the average number of species in any  COUNTRY, the species of these genera have more than
0063  quickly become so inordinately great that no   COUNTRY  could support the product. Hence, as more in
0066  and the ostrich a score, and yet in the same   COUNTRY  the condor may be the more numerous of the t
0071  , which have to struggle together in the same  COUNTRY. I will give only a single instance, which,
0078  hat when a plant or animal is placed in a new  COUNTRY  amongst new competitors, though the climate
0078  way to what we should have done in its native  COUNTRY; for we should have to give it some advantag
0081  of natural selection by taking the case of a  COUNTRY  undergoing some physical change, for instanc
0081  mplex manner in which the inhabitants of each  COUNTRY  are bound together, that any change in the n
0081  t seriously affect many of the others. If the  COUNTRY  were open on its borders, new forms would ce
0081  to be. But in the case of an island, or of a  COUNTRY  partly surrounded by barriers, into which ne
0082  habitants. For as all the inhabitants of each  COUNTRY  are struggling together with nicely balanced
0082  ften still further increase the advantage. No  COUNTRY  can be named in which all the native inhabit
0083  eeps the natives of many climates in the same  COUNTRY; he seldom exercises each selected character
0090  deer for instance, had from any change in the  COUNTRY  increased in numbers, or that other prey had
0095  in, if humble bees were to become rare in any  COUNTRY, it might be a great advantage to the red cl
0100  n other plants, I find to be the case in this  COUNTRY; and at my request Dr. Hooker tabulated the
0104  thus new places in the natural economy of the  COUNTRY  are left open for the old inhabitants to str
0108  er occupied by some of the inhabitants of the  COUNTRY  undergoing modification of some kind. The ex
0113  which the number that can be supported in any  COUNTRY  has long ago arrived at its full average. If
0113  wed to act, it can succeed in increasing (the  COUNTRY  not undergoing any change in its conditions)
0115  s specially created and adapted for their own  COUNTRY. It might, also, perhaps have been expected
0115  uggled successfully with the indigenes of any  COUNTRY, and have there become naturalised, we can g
0116  esent the species of a genus large in its own  COUNTRY; these species are supposed to resemble each
0117  pecies, belonging to a genus large in its own  COUNTRY. The little fan of diverging dotted lines of
0118  han most of the other inhabitants of the same  COUNTRY; they will likewise partake of those more ge
0118  nt species belonged, a large genus in its own  COUNTRY. And these circumstances we know to be favou
0121  d an important part. As in each fully stocked  COUNTRY  natural selection necessarily acts by the se
0122  offspring of a species get into some distinct  COUNTRY, or become quickly adapted to some quite new
0122  elated places in the natural economy of their  COUNTRY. It seems, therefore, to me extremely probab
0125  and unknown genus. We have seen that in each  COUNTRY  it is the species of the larger genera which
0139  ry closely allied to those of the surrounding  COUNTRY. It would be most difficult to give any rati
0140  t heights on the Himalaya, were found in this  COUNTRY  to possess different constitutional powers o
0151  of distinct genera. As birds within the same  COUNTRY  vary in a remarkably small degree, I have pa
0172  s, each new form will tend in a fully stocked  COUNTRY  to take the place of, and finally to extermi
0175  see that the range of the inhabitants of any  COUNTRY  by no means exclusively depends on insensibl
0177  ieve, accounts for the common species in each  COUNTRY, as shown in the second chapter, presenting
0178  nd until a place in the natural polity of the  COUNTRY  can be better filled by some modification of
0180  is of use to each kind of squirrel in its own  COUNTRY, by enabling it to escape birds or beasts of
0183  shing like a kingfisher at a fish. In our own  COUNTRY  the larger titmouse (Parus major) may be see
0184  pted competitors did not already exist in the  COUNTRY, I can see no difficulty in a race of bears
0186  n advantage over some other inhabitant of the  COUNTRY, it will seize on the place of that inhabita
0198  differ from lowland breeds; and a mountainous  COUNTRY  would probably affect the hind limbs from ex
0201  rfect than, the other inhabitants of the same  COUNTRY  with which it has to struggle for existence.
0205  owner. Natural selection in each well stocked  COUNTRY, must act chiefly through the competition of
0205  life, only according to the standard of that  COUNTRY. Hence the inhabitants of one country, gener
0205  of that country. Hence the inhabitants of one  COUNTRY, generally the smaller one, will often yield
0205  e inhabitants of another and generally larger  COUNTRY. For in the larger country there will have e
0206  d generally larger country. For in the larger  COUNTRY  there will have existed more individuals, an
0212  osen and on the nature and temperature of the  COUNTRY  inhabited, but often from causes wholly unkn
0223  laves and larvae. So that the masters in this  COUNTRY  receive much less service from their slaves
0234  umbers of a humble bee which could exist in a  COUNTRY; and let us further suppose that the communi
0245  t first probable, for species within the same  COUNTRY  could hardly have kept distinct had they bee
0253  d in distinct genera, have often bred in this  COUNTRY  with either pure parent, and in one single i
0253  rossed geese are kept in various parts of the  COUNTRY; and as they are kept for profit, where neit
0264  re and after birth: when born and living in a  COUNTRY  where their two parents can live, they are g
0289  k in Russia, what wide gaps there are in that  COUNTRY  between the superimposed formations; so it i
0289  eriods which were blank and barren in his own  COUNTRY, great piles of sediment, charged with new a
0295  ion, like the whole pile of formations in any  COUNTRY, has generally been intermittent in its accu
0298  ance of discovering in a formation in any one  COUNTRY  all the early stages of transition between a
0314  development, causing all the inhabitants of a  COUNTRY  to change abruptly, or simultaneously, or to
0314  he slowly changing physical conditions of the  COUNTRY, and more especially on the nature of the ot
0314  er chapter. When many of the inhabitants of a  COUNTRY  have become modified and improved, we can un
0318  to south America, has run wild over the whole  COUNTRY  and has increased in numbers at an unparalle
0322  and not another can be naturalised in a given  COUNTRY; then, and not till then, we may justly feel
0325  e advantage over the other forms in their own  COUNTRY, would naturally oftener give rise to new v
0351  in a body into a new and afterwards isolated  COUNTRY, they will be little liable to modification;
0359  nclude that the seeds of 14/100 plants of any  COUNTRY  might be floated by sea currents during 28 d
```

Page **(Key Word)**

Page **************************************(Key Word)**************************************

0360 , the seeds of 14/100 plants belonging to one COUNTRY might be floated across 924 miles of sea to
0360 be floated across 924 miles of sea to another COUNTRY; and when stranded, if blown to a favourable
0368 ains of Siberia to the arctic regions of that COUNTRY. These views, grounded as they are on the pe
0378 ty five degrees of latitude from their native COUNTRY and covered the land at the foot of the Pyre
0383 would not have ranged widely within the same COUNTRY, and as the sea is apparently a still more i
0385 ecies have rapidly spread throughout the same COUNTRY. But two facts, which I have observed, and n
0388 ntly an intruder from the waters of a foreign COUNTRY, would have a better chance of seizing on a
0400 r of considering the physical conditions of a COUNTRY as the most important for its inhabitants; w
0472 ompetition, it adapts the inhabitants of each COUNTRY only in relation to the degree of perfection
0472 eel no surprise at the inhabitants of any one COUNTRY, although on the ordinary view supposed to h
0472 e/been specially created and adapted for that COUNTRY, being beaten and supplanted by the naturali
0476 cases, is intelligible, for within a confined COUNTRY, the recent and the extinct will naturally b
0389 reat a difference in number. Even the uniform COUNTY of Cambridge has 847 plants, and the little i
0066 t, Let it be ever so large. The condor lays a COUPLE of eggs and the ostrich a score, and yet in t
0088 victor to breed might surely give indomitable COURAGE, Length to the spur, and strength to the win
0214 bulldog has affected for many generations the COURAGE and obstinacy of greyhounds; and a cross wit
0222 attacked their big neighbours with surprising COURAGE. Now I was curious to ascertain whether F. s
0219 sterile females, though most energetic and COURAGEOUS in capturing slaves, do no other work. They
0222 th. Although so small a species, it is very COURAGEOUS, and I have seen it ferociously attack othe
0014 the Laws of embryology. These remarks are of COURSE confined to the first appearance of the pecul
0036 an excellent illustration of the effects of a COURSE of selection, which may be considered as unco
0045 good authority, as I have collected, during a COURSE of years. It should be remembered that system
0068 reely. The amount of food for each species of COURSE gives the extreme limit to which each can inc
0075 nts and animals which have determined, in the COURSE of centuries, the proportional numbers and ki
0080 battle of life, should sometimes occur in the COURSE of thousands of generations? If such do occur
0081 orphic. We shall best understand the probable COURSE of natural selection by taking the case of a
0082 case, every slight modification, which in the COURSE of ages chanced to arise, and which in any wa
0096 mals and plants with separate sexes, it is of COURSE obvious that two individuals must always unit
0106 r the production of new species, yet that the COURSE of modification will generally have been more
0109 ns of life, which may be effected in the long COURSE of time by nature's power of selection. Extin
0110 nevitably follows, that as new species in the COURSE of time are formed through natural selection,
0112 arly differences would be very slight; in the COURSE of time, from the continued selection of swif
0114 ers. Consequently, I cannot doubt that in the COURSE of many thousands of generations, the most di
0126 t day. Summary of Chapter. If during the long COURSE of ages and under varying conditions of life,
0131 e of nature, had been due to chance. This, of COURSE, is a wholly incorrect expression, but it ser
0154 nd variability on the other hand, will in the COURSE of time cease: and that the most abnormally d
0191 ts on the sides of the neck and the loop like COURSE of the arteries still marking in the embryo t
0204 ects; consequently the two Latter, during the COURSE of further modification, from existing in gre
0234 element of success in any family of bees. Of COURSE the success of any species of bee may be depe
0235 long the planes of intersection. The bees, of COURSE, no more knowing that they swept their sphere
0246 rids produced from them. Pure species have of COURSE their organs of reproduction in a perfect con
0258 ids raised from reciprocal crosses, though of COURSE compounded of the very same two species, the
0269 some eminent naturalists believe that a long COURSE of domestication tends to eliminate sterility
0279 lly be beaten out and exterminated during the COURSE of further modification and improvement. The
0286 requisite to have denuded the Weald. This, of COURSE, cannot be done: but we may, in order to form
0326 e dominant species, and then their triumphant COURSE, or even their existence, would cease. We kno
0326 rom the other might be victorious. But in the COURSE of time, the forms dominant in the highest de
0330 roups which have undergone much change in the COURSE of geological ages; and it would be difficult
0332 t directly, but only by a long and circuitous COURSE through many widely different forms. If many
0337 ain were set free in New Zealand, that in the COURSE of time a multitude of British forms would be
0338 yo almost unaltered, continually adds, in the COURSE of successive generations, more and more diff
0345 re intermediate only by a long and circuitous COURSE through many extinct and very different forms
0355 tinent, would probably receive from it in the COURSE of time a few colonists, and their descendant
0357 ded from a single birthplace, and when in the COURSE of time we know something definite about the
0361 even the digestive organs of a turkey. In the COURSE of two months, I picked up in my garden 12 ki
0364 now see them to be. The currents, from their COURSE, would never bring seeds from North America t
0377 n crossed the equator. The invasion would, of COURSE, have been greatly favoured by high land, and
0389 ommon birthplace, notwithstanding that in the COURSE of time they have come to inhabit distant poi
0396 ort having been largely efficient in the long COURSE of time, than with the view of all our oceani
0408 not insuperable in admitting that in the long COURSE of time the individuals of the same species,
0418 ed from the adult, for our classifications of COURSE include all ages of each species. But it is b
0425 cies of kangaroo had been produced, by a long COURSE of modification, from a bear? Ought we to ran
0425 with the other species? The supposition is of COURSE preposterous; and I might answer by the argum
0438 at natural selection, during a long continued COURSE of modification, should have seized on a cert
0438 they are far from meaning that during a long COURSE of descent, primordial organs of any kind, ve
0439 y had really been metamorphosed during a long COURSE of descent from true legs, or from some simpl
0440 ryos of the vertebrata the peculiar loop like COURSE of the arteries near the branchial slits are
0441 ges barely distinguishable. The embryo in the COURSE of development generally rises in organisatio
0447 in the parent species, may become, by a long COURSE of modification, adapted in one descendant to
0448 tingencies: firstly, from the young, during a COURSE of modification carried on for many generatio
0449 either by the successive variations in a long COURSE of modification having supervened at a very e
0452 t some Compositae, the male florets, which of COURSE cannot be fecundated, have a pistil, which is
0453 ses round the sun, satellites follow the same COURSE round the planets, for the sake of symmetry,
0458 on parents, and have all been modified in the COURSE of descent, that I should without hesitation
0461 he world they are now found, they must in the COURSE of successive generations have passed from so
0470 ny one species. Hence during a long continued COURSE of modification, the slight differences, char
0476 we admit that there has been during the long COURSE of ages much migration from one part of the w
0477 ous periods and in different proportions, the COURSE of modification in the two areas will inevita
0481 be proved that the amount of variation in the COURSE of long ages is a limited quantity; no clear
0481 multitude of facts all viewed, during a long COURSE of years, from a point of view directly oppos
0453 say that because planets revolve in elliptic COURSES round the sun, satellites follow the same co
0028 r less than 20,000 pigeons were taken with the COURT. The monarchs of Iran and Turan sent him some
0028 him some very rare birds; and, continues the COURTLY historian, His Majesty by crossing the breed
0421 ame degree; they may metaphorically be called COUSINS to the same millionth degree; yet they diffe
0380 esent day, that very many European productions COVER the ground in La Plata, and in a lesser degre
0064 hat if not destroyed, the earth would soon be COVERED by the progeny of a single pair. Even slow b

Page **************************************(Key Word)**************************************

Page **(Key Word)**

```
0099  uture chapter. In the case of a gigantic tree  COVERED  with innumerable flowers, it may be objected
0137  imentary in size, and in some cases are quite  COVERED  up by skin and fur. This state of the eyes i
0286  eself the great dome of rocks which must have  COVERED  up the Weald within so limited a period as s
0288  ny cases on record of a formation conformably  COVERED, after an enormous interval of time, by anot
0296  een upraised, denuded, submerged, and then re  COVERED  by the upper beds of the same formation; fac
0366  they would perish. The mountains would become  COVERED  with snow and ice, and their former Alpine i
0367  egions of the United States would likewise be  COVERED  by arctic plants and animals, and these woul
0378  ees of latitude from their native country and  COVERED  the land at the foot of the Pyrenees. At thi
0381  glacial period, when the antarctic lands, now  COVERED  with ice, supported a highly peculiar and is
0385  ect. When a duck suddenly emerges from a pond  COVERED  with duck weed, I have twice seen these litt
0386  when dry weighed only 6 3/4 ounces: I kept it  COVERED  up in my study for six months, pulling up an
0416  e vertebrata. If the Ornithorhynchus had been  COVERED  with feathers instead of hair, this external
0426  insect's wing is folded, whether the skin be  COVERED  by hair or feathers, if it prevail throughou
0473  hich are habitually blind and have their eyes  COVERED  with skin! or when we look at the blind anim
0144  alia which are most abnormal in their dermal  COVERING, viz. Cetacea (whales) and Edentata (armadil
0231  i was able practically to show this fact, by  COVERING  the edges of the hexagonal walls of a single
0366  hould have a uniform arctic fauna and flora,  COVERING  the central parts of Europe, as far south as
0416  flower in grasses, the nature of the dermal  COVERING, as hair or feathers, in the Vertebrata. If
0130  d broken branches the crust of the earth, and  COVERS  the surface with its ever branching and beaut
0480  the shrivelled wings under the soldered wing  COVERS  of some beetles, should thus so frequently be
0014  as in the crossed offspring from a short horned  COW  by a long horned bull, the greater length of h
0147  to another part; thus it is difficult to get a  COW  to give milk and to fatten readily. The same v
0011  eat and inherited development of the udders in  COWS  and goats in countries where they are habitual
0237  , in comparison with the horns of the bulls or  COWS  of these same breeds. Hence I can see no real
0238  carefully watching which individual bulls and  COWS, when matched, produced oxen with the longest
0451  e udders of the genus Bos, but in our domestic  COWS  the two sometimes become developed and give mi
0049  tance, the well known one of the primrose and  COWSLIP, or Primula veris and elatior. These plants
0247  ral years repeatedly crossed the primrose and  COWSLIP, which we have such good reason to believe t
0268  the blue and red pimpernel, the primrose and  COWSLIP, which are considered by many of our best bo
0424  e descended from it. He who believes that the  COWSLIP  is descended from the primrose, or conversel
0485  y of specific names, as with the primrose and  COWSLIP; and in this case scientific and common lang
0349  we do not find the beaver or musk rat, but the  COYPU  and capybara, rodents of the American type. I
0348  e more distinct, with hardly a fish, shell, or  CRAB  in common, than those of the eastern and weste
0348  inct marine faunas. Although hardly one shell,  CRAB  or fish is common to the above named three app
0439  wonderful fact of the jaws, for instance, of a  CRAB  retaining numerous characters, which they woul
0479  h different purpose, in the jaws and legs of a  CRAB, in the petals, stamens, and pistils of a flow
0137  ria and of Kentucky, are blind. In some of the  CRABS  the foot stalk for the eye remains, though th
0438  formed of metamorphosed vertebrae? the jaws of  CRABS  as metamorphosed legs! the stamens and pistil
0285  pth of thousands of feet; for since the crust  CRACKED, the surface of the land has been so complet
0285  till more plainly told by faults, those great  CRACKS  along which the strata have been upheaved on
0090  ich preys on various animals, securing some by  CRAFT, some by strength, and some by fleetness; and
0285  vast dislocations is externally visible. The  CRAVEN  fault, for instance, extends for upwards of 3
0213  tand motionless like a statue, and then slowly  CRAWL  forward with a peculiar gait; and another kin
0231  strong coping of wax, the bees can cluster and  CRAWL  over the comb without injuring the delicate h
0397  tal. Would the just hatched young occasionally  CRAWL  on and adhere to the feet of birds roosting o
0222  shortly after all the little yellow ants had  CRAWLED  away, they took heart and carried off the pu
0385  the extremely minute and just hatched shells  CRAWLED  on the feet, and clung to them so firmly tha
0397  rteen days in sea water, and it recovered and  CRAWLED  away: but more experiments are wanted on thi
0489  rious insects flitting about, and with worms  CRAWLING  through the damp earth, and to reflect that
0164  between brown and black to a close approach to  CREAM  colour. I am aware that Colonel Hamilton Smit
0003  that each species had not been independently  CREATED, but had descended, like varieties, from oth
0006  ely, that each species has been independently  CREATED, is erroneous. I am fully convinced that spe
0059  icable if each species has been independently  CREATED. We have, also, seen that it is the most flo
0115  for these are commonly looked at as specially  CREATED  and adapted for their own country. It might,
0129  view that each species has been independently  CREATED, I can see no explanation of this great fact
0133  ave to say that this shell, for instance, was  CREATED  with bright colours for a warm sea: but that
0138  w of the blind animals having been separately  CREATED  for the American and European caverns, close
0152  view that each species has been independently  CREATED, with all its parts as we now see them, I ca
0155  iew of each species having been independently  CREATED, why should that part of the structure, whic
0155  ers from the same part in other independently  CREATED  species of the same genus, be more variable
0159  iew of each species having been independently  CREATED, we should have to attribute this similarity
0162  ll these forms be considered as independently  CREATED  species, that the one in varying has assumed
0167  es that each equine species was independently  CREATED, will, I presume, assert that each species h
0167  i presume, assert that each species has been  CREATED  with a tendency to vary, both under nature a
0167  species of the genus? and that each has been  CREATED  with a strong tendency, when crossed with sp
0167  t fossil shells had never lived, but had been  CREATED  in stone so as to mock the shells now living
0185  ter. He who believes that each being has been  CREATED  as we now see it, must occasionally have fel
0194  beings, each supposed to have been separately  CREATED  for its proper place in nature, be so invari
0199  y believe that very many structures have been  CREATED  for beauty in the eyes of man, or for mere v
0244  of caterpillars, not as specially endowed or  CREATED  instincts, but as small consequences of one
0275  f we look at species as having been specially  CREATED, and at varieties as having been produced by
0303  ation such species will appear as if suddenly  CREATED. I may here recall a remark formerly made, n
0352  aturalists, namely, whether species have been  CREATED  at one or more points of the earth's surface
0355  suppose, from many individuals simultaneously  CREATED. With those organic beings which never inter
0365  the same species must have been independently  CREATED  at several distinct points: and we might hav
0372  f creation. We cannot say that they have been  CREATED  alike, in correspondence with the nearly sim
0390  best adapted plants and animals have not been  CREATED  on oceanic islands! for man has unintentiona
0393  theory of creation, they should not have been  CREATED  there, it would be very difficult to explain
0398  ld as distinct species, supposed to have been  CREATED  here! yet the close affinity of most of thes
0398  d the species which are supposed to have been  CREATED  in the Galapagos Archipelago, and nowhere el
0398  e, bear so plain a stamp of affinity to those  CREATED  in America? There is nothing in the conditio
0437  of birds. Why should similar bones have been  CREATED  in the formation of the wing and leg of a ba
0453  entary organs are generally said to have been  CREATED  for the sake of symmetry, or in order to com
0465  al formation, they will appear as if suddenly  CREATED  there, and will be simply classed as new spe
0470  iew of each species having been independently  CREATED, but are intelligible if all species first e
0471  nature if each species has been independently  CREATED, no man can explain. Many other facts are, a
```

Page **(Key Word)**

created

Page **(Key Word)**

```
0471  nder the form of woodpecker, should have been  CREATED  to prey on insects on the ground; that uplan
0471  , which never or rarely swim should have been  CREATED  with webbed feet; that a thrush should have
0471  h webbed feet; that a thrush should have been  CREATED  to dive and feed on sub aquatic insects; and
0471  c insects; and that a petrel should have been  CREATED  with habits and structure fitting it for the
0472  ordinary view supposed to have been specially  CREATED  and adapted for that country, being beaten a
0473  iew of each species having been independently  CREATED, why should the specific characters, or thos
0473  , if the other species, supposed to have been  CREATED  independently, have differently coloured flo
0475  ange facts if species have been independently  CREATED, and varieties have been produced by seconda
0480  and each separate organ having been specially  CREATED, how utterly inexplicable it is that parts,
0482  es are real, that is, have been independently  CREATED. This seems to me a strange conclusion to ar
0482  can define, or even conjecture, which are the  CREATED  forms of life, and which are those produced
0483  finitely numerous kinds of animals and plants  CREATED  as eggs or seed, or as full grown? and in th
0483  grown? and in the case of mammals, were they  CREATED  bearing the false marks of nourishment from
0488  numerable extinct and living descendants, was  CREATED. In the distant future I see open fields for
0488  view that each species has been independently  CREATED. To my mind it accords better with what we k
0003  plant itself. The author of the Vestiges of  CREATION  would, I presume, say that after a certain u
0044  des the unknown element of a distinct act of  CREATION. The term variety is almost equally difficul
0055  we look at each species as a special act of  CREATION, there is no apparent reason why more variet
0095  rinciple, banish the belief of the continued  CREATION  of new organic beings, or of any great and s
0133  o have fleshy leaves. He who believes in the  CREATION  of each species, will have to say that this
0139  ts on the ordinary view of their independent  CREATION. That several of the inhabitants of the cave
0159  o three separate yet closely related acts of  CREATION. With pigeons, however, we have another case
0185  believes in separate and innumerable acts of  CREATION  will say, that in these cases it has pleased
0194  niggard in innovation. Why, on the theory of  CREATION, should this be so? Why should all the parts
0203  , which on the theory of independent acts of  CREATION  are utterly obscure. We have seen that speci
0315  iew, does not mark a new and complete act of  CREATION, but only an occasional scene, taken almost
0346  roductions of the same continent. Centres of  CREATION. Means of dispersal, by changes of climate a
0355  e, inexplicable on the theory of independent  CREATION. This view of the relation of species in one
0355  us remarks on single and multiple centres of  CREATION  do not directly bear on another allied quest
0356  ifficulty on the theory of single centres of  CREATION, I must say a few words on the means of disp
0372  nd europe, are inexplicable on the theory of  CREATION. We cannot say that they have been created a
0389  the truth of the two theories of independent  CREATION  and of descent with modification. The specie
0390  oductions. He who admits the doctrine of the  CREATION  of each separate species, will have to admit
0393  ny oceanic island. But why, on the theory of  CREATION, they should not have been created there, it
0394  . it cannot be said, on the ordinary view of  CREATION, that there has not been time for the creati
0394  eation, that there has not been time for the  CREATION  of mammals; many volcanic islands are suffic
0396  relation on the view of independent acts of  CREATION. All the foregoing remarks on the inhabitant
0398  lanation on the ordinary view of independent  CREATION; whereas on the view here maintained, it is
0406  able on the ordinary view of the independent  CREATION  of each species, but are explicable on the v
0407  s under the designation of single centres of  CREATION, by some general considerations, more especi
0414  sification either gives some unknown plan of  CREATION, or is simply a scheme for enunciating gener
0420  iously seeking, and not some unknown plan of  CREATION, or the enunciation of general propositions,
0434  iew, and do not look to some unknown plan of  CREATION, we may hope to make sure but slow progress.
0435  mbs. On the ordinary view of the independent  CREATION  of each being, we can only say that so it is
0437  able are these facts on the ordinary view of  CREATION! Why should the brain be enclosed in a box c
0456  hey assuredly do on the ordinary doctrine of  CREATION, might even have been anticipated, and can b
0469  sed to have been produced by special acts of  CREATION, and varieties which are acknowledged to hav
0471  to me utterly inexplicable on the theory of  CREATION. As natural selection acts solely by accumul
0473  ers occur. How inexplicable on the theory of  CREATION  is the occasional appearance of stripes on t
0474  period. It is inexplicable on the theory of  CREATION  why a part developed in a very unusual manne
0478  licable on the theory of independent acts of  CREATION. The existence of closely allied or represen
0478  acts receive no explanation on the theory of  CREATION. The fact, as we have seen, that all past an
0481  sterility is a special endowment and sign of  CREATION. The belief that species were immutable prod
0482  orance under such expressions as the plan of  CREATION, unity of design, etc., and to think that we
0483  seem no more startled at a miraculous act of  CREATION  than at an ordinary birth. But do they reall
0483  do they believe that at each supposed act of  CREATION  one individual or many were produced? Were a
0486  en truly give what may be called the plan of  CREATION. The rules for classifying will no doubt bec
0487  isting causes, and not by miraculous acts of  CREATION  and by catastrophes; and as the most importa
0482  lately they themselves thought were special  CREATIONS, and which are still thus looked at by the
0488  idual. When I view all beings not as special  CREATIONS, but as the lineal descendants of some few
0394  bals. Why, it may be asked, has the supposed  CREATIVE  force produced bats and no other mammals on
0186  l say, that in these cases it has pleased the  CREATOR  to cause a being of one type to take the pla
0188  mptuous? Have we any right to assume that the  CREATOR  works by intellectual powers like those of m
0189  superior to one of glass, as the works of the  CREATOR  are to those of man? If it could be demonstr
0413  they believe that it reveals the plan of the  CREATOR; but unless it be specified whether order in
0413  ace, or what else is meant by the plan of the  CREATOR, it seems to me that nothing is thus added t
0435  say that so it is; that it has so pleased the  CREATOR  to construct each animal and plant. The expl
0488  e know of the laws impressed on matter by the  CREATOR, that the production and extinction of the p
0184  abits, with larger and larger mouths, till a  CREATURE  was produced as monstrous as a whale. As we
0200  ce every detail of structure in every living  CREATURE  (making some little allowance for the direct
0269  ganisation, in any way which may be for each  CREATURE's own good; and thus she may, either directl
0416  ining the degree of affinity of this strange  CREATURE  to birds and reptiles, as an approach in str
0469  constitution, structure, and habits of each  CREATURE, favouring the good and rejecting the bad? I
0480  use and selection will generally act on each  CREATURE, when it has come to maturity and has to pla
0488  the ages which have elapsed since the first  CREATURE, the progenitor of innumerable extinct and l
0307  riods of time, the world swarmed with living  CREATURES. To the question why we do not find records
0074  ats that destroy the mice. Hence it is quite  CREDIBLE  that the presence of a feline animal in larg
0184  may be seen climbing branches; almost like a  CREEPER. It often, like a shrike, kills small birds
0002  ence in my accuracy. No doubt errors will have  CREPT  in, though I hope I have always been cautious
0403  sentative land shells, some of which live in  CREVICES  of stone; and although large quantities of s
0250  example, namely, that every ovule in a pod of  CRINUM  capense fertilised by C. revolutum produced a
0250  etween two distinct species. This case of the  CRINUM  leads me to refer to a most singular fact, na
0056  ce. We have seen that there is no infallible  CRITERION  by which to distinguish species and well ma
0056  e amount of difference is one very important  CRITERION  in settling whether two forms should be ran
0270  es consider fertility and sterility as safe  CRITERIONS  of specific distinction. Gartner kept durin
0313  g instance of a similar fact, in an existing  CROCODILE  associated with many strange and lost mamma
```

Page **(Key Word)**

```
0021  wings, and legs; and its enormously developed  CROP, which it glories in inflating, may well excit
0022  tion with the length of beak), the size of the  CROP and of the upper part of the oesophagus, the d
0023  t stocks possessed the characteristic enormous  CROP? The supposed aboriginal stocks must all have
0024  rsed feathers like those of the 'acobin; for a  CROP like that of the pouter; for tail feathers lik
0039  anner, or a pouter till he saw a pigeon with a  CROP of somewhat unusual size: and the more abnorma
0039  ps the first pouter pigeon did not inflate its  CROP much more than the turbit now does the upper p
0153  s with the wattle of carriers and the enlarged  CROP of pouters. Now let us turn to nature. When a
0280  nce, combining a tail somewhat expanded with a  CROP somewhat enlarged, the characteristic features
0361  ersion in sea water; but some taken out of the  CROP of a pigeon, which had·floated on artificial s
0361  ately devoured; and seeds of many kinds in the  CROPS of floating birds long retain their vitality;
0361  but the following fact is more important: the   CROPS of birds do not secrete gastric juice, and do
0362  or tired birds, and the contents of their torn  CROPS might thus readily get scattered. Mr. Brent i
0364  a water; nor could they be long carried in the  CROPS or intestines of birds. These means, however,
0020  ject, and failed. The offspring from the first  CROSS between two pure breeds is tolerably and some
0026  o reversion to any character derived from such  CROSS will naturally become less and less, as in ea
0026  the foreign blood; but when there has been no  CROSS with a distinct breed, and there is a tendenc
0032  amongst closely allied sub breeds. And when a  CROSS has been made, the closest selection is far m
0034  arly period attended to. Savages now sometimes  CROSS their dogs with wild canine animals, to impro
0091  might slowly be formed. These varieties would  CROSS and blend where they met; but to this subject
0092  devouring insects from flower to flower, and a  CROSS thus effected, although nine tenths of the po
0096  ef of breeders, that with animals and plants a  CROSS between different varieties, or between indiv
0097  elf for an eternity of generations; but that a  CROSS with another individual is occasionally, perh
0097  y exposed to the weather; but if an occasional  CROSS be indispensable, the fullest freedom for the
0098  e pure seedlings, so largely do they naturally  CROSS. In many other cases, far from there being an
0099  e facts explained on the view of an occasional  CROSS with a distinct individual being advantageous
0100  rrestrial plants, on the view of an occasional  CROSS being indispensable, by considering the mediu
0100  in the case of plants, by which an occasional  CROSS could be effected with terrestrial animals wi
0100  water offer an obvious means for an occasional  CROSS. And, as in the case of flowers, I have as ye
0101  self fertilising hermaphrodites, do sometimes  CROSS. It must have struck most naturalists as a st
0101  we may conclude that in many organic beings, a  CROSS between two individuals is an obvious necessi
0103  be the case. In hermaphrodite organisms which  CROSS only occasionally, and likewise in animals wh
0214  s of dogs are crossed. Thus it is known that a  CROSS with a bulldog has affected for many generati
0214  the courage and obstinacy of greyhounds; and a  CROSS with a greyhound has given to a whole family
0248  r some hybrids, carefully guarding them from a  CROSS with either pure parent, for six or seven, an
0249  ty, and, on the other hand, that an occasional  CROSS with a distinct individual or variety increas
0249  er itself which is to be fertilised; so that a  CROSS between two flowers, though probably on the s
0249  d this would have insured in each generation a  CROSS with the pollen from a distinct flower, eithe
0250  re than commonly perfect, fertility in a first  CROSS between two distinct species. This case of th
0252  the same hybrid variety are allowed to freely  CROSS with each other, and the injurious influence
0253  geese) from one nest. In India, however, these  CROSS bred geese must be far more fertile: for I am
0256  s from two species which are very difficult to  CROSS, and which rarely produce any offspring, are
0256  elism between the difficulty of making a first  CROSS, and the sterility of the hybrids thus produc
0258  an often be crossed with ease. By a reciprocal  CROSS between two species, I mean the case, for ins
0258  prove that the capacity in any two species to  CROSS is often completely independent of their syst
0259  o means always the same in degree in the first  CROSS and in the hybrids produced from this cross.
0259  st cross and in the hybrids produced from this  CROSS. That the fertility of hybrids is not related
0259  nd lastly, that the facility of making a first  CROSS between any two species is not always governe
0260  s of the same species. Why should some species  CROSS with facility, and yet produce very sterile h
0260  roduce very sterile hybrids; and other species  CROSS with extreme difficulty, and yet produce fair
0260  eat a difference in the result of a reciprocal  CROSS between the same two species? Why, it may eve
0267  nd animals, there is abundant evidence, that a  CROSS between very distinct individuals of the same
0268  differ from each other in external appearance,  CROSS with perfect facility, and yield perfectly fe
0268  erican indigenous domestic dogs do not readily  CROSS with European dogs, the explanation which wil
0273  egree of variability in hybrids from the first  CROSS or in the first generation, in contrast with
0274  iety. Hybrid plants produced from a reciprocal  CROSS, generally resemble each other closely; and s
0274  ; and so it is with mongrels from a reciprocal  CROSS. Both hybrids and mongrels can be reduced to
0275  looking to the cases which I have collected of  CROSS bred animals closely resembling one parent, t
0276  s. it is not always equal in degree in a first  CROSS and in the hybrid produced from this cross. I
0276  rst cross and in the hybrid produced from this  CROSS. In the same manner as in grafting trees, the
0277  prising that the facility of effecting a first  CROSS, the fertility of the hybrids produced, and t
0363  oment on the millions of quails which annually  CROSS the Mediterranean; and can we doubt that the
0443  of one parent. Or again, as when the horns of  CROSS bred cattle have been affected by the shape o
0477  can clearly see why those animals which cannot  CROSS wide spaces of ocean, as frogs and terrestria
0014  elementi in nearly the same manner as in the   CROSSED offspring from a short horned cow by a long
0016  at of the perfect fertility of varieties when  CROSSED, a subject hereafter to be discussed), domes
0020  ms simple enough; but when these mongrels are  CROSSED one with another for several generations, ha
0025  wo birds belonging to two distinct breeds are  CROSSED, neither of which is blue or has any of the
0025  to acquire these characters; for instance, I  CROSSED some uniformly white fantails with some unif
0025  mottled brown and black birds; these I again  CROSSED together, and one grandchild of the pure whi
0026  at most, within a score of generations, been  CROSSED by the rock pigeon: I say within a dozen or
0026  ber of generations. In a breed which has been  CROSSED only once with some distinct breed, the tend
0027  all the breeds, both when kept pure and when  CROSSED; the mongrel offspring being perfectly ferti
0050  at most careful observer Gartner, they can be  CROSSED only with much difficulty. We could hardly w
0092  ndividuals of the same species would thus get  CROSSED; and the act of crossing, we have good reaso
0092  est visited by insects, and would·be oftenest  CROSSED; and so in the long run would gain the upper
0099  fact separated sexes, and must habitually be  CROSSED. How strange are these facts! How strange th
0099  f the same species. When distinct species are  CROSSED the case is directly the reverse, for a plan
0160  d marks are eminently liable to appear in the  CROSSED offspring of two distinct and differently co
0160  eds of generations. But when a breed has been  CROSSED only once by some other breed, the offspring
0160  foreign blood. In a breed which has not been  CROSSED, but in which both parents have lost some ch
0162  g when distinct breeds of diverse colours are  CROSSED. Hence, though under nature it must generall
0166  dest and truest breeds of various colours are  CROSSED, we see a strong tendency for the blue tint
0167  has been created with a strong tendency, when  CROSSED with species inhabiting distant quarters of
0172  ourthly, how can we account for species, when  CROSSED, being sterile and producing sterile offspri
0172  terile offspring, whereas, when varieties are  CROSSED their fertility is unimpaired? The two first
0214  well shown when different breeds of dogs are  CROSSED. Thus it is known that a cross with a bulldo
```

Page **************************************(Key Word)**************************************

```
0245 ife and crossing. Fertility of varieties when CROSSED and of their mongrel offspring not universal
0246 mely, the sterility of two species when first CROSSED, and the sterility of the hybrids produced f
0246 ies. First, for the sterility of species when CROSSED and of their hybrid offspring. It is impossi
0247 number of seeds produced by two species when CROSSED and by their hybrid offspring, with the aver
0247 r, as Gartner during several years repeatedly CROSSED the primrose and cowslip, which we have such
0248 d, that the sterility of various species when CROSSED is so different in degree and graduates away
0251 e lesser or greater fertility of species when CROSSED, in comparison with the same species when se
0251 laria, Petunia, Rhododendron, etc., have been CROSSED, yet many of these hybrids seed freely. For
0252 ted in the scale of nature can be more easily CROSSED than in the case of plants; but the hybrids
0252 tried: for instance, the canary bird has been CROSSED with nine other finches, but as not one of t
0253 trary, brothers and sisters have usually been CROSSED in each successive generation, in opposition
0253 and Capt. Hutton, that whole flocks of these CROSSED geese are kept in various parts of the count
0254 sterility of distinct species of animals when CROSSED: or we must look at sterility, not as an ind
0256 he other hand, there are species which can be CROSSED very rarely, or with extreme difficulty, but
0256 always the same when the same two species are CROSSED under the same circumstances, but depends in
0257 n which very many species can most readily be CROSSED; and another genus, as Silene, in which the
0257 y species of Nicotiana have been more largely CROSSED than the species of almost any other genus:
0257 , in the fruit, and in the cotyledons, can be CROSSED. Annual and perennial plants, deciduous and
0257 or extremely different climates, can often be CROSSED with ease. By a reciprocal cross between two
0258 for instance, of a stallion horse being first CROSSED with a female ass, and then a male ass with
0258 es may then be said to have been reciprocally CROSSED. There is often the widest possible differen
0260 different in degree, when various species are CROSSED, all of which we must suppose it would be ea
0260 eproductive systems, of the species which are CROSSED. The differences being of so peculiar and li
0267 rinciple of life. Fertility of varieties when CROSSED, and of their Mongrel offspring. It may be u
0268 said by Gartner not to be quite fertile when CROSSED, and he consequently ranks them as undoubted
0269 ir greater or lesser degree of sterility when CROSSED; and we may apply the same rule to domestic
0270 ts have separated sexes, they never naturally CROSSED. He then fertilised thirteen flowers of the
0270 specifically distinct. Girou de Buzareingues CROSSED three varieties of gourd, which like the mai
0271 yellow and white varieties of one species are CROSSED with yellow and white varieties of a distinc
0271 y of the common tobacco is more fertile, when CROSSED with a widely distinct species, than are the
0271 es, when used either as father or mother, and CROSSED with the Nicotiana glutinosa, always yielded
0271 e produced from the four other varieties when CROSSED with N. glutinosa. Hence the reproductive sy
0272 e reproductive systems of the forms which are CROSSED. Hybrids and Mongrels compared, independentl
0272 n of fertility, the offspring of species when CROSSED and of varieties when crossed may be compare
0272 of species when crossed and of varieties when CROSSED may be compared in several other respects. G
0274 though most closely allied to each other, are CROSSED with a third species, the hybrids are widely
0274 wo very distinct varieties of one species are CROSSED with another species, the hybrids do not dif
0274 o gartner the same laws. When two species are CROSSED, one has sometimes a prepotent power of impr
0274 x than in the other, both when one species is CROSSED with another, and when one variety is crosse
0274 crossed with another, and when one variety is CROSSED with another variety. For instance, I think
0275 eral and close similarity in the offspring of CROSSED species, and of crossed varieties. If we loo
0275 y in the offspring of crossed species, and of CROSSED varieties. If we look at species as having b
0277 ce of some kind between the species which are CROSSED. Nor is it surprising that the facility of e
0377 ted the native ranks and have reached or even CROSSED the equator. The invasion would, of course,
0378 t that some temperate productions entered and CROSSED even the lowlands of the tropics at the peri
0378 into the intertropical regions, and some even CROSSED the equator. As the warmth returned, these t
0379 s; but the forms, chiefly northern, which had CROSSED the equator, would travel still further from
0382 rom the north and from the south, and to have CROSSED at the equator; but to have flowed with grea
0460 ost universal sterility of species when first CROSSED, which forms so remarkable a contrast with t
0460 almost universal fertility of varieties when CROSSED, I must refer the reader to the recapitulati
0460 in the result, when the same two species are CROSSED reciprocally; that is, when one species is f
0461 odified forms or varieties acquire from being CROSSED increased vigour and fertility. So that, on
0475 anent varieties, we can at once see why their CROSSED offspring should follow the same complex law
0475 crosses, and in other such points, as do the CROSSED offspring of acknowledged varieties. On the
0020 ubt that a race may be modified by occasional CROSSES, if aided by the careful selection of those
0035 has, it is believed, been chiefly effected by CROSSES with the fox hound; but what concerns us is,
0042 s with separate sexes, facility in preventing CROSSES is an important element of success in the fo
0142 ew survivors, with care to prevent accidental CROSSES, and then again get seed from these seedling
0165 described appearances are all due to ancient CROSSES with the dun stock. But I am not at all sati
0245 m. distinction between the sterility of first CROSSES and of hybrids. Sterility various in degree,
0245 differences. Causes of the sterility of first CROSSES and of hybrids. Parallelism between the effe
0251 he degree of fertility of some of the complex CROSSES of Rhododendrons, and I am assured that many
0252 nt, we have no right to expect that the first CROSSES between them and the canary, or that their h
0254 that some degree of sterility, both in first CROSSES and in hybrids, is an extremely general resu
0254 versal. Laws governing the Sterility of first CROSSES and of Hybrids. We will now consider a littl
0255 es and rules governing the sterility of first CROSSES and of hybrids. Our chief object will be to
0255 , that the degree of fertility, both of first CROSSES and of hybrids, graduates from zero to perfe
0256 ite cases occur. The fertility, both of first CROSSES and of hybrids, is more easily affected by u
0256 he allied species. Now the fertility of first CROSSES between species; and of the hybrids produced
0258 fference in the facility of making reciprocal CROSSES. Such cases are highly important, for they p
0258 . this difference in the result of reciprocal CROSSES between the same two species was long ago ob
0258 s difference of facility in making reciprocal CROSSES is extremely common in a lesser degree. He h
0258 ble fact, that hybrids raised from reciprocal CROSSES, though of course compounded of the very sam
0259 ow given, which govern the fertility of first CROSSES and of hybrids, we see that when forms, whic
0260 ter statement is clearly proved by reciprocal CROSSES between the same two species, for according
0260 e hybrids, moreover, produced from reciprocal CROSSES often differ in fertility. Now do these comp
0260 to indicate that the sterility both of first CROSSES and of hybrids is simply incidental or depen
0261 iar and limited a nature, that, in reciprocal CROSSES between two species the male sexual element
0262 s in being grafted together. As in reciprocal CROSSES, the facility of effecting an union is often
0263 complex laws governing the facility of first CROSSES, are incidental on unknown differences, chie
0263 eir welfare. Causes of the Sterility of first CROSSES and of Hybrids. We may now look a little clo
0263 the probable causes of the sterility of first CROSSES and of hybrids. These two cases are fundamen
0263 in hybrids they are imperfect. Even in first CROSSES, the greater or lesser difficulty in effecti
0264 s a very frequent cause of sterility in first CROSSES. I was at first very unwilling to believe in
0266 fertility of hybrids produced from reciprocal CROSSES; or the increased sterility in those hybrids
0267 ic beings, and on the other hand, that slight CROSSES; that is crosses between the males and femal
```

Page **************************************(Key Word)**************************************

crosses

```
0267  the other hand, that slight crosses, that is  CROSSES between the males and females of the same sp
0267  ings in some degree sterile; and that greater  CROSSES, that is crosses between males and females w
0267  ee sterile; and that greater crosses, that is  CROSSES between males and females which have become
0271  istinct species, more seed is produced by the  CROSSES between the same coloured flowers, than betw
0271  by the severest trial, namely, by reciprocal   CROSSES, and he found their mongrel offspring perfec
0272  neral, but not invariable, sterility of first  CROSSES and of hybrids, namely, that it is not a spe
0274  duced to either pure parent form, by repeated  CROSSES in successive generations with either parent
0276  cies and varieties. Summary of Chapter. First  CROSSES between forms sufficiently distinct to be ra
0276  and sometimes widely different, in reciprocal  CROSSES between the same two species. It is not alwa
0276  arched in our forests. The sterility of first  CROSSES between pure species, which have their repro
0277  nds of resemblance between all species. First  CROSSES between forms known to be varieties, or suff
0461  s is a very different case from that of first  CROSSES, for their reproductive organs are more or l
0461  less functionally impotent; whereas in first   CROSSES the organs on both sides are in a perfect co
0461  derable changes in the conditions of life and  CROSSES between greatly modified forms, lessen ferti
0461  lesser changes in the conditions of life and   CROSSES between less modified forms, increase fertil
0475  being absorbed into each other by successive   CROSSES, and in other such points, as do the crossed
0019  our races of dogs have been produced by the    CROSSING of a few aboriginal species; but by crossing
0019  crossing of a few aboriginal species; but by   CROSSING we can get only forms in some degree interme
0020  the possibility of making distinct races by    CROSSING has been greatly exaggerated. There can be n
0023  e to make the present domestic breeds by the   CROSSING of any lesser number: how, for instance, cou
0023  for instance, could a pouter be produced by    CROSSING two breeds unless one of the parent stocks p
0028  tinues the courtly historian, His Majesty by   CROSSING the breeds, which method was never practised
0031  improvement is by no means generally due to    CROSSING different breeds; all the best breeders are
0037  ies or races have become blended together by   CROSSING, may plainly be recognised in the increased
0042  om them, then, there appeared (aided by some   CROSSING with distinct species) those many admirable
0043  of new sub breeds; but the importance of the   CROSSING of varieties has, I believe, been greatly ex
0043  cuttings, buds, etc., the importance of the    CROSSING both of distinct species and of varieties is
0092  ecies would thus get crossed; and the act of   CROSSING, we have good reason to believe (as will her
0102  selection, notwithstanding a large amount of   CROSSING with inferior animals. Thus it will be in na
0160  rks, beyond the influence of the mere act of   CROSSING on the laws of inheritance. No doubt it is a
0165  the world. Now let us turn to the effects of   CROSSING the several species of the horse genus. Roll
0174  of new species, more especially with freely   CROSSING and wandering animals. In looking at species
0214  hese domestic instincts, when thus tested by   CROSSING, resemble natural instincts, which in a like
0245  he effects of changed conditions of life and  CROSSING. Fertility of varieties when crossed and of
0245  have kept distinct had they been capable of   CROSSING freely. The importance of the fact that hybr
0255  with this quality, in order to prevent their  CROSSING and blending together in utter confusion. Th
0257  ween systematic affinity and the facility of  CROSSING is by no means strict. A multitude of cases
0257  aracter is sufficient to prevent two species  CROSSING. It can be shown that plants most widely dif
0258  ese cases clearly show that the capacity for  CROSSING is connected with constitutional differences
0259  nce, some species have a remarkable power of  CROSSING with other species: other species of the sam
0262  erent individuals of the same two species in  CROSSING: so Sagaret believes this to be the case wit
0263  arallelism in the results of grafting and of  CROSSING distinct species. And as we must look at the
0263  r or lesser difficulty of either grafting or  CROSSING together various species has been a special
0263  a special endowment: although in the case of  CROSSING, the difficulty is as important for the endu
0265  d when hybrids are produced by the unnatural  CROSSING of two species, the reproductive system, ind
0267  our fourth chapter, that a certain amount of  CROSSING is indispensable even with hermaphrodites; a
0273  r added to that arising from the mere act of  CROSSING. The slight degree of variability in hybrids
0276  fferences in their vegetative systems, so in  CROSSING, the greater or less facility of one species
0276  various degrees of sterility to prevent them  CROSSING and blending in nature, than to think that t
0277  arallelism of another kind: namely, that the  CROSSING of forms only slightly different is favourab
0317  ted by a vertical line of varying thickness,  CROSSING the successive geological formations in whic
0373  mound of detritus, about 800 feet in height,  CROSSING a valley of the Andes; and this I now feel c
0405  species which apparently has the capacity of  CROSSING barriers and ranging widely, as in the case
0405  o range widely implies not only the power of  CROSSING barriers, but the more important power of be
0451  ent of a pistil; and Kolreuter found that by  CROSSING such male plants with an hermaphrodite speci
0212  n england, is tame in Norway, as is the hooded CROW in Egypt. That the general disposition of indi
0171  fore having arrived at this part of my work,  a CROWD of difficulties will have occurred to the rea
0224  e form, though this is perfectly effected by a CROWD of bees working in a dark hive. Grant whateve
0070  art through facility of diffusion amongst the  CROWDED animals, been disproportionably favoured: an
0110  f good Hope, where more species of plants are  CROWDED together than in any other quarter of the wo
0377  as many species as we see at the present day  CROWDED together at the Cape of Good Hope, and in pa
0162  he rock pigeon was not feather footed or turn  CROWNED, we could not have told, whether these chara
0231  thin wall steadily growing upward: but always  CROWNED by a gigantic coping. From all the cells, bo
0231  ust commenced and those completed, being thus  CROWNED by a strong coping of wax, the bees can clus
0452  hich is in a rudimentary state, for it is not  CROWNED with a stigma: but the style remains well de
0115  ave there become naturalised, we can gain some CRUDE idea in what manner some of the natives would
0286  not be done; but we may, in order to form some CRUDE notion on the subject, assume that the sea wo
0124  ection of the successive strata of the earth's CRUST including extinct remains. We shall, when we
0130  ch fills with its dead and broken branches the CRUST of the earth, and covers the surface with its
0172  find them embedded in countless numbers in the CRUST of the earth? It will be much more convenient
0173  will be blanks in our geological history. The CRUST of the earth is a vast museum: but the natura
0284  d measure of the degradation which the earth's CPUST has elsewhere suffered. And what an amount of
0285  t or depth of thousands of feet: for since the CRUST cracked, the surface of the land has been so
0487  om the extreme imperfection of the record. The CRUST of the earth with its embedded remains must n
0182  mbers of such water breathing classes as the   CRUSTACEA and Mollusca are adapted to live on the lan
0376  land should have a closer resemblance in its   CRUSTACEA to Great Britain, its antipode, than to any
0431  ess in some other classes, as in that of the   CRUSTACEA, for here the most wonderfully diverse form
0306  ian trilobites have descended from some one    CRUSTACEAN, which must have lived long before the Silu
0437  totally different purposes? Why should one     CRUSTACEAN, which has an extremely complex mouth forme
0440  that a barnacle was, as it certainly is, a     CRUSTACEAN: but a glance at the larva shows this to be
0187  rately high stage of perfection. In certain    CRUSTACEANS, for instance, there is a double cornea, t
0187  h there is a lens shaped swelling. In other    CRUSTACEANS the transparent cones which are coated by
0188  h graduated diversity in the eyes of living    CRUSTACEANS, and bearing in mind how small the number
0313  of the other Silurian Molluscs and all the     CRUSTACEANS have changed greatly. The productions of t
0337  ee no way of testing this sort of progress.    CRUSTACEANS, for instance, not the highest in their ow
0372  l more striking case of many closely allied    CRUSTACEANS (as described in Dana's admirable work), o
```

Page **************************************(Key Word)**************************************

0419 ers common to all birds: but in the case of CRUSTACEANS, such definition has hitherto been found i
0419 s hitherto been found impossible. There are CRUSTACEANS at the opposite ends of the series, which
0435 the construction of the mouths and limbs of CRUSTACEANS. So it is with the flowers of plants. Noth
0436 rds, and in the mouths of certain suctorial CRUSTACEANS, the general pattern seems to have been th
0436 ng the wonderfully complex jaws and legs in CRUSTACEANS. It is familiar to almost every one, that
0436 other: and we can actually see in embryonic CRUSTACEANS and in many other animals, and in flowers,
0441 e than the larva, as with certain parasitic CRUSTACEANS. To refer once again to cirripedes: the la
0449 r larvae as belonging to the great class of CRUSTACEANS. As the embryonic state of each species an
0238 and instincts extraordinarily different: in CRYPTOCERUS, the workers of one caste alone carry a wo
0137 erica, a burrowing rodent, the tuco tuco, or CTENOMYS, is even more subterranean in its habits tha
0091 bers of the animals on which our wolf preyed, a CUB might be born with an innate tendency to pursu
0207 incts, their origin. Natural instincts of the CUCKOO, ostrich, and parasitic bees. Slave making an
0207 ant, when it is said that instinct impels the CUCKOO to migrate and to lay her eggs in other birds
0216 re work, namely, the instinct which leads the CUCKOO to lay her eggs in other birds' nests: the sl
0216 hat the more immediate and final cause of the CUCKOO's instinct is, that she lays her eggs, not da
0217 to be fed by the male alone. But the American CUCKOO is in this predicament: for she makes her own
0217 time. It has been asserted that the American CUCKOO occasionally lays her eggs in other birds'.ne
0217 e that the ancient progenitor of our European CUCKOO had the habits of the American cuckoo; but th
0217 uropean cuckoo had the habits of the American CUCKOO; but that occasionally she laid an egg in ano
0217 e, i believe that the strange instinct of our CUCKOO could be, and has been, generated. I may add
0218 ray and to some other observers, the European CUCKOO has not utterly lost all maternal love and ca
0218 rge number of eggs; but as in the case of the CUCKOO, at intervals of two or three days. This inst
0218 this case is more remarkable than that of the CUCKOO; for these bees have not only their instincts
0218 n this case, as with the supposed case of the CUCKOO, I can see no difficulty in natural selection
0244 actory to look at such instincts as the young CUCKOO ejecting its foster brothers, ants making sla
0030 ti when we compare the host of agricultural, CULINARY, orchard, and flower garden races of plants
0086 e, in the seeds of the many varieties of our CULINARY and agricultural plants; in the caterpillar
0015 we could succeed in naturalising, or were to CULTIVATE, during many generations, the several races
0004 areful study of domesticated animals and of CULTIVATED plants would offer the best chance of makin
0007 he same variety or sub variety of our older CULTIVATED plants and animals, one of the first points
0007 y of the plants and animals which have been CULTIVATED, and which have varied during all ages unde
0008 o be variable under cultivation. Our oldest CULTIVATED plants, such as wheat, still often yield ne
0008 ributed to vitiated instincts: but how many CULTIVATED plants display the utmost vigour, and yet r
0012 eral treatises published on some of our old CULTIVATED plants, as on the hyacinth, potato, even th
0030 e various breeds of sheep fitted either for CULTIVATED Land or mountain pasture, with the wool of
0037 come from a garden stock. The pear, though CULTIVATED in classical times, appears, from Pliny's d
0037 the gardeners of the classical period, who CULTIVATED the best pear they could procure, never tho
0037 where find. A large amount of change in our CULTIVATED plants, thus slowly and unconsciously accum
0037 tocks of the plants which have been longest CULTIVATED in our flower and kitchen gardens. If it ha
0041 e strawberry had always varied since it was CULTIVATED, but the slight varieties had been neglecte
0142 ve countries. In treatises on many kinds of CULTIVATED plants, certain varieties are said to withs
0272 t hybrids from species which have long been CULTIVATED even often variable in the first generation:
0273 escended from species (excluding those long CULTIVATED) which have not had their reproductive syst
0037 t unconsciously. It has consisted in always CULTIVATING the best known variety, sowing its seeds,
0085 ight differences make a great difference in CULTIVATING the several varieties, assuredly, in a sta
0008 variable being ceasing to be variable under CULTIVATION. Our oldest cultivated plants, such as whe
0008 the remarkable effect which confinement or CULTIVATION has on the functions of the reproductive s
0009 imals and plants withstand domestication or CULTIVATION, and vary very slightly, perhaps hardly mo
0010 rare under nature, but far from rare under CULTIVATION; and in this case we see that the treatmen
0131 cy of monstrosities, under domestication or CULTIVATION, than under nature, leads me to believe th
0159 ral botanists rank as varieties produced by CULTIVATION from a common parent: if this be not so, t
0043 pecies and of varieties is immense: for the CULTIVATOR here quite disregards the extreme variabili
0028 ers of the various domestic animals and the CULTIVATORS of plants, with whom I have ever conversed
0038 sed man, has afforded us a single plant worth CULTURE. It is not that these countries, so rich in
0094 r flower or on another plant. In plants under CULTURE and placed under new conditions of life, som
0265 er confinement or any plant seed freely under CULTURE; nor can he tell, till he tries, whether any
0387 the viscid mud was all contained in a breakfast CUP! Considering these facts, I think it would be
0085 nned fruits suffer far more from a beetle, a CURCULIO, than those with down: that purple plums suf
0215 ake an attack, and are then beaten; and if not CURED, they are destroyed: so that habit, with some
0350 conditions. The naturalist must feel little CURIOSITY, who is not led to inquire what this bond i
0009 opious details which I have collected on this CURIOUS subject; but to show how singular the laws a
0011 lations between quite distinct parts are very CURIOUS; and many instances are given in Isidore Geo
0072 e of cattle. Perhaps Paraguay offers the most CURIOUS instance of this: for here neither cattle no
0097 perhaps in all, such flowers, there is a very CURIOUS adaptation between the structure of the flow
0110 the selection of improved forms by man. Many CURIOUS instances could be given showing how quickly
0124 o the other fourteen new species will be of a CURIOUS and circuitous nature. Having descended from
0163 me very remarkable. I will, however, give one CURIOUS and complex case, not indeed as affecting an
0193 ld be given: for instance in plants, the very CURIOUS contrivance of a mass of pollen grains, born
0213 ualities of our domestic animals. A number of CURIOUS and authentic instances could be given of th
0222 neighbours with surprising courage. Now I was CURIOUS to ascertain whether F. sanguinea could dist
0231 sides. The manner in which the bees build is CURIOUS: they always make the first rough wall from
0232 rsection between these spheres. It was really CURIOUS to note in cases of difficulty, as when two
0255 rfect fertility. It is surprising in how many CURIOUS ways this gradation can be shown to exist: b
0263 distinct species. And as we must look at the CURIOUS and complex laws governing the facility with
0273 riability in the succeeding generations, is a CURIOUS fact and deserves attention. For it bears on
0276 stematic affinity: but is governed by several CURIOUS and complex laws. It is generally different,
0328 those in France, although he finds in both a CURIOUS accordance in the numbers of the species bel
0407 ignorant we are with respect to the many and CURIOUS means of occasional transport, a subject whi
0434 be said to be its very soul. What can be more CURIOUS than that the hand of a man, formed for gras
0434 y long spiral proboscis of a sphinx moth, the CURIOUS folded one of a bee or bug, and the great ja
0436 extent obscured. There is another and equally CURIOUS branch of the present subject; namely, the c
0445 scended from the same wild stock: hence I was CURIOUS to see how far their puppies differed from e
0450 the cases of rudimentary organs are extremely CURIOUS: for instance, the presence of teeth in foet
0460 ory of natural selection: and one of the most CURIOUS of these is the existence of two or three be
0482 he day will come when this will be given as a CURIOUS illustration of the blindness of preconceive
0098 shown to be the case with the barberry: and CURIOUSLY in this very genus, which seems to have a s

Page **************************************(Key Word)**************************************

0214 its, and dispositions are inherited, and how CURIOUSLY they become mingled, is well shown when dif
0214 ral instincts, which in a like manner become CURIOUSLY blended together, and for a long period exh
0201 e. i would almost as soon believe that the cat CURLS the end of its tail when preparing to spring,
0262 berry, for instance, cannot be grafted on the CURRANT, whereas the currant will take, though with
0262 cannot be grafted on the currant, whereas the CURRANT will take, though with difficulty, on the go
0100 ny self fertilising hermaphrodites; but here CURRENTS in the water offer an obvious means for an o
0284 ces very fine sediment is transported by the CURRENTS of the sea, the process of accumulation in a
0295 as been much interrupted, as a change in the CURRENTS of the sea and a supply of sediment of a dif
0325 s, cannot be owing to mere changes in marine CURRENTS or other causes more or less local and tempo
0325 , indeed, quite futile to look to changes of CURRENTS, climate, or other physical conditions; as t
0359 lants of any country might be floated by sea CURRENTS during 28 days, and would retain their power
0359 as, the average rate of the several Atlantic CURRENTS is 33 miles per diem (some currents running
0359 atlantic currents is 33 miles per diem (some CURRENTS running at the rate of 60 miles per diem); o
0364 ental, but this is not strictly correct: the CURRENTS of the sea are not accidental, nor is the di
0364 in as distinct as we now see them to be. The CURRENTS, from their course, would never bring seeds
0399 tones on icebergs, drifted by the prevailing CURRENTS, this anomaly disappears. New Zealand in its
0401 former period been continuously united. The CURRENTS of the sea are rapid and sweep across the ar
0022 bones of the face in length and breadth and CURVATURE differs enormously. The shape, as well as t
0094 in the size and form of the body, or in the CURVATURE and length of the proboscis, etc., far too
0141 ed, is a very obscure question. That habit or CUSTOM has some influence I must believe, both from
0074 y one has heard that when an American forest is CUT down, a very different vegetation springs up;
0149 lone. In the same way that a knife which has to CUT all sorts of things may be of almost any shape
0231 in the middle; the masons always piling up the CUT away cement, and adding fresh cement, on the s
0371 he polar Circle, they must have been completely CUT off from each other. This separation, as far a
0450 r heads; and the presence of teeth, which never CUT through the gums, in the upper jaws of our unb
0480 for instance, has inherited teeth, which never CUT through the gums of the upper jaw, from an ear
0246 olreuter makes the rule universal; but then he CUTS the knot, for in ten cases in which he found t
0357 ently been united to some continent. This view CUTS the Gordian knot of the dispersal of the same
0094 re to save time; for instance, their habit of CUTTING holes and sucking the nectar at the bases of
0231 up a broad ridge of cement, and then to begin CUTTING it away equally on both sides near the groun
0043 n plants which are temporarily propagated by CUTTINGS, buds, etc., the importance of the crossing
0442 he adult: thus Owen has remarked in regard to CUTTLE fish, there is no metamorphosis: the cephalop
0448 t in certain whole groups of animals, as with CUTTLE fish and spiders, and with a few members of t
0206 ence, so often insisted on by the illustrious CUVIER, is fully embraced by the principle of natura
0208 ls very low in the scale of nature. Frederick CUVIER and several of the older metaphysicians have
0303 nearly the commencement of this great series. CUVIER used to urge that no monkey occurred in any t
0310 all the most eminent palaeontologists, namely CUVIER, Owen, Agassiz, Barrande, Falconer, E. Forbes
0329 inct animals fall in between existing groups. CUVIER ranked the Ruminants and Pachyderms, as the t
0440 a good instance of this: even the illustrious CUVIER did not perceive that a barnacle was, as it c
0129 und other points, and so on in almost endless CYCLES. On the view that each species has been indep
0382 the world has recently felt one of his great CYCLES of change; and that on this view, combined wi
0490 to one; and that, whilst this planet has gone CYCLING on according to the fixed law of gravity, fr
0253 ybrids from the common and Chinese geese (A. CYGNOIDES), species which are so different that they
0233 more materials would be required than for a CYLINDER. As natural selection acts only by the accum
0227 in the same way as the rude humble bee adds CYLINDERS of wax to the circular mouths of her old co
0233 of the cells just begun, sweeping spheres or CYLINDERS, and building up intermediate planes. It is
0225 er: it forms a nearly regular waxen comb of CYLINDRICAL cells, in which the young are hatched, and
0227 a certain extent, and seeing what perfectly CYLINDRICAL burrows in wood many insects can make, app
0227 ls in level layers, as she already does her CYLINDRICAL cells; and we must further suppose, and th
0012 plants, as on the hyacinth, potato, even the DAHLIA, etc.; and it is really surprising to note th
0037 rieties of the heartsease, rose, pelargonium, DAHLIA, and other plants, when compared with the old
0037 ever expect to get a first rate heartsease or DAHLIA from the seed of a wild plant. No one would e
0084 ship? It may be said that natural selection is DAILY and hourly scrutinising, throughout the world
0217 koo's instinct is, that she lays her eggs, not DAILY, but at intervals of two or three days; so th
0387 her birds, century after century, have gone on DAILY devouring fish; they then take flight and go
0145 ray and central florets of, for instance, the DAISY, and this difference is often accompanied wit
0068 res to attack a young elephant protected by its DAM. Climate plays an important part in determinin
0445 but having had careful measurements made of the DAM and of a three days old colt of a race and hea
0069 when we travel from south to north, or from a DAMP region to a dry, we invariably see some specie
0139 o again, many succulent plants cannot endure a DAMP climate. But the degree of adaptation of speci
0198 ection. Careful observers are convinced that a DAMP climate affects the growth of the hair, and th
0378 te; for Dr. Falconer informs me that it is the DAMP with the heat of the tropics which is so destr
0385 their nature, survived on the duck's feet, in DAMP air, from twelve to twenty hours; and in this
0489 ing about, and with worms crawling through the DAMP earth, and to reflect that these elaborately c
0078 ere it ranges into slightly hotter or colder, DAMPER or drier districts. In this case we can clear
0078 duals of the same species, for the warmest or DAMPEST spots. Hence, also, we can see that when a p
0077 y well withstand a little more heat or cold, DAMPNESS or dryness, for elsewhere it ranges into sli
0139 merican cave animals, as I hear from Professor DANA; and some of the European cave insects are ver
0372 ny closely allied crustaceans (as described in DANA's admirable work), of some fish and other mari
0376 quote a remark by the highest authority, Prof. DANA, that it is certainly a wonderful fact that Ne
0088 ing, and whirling round, like Indians in a war DANCE, for the possession of the females; male salm
0077 y. but in the beautifully plumed seed of the DANDELION, and in the flattened and fringed legs of t
0011 r, from the animals not being much alarmed by DANGER, seems probable. There are many laws regulati
0084 ese birds and insects in preserving them from DANGER. Grouse, if not destroyed at some period of t
0134 ing birds seldom take flight except to escape DANGER, I believe that the nearly wingless condition
0134 indeed inhabits continents and is exposed to DANGER from which it cannot escape by flight, but by
0180 there is reason to believe, by lessening the DANGER from occasional falls. But it does not follow
0216 ly of dogs and cats, for if the hen gives the DANGER chuckle, they will run (more especially young
0075 re the same food, and are exposed to the same DANGERS. In the case of varieties of the same specie
0454 n earless breeds, the reappearance of minute DANGLING horns in hornless breeds of cattle, more esp
0197 w that there were many black and pied kinds, I DARE say that we should have thought that the green
0240 lly lesser size; and I fully believe, though I DARE not assert so positively, that the workers of
0068 s of prey: even the tiger in India most rarely DARES to attack a young elephant protected by its d
0025 nd, having it bluish); the tail has a terminal DARK bar, with the bases of the outer feathers exte
0139 competition to which the inhabitants of these DARK abodes will probably have been exposed. Acclim
0163 times very obscure, or actually quite lost, in DARK coloured asses. The koulan of Pallas is said t

Page **(Key Word)**

```
0224 ectly effected by a crowd of bees working in a DARK hive. Grant whatever instincts you please, and
0454 in the case of the eyes of animals inhabiting DARK caverns, and of the wings of birds inhabiting
0473 en we look at the blind animals inhabiting the DARK caves of America and Europe. In both varieties
0145 lower of the truss often loses the patches of DARKER colour in the two upper petals; and that when
0137 be in any way injurious to animals living in DARKNESS, I attribute their loss wholly to disuse. In
0138 forms, prepare the transition from light to DARKNESS. Next follow those that are constructed for
0138 ; and, last of all, those destined for total DARKNESS. By the time that an animal had reached, aft
0183 g stationary on the margin of water, and then DASHING like a kingfisher at a fish. In our own coun
0287 an ample allowance. At this rate, on the above DATA, the denudation of the Weald must have require
0323 ant parts of the world: we have not sufficient DATA to judge whether the productions of the land a
0203 er instantly to destroy the young queens her DAUGHTERS as soon as born, or to perish herself in th
0472 hatred of the queen bee for her own fertile DAUGHTERS; at ichneumonidae feeding within the live b
0307 nic remains of the lowest Silurian stratum the DAWN of life on this planet. Other highly competent
0318 e seen, having endured from the earliest known DAWN of life to the present day: some having disapp
0411 heir origin explained. Summary. From the first DAWN of life, all organic beings are found to resem
0488 f change was probably slower: and at the first DAWN of life, when very few forms of the simplest s
0296 have been preserved: thus, Messrs. Lyell and DAWSON found carboniferous beds 1400 feet thick in N
0001 to me probable: from that period to the present DAY I have steadily pursued the same object. I hop
0032 rists' flowers, when the flowers of the present DAY are compared with drawings made only twenty or
0088 males: male salmons have been seen fighting all DAY long; male stag beetles often bear wounds from
0107 iving fossils; they have endured to the present DAY, from having inhabited a confined area, and fr
0126 amilies, orders, and classes, as at the present DAY. Summary of Chapter. If during the long course
0153 breed true. In the long run selection gains the DAY, and we do not expect to fail so far as to bre
0156 t probable that they should vary at the present DAY. On the other hand, the points in which specie
0171 reader. Some of them are so grave that to this DAY I can never reflect on them without being stag
0182 re will seldom continue to exist to the present DAY, for they will have been supplanted by the ver
0218 ggs lie strewed over the plains, so that in one DAY's hunting I picked up no less than twenty lost
0221 ter numbers in Switzerland than in England. One DAY I fortunately chanced to witness a migration f
0221 described, their slaves in their jaws. Another DAY my attention was struck by about a score of th
0279 bsence of intermediate varieties at the present DAY. On the nature of extinct intermediate varieti
0279 such links do not commonly occur at the present DAY, under the circumstances apparently most favou
0282 he varieties of the same species at the present DAY; and these parent species, now generally extin
0283 reach the cliffs only for a short time twice a DAY, and the waves eat into them only when they ar
0295 ng from before the glacial epoch to the present DAY. In order to get a perfect gradation between t
0298 it should not be forgotten, that at the present DAY, with perfect specimens for examination, two f
0305 h to north as they are at present. Even at this DAY, if the Malay Archipelago were converted into
0307 e interval from the Silurian age to the present DAY; and that during these vast, yet quite unknown
0316 from the lowest Silurian stratum to the present DAY. We have seen in the last chapter that the spe
0318 the earliest known dawn of life to the present DAY; some having disappeared before the close of t
0324 ll the marine animals which live at the present DAY in Europe, and all those that lived in Europe
0324 hat fossiliferous beds deposited at the present DAY on the shores of North America would hereafter
0330 ty could not be named, asserts that he is every DAY taught that palaeozoic animals, though belongi
0330 ies, or genera with those living at the present DAY, were not at this early epoch limited in such
0330 pposing them to be distinguished at the present DAY from each other by a dozen characters, the anc
0332 eriods, and some to have endured to the present DAY. By looking at the diagram we can see that if
0338 pical members of the same groups at the present DAY, it would be vain to look for animals having t
0357 it is fully admitted, as I believe it will some DAY be, that each species has proceeded from a sin
0369 nd the polar regions as they are at the present DAY. But the foregoing remarks on distribution app
0369 mencement of the Glacial period. At the present DAY, the sub arctic and northern temperate product
0370 now, the climate was warmer than at the present DAY. Hence we may suppose that the organisms now l
0377 ported as many species as we see at the present DAY crowded together at the Cape of Good Hope, and
0379 ust in the same manner as we see at the present DAY, that very many European productions cover the
0394 an fly across. Bats have been seen wandering by DAY far over the Atlantic Ocean; and two North Ame
0399 o the plants, and will, I do not doubt, be some DAY explained. The law which causes the inhabitant
0420 transmitted modified descendants to the present DAY, represented by the fifteen genera (a14 to z14
0429 having occasionally transmitted to the present DAY descendants but little modified, will give to
0480 n inherited from a remote period to the present DAY. On the view of each organic being and each se
0482 assigning any distinction in the two cases. The DAY will come when this will be given as a curious
0485 wn, or believed, to be connected at the present DAY by intermediate gradations, whereas species we
0093 ion of pollen. As the wind had set for several DAYS from the female to the male tree, the pollen c
0137 an thought that it regained; after living some DAYS in the light, some slight power of vision. In
0217 s, not daily, but at intervals of two or three DAYS: so that, if she were to make her own nest and
0218 se of the cuckoo, at intervals of two or three DAYS. This instinct, however, of the American ostri
0234 reover, many bees have to remain idle for many DAYS during the process of secretion. A large store
0251 t flowers soon ceased to grow, and after a few DAYS perished entirely; whereas the pod impregnated
0358 kinds, 64 germinated after an immersion of 28 DAYS, and a few survived an immersion of 137 days.
0358 8 days, and a few survived an immersion of 137 DAYS. For convenience sake I chiefly tried small se
0359 le or fruit; and as all of these sank in a few DAYS, they could not be floated across wide spaces
0359 mediately; but when dried, they floated for 90 DAYS and afterwards when planted they germinated: a
0359 paragus plant with ripe berries floated for 23 DAYS, when dried it floated for 85 days, and the se
0359 ated for 23 days, when dried it floated for 85 DAYS, and the seeds afterwards germinated: the ripe
0359 ed: the ripe seeds of Helosciadium sank in two DAYS, when dried they floated for above 90 days, an
0359 two days, when dried they floated for above 90 DAYS, and afterwards germinated. Altogether out of
0359 + the 94 dried plants, 18 floated for above 28 DAYS, and some of the 18 floated for a very much lo
0359 4/87 seeds germinated after an immersion of 28 DAYS; and as 18/94 plants with ripe fruit (but not
0359 ment) floated, after being dried, for above 28 DAYS; as far as we may infer anything from these sc
0359 try might be floated by sea currents during 28 DAYS, and would retain their power of germination.
0360 ult was that 18/98 of his seeds floated for 42 DAYS, and were then capable of germination. But I d
0361 etches, for instance, are killed by even a few DAYS immersion in sea water: but some taken out of
0361 ch had floated on artificial salt water for 30 DAYS, to my surprise nearly all germinated. Living
0362 t grew after having been thus retained for two DAYS and fourteen hours. Fresh water fish, I find,
0397 injured an immersion in sea water during seven DAYS: one of these shells was the Helix pomatia, an
0397 in hybernated I put it in sea water for twenty DAYS, and it perfectly recovered. As this species h
0397 new membranous one, I immersed it for fourteen DAYS in sea water, and it recovered and crawled awa
0445 actually measuring the old dogs and their six DAYS old puppies, I found that the puppies had not
0445 ul measurements made of the dam and of a three DAYS old colt of a race and heavy cart horse, I fin
```

Page **(Key Word)**

days

Page **(Key Word)**

```
0053 her questions, hereafter to be discussed. Alph. DE Candolle and others have shown that plants whic
0062 ell deserves, at much greater length. The elder DE Candolle and Lyell have largely and philosophic
0115 omes. But the case is very different; and Alph. DE Candolle has well remarked in his great and adm
0146 spermous in the central flowers, that the elder DE Candolle founded his main divisions of the orde
0146 natural selection can act. For instance, Alph. DE Candolle has remarked that winged seeds are nev
0175 s it is quite remarkable how abruptly, as Alph. DE Candolle has observed, a common alpine species
0270 e two varieties as specifically distinct. Girou DE Buzareingues crossed three varieties of gourd,
0317 lly given up, even by those geologists, as Elie DE Beaumont, Murchison, Barrande, etc., whose gene
0325 s greatly struck those admirable observers, MM. DE Verneuil and d'Archiac. After referring to the
0360 ly be transported by any other means; and Alph. DE Candolle has shown that such plants generally h
0379 on by Hooker in regard to America, and by Alph. DE Candolle in regard to Australia, that many more
0386 this is strikingly shown, as remarked by Alph. DE Candolle, in large groups of terrestrial plants
0387 water lily the Nelumbium, and remembered Alph. DE Candolle's remarks on this plant, I thought tha
0389 ed with those on equal continental areas: Alph. DE Candolle admits this for plants, and Wollaston
0392 de only herbaceous species: now trees, as Alph. DE Candolle has shown, generally have, whatever th
0398 of the islands, between the Galapagos and Cape DE Verde Archipelagos: but what an entire and abso
0398 their inhabitants! The inhabitants of the Cape DE Verde Islands are related to those of Africa, l
0399 rly continuous land, from America; and the Cape DE Verde islands from Africa: and that such coloni
0402 a large proportion of cases, as shown by Alph. DE Candolle, to distinct genera. In the Galapagos
0406 ch has lately been admirably discussed by Alph. DE Candolle in regard to plants, namely, that the
0417 fect and degraded flowers; in the latter, as A. DE Jussieu has remarked, the greater number of the
0430 ted to habits like those of a Rodent. The elder DE Candolle has made nearly similar observations o
0478 ouring American mainland; and those of the Cape DE Verde archipelago and other African islands to
0113 g enabled to feed on new kinds of prey, either DEAD or alive; some inhabiting new stations, climbi
0130 h the great Tree of Life, which fills with its DEAD and broken branches the crust of the earth, an
0185 er hand, the acutest observer by examining the DEAD body of the water ouzel would never have suspe
0221 illed their small opponents, and carried their DEAD bodies as food to their nest, twenty nine yard
0222 f these ants entering their nest, carrying the DEAD bodies of F. fusca (showing that it was not a
0339 relationship in the same continent between the DEAD and the living. Professor Owen has subsequentl
0362 orced many kinds of seeds into the stomachs of DEAD fish, and then gave their bodies to fishing ea
0012 sical: thus cats with blue eyes are invariably DEAF; colour and constitutional peculiarities go to
0144 ular than the relation between blue eyes and DEAFNESS in cats, and the tortoise shell colour with
0062 with more spirit and ability than W. Herbert, DEAN of Manchester, evidently the result of his gre
0036 ng and devouring their old women, in times of DEARTH, as of less value than their dogs. In plants
0062 ving progeny. Two canine animals in a time of DEARTH, may be truly said to struggle with each othe
0195 o disease, or not so well enabled in a coming DEARTH to search for food, or to escape from beasts
0079 e is not incessant, that no fear is felt, that DEATH is generally prompt, and that the vigorous, t
0088 r possession of the females; the result is not DEATH to the unsuccessful competitor, but few or no
0157 than ordinary selection, as it does not entail DEATH, but only gives fewer offspring to the less f
0194 ortance. As natural Selection acts by Life and DEATH, by the preservation of individuals with any
0202 kward serratures, and so inevitably causes the DEATH of the insect by tearing out its viscera? If
0202 e sting should so often cause the insect's own DEATH: for if on the whole the power of stinging be
0202 of natural selection, though it may cause the DEATH of some few members. If we admire the truly w
0264 hybridising gallinaceous birds, that the early DEATH of the embryo is a very frequent cause of ste
0277 cumstances; in some cases largely on the early DEATH of the embryo. The sterility of hybrids, whic
0320 ickness in the individual is the forerunner of DEATH, to feel no surprise at sickness, but when th
0472 at the sting of the bee causing the bee's own DEATH; at drones being produced in such vast number
0488 y causes, like those determining the birth and DEATH of the individual. When I view all beings not
0490 thus; from the war of nature, from famine and DEATH, the most exalted object which we are capable
0288 y soft can be preserved. Shells and bones will DECAY and disappear when left on the bottom of the
0292 d preserve the remains before they had time to DECAY. On the other hand, as long as the bed of the
0300 sufficient rate to protect organic bodies from DECAY, no remains could be preserved. In our archip
0129 rowth of the tree, many a limb and branch has DECAYED and dropped off; and these lost branches of
0484 rom some one prototype. But analogy may be a DECEITFUL guide. Nevertheless all living things have
0105 ction of new species. But we may thus greatly DECEIVE ourselves; for to ascertain whether a small
0420 sification are explained, if I do not greatly DECEIVE myself, on the view that the natural system
0167 it makes the works of God a mere mockery and DECEPTION: I would almost as soon believe with the ol
0047 e to follow. We must, however, in many cases, DECIDE by a majority of naturalists, for few well ma
0464 severe, and we sometimes see the contest soon DECIDED: for instance, if several varieties of wheat
0484 es will cease. Systematists will have only to DECIDE (not that this will be easy) whether any form
0011 ickness of fur from climate. Habit also has a DECIDED influence, as in the period of flowering wit
0075 severe, and we sometimes see the contest soon DECIDED: for instance, if several varieties of wheat
0148 as, though effected by slow steps, would be a DECIDED advantage to each successive individual of t
0183 age of perfection, so as to have given them a DECIDED advantage over other animals in the battle f
0301 onfined to one place, but if possessed of any DECIDED advantage, or when further modified and impr
0249 quent ill effects of manipulation, sometimes DECIDEDLY increases, and goes on increasing. Now, in
0047 will not here enumerate, sometimes occur in DECIDING whether or not to rank one form as a variety
0257 can be crossed. Annual and perennial plants, DECIDUOUS and evergreen trees, plants inhabiting diff
0261 erbaceous, one being evergreen and the other DECIDUOUS, and adaptation to widely different climate
0001 ow that I have not been hasty in coming to a DECISION. My work is now nearly finished; but as it w
0014 have in vain endeavoured to discover on what DECISIVE facts the above statement has so often and s
0127 cted in the world's history, geology plainly DECLARES. Natural selection, also, leads to divergenc
0465 ng enough intervals of time, geology plainly DECLARES that all species have changed; and they have
0056 arge genera have often come to their maxima, DECLINED, and disappeared. All that we want to show i
0069 ked with inhabitants, the other species will DECREASE. When we travel southward and see a species
0072 ere to increase in Paraguay, the flies would DECREASE, then cattle and horses would become feral,
0076 on, otherwise the weaker kinds will steadily DECREASE in numbers and disappear. So again with the
0076 of one species of swallow having caused the DECREASE of another species. The recent increase of t
0076 l thrush in parts of Scotland has caused the DECREASE of the song thrush. How frequently we hear o
0109 s in number, so will the less favoured forms DECREASE and become rare. Rarity, as geology tells us
0181 ve that some at least of the squirrels would DECREASE in numbers or become exterminated, unless th
0292 nhabited area and number of inhabitants will DECREASE (excepting the productions on the shores of
0317 ely thins out in the upper beds, marking the DECREASE and final extinction of the species. This gr
0467 es shall increase in number, and which shall DECREASE, or finally become extinct. As the individua
0090 increased in numbers, or that other prey had DECREASED in numbers, during that season of the year
0248 ility never increased, but generally greatly DECREASED. I do not doubt that this is usually the ca
```

Page **(Key Word)**

Page **************************************(Key Word)**************************************

```
0069  of all kinds, and therefore of competitors, DECREASES northwards; hence in going northward, or in
0248  case, and that the fertility often suddenly DECREASES in the first few generations. Nevertheless
0316  mum, and then, sooner or later, it gradually DECREASES. If the number of the species of a genus, o
0069  when we travel southward and see a species DECREASING in numbers, we may feel sure that the cause
0105  f new species through natural selection, by DECREASING the chance of the appearance of favourable
0252  . had hybrids, when fairly treated, gone on DECREASING in fertility in each successive generation,
0341  failing orders, with the genera and species DECREASING in numbers, as apparently is the case of th
0371  ion of their inhabitants. During the slowly DECREASING warmth of the Pliocene period, as soon as t
0015  tructure, in such case, I grant that we could DEDUCE nothing from domestic varieties in regard to
0077  a corollary of the highest importance may be DEDUCED from the foregoing remarks, namely, that the
0243  known bird. Finally, it may not be a logical DEDUCTION, but to my imagination it is far more satis
0014  al stocks. Hence it has been argued that no DEDUCTIONS can be drawn from domestic races to species
0320  er and to suspect that he died by some unknown DEED of violence. The theory of natural selection i
0003  et this opportunity pass without expressing my DEEP obligations to Dr. Hooker, who for the last fi
0048  rieties. Mr. H. C. Watson, to whom I lie under DEEP obligation for assistance of all kinds, has ma
0138  o imagine conditions of life more similar than DEEP limestone caverns under a nearly similar clima
0288  a single Mediterranean species, which inhabits DEEP water and has been found fossil in Sicily, whe
0350  the American seas. We see in these facts some DEEP organic bond, prevailing throughout space and
0395  which is traversed near Celebes by a space of DEEP ocean; and this space separates two widely dis
0395  er side the islands are situated on moderately DEEP submarine banks, and they are inhabited by cl
0401  hough in sight of each other, are separated by DEEP arms of the sea, in most cases wider than the
0230  site sides, always working circularly as they DEEPEN each cell. They do not make the whole three s
0228  vate minute circular pits in it; and as they DEEPENED these little pits, they make them wider and
0229  he basins, as soon as they had been a little DEEPENED, came to have flat bottoms: and these flat b
0229  lion wax, as they circularly gnawed away and DEEPENED the basins on both sides, in order to have s
0133  hter coloured than those confined to cold and DEEPER seas. The birds which are confined to confine
0138  ive generations from the outer world into the DEEPER and deeper recesses of the Kentucky caves, as
0138  ions from the outer world into the deeper and DEEPER recesses of the Kentucky caves, as did Europe
0396  iod to the mainland than islands separated by DEEPER channels, we can understand the frequent rela
0138  ad reached, after numberless generations, the DEEPEST recesses, disuse will on this view have more
0095  e to the red clover to have a shorter or more DEEPLY divided tube to its corolla, so that the hive
0185  oh not in structure. In the frigate bird, the DEEPLY scooped membrane between the toes shows that
0246  ed their lives to this subject, without being DEEPLY impressed with the high generality of some de
0283  e sea on our shores, will, I believe, be most DEEPLY impressed with the slowness with which rocky
0287  thus have escaped the action of the sea: when DEEPLY submerged for perhaps equally long periods, i
0395  australia. The West Indian Islands stand on a DEEPLY submerged bank, nearly 1000 fathoms in depth,
0400  ficulty: but it arises in chief part from the DEEPLY seated error of considering the physical cond
0017  small power of endurance of warmth by the rein DEER, or of cold by the common camel, prevented the
0090  ; and let us suppose that the fleetest prey, a DEER for instance, had from any change in the count
0091  ith a light greyhound like form, which pursues DEER, and the other more bulky, with shorter legs,
0213  f wolf rushing round, instead of at, a herd of DEER, and driving them to a distant point, we shoul
0088  hough to them and to others, special means of DEFENCE may be given through means of sexual selecti
0090  over other males, in their weapons, means of DEFENCE, or charms; and have transmitted these advan
0201  the rattlesnake has a poison fang for its own DEFENCE and for the destruction of its prey; but som
0468  depend on having special weapons or means of DEFENCE, or on the charms of the males; and the slig
0134  annot escape by flight, but by kicking it can DEFEND itself from enemies, as well as any of the sm
0195  so that individuals which could by any means DEFEND themselves from these small enemies, would be
0220  and like their masters are much agitated and DEFEND the nest: when the nest is much disturbed and
0293  tologists, whose opinions are worthy of much DEFERENCE, namely Bronn and Woodward, have concluded
0275  denly appeared, such as albinism, melanism, DEFICIENCY of tail or horns, or additional fingers and
0135  ed beetle of the Egyptians, they are totally DEFICIENT. There is not sufficient evidence to induce
0135  e 550 species inhabiting Madeira, are so far DEFICIENT in wings that they cannot fly: and that of
0391  alogous facts. Oceanic islands are sometimes DEFICIENT in certain classes, and their places are ap
0484  r forms, to be capable of definition; and if DEFINABLE, whether the differences be sufficiently im
0044  e term variety is almost equally difficult to DEFINE, but here community of descent is almost univ
0419  of affinities. Nothing can be easier than to DEFINE a number of characters common to all birds; b
0432  blend together. We could not, as I have said, DEFINE the several groups; but we could pick out typ
0433  to types, whether or not we can separate and DEFINE the groups to which such types belong. Finall
0441  ough I am aware that it is hardly possible to DEFINE clearly what is meant by the organisation bei
0482  evertheless they do not pretend that they can DEFINE, or even conjecture, which are the created to
0022  certainly, I think, be ranked by him as well DEFINED species. Moreover, I do not believe that any
0055  g at species as only strongly marked and well DEFINED varieties, I was led to anticipate that the
0111  e well marked species, present slight and ill DEFINED differences. Mere chance, as we may call it,
0120  mount, to convert these three forms into well DEFINED species: thus the diagram illustrates the st
0171  ad of the species being, as we see them, well DEFINED? Secondly, is it possible that an animal hav
0174  difficulty: for I believe that many perfectly DEFINED species have been formed on strictly continu
0175  competition; and as these species are already DEFINED objects (however they may have become so), n
0175  the range of others, will tend to be sharply DEFINED. Moreover, each species on the confines of t
0175  ical range will come to be still more sharply DEFINED. If I am right in believing that allied or r
0177  elieve that species come to be tolerably well DEFINED objects, and do not at any one period presen
0238  duate into each other, but are perfectly well DEFINED; being as distinct from each other, as are a
0241  believe, the wonderful fact of two distinctly DEFINED castes of sterile workers existing in the sa
0279  rtant manner on the presence of other already DEFINED organic forms, than on climate: and, therefo
0336  this subject, for naturalists have not as yet DEFINED to each other's satisfaction what is meant b
0345  ; and this may account for that vague yet ill DEFINED sentiment, felt by many palaeontologists, th
0348  and over this vast space we meet with no well DEFINED and distinct marine faunas. Although hardly
0417  fling points of resemblance, too slight to be DEFINED. Certain plants, belonging to the Malpighiac
0460  ous of these is the existence of two or three DEFINED castes of workers or sterile females in the
0470  d thus species are rendered to a large extent DEFINED and distinct objects. Dominant species belon
0411  mplex and radiating. Extinction separates and DEFINES groups. Morphology, between members of the s
0418  al value of the characters which they use in DEFINING a group, or in allocating any particular spe
0431  rth chapter, has played an important part in DEFINING and widening the intervals between the sever
0017  i do not think it is possible to come to any DEFINITE conclusion, whether they have descended from
0040  a language, can hardly be said to have had a DEFINITE origin. A man preserves and breeds from an i
0046  ic genera some of the species have fixed and DEFINITE characters. Genera which are polymorphic in
0046  quently have not been seized on and rendered DEFINITE by natural selection, as hereafter will be e
```

Page **************************************(Key Word)**************************************

Page **(Key Word)**

```
0052 plained) differences of structure in certain DEFINITE directions. Hence I believe a well marked va
0075 and all must fall to the ground according to DEFINITE laws; but how simple is this problem compare
0102 odical selection, a breeder selects for some DEFINITE object, and free intercrossing will wholly s
0318 e even supposed that as the individual has a DEFINITE length of life, so have species a definite d
0318 a definite length of life, so have species a DEFINITE duration. No one I think can have marvelled
0357 when in the course of time we know something DEFINITE about the means of distribution; we shall be
0486 will no doubt become simpler when we have a DEFINITE object in view. We possess no pedigrees or a
0044 have been given of the term species. No one DEFINITION has as yet satisfied all naturalists, yet e
0049 tly called species or varieties, before any DEFINITION of these terms has been generally accepted,
0207 thin the same class. I will not attempt any DEFINITION of instinct. It would be easy to show that
0416 sally admitted as highly serviceable in the DEFINITION of whole groups. For instance, whether or n
0419 birds; but in the case of crustaceans, such DEFINITION has hitherto been found impossible. There a
0424 her as a single species, and gives a single DEFINITION by which the several members of the several
0432 se it would be quite impossible to give any DEFINITION by which the several members of the several
0484 distinct from other forms, to be capable of DEFINITION; and if definable, whether the differences
0044 work. Nor shall I here discuss the various DEFINITIONS which have been given of the term species.
0432 though it would be quite impossible to give DEFINITIONS by which each group could be distinguished
0341 he sloth, armadillo, and anteater, as their DEGENERATE descendants. This cannot for an instant be
0152 racter. The breed will then be said to have DEGENERATED. In rudimentary organs, and in those which
0173 e to withstand an enormous amount of future DEGRADATION; and such fossiliferous masses can be accu
0283 erately hard rocks, and mark the process of DEGRADATION. The tides in most cases reach the cliffs
0283 ny line of rocky cliff, which is undergoing DEGRADATION, we find that it is only here and there, a
0284 ormations are the result and measure of the DEGRADATION which the earth's crust has elsewhere suff
0284 s elsewhere suffered. And what an amount of DEGRADATION is implied by the sedimentary deposits of
0286 he most exposed coasts; though no doubt the DEGRADATION of a lofty cliff would be more rapid from
0286 ten or twenty miles in length, ever suffers DEGRADATION at the same time along its whole indented
0290 for ages have been great, from the enormous DEGRADATION of the coast rocks and from muddy streams
0291 ugh, when upraised, to resist any amount of DEGRADATION, may be formed. I am convinced that all ou
0291 which has been bulky enough to resist such DEGRADATION as it has as yet suffered, but which will
0291 ly thick and extensive to resist subsequent DEGRADATION, may have been formed over wide spaces dur
0342 rous deposits thick enough to resist future DEGRADATION, enormous intervals of time have elapsed b
0394 ciently ancient, as shown by the stupendous DEGRADATION which they have suffered and by their tert
0465 s formations, thick enough to resist future DEGRADATION, can be accumulated only where much sedime
0284 pendently of the rate of accumulation of the DEGRADED matter, probably offers the best evidence of
0417 nging to the Malpighiaceae, bear perfect and DEGRADED flowers; in the latter, as A. de Jussieu has
0417 oduced in France, during several years, only DEGRADED flowers, departing so wonderfully in a numbe
0012 me plastic, and tends to depart in some small DEGREE from that of the parental type. Any variation
0013 tted either exclusively, or in a much greater DEGREE, to males alone. A much more important rule,
0016 ing respects; they often differ in an extreme DEGREE in some one part, both when compared one with
0016 ame manner as, only in most cases in a lesser DEGREE than, do closely allied species of the same g
0018 horner's researches have rendered it in some DEGREE probable that man sufficiently civilized to h
0019 but by crossing we can get only forms in some DEGREE intermediate between their parents; and if we
0022 in the sternum are highly variable: so is the DEGREE of divergence and relative size of the two ar
0022 s and females have come to differ to a slight DEGREE from each other. Altogether at least a score
0023 which have led me to this belief are in some DEGREE applicable in other cases, I will here briefl
0024 ngencies seem to me improbable in the highest DEGREE. Some facts in regard to the colouring of pig
0033 sult, I may add, has been, in a corresponding DEGREE, rapid and important. But it is very far from
0037 ugh we owe our excellent fruit, in some small DEGREE, to their having naturally chosen and preserv
0039 s which are first given to him in some slight DEGREE by nature. No man would ever try to make a fa
0039 a pigeon with a tail developed in some slight DEGREE in an unusual manner, or a pouter till he saw
0040 reverse, to man's power of selection. A high DEGREE of variability is obviously favourable, as fr
0046 s: yet quite recently Mr. Lubbock has shown a DEGREE of variability in these main nerves in Coccus
0047 hose forms which possess in some considerable DEGREE the character of species, but which are so cl
0051 ory. And I look at varieties which are in any DEGREE more distinct and permanent, as steps leading
0054 for, as varieties, in order to become in any DEGREE permanent, necessarily have to struggle with
0054 yield offspring which, though in some slight DEGREE modified, will still inherit those advantages
0061 om whatever cause proceeding, if it be in any DEGREE profitable to an individual of any species, i
0065 nly and temporarily increased in any sensible DEGREE. The obvious explanation is that the conditio
0069 these enemies or competitors be in the least DEGREE favoured by any slight change of climate, the
0069 we travel northward; but in a somewhat lesser DEGREE, for the number of species of all kinds, and
0074 umble bees in any district depends in a great DEGREE on the number of field mice, which destroy th
0080 es our domestic productions, and, in a lesser DEGREE, those under nature, vary; and how strong the
0080 d that the whole organisation becomes in some DEGREE plastic. Let it be borne in mind how infinite
0081 may feel sure that any variation in the least DEGREE injurious would be rigidly destroyed. This pr
0082 hysical change, as of climate, or any unusual DEGREE of isolation to check immigration, is actuall
0086 ent period of life, shall not be in the least DEGREE injurious: for if they became so, they would
0092 ts which visited them, so as to favour in any DEGREE the transportal of their pollen from flower t
0094 we suppose this to occur in ever so slight a DEGREE under nature, then as pollen is already carri
0102 come modified and improved in a corresponding DEGREE with its competitors, it will soon be extermi
0104 nditions of life will generally be in a great DEGREE uniform; so that natural selection will tend
0106 s will have to be improved in a corresponding DEGREE or they will be exterminated. Each new form,
0111 t in the very same character and in a greater DEGREE; but this alone would never account for so ha
0123 at an earlier period from a5, will be in some DEGREE distinct from the three first named species;
0124 either unaltered or altered only in a slight DEGREE. In this case, its affinities to the other fo
0124 to be extinct and unknown, it will be in some DEGREE intermediate in character between the two gro
0124 with those now living, yet are often, in some DEGREE, intermediate in character between existing g
0130 orhynchus or Lepidosiren, which in some small DEGREE connects by its affinities two large branches
0131 c beings under domestication, and in a lesser DEGREE in those in a state of nature, had been due t
0132 near the sea shore have their leaves in some DEGREE fleshy, though not elsewhere fleshy. Several
0133 her species, often acquiring in a very slight DEGREE some of the characters of such species, accor
0139 plants cannot endure a damp climate. But the DEGREE of adaptation of species to the climates unde
0148 rt to be largely developed in a corresponding DEGREE. And, conversely, that natural selection may
0150 developed in any species in an extraordinary DEGREE or manner, in comparison with the same part i
0151 n the same country vary in a remarkably small DEGREE, I have particularly attended to them, and th
0151 e any part or organ developed in a remarkable DEGREE or manner in any species, the fair presumptio
0153 s period will seldom be remote in any extreme DEGREE, as species very rarely endure for more than
```

Page **(Key Word)**

Page ***(Key Word)***

```
0154  as it may be called, still present in a high DEGREE. For in this case the variability will seldom
0154  ndividuals varying in the required manner and DEGREE, and by the continued rejection of those tend
0156  ntly have not varied or come to differ in any DEGREE, or only in a slight degree, it is not probab
0156  to differ in any degree, or only in a slight DEGREE, it is not probable that they should vary at
0156  bable that they should still often be in some DEGREE variable, at least more variable than those p
0158  same part in its congeners; and the not great DEGREE of variability in a part, however extraordina
0162  ccasionally varying so as to acquire, in some DEGREE, the character of the same part or organ in a
0169  ould often still be variable in a much higher DEGREE than other parts; for variation is a long con
0178  nts, and, probably, in a still more important DEGREE, on some of the old inhabitants becoming slow
0178  ing slight modifications of structure in some DEGREE permanent; and this assuredly we do see. Seco
0186  reely confess, absurd in the highest possible DEGREE. Yet reason tells me, that if numerous gradat
0188  ficulty far too keenly to be surprised at any DEGREE of hesitation in extending the principle of n
0189  ried circumstances, may in any way, or in any DEGREE, tend to produce a distincter image. We must
0201  le for existence. And we see that this is the DEGREE of perfection attained under nature. The ende
0204  ibility in the acquirement of any conceivable DEGREE of perfection through natural selection. In t
0211  nsable, they may be here passed over. As some DEGREE of variation in instincts under a state of na
0213  re taken out; retrieving is certainly in some DEGREE inherited by retrievers; and a tendency to ru
0215  they are destroyed; so that habit, with some DEGREE of selection, has probably concurred in civil
0225  , is that these cells are always made at that DEGREE of nearness to each other, that they would ha
0238  each other, sometimes to an almost incredible DEGREE, and are thus divided into two or even three
0242  ection could have been efficient in so high a DEGREE, had not the case of these neuter insects con
0243  given in this chapter strengthen in any great DEGREE my theory; but none of the cases of difficult
0245  crosses and of hybrids. Sterility various in DEGREE, not universal, affected by close interbreedi
0246  ly impressed with the high generality of some DEGREE of sterility. Kolreuter makes the rule univer
0247  the seeds, in order to show that there is any DEGREE of sterility. He always compares the maximum
0247  ese twenty plants had their fertility in some DEGREE impaired. Moreover, as Gartner during several
0248  rious species when crossed is so different in DEGREE and graduates away so insensibly, and, on the
0248  e graduates away, and is doubtful in the same DEGREE as is the evidence derived from other constit
0250  ecies, as are Kolreuter and Gartner that some DEGREE of sterility between distinct species is a un
0251  ile. I have taken some pains to ascertain the DEGREE of fertility of some of the complex crosses o
0254  ts and animals, it may be concluded that some DEGREE of sterility, both in first crosses and in hy
0255  gdoms. It has been already remarked, that the DEGREE of fertility, both of first crosses and of hy
0256  of incipient fertilisation. From this extreme DEGREE of sterility we have self fertilised hybrids
0256  han is the fertility of pure species. But the DEGREE of fertility is likewise innately variable; f
0256  experiment. So it is with hybrids, for their DEGREE of fertility is often found to differ greatly
0258  rocal crosses is extremely common in a lesser DEGREE. He has observed it even between forms so clo
0258  lity in a small, and occasionally in a higher DEGREE. Several other singular rules could be given
0259  eed from the same capsule have a considerable DEGREE of fertility. These facts show how completely
0259  le. That it is by no means always the same in DEGREE in the first cross and in the hybrids produce
0259  he fertility of hybrids is not related to the DEGREE in which they resemble in external appearance
0260  ways governed by their systematic affinity or DEGREE of resemblance to each other. This latter sta
0260  ld the sterility be so extremely different in DEGREE, when various species are crossed, all of whi
0260  o keep from blending together? Why should the DEGREE of sterility be innately variable in the indi
0262  act of reproduction, yet that there is a rude DEGREE of parallelism in the results of grafting and
0264  ther, it may be exposed to conditions in some DEGREE unsuitable, and consequently be liable to per
0265  g been specially affected, though in a lesser DEGREE than when sterility ensues. So it is with hyb
0265  e been disturbed, though often in so slight a DEGREE as to be inappreciable by us; in the other ca
0266  urprised that their sterility, though in some DEGREE variable, rarely diminishes. It must, however
0267  r nature, often render organic beings in some DEGREE sterile; and that greater crosses, that is cr
0267  e hybrids which are generally sterile in some DEGREE. I cannot persuade myself that this paralleli
0268  wo hitherto reputed varieties be found in any DEGREE sterile together, they are at once ranked by
0269  es does not determine their greater or lesser DEGREE of sterility when crossed; and we may apply t
0271  ety must have been in some manner and in some DEGREE modified. From these facts; from the great di
0271  e, for a supposed variety if infertile in any DEGREE would generally be ranked as species; from ma
0272  es; and this shows that the difference in the DEGREE of variability graduates away. When mongrels
0273  ing from the mere act of crossing. The slight DEGREE of variability in hybrids from the first cros
0273  it be true, is certainly only a difference in DEGREE. Gartner further insists that when any two sp
0276  f favourable and unfavourable conditions. The DEGREE of sterility does not strictly follow systema
0276  e same two species. It is not always equal in DEGREE in a first cross and in the hybrid produced f
0277  organic beings. It is not surprising that the DEGREE of difficulty in uniting two species, and the
0277  of difficulty in uniting two species, and the DEGREE of sterility of their hybrid offspring should
0289  n fossil remains is fragmentary in an extreme DEGREE. For instance, not a land shell is known belo
0298  n modified and perfected in some considerable DEGREE. According to this view, the chance of discov
0299  success seems to me improbable in the highest DEGREE. Geological research, though it has added num
0301  arly uniform, though perhaps extremely slight DEGREE, they would, according to the principles foll
0301  and distinct species. If then, there be some DEGREE of truth in these remarks, we have no right t
0306  ng, for they do not present characters in any DEGREE intermediate between them. If, moreover, they
0310  ho think the natural geological record in any DEGREE perfect, and who do not attach much weight to
0313  not changed at the same rate, or in the same DEGREE. In the oldest tertiary beds a few living she
0313  life have seldom changed in exactly the same DEGREE. Yet if we compare any but the most closely r
0314  e abruptly, or simultaneously, or to an equal DEGREE. The process of modification must be extremel
0314  , that any form which does not become in some DEGREE modified and improved, will be liable to be e
0316  e or less quickly, and in a greater or lesser DEGREE. A group does not reappear after it has once
0319  had this horse been still living, but in some DEGREE rare, no naturalist would have felt the least
0319  what period of the horse's life, and in what DEGREE, they severally acted. If the conditions had
0323  ains in certain beds present an unmistakeable DEGREE of resemblance to those of the Chalk. It is n
0324  ralia, from containing fossil remains in some DEGREE allied, and from not including those forms wh
0325  e latter must be victorious in a still higher DEGREE in order to be preserved and to survive. We h
0326  find, as we apparently do find, a less strict DEGREE of parallel succession in the productions of
0326  g period favourably circumstanced in an equal DEGREE, whenever their inhabitants met, the battle w
0327  se of time, the forms dominant in the highest DEGREE, wherever produced, would tend everywhere to
0333  t and extinct genera are often in some slight DEGREE intermediate in character between their modif
0335  ns from an intermediate formation are in some DEGREE intermediate in character, is the fact, insis
0338  ogical succession of extinct forms is in some DEGREE parallel to the embryological development of
0338  hough some of these old forms are in a slight DEGREE less distinct from each other than are the ty
0340  period of time, closely allied though in some DEGREE modified descendants. If the inhabitants of o
```

Page ***(Key Word)***

Page ***(Key Word)***

```
0340  ts still differ in nearly the same manner and  DEGREE. But after very long intervals of time and af
0343  together, or at the same rate, or in the same   DEGREE: yet in the long run that all undergo modific
0344  n, apparently, it displays characters in some   DEGREE intermediate between groups now distinct; for
0347  , having conditions peculiar in only a slight   DEGREE; for instance, small areas in the Old World c
0348  tinents, the differences are ever inferior in   DEGREE to those characteristic of distinct continent
0350  natural selection, and in a quite subordinate   DEGREE to the direct influence of different physical
0350  fluence of different physical conditions. The   DEGREE of dissimilarity will depend on the migration
0351  dual in its complex struggle for life, so the   DEGREE of modification in different species will be
0351  ew relations with each other, and in a lesser   DEGREE with the surrounding physical conditions. As
0358  seas to their present inhabitants, a certain   DEGREE of relation (as we shall hereafter see) betwe
0364  not by such means become mingled in any great   DEGREE: but would remain as distinct as we now see t
0368  ince the Glacial period, has ever been in any   DEGREE warmer than at present (as some geologists in
0369  mutual relations will thus have been in some   DEGREE disturbed; consequently they will have been l
0369  sonably asked how I account for the necessary   DEGREE of uniformity of the sub arctic and northern
0380  cover the ground in La Plata, and in a lesser   DEGREE in Australia, and have to a certain extent be
0381  heir subsequent modification to the necessary   DEGREE. The facts seem to me to indicate that peculi
0388  retaining the same identical form or in some   DEGREE modified, I believe mainly depends on the wid
0396  odification in all cases depends to a certain   DEGREE on the lapse of time, and as during changes o
0396  relation between the depth of the sea and the   DEGREE of affinity of the mammalian inhabitants
0398  s. on the other hand, there is a considerable   DEGREE of resemblance in the volcanic nature of the
0400  stinct, are related in an incomparably closer   DEGREE to each other than to the inhabitants of any
0400  differently modified, though only in a small   DEGREE. This long appeared to me a great difficulty:
0401  of the Galapagos Archipelago, and in a lesser   DEGREE in some analogous instances, is that the new
0404  strikingly displayed in Bats, and in a lesser   DEGREE in the Felidae and Canidae. We see it, if we
0416  alists as important an aid in determining the   DEGREE of affinity of this strange creature to birds
0420  branches or groups, though allied in the same   DEGREE in blood to their common progenitor, may diff
0421  ed as related in blood or descent to the same   DEGREE; they may metaphorically be called cousins to
0421  cally be called cousins to the same millionth   DEGREE; yet they differ widely and in different degr
0421  beings all related to each other in the same   DEGREE in blood, has come to be wicely different. Me
0425  se cases the modification has been greater in   DEGREE, and has taken a longer time to complete? I b
0429  life often present characters in some slight   DEGREE intermediate between existing groups. A few o
0430  pial, which will have had a character in some   DEGREE intermediate with respect to all existing Mar
0438  iscovering in such parts or organs, a certain   DEGREE of fundamental resemblance, retained by the s
0446  h not proved true, can be shown to be in some   DEGREE probable, to species in a state of nature. Le
0447  remain unmodified, or be modified in a lesser   DEGREE, by the effects of use and disuse. In certain
0448  ted the young to follow habits of life in any   DEGREE different from those of their parent, and con
0449  ly similar parents, and are therefore in that   DEGREE closely related. Thus, community in embryonic
0452  f the same species are very liable to vary in   DEGREE of development and in other respects. Moreove
0452  cts. Moreover, in closely allied species, the   DEGREE to which the same organ has been rendered rud
0453  imentary, or even cannot be said to be in any   DEGREE rudimentary. Hence, also, a rudimentary organ
0459  organs and instincts are, in ever so slight a   DEGREE, variable, and, lastly, that there is a strug
0461  t organisms of all kinds are rendered in some   DEGREE sterile from their constitutions having been
0461  ed not feel surprise at hybrids being in some   DEGREE sterile, for their constitutions can hardly f
0462  usion of species of the same genus is in some   DEGREE lessened. As on the theory of natural selecti
0465  l will admit; but that it is imperfect to the   DEGREE which I require, few will be inclined to admi
0468  ion, or better adaptation in however slight a   DEGREE to the surrounding physical conditions, will
0470  ies or incipient species, retain to a certain   DEGREE the character of varieties: for they differ f
0472  tants of each country only in relation to the   DEGREE of perfection of their associates; so that we
0473  of which the characters have become in a high   DEGREE permanent, we can understand this fact: for t
0475  geological record is imperfect in an extreme   DEGREE, then such facts as the record gives; support
0475  ossil remains of each formation being in some   DEGREE intermediate in character between the fossils
0476  nt a fossil is, the oftener it stands in some   DEGREE intermediate between existing and allied grou
0486  logy will reveal to us the structure, in some   DEGREE obscured, or the prototypes of each great cla
0488  te of change may have been slow in an extreme   DEGREE. The whole history of the world, as at presen
0043  ffects of variability are modified by various   DEGREES of inheritance and of reversion. Variability
0046  nature could have been effected only by slow   DEGREES: yet quite recently Mr. Lubbock has shown a
0102  g in the right direction, though in different   DEGREES, so as better to fill up the unoccupied plac
0116  re supposed to resemble each other in unequal   DEGREES, as is so generally the case in nature, and
0122  re supposed to resemble each other in unequal   DEGREES, as is so generally the case in nature; spec
0128  sely related, and genera related in different   DEGREES, forming sub families, families, orders, sub
0151  ticularly difficult to compare their relative   DEGREES of variability. When we see any part or orga
0235  r instincts: natural selection having by slow   DEGREES, more and more perfectly, led the bees to sw
0245  ntinued preservation of successive profitable   DEGREES of sterility. I hope, however, to be able to
0260  o stop their further propagation by different   DEGREES of sterility, not strictly related to the fa
0262  ent varieties of the pear take with different   DEGREES of facility on the quince; so do different v
0265  n both cases, the sterility occurs in various   DEGREES: in both, the male element is the most liabl
0276  universally, sterile. The sterility is of all   DEGREES, and is often so slight that the two most ca
0276  cies have been specially endowed with various   DEGREES of sterility to prevent them crossing and bl
0276  y endowed with various and somewhat analogous   DEGREES of difficulty in being grafted together in o
0333  time, and will have been modified in various   DEGREES. As we possess only the last volume of the g
0347  d western South America, between latitudes 25   DEGREES and 35 degrees, we shall find parts extremel
0347  america, between latitudes 25 degrees and 35   DEGREES, we shall find parts extremely similar in al
0347  productions of South America south of lat. 35   DEGREES with those north of 25 degrees, which conseq
0347  uth of lat. 35 degrees with those north of 25   DEGREES, which consequently inhabit a considerably d
0370  s now living under the climate of latitude 60   DEGREES, during the Pliocene period lived further no
0370  north under the polar circle, in latitude 66   DEGREES 67 degrees; and that the strictly arctic pro
0370  r the polar circle, in latitude 66 degrees 67   DEGREES: and that the strictly arctic productions th
0373  d on the eastern side as far south as lat. 36   DEGREES to 37 degrees, and on the shores of the Paci
0373  rn side as far south as lat. 36 degrees to 37   DEGREES, and on the shores of the Pacific, where the
0373  is now so different, as far south as lat. 46   DEGREES: erratic boulders have, also, been noticed o
0373  on both sides of the continent, from lat. 41   DEGREES to the southernmost extremity, we have the c
0378  en arctic forms had migrated some twenty five   DEGREES of latitude from their native country and co
0411  re found to resemble each other in descending   DEGREES, so that they can be classed in groups under
0413  ings, is the bond, hidden as it is by various   DEGREES of modification, which is partially revealed
0420  ay differ greatly, being due to the different   DEGREES of modification which they have undergone: a
0421  gree: yet they differ widely and in different   DEGREES from each other. The forms descended from A,
```

Page ***(Key Word)***

```
0422  in its arrangement, like a pedigree; but the DEGREES of modification which the different groups h
0422  many new languages and dialects. The various DEGREES of difference in the languages from the same
0423  es, namely, closeness of descent with various DEGREES of modification. Nearly the same rules are f
0431  its characters, modified in various ways and DEGREES, to all; and the several species will conseq
0475  should follow the same complex laws in their DEGREES and kinds of resemblance to their parents, i
0018  ancient periods, savages, like those of Tierra DEL Fuego or Australia, who possess a semi domesti
0036  set on animals even by the barbarians of Tierra DEL Fuego, by their killing and devouring their ol
0184  nic of birds, yet in the quiet Sounds of Tierra DEL Fuego, the Puffinuria berardi, in its general
0215  home as puppies from countries, such as Tierra DEL Fuego and Australia, where the savages do not
0323  america, in equatorial South America, in Tierra DEL Fuego, at the Cape of Good Hope, and in the pe
0374  rty and fifty of the flowering plants of Tierra DEL Fuego, forming no inconsiderable part of its s
0378  ts, about forty six in number, common to Tierra DEL Fuego and to Europe still exist in North Ameri
0006  e, i can entertain no doubt, after the most DELIBERATE study and dispassionate judgment of which I
0020  to study some special group, I have, after DELIBERATION, taken up domestic pigeons. I have kept ev
0057  and most experienced observers, and, after DELIBERATION, they concur in this view. In this respect
0087  weak beaks would inevitably perish: or, more DELICATE and more easily broken shells might be selec
0231  and crawl over the comb without injuring the DELICATE hexagonal walls, which are only about one fo
0232  i invariably found that the colour was most DELICATELY diffused by the bees, as delicately as a pa
0232  as most delicately diffused by the bees, as DELICATELY as a painter could have done with his brush
0213  nner by each individual, performed with eager DELIGHT by each breed, and without the end being kno
0483  er's womb? Although naturalists very properly DEMAND a full explanation of every difficulty from t
0051  ise to a climax. Certainly no clear line of DEMARCATION has as yet been drawn between species and
0469  riety, we can see why it is that no line of DEMARCATION can be drawn between species, commonly sup
0189  reator are to those of man? If it could be DEMONSTRATED that any complex organ existed, which coul
0450  r a long period, or for ever, incapable of DEMONSTRATION. Thus, as it seems to me, the leading fac
0242  am surprised that no one has advanced this DEMONSTRATIVE case of neuter insects, against the well
0462  ies in the intermediate regions. It cannot be DENIED that we are as yet very ignorant of the full
0058  that of its supposed parent species, their DENOMINATIONS ought to be reversed. But there is also r
0203  ly perfect the elaboration by our fir trees of DENSE clouds of pollen, in order that a few granule
0145  as dr. Hooker informs me, in species with the DENSEST heads that the inner and outer flowers most
0189  , so as to separate into layers of different DENSITIES and thicknesses, placed at different distan
0189  is layer to be continually changing slowly in DENSITY, so as to separate into layers of different
0279  inferred from the rate of deposition and of DENUDATION. On the poorness of our palaeontological co
0284  a must be extremely slow. But the amount of DENUDATION which the strata have in many places suffer
0284  aving been much struck with the evidence of DENUDATION, when viewing volcanic islands, which have
0285  e one other case, the well known one of the DENUDATION of the Weald. Though it must be admitted th
0285  weald. Though it must be admitted that the DENUDATION of the Weald has been a mere trifle, in com
0286  lude that for a cliff 500 feet in height, a DENUDATION of one inch per century for the whole lenot
0287  wance. At this rate, on the above data, the DENUDATION of the Weald must have required 306,662,400
0307  t ancient beds had been wholly worn away by DENUDATION, or obliterated by metamorphic action, we o
0308  , the more it has suffered the extremity of DENUDATION and metamorphism. The case at present must
0286  , we could measure the time requisite to have DENUDED the Weald. This, of course, cannot be done;
0296  wer beds of a formation having been upraised, DENUDED, submerged, and then re covered by the upper
0025  have descended from the rock pigeon. But if we DENY this, we must make one of the two following hi
0396  the relation of organism to organism. I do not DENY that there are many and grave difficulties in
0459  dification through natural selection, I do not DENY. I have endeavoured to give to them their full
0012  on seems to have become plastic, and tends to DEPART in some small degree from that of the parenta
0104  hrough natural selection destroying any which DEPART from the proper type; but if their conditions
0152  on, and frequently individuals are born which DEPART widely from the standard. There may be truly
0417  ce, as has often been remarked, a species may DEPART from its allies in several characters, both o
0426  stand why a species or a group of species may DEPART, in several of its most important characteris
0417  during several years, only degraded flowers, DEPARTING so wonderfully in a number of the most impo
0434  er of morphology. This is the most interesting DEPARTMENT of natural history, and may be said to be i
0485  he term species. The other and more general DEPARTMENTS of natural history will rise greatly in in
0066  is of some importance to those species, which DEPEND on a rapidly fluctuating amount of food, for
0088  most progeny. But in many cases, victory will DEPEND not on general vigour, but on having special
0108  kind. The existence of such places will often DEPEND on physical changes, which are generally very
0108  natural selection will probably still oftener DEPEND on some of the inhabitants becoming slowly mo
0119  ectly occupied by other beings; and this will DEPEND on infinitely complex relations. But as a gen
0136  tion to enlarge or to reduce the wings, would DEPEND on whether a greater number of individuals we
0178  of its inhabitants. And such new places will DEPEND on slow changes of climate, or on the occasio
0234  d the security of the hive is known mainly to DEPEND on a large number of bees being supported. He
0277  their reproductive systems perfect, seems to DEPEND on several circumstances; in some cases large
0277  pond, though due to distinct causes; for both DEPEND on the amount of difference of some kind betw
0325  causes more or less local and temporary, but DEPEND on general laws which govern the whole animal
0350  conditions. The degree of dissimilarity will DEPEND on the migration of the more dominant forms o
0390  s, are peculiar; and this difference seems to DEPEND on the species which do not become modified h
0468  ly leave most progeny. But success will often DEPEND on having special weapons or means of defence
0094  be a common plant; and that certain insects DEPENDED in main part on its nectar for food. I could
0487  erous formation will be recognised as having DEPENDED on an unusual concurrence of circumstances,
0062  n a large and metaphorical sense, including DEPENDENCE of one being on another, and including (whi
0212  th the nests of birds, which vary partly in DEPENDENCE on the situations chosen and on the nature
0075  es now growing on the old Indian ruins! The DEPENDENCY of one organic being on another, as of a pa
0062  though more properly it should be said to be DEPENDENT on the moisture. A plant which annually pro
0063  already clothe the ground. The missletoe is DEPENDENT on the apple and a few other trees, but can
0074  r england. Now the number of mice is largely DEPENDENT, as every one knows, on the number of cats;
0219  is celebrated father. This ant is absolutely DEPENDENT on its slaves; without their aid, the speci
0223  nd cannot even feed itself: it is absolutely DEPENDENT on its numerous slaves. Formica sanguinea,
0224  species, until an ant was formed as abjectly DEPENDENT on its slaves as is the Formica rufescens.
0234  rse the success of any species of bee may be DEPENDENT on the number of its parasites or other ene
0260  osses and of hybrids is simply incidental or DEPENDENT on unknown differences, chiefly in the repr
0326  s of diffusion may often be very slow, being DEPENDENT on climatal and geographical changes, or on
0489  ted forms, so different from each other, and DEPENDENT on each other in so complex a manner, have
0175  le gradations, the range of any one species, DEPENDING as it does on the range of others, will ten
0063  letoe is disseminated by birds, its existence DEPENDS on birds; and it may metaphorically be said
0066  es, the average number of any animal or plant DEPENDS only indirectly on the number of its eggs or
```

Page ***(Key Word)**

```
0068 ridges, grouse, and hares on any large estate DEPENDS chiefly on the destruction of vermin. If not
0074 ar. The number of humble bees in any district DEPENDS in a great degree on the number of field mic
0088 w words on what I call Sexual Selection. This DEPENDS, not on a struggle for existence, but on a s
0095 periment that the fertility of clover greatly DEPENDS on bees visiting and moving parts of the cor
0108 h extreme slowness, I fully admit. Its action DEPENDS on there being places in the polity of natur
0175 itants of any country by no means exclusively DEPENDS on insensibly changing physical conditions,
0175 on the presence of other species, on which it DEPENDS, or by which it is destroyed, or with it com
0195 and other animals in South America absolutely DEPENDS on their power of resisting the attacks of i
0221 ers and slaves in the two countries, probably DEPENDS merely on the slaves being captured in great
0251 same species when self fertilised, sometimes DEPENDS. The practical experiments of horticulturist
0256 are crossed under the same circumstances, but DEPENDS in part upon the constitution of the individ
0263 er difficulty in effecting a union apparently DEPENDS on several distinct causes. There must somet
0277 gether, though this latter capacity evidently DEPENDS on widely different circumstances, should al
0279 voured to show, that the life of each species DEPENDS in a more important manner on the presence o
0279 not now occuring everywhere throughout nature DEPENDS on the very process of natural selection, th
0314 mount of modification in the varying species, DEPENDS on many complex contingencies, on the variab
0315 ion of long enduring fossiliferous formations DEPENDS on great masses of sediment having been depo
0322 ncies, on which the existence of each species DEPENDS. If we forget for an instant, that each spec
0331 e of character is a necessary contingency; it DEPENDS solely on the descendants from a species bei
0343 cess of modification is necessarily slow, and DEPENDS on many complex contingencies. The dominant
0388 or in some degree modified, I believe mainly DEPENDS on the wide dispersal of their seeds and egg
0395 t. as the amount of modification in all cases DEPENDS to a certain degree on the lapse of time, an
0414 gans of vegetation, on which their whole life DEPENDS, are of little signification, excepting in t
0415 eir importance for classification, I believe, DEPENDS on their greater constancy throughout large
0415 t large groups of species; and this constancy DEPENDS on such organs having generally been subject
0417 lassification, of trifling characters, mainly DEPENDS on their being correlated with several other
0294 few still existing species are common in the DEPOSIT, but have become extinct in the immediately
0294 ea, but are rare or absent in this particular DEPOSIT. It is an excellent lesson to reflect on the
0295 er and lower parts of the same formation, the DEPOSIT must have gone on accumulating for a very lo
0295 for the slow process of variation; hence the DEPOSIT will generally have to be a very thick one;
0410 t having as yet discovered in an intermediate DEPOSIT the forms which are therein absent, but whic
0173 n be accumulated only where much sediment is DEPOSITED on the shallow bed of the sea, whilst it sl
0173 ising, or when very little sediment is being DEPOSITED, there will be blanks in our geological his
0200 itor of the ichneumon, by which its eggs are DEPOSITED in the living bodies of other insects. If i
0284 od observers have estimated that sediment is DEPOSITED by the great Mississippi river at the rate
0288 ly admit to ourselves that sediment is being DEPOSITED over nearly the whole bed of the sea, at a
0291 t to a distant geological age, was certainly DEPOSITED during a downward oscillation of level, and
0293 earing before the uppermost layers have been DEPOSITED, it would be equally rash to suppose that i
0294 ot, for instance, probable that sediment was DEPOSITED during the whole of the glacial period near
0294 o this space of time. When such beds as were DEPOSITED in shallow water near the mouth of the Miss
0307 that before the lowest Silurian stratum was DEPOSITED, long periods elapsed, as long as, or proba
0308 surface near land, on which sediment was not DEPOSITED, or again as the bed of an open and unfatho
0309 strata, supposing such to have been formerly DEPOSITED; for it might well happen that strata which
0315 ends on great masses of sediment having been DEPOSITED on areas whilst subsiding, our formations h
0324 be so, it is evident that fossiliferous beds DEPOSITED at the present day on the shores of North A
0327 ll our greater fossiliferous formations were DEPOSITED during periods of subsidence; and that blan
0328 ame movements. When two formations have been DEPOSITED in two regions during nearly, but not exact
0328 al formations in these regions have not been DEPOSITED during the same exact periods, a formation
0342 h single formation has not been continuously DEPOSITED; that the duration of each formation is, pe
0343 ore the first bed of the Silurian system was DEPOSITED; I can answer this latter question only hyp
0464 theory such strata must somewhere have been DEPOSITED at these ancient and utterly unknown epochs
0465 be accumulated only where much sediment is DEPOSITED on the subsiding bed of the sea. During the
0489 ore the first bed of the Silurian system was DEPOSITED, they seem to me to become ennobled. Judgin
0279 ore the first bed of time, as inferred from the rate of DEPOSITION and of denudation. On the poorness of our p
0293 bly required a vast number of years for its DEPOSITION, I can see several reasons why each should
0295 may reasonably suspect that the process of DEPOSITION has been much interrupted, as a change in t
0296 rmation give any idea of the time which its DEPOSITION has consumed. Many instances could be given
0296 and changes of level during the process of DEPOSITION, which would never even have been suspected
0296 on the same spot during the whole period of DEPOSITION, but have disappeared and reappeared, perha
0066 the most numerous bird in the world. One fly DEPOSITS hundreds of eggs, and another, like the hipp
0283 bly formed at a quicker rate than many other DEPOSITS, yet, from being formed of worn and rounded
0284 of degradation is implied by the sedimentary DEPOSITS of many countries! Professor Ramsay has give
0286 he flanks of which the overlying sedimentary DEPOSITS might have accumulated in thinner masses tha
0289 discovered either in caves or in lacustrine DEPOSITS; and that not a cave or true lacustrine bed
0290 ecent period, than the absence of any recent DEPOSITS sufficiently extensive to last for even a sh
0290 oubt, is, that the littoral and sub littoral DEPOSITS are continually worn away, as soon as they a
0292 he bed of the sea remained stationary, thick DEPOSITS could not have been accumulated in the shall
0292 e very periods of subsidence, that our great DEPOSITS rich in fossils have been accumulated. Natur
0294 o the European seas. In examining the latest DEPOSITS of various quarters of the world, it has eve
0294 her in any quarter of the world, sedimentary DEPOSITS, including fossil remains, have gone on accu
0295 re than one palaeontologist, that very thick DEPOSITS are usually barren of organic remains, excep
0308 iptions which we now possess of the Silurian DEPOSITS over immense territories in Russia and in No
0313 st mammals and reptiles in the sub Himalayan DEPOSITS. The Silurian Lingula differs but little fro
0323 several European and North American tertiary DEPOSITS. Even if the new fossil species which are co
0324 which are only found in the older underlying DEPOSITS, would be correctly ranked as simultaneous i
0328 wich, in his admirable Memoirs on the eocene DEPOSITS of England and France, is able to draw a clo
0328 neral parallelism in the successive Silurian DEPOSITS of Bohemia and Scandinavia: nevertheless he
0342 essary for the accumulation of fossiliferous DEPOSITS thick enough to resist future degradation, e
0175 ht to cause surprise, as climate and height or DEPTH graduate away insensibly. But when we bear in
0228 ut the width of an ordinary cell), and were in DEPTH about one sixth of the diameter of the sphere
0228 basins, if they had been excavated to the same DEPTH as in the former experiment, would have broke
0285 or thrown down on the other, to the height or DEPTH of thousands of feet; for since the crust cra
0294 mouth of the Mississippi, within that limit of DEPTH at which marine animals can flourish; for we
0295 during a period of subsidence; and to keep the DEPTH approximately the same, which is necessary in
0358 e) between the distribution of mammals and the DEPTH of the sea, these and other such facts seem t
0395 in extent independent of distance, between the DEPTH of the sea separating an island from the neig
```

Page ***(Key Word)**

```
0395   deeply submerged bank, nearly 1000 fathoms in DEPTH, and here we find American forms, but the spe
0396   n understand the frequent relation between the DEPTH of the sea and the degree of affinity of the
0409   in a more or less modified condition, and the DEPTH of the sea between an island and the mainland
0132   he same species further north or from greater DEPTHS. Gould believes that birds of the same specie
0173   due to organic beings not inhabiting profound DEPTHS of the sea, and to their remains being embedd
0175   ct has been noticed by Forbes in sounding the DEPTHS of the sea with the dredge. To those who look
0291   ay be formed in two ways: either, in profound DEPTHS of the sea, in which case, judging from the r
0304   s to the equator, inhabiting various zones of DEPTHS from the upper tidal limits to 50 fathoms; fr
0308   hich do not appear to have inhabited profound DEPTHS, in the several formations of Europe and of t
0029   they not learn a lesson of caution, when they DERIDE the idea of species in a state of nature bein
0455   ut which serve as a clue in seeking for its DERIVATION. On the view of descent with modification,
0019   ed in Europe: for whence could they have been DERIVED, as these several countries do not possess a
0026   d, the tendency to reversion to any character DERIVED from such cross will naturally become less a
0035   ies are convinced that the setter is directly DERIVED from the spaniel, and has probably been slow
0099   this is part of the general law of good being DERIVED from the intercrossing of distinct individua
0116   see how this principle of great benefit being DERIVED from divergence of character, combined with
0117   importance of the principle of benefit being DERIVED from divergence of character comes in: for t
0233   case has been observed: nor would any good be DERIVED from a single hexagon being built, as in its
0248   oubtful in the same degree as is the evidence DERIVED from other constitutional and structural dif
0267   animals, we plainly see that great benefit is DERIVED from almost any change in the habits of life
0270   multitude of species. The evidence is, also, DERIVED from hostile witnesses, who in all other cas
0290   rrounding lands, whence the sediment has been DERIVED, accords with the belief of vast intervals o
0295   tend to sink the area whence the sediment is DERIVED, and thus diminish the supply whilst the dow
0308   ds or tracts of land, whence the sediment was DERIVED, occurred in the neighbourhood of the existi
0308   obability have been accumulated from sediment DERIVED from their wear and tear; and would have bee
0403   whence colonists could most readily have been DERIVED, the colonists having been subsequently modi
0409   other source whence immigrants were probably DERIVED. We can see why in two areas, however distan
0416   merous instances could be given of characters DERIVED from parts which must be considered of very
0418   subordinate value. We can see why characters DERIVED from the embryo should be of equal importanc
0418   bryo should be of equal importance with those DERIVED from the adult, for our classifications of c
0418   ain divisions have been founded on characters DERIVED from the embryo, on the number and position
0437   es of bone? As Owen has remarked, the benefit DERIVED from the yielding of the separate pieces in
0478   rest source whence immigrants might have been DERIVED. We see this in nearly all the plants and an
0479   importance in classification; why characters DERIVED from rudimentary parts, though of no service
0144   of mammalia which are most abnormal in their DERMAL covering, viz. Cetacea (whales) and Edentata
0416   s of the flower in grasses, the nature of the DERMAL covering, as hair or feathers, in the Vertebr
0050   t of experimental evidence, showing that they DESCEND from common parents, and consequently must b
0187   n this great class we should probably have to DESCEND far beneath the lowest known fossiliferous s
0366   ce, and their former Alpine inhabitants would DESCEND to the plains. By the time that the cold had
0119   y or may not produce more than one modified DESCENDANT; for natural selection will always act acco
0447   long course of modification, adapted in one DESCENDANT to act as hands, in another as paddles, in
0006   what are called the same genera are lineal DESCENDANTS of some other and generally extinct specie
0006   legged varieties of any one species are the DESCENDANTS of that species. Furthermore, I am convinc
0016   eties, and by other competent judges as the DESCENDANTS of aboriginally distinct species. If any m
0029   f species in a state of nature being lineal DESCENDANTS of other species? Selection. Let us now br
0039   lightly larger tail, never dreamed what the DESCENDANTS of that pigeon would become through long c
0059   inant by leaving many modified and dominant DESCENDANTS. But by steps hereafter to be explained, t
0080   ergence of Character and Extinction, on the DESCENDANTS from a common parent. Explains the groupin
0094   have a better chance of living and leaving DESCENDANTS. Its descendants would probably inherit a
0094   ance of living and leaving descendants. Its DESCENDANTS would probably inherit a tendency to a sim
0110   beaten in the race for life by the modified DESCENDANTS of the commoner species. From these severa
0112   circumstance that the more diversified the DESCENDANTS from any one species become in structure,
0113   ange in its conditions) only by its varying DESCENDANTS seizing on places at present occupied by o
0113   ore diversified in habits and structure the DESCENDANTS of our carnivorous animal became, the more
0113   is species of grass, including its modified DESCENDANTS, would succeed in living on the same piece
0116   , we may, I think, assume that the modified DESCENDANTS of any one species will succeed by so much
0118   produce any. Thus the varieties or modified DESCENDANTS, proceeding from the common parent (A), wi
0119   rule, the more diversified in structure the DESCENDANTS from any one species can be rendered, the
0119   of divergent variation. As all the modified DESCENDANTS from a common and widely diffused species,
0119   ngle line of descent, and the number of the DESCENDANTS will not be increased: although the amount
0121   to produce the greatest number of modified DESCENDANTS; for these will have the best chance of fi
0121   long period continue transmitting unaltered DESCENDANTS; and this is shown in the diagram by the d
0121   will be a constant tendency in the improved DESCENDANTS of any one species to supplant and extermi
0122   other species of the genus. Their modified DESCENDANTS, fourteen in number at the fourteen thousa
0122   ther nine original species, has transmitted DESCENDANTS to this late stage of descent. The new spe
0123   in a widely different manner. Of the eight DESCENDANTS from (A) the three marked a14, q14, p14, w
0123   sub genus or even a distinct genus. The six DESCENDANTS from (I) will form two sub genera or even
0123   treme points of the original genus, the six DESCENDANTS from (I) will, owing to inheritance, diffe
0123   ritance, differ considerably from the eight DESCENDANTS from (A); the two groups, moreover, are su
0123   e, excepting (F), extinct, and have left no DESCENDANTS. Hence the six new species descended from
0125   ogle for the production of new and modified DESCENDANTS, will mainly lie between the larger groups
0126   come utterly extinct, and leave no modified DESCENDANTS; and consequently that of the species livi
0126   any one period, extremely few will transmit DESCENDANTS to a remote futurity. I shall have to retu
0126   the more ancient species having transmitted DESCENDANTS, and on the view of all the descendants of
0126   ted descendants, and on the view of all the DESCENDANTS of the same species making a class, we can
0126   nt species may now have living and modified DESCENDANTS, yet at the most remote geological period,
0128   s. therefore during the modification of the DESCENDANTS of any one species, and during the incessa
0128   ease in numbers, the more diversified these DESCENDANTS become, the better will he their chance of
0129   iods, very few now have living and modified DESCENDANTS. From the first growth of the tree, many a
0146   having been transmitted to a whole group of DESCENDANTS with diverse habits, would naturally be th
0154   imately the same condition to many modified DESCENDANTS, as in the case of the wing of the bat, it
0157   e of the common progenitor, or of its early DESCENDANTS, became variable: variations of this part
0169   rgan has become the parent of many modified DESCENDANTS, which on my view must be a very slow proc
0187   f the same group, that is to the collateral DESCENDANTS from the same original parent-form, in ord
0193   ectric organs, which most of their modified DESCENDANTS have lost. The presence of luminous organs
0196   quently have been taken advantage of by the DESCENDANTS of the species under new conditions of lif
```

Page **(Key Word)**

0200 form, or as being now of special use to the DESCENDANTS of this form, either directly, or indirect
0205 advantage of by the still further modified DESCENDANTS of this species. We may, also, believe tha
0205 ail of an aquatic animal by its terrestrial DESCENDANTS), though it has become of such small impor
0242 s of the fertile members, which alone leave DESCENDANTS. I am surprised that no one has advanced t
0269 dify the reproductive system in the several DESCENDANTS from any one species. Seeing this differen
0273 ctive systems seriously affected, and their DESCENDANTS are highly variable. But to return to our
0280 ered in some respects from all its modified DESCENDANTS. To give a simple illustration: the fantai
0281 ure of the parent with that of its modified DESCENDANTS, unless at the same time we had a nearly p
0281 or a very long period unaltered, whilst its DESCENDANTS had undergone a vast amount of change; and
0297 the parent species and its several modified DESCENDANTS from the lower and upper beds of a formati
0302 have lived long ages before their modified DESCENDANTS. But we continually over rate the perfecti
0307 exterminated by their numerous and improved DESCENDANTS. Consequently, if my theory be true, it is
0321 l respects. Hence the improved and modified DESCENDANTS of a species will generally cause the exte
0331 ssary contingency; it depends solely on the DESCENDANTS from a species being thus enabled to seize
0333 mediate in character their modified DESCENDANTS, or between their collateral relations. In
0340 osely allied though in some degree modified DESCENDANTS. If the inhabitants of one continent forme
0340 f another continent, so will their modified DESCENDANTS still differ in nearly the same manner and
0341 rmadillo, and anteater, as their degenerate DESCENDANTS. This cannot for an instant be admitted. T
0341 h of the six older genera has left modified DESCENDANTS, constituting the six new genera. The othe
0341 a and species will have left modified blood DESCENDANTS. Summary of the preceding and present Chap
0344 dominant groups tend to leave many modified DESCENDANTS, and thus new sub groups and groups are fo
0344 ry slow process, from the survival of a few DESCENDANTS, lingering in protected and isolated situa
0344 people the world with allied, but modified, DESCENDANTS; and these will generally succeed in takin
0350 orious, and will produce groups of modified DESCENDANTS. On this principle of inheritance with mod
0355 e course of time a few colonists, and their DESCENDANTS, though modified, would still be plainly r
0371 ial period. We now see, as I believe, their DESCENDANTS, mostly in a modified condition, in the ce
0390 , and will often produce groups of modified DESCENDANTS. But it by no means follows, that, because
0412 her attempted to show that from the varying DESCENDANTS of each species trying to occupy as many a
0412 the inevitable result is that the modified DESCENDANTS proceeding from a progenitor become brok
0420 era (A, F, and I) have transmitted modified DESCENDANTS to the present day, represented by the fif
0420 ost horizontal line. Now all these modified DESCENDANTS from a single species, are represented as
0421 cessive period of descent. All the modified DESCENDANTS from A will have inherited something in co
0421 n from their common parent, as will all the DESCENDANTS from I; so will it be with each subordinat
0421 will it be with each subordinate branch of DESCENDANTS, at each successive period. If, however, w
0421 wever, we choose to suppose that any of the DESCENDANTS of A or of I have been so much modified as
0421 e occurred with existing organisms. All the DESCENDANTS of the genus F, along its whole line of de
0428 tions have probably arisen. As the modified DESCENDANTS of dominant species, belonging to the larg
0429 occasionally transmitted to the present day DESCENDANTS but little modified, will give to us our s
0432 hich have produced large groups of modified DESCENDANTS. Every intermediate link between these ele
0432 link in each branch and sub branch of their DESCENDANTS, may be supposed to be alive; and the link
0433 ion and divergence of character in the many DESCENDANTS from one dominant parent species, explains
0433 , with the grades of difference between the DESCENDANTS from a common parent, expressed by the ter
0447 he fore limbs in the embryos of the several DESCENDANTS of the parent species will still resemble
0449 f life should resemble the embryos of their DESCENDANTS, our existing species. Agassiz believes th
0450 ht modifications not appearing, in the many DESCENDANTS from some one ancient progenitor, at a ver
0470 inordinately in number; and as the modified DESCENDANTS of each species will be enabled to increas
0472 on always ready to adapt the slowly varying DESCENDANTS of each to any unoccupied or ill occupied
0475 forms, with the slow modification of their DESCENDANTS, causes the forms of life, after long inte
0476 in character, the progenitor with its early DESCENDANTS will often be intermediate in character in
0476 e in character in comparison with its later DESCENDANTS; and thus we can see why the more ancient
0477 plainly related; for they will generally be DESCENDANTS of the same progenitors and early colonist
0483 s an enormous amount of modification in the DESCENDANTS. Throughout whole classes various structur
0488 rogenitor of innumerable extinct and living DESCENDANTS, was created. In the distant future I see
0488 not as special creations, but as the lineal DESCENDANTS of some few beings which lived long before
0489 ll the species of many genera, have left no DESCENDANTS, but have become utterly extinct. We can s
0489 all the living forms of life are the lineal DESCENDANTS of those which lived long before the Silur
0003 had not been independently created, but had DESCENDED, like varieties, from other species. Nevert
0016 in doubt, from not knowing whether they have DESCENDED from one or several parent species. This po
0017 shall presently see, that all our dogs have DESCENDED from any one wild species; but, in the case
0017 o any definite conclusion, whether they have DESCENDED from one or several species. The argument m
0018 highly probable that our domestic dogs have DESCENDED from several wild species. In regard to she
0018 of the humped Indian cattle, that these had DESCENDED from a different aboriginal stock from our
0018 to several authors, that all the races have DESCENDED from one wild stock. Mr. Blyth, whose opini
0019 structure, I do not doubt that they all have DESCENDED from the common wild duck and rabbit. The d
0019 ole world, which I fully admit have probably DESCENDED from several wild species, I cannot doubt t
0023 aturalists is correct, namely, that all have DESCENDED from the rock pigeon (Columba livia), inclu
0023 oceeded from the rock pigeon, they must have DESCENDED from at least seven or eight aboriginal sto
0025 characters, if all the domestic breeds have DESCENDED from the rock pigeon. But if we deny this,
0027 l no doubt that all our domestic breeds have DESCENDED from the Columba livia with its geographica
0028 culty in believing that they could ever have DESCENDED from a common parent, as any naturalist cou
0028 veral breeds to which each has attended, are DESCENDED from so many aboriginally distinct species.
0029 rd cattle, whether his cattle might not have DESCENDED from long horns, and he will laugh you to s
0029 not fully convinced that each main breed was DESCENDED from a distinct species. Van Mons, in his t
0029 t admit that many of our domestic races have DESCENDED from the same parents, may they not learn a
0064 re would be alive fifteen million elephants, DESCENDED from the first pair. But we have better evi
0120 rked by the letters between a14 and m14, all DESCENDED from (A). Thus, as I believe, species are m
0123 e of descent. The new species in our diagram DESCENDED from the original eleven species, will now
0123 ft no descendants. Hence the six new species DESCENDED from (I), and the eight descended from (A),
0123 ew species descended from (I), and the eight DESCENDED from (A), will have to be ranked as very di
0124 or more parent species are supposed to have DESCENDED from some one species of an earlier genus.
0124 e of a curious and circuitous nature. Having DESCENDED from a form which stood between the two par
0124 rmediate in character between the two groups DESCENDED from these species. But as these two groups
0125 we shall also have two very distinct genera DESCENDED from (I); and as these latter two genera, b
0125 nt, will differ widely from the three genera DESCENDED from (A), the two little groups of genera w
0125 d the two new families, or orders, will have DESCENDED from two species of the original genus; and
0125 ; and these two species are supposed to have DESCENDED from one species of a still more ancient an

Page **(Key Word)**

Page ***************************************(Key Word)***************************************

```
0139  that all the species of the same genus have DESCENDED from a single parent, if this view be corre
0152  but on the view that groups of species have DESCENDED from other species, and have been modified
0157  ies of the same genus as having as certainly DESCENDED from the same progenitor, as have the two s
0158  due to the species of the same group having DESCENDED from a common progenitor, from whom they ha
0161  me genus are supposed, on my theory, to have DESCENDED from a common parent, it might be expected
0164  es that the several breeds of the horse have DESCENDED from several aboriginal species, one of whi
0166  e of the several breeds of pigeons: they are DESCENDED from a pigeon (including two or three sub s
0167  of our domestic horse, whether or not it be DESCENDED from one or more wild stocks, or the ass, t
0171  ollowing heads: Firstly, why if species have DESCENDED from other species by insensibly fine grada
0172  n hand. Hence, if we look at each species as DESCENDED from some other unknown form, both the pare
0173  each. By my theory these allied species have DESCENDED from a common parent; and during the proces
0191  ll vertebrate animals having true lungs have DESCENDED by ordinary generation from an ancient prot
0246  at is of the forms known or believed to have DESCENDED from common parents, when intercrossed, and
0250  ertilised by the pollen of a compound hybrid DESCENDED from three other and distinct species: the
0254  mely, that most of our domestic animals have DESCENDED from two or more aboriginal species, since
0254  i believe, for instance, that our dogs have DESCENDED from several wild stocks; yet, with perhaps
0268  obably the true one, is that these dogs have DESCENDED from several aboriginally distinct species.
0273  orm. Now hybrids in the first generation are DESCENDED from species (excluding those long cultivat
0275  ore likely to occur with mongrels, which are DESCENDED from varieties often suddenly produced and
0275  s in character, than with hybrids, which are DESCENDED from species slowly and naturally produced.
0280  on: the fantail and pouter pigeons have both DESCENDED from the rock pigeon; if we possessed all t
0281  th that of the rock pigeon, whether they had DESCENDED from this species or from some other allied
0281  ory, that one of two living forms might have DESCENDED from the other: for instance, a horse from
0299  eeds of cattle, sheep, horses, and dogs have DESCENDED from a single stock or from several aborigi
0302  pment of a group of forms, all of which have DESCENDED from some one progenitor, must have been an
0306  the existing species of the same group have DESCENDED from one progenitor, apply with nearly equa
0306  doubt that all the Silurian trilobites have DESCENDED from some one crustacean, which must have l
0316  or as all the species of the same group have DESCENDED from one species, it is clear that as long
0321  thus, as I believe, a number of new species DESCENDED from one species, that is a new genus, come
0332  f14. All the many forms, extinct and recent, DESCENDED from A, make, as before remarked, one order
0341  eory, all the species of the same genus have DESCENDED from some one species; so that if six gener
0351  proceeded from the same source, as they have DESCENDED from the same progenitor. In the case of th
0354  cies of a genus, which on my theory have all DESCENDED from a common progenitor, can have migrated
0355  all the individuals of the same species have DESCENDED from a single pair, or single hermaphrodite
0355  exist), the species, on my theory, must have DESCENDED from a succession of improved varieties, wh
0355  ll the individuals of each variety will have DESCENDED from a single parent. But in the majority o
0385  and allied species, which, on my theory, are DESCENDED from a common parent and must have proceede
0389  both of the same and of allied species have DESCENDED from a single parent; and therefore have al
0405  he view of all the species of a genus having DESCENDED from a single parent, though now distribute
0407  of the same species, wherever located, have DESCENDED from the same parents; are not insuperable.
0412  s line form together one class, for all have DESCENDED from one ancient but unseen parent; and, co
0412  still earlier period. And all these genera, DESCENDED from (A), form an order distinct from the g
0413  (a), form an order distinct from the genera DESCENDED from (I). So that we here have many species
0413  from (I). So that we here have many species DESCENDED from a single progenitor grouped into gener
0420  ed during the Silurian epoch, and these have DESCENDED from a species which existed at an unknown
0421  different degrees from each other. The forms DESCENDED from A, now broken up into two or three fam
0421  lies, constitute a distinct order from those DESCENDED from I, also broken up into two families. N
0421  two families. Nor can the existing species, DESCENDED from A, be ranked in the same genus with th
0421  cter between A and I, and the several genera DESCENDED from these two genera will have inherited t
0422  states of civilisation of the several races, DESCENDED from a common race) had altered much, and h
0423  rieties, which are believed or known to have DESCENDED from one species. These are grouped under s
0424  if it could be proved that the Hottentot had DESCENDED from the Negro, I think he would be classed
0424  semble the parent form, but because they are DESCENDED from it. He who believes that the cowslip i
0424  from it. He who believes that the cowslip is DESCENDED from the primrose, or conversely, ranks the
0427  and isolated region, have in all probability DESCENDED from the same parents. We can understand, o
0430  adual divergence in character of the species DESCENDED from a common parent, together with their r
0432  the principle of inheritance, all the forms DESCENDED from A, or from I, would have something in
0444  eties most closely allied, and have probably DESCENDED from the same wild stock: hence I was curio
0445  t the several domestic breeds of Pigeon have DESCENDED from one wild species, I compared young pig
0446  ate of nature. Let us take a genus of birds, DESCENDED on my theory from some one parent species,
0449  ges, we may feel assured that they have both DESCENDED from the same or nearly similar parents, an
0458  , with which this world is peopled, have all DESCENDED, each within its own class or group, from c
0461  same genus, or even higher group, must have DESCENDED from common parents; and therefore, in howe
0473  lained if we believe that these species have DESCENDED from a striped progenitor, in the same mann
0473  s the several domestic breeds of pigeon have DESCENDED from the blue and barred rock pigeon! On th
0474  of all the species of the same genus having DESCENDED from a common parent, and having inherited
0476  of common parents. As the groups which have DESCENDED from an ancient progenitor have generally d
0484  the same class. I believe that animals have DESCENDED from at most only four or five progenitors,
0484  the belief that all animals and plants have DESCENDED from some one prototype. But analogy may be
0484  ngs which have ever lived on this earth have DESCENDED from some one primordial form, into which l
0486  genera, have within a not very remote period DESCENDED from one parent, and have migrated from som
0069  , than we do in proceeding southwards or in DESCENDING a mountain. When we reach the Arctic region
0411  beings are found to resemble each other in DESCENDING degrees, so that they can be classed in gro
0088  ow in the scale of nature this law of battle DESCENDS, I know not; male alligators have been descr
0029  f the intermediate links in the long lines of DESCENT is almost universally implied, though it can
0044  ly difficult to define, but here community of DESCENT is almost universally implied, though it can
0119  more highly improved branches in the lines of DESCENT, will, it is probable, often take the place
0119  fication will be confined to a single line of DESCENT, and the number of the descendants will not
0121  to supplant and exterminate in each stage of DESCENT their predecessors and their original parent
0121  y will be with many whole collateral lines of DESCENT, which will be conquered by later and improv
0121  l be conquered by later and improved lines of DESCENT. If, however, the modified offspring of a sp
0122  oved in a diversified manner at each stage of DESCENT, so as to have become adapted to many relate
0122  transmitted descendants to this late stage of DESCENT. The new species in our diagram descended fr
0123  ieve, that two or more genera are produced by DESCENT, with modification, from two or more species
0125  ry ancient epochs when the branching lines of DESCENT had diverged less. I see no reason to limit
0139  ust be readily effected during long continued DESCENT. It is notorious that each species is adapte
```

Page ***************************************(Key Word)***************************************

Page **(Key Word)**

0159 plants, not to the vera causa of community of DESCENT, and a consequent tendency to vary in a like
0171 ties on Theory. Difficulties on the theory of DESCENT with modification. Transitions. Absence or r
0187 g been transmitted from the earlier stages of DESCENT, in an unaltered or little altered condition
0188 explicable, can be explained by the theory of DESCENT, ought not to hesitate to go further, and to
0206 heory, unity of type is explained by unity of DESCENT. The expression of conditions of existence,
0210 t we ought to find in the collateral lines of DESCENT some evidence of such gradations; or we ough
0302 nce, the fact would be fatal to the theory of DESCENT with slow modification through natural selec
0312 their slow and gradual modification, through DESCENT and natural selection. New species have appe
0329 fact is at once explained on the principle of DESCENT. The more ancient any form is, the more, as
0331 acts and inferences accord with the theory of DESCENT with modification. As the subject is somewha
0331 e many extinct genera on the several lines of DESCENT diverging from the parent form A, will form
0333 ber of characters; for at this early stage of DESCENT they have not diverged in character from the
0333 uently to be the case. Thus, on the theory of DESCENT with modification, the main facts with respe
0334 ecies which lived at the sixth great stage of DESCENT in the diagram are the modified offspring of
0336 the sea have been affected. On the theory of DESCENT, the full meaning of the fact of fossil rema
0340 tribution of marine animals. On the theory of DESCENT with modification, the great law of the long
0343 seem to me simply to follow on the theory of DESCENT with modification through natural selection.
0356 ion will not have been due, at each stage, to DESCENT from a single parent. To illustrate what I m
0356 o not owe their difference and superiority to DESCENT from any single pair, but to continued care
0373 ciers have left the marks of their former low DESCENT; and in Sikkim, Dr. Hooker saw maize growing
0381 the southern hemisphere, is, on my theory of DESCENT with modification, a far more remakable case
0389 e two theories of independent creation and of DESCENT with modification. The species of all kinds
0411 in classification, explained on the theory of DESCENT with modification. Classification of varieti
0411 th modification. Classification of varieties. DESCENT always used in classification. Analogical or
0412 ed from a common parent at the fifth stage of DESCENT. These five genera have also much, though le
0413 ing more is included; and that propinquity of DESCENT, the only known cause of the similarity of o
0419 classification tacitly including the idea of DESCENT. Our classifications are often plainly influ
0420 he view that the natural system is founded on DESCENT with modification: that the characters which
0420 sification is genealogical: that community of DESCENT is the hidden bond which naturalists have be
0420 ecies, are represented as related in blood or DESCENT to the same degree; they may metaphorically
0421 resent time, but at each successive period of DESCENT. All the modified descendants from A will ha
0421 dants of the genus F, along its whole line of DESCENT, are supposed to have been but little modifi
0423 rieties as with species; namely, closeness of DESCENT with various degrees of modification. Nearly
0424 nature, every naturalist has in fact brought DESCENT into his classification; for he includes in
0425 preposterous; for where there has been close DESCENT in common, there will certainly be close res
0425 ertainly be close resemblance or affinity. As DESCENT has universally been used in classing togeth
0425 of modification, may not this same element of DESCENT have been unconsciously used in grouping spe
0425 n pedigrees; we have to make out community of DESCENT by resemblances of any kind. Therefore we ch
0426 ts, we may feel almost sure, on the theory of DESCENT, that these characters have been inherited f
0426 tant, betrays the hidden bond of community of DESCENT. Let two forms have not a single character i
0426 oups, we may at once infer their community of DESCENT, and we put them all into the same class. As
0427 classification, only in so far as they reveal DESCENT, we can clearly understand why analogical or
0427 mals, belonging to two most distinct lines of DESCENT, may readily become adapted to similar condi
0427 r blood relationship to their proper lines of DESCENT. We can also understand the apparent paradox
0433 n in group under group. We use the element of DESCENT in classing the individuals of both sexes an
0433 aracters in common, under one species; we use DESCENT in classing acknowledged varieties, however
0433 m their parent; and I believe this element of DESCENT is the hidden bond of connexion which natura
0438 far from meaning that during a long course of DESCENT, primordial organs of any kind, vertebrae in
0439 ly been metamorphosed during a long course of DESCENT from true legs, or from some simple appendag
0443 can be explained, as follows, on the view of DESCENT with modification. It is commonly assumed, p
0449 possible arrangement, would be genealogical. DESCENT being on my view the hidden bond of connexio
0449 y in embryonic structure reveals community of DESCENT. It will reveal this community of descent, h
0449 of descent. It will reveal this community of DESCENT, however much the structure of the adult may
0454 e were formed for this purpose. On my view of DESCENT with modification, the origin of rudimentary
0455 in seeking for its derivation. On the view of DESCENT with modification, we may conclude that the
0456 t should be borne in mind that the element of DESCENT has been universally used in ranking togethe
0456 ture. If we extend the use of this element of DESCENT, the only certainly known cause of similarit
0456 es, orders, and classes. On this same view of DESCENT with modification, all the great facts in Mo
0458 , and have all been modified in the course of DESCENT, that I should without hesitation adopt this
0459 ections may be advanced against the theory of DESCENT with modification through natural selection,
0461 the difficulties encountered on the theory of DESCENT with modification are grave enough. All the
0466 judgment they do not overthrow the theory of DESCENT with modification. Now let us turn to the ot
0475 ts as the record gives, support the theory of DESCENT with modification. New species have come on
0476 y their intermediate position in the chain of DESCENT. The grand fact that all extinct organic bei
0476 t and the extinct will naturally be allied by DESCENT. Looking to geographical distribution, if we
0476 sal, then we can understand, on the theory of DESCENT with modification, most of the great leading
0478 s in any two areas, implies, on the theory of DESCENT with modification, that the same parents for
0479 beings are due to inheritance or community of DESCENT. The natural system is a genealogical arrang
0479 nt, in which we have to discover the lines of DESCENT by the most permanent characters, however sl
0479 , at once explain themselves on the theory of DESCENT with slow and slight successive modification
0483 therefore I cannot doubt that the theory of DESCENT with modification embraces all the members o
0486 iscover and trace the many diverging lines of DESCENT in our natural genealogies, by characters of
0047 the most common; but sometimes the one first DESCRIBED, as the species, and the other as the varie
0054 iots. If the plants inhabiting a country and DESCRIBED in any Flora be divided into two equal mass
0088 cends, I know not; male alligators have been DESCRIBED as fighting, bellowing, and whirling round,
0089 ferences and dislikes: thus Sir R. Heron has DESCRIBED how one pied peacock was eminently attracti
0135 so habitually lost, that the insect has been DESCRIBED as not having them. In some other genera th
0145 th its accessory parts, differs, as has been DESCRIBED by Cassini. These differences have been att
0163 ne. A white ass, but not an albino, has been DESCRIBED without either spinal or shoulder stripe; a
0164 ch, the dun, was striped; and that the above DESCRIBED appearances are all due to ancient crosses
0221 the masters carefully carrying, as Huber has DESCRIBED, their slaves in their jaws. Another day my
0222 though rarely, made into slaves, as has been DESCRIBED by Mr. Smith. Although so small a species,
0225 of the Mexican Melipona domestica, carefully DESCRIBED and figured by Pierre Huber. The Melipona i
0226 tly correct: If a number of equal spheres be DESCRIBED with their centres placed in two parallel l
0372 case of many closely allied crustaceans (as DESCRIBED in Dana's admirable work), of some fish and
0378 at the base of the Himalaya, as graphically DESCRIBED by Hooker. Thus, as I believe, a considerab

Page **(Key Word)**

Page ***(Key Word)**

```
0214  tincts of either parent: for example, Le Roy DESCRIBES a dog, whose great grandfather was a wolf,
0227  nabled to judge of distance, that she always DESCRIBES her spheres so as to intersect largely: and
0431  culty which naturalists have experienced in DESCRIBING, without the aid of a diagram, the various
0037  d in classical times, appears, from Pliny's DESCRIPTION, to have been a fruit of very inferior qua
0191  s 1 infer from Professor Owen's interesting DESCRIPTION of these parts, understand the strange fac
0413  nd then by adding a single sentence, a full DESCRIPTION is given of each kind of dog. The ingenuit
0308  ally in a metamorphosed condition. But the DESCRIPTIONS which we now possess of the Silurian depos
0062  t food and live. But a plant on the edge of a DESERT is said to struggle for life against the drou
0078  arctic regions or on the borders of an utter DESERT, will competition cease. The land may be extr
0212  lsewhere shown, by various animals inhabiting DESERT islands; and we may see an instance of this e
0069  regions, or snow capped summits, or absolute DESERTS, the struggle for life is almost exclusively
0346  ed conditions; the most humid districts, arid DESERTS, lofty mountains, grassy plains, forests, ma
0348  and continuous mountain ranges, and of great DESERTS, and sometimes even of large rivers, we find
0348  erent productions; though as mountain chains, DESERTS, etc., are not as impassable, or likely to h
0408  s and mountains, of the forests, marshes, and DESERTS, are in so mysterious a manner linked togeth
0477  er heat and cold, on mountain and lowland, on DESERTS and marshes, most of the inhabitants within
0025  ts in regard to the colouring of pigeons well DESERVE consideration. The rock pigeon is of a slaty
0049  gly marked varieties or doubtful species well DESERVE consideration; for several interesting lines
0056  rge genera and their recorded varieties which DESERVE notice. We have seen that there is no infall
0251  s, though not made with scientific precision, DESERVE some notice. It is notorious in how complica
0484  the differences be sufficiently important to DESERVE a specific name. This latter point will beco
0062  rk this subject shall be treated, as it well DESERVES, at much greater length. The elder De Candol
0153  tructure undergoing modification. It further DESERVES notice that these variable characters, produ
0236  annot propagate their kind. The subject well DESERVES to be discussed at great length, but I will
0273  ucceeding generations, is a curious fact and DESERVES attention. For it bears on and corroborates
0466  ing many years to doubt their weight. But it DESERVES especial notice that the more important obje
0482  expressions as the plan of creation, unity of DESIGN, etc., and to think that we give an explanati
0111  e of Character. The principle, which I have DESIGNATED by this term, is of high importance on my t
0407  en arrived at by many naturalists under the DESIGNATION of single centres of creation, by some gen
0030  rrelsome; with everlasting layers which never DESIRE to sit, and with the bantam so small and eleg
0020  those individual mongrels, which present any DESIRED character; but that a race could be obtained
0041  a large amount of modification in almost any DESIRED direction. But as variations manifestly usef
0237  l as to the individual, and may thus gain the DESIRED end. Thus, a well flavoured vegetable is coo
0239  , all the neuters ultimately came to have the DESIRED character. On this view we ought occasionall
0446  ly grown up: they are indifferent whether the DESIRED qualities and structures have been acquired
0467  im by nature, and thus accumulate them in any DESIRED manner. He thus adapts animals and plants fo
0073  o not see the cause, we invoke cataclysms to DESOLATE the world, or invent laws on the duration of
0222  rying a pupa; but I was not able to find the DESOLATED nest in the thick heath. The nest, however,
0489  once been broken, and that no cataclysm has DESOLATED the whole world. Hence we may look with som
0138  ructed for twilight; and, last of all, those DESTINED for total darkness. By the time that an anim
0236  , as in the shape of the thorax and in being DESTITUTE of wings and sometimes of eyes, and in inst
0441  e sack, which lives for a short time, and is DESTITUTE of mouth, stomach, or other organs of impor
0074  eat degree on the number of field mice, which DESTROY their combs and nests; and Mr. H. Newman, wh
0074  which I attribute to the number of cats that DESTROY the mice. Hence it is quite credible that 'th
0083  struggle for the females. He does not rigidly DESTROY all inferior animals, but protects during ea
0085  essential it is in a flock of white sheep to DESTROY every lamb with the faintest trace of black.
0098  ffect, that it will invariably and completely DESTROY, as has been shown by Gartner, any influence
0119  is probable, often take the place of, and so DESTROY, the earlier and less improved branches: thi
0126  nature, will constantly tend to supplant and DESTROY the earlier and less improved sub groups. Sm
0202  f the queen bee, which urges her instantly to DESTROY the young queens her daughters as soon as bo
0062  ters, or their eggs, or their nestlings, are DESTROYED by birds and beasts of prey: we do not alwa
0064  lly increases at so high a rate, that if not DESTROYED, the earth would soon be covered by the pro
0066  fully kept up; but if many eggs or young are DESTROYED, many must be produced, or the species will
0066  d years, supposing that this seed were never DESTROYED, and could be ensured to germinate in a fit
0067  cked with other plants. Seedlings, also, are DESTROYED in vast numbers by various enemies; for ins
0067  up, and out of the 357 no less than 295 were DESTROYED, chiefly by slugs and insects. If turf whic
0068  and, and at the same time, if no vermin were DESTROYED, there would, in all probability, be less g
0068  s with the elephant and rhinoceros, none are DESTROYED by beasts of prey: even the tiger in India
0068  ecks. I estimated that the winter of 1854 55 DESTROYED four fifths of the birds in my own grounds;
0074  s that more than two thirds of them are thus DESTROYED all over England. Now the number of mice is
0078  lants or animals range so far, that they are DESTROYED by the rigour of the climate alone. Not unt
0081  the least degree injurious would be rigidly DESTROYED. This preservation of favourable variations
0084  preserving them from danger. Grouse, if not DESTROYED at some period of their lives, would increa
0092  ted, although nine tenths of the pollen were DESTROYED, it might still be a great gain to the plan
0136  st have been blown to sea and thus have been DESTROYED. The insects in Madeira which are not groun
0142  ns so early that a very large proportion are DESTROYED by frost, and then collect seed from the fe
0175  cies, on which it depends, or by which it is DESTROYED, or with it comes into competition; and as
0189  ter be produced, and then the old ones to be DESTROYED. In living bodies, variation will cause the
0195  not that the larger quadrupeds are actually DESTROYED (except in some rare cases) by the flies, b
0215  are then beaten; and if not cured, they are DESTROYED; so that habit, with some degree of selecti
0237  d vegetable is cooked, and the individual is DESTROYED; but the horticulturist sows seeds of the s
0292  s which were then accumulated will have been DESTROYED by being upraised and brought within the li
0300  ising, each fossiliferous formation would be DESTROYED, almost as soon as accumulated, by the ince
0315  st possible, if our fantail pigeons were all DESTROYED, that fanciers, by striving during long age
0315  ail; but if the parent rock pigeon were also DESTROYED, and in nature we have every reason to beli
0362  o england, as the hawks on the English coast DESTROYED so many on their arrival. Some hawks and ow
0062  n insects or seeds, and are thus constantly DESTROYING life; or we forget how largely these songst
0104  inheritance, and through natural selection DESTROYING any which depart from the proper type; but
0034  re passed to prevent their exportation: the DESTRUCTION of horses under a certain size was ordered
0063  produces several eggs or seeds; must suffer DESTRUCTION during some period of its life, and during
0065  , and that there has consequently been less DESTRUCTION of the old and young, and that nearly all
0065  cal tendency to increase must be checked by DESTRUCTION at some period of life. Our familiarity wi
0065  ds, I think, to mislead us: we see no great DESTRUCTION falling on them, and we forget that thousa
0066  ber of eggs or seeds is to make up for much DESTRUCTION at some period of life; and this period in
0066  ggle at some period of its life: that heavy DESTRUCTION inevitably falls either on the young or ol
0067  intervals. Lighten any check, mitigate the DESTRUCTION ever so little, and the number of the spec
```

Page ***(Key Word)**

0067 iably the case. With plants there is a vast DESTRUCTION of seeds, but, from some observations whic
0068 on any large estate depends chiefly on the DESTRUCTION of vermin. If not one head of game were sh
0068 in my own grounds: and this is a tremendous DESTRUCTION, when we remember that ten per cent. is an
0069 t abounds, is constantly suffering enormous DESTRUCTION at some period of its life, from enemies o
0069 compete with our native plants, nor resist DESTRUCTION by our native animals. When a species, owi
0070 gether, and thus save each other from utter DESTRUCTION. I should add that the good effects of fre
0079 s to struggle for life, and to suffer great DESTRUCTION. When we reflect on this struggle, we may
0085 white pigeons, as being the most liable to DESTRUCTION. Hence I can see no reason to doubt that n
0085 . nor ought we to think that the occasional DESTRUCTION of an animal of any particular colour woul
0092 d for the sole object of fertilisation, its DESTRUCTION appears a simple loss to the plant; yet if
0194 s with any favourable variation, and by the DESTRUCTION of those with any unfavourable deviation o
0201 poison fang for its own defence and for the DESTRUCTION of its prey; but some authors suppose that
0365 itants. On almost bare land, with few or no DESTRUCTIVE insects or birds living there, nearly ever
0378 mp with the heat of the tropics which is so DESTRUCTIVE to perennial plants from a temperate clima
0002 o of the necessity of hereafter publishing in DETAIL all the facts, with references, on which my c
0062 of Art. We will now discuss in a little more DETAIL the struggle for existence. In my future work
0173 absolutely distinct from each other in every DETAIL of structure as are specimens taken from the
0199 , against the utilitarian doctrine that every DETAIL of structure has been produced for the good o
0200 sion, correlation of growth, etc. Hence every DETAIL of structure in every living creature (making
0212 se general statements, without facts given in DETAIL, can produce but a feeble effect on the reade
0220 ions which I have myself made, in some little DETAIL. I opened fourteen nests of F. sanguinea, and
0254 ybrids. We will now consider a little more in DETAIL the circumstances and rules governing the ste
0289 their preservation, far better than pages of DETAIL. Nor is their rarity surprising, when we reme
0356 to discuss this branch of the subject in some DETAIL. Changes of level in the Land must also have
0001 may be excused for entering on these personal DETAILS, as I give them to show that I have not been
0008 ts a seed. I cannot here enter on the copious DETAILS which I have collected on this curious subje
0018 rtheless, I may, without here entering on any DETAILS, state that, from geographical and other con
0089 rently weak means: I cannot here enter on the DETAILS necessary to support this view: but if man c
0096 duction? As it is impossible here to enter on DETAILS, I must trust to some general considerations
0156 k it will be admitted, without my entering on DETAILS, that secondary sexual characters are very v
0164 rown animal. Without here entering on further DETAILS, I may state that I have collected cases of
0199 ar kind, but without here entering on copious DETAILS my reasoning would appear frivolous. The for
0211 t be considered as absolutely perfect: but as DETAILS on this and other such points are not indisp
0224 the Hive Bee. I will not here enter on minute DETAILS on this subject, but will merely give an out
0230 od of wax; but I will not here enter on these DETAILS. We see how important a part excavation play
0248 ompare; but I have not space here to enter on DETAILS, the evidence advanced by our best botanists
0445 sured the proportions (but will not here give DETAILS) of the beak, width of mouth, length of nost
0093 thers, in which not a grain of pollen can be DETECTED. Having found a female tree exactly sixty ya
0193 some fundamental difference can generally be DETECTED. I am inclined to believe that in nearly the
0255 hese cases a first trace of fertility may be DETECTED, by the pollen of one of the pure parent spe
0307 d peculiar species. Traces of life have been DETECTED in the longmynd beds beneath Barrande's so c
0419 been done, not because further research has DETECTED important structural differences, at first o
0445 that of the width of mouth, could hardly be DETECTED in the young. But there was one remarkable e
0450 ood authority that ruciments of teeth can be DETECTED in the beaks of certain embryonic birds. Not
0452 r jaws of whales and ruminants, can often be DETECTED in the embryo, but afterwards wholly disappe
0056 ms, naturalists are compelled to come to a DETERMINATION by the amount of difference between them,
0006 ions are of the highest importance, for they DETERMINE the present welfare, and, as I believe, the
0008 er at some particular period of growth, will DETERMINE whether or not the plant sets a seed. I can
0009 but to show how singular the laws are which DETERMINE the reproduction of animals under confineme
0015 ection, as will hereafter be explained, will DETERMINE how far the new characters thus arising sha
0049 have been brought to bear on the attempt to DETERMINE their rank. I will here give only a single
0050 own to him, he is at first much perplexed to DETERMINE what differences to consider as specific, a
0066 , a single one: but this difference does not DETERMINE how many individuals of the two species can
0072 for food. Here we see that cattle absolutely DETERMINE the existence of the Scotch fir; but in sev
0072 ri: but in several parts of the world insects DETERMINE the existence of cattle. Perhaps Paraguay o
0074 animal in large numbers in a district might DETERMINE, through the intervention first of mice and
0114 nying differences of habit and constitution, DETERMINE the inhabitants, which thus jostle eac
0144 ape of the body and the manner of swallowing DETERMINE the position of several of the most importa
0219 they have to migrate, it is the slaves which DETERMINE the migration, and actually carry their mas
0223 latter does not build its own nest, does not DETERMINE its own migrations, does not collect food f
0223 st of the summer extremely few. The masters DETERMINE when and where a new nest shall be formed,
0234 stance determined, as it probably often does DETERMINE, the numbers of a humble bee which could ex
0269 l dissimilarity between two species does not DETERMINE their greater or lesser degree of sterility
0318 the palaeozoic period. No fixed law seems to DETERMINE the length of time during which any single
0405 vary and give rise to new forms will largely DETERMINE their average range. For instance, two vari
0415 hysiological importance of an organ does not DETERMINE its classificatory value, is almost shown b
0467 ossibly survive. A grain in the balance will DETERMINE which individual shall live and which shall
0075 e innumerable plants and animals which have DETERMINED, in the course of centuries, the proportion
0234 et us suppose that this latter circumstance DETERMINED, as it probably often does determine, the n
0281 in, it would not have been possible to have DETERMINED from a mere comparison of their structure w
0414 ht) that those parts of the structure which DETERMINED the habits of life, and the general place o
0068 the serving as prey to other animals, which DETERMINES the average numbers of a species. Thus, the
0403 als, birds, and plants. The principle which DETERMINES the general character of the fauna and flor
0016 rect: but naturalists differ most widely in DETERMINING what characters are of generic value; all
0047 f doubt and conjecture is opened. Hence, in DETERMINING whether a form should be ranked as a speci
0068 its dam. Climate plays an important part in DETERMINING the average numbers of a species, and peri
0074 ally the most potent, but all concurring in DETERMINING the average number or even the existence o
0195 it and the colour of the flesh, which, from DETERMINING the attacks of insects or from being corre
0416 dered by naturalists as important an aid in DETERMINING the degree of affinity of this strange cre
0488 ve been due to secondary causes, like those DETERMINING the birth and death of the individual. Whe
0168 of the structure which can be saved without DETRIMENT to the individual, will be saved. Changes o
0373 tonished at the structure of a vast mound of DETRITUS, about 800 feet in height; crossing a valley
0005 n can be changed and perfected into a highly DEVELOPED being or elaborately constructed organ: sec
0021 ed body, wings, and legs: and its enormously DEVELOPED crop, which it glories in inflating, may we
0025 er tail feathers, sometimes concur perfectly DEVELOPED. Moreover, when two birds belonging to two
0039 a fantail, till he saw a pigeon with a tail DEVELOPED in some slight degree in an unusual manner,

Page **************************************(Key Word)**************************************

```
0126  that many groups, formerly most extensively  DEVELOPED, have now become extinct. Looking still mor
0131  d lowly organised structures variable. Parts  DEVELOPED in an unusual manner are highly variable: s
0134  s, however slight, until they become plainly  DEVELOPED and appreciable by us. Effects of Use and D
0136  gs having been ever so little less perfectly  DEVELOPED or from indolent habit, will have had the b
0147  ts, on the one hand, of a part being largely  DEVELOPED through natural selection and another and a
0148  ant anterior segments of the head enormously  DEVELOPED, and furnished with great nerves and muscle
0148  means causing some other part to be largely  DEVELOPED in a corresponding degree. And, conversely,
0150  se, and to the tendency to reversion. A part  DEVELOPED in any species in an extraordinary degree o
0150  unusually developed, unless it be unusually  DEVELOPED, unless it be unusually developed in compar
0150  means applies to any part, however unusually  DEVELOPED in comparison with the same part in closely
0150  nly if some one species of bat had its wings  DEVELOPED in some remarkable manner in comparison wit
0151  f variability. When we see any part or organ  DEVELOPED in a remarkable degree or manner in any spe
0153  let us turn to nature. When a part has been  DEVELOPED in an extraordinary manner in any one speci
0153  ut as the variability of the extraordinarily  DEVELOPED part or organ has been so great and long co
0154  of time cease: and that the most abnormally  DEVELOPED organs may be made constant, I can see no r
0158  ent extreme variability of any part which is  DEVELOPED in a species in an extraordinary manner in
0158  in a part, however extraordinarily it may be  DEVELOPED, if it be common to a whole group of specie
0168  age will generally affect parts subsequently  DEVELOPED; and there are very many other correlations
0168  and internal parts. When one part is largely  DEVELOPED, perhaps it tends to draw nourishment from
0169  species of the same genus. Any part or organ  DEVELOPED to an extraordinary size or in an extraordi
0169  but when a species with any extraordinarily  DEVELOPED organ has become the parent of many modifie
0169  n however extraordinary a manner it may be  DEVELOPED. Species inheriting nearly the same constit
0183  fferent habits of life will rarely have been  DEVELOPED at an early period in great numbers and und
0183  ishes capable of true flight would have been  DEVELOPED under many subordinate forms, for taking pr
0183  bers, than in the case of species with fully  DEVELOPED structures. I will now give two or three in
0189  all the many members of the class have been  DEVELOPED; and in order to discover the early transit
0196  n, may perhaps be thus accounted for. A well  DEVELOPED tail having been formed in an aquatic anima
0223  pupae originally stored as food might become  DEVELOPED; and the ants thus unintentionally reared w
0239  f another caste, and they have an enormously  DEVELOPED abdomen which secretes a sort of honey, sup
0240  male and female ants of this genus have well  DEVELOPED ocelli. I may give one other case: so confi
0246  n the second case they are either not at all  DEVELOPED, or are imperfectly developed. This distinc
0246  her not at all developed, or are imperfectly  DEVELOPED. This distinction is important, when the ca
0264  but be incapable of causing an embryo to be  DEVELOPED, as seems to have been the case with some o
0264  grafted on others. Lastly, an embryo may be  DEVELOPED, and then perish at an early period. This l
0264  in which the sexual elements are imperfectly  DEVELOPED, the case is very different. I have more th
0290  marine fauna, tertiary beds are so scantily  DEVELOPED, that no record of several successive and p
0304  uded that this great group had been suddenly  DEVELOPED at the commencement of the tertiary series.
0305  onfined range, and after having been largely  DEVELOPED in some one sea, might have spread widely.
0321  ent species: and if many new forms have been  DEVELOPED from any one species, the nearest allies of
0321  s extermination: and if many allied forms be  DEVELOPED from the successful intruder, many will hav
0336  cussion whether recent forms are more highly  DEVELOPED than ancient. I will not here enter on this
0392  ccessfully competing in stature with a fully  DEVELOPED tree, when established on an island and hav
0396  ders of plants, herbaceous forms having been  DEVELOPED into trees, etc., seem to me to accord bett
0408  eatly, and some only slightly modified, some  DEVELOPED in great force, some existing in scanty num
0441  on at this stage is, to search by their well  DEVELOPED organs of sense, and to reach by their acti
0441  dition. But in some genera the larvae become  DEVELOPED either into hermaphrodites having the ordin
0443  ion than the mature animal, into which it is  DEVELOPED. I believe that all these facts can be expl
0451  etain their potentiality, and are merely not  DEVELOPED: this seems to be the case with the mammae
0451  on record of these organs having become well  DEVELOPED in full grown males, and having secreted mi
0451  reted milk. So again there are normally four  DEVELOPED and two rudimentary teats in the udders of
0451  n our domestic cows the two sometimes become  DEVELOPED and give milk. In individual plants of the
0451  r as mere rudiments, and sometimes in a well  DEVELOPED state. In plants with separated sexes, the
0452  ed with a stigma: but the style remains well  DEVELOPED and is clothed with hairs as in other compo
0474  licable on the theory of creation why a part  DEVELOPED in a very unusual manner in any one species
0474  ally to be still variable. But a part may be  DEVELOPED in the most unusual manner, like the wing o
0479  e air dissolved in water, by the aid of well  DEVELOPED branchiae. Disuse, aided sometimes by natur
0480  er jaw, from an early progenitor having well  DEVELOPED teeth: and we may believe, that the teeth i
0483  an early progenitor had the organ in a fully  DEVELOPED state: and this in some instances necessari
0148  g itself, by less nutriment being wasted in  DEVELOPING a structure now become useless. Thus, as I
0148  ction may perfectly well succeed in largely  DEVELOPING any organ, without requiring as a necessary
0008  whether during the early or late period of  DEVELOPMENT of the embryo, or at the instant of concep
0011  an its wild parent. The great and inherited  DEVELOPMENT of the udders in cows and goats in countri
0021  bird, is also remarkable from the wonderful  DEVELOPMENT of the carunculated skin about the head, a
0022  in the skeletons of the several breeds, the  DEVELOPMENT of the bones of the face in length and bre
0022  nd of the upper part of the oesophagus, the  DEVELOPMENT and abortion of the oil gland: the number
0022  ti the number of scutellae on the toes, the  DEVELOPMENT of skin between the toes, are all points o
0116  ication in an early and incomplete stage of  DEVELOPMENT. After the foregoing discussion, which oug
0143  n is so tied together during its growth and  DEVELOPMENT, that when slight variations in any one pa
0145  differ. It might have been thought that the  DEVELOPMENT of the ray petals by drawing nourishment f
0148  ful, any diminution, however slight, in its  DEVELOPMENT, will be seized on by natural selection, f
0266  e, without some disturbance occuring in the  DEVELOPMENT, or periodical action, or mutual relation
0302  fication through natural selection. For the  DEVELOPMENT of a group of forms, all of which have des
0312  ther and to living species. On the state of  DEVELOPMENT of ancient forms. On the succession of the
0314  ith my theory. I believe in no fixed law of  DEVELOPMENT, causing all the inhabitants of a country
0322  by sudden immigration or by unusually rapid  DEVELOPMENT, many species of a new group have taken po
0336  s right to expect to find. On the state of  DEVELOPMENT of Ancient Forms. There has been much disc
0338  n some degree parallel to the embryological  DEVELOPMENT of recent forms. I must follow Pictet and
0351  in the last chapter, in no law of necessary  DEVELOPMENT. As the variability of each species is an
0410  ups of species have their points of maximum  DEVELOPMENT. Groups of species, belonging either to a
0419  ic leaves or cotyledons, and on the mode of  DEVELOPMENT of the plumule and radicle. In our discuss
0441  istinguishable. The embryo in the course of  DEVELOPMENT generally rises in organisation: I use thi
0441  complemental males: and in the latter, the  DEVELOPMENT has assuredly been retrograde: for the mal
0442  l pass through a similar worm like stage of  DEVELOPMENT; but in some few cases, as in that of Aphi
0442  mirable drawings by Professor Huxley of the  DEVELOPMENT of this insect, we see no trace of the ver
0448  form. We have seen that this is the rule of  DEVELOPMENT in certain whole groups of animals, as wit
0448  or their own wants at a very early stage of  DEVELOPMENT, and secondly from their following exactly
```

Page **************************************(Key Word)**************************************

developed

```
0448 become correlated with successive stages of DEVELOPMENT; so that the Larvae, in the first stage, m
0452 pecies are very liable to vary in degree of DEVELOPMENT and in other respects. Moreover, in closel
0032 l up the rogues, as they call the plants that DEVIATE from the proper standard. With animals this
0036 subject that the owner of either of them has DEVIATED in any one instance from the pure blood of M
0013 inciple by theoretical writers alone. When a DEVIATION appears not unfrequently, and we see it in
0013 xposed to the same conditions, any very rare DEVIATION, due to some extraordinary combination of c
0038 ly select, or only with much difficulty, any DEVIATION of structure excepting such as is externall
0039 breed. Nor let it be thought that some great DEVIATION of structure would be necessary to catch th
0040 d breeds from an individual with some slight DEVIATION of structure, or takes more care than usual
0041 tention should be paid to even the slightest DEVIATION in the qualities or structure of each indiv
0044 trosity I presume is meant some considerable DEVIATION of structure in one part, either injurious
0094 an see no reason to doubt that an accidental DEVIATION in the size and form of the body, or in the
0094 bably inherit a tendency to a similar slight DEVIATION of structure. The tubes of the corollas of
0132 sure that there must be some cause for each DEVIATION of structure, however slight. How much dire
0149 hould have preserved or rejected each little DEVIATION of form less carefully than when the part h
0194 e destruction of those with any unfavourable DEVIATION of structure, I have sometimes felt much di
0459 ading to the preservation of each profitable DEVIATION of structure or instinct. The truth of thes
0012 but the number and diversity of inheritable DEVIATIONS of structure, both those of slight and thos
0013 ers of the same family. If strange and rare DEVIATIONS of structure are truly inherited, less stra
0013 truly inherited, less strange and commoner DEVIATIONS may be freely admitted to be inheritable. P
0015 ght check, by blending together, any slight DEVIATIONS of structure, in such case, I grant that we
0039 st pigeons, which are rejected as faults or DEVIATIONS from the standard of perfection of each bre
0095 s presenting mutual and slightly favourable DEVIATIONS of structure. I am well aware that this doc
0131 duce individual differences, or very slight DEVIATIONS of structure, as to make the child like its
0131 than under nature, leads me to believe that DEVIATIONS of structure are in some way due to the nat
0149 natural selection having no power to check DEVIATIONS in their structure. Thus rudimentary parts
0196 very slight use; and any actually injurious DEVIATIONS in their structure will always have been ch
0209 he same unknown causes which produce slight DEVIATIONS of bodily structure. No complex instinct ca
0334 mely, the manner in which the fossils of the DEVONIAN system, when this system was first discovere
0004 turalists. From these considerations, I shall DEVOTE the first chapter of this Abstract to Variati
0031 cessary to read several of the many treatises DEVOTED to this subject, and to inspect the animals.
0246 observers, Kolreuter and Gartner, who almost DEVOTED their lives to this subject, without being d
0032 es, and he studies his subject for years, and DEVOTES his lifetime to it with indomitable persever
0063 it bearing plants, in order to tempt birds to DEVOUR and thus disseminate its seeds rather than th
0182 lusively, as far as we know, to escape being DEVOURED by other fish? When we see any structure hig
0211 impid drop of sweet juice, which was eagerly DEVOURED by the ant. Even the quite young aphides beh
0361 the sea, sometimes escape being immediately DEVOURED; and seeds of many kinds in the crops of flo
0362 ion of seeds; now after a bird has found and DEVOURED a large supply of food, it is positively ass
0362 y land and water plants: fish are frequently DEVOURED by birds, and thus the seeds might be transp
0036 ns of Tierra del Fuego, by their killing and DEVOURING their old women, in times of dearth, as of
0092 asionally and then habitually, by the pollen DEVOURING insects from flower to flower, and a cross
0387 s, century after century, have gone on daily DEVOURING fish; they then take flight and go to other
0136 of wingless beetles is larger on the exposed DEZERTAS than in Madeira itself; and especially the e
0116 xtinction, will tend to act. The accompanying DIAGRAM will aid us in understanding this rather per
0116 the case in nature, and is represented in the DIAGRAM by the letters standing at unequal distances
0117 intervals between the horizontal lines in the DIAGRAM, may represent each a thousand generations;
0118 this interval, variety a1 is supposed in the DIAGRAM to have produced variety a2, which will, owi
0118 in number and diverging in character. In the DIAGRAM the process is represented up to the ten tho
0118 goes on so regularly as is represented in the DIAGRAM, though in itself made somewhat irregular. I
0119 ir modified progeny will be increased. In our DIAGRAM the line of succession is broken at regular
0119 ging in character: this is represented in the DIAGRAM by the several divergent branches proceeding
0119 improved branches: this is represented in the DIAGRAM by some of the lower branches not reaching t
0120 ations. This case would be represented in the DIAGRAM, if all the lines proceeding from (A) were r
0120 of change between each horizontal line in our DIAGRAM to be excessively small, these three forms m
0120 ree forms into well defined species: thus the DIAGRAM illustrates the steps by which the small dif
0120 reater number of generations (as shown in the DIAGRAM in a condensed and simplified manner), we ge
0120 that more than one species would vary. In the DIAGRAM I have assumed that a second species (I) has
0121 places in the polity of nature: hence in the DIAGRAM I have chosen the extreme species (A), and t
0121 altered descendants; and this is shown in the DIAGRAM by the dotted lines not prolonged far upward
0121 e process of modification, represented in the DIAGRAM, another of our principles, namely that of e
0122 tion, both may continue to exist. If then our DIAGRAM be assumed to represent a considerable amoun
0123 late stage of descent. The new species in our DIAGRAM descended from the original eleven species,
0124 some one species of an earlier genus. In our DIAGRAM, this is indicated by the broken lines, bene
0124 bring some such case before his mind. In the DIAGRAM, each horizontal line has hitherto been supp
0124 bject, and I think we shall then see that the DIAGRAM throws light on the affinities of extinct be
0125 to the formation of genera alone. If, in our DIAGRAM, we suppose the amount of change represented
0125 odification supposed to be represented in the DIAGRAM. And the two new families, or orders, will h
0129 character, as we have seen illustrated in the DIAGRAM. The affinities of all the beings of the sam
0331 lex, I must request the reader to turn to the DIAGRAM in the fourth chapter. We may suppose that t
0331 ging from them the species in each genus. The DIAGRAM is much too simple, too few genera and too f
0331 acter, which was formerly illustrated by this DIAGRAM, the more recent any form is, the more it wi
0332 l characteristics. This is represented in the DIAGRAM by the letter F14. All the many forms, extin
0332 endured to the present day. By looking at the DIAGRAM we can see that if many of the extinct forms
0333 r more complicated than is represented in the DIAGRAM; for the groups will have been more numerous
0334 ed at the sixth great stage of descent in the DIAGRAM are the modified offspring of those which li
0412 ng forms. I request the reader to turn to the DIAGRAM illustrating the action, as formerly explain
0412 up into groups subordinate to groups. In the DIAGRAM each letter on the uppermost line may repres
0420 he will take the trouble of referring to the DIAGRAM in the fourth chapter. We will suppose the l
0422 shown, as far as is possible on paper, in the DIAGRAM, but in much too simple a manner. If a branc
0422 t in much too simple a manner. If a branching DIAGRAM had not been used, and only the names of the
0431 ity of various lengths (as may be seen in the DIAGRAM so often referred to), mounting up through m
0431 perienced in describing, without the aid of a DIAGRAM, the various affinities which they perceive
0432 possible. We shall see this by turning to the DIAGRAM: the letters, A to L, may represent eleven S
0432 rogenitor. Yet the natural arrangement in the DIAGRAM would still hold good; and, on the principle
0040 omestic breeds. But, in fact, a breed, like a DIALECT of a language, can hardly be said to have ha
0310 d imperfectly kept, and written in a changing DIALECT; of this history we possess the last volume
```

Page **(Key Word)**

```
0422 es, and all intermediate and slowly changing DIALECTS, had to be included, such an arrangement wou
0422 and had given rise to many new languages and DIALECTS. The various degrees of difference in the la
0228 true or parts of a sphere, and of about the DIAMETER of a cell. It was most interesting to me to
0228 l), and were in depth about one sixth of the DIAMETER of the sphere of which they formed a part, t
0248 reuter and Gartner, should have arrived at DIAMETRICALLY opposite conclusions in regard to the ver
0276 ntalists who have ever lived, have come to DIAMETRICALLY opposite conclusions in ranking forms by
0256 he limits of the same genus, for instance in DIANTHUS, these two opposite cases occur. The fertili
0257 in the same family there may be a genus, as DIANTHUS, in which very many species can most readily
0397 ls, when hybernating and having a membranous DIAPHRAGM over the mouth of the shell, might be float
0361 wood in an oak about 50 years old, three DICOTYLEDONOUS plants germinated: I am certain of the ac
0063 tes grow on the same tree, it will languish and DIE. But several seedling misstletoes, growing clos
0244 y, vary, let the strongest live and the weakest DIE. Chapter VIII. Hybridism. Distinction between
0467 ine which individual shall live and which shall DIE, which variety or species shall increase in nu
0320 ick man dies, to wonder and to suspect that he DIED by some unknown deed of violence. The theory o
0341 other seven species of the old genera have all DIED out and have left no progeny. Or, which would
0359 the several Atlantic currents is 33 miles per DIEM (some currents running at the rate of 60 miles
0360 e currents running at the rate of 60 miles per DIEM); on this average, the seeds of 14/100 plants
0320 no surprise at sickness, but when the sick man DIES, to wonder and to suspect that he died by some
0007 nts which strikes us, is, that they generally DIFFER much more from each other, than do the indivi
0010 , and the young of the same litter, sometimes DIFFER considerably from each other, though both the
0012 tion in which the varieties and sub varieties DIFFER slightly from each other. The whole organisat
0016 nus, in several trifling respects, they often DIFFER in an extreme degree in some one part, both w
0016 iscussed), domestic races of the same species DIFFER from each other in the same manner as, only i
0016 often been stated that domestic races do not DIFFER from each other in characters of. generic valu
0016 statement is hardly correct: but naturalists DIFFER most widely in determining what characters ar
0019 ard to ducks and rabbits, the breeds of which DIFFER considerably from each other in structure, I
0022 in breeds, the males and females have come to DIFFER to a slight degree from each other. Altogethe
0023 eral geographical races or sub species, which DIFFER from each other in the most trifling respects
0033 fruit of the different kinds of gooseberries DIFFER in size, colour, shape, and hairiness, and ye
0033 fferences. It is not that the varieties which DIFFER largely in some one point do not differ at al
0033 which differ largely in some one point do not DIFFER at all in other points: this is hardly ever,
0036 hich they have insensibly passed, and come to DIFFER so greatly from the rock pigeon. Youatt gives
0039 nce the Thoulouse and the common breed, which DIFFER only in colour, that most fleeting of charact
0048 nd insects in North America and Europe, which DIFFER very slightly from each other, have been rank
0049 p, or Primula veris and elatior. These plants DIFFER considerably in appearance; they have a diffe
0052 each other, and that it does not essentially DIFFER from the term variety, which is given to less
0057 a certain extent resemble varieties, for they DIFFER from each other by a less than usual amount o
0061 stinct species, which in most cases obviously DIFFER from each other far more than do the varietie
0061 te what are called distinct genera, and which DIFFER from each other more than do the species of t
0089 mal have the same general habits of life, but DIFFER in structure, colour, or ornament, such diffe
0094 ncarnatum) do not on a hasty glance appear to DIFFER in length: yet the hive bee can easily suck t
0111 in many cases how to rank them, yet certainly DIFFER from each other far less than do good and dis
0111 as we may call it, might cause one variety to DIFFER in some character from its parents, and the o
0111 s, and the offspring of this variety again to DIFFER from its parent in the very same character an
0113 ner as distinct species and genera of grasses DIFFER from each other, a greater number of individu
0115 ants are of a highly diversified nature. They DIFFER, moreover, to a large extent from the indigen
0118 h will, owing to the principle of divergence, DIFFER more from (A) than did variety al. Variety m1
0120 the successive generations, will have come to DIFFER largely, but perhaps unequally, from each oth
0123 cendants from (I) will, owing to inheritance, DIFFER considerably from the eight descendants from
0125 rom inheritance from a different parent, will DIFFER widely from the three genera descended from (
0132 in its earliest condition does not apparently DIFFER essentially from an ovule, is alone affected.
0145 t, in some Compositous plants, the seeds also DIFFER in shape and sculpture; and even the ovary it
0145 t the inner and outer flowers most frequently DIFFER. It might have been thought that the developm
0151 the word, very important structures, and they DIFFER extremely little even in different genera: bu
0151 s no exaggeration to state that the varieties DIFFER more from each other in the characters of the
0155 ly recent period, and which have thus come to DIFFER. Or to state the case in another manner: the
0155 genus resemble each other, and in which they DIFFER from the species of some other genus, are cal
0156 , and subsequently have not varied or come to DIFFER in any degree, or only in a slight degree, it
0156 n the other hand, the points in which species DIFFER from other species of the same genus, are cal
0156 e specific characters have varied and come to DIFFER within the period of the branching off of the
0156 ll be admitted that species of the same group DIFFER from each other more widely in their secondar
0157 which the different species of the same genus DIFFER from each other. Of this fact I will give in
0168 s, that is, the characters which have come to DIFFER since the several species of the same genus b
0168 se they have recently varied and thus come to DIFFER; but we have also seen in the second Chapter
0169 ters are highly variable, and such characters DIFFER much in the species of the same group. Variab
0175 rarer: then, as varieties do not essentially DIFFER from species, the same rule will probably app
0199 horns are correlated. Mountain breeds always DIFFER from lowland breeds; and a mountainous countr
0213 the cabbage, I cannot see that these actions DIFFER essentially from true instincts. If we were t
0230 walls are commenced. Some of these statements DIFFER from those made by the justly celebrated elde
0236 n irsect communities: for these neuters often DIFFER widely in instinct and in structure from both
0238 ly, the fact that the neuters of several ants DIFFER, not only from the fertile females and males,
0239 prisingly the neuters of several British ants DIFFER from each other in size and sometimes in colo
0256 r their degree of fertility is often found to DIFFER greatly in the several individuals raised fro
0256 re of high physiological importance and which DIFFER little in the allied species. Now the fertili
0258 the father and then as the mother, generally DIFFER in fertility in a small, and occasionally in
0260 eover, produced from reciprocal crosses often DIFFER in fertility. Now do these complex and singul
0268 inasmuch as varieties, however much they may DIFFER from each other in external appearance, cross
0274 ssed with another species, the hybrids do not DIFFER much. But this conclusion, as far as I can ma
0275 parents are the same, whether the two parents DIFFER much or little from each other, namely in the
0281 erably from both, even perhaps more than they DIFFER from each other. Hence in all such cases, we
0301 eturned to their ancient homes, as they would DIFFER from their former state, in a nearly uniform,
0306 imals, as the Nautilus, Lingula, etc., do not DIFFER much from living species; and it cannot on my
0310 bts on this subject. I feel how rash it is to DIFFER from these great authorities; to whom, with o
0314 oleopterous insects of Madeira having come to DIFFER considerably from their nearest allies on the
0328 o the same genera, yet the species themselves DIFFER in a manner very difficult to account for, co
0331 ecent any form is, the more it will generally DIFFER from its ancient progenitor. Hence we can und
```

Page **(Key Word)**

```
0331  rstand the rule that the most ancient fossils  DIFFER  most from existing forms. We must not, howeve
0333  se the existing genera of the two families to  DIFFER  from each other by a dozen characters, in thi
0333  genera, at the early period marked VI., would  DIFFER  by a lesser number of characters: for at this
0333   each other; so that the older members should  DIFFER  less from each other in some of their charact
0340  ent, so will their modified descendants still  DIFFER  in nearly the same manner and degree. But aft
0349  off the American shore, however much they may  DIFFER  in geological structure, the inhabitants, tho
0355  s in one region to those in another, does not  DIFFER  much (by substituting the word variety for sp
0356  lustrate what I mean: our English race horses  DIFFER  slightly from the horses of every other breed
0410  ughout the world, we find that some organisms  DIFFER  little, whilst others belonging to a differen
0410  only to a different family of the same order,  DIFFER  greatly. In both time and space the lower mem
0415  r work he says, the genera of the Connaraceae  DIFFER  in having one or more ovaria, in the existenc
0416  nstant in structure: in another division they  DIFFER  much, and the differences are of quite subord
0420  gree in blood to their common progenitor, may  DIFFER  greatly, being due to the different degrees o
0421  ousins to the same millionth degree; yet they  DIFFER  widely and in different degrees from each oth
0423  in tumbler pigeons, though some sub varieties  DIFFER  from the others in the important character of
0424   under the Negro group, however much he might  DIFFER  in colour and other important characters from
0424  two sexes, and how enormously these sometimes  DIFFER  in the most important characters, is known to
0424  of the same individual, however much they may  DIFFER  from each other and from the adult; as he lik
0426  her group or section of a group, are found to  DIFFER  much, we at once value them less in our class
0440  rripedes, the pedunculated and sessile, which  DIFFER  widely in external appearance, have larvae in
0442  her groups, the embryo does not at any period  DIFFER  widely from the adult: thus Owen has remarked
0445  mblers. Now some of these birds, when mature,  DIFFER  so extraordinarily in length and form of beak
0447  al new species, the embryonic fore limbs will  DIFFER  greatly from the fore limbs in the mature and
0448  so that the larvae, in the first stage, might  DIFFER  greatly from the larvae in the second stage,
0449  s of animal, however much they may at present  DIFFER  from each other in structure and habits, if t
0469  d species. Let it be observed how naturalists  DIFFER  in the rank which they assign to the many rep
0470  n degree the character of varieties; for they  DIFFER  from each other by a less amount of differenc
0473   those by which the species of the same genus  DIFFER  from each other, be more variable than the ge
0113  se varieties were continually selected which  DIFFERED  from each other in at all the same manner as
0114  ht orders, which shows how much these plants  DIFFERED  from each other. So it is with the plants an
0123  even genera. But as the original species (I)  DIFFERED  largely from (A), standing nearly at the ext
0155  tant throughout large groups of species, has  DIFFERED  considerably in closely allied species, that
0168  which have long been inherited, and have not  DIFFERED  within this same period. In these remarks we
0241  er, of the working ants of the several sizes  DIFFERED  wonderfully in shape, and in the form and nu
0280  itor; and the progenitor will generally have  DIFFERED  in some respects from all its modified desce
0281  se; but in some points of structure may have  DIFFERED  considerably from both, even perhaps more th
0306   before the Silurian age, and which probably  DIFFERED  greatly from any known animal. Some of the m
0340  if the inhabitants of one continent formerly  DIFFERED  greatly from those of another continent, so
0445  e I was curious to see how far their puppies  DIFFERED  from each other: I was told by breeders that
0445  each other: I was told by breeders that they  DIFFERED  just as much as their parents, and this, jud
0445  told that the foals of cart and race horses  DIFFERED  as much as the full grown animals: and this
0446  le, for the young of the short faced tumbler  DIFFERED  from the young of the wild rock pigeon and o
0010  st physiologists that there is no essential  DIFFERENCE  between a bud and an ovule in their earlies
0016  ttempt to estimate the amount of structural  DIFFERENCE  between the domestic races of the same spec
0021   short faced tumbler, and see the wonderful  DIFFERENCE  in their beaks, entailing corresponding dif
0036   blood of Mr. Bakewell's flock, and yet the  DIFFERENCE  between the sheep possessed by these two ge
0048  cies, whereas Mr. Bentham gives only 112, a  DIFFERENCE  of 13° doubtful forms! Amongst animals whic
0051  ncier before alluded to, with the amount of  DIFFERENCE  in the forms which he is continually studyi
0052  d to species. The passage from one stage of  DIFFERENCE  to another and higher stage may be, in some
0056  to come to a determination by the amount of  DIFFERENCE  between them, judging by analogy whether or
0056  to the rank of species. Hence the amount of  DIFFERENCE  is one very important criterion in settling
0057  insects, that in large genera the amount of  DIFFERENCE  between the species is often exceedingly sm
0057  m each other by a less than usual amount of  DIFFERENCE. Moreover, the species of the large genera
0057  btedly there is one most-important point of  DIFFERENCE  between varieties and species; namely, that
0057  ies and species; namely, that the amount of  DIFFERENCE  between varieties, when compared with each
0059  nd except, secondly, by a certain amount of  DIFFERENCE, for two forms, if differing very little, a
0059  have not been discovered; but the amount of  DIFFERENCE  considered necessary to give to two forms t
0065  ld have somehow to be disposed of. The only  DIFFERENCE  between organisms which annually produce eg
0066  like the hippobosca, a single one; but this  DIFFERENCE  does not determine how many individuals of
0083  nal organ, or every shade of constitutional  DIFFERENCE, on the whole machinery of life. Man select
0093  useful to him. Under nature, the slightest  DIFFERENCE  of structure or constitution may well turn
0085  art, these slight differences make a great  DIFFERENCE  in cultivating the several varieties, assur
0095  succeed. In looking at many small points of  DIFFERENCE  between species, which, as far as our ignor
0101  ally intercross with other individuals, the  DIFFERENCE  between hermaphrodites and unisexual specie
0111  cipient species. How, then, does the lesser  DIFFERENCE  between varieties become augmented into the
0111  varieties become augmented into the greater  DIFFERENCE  between species? That this does habitually
0111  ount for so habitual and large an amount of  DIFFERENCE  as that between varieties of the same speci
0123  of natural selection, the extreme amount of  DIFFERENCE  in character between species a14 and z14 wi
0132  ure, however slight. How much direct effect  DIFFERENCE  of climate, food, etc., produces on any bei
0133  ve lived; but who can tell how much of this  DIFFERENCE  may be due to the warmest clad individuals
0144  ore, of natural selection, than that of the  DIFFERENCE  between the outer and inner flowers in some
0145  d umbelliferous plants. Every one knows the  DIFFERENCE  in the ray and central florets of, for inst
0145  orets of, for instance, the daisy, and this  DIFFERENCE  is often accompanied with the abortion of p
0145  abortion; but in some Compositae there is a  DIFFERENCE  in the seeds of the outer and inner florets
0145   of the outer and inner florets without any  DIFFERENCE  in the corolla. Possibly, these several dif
0145  eral differences may be connected with some  DIFFERENCE  in the flow of nutriment towards the centra
0145  is only much shortened. With respect to the  DIFFERENCE  in the corolla of the central and exterior
0152  the pigeon; see what a prodigious amount of  DIFFERENCE  there is in the beak of the different tumbl
0156  ation; compare, for instance, the amount of  DIFFERENCE  between the males of gallinaceous birds, in
0156  are strongly displayed, with the amount of  DIFFERENCE  between their females; and the truth of thi
0157  ecies of the same group a greater amount of  DIFFERENCE  in their sexual characters, than in other p
0158  sexual characters, and the great amount of  DIFFERENCE  in these same characters between closely al
0170  e, whatever the cause may be of each slight  DIFFERENCE  in the offspring from their parents, and a
0193  organ may be the same, yet some fundamental  DIFFERENCE  can generally be detected. I am inclined to
0221  cipal office is to search for aphides. This  DIFFERENCE  in the usual habits of the masters and slav
0236  instinct alone is concerned, the prodigious  DIFFERENCE  in this respect between the workers and the
```

Page ***(Key Word)***

```
0238 n repeated, until that prodigious amount of DIFFERENCE between the fertile and sterile females of
0240 will perhaps best appreciate the amount of DIFFERENCE in these workers, by my giving not the actu
0240 , but a strictly accurate illustration: the DIFFERENCE was the same as if we were to see a set of
0250 f the very same species as did Gartner. The DIFFERENCE in their results may, I think, be in part a
0257 s of the same genus, we meet with this same DIFFERENCE; for instance, the many species of Nicotian
0257 to point out what kind, or what amount, of DIFFERENCE in any recognisable character is sufficient
0258 crossed. There is often the widest possible DIFFERENCE in the facility of making reciprocal crosse
0258 systematic affinity, or of any recognisable DIFFERENCE in their whole organisation. On the other h
0258 d confined to the reproductive system. This DIFFERENCE in the result of reciprocal crosses between
0258 or fuci. Gartner, moreover, found that this DIFFERENCE of facility in making reciprocal crosses is
0260 ther or the mother, there is generally some DIFFERENCE, and occasionally the widest possible diffe
0260 rence, and occasionally the widest possible DIFFERENCE, in the facility of effecting an union. The
0260 brids? Why should there often be so great a DIFFERENCE in the result of a reciprocal cross between
0262 er found that there was sometimes an innate DIFFERENCE in different individuals of the same two sp
0262 t although there is a clear and fundamental DIFFERENCE between the mere adhesion of grafted stocks
0269 scendants from any one species. Seeing this DIFFERENCE in the process of selection, as carried on
0269 nd nature, we need not be surprised at some DIFFERENCE in the result. I have as yet spoken as if t
0271 ese varieties of Verbascum present no other DIFFERENCE besides the mere colour of the flower; and
0272 y distinct species; and this shows that the DIFFERENCE in the degree of variability graduates away
0273 ut this, if it be true, is certainly only a DIFFERENCE in degree. Gartner further insists that whe
0277 ct causes; for both depend on the amount of DIFFERENCE of some kind between the species which are
0297 they meet with a somewhat greater amount of DIFFERENCE between any two forms, they rank both as sp
0328 evertheless he finds a surprising amount of DIFFERENCE in the species. If the several formations i
0330 lves by fine gradations the apparently wide DIFFERENCE between the pig and the camel. In regard to
0338 se of successive generations, more and more DIFFERENCE to the adult. Thus the embryo comes to be l
0347 f various regions. We see this in the great DIFFERENCE of nearly all the terrestrial productions o
0347 ductions. We see the same fact in the great DIFFERENCE between the inhabitants of Australia, Afric
0356 very other breed; but they do not owe their DIFFERENCE and superiority to descent from any single
0358 al facts in distribution, such as the great DIFFERENCE in the marine faunas on the opposite sides
0359 ed for a long time. It is well known what a DIFFERENCE there is in the buoyancy of green and seaso
0367 tle earlier or later; but this will make no DIFFERENCE in the final result. As the warmth returned
0384 sting land and fresh water shells. The wide DIFFERENCE of the fish on opposite sides of continuous
0389 t that something quite independently of any DIFFERENCE in physical conditions has caused so great
0389 n physical conditions has caused so great a DIFFERENCE in number. Even the uniform county of Cambr
0390 n of the same class, are peculiar; and this DIFFERENCE seems to depend on the species which do not
0398 chipelagos: but what an entire and absolute DIFFERENCE in their inhabitants! The inhabitants of th
0401 arrival), we find a considerable amount of DIFFERENCE in the several islands. This difference mig
0401 of cifference in the several islands. This DIFFERENCE might indeed have been expected on the view
0419 species, with slightly different grades of DIFFERENCE, have been subsequently discovered. All the
0420 order to be natural; but that the amount of DIFFERENCE in the several branches or groups, though a
0422 guages and dialects. The various degrees of DIFFERENCE in the languages from the same stock, would
0423 mestic productions, several other grades of DIFFERENCE are requisite, as we have seen with pigeons
0433 ical in its arrangement, with the grades of DIFFERENCE between the descendants from a common paren
0442 namely the very general, but not universal DIFFERENCE in structure between the embryo and the adu
0445 acquired their full amount of proportional DIFFERENCE. So, again, I was told that the foals of ca
0445 me greatly, as I think it probable that the DIFFERENCE between these two breeds has been wholly ca
0445 acquired their full amount of proportional DIFFERENCE. As the evidence appears to me conclusive,
0445 grown birds. Some characteristic points of DIFFERENCE, for instance, that of the width of mouth,
0456 ed arrangement, with the grades of acquired DIFFERENCE marked by the terms varieties, species, gen
0460 ee the truth of this conclusion in the vast DIFFERENCE in the result, when the same two species ar
0470 differ from each other by a less amount of DIFFERENCE than do the species of smaller genera. The
0485 ly and to value higher the actual amount of DIFFERENCE between them. It is quite possible that for
0007 ne or more Species. Domestic Pigeons, their DIFFERENCES and Origin. Principle of Selection ancient
0016 right to expect often to meet with generic DIFFERENCES in our domesticated productions. When we a
0021 nce in their beaks, entailing corresponding DIFFERENCES in their skulls. The carrier, more especia
0023 lled them, could be shown him. Great as the DIFFERENCES are between the breeds of pigeons, I am fu
0029 study they are strongly impressed with the DIFFERENCES between the several races; and though they
0029 y win their prizes by selecting such slight DIFFERENCES, yet they ignore all general arguments, an
0029 and refuse to sum up in their minds slight DIFFERENCES accumulated during many successive generat
0029 who would account by such agencies for the DIFFERENCES of a dray and cart horse, a greyhound and
0032 irection, during successive generations, of DIFFERENCES absolutely inappreciable by an uneducated
0032 olutely inappreciable by an uneducated eye, DIFFERENCES which I for one have vainly attempted to a
0033 ss, and yet the flowers present very slight DIFFERENCES. It is not that the varieties which differ
0033 hould never be overlooked, will ensure some DIFFERENCES; but, as a general rule, I cannot doubt th
0038 r of our domestic races, and likewise their DIFFERENCES being so great in external characters and
0039 fancier's eye: he perceives extremely small DIFFERENCES, and it is in human nature to value any no
0039 e which would formerly be set on any slight DIFFERENCES in the individuals of the same species, he
0039 e once fairly been established. Many slight DIFFERENCES might, and indeed do now, arise amongst pi
0040 ection to work on; not that mere individual DIFFERENCES are not amply sufficient, with extreme car
0044 ation Under Nature. Variability. Individual DIFFERENCES. Doubtful species. Wide ranging, much diff
0045 alled a variety. Again, we have many slight DIFFERENCES which may be called individual differences
0045 differences which may be called individual DIFFERENCES, such as are known frequently to appear in
0045 st in the very same mould. These individual DIFFERENCES are highly important for us, as they affor
0045 ccumulate in any given direction individual DIFFERENCES in his domesticated productions. These ind
0045 domesticated productions. These individual DIFFERENCES generally affect what naturalists consider
0046 here is one point connected with individual DIFFERENCES, which seems to me extremely perplexing: I
0050 s at first much perplexed to determine what DIFFERENCES to consider as specific, and what as varie
0051 or between lesser varieties and individual DIFFERENCES. These differences blend into each other i
0051 varieties and individual differences. These DIFFERENCES blend into each other in an insensible ser
0051 actual passage. Hence I look at individual DIFFERENCES, though of small interest to the systemati
0052 (as will hereafter be more fully explained) DIFFERENCES of structure in certain definite direction
0052 , again, in comparison with mere individual DIFFERENCES, is also applied arbitrarily, and for mere
0058 w this may be explained, and how the lesser DIFFERENCES between varieties will tend to increase in
0058 ties will tend to increase into the greater DIFFERENCES between species. There is one other point
0082 g up in any given direction mere individual DIFFERENCES, so could Nature, but far more easily, fro
0085 if, with all the aids of art, these slight DIFFERENCES make a great difference in cultivating the
```

Page ***(Key Word)***

Page **************************************(Key Word)**************************************

```
0085 ther trees and with a host of enemies, such DIFFERENCES would effectually settle which variety, wh
0089 fer in structure, colour, or ornament, such DIFFERENCES have been mainly caused by sexual selectio
0090 would not wish to attribute all such sexual DIFFERENCES to this agency: for we see peculiarities a
0091 very improbable; for we often observe great DIFFERENCES in the natural tendencies of our domestic
0102 s favourable, but I believe mere individual DIFFERENCES suffice for the work. A large number of in
0111 es throughout nature presenting well marked DIFFERENCES: whereas varieties, the supposed prototype
0111 ked species, present slight and ill defined DIFFERENCES. Mere chance, as we may call it, might cau
0112 r stronger and more bulky horses. The early DIFFERENCES would be very slight; in the course of tim
0112 eeders, and of stronger ones by others, the DIFFERENCES would become greater, and would be noted a
0112 ell established and distinct breeds. As the DIFFERENCES slowly become greater, the inferior animal
0112 called the principle of divergence, causing DIFFERENCES, at first barely appreciable, steadily to
0114 ication of structure, with the accompanying DIFFERENCES of habit and constitution, determine that
0115 tion of structure, amounting to new generic DIFFERENCES, would have been profitable to them. The a
0120 am illustrates the steps by which the small DIFFERENCES distinguishing varieties are increased int
0120 ing varieties are increased into the larger DIFFERENCES distinguishing species. By continuing the
0128 eding in the battle of life. Thus the small DIFFERENCES distinguishing varieties of the same speci
0128 ncrease till they come to equal the greater DIFFERENCES between species of the same genus, or even
0131 e reproductive system to produce individual DIFFERENCES, or very slight deviations of structure, a
0142 ricin, they cannot owe their constitutional DIFFERENCES to habit. The case of the Jerusalem artich
0142 even tried. Nor let it be supposed that no DIFFERENCES in the constitution of seedling kidney bea
0143 astered by, the natural selection of innate DIFFERENCES. Correlation of Growth. I mean by this exp
0145 rs, as has been described by Cassini. These DIFFERENCES have been attributed by some authors to pr
0145 nce in the corolla. Possibly, these several DIFFERENCES may be connected with some difference in t
0146 y have come into play. But in regard to the DIFFERENCES both in the internal and external structur
0146 s, which are not always correlated with any DIFFERENCES in the flowers, it seems impossible that t
0146 to the plant: yet in the Umbelliferae these DIFFERENCES are of such apparent importance, the seeds
0146 is main divisions of the order on analogous DIFFERENCES. Hence we see that modifications of struct
0157 remarkable fact, that the secondary sexual DIFFERENCES between the two sexes of the same species
0157 hich happen to stand on my list; and as the DIFFERENCES in these cases are of a very unusual natur
0158 that secondary sexual and ordinary specific DIFFERENCES are generally displayed in the same parts
0167 ppear to have acted in producing the lesser DIFFERENCES between varieties of the same species, and
0167 ieties of the same species, and the greater DIFFERENCES between species of the same genus. The ext
0167 ications. Habit in producing constitutional DIFFERENCES, and use in strengthening, and disuse in w
0169 ken advantage of in giving secondary sexual DIFFERENCES to the sexes of the same species, and spec
0169 the sexes of the same species, and specific DIFFERENCES to the several species of the same genus.
0170 ulation, through natural selection, of such DIFFERENCES, when beneficial to the individual, that g
0195 r from being correlated with constitutional DIFFERENCES, might assuredly be acted on by natural se
0198 made conscious of this by reflecting on the DIFFERENCES in the breeds of our domesticated animals
0198 re unable to account for the characteristic DIFFERENCES of our domestic breeds, which nevertheless
0199 f the precise cause of the slight analogous DIFFERENCES between species. I might have adduced for
0199 ight have adduced for this same purpose the DIFFERENCES between the races of man, which are so str
0199 apparently be thrown on the origin of these DIFFERENCES, chiefly through sexual selection of a par
0212 dubon has given several remarkable cases of DIFFERENCES in nests of the same species in the northe
0237 those in a state of nature, of all sorts of DIFFERENCES of structure which have become correlated
0237 to certain ages; and to either sex. We have DIFFERENCES correlated not only to one sex, but to tha
0237 aws of the male salmon. We have even slight DIFFERENCES in the horns of different breeds of cattle
0245 special endowment, but incidental on other DIFFERENCES. Causes of the sterility of first crosses
0245 uality, but is incidental on other acquired DIFFERENCES. In treating this subject, two classes of
0248 ed from other constitutional and structural DIFFERENCES. In regard to the sterility of hybrids in
0257 eral appearance, and having strongly marked DIFFERENCES in every part of the flower, even in the p
0258 r crossing is connected with constitutional DIFFERENCES imperceptible by us, and confined to the r
0260 s simply incidental or dependent on unknown DIFFERENCES, chiefly in the reproductive systems, of t
0260 tems, of the species which are crossed. The DIFFERENCES being of so peculiar and limited a nature,
0261 mean by sterility being incidental on other DIFFERENCES, and not a specially endowed quality. As t
0261 ty, but will admit that it is incidental on DIFFERENCES in the laws of growth of the two plants. W
0261 why one tree will not take on another, from DIFFERENCES in their rate of growth, in the hardness o
0263 fted on each other as incidental on unknown DIFFERENCES in their vegetative systems, so I believe
0263 of first crosses, are incidental on unknown DIFFERENCES, chiefly in their reproductive systems. Th
0263 hiefly in their reproductive systems. These DIFFERENCES, in both cases, follow to a certain extent
0269 wishes to select, nor could select, slight DIFFERENCES in the reproductive system, or other const
0269 eproductive system, or other constitutional DIFFERENCES correlated with the reproductive system. H
0270 sation is by so much the less easy as their DIFFERENCES are greater. How far these experiments may
0271 ng able to produce recondite and functional DIFFERENCES in the reproductive system; from these sev
0272 w ard, as it seems to me, quite unimportant DIFFERENCES between the so called hybrid offspring of
0274 kolreuter. These alone are the unimportant DIFFERENCES, which Gartner is able to point out, betwe
0276 another, is incidental on generally unknown DIFFERENCES in their vegetative systems, so in crossin
0276 nite with another, is incidental on unknown DIFFERENCES in their reproductive systems. There is no
0278 stication by the selection of mere external DIFFERENCES, and not of differences in the reproductiv
0278 on of mere external differences, and not of DIFFERENCES in the reproductive system. In all other r
0281 d with the parent species of each genus, by DIFFERENCES not greater than we see between the variet
0297 it is notorious on what excessively slight DIFFERENCES many palaeontologists have founded their s
0316 s new progenitor some slight characteristic DIFFERENCES. Groups of species, that is, genera and fa
0346 ent distribution cannot be accounted for by DIFFERENCES in physical conditions. Importance of barr
0347 ated in a close and important manner to the DIFFERENCES between the productions of various regions
0348 ng as the oceans separating continents, the DIFFERENCES are very inferior in degree to those chara
0409 ith those governing at the present time the DIFFERENCES in different areas. We see this in many fa
0416 another division they differ much, and the DIFFERENCES are of quite subordinate value in classifi
0419 research has detected important structural DIFFERENCES, at first overlooked, but because numerous
0421 genera. So that the amount or value of the DIFFERENCES between organic beings all related to each
0432 hus give a general idea of the value of the DIFFERENCES between them. This is what we should be dr
0442 roduction. We are so much accustomed to see DIFFERENCES in structure between the embryo and the ad
0445 hed from each other, yet their proportional DIFFERENCES in the above specified several points were
0446 geons, seem to show that the characteristic DIFFERENCES which give value to each breed, and which
0446 universal rule: for here the characteristic DIFFERENCES must either have appeared at an earlier pe
0446 rlier period than usual, or, if not so, the DIFFERENCES must have been inherited, not at the corre
0448 conceivable extent from their parents. Such DIFFERENCES might, also, become correlated with succes
```

Page **************************************(Key Word)**************************************

0460 but that it is incidental on constitutional DIFFERENCES in the reproductive systems of the intercr
0467 , in each successive generation, individual DIFFERENCES so slight as to be quite inappreciable by
0468 a great result by adding up mere individual DIFFERENCES in his domestic productions; and every one
0468 e admits that there are at least individual DIFFERENCES in species under nature. But, besides such
0468 in species under nature. But, besides such DIFFERENCES, all naturalists have admitted the existen
0469 aw any clear distinction between individual DIFFERENCES and slight varieties; or between more plai
0470 ontinued course of modification, the slight DIFFERENCES, characteristic of varieties of the same s
0470 cies, tend to be augmented into the greater DIFFERENCES characteristic of species of the same genu
0484 f definition; and if definable, whether the DIFFERENCES be sufficiently important to deserve a spe
0485 al consideration than it is at present; for DIFFERENCES, however slight, between any two forms, if
0487 le world. Even at present, by comparing the DIFFERENCES of the inhabitants of the sea on the oppos
0007 h have varied during all ages under the most DIFFERENT climates and treatment, I think we are driv
0007 ions of life not so uniform as, and somewhat DIFFERENT from, those to which the parent species hav
0009 ch suddenly assumes a new and sometimes very DIFFERENT character from that of the rest of the plan
0013 ; no one can say why the same peculiarity in DIFFERENT individuals of the same species, and in ind
0013 s of the same species, and in individuals of DIFFERENT species, is sometimes inherited and sometim
0017 for instance, of the many foxes, inhabiting DIFFERENT quarters of the world. I do not believe, as
0018 dian cattle, that these had descended from a DIFFERENT aboriginal stock from our European cattle:
0020 ed nearly intermediate between two extremely DIFFERENT races or species, I can hardly believe. Sir
0020 on. C. Murray from Persia. Many treatises in DIFFERENT languages have been published on pigeons, a
0021 augher, as their names express, utter a very DIFFERENT coo from the other breeds. The fantail has
0028 geons can be easily mated for life; and thus DIFFERENT breeds can be kept together in the same avi
0030 ny breeds of dogs, each good for man in very DIFFERENT ways; when we compare the game cock, so per
0030 garden races of plants most useful to man at DIFFERENT seasons and for different purposes, or so b
0030 t useful to man at different seasons and for DIFFERENT purposes, or so beautiful in his eyes, we m
0031 ent is by no means generally due to crossing DIFFERENT breeds; all the best breeders are strongly
0033 by comparing the diversity of flowers in the DIFFERENT varieties of the same species in the flower
0033 lowers of the same set of varieties. See how DIFFERENT the leaves of the cabbage are, and how extr
0033 alike the leaves: how much the fruit of the DIFFERENT kinds of gooseberries differ in size, colou
0036 that they have the appearance of being quite DIFFERENT varieties. If there exist savages so barbar
0038 viduals of the same species, having slightly DIFFERENT constitutions or structure, would often suc
0048 france or of the United States, drawn up by DIFFERENT botanists, and see what a surprising number
0049 ffer considerably in appearance: they have a DIFFERENT flavour and emit a different odour; they fl
0049 ce: they have a different flavour and emit a DIFFERENT odour: they flower at slightly different pe
0049 t a different odour: they flower at slightly DIFFERENT periods: they grow in somewhat different st
0049 tly different periods: they grow in somewhat DIFFERENT stations: they ascend mountains to differen
0049 different stations; they ascend mountains to DIFFERENT heights; they have different geographical r
0049 nd mountains to different heights; they have DIFFERENT geographical ranges; and lastly, according
0052 , due merely to the long continued action of DIFFERENT physical conditions in two different region
0052 tion of different physical conditions in two DIFFERENT regions; but I have not much faith in this
0053 , is a far more important circumstance) with DIFFERENT sets of organic beings. But my tables furth
0053 used within their own country (and this is a DIFFERENT consideration from wide range, and to a cer
0071 is generally seen in passing from one quite DIFFERENT soil to another: not only the proportional
0074 district! In the case of many species, many DIFFERENT checks, acting at different periods of life
0074 ny species, many different checks, acting at DIFFERENT periods of life, and during different seaso
0074 ing at different periods of life, and during DIFFERENT seasons or years, probably come into play;
0074 s. in some cases it can be shown that widely DIFFERENT checks act on the same species in different
0074 different checks act on the same species in DIFFERENT districts. When we look at the plants and b
0074 when an American forest is cut down, a very DIFFERENT vegetation springs up; but it has been obse
0076 me result has followed from keeping together DIFFERENT varieties of the medicinal leech. It may ev
0076 the place of another species under the most DIFFERENT climates. In Russia the small Asiatic cockr
0078 s new home, we should have to modify it in a DIFFERENT way to what we should have done in its nati
0078 should have to give it some advantage over a DIFFERENT set of competitors or enemies. It is good t
0084 , that we only see that the forms of are now DIFFERENT from what they formerly were. Although natu
0086 n insect to a score of contingencies, wholly DIFFERENT from those which concern the mature insect.
0086 tions consequent on other modifications at a DIFFERENT period of life, shall not be in the least d
0087 s to the other sex, or in relation to wholly DIFFERENT habits of life in the two sexes, as is some
0091 lowlands, would naturally be forced to hunt DIFFERENT prey; and from the continued preservation o
0093 ut the stigmas of twenty flowers, taken from DIFFERENT branches, under the microscope, and on all,
0096 that with animals and plants a cross between DIFFERENT varieties, or between individuals of the sa
0099 ed 233 seedling cabbages from some plants of DIFFERENT varieties growing near each other, and of t
0102 ls varying in the right direction, though in DIFFERENT degrees, so as better to fill up the unoccu
0102 eral districts will almost certainly present DIFFERENT conditions of life; and then if natural sel
0103 imal can long remain distinct, from haunting DIFFERENT stations, from breeding at slightly differe
0103 ifferent stations, from breeding at slightly DIFFERENT seasons, or from varieties of the same kind
0114 y a rotation of plants belonging to the most DIFFERENT orders: nature follows what may be called a
0114 l, as a general rule, belong to what we call DIFFERENT genera and orders. The same principle is se
0115 ons in their new homes. But the case is very DIFFERENT; and Alph. de Candolle has well remarked in
0116 y the animals and plants are diversified for DIFFERENT habits of life, so will a greater number of
0117 in; for this will generally lead to the most DIFFERENT or divergent variations (represented by the
0121 us, the species, which are already extremely DIFFERENT in character, will generally tend to produc
0121 ve the best chance of filling new and widely DIFFERENT places in the polity of nature: hence in th
0123 l be much greater than that between the most DIFFERENT of the original eleven species. The new spe
0123 er, will be allied to each other in a widely DIFFERENT manner. Of the eight descendants from (A) t
0123 the process of modification, will be widely DIFFERENT from the other five species, and may consti
0123 r, are supposed to have gone on diverging in DIFFERENT directions. The intermediate species, also
0125 nce of character and from inheritance from a DIFFERENT parent, will differ widely from the three g
0128 less closely related, and genera related in DIFFERENT degrees, forming sub families, families, or
0133 y being produced under conditions of life as DIFFERENT as can well be conceived; and, on the other
0133 ell he conceived; and, on the other hand, of DIFFERENT varieties being produced from the same spec
0137 that several animals, belonging to the most DIFFERENT classes, which inhabit the caves of Styria
0140 to a certain extent, naturally habituated to DIFFERENT temperatures, or becoming acclimatised: thu
0140 ollected by Dr. Hooker from trees growing at DIFFERENT heights on the Himalaya, were found in this
0140 alaya, were found in this country to possess DIFFERENT constitutional powers of resisting cold. Mr
0140 ic animals of not only withstanding the most DIFFERENT climates but of being perfectly fertile (a
0141 ture, could easily be brought to bear widely DIFFERENT climates. We must not, however, push the fo

Page **(Key Word)**

```
0141  this view, the capacity of enduring the most  DIFFERENT  climates by man himself and by his domestic
0141  to the natural selection of varieties having  DIFFERENT  innate constitutions, and how much to both
0151  es, and they differ extremely little even in  DIFFERENT  genera; but in the several species of one g
0151  iversification: the homologous valves in the  DIFFERENT  species being sometimes wholly unlike in sh
0152  nt of difference there is in the beak of the  DIFFERENT  tumblers, in the beak and wattle of the dif
0152  rent tumblers, in the beak and wattle of the  DIFFERENT  carriers, in the carriage and tail of our f
0155  at the more an organ normally differs in the  DIFFERENT  species of the same group, the more subject
0156  veral species, fitted to more or less widely  DIFFERENT  habits, in exactly the same manner: and as
0157  same parts of the organisation in which the   DIFFERENT  species of the same genus differ from each
0157  certain genera the neuration differs in the   DIFFERENT  species, and likewise in the two sexes of t
0158  ch other, or to fit the males and females to  DIFFERENT  habits of life, or the males to struggle wi
0164  f leg and shoulder stripes in horses of very  DIFFERENT  breeds, in various countries from Britain t
0171  the same species. Species with habits widely  DIFFERENT  from those of their allies. Organs of extre
0171  the modification of some animal with wholly   DIFFERENT  habits? Can we believe that natural selecti
0178  en two or more varieties have been formed in  DIFFERENT  portions of a strictly continuous area, int
0180  ther polecats on mice and land animals. If a  DIFFERENT  case had been taken, and it had been asked
0183  al grades between structures fitted for very  DIFFERENT  habits of life will rarely have been develo
0183  abits, or exclusively for one of its several  DIFFERENT  habits. But it is difficult to tell, and im
0184  viduals of a species following habits widely  DIFFERENT  from those both of their own species and of
0186  ize on the place of that inhabitant, however  DIFFERENT  it may be from its own place. Hence it will
0186  able contrivances for adjusting the focus to  DIFFERENT  distances, for admitting different amounts
0186  focus to different distances, for admitting   DIFFERENT  amounts of light, and for the correction of
0187  tructure, branching off in two fundamentally  DIFFERENT  lines, can be shown to exist, until we reac
0189  in density, so as to separate into layers of  DIFFERENT  densities and thicknesses, placed at differ
0189  fferent densities and thicknesses, placed at  DIFFERENT  distances from each other, and with the sur
0190  tion, may be converted into one for a wholly  DIFFERENT  purpose, namely respiration. The swimbladde
0193  class, especially if in members having very   DIFFERENT  habits of life, we may attribute its presen
0193  minous organs in a few insects, belonging to  DIFFERENT  families and orders, offers a parallel case
0195  es felt as much difficulty, though of a very  DIFFERENT  kind, on this head, as in the case of an or
0198  in the breeds of our domesticated animals in  DIFFERENT  countries, more especially in the less civi
0198  d come into play. Animals kept by savages in  DIFFERENT  countries often have to struggle for their
0198  ral selection, and individuals with slightly  DIFFERENT  constitutions would succeed best under diff
0198  erent constitutions would succeed best under  DIFFERENT  climates; and there is reason to believe th
0204  ous we should be in concluding that the most  DIFFERENT  habits of life could not graduate into each
0205  e organ having performed simultaneously very  DIFFERENT  functions, and then having been specialised
0207  tinct. Instincts comparable with habits, but  DIFFERENT  in their origin. Instincts graduated. Aphid
0210  es be facilitated by the same species having  DIFFERENT  instincts at different periods of life, or
0210  e same species having different instincts at  DIFFERENT  periods of life, or at different seasons of
0210  nstincts at different periods of life, or at  DIFFERENT  seasons of the year, or when placed under d
0210  nt seasons of the year, or when placed under  DIFFERENT  circumstances, etc.: in which case either o
0214  usly they become mingled, is well shown when  DIFFERENT  breeds of dogs are crossed. Thus it is know
0217  d, or there would be eggs and young birds of  DIFFERENT  ages in the same nest. If this were the cas
0217  ardly fail to be by having eggs and young of  DIFFERENT  ages at the same time; then the old birds o
0224  ngthened and rendered permanent for the very  DIFFERENT  purpose of raising slaves. When the instinc
0232  bees would entirely pull down and rebuild in  DIFFERENT  ways the same cell, sometimes recurring to
0237  have even slight differences in the horns of  DIFFERENT  breeds of cattle in relation to an artifici
0238  ers, with jaws and instincts extraordinarily  DIFFERENT: in Cryptocerus, the workers of one caste a
0239  election, as I believe to be quite possible,  DIFFERENT  from the fertile males and females, in this
0240  in important points of structure between the  DIFFERENT  castes of neuters in the same species, that
0241  gh the workers can be grouped into castes of  DIFFERENT  sizes, yet they graduate insensibly into ea
0241  sensibly into each other, as does the widely  DIFFERENT  structure of their jaws. I speak confidentl
0241  all of small size with jaws having a widely   DIFFERENT  structure; or lastly, and this is our clima
0241  d simultaneously another set of workers of a  DIFFERENT  size and structure; a graduated series havi
0241  rkers existing in the same nest, both widely  DIFFERENT  from each other and from their parents, has
0243  s of the world and living under considerably  DIFFERENT  conditions of life, yet often retaining nea
0245  es of facts, to a large extent fundamentally  DIFFERENT, have generally been confounded together; n
0248  rility of various species when crossed is so  DIFFERENT  in degree and graduates away so insensibly,
0248  with the evidence from fertility adduced by   DIFFERENT  hybridisers, or by the same author, from ex
0248  he same author, from experiments made during  DIFFERENT  years. It can thus be shown that neither st
0253  ybrid have been raised at the same time from  DIFFERENT  parents, so as to avoid the ill effects of
0253  e geese (A. cygnoides), species which are so  DIFFERENT  that they are generally ranked in distinct
0253  d two hybrids from the same parents but from  DIFFERENT  hatches; and from these two birds he raised
0255  is absolute zero of fertility, the pollen of  DIFFERENT  species of the same genus applied to the st
0257  ing. It can be shown that plants most widely  DIFFERENT  in habit and general appearance, and having
0257  duous and evergreen trees, plants inhabiting  DIFFERENT  stations and fitted for extremely different
0257  different stations and fitted for extremely   DIFFERENT  climates, can often be crossed with ease. B
0260  for why should the sterility be so extremely  DIFFERENT  in degree, when various species are crossed
0260  nd then to stop their further propagation by  DIFFERENT  degrees of sterility, not strictly related
0261  he other deciduous, and adaptation to widely  DIFFERENT  climates, does not always prevent the two g
0261  e, which is a member of the same genus. Even  DIFFERENT  varieties of the pear take with different d
0262  en different varieties of the pear take with  DIFFERENT  degrees of facility on the quince; so do di
0262  ent degrees of facility on the quince; so do  DIFFERENT  varieties of the apricot and peach on certa
0262  there was sometimes an innate difference in   DIFFERENT  individuals of the same two species in cros
0262  so sagaret believes this to be the case with  DIFFERENT  individuals of the same two species in bein
0262  organs in an imperfect condition, is a very  DIFFERENT  case from the difficulty of uniting two pur
0263  f hybrids. These two cases are fundamentally  DIFFERENT. I have more than once alluded to a large b
0264  are imperfectly developed, the case is very  DIFFERENT  structures and constitutions having been bl
0266  t the organisation has been disturbed by two  DIFFERENT  parts and organs one to another, or to the
0266  periodical action, or mutual relation of the  DIFFERENT  class of facts. It is an old and almost uni
0266  ar parallelism extends to an allied yet very  DIFFERENT  strains or sub breeds, gives vigour and fer
0267  the same species, that is between members of  DIFFERENT, give vigour and fertility to the offspring
0267  pecies which have varied and become slightly  DIFFERENT, produce hybrids which are generally steril
0267  les which have become widely or specifically  DIFFERENT  from each other; whereas if two very distin
0274  with a third species, the hybrids are widely  DIFFERENT  varieties, or of distinct species. Laying a
0275  on of individuals of the same variety, or of  DIFFERENT, and sometimes widely different, in recipro
0276  al curious and complex laws. It is generally  DIFFERENT, and sometimes widely different, in recipro
```

Page **(Key Word)**

```
0276  is generally different, and sometimes widely DIFFERENT, in reciprocal crosses between the same two
0277  ly, that the crossing of forms only slightly DIFFERENT is favourable to the vigour and fertility o
0277  latter capacity evidently depends on widely  DIFFERENT circumstances, should all run, to a certain
0282  of geology, or to read special treatises by  DIFFERENT observers on separate formations, and to ma
0284  ew cases from estimate, of each formation in DIFFERENT parts of Great Britain; and this is the res
0294  will probably first appear and disappear at  DIFFERENT levels, owing to the migration of species a
0295  en the case, a formation composed of beds of DIFFERENT mineralogical composition, we may reasonabl
0295  nts of the sea and a supply of sediment of a DIFFERENT nature will generally have been due to geog
0296  above the other, at no less than sixty eight DIFFERENT levels. Hence, when the same species occur
0297  the more readily if the specimens come from  DIFFERENT sub stages of the same formation. Some expe
0297  gh almost universally ranked as specifically  DIFFERENT, yet are far more closely allied to each ot
0299  uture period will be able to prove, that our DIFFERENT breeds of cattle, sheep, horses, and dogs h
0301  inks, let them be ever so close, if found in DIFFERENT stages of the same formation, would, by mos
0311  s supposed to be written, being more or less DIFFERENT in the interrupted succession of chapters,
0312  ccessive appearance of new species. On their DIFFERENT rates of change. Species once lost do not r
0313  neous in each separate formation. Species of DIFFERENT genera and classes have not changed at the
0315  ame; for both would almost certainly inherit DIFFERENT characters from their distinct progenitors.
0322  y distant parts of the world, under the most DIFFERENT climates, where not a fragment of the miner
0325  of life throughout the world, under the most DIFFERENT climates. We must, as Barrande has remarked
0331  cies being thus enabled to seize on many and DIFFERENT places in the economy of nature. Therefore
0332  me of which are supposed to have perished at DIFFERENT periods, and some to have endured to the pr
0332  ng and circuitous course through many widely DIFFERENT forms. If many extinct forms were to be dis
0336  hout the world, and therefore under the most DIFFERENT climates and conditions. Consider the prodi
0340  ribution of terrestrial mammals was formerly DIFFERENT from what it now is. North America formerly
0343  eriod, the world may have presented a wholly DIFFERENT aspect: and that the older continents, form
0343  e in slowly and successively; how species of DIFFERENT classes do not necessarily change together,
0345  cuitous course through many extinct and very DIFFERENT forms. We can clearly see why the organic r
0347  itions of the Old and New Worlds, how widely DIFFERENT are their living productions! In the southe
0347  s, which consequently inhabit a considerably DIFFERENT climate, and they will be found incomparabl
0347  nd almost joins, and where, under a slightly DIFFERENT climate, there might have been free migrati
0348  and sometimes even of large rivers, we find  DIFFERENT productions; though as mountain chains, des
0349  hough the species themselves are distinct at DIFFERENT points and stations. It is a law of the wid
0350  her. The dissimilarity of the inhabitants of DIFFERENT regions may be attributed to modification t
0350  ubordinate degree to the direct influence of DIFFERENT physical conditions. The degree of dissimil
0351  e for life, so the degree of modification in DIFFERENT species will be no uniform quantity. If, fo
0356  on migration: a region when its climate was  DIFFERENT may have been a high road for migration, bu
0360  y floating plants. He tried 98 seeds, mostly DIFFERENT from mine; but he chose many large fruits a
0362  welve to twenty one hours in the stomachs of DIFFERENT birds of prey; and two seeds of beet grew a
0368  mountains, the case will have been somewhat  DIFFERENT; for it is not likely that all the same arc
0369  ey will, also, have been exposed to somewhat DIFFERENT climatal influences. Their mutual relations
0373  of the Pacific, where the climate is now so  DIFFERENT, as far south as lat. 46 degrees; erratic b
0379  ion northward, the case may have been wholly DIFFERENT with those intruding forms which settled th
0383  many fresh water species, belonging to quite DIFFERENT classes, an enormous range, but allied spec
0384  stems will have some fish in common and some DIFFERENT. A few facts seem to favour the possibility
0386  february three tablespoons of mud from three DIFFERENT points, beneath water, on the edge of a lit
0391  three or four hundred miles of open sea. The DIFFERENT orders of insects in Madeira apparently pre
0391  s shown that the proportional numbers of the DIFFERENT orders are very different from what they ar
0391  nal numbers of the different orders are very DIFFERENT from what they are elsewhere. Such cases ar
0401  another, it would undoubtedly be exposed to  DIFFERENT conditions of life in the different islands
0401  posed to different conditions of life in the DIFFERENT islands; for it would have to compete with
0401  t islands, for it would have to compete with DIFFERENT sets of organisms: a plant, for instance, w
0401  would be exposed to the attacks of somewhat  DIFFERENT enemies. If then it varied, natural selecti
0401  ied, natural selection would probably favour DIFFERENT varieties in the different islands. Some sp
0401  d probably favour different varieties in the DIFFERENT islands. Some species, however, might sprea
0404  present time or at some former period under  DIFFERENT physical conditions, and the existence at r
0406  and dry lands, though these stations are so  DIFFERENT, the very close relation of the distinct sp
0408  enera and families; and how it is that under DIFFERENT latitudes, for instance in South America, t
0408  conditions should often be inhabited by very DIFFERENT forms of life; for according to the length
0408  g more or less rapidly, there would ensue in DIFFERENT regions, independently of their physical co
0409  rce, some existing in scanty numbers, in the DIFFERENT great geographical provinces of the world.
0409  rning at the present time the differences in DIFFERENT areas. We see this in many facts. The endur
0410  for by migration at some former period under DIFFERENT conditions or by occasional means of transp
0410  differ little, whilst others belonging to a  DIFFERENT class, or to a different order, or even onl
0410  hers belonging to a different class, or to a DIFFERENT order, or even only to a different family o
0410  , or to a different order, or even only to a DIFFERENT family of the same order, differ greatly. I
0411  le matter, and so on: but the case is widely DIFFERENT in nature; for it is notorious how commonly
0411  only members of even the same sub group have DIFFERENT habits. In our second and fourth chapters,
0412  each species trying to occupy as many and as DIFFERENT places as possible in the economy of nature
0415  al value, its classificatory value is widely DIFFERENT. No naturalist can have worked at any group
0419  cause numerous allied species, with slightly DIFFERENT grades of difference, have been subsequentl
0420  enitor, may differ greatly, being due to the DIFFERENT degrees of modification which they have und
0420  is expressed by the forms being ranked under DIFFERENT genera, families, sections, or orders. The
0421  lionth degree! yet they differ widely and in DIFFERENT degrees from each other. The forms descende
0421  same degree in blood, has come to be widely  DIFFERENT. Nevertheless their genealogical arrangemen
0422  ei but the degrees of modification which the DIFFERENT groups have undergone, have to be expressed
0422  , have to be expressed by ranking them under DIFFERENT so called genera, sub families, families, s
0425  d females and larvae are sometimes extremely DIFFERENT; and as it has been used in classing variet
0426  feathers, if it prevail throughout many and  DIFFERENT species, especially those having very diffe
0426  ferent species, especially those having very DIFFERENT habits of life, it assumes high value; for
0426  for its presence in so many forms with such  DIFFERENT habits, only by its inheritance from a comm
0426  er throughout a large group of beings having DIFFERENT habits, we may feel almost sure, on the the
0433  in classing acknowledged varieties, however  DIFFERENT they may be from their parent: and I believ
0434  ing that the several parts and organs in the DIFFERENT species of the class are homologous. The wh
0434  n be given to the homologous bones in widely DIFFERENT animals. We see the same great law in the c
0434  n of the mouths of insects: what can be more DIFFERENT than the immensely long spiral proboscis of
0434  etle? yet all these organs, serving for such DIFFERENT purposes, are formed by infinitely numerous
```

0436 mely, the comparison not of the same part in DIFFERENT members of a class, but of the different pa
0436 in different members of a class, but of the DIFFERENT parts or organs in the same individual. Mos
0437 t organs, which when mature become extremely DIFFERENT, are at an early stage of growth exactly al
0437 of a bat, used as they are for such totally DIFFERENT purposes? Why should one crustacean, which
0437 vidual flower, though fitted for such widely DIFFERENT purposes, be all constructed on the same pa
0439 individual, which when mature become widely DIFFERENT and serve for different purposes, are in th
0439 mature become widely different and serve for DIFFERENT purposes, are in the embryo exactly alike.
0439 of structure, in which the embryos of widely DIFFERENT animals of the same class resemble each oth
0440 hich they are exposed. The case, however, is DIFFERENT when an animal during any part of its embry
0442 a close similarity in the embryos of widely DIFFERENT animals within the same class, that we migh
0442 early period of growth alike; of embryos of DIFFERENT species within the same class, generally, b
0444 greyhound and bull dog, though appearing so DIFFERENT, are really varieties most closely allied,
0448 young to follow habits of life in any degree DIFFERENT from those of their parent, and consequentl
0448 consequently to be constructed in a slightly DIFFERENT manner, then, on the principle of inheritan
0448 ight easily be rendered by natural selection DIFFERENT to any conceivable extent from their parent
0452 n tracing the homologies of the same part in DIFFERENT members of a class, nothing is more common,
0456 edged varieties of the same species, however DIFFERENT they may be in structure. If we extend the
0457 organs, to whatever purpose applied, of the DIFFERENT species of a class; or to the homologous pa
0457 parts, which when matured will become widely DIFFERENT from each other in structure and function;
0457 ructure and function; and the resemblance in DIFFERENT species of a class of the homologous parts
0457 fitted in the adult members for purposes as DIFFERENT as possible. Larvae are active embryos, whi
0461 terility. The sterility of hybrids is a very DIFFERENT case from that of first crosses, for their
0461 stitutions having been disturbed by slightly DIFFERENT and new conditions of life, we need not fee
0470 as to be enabled to seize on many and widely DIFFERENT places in the economy of nature, there will
0474 nature moves by graduated steps in endowing DIFFERENT animals of the same class with their severa
0474 lied species, when placed under considerably DIFFERENT conditions of life, yet should follow nearl
0475 ge, after equal intervals of time, is widely DIFFERENT in different groups. The extinction of spec
0475 al intervals of time, is widely different in DIFFERENT groups. The extinction of species and of wh
0477 n the most distant mountains, under the most DIFFERENT climates; and likewise the close alliance o
0477 o surprise at their inhabitants being widely DIFFERENT, if they have been for a long period comple
0477 r from each other, at various periods and in DIFFERENT proportions, the course of modification in
0477 fication in the two areas will inevitably be DIFFERENT. On this view of migration, with subsequent
0479 wing and leg of a bat, though used for such DIFFERENT purpose, in the jaws and legs of a crab, in
0482 end the same view to other and very slightly DIFFERENT forms. Nevertheless they do not pretend tha
0489 that these elaborately constructed forms, so DIFFERENT from each other, and dependent on each othe
0169 there has been much former variation and DIFFERENTIATION, or where the manufactory of new specifi
0012 r, it appears that white sheep and pigs are DIFFERENTLY affected from coloured individuals by cert
0038 certain seasons. And in two countries very DIFFERENTLY circumstanced, individuals of the same spe
0095 o the hive bee to have a slightly longer or DIFFERENTLY constructed proboscis. On the other hand,
0104 se would have inhabited the surrounding and DIFFERENTLY circumstanced districts, will be prevented
0160 n the crossed offspring of two distinct and DIFFERENTLY coloured breeds; and in this case there is
0167 ed like a zebra, but perhaps otherwise very DIFFERENTLY constructed, the common parent of our dome
0192 hat of neuter insects, which are often very DIFFERENTLY constructed from either the males or ferti
0220 in number, I thought that they might behave DIFFERENTLY when more numerous; but Mr. Smith informs
0264 e of the common mule. Hybrids, however, are DIFFERENTLY circumstanced before and after birth: when
0271 oured flowers, than between those which are DIFFERENTLY coloured. Yet these varieties of Verbascum
0400 hat many of the immigrants should have been DIFFERENTLY modified, though only in a small degree. T
0473 ed to have been created independently, have DIFFERENTLY coloured flowers, than if all the species
0016 s character; by which I mean, that, although DIFFERING from each other, and from the other species
0033 he flowers, or the fruit, will produce races DIFFERING from each other chiefly in these characters
0059 tain amount of difference, for two forms, if DIFFERING very little, are generally ranked as variet
0116 an marsupials, which are divided into groups DIFFERING but little from each other, and feebly repr
0118 ve produced two varieties, namely m2 and s2, DIFFERING from each other, and more considerably from
0236 he great difficulty lies in the working ants DIFFERING widely from both the males and the fertile
0237 . but with the working ant we have an insect DIFFERING greatly from its parents, yet absolutely st
0240 bodies of sterile workers in the same nest, DIFFERING not only in size, but in their organs of vi
0268 ect fertility of so many domestic varieties, DIFFERING widely from each other in appearance, for i
0440 of two species, or of two groups of species, DIFFERING quite as much, or even more, from each othe
0022 the face in length and breadth and curvature DIFFERS enormously. The shape, as well as the breadt
0022 d size of the eggs vary. The manner of flight DIFFERS remarkably; as does in some breeds the voice
0027 ugh an English carrier or short faced tumbler DIFFERS immensely in certain characters from the roc
0052 assage of a variety, from a state in which it DIFFERS very slightly from its parent to one in whic
0052 y slightly from its parent to one in which it DIFFERS more, to the action of natural selection in
0145 n the ovary itself, with its accessory parts, DIFFERS, as has been described by Cassini. These dif
0155 ain no doubt, that the more an organ normally DIFFERS in the different species of the same group,
0155 why should that part of the structure, which DIFFERS from the same part in other independently cr
0157 umber varies greatly; and the number likewise DIFFERS in the two sexes of the same species: again
0157 e groups; but in certain genera the neuration DIFFERS in the different species, and likewise in th
0167 nd to assign any reason why this or that part DIFFERS, more or less, from the same part in the par
0313 sub Himalayan deposits. The Silurian Lingula DIFFERS but little from the living species of this g
0329 any form is, the more, as a general rule, it DIFFERS from living forms. But, as Buckland long ago
0338 hapter I shall attempt to show that the adult DIFFERS from its embryo, owing to variations superve
0344 more ancient a form is, the more it generally DIFFERS from those now living. Why ancient and extin
0452 an has been rendered rudimentary occasionally DIFFERS much. This latter fact is well exemplified i
0180 uld have given no answer. Yet I think such DIFFICULTIES have very little weight. Here, as on othe
0008 than to tame an animal, and few things more DIFFICULT than to get it to breed freely under confin
0010 f heat, moisture, light, food, etc., is most DIFFICULT: my impression is, that with animals such a
0024 all recent experience shows that it is most DIFFICULT to get any wild animal to breed freely unde
0026 made on the most distinct breeds. Now, it is DIFFICULT, perhaps impossible, to bring forward one c
0044 creation. The term variety is almost equally DIFFICULT to define, but here community of descent is
0062 of the universal struggle for life, or more DIFFICULT, at least I have found it so, than constant
0078 a conviction as necessary, as it seems to be DIFFICULT to acquire. All that we can do, is to keep
0137 ope with its glasses has been lost. As it is DIFFICULT to imagine that eyes, though useless, could
0138 by itself seems to have done its work. It is DIFFICULT to imagine conditions of life more similar
0139 of the surrounding country. It would be most DIFFICULT to give any rational explanation of the aff
0147 least in excess, to another part; thus it is DIFFICULT to get a cow to give milk and to fatten rea

Page ***(Key Word)***

```
0151 t variability in plants made it particularly DIFFICULT to compare their relative degrees of variab
0152 n the short faced tumbler, it is notoriously DIFFICULT to breed them nearly to perfection, and fre
0180 g bat, the question would have been far more DIFFICULT, and I could have given no answer. Yet I th
0183 e of its several different habits. But it is DIFFICULT to tell, and immaterial for us, whether hab
0194 ancestor. Although in many cases it is most DIFFICULT to conjecture by what transitions an organ
0202 r industrious and sterile sisters? It may be DIFFICULT, but we ought to admire the savage instinct
0215 n some cases has sufficed; no animal is more DIFFICULT to tame than the young of the wild rabbit:
0224 tting tools and measures, would find it very DIFFICULT to make cells of wax of the true form, thou
0233 s of the hive bee? I think the answer is not DIFFICULT: it is known that bees are often hard press
0235 r existence. No doubt many instincts of very DIFFICULT explanation could be opposed to the theory
0248 , that for all practical purposes it is most DIFFICULT to say where perfect fertility ends and ste
0256 ity. Hybrids from two species which are very DIFFICULT to cross, and which rarely produce any offs
0280 eory, have formerly existed. I have found it DIFFICULT, when looking at any two species, to avoid
0289 rks, or when we follow them in nature, it is DIFFICULT to avoid believing that they are closely co
0292 ention to any one formation, it becomes more DIFFICULT to understand, why we do not therein find c
0319 some more successful competitor. It is most DIFFICULT always to remember that the increase of eve
0328 e species themselves differ in a manner very DIFFICULT to account for, considering the proximity o
0330 e course of geological ages; and it would be DIFFICULT to prove the truth of the proposition, for
0364 ot, as far as is known (and it would be very DIFFICULT to prove this), received within the last fe
0374 of the continent. If this be admitted, it is DIFFICULT to avoid believing that the temperature of
0393 ot have been created there, it would be very DIFFICULT to explain. Mammals offer another and simil
0404 s rule is generally true, though it would be DIFFICULT to prove it. Amongst mammals, we see it str
0431 nting up through many predecessors. As it is DIFFICULT to show the blood relationship between the
0459 full force. Nothing at first can appear more DIFFICULT to believe than that the more complex organ
0460 ink, be disputed. It is, no doubt, extremely DIFFICULT even to conjecture by what gradations many
0472 seem to have produced some effect: for it is DIFFICULT to resist this conclusion when we look, for
0483 the modification of species. The question is DIFFICULT to answer, because the more distinct the fo
0005 ng chapters, the most apparent and gravest DIFFICULTIES on the theory will be given: namely, first
0005 e theory will be given: namely, first, the DIFFICULTIES of transitions, or in understanding how a
0051 trust almost entirely to analogy, and his DIFFICULTIES will rise to a climax. Certainly no clear
0053 ct, soon convinced me that there were many DIFFICULTIES, as did subsequently Dr. Hooker, even in s
0053 for my future work the discussion of these DIFFICULTIES, and the tables themselves of the proporti
0171 d the best adapted to survive. Chapter VI. DIFFICULTIES on Theory. Difficulties on the theory of d
0171 rvive. Chapter VI. Difficulties on Theory. DIFFICULTIES on the theory of descent with modification
0171 rrived at this part of my work, a crowd of DIFFICULTIES will have occurred to the reader. Some of
0171 re not, I think, fatal to my theory. These DIFFICULTIES and objections may be classed under the fo
0203 have in this chapter discussed some of the DIFFICULTIES and objections which may be urged against
0207 ants. Hive bee, its cell making instinct. DIFFICULTIES on the theory of the Natural Selection of
0268 e, we are immediately involved in hopeless DIFFICULTIES; for if two hitherto reputed varieties be
0293 uration of specific forms. But insuperable DIFFICULTIES, as it seems to me, prevent us coming to a
0310 etely metamorphosed condition. The several DIFFICULTIES here discussed, namely our not finding in
0311 y separated, formations. On this view, the DIFFICULTIES above discussed are greatly diminished, or
0343 buried under the ocean. Passing from these DIFFICULTIES, all the other great leading facts in pala
0380 s agency. I am far from supposing that all DIFFICULTIES are removed on the view here given in rega
0380 ns of the intertropical regions. Very many DIFFICULTIES remain to be solved. I do not pretend to i
0381 hin their own homes. I have said that many DIFFICULTIES remain to be solved: some of the most rema
0389 ome continent. This view would remove many DIFFICULTIES, but it would not, I think, explain all th
0396 do not deny that there are many and grave DIFFICULTIES in understanding how several of the inhabi
0407 ct in the intermediate tracts, I think the DIFFICULTIES in believing that all the individuals of t
0407 r their migration, I do not think that the DIFFICULTIES are insuperable: though they often are in
0408 spersal of fresh water productions. If the DIFFICULTIES be not insuperable in admitting that in th
0411 inate to groups. Natural system. Rules and DIFFICULTIES in classification, explained on the theory
0414 rules followed in classification, and the DIFFICULTIES which are encountered on the view that cla
0420 ered. All the foregoing rules and aids and DIFFICULTIES in classification are explained, if I do n
0456 e grand system; the rules followed and the DIFFICULTIES encountered by naturalists in their classi
0457 al abortion, present to us no inexplicable DIFFICULTIES; on the contrary, their presence might hav
0459 tion and Conclusion. Recapitulation of the DIFFICULTIES on the theory of Natural Selection. Recapi
0461 turning to geographical distribution, the DIFFICULTIES encountered on the theory of descent with
0465 he sum of the several chief objections and DIFFICULTIES which may justly be urged against my theor
0465 ch can be given to them. I have felt these DIFFICULTIES far too heavily during many years to doubt
0466 ological Record is. Grave as these several DIFFICULTIES are, in my judgment they do not overthrow
0469 tulated, as fairly as I could, the opposed DIFFICULTIES and objections: now let us turn to the spe
0482 s him to attach more weight to unexplained DIFFICULTIES than to the explanation of a certain numbe
0007 heritance. Character of Domestic Varieties. DIFFICULTY of distinguishing between Varieties and Spe
0014 d so boldly been made. There would be great DIFFICULTY in proving its truth: we may safely conclud
0020 of them will be alike; and then the extreme DIFFICULTY, or rather utter hopelessness, of the task
0028 ll how true they bred, I felt fully as much DIFFICULTY in believing that they could ever have desc
0038 s. man can hardly select, or only with much DIFFICULTY, any deviation of structure excepting such
0042 g been brought into play: in cats, from the DIFFICULTY in pairing them; in donkeys, from only a fe
0047 he other as the variety. But cases of great DIFFICULTY, which I will not here enumerate, sometimes
0047 of the intermediate links always remove the DIFFICULTY. In very many cases, however, one form is r
0050 gartner, they can be crossed only with much DIFFICULTY. We could hardly wish for better evidence o
0051 servations, he will meet with more cases of DIFFICULTY; for he will encounter a greater number of
0086 seminated by the wind, I can see no greater DIFFICULTY in this being effected through natural sele
0101 eared to me to present a case of very great DIFFICULTY under this point of view: but I have been e
0101 that there are, on this view, many cases of DIFFICULTY, some of which I am trying to investigate.
0162 case in nature. A considerable part of the DIFFICULTY in recognising a variable species in our sy
0171 e perfection. Means of transition. Cases of DIFFICULTY. Natura non facit saltum. Organs of small i
0174 close linking intermediate varieties? This DIFFICULTY for a long time quite confounded me. But I
0174 ill pass over this way of escaping from the DIFFICULTY: for I believe that many perfectly defined
0180 t of such cases is sufficient to lessen the DIFFICULTY in any particular case like that of the bat
0181 rresponding manner. Therefore, I can see no DIFFICULTY, more especially under changing conditions
0181 with the other Lemuridae, yet I can see no DIFFICULTY in supposing that such links formerly exist
0181 ts possessor. Nor can I see any insuperable DIFFICULTY in further believing it possible that the m
0184 already exist in the country, I can see no DIFFICULTY in a race of bears being rendered, by natur
0186 under changing conditions of life, then the DIFFICULTY of believing that a perfect and complex eye
```

Page ***(Key Word)***

Page ***(Key Word)***

```
0188  ave become extinct, I can see no very great  DIFFICULTY (not more than in the case of many other st
0188  uer his imagination; though I have felt the  DIFFICULTY far too keenly to be surprised at any degre
0191  als: hence there seems to me to be no great  DIFFICULTY in believing that natural selection has act
0192  radations, yet, undoubtedly, grave cases of  DIFFICULTY occur, some of which will be discussed in m
0192  ans of fishes offer another case of special  DIFFICULTY; it is impossible to conceive by what steps
0193  organs offer another and even more serious  DIFFICULTY; for they occur in only about a dozen fishe
0193  ilies and orders, offers a parallel case of  DIFFICULTY. Other cases could be given: for instance i
0194  on of structure, I have sometimes felt much  DIFFICULTY in understanding the origin of simple parts
0195  incividuals. I have sometimes felt as much  DIFFICULTY, though of a very different kind, on this h
0207  robably have occurred to many readers, as a  DIFFICULTY sufficient to overthrow my whole theory. I
0209  s do vary ever so little, then I can see no  DIFFICULTY in natural selection preserving and continu
0219  e supposed case of the cuckoo, I can see no  DIFFICULTY in natural selection making an occasional h
0224  e same species in Switzerland, I can see no  DIFFICULTY in natural selection increasing and modifyi
0224  ceive when they are correctly made. But the  DIFFICULTY is not nearly so great as it at first appea
0227  t further suppose, and this is the greatest  DIFFICULTY, that she can somehow judge accurately at w
0227  we have further to suppose, but this is no  DIFFICULTY, that after hexagonal prisms have been form
0229  thin wax is, I do not see that there is any  DIFFICULTY in the bees, whilst at work on the two side
0231  omy of wax. It seems at first to add to the  DIFFICULTY of understanding how the cells are made, th
0232  . it was really curious to note in cases of  DIFFICULTY, as when two pieces of comb met at an angle
0233  ubject. Nor does there seem to me any great  DIFFICULTY in a single insect (as in the case of a que
0236  ses, but will confine myself to one special  DIFFICULTY, which at first appeared to me insuperable,
0236  the workers have been rendered sterile is a  DIFFICULTY; but not much greater than that of any othe
0236  ble of procreation, I can see no very great  DIFFICULTY in this being effected by natural selection
0236  tion. But I must pass over this preliminary  DIFFICULTY. The great difficulty lies in the working a
0236  over this preliminary difficulty. The great  DIFFICULTY lies in the working ants differing widely f
0237  these same breeds. Hence I can see no real  DIFFICULTY in any character having become correlated w
0237  certain members of insect communities: the  DIFFICULTY lies in understanding how such correlated m
0237  owly accumulated by natural selection. This  DIFFICULTY, though appearing insuperable, is lessened,
0238  ave not as yet touched on the climax of the  DIFFICULTY; namely, the fact that the neuters of sever
0241  cture; or lastly, and this is our climax of  DIFFICULTY, one set of workers of one size and structu
0242  use this is by far the most serious special  DIFFICULTY, which my theory has encountered. The case,
0243  ance to each animal. Therefore I can see no  DIFFICULTY, under changing conditions of life, in natu
0243  degree my theory; but none of the cases of  DIFFICULTY, to the best of my judgment, annihilate it.
0247  uminosae, in which there is an acknowledged  DIFFICULTY in the manipulation) half of these twenty p
0256  ry sterile: but the parallelism between the  DIFFICULTY of making a first cross, and the sterility
0256  can be crossed very rarely, or with extreme  DIFFICULTY, but the hybrids, when at last produced, ar
0257  which will not unite, or only with extreme  DIFFICULTY; and on the other hand of very distinct spe
0260  brids; and other species cross with extreme  DIFFICULTY, and yet produce fairly fertile hybrids? Wh
0262  whereas the currant will take, though with  DIFFICULTY, on the gooseberry. We have seen that the s
0262  ondition, is a very different case from the  DIFFICULTY of uniting two pure species, which have the
0262  s, and which could be grafted with no great  DIFFICULTY on another species, when thus grafted were
0263  o me to indicate that the greater or lesser  DIFFICULTY of either grafting or crossing together var
0263  ment; although in the case of crossing, the  DIFFICULTY is as important for the endurance and stabi
0263  ven in first crosses, the greater or lesser  DIFFICULTY in effecting a union apparently depends on
0271  modified. From these facts: from the great  DIFFICULTY of ascertaining the infertility of varietie
0276  h various and somewhat analogous degrees of  DIFFICULTY in being grafted together in order to preve
0277  gs. It is not surprising that the degree of  DIFFICULTY in uniting two species, and the degree of s
0279  rable transitional links, is a very obvious  DIFFICULTY. I assigned reasons why such links do not c
0302  d geological section presented, had not the  DIFFICULTY of our not discovering innumerable transiti
0305  cannot see that it would be an insuperable  DIFFICULTY on my theory, unless it could likewise be s
0306  iferous strata. There is another and allied  DIFFICULTY, which is much graver. I allude to the mann
0307  existence of life at these periods. But the  DIFFICULTY of understanding the absence of vast piles
0351  but little modification, there is not much  DIFFICULTY in believing that they may have migrated fr
0351  comparatively recent times, there is great  DIFFICULTY on this head. It is also obvious that the f
0352  btedly there are very many cases of extreme  DIFFICULTY, in understanding how the same species coul
0352  ts of the world. No geologist will feel any  DIFFICULTY in such cases as Great Britain having been
0356  lected as presenting the greatest amount of  DIFFICULTY on the theory of single centres of creation
0357  the most distant points, and removes many a  DIFFICULTY: but to the best of my judgment we are not
0370  wider spaces of ocean. I believe the above  DIFFICULTY may be surmounted by looking to still earli
0381  modification, a far more remakable case of  DIFFICULTY. For some of these species are so distinct,
0389  lected as presenting the greatest amount of  DIFFICULTY, on the view that all the individuals both
0392  nd fur of quadrupeds. This case presents no  DIFFICULTY on my view, for a hooked seed might be tran
0393  y view we can see that there would be great  DIFFICULTY in their transportal across the sea, and th
0395  n this great archipelago, and there is much  DIFFICULTY in forming a judgment in some cases owing t
0397  ve a single instance of one of the cases of  DIFFICULTY. Almost all oceanic islands, even the most
0399  that the fact becomes an anomaly. But this  DIFFICULTY almost disappears on the view that both New
0400  ll degree. This long appeared to me a great  DIFFICULTY: but it arises in chief part from the deepl
0431  is aid, we can understand the extraordinary  DIFFICULTY which naturalists has experienced in descr
0456  uite aborted, far from presenting a strange  DIFFICULTY, as they assuredly do on the ordinary doctr
0459  he individual possessor. Nevertheless, this  DIFFICULTY, though appearing to our imagination insupe
0460  are, it must be admitted, cases of special  DIFFICULTY on the theory of natural selection; and one
0460  ants: but I have attempted to show how this  DIFFICULTY can be mastered. With respect to the almost
0462  ng a very long period; and consequently the  DIFFICULTY of the wide diffusion of species of the sam
0466  physical conditions of life. There is much  DIFFICULTY in ascertaining how much modification our d
0474  rvellous as some are, they offer no greater  DIFFICULTY than does corporeal structure on the thoery
0481  h we do not see the intermediate steps. The  DIFFICULTY is the same as that felt by so many geologi
0483  properly demand a full explanation of every  DIFFICULTY from those who believe in the mutabliity of
0044  rences. Doubtful species. Wide ranging, much  DIFFUSED, and common species vary most. Species of th
0053  duals, and the species which are most widely  DIFFUSED within their own country (and this is a diff
0054  ch range widely over the world, are the most  DIFFUSED in their own country, and are the most numer
0054  at larger number of the very common and much  DIFFUSED or dominant species will be found on the sid
0054  have generally very wide ranges and are most  DIFFUSED, but this seems to be connected with the nat
0055  organisation are generally much more widely  DIFFUSED than plants higher in the scale: and here ag
0117  which are the commonest and the most widely  DIFFUSED, vary more than rare species with restricted
0117  stricted ranges. Let (A) be a common, widely  DIFFUSED, and varying species, belonging to a genus l
0119  odified descendants from a common and widely  DIFFUSED species, belonging to a large genus, will te
```

Page ***(Key Word)***

diffused

Page **************************************(Key Word)**************************************

```
0122 e also supposed to be very common and widely DIFFUSED species, so that they must originally have h
0128 have seen that it is the common, the widely DIFFUSED, and widely ranging species, belonging to th
0232 ly found that the colour was most delicately DIFFUSED by the bees, as delicately as a painter coul
0325 nest in their own homes, and are most widely DIFFUSED, having produced the greatest number of new
0411 show that it is the widely ranging, the much DIFFUSED and common, that is the dominant species bel
0065 the extraordinarily rapid increase and wide DIFFUSION of naturalised productions in their new hom
0070 cause, possibly in part through facility of DIFFUSION amongst the crowded animals, been dispropor
0326 to new varieties and species. The process of DIFFUSION may often be very slow, being dependent on
0326 rms will generally succeed in spreading. The DIFFUSION would, it is probable, be slower with the t
0405 e widely, undergoing modification during its DIFFUSION, and should place itself under diverse cond
0462 ought not to be laid on the occasional wide DIFFUSION of the same species: for during very long p
0462 and consequently the difficulty of the wide DIFFUSION of species of the same genus is in some deg
0475 generation has once been broken. The gradual DIFFUSION of dominant forms, with the slow modificati
0116 ist doubts that a stomach by being adapted to DIGEST vegetable matter alone, or flesh alone, draws
0190 nside out, and the exterior surface will then DIGEST and the stomach respire. In such cases natura
0361 f fruit will pass uninjured through even the DIGESTIVE organs of a turkey. In the course of two mo
0190 unctions; thus the alimentary canal respires, DIGESTS, and excretes in the larva of the dragonfly
0434 man, formed for grasping, that of a mole for DIGGING, the leg of the horse, the paddle of the por
0450 rd wing in birds may be safely considered as a DIGIT in a rudimentary state: in very many snakes o
0186 this seems to me only restating the fact in DIGNIFIED language. He who believes in the struggle f
0240 w members in an intermediate condition. I may DIGRESS by adding, that if the smaller workers had b
0096 individuals. I must here introduce a short DIGRESSION. In the case of animals and plants with sep
0095 the excavation of a great valley by a single DILUVIAL wave, so will natural selection, if it be a
0295 rea whence the sediment is derived, and thus DIMINISH the supply whilst the downward movement cont
0096 ir for each birth. Modern research has much DIMINISHED the number of supposed hermaphrodites, and
0097 elsewhere, that their fertility is greatly DIMINISHED if these visits be prevented. Now, it is sc
0147 s on the head is generally accompanied by a DIMINISHED comb, and a large beard by diminished wattl
0147 by a diminished comb, and a large beard by DIMINISHED wattles. With species in a state of nature
0209 and are increased by, use or habit, and are DIMINISHED or lost by disuse; so I do not doubt it has
0248 ll these experiments the fertility has been DIMINISHED by an independent cause, namely, from close
0311 he difficulties above discussed are greatly DIMINISHED, or even disappear. Chapter X. On the Geolo
0345 the theory of natural selection are greatly DIMINISHED or disappear. On the other hand, all the ch
0096 on the other hand, that close interbreeding DIMINISHES vigour and fertility; that these facts alon
0134 hens and enlarges certain parts, and disuse DIMINISHES them; and that such modifications are inher
0266 ity, though in some degree variable, rarely DIMINISHES. It must, however, be confessed that we can
0168 strengthening, and disuse in weakening and DIMINISHING organs, seem to have been more potent in t
0148 ture before useful becomes less useful, any DIMINUTION, however slight, in its development, will b
0011 regulating variation, some few of which can be DIMLY seen, and will be hereafter briefly mentioned
0012 . the result of the various, quite unknown, or DIMLY seen laws of variation is infinitely complex
0062 abundance, extinction, and variation, will be DIMLY seen or quite misunderstood. We behold the fa
0076 upplant another, and so in other cases. We can DIMLY see why the competition should be most severe
0132 ignorant. Nevertheless, we can here and there DIMLY catch a faint ray of light, and we may feel s
0484 analogous views are generally admitted, we can DIMLY foresee that there will be a considerable rev
0030 l contrivance, is only a variety of the wild DIPSACUS; and this amount of change may have suddenly
0010 s of life; and this shows how unimportant the DIRECT effects of the conditions of life are in comp
0010 ce; for had the action of the conditions been DIRECT, if any of the young had varied, all would pr
0010 of any variation, we should attribute to the DIRECT action of heat, moisture, light, food, etc.,
0010 imals such agencies have produced very little DIRECT effect, though apparently more in the case of
0011 of change may, I think, be attributed to the DIRECT action of the conditions of life, as, in some
0015 ome effect would have to be attributed to the DIRECT action of the poor soil), that they would to
0029 tle effect may, perhaps, be attributed to the DIRECT action of the external conditions of life, an
0043 of growth. Something may be attributed to the DIRECT action of the conditions of life. Something m
0069 tempted to attribute the whole effect to its DIRECT action. But this is a very false view: we for
0085 food, etc., probably produce some slight and DIRECT effect. It is, however, far more necessary to
0132 iation of structure, however slight. How much DIRECT effect difference of climate, food, etc., pro
0133 during many generations, and how much to the DIRECT action of the severe climate? for it would ap
0133 te? for it would appear that climate has some DIRECT action on the hair of our domestic quadrupeds
0134 e incline me to lay very little weight on the DIRECT action of the conditions of life. Indirectly,
0196 limate, food, etc., probably have some little DIRECT influence on the organisation; that character
0197 head of a vulture is generally looked at as a DIRECT adaptation for wallowing in putridity; and so
0197 o it may be; or it may possibly be due to the DIRECT action of putrid matter; but we should be ver
0199 i fully admit that many structures are of no DIRECT use to their possessors. Physical conditions
0199 iled on other parts diversified changes of no DIRECT use. So again characters which formerly were
0199 r from the law of reversion, though now of no DIRECT use. The effects of sexual selection, when di
0199 place in nature, many structures now have no DIRECT relation to the habits of life of each specie
0200 reature (making some little allowance for the DIRECT action of physical conditions) may be viewed,
0200 can and does often produce structures for the DIRECT injury of other species, as we see in the fan
0206 se and disuse, being slightly affected by the DIRECT action of the external conditions of life, an
0208 l actions are performed, indeed not rarely in DIRECT opposition to our conscious will! yet they ma
0254 believe in its truth, although it rests on no DIRECT evidence. I believe, for instance, that our d
0281 tance, a horse from a tapir; and in this case DIRECT intermediate links will have existed between
0350 ion, and in a quite subordinate degree to the DIRECT influence of different physical conditions. T
0351 instance, a number of species, which stand in DIRECT competition with each other, migrate in a bod
0373 moraines. South of the equator, we have some DIRECT evidence of former glacial action in New Zeal
0373 ch has been published can be trusted, we have DIRECT evidence of glacial action in the south easte
0408 hich entered happened to come in more or less DIRECT competition with each other and with the abor
0436 in a spire. In monstrous plants, we often get DIRECT evidence of the possibility of one organ bein
0439 same class resemble each other, often have no DIRECT relation to their conditions of existence. We
0444 at a not very early period of life; and some DIRECT evidence from our domestic animals supports t
0466 tion of growth, by use and disuse; and by the DIRECT action of the physical conditions of life. Th
0472 l conditions seem to have produced but little DIRECT effect; yet when varieties enter any zone, th
0486 wth, on the effects of use and disuse, on the DIRECT action of external conditions, and so forth.
0489 production; Variability from the indirect and DIRECT action of the external conditions of life, an
0330 lepidosiren, is discovered having affinities DIRECTED towards very distinct groups. Yet if we comp
0032 t effect produced by the accumulation in one DIRECTION, during successive generations, of differen
0041 amount of modification in almost any desired DIRECTION. But as variations manifestly useful or ple
```

Page **************************************(Key Word)**************************************

```
0045 me manner as man can accumulate in any given DIRECTION individual differences in his domesticated
0082 duce great results by adding up in any given DIRECTION mere individual differences, so could Natur
0102 rve all the individuals varying in the right DIRECTION, though in different degrees, so as better
0212 , the migratory instinct, both in extent and DIRECTION, and in its total loss. So it is with the n
0243 ons of instinct to any extent, in any useful DIRECTION. In some cases habit or use and disuse have
0261 element of the other, but not in a reversed DIRECTION. It will be advisable to explain a little m
0364 ts of the sea are not accidental, nor is the DIRECTION of prevalent gales of wind. It should be ob
0379 m the north to the south, than in a reversed DIRECTION. We see, however, a few southern vegetable
0433 , in certain classes, we are tending in this DIRECTION; and Milne Edwards has lately insisted, in
0030 ive variations; man adds them up in certain DIRECTIONS useful to him. In this sense he may be said
0052 ifferences of structure in certain definite DIRECTIONS. Hence I believe a well marked variety may
0123 osed to have gone on diverging in different DIRECTIONS. The intermediate species, also (and this I
0430 ce undergone much modification in divergent DIRECTIONS. On either view we may suppose that the biz
0002 ced, often apparently leading to conclusions DIRECTLY opposite to those at which I have arrived. A
0010 same way, the change at first appears to be DIRECTLY due to such conditions; but in some cases it
0035 authorities are convinced that the setter is DIRECTLY derived from the spaniel, and has probably b
0044 technical sense, as implying a modification DIRECTLY due to the physical conditions of life, and
0069 hen climate, for instance extreme cold, acts DIRECTLY, it will be the least vigorous, or those whi
0069 oftener meet with stunted forms, due to the DIRECTLY injurious action of climate, than we do in p
0099 hen distinct species are crossed the case is DIRECTLY the reverse, for a plant's own pollen is alw
0124 r parents, the new species (F14) will not be DIRECTLY intermediate between them, but rather betwee
0150 s which are attached to one sex, but are not DIRECTLY connected with the act of reproduction. The
0175 in short, that each organic being is either DIRECTLY or indirectly related in the most important
0200 use to the descendants of this form, either DIRECTLY, or indirectly through the complex laws of g
0232 inc, for instance, on a slip of wood, placed DIRECTLY under the middle of a comb growing downwards
0269 reature's own good; and thus she may, either DIRECTLY, or more probably indirectly, through correl
0274 is founded on a single experiment; and seems DIRECTLY opposed to the results of several experiment
0280 species, to avoid picturing to myself, forms DIRECTLY intermediate between them. But this is a who
0280 rock pigeon; but we should have no varieties DIRECTLY intermediate between the fantail and pouter;
0281 no reason to suppose that links ever existed DIRECTLY intermediate between them, but between each
0330 his term it is meant that an extinct form is DIRECTLY intermediate in all its characters between t
0332 be justified, as they are intermediate, not DIRECTLY, but only by a long and circuitous course th
0345 e widely divergent. Extinct forms are seldom DIRECTLY intermediate between existing forms; but are
0353 s, to the individuals of the same species, a DIRECTLY opposite rule prevailed; and species were no
0355 ngle and multiple centres of creation do not DIRECTLY bear on another allied question, namely whet
0461 lelism is supported by another parallel, but DIRECTLY opposite, class of facts; namely, that the v
0462 expect (excepting in rare cases) to discover DIRECTLY connecting links between them, but only betw
0481 a long course of years, from a point of view DIRECTLY opposite to mine. It is so easy to hide our
0490 amely, the production of the higher animals, DIRECTLY follows. There is grandeur in this view of l
0364 these wanderers only by one means, namely, in DIRT sticking to their feet, which is in itself a r
0386 alight on the surface of the sea, so that the DIRT would not be washed off their feet; when makin
0163 ; but here, as before, I lie under a great DISADVANTAGE in not being able to give them. I can only
0180 as on other occasions, I lie under a heavy DISADVANTAGE, for out of the many striking cases which
0074 red clover would become very rare, or wholly DISAPPEAR. The number of humble bees in any district
0076 kinds will steadily decrease in numbers and DISAPPEAR. So again with the varieties of sheep: it h
0112 have been neglected, and will have tended to DISAPPEAR. Here, then, we see in man's productions th
0126 n groups and sub groups will finally tend to DISAPPEAR. Looking to the future, we can predict that
0164 re plainest in the foal; and sometimes quite DISAPPEAR in old horses. Colonel Poole has seen both
0176 general rule they should be exterminated and DISAPPEAR, sooner than the forms which they originall
0234 ve seen, the spherical surfaces would wholly DISAPPEAR, and would all be replaced by plane surface
0288 e preserved. Shells and bones will decay and DISAPPEAR when left on the bottom of the sea, where s
0294 ganic remains will probably first appear and DISAPPEAR at different levels, owing to the migration
0311 ve discussed are greatly diminished, or even DISAPPEAR. Chapter X. On the Geological Succession of
0317 that species and groups of species gradually DISAPPEAR, one after another, first from one spot, th
0327 heritance, whole groups would tend slowly to DISAPPEAR; though here and there a single member migh
0327 spread throughout the world, old groups will DISAPPEAR from the world; and the succession of forms
0345 natural selection are greatly diminished or DISAPPEAR. On the other hand, all the chief laws of p
0394 inents it is thought that mammals appear and DISAPPEAR at a quicker rate than other and lower anim
0417 , to the genus, to the family, to the class, DISAPPEAR, and thus laugh at our classification. But
0452 etected in the embryo, but afterwards wholly DISAPPEAR. It is also, I believe, a universal rule, t
0312 same general rules in their appearance and DISAPPEARANCE as do single species. On Extinction. On s
0313 n has remarked, neither the appearance nor DISAPPEARANCE of their many now extinct species has bee
0316 same general rules in their appearance and DISAPPEARANCE as do single species, changing more or le
0317 ave as yet spoken only incidentally of the DISAPPEARANCE of species and of groups of species. On t
0318 an their production: if the appearance and DISAPPEARANCE of a group of species be represented, as
0320 . thus the appearance of new forms and the DISAPPEARANCE of old forms, both natural and artificial
0335 hat the record of the first appearance and DISAPPEARANCE of the species was perfect, we have no re
0335 on, and still less with the order of their DISAPPEARANCE; for the parent rock pigeon now lives; an
0056 e often come to their maxima, declined, and DISAPPEARED. All that we want to show is, that where m
0191 higher Vertebrata the branchiae have wholly DISAPPEARED, the slits on the sides of the neck and th
0296 ng the whole period of deposition, but have DISAPPEARED and reappeared, perhaps many times, during
0313 ergone some change. When a species has once DISAPPEARED from the face of the earth, we have reason
0316 a group does not reappear after it has once DISAPPEARED; or its existence, as long as it lasts, is
0318 awn of life to the present day; some having DISAPPEARED before the close of the palaeozoic period.
0343 can understand why when a species has once DISAPPEARED it never reappears. Groups of species incr
0344 ed situations. When a group has once wholly DISAPPEARED, it does not reappear; for the link of gen
0069 ually getting rarer and rarer, and finally DISAPPEARING; and the change of climate being conspicuo
0174 rer and rarer on the confines, and finally DISAPPEARING. Hence the neutral territory between two r
0269 xpect to find sterility both appearing and DISAPPEARING under nearly the same conditions of life.
0293 y existed. So again when we find a species DISAPPEARING before the uppermost layers have been depo
0175 dolle has observed, a common alpine species DISAPPEARS. The same fact has been noticed by Forbes i
0237 insuperable, is lessened, or, as I believe, DISAPPEARS, when it is remembered that selection may b
0399 ed by the prevailing currents, this anomaly DISAPPEARS. New Zealand in its endemic plants is much
0399 omes an anomaly. But this difficulty almost DISAPPEARS on the view that both New Zealand, South Am
0004 out this obscure problem. Nor have I been DISAPPOINTED: in this and in all other perplexing cases
0342 stages of the same great formation. He may DISBELIEVE in the enormous intervals of time which hav
```

Page **(Key Word)**

```
0029  e on pears and apples, shows how utterly he DISBELIEVES that the several sorts, for instance a Rib
0193  ratus, and yet do not, as Matteucci asserts, DISCHARGE any electricity, we must own that we are fa
0353  l times, must have interrupted or rendered DISCONTINUOUS the formerly continuous range of many spe
0014  ate of nature. I have in vain endeavoured to DISCOVER on what decisive facts the above statement h
0101  st authorities, namely, Professor Huxley, to DISCOVER a single case of an hermaphrodite animal wit
0187  th the lowest known fossiliferous stratum to DISCOVER the earlier stages, by which the eye has bee
0189  e class have been developed; and in order to DISCOVER the early transitional grades through which
0325  rn our attention to North America, and there DISCOVER a series of analogous phenomena, it will app
0422  , on a flat surface, the affinities which we DISCOVER in nature amongst the beings of the same gro
0462  right to expect (excepting in rare cases) to DISCOVER directly connecting links between them, but
0464  se many links we could hardly ever expect to DISCOVER, owing to the imperfection of the geological
0479  enealogical arrangement, in which we have to DISCOVER the lines of descent by the most permanent c
0486  digrees or armorial bearings; and we have to DISCOVER and trace the many diverging lines of descen
0059  at intermediate linking forms have not been DISCOVERED; but the amount of difference considered ne
0135  due to natural selection. Mr. Wollaston has DISCOVERED the remarkable fact that 200 beetles, out o
0210  ading to the most complex instincts, can be DISCOVERED. The canon of Natura non facit saltum appli
0211  , to be well aware what a rich flock it had DISCOVERED; it then began to play with its antennae on
0219  nstinct. This remarkable instinct was first DISCOVERED in the Formica (Polyerges) rufescens by Pie
0219  ected. Formica sanguinea was likewise first DISCOVERED by P. Huber to be a slave making ant. This
0289  r of these vast periods, with one exception DISCOVERED by Sir C. Lyell in the carboniferous strata
0289  of the bones of tertiary mammals have been DISCOVERED either in caves or in lacustrine deposits;
0303  ondary series; and one true mammal has been DISCOVERED in the new red sandstone at nearly the comm
0303  stratum: but now extinct species have been DISCOVERED in India, South America, and in Europe even
0303  of not a single bone of a whale having been DISCOVERED in any secondary formation, seemed fully to
0304  hey would certainly have been preserved and DISCOVERED; and as not one species had been discovered
0304  discovered; and as not one species had been DISCOVERED in beds of this age, I concluded that this
0329  st distinct orders of mammals; but Owen has DISCOVERED so many fossil links, that he has had to al
0330  ven a living animal, as the lepidosiren, is DISCOVERED having affinities directed towards very dis
0332  embedded in the successive formations, were DISCOVERED at several points low down in the series, t
0332  ent forms. If many extinct forms were to be DISCOVERED above one of the middle horizontal lines or
0334  devorian system, when this system was first DISCOVERED, were at once recognised by palaeontologist
0338  far beneath the lowest Silurian strata are DISCOVERED, a discovery of which the chance is very sm
0363  her means, which without doubt remain to be DISCOVERED, have been in action year after year, for c
0375  ropean forms are found, which have not been DISCOVERED in the intertropical parts of Africa. On th
0375  n mountains of Australia, Dr. F. Muller has DISCOVERED several European species; other species, no
0410  irly be attributed to our not having as yet DISCOVERED in an intermediate deposit the forms which
0418  though no apparent bond of connexion can be DISCOVERED between them, especial value is set on them
0419  rades of difference, have been subsequently DISCOVERED. All the foregoing rules and aids and diffi
0464  in future ages so many fossil links will be DISCOVERED, that naturalists will be able to decide, o
0464  if any one link or intermediate variety be DISCOVERED, it will simply be classed as another and d
0465  d and improved; and when they do spread, if DISCOVERED in a geological formation, they will appear
0172  lls, which have practically anticipated the DISCOVERIES of profound mathematicians? Fourthly, how
0288  part with sufficient care, as the important DISCOVERIES made every year in Europe prove. No organi
0306  ontological ideas on many points, which the DISCOVERIES of even the last dozen years have effected
0340  manner we know from Falconer and Cautley's DISCOVERIES, that Northern India was formerly more clo
0183  in the battle for life. Hence the chance of DISCOVERING species with transitional grades of struct
0298  gree. According to this view, the chance of DISCOVERING in a formation in any one country all the
0299  ld be effected only by the future geologist DISCOVERING in a fossil state numerous intermediate gr
0302  resented, had not the difficulty of our not DISCOVERING innumerable transitional links between the
0438  ght successive steps, we need not wonder at DISCOVERING in such parts or organs, a certain degree
0033  far from true that the principle is a modern DISCOVERY. I could give several references to the ful
0058  uished from species, except, firstly, by the DISCOVERY of intermediate linking forms, and the occu
0065  ch have been imported from America since its DISCOVERY. In such cases, and endless instances could
0292  be said to have guarded against the frequent DISCOVERY of her transitional or linking forms. From
0322  out the World. Scarcely any palaeontological DISCOVERY is more striking than the fact, that the fo
0333  ct from each other than they were before the DISCOVERY of the fossils. If, for instance, we suppos
0338  the lowest Silurian strata are discovered, a DISCOVERY of which the chance is very small. On the S
0429  ut the world, the fact is striking, that the DISCOVERY of Australia has not added a single insect
0452  common, or more necessary, than the use and DISCOVERY of rudiments. This is well shown in the dra
0464  en at first local, both causes rendering the DISCOVERY of intermediate links less likely. Local va
0001  es of facts. We shall, however, be enabled to DISCUSS what circumstances are most favourable to va
0005  nce of Character. In the next chapter I shall DISCUSS the complex and little known laws of variati
0044  beings in a state of nature, we must briefly DISCUSS whether these latter are subject to any vari
0044  reserve for my future work. Nor shall I here DISCUSS the various definitions which have been give
0049  etent judges as good and true species. But to DISCUSS whether they are rightly called species or v
0057  pecies of the same genus. But when we come to DISCUSS the principle, as I call it, of Divergence o
0062  ks of Nature are to those of Art. We will now DISCUSS in a little more detail the struggle for exi
0067  eral authors, and I shall, in my future work, DISCUSS some of the checks at considerable length, m
0172  the earth? It will be much more convenient to DISCUSS this question in the chapter on the Imperfec
0216  ree, out of the several which I shall have to DISCUSS in my future work, namely, the instinct whic
0272  in very many important respects. I shall here DISCUSS this subject with extreme brevity. The most
0306  me one barren point in Australia, and then to DISCUSS the number and range of its productions. On
0354  s it could. It would be hopelessly tedious to DISCUSS all the exceptional cases of the same specie
0354  s. but after some preliminary remarks, I will DISCUSS a few of the most striking classes of facts;
0356  passable; I shall, however, presently have to DISCUSS this branch of the subject in some detail. C
0002  m well aware that scarcely a single point is DISCUSSED in this volume on which facts cannot be add
0016  ties when crossed, a subject hereafter to be DISCUSSED), domestic races of the same species differ
0028  be kept together in the same aviary. I have DISCUSSED the probable origin of domestic pigeons at
0053  racter, and other questions, hereafter to be DISCUSSED. Alph. de Candolle and others have shown th
0055  owly organised plants ranging widely will be DISCUSSED in our chapter on geographical distribution
0080  beings. How will the struggle for existence, DISCUSSED too briefly in the last chapter, act in reg
0109  extinction. This subject will be more fully DISCUSSED in our chapter on Geology: but it must be h
0172  nimpaired? The two first heads shall be here DISCUSSED, Instinct and Hybridism in separate chapter
0192  s of difficulty occur, some of which will be DISCUSSED in my future work. One of the gravest is th
0203  summary of Chapter. We have in this chapter DISCUSSED some of the difficulties and objections whi
0236  their kind. The subject well deserves to be DISCUSSED at great length, but I will here take only
```

Page **(Key Word)**

Page ***(Key Word)***

```
0242  convinced me of the fact. I have, therefore, DISCUSSED this case, at some little but wholly insuff
0279  d in this volume. Most of them have now been DISCUSSED. One, namely the distinctness of specific f
0310  sed condition. The several difficulties here DISCUSSED, namely our not finding in the successive f
0311  ations. On this view, the difficulties above DISCUSSED are greatly diminished, or even disappear.
0352  ought to the question which has been largely DISCUSSED by naturalists, namely, whether species hav
0381  the antarctic regions. These cannot be here DISCUSSED. I will only say that as far as regards the
0406  bserved, and which has lately been admirably DISCUSSED by Alph. de Candolle in regard to plants, n
0406  idely it is apt to range. The relations just DISCUSSED, namely, low and slowly changing organisms
0407  re the means of occasional transport, I have DISCUSSED at some little length the means of dispersa
0354  aw, seems to me incomparably the safest. In  DISCUSSING this subject, we shall be enabled at the sa
0356  individuals during many generations. Before  DISCUSSING the three classes of facts, which I have se
0053  rms. I shall reserve for my future work the  DISCUSSION of these difficulties, and the tables thems
0096  i have the materials prepared for an ample   DISCUSSION. All vertebrate animals, all insects, and s
0116  e stage of development. After the foregoing  DISCUSSION, which ought to have been much amplified, w
0203  hem are very grave; but I think that in the  DISCUSSION light has been thrown on several facts, whi
0336  pment of Ancient Forms. There has been much  DISCUSSION whether recent forms are more highly develo
0419  elopment of the plumule and radicle. In our  DISCUSSION on embryology, we shall see why such charac
0085  t purple plums suffer far more from a certain DISEASE than yellow plums; whereas another disease a
0085  in disease than yellow plums; whereas another DISEASE attacks yellow fleshed peaches far more than
0195  cth reduced, so that they are more subject to DISEASE, or not so well enabled in a coming dearth t
0443  ay appear late in life; as when an hereditary DISEASE, which appears in old age alone, has been co
0014  caterpillar or cocoon stage. But hereditary  DISEASES and some other facts make me believe that th
0434  es of affinities. We shall never, probably,  DISENTANGLE the inextricable web of affinities between
0362  an interval of from twelve to twenty hours,  DISGORGE pellets, which, as I know from experiments m
0332  he genera a1, a5, a10, f8, m3, m6, m9, were  DISINTERRED, these three families would be so closely
0372  out identity, of the inhabitants of seas now DISJOINED, and likewise of the past and present inhab
0089  t they often take individual preferences and DISLIKES: thus Sir R. Heron has described how one pie
0285  on of the sea, that no trace of these vast   DISLOCATIONS is externally visible. The Craven fault, f
0006  doubt, after the most deliberate study and   DISPASSIONATE judgment of which I am capable, that the
0147  ge over those producing seed less fitted for DISPERSAL; and this process could not possibly go on
0346  ame continent. Centres of creation. Means of DISPERSAL, by changes of climate and of the level of
0346  level of the land, and by occasional means.  DISPERSAL during the Glacial period co extensive with
0353  reas some plants, from their varied means of DISPERSAL, have migrated across the vast and broken i
0356  tion, I must say a few words on the means of DISPERSAL. Means of Dispersal. Sir C. Lyell and other
0356  ew words on the means of dispersal. Means of DISPERSAL. Sir C. Lyell and other authors have ably t
0357  nent. This view cuts the Gordian knot of the DISPERSAL of the same species to the most distant pqi
0365  ive, would be sure to germinate and survive. DISPERSAL during the Glacial period. The identity of
0381  lyell, have been largely concerned in their  DISPERSAL. But the existence of several quite distinc
0383  from stream to stream; and liability to wide DISPERSAL would follow from this capacity as an almos
0384  he water. But I am inclined to attribute the DISPERSAL of fresh water fish mainly to slight change
0386  very wide range. I think favourable means of DISPERSAL explain this fact. I have before mentioned
0388  dified, I believe mainly depends on the wide DISPERSAL of their seeds and eggs by animals, more es
0389  l not confine myself to the mere question of DISPERSAL; but shall consider some other facts, which
0407  discussed at some little length the means of DISPERSAL of fresh water productions. If the difficul
0476  to the many occasional and unknown means of  DISPERSAL, then we can understand, on the theory of d
0382  y the Glacial epoch, a few forms were widely DISPERSED to various points of the southern hemispher
0391  ow, though we do not know how sea shells are DISPERSED, yet we can see that their eggs or larvae,
0111  lly known that the ancient black cattle were DISPLACED by the long horns, and that these were swep
0285  30 miles, and along this line the vertical   DISPLACEMENT of the strata has varied from 600 to 3000
0008  ted instincts; but how many cultivated plants DISPLAY the utmost vigour, and yet rarely or never s
0042  ally from no pleasure having been felt in the DISPLAY of distinct breeds. To sum up on the origin
0074  indian mounds, in the Southern United States, DISPLAY the same beautiful diversity and proportion
0089  some others, congregate; and successive males DISPLAY their gorgeous plumage and perform strange a
0287  richest geological museums; and what a paltry DISPLAY we behold! On the poorness of our Palaeontol
0349  o the same order of Rodents, but they plainly DISPLAY an American type of structure. We ascend the
0150  he case of secondary sexual characters, when  DISPLAYED in any unusual manner. The term, secondary
0151  iability of these characters, whether or not  DISPLAYED in any unusual manner, of which fact I thin
0156  ich secondary sexual characters are strongly  DISPLAYED, with the amount of difference between thei
0157  two sexes of the same species are generally   DISPLAYED in the very same parts of the organisation
0158  ordinary specific differences are generally   DISPLAYED in the same parts of the organisation, ane
0166  his tendency to become striped most strongly  DISPLAYED in hybrids from between several of the most
0199  t use. The effects of sexual selection, when  DISPLAYED in beauty to charm the females, can be call
0214  re terrier. When the first tendency was once  DISPLAYED, methodical selection and the inherited eff
0339  n of most genera of molluscs, it is not well  DISPLAYED by them. Other cases could be added, as the
0400  e of the nearest continent, we sometimes see  DISPLAYED on a small scale, yet in a most interestinc
0404  ve it. Amongst mammals, we see it strikingly  DISPLAYED in Bats, and in a lesser degree in the Feli
0443  been caused; but at what period it is fully   DISPLAYED. The cause may have acted, and I believe ge
0457  ligible, whether we look to the same pattern  DISPLAYED in the homologous organs, to whatever purpo
0182  flight, we should bear in mind that animals   DISPLAYING early transitional grades of the structure
0344  nt a form is, the more often, apparently, it DISPLAYS characters in some degree intermediate betwe
0082  from having incomparably longer time at her   DISPOSAL. Nor do I believe that any great physical ch
0065  ure an equal number would have somehow to be  DISPOSED of. The only difference between organisms wh
0022  kably; as does in some breeds the voice and   DISPOSITION. Lastly, in certain breeds, the males and
0212  the hooded crow in Egypt. That the general   DISPOSITION of individuals of the same species, born i
0213  e given of the inheritance of all shades of   DISPOSITION and tastes, and likewise of the oddest tri
0482  when we only restate a fact. Any one whose   DISPOSITION leads him to attach more weight to unexpla
0213  ngly these domestic instincts, habits, and   DISPOSITIONS are inherited, and how curiously they beco
0070  fusion amongst the crowded animals, been DISPROPORTIONABLY favoured: and here comes in a sort of s
0017  ewise to withstand diverse climates. I do not DISPUTE that these capacities have added largely to
0192  rge folded branchiae. Now I think no one will DISPUTE that the ovigerous frena in the one family a
0243  ry slightly in a state of nature. No one will DISPUTE that instincts are of the highest importance
0307  tent judges, as Lyell and the late E. Forbes, DISPUTE this conclusion. We should not forget that o
0357  rom one to the other. No other geologist will DISPUTE that great mutations of level, have occurred
0416  of high value in classification. No one will DISPUTE that the rudimentary teeth in the upper jaws
0441  inc higher or lower. But no one probably will DISPUTE that the butterfly is higher than the caterp
0008  pid improvement or modification. It has been  DISPUTED at what period of life the causes of variabi
```

Page ***(Key Word)***

disputed

Page **(Key Word)**

```
0048  btful nature are far from uncommon cannot be DISPUTED. Compare the several floras of Great Britain
0051  nd the truth of this admission will often be DISPUTED by other naturalists. When, moreover, he com
0060  ndeed I am not aware that this has ever been DISPUTED. It is immaterial for us whether a multitude
0127  eir organisation, and I think this cannot be DISPUTED; if there be, owing to the high geometrical
0127  uggle for life, and this certainly cannot be DISPUTED; then, considering the infinite complexity o
0329  ting genera, families, and orders, cannot be DISPUTED. For if we confine our attention either to t
0400  inhabitants; whereas it cannot, I think, be DISPUTED that the nature of the other inhabitants, wi
0459  th of these propositions cannot, I think, be DISPUTED. It is, no doubt, extremely difficult even t
0247  so, makes the rule equally universal; and he DISPUTES the entire fertility of Kolreuter's ten case
0484  ience, will be no slight relief. The endless DISPUTES whether or not some fifty species of British
0039  er part of its oesophagus, a habit which is DISREGARDED by all fanciers, as it is not one of the p
0040  only slightly valued, their history will be DISREGARDED. When further improved by the same slow an
0168  mentary organs, from being useless, will be DISREGARDED by natural selection, and hence probably a
0043  s is immense; for the cultivator here quite DISREGARDS the extreme variability both of hybrids and
0240  e their ocelli rudimentary. Having carefully DISSECTED several specimens of these workers, I can a
0241  th the camera lucida of the jaws which I had DISSECTED from the workers of the several sizes. With
0137  n this condition, the cause, as appeared on DISSECTION, having been inflammation of the nictitatin
0063  in order to tempt birds to devour and thus DISSEMINATE its seeds rather than those of other plant
0063  uggle with each other. As the missletoe is DISSEMINATED by birds, its existence depends on birds;
0086  ant to have its seeds more and more widely DISSEMINATED by the wind, I can see no greater difficul
0358  plant is stated to be ill adapted for wide DISSEMINATION; but for transport across the sea, the gr
0046  nts of structure which are of no service or DISSERVICE to the species, and which consequently have
0251  ifolia and plantaginea, species most widely DISSIMILAR in general habit, reproduced itself as perf
0339  uld attempt to account, on the one hand, by DISSIMILAR physical conditions for the dissimilarity o
0347  nt out three faunas and floras more utterly DISSIMILAR. Or again we may compare the productions of
0372  ditions, but with their inhabitants utterly DISSIMILAR. But we must return to our more immediate s
0263  ty, by which every kind of resemblance and DISSIMILARITY between organic beings is attempted to be
0269  place, be clearly shown that mere external DISSIMILARITY between two species does not determine th
0339  by dissimilar physical conditions for the DISSIMILARITY of the inhabitants of these two continent
0346  us is, that neither the similarity nor the DISSIMILARITY of the inhabitants of various regions can
0350  e of varieties nearly like each other. The DISSIMILARITY of the inhabitants of different regions m
0350  fferent physical conditions. The degree of DISSIMILARITY will depend on the migration of the more
0383  sh water insects, shells, etc., and at the DISSIMILARITY of the surrounding terrestrial beings, co
0398  can coast: in fact there is a considerable DISSIMILARITY in all these respects. On the other hand,
0400  ginal source, or from each other. But this DISSIMILARITY between the endemic inhabitants of the is
0190  with gills or branchiae that breathe the air DISSOLVED in the water, at the same time that they br
0288  will when the beds are upraised generally be DISSOLVED by the percolation of rain water. I suspect
0479  those in a fish which has to breathe the air DISSOLVED in water, by the aid of well developed bran
0329  me sub order with ruminants: for example, he DISSOLVES the apparently wide diff
0049  ed species peculiar to Great Britain. A wide DISTANCE between the homes of two doubtful forms lead
0049  s to rank both as distinct species; but what DISTANCE, it has been well asked, will suffice? If th
0180  m to glide through the air to an astonishing DISTANCE from tree to tree. We cannot doubt that each
0226  ayers: with the centre of each sphere at the DISTANCE of radius x sq. Rt. 2, or radius x 1.41421 (
0226  meliopna had made its spheres at some given DISTANCE from each other, and had made them of equal
0226  t. 2, or radius x 1.41421 (or at some lesser DISTANCE) from the centres of the six surrounding sph
0226  g spheres in the same layer: and at the same DISTANCE from the centres of the adjoining spheres in
0227  hat she can somehow judge accurately at what DISTANCE to stand from her fellow labourers when seve
0227  ut she is already so far enabled to judge of DISTANCE, that she always describes her spheres so as
0228  ogether, they had begun their work at such a DISTANCE from each other, that by the time the basins
0230  the proper shape, by standing at the proper DISTANCE from each other, by excavating at the same r
0232  instinctively standing at the same relative DISTANCE from each other, all trying to sweep equal s
0233  time, always standing at the proper relative DISTANCE from the parts of the cells just begun, swee
0235  d the bees to sweep equal spheres at a given DISTANCE from each other in a double layer, and to bu
0235  t they swept their spheres at one particular DISTANCE from each other, than they know what are the
0285  uth Downs: for, remembering that at no great DISTANCE to the west the northern and southern escarp
0286  the latter part of the Chalk formation. The DISTANCE from the northern to the southern Downs is a
0355  nd, for instance, upheaved and formed at the DISTANCE of a few hundreds of miles from a continent,
0362  n this interval might easily be blown to the DISTANCE of 500 miles, and hawks are known to look ou
0371  ip which is most remarkable, considering the DISTANCE of the two areas, and their separation by th
0391  the other hand, which lies at about the same DISTANCE from North America as the Galapagos Islands
0393  nd; and many islands situated at a much less DISTANCE are equally barren. The Falkland Islands, wh
0394  ularly or occasionally visit Bermuda, at the DISTANCE of 600 miles from the mainland. I hear from
0395  relation, to a certain extent independent of DISTANCE, between the depth of the sea separating an
0116  e diagram to the letters standing at unequal DISTANCES. I have said a large genus, because we have
0186  ivances for adjusting the focus to different DISTANCES, for admitting different amounts of light,
0189  nsities and thicknesses, placed at different DISTANCES from each other, and with the surfaces of e
0232  be enabled to stand at their proper relative DISTANCES from each other and from the walls of the l
0233  to five other points, at the proper relative DISTANCES from the central point and from each other,
0361  rds of many kinds are blown by gales to vast DISTANCES across the ocean. We may I think safely ass
0363  s might occasionally be transported to great DISTANCES; for many facts could be given showing that
0364  f transport would carry seeds for very great DISTANCES: for seeds do not retain their vitality whe
0387  port the seeds of fresh water plants to vast DISTANCES, and if consequently the range of these pla
0027  se breeds, more especially those brought from DISTANT countries, we can make an almost perfect ser
0071  , with a few clumps of old Scotch firs on the DISTANT hilltops: within the last ten years large sp
0072  ne square yard, at a point some hundred yards DISTANT from one of the old clumps, I counted thirty
0161  es after an ancestor some hundred generations DISTANT, but that in each successive generation ther
0165  anky Kattywar race, etc., inhabiting the most DISTANT parts of the world. Now let us turn to the e
0167  endency, when crossed with species inhabiting DISTANT quarters of the world, to produce hybrids re
0213  of at, a herd of deer, and driving them to a DISTANT point, we should assuredly call these action
0221  o a tall Scotch fir tree, twenty five yards DISTANT, which they ascended together probably in se
0221  dies as food to their nest, twenty nine yards DISTANT, but they were prevented from getting any pu
0243  certainly distinct, species, when inhabiting DISTANT parts of the world and living under consider
0285  o stand on the North Downs and to look at the DISTANT South Downs; for, remembering that at no gre
0290  marine faunas will probably be preserved to a DISTANT age. A little reflection will explain why at
0291  yet suffered, but which will hardly last to a DISTANT geological age, was certainly deposited duri
0294  ecies and to geographical changes. And in the DISTANT future, a geologist examining these beds, mi
```

Page **(Key Word)**

Page ***************************************(Key Word)***************************************

```
0300  dded in gravel or sand, would not endure to a  DISTANT  epoch. Wherever sediment did not accumulate
0300  of sufficient thickness to last to an age, as   DISTANT  in futurity as the secondary formations lie
0306  africa or Australia, and thus reach other and   DISTANT  seas. From these and similar considerations,
0321  eculiar line of life, or from inhabiting some    DISTANT  and isolated station, where they have escape
0322  ean Chalk formation can be recognised in many   DISTANT  parts of the world, under the most different
0323  , and in the peninsula of India. For at these   DISTANT  points, the organic remains in certain beds
0323  above or below, are similarly absent at these   DISTANT  points of the world. In the several successi
0323  however, relate to the marine inhabitants of    DISTANT  parts of the world: we have not sufficient d
0323  ions of the land and of fresh water change at   DISTANT  points in the same parallel manner. We may d
0324  simultaneously, in the above large sense, at    DISTANT  parts of the world, has greatly struck those
0351  of the same genus, though inhabiting the most   DISTANT  quarters of the world, must originally have
0352  ls of the same species, though now inhabiting   DISTANT  and isolated regions, must have proceeded fr
0352  e migrated from some one point to the several   DISTANT  and isolated points, where now found. Nevert
0352  when a plant or animal inhabits two points so   DISTANT  from each other, or with an interval of such
0352  xplicable cases of the same mammal inhabiting   DISTANT  points of the world. No geologist will feel
0353  inal plants are identically the same at these   DISTANT  points of the northern and southern hemisphe
0354  onal cases of the same species, now living at   DISTANT  and separated points: nor do I for a moment
0354  istence of the same species on the summits of   DISTANT  mountain ranges, and at distant points in th
0354  he summits of distant mountain ranges, and at   DISTANT  points in the arctic and antarctic regions:
0354  sea. If the existence of the same species at   DISTANT  and isolated points of the earth's surface,
0357  the dispersal of the same species to the most   DISTANT  points, and removes many a difficulty: but t
0364  nt to a neighbouring island, but not from one   DISTANT  continent to another. The floras of distant
0364  e distant continent to another. The floras of   DISTANT  continents would not by such means become mi
0365  ing cases known of the same species living at   DISTANT  points, without the apparent possibility of
0367  old and New Worlds, would be left isolated on   DISTANT  mountain summits (having been exterminated o
0368  in other regions we find the same species on   DISTANT  mountain summits, we may almost conclude wit
0369  pecies will have been left on mountain ranges   DISTANT  from each other, and have survived there eve
0373  as strictly simultaneous at these several far   DISTANT  points on opposite sides of the world. But w
0381  ely confined to the south, at these and other   DISTANT  points of the southern hemisphere, is, on my
0383  rrier, that they never would have extended to   DISTANT  countries. But the case is exactly the rever
0384  me species never occur in the fresh waters of   DISTANT  continents. But on the same continent the sp
0384  to allied fresh water fish occurring at very   DISTANT  points of the world, no doubt there are many
0385  modified and adapted to the fresh waters of a   DISTANT  land. Some species of fresh water shells hav
0385  ross sea to an oceanic island or to any other   DISTANT  point. Sir Charles Lyell also informs me tha
0386  ew on board the Beagle, when forty five miles   DISTANT  from the nearest land: how much farther it m
0388  aturally travel from one to another and often   DISTANT  piece of water. Nature, like a careful garde
0389  the course of time they have come to inhabit   DISTANT  points of the globe. I have already stated t
0392  . hence trees would be little likely to reach   DISTANT  oceanic islands: and an herbaceous plant, th
0394  anges, and are found on continents and on far   DISTANT  islands. Hence we have only to suppose that
0398  nts of these volcanic islands in the Pacific,   DISTANT  several hundred miles from the continent, ye
0399  lly stocked from a nearly intermediate though  DISTANT. point, namely from the antarctic islands, wh
0404  wherever in two regions, let them be ever so   DISTANT, many closely allied or representative speci
0405  e more important power of being victorious in   DISTANT  lands in the struggle for life with foreign
0406  forms being very minute and better fitted for   DISTANT  transportation, probably accounts for a law
0409  derived. We can see why in two areas, however   DISTANT  from each other, there should be a correlati
0410  long succession of ages, as in now looking to   DISTANT  provinces throughout the world, we find that
0410  which have changed after having migrated into   DISTANT  quarters, in both cases the forms within eac
0461  rom common parents: and therefore, in however  DISTANT  and isolated parts of the world they are now
0462  nct species of the same genus inhabiting very   DISTANT  and isolated regions, as the process of modi
0464  ocal varieties will not spread into other and   DISTANT  regions until they are considerably modified
0465  h other, than are the fossils from formations   DISTANT  from each other in time. Such is the sum of
0477  he close alliance of many others, on the most   DISTANT  mountains, under the most different climates
0477  cean, should so often be found on islands far  DISTANT  from any continent. Such facts as the presen
0488  t and living descendants, was created. In the  DISTANT  future I see open fields for far more import
0489  ies will transmit its unaltered likeness to a  DISTANT  futurity. And of the species now living very
0489  ew will transmit progeny of any kind to a far  DISTANT  futurity: for the manner in which all organi
0263  of one species is placed on the stigma of a    DISTANTLY  allied species, though the pollen tubes pro
0301  or a few links, some more closely, some more   DISTANTLY  related to each other: and these links, let
0003  this parasite, with its relations to several   DISTINCT  organic beings, by the effects of external c
0011  onstrosities, the correlations between quite   DISTINCT  parts are very curious: and many instances a
0016  nt judges as the descendants of aboriginally   DISTINCT  species. If any marked distinction existed b
0019  rdly one peculiar mammal, and France but few   DISTINCT  from those of Germany and conversely, and so
0019  not possess a number of peculiar species as   DISTINCT  parent stocks? So it is in India. Even in th
0020  d state. Moreover, the possibility of making   DISTINCT  races by crossing has been greatly exaggerat
0020  ainly, a breed intermediate between two very   DISTINCT  breeds could not be got without extreme care
0022  l gland is quite aborted. Several other less   DISTINCT  breeds might have been specified. In the ske
0025  d. moreover, when two birds belonging to two   DISTINCT  breeds are crossed; neither of which is blue
0026  d which has been crossed only once with some   DISTINCT  breed, the tendency to reversion to any char
0026  ood: but when there has been no cross with a   DISTINCT  breed, and there is a tendency in both paren
0026  indefinite number of generations. These two   DISTINCT  cases are often confounded in treatises on i
0026  own observations, purposely made on the most  DISTINCT  breeds. Now, it is difficult, perhaps imposs
0026  the hybrid offspring of two animals clearly   DISTINCT  being themselves perfectly fertile. Some aut
0026  as to suppose that species, aboriginally as   DISTINCT  as carriers, tumblers, pouters, and fantails
0028  avourable circumstance for the production of   DISTINCT  breeds, that male and female pigeons can be
0028  ded, are descended from so many aboriginally  DISTINCT  species. Ask, as I have asked, a celebrated
0029  ed that each main breed was descended from a  DISTINCT  species. Van Mons, in his treatise on pears
0032  ion consisted merely in separating some very  DISTINCT  variety, and breeding from it, the principle
0034  breeders try by methodical selection, with a  DISTINCT  object in view, to make a new strain or sub
0036  which ensued, namely, the production of two   DISTINCT  strains. The two flocks of Leicester sheep k
0036  est individuals, whether or not sufficiently  DISTINCT  to be ranked at their first appearance as di
0037  ct to be ranked at their first appearance as  DISTINCT  varieties, and whether or not two or more sp
0039  va fantail, or like individuals of other and  DISTINCT  breeds, in which as many as seventeen tail f
0040  of characters, have lately been exhibited as  DISTINCT  at our poultry shows. I think these views fu
0040  bourhood. But as yet they will hardly have a  DISTINCT  name, and from being only slightly valued, t
0040  widely, and will get recognised as something  DISTINCT  and valuable, and will then probably first r
0042  there appeared (aided by some crossing with   DISTINCT  species) those many admirable varieties of t
```

Page ***************************************(Key Word)***************************************

Page **(Key Word)**

```
0042  by women and children, we hardly ever see a  DISTINCT  breed kept up; such breeds as we do sometime
0042  ss than others, yet the rarity or absence of   DISTINCT  breeds of the cat, the donkey, peacock, goos
0042  pleasure having been felt in the display of    DISTINCT  breeds. To sum up on the origin of our Domes
0043  t the intercrossing of species, aboriginally   DISTINCT, has played an important part in the origin
0043  etc., the importance of the crossing both of   DISTINCT  species and of varieties is immense: for the
0044  y term includes the unknown element of a       DISTINCT  act of creation. The term variety is almost
0047  that naturalists do not like to rank them as   DISTINCT  species, are in several respects the most im
0049  hich it cannot be doubted would be ranked as   DISTINCT  species by many entomologists. Even Ireland
0049  forms leads many naturalists to rank both as   DISTINCT  species: but what distance, it has been well
0050  evidence of the two forms being specifically   DISTINCT. On the other hand, they are united by many
0050  le and pedunculated oaks are either good and   DISTINCT  species or mere varieties. When a young natu
0051  ok at varieties which are in any degree more   DISTINCT  and permanent, as steps leading to more stro
0052  rom the term variety, which is given to less   DISTINCT  and more fluctuating forms. The term variety
0057  that all the species of a genus are equally    DISTINCT  from each other; they may generally be divid
0059  r see, tend to become converted into new and   DISTINCT  species. The larger genera thus tend to beco
0060  t, and to the conditions of life, and of one   DISTINCT  organic being to another being, been perfect
0061  s, become ultimately converted into good and   DISTINCT  species, which in most cases obviously diffe
0061  of species, which constitute what are called   DISTINCT  genera, and which differ from each other mor
0063  the same species, or with the individuals of   DISTINCT  species, or with the physical conditions of
0068  n the individuals, whether of the same or of   DISTINCT  species, which subsist on the same kind of f
0071  and the heath was frequented by two or three   DISTINCT  insectivorous birds. Here we see how potent
0076  ion with each other, than between species of   DISTINCT  genera. We see this in the recent extension
0092  stigma of another flower. The flowers of two   DISTINCT  individuals of the same species would thus g
0098  thus produce a multitude of hybrids between    DISTINCT  species; for if you bring on the same brush
0099  ed on the view of an occasional cross with a   DISTINCT  individual being advantageous or indispensab
0099  pect that it must arise from the pollen of a   DISTINCT  variety having a prepotent effect over a flo
0099  good being derived from the intercrossing of   DISTINCT  individuals of the same species. When distin
0099  stinct individuals of the same species. When  DISTINCT  species are crossed the case is directly the
0100  lowers on the same tree can be considered as   DISTINCT  individuals only in a limited sense. I belie
0101  om without and the occasional influence of a   DISTINCT  individual can be shown to be physically imp
0101  al kingdoms, an occasional intercross with a   DISTINCT  individual is a law of nature. I am well awa
0103  varieties of the same animal can long remain  DISTINCT, from haunting different stations, from bree
0111  er from each other far less than do good and   DISTINCT  species. Nevertheless, according to my view,
0112  come converted into two well established and   DISTINCT  breeds. As the differences slowly become gre
0113  ass, and a similar plot be sown with several   DISTINCT  genera of grasses, a greater number of plant
0113  from each other in at all the same manner as   DISTINCT  species and genera of grasses differ from ea
0114  e of many thousands of generations, the most  DISTINCT  varieties of any one species of grass would
0114  in numbers, and thus of supplanting the less  DISTINCT  varieties; and varieties, when rendered very
0114  varieties, and varieties, when rendered very  DISTINCT  from each other, take the rank of species. T
0119  cessive forms which have become sufficiently  DISTINCT  to be recorded as varieties. But these break
0122  odified offspring of a species get into some   DISTINCT  country, or become quickly adapted to some o
0123  rlier period from a5, will be in some degree   DISTINCT  from the three first named species; and last
0123  es, and may constitute a sub genus or even a   DISTINCT  genus. The six descendants from (I) will for
0123  ded from (A), will have to be ranked as very   DISTINCT  genera, or even as distinct sub families. Th
0123  e ranked as very distinct genera, or even as   DISTINCT  sub families. Thus it is, as I believe, that
0125  hose marked c14 to m14, will form three very   DISTINCT  genera. We shall also have two very distinct
0125  distinct genera. We shall also have two very   DISTINCT  genera descended from (I); and as these latt
0125  he two little groups of genera will form two   DISTINCT  families, or even orders, according to the a
0128  etween species of the same genus, or even of   DISTINCT  genera. We have seen that it is the common,
0128  forming sections and sub genera; species of   DISTINCT  genera much less closely related, and genera
0151  se important valves than do other species of   DISTINCT  genera. As birds within the same country var
0159  sexual, and for ordinary specific purposes.    DISTINCT  species present analogous variations; and a
0159  d by looking to our domestic races. The most  DISTINCT  breeds of pigeons, in countries most widely
0159  then are analogous variations in two or more  DISTINCT  races. The frequent presence of fourteen or
0159  one of analogous variation in two so called   DISTINCT  species; and to these a third may be added,
0160  le to appear in the crossed offspring of two  DISTINCT  and differently coloured breeds; and in this
0162  lue colour and marks so often appearing when  DISTINCT  breeds of diverse colours are crossed. Hence
0163  ly of reversion. The ass not rarely has very  DISTINCT  transverse bars on its legs, like those on t
0163  dr. Gray has figured one specimen with very   DISTINCT  zebra like bars on the hocks. With respect t
0163  d of the spinal stripe in horses of the most  DISTINCT  breeds, and of all colours; transverse bars
0165  and should be loth to apply it to breeds as   DISTINCT  as the heavy Belgian cart horse, welch ponie
0166  to these several facts? We see several very   DISTINCT  species of the horse genus becoming, by simp
0166  in hybrids from between several of the most   DISTINCT  species. Now observe the case of the several
0173  ntermingle, they are generally as absolutely  DISTINCT  from each other in every detail of structure
0174  later tertiary periods; and in such islands   DISTINCT  species might have been separately formed wi
0176  heory to be converted and perfected into two  DISTINCT  species, the two which exist in larger numbe
0178  y have separately been rendered sufficiently  DISTINCT  to rank as representative species. In this c
0190  ame organ performing at the same time wholly  DISTINCT  functions; thus the alimentary canal respire
0190  y change its nature by insensible steps. Two  DISTINCT  organs sometimes perform simultaneously the
0190  n might be modified for some other and quite  DISTINCT  purpose, or be quite obliterated. The illust
0191  orked in by natural selection for some quite  DISTINCT  purpose: in the same manner as, on the view
0193  ering plants. In all these cases of two very  DISTINCT  species furnished with apparently the same a
0197  o doubt that the colour is due to some quite  DISTINCT  cause, probably to sexual selection. A trail
0205  n specialised for one function: and two very  DISTINCT  organs having performed at the same time the
0207  tinct. It would be easy to show that several  DISTINCT  mental actions are commonly embraced by this
0211  ction for the exclusive good of another of a  DISTINCT  species; yet each species tries to take adva
0218  ther birds' nests either of the same or of a  DISTINCT  species, is not very uncommon with the Galli
0234  its parasites or other enemies, or on quite  DISTINCT  causes, and so be altogether independent of
0238  er, but are perfectly well defined; being as  DISTINCT  from each other, as are any two species of t
0243  common case of closely allied, but certainly  DISTINCT, species, when inhabiting distant parts of t
0243  ck nests, to roost in, like the males of our  DISTINCT  Kitty wrens, a habit wholly unlike that of a
0245  thin the same country could hardly have kept  DISTINCT  had they been capable of crossing freely. Th
0246  und two forms, considered by most authors as  DISTINCT  species, quite fertile together, he unhesita
0249  other hand, that an occasional cross with a   DISTINCT  individual or variety increases fertility, t
0249  ch generation a cross with the pollen from a  DISTINCT  flower, either from the same plant or from a
0250  artner that some degree of sterility between  DISTINCT  species is a universal law of nature. He exp
```

Page **(Key Word)**

distinct

Page ***(Key Word)***

```
0250 fect, fertility in a first cross between two DISTINCT species. This case of the Crinum leads me to
0250 sily fertilised by the pollen of another and DISTINCT species, than by their own pollen. For these
0250  been found to yield seed to the pollen of a DISTINCT species, though quite sterile with their own
0250 ound to be perfectly good, for it fertilised DISTINCT species. So that certain individual plants a
0250 mpound hybrid descended from three other and DISTINCT species: the result was that the ovaries of
0252 ted, that is if the genera of animals are as DISTINCT from each other, as are the genera of plants
0253  different that they are generally ranked in DISTINCT genera, have often bred in this country with
0254 r. blyth, I think they must be considered as DISTINCT species. On this view of the origin of many
0254  belief of the almost universal sterility of DISTINCT species of animals when crossed; or we must
0255 mily is placed on the stigma of a plant of a DISTINCT family, it exerts no more influence than so
0257 ed between species ranked by systematists in DISTINCT families; and on the other hand, by very clo
0257 me difficulty; and on the other hand of very DISTINCT species which unite with the utmost facility
0257 at n. acuminata, which is not a particularly DISTINCT species, obstinately failed to fertilise, or
0259  forms, which must be considered as good and DISTINCT species, are united, their fertility graduat
0261 e to graft trees together belonging to quite DISTINCT families; and, on the other hand, closely al
0261 verned by systematic affinity. Although many DISTINCT genera within the same family have been graf
0261  readily on the quince, which is ranked as a DISTINCT genus, than on the apple, which is a member
0262 r reproductive organs perfect; yet these two DISTINCT cases run to a certain extent parallel. Some
0262 re freely when fertilized with the pollen of DISTINCT species, than when self fertilised with thei
0263 m in the results of grafting and of crossing DISTINCT species. And as we must look at the curious
0263 ecting a union apparently depends on several DISTINCT causes. There must sometimes be a physical i
0267 abundant evidence, that a cross between very DISTINCT individuals of the same species, that is bet
0268 ogs have descended from several aboriginally DISTINCT species. Nevertheless the perfect fertility
0270  suspected that these varieties of maize are DISTINCT species; and it is important to notice that
0270 o consider the two varieties as specifically DISTINCT. Girou de Buzareingues crossed three varieti
0271 crossed with yellow and white varieties of a DISTINCT species, more seed is produced by the crosse
0271  is more fertile, when crossed with a widely DISTINCT species, than are the other varieties. He ex
0271 nal characters in the production of the most DISTINCT domestic varieties, and from not wishing or
0272 ecies are more variable than those from very DISTINCT species; and this shows that the difference
0274 fferent from each other; whereas if two very DISTINCT varieties of one species are crossed with an
0275 me variety, or of different varieties, or of DISTINCT species. Laying aside the question of fertil
0276 er. First crosses between forms sufficiently DISTINCT to be ranked as species, and their hybrids,
0277 isation disturbed by being compounded of two DISTINCT species, seems closely allied to that steril
0277 o should generally correspond, though due to DISTINCT causes: for both depend on the amount of dif
0281 th natural species, if we look to forms very DISTINCT, for instance to the horse and tapir, we hav
0293 are on record of the same species presenting DISTINCT varieties in the upper and lower parts of th
0297 c, it would simply be ranked as a third and DISTINCT species, unless at the same time it could be
0297 onsequently be compelled to rank them all as DISTINCT species. It is notorious on what excessively
0297 e look to rather wider intervals, namely, to DISTINCT but consecutive stages of the same great for
0299 a, which are ranked by some conchologists as DISTINCT species from their European representatives,
0299 ieties or are, as it is called, specifically DISTINCT. This could be effected only by the future g
0301 nal varieties would merely appear as so many DISTINCT species. It is, also, probable that each gre
0301  many palaeontologists, be ranked as new and DISTINCT species. If then, there be some degree of tr
0302 oulc, by most palaeontologists, be ranked as DISTINCT species. But I do not pretend that I should
0304 ly to justify the belief that this great and DISTINCT order had been suddenly produced in the inte
0313 t it is a case of temporary migration from a DISTINCT geographical province, seems to me satisfact
0315 inly inherit different characters from their DISTINCT progenitors. For instance, it is just possib
0321 e place occupied by a species belonging to a DISTINCT group, and thus caused its extermination; an
0321  it be species belonging to the same or to a DISTINCT class, which yield their places to other spe
0325 rder to be preserved and to survive. We have DISTINCT evidence on this head, in the plants which a
0326 e slower with the terrestrial inhabitants of DISTINCT continents than with the marine inhabitants
0328 t an isthmus separated two seas inhabited by DISTINCT, but contemporaneous, faunas. Lyell has made
0329 he ruminants and Pachyderms, as the two most DISTINCT orders of mammals; but Owen has discovered s
0330 were not at this early epoch limited in such DISTINCT groups as they now are. Some writers have ob
0330 ing genera, even between genera belonging to DISTINCT families. The most common case, especially w
0330 common case, especially with respect to very DISTINCT groups, such as fish and reptiles, seems to
0330 o that the two groups, though formerly quite DISTINCT, at that period made some small approach to
0330 ered having affinities directed towards very DISTINCT groups. Yet if we compare the older Reptiles
0332 on the uppermost line would be rendered less DISTINCT from each other. If, for instance, the gener
0332 ive genera, and o14 to m14) would yet remain DISTINCT. These two families, however, would be less
0332 . these two families, however, would be less DISTINCT from each other than they were before the di
0333 ervals in the natural system, and thus unite DISTINCT families or orders. All that we have a right
0335  the chalk formation, though the species are DISTINCT in each stage. This fact alone, from its gen
0335  to account for the close resemblance of the DISTINCT species in closely consecutive formations, b
0336 ely consecutive formations, though ranked as DISTINCT species, being closely related, is obvious.
0338  these old forms are in a slight degree less DISTINCT from each other than are the typical members
0344 es blending two groups previously classed as DISTINCT into one; but more commonly only bringing th
0344  some degree intermediate between groups now DISTINCT; for the more ancient a form is, the more ne
0348 nferior in degree to those characteristic of DISTINCT continents. Turning to the sea, we find the
0348  the same law. No two marine faunas are more DISTINCT, with hardly a fish, shell, or crab in commo
0348 nds of the Pacific, with another and totally DISTINCT fauna. So that here three marine faunas rang
0348  either of land or open sea, they are wholly DISTINCT. On the other hand, proceeding still further
0348  vast space we meet with no well defined and DISTINCT marine faunas. Although hardly one shell, cr
0349 nt or sea, though the species themselves are DISTINCT at different points and stations. It is a la
0349 ch successive groups of beings, specifically DISTINCT, yet clearly related, replace each other. He
0349 ach other. He hears from closely allied, yet DISTINCT kinds of birds, notes nearly similar, and se
0352  natural selection from parents specifically DISTINCT. We are thus brought to the question which h
0353  local, but had been produced in two or more DISTINCT areas! Hence it seems to me, as it has to ma
0354 mportant for us, namely, whether the several DISTINCT species of a genus, which on my theory have
0364 led in any great degree; but would remain as DISTINCT as we now see them to be. The currents, from
0365 t have been independently created at several DISTINCT points; and we might have remained in this s
0369 e ranked as doubtful forms, and some few are DISTINCT yet closely allied or representative species
0372 ts rank as geographical races, and others as DISTINCT species; and a host of closely allied or rep
0372 re ranked by all naturalists as specifically DISTINCT. As on the land, so in the waters of the sea
0374 ultaneous throughout the world. Without some DISTINCT evidence to the contrary, we may at least ad
0376 same; but they are much oftener specifically DISTINCT, though related to each other in a most rema
```

Page ***(Key Word)***

Page ***************************************(Key Word)***************************************

```
0376  g ranked by some naturalists as specifically DISTINCT, by others as varieties: but some are certai
0376  related to northern forms, must be ranked as DISTINCT species. Now let us see what light can be th
0379  eir new homes as well marked varieties or as DISTINCT species. It is a remarkable fact, strongly f
0381  ispersal. But the existence of several quite DISTINCT species, belonging to genera exclusively con
0381  difficulty. For some of these species are so DISTINCT, that we cannot suppose that there has been
0381  eem to me to indicate that peculiar and very DISTINCT species have migrated in radiating lines fro
0395  p ocean; and this space separates two widely DISTINCT mammalian faunas. On either side the islands
0395  rms, but the species and even the genera are DISTINCT. As the amount of modification in all cases
0398  nty five of these are ranked by Mr. Gould as DISTINCT species, supposed to have been created here;
0400  tants of an archipelago, though specifically DISTINCT, to be closely allied to those of the neares
0400  tants of each separate island, though mostly DISTINCT, are related in an incomparably closer degre
0401  est fitted ground more perfectly occupied by DISTINCT plants in one island than in another, and it
0402   to the aboriginal inhabitants, but are very DISTINCT species, belonging in a large proportion of
0402   of cases, as shown by Alph. de Candolle, to DISTINCT genera. In the Galapagos Archipelago, many e
0402  dapted for flying from island to island, are DISTINCT on each; thus there are three closely allied
0403   adjoining islet of Porto Santo possess many DISTINCT but representative land shells, some of whic
0403   , yet they are inhabited by a vast number of DISTINCT mammals, birds, and plants. The principle wh
0404  nd many forms which some naturalists rank as DISTINCT species, and some as varieties: these doubtf
0405   the two varieties would have been ranked as DISTINCT species, and the common range would have bee
0406  so different, the very close relation of the DISTINCT species which inhabit the islets of the same
0407  a, genera, and families. With respect to the DISTINCT species of the same genus, which on my theor
0409  tants of an archipelago, though specifically DISTINCT on the several islets, should be closely rel
0409  s, of varieties, of doubtful species, and of DISTINCT but representative species. As the late Edwa
0411  become converted, as I believe, into new and DISTINCT species; and these, on the principle of inhe
0412  iple, much in common, and form a sub family, DISTINCT from that including the next two genera on t
0412  ough less, in common; and they form a family DISTINCT from that including the three genera still f
0412  se genera, descended from (A), form an order DISTINCT from the genera descended from (I). So that
0421   up into two or three families, constitute a DISTINCT order from those descended from I, also brok
0424   ), which had previously been ranked as three DISTINCT genera, were known to be sometimes produced
0427  he species of the same genus, inhabiting any DISTINCT and isolated region, have in all probability
0427  have been drawn by some authors between very DISTINCT animals. On my view of characters being of r
0427  tematist. For animals, belonging to two most DISTINCT lines of descent, may readily become adapted
0428  ncestor. So it is with fishes. As members of DISTINCT classes have often been adapted by successiv
0428  imes been observed between the sub groups in DISTINCT classes. A naturalist, struck by a paralleli
0429   such species as do occur are generally very DISTINCT from each other, which again implies extinct
0430  p of animals exhibits an affinity to a quite DISTINCT group, this affinity in most cases is genera
0430  s on the general nature of the affinities of DISTINCT orders of plants. On the principle of the mu
0431  of species, now broken up by extinction into DISTINCT groups and sub groups, will have transmitted
0433  portance: why, in comparing one group with a DISTINCT group, we summarily reject analogical or ada
0434  members of any one class; but when we have a DISTINCT object in view, and do not look to some unkn
0438  rts of one species with those of another and DISTINCT species, we can indicate but few serial homo
0439   embryo exactly alike. The embryos, also, of DISTINCT animals within the same class are often stri
0445  hat they would, I cannot doubt, be ranked in DISTINCT genera, had they been natural productions. B
0452  ry for its proper purpose, and be used for a DISTINCT object: in certain fish the swim bladder see
0461   been disturbed from being compounded of two DISTINCT organisations. This parallelism is supported
0462  casional means of transport. With respect to DISTINCT species of the same genus inhabiting very di
0464  ed, it will simply be classed as another and DISTINCT species. Only a small portion of the world h
0467  e great agency in the production of the most DISTINCT and useful domestic breeds. That many of the
0469   of varieties, which they think sufficiently DISTINCT to be worthy of record in systematic works.
0470  s are rendered to a large extent defined and DISTINCT objects. Dominant species belonging to the l
0473  , by which they have come to be specifically DISTINCT from each other; and therefore these same ch
0478  till exist. Wherever many closely allied yet DISTINCT species occur, many doubtful forms and and v
0481  hat one species has given birth to other and DISTINCT species, is that we are always slow in admit
0483  ion is difficult to answer, because the more DISTINCT the forms are which we may consider, by so m
0484  hether any form be sufficiently constant and DISTINCT from other forms, to be capable of definitio
0189  ny way, or in any degree, tend to produce a  DISTINCTER image. We must suppose each new state of th
0008  es cannot be separated by any clear line of  DISTINCTION from mere variations. But I am strongly in
0016  boriginally distinct species. If any marked  DISTINCTION existed between domestic races and species
0048  uck how entirely vague and arbitrary is the  DISTINCTION between species and varieties. On the isle
0245  d the weakest die. Chapter VIII. Hybridism.  DISTINCTION between the sterility of first crosses and
0246  veloped, or are imperfectly developed. This  DISTINCTION is important, when the cause of the steril
0246  to the two cases, has to be considered. The  DISTINCTION has probably been slurred over, owing to t
0246  ies; for it seems to make a broad and clear  DISTINCTION between varieties and species. First, for
0248  r sterility nor fertility affords any clear  DISTINCTION between species and varieties; but that th
0268  argument, that there must be some essential  DISTINCTION between species and varieties; and that th
0270  nd sterility as safe criterions of specific  DISTINCTION. Gartner kept during several years a dwarf
0272  iversal occurence, or to form a fundamental  DISTINCTION between varieties and species. The general
0272  se strong wish was to draw a marked line of  DISTINCTION between species and varieties, could find
0272  ct with extreme brevity. The most important  DISTINCTION is, that in the first generation mongrels
0276  ly with the view that there is no essential  DISTINCTION between species and varieties. Summary of
0278  port the view, that there is no fundamental  DISTINCTION between species and varieties. Chapter IX.
0299  done scarcely anything in breaking down the  DISTINCTION between species, by connecting them togeth
0427  erstand, on these views, the very important  DISTINCTION between real affinities and analogical or
0427  ces. Lamarck first called attention to this  DISTINCTION, and he has been ably followed by Macleay
0469  systematic works. No one can draw any clear  DISTINCTION between individual differences and slight
0481  f long ages is a limited quantity; no clear  DISTINCTION has been, or can be, drawn between species
0482  reject it in another, without assigning any  DISTINCTION in the two cases. The day will come when t
0485  l be compelled to acknowledge that the only  DISTINCTION between species and well marked varieties
0019   that every race which breeds true, let the  DISTINCTIVE characters be ever so slight, has had its
0027   thirdly, those characters which are mainly  DISTINCTIVE of each breed, for instance the wattle and
0034  rserymen. The principle of selection I find  DISTINCTLY given in an ancient Chinese encyclopaedia.
0241  us, as I believe, the wonderful fact of two  DISTINCTLY defined castes of sterile workers existing
0352  apacity of migrating across the sea is more  DISTINCTLY limited in terrestrial mammals, than perhap
0279  m have now been discussed. One, namely the   DISTINCTNESS of specific forms, and their not being be
0431  ch class. We may thus account even for the   DISTINCTNESS of whole classes from each other, for inst
0439  otted in lines; and stripes can be plainly   DISTINGUISED in the whelp of the lion. We occasionally
```

Page ***************************************(Key Word)***************************************

0056 here is no infallible criterion by which to DISTINGUISH species and well marked varieties; and in
0158 lity of specific characters, or those which DISTINGUISH species from species, than of generic char
0161 he common ancestor of a group, we could not DISTINGUISH these two cases: if, for instance, we did
0222 ous to ascertain whether F. sanguinea could DISTINGUISH the pupae of F. fusca, which they habitual
0222 e, and it was evident that they did at once DISTINGUISH them: for we have seen that they eagerly a
0297 naturalists have no golden rule by which to DISTINGUISH species and varieties; they grant some lit
0315 ame object, might make a new breed hardly DISTINGUISHABLE from our present fantail; but if the par
0440 larvae in all their several stages barely DISTINGUISHABLE. The embryo in the course of development
0058 characters as species, for they cannot be DISTINGUISHED from species, except, firstly, by the dis
0209 an instinct becomes so close as not to be DISTINGUISHED. If Mozart, instead of playing the pianof
0240 ocelli), which though small can be plainly DISTINGUISHED, whereas the smaller workers have their o
0330 es, seems to be, that supposing them to be DISTINGUISHED at the present day from each other by a d
0330 nt members of the same two groups would be DISTINGUISHED by a somewhat lesser number of characters
0432 e definitions by which each group could be DISTINGUISHED from other groups, as all would blend tog
0432 ral members of the several groups could be DISTINGUISHED from their more immediate parents; or tho
0445 ced in a row, though most of them could be DISTINGUISHED from each other, yet their proportional d
0416 acter, according to Owen, which absolutely DISTINGUISHES fishes and reptiles, the inflection of th
0007 cter of Domestic Varieties. Difficulty of DISTINGUISHING between Varieties and Species. Origin of
0120 the steps by which the small differences DISTINGUISHING varieties are increased into the larger d
0120 are increased into the larger differences DISTINGUISHING species. By continuing the same process f
0128 attle of life. Thus the small differences DISTINGUISHING varieties of the same species, will stead
0147 ny instances, for I see hardly any way of DISTINGUISHING between the effects, on the one hand, of
0077 her plants; so that the seeds may be widely DISTRIBUTED and fall on unoccupied ground. In the wate
0174 mals. In looking at species as they are now DISTRIBUTED over a wide area, we generally find them t
0175 abiting a continuous area, are generally so DISTRIBUTED that each has a wide range, with a compara
0405 descended from a single parent, though now DISTRIBUTED to the most remote points of the world, we
0427 ully into play in classing large and widely DISTRIBUTED genera, because all the species of the sam
0001 was much struck with certain facts in the DISTRIBUTION of the inhabitants of South America and in
0003 mbryological relations, their geographical DISTRIBUTION, geological succession, and other such fac
0005 e eleventh and twelfth, their geographical DISTRIBUTION throughout space: in the thirteenth, their
0049 stinc lines of argument, from geographical DISTRIBUTION, analogical variation, hybridism, etc., ha
0055 e discussed in our chapter on geographical DISTRIBUTION. From looking at species as only strongly
0062 hole economy of nature, with every fact on DISTRIBUTION, rarity, abundance, extinction, and variat
0105 e shall see in our chapter on geographical DISTRIBUTION; yet of these species a very large proport
0106 alluded to in our chapter on geographical DISTRIBUTION; for instance, that the productions of wha
0175 s of life as the all important elements of DISTRIBUTION, these facts ought to cause surprise, as c
0178 ed (namely from what we know of the actual DISTRIBUTION of closely allied or representative specie
0195 ve even in this case, for we know that the DISTRIBUTION and existence of cattle and other animals
0314 less. We see the same fact in geographical DISTRIBUTION; for instance, in the land shells and cole
0325 more clearly when we treat of the present DISTRIBUTION of organic beings, and find how slight is
0335 of species. He who is acquainted with the DISTRIBUTION of existing species over the globe, will n
0339 ds good with sea shells, but from the wide DISTRIBUTION of most genera of molluscs, it is not well
0340 ove alluded to, that in America the law of DISTRIBUTION of terrestrial mammals was formerly differ
0340 us facts could be given in relation to the DISTRIBUTION of marine animals. On the theory of descen
0340 immutable in the laws of past and present DISTRIBUTION. It may be asked in ridicule, whether I su
0346 atural Selection. Chapter XI. Geographical DISTRIBUTION. Present distribution cannot be accounted
0346 ter XI. Geographical Distribution. Present DISTRIBUTION cannot be accounted for by differences in
0346 tensive with the world. In considering the DISTRIBUTION of organic beings over the face of the glo
0346 most fundamental divisions in geographical DISTRIBUTION is that between the New and Old Worlds; ve
0353 e which barriers of every kind have had on DISTRIBUTION, is intelligible only on the view that the
0354 ondly (in the following chapter), the wide DISTRIBUTION of fresh water productions; and thirdly, t
0357 know something definite about the means of DISTRIBUTION, we shall be enabled to speculate with sec
0358 existing oceanic islands. Several facts in DISTRIBUTION, such as the great difference in the marin
0358 on (as we shall hereafter see) between the DISTRIBUTION of mammals and the depth of the sea, these
0358 operly might be called occasional means of DISTRIBUTION. I shall here confine myself to plants. In
0366 er influence of the glacial climate on the DISTRIBUTION of the inhabitants of Europe, as explained
0368 in in so satisfactory a manner the present DISTRIBUTION of the Alpine and Arctic productions of Eu
0368 ve to have been the case, chiefly from the DISTRIBUTION of the fossil Gnathodon), then the arctic
0369 present day. But the foregoing remarks on DISTRIBUTION apply not only to strictly arctic forms, b
0374 e, much light can be thrown on the present DISTRIBUTION of identical and allied species. In Americ
0376 ctly analogous facts could be given on the DISTRIBUTION of terrestrial animals. In marine producti
0382 at alternations of climate on geographical DISTRIBUTION. I believe that the world has recently fel
0382 ction, a multitude of facts in the present DISTRIBUTION both of the same and of allied forms of li
0383 Lowlands. Chapter XII. Geographical DISTRIBUTION, continued. Distribution of fresh water pr
0383 xii. Geographical Distribution, continued. DISTRIBUTION of fresh water productions. On the inhabit
0385 ource, prevail throughout the world. Their DISTRIBUTION at first perplexed me much, as their ova a
0387 remarks on this plant, I thought that its DISTRIBUTION must remain quite inexplicable; but Audubo
0387 ped. In considering these several means of DISTRIBUTION, it should be remembered that when a pond
0388 inct in intermediate regions. But the wide DISTRIBUTION of fresh water plants and of the lower ani
0404 and Canidae. We see it, if we compare the DISTRIBUTION of butterflies and beetles. So it is with
0405 into new species. In considering the wide DISTRIBUTION of certain genera, we should bear in mind
0407 rtance of barriers and from the analogical DISTRIBUTION of sub genera, genera, and families. With
0407 lifying the effects of climatal changes on DISTRIBUTION, I have attempted to show how important ha
0408 ll the grand leading facts of geographical DISTRIBUTION are explicable on the theory of migration
0419 no other class of Articulata. Geographical DISTRIBUTION has often been used, though perhaps not qu
0427 gh classificatory importance. Geographical DISTRIBUTION may sometimes be brought usefully into pla
0461 ncrease fertility. Turning to geographical DISTRIBUTION, the difficulties encountered on the theor
0462 the influence of the Glacial period on the DISTRIBUTION both of the same and of representative spe
0466 ended that we know all the varied means of DISTRIBUTION during the long lapse of years, or that we
0476 allied by descent. Looking to geographical DISTRIBUTION, if we admit that there has been during th
0476 cation, most of the great leading facts in DISTRIBUTION. We can see why there should be so strikin
0476 should be so striking a parallelism in the DISTRIBUTION of organic beings throughout space, and in
0040 ses or falls in fashion, perhaps more in one DISTRICT than in another, according to the state of c
0066 people, under favourable conditions, a whole DISTRICT, let it be ever so large. The condor lays a
0066 als of the two species can be supported in a DISTRICT. A large number of eggs is of some importanc
0074 disappear. The number of humble bees in any DISTRICT depends in a great degree on the number of f

```
0074  nce of a feline animal in large numbers in a DISTRICT might determine, through the intervention fi
0074  es, the frequency of certain flowers in that DISTRICT! In the case of many species, many different
0091  , again, the wolves inhabiting a mountainous DISTRICT, and those frequenting the lowlands, would n
0103  ending to modify all the individuals in each DISTRICT in exactly the same manner to the conditions
0103   generally graduate away insensibly from one DISTRICT to another. The intercrossing will most affe
0142  ery cautious in transposing animals from one DISTRICT to another; for it is not likely that man sh
0168  le applies to the whole individual; for in a DISTRICT where many species of any genus are found, t
0286  e as applied to the western extremity of the DISTRICT. If, then, we knew the rate at which the sea
0287  f fresh water on the gently inclined Wealden DISTRICT, when upraised, could hardly have been great
0390  g after long intervals in a new and isolated DISTRICT, and having to compete with new associates,
0463  ns of life change insensibly in going from a DISTRICT occupied by one species into another distric
0463  istrict occupied by one species into another DISTRICT occupied by a closely allied species, we hav
0035  he same breed may be found in less civilised DISTRICTS, where the breed has been less improved. Th
0055  untries, and the coleopterous insects of two DISTRICTS, into two nearly equal masses, the species
0074  checks act on the same species in different DISTRICTS. When we look at the plants and bushes clot
0075  the same species, for they frequent the same DISTRICTS, require the same food, and are exposed to
0078  o slightly hotter or colder, damper or drier DISTRICTS. In this case we can clearly see that if we
0102  place. But if the area be large, its several DISTRICTS will almost certainly present different con
0102  fying and improving a species in the several DISTRICTS, there will be intercrossing with the other
0103  ed might subsequently slowly spread to other DISTRICTS. On the above principle, nurserymen always
0104  he surrounding and differently circumstanced DISTRICTS, will be prevented. But isolation probably
0142  constitutions specially fitted for their own DISTRICTS: the result must, I think, be due to habit.
0203  nother. When two varieties are formed in two DISTRICTS of a continuous area, an intermediate varie
0335  terrestrial productions inhabiting separated DISTRICTS. To compare small things with great: if the
0346  most diversified conditions; the most humid DISTRICTS, arid deserts, lofty mountains, grassy plai
0378  e other hand, the the most humid and hottest DISTRICTS will have afforded an asylum to the tropica
0403  . in many other instances, as in the several DISTRICTS of the same continent, pre occupation has p
0081  inly immigrate, and this also would seriously DISTURB the relations of some of the former inhabita
0266  should be compounded into one, without some DISTURBANCE occurring in the development, or periodical
0108   of many of the other inhabitants being thus DISTURBED. Nothing can be effected, unless favourable
0132  life; and to this system being functionally DISTURBED in the parents, I chiefly attribute the var
0132  but why, because the reproductive system is DISTURBED, this or that part should vary more or less
0220  nce is very great. When the nest is slightly DISTURBED, the slaves occasionally come out, and like
0220  d and defend the nest: when the nest is much DISTURBED and the larvae and pupae are exposed, the s
0222  ng f. sanguinea; and when I had accidentally DISTURBED both nests, the little ants attacked their
0265  e one case, the conditions of life have been DISTURBED, though often in so slight a degree as to b
0266  ined the same, but the organisation has been DISTURBED by two different structures and constitutio
0266  case from the conditions of life having been DISTURBED, in the other case from the organisation ha
0266  other case from the organisation having been DISTURBED by two organisations having been compounded
0277  had this system and their whole organisation DISTURBED by being compounded of two distinct species
0277  n their natural conditions of life have been DISTURBED. This view is supported by a parallelism of
0368  eir mutual relations will not have been much DISTURBED, and, in accordance with the principles inc
0369  relations will thus have been in some degree DISTURBED; consequently they will have been liable to
0390  at their mutual relations have not been much DISTURBED. Thus in the Galapagos Islands nearly every
0461  sterile from their constitutions having been DISTURBED by slightly different and new conditions of
0461  r constitutions can hardly fail to have been DISTURBED from being compounded of two distinct organ
0011  some authors, that the drooping is due to the DISUSE of the muscles of the ear, from the animals n
0043  life. Something must be attributed to use and DISUSE. The final result is thus rendered infinitely
0131  tion. Effects of external conditions, and use and DISUSE, combined with natural selection; organs of f
0134  ped and appreciable by us. Effects of Use and DISUSE. From the facts alluded to in the first chapt
0134  s strengthens and enlarges certain parts, and DISUSE diminishes them: and that such modifications
0134  judge of the effects of long continued use or DISUSE, for we know not the parent forms; but many a
0134  ures which can be explained by the effects of DISUSE. As Professor Owen has remarked, there is no
0134  anted by no beast of prey, has been caused by DISUSE. The ostrich indeed inhabits continents and i
0135  ther genera, by the long continued effects of DISUSE in their progenitors; for as the tarsi are al
0135  ts. In some cases we might easily put down to DISUSE modifications of structure which are wholly,
0136  natural selection, but combined probably with DISUSE. For during thousands of successive generatio
0137  yes is probably due to gradual reduction from DISUSE, but aided perhaps by natural selection. In S
0137  selection would constantly aid the effects of DISUSE. It is well known that several animals, belon
0137  in darkness, I attribute their loss wholly to DISUSE. In one of the blind animals, namely, the cav
0137  reduced by natural selection aided by use and DISUSE, so in the case of the cave rat natural selec
0138  with all the other inhabitants of the caves, DISUSE by itself seems to have done its work. It is
0138  numberless generations, the deepest recesses, DISUSE will on this view have more or less perfectly
0143  i think we may conclude that habit, use, and DISUSE, have, in some cases, played a considerable p
0143  rious organs; but that the effects of use and DISUSE have often been largely combined with, and so
0147  part being reduced by this same process or by DISUSE, and, on the other hand, the actual withdrawa
0150  s of growth, to the effects of long continued DISUSE, and to the tendency to reversion. A part dev
0168  al differences, and use in strengthening, and DISUSE in weakening and diminishing organs, seem to
0182  to, which perhaps may all have resulted from DISUSE, indicate the natural steps by which birds ha
0193  ce in some of the members to its loss through DISUSE or natural selection. But if the electric org
0206  ptations being aided in some cases by use and DISUSE, being slightly affected by the direct action
0209  , use or habit, and are diminished or lost by DISUSE, so I do not doubt it has been with instincts
0216  cation, for the mother hen has almost lost by DISUSE the power of flight. Hence we may conclude, t
0243  ful direction. In some cases habit or use and DISUSE have probably come into play. I do not preten
0447  ontinued exercise or use on the one hand, and DISUSE on the other, may have in modifying an organ,
0447  in a lesser degree, by the effects of use and DISUSE. In certain cases the successive steps of var
0454  e ever undergo abrupt changes. I believe that DISUSE has been the main agency; that it has led in
0455  natural selection. At whatever period of life DISUSE or selection reduces an organ, and this will
0457   when organs are reduced in size, either from DISUSE or selection, it will generally be at that pe
0466  ex laws, by correlation of growth, by use and DISUSE, and by the direct action of the physical con
0472  zore. In both varieties and species, use and DISUSE seem to have produced some effect: for it is
0479  ater, by the aid of well developed branchiae. DISUSE, aided sometimes by natural selection, will o
0480  s view the meaning of rudimentary organs. But DISUSE and selection will generally act on each crea
0480  re reduced, during successive generations, by DISUSE or by the tongue and palate having been fitte
0480  eeth have been left untouched by selection or DISUSE, and on the principle of inheritance at corre
0486  relation of growth, on the effects of use and DISUSE, on the direct action of external conditions,
```

0490 external conditions of life, and from use and DISUSE; a Ratio of Increase so high as to lead to a
0471 eet; that a thrush should have been created to DIVE and feed on sub aquatic insects; and that a pe
0112 able, steadily to increase, and the breeds to DIVERGE in character both from each other and from t
0128 n be supported on the same area the more they DIVERGE in structure, habits, and constitution, of w
0412 is a constant tendency in their characters to DIVERGE. This conclusion was supported by looking at
0120 forms, a10, f10, and m10, which, from having DIVERGED in character during the successive generatio
0123 nched off from a10; b14 and f14, from having DIVERGED at an earlier period from a5, will be in som
0123 ly related one to the other, but from having DIVERGED at the first commencement of the process of
0124 w species f14, which is supposed not to have DIVERGED much in character, but to have retained the
0125 ochs when the branching lines of descent had DIVERGED less. I see no reason to limit the process o
0333 at this early stage of descent they have not DIVERGED in character from the common progenitor of t
0333 e order, nearly so much as they subsequently DIVERGED. Thus it comes that ancient and extinct gene
0412 the next two genera on the right hand, which DIVERGED from a common parent at the fifth stage of d
0412 enera still further to the right hand, which DIVERGED at a still earlier period. And all these gen
0429 the principle of each group having generally DIVERGED much in character during the long continued
0476 ed from an ancient progenitor have generally DIVERGED in character, the progenitor with its early
0005 rms of life, and induces what I have called DIVERGENCE of Character. In the next chapter I shall d
0022 um are highly variable; so is the degree of DIVERGENCE and relative size of the two arms of the fu
0053 t be avoided to the struggle for existence, DIVERGENCE of character, and other questions, hereafte
0057 to discuss the principle, as I call it, of DIVERGENCE of Character, we shall see how this may be
0080 on. Extinction caused by Natural Selection. DIVERGENCE of Character, related to the diverstiy of i
0080 ation. Action of Natural Selection, through DIVERGENCE of Character and Extinction, on the descend
0111 writer) as if by some murderous pestilence. DIVERGENCE of Character. The principle, which I have d
0112 tion of what may be called the principle of DIVERGENCE, causing differences, at first barely appre
0116 inciple of great benefit being derived from DIVERGENCE of character, combined with the principles
0117 the principle of benefit being derived from DIVERGENCE of character comes in; for this will genera
0118 y a2, which will, owing to the principle of DIVERGENCE, differ more from (A) than did variety a1.
0125 hese latter two genera, both from continued DIVERGENCE of character and from inheritance from a di
0127 declares. Natural selection, also, leads to DIVERGENCE of character; for more living beings can be
0128 ection, as has just been remarked, leads to DIVERGENCE of character and to much extinction of the
0129 natural selection, entailing extinction and DIVERGENCE of character, as we have seen illustrated i
0331 the principle of the continued tendency to DIVERGENCE of character, which was formerly illustrate
0331 ng forms. We must not, however, assume that DIVERGENCE of character is a necessary contingency; it
0332 rom the continued effects of extinction and DIVERGENCE of character, has become divided into sever
0344 understand, from the continued tendency to DIVERGENCE of character, why the more ancient a form i
0430 principle of the multiplication and gradual DIVERGENCE in character of the species descended from
0433 ch almost inevitably induces extinction and DIVERGENCE of character in the many descendants from o
0456 n, with its contingencies of extinction and DIVERGENCE of character. In considering this view of c
0478 on with its contingencies of extinction and DIVERGENCE of character. On these same principles we s
0490 consequence to Natural Selection, entailing DIVERGENCE of Character and the Extinction of less imp
0117 will generally lead to the most different or DIVERGENT variations (represented by the outer dotted
0118 n, these two varieties be variable, the most DIVERGENT of their variations will generally be prese
0119 egular. I am far from thinking that the most DIVERGENT varieties will invariably prevail and multi
0119 the accumulation of a considerable amount of DIVERGENT variation. As all the modified descendants
0119 is represented in the diagram by the several DIVERGENT branches proceeding from (A). The modified
0120 ill not be increased; although the amount of DIVERGENT modification may have been increased in the
0123 will now be fifteen in number. Owing to the DIVERGENT tendency of natural selection, the extreme
0125 , or even orders, according to the amount of DIVERGENT modification supposed to be represented in
0303 hort time would be necessary to produce many DIVERGENT forms, which would be able to spread rapidl
0345 on progenitor of groups; since become widely DIVERGENT. Extinct forms are seldom directly intermed
0412 racter, to supplant and exterminate the less DIVERGENT, the less improved, and preceding forms. T
0430 ps have since undergone much modification in DIVERGENT directions. On either view we may suppose t
0470 cy in natural selection to preserve the most DIVERGENT offspring of any one species. Hence during
0471 come still larger, and at the same time more DIVERGENT in character. But as all groups cannot thus
0117 large in its own country. The little fan of DIVERGING dotted lines of unequal lengths proceeding
0118 ill generally go on increasing in number and DIVERGING in character. In the diagram the process is
0119 rally go on multiplying in number as well as DIVERGING in character: this is represented in the di
0120 pointer have apparently both gone on slowly DIVERGING in character from their original stocks, wi
0123 oups, moreover, are supposed to have gone on DIVERGING in different directions. The intermediate s
0124 pecies. But as these two groups have gone on DIVERGING from the type of their parents
0125 ange represented by each successive group of DIVERGING dotted lines to be very great, the forms ma
0331 tters represent genera, and the dotted lines DIVERGING from them the species in each genus. The di
0331 tinct genera on the several lines of descent DIVERGING from the parent form A, will form an order;
0412 the forms which are increasing in number and DIVERGING in character, to supplant and exterminate t
0471 large groups to go on increasing in size and DIVERGING in character, together with the almost inev
0486 ; and we have to discover and trace the many DIVERGING lines of descent in our natural genealogies
0017 t tendency to vary, and likewise to withstand DIVERSE climates. I do not dispute that these capaci
0017 ticated productions, and belonging to equally DIVERSE classes and countries, were taken from a sta
0053 have been expected, as they become exposed to DIVERSE physical conditions, and as they come into c
0146 nsmitted to a whole group of descendants with DIVERSE habits, would naturally be thought to be cor
0161 by natural selection, in accordance with the DIVERSE habits of the species, and will not be left
0162 ks so often appearing when distinct breeds of DIVERSE colours are crossed. Hence, though under nat
0405 its diffusion; and should place itself under DIVERSE conditions favourable for the conversion of
0426 e which serve to preserve life under the most DIVERSE conditions of existence, are generally the m
0431 the Crustacea, for here the most wonderfully DIVERSE forms are still tied together by a long, but
0438 s repeated, and have adapted them to the most DIVERSE purposes. And as the whole amount of modific
0442 arvae of insects, whether adapted to the most DIVERSE and active habits, or quite inactive, being
0442 h ultimately become very unlike and serve for DIVERSE purposes, being at this early period of grow
0446 gh natural selection in accordance with their DIVERSE habits. Then, from the many slight successiv
0477 y, that on the same continent, under the most DIVERSE conditions, under heat and cold, on mountain
0114 amount of life can be supported by great DIVERSIFICATION of structure, is seen under many natural
0114 tition with each other, the advantages of DIVERSIFICATION of structure, with the accompanying diff
0115 may, I think, at least safely infer that DIVERSIFICATION of structure, amounting to new generic d
0115 been profitable to them. The advantage of DIVERSIFICATION in the inhabitants of the same region is
0116 australian mammals, we see the process of DIVERSIFICATION in an early and incomplete stage of deve
0151 ese valves present a marvellous amount of DIVERSIFICATION: the homologous valves in the different

Page ***(Key Word)**

```
0012 laws of variation is infinitely complex and DIVERSIFIED. It is well worth while carefully to study
0102 subject. A large amount of inheritable and DIVERSIFIED variability is favourable, but I believe m
0112 from the simple circumstance that the more DIVERSIFIED the descendants from any one species becom
0112 better enabled to seize on many and widely DIVERSIFIED places in the polity of nature, and so be
0113 perhaps becoming less carnivorous. The more DIVERSIFIED in habits and structure the descendants of
0115 at these naturalised plants are of a highly DIVERSIFIED nature. They differ, moreover, to a large
0116 ly and perfectly the animals and plants are DIVERSIFIED for different habits of life, so will a or
0116 animals, with their organisation but little DIVERSIFIED, could hardly compete with a set more perf
0116 ld hardly compete with a set more perfectly DIVERSIFIED in structure. It may be doubted, for insta
0116 d by so much the better as they become more DIVERSIFIED in structure, and are thus enabled to encr
0117 sed to be extremely slight, but of the most DIVERSIFIED nature; they are not supposed all to appea
0119 relations. But as a general rule, the more DIVERSIFIED in structure the descendants from any one
0122 y have also been modified and improved in a DIVERSIFIED manner at each stage of descent, so as to
0128 ll species to increase in numbers, the more DIVERSIFIED these descendants become, the better will
0149 and as long as the same part has to perform DIVERSIFIED work, we can perhaps see why it should rem
0171 l varieties. Transitions in habits of life. DIVERSIFIED habits in the same species. Species with h
0180 ly allied species of the same genus: and of DIVERSIFIED habits, either constant or occasional, and
0182 ght: but they serve, at least, to show what DIVERSIFIED means of transition are possible. Seeing t
0182 rds and mammals, flying insects of the most DIVERSIFIED types, and formerly had flying reptiles, i
0183 . i will now give two or three instances of DIVERSIFIED and of changed habits in the individuals o
0183 or exclusively on artificial substances. Of DIVERSIFIED habits innumerable instances could be give
0199 art will often have entailed on other parts DIVERSIFIED changes of no direct use. So again charact
0200 now are to these animals having such widely DIVERSIFIED habits. Therefore we may infer that these
0204 nditions of life change its habits, or have DIVERSIFIED habits, with some habits very unlike those
0206 ill have existed more individuals, and more DIVERSIFIED forms, and the competition will have been
0212 es, born in a state of nature, is extremely DIVERSIFIED, can be shown by a multitude of facts. Sev
0346 treme southern point, we meet with the most DIVERSIFIED conditions: the most hum*d districts, arid
0407 east great meridional belts. As showing how DIVERSIFIED are the means of occasional transport, I h
0408 ly of their physical conditions, infinitely DIVERSIFIED conditions of life, there would be an almo
0470 ase by so much the more as they become more DIVERSIFIED in habits and structure, so as to be enabl
0207 life itself. We are concerned only with the DIVERSITIES of instinct and of the other mental qualit
0007 state of nature. When we reflect on the vast DIVERSITY of the plants and animals which have been c
0012 ed is unimportant for us. But the number and DIVERSITY of inheritable deviations of structure, bot
0018 e especially on the monuments of Egypt, much DIVERSITY in the breeds: and that some of the breeds
0021 to join two of the London Pigeon Clubs. The DIVERSITY of the breeds is something astonishing. Com
0033 ffects of selection namely, by comparing the DIVERSITY of flowers in the different varieties of th
0033 f the same species in the flower garden: the DIVERSITY of leaves, pods, or tubers, or whatever par
0033 h the flowers of the same varieties: and the DIVERSITY of fruit of the same species in the orchard
0074 rn united States, display the same beautiful DIVERSITY and proportion of kinds as in the surroundi
0114 ividual must be severe, we always find great DIVERSITY in its inhabitants. For instance, I found t
0127 conditions of existence, causing an infinite DIVERSITY in structure, constitution, and habits, to
0144 rts: it is believed by some authors that the DIVERSITY in the shape of the pelvis in birds causes
0144 of the pelvis in birds causes the remarkable DIVERSITY in the shape of their kidneys. Others belie
0169 ons will add to the beautiful and harmonious DIVERSITY of nature. Whatever the cause may be of eac
0188 ven, which show that there is much graduated DIVERSITY in the eyes of living crustaceans, and bear
0210 by natural selection. And such instances of DIVERSITY of instinct in the same species can be show
0261 ases we can assign no reason whatever. Great DIVERSITY in the size of two plants, one being woody
0412 lusion was supported by looking at the great DIVERSITY of the forms of life which, in any small ar
0436 ural selection will account for the infinite DIVERSITY in structure and function of the mouths of
0080 ion. Divergence of Character, related to the DIVERSTIY of inhabitants of any small area, and to na
0061 f a bird: in the structure of the beetle which DIVES through the water: in the plumed seed which i
0179 s, and form of tail; during summer this animal DIVES for and preys on fish, but during the long wi
0054 iting a country and described in any Flora be DIVIDED into two equal masses, all those in the larg
0057 stinct from each other: they may generally be DIVIDED into sub genera, or sections, or lesser grou
0058 .9 of the provinces into which Mr. Watson has DIVIDED Great Britain. Now, in this same catalogue,
0059 forms of life throughout the universe become DIVIDED into groups subordinate to groups. Chapter I
0095 e red clover to have a shorter or more deeply DIVIDED tube to its corolla, so that the hive bee co
0116 whether the Australian marsupials, which are DIVIDED into groups differing but little from each o
0129 ecies in the great battle for life. The limbs DIVIDED into great branches, and these into lesser a
0187 ance, there is a double cornea, the inner one DIVIDED into facets, within each of which there is a
0190 ductus pneumaticus for its supply, and being DIVIDED by highly vascular partitions. In these case
0238 to an almost incredible degree, and are thus DIVIDED into two or even three castes. The castes, m
0332 ction and divergence of character, has become DIVIDED into several sub families and families, some
0437 ppendages: in the articulata, we see the body DIVIDED into a series of segments, bearing external
0439 of the phyllodineous acaceas, are pinnate or DIVIDED like the ordinary leaves of the Leguminosae.
0077 he structure of its legs, so well adapted for DIVING, allows it to compete with other aquatic inse
0185 s general habits, in its astonishing power of DIVING, its manner of swimming, and of flying when u
0185 terrestrial thrush family wholly subsists by DIVING, grasping the stones with its feet and using
0186 where not a tree grows: that there should be DIVING thrushes, and petrels with the habits of auks
0204 d geese with webbed feet, ground woodpeckers, DIVING thrushes, and petrels with the habits of auks
0055 nera. Both these results follow when another DIVISION is made, and when all the smallest genera, w
0093 ge of what has been called the physiological DIVISION of labour: hence we may believe that it woul
0094 ould be advantageous on the principle of the DIVISION of labour, individuals with this tendency mo
0115 fact, the same as that of the physiological DIVISION of labour in the organs of the same individu
0126 here exist but very few classes in each main DIVISION of the animal and vegetable kingdoms. Althou
0241 y of insects, on the same principle that the DIVISION of labour is useful to civilised man. As ant
0242 edge and manufactured instruments, a perfect DIVISION of labour could be effected with them only b
0242 e has, as I believe, effected this admirable DIVISION of labour in the communities of ants, by the
0415 ive an example amongst insects, in one great DIVISION of the Hymenoptera, the antennae, as Westwoo
0416 , are most constant in structure: in another DIVISION they differ much, and the differences are of
0146 that the elder De Candolle founded his main DIVISIONS of the order on analogous differences. Henc
0346 thors agree that one of the most fundamental DIVISIONS in geographical distribution is that betwee
0414 e signification, excepting in the first main DIVISIONS: whereas the organs of reproduction, with t
0416 ably will say that the antennae in these two DIVISIONS of the same order are of unequal physiologi
0418 with flowering plants, of which the two main DIVISIONS have been founded on characters derived fro
0440 unmistakeable manner. So again the two main DIVISIONS of cirripedes, the pedunculated and sessile
```

Page ***(Key Word)**

Page ***(Key Word)***

```
0210  s from a group of about a dozen aphides on a DOCKPLANT, and prevented their attendance during seve
0005  of increase, will be treated of. This is the DOCTRINE of Malthus, applied to the whole animal and
0013  als, and it reappears in the child, the mere DOCTRINE of chances almost compels us to attribute it
0019  ed from the common wild duck and rabbit. The DOCTRINE of the origin of our several domestic races
0063  h the physical conditions of life. It is the DOCTRINE of Malthus applied with manifold force to th
0095  ions of structure. I am well aware that this DOCTRINE of natural selection, exemplified in the abo
0199  by some naturalists, against the utilitarian DOCTRINE that every detail of structure has been prod
0199  n the eyes of man, or for mere variety. This DOCTRINE, if true, would be absolutely fatal to my th
0242  se of neuter insects, against the well known DOCTRINE of Lamarck. Summary. I have endeavoured brie
0253  ts, they must certainly be highly fertile. A DOCTRINE which originated with Pallas, has been Large
0338  nd huxley in thinking that the truth of this DOCTRINE is very far from proved. Yet I fully expect
0338  within comparatively recent times. For this DOCTRINE of Agassiz accords well with the theory of n
0390  d many native productions. He who admits the DOCTRINE of the creation of each separate species, wi
0418  y in the classification of animals; and this DOCTRINE has very generally been admitted as true. Th
0435  bers of the same class, by utility or by the DOCTRINE of final causes. The hopelessness of the att
0456  iculty, as they assuredly do on the ordinary DOCTRINE of creation, might even have been anticipate
0463  run, be supplanted and exterminated. On this DOCTRINE of the extermination of an infinitude of con
0483  ilence. It may be asked how far I extend the DOCTRINE of the modification of species. The question
0017  eyhound, bloodhound, terrier, spaniel, and bull DOG, which we all know propagate their kind so tru
0018  fuego or Australia, who possess a semi domestic DOG, may not have existed in Egypt? The whole subj
0019  the Italian greyhound, the bloodhound, the bull DOG, or Blenheim spaniel, etc., so unlike all wild
0020  rms, as the Italian greyhound, bloodhound, bull DOG, etc., in the wild state. Moreover, the possib
0026  tendency to sterility: from the history of the DOG I think there is some probability in this hypo
0030  line. So it has probably been with the turnspit DOG; and this is known to have been the case with
0035  s not seen, as I am informed by him, any native DOG in Spain like our pointer. By a similar proces
0141  instance, of a tropical and arctic wolf or wild DOG may perhaps be mingled in our domestic breeds.
0144  between the hair and teeth in the naked Turkish DOG, though here probably homology comes into play
0196  rehension, or as an aid in turning, as with the DOG, though the aid must be slight, for the hare,
0214  either parent: for example, Le Roy describes a DOG, whose great grandfather was a wolf, and this
0214  o, whose great grandfather was a wolf, and this DOG showed a trace of its wild parentage only in o
0214  hether any one would have thought of training a DOG to point, had not some one dog naturally shown
0214  ht of training a dog to point, had not some one DOG naturally shown a tendency in this line; and t
0215  t the love of man has become instinctive in the DOG. All wolves, foxes, jackals, and species of th
0215  ns have lost, wholly by habit, that fear of the DOG and cat which no doubt was originally instinct
0268  is stated, for instance, that the German Spitz DOG unites more easily than other dogs with foxes,
0413  o all carnivora, by another those common to the DOG genus, and then by adding a single sentence, a
0413  ce, a full description is given of each kind of DOG. The ingenuity and utility of this system are
0444  n on Dogs, maintain that the greyhound and bull DOG, though appearing so different, are really var
0306  it seems to me to be about as rash in us to DOGMATIZE on the succession of organic beings through
0012  viduals by certain vegetable poisons. Hairless DOGS have imperfect teeth; long haired and coarse h
0017  lieve, as we shall presently see, that all our DOGS have descended from any one wild species: but,
0018  , i think it highly probable that our domestic DOGS have descended from several wild species. In r
0019  is in India. Even in the case of the domestic DOGS of the whole world, which I fully admit have p
0019  often been loosely said that all our races of DOGS have been produced by the crossing of a few ab
0030  er purpose; when we compare the many breeds of DOGS, each good for man in very different ways; whe
0034  attended to. Savages now sometimes cross their DOGS with wild canine animals, to improve the breed
0034  ur, as do some of the Escuimaux their teams of DOGS. Livingstone shows how much good domestic bree
0034  eeping pointers naturally tries to get as good DOGS as he can, and afterwards breeds from his own
0034  e can, and afterwards breeds from his own best DOGS; but he has no wish or expectation of permanen
0036  n times of dearth, as of less value than their DOGS. In plants the same gradual process of improve
0072  f this; for here neither cattle nor horses nor DOGS have ever run wild, though they swarm southwar
0091  results from each man trying to keep the best DOGS without any thought to modifying the breed. Ev
0213  to the familiar case of the several breeds of DOGS: it cannot be doubted that young pointers (I h
0213  ance) will sometimes point and even back other DOGS the very first time that they are taken out: r
0213  , instead of at, a flock of sheep; by shepherd DOGS. I cannot see that these actions, performed wi
0214  ingled, is well shown when different breeds of DOGS are crossed. Thus it is known that a cross wit
0214  yhound has given to a whole family of shepherd DOGS a tendency to hunt hares. These domestic insti
0215  ocure, without intending to improve the breed, DOGS which will stand and hunt best. On the other h
0215  and this tendency have been found incurable in DOGS which have been brought home as puppies from c
0215  ow rarely, on the other hand, do our civilised DOGS, even when quite young, require to be taught w
0215  bly concurred in civilising by inheritance our DOGS. On the other hand, young chickens have lost,
0216  chickens have lost all fear, but fear only of DOGS and cats, for if the hen gives the danger chuc
0254  ct evidence. I believe, for instance, that our DOGS have descended from several wild stocks; yet,
0254  s the exception of certain indigenous domestic DOGS of South America, all are quite fertile togeth
0268  german Spitz dog unites more easily than other DOGS with foxes, or that certain South American ind
0268  hat certain South American indigenous domestic DOGS do not readily cross with European dogs, the e
0268  mestic dogs do not readily cross with European DOGS, the explanation which will occur to everyone,
0268  yone, and probably the true one, is that these DOGS have descended from several aboriginally disti
0299  different breeds of cattle, sheep, horses, and DOGS have descended from a single stock or from sev
0444  ic varieties. Some authors who have written on DOGS, maintain that the greyhound and bull dog, tho
0445  be the case; but on actually measuring the old DOGS and their six days old puppies, I found that t
0446  estic varieties. Fanciers select their horses, DOGS, and pigeons, for breeding, when they are near
0286  e, one can safely picture to oneself the great DOME of rocks which must have covered up the Weald
0007  elation of Growth. Inheritance. Character of DOMESTIC Varieties. Difficulty of distinguishing betw
0007  ing between Varieties and Species. Origin of DOMESTIC Varieties from one or more Species. Domestic
0007  domestic Varieties from one or more Species. DOMESTIC Pigeons, their Differences and Origin. Princ
0007  unconscious Selection. Unknown Origin of our DOMESTIC Productions. Circumstances favourable to Man
0007  his greater variability is simply due to our DOMESTIC productions having been raised under conditi
0011  e marked effect; for instance, I find in the DOMESTIC duck that the bones of the wing weigh less a
0011  this change may be safely attributed to the DOMESTIC duck flying much less, and walking more, tha
0011  instance of the effect of use. Not a single DOMESTIC animal can be named which has not in some co
0013  peculiarities appearing in the males of our DOMESTIC breeds are often transmitted either exclusiv
0014  often made by naturalists, namely, that our DOMESTIC varieties, when run wild, gradually but cert
0014  argued that no deductions can be drawn from DOMESTIC races to species in a state of nature. I hav
0014  e that very many of the most strongly marked DOMESTIC varieties could not possibly live in a wild
0015  e are changed. If it could be shown that our DOMESTIC varieties manifested a strong tendency to re
```

Page ***(Key Word)***

Page ***(Key Word)**

```
0015  e, i grant that we could deduce nothing from  DOMESTIC  varieties in regard to species. But there is
0015  to the hereditary varieties or races of our   DOMESTIC  animals and plants, and compare them with sp
0015  lied together, we generally perceive in each  DOMESTIC  race, as already remarked, less uniformity o
0015  niformity of character than in true species.  DOMESTIC  races of the same species, also, often have
0016  ossed, a subject hereafter to be discussed),  DOMESTIC  races of the same species differ from each o
0016  tted, when we find that there are hardly any  DOMESTIC  races, either amongst animals or plants, whi
0016  s. if any marked distinction existed between  DOMESTIC  races and species, this source of doubt coul
0016  etually recur. It has often been stated that  DOMESTIC  races do not differ from each other in chara
0016  amount of structural difference between the   DOMESTIC  races of the same species, we are soon invol
0017  wild species: but, in the case of some other  DOMESTIC  races, there is presumptive, or even strong,
0018  se who believe in the multiple origin of our  DOMESTIC  animals is, that we find in the most ancient
0018  a del Fuego or Australia, who possess a semi  DOMESTIC  dog, may not have existed in Egypt? The whol
0018  rations, I think it highly probable that our  DOMESTIC  dogs have descended from several wild specie
0019  t. the doctrine of the origin of our several  DOMESTIC  races from several aboriginal stocks, has be
0019  cattle, sheep, etc., we must admit that many  DOMESTIC  breeds have originated in Europe: for whence
0019  ? so it is in India. Even in the case of the  DOMESTIC  dogs of the whole world, which I fully admit
0020  r parents: and if we account for our several  DOMESTIC  races by this process, we must admit the for
0020  aving been thus formed. On the Breeds of the  DOMESTIC  Pigeon. Believing that it is always best to
0020  group, I have, after deliberation, taken up   DOMESTIC  pigeons. I have kept every breed which I cou
0023  ks: for it is impossible to make the present  DOMESTIC  breeds by the crossing of any lesser number:
0023  these have not any of the characters of the  DOMESTIC  breeds. Hence the supposed aboriginal stocks
0023  k pigeon, which has the same habits with the  DOMESTIC  breeds, has not been exterminated even on se
0025  te: the wings have two black bars; some semi  DOMESTIC  breeds and some apparently truly wild breeds
0025  f the whole family. Now, in every one of the  DOMESTIC  breeds, taking thoroughly well-bred birds, a
0025  eversion to ancestral characters, if all the  DOMESTIC  breeds have descended from the rock pigeon.
0026  the hybrids or mongrels from between all the  DOMESTIC  breeds of pigeons are perfectly fertile. I c
0027  n together, I can feel no doubt that all our  DOMESTIC  breeds have descended from the Columba livia
0027  t number of points of structure with all the DOMESTIC  breeds. Secondly, although an English carrie
0028  ary. I have discussed the probable origin of  DOMESTIC  pigeons at some, yet quite insufficient, len
0028  namely, that all the breeders of the various  DOMESTIC  animals and the cultivators of plants, with
0029  lines of descent, yet admit that many of our  DOMESTIC  races have been descended from the same parents.
0029  t us now briefly consider the steps by which  DOMESTIC  races have been produced, either from one or
0034  s in Genesis, it is clear that the colour of  DOMESTIC  animals was at that early period attended to
0034  ams of dogs. Livingstone shows how much good  DOMESTIC  breeds are valued by the negroes of the inte
0034  election, but they show that the breeding of  DOMESTIC  animals was carefully attended to in ancient
0036  nherited character of the offspring of their  DOMESTIC  animals, yet any one animal particularly use
0038  ntries anciently civilised. In regard to our  DOMESTIC  animals kept by uncivilised man, it should n
0038  becomes at once obvious, how it is that our   DOMESTIC  races show adaptation in their structure or
0038  and the frequently abnormal character of our  DOMESTIC  races, and likewise their differences being
0040  ng about the origin or history of any of our  DOMESTIC  breeds. But, in fact, a breed' like a dialec
0042  m islands. Although I do not doubt that some  DOMESTIC  animals vary less than others, yet the rarit
0043  tinct breeds. To sum up on the origin of our  DOMESTIC  Races of animals and plants. I believe that
0043  layed an important part in the origin of our  DOMESTIC  productions. When in any country several dom
0043  tic productions. When in any country several  DOMESTIC  breeds have once been established, their occ
0064  still more striking is the evidence from our  DOMESTIC  animals of many kinds which have run wild in
0065  iod of life. Our familiarity with the larger  DOMESTIC  animals tends, I think, to mislead us: we se
0076  bted whether the varieties of any one of our  DOMESTIC  plants or animals have so exactly the same s
0080  endless number of strange peculiarities our   DOMESTIC  productions, and, in a lesser degree, those
0090  and becoming attached to the male sex in our  DOMESTIC  animals (as the wattle in male carriers, hor
0091  differences in the natural tendencies of our  DOMESTIC  animals: one cat, for instance, taking to ca
0112  ice, let us seek light on this head from our  DOMESTIC  productions. We shall here find something an
0133  te has some direct action on the hair of our  DOMESTIC  quadrupeds. Instances could be given of the
0134  nk there can be little doubt that use in our  DOMESTIC  animals strengthens and enlarges certain par
0134  ts wings in nearly the same condition as the  DOMESTIC  Aylesbury duck. As the larger ground feeding
0140  ed to their new homes. As I believe that our  DOMESTIC  animals were originally chosen by uncivilise
0140  the common and extraordinary capacity in our  DOMESTIC  animals of not only withstanding the most di
0141  ccount of the probable origin of some of our  DOMESTIC  animals from several wild stocks: the blood,
0141  lf or wild dog may perhaps be mingled in our  DOMESTIC  breeds. The rat and mouse cannot be consider
0141  s. the 'rat and mouse cannot be considered as DOMESTIC  animals, but they have been transported by m
0141  different climates by man himself and by his  DOMESTIC  animals, and such facts as that former speci
0147  this holds true to a certain extent with our  DOMESTIC  productions: if nourishment flows to one par
0152  on, I think we can obtain some light. In our  DOMESTIC  animals, if any part, or the whole animal, b
0152  more especially concerns us is, that in our  DOMESTIC  animals those points, which at the present t
0159  be most readily understood by looking to our  DOMESTIC  races. The most distinct breeds of pigeons,
0162  t have told, whether these characters in our  DOMESTIC  breeds were reversions or only analogous var
0167  rently constructed, the common parent of our  DOMESTIC  horse, whether or not it be descended from o
0198  nt for the characteristic differences of our  DOMESTIC  breeds, which nevertheless we generally admi
0207  uated. Aphides and ants. Instincts variable.  DOMESTIC  instincts, their origin. Natural instincts o
0213  yed in modifying the mental qualities of our  DOMESTIC  animals. A number of curious and authentic i
0213  ld assuredly call these actions instinctive.  DOMESTIC  instincts, as they may be called, are certai
0213  fixed conditions of life. How stongly these   DOMESTIC  instincts, habits, and dispositions are inhe
0214  hepherd dogs a tendency to hunt hares. These  DOMESTIC  instincts, when thus tested by crossing, res
0214  n a straight line to his master when called.  DOMESTIC  instincts are sometimes spoken of as actions
0215  f the tame rabbit; but I do not suppose that  DOMESTIC  rabbits have ever been selected for tameness
0215  how universally and largely the minds of our  DOMESTIC  animals have been modified by domestication.
0215  stralia, where the savages do not keep these  DOMESTIC  animals. How rarely, on the other hand, do o
0216  power of flight. Hence we may conclude, that  DOMESTIC  instincts have been acquired and natural ins
0237  t we have innumerable instances, both in our  DOMESTIC  productions and in those in a state of natur
0239  lace of that excreted by the aphides, or the  DOMESTIC  cattle as they may be called, which our Euro
0242  ter to show that the mental qualities of our  DOMESTIC  animals vary, and that the variations are in
0254  modern naturalists: namely, that most of our  DOMESTIC  animals have descended from two or more abor
0254  perhaps the exception of certain indigenous  DOMESTIC  dogs of South America, all are quite fertile
0254  s. on this view of the origin of many of our  DOMESTIC  animals, we must either give up the belief o
0268  s, or that certain South American indigenous  DOMESTIC  dogs do not readily cross with European dogs
0268  evertheless the perfect fertility of so many  DOMESTIC  varieties, differing widely from each other
0268  iderations, however, render the fertility of  DOMESTIC  varieties less remarkable than at first appe
```

Page ***(Key Word)**

Page **(Key Word)**

```
0269 n crossed; and we may apply the same rule to DOMESTIC varieties. In the second place, some eminent
0271 cters in the production of the most distinct DOMESTIC varieties, and from not wishing or being abl
0273 arents of mongrels are varieties, and mostly DOMESTIC varieties (very few experiments having been
0316 from the other well established races of the DOMESTIC pigeon, for the newly formed fantail would b
0319 rom the history of the naturalisation of the DOMESTIC horse in South America, that under more favo
0320 inevitably follows. It is the same with our DOMESTIC productions: when a new and slightly improve
0335 he principal living and extinct races of the DOMESTIC pigeon were arranged as well as they could b
0423 sub varieties under varieties; and with our DOMESTIC productions, several other grades of differe
0427 e see something of the same kind even in our DOMESTIC varieties, as in the thickened stems of the
0444 d of life; and some direct evidence from our DOMESTIC animals supports this view. But in other cas
0444 irst let us look at a few analogous cases in DOMESTIC varieties. Some authors who have written on
0445 e appears to me conclusive, that the several DOMESTIC breeds of Pigeon have descended from one wil
0446 regard to the later embryonic stages of our DOMESTIC varieties. Fanciers select their horses, dog
0451 s in the udders of the genus Bos, but in our DOMESTIC cows the two sometimes become developed and
0454 plenty of cases of rudimentary organs in our DOMESTIC productions, as the stump of a tail in taill
0466 ty in ascertaining how much modification our DOMESTIC productions have undergone; but we may safel
0467 e production of the most distinct and useful DOMESTIC breeds. That many of the breeds produced by
0468 adding up mere individual differences in his DOMESTIC productions; and every one admits that there
0473 ight, in nearly the same condition as in the DOMESTIC duck; or when we look at the burrowing tucut
0473 rogenitor, in the same manner as the several DOMESTIC breeds of pigeon have descended from the blu
0486 ernal conditions, and so forth. The study of DOMESTIC productions will rise immensely in value. A
0225 e, we have the cells of the Mexican Melipona DOMESTICA, carefully described and figured by Pierre
0004 med to me probable that a careful study of DOMESTICATED animals and of cultivated plants would off
0008 till often yield new varieties; our oldest DOMESTICATED animals are still capable of rapid improve
0009 ile hybrids. When, on the one hand, we see DOMESTICATED animals and plants, though often weak and
0016 en to meet with generic differences in our DOMESTICATED productions. When we attempt to estimate t
0017 added largely to the value of most of our DOMESTICATED productions; but how could a savage possib
0017 animals and plants, equal in number to our DOMESTICATED productions, and belonging to equally dive
0017 gely as the parent species of our existing DOMESTICATED productions have varied. In the case of mo
0017 ried. In the case of most of our anciently DOMESTICATED animals and plants, I do not think it is p
0023 n the countries where they were originally DOMESTICATED, and yet be unknown to ornithologists; and
0024 umption. Moreover, the several above named DOMESTICATED breeds have been transported to all parts
0024 seven or eight species were so thoroughly DOMESTICATED in ancient times by half civilized man, as
0027 , and loved by many people. They have been DOMESTICATED for thousands of years in several quarters
0029 one of the most remarkable features in our DOMESTICATED races is that we see in them adaptation, n
0045 en direction individual differences in his DOMESTICATED productions. These individual differences
0110 same process of extermination amongst our DOMESTICATED productions, through the selection of impr
0198 no on the differences in the breeds of our DOMESTICATED animals in different countries, more espec
0393 doubt, of a terrestrial mammal (excluding DOMESTICATED animals kept by the natives) inhabiting an
0466 ccasionally produced by our most anciently DOMESTICATED productions. Man does not actually produce
0024 half civilized man succeeded in thoroughly DOMESTICATING several species, but that he intentionall
0004 imperfect though it be, of variation under DOMESTICATION, afforded the best and safest clue. I may
0004 hapter of this Abstract to Variation under DOMESTICATION. We shall thus see that a large amount of
0007 ive means of modification. Variation Under DOMESTICATION Chapter I. Causes of Variability. Effects
0009 so will some animals and plants withstand DOMESTICATION or cultivation, and vary very slightly, p
0017 often been assumed that man has chosen for DOMESTICATION animals and plants having an extraordinar
0017 cold by the common camel, prevented their DOMESTICATION? I cannot doubt that if other animals and
0017 d for an equal number of generations under DOMESTICATION, they would vary on an average as largely
0024 get any wild animal to breed freely under DOMESTICATION; yet on the hypothesis of the multiple or
0026 . some authors believe that long continued DOMESTICATION eliminates this strong tendency to steril
0027 d species of pigeons to breed freely under DOMESTICATION; these supposed species being quite unkno
0027 the rock pigeon, has been found capable of DOMESTICATION in Europe and in India; and that it agree
0080 w strong the hereditary tendency is. Under DOMESTICATION, it may be truly said that the whole orga
0086 s we see that those variations which under DOMESTICATION appear at any particular period of life,
0087 asmuch as peculiarities often appear under DOMESTICATION in one sex and become hereditarily attach
0090 bird; indeed, had the tuft appeared under DOMESTICATION, it would have been called a monstrosity.
0131 mmon and multiform in organic beings under DOMESTICATION, and in a lesser degree in those in a sta
0131 greater frequency of monstrosities, under DOMESTICATION or cultivation, than under nature, leads
0163 al species of the same genus, partly under DOMESTICATION and partly under nature. It is a case app
0167 dency to vary, both under nature and under DOMESTICATION, in this particular manner, so as often t
0213 d by briefly considering a few cases under DOMESTICATION. We shall thus also be enabled to see the
0215 finement. Natural instincts are lost under DOMESTICATION: a remarkable instance of this is seen in
0215 our domestic animals have been modified by DOMESTICATION. It is scarcely possible to doubt that th
0216 d by our chickens has become useless under DOMESTICATION, for the mother hen has almost lost by di
0245 ffected by close interbreeding, removed by DOMESTICATION. Laws governing the sterility of hybrids.
0254 subsequent generations quite fertile under DOMESTICATION. This latter alternative seems to me the
0254 ic, but as one capable of being removed by DOMESTICATION. Finally, looking to all the ascertained
0265 ed. This, in fact, is the great bar to the DOMESTICATION of animals. Between the sterility thus su
0268 , or supposed to have been produced, under DOMESTICATION, we are still involved in doubt. For when
0269 naturalists believe that a long course of DOMESTICATION tends to eliminate sterility in the succe
0269 s of animals and plants are produced under DOMESTICATION by man's methodical and unconscious power
0277 mber of varieties have been produced under DOMESTICATION by the selection of mere external differe
0445 has been wholly caused by selection under DOMESTICATION; but having had careful measurements made
0461 experimentised on have been produced under DOMESTICATION; and as domestication apparently tends to
0461 been produced under domestication; and as DOMESTICATION apparently tends to eliminate sterility,
0466 n to the other side of the argument. Under DOMESTICATION we see much variability. This seems to be
0467 ples which have acted so efficiently under DOMESTICATION should not have acted under nature. In th
0468 ave varied under the changed conditions of DOMESTICATION. And if there be any variability under na
0053 flourishing, or, as they may be called, the DOMINANT species, those which range widely over the w
0054 f the country, the species which are already DOMINANT will be the most likely to yield offspring w
0054 antages that enabled their parents to become DOMINANT over their compatriots. If the plants inhabi
0054 mber of the very common and much diffused or DOMINANT species will be found on the side of the lar
0054 many species, a large proportional number of DOMINANT species. But so many causes tend to obscure
0059 so, seen that it is the most flourishing and DOMINANT species of the larger genera which on an ave
0059 ghout nature the forms of life which are now DOMINANT tend to become still more dominant by leavin
0059 h are now dominant tend to become still more DOMINANT by leaving many modified and dominant descen
```

Page **(Key Word)**

Page **************************************(Key Word)**************************************

```
0059 l more dominant by leaving many modified and DOMINANT descendants. But by steps hereafter to be ex
0128 spring that superiority which now makes them DOMINANT in their own countries. Natural selection, a
0325 er forms; and those forms, which are already DOMINANT, or have some advantage over the other forms
0325 idence on this head, in the plants which are DOMINANT, that is, which are commonest in their own h
0325 f new varieties. It is also natural that the DOMINANT, varying, and far spreading species, which a
0326 n strange accidents, but in the long run the DOMINANT forms will generally succeed in spreading. T
0326 the productions of the land than of the sea. DOMINANT species spreading from any region might enco
0326 g from any region might encounter still more DOMINANT species, and then their triumphant course, o
0326 favourable for the multiplication of new and DOMINANT species; but we can, I think, clearly see th
0326 ost favourable for the production of new and DOMINANT species on the land, and another for those i
0327 orious. But in the course of time, the forms DOMINANT in the highest degree, wherever produced, wo
0327 inciple of new species having been formed by DOMINANT species spreading widely and varying; the ne
0327 e new species thus produced being themselves DOMINANT owing to inheritance, and to having already
0340 igration, the feebler will yield to the more DOMINANT forms; and there will be nothing immutable i
0343 d depends on many complex contingencies. The DOMINANT species of the larger dominant groups tend t
0343 ngencies. The dominant species of the larger DOMINANT groups tend to leave many modified descendan
0344 . we can understand how the spreading of the DOMINANT forms of life, which are those that oftenest
0350 ity will depend on the migration of the more DOMINANT forms of life from one region into another h
0377 at a certain number of the more vigorous and DOMINANT temperate forms might have penetrated the na
0380 lmost everywhere largely yielded to the more DOMINANT forms, generated in the larger areas and mor
0408 e theory of migration (generally of the more DOMINANT forms of life), together with subsequent mod
0411 o, the much diffused and common, that is the DOMINANT species belonging to the larger genera, were
0411 f inheritance, tend to produce other new and DOMINANT species. Consequently the groups which are n
0412 now large, and which generally include many DOMINANT species, tend to go on increasing indefinite
0428 bably arisen. As the modified descendants of DOMINANT species, belonging to the larger genera, ten
0428 to which they belong large and their parents DOMINANT, they are almost sure to spread widely, and
0428 n the economy of nature. The larger and more DOMINANT groups thus tend to go on increasing in size
0433 f character in the many descendants from one DOMINANT parent species, explains that great and univ
0470 a large extent defined and distinct objects. DOMINANT species belonging to the larger groups tend
0470 larger groups tend to give birth to new and DOMINANT forms; so that each large group tends to bec
0471 for the world would not hold them, the more DOMINANT groups beat the less dominant. This tendency
0471 them, the more dominant groups beat the less DOMINANT. This tendency in the large groups to go on
0475 s once been broken. The gradual diffusion of DOMINANT forms, with the slow modification of their d
0489 spread species, belonging to the larger and DOMINANT groups, which will ultimately prevail and pr
0489 ill ultimately prevail and procreate new and DOMINANT species. As all the living forms of life are
0379 petition to a higher stage of perfection or DOMINATING power, than the southern forms. And thus, w
0042 or absence of distinct breeds of the cat, the DONKEY, peacock, goose, etc., may be attributed in m
0042 cats, from the difficulty in pairing them; in DONKEYS, from only a few being kept by poor people,
0201 hen preparing to spring, in order to warn the DOOMED mouse. But I have not space here to enter on
0047 or may formerly have existed: and here a wide DOOR for the entry of doubt and conjecture is expe
0221 ng the nest, and they alone open and close the DOORS in the morning and evening: and as Huber expr
0152 ied, that part (for instance, the comb in the DORKING fowl) or the whole breed will cease to have
0191 ed by some naturalists that the branchiae and DORSAL scales of Annelids are homologous with the wi
0208 characters of instinct are universal. A little DOSE, as Pierre Huber expresses it, of judgment or
0117 its own country. The little fan of diverging DOTTED lines of unequal lengths proceeding from (A),
0117 ivergent variations (represented by the outer DOTTED lines) being preserved and accumulated by nat
0117 and accumulated by natural selection. When a DOTTED line reaches one of the horizontal lines, and
0121 ants; and this is shown in the diagram by the DOTTED lines not prolonged far upwards from want of
0125 esented by each successive group of diverging DOTTED lines to be very great, the forms marked a14
0331 he numbered letters represent genera, and the DOTTED lines diverging from them the species in each
0025 beautiful a blue colour, with the white rump, DOUBLE black wing bar, and barred and white edged ta
0077 nt in the midst of its range, why does it not DOUBLE or quadruple its numbers? We know that it can
0163 that the stripe on each shoulder is sometimes DOUBLE. The shoulder stripe is certainly very variab
0163 an of Pallas is said to have been seen with a DOUBLE shoulder stripe. The hemionus has no shoulder
0164 tch for me of a dun Belgian cart horse with a DOUBLE stripe on each shoulder and with leg stripes:
0164 ; and the shoulder stripe, which is sometimes DOUBLE and sometimes treble; is common; the side of
0165 t of the body; and in one of them there was a DOUBLE shoulder stripe. In Lord Moreton's famous hyb
0187 certain crustaceans, for instance, there is a DOUBLE cornea, the inner one divided into facets, wi
0196 ight, for the hare, with hardly any tail, can DOUBLE quickly enough. In the second place, we may s
0225 e have the cells of the hive bee, placed in a DOUBLE layer: each cell, as is well known, is an hex
0226 lat walls between the adjoining cells are not DOUBLE, but are of the same thickness as the outer s
0226 izes and had arranged them symmetrically in a DOUBLE layer, the resulting structure would probably
0227 in both layers be formed, there will result a DOUBLE layer of hexagonal prisms united together by
0235 eres at a given distance from each other in a DOUBLE layer, and to build up and excavate the wax a
0306 pted to a cooler climate, and were enabled to DOUBLE the southern capes of Africa or Australia, an
0064 of a single pair. Even slow breeding man has DOUBLED in twenty five years, and at this rate, in a
0436 oldering together of other parts; and by the DOUBLING or multiplication of others, variations whic
019 usively for respiration. I can, indeed, hardly DOUBT that all vertebrate animals having true lungs
0002 er reposing some confidence in my accuracy. No DOUBT errors will have crept in, though I hope I ha
0006 d will long remain obscure, I can entertain no DOUBT, after the most deliberate study and dispassi
0016 een domestic races and species, this source of DOUBT could not so perpetually recur. It has often
0016 s of the same species, we are soon involved in DOUBT, from not knowing whether they have descended
0017 uch facts would have great weight in making us DOUBT about the immutability of the many very close
0017 camel, prevented their domestication? I cannot DOUBT that if other animals and plants, equal in nu
0019 derably from each other in structure, I do not DOUBT that they all have descended from the common
0019 descended from several wild species, I cannot DOUBT that there has been an immense amount of inhe
0020 has been greatly exaggerated. There can be no DOUBT that a race may be modified by occasional cro
0027 several reasons, taken together, I can feel no DOUBT that all our domestic breeds have descended f
0033 differences; but, as a general rule, I cannot DOUBT that the continued selection of slight variat
0035 ntly altering the breed. Nevertheless I cannot DOUBT that this process, continued during centuries
0037 rom such poor materials: but the art, I cannot DOUBT, has been simple, and, as far as the final re
0039 ion as trying to make a fantail, is, I have no DOUBT, in most cases, utterly incorrect. The man wh
0041 ners began to attend closely to this plant. No DOUBT the strawberry had always varied since it was
0042 country, often from islands. Although I do not DOUBT that some domestic animals vary less than oth
0043 ed infinitely complex. In some cases, I do not DOUBT that the intercrossing of species, aboriginal
0043 ercrossing, with the aid of selection, has, no DOUBT, largely aided in the formation of new sub br
```

Page **************************************(Key Word)**************************************

doubt

Page **************************************(Key Word)***************************************

```
0047 existed; and here a wide door for the entry of DOUBT and conjecture is opened. Hence, in determini
0068 s of a species. Thus, there seems to be little DOUBT that the stock of partridges, grouse, and har
0073 not reach the nectar. Hence I have very little DOUBT, that if the whole genus of humble bees becam
0077 nd water. Yet the advantage of plumed seeds no DOUBT stands in the closest relation to the land be
0080 sands of generations? If such do occur, can we DOUBT (remembering that many more individuals are b
0085 e to destruction. Hence I can see no reason to DOUBT that natural selection might be most effectiv
0086 the mature insect. These modifications will no DOUBT affect, through the laws of correlation, the
0089 tandard of beauty, I can see no good reason to DOUBT that female birds, by selecting, during thous
0090 can under such circumstances see no reason to DOUBT that the swiftest and slimmest wolves would h
0090 on other animals. I can see no more reason to DOUBT this, than that man can improve the fleetness
0094 ing such facts in mind, I can see no reason to DOUBT that an accidental deviation in the size and
0098 ed solely to ensure self fertilisation; and no DOUBT it is useful for this end: but, the agency of
0105 s we are incapable of doing. Although I do not DOUBT that isolation is of considerable importance
0114 o increase its numbers. Consequently, I cannot DOUBT that in the course of many thousands of gener
0119 upper horizontal lines. In some cases I do not DOUBT that the process of modification will be conf
0134 the first chapter, I think there can be little DOUBT that use in our domestic animals strengthens
0142 bit. On the other hand, I can see no reason to DOUBT that natural selection will continually tend
0143 ous with the limbs. These tendencies, I do not DOUBT, may be mastered more or less completely by n
0146 in some necessary manner. So, again, I do not DOUBT that some apparent correlations, occurring th
0151 ner, of which fact I think there can be little DOUBT. But that our rule is not confined to seconda
0154 s may be made constant, I can see no reason to DOUBT. Hence when an organ, however abnormal it may
0155 is. Geoffroy St. Hilaire seems to entertain no DOUBT, that the more an organ normally differs in t
0159 race, the fantail. I presume that no one will DOUBT that all such analogous variations are due to
0160 parent rock pigeon. I presume that no one will DOUBT that this is a case of reversion, and not of
0160 act of crossing on the laws of inheritance. No DOUBT it is a very surprising fact that characters
0174 on strictly continuous areas; though I do not DOUBT that the formerly broken condition of areas n
0180 onishing distance from tree to tree. We cannot DOUBT that each structure is of use to each kind of
0189 reak down. But I can find out no such case. No DOUBT many organs exist of which we do not know the
0192 y graduate into each other. Therefore I do not DOUBT that little folds of skin, which originally s
0197 through natural selection; as it is, I have no DOUBT that the colour is due to some quite distinct
0197 ends of the branches, and this contrivance, no DOUBT, is of the highest service to the plant: but
0197 iful adaptation for aiding parturition, and no DOUBT they facilitate, or may be indispensable for
0199 good thus gained. Correlation of growth has no DOUBT played a most important part, and a useful mo
0200 goose and of the frigate bird, webbed feet no DOUBT were as useful as they now are to the most aq
0209 are diminished or lost by disuse, so I do not DOUBT it has been with instincts. But I believe tha
0215 d by domestication. It is scarcely possible to DOUBT that the love of man has become instinctive i
0215 ght not to attack poultry, sheep, and pigs! No DOUBT they occasionally do make an attack, and are
0215 f habit, that fear of the dog and cat which no DOUBT was originally instinctive in them, in the sa
0234 a store of honey: there can in this case be no DOUBT that it would be an advantage to our humble b
0235 f succeeding in the struggle for existence. No DOUBT many instincts of very difficult explanation
0238 aith in the powers of selection, that I do not DOUBT that a breed of cattle, always yielding oxen
0247 t seems to me that we may well be permitted to DOUBT whether many other species are really so ster
0248 sed, but generally greatly decreased. I do not DOUBT that this is usually the case, and that the f
0249 or variety increases fertility, that I cannot DOUBT the correctness of this almost universal beli
0252 brics themselves are, I think, more sterile. I DOUBT whether any case of a perfectly fertile hybri
0254 fertile together; and analogy makes me greatly DOUBT, whether the several aboriginal species would
0268 under domestication, we are still involved in DOUBT. For when it is stated, for instance, that th
0286 stimate would be erroneous; but this source of DOUBT probably would not greatly affect the estimat
0286 n one yard in nearly every twenty two years. I DOUBT whether any rock, even as soft as chalk, woul
0286 xcepting on the most exposed coasts; though no DOUBT the degradation of a lofty cliff would be mor
0290 streams entering the sea. The explanation, no DOUBT, is, that the littoral and sub littoral depos
0306 earliest known species. For instance, I cannot DOUBT that all the Silurian trilobites have descend
0307 of fossiliferous strata, which on my theory no DOUBT were somewhere accumulated before the Siluria
0309 arge tracts of land have existed, subjected no DOUBT to great oscillations of level, since the ear
0315 spring of one species might be adapted (and no DOUBT this has occurred in innumerable instances) t
0323 ant points in the same parallel manner. We may DOUBT whether they have thus changed: if the Megath
0324 ly future epoch, there can, I think, be little DOUBT that all the more modern marine formations, n
0330 dely separated from each other. This remark no DOUBT must be restricted to those groups which have
0337 alaeozoic fauna by a secondary fauna. I do not DOUBT that this process of improvement has affected
0337 ving become wild in any part of Europe, we may DOUBT, if all the productions of New Zealand were s
0360 were then capable of germination. But I do not DOUBT that plants exposed to the waves would float
0363 h annually cross the Mediterranean; and can we DOUBT that the earth adhering to their feet would s
0363 nes, and the nest of a land bird, I can hardly DOUBT that they must occasionally have transported
0363 t, and that several other means, which without DOUBT remain to be discovered, have been in action
0365 t receive colonists by similar means. I do not DOUBT that out of twenty seeds or animals transport
0378 t have lain on the line of march. But I do not DOUBT that some temperate productions entered and c
0380 ve occurred on the intertropical mountains: no DOUBT before the Glacial period they were stocked w
0384 urring at very distant points of the world, no DOUBT there are many cases which cannot at present
0385 . but two facts, which I have observed, and no DOUBT many others remain to be observed, throw some
0393 i have not found a single instance, free from DOUBT, of a terrestrial mammal (excluding domestica
0395 by closely allied or identical quadrupeds. No DOUBT some few anomalies occur in this great archip
0399 is confined to the plants, and will, I do not DOUBT, be some day explained. The law which causes
0403 olorised by some European land shells which no DOUBT had some advantage over the indigenous specie
0404 he species have very wide ranges. I can hardly DOUBT that this rule is generally true, though it w
0415 of high vital or physiological importance. No DOUBT this view of the classificatory importance of
0417 t universal prevalence, and yet leave us in no DOUBT where it should be ranked. Hence, also, it ha
0428 ny characters, great and small, that we cannot DOUBT that they have inherited their general shape
0445 th and form of beak, that they would, I cannot DOUBT, be ranked in distinct genera, had they been
0451 ents of membrane; and here it is impossible to DOUBT, that the rudiments represent wings. Rudiment
0454 rudiments of various parts in monsters. But I DOUBT whether any of these cases throw light on the
0454 showing that rudiments can be produced; for I DOUBT whether species under nature ever undergo abr
0460 itions cannot, I think, be disputed. It is, no DOUBT, extremely difficult even to conjecture by wh
0466 ficulties far too heavily during many years to DOUBT their weight. But it deserves especial notice
0469 always ready to act and select, why should we DOUBT that variations in any way useful to beings,
0474 architectural powers of the hive bee. Habit no DOUBT sometimes comes into play in modifying instin
0482 ibility of mind, and who have already begun to DOUBT on the immutability of species, may be influe
```

Page **************************************(Key Word)***************************************

Page ***********************************(Key Word)**

0483 losely resemble each other. Therefore I cannot DOUBT that the theory of descent with modification
0484 will not be incessantly haunted by the shadowy DOUBT whether this or that form be in essence a spe
0486 of creation. The rules for classifying will no DOUBT become simpler when we have a definite object
0049 ston's admirable work, but which it cannot be DOUBTED would be ranked as distinct species by many
0076 ieties of the medicinal leech. It may even be DOUBTED whether the varieties of any one of our dome
0116 perfectly diversified in structure. It may be DOUBTED, for instance, whether the Australian marsup
0213 e of the several breeds of dogs: it cannot be DOUBTED that young pointers (I have myself seen a st
0214 high without going head over heels. It may be DOUBTED whether any one would have thought of traini
0247 jurious to the fertility of a plant cannot be DOUBTED: for Gartner gives in his table about a scor
0292 rom the foregoing considerations it cannot be DOUBTED that the geological record, viewed as a whol
0294 ithin this same glacial period. Yet it may be DOUBTED whether in any quarter of the world, sedimen
0300 record would then be least perfect. It may be DOUBTED whether the duration of any one great period
0044 nature. Variability. Individual differences. DOUBTFUL species. Wide ranging, much diffused, and co
0047 e every reason to believe that many of these DOUBTFUL and closely allied forms have permanently in
0048 ome competent judges. That varieties of this DOUBTFUL nature are far from uncommon cannot be dispu
0048 bentham gives only 112, a difference of 139 DOUBTFUL forms! Amongst animals which unite for each
0048 each birth, and which are highly locomotive, DOUBTFUL forms, ranked by one zoologist as a species
0049 in. A wide distance between the homes of two DOUBTFUL forms leads many naturalists to rank both as
0049 of the cases of strongly marked varieties or DOUBTFUL species well deserve consideration: for seve
0050 d by many intermediate links, and it is very DOUBTFUL whether these links are hybrids; and there i
0050 ring naturalists to an agreement how to rank DOUBTFUL forms. Yet it must be confessed, that it is
0050 that we find the greatest number of forms of DOUBTFUL value. I have been struck with the fact, tha
0050 oon make up his mind how to rank most of the DOUBTFUL forms. His general tendency will be to make
0051 e to find the intermediate links between his DOUBTFUL forms, he will have to trust almost entirely
0056 termediate links have not been found between DOUBTFUL forms, naturalists are compelled to come to
0058 closely allied to other species as to be of DOUBTFUL value: these 63 reputed species range on an
0058 allied forms, marked for me by Mr. Watson as DOUBTFUL species, but which are almost universally ra
0060 is immaterial for us whether a multitude of DOUBTFUL forms be called species or sub species or va
0060 rank, for instance, the two or three hundred DOUBTFUL forms of British plants are entitled to hold
0120 d varieties; or they may have arrived at the DOUBTFUL category of sub species; but we have only to
0132 ody, etc., produces on any being is extremely DOUBTFUL. My impression is, that the effect is extrem
0162 hough under nature it must generally be left DOUBTFUL, what cases are reversions to an anciently e
0248 st botanists on the question whether certain DOUBTFUL forms should be ranked as species or varieti
0248 ence from this source graduates away, and is DOUBTFUL in the same degree as is the evidence derive
0369 , some present varieties, some are ranked as DOUBTFUL forms, and some few are distinct yet closely
0376 earth and in the southern hemisphere are of DOUBTFUL value, being ranked by some naturalists as s
0392 ut this explanation seems to me not a little DOUBTFUL. Facility of immigration, I believe, has bee
0404 stinct species, and some as varieties; these DOUBTFUL forms showing us the steps in the process of
0409 sence of identical species, of varieties, of DOUBTFUL species, and of distinct but representative
0464 of the geological record. Numerous existing DOUBTFUL forms could be named which are probably vari
0464 de, on the common view, whether or not these DOUBTFUL forms are varieties? As long as most of the
0478 sely allied yet distinct species occur, many DOUBTFUL forms and varieties of the same species
0018 from reasons which I cannot give here, I am DOUBTFULLY inclined to believe, in opposition to sever
0162 n two other forms, which themselves must be DOUBTFULLY ranked as either varieties or species; and
0220 of mind, as any one may well be excused for DOUBTING the truth of so extraordinary and odious an
0012 lest and the best on this subject. No breeder DOUBTS how strong is the tendency to inheritance: li
0012 like produces like is his fundamental belief: DOUBTS have been thrown on this principle by theoret
0093 another process might commence. No naturalist DOUBTS the advantage of what has been called the phy
0111 acter of species, as is shown by the hopeless DOUBTS in many cases how to rank them, yet certainly
0116 elucidated by Milne Edwards. No physiologist DOUBTS that a stomach by being adapted to digest veg
0310 yell, from further reflexion entertains grave DOUBTS on this subject. I feel how rash it is to dif
0467 natural species, is shown by the inextricable DOUBTS whether very many of them are varieties or ab
0024 one has ever become wild or feral, though the DOVECOT pigeon, which is the rock pigeon in a very s
0085 yet we hear from an excellent horticulturist, DOWNING, that in the United States smooth skinned fr
0285 t is an admirable lesson to stand on the North DOWNS and to look at the distant South Downs: for,
0285 e north Downs and to look at the distant South DOWNS: for, remembering that at no great distance t
0286 the distance from the northern to the southern DOWNS is about 22 miles, and the thickness of the s
0285 . prof. Ramsay has published an account of a DOWNTHROW in Anglesea of 2300 feet; and he informs me
0291 ogical age, was certainly deposited during a DOWNWARD oscillation of level, and thus gained consid
0295 ved, and thus diminish the supply whilst the DOWNWARD movement continues. In fact, this nearly exa
0124 capital letters, converging in sub branches DOWNWARDS towards a single point; this point represen
0232 directly under the middle of a comb growing DOWNWARDS so that the comb has to be built over one f
0085 ally settle which variety, whether a smooth or DOWNY, a yellow or purple fleshed fruit; should suc
0026 that each breed, even the purest, has within a DOZEN or, at most, within a score of generations, b
0026 een crossed by the rock pigeon: I say within a DOZEN or twenty generations, for we know of no fact
0050 studied; yet a German author makes more than a DOZEN species out of forms, which are very generall
0076 s of a mixed stock could be kept up for half a DOZEN generations, if they were allowed to struggle
0160 gn breed for many generations, some say, for a DOZEN or even a score of generations. After twelve
0181 ugh the air rather than for flight. If about a DOZEN genera of birds had become extinct or were un
0193 ous difficulty; for they occur in only about a DOZEN fishes, of which several are widely remote in
0210 i removed all the ants from a group of about a DOZEN aphides on a dockplant, and prevented their a
0306 points, which the discoveries of even the last DOZEN years have effected, it seems to me to be abo
0330 uished at the present day from each other by a DOZEN characters, the ancient members of the same t
0333 he two families to differ from each other by a DOZEN characters, in this case the genera, at the e
0389 ascension aboriginally possessed under half a DOZEN flowering plants; yet many have become natura
0429 n less aberrant had each been represented by a DOZEN species instead of a single one; but such ric
0190 s, digests, and excretes in the larva of the DRAGONFLY and in the fish Cobites. In the Hydra, the
0445 ld stock, in pouters, fantails, runts, barbs, DRAGONS, carriers, and tumblers. Now some of these b
0315 , taken almost at hazard, in a slowly changing DRAMA. We can clearly understand why a species when
0034 liny. The savages in South Africa match their DRAUGHT cattle by colour, as do some of the Esquimau
0168 part is largely developed, perhaps it tends to DRAW nourishment from the adjoining parts; and ever
0272 er respects. Gartner, whose strong wish was to DRAW a marked line of distinction between species a
0328 ene deposits of England and France, is able to DRAW a close general parallelism between the succes
0469 rthy of record in systematic works. No one can DRAW any clear distinction between individual diffe
0145 ght that the development of the ray petals by DRAWING nourishment from certain other parts of the
0197 rid matter: but we should be very cautious in DRAWING any such inference, when we see that the ski

Page ***********************************(Key Word)**

Page ***(Key Word)***

```
0304 kilful palaeontologist, M. Bosquet, sent me a DRAWING of a perfect specimen of an unmistakeable se
0032 flowers of the present day are compared with DRAWINGS made only twenty or thirty years ago. When a
0035 gnised unless actual measurements or careful DRAWINGS of the breeds in question hac been made long
0165 a figure of a similar mule. In four coloured DRAWINGS, which I have seen, of hybrids between the a
0241 ly on this latter point, as Mr. Lubbock made DRAWINGS for me with the camera lucida of the jaws wh
0442 n that of Aphis, if we look to the admirable DRAWINGS by Professor Huxley of the development of th
0452 very of rudiments. This is well shown in the DRAWINGS given by Owen of the bones of the leg of the
0014 e it has been argued that no deductions can be DRAWN from domestic races to species in a state of
0048 at britain, of France or of the United States, DRAWN up by different botanists; and see what a sur
0051 y no clear line of demarcation has as yet been DRAWN between species and sub species, that is, the
0226  has kindly read over the following statement, DRAWN up from his information, and tells me that it
0255 he following rules and conclusions are chiefly DRAWN up from Gartner's admirable work on the hybri
0427 re fanciful than the analogies which have been DRAWN by some authors between very distinct animals
0469 e why it is that no line of demarcation can be DRAWN between species, commonly supposed to have be
0481 ity; no clear distinction has been, or can be, DRAWN between species and well marked varieties. It
0003 of trees. In the case of the misseltoe, which DRAWS its nourishment from certain trees, which has
0116 digest vegetable matter alone, or flesh alone, DRAWS most nutriment from these substances. So in t
0029 ount by such agencies for the differences of a DRAY and race horse, a greyhound and bloodhound, a
0030  with the ancon sheep. But when we compare the DRAY horse and race horse, the dromedary and camel,
0039 d a pigeon with a slightly larger tail, never DREAMED what the descendants of that pigeon would be
0424 ertain cirripedes, when adult, and yet no one DREAMS of separating them. The naturalist includes a
0175 es in sounding the depths of the sea with the DREDGE. To those who look at climate and the physica
0001 llowed myself to speculate on the subject, and DREW up some short notes; these I enlarged in 1844
0359 wn plants or branches; and that these might be DRIED on the banks, and then by a fresh rise in the
0359 ilst green floated for a very short time, when DRIED floated much longer; for instance, ripe hazel
0359 ce, ripe hazel nuts sank immediately, but when DRIED, they floated for 90 days and afterwards when
0359 nt with ripe berries floated for 23 days, when DRIED it floated for 85 days, and the seeds afterwa
0359 e seeds of Helosciadium sank in two days, when DRIED they floated for above 90 days, and afterward
0359 fterwards germinated. Altogether out of the 94 DRIED plants, 18 floated for above 28 days, and som
0359 the foregoing experiment) floated, after being DRIED, for above 28 days, as far as we may infer an
0360 ut 10/100 plants of a flora, after having been DRIED, could be floated across a space of sea 900 m
0078 nges into slightly hotter or colder, damper or DRIER districts. In this case we can clearly see th
0360 be occasionally transported in another manner. DRIFT timber is thrown up on most islands, even on
0382 ly inundated the south. As the tide leaves its DRIFT in horizontal lines, though rising higher on
0382 t, so have the living waters left their living DRIFT on our mountain summits, in a line gently ris
0361 nes for their tools, solely from the roots of DRIFTED trees, these stones being a valuable royal t
0366 states, erratic boulders, and rocks scored by DRIFTED icebergs and coast ice, plainly reveal a for
0397 h of the shell, might be floated in chinks of DRIFTED timber across moderately wide arms of the se
0399 ds brought with earth and stones on icebergs, DRIFTED by the prevailing currents, this anomaly dis
0191 e strange fact that every particle of food and DRINK which we swallow has to pass over the orifice
0007 ferent climates and treatment, I think we are DRIVEN to conclude that this greater variability is
0067 ousand sharp wedges packed close together and DRIVEN inwards by incessant blows, sometimes one wed
0076 ia the small Asiatic cockroach has everywhere DRIVEN before it its great congener. One species of
0369 its coldest period will have been temporarily DRIVEN down to the plains; they will, also, have bee
0382 ded may be compared with savage races of man, DRIVEN up and surviving in the mountian fastnesses o
0432 ences between them. This is what we should be DRIVEN to, if we were ever to succeed in collecting
0240  numerous specimens from the same nest of the DRIVER ant (Anomma) of West Africa. The reader will
0241 ving been first formed, as in the case of the DRIVER ant, and then the extreme forms, from being t
0195 tter and better, for so trifling an object as DRIVING away flies; yet we should pause before being
0213 ing round, instead of at, a herd of deer, and DRIVING them to a distant point, we should assuredly
0030 e compare the dray horse and race horse, the DROMEDARY and camel, the various breeds of sheep fitt
0202 ction for this single purpose of thousands of DRONES, which are utterly useless to the community f
0472 ng of the bee causing the bee's own death; at DRONES being produced in such vast numbers for one s
0011 l can be named which has not in some country DROOPING ears; and the view suggested by some authors
0011 the view suggested by some authors, that the DROOPING is due to the disuse of the muscles of the e
0211 ly lifted up its abdomen and excreted a limpid DROP of sweet juice, which was eagerly devoured by
0385 ewhat more advanced age they would voluntarily DROP off. These just hatched molluscs, though aquat
0129 tree; many a limb and branch has decayed and DROPPED off; and these lost branches of various size
0384 means; like that of the live fish not rarely DROPPED by whirlwinds in India, and the vitality of
0387 e nelumbium undigested; or the seeds might be DROPPED by the bird whilst feeding its young, in the
0387 he same way as fish are known sometimes to be DROPPED. In considering these several means of distr
0388 r seeds from a bed of a particular nature, and DROPS them in another equally well fitted for them.
0062 sert is said to struggle for life against the DROUGHT, though more properly it should be said to b
0068 es, and periodical seasons of extreme cold or DROUGHT, I believe to be the most effective of all c
0044 is subject at all properly, a long catalogue of DRY facts should be given, but these I shall reser
0069 from south to north, or from a damp region to a DRY, we invariably see some species gradually gett
0078 tition cease. The land may be extremely cold or DRY, yet there will be competition between some fe
0113 reater number of plants and a greater weight of DRY herbage can thus be raised. The same has been
0186 te birds with webbed feet, either living on the DRY land or most rarely alighting on the water; th
0233 t no less than from twelve to fifteen pounds of DRY sugar are consumed by a hive of bees for the s
0308 united States during these intervals existed as DRY land, or as a submarine surface near land, on
0359 ream be washed into the sea. Hence I was led to DRY stems in branches of 94 plants with ripe fruit
0360 water. On the other hand he did not previously DRY the plants or branches with the fruit; and thi
0362 in one instance I removed twenty two grains of DRY argillaceous earth from one foot of a partridg
0378 greatly favoured by high land, and perhaps by a DRY climate; for Dr. Falconer informs me that it i
0386 er, on the edge of a little pond; this mud when DRY weighed only 6 3/4 ounces; I kept it covered u
0406 fied) to those on the surrounding low lands and DRY lands, though these stations are so different,
0077 stand a little more heat or cold, dampness or DRYNESS, for elsewhere it ranges into slightly hotte
0011 d effect; for instance, I find in the domestic DUCK that the bones of the wing weigh less and the
0011 e skeleton, than do the same bones in the wild DUCK; and I presume that this change may be safely
0011 hange may be safely attributed to the domestic DUCK flying much less, and walking more, than its w
0019 t they all have descended from the common wild DUCK and rabbit. The doctrine of the origin of our
0029 orn. I have never met a pigeon, or poultry, or DUCK, or rabbit fancier, who was not fully convince
0134 re are several in this state. The loggerheaded DUCK of South America can only flap along the surfa
0134 y the same condition as the domestic Aylesbury DUCK. As the larger ground feeding birds seldom tak
0182 ngs solely as flappers, like the logger headed DUCK (Micropterus of Eyton); as fins in the water a
```

Page ***(Key Word)***

Page ***(Key Word)***

```
0385 rved, throw some light on this subject. When a DUCK suddenly emerges from a pond covered with duck
0385 duck suddenly emerges from a pond covered with DUCK weed, I have twice seen these little plants ad
0385 nd it has happened to me, in removing a little DUCK weed from one aquarium to another, that I have
0385 gency is perhaps more effectual: I suspended a DUCK's feet, which might represent those of a bird
0385 hough aquatic in their nature, survived on the DUCK's feet, in damp air, from twelve to twenty hou
0385 to twenty hours; and in this length of time a DUCK or heron might fly at least six or seven hundr
0473 en we look, for instance, at the logger headed DUCK, which has wings incapable of flight, in nearl
0473 n nearly the same condition as in the domestic DUCK; or when we look at the burrowing tucutucu, wh
0019 ild Indian fowl (Gallus bankiva). In regard to DUCKS and rabbits, the breeds of which differ consi
0185 at can be plainer than that the webbed feet of DUCKS and geese are formed for swimming? yet there
0230 nd in the opposed cells and push and bend the DUCTILE and warm wax (which as I have tried is easil
0190 heir swimbladders, this latter organ having a DUCTUS pneumaticus for its supply, and being divided
0006 he origin of species and varieties, if he makes DUE allowance for our profound ignorance in regard
0007 onclude that this greater variability is simply DUE to our domestic productions having been raised
0010 way, the change at first appears to be directly DUE to such conditions; but in some cases it can b
0011 suggested by some authors, that the drooping is DUE to the disuse of the muscles of the ear, from
0013 and child, we cannot tell whether it may not be DUE to the same original cause acting on both; but
0013 o the same conditions; any very rare deviation, DUE to some extraordinary combination of circumsta
0014 horn, though appearing late in life, is clearly DUE to the male element. Having alluded to the sub
0031 world. The improvement is by no means generally DUE to crossing different breeds; all the best bre
0044 ical sense, as implying a modification directly DUE to the physical conditions of life, and variat
0052 another and higher stage may be, in some cases, DUE merely to the long continued action of differe
0069 untain; we far oftener meet with stunted forms, DUE to the directly injurious action of climate, t
0070 some of these so called epidemics appear to be DUE to parasitic worms, which have from some cause
0075 arvested separately, and the seed then mixed in DUE proportion, otherwise the weaker kinds will st
0131 degree in those in a state of nature, had been DUE to chance. This, of course, is a wholly incorr
0131 ve that deviations of structure are in some way DUE to the nature of the conditions of life, to wh
0133 who can tell how much of this difference may be DUE to the warmest clad individuals having been fa
0135 tions of structure which are wholly, or mainly, DUE to natural selection. Mr. Wollaston has discov
0136 condition of so many Madeira beetles is mainly DUE to the action of natural selection, but combin
0137 kin and fur. This state of the eyes is probably DUE to gradual reduction from disuse, but aided pe
0141 atisation of species to any peculiar climate is DUE to mere habit, and how much to the natural sel
0142 eir own districts: the result must, I think, be DUE to habit. On the other hand, I can see no reas
0146 by systematists as of high value, may be wholly DUE to unknown laws of correlated growth, and with
0146 roups of species, and which in truth are simply DUE to inheritance; for an ancient progenitor may
0146 occurring throughout whole orders, are entirely DUE to the manner alone in which natural selection
0150 al causes of error, but I hope that I have made DUE allowance for them. It should be understood th
0151 the case of secondary sexual characters, may be DUE to the great variability of these characters,
0158 es closely connected together. All being mainly DUE to the species of the same group having descen
0159 ll doubt that all such analogous variations are DUE to the several races of the pigeon having inhe
0162 a variable species in our systematic works, is DUE to its varieties mocking, as it were, some of
0164 nd that the above described appearances are all DUE to ancient crosses with the dun stock. But I a
0172 d; the imperfection of the record being chiefly DUE to organic beings not inhabiting profound dept
0197 n; as it is, I have no doubt that the colour is DUE to some quite distinct cause, probably to sexu
0197 ridity; and so it may be, or it may possibly be DUE to the direct action of putrid matter; but we
0199 rt of the organisation of every being is simply DUE to inheritance; and consequently, though each
0205 idently believe that many modifications, wholly DUE to the laws of growth, and at first in no way
0228 to happen, and they stopped their excavation in DUE time; so that the basins, as soon as they had
0265 em, they are extremely liable to vary, which is DUE, as I believe, to their reproductive systems h
0273 use of ordinary variability; namely, that it is DUE to the reproductive system being eminently sen
0277 d offspring should generally correspond, though DUE to distinct causes; for both depend on the amo
0285 ight; for the gentle slope of the lava streams, DUE to their formerly liquid state, showed at a ql
0293 lived; but I can by no means pretend to assign DUE proportional weight to the following considera
0296 of a different nature will generally have been DUE to geographical changes requiring much time. N
0302 europe and of the United States. We do not make DUE allowance for the enormous intervals of time,
0356 whole amount of modification will not have been DUE, at each stage, to descent from a single paren
0367 e especially related to the arctic forms living DUE north or nearly due north of them: for the mig
0367 to the arctic forms living due north or nearly DUE north of them: for the migration as the cold c
0367 the returning warmth, will generally have been DUE south and north. The Alpine plants, for exampl
0379 s preponderant migration from north to south is DUE to the greater extent of land in the north, an
0406 ers I have endeavoured to show that if we make DUE allowance for our ignorance of the full effect
0420 arrangement of the groups within each class, in DUE subordination and relation to the other groups
0420 ir common progenitor, may differ greatly, being DUE to the different degrees of modification which
0430 ed to be real and not merely adaptive, they are DUE on my theory to inheritance in common. Therefo
0443 the embryo is formed; and the variation may be DUE to the male and female sexual elements having
0455 . as the presence of rudimentary organs is thus DUE to the tendency in every part of the organisat
0466 e see much variability. This seems to be mainly DUE to the reproductive system being eminently sus
0479 . the real affinities of all organic beings are DUE to inheritance or community of descent. The na
0488 esent inhabitants of the world should have been DUE to secondary causes, like those determining th
0067 a piece of ground three feet long and two wide, DUG and cleared, and where there could be no choki
0221 rom getting any pupae to rear as slaves. I then DUG up a small parcel of the pupae of F. fusca fro
0414 ternal similarity of a mouse to a shrew, of a DUGONG to a whale, of a whale to a fish, as of any f
0414 ion. As an instance: Owen, in speaking of the DUGONG, says, The generative organs being those whic
0427 d in the fin like anterior limbs, between the DUGONG, which is a pachydermatous animal, and the wh
0224 clusions at which I have arrived. He must be a DULL man who can examine the exquisite structure of
0164 de a careful examination and sketch for me of a DUN Belgian cart horse with a double stripe on eac
0164 n implicitly trust, has examined for me a small DUN Welch pony with three short parellel stripes o
0164 ar oftenest in duns and mouse duns; by the term DUN a large range of colour is included, from one
0164 m several aboriginal species, one of which, the DUN, was striped; and that the above described app
0165 arances are all due to ancient crosses with the DUN stock. But I am not at all satisfied with this
0165 three short shoulder stripes, like those on the DUN Welch pony, and even had some zebra like strip
0166 he horse we see this tendency strong whenever a DUN tint appears, a tint which approaches to that
0135 that the anterior tarsi, or feet, of many male DUNG feeding beetles are very often broken off; he
0135 or as the tarsi are almost always lost in many DUNG feeding beetles, they must be lost early in li
0163 s; transverse bars on the legs are not rare in DUNS, mouse duns, and in one instance in a chestnut
0163 e bars on the legs are not rare in duns, mouse DUNS, and in one instance in a chestnut: a faint sh
```

Page ***(Key Word)**

duns

```
0163 faint shoulder stripe may sometimes be seen in DUNS, and I have see a trace in a bay horse. My son
0164  the world these stripes occur far oftenest in DUNS and mouse duns; by the term dun a large range
0164 e stripes occur far oftenest in duns and mouse DUNS; by the term dun a large range of colour is in
0073 to desolate the world, or invent laws on the DURATION of the forms of life! I am tempted to give o
0178 es, but they will generally have had a short DURATION. For these intermediate varieties will, from
0282 r attempts to give an inadequate idea of the DURATION of each formation or even each stratum. A ma
0293 nd woodward, have concluded that the average DURATION of each formation is twice or thrice as long
0293 on is twice or thrice as long as the average DURATION of specific forms. But insuperable difficult
0294 ight be tempted to conclude that the average DURATION of life of the embedded fossils had been les
0300 least perfect. It may be doubted whether the DURATION of any one great period of subsidence over t
0300 lation of sediment, would exceed the average DURATION of the same specific forms; and these contin
0318 e length of life, so have species a definite DURATION. No one I think can have marvelled more at t
0327 subsidence; and that blank intervals of vast DURATION occurred during the periods when the bed of
0342 as not been continuously deposited; that the DURATION of each formation is, perhaps, short compare
0342 is, perhaps, short compared with the average DURATION of specific forms; that migration has played
0465 ntermittent in their accumulation; and their DURATION, I am inclined to believe, has been shorter
0465 o believe, has been shorter than the average DURATION of specific forms. Successive formations are
0481 tory of the world was thought to be of short DURATION; and now that we have acquired some idea of
0487 the successive stages as having been of vast DURATION. But we shall be able to gauge with some sec
0487 hall be able to gauge with some security the DURATION of these intervals by a comparison of the pr
0255 xerts no more influence than so much inorganic DUST. From this absolute zero of fertility, the pol
0092  case insects in seeking the nectar would get DUSTED with pollen, and would certainly often transp
0093 ectually fertilised by the bees, accidentally DUSTED with pollen, having flown from tree to tree i
0028 them astonishingly. About this same period the DUTCH were as eager about pigeons as were the old R
0270 stinction. Gartner kept during several years a DWARF kind of maize with yellow seeds, and a tall v
0044 not to be inherited; but who can say that the DWARFED condition of shells in the brackish waters o
0044 ells in the brackish waters of the Baltic, or DWARFED plants on Alpine summits, or the thicker fur
0027 n record of pigeons is in the fifth Aegyptian DYNASTY, about 3000 B.C., as was pointed out to me b
0028 s are given in a bill of fare in the previous DYNASTY. In the time of the Romans, as we hear from
0386 int. Sir Charles Lyell also informs me that a DYTICUS has been caught with an Ancylus (a fresh wat
0028 ngly. About this same period the Dutch were as EAGER about pigeons as were the old Romans. The par
0211 visit them, and it immediately seemed, by its EAGER way of running about, to be well aware what a
0213 same manner by each individual, performed with EAGER delight by each breed, and without the end be
0215 ies of the cat genus, when kept tame, are most EAGER to attack poultry, sheep and pigs; and this t
0211 reted a limpid drop of sweet juice, which was EAGERLY devoured by the ant. Even the quite young ap
0221 bare spot near the place of combat; they were EAGERLY seized, and carried off by the tyrants, who
0222 distinguish them: for we have seen that they EAGERLY and instantly seized the pupae of F. fusca,
0188 t a structure even as perfect as the eye of an EAGLE might be formed by natural selection, althoug
0362 d fish, and then gave their bodies to fishing EAGLES, storks, and pelicans; these birds after an i
0011 pinc is due to the disuse of the muscles of the EAR, from the animals not being much alarmed by da
0454 of a tail in tailless breeds, the vestige of an EAR in earless breeds, the reappearance of minute
0395 a more or less modified condition. Mr. Windsor EARL has made some striking observations on this he
0454 in tailless breeds, the vestige of an ear in EARLESS breeds, the reappearance of minute dangling
0013 ring at a corresponding age, though sometimes EARLIER. In many cases this could not be otherwise:
0041 d out individual plants with slightly larger, EARLIER, or better fruit, and raised seedlings from
0119 often take the place of, and so destroy, the EARLIER and less improved branches: this is represen
0121 hence all the intermediate forms between the EARLIER and later states, that is between the less a
0122 ount of modification, species (A) and all the EARLIER varieties will have become extinct, having b
0123 a10; b14 and f14, from having diverged at an EARLIER period from a5, will be in some degree disti
0124 to have descended from some one species of an EARLIER genus. In our diagram, this is indicated by
0126 l constantly tend to supplant and destroy the EARLIER and less improved sub groups. Small and brok
0187 e gradations having been transmitted from the EARLIER stages of descent, in an unaltered or little
0187 t known fossiliferous stratum to discover the EARLIER stages, by which the eye has been perfected.
0255 es causing the flower of the hybrid to wither EARLIER than it otherwise would have done; and the e
0294 tance, that several species appeared somewhat EARLIER in the palaeozoic beds of North America than
0335 ortant character of length of beak originated EARLIER than short beaked tumblers, which are at the
0367 pose that the Glacial period came on a little EARLIER or later in North America than in Europe, so
0367 e southern migration there have been a little EARLIER or later; but this will make no difference i
0370 ficulty may be surmounted by looking to still EARLIER changes of climate of an opposite nature. We
0370 the above view, and to infer that during some EARLIER and still warmer period, such as the older P
0372 which during the Pliocene or even a somewhat EARLIER period, was nearly uniform along the continu
0374 t. the cold may have come on, or have ceased, EARLIER at one point of the globe than at another, b
0412 to the right hand, which diverged at a still EARLIER period. And all these genera, descended from
0440 or itself. The period of activity may come on EARLIER or later in life; but whenever it comes on,
0444 of its characters are fully acquired a little EARLIER or later in life. It would not signify, for
0444 for all that we can see, might have appeared EARLIER or later in life, tend to appear at a corres
0444 he largest sense) which have supervened at an EARLIER age in the child than in the parent. These t
0446 d qualities and structures have been acquired EARLIER or later in life, if the full grown animal p
0446 c differences must either have appeared at an EARLIER period than usual, or, if not so, the differ
0446 nherited, not at the corresponding, but at an EARLIER age. Now let us apply these facts and the ab
0447 f life, or each step might be inherited at an EARLIER period than that at which it first appeared.
0450 by the variations having been inherited at an EARLIER period than that at which they first appeare
0010 fference between a bud and an ovule in their EARLIEST stages of formation; so that, in fact, sport
0027 years in several quarters of the world; the EARLIEST known record of pigeons is in the fifth Aegy
0132 se of sporting plants, the bud, which in its EARLIEST condition does not apparently differ essenti
0304 he interval between the latest secondary and EARLIEST tertiary formation. But now we may read in t
0306 enitor, apply with nearly equal force to the EARLIEST known species. For instance, I cannot doubt
0309 bt to great oscillations of level, since the EARLIEST silurian period. The coloured map appended t
0318 ps, as we have seen, having endured from the EARLIEST known dawn of life to the present day; some
0448 closely resembling their parents from their EARLIEST age, we can see that this would result from
0450 e life of each, though perhaps caused at the EARLIEST, and being inherited at a corresponding not
0008 they may be, generally acti; whether during the EARLY or late period of development of the embryo,
0034 hat the colour of domestic animals was at that EARLY period attended to. Savages now sometimes cro
0035 tle of England have increased in weight and in EARLY maturity, compared with the stock formerly ke
0066 is period in the great majority of cases is an EARLY one. If an animal can in any way protect its
0112 horter beaks. Again, we may suppose that at an EARLY period one man preferred swifter horses; anot
```

Page ***(Key Word)***

```
0112  s; another stronger and more bulky horses. The EARLY differences would be very slight; in the cour
0116  s, we see the process of diversification in an EARLY and incomplete stage of development. After th
0135  he smaller quadrupeds. We may imagine that the EARLY progenitor of the ostrich had habits like tho
0135  n many dung feeding beetles, they must be lost EARLY in life, and therefore cannot be much used by
0142  ng a score of generations, his kidney beans so EARLY that a very large proportion are destroyed by
0143  me manner as any malconformation affecting the EARLY embryo, seriously affects the whole organisat
0143  he body which are homologous, and which, at an EARLY embryonic period, are alike, seem liable to v
0157  structure of the common progenitor, or of its EARLY descendants, became variable; variations of t
0159  es, or reverts to some of the characters of an EARLY progenitor. These propositions will be most r
0168  ual, will be saved. Changes of structure at an EARLY age will generally affect parts subsequently
0182  ected, who would have ever imagined that in an EARLY transitional state they had been inhabitants
0182  we should bear in mind that animals displaying EARLY transitional grades of the structure will sel
0183  of life will rarely have been developed at an EARLY period in great numbers and under many subord
0189  e been developed; and in order to discover the EARLY transitional grades through which the organ h
0195  ly in some cases been of high importance to an EARLY progenitor, and, after having been slowly per
0217  ore especially as she has to migrate at a very EARLY period; and the first hatched young would pro
0223  hand, possesses much fewer slaves, and in the EARLY part of the summer extremely few. The masters
0255  ier than it otherwise would have done; and the EARLY withering of the flower is well known to be a
0264  embryo may be developed, and then perish at an EARLY period. This latter alternative has not been
0264  ce in hybridising gallinaceous birds, that the EARLY death of the embryo is a very frequent cause
0264  le, and consequently be liable to perish at an EARLY period; more especially as all very young bei
0277  al circumstances; in some cases largely on the EARLY death of the embryo. The sterility of hybrids
0298  ring in a formation in any one country all the EARLY stages of transition between any two forms, i
0330  se living at the present day, were not at this EARLY epoch limited in such distinct groups as they
0333  en characters, in this case the genera, at the EARLY period marked VI., would differ by a lesser n
0333  by a lesser number of characters; for at this EARLY stage of descent they have not diverged in ch
0338  bryo, owing to variations supervening at a not EARLY age, and being inherited at a corresponding a
0384  s of continuous mountain ranges, which from an EARLY period must have parted river systems and com
0411  explained by variations not supervening at an EARLY age, and being inherited at a corresponding a
0430  character of their common progenitor, or of an EARLY member of the group. On the other hand, of al
0431  life have been utterly lost, through which the EARLY progenitors of birds were formerly connected
0431  tors of birds were formerly connected with the EARLY progenitors of the other vertebrate classes.
0437  n mature become extremely different, are at an EARLY stage of growth exactly alike. How inexplicab
0442  and serve for diverse purposes, being at this EARLY period of growth alike; of embryos of differe
0443  trosities often affecting the embryo at a very EARLY period, that slight variations necessarily ap
0443  ht variations necessarily appear at an equally EARLY period. But we have little evidence on this h
0443  . nevertheless an effect thus caused at a very EARLY period, even before the formation of the embr
0444  t structure, may have supervened at a not very EARLY period of life; and some direct evidence from
0444  ost of them, may have appeared at an extremely EARLY period. I have stated in the first chapter, t
0446  ction, have not generally first appeared at an EARLY period of life, and have been inherited by th
0446  erited by the offspring at a corresponding not EARLY period. But the case of the short faced tumbl
0447  ses of which we are wholly ignorant, at a very EARLY period of life, or each step might be inherit
0448  aving to provide for their own wants at a very EARLY stage of development, and secondly from their
0448  s, that the child should be modified at a very EARLY age in the same manner with its parents, in a
0450  se of modification having supervened at a very EARLY age, or by the variations having been inherit
0450  ts from some one ancient progenitor, at a very EARLY period in the life of each; though perhaps ca
0450  st, and being inherited at a corresponding not EARLY period. Embryology rises greatly in interest,
0453  , than in the adult; so that the organ at this EARLY age is less rudimentary, or even cannot be sa
0455  at the corresponding age, but at an extremely EARLY period of life (as we have good reason to bel
0457  necessarily or generally supervening at a very EARLY period of life, and being inherited at a corr
0476  diverged in character, the progenitor with its EARLY descendants will often be intermediate in cha
0477  lly be descendants of the same progenitors and EARLY colonists. On this same principle of former m
0479  on of parts or organs, which were alike in the EARLY progenitor of each class. On the principle of
0479  essive variations not always supervening at an EARLY age, and being inherited at a corresponding n
0479  ge, and being inherited at a corresponding not EARLY period of life, we can clearly see why the em
0480  have little power of acting on an organ during EARLY life; hence the organ will not be much reduce
0480  e much reduced or rendered rudimentary at this EARLY age. The calf, for instance, has inherited an
0480  cut through the gums of the upper jaw, from an EARLY progenitor having well developed teeth; and w
0483  n a rudimentary condition plainly show that an EARLY progenitor had the organ in a fully developed
0488  of organic change as a measure of time. During EARLY periods of the earth's history, when the form
0011  c named which has not in some country drooping EARS; and the view suggested by some authors, that
0064  at so high a rate, that if not destroyed, the EARTH would soon be covered by the progeny of a sin
0084  of heather, and the black grouse that of peaty EARTH, we must believe that these tints are of serv
0095  ell's noble views on the modern changes of the EARTH, as illustrative of geology; but we now very
0100  are some hermaphrodites, as land mollusca and EARTH worms; but these all pair. As yet I have not
0124  wise a section of the successive strata of the EARTH's crust including extinct remains. We shall,
0126  yet at the most remote geological period, the EARTH may have been as well peopled with many speci
0130  its dead and broken branches the crust of the EARTH, and covers the surface with its ever branchi
0170  ich the innumerable beings on the face of this EARTH are enabled to struggle with each other, and
0172  edded in countless numbers in the crust of the EARTH? It will be much more convenient to discuss t
0173  ks in our geological history. The crust of the EARTH is a vast museum; but the natural collections
0222  d when they came across the pupae, or even the EARTH from the nest of F. flava, and quickly ran aw
0280  varieties, which have formerly existed on the EARTH, be truly enormous. Why then is not every geo
0282  this theory be true, such have lived upon this EARTH. On the lapse of Time. Independently of our n
0284  esult and measure of the degradation which the EARTH's crust has elsewhere suffered. And what an a
0287  ot. Only a small portion of the surface of the EARTH has been geologically explored, and no part w
0309  ubsided some miles nearer to the centre of the EARTH, and which had been pressed on by an enormous
0312  lly, or, as far as we know, on the face of the EARTH. If we may trust the observations of Philippi
0313  cies has once disappeared from the face of the EARTH, we have reason to believe that the same iden
0317  . the old notion of all the inhabitants of the EARTH having been swept away at successive periods
0333  dent that the fauna of any great period in the EARTH's history will be intermediate in general cha
0344  leave no modified offspring on the face of the EARTH. But the utter extinction of a whole group of
0352  have been created at one or more points of the EARTH's surface. Undoubtedly there are very many ca
0354  species at distant and isolated points of the EARTH's surface, can in many instances be explained
0361  bedded in the roots of trees, small parcels of EARTH are very frequently enclosed in their interst
0361  longest transport: out of one small portion of EARTH thus completely enclosed by wood in an oak ab
```

Page ***(Key Word)***

Page ********************************(Key Word)********************************

```
0362  rds are generally quite clean, I can show that  EARTH  sometimes adheres to them: in one instance I
0362  removed twenty two grains of dry argillaceous  EARTH  from one foot of a partridge, and in this ear
0362  arth from one foot of a partridge, and in this  EARTH  there was a pebble quite as large as the seed
0363  s the Mediterranean: and can we doubt that the  EARTH  adhering to their feet would sometimes includ
0363  icebergs are known to be sometimes loaded with  EARTH  and stones, and have even carried brushwood,
0376  on the mountians of the warmer regions of the  EARTH  and in the southern hemisphere are of doubtfu
0386  xplain this fact. I have before mentioned that  EARTH  occasionally, though rarely, adheres in some
0399  has been mainly stocked by seeds brought with  EARTH  and stones on icebergs, drifted by the prevai
0432  for if every form which has ever lived on this  EARTH  were suddenly to reappear, though it would be
0449  inct and recent, which have ever lived on this  EARTH  have to be classed together, and as all have
0462  d geographical changes which have affected the  EARTH  during modern periods: and such changes will
0483  lly believe that at innumerable periods in the  EARTH's history certain elemental atoms have been c
0484  e organic beings which have ever lived on this  EARTH  have descended from some one primordial form,
0487  e imperfection of the record. The crust of the  EARTH  with its embedded remains must not be looked
0488  a measure of time. During early periods of the  EARTH's history, when the forms of life were probab
0489  bout, and with worms crawling through the damp  EARTH, and to reflect that these elaborately constr
0190  these cases, one of the two organs might with  EASE  be modified and perfected so as to perform all
0257  different climates, can often be crossed with  EASE. By a reciprocal cross between two species, I
0261  n usually, but not invariably, be grafted with  EASE. But this capacity, as in hybridisation, is by
0304  reserved in the oldest tertiary beds: from the  EASE  with which even a fragment of a valve can be r
0350  another having been effected with more or less  EASE, at periods more or less remote: on the nature
0062  his great horticultural knowledge. Nothing is  EASIER  than to admit in words the truth of the unive
0419  enced by chains of affinities. Nothing can be  EASIER  than to define a number of characters common
0009  than in a state of nature. A long list could  EASILY  be given of sporting plants: by this term gar
0028  t breeds, that male and female pigeons can be  EASILY  mated for life: and thus different breeds can
0042  ir breeding: in peacocks, from not being very  EASILY  reared and a large stock not kept: in geese,
0070  y necessary for its preservation. Thus we can  EASILY  raise plenty of corn and rape seed, etc., in
0082  al differences, so could Nature, but far more  EASILY, from having incomparably longer time at her
0087  inevitably perish: or, more delicate and more  EASILY  broken shells might be selected, the thickne
0093  at they can most effectually do this, I could  EASILY  show by many striking instances. I will give
0094  ear to differ in length: yet the hive bee can  EASILY  suck the nectar out of the incarnate clover,
0127  ages, can modify the egg, seed, or young, as  EASILY  as the adult. Amongst many animals, sexual se
0135  used by these insects. In some cases we might  EASILY  put down to disuse modifications of structure
0141  ther animals, now in a state of nature, could  EASILY  be brought to bear widely different climates.
0190  espire. In such cases natural selection might  EASILY  specialise, if any advantage were thus gained
0202  ure, this same reason tells us, though we may  EASILY  err on both sides, that some other contrivanc
0208  may be modified by the will or reason. Habits  EASILY  become associated with other habits, and with
0230  uctile and warm wax (which as I have tried is  EASILY  done) into its proper intermediate plane, and
0248  and, that the fertility of pure species is so  EASILY  affected by various circumstances, that for a
0250  the genus Hippeastrum, which can be far more  EASILY  fertilised by the pollen of another and disti
0252  separated in the scale of nature can be more  EASILY  crossed than in the case of plants: but the h
0256  both of first crosses and of hybrids, is more  EASILY  affected by unfavourable conditions, than is
0258  r. to give an instance: Mirabilis jalappa can  EASILY  be fertilised by the pollen of M. longiflora,
0268  stance, that the German Spitz dog unites more  EASILY  than other dogs with foxes, or that certain S
0296  same formation: facts, showing what wide, yet  EASILY  overlooked, intervals have occurred in its ac
0323  manifest, and the several formations could be  EASILY  correlated. These observations, however, rela
0352  of such a nature, that the space could not be  EASILY  passed over by migration, the fact is given a
0362  even 18 hours. A bird in this interval might  EASILY  be blown to the distance of 500 miles, and ha
0391  rine birds could arrive at these islands more  EASILY  than land birds. Bermuda, on the other hand,
0391  f wading birds, might be transported far more  EASILY  than land shells, across three or four hundre
0394  remote islands? On my view this question can  EASILY  be answered: for no terrestrial mammal can be
0397  now it is notorious that land shells are very  EASILY  killed by salt: their eggs, at least such as
0428  valuation has hitherto been arbitrary), could  EASILY  extend the parallelism over a wide range: and
0448  onding ages, the active young or larvae might  EASILY  be rendered by natural selection different to
0454  , useless or injurious for one purpose, might  EASILY  be modified and used for another purpose. Or
0403  r the same conditions of life. Thus, the south  EAST  and south west corners of Australia have nearl
0164  breeds, in various countries from Britain to  EASTERN  China: and from Norway in the north to the M
0348  , shell, or crab in common, than those of the  EASTERN  and western shores of South and Central Amer
0348  and as soon as this is passed we meet in the  EASTERN  islands of the Pacific, with another and tot
0348  d, proceeding still further westward we meet in the  EASTERN  islands of the tropical parts of the Pacific
0348  o the above named three approximate faunas of  EASTERN  and Western America and the eastern Pacific
0348  faunas of Eastern and Western America and the  EASTERN  Pacific Islands, yet many fish range from th
0348  dian Ocean, and many shells are common to the  EASTERN  islands of the Pacific and the eastern shore
0349  to the eastern islands of the Pacific and the  EASTERN  shores of Africa, on almost exactly opposite
0370  land from western Europe, through Siberia, to  EASTERN  America. And to this continuity of the circu
0372  ting and tertiary representative forms on the  EASTERN  and western shores of temperate North Americ
0373  irect evidence of glacial action in the south  EASTERN  corner of Australia. Looking to America: in
0373  e fragments of rock have been observed on the  EASTERN  side as far south as lat. 36 degrees to 37 d
0374  at the glacial action was simultaneous on the  EASTERN  and western sides of North America, in the C
0008  nge in the conditions of life. Nothing is more  EASY  than to tame an animal, and few things more di
0179  transitional state have subsisted? It would be  EASY  to show that within the same group carnivorous
0183  species. When either case occurs, it would be  EASY  for natural selection to fit the animal, by so
0207  ttempt any definition of instinct. It would be  EASY  to show that several distinct mental actions a
0270  ir mutual fertilisation is by so much the less  EASY  as their differences are greater. How far thes
0481  nt of view directly opposite to mine. It is so  EASY  to hide our ignorance under such expressions a
0484  ill have only to decide (not that this will be  EASY) whether any form be sufficiently constant and
0037  re, never thought what splendid fruit we should  EAT: though we owe our excellent fruit, in some sm
0283  nly for a short time twice a day, and the waves  EAT  into them only when they are charged with sand
0286  otion on the subject, assume that the sea would  EAT  into cliffs 500 feet in height at the rate of
0362  s and fourteen hours. Fresh water fish, I find,  EAT  seeds of many land and water plants: fish are
0387  yed a part. I have stated that fresh water fish  EAT  some kinds of seeds, though they reject many o
0286  ere to assume a cliff one yard in height to be  EATEN  back along a whole line of coast at the rate
0084  tance, may thus be acted on. When we see leaf  EATING  insects green, and bark feeders mottled grey:
0238  as any two genera of the same family. Thus in  ECITON, there are working and soldier neuters, with
0235  ansmitted by inheritance its newly acquired  ECONOMICAL  instinct to new swarms, which in their turn
0147  er to spend on one side, nature is forced to  ECONOMISE  on the other side. I think this holds true
```

Page ********************************(Key Word)********************************

Page **************************************(Key Word)**************************************

0147 t natural selection is continually trying to ECONOMISE in every part of the organisation. If under
0235 far as we can see, is absolutely perfect in ECONOMISING wax. Thus, as I believe, the most wonderfu
0062 ed in the mind, I am convinced that the whole ECONOMY of nature, with every fact on distribution,
0076 orms, which fill nearly the same place in the ECONOMY of nature; but probably in no one case could
0081 eely enter, we should then have places in the ECONOMY of nature which would assuredly be better fi
0102 it may be said, to seize on each place in the ECONOMY of nature, if any one species does not becom
0104 and, etc.; and thus new places in the natural ECONOMY of the country are left open for the old inh
0116 ment from these substances. So in the general ECONOMY of any land, the more widely and perfectly t
0122 adapted to many related places in the natural ECONOMY of their country. It seems, therefore, to me
0131 tion. Correlation of growth. Compensation and ECONOMY of growth. False correlations. Multiple, rud
0158 everal species to their several places in the ECONOMY of nature, and likewise to fit the two sexes
0173 filling nearly the same place in the natural ECONOMY of the land. These representative species of
0195 are much too ignorant in regard to the whole ECONOMY of any one organic being, to say what slight
0231 y given to the comb, with the utmost ultimate ECONOMY of wax. It seems at first to add to the diff
0235 the process of natural selection having been ECONOMY of wax; that individual swarm which wasted l
0315 ill the exact place of another species in the ECONOMY of nature, and thus supplant it; yet the two
0322 action, yet seldom perceived by us, the whole ECONOMY of nature will be utterly obscured. Whenever
0331 to seize on many and different places in the ECONOMY of nature. Therefore it is quite possible, a
0412 ny and as different places as possible in the ECONOMY of nature, there is a constant tendency in t
0414 e, and the general place of each being in the ECONOMY of nature, would be of very high importance
0418 adult, which alone plays its full part in the ECONOMY of nature. Yet it has been strongly urged by
0428 , and to seize on more and more places in the ECONOMY of nature. The larger and more dominant grou
0455 of complete abortion. The principle, also, of ECONOMY, explained in a former chapter, by which the
0470 ze on many and widely different places in the ECONOMY of nature, there will be a constant tendency
0144 r dermal covering, viz. Cetacea (whales) and EDENTATA (armadilloes, scaly anteaters, etc.), that t
0340 fly or solely produced in Australia; or that EDENTATA and other American types should have been so
0341 in numbers, as apparently is the case of the EDENTATA of South America, still fewer genera and spe
0062 ch shall get food and live. But a plant on the EDGE of a desert is said to struggle for life again
0228 h hexagonal prism was built upon the festooned EDGE of a smooth basin, instead of on the straight
0230 bees, as may be clearly seen by examining the EDGE of a growing comb, do make a rough, circumfere
0386 three different points, beneath water, on the EDGE of a little pond; this mud when dry weighed on
0025 ith the bases of the outer feathers externally EDGED with white; the wings have two black bars; so
0025 p, double black wing bar, and barred and white EDGED tail feathers, as any wild rock pigeon! We ca
0160 f the tail, with the outer feathers externally EDGED near their bases with white. As all these mar
0228 square piece of wax, a thin and narrow, knife EDGED ridge, coloured with vermilion. The bees inst
0225 l known, is an hexagonal prism, with the basal EDGES of its six sides bevelled so as to join on to
0228 of a smooth basin, instead of on the straight EDGES of a three sided pyramid as in the case of or
0230 case may be; and they never complete the upper EDGES of the rhombic plates, until the hexagonal wa
0231 practically to show this fact, by covering the EDGES of the hexagonal walls of a single cell, or t
0232 t had been placed, and worked into the growing EDGES of the cells all round. The work of construct
0386 birds. Wading birds, which frequent the muddy EDGES of ponds, if suddenly flushed, would be the m
0025 birds, all the above marks, even to the white EDGING of the outer tail feathers, sometimes concur
0058 e well sifted London Catalogue of plants (4th EDITION) 63 plants which are therein ranked as speci
0115 ecies. To give a single instance: in the last EDITION of Dr. Asa Gray's Manual of the Flora of the
0287 k of that admirable palaeontologist, the late EDWARD Forbes, should not be forgotten, namely, that
0357 rred within the period of existing organisms. EDWARD Forbes insisted that all the islands in the A
0366 pe, as explained with remarkable clearness by EDWARD Forbes, is substantially as follows. But we s
0409 tinct but representative species. As the late EDWARD Forbes often insisted, there is a striking pa
0116 l body, a subject so well elucidated by Milne EDWARDS. No physiologist doubts that a stomach by ta
0164 spect, from information given me by Mr. W. W. EDWARDS, that with the English race horse the spinal
0194 st every experienced naturalist; or, as Milne EDWARDS has well expressed it, nature is prodigal in
0418 ongly urged by those great naturalists, Milne EDWARDS and Agassiz, that embryonic characters are t
0433 , we are tending in this direction; and Milne EDWARDS has lately insisted, in an able paper, on th
0008 in this; but the chief one is the remarkable EFFECT which confinement or cultivation has on the f
0010 uch agencies have produced very little direct EFFECT, though apparently more in the case of plants
0011 e to another. In animals it has a more marked EFFECT; for instance, I find in the domestic duck th
0011 n other countries, is another instance of the EFFECT of use. Not a single domestic animal can be n
0015 very poor soil (in which case, however, some EFFECT would have to be attributed to the direct act
0029 e or from several allied species. Some little EFFECT may, perhaps, be attributed to the direct act
0031 space I could quote numerous passages to this EFFECT from highly competent authorities. Youatt, wh
0032 ice; but its importance consists in the great EFFECT produced by the accumulation in one direction
0069 icuous, we are tempted to attribute the whole EFFECT to its direct action. But this is a very fals
0071 s, which could not be found on the heath. The EFFECT on the insects must have been still greater,
0071 us birds. Here we see how potent has been the EFFECT of the introduction of a single tree, nothing
0083 cious means of selection, what may not nature EFFECT? Man can act only on external and visible cha
0085 of any particular colour would produce little EFFECT: we should remember how essential it is in a
0085 etc., probably produce some slight and direct EFFECT. It is, however, far more necessary to bear l
0087 nother species; and though statements to this EFFECT may be found in works of natural history, I c
0089 irds. It may appear childish to attribute any EFFECT to such apparently weak means: I cannot here
0089 ir standard of beauty, might produce a marked EFFECT. I strongly suspect that some well known laws
0098 pecies, the former will have such a prepotent EFFECT, that it will invariably and completely destr
0099 llen of a distinct variety having a prepotent EFFECT over a flower's own pollen; and that this is
0132 of structure, however slight. How much direct EFFECT difference of climate, food, etc., produces o
0132 xtremely doubtful. My impression is, that the EFFECT is extremely small in the case of animals, bu
0150 uch struck with a remark, nearly to the above EFFECT, published by Mr. Waterhouse. I infer also fr
0199 ical conditions probably have had some little EFFECT on structure, quite independently of any good
0201 be found in works on natural history to this EFFECT, I cannot find even one which seems to me of
0212 cts given in detail, can produce but a feeble EFFECT on the reader's mind. I can only repeat my as
0283 here is reason to believe that pure water can EFFECT little of nothing in wearing away rock. At la
0297 e reasons just assigned we can seldom hope to EFFECT in any one geological section. Supposing B an
0325 s made forcible remarks to precisely the same EFFECT. It is, indeed, quite futile to look to chang
0443 ancestors, have been exposed. Nevertheless an EFFECT thus caused at a very early period, even befo
0472 tions seem to have produced but little direct EFFECT; yet when varieties enter any zone, they occa
0472 es, use and disuse seem to have produced some EFFECT; for it is difficult to resist this conclusio
0031 reeding. What English breeders have actually EFFECTED is proved by the enormous prices given for a
0035 the change has, it is believed, been chiefly EFFECTED by crosses with the fox hound; but what conc

Page **************************************(Key Word)**************************************

Page ***(Key Word)***

```
0035  hat concerns us is, that the change has been  EFFECTED unconsciously and gradually, and yet so effe
0041  unless such attention be paid nothing can be  EFFECTED. I have seen it gravely remarked, that it wa
0045  that changes of this nature could have been  EFFECTED only by slow degrees: yet quite recently Mr.
0086  can see no greater difficulty in this being  EFFECTED through natural selection, than in the cotto
0092  something injurious from their sap: this is  EFFECTED by glands at the base of the stipules in som
0092  ects from flower to flower, and a cross thus  EFFECTED, although nine tenths of the pollen were des
0100  lants, by which an occasional cross could be  EFFECTED with terrestrial animals without the concurr
0108  bitants being thus disturbed. Nothing can be  EFFECTED, unless favourable variations occur, and var
0109  ir physical conditions of life, which may be  EFFECTED in the long course of time by nature's power
0139  eyes, and natural selection will often have  EFFECTED other changes, such as an increase in the le
0139  be correct, acclimatisation must be readily  EFFECTED during long continued descent. It is notorio
0142  s, as proving that acclimatisation cannot be  EFFECTED! The case, also, of the kidney bean has been
0148  parasitic habits of the Proteolepas, though  EFFECTED by slow steps, would be a decided advantage
0182  o perfectly winged animals. If this had been  EFFECTED, who would have ever imagined that in an ear
0224  x of the true form, though this is perfectly  EFFECTED by a crowd of bees working in a dark hive. G
0229  e little rhombic plate, that they could have  EFFECTED this by gnawing away the convex side; and I
0236  n see no very great difficulty in this being  EFFECTED by natural selection. But I must pass over t
0242  ments, a perfect division of labour can be  EFFECTED with them only by the workers being sterile;
0242  ecome blended. And nature has, as I believe,  EFFECTED this admirable division of labour in the com
0242  y amount of modification in structure can be  EFFECTED by the accumulation of numerous, slight, and
0249  gh probably on the same plant, would be thus  EFFECTED. Moreover, whenever complicated experiments
0253  e instance they have bred inter se. This was  EFFECTED by Mr. Eyton, who raised two hybrids from th
0282  t of organic change, all changes having been  EFFECTED very slowly through natural selection. It is
0298  case of fossil species this could rarely be  EFFECTED by palaeontologists. We shall, perhaps, best
0299  called, specifically distinct. This could be  EFFECTED only by the future geologist discovering in
0299  rmediate varieties; and this not having been  EFFECTED, is probably the gravest and most obvious of
0303  through the air; but that when this had been  EFFECTED, and a few species had thus acquired a great
0306  iscoveries of even the last dozen years have  EFFECTED, it seems to me to be about as rash in us to
0350  ife from one region into another having been  EFFECTED with more or less ease, at periods more or l
0365  , is no valid argument against what would be  EFFECTED by occasional means of transport, during the
0438  whole amount of modification will have been  EFFECTED by slight successive steps, we need not wond
0454  entary. Any change in function, which can be  EFFECTED by insensibly small steps, is within the pow
0462  even to conjecture how this could have been  EFFECTED. Yet, as we have reason to believe that some
0463  t any one period; and all changes are slowly  EFFECTED. I have also shown that the intermediate var
0260  dest possible difference, in the facility of  EFFECTING an union. The hybrids, moreover, produced f
0262  r. as in reciprocal crosses, the facility of  EFFECTING an union is often very far from equal, so i
0263  crosses, the greater or lesser difficulty in  EFFECTING a union apparently depends on several disti
0277  d. nor is it surprising that the facility of  EFFECTING a first cross, the fertility of the hybrids
0068  me cold or drought, I believe to be the most  EFFECTIVE of all checks. I estimated that the winter
0085  o doubt that natural selection might be most  EFFECTIVE in giving the proper colour to each kind of
0361  d. living birds can hardly fail to be highly  EFFECTIVE agents in the transportation of seeds. I co
0003  ns to several distinct organic beings, by the  EFFECTS of external conditions, or of habit, or of t
0007  mestication Chapter I. Causes of Variability.  EFFECTS of Habit. Correlation of Growth. Inheritance
0007  rinciple of Selection anciently followed, its  EFFECTS. Methodical and Unconscious Selection. Unkno
0010  fe; and this shows how unimportant the direct  EFFECTS of the conditions of life are in comparison
0014  d be quite necessary, in order to prevent the  EFFECTS of intercrossing, that only a single variety
0033  is another means of observing the accumulated  EFFECTS of selection namely, by comparing the divers
0036  youatt gives an excellent illustration of the  EFFECTS of a course of selection, which may be consi
0043  nic beings, as some authors have thought. The  EFFECTS of variability are modified by various degre
0060  he checks to increase. Competition universal.  EFFECTS of climate. Protection from the number of in
0070  utter destruction. I should add that the good  EFFECTS of frequent intercrossing, and the ill effec
0070  ffects of frequent intercrossing, and the ill  EFFECTS of close interbreeding, probably come into p
0102  on the confines of each. And in this case the  EFFECTS of intercrossing can hardly be counterbalanc
0103  nite for each birth, we must not overrate the  EFFECTS of intercrosses in retarding natural selecti
0106  isted in a broken condition, so that the good  EFFECTS of isolation will generally, to a certain ex
0131              Chapter V. Laws of Variation.  EFFECTS of external conditions. Use and disuse, comb
0134  come plainly developed and appreciable by us.  EFFECTS of Use and Disuse. From the facts alluded to
0134  ndard of comparison, by which to judge of the  EFFECTS of long continued use or disuse; for we know
0134  have structures which can be explained by the  EFFECTS of disuse. As Professor Owen has remarked, t
0135  n in some other genera, by the long continued  EFFECTS of disuse in their progenitors; for as the t
0137  o, natural selection would constantly aid the  EFFECTS of disuse. It is well known that several ani
0143  the structure of various organs; but that the  EFFECTS of use and disuse have often been largely co
0147  hardly any way of distinguishing between the  EFFECTS, on the one hand, of a part being largely de
0150  ee play of the various laws of growth, to the  EFFECTS of long continued disuse, and to the tendenc
0165  nt parts of the world. Now let us turn to the  EFFECTS of crossing the several species of the horse
0168  rgans, seem to have been more potent in their  EFFECTS. Homologous parts tend to vary in the same w
0181  ch being propagated, until by the accumulated  EFFECTS of this process of natural selection, a perf
0199  f reversion, though now of no direct use. The  EFFECTS of sexual selection, when displayed in beaut
0209  s been with instincts. But I believe that the  EFFECTS of habit are of quite subordinate importance
0209  it are of quite subordinate importance to the  EFFECTS of the natural selection of what may be call
0214  layed, methodical selection and the inherited  EFFECTS of compulsory training in each successive ge
0245  osses and of hybrids. Parallelism between the  EFFECTS of changed conditions of life and crossing.
0249  r fertility, notwithstanding the frequent ill  EFFECTS of manipulation, sometimes decidedly increas
0253  rom different parents, so as to avoid the ill  EFFECTS of close interbreeding. On the contrary, bro
0332  one order; and this order, from the continued  EFFECTS of extinction and divergence of character, h
0382  n language almost identical with mine, on the  EFFECTS of great alternations of climate on geograph
0406  e due allowance for our ignorance of the full  EFFECTS of all the changes of climate and of the lev
0407  species, extremely grave. As exemplifying the  EFFECTS of climatal changes on distribution, I have
0447  ivity and has to gain its own living; and the  EFFECTS thus produced will be inherited at a corresp
0447  ed, or be modified in a lesser degree, by the  EFFECTS of use and disuse. In certain cases the succ
0459  theory of natural selection may be extended.  EFFECTS of its adoption on the study of Natural hist
0467  of increase is proved by calculation, by the  EFFECTS of a succession of peculiar seasons; and by
0474  nsects, which leave no progeny to inherit the  EFFECTS of long continued habit. On the view of all
0481  years; it cannot add up and perceive the full  EFFECTS of many slight variations, accumulated durin
0486  f variation, on correlation of growth, on the  EFFECTS of use and disuse, on the direct action of e
0385  he other. But another agency is perhaps more  EFFECTUAL: I suspended a duck's feet, which might rep
```

Page ***(Key Word)***

Page **(Key Word)***

```
0035 ted unconsciously and gradually, and yet so EFFECTUALLY, that, though the old Spanish pointer cert
0041 enerally be allowed to breed, and this will EFFECTUALLY prevent selection. But probably the most i
0072 gined that cattle would have so closely and EFFECTUALLY searched it for food. Here we see that cat
0080 ? i think we shall see that it can act most EFFECTUALLY. Let it be borne in mind in what an endles
0085 h a host of enemies, such differences would EFFECTUALLY settle which variety, whether a smooth or
0093 om flower to flower; and that they can most EFFECTUALLY do this, I could easily show by many strik
0093 ery female flower which I examined had been EFFECTUALLY fertilised by the bees, accidentally duste
0098 prengel and from my own observations, which EFFECTUALLY prevent the stigma receiving pollen from i
0401 rily rare; so that the islands are far more EFFECTUALLY separated from each other than they appear
0252 any one may readily convince himself of the EFFICIENCY of insect agency by examining the flowers o
0242 pated that natural selection could have been EFFICIENT in so high a degree, had not the case of th
0380 orms, generated in the larger areas and more EFFICIENT workshops of the north. In many islands the
0396 ional means of transport having been largely EFFICIENT in the long course of time, than with the v
0397 ust be, on my view, some unknown, but highly EFFICIENT means for their transportal. Would the just
0451 more important purpose; and remain perfectly EFFICIENT for the other. Thus in plants, the office o
0043 or unconsciously and more slowly, but more EFFICIENTLY, is by far the predominant Power. Chapter
0103 racter. It will obviously thus act far more EFFICIENTLY with those animals which unite for each bi
0104 prevented. But isolation probably acts more EFFICIENTLY in checking the immigration of better adap
0112 ature? I believe it can and does apply most EFFICIENTLY, from the simple circumstance that the mor
0467 ason why the principles which have acted so EFFICIENTLY under domestication should not have acted
0061 d is as immeasurably superior to man's feeble EFFORTS, as the works of Nature are to those of Art.
0084 be preserved. How fleeting are the wishes and EFFORTS of man! how short his time! and consequently
0188 ment has been perfected by the long continued EFFORTS of the highest human intellects; and we natu
0257 us, as Silene, in which the most perservering EFFORTS have failed to produce between extremely clo
0066 rous of the two: the Fulmar petrel lays but one EGG, yet it is believed to be the most numerous bi
0087 e beak of nestling birds, used for breaking the EGG. It has been asserted, that of the best short
0087 short beaked tumbler pigeons more perish in the EGG than are able to get out of it; so that fancie
0087 igorous selection of the young birds within the EGG, which had the most powerful and hardest peaks
0127 inherited at corresponding ages, can modify the EGG, seed, or young, as easily as the adult. Among
0197 ptiles, which have only to escape from a broken EGG, we may infer that this structure has arisen f
0217 rican cuckoo; but that occasionally she laid an EGG in another bird's nest.If the old bird profite
0264 ourished within its mother's womb or within the EGG or seed produced by the mother, it may be expo
0388 et, it will be unoccupied; and a single seed or EGG will have a good chance of succeeding. Althoug
0440 is nourished in the womb of its mother, in the EGG of the bird which is hatched in a nest, and in
0443 g as it remains in its mother's womb, or in the EGG, or as long as it is nourished and protected b
0009 the rarest exceptions, hardly ever lay fertile EGGS. Many exotic plants have pollen utterly worthl
0022 lothed when hatched. The shape and size of the EGGS vary. The manner of flight differs remarkably;
0062 e forget how largely th se songsters, or their EGGS, or their nestlings, are destroyed by birds an
0063 h during its natural lifetime produces several EGGS or seeds, must suffer destruction during some
0065 rence between organisms which annually produce EGGS or seeds by the thousand, and those which prod
0066 be ever so large. The condor lays a couple of EGGS and the ostrich a score, and yet in the same c
0066 ird in the world. One fly deposits hundreds of EGGS, and another, like the hippobosca, a single on
0066 be supported in a district. A large number of EGGS is of some importance to those species, which
0066 . but the real importance of a large number of EGGS or seeds is to make up for much destruction at
0066 e. if an animal can in any way protect its own EGGS or young, a small number may be produced, and
0066 he average stock be fully kept up; but if many EGGS or young are destroyed, many must be produced,
0066 t depends only indirectly on the number of its EGGS or seeds. In looking at Nature, it is most nec
0067 to the reader's mind some of the chief points. EGGS or very young animals seem generally to suffer
0072 r in Paraguay of a certain fly, which lays its EGGS in the navels of these animals when first born
0086 tages of the varieties of the silkworm; in the EGGS of poultry, and in the colour of the down of t
0192 the means of a sticky secretion, to retain the EGGS until they are hatched within the sack. These
0192 n the other hand, have no ovigerous frena, the EGGS lying loose at the bottom of the sack, in the
0200 the ovipositor of the ichneumon, by which its EGGS are deposited in the living bodies of other in
0207 ct impels the cuckoo to migrate and to lay her EGGS in other birds' nests. An action, which we our
0213 han the white butterfly knows why she lays her EGGS on the leaf of the cabbage, I cannot see that
0215 me broody, that is, never wish to sit on their EGGS. Familiarity alone prevents our seeing how uni
0216 the instinct which leads the cuckoo to lay her EGGS in other birds' nests; the slave making instin
0217 of the cuckoo's instinct is, that she lays her EGGS, not daily, but at intervals of two or three d
0217 e were to make her own nest and sit on her own EGGS, those first laid would have to be left for so
0217 t for some time unincubated, or there would be EGGS and young birds of different ages in the same
0217 redicament; for she makes her own nest and has EGGS and young successively hatched, all at the sam
0217 that the American cuckoo occasionally lays her EGGS in other birds' nests; but I hear on the high
0217 hich have been known occasionally to lay their EGGS in other bird's nests. Now let us suppose that
0217 umbered as she can hardly fail to be by having EGGS and young of different ages at the same time;
0217 er and in their turn would be apt to lay their EGGS in other birds' nests, and thus be successful
0218 ng. The occasional habit of birds laying their EGGS in one nest and then in another; and these are
0218 he american species, unite and lay first a few EGGS in one nest and then in another; and these are
0218 the fact of the hens laying a large number of EGGS; but as in the case of the cuckoo, at interval
0218 yet been perfected; for a surprising number of EGGS lie strewed over the plains, so that in one da
0218 picked up no less than twenty lost and wasted EGGS. Many bees are parasitic, and always lay their
0218 many bees are parasitic, and always lay their EGGS in the nests of bees of other kinds. This case
0349 milarly constructed, but not quite alike, with EGGS coloured in nearly the same manner. The plains
0387 e same agency may have come into play with the EGGS of some of the smaller fresh water animals. Ot
0398 pends on the wide dispersal of their seeds and EGGS by animals, more especially by fresh water bir
0391 hells are dispersed, yet we can see that their EGGS or larvae, perhaps attached to seaweed or floa
0397 d shells are very easily killed by salt; their EGGS, at least such as I have tried, sink in sea wa
0402 tocked with its own species, for annually more EGGS are laid there than can possibly be reared; an
0406 racter. This fact, together with the seeds and EGGS of many low forms being very minute and better
0483 umerous kinds of animals and plants created as EGGS or seed, or as full grown? and in the case of
0018 t records, more especially on the monuments of EGYPT, much diversity in the breeds; and that some
0018 s a semi domestic dog, may not have existed in EGYPT? The whole subject must, I think, remain vagu
0212 d, is tame in Norway, as is the hooded crow in EGYPT. That the general disposition of individuals
0135 ion. In the Ateuchus or sacred beetle of the EGYPTIANS, they are totally deficient. There is not s
0023 hey must have descended from at least seven or EIGHT aboriginal stocks; for it is impossible to ma
0024 ons, it must be assumed that at least seven or EIGHT species were so thoroughly domesticated in an
0026 robability of man having formerly got seven or EIGHT supposed species of pigeons to breed freely u
```

Page **(Key Word)***

eight

Page **(Key Word)**

```
0114  , and these belonged to eighteen genera and to EIGHT    orders, which shows how much these plants dif
0120  in a condensed and simplified manner), we get   EIGHT    species, marked by the letters between a14 an
0122  l have become extinct, having been replaced by  EIGHT    new species (a14 to m14); and (I) will have b
0123  ach other in a widely different manner. Of the  EIGHT    descendants from (A) the three marked a14, q1
0123  g to inheritance, differ considerably from the  EIGHT    descendants from (A); the two groups, moreove
0123  he six new species descended from (I), and the  EIGHT    descended from (A), will have to be ranked as
0253  nd from these two birds he raised no less than  EIGHT    hybrids (grandchildren of the pure geese) fro
0257  ertilise, or to be fertilised by, no less than  EIGHT    other species of Nicotiana. Very many analogo
0258  uter tried more than two hundred times, during  EIGHT    following years, to fertilise reciprocally M.
0296  ta, one above the other, at no less than sixty  EIGHT    different levels. Hence, when the same specie
0341  ne species; so that if six genera, each having  EIGHT    species, be found in one geological formation
0114  nty species of plants, and these belonged to EIGHTEEN  genera and to eight orders, which shows how
0214  , as I hear from Mr. Brent, which cannot fly  EIGHTEEN  inches high without going head over heels. I
0460  tulation of the facts given at the end of the  EIGHTH    chapter, which seem to me conclusively to sho
0013  of our domestic breeds are often transmitted   EITHER    exclusively, or in a much greater degree, to
0016  ind that there are hardly any domestic races,  EITHER    amongst animals or plants, which have not bee
0023  ds. Hence the supposed aboriginal stocks must  EITHER    still exist in the countries where they were
0025  two following highly improbable suppositions,  EITHER,   firstly, that all the several imagined abori
0029  s by which domestic races have been produced,  EITHER    from one or from several allied species. Some
0030  and camel, the various breeds of sheep fitted  EITHER    for cultivated land or mountain pasture, with
0033  the continued selection of slight variations,  EITHER    in the leaves, the flowers, or the fruit, wil
0036  acquainted with the subject that the owner of  EITHER    of them has deviated in any one instance from
0044  siderable deviation of structure in one part,  EITHER    injurious to or not useful to the species, an
0047  because analogy leads the observer to suppose  EITHER    that they do now somewhere exist, or may form
0050  ow that the sessile and pedunculated oaks are  EITHER    good and distinct species or mere varieties.
0063  st in every case be a struggle for existence,  EITHER    one individual with another of the same speci
0066  life; that heavy destruction inevitably falls  EITHER    on the young or old, during each generation o
0089  o inherited at corresponding ages or seasons,  EITHER    by the males alone, or by the males and femal
0090  in fowls, etc.) which we cannot believe to be  EITHER    useful to the males in battle, or attractive
0090  reast of the turkey cock, which can hardly be  EITHER    useful or ornamental to this bird; indeed, ha
0091  ss, a new variety might be formed which would  EITHER    supplant or coexist with the parent form of w
0095  d how a flower and a bee might slowly become,  EITHER    simultaneously or one after the other, modifi
0096  that with all hermaphrodites two individuals,  EITHER    occasionally or habitually, concur for the re
0097  ees suck the nectar; for, in doing this, they  EITHER    push the flowers own pollen on the stigma, or
0099  c. Sprengel has shown, and as I can confirm,  EITHER    the anthers burst before the stigma is ready
0105  if, however, an isolated area be very small,  EITHER    from being surrounded by barriers, or from ha
0113  , being enabled to feed on new kinds of prey,  EITHER    dead or alive; some inhabiting new stations,
0119  cording to the nature of the places which are  EITHER    unoccupied or not perfectly occupied by other
0120  character from their original stocks, without  EITHER    having given off any fresh branches or races.
0120  logous steps, after ten thousand generations,  EITHER    two well marked varieties (w10 and z10) or tw
0124  racter, but to have retained the form of (F),  EITHER    unaltered or altered only in a slight degree.
0136  ions each individual beetle which flew least,  EITHER    from its wings having been ever so little les
0152  ral case; for in such cases natural selection  EITHER    has not or cannot come into full play, and th
0162  o offspring of a species assuming characters  (EITHER    from reversion or from analogous variation)'w
0162  which themselves must be doubtfully ranked as  EITHER    varieties or species; and this shows, unless
0163  but not an albino, has been described without  EITHER    spinal or shoulder stripe; and these stripes
0166  al species of the horse genus the stripes are  EITHER    plainer or appear more commonly in the young
0175  for other competing species; that nearly all  EITHER    prey on or serve as prey for others; in short
0175  others; in short, that each organic being is  EITHER    directly or indirectly related to the most im
0180  of the same genus; and of diversified habits,  EITHER    constant or occasional, in the same species.
0183  in the individuals of the same species. When  EITHER    case occurs, it would be easy for natural sel
0184  ng anomalous habits, and with their structure  EITHER    slightly or considerably modified from that o
0186  nd that if any one being vary ever so little,  EITHER    in habits or structure, and thus gain an adva
0186  be geese and frigate birds with webbed feet,  EITHER    living on the dry land or most rarely alighti
0192  h are often very differently constructed from  EITHER    the males or fertile females; but this case w
0200  action of physical conditions) may be viewed,  EITHER    as having been of special use to some ancestr
0200  special use to the descendants of this form,  EITHER    directly, or indirectly through the complex l
0206  ural selection. For natural selection acts by  EITHER    now adapting the varying parts of each being
0210  different circumstances, etc.; in which case  EITHER    one or the other instinct might be preserved
0214  ong period exhibit traces of the instincts of  EITHER    parent: for example, Le Roy describes a dog,
0218  birds laying their eggs in other birds' nests  EITHER    of the same or of a distinct species, is not
0220  s in Surrey and Sussex, and never saw a slave  EITHER    leave or enter a nest. As, during these month
0220  s, though present in large numbers in August,  EITHER    leave or enter the nest. Hence he considers t
0237  ave become correlated to certain ages, and to  EITHER    sex. We have differences correlated not only
0241  ecies which should regularly produce neuters,  EITHER    all of large size with one form of jaw, or al
0246  condition, yet when intercrossed they produce  EITHER    few or no offspring. Hybrids, on the other ha
0246  bryo are perfect; in the second case they are  EITHER    not at all developed, or are imperfectly deve
0248  ds, carefully guarding them from a cross with  EITHER    pure parent, for six or seven, and in one cas
0249  cross with the pollen from a distinct flower,  EITHER    from the same plant or from another plant of
0253  genera, have often been bred in this country with  EITHER  pure parent, and in one single instance they
0254  ng. On this view, the aboriginal species must  EITHER    at first have produced quite fertile hybrids,
0254  igin of many of our domestic animals, we must  EITHER    give up the belief of the almost universal st
0255  never would produce, even with the pollen of  EITHER    pure parent, a single fertile seed: but in so
0259  is independent of its external resemblance to  EITHER    pure parent. Considering the several rules no
0259  in which they resemble in external appearance  EITHER    parent. And lastly, that the facility of maki
0263  cate that the greater or lesser difficulty of  EITHER    grafting or crossing together various species
0266  casionally and exceptionally resemble closely  EITHER    pure parent. Nor do I pretend that the forego
0269  r each creature's own good; and thus she may,  EITHER    directly, or more probably indirectly, throug
0270  when intercrossed produce less seed, than do  EITHER    coloured varieties when fertilised with polle
0271  e. but one of these five varieties, when used  EITHER    as father or mother, and crossed with the Nic
0273  conditions of life, being thus often rendered  EITHER    impotent or at least incapable of its proper
0273  els are more liable than hybrids to revert to  EITHER    parent form; but this, if it be true, is cert
0274  . both hybrids and mongrels can be reduced to  EITHER    pure parent form, by repeated crosses in succ
0274  peated crosses in successive generations with  EITHER    parent. These several remarks are apparently
0275  sudden reversions to the perfect character of  EITHER    parent would be more likely to occur with mon
0289  tance, not a land shell is known belonging to  EITHER    of these vast periods, with one exception dis
```

Page **(Key Word)**

```
0289 ones of tertiary mammals have been discovered EITHER in caves or in lacustrine deposits: and that
0290 ations of sediment may be formed in two ways: EITHER, in profound depths of the sea, in which case
0297 time it could be most closely connected with EITHER one or both forms by intermediate varieties.
0300 ous intervals, during which the area would be EITHER stationary or rising: whilst rising, each fos
0312 rms, having here appeared for the first time, EITHER locally, or, as far as we know, on the face o
0316 tinuously existed, in order to have generated EITHER new and modified or the same old and unmodifi
0320 h those animals which have been exterminated, EITHER locally or wholly, through man's agency. I ma
0323 of Europe, but which occur in the formations EITHER above or below, are similarly absent at these
0327 uring the periods when the bed of the sea was EITHER stationary or rising, and likewise when sedim
0329 long ago remarked, all fossils can be classed EITHER in still existing groups, or between them. Th
0329 be disputed. For if we confine our attention EITHER to the living or to the extinct alone, the se
0348 rated from each other by impassable barriers, EITHER of land or open sea, they are wholly distinct
0362 these birds after an interval of many hours, EITHER rejected the seeds in pellets or passed them
0375 he volcanic cones of Java, many plants occur, EITHER identically the same or representing each oth
0391 , during their great annual migrations, visit EITHER periodically or occasionally this island. Mad
0394 tlantic Ocean: and two North American species EITHER regularly or occasionally visit Bermuda, at t
0395 ates two widely distinct mammalian faunas. On EITHER side the islands are situated on moderately d
0404 e power and extent of migration of a species, EITHER at the present time or at some former period
0408 llowed certain forms and not others to enter, EITHER in greater or lesser numbers: according or no
0410 mum development. Groups of species, belonging EITHER to a certain period of time, or to a certain
0414 e encountered on the view that classification EITHER gives some unknown plan of creation, or is si
0430 eritance in common. Therefore we must suppose EITHER that all Rodents, including the bizcacha, bra
0430 much modification in divergent directions. On EITHER view we may suppose that the bizcacha has ret
0441 pleted state, cirripedes may be considered as EITHER more highly or more lowly organised then they
0441 ut in some genera the larvae become developed EITHER into hermaphrodites having the ordinary struc
0443 ving been affected by the conditions to which EITHER parent, or their ancestors, have been exposed
0443 ve been affected by the shape of the horns of EITHER parent. For the welfare of a very young anima
0446 for here the characteristic differences must EITHER have appeared at an earlier period than usual
0447 riod than that at which it first appeared. In EITHER case (as with the short faced tumbler) the yo
0449 ed in many embryos, has not been obliterated, EITHER by the successive variations in a long course
0457 n mind, that when organs are reduced in size, EITHER from disuse or selection, it will generally b
0459 any organ or instinct, which we may consider, EITHER do now exist or could have existed, each good
0460 g when we remember that it is not likely that EITHER their constitutions or their reproductive sys
0463 iable to be supplanted by the allied forms on EITHER hand: and the latter, from existing in greate
0476 o the same system with recent beings, falling EITHER into the same or into intermediate groups, fo
0244 o look at such instincts as the young cuckoo EJECTING its foster brothers, ants making slaves, the
0098 lia fulgens, there is a really beautiful and ELABORATE contrivance by which every one of the infin
0005 perfected into a highly developed being or ELABORATELY constructed organ: seconcly, the subject o
0489 h the damp earth, and to reflect that these ELABORATELY constructed forms, so different from each
0203 ncy, can we consider as equally perfect the ELABORATION by our fir trees of dense clouds of pollen
0283 the vegetation show that elsewhere years have ELAPSED since the waters washed their base. He who m
0284 but an inadequate idea of the time which has ELAPSED during their accumulation: yet what time thi
0287 far longer period than 300 million years has ELAPSED since the latter part of the Secondary perio
0289 can be formed of the length of time which has ELAPSED between the consecutive formations, we may i
0290 h the belief of vast intervals of time having ELAPSED between each formation. But we can, I think,
0303 ormous intervals of time, which have probably ELAPSED between our consecutive formations, longer p
0307 silurian stratum was deposited, long periods ELAPSED, as long as, or probably far longer than, th
0334 termediate, as unequal intervals of time have ELAPSED between consecutive formations. It is no rea
0342 degradation, enormous intervals of time have ELAPSED between the successive formations: that ther
0342 in the enormous intervals of time which have ELAPSED between our consecutive formations: he may o
0408 for according to the length of time which has ELAPSED since new inhabitants entered one region: ac
0488 nt of time, compared with the ages which have ELAPSED since the first creature, the progenitor of
0049 he primrose and cowslip, or Primula veris and ELATIOR. These plants differ considerably in appeara
0062 it well deserves, at much greater length. The ELDER De Candolle and Lyell have largely and philos
0146 coelospermous in the central flowers, that the ELDER De Candolle founded his main divisions of the
0147 ossibly go on in fruit which did not open. The ELDER Geoffroy and Goethe propounced, at about the
0230 iffer from those made by the justly celebrated ELDER Huber, but I am convinced of their accuracy:
0430 adapted to habits like those of a Rodent. The ELDER De Candolle have made nearly similar observati
0192 will be treated of in the next chapter. The ELECTRIC organs of fishes offer another case of speci
0193 rays have an organ closely analogous to the ELECTRIC apparatus, and yet do not, as Matteuchi asse
0193 t no transition of any kind is possible. The ELECTRIC organs offer another and even more serious d
0193 ough disuse or natural selection. But if the ELECTRIC organs had been inherited from one ancient p
0193 us provided, we might have expected that all ELECTRIC fishes would have been specially related to
0193 to the belief that formerly most fisnes had ELECTRIC organs, which most of their modified descend
0193 do not, as Matteuchi asserts, discharge any ELECTRICITY, we must own that we are far too ignorant
0030 sire to sit, and with the bantam so small and ELEGANT; when we compare the host of agricultural, c
0089 his view: but if man can in a short time give ELEGANT carriage and beauty to his bantams, accordin
0014 e, which may have acted on the ovules or male ELEMENT; in nearly the same manner as in the crossed
0014 ring late in life, is clearly due to the male ELEMENT. Having alluded to the subject of reversion,
0042 acility in preventing crosses is an important ELEMENT of success in the formation of new races, at
0044 cies. Generally the term includes the unknown ELEMENT of a distinct act of creation. The term vari
0071 cattle could not enter. But how important an ELEMENT enclosure is, I plainly saw near Farnham, in
0081 al selection, and would be left a fluctuating ELEMENT, as perhaps we see in the species called pol
0100 imals live, and the nature of the fertilising ELEMENT; for we know of no means, analogous to the f
0102 at, and is, I believe, an extremely important ELEMENT of success. Though nature grants vast period
0104 variations. Isolation, also, is an important ELEMENT in the process of natural selection. In a co
0234 largely saving honey must be a most important ELEMENT of success in any family of bees. Of course
0246 may be clearly seen in the state of the male ELEMENT in both plants and animals: though the organ
0261 l crosses between two species the male sexual ELEMENT of the one will often freely act on the fema
0261 ne will often freely act on the female sexual ELEMENT of the other, but not in a reversed directio
0263 times be a physical impossibility in the male ELEMENT reaching the ovule, as would be the case wit
0264 etrate the stigmatic surface. Again, the male ELEMENT may reach the female element, but be incapab
0264 again, the male element may reach the female ELEMENT, but be incapable of causing an embryo to be
0265 occurs in various degrees: in both, the male ELEMENT is the most liable to be affected: but somet
0400 important, and generally a far more important ELEMENT of success. Now if we look to those inhabita
0425 ble amount of modification, may not this same ELEMENT of descent have been unconsciously used in g
```

```
0433  ubordination in group under group. We use the  ELEMENT of descent in classing the individuals of bo
0433  may be from their parent; and I believe this   ELEMENT of descent is the hidden bond of connexion w
0438  not one from the other, but from some common   ELEMENT. Naturalists, however, use such language onl
0443  icated to the offspring from the reproductive  ELEMENT of one parent. Or again, as when the horns o
0456  fication, it should be borne in mind that the  ELEMENT of descent has been universally used in rank
0456  be in structure. If we extend the use of this  ELEMENT of descent, the only certainly known cause o
0436  n number and in relative connexion with, the   ELEMENTAL parts of a certain number of vertebrae. The
0483  rable periods in the earth's history certain   ELEMENTAL atoms have been commanded suddenly to flash
0008  tributed to the male and female reproductive   ELEMENTS having been affected prior to the act of con
0069  ggle for life is almost exclusively with the   ELEMENTS. That climate acts in main part indirectly b
0077  the relation seems at first confinec to the    ELEMENTS of air and water. Yet the advantage of plume
0132  of the offspring. The male and female sexual   ELEMENTS seem to be affected before that union takes
0175  ical conditions of life as the all important   ELEMENTS of distribution, these facts ought to cause
0246  pe reveals. In the first case the two sexual   ELEMENTS which go to form the embryo are perfect; in
0262  stocks, and the union of the male and female   ELEMENTS in the act of reproduction, yet that there i
0263  two pure species the male and female sexual    ELEMENTS are perfect, whereas in hybrids they are imp
0264  he sterility of hybrids, in which the sexual   ELEMENTS are imperfectly developed, the case is very
0428  umstances, to inhabit for instance the three   ELEMENTS of land, air, and water, we can perhaps unde
0438  a certain number of the primordially similar   ELEMENTS, many times repeated, and have adapted them
0443  ion may be due to the male and female sexual   ELEMENTS having been affected by the conditions to wh
0064  y years there would be a million plants. The   ELEPHANT is reckoned to be the slowest breeder of all
0068  n the other hand, in some cases, as with the   ELEPHANT and rhinoceros, none are destroyed by beasts
0068  in India most rarely dares to attack a young   ELEPHANT protected by its dam. Climate plays an impor
0141  and such facts as that former species of the   ELEPHANT and rhinoceros were capable of enduring a gl
0319  all other mammals, even of the slow breeding   ELEPHANT, and from the history of the naturalisation
0479  e forming the neck of the giraffe and of the   ELEPHANT, and innumerable other such facts, at once e
0064  century there would be alive fifteen million   ELEPHANTS, descended from the first pair. But we have
0334  ons to the rule. For instance, mastodons and   ELEPHANTS, when arranged by Dr. Falconer in two serie
0104  r any physical change, such as of climate or   ELEVATION of the land, etc.; and thus new places in t
0108  ell modified and perfected. When, by renewed   ELEVATION, the islands shall be re converted into a c
0292  ave happened during the alternate periods of   ELEVATION; or, to speak more accurately, the beds whi
0292  re worth a passing notice. During periods of   ELEVATION the area of the land and of the adjoining s
0300  ch extinction of life; during the periods of   ELEVATION, there would be much variation, but the geo
0309  ations of level, and the continents areas of   ELEVATION. But have we any right to assume that thing
0309  many oscillations of level, of the force of   ELEVATION; but may not the areas of preponderant move
0342  ce, and more variation during the periods of   ELEVATION, and during the latter the record will have
0465  of the sea. During the alternate periods of   ELEVATION and of stationary level the record will be
0019  that there formerly existed in Great Britain   ELEVEN wild species of sheep peculiar to it! When we
0123  es in our diagram descended from the original   ELEVEN species, will now be fifteen in number. Owing
0123  at between the most different of the original   ELEVEN species. The new species, moreover, will be a
0390  arly every land bird, but only two out of the   ELEVEN marine birds, are peculiar; and it is obvious
0432  e diagram: the letters, A to L, may represent   ELEVEN Silurian genera, some of which have produced
0432  ndants. Every intermediate link between these   ELEVEN genera and there primordial parent, and every
0005  on of organic beings throughout time; in the   ELEVENTH and twelfth, their geographical distribution
0317  nerally given up, even by those geologists, as  ELIE de Beaumont, Murchison, Barrande, etc., whose
0269  that a long course of domestication tends to   ELIMINATE sterility in the successive generations of
0461  on; and as domestication apparently tends to   ELIMINATE sterility, we ought not to expect it also t
0026  s believe that long continued domestication   ELIMINATES this strong tendency to sterility: from the
0091  e a sweet juice, apparently for the sake of   ELIMINATING something injurious from their sap: this i
0020  of the world, more especially by the Hon. W.   ELLIOT from India, and by the Hon. C. Murray from Pe
0453  cient to say that because planets revolve in   ELLIPTIC courses round the sun, satellites follow the
0011  ng limbs are almost always accompanied by an   ELONGATED head. Some instances of correlation are qui
0021  the head, and this is accompanied by greatly   ELONGATED eyelids, very large external orifices to th
0021  rt and very broad one. The pouter has a much   ELONGATED body, wings, and legs; and its enormously d
0021  and it has, proportionally to its size, much   ELONGATED wing and tail feathers. The trumpeter and l
0181  to the tail, and including the limbs and the   ELONGATED fingers: the flank membrane is, also, furni
0184  which feed largely on fruit, and others with   ELONGATED wings which chase insects on the wing; and
0071  ntroduction of a single tree, nothing whatever  ELSE having been done, with the exception that the
0105  that is, have been produced there, and nowhere  ELSE. Hence an oceanic island at first sight seems
0390  n of endemic species (i.e. those found nowhere  ELSE in the world) is often extremely large. If we
0394  . new Zealand possesses two bats found nowhere  ELSE in the world: Norfolk Island, the Viti Archipe
0398  ated in the Galapagos Archipelago, and nowhere  ELSE, bear so plain a stamp of affinity to those cr
0413  cified whether order in time or space, or what  ELSE is meant by the plan of the Creator, it seems
0074  the nests of humble bees more numerous than   ELSEWHERE, which I attribute to the number of cats th
0077  more heat or cold, dampness or dryness, for   ELSEWHERE it ranges into slightly hotter or colder, d
0097  that I have found, by experiments published   ELSEWHERE, that their fertility is greatly diminished
0101  i have been enabled, by a fortunate chance,   ELSEWHERE to prove that two individuals, though both
0107  productions will have been less severe than   ELSEWHERE; new forms will have been more slowly forme
0132  eir leaves in some degree fleshy, though not   ELSEWHERE fleshy. Several other such cases could be g
0136  absence of certain large groups of beetles,   ELSEWHERE excessively numerous, and which groups have
0212  ut fear of man is slowly acquired, as I have   ELSEWHERE shown, by various animals inhabiting desert
0283  of the surface and the vegetation show that   ELSEWHERE years have elapsed since the waters washed
0284  the degradation which the earth's crust has   ELSEWHERE suffered. And what an amount of degradation
0286  ight have accumulated in thinner masses than   ELSEWHERE, the above estimate would be erroneous; but
0289  ged with new and peculiar forms of life, had   ELSEWHERE been accumulated. And if in each separate t
0293  rash in the extreme to infer that it had not   ELSEWHERE previously existed. So again when we find a
0296  feet in thickness, representing formations,   ELSEWHERE thousands of feet in thickness; and which m
0302  amined; we forget that groups of species may   ELSEWHERE have long existed and have slowly multiplie
0335  ht occasionally last much longer than a form   ELSEWHERE subsequently produced, especially in the ca
0392  orders are very different from what they are   ELSEWHERE. Such cases are generally accounted for by
0392  ss trees or bushes belonging to orders which   ELSEWHERE include only herbaceous species; now trees,
0397  emic species, but sometimes by species found   ELSEWHERE. Dr. Aug. A. Gould has given several intere
0400  alapagos Archipelago are tenanted, as I have   ELSEWHERE shown, in a quite marvellous manner, by ver
0115  the same individual body, a subject so well   ELUCIDATED by Milne Edwards. No physiologist doubts th
0392  e, as the shrivelled wings under the soldered  ELYTRA of many insular beetles. Again, islands often
0208  om feeling the benefit of this, it was much   EMBARRASSED, and, in order to complete its hammock; se
```

0288 ed of the sea, at a rate sufficiently quick to EMBED and preserve fossil remains. Throughout an en
0291 was sufficient to keep the sea shallow and to EMBED and preserve the remains before they had time
0327 sediment was not thrown down quickly enough to EMBED and preserve organic remains. During these lo
0172 s must have existed, why do we not find them EMBEDDED in countless numbers in the crust of the ear
0173 epths of the sea, and to their remains being EMBEDDED and preserved to a future age only in masses
0174 ugh they must have existed there, and may be EMBEDDED there in a fossil condition. But in the inte
0288 tered condition. The remains which do become EMBEDDED, if in sand or gravel, will when the beds ar
0295 ude that the average duration of life of the EMBEDDED fossils had been less than that of the glaci
0297 f the same great formation, we find that the EMBEDDED fossils, though almost universally ranked as
0300 ich lived on naked submarine rocks, would be EMBEDDED; and those embedded in gravel or sand, would
0300 ubmarine rocks, would be embedded; and those EMBEDDED in gravel or sand, would not endure to a dis
0315 t of organic change exhibited by the fossils EMBEDDED in consecutive formations is not equal. Each
0318 hen I found in La Plata the tooth of a horse EMBEDDED with the remains of Mastodon, Megatherium, T
0332 if many of the extinct forms, supposed to be EMBEDDED in the successive formations, were discovere
0361 ion, that when irregularly shaped stones are EMBEDDED in the roots of trees, small parcels of eart
0487 the record. The crust of the earth with its EMBEDDED remains must not be looked at as a well fill
0171 y of Type and of the Conditions of Existence EMBRACED by the theory of Natural Selection. Long bef
0206 isted on by the illustrious Cuvier, is fully EMBRACED by the principle of natural selection. For n
0207 several distinct mental actions are commonly EMBRACED by this term; but every one understands what
0484 that the theory of descent with modification EMBRACES all the members of the same class. I believe
0008 he early or late period of development of the EMBRYO, or at the instant of conception. Geoffroy St
0008 eriments show that unnatural treatment of the EMBRYO causes monstrosities; and monstrosities canno
0011 lled correlation of growth. Any change in the EMBRYO or larva will almost certainly entail changes
0143 er as any malconformation affecting the early EMBRYO, seriously affects the whole organisation of
0191 e course of the arteries still marking in the EMBRYO their former position. But it is, conceivable
0246 the two sexual elements which go to form the EMBRYO are perfect; in the second case they are eith
0264 emale element, but be incapable of causing an EMBRYO to be developed, as seems to have been the ca
0264 trees cannot be grafted on others. Lastly, an EMBRYO may be developed, and then perish at an early
0264 llinaceous birds, that the early death of the EMBRYO is a very frequent cause of sterility in firs
0277 some cases largely on the early death of the EMBRYO. The sterility of hybrids, which have their r
0338 tempt to show that the adult differs from its EMBRYO, owing to variations supervening at a not ear
0338 nding age. This process, whilst it leaves the EMBRYO almost unaltered, continually adds, in the co
0338 re and more difference to the adult. Thus the EMBRYO comes to be left as a sort of picture, preser
0418 e. we can see why characters derived from the EMBRYO should be of equal importance with those deri
0418 n the ordinary view, why the structure of the EMBRYO should be more important for this purpose tha
0418 e been founded on characters derived from the EMBRYO, on the number and position of the embryonic
0439 and serve for different purposes, are in the EMBRYO exactly alike. The embryos, also, of distinct
0439 , namely, that having forgotten to ticket the EMBRYO of some vertebrate animal, he cannot now tell
0441 ir several stages barely distinguishable. The EMBRYO in the course of development generally rises
0442 d to see differences in structure between the EMBRYO and the adult, and likewise a close similarit
0442 s soon as any structure became visible in the EMBRYO. And in some whole groups of animals and in c
0442 s and in certain members of other groups, the EMBRYO does not at any period differ widely from the
0442 er is manifested long before the parts of the EMBRYO are completed; and again in spiders, there is
0442 universal difference in structure between the EMBRYO and the adult; of parts in the same individid
0442 the adult; of parts in the same individividual EMBRYO, which ultimately become very unlike and serv
0442 esembling each other; of the structure of the EMBRYO not being closely related to its conditions o
0443 its conditions of existence, except when the EMBRYO becomes at any period of life active and has
0443 active and has to provide for itself; of the EMBRYO apparently having sometimes a higher organisa
0443 erhaps from monstrosities often affecting the EMBRYO at a very early period, that slight variation
0443 believe generally has acted, even before the EMBRYO is formed; and the variation may be due to th
0443 arly period, even before the formation of the EMBRYO, may appear late in life; as when an heredita
0447 as with the short faced tumbler) the young or EMBRYO would closely resemble the mature parent form
0448 ts. Some further explanation, however, of the EMBRYO not undergoing any metamorphosis is perhaps r
0449 yes of most naturalists, the structure of the EMBRYO is even more important for classification tha
0449 lassification than that of the adult. For the EMBRYO is the animal in its less modified state; and
0450 greatly in interest, when we thus look at the EMBRYO as a picture, more or less obscured, of the c
0452 s and ruminants, can often be detected in the EMBRYO, but afterwards wholly disappear. It is also,
0453 size relatively to the adjoining parts in the EMBRYO, than in the adult; so that the organ at this
0455 uently will seldom affect or reduce it in the EMBRYO. Thus we can understand the greater relative
0455 er relative size of rudimentary organs in the EMBRYO, and their lesser relative size in the adult.
0457 ogy; namely, the resemblance in an individual EMBRYO of the homologous parts, which when matured w
0479 e adult forms. We may cease marvelling at the EMBRYO of an air breathing mammal or bird having bra
0003 ual affinities of organic beings, on their EMBRYOLOGICAL relations, their geographical distributio
0338 ct forms is in some degree parallel to the EMBRYOLOGICAL development of recent forms. I must follo
0338 vain to look for animals having the common EMBRYOLOGICAL character of the Vertebrata, until beds f
0426 shall hereafter, I think, clearly see why EMBRYOLOGICAL characters are of such high classificator
0457 e been even anticipated. The importance of EMBRYOLOGICAL characters and of rudimentary organs in c
0479 ften of high classificatory value; and why EMBRYOLOGICAL characters are the most valuable of all.
0014 ighest importance in explaining the laws of EMBRYOLOGY. These remarks are of course confined to th
0411 l affinities of Organic Beings. Morphology. EMBRYOLOGY. Rudimentary Organs. Classification, groups
0411 lass, between parts of the same individual. EMBRYOLOGY, laws of, explained by variations not super
0419 e plumule and radicle. In our discussion on EMBRYOLOGY, we shall see why such characters are so va
0439 r from some simple appendage, is explained. EMBRYOLOGY. It has already been casually remarked that
0442 then, can we explain these several facts in EMBRYOLOGY, namely the very general, but not universal
0444 in all the above specified leading facts in EMBRYOLOGY. But first let us look at a few analogous c
0450 us, as it seems to me, the leading facts in EMBRYOLOGY, which are second in importance to none in
0450 erited at a corresponding not early period. EMBRYOLOGY rises greatly in interest, when we thus loo
0457 e can understand the great leading facts in EMBRYOLOGY; namely, the resemblance in an individual e
0486 ing a picture of the ancient forms of life. EMBRYOLOGY will reveal to us the structure, in some de
0005 utual affinities, both when mature and in an EMBRYONIC condition. In the last chapter I shall give
0143 which are homologous, and which, at an early EMBRYONIC period, are alike, seem liable to vary in a
0418 naturalists, Milne Edwards and Agassiz, that EMBRYONIC characters are the most important of any in
0418 he embryo, on the number and position of the EMBRYONIC leaves or cotyledons, and on the mode of de
0436 med into another; and we can actually see in EMBRYONIC crustaceans and in many other animals, and
0439 special lines of life. A trace of the law of EMBRYONIC resemblance, sometimes lasts till a rather
0439 e something of this kind in plants: thus the EMBRYONIC leaves of the ulex or furze, and the first

embryonic

Page **********************************(Key Word)**********************************

0440 ferent when an animal during any part of its EMBRYONIC career is active, and has to provide for it
0440 obey more or less closely the law of common EMBRYONIC resemblance. Cirripedes afford a good insta
0446 o explain these facts in regard to the later EMBRYONIC stages of our domestic varieties. Fanciers
0447 ied. But in each individual new species, the EMBRYONIC fore limbs will differ greatly from the for
0449 ts, if they pass through the same or similar EMBRYONIC stages, we may feel assured that they have
0449 t degree closely related. Thus, community in EMBRYONIC structure reveals community of descent. It
0449 ng to the great class of crustaceans. As the EMBRYONIC state of each species and group of species
0450 resemblance of ancient forms of life to the EMBRYONIC stages of recent forms, may be true, but ye
0451 eeth can be detected in the beaks of certain EMBRYONIC birds. Nothing can be plainer than that win
0453 he adult, is often said to have retained its EMBRYONIC condition. I have now given the leading fac
0453 can be of any service to the rapidly growing EMBRYONIC calf by the excretion of precious phosphate
0480 able it is that parts, like the teeth in the EMBRYONIC calf or like the shrivelled wings under the
0483 es are formed on the same pattern, and at an EMBRYONIC age the species closely resemble each other
0338 ient animals resemble to a certain extent the EMBRYOS of recent animals of the same classes; or th
0345 ient animals resemble to a certain extent the EMBRYOS of more recent animals of the same class, th
0439 urposes, are in the embryo exactly alike. The EMBRYOS, also, of distinct animals within the same c
0439 ature insects; but in the case of larvae, the EMBRYOS are active, and have been adapted for specia
0439 inosae. The points of structure, in which the EMBRYOS of widely different animals of the same clas
0440 we cannot, for instance, suppose that in the EMBRYOS of the vertebrata the peculiar loop like cou
0440 tions, the similarity of the larvae or active EMBRYOS of allied animals is sometimes much obscured
0442 adult, and likewise a close similarity in the EMBRYOS of widely different animals within the same
0442 eing at this early period of growth alike; of EMBRYOS of different species within the same class,
0447 corresponding late age, the fore limbs in the EMBRYOS of the several descendants of the parent spe
0449 and extinct forms of life should resemble the EMBRYOS of their descendants, our existing species.
0449 state, now supposed to be represented in many EMBRYOS, has not been obliterated, either by the suc
0457 s as different as possible. Larvae are active EMBRYOS, which have become specially modified in rel
0479 ly period of life, we can clearly see why the EMBRYOS of mammals, birds, reptiles, and fishes shou
0222 nce I saw the last individual of F. sanguinea EMERGE, carrying a pupa; but I was not able to find
0385 e light on this subject. When a duck suddenly EMERGES from a pond covered with duck weed, I have t
0349 f the same genus; and not by a true ostrich or EMEU, like those found in Africa and Australia unde
0348 s, with not an island as a halting place for EMIGRANTS; here we have a barrier of another kind, an
0488 man and his history. Authors of the highest EMINENCE seem to be fully satisfied with the view tha
0020 ble antiquity. I have associated with several EMINENT fanciers, and have been permitted to join tw
0030 pothetical. It is certain that several of our EMINENT breeders have, even within a single lifetime
0032 y of eye and judgment sufficient to become an EMINENT breeder. If gifted with these qualities, and
0034 qualities is so obvious. At the present time, EMINENT breeders try by methodical selection, with a
0048 htly from each other, have been ranked by one EMINENT naturalist as undoubted species, and by anot
0269 domestic varieties. In the second place, some EMINENT naturalists believe that a long course of do
0307 e no satisfactory answer. Several of the most EMINENT geologists, with Sir R. Murchison at their h
0310 plainest manner by the fact that all the most EMINENT palaeontologists, namely Cuvier, Owen, Agass
0453 try, and to complete the scheme of nature? An EMINENT physiologist accounts for the presence of ru
0480 ions. Why, it may be asked, have all the most EMINENT living naturalists and geologists rejected t
0482 is subject is overwhelmed be removed. Several EMINENT naturalists have of late published their bel
0027 l feathers in the fantail, are in each breed EMINENTLY variable; and the explanation of this fact
0089 heron has described how one pied peacock was EMINENTLY attractive to all his hen birds. It may app
0131 the remark, that the reproductive system is EMINENTLY susceptible to changes in the conditions of
0152 ecies; nevertheless the part in this case is EMINENTLY liable to variation. Why should this be so?
0152 apid change by continued selection, are also EMINENTLY liable to variation. Look at the breeds of
0160 e, as we have seen, these coloured marks are EMINENTLY liable to appear in the crossed offspring o
0166 whether such face stripes ever occur in the EMINENTLY striped Kattywar breed of horses, and was,
0176 rticular case the intermediate form would be EMINENTLY liable to the inroads of closely allied for
0185 sea. On the other hand, grebes and coots are EMINENTLY aquatic, although their toes are only borde
0253 be far more fertile; for I am assured by two EMINENTLY capable judges, namely Mr. Blyth and Capt.
0259 excess. That their fertility, besides being EMINENTLY susceptible to favourable and unfavourable
0264 ore especially as all very young beings seem EMINENTLY sensitive to injurious or unnatural conditi
0265 s, for hybrids in successive generations are EMINENTLY liable to vary, as every experimentalist ha
0273 t it is due to the reproductive system being EMINENTLY sensitive to any change in the conditions o
0276 e in individuals of the same species, and is EMINENTLY susceptible of favourable and unfavourable
0390 ving to compete with new associates, will be EMINENTLY liable to modification, and will often prod
0437 erly seen that parts many times repeated are EMINENTLY liable to vary in number and structure; con
0466 mainly due to the reproductive system being EMINENTLY susceptible to changes in the conditions of
0474 f great importance to the species, should be EMINENTLY liable to variation; but, on my view, this
0049 appearance; they have a different flavour and EMIT a different odour; they flower at slightly dif
0250 mely, the Hon. and Rev. W. Herbert. He is as EMPHATIC in his conclusion that some hybrids are perf
0016 value; all such valuations being at present EMPIRICAL. Moreover, on the view of the origin of gen
0438 occurred, that naturalists can hardly avoid EMPLOYING language having this plain signification. O
0207 ich we ourselves should require experience to ENABLE us to perform, when performed by an animal, m
0295 tely the same, which is necessary in order to ENABLE the same species to live on the same space, t
0004 ng catalogues of facts. We shall, however, be ENABLED to discuss what circumstances are most favou
0051 ely extended, he will in the end generally be ENABLED to make up his own mind which to call variet
0054 ied, will still inherit those advantages that ENABLED their parents to become dominant over their
0065 oung, and that nearly all the young have been ENABLED to breed. In such cases the geometrical rati
0086 a state of nature, natural selection will be ENABLED to act on and modify organic beings at any
0101 lty under this point of view; but I have been ENABLED, by a fortunate chance, elsewhere to prove t
0108 e most favoured or improved varieties will be ENABLED to spread: there will be much extinction of
0112 n, and habits, by so much will they be better ENABLED to seize on many and widely diversified plac
0112 ied places in the polity of nature, and so be ENABLED to increase in numbers. We can clearly see t
0113 er animals: some of them, for instance, being ENABLED to feed on new kinds of prey, either dead or
0113 animal became, the more places they would be ENABLED to occupy. What applies to one animal will a
0116 e more diversified in structure, and are thus ENABLED to encroach on places occupied by other bein
0119 can be rendered, the more places they will be ENABLED to seize on, and the more their modified pro
0170 umerable beings on the face of this earth are ENABLED to struggle with each other, and the best ad
0195 y are more subject to disease, or not so well ENABLED in a coming dearth to search for food, or to
0205 far too ignorant, in almost every case, to be ENABLED to assert that any part or organ is so unimp
0213 es under domestication. We shall thus also be ENABLED to see the respective parts which habit and
0227 king their spheres; but she is already so far ENABLED to judge of distance, that she always descri

Page **********************************(Key Word)**********************************

Page ***(Key Word)***

```
0232  ed cells. It suffices that the bees should be ENABLED to stand at their proper relative distances
0248  in successive generations; though Gartner was ENABLED to rear some hybrids, carefully guarding the
0297  s, they rank both as species, unless they are ENABLED to connect them together by close intermedia
0298  best perceive the improbability of our being ENABLED to connect species by numerous, fine, interm
0306  became adapted to a cooler climate, and were ENABLED to double the southern capes of Africa or Au
0327  here and there a single member might long be ENABLED to survive. Thus, as it seems to me, the par
0331  on the descendants from a species being thus ENABLED to seize on many and different places in the
0337  ain, whether any considerable number would be ENABLED to seize on places now occupied by our nativ
0354  fest. In discussing this subject, we shall be ENABLED at the same time to consider a point equally
0357  about the means of distribution, we shall be ENABLED to speculate with security on the former ext
0379  g the Glacial period, the northern forms were ENABLED to beat the less powerful southern forms. Ju
0438  few serial homologies: that is, we are seldom ENABLED to say that one part or organ is homologous
0470  modified descendants of each species will be ENABLED to increase by so much the more as they beco
0470  ersified in habits and structure, so as to be ENABLED to seize on many and widely different places
0487  of the level of the land, we shall surely be ENABLED to trace in an admirable manner the former m
0031  s of the principle of selection as that which ENABLES the agriculturist, not only to modify the ch
0180  each kind of squirrel in its own country, by ENABLING it to escape birds or beasts of prey, or to
0071  ed acres of exactly the same nature had been ENCLOSED twenty five years previously and planted wit
0071  e, with the exception that the land had been ENCLOSED, so that cattle could not enter. But how imp
0071  in the last ten years large spaces have been ENCLOSED, and self sown firs are now springing up in
0072  led. No wonder that, as soon as the land was ENCLOSED, it became thickly clothed with vigorously g
0097  have their organs of fructification closely ENCLOSED, as in the great papilionaceous or pea famil
0101  with the organs of reproduction so perfectly ENCLOSED within the body, that access from without an
0192  loose at the bottom of the sack, in the well ENCLOSED shell; but they have large folded branchiae.
0305  ndian Ocean would form a large and perfectly ENCLOSED basin, in which any great group of marine an
0361  , small parcels of earth are very frequently ENCLOSED in their interstices and behind them, so per
0361  f one small portion of earth thus completely ENCLOSED by wood in an oak about 50 years old, three
0437  ry view of creation! Why should the brain be ENCLOSED in a box composed of such numerous and such
0042  dy stocked with other races. In this respect ENCLOSURE of the land plays a part. Wandering savages
0071  ould not enter. But how important an element ENCLOSURE is, I plainly saw near Farnham, in Surrey.
0051  t with more cases of difficulty: for he will ENCOUNTER a greater number of closely allied forms. B
0326  nant species spreading from any region might ENCOUNTER still more dominant species, and then their
0348  nds of the tropical parts of the Pacific, we ENCOUNTER no impassable barriers, and we have innumer
0242  ous special difficulty, which my theory has ENCOUNTERED. The case, also, is very interesting, as i
0414  ssification, and the difficulties which are ENCOUNTERED on the view that classification either giv
0456  emi the rules followed and the difficulties ENCOUNTERED by naturalists in their classifications: t
0461  geographical distribution, the difficulties ENCOUNTERED on the theory of descent with modification
0116  sified in structure, and are thus enabled to ENCROACH on places occupied by other beings. Now let
0217  ther bird, than by their own mother's care, ENCUMBERED as she can hardly fail to be by having eggs
0034  ind distinctly given in an ancient Chinese ENCYCLOPAEDIA. Explicit rules are laid down by some of
0141  n agricultural works, even in the ancient ENCYCLOPAEDIAS of China, to be very cautious in transpos
0051  observations be widely extended, he will in the END generally be enabled to make up his own mind w
0064  f young in this interval: if this be so, at the END of the fifth century there would be alive fift
0098  rtilisation; and no doubt it is useful for this END: but, the agency of insects is often required
0160  k bars on the wings, a white rump, a bar at the END of the tail, with the outer feathers externall
0193  orne on a foot stalk with a sticky gland at the END, is the same in Orchis and Asclepias, genera a
0201  d almost as soon believe that the cat curls the END of its tail when preparing to spring, in order
0202  utterly useless to the community for any other END, and which are ultimately slaughtered by their
0213  th eager delight by each breed, and without the END being known, for the young pointer can no more
0224  ucture of a comb, so beautifully adapted to its END, without enthusiastic admiration. We hear from
0225  oes not reveal to us her method of work. At one END of a short series we have humble bees, which u
0225  ry irregular rounded cells of wax. At the other END of the series we have the cells of the hive be
0237  o the individual, and may thus gain the desired END. Thus, a well flavoured vegetable is cooked, a
0317  sometimes falsely appear to begin at its lower END, not in a sharp point, but abruptly: it then o
0318  e is found to taper more gradually at its upper END, which marks the progress of extermination, th
0318  he progress of extermination, than at its lower END, which marks the first appearance and increase
0335  hort beaked tumblers, which are at the opposite END of the series in this same respect. Closely co
0460  to the recapitulation of the facts given at the END of the eighth chapter, which seem to me conclu
0285  d almost in the same manner as does the vain ENDEAVOUR to grapple with the idea of eternity. I am
0014  pecies in a state of nature. I have in vain ENDEAVOURED to discover on what decisive facts the abo
0057  species is often exceedingly small. I have ENDEAVOURED to test this numerically by averages, and,
0242  known doctrine of Lamarck. Summary. I have ENDEAVOURED briefly in this chapter to show that the m
0279  area with graduated physical conditions. I ENDEAVOURED to show, that the life of each species dep
0279  y quite insensibly like heat or moisture. I ENDEAVOURED, also, to show that intermediate varieties
0406  present Chapters. In these chapters I have ENDEAVOURED to show, that if we make due allowance for
0409  me principles, we can understand, as I have ENDEAVOURED to show, why oceanic islands should have f
0459  gh natural selection, I do not deny. I have ENDEAVOURED to give to them their full force. Nothing
0186  dge that every organic being is constantly ENDEAVOURING to increase in numbers; and that if any on
0230  er, by excavating at the same rate, and by ENDEAVOURING to make equal spherical hollows, but never
0073  his series by insectivorous birds, and we have ENDED with them. Not that in nature the relations c
0105  of these species a very large proportion are ENDEMIC, that is, have been produced there, and nowh
0135  they cannot fly; and that of the twenty nine ENDEMIC genera, no less than twenty three genera hav
0201  gree of perfection attained under nature. The ENDEMIC productions of New Zealand, for instance, ar
0380  ore the Glacial period they were stocked with ENDEMIC Alpine forms: but these have almost everywhe
0390  s of inhabitants is scanty, the proportion of ENDEMIC species (i.e. those found nowhere else in th
0390  f we compare; for instance, the number of the ENDEMIC land shells in Madeira, or of the endemic bi
0390  the endemic land shells in Madeira, or of the ENDEMIC birds in the Galapagos Archipelago, with the
0391  as a very peculiar soil, does not possess one ENDEMIC land bird: and we know from Mr. J. M. Jones'
0392  islands not tenanted by mammals, some of the ENDEMIC plants have beautifully hooked seeds; yet fe
0392  ill retaining its hooked seeds, would form an ENDEMIC species, having as useless an appendage as a
0394  here has also been time for the production of ENDEMIC species belonging to other classes: and on c
0395  sition, and we can understand the presence of ENDEMIC bats on islands, with the absence of all ter
0396  amely, the scarcity of kinds, the richness in ENDEMIC forms in particular classes or sections of c
0397  t, are inhabited by land shells, generally by ENDEMIC species, but sometimes by species found else
0399  ndeed it is an almost universal rule that the ENDEMIC productions of islands are related to those
0399  , this anomaly disappears. New Zealand in its ENDEMIC plants is much more closely related to Austr
```

Page ***(Key Word)***

```
0400  ach other. But this dissimilarity between the  ENDEMIC inhabitants of the islands may be used as an
0400  world (laying on one side for the moment the  ENDEMIC species, which cannot be here fairly include
0403  ons I think we need not greatly marvel at the  ENDEMIC and representative species, which inhabit th
0409  itants, but of these a great number should be  ENDEMIC or peculiar; and why, in relation to the mea
0409  n the same class, should have all its species  ENDEMIC, and another group should have all its speci
0476  me continent, of marsupials in Australia, of  ENDENTATA in America, and other such cases, is intell
0012  etc.; and it is really surprising to note the  ENDLESS points in structure and constitution in whic
0012  e of considerable physiological importance is  ENDLESS. Dr. Prosper Lucas's treatise, in two large
0065  erica since its discovery. In such cases, and  ENDLESS instances could be given, no one supposes th
0080  fectually. Let it be borne in mind in what an  ENDLESS number of strange peculiarities our domestic
0129  these round other points, and so on in almost  ENDLESS cycles. On the view that each species has be
0161  produce any character being inherited for an  ENDLESS number of generations, than in quite useless
0408  conditions of life, there would be an almost  ENDLESS amount of organic action and reaction, and w
0471  for the life of an auk or grebe! and so on in  ENDLESS other cases. But on the view of each species
0484  ter experience, will be no slight relief. The  ENDLESS disputes whether or not some fifty species o
0490  ed law of gravity, from so simple a beginning  ENDLESS forms most beautiful and most wonderful have
0244  live bodies of caterpillars, not as specially  ENDOWED or created instincts, but as small consequen
0245  ecies, when intercrossed, have been specially  ENDOWED with the quality of sterility, in order to p
0245  that sterility is not a specially acquired or  ENDOWED quality, but is incidental on other acquired
0255  les indicate that species have specially been  ENDOWED with this quality, in order to prevent their
0260  ingular rules indicate that species have been  ENDOWED with sterility simply to prevent their becom
0261  tal on other differences, and not a specially  ENDOWED quality. As the capacity of one plant to be
0261  ill suppose that this capacity is a specially  ENDOWED quality, but will admit that it is incidenta
0276  son to think that species have been specially  ENDOWED with various degrees of sterility to prevent
0276  than to think that trees have been specially  ENDOWED with various and somewhat analogous degrees
0482  ertainly reject my theory. A few naturalists,  ENDOWED with much flexibility of mind, and who have
0474  stand why nature moves by graduated steps in  ENDOWING different animals of the same class with the
0245  terility of hybrids. Sterility not a special  ENDOWMENT, but incidental on other differences. Cause
0246  y in both cases being looked on as a special  ENDOWMENT, beyond the province of our reasoning power
0263  together various species has been a special  ENDOWMENT; although in the case of crossing, the diff
0272  of hybrids, namely, that it is not a special  ENDOWMENT, but is incidental on slowly acquired modif
0460  how that this sterility is no more a special  ENDOWMENT than is the incapacity of two trees to be g
0481  ably fertile; or that sterility is a special  ENDOWMENT and sign of creation. The belief that speci
0489  ood of each being, all corporeal and mental  ENDOWMENTS will tend to progress towards perfection. I
0187  al pencils of light, are convex at their upper  ENDS and must act by convergence; and at their lowe
0188  nd must act by convergence; and at their lower  ENDS there seems to be an imperfect vitreous substa
0197  isitely constructed hooks clustered around the  ENDS of the branches; and this contrivance, no doub
0248  most difficult to say where perfect fertility  ENDS and sterility begins. I think no better eviden
0419  ossible. There are crustaceans at the opposite  ENDS of the series, which have hardly a character i
0419  a character in common; yet the species at both  ENDS, from being plainly allied to others, and thes
0017  he ass or guinea fowl, or the small power of  ENDURANCE of warmth by the rein deer, or of cold by t
0043  ed are of little importance to us, for their  ENDURANCE is only temporary. Over all these causes of
0263  sinc, the difficulty is as important for the  ENDURANCE and stability of specific forms, as in the
0409  ferent areas. We see this in many facts. The  ENDURANCE of each species and group of species is con
0476  ruggle for life. Lastly, the law of the long  ENDURANCE of allied forms on the same continent, of m
0017  succeeding generations, and whether it would  ENDURE other climates? Has the little variability of
0052  s incipient state become extinct, or they may  ENDURE as varieties for very long periods, as has be
0069  lants in our gardens which can perfectly well  ENDURE our climate, but which never become naturalis
0107  oduction of many new forms of life, likely to  ENDURE long and to spread widely. For the area will
0109  in some way advantageous, which consequently  ENDURE. But as from the high geometrical powers of i
0117  tervals of time; nor are they all supposed to  ENDURE for equal periods. Only those variations whic
0119  il and multiply: a medium form may often long  ENDURE, and may or may not produce more than one mod
0139  arctic or even from a temperate region cannot  ENDURE a tropical climate, or conversely. So again,
0139  rsely. So again, many succulent plants cannot  ENDURE a damp climate. But the degree of adaptation
0140  predict whether or not an imported plant will  ENDURE our climate, and from the number of plants an
0153  in any extreme degree, as species very rarely  ENDURE for more than one geological period. An extra
0176  erstand why intermediate varieties should not  ENDURE for very long periods; why as a general rule
0300  d those embedded in gravel or sand, would not  ENDURE to a distant epoch. Wherever sediment did not
0335  that forms successively produced necessarily  ENDURE for corresponding lengths of time: a very anc
0343  ps of species increase in numbers slowly, and  ENDURE for unequal periods of time; for the process
0364  ng an immersion in salt water, they could not  ENDURE our climate. Almost every year, one or two la
0107  ay almost be called living fossils; they have  ENDURED to the present day; from having inhabited a
0318  periods; some groups, as we have seen, having  ENDURED from the earliest known dawn of life to the
0332  rished at different periods, and some to have  ENDURED to the present day. By looking at the diagra
0333  will have been more numerous, they will have  ENDURED for extremely unequal lengths of time, and w
0348  tc., are not as impassable, or likely to have  ENDURED so long as the oceans separating continents,
0374  d. we have, also, excellent evidence, that it  ENDURED for an enormous time, as measured by years,
0374  the globe than at another, but seeing that it  ENDURED for long at each, and that it was contempora
0318  which any single species or any single genus  ENDURES. There is reason to believe that the complet
0105  tion of species, which will prove capable of  ENDURING for a long period, and of spreading widely.
0141  most animals. On this view, the capacity of  ENDURING the most different climates by man himself a
0141  the elephant and rhinoceros were capable of  ENDURING a glacial climate, whereas the living specie
0315  ly the same: but as the accumulation of long  ENDURING fossiliferous formations depends on great ma
0340  with modification, the great law of the long  ENDURING, but not immutable, succession of the same t
0067  lso, are destroyed in vast numbers by various  ENEMIES; for instance, on a piece of ground three fe
0069  destruction at some period of its life, from  ENEMIES or from competitors for the same place and f
0069  ors for the same place and food: and if these  ENEMIES or competitors be in the least degree favour
0070  ame species, relatively to the numbers of its  ENEMIES, is absolutely necessary for its preservatio
0078  antage over a different set of competitors or  ENEMIES. It is good thus to try in our imagination t
0085  struggle with other trees and with a host of  ENEMIES, such differences would effectually settle w
0109  ations in the seasons or in the number of its  ENEMIES, run a good chance of utter extinction. But
0134  ght; but by kicking it can defend itself from  ENEMIES, as well as any of the smaller quadrupeds. W
0175  ill, during fluctuations in the number of its  ENEMIES or of its prey, or in the seasons, be extrem
0195  any means defend themselves from these small  ENEMIES, would be able to range into new pastures an
0197  n to hide this tree frequenting bird from its  ENEMIES; and consequently that it was a character of
0234  ndent on the number of its parasites or other  ENEMIES; or on quite distinct causes, and so be alto
```

0401 exposed to the attacks of somewhat different ENEMIES. If then it varied, natural selection would
0212 southern United States. Fear of any particular ENEMY is certainly an instinctive quality, as may b
0212 perience, and by the sight of fear of the same ENEMY in other animals. But fear of man is slowly a
0219 the workers or sterile females, though most ENERGETIC and courageous in capturing slaves, do no o
0220 vae and pupae are exposed, the slaves work ENERGETICALLY with their masters in carrying them away
0157 n to very large groups of beetles, but in the ENGIDAE, as Westwood has remarked, the number varies
0035 encer and others have shown how the cattle of ENGLAND have increased in weight and in early maturi
0068 ame were shot during the next twenty years in ENGLAND, and at the same time, if no vermin were des
0073 t the exotic Lobelia fulgens, in this part of ENGLAND, is never visited by insects, and consequent
0074 of humble bees became extinct or very rare in ENGLAND, the heartsease and red clover would become
0074 wo thirds of them are thus destroyed all over ENGLAND. Now the number of mice is largely dependent
0140 species of plants brought from the Azores to ENGLAND. In regard to animals, several authentic cas
0163 spect to the horse, I have collected cases in ENGLAND of the spinal stripe in horses of the most d
0212 si and we may see an instance of this even in ENGLAND, in the greater wildness of all our large bi
0212 earful than small; and the magpie, so wary in ENGLAND, is tame in Norway, as is the hooded crow in
0219 his species is found in the southern parts of ENGLAND, and its habits have been attended to by Mr.
0221 red in greater numbers in Switzerland than in ENGLAND. One day I fortunately chanced to witness a
0223 ers carry the slaves. Both in Switzerland and ENGLAND the slaves seem to have the exclusive care o
0223 thus both collect food for the community. In ENGLAND the masters alone usually leave the nest to
0284 of these formations, which are represented in ENGLAND by thin beds, are thousands of feet in thick
0328 s admirable Memoirs on the eocene deposits of ENGLAND and France, is able to draw a close general
0328 tries; but when he compares certain stages in ENGLAND with those in France, although he finds in b
0362 give up flying carrier pigeons from France to ENGLAND, as the hawks on the English coast destroyed
0364 america to the western shores of Ireland and ENGLAND; but seeds could be transported by these wan
0021 breeds is something astonishing. Compare the ENGLISH carrier and the short faced tumbler, and see
0023 elieve that any ornithologist would place the ENGLISH carrier, the short faced tumbler, the runt,
0024 ily of Columbidae for a beak like that of the ENGLISH carrier, or that of the short faced tumbler,
0027 ll the domestic breeds. Secondly, although an ENGLISH carrier or short faced tumbler differs immen
0031 may ultimately be selected for breeding. What ENGLISH breeders have actually effected is proved by
0034 h antiquity. In rude and barbarous periods of ENGLISH history choice animals were often imported,
0035 slowly altered from it. It is known that the ENGLISH pointer has been greatly changed within the
0035 n, and by careful training, the whole body of ENGLISH racehorses have come to surpass in fleetness
0120 a1 to a10. In the same way, for instance, the ENGLISH race horse and English pointer have apparent
0120 way, for instance, the English race horse and ENGLISH pointer have apparently both gone on slowly
0152 se being the points now mainly attended to by ENGLISH fanciers. Even in the sub breeds, as in the
0164 given me by Mr. W. W. Edwards, that with the ENGLISH race horse the spinal stripe is much commone
0356 single parent. To illustrate what I mean: our ENGLISH race horses differ slightly from the horses
0362 s from France to England, as the hawks on the ENGLISH coast destroyed so many on their arrival. So
0062 clusion in mind. Yet unless it be thoroughly ENGRAINED in the mind, I am convinced that the whole
0140 imals brought from warmer countries which here ENJOY good health. We have reason to believe that s
0071 but on this intricate subject I will not here ENLARGE. Many cases are on record showing how comple
0136 island, the tendency of natural selection to ENLARGE or to reduce the wings, would depend on whet
0001 bject, and drew up some short notes; these I ENLARGED in 1844 into a sketch of the conclusions, wh
0136 ts, their wings not at all reduced, but even ENLARGED. This is quite compatible with the action of
0137 a the wings of some of the insects have been ENLARGED, and the wings of others have been reduced b
0153 sex, as with the wattle of carriers and the ENLARGED crop of pouters. Now let us turn to nature.
0159 e have a case of analogous variation, in the ENLARGED stems, or roots as commonly called, of the S
0159 uld have to attribute this similarity in the ENLARGED stems of these three plants, not to the vera
0280 tail somewhat expanded with a crop somewhat ENLARGED, the characteristic features of these two br
0134 use in our domestic animals strengthens and ENLARGES certain parts, and disuse diminishes them; a
0489 tem was deposited, they seem to me to become ENNOBLED. Judging from the past, we may safely infer
0023 e parent stocks possessed the characteristic ENORMOUS crop? The supposed aboriginal stocks must al
0031 ders have actually effected is proved by the ENORMOUS prices given for animals with a good pedigre
0069 ere it most abounds, is constantly suffering ENORMOUS destruction at some period of its life, from
0173 iciently thick and extensive to withstand an ENORMOUS amount of future degradation; and such fossi
0275 h dr. Prosper Lucas, who, after arranging an ENORMOUS body of facts with respect to animals, comes
0280 his process of extermination has acted on an ENORMOUS scale, so must the number of intermediate va
0280 have formerly existed on the earth, be truly ENORMOUS. Why then is not every geological formation
0288 of a formation conformably covered, after an ENORMOUS interval of time, by another and later forma
0290 ment must for ages have been great, from the ENORMOUS degradation of the coast rocks and from mudd
0296 n thickness, and which must have required an ENORMOUS period for their accumulation: yet no one is
0300 idence would be separated from each other by ENORMOUS intervals, during which the area would be ei
0302 states. We do not make due allowance for the ENORMOUS intervals of time, which have probably elaps
0309 e earth, and which had been pressed on by an ENORMOUS weight of superincumbent water, might have u
0342 s thick enough to resist future degradation, ENORMOUS intervals of time have elapsed between the s
0342 me great formation. He may disbelieve in the ENORMOUS intervals of time which have elapsed between
0343 where our oceans now extend they have for an ENORMOUS period extended, and where our oscillating c
0357 ment we are not authorized in admitting such ENORMOUS geographical changes within the period of ex
0374 , excellent evidence, that it endured for an ENORMOUS time, as measured by years, at each point. T
0383 es, belonging to quite different classes, an ENORMOUS range, but allied species prevail in a remar
0386 spect to plants, it has long been known what ENORMOUS ranges many fresh water and even marsh speci
0394 s family, that many of the same species have ENORMOUS ranges, and are found on continents and on f
0404 the world, and many individual species have ENORMOUS ranges. It is not meant that in world rangin
0407 that some forms of life change most slowly, ENORMOUS periods of time being thus granted for their
0465 formations are separated from each other by ENORMOUS blank intervals of time: for fossiliferous f
0474 ch have been inherited without change for an ENORMOUS period. It is inexplicable on the theory of
0483 his in some instances necessarily implies an ENORMOUS amount of modification in the descendants. T
0021 ch elongated body, wings, and legs; and its ENORMOUSLY developed crop, which it glories in inflati
0022 in length and breadth and curvature differs ENORMOUSLY. The shape, as well as the breadth and leng
0148 hly important anterior segments of the head ENORMOUSLY developed, and furnished with great nerves
0173 ngencies will concur only rarely, and after ENORMOUSLY long intervals. Whilst the bed of the sea i
0239 workers of another caste, and they have an ENORMOUSLY developed abdomen which secretes a sort of
0284 we have, in the opinion of most geologists, ENORMOUSLY long blank periods. So that the lofty pile
0288 and preserve fossil remains. Throughout an ENORMOUSLY large proportion of the ocean, the bright b
0309 conclude must have intervened during these ENORMOUSLY long periods. If then we may infer anything
0324 in europe during the pleistocene period (an ENORMOUSLY remote period as measured by years, includi

Page **(Key Word)**

0351 retained nearly the same character from an ENORMOUSLY remote geological period, so certain specie
0374 of its scanty flora, are common to Europe, ENORMOUSLY remote as these two points are: and there a
0381 ccurrence of identical species at points so ENORMOUSLY remote as Kerguelen Land, New Zealand, and
0399 although the next nearest continent, is so ENORMOUSLY remote, that the fact becomes an anomaly. B
0424 r that of a species, the two sexes: and how ENORMOUSLY these sometimes differ in the most importan
0462 e same specific form for very long periods, ENORMOUSLY long as measured by years, too much stress
0070 enerally to occur with our game animals, often ENSUE: and here we have a limiting check independen
0408 e of varying more or less rapidly, there would ENSUE in different regions, independently of their
0014 l whether or not nearly perfect reversion had ENSUED. It would be quite necessary, in order to pre
0036 have wished to have produced the result which ENSUED, namely, the production of two distinct strai
0265 though in a lesser degree than when sterility ENSUES. So it is with hybrids, for hybrids in succes
0033 nce of which should never be overlooked, will ENSURE some differences: but, as a general rule, I c
0086 arva: but in all cases natural selection will ENSURE that modifications consequent on other modifi
0097 the stigma of another with the same brush to ENSURE fertilisation: but it must not be supposed th
0098 s it, the contrivance seems adapted solely to ENSURE self fertilisation: and no doubt it is useful
0066 this seed were never destroyed, and could be ENSURED to germinate in a fitting place. So that in
0011 in the embryo or larva will almost certainly ENTAIL changes in the mature animal. In monstrositie
0157 ction than ordinary selection, as it does not ENTAIL death, but only gives fewer offspring to the
0199 ful modification of one part will often have ENTAILED on other parts diversified changes of no dir
0021 see the wonderful difference in their beaks, ENTAILING corresponding differences in their skulls.
0129 and the complex action of natural selection, ENTAILING extinction and divergence of character, as
0487 sm to organism, the improvement of one being ENTAILING the improvement or the extermination of of
0490 , and as a consequence to Natural Selection, ENTAILING Divergence of Character and the Extinction
0127 following chapters. But we already see how it ENTAILS extinction: and how largely extinction has a
0074 we look at the plants and bushes clothing an ENTANGLED bank, we are tempted to attribute their pro
0489 fection. It is interesting to contemplate an ENTANGLED bank, clothed with many plants of many kind
0008 er or not the plant sets a seed. I cannot here ENTER on the copious details which I have collected
0071 nd had been enclosed, so that cattle could not ENTER. But how important an element enclosure is, I
0081 new and better adapted forms could not freely ENTER, we should then have places in the economy of
0089 t to such apparently weak means: I cannot here ENTER on the details necessary to support this view
0089 ales and females: but I have not space here to ENTER on this subject. Thus it is, as I believe, th
0094 ich they can, with a very little more trouble, ENTER by the mouth. Bearing such facts in mind, I c
0096 r in reproduction? As it is impossible here to ENTER on details, I must trust to some general cons
0201 the doomed mouse. But I have not space here to ENTER on this and other such cases. Natural selecti
0220 sussex, and never saw a slave either leave or ENTER a nest. As, during these months, the slaves a
0220 nt in large numbers in August, either leave or ENTER the nest. Hence he considers them as strictly
0224 king instinct of the Hive Bee. I will not here ENTER on minute details on this subject, but will m
0225 base of a single cell on one side of the comb, ENTER into the composition of the bases of three ad
0226 ree plane surfaces in any one cell necessarily ENTER into the construction of three adjoining cell
0230 been a little hood of wax: but I will not here ENTER on these details. We see how important a part
0233 rictly hexagonal: but I have not space here to ENTER on this subject. Nor does there seem to me an
0236 ent acts of natural selection. I will not here ENTER on these several cases, but will confine myse
0248 ctive to compare, but I have not space here to ENTER on details, the evidence advanced by our best
0336 highly developed than ancient. I will not here ENTER on this subject, for naturalists have not as
0408 which allowed certain forms and not others to ENTER, either in greater or lesser numbers: accordi
0472 d but little direct effect: yet when varieties ENTER any zone, they occasionally assume some of th
0378 do not doubt that some temperate productions ENTERED and crossed even the lowlands of the tropics
0408 time which has elapsed since new inhabitants ENTERED one region: according to the nature of the c
0408 ser numbers: according or not, as those which ENTERED happened to come in more or less direct comp
0001 ame object. I hope that I may be excused for ENTERING on these personal details, as I give them to
0018 ain vague: nevertheless, I may, without here ENTERING on any details, state that, from geographica
0060 the most important of all relations. Before ENTERING on the subject of this chapter, I must make
0156 rks. I think it will be admitted, without my ENTERING on details, that secondary sexual characters
0164 than in the full grown animal. Without here ENTERING on further details, I may state that I have
0199 ction of a particular kind, but without here ENTERING on copious details my reasoning would appear
0222 sanguinea, and found a number of these ants ENTERING their nest, carrying the dead bodies of F. f
0290 on of the coast rocks and from muddy streams ENTERING the sea. The explanation, no doubt, is, that
0006 obscure, and will long remain obscure, I can ENTERTAIN no doubt, after the most celiberate study a
0006 apable, that the view which most naturalists ENTERTAIN, and which I formerly entertained, namely,
0155 : at least Is. Geoffroy St. Hilaire seems to ENTERTAIN no doubt, that the more an organ normally p
0006 naturalists entertain, and which I formerly ENTERTAINED, namely, that each species has been indepe
0191 purpose: in the same manner as, on the view ENTERTAINED by some naturalists that the branchiae and
0245 heir fertility. Summary. The view generally ENTERTAINED by naturalists is that species, when inter
0308 as a valid argument against the views here ENTERTAINED. To show that it may hereafter receive som
0484 ich life was first breathed. When the views ENTERTAINED in this volume on the origin of species, o
0310 , sir Charles Lyell, from further reflexion ENTERTAINS grave doubts on this subject. I feel how ra
0202 eye. If our reason leads us to admire with ENTHUSIASM a multitude of inimitable contrivances in n
0224 so beautifully adapted to its end, without ENTHUSIASTIC admiration. We hear from mathematicians th
0135 inherited: and I should prefer explaining the ENTIRE absence of the anterior tarsi in Ateuchus, an
0136 y insisted on by Mr. Wollaston, of the almost ENTIRE absence of certain large groups of beetles, e
0247 e rule equally universal: and he disputes the ENTIRE fertility of Kolreuter's ten cases. But in th
0310 appear in our European formations: the almost ENTIRE absence, as at present known, of fossiliferou
0334 ve and below. We must, however, allow for the ENTIRE extinction of some preceding forms, and for t
0398 s and Cape de Verde Archipelagos: but what an ENTIRE and absolute difference in their inhabitants!
0431 other vertebrate classes. There has been less ENTIRE extinction of the forms of life which once co
0455 me into play: and this will tend to cause the ENTIRE obliteration of a rudimentary organ. As the p
0048 ked by some botanists as species, and he has ENTIRELY omitted several highly polymorphic genera. U
0048 the American mainland, I was much struck how ENTIRELY vague and arbitrary is the distinction betwe
0051 doubtful forms, he will have to trust almost ENTIRELY to analogy, and his difficulties will rise t
0146 ions, occurring throughout whole orders, are ENTIRELY due to the manner alone in which natural sel
0232 mb met at an angle, how often the bees would ENTIRELY pull down and rebuild in different ways the
0251 eased to grow, and after a few days perished ENTIRELY, whereas the pod impregnated by the pollen o
0261 ant to be grafted or budded on another is so ENTIRELY unimportant for its welfare in a state of na
0275 lowly and naturally produced. On the whole I ENTIRELY agree with Dr. Prosper Lucas, who, after arr
0415 very unequal, and in some cases seems to be ENTIRELY lost. Again in another work he says, the gen
0060 hundred doubtful forms of British plants are ENTITLED to hold, if the existence of any well marked

Page **(Key Word)**

Page **(Key Word)**

```
0311 e apparently abruptly changed forms of life, ENTOMBED in our consecutive, but widely separated, fo
0049 ould be ranked as distinct species by many ENTOMOLOGISTS. Even Ireland has a few animals, now gene
0419 birds; and it has been followed by several ENTOMOLOGISTS and botanists. Finally, with respect to t
0097 e indispensable, the fullest freedom for the ENTRANCE of pollen from another individual will expla
0047 rly have existed; and here a wide door for the ENTRY of doubt and conjecture is opened. Hence, in
0047 s of great difficulty, which I will not here ENUMERATE, sometimes occur in deciding whether or not
0115 n united States, 260 naturalised plants are ENUMERATED, and these belong to 162 genera. We thus se
0279 ossiliferous strata. In the sixth chapter I ENUMERATED the chief objections which might be justly
0413 most unlike; or as an artificial means for ENUNCIATING, as briefly as possible, general propositi
0414 plan of creation, or is simply a scheme for ENUNCIATING general propositions and of placing togeth
0420 d not some unknown plan of creation, or the ENUNCIATION of general propositions, and the mere putt
0435 widened to any extent, and become gradually ENVELOPED in thick membrane, so as to serve as a fin;
0303 merica, and in Europe even as far back as the EOCENE stage. The most striking case, however, is th
0328 r. prestwich, in his admirable Memoirs on the EOCENE deposits of England and France, is able to dr
0331 he older Fish, the older Cephalopods, and the EOCENE mammals, with the more recent members of the
0337 forms. If under a nearly similar climate, the EOCENE inhabitants of one quarter of the world were
0337 itants of the same or some other quarter, the EOCENE fauna or flora would certainly be beaten and
0337 xterminated; as would a secondary fauna by an EOCENE, and a palaeozoic fauna by a secondary fauna.
0068 is an extraordinarily severe mortality from EPIDEMICS with man. The action of climate seems at fi
0070 es inordinately in numbers in a small tract, EPIDEMICS, at least, this seems generally to occur wi
0070 e for life. But even some of these so called EPIDEMICS appear to be due to parasitic worms, which
0295 ter, that is extending from before the glacial EPOCH to the present day. In order to get a perfect
0300 gravel or sand, would not endure to a distant EPOCH. Wherever sediment did not accumulate on the
0307 were somewhere accumulated before the glacial EPOCH, is very great. If these most ancient beds ha
0309 eriod immeasurably antecedent to the silurian EPOCH, continents may have existed where oceans are
0310 many formations long anterior to the silurian EPOCH in a completely metamorphosed condition. The
0324 measured by years, including the whole glacial EPOCH), were to be compared with those now living i
0324 ds. Nevertheless, looking to a remotely future EPOCH, there can, I think, be little doubt that all
0330 ing at the present day, were not at this early EPOCH limited in such distinct groups as they now a
0343 stand they have stood ever since the Silurian EPOCH; but that long before that period, the world
0363 stocked by ice borne seeds, during the glacial EPOCH. At my request Sir C. Lyell wrote to M. Hartu
0369 untains before the commencement of the Glacial EPOCH, and which during its coldest period will hav
0370 the newer Pliocene period, before the Glacial EPOCH, and whilst the majority of the inhabitants o
0370 ew worlds, at a period anterior to the Glacial EPOCH. Believing, from reasons before alluded to, t
0373 parent source. We do not know that the Glacial EPOCH was strictly simultaneous at these several fa
0373 e good evidence in almost every case, that the EPOCH was included within the latest geological per
0381 ore this flora was exterminated by the Glacial EPOCH, a few forms were widely dispersed to various
0403 throughout the world during the recent Glacial EPOCH, are related to those of the surrounding Lowl
0405 branched off from a common parent at a remote EPOCH; so that in such cases there will have been a
0420 allied genera, which lived during the Silurian EPOCH, and these have descended from a species whic
0489 of those which lived long before the Silurian EPOCH, we may feel certain that the ordinary succes
0006 of the world during the many past geological EPOCHS in its history. Although much remains obscure
0125 for the extinct species lived at very ancient EPOCHS when the branching lines of descent had diver
0464 eposited at these ancient and utterly unknown EPOCHS in the world's history. I can answer these qu
0017 cannot doubt that if other animals and plants, EQUAL in number to our domesticated productions, an
0017 e of nature, and could be made to breed for an EQUAL number of generations under domestication, th
0054 and described in any Flora be divided into two EQUAL masses, all those in the larger genera being
0055 rous insects of two districts, into two nearly EQUAL masses, the species of the larger genera on o
0065 red for food, and that in a state of nature an EQUAL number would have somehow to be disposed of.
0105 forms, we ought to make the comparison within EQUAL times; and this we are incapable of doing. Al
0113 ral mixed varieties of wheat have been sown on EQUAL spaces of ground. Hence, if any one species o
0117 time; nor are they all supposed to endure for EQUAL periods. Only those variations which are in s
0128 ll steadily tend to increase till they come to EQUAL the greater differences between species of th
0177 ; and that the inhabitants are all trying with EQUAL steadiness and skill to improve their stocks
0210 of natura non facit saltum applies with almost EQUAL force to instincts as to bodily organs. Chang
0225 atter cells are nearly spherical and of nearly EQUAL sizes, and are aggregated into an irregular m
0226 distance from each other, and had made them of EQUAL sizes and had arranged them symmetrically in
0226 me that it is strictly correct: If a number of EQUAL spheres be described with their centres place
0227 pona to make her cells truly spherical, and of EQUAL sizes; and this would not be very surprising,
0230 at the same rate, and by endeavouring to make EQUAL spherical hollows, but never allowing the oph
0232 distance from each other, all trying to sweep EQUAL spheres, and then building up, or leaving ung
0235 more and more perfectly, led the bees to sweep EQUAL spheres at a given distance from each other i
0246 their mongrel offspring, is, on my theory, of EQUAL importance with the sterility of species; for
0262 y of effecting an union is often very far from EQUAL, so it sometimes is in grafting; the common g
0276 between the same two species. It is not always EQUAL in degree in a first cross and in the hybrid
0306 scended from one progenitor, apply with nearly EQUAL force to these several known species. For inst
0314 o change abruptly, or simultaneously, or to an EQUAL degree. The process of modification must be e
0315 the average amount of change, during long and EQUAL periods of time, may, perhaps, be nearly the
0315 sils embedded in consecutive formations is not EQUAL. Each formation, on this view, does not mark
0317 kens upwards, sometimes keeping for a space of EQUAL thickness, and ultimately thins out in the up
0326 r a long period favourably circumstanced in an EQUAL degree, whenever their inhabitants met, the b
0389 lands are few in number compared with those on EQUAL continental areas: Alph. de Candolle admits t
0389 g plants, only 750 in number, with those on an EQUAL area at the Cape of Good Hope or in Australia
0418 haracters derived from the embryo should be of EQUAL importance with those derived from the adult,
0453 ly adapted for certain purposes, tells us with EQUAL plainness that these rudimentary or atrophied
0475 ive intervals; and the amount of change, after EQUAL intervals of time, is widely different in dif
0484 y four or five progenitors, and plants from an EQUAL or lesser number. Analogy would lead me one s
0380 ny islands the native productions are nearly EQUALLED or even outnumbered by the naturalised: and
0003 ng pollen from one flower to the other, it is EQUALLY preposterous to account for the structure of
0004 dification is at least possible; and, what is EQUALLY or more important, we shall see how great is
0017 ur domesticated productions, and belonging to EQUALLY diverse classes and countries, were taken fr
0044 t act of creation. The term variety is almost EQUALLY difficult to define; but here community of d
0057 pretends that all the species of a genus are EQUALLY distinct from each other; they may generally
0075 pecies, the struggle will generally be almost EQUALLY severe, and we sometimes see the contest soo
0203 sed through insect agency, can we consider as EQUALLY perfect the elaboration by our fir trees of
0231 of cement, and then to begin cutting it away EQUALLY on both sides near the ground, till a smooth
```

Page **(Key Word)**

equally

Page **(Key Word)**

```
0247 m as varieties. Gartner, also, makes the rule EQUALLY universal; and he disputes the entire fertil
0258 m. jalappa, and utterly failed. Several other EQUALLY striking cases could be given. Thuret has ob
0260 sed, all of which we must suppose it would be EQUALLY important to keep from blending together? Wh
0287 of the sea: when deeply submerged for perhaps EQUALLY long periods, it would, likewise, have escap
0293 rmost layers have been deposited, it would be EQUALLY rash to suppose that it then became wholly e
0317 into species, which in their turn produce by EQUALLY slow steps other species; and so on, like th
0354 enabled at the same time to consider a point EQUALLY important for us, namely, whether the severa
0388 particular nature, and drops them in another EQUALLY well fitted for them. On the Inhabitants of
0393 islands situated at a much less distance are EQUALLY barren. The Falkland Islands, which are inha
0396 d, all the forms of life would have been more EQUALLY modified, in accordance with the paramount i
0402 holly or in part supplant it; but if both are EQUALLY well fitted for their own places in nature,
0436 certain extent obscured. There is another and EQUALLY curious branch of the present subject; namel
0443 at slight variations necessarily appear at an EQUALLY early period. But we have little evidence on
0468 e most severe between them; it will be almost EQUALLY severe between the varieties of the same spe
0489 ok with some confidence to a secure future of EQUALLY inappreciable length. And as natural selecti
0299 an, and from Britain to Russia; and therefore EQUALS all the geological formations which have been
0304 ver the world, from the Arctic regions to the EQUATOR, inhabiting various zones of depths from the
0305 y any fossil fish are known from south of the EQUATOR; and by running through Pictet's Palaeontolo
0373 nc on gigantic ancient moraines. South of the EQUATOR, we have some direct evidence of former glac
0374 of north America, in the Cordillera under the EQUATOR and under the warmer temperate zones, and on
0377 ll have retreated from both sides towards the EQUATOR, followed in the rear by the temperate produ
0377 te productions, after migrating nearer to the EQUATOR, though they will have been placed under som
0377 ve ranks and have reached or even crossed the EQUATOR. The invasion would, of course, have been or
0378 me cold, I believe that the climate under the EQUATOR at the level of the sea was about the same w
0378 ertropical regions, and some even crossed the EQUATOR. As the warmth returned, these temperate for
0378 the lowlands: those which had not reached the EQUATOR, would re migrate northward or southward tow
0379 orms, chiefly northern, which had crossed the EQUATOR, would travel still further from their homes
0382 nd from the south, and to have crossed at the EQUATOR; but to have flowed with greater force from
0382 e arctic lowlands to a great height under the EQUATOR. The various beings thus left stranded may b
0397 the Galapagos Archipelago, situated under the EQUATOR, between 500 and 600 miles from the shores o
0323 can be found; namely, in North America, in EQUATORIAL South America, in Tierra del Fuego, at the
0372 y a continent and by nearly a hemisphere of EQUATORIAL ocean. These cases of relationship, without
0373 n the Rocky Mountains. In the Cordillera of EQUATORIAL South America, glaciers once extended far b
0374 y allied species. On the lofty mountains of EQUATORIAL America a host of peculiar species belongin
0376 ly remarked, In receding from polar towards EQUATORIAL latitudes, the Alpine or mountain floras re
0167 quagga, and zebra. He who believes that each EQUINE species was independently created, will, I pr
0021 feathers are kept expanded, and are carried so ERECT that in good birds the head and tail touch; t
0202 this same reason tells us, though we may easily ERR on both sides, that some other contrivances ar
0426 by its inheritance from a common parent. We may ERR in this respect in regard to single points of
0363 m. hartung to inquire whether he had observed ERRATIC boulders on these islands, and he answered t
0366 throughout a large part of the United States, ERRATIC boulders, and rocks scored by drifted iceber
0373 o different, as far south as lat. 46 degrees: ERRATIC boulders have, also, been noticed on the Roc
0006 h species has been independently created, is ERRONEOUS. I am fully convinced that species are not
0284 d thousand years. This estimate may be quite ERRONEOUS; yet, considering over what wide spaces ver
0286 than elsewhere, the above estimate would be ERRONEOUS; but this source of doubt probably would no
0288 . i believe we are continually taking a most ERRONEOUS view, when we tacitly admit to ourselves th
0402 o the others. But we often take, I think, an ERRONEOUS view of the probability of closely allied s
0150 gh generality. I am aware of several causes of ERROR, but I hope that I have made due allowance fo
0209 nstinctively. But it would be the most serious ERROR to suppose that the greater number of instinc
0230 truction of the cells; but it would be a great ERROR to suppose that the bees cannot build up a ro
0247 s in a state of nature. But a serious cause of ERROR seems to me to be here introduced: a plant to
0268 ies and varieties, and that there must be some ERROR in all the foregoing remarks, inasmuch as var
0303 hese remarks; and to show how liable we are to ERROR in supposing that whole groups of species hav
0364 nd coming to maturity! But it would be a great ERROR to argue that because a well stocked island,
0400 it arises in chief part from the deeply seated ERROR of considering the physical conditions of a c
0002 sing some confidence in my accuracy. No doubt ERRORS will have crept in, though I hope I have alwa
0077 or food or residence, or from which it has to ESCAPE, or on which it preys. This is obvious in the
0077 tic insects, to hunt for its own prey, and to ESCAPE serving as prey to other animals. The store o
0134 nd feeding birds seldom take flight except to ESCAPE danger, I believe that the nearly wingless co
0134 and is exposed to danger from which it cannot ESCAPE by flight, but by kicking it can defend itsel
0180 quirrel in its own country, by enabling it to ESCAPE birds or beasts of prey, or to collect food m
0182 of flight exclusively, as far as we know, to ESCAPE being devoured by other fish? When we see any
0195 in a coming dearth to search for food, or to ESCAPE from beasts of prey. Organs now of trifling i
0197 young birds and reptiles, which have only to ESCAPE from a broken egg, we may infer that this str
0201 r its own injury, namely, to warn its prey to ESCAPE. I would almost as soon believe that the cat
0361 of birds, when floating on the sea, sometimes ESCAPE being immediately devoured; and seeds of many
0287 for millions of years as land, and thus have ESCAPED the action of the sea: when deeply submerged
0287 qually long periods, it would, likewise, have ESCAPED the action of the coast waves. So that in al
0321 distant and isolated station, where they have ESCAPED severe competition. For instance, a single s
0377 considerable amount of cold, many might have ESCAPED extermination during a moderate fall of temp
0174 at present. But I will pass over this way of ESCAPING from the difficulty; for I believe that many
0377 rate fall of temperature, more especially by ESCAPING into the warmest spots. But the great fact t
0285 tance to the west the northern and southern ESCARPMENTS meet and close, one can safely picture to
0015 d cattle, and poultry of various breeds, and ESCULENT vegetables, for an almost infinite number of
0423 dish and common turnips together, though the ESCULENT and thickened stems are so similar. Whatever
0418 of connexion can be discovered between them, ESPECIAL value is set on them. As in most groups of a
0426 uch correlated or aggregated characters have ESPECIAL value in classification. We can understand w
0426 , are generally the most constant, we attach ESPECIAL value to them; but if these same organs, in
0466 years to doubt their weight. But it deserves ESPECIAL notice that the more important objections re
0001 urged to publish this Abstract. I have more ESPECIALLY been induced to do this, as Mr. Wallace, wh
0016 th when compared one with another, and more ESPECIALLY when compared with all the species in natur
0018 t we find in the most ancient records, more ESPECIALLY on the monuments of Egypt, much diversity i
0020 ns from several quarters of the world, more ESPECIALLY by the Hon. W. Elliot from India, and by th
0021 ferences in their skulls. The carrier, more ESPECIALLY the male bird, is also remarkable from the
0023 pouter, and fantail in the same genus; more ESPECIALLY as in each of these breeds several truly in
0027 he several sub breeds of these breeds, more ESPECIALLY those brought from distant countries, we ca
```

Page **(Key Word)**

Page **(Key Word)**

0042 r two purposes, food and feathers, and more ESPECIALLY from no pleasure having been felt in the di
0043 lity is governed by many unknown laws, more ESPECIALLY by that of correlation of growth. Something
0056 find the manufactory still in action, more ESPECIALLY as we have every reason to believe the proc
0067 of the checks at considerable length, more ESPECIALLY in regard to the feral animals of South Ame
0097 l will explain this state of exposure, more ESPECIALLY as the plant's own anthers and pistil gener
0105 rgeness of area is of more importance, more ESPECIALLY in the production of species, which will pr
0114 circumstances. In an extremely small area, ESPECIALLY if freely open to immigration, and where th
0115 ts would have belonged to a few groups more ESPECIALLY adapted to certain stations in their new ho
0136 xposed Dezertas than in Madeira itself; and ESPECIALLY the extraordinary fact, so strongly insiste
0147 application; but many good observers, more ESPECIALLY botanists, believe in its truth. I will not
0152 fluctuating condition. But what here more ESPECIALLY concerns us is, that in our domestic animal
0162 appear together from simple variation. More ESPECIALLY we might have inferred this, from the blue
0174 part in the formation of new species, more ESPECIALLY with freely crossing and wandering animals.
0178 solated portions, in which many forms, more ESPECIALLY amongst the classes which unite for each bi
0181 r. therefore, I can see no difficulty, more ESPECIALLY under changing conditions of life, in the c
0189 e do not know the transitional grades, more ESPECIALLY if we look to much isolated species, round
0193 pears in several members of the same class, ESPECIALLY if in members having very different habits
0198 icated animals in different countries, more ESPECIALLY in the less civilized countries where there
0207 convenient to treat the subject separately, ESPECIALLY as so wonderful an instinct as that of the
0207 perform, when performed by an animal, more ESPECIALLY by a very young one, without any experience
0216 ves the danger chuckle, they will run (more ESPECIALLY young turkeys) from under her, and conceal
0217 hatching might be inconveniently long, more ESPECIALLY as she has to migrate at a very early perio
0245 the theory of natural selection the case is ESPECIALLY important, inasmuch as the sterility of hyb
0256 cies in structure and in constitution, more ESPECIALLY in the structure of parts which are of high
0264 e liable to perish at an early period; more ESPECIALLY as all very young beings seem eminently sen
0267 generations between the nearest relations, ESPECIALLY if these be kept under the same conditions
0268 of the cabbage, is a remarkable fact; more ESPECIALLY when we reflect how many species there are,
0272 ntal on slowly acquired modifications, more ESPECIALLY in the reproductive systems of the forms wh
0274 n hybrids to their respective parents, more ESPECIALLY in hybrids produced from nearly related spe
0274 ce of secondary sexual characters; but more ESPECIALLY owing to prepotency in transmitting likenes
0314 hysical conditions of the country, and more ESPECIALLY on the nature of the other inhabitants with
0330 to distinct families. The most common case, ESPECIALLY with respect to very distinct groups, such
0335 han a form elsewhere subsequently produced, ESPECIALLY in the case of terrestrial productions inha
0367 pine plants of each mountain range are more ESPECIALLY related to the arctic forms living due nort
0368 e pyrenees, as remarked by Ramond, are more ESPECIALLY allied to the plants of northern Scandinavi
0368 osed to nearly the same climate, and, as is ESPECIALLY to be noticed, they will have kept in a bod
0377 during a moderate fall of temperature, more ESPECIALLY by escaping into the warmest spots. But the
0388 al of their seeds and eggs by animals, more ESPECIALLY by fresh water birds, which have large powe
0406 bit the islets of the same archipelago, and ESPECIALLY the striking relation of the inhabitants of
0407 ation, by some general considerations, more ESPECIALLY from the importance of barriers and from th
0419 ot quite logically, in classification, more ESPECIALLY in very large groups of closely allied form
0426 vail throughout many and different species, ESPECIALLY those having very different habits of life,
0446 sesses them. And the cases just given, more ESPECIALLY that of pigeons, seem to show that the char
0454 ng horns in hornless breeds of cattle, more ESPECIALLY, according to Youatt, in young animals, and
0460 s many structures have been perfected, more ESPECIALLY amongst broken and failing groups of organi
0034 draught cattle by colour, as do some of the ESQUIMAUX their teams of dogs. Livingstone shows how
0484 shadowy doubt whether this or that form be in ESSENCE a species. This I feel sure, and I speak aft
0485 earch for the undiscovered and undiscoverable ESSENCE of the term species. The other and more gene
0010 inion of most physiologists that there is no ESSENTIAL difference between a bud and an ovule in th
0077 every organic being is related, in the most ESSENTIAL yet often hidden manner, to that of all oth
0078 of its life will generally be changed in an ESSENTIAL manner. If we wished to increase its averag
0085 roduce little effect: we should remember how ESSENTIAL it is in a flock of white sheep to destroy
0184 grows, there is a woodpecker, which in every ESSENTIAL part of its organisation, even in its colou
0268 t forcible argument, that there must be some ESSENTIAL distinction between species and varieties,
0276 ses perfectly with the view that there is no ESSENTIAL distinction between species and varieties.
0414 e organs to mistake a merely adaptive for an ESSENTIAL character. So with plants, how remarkable i
0484 me. This latter point will become a far more ESSENTIAL consideration than it is at present; for di
0052 resembling each other, and that it does not ESSENTIALLY differ from the term variety, which is giv
0132 rliest condition does not apparently differ ESSENTIALLY from an ovule, is alone affected. But why,
0175 rarer and rarer; then, as varieties do not ESSENTIALLY differ from species, the same rule will pr
0185 ne for an auk or grebe; nevertheless, it is ESSENTIALLY a petrel, but with many parts of its organ
0213 age, I cannot see that these actions differ ESSENTIALLY from true instincts. If we were to see one
0267 r by some common but unknown bond, which is ESSENTIALLY related to the principle of life. Fertilit
0349 hough they may be all peculiar species, are ESSENTIALLY American. We may look back to past ages, a
0451 hat the rudiment and the perfect pistil are ESSENTIALLY alike in nature. An organ serving for two
0032 . when a race of plants is once pretty well ESTABLISHED, the seed raisers do not pick out the best
0039 after several breeds have once fairly been ESTABLISHED. Many slight differences might, and indeed
0043 ntry several domestic breeds have once been ESTABLISHED, their occasional intercrossing, with the
0053 at the following statements are fairly well ESTABLISHED. The whole subject, however, treated as it
0112 breeds would become converted into two well ESTABLISHED and distinct breeds. As the differences sl
0239 i do not admit that such wonderful and well ESTABLISHED facts at once annihilate my theory. In the
0316 cies of pigeon, or even from the other well ESTABLISHED races of the domestic pigeon, for the newl
0392 n stature with a fully developed tree, when ESTABLISHED on an island and having to compete with he
0402 n mocking thrush: why should it succeed in ESTABLISHING itself there? We may safely infer that Cha
0068 of partridges, grouse, and hares on any large ESTATE depends chiefly on the destruction of vermin.
0071 , has interested me. In Staffordshire, on the ESTATE of a relation where I had ample means of inve
0016 domesticated productions. When we attempt to ESTIMATE the amount of structural difference between
0064 nown animals, and I have taken some pains to ESTIMATE its probable minimum rate of natural increas
0284 from actual measurement, in a few cases from ESTIMATE, of each formation in different parts of Gre
0284 y 600 feet in a hundred thousand years. This ESTIMATE may be quite erroneous; yet, considering.ove
0286 in thinner masses than elsewhere, the above ESTIMATE would be erroneous; but this source of doubt
0286 doubt probably would not greatly affect the ESTIMATE as applied to the western extremity of the d
0287 reat, but it would somewhat reduce the above ESTIMATE. On the other hand, during oscillations of l
0361 uri and some authors have given a far higher ESTIMATE. I have never seen an instance of nutritious
0068 ve to be the most effective of all checks. I ESTIMATED that the winter of 1854 55 destroyed four f
0284 this must have consumed! Good observers have ESTIMATED that sediment is deposited by the great Mis

Page **(Key Word)**

Page **(Key Word)***

```
0097  organic being self fertilises itself for an  ETERNITY of generations; but that a cross with anothe
0285  e vain endeavour to grapple with the idea of  ETERNITY. I am tempted to give one other case, the we
0309  o assume that things have thus remained from  ETERNITY? Our continents seem to to have been formed
0106  are now yielding, before those of the larger  EUROPAEO Asiatic area. Thus, also, it is that contine
0019  d cattle, as many sheep, and several goats in  EUROPE alone, and several even within Great Britain.
0019  that many domestic breeds have originated in  EUROPE; for whence could they have been derived, as
0027  n, has been found capable of domestication in  EUROPE and in India; and that it agrees in habits an
0048  those birds and insects in North America and  EUROPE, which differ very slightly from each other,
0049  ed, will suffice? If that between America and  EUROPE is ample, will that between the Continent and
0065  f all other plants, have been introduced from  EUROPE; and there are plants which now range in Indi
0107  heer, resembles the extinct tertiary flora of  EUROPE. All fresh water basins, taken together, make
0138  of the other inhabitants of North America and  EUROPE. On my view we must suppose that American ani
0138  es, as did European animals into the caves of  EUROPE. We have some evidence of this gradation of h
0138  nhabitants of that continent, and in those of  EUROPE, to the inhabitants of the European continent
0139  ind Proteus with reference to the reptiles of  EUROPE, I am only surprised that more wrecks of anci
0202  legions of plants and animals introduced from  EUROPE. Natural selection will not produce absolute
0210  als having been but little observed except in  EUROPE and North America, and for no instinct being
0239  en, considering how few neuter insects out of  EUROPE have been carefully examined. Mr. F. Smith ha
0288  the important discoveries made every year in  EUROPE prove. No organism wholly soft can be preserv
0293  olly extinct. We forget how small the area of  EUROPE is compared with the rest of the world; nor h
0293  veral stages of the same formation throughout  EUROPE been correlated with perfect accuracy. With m
0294  eozoic beds of North America than in those of  EUROPE; time having apparently been required for the
0294  ned amount of migration of the inhabitants of  EUROPE during the Glacial period, which forms only a
0298  Limits of the known geological formations of  EUROPE, which have oftenest given rise, first to loc
0299  the Malay Archipelago is of about the size of  EUROPE from the North Cape to the Mediterranean, and
0299  seas, probably represents the former state of  EUROPE, when most of our formations were accumulatin
0302  ore they invaded the ancient archipelagoes of  EUROPE and of the United States. We do not make due
0303  en discovered in India, South America, and in  EUROPE even as far back as the eocene stage. The mos
0305  species are known from several formations in  EUROPE. Some few families of fish now have a confine
0306  ogy of other countries beyond the confines of  EUROPE and the United States; and from the revolutio
0308  profound depths, in the several formations of  EUROPE and of the United States; and from the amount
0308  e neighbourhood of the existing continents of  EUROPE and North America. But we do not know what wa
0308  ls between the successive formations; whether  EUROPE and the United States during these intervals
0314  from their nearest allies on the continent of  EUROPE, whereas the marine shells and birds have rem
0323  er forms, which are not found in the Chalk of  EUROPE, but which occur in the formations either abo
0323  sive palaeozoic formations of Russia, Western  EUROPE and North America, a similar parallelism in t
0323  macrauchenia, and Toxodon had been brought to  EUROPE from La Plata, without any information in reg
0324  rine animals which live at the present day in  EUROPE, and all those that lived in Europe during th
0324  nt day in Europe, and all those that lived in  EUROPE during the pleistocene period (an enormously
0324  he existing or the pleistocene inhabitants of  EUROPE resembled most closely those of the southern
0324  more closely related to those which lived in  EUROPE during certain later tertiary stages, than to
0324  the pleistocene and strictly modern beds, of  EUROPE, North and South America, and Australia, from
0325  palaeozoic forms of life in various parts of  EUROPE, they add, if struck by this strange sequence
0328  ct that cases of this nature have occurred in  EUROPE. Mr. Prestwich, in his admirable Memoirs on t
0337  hemisphere having become wild in any part of  EUROPE, we may doubt, if all the productions of New
0340  y produced in South America. For we know that  EUROPE in ancient times was peopled by numerous mars
0343  ons of any one great region alone, as that of  EUROPE, are considered; he may urge the apparent, bu
0352  great Britain having been formerly united to  EUROPE, and consequently possessing the same quadrup
0352  why do we not find a single mammal common to  EUROPE and Australia or South America? The condition
0357  lantic must recently have been connected with  EUROPE or Africa, and Europe likewise with America.
0357  ave been connected with Europe or Africa, and  EUROPE likewise with America. Other authors have thu
0363  rge number of the species of plants common to  EUROPE, in comparison with the plants of other ocean
0365  ccasional means of transport, immigrants from  EUROPE or any other continent, that a poorly stocked
0365  yrenees, and in the extreme northern parts of  EUROPE; but it is far more remarkable, that the plan
0365  gray, with those on the loftiest mountains of  EUROPE. Even as long ago as 1747, such facts led Gme
0366  thin a very recent geological period, central  EUROPE and North America suffered under an Arctic cl
0366  lately filled. So greatly has the climate of  EUROPE changed, that in Northern Italy, gigantic mor
0366  ate on the distribution of the inhabitants of  EUROPE, as explained with remarkable clearness by Ed
0366  auna and flora, covering the central parts of  EUROPE, as far south as the Alps and Pyrenees, and e
0367  these would be nearly the same with those of  EUROPE; for the present circumpolar inhabitants, whi
0367  tter earlier or later in North America than in  EUROPE, so will the southern migration there have be
0367  on the mountains of the United States and of  EUROPE. We can thus also understand the fact that th
0368  ution of the Alpine and Arctic productions of  EUROPE and America, that when in other regions we fi
0369  ntains and on the plains of North America and  EUROPE; and it may be reasonably asked how I account
0370  there is almost continuous land from western  EUROPE, through Siberia, to eastern America. And to
0371  a modified condition, in the central parts of  EUROPE and the United States. On this view we can un
0371  between the productions of North America and  EUROPE, a relationship which is most remarkable, con
0371  by several observers, that the productions of  EUROPE and America during the later tertiary stages
0372  s of the temperate lands of North America and  EUROPE, are inexplicable on the theory of creation.
0373  hat Forbes's view may be largely extended. In  EUROPE we have the plainest evidence of the cold per
0374  rable part of its scanty flora, are common to  EUROPE, enormously remote as these two points are; a
0375  , and at the same time representing plants of  EUROPE, not found in the intervening hot lowlands. A
0375  s a picture of a collection made on a hill in  EUROPE! Still more striking is the fact that souther
0376  ies of Algae are common to New Zealand and to  EUROPE, but have not been found in the intermediate
0378  in number, common to Tierra del Fuego and to  EUROPE still exist in North America, which must have
0380  forms have become naturalised in any part of  EUROPE, though hides, wool, and other objects likely
0380  o carry seeds have been largely imported into  EUROPE during the last two or three centuries from L
0395  e britain separated by a shallow channel from  EUROPE, and the mammals are the same on both sides;
0404  nimals inhabiting the caves of America and of  EUROPE. Other analogous facts could be given. And it
0405  eties of the same species inhabit America and  EUROPE, and the species thus has an immense range; b
0469  ey assign to the many representative forms in  EUROPE and North America. If then we have under natu
0473  mals inhabiting the dark caves of America and  EUROPE. In both varieties and species correlation of
0018  d from a different aboriginal stock from our  EUROPEAN cattle; and several competent judges believe
0138  been separately created for the American and  EUROPEAN caverns, close similarity in their organisat
0138  eeper recesses of the Kentucky caves, as did  EUROPEAN animals into the caves of Europe. We have so
0138  n those of Europe, to the inhabitants of the  EUROPEAN continent. And this is the case with some of
```

Page **(Key Word)***

Page **(Key Word)**

```
0139  i hear from Professor Dana; and some of the EUROPEAN cave insects are very closely allied to thos
0140  ations have been made by Mr. H. C. Watson on EUROPEAN species of plants brought from the Azores to
0217  s suppose that the ancient progenitor of our EUROPEAN cuckoo had the habits of the American cuckoo
0218  to dr. Gray and to some other observers, the EUROPEAN cuckoo has not utterly lost all maternal lov
0239  stic cattle as they may be called, which our EUROPEAN ants guard or imprison. It will indeed be th
0254  so again there is reason to believe that our EUROPEAN and the humped Indian cattle are quite ferti
0268  nous domestic dogs do not readily cross with EUROPEAN dogs, the explanation which will occur to ev
0294  for their migration from the American to the EUROPEAN seas. In examining the latest deposits of va
0299  conchologists as distinct species from their EUROPEAN representatives, and by other conchologists
0310  which whole groups of species appear in our  EUROPEAN formations; the almost entire absence, as at
0322  imultaneously throughout the world. Thus our  EUROPEAN Chalk formation can be recognised in many di
0323  it is, according to Lyell, with the several  EUROPEAN and North American tertiary deposits. Even i
0324  be liable to be classed with somewhat older  EUROPEAN beds. Nevertheless, looking to a remotely fu
0337  uses. From the extraordinary manner in which EUROPEAN productions have recently spread over New Ze
0353  are nearly the same, so that a multitude of  EUROPEAN animals and plants have become naturalised i
0369  pine plants and animals of the several great EUROPEAN mountain ranges, though very many of the spe
0374  rica a host of peculiar species belonging to EUROPEAN genera occur. On the highest mountains of Br
0374  on the highest mountains of Brazil, some few EUROPEAN genera were found by Gardner, which do not e
0375  lera. On the mountains of Abyssinia, several EUROPEAN forms and some few representatives of the pe
0375  e occur. At the Cape of Good Hope a very few EUROPEAN species, believed not to have been introduce
0375  nd on the mountains, some few representative EUROPEAN forms are found, which have not been discove
0375  tralia, Dr. F. Muller has discovered several EUROPEAN species; other species, not introduced by ma
0375  be given, as I am informed by Dr. Hooker, of EUROPEAN genera, found in Australia, but not in the i
0380  as we see at the present day, that very many EUROPEAN productions cover the ground in La Plata, an
0391  does not possess one peculiar bird, and many EUROPEAN and African birds are almost every year blow
0403  ess both islands have been colonised by some EUROPEAN land shells which no doubt had some advantag
0034  erior of Africa who have not associated with EUROPEANS. Some of these facts do not show actual sel
0221  e open and close the doors in the morning and EVENING; and as Huber expressly states, their princi
0222  hey took heart and carried off the pupae. One EVENING I visited another community of F. sanguinea,
0281  child and parent, will render this a very rare EVENT: for in all cases the new and improved forms
0319  nd we know that this has been the progress of EVENTS with those animals which have been exterminat
0257  . annual and perennial plants, deciduous and EVERGREEN trees, plants inhabiting different stations
0261  ng woody and the other herbaceous, one being EVERGREEN and the other deciduous, and adaptation to
0030  th other breeds so little quarrelsome, with  EVERLASTING layers which never desire to sit, and with
0268  an dogs, the explanation which will occur to EVERYONE, and probably the true one, is that these do
0020  found with pigeons) extremely uniform, and  EVERYTHING seems simple enough; but when these mongrel
0371  of the Old World. Consequently we have here  EVERYTHING favourable for much modification, for far m
0061  eze; in short, we see beautiful adaptations  EVERYWHERE and in every part of the organic world. Aga
0076  . in Russia the small Asiatic cockroach has  EVERYWHERE driven before it its great congener. One sp
0083  on of the land. And as foreigners have thus  EVERYWHERE beaten some of the natives, we may safely c
0106  so, it is that continental productions have  EVERYWHERE become so largely naturalised on islands. O
0128  ubordinate to group, in the manner which we  EVERYWHERE behold, namely, varieties of the same speci
0132  one organic being and another, which we see  EVERYWHERE throughout nature. Some little influence ma
0171  es by insensibly fine gradations, do we not  EVERYWHERE see innumerable transitional forms? Why is
0206  mited faculties, can absolute perfection be  EVERYWHERE found. On the theory of natural selection w
0279  merable intermediate links not now occuring EVERYWHERE throughout nature depends on the very proce
0294  ts of various quarters of the world, it has  EVERYWHERE been noted, that some few still existing sp
0327  ghest degree, wherever produced, would tend  EVERYWHERE to prevail. As they prevailed, they would c
0327  d the succession of forms in both ways will  EVERYWHERE tend to correspond. There is one other rema
0363  cts could be given showing that soil almost  EVERYWHERE is charged with seeds. Reflect for a moment
0367  polar inhabitants, which we suppose to have  EVERYWHERE travelled southward, are remarkably uniform
0380  endemic Alpine forms; but these have almost  EVERYWHERE largely yielded to the more dominant forms,
0380  way as the productions of real islands have EVERYWHERE lately yielded to continental forms, natura
0471  ithin a few great classes, which we now see EVERYWHERE around us, and which has prevailed througho
0015  ard to species. But there is not a shadow of EVIDENCE in favour of this view: to assert that we co
0017  races, there is presumptive, or even strong, EVIDENCE in favour of this view. It has often been as
0050  difficulty. We could hardly wish for better EVIDENCE of the two forms being specifically distinct
0050  o me, an overwhelming amount of experimental EVIDENCE, showing that they descend from common paren
0064  nded from the first pair. But we have better EVIDENCE on this subject than mere theoretical calcul
0064  ollowing seasons. Still more striking is the EVIDENCE from our domestic animals of many kinds whic
0110  given period favourable variations. We have EVIDENCE of this, in the facts given in the second ch
0127  dged of by the general tenour and balance of EVIDENCE given in the following chapters. But we alre
0135  e totally deficient. There is not sufficient EVIDENCE to induce us to believe that mutilations are
0138  imals into the caves of Europe. We have some EVIDENCE of this gradation of habit; for, as Schiodte
0140  adaptation be generally very close, we have  EVIDENCE, in the case of some few plants, of their be
0155  on. It would be almost superfluous to adduce EVIDENCE in support of the above statement, that spec
0162  produce the intermediate form. But the best  EVIDENCE is afforded by parts or organs of an importa
0179  rms and the intermediate links. Consequently EVIDENCE of their former existence could be found onl
0210  find in the collateral lines of descent some EVIDENCE of such gradations; or we ought at least to
0212  assurance, that I do not speak without good  EVIDENCE. The possibility, or even probability, of in
0248  ends and sterility begins. I think no better EVIDENCE of this can be required than that the two mo
0248  have not space here to enter on details, the EVIDENCE advanced by our best botanists on the questi
0248  be ranked as species or varieties, with the  EVIDENCE from fertility adduced by different hybridis
0248  between species and varieties; but that the  EVIDENCE from this source graduates away, and is doub
0248  and is doubtful in the same degree as is the EVIDENCE derived from other constitutional and struct
0254  in its truth, although it rests on no direct EVIDENCE. I believe, for instance, that our dogs have
0266  founded, I think, on a considerable body of  EVIDENCE, that slight changes in the conditions of li
0267  h with plants and animals, there is abundant EVIDENCE, that a cross between very distinct individu
0269  but it seems to me impossible to resist the  EVIDENCE of the existence of a certain amount of ster
0269  ng cases, which I will briefly abstract. The EVIDENCE is at least as good as that from which we be
0270  the sterility of a multitude of species. The EVIDENCE is, also, derived from hostile witnesses, wh
0280  ed, that if we had no historical or indirect EVIDENCE regarding their origin, it would not have be
0284  he degraded matter, probably offers the best EVIDENCE of the lapse of time. I remember having been
0284  i remember having been much struck with the  EVIDENCE of denudation, when viewing volcanic islands
0288  periods, it is superfluous to state that our EVIDENCE from fossil remains is fragmentary in an ext
0296  ulation. In other cases we have the plainest EVIDENCE in great fossilised trees, still standing up
```

Page **(Key Word)**

Page **(Key Word)***

```
0297  ies; and on this view we do find the kind of EVIDENCE of change which on my theory we ought to fin
0304  to Lyell's Manual, published in 1858, clear EVIDENCE of the existence of whales in the upper gree
0312  own that it is hardly possible to resist the EVIDENCE on this head in the case of the several tert
0325  e preserved and to survive. We have distinct EVIDENCE on this head, in the plants which are domina
0333  the same groups; and this by the concurrent EVIDENCE of our best palaeontologists seems frequentl
0336  e assuredly do find. We find, in short, such EVIDENCE of the slow and scarcely sensible mutation o
0357  pecies. It seems to me that we have abundant EVIDENCE of great oscillations of level in our contin
0366  a simple explanation of these facts. We have EVIDENCE of almost every conceivable kind, organic an
0368  ummits, we may almost conclude without other EVIDENCE, that a colder climate permitted their forme
0368  t homes; but I have met with no satisfactory EVIDENCE with respect to this intercalated slightly w
0373  ely extended. In Europe we have the plainest EVIDENCE of the cold period, from the western whores
0373  s. south of the equator, we have some direct EVIDENCE of former glacial action in New Zealand; and
0373  een published can be trusted, we have direct EVIDENCE of glacial action in the south eastern corne
0373  southernmost extremity, we have the clearest EVIDENCE of former glacial action, in huge boulders t
0373  pposite sides of the world. But we have good EVIDENCE in almost every case, that the epoch was inc
0374  geological period. We have, also, excellent EVIDENCE, that it endured for an enormous time, as me
0374  throughout the world. Without some distinct EVIDENCE to the contrary, we may at least admit as or
0376  orted as it is by a large body of geological EVIDENCE, that the whole world, or a large part of it
0379  gh we have reason to believe from geological EVIDENCE that the whole body of arctic shells underwe
0384  floods, without any change of level. We have EVIDENCE in the loess of the Rhine of considerable ch
0389  me other respects is not quite fair. We have EVIDENCE that the barren island of Ascension aborigin
0406  also, some reason to believe from geological EVIDENCE that organisms low in the scale within each
0429  erminated and utterly lost. And we have some EVIDENCE of aberrant forms having suffered severely f
0436  re. In monstrous plants, we often get direct EVIDENCE of the possibility of one organ being transf
0443  an equally early period. But we have little EVIDENCE on this head, indeed the evidence rather poi
0443  ave little evidence on this head, indeed the EVIDENCE rather points the other way; for it is notor
0444  t very early period of life; and some direct EVIDENCE from our domestic animals supports this view
0444  ted in the first chapter, that there is some EVIDENCE to render it probable, that at whatever age
0445  ll amount of proportional difference. As the EVIDENCE appears to me conclusive, that the several d
0463  ry collection of fossil remains afford plain EVIDENCE of the gradation and mutation of the forms o
0463  n of the forms of life? We meet with no such EVIDENCE, and this is the most obvious and forcible o
0466  er of generations. On the other hand we have EVIDENCE that variability, when it has once come into
0481  perfect that it would have afforded us plain EVIDENCE of the mutation of species, if they had unde
0222  .flava, which they rarely capture, and it was EVIDENT that they did at once distinguish them: for
0324  which now live here; and if this be so, it is EVIDENT that fossiliferous beds deposited at the pre
0333  on any other view. On this same theory, it is EVIDENT that the fauna of any great period in the ea
0417  indeed of an aggregate of characters is very EVIDENT in natural history. Hence, as has often been
0062  ability than W. Herbert, Dean of Manchester, EVIDENTLY the result of his great horticultural knowl
0173  th closely allied or representative species, EVIDENTLY filling nearly the same place in the natura
0216  e surrounding grass or thickets; and this is EVIDENTLY done for the instinctive purpose of allowin
0221  the slave makers haunting the same spot, and EVIDENTLY not in search of food; they approached and
0277  rafted together, though this latter capacity EVIDENTLY depends on widely different circumstances,
0411  groups under groups. This classification is EVIDENTLY not arbitrary like the grouping of the star
0201  a fair balance be struck between the good and EVIL caused by each part, each will be found on the
0490  and most wonderful have been, and are being, EVOLVED.
0009  nts have pollen utterly worthless, in the same EXACT condition as in the most sterile hybrids. Whe
0032  fs that this is not so in some cases, in which EXACT records have been kept; thus, to give a very
0161  ral characters. As, however, we never know the EXACT character of the common ancestor of a group,
0295  nward movement continues. In fact, this nearly EXACT balancing between the supply of sediment and
0315  occurred in innumerable instances) to fill the EXACT place of another species in the economy of na
0328  egions have not been deposited during the same EXACT periods, a formation in one region often corr
0380  to be solved. I do not pretend to indicate the EXACT lines and means of migration, or the reason w
0002  the Malay archipelago, has arrived at almost EXACTLY the same general conclusions that I have on
0010  has remarked, have apparently been exposed to EXACTLY the same conditions of life; and this shows
0067  rease be still further increased. We know not EXACTLY what the checks are in even one single insta
0071  the hand of man; but several hundred acres of EXACTLY the same nature had been enclosed twenty fiv
0076  one of our domestic plants or animals have so EXACTLY the same strength, habits, and constitution,
0078  st new competitors, though the climate may be EXACTLY the same as in its former home, yet the cond
0093  n can be detected. Having found a female tree EXACTLY sixty yards from a male tree, I put the stig
0103  odify all the individuals in each district in EXACTLY the same manner to the conditions of each; f
0114  ize, which had been exposed for many years to EXACTLY the same conditions, supported twenty specie
0156  d to more or less widely different habits, in EXACTLY the same manner: and as these so called gene
0166  ave bred true for centuries; species; and how EXACTLY parallel is the case with that of the horse g
0229  were situated, as far as the eye could judge, EXACTLY along the planes of imaginary intersection b
0229  the bees do not always succeed in working at EXACTLY the same rate from the opposite sides; for I
0240  of intermediate size have their ocelli in an EXACTLY intermediate condition. So that we here have
0256  m seed out of the same capsule and exposed to EXACTLY the same conditions. By the term systematic
0313  ons, the forms of life have seldom changed in EXACTLY the same degree. Yet if we compare any but t
0328  posited in two regions during nearly, but not EXACTLY the same period, we should find in both, fro
0328  the forms of life; but the species would not EXACTLY correspond; for there will have been a littl
0334  ian system. Put each fauna is not necessarily EXACTLY intermediate, as unequal intervals of time h
0349  c and the eastern shores of Africa, on almost EXACTLY opposite meridians of longitude. A third gre
0383  xtended to distant countries. But the case is EXACTLY the reverse. Not only have many fresh water
0437  ly different, are at an early stage of growth EXACTLY alike. How inexplicable are these facts on t
0439  rve for different purposes, are in the embryo EXACTLY alike. The embryos, also, of distinct animal
0446  other breeds, in all its proportions, almost EXACTLY as much as in the adult state. The two princ
0448  evelopment, and secondly from their following EXACTLY the same habits of life with their parents;
0466  ndered impotent, fails to reproduce offspring EXACTLY like the parent form. Variability is governe
0020  distinct races by crossing has been greatly EXAGGERATED. There can be no doubt that a race may be
0043  g of varieties has, I believe, been greatly EXAGGERATED, both in regard to animals and to those pl
0151  of the species is so great, that it is no EXAGGERATION to state that the varieties differ more fr
0490  ar of nature, from famine and death, the most EXALTED object which we are capable of conceiving, n
0164  trace in a bay horse. My son made a careful EXAMINATION and sketch for me of a dun Belgian cart ho
0298  the present day, with perfect specimens for EXAMINATION, two forms can seldom be connected by inte
0337  nd. Yet the most skilful naturalist from an EXAMINATION of the species of the two countries could
0361  tones being a valuable royal tax. I find on EXAMINATION, that when irregularly shaped stones are e
```

Page ***(Key Word)**

Page **(Key Word)**

```
0045 t there are not many men who will laboriously EXAMINE internal and important organs, and compare t
0072 ent to several points of view, whence I could EXAMINE hundreds of acres of the unenclosed heath, a
0224 i have arrived. He must be a dull man who can EXAMINE the exquisite structure of a comb, so beauti
0282 on or even each stratum. A man must for years EXAMINE for himself great piles of superimposed stra
0283 ve. With the mind thus impressed, let any one EXAMINE beds of conglomerate many thousand feet in t
0464 rent of any one or more species if we were to EXAMINE them ever so closely, unless we likewise pos
0053 ter having carefully read my manuscript, and  EXAMINED the tables, he thinks that the following sta
0093 es, nevertheless every female flower which I  EXAMINED had been effectually fertilised by the bees,
0135 eeding beetles are very often broken off; he  EXAMINED seventeen specimens in his own collection, a
0164 and a man, whom I can implicitly trust, has   EXAMINED for me a small dun Welch pony with three sho
0164 ped, that, as I hear from Colonel Poole, who  EXAMINED the breed for the Indian Government, a horse
0229 to go on working for a short time, and again  EXAMINED the cell, and I found that the rhombic plate
0239 er insects out of Europe have been carefully  EXAMINED. Mr. F. Smith has shown how surprisingly the
0299 ll the geological formations which have been  EXAMINED with any accuracy, excepting those of the Un
0302 ur geological formations have been carefully  EXAMINED: we forget that groups of species may elsewh
0148 and a fact with which I was much struck when  EXAMINING cirripedes; and of which many other instanc
0185 . on the other hand, the acutest observer by  EXAMINING the dead body of the water ouzel would neve
0230 h other. Now bees, as may be clearly seen by  EXAMINING the edge of a growing comb, do make a rough
0252 imself of the efficiency of insect agency by  EXAMINING the flowers of the more sterile kinds of hy
0290 uence. Scarcely any fact struck me more when  EXAMINING many hundred miles of the South American co
0294 n from the American to the European seas. In  EXAMINING the latest deposits of various quarters of
0294 nges. And in the distant future, a geologist  EXAMINING these beds, might be tempted to conclude th
0154 variable than generic. To explain by a simple EXAMPLE what is meant. If some species in a large ge
0154 more unusual circumstance. I have chosen this EXAMPLE because an explanation is not in this case a
0214 traces of the instincts of either parent: for EXAMPLE, Le Roy describes a dog, whose great grandfa
0228 ry can be tested by experiment. Following the EXAMPLE of Mr. Tegetmeier, I separated two combs, an
0250 ents I will here give only a single one as an EXAMPLE, namely, that every ovule in a pod of Crinum
0261 dvisable to explain a little more fully by an EXAMPLE what I mean by sterility being incidental on
0329 rms in the same sub order with ruminants: for EXAMPLE, he dissolves by fine gradations the apparen
0367 n due south and north. The Alpine plants, for EXAMPLE, of Scotland, as remarked by Mr. H. C. Watso
0376 arine productions, similar cases occur; as an EXAMPLE, I may quote a remark by the highest authori
0415 to separate Cnestis from Connarus. To give an EXAMPLE amongst insects, in one great division of th
0429 e genera Ornithorhynchus and Lepidosiren, for EXAMPLE, would not have been less aberrant had each
0462 sly have greatly facilitated migration. As an EXAMPLE, I have attempted to show how potent has bee
0029 he seeds of the same tree. Innumerable other  EXAMPLES could be given. The explanation, I think, is
0141 to be looked at as anomalies, but merely as   EXAMPLES of a very common flexibility of constitution
0303 throughout the world. I will now give a few   EXAMPLES to illustrate these remarks; and to show how
0320 re, as formerly explained and illustrated by  EXAMPLES, between the forms which are most like each
0228 re strip of wax: the bees instantly began to  EXCAVATE minute circular pits in it; and as they deep
0228 erve that wherever several bees had begun to  EXCAVATE these basins near together, they had begun t
0228 as soon as this occurred, the bees ceased to  EXCAVATE, and began to build up flat walls of wax on
0228 n. the bees instantly began on both sides to  EXCAVATE little basins near to each other, in the san
0235 other in a double layer, and to build up and  EXCAVATE the wax along the planes of intersection. Th
0228 the bottoms of the basins, if they had been  EXCAVATED to the same depth as in the former experime
0229 one side, where I suppose that the bees had  EXCAVATED too quickly, and convex on the opposed side
0230 uber's statement that the very first cell is  EXCAVATED out of a little parallel siced wall of wax,
0481 nd cliffs had been formed, and great valleys  EXCAVATED, by the slow action of the coast waves. The
0230 at the proper distance from each other, by   EXCAVATING at the same rate, and by endeavouring to ma
0095 nd insignificant cause, when applied to the   EXCAVATION of gigantic valleys or to the formation of
0095 ology has almost banished such views as the   EXCAVATION of a great valley by a single diluvial wave
0228 ffer this to happen, and they stopped their   EXCAVATION in due time; so that the basins, as soon as
0230 these details. We see how important a part    EXCAVATION plays in the construction of the cells; but
0052 deira. If a variety were to flourish so as to EXCEED in numbers the parent species, it would then
0300 ntemporaneous accumulation of sediment, would EXCEED the average duration of the same specific for
0298 y those which have had the widest range, far  EXCEEDING the limits of the known geological formatio
0057 of difference between the species is often   EXCEEDINGLY small. I have endeavoured to test this num
0002 it advisable to publish, with Mr. Wallace's   EXCELLENT memoir, some brief extracts from my manuscr
0003 way by his large stores of knowledge and his  EXCELLENT judgment. In considering the Origin of Spec
0036 reatly from the rock pigeon. Youatt gives an  EXCELLENT illustration of the effects of a course of
0037 endid fruit we should eat: though we owe our  EXCELLENT fruit, in some small degree, to their havin
0083 ost trifling importance: yet we hear from an  EXCELLENT horticulturist, Downing, that in the United
0165 uct of a zebra; and Mr. W. C. Martin, in his  EXCELLENT treatise on the horse, has given a figure o
0283 ons on this head by Hugh Miller, and by that  EXCELLENT observer Mr. Smith of Jordan Hill, are most
0294 absent in this particular deposit. It is an   EXCELLENT lesson to reflect on the ascertained amount
0374 the latest geological period. We have, also, EXCELLENT evidence, that it endured for an enormous t
0418 e one; and by none more clearly than by that  EXCELLENT botanist, Aug. St. Hilaire. If certain char
0032 eeders are strongly opposed to this practice, EXCEPT sometimes amongst closely allied sub breeds.
0058 or they cannot be distinguished from species, EXCEPT, firstly, by the discovery of intermediate li
0058 aracters of the forms which they connect; and EXCEPT, secondly, by a certain amount of difference,
0072 iterally I could not see a single Scotch fir, EXCEPT the old planted clumps. But on looking closel
0083 acters: nature cares nothing for appearances, EXCEPT in so far as they may be useful to any being.
0134 arger ground feeding birds seldom take flight EXCEPT to escape danger, I believe that the nearly w
0146 ally become winged through natural selection, EXCEPT in fruits which opened; so that the individua
0185 rarely or never go near the water; and no one EXCEPT Audubon has seen the frigate bird, which has
0195 the larger quadrupeds are actually destroyed (EXCEPT in some rare cases) by the flies, but they ar
0210 ssibly be produced through natural selection, EXCEPT by the slow and gradual accumulation of numer
0210 ts of animals having been but little observed EXCEPT in Europe and North America, and for no insti
0295 posits are usually barren of organic remains, EXCEPT near their upper or lower limits. It would se
0333 broken condition, we have no right to expect, EXCEPT in very rare cases, to fill up wide intervals
0442 osely related to its conditions of existence, EXCEPT when the embryo becomes at any period of life
0038 much difficulty, any deviation of structure  EXCEPTING such as is externally visible: and indeed h
0038 is internal. He can never act by selection,  EXCEPTING on variations which are first given to him
0120 the lines proceeding from (A) were removed,   EXCEPTING that from a1 to .a10. In the same way, for i
0123 iginal species (A) and (I), have all become,  EXCEPTING (F), extinct, and have left no descendants.
0266 ver, be confessed that we cannot understand,  EXCEPTING on vague hypotheses, several facts with res
0286 n as soft as chalk, would yield at this rate  EXCEPTING on the most exposed coasts; though no doubt
```

Page **(Key Word)**

Page **(Key Word)**

```
0292  rea and number of inhabitants will decrease (EXCEPTING the productions on the shores of a continen
0299   which have been examined with any accuracy, EXCEPTING those of the United States of America. I fu
0347  trial productions of the New and Old Worlds, EXCEPTING in the northern parts, where the land almos
0403  in every lake and marsh. For Alpine species, EXCEPTING in so far as the same forms, chiefly of pla
0404  s with the inhabitants of lakes and marshes, EXCEPTING in so far as great facility of transport ha
0414  e life depends, are of little signification, EXCEPTING in the first main divisions; whereas the or
0441  uth, stomach, or other organs of importance, EXCEPTING for reproduction. We are so much accustomed
0462  ld remember that we have no right to expect (EXCEPTING in rare cases) to discover directly connect
0009  ry pretty freely under confinement, with the EXCEPTION of the plantigrades or bear family; whereas
0064  r the world would not hold them. There is no EXCEPTION to the rule that every organic being natura
0071  ing whatever else having been done, with the EXCEPTION that the land had been enclosed, so that ca
0093  s, under the microscope, and on all, without EXCEPTION, there were pollen grains; and on some a pr
0254  m several wild stocks: yet, with perhaps the EXCEPTION of certain indigenous domestic dogs of Sout
0288  rs: they are all strictly littoral, with the EXCEPTION of a single Mediterranean species, which in
0289  ng to either of these vast periods, with one EXCEPTION discovered by Sir C. Lyell in the carbonife
0313  form never reappears. The strongest apparent EXCEPTION to this latter rule, is that of the so call
0793  land of New Zealand: but I suspect that this EXCEPTION (if the information be correct) may be expl
0393  bited by a wolf like fox, come nearest to an EXCEPTION; but this group cannot be considered as oce
0446  d in the young. But there was one remarkable EXCEPTION to this rule, for the young of the short fa
0259  mediate in structure between their parents, EXCEPTIONAL and abnormal individuals sometimes are bor
0316  l to my views. But such cases are certainly EXCEPTIONAL: the general rule being a gradual increase
0352  e fact is given as something remarkable and EXCEPTIONAL. The capacity of migrating across the sea
0354  ld be hopelessly tedious to discuss all the EXCEPTIONAL cases of the same species, now living at d
0266  ty in those hybrids which occasionally and EXCEPTIONALLY resemble closely either pure parent. Nor
0009  whereas, carnivorous birds, with the rarest EXCEPTIONS, hardly ever lay fertile eggs. Many exotic
0016  o which they are nearest allied. With these EXCEPTIONS (and with that of the perfect fertility of
0046  ic in one country seem to be, with some few EXCEPTIONS, polymorphic in other countries, and likewi
0259  of their pure parent species, are with rare EXCEPTIONS extremely sterile. So again amongst hybrids
0313  y than those that are low: though there are EXCEPTIONS to this rule. The amount of organic change,
0316  us. I am aware that there are some apparent EXCEPTIONS to this rule, but the exceptions are surpri
0316  e apparent exceptions to this rule, but the EXCEPTIONS are surprisingly few, so few, that E. Forbe
0334  ucceeding faunas, that certain genera offer EXCEPTIONS to the rule. For instance, mastodons and el
0353  that we are reduced to consider whether the EXCEPTIONS to continuity of range are so numerous and
0399  st continent, or of other near islands. The EXCEPTIONS are few, and most of them can be explained.
0406  d marsh productions being related (with the EXCEPTIONS before specified) to those on the surroundi
0409  p of species is continuous in time: for the EXCEPTIONS to the rule are so few, that they may fairl
0410  a group of species, is continuous: and the EXCEPTIONS, which are not rare, may, as I have attempt
0410  higher; but there are in both cases marked EXCEPTIONS to the rule. On my theory these several rel
0007  this variability may be partly connected with EXCESS of food. It seems pretty clear that organic b
0070  in our fields, because the seeds are in great EXCESS compared with the number of birds which feed
0147  if nourishment flows to one part or organ in EXCESS, it rarely flows, at least in excess, to anot
0147  organ in excess, it rarely flows, at least in EXCESS, to another part; thus it is difficult to get
0147  rawal of nutriment from one part owing to the EXCESS of growth in another and adjoining part. I su
0255  e seen, in certain abnormal cases, even to an EXCESS of fertility, beyond that which the plant's o
0259  even to fertility under certain conditions in EXCESS. That their fertility, besides being eminentl
0265  f general health, and is often accompanied by EXCESS of size or great luxuriance. In both cases, t
0453  upposing that they serve to excrete matter in EXCESS, or injurious to the system: but can we suppo
0120  n each horizontal line in our diagram to be EXCESSIVELY small, these three forms may still be only
0136   certain large groups of beetles; elsewhere EXCESSIVELY numerous, and which groups have habits of
0153  reat and long continued within a period not EXCESSIVELY remote, we might, as a general rule, expec
0231  l from ten to twenty times thicker than the EXCESSIVELY thin finished wall of the cell, which will
0274  licable to animals; but the subject is here EXCESSIVELY complicated, partly owing to the existence
0297  s distinct species. It is notorious on what EXCESSIVELY slight differences many palaeontologists h
0300  of the archipelago would be preserved in an EXCESSIVELY imperfect manner in the formations which w
0431  characters in common, we can understand the EXCESSIVELY complex and radiating affinities by which
0469  ns in any way useful to beings, under their EXCESSIVELY complex relations of life, would be preser
0267  n by farmers and gardeners in their frequent EXCHANGES of seed, tubers, etc, from one soil or clim
0021  crop, which it glories in inflating, may well EXCITE astonishment and even laughter. The turbit ha
0003   structure and coadaptation which most justly EXCITES our admiration. Naturalists continually refe
0108  tly retarded by free intercrossing. Many will EXCLAIM that these several causes are amply sufficie
0346   almost suffice to prove its truth: for if we EXCLUDE the northern parts where the circumpolar lan
0055  rom only one to four species, are absolutely EXCLUDED from the tables. These facts are of plain si
0187  d by pigment, and which properly act only by EXCLUDING lateral pencils of light, are convex at the
0247  ially fertilised with their own pollen; and (EXCLUDING all cases such as the Leguminosae, in which
0273  first generation are descended from species (EXCLUDING those long cultivated) which have not had t
0278  reproductive system. In all other respects, EXCLUDING fertility, there is a close general resembl
0393  e, free from doubt, of a terrestrial mammal (EXCLUDING domesticated animals kept by the natives) i
0065  hing square leagues of surface almost to the EXCLUSION of all other plants, have been introduced f
0006   natural Selection has been the main but not EXCLUSIVE means of modification. Variation Under Dome
0201  e of any one species had been formed for the EXCLUSIVE good of another species, it would annihilat
0205   will produce nothing in one species for the EXCLUSIVE good or injury of another: though it may we
0210  s far as we can judge, been produced for the EXCLUSIVE good of others. One of the strongest instan
0211  imal in the world performs an action for the EXCLUSIVE good of another of a distinct species, yet
0223  land and England the slaves seem to have the EXCLUSIVE care of the larvae, and the masters alone g
0243  : that no instinct has been produced for the EXCLUSIVE good of other animals, but that each animal
0013  or to one sex alone, more commonly but not EXCLUSIVELY to the like sex. It is a fact of some litt
0013  omestic breeds are often transmitted either EXCLUSIVELY, or in a much greater degree, to males alo
0069  te deserts, the struggle for life is almost EXCLUSIVELY with the elements. That climate acts in ma
0087  jaws possessed by certain insects, and used EXCLUSIVELY for opening the cocoon, or the hard tip to
0133  re often brassy or lurid. Plants which live EXCLUSIVELY on the sea side are very apt to have flesh
0175  the inhabitants of any country by no means EXCLUSIVELY depends on insensibly changing physical co
0182  d had used their incipient organs of flight EXCLUSIVELY, as far as we know, to escape being devour
0183  f its structure, for its changed habits, or EXCLUSIVELY for one of its several different habits. B
0183  insects which now feed on exotic plants, or EXCLUSIVELY on artificial substances. Of diversified h
0187  pecies has been perfected, we ought to look EXCLUSIVELY to its lineal ancestors; but this is scarc
0191  ed a swimbladder into a lung, or organ used EXCLUSIVELY for respiration. I can, indeed, hardly dou
```

Page **(Key Word)**

Page **************************************(Key Word)**************************************

0200 produce any modification in any one species EXCLUSIVELY for the good of another species; though th
0289 kilful geologist, if his attention had been EXCLUSIVELY confined to these large territories, would
0381 quite distinct species, belonging to genera EXCLUSIVELY confined to the south, at these and other
0384 is hardly a single group of fishes confined EXCLUSIVELY to fresh water, so that we may imagine tha
0411 simple signification, if one group had been EXCLUSIVELY fitted to inhabit the land, and another th
0361 p in my garden 12 kinds of seeds, out of the EXCREMENT of small birds, and these seemed perfect, a
0362 the seeds in pellets or passed them in their EXCREMENT; and several of these seeds retained their
0387 germination, when rejected in pellets or in EXCREMENT, many hours afterwards. When I saw the grea
0091 now take a more complex case. Certain plants EXCRETE a sweet juice, apparently for the sake of el
0211 l, i felt sure that the aphides would want to EXCRETE. I watched them for some time through a lens
0211 ore probably the aphides do not instinctively EXCRETE for the sole good of the ants. Although I do
0453 ntary organs, by supposing that they serve to EXCRETE matter in excess, or injurious to the system
0454 from unknown laws of growth, but in order to EXCRETE horny matter, as that the rudimentary nails
0092 suppose a little sweet juice or nectar to be EXCRETED by the inner bases of the petals of a flower
0092 d the largest glands or nectaries, and which EXCRETED most nectar, would be oftenest visited by in
0211 em for some time through a lens, but not one EXCRETED; I then tickled and stroked them with a hair
0211 the ants do with their antennae; but not one EXCRETED. Afterwards I allowed an ant to visit them,
0211 ennae, immediately lifted up its abdomen and EXCRETED a limpid drop of sweet juice, which was eage
0239 a sort of honey, supplying the place of that EXCRETED by the aphides, or the domestic cattle as th
0190 the alimentary canal respires, digests, and EXCRETES in the larva of the dragonfly and in the fis
0092 seedlings would probably inherit the nectar EXCRETING power. Those individual flowers which had t
0210 of aphides voluntarily yielding their sweet EXCRETION to ants: that they do so voluntarily, the f
0211 and not the result of experience. But as the EXCRETION is extremely viscid, it is probably a conve
0453 to the rapidly growing embryonic calf by the EXCRETION of precious phosphate of lime? When a man's
0205 ough it may well produce parts, organs, and EXCRETIONS highly useful or even indispensable, or hig
0001 pursued the same object. I hope that I may be EXCUSED for entering on these personal details, as I
0220 eptical frame of mind, as any one may well be EXCUSED for doubting the truth of so extraordinary a
0095 re that this doctrine of natural selection, EXEMPLIFIED in the above imaginary instances, is open
0236 perfect females, would have been far better EXEMPLIFIED by the hive bee. If a working ant or other
0452 ally differs much. This latter fact is well EXEMPLIFIED in the state of the wings of the female mo
0407 s of the same species, extremely grave. As EXEMPLIFYING the effects of climatal changes on distrib
0083 beaked pigeon on the same food; he does not EXERCISE a long backed or long legged quadruped in an
0242 which are in any manner profitable, without EXERCISE or habit having come into play. For no amoun
0242 abit having come into play. For no amount of EXERCISE, or habit, or volition, in the utterly steri
0447 or wings. Whatever influence long continued EXERCISE or use on the one hand, and disuse on the ot
0083 she tends. Every selected character is fully EXERCISED by her; and the being is placed under well
0083 many climates in the same country; he seldom EXERCISES each selected character in some peculiar an
0198 y would probably affect the hind limbs from EXERCISING them more, and possibly even the form of th
0255 he stigma of a plant of a distinct family, it EXERTS no more influence than so much inorganic dust
0161 species of the same genus would occasionally EXHIBIT reversions to lost ancestral characters. As,
0214 ously blended together, and for a long period EXHIBIT traces of the instincts of either parent: fo
0040 ost fleeting of characters, have lately been EXHIBITED as distinct at our poultry shows. I think t
0315 si consequently the amount of organic change EXHIBITED by the fossils embedded in consecutive form
0416 bones of the leg, are highly serviceable in EXHIBITING the close affinity between Ruminants and Pa
0428 body and fin like limbs serve as characters EXHIBITING true affinity between the several members o
0430 n a member belonging to one group of animals EXHIBITS an affinity to a quite distinct group, this
0023 e supposed aboriginal stocks must either still EXIST in the countries where they were originally d
0036 e of being quite different varieties. If there EXIST savages so barbarous as never to think of the
0047 r to suppose either that they do now somewhere EXIST, or may formerly have existed: and here a wid
0052 terminate the parent species; or both might co EXIST, and both rank as independent species. But we
0065 tock every station in which they could any how EXIST, and that the geometrical tendency to increas
0070 such cases, we may believe, that a plant could EXIST only where the conditions of its life were so
0070 of its life were so favourable that many could EXIST together, and thus save each other from utter
0104 ven at rare intervals, will be great. If there EXIST organic beings which never intercross, unifor
0107 llations of level, and which consequently will EXIST for long periods in a broken condition, will
0107 into large separate islands, there will still EXIST many individuals of the same species on each
0122 ot come into competition, both may continue to EXIST. If then our diagram be assumed to represent
0126 class, we can understand how it is that there EXIST but very few classes in each main division of
0170 from their parents, and a cause for each must EXIST, it is the steady accumulation, through natur
0176 . the intermediate variety, consequently, will EXIST in lesser numbers from inhabiting a narrow an
0176 ected into two distinct species, the two which EXIST in larger numbers from inhabiting larger area
0177 n to seize or, than will the rarer forms which EXIST in lesser numbers. Hence, the more common for
0178 natural selection, so that they will no longer EXIST in a living state. Thirdly, when two or more
0178 cies, and likewise of acknowledged varieties), EXIST in the intermediate zones in lesser numbers t
0179 that within the same group carnivorous animals EXIST having every intermediate grade between truly
0182 rades of the structure will seldom continue to EXIST to the present day, for they will have been s
0184 if better adapted competitors did not already EXIST in the country, I can see no difficulty in a
0186 being useful to its possessor, can be shown to EXIST; if further, the eye does vary ever so slight
0187 fundamentally different lines, can be shown to EXIST, until we reach a moderately high stage of pe
0189 an find out no such case. No doubt many organs EXIST of which we do not know the transitional grad
0203 ssigned, the intermediate variety will usually EXIST in lesser numbers than the two forms which it
0234 rmine, the numbers of a humble bee which could EXIST in a country; and let us further suppose that
0235 which no intermediate gradations are known to EXIST; cases of instinct of apparently such triflin
0255 ny curious ways this gradation can be shown to EXIST; but only the barest outline of the facts can
0302 und beneath a certain stage, that they did not EXIST before that stage. We continually forget how
0309 d on the other hand, that where continents now EXIST, large tracts of land have existed, subjected
0310 ional links between the many species which now EXIST or have existed; the sudden manner in which w
0320 s, and yet to marvel greatly when it ceases to EXIST, is much the same as to admit that sickness i
0355 organic beings which never intercross (if such EXIST), the species, on my theory, must have descen
0365 s, where the Alpine species could not possibly EXIST, is one of the most striking cases known of t
0374 ean genera were found by Gardner, which do not EXIST in the wide intervening hot countries. So on
0378 common to Tierra del Fuego and to Europe still EXIST in North America, which must have lain on the
0379 n of the northern or southern hemispheres, now EXIST in their new homes as well marked varieties o
0393 across the sea, and therefore why they do not EXIST on any oceanic island. But why, on the theory
0459 instinct, which we may consider, either do now EXIST or could have existed, each good of its kind,
0463 mediate varieties which will at first probably EXIST in the intermediate zones, will be liable to

Page **************************************(Key Word)**************************************

exist

Page ***(Key Word)***

```
0463  er rate than the intermediate varieties, which EXIST in lesser numbers; so that the intermediate v
0478  s, some identical species common to both still EXIST. Wherever many closely allied yet distinct sp
0016  y distinct species. If any marked distinction EXISTED between domestic races and species, this sou
0018  iently civilized to have manufactured pottery EXISTED in the valley of the Nile thirteen or fourte
0018  who possess a semi domestic dog, may not have EXISTED in Egypt? The whole subject must, I think, r
0019  wild prototype. At this rate there must have EXISTED at least a score of species of wild cattle,
0019  tain. One author believes that there formerly EXISTED in Great Britain eleven wild species of shee
0019  aniel, etc., so unlike all wild Canidae, ever EXISTED freely in a state of nature? It has often be
0047  do now somewhere exist, or may formerly have EXISTED; and here a wide door for the entry of doubt
0059  erstand these analogies, if species have once EXISTED as varieties, and have thus originated: wher
0106  cillations of level, will often have recently EXISTED in a broken condition, so that the good effe
0107  o spread widely. For the area will first have EXISTED as a continent, and the inhabitants, at this
0143  atural selection: thus a family of stags once EXISTED with an antler only on one side; and if this
0154  the case of the wing of the bat, it must have EXISTED, according to my theory, for an immense peri
0172  eory innumerable transitional forms must have EXISTED, why do we not find them embedded in countle
0174  rieties in each region, though they must have EXISTED there, and may be embedded there in a fossil
0174  , marine areas now continuous must often have EXISTED within recent times in a far less continuous
0176  g two other varieties together have generally EXISTED in lesser numbers than the forms which they
0177  edi and thus the two breeds, which originally EXISTED in greater numbers, will come into close con
0178  econdly, areas now continuous must often have EXISTED within the recent period in isolated portion
0178  s and their common parent, must formerly have EXISTED in each broken portion of the land, but thes
0179  the same group together, must assuredly have EXISTED; but the very process of natural selection c
0181  ficulty in supposing that such links formerly EXISTED, and that each had been formed by the same s
0182  ntured to have surmised that birds might have EXISTED which used their wings solely as flappers, l
0183  dition will always be less, from their having EXISTED in lesser numbers, than in the case of speci
0189  could be demonstrated that any complex organ EXISTED, which could not possibly have been formed b
0192  ranchiae in this latter family had originally EXISTED as organs for preventing the ova from being
0197  atter remarks. If green woodpeckers alone had EXISTED, and we did not know that there were many bl
0202  at the sting of the bee, as having originally EXISTED in a remote progenitor as a boring and serra
0204  y cautious in concluding that none could have EXISTED, for the homologies of many organs and their
0206  ry. For in the larger country there will have EXISTED more individuals, and more diversified forms
0280  f intermediate varieties, which have formerly EXISTED on the earth, be truly enormous. Why then is
0280  diate forms must, on my theory, have formerly EXISTED. I have found it difficult, when looking at
0280  ll the intermediate varieties which have ever EXISTED, we should have an extremely close series be
0281  we have no reason to suppose that links ever EXISTED directly intermediate between them, but betw
0281  this case direct intermediate links will have EXISTED between them. But such a case would imply th
0287  this area has undergone, the surface may have EXISTED for millions of years as land, and thus have
0288  yet it is now known that the genus Chthamalus EXISTED during the chalk period. The molluscan genus
0291  erfect record of the forms of life which then EXISTED; or, sediment may be accumulated to any thic
0293  to infer that it had not elsewhere previously EXISTED. So again when we find a species disappearin
0296  diate gradations which must on my theory have EXISTED between them, but abrupt, though perhaps ver
0302  hat groups of species may elsewhere have long EXISTED and have slowly multiplied before they invad
0304  ances, I inferred that had sessile cirripedes EXISTED during the secondary periods, they would cer
0305  e now positively know that sessile cirripedes EXISTED during the secondary period; and these cirri
0308  and the United States during these intervals EXISTED as dry land, or as a submarine surface near
0308  s, neither continents nor continental islands EXISTED where our oceans now extend; for had they ex
0308  ted where our oceans now extend: for had they EXISTED there, palaeozoic and secondary formations w
0309  ntinents now exist, large tracts of land have EXISTED, subjected no doubt to great oscillations of
0309  nt to the silurian epoch, continents may have EXISTED where oceans are now spread out; and clear a
0309  pread out; and clear and open oceans may have EXISTED where our continents now stand. Nor should w
0310  ween the many species which now exist or have EXISTED; the sudden manner in which whole groups of
0316  s, so long must its members have continuously EXISTED, in order to have generated either new and m
0316  lingula, for instance, must have continuously EXISTED by an unbroken succession of generations, fr
0318  don, and other extinct monsters, which all co EXISTED with still living shells at a very late geol
0343  infinitely numerous organisms which must have EXISTED long before the first bed of the Silurian sy
0358  of sunken continents; if they had originally EXISTED as mountain ranges on the land, some at leas
0369  with ancient Alpine species, which must have EXISTED on the mountains before the commencement of
0379  n the north, and to the northern forms having EXISTED in their own homes in greater numbers, and h
0396  s. but the probability of many islands having EXISTED as halting places, of which not a wreck now
0420  and these have descended from a species which EXISTED at an unknown anterior period. Species of th
0455  very part of the organisation, which has long EXISTED, to be inherited, we can understand, on the
0459  y consider, either do now exist or could have EXISTED, each good of its kind, that all organs and
0462  inable number of intermediate forms must have EXISTED, linking together all the species in each gr
0464  ons of countless species which certainly have EXISTED. We should not be able to recognise a specie
0469  manent varieties, and that each species first EXISTED as a variety, we can see why it is that no l
0470  ed, but are intelligible if all species first EXISTED as varieties. As each species tends by its g
0488  when very few forms of the simplest structure EXISTED, the rate of change may have been slow in an
0004  iation. In the next chapter the Struggle for EXISTENCE amongst all organic beings throughout the w
0005  there is a frequently recurring struggle for EXISTENCE, it follows that any being, if it vary howe
0020  es by this process, we must admit the former EXISTENCE of the most extreme forms; as the Italian g
0031  orm perfect in itself, and then had given it EXISTENCE. That most skilful breeder, Sir John Sebrig
0053  usions cannot be avoided to the struggle for EXISTENCE, divergence of character, and other questio
0060  rdinate to groups. Chapter III. Struggle For EXISTENCE. Bears on natural selection. The term used
0060  minary remarks, to show how the struggle for EXISTENCE bears on Natural Selection. It has been see
0060  british plants are entitled to hold, if the EXISTENCE of any well marked varieties be admitted. B
0060  l marked varieties be admitted. But the mere EXISTENCE of individual variability and of some few w
0062  uss in a little more detail the struggle for EXISTENCE. In my future work this subject shall be tr
0062  uld premise that I use the term Struggle for EXISTENCE in a large and metaphorical sense, includin
0063  the missletoe is disseminated by birds, its EXISTENCE depends on birds; and it may metaphorically
0063  nience sake the general term of struggle for EXISTENCE. A struggle for existence inevitably follow
0063  rm of struggle for existence. A struggle for EXISTENCE inevitably follows from the high rate at wh
0063  , there must in every case be a struggle for EXISTENCE, either one individual with another of the
0068  to be quite independent of the struggle for EXISTENCE; but in so far as climate chiefly acts in r
0072  we see that cattle absolutely determine the EXISTENCE of the Scotch fir; but in several parts of
0072  ral parts of the world insects determine the EXISTENCE of cattle. Perhaps Paraguay offers the most
0074  n determining the average number or even the EXISTENCE of the species. In some cases it can be sho
```

Page ***(Key Word)***

Page **(Key Word)**

```
0075 ctly be said to struggle with each other for  EXISTENCE, as in the case of locusts and grass feedin
0080 ll organic beings. How will the struggle for  EXISTENCE, discussed too briefly in the last chapter,
0088 lection. This depends, not on a struggle for  EXISTENCE, but on a struggle between the males for po
0108 ry undergoing modification of some kind. The  EXISTENCE of such places will often depend on physica
0110 have seen in the chapter on the Struggle for  EXISTENCE that it is the most closely allied forms, v
0125 vantage over other forms in the struggle for  EXISTENCE, it will chiefly act on those which already
0127 ngs to each other and to their conditions of  EXISTENCE, causing an infinite diversity in structure
0171 aw of Unity of Type and of the Conditions of  EXISTENCE embraced by the theory of Natural Selection
0179 links. Consequently evidence of their former  EXISTENCE could be found only amongst fossil remains,
0186 anguage. He who believes in the struggle for  EXISTENCE and in the principle of natural selection,
0195 case, for we know that the distribution and  EXISTENCE of cattle and other animals in South Americ
0201 me country with which it has to struggle for  EXISTENCE. And we see that this is the degree of perf
0206 t laws, Unity of Type, and the Conditions of  EXISTENCE. By unity of type is meant that fundamental
0206 of descent. The expression of conditions of  EXISTENCE, so often insisted on by the illustrious Cu
0206 hence, in fact, the law of the Conditions of  EXISTENCE is the higher law: as it includes, through
0235 est chance of succeeding in the struggle for  EXISTENCE. No doubt many instincts of very difficult
0269 me impossible to resist the evidence of the  EXISTENCE of a certain amount of sterility in the few
0274 excessively complicated, partly owing to the  EXISTENCE of secondary sexual characters: but more es
0304 al, published in 1858, clear evidence of the  EXISTENCE of whales in the upper greensand, some time
0305 . Lately, Professor Pictet has carried their  EXISTENCE one sub stage further back: and some palaeo
0307 t azoic rocks, probably indicates the former  EXISTENCE of life at these periods. But the difficult
0316 appear after it has once disappeared; or its  EXISTENCE, as long as it lasts, is continuous. I am a
0322 the many complex contingencies, on which the  EXISTENCE of each species depends. If we forget for a
0326 then their triumphant course, or even their  EXISTENCE, would cease. We know not at all precisely
0334 ities and then according to their periods of  EXISTENCE, do not accord in arrangement. The species
0344 hich are their inferiors in the struggle for  EXISTENCE. Hence, after long intervals of time, the p
0354 most striking classes of facts; namely, the  EXISTENCE of the same species on the summits of dista
0354 ted by hundreds of miles of open sea. If the  EXISTENCE of the same species at distant and isolated
0355 concludes, that every species has come into  EXISTENCE coincident both in space and time with a pr
0357 g oceanic islands. I freely admit the former  EXISTENCE of many islands, now buried beneath the sea
0368 ning tracts, since become too warm for their  EXISTENCE. If the climate, since the Glacial period,
0381 argely concerned in their dispersal. But the  EXISTENCE of several quite distinct species, belongin
0404 under different physical conditions, and the  EXISTENCE at remote points of the world of other spec
0411 grouping of the stars in constellations. The  EXISTENCE of groups would have been of simple signifi
0415 differ in having one or more ovaria, in the  EXISTENCE or absence of albumen, in the imbricate or
0423 we have seen with pigeons. The origin of the  EXISTENCE of groups subordinate to groups, is the sam
0426 ve life under the most diverse conditions of  EXISTENCE, are generally the most constant, we attach
0433 lection, which results from the struggle for  EXISTENCE, and which almost inevitably induces extinc
0440 ve no direct relation to their conditions of  EXISTENCE. We cannot, for instance, suppose that in t
0442 t being closely related to its conditions of  EXISTENCE, except when the embryo becomes at any peri
0448 this case, it would be indispensable for the  EXISTENCE of the species, that the child should be mo
0455 with modification, we may conclude that the  EXISTENCE of organs in a rudimentary, imperfect, and
0459 e, and, lastly, that there is a struggle for  EXISTENCE leading to the preservation of each profita
0460 and one of the most curious of these is the  EXISTENCE of two or three defined castes of workers o
0467 during the constantly recurrent Struggle for  EXISTENCE, we see the most powerful and ever acting m
0467 acting means of selection. The struggle for  EXISTENCE inevitably follows from the high geometrica
0469 fferences, all naturalists have admitted the  EXISTENCE of varieties, which they think sufficiently
0478 theory of independent acts of creation. The  EXISTENCE of closely allied or representative species
0480 as to play its full part in the struggle for  EXISTENCE, and will thus have little power of acting
0485 e rejecting the consideration of the present  EXISTENCE of intermediate gradations between any two
0017 rage as largely as the parent species of our  EXISTING domesticated productions have varied. In the
0018 ble, perhaps are identical with, those still  EXISTING. Even if this latter fact were found more st
0025 rked like the rock pigeon, although no other  EXISTING species is thus coloured and marked, so that
0034 ew strain or sub breed, superior to anything  EXISTING in the country. But, for our purpose, a kind
0036 rriers and tumblers with these breeds as now  EXISTING in Britain, India, and Persia, we can, I thi
0036 rds of fifty years. There is not a suspicion  EXISTING in the mind of any one at all acquainted wit
0106 ely complex from the large number of already  EXISTING species: and if some of these many species b
0124 me degree, intermediate in character between  EXISTING groups: and we can understand this fact, for
0129 h. the green and budding twigs may represent  EXISTING species: and those produced during each form
0162 l, what cases are reversions to an anciently  EXISTING character and what are new but analogous var
0174 ut the possibility of intermediate varieties  EXISTING in the intermediate zones. By changes in the
0176 hey originally linked together. For any form  EXISTING in lesser numbers would, as already remarked
0176 reater chance of being exterminated than one  EXISTING in large numbers: and in this particular cas
0176 iable to the inroads of closely allied forms  EXISTING on both sides of it. But a far more importan
0177 in a narrow and intermediate zone. For forms  EXISTING in larger numbers will always have a better
0179 the forms which they connect: for these from  EXISTING in greater numbers will, in the aggregate, p
0187 altered or little altered condition. Amongst  EXISTING Vertebrata, we find but a small amount of gr
0200 seful as they now are to the most aquatic of  EXISTING birds. So we may believe that the progenitor
0204 ing the course of further modification, from  EXISTING in greater numbers, will have a great advant
0241 distinctly defined castes of sterile workers  EXISTING in the same nest, both widely different from
0279 o, to show that intermediate varieties, from  EXISTING in lesser numbers than the forms which they
0294 s everywhere been noted, that some few still  EXISTING species are common in the deposit, but have
0299 rch, though it has added numerous species to  EXISTING and extinct genera, and has made the interva
0304 edes, I have stated that, from the number of  EXISTING and extinct tertiary species: from the extra
0305 een the progenitors of our many tertiary and  EXISTING species. The case most frequently insisted o
0305 d. this group includes the large majority of  EXISTING species. Lately, Professor Pictet has carrie
0306 guments which have convinced me that all the  EXISTING species of the same group have descended fro
0308 erived, occurred in the neighbourhood of the  EXISTING continents of Europe and North America. But
0308 an open and unfathomable sea. Looking to the  EXISTING oceans, which are thrice as extensive as the
0313 a striking instance of a similar fact, in an  EXISTING crocodile associated with many strange and l
0313 f an older formation, and then allow the pre  EXISTING fauna to reappear: but Lyell's explanation,
0316 ncredible that a fantail, identical with the  EXISTING breed, could be raised from any other specie
0319 d that the tooth, though so like that of the  EXISTING horse, belonged to an extinct species. Had t
0319 on the supposition of the fossil horse still  EXISTING as a rare species, we might have felt certai
0324 list would hardly be able to say whether the  EXISTING or the pleistocene inhabitants of Europe res
0324 highly competent observers believe that the  EXISTING productions of the United States are more cl
```

Page **(Key Word)**

Page **(Key Word)**

```
0326  nd that severe competition with many already  EXISTING  forms, would be highly favourable, as would
0329  , all fossils can be classed either in still  EXISTING  groups, or between them. That the extinct fo
0329  e help to fill up the wide intervals between  EXISTING  genera, families, and orders, cannot be disp
0329  showing how extinct animals fall in between  EXISTING  groups. Cuvier ranked the Ruminants and Pach
0331  line may be considered as extinct. The three  EXISTING  genera, a14, q14, p14, will form a small fam
0331  at the most ancient fossils differ most from  EXISTING  forms. We must not, however, assume that div
0332  ral points low down in the series, the three  EXISTING  families on the uppermost line would be rend
0333  he fossils. If, for instance, we suppose the  EXISTING  genera of the two families to differ from ea
0333  ther in some of their characters than do the  EXISTING  members of the same groups: and this by the
0335  e who is acquainted with the distribution of  EXISTING  species over the globe, will not attempt to
0337  the world were put into competition with the  EXISTING  inhabitants of the same or some other quarte
0342  ies, connecting together all the extinct and  EXISTING  forms of life by the finest graduated steps.
0344  nct forms often tend to fill up gaps between  EXISTING  forms, sometimes blending two groups previou
0345  rms are seldom directly intermediate between  EXISTING  forms: but are intermediate only by a long a
0355  coincident both in space and time with a pre  EXISTING  closely allied species. And I now know from
0357  of level, have occurred within the period of  EXISTING  organisms. Edward Forbes insisted that all t
0357  us geographical changes within the period of  EXISTING  species. It seems to me that we have abundan
0358  y, united with each other, and with the many  EXISTING  oceanic islands. Several facts in distributi
0372  hink, we can understand the presence of many  EXISTING  and tertiary representative forms on the eas
0373  d was a gigantic moraine, left far below any  EXISTING  glacier. Further south on both sides of the
0382  sport, and by the aid, as halting places, of  EXISTING  and now sunken islands, and perhaps at the c
0384  period, and when the surface was peopled by  EXISTING  land and fresh water shells. The wide differ
0389  the belief that within the recent period all  EXISTING  islands have been nearly or quite joined to
0409  odified, some developed in great force, some  EXISTING  in scanty numbers, in the different great ge
0421  lso broken up into two families. Nor can the  EXISTING  species, descended from A, be ranked in the
0421  or those from I, with the parent I. But the  EXISTING  genus F14 may be supposed to have been but s
0421  st, as sometimes seems to have occurred with  EXISTING  organisms. All the descendants of the genus
0429  s in some slight degree intermediate between  EXISTING  groups. A few old and intermediate parent fo
0430  some degree intermediate with respect to all  EXISTING  Marsupials: or that both Rodents and Marsupi
0430  it will not be specially related to any one  EXISTING  Marsupial, but indirectly to all or nearly a
0432  by steps as fine as those between the finest  EXISTING  varieties, nevertheless a natural classifica
0435  ll mammals, had its limbs constructed on the  EXISTING  general pattern, for whatever purpose they s
0449  semble the embryos of their descendants, our  EXISTING  species. Agassiz believes this to be a law o
0462  er in an inextricable chaos? With respect to  EXISTING  forms, we should remember that we have no ri
0463  d forms on either hand: and the latter, from  EXISTING  in greater numbers, will generally be modifi
0464  erfection of the geological record. Numerous  EXISTING  doubtful forms could be named which are prob
0476  t stands in some degree intermediate between  EXISTING  and allied groups. Recent forms are generall
0483  tend to fill up very wide intervals between  EXISTING  orders. Organs in a rudimentary condition pl
0487  and exterminated by slowly acting and still  EXISTIMG  causes; and not by miraculous acts of creati
0175  pecies on the confines of its range, where it  EXISTS  in lessened numbers, will, during fluctuation
0176  dvantage over the intermediate variety, which  EXISTS  in smaller numbers in a narrow and intermedia
0179  and strictly terrestrial habits: and as each  EXISTS  by a struggle for life, it is clear that each
0253  or profit, where neither pure parent species  EXISTS, they must certainly be highly fertile. A doc
0357  ust be admitted that scarcely a single island  EXISTS  which has not recently been united to some co
0393  ue. I have, however, been assured that a frog  EXISTS  on the mountains of the great island of New Z
0009  xceptions, hardly ever lay fertile eggs. Many  EXOTIC  plants have pollen utterly worthless, in the
0073  hall hereafter have occasion to show that the  EXOTIC  Lobelia fulgens, in this part of England, is
0183  of the many British insects which now feed on  EXOTIC  plants, or exclusively on artificial substanc
0021  t pigeon family: and these feathers are kept  EXPANDED, and are carried so erect that in good birds
0039  ails had only fourteen tailfeathers somewhat  EXPANDED, like the present Java fantail, or like indi
0280  one, for instance, combining a tail somewhat  EXPANDED  with a crop somewhat enlarged, the character
0021  breast: and it has the habit of continually  EXPANDING  slightly the upper part of the oesophagus.
0180  d even the base of the tail united by a broad  EXPANSE  of skin, which serves as a parachute and all
0016  h I shall presently give, we have no right to  EXPECT  often to meet with generic differences in our
0037  r with their parent stocks. No one would ever  EXPECT  to get a first rate heartsease or dahlia from
0037  a from the seed of a wild plant. No one would  EXPECT  to raise a first rate melting pear from the s
0055  now forming. Where many large trees grow, we  EXPECT  to find saplings. Where many species of a gen
0055  favourable for variation; and hence we might  EXPECT  that the circumstances would generally be sti
0138  notwithstanding such modifications, we might  EXPECT  still to see in the cave animals of America,
0139  ew worlds should be closely related, we might  EXPECT  from the well known relationship of most of t
0153  ng run selection gains the day, and we do not  EXPECT  to fail so far as to breed a bird as coarse a
0153  essively remote, we might, as a general rule,  EXPECT  still to find more variability in such parts
0155  y marked and fixed varieties, we might surely  EXPECT  to find them still often continuing to vary i
0173  ast and present states. Hence we ought not to  EXPECT  at the present time to meet with numerous tra
0184  the other species of the same genus, we might  EXPECT, on my theory, that such individuals would oc
0240  may give one other case: so confidently did I  EXPECT  to find gradations in important points of str
0252  ds freely in confinement, we have no right to  EXPECT  that the first crosses between them and the c
0269  lei and if this be so, we surely ought not to  EXPECT  to find sterility both appearing and disappea
0273  en recent variability; and therefore we might  EXPECT  that such variability would often continue an
0301  f truth in these remarks, we have no right to  EXPECT  to find in our geological formations, an infi
0326  nts of the continuous sea. We might therefore  EXPECT  to find, as we apparently do find, a less str
0333  a very broken condition, we have no right to  EXPECT, except in very rare cases, to fill up wide i
0333  milies or orders. All that we have a right to  EXPECT, is that those groups, which have within know
0336  etween successive formations, we ought not to  EXPECT  to find, as I attempted to show in the last c
0336  of specific forms, as we have a just right to  EXPECT  to find. On the state of Development of Ancie
0338  doctrine is very far from proved. Yet I fully  EXPECT  to see it hereafter confirmed, at least in re
0452  e of an organ, which analogy would lead us to  EXPECT  to find, and which is occasionally found in m
0461  tends to eliminate sterility, we ought not to  EXPECT  it also to produce sterility. The sterility o
0462  , we should remember that we have no right to  EXPECT  (excepting in rare cases) to discover directl
0463  sely allied species, we have no just right to  EXPECT  often to find intermediate varieties in the i
0464  es: and these many links we could hardly ever  EXPECT  to discover, owing to the imperfection of the
0470  ufactory of species has been active, we might  EXPECT, as a general rule, to find it still in actio
0474  lity and modification, and therefore we might  EXPECT  this part generally to be still variable. But
0481  under the form of an abstract, I by no means  EXPECT  to convince experienced naturalists whose min
0034  om his own best dogs; but he has no wish or  EXPECTATION  of permanently altering the breed. Neverth
0036  in so far that the breeders could never have  EXPECTED  or even have wished to have produced the res
```

Page **(Key Word)**

```
0045  ens of the same species. I should never have  EXPECTED  that the branching of the main nerves close
0045  variable in the same species; I should have   EXPECTED  that changes of this nature could have been
0053  present varieties; and this might have been    EXPECTED, as they become exposed to diverse physical
0054  the genus; and, consequently, we might have    EXPECTED  to have found in the larger genera, or those
0086  cause other modifications, often of the most   EXPECTED  nature. As we see that those variations whic
0115  agency in foreign lands. It might have been    EXPECTED  that the plants which have succeeded in beco
0115  n country. It might, also, perhaps have been   EXPECTED  that naturalised plants would have belonged
0125  pient species. This, indeed, might have been   EXPECTED; for as natural selection acts through one f
0138  organisation and affinities might have been    EXPECTED; but, as Schiodte and others have remarked,
0153  ion is rapidly going on, there may always be   EXPECTED  to be much variability in the structure unde
0161  descended from a common parent, it might be    EXPECTED  that they would occasionally vary in an anal
0161  inherited constitution. It might further be    EXPECTED  that the species of the same genus would occ
0193  ient progenitor thus provided, we might have   EXPECTED  that all electric fishes would have been spe
0263  llow to a certain extent, as might have been   EXPECTED, systematic affinity, by which every kind of
0390  that this is true. This fact might have been   EXPECTED  on my theory; for, as already explained, spe
0399  her region: and this is what might have been   EXPECTED; but it is also plainly related to South Ame
0400  world. And this is just what might have been   EXPECTED  on my view, for the islands are situated so
0401  ands. This difference might indeed have been   EXPECTED  on the view of the islands having been stock
0468  ergone great physical changes, we might have   EXPECTED  that organic beings would have varied under
0237  sows seeds of the same stock, and confidently  EXPECTS  to get nearly the same variety; breeders of
0223  e, and the masters alone go on slave making    EXPEDITIONS. In Switzerland the slaves and masters wor
0051  h species; but he will succeed in this at the  EXPENSE  of admitting much variation, and the truth o
0015  ber of generations, would be opposed to all    EXPERIENCE. I may add, that when under nature the cond
0024  feral in several places. Again, all recent     EXPERIENCE  shows that it is most difficult to get any
0047  naturalists having sound judgment and wide     EXPERIENCE  seems the only guide to follow. We must, ho
0207  n action, which we ourselves should require    EXPERIENCE  to enable us to perform, when performed by
0207  especially by a very young one, without any    EXPERIENCE, and when performed by many individuals in
0211  tion was instinctive, and not the result of    EXPERIENCE. But as the excretion is extremely viscid,
0212  stling birds, though it is strengthened by     EXPERIENCE, and by the sight of fear of the same enemy
0213  t see that these actions, performed without    EXPERIENCE  by the young, and in nearly the same manner
0249  ften taken by chance (as I know from my own    EXPERIENCE) from the anthers of another flower, as fro
0264  ated to me by Mr. Hewitt, who has had great    EXPERIENCE  in hybridising gallinaceous birds, that the
0428  of the groups in other classes (and all our    EXPERIENCE  shows that this valuation has hitherto been
0484  pecies. This I feel sure, and I speak after    EXPERIENCE, will be no slight relief. The endless disp
0486  ention as the summing up of the labour, the    EXPERIENCE, the reason, and even the blunders of numer
0486  ing, how far more interesting, I speak from    EXPERIENCE, will the study of natural history become!
0045  same species. I am convinced that the most    EXPERIENCED  naturalist would be surprised at the numbe
0049  as species by some zoologists. Several most    EXPERIENCED  ornithologists consider our British red gr
0057  have also consulted some sagacious and most   EXPERIENCED  observers; and, after deliberation, they c
0194  s admission in the writings of almost every   EXPERIENCED  naturalist; or, as Milne Edwards has well
0248  this can be required than that the two most   EXPERIENCED  observers who have ever lived, namely, Kol
0249  to the results arrived at by the third most   EXPERIENCED  hybridiser, namely, the Hon. and Rev. W. H
0297  rent sub stages of the same formation. Some   EXPERIENCED  conchologists are now sinking many of the
0431  aordinary difficulty which naturalists have   EXPERIENCED  in describing, without the aid of a diagra
0481  abstract, I by no means expect to convince    EXPERIENCED  naturalists whose minds are stocked with a
0015  e wild aboriginal stock. Whether or not the   EXPERIMENT  would succeed, is not of great importance f
0015  rtance for our line of argument; for by the   EXPERIMENT  itself the conditions of life are changed.
0026  ether, though it is unsupported by a single   EXPERIMENT. But to extend the hypothesis so far as to
0095  oboscis. On the other hand, I have found by   EXPERIMENT  that the fertility of clover greatly depend
0142  e seedlings, with the same precautions, the   EXPERIMENT  cannot be said to have been even tried. Nor
0228  al powers. But this theory can be tested by   EXPERIMENT. Following the example of Mr. Tegetmeier, I
0228  xcavated to the same depth as in the former   EXPERIMENT, would have broken into each other from the
0230  ediate plane, and thus flatten it. From the   EXPERIMENT  of the ridge of vermilion wax, we can clear
0251  herbert told me that he had then tried the   EXPERIMENT  during five years, and he continued to try
0256  ls which happen to have been chosen for the   EXPERIMENT. So it is with hybrids, for their degree of
0274  r as I can make out, is founded on a single   EXPERIMENT; and seems directly opposed to the results
0277  ffinity of the forms which are subjected to   EXPERIMENT; for systematic affinity attempts to expres
0359  ot all the same species as in the foregoing   EXPERIMENT) floated, after being dried, for above 28 d
0050  it seems to me, an overwhelming amount of     EXPERIMENTAL  evidence, showing that they descend from c
0265  ns are eminently liable to vary, as every     EXPERIMENTALIST  has observed. Thus we see that when orga
0249  t breeders. Hybrids are seldom raised by      EXPERIMENTALISTS  in great numbers; and as the parent spec
0276  ften so slight that the two most careful      EXPERIMENTALISTS  who have ever lived, have come to diamet
0113  g. so it will be with plants. It has been     EXPERIMENTALLY  proved, that if a plot of ground be sown
0233  formed by Mr. Tegetmeier that it has been     EXPERIMENTALLY  found that no less than from twelve to fi
0020  hardly believe. Sir J. Sebright expressly     EXPERIMENTISED  for this object, and failed. The offsprin
0247  from other plants. Nearly all the plants      EXPERIMENTISED  on by Gartner were potted, and apparently
0250  species is a universal law of nature. He     EXPERIMENTISED  on some of the very same species as did G
0270  may be trusted, I know not; but the forms     EXPERIMENTISED  on, are ranked by Sageret, who mainly fou
0271  species, than are the other varieties. He    EXPERIMENTISED  on five forms, which are commonly reputed
0407  bject which has hardly ever been properly     EXPERIMENTISED  on; if we bear in mind how often a specie
0461  er, most of the varieties which have been     EXPERIMENTISED  on have been produced under domestication
0008  stant of conception. Geoffroy St. Hilaire's   EXPERIMENTS  show that unnatural treatment of the embry
0010  er this point of view, Mr. Buckman's recent   EXPERIMENTS  on plants seem extremely valuable. When al
0049  ges; and lastly, according to very numerous   EXPERIMENTS  made during several years by that most car
0073  r other bees do not visit this flower. From   EXPERIMENTS  which I have tried, I have found that the
0097  pilionaceous flowers, that I have found, by   EXPERIMENTS  published elsewhere, that their fertility
0248  nt hybridisers, or by the same author, from  EXPERIMENTS  made during different years. It can thus b
0248  s. nevertheless I believe that in all these  EXPERIMENTS  the fertility has been diminished by an in
0249  us effected. Moreover, whenever complicated  EXPERIMENTS  are in progress, so careful an observer as
0251  and Verbascum. Although the plants in these  EXPERIMENTS  appeared perfectly healthy, and although b
0251  ertilised, sometimes depends. The practical  EXPERIMENTS  of horticulturists, though not made with s
0252  r flowers. In regard to animals, much fewer  EXPERIMENTS  have been carefully tried than with plants
0252  mals breeding freely under confinement, few  EXPERIMENTS  have been fairly tried: for instance, the
0264  to have been the case with some of Thuret's  EXPERIMENTS  on Fuci. No explanation can be given of th
0270  heir differences are greater. How far these  EXPERIMENTS  may be trusted, I know not; but the forms
0270  t is the result of an astonishing number of  EXPERIMENTS  made during many years on nine species of
```

Page **************************************(Key Word)**************************************

```
0273  es, and mostly domestic varieties (very few EXPERIMENTS having been tried on natural varieties), a
0274  directly opposed to the results of several EXPERIMENTS made by Kolreuter. These alone are the uni
0358  til I tried, with Mr. Berkeley's aid, a few EXPERIMENTS, it was not even known how far seeds could
0360  e, they would germinate. Subsequently to my EXPERIMENTS, M. Martens tried similar ones, but in a m
0360  e protected from violent movement as in our EXPERIMENTS. Therefore it would perhaps be safer to as
0362  rs, disgorge pellets, which, as I know from EXPERIMENTS made in the Zoological Gardens, include se
0386  is with seeds: I have tried several little EXPERIMENTS, but will here give only the most striking
0397  and it recovered and crawled away: but more EXPERIMENTS are wanted on this head. The most striking
0006  all the beings which live around us. Who can EXPLAIN why one species ranges widely and is very nu
0040  ur poultry shows. I think these views further EXPLAIN what has sometimes been noticed, namely, tha
0097  trance of pollen from another individual will EXPLAIN this state of exposure, more especially as t
0146  r found in fruits which do not open: I should EXPLAIN the rule by the fact that seeds could not gr
0154  characters are more variable than generic. To EXPLAIN by a simple example what is meant. If some s
0261  a reversed direction. It will be advisable to EXPLAIN a little more fully by an example what I mea
0290  ed to a distant age. A little reflection will EXPLAIN why along the rising coast of the western si
0342  tremely imperfect, and will to a large extent EXPLAIN why we do not find interminable varieties, c
0353  oubtedly many cases occur, in which we cannot EXPLAIN how the same species could have passed from
0368  nce of a former Glacial period, seem to me to EXPLAIN in so satisfactory a manner the present dist
0381  rs have remained unaltered. We cannot hope to EXPLAIN such facts, until we can say why one species
0386  range. I think favourable means of dispersal EXPLAIN this fact. I have before mentioned that eart
0389  many difficulties, but it would not, I think, EXPLAIN all the facts in regard to insular productio
0393  created there, it would be very difficult to EXPLAIN. Mammals offer another and similar case. I h
0420  rating objects more or less alike. But I must EXPLAIN my meaning more fully. I believe that the ar
0435  thing can be more hopeless than to attempt to EXPLAIN this similarity of pattern in members of the
0437  t of parturition of mammals, will by no means EXPLAIN the same construction in the skulls of birds
0442  ace of the vermiform stage. How, then, can we EXPLAIN these several facts in embryology, namely th
0444  if their truth be admitted, will, I believe, EXPLAIN all the above specified leading facts in emb
0446  the two principles above given seem to me to EXPLAIN these facts in regard to the later embryonic
0471  es has been independently created, no man can EXPLAIN. Many other facts are, as it seems to me, ex
0479  nt, and innumerable other such facts, at once EXPLAIN themselves on the theory of descent with slo
0015  but natural selection, as will hereafter be EXPLAINED, will determine how far the new characters
0038  l selection, as will hereafter be more fully EXPLAINED, two sub breeds might be formed. This, perh
0046  e by natural selection, as hereafter will be EXPLAINED. Those forms which possess in some consider
0052  ccumulated (as will hereafter be more fully EXPLAINED) differences of structure in certain defini
0058  e of Character, we shall see how this may be EXPLAINED, and how the lesser differences between var
0059  nt descendants. But by steps hereafter to be EXPLAINED, the larger genera also tend to break up in
0089  arison with the plumage of the young, can be EXPLAINED on the view of plumage having been chiefly
0099  ss to each other! How simply are these facts EXPLAINED on the view of an occasional cross with a d
0125  to limit the process of modification, as now EXPLAINED, to the formation of genera alone. If, in o
0128  the affinities of all organic beings may be EXPLAINED. It is a truly wonderful fact, the wonder o
0129  ings: but, to the best of my judgment, it is EXPLAINED through inheritance and the complex action
0134  ut many animals have structures which can be EXPLAINED by the effects of disuse. As Professor Owen
0174  nded me. But I think it can be in large part EXPLAINED. In the first place we should be extremely
0188  ies of facts, otherwise inexplicable, can be EXPLAINED by the theory of descent, ought not to hesi
0206  bits of life. On my theory, unity of type is EXPLAINED by unity of descent. The expression of cond
0235  nown instincts, that of the hive bee, can be EXPLAINED by natural selection having taken advantage
0292  circumstances most favourable, as previously EXPLAINED, for the formation of new varieties and spe
0297  eties. Nor should it be forgotten, as before EXPLAINED, that A might be the actual progenitor of B
0314  organic and inorganic conditions of life, as EXPLAINED in a former chapter. When many of the inhab
0320  n will generally be most severe, as formerly EXPLAINED and illustrated by examples, between the fo
0326  would probably be also favourable, as before EXPLAINED. One quarter of the world may have been mos
0328  iod, we should find in both, from the causes EXPLAINED in the foregoing paragraphs, the same gener
0329  and natural system: and this fact is at once EXPLAINED on the principle of descent. The more ancie
0333  o each other and to living forms, seem to me EXPLAINED in a satisfactory manner. And they are whol
0340  same types within the same areas, is at once EXPLAINED; for the inhabitants of each quarter of the
0345  riods ceases to be mysterious, and is simply EXPLAINED by inheritance. If then the geological reco
0352  e their parents were first produced: for, as EXPLAINED in the last chapter, it is incredible that
0354  he earth's surface, can in many instances be EXPLAINED on the view of each species having migrated
0366  istribution of the inhabitants of Europe, as EXPLAINED with remarkable clearness by Edward Forbes.
0382  the same and of allied forms of life can be EXPLAINED. The living waters may be said to have flow
0383  o unexpected, can, I think, in most cases be EXPLAINED by their having become fitted, in a manner
0384  re are many cases which cannot at present be EXPLAINED: but some fresh water fish belong to very a
0390  been expected on my theory, for, as already EXPLAINED, species occasionally arriving after long i
0393  ption (if the information be correct) may be EXPLAINED through glacial agency. This general absenc
0399  exceptions are few, and most of them can be EXPLAINED. Thus the plants of Kerguelen Land, though
0399  lants, and will, I do not doubt, be some day EXPLAINED. The law which causes the inhabitants of an
0411  m. rules and difficulties in classification, EXPLAINED on the theory of descent with modification.
0411  of the same individual. Embryology, Laws of; EXPLAINED by variations not supervening at an early a
0411  onding age. Rudimentary Organs: their origin EXPLAINED. Summary. From the first dawn of life, all
0412  diagram illustrating the, as formerly EXPLAINED, of these several principles: and he will s
0413  ficiently strike us, is in my judgment fully EXPLAINED. Naturalists try to arrange the species, ge
0420  aids and difficulties in classification are EXPLAINED, if I do not greatly deceive myself, on the
0439  true legs, or from some simple appendage, is EXPLAINED. Embryology. It has already been casually r
0443  loped. I believe that all these facts can be EXPLAINED, as follows, on the view of descent with mo
0450  n importance to none in natural history, are EXPLAINED on the principle of slight modifications no
0455  e abortion. The principle, also, of economy, EXPLAINED in a former chapter, by which the materials
0467  ns, and by the results of naturalisation, as EXPLAINED in the third chapter. More individuals are
0473  nd in their hybrids! How simply is this fact EXPLAINED if we believe that these species have desce
0476  in the formations above and below, is simply EXPLAINED by their intermediate position in the chain
0014  his rule to be of the highest importance in EXPLAINING the laws of embryology. These remarks are o
0028  mount importance of these considerations in EXPLAINING the immense amount of variation which pigeo
0135  ons are ever inherited; and I should prefer EXPLAINING the entire absence of the anterior tarsi in
0037  , thus slowly and unconsciously accumulated, EXPLAINS, as I believe, the well known fact, that in
0038  reeds might be formed. This, perhaps, partly EXPLAINS what has been remarked by some authors, name
0065  f which never fails to be surprising, simply EXPLAINS the extraordinarily rapid increase and wide
0070  ck of the same species for its preservation, EXPLAINS, I believe, some singular facts in nature, s
```

Page **************************************(Key Word)**************************************

Page ***************************************(Key Word)***************************************

```
0080  on, on the descendants from a common parent. EXPLAINS the grouping of all organic beings. How will
0111  erm, is of high importance on my theory, and EXPLAINS, as I believe, several important facts. In t
0218  ommon with the Gallinaceae; and this perhaps EXPLAINS the origin of a singular instinct in the all
0417  racters, even when none are important, alone EXPLAINS, I think, that saying of Linnaeus, that the
0433  escandants from one dominant parent species; EXPLAINS that great and universal feature in the affi
0471  t inevitable contingency of much extinction, EXPLAINS the arrangement of all the forms of life, in
0004  m; but this assumption seems to me to be no EXPLANATION, for it leaves the case of the coadaptatio
0027  e in each breed eminently variable; and the EXPLANATION of this fact will be obvious when we come
0029  umerable other examples could be given. The EXPLANATION, I think, is simple: from long continued s
0065  creased in any sensible degree. The obvious EXPLANATION is that the conditions of life have been v
0129  as been independently created, I can see no EXPLANATION of this great fact in the classification a
0139  ould be most difficult to give any rational EXPLANATION of the affinities of the blind cave animal
0152  its parts as we now see them, I can see no EXPLANATION. But on the view that groups of species ha
0154  ance. I have chosen this example because an EXPLANATION is not in this case applicable, which most
0154  ly used for classing genera. I believe this EXPLANATION is partly, yet only indirectly, true; I sh
0155  the several species? I do not see that any EXPLANATION can be given. But on the view of species b
0235  . no doubt many instincts of very difficult EXPLANATION could be opposed to the theory of natural
0264  th some of Thuret's experiments on Fuci. No EXPLANATION can be given of these facts, any more than
0266  ng remarks go to the root of the matter: no EXPLANATION is offered why an organism, when placed un
0268  o not readily cross with European dogs, the EXPLANATION which will occur to everyone, and probably
0280  n which can be urged against my theory. The EXPLANATION lies, as I believe, in the extreme imperfe
0290  nd from muddy streams entering the sea. The EXPLANATION, no doubt, is, that the littoral and sub l
0308  to show that it may hereafter receive some EXPLANATION, I will give the following hypothesis. Fro
0310  always seemed to me to require some special EXPLANATION; and we may perhaps believe that we see in
0313  pre existing fauna to reappear; but Lyell's EXPLANATION, namely,that it is a case of temporary mig
0316  n abruptly; and I have attempted to give an EXPLANATION of this fact, which if true would have bee
0354  nts; nor do I for a moment pretend that any EXPLANATION could be offered of many such cases. But a
0366  we shall immediately see, affords a simple EXPLANATION of these facts. We have evidence of almost
0392  hysical conditions of the islands; but this EXPLANATION seems to me not a little doubtful. Facilit
0398  ieve this grand fact can receive no sort of EXPLANATION on the ordinary view of independent creati
0435  tor to construct each animal and plant. The EXPLANATION is manifest on the theory of the natural s
0448  nce with their similar habits. Some further EXPLANATION, however, of the embryo not undergoing any
0453  e scheme of nature; but this seems to me no EXPLANATION, merely a restatement of the fact. Would I
0478  ust be admitted that these facts receive no EXPLANATION on the theory of creation. The fact, as we
0482  design, etc., and to think that we give an EXPLANATION when we only restate a fact. Any one whose
0482  cht to unexplained difficulties than to the EXPLANATION of a certain number of facts will certainl
0483  ugh naturalists very properly demand a full EXPLANATION of every difficulty from those who believe
0465  now briefly recapitulated the answers and EXPLANATIONS which can be given to them. I have felt t
0243  l as to corporeal structure, and is plainly EXPLICABLE on the foregoing views, but is otherwise in
0288  red in the interval any wear and tear, seem EXPLICABLE only on the view of the bottom of the sea n
0325  the forms of life throughout the world, is EXPLICABLE on the theory of natural selection. New spe
0406  dependent creation of each species; but are EXPLICABLE on the view of colonisation from the neares
0408  ding facts of geographical distribution are EXPLICABLE on the theory of migration (generally of th
0471  n. many other facts are, as it seems to me, EXPLICABLE on this theory. How strange it is that a bi
0034  y given in an ancient Chinese encyclopaedia. EXPLICIT rules are laid down by some of'the Roman cla
0287  e surface of the earth has been geologically EXPLORED, and no part with sufficient care, as the im
0341  l portion of the globe has been geologically EXPLORED with care; that only certain classes of orga
0464  l portion of the world has been geologically EXPLORED. Only organic beings of certain classes can
0034  rted, and laws were passed to prevent their EXPORTATION: the destruction of horses under a certain
0031  ith a good pedigree; and these have now been EXPORTED to almost every quarter of the world. The im
0007  , those to which the parent species have been EXPOSED under nature. There is, also, I think, some
0007  eems pretty clear that organic beings must be EXPOSED during several generations to the new condit
0010  as Muller has remarked, have apparently been EXPOSED to exactly the same conditions of life; and
0010  uable. When all or nearly all the individuals EXPOSED to certain conditions are affected in the sa
0013  othi but when amongst individuals, apparently EXPOSED to the same conditions, any very rare deviat
0053  this might have been expected, as they become EXPOSED to diverse physical conditions, and as they
0062  osophically shown that all organic beings are EXPOSED to severe competition. In regard to plants,
0075  ame districts, require the same food, and are EXPOSED to the same dangers. In the case of varietie
0097  flowers have their anthers and stigmas fully EXPOSED to the weather! but if an occasional cross b
0107  ed a confined area, and from having thus been EXPOSED to less severe competition. To sum up the ci
0114  f, three feet by four in size, which had been EXPOSED for many years to exactly the same condition
0117  e two varieties will generally continue to be EXPOSED to the same conditions which made their pare
0131  nts and their more remote ancestors have been EXPOSED during several generations. I have remarked
0134  the ostrich indeed inhabits continents and is EXPOSED to danger from which it cannot escape by fli
0136  oportion of wingless beetles is larger on the EXPOSED Dezertas than in Madeira itself; and especia
0139  of these dark abodes will probably have been EXPOSED. Acclimatisation. Habit is hereditary with p
0148  he struggle for life to which every animal is EXPOSED, each individual Proteolepas would have a be
0169  he same constitution from a common parent and EXPOSED to similar influences will naturally tend to
0182  , under the conditions of life to which it is EXPOSED, for each has to live by a struggle; but it
0198  uggle for their own subsistence, and would be EXPOSED to a certain extent to natural selection, an
0220  s much disturbed and·the larvae and pupae are EXPOSED, the slaves work energetically with their ma
0256  raised from seed out of the same capsule and EXPOSED to exactly the same conditions. By the term
0264  egg or seed produced by the mother, it may be EXPOSED to conditions in some degree unsuitable, and
0286  ould yield at this rate excepting on the most EXPOSED coasts; though no doubt the degradation of a
0350  ew countries. In their new homes they will be EXPOSED to new conditions, and will frequently under
0360  al sea, so that they were alternately wet and EXPOSED to the air like really floating plants. He t
0360  f germination. But I do not doubt that plants EXPOSED to the waves would float for a less time tha
0364  ; for seeds do not retain their vitality when EXPOSED for a great length of time to the action of
0368  on and re migration northward, will have been EXPOSED to nearly the same climate, and, as is espec
0369  own to the plains; they will, also, have been EXPOSED to somewhat different climatal influences. T
0401  ne island to another, it would undoubtedly be EXPOSED to different conditions of life in the diffe
0401  n one island than in another, and it would be EXPOSED to the attacks of somewhat different enemies
0425  life to which each species has been recently EXPOSED. Rudimentary structures on this view are as
0440  e related to the conditions to which they are EXPOSED. The case, however, is different when an ani
0443  either parent, or their ancestors, have been EXPOSED. Nevertheless an effect thus caused at a ver
0454  with the wings of beetles living on small and EXPOSED islands; and in this case natural selection
```

Page ***************************************(Key Word)***************************************

Page **(Key Word)**

```
0083  g legged quadruped in any peculiar manner; he EXPOSES sheep with long and short wool to the same c
0467  produce variability; he only unintentionally EXPOSES organic beings to new conditions of life, an
0097  ble. Every hybridizer knows how unfavourable EXPOSURE to wet is to the fertilisation of a flower,
0097  nother individual will explain this state of EXPOSURE, more especially as the plant's own anthers
0004  ed the best and safest clue. I may venture to EXPRESS my conviction of the high value of such stud
0021  rs. The trumpeter and laugher, as their names EXPRESS, utter a very different coo from the other b
0277  periment: for systematic affinity attempts to EXPRESS all kinds of resemblance between all species
0037  inferior quality. I have seen great surprise  EXPRESSED in horticultural works at the wonderful ski
0147  tion or balancement of growth; or, as Goethe  EXPRESSED it, in order to spend on one side, nature i
0194  ed naturalist; or, as Milne Edwards has well  EXPRESSED it, nature is prodigal in variety, but nicg
0263  ty between organic beings is attempted to be  EXPRESSED. The facts by no means seem to me to indica
0420  ation which they have undergone; and this is  EXPRESSED by the forms being ranked under different g
0422  different groups have undergone, have to be   EXPRESSED by ranking them under different so called g
0422  guages from the same stock, would have to be  EXPRESSED by groups subordinate to groups; but the pr
0433  etween the descendants from a common parent,  EXPRESSED by the terms genera, families, orders, etc.
0434  heir organisation. This resemblance is often  EXPRESSED by the term unity of type; or by saying tha
0208  re universal. A little dose, as Pierre Huber  EXPRESSES it, of judgment or reason, often comes into
0003  however, let this opportunity pass without    EXPRESSING my deep obligations to Dr. Hooker, who for
0482  ble will do good service by conscientiously   EXPRESSING his conviction; for only thus can the load
0039  to catch his attention. But to use such an    EXPRESSION as trying to make a fantail, is, I have no
0056  ve been formed, or where, if we may use the   EXPRESSION, the manufactory of species has been active
0131  nce. This, of course, is a wholly incorrect   EXPRESSION, but it serves to acknowledge plainly our i
0143  nces. Correlation of Growth. I mean by this   EXPRESSION that the whole organisation is so tied toge
0149  vegetative repetition, to use Prof. Owen's    EXPRESSION, seems to be a sign of low organisation; th
0160  s, the proportion of blood, to use a common   EXPRESSION, of any one ancestor, is only 1 in 2048; an
0206  type is explained by unity of descent. The    EXPRESSION of conditions of existence, so often insist
0324  he world, it must not be supposed that this   EXPRESSION relates to the same thousandth or hundred t
0441  generally rises in organisation: I use this   EXPRESSION, though I am aware that it is hardly possib
0413  othing is thus added to our knowledge. Such   EXPRESSIONS as that famous one of Linnaeus, and which
0482  is so easy to hide our ignorance under such   EXPRESSIONS as the plan of creation, unity of design,
0020  ecies, I can hardly believe. Sir J. Sebright  EXPRESSLY experimentised for this object, and failed.
0221  ors in the morning and evening; and as Huber  EXPRESSLY states, their principal office is to search
0435  es. The hopelessness of the attempt has been  EXPRESSLY admitted by Owen in his most interesting wo
0060  species arise in nature. How have all those   EXQUISITE adaptations of one part of the organisation
0224  d. he must be a dull man who can examine the  EXQUISITE structure of a comb, so beautifully adapted
0197  ago climbs the loftiest trees by the aid of   EXQUISITELY constructed hooks clustered around the end
0453  s us plainly that most parts and organs are   EXQUISITELY adapted for certain purposes, tells us wit
0026  is unsupported by a single experiment. But to EXTEND the hypothesis so far as to suppose that spec
0308  tinental islands existed where our oceans now EXTEND: for had they existed there, palaeozoic and s
0309  facts, we may infer that where our oceans now EXTEND, oceans have extended from the remotest perio
0343  at as far as we can see, where our oceans now EXTEND they have for an enormous period extended, an
0370  illations of level, I am strongly inclined to EXTEND the above view, and to infer that during some
0375  australian forms, as I hear from Dr. Hooker,  EXTEND along the heights of the peninsula of Malacca
0428  on has hitherto been arbitrary), could easily EXTEND the parallelism over a wide range; and thus t
0447  s we have seen in the case of pigeons. We may EXTEND this view to whole families or even classes.
0456  ver different they may be in structure. If we EXTEND the use of this element of descent, the only
0482  een produced by variation, but they refuse to EXTEND the same view to other and very slightly diff
0483  r reverent silence. It may be asked how far I EXTEND the doctrine of the modification of species.
0483  ce. But some arguments of the greatest weight EXTEND very far. All the members of whole classes ca
0051  ied forms. But if his observations be widely  EXTENDED, he will in the end generally be enabled to
0140  ecies within historical times having largely  EXTENDED their range from warmer to cooler latitudes,
0140  they were subsequently found capable of far   EXTENDED transportation, I think the common and extra
0154  e principle included in these remarks may be  EXTENDED. It is notorious that specific characters ar
0181  a bat. In bats which have the wing membrane   EXTENDED from the top of the shoulder to the tail, in
0285  glance how far the hard, rocky beds had once  EXTENDED into the open ocean. The same story is still
0309  hat where our oceans now extend, oceans have  EXTENDED from the remotest period of which we have an
0339  the living. Professor Owen has subsequently   EXTENDED the same generalisation to the mammals of th
0343  now extend they have for an enormous period   EXTENDED, and where our oscillating continents now st
0350  ed over many competitors in their own widely  EXTENDED homes will have the best chance of seizing o
0373  convinced that Forbes's view may be largely   EXTENDED. In Europe we have the plainest evidence of
0373  a of Equatorial South America, glaciers once  EXTENDED far below their present level. In central Ch
0383  passable barrier; that they never would have  EXTENDED to distant countries. But the case is exactl
0459  w far the theory of natural selection may be  EXTENDED. Effects of its adoption on the study of Nat
0188  be surprised at any degree of hesitation in   EXTENDING the principle of natural selection to such
0295  d of having been really far greater, that is  EXTENDING from before the glacial epoch to the presen
0389  rge size and varied stations of New Zealand,  EXTENDING over 780 miles of latitude, and compare its
0450  but yet, owing to the geological record not   EXTENDING far enough back in time, may remain for a l
0051  which to correct his first impressions. As he EXTENDS the range of his observations, he will meet
0266  ful, but I suspect that a similar parallelism EXTENDS to an allied yet very different class of fac
0285  ally visible. The Craven fault, for instance, EXTENDS for upwards of 30 miles, and along this line
0348  shores of America, a wide space of open ocean EXTENDS, with not an island as a halting place for e
0356  may formerly have blended: where the sea now  EXTENDS, land may at a former period have connected
0014  ts make me believe that the rule has a wider  EXTENSION, and that when there is no apparent reason
0076  f distinct genera. We see this in the recent  EXTENSION over parts of the United States of one spec
0357  t of such vast changes in their position and  EXTENSION, as to have united them within the recent p
0357  led to speculate with security on the former  EXTENSION of the land. But I do not believe that it w
0389  honestly admit Forbes's view on continental  EXTENSIONS, which, if legitimately followed out, would
0071  saw near Farnham, in Surrey. Here there are   EXTENSIVE heaths, with a few clumps of old Scotch fir
0072  yet the heath was so extremely barren and so  EXTENSIVE that no one would ever have imagined that c
0173  in masses of sediment sufficiently thick and  EXTENSIVE to withstand an enormous amount of future d
0177  eties of sheep to be kept, one adapted to an  EXTENSIVE mountainous region; a second to a comparati
0279  favourable for their presence, namely on an   EXTENSIVE and continous area with graduated physical
0290  absence of any recent deposits sufficiently   EXTENSIVE to last for even a short geological period.
0290  ast of the western side of South America, no  EXTENSIVE formations with recent or tertiary remains
0290  be accumulated in extremely thick, solid, or  EXTENSIVE masses, in order to withstand the incessant
0290  equent oscillations of level. Such thick and  EXTENSIVE accumulations of sediment may be formed in
```

Page **(Key Word)**

```
0291  s rich in fossils and sufficiently thick and EXTENSIVE to resist subsequent degradation, may have
0308  to the existing oceans, which are thrice as EXTENSIVE as the land, we see them studded with many
0346  eans. Dispersal during the Glacial period co EXTENSIVE with the world. In considering the distribu
0126  e well know that many groups, formerly most EXTENSIVELY developed, have now become extinct. Lookin
0181  e flank membrane is, also, furnished with an EXTENSOR muscle. Although no graduated links of struc
0015  of the poor soil), that they would to a large EXTENT, or even wholly, revert to the wild aborigina
0031  within a single lifetime, modified to a large EXTENT some breeds of cattle and sheep. In order ful
0035  el has been unconsciously modified to a large EXTENT since the time of that monarch. Some highly c
0053  nsideration from wide range, and to a certain EXTENT from commonness), often give rise to varietie
0057  ecies already manufactured still to a certain EXTENT resemble varieties, for they differ from each
0087  gh importance to it, might be modified to any EXTENT by natural selection; for instance, the great
0106  cts of isolation will generally, to a certain EXTENT, have concurred. Finally, I conclude that, al
0107  en, which, like fossils, connect to a certain EXTENT orders now widely separated in the natural sc
0115  ied nature. They differ, moreover, to a large EXTENT from the indigenes, for out of the 162 genera
0140  e few plants, of their becoming, to a certain EXTENT, naturally habituated to different temperatur
0147  er side. I think this holds true to a certain EXTENT with our domestic productions: if nourishment
0198  ubsistence, and would be exposed to a certain EXTENT to natural selection, and individuals with sl
0209  ly accumulating variations of instinct to any EXTENT that may be profitable. It is thus, as I beli
0212  for instance, the migratory instinct, both in EXTENT and direction, and in its total loss. So it i
0224  once acquired, if carried out to a much less EXTENT even than in our British F. sanguinea, which,
0227  seeing that she already does so to a certain EXTENT, and seeing what perfectly cylindrical burrow
0243  ating slight modifications of instinct to any EXTENT, in any useful direction. In some cases habit
0245  his subject, two classes of facts, to a large EXTENT fundamentally different, have generally been
0262  yet these two distinct cases run to a certain EXTENT parallel. Something analogous occurs in graft
0263  fferences, in both cases, follow to a certain EXTENT, as might have been expected, systematic affi
0265  male. In both, the tendency goes to a certain EXTENT with systematic affinity, for whole groups of
0277  t circumstances, should all run, to a certain EXTENT, parallel with the systematic affinity of the
0283  ell's profound remark, that the thickness and EXTENT of sedimentary formations are the result and
0291  iment may be accumulated to any thickness and EXTENT over a shallow bottom, if it continue slowly
0326  cies, which already have invaded to a certain EXTENT the territories of other species, should be t
0338  ts that ancient animals resemble to a certain EXTENT the embryos of recent animals of the same cla
0342  cord extremely imperfect, and will to a large EXTENT explain why we do not find interminable varie
0343  ong run that all undergo modification to some EXTENT. The extinction of old forms is the almost in
0345  ed that ancient animals resemble to a certain EXTENT the embryos of more recent animals of the sam
0377  l productions will have suffered to a certain EXTENT. On the other hand, the temperate productions
0379  ion from north to south is due to the greater EXTENT of land in the north, and to the northern for
0380  er degree in Australia, and have to a certain EXTENT beaten the natives; whereas extremely few sou
0395  nents, there is also a relation, to a certain EXTENT independent of distance, between the depth of
0404  fication. This relation between the power and EXTENT of migration of a species, either at the pres
0422  e two genera will have inherited to a certain EXTENT their characters. This natural arrangement is
0434  us organs: the parts may change to almost any EXTENT in form and size, and yet they always remain
0435  f a limb might be shortened or widened to any EXTENT, and become gradually enveloped in thick memb
0435  ts bones, or certain bones, lengthened to any EXTENT, and the membrane connecting them increased t
0435  the membrane connecting them increased to any EXTENT, so as to serve as a wing: yet in all this gr
0436  pattern seems to have been thus to a certain EXTENT obscured. There is another and equally curiou
0448  atural selection different to any conceivable EXTENT from their parents. Such differences might, a
0462  that we are as yet very ignorant of the full EXTENT of the various climatal and geographical chan
0467  of the breeds produced by man have to a large EXTENT the character of natural species, is shown by
0470  ies; and thus species are rendered to a large EXTENT defined and distinct objects. Dominant specie
0145  difference in the corolla of the central and EXTERIOR flowers of a head or umbel, I do not feel at
0146  s, according to Tausch, orthospermous in the EXTERIOR flowers and coelospermous in the central flo
0190  the animal may be turned inside out, and the EXTERIOR surface will then digest and the stomach res
0052  e variety; or it might come to supplant and EXTERMINATE the parent species; or both might co exist
0110  hardest on its nearest kindred, and tend to EXTERMINATE them. We see the same process of extermina
0121  cendants of any one species to supplant and EXTERMINATE in each stage of descent their predecessor
0172  ountry to take the place of, and finally to EXTERMINATE, its own less improved parent or other les
0179  ly tends, as has been so often remarked, to EXTERMINATE the parent forms and the intermediate link
0280  arieties continually take the places of and EXTERMINATE their parent forms. But just in proportion
0337  ome thoroughly naturalized there, and would EXTERMINATE many of the natives. On the other hand, fr
0412  and diverging in character, to supplant and EXTERMINATE the less divergent, the less improved, and
0470  oved varieties will inevitably supplant and EXTERMINATE the older, less improved and intermediate
0023  pices, and good fliers, are unlikely to be EXTERMINATED, and the common rock pigeon, which has the
0023  its with the domestic breeds, has not been EXTERMINATED even on several of the smaller British isl
0102  gree with its competitors, it will soon be EXTERMINATED. In man's methodical selection, a breeder
0106  in a corresponding degree or they will be EXTERMINATED. Each new form, also, as soon as it has be
0107  e slowly formed, and old forms more slowly EXTERMINATED. And it is in fresh water that we find sev
0122  ey will have taken the places of, and thus EXTERMINATED, not only their parents (A) and (I), but l
0172  itional varieties will generally have been EXTERMINATED by the very process of formation and perfe
0173  of its own region, and has supplanted and EXTERMINATED its original parent and all the transition
0176  iods; why as a general rule they should be EXTERMINATED and disappear, sooner than the forms which
0176  dy remarked, run a greater chance of being EXTERMINATED than one existing in large numbers; and in
0178  these links will have been supplanted and EXTERMINATED during the process of natural selection, s
0181  irrels would decrease in numbers or become EXTERMINATED, unless they also became modified and impr
0219  thus feloniously appropriated, be not thus EXTERMINATED. Slave making instinct. This remarkable in
0279  connect, will generally be beaten out and EXTERMINATED during the course of further modification
0307  ertainly have been long ago supplanted and EXTERMINATED by their numerous and improved descendants
0315  odified and improved, will be liable to be EXTERMINATED. Hence we can see why all the species in t
0316  rent form will generally be supplanted and EXTERMINATED by its improved offspring, it is quite inc
0318  l asked myself what could so recently have EXTERMINATED the former horse under conditions of life
0320  events with those animals which have been EXTERMINATED, either locally or wholly, through man's a
0320  than that of the old forms which have been EXTERMINATED: but we know that the number of species ha
0322  n possession of a new area, they will have EXTERMINATED in a correspondingly rapid manner many of
0337  una or flora would certainly be beaten and EXTERMINATED; as would a secondary fauna by an eocene;
0367  d on distant mountain summits (having been EXTERMINATED on all lesser heights) and in the artic re
0378  turally ascend the higher mountains, being EXTERMINATED on the lowlands; those which had not reach
0380  and if the natives have not been actually EXTERMINATED, their numbers have been greatly reduced,
```

Page ***(Key Word)***

```
0381 lora. I suspect that before this flora was EXTERMINATED by the Glacial epoch, a few forms were wid
0390 ed plants and animals have nearly or quite EXTERMINATED many native productions. He who admits the
0429 necting forms which on my theory have been EXTERMINATED and utterly lost. And we have some evidenc
0463 s will, in the long run, be supplanted and EXTERMINATED. On this doctrine of the extermination of
0487 forms of life. As species are produced and EXTERMINATED by slowly acting and still existing causes
0204 thus generally succeed in supplanting and EXTERMINATING it. We have have seen in this chapter how
0024 s of the Mediterranian. Hence the supposed EXTERMINATION of so many species having similar habits
0106 will have been less modification and less EXTERMINATION. Hence, perhaps, it comes that the flora
0110 terminate them. We see the same process of EXTERMINATION amongst our domesticated productions, thr
0175 the seasons, be extremely liable to utter EXTERMINATION; and thus its geographical range will com
0179 ate varieties will be liable to accidental EXTERMINATION; and during the process of further modifi
0280 but just in proportion as this process of EXTERMINATION has acted on an enormous scale, so must t
0318 its upper end, which marks the progress of EXTERMINATION, than at its lower end, which marks the f
0318 f the species. In some cases, however, the EXTERMINATION of whole groups of beings, as of ammonite
0321 ants of a species will generally cause the EXTERMINATION of the parent species; and if many new fo
0321 the same genus, will be the most liable to EXTERMINATION. Thus, as I believe, a number of new spec
0321 g to a distinct group, and thus caused its EXTERMINATION; and if many allied forms be developed fr
0321 ion. With respect to the apparently sudden EXTERMINATION of whole families or orders, as of Trilob
0322 se intervals there may have been much slow EXTERMINATION. Moreover, when by sudden immigration or
0377 le amount of cold, many might have escaped EXTERMINATION during a moderate fall of temperature, mo
0463 and exterminated. On this doctrine of the EXTERMINATION of an infinitude of connecting links, bet
0487 one being entailing the improvement or the EXTERMINATION of of others; it follows, that the amount
0003 admiration. Naturalists continually refer to EXTERNAL conditions, such as climate, food, etc., as
0003 but it is preposterous to attribute to mere EXTERNAL conditions, the structure, for instance, of
0003 l distinct organic beings, by the effects of EXTERNAL conditions, or of habit, or of the volition
0021 ied by greatly elongated eyelids, very large EXTERNAL orifices to the nostrils, and a wide gape of
0029 s, be attributed to the direct action of the EXTERNAL conditions of life, and some little to habit
0038 likewise their differences being so great in EXTERNAL characters and relatively so slight in inter
0061 lex relations to other organic beings and to EXTERNAL nature, will tend to the preservation of tha
0083 t may not nature effect? Man can act only on EXTERNAL and visible characters: nature cares nothing
0131 Chapter V. Laws of Variation. Effects of EXTERNAL conditions. Use and disuse, combined with na
0145 he flow of nutriment towards the central and EXTERNAL flowers: we know, at least, that in irregula
0146 to the differences both in the internal and EXTERNAL structure of the seeds, which are not always
0160 ds; and in this case there is nothing in the EXTERNAL conditions of life to cause the reappearance
0167 ences between species of the same genus. The EXTERNAL conditions of life, as climate and food, etc
0168 o cohere. Modifications in hard parts and in EXTERNAL parts sometimes affect softer and internal p
0196 lection will often have largely modified the EXTERNAL characters of animals having a will, to give
0206 lightly affected by the direct action of the EXTERNAL conditions of life, and being in all cases s
0259 ertility in the hybrid is independent of its EXTERNAL resemblance to either pure parent. Consideri
0259 ated to the degree in which they resemble in EXTERNAL appearance either parent. And lastly, that t
0266 ; in the other case, or that of hybrids, the EXTERNAL conditions have remained the same, but the o
0268 ever much they may differ from each other in EXTERNAL appearance, cross with perfect facility, and
0269 the first place, be clearly shown that mere EXTERNAL dissimilarity between two species does not d
0271 e ranked as species; from man selecting only EXTERNAL characters in the production of the most dis
0278 under domestication by the selection of mere EXTERNAL differences, and not of differences in the r
0414 othing can be more false. No one regards the EXTERNAL similarity of a mouse to a shrew, of a dugon
0416 covered with feathers instead of hair, this EXTERNAL and trifling character would, I think, have
0427 mmerable instances: thus Linnaeus, misled by EXTERNAL appearances, actually classed an homopterous
0427 similar conditions, and thus assume a close EXTERNAL resemblance; but such resemblances will not
0437 y divided into a series of segments, bearing EXTERNAL appendages; and in flowering plants, we see
0440 nculated and sessile, which differ widely in EXTERNAL appearance, have larvae in all their several
0468 ctly limited quantity. Man, though acting on EXTERNAL characters alone and often capriciously, can
0482 turalists, and which consequently have every EXTERNAL characteristic feature of true species, they
0486 s of use and disuse, on the direct action of EXTERNAL conditions, and so forth. The study of domes
0489 y from the indirect and direct action of the EXTERNAL conditions of life, and from use and disuse;
0025 k bar, with the bases of the outer feathers EXTERNALLY edged with white; the wings have two black
0038 deviation of structure excepting such as is EXTERNALLY visible; and indeed he rarely cares for wha
0160 he end of the tail, with the outer feathers EXTERNALLY edged near their bases with white. As all t
0259 emble one of them; and such hybrids, though EXTERNALLY so like one of their pure parent species, a
0285 that no trace of these vast dislocations is EXTERNALLY visible. The Craven fault, for instance, ex
0006 ineal descendants of some other and generally EXTINCT species, in the same manner as the acknowled
0023 ems very improbable; or they must have become EXTINCT in the wild state. But birds breeding on pre
0024 that these very species have since all become EXTINCT or unknown. So many strange contingencies se
0052 hey may whilst in this incipient state become EXTINCT, or they may endure as varieties for very lo
0066 must be produced, or the species will become EXTINCT. It would suffice to keep up the full number
0074 that if the whole genus of humble bees became EXTINCT or very rare in England, the heartsease and
0081 dergo a change, and some species might become EXTINCT. We may conclude, from what we have seen of
0107 eira, according to Oswald Heer, resembles the EXTINCT tertiary flora of Europe. All fresh water ba
0109 ly increasing, numbers inevitably must become EXTINCT. That the number of specific forms has not i
0110 hers will become rarer and rarer, and finally EXTINCT. The forms which stand in closest competitio
0121 species itself, will generally tend to become EXTINCT. So it probably will be with many whole coll
0122 nd all the earlier varieties will have become EXTINCT, having been replaced by eight new species (
0123 (a) and (I), have all become, excepting (F), EXTINCT, and have left no descendants. Hence the six
0124 arent species (A) and (I), now supposed to be EXTINCT and unknown, it will be in some degree inter
0124 cessive strata of the earth's crust including EXTINCT remains. We shall, when we come to our chapt
0124 the diagram throws light on the affinities of EXTINCT beings, which, though generally belonging to
0125 ups; and we can understand this fact, for the EXTINCT species lived at very ancient epochs when th
0126 y most extensively developed, have now become EXTINCT. Looking still more remotely to the future,
0126 ltitude of smaller groups will become utterly EXTINCT, and leave no modified descendants; and cons
0129 mer year may represent the long succession of EXTINCT species. At each period of growth all the gr
0129 may well represent the classification of all EXTINCT and living species in groups subordinate to
0181 . if about a dozen genera of birds had become EXTINCT or were unknown, who would have ventured to
0188 s is in proportion to those which have become EXTINCT, I can see no very great difficulty (not mor
0190 ry ancient ancestral forms, long since become EXTINCT. We should be extremely cautious in concludi
0192 ds. If all pedunculated cirripedes had become EXTINCT, and they have already suffered far more ext
0194 e proportion of living and known forms to the EXTINCT and unknown is very small, I have been aston
```

Page ***(Key Word)***

extinct

Page **(Key Word)**

```
0201  ed; or if it be not so, the being will become  EXTINCT, as myriads have become extinct. Natural sel
0201  g will become extinct, as myriads have become  EXTINCT. Natural selection tends only to make each o
0210  rica, and for no instinct being known amongst  EXTINCT species, how very generally gradations, lead
0219  their aid, the species would certainly become  EXTINCT in a single year. The males and fertile fema
0279  arieties at the present day. On the nature of  EXTINCT intermediate varieties; on their number. On
0282  day; and these parent species, now generally  EXTINCT, have in their turn been similarly connected
0282  nd transitional links, between all living and  EXTINCT species, must have been inconceivably great.
0293  ly rash to suppose that it then became wholly  EXTINCT. We forget how small the area of Europe is c
0294  es are common in the deposit, but have become  EXTINCT in the immediately surrounding sea; or, conv
0299  it has added numerous species to existing and  EXTINCT genera, and has made the intervals between s
0303  key occurred in any tertiary stratum; but now  EXTINCT species have been discovered in India, South
0304  stated that, from the number of existing and  EXTINCT tertiary species; from the extraordinary abu
0312  fe throughout the world. On the affinities of  EXTINCT species to each other and to living species.
0313  ppearance nor disappearance of their many now  EXTINCT species has been simultaneous in each separa
0313  still be found in the midst of a multitude of  EXTINCT forms. Falconer has given a striking instanc
0315  ed; for those which do not change will become  EXTINCT. In members of the same class the average an
0318  of Mastodon, Megatherium, Toxodon, and other  EXTINCT monsters, which all co existed with still li
0319  ke that of the existing horse, belonged to an  EXTINCT species. Had this horse been still living, b
0319  inly have become rarer and rarer, and finally  EXTINCT; its place being seized on by some more succ
0320  cies generally become rare before they become  EXTINCT, to feel no surprise at the rarity of a spec
0321  as; and a few members of the great and almost  EXTINCT group of Ganoid fishes still inhabit our fre
0322  le species and whole groups of species become  EXTINCT, accords well with the theory of natural sel
0329  ages in the two regions. On the Affinities of  EXTINCT Species to each other, and to living forms.
0329  . Let us now look to the mutual affinities of  EXTINCT and living species. They all fall into one g
0329  ll existing groups, or between them. That the  EXTINCT forms of life help to fill up the wide inter
0329  our attention either to the living or to the  EXTINCT alone, the series is far less perfect than i
0329  our great palaeontologist, Owen, showing how  EXTINCT animals fall in between existing groups. Cuv
0330  ey now are. Some writers have objected to any  EXTINCT species or group of species being considered
0330  r groups. If by this term it is meant that an  EXTINCT form is directly intermediate in all its cha
0330  ave to stand between living species, and some  EXTINCT genera between living genera, even between g
0331  neath the uppermost line may be considered as  EXTINCT. The three existing genera, a14, o14, p14, w
0331  these three families, together with the many  EXTINCT genera on the several lines of descent diver
0332  iagram by the letter F14. All the many forms,  EXTINCT and recent, descended from A, make, as befor
0332  at the diagram we can see that if many of the  EXTINCT forms, supposed to be embedded in the succes
0332  d pachyderms. Yet he who objected to call the  EXTINCT genera, which thus linked the living genera
0332  through many widely different forms. If many  EXTINCT forms were to be discovered above one of the
0333  ntly diverged. Thus it comes that ancient and  EXTINCT genera are often in some slight degree inter
0333  with respect to the mutual affinities of the  EXTINCT forms of life to each other and to living fo
0335  hings with great: if the principal living and  EXTINCT races of the domestic pigeon were arranged a
0335  n the rock pigeon and the carrier have become  EXTINCT; and carriers which are extreme in the impor
0338  classes; or that the geological succession of  EXTINCT forms is in some degree parallel to the embr
0339  same law in this author's restorations of the  EXTINCT and gigantic birds of New Zealand. We see it
0339  s could be added, as the relation between the  EXTINCT and living land shells of Madeira; and betwe
0339  iving land shells of Madeira; and between the  EXTINCT and living brackish water shells of the Aral
0341  mitted. These huge animals have become wholly  EXTINCT, and have left no progeny. But in the caves
0341  y. but in the caves of Brazil, there are many  EXTINCT species which are closely allied in size and
0341  the other whole genera having become utterly  EXTINCT. In failing orders, with the genera and spec
0342  inable varieties, connecting together all the  EXTINCT and existing forms of life by the finest gra
0344  ited from a common progenitor, tend to become  EXTINCT together, and to leave no modified offspring
0344  iffers from those now living. Why ancient and  EXTINCT forms often tend to fill up gaps between exi
0345  tor of groups, since become widely divergent.  EXTINCT forms are seldom directly intermediate betwe
0345  by a long and circuitous course through many  EXTINCT and very different forms. We can clearly see
0388  immense areas, and having subsequently become  EXTINCT in intermediate regions. But the wide distri
0407  uously over a wide area, and then have become  EXTINCT in the intermediate tracts, I think the diff
0408  r by affinity, and are likewise linked to the  EXTINCT beings which formerly inhabited the same con
0410  f transport, and by the species having become  EXTINCT in the intermediate tracts. Both in time and
0422  s now spoken throughout the world; and if all  EXTINCT languages, and all intermediate and slowly c
0423  , as it would connect together all languages,  EXTINCT and modern, by the closest affinities, and w
0428  t for the fact that all organisms, recent and  EXTINCT, are included under a few great orders, unde
0431  ich they perceive between the many living and  EXTINCT members of the same great natural class. Ext
0433  can clearly see how it is that all living and  EXTINCT forms can be grouped together in one great s
0436  limits of possibility. In the paddles of the  EXTINCT gigantic sea lizards, and in the mouths of c
0448  to be retrograde. As all the organic beings,  EXTINCT and recent, which have ever lived on this ea
0449  ogenitors, we can clearly see why ancient and  EXTINCT forms of life should resemble the embryos of
0456  of the relationship, by which all living and  EXTINCT beings are united by complex, radiating, and
0463  between them, but only between each and some  EXTINCT and supplanted form. Even on a wide area, wh
0463  e of connecting links, between the living and  EXTINCT inhabitants of the world, and at each succes
0463  ld, and at each successive period between the  EXTINCT and still older species, why is not every ge
0467  , and which shall decrease, or finally become  EXTINCT. As the individuals of the same species come
0476  the chain of descent. The grand fact that all  EXTINCT organic beings belong to the same system wit
0476  diate groups, follows from the living and the  EXTINCT being the offspring of common parents. As th
0476  in some vague sense, higher than ancient and  EXTINCT forms; and they are in so far higher as the
0476  within a confined country, the recent and the  EXTINCT will naturally be allied by descent. Looking
0478  em, with group subordinate to group, and with  EXTINCT groups often falling in between recent group
0488  first creature, the progenitor of innumerable  EXTINCT and living descendants, was created. In the
0489  left no descendants, but have become utterly  EXTINCT. We can so far take a prophetic glance into
0005  ral Selection almost inevitably causes much  EXTINCTION of the less improved forms of life, and ind
0062  ry fact on distribution, rarity, abundance,  EXTINCTION, and variation, will be dimly seen or quite
0073  umption, that we marvel when we hear of the  EXTINCTION of an organic being; and as we do not see t
0080  lation, number of individuals. Slow action.  EXTINCTION caused by Natural Selection. Divergence of
0080  ection, through Divergence of Character and  EXTINCTION, on the descendants from a common parent. E
0086  for if they became so, they would cause the  EXTINCTION of the species. Natural selection will modi
0108  ll be enabled to spread: there will be much  EXTINCTION of the less improved forms, and the relativ
0109  rse of time by nature's power of selection.  EXTINCTION. This subject will be more fully discussed
0109  ys, as geology tells us, is the precursor to  EXTINCTION. We can, also, see that any form represente
0109  of its enemies, run a good chance of utter  EXTINCTION. But we may go further than this! for as ne
```

Page **(Key Word)**

Page ***(Key Word)***

0110 ed, without causing, as far as we know, the EXTINCTION of any natives. Furthermore, the species wh
0116 the principles of natural selection and of EXTINCTION, will tend to act. The accompanying diagram
0121 , another of our principles, namely that of EXTINCTION, will have played an important part. As in
0126 , that is, which as yet have suffered least EXTINCTION, will for a long period continue to increas
0127 chapters. But we already see how it entails EXTINCTION; and how largely extinction has acted in th
0127 how it entails extinction; and how largely EXTINCTION has acted in the world's history, geology p
0128 eads to divergence of character and to much EXTINCTION of the less improved and intermediate forms
0129 plex action of natural selection, entailing EXTINCTION and divergence of character, as we have see
0172 with which it comes into competition. Thus EXTINCTION and natural selection will, as we have seen
0189 according to my theory, there has been much EXTINCTION. Or again, if we look to an organ common to
0192 ct, and they have already suffered far more EXTINCTION than have sessile cirripeces, who would eve
0203 lmost implies the continual supplanting and EXTINCTION of preceding and intermediate gradations. C
0292 uring subsidence, though there will be much EXTINCTION, fewer new varieties or species will be for
0300 of subsidence there would probably be much EXTINCTION of life; during the periods of elevation, t
0312 and disappearance as do single species. On EXTINCTION. On simultaneous changes in the forms of li
0317 upper beds, marking the decrease and final EXTINCTION of the species. This gradual increase in nu
0317 ngle stem, till the group becomes large. On EXTINCTION. We have as yet spoken only incidentally of
0317 ies. On the theory of natural selection the EXTINCTION of old forms and the production of new and
0318 here is reason to believe that the complete EXTINCTION of the species of a group is generally a sl
0318 onderfully sudden. The whole subject of the EXTINCTION of species has been involved in the most gr
0318 one I think can have marvelled more at the EXTINCTION of species, than I have done. When I found
0319 ply sufficient to cause rarity, and finally EXTINCTION. We see in many cases in the more recent te
0319 t tertiary formations, that rarity precedes EXTINCTION; and we know that this has been the progres
0320 comes into competition; and the consequent EXTINCTION of less favoured forms almost inevitably fo
0320 the production of new forms has caused the EXTINCTION of about the same number of old forms. The
0321 habit our fresh waters. Therefore the utter EXTINCTION of a group is generally, as we have seen, a
0322 of natural selection. We need not marvel at EXTINCTION; if we must marvel, let it be at our presum
0322 feel surprise why we cannot account for the EXTINCTION of this particular species or group of spec
0325 t all these modifications of species, their EXTINCTION, and the introduction of new ones, cannot b
0327 il. As they prevailed, they would cause the EXTINCTION of other and inferior forms; and as these i
0327 t a considerable amount of modification and EXTINCTION, and that there was much migration from oth
0328 region than in the other for modification, EXTINCTION, and immigration. I suspect that cases of t
0332 d this order, from the continued effects of EXTINCTION and divergence of character, has become div
0334 low. We must, however, allow for the entire EXTINCTION of some preceding forms, and for the coming
0342 rmations; that there has probably been more EXTINCTION during the periods of subsidence, and more
0343 ll undergo modification to some extent. The EXTINCTION of old forms is the almost inevitable conse
0344 ing on the face of the earth. But the utter EXTINCTION of a whole group of species may often be a
0377 the tropical plants probably suffered much EXTINCTION; how much no one can say; perhaps formerly
0380 educed, and this is the first stage towards EXTINCTION. A mountain is an island on the land; and t
0411 affinities, general, complex and radiating. EXTINCTION separates and defines groups. Morphology, b
0429 berrant forms having suffered severely from EXTINCTION, for they are generally represented by extr
0429 stinct from each other, which again implies EXTINCTION. The genera Ornithorhynchus and Lepidosiren
0431 a whole family of species, now broken up by EXTINCTION into distinct groups and sub groups, will h
0431 ct members of the same great natural class. EXTINCTION, as we have seen in the fourth chapter, has
0431 tebrate classes. There has been less entire EXTINCTION of the forms of life which once connected f
0432 by a long, but broken, chain of affinities. EXTINCTION has only separated groups: it has by no mea
0433 stence, and which almost inevitably induces EXTINCTION and divergence of character in the many des
0456 atural selection, with its contingencies of EXTINCTION and divergence of character. In considering
0462 ted range may often be accounted for by the EXTINCTION of the species in the intermediate regions.
0465 of life; during periods of subsidence, more EXTINCTION. With respect to the absence of fossilifero
0471 h the almost inevitable contingency of much EXTINCTION, explains the arrangement of all the forms
0475 s widely different in different groups. The EXTINCTION of species and of whole groups of species,
0478 natural selection with its contingencies of EXTINCTION and divergence of character. On these same
0488 ter by the Creator, that the production and EXTINCTION of the past and present inhabitants of the
0490 , entailing Divergence of Character and the EXTINCTION of less improved forms. Thus, from the war
0304 able sessile cirripede, which he had himself EXTRACTED from the chalk of Belgium. And, as if to ma
0002 h mr. Wallace's excellent memoir, some brief EXTRACTS from my manuscripts. This Abstract, which I
0024 he intentionally or by chance picked out EXTRAORDINARILY abnormal species; and further, that thes
0065 ils to be surprising, simply explains the EXTRAORDINARILY rapid increase and wide diffusion of nat
0068 when we remember that ten per cent. is an EXTRAORDINARILY severe mortality from epidemics with man
0153 he species. But as the variability of the EXTRAORDINARILY developed part or organ has been so grea
0154 ication has been comparatively recent and EXTRAORDINARILY great that we ought to find the generati
0158 degree of variability in a part, however EXTRAORDINARILY it may be developed, if it be common to
0169 dified state. But when a species with any EXTRAORDINARILY developed organ has become the parent of
0238 reed of cattle, always yielding oxen with EXTRAORDINARILY long horns, could be slowly formed by ca
0238 soldier neuters, with jaws and instincts EXTRAORDINARILY different: in Cryptocerus, the workers o
0401 ss the archipelago, and gales of wind are EXTRAORDINARILY rare; so that the islands are far more e
0437 a box composed of such numerous and such EXTRAORDINARILY shaped pieces of bone? As Owen has remar
0445 me of these birds, when mature, differ so EXTRAORDINARILY in length and form of beak, that they wo
0013 ions, any very rare deviation, due to some EXTRAORDINARY combination of circumstances, appears in
0017 domestication animals and plants having an EXTRAORDINARY inherent tendency to vary, and likewise t
0127 ageous to them, I think it would be a most EXTRAORDINARY fact if no variation ever had occurred un
0136 than in Madeira itself; and especially the EXTRAORDINARY fact, so strongly insisted on by Mr. Woll
0140 ded transportation, I think the common and EXTRAORDINARY capacity in our domestic animals of not o
0148 he case with the male Ibla, and in a truly EXTRAORDINARY manner with the Proteolepas: for the cara
0150 ion. A part developed in any species in an EXTRAORDINARY degree or manner, in comparison with the
0153 ture. When a part has been developed in an EXTRAORDINARY manner in any one species, compared with
0153 y conclude that this part has undergone an EXTRAORDINARY amount of modification, since the period
0153 re for more than one geological period. An EXTRAORDINARY amount of modification implies an unusual
0158 part which is developed in a species in an EXTRAORDINARY manner in comparison with the same part i
0169 e genus. Any part or organ developed to an EXTRAORDINARY size or in an extraordinary manner, in co
0169 eveloped to an extraordinary size or in an EXTRAORDINARY manner, in comparison with the same part
0169 allied species, must have gone through an EXTRAORDINARY amount of modification since the genus ar
0169 a fixed character to the organ, in however EXTRAORDINARY a a manner it may be developed. Species i
0219 e, and put all to rights. What can be more EXTRAORDINARY than these well ascertained facts? If we
0220 ll be excused for doubting the truth of so EXTRAORDINARY and odious an instinct as that of making

Page ***(Key Word)***

Page **(Key Word)**

```
0262 we are reminded by this latter fact of the EXTRAORDINARY case of Hippeastrum, Lobelia, etc., which
0304 ing and extinct tertiary species; from the EXTRAORDINARY abundance of the individuals of many spec
0337 have beaten the highest molluscs. From the EXTRAORDINARY manner in which European productions have
0431 is without this aid, we can understand the EXTRAORDINARY difficulty which naturalists have experie
0016 al trifling respects, they often differ in an EXTREME degree in some one part, both when compared
0019 riginal stocks, has been carried to an absurd EXTREME by some authors. They believe that every rac
0020 e must admit the former existence of the most EXTREME forms, as the Italian greyhound, bloodhound,
0020 ardly two of them will be alike, and then the EXTREME difficulty, or rather utter hopelessness, of
0020 very distinct breeds could not be got without EXTREME care and long continued selection; nor can I
0026 ly fertile, inter se, seems to me rash in the EXTREME. From these several reasons, namely, the imp
0041 al differences are not amply sufficient, with EXTREME care, to allow of the accumulation of a larg
0043  for the cultivator here quite disregards the EXTREME variability both of hybrids and mongrels, an
0068  of food for each species of course gives the EXTREME limit to which each can increase; but very f
0068 mbers of a species, and periodical seasons of EXTREME cold or drought, I believe to be the most ef
0068 kind of food. Even when climate, for instance EXTREME cold, acts directly, it will be the least vi
0070 hat is, abounding in individuals, even on the EXTREME confines of their range. For in such cases,
0078  of the climate alone. Not until we reach the EXTREME confines of life, in the arctic regions or o
0082 n can do nothing. Not that, as I believe, any EXTREME amount of variability is necessary; as man c
0096 tance; but I must here treat the subject with EXTREME brevity, though I have the materials prepare
0107 avourable to natural selection, as far as the EXTREME intricacy of the subject permits. I conclude
0108 . that natural selection will always act with EXTREME slowness, I fully admit. Its action depends
0121 ature: hence in the diagram I have chosen the EXTREME species (A), and the nearly extreme species
0121 hosen the extreme species (A), and the nearly EXTREME species (I), as those which have largely var
0123 divergent tendency of natural selection, the EXTREME amount of difference in character between sp
0123 ered largely from (A), standing nearly at the EXTREME points of the original genus, the six descen
0153 nus. This period will seldom be remote in any EXTREME degree, as species very rarely endure for mo
0158  species possess in common; that the frequent EXTREME variability of any part which is developed i
0171 fferent from those of their allies. Organs of EXTREME perfection. Means of transition. Cases of di
0184 ike a whale, insects in the water. Even in so EXTREME a case as this, if the supply of insects wer
0186 nd petrels with the habits of auks. Organs of EXTREME perfection and complication. To suppose that
0215 ribute the whole of the inherited change from EXTREME wildness to extreme tameness, simply to habi
0215 the inherited change from extreme wildness to EXTREME tameness, simply to habit and long continued
0225  the opposite side. In the series between the EXTREME perfection of the cells of the hive bee and
0229 flat: it was absolutely impossible, from the EXTREME thinness of the little rhombic plate, that t
0230 nly the one rhombic plate which stands on the EXTREME growing margin, or the two plates, as the ca
0232 the hexagonal walls of a single cell, or the EXTREME margin of the circumferential rim of a growi
0233 regoing theory; namely, that the cells on the EXTREME margin of wasp combs are sometimes strictly
0239 in size and sometimes in colour; and that the EXTREME forms can sometimes be perfectly linked toge
0241 s in the case of the driver ant; and then the EXTREME forms, from being the most useful to the com
0256  a sign of incipient fertilisation. From this EXTREME degree of sterility we have self fertilised
0256 ies which can be crossed very rarely, or with EXTREME difficulty, but the hybrids, when at last pr
0257 ed species which will not unite, or only with EXTREME difficulty; and on the other hand of very di
0260 sterile hybrids; and other species cross with EXTREME difficulty, and yet produce fairly fertile h
0272 pects. I shall here discuss this subject with EXTREME brevity. The most important distinction is,
0272 ids are propagated for several generations an EXTREME amount of variability in their offspring is
0273  the first generation, in contrast with their EXTREME variability in the succeeding generations, i
0280 y. the explanation lies, as I believe, in the EXTREME imperfection of the geological record. In th
0289 ence from fossil remains is fragmentary in an EXTREME degree. For instance, not a land shell is kn
0293 dle of any formation, it would be rash in the EXTREME to infer that it had not elsewhere previousl
0334 ce, do not accord in arrangement. The species EXTREME in character are not the oldest, or the most
0335 r have become extinct; and carriers which are EXTREME in the important character of length of beak
0346 the central parts of the United States to its EXTREME southern point, we meet with the most divers
0352 ace. Undoubtedly there are very many cases of EXTREME difficulty, in understanding how the same sp
0365 y regions of the Alps or Pyrenees, and in the EXTREME northern parts of Europe; but it is far more
0369 m each other by the Atlantic Ocean and by the EXTREME northern part of the Pacific. During the Gla
0378 t the foot of the Pyrenees. At this period of EXTREME cold, I believe that the climate under the e
0426 ot a single character in common, yet if these EXTREME forms are connected together by a chain of i
0475 that the geological record is imperfect in an EXTREME degree, then such facts as the record gives,
0487 noble science of Geology loses glory from the EXTREME imperfection of the record. The crust of the
0488 , the rate of change may have been slow in an EXTREME degree. The whole history of the world, as a
0010 tc., and sometimes by seed. These sports are EXTREMELY rare under nature, but far from rare under
0010 buckman's recent experiments on plants seem EXTREMELY valuable. When all or nearly all the indivi
0020  be obtained nearly intermediate between two EXTREMELY different races or species, I can hardly be
0020 and sometimes (as I have found with pigeons) EXTREMELY uniform, and everything seems simple enough
0033 erent the leaves of the cabbage are, and how EXTREMELY alike the flowers, how unlike the flowers o
0039 ary to catch the fancier's eye: he perceives EXTREMELY small differences, and it is in human natur
0046 th individual differences, which seems to me EXTREMELY perplexing: I refer to those genera which h
0065 eds by the thousand, and those which produce EXTREMELY few, is, that the slow breeders would requi
0070  as that of very rare plants being sometimes EXTREMELY abundant in the few spots where they do occ
0071 eans of investigation, there was a large and EXTREMELY barren heath, which had never been touched
0072 sly crowing young firs. Yet the heath was so EXTREMELY barren and so extensive that no one would e
0075 eties. To keep up a mixed stock of even such EXTREMELY close varieties as the variously coloured s
0078 ert, will competition cease. The land may be EXTREMELY cold or dry, yet there will be competition
0082 agling together with nicely balanced forces, EXTREMELY slight modifications in the structure or ha
0102  favourable to Natural Selection. This is an EXTREMELY intricate subject. A large amount of inheri
0102 ty in each individual, and is, I believe, an EXTREMELY important element of success. Though nature
0114 seen under many natural circumstances. In an EXTREMELY small area, especially if freely open to im
0117 offspring. The variations are supposed to be EXTREMELY slight, but of the most diversified nature;
0121 n each genus, the species, which are already EXTREMELY different in character, will generally tend
0122 of their country. It seems, therefore to me EXTREMELY probable that they will have taken the plac
0126 hat of the species living at any one period, EXTREMELY few will transmit descendants to a remote f
0126 fication, but I may add that on this view of EXTREMELY few of the more ancient species having tran
0126 the animal and vegetable kingdoms. Although EXTREMELY few of the most ancient species may now hav
0132 limate, food, etc., produces on any being is EXTREMELY doubtful. My impression is, that the effect
0132 ubtful. My impression is, that the effect is EXTREMELY small in the case of animals, but perhaps r
0139 say a few words on acclimatisation. As it is EXTREMELY common for species of the same genus to inh
```

Page **(Key Word)**

Page **(Key Word)**

```
0151  , very important structures, and they differ  EXTREMELY  little even in different genera: but in the
0174  t explained. In the first place we should be    EXTREMELY  cautious in inferring, because an area is n
0175  nemies or of its prey, or in the seasons, be    EXTREMELY  liable to utter extermination: and thus its
0179  l in a future chapter attempt to show, in an    EXTREMELY  imperfect and intermittent record. On the o
0181  y was falsely ranked amongst bats. It has an    EXTREMELY  wide flank membrane, stretching from the co
0189  the organ must have been first formed at an     EXTREMELY  remote period, since which all the many mem
0190  rms, long since become extinct. We should be    EXTREMELY  cautious in concluding that an organ could
0192  washed out of the sack? Although we must be     EXTREMELY  cautious in concluding that any organ could
0211  esult of experience. But as the excretion is    EXTREMELY  viscid, it is probably a convenience to the
0212  same species, born in a state of nature, is     EXTREMELY  diversified, can be shown by a multitude of
0223  slaves, and in the early part of the summer     EXTREMELY  few. The masters determine when and where a
0232  rcumferential rim of a growing comb, with an    EXTREMELY  thin layer of melted vermilion wax; and I i
0254  both in first crosses and in hybrids, is an     EXTREMELY  general result: but that it cannot, under o
0257  ering efforts have failed to produce between    EXTREMELY  close species a single hybrid. Even within
0257  inhabiting different stations and fitted for    EXTREMELY  different climates, can often be crossed wi
0258  of facility in making reciprocal crosses is     EXTREMELY  common in a lesser degree. He has observed
0259  ure parent species, are with rare exceptions    EXTREMELY  sterile. So again amongst hybrids which are
0260  hink not. For why should the sterility be so    EXTREMELY  different in degree, when various species a
0264  oved from their natural conditions, they are    EXTREMELY  liable to have their reproductive systems s
0265  der conditions not natural to them, they are    EXTREMELY  liable to vary, which is due, as T believe,
0280  s which have ever existed, we should have an    EXTREMELY  close series between both and the rock pige
0284  cess of accumulation in any one area must be    EXTREMELY  slow. But the amount of denudation which th
0290  onclude that sediment must be accumulated in    EXTREMELY  thick, solid, or extensive masses, in order
0291  onclude that the bottom will be inhabited by    EXTREMELY  few animals, and the mass when upraised wil
0292  the geological record, viewed as a whole, is    EXTREMELY  imperfect: but if we confine our attention
0301  r state, in a nearly uniform, though perhaps    EXTREMELY  slight degree, they would, according to the
0302  from some one progenitor, must have been an     EXTREMELY  slow process: and the progenitors must have
0314  degree. The process of modification must be    EXTREMELY  slow. The variability of each species is qu
0333  en more numerous, they will have endured for    EXTREMELY  unequal lengths of time, and will have been
0341  empted to show that the geological record is    EXTREMELY  imperfect; that only a small portion of the
0342  st have tended to make the geological record    EXTREMELY  imperfect, and will to a large extent expla
0347  degrees and 35 degrees, we shall find parts    EXTREMELY  similar in all their conditions, yet it wou
0380  a certain extent beaten the natives: whereas    EXTREMELY  few southern forms have become naturalised
0385  re hatching; and I found that numbers of the    EXTREMELY  minute and just hatched shells crawled on t
0390  se found nowhere else in the world) is often    EXTREMELY  large. If we compare, for instance, the num
0405  genera, we should bear in mind that some are    EXTREMELY  ancient, and must have branched off from a
0407  that of the individuals of the same species,    EXTREMELY  grave. As exemplifying the effects of clima
0425  e males and females and larvae are sometimes    EXTREMELY  different; and as it has been used in class
0429  ction, for they are generally represented by    EXTREMELY  few species: and such species as do occur a
0437  owers; that organs, which when mature become    EXTREMELY  different, are at an early stage of growth
0437  ses? Why should one crustacean, which has an    EXTREMELY  complex mouth formed of many parts, consequ
0441  gs, a pair of magnificent compound eyes, and    EXTREMELY  complex antennae; but they have a closed an
0444  on, or most of them, may have appeared at an    EXTREMELY  early period. I have stated in the first ch
0450  ndition, bearing the stamp of inutility, are    EXTREMELY  common throughout nature. For instance, rud
0450  some of the cases of rudimentary organs are    EXTREMELY  curious: for instance, the presence of teet
0455  ted, not at the corresponding age, but at an    EXTREMELY  early period of life (as we have good reaso
0460  nnot, I think, be disputed. It is, no doubt,    EXTREMELY  difficult even to conjecture by what gradat
0460  natura non facit saltum, that we ought to be    EXTREMELY  cautious in saying that any organ or instin
0027  an make an almost perfect series between the   EXTREMES  of structure. Thirdly, those characters whic
0112  will not admire a medium standard, but like   EXTREMES, they both go on (as has actually occurred w
0286  ffect the estimate as applied to the western  EXTREMITY  of the district. If, then, we knew the rate
0308  a formation is, the more it has suffered the  EXTREMITY  of denudation and metamorphism. The case at
0373  nt, from lat. 41 degrees to the southernmost  EXTREMITY, we have the clearest evidence of former gl
0374  ate zones, and on both sides of the southern  EXTREMITY  of the continent. If this be admitted, it i
0032  ences absolutely inappreciable by an uneducated  EYE, differences which I for one have vainly attem
0032  iate. Not one man in a thousand has accuracy of  EYE  and judgment sufficient to become an eminent b
0039  cture would be necessary to catch the fancier's  EYE: he perceives extremely small differences, and
0083  some modification prominent enough to catch his  EYE, or to be plainly useful to him. Under nature,
0137  nd. In some of the crabs the foot stalk for the  EYE  remains, though the eye is gone; the stand for
0137  the foot stalk for the eye remains, though the  EYE  is gone; the stand for the telescope is there,
0172  and, organs of such wonderful structure, as the  EYE, of which we hardly as yet fully understand th
0186  erfection and complication. To suppose that the  EYE, with all its inimitable contrivances for adju
0186  numerous gradations from a perfect and complex  EYE  to one very imperfect and simple, each grade b
0186  ssessor, can be shown to exist; if further, the  EYE  does vary ever so slightly, and the variations
0186  ficulty of believing that a perfect and complex  EYE  could be formed by natural selection, though i
0187  all amount of gradation in the structure of the  EYE, and from fossil species we can learn nothing
0187  um to discover the earlier stages, by which the  EYE  has been perfected. In the Articulata we can c
0188  o admit that a structure even as perfect as the  EYE  of an eagle might be formed by natural selecti
0188  it is scarcely possible to avoid comparing the   EYE  to a telescope. We know that this instrument h
0188  man intellects; and we naturally infer that the  EYE  has been formed by a somewhat analogous proces
0188  owers like those of man? If we must compare the  EYE  to an optical instrument, we ought in imaginat
0195  case of an organ as perfect and complex as the  EYE. In the first place, we are much too ignorant
0202  be perfect even in that most perfect organ, the  EYE. If our reason leads us to admire with enthusi
0204  ough the belief that an organ so perfect as the  EYE  could have been formed by natural selection, i
0228  converted into shallow basins, appearing to the  EYE  perfectly true or parts of a sphere, and of ab
0229  een left ungnawed, were situated, as far as the  EYE  could judge, exactly along the planes of imagi
0339  relationship is manifest, even to an uneducated  EYE, in the gigantic pieces of armour like those o
0441  have three pairs of legs, a very simple single   EYE, and a proprosciform mouth, with which they f
0441  onverted into a minute, single, and very simple  EYE  spot. In this last and completed state, cirrip
0445  much as their parents; and this, judging by the  EYE, seemed almost to be the case: but on actually
0467  t as to be quite inappreciable by an uneducated  EYE. This process of selection has been the great
0445  eak, width of mouth, length of nostril and of   EYELID, size of feet and length of leg, in the wild
0021  and this is accompanied by greatly elongated    EYELIDS, very large external orifices to the nostril
0022  gape of mouth, the proportional length of the   EYELIDS, of the orifice of the nostrils, of the tong
0137  uction in their size with the adhesion of the   EYELIDS  and growth of fur over them, might in such c
0012  ation are quite whimsical: thus cats with blue  EYES  are invariably deaf; colour and constitutional
```

Page ***(Key Word)***

Page **(Key Word)**

```
0030  for different purposes, or so beautiful in his  EYES, we must, I think, look further than to mere v
0137  to swim at all and had stuck to the wreck. The  EYES of moles and of some burrowing rodents are rud
0137  covered up by skin and fur. This state of the   EYES is probably due to gradual reduction from disu
0137  ting membrane. As frequent inflammation of the  EYES must be injurious to any animal, and as eyes a
0137  e eyes must be injurious to any animal, and as  EYES are certainly not indispensable to animals wit
0137  been lost. As it is difficult to imagine that  EYES, though useless, could be in any way injurious
0137  f the blind animals, namely, the cave rat, the  EYES are of immense size; and Professor Silliman th
0138  of light and to have increased the size of the  EYES; whereas with all the other inhabitants of the
0138  ew have more or less perfectly obliterated its  EYES, and natural selection will often have effecte
0144  e more singular than the relation between blue  EYES and deafness in cats; and the tortoise shell c
0188  that there is much graduated diversity in the  EYES of living crustaceans, and bearing in mind how
0199  structures have been created for beauty in the  EYES of man, or for mere variety. This doctrine, if
0236  d in being destitute of wings and sometimes of  EYES, and in instinct. As far as instinct alone is
0240  h has observed, the larger workers have simple  EYES (ocelli), which though small can be plainly di
0240  cimens of these workers, I can affirm that the  EYES are far more rudimentary in the smaller worker
0304  nstance, which from having passed under my own  EYES has much struck me. In a memoir on Fossil Sess
0441  natatory legs, a pair of magnificent compound  EYES, and extremely complex antennae; but they have
0441  outh; but they have no antennae, and their two  EYES are now reconverted into a minute, single, and
0449  view we can understand how it is that, in the   EYES of most naturalists, the structure of the embr
0454  they become rudimentary, as in the case of the  EYES of animals inhabiting dark caverns, and of the
0473  les, which are habitually blind and have their  EYES covered with skin; or when we look at the blin
0084  birds of prey; and hawks are guided by their   EYESIGHT to their prey, so much so, that on parts of
0182  s, like the logger headed duck (Micropterus of  EYTON); as fins in the water and front legs on the
0253  y have bred inter se. This was effected by Mr.  EYTON, who raised two hybrids from the same parents
0218  nsects) are parasitic on other species; and M.  FABRE has lately shown good reason for believing th
0022  al breeds, the development of the bones of the_ FACE in length and breadth and curvature differs en
0062  mly seen or quite misunderstood. We behold the  FACE of nature bright with gladness, we often see s
0067  st instantaneously increase to any amount. The  FACE of Nature may be compared to a yielding surfac
0073  un the forces are so nicely balanced, that the  FACE of nature remains uniform for long periods of
0164  d sometimes treble, is common; the side of the  FACE, moreover, is sometimes striped. The stripes a
0165  ad some zebra like stripes on the sides of its  FACE. With respect to this last fact, I was so conv
0165  hat I was led solely from the occurence of the  FACE stripes on this hybrid from the ass and hemion
0166  nd hemionus, to ask Colonel Poole whether such  FACE stripes ever occur in the eminently striped Ka
0170  ucture, by which the innumerable beings on the  FACE of this earth are enabled to struggle with eac
0232  ards so that the comb has to be built over one  FACE of the slip, in this case the bees can lay the
0312  either locally, or, as far as we know, on the  FACE of the earth. If we may trust the observations
0313  . when a species has once disappeared from the  FACE of the earth, we have reason to believe that t
0344  her, and to leave no modified offspring on the  FACE of the earth. But the utter extinction of a wh
0346  ng the distribution of organic beings over the  FACE of the globe, the first great fact which strik
0021  ing. Compare the English carrier and the short  FACED tumbler, and see the wonderful difference in
0021  nostrils, and a wide gape of mouth. The short   FACED tumbler has a beak in outline almost like tha
0023  ist would place the English carrier, the short  FACED tumbler, the runt, the barb, pouter, and fant
0024  t of the English carrier, or that of the short  FACED tumbler, or barb; for reversed feathers like
0027  secondly, although an English carrier or short  FACED tumbler differs immensely in certain characte
0152  ciers. Even in the sub breeds, as in the short  FACED tumbler, it is notoriously cifficult to breed
0153  s coarse as a common tumbler from a good short  FACED strain. But as long as selection is rapidly g
0424  ng the common habit of tumbling; but the short  FACED breed has nearly or quite lost this habit; ne
0446  ption to this rule, for the young of the short  FACED tumbler differed from the young of the wild r
0446  ng not early period. But the case of the short  FACED tumbler, which when twelve hours old had acqu
0447  st appeared. In either case (as with the short  FACED tumbler) the young or embryo would closely re
0187  s a double cornea, the inner one divided into   FACETS, within each of which there is a lens shaped
0197  n for aiding parturition, and no doubt they    FACILITATE, or may be indispensable for this act; but
0205  aided by the other, must often have largely    FACILITATED transitions. We are far too ignorant, in a
0210  rgans. Changes of instinct may sometimes be    FACILITATED by the same species having different insti
0462  nd such changes will obviously have greatly    FACILITATED migration. As an example, I have attempted
0358  ansport across the sea, the greater or less    FACILITIES may be said to be almost wholly unknown. Un
0042  in the case of animals with separate sexes,    FACILITY in preventing crosses is an important elemen
0070  ve from some cause, possibly in part through    FACILITY of diffusion amongst the crowded animals, be
0256  two pure species can be united with unusual    FACILITY, and produce numerous hybrid offspring, yet
0257  losely allied species generally uniting with    FACILITY. But the correspondence between systematic a
0257  pondence between systematic affinity and the    FACILITY of crossing is by no means strict. A multitu
0257  distinct species which unite with the utmost   FACILITY. In the same family there may be a genus, as
0258  often the widest possible difference in the    FACILITY of making reciprocal crosses. Such cases are
0258  ner, moreover, found that this difference of    FACILITY in making reciprocal crosses is extremely co
0259  pearance either parent. And lastly, that the   FACILITY of making a first cross between any two spec
0260  nally the widest possible difference, in the   FACILITY of effecting an union. The hybrids, moreover
0260  species? Why should some species cross with    FACILITY, and yet produce very sterile hybrids; and o
0260  es of sterility, not strictly related to the   FACILITY of the first union between their parents; I
0262  s of the pear take with different degrees of   FACILITY on the quince; so do different varieties of
0262  fted together. As in reciprocal crosses, the   FACILITY of effecting an union is often very far from
0263  t the curious and complex laws governing the   FACILITY with which trees can be grafted on each othe
0263  at the still more complex laws governing the   FACILITY of first crosses, are incidental on unknown
0268  r in external appearance, cross with perfect   FACILITY, and yield perfectly fertile offspring. I fu
0276  systems, so in crossing, the greater or less   FACILITY of one species to unite with another, is inc
0277  h are crossed. Nor is it surprising that the   FACILITY of effecting a first cross, the fertility of
0390  o not become modified having immigrated with   FACILITY and in a body, so that their mutual relation
0392  planation seems to me not a little doubtful.   FACILITY of immigration, I believe, has been at least
0404  es and marshes, excepting in so far as great   FACILITY of transport has given the same general form
0405  me of the species range very widely; for the  FACILITY with which widely ranging species vary and g
0171  of transition. Cases of difficulty. Natura non FACIT saltum. Organs of small importance. Organs no
0194  hat old canon in natural history of Natura non FACIT saltum. We meet with this admission in the wr
0206  that old canon in natural history, Natura non FACIT saltum. This canon if we look only to the pre
0210  ts, can be discovered. The canon of Natura non FACIT saltum applies with almost equal force to ins
0243  at the canon in natural history, of natura non FACIT saltum is applicable to instincts as well as
0460  ure, as is proclaimed by the canon, Natura non FACIT saltum, that we ought to be extremely cautiou
0471  and slow steps. Hence the canon of Natura non FACIT saltum, which every fresh addition to our kno
```

Page **(Key Word)**

Page **(Key Word)**

```
0010  heir earliest stages of formation; so that, in FACT, sports support my view, that variability may
0013  y but not exclusively to the like sex. It is a FACT of some little importance to us, that peculiar
0018  ith, those still existing. Even if this latter FACT were found more strictly and generally true th
0026  dozen or twenty generations, for we know of no FACT countenancing the belief that the child ever r
0027  minently variable; and the explanation of this FACT will be obvious when we come to treat of selec
0033  rd. With animals this kind of selection is, in FACT, also followed; for hardly any one is so carel
0034  savages. It would, indeed, have been a strange FACT, had attention not been paid to breeding, for
0037  ulated, explains, as I believe, the well known FACT, that in a vast number of cases we cannot reco
0040  history of any of our domestic breeds. But, in FACT, a breed, like a dialect of a language, can ha
0050  of doubtful value. I have been struck with the FACT, that if any animal or plant in a state of nat
0054  ain, might have been anticipated; for the mere FACT of many species of the same genus inhabiting a
0062  d that the whole economy of nature, with every FACT on distribution, rarity, abundance, extinction
0087  me hereditarily attached to that sex, the same FACT probably occurs under nature, and if so, natur
0099  flower is ready, so that these plants have in FACT separated sexes, and must habitually be crosse
0100  ises itself. We can understand this remarkable FACT, which offers so strong a contrast with terres
0101  odites, and some of them unisexual. But if, in FACT, all hermaphrodites do occasionally intercross
0115  n in the inhabitants of the same region is, in FACT, the same as that of the physiological divisio
0124  en existing groups; and we can understand this FACT, for the extinct species lived at very ancient
0127  them, I think it would be a most extraordinary FACT if no variation ever had occurred useful to ea
0128  ings may be explained. It is a truly wonderful FACT, the wonder of which we are apt to overlook fr
0129  reated, I can see no explanation of this great FACT in the classification of all organic beings; b
0132  . several other such cases could be given. The FACT of varieties of one species, when they range i
0135  rby has remarked (and I have observed the same FACT) that the anterior tarsi, or feet, of many mal
0135  n. mr. Wollaston has discovered the remarkable FACT that 200 beetles, out of the 550 species inhab
0136  deira itself; and especially the extraordinary FACT, so strongly insisted on by Mr. Wollaston, of
0146  do not open; I should explain the rule by the FACT that seeds could not gradually become winged t
0148  seless structure. I can thus only understand a FACT with which I was much struck when examining ci
0151  not displayed in any unusual manner, of which FACT I think there can be little doubt. But that ou
0155  e individuals of some of the species. And this FACT shows that a character, which is generally of
0157  r parts of their structure. It is a remarkable FACT, that the secondary sexual differences between
0157  the same genus differ from each other. Of this FACT I will give in illustration two instances, the
0160  inheritance. No doubt it is a very surprising FACT that characters should reappear after having b
0165  e sides of its face. With respect to this last FACT, I was so convinced that not even a stripe of
0174  the territory proper to each. We see the same FACT in ascending mountains, and sometimes it is qu
0175  , a common alpine species disappears. The same FACT has been noticed by Forbes in sounding the dep
0180  sional falls. But it does not follow from this FACT that the structure of each squirrel is the bes
0186  type; but this seems to me only restating the FACT in dignified language. He who believes in the
0190  cause it shows us clearly the highly important FACT that an organ originally constructed for one p
0191  ription of these parts, understand the strange FACT that every particle of food and drink which we
0206  ected to the several laws of growth. Hence, in FACT, the Law of the Conditions of Existence is the
0218  instinct may probably be accounted for by the FACT of the hens laying a large number of eggs; but
0231  irst cell. I was able practically to show this FACT, by covering the edges of the hexagonal walls
0233  ommenced cells, is important, as it bears on a FACT, which seems at first quite subversive of the
0238  d on the climax of the difficulty; namely, the FACT that the neuters of several ants differ, not o
0241  orm and number of the teeth. But the important FACT for us is, that though the workers can be grou
0241  re produced. Thus, as I believe, the wonderful FACT of two distinctly defined castes of sterile wo
0242  se of these neuter insects convinced me of the FACT. I have, therefore, discussed this case, at so
0243  udgment, annihilate it. On the other hand, the FACT that instincts are not always absolutely perfe
0245  able of crossing freely. The importance of the FACT that hybrids are very generally sterile, has,
0249  the same hybrid nature. And thus, the strange FACT of the increase of fertility in the successive
0250  he crinum leads me to refer to a most singular FACT, namely, that there are individual plants, as
0252  ation, as Gartner believes to be the case, the FACT would have been notorious to nurserymen. Horti
0258  s could be given. Thuret has observed the same FACT with certain sea weeds or Fuci. Gartner, moreo
0258  hem only as varieties. It is also a remarkable FACT, that hybrids raised from reciprocal crosses,
0262  heir own roots. We are reminded by this latter FACT of the extraordinary case of Hippeastrum, Lobe
0264  roductive systems seriously affected. This, in FACT, is the great bar to the domestication of anim
0268  the pigeon or of the cabbage, is a remarkable FACT; more especially when we reflect how many spec
0271  subsequent observer, has proved the remarkable FACT, that one variety of the common tobacco is mor
0272  i have myself seen striking instances of this FACT. Gartner further admits that hybrids between v
0273  ty in the succeeding generations, is a curious FACT and deserves attention. For it bears on and co
0275  has been laid by some authors on the supposed FACT, that mongrel animals alone are found closely l
0276  laws, this similarity would be an astonishing FACT. But it harmonises perfectly with the view tha
0290  wed each other in close sequence. Scarcely any FACT struck me more when examining many hundred mil
0295  ply whilst the downward movement continues. In FACT, this nearly exact balancing between the suppl
0296  heir accumulation; yet no one ignorant of this FACT would have suspected the vast lapse of time re
0302  have really started into life all at once, the FACT would be fatal to the theory of descent with s
0303  nly been produced. I may recall the well known FACT that in geological treatises, published not ma
0303  nes, are marine, and range over the world, the FACT of not a single bone of a whale having been di
0305  t the commencement of the chalk formation, the FACT would certainly be highly remarkable; but I ca
0310  ure. We see this in the plainest manner by the FACT that all the most eminent palaeontologists, na
0313  ner has given a striking instance of a similar FACT, in an existing crocodile associated with many
0314  g, that it should change less. We see the same FACT in geographical distribution; for instance, in
0316  have attempted to give an explanation of this FACT, which if true would have been fatal to my vie
0319  le, we assuredly should not have perceived the FACT, yet the fossil horse would certainly have bec
0322  ntological discovery is more striking than the FACT, that the forms of life change almost simultan
0324  ked as simultaneous in a geological sense. The FACT of the forms of life changing simultaneously,
0325  nd the nature of their inhabitants. This great FACT of the parallel succession of the forms of lif
0329  l fall into one grand natural system; and this FACT is at once explained on the principle of desce
0335  some degree intermediate in character, is the FACT, insisted on by all palaeontologists, that fos
0335  h the species are distinct in each stage. This FACT alone, from its generality, seems to have shak
0336  the theory of descent, the full meaning of the FACT of fossil remains from closely consecutive for
0345  of more recent animals of the same class, the FACT will be intelligible. The succession of the sa
0346  gs over the face of the globe, the first great FACT which strikes us is, that neither the similari
0347  to the inhabitants of the sea. A second great FACT which strikes us in our general review is, tha
0347  e strictly arctic productions. We see the same FACT in the great difference between the inhabitant
0347  ible. On each continent, also, we see the same FACT; for on the opposite sides of lofty and contin
```

Page **(Key Word)**

0349 opposite meridians of longitude. A third great FACT, partly included in the foregoing statements,
0352 ld not be easily passed over by migration, the FACT is given as something remarkable and exception
0360 miles in width, and would then germinate. The FACT of the larger fruits often floating longer tha
0361 , which I tried, germinated. But the following FACT is more important: the crops of birds do not s
0364 nds of years, it would I think be a marvellous FACT if many plants had not thus become widely tran
0365 om one to the other. It is indeed a remarkable FACT to see so many of the same plants living on th
0367 and of Europe. We can thus also understand the FACT that the Alpine plants of each mountain range
0371 ocean. We can further understand the singular FACT remarked on by several observers, that the pro
0375 n a hill in Europe! Still more striking is the FACT that southern Australian forms are clearly rep
0376 , prof. Dana, that it is certainly a wonderful FACT that New Zealand should have a closer resembla
0377 escaping into the warmest spots. But the great FACT to bear in mind is, that all tropical producti
0378 great lines of invasion: and it is a striking FACT, lately communicated to me by Dr. Hooker, that
0379 ies or as distinct species. It is a remarkable FACT, strongly insisted on by Hooker in regard to A
0386 ink favourable means of dispersal explain this FACT. I have before mentioned that earth occasional
0387 a heron's stomach: although I do not know the FACT, yet analogy makes me believe that a heron fly
0390 ontinent, we shall see that this is true. This FACT might have been expected on my theory, for, as
0397 on this head. The most striking and important FACT for us in regard to the inhabitants of islands
0397 ies. Numerous instances could be given of the FACT. I will give only one, that of the Galapagos A
0398 the conditions of the South American coast: in FACT there is a considerable dissimilarity in all t
0398 the Galapagos to America. I believe this grand FACT can receive no sort of explanation on the ordi
0399 t continent, is so enormously remote, that the FACT becomes an anomaly. But this difficulty almost
0401 and remaining the same. The really surprising FACT in this case of the Galapagos Archipelago, and
0402 st any length of time. Being familiar with the FACT that many species, naturalised through man's a
0402 Wollaston have communicated to me a remarkable FACT bearing on this subject; namely, that Madeira
0406 ll retaining the same specific character. This FACT, together with the seeds and eggs of many low
0413 rs, all united into one class. Thus, the grand FACT in natural history of the subordination of gro
0415 assificatory value, is almost shown by the one FACT, that in allied groups, in which the same orga
0415 ed at any group without being struck with this FACT; and it has been most fully acknowledged in th
0416 rms. Robert Brown has strongly insisted on the FACT that the rudimentary florets are of the highes
0418 very generally been admitted as true. The same FACT holds good with flowering plants, of which the
0424 in a state of nature, every naturalist has in FACT brought descent into his classification; for h
0424 s known to every naturalist: scarcely a single FACT can be predicated in common of the males and f
0428 nd feebler groups. Thus we can account for the FACT that all organisms, recent and extinct, are in
0429 dely spread they are throughout the world, the FACT is striking, that the discovery of Australia h
0429 rant genera. We can, I think, account for this FACT only by looking at aberrant forms as failing g
0438 he same individual. And we can understand this FACT; for in molluscs, even in the lowest members o
0439 terms may be used literally; and the wonderful FACT of the jaws, for instance, of a crab retaining
0452 mentary occasionally differs much. This latter FACT is well exemplified in the state of the wings
0452 horse, ox, and rhinoceros. It is an important FACT that rudimentary organs, such as teeth in the
0453 me no explanation, merely a restatement of the FACT. Would it be thought sufficient to say that be
0468 ity under nature, it would be an unaccountable FACT if natural selection had not come into play. I
0471 h has prevailed throughout all time. The grand FACT of the grouping of all organic beings seems to
0473 genus and in their hybrids! How simply is this FACT explained if we believe that these species hav
0473 high degree permanent, we can understand this FACT; for they have already varied since they branc
0475 anged simultaneously throughout the world. The FACT of the fossil remains of each formation being
0476 te position in the chain of descent. The grand FACT that all extinct organic beings belong to the
0477 same. We see the full meaning of the wonderful FACT, which must have struck every traveller, namel
0478 no explanation on the theory of creation. The FACT, as we have seen, that all past and present or
0482 we give an explanation when we only restate a FACT. Any one whose disposition leads him to attach
0001 as naturalist, I was much struck with certain FACTS in the distribution of the inhabitants of Sou
0001 the past inhabitants of that continent. These FACTS seemed to me to throw some light on the origi
0001 ly accumulating and reflecting on all sorts of FACTS which could possibly have any bearing on it.
0002 onclusions at which I have arrived, with a few FACTS in illustration, but which, I hope, in most c
0002 sity of hereafter publishing in detail all the FACTS, with references, on which my conclusions hav
0002 gle point is discussed in this volume on which FACTS cannot be adduced, often apparently leading t
0002 tained only by fully stating and balancing the FACTS and arguments on both sides of each question;
0003 ibution, geological succession, and other such FACTS, might come to the conclusion that each speci
0004 ted properly only by giving long catalogues of FACTS. We shall, however, be enabled to discuss wha
0012 be given amongst animals and plants. From the FACTS collected by Heusinger, it appears that white
0014 stage. But hereditary diseases and some other FACTS make me believe that the rule has a wider ext
0014 vain endeavoured to discover on what decisive FACTS the above statement has so often and so bold l
0017 the offspring of any single species, then such FACTS would have great weight in making us doubt ab
0018 ts I can form no opinion. I should think, from FACTS communicated to me by Mr. Blyth, on the habit
0025 m to me improbable in the highest degree. Some FACTS in regard to the colouring of pigeons well de
0025 any wild rock pigeon! We can understand these FACTS, on the well known principle of reversion to
0034 e not associated with Europeans. Some of these FACTS do not show actual selection, but they show t
0044 bject at all properly, a long catalogue of dry FACTS should be given, but these I shall reserve fo
0045 parts; but I could show by a long catalogue of FACTS, that parts which must be called important, w
0046 iopod shells, at former periods of time. These FACTS seem to be very perplexing, for they seem to
0052 judged of by the general weight of the several FACTS and views given throughout this work. It need
0056 are absolutely excluded from the tables. These FACTS are of plain signification on the view that s
0070 eservation, explains, I believe, some singular FACTS in nature, such as that of very rare plants b
0094 part on its nectar for food. I could give many FACTS, showing how anxious bees are to save time; f
0094 more trouble, enter by the mouth. Bearing such FACTS in mind, I can see no reason to doubt that an
0096 rst place, I have collected so large a body of FACTS, showing, in accordance with the almost unive
0097 ng diminishes vigour and fertility; that these FACTS alone incline me to believe that it is a gene
0097 , I think, understand several large classes of FACTS, such as the following, which on any other vi
0099 t habitually be crossed. How strange are these FACTS! How strange that the pollen and stigmatic su
0099 ly useless to each other! How simply are these FACTS explained on the view of an occasional cross
0101 veral considerations and from the many special FACTS which I have collected, but which I am not he
0103 nt; for I can bring a considerable catalogue of FACTS, showing that within the same area, varieties
0106 car, perhaps, on these views, understand some FACTS which will be again alluded to in our chapter
0110 e variations. We have evidence of this, in the FACTS given in the second chapter, showing that if
0111 and explains, as I believe, several important FACTS. In the first place, varieties, even strongly
0131 in the first chapter, but a long catalogue of FACTS which cannot be here given would be necessary
0133 e same species under the same conditions. Such FACTS show how indirectly the conditions of life mu

Page **(Key Word)**

0134 ble by us. Effects of Use and Disuse. From the FACTS alluded to in the first chapter, I think ther
0135 e all their species in this condition! Several FACTS, namely, that beetles in many parts of the wo
0140 waites informs me that he has observed similar FACTS in Ceylon, and analogous observations have be
0141 himself and by his domestic animals, and such FACTS as that former species of the elephant and rh
0147 ch have been advanced, and likewise some other FACTS, may be merged under a more general principle
0150 s proposition without giving the long array of FACTS which I have collected, and which cannot poss
0161 s hypothetical, but could be supported by some FACTS; and I can see no more abstract improbability
0166 ative. What now are we to say to these several FACTS? We see several very distinct species of the
0175 all important elements of distribution, these FACTS ought to cause surprise, as climate and heigh
0176 which they connect. Now, if we may trust these FACTS and inferences, and therefore conclude that v
0187 irst originated; but I may remark that several FACTS make me suspect that any sensitive nerve may
0188 be an imperfect vitreous substance. With these FACTS, here far too briefly and imperfectly given,
0188 n finishing this treatise that large bodies of FACTS, otherwise inexplicable, can be explained by
0203 he discussion light has been thrown on several FACTS, which on the theory of independent acts of c
0210 ts: that they do so voluntarily, the following FACTS show. I removed all the ants from a group of
0212 ly diversified, can be shown by a multitude of FACTS. Several cases also, could be given, of occas
0212 l aware that these general statements, without FACTS given in detail, can produce but a feeble eff
0219 more extraordinary than these well ascertained FACTS? If we had not known of any other slave makin
0223 y of heath over its ravaged home. Such are the FACTS, though they did not need confirmation by me,
0239 admit that such wonderful and well established FACTS at once annihilate my theory. In the simpler
0241 m the workers of the several sizes. With these FACTS before me, I believe that natural selection,
0243 ably come into play. I do not pretend that the FACTS given in this chapter strengthen in any great
0243 heory is, also, strengthened by some few other FACTS in regard to instincts; as by that common cas
0245 nces. In treating this subject, two classes of FACTS, to a large extent fundamentally different, h
0248 rbreeding. I have collected so large a body of FACTS, showing that close interbreeding lessens fer
0251 were in an-unnatural state. Nevertheless these FACTS show on what slight and mysterious causes the
0254 an cattle are quite fertile together; but from FACTS communicated to me by Mr. Blyth, I think they
0254 ation. Finally, looking to all the ascertained FACTS on the intercrossing of plants and animals, i
0255 n to exist; but only the barest outline of the FACTS can here be given. When pollen from a plant o
0256 y of the hybrids thus produced, two classes of FACTS which are generally confounced together, is b
0257 ther species of Nicotiana. Very many analogous FACTS could be given. No one has been able to point
0259 have a considerable degree of fertility. These FACTS show how completely fertility in the hybrid i
0260 a strange arrangement. The foregoing rules and FACTS, on the other hand, appear to me clearly to i
0263 ganic beings is attempted to be expressed. The FACTS by no means seem to me to indicate that the g
0264 on Fuci. No explanation can be given of these FACTS, any more than why certain trees cannot be gr
0264 have more than once alluded to a large body of FACTS, which I have collected, showing that when an
0266 rstand, excepting on vague hypotheses, several FACTS with respect to the sterility of hybrids; for
0266 tends to an allied yet very different class of FACTS. It is an old and almost universal belief, fo
0267 to the offspring. I believe, indeed, from the FACTS alluded to in our fourth chapter, that a cert
0267 is an accident or an illusion. Both series of FACTS seem to be connected together by some common
0271 nclined to suspect that they present analogous FACTS. Kolreuter, whose accuracy has been confirmed
0271 manner and in some degree modified. From these FACTS; from the great difficulty of ascertaining th
0271 system; from these several considerations and FACTS, I do not think that the very general fertili
0275 ucas; who, after arranging an enormous body of FACTS with respect to animals, comes to the conclus
0278 tween hybrids and mongrels. Finally, then, the FACTS briefly given in this chapter do not seem to
0282 der, who may not be a practical geologist, the FACTS leading the mind feebly to comprehend the lap
0285 moothly swept away. The consideration of these FACTS impresses my mind almost in the same manner a
0291 gained considerable thickness. All geological FACTS tell us plainly that each area has undergone
0296 vered by the upper beds of the same formation; FACTS, showing what wide, yet easily overlooked, in
0309 iods. If then we may infer anything from these FACTS, we may infer that where our oceans now exten
0310 fect, and who do not attach much weight to the FACTS and arguments of other kinds given in this vo
0312 t chapters. Let us now see whether the several FACTS and rules relating to the geological successi
0314 vince, seems to me satisfactory. These several FACTS accord well with my theory. I believe in no f
0331 n the remark. Let us see how far these several FACTS and inferences accord with the theory of desc
0333 theory of descent with modification, the main FACTS with respect to the mutual affinities of the
0339 of Brazil. I was so much impressed with these FACTS that I strongly insisted, in 1839 and 1845, o
0340 rica than it is at the present time. Analogous FACTS could be given in relation to the distributio
0343 hese difficulties, all the other great leading FACTS in palaeontology seem to me simply to follow
0347 frica under nearly the same climate. Analogous FACTS could be given with respect to the inhabitant
0350 nent and in the American seas. We see in these FACTS some deep organic bond, prevailing throughout
0354 discuss a few of the most striking classes of FACTS; namely, the existence of the same species on
0356 ations. Before discussing the three classes of FACTS, which I have selected as presenting the grea
0356 ly the briefest abstract of the more important FACTS. Change of climate must have had a powerful i
0358 ith the many existing oceanic islands. Several FACTS in distribution, such as the great difference
0358 and the depth of the sea, these and other such FACTS seem to me opposed to the admission of such p
0359 far as we may infer anything from these scanty FACTS, we may conclude that the seeds of 14/100 pla
0361 the transportation of seeds. I could give many FACTS showing how frequently birds of many kinds ar
0363 ly be transported to great distances; for many FACTS could be given showing that soil almost every
0365 ains of Europe. Even as long ago as 1747, such FACTS led Gmelin to conclude that the same species
0366 ely see, affords a simple explanation of these FACTS. We have evidence of almost every conceivable
0375 zealand, by Dr. Hooker, analogous and striking FACTS are given in regard to the plants of that lar
0376 plies to plants alone: some strictly analogous FACTS could be given on the distribution of terrest
0376 see what light can be thrown on the foregoing FACTS, on the belief, supported as it is by a large
0381 ined unaltered. We cannot hope to explain such FACTS, until we can say why one species and not ano
0381 uent modification to the necessary degree. The FACTS seem to me to indicate that peculiar and very
0382 tion through natural selection, a multitude of FACTS in the present distribution both of the same
0384 some fish in common and some different. A few FACTS seem to favour the possibility of their occas
0385 ly spread throughout the same country. But two FACTS, which I have observed, and no doubt many oth
0387 ontained in a breakfast cup! Considering these FACTS, I think it would be an inexplicable circumst
0388 e now come to the last of the three classes of FACTS, which I have selected as presenting the grea
0389 es, but it would not, I think, explain all the FACTS in regard to insular productions. In the foll
0389 on of dispersal; but shall consider some other FACTS, which bear on the truth of the two theories
0391 nsects in Madeira apparently present analogous FACTS. Oceanic islands are sometimes deficient in c
0392 ture of the conditions. Many remarkable little FACTS could be given with respect to the inhabitant
0395 the same on both sides; we meet with analogous FACTS on many islands separated by similar channels
0399 ying their original birthplace. Many analogous FACTS could be given: indeed it is an almost univer

Page **(Key Word)**

Page ***(Key Word)***

```
0402  several islands, and we may infer from certain FACTS that these have probably spread from some one
0404  aves of America and of Europe. Other analogous FACTS could be given. And it will, I believe, be un
0406  ranging genera themselves ranging widely, such FACTS, as alpine, lacustrine, and marsh productions
0408  one source; then I think all the grand leading FACTS of geographical distribution are explicable o
0409  rences in different areas. We see this in many FACTS. The endurance of each species and group of s
0412  closest competition, and by looking to certain FACTS in naturalisation. I attempted also to show t
0437  owth exactly alike. How inexplicable are these FACTS on the ordinary view of creation! Why should
0442  e class, that we might be led to look at these FACTS as necessarily contingent in some manner on g
0442  stage. How, then, can we explain these several FACTS in embryology, namely the very general, but n
0443  hich it is developed. I believe that all these FACTS can be explained, as follows, on the view of
0444  lieve, explain all the above specified leading FACTS in embryology. But first let us look at a few
0446  ciples above given seem to me to explain these FACTS in regard to the later embryonic stages of ou
0446  but at an earlier age. Now let us apply these FACTS and the above two principles, which latter, t
0450  stration. Thus, as it seems to me, the leading FACTS in embryology, which are second in importance
0453  ryonic condition. I have now given the leading FACTS with respect to rudimentary organs. In reflec
0456  ew of descent with modification, all the great FACTS in Morphology become intelligible, whether we
0457  ng period, we can understand the great leading FACTS in Embryology; namely, the resemblance in an
0457  genealogical. Finally, the several classes of FACTS which have been considered in this chapter, s
0458  his view, even if it were unsupported by other FACTS or arguments. Chapter XIV. Recapitulation and
0459  e convenient to the reader to have the leading FACTS and inferences briefly recapitulated. That ma
0460  refer the reader to the recapitulation of the FACTS given at the end of the eighth chapter, which
0461  ther parallel, but directly opposite, class of FACTS: namely, that the vigour and fertility of all
0469  and objections: now let us turn to the special FACTS and arguments in favour of the theory. On the
0471  dently created, no man can explain. Many other FACTS are, as it seems to me, explicable on this th
0472  ccupied or ill occupied place in nature, these FACTS cease to be strange, or perhaps might even ha
0475  ies. On the other hand, these would be strange FACTS if species have been independently created, a
0475  d is imperfect in an extreme degree, then such FACTS as the record gives, support the theory of de
0476  t with modification, most of the great leading FACTS in Distribution. We can see why there should
0477  n islands far distant from any continent. Such FACTS as the presence of peculiar species of bats,
0478  rican mainland. It must be admitted that these FACTS receive no explanation on the theory of creat
0479  nd of the elephant, and innumerable other such FACTS, at once explain themselves on the theory of
0480  understand. I have now recapitulated the chief FACTS and considerations which have thoroughly conv
0481  ts whose minds are stocked with a multitude of FACTS all viewed, during a long course of years, fr
0482  than to the explanation of a certain number of FACTS will certainly reject my theory. A few natura
0206  ; nor, as far as we can judge by our limited FACULTIES, can absolute perfection be everywhere foun
0009  seriously affected by unperceived causes as to FAIL in acting, we need not be surprised at this sy
0032  ants any of these qualities, he will assuredly FAIL. Few would readily believe in the natural capa
0153  lection gains the day, and we do not expect to FAIL so far as to breed a bird as coarse as a commo
0217  wn mother's care, encumbered as she can hardly FAIL to be by having eggs and young of different ag
0334  at the seventh stage; hence they could hardly FAIL to be nearly intermediate in character between
0361  nearly all germinated. Living birds can hardly FAIL to be highly effective agents in the transport
0461  ee sterile, for their constitutions can hardly FAIL to have been disturbed from being compounded o
0469  riations most useful to himself, should nature FAIL in selecting variations useful,,under changing
0020  expressly experimentised for this object, and FAILED. The offspring from the first cross between t
0072  ts head above the stems of the heath, and had FAILED. No wonder that, as soon as the land was encl
0101  and, as in the case of flowers, I have as yet FAILED, after consultation with one of the highest a
0257  , in which the most persevering efforts have FAILED to produce between extremely close species a
0257  a particularly distinct species, obstinately FAILED to fertilise, or to be fertilised by, no less
0258  ra with the pollen of M. jalappa, and utterly FAILED. Several other equally striking cases could b
0417  er, however important that may be, has always FAILED: for no part of the organisation is universal
0118  me producing two or three varieties, and some FAILING to produce any. Thus the varieties or modifi
0341  hole genera having become utterly extinct. In FAILING orders, with the genera and species decreasi
0429  his fact only by looking at aberrant forms as FAILING groups conquered by more successful competit
0460  perfected, more especially amongst broken and FAILING groups of organic beings; but we see so many
0065  l ratio of increase, the result of which never FAILS to be surprising, simply explains the extraor
0349  lling, for instance, from north to south never FAILS to be struck by the manner in which successiv
0466  that this system, when not rendered impotent, FAILS to reproduce offspring exactly like the paren
0132  ertheless, we can here and there dimly catch a FAINT ray of light, and we may feel sure that there
0163  use duns, and in one instance in a chestnut: a FAINT shoulder stripe may sometimes be seen in duns
0085  f white sheep to destroy every lamb with the FAINTEST trace of black. In plants the down of the fr
0163  species are generally striped on the legs and FAINTLY on the shoulder. The quagga, though so plain
0002  y opposite to those at which I have arrived. A FAIR result can be obtained only by fully stating a
0108  ll again be changed: and again there will be a FAIR field for natural selection to improve still f
0151  emarkable degree or manner in any species, the FAIR presumption is that it is of high importance t
0201  or for doing an injury to its possessor. If a FAIR balance be struck between the good and evil ca
0389  comparison in some other respects is not quite FAIR. We have evidence that the barren island of As
0488  of consecutive formations probably serves as a FAIR measure of the lapse of actual time. A number
0039  e set on them, after several breeds have once FAIRLY been established. Many slight differences mig
0053  , he thinks that the following statements are FAIRLY well established. The whole subject, however,
0117  sed to have been accumulated to have formed a FAIRLY well marked variety, such as would be thought
0117  species (A) is supposed to have produced two FAIRLY well marked varieties, namely a1 and m1. Thes
0252  it is possible to imagine. Had hybrids, when FAIRLY treated, gone on decreasing in fertility in e
0252  beds of the same hybrids, and such alone are FAIRLY treated, for by insect agency the several ind
0252  under confinement, few experiments have been FAIRLY tried: for instance, the canary bird has been
0260  ross with extreme difficulty, and yet produce FAIRLY fertile hybrids? Why should there often be so
0309  ed by the oscillations of level, which we may FAIRLY conclude must have intervened during these en
0400  ent the endemic species, which cannot be here FAIRLY included, as we are considering how they have
0410  eptions to the rule are so few, that they may FAIRLY be attributed to our not having as yet discov
0469  lf probable. I have already recapitulated, as FAIRLY as I could, the opposed difficulties and obje
0052  in two different regions; but I have not much FAITH in this view; and I attribute the passage of
0238  ith confidence to the same family. I have all FAITH in the powers of selection, that I do not dou
0242  . but I am bound to confess, that, with all my FAITH in this principle, I should never have antici
0065  which now range in India, as I hear from Dr. FALCONER, from Cape Comorin to the Himalaya, which ha
0310  sts, namely Cuvier, Owen, Agassiz, Barrande, FALCONER, E. Forbes, etc., and all our greatest geolo
0313  n the midst of a multitude of extinct forms. FALCONER has given a striking instance of a similar f
0334  astodons and elephants, when arranged by Dr. FALCONER in two series, first according to their mutu
```

Page ***(Key Word)***

Page ***(Key Word)***

```
0340  thern half. In a similar manner we know from FALCONER and Cautley's discoveries, that Northern Ind
0378  land, and perhaps by a dry climate; for Dr. FALCONER informs me that it is the damp with the heat
0393  a much less distance are equally barren. The FALKLAND Islands, which are inhabited by a wolf like
0141  old climate of Faroe in the north and of the FALKLANDS in the south, and on many islands in the to
0075  ! throw up a handful of feathers, and all must FALL to the ground according to definite laws; but
0077  o that the seeds may be widely distributed and FALL on unoccupied ground. In the water beetle, the
0283  ase of the cliff is undermined, huge fragments FALL down, and these remaining fixed, have to be wo
0329  nities of extinct and living species. They all FALL into one grand natural system; and this fact i
0329  eontologist, Owen, showing how extinct animals FALL in between existing groups. Cuvier ranked the
0377  t have escaped extermination during a moderate FALL of temperature, more especially by escaping in
0429  nd after some investigation, does not commonly FALL to the lot of aberrant genera. We can, I think
0483  hich we may consider, by so much the arguments FALL away in force. But some arguments of the great
0286  would be more rapid from the breakage of the FALLEN fragments. On the other hand, I do not believ
0065  k, to mislead us: we see no great destruction FALLING on them, and we forget that thousands are an
0191  the orifice of the trachea, with some risk of FALLING into the lungs, notwithstanding the beautifu
0364  case; how small would the chance be of a seed FALLING on favourable soil, and coming to maturity!
0476  belong to the same system with recent beings, FALLING either into the same or into intermediate gr
0478  inate to group, and with extinct groups often FALLING in between recent groups, is intelligible on
0040  period than at another, as the breed rises or FALLS in fashion, perhaps more in one district than
0066  of its life; that heavy destruction inevitably FALLS either on the young or old, during each gener
0180  lieve, by lessening the danger from occasional FALLS. But it does not follow from this fact that t
0069  ffect to its direct action. But this is a very FALSE view: we forget that each species, even where
0074  bers and kinds to what we call chance. But how FALSE a view is this! Every one has heard that when
0131  of growth. Compensation and economy of growth. FALSE correlations. Multiple, rudimentary, and lowl
0280  ntermediate between them. But this is a wholly FALSE view; we should always look for forms interme
0414  ortance in classification. Nothing can be more FALSE. No one regards the external similarity of a
0483  case of mammals, were they created bearing the FALSE marks of nourishment from the mother's womb?
0146  lightest service to the species. We may often FALSELY attribute to correlation of growth, structur
0181  opithecus or flying Lemur, which formerly was FALSELY ranked amongst bats. It has an extremely wid
0302  the perfection of the geological record, and FALSELY infer, because certain genera or families ha
0316  chapter that the species of a group sometimes FALSELY appear to have come in abruptly; and I have
0317  he species are found, the line will sometimes FALSELY appear to begin at its lower end, not in a s
0329  sion of the form of life, and the order would FALSELY appear to be strictly parallel; nevertheless
0343  nsidered; he may urge the apparent, but often FALSELY apparent, sudden coming in of whole groups o
0463  d species appear, though certainly they often FALSELY appear, to have come in suddenly on the seve
0213  d or periods of time. But let us look to the FAMILIAR case of the several breeds of dogs: it canno
0402  eparate for almost any length of time. Being FAMILIAR with the fact that many species, naturalised
0436  complex jaws and legs in crustaceans. It is FAMILIAR to almost every one, that in a flower the re
0065  by destruction at some period of life. Our FAMILIARITY with the larger domestic animals tends, I
0128  wonder of which we are apt to overlook from FAMILIARITY, that all animals and all plants throughou
0215  , that is, never wish to sit on their eggs. FAMILIARITY alone prevents our seeing how universally
0413  ation of group under group, which, from its FAMILIARITY, does not always sufficiently strike us, i
0123  ery distinct genera, or even as distinct sub FAMILIES. Thus it is, as I believe, that two or more
0124  h generally belonging to the same orders, or FAMILIES, or genera, with those now living, yet are o
0125  ttle groups of genera will form two distinct FAMILIES, or even orders, according to the amount of
0125  represented in the diagram. And the two new FAMILIES, or orders, will have descended from two spe
0126  ll peopled with many species of many genera, FAMILIES, orders, and classes, as at the present day.
0128  ra related in different degrees, forming sub FAMILIES, families, orders, sub classes, and classes.
0128  in different degrees, forming sub families, FAMILIES, orders, sub classes, and classes. The sever
0129  ious sizes may represent those whole orders, FAMILIES, and genera which have now no living represe
0193  ans in a few insects, belonging to different FAMILIES and orders, offers a parallel case of diffic
0253  s, I hardly know of an instance in which two FAMILIES of the same hybrid have been raised at the s
0257  n species ranked by systematists in distinct FAMILIES; and on the other hand, by very closely alli
0261  t trees together belonging to quite distinct FAMILIES; and, on the other hand, closely allied spec
0302  ous species, belonging to the same genera or FAMILIES, have really started into life all at once,
0302  and falsely infer, because certain genera or FAMILIES have not been found beneath a certain stage,
0305  from several formations in Europe. Some few FAMILIES of fish now have a confined range; the teleo
0316  nces. Groups of species, that is, genera and FAMILIES, follow the same general rules in their appe
0321  the apparently sudden extermination of whole FAMILIES or orders, as of Trilobites at the close of
0323  ically the same, but they belong to the same FAMILIES, genera, and sections of genera, and sometim
0329  the wide intervals between existing genera, FAMILIES, and orders, cannot be disputed. For if we c
0330  nimals, though belonging to the same orders, FAMILIES, or genera with those living at the present
0330  a, even between genera belonging to distinct FAMILIES. The most common case, especially with respe
0331  d o14, e14, m14, a third family. These three FAMILIES, together with the many extinct genera on th
0332  aracter, has become divided into several sub FAMILIES and families, some of which are supposed to
0332  become divided into several sub families and FAMILIES, some of which are supposed to have perished
0332  s low down in the series, the three existing FAMILIES on the uppermost line would be rendered less
0332  8, m3, m6, m9, were disinterred, these three FAMILIES would be so closely linked together that the
0332  which thus linked the living genera of three FAMILIES together, intermediate in character, would b
0332  ne from beneath this line, then only the two FAMILIES on the left hand (namely, a14, etc., and b14
0332  be united into one family; and the two other FAMILIES (namely, a14 to f14 now including five gener
0332  to m14) would yet remain distinct. These two FAMILIES, however, would be less distinct from each o
0333  e, we suppose the existing genera of the two FAMILIES to differ from each other by a dozen charact
0333  the natural system, and thus unite distinct FAMILIES or orders. All that we have a right to expec
0351  t sections of genera, whole genera, and even FAMILIES are confined to the same areas, as is so com
0353  able to migrate to the other side. Some few FAMILIES, many sub families, very many genera, and a
0353  the other side. Some few families, many sub FAMILIES, very many genera, and a still greater numbe
0407  ical distribution of sub genera, genera, and FAMILIES. With respect to the distinct species of the
0408  d the localisation of sub genera, genera and FAMILIES; and how it is that under different latitude
0413  nera are included in, or subordinate to, sub FAMILIES, families, and orders, all united into one c
0413  ncluded in, or subordinate to, sub families, FAMILIES, and orders, all united into one class. Thus
0413  ists try to arrange the species, genera, and FAMILIES in each class, on what is called the Natural
0419  oups of species, such as orders, sub orders, FAMILIES, sub families, and genera, they seem to be,
0419  s, such as orders, sub orders, families, sub FAMILIES, and genera, they seem to be, at least at pr
0420  e forms being ranked under different genera, FAMILIES, sections, or orders. The reader will best u
0421  nded from A, now broken up into two or three FAMILIES, constitute a distinct order from those desc
```

Page ***(Key Word)***

Page ∗∗∗(Key Word)∗∗∗

```
0421  se descended from I, also broken up into two FAMILIES. Nor can the existing species, descended fro
0422  g them under different so called genera, sub FAMILIES, families, sections, orders, and classes. It
0422  er different so called genera, sub families, FAMILIES, sections, orders, and classes. It may be wo
0433  ommon parent, expressed by the terms genera, FAMILIES, orders, etc., we can understand the rules w
0447  of pigeons. We may extend this view to whole FAMILIES or even classes. The fore limbs, for instanc
0456  ked by the terms varieties, species, genera, FAMILIES, orders, and classes. On this same view of d
0457  ly, that the inumerable species, genera, and FAMILIES of organic beings, with which this world is
0009  ith the exception of the plantigrades or bear FAMILY; whereas, carnivorous birds, with the rarest
0013  tc., appearing in several members of the same FAMILY. If strange and rare deviations of structure
0021  mal number in all members of the great pigeon FAMILY; and these feathers are kept expanded, and ar
0024  e may look in vain throughout the whole great FAMILY of Columbidae for a beak like that of the Eng
0025  ur together in any other species of the whole FAMILY. Now, in every one of the domestic breeds, ta
0097  closed, as in the great papilionaceous or pea FAMILY; but in several, perhaps in all, such flowers
0101  both animals and plants, species of the same FAMILY and even of the same genus, though agreeing c
0143  less completely by natural selection: thus a FAMILY of stags once existed with an antler only on
0180  icular case like that of the bat. Look at the FAMILY of squirrels; here we have the finest gradati
0185  ous member of the strictly terrestrial thrush FAMILY wholly subsists by diving, grasping the stone
0192  l dispute that the ovigerous frena in the one FAMILY are strictly homologous with the branchiae of
0192  ly homologous with the branchiae of the other FAMILY; indeed, they graduate into each other. There
0192  ve imagined that the branchiae in this latter FAMILY had originally existed as organs for preventi
0214  a cross with a greyhound has given to a whole FAMILY of shepherd dogs a tendency to hunt hares. Th
0234  be a most important element of success in any FAMILY of bees. Of course the success of any species
0237  membered that selection may be applied to the FAMILY, as well as to the individual, and may thus g
0238  the breeder goes with confidence to the same FAMILY. I have such faith in the powers of selection
0238  enus, or rather as any two genera of the same FAMILY. Thus in Eciton, there are working and soldie
0255  ere be given. When pollen from a plant of one FAMILY is placed on the stigma of a plant of a disti
0255  placed on the stigma of a plant of a distinct FAMILY, it exerts no more influence than so much ino
0257  h unite with the utmost facility. In the same FAMILY there may be a genus, as Dianthus, in which v
0261  although many distinct genera within the same FAMILY have been grafted together, in other cases sp
0288  he several species of the Chthamalinae (a sub FAMILY of sessile cirripedes) coat the rocks all ove
0303  striking case, however, is that of the Whale FAMILY: as these animals have huge bones, are marine
0317  of a genus, or the number of the genera of a FAMILY, be represented by a vertical line of varying
0317  of the same genus, and the genera of the same FAMILY, can increase only slowly and progressively;
0321  supplant an old genus, belonging to the same FAMILY. But it must often have happened that a new s
0331  ting genera, a14, o14, will form a small FAMILY; b14 and f14 a closely allied family or sub f
0331  a small family; b14 and f14 a closely allied FAMILY or sub family; and o14, e14, m14, a third fam
0331  y; b14 and f14 a closely allied family or sub FAMILY; and o14, e14, m14, a third family. These thr
0331  ily or sub family; and o14, e14, m14, a third FAMILY. These three families, together with the many
0332  obably would have to be united into one great FAMILY, in nearly the same manner as has occurred wi
0332  d b14, etc.) would have to be united into one FAMILY; and the two other families (namely, a14 to f
0386  dhering to it; and a water beetle of the same FAMILY, a Colymbetes, once flew on board the Beagle,
0394  rom Mr. Tomes, who has specially studied this FAMILY, that many of the same species have enormous
0410  different order, or even only to a different FAMILY of the same order, differ greatly. In both ti
0412  ame principle, much in common, and form a sub FAMILY, distinct from that including the next two ge
0412  much, though less, in common; and they form a FAMILY distinct from that including the three genera
0415  in this but, as I apprehend, in every natural FAMILY, is very unequal, and in some cases seems to
0417  s proper to the species, to the genus, to the FAMILY, to the class, disappear, and thus laugh at o
0419  a genus, and then raised to the rank of a sub FAMILY or family; and this has been done, not becaus
0419  nd then raised to the rank of a sub family or FAMILY; and this has been done, not because further
0425  suredly all the other species of the kangaroo FAMILY would have to be classed under the bear genus
0428  nity between the several members of the whale FAMILY: for these cetaceans agree in so many charact
0431  finities by which all the members of the same FAMILY or higher group are connected together. For t
0431  ed together. For the common parent of a whole FAMILY of species, now broken up by extinction into
0431  the numerous kindred of any ancient and noble FAMILY, even by the aid of a genealogical tree, and
0490  ved forms. Thus, from the war of nature, from FAMINE and death, the most exalted object which we a
0036  purpose, would be carefully preserved during FAMINES and other accidents, to which savages are so
0165  s a double shoulder stripe. In lord Moreton's FAMOUS hybrid from a chestnut mare and male quagga,
0413  ed to our knowledge. Such expressions as that FAMOUS one of Linnaeus, and which we often meet with
0117  to a genus large in its own country. The little FAN of diverging dotted lines of unequal lengths p
0221  , and carried off by the tyrants, who perhaps FANCIED that, after all, they had been victorious in
0029  met a pigeon, or poultry, or duck, or rabbit FANCIER, who was not fully convinced that each main
0032  ice requisite to become even a skilful pigeon FANCIER. The same principles are followed by horticu
0039  of structure would be necessary to catch the FANCIER's eye: he perceives extremely small differen
0042  life, and this is a great convenience to the FANCIER, for thus many races may be kept true, thoug
0051  me impressed, just like the pigeon or poultry FANCIER before alluded to, with the amount of differ
0112  ns. We shall here find something analogous. A FANCIER is struck by a pigeon having a slightly shor
0112  igeon having a slightly shorter beak; another FANCIER is struck by a pigeon having a rather longer
0020  uity. I have associated with several eminent FANCIERS, and have been permitted to join two of the
0039  ophagus, a habit which is disregarded by all FANCIERS, as it is not one of the points of the breed
0087  egg than are able to get out of it; so that FANCIERS assist in the act of hatching. Now, if natur
0112  beak; and on the acknowledged principle that FANCIERS do not and will not admire a medium standard
0152  the points now mainly attended to by English FANCIERS. Even in the sub breeds, as in the short fac
0315  our fantail pigeons were all destroyed, that FANCIERS, by striving during long ages for the same o
0446  embryonic stages of our domestic varieties. FANCIERS select their horses, dogs, and pigeons, for
0038  tructure or in their habits to man's wants or FANCIES. We can, I think, further understand the fre
0266  having been compounded into one. It may seem FANCIFUL, but I suspect that a similar parallelism ex
0427  f the greyhound and racehorse is hardly more FANCIFUL than the analogies which have been drawn by
0486  s, which are called aberrant, and which may FANCIFULLY be called living fossils, will aid us in fo
0030  mal's or plant's own good, but to man's use or FANCY. Some variations useful to him have probably
0443  s that breeders of cattle, horses, and various FANCY animals, cannot positively tell, until some t
0200  rect injury of other species, as we see in the FANG of the adder, and in the ovipositor of the ich
0201  is admitted that the rattlesnake has a poison FANG for its own defence and for the destruction of
0021  very different coo from the other breeds. The FANTAIL has thirty or even forty tail feathers, inst
0023  aced tumbler, the runt, the barb, pouter, and FANTAIL in the same genus: more especially as in eac
0024  e pouter: for tail feathers like those of the FANTAIL. Hence it must be assumed not only that half
0025  ogether, and one grandchild of the pure white FANTAIL and pure black barb was of as beautiful a bl
```

Page ∗∗∗(Key Word)∗∗∗

Page ***(Key Word)***

```
0027 mbler, and the number of tail feathers in the FANTAIL, are in each breed eminently variable: and t
0039 ee by nature. No man would ever try to make a FANTAIL, till he saw a pigeon with a tail developed
0039 to use such an expression as trying to make a FANTAIL, is, I have no doubt, in most cases, utterly
0039 hers somewhat expanded, like the present Java FANTAIL, or like individuals of other and distinct b
0159 ing the normal structure of another race, the FANTAIL. I presume that no one will doubt that all s
0280 scendants. To give a simple illustration: the FANTAIL and pouter pigeons have both descended from
0280 o varieties directly intermediate between the FANTAIL and pouter: none, for instance, combining a
0315 rs. For instance, it is just possible, if our FANTAIL pigeons were all destroyed, that fanciers, b
0315 breed hardly distinguishable from our present FANTAIL; but if the parent rock pigeon were also des
0316 oved offspring, it is quite incredible that a FANTAIL, identical with the existing breed, could be
0316 of the domestic pigeon, for the newly formed FANTAIL would be almost sure to inherit from its new
0025 for instance, I crossed some uniformly white FANTAILS with some uniformly black barbs, and they pr
0026 distinct as carriers, tumblers, pouters, and FANTAILS now are, should yield offspring perfectly fe
0039 al selection. Perhaps the parent bird of all FANTAILS had only fourteen tailfeathers somewhat expa
0152 nt carriers, in the carriage and tail of our FANTAILS, etc., these being the points now mainly att
0445 ength of leg, in the wild stock, in pouters, FANTAILS, runts, barbs, dragons, carriers, and tumble
0028 informs me that pigeons are given in a bill of FARE in the previous dynasty. In the time of the Ro
0114 islets: and so in small ponds of fresh water. FARMERS find that they can raise most food by a rota
0267 to all living things. We see this acted on by FARMERS and gardeners in their frequent exchanges of
0071 t an element enclosure is, I plainly saw near FARNHAM, in Surrey. Here there are extensive heaths,
0141 rodent, living free under the cold climate of FAROE in the north and of the Falklands in the sout
0386 miles distant from the nearest land: how much FARTHER it might have flown with a favouring gale no
0040 an at another, as the breed rises or falls in FASHION, perhaps more in one district than in anothe
0382 an, driven up and surviving in the mountian FASTNESSES of almost every land, which serve as a reco
0238 variety: breeders of cattle wish the flesh and FAT to be well marbled together: the animal has be
0056 g; for if this had been so, it would have been FATAL to my theory: inasmuch as geology plainly tel
0130 life, and which has apparently been saved from FATAL competition by having inhabited a protected s
0171 ent, and those that are real are not, I think, FATAL to my theory. These difficulties and objectio
0199 y. this doctrine, if true, would be absolutely FATAL to my theory. Yet I fully admit that many str
0236 first appeared to me insuperable, and actually FATAL to my whole theory. I allude to the neuters o
0302 more forcibly than by Professor Sedgwick, as a FATAL objection to the belief in the transmutation
0302 arted into life all at once, the fact would be FATAL to the theory of descent with slow modificati
0316 on of this fact, which if true would have been FATAL to my views. But such cases are certainly exc
0013 ppears not unfrequently, and we see it in the FATHER and child, we cannot tell whether it may not
0219 r, a better observer even than his celebrated FATHER. This ant is absolutely dependent on its slav
0258 the one species having first been used as the FATHER and then as the mother, generally differ in f
0260 s the one species or the other is used as the FATHER or the mother, there is generally some differ
0271 of these five varieties, when used either as FATHER or mother, and crossed with the Nicotiana glu
0460 hat is, when one species is first used as the FATHER and then as the mother. The fertility of vari
0304 s of depths from the upper tidal limits to 50 FATHOMS; from the perfect manner in which specimens
0395 stand on a deeply submerged bank, nearly 1000 FATHOMS in depth, and here we find American forms, b
0147 is difficult to get a cow to give milk and to FATTEN readily. The same varieties of the cabbage do
0285 dislocations is externally visible. The Craven FAULT, for instance, extends for upwards of 30 mile
0039 arise amongst pigeons, which are rejected as FAULTS or deviations from the standard of perfection
0285 the same story is still more plainly told by FAULTS, those great cracks along which the strata ha
0290 coast, which is inhabited by a peculiar marine FAUNA, tertiary beds are so scantily developed, tha
0313 der formation, and then allow the pre existing FAUNA to reappear: but Lyell's explanation, namely,
0333 w. on this same theory, it is evident that the FAUNA of any great period in the earth's history wi
0334 e formations. Subject to these allowances, the FAUNA of each geological period undoubtedly is inte
0334 rous, and underlying Silurian system. But each FAUNA is not necessarily exactly intermediate, as u
0334 ection to the truth of the statement, that the FAUNA of each period as a whole is nearly intermedi
0337 of the same or some other quarter, the eocene FAUNA or flora would certainly be beaten and exterm
0337 beaten and exterminated: as would a secondary FAUNA by an eocene, and a palaeozoic fauna by a sec
0337 secondary fauna by an eocene, and a palaeozoic FAUNA by a secondary fauna. I do not doubt that thi
0337 eocene, and a palaeozoic fauna by a secondary FAUNA. I do not doubt that this process of improvem
0347 rld, yet these are not inhabited by a peculiar FAUNA or flora. Notwithstanding this parallelism in
0348 the Pacific, with another and totally distinct FAUNA. So that here three marine faunas range far n
0366 d its maximum, we should have a uniform arctic FAUNA and flora, covering the central parts of Euro
0372 the sea, a slow southern migration of a marine FAUNA, which during the Pliocene or even a somewhat
0403 which determines the general character of the FAUNA and flora of oceanic islands, namely, that th
0290 ord of several successive and peculiar marine FAUNAS will probably be preserved to a distant age.
0328 s inhabited by distinct, but contemporaneous, FAUNAS. Lyell has made similar observations on some
0334 aracter, between the preceding and succeeding FAUNAS. I need give only one instance, namely, the m
0334 haracter between the preceding and succeeding FAUNAS, that certain genera offer exceptions to the
0347 t it would not be possible to point out three FAUNAS and floras more utterly dissimilar. Or again
0348 the sea, we find the same law. No two marine FAUNAS are more distinct, with hardly a fish, shell,
0348 of south and Central America: yet these great FAUNAS are separated only by the narrow, but impassa
0348 lly distinct fauna. So that here three marine FAUNAS range far northward and southward, in paralle
0348 meet with no well defined and distinct marine FAUNAS. Although hardly one shell, crab or fish is c
0348 s common to the above named three approximate FAUNAS of Eastern and Western America and the easter
0356 al: a narrow isthmus now separates two marine FAUNAS; submerge it, or let it formerly have been su
0356 it formerly have been submerged, and the two FAUNAS will now blend or may formerly have blended:
0358 n, such as the great difference in the marine FAUNAS on the opposite sides of almost every contine
0395 space separates two widely distinct mammalian FAUNAS. On either side the islands are situated on m
0161 er in question, which at last, under unknown FAVORABLE conditions, gains an ascendancy. For instan
0015 ies. But there is not a shadow of evidence in FAVOUR of this view: to assert that we could not bre
0017 e is presumptive, or even strong, evidence in FAVOUR of this view. It has often been assumed that
0027 a livia with its geographical sub species. In FAVOUR of this view, I may add, firstly, that C. liv
0077 chief use of the nutriment in the seed is to FAVOUR the growth of the young seedling, whilst stru
0092 rticular insects which visited them, so as to FAVOUR in any degree the transportal of their pollen
0177 the chances in this case will be strongly in FAVOUR of the great holders on the mountains or on t
0358 their almost universally volcanic composition FAVOUR the admission that they are the wrecks of sun
0384 ommon and some different. A few facts seem to FAVOUR the possibility of their occasional transport
0401 n it varied, natural selection would probably FAVOUR different varieties in the different islands.
0459 the general and special circumstances in its FAVOUR. Causes of the general belief in the immutabi
0469 us turn to the special facts and arguments in FAVOUR of the theory. On the view that species are o
```

Page ***(Key Word)***

Page ***(Key Word)***

```
0004 bled to discuss what circumstances are most FAVOURABLE to variation. In the next chapter the Strug
0007 of our Domestic Productions. Circumstances  FAVOURABLE to Man's power of Selection. When we look t
0028 what monstrous character. It is also a most  FAVOURABLE circumstance for the production of distinct
0040 t now say a few words on the circumstances,  FAVOURABLE, or the reverse, to man's power of selectio
0040 . a high degree of variability is obviously  FAVOURABLE, as freely giving the materials for selecti
0041 res that the species should be placed under  FAVOURABLE conditions of life, so as to breed freely i
0054 nic or inorganic conditions of that country  FAVOURABLE to the genus; and, consequently, we might h
0055 through variation, circumstances have been   FAVOURABLE for variation; and hence we might expect th
0055 the circumstances would generally be still   FAVOURABLE to variation. On the other hand, if we look
0064 ate of nature, when circumstances have been  FAVOURABLE to them during two or three following seaso
0065 that the conditions of life have been very   FAVOURABLE, and that there has consequently been less
0065 d require a few more years to people, under  FAVOURABLE conditions, a whole district, let it be eve
0070 ve animals. When a species, owing to highly  FAVOURABLE circumstances, increases inordinately in nu
0070 ly where the conditions of its life were so  FAVOURABLE that many could exist together, and thus sa
0080 ividuals of the same species. Circumstances  FAVOURABLE and unfavourable to Natural Selection, name
0081 be rigidly destroyed. This preservation of   FAVOURABLE variations and the rejection of injurious v
0082 gone a change, and this would manifestly be  FAVOURABLE to natural selection, by giving a better ch
0093 been cold and boisterous, and therefore not  FAVOURABLE to bees, nevertheless every female flower w
0095 individuals presenting mutual and slightly   FAVOURABLE deviations of structure. I am well aware th
0101 isation go on for perpetuity. Circumstances  FAVOURABLE to Natural Selection. This is an extremely
0102 inheritable and diversified variability is   FAVOURABLE, but I believe mere individual differences
0104 ly by natural selection preserving the same  FAVOURABLE variations. Isolation, also, is an importan
0105 decreasing the chance of the appearance of   FAVOURABLE variations. If we turn to nature to test th
0105 nd at first sight seems to have been highly  FAVOURABLE for the production of new species. But we m
0105 e open area like a continent, has been most  FAVOURABLE for the production of new organic forms, we
0105 , not only will there be a better chance of  FAVOURABLE variations arising from the large number of
0106 probably have been in some respects highly   FAVOURABLE for the production of new species, yet that
0107 re competition. To sum up the circumstances  FAVOURABLE and unfavourable to natural selection, as f
0107 ods in a broken condition, will be the most-FAVOURABLE for the production of many new forms of lif
0108 disturbed. Nothing can be effected, unless   FAVOURABLE variations occur, and variation itself is a
0110 chance of producing within any given period  FAVOURABLE variations. We have evidence of this, in th
0118 ntry. And these circumstances we know to be  FAVOURABLE to the production of new varieties. If, the
0177 hin any given period, of presenting further  FAVOURABLE variations for natural selection to seize o
0177 and natural selection can do nothing until  FAVOURABLE variations chance to occur, and until a pla
0194 by the preservation of individuals with any  FAVOURABLE variation, and by the destruction of those
0259 ity, besides being eminently susceptible to  FAVOURABLE and unfavourable conditions, is innately va
0276 me species, and is eminently susceptible of  FAVOURABLE and unfavourable conditions. The degree of
0277 rossing of forms only slightly different is  FAVOURABLE to the vigour and fertility of their offspr
0277 es in the conditions of life are apparently  FAVOURABLE to the vigour and fertility of all organic
0279 ay, under the circumstances apparently most  FAVOURABLE for their presence, namely on an extensive
0291 each other, the sea will remain shallow and  FAVOURABLE for life, and thus a fossiliferous formatio
0292 ed in the shallow parts, which are the most  FAVOURABLE to life. Still less could this have happene
0292 ill often be formed: all circumstances most  FAVOURABLE, as previously explained, for the formation
0318 orse under conditions of life apparently so  FAVOURABLE. But how utterly groundless was my astonish
0319 tic horse in South America, that under more  FAVOURABLE conditions it would in a very few years hav
0319 on, however slowly, becoming less and less   FAVOURABLE, we assuredly should not have perceived the
0326 precisely what are all the conditions most  FAVOURABLE for the multiplication of new and dominant
0326 giving a better chance of the appearance of  FAVOURABLE variations, and that severe competition wit
0326 any already existing forms, would be highly  FAVOURABLE, as would be the power of spreading into ne
0326 o intervals of time, would probably be also  FAVOURABLE, as before explained. One quarter of the wo
0326 one quarter of the world may have been most  FAVOURABLE for the production of new and dominant spec
0360 r country; and when stranded, if blown to a  FAVOURABLE spot by an inland gale, they would germinat
0364 ll would the chance be of a seed falling on  FAVOURABLE soil, and coming to maturity! But it would
0370 ent freedom for intermigration under a more  FAVOURABLE climate, I attribute the necessary amount o
0371 world. Consequently we have here everything  FAVOURABLE for much modification, for far more modific
0386 in consequence, a very wide range. I think   FAVOURABLE means of dispersal explain this fact. I hav
0405 hould place itself under diverse conditions  FAVOURABLE for the conversion of its offspring, firstl
0429 rs preserved by some unusual coincidence of  FAVOURABLE circumstances. Mr. Waterhouse has remarked
0471 solely by accumulation slight, successive,  FAVOURABLE variations, it can produce no great or sudd
0480 ation and accumulation of successive slight  FAVOURABLE variations. Why, it may be asked, have all
0326 wo great regions had been for a long period  FAVOURABLY circumstanced in an equal degree, whenever
0020 urchase or obtain, and have been most kindly  FAVOURED with skins from several quarters of the worl
0035 the regulations for the Goodwood Races, are  FAVOURED in the weights they carry. Lord Spencer and
0042 ary, and this circumstance must have largely  FAVOURED the improvement and formation of new breeds.
0069 nemies or competitors be in the least degree  FAVOURED by any slight change of climate, they will i
0069 se lies quite as much in other species being  FAVOURED, as in this one being hurt. So it is when we
0070 the crowded animals, been disproportionably  FAVOURED: and here comes in a sort of struggle betwee
0082 ages chanced to arise, and which in any way  FAVOURED the individuals of any of the species, by be
0092 len from flower to flower, would likewise be  FAVOURED or selected. We might have taken the case of
0094 ore and more increased, would be continually  FAVOURED or selected, until at last a complete separa
0108 e will again be severe competition: the most  FAVOURED or improved varieties will be enabled to spr
0109 itants, it follows that as each selected and  FAVOURED form increases in number, so will the less f
0109 d form increases in number, so will the less  FAVOURED forms decrease and become rare. Rarity, as g
0130 in a tree, and which by some chance has been  FAVOURED and is still alive on its summit, so we occa
0133 to the warmest clad individuals having been  FAVOURED and preserved during many generations, and h
0157 , but only gives fewer offspring to the less  FAVOURED males. Whatever the cause may be of the vari
0172 , its own less improved parent or other less  FAVOURED forms with which it comes into competition.
0320 ition; and the consequent extinction of less  FAVOURED forms almost inevitably follows. It is the s
0360 which live near the sea; and this would have  FAVOURED the average length of their flotation and of
0377 invasion would, of course, have been greatly  FAVOURED by high land, and perhaps by a dry climate:
0467 e acted under nature. In the preservation of  FAVOURED individuals and races, during the constantly
0069 that climate acts in main part indirectly by  FAVOURING other species, we may clearly see in the pr
0386 how much farther it might have flown with a  FAVOURING gale no one can tell. With respect to plant
0469 ion, structure, and habits of each creature,  FAVOURING the good and rejecting the bad? I can see n
0079 at the war of nature is not incessant; that no FEAR is felt, that death is generally prompt, and t
0212 es in the northern and southern United States. FEAR of any particular enemy is certainly an instin
```

Page ***(Key Word)***

Page **(Key Word)**

```
0212  renghthened by experience, and by the sight of FEAR of the same enemy in other animals. But fear o
0212  f fear of the same enemy in other animals. But FEAR of man is slowly acquired, as I have elsewhere
0215  oung chickens have lost, wholly by habit, that FEAR of the dog and cat which no doubt was original
0216  r a hen. It is not that chickens have lost all FEAR, but fear only of dogs and cats, for if the he
0216  t is not that chickens have lost all fear, but FEAR only of dogs and cats, for if the hen gives th
0212  uninhabited islands large birds are not more FEARFUL than small; and the magpie, so wary in Engla
0031  t to pigeons, that he would produce any given FEATHER in three years; but it would take him six ye
0162  we did not know that the rock pigeon was not FEATHER footed or turn crowned, we could not have to
0012  s asserted, long or many horns; pigeons with FEATHERED feet have skin between their outer toes; pi
0144  rtoise shell colour with the female sex; the FEATHERED feet and skin between the outer toes in pig
0021  rt and conical beak, with a line of reversed FEATHERS down the breast; and it has the habit of con
0021  part of the oesophagus. The Jacobin has the FEATHERS so much reversed along the back of the neck
0021  ly to its size, much elongated wing and tail FEATHERS. The trumpeter and laugher, as their names e
0021  s. the fantail has thirty or even forty tail FEATHERS, instead of twelve or fourteen, the normal n
0021  embers of the great pigeon family; and these FEATHERS are kept expanded, and are carried so erect
0022  d; the number of the primary wing and caudal FEATHERS; the relative length of wing and tail to eac
0024  e short faced tumbler, or barb; for reversed FEATHERS like those of the jacobin; for a crop like t
0024  for a crop like that of the pouter; for tail FEATHERS like those of the fantail. Hence it must be
0025  rminal dark bar, with the bases of the outer FEATHERS externally edged with white; the wings have
0025  , even to the white edging of the outer tail FEATHERS, sometimes concur perfectly developed. Moreo
0025  ck wing bar, and barred and white edged tail FEATHERS, as any wild rock pigeon! We can understand
0027  that of the tumbler, and the number of tail FEATHERS in the fantail, are in each breed eminently
0039  t breeds, in which as many as seventeen tail FEATHERS have been counted. Perhaps the first pouter
0042  ing valuable only for two purposes, food and FEATHERS, and more especially from no pleasure having
0061  which clings to the hairs of a quadruped or FEATHERS of a bird; in the structure of the beetle wh
0075  e growth of the trees! Throw up a handful of FEATHERS, and all must fall to the ground according t
0147  and quality. In our poultry, a large tuft of FEATHERS on the head is generally accompanied by a di
0159  t apart, present sub varieties with reversed FEATHERS on the head and feathers on the feet, charac
0159  eties with reversed feathers on the head and FEATHERS on the feet, characters not possessed by the
0159  nt presence of fourteen or even sixteen tail FEATHERS in the pouter, may be considered as a variat
0160  a bar at the end of the tail; with the outer FEATHERS externally edged near their bases with white
0416  he nature of the dermal covering, as hair or FEATHERS, in the Vertebrata. If the Ornithorhynchus h
0416  if the Ornithorhynchus had been covered with FEATHERS instead of hair, this external and trifling
0426  lded, whether the skin be covered by hair or FEATHERS, if it prevail throughout many and different
0439  and second plumage; as we see in the spotted FEATHERS in the thrush group. In the cat tribe, most
0433  nt species, explains that great and universal FEATURE in the affinities of all organic beings, nam
0482  nsequently have every external characteristic FEATURE of true species, they admit that these have
0029  d tumbler pigeon. One of the most remarkable FEATURES in our domesticated races is that we see in
0040  bitants, slowly to add to the characteristic FEATURES of the breed, whatever they may be. But the
0280  a crop somewhat enlarged, the characteristic FEATURES of these two breeds. These two breeds, moreo
0443  d will be tall or short, or what its precise FEATURES will be. The question is not, at what period
0386  give only the most striking case; I took in FEBRUARY three tablespoons of mud from three differen
0452  the male florets, which of course cannot be FECUNDATED, have a pistil, which is in a rudimentary s
0250  never saw to occur in a case of its natural FECUNDATION. So that we here have perfect, or even mor
0217  e first hatched young would probably have to be FED by the male alone. But the American cuckoo is
0219  lave (F. fusca), and she instantly set to work, FED and saved the survivors; made some cells and t
0239  ers of one caste never leave the nest; they are FED by the workers of another caste, and they have
0442  rse and active habits, or quite inactive, being FED by their parents or placed in the midst of pro
0444  ak of this particular length, as long as it was FED by its parents. Hence, I conclude, that it is
0061  ion, and is as immeasurably superior to man's FEEBLE efforts, as the works of Nature are to those
0109  ow though the process of selection may be, if FEEBLE man can do much by his powers of artificial s
0212  hout facts given in detail, can produce but a FEEBLE effect on the reader's mind. I can only repea
0399  e glacial period. The affinity, which, though FEEBLE, I am assured by Dr. Hooker is real, between
0130  s, branch out and overtop on all sides many a FEEBLER branch, so by generation I believe it has be
0340  changes, permitting much inter migration, the FEEBLER will yield to the more dominant forms, and t
0428  d they consequently supplant many smaller and FEEBLER groups. Thus we can account for the fact tha
0116  ups differing but little from each other, and FEEBLY representing, as Mr. Waterhouse and others ha
0282  actical geologist, the facts leading the mind FEEBLY to comprehend the lapse of time. He who can r
0070  excess compared with the number of birds which FEED on them; nor can the birds, though having a su
0086  ich live only for a few hours, and which never FFED, a large part of their structure is merely the
0113  : some of them, for instance, being enabled to FEED on new kinds of prey, either dead or alive; so
0183  to that of the many British insects which now FEED on exotic plants, or exclusively on artificial
0184  t in North America there are woodpeckers which FEED largely on fruit, and others with elongated wi
0218  s it with paralysed prey for its own larvae to FEED on, yet that when this insect finds a burrow a
0219  to work, they did nothing; they could not even FEED themselves, and many perished of hunger. Huber
0223  food for itself or its young, and cannot even FEED itself: it is absolutely dependent on its nume
0411  nhabit the land, and another the water; one to FEED on flesh, another on vegetable matter, and so
0441  e, and a probosciformed mouth, with which they FEED largely, for they increase much in size. In th
0441  have a closed and imperfect mouth, and cannot FEED: their function at this stage is, to search by
0471  a thrush should have been created to dive and FEED on sub aquatic insects; and that a petrel shou
0094  en we see leaf eating insects green, and bark FEEDERS mottled grey; the alpine ptarmigan white in
0136  . the insects in Madeira which are not ground FEEDERS, and which, as the flower feeding coleoptera
0075  ts of prey, all striving to increase, and all FEEDING on each other or on the trees or their seeds
0075  xistence, as in the case of locusts and grass FEEDING quadrupeds. But the struggle almost invariab
0094  ld be affected. Let us now turn to the nectar FEEDING insects in our imaginary case: we may suppos
0134  domestic Aylesbury duck. As the larger ground FEEDING birds seldom take flight except to escape da
0135  he anterior tarsi, or feet, of many male dung FEEDING beetles are very often broken off; he examin
0135  the tarsi are almost always lost in many dung FEEDING beetles, they must be lost early in life, an
0136  not ground feeders, and which, as the flower FEEDING coleoptera and lepidoptera, must habitually
0197  we see that the skin on the head of the clean FEEDING male turkey is likewise naked. The sutures i
0219  re incapable of making their own nests; or of FEEDING their own larvae. When the old nest is found
0244  ts making slaves, the larvae of ichneumonidae FEEDING within the live bodies of caterpillars, not
0387  the seeds might be dropped by the bird whilst FEEDING its young, in the same way as fish are known
0472  r her own fertile daughters; at ichneumonidae FEEDING within the live bodies of caterpillars; and
0083  racter in some peculiar and fitting manner; he FEEDS a long and a short beaked pigeon on the same
0002  i hope, in most cases will suffice. No one can FEEL more sensible than I do of the necessity of he
```

Page **(Key Word)**

Page **(Key Word)**

```
0006  and a few concluding remarks. No one ought to FEEL surprise at much remaining as yet unexplained
0027  m these several reasons, taken together, I can FEEL no doubt that all our domestic breeds have des
0069  nd see a species decreasing in numbers, we may FEEL sure that the cause lies quite as much in othe
0081  creating their kind? On the other hand, we may FEEL sure that any variation in the least degree in
0132  e dimly catch a faint ray of light, and we may FEEL sure that there must be some cause for each de
0145   exterior flowers of a head or umbel, I do not FEEL at all sure that C. C. Sprengel's idea that th
0220  of safety. Hence, it is clear, that the slaves FEEL quite at home. During the months of June and J
0292  almost necessarily be rendered intermittent. I FEEL much confidence in the truth of these views, f
0310  ion entertains grave doubts on this subject. I FEEL how rash it is to differ from these great auth
0320  lly become rare before they become extinct, to FEEL no surprise at the rarity of a species, and ye
0320   the individual is the forerunner of death, to FEEL no surprise at sickness, but when the sick man
0322  ountry; then, and not till then, we may justly FEEL surprise why we cannot account for the extinct
0350  their physical conditions. The naturalist must FEEL little curiosity, who is not led to inquire wh
0352  distant points of the world. No geologist will FEEL any difficulty in such cases as Great Britain
0373  crossing a valley of the Andes; and this I now FEEL convinced was a gigantic moraine, left far bel
0423  s been attempted by some authors. For we might FEEL sure, whether there had been more or less modi
0426  roup of beings having different habits, we may FEEL almost sure, on the theory of descent, that th
0449  h the same or similar embryonic stages, we may FEEL assured that they have both descended from the
0461  ferent and new conditions of life, we need not FEEL surprise at hybrids being in some degree steri
0472  erfection of their associates; so that we need FEEL no surprise at the inhabitants of any one coun
0477   the same physical conditions of life, we need FEEL no surprise at their inhabitants being widely
0484  s or that form be in essence a species. This I FEEL sure, and I speak after experience, will be no
0486  he prototypes of each great class. When we can FEEL assured that all the individuals of the same s
0489  h lived long before the Silurian epoch, we may FEEL certain that the ordinary succession by genera
0139   of most of their other productions. Far from FEELING any surprise that some of the cave animals a
0208  of its work was already done for it, far from FEELING the benefit of this, it was much embarrassed
0383  rst collecting in the fresh waters of Brazil, FEELING much surprise at the similarity of the fresh
0398  several hundred miles from the continent, yet FEELS that he is standing on American land. Why sho
0003  ure, for instance, of the woodpecker, with its FEET, tail, beak, and tongue, so admirably adapted
0012  ed, long or many horns; pigeons with feathered FEET have skin between their outer toes; pigeons wi
0012  uter toes; pigeons with short beaks have small FEET; and those with long beaks large feet. Hence,
0012  ve small feet; and those with long beaks large FEET. Hence, if man goes on selecting, and thus aug
0021  great size, with long, massive beak and large FEET; some of the sub breeds of runts have very lon
0022  he body; the relative length of leg and of the FEET; the number of scutellae on the toes, the deve
0067  mies; for instance, on a piece of ground three FEET long and two wide, dug and cleared, and where
0068  pecies growing on a little plot of turf (three FEET by four) nine species perished from the other
0114  instance, I found that a piece of turf, three FEET by four in size, which had been exposed for ma
0135  ved the same fact) that the anterior tarsi, or FEET, of many male dung feeding beetles are very of
0144  hell colour with the female sex; the feathered FEET and skin between the outer toes in pigeons, an
0159  ersed feathers on the head and feathers on the FEET, characters not possessed by the aboriginal ro
0179  stela vison of North America, which has webbed FEET and which resembles an otter in its fur, short
0185  bsists by diving, grasping the stones with its FEET and using its wings under water. He who believ
0185  ment. What can be plainer than that the webbed FEET of ducks and geese are formed for swimming? ye
0185  imming? yet there are upland geese with webbed FEET which rarely or never go near the water; and n
0185   corresponding change of structure. The webbed FEET of the upland goose may be said to have become
0186   should be geese and frigate birds with webbed FEET, either living on the dry land or most rarely
0199  s. thus, we can hardly believe that the webbed FEET of the upland goose or of the frigate bird are
0200  e upland goose and of the frigate bird, webbed FEET no doubt were as useful as they now are to the
0204  arisen that there are upland geese with webbed FEET, ground woodpeckers, diving thrushes, and petr
0241  orkmen building a house of whom many were five FEET four inches high, and many sixteen feet high;
0241  e five feet four inches high, and many sixteen FEET high; but we must suppose that the larger work
0283  one examine beds of conglomerate many thousand FEET in thickness, which, though probably formed at
0284  strata (not including igneous beds).....57,154 FEET. Secondary strata.........................
0284  trata.........................................13,190 FEET. Tertiary strata...................
0284  rata..................................... 2,240 FEET. making altogether.....................
0284  gether....................................72,584 FEET; that is, very nearly thirteen and three quart
0284  nted in England by thin beds, are thousands of FEET in thickness on the Continent. Moreover, betwe
0284  reat Mississippi river at the rate of only 600 FEET in a hundred thousand years. This estimate may
0285  to perpendicular cliffs of one or two thousand FEET in height; for the gentle slope of the lava st
0285   other, to the height or depth of thousands of FEET; for since the crust cracked, the surface of t
0285  ment of the strata has varied from 600 to 3000 FEET. Prof. Ramsay has published an account of a do
0285  an account of a downthrow in Anglesea of 2300 FEET; and he informs me that he fully believes ther
0285  ieves there is one in Merionethshire of 12,000 FEET; yet in these cases there is nothing on the su
0285  f our palaeozoic strata, in parts ten thousand FEET in thickness, as shown in Prof. Ramsay's maste
0286  several formations is on an average about 1100 FEET, as I am informed by Prof. Ramsay. But if, as
0286  assume that the sea would eat into cliffs 500 FEET in height at the rate of one inch in a century
0286  circumstances, I conclude that for a cliff 500 FEET in height, a denudation of one inch per centur
0290  asts, which have been upraised several hundred FEET within the recent period, than the absence of
0296  ny instances could be given of beds only a few FEET in thickness, representing formations, elsewhe
0296  epresenting formations, elsewhere thousands of FEET in thickness, and which must have required an
0296  Lyell and Dawson found carboniferous beds 1400 FEET thick in Nova Scotia, with ancient root bearin
0362  killed by this process. Although the beaks and FEET of birds are generally quite clean, I can show
0363   can we doubt that the earth adhering to their FEET would sometimes include a few minute seeds? Bu
0364  y one means, namely, in dirt sticking to their FEET, which is in itself a rare accident. Even in t
0373  ructure of a vast mound of detritus, about 800 FEET in height, crossing a valley of the Andes; and
0378  t there at the height of six or seven thousand FEET. During this the coldest period, I suppose tha
0385  s perhaps more effectual: I suspended a duck's FEET, which might represent those of a bird sleepin
0385  minute and just hatched shells crawled on the FEET, and clung to them so firmly that when taken o
0385  quatic in their nature, survived on the duck's FEET, in damp air, from twelve to twenty hours; and
0386  though rarely, adheres in some quantity to the FEET and beaks of birds. Wading birds, which freque
0386  lushed, would be the most likely to have muddy FEET. Birds of this order I can show are the greate
0386  so that the dirt would not be washed off their FEET; then making land, they would be sure to fly t
0391  ached to seaweed or floating timber, or to the FEET of wading birds, might be transported far more
0397  young occasionally crawl on and adhere to the FEET of birds roosting on the ground, and thus get
0445  outh, length of nostril and of eyelid, size of FEET and length of leg, in the wild stock, in poute
0471  rely swim should have been created with webbed FEET; that a thrush should have been created to div
```

Page **(Key Word)**

266   feet

Page **(Key Word)**

```
0404 played in Bats, and in a lesser degree in the FELIDAE and Canidae. We see it, if we compare the di
0074 e it is quite credible that the presence of a FELINE animal in large numbers in a district might d
0227 accurately at what distance to stand from her FELLOW labourers when several are making their spher
0219 insect whose nest and stored food are thus FELONIOUSLY appropriated, be not thus exterminated. Sl
0028 eral kinds, knowing well how true they bred, I FELT fully as much difficulty in believing that the
0042 d more especially from no pleasure having been FELT in the display of distinct breeds. To sum up o
0079 ar of nature is not incessant, that no fear is FELT, that death is generally prompt, and that the
0185 eated as we now see it, must occasionally have FELT surprise when he has met with an animal having
0188 ucht to conquer his imagination; though I have FELT the difficulty far too keenly to be surprised
0194 rable deviation of structure, I have sometimes FELT much difficulty in understanding the origin of
0195 essively varying individuals. I have sometimes FELT as much difficulty, though of a very different
0211 e during several hours. After this interval, I FELT sure that the aphides would want to excrete. I
0211 then of another; and each aphis, as soon as it FELT the antennae, immediately lifted up its abdome
0319 in some degree rare, no naturalist would have FELT the least surprise at its rarity; for rarity i
0319 till existing as a rare species, we might have FELT certain from the analogy of all other mammals,
0345 ount for that vague yet ill defined sentiment, FELT by many palaeontologists, that organisation on
0378 el of the sea was about the same with that now FELT there at the height of six or seven thousand f
0382 ibution. I believe that the world has recently FELT one of his great cycles of change; and that on
0465 xplanations which can be given to them. I have FELT these difficulties far too heavily during many
0481 iate steps. The difficulty is the same as that FELT by so many geologists, when Lyell first insist
0008 variability may be attributed to the male and FEMALE reproductive elements having been affected pr
0008 ent, even in the many cases when the male and FEMALE unite. How many animals there are which will
0028 production of distinct breeds, that male and FEMALE pigeons can be easily mated for life; and thu
0089 eauty, I can see no good reason to doubt that FEMALE birds, by selecting, during thousands of gene
0089 laws with respect to the plumage of male and FEMALE birds, in comparison with the plumage of the
0093 dimentary pistil: other holly trees bear only FEMALE flowers; these have a full sized pistil, and
0093 ain of pollen can be detected. Having found a FEMALE tree exactly sixty yards from a male tree, I
0093 as the wind had set for several days from the FEMALE to the male tree, the pollen could not thus h
0093 re not favourable to bees, nevertheless every FEMALE flower which I examined had been effectually
0094 , sometimes the male organs and sometimes the FEMALE organs become more or less impotent; now if w
0100 he sexes are separated, although the male and FEMALE flowers may be produced on the same tree, we
0132 stic condition of the offspring. The male and FEMALE sexual elements seem to be affected before th
0144 cats, and the tortoise shell colour with the FEMALE sex: the feathered feet and skin between the
0240 even rudiments of ocelli, though the male and FEMALE ants of this genus have well developed ocelli
0258 f a stallion horse being first crossed with a FEMALE ass, and then a male ass with a mare: these t
0261 ement of the one will often freely act on the FEMALE sexual element of the other, but not in a rev
0262 grafted stocks, and the union of the male and FEMALE elements in the act of reproduction, yet that
0263 in the union of two pure species the male and FEMALE sexual elements are perfect; whereas in hybri
0264 urface. Again, the male element may reach the FEMALE element, but be incapable of causing an embry
0265 most liable to be affected: but sometimes the FEMALE more than the male. In both, the tendency goe
0275 uns more strongly in the male ass than in the FEMALE, so that the mule, which is the offspring of
0275 n is the hinny, which is the offspring of the FEMALE ass and stallion. Much stress has been laid b
0443 and the variation may be due to the male and FEMALE sexual elements having been affected by the c
0452 exemplified in the state of the wings of the FEMALE moths in certain groups. Rudimentary organs m
0022 ion. Lastly, in certain breeds, the males and FEMALES have come to differ to a slight degree from
0083 w the most vigorous males to struggle for the FEMALES. He does not rigidly destroy all inferior an
0088 uggle between the males for possession of the FEMALES; the result is not death to the unsuccessful
0088 ans in a war dance, for the possession of the FEMALES: male salmons have been seen fighting all da
0089 les of many species to attract by singing the FEMALES. The rock thrush of Guiana, birds of Paradis
0089 plumage and perform strange antics before the FEMALES, which standing by as spectators, at last ch
0089 ither by the males alone, or by the males and FEMALES: but I have not space here to enter on this
0089 it is, as I believe, that when the males and FEMALES of any animal have the same general habits o
0090 to the males in battle, or attractive to the FEMALES. We see analogous cases under nature, for in
0150 f reproduction. The rule applies to males and FEMALES: but as females more rarely offer remarkable
0150 the rule applies to males and females; but as FEMALES more rarely offer remarkable secondary sexua
0156 , with the amount of difference between their FEMALES; and the truth of this proposition will be g
0158 pecies to each other, or to fit the males and FEMALES to different habits of life, or the males to
0158 le with other males for the possession of the FEMALES. Finally, then, I conclude that the greater
0192 constructed from either the males or fertile FEMALES; but this case will be treated of in the nex
0196 e in fighting with another or in charming the FEMALES. Moreover when a modification of structure h
0199 ection, when displayed in beauty to charm the FEMALES, can be called useful only in rather a force
0202 by which the males of many insects find their FEMALES, can we admire the production for this singl
0219 tinct in a single year. The males and fertile FEMALES do no work. The workers or sterile females,
0219 le females do no work. The workers or sterile FEMALES, though most energetic and courageous in cap
0220 found a few slaves in all. Males and fertile FEMALES of the slave species are found only in their
0236 le theory. I allude to the neuters or sterile FEMALES in insect communities: for these neuters oft
0236 in structure from both the males and fertile FEMALES, and yet, from being sterile, they cannot pr
0236 ng widely from both the males and the fertile FEMALES in structure, as in the shape of the thorax
0236 s respect between the workers and the perfect FEMALES, would have been far better exemplified by t
0238 community: consequently the fertile males and FEMALES of the same community flourished, and transm
0238 of difference between the fertile and sterile FEMALES of the same species has been produced, which
0238 everal ants differ, not only from the fertile FEMALES and males, but from each other, sometimes to
0239 ossible, different from the fertile males and FEMALES, in this case, we may safely conclude from t
0240 useful to the community, and those males and FEMALES had been continually selected, which produce
0267 rosses, that is crosses between the males and FEMALES of the same species which have varied and be
0267 er crosses, that is crosses between males and FEMALES which have become widely or specifically dif
0425 als of the same species, though the males and FEMALES and larvae are sometimes extremely different
0460 or three defined castes of workers or sterile FEMALES in the same community of ants: but I have at
0468 struggle between males for possession of the FEMALES. The most vigorous individuals, or those whi
0024 e country; but not one has ever become wild or FERAL, though the dovecot pigeon, which is the rock
0024 n in a very slightly altered state, has become FERAL in several places. Again, all recent experien
0027 wn in a wild state, and their becoming nowhere FERAL; these species having very abnormal character
0067 rable length, more especially in regard to the FERAL animals of South America. Here I will make on
0072 though they swarm southward and northward in a FERAL state; and Azara and Rengger have shown that
0072 decrease, then cattle and horses would become FERAL, and this would certainly greatly alter (as i
0478 nimals of the Galapagos archipelago, of Juan FERNANDEZ, and of the other American islands being re
```

Page **(Key Word)**

0389 d the little island of Anglesea 764, but a few FERNS and a few introduced plants are included in
0222 , it is very courageous, and I have seen it FEROCIOUSLY attack other ants. In one instance I foun
0009 ural conditions (for instance, the rabbit and FERRET kept in hutches), showing that their reprodu
0009 , with the rarest exceptions, hardly ever lay FERTILE eggs. Many exotic plants have pollen utterl;
0026 the domestic breeds of pigeons are perfectly FERTILE. I can state this from my own observations,
0026 s clearly distinct being themselves perfectly FERTILE. Some authors believe that long continued d
0026 ils now are, should yield offspring perfectly FERTILE; inter se, seems to me rash in the extreme.
0027 rossed; the mongrel offspring being perfectly FERTILE; from these several reasons, taken together,
0075 he soil or climate, or are naturally the most FERTILE, will beat the others and so yield more see
0141 ost different climates but of being perfectly FERTILE (a far severer test) under them, may be use
0192 ferently constructed from either the males or FERTILE females; but this case will be treated of in
0219 ecome extinct in a single year. The males and FERTILE females do no work. The workers or sterile
0220 nea, and found a few slaves in all. Males and FERTILE females of the slave species are found only
0236 inct and in structure from both the males and FERTILE females, and yet, from being sterile, they
0236 differing widely from both the males and the FERTILE females in structure, as in the shape of the
0238 vantageous to the community: consequently the FERTILE males and females of the same community flou
0238 ommunity flourished, and transmitted to their FERTILE offspring a tendency to produce sterile memb
0238 t prodigious amount of difference between the FERTILE and sterile females of the same species. Man
0238 ers of several ants differ, not only from the FERTILE females and males, but from each other, some
0239 ieve to be quite possible, different from the FERTILE males and females, in this case, we may safe
0239 d that by the long continued selection of the FERTILE parents which produced most neuters with the
0241 ieve that natural selection, by acting on the FERTILE parents, could form a species which should r
0242 the workers being sterile; for had they been FERTILE, they would have intercrossed, and their ins
0242 ve affected the structure or instincts of the FERTILE members, which alone leave descendants. I am
0246 ed by most authors as distinct species, quite FERTILE together, he unhesitatingly ranks them as va
0247 , and only once or twice succeeded in getting FERTILE seed; as he found the common red and blue pi
0249 ade by Gartner, namely, that if even the less FERTILE hybrids be artificially fertilised with hybr
0250 is conclusion that some hybrids are perfectly FERTILE, as fertile as the pure parent species, as a
0250 n that some hybrids are perfectly fertile, as FERTILE as the pure parent species, as are Kolreuter
0251 I am assured that many of them are perfectly FERTILE. Mr. C. Noble, for instance, informs me that
0252 rile. I doubt whether any case of a perfectly FERTILE hybrid animal can be considered as thoroughl
0252 y, or that their hybrids, should be perfectly FERTILE. Again, with respect to the fertility in suc
0252 rtility in successive generations of the more FERTILE hybrid animals, I hardly know of an instance
0253 rouchly well authenticated cases of perfectly FERTILE hybrid animals, I have some reason to believ
0253 orouatus and with P. versicolor are perfectly FERTILE. The hybrids from the common and Chinese gee
0253 ever, these cross bred geese must be far more FERTILE; for I am assured by two eminently capable j
0253 species exists, they must certainly be highly FERTILE. A doctrine which originated with Pallas, ha
0254 cies must either at first have produced quite FERTILE hybrids, or the hybrids must have become in
0254 t have become in subsequent generations quite FERTILE under domestication. This latter alternative
0254 domestic dogs of South America, all are quite FERTILE together; and analogy makes me greatly doubt
0254 freely bred together and have produced quite FERTILE hybrids. So again there is reason to believe
0254 ropean and the humped Indian cattle are quite FERTILE together; but from facts communicated to me
0255 th the pollen of either pure parent, a single FERTILE seed: but in some of these cases a first tra
0256 the hybrids, when at last produced, are very FERTILE. Even within the limits of the same genus, f
0258 nd the hybrids thus produced are sufficiently FERTILE; but Kolreuter tried more than two hundred t
0260 th extreme difficulty, and yet produce fairly FERTILE hybrids? Why should there often be so great
0265 ain species in a group will produce unusually FERTILE hybrids. No one can tell, till he tries, whe
0268 ss with perfect facility, and yield perfectly FERTILE offspring. I fully admit that this is almost
0268 arieties, are said by Gartner not to be quite FERTILE when crossed, and he consequently ranks them
0269 varieties of the same species were invariably FERTILE when intercrossed. But it seems to me imposs
0270 plants thus raised were themselves perfectly FERTILE; so that even Gartner did not venture to con
0271 hat one variety of the common tobacco is more FERTILE, when crossed with a widely distinct species
0271 nd he found their mongrel offspring perfectly FERTILE. But one of these five varieties, when used
0272 ty graduates away. When mongrels and the more FERTILE hybrids are propagated for several generatio
0277 re very generally, but not quite universally, FERTILE. Nor is this nearly general and perfect fert
0472 stinctive hatred of the queen bee for her own FERTILE daughters; at ichneumonidae feeding within t
0481 invariably sterile, and varieties invariably FERTILE; or that sterility is a special endowment an
0073 hat humble bees are indispensable to the FERTILISATION of the heartsease (Viola tricolor), for o
0073 ble, are at least highly beneficial to the FERTILISATION of our clovers; but humble bees alone vis
0092 as pollen is formed for the sole object of FERTILISATION, its destruction appears a simple loss to
0097 how unfavourable exposure to wet is to the FERTILISATION of a flower, yet what a multitude of flow
0097 enerally stand so close together that self FERTILISATION seems almost inevitable. Many flowers, on
0097 a of another with the same brush to ensure FERTILISATION; but it must not be supposed that bees wo
0098 ivance seems adapted solely to ensure self FERTILISATION; and no doubt it is useful for this end:
0098 ems to have a special contrivance for self FERTILISATION, it is well known that if very closely al
0098 es, far from there being any aids for self FERTILISATION, there are special contrivances, as I cou
0099 thers burst before the stigma is ready for FERTILISATION, or the stigma is ready before the pollen
0099 gether, as if for the very purpose of self FERTILISATION, should in so many cases be mutually usel
0101 rvals; but in none, as I suspect, can self FERTILISATION go on for perpetuity. Circumstances favou
0104 ver the offspring from long continued self FERTILISATION, that they will have a better chance of s
0145 whose agency is highly advantageous in the FERTILISATION of plants of these two orders, is so far
0249 and goes on increasing. Now, in artificial FERTILISATION pollen is as often taken by chance (as I
0256 er is well known to be a sign of incipient FERTILISATION. From this extreme degree of sterility we
0270 ed sexes, and he asserts that their mutual FERTILISATION is by so much the less easy as their diff
0073 hs to remove their pollen masses and thus to FERTILISE them. I have, also, reason to believe that
0257 arly distinct species, obstinately failed to FERTILISE, or to be fertilised by, no less than eight
0258 dred times, during eight following years, to FERTILISE reciprocally M. longiflora with the pollen
0093 lower which I examined had been effectually FERTILISED by the bees, accidentally dusted with polle
0203 of the orchis and of many other plants are FERTILISED through insect agency, can we consider as e
0247 plants which he castrated, and artificially FERTILISED with their own pollen, and (excluding all c
0249 ing season: hence hybrids will generally be FERTILISED during each generation by their own individ
0249 en the less fertile hybrids are artificially FERTILISED with hybrid pollen of the same kind, their
0249 anthers of the flower itself which is to be FERTILISED: so that a cross between two flowers, thoug
0249 the successive generations of artificially FERTILISED hybrids may, I believe, be accounted for by
0250 that every ovule in a pod of Crinum capense FERTILISED by C. revolutum produced a plant, which (he
0250 s hippeastrum, which can be far more easily FERTILISED by the pollen of another and distinct speci

```
0250  llen was found to be perfectly good, for it  FERTILISED  distinct species. So that certain individua
0250  sed much more readily than they can be self  FERTILISED! For instance, a bulb of Hippeastrum aulicu
0250  m aulicum produced four flowers; three were  FERTILISED  by Herbert with their own pollen, and the f
0250  own pollen, and the fourth was subsequently  FERTILISED  by the pollen of a compound hybrid descende
0251  comparison with the same species when self  FERTILISED, sometimes depends. The practical experimen
0256  is extreme degree of sterility we have self  FERTILISED  hybrids producing a greater and greater num
0257  , obstinately failed to fertilise, or to be  FERTILISED  by, no less than eight other species of Nic
0258  n instance: Mirabilis jalappa can easily be  FERTILISED  by the pollen of M. longiflora, and the hyb
0262  pollen of distinct species, than when self  FERTILISED  with their own pollen. We thus see, that al
0270  exes, they never naturally crossed. He then  FERTILISED  thirteen flowers of the one with the pollen
0270  eed, than do either coloured varieties when  FERTILISED  with pollen from their own coloured flowers
0097  ning of the law) that no organic being self  FERTILISES  itself for an eternity of generations; but
0100  a single case of a terrestrial animal which  FERTILISES  itself. We can understand this remarkable f
0100  estrial animals live, and the nature of the  FERTILISING  element; for we know of no means, analogou
0100  ls. Of aquatic animals, there are many self  FERTILISING  hermaphrodites; but here currents in the w
0101  that two individuals, though both are self  FERTILISING  hermaphrodites, do sometimes cross. It mus
0005  idism, or the infertility of species and the  FERTILITY  of varieties when intercrossed; and fourthl
0016  ese exceptions (and with that of the perfect  FERTILITY  of varieties when crossed, a subject hereaf
0065  ces could be given, no one supposes that the  FERTILITY  of these animals or plants has been suddenl
0095  er hand, I have found by experiment that the  FERTILITY  of clover greatly depends on bees visiting
0096  iety but of another strain, gives vigour and  FERTILITY  to the offspring; and on the other hand, th
0096  at close interbreeding diminishes vigour and  FERTILITY; that these facts alone incline me to belie
0097  experiments published elsewhere, that their  FERTILITY  is greatly diminished if these visits be pr
0104  hus produced will gain so much in vigour and  FERTILITY  over the offspring from long continued self
0172  g, whereas, when varieties are crossed their  FERTILITY  is unimpaired? The two first heads shall be
0245  of changed conditions of life and crossing.  FERTILITY  of varieties when crossed and of their mong
0245  and mongrels compared independently of their  FERTILITY. Summary. The view generally entertained by
0246  nd the province of our reasoning powers. The  FERTILITY  of varieties, that is of the forms known or
0246  parents, when intercrossed, and likewise the  FERTILITY  of their mongrel offspring, is, on my theor
0247  qually universal; and he disputes the entire  FERTILITY  of Kolreuter's ten cases. But in these and
0247  t these processes are often injurious to the  FERTILITY  of a plant cannot be doubted; for Gartner g
0247  ation) half of these twenty plants had their  FERTILITY  in some degree impaired. Moreover, as Gartn
0248  insensibly, and, on the other hand, that the  FERTILITY  of pure species is so easily affected by va
0248  es it is most difficult to say where perfect  FERTILITY  ends and sterility begins. I think no bette
0248  species or varieties, with the evidence from  FERTILITY  adduced by different hybridisers, or by the
0248  can thus be shown that neither sterility nor  FERTILITY  affords any clear distinction between speci
0248  ations, yet he asserts positively that their  FERTILITY  never increased, but generally greatly decr
0248  that this is usually the case, and that the  FERTILITY  often suddenly decreases in the first few g
0248  i believe that in all these experiments the  FERTILITY  has been diminished by an independent cause
0249  ts, showing that close interbreeding lessens  FERTILITY, and, on the other hand, that an occasional
0249  h a distinct individual or variety increases  FERTILITY, that I cannot doubt the correctness of thi
0249  vinced that this would be injurious to their  FERTILITY, already lessened by their hybrid origin. I
0249  d with hybrid pollen of the same kind, their  FERTILITY, notwithstanding the frequent ill effects o
0249  nd thus, the strange fact of the increase of  FERTILITY  in the successive generations of artificial
0250  perfect, or even more than commonly perfect,  FERTILITY  in a first cross between two distinct speci
0251  and mysterious causes the lesser or greater  FERTILITY  of species when crossed, in comparison with
0251  taken some pains to ascertain the degree of  FERTILITY  of some of the complex crosses of Rhododend
0252  , when fairly treated, gone on decreasing in  FERTILITY  in each successive generation, as Gartner b
0252  erfectly fertile. Again, with respect to the  FERTILITY  in successive generations of the more ferti
0255  as been already remarked, that the degree of  FERTILITY, both of first crosses and of hybrids, grad
0255  d of hybrids, graduates from zero to perfect  FERTILITY. It is surprising in how many curious ways
0255  h inorganic dust. From this absolute zero of  FERTILITY, the pollen of different species of the sam
0255  up to nearly complete or even quite complete  FERTILITY; and, as we have seen, in certain abnormal
0255  certain abnormal cases, even to an excess of  FERTILITY, beyond that which the plant's own pollen w
0255  but in some of these cases a first trace of  FERTILITY  may be detected, by the pollen of one of th
0256  er and greater number of seeds up to perfect  FERTILITY. Hybrids from two species which are very di
0256  ianthus, these two opposite cases occur. The  FERTILITY, both of first crosses and of hybrids, is m
0256  cted by unfavourable conditions, than is the  FERTILITY  of pure species. But the degree of fertilit
0256  fertility of pure species. But the degree of  FERTILITY  is likewise innately variable; for it is no
0256  . so it is with hybrids, for their degree of  FERTILITY  is often found to differ greatly in the sev
0256  differ little in the allied species. Now the  FERTILITY  of first crosses between species, and of th
0258  and then as the mother, generally differ in  FERTILITY  in a small, and occasionally in a higher de
0259  e same capsule have a considerable degree of  FERTILITY. These facts show how completely fertility
0259  f fertility. These facts show how completely  FERTILITY  in the hybrid is independent of its externa
0259  he several rules now given, which govern the  FERTILITY  of first crosses and of hybrids, we see tha
0259  good and distinct species, are united, their  FERTILITY  graduates from zero to perfect fertility, o
0259  eir fertility graduates from zero to perfect  FERTILITY, or even to fertility under certain conditi
0259  s from zero to perfect fertility, or even to  FERTILITY  under certain conditions in excess. That th
0259  der certain conditions in excess. That their  FERTILITY, besides being eminently susceptible to fav
0259  e hybrids produced from this cross. That the  FERTILITY  of hybrids is not related to the degree in
0260  uced from reciprocal crosses often differ in  FERTILITY. Now do these complex and singular rules in
0265  great changes of conditions with unimpaired  FERTILITY; and certain species in a group will produc
0266  rility of hybrids; for instance, the unequal  FERTILITY  of hybrids produced from reciprocal crosses
0267  rent strains or sub breeds, gives vigour and  FERTILITY  to the offspring. I believe, indeed, from t
0267  d become slightly different, give vigour and  FERTILITY  to the offspring. But we have seen that gre
0267  ssentially related to the principle of life.  FERTILITY  of varieties when crossed, and of their Mon
0268  d species. If we thus argue in a circle, the  FERTILITY  of all varieties produced under nature will
0268  y distinct species. Nevertheless the perfect  FERTILITY  of so many domestic varieties, differing wi
0268  several considerations, however, render the  FERTILITY  of domestic varieties less remarkable than
0270  e witnesses, who in all other cases consider  FERTILITY  and sterility as safe criterions of specifi
0272  facts, I do not think that the very general  FERTILITY  of varieties can be proved to be of univers
0272  n between varieties and species. The general  FERTILITY  of varieties does not seem to me sufficient
0272  nd mongrels compared, independently of their  FERTILITY. Independently of the question of fertility
0272  fertility. Independently of the question of  FERTILITY, the offspring of species when crossed and
0275  stinct species. Laying aside the question of  FERTILITY  and sterility, in all other respects there
0277  ly different is favourable to the vigour and  FERTILITY  of their offspring; and that slight changes
```

0277 are apparently favourable to the vigour and FERTILITY of all organic beings. It is not surprising
0277 the facility of effecting a first cross, the FERTILITY of the hybrids produced, and the capacity o
0277 tile. Nor is this nearly general and perfect FERTILITY surprising, when we remember how liable we
0278 ive system. In all other respects, excluding FERTILITY, there is a close general resemblance betwe
0460 arkable a contrast with the almost universal FERTILITY of varieties when crossed, I must refer the
0460 ed as the father and then as the mother. The FERTILITY of varieties when intercrossed and of their
0460 ered as universal; nor is their very general FERTILITY surprising when we remember that it is not
0461 class of facts; namely, that the vigour and FERTILITY of all organic beings are increased by slig
0461 uire from being crossed increased vigour and FERTILITY. So that, on the one hand, considerable cha
0461 osses between greatly modified forms, lessen FERTILITY; and on the other hand, lesser changes in t
0461 rosses between less modified forms, increase FERTILITY. Turning to geographical distribution, the
0262 a, etc., which seeded much more freely when FERTILIZED with the pollen of distinct species, than w
0228 that each hexagonal prism was built upon the FESTOONED edge of a smooth basin, instead of on the s
0063 and a few other trees, but can only in a far FETCHED sense be said to struggle with these trees,
0146 tion of plants of these two orders, is so far FETCHED, as it may at first appear: and if it be adv
0157 n, as it does not entail death, but only gives FEWER offspring to the less favoured males. Whateve
0223 a sanguinea, on the other hand, possesses much FEWER slaves, and in the early part of the summer e
0252 from other flowers. In regard to animals, much FEWER experiments have been carefully tried than wi
0292 sidence, though there will be much extinction, FEWER new varieties or species will be formed: and
0341 e case of the Edentata of South America, still FEWER genera and species will have left modified bl
0429 included under a few great orders, under still FEWER classes, and all in one great natural system.
0437 formed of many parts, consequently always have FEWER legs; or conversely, those with many legs hav
0488 history, when the forms of life were probably FEWER and simpler, the rate of change was probably
0105 ted on it will necessarily be very small; and FEWNESS of individuals will greatly retard the produ
0074 ict depends in a great degree on the number of FIELD mice, which destroy their combs and nests: an
0108 ain be changed: and again there will be a fair FIELD for natural selection to improve still furthe
0486 l history become! A grand and almost untrodden FIELD of inquiry will be opened, on the causes and
0070 se plenty of corn and rape seed, etc., in our FIELDS, because the seeds are in great excess compar
0095 s visited by humble bees alone: so that whole FIELDS of the red clover offer in vain an abundant s
0488 was created. In the distant future I see open FIELDS for far more important researches. Psychology
0003 p obligations to Dr. Hooker, who for the last FIFTEEN years has aided me in every possible way by
0064 end of the fifth century there would be alive FIFTEEN million elephants, descended from the first
0123 from the original eleven species, will now be FIFTEEN in number. Owing to the divergent tendency o
0233 ntally found that no less than from twelve to FIFTEEN pounds of dry sugar are consumed by a hive o
0420 ndants to the present day, represented by the FIFTEEN genera (a14 to z14) on the uppermost horizon
0027 the earliest known record of pigeons is in the FIFTH Aegyptian dynasty, about 3000 B.C., as was po
0064 his interval: if this be so, at the end of the FIFTH century there would be alive fifteen million
0161 ommon snapdragon (Antirrhinum) a rudiment of a FIFTH stamen so often appears, that this plant must
0208 he caterpillar simply re performed the fourth, FIFTH, and sixth stages of construction. If however
0334 modified offspring of those which lived at the FIFTH stage, and are the parents of those which bec
0412 nd, which diverged from a common parent at the FIFTH stage of descent. These five genera have also
0452 inum) we generally do not find a rudiment of a FIFTH stamen; but this may sometimes be seen. In tr
0068 ted that the winter of 1854 55 destroyed four FIFTHS of the birds in my own grounds; and this is a
0036 original stock of Mr. Bakewell for upwards of FIFTY years. There is not a suspicion existing in t
0374 a, dr. Hooker has shown that between forty and FIFTY of the flowering plants of Tierra del Fuego,
0484 lief. The endless disputes whether or not some FIFTY species of British brambles are true species
0088 not: male alligators have been described as FIGHTING, bellowing, and whirling round, like Indians
0088 of the females: male salmons have been seen FIGHTING all day long: male stag beetles often bear w
0196 ing a will, to give one male an advantage in FIGHTING with another or in charming the females. Mor
0165 excellent treatise on the horse, has given a FIGURE of a similar mule. In four coloured drawings,
0163 is without bars on the legs: but Dr. Gray has FIGURED one specimen with very distinct zebra like b
0165 other most remarkable case, a hybrid has been FIGURED by Dr. Gray (and he informs me that he knows
0225 n melipona domestica, carefully described and FIGURED by Pierre Huber. The Melipona itself is inte
0129 oups in any class cannot be ranked in a single FILE, but seem rather to be clustered round points,
0222 on) and numerous pupae. I traced the returning FILE burthened with booty, for about forty yards, t
0423 y the closest affinities, and would give the FILIATION and origin of each tongue. In confirmation
0076 uld be most severe between allied forms, which FILL nearly the same place in the economy of nature
0082 and unoccupied places for natural selection to FILL up by modifying and improving some of the vary
0102 , though in different degrees, so as better to FILL up the unoccupied place. But if the area be la
0106 places will be formed, and the competition to FILL them will be more severe, on a large than on a
0312 veral tertiary stages; and every other tends to FILL up the blanks between them, and to make the pe
0315 this has occurred in innumerable instances) to FILL the exact place of another species in the econ
0329 n them. That the extinct forms of life help to FILL up the wide intervals between existing genera,
0333 right to expect, except in very rare cases, to FILL up wide intervals in the natural system, and t
0344 g. why ancient and extinct forms often tend to FILL up gaps between existing forms, sometimes blen
0483 te to groups. Fossil remains sometimes tend to FILL up very wide intervals between existing orders
0081 omy of nature which would assuredly be better FILLED up, if some of the original inhabitants were
0108 in the polity of each island will have to be FILLED up by modifications of the old inhabitants; a
0178 e natural polity of the country can be better FILLED by some modification of some one or more of i
0318 hells at a very late geological period, I was FILLED with astonishment: for seeing that the horse,
0329 spect to the Vertebrata, whole pages could be FILLED with striking illustrations from our great pa
0366 streams with which their valleys were lately FILLED. So greatly has the climate of Europe changed
0487 edded remains must not be looked at as a well FILLED museum, but as a poor collection made at haza
0121 dants: for these will have the best chance of FILLING new and widely different places in the polit
0173 y allied or representative species, evidently FILLING nearly the same place in the natural economy
0130 it has been with the great Tree of Life, which FILLS with its dead and broken branches the crust o
0427 esemblance, in the shape of the body and in the FIN like anterior limbs, between the dugong, which
0428 ne with another: thus the shape of the body and FIN like limbs are only analogical when whales are
0428 hrough the water: but the shape of the body and FIN like limbs serve as characters exhibiting true
0435 nveloped in thick membrane, so as to serve as a FIN; or a webbed foot might have all its bones, or
0440 bones in the hand of a man, wing of a bat, and FIN of a porpoise, are related to similar conditio
0442 on why, for instance, the wing of a bat, or the FIN of a porpoise, should not have been sketched o
0454 ny matter, as that the rudimentary nails on the FIN of the manatee were formed for this purpose. O
0479 ing the same in the hand of a man, wing of bat, FIN of the porpoise, and leg of the horse, the sam
0037 not doubt, has been simple, and, as far as the FINAL result is concerned, has been followed almost
0043 hing must be attributed to use and disuse. The FINAL result is thus rendered infinitely complex. I

Page **(Key Word)**

```
0216  commonly admitted that the more immediate and FINAL cause of the cuckoo's instinct is, that she l
0317  ut in the upper beds, marking the decrease and FINAL extinction of the species. This gradual incre
0367  later; but this will make no difference in the FINAL result. As the warmth returned, the arctic fo
0435  e same class, by utility or by the doctrine of FINAL causes. The hopelessness of the attempt has b
0441  which to become attached and to undergo their FINAL metamorphosis. When this is completed they ar
0448  of insects, as with Aphis. With respect to the FINAL cause of the young in these cases not undergo
0448  , etc., would be useless; and in this case the FINAL metamorphosis would be said to be retrograde.
0457  the occurrence of rudimentary organs and their FINAL abortion, present to us no inexplicable diffi
0058  y british botanists as good and true species.  FINALLY, then, varieties have the same general chara
0069  pecies gradually getting rarer and rarer, and  FINALLY disappearing; and the change of climate bein
0101  ty, some of which I am trying to investigate.  FINALLY then, we may conclude that in many organic b
0106  nerally, to a certain extent, have concurred.  FINALLY, I conclude that, although small isolated ar
0110  tion, others will become rarer and rarer, and  FINALLY extinct. The forms which stand in closest co
0112  and would be noted as forming two sub breeds;  FINALLY, after the lapse of centuries, the sub breed
0126  . small and broken groups and sub groups will  FINALLY tend to disappear. Looking to the future, we
0158  ther males for the possession of the females.  FINALLY, then, I conclude that the greater variabili
0172  lly stocked country to take the place of, and  FINALLY to exterminate, its own less improved parent
0174  abruptly rarer and rarer on the confines, and  FINALLY disappearing. Hence the neutral territory be
0196  nfluence in modifying various structures; and  FINALLY, that sexual selection will often have large
0243  t wholly unlike that of any other known bird.  FINALLY, it may not be a logical deduction, but to m
0254  ne capable of being removed by domestication.  FINALLY, looking to all the ascertained facts on the
0278  ral resemblance between hybrids and mongrels.  FINALLY, then, the facts briefly given in this chapt
0318  r, first from one spot, the from another, and  FINALLY from the world. Both single species and whol
0319  ld certainly have become rarer and rarer, and  FINALLY extinct; its place being seized on by some m
0319  ies are amply sufficient to cause rarity, and  FINALLY extinction. We see in many cases in the more
0419  lowed by several entomologists and botanists.  FINALLY, with respect to the comparative value of th
0433  define the groups to which such types belong.  FINALLY, we have seen that natural selection, which
0436  organ might become so much obscured as to be  FINALLY lost, by the atrophy and ultimately by the c
0457  is only so far natural as it is genealogical.  FINALLY, the several classes of facts which have bee
0467  rease in number, and which shall decrease, or  FINALLY become extinct. As the individuals of the sa
0021  er has a beak in outline almost like that of a FINCH; and the common tumbler has the singular and
0028  r conclusion in regard to the many species of  FINCHES, or other large groups of birds, in nature.
0252  canary bird has been crossed with nine other  FINCHES; but as not one of these nine species breeds
0011  s it has a more marked effect; for instance, I FIND in the domestic duck that the bones of the win
0016  nature. I think this must be admitted, when we FIND that there are hardly any domestic races, eith
0018  ple origin of our domestic animals is, that we FIND in the most ancient records, more especially o
0020  e care and long continued selection; nor can I FIND a single case on record of a permanent race ha
0034  ts by nurserymen. The principle of selection I FIND distinctly given in an ancient Chinese encyclo
0037  eserved the best varieties they could anywhere FIND. A large amount of change in our cultivated pl
0050  that it is in the best known countries that we FIND the greatest number of forms of doubtful value
0051  ontinuous, in which case he can hardly hope to FIND the intermediate links between his doubtful fo
0055  ing. Where many large trees grow, we expect to FIND saplings. Where many species of a genus have b
0056  species has been active, we ought generally to FIND the manufactory still in action, more especial
0087  be found in works of natural history, I cannot FIND one case which will bear investigation. A stru
0100  exes more often separated than other plants, I FIND to be the case in this country; and at my requ
0107  exterminated. And it is in fresh water that we FIND seven genera of Ganoid fishes, remnants of a o
0107  once proponderant order: and in fresh water we FIND some of the most anomalous forms now known in
0112  d from our domestic productions. We shall here FIND something analogous. A fancier is struck by a
0114  idual and individual must be severe, we always FIND great diversity in its inhabitants. For instan
0114  and so in small ponds of fresh water. Farmers FIND that they can raise most food by a rotation of
0153  , we might, as a general rule, expect still to FIND more variability in such parts than in other p
0154  ent and extraordinarily great that we ought to FIND the generative variability, as it may be calle
0155  and fixed varieties, we might surely expect to FIND them still often continuing to vary in those p
0162  ions, yet we ought, on my theory, sometimes to FIND the varying offspring of a species assuming ch
0169  actively at work, there, on an average, we now FIND most varieties or incipient species. Secondary
0172  itional forms must have existed, why do we not FIND them embedded in countless numbers in the crus
0173  inhabit the same territory we surely ought to  FIND at the present time many transitional forms. L
0174  rmediate conditions of life, why do we not now FIND close linking intermediate varieties? This dif
0174  now distributed over a wide area, we generally FIND them tolerably numerous over a large territory
0187  red condition. Amongst existing Vertebrata, we FIND but a small amount of gradation in the structu
0188  ticulate class. He who will go thus far, if he FIND on finishing this treatise that large bodies o
0189  theory would absolutely break down. But I can  FIND out no such case. No doubt many organs exist o
0201  ks on natural history to this effect, I cannot FIND even one which seems to me of any weight. It i
0202  er of scent by which the males of many insects FIND their females, can we admire the production fo
0210  the case of corporeal structures, we ought to  FIND in nature, not the actual transitional gradati
0210  eal ancestors of each species; but we ought to FIND in the collateral lines of descent some eviden
0210  we certainly can do. I have been surprised to  FIND, making allowance for the instincts of animals
0222  emerge, carrying a pupa; but I was not able to FIND the desolated nest in the thick heath. The nes
0224  orkman, with fitting tools and measures, would FIND it very difficult to make cells of wax of the
0239  aracter. On this view we ought occasionally to FIND neuter insects of the same species, in the sam
0239  enting gradations of structure; and this we do FIND, even often, considering how few neuter insect
0240  one other case: so confidently did I expect to FIND gradations in important points of structure be
0252  dendrons, which produce no pollen, for he will FIND on their stigmas plenty of pollen brought from
0255  rd to hybrid animals, I have been surprised to FIND how generally the same rules apply to both kin
0269  f this be so, we surely ought not to expect to FIND sterility both appearing and disappearing unde
0272  stinction between species and varieties, could FIND very few and, as it seems to me, quite unimpor
0283  cky cliff, which is undergoing degradation, we FIND that it is only here and there, along a short
0293  difficult to understand, why we do not therein FIND closely graduated varieties between the allied
0293  elsewhere previously existed. So again when we FIND a species disappearing before the uppermost la
0297  the rank of varieties; and on this view we do  FIND the kind of evidence of change which on my the
0297  dence of change which on my theory we ought to FIND. Moreover, if we look to rather wider interval
0297  ecutive stages of the same great formation, we FIND that the embedded fossils, though almost unive
0301  n these remarks, we have no right to expect to FIND in our geological formations, an infinite numb
0307  iving creatures. To the question why we do not FIND records of these vast primordial periods, I ca
0307  obliterated by metamorphic action, we ought to FIND only small remnants of the formations next suc
0309  ow converted into a continent, we should there FIND formations older than the Silurian strata, sup
```

Page **(Key Word)**

Page **************************************(Key Word)**************************************

```
0325 he present distribution of organic beings, and FIND how slight is the relation between the physica
0326 e continuous sea. We might therefore expect to FIND, as we apparently do find, a less strict degre
0326  therefore expect to find, as we apparently do FIND, a less strict degree of parallel succession i
0328 ly, but not exactly the same period, we should FIND in both, from the causes explained in the fore
0336 ccessive formations, we ought not to expect to FIND, as I attempted to show in the last chapter, i
0336 nt and close of these periods; but we ought to FIND after intervals, very long as measured by year
0336 resentative species; and these we assuredly do FIND. We find, in short, such evidence of the slow
0336 ve species; and these we assuredly do find. We FIND, in short, such evidence of the slow and scarc
0336 ic forms, as we have a just right to expect to FIND. On the state of Development of Ancient Forms.
0342 d will to a large extent explain why we do not FIND interminable varieties, connecting together al
0346 nerally require; for it is a most rare case to FIND a group of organisms confined to any small spo
0347 latitudes 25 degrees and 35 degrees, we shall FIND parts extremely similar in all their condition
0348 eserts, and sometimes even of large rivers, we FIND different productions; though as mountain chai
0348 of distinct continents. Turning to the sea, we FIND the same law. No two marine faunas are more di
0349 scend the lofty peaks of the Cordillera and we FIND an alpine species of bizcacha; we look to the
0349 bizcacha; we look to the waters, and we do not FIND the beaver or musk rat, but the coypu and capy
0349 ast ages, as shown in the last chapter, and we FIND American types then prevalent on the American
0352 any other organic beings; and, accordingly, we FIND no inexplicable cases of the same mammal inhab
0352 procuced at two separate points, why do we not FIND a single mammal common to Europe and Australia
0361 es, these stones being a valuable royal tax. I FIND on examination, that when irregularly shaped s
0362 o days and fourteen hours. Fresh water fish, I FIND, eat seeds of many land and water plants: fish
0368 ope and America, that when in other regions we FIND the same species on distant mountain summits,
0369  have been liable to modification; and this we FIND has been the case; for if we compare the prese
0371 emperate regions of the New and Old Worlds, we FIND very few identical species (though Asa Gray ha
0372 identical than was formerly supposed), but we FIND in every great class many forms, which some na
0395 ank, nearly 1000 fathoms in depth, and here we FIND American forms, but the species and even the g
0400 e come to be modified since their arrival), we FIND a considerable amount of difference in the sev
0401 ets of organisms: a plant, for instance, would FIND the best fitted ground more perfectly occupied
0405 e most remote points of the world, we ought to FIND, and I believe as a general rule we do find, t
0405 to find, and I believe as a general rule we do FIND, that some at least of the species range very
0408 of organic action and reaction, and we should FIND, as we do find, some groups of beings greatly,
0408 ion and reaction, and we should find, as we do FIND, some groups of beings greatly, and some only
0410 to distant provinces throughout the world, we FIND that some organisms differ little, whilst othe
0418 in allocating any particular species. If they FIND a character nearly uniform, and common to a gr
0426 and we put them all into the same class. As we FIND organs of high physiological importance, those
0429 single one; but such richness in species, as I FIND after some investigation, does not commonly fa
0434 connected together in the same order. We never FIND, for instance, the bones of the arm and forear
0438 in the lowest members of the class, we do not FIND nearly so much indefinite repetition of any on
0438 h indefinite repetition of any one part, as we FIND in the other great classes of the animal and v
0445 ays old colt of a race and heavy cart horse, I FIND that the colts have by no means acquired their
0452 be utterly aborted; and this implies, that we FIND in an animal or plant no trace of an organ, wh
0452 rgan, which analogy would lead us to expect to FIND, and which is occasionally found in monstrous
0452 e snapdragon (antirrhinum) we generally do not FIND a rudiment of a fifth stamen: but this may sóm
0463 cies, we have no just right to expect often to FIND intermediate varieties in the intermediate zon
0463 n the several geological stages? Why do we not FIND great piles of strata beneath the Silurian sys
0470 active, we might expect, as a general rule, to FIND it still in action; and this is the case if va
0478 inhabited both areas; and we almost invariably FIND that wherever many closely allied species inha
0045 red that systematists are far from pleased at FINDING variability in important characters, and tha
0282 n the lapse of Time. Independently of our not FINDING fossil remains of such infinitely numerous c
0310 l difficulties here discussed; namely our not FINDING in the successive formations infinitely nume
0100 d, dr. Hooker has recently informed me that he FINDS that the rule does not hold in Australia; and
0218 n larvae to feed on, yet that when this insect FINDS a burrow already made and stored by another s
0328 s in England with those in France, although he FINDS in both a curious accordance in the numbers o
0328 ts of Bohemia and Scandinavia; nevertheless he FINDS a surprising amount of difference in the spec
0171 ave descended from other species by insensibly FINE gradations, do we not everywhere see innumerab
0284 si; yet, considering over what wide spaces very FINE sediment is transported by the currents of the
0296 , a section would not probably include all the FINE intermediate gradations which must on my theor
0297 conchologists are now sinking many of the very FINE species of D'Orbigny and others into the rank
0298 being enabled to connect species by numerous, FINE, intermediate, fossil links, by asking ourselv
0299 cies, by connecting them together by numerous, FINE, intermediate varieties; and this not having b
0301 ogical formations, an infinite number of those FINE transitional forms, which on my theory assured
0329 r with ruminants: for example, he dissolves by FINE gradations the apparently wide difference betw
0387 when I saw the great size of the seeds of that FINE water lily the Nelumbium, and remembered Alph.
0432 roups, as all would blend together by steps as FINE as those between the finest existing varieties
0432 e supposed to be alive; and the links to be as FINE as those between the finest existing varieties
0462 all the species in each group by gradations as FINE as our present varieties, it may be asked, Why
0280 s? geology assuredly does not reveal any such FINELY graduated organic chain; and this, perhaps, i
0190 at the family of squirrels: here we have the FINEST gradation from animals with their tails only
0342 the extinct and existing forms of life by the FINEST graduated steps. He who rejects these views c
0432 ogether by steps as fine as those between the FINEST existing varieties, nevertheless a natural cl
0432 the links to be as fine as those between the FINEST varieties. In this case it would be quite imp
0449 gether, and as all have been connected by the FINEST gradations, the best, or indeed, if our colle
0181 il, and including the limbs and the elongated FINGERS: the flank membrane is, also, furnished with
0181 eving it possible that the membrane connected FINGERS and fore arm of the Galeopithecus might be g
0275 m, deficiency of tail or horns, or additional FINGERS and toes; and do not relate to characters wh
0453 n of precious phosphate of lime? When a man's FINGERS have been amputated, imperfect nails sometim
0232 far as I have seen, they never gnaw away and FINISH off the angles of a cell till a large part bo
0001 coming to a decision. My work is now nearly FINISHED; but as it will take me two or three more ye
0208 e, to the third stage, and were put into one FINISHED up to the sixth stage, so that much of its w
0208 off, and thus tried to complete the already FINISHED work. If we suppose any habitual action to b
0231 he rough wall of wax has in every case to be FINISHED off, by being largely gnawed away on both si
0231 enty times thicker than the excessively thin FINISHED wall of the cell, which will ultimately be l
0393 ly searched the oldest voyages, but have not FINISHED my search; as yet I have not found a single
0188 lass. He who will go thus far, if he find on FINISHING this treatise that large bodies of facts, o
0182 logger headed duck (Micropterus of Eyton); as FINS in the water and front legs on the land, like
0182 ing and turning by the aid of their fluttering FINS, might have been modified into perfectly winge
```

Page **************************************(Key Word)**************************************

Page **(Key Word)**

```
0071  y five years previously and planted with Scotch FIR. The change in the native vegetation of the pl
0072  , and literally I could not see a single Scotch FIR, except the old planted clumps. But on looking
0072  bsolutely determine the existence of the Scotch FIR; but in several parts of the world insects det
0203  sider as equally perfect the elaboration by our FIR trees of dense clouds of pollen, in order that
0221  d marching along the same road to a tall Scotch FIR tree, twenty five yards distant, which they as
0472  ters; at the astonishing waste of pollen by our FIR trees; at the instinctive hatred of the queen
0366  arctic climate. The ruins of a house burnt by FIRE do not tell their tale more plainly, than do t
0083  ons, that they have allowed foreigners to take FIRM possession of the land. And as foreigners have
0335  , seems to have shaken Professor Pictet in his FIRM belief in the immutability of species. He who
0377  suffering state and could not have presented a FIRM front against intruders, that a certain number
0028  onversed, or whose treatises I have read, are FIRMLY convinced that the several breeds to which ea
0385  lls crawled on the feet, and clung to them so FIRMLY that when taken out of the water they could n
0386  n ancylus (a fresh water shell like a limpet) FIRMLY adhering to it; and a water beetle of the sam
0451  light, and not rarely lying under wing cases, FIRMLY soldered together! The meaning of rudimentary
0071  ensive heaths, with a few clumps of old Scotch FIRS on the distant hilltops: within the last ten y
0071  large spaces have been enclosed, and self sown FIRS are now springing up in multitudes, so close t
0072  thickly clothed with vigorously growing young FIRS. Yet the heath was so extremely barren and so
0004  from these considerations, I shall devote the FIRST chapter of this Abstract to Variation under D
0005  ficulties on the theory will be given: namely, FIRST, the difficulties of transitions, or in under
0007  lder cultivated plants and animals, one of the FIRST points which strikes us, is, that they genera
0010  ns are affected in the same way, the change at FIRST appears to be directly due to such conditions
0013  that, at whatever period of life a peculiarity FIRST appears, it tends to appear in the offspring
0014  n the offspring at the same period at which it FIRST appeared in the parent. I believe this rule t
0014  y. these remarks are of course confined to the FIRST appearance of the peculiarity, and not to its
0017  but how could a savage possibly know, when he FIRST tamed an animal, whether it would vary in suc
0020  his object, and failed. The offspring from the FIRST cross between two pure breeds is tolerably un
0028  yet quite insufficient, length: because when I FIRST kept pigeons and watched the several kinds, k
0037  ot sufficiently distinct to be ranked at their FIRST appearance as distinct varieties, and whether
0037  rent stocks. No one would ever expect to get a FIRST rate heartsease or dahlia from the seed of a
0037  f a wild plant. No one would expect to raise a FIRST rate melting pear from the seed of the wild p
0039  y selection, excepting on variations which are FIRST given to him in some slight degree by nature.
0039  abnormal or unusual any character was when it FIRST appeared, the more likely it would be to catc
0039  in most cases, utterly incorrect. The man who FIRST selected a pigeon with a slightly larger tail
0039  n tail feathers have been counted. Perhaps the FIRST pouter pigeon did not inflate its crop much m
0040  distinct and valuable, and will then probably FIRST receive a provincial name. In semi civilised
0047  ranking the most common, but sometimes the one FIRST described, as the species, and the other as t
0050  up of organisms quite unknown to him, he is at FIRST much perplexed to determine what differences
0051  nd in other countries, by which to correct his FIRST impressions. As he extends the range of his o
0051  st, as of high importance for us, as being the FIRST step towards such slight varieties as are bar
0053  he varieties in several well worked floras. At FIRST this seemed a simple task: but Mr. H. C. Wats
0064  fifteen million elephants, descended from the FIRST pair. But we have better evidence on this sub
0068  emics with man. The action of climate seems at FIRST sight to be quite independent of the struggle
0072  s its eggs in the navels of these animals when FIRST born. The increase of these flies, numerous a
0074  rict might determine, through the intervention FIRST of mice and then of bees, the frequency of ce
0075  ds and seedlings, or on the other plants which FIRST clothed the ground and thus checked the growt
0077  egs of the water beetle, the relation seems at FIRST confined to the elements of air and water. Ye
0077  id up within the seeds of many plants seems at FIRST sight to have no sort of relation to other pl
0082  t. we have reason to believe, as stated in the FIRST chapter, that a change in the conditions of l
0092  plant: yet if a little pollen were carried, at FIRST occasionally and then habitually, by the poll
0095  , is open to the same objections which were at FIRST urged against Sir Charles Lyell's noble views
0096  ction of their kind. This view, I may add, was FIRST suggested by Andrew Knight. We shall presentl
0096  t to some general considerations alone. In the FIRST place, I have collected so large a body of fa
0105  , and nowhere else. Hence an oceanic island at FIRST sight seems to have been highly favourable fo
0107  e long and to spread widely. For the area will FIRST have existed as a continent, and the inhabita
0111  as I believe, several important facts. In the FIRST place, varieties, even strongly marked ones,
0112  inciple of divergence, causing differences, at FIRST barely appreciable, steadily to increase, and
0113  sed. The same has been found to hold good when FIRST one variety and then several mixed varieties
0123  will be in some degree distinct from the three FIRST named species; and lastly, o14, e14, and m14,
0123  to the other, but from having diverged at the FIRST commencement of the process of modification,
0129  have living and modified descendants. From the FIRST growth of the tree, many a limb and branch ha
0131  ng several generations. I have remarked in the FIRST chapter, but a long catalogue of facts which
0134  e and Disuse. From the facts alluded to in the FIRST chapter, I think there can be little doubt th
0136  on of natural selection. For when a new insect FIRST arrived on the island, the tendency of natura
0144  e of more or less down on the young birds when FIRST hatched, with the future colour of their plum
0146  se two orders, is so far fetched, as it may at FIRST appear: and if it be advantageous, natural se
0156  ote period, since that period when the species FIRST branched off from their common progenitor, an
0157  i will give in illustration two instances, the FIRST which happen to stand on my list: and as the
0164  both gray and bay Kattywar horses striped when FIRST foaled. I have, also, reason to suspect, from
0165  with its legs so much striped that any one at FIRST would have thought that it must have been the
0172  crossed their fertility is unimpaired? The two FIRST heads shall be here discussed, Instinct and H
0174  hink it can be in large part explained. In the FIRST place we should be extremely cautious in infe
0178  ntermediate varieties will, it is probable, at FIRST have been formed in the intermediate zones, b
0183  terial for us, whether habits generally change FIRST and structure afterwards: or whether slight m
0187  , hardly concerns us more than how life itself FIRST originated: but I may remark that several fac
0189  r in this latter case the organ must have been FIRST formed at an extremely remote period, since w
0195  rgan as perfect and complex as the eye. In the FIRST place, we are much too ignorant in regard to
0195  ially constructed fly flapper; and it seems at FIRST incredible that this could have been adapted
0196  m the above or other unknown causes, it may at FIRST have been of no advantage to the species, but
0204  d by natural selection from an animal which at FIRST could only glide through the air. We have see
0205  ions, wholly due to the laws of growth, and at FIRST in no way advantageous to a species, have bee
0211  began to play with its antennae on the abdomen FIRST of one aphis and then of another: and each ap
0213  etimes point and even back other dogs the very FIRST time that they are taken out: retrieving is c
0214  pen, as I once saw in a pure terrier. When the FIRST tendency was once displayed, methodical selec
0216  , peculiar mental habits and actions, which at FIRST appeared from what we must in our ignorance c
0217  ke her own nest and sit on her own eggs, those FIRST laid would have to be left for some time unin
0217  has to migrate at a very early period: and the FIRST hatched young would probably have to be fed b
```

Page **(Key Word)**

Page **************************************(Key Word)***************************************

```
0218  he case of the American species, unite and lay  FIRST  a few eggs in one nest and then in another; a
0219  making instinct. This remarkable instinct was   FIRST  discovered in the Formica (Polyerges) rufesce
0219  been perfected. Formica sanguinea was likewise  FIRST  discovered by P. Huber to be a slave making a
0224  whatever instincts you please, and it seems at  FIRST  quite inconceivable how they can make all the
0224  the difficulty is not nearly so great as it at  FIRST  appears: all this beautiful work can be shown
0230  ith my theory. Huber's statement that the very  FIRST  cell is excavated out of a little parallel si
0230  , as far as I have seen, strictly correct; the  FIRST  commencement having always been a little hood
0231  he bees build is curious; they always make the  FIRST  rough wall from ten to twenty times thicker t
0231  understand how they work, by supposing masons   FIRST  to pile up a broad ridge of cement, and then
0231  he utmost ultimate economy of wax. It seems at  FIRST  to add to the difficulty of understanding how
0231  dividuals work even at the commencement of the  FIRST  cell. I was able practically to show this fac
0232  metimes recurring to a shape which they had at  FIRST  rejected. When bees have a place on which the
0233  h to commence a cell, and then moving outside,  FIRST  quite subversive of the foregoing theory; nam
0233  portant, as it bears on a fact, which seems at  FIRST  to one point; and then to five other points,
0236  ine myself to one special difficulty, which at  FIRST  appeared to me insuperable, and actually fata
0237  his case with the theory of natural selection?  FIRST, let it be remembered that we have innumerabl
0239  t, profitable modification did not probably at  FIRST  appear in all the individual neuters in the s
0241  and structure; a graduated series having been   FIRST  formed, as in the case of the driver ant, and
0245  ybridism. Distinction between the sterility of  FIRST  crosses and of hybrids. Sterility various in
0245  other differences. Causes of the sterility of  FIRST  crosses and of hybrids. Parallelism between t
0245  ll organic forms. This view certainly seems at  FIRST  probable, for species within the same country
0246  heri namely, the sterility of two species when  FIRST  crossed, and the sterility of the hybrids pro
0246  ture, as far as the microscope reveals. In the  FIRST  case the two sexual elements which go to form
0246  ear distinction between varieties and species.  FIRST, for the sterility of species when crossed an
0248  the fertility often suddenly decreases in the   FIRST  few generations. Nevertheless I believe that
0250  ven more than commonly perfect, fertility in a  FIRST  cross between two distinct species. This case
0250  : the result was that the ovaries of the three  FIRST  flowers soon ceased to grow, and after a few
0252  nfinement, we have no right to expect that the  FIRST  crosses between them and the canary, or that
0254  is view, the aboriginal species must either at  FIRST  have produced quite fertile hybrids, or the h
0254  hether the several aboriginal species would at  FIRST  have freely bred together and have produced o
0254  ncluded that some degree of sterility, both in  FIRST  crosses and in hybrids, is an extremely gener
0254  ely universal. Laws governing the Sterility of  FIRST  Crosses and of Hybrids. We will now consider
0255  umstances and rules governing the sterility of  FIRST  crosses and of hybrids. Our chief object will
0255  emarked, that the degree of fertility, both of  FIRST  crosses and of hybrids, graduates from zero t
0255  gle fertile seed: but in some of these cases a  FIRST  trace of fertility may be detected, by the po
0256  parallelism between the difficulty of making a  FIRST  cross, and the sterility of the hybrids thus
0256  o opposite cases occur. The fertility, both of  FIRST  crosses and of hybrids, is more easily affect
0256  le in the allied species. Now the fertility of  FIRST  crosses between species, and of the hybrids p
0258  case, for instance, of a stallion horse being  FIRST  crossed with a female ass, and then a male as
0258  very same two species, the one species having  FIRST  been used as the father and then as the mothe
0259  rules now given, which govern the fertility of  FIRST  crosses and of hybrids, we see that when form
0259  s by no means always the same in degree in the  FIRST  cross and in the hybrids produced from this c
0259  ent. And lastly, that the facility of making a  FIRST  cross between any two species is not always g
0260  y, not strictly related to the facility of the  FIRST  union between their parents, seems to be a st
0260  clearly to indicate that the sterility both of  FIRST  crosses and of hybrids is simply incidental o
0263  ll more complex laws governing the facility of  FIRST  crosses, are incidental on unknown difference
0263  for their welfare. Causes of the Sterility of  FIRST  Crosses and of Hybrids. We may now look a lit
0263  ser at the probable causes of the sterility in  FIRST  crosses and of hybrids. These two cases are f
0263  whereas in hybrids they are imperfect. Even in  FIRST  crosses, the greater or lesser difficulty in
0264  mbryo is a very frequent cause of sterility in  FIRST  crosses. I was at first very unwilling to bel
0264  cause of sterility in first crosses. I was at  FIRST  very unwilling to believe in this view: as th
0269  of domestic varieties less remarkable than at  FIRST  appears. It can, in the first place, be clear
0269  markable than at first appears. It can, in the  FIRST  place, be clearly shown that mere external di
0269  ccessive generations of hybrids, which were at  FIRST  only slightly sterile; and if this be so, we
0270  wing case is far more remarkable, and seems at  FIRST  quite incredible; but it is the result of an
0272  very general, but not invariable, sterility of  FIRST  crosses and of hybrids, namely, that it is no
0272  the most important distinction is, that in the  FIRST  generation mongrels are more variable than hy
0272  long been cultivated are often variable in the  FIRST  generation; and I have myself seen striking i
0273  ight degree of variability in hybrids from the  FIRST  cross or in the first generation, in contrast
0273  lity in hybrids from the first cross or in the  FIRST  generation, in contrast with their extreme va
0273  tical with the parent form. Now hybrids in the  FIRST  generation are descended from species (exclud
0276  een species and varieties. Summary of Chapter.  FIRST  crosses between forms sufficiently distinct t
0276  species. It is not always equal in degree in a  FIRST  cross and in the hybrid produced from this cr
0276  ming inarched in our forests. The sterility of  FIRST  crosses between pure species, which have thei
0277  it surprising that the facility of effecting a  FIRST  cross, the fertility of the hybrids produced,
0277  all kinds of resemblance between all species.  FIRST  crosses between forms known to be varieties,
0280  imperfection of the geological record. In the  FIRST  place it should always be borne in mind what
0286  he rate of one inch in a century. This will at  FIRST  appear much too small an allowance; but it is
0290  hstand the incessant action of the waves, when  FIRST  upraised and during subsequent oscillations o
0292  productions on the shores of a continent when  FIRST  broken up into an archipelago), and consequen
0293  conclusion on this head. When we see a species  FIRST  appearing in the middle of any formation, the
0293  l and other changes; and when we see a species  FIRST  appearing in any formation, the probability i
0294  ormation, the probability is that it only then  FIRST  immigrated into that area. It is well known,
0294  e been upraised, organic remains will probably  FIRST  appear and disappear at different levels, owi
0298  ly seen, that their varieties are generally at  FIRST  local; and that such local varieties do not s
0298  ons of Europe, which have oftenest given rise,  FIRST  to local varieties and ultimately to new spec
0301  duce new varieties; and the varieties would at  FIRST  generally be local or confined to one place,
0308  ormations are composed, we may infer that from  FIRST  to last large islands or tracts of land, when
0312  wo are new forms, having here appeared for the  FIRST  time, either locally, or, as far as we know,
0317  t be slow and gradual, one species giving rise  FIRST  to two or three varieties, these being slowly
0317  pecies gradually disappear, one after another,  FIRST  from one spot, the from another, and finally
0318  nation, than at its lower end, which marks the  FIRST  appearance and increase in numbers of the spe
0320  ightly improved variety has been raised, it at  FIRST  supplants the less improved varieties in the
0334  s of the Devonian system, when this system was  FIRST  discovered, were at once recognised by palaeo
0334  , when arranged by Dr. Falconer in two series,  FIRST  according to their mutual affinities and then
0335  s and other such cases, that the record of the  FIRST  appearance and disappearance of the species w
```

Page **************************************(Key Word)***************************************

Page **(Key Word)**

```
0342   migration has played an important part in the FIRST appearance of new forms in any one area and f
0342   ise to new species; and that varieties have at FIRST often been local. All these causes taken conj
0343   anisms which must have existed long before the FIRST bed of the Silurian system was deposited: I c
0346   organic beings over the face of the globe, the FIRST great fact which strikes us is, that neither
0352   ceeded from one spot, where their parents were FIRST produced: for, as explained in the last chapt
0352   e simplicity of the view that each species was FIRST produced within a single region captivates th
0368   lated from the moment of the returning warmth, FIRST at the bases and ultimately on the summits of
0380   ers have been greatly reduced, and this is the FIRST stage towards extinction. A mountain is an is
0383   er throughout the world. I well remember, when FIRST collecting in the fresh waters of Brazil, fee
0385   il throughout the world. Their distribution at FIRST perplexed me much, as their ova are not likel
0388   ld be remembered that when a pond or stream is FIRST formed, for instance, on a rising islet, it w
0392   ver order they belonged, and thus convert them FIRST into bushes and ultimately into trees. With r
0411   ans; their origin explained. Summary. From the FIRST dawn of life, all organic beings are found to
0414    are of little signification, excepting in the FIRST main divisions; whereas the organs of reprodu
0419   ongst plants and insects, of a group of forms, FIRST ranked by practised naturalists as only a gen
0419    detected important structural differences, at FIRST overlooked, but because numerous allied speci
0427   d analogical or adaptive resemblances. Lamarck FIRST called attention to this distinction, and he
0439   ied genera, often resemble each other in their FIRST and second plumage: as we see in the spotted
0439   embryonic leaves of the ulex or furze, and the FIRST leaves of the phyllodineous acaceas, are pinn
0441   er once again to cirripedes: the larvae in the FIRST stage have three pairs of legs, a very simple
0444   n extremely early period. I have stated in the FIRST chapter, that there is some evidence to rende
0444   t probable, that at whatever age any variation FIRST appears in the parent, it tends to reappear a
0444   ove specified leading facts in embryology. But FIRST let us look at a few analogous cases in domes
0446   mulated by man's selection, have not generally FIRST appeared at an early period of life, and have
0447   ted at an earlier period than that at which it FIRST appeared. In either case (as with the short f
0448   ges of development: so that the larvae, in the FIRST stage, might differ greatly from the larvae i
0450   d at an earlier period than that at which they FIRST appeared. It should also be borne in mind, th
0459   d to give to them their full force. Nothing at FIRST can appear more difficult to believe than tha
0460   the almost universal sterility of species when FIRST crossed, which forms so remarkable a contrast
0460   sed reciprocally; that is, when one species is FIRST used as the father and then as the mother. Th
0461   hybrids is a very different case from that of FIRST crosses, for their reproductive organs are mo
0461   more or less functionally impotent: whereas in FIRST crosses the organs on both sides are in a per
0463    that the intermediate varieties which will at FIRST probably exist in the intermediate zones, wil
0464   species vary most, and varieties are often at FIRST local, both causes rendering the discovery of
0469   and permanent varieties, and that each species FIRST existed as a variety, we can see why it is th
0470   y created, but are intelligible if all species FIRST existed as varieties. As each species tends b
0481   as that felt by so many geologists, when Lyell FIRST insisted that long lines of inland cliffs had
0483    own side they ignore the whole subject of the FIRST appearance of species in what they consider r
0484   some one primordial form, into which life was FIRST breathed. When the views entertained in this
0488   rate of change was probably slower; and at the FIRST dawn of life, when very few forms of the simp
0488   red with the ages which have elapsed since the FIRST creature, the progenitor of innumerable extin
0489   of some few beings which lived long before the FIRST bed of the Silurian system was deposited, the
0025   owing highly improbable suppositions. Either, FIRSTLY, that all the several imagined aboriginal st
0027   b species. In favour of this view, I may add, FIRSTLY, that C. livia, or the rock pigeon, has been
0058   cannot be distinguished from species, except, FIRSTLY, by the discovery of intermediate linking o
0171   ons may be classed under the following heads: FIRSTLY, why if species have descended from other sp
0177   able chaos of varying and intermediate links: FIRSTLY, because new varieties are very slowly forme
0405   vourable for the conversion of its offspring, FIRSTLY into new varieties and ultimately into new s
0448    result from the two following contingencies; FIRSTLY, for the young, during a course of modifica
0139   as agassiz has remarked in regard to the blind FISH, the Amblyopsis, and as is the case with the b
0179   ring summer this animal dives for and preys on FISH, but during the long winter it leaves the froz
0182   flying reptiles, it is conceivable that flying FISH, which now glide far through the air, slightly
0182    as we know, to escape being devoured by other FISH? When we see any structure highly perfected fo
0183   rn to our imaginary illustration of the flying FISH, it does not seem probable that fishes capable
0183   water, and then dashing like a kingfisher at a FISH. In our own country the larger titmouse (Parus
0190   retes in the larva of the dragonfly and in the FISH Cobites. In the Hydra, the animal may be turne
0190   me individual; to give one instance, there are FISH with gills or branchiae that breathe the air d
0190   an accessory to the auditory organs of certain FISH, or, for I do not know which view is now gener
0305   t superfluous to remark that hardly any fossil FISH are known from south of the equator; and by ru
0305   ral formations in Europe. Some few families of FISH now have a confined range; the teleostean fish
0305   fish now have a confined range; the teleostean FISH might formerly have had a similarly confined r
0330   with respect to very distirct groups, such as FISH and reptiles, seems to be, that supposing them
0331   the older Reptiles and Batrachians, the older FISH, the older Cephalopods, and the Eocene mammals
0338   , that the oldest known mammals, reptiles, and FISH strictly belong to their own proper classes, t
0348   marine faunas. Although hardly one shell, crab FISH, shell, or crab in common, than those of the e
0348   ine faunas. Although hardly one shell, crab or FISH is common to the above named three approximate
0348   rica and the eastern Pacific Islands, yet many FISH range from the Pacific into the Indian Ocean,
0362   d for two days and fourteen hours. Fresh water FISH, I find, eat seeds of many land and water plan
0362   find, eat seeds of many land and water plants: FISH are frequently devoured by birds, and thus the
0362   many kinds of seeds into the stomachs of dead FISH, and then gave their bodies to fishing eagles,
0372   s described in Dana's admirable work), of some FISH and other marine animals, in the Mediterranean
0376   zealand, Tasmania, etc., of northern forms of FISH. Dr. Hooker informs me that twenty five specie
0384   n here consider only a few cases. In regard to FISH, I believe that the same species never occur i
0384   iciously; for two river systems will have some FISH in common and some different. A few facts seem
0384   ort by accidental means: like that of the live FISH not rarely dropped by whirlwinds in India, and
0384   ined to attribute the dispersal of fresh water FISH mainly to slight changes within the recent per
0384   fresh water shells. The wide difference of the FISH on opposite sides of continuous mountain range
0384   conclusion. With respect to allied fresh water FISH occurring at very distant points of the world,
0384   at present be explained: but some fresh water FISH belong to very ancient forms, and in such case
0384   uch migration. In the second place, salt water FISH can with care be slowly accustomed to live in
0387   played a part. I have stated that fresh water FISH eat some kinds of seeds, though they reject ma
0387   kinds after having swallowed them; even small FISH swallow seeds of moderate size, as of the yell
0387   ry after century, have gone on daily devouring FISH; they then take flight and go to other waters,
0387   g to another pond and getting a hearty meal of FISH, would probably reject from its stomach a pell
0387   d whilst feeding its young, in the same way as FISH are known sometimes to be dropped. In consider
0414   shrew, of a dugong to a whale, of a whale to a FISH, as of any importance. These resemblances, tho
```

Page **(Key Word)**

Page ***************************************(Key Word)***************************************

```
0442  lt: thus Owen has remarked in regard to cuttle FISH, there is no metamorphosis; the cephalopodic c
0448  ertain whole groups of animals, as with cuttle FISH and spiders, and with a few members of the gre
0452   and be used for a distinct object: in certain FISH the swim bladder seems to be rudimentary for t
0479  and arteries running in loops, like those in a FISH which has to breathe the air dissolved in wate
0107  esh water that we find seven genera of Ganoid FISHES, remnants of a once proponderant order: and t
0183  e flying fish, it does not seem probable that FISHES capable of true flight would have been develo
0190  rated. The illustration of the swimbladder in FISHES is a good one, because it shows us clearly th
0192  f in the next chapter. The electric organs of FISHES offer another case of special difficulty; it
0193  ficulty; for they occur in only about a dozen FISHES, of which several are widely remote in their
0193  ded, we might have expected that all electric FISHES would have been specially related to each oth
0193  at all lead to the belief that formerly most FISHES had electric organs, which most of their modi
0305  e group of species, is that of the teleostean FISHES, low down in the Chalk period. This group inc
0305  aeontologists believe that certain much older FISHES, of which the affinities are as yet imperfect
0321   the great and almost extinct group of Ganoid FISHES still inhabit our fresh waters. Therefore the
0384  lenciennes, there is hardly a single group of FISHES confined exclusively to fresh water, so that
0416  rding to Owen, which absolutely distinguishes FISHES and reptiles, the inflection of the angle of
0427  the whale, and between both these mammals and FISHES, is analogical. Amongst insects there are inn
0428  only analogical when whales are compared with FISHES, being adaptations in both classes for swimmi
0428  f limbs from a common ancestor. So it is with FISHES. As members of distinct classes have often be
0431  ion of the forms of life which once connected FISHES with batrachians. There has been still less t
0479   the embryos of mammals, birds, reptiles, and FISHES should be so closely alike, and should be so
0362  s of dead fish, and then gave their bodies to FISHING eagles, storks, and pelicans; these birds af
0158  of by natural and sexual selection, in order to FIT the several species to their several places in
0158  laces in the economy of nature, and likewise to FIT the two sexes of the same species to each othe
0158  sexes of the same species to each other, or to FIT the males and females to different habits of l
0183  curs, it would be easy for natural selection to FIT the animal, by some modification of its struct
0472   if some of them be abhorrent to our ideas of FITNESS. We need not marvel at the sting of the bee
0030  medary and camel, the various breeds of sheep FITTED either for cultivated land or mountain pastur
0088  the most vigorous males, those which are best FITTED for their places in nature, will leave most p
0091  ontinued preservation of the individuals best FITTED for the two sites, two varieties might slowly
0142  s and sub breeds with constitutions specially FITTED for their own districts: the result must, I t
0147  ts producing seeds which were a little better FITTED to be wafted further, might get an advangage
0147  t an advangage over those producing seed less FITTED for dispersal; and this process could not pos
0156  selection will have modified several species, FITTED to more or less widely different habits, in e
0181  le. Although no graduated links of structure, FITTED for gliding through the air, now connect the
0183  e that transitional grades between structures FITTED for very different habits of life will rarely
0199  equently, though each being assuredly is well FITTED for its place in nature, many structures now
0200   had not a flipper, but a foot with five toes FITTED for walking or grasping; and we may further v
0203  an intermediate variety will often be formed, FITTED for an intermediate zone; but from reasons as
0257  ees, plants inhabiting different stations and FITTED for extremely different climates, can often b
0321  erers may often long be preserved, from being FITTED to some peculiar line of life, or from inhabi
0365  tain, scarcely more than one would be so well FITTED to its new home, as to become naturalised. Bu
0366  ame on, and as each more southern zone became FITTED for arctic beings and ill fitted for their fo
0366   zone became fitted for arctic beings and ill FITTED for their former more temperate inhabitants,
0383  ost cases be explained by their having become FITTED, in a manner highly useful to them, for short
0388  ature, and drops them in another equally well FITTED for them. On the Inhabitants of Oceanic Islan
0393  eed it seems that islands are peculiarly well FITTED for these animals: for frogs have been introd
0401  s: a plant, for instance, would find the best FITTED ground more perfectly occupied by distinct pl
0402  art supplant it; but if both are equally well FITTED for their own places in nature, both probably
0402  eculiar to Charles Island is at least as well FITTED for its home as is the species peculiar to Ch
0403   having been subsequently modified and better FITTED to their new homes, is of the widest applicat
0406  f many low forms being very minute and better FITTED for distant transportation, probably accounts
0411  nification, if one group had been exclusively FITTED to inhabit the land, and another the water: o
0437   and pistils in any individual flower, though FITTED for such widely different purposes, be all co
0448  case with cirripedes. The adult might become FITTED for sites or habits, in which organs of locom
0457  ass of the homologous parts or organs, though FITTED in the adult members for purposes as differen
0480  isuse or by the tongue and palate having been FITTED by natural selection to browse without their
0066  royed, and could be ensured to germinate in a FITTING place. So that in all cases, the average num
0080  orne in mind how infinitely complex and close FITTING are the mutual relations of all organic bein
0083  each selected character in some peculiar and FITTING manner; he feeds a long and a short beaked p
0224  as been remarked that a skilful workman, with FITTING tools and measures, would find it very diffi
0471  d have been created with habits and structure FITTING it for the life of an auk or grebe! and so o
0001  h could possibly have any bearing on it. After FIVE years' work I allowed myself to speculate on t
0018  t some of our breeds originated there, four or FIVE thousand years ago? But Mr. Horner's researche
0064  . even slow breeding man has doubled in twenty FIVE years, and at this rate, in a few thousand yea
0071  actly the same nature had been enclosed twenty FIVE years previously and planted with Scotch fir.
0123  ation, will be widely different from the other FIVE species, and may constitute a sub genus or eve
0200  of the seal had not a flipper, but a foot with FIVE toes fitted for walking or grasping; and we ma
0221  he same road to a tall Scotch fir tree, twenty FIVE yards distant, which they ascended together pr
0233  oving outside, first to one point, and then to FIVE other points, at the proper relative distances
0241  of workmen building a house of whom many were FIVE feet four inches high, and many sixteen feet h
0241  g as those of the smaller men, and jaws nearly FIVE times as big. The jaws, moreover, of the worki
0251  e that he had then tried the experiment during FIVE years, and he continued to try it during sever
0270  uced any seed, and this one head produced only FIVE grains. Manipulation in this case could not ha
0271   are the other varieties. He experimentised on FIVE forms, which are commonly reputed to be variet
0271  offspring perfectly fertile. But one of these FIVE varieties, when used either as father or mothe
0306  d, as it would be for a naturalist to land for FIVE minutes on some one barren point in Australia,
0332  her families (namely, a14 to f14 now including FIVE genera, and o14 to m14) would yet remain disti
0376  rms of fish. Dr. Hooker informs me that twenty FIVE species of Algae are common to New Zealand and
0378  se, when arctic forms had migrated some twenty FIVE degrees of latitude from their native country
0386  tes, once flew on board the Beagle, when forty FIVE miles distant from the nearest land: how much
0398  t. there are twenty six land birds, and twenty FIVE of these are ranked by Mr. Gould as distinct s
0412  on parent at the fifth stage of descent. These FIVE genera have also much, though less, in communi
0484  imals have descended from at most only four or FIVE progenitors, and plants from an equal or lesse
0046  st polymorphic genera some of the species have FIXED and definite characters. Genera which are pol
0154  e the variability will seldom as yet have been FIXED by the continued selection of the individuals
```

Page **************************************(Key Word)***************************************

Page ***(Key Word)***

```
0155  view of species being only strongly marked and FIXED varieties, we might surely expect to find the
0169  lection may readily have succeeded in giving a FIXED character to the organ, in however extraordin
0213  as they may be called, are certainly far less FIXED or invariable than natural instincts; but the
0213  for an incomparably shorter period, under less FIXED conditions of life. How stongly these domesti
0227  cts can make, apparently by turning round on a FIXED point. We must suppose the Melipona to arrang
0283  huge fragments fall down, and these remaining FIXED, have to be worn away, atom by atom, until re
0314  ts accord well with my theory. I believe in no FIXED law of development, causing all the inhabitan
0318  before the close of the palaeozoic period. No FIXED law seems to determine the length of time dur
0441  metamorphosis. When this is completed they are FIXED for life: their legs are now converted into p
0490  is planet has gone cycling on according to the FIXED law of gravity, from so simple a beginning en
0233  is even conceivable that an insect might, by FIXING on a point at which to commence a cell, and t
0181  ervation of individuals with fuller and fuller FLANK membranes, each modification being useful, ea
0181  uding the limbs and the elongated fingers: the FLANK membrane is, also, furnished with an extensor
0180  bodies rather wide and with the skin on their FLANKS rather full, to the so called flying squirrel
0286  ge of older rocks underlies the Weald, on the FLANKS of which the overlying sedimentary deposits m
0366  ains of Scotland and Wales, with their scored FLANKS, polished surfaces, and perched boulders, of
0134  he loggerheaded duck of South America can only FLAP along the surface of the water, and has its wi
0171  the tail of a giraffe, which serves as a fly FLAPPER, and on the other hand, organs of such wonde
0195  fe looks like an artificially constructed fly FLAPPER; and it seems at first incredible that this
0196  worked in for all sorts of purposes, as a fly FLAPPER, an organ of prehension, or as an aid in tur
0182  ave existed which used their wings solely as FLAPPERS, like the logger headed duck (Micropterus of
0483  lemental atoms have been commanded suddenly to FLASH into living tissues? Do they believe that at
0226  s never permitted, the bees building perfectly FLAT walls of wax between the spheres which thus te
0226  l portion and of two, three, or more perfectly FLAT surfaces, according as the cell adjoins two, t
0226  frequently and necessarily the case, the three FLAT surfaces are united into a pyramid; and this p
0226  saves wax by this manner of building; for the FLAT walls between the adjoining cells are not doub
0226  as the outer spherical portions, and yet each FLAT portion forms a part of two cells. Reflecting
0227  unites the points of intersection by perfectly FLAT surfaces. We have further to suppose, but this
0228  bees ceased to excavate, and began to build up FLAT walls of wax on the lines of intersection betw
0229  they had been a little deepened, came to have FLAT bottoms; and these flat bottoms, formed by thi
0229  deepened, came to have flat bottoms: and these FLAT bottoms, formed by thin little plates of the v
0229  es, in order to have succeeded in thus leaving FLAT plates between the basins, by stopping work al
0229  e had been completed, and had become perfectly FLAT: it was absolutely impossible, from the extrem
0422  ly not possible to represent in a series, on a FLAT surface, the affinities which we discover in n
0230  into its proper intermediate plane, and thus FLATTEN it. From the experiment of the ridge of verm
0077  lly plumed seed of the dandelion, and in the FLATTENED and fringed legs of the water beetle, the r
0180  from animals with their tails only slightly FLATTENED, and from others, as Sir J. Richardson has
0222  all parcel of the pupae of another species, F. FLAVA, with a few of these little yellow ants still
0222  to my surprise an independent community of F. FLAVA under a stone beneath a nest of the slave mak
0222  slaves, from those of the little and furious F.FLAVA, which they rarely capture, and it was eviden
0222  e pupae, or even the earth from the nest of F. FLAVA, and quickly ran away: but in about a quarter
0239  n intermediate size scanty in numbers, Formica FLAVA has larger and smaller workers, with some of
0049  iderably in appearance: they have a different FLAVOUR and emit a different odour: they flower at s
0237  may thus gain the desired end. Thus, a well FLAVOURED vegetable is cooked, and the individual is
0090  me by fleetness; and let us suppose that the FLEETEST prey, a deer for instance, had from any chan
0040  reed, which differ only in colour, that most FLEETING of characters, have lately been exhibited as
0084  struggle for life, and so be preserved. How FLEETING are the wishes and efforts of man! how short
0035  f english racehorses have come to surpass in FLEETNESS and size the parent Arab stock, so that the
0090  some by craft, some by strength, and some by FLEETNESS; and let us suppose that the fleetest prey,
0090  to doubt this, than that man can improve the FLEETNESS of his greyhounds by careful and methodical
0085  ts the down of the fruit and the colour of the FLESH are considered by botanists as characters of
0085  eaches far more than those with other coloured FLESH. If, with all the aids of art, these slight d
0116  g adapted to digest vegetable matter alone, or FLESH alone, draws most nutriment from these substa
0195  uch as the down on fruit and the colour of the FLESH, which, from determining the attacks of insec
0238  the same variety; breeders of cattle wish the FLESH and fat to be well marbled together: the anim
0411  he land, and another the water: one to feed on FLESH, another on vegetable matter, and so on: but
0085  plums; whereas another disease attacks yellow FLESHED peaches far more than those with other colou
0085  whether a smooth or downy, a yellow or purple FLESHED fruit, should succeed. In looking at many sm
0132  he sea shore have their leaves in some degree FLESHY, though not elsewhere fleshy. Several other s
0132  s in some degree fleshy, though not elsewhere FLESHY. Several other such cases could be given. The
0133  lusively on the sea side are very apt to have FLESHY leaves. He who believes in the creation of ea
0136  ssive generations each individual beetle which FLEW least, either from its wings having been ever
0386  beetle of the same family, a Colymbetes, once FLEW on board the Beagle, when forty five miles dis
0141  a quality readily grafted on an innate wide FLEXIBILITY of constitution, which is common to most a
0141  es, but merely as examples of a very common FLEXIBILITY of constitution, brought, under peculiar c
0482  heory. A few naturalists, endowed with much FLEXIBILITY of mind, and who have already begun to dou
0229  s or planes of intersection. Considering how FLEXIBLE thin wax is, I do not see that there is any
0231  ial rim or wall of wax round a growing comb, FLEXURES may sometimes be observed, corresponding in
0023  e. but birds breeding on precipices; and good FLIERS, are unlikely to be exterminated, and the com
0072  animals when first born. The increase of these FLIES, numerous as they are, must be habitually che
0072  sts of prey) were to increase in Paraguay, the FLIES would decrease, then cattle and horses would
0195  ter, for so trifling an object as driving away FLIES; yet we should pause before being too positiv
0195  y destroyed (except in some rare cases) by the FLIES, but they are incessantly harassed and their
0198  hat in cattle susceptibility to the attacks of FLIES is correlated with colour, as is the liabilit
0439  rd, or reptile. The vermiform larvae of moths, FLIES, beetles, etc., resemble each other much more
0022  hape and size of the eggs vary. The manner of FLIGHT differs remarkably: as does in some breeds th
0131  e, combined with natural selection; organs of FLIGHT and of vision. Acclimatisation. Correlation o
0134  s the larger ground feeding birds seldom take FLIGHT except to escape danger, I believe that the n
0134  osed to danger from which it cannot escape by FLIGHT, but by kicking it can defend itself from ene
0135  ts wings less, until they became incapable of FLIGHT. Kirby has remarked (and I have observed the
0136  habits of life almost necessitating frequent FLIGHT; these several considerations have made me be
0136  and, those beetles which most readily took to FLIGHT will oftenest have been blown to sea and thus
0181  selection; and this, as far as the organs of FLIGHT are concerned, would convert it into a bat. I
0181  d for gliding through the air rather than for FLIGHT. If about a dozen genera of birds had become
0182  ch birds have acquired their perfect power of FLIGHT; but they serve, at least, to show what diver
```

Page ***(Key Word)***

0182 ocean, and had used their incipient organs of FLIGHT exclusively, as far as we know, to escape bei
0182 particular habit, as the wings of a bird for FLIGHT, we should bear in mind that animals displayi
0183 not seem probable that fishes capable of true FLIGHT would have been developed under many subordin
0183 land and in the water, until their organs of FLIGHT had come to a high stage of perfection, so as
0184 n the harsh tone of its voice, and undulatory FLIGHT, told me plainly of its close blood relations
0185 ming, and of flying when unwillingly it takes FLIGHT, would be mistaken by any one for an auk or g
0191 n have been actually converted into organs of FLIGHT. In considering transitions of organs, it is
0216 er hen has almost lost by disuse the power of FLIGHT. Hence we may conclude, that domestic instinc
0361 e that under such circumstances their rate of FLIGHT would often be 35 miles an hour; and some aut
0387 gone on daily devouring fish; they then take FLIGHT and go to other waters, or are blown across t
0388 fresh water birds, which have large powers of FLIGHT, and naturally travel from one to another and
0451 can be plainer than that wings are formed for FLIGHT, yet in how many insects do we see wings so r
0451 reduced in size as to be utterly incapable of FLIGHT, and not rarely lying under wing cases, firml
0454 slands, which have seldom been forced to take FLIGHT, and have ultimately lost the power of flying
0473 ger headed duck, which has wings incapable of FLIGHT, in nearly the same condition as in the domes
0200 the horse, in the wing of the bat, and in the FLIPPER of the seal, are of special use to these ani
0200 eve that the progenitor of the seal had not a FLIPPER, but a foot with five toes fitted for walkin
0489 singing on the bushes, with various insects FLITTING about, and with worms crawling through the d
0360 t doubt that plants exposed to the waves would FLOAT for a less time than those protected from vio
0359 f these sank in a few days, they could not be FLOATED across wide spaces of the sea, whether or no
0359 ger fruits, capsules, etc., and some of these FLOATED for a long time. It is well known what a dif
0359 ity sank quickly, but some which whilst green FLOATED for a very short time, when dried floated mu
0359 een floated for a very short time, when dried FLOATED much longer; for instance, ripe hazel nuts s
0359 l nuts sank immediately, but when dried, they FLOATED for 90 days and afterwards when planted they
0359 minated; an asparagus plant with ripe berries FLOATED for 23 days, when dried it floated for 85 da
0359 pe berries floated for 23 days, when dried it FLOATED for 85 days, and the seeds afterwards germin
0359 elosciadium sank in two days, when dried they FLOATED for above 90 days, and afterwards germinated
0359 ed. Altogether out of the 94 dried plants, 18 FLOATED for above 28 days, and some of the 18 floate
0359 floated for above 28 cays, and some of the 18 FLOATED for a very much longer period. So that as 64
0359 same species as in the foregoing experiment) FLOATED, after being dried, for above 28 days, as fa
0359 eeds of 14/100 plants of any country might be FLOATED by sea currents during 28 cays, and would re
0360 /100 plants belonging to one country might be FLOATED across 924 miles of sea to another country;
0360 seen, would have caused some of them to have FLOATED much longer. The result was that 18/98 of hi
0360 onger. The result was that 18/98 of his seeds FLOATED for 42 days, and were then capable of germin
0360 of a flora, after having been dried, could be FLOATED across a space of sea 900 miles in width, an
0361 taken out of the crop of a pigeon, which had FLOATED on artificial salt water for 30 days, to my
0397 aphragm over the mouth of the shell, might be FLOATED in chinks of drifted timber across moderatel
0185 tores are formed for walking over swamps and FLOATING plants, yet the water hen is nearly as aquat
0191 , of which we know nothing, furnished with a FLOATING apparatus or swimbladder. We can thus, as I
0360 ately wet and exposed to the air like really FLOATING plants. He tried 98 seeds, mostly different
0360 rminate. The fact of the larger fruits often FLOATING longer than the small, is interesting; as pl
0361 i can show that the carcasses of birds, when FLOATING on the sea, sometimes escape being immediate
0361 red; and seeds of many kinds in the crops of FLOATING birds long retain their vitality: peas and v
0391 gs or larvae, perhaps attached to seaweed or FLOATING timber, or to the feet of wading birds, migh
0021 habit of flying at a great height in a compact FLOCK, and tumbling in the air head over heels. The
0031 urist, not only to mocify the character of his FLOCK, but to change it altogether. It is the magic
0036 instance from the pure blood of Mr. Bakewell's FLOCK, and yet the difference between the sheep pos
0085 t: we should remember how essential it is in a FLOCK of white sheep to destroy every lamb with the
0211 of running about, to be well aware what a rich FLOCK it had discovered; it then began to play with
0213 and a tendency to run round, instead of at, a FLOCK of sheep, by shepherd dogs. I cannot see that
0036 e production of two distinct strains. The two FLOCKS of Leicester sheep kept by Mr. Buckley and Mr
0091 which more frequently attacks the shepherd's FLOCKS. Let us now take a more complex case. Certain
0253 namely Mr. Blyth and Capt. Hutton, that whole FLOCKS of these crossed geese are kept in various pa
0359 d seasoned timber; and it occurred to me that FLOODS might wash down plants or branches, and that
0384 could be given of this having occurred during FLOODS, without any change of level. We have evidenc
0054 ants inhabiting a country and described in any FLORA be divided into two equal masses, all those i
0107 termination. Hence, perhaps, it comes that the FLORA of Madeira, according to Oswald Heer, resembl
0107 to oswald Heer, resembles the extinct tertiary FLORA of Europe. All fresh water basins, taken toge
0115 e last edition of Dr. Asa Gray's Manual of the FLORA of the Northern United States, 260 naturalise
0337 ame or some other quarter, the eocene fauna or FLORA would certainly be beaten and exterminated; a
0347 these are not inhabited by a peculiar fauna or FLORA. Notwithstanding this parallelism in the cond
0360 ume that the seeds of about 10/100 plants of a FLORA, after having been dried, could be floated ac
0363 n) from the somewhat northern character of the FLORA in comparison with the latitude, I suspected
0366 mum, we should have a uniform arctic fauna and FLORA, covering the central parts of Europe, as far
0374 , forming no inconsiderable part of its scanty FLORA, are common to Europe, enormously remote as t
0375 s and some few representatives of the peculiar FLORA of the Cape of Good Hope occur. At the Cape o
0375 regions. In the admirable Introduction to the FLORA of New Zealand, by Dr. Hooker, analogous and
0381 ice, supported a highly peculiar and isolated FLORA. I suspect that before this flora was extermi
0381 and isolated flora. I suspect that before this FLORA was exterminated by the Glacial epoch, a few
0398 n by Dr. Hooker in his admirable memoir on the FLORA of this archipelago. The naturalist, looking
0399 am assured by Dr. Hooker is real, between the FLORA of the south western corner of Australia and
0403 ermines the general character of the fauna and FLORA of oceanic islands, namely, that the inhabita
0048 ommon cannot be disputed. Compare the several FLORAS of Great Britain, of France or of the United
0053 ting all the varieties in several well worked FLORAS. At first this seemed a simple task; but Mr.
0115 emarked in his great and admirable work, that FLORAS gain by naturalisation, proportionally with t
0347 not be possible to point out three faunas and FLORAS more utterly dissimilar. Or again we may comp
0364 ot from one distant continent to another. The FLORAS of distant continents would not by such means
0376 equatorial latitudes, the Alpine or mountain FLORAS really become less and less arctic. Many of t
0145 e knows the difference in the ray and central FLORETS of, for instance, the daisy, and this differ
0145 essure, and the shape of the seeds in the ray FLORETS in some Compositae countenances this idea; b
0145 ifference in the seeds of the outer and inner FLORETS without any difference in the corolla. Possi
0145 sure that C. C. Sprengel's idea that the ray FLORETS serve to attract insects, whose agency is hi
0416 gly insisted on the fact that the rudimentary FLORETS are of the highest importance in the classif
0452 d on the style; but some Compositae, the male FLORETS, which of course cannot be fecundated, have
0032 d. we see an astonishing improvement in many FLORISTS' flowers, when the flowers of the present da
0190 iginally constructed for one purpose, namely FLOTATION, may be converted into one for a wholly dif

Page ***(Key Word)***

```
0360  ld have favoured the average length of their FLOTATION and of their resistance to the injurious ac
0052  land shells in Madeira. If a variety were to FLOURISH so as to exceed in numbers the parent specie
02º4  t limit of depth at which marine animals can FLOURISH: for we know what vast geographical changes
0470  genus have been produced, and where they now FLOURISH, these same species should present many vari
0071  f plants (not counting grasses and carices) FLOURISHED in the plantations, which could not be foun
0129  ordinate to groups. Of the many twigs which FLOURISHED when the tree was a mere bush, only two or
0238  ile males and females of the same community FLOURISHED, and transmitted to their fertile offspring
0053  ed in botanical works. Hence it is the most FLOURISHING, or, as they may be called, the dominant s
0059  ed. We have, also, seen that it is the most FLOURISHING and dominant species of the larger genera
0092  consequently would have the best chance of FLOURISHING and surviving. Some of these seedlings wou
0320  artificial, are bound together. In certain FLOURISHING groups, the number of new specific forms w
0145  s may be connected with some difference in the FLOW of nutriment towards the central and external
0261  e hardness of their wood, in the period of the FLOW or nature of their sap, etc., but in a multitu
0384  the level of the land, having caused rivers to FLOW into each other. Instances, also, could be giv
0382  lained. The living waters may be said to have FLOWED during one short period from the north and fr
0382  d to have crossed at the equator; but to have FLOWED with greater force from the north so as to ha
0003  y of certain insects to bring pollen from one FLOWER to the other, it is equally preposterous to a
0030  host of agricultural, culinary, orchard, and FLOWER garden races of plants most useful to man at
0033  ifferent varieties of the same species in the FLOWER garden; the diversity of leaves, pods, or tub
0037  nts which have been longest cultivated in our FLOWER and kitchen gardens. If it has taken centurie
0049  rent flavour and emit a different odour; they FLOWER at slightly different periods; they grow in s
0073  a tricolor), for other bees do not visit this FLOWER. From experiments which I have tried, I have
0092  xcreted by the inner bases of the petals of a FLOWER. In this case insects in seeking the nectar w
0092  certainly often transport the pollen from one FLOWER to the stigma of another flower. The flowers
0092  llen from one flower to the stigma of another FLOWER. The flowers of two distinct individuals of t
0092  y degree the transportal of their pollen from FLOWER to flower, would likewise be favoured or sele
0092  he transportal of their pollen from flower to FLOWER, would likewise be favoured or selected. We m
0092  itually, by the pollen devouring insects from FLOWER to flower, and a cross thus effected, althoug
0092  y the pollen devouring insects from flower to FLOWER, and a cross thus effected, although nine ten
0093  ly on their part, regularly carry pollen from FLOWER to flower; and that they can most effectually
0093  r part, regularly carry pollen from flower to FLOWER; and that they can most effectually do this,
0093  favourable to bees, nevertheless every female FLOWER which I examined had been effectually fertili
0093  nsects that pollen was regularly carried from FLOWER to flower, another process might commence. No
0093  t pollen was regularly carried from flower to FLOWER, another process might commence. No naturalis
0093  us to a plant to produce stamens alone in one FLOWER or on one whole plant, and pistils alone in a
0094  one whole plant, and pistils alone in another FLOWER or on another plant. In plants under culture
0094  n as pollen is already carried regularly from FLOWER to flower, and as a more complete separation
0094  n is already carried regularly from flower to FLOWER, and as a more complete separation of the sex
0095  isit its flowers. Thus I can understand how a FLOWER and a bee might slowly become, either simulta
0097  exposure to wet is to the fertilisation of a FLOWER, yet what a multitude of flowers have their a
0097  rious adaptation between the structure of the FLOWER and the manner in which bees suck the nectar;
0097  n on the stigma, or bring pollen from another FLOWER. So necessary are the visits of bees to papil
0097  s scarcely possible that bees should fly from FLOWER to flower, and not carry pollen from one to t
0097  possible that bees should fly from flower to FLOWER, and not carry pollen from one to the other,
0097  e sufficient just to touch the anthers of one FLOWER and then the stigma of another. With the same
0098  rom the foreign pollen. When the stamens of a FLOWER suddenly spring towards the pistil, or slowly
0098  vent the stigma receiving pollen from its own FLOWER: for instance, in Lobelia fulgens, there is a
0098  re swept out of the conjoined anthers of each FLOWER, before the stigma of that individual flower
0098  flower, before the stigma of that individual FLOWER is ready to receive them: and as this flower
0098  flower is ready to receive them: and as this FLOWER is never visited, at least in my garden, by i
0098  ets a seed, though by placing pollen from one FLOWER on the stigma of another, I raised plenty of whi
0098  anical contrivance to prevent the stigma of a FLOWER receiving its own pollen, yet, as C. C. Spren
0099  the stigma is ready before the pollen of that FLOWER is ready, so that these plants have in fact s
0099  the pollen and stigmatic surface of the same FLOWER, though placed so close together, as if for t
0099  erfectly true. Yet the pistil of each cabbage FLOWER is surrounded not only by its own six stamens
0099  inct variety having a prepotent effect over a FLOWER's own pollen; and that this is part of the ge
0099  ried from tree to tree, and at most only from FLOWER to flower on the same tree, and that flowers
0100  tree to tree, and at most only from flower to FLOWER on the same tree, and that flowers on the sam
0100  ee that pollen must be regularly carried from FLOWER to flower; and this will give a better chance
0100  llen must be regularly carried from flower to FLOWER; and this will give a better chance of pollen
0136  ich are not ground feeders, and which, as the FLOWER feeding coleoptera and lepidoptera, must habi
0145  accompanied with the abortion of parts of the FLOWER. But, in some Compositous plants, the seeds a
0145  g nourishment from certain other parts of the FLOWER had caused their abortion; but in some Compos
0145  in some garden pelargoniums, that the central FLOWER of the truss often loses the patches of darke
0249  y own experience) from the anthers of another FLOWER, as from the anthers of the flower itself whi
0249  of another flower, as from the anthers of the FLOWER itself which is to be fertilised: so that a c
0249  ation a cross with the pollen from a distinct FLOWER, either from the same plant or from another p
0251  though both the ovules and pollen of the same FLOWER were perfectly good with respect to other spe
0255  of one of the pure parent species causing the FLOWER of the hybrid to wither earlier than it other
0255  uld have done; and the early withering of the FLOWER is well known to be a sign of incipient ferti
0257  ongly marked differences in every part of the FLOWER, even in the pollen, in the fruit, and in the
0271  her difference besides the mere colour of the FLOWER: and one variety can sometimes be raised from
0416  ertain Algae, mere pubescence on parts of the FLOWER in grasses, the nature of the dermal covering
0436  it is familiar to almost every one, that in a FLOWER the relative position of the sepals, petals,
0437  etals, stamens, and pistils in any individual FLOWER, though fitted for such widely different purp
0454  in young animals, and the state of the whole FLOWER in the cauliflower. We often see rudiments of
0473  ee? Why, for instance, should the colour of a FLOWER be more likely to vary in any one species of
0479  rab, in the petals, stamens, and pistils of a FLOWER, is likewise intelligible on the view of the
0011  has a decided influence, as in the period of FLOWERING with plants when transported from one clima
0139  hereditary with plants, as in the period of FLOWERING, in the amount of rain requisite for seeds
0193  genera almost as remote as possible amongst FLOWERING plants. In all these cases of two very dist
0249  sects must be carefully prevented during the FLOWERING season: hence hybrids will generally be fer
0374  as shown that between forty and fifty of the FLOWERING plants of Tierra del Fuego, forming no inco
0378  mmunicated to me by Dr. Hooker, that all the FLOWERING plants, about forty six in number, common t
0389  over 780 miles of latitude, and compare its FLOWERING plants, only 750 in number, with those on a
0389  on aboriginally possessed under half a dozen FLOWERING plants; yet many have become naturalised on
```

Page ***(Key Word)***

Page **(Key Word)**

```
0418  itted as true. The same fact holds good with  FLOWERING plants, of which the two main divisions hav
0437  egments, bearing external appendages; and in  FLOWERING plants, we see a series of successive spira
0437  many segments; and the unknown progenitor of  FLOWERING plants, many spiral whorls of leaves. We ha
0003  e transported by certain birds, and which has  FLOWERS with separate sexes absolutely requiring the
0032  an astonishing improvement in many florists'  FLOWERS, when the flowers of the present day are com
0032  provement in many florists' flowers, when the  FLOWERS of the present day are compared with drawing
0033  lection namely, by comparing the diversity of  FLOWERS in the different varieties of the same speci
0033  in the kitchen garden, in comparison with the  FLOWERS of the same varieties; and the diversity of
0033  he orchard, in comparison with the leaves and  FLOWERS of the same set of varieties. See how differ
0033  the cabbage are, and how extremely alike the   FLOWERS, how unlike the flowers of the heartsease ar
0033  w extremely alike the flowers, how unlike the  FLOWERS of the heartsease are, and how alike the lea
0033  ze, colour, shape, and hairiness, and yet the  FLOWERS present very slight differences. It is not t
0033  slight variations, either in the leaves, the   FLOWERS, or the fruit, will produce races differing
0074  ce and then of bees, the frequency of certain  FLOWERS in that district! In the case of many specie
0092  e flower to the stigma of another flower. The  FLOWERS of two distinct individuals of the same spec
0092  the nectar excreting power. Those individual   FLOWERS which had the largest glands or nectaries, a
0092  the long run would gain the upper hand. Those  FLOWERS, also, which had their stamens and pistils p
0092  might have taken the case of insects visiting  FLOWERS for the sake of collecting pollen instead of
0093  natural selection of more and more attractive  FLOWERS, had been rendered highly attractive to inse
0093  e alluded to. Some holly trees bear only male  FLOWERS, which have four stamens producing rather a
0093  ry pistil; other holly trees bear only female  FLOWERS; these have a full sized pistil, and four st
0093  from a male tree, I put the stigmas of twenty  FLOWERS, taken from different branches, under the mi
0094  nd sucking the nectar at the bases of certain  FLOWERS, which they can, with a very little more tro
0095  corolla, so that the hive bee could visit its  FLOWERS. Thus I can understand how a flower and a be
0097  lisation of a flower, yet what a multitude of  FLOWERS have their anthers and stigmas fully exposed
0097  f fertilisation seems almost inevitable. Many  FLOWERS, on the other hand, have their organs of fru
0097  family; but in several, perhaps in all, such   FLOWERS, there is a very curious adaptation between
0097  tar: for, in doing this, they either push the  FLOWERS own pollen on the stigma, or bring pollen fr
0097  sary are the visits of bees to papilionaceous  FLOWERS, that I have found, by experiments published
0099  n six stamens; but by those of the many other  FLOWERS on the same plant. How, then, comes it that
0099  e of a gigantic tree covered with innumerable  FLOWERS, it may be objected that pollen could seldom
0100  m flower to flower on the same tree, and that  FLOWERS on the same tree can be considered as distin
0100  by giving to trees a strong tendency to bear  FLOWERS with separated sexes. When the sexes are sep
0100  s are separated, although the male and female  FLOWERS may be produced on the same tree, we can see
0100  r an occasional cross. And, as in the case of  FLOWERS, I have as yet failed, after consultation wi
0111  e, sheep, and other animals, and varieties of  FLOWERS, take the place of older and inferior kinds.
0144  of the difference between the outer and inner  FLOWERS in some Compositous and Umbelliferous plants
0145  th the densest heads that the inner and outer  FLOWERS most frequently differ. It might have been t
0145  of nutriment towards the central and external  FLOWERS: we know, at least, that in irregular flower
0145  flowers: we know, at least, that in irregular  FLOWERS, those nearest to the axis are oftenest subj
0145  ce in the corolla of the central and exterior  FLOWERS of a head or umbel, I do not feel at all sur
0146  always correlated with any differences in the  FLOWERS, it seems impossible that they can be in any
0146  ding to Tausch, orthospermous in the exterior  FLOWERS and coelospermous in the central flowers, th
0146  rior flowers and coelospermous in the central  FLOWERS, that the elder De Candolle founded his main
0149  rae in snakes; and the stamens in polyandrous  FLOWERS) the number is variable; whereas the number
0154  e species in a large genus of plants had blue  FLOWERS and some had red, the colour would by only a
0154  r conversely; but if all the species had blue  FLOWERS, the colour would become a generic character
0203  several ingenious contrivances, by which the   FLOWERS of the orchis and of many other plants are f
0249  to be fertilised: so that a cross between two  FLOWERS, though probably on the same plant, would be
0250  , a bulb of Hippeastrum aulicum produced four  FLOWERS; three were fertilised by Herbert with their
0250  esult was that the ovaries of the three first  FLOWERS soon ceased to grow; and after a few days we
0252  efficiency of insect agency by examining the   FLOWERS of the more sterile kinds of hybrid rhododen
0252  r stigmas plenty of pollen brought from other  FLOWERS. In regard to animals, much fewer experiment
0270  aturally crossed. He then fertilised thirteen  FLOWERS of the one with the pollen of the other; but
0270  ertilised with pollen from their own coloured  FLOWERS. Moreover, he asserts that when yellow and w
0271  uced by the crosses between the same coloured  FLOWERS, than between those which are differently co
0417  the Malpighiaceae, bear perfect and degraded  FLOWERS; in the latter, as A. de Jussieu has remarke
0417  n france, during several years, only degraded  FLOWERS, departing so wonderfully in a number of the
0435  s and limbs of crustaceans. So it is with the  FLOWERS of plants. Nothing can be more hopeless than
0436  crustaceans and in many other animals, and in  FLOWERS, that organs, which when mature become extre
0438  etamorphosed legs; the stamens and pistils of  FLOWERS as metamorphosed leaves; but it would in the
0451  ate. In plants with separated sexes, the male  FLOWERS often have a rudiment of a pistil; and Kolre
0453  la, which often represents the pistil in male  FLOWERS, and which is formed merely of cellular tiss
0473  ated independently, have differently coloured  FLOWERS, than if all the species of the genus have t
0473  e species of the genus have the same coloured  FLOWERS? If species are only well marked varieties,
0093  bees, accidentally dusted with pollen, having  FLOWN from tree to tree in search of nectar. But to
0386  e nearest land: how much farther it might have  FLOWN with a favouring gale no one can tell. With r
0147  with our domestic productions: if nourishment  FLOWS to one part or organ in excess, it rarely flo
0147  lows to one part or organ in excess, it rarely  FLOWS, at least in excess, to another part; thus it
0052  y, which is given to. less distinct and more   FLUCTUATING forms. The term variety, again, in compari
0066  to those species, which depend on a rapidly   FLUCTUATING amount of food, for it allows them rapidly
0081  d by natural selection, and would be left a   FLUCTUATING element, as perhaps we see in the species
0109  epresented by few individuals will, during   FLUCTUATIONS in the seasons or in the number of its ind
0175  t exists in lessened numbers, will, during   FLUCTUATIONS in the number of its enemies or of its pre
0152  ay, and thus the organisation is left in a   FLUCTUATUING condition. But what here more especially c
0233  pound of wax: so that a prodigious quantity of  FLUID nectar must be collected and consumed by the
0386  requent the muddy edges of ponds, if suddenly  FLUSHED, would be the most likely to have muddy feet
0182  htly rising and turning by the aid of their   FLUTTERING fins, might have been modified into perfect
0066  to be the most numerous bird in the world. One  FLY deposits hundreds of eggs, and another, like t
0072  by the greater number in Paraguay of a certain  FLY, which lays its eggs in the navels of these an
0097  . now, it is scarcely possible that bees should  FLY from flower to flower, and not carry pollen fr
0134  eater anomaly in nature than a bird that cannot  FLY: yet there are several in this state. The logg
0135  are so far deficient in wings that they cannot  FLY; and that of the twenty nine endemic genera, n
0171  uch as the tail of a giraffe, which serves as a  FLY flapper, and on the other hand, organs of such
0195  giraffe looks like an artificially constructed  FLY flapper; and it seems at first incredible that
0196  to be worked in for all sorts of purposes, as a  FLY flapper, an organ of prehension, or as an aid
```

Page **(Key Word)**

Page **(Key Word)**

```
0214  umblers, as I hear from Mr. Brent, which cannot FLY eighteen inches high without going head over h
0216  as we see in wild ground birds, their mother to FLY away. But this instinct retained by our chicke
0303  new and peculiar line of life, for instance to FLY through the air; but that when this had been e
0385  nd in this length of time a duck or heron might FLY at least six or seven hundred miles, and would
0386  r feet; when making land, they would be sure to FLY to their natural fresh water haunts. I do not
0394  ported across a wide space of sea, but bats can FLY across. Bats have been seen wandering by day f
0484  nimals; or that the poison secreted by the gall FLY produces monstrous growths on the wild rose or
0183  uld be given: I have often watched a tyrant FLYCATCHER (Saurophagus sulphuratus) in South America,
0011  may be safely attributed to the domestic duck FLYING much less, and walking more, than its wild pa
0021  the singular and strictly inherited habit of FLYING at a great height in a compact flock, and tum
0136  by giving up the attempt and rarely or never FLYING. As with mariners shipwrecked near a coast, i
0180  ped could possibly have been converted into a FLYING bat, the question would have been far more di
0190  on their flanks rather full, to the so called FLYING squirrels; and flying squirrels have their li
0180  full, to the so called flying squirrels; and FLYING squirrels have their limbs and even the base
0181  ess of natural selection, a perfect so called FLYING squirrel was produced. Now look at the Galeop
0181  as produced. Now look at the Galeopithecus or FLYING Lemur, which formerly was falsely ranked amon
0182  to live on the land, and seeing that we have FLYING birds and mammals, flying insects of the most
0182  seeing that we have flying birds and mammals, FLYING insects of the most diversified types, and fo
0182  the most diversified types, and formerly had FLYING reptiles, it is conceivable that flying fish,
0182  y had flying reptiles, it is conceivable that FLYING fish, which now glide far through the air, sl
0183  o return to our imaginary illustration of the FLYING fish, it does not seem probable that fishes c
0185  wer of diving, its manner of swimming, and of FLYING when unwillingly it takes flight, would be mi
0362  nforms me that a friend of his had to give up FLYING carrier pigeons from France to England, as th
0387  ct, yet analogy makes me believe that a heron FLYING to another pond and getting a hearty meal of
0402  even of the birds, though so well adapted for FLYING from island to island, are distinct on each;
0454  flight, and have ultimately lost the power of FLYING. Again, an organ useful under certain conditi
0163  s been asserted that these are plainest in the FOAL, and from inquiries which I have made, I belie
0164  times striped. The stripes are plainest in the FOAL; and sometimes quite disappear in old horses.
0164  orse the spinal stripe is much commoner in the FOAL than in the full grown animal. Without here en
0164  ay and bay Kattywar horses striped when first FOALED. I have, also, reason to suspect, from inform
0163  i have been informed by Colonel Poole that the FOALS of this species are generally striped on the
0445  nal difference. So, again, I was told that the FOALS of cart and race horses differed as much as t
0186  its inimitable contrivances for adjusting the FOCUS to different distances, for admitting differe
0450  rious; for instance, the presence of teeth in FOETAL whales, which when grown up have not a tooth
0192  the well enclosed shell; but they have large FOLDED branchiae. Now I think no one will dispute th
0416  the manner in which the wings of insects are FOLDED, mere colour in certain Algae, mere pubescenc
0426  jaw, the manner in which an insect's wing is FOLDED, whether the skin be covered by hair or feath
0434  piral proboscis of a sphinx moth; the curious FOLDED one of a bee or bug, and the great jaws of a
0191  tance. Pedunculated cirripedes have two minute FOLDS of skin, called by me the ovigerous frena, wh
0192  ch other. Therefore I do not doubt that little FOLDS of skin, which originally served as ovigerous
0147  cabbage do not yield abundant and nutritious FOLIAGE and a copious supply of oil bearing seeds. W
0031  merino sheep is so fully recognised, that men FOLLOW it as a trade: the sheep are placed on a tabl
0047  t and wide experience seems the only guide to FOLLOW. We must, however, in many cases, decide by a
0055  ecies of the small genera. Both these results FOLLOW when another division is made, and when all t
0061  we shall more fully see in the next chapter, FOLLOW inevitably from the struggle for life. Owing
0102  mprovement and modification surely but slowly FOLLOW from this unconscious process of selection; n
0138  e the transition from light to darkness. Next FOLLOW those that are constructed for twilight; and,
0180  danger from occasional falls. But it does not FOLLOW from this fact that the structure of each squ
0217  e, that the young thus reared would be apt to FOLLOW by inheritance the occasional and aberrant ha
0223  e ants thus unintentionally reared would then FOLLOW their proper instincts, and do what work they
0224  this beautiful work can be shown, I think, to FOLLOW from a few very simple instincts. I was led t
0263  ve systems. These differences, in both cases, FOLLOW to a certain extent, as might have been expec
0276  ns. The degree of sterility does not strictly FOLLOW systematic affinity, but is governed by sever
0283  seldom they are rolled about! Moreover, if we FOLLOW for a few miles any line of rocky cliff, whic
0289  ations tabulated in written works, or when we FOLLOW them in nature, it is difficult to avoid beli
0312  once lost do not reappear. Groups of species FOLLOW the same general rules in their appearance an
0316  ups of species, that is, genera and families, FOLLOW the same general rules in their appearance an
0338  yological development of recent forms. I must FOLLOW Pictet and Huxley in thinking that the truth
0343  g facts in palaeontology seem to me simply to FOLLOW on the theory of descent with modification th
0366  es, is substantially as follows. But we shall FOLLOW the changes more readily, by supposing a new
0383  stream; and liability to wide dispersal would FOLLOW from this capacity as an almost necessary con
0395  of mr. Wallace. I have not as yet had time to FOLLOW up this subject in all other quarters of the
0433  nderstand the rules which we are compelled to FOLLOW in our classification. We can understand why
0448  , on the other hand, it profited the young to FOLLOW habits of life in any degree different from t
0453  in elliptic courses round the sun, satellites FOLLOW the same course round the planets, for the sa
0456  affinity; and other such rules; all naturally FOLLOW on the view of the common parentage of those
0475  ably different conditions of life, yet should FOLLOW nearly the same instincts; why the thrush of
0475  t once see why their crossed offspring should FOLLOW the same complex laws in their degrees and ki
0007  and Origin. Principle of Selection anciently FOLLOWED, its Effects. Methodical and Unconscious Sel
0032  lful pigeon fancier. The same principles are FOLLOWED by horticulturists; but the variations are h
0033  als this kind of selection is, in fact, also FOLLOWED; for hardly any one is so careless as to all
0036  on, which may be considered as unconsciously FOLLOWED, in so far that the breeders could never hav
0037  r as the final result is concerned, has been FOLLOWED almost unconsciously. It has consisted in al
0076  cannot be kept together. The same result has FOLLOWED from keeping together different varieties of
0290  t invariably intermittent; that is, have not FOLLOWED each other in close sequence. Scarcely any f
0301  ree, they would, according to the principles FOLLOWED by many palaeontologists, be ranked as new a
0367  rctic forms would retreat northward, closely FOLLOWED up in their retreat by the productions of th
0377  treated from both sides towards the equator, FOLLOWED in the rear by the temperate productions, an
0389  ntinental extensions, which, if legitimately FOLLOWED out, would lead to the belief that within th
0414  ssifications. Let us now consider the rules FOLLOWED in classification, and the difficulties whic
0419  in certain groups of birds; and it has been FOLLOWED by several entomologists and botanists. Fina
0423  s of modification. Nearly the same rules are FOLLOWED in classifying varieties, as with species. A
0425  the several rules and guides which have been FOLLOWED by our best systematists. We have no written
0427  on to this distinction, and he has been ably FOLLOWED by Macleay and others. The resemblance, in t
0456  affinities into one grand system; the rules FOLLOWED and the difficulties encountered by naturali
0358  advanced by Forbes and admitted by his many FOLLOWERS. The nature and relative proportions of the
```

Page **(Key Word)**

Page **********************************(Key Word)**********************************

```
0025 if we deny this, we must make one of the two FOLLOWING highly improbable suppositions. Either, fir
0053 and examined the tables, he thinks that the FOLLOWING statements are fairly well established. The
0064 been favourable to them during two or three FOLLOWING seasons. Still more striking is the evidenc
0097 several large classes of facts, such as the FOLLOWING, which on any other view are inexplicable.
0127 tenour and balance of evidence given in the FOLLOWING chapters. But we already see how it entails
0171 ties and objections may be classed under the FOLLOWING heads: Firstly, why if species have descend
0184 as we sometimes see individuals of a species FOLLOWING habits widely different from those both of
0210 on to ants: that they do so voluntarily, the FOLLOWING facts show. I removed all the ants from a g
0225 to the presence of adjoining cells; and the FOLLOWING view may, perhaps, be considered only as a
0226 , and this geometer has kindly read over the FOLLOWING statement, drawn up from his information, a
0228 but this theory can be tested by experiment. FOLLOWING the example of Mr. Tegetmeier, I separated
0255 nd blending together in utter confusion. The FOLLOWING rules and conclusions are chiefly drawn up
0258 ed more than two hundred times, during eight FOLLOWING years, to fertilise reciprocally M. longifl
0269 of a certain amount of sterility in the few FOLLOWING cases, which I will briefly abstract. The e
0270 y the test of infertility, as varieties. The FOLLOWING case is far more remarkable, and seems at f
0293 end to assign due proportional weight to the FOLLOWING considerations. Although each formation may
0298 o this subject I shall have to return in the FOLLOWING chapter. One other consideration is worth n
0308 er receive some explanation, I will give the FOLLOWING hypothesis. From the nature of the organic
0310 tedly at once reject my theory. For my part, FOLLOWING out Lyell's metaphor, I look at the natural
0354 and antarctic regions; and secondly (in the FOLLOWING chapter), the wide distribution of fresh wa
0361 of them, which I tried, germinated. But the FOLLOWING fact is more important: the crops of birds
0389 cts in regard to insular productions. In the FOLLOWING remarks I shall not confine myself to the m
0448 can see that this would result from the two FOLLOWING contingencies: firstly, from the young, dur
0448 tage of development, and secondly from their FOLLOWING exactly the same habits of life with their
0459 t, cannot be considered real if we admit the FOLLOWING propositions, namely, that gradations in th
0004 beings throughout the world, which inevitably FOLLOWS from their high geometrical powers of increa
0005 equently recurring struggle for existence, it FOLLOWS that any being, if it vary however slightly
0063 xistence. A struggle for existence inevitably FOLLOWS from the high rate at which all organic bein
0109 is already fully stocked with inhabitants, it FOLLOWS that as each selected and favoured form incr
0110 several considerations I think it inevitably FOLLOWS, that as new species in the course of time a
0114 elonging to the most different orders: nature FOLLOWS what may be called a simultaneous rotation.
0208 well known song, so in instincts, one action FOLLOWS another by a sort of rhythm; if a person be
0274 hybrids produced from nearly related species, FOLLOWS according to Gartner the same laws. When two
0320 tion of less favoured forms almost inevitably FOLLOWS. It is the same with our domestic production
0366 earness by Edward Forbes, is substantially as FOLLOWS. But we shall follow the changes more readil
0390 s of modified descendants. But it by no means FOLLOWS, that, because in an island nearly all the s
0443 eve that all these facts can be explained, as FOLLOWS, on the view of descent with modification. I
0467 ection. The struggle for existence inevitably FOLLOWS, from the high geometrical ratio of increase
0475 story of the organic world, almost inevitably FOLLOWS on the principle of natural selection; for o
0476 er into the same or into intermediate groups, FOLLOWS from the living and the extinct being the of
0488 ovement or the extermination of of others; it FOLLOWS, that the amount of organic change in the fo
0490 he production of the higher animals, directly FOLLOWS. There is grandeur in this view of life, wit
0003 refer to external conditions, such as climate, FOOD, etc., as the only possible cause of variation
0007 ability may be partly connected with excess of FOOD. It seems pretty clear that organic beings mus
0010 to the direct action of heat, moisture, light, FOOD, etc., is most difficult: my impression is, th
0011 , in some cases, increased size from amount of FOOD, colour from particular kinds of food and from
0011 mount of food, colour from particular kinds of FOOD and from light, and perhaps the thickness of f
0038 y almost always have to struggle for their own FOOD, at least during certain seasons. And in two c
0042 freely rejected, as when killed they serve for FOOD. On the other hand, cats, from their nocturnal
0042 se, from being valuable only for two purposes, FOOD and feathers, and more especially from no plea
0062 with gladness, we often see superabundance of FOOD; we do not see, or we forget, that the birds w
0062 ey; we do not always bear in mind, that though FOOD may be now superabundant, it is not so at all
0062 id to struggle with each other which shall get FOOD and live. But a plant on the edge of a desert
0063 is case there can be no artificial increase of FOOD, and no prudential restraint from marriage. Al
0065 et that thousands are annually slaughtered for FOOD, and that in a state of nature an equal number
0066 hich depend on a rapidly fluctuating amount of FOOD, for it allows them rapidly to increase in num
0068 being allowed to grow up freely. The amount of FOOD for each species of course gives the extreme l
0068 e; but very frequently it is not the obtaining FOOD, but the serving as prey to other animals, whi
0068 in so far as climate chiefly acts in reducing FOOD, it brings on the most severe struggle between
0068 nct species, which subsist on the same kind of FOOD. Even when climate, for instance extreme cold,
0069 least vigorous, or those which have got least FOOD through the advancing winter, which will suffe
0069 ies or from competitors for the same place and FOOD; and if these enemies or competitors be in the
0070 the birds, though having a super abundance of FOOD at this one season, increase in number proport
0072 ave so closely and effectually searched it for FOOD. Here we see that cattle absolutely determine
0075 frequent the same districts, require the same FOOD, and are exposed to the same dangers. In the c
0077 ings, with which it comes into competition for FOOD or residence, or from which it has to escape,
0083 s a long and a short beaked pigeon on the same FOOD; he does not exercise a long backed or long le
0085 unimportant, we must not forget that climate, FOOD, etc., probably produce some slight and direct
0090 the year when the wolf is hardest pressed for FOOD. I can under such circumstances see no reason
0094 nsects depended in main part on its nectar for FOOD. I could give many facts, showing how anxious
0094 l so characterised would be able to obtain its FOOD more quickly, and so have a better chance of
0114 h water. Farmers find that they can raise most FOOD by a rotation of plants belonging to the most
0132 how much direct effect difference of climate, FOOD, etc., produces on any being is extremely doub
0132 little influence may be attributed to climate, FOOD, etc.: thus, E. Forbes speaks confidently that
0167 he external conditions of life, as climate and FOOD, etc., seem to have induced some slight modifi
0180 escape birds or beasts of prey, or to collect FOOD more quickly, or as there is reason to believe
0191 rstand the strange fact that every particle of FOOD and drink which we swallow has to pass over th
0195 well enabled in a coming dearth to search for FOOD, or to escape from beasts of prey. Organs now
0196 al selection. We should remember that climate, FOOD, etc., probably have some little direct influe
0218 which would be necessary if they had to store FOOD for their own young. Some species, likewise, o
0219 ecies, and if the insect whose nest and stored FOOD are thus feloniously appropriated, be not thus
0219 f them without a slave, but with plenty of the FOOD which they like best, and with their larvae an
0220 y seen bringing in materials for the nest, and FOOD of all kinds. During the present year, however
0221 the same spot, and evidently not in search of FOOD; they approached and were vigorously repulsed
0221 ll opponents, and carried their dead bodies as FOOD to their nest, twenty nine yards distant, but
0223 determine its own migrations, does not collect FOOD for itself or its young, and cannot even feed
```

Page **********************************(Key Word)**********************************

Page **(Key Word)**

```
0223  e called, their aphides: and thus both collect FOOD for the community. In England the masters alon
0223  ave the nest to collect building materials and FOOD for themselves, their slaves and larvae. So th
0223  it is possible that pupae originally stored as FOOD might become developed: and the ants thus unin
0224  , the habit of collecting pupae originally for FOOD might by natural selection be strengthened and
0269  e supplies his several varieties with the same FOOD: treats them in nearly the same manner, and do
0362  bird has found and devoured a large supply of FOOD, it is positively asserted that all the grains
0414  ch are most remotely related to the habits and FOOD of an animal, I have always regarded as afford
0444  fy, for instance, to a bird which obtained its FOOD best by having a long beak, whether or not it
0137  kentucky, are blind. In some of the crabs the FOOT stalk for the eye remains, though the eye is g
0193  rivance of a mass of pollen grains, borne on a FOOT stalk with a sticky gland at the end, is the s
0200  rogenitor of the seal had not a flipper, but a FOOT with five toes fitted for walking or grasping:
0362  two grains of dry argillaceous earth from one FOOT of a partridge, and in this earth there was a
0378  eir native country and covered the land at the FOOT of the Pyrenees. At this period of extreme col
0435  membrane, so as to serve as a fin: or a webbed FOOT might have all its bones, or certain bones, le
0162  not know that the rock pigeon was not feather FOOTED or turn crowned, we could not have told, whet
0132  e attributed to climate, food, etc.: thus, F. FORBES speaks confidently that shells at their south
0175  disappears. The same fact has been noticed by FORBES in sounding the depths of the sea with the dr
0287  at admirable palaeontologist, the late Edward FORBES, should not be forgotten, namely, that number
0291  which case, judging from the researches of E. FORBES, we may conclude that the bottom will be inha
0292  principles inculcated by Sir C. Lyell; and E. FORBES independently arrived at a similar conclusion
0307  ly competent judges, as Lyell and the late E. FORBES, dispute this conclusion. We should not forge
0310  cuvier, Owen, Agassiz, Barrande, Falconer, E. FORBES, etc., and all our greatest geologists, as ly
0316  eptions are surprisingly few, so few, that E. FORBES, Pictet, and Woodward (though all strongly op
0357  thin the period of existing organisms. Edward FORBES insisted that all the islancs in the Atlantic
0357  ome mainland. If indeed the arguments used by FORBES are to be trusted, it must be admitted that s
0358  , as are necessitated on the view advanced by FORBES and admitted by his many followers. The natur
0366  explained with remarkable clearness by Edward FORBES, is substantially as follows. But we shall fo
0372  ject, the Glacial period. I am convinced that FORBES's view may be largely extended. In Europe we
0389  e already stated that I cannot honestly admit FORBES's view on continental extensions, which, if l
0409  ut representative species. As the late Edward FORBES often insisted, there is a striking parelleli
0063  the doctrine of Malthus applied with manifold FORCE to the whole animal and vegetable kingdoms: f
0067  ge being struck, and then another with greater FORCE. What checks the natural tendency of each spe
0210  ura non facit saltum applies with almost equal FORCE to instincts as to bodily organs. Changes of
0306  d from one progenitor, apply with nearly equal FORCE to the earliest known species. For instance,
0309  nce, during many oscillations of level, of the FORCE of elevation: but may not the areas of prepon
0382  t the equator; but to have flowed with greater FORCE from the north so as to have freely inundated
0394  hy, it may be asked, has the supposed creative FORCE produced bats and no other mammals on remote
0409  nly slightly modified, some developed in great FORCE, some existing in scanty numbers, in the diff
0459  i have endeavoured to give to them their full FORCE. Nothing at first can appear more difficult t
0483  onsider, by so much the arguments fall away in FORCE. But some arguments of the greatest weight ex
0091  frequenting the lowlands, would naturally be FORCED to hunt different prey: and from the continue
0147  it, in order to spend on one side, nature is FORCED to economise on the other side. I think this
0187  ut this is scarcely ever possible, and we are FORCED in each case to look to species of the same g
0199  emales, can be called useful only in rather a FORCED sense. But by far the most important consider
0208  n repeating anything by rote, he is generally FORCED to go back to recover the habitual train of t
0208  and, in order to complete its hammock, seemed FORCED to start from the third stage, where it had l
0362  s might be transported from place to place. I FORCED many kinds of seeds into the stomachs of dead
0454  iting oceanic islands, which have seldom been FORCED to take flight, and have ultimately lost the
0073  varying success; and yet in the long run the FORCES are so nicely balanced, that the face of natu
0082  are struggling together with nicely balanced FORCES, extremely slight modifications in the struct
0267  ongrel offspring. It may be urged, as a most FORCIBLE argument, that there must be some essential
0325  e whole animal kingdom. M. Barrande has made FORCIBLE remarks to precisely the same effect. It is,
0463  h evidence, and this is the most obvious and FORCIBLE of the many objections which may be urged ag
0144  ite obscure. M. Is. Geoffroy St. Hilaire has FORCIBLY remarked, that certain malconformations are
0302  stance, by Agassiz, Pictet, and by none more FORCIBLY than by Professor Sedgwick, as a fatal objec
0181  ssible that the membrane connected fingers and FORE arm of the Galeopithecus might be greatly leng
0200  he same bones in the arm of the monkey, in the FORE leg of the horse, in the wing of the bat, and
0447  is view to whole families or even classes. The FORE limbs, for instance, which served as legs in t
0447  ing inherited at a corresponding late age, the FORE limbs in the embryos of the several descendant
0447  in each individual new species, the embryonic FORE limbs will differ greatly from the fore limbs
0447  ryonic fore limbs will differ greatly from the FORE limbs in the mature animal; the limbs in the l
0434  find, for instance, the bones of the arm and FOREARM, or of the thigh and leg, transposed. Hence
0066  at Nature, it is most necessary to keep the FOREGOING considerations always in mind, never to for
0077  e highest importance may be deduced from the FOREGOING remarks, namely, that the structure of ever
0082  causes or increases variability: and in the FOREGOING case the conditions of life are supposed to
0116  d incomplete stage of development. After the FOREGOING discussion, which ought to have been much a
0141  ent climates. We must not, however, push the FOREGOING argument too far, on account of the probabl
0149  seems to be a sign of low organisation: the FOREGOING remark seems connected with the very genera
0199  ils my reasoning would appear frivolous. The FOREGOING remarks lead me to say a few words on the p
0233  which seems at first cuite subversive of the FOREGOING theory: namely, that the cells on the extre
0243  structure, and is plainly explicable on the FOREGOING views: but is otherwise inexplicable, all t
0260  ents, seems to be a strange arrangement. The FOREGOING rules and facts, on the other hand, appear
0266  ither pure parent. Nor do I pretend that the FOREGOING remarks go to the root of the matter: no ex
0268  and that there must be some error in all the FOREGOING remarks, inasmuch as varieties, however muc
0289  her and more important cause than any of the FOREGOING: namely, from the several formations being
0292  her transitional or linking forms. From the FOREGOING considerations it cannot be doubted that th
0299  . hence it will be worth while to sum up the FOREGOING remarks, under an imaginary illustration. T
0328  nd in both, from the causes explained in the FOREGOING paragraphs, the same general succession in
0349  . a third great fact, partly included in the FOREGOING statements, is the affinity of the producti
0359  ruit (but not all the same species as in the FOREGOING experiment) floated, after being dried, for
0369  ions as they are at the present day. But the FOREGOING remarks on distribution apply not only to s
0376  w let us see what light can be thrown on the FOREGOING facts, on the belief, supported as it is by
0396  iew of independent acts of creation. All the FOREGOING remarks on the inhabitants of oceanic islan
0404  cal species, showing, in accordance with the FOREGOING view, that at some former period there has
0420  , have been subsequently discovered. All the FOREGOING rules and aids and difficulties in classifi
0026  cceeding generation there will be less of the FOREIGN blood: but when there has been no cross with
```

Page **(Key Word)**

Page **************************************(Key Word)***************************************

```
0098  been shown by Gartner, any influence from the FOREIGN pollen. When the stamens of a flower suddenl
0099  a plant's own pollen is always prepotent over FOREIGN pollen; but to this subject we shall return
0110  than in any other quarter of the world, some FOREIGN plants have become naturalised, without caus
0115  uralisation of plants through man's agency in FOREIGN lands. It might have been expected that the
0160  show a tendency to revert in character to the FOREIGN breed for many generations, some say, for a
0160  is retained by this very small proportion of FOREIGN blood. In a breed which has not been crossed
0381  ther becomes naturalised by man's agency in a FOREIGN land; why one ranges twice or thrice as far,
0388  consequently an intruder from the waters of a FOREIGN country, would have a better chance of seizi
0405  n distant lands in the struggle for life with FOREIGN associates. But on the view of all the speci
0488  ew countries and coming into competition with FOREIGN associates, might become modified; so that w
0083  ralised productions, that they have allowed FOREIGNERS to take firm possession of the land. And as
0083  to take firm possession of the land. And as FOREIGNERS have thus everywhere beaten some of the nat
0320  dmit that sickness in the individual is the FORERUNNER of death, to feel no surprise at sickness,
0484  us views are generally admitted, we can dimly FORESEE that there will be a considerable revolution
0338  species of the two countries could not have FORESEEN this result. Agassiz insists that ancient an
0074  is! Every one has heard that when an American FOREST is cut down, a very different vegetation spri
0074  portion of kinds as in the surrounding virgin FORESTS. What a struggle between the several kinds o
0276  rder to prevent them becoming inarched in our FORESTS. The sterility of first crosses between pure
0346  arid deserts, lofty mountains, grassy plains, FORESTS, marshes, lakes, and great rivers, under alm
0408  habitants of the plains and mountains, of the FORESTS, marshes, and deserts, are in so mysterious
0489  r take a prophetic glance into futurity as to FORETEL that it will be the common and widely spread
0062  superabundance of food: we do not see, or we FORGET, that the birds which are idly singing round
0062  nd are thus constantly destroying life; or we FORGET how largely these songsters, or their eggs, o
0065  no great destruction falling on them; and we FORGET that thousands are annually slaughtered for f
0066  coing considerations always in mind, never to FORGET that every single organic being around us may
0069  ect action. But this is a very false view: we FORGET that each species, even where it most abounds
0085  ge, seem to be quite unimportant, we must not FORGET that climate, food, etc., probably produce so
0293  uppose that it then became wholly extinct. We FORGET how small the area of Europe is compared with
0302  d not exist before that stage. We continually FORGET how large the world is, compared with the are
0302  l formations have been carefully examined; we FORGET that groups of species may elsewhere have lon
0307  orbes, dispute this conclusion. We should not FORGET that only a small portion of the world is kno
0322  the existence of each species depends. If we FORGET for an instant, that each species tends to in
0388  on of the same aquatic species, we should not FORGET the probability of many species having former
0405  necessarily range widely; for we should never FORGET that to range widely implies not only the pow
0149  shape. Natural selection, it should never be FORGOTTEN, can act on each part of each being, solely
0287  ogist, the late Edward Forbes, should not be FORGOTTEN, namely, that numbers of our fossil species
0297  by intermediate varieties. Nor should it be FORGOTTEN, as before explained, that A might be the a
0298  y one geological formation. It should not be FORGOTTEN, that at the present day, with perfect spec
0341  rogenitors of living species. It must not be FORGOTTEN that, on my theory, all the species of the
0439  ce mentioned by Agassiz, namely, that having FORGOTTEN to ticket the embryo of some vertebrate ani
0130  see a thin straggling branch springing from a FORK low down in a tree, and which by some chance h
0432  cify this or that branch, though at the actual FORK the two unite and blend together. We could not
0005  ty will tend to propagate its new and modified FORM. This fundamental subject of Natural Selection
0018  ld species. In regard to sheep and goats I can FORM no opinion. I should think, from facts communi
0021  reversed along the back of the neck that they FORM a hood, and it has, proportionally to its size
0031  eans of which he may summon into life whatever FORM and mould he pleases. Lord Somerville, speakin
0031  seem as if they had chalked out upon a wall a FORM perfect in itself, and then had given it exist
0045  nerations? and in this case I presume that the FORM would be called a variety. Again, we have many
0047  s occur in deciding whether or not to rank one FORM as a variety of another, even when they are cl
0047  e difficulty. In very many cases, however, one FORM is ranked as a variety of another, not because
0047  ure is opened. Hence, in determining whether a FORM should be ranked as a species or a variety, th
0078  ood thus to try in our imagination to give any FORM some advantage over another. Probably in no si
0083  en begins his selection by some half monstrous FORM; or at least by some modification prominent en
0091  uld either supplant or coexist with the parent FORM of wolf. Or, again, the wolves inhabiting a mo
0091  united States, one with a light greyhound like FORM, which pursues deer, and the other more bulky,
0094  t that an accidental deviation in the size and FORM of the body, or in the curvature and length of
0106  degree or they will be exterminated. Each new FORM, also, as soon as it has been much improved, w
0109  it follows that as each selected and favoured FORM increases in number, so will the less favoured
0109  rsor to extinction. We can, also, see that any FORM represented by few individuals will, during fl
0118  neration, and under a condensed and simplified FORM up to the fourteen thousandth generation. But
0119  will invariably prevail and multiply: a medium FORM may often long endure, and may or may not prod
0121  ral selection necessarily acts by the selected FORM having some advantage in the struggle for life
0123  tinct genus. The six descendants from (I) will FORM two sub genera or even genera. But as the orig
0124  ed much in character, but to have retained the FORM of (F), either unaltered or altered only in a
0124  and circuitous nature. Having descended from a FORM which stood between the two parent species (A)
0125  b14 and f14, and those marked o14 to m14, will FORM three very distinct genera. We shall also have
0125  from (A), the two little groups of genera will FORM two distinct families; or even orders, accordi
0125  ted; for as natural selection acts through one FORM having some advantage over other forms in the
0132  cted before that union takes place which is to FORM a new being. In the case of sporting plants, t
0144  lla into a tube. Hard parts seem to affect the FORM of adjoining soft parts; it is believed by som
0149  preserved or rejected each little deviation of FORM less carefully than when the part has to serve
0162  f the other, so as to produce the intermediate FORM. But the best evidence is afforded by parts or
0166  he stripes is not accompanied by any change of FORM or by any other new character. We see this ten
0166  ably reappear; but without any other change of FORM or character. When the oldest and truest breed
0172  ervation of profitable modifications, each new FORM will tend in a fully stocked country to take t
0172  h species as descended from some other unknown FORM, both the parent and all the transitional vari
0172  process of formation and perfection of the new FORM. But, as by this theory innumerable transition
0174  g in the intermediate zones. By changes in the FORM of the land and of climate, marine areas now c
0176  which they originally linked together. For any FORM existing in lesser numbers would, as already r
0176  ; and in this particular case the intermediate FORM would be eminently liable to the inroads of cl
0179  resembles an otter in its fur, short legs, and FORM of tail; during summer this animal dives for a
0187  eral descendants from the same original parent FORM, in order to see what gradations are possible,
0189  the surfaces of each layer slowly changing in FORM. Further we must suppose that there is a power
0198  om exercising them more, and possibly even the FORM of the pelvis; and then by the law of homologo
0200  s having been of special use to some ancestral FORM, or as being now of special use to the descend
0200  now of special use to the descendants of this FORM, either directly, or indirectly through the co
```

Page **************************************(Key Word)***************************************

Page ***(Key Word)***

```
0224 ery difficult to make cells of wax of the true FORM, though this is perfectly effected by a crowd
0225 ject by Mr. Waterhouse, who has shown that the FORM of the cell stands in close relation to the pr
0225 hombs have certain angles, and the three which FORM the pyramidal base of a single cell on one sid
0241 izes differed wonderfully in shape, and in the FORM and number of the teeth. But the important fac
0241 ction, by acting on the fertile parents, could FORM a species which should regularly produce neute
0241 uce neuters, either all of large size with one FORM of jaw, or all of small size with jaws having
0246 first case the two sexual elements which go to FORM the embryo are perfect; in the second case the
0272 be proved to be of universal occurence, or to FORM a fundamental distinction between varieties an
0273 producing offspring identical with the parent FORM. Now hybrids in the first generation are desce
0273 liable than hybrids to revert to either parent FORM; but this, if it be true, is certainly only a
0274 mongrels can be reduced to either pure parent FORM, by repeated crosses in successive generations
0281 s, we should be unable to recognise the parent FORM of any two or more species, even if we closely
0281 een them. But such a case would imply that one FORM had remained for a very long period unaltered,
0286 ourse, cannot be done; but we may, in order to FORM some crude notion on the subject, assume that
0286 r nodules, which from long resisting attrition FORM a breakwater at the base. Hence, under ordinar
0296 abrupt, though perhaps very slight, changes of FORM. It is all important to remember that naturali
0305 , the tropical parts of the Indian Ocean would FORM a large and perfectly enclosed basin, in which
0313 have reason to believe that the same identical FORM never reappears. The strongest apparent except
0314 t one species should retain the same identical FORM much longer than others; or, if changing, that
0314 nt relations of organism to organism, that any FORM which does not become in some degree modified
0315 e have every reason to believe that the parent FORM will generally be supplanted and exterminated
0329 accordance with the general succession of the FORM of life, and the order would falsely appear to
0329 the principle of descent. The more ancient any FORM is, the more, as a general rule, it differs fr
0330 s. if by this term it is meant that an extinct FORM is directly intermediate in all its characters
0330 it is a common belief that the more ancient a FORM is, by so much the more it tends to connect by
0331 the three existing genera, a14, a14, p14, will FORM a small family; b14 and f14 a closely allied f
0331 ral lines of descent diverging from the parent FORM A, will form an order; for all will have inher
0331 descent diverging from the parent form A, will FORM an order; for all will have inherited somethin
0331 lustrated by this diagram, the more recent any FORM is, the more it will generally differ from its
0335 corresponding lengths of time: a very ancient FORM might occasionally last much longer than a for
0335 orm might occasionally last much longer than a FORM elsewhere subsequently produced, especially in
0344 ivergence of character, why the more ancient a FORM is, the more it generally differs from those n
0344 m a little closer together. The more ancient a FORM is, the more often, apparently, it displays ch
0344 en groups now distinct; for the more ancient a FORM is, the more nearly it will be related to, and
0388 animals, whether retaining the same identical FORM or in some degree modified, I believe mainly d
0392 d, but still retaining its hooked seeds, would FORM an endemic species, having as useless an appen
0396 nds, whether still retaining the same specific FORM or modified since their arrival, could have re
0412 veral species; and all the genera on this line FORM together one class, for all have descended fro
0412 e, on this same principle, much in common, and FORM a sub family, distinct from that including the
0412 ve also much, though less, in common; and they FORM a family distinct from that including the thre
0412 iod. And all these genera, descended from (A), FORM an order distinct from the genera descended fr
0413 we often meet with in a more or less concealed FORM, that the characters do not make the genus, bu
0421 to have been but little modified, and they yet FORM a single genus. But this genus, though much is
0424 olely because they closely resemble the parent FORM, but because they are descended from it. He wh
0429 lant or aberrant groups. The more aberrant any FORM is, the greater must be the number of connecti
0432 ps: it has by no means made them; for if every FORM which has ever lived on this earth were sudden
0434 : the parts may change to almost any extent in FORM and size, and yet they always remain connected
0435 n being profitable in some way to the modified FORM, but often affecting by correlation of growth
0436 llae, these parts being perhaps very simple in FORM; and then natural selection will account for t
0443 r the animal has been born, what its merits or FORM will ultimately turn out. We see this plainly
0445 ature, differ so extraordinarily in length and FORM of beak, that they would, I cannot doubt, be r
0448 mbryo would closely resemble the mature parent FORM. We have seen that this is the rule of develop
0450 e, more or less obscured, of the common parent FORM of each great class of animals. Rudimentary, a
0462 t some species have retained the same specific FORM for very long periods, enormously long as meas
0463 y between each and some extinct and supplanted FORM. Even on a wide area, which has during a long
0466 to reproduce offspring exactly like the parent FORM. Variability is governed by many complex laws,
0469 power, in slowly and beautifully adapting each FORM to the most complex relations of life. The the
0471 eory. How strange it is that a bird, under the FORM of woodpecker, should have been created to pre
0481 th of the views given in this volume under the FORM of an abstract, I by no means expect to convin
0484 earth have descended from some one primordial FORM, into which life was first breathed. When the
0484 nted by the shadowy doubt whether this or that FORM be in essence a species. This I feel sure, and
0484 ecide (not that this will be easy) whether any FORM be sufficiently constant and distinct from oth
0010 bud and an ovule in their earliest stages of FORMATION; so that, in fact, sports support my view,
0042 es is an important element of success in the FORMATION of new races, at least, in a country which
0042 st have largely favoured the improvement and FORMATION of new breeds. Pigeons, I may add, can be p
0043 lection, has, no doubt, largely aided in the FORMATION of new sub breeds; but the importance of th
0095 the excavation of gigantic valleys or to the FORMATION of the longest lines of inland cliffs. Natu
0110 riety or species, during the progress of its FORMATION, will generally press hardest on its neares
0111 few, varieties are species in the process of FORMATION, or are, as I have called them, incipient s
0125 ss of modification, as now explained, to the FORMATION of genera alone. If, in our diagram, we sup
0172 ave been exterminated by the very process of FORMATION and perfection of the new form. But, as by
0174 ontinous has played an important part in the FORMATION of new species, more especially with freely
0279 bsence of intermediate varieteies in any one FORMATION. On the sudden appearance of groups of spec
0280 y enormous. Why then is not every geological FORMATION and every stratum full of such intermediate
0282 e an inadequate idea of the duration of each FORMATION or even each stratum. A man must for years
0284 ement, in a few cases from estimate, of each FORMATION in different parts of Great Britain; and th
0284 continent. Moreover, between each successive FORMATION, we have, in the opinion of most geologists
0286 period as since the latter part of the Chalk FORMATION. The distance from the northern to the sout
0288 ks its purity. The many cases on record of a FORMATION conformably covered, after an enormous inte
0288 rmous interval of time, by another and later FORMATION, without the underlying bed having suffered
0288 cies has hitherto been found in any tertiary FORMATION: yet it is now known that the genus Chthama
0290 ntervals of time having elapsed between each FORMATION. But we can, I think, see why the geologica
0291 avourable for life, and thus a fossiliferous FORMATION thick enough, when upraised, to resist any
0291 er author, in treating of this or that great FORMATION, has come to the conclusion that it was acc
0291 e. i may add, that the only ancient tertiary FORMATION on the west coast of South America, which h
0292 favourable, as previously explained, for the FORMATION of new varieties and species; but during su
```

Page ***(Key Word)***

```
0292 ; but if we confine our attention to any one  FORMATION, it becomes more difficult to understand, w
0293 ies in the upper and lower parts of the same   FORMATION, but, as they are rare, they may be here pa
0293 they may be here passed over. Although each    FORMATION has indisputably required a vast number of
0293 the following considerations. Although each    FORMATION may mark a very long lapse of years, each p
0293 concluded.that the average duration of each    FORMATION is twice or thrice as long as the average d
0293 species first appearing in the middle of any   FORMATION, it would be rash in the extreme to infer t
0293 rld; nor have the several stages of the same   FORMATION throughout Europe been correlated with perf
0293 when we see a species first appearing in any   FORMATION, the probability is that it only then first
0295 rms in the upper and lower parts of the same   FORMATION, the deposit must have gone on accumulating
0295 but we have seen that a thick fossiliferous    FORMATION can only be accumulated during a period of
0295 wer limits. It would seem that each separate   FORMATION, like the whole pile of formations in any c
0295 ion. When we see, as is so often the case, a   FORMATION composed of beds of different mineralogical
0296 h time. Nor will the closest inspection of a   FORMATION give any idea of the time which its deposit
0296 ast lapse of time represented by the thinner   FORMATION. Many cases could be given of the lower bed
0296 cases could be given of the lower beds of a    FORMATION having been upraised, denuded, submerged, a
0296 hen re covered by the upper beds of the same   FORMATION; facts, showing what wide, yet easily overl
0296 es occur at the bottom, middle, and top of a   FORMATION, the probability is that they have not live
0297 scendants from the lower and upper beds of a   FORMATION, and unless we obtained numerous transition
0297 s come from different sub stages of the same   FORMATION. Some experienced conchologists are now sin
0297 nct but consecutive stages of the same great   FORMATION, we find that the embedded fossils, though
0298 to this view, the chance of discovering in a   FORMATION in any one country all the early stages of
0298 e stages of transition in any one geological   FORMATION. It should not be forgotten, that at the pr
0300 or rising; whilst rising, each fossiliferous   FORMATION would be destroyed, almost as soon as accum
0301 modifications could be preserved in any one    FORMATION. Very many of the marine inhabitants of the
0301 se, if found in different stages of the same   FORMATION, would, by most palaeontologists, be ranked
0302 peared at the commencement and close of each   FORMATION, pressed so hardly on my theory. On the sud
0303 e time required for the accumulation of each   FORMATION. These intervals will have given time for t
0303 some few parent forms; and in the succeeding   FORMATION such species will appear as if suddenly cre
0304 hale having been discovered in any secondary   FORMATION, seemed fully to justify the belief that th
0304 n the latest secondary and earliest tertiary  FORMATION. But now we may read in the Supplement to L
0305 z believes, at the commencement of the chalk   FORMATION, the fact would certainly be highly remarka
0308 a, do not support the view, that the older a   FORMATION is, the more it has suffered the extremity
0308 ven a remnant of any palaeozoic or secondary   FORMATION. Hence we may perhaps infer, that during th
0313 ecies has been simultaneous in each separate   FORMATION. Species of different genera and classes ha
0313 ntrude for a period in the midst of an older   FORMATION, and then allow the pre existing fauna to r
0315 in consecutive formations is not equal. Each   FORMATION, on this view, does not mark a new and comp
0322 hroughout the world. Thus our European Chalk   FORMATION can be recognised in many distant parts of
0329 n deposited during the same exact periods, a   FORMATION in one region often corresponding with a bl
0335 hat the organic remains from an intermediate   FORMATION are in some degree intermediate in characte
0335 remains from the several stages of the chalk  FORMATION, though the species are distinct in each st
0336 ted, is obvious. As the accumulation of each  FORMATION has often been interrupted, and as long bla
0341 ng eight species, be found in one geological  FORMATION, and in the next succeeding formation there
0341 ogical formation, and in the next succeeding  FORMATION there be six other allied or representative
0342 h must have passed away even during a single  FORMATION; that, owing to subsidence being necessary
0342 been least perfectly kept; that each single   FORMATION has not been continuously deposited; that t
0342 nuously deposited; that the duration of each  FORMATION is, perhaps, short compared with the averag
0342 appearance of new forms in any one area and   FORMATION; that widely ranging species are those whic
0342 ound in the several stages of the same great  FORMATION. He may disbelieve in the enormous interval
0345 early see why the remains of an intermediate  FORMATION are intermediate in character. The inhabita
0437 hould similar bones have been created in the  FORMATION of the wing and leg of a bat, used as they
0443 used at a very early period, even before the  FORMATION of the embryo, may appear late in life; as
0453 issue, can thus act? Can we suppose that the  FORMATION of rudimentary teeth which are subsequently
0463 l older species, why is not every geological  FORMATION charged with such links? Why does not every
0465 hey do spread, if discovered in a geological  FORMATION, they will appear as if suddenly created th
0475 orld. The fact of the fossil remains of each  FORMATION being in some degree intermediate in charac
0487 the accumulation of each great fossiliferous  FORMATION will be recognised as having depended on an
0279 ections. On the intermittence of geological   FORMATIONS. On the absence of intermediate varieteies
0282 reatises by different observers on separate   FORMATIONS, and to mark how each author attempts to gi
0283 hat the thickness and extent of sedimentary   FORMATIONS are the result and measure of the degradati
0284 three quarters British miles. Some of these   FORMATIONS, which are represented in England by from 1
0286 22 miles, and the thickness of the several   FORMATIONS io on an average about 1100 feet, as I am i
0289 g to the age of our secondary or palaeozoic   FORMATIONS. But the imperfection in the geological rec
0289 of the foregoing; namely, from the several   FORMATIONS being separated from each other by wide int
0289 by wide intervals of time. When we see the   FORMATIONS tabulated in written works, or when we foll
0289 re in that country between the superimposed  FORMATIONS; so it is in North America, and in many oth
0289 e which has elapsed between the consecutive  FORMATIONS, we may infer that this could nowhere be as
0290 he mineralogical composition of consecutive  FORMATIONS, generally implying great changes in the ge
0290 but we can, I think, see why the geological  FORMATIONS of each region are almost invariably interm
0290 western side of South America, no extensive  FORMATIONS with recent or tertiary remains can anywher
0291 formed. I am convinced that all our ancient  FORMATIONS, which are rich in fossils, have thus been
0291 ons have affected wide spaces. Consequently  FORMATIONS rich in fossils and sufficiently thick and
0295 separate formation, like the whole pile of  FORMATIONS in any country, has generally been intermit
0296 only a few feet in thickness, representing   FORMATIONS, elsewhere thousands of feet in thickness,
0298 the species found in more widely separated  FORMATIONS; so to this subject I shall have to return
0298 xceeding the limits of the known geological  FORMATIONS of Europe, which have oftenest given rise,
0299 ia; and therefore equals all the geological  FORMATIONS which have been examined with any accuracy,
0299 he former state of Europe, when most of our  FORMATIONS were accumulating. The Malay Archipelago is
0300 d in an excessively imperfect manner in the  FORMATIONS which we suppose to be there accumulating.
0300 r archipelago, I believe that fossiliferous  FORMATIONS could be formed of sufficient thickness to
0300 ge, as distant in futurity as the secondary  FORMATIONS lie in the past, only during periods of sub
0301 o right to expect to find in our geological  FORMATIONS, an infinite number of those fine transitio
0302 roups of species suddenly appear in certain  FORMATIONS, has been urged by several palaeontologists
0302 red with the area over which our geological  FORMATIONS have been carefully examined; we forget tha
0303 ve probably elapsed between our consecutive  FORMATIONS, longer perhaps in some cases than the time
0305 hat very few species are known from several  FORMATIONS in Europe. Some few families of fish now ha
0307 we ought to find only small remnants of the  FORMATIONS next succeeding them in age, and these ough
```

```
0308  e inhabited profound depths, in the several  FORMATIONS of Europe and of the United States; and fro
0308  sediment, miles in thickness, of which the    FORMATIONS are composed, we may infer that from first
0308  nos in the intervals between the successive   FORMATIONS; whether Europe and the United States durin
0308  hey existed there, palaeozoic and secondary   FORMATIONS would in all probability have been accumula
0309  rted into a continent, we should there find   FORMATIONS older than the Silurian strata, supposing s
0310  that we see in these large areas, the many    FORMATIONS long anterior to the silurian epoch in a co
0310  d, namely our not finding in the successive   FORMATIONS infinitely numerous transitional links betw
0310  le groups of species appear in our European   FORMATIONS; the almost entire absence, as at present k
0310  ence, as at present known, of fossiliferous   FORMATIONS beneath the Silurian strata, are all undoub
0311  d in our consecutive, but widely separated.   FORMATIONS. On this view, the difficulties above discu
0312  e been many and most gradual. The secondary   FORMATIONS are more broken; but, as Bronn has remarked
0313  spond with the succession of our geological   FORMATIONS; so that between each two consecutive forma
0313  tions; so that between each two consecutive   FORMATIONS, the forms of life have seldom changed in e
0313  we compare any but the most closely related   FORMATIONS, all the species will be found to have unde
0315  accumulation of long enduring fossiliferous   FORMATIONS depends on great masses of sediment having
0315  en deposited on areas whilst subsiding, our   FORMATIONS have been almost necessarily accumulated at
0315  ited by the fossils embedded in consecutive   FORMATIONS is not equal. Each formation, on this view,
0317  ickness, crossing the successive geological   FORMATIONS in which the species are founc, the line wi
0317  to believe, from the study of the tertiary   FORMATIONS, that species and groups of species gradual
0319  e in many cases in the more recent tertiary   FORMATIONS, that rarity precedes extinction; and we kn
0321  a, a great genus of shells in the secondary   FORMATIONS, survives in the Australian seas; and a few
0322  e intervals of time between our consecutive   FORMATIONS; and in these intervals there may have been
0323  the Chalk of Europe, but which occur in the   FORMATIONS either above or below, are similarly absent
0323  world. In the several successive palaeozoic   FORMATIONS of Russia, Western Europe and North America
0323  s, would still be manifest, and the several   FORMATIONS could be easily correlated. These observati
0324  ittle doubt that all the more modern marine   FORMATIONS, namely, the upper pliocene, the pleistocen
0327  elieving that all our greater fossiliferous   FORMATIONS were deposited during periods of subsidence
0328  t is probable that strictly contemporaneous   FORMATIONS have often been accumulated over very wide
0328  en affected by the same movements. When two   FORMATIONS have been deposited in two regions during n
0328  observations on some of the later tertiary   FORMATIONS. Barrande, also, shows that there is a stri
0328  f difference in the species. If the several   FORMATIONS in these regions have not been deposited du
0329  ging during the accumulation of the several   FORMATIONS and during the long intervals of time betwe
0329  ime between them; in this case, the several   FORMATIONS in the two regions could be arranged in the
0331  l lines may represent successive geological   FORMATIONS, and all the forms beneath the uppermost li
0332  , supposed to be embedded in the successive   FORMATIONS, were discovered at several points low down
0332  f the middle horizontal lines or geological   FORMATIONS, for instance, above No. VI., but none from
0333  gone much modification, should in the older   FORMATIONS make some slight approach to each other; so
0334  and blank intervals between the successive   FORMATIONS. Subject to these allowances, the fauna of
0334  ls of time have elapsed between consecutive   FORMATIONS. It is no real objection to the truth of th
0335  ologists, that fossils from two consecutive   FORMATIONS are far more closely related to each other,
0335  other, than are the fossils from two remote   FORMATIONS. Pictet gives as a well known instance, the
0336  the distinct species in closely consecutive   FORMATIONS, by the physical conditions of the ancient
0336  of fossil remains from closely consecutive   FORMATIONS, though ranked as distinct species, being c
0336  ntervals have intervened between successive   FORMATIONS, we ought not to expect to find, as I attem
0336  show in the last chapter, in any one or two   FORMATIONS all the intermediate varieties between the
0342  of time have elapsed between the successive   FORMATIONS; that there has probably been more extincti
0342  which have elapsed between our consecutive   FORMATIONS; he may overlook how important a part migra
0343  a part migration must have played, when the   FORMATIONS of any one great region alone, as that of E
0343  t; and that the older continents, formed of   FORMATIONS older than any known to us, may now all be
0345  the organic remains of closely consecutive   FORMATIONS are more closely allied to each other, than
0345  ied to each other, than are those of remote   FORMATIONS; for the forms are more closely linked toge
0465  will be simply classed as new species. Most   FORMATIONS have been intermittent in their accumulati
0465  rage duration of specific forms. Successive   FORMATIONS are separated from each other by enormous b
0465  blank intervals of time; for fossiliferous   FORMATIONS, thick enough to resist future degradation,
0465  ith respect to the absence of fossiliferous   FORMATIONS beneath the lowest Silurian strata, I can o
0465  this in the fossil remains from consecutive   FORMATIONS invariably being much more closely related
0465  ed to each other, than are the fossils from   FORMATIONS distant from each other in time. Such is th
0476  ate in character between the fossils in the   FORMATIONS above and below, is simply explained by the
0487  o correlate as strictly contemporaneous two   FORMATIONS, which include few identical species, by th
0488  rganic change in the fossils of consecutive   FORMATIONS probably serves as a fair measure of the la
0020  n record of a permanent race having been thus  FORMED. On the Breeds of the Domestic Pigeon. Believ
0038  more fully explained, two sub breeds might be  FORMED. This, perhaps, partly explains what has been
0055  es (i.e. species of the same genus) have been  FORMED, many varieties or incipient species ought, a
0055  ings. Where many species of a genus have been  FORMED through variation, circumstances have been fa
0056  ever many species of the same genus have been  FORMED, or where, if we may use the expression, the
0056  t, wherever many species of a genus have been  FORMED, the species of that genus present a number o
0056  that where many species of a genus have been  FORMED, on an average many are still forming; and th
0091  ition of this process, a new variety might be  FORMED which would either supplant or coexist with t
0091  the two sites, two varieties might slowly be  FORMED. These varieties would cross and blend where
0092  ng pollen instead of nectar; and as pollen is  FORMED for the sole object of fertilisation, its des
0103  , a new and improved variety might be quickly  FORMED on any one spot, and might there maintain its
0103  e new variety. A local variety when once thus  FORMED might subsequently slowly spread to other dis
0106  th many others. Hence more new places will be  FORMED, and the competition to fill them will be mor
0107  sewhere; new forms will have been more slowly  FORMED, and old forms more slowly exterminated. And
0110  that as new species in the course of time are  FORMED through natural selection, others will become
0117  is supposed to have been accumulated to have  FORMED a fairly well marked variety, such as would b
0120  elieve, species are multiplied and genera are  FORMED. In a large genus it is probable that more th
0171  tructure and habits of a bat, could have been  FORMED by the modification of some animal with wholl
0174  s distinct species might have been separately  FORMED without the possibility of intermediate varie
0174  that many perfectly defined species have been  FORMED on strictly continuous areas; though I do not
0177  irstly, because new varieties are very slowly  FORMED, for variation is a very slow process, and na
0178  thirdly, when two or more varieties have been  FORMED in different portions of a strictly continuou
0178  ties will, it is probable, at first have been  FORMED in the intermediate zones, but they will gene
0181  inks formerly existed, and that each had been  FORMED by the same steps as in the case of the less
0185  n that the webbed feet of ducks and geese are  FORMED for swimming? yet there are upland geese with
0185  er than that the long toes of grallatores are  FORMED for walking over swamps and floating plants,
```

Page ********************************(Key Word)********************************

```
0186  cal and chromatic aberration, could have been  FORMED  by natural selection, seems, I freely confess
0186  eving that a perfect and complex eye could be   FORMED  by natural selection, though insuperable by o
0188  en as perfect as the eye of an eagle might be   FORMED  by natural selection, although in this case h
0188  and we naturally infer that the eye has been    FORMED  by a somewhat analogous process. But may not
0189  hat a living optical instrument might thus be   FORMED  as superior to one of glass, as the works of
0189  n existed, which could not possibly have been   FORMED  by numerous, successive, slight modifications
0189  is latter case the organ must have been first   FORMED  at an extremely remote period, since which al
0190  concluding that an organ could not have been    FORMED  by transitional gradations of some kind. Nume
0196  ounted for. A well developed tail having been   FORMED  in an aquatic animal, it might subsequently c
0201  of the structure of any one species had been    FORMED  for the exclusive good of another species, it
0201  by and for the good of each. No organ will be   FORMED, as Paley has remarked, for the purpose of ca
0203  ng on a continuous area, must often have been   FORMED  when the area was not continuous, and when th
0203  m one part to another. When two varieties are   FORMED  in two districts of a continuous area, an int
0203  s area, an intermediate variety will often be   FORMED, fitted for an intermediate zone; but from re
0204  that a bat, for instance, could not have been   FORMED  by natural selection from an animal which at
0204  n organ so perfect as the eye could have been   FORMED  by natural selection, is more than enough to
0206  cknowledged that all organic beings have been   FORMED  on two great laws, Unity of Type, and the Con
0223  determine when and where a new nest shall be    FORMED, and when they migrate, the masters carry the
0224  to be of use to the species, until an ant was   FORMED  as abjectly dependent on its slaves as is the
0225  sides bevelled so as to join on to a pyramid,   FORMED  of three rhombs. These rhombs have certain an
0227  between the several spheres in both layers be   FORMED, there will result a double layer of hexagona
0227  nal prisms united together by pyrimidal bases   FORMED  of three rhombs; and the rhombs and the sides
0227  iculty, that after hexagonal prisms have been   FORMED  by the intersection of adjoining spheres in t
0228  h of the diameter of the sphere of which they   FORMED  a part, the rims of the basins intersected or
0229  to have flat bottoms; and these flat bottoms,   FORMED  by thin little plates of the vermilion wax ha
0238  h extraordinarily long horns, could be slowly   FORMED  by carefully watching which individual bulls
0241  ructure; a graduated series having been first   FORMED, as in the case of the driver ant, and then t
0282  good to wander along lines of sea coast, when   FORMED  of moderately hard rocks, and mark the proces
0283  and feet in thickness, which, though probably   FORMED  at a quicker rate than many other deposits, y
0283  ate than many other deposits, yet, from being   FORMED  of worn and rounded pebbles, each of which be
0289  ch separate territory, hardly any idea can be   FORMED  of the length of time which has elapsed betwe
0290  nd extensive accumulations of sediment may be   FORMED  in two ways: either, in profound depths of th
0291  , to resist any amount of degradation, may be   FORMED. I am convinced that all our ancient formatio
0291  ns, which are rich in fossils, have thus been   FORMED  during subsidence. Since publishing my views
0291  resist subsequent degradation, may have been   FORMED  over wide spaces during periods of subsidence
0292  be increased, and new stations will often be   FORMED; all circumstances most favourable, as previo
0292  ction, fewer new varieties or species will be   FORMED; and it is during these very periods of subsi
0300  elieve that fossiliferous formations could be   FORMED  of sufficient thickness to last to an age, as
0309  eternity? Our continents seem to to have been   FORMED  by a preponderance, during many oscillations
0316  d races of the domestic pigeon, for the newly   FORMED  fantail would be almost sure to inherit from
0325  theory of natural selection. New species are   FORMED  by new varieties arising, which have some adv
0327  with the principle of new species having been   FORMED  by dominant species spreading widely and vary
0337  han the more ancient: for each new species is   FORMED  by having had some advantage in the struggle,
0343  ferent aspect: and that the older continents,   FORMED  of formations older than any known to us, may
0344  dants, and thus new sub groups and groups are   FORMED. As these are formed, the species of the less
0344  ub groups and groups are formed. As these are   FORMED, the species of the less vigorous groups, fro
0355  a volcanic island, for instance, upheaved and   FORMED  at the distance of a few hundreds of miles fr
0358  some at least of the islands would have been    FORMED, like other mountain summits, of granite, met
0365  time, whilst an island was being upheaved and   FORMED, and before it had become fully stocked with
0388  emembered that when a pond or stream is first   FORMED, for instance, on a rising islet, it will be
0397  eous operculum, I removed it, and when it had   FORMED  a new membranous one, I immersed it for fourt
0401  analogous instances, is that the new species   FORMED  in the separate islands have not quickly spre
0434  be more curious than that the hand of a man,   FORMED  for grasping, that of a mole for digging, the
0435  ans, serving for such different purposes, are   FORMED  by infinitely numerous modifications of an up
0437  stacean, which has an extremely complex mouth   FORMED  of many parts, consequently always have fewer
0438  naturalists frequently speak of the skull as   FORMED  of metamorphosed vertebrae: the jaws of crabs
0443  enerally has acted, even before the embryo is   FORMED; and the variation may be due to the male and
0451  s. nothing can be plainer than that wings are   FORMED  for flight, yet in how many insects do we see
0453  ents the pistil in male flowers, and which is   FORMED  merely of cellular tissue, can thus act? Can
0454  imertary nails on the fin of the manatee were   FORMED  for this purpose. On my view of descent with
0401  ted that long lines of inland cliffs had been   FORMED, and great valleys excavated, by the slow act
0483  roughout whole classes various structures are   FORMED  on the same pattern, and at an embryonic age
0020  stic races by this process, we must admit the   FORMER  existence of the most extreme forms, as the I
0026  a character, which has been lost during some   FORMER  generation, this tendency, for all that we ca
0046  likewise, judging from Brachiopod shells, at   FORMER  periods of time. These facts seem to be very
0078  the climate may be exactly the same as in its   FORMER  home, yet the conditions of its life will oen
0081  eriously disturb the relations of some of the   FORMER  inhabitants. Let it be remembered how powerfu
0098  n pollen and pollen from another species, the   FORMER  will have such a prepotent effect, that it wi
0129  sting species; and those produced during each   FORMER  year may represent the long succession of ext
0129  all, budding twigs; and this connexion of the   FORMER  and present buds by ramifying branches may we
0141  his domestic animals, and such facts as that   FORMER  species of the elephant and rhinoceros were c
0154  ued rejection of those tending to revert to a   FORMER  and less modified condition. The principle in
0168  are found, that is, where there has been much   FORMER  variation and differentiation, or where the m
0179  mediate links. Consequently evidence of the    FORMER  existence could be found only amongst fossil
0191  he arteries still marking in the embryo their   FORMER  position. But it is conceivable that the now
0195  fications would be of importance or not. In a   FORMER  chapter I have given instances of most trifli
0196  and, after having been slowly perfected at a   FORMER  period, have been transmitted in nearly the s
0206  wi as it includes, through the inheritance of   FORMER  adaptations, that of Unity of Type. Chapter V
0228  ad been excavated to the same depth as in the   FORMER  experiment, would have broken into each other
0299  ide and shallow seas, probably represents the   FORMER  state of Europe, when most of our formations
0301  ncient homes, as they would differ from their   FORMER  state, in a nearly uniform, though perhaps ex
0307  he lowest azoic rocks, probably indicates the   FORMER  existence of life at these periods. But the d
0314  organic conditions of life, as explained in a   FORMER  chapter. When many of the inhabitants of a co
0318  what could so recently have exterminated the   FORMER  horse under conditions of life apparently so
0350  less remote: on the nature and number of the   FORMER  immigrants: and on their action and reaction,
0354  en, considering our ignorance with respect to   FORMER  climatal and geographical changes and various
```

Page ********************************(Key Word)********************************

Page **(Key Word)**

```
0355  second region, has probably received at some  FORMER  period immigrants from this other region, my
0356  ded: where the sea now extends, land may at a  FORMER  period have connected islands or possibly eve
0357  tervening oceanic islands. I freely admit the  FORMER  existence of many islands, now buried beneath
0357  be enabled to speculate with security on the   FORMER  extension of the land. But I do not believe t
0358  ise seem to me opposed to the belief of their  FORMER  continuity with continents. Nor does their al
0366  fted icebergs and coast ice, plainly reveal a  FORMER  cold period. The former influence of the glac
0366  ice, plainly reveal a former cold period. The  FORMER  influence of the glacial climate on the distr
0366  ed for arctic beings and ill fitted for their  FORMER  more temperate inhabitants, the latter would
0366  d become covered with snow and ice, and their  FORMER  Alpine inhabitants would descend to the plain
0368  he perfectly well ascertained occurrence of a  FORMER  Glacial period, seem to me to explain in so s
0368  idence, that a colder climate permitted their  FORMER  migration across the low intervening tracts,
0373  apart, glaciers have left the marks of their   FORMER  low descent; and in Sikkim, Dr. Hooker saw ma
0373  the equator, we have some direct evidence of   FORMER  glacial action in New Zealand; and the same p
0373  t extremity, we have the clearest evidence of  FORMER  glacial action, in huge boulders transported
0378  migrate northward or southward towards their   FORMER  homes: but the forms, chiefly northern, which
0381  southern, as in the northern hemisphere, to a  FORMER  and warmer period, before the commencement of
0382  e as a record, full of interest to us, of the  FORMER  inhabitants of the surrounding lowlands. Chap
0391  or long ages have struggled together in their  FORMER  homes, and have become mutually adapted to ea
0401  nother plant to another island. Hence when in  FORMER  times an immigrant settled on any one or more
0401  is no reason to suppose that they have at any  FORMER  period been continuously united. The currents
0404  ordance with the foregoing view, that at some  FORMER  period there has been intercommunication or m
0404  pecies, either at the present time or at some  FORMER  period under different physical conditions, a
0410  o show, be accounted for by migration at some  FORMER  period under different conditions or by occas
0455  organ might be retained for one alone of its   FORMER  functions. An organ, when rendered useless, m
0455  e principle, also, of economy, explained in a  FORMER  chapter, by which the materials forming any p
0476  om one part of the world to another, owing to  FORMER  climatal and geographical changes and to the
0477  nd early colonists. On this same principle of  FORMER  migration, combined in most cases with modifi
0487  gy now throws, and will continue to throw, on  FORMER  changes of climate and of the level of the la
0487  e enabled to trace in an admirable manner the  FORMER  migrations of the inhabitants of the whole wo
0006  hich most naturalists entertain, and which I   FORMERLY  entertained, namely, that each species has b
0019  reat Britain. One author believes that there  FORMERLY  existed in Great Britain eleven wild species
0026  ons, namely, the improbability of man having  FORMERLY  got seven or eight supposed species of pigeo
0034  nine animals, to improve the breed, and they  FORMERLY  did so, as is attested by passages in Pliny.
0035  d in early maturity, compared with the stock  FORMERLY  kept in this country. By comparing the accou
0039  n possession. Nor must the value which would  FORMERLY  be set on any slight differences in the indi
0047  her that they do now somewhere exist, or may  FORMERLY  have existed; and here a wide door for the e
0084  he forms of are now different from what they  FORMERLY  were. Although natural selection can act onl
0106  s of the smaller continent of Australia have  FORMERLY  yielded, and apparently are now yielding, to
0126  predict: for we well know that many groups,  FORMERLY  most extensively developed, have now become
0160  eproduce the lost character might be, as was  FORMERLY  remarked, for all that we can see to the con
0174  inuous areas: though I do not doubt that the  FORMERLY  broken condition of areas now continous has
0178  tative species and their common parent, must  FORMERLY  have existed in each broken portion of the l
0181  at the Galeopithecus or flying Lemur, which   FORMERLY  was falsely ranked amongst bats. It has an e
0181  e no difficulty in supposing that such links  FORMERLY  existed, and that each had been formed by th
0182  g insects of the most diversified types, and  FORMERLY  had flying reptiles, it is conceivable that
0193  does geology at all lead to the belief that  FORMERLY  most fishes had electric organs, which most
0199  of no direct use. So again characters which  FORMERLY  were useful, or which formerly had arisen fr
0199  racters which formerly were useful, or which  FORMERLY  had arisen from correlation of growth, or fr
0200  een inherited from a common progenitor, were  FORMERLY  of more special use to that progenitor, or i
0200  cquired through natural selection, subjected  FORMERLY, as now, to the several laws of inheritance,
0205  s species. We may, also, believe that a part  FORMERLY  of high importance has often been retained (
0280  number of intermediate varieties, which have  FORMERLY  existed on the earth, be truly enormous. Why
0280  intermediate forms must, on my theory, have  FORMERLY  existed. I have found it difficult, when loo
0285  ntle slope of the lava streams, due to their  FORMERLY  liquid state, showed at a glance how far the
0298  tive, there is reason to suspect, as we have  FORMERLY  seen, that their varieties are generally at
0303  suddenly created. I may here recall a remark  FORMERLY  made, namely that it might require a long su
0305  a confined range: the teleostean fish might  FORMERLY  have had a similarly confined range, and aft
0309  silurian strata, supposing such to have been  FORMERLY  deposited; for it might well happen that str
0320  ompetition will generally be most severe, as  FORMERLY  explained and illustrated by examples, betwe
0330  f characters; so that the two groups, though  FORMERLY  quite distinct, at that period made some sma
0331  ndency to divergence of character, which was  FORMERLY  illustrated by this diagram, the more recent
0340  w of distribution of terrestrial mammals was  FORMERLY  different from what it now is. North America
0340  different from what it now is. North America  FORMERLY  partook strongly of the present character of
0340  of the continent; and the southern half was  FORMERLY  more closely allied, than it is at present,
0340  utley's discoveries, that Northern India was  FORMERLY  more closely related in its mammals to Afric
0340  endants. If the inhabitants of one continent  FORMERLY  differed greatly from those of another conti
0342  the numberless transitional links which must  FORMERLY  have connected the closely allied or represe
0352  y in such cases as Great Britain having been  FORMERLY  united to Europe, and consequently possessin
0353  ve interrupted or rendered discontinuous the  FORMERLY  continuous range of many species. So that we
0356  es two marine faunas: submerge it, or let it  FORMERLY  have been submerged, and the two faunas will
0356  ed, and the two faunas will now blend or may  FORMERLY  have blended: where the sea now extends, lan
0363  ago. Hence we may safely infer that icebergs  FORMERLY  landed their rocky burthens on the shores of
0366  od to come slowly on, and then pass away, as  FORMERLY  occurred. As the cold came on, and as each m
0372  hown that more plants are identical than was  FORMERLY  supposed), but we find in every great class
0377  extinction: how much no one can say: perhaps  FORMERLY  the tropics supported as many species as we
0388  orget the probability of many species having  FORMERLY  ranged as continuously as fresh water produc
0393  nected with the mainland; moreover, icebergs  FORMERLY  brought boulders to its western shores, and
0394  ers to its western shores, and they may have  FORMERLY  transported foxes, as so frequently now happ
0396  view of all our oceanic islands having been  FORMERLY  connected by continuous land with the neares
0399  ether by occasional means of transport or by  FORMERLY  continuous land, from America; and the Cape
0408  likewise linked to the extinct beings which  FORMERLY  inhabited the same continent. Bearing in min
0412  n to the diagram illustrating the action, as  FORMERLY  explained, of these several principles; and
0431  gh which the early progenitors of birds were  FORMERLY  connected with the early progenitors of the
0437  lants, many spiral whorls of leaves. We have  FORMERLY  seen that parts many times repeated are emin
0478  ent with modification, that the same parents  FORMERLY  inhabited both areas; and we almost invariab
0485  ntermediate gradations, whereas species were  FORMERLY  thus connected. Hence, without quite rejecti
```

Page **(Key Word)**

Page ***(Key Word)***

0219 markable instinct was first discovered in the FORMICA (Polyerges) rufescens by Pierre Huber, a bet
0219 derful an instinct could have been perfected. FORMICA sanguinea was likewise first discovered by P
0223 absolutely dependent on its numerous slaves. FORMICA sanguinea, on the other hand, possesses much
0224 as abjectly dependent on its slaves as is the FORMICA rufescens. Cell making instinct of the Hive
0239 se of an intermediate size scanty in numbers, FORMICA flava has larger and smaller workers, with s
0055 t species ought, as a general rule, to be now FORMING. Where many large trees grow, we expect to f
0056 ave been formed, on an average many are still FORMING; and this holds good. There are other relati
0059 o be closely, but unequally, allied together, FORMING little clusters round certain species. Speci-
0112 s would become greater, and would be noted as FORMING two sub breeds: finally, after the lapse of
0128 less closely and unequally related together, FORMING sections and sub genera, species of distinct
0128 ted, and genera related in different degrees, FORMING sub families, families, orders, sub classes,
0374 of the flowering plants of Tierra del Fuego, FORMING no inconsiderable part of its scanty flora,
0395 archipelago, and there is much difficulty in FORMING a judgment in some cases owing to the probab
0455 d in a former chapter, by which the materials FORMING any part or structure, if not useful to the
0479 eg of the horse, the same number of vertebrae FORMING the neck of the giraffe and of the elephant,
0486 ully be called living fossils, will aid us in FORMING a picture of the ancient forms of life. Embr
0005 ly causes much Extinction of the less improved FORMS of life, and induces what I have called Diver
0015 evert in some of their characters to ancestral FORMS, it seems to me not improbable, that if we co
0019 ginal species; but by crossing we can get only FORMS in some degree intermediate between their par
0020 admit the former existence of the most extreme FORMS, as the Italian greyhound, bloodhound, bull d
0035 y modify, even during their own lifetimes, the FORMS and qualities of their cattle. Slow and insen
0046 on; and hardly two naturalists can agree which FORMS to rank as species and which as varieties. We
0047 lection, as hereafter will be explained. Those FORMS which possess in some considerable degree the
0047 but which are so closely similar to some other FORMS, or are so closely linked to them by intermed
0047 that many of these doubtful and closely allied FORMS have permanently retained their characters in
0047 . practically, when a naturalist can unite two FORMS together by others having intermediate charac
0048 botanists, and see what a surprising number of FORMS have been ranked by one botanist as good spec
0048 . under genera, including the most polymorphic FORMS, Mr. Babington gives 251 species; whereas Mr.
0048 m gives only 112, a difference of 139 doubtful FORMS! Amongst animals which unite for each birth,
0048 rth, and which are highly locomotive, doubtful FORMS, ranked by one zoologist as a species and by
0049 ide distance between the homes of two doubtful FORMS leads many naturalists to rank both as distin
0049 , be sufficient? It must be admitted that many FORMS, considered by highly competent judges as var
0050 uld hardly wish for better evidence of the two FORMS being specifically distinct. On the other han
0050 turalists to an agreement how to rank doubtful FORMS. Yet it must be confessed, that it is in the
0050 countries that we find the greatest number of FORMS of doubtful value. I have been struck with th
0050 author makes more than a dozen species out of FORMS, which are very generally considered as varie
0050 e up his mind how to rank most of the doubtful FORMS. His general tendency will be to make many sp
0051 luded to, with the amount of difference in the FORMS which he is continually studying; and he has
0051 l encounter a greater number of closely allied FORMS. But if his observations be widely extended,
0051 ists. When, moreover, he comes to study allied FORMS brought from countries not now continuous, in
0051 nd the intermediate links between his doubtful FORMS, he will have to trust almost entirely to ana
0051 between species and sub species, that is, the FORMS which in the opinion of some naturalists come
0052 is given to less distinct and more fluctuating FORMS. The term variety, again, in comparison with
0056 ate links have not been found between doubtful FORMS, naturalists are compelled to come to a deter
0056 ry important criterion in settling whether two FORMS should be ranked as species or varieties. Now
0057 species. And what are varieties but groups of FORMS, unequally related to each other, and cluster
0057 ted to each other, and clustered round certain FORMS, that is, round their parent species? Undoubt
0058 erage range, as have those very closely allied FORMS, marked for me by Mr. Watson as doubtful spec
0058 stly, by the discovery of intermediate linking FORMS, and the occurrence of such links cannot affe
0058 nks cannot affect the actual characters of the FORMS which they connect; and except, secondly, by
0059 ly, by a certain amount of difference, for two FORMS, if differing very little, are generally rank
0059 ies, notwithstanding that intermediate linking FORMS have not been discovered; but the amount of d
0059 difference considered necessary to give to two FORMS the rank of species is quite indefinite. In g
0059 nd to become larger; and throughout nature the FORMS of life which are now dominant tend to become
0059 to break up into smaller genera. And thus, the FORMS of life throughout the universe become divide
0060 aterial for us whether a multitude of doubtful FORMS be called species or sub species or varieties
0060 or instance, the two or three hundred doubtful FORMS of British plants are entitled to hold, if th
0069 g a mountain, we far oftener meet with stunted FORMS, due to the directly injurious action of clim
0073 e world, or invent laws on the duration of the FORMS of life! I am tempted to give one more instan
0076 mpetition should be most severe between allied FORMS, which fill nearly the same place in the econ
0081 . if the country were open on its borders, new FORMS would certainly immigrate, and this also woul
0081 by barriers, into which new and better adapted FORMS could not freely enter, we should then have p
0084 ast geological ages, that we only see that the FORMS of are now different from what they formerly
0098 , it is well known that if very closely allied FORMS or varieties are planted near each other, it
0105 t favourable for the production of new organic FORMS, we ought to make the comparison within equal
0106 reas; and what is more important, that the new FORMS produced on large areas, which already have b
0107 will have been less severe than elsewhere; new FORMS will have been more slowly formed, and old fo
0107 rms will have been more slowly formed, and old FORMS more slowly exterminated. And it is in fresh
0107 fresh water we find some of the most anomalous FORMS now known in the world, as the Ornithorhynchu
0107 eparated in the natural scale. These anomalous FORMS may almost be called living fossils; they hav
0107 most favourable for the production of many new FORMS of life, likely to endure long and to spread
0108 e will be much extinction of the less improved FORMS, and the relative proportional numbers of the
0108 slow, and on the immigration of better adapted FORMS having been checked. But the action of natura
0109 increases in number, so will the less favoured FORMS decrease and become rare. Rarity, as geology
0109 n. but we may go further than this; for as new FORMS are continually and slowly being produced, un
0109 unless we believe that the number of specific FORMS goes on perpetually and almost indefinitely i
0109 st become extinct. That the number of specific FORMS has not indefinitely increased, geology shows
0110 come rarer and rarer, and finally extinct. The FORMS which stand in closest competition with those
0110 r existence that it is the most closely allied FORMS, varieties of the same species, and species o
0110 productions, through the selection of improved FORMS by man. Many curious instances could be given
0118 se two varieties, being only slightly modified FORMS, will tend to inherit those advantages which
0119 small numbered letters marking the successive FORMS which have become sufficiently distinct to be
0120 species (A) is supposed to have produced three FORMS, a10, f10, and m10, which, from having diverg
0120 r diagram to be excessively small, these three FORMS may still be only well marked varieties; or t
0120 s or greater in amount, to convert these three FORMS into well defined species: thus the diagram i
0121 advantage in the struggle for life over other FORMS, there will be a constant tendency in the imp

Page ***(Key Word)***

forms

Page **(Key Word)***

```
0121 on will generally be most severe between those FORMS which are most nearly related to each other i
0121 ion, and structure. Hence all the intermediate FORMS between the earlier and later states, that is
0125 f diverging dotted lines to be very great, the FORMS marked a14 to p14, those marked b14 and f14,
0125 ough one form having some advantage over other FORMS in the struggle for existence, it will chiefl
0127 nature, in modifying and adapting the various FORMS of life to their several conditions and stati
0128 tinction of the less improved and intermediate FORMS of life. On these principles, I believe, the
0134 nued use or disuse, for we know not the parent FORMS; but many animals have structures which can b
0138 remarks, animals not far remote from ordinary FORMS, prepare the transition from light to darknes
0162 onsiderable catalogue, also, could be given of FORMS intermediate between two other forms, which t
0162 given of forms intermediate between two other FORMS, which themselves must be doubtfully ranked a
0162 s or species; and this shows, unless all these FORMS be considered as independently created specie
0169 tion, or where the manufactury of new specific FORMS has been actively at work, there, on an avera
0171 we not everywhere see innumerable transitional FORMS? Why is not all nature in confusion instead o
0172 wn less improved parent or other less favoured FORMS with which it comes into competition. Thus ex
0172 ut, as by this theory innumerable transitional FORMS must have existed, why do we not find them em
0173 to find at the present time many transitional FORMS. Let us take a simple case: in travelling fro
0176 when varieties intermediate between two other FORMS occur, they are much rarer numerically than t
0176 ccur, they are much rarer numerically than the FORMS which they connect. Now, if we may trust thes
0176 e generally existed in lesser numbers than the FORMS which they connect, then, I think, we can und
0176 be exterminated and disappear, sooner than the FORMS which they originally linked together. For an
0176 nently liable to the inroads of closely allied FORMS existing on both sides of it. But a far more
0177 numbers in a narrow and intermediate zone. For FORMS existing in larger numbers will always have a
0177 ral selection to seize on, than will the rarer FORMS which exist in lesser numbers. Hence, the mor
0177 xist in lesser numbers. Hence, the more common FORMS, in the race for life, will tend to beat and
0177 will tend to beat and supplant the less common FORMS, for these will be more slowly modified and i
0178 bitants becoming slowly modified, with the new FORMS thus produced and the old ones acting and rea
0178 ent period in isolated portions, in which many FORMS, more especially amongst the classes which un
0179 most certainly be beaten and supplanted by the FORMS which they connect; for these from existing i
0179 n so often remarked, to exterminate the parent FORMS and the intermediate links. Consequently evid
0183 od in great numbers and under many subordinate FORMS. Thus, to return to our imaginary illustratio
0183 uld have been developed under many subordinate FORMS, for taking prey of many kinds in many ways,
0190 should have to look to very ancient ancestral FORMS, long since become extinct. We should be extr
0194 dering that the proportion of living and known FORMS to the extinct and unknown is very small, I h
0203 will act, at any one time, only on a very few FORMS; and partly because the very process of natur
0204 l usually exist in lesser numbers than the two FORMS which it connects; consequently the two latte
0206 existed more individuals, and more diversified FORMS, and the competition will have been severer,
0225 bee, but more nearly related to the latter: it FORMS a nearly regular waxen comb of cylindrical ce
0226 spherical portions, and yet each flat portion FORMS a part of two cells. Reflecting on this case,
0239 and sometimes in colour; and that the extreme FORMS can sometimes be perfectly linked together by
0241 e case of the driver ant, and then the extreme FORMS, from being the most useful to the community,
0245 order to prevent the confusion of all organic FORMS. This view certainly seems at first probable,
0246 rs. The fertility of varieties, that is of the FORMS known or believed to have descended from comm
0246 e knot, for in ten cases in which he found two FORMS, considered by most authors as distinct speci
0248 nists on the question whether certain doubtful FORMS should be ranked as species or varieties, wit
0258 lesser degree. He has observed it even between FORMS so closely related (as Matthiola annua and gl
0259 first crosses and of hybrids, we see that when FORMS, which must be considered as good and distinc
0263 nt for the endurance and stability of specific FORMS, as in the case of grafting it is unimportant
0270 xperiments may be trusted, I know not; but the FORMS experimentised on, are ranked by Sageret, who
0271 the other varieties. He experimentised on five FORMS, which are commonly reputed to be varieties,
0272 especially in the reproductive systems of the FORMS which are crossed. Hybrids and Mongrels compa
0276 ies. Summary of Chapter. First crosses between FORMS sufficiently distinct to be ranked as species
0276 diametrically opposite conclusions in ranking FORMS by this test. The sterility is innately varia
0277 of another kind: namely, that the crossing of FORMS only slightly different is favourable to the
0277 , parallel with the systematic affinity of the FORMS which are subjected to experiment; for system
0277 nce between all species. First crosses between FORMS known to be varieties, or sufficiently alike
0279 ssed. One, namely the distinctness of specific FORMS, and their not being blended together by innu
0279 the presence of other already defined organic FORMS, than on climate; and, therefore, that the re
0279 ties, from existing in lesser numbers than the FORMS which they connect, will generally be beaten
0280 ake the places of and exterminate their parent FORMS. But just in proportion as this process of ex
0280 ays be borne in mind what sort of intermediate FORMS must, on my theory, have formerly existed. I
0280 any two species, to avoid picturing to myself, FORMS directly intermediate between them. But this
0280 a wholly false view; we should always look for FORMS intermediate between each species and a commo
0281 oenas. So with natural species, if we look to FORMS very distinct, for instance to the horse and
0281 possible by my theory, that one of two living FORMS might have descended from the other; for inst
0281 e event; for in all cases the new and improved FORMS of life will tend to supplant the old and uni
0281 e will tend to supplant the old and unimproved FORMS. By the theory of natural selection all livin
0287 the water has been peopled by hosts of living FORMS. What an infinite number of generations, whic
0289 les of sediment, charged with new and peculiar FORMS of life, had elsewhere been accumulated. And
0291 aised will give a most imperfect record of the FORMS of life which then existed; or, sediment may
0292 quent discovery of her transitional or linking FORMS. From the foregoing considerations it cannot
0293 ce as long as the average duration of specific FORMS. But insuperable difficulties, as it seems to
0294 nts of Europe during the Glacial period, which FORMS only a part of one whole geological period; a
0295 n order to get a perfect gradation between two FORMS in the upper and lower parts of the same form
0297 t greater amount of difference between any two FORMS, they rank both as species, unless they are e
0297 most closely connected with either one or both FORMS by intermediate varieties. Nor should it be f
0298 do not spread widely and supplant their parent FORMS until they have been modified and perfected i
0298 the early stages of transition between any two FORMS, is small, for the successive changes are sup
0298 y, with perfect specimens for examination, two FORMS can seldom be connected by intermediate varie
0300 ceed the average duration of the same specific FORMS; and these contingencies are indispensable fo
0301 would slowly spread and supplant their parent FORMS. When such varieties returned to their ancien
0301 an infinite number of those fine transitional FORMS, which on my theory assuredly have connected
0302 l selection. For the development of a group of FORMS, all of which have descended from some one pr
0303 on of species from some one or some few parent FORMS; and in the succeeding formation such species
0303 e would be necessary to produce many divergent FORMS, which would be able to spread rapidly and wi
0311 may represent the apparently abruptly changed FORMS of life, entombed in our consecutive, but wid
0312 on Extinction. On simultaneous changes in the FORMS of life throughout the world. On the affiniti
```

Page **(Key Word)***

Page ***(Key Word)***

```
0312  pecies. On the state of development of ancient  FORMS. On the succession of the same types within t
0312  to make the percentage system of lost and new   FORMS more gradual. In some of the most recent beds
0312  red by years, only one or two species are lost   FORMS, and only one or two are new forms, having he
0312  es are lost forms, and only one or two are new   FORMS, having here appeared for the first time, eit
0313  e found in the midst of a multitude of extinct   FORMS. Falconer has given a striking instance of a
0313  t between each two consecutive formations, the   FORMS of life have seldom changed in exactly the sa
0315  y of nature, and thus supplant it; yet the two   FORMS, the old and the new, would not be identical
0316  ew and modified or the same old and unmodified   FORMS. Species of the genus Lingula, for instance,
0317  ation and the production of a number of allied   FORMS must be slow and gradual, one species giving
0317  ory of natural selection the extinction of old   FORMS and the production of new and improved forms
0317  d forms and the production of new and improved   FORMS are intimately connected together. The old no
0320  and the consequent extinction of less favoured   FORMS almost inevitably follows. It is the same wit
0320  in other countries. Thus the appearance of new   FORMS and the disappearance of old forms, both natu
0320  ance of new forms and the disappearance of old   FORMS, both natural and artificial, are bound toget
0320  flourishing groups, the number of new specific   FORMS which have been produced within a given time
0320  time is probably greater than that of the old   FORMS which have been exterminated; but we know tha
0320  imes we may believe that the production of new   FORMS has caused the extinction of about the same n
0320  the extinction of about the same number of old   FORMS. The competition will generally be most sever
0320  ained and illustrated by examples, between the   FORMS which are most like each other in all respect
0321  ination of the parent species; and if many new   FORMS have been developed from any one species, the
0321  s caused its extermination; and if many allied   FORMS be developed from the successful intruder, ma
0321  their places; and it will generally be allied   FORMS, which will suffer from some inherited inferi
0322  id manner many of the old inhabitants; and the   FORMS which thus yield their places will commonly b
0322  particular species or group of species. On the   FORMS of Life changing almost simultaneously throug
0322  overy is more striking than the fact, that the   FORMS of life change almost simultaneously througho
0323  as mere superficial sculpture. Moreover other   FORMS, which are not found in the Chalk of Europe,
0323  nd north America, a similar parallelism in the   FORMS of life has been observed by several authors:
0323  iew, the general parallelism in the successive   FORMS of life, in the stages of the widely separate
0324  of the later tertiary stages. When the marine   FORMS of life are spoken of as having changed simul
0324  me degree allied, and from not including those   FORMS which are only found in the older underlying
0324  taneous in a geological sense. The fact of the   FORMS of life changing simultaneously, in the above
0325  referring to the parallelism of the palaeozoic   FORMS of life in various parts of Europe, they add,
0325  , as the cause of these great mutations in the   FORMS of life throughout the world, under the most
0325  s great fact of the parallel succession of the   FORMS of life throughout the world, is explicable o
0325  arising, which have some advantage over older   FORMS; and those forms, which are already dominant,
0325  ave some advantage over older forms; and those   FORMS, which are already dominant, or have some adv
0325  ominant, or have some advantage over the other   FORMS in their own country, would naturally oftenes
0326  ge accidents, but in the long run the dominant   FORMS will generally succeed in spreading. The diff
0326  severe competition with many already existing   FORMS, would be highly favourable, as would be the
0327  be victorious. But in the course of time, the   FORMS dominant in the highest degree, wherever prod
0327  uld cause the extinction of other and inferior   FORMS; and as these inferior forms would be allied
0327  ther and inferior forms; and as these inferior   FORMS would be allied in groups by inheritance, who
0327  ge sense, simultaneous, succession of the same   FORMS of life throughout the world, accords well wi
0327  ading, varying, and producing new species. The   FORMS which are beaten and which yield their places
0327  h yield their places to the new and victorious   FORMS, will generally be allied in groups, from inh
0327  isappear from the world; and the succession of   FORMS in both ways will everywhere tend to correspo
0328  paragraphs, the same general succession in the   FORMS of life; but the species would not exactly co
0329  f extinct Species to each other, and to living   FORMS. Let us now look to the mutual affinities of
0329  ore, as a general rule, it differs from living   FORMS. But, as Buckland long ago remarked, all foss
0329  ting groups, or between them. That the extinct   FORMS of life help to fill up the wide intervals be
0330  diate in all its characters between two living   FORMS, the objection is probably valid. But I appre
0331  successive geological formations, and all the   FORMS beneath the uppermost line may be considered
0331  most ancient fossils differ most from existing   FORMS. We must not, however, assume that divergence
0331  , as we have seen in the case of some Silurian   FORMS, that a species might go on being slightly mo
0332  in the diagram by the letter F14. All the many   FORMS, extinct and recent, descended from A, make,
0332  diagram we can see that if many of the extinct   FORMS, supposed to be embedded in the successive fo
0332  ircuitous course through many widely different   FORMS. If many extinct forms were to be discovered
0332  h many widely different forms. If many extinct   FORMS were to be discovered above one of the middle
0333  espect to the mutual affinities of the extinct   FORMS of life to each other and to living forms, se
0333  inct forms of life to each other and to living   FORMS, seem to me explained in a satisfactory manne
0334  e nearly intermediate in character between the   FORME of life above and below. We must, however, al
0334  ow for the entire extinction of some preceding   FORMS, and for the coming in of quite new forms by
0334  ding forms, and for the coming in of quite new   FORMS by immigration, and for a large amount of mod
0335  was perfect, we have no reason to believe that   FORMS successively produced necessarily endure for
0336  nearly the same. Let it be remembered that the   FORMS of life, at least those inhabiting the sea, h
0336  acial period, and note how little the specific   FORMS of the inhabitants of the sea have been affec
0336  long as measured geologically, closely allied   FORMS, or, as they have been called by some authors
0336  low and scarcely sensible mutation of specific   FORMS, as we have a just right to expect to find. O
0336  o find. On the state of Development of Ancient   FORMS. There has been much discussion whether recen
0336  there has been much discussion whether recent   FORMS are more highly developed than ancient. I wil
0336  r's satisfaction what is meant by high and low   FORMS. But in one particular sense the more recent
0337  s. but in one particular sense the more recent   FORMS must, on my theory, be higher than the more a
0337  the struggle for life over other and preceding   FORMS. If under a nearly similar climate, the eocen
0337  organisation of the more recent and victorious   FORMS of life, in comparison with the ancient and b
0337  ife, in comparison with the ancient and beaten   FORMS; but I can see no way of testing this sort of
0337  t in the course of time a multitude of British   FORMS would become thoroughly naturalized there, an
0338  ; or that the geological succession of extinct   FORMS is in some degree parallel to the embryologic
0338  lel to the embryological development of recent   FORMS. I must follow Pictet and Huxley in thinking
0338  r own proper classes, though some of these old   FORMS are in a slight degree less distinct from eac
0340  n, the feebler will yield to the more dominant   FORMS, and there will be nothing immutable in the l
0342  compared with the average duration of specific   FORMS; that migration has played an important part
0342  important part in the first appearance of new   FORMS in any one area and formation; that widely ra
0342  nnecting together all the extinct and existing   FORMS of life by the finest graduated steps. He who
0343  fication to some extent. The extinction of old   FORMS is the almost inevitable consequence of the p
0343  nevitable consequence of the production of new   FORMS. We can understand why when a species has onc
0344  n understand how the spreading of the dominant   FORMS of life, which are those that oftenest vary,
```

Page ***(Key Word)***

```
0344  usly. We can understand how it is that all the  FORMS of life, ancient and recent, make together on
0344  from those now living. Why ancient and extinct  FORMS often tend to fill up gaps between existing f
0344  ms often tend to fill up gaps between existing  FORMS, sometimes blending two groups previously cla
0345  groups, since become widely divergent. Extinct  FORMS are seldom directly intermediate between exis
0345  seldom directly intermediate between existing  FORMS: but are intermediate only by a long and circ
0345  course through many extinct and very different  FORMS. We can clearly see why the organic remains o
0345  , than are those of remote formations: for the  FORMS are more closely linked together by generatio
0345  have been produced by ordinary generation: old  FORMS having been supplanted by new and improved fo
0345  rms having been supplanted by new and improved  FORMS of life, produced by the laws of variation st
0347  been free migration for the northern temperate  FORMS, as there now is for the strictly arctic prod
0350  l depend on the migration of the more dominant  FORMS of life from one region into another having b
0351  as we have seen in the last chapter that some  FORMS have retained nearly the same character from
0367  nal result. As the warmth returned, the arctic  FORMS would retreat northward, closely followed up
0367  ed from the bases of the mountains, the arctic  FORMS would seize on the cleared and thawed ground,
0367  ange are more especially related to the arctic  FORMS living due north or nearly due north of them:
0368  r period, since the Glacial period. The arctic  FORMS, during their long southern migration and re
0369  present varieties, some are ranked as doubtful  FORMS, and some few are distinct yet closely allied
0369  distribution apply not only to strictly arctic  FORMS, but also to many sub arctic and to some few
0369  sub arctic and to some few northern temperate  FORMS, for some of these are the same on the lower
0369  rmity of the sub arctic and northern temperate  FORMS round the world, at the commencement of the G
0372  pposed), but we find in every great class many  FORMS, which some naturalists rank as geographical
0372  and a host of closely allied or representative  FORMS which are ranked by all naturalists as specif
0372  heory of modification, for many closely allied  FORMS now living in areas completely sundered. Thus
0372  e of many existing and tertiary representative  FORMS on the eastern and western shores of temperat
0375  n the mountains of Abyssinia, several European  FORMS and some few representatives of the peculiar
0375  he mountains, some few representative European  FORMS are found, which have not been discovered in
0375  striking is the fact that southern Australian  FORMS are clearly represented by plants growing on
0375  mountains of Borneo. Some of these Australian  FORMS, as I hear from Dr. Hooker, extend along the
0376  es of New Zealand, Tasmania, etc., of northern  FORMS of fish. Dr. Hooker informs me that twenty fi
0376  ould be observed that the northern species and  FORMS found in the southern parts of the southern h
0376  eally become less and less arctic. Many of the  FORMS living on the mountains of the warmer regions
0376  , and many, though closely related to northern  FORMS, must be ranked as distinct species. Now let
0377  er of the more vigorous and dominant temperate  FORMS might have penetrated the native ranks and ha
0378  od when the cold was most intense, when arctic  FORMS had migrated some twenty five degrees of lati
0378  uator. As the warmth returned, these temperate  FORMS would naturally ascend the higher mountains,
0379  southward towards their former homes: but the  FORMS, chiefly northern, which had crossed the equa
0379  ave been wholly different with those intruding  FORMS which settled themselves on the intertropical
0379  rangers will have had to compete with many new  FORMS of life: and it is probable that selected mod
0379  ia, that many more identical plants and allied  FORMS have apparently migrated from the north to th
0379  ion. We see, however, a few southern vegetable  FORMS on the mountains of Borneo and Abyssinia. I s
0379  tent of land in the north, and to the northern  FORMS having existed in their own homes in greater
0379  fection or dominating power, than the southern  FORMS. And thus, when they became commingled during
0379  ingled during the Glacial period, the northern  FORMS were enabled to beat the less powerful southe
0379  ere enabled to beat the less powerful southern  FORMS. Just in the same manner as we see at the pre
0380  en the natives: whereas extremely few southern  FORMS have become naturalised in any part of Europe
0380  l period they were stocked with endemic Alpine  FORMS: but these have almost everywhere largely yie
0380  verywhere largely yielded to the more dominant  FORMS, generated in the larger areas and more effic
0380  have everywhere lately yielded to continental  FORMS, naturalised by man's agency. I am far from s
0381  modified and have given rise to new groups of  FORMS, and others have remained unaltered. We canno
0381  a was exterminated by the Glacial epoch, a few  FORMS were widely dispersed to various points of th
0382  ve become slightly tinted by the same peculiar  FORMS of vegetable life. Sir C. Lyell in a striking
0382  nt distribution both of the same and of allied  FORMS of life can be explained. The living waters m
0384  t some fresh water fish belong to very ancient  FORMS, and in such cases there will have been ample
0395  00 fathoms in depth, and here we find American  FORMS, but the species and even the genera are dist
0396  the scarcity of kinds, the richness in endemic  FORMS in particular classes or sections of classes,
0396  rtions of certain orders of plants, herbaceous  FORMS having been developed into trees, etc., seem
0396  lete: and if modification be admitted, all the  FORMS of life would have been more equally modified
0402  d thus spread: but we should remember that the  FORMS which become naturalised in new countries are
0403  lpine species, excepting in so far as the same  FORMS, chiefly of plants, have spread widely throug
0403  alpine plants, etc., all of strictly American  FORMS, and it is obvious that a mountain, as it bec
0404  cility of transport has given the same general  FORMS to the whole world. We see this same principl
0404  allied species occur, there will be found many  FORMS which some naturalists rank as distinct speci
0404  species, and some as varieties: these doubtful  FORMS showing us the steps in the process of modifi
0405  dely ranging species vary and give rise to new  FORMS will largely determine their average range. F
0406  erally change at a slower rate than the higher  FORMS: and consequently the lower forms will have h
0406  n the higher forms: and consequently the lower  FORMS will have had a better chance of ranging wide
0406  , together with the seeds and eggs of many low  FORMS being very minute and better fitted for dista
0407  fore for our ignorance, and remember that some  FORMS of life change most slowly, enormous periods
0408  y of migration (generally of the more dominant  FORMS of life), together with subsequent modificati
0408  ent modification and the multiplication of new  FORMS. We can thus understand the high importance o
0408  ns should often be inhabited by very different  FORMS of life: for according to the length of time
0408  ure of the communication which allowed certain  FORMS and not others to enter, either in greater or
0409  nd space: the laws governing the succession of  FORMS in past times being nearly the same with thos
0410  yet discovered in an intermediate deposit the  FORMS which are therein absent, but which occur abo
0410  e are intelligible: for whether we look to the  FORMS of life which have changed during successive
0410  rated into distant quarters, in both cases the  FORMS within each class have been connected by the
0410  dinary generation: and the more nearly any two  FORMS are related in blood, the nearer they will ge
0412  orted by looking at the great diversity of the  FORMS of life which, in any small area, come into t
0412  show that there is a constant tendency in the  FORMS which are increasing in number and diverging
0412  ss divergent, the less improved, and preceding  FORMS. I request the reader to turn to the diagram
0414  neral propositions and of placing together the  FORMS most like each other. It might have been thou
0418  early uniform, and common to a great number of  FORMS, and not common to others, they use it as one
0419  ecially in very large groups of closely allied  FORMS. Temminck insists on the utility or even nece
0419  iven amongst plants and insects, of a group of  FORMS, first ranked by practised naturalists as onl
0420  y have undergone: and this is expressed by the  FORMS being ranked under different genera, families
0421  and in different degrees from each other. The  FORMS descended from A, now broken up into two or t
```

Page **************************************(Key Word)**************************************

```
0423 n, the principle of inheritance would keep the FORMS together which were allied in the greatest nu
0424 single definition. As soon as three Orchidean   FORMS (Monochanthus, Myanthus, and Catasetum), whic
0426 for we can account for its presence in so many  FORMS with such different habits, only by its inher
0426 e hidden bond of community of descent. Let two  FORMS have not a single character in common, yet if
0426 ngle character in common, yet if these extreme  FORMS are connected together by a chain of intermed
0429 mocification, how it is that the more ancient   FORMS of life often present characters in some slig
0429 ting groups. A few old and intermediate parent  FORMS having occasionally transmitted to the presen
0429 , the greater must be the number of connecting  FORMS which on my theory have been exterminated and
0429 ly lost. And we have some evidence of aberrant  FORMS having suffered severely from extinction, for
0429 ount for this fact only by looking at aberrant  FORMS as failing groups conquered by more successfu
0431 brate animals, by the belief that many ancient FORMS of life have been utterly lost, through which
0431 . there has been less entire extinction of the FORMS of life which once connected fishes with batr
0431 ustacea, for here the most wonderfully diverse FORMS are still tied together by a long, but broken
0432 and, on the principle of inheritance, all the  FORMS descended from A, or from I, would have somet
0432 everal groups; but we could pick out types, or FORMS, representing most of the characters of each
0432 we were ever to succeed in collecting all the  FORMS in any class which have lived throughout all
0433 arly see how it is that all living and extinct FORMS can be grouped together in one great system;
0437 en has observed) of all low or little modified FORMS; therefore we may readily believe that the un
0449 rs, we can clearly see why ancient and extinct FORMS of life should resemble the embryos of their
0450 hat the supposed law of resemblance of ancient FORMS of life to the embryonic stages of recent for
0450 orms of life to the embryonic stages of recent FORMS, may be true, but yet, owing to the geologica
0456 w on the view of the common parentage of those FORMS which are considered by naturalists as allied
0460 sterility of species when first crossed, which FORMS so remarkable a contrast with the almost univ
0461 e, and that the offspring of slightly modified FORMS or varieties acquire from being crossed incre
0461 s of life and crosses between greatly modified FORMS, lessen fertility; and on the other hand, les
0461 ions of life and crosses between less modified FORMS, increase fertility. Turning to geographical
0462 lection an interminable number of intermediate FORMS must have existed, linking together all the s
0462 may be asked, Why do we not see these linking  FORMS all around us? Why are not all organic beings
0462 n inextricable chaos? With respect to existing FORMS, we should remember that we have no right to
0463 will be liable to be supplanted by the allied  FORMS on either hand; and the latter, from existing
0463 evidence of the gradation and mutation of the  FORMS of life? We meet with no such evidence, and t
0464 geological record. Numerous existing doubtful  FORMS could be named which are probably varieties;
0464 the common view, whether or not these doubtful FORMS are varieties? As long as most of the links b
0465 shorter than the average duration of specific  FORMS. Successive formations are separated from eac
0465 there will probably be more variability in the FORMS of life; during periods of subsidence, more e
0469 k which they assign to the many representative  FORMS in Europe and North America. If then we have
0471 oroups tend to give birth to new and dominant  FORMS; so that each large group tends to become sti
0471 xtinction, explains the arrangement of all the FORMS of life, in groups subordinate to groups, all
0472 governed the production of so called dominant  FORMS. In both cases physical conditions seem to ha
0474 ure, if the part be common to so many subordinate FORMS, that is, if it has been inherited for a very
0475 on the principle of natural selection; for old FORMS will be supplanted by new and improved forms.
0475 d forms will be supplanted by new and improved FORMS. Neither single species nor groups of species
0475 been broken. The gradual diffusion of dominant FORMS, with the slow modification of their descenda
0475 modification of their descendants, causes the  FORMS of life, after long intervals of time, to app
0476 ate between existing and allied groups. Recent FORMS are generally looked at as being, in some vag
0476 e vague sense, higher than ancient and allied  FORMS; and they are in so far higher as the later a
0476 n so far higher as the later and more improved FORMS have conquered the older and less improved or
0476 astly, the law of the long endurance of allied FORMS on the same continent, of marsupials in Austr
0478 lied yet distinct species occur, many doubtful FORMS and and varieties of the same species likewis
0479 osely alike, and should be so unlike the adult FORMS. We may cease marvelling at the embryo of an
0482 n to arrive at. They admit that a multitude of FORMS, which t'll lately they themselves thought we
0482 same view to other and very slightly different FORMS. Nevertheless they do not pretend that they c
0482 ine, or even conjecture, which are the created FORMS of life, and which are those produced by seco
0483 icult to answer, because the more distinct the FORMS are which we may consider, by so much the arg
0484 sufficiently constant and distinct from other  FORMS, to be capable of definition; and if definabl
0485 r differences, however slight, between any two FORMS, if not blended by intermediate gradations, a
0485 y most naturalists as sufficient to raise both FORMS to the rank of species. Hereafter we shall be
0485 nce of intermediate gradations between any two FORMS, we shall be led to weigh more carefully and
0485 erence between them. It is quite possible that FORMS now generally acknowledged to be merely varie
0486 ill aid us in forming a picture of the ancient FORMS of life. Embryology will reveal to us the str
0487 arison of the preceding and succeeding organic FORMS. We must be cautious in attempting to correla
0487 al species, by the general succession of their FORMS of life. As species are produced and extermin
0488 early periods of the earth's history, when the FORMS of life were probably fewer and simpler, the
0488 ; and at the first dawn of life, when very few FORMS of the simplest structure existed, the rate o
0489 te new and dominant species. As all the living FORMS of life are the lineal descendants of those w
0489 to reflect that these elaborately constructed  FORMS, so different from each other, and dependent
0490 character and the Extinction of less improved  FORMS. Thus, from the war of nature, from famine an
0490 rs, having been originally breathed into a few FORMS or into one; and that, whilst this planet has
0490 of gravity, from so simple a beginning endless FORMS most beautiful and most wonderful have been,
0064 as on breeding till ninety years old, bringing FORTH three pair of young in this interval; if this
0486 e direct action of external conditions, and so FORTH. The study of domestic productions will rise
0041 s seen it gravely remarked, that it was most   FORTUNATE that the strawberry began to vary just when
0101 point of view; but I have been enabled, by a   FORTUNATE chance, elsewhere to prove that two individ
0203 love or maternal hatred, though the latter     FORTUNATELY is most rare, is all the same to the inexo
0221 s in Switzerland than in England. One day I    FORTUNATELY chanced to witness a migration from one ne
0021 e other breeds. The fantail has thirty or even FORTY tail feathers, instead of twelve or fourteen,
0042 ich have been raised during the last thirty or FORTY years. In the case of animals with separate s
0222 returning file burthened with booty, for about FORTY yards, to a very thick clump of heath, whence
0374 in America, Dr. Hooker has shown that between  FORTY and fifty of the flowering plants of Tierra d
0378 . hooker, that all the flowering plants, about FORTY six in number, common to Tierra del Fuego and
0380 s from La Plata, and during the last thirty or FORTY years from Australia. Something of the same k
0386 olymbetes, once flew on board the Beagle, when FORTY five miles distant from the nearest land; how
0002 on this subject, with a request that I would   FORWARD it to Sir Charles Lyell, who sent it to the
0026 it is difficult, perhaps impossible, to bring  FORWARD one case of the hybrid offspring of two anim
0098 often required to cause the stamens to spring  FORWARD, as Kolreuter has shown to be the case with
0213 tionless like a statue, and then slowly crawl  FORWARD with a peculiar gait; and another kind of wo
```

Page **************************************(Key Word)**************************************

forward

Page ***************************************(Key Word)***************************************

```
0052  y mr. Wollaston with the varieties of certain  FOSSIL  land shells in Madeira. If a variety were to
0130  known to us only from having been found in a   FOSSIL  state. As we here and there see a thin stragg
0167  with the old and ignorant cosmogonists, that   FOSSIL  shells had never lived, but had been created
0174  existed there, and may be embedded there in a  FOSSIL  condition. But in the intermediate region, ha
0179  former existence could be found only amongst   FOSSIL  remains, which are preserved, as we shall in
0183  es with transitional grades of structure in a  FOSSIL  condition will always be less, from their hav
0187  adation in the structure of the eye, and from  FOSSIL  species we can learn nothing on this head. In
0282  pse of Time. Independently of our not finding  FOSSIL  remains of such infinitely numerous connectin
0287  not be forgotten, namely, that numbers of our  FOSSIL  species are known and named from single and o
0288  rate sufficiently quick to embed and preserve  FOSSIL  remains. Throughout an enormously large propo
0288  which inhabits deep water and has been found   FOSSIL  in Sicily, whereas not one other species has
0288  s superfluous to state that our evidence from  FOSSIL  remains is fragmentary in an extreme degree.
0294  of the world, sedimentary deposits, including  FOSSIL  remains, have gone on accumulating within the
0298  ollected from many places; and in the case of  FOSSIL  species this could rarely be effected by pala
0298  nect species by numerous, fine, intermediate,  FOSSIL  links, by asking ourselves whether, for insta
0299  only by the future geologist discovering in a  FOSSIL  state numerous intermediate gradations; and s
0303  now one of the richest known accumulations of  FOSSIL  mammals belongs to the middle of the secondar
0304  y own eyes has much struck me. In a memoir on  FOSSIL  Sessile Cirripedes, I have stated that, from
0305  almost superfluous to remark that hardly any   FOSSIL  fish are known from south of the equator; and
0319  n hardly ever tell. On the supposition of the  FOSSIL  horse still existing as a rare species, we mi
0319  y should not have perceived the fact, yet the  FOSSIL  horse would certainly have become rarer and r
0323  h american tertiary deposits. Even if the new  FOSSIL  species which are common to the Old and New W
0324  south America, and Australia, from containing  FOSSIL  remains in some degree allied, and from not i
0329  s of mammals; but Owen has discovered so many  FOSSIL  links, that he has had to alter the whole cla
0330  at in a perfectly natural classification many  FOSSIL  species would have to stand between living sp
0336  y of descent, the full meaning of the fact of  FOSSIL  remains from closely consecutive formations,
0339  ods. Mr. Clift many years ago showed that the  FOSSIL  mammals from the Australian caves were closel
0339  in the most striking manner that most of the   FOSSIL  mammals, buried there in such numbers, are re
0339  e clearly seen in the wonderful collection of  FOSSIL  bones made by MM. Lund and Clausen in the cav
0342  ganic beings have been largely preserved in a  FOSSIL  state; that the number both of specimens and
0368  he case, chiefly from the distribution of the  FOSSIL  Gnathodon), then the arctic and temperate pro
0463  such links? Why does not every collection of   FOSSIL  remains afford plain evidence of the gradatio
0464  who will pretend that in future ages so many   FOSSIL  links will be discovered, that naturalists wi
0464  ings of certain classes can be preserved in a  FOSSIL  condition, at least in any great number. Wide
0465  graduated manner. We clearly see this in the   FOSSIL  remains from consecutive formations invariabl
0475  neously throughout the world. The fact of the  FOSSIL  remains of each formation being in some degre
0476  si and thus we can see why the more ancient a  FOSSIL  is, the oftener it stands in some degree inte
0483  principle, in groups sub ordinate to groups.   FOSSIL  remains sometimes tend to fill up very wide i
0173  ous amount of future degradation; and such     FOSSILIFEROUS  masses can be accumulated only where much
0187  ve to descend far beneath the lowest known     FOSSILIFEROUS  stratum to discover the earlier stages, b
0279  heir sudden appearance in the lowest known     FOSSILIFEROUS  strata. In the sixth chapter I enumerated
0291  hallow and favourable for life, and thus a     FOSSILIFEROUS  formation thick enough, when upraised, to
0295  whole time. But we have seen that a thick      FOSSILIFEROUS  formation can only be accumulated during
0300  served. In our archipelago, I believe that    FOSSILIFEROUS  formations could be formed of sufficient
0300  stationary or rising; whilst rising, each      FOSSILIFEROUS  formation would be destroyed, almost as s
0306  oups of Allied Species in the lowest known     FOSSILIFEROUS  strata. There is another and allied diffi
0306  group, suddenly appear in the lowest known     FOSSILIFEROUS  rocks. Most of the arguments which have c
0307  understanding the absence of vast piles of     FOSSILIFEROUS  strata, which on my theory no doubt were
0310  st entire absence, as at present known, of     FOSSILIFEROUS  formations beneath the Silurian strata, a
0315  ; but as the accumulation of long enduring     FOSSILIFEROUS  formations depends on great masses of sed
0324  erei and if this be so, it is evident that     FOSSILIFEROUS  beds deposited at the present day on the
0327  reasons for believing that all our greater     FOSSILIFEROUS  formations were deposited during periods
0342  ce being necessary for the accumulation of     FOSSILIFEROUS  deposits thick enough to resist future de
0358  mits, of granite, metamorphic schists, old     FOSSILIFEROUS  or other such rocks, instead of consistin
0465  r by enormous blank intervals of time; for     FOSSILIFEROUS  formations, thick enough to resist future
0465  extinction. With respect to the absence of     FOSSILIFEROUS  formations beneath the lowest Silurian st
0487  intervals. The accumulation of each great     FOSSILIFEROUS  formation will be recognised as having de
0296  ases we have the plainest evidence in great    FOSSILISED  trees, still standing upright as they grew,
0107  ornithorhynchus and Lepidosiren, which, like   LIVING  FOSSILS, connect to a certain extent orders now wide
0107  e anomalous forms may almost be called living  FOSSILS: they have endured to the present day, from
0291  all our ancient formations, which are rich in  FOSSILS, have thus been formed during subsidence. Si
0291  wide spaces. Consequently formations rich in   FOSSILS and sufficiently thick and extensive to resi
0292  f subsidence, that our great deposits rich in  FOSSILS have been accumulated. Nature may almost be
0295  the average duration of life of the embedded   FOSSILS had been less than that of the glacial perio
0297  me great formation, we find that the embedded  FOSSILS, though almost universally ranked as specifi
0315  the amount of organic change exhibited by the  FOSSILS embedded in consecutive formations is not eq
0329  orms. But, as Buckland long ago remarked, all  FOSSILS can be classed either in still existing grou
0331  can understand the rule that the most ancient  FOSSILS differ most from existing forms. We must not
0333  er than they were before the discovery of the  FOSSILS. If, for instance, we suppose the existing g
0334  one instance, namely, the manner in which the  FOSSILS of the Devonian system, when this system was
0335  ct, insisted on by all palaeontologists, that  FOSSILS from two consecutive formations are far more
0335  e closely related to each other, than are the  FOSSILS from two remote formations. Pictet gives as
0341  ll living in South America; and some of these  FOSSILS may be the actual progenitors of living spec
0463  of the progenitors of the Silurian groups of  FOSSILS? For certainly on my theory such strata must
0465  e closely related to each other, than are the  FOSSILS from formations distant from each other in t
0476  degree intermediate in character between the   FOSSILS in the formations above and below, is simply
0486  nt, and which may fancifully be called living  FOSSILS, will aid us in forming a picture of the anc
0488  ows, that the amount of organic change in the  FOSSILS of consecutive formations probably serves as
0157  the two sexes of the same species: again in   FOSSORIAL  hymenoptera, the manner of neuration of the
0244  ch instincts as the young cuckoo ejecting its  FOSTER  brothers, ants making slaves, the larvae of i
0217  at the same time; then the old birds or the    FOSTERED  young would gain an advantage. And analogy w
0004  n all other perplexing cases I have invariably  FOUND  that our knowledge, imperfect though it be, o
0008  never seed! In some few such cases it has been  FOUND  out that very trifling changes, such as a lit
0018  still existing. Even if this latter fact were  FOUND  more strictly and generally true than appears
0020  e breeds is tolerably and sometimes (as I have  FOUND  with pigeons) extremely uniform, and everythi
0027  y, that C. livia, or the rock pigeon, has been  FOUND  capable of domestication in Europe and in Ind
```

Page ***************************************(Key Word)***************************************

Page **(Key Word)**

```
0035 e changed individuals of the same breed may be FOUND in less civilised districts, where the breed
0046 ance of an important part varying will ever be FOUND: but under any other point of view many insta
0047 ause the intermediate links have actually been FOUND, but because analogy leads the observer to su
0048 ies and by another as a variety, can rarely be FOUND within the same country, but are common in se
0050 on, varieties of it will almost universally be FOUND recorded. These varieties, moreover, will be
0054 and much diffused or dominant species will be FOUND on the side of the larger genera. This, again
0054 , consequently, we might have expected to have FOUND in the larger genera, or those including many
0056 ases in which intermediate links have not been FOUND between doubtful forms, naturalists are compe
0058 cely more than a truism, for if a variety were FOUND to have a wider range than that of its suppos
0062 e for life, or more difficult, at least I have FOUND it so, than constantly to bear this conclusio
0071 urished in the plantations, which could not be FOUND on the heath. The effect on the insects must
0072 king closely between the stems of the heath, I FOUND a multitude of seedlings and little trees, wh
0073 r. from experiments which I have tried, I have FOUND that the visits of bees, if not indispensable
0074 man says, near villages and small towns I have FOUND the nests of humble bees more numerous than e
0087 s: and though statements to this effect may be FOUND in works of natural history, I cannot find on
0093 not a grain of pollen can be detected. Having FOUND a female tree exactly sixty yards from a male
0095 nstructed proboscis. On the other hand, I have FOUND by experiment that the fertility of clover gr
0097 of bees to papilionaceous flowers, that I have FOUND, by experiments published elsewhere, that the
0099 d near each other, a large majority, as I have FOUND, of the seedlings thus raised will turn out m
0100 h worms: but these all pair. As yet I have not FOUND a single case of a terrestrial animal which f
0105 l number of the species inhabiting it, will be FOUND to be small, as we shall see in our chapter o
0113 herbage can thus be raised. The same has been FOUND to hold good when first one variety and then
0114 diversity in its inhabitants. For instance, I FOUND that a piece of turf, three feet by four in s
0130 nd which are known to us only from having been FOUND in a fossil state. As we here and there see a
0140 ing at different heights on the Himalaya, were FOUND in this country to possess different constitu
0140 nement, and not because they were subsequently FOUND capable of far extended transportation, I thi
0146 dolle has remarked that winged seeds are never FOUND in fruits which do not open: I should explain
0168 a district where many species of any genus are FOUND, that is, where there has been much former va
0179 ly evidence of their former existence could be FOUND only amongst fossil remains, which are preser
0201 ral selection. Although many statements may be FOUND in works on natural history to this effect, I
0201 ood and evil caused by each part, each will be FOUND on the whole advantageous. After the lapse of
0206 culties, can absolute perfection be everywhere FOUND. On the theory of natural selection we can cl
0208 ver the habitual train of thought: so P. Huber FOUND it was with a caterpillar, which makes a very
0210 instinct has been acquired, for these could be FOUND only in the lineal ancestors of each species,
0215 ry, sheep and pigs: and this tendency has been FOUND incurable in dogs which have been brought hom
0219 feeding their own larvae. When the old nest is FOUND inconvenient, and they have to migrate, it is
0219 uber to be a slave making ant. This species is FOUND in the southern parts of England, and its hab
0220 . I opened fourteen nests of F. sanguinea, and FOUND a few slaves in all. Males and fertile female
0220 s and fertile females of the slave species are FOUND only in their own proper communities, and hav
0222 rociously attack other ants. In one instance I FOUND to my surprise an independent community of F.
0222 visited another community of F. sanguinea, and FOUND a number of these ants entering their nest, c
0229 short time, and again examined the cell, and I FOUND that the rhombic plate had been, completed, an
0232 ayer of melted vermilion wax; and I invariably FOUND that the colour was most delicately diffused
0233 mr. Tegetmeier that it has been experimentally FOUND that no less than from twelve to fifteen poun
0246 he cuts the knot, for in ten cases in which he FOUND two forms, considered by most authors as dist
0247 twice succeeded in getting fertile seed: as he FOUND the common red and blue pimpernels (Anagallis
0250 y their own pollen. For these plants have been FOUND to yield seed to the pollen of a distinct spe
0250 len, notwithstanding that their own pollen was FOUND to be perfectly good, for it fertilised disti
0256 ybrids, for their degree of fertility is often FOUND to differ greatly in the several individuals
0257 species of almost any other genus: but Gartner FOUND that N. acuminata, which is not a particularl
0258 certain sea weeds or Fuci. Gartner, moreover, FOUND that this difference of facility in making re
0262 h on certain varieties of the plum. As Gartner FOUND that there was sometimes an innate difference
0262 thing analogous occurs in grafting; for Thouin FOUND that three species of Robinia, which seeded f
0268 ties: for if two hitherto reputed varieties be FOUND in any degree sterile together, they are at o
0271 t trial, namely, by reciprocal crosses, and he FOUND their mongrel offspring perfectly fertile. Bu
0280 t, on my theory, have formerly existed. I have FOUND it difficult, when looking at any two species
0288 pecies, which inhabits deep water and has been FOUND fossil in Sicily, whereas not one other speci
0288 hereas not one other species has hitherto been FOUND in any tertiary formation: yet it is now know
0290 ith recent or tertiary remains can anywhere be FOUND, though the supply of sediment must for ages
0296 been preserved: thus, Messrs. Lyell and Dawson FOUND carboniferous beds 1400 feet thick in Nova Sc
0297 and C to be two species, and a third, A, to be FOUND in an underlying bed: even if A were strictly
0298 sely allied to each other than are the species FOUND in more widely separated formations; but to t
0301 and these links, let them be ever so close, if FOUND in different stages of the same formation, wo
0302 cause certain genera or families have not been FOUND beneath a certain stage, that they did not ex
0304 nus, of which not one specimen has as yet been FOUND even in any tertiary stratum. Hence we now po
0313 tertiary beds a few living shells may still be FOUND in the midst of a multitude of extinct forms.
0313 ly related formations, all the species will be FOUND to have undergone some change. When a species
0317 geological formations in which the species are FOUND, the line will sometimes falsely appear to be
0318 ertical line of varying thickness, the line is FOUND to taper more gradually at its upper end, whi
0318 xtinction of species, than I have done. When I FOUND in La Plata the tooth of a horse embedded wit
0322 a fragment of the mineral chalk itself can be FOUND; namely, in North America, in equatorial Sout
0323 sculpture. Moreover other forms, which are not FOUND in the Chalk of Europe, but which occur in th
0324 from not including those forms which are only FOUND in the older underlying deposits, would be co
0339 pieces of armour like those of the armadillo, FOUND in several parts of La Plata; and Professor O
0341 t if six genera, each having eight species, be FOUND in one geological formation, and in the next
0342 the closely allied or representative species, FOUND in the several stages of the same great forma
0347 nsiderably different climate, and they will be FOUND incomparably more closely related to each oth
0349 and not by a true ostrich or emeu, like those FOUND in Africa and Australia under the same latitu
0352 several distant and isolated points, where now FOUND. Nevertheless the simplicity of the view that
0358 njurious action of sea water. To my surprise I FOUND that out of 87 kinds, 64 germinated after an
0362 the germination of seeds: now after a bird has FOUND and devoured a large supply of food, it is po
0363 on these islands, and he answered that he had FOUND large fragments of granite and other rocks, w
0373 al action in New Zealand: and the same plants, FOUND on widely separated mountains in this island,
0374 tains of Brazil, some few European genera were FOUND by Gardner, which do not exist in the wide in
0374 of Caraccas the illustrious Humboldt long ago FOUND species belonging to genera characteristic of
0375 ns, some few representative European forms are FOUND, which have not been discovered in the intert
```

Page **(Key Word)**

Page **(Key Word)**

```
0375  e same time representing plants of Europe, not  FOUND  in the intervening hot lowlands. A list of th
0375  am informed by Dr. Hooker, of European genera,  FOUND  in Australia, but not in the intermediate tor
0376  o new Zealand and to Europe, but have not been  FOUND  in the intermediate tropical seas. It should
0376  e observed that the northern species and forms  FOUND  in the southern parts of the southern hemisph
0385  ova of fresh water shells were hatching: and I  FOUND  that numbers of the extremely minute and just
0386  e the greatest wanderers; and are occasionally  FOUND  on the most remote and barren islands in the
0387  quite inexplicable; but Audubon states that he  FOUND  the seeds of the great southern water lily (p
0390  the proportion of endemic species (i.e. those  FOUND  nowhere else in the world) is often extremely
0390  in the Galapagos Archipelago, with the number  FOUND  on any continent, and then compare the area o
0393  rachians (frogs, toads, newts) have never been  FOUND  on any of the many islands with which the gre
0393  ken pains to verify this assertion, and I have  FOUND  it strictly true. I have, however, been assur
0393  have not finished my search: as yet I have not  FOUND  a single instance, free from doubt, of a terr
0394  t every island. New Zealand possesses two bats  FOUND  nowhere else in the world: Norfolk Island, th
0394  the same species have enormous ranges, and are  FOUND  on continents and on far distant islands. Hen
0397  y by endemic species, but sometimes by species  FOUND  elsewhere. Dr. Aug. A. Gould has given severa
0397  across moderately wide arms of the sea. And I  FOUND  that several species did in this state withst
0400  bitants of the Galapagos Archipelago which are  FOUND  in other parts of the world (laying on one si
0401  . nevertheless a good many species, both those  FOUND  in other parts of the world and those confine
0404  given. And it will, I believe, be universally  FOUND  to be true, that wherever in two regions, let
0404  entative species occur, there will likewise be  FOUND  some identical species, showing, in accordanc
0404  ny closely allied species occur, there will be  FOUND  many forms which some naturalists rank as dis
0411  the first dawn of life, all organic beings are  FOUND  to resemble each other in descending degrees,
0417  it should be ranked. Hence, also, it has been  FOUND, that a classification founded on any single
0418  st. Hilaire. If certain characters are always  FOUND  correlated with others, though no apparent bo
0418  inc it, or those for propagating the race, are  FOUND  nearly uniform, they are considered as highly
0418  ll these, the most important vital organs, are  FOUND  to offer characters of quite subordinate valu
0419  crustaceans, such definition has hitherto been  FOUND  impossible. There are crustaceans at the oppo
0423  ickened stems are so similar. Whatever part is  FOUND  to be most constant, is used in classing vari
0426  s, in another group or section of a group, are  FOUND  to differ much, we at once value them less in
0445  the old dogs and their six days old puppies, I  FOUND  that the puppies had not nearly acquired thei
0451  ten have a rudiment of a pistil; and Kolreuter  FOUND  that by crossing such male plants with an her
0452  s to expect to find, and which is occasionally  FOUND  in monstrous individuals of the species. Thus
0455  assification, how it is that systematists have  FOUND  rudimentary parts as useful as, or even somet
0461  t and isolated parts of the world they are now  FOUND, they must in the course of successive genera
0477  ich can traverse the ocean, should so often be  FOUND  on islands far distant from any continent. Su
0060  l marked varieties, though necessary as the  FOUNDATION  for the work, helps us but little in unders
0488  searches. Psychology will be based on a new  FOUNDATION, that of the necessary acquirement of each
0232  the slip, in this case the bees can lay the  FOUNDATIONS  of one wall of a new hexagon, in its stric
0003  nevertheless, such a conclusion, even if well  FOUNDED, would be unsatisfactory, until it could be
0146  e central flowers, that the elder De Candolle  FOUNDED  his main divisions of the order on analogous
0266  ts. It is an old and almost universal belief,  FOUNDED, I think, on a considerable body of evidence
0274  this conclusion, as far as I can make out, is  FOUNDED  on a single experiment; and seems directly o
0297  slight differences many palaeontologists have  FOUNDED  their species: and they do this the more rea
0417  lso, it has been found, that a classification  FOUNDED  on any single character, however important t
0417  s gives the characters; for this saying seems  FOUNDED  on an appreciation of many trifling points o
0417  our classifications are sometimes necessarily  FOUNDED. Practically when naturalists are at work, t
0418  ts, of which the two main divisions have been  FOUNDED  on characters derived from the embryo, on th
0420  yself, on the view that the natural system is  FOUNDED  on descent with modification; that the chara
0270  entised on, are ranked by Sagaret, who mainly  FOUNDS  his classification by the test of infertility
0018  xisted in the valley of the Nile thirteen or  FOURTEEN  thousand years ago, and who will pretend to
0021  en forty tail feathers, instead of twelve or  FOURTEEN, the normal number in all members of the gre
0039  aps the parent bird of all fantails had only  FOURTEEN  tailfeathers somewhat expanded, like the pre
0118  er a condensed and simplified form up to the  FOURTEEN  thousandth generation. But I must here remar
0121  resented between the horizontal lines. After  FOURTEEN  thousand generations, six new species, marke
0122  es of the genus. Their modified descendants,  FOURTEEN  in number at the fourteen thousandth generat
0122  ified descendants, fourteen in number at the  FOURTEEN  thousandth generation, will probably have in
0122  ecies will have transmitted offspring to the  FOURTEEN  thousandth generation. We may suppose that o
0124  e. in this case, its affinities to the other  FOURTEEN  new species will be of a curious and circuit
0159  ore distinct races. The frequent presence of  FOURTEEN  or even sixteen tail feathers in the pouter,
0220  myself made, in some little detail. I opened  FOURTEEN  nests of F. sanguinea, and found a few slave
0362  r having been thus retained for two days and  FOURTEEN  hours. Fresh water fish, I find, eat seeds o
0397  rmed a new membranous one, I immersed it for  FOURTEEN  days in sea water, and it recovered and craw
0005  lection will be treated at some length in the  FOURTH  chapter; and we shall then see how Natural Se
0208  tage, the caterpillar simply re performed the  FOURTH, fifth, and sixth stages of construction. If
0250  sed by Herbert with their own pollen, and the  FOURTH  was subsequently fertilised by the pollen of
0267  eve, indeed, from the facts alluded to in our  FOURTH  chapter, that a certain amount of crossing is
0331  uest the reader to turn to the diagram in the  FOURTH  chapter. We may suppose that the numbered let
0411  roup have different habits. In our second and  FOURTH  chapters, on Variation and on Natural Selecti
0420  he trouble of referring to the diagram in the  FOURTH  chapter. We will suppose the letters A to L t
0431  ral class. Extinction, as we have seen in the  FOURTH  chapter, has played an important part in defi
0005  ertility of varieties when intercrossed; and  FOURTHLY, the imperfection of the Geological Record.
0027  obvious when we come to treat of selection.  FOURTHLY, pigeons have been watched, and tended with
0172  the discoveries of profound mathematicians?  FOURTHLY, how can we account for species, when crosse
0017  as the little variability of the ass or guinea  FOWL, or the small power of endurance of warmth by
0019  try have proceeded from the common wild Indian  FOWL (Gallus bankiva). In regard to ducks and rabbi
0152  at part (for instance, the comb in the Dorking  FOWL) or the whole breed will cease to have a nearl
0090  orn like protuberances in the cocks of certain  FOWLS, etc.) which we cannot believe to be either u
0215  le instance of this is seen in those breeds of  FOWLS  which very rarely or never become broody, tha
0035  eved, been chiefly effected by crosses with the  FOX  hound; but what concerns us is, that the chang
0393  and Islands, which are inhabited by a wolf like  FOX, come nearest to an exception; but this group
0017  and natural species, for instance, of the many  FOXES, inhabiting different quarters of the world.
0215  has become instinctive in the dog. All wolves,  FOXES, jackals, and species of the cat genus, when
0268  tz cog unites more easily than other dogs with  FOXES, or that certain South American indigenous do
0394  shores, and they may have formerly transported  FOXES, as so frequently now happens in the arctic r
0304  rtiary beds: from the ease with which even a  FRAGMENT  of a valve can be recognised; from all these
0322  der the most different climates, where not a  FRAGMENT  of the mineral chalk itself can be found; na
```

Page **(Key Word)**

```
0488 y us, will hereafter be recognised as a mere FRAGMENT of time, compared with the ages which have e
0289 te that our evidence from fossil remains is FRAGMENTARY in an extreme degree. For instance, not a
0222 ese little yellow ants still clinging to the FRAGMENTS of the nest. This species is sometimes, tho
0283 st the base of the cliff is undermined, huge FRAGMENTS fall down, and these remaining fixed, have
0286 e more rapid from the breakage of the fallen FRAGMENTS. On the other hand, I do not believe that a
0363 nds, and he answered that he had found large FRAGMENTS of granite and other rocks, which do not oc
0373 to America; in the northern half, ice borne FRAGMENTS of rock have been observed on the eastern s
0208 , i think, a remarkably accurate notion of the FRAME of mind under which an instinctive action is
0220 i tried to approach the subject in a sceptical FRAME of mind, as any one may well be excused for d
0213 of the oddest tricks, associated with certain FRAMES of mind or periods of time. But let us look t
0435 ation there will be no tendency to alter the FRAMEWORK of bones or the relative connexion of the s
0479 er slight their vital importance may be. The FRAMEWORK of bones being the same in the hand of a ma
0019 itain has now hardly one peculiar mammal, and FRANCE but few distinct from those of Germany and co
0048 mpare the several floras of Great Britain, of FRANCE or of the United States, drawn up by differen
0328 memoirs on the eocene deposits of England and FRANCE, is able to draw a close general parallelism
0328 pares certain stages in England with those in FRANCE, although he finds in both a curious accordan
0362 is had to give up flying carrier pigeons from FRANCE to England, as the hawks on the English coast
0417 lassification. But when Aspicarpa produced in FRANCE, during several years, only degraded flowers,
0208 in animals very low in the scale of nature. FREDERICK Cuvier and several of the older metaphysici
0015 nd whilst kept in a considerable body, so that FREE intercrossing might check, by blending togethe
0040 name. In semi civilised countries, with little FREE communication, the spreading and knowledge of
0082 eserved; and natural selection would thus have FREE scope for the work of improvement. We have rea
0102 breeder selects for some definite object, and FREE intercrossing will wholly stop his work. But w
0108 the process will often be greatly retarded by FREE intercrossing. Many will exclaim that these se
0134 d that such modifications are inherited. Under FREE nature, we can have no standard of comparison,
0141 far wider range than any other rodent, living FREE under the cold climate of Faroe in the north a
0150 ucture. Thus rudimentary parts are left to the FREE play of the various laws of growth, to the eff
0190 the water, at the same time that they breathe FREE air in their swimbladders, this latter organ h
0337 e animals and plants of Great Britain were set FREE in New Zealand, that in the course of time a m
0337 if all the productions of New Zealand were set FREE in Great Britain, whether any considerable num
0347 is, that barriers of any kind, or obstacles to FREE migration, are related in a close and importan
0347 ghtly different climate, there might have been FREE migration for the northern temperate forms, as
0393 chi as yet I have not found a single instance, FREE from doubt, of a terrestrial mammal (excluding
0402 invading each other's territory, when put into FREE intercommunication. Undoubtedly if one species
0485 eering prospect; but we shall have at least be FREED from the vain search for the undiscovered and
0097 ccasional cross be indispensable, the fullest FREEDOM for the entrance of pollen from another indi
0370 f the circumpolar land, and to the consequent FREEDOM for intermigration under a more favourable c
0008 things more difficult than to get it to breed FREELY under confinement, even in the many cases whe
0009 rom the tropics, breed in this country pretty FREELY under confinement, with the exception of the
0009 ugh often weak and sickly, yet breeding quite FREELY under confinement; and when, on the other han
0009 y add, that as some organisms will breed most FREELY under the most unnatural conditions (for inst
0013 , less strange and commoner deviations may be FREELY admitted to be inheritable. Perhaps the corre
0019 tc., so unlike all wild Canidae, ever existed FREELY in a state of nature? It has often been loose
0024 ost difficult to get any wild animal to breed FREELY under domestication; yet on the hypothesis of
0027 or eight supposed species of pigeons to breed FREELY under domestication; these supposed species b
0040 ee of variability is obviously favourable, as FREELY giving the materials for selection to work on
0041 favourable conditions of life, so as to breed FREELY in that country. When the individuals of any
0042 a very quick rate, and inferior birds may be FREELY rejected, as when killed they serve for food.
0068 om the other species being allowed to grow up FREELY. The amount of food for each species of cours
0081 which new and better adapted forms could not FREELY enter, we should then have places in the econ
0098 ing close by, which is visited by bees, seeds FREELY. In very many other cases, though there be no
0114 es. In an extremely small area, especially if FREELY open to immigration, and where the contest be
0174 ormation of new species, more especially with FREELY crossing and wandering animals. In looking at
0186 ve been formed by natural selection, seems, I FREELY confess, absurd in the highest possible degre
0245 pt distinct had they been capable of crossing FREELY. The importance of the fact that hybrids are
0251 maturity, and bore good seed, which vegetated FREELY. In a letter to me, in 1839, Mr. Herbert told
0251 been crossed, yet many of these hybrids seed FREELY. For instance, Herbert asserts that a hybrid
0252 nd catawbiense, and that this hybrid seeds as FREELY as it is possible to imagine. Had hybrids, wh
0252 als of the same hybrid variety are allowed to FREELY cross with each other, and the injurious infl
0252 e in mind that, owing to few animals breeding FREELY under confinement, few experiments have been
0252 , but as not one of these nine species breeds FREELY in confinement, we have no right to expect th
0254 everal aboriginal species would at first have FREELY bred together and have produced quite fertile
0261 the male sexual element of the one will often FREELY act on the female sexual element of the other
0262 d that three species of Robinia, which seeded FREELY on their own roots, and which could be grafte
0262 astrum, Lobelia, etc., which seeded much more FREELY when fertilized with the pollen of distinct s
0265 ill breed under confinement or any plant seed FREELY under culture; nor can he tell, till he tries
0305 hat the seas of the world have always been so FREELY open from south to north as they are at prese
0357 to the several intervening oceanic islands. I FREELY admit the former existence of many islands, n
0382 th greater force from the north so as to have FREELY inundated the south. As the tide leaves its d
0192 nute folds of skin, called by me the ovigerous FRENA, which serve, through the means of a sticky s
0192 face of the body and sack, including the small FRENA, serving for respiration. The Balanidae or se
0192 rripedes, on the other hand, have no ovigerous FRENA, the eggs lying loose at the bottom of the sa
0192 i think no one will dispute that the ovigerous FRENA in the one family are strictly homologous wit
0192 of skin, which originally served as ovigerous FRENA, but which, likewise, very slightly aided the
0074 rvention first of mice and then of bees, the FREQUENCY of certain flowers in that district! In the
0131 greater variability, as well as the greater FREQUENCY of monstrosities, under domestication or cu
0008 m strongly inclined to suspect that the most FREQUENT cause of variability may be attributed to th
0043 bility both of hybrids and mongrels, and the FREQUENT sterility of hybrids; but the cases of plant
0070 ction. I should add that the good effects of FREQUENT intercrossing, and the ill effects of close
0075 he individuals of the same species, for they FREQUENT the same districts, require the same food, a
0136 ups have habits of life almost necessitating FREQUENT flight; these several considerations have ma
0137 inflammation of the nictitating membrane. As FREQUENT inflammation of the eyes must be injurious t
0140 often overrated. We may infer this from our FREQUENT inability to predict whether or not an impor
0158 hich the species possess in common; that the FREQUENT extreme variability of any part which is dev
0159 ariations in two or more distinct races. The FREQUENT presence of fourteen or even sixteen tail fe
0173 r and rarer, the other becomes more and more FREQUENT, till the one replaces the other. But if we
```

Page ***(Key Word)**

```
0249  e kind, their fertility, notwithstanding the  FREQUENT    ill effects of manipulation, sometimes decid
0264  that the early death of the embryo is a very   FREQUENT    cause of sterility in first crosses. I was a
0267  s acted on by farmers and gardeners in their   FREQUENT    exchanges of seed, tubers, etc, from one soi
0289  that this could nowhere be ascertained. The    FREQUENT    and great changes in the mineralogical compo
0292  y almost be said to have guarded against the   FREQUENT    discovery of her transitional or linking for
0383  manner highly useful to them, for short and    FREQUENT    migrations from pond to pond, or from stream
0386  feet and beaks of birds. Wading birds, which   FREQUENT    the muddy edges of ponds, if suddenly flushe
0396  ed by deeper channels, we can understand the   FREQUENT    relation between the depth of the sea and th
0071  to be seen on the heath; and the heath was     FREQUENTED  by two or three distinct insectivorous bird
0091  nhabiting a mountainous district, and those    FREQUENTING the lowlands, would naturally be forced to
0113  me inhabiting new stations, climbing trees,    FREQUENTING water, and some perhaps becoming less carn
0197  as a beautiful adaptation to hide this tree    FREQUENTING bird from its enemies; and consequently th
0005  y survive; and as, consequently, there is a    FREQUENTLY  recurring struggle for existence, it follow
0038  es. We can, I think, further understand the    FREQUENTLY  abnormal character of our domestic races, a
0045  d individual differences, such as are known    FREQUENTLY  to appear in the offspring from the same pa
0045  may be presumed to have thus arisen, being     FREQUENTLY  observed in the individuals of the same spe
0068  limit to which each can increase: but very     FREQUENTLY  it is not the obtaining food, but the servi
0076  caused the decrease of the song thrush. How    FREQUENTLY  we hear of one species of rat taking the pl
0091  r more bulky, with shorter legs, which more    FREQUENTLY  attacks the shepherd's flocks. Let us now t
0135  beetles in many parts of the world are very    FREQUENTLY  blown to sea and perish; that the beetles i
0137  , who had often caught them, that they were    FREQUENTLY  blind; one which I kept alive was certainly
0144  e nature of the bond of correlation is very    FREQUENTLY  quite obscure. M. Is. Geoffroy St. Hilaire
0144  emarked, that certain malconformations very    FREQUENTLY, and that others rarely coexist, without ou
0145  heads that the inner and outer flowers most    FREQUENTLY  differ. It might have been thought that the
0152  ult to breed them nearly to perfection, and    FREQUENTLY  individuals are born which depart widely fr
0226  eres being nearly of the same size, is very    FREQUENTLY  and necessarily the case, the three flat su
0275  s occur with hybrids: yet I grant much less    FREQUENTLY  with hybrids than with mongrels. Looking to
0277  s closely allied to that sterility which so    FREQUENTLY  affects pure species, when their natural co
0305  ertiary and existing species. The case most   FREQUENTLY  insisted on by palaeontologists of the appa
0333  evidence of our best palaeontologists seems   FREQUENTLY  to be the case. Thus, on the theory of desc
0350  will be exposed to new conditions, and will   FREQUENTLY  undergo further modification and improvemen
0361  s of trees, small parcels of earth are very   FREQUENTLY  enclosed in their interstices and behind th
0361  seeds. I could give many facts showing how    FREQUENTLY  birds of many kinds are blown by gales to v
0362  eds of many land and water plants: fish are   FREQUENTLY  devoured by birds, and thus the seeds might
0394  may have formerly transported foxes, as so    FREQUENTLY  now happens in the arctic regions. Yet it c
0415  tion. Any one of these characters singly is   FREQUENTLY  of more than generic importance, though her
0438  animal and vegetable kingdoms. Naturalists    FREQUENTLY  speak of the skull as formed of metamorphos
0480  wing covers of some beetles; should thus so   FREQUENTLY  bear the plain stamp of inutility! Nature m
0054  l here allude to only two causes of obscurity. FRESH      water and salt loving plants have generally v
0107  bles the extinct tertiary flora of Europe. All FRESH      water basins, taken together, make a small ar
0107  ndi and, consequently, the competition between FRESH      water productions will have been less severe
0107  d forms more slowly exterminated. And it is in  FRESH      water that we find seven genera of Ganoid fis
0107  remnants of a once proponderant order: and in   FRESH      water we find some of the most anomalous form
0114  l and uniform islets: and so in small ponds of  FRESH      water. Farmers find that they can raise most
0120  al stocks, without either having given off any  FRESH      branches or races. After ten thousand generat
0130  tected station. As buds give rise by growth to  FRESH      buds, and these, if vigorous, branch out and
0231  ways piling up the cut away cement, and adding  FRESH      cement, on the summit of the ridge. We shall
0282  sea at work grinding down old rocks and making  FRESH      sediment, before he can hope to comprehend an
0287  say three hundred million years. The action of  FRESH      water on the gently inclined Wealden district
0321  tinct group of Ganoid fishes still inhabit our  FRESH      waters. Therefore the utter extinction of a g
0323  dge whether the productions of the land and of  FRESH      water change at distant points in the same pa
0354  e following chapter), the wide distribution of  FRESH      water productions; and thirdly, the occurrenc
0359  ese might be dried on the banks, and then by a  FRESH      rise in the stream be washed into the sea. He
0362  thus retained for two days and fourteen hours.  FRESH      water fish, I find, eat seeds of many land an
0383  hical Distribution, continued. Distribution of  FRESH      water productions. On the inhabitants of ocea
0383  riers of land, it might have been thought that  FRESH      water productions would not have ranged widel
0383  ase is exactly the reverse. Not only have many  FRESH      water species, belonging to quite different c
0383  i well remember, when first collecting in the   FRESH      waters of Brazil, feeling much surprise at th
0383  feeling much surprise at the similarity of the  FRESH      water insects, shells, etc., and at the dissi
0383  pared with those of Britain. But this power in  FRESH      water productions of ranging widely, though s
0384  lieve that the same species never occur in the  FRESH      waters of distant continents. But on the same
0384  ut I am inclined to attribute the dispersal of  FRESH      water fish mainly to slight changes within th
0384  n the surface was peopled by existing land and  FRESH      water shells. The wide difference of the fish
0384  o this same conclusion. With respect to allied  FRESH      water fish occurring at very distant points o
0384  which cannot at present be explained: but some  FRESH      water fish belong to very ancient forms, and
0384  can with care be slowly accustomed to live in   FRESH      water; and, according to Valenciennes, there
0384  single group of fishes confined exclusively to  FRESH      water, so that we may imagine that a marine m
0384  that we may imagine that a marine member of a   FRESH      water group might travel far along the shores
0385  ubsequently become modified and adapted to the  FRESH      waters of a distant land. Some species of fre
0385  resh waters of a distant land. Some species of  FRESH      water shells have a very wide range, and alli
0385  ave quite unintentionally stocked the one with  FRESH      water shells from the other. But another agen
0385  atural pond, in an aquarium, where many ova of  FRESH      water shells were hatching; and I found that
0386  t a Dyticus has been caught with an Ancylus (a  FRESH      water shell like a limpet) firmly adhering to
0386  has long been known what enormous ranges many   FRESH      water and even marsh species have, both over
0386  nd, they would be sure to fly to their natural  FRESH      water haunts. I do not believe that botanists
0387  if water birds did not transport the seeds of   FRESH      water plants to vast distances, and if consec
0387  into play with the eggs of some of the smaller  FRESH      water animals. Other and unknown agencies pro
0387  ly have also played a part. I have stated that  FRESH      water fish eat some kinds of seeds, though th
0388  hould, also, remember that some, perhaps many,  FRESH      water productions are low in the scale of nat
0388  cies having formerly ranged as continuously as  FRESH      water productions ever can range, over immens
0388  rmediate regions. But the wide distribution of  FRESH      water plants and of the lower animals, whethe
0388  seeds and eggs by animals, more especially by   FRESH      water birds, which have large powers of fligh
0404  of butterflies and beetles. So it is with most  FRESH      water productions, in which so many genera ra
0407  t some little length the means of dispersal of  FRESH      water productions. If the difficulties be not
0471  canon of Natura non facit saltum, which every   FRESH      addition to our knowledge tends to make more
0362  ly get scattered. Mr. Brent informs me that a   FRIEND     of his had to give up flying carrier pigeons
```

Page ***(Key Word)**

```
0057  should he ranked as species or varieties. Now FRIES has remarked in regard to plants, and Westwoc
0057  sub genera, or sections, or lesser groups. As FRIES has well remarked, little groups of species a
0185  water; and no one except Audubon has seen the FRIGATE bird, which has all its four toes webbed, al
0185  in function, though not in structure. In the  FRIGATE bird, the deeply scooped membrane between th
0186  im no surprise that there should be geese and  FRIGATE birds with webbed feet, either living on the
0200  the webbed feet of the upland goose or of the  FRIGATE bird are of special use to these birds; we c
0200  the progenitor of the upland goose and of the  FRIGATE bird, webbed feet no doubt were as useful as
0077  ed of the dandelion, and in the flattened and  FRINGED legs of the water beetle, the relation seems
0199  on copious details my reasoning would appear   FRIVOLOUS. The foregoing remarks lead me to say a fe
0393  tly true. I have, however, been assured that a FROG exists on the mountains of the great island of
0440  ch is hatched in a nest, and in the spawn of a FROG under water. We have no more reason to believe
0393  t. vincent long ago remarked that Batrachians  (FROGS, toads, newts) have never been found on any o
0393  hrough glacial agency. This general absence of FROGS, toads, and newts on so many oceanic islands
0393  peculiarly well fitted for these animals; for  FROGS have been introduced into Madeira, the Azores
0477  ls which cannot cross wide spaces of ocean, as FROGS and terrestrial mammals, should not inhabit o
0143  of the body varying in the same manner; in the FRONT and hind legs, and even in the jaws and limbs
0182  icropterus of Eyton); as fins in the water and FRONT legs on the land, like the penguin; as sails,
0198  d then by the law of homologous variation, the FRONT limbs and even the head would probably be aff
0377  ring state and could not have presented a firm FRONT against intruders, that a certain number of t
0142  that a very large proportion are destroyed by  FROST, and then collect seed from the few survivors
0180  ish, but during the long winter it leaves the  FROZEN waters, and preys like other polecats on mice
0373  hward to the Pyrenees. We may infer, from the  FROZEN mammals and nature of the mountain vegetation
0097  , on the other hand, have their organs of      FRUCTIFICATION closely enclosed, as in the great papilio
0010  the act of generation. Seedlings from the same FRUIT, and the young of the same litter, sometimes
0033  rs of the same varieties; and the diversity of FRUIT of the same species in the orchard, in compar
0033  se are, and how alike the leaves; how much the FRUIT of the different kinds of gooseberries differ
0033  ons, either in the leaves, the flowers, or the FRUIT, will produce races differing from each other
0037  ears, from Pliny's description, to have been a FRUIT of very inferior quality. I have seen great s
0037  hey could procure, never thought what splendid FRUIT we should eat; though we owe our excellent fr
0037  uit we should eat; though we owe our excellent FRUIT, in some small degree, to their having natura
0041  lants with slightly larger, earlier, or better FRUIT, and raised seedlings from them, and again pi
0063  metaphorically be said to struggle with other  FRUIT bearing plants, in order to tempt birds to de
0085  test trace of black. In plants the down of the FRUIT and the colour of the flesh are considered by
0085  a smooth or downy, a yellow or purple fleshed  FRUIT, should succeed. In looking at many small poi
0142  ers: this is very strikingly shown in works on FRUIT trees published in the United States, in whic
0147  ; and this process could not possibly go on in FRUIT which did not open. The elder Geoffroy and Go
0147  the seeds in our fruits become atrophied, the  FRUIT itself gains largely in size and quality. In
0184  ca there are woodpeckers which feed largely on FRUIT, and others with elongated wings which chase
0195  most trifling characters, such as the down on  FRUIT and the colour of the flesh, which, from dete
0257  part of the flower, even in the pollen, in the FRUIT, and in the cotyledons, can be crossed. Annua
0262  rafted on other species, yielded twice as much FRUIT as when on their own roots. We are reminded b
0359  efly tried small seeds, without the capsule or FRUIT; and as all of these sank in a few days, they
0359  o dry stems in branches of 94 plants with ripe FRUIT, and to place them on sea water. The majority
0359  sion of 28 days; and as 18/94 plants with ripe FRUIT (but not all the same species as in the foreg
0360  previously dry the plants or branches with the FRUIT: and this, as we have seen, would have caused
0360  is interesting; as plants with large seeds or  FRUIT could hardly be transported by any other mean
0361  gh the intestines of a bird: but hard seeds of FRUIT will pass uninjured through even the digestiv
0423  the pine apple together, merely because their  FRUIT, though the most important part, happens to b
0085  ing, that in the United States smooth skinned  FRUITS suffer far more from a beetle, a curculio, th
0146  remarked that winged seeds are never found in  FRUITS which do not open: I should explain the rule
0146  e winged through natural selection, except in  FRUITS which opened; so that the individual plants p
0147  y of oil bearing seeds. When the seeds in our  FRUITS become atrophied, the fruit itself gains larg
0359  he salt water. Afterwards I tried some larger  FRUITS, capsules, etc., and some of these floated fo
0360  different from mine; but he chose many large   FRUITS and likewise seeds from plants which live nea
0360  would then germinate. The fact of the larger   FRUITS often floating longer than the small, is inte
0251  plicated a manner the species of Pelargonium,  FUCHSIA, Calceolaria, Petunia, Rhododendron, etc., h
0258  served the same fact with certain sea weeds or FUCI. Gartner, moreover, found that this difference
0264  the case with some of Thuret's experiments on  FUCI. No explanation can be given of these facts, a
0381  ly remote as Kerguelen Land, New Zealand, and  FUEGIA, I believe that towards the close of the Glac
0018  ent periods, savages, like those of Tierra del FUEGO or Australia, who possess a semi domestic dog
0030  n animals even by the barbarians of Tierra del FUEGO, by their killing and devouring their old wom
0184  f birds, yet in the quiet Sounds of Tierra del FUEGO, the Puffinuria berard?, in its general habit
0215  as puppies from countries; such as Tierra del  FUEGO and Australia, where the savages do not keep
0323  ca, in equatorial South America, in Tierra del FUEGO, at the Cape of Good Hope, and in the peninsu
0374  nd fifty of the flowering plants of Tierra del FUEGO, forming no inconsiderable part of its scanty
0378  bout forty six in number, common to Tierra del FUEGO and to Europe still exist in North America, w
0202  stinging be useful to the community; it will   FULFIL all the requirements of natural selection, th
0073  have occasion to show that the exotic Lobelia  FULGENS, in this part of England, is never visited b
0098  from its own flower: for instance, in Lobelia  FULGENS, there is a really beautiful and elaborate c
0033  covery. I could give several references to the FULL acknowledgement of the importance of the princi
0066  ecome extinct. It would suffice to keep up the FULL number of a tree, which lived on an average fo
0079  is struggle, we may console ourselves with the FULL belief, that the war of nature is not incessan
0087  hing. Now, if nature had to make the beak of a FULL grown pigeon very short for the bird's own adv
0093  y trees bear only female flowers; these have a FULL sized pistil, and four stamens with shrivelled
0113  ted in any country has long ago arrived at its FULL average. If its natural powers of increase be
0152  l selection either has not or cannot come into FULL play, and thus the organisation is left in a f
0164  tripe is much commoner in the foal than in the FULL grown animal. Without here entering on further
0180  wide and with the skin on their flanks rather  FULL, to the so called flying squirrels; and flying
0206  atural selection we can clearly understand the FULL meaning of that old canon in natural history,
0280  t every geological formation and every stratum FULL of such intermediate links? Geology assuredly
0336  e been affected. On the theory of descent, the FULL meaning of the fact of fossil remains from clo
0338  ay be true, and yet it may never be capable of FULL proof. Seeing, for instance, that the oldest k
0382  of almost every land, which serve as a record, FULL of interest to us, of the former inhabitants o
0406  we make due allowance for our ignorance of the FULL effects of all the changes of climate and of t
0413  genus, and then by adding a single sentence, a FULL description is given of each kind of dog. The
0418  than that of the adult, which alone plays its  FULL part in the economy of nature. Yet it has been
```

Page ***(Key Word)***

```
0444  e silk moth; or, again, in the horns of almost  FULL  grown cattle. But further than this, variation
0445  that the puppies had not nearly acquired their  FULL  amount of proportional difference. So, again,
0445  f cart and race horses differed as much as the  FULL  grown animals: and this surprised me greatly,
0445  that the colts have by no means acquired their  FULL  amount of proportional difference. As the evid
0445  eral points were incomparably less than in the  FULL  grown birds. Some characteristic points of dif
0446  been acquired earlier or later in life, if the  FULL  grown animal possesses them. And the cases jus
0447  ffect the mature animal, which has come to its  FULL  powers of activity and has to gain its own liv
0451  losely in all respects, one of which will have  FULL  sized wings, and another mere rudiments of mem
0451  f these organs having become well developed in  FULL  grown males, and having secreted milk. So agai
0455  when the being has come to maturity and to its  FULL  powers of action, the principle of inheritance
0459  deny. I have endeavoured to give to them their  FULL  force. Nothing at first can appear more diffic
0462  denied that we are as yet very ignorant of the  FULL  extent of the various climatal and geographica
0477  of modification have been the same. We see the  FULL  meaning of the wonderful fact, which must have
0480  en it has come to maturity and has to play its  FULL  part in the struggle for existence, and will t
0481  oast waves. The mind cannot possibly grasp the  FULL  meaning of the term of a hundred million years
0481  llion years; it cannot add up and perceive the  FULL  effects of many slight variations, accumulated
0483  mals and plants created as eggs or seed, or as  FULL  grown? and in the case of mammals, were they c
0483  b? although naturalists very properly demand a  FULL  explanation of every difficulty from those who
0030  any botanists, for instance, believe that the  FULLER's teazle, with its hooks, which cannot be riv
0181  he continued preservation of individuals with  FULLER  and fuller flank membranes, each modification
0181  d preservation of individuals with fuller and  FULLER  flank membranes, each modification being usef
0012  ucas's treatise, in two large volumes, is the  FULLEST  and the best on this subject. No breeder dou
0097  if an occasional cross be indispensable, the  FULLEST  freedom for the entrance of pollen from anot
0002  arrived. A fair result can be obtained only by  FULLY  stating and balancing the facts and arguments
0006  been independently created, is erroneous. I am  FULLY  convinced that species are not immutable; but
0019  the domestic dogs of the whole world, which I  FULLY  admit have probably descended from several wi
0023  rences are between the breeds of pigeons, I am  FULLY  convinced that the common opinion of naturali
0028  kinds, knowing well how true they bred, I felt  FULLY  as much difficulty in believing that they cou
0029  ultry, or duck, or rabbit fancier, who was not  FULLY  convinced that each main breed was descended
0031  tent some breeds of cattle and sheep. In order  FULLY  to realise what they have done, it is almost
0031  e of selection in regard to merino sheep is so  FULLY  recognised, that men follow it as a trade: th
0038  f natural selection, as will hereafter be more  FULLY  explained, two sub breeds might be formed. Th
0040  points of value of the new sub breed are once  FULLY  acknowledged, the principle, as I have called
0052  ion in accumulating (as will hereafter be more  FULLY  explained) differences of structure in certai
0061  us, arise? All these results, as we shall more  FULLY  see in the next chapter, follow inevitably fr
0066  may be produced; and yet the average stock is  FULLY  kept up; but if many eggs or young are destro
0068  lants gradually kill the less vigorous, though  FULLY  grown, plants: thus out of twenty species gro
0069  rease in numbers; and, as each area is already  FULLY  stocked with inhabitants, the other species w
0083  g which she tends. Every selected character is  FULLY  exercised by her; and the being is placed und
0092  d reason to believe (as will hereafter be more  FULLY  alluded to), would produce very vigorous seed
0097  tude of flowers have their anthers and stigmas  FULLY  exposed to the weather! but if an occasional
0108  ction will always act with extreme slowness; I  FULLY  admit. Its action depends on there being plac
0109  lection. Extinction. This subject will be more  FULLY  discussed in our chapter on Geology; but it m
0109  se of all organic beings, each area is already  FULLY  stocked with inhabitants, it follows that as
0110  ximum of species. Probably no region is as yet  FULLY  stocked, for at the Cape of Good Hope, where
0121  will have played an important part. As in each  FULLY  stocked country natural selection necessarily
0151  ark, whilst investigating this Order, and I am  FULLY  convinced that the rule almost invariably hol
0172  ructure, as the eye, of which we hardly as yet  FULLY  understand the inimitable perfection? Thirdly
0172  le modifications, each new form will tend in a  FULLY  stocked country to take the place of, and fin
0183  sser numbers, than in the case of species with  FULLY  developed structures. I will now give two or
0199  would be absolutely fatal to my theory. Yet I  FULLY  admit that many structures are of no direct u
0206  ften insisted on by the illustrious Cuvier, is  FULLY  embraced by the principle of natural selectio
0220  formation on this and other subjects. Although  FULLY  trusting to the statements of Huber and Mr. S
0240  ely by their proportionally lesser size; and I  FULLY  believe, though I dare not assert so positive
0261  it will be advisable to explain a little more  FULLY  by an example what I mean by sterility being
0268  lity, and yield perfectly fertile offspring. I  FULLY  admit that this is almost invariably the case
0285  glesea of 2300 feet; and he informs me that he  FULLY  believes there is one in Merionethshire of 12
0299  pting those of the United States of America. I  FULLY  agree with Mr. Godwin Austen, that the presen
0301  o or more species. If such gradations were not  FULLY  preserved, transitional varieties would merel
0304  discovered in any secondary formation, seemed  FULLY  to justify the belief that this great and dis
0338  f this doctrine is very far from proved. Yet I  FULLY  expect to see it hereafter confirmed, at leas
0355  re common, and are, as we shall hereafter more  FULLY  see, inexplicable on the theory of independen
0357  l or atolls standing over them. Whenever it is  FULLY  admitted, as I believe it will some day be, t
0365  upheaved and formed, and before it had become  FULLY  stocked with inhabitants. On almost bare land
0367  r northern journey. Hence, when the warmth had  FULLY  returned, the same arctic species, which had
0390  lly stocked them from various sources far more  FULLY  and perfectly than has nature. Although in oc
0392  ce cf successfully competing in stature with a  FULLY  developed tree, when established on an island
0407  uence of the modern Glacial period, which I am  FULLY  convinced simultaneously affected the whole w
0413  ways sufficiently strike us, is in my judgment  FULLY  explained. Naturalists try to arrange the spe
0415  ng struck with this fact; and it has been most  FULLY  acknowledged in the writings of almost every
0420  less alike. But I must explain my meaning more  FULLY.  I believe that the arrangement of the groups
0443  tion has been caused, but at what period it is  FULLY  displayed. The cause may have acted, and it
0443  unimportant whether most of its characters are  FULLY  acquired a little earlier or later in life. I
0481  infinite number of generations. Although I am  FULLY  convinced of the truth of the views given in
0483  ow that an early progenitor had the organ in a  FULLY  developed state; and this in some instances n
0488  ry. Authors of the highest eminence seem to be  FULLY  satisfied with the view that each species has
0066  ndor may be the more numerous of the two: the  FULMAR  petrel lays but one egg, yet it is believed t
0101  maphrodites and unisexual species, as far as the  FUNCTION  is concerned, becomes very small. From these
0131  n. some authors believe it to be as much the  FUNCTION  of the reproductive system to produce indivi
0168  g been closely specialised to any particular  FUNCTION,  so that their modifications have not been c
0185  se may be said to have become rudimentary in  FUNCTION,  though not in structure. In the frigate bir
0190  , which had performed two functions, for one  FUNCTION  alone, and thus wholly change its nature by
0190  ns sometimes perform simultaneously the same  FUNCTION  in the same individual; to give one instance
0191  mind the probability of conversion from one  FUNCTION  to another, that I will give one more instan
0193  ed that, although the general appearance and  FUNCTION  of the organ may be the same, yet some funda
0204  states show that wonderful metamorphoses in  FUNCTION  are at least possible. For instance, a swim
```

Page ***(Key Word)***

Page ***************************************(Key Word)***

```
0205  ns, and then having been specialised for one FUNCTION; and two very distinct organs having perform
0205  s having performed at the same time the same FUNCTION, the one having been perfected whilst aided
0273  impotent or at least incapable of its proper FUNCTION of producing offspring identical with the pa
0436  for the infinite diversity in structure and FUNCTION of the mouths of insects. Nevertheless, it i
0441  and imperfect mouth, and cannot feed: their FUNCTION at this stage is, to search by their well de
0452  adder seems to be rudimentary for its proper FUNCTION of giving buoyancy, but has become converted
0454  ered harmless and rudimentary. Any change in FUNCTION, which can be effected by insensibly small s
0457  y different from each other in structure and FUNCTION; and the resemblance in different species of
0087  ction will be able to modify one sex in its FUNCTIONAL relations to the other sex, or in relation
0271  hing or being able to produce recondite and FUNCTIONAL differences in the reproductive system; fro
0132  nditions of life; and to this system being FUNCTIONALLY disturbed in the parents, I chiefly attrib
0182  e penguin; as sails, like the ostrich; and FUNCTIONALLY for no purpose like the Apteryx. Yet the s
0246  other hand, have their reproductive organs FUNCTIONALLY impotent, as may be clearly seen in the st
0251  respect to other species, yet as they were FUNCTIONALLY imperfect in their mutual self action, we
0461  their reproductive organs are more or less FUNCTIONALLY impotent; whereas in first crosses the org
0008  which confinement or cultivation on the FUNCTIONS of the reproductive system; this system app
0149  e been but little specialised for particular FUNCTIONS; and as long as the same part has to perfor
0190  performing at the same time wholly distinct FUNCTIONS; thus the alimentary canal respires, digest
0190  ed, a part or organ, which had performed two FUNCTIONS, for one function alone, and thus wholly ch
0205  ving performed simultaneously very different FUNCTIONS, and then having been specialised for one f
0455  ight be retained for one alone of its former FUNCTIONS. An organ, when rendered useless, may well
0005  o propagate its new and modified form. This FUNDAMENTAL subject of Natural Selection will be treat
0012  y to inheritance: like produces like is his FUNDAMENTAL belief: doubts have been thrown on this pr
0193  tion of the organ may be the same, yet some FUNDAMENTAL difference can generally be detected. I am
0206  f existence. By unity of type is meant that FUNDAMENTAL agreement in structure, which we see in or
0262  hus see, that although there is a clear and FUNDAMENTAL difference between the mere adhesion of gr
0272  to be of universal occurence, or to form a FUNDAMENTAL distinction between varieties and species.
0278  ather to support the view, that there is no FUNDAMENTAL distinction between species and varieties.
0346  ous, all authors agree that one of the most FUNDAMENTAL divisions in geographical distribution is
0438  n such parts or organs, a certain degree of FUNDAMENTAL resemblance, retained by the strong princi
0187  dations of structure, branching off in two FUNDAMENTALLY different lines, can be shown to exist, u
0245  t, two classes of facts, to a large extent FUNDAMENTALLY different, have generally been confounded
0263  rosses and of hybrids. These two cases are FUNDAMENTALLY different, for, as just remarked, in the
0011  od and from light, and perhaps the thickness of FUR from climate. Habit also has a decided influen
0045  warfed plants on Alpine summits, or the thicker FUR of an animal from far northwards, would not in
0133  als of the same species have thicker and better FUR the more severe the climate is under which the
0137  in some cases are quite covered up by skin and FUR. This state of the eyes is probably due to gra
0137  with the adhesion of the eyelids and growth of FUR over them, might in such case be an advantage;
0179  webbed feet and which resembles an otter in its FUR, short legs, and form of tail: during summer t
0392  of hooked seeds for transportal by the wool and FUR of quadrupeds. This case presents no difficult
0022  ence and relative size of the two arms of the FURCULA. The proportional width of the gape of mouth
0222  ake into slaves, from those of the little and FURIOUS F.flava, which they rarely capture, and it w
0148  gments of the head enormously developed, and FURNISHED with great nerves and muscles; but in the p
0181  ngated fingers: the flank membrane is, also, FURNISHED with an extensor muscle. Although no gradua
0191  ancient prototype, of which we know nothing, FURNISHED with a floating apparatus or swimbladder. W
0193  all these cases of two very distinct species FURNISHED with apparently the same anomalous organ, i
0201  suppose that at the same time this snake is FURNISHED with a rattle for its own injury, namely, t
0133  onditions of life. Thus, it is well known to FURRIERS that animals of the same species have thicke
0024  ked out extraordinarily abnormal species; and FURTHER, that these very species have since all beco
0030  beautiful in his eyes, we must, I think, look FURTHER than to mere variability. We cannot suppose
0038  s to man's wants or fancies. We can, I think, FURTHER understand the frequently abnormal character
0040  nct at our poultry shows. I think these views FURTHER explain what has sometimes been noticed, nam
0040  lued, their history will be disregarded. When FURTHER improved by the same slow and gradual proces
0053  fferent sets of organic beings. But my tables FURTHER show that, in any limited country, the speci
0067  o much will its tendency to increase be still FURTHER increased. We know not exactly what the chec
0082  n give it an advantage over others; and still FURTHER modifications of the same kind would often s
0082  ifications of the same kind would often still FURTHER increase the advantage. No country can be na
0108  field for natural selection to improve still FURTHER the inhabitants, and thus produce new specie
0108  itants of the same region at the same time. I FURTHER believe, that this very slow, intermittent a
0109  ood chance of utter extinction. But we may go FURTHER than this: for as new forms are continually
0122  y oix (n1^ to z1^) now species. But we may go FURTHER than this. The original species of our genus
0125  ce its numbers, and thus lessen its chance of FURTHER variation and improvement. Within the same l
0132  ghtly coloured than those of the same species FURTHER north or from greater depths. Gould believes
0136  swimmers if they had been able to swim still FURTHER, whereas it would have been better for the b
0147  hich were a little better fitted to be wafted FURTHER, might get an advangage over those producing
0149  tant. The same author and some botanists have FURTHER remarked that multiple parts are also very l
0152  ified state, as well as an innate tendency to FURTHER variability of all kinds and on the other ha
0153  in the structure undergoing modification. It FURTHER deserves notice that these variable characte
0158  overmastered the tendency to reversion and to FURTHER variability, to sexual selection being less
0161  of a similar inherited constitution. It might FURTHER be expected that the species of the same gen
0164  e full grown animal. Without here entering on FURTHER details, I may state that I have collected c
0169  yet have had time to overcome the tendency to FURTHER variability and to reversion to a less modif
0176  as I believe, is that, during the process of FURTHER modification, by which two varieties are sup
0177  hance, within any given period, of presenting FURTHER favourable variations for natural selection
0179  ntal extermination; and during the process of FURTHER modification through natural selection, they
0179  ggregate, present more variation, and thus be FURTHER improved through natural selection and gain
0179  r improved through natural selection and gain FURTHER advantages. Lastly, looking not to any one t
0181  . nor can I see any insuperable difficulty in FURTHER believing it possible that the membrane conn
0186  l to its possessor, can be shown to exist! if FURTHER, the eye does vary ever so slightly, and the
0188  heory of descent, ought not to hesitate to go FURTHER, and to admit that a structure even as perfe
0189  rfaces of each layer slowly changing in form. FURTHER we must suppose that there is a power always
0197  ly taken advantage of by the plant undergoing FURTHER modification and becoming a climber. The nak
0200  es fitted for walking or grasping; and we may FURTHER venture to believe that the several bones in
0204  equently the two latter, during the course of FURTHER modification, from existing in greater numbe
0205  subsequently taken advantage of by the still FURTHER modified descendants of this species. We may
0227  ready does her cylindrical cells; and we must FURTHER suppose, and this is the greatest difficulty
```

Page ***************************************(Key Word)***

Page **(Key Word)**

```
)227  ersection by perfectly flat surfaces. We have  FURTHER  to suppose, but this is no difficulty, that
)234  ee which could exist in a country; and let us   FURTHER  suppose that the community lived throughout
)260  of producing hybrids, and then to stop their   FURTHER  propagation by different degrees of sterilit
)272  seen striking instances of this fact. Gartner  FURTHER  admits that hybrids between very closely all
)273  ertainly only a difference in degree. Gartner  FURTHER  insists that when any two species, although
)279  ten out and exterminated during the course of  FURTHER  modification and improvement. The main cause
)301  f possessed of any decided advantage, or when  FURTHER  modified and improved; they would slowly spr
)305  tet has carried their existence one sub stage  FURTHER  back; and some palaeontologists believe that
)310  one great authority, Sir Charles Lyell, from  FURTHER  reflexion entertains grave doubts on this su
)326  would have the best chance of spreading still  FURTHER, and of giving rise in new countries to new
)348  distinct. On the other hand, proceeding still  FURTHER  westward from the eastern islands of the tro
)350  o new conditions, and will frequently undergo  FURTHER  modification and improvement; and thus they
)350  improvement; and thus they will become still  FURTHER  victorious, and will produce groups of modif
)368  l at a very late period have marched a little  FURTHER  north, and subsequently have retreated to th
)370  e inhabitants of the Old and New Worlds lived  FURTHER  southwards than at present, they must have b
)370  60 degrees, during the Pliocene period lived  FURTHER  north under the polar circle, in latitude 66
)371  heir separation by the Atlantic Ocean. We can  FURTHER  understand the singular fact remarked on by
)373  moraine, left far below any existing glacier.  FURTHER  south on both sides of the continent, from l
)379  h had crossed the equator, would travel still  FURTHER  from their homes into the most temperate lat
)412  d to go on increasing indefinitely in size. I  FURTHER  attempted to show that from the varying desc
)412  ct from that including the three genera still  FURTHER  to the right hand, which diverged at a still
)419  r family; and this has been done, not because  FURTHER  research has detected important structural d
)444  in the horns of almost full grown cattle. But  FURTHER  than this, variations which, for all that we
)448  in accordance with their similar habits. Some  FURTHER  explanation, however, of the embryo not unde
)454  n of rudimentary organs in a state of nature,  FURTHER  than by showing that rudiments can be produc
)469  ry of natural selection, even if we looked no  FURTHER  than this, seems to me to be in itself proba
)484  lesser number. Analogy would lead me one step  FURTHER, namely, to the belief that all animals and
0006  pecies are the descendants of that species.  FURTHERMORE, I am convinced that Natural Selection has
0110  as we know, the extinction of any natives.  FURTHERMORE, the species which are most numerous in in
0182  ss of perfection through natural selection.  FURTHERMORE, we may conclude that transitional grades
0439  ants: thus the embryonic leaves of the ulex or FURZE, and the first leaves of the phyllodineous ac
0219  nger. Huber then introduced a single slave (F. FUSCA), and she instantly set to work, fed and save
0221  independent community of the slave species (F. FUSCA) sometimes as many as three of these ants cli
0221  then dug up a small parcel of the pupae of F.  FUSCA from another nest, and put them down on a bar
0222  f. sanguinea could distinguish the pupae of F. FUSCA, which they habitually make into slaves, from
0222  y eagerly and instantly seized the pupae of F. FUSCA, whereas they were much terrified when they c
0222  ing their nest, carrying the dead bodies of F. FUSCA (showing that it was not a migration) and num
0222  se at hand, for two or three individuals of F. FUSCA were rushing about in the greatest agitation,
0325  ecisely the same effect. It is, indeed, quite  FUTILE to look to changes of currents, climate, or o
0002  nclusions have been grounded: and I hope in a  FUTURE work to do this. For I am well aware that sca
0006  e the present welfare; and, as I believe, the  FUTURE success and modification of every inhabitant
0044  ld be given, but these I shall reserve for my  FUTURE work. Nor shall I here discuss the various de
0053  ven in stronger terms. I shall reserve for my  FUTURE work the discussion of these difficulties, an
0062  more detail the struggle for existence. In my  FUTURE work this subject shall be treated, as it wel
0067  reated by several authors; and I shall, in my  FUTURE work, discuss some of the checks at considera
0099  len: but to this subject we shall return in a  FUTURE chapter. In the case of a gigantic tree cover
0107  e subject permits. I conclude, looking to the  FUTURE, that for terrestrial productions a large con
0111  eties, the supposed prototypes and parents of  FUTURE well marked species, present slight and ill d
0126  ill finally tend to disappear. Looking to the  FUTURE, we can predict that the groups of organic be
0126  e extinct. Looking still more remotely to the  FUTURE, we may predict that, owing to the continued
0144  the young birds when first hatched, with the  FUTURE colour of their plumage; or, again, the relat
0151  ly holds good with cirripedes. I shall, in my  FUTURE work, give a list of the more remarkable case
0173  eir remains being embedded and preserved to a  FUTURE age only in masses of sediment sufficiently t
0173  extensive to withstand an enormous amount of  FUTURE degradation; and such fossiliferous masses ca
0179  emains, which are preserved, as we shall in a  FUTURE chapter attempt to show, in an extremely impe
0192  occur, some of which will be discussed in my  FUTURE work. One of the gravest is that of neuter in
0216  e several which I shall have to discuss in my  FUTURE work, namely, the instinct which leads the cu
0231  to the planes of the rhombic basal plates of  FUTURE cells. But the rough wall of wax has in every
0282  work on the Principles of Geology, which the  FUTURE historian will recognise as having produced a
0294  d to geographical changes. And in the distant  FUTURE, a geologist examining these beds, might be t
0299  ves whether, for instance, geologists at some  FUTURE period will be able to prove, that our differ
0299  distinct. This could be effected only by the  FUTURE geologist discovering in a fossil state numer
0324  ean beds. Nevertheless, looking to a remotely  FUTURE epoch, there can, I think, be little doubt th
0338  ll with the theory of natural selection. In a  FUTURE chapter I shall attempt to show that the adul
0342  fossiliferous deposits thick enough to resist  FUTURE degradation, enormous intervals of time have
0464  bably varieties: but who will pretend that in  FUTURE ages so many fossil links will be discovered,
0465  siliferous formations, thick enough to resist  FUTURE degradation, can be accumulated only where mu
0482  his volume: but I look with confidence to the  FUTURE, to young and rising naturalists, who will be
0488  ving descendants, was created. In the distant  FUTURE I see open fields for far more important rese
0489  we may look with some confidence to a secure  FUTURE of equally inappreciable length. And as natur
0126  ly few will transmit descendants to a remote  FUTURITY. I shall have to return to this subject in t
0300  l thickness to last to an age, as distant in  FUTURITY as the secondary formations lie in the past,
0489  transmit its unaltered likeness to a distant  FUTURITY. And of the species now living very few will
0489  ransmit progeny of any kind to a far distant  FUTURITY; for the manner in which all organic beings
0489  . we can so far take a prophetic glance into  FUTURITY as to foretel that it will be the common and
0004  it is, therefore, of the highest importance to  GAIN  a clear insight into the means of modification
0092  oftenest crossed: and so in the long run would  GAIN  the upper hand. Those flowers, also, which had
0092  llen were destroyed, it might still be a great  GAIN  to the plant: and those individuals which prod
0104  ance imagined that the young thus produced will GAIN  so much in vigour and fertility over the offsp
0115  d in his great and admirable work, that floras  GAIN  by naturalisation, proportionally with the num
0115  try, and have there become naturalised, we can  GAIN  some crude idea in what manner some of the nat
0136  epidoptera, must habitually use their wings to  GAIN  their subsistence, have, as Mr. Wollaston susp
0179  further improved through natural selection and  GAIN  further advantages. Lastly, looking not to any
0186  ittle, either in habits or structure, and thus  GAIN  an advantage over some other inhabitant of the
0195  ld be able to range into new pastures and thus  GAIN  a great advantage. It is not that the larger q
0217  then the old birds or the fostered young would  GAIN  an advantage. And analogy would lead me to bel
```

Page ***(Key Word)***

Page **(Key Word)**

```
0237  ly, as well as to the individual, and may thus GAIN the desired end. Thus, a well flavoured vegeta
0287  marks because it is highly important for us to GAIN some notion, however imperfect, of the lapse c
0392  te with herbaceous plants alone, might readily GAIN an advantage by growing taller and taller and
0447  come to its full powers of activity and has to GAIN its own living; and the effects thus produced
0115  uld have had to be modified, in order to have GAINED an advantage over the other natives; and we m
0161  ed and permanent variety. But characters thus GAINED would probably be of an important nature, for
0190  easily specialise, if any advantage were thus GAINED, a part or organ, which had performed two fur
0199  ructure, quite independently of any good thus GAINED. Correlation of growth has no doubt played a
0291  ing a downward oscillation of level, and thus GAINED considerable thickness. All geological facts
0147  our fruits become atrophied, the fruit itself GAINS largely in size and quality. In our poultry,
0153  keep the breed true. In the long run selection GAINS the day, and we do not expect to fail so far
0161  h at last, under unknown favorable conditions, GAINS an ascendancy. For instance, it is probable t
0213  and then slowly crawl forward with a peculiar GAIT; and another kind of wolf rushing round, inste
0048  , the birds from the separate islands of the GALAPAGOS Archipelago, both one with another, and wit
0390  s in Madeira, or of the endemic birds in the GALAPAGOS Archipelago, with the number found on any c
0390  ns have not been much disturbed. Thus in the GALAPAGOS Islands nearly every land bird, but only tw
0391  the same distance from North America as the GALAPAGOS Islands do from South America, and which ha
0391  ly occupied by the other inhabitants; in the GALAPAGOS Islands reptiles, and in New Zealand gigant
0391  e the place of mammals. In the plants of the GALAPAGOS Islands, Dr. Hooker has shown that the prop
0397  this fact. I will give only one, that of the GALAPAGOS Archipelago, situated under the equator, be
0398  ich are supposed to have been created in the GALAPAGOS Archipelago, and nowhere else, bear so plai
0398  height, and size of the islands, between the GALAPAGOS and Cape de Verde Archipelagos: but what ar
0398  elated to those of Africa, like those of the GALAPAGOS to America. I believe this grand fact can r
0398  view here maintained, it is obvious that the GALAPAGOS Islands would be likely to receive colonist
0400  archipelago. Thus the several islands of the GALAPAGOS Archipelago are tenanted, as I have elsewhe
0400  . now if we look to those inhabitants of the GALAPAGOS Archipelago which are found in other parts
0401  e really surprising fact in this case of the GALAPAGOS Archipelago, and in a lesser degree in some
0402  lph. de Candolle, to distinct genera. In the GALAPAGOS Archipelago, many even of the .birds, though
0403  es, which inhabit the several islands of the GALAPAGOS Archipelago, not having universally spread
0478  in nearly all the plants and animals of the GALAPAGOS archipelago, of Juan Fernandez, and of the
0360  ed, if blown to a favourable spot by an inland GALE, they would germinate. Subsequently to my expe
0386  h farther it might have flown with a favouring GALE no one can tell. With respect to plants, it ha
0181  ing squirrel was produced. Now look at the GALEOPITHECUS or flying Lemur, which formerly was false
0181  r gliding through the air, now connect the GALEOPITHECUS with the other Lemuridae, yet I can see r
0181  rane connected fingers and fore arm of the GALEOPITHECUS might be greatly lengthened by natural se
0361  ow frequently birds of many kinds are blown by GALES to vast distances across the ocean. We may I
0364  accidental, nor is the direction of prevalent GALES of wind. It should be observed that scarcely
0401  re rapid and sweep across the archipelago; and GALES of wind are extraordinarily rare; so that the
0484  nd animals; or that the poison secreted by the GALL fly produces monstrous growths on the wild ros
0218  inct species, is not very uncommon with the GALLINACEAE; and this perhaps explains the origin of a
0156  amount of difference between the males of GALLINACEOUS birds, in which secondary sexual character
0264  ho has had great experience in hybridising GALLINACEOUS birds, that the early death of the embryo
0202  e, with the poison originally adapted to cause GALLS subsequently intensified, we can perhaps unde
0019  e proceeded from the common wild Indian fowl (GALLUS bankiva). In regard to ducks and rabbits, the
0030  an in very different ways; when we compare the GAME cock, so pertinacious in battle, with other br
0068  the destruction of vermin. If not one head of GAME were shot during the next twenty years in Engl
0068  oyec, there would, in all probability, be less GAME than at present, although hundreds of thousand
0068  at present, although hundreds of thousands of GAME animals are now annually killec. On the other
0070  least, this seems generally to occur with our GAME animals, often ensue: and here we have a limit
0091  ccording to Mr. St. John, bringing home winged GAME, another hares or rabbits, and another hunting
0045  f the main nerves close to the great central GANGLION of an insect would have been variable in the
0107  s in fresh water that we find seven genera of GANOID fishes, remnants of a once preponderant order
0321  bers of the great and almost extinct group of GANOID fishes still inhabit our fresh waters. Theref
0021  external orifices to the nostrils, and a wide GAPE of mouth. The short faced tumbler has a beak t
0022  of the furcula. The proportional width of the GAPE of mouth, the proportional length of the eyeli
0289  r. murchison's great work in Russia, what wide GAPS there are in that country between the superimp
0344  ncient and extinct forms often tend to fill up GAPS between existing forms, sometimes blending two
0009  source of all the choicest productions of the GARDEN. I may add, that as some organisms will breed
0030  f agricultural, culinary, orchard, and flower GARDEN races of plants most useful to man at differe
0033  t varieties of the same species in the flower GARDEN; the diversity of leaves, pods, or tubers, or
0033  s, or whatever part is valued, in the kitchen GARDEN, in comparison with the flowers of the same v
0037  seedling growing wild, if it had come from a GARDEN stock. The pear, though cultivated in classic
0070  ed from a few wheat or other such plants in a GARDEN; I have in this case lost every single seed.
0098  this flower is never visited, at least in my GARDEN, by insects, it never sets a seed, though by
0145  lation, that I have recently observed in some GARDEN pelargoniums, that the central flower of the
0249  er allied hybrids, generally grow in the same GARDEN, the visits of insects must be carefully prev
0270  ith red seeds, growing near each other in his GARDEN; and although these plants have separated sex
0361  n the course of two months, I picked up in my GARDEN 12 kinds of seeds, out of the excrement of sm
0388  stant piece of water. Nature, like a careful GARDENER, thus takes her seeds from a bed of a partic
0009  ly be given of sporting plants; by this term GARDENERS mean a single bud or offset, which suddenly
0037  orticultural works at the wonderful skill of GARDENERS, in having produced such splendid results f
0037  ppear, selecting it, and so onwards. But the GARDENERS of the classical period, who cultivated the
0041  that the strawberry began to vary just when GARDENERS began to attend closely to this plant. No o
0041  ies had been neglected. As soon, however, as GARDENERS picked out individual plants with slightly
0267  things. We see this acted on by farmers and GARDENERS in their frequent exchanges of seed, tubers
0037  longest cultivated in our flower and kitchen GARDENS. If it has taken centuries or thousands of y
0069  see in the prodigious number of plants in our GARDENS which can perfectly well endure our climate,
0362  know from experiments made in the Zoological GARDENS, include seeds capable of germination. Some
0374  razil, some few European genera were found by GARDNER, which do not exist in the wide intervening
0050  g several years by that most careful observer GARTNER, they can be crossed only with much difficul
0098  and completely destroy, as has been shown by GARTNER, any influence from the foreign pollen. When
0246  ntious and admirable observers, Kolreuter and GARTNER, who almost devoted their lives to this subj
0247  r, he unhesitatingly ranks them as varieties. GARTNER, also, makes the rule equally universal; and
0247  cases. But in these and in many other cases, GARTNER is obliged carefully to count the seeds, in
0247  s. nearly all the plants experimentised on by GARTNER were potted, and apparently were kept in a c
0247  e fertility of a plant cannot be doubted; for GARTNER gives in his table about a score of cases of
```

Page **(Key Word)**

Page **(Key Word)**

```
0247 rtility in some degree impaired. Moreover, as GARTNER during several years repeatedly crossed the
0247 are really so sterile, when intercrossed, as GARTNER believes. It is certain, on the one hand, th
0248 rs who have ever lived, namely, Kolreuter and GARTNER, should have arrived at diametrically opposi
0248 of hybrids in successive generations; though GARTNER was enabled to rear some hybrids, carefully
0249 by a remarkable statement repeatedly made by GARTNER, namely, that if even the less fertile hybri
0249 ts are in progress, so careful an observer as GARTNER would have castrated his hybrids, and this w
0250 the pure parent species, as are Kolreuter and GARTNER that some degree of sterility between distin
0250 tised on some of the very same species as did GARTNER. The difference in their results may, I thin
0252 n fertility in each successive generation, as GARTNER believes to be the case, the fact would have
0255 les and conclusions are chiefly drawn up from GARTNER's admirable work on the hybridisation of pla
0257 an the species of almost any other genus; but GARTNER found that N. acuminata, which is not a part
0258 the same fact with certain sea weeds or Fuci. GARTNER, moreover, found that this difference of fac
0259 eral other singular rules could be given from GARTNER: for instance, some species have a remarkabl
0262 nd peach on certain varieties of the plum. As GARTNER found that there was sometimes an innate dif
0268 our best botanists as varieties, are said by GARTNER not to be quite fertile when crossed, and he
0270 y as safe criterions of specific distinction. GARTNER kept during several years a dwarf kind of ma
0270 re themselves perfectly fertile; so that even GARTNER did not venture to consider the two varietie
0270 good an observer and so hostile a witness, as GARTNER: namely, that yellow and white varieties of
0272 ed may be compared in several other respects. GARTNER, whose strong wish was to draw a marked line
0272 mongrels are more variable than hybrids; but GARTNER admits that hybrids from species which have
0272 myself seen striking instances of this fact. GARTNER further admits that hybrids between very clo
0273 rn to our comparison of mongrels and hybrids. GARTNER states that mongrels are more liable than hy
0273 ue, is certainly only a difference in degree. GARTNER further insists that when any two species, a
0274 alone are the unimportant differences, which GARTNER is able to point out, between hybrid and mon
0274 nearly related species, follows according to GARTNER the same laws. When two species are crossed,
0361 important: the crops of birds do not secrete GASTRIC juice, and do not in the least injure, as I
0487 been of vast duration. But we shall be able to GAUGE with some security the duration of these inte
0042 y easily reared and a large stock not kept; in GEESE, from being valuable only for two purposes, f
0185 plainer than that the webbed feet of ducks and GEESE are formed for swimming? yet there are upland
0185 are formed for swimming? yet there are upland GEESE with webbed feet which rarely or never go nea
0186 ill cause him no surprise that there should be GEESE and frigate birds with webbed feet, either li
0204 live, how it has arisen that there are upland GEESE with webbed feet, ground woodpeckers, diving
0253 rtile. The hybrids from the common and Chinese GEESE (A. cygnoides), species which are so differen
0253 than eight hybrids (grandchildren of the pure GEESE) from one nest. In India, however, these cros
0253 one nest. In India, however, these cross bred GEESE must be far more fertile; for I am assured by
0253 pt. Hutton, that whole flocks of these crossed GEESE are kept in various parts of the country; and
0471 to prey on insects on the ground; that upland GEESE, which never or rarely swim should have been
0420 and, in so far, all true classification is GENEALOGICAL: that community of descent is the hidden b
0420 tion to the other groups, must be strictly GENEALOGICAL in order to be natural; but that the amoun
0421 to be widely different. Nevertheless their GENEALOGICAL arrangement remains strictly true, not onl
0422 e view which I hold, the natural system is GENEALOGICAL in its arrangement, like a pedigree; but t
0422 possessed a perfect pedigree of mankind, a GENEALOGICAL arrangement of the races of man would affo
0422 n only possible arrangement would still be GENEALOGICAL; and this would be strictly natural, as it
0423 , i apprehend if we had a real pedigree, a GENEALOGICAL classification would be universally prefer
0431 ent and noble family, even by the aid of a GENEALOGICAL tree, and almost impossible to do this wit
0433 being, in so far as it has been perfected, GENEALOGICAL in its arrangement, with the grades of dif
0449 t, the only possible arrangement, would be GENEALOGICAL. Descent being on my view the hidden bond
0455 to be inherited, we can understand, on the GENEALOGICAL view of classification, how it is that sys
0456 what is meant by the natural system: it is GENEALOGICAL in its attempted arrangement, with the gra
0457 rrangement is only so far natural as it is GENEALOGICAL. Finally, the several classes of facts whi
0479 munity of descent. The natural system is a GENEALOGICAL arrangement, in which we have to discover
0486 come to be, as far as they can be so made, GENEALOGIES; and will then truly give what may be call
0486 y diverging lines of descent in our natural GENEALOGIES, by characters of any kind which have long
0006 t those belonging to what are called the same GENERA are lineal descendants of some other and gene
0016 rical. Moreover, on the view of the origin of GENERA which I shall presently give, we have no righ
0044 mmon species vary most. Species of the larger GENERA in any country vary more than the species of
0044 try vary more than the species of the smaller GENERA. Many of the species of the larger genera res
0044 ler genera. Many of the species of the larger GENERA resemble varieties in being very closely, but
0046 to me extremely perplexing: I refer to those GENERA which have sometimes been called protean or p
0046 , rosa, and Hieracium amongst plants, several GENERA of insects, and several genera of Brachiopod
0046 lants, several genera of insects, and several GENERA of Brachiopod shells. In most polymorphic gen
0046 era of Brachiopod shells. In most polymorphic GENERA some of the species have fixed and definite c
0046 e species have fixed and definite characters. GENERA which are polymorphic in one country seem to
0046 d to suspect that we see in these polymorphic GENERA variations in points of structure which are o
0048 s entirely omitted several highly polymorphic GENERA. Under genera, including the most polymorphic
0048 tted several highly polymorphic genera. Under GENERA, including the most polymorphic forms, Mr. Ba
0054 nto two equal masses, all those in the larger GENERA being placed on one side, and all those in th
0054 ced on one side, and all those in the smaller GENERA on the other side, a somewhat larger number o
0054 ecies will be found on the side of the larger GENERA. This, again, might have been anticipated; fo
0054 ght have expected to have found in the larger GENERA, or those including many species, a large pro
0054 en a small majority on the side of the larger GENERA. I will here allude to only two causes of obs
0054 has little or no relation to the size of the GENERA to which the species belong. Again, plants lo
0055 there is no close relation to the size of the GENERA. The cause of lowly organised plants ranging
0055 to anticipate that the species of the larger GENERA in each country would oftener present varieti
0055 nt varieties, than the species of the smaller GENERA; for wherever many closely related species (i
0055 early equal masses, the species of the larger GENERA on one side, and those of the smaller genera
0055 genera on one side, and those of the smaller GENERA on the other side, and it has invariably prov
0055 tion of the species on the side of the larger GENERA present varieties, than on the side of the sm
0055 nt varieties, than on the side of the smaller GENERA. Moreover, the species of the large genera wh
0055 er genera. Moreover, the species of the large GENERA which present any varieties, invariably prese
0055 of varieties than do the species of the small GENERA. Both these results follow when another divis
0055 r division is made, and when all the smallest GENERA, with from only one to four species, are abso
0056 beyond the average. It is not that all large GENERA are now varying much, and are thus increasing
0056 the number of their species, or that no small GENERA are now varying and increasing; for if this h
0056 asmuch as geology plainly tells us that small GENERA have in the lapse of time often increased gre
0056 ten increased greatly in size; and that large GENERA have often come to their maxima, declined, an
```

Page **(Key Word)**

Page ***************************************(Key Word)**

```
0056   other relations between the species of large GENERA and their recorded varieties which deserve n
0057   westwood in regard to insects, that in large GENERA the amount of difference between the species
0057   respect, therefore, the species of the larger GENERA resemble varieties, more than do the species
0057   ties, more than do the species of the smaller GENERA. Or the case may be put in another way, and
0057   r way, and it may be said, that in the larger GENERA, in which a number of varieties or incipient
0057   ifference. Moreover, the species of the large GENERA are related to each other, in the same manner
0057   other; they may generally be divided into sub GENERA, or sections, or lesser groups. As Fries has
0059   s the rank of species is quite indefinite. In GENERA having more than the average number of specie
0059   species in any country, the species of these GENERA have more than the average number of varietie
0059   han the average number of varieties. In large GENERA the species are apt to be closely, but unequa
0059   l these several respects the species of large GENERA present a strong analogy with varieties. And
0059   lourishing and dominant species of the larger GENERA which on an average vary most; and varieties
0059   ted into new and distinct species. The larger GENERA thus tend to become larger; and throughout na
0059   y steps hereafter to be explained, the larger GENERA also tend to break up into smaller genera. Ar
0059   ger genera also tend to break up into smaller GFNERA. And thus, the forms of life throughout the
0061   es, which constitute what are called distinct GENERA, and which differ from each other more than d
0076   each other, than between species of distinct GENERA. We see this in the recent extension over pa
0107   . and it is in fresh water that we find seven GENERA of Ganoid fishes, remnants of a once proponde
0110   , and species of the same genus or of related GENERA, which, from having nearly the same structure
0113   a similar plot be sown with several distinct GENERA of grasses, a greater number of plants and a
0113   t all the same manner as distinct species and GENERA of grasses differ from each other, a greater
0114   ies of plants, and these belonged to eighteen GENERA and to eight orders, which shows how much the
0114   eneral rule, belong to what we call different GENERA and orders. The same principle is seen in the
0115   proportionally with the number of the native GENERA and species, far more in new genera than in r
0115   he native genera and species, far more in new GENERA than in new species. To give a single instanc
0115   lants are enumerated, and these belong to 162 GENERA. We thus see that these naturalised plants a
0115   extent from the indigenes, for out of the 162 GENERA, no less than 100 genera are not there indige
0115   , for out of the 162 genera, no less than 100 GENERA are not there indigenous, and thus a large pr
0115   a large proportional addition is made to the GENERA of these States. By considering the nature o
0117   at on an average more of the species of large GENERA vary than of small genera; and the varying sp
0117   he species of large genera vary than of small GENERA; and the varying species of the large genera
0117   genera; and the varying species of the large GENERA present a greater number of varieties. We hav
0120   hus, as I believe, species are multiplied and GENERA are formed. In a large genus it is probable
0123   he six descendants from (I) will form two sub GENERA or even genera. But as the original species
0123   nts from (I) will form two sub genera or even GENERA. But as the original species (I) differed far
0123   (a), will have to be ranked as very cistinct GENERA, or even as distinct sub families. Thus it is
0123   s. thus it is, as I believe, that two or more GENERA are produced by descent, with modification,
0124   supposed single parent of our several new sub GENERA and genera. It is worth while to reflect for
0124   ngle parent of our several new sub genera and GENERA. It is worth while to reflect for a moment on
0124   belonging to the same orders, or families, or GENERA, with those now living, yet are often, in som
0125   cation, as now explained, to the formation of GENERA alone. If, in our diagram, we suppose the amo
0125   ked o14 to m14, will form three very distinct GENERA. We shall also have two very distinct genera
0125   genera. We shall also have two very distinct GENERA descended from (I); and as these latter two g
0125   a descended from (I); and as these latter two GENERA, both from continued divergence of character
0125   ent parent, will differ widely from the three GENERA descended from (A), the two little groups of
0125   descended from (A), the two little groups of GENERA will form two distinct families, or even orde
0125   each country it is the species of the larger GENERA which oftenest present varieties or incipient
0126   een as well peopled with many species of many GENERA, families, orders, and classes, as at the pre
0128   pecies of the same genus, or even of distinct GENERA. We have seen that it is the common, the wide
0128   dely ranging species, belonging to the larger GENERA, which vary most; and these will tend to tran
0128   ly related together, forming sections and sub GENERA, species of distinct genera much less closely
0128   sections and sub genera, species of distinct GENERA much less closely related, and genera related
0128   istinct genera much less closely related, and GENERA related in different degrees, forming sub fam
0129   y represent those whole orders, families, and GENERA which have now no living representatives, and
0135   n described as not having them. In some other GENERA they are present, but in a rudimentary condit
0135   and their rudimentary condition in some other GENERA, by the long continued effects of disuse in t
0135   nnot fly; and that of the twenty nine endemic GENERA, no less than twenty three genera have all th
0135   ine endemic genera, no less than twenty three GENERA had all their species in this condition! Sev
0151   hey differ extremely little even in different GENERA. As birds within the same country vary in a r
0151   tant valves than do other species of distinct GENERA; but in the several species of one genus, Pyr
0154   ortance than those commonly used for classing GENERA. I believe this explanation is partly, yet or
0157   ecause common to large groups; but in certain GENERA the neuration differs in the different specie
0181   air rather than for flight. If about a dozen GENERA of birds had become extinct or were unknown,
0193   the end, is the same in Orchis and Asclepias, GENERA almost as remote as possible amongst flowerir
0238   ecies of the same genus, or rather as any two GENERA of the same family. Thus in Eciton, there are
0251   rvers in the case of Hippeastrum with its sub GENERA, and in the case of some other genera, as Lob
0251   its sub genera, and in the case of some other GENERA, as Lobelia, Passiflora and Verbascum. Althou
0252   c arrangements can be trusted, that is if the GENERA of animals are as distinct from each other, a
0252   s are as distinct from each other, as are the GENERA of plants, then we may infer that animals mor
0253   nt that they are generally ranked in distinct GENERA, have often bred in this country with either
0261   y systematic affinity. Although many distinct GENERA within the same family have been grafted toge
0299   dded numerous species to existing and extinct GENERA, and has made the intervals between some few
0302   s. if numerous species, belonging to the same GENERA or families, have really started into life al
0302   al record, and falsely infer, because certain GENERA or families have not been found beneath a cer
0313   each separate formation. Species of different GENERA and classes have not changed at the same rate
0316   stic differences. Groups of species, that is, GENERA and families, follow the same general rules
0317   the species of a genus, or the number of the GENERA of a family, be represented by a vertical lir
0317   ry; as the species of the same genus, and the GENERA of the same family, can increase only slowly
0323   e same; but they belong to the same families, GENERA, and sections of genera, and sometimes are st
0323   to the same families, genera, and sections of GENERA, and sometimes are similarly characterised ir
0328   numbers of the species belonging to the same GENERA, yet the species themselves differ in a manne
0329   o fill up the wide intervals between existing GENERA, families, and orders, cannot be disputed. Fo
0330   gh belonging to the same orders, families, or GENERA with those living at the present day, were no
0330   tand between living species, and some extinct GENERA between living genera, even between genera be
0330   ecies, and some extinct genera between living GENERA, even between genera belonging to distinct fa
0330   ct genera between living genera, even between GENERA belonging to distinct families. The most comm
```

Page ***************************************(Key Word)**

Page **(Key Word)**

```
0331  y suppose that the numbered letters represent  GENERA, and the dotted lines diverging from them the
0331  enus. The diagram is much too simple, too few   GENERA and too few species being given, but this is
0331  be considered as extinct. The three existing    GENERA, a14, q14, p14, will form a small family: b14
0331  hree families, together with the many extinct   GENERA on the several lines of descent diverging fro
0332  stinct from each other. If, for instance, the   GENERA a1, a5, a10, f8, m3, m6, m9, were disinterred
0332  erms. Yet he who objected to call the extinct   GENERA, which thus linked the living genera of three
0332  extinct genera, which thus linked the living    GENERA of three families together, intermediate in c
0332  milies (namely, a14 to f14 now including five    GENERA, and o14 to m14) would yet remain distinct. T
0333  ls. If, for instance, we suppose the existing    GENERA of the two families to differ from each other
0333  other by a dozen characters, in this case the    GENERA, at the early period marked VI., would differ
0333  erged. Thus it comes that ancient and extinct    GENERA are often in some slight degree intermediate
0334  preceding and succeeding faunas, that certain    GENERA offer exceptions to the rule. For instance, m
0339  hells, but from the wide distribution of most    GENERA of molluscs, it is not well displayed by them
0341  scended from some one species; so that if six    GENERA, each having eight species, be found in one g
0341  n there be six other allied or representative    GENERA with the same number of species, then we may
0341  hat only one species of each of the six older    GENERA has left modified descendants, constituting t
0341  odified descendants, constituting the six new    GENERA. The other seven species of the old genera ha
0341  ew genera. The other seven species of the old    GENERA have all died out and have left no progeny. O
0341  pecies of two or three alone of the six older    GENERA will have been the parents of the six new gen
0341  era will have been the parents of the six new    GENERA: the other old species and the other whole ge
0341  ra; the other old species and the other whole    GENERA having become utterly extinct. In failing ord
0341  utterly extinct. In failing orders, with the    GENERA and species decreasing in numbers, as apparen
0341  of the Edentata of South America, still fewer    GENERA and species will have left modified blood des
0350  we can understand how it is that sections of    GENERA, whole genera, and even families are confined
0350  tand how it is that sections of genera, whole    GENERA, and even families are confined to the same a
0353  me few families, many sub families, very many    GENERA, and a still greater number of sections of ge
0353  ra, and a still greater number of sections of    GENERA are confined to a single region: and it has b
0353  by several naturalists, that the most natural    GENERA, or those genera in which the species are mos
0353  lists, that the most natural genera, or those    GENERA in which the species are most closely related
0354  are closely related to, or belong to the same    GENERA with the species of a second region, has prob
0374  ost of peculiar species belonging to European    GENERA occur. On the highest mountains of Brazil, so
0374  ighest mountains of Brazil, some few European    GENERA were found by Gardner, which do not exist in
0375  humboldt long ago found species belonging to    GENERA characteristic of the Cordillera. On the moun
0375  n the intervening hot lowlands. A list of the    GENERA collected on the loftier peaks of Java raises
0375  , as I am informed by Dr. Hooker, of European    GENERA, found in Australia, but not in the intermedi
0381  several quite distinct species, belonging to    GENERA exclusively confined to the south, at these a
0395  american forms, but the species and even the    GENERA are distinct. As the amount of modification i
0402  s, as shown by Alph. de Candolle, to distinct    GENERA. In the Galapagos Archipelago, many even of t
0404  gould remarked to me long ago, that in those    GENERA of birds which range over the world, many of
0404  ost fresh water productions, in which so many    GENERA range over the world, and many individual spe
0405  ranges. It is not meant that in world ranging    GENERA all the species have a wide range, or even th
0405  considering the wide distribution of certain    GENERA, we should bear in mind that some are extreme
0406  e high, some of the species of widely ranging    GENERA themselves ranging widely, such facts, as alp
0407  s and from the analogical distribution of sub    GENERA, genera, and families. With respect to the di
0407  om the analogical distribution of sub genera,    GENERA, and families. With respect to the distinct s
0408  e can thus understand the localisation of sub    GENERA, genera and families; and how it is that unde
0408  us understand the localisation of sub genera,    GENERA and families; and how it is that under differ
0411  the dominant species belonging to the larger    GENERA, which vary most. The varieties, or incipient
0412  genus including several species; and all the    GENERA on this line form together one class, for all
0412  inherited something in common. But the three    GENERA on the left hand have, on this same principle
0412  ly, distinct from that including the next two    GENERA on the right hand, which diverged from a comm
0412  ent at the fifth stage of descent. These five    GENERA have also much, though less, in common; and t
0412  family distinct from that including the three    GENERA still further to the right hand, which diverg
0412  rged at a still earlier period. And all these    GENERA, descended from (A), form an order distinct f
0413  ded from (A), form an order distinct from the    GENERA descended from (I). So that we here have many
0413  scended from a single progenitor grouped into    GENERA; and the genera are included in, or subordina
0413  ingle progenitor grouped into genera; and the    GENERA are included in, or subordinate to, sub famil
0413  ined. Naturalists try to arrange the species,    GENERA, and families in each class, on what is calle
0415  rely lost. Again in another work he says, the    GENERA of the Connaraceae differ in having one or mo
0419  ders, sub orders, families, sub families, and    GENERA, they seem to be, at least at present, almost
0420  sed by the forms being ranked under different    GENERA, families, sections, or orders. The reader wi
0420  uppose the letters A to L to represent allied    GENERA, which lived during the Silurian epoch, and t
0420  wn anterior period. Species of three of these    GENERA (A, F, and I) have transmitted modified desce
0420  o the present day, represented by the fifteen    GENERA (a14 to z14) on the uppermost horizontal line
0421  till living organic beings belong to Silurian    GENERA. So that the amount or value of the differenc
0421  in character between A and I, and the several    GENERA descended from these two genera will have inh
0421  d the several genera descended from these two    GENERA will have inherited to a certain extent their
0422  sed by ranking them under different so called    GENERA, sub families, families, sections, orders, an
0424  had previously been ranked as three distinct    GENERA, were known to be sometimes produced on the s
0425  unconsciously used in grouping species under    GENERA, and genera under higher groups, though in th
0425  ly used in grouping species under genera, and    GENERA under higher groups, though in these cases th
0427  play in classing large and widely distributed    GENERA, because all the species of the same genus, i
0428  of dominant species, belonging to the larger    GENERA, tend to inherit the advantages, which made t
0429  ch other, which again implies extinction. The    GENERA Ornithorhynchus and Lepidosiren, for example,
0429  does not commonly fall to the lot of aberrant    GENERA. We can, I think, account for this fact only
0432  etters, A to L, may represent eleven Silurian    GENERA, some of which have produced large groups of
0432  every intermediate link between these eleven    GENERA and there primordial parent, and every interm
0433  from a common parent, expressed by the terms    GENERA, families, orders, etc., we can understand th
0439  irds of the same genus, and of closely allied    GENERA, often resemble each other in their first and
0441  hey were in the larval condition. But in some    GENERA the larvae become developed either into herma
0445  would, I cannot doubt, be ranked in distinct    GENERA, had they been natural productions. But when
0456  rence marked by the terms varieties, species,    GENERA, families, orders, and classes. On this same
0457  laim so plainly, that the inumerable species,    GENERA, and families of organic beings, with which t
0470  species. Moreover, the species of the larger    GENERA, which afford the greater number of varieties
0470  of difference than do the species of smaller    GENERA. The closely allied species also of the large
0470  the closely allied species also of the larger    GENERA apparently have restricted ranges, and they a
```

Page **(Key Word)**

Page **(Key Word)**

0478 that the mutual affinities of the species and GENERA within each class are so complex and circuito
0485 in the same manner as those naturalists treat GENERA, who admit that genera are merely artificial
0485 hose naturalists treat genera, who admit that GENERA are merely artificial combinations made for c
0486 s, and all the closely allied species of most GENERA, have within a not very remote period descend
0489 es of each genus, and all the species of many GENERA, have left no descendants, but have become ut
0002 elago, has arrived at almost exactly the same GENERAL conclusions that I have on the origin of spe
0002 d authorities alone. I can here give only the GENERAL conclusions at which I have arrived, with a
0029 such slight differences, yet they ignore all GENERAL arguments, and refuse to sum up in their min
0033 oked, will ensure some differences; but, as a GENERAL rule, I cannot doubt that the continued sele
0051 d 'how to rank most of the doubtful forms. His GENERAL tendency will be to make many species, for h
0051 he is continually studying; and he has little GENERAL knowledge of analogical variation in other g
0052 elief be justifiable must be judged of by the GENERAL weight of the several facts and views given
0055 ny varieties or incipient species ought, as a GENERAL rule, to be now forming. Where many large tr
0056 ient species; for my tables clearly show as a GENERAL rule that, wherever many species of a genus
0058 ecies. Finally, then, varieties have the same GENERAL characters as species, for they cannot be di
0063 to each other, I use for convenience sake the GENERAL term of struggle for existence. A struggle f
0088 but in many cases, victory will depend not on GENERAL vigour, but on having special weapons, confi
0089 males and females of any animal have the same GENERAL habits of life, but differ in structure, col
0096 ere to enter on details, I must trust to some GENERAL considerations alone. In the first place, I
0097 acts alone incline me to believe that it is a GENERAL law of nature (utterly ignorant though we be
0099 er's own pollen: and that this is part of the GENERAL law of good being derived from the intercros
0114 s jostle each other most closely, shall, as a GENERAL rule, belong to what we call different gener
0116 st nutriment from these substances. So in the GENERAL economy of any land, the more widely and per
0118 try: they will likewise partake of those more GENERAL advantages which made the genus to which the
0119 end on infinitely complex relations. But as a GENERAL rule, the more diversified in structure the
0127 itions and stations, must be judged of by the GENERAL tenour and balance of evidence given in the
0138 ied than might have been anticipated from the GENERAL resemblance of the other inhabitants of Nort
0147 some other facts, may be merged under a more GENERAL principle, namely, that natural selection is
0149 oregoing remark seems connected with the very GENERAL opinion of naturalists, that beings low in t
0149 ighly variable. We shall have to recur to the_GENERAL subject of rudimentary and aborted organs: a
0153 period not excessively remote, we might, as a GENERAL rule, expect still to find more variability
0166 pears, a tint which approaches to that of the GENERAL colouring of the other species of the genus.
0176 ld not endure for very long periods: why as a GENERAL rule they should be exterminated and disappe
0185 rra del Fuego, the Puffinuria berardi, in its GENERAL habits, in its astonishing power of diving,
0193 can, it should be observed that, although the GENERAL appearance and function of the organ may be
0196 tion the tail is in most aquatic animals, its GENERAL presence and use for many purposes in so man
0212 way, as is the hooded crow in Egypt. That the GENERAL disposition of individuals of the same speci
0212 new instincts. But I am well aware that these GENERAL statements, without facts given in detail, c
0244 d instincts, but as small consequences of one GENERAL law, leading to the advancement of all organ
0251 lantaginea, species most widely dissimilar in GENERAL habit, reproduced itself as perfectly as if
0254 first crosses and in hybrids, is an extremely GENERAL result: but that it cannot, under our presen
0257 hat plants most widely different in habit and GENERAL appearance, and having strongly marked diffe
0265 in both cases the sterility is independent of GENERAL health, and is often accompanied by excess o
0265 the reproductive system, independently of the GENERAL state of health, is affected by sterility in
0269 same manner, and does not wish to alter their GENERAL habits of life. Nature acts uniformly and sl
0271 tions and facts, I do not think that the very GENERAL fertility of varieties can be proved to be o
0272 istinction between varieties and species. The GENERAL fertility of varieties does not seem to me s
0272 w which I have taken with respect to the very GENERAL, but not invariable, sterility of first cros
0275 ty, in all other respects there seems to be a GENERAL and close similarity in the offspring of cro
0277 uite universally, fertile. Nor is this nearly GENERAL and perfect fertility surprising, when we re
0278 spects, excluding fertility, there is a close GENERAL resemblance between hybrids and mongrels. Fi
0281 will have had in its whole organisation much GENERAL resemblance to the tapir and the horse; but
0292 s, for they are in strict accordance with the GENERAL principles inculcated by Sir C. Lyell; and E
0312 t reappear. Groups of species follow the same GENERAL rules in their appearance and disappearance
0316 that is, genera and families, follow the same GENERAL rules in their appearance and disappearance
0316 but such cases are certainly exceptional; the GENERAL rule being a gradual increase in number, til
0317 de beaumont, Murchison, Barrande, etc., whose GENERAL views would naturally lead them to this conc
0323 nd new Worlds be kept wholly out of view, the GENERAL parallelism in the successive forms of life,
0325 re or less local and temporary, but depend on GENERAL laws which govern the whole animal kingdom.
0328 plained in the foregoing paragraphs, the same GENERAL succession in the forms of life; but the spe
0328 f england and France, is able to draw a close GENERAL parallelism between the successive stages in
0328 arrade, also, shows that there is a striking GENERAL parallelism in the successive Silurian depos
0329 ged in the same order, in accordance with the GENERAL succession of the form of life, and the orde
0329 the more ancient any form is, the more, as a GENERAL rule, it differs from living forms. But, as
0329 less perfect than if we combine both into one GENERAL system. With respect to the Vertebrata, whol
0332 yet retain throughout a vast period the same GENERAL characteristics. This is represented in the
0334 n the earth's history will be intermediate in GENERAL character between that which preceded and th
0335 s. pictet gives as a well known instance, the GENERAL resemblance of the organic remains from the
0347 . a second great fact which strikes us in our GENERAL review is, that barriers of any kind, or obs
0354 t to give up the belief, rendered probable by GENERAL considerations, that each species has been p
0393 may be explained through glacial agency. This GENERAL absence of frogs, toads, and newts on so man
0403 nd plants. The principle which determines the GENERAL character of the fauna and flora of oceanic
0404 reat facility of transport has given the same GENERAL forms to the whole world. We see this same p
0404 es allied to it, is shown in another and more GENERAL way. Mr. Gould remarked to me long ago, that
0405 e world, we ought to find; and I believe as a GENERAL rule we do find, that some at least of the s
0407 nation of single centres of creation, by some GENERAL considerations, more especially from the imp
0410 e and below: so in space, it certainly is the GENERAL rule that the area inhabited by a single spe
0411 nalogical or adaptive characters. Affinities, GENERAL, complex and radiating. Extinction separates
0413 eans for enunciating, as briefly as possible, GENERAL propositions, that is, by one sentence to gi
0414 eation, or is simply a scheme for enunciating GENERAL propositions and of placing together the for
0414 which determined the habits of life, and the GENERAL place of each being in the economy of nature
0414 hall have to recur. It may even be given as a GENERAL rule, that the less any part of the organisa
0420 known plan of creation, or the enunciation of GENERAL propositions, and the mere putting together
0428 e cannot doubt that they have inherited their GENERAL shape of body and structure of limbs from a
0430 istinct group, this affinity in most cases is GENERAL and not special; thus, according to Mr. Wate
0430 h it approaches this order, its relations are GENERAL, and not to any one marsupial species more t

Page **(Key Word)**

Page ***(Key Word)***************************************

```
0430  les most nearly, not any one species, but the  GENERAL order of Rodents. In this case, however, it
0430  e 'has made nearly similar observations on the  GENERAL nature of the affinities of distinct orders
0432  roup, whether large or small, and thus give a   GENERAL idea of the value of the differences between
0434  ir habits of life, resemble each other in the   GENERAL plan of their organisation. This resemblance
0434  cous. The whole subject is included under the   GENERAL name of Morphology. This is the most interes
0435  ls, had its limbs constructed on the existing   GENERAL pattern, for whatever purpose they served, w
0436  cts. Nevertheless, it is conceivable that the   GENERAL pattern of an organ might become so much obs
0436  mouths of certain suctorial crustaceans, the    GENERAL pattern seems to have been thus to a certain
0442  several facts in embryology, namely the very    GENERAL, but not universal difference in structure b
0450  re. For instance, rudimentary mammae are very   GENERAL in the males of mammals: I presume that the
0459  y of Natural Selection. Recapitulation of the   GENERAL and special circumstances in its favour. Cau
0459  al circumstances in its favour. Causes of the   GENERAL belief in the immutability of species. How f
0460  be considered as universal: nor is their very   GENERAL fertility surprising when we remember that i
0470  pecies has been active, we might expect, as a   GENERAL rule, to find it still in action: and this i
0485  sence of the term species. The other and more   GENERAL departments of natural history will rise gre
0487  , which include few identical species, by the   GENERAL succession of their forms of life. As specie
0339  r owen has subsequently extended the same       GENERALISATION to the mammals of the Old World. We see t
0080  and on both sexes. Sexual Selection. On the     GENERALITY of intercrosses between individuals of the
0150  ate my conviction that it is a rule of high     GENERALITY. I am aware of several causes of error, but
0246  ithout being deeply impressed with the high     GENERALITY of some degree of sterility. Kolreuter make
0335  ct in each stage. This fact alone, from its     GENERALITY, seems to have shaken Professor Pictet in h
0349  nts and stations. It is a law of the widest     GENERALITY, and every continent offers innumerable ins
0478  pecies likewise occur. It is a rule of high     GENERALITY that the inhabitants of each area are relat
0006  era are lineal descendants of some other and    GENERALLY extinct species, in the same manner as the
0007  first points which strikes us, is, that they    GENERALLY differ much more from each other, than do t
0007  the organisation has once begun to vary, it     GENERALLY continues to vary for many generations. No
0008  causes of variability, whatever they may be,    GENERALLY act: whether during the early or late perio
0008  confinement in their native country! This is    GENERALLY attributed to vitiated instincts: but how m
0015  hem with species closely allied together, we    GENERALLY perceive in each domestic race, as already
0018  his latter fact were found more strictly and    GENERALLY true than seems to me to be the case, what
0024  the above specified breeds, though agreeing     GENERALLY in constitution, habits, voice, colouring,
0031  of the world. The improvement is by no means    GENERALLY due to crossing different breeds: all the b
0036  o liable, and such choice animals would thus    GENERALLY leave more offspring than the inferior ones
0041  he sheep of parts of Yorkshire, that as they    GENERALLY belong to poor people, and are mostly in sm
0041  raising large stocks of the same plants, are    GENERALLY far more successful than amateurs in gettin
0041  viduals, whatever their quality may be, will    GENERALLY be allowed to breed, and this will effectua
0044  y what he means when he speaks of a species.    GENERALLY the term includes the unknown element of a
0044  ous to or not useful to the species, and not    GENERALLY propagated. Some authors use the term varia
0045  ed productions. These individual differences    GENERALLY affect what naturalists consider unimportan
0048  marked for me 182 British plants, which are     GENERALLY considered as varieties, but which have all
0049  logists. Even Ireland has a few animals, now    GENERALLY regarded as varieties, but which have been
0049  efore any definition of these terms has been    GENERALLY accepted, is vainly to beat the air. Many o
0050  a dozen species out of forms, which are very    GENERALLY considered as varieties, and in this countr
0050  subject: and this shows, at least, how very     GENERALLY there is some variation. But if he confine
0051  tions be widely extended, he will in the end    GENERALLY be enabled to make up his own mind which to
0053  hown that plants which have very wide ranges    GENERALLY present varieties: and this might have been
0054  ity. Fresh water and salt loving plants have    GENERALLY very wide ranges and are much diffused, but
0055  plants low in the scale of organisation are    GENERALLY much more widely diffused than plants highe
0055  we might expect that the circumstances would    GENERALLY be still favourable to variation. On the ot
0056  factory of species has been active, we ought    GENERALLY to find the manufactory still in action, mo
0057  e equally distinct from each other: they may    GENERALLY be divided into sub genera, or sections, or
0057  well remarked, little groups of species are     GENERALLY clustered like satellites around certain ot
0058  nt which seems to me worth notice. Varieties    GENERALLY have much restricted ranges: this statement
0059  for two forms, if differing very little, are    GENERALLY ranked as varieties, notwithstanding that i
0061  he preservation of that individual, and will    GENERALLY be inherited by its offspring. The offsprin
0067  hief points. Eggs or very young animals seem    GENERALLY to suffer most, but this is not invariably
0070  small tract, epidemics, at least, this seems    GENERALLY to occur with our game animals, often ensue
0071  the heath was most remarkable, more than is    GENERALLY seen in passing from one quite different so
0074  into play: some one check or some few being     GENERALLY the most potent, but all concurring in dete
0075  another, as of a parasite on its prey, lies    GENERALLY between beings remote in the scale of natur
0075  eties of the same species, the struggle will    GENERALLY be almost equally severe, and we sometimes
0076  , and always in structure, the struggle will    GENERALLY be more severe between species of the same
0078  er home, yet the conditions of its life will    GENERALLY be changed in an essential manner. If we wi
0079  cessant, no fear is felt, that death is        GENERALLY prompt, and that the vigorous, the healthy,
0088  efore, less rigorous than natural selection.    GENERALLY, the most vigorous males, those which are b
0097  cially as the plant's own anthers and pistil    GENERALLY stand so close together that self fertilisa
0103  or in a continuous area, the conditions will    GENERALLY graduate away insensibly from one district
0103  ature, for instance in birds, varieties will    GENERALLY be confined to separated countries: and thi
0104  rganic and inorganic conditions of life will    GENERALLY be in a great degree uniform: so that natur
0106  , so that the good effects of isolation will    GENERALLY, to a certain extent, have concurred. Final
0106  es, yet that the course of modification will    GENERALLY have been more rapid on large areas: and wh
0108  often depend on physical changes, which are    GENERALLY very slow, and on the immigration of better
0108  y, often only at long intervals of time, and    GENERALLY on only a very few of the inhabitants of th
0110  he same structure, constitution, and habits,    GENERALLY come into the severest competition with eac
0110  , during the progress of its formation, will    GENERALLY press hardest on its nearest kindred, and t
0115  ed in becoming naturalised in any land would    GENERALLY have been closely allied to the indigenes:
0116  mble each other in unequal degrees, as is so    GENERALLY the case in nature, and is represented in t
0117  ergence of character comes in: for this will    GENERALLY lead to the most different or divergent var
0117  , namely a1 and m1. These two varieties will    GENERALLY continue to be exposed to the same conditio
0118  ry, consequently they will tend to vary, and    GENERALLY to vary in nearly the same manner as their
0118  the most divergent of their variations will    GENERALLY be preserved during the next thousand gener
0118  proceeding from the common parent (A), will    GENERALLY go on increasing in number and diverging in
0119  e their parent successful in life, they will    GENERALLY go on multiplying in number as well as dive
0121  ready extremely different in character, will    GENERALLY tend to produce the greatest number of modi
0121  ould be remembered that the competition will    GENERALLY be most severe between those forms which ar
0121  as the original parent species itself, will    GENERALLY tend to become extinct. So it probably will
```

Page ***(Key Word)***************************************

Page ********************************(Key Word)********************************

```
0122  mble each other in unequal degrees, as is so  GENERALLY  the case in nature; species (A) being more
0124  affinities of extinct beings, which, though    GENERALLY  belonging to the same orders, or families,
0133  re confined to tropical and shallow seas are   GENERALLY  brighter coloured than those confined to co
0140  imates. But whether or not the adaptation be   GENERALLY  very close, we have evidence, in the case o
0147  try, a large tuft of feathers on the head is   GENERALLY  accompanied by a diminished comb, and a lar
0153  wn to us, more to one sex than to the other,   GENERALLY  to the male sex, as with the wattle of carr
0155  that some important organ or part, which is    GENERALLY  very constant throughout large groups of sp
0155  d this fact shows that a character, which is   GENERALLY  of generic value, when it sinks in value an
0157  etween the two sexes of the same species are   GENERALLY  displayed in the very same parts of the org
0157  number of joints in the tarsi is a character   GENERALLY  common to very large groups of beetles, but
0158  sexual and ordinary specific differences are  GENERALLY  displayed in the same parts of the organisa
0160  r, is only 1 in 2048; and yet, we see, it is   GENERALLY  believed that a tendency to reversion is re
0162  crossed. Hence, though under nature it must    GENERALLY  be left doubtful, what cases are reversions
0163  nel Poole that the foals of this species are   GENERALLY  striped on the legs and faintly on the shou
0164  of India the Kattywar breed of horses is so    GENERALLY  striped, that, as I hear from Colonel Poole
0164  d. the spine is always striped; the legs are   GENERALLY  barred; and the shoulder stripe, which is s
0168  d. changes of structure at an early age will   GENERALLY  affect parts subsequently developed; and th
0169  ty in the same parts of the organisation has   GENERALLY  been taken advantage of in giving secondary
0172  rent and all the transitional varieties will   GENERALLY  have been exterminated by the very process
0172  cord being incomparably less perfect than is   GENERALLY  supposed; the imperfection of the record be
0173  ing from north to south over a continent, we   GENERALLY  meet at successive intervals with closely a
0173  ese species where they intermingle, they are   GENERALLY  as absolutely distinct from each other in e
0174  hey are now distributed over a wide area, we   GENERALLY  find them tolerably numerous over a large t
0174  ritory between two representative species is   GENERALLY  narrow in comparison with the territory pro
0175  cies, when inhabiting a continuous area, are   GENERALLY  so distributed that each has a wide range,
0176  atson, Dr. Asa Gray, and Mr. Wollaston, that  GENERALLY  when varieties intermediate between two oth
0176  es linking two other varieties together have  GENERALLY  existed in lesser numbers than the forms wh
0178  med in the intermediate zones, but they will  GENERALLY  have had a short duration. For these interm
0183  tell, and immaterial for us, whether habits   GENERALLY  change first and structure afterwards; or w
0191  ish, or, for I do not know which view is now   GENERALLY  held, a part of the auditory apparatus has
0193  veral are widely remote in their affinities.  GENERALLY  when the same organ appears in several memb
0193  he same, yet some fundamental difference can   GENERALLY  be detected. I am inclined to believe that
0197  . the naked skin on the head of a vulture is   GENERALLY  looked at as a direct adaptation for wallow
0198  f our domestic breeds, which nevertheless we  GENERALLY  admit to have arisen through ordinary gener
0204  numerous intermediate variety, and will thus  GENERALLY  succeed in supplanting and exterminating it
0205  untry. Hence the inhabitants of one country,  GENERALLY  the smaller one, will often yield, as we se
0205  do yield, to the inhabitants of another and   GENERALLY  larger country. For in the larger country t
0206  it must by my theory be strictly true. It is  GENERALLY  acknowledged that all organic beings have b
0208  once, or in repeating anything by rote, he is  GENERALLY  forced to go back to recover the habitual t
0210  eing known amongst extinct species, how very  GENERALLY  gradations, leading to the most complex ins
0216  he hive bee: these two latter instincts have  GENERALLY, and most justly, been ranked by naturalist
0218  r believing that although the Tachytes nigra  GENERALLY  makes its own burrow and stores it with par
0238  n three castes. The castes, moreover, do not  GENERALLY  graduate into each other, but are perfectly
0245  dently of their fertility. Summary. The view  GENERALLY  entertained by naturalists is that species,
0245  importance of the fact that hybrids are very  GENERALLY  sterile, has, I think, been much underrated
0245  a large extent fundamentally different, have  GENERALLY  been confounded together; namely, the steri
0248  ly that their fertility never increased, but  GENERALLY  greatly decreased. I do not doubt that this
0249  the parent species, or other allied hybrids,  GENERALLY  grow in the same garden, the visits of inse
0249  ing the flowering season: hence hybrids will  GENERALLY  be fertilised during each generation by the
0253  species which are so different that they are  GENERALLY  ranked in distinct genera, have often bred
0255  d animals, I have been surprised to find how  GENERALLY  the same rules apply to both kingdoms. It h
0256  and which rarely produce any offspring, are   GENERALLY  very sterile; but the parallelism between t
0256  hus produced, two classes of facts which are  GENERALLY  confounded together, is by no means strict.
0257  e other hand, by very closely allied species  GENERALLY  uniting with facility. But the corresponden
0258  n used as the father and then as the mother,  GENERALLY  differ in fertility in a small, and occasio
0260  s used as the father or the mother, there is  GENERALLY  some difference, and occasionally the wides
0264  n this view: as hybrids, when once born, are  GENERALLY  healthy and long lived, as we see in the ca
0264  y where their two parents can live, they are  GENERALLY  placed under suitable conditions of life. B
0267  fically different, produce hybrids which are  GENERALLY  sterile in some degree. I cannot persuade m
0271  sed variety if infertile in any degree would  GENERALLY  be ranked as species; from man selecting on
0274  rid plants produced from a reciprocal cross,  GENERALLY  resemble each other closely; and so it is w
0276  nked as species, and their hybrids, are very  GENERALLY, but not universally, sterile. The sterilit
0276  d by several curious and complex laws. It is  GENERALLY  different, and sometimes widely different,
0276  variety to take on another, is incidental on  GENERALLY  unknown differences in their vegetative sys
0277  f sterility of their hybrid offspring should  GENERALLY  correspond, though due to distinct causes;
0277  eties, and their mongrel offspring, are very  GENERALLY, but not quite universally, fertile. Nor is
0279  bers than the forms which they connect, will  GENERALLY  be beaten out and exterminated during the c
0280  unknown progenitor; and the progenitor will  GENERALLY  have differed in some respects from all its
0282  e present day: and these parent species, now  GENERALLY  extinct, have in their turn been similarly
0288  d or gravel, will when the beds are upraised  GENERALLY  be dissolved by the percolation of rain wat
0290  gical composition of consecutive formations,  GENERALLY  implying great changes in the geography of
0292  species; but during such periods there will  GENERALLY  be a blank in the geological record. On the
0295  process of variation; hence the deposit will  GENERALLY  have to be a very thick one; and the specie
0295  whole pile of formations in any country, has  GENERALLY  been intermittent in its accumulation. When
0296  upply of sediment of a different nature will  GENERALLY  have been due to geographical changes requi
0298  have formerly seen, that their varieties are  GENERALLY  at first local; and that such local varieti
0301  varieties; and the varieties would at first   GENERALLY  be local or confined to one place; but if p
0307  ding them in age, and these ought to be very  GENERALLY  in a metamorphosed condition. But the descr
0315  reason to believe that the parent form will   GENERALLY  be supplanted and exterminated by its impro
0317  successive periods by catastrophes, is very   GENERALLY  given up, even by those geologists, as Elie
0318  lete extinction of the species of a group is  GENERALLY  a slower process than their production: if
0320  in 1845, namely, that to admit that species   GENERALLY  become rare before they become extinct, to
0320  me number of old forms. The competition will  GENERALLY  be most severe, as formerly explained and i
0321  d and modified descendants of a species will  GENERALLY  cause the extermination of the parent speci
0321  will have to yield their places: and it will  GENERALLY  be allied forms, which will suffer from som
0321  therefore the utter extinction of a group is  GENERALLY, as we have seen, a slower process than its
```

Page ********************************(Key Word)********************************

Page **(Key Word)***

```
0326   but in the long run the dominant forms will   GENERALLY succeed in spreading. The diffusion would,
0327   places to the new and victorious forms, will   GENERALLY be allied in groups, from inheriting some i
0331   he more recent any form is, the more it will   GENERALLY differ from its ancient progenitor. Hence w
0344   d, but modified, descendants: and these will   GENERALLY succeed in taking the places of those group
0344   why the more ancient a form is, the more it   GENERALLY differs from those now living. Why ancient
0346   new, at least as closely as the same species   GENERALLY require: for it is a most rare case to find
0353   are most closely related to each other, are   GENERALLY local, or confined to one area. What a stra
0360   alph. de Candolle has shown that such plants   GENERALLY have restricted ranges. But seeds may be oc
0362   ss. Although the beaks and feet of birds are   GENERALLY quite clean, I can show that earth sometime
0367   e re migration on the returning warmth, will   GENERALLY have been due south and north. The Alpine p
0392   from what they are elsewhere. Such cases are   GENERALLY accounted for by the physical conditions of
0392   : now trees, as Alph. de Candolle has shown,   GENERALLY have, whatever the cause may be, confined r
0395   rld: but as far as I have gone, the relation   GENERALLY holds good. We see Britain separated by a s
0397   and smallest, are inhabited by land shells,   GENERALLY by endemic species, but sometimes by specie
0400   as to compete, is at least as important, and   GENERALLY a far more important element of success. No
0402   become naturalised in new countries are not   GENERALLY closely allied to the aboriginal inhabitant
0404   ranges. I can hardly doubt that this rule is   GENERALLY true, though it would be difficult to prove
0406   ms low in the scale within each great class,   GENERALLY change at a slower rate than the higher for
0408   n are explicable on the theory of migration (GENERALLY of the more dominant forms of life), togeth
0410   me and space the lower members of each class   GENERALLY change less than the higher: but there are
0410   s are related in blood, the nearer they will   GENERALLY stand to each other in time and space: in b
0412   ly the groups which are now large, and which   GENERALLY include many dominant species, tend to go o
0415   importance of organs which are important is   GENERALLY, but by no means always, true. But their im
0415   this constancy depends on such organs having   GENERALLY been subjected to less change in the adapta
0418   ation of animals: and this doctrine has very   GENERALLY been admitted as true. The same fact holds
0426   he most diverse conditions of existence, are   GENERALLY the most constant, we attach especial value
0429   show, on the principle of each group having   GENERALLY diverged much in character during the long
0429   fered severely from extinction, for they are   GENERALLY represented by extremely few species: and s
0429   ew species: and such species as do occur are   GENERALLY very distinct from each other, which again
0441   ble. The embryo in the course of development   GENERALLY rises in organisation: I use this expressio
0441   in some cases, however, the mature animal is   GENERALLY considered as lower in the scale than the l
0442   of different species within the same class,   GENERALLY, but not universally, resembling each other
0443   yed. The cause may have acted, and I believe   GENERALLY has acted, even before the embryo is formed
0446   een accumulated by man's selection, have not   GENERALLY first appeared at an early period of life,
0452   ies. Thus in the snapdragon (antirrhinum) we   GENERALLY do not find a rudiment of a fifth stamen: b
0453   ks on natural history rudimentary organs are   GENERALLY said to have been created for the sake of s
0455   or selection reduces an organ, and this will   GENERALLY be when the being has come to maturity and
0457   essive slight variations, not necessarily or   GENERALLY supervening at a very early period of life,
0457   ze, either from disuse or selection, it will   GENERALLY be at that period of life when the being ha
0463   tter, from existing in greater numbers, will   GENERALLY be modified and improved at a quicker rate
0468   mpetition with each other, the struggle will   GENERALLY be most severe between them: it will be alm
0468   truggled with their conditions of life, will   GENERALLY leave most progeny. But success will often
0468   varied under nature, in the same way as they   GENERALLY have varied under the changed conditions of
0474   ion, and therefore we might expect this part   GENERALLY to be still variable. But a part may be dev
0476   ve descended from an ancient progenitor have   GENERALLY diverged in character, the progenitor with
0476   existing and allied groups. Recent forms are   GENERALLY looked at as being, in some vague sense, hi
0477   eat class are plainly related: for they will   GENERALLY be descendants of the same progenitors and
0480   entary organs. But disuse and selection will   GENERALLY act on each creature, when it has come to m
0484   igin of species, or when analogous views are   GENERALLY admitted, we can dimly foresee that there w
0485   en them. It is quite possible that forms now   GENERALLY acknowledged to be merely varieties may her
0218   stinct of our cuckoo could be, and has been,   GENERATED. I may add that, according to Dr. Gray and
0241   h the natural selection of the parents which   GENERATED them: until none with an intermediate struc
0316   have continuously existed, in order to have   GENERATED either new and modified or the same old and
0380   largely yielded to the more dominant forms,   GENERATED in the larger areas and more efficient work
0010   some authors have supposed, with the act of   GENERATION. Seedlings from the same fruit, and the you
0026   become less and less, as in each succeeding   GENERATION there will be less of the foreign blood: bu
0026   ter, which has been lost during some former   GENERATION, this tendency, for all that we can see to
0066   lls either on the young or old, during each   GENERATION or at recurrent intervals. Lighten any chec
0079   during some season of the year, during each   GENERATION or at intervals, has to struggle for life,
0118   ess is represented up to the ten thousandth   GENERATION, and under a condensed and simplified form
0118   mplified form up to the fourteen thousandth   GENERATION. But I must here remark that I do not suppo
0122   urteen in number at the fourteen thousandth   GENERATION, will probably have inherited some of the s
0122   mitted offspring to the fourteen thousandth   GENERATION. We may suppose that only one (F), of the t
0130   p on all sides many a feebler branch, so by   GENERATION I believe it has been with the great Tree o
0161   ations distant, but that in each successive   GENERATION there has been a tendency to reproduce the
0161   . for instance, it is probable that in each   GENERATION of the barb pigeon, which produces most rar
0161   red bird, there has been a tendency in each   GENERATION in the plumage to assume this colour. This
0166   a tendency in the young of each successive   GENERATION to produce the long lost character, and tha
0189   ariation will cause the slight alterations,   GENERATION will multiply them almost infinitely, and n
0191   aving true lungs have descended by ordinary   GENERATION from an ancient prototype, of which we know
0198   rally admit to have arisen through ordinary   GENERATION, we ought not to lay too much stress on our
0209   nstincts have been acquired by habit in one   GENERATION, and then transmitted by inheritance to suc
0214   s of compulsory training in each successive   GENERATION would soon complete the work: and unconscio
0249   ds will generally be fertilised during each   GENERATION by their own individual pollen: and I am co
0249   ybrids, and this would have insured in each   GENERATION a cross with the pollen from a distinct flo
0252   decreasing in fertility in each successive   GENERATION, as Gartner believes to be the case, the fa
0253   ave usually been crossed in each successive   GENERATION, in opposition to the constantly repeated a
0266   r se, they transmit to their offspring from   GENERATION to generation the same compounded organisat
0266   nsmit to their offspring from generation to   GENERATION the same compounded organisation, and hence
0272   important distinction is, that in the first   GENERATION mongrels are more variable than hybrids: bu
0272   cultivated are often variable in the first   GENERATION: and I have myself seen striking instances
0273   ybrids from the first cross or in the first   GENERATION, in contrast with their extreme variability
0273   h the parent form. Now hybrids in the first   GENERATION are descended from species (excluding those
0344   ared, it does not reappear: for the link of   GENERATION has been broken. We can understand how the
0344   one grand system: for all are connected by   GENERATION. We can understand, from the continued tend
0345   e forms are more closely linked together by   GENERATION: we can clearly see why the remains of an i
```

Page **(Key Word)***

generation

0345	that species have been produced by ordinary	GENERATION: old forms having been supplanted by new an
0352	ects it, rejects the vera causa of ordinary	GENERATION with subsequent migration, and calls in the
0355	nce, that this coincidence he attributes to	GENERATION with modification. The previous remarks rela
0410	been connected by the same bond of ordinary	GENERATION; and the more nearly any two forms are rela
0467	of a breed by selecting, in each successive	GENERATION, individual differences so slight as to be
0475	species reappear when the chain of ordinary	GENERATION has once been broken. The gradual diffusion
0476	have been connected by the bond of ordinary	GENERATION, and the means of modification have been th
0489	eel certain that the ordinary succession by	GENERATION has never once been broken, and that no cat
0004	say that after a certain unknown number of	GENERATIONS, some bird had given birth to a woodpecker
0007	ganic beings must be exposed during several	GENERATIONS to the new conditions of life to cause any
0007	ry, it generally continues to vary for many	GENERATIONS. No case is on record of a variable being
0015	ralising, or were to cultivate, during many	GENERATIONS, the several races, for instance, of the c
0015	egetables, for an almost infinite number of	GENERATIONS, would be opposed to all experience. I may
0017	animal, whether it would vary in succeeding	GENERATIONS, and whether it would endure other climate
0017	uld be made to breed for an equal number of	GENERATIONS under domestication, they would vary on an
0020	ls are crossed one with another for several	GENERATIONS, hardly two of them will be alike, and the
0026	thin a dozen or, at most, within a score of	GENERATIONS, been crossed by the rock pigeon: I say wi
0026	rock pigeon: I say within a dozen or twenty	GENERATIONS, for we know of no fact countenancing the
0026	ne ancestor, removed by a greater number of	GENERATIONS. In a breed which has been crossed only on
0026	ed undiminished for an indefinite number of	GENERATIONS. These two distinct cases are often confou
0029	ferences accumulated during many successive	GENERATIONS. May not those naturalists who, knowing fa
0032	ulation in one direction, during successive	GENERATIONS? and in this case I presume that the form
0045	me cases be inherited for at least some few	GENERATIONS, of differences absolutely inappreciable b
0076	xed stock could be kept up for half a dozen	GENERATIONS, if they were allowed to struggle together
0080	metimes occur in the course of thousands of	GENERATIONS? If such do occur, can we doubt (rememberi
0089	le birds, by selecting, during thousands of	GENERATIONS, the most melodious or beautiful males, ac
0089	s, individual males have had, in successive	GENERATIONS, some slight advantage over other males, i
0097	g self fertilises itself for an eternity of	GENERATIONS; but that a cross with another individual
0114	ubt that in the course of many thousands of	GENERATIONS, the most distinct varieties of any one sp
0117	the diagram, may represent each a thousand	GENERATIONS; but it would have been better if each had
0117	better if each had represented ten thousand	GENERATIONS. After a thousand generations, species (A)
0117	ten thousand generations. After a thousand	GENERATIONS, species (A) is supposed to have produced
0118	rally be preserved during the next thousand	GENERATIONS. And after this interval, variety a1 is su
0118	some of the varieties, after each thousand	GENERATIONS, producing only a single variety, but in a
0120	n may have been increased in the successive	GENERATIONS. This case would be represented in the dia
0120	fresh branches or races. After ten thousand	GENERATIONS, species (A) is supposed to have produced
0120	diverged in character during the successive	GENERATIONS, will have come to differ largely, but per
0120	ng the same process for a greater number of	GENERATIONS (as shown in the diagram in a condensed an
0120	ced, by analogous steps, after ten thousand	GENERATIONS, either two well marked varieties (w10 and
0121	e horizontal lines. After fourteen thousand	GENERATIONS, six new species, marked by the letters n1
0124	herto been supposed to represent a thousand	GENERATIONS, but each may represent a million or a hun
0124	ay represent a million or a hundred million	GENERATIONS, and likewise a section of the successive
0131	ancestors have been exposed during several	GENERATIONS. I have remarked in the first chapter, but
0133	ing been favoured and preserved during many	GENERATIONS, and how much to the direct action of the
0135	s natural selection increased in successive	GENERATIONS the size and weight of its body, its legs
0136	disuse. For during thousands of successive	GENERATIONS each individual beetle which flew least, e
0138	rs of vision, slowly migrated by successive	GENERATIONS from the outer world into the deeper and d
0138	hat an animal had reached, after numberless	GENERATIONS, the deepest recesses, disuse will on this
0142	until some one will sow, during a score of	GENERATIONS, his kidney beans so early that a very lar
0146	ation in structure; and, after thousands of	GENERATIONS, some other and independent modification;
0160	been lost for many, perhaps for hundreds of	GENERATIONS. But when a breed has been crossed only on
0160	in character to the foreign breed for many	GENERATIONS, some say, for a dozen or even a score of
0160	s, some say, for a dozen or even a score of	GENERATIONS. After twelve generations, the proportion
0160	r even a score of generations. After twelve	GENERATIONS, the proportion of blood, to use a common
0160	trary, transmitted for almost any number of	GENERATIONS. When a character which has been lost in a
0160	a breed, reappears after a great number of	GENERATIONS, the most probable hypothesis is, not that
0160	ddenly takes after an ancestor some hundred	GENERATIONS distant, but that in each successive gener
0161	er being inherited for an endless number of	GENERATIONS, than in quite useless or rudimentary orga
0167	ntly to look back thousands on thousands of	GENERATIONS, and I see an animal striped like a zebra,
0209	en transmitted by inheritance to succeeding	GENERATIONS. It can be clearly shown that the most won
0214	cross with a bulldog has affected for many	GENERATIONS the courage and obstinacy of greyhounds; a
0214	ction of the best individuals in successive	GENERATIONS made tumblers what they now are: and near
0216	electing and accumulating during successive	GENERATIONS, peculiar mental habits and actions, which
0248	d to the sterility of hybrids in successive	GENERATIONS; though Gartner was enabled to rear some h
0248	, for six or seven, and in one case for ten	GENERATIONS, yet he asserts positively that their fert
0248	y often suddenly decreases in the first few	GENERATIONS. Nevertheless I believe that in all these
0249	the increase of fertility in the successive	GENERATIONS of artificially fertilised hybrids may, I
0252	with respect to the fertility in successive	GENERATIONS of the more fertile hybrid animals, I hard
0253	breed would assuredly be lost in a very few	GENERATIONS. Although I do not know of any thoroughly
0254	the hybrids must have become in subsequent	GENERATIONS quite fertile under domestication. This la
0265	en organic beings are placed during several	GENERATIONS under conditions not natural to them, they
0265	is with hybrids, for hybrids in successive	GENERATIONS are eminently liable to vary, as every exp
0267	lose interbreeding continued during several	GENERATIONS between the nearest relations, especially
0269	ds to eliminate sterility in the successive	GENERATIONS of hybrids, which were at first only sligh
0272	fertile hybrids are propagated for several	GENERATIONS an extreme amount of variability in their
0273	the variability, however, in the successive	GENERATIONS of mongrels is, perhaps, greater than in h
0273	their extreme variability in the succeeding	GENERATIONS, is a curious fact and deserves attention.
0274	ent form, by repeated crosses in successive	GENERATIONS with either parent. These several remarks
0287	of living forms. What an infinite number of	GENERATIONS, which the mind cannot grasp, must have su
0316	uously existed by an unbroken succession of	GENERATIONS, from the lowest Silurian stratum to the p
0338	ntinually adds, in the course of successive	GENERATIONS, more and more difference to the adult. Th
0342	ng compared with the incalculable number of	GENERATIONS which must have passed away even during a
0356	g and training many individuals during many	GENERATIONS. Before discussing the three classes of fa
0424	e likewise includes the so called alternate	GENERATIONS of Steenstrup, which can only in a technic
0448	course of modification carried on for many	GENERATIONS, having to provide for their own wants at
0454	main agency: that it has led in successive	GENERATIONS to the gradual reduction of various organs

0461 ound, they must in the course of successive GENERATIONS have passed from some one part to the othe
0464 tely as nothing compared with the countless GENERATIONS of countless species which certainly have
0466 , which has already been inherited for many GENERATIONS, may continue to be inherited for an almos
0466 inherited for an almost infinite number of GENERATIONS. On the other hand we have evidence that v
0480 ture animal were reduced, during successive GENERATIONS, by disuse or by the tongue and palate hav
0481 mulated during an almost infinite number of GENERATIONS. Although I am fully convinced of the trut
0154 aordinarily great that we ought to find the GENERATIVE variability, as it may be called, still pre
0414 owen, in speaking of the dugong, says, The GENERATIVE organs being those which are most remotely
0016 o not differ from each other in characters of GENERIC value. I think it could be shown that this s
0016 widely in determining what characters are of GENERIC value; all such valuations being at present
0016 we have no right to expect often to meet with GENERIC differences in our domesticated productions.
0115 iversification of structure, amounting to new GENERIC differences, would have been profitable to t
0131 iable: specific characters more variable than GENERIC: secondary sexual characters variable. Speci
0154 at specific characters are more variable than GENERIC. To explain by a simple example what is mean
0154 s had blue flowers, the colour would become a GENERIC character, and its variation would be a more
0154 at specific characters are more variable than GENERIC, because they are taken from parts of less p
0155 at specific characters are more variable than GENERIC; but I have repeatedly noticed in works on n
0155 shows that a character, which is generally of GENERIC value, when it sinks in value and becomes on
0155 m the species of some other genus, are called GENERIC characters; and these characters in common I
0156 actly the same manner: and as these so called GENERIC characters have been inherited from a remote
0158 ich distinguish species from species, than of GENERIC characters, or those which the species posse
0168 from a common parent, are more variable than GENERIC characters, or those which have long been in
0415 certain organs in the Proteaceae, says their GENERIC importance, like that of all their parts, no
0415 characters singly is frequently of more than GENERIC importance, though here even when all taken
0473 er from each other, be more variable than the GENERIC characters in which they all agree? Why, for
0474 be more likely still to be variable than the GENERIC characters which have been inherited without
0002 having the satisfaction of acknowledging the GENEROUS assistance which I have received from very m
0034 the Roman classical writers. From passages in GENESIS, it is clear that the colour of domestic ani
0285 f one or two thousand feet in height; for the GENTLE slope of the lava streams, due to their forme
0036 nce between the sheep possessed by these two GENTLEMEN is so great that they have the appearance o
0061 ri in the plumed seed which is wafted by the GENTLEST breeze; in short, we see beautiful adaptatio
0287 llion years. The action of fresh water on the GENTLY inclined Wealden district, when upraised, cou
0382 ving drift on our mountain summits, in a line GENTLY rising from the arctic lowlands to a great he
0016 other, and from the other species of the same GENUS, in several trifling respects, they often dif
0016 ee than, do closely allied species of the same GENUS in a state of nature. I think this must be ad
0023 unt, the barb, pouter, and fantail in the same GENUS: more especially as in each of these breeds s
0054 for the mere fact of many species of the same GENUS inhabiting any country, shows that there is s
0054 c conditions of that country favourable to the GENUS: and, consequently, we might have expected to
0055 sely related species (i.e. species of the same GENUS) have been formed, many varieties or incipien
0055 pect to find saplings. Where many species of a GENUS have been formed through variation, circumsta
0056 rieties; for wherever many species of the same GENUS have been formed, or where, if we may use the
0056 general rule that, wherever many species of a GENUS have been formed, the species of that genus p
0056 a genus have been formed, the species of that GENUS present a number of varieties, that is of inc
0056 want to show is, that where many species of a GENUS have been formed, on an average many are stil
0057 naturalist pretends that all the species of a GENUS are equally distinct from each other; they ma
0057 less than that between the species of the same GENUS. But when we come to discuss the principle, a
0060 cies; often severe between species of the same GENUS. The relation of organism to organism the mos
0061 ach other more than do the species of the same GENUS, arise? All these results, as we shall more f
0073 ce i have very little doubt, that if the whole GENUS of humble bees became extinct or very rare in
0076 re not annually sorted. As species of the same GENUS have usually, though by no means invariably,
0076 lly be more severe between species of the same GENUS, when they come into competition with each ot
0098 with the barberry; and curiously in this very GENUS, which seems to have a special contrivance fo
0101 pecies of the same family and even of the same GENUS, though agreeing closely with each other in a
0110 s of the same species, and species of the same GENUS or of related genera, which, from having near
0111 es of the same species and species of the same GENUS. As has always been my practice, let us seek
0116 subject. Let A to L represent the species of a GENUS large in its own country; these species are s
0116 ding at unequal distances. I have said a large GENUS, because we have seen in the second chapter,
0117 diffused, and varying species, belonging to a GENUS large in its own country. The little fan of d
0118 f those more general advantages which made the GENUS to which the parent species belonged, a large
0118 to which the parent species belonged, a large GENUS in its own country. And these circumstances w
0119 widely diffused species, belonging to a large GENUS, will tend to partake of the same advantages
0120 e multiplied and genera are formed. In a large GENUS it is probable that more than one species wou
0121 4, are supposed to have been produced. In each GENUS, the species, which are already extremely dif
0121 es (marked by capital letters) of our original GENUS, may for a long period continue transmitting
0122 further than this. The original species of our GENUS were supposed to resemble each other in unequ
0122 dvantage over most of the other species of the GENUS. Their modified descendants, fourteen in numb
0123 e other five species, and may constitute a sub GENUS or even a distinct genus. The six descendants
0123 may constitute a sub genus or even a distinct GENUS. The six descendants from (I) will form two s
0123 g nearly at the extreme points of the original GENUS, the six descendants from (I) will, owing to
0123 fication, from two or more species of the same GENUS. And the two or more parent species are suppo
0124 descended from some one species of an earlier GENUS. In our diagram, this is indicated by the bro
0125 ave descended from two species of the original GENUS; and these two species are supposed to have d
0125 ne species of a still more ancient and unknown GENUS. We have seen that in each country it is the
0128 reater differences between species of the same GENUS, or even of distinct genera. We have seen tha
0128 closely related together, species of the same GENUS less closely and unequally related together,
0131 exual characters variable. Species of the same GENUS vary in an analogous manner. Reversions to lo
0139 it is extremely common for species of the same GENUS to inhabit very hot and very cold countries,
0139 as I believe that all the species of the same GENUS have descended from a single parent, if this
0150 comparison with the other species of the same GENUS. The rule applies very strongly in the case o
0151 rent genera; but in the several species of one GENUS, Pyrgoma, these valves present a marvellous a
0153 s, compared with the other species of the same GENUS, we may conclude that this part has undergone
0153 branched off from the common progenitor of the GENUS. This period will seldom be remote in any ext
0154 mple what is meant. If some species in a large GENUS of plants had blue flowers and some had red,
0155 ther independently created species of the same GENUS, be more variable than those parts which are
0155 nner: the points in which all the species of a GENUS resemble each other, and in which they differ
0155 ich they differ from the species of some other GENUS, are called generic characters; and these cha

Page ***(Key Word)***

```
0156  species differ from other species of the same  GENUS, are called specific characters; and as these
0157  ion in which the different species of the same   GENUS differ from each other. Of this fact I will g
0157  subject: I look at all the species of the same   GENUS as having as certainly descended from the sam
0161  to produce it. As all the species of the same    GENUS are supposed, on my theory, to have descended
0161  rther be expected that the species of the same   GENUS would occasionally exhibit reversions to lost
0162  it were, some of the other species of the same   GENUS. A considerable catalogue, also, could be giv
0163  from occurring in several species of the same    GENUS, partly under domestication and partly under
0165  s of crossing the several species of the horse   GENUS. Rollin asserts, that the common mule from th
0166  see several very distinct species of the horse   GENUS becoming, by simple variation, striped on the
0166  general colouring of the other species of the    GENUS. The appearance of the stripes is not accompa
0166  just seen that in several species of the horse   GENUS the stripes are either plainer or appear more
0166  tly parallel is the case with that ofthe horse   GENUS! For myself, I venture confidently to look ba
0167  en to become striped like other species of the   GENUS! and that each has been created with a strong
0167  ot their own parents, but other species of the   GENUS. To admit this view is, as it seems to me, to
0167  reater differences between species of the same   GENUS. The external conditions of life, as climate
0168  o differ since the several species of the same   GENUS branched off from a common parent, are more v
0168  li; for in a district where many species of any  GENUS are found, that is, where there has been much
0169  differences to the several species of the same   GENUS. Any part or organ developed to an extraordin
0169  extraordinary amount of modification since the   GENUS arose; and thus we can understand why it shou
0176  ermediate between well marked varieties in the   GENUS Balanus. And it would appear from information
0180  ructures in closely allied species of the same   GENUS; and of diversified habits, either constant o
0184  n species and of the other species of the same   GENUS, we might expect, on my theory, that such ind
0215  wolves, foxes, jackals, and species of the cat   GENUS, when kept tame, are most eager to attack pou
0238  each other, as are any two species of the same   GENUS, or rather as any two genera of the same fami
0240  celli, though the male and female ants of this   GENUS have well developed ocelli. I may give one ot
0250  es of Lobelia, and with all the species of the   GENUS Hippeastrum, which can be far more easily fer
0255  y, the pollen of different species of the same   GENUS applied to the stigma of some one species, yi
0256  ry fertile. Even within the limits of the same   GENUS, for instance in Dianthus, these two opposite
0257  st facility. In the same family there may be a   GENUS, as Dianthus, in which very many species can
0257  ecies can most readily be crossed; and another  GENUS, as Silene, in which the most persevering ef
0257  gle hybrid. Even within the limits of the same   GENUS, we meet with this same difference: for insta
0257  y crossed than the species of almost any other  GENUS: but Gartner found that N. acuminata, which i
0259  with other species; other species of the same   GENUS have a remarkable power of impressing their l
0261  d together, in other cases species of the same   GENUS will not take on each other. The pear can be
0261  y on the quince, which is ranked as a distinct   GENUS, than on the apple, which is a member of the
0261  an on the apple, which is a member of the same   GENUS. Even different varieties of the pear take wi
0265  l, till he tries, whether any two species of a   GENUS will produce more or less sterile hybrids. La
0281  been connected with the parent species of each   GENUS, by differences not greater than we see betwe
0288  rtiary formation: yet it is now known that the   GENUS Chthamalus existed during the chalk period. T
0288  existed during the chalk period. The molluscan  GENUS Chiton offers a partially analogous case. Wit
0304  thamalus, a very commom, large, and ubiquitous  GENUS, of which not one specimen has as yet been fo
0313  ers but little from the living species of this   GENUS; whereas most of the other Silurian Molluscs
0316  same old and unmodified forms. Species of the   GENUS Lingula, for instance, must have continuously
0317  y decreases. If the number of the species of a  GENUS, or the number of the genera of a family, be
0317  ble with my theory; as the species of the same  GENUS, and the genera of the same family, can incre
0318  during which any single species or any single   GENUS endures. There is reason to believe that the
0321  of that species; i.e. the species of the same   GENUS, will be the most liable to extermination. Th
0321  cies descended from one species, that is a new   GENUS, comes to supplant an old genus, belonging to
0321  that is a new genus, comes to supplant an old   GENUS, belonging to the same family. But it must of
0321  nstance, a single species of Trigonia, a great  GENUS of shells in the secondary formations, surviv
0331  lines diverging from them the species in each   GENUS. The diagram is much too simple, too few gene
0341  hat, on my theory, all the species of the same  GENUS have descended from some one species; so that
0349  ins of La Plata by another species of the same  GENUS: and not by a true ostrich or emeu, like thos
0351  obvious, that the several species of the same   GENUS, though inhabiting the most distant quarters
0351  e have reason to believe that the species of a  GENUS have been produced within comparatively recen
0354  ely, whether the several distinct species of a  GENUS, which on my theory have all descended from a
0405  iates. But on the view of all the species of the  GENUS having descended from a single parent, though
0407  th respect to the distinct species of the same  GENUS, which on my theory must have spread from one
0412  h letter on the uppermost line may represent a  GENUS including several species; and all the genera
0413  carnivora, by another those common to the dog   GENUS, and then by adding a single sentence, a full
0413  aled form, that the characters do not make the  GENUS, but that the genus gives the characters, see
0413  characters do not make the genus, but that the  GENUS gives the characters, seem to imply that some
0417  linnaeus, that the characters do not give the   GENUS, but the genus gives the characters: for this
0417  the characters do not give the genus, but the   GENUS gives the characters; for this saying seems f
0417  f the characters proper to the species, to the  GENUS, to the family, to the class, disappear, and
0417  agaciously saw, as Jussieu observes, that this  GENUS should still be retained amongst the Malpighi
0419  irst ranked by practised naturalists as only a  GENUS, and then raised to the rank of a sub family
0421  ecies, descended from A, be ranked in the same  GENUS with the parent A; or those from I, with the
0421  se from I, with the parent I. But the existing  GENUS F14 may be supposed to have been but slightly
0421  odified; and it will then rank with the parent  GENUS F; just as some few still living organic bein
0421  existing organisms. All the descendants of the  GENUS F, along its whole line of descent, are suppo
0421  ut little modified, and they yet form a single  GENUS. But this genus, though much isolated, will s
0421  ed, and they yet form a single genus. But this  GENUS, though much isolated, will still occupy its
0425  family would have to be classed under the bear  GENUS. The whole case is preposterous; for where th
0427  ed genera, because all the species of the same  GENUS, inhabiting any distinct and isolated region,
0439  till a rather late age: thus birds of the same  GENUS, and of closely allied genera, often resemble
0446  to species in a state of nature. Let us take a  GENUS of birds, descended on my theory from some on
0447  , the young of the new species of our supposed  GENUS will manifestly tend to resemble each other m
0451  le: for instance there are beetles of the same  GENUS (and even of the same species) resembling eac
0451  and two rudimentary teats in the udders of the  GENUS Bos, but in our domestic cows the two sometim
0461  same species, and all the species of the same   GENUS, or even higher group, must have descended fr
0462  . with respect to distinct species of the same  GENUS inhabiting very distant and isolated regions,
0462  y of the wide diffusion of species of the same  GENUS is in some degree lessened. As on the theory
0468  xt in severity between the species of the same  GENUS. But the struggle will often be very severe b
0470  is that in each region where many species of a  GENUS have been produced, and where they now flouri
0470  ferences characteristic of species of the same  GENUS. New and improved varieties will inevitably s
```

Page ***(Key Word)***

Page ***(Key Word)***

```
0473  r and legs of the several species of the horse GENUS and in their hybrids! How simply is this fact
0473  ers, or those by which the species of the same GENUS differ from each other, be more variable than
0473  be more likely to vary in any one species of a GENUS, if the other species, supposed to have been
0473  loured flowers, than if all the species of the GENUS have the same coloured flowers? If species ar
0474  a very unusual manner in any one species of a GENUS, and therefore, as we may naturally infer, of
0474  it. On the view of all the species of the same GENUS having descended from a common parent, and ha
0482  ef that a multitude of reputed species in each GENUS are not real species; but that other species
0489  ows that the greater number of species of each GENUS, and all the species of many genera, have lef
0008  the embryo, or at the instant of conception. GEOFFROY St. Hilaire's experiments show that unnatura
0011  ous; and many instances are given in Isidore GEOFFROY St. Hilaire's great work on this subject. Br
0144  ion is very frequently quite obscure. M. Is. GEOFFROY St. Hilaire has forcibly remarked, that cert
0147  go on in fruit which did not open. The elder GEOFFROY and Goethe propounded, at about the same per
0149  t. it seems to be a rule, as remarked by Is. GEOFFROY St. Hilaire, both in varieties and in specie
0155  kind applies to monstrosities: at least Is. GEOFFROY St. Hilaire seems to entertain no doubt, tha
0434  same bones, in the same relative positions? GEOFFROY St. Hilaire has insisted strongly on the hig
0003  s, on their embryological relations, their GEOGRAPHICAL distribution, geological succession, and o
0005  t time; in the eleventh and twelfth, their GEOGRAPHICAL distribution throughout space; in the thir
0018  entering on any details, state that, from GEOGRAPHICAL and other considerations, I think it highl
0023  livia), including under this term several GEOGRAPHICAL races or sub species, which differ from ea
0023  o on trees. But besides C. Livia, with its GEOGRAPHICAL sub species, only two or three other speci
0027  descended from the Columba livia with its GEOGRAPHICAL sub species. In favour of this view, I may
0048  arieties, or, as they are often called, as GEOGRAPHICAL races! Many years ago, when comparing, and
0049  everal interesting lines of argument, from GEOGRAPHICAL distribution, analogical variation, hybrid
0049  to different heights; they have different GEOGRAPHICAL ranges; and lastly, according to very nume
0055  widely will be discussed in our chapter on GEOGRAPHICAL distribution. From looking at species as o
0078  which preyed on it. On the confines of its GEOGRAPHICAL range, a change of constitution with respe
0105  e small, as we shall see in our chapter on GEOGRAPHICAL distribution; yet of these species a very
0106  will be again alluded to in our chapter on GEOGRAPHICAL distribution; for instance, that the produ
0166  eon (including two or three sub species or GEOGRAPHICAL races) of a bluish colour, with certain ba
0175  iable to utter extermination; and thus its GEOGRAPHICAL range will come to be still more sharply d
0294  nimals can flourish; for we know what vast GEOGRAPHICAL changes occurred in other parts of America
0294  , owing to the migration of species and to GEOGRAPHICAL changes. And in the distant future, a geol
0296  ent nature will generally have been due to GEOGRAPHICAL changes requiring much time. Nor will the
0313  ase of temporary migration from a distinct GEOGRAPHICAL province, seems to me satisfactory. These
0314  hould change less. We see the same fact in GEOGRAPHICAL distribution; for instance, in the land sh
0326  very slow, being dependent on climatal and GEOGRAPHICAL changes, or on strange accidents, but in t
0340  ery long intervals of time and after great GEOGRAPHICAL changes, permitting much inter migration,
0346  reserved by Natural Selection. Chapter XI. GEOGRAPHICAL Distribution. Present distribution cannot
0346  t one of the most fundamental divisions in GEOGRAPHICAL distribution is that between the New and O
0351  from the same region; for during the vast GEOGRAPHICAL and climatal changes which will have super
0353  assed from one point to the other. But the GEOGRAPHICAL and climatal changes, which have certainly
0354  orance with respect to former climatal and GEOGRAPHICAL changes and various occasional means of tr
0357  not authorized in admitting such enormous GEOGRAPHICAL changes within the period of existing spec
0358  pposed to the admission of such prodigious GEOGRAPHICAL revolutions within the recent period, as a
0372  many forms, which some naturalists rank as GEOGRAPHICAL races, and others as distinct species; and
0382  ffects of great alternations of climate on GEOGRAPHICAL distribution. I believe that the world has
0383                      Lowlands. Chapter XII. GEOGRAPHICAL Distribution, continued. Distribution of f
0384  there will have been ample time for great GEOGRAPHICAL changes, and consequently time and means f
0405  ave been ample time for great climatal and GEOGRAPHICAL changes and for accidents of transport; an
0408  hen I think all the grand leading facts of GEOGRAPHICAL distribution are explicable on the theory
0409  in scanty numbers, in the different great GEOGRAPHICAL provinces of the world. On these same prin
0419  this, and to no other class of Articulata. GEOGRAPHICAL distribution has often been used, though p
0427  re of such high classificatory importance. GEOGRAPHICAL distribution may sometimes be brought usef
0461  fied forms, increase fertility. Turning to GEOGRAPHICAL distribution, the difficulties encountered
0462  he full extent of the various climatal and GEOGRAPHICAL changes which have affected the earth duri
0476  naturally be allied by descent. Looking to GEOGRAPHICAL distribution, if we admit that there has b
0476  d to another, owing to former climatal and GEOGRAPHICAL changes and to the many occasional and unk
0290  ons, generally implying great changes in the GEOGRAPHY of the surrounding lands, whence the sedime
0487  gration, some light can be thrown on ancient GEOGRAPHY. The noble science of Geology loses glory f
0001  the inhabitants of South America and in the GEOLOGICAL relations of the present to the past inhabi
0003  relations, their geographical distribution, GEOLOGICAL succession, and other such facts, might com
0005  ssed; and fourthly, the imperfection of the GEOLOGICAL Record. In the next chapter I shall conside
0005  d. in the next chapter I shall consider the GEOLOGICAL succession of organic beings throughout tim
0006  habitants of the world during the many past GEOLOGICAL epochs in its history. Although much remain
0084  th those accumulated by nature during much GEOLOGICAL periods. Can we wonder, then, that nature's
0084  hen so imperfect is our view into long past GEOLOGICAL ages, that we only see that the forms of ar
0126  odified descendants, yet at the most remote GEOLOGICAL period, the earth may have been as well peo
0129  th the species which lived during long past GEOLOGICAL periods, very few now have living and modif
0153  pecies very rarely endure for more than one GEOLOGICAL period. An extraordinary amount of modifica
0172  n in the chapter on the Imperfection of the GEOLOGICAL record; and I will here only state that I b
0173  eing deposited, there will be blanks in our GEOLOGICAL history. The crust of the earth is a vast m
0279  ies. Chapter IX. On the Imperfection of the GEOLOGICAL Record. On the absence of intermediate vari
0279  ogical collections. On the intermittence of GEOLOGICAL formations. On the absence of intermediate
0280  h, be truly enormous. Why then is not every GEOLOGICAL formation and every stratum full of such in
0280  believe, in the extreme imperfection of the GEOLOGICAL record. In the first place it should always
0287  long roll of years! Now turn to our richest GEOLOGICAL museums, and what a paltry display we behol
0289  oic formations. But the imperfection in the GEOLOGICAL record mainly results from another and more
0290  formation. But we can, I think, see why the GEOLOGICAL formations of each region are almost invari
0290  iciently extensive to last for even a short GEOLOGICAL period. Along the whole west coast, which i
0291  ed, but which will hardly last to a distant GEOLOGICAL age, was certainly deposited during a downw
0291  and thus gained considerable thickness. All GEOLOGICAL facts tell us plainly that each area has un
0292  in the limits of the coast action. Thus the GEOLOGICAL record will almost necessarily be rendered
0292  iods there will generally be a blank in the GEOLOGICAL record. On the other hand, during subsidenc
0292  onsiderations it cannot be doubted that the GEOLOGICAL record, viewed as a whole, is extremely imp
0294  eriod, which forms only a part of one whole GEOLOGICAL period; and likewise to reflect on the grea
0296  peared, perhaps many times, during the same GEOLOGICAL period. So that if such species were to und
```

Page ***(Key Word)***

Page **************************************(Key Word)**************************************

0296	rable amount of modification during any one	GEOLOGICAL	period, a section would not probably includ
0297	ned we can seldom hope to effect in any one	GEOLOGICAL	section. Supposing B and C to be two specie
0298	ance, far exceeding the limits of the known	GEOLOGICAL	formations of Europe, which have oftenest g
0298	o trace the stages of transition in any one	GEOLOGICAL	formation. It should not be forgotten, that
0299	ems to me improbable in the highest degree.	GEOLOGICAL	research, though it has added numerous spec
0299	ain to Russia; and therefore equals all the	GEOLOGICAL	formations which have been examined with an
0300	ion, there would be much variation, but the	GEOLOGICAL	record would then be least perfect. It may
0301	, we have no right to expect to find in our	GEOLOGICAL	formations, an infinite number of those fin
0302	f the mutations of life, the best preserved	GEOLOGICAL	section presented, had not the difficulty o
0302	continually over rate the perfection of the	GEOLOGICAL	record, and falsely infer, because certain
0302	d is, compared with the area over which our	GEOLOGICAL	formations have been carefully examined; we
0303	d. i may recall the well known fact that in	GEOLOGICAL	treatises, published not many years ago, th
0310	our knowledge. Those who think the natural	GEOLOGICAL	record in any degree perfect, and who do no
0310	out Lyell's metaphor, I look at the natural	GEOLOGICAL	record, as a history of the world imperfect
0312	ished, or even disappear. Chapter X. On the	GEOLOGICAL	Succession of Organic Beings. On the slow a
0312	the several facts and rules relating to the	GEOLOGICAL	succession of organic beings, better accord
0313	ictly correspond with the succession of our	GEOLOGICAL	formations: so that between each two consec
0317	varying thickness, crossing the successive	GEOLOGICAL	formations in which the species are found,
0318	ted with still living shells at a very late	GEOLOGICAL	period, I was filled with astonishment; for
0320	itely increasing, at least during the later	GEOLOGICAL	periods, so that looking to later times we
0323	without any information in regard to their	GEOLOGICAL	position, no one would have suspected that
0324	dth year, or even that it has a very strict	GEOLOGICAL	sense: for if all the marine animals which
0324	ld be correctly ranked as simultaneous in a	GEOLOGICAL	sense. The fact of the forms of life changi
0330	have undergone much change in the course of	GEOLOGICAL	ages; and it would be difficult to prove th
0331	e horizontal lines may represent successive	GEOLOGICAL	formations, and all the forms beneath the u
0332	above one of the middle horizontal lines or	GEOLOGICAL	formations, for instance, above No. VI., bu
0333	. as we possess only the last volume of the	GEOLOGICAL	record, and that in a very broken condition
0333	that those groups, which have within known	GEOLOGICAL	periods undergone much modification, should
0334	ject to these allowances, the fauna of each	GEOLOGICAL	period uncoubtedly is intermediate in chara
0338	nt animals of the same classes; or that the	GEOLOGICAL	succession of extinct forms is in some degr
0341	each having eight species; be found in one	GEOLOGICAL	formation, and in the next succeeding forma
0341	chapters. I have attempted to show that the	GEOLOGICAL	record is extremely imperfect; that only a
0342	en conjointly, must have tended to make the	GEOLOGICAL	record extremely imperfect, and will to a l
0342	ho rejects these views on the nature of the	GEOLOGICAL	record, will rightly reject my whole theory
0345	ture within the same areas during the later	GEOLOGICAL	periods ceases to be mysterious, and is sim
0345	imply explained by inheritance. If then the	GEOLOGICAL	record be as imperfect as I believe it to b
0349	ican shore, however much they may differ in	GEOLOGICAL	structure, the inhabitants, though they may
0351	he same character from an enormously remote	GEOLOGICAL	period, so certain species have migrated ov
0351	species, which have undergone during whole	GEOLOGICAL	periods but little modification, there is n
0353	which have certainly occurred within recent	GEOLOGICAL	times, must have interrupted or rendered di
0365	eans of transport, during the long lapse of	GEOLOGICAL	time, whilst an island was being upheaved a
0366	ic and inorganic, that within a very recent	GEOLOGICAL	period, central Europe and North America su
0374	at the epoch was included within the latest	GEOLOGICAL	period. We have, also, excellent evidence,
0374	each, and that it was contemporaneous in a	GEOLOGICAL	sense, it seems to me probable that it was,
0376	lief, supported as it is by a large body of	GEOLOGICAL	evidence, that the whole world, or a large
0379	re. Although we have reason to believe from	GEOLOGICAL	evidence that the whole body of arctic shel
0384	s of level in the land within a very recent	GEOLOGICAL	period, and when the surface was peopled by
0398	s nothing in the conditions of life, in the	GEOLOGICAL	nature of the islands, in their height or c
0400	within sight of each other, having the same	GEOLOGICAL	nature, the same height, climate, etc., tha
0406	there is, also, some reason to believe from	GEOLOGICAL	evidence that organisms low in the scale wi
0429	ree orders of small size. In the chapter on	GEOLOGICAL	succession I attempted to show, on the prin
0450	t forms, may be true, but yet, owing to the	GEOLOGICAL	record not extending far enough back in tim
0463	t and still older species, why is not every	GEOLOGICAL	formation charged with such links? Why does
0463	ar, to have come in suddenly on the several	GEOLOGICAL	stages? Why do we not find great piles of s
0464	objections only on the supposition that the	GEOLOGICAL	record is far more imperfect than most geol
0464	discover, owing to the imperfection of the	GEOLOGICAL	record. Numerous existing doubtful forms co
0465	and when they do spread, if discovered in a	GEOLOGICAL	formation, they will appear as if suddenly
0465	thesis given in the ninth chapter. That the	GEOLOGICAL	record is imperfect all will admit; but tha
0466	of years, or that we know how imperfect the	GEOLOGICAL	Record is. Grave as these several difficult
0475	ced by secondary laws. If we admit that the	GEOLOGICAL	record is imperfect in an extreme degree, t
0476	ganic beings throughout space, and in their	GEOLOGICAL	succession throughout time: for in both cas
0481	too apt to assume, without proof, that the	GEOLOGICAL	record is so perfect that it would have aff
0287	rtion of the surface of the earth has been	GEOLOGICALLY	explored, and no part with sufficient care
0336	ears, but only moderately long as measured	GEOLOGICALLY,	closely allied forms, or, as they have be
0341	only a small portion of the globe has been	GEOLOGICALLY	explored with care; that only certain clas
0464	only a small portion of the world has been	GEOLOGICALLY	explored. Only organic beings of certain c
0282	ll to the reader, who may not be a practical	GEOLOGIST,	the facts leading the mind feebly to compr
0289	y other parts of the world. The most skilful	GEOLOGIST,	if his attention had been exclusively conf
0294	phical changes. And in the distant future, a	GEOLOGIST	examining these beds, might be tempted to c
0299	t. this could be effected only by the future	GEOLOGIST	discovering in a fossil state numerous inte
0352	l inhabiting distant points of the world. No	GEOLOGIST	will feel any difficulty in such cases as G
0357	ions to pass from one to the other. No other	GEOLOGIST	will dispute that great mutations of level,
0284	formation, we have, in the opinion of most	GEOLOGISTS,	enormously long blank periods. So that the
0286	m informed by Prof. Ramsay. But if, as some	GEOLOGISTS	suppose, a range of older rocks underlies t
0299	by asking ourselves whether, for instance	GEOLOGISTS	at some future period will be able to prove
0307	factory answer. Several of the most eminent	GEOLOGISTS,	with Sir R. Murchison at their head, are c
0310	oner, E. Forbes, etc., and all our greatest	GEOLOGISTS,	as Lyell, Murchison, Sedgwick, etc., have
0317	, is very generally given up, even by those	GEOLOGISTS,	as Elie de Beaumont, Murchison, Barrande,
0368	any degree warmer than at present (as some	GEOLOGISTS	in the United States believe to have been t
0464	ical record is far more imperfect than most	GEOLOGISTS	believe. It cannot be objected that there h
0480	all the most eminent living naturalists and	GEOLOGISTS	rejected this view of the mutability of spe
0481	ficulty is the same as that felt by so many	GEOLOGISTS,	when Lyell first insisted that long lines
0056	uld have been fatal to my theory; inasmuch as	GEOLOGY	plainly tells us that small genera have in t
0095	dern changes of the earth; as illustrative of	GEOLOGY;	but we now very seldom hear the action, for
0095	fitable to the preserved being; and as modern	GEOLOGY	has almost banished such views as the excava
0109	al selection accords perfectly well with what	GEOLOGY	tells us of the rate and manner at which the

Page **************************************(Key Word)**************************************

Page **(Key Word)***

```
0109  ill be more fully discussed in our chapter on  GEOLOGY: but it must be here alluded to from being i
0109  ed forms decrease and become rare. Rarity, as  GEOLOGY tells us, is the precursor to extinction. We
0109  pecific forms has not indefinitely increased,  GEOLOGY shows us plainly: and indeed we can see reas
0124  ins. We shall, when we come to our chapter on  GEOLOGY, have to refer again to this subject, and I
0127  extinction has acted in the world's history,  GEOLOGY plainly declares. Natural selection, also, l
0174  it has been continuous during a long period.  GEOLOGY would lead us to believe that almost every c
0193  een specially related to each other. Nor does  GEOLOGY at all lead to the belief that formerly most
0280  very stratum full of such intermediate links?  GEOLOGY assuredly does not reveal any such finely or
0282  arles Lyell's grand work on the Principles of  GEOLOGY, which the future historian will recognise a
0282  t that it suffices to study the principles of  GEOLOGY, or to read special treatises by different o
0291  bject in 1845, I have watched the progress of  GEOLOGY, and have been surprised to note how author
0306  ations, but chiefly from our ignorance of the  GEOLOGY of other countries beyond the confines of Eu
0465  if we look to long enough intervals of time,  GEOLOGY plainly declares that all species have chang
0468  slightest advantage will lead to victory. As  GEOLOGY plainly proclaims that each land has undergo
0487  means of migration, then, by the light which  GEOLOGY now throws, and will continue to throw, on f
0487  wn on ancient geography. The noble science of  GEOLOGY loses glory from the extreme imperfection of
0226  to Professor Miller, of Cambridge, and this  GEOMETER has kindly read over the following statement
0004  d, which inevitably follows from their high  GEOMETRICAL powers of increase, will be treated of. Th
0060  l selection. The term used in a wide sense.  GEOMETRICAL powers of increase. Rapid increase of natu
0063  sional year, otherwise, on the principle of  GEOMETRICAL increase, its numbers would quickly become
0065  ve been enabled to breed. In such cases the  GEOMETRICAL ratio of increase, the result of which nev
0065  ts and animals are tending to increase at a  GEOMETRICAL ratio, that all would most rapidly stock e
0065  hich they could any how exist, and that the  GEOMETRICAL tendency to increase must be checked by de
0078  organic being is striving to increase at a  GEOMETRICAL ratio: that each at some period of its lif
0109  h consequently endure. But as from the high  GEOMETRICAL powers of increase of all organic beings,
0127  be disputed; if there be, owing to the high  GEOMETRICAL powers of increase of each species, at som
0467  existence inevitably follows from the high  GEOMETRICAL ratio of increase which is common to all o
0470  as varieties. As each species tends by its  GEOMETRICAL ratio of reproduction to increase inordina
0050  n ozk, how closely it has been studied; yet a  GERMAN author makes more than a dozen species out of
0268  for when it is stated, for instance, that the  GERMAN Spitz dog unites more easily than other dogs
0019  al, and France but few distinct from those of  GERMANY and conversely, and so with Hungary, Spain,
0484  common, in their chemical compositior, their  GERMINAL vesicles, their cellular structure, and thei
0066  ere never destroyed, and could be ensured to  GERMINATE in a fitting place. So that in all cases, t
0139  in the amount of rain requisite for seeds to  GERMINATE, in the time of sleep, etc., and this leads
0360  avourable spot by an inland gale, they would  GERMINATE. Subsequently to my experiments, M. Martens
0360  ce of sea 900 miles in width, and would then  GERMINATE. The fact of the larger fruits often floati
0365  d, which chanced to arrive, would be sure to  GERMINATE and survive. Dispersal during the Glacial p
0358  y surprise I found that out of 87 kinds, 64  GERMINATED after an immersion of 28 days, and a few su
0359  or 90 days and afterwards when planted they  GERMINATED: an asparagus plant with ripe berries float
0359  oated for 85 days, and the seeds afterwards  GERMINATED: the ripe seeds of Helosciadium sank in two
0359  y floated for above 90 days, and afterwards  GERMINATED. Altogether out of the 94 dried plants, 18
0359  much longer period. So that as 64/87 seeds  GERMINATED after an immersion of 28 days; and as 18/94
0361  t 50 years old, three dicotyledonous plants  GERMINATED: I am certain of the accuracy of this obser
0361  ater for 30 days; to my surprise nearly all  GERMINATED. Living birds can hardly fail to be highly
0361  d perfect, and some of them, which I tried,  GERMINATED. But the following fact is more important:
0362  eat, millet, canary, hemp, clover, and beet  GERMINATED after having been from twelve to twenty one
0067  it is the seedlings which suffer most from  GERMINATING in ground already thickly stocked with oth
0359  ng 28 days, and would retain their power of  GERMINATION. In Johnston's Physical Atlas, the average
0360  oated for 42 days, and were then capable of  GERMINATION. But I do not doubt that plants exposed to
0362  n the least injure, as I know by trial, the  GERMINATION of seeds: now after a bird has found and d
0362  oological Gardens, include seeds capable of  GERMINATION. Some seeds of the oat, wheat, millet, can
0362  eral of these seeds retained their power of  GERMINATION. Certain seeds, however, were always kille
0387  have seen that seeds retain their power of  GERMINATION, when rejected in pellets or in excrement,
0398  species in every character, in their habits,  GESTURES, and tones of voice, was manifest. So it is
0032  t sufficient to become an eminent breeder. If  GIFTED with these qualities, and he studies his subj
0095  ant cause, when applied to the excavation of  GIGANTIC valleys or to the formation of the longest l
0099  return in a future chapter. In the case of a  GIGANTIC tree covered with innumerable flowers, it ma
0231  dily growing upward; but always crowned by a  GIGANTIC coping. From all the cells, both those just
0339  manifest, even to an uneducated eye, in the  GIGANTIC pieces of armour like those of the armadillo
0339  his author's restorations of the extinct and  GIGANTIC birds of New Zealand. We see it also in the
0366  e of Europe changed, that in Northern Italy,  GIGANTIC moraines, left by old glaciers, are now clot
0373  d in Sikkim, Dr. Hooker saw maize growing on  GIGANTIC ancient moraines. South of the equator, we h
0373  e andes: and this I now feel convinced was a  GIGANTIC moraine, left far below any existing glacier
0391  lapagos Islands reptiles, and in New Zealand  GIGANTIC wingless birds, take the place of mammals. I
0436  f possibility. In the paddles of the extinct  GIGANTIC sea lizards, and in the mouths of certain su
0190  uali; to give one instance; there are fish with  GILLS or branchiae that breathe the air dissolved i
0171  of trifling importance, such as the tail of a  GIRAFFE, which serves as a fly flapper, and on the o
0195  cted on by natural selection. The tail of the  GIRAFFE looks like an artificially constructed fly f
0479  e number of vertebrae forming the neck of the  GIRAFFE and of the elephant, and innumerable other s
0270  er the two varieties as specifically distinct.  GIROU de Buzareingues crossed three varieties of go
0362  rted that all the grains do not pass into the  GIZZARD for 12 or even 18 hours. A bird in this inte
0258  ms so closely related (as Matthiola annua and  GLABRA) that many botanists rank them only as variet
0141  ant and rhinoceros were capable of enduring a  GLACIAL climate, whereas the living species are now
0294  ation of the inhabitants of Europe during the  GLACIAL period, which forms only a part of one whole
0294  lapse of time, all included within this same  GLACIAL period. Yet it may be doubted whether in any
0294  ediment was deposited during the whole of the  GLACIAL period near the mouth of the Mississippi, wi
0294  th of the Mississippi during some part of the  GLACIAL period shall have been upraised, organic rem
0295  bedded fossils had been less than that of the  GLACIAL period, instead of having been really far or
0295  ar greater; that is extending from before the  GLACIAL epoch to the present day. In order to get a
0324  iod as measured by years, including the whole  GLACIAL epoch), were to be compared with those now l
0336  pleistocene period, which includes the whole  GLACIAL period, and note how little the specific for
0346  and by occasional means. Dispersal during the  GLACIAL period co extensive with the world. In consi
0363  egions, as suggested by Lyell: and during the  GLACIAL period from one part of the now temperate re
0363  partly stocked by ice borne seeds, during the  GLACIAL epoch. At my request Sir C. Lyell wrote to M
0365  o germinate and survive. Dispersal during the  GLACIAL period. The identity of many plants and anim
0366  ssiz and others called vivid attention to the  GLACIAL Period, which, as we shall immediately see,
```

Page **(Key Word)***

Page ***(Key Word)***

```
0366  rmer cold period. The former influence of the  GLACIAL  climate on the distribution of the inhabitan
0366  the changes more readily, by supposing a new    GLACIAL  period to come slowly on, and then pass away
0367  form round the world. We may suppose that the   GLACIAL  period came on a little earlier or later in
0368  ectly well ascertained occurrence of a former   GLACIAL  period, seem to me to explain in so satisfac
0368  or their existence. If the climate, since the   GLACIAL  period, has ever been in any degree warmer t
0368  ntercalated slightly warmer period, since the   GLACIAL  period. The arctic forms, during their long
0369  the mountains before the commencement of the    GLACIAL  epoch, and which during its coldest period w
0369  as I believe, actually took place during the    GLACIAL  period, I assumed that at its commencement t
0369  s round the world, at the commencement of the   GLACIAL  period. At the present day, the sub arctic a
0369  reme northern part of the Pacific. During the    GLACIAL  period, when the inhabitants of the Old and
0370  during the newer Pliocene period, before the    GLACIAL  epoch, and whilst the majority of the inhabi
0370  d and New Worlds, at a period anterior to the    GLACIAL  epoch. Believing, from reasons before allude
0371  ess warm, long before the commencement of the   GLACIAL  period. We now see, as I believe, their desc
0372  ust return to our more immediate subject, the   GLACIAL  period. I am convinced that Forbes's view ma
0373  uator, we have some direct evidence of former   GLACIAL  action in New Zealand; and the same plants,
0373  ed can be trusted, we have direct evidence of   GLACIAL  action in the south eastern corner of Austra
0373  mity, we have the clearest evidence of former   GLACIAL  action, in huge boulders transported far fro
0373  their parent source. We do not know that the    GLACIAL  epoch was strictly simultaneous at these sev
0374  y, we may at least admit as probable that the   GLACIAL  action was simultaneous on the eastern and w
0376  world, or a large part of it, was during the    GLACIAL  period simultaneously much colder than at pr
0377  multaneously much colder than at present. The   GLACIAL  period, as measured by years, must have been
0378  some marine productions, migrated during the    GLACIAL  period from the northern and southern temper
0379  thus, when they became commingled during the    GLACIAL  period, the northern forms were enabled to b
0380  intertropical mountains: no doubt before the    GLACIAL  period they were stocked with endemic Alpine
0380  d; and the intertropical mountains before the   GLACIAL  period must have been completely isolated; a
0381  egia, I believe that towards the close of the   GLACIAL  period, icebergs, as suggested by Lyell, hav
0381  e has been time since the commencement of the   GLACIAL  period for their migration, and for their su
0381  warmer period, before the commencement of the   GLACIAL  period, when the antarctic lands, now covere
0381  hat before this flora was exterminated by the   GLACIAL  epoch, a few forms were widely dispersed to
0382  lands, and perhaps at the commencement of the   GLACIAL  period, by icebergs. By these means, as I be
0393  ormation be correct) may be explained through   GLACIAL  agency. This general absence of frogs, toads
0399  th vegetation, before the commencement of the   GLACIAL  period. The affinity, which, though feeble,
0403  widely throughout the world during the recent   GLACIAL  epoch, are related to those of the surroundi
0407  mportant has been the influence of the modern   GLACIAL  period, which I am fully convinced simultane
0462  show how potent has been the influence of the   GLACIAL  period on the distribution both of the same
0477  ication, we can understand, by the aid of the   GLACIAL  period, the identity of some few plants, and
0373  gigantic moraine, left far below any existing   GLACIER.  Further south on both sides of the continen
0366  rthern Italy, gigantic moraines, left by old    GLACIERS,  are now clothed by the vine and maize. Thro
0373  ong the Himalaya, at points 900 miles apart,   GLACIERS  have left the marks of their former low desc
0373  the Cordillera of Equatorial South America,    GLACIERS  once extended far below their present level.
0240  castes of neuters in the same species, that I   GLADLY  availed myself of Mr. F. Smith's offer of num
0062  od. We behold the face of nature bright with    GLADNESS, we often see superabundance of food; we do
0094  um pratense and incarnatum) do not on a hasty   GLANCE  appear to differ in length; yet the hive bee
0285  e to their formerly liquid state, showed at a   GLANCE  how far the hard, rocky beds had once extende
0289  a. In regard to mammiferous remains, a single   GLANCE  at the historical table published in the Supp
0423  tongue. In confirmation of this view, let us    GLANCE  at the classification of varieties, which are
0440  was, as it certainly is, a crustacean; but a   GLANCE  at the larva shows this to be the case in an
0489  terly extinct. We can so far take a prophetic   GLANCE  into futurity so to foretel that it will be t
0474  onstant by long continued natural selection.   GLANCING  at instincts, marvellous as some are, they o
0022  in good birds the head and tail touch; the oil  GLAND  is quite aborted. Several other less distinct
0022  hagus, the development and abortion of the oil  GLAND; the number of the primary wing and caudal fe
0193  en grains, borne on a foot stalk with a sticky  GLAND  at the end, is the same in Orchis and Asclepi
0092  injurious from their sap: this is effected by   GLANDS  at the base of the stipules in some Leguminos
0092  hose individual flowers which had the largest   GLANDS  or nectaries, and which excreted most nectar,
0192  r size and the obliteration of their adhesive   GLANDS. If all pedunculated cirripedes had become ex
0214  ons made tumblers what they now are: and near   GLASGOW  there are house tumblers, as I hear from Mr.
0189  ent might thus be formed as superior to one of  GLASS, as the works of the Creator are to those of
0137  scope is there, though the telescope with its   GLASSES  has been lost. As it is difficult to imagine
0180  which serves as a parachute and allows them to  GLIDE  through the air to an astonishing distance fr
0182  it is conceivable that flying fish, which now   GLIDE  far through the air, slightly rising and turn
0204  ction from an animal which at first could only  GLIDE  through the air. We have seen that a species
0181  h no graduated links of structure, fitted for  GLIDING  through the air, now connect the Galeopithec
0181  me steps as in the case of the less perfectly  GLIDING  squirrels; and that each grace of structure
0181  es of an apparatus originally constructed for  GLIDING  through the air rather than for flight. If a
0335  the distribution of existing species over the  GLOBE, will not attempt to account for the close re
0341  ly imperfect; that only a small portion of the  GLOBE  has been geologically explored with care; tha
0346  ibution of organic beings over the face of the  GLOBE, the first great fact which strikes us is, th
0370  still nearer to the pole. Now if we look at a   GLOBE, we shall see that under the Polar Circle ver
0374  n, or have ceased, earlier at one point of the  GLOBE  than at another; but seeing that it endured f
0389  hey have come to inhabit distant points of the  GLOBE. I have already stated that I cannot honestly
0021  ; and its enormously developed crop, which it   GLORIES  in inflating, may well excite astonishment a
0487  geography. The noble science of Geology loses   GLORY  from the extreme imperfection of the record.
0191  anding the beautiful contrivance by which the   GLOTTIS  is closed. In the higher Vertebrata the bran
0271  er or mother, and crossed with the Nicotiana   GLUTINOSA, always yielded hybrids not so sterile as t
0271  he four other varieties when crossed with N.    GLUTINOSA. Hence the reproductive system of this one
0365  ope. Even as long ago as 1747, such facts led   GMELIN  to conclude that the same species must have b
0368  chiefly from the distribution of the fossil    GNATHODON), then the arctic and temperate productions
0230  ntial wall or rim all round the comb: and they  GNAW  into this from the opposite sides, always work
0232  pheres; but, as far as I have seen, they never  GNAW  away and finish off the angles of a cell till
0229  he ridge of vermilion wax, as they circularly   GNAWED  away and deepened the basins on both sides, i
0229  of a strip of wax, perceiving when they have    GNAWED  the wax away to the proper thinness, and then
0231  ery case to be finished off, by being largely   GNAWED  away on both sides. The manner in which the b
0230  plate, that they could have effected this by    GNAWING  away the convex side; and I suspect that the
0011  nherited development of the udders in cows and  GOATS  in countries where they are habitually milked
0018  m several wild species. In regard to sheep and  GOATS  I can form no opinion. I should think, from f
0019  ies of wild cattle, as many sheep, and several  GOATS  in Europe alone, and several even within Grea
```

Page ***(Key Word)***

Page **(Key Word)**

```
0167  st for an unknown, cause. It makes the works of  GOD  a mere mockery and deception! I would almost a
0299  ted States of America. I fully agree with Mr.  GODWIN  Austen, that the present condition of the Mal
0147  it which did not open. The elder Geoffroy and  GOETHE  propounded, at about the same period, their l
0147  compensation or balancement of growth; or, as  GOETHE  expressed it, in order to spend on one side,
0297  mportant to remember that naturalists have no  GOLDEN  rule by which to distinguish species and vari
0002  one I have always been cautious in trusting to  GOOD  authorities alone. I can here give only the ge
0021  ent expanded, and are carried so erect that in  GOOD  birds the head and tail touch; the oil gland i
0023  d state. But birds breeding on precipices, and  GOOD  fliers, are unlikely to be exterminated, and t
0030  ion, not indeed to the animal's or plant's own  GOOD, but to man's use or fancy. Some variations us
0030  r mountain pasture, with the wool of one breed  GOOD  for one purpose, and that of another breed for
0030  when we compare the many breeds of dogs, each  GOOD  for man in very different ways; when we compar
0031  y other individual, and who was himself a very  GOOD  judge of an animal, speaks of the principle of
0031  y the enormous prices given for animals with a  GOOD  pedigree; and these have now been exported to
0034  heir teams of dogs. Livingstone shows how much  GOOD  domestic breeds are valued by the negroes of t
0034  been paid to breeding, for the inheritance of  GOOD  and bad qualities is so obvious. At the presen
0034  nds keeping pointers naturally tries to get as  GOOD  dogs as he can, and afterwards breeds from his
0038  how it is that neither Australia, the Cape of  GOOD  Hope, nor any other region inhabited by quite
0045  parts of structure, which he could collect on  GOOD  authority, as I have collected, during a cours
0047  time; for as long, as far as we know, as have  GOOD  and true species. Practically, when a naturali
0048  r of forms have been ranked by one botanist as  GOOD  species, and by another as mere varieties. Mr.
0049  are ranked by other highly competent judges as  GOOD  and true species. But to discuss whether they
0050  t the sessile and pedunculated oaks are either  GOOD  and distinct species or mere varieties. When a
0056  average many are still forming; and this holds  GOOD. There are other relations between the species
0058  ost universally ranked by British botanists as  GOOD  and true species. Finally, then, varieties hav
0061  ient species, become ultimately converted into  GOOD  and distinct species, which in most cases obvi
0070  from utter destruction. I should add that the  GOOD  effects of frequent intercrossing, and the ill
0078  different set of competitors or enemies. It is  GOOD  thus to try in our imagination to give any for
0083  achinery of life. Man selects only for his own  GOOD; Nature only for that of the being which she t
0084  h is bad, preserving and adding up all that is  GOOD; silently and insensibly working, whenever and
0084  ral selection can act only through and for the  GOOD  of each being, yet characters and structures,
0086  s are accumulated by natural selection for the  GOOD  of the being, will cause other modifications,
0087  cies, without giving it any advantage, for the  GOOD  of another species; and though statements to t
0089  ording to his standard of beauty, I can see no  GOOD  reason to doubt that female birds, by selectin
0092  get crossed; and the act of crossing, we have  GOOD  reason to believe (as will hereafter be more f
0097  rry pollen from one to the other, to the great  GOOD, as I believe, of the plant. Bees will act lik
0099  ni and that this is part of the general law of  GOOD  being derived from the intercrossing of distin
0106  tly existed in a broken condition; so that the  GOOD  effects of isolation will generally, to a cert
0109  seasons or in the number of its enemies, run a  GOOD  chance of utter extinction. But we may go furt
0110  on is as yet fully stocked, for at the Cape of  GOOD  Hope, where more species of plants are crowded
0111  tainly differ from each other far less than do  GOOD  and distinct species. Nevertheless, according
0113  hus be raised. The same has been found to hold  GOOD  when first one variety and then several mixed
0136  ear a coast, it would have been better for the  GOOD  swimmers if they had been able to swim still f
0140  brought from warmer countries which here enjoy  GOOD  health. We have reason to believe that species
0143  that modifications accumulated solely for the  GOOD  of the young or larva, will, it may safely be
0147  the law is of universal application; but many  GOOD  observers, more especially botanists, believe
0151  onvinced that the rule almost invariably holds  GOOD  with cirripedes. I shall, in my future work, g
0151  em, and the rule seems to me certainly to hold  GOOD  in this class. I cannot make out that it appli
0153  ed a bird as coarse as a common tumbler from a  GOOD  short faced strain. But as long as selection i
0176  lly, as far as I can make out, this rule holds  GOOD  with varieties in a state of nature. I have me
0182  x. yet the structure of each of these birds is  GOOD  for it, under the conditions of life to which
0190  illustration of the swimbladder in fishes is a  GOOD  one, because it shows us clearly the highly im
0194  vention, so natural selection, working for the  GOOD  of each being and taking advantage of analogou
0198  that constitution and colour are correlated. A  GOOD  observer, also, states that in cattle suscepti
0199  detail of structure has been produced for the  GOOD  of its possessor. They believe that very many
0199  ffect on structure, quite independently of any  GOOD  thus gained. Correlation of growth has no doub
0200  ication in any one species exclusively for the  GOOD  of another species; though throughout nature o
0201  one species had been formed for the exclusive  GOOD  of another species, it would annihilate my the
0201  r natural selection acts solely by and for the  GOOD  of each. No organ will be formed, as Paley has
0201  essor. If a fair balance be struck between the  GOOD  and evil caused by each part, each will be fou
0203  in the combat; for undoubtedly this is for the  GOOD  of the community; and maternal love or materna
0204  a long series of gradations in complexity each  GOOD  for its possessor, then, under changing condit
0205  oduce nothing in one species for the exclusive  GOOD  or injury of another; though it may well produ
0210  ith my theory, the instinct of each species is  GOOD  for itself, but has never, as far as we can ju
0210  we can judge, been produced for the exclusive  GOOD  of others. One of the strongest instances of a
0210  l apparently performing an action for the sole  GOOD  of another, with which I am acquainted, is tha
0211  ides do not instinctively excrete for the sole  GOOD  of the ants. Although I do not believe that an
0211  the world performs an action for the exclusive  GOOD  of another of a distinct species, yet each spe
0212  peat my assurance, that I do not speak without  GOOD  evidence. The possibility, or even probability
0218  n other species; and M. Fabre has lately shown  GOOD  reason for believing that although the Tachyte
0233  any such case has been observed; nor would any  GOOD  be derived from a single hexagon being built,
0243  o instinct has been produced for the exclusive  GOOD  of other animals; but that each animal takes a
0247  d the primrose and cowslip, which we have such  GOOD  reason to believe to be varieties, and only on
0250  hat their own pollen was found to be perfectly  GOOD, for it fertilised distinct species. So that c
0251  rowth and rapid progress to maturity, and bore  GOOD  seed, which vegetated freely. In a letter to m
0251  s and pollen of the same flower were perfectly  GOOD  with respect to other species, yet as they wer
0259  e that when forms, which must be considered as  GOOD  and distinct species, are united, their fertil
0269  n any way which may be for each creature's own  GOOD; and thus she may, either directly, or more pr
0269  briefly abstract. The evidence is at least as  GOOD  as that from which we believe in the sterility
0270  many years on nine species of Verbascum, by so  GOOD  an observer and so hostile a witness, as Gartn
0282  the monuments of which we see around us. It is  GOOD  to wander along lines of sea coast, when forme
0283  es, each of which bears the stamp of time, are  GOOD  to show how slowly the mass has been accumulat
0284  lation; yet what time this must have consumed!  GOOD  observers have estimated that sediment is depo
0323  h america, in Tierra del Fuego, at the Cape of  GOOD  Hope, and in the peninsula of India. For at th
0339  mr. Woodward has shown that the same law holds  GOOD  with sea shells; but from the wide distributio
0370  nges of climate of an opposite nature. We have  GOOD  reason to believe that during the newer Plioce
0373  ts or opposite sides of the world. But we have  GOOD  evidence in almost every case, that the epoch
```

Page **(Key Word)**

Page **(Key Word)**

```
0375 entatives of the peculiar flora of the Cape of  GOOD Hope occur. At the Cape of Good Hope a very fe
0375 of the Cape of Good Hope occur. At the Cape of  GOOD Hope a very few European species, believed not
0377 he present day crowded together at the Cape of  GOOD Hope, and in parts of temperate Australia. As
0388 occupied; and a single seed or egg will have a  GOOD chance of succeeding. Although there will alwa
0389 er, with those on an equal area at the Cape of  GOOD Hope or in Australia, we must, I think, admit
0395 r as I have gone, the relation generally holds  GOOD. We see Britain separated by a shallow channel
0399 western corner of Australia and of the Cape of  GOOD Hope, is a far more remarkable case, and is at
0401 han they appear to be on a map. Nevertheless a  GOOD many species, both those found in other parts
0418 lly been admitted as true. The same fact holds  GOOD with flowering plants, of which the two main d
0425 ed. Rudimentary structures on this view are as  GOOD as, or even sometimes better than, other parts
0432 al arrangement in the diagram would still hold  GOOD; and, on the principle of inheritance, all the
0440 mon embryonic resemblance. Cirripedes afford a  GOOD instance of this: even the illustrious Cuvier
0444 his is invariably the case; and I could give a  GOOD many cases of variations (taking the word in t
0450 our unborn calves. It has even been stated on  GOOD authority that rudiments of teeth can be detec
0455 an extremely early period of life (as we have  GOOD reason to believe to be possible) the rudiment
0459 ulation of innumerable slight variations, each  GOOD for the individual possessor. Nevertheless, th
0459 ither do now exist or could have existed, each  GOOD of its kind, that all organs and instincts are
0462 ry long periods of time there will always be a  GOOD chance for wide migration by many means. A bro
0469 re, and habits of each creature, favouring the  GOOD and rejecting the bad? I can see no limit to t
0482 ed to believe that species are mutable will do  GOOD service by conscientiously expressing his conv
0489 natural selection works solely by and for the  GOOD of each being, all corporeal and mental endowm
0035 that the latter, by the regulations for the  GOODWOOD Races, are favoured in the weights they carr
0039 andard of perfection of each breed. The common  GOOSE has not given rise to any marked varieties: h
0042 stinct breeds of the cat, the donkey, peacock,  GOOSE, etc., may be attributed in main part to sele
0185 ge of structure. The webbed feet of the upland  GOOSE may be said to have become rudimentary in fun
0200 dly believe that the webbed feet of the upland  GOOSE or of the frigate bird are of special use to
0200 heritance. But to the progenitor of the upland  GOOSE and of the frigate bird, webbed feet no doubt
0033 w much the fruit of the different kinds of  GOOSEBERRIES differ in size, colour, shape, and hairine
0032 the steadily increasing size of the common  GOOSEBERRY may be quoted. We see an astonishing improv
0262 so it sometimes is in grafting; the common  GOOSEBERRY, for instance, cannot be grafted on the cur
0262 t will take, though with difficulty, on the  GOOSEBERRY. We have seen that the sterility of hybrids
0357 united to some continent. This view cuts the  GORDIAN knot of the dispersal of the same species to
0089 ngregate; and successive males display their  GORGEOUS plumage and perform strange antics before th
0132 species further north or from greater depths.  GOULD believes that birds of the same species are m
0133 e confined to continents are, according to Mr.  GOULD, brighter coloured than those of islands. The
0397 etimes by species found elsewhere. Dr. Aug. A.  GOULD has given several interesting cases in regard
0398 ds, and twenty five of these are ranked by Mr.  GOULD as distinct species, supposed to have been cr
0404 is shown in another and more general way. Mr.  GOULD remarked to me long ago, that in those genera
0270 rou de Buzareingues crossed three varieties of  GOURD, which like the maize has separated sexes, an
0259 onsidering the several rules now given, which  GOVERN the fertility of first crosses and of hybrids
0325 d temporary, but depend on general laws which  GOVERN the whole animal kingdom. M. Barrande has mad
0435 es, and two pairs of maxillae. Analogous laws  GOVERN the construction of the mouths and limbs of c
0043 inheritance and of reversion. Variability is  GOVERNED by many unknown laws, more especially by tha
0161 presence of all important characters will be  GOVERNED by natural selection, in accordance with the
0257 f the hybrids produced from them, is largely  GOVERNED by their systematic affinity. This is clearl
0259 cross between any two species is not always  GOVERNED by their systematic affinity or degree of re
0261 in hybridisation, is by no means absolutely  GOVERNED by systematic affinity. Although many distin
0276 strictly follow systematic affinity, but is  GOVERNED by several curious and complex laws. It is g
0466 exactly like the parent form. Variability is  GOVERNED by many complex laws, by correlation of grow
0472 far as we can see, with the laws which have  GOVERNED the production of so called specific forms.
0013 and non inheritance as the anomaly. The Laws  GOVERNING inheritance are quite unknown: no one can s
0245 nterbreeding, removed by domestication. Laws  GOVERNING the sterility of hybrids. Sterility not a s
0254 be considered as absolutely universal. Laws  GOVERNING the Sterility of first Crosses and of Hybri
0255 e more in detail the circumstances and rules  GOVERNING the sterility of first crosses and of hybri
0263 we must look at the curious and complex laws  GOVERNING the facility with which trees can be grafte
0263 o i believe that the still more complex laws  GOVERNING the facility of first crosses, are incident
0279 on climate; and, therefore, that the really  GOVERNING conditions of life do not graduate away qui
0409 of life throughout time and space; the laws  GOVERNING the succession of forms in past times being
0409 past times being nearly the same with those  GOVERNING at the present time the differences in diff
0472 observed. The complex and little known laws  GOVERNING variation are the same, as far as we can se
0164 oole, who examined the breed for the Indian  GOVERNMENT, a horse without stripes is not considered
0138 ves of Europe. We have some evidence of this  GRADATION of habit; for, as Schiodte remarks, animals
0180 family of squirrels; here we have the finest  GRADATION from animals with their tails only slightly
0187 ng vertebrata, we find but a small amount of  GRADATION in the structure of the eye, and from fossi
0225 heory. Let us look to the great principle of  GRADATION, and see whether Nature does not reveal to
0255 is surprising in how many curious ways this  GRADATION can be shown to exist; but only the barest
0255 stigma of some one species, yields a perfect  GRADATION in the number of seeds produced, up to near
0255 o the present day. In order to get a perfect  GRADATION between two forms in the upper and lower pa
0463 fossil remains afford plain evidence of the  GRADATION and mutation of the forms of life? We meet
0474 pted to show how much light the principle of  GRADATION throws on the admirable architectural power
0488 irement of each mental power and capacity by  GRADATION. Light will be thrown on the origin of man
0047 e so closely linked to them by intermediate  GRADATIONS, that naturalists do not like to rank them
0171 ended from other species by insensibly fine  GRADATIONS, do we not everywhere see innumerable trans
0175 not blending one into another by insensible  GRADATIONS, the range of any one species, depending as
0186 gree. Yet reason tells me, that if numerous  GRADATIONS from a perfect and complex eye to one very
0187 air which produce sound. In looking for the  GRADATIONS by which an organ in any species has been p
0187 original parent form, in order to see what  GRADATIONS are possible, and for the chance of some gr
0187 ns are possible, and for the chance of some  GRADATIONS having been transmitted from the earlier st
0187 echanism; and from this low stage, numerous  GRADATIONS of structure, branching off in two fundamen
0190 could not have been formed by transitional  GRADATIONS of some kind. Numerous cases could be given
0192 ve been produced by successive transitional  GRADATIONS, yet, undoubtedly, grave cases of difficult
0203 ked together by a multitude of intermediate  GRADATIONS, partly because the process of natural sele
0203 nd extinction of preceding and intermediate  GRADATIONS. Closely allied species, now living on a co
0204 f any organ, if we know of a long series of  GRADATIONS in complexity each good for its possessor,
0210 find in nature, not the actual transitional  GRADATIONS by which each complex instinct has been acc
0210 eral lines of descent some evidence of such  GRADATIONS; or we ought at least to be able to show th
```

Page **(Key Word)**

Page ***(Key Word)***

```
0210  r we ought at least to be able to show that  GRADATIONS of some kind are possible: and this we cert
0210  amongst extinct species, how very generally  GRADATIONS, leading to the most complex instincts, can
0235  originated: cases, in which no intermediate  GRADATIONS are known to exist: cases of instinct of ap
0239  same species, in the same nest, presenting   GRADATIONS of structure: and this we do find, even oft
0239  e same nest: I have myself compared perfect  GRADATIONS of this kind. It often happens that the lar
0240  r case: so confidently did I expect to find  GRADATIONS in important points of structure between th
0296  probably include all the fine intermediate   GRADATIONS which must on my theory have existed betwee
0297  connect them together by close intermediate  GRADATIONS. And this from the reasons just assigned we
0297  nd unless we obtained numerous transitional  GRADATIONS, we should not recognise their relationship
0299  ing in a fossil state numerous intermediate  GRADATIONS: and such success seems to me improbable in
0301  or the preservation of all the transitional  GRADATIONS between any two or more species. If such or
0301  ns between any two or more species. If such   GRADATIONS were not fully preserved, transitional vari
0329  uminants: for example, he dissolves by fine  GRADATIONS the apparently wide difference between the
0449  nd as all have been connected by the finest  GRADATIONS, the best, or indeed, if our collections we
0459  it the following propositions, namely, that  GRADATIONS in the perfection of any organ or instinct,
0460  remely difficult even to conjecture by what  GRADATIONS many structures have been perfected, more e
0460  organic beings: but we see so many strange   GRADATIONS in nature, as is proclaimed by the canon, N
0462  o together all the species in each group by  GRADATIONS as fine as our present varieties, it may be
0466  e do not know all the possible transitional  GRADATIONS between the simplest and the most perfect o
0485  y two forms, if not blended by intermediate  GRADATIONS, are looked at by most naturalists as suffi
0485  onnected at the present day by intermediate  GRADATIONS, whereas species were formerly thus connect
0485  on of the present existence of intermediate  GRADATIONS between any two forms, we shall be led to w
0179  vorous animals exist having every intermediate  GRADE between truly aquatic and strictly terrestria
0181  ess perfectly gliding squirrels: and that each  GRADE of structure had been useful to its possessor
0186  lex eye to one very imperfect and simple, each  GRADE being useful to its possessor, can be shown t
0194  an can be named, towards which no transitional  GRADE is known to lead. The truth of this remark is
0424  ctassification: for he includes in his lowest  GRADE, or that of a species, the two sexes: and how
0182  e inferred from these remarks that any of the  GRADES of wing structure here alluded to, which perh
0182  nd that animals displaying early transitional  GRADES of the structure will seldom continue to exis
0183  urthermore, we may conclude that transitional  GRADES between structures fitted for very different
0183  ance of discovering species with transitional  GRADES of structure in a fossil condition will alway
0188  case he does not know any of the transitional  GRADES. His reason ought to conquer his imagination:
0189  xist of which we do not know the transitional  GRADES, more especially if we look to much isolated
0189  d in order to discover the early transitional  GRADES through which the organ has passed, we should
0419  erous allied species, with slightly different  GRADES of difference, have been subsequently discove
0423  with our domestic productions, several other  GRADES of difference are requisite, as we have seen
0433  ed, genealogical in its arrangement, with the  GRADES of difference between the descendants from a
0456  ogical in its attempted arrangement, with the  GRADES of acquired difference marked by the terms va
0036  ess value than their dogs. In plants the same  GRADUAL process of improvement, through the occasion
0040  d. when further improved by the same slow and  GRADUAL process, they will spread more widely, and w
0137  ur. This state of the eyes is probably due to  GRADUAL reduction from disuse, but aided perhaps by
0210  ugh natural selection, except by the slow and  GRADUAL accumulation of numerous, slight, yet profit
0290  s soon as they are brought up by the slow and  GRADUAL rising of the land within the grinding actio
0312  ty of species, or with that of their slow and  GRADUAL modification, through descent and natural se
0312  percentage system of lost and new forms more  GRADUAL. In some of the most recent beds, though und
0312  itants of that island have been many and most  GRADUAL. The secondary formations are more broken: b
0316  rtainly exceptional: the general rule being a  GRADUAL increase in number, till the group reaches i
0317  ase and final extinction of the species. This  GRADUAL increase in number of the species of a group
0317  of a number of allied forms must be slow and  GRADUAL, one species giving rise first to two or thr
0430  s. on the principle of the multiplication and  GRADUAL divergence in character of the species desce
0454  t it has led in successive generations to the  GRADUAL reduction of various organs, until they beco
0475  ordinary generation has once been broken. The  GRADUAL diffusion of dominant forms, with the slow m
0479  , is likewise intelligible on the view of the  GRADUAL modification of parts or organs, which were
0014  that our domestic varieties, when run wild,  GRADUALLY but certainly revert in character to their
0035  e change has been effected unconsciously and  GRADUALLY, and yet so effectually, that, though the o
0068  ds, be let to grow, the more vigorous plants  GRADUALLY kill the less vigorous, though fully grown,
0069  ion to a dry, we invariably see some species  GRADUALLY getting rarer and rarer, and finally disapp
0146  in the rule by the fact that seeds could not  GRADUALLY become winged through natural selection, ex
0191  e now utterly lost branchiae might have been  GRADUALLY worked in by natural selection for some qui
0192  htly aided the act of respiration, have been  GRADUALLY converted by natural selection into branchi
0316  s its maximum, and then, sooner or later, it  GRADUALLY decreases. If the number of the species of
0317  not in a sharp point, but abruptly: it then  GRADUALLY thickens upwards, sometimes keeping for a s
0317  rmations, that species and groups of species  GRADUALLY disappear, one after another, first from on
0318  g thickness, the line is found to taper more  GRADUALLY at its upper end, which marks the progress
0435  ortened or widened to any extent, and become  GRADUALLY enveloped in thick membrane, so as to serve
0044  also what are called monstrosities: but they  GRADUATE into varieties. By a monstrosity I presume i
0103  ntinuous area, the conditions will generally  GRADUATE away insensibly from one district to another
0175  use surprise, as climate and height or depth  GRADUATE away insensibly. But when we bear in mind th
0192  branchiae of the other family: indeed, they  GRADUATE into each other. Therefore I do not doubt th
0203  en the conditions of life did not insensibly  GRADUATE away from one part to another. When two vari
0204  the most different habits of life could not  GRADUATE into each other: that a bat, for instance, c
0238  stes. The castes, moreover, do not generally  GRADUATE into each other, but are perfectly well defi
0241  ped into castes of different sizes, yet they  GRADUATE insensibly into each other, as does the wide
0279  e really governing conditions of life do not  GRADUATE away quite insensibly like heat or moisture.
0181  rnished with an extensor muscle. Although no  GRADUATED links of structure, fitted for gliding thro
0188  rfectly given, which show that there is much  GRADUATED diversity in the eyes of living crustaceans
0194  nature, be so invariably linked together by  GRADUATED steps? Why should not Nature have taken a l
0207  ts, but different in their origin. Instincts  GRADUATED. Aphides and ants. Instincts variable. Dome
0233  , it may reasonably be asked, how a long and  GRADUATED succession of modified architectural instin
0241  workers of a different size and structure: a  GRADUATED series having been first formed, as in the
0279  mely on an extensive and continous area with  GRADUATED physical conditions. I endeavoured to show,
0280  gy assuredly does not reveal any such finely  GRADUATED organic chain: and this, perhaps, is the mo
0293  derstand, why we do not therein find closely  GRADUATED varieties between the allied species which
0293  everal reasons why each should not include a  GRADUATED series of links between the species which t
0342  nct and existing forms of life by the finest  GRADUATED steps. He who rejects these views on the na
0460  ot have arrived at its present state by many  GRADUATED steps. There are, it must be admitted, case
```

Page ***(Key Word)***

Page ***(Key Word)***

```
0465  uires, for they have changed slowly and in a GRADUATED manner. We clearly see this in the fossil r
0474  . we can thus understand why nature moves by GRADUATED steps in endowing different animals of the
0248  s when crossed is so different in degree and GRADUATES away so insensibly; and, on the other hand,
0248  ties; but that the evidence from this source GRADUATES away, and is doubtful in the same degree as
0255  ility, both of first crosses and of hybrids, GRADUATES from zero to perfect fertility. It is surpr
0259  istinct species, are united, their fertility GRADUATES from zero to perfect fertility, or even to
0272   the difference in the degree of variability GRADUATES away. When mongrels and the more fertile hy
0261  stematic affinity, for no one has been able to GRAFT trees together belonging to quite distinct fa
0141  n to any special climate as a quality readily GRAFTED on an innate wide flexibility of constitutio
0261  d quality. As the capacity of one plant to be GRAFTED or budded on another is so entirely unimport
0261  species, can usually, but not invariably, be GRAFTED with ease. But this capacity, as in hybridis
0261  tinct genera within the same family have been GRAFTED together, in other cases species of the same
0261  will not take on each other. The pear can be GRAFTED far more readily on the quince, which is ran
0262  individuals of the same two species in being GRAFTED together. As in reciprocal crosses, the faci
0262  he common gooseberry, for instance, cannot be GRAFTED on the currant, whereas the currant will tak
0262  freely on their own roots; and which could be GRAFTED with no great difficulty on another species,
0262  reat difficulty on another species, when thus GRAFTED were rendered barren. On the other hand, cer
0262  e other hand, certain species of Sorbus, when GRAFTED on other species, yielded twice as much frui
0262  ental difference between the mere adhesion of GRAFTED stocks, and the union of the male and female
0263  overning the facility with which trees can be GRAFTED on each other as incidental on unknown diffe
0264  ts, any more than why certain trees cannot be GRAFTED on others. Lastly, an embryo may be develope
0276  what analogous degrees of difficulty in being GRAFTED together in order to prevent them becoming i
0277  e hybrids produced, and the capacity of being GRAFTED together, though this latter capacity eviden
0460  ent than is the incapacity of two trees to be GRAFTED together; but that it is incidental on const
0010  of the plant. Such buds can be propagated by GRAFTING, etc., and sometimes by seed. These sports a
0251  stance, informs me that he raises stocks for GRAFTING from a hybrid between Rhod. Ponticum and Cat
0261  nt climates, does not always prevent the two GRAFTING together. As in hybridisation, so with graft
0261  fting together. As in hybridisation, so with GRAFTING, the capacity is limited by systematic affin
0262  n very far from equal, so it sometimes is in GRAFTING: the common gooseberry, for instance, cannot
0262  tent parallel. Something analogous occurs in GRAFTING: for Thouin found that three species of Robi
0262  rude degree of parallelism in the results of GRAFTING and of crossing distinct species. And as we
0263  t the greater or lesser difficulty of either GRAFTING or crossing together various species has bee
0263  ability of specific forms, as in the case of GRAFTING it is unimportant for their welfare. Causes
0276  ed from this cross. In the same manner as in GRAFTING trees, the capacity of one species or variet
0093  tamens with shrivelled anthers, in which not a GRAIN of pollen can be detected. Having found a fem
0467  ividuals are born than can possibly survive. A GRAIN in the balance will determine which individua
0093  on all, without exception, there were pollen GRAINS, and on some a profusion of pollen. As the wi
0193  very curious contrivance of a mass of pollen GRAINS, borne on a foot stalk with a sticky gland at
0270  ny seed, and this one head produced only five GRAINS. Manipulation in this case could not have bee
0362  food, it is positively asserted that all the GRAINS do not pass into the gizzard for 12 or even 1
0362  to them: in one instance I removed twenty two GRAINS of dry argillaceous earth from one foot of a
0185  at seems plainer than that the long toes of GRALLATORES are formed for walking over swamps and flo
0282  e of time. He who can read Sir Charles Lyell's GRAND work on the Principles of Geology, which the
0329  nct and living species. They all fall into one GRAND natural system; and this fact is at once expl
0344  of life, ancient and recent, make together one GRAND system; for all are connected by generation.
0398  se of the Galapagos to America. I believe this GRAND fact can receive no sort of explanation on th
0408  ded from some one source; then I think all the GRAND leading facts of geographical distribution ar
0413  d orders, all united into one class. Thus, the GRAND fact in natural history of the subordination
0456  g, and circuitous lines of affinities into one GRAND system; the rules followed and the difficulti
0471  d which has prevailed throughout all time. The GRAND fact of the grouping of all organic beings se
0476  rmediate position in the chain of descent. The GRAND fact that all extinct organic beings belong t
0478  past and present organic beings constitute one GRAND natural system, with group subordinate to gro
0486  e, will the study of natural history become! A GRAND and almost untrodden field of inquiry will be
0025  ds; these I again crossed together, and one GRANDCHILD of the pure white fantail and pure black ba
0253  irds he raised no less than eight hybrids (GRANDCHILDREN of the pure geese) from one nest. In Indi
0490  e higher animals, directly follows. There is GRANDEUR in this view of life, with its several power
0013  often reverts in certain characters to its GRANDFATHER or grandmother or other much more remote a
0214  xample, Le Roy describes a dog, whose great GRANDFATHER was a wolf, and this dog showed a trace of
0013  in certain characters to its grandfather or GRANDMOTHER or other much more remote ancestor: why a
0358  been formed, like other mountain summits, of GRANITE, metamorphic schists, old fossiliferous or o
0363  answered that he had found large fragments of GRANITE and other rocks, which do not occur in the a
0015  light deviations of structure, in such case, I GRANT that we could deduce nothing from domestic va
0102  or the work of natural selection, she does not GRANT an indefinite period: for as all organic bein
0224  ted by a crowd of bees working in a dark hive. GRANT whatever instincts you please, and it seems a
0260  s the production of hybrids been permitted? to GRANT to species the special power of producing hyb
0275  this does sometimes occur with hybrids: yet I GRANT much less frequently with hybrids than with m
0297  ich to distinguish species and varieties; they GRANT some little variability to each species, but
0156  es; and the truth of this proposition will be GRANTED. The cause of the original variability of a
0268  oduced under nature will assuredly have to be GRANTED. If we turn to varieties, produced, or suppo
0407  t slowly, enormous periods of time being thus GRANTED for their migration, I do not think that the
0102  y important element of success. Though nature GRANTS vast periods of time for the work of natural
0098  every one of the infinitely numerous pollen GRANULES are swept out of the conjoined anthers of ea
0203  dense clouds of pollen, in order that a few GRANULES may be wafted by a chance breeze on to the o
0378  luxuriance at the base of the Himalaya, as GRAPHICALLY described by Hooker. Thus, as I believe, a
0285  the same manner as does the vain endeavour to GRAPPLE with the idea of eternity. I am tempted to g
0287  e number of generations, which the mind cannot GRASP, must have succeeded each other in the long r
0481  n of the coast waves. The mind cannot possibly GRASP the full meaning of the term of a hundred mil
0185  ial thrush family wholly subsists by diving, GRASPING the stones with its feet and using its wings
0200   a foot with five toes fitted for walking or GRASPING; and we may further venture to believe that
0434  ious than that the hand of a man, formed for GRASPING, that of a mole for digging, the leg of the
0075  r for existence, as in the case of locusts and GRASS feeding quadrupeds. But the struggle almost i
0077  eas and beans), when sown in the midst of long GRASS, I suspect that the chief use of the nutrimen
0113  f a plot of ground be sown with one species of GRASS, and a similar plot be sown with several dist
0113  spaces of ground. Hence, if any one species of GRASS were to go on varying, and those varieties we
0113  number of individual plants of this species of GRASS, including its modified descendants, would su
0113  ell know that each species and each variety of GRASS is annually sowing almost countless seeds; an
```

Page ***(Key Word)***

Page **(Key Word)**

```
0114  most distinct varieties of any one species of  GRASS would always have the best chance of succeed1
0216  her, and conceal themselves in the surrounding  GRASS or thickets: and this is evidently done for t
0071  d, but twelve species of plants (not counting   GRASSES and carices) flourished in the plantations,
0113  plot be sown with several distinct genera of    GRASSES, a greater number of plants and a greater we
0113  same manner as distinct species and genera of   GRASSES differ from each other, a greater number of
0416  ghest importance in the classification of the   GRASSES. Numerous instances could be given of charac
0416  oc, mere pubescence on parts of the flower in   GRASSES, the nature of the dermal covering, as hair
0346  mid districts, arid deserts, lofty mountains,   GRASSY plains, forests, marshes, lakes, and great ri
0318  on of species has been involved in the most     GRATUITOUS mystery. Some authors have even supposed th
0171  ve occurred to the reader. Some of them are so   GRAVE that to this day I can never reflect on them
0192  ive transitional gradations, yet, undoubtedly,  GRAVE cases of difficulty occur, some of which will
0203  urged against my theory. Many of them are very  GRAVE: but I think that in the discussion light has
0310  arles Lyell, from further reflexion entertains  GRAVE doubts on this subject. I feel how rash it is
0354  continuity of range are so numerous and of so   GRAVE a nature, that we ought to give up the belief
0396  rganism. I do not deny that there are many and   GRAVE difficulties in understanding how several of
0407  the individuals of the same species, extremely  GRAVE. As exemplifying the effects of climatal chan
0459  nferences briefly recapitulated. That many and  GRAVE objections may be advanced against the theory
0461  on the theory of descent with modification are  GRAVE enough. All the individuals of the same speci
0464  ld's history. I can answer these questions and  GRAVE objections only on the supposition that the g
0466  e know how imperfect the Geological Record is.  GRAVE as these several difficulties are, in my judg
0288  mains which do become embedded, if in sand or   GRAVEL, will when the beds are upraised generally be
0300  cks, would be embedded; and those embedded in   GRAVEL or sand, would not endure to a distant epoch.
0041  paid nothing can be effected. I have seen it    GRAVELY remarked, that it was most fortunate that th
0306  another and allied difficulty, which is much    GRAVER. I allude to the manner in which numbers of s
0005  ur succeeding chapters, the most apparent and   GRAVEST difficulties on the theory will be given: na
0192  ll be discussed in my future work. One of the   GRAVEST is that of neuter insects, which are often v
0280  ni and this, perhaps, is the most obvious and   GRAVEST objection which can be urged against my theo
0299  his not having been effected, is probably the   GRAVEST and most obvious of all the many objections
0310  e silurian strata, are all undoubtedly of the   GRAVEST nature. We see this in the plainest manner b
0490  gone cycling on according to the fixed law of   GRAVITY, from so simple a beginning endless forms mo
0100  abulated the trees of New Zealand, and Dr. Asa  GRAY those of the United States, and the result was
0115  ingle instance: in the last edition of Dr. Asa  GRAY's Manual of the Flora of the Northern United S
0163  the body, is without bars on the legs: but Dr.  GRAY has figured one specimen with very distinct ze
0164  ear in old horses. Colonel Poole has seen both  GRAY and bay Kattywar horses striped when first foa
0165  arkable case, a hybrid has been figured by Dr.  GRAY (and he informs me that he knows of a second c
0176  om information given me by Mr. Watson, Dr. Asa  GRAY, and Mr. Wollaston, that generally when variet
0218  n, generated. I may add that, according to Dr.  GRAY and to some other observers, the European cuck
0365  , and nearly all the same, as we hear from Asa  GRAY, with those on the loftiest mountains of Europ
0372  we find very few identical species (though Asa  GRAY has lately shown that more plants are identica
0004  is equally or more important, we shall see how  GREAT is the power of man in accumulating by his Se
0011  s, and walking more, than its wild parent. We   GREAT and inherited development of the udders in co
0011  es are given in Isidore Geoffroy St. Hilaire's  GREAT work on this subject. Breeders believe that l
0014  often and so boldly been made. There would be   GREAT difficulty in proving its truth: we may safel
0015  or not the experiment would succeed, is not of  GREAT importance for our line of argument: for by t
0017  any single species, then such facts would have  GREAT weight in making us doubt about the immutabil
0019  goats in Europe alone, and several even within  GREAT Britain. One author believes that there forme
0019  author believes that there formerly existed in  GREAT Britain eleven wild species of sheep peculiar
0021  ar and strictly inherited habit of flying at a  GREAT height in a compact flock, and tumbling in th
0021  the air head over heels. The runt is a bird of  GREAT size, with long, massive beak and large feet:
0021  rteen, the normal number in all members of the  GREAT pigeon family: and these feathers are kept ex
0023  he might have called them, could be shown him.  GREAT as the differences are between the breeds of
0024  onfinement. An argument, as it seems to me, of  GREAT weight, and applicable in several other cases
0024  ture, we may look in vain throughout the whole  GREAT family of Columbidae for a beak like that of
0027  n India: and that it agrees in habits and in a  GREAT number of points of structure with all the do
0030  be said to make for himself useful breeds. The  GREAT power of this principle of selection is not h
0032  rth notice: but its importance consists in the  GREAT effect produced by the accumulation in one di
0032  le perseverance, he will succeed, and may make  GREAT improvements: if he wants any of these qualit
0036  e sheep possessed by these two gentlemen is so  GREAT that they have the appearance of being quite
0037  a fruit of very inferior quality. I have seen   GREAT surprise expressed in horticultural works at
0038  races, and likewise their differences being so  GREAT in external characters and relatively so slig
0039  of the breed. Nor let it be thought that some   GREAT deviation of structure would be necessary to
0042  . pigeons can be mated for life, and this is a  GREAT convenience to the fancier, for thus many rac
0042  eeds. Pigeons, I may add, can be propagated in  GREAT numbers and at a very quick rate, and inferio
0045  the branching of the main nerves close to the  GREAT central ganglion of an insect would have been
0047  es, and the other as the variety. But cases of  GREAT difficulty, which I will not here enumerate,
0048  not be disputed. Compare the several floras of  GREAT Britain, of France or of the United States, d
0049  er rank it as an undoubted species peculiar to  GREAT Britain. A wide distance between the homes of
0058  he provinces into which Mr. Watson has divided  GREAT Britain. Now, in this same catalogue, 53 ackn
0061  en that man by selection can certainly produce  GREAT results, and can adapt organic beings to his
0062  ean of Manchester, evidently the result of his  GREAT horticultural knowledge. Nothing is easier th
0063  s numbers would quickly become so inordinately  GREAT that no country could support the product. He
0065  imals tends, I think, to mislead us: we see no  GREAT destruction falling on them, and we forget th
0066  at some period of life: and this period in the  GREAT majority of cases is an early one. If an anim
0070  etc., in our fields, because the seeds are in   GREAT excess compared with the number of birds whic
0074  er of humble beings in any district depends in  GREAT degree on the number of field mice, which des
0076  cockroach has everywhere driven before it its   GREAT congener. One species of charlock will suppla
0076  pecies has been victorious over another in the  GREAT battle of life. A corollary of the highest im
0079  rvals, has to struggle for life, and to suffer  GREAT destruction. When we reflect on this struggle
0080  ations useful in some way to each being in the  GREAT and complex battle of life, should sometimes
0082  ity is necessary: as man can certainly produce  GREAT results by adding up in any given direction m
0082  ime at her disposal. Nor do I believe that any  GREAT physical change, as of climate, or any unusua
0083  s man can produce and certainly has produced a  GREAT result by his methodical and unconscious mean
0085  e aids of art, these slight differences make a  GREAT difference in cultivating the several varieti
0087  extent by natural selection: for instance, the  GREAT jaws possessed by certain insects, and used e
0091  thought very improbable: for we often observe   GREAT differences in the natural tendencies of our
0092  the pollen were destroyed, it might still be a  GREAT gain to the plant: and those individuals whic
```

Page **(Key Word)**

Page ∗∗∗(Key Word)∗∗

```
0095  ous nectar to the hive bee. Thus it might be a  GREAT  advantage to the hive bee to have a slightly
0095  e to become rare in any country, it might be a  GREAT  advantage to the red clover to have a shorter
0095  ost banished such views as the excavation of a  GREAT  valley by a single diluvial wave, so will nat
0096  nued creation of new organic beings, or of any  GREAT  and sudden modification in their structure. O
0097  of fructification closely enclosed, as in the  GREAT  papilionaceous or pea family; but in several,
0097  not carry pollen from one to the other, to the  GREAT  good, as I believe, of the plant. Bees will a
0101  long appeared to me to present a case of very  GREAT  difficulty under this point of view: but I ha
0104  f intercrosses even at rare intervals, will be  GREAT. If there exist organic beings which never b
0104  anic conditions of life will generally be in a  GREAT  degree uniform; so that natural selection wil
0105  period, and of spreading widely. Throughout a  GREAT  and open area, not only will there be a bette
0106  e than on a small and isolated area. Moreover,  GREAT  areas, though now continuous, owing to oscill
0109  es in the polity of nature is not indefinitely  GREAT, not that we have any means of knowing that a
0114  he greatest amount of life can be supported by  GREAT  diversification of structure, is seen under m
0114  and individual must be severe, we always find  GREAT  diversity in its inhabitants. For instance, I
0115  and Alph. de Candolle has well remarked in his  GREAT  and admirable work, that floras gain by natur
0116  r beings. Now let us see how this principle of  GREAT  benefit being derived from divergence of char
0125  ive group of diverging dotted lines to be very  GREAT, the forms marked a14 to p14, those marked b1
0129  ntly created, I can see no explanation of this  GREAT  fact in the classification of all organic bei
0129  ame class have sometimes been represented by a  GREAT  tree. I believe this simile largely speaks th
0129  have tried to overmaster other species in the  GREAT  battle for life. The limbs divided into great
0129  great battle for life. The limbs divided into  GREAT  branches, and these into lesser and lesser br
0129  a mere bush, only two or three, now grown into  GREAT  branches, yet survive and bear all the other
0130  o by generation I believe it has been with the  GREAT  Tree of Life, which fills with its dead and b
0143  only on one side; and if this had been of any  GREAT  use to the breed it might probably have been
0148  head enormously developed, and furnished with  GREAT  nerves and muscles; but in the parasitic and
0151  secondary sexual characters, may be due to a  GREAT  variability of these characters, whether or n
0151  he individuals of several of the species is so  GREAT, that it is no exaggeration to state that the
0151  ave shaken my belief in its truth, had not the  GREAT  variability in plants made it particularly di
0153  ordinarily developed part or organ has been so  GREAT  and long continued within a period not excess
0154  been comparatively recent and extraordinarily  GREAT  that we ought to find the generative variabil
0158  th the same part in its congeners; and the not  GREAT  degree of variability in a part, however extr
0158  e common to a whole group of species; that the  GREAT  variability of secondary sexual characters, a
0158  bility of secondary sexual characters, and the  GREAT  amount of difference in these same characters
0160  ch has been lost in a breed, reappears after a  GREAT  number of generations, the most probable hypo
0163  such cases; but here, as before, I lie under a  GREAT  disadvantage in not being able to give them.
0176  bers from inhabiting larger areas, will have a  GREAT  advantage over the intermediate variety, whic
0177  in this case will be strongly in favour of the  GREAT  holders on the mountains or on the plains imp
0183  rely have been developed at an early period in  GREAT  numbers and under many subordinate forms. Thu
0187  ies we can learn nothing on this head. In this  GREAT  class we should probably have to descend far
0188  e which have become extinct, I can see no very  GREAT  difficulty (not more than in the case of many
0188  s perfect as is possessed by any member of the  GREAT  Articulate class. He who will go thus far, if
0191  rate animals: hence there seems to me to be no  GREAT  difficulty in believing that natural selectio
0195  ble to range into new pastures and thus gain a  GREAT  advantage. It is not that the larger quadrupe
0202  ment, like that in so many members of the same  GREAT  order, and which has been modified but not pe
0204  from existing in greater numbers, will have a  GREAT  advantage over the less numerous intermediate
0206  hat all organic beings have been formed on two  GREAT  laws, Unity of Type, and the Conditions of Ex
0214  nt: for example, Le Roy describes a dog, whose  GREAT  grandfather was a wolf, and this dog showed a
0220  that the contrast in their appearance is very  GREAT. When the nest is slightly disturbed, the sla
0224  ctly made. Put the difficulty is not nearly so  GREAT  as it at first appears: all this beautiful wo
0225  modification of his theory. Let us look to the  GREAT  principle of gradation, and see whether Natur
0230  e construction of the cells: but it would be a  GREAT  error to suppose that the bees cannot build u
0233  on this subject. Nor does there seem to me any  GREAT  difficulty in a single insect (as in the case
0236  . the subject well deserves to be discussed at  GREAT  length, but I will here take only a single ca
0236  ut incapable of procreation, I can see no very  GREAT  difficulty in this being effected by natural
0236  ust pass over this preliminary difficulty. The  GREAT  difficulty lies in the working ants differing
0243  facts given in this chapter strengthen in any  GREAT  degree my theory: but none of the cases of di
0249  brids are seldom raised by experimentalists in  GREAT  numbers; and as the parent species, or other
0250  i think, be in part accounted for by Herbert's  GREAT  horticultural skill, and by his having hothou
0260  fertile hybrids? Why should there often be so  GREAT  a difference in the result of a reciprocal cr
0261  ude of cases we can assign no reason whatever.  GREAT  diversity in the size of two plants, one bein
0262  own roots, and which could be grafted with no  GREAT  difficulty on another species, when thus graf
0264  communicated to me by Mr. Hewitt, who has had  GREAT  experience in hybridising gallinaceous birds,
0265  tems seriously affected. This, in fact, is the  GREAT  bar to the domestication of animals. Between
0265  and is often accompanied by excess of size or  GREAT  luxuriance. In both cases, the sterility occu
0265  , one species in a group will sometimes resist  GREAT  changes of conditions with unimpaired fertili
0267  convalescence of animals, we plainly see that  GREAT  benefit is derived from almost any change in
0271  me degree modified. From these facts; from the  GREAT  difficulty of ascertaining the infertility of
0282  ways converging to the common ancestor of each  GREAT  class. So that the number of intermediate and
0282  extinct species, must have been inconceivably  GREAT. But assuredly, if this theory be true, such
0282  ected, that time will not have sufficed for so  GREAT  an amount of organic change, all changes havi
0282  atum. A man must for years examine for himself  GREAT  piles of superimposed strata, and watch the s
0284  imate, of each formation in different parts of  GREAT  Britain: and this is the result: Palaeozoic s
0284  ve estimated that sediment is deposited by the  GREAT  Mississippi river at the rate of only 600 fee
0285  ry is still more plainly told by faults, those  GREAT  cracks along which the strata have been uphea
0285  stant South Downs; for, remembering that at no  GREAT  distance to the west the northern and souther
0286  d close, one can safely picture to oneself the  GREAT  dome of rocks which must have covered up the
0287  istrict, when upraised, could hardly have been  GREAT, but it would somewhat reduce the above estim
0289  we know, for instance, from Sir R. Murchison's  GREAT  work in Russia, what wide gaps there are in t
0289  hich were blank and barren in his own country,  GREAT  piles of sediment, charged with new and pecul
0290  could nowhere be ascertained. The frequent and  GREAT  changes in the mineralogical composition of. c
0290  of consecutive formations, generally implying  GREAT  changes in the geography of the surrounding l
0290  the supply of sediment must for ages have been  GREAT, from the enormous degradation of the coast r
0291  thor after author, in treating of this or that  GREAT  formation, has come to the conclusion that it
0292  ing these very periods of subsidence, that our  GREAT  deposits rich in fossils have been accumulate
0294  logical period; and likewise to reflect on the  GREAT  changes of level, on the inordinately great c
0294  he great changes of level, on the inordinately  GREAT  change of climate, on the prodigious lapse of
```

Page ∗∗∗(Key Word)∗∗

Page ***(Key Word)***

```
0296 n other cases we have the plainest evidence in GREAT fossilised trees, still standing upright as t
0297 to distinct but consecutive stages of the same GREAT formation, we find that the embedded fossils,
0300 may be doubted whether the duration of any one GREAT period of subsidence over the whole or part o
0301 tinct species. It is, also, probable that each GREAT period of subsidence would be interrupted by
0303 ffected, and a few species had thus acquired a GREAT advantage over other organisms, a comparative
0303 l treatises, published not many years ago, the GREAT class of mammals was always spoken of as havi
0303 d sandstone at nearly the commencement of this GREAT series. Cuvier used to urge that no monkey oc
0304 , seemed fully to justify the belief that this GREAT and distinct order had been suddenly produced
0304 red in beds of this age, I concluded that this GREAT group had been suddenly developed at the comm
0304 ne more instance of the abrupt appearance of a GREAT group of species. But my work had hardly been
0305 rge and perfectly enclosed basin, in which any GREAT group of marine animals might be multiplied:
0307 accumulated before the Silurian epoch, is very GREAT. If these most ancient beds had been wholly w
0309 ts of land have existed, subjected no doubt to GREAT oscillations of level, since the earliest sil
0309 me on Coral Reefs, led me to conclude that the GREAT oceans are still mainly areas of subsidence,
0309 eans are still mainly areas of subsidence, the GREAT archipelagoes still areas of oscillations of
0310 rphic rocks, which must have been heated under GREAT pressure, have always seemed to me to require
0310 species. But I have reason to believe that one GREAT authority, Sir Charles Lyell, from further re
0310 ct. I feel how rash it is to differ from these GREAT authorities, to whom, with others, we owe all
0315 g enduring fossiliferous formations depends on GREAT masses of sediment having been deposited on a
0317 er species, and so on, like the branching of a GREAT tree from a single stem, till the group becom
0321 for instance, a single species of Trigonia, a GREAT genus of shells in the secondary formations,
0321 the Australian seas; and a few members of the GREAT and almost extinct group of Ganoid fishes sti
0325 her physical conditions, as the cause of these GREAT mutations in the forms of life throughout the
0325 ies, and the nature of their inhabitants. This GREAT fact of the parallel succession of the forms
0326 her for those in the waters of the sea. If two GREAT regions had been for a long period favourably
0329 be filled with striking illustrations from our GREAT palaeontology, Owen, showing how extinct an
0332 they probably would have to be united into one GREAT family, in nearly the same manner as has occu
0333 me theory, it is evident that the fauna of any GREAT period in the earth's history will be interme
0334 it. Thus, the species which lived at the sixth GREAT stage of descent in the diagram are the modif
0335 arated districts. To compare small things with GREAT: if the principal living and extinct races of
0337 may believe, if all the animals and plants of GREAT Britain were set free in New Zealand, that in
0337 he productions of New Zealand were set free in GREAT Britain, whether any considerable number woul
0337 . under this point of view, the productions of GREAT Britain may be said to be higher than those o
0340 n the theory of descent with modification, the GREAT law of the long enduring, but not immutable,
0340 ut after very long intervals of time and after GREAT geographical changes, permitting much inter m
0342 ecies, found in the several stages of the same GREAT formation. He may disbelieve in the enormous
0343 st have played, when the formations of any one GREAT region alone, as that of Europe, are consider
0343 passing from these difficulties, all the other GREAT leading facts in palaeontology seem to me sim
0346 c beings over the face of the globe, the first GREAT fact which strikes us is, that neither the si
0346 s, grassy plains, forests, marshes, lakes, and GREAT rivers, under almost every temperature. There
0347 espect to the inhabitants of the sea. A second GREAT fact which strikes us in our general review i
0347 uctions of various regions. We see this in the GREAT difference of nearly all the terrestrial prod
0347 rctic productions. We see the same fact in the GREAT difference between the inhabitants of Austral
0348 f lofty and continuous mountain ranges, and of GREAT deserts, and sometimes even of large rivers,
0348 shores of South and Central America; yet these GREAT faunas are separated only by the narrow, but
0349 actly opposite meridians of longitude. A third GREAT fact, partly included in the foregoing statem
0351 ed within comparatively recent times, there is GREAT difficulty on this head. It is also obvious t
0352 gist will feel any difficulty in such cases as GREAT Britain having been formerly united to Europe
0353 ted across the vast and broken interspace. The GREAT and striking influence which barriers of ever
0353 ion, is intelligible only on the view that the GREAT majority of species have been produced on one
0357 he other. No other geologist will dispute that GREAT mutations of level, have occurred within the
0357 seems to me that we have abundant evidence of GREAT oscillations of level in our continents; but
0358 ds. Several facts in distribution, such as the GREAT difference in the marine faunas on the opposi
0363 hus seeds might occasionally be transported to GREAT distances; for many facts could be given show
0364 means of transport would carry seeds for very GREAT distances; for seeds do not retain their vita
0364 o not retain their vitality when exposed for a GREAT length of time to the action of sea water; no
0364 would not by such means become mingled in any GREAT degree; but would remain as distinct as we no
0364 oil, and coming to maturity! But it would be a GREAT error to argue that because a well stocked is
0364 argue that because a well stocked island, like GREAT Britain, has not, as far as is known (and it
0369 esent Alpine plants and animals of the several GREAT European mountain ranges, though very many of
0371 ward, they will have become mingled in the one GREAT region with the native American productions,
0371 ave had to compete with them; and in the other GREAT region, with those of the Old World. Conseque
0372 n was formerly supposed), but we find in every GREAT class many forms, which some naturalists rank
0376 have a closer resemblance in its crustacea to GREAT Britain, its antipode, than to any other part
0377 ly by escaping into the warmest spots. But the GREAT fact to bear in mind is, that all tropical pr
0378 e of the Cordillera, seem to have afforded two GREAT lines of invasion: and it is a striking fact,
0382 almost identical with mine, on the effects of GREAT alternations of climate on geographical distr
0382 ve that the world has recently felt one of his GREAT cycles of change: and that on this view, comb
0382 ne gently rising from the arctic lowlands from GREAT height under the equator. The various beings
0384 such cases there will have been ample time for GREAT geographical changes, and consequently time a
0387 quently the range of these plants was not very GREAT. The same agency may have come into play with
0387 crement, many hours afterwards. When I saw the GREAT size of the seeds of that fine water lily (probably
0387 audubon states that he found the seeds of the GREAT southern water lily (probably, according to D
0389 ifference in physical conditions has caused so GREAT a difference in number. Even the uniform coun
0391 t very many North American birds, during their GREAT annual migrations, visit either periodically
0393 ound on any of the many islands with which the GREAT oceans are studded. I have taken pains to ver
0393 red that a frog exists on the mountains of the GREAT island of New Zealand; but I suspect that thi
0393 ter, on my view we can see that there would be GREAT difficulty in their transportal across the se
0393 d situated above 300 miles from a continent or GREAT continental island; and many islands situated
0395 ing observations on this head in regard to the GREAT Malay Archipelago, which is traversed near Ce
0395 eds. No doubt some few anomalies occur in this GREAT archipelago, and there is much difficulty in
0400 in a small degree. This long appeared to me a GREAT difficulty: but it arises in chief part from
0404 s of lakes and marshes, excepting in so far as GREAT facility of transport has given the same gene
0405 such cases there will have been ample time for GREAT climatal and geographical changes and for acc
0406 ce that organisms low in the scale within each GREAT class, generally change at a slower rate than
0407 aneously affected the whole world, or at least GREAT meridional belts. As showing how diversified
```

Page ***(Key Word)***

Page **(Key Word)**

0409	some only slightly modified, some developed in	GREAT	force, some existing in scanty numbers, in th
0409	e existing in scanty numbers, in the different	GREAT	geographical provinces of the world. On these
0409	ds should have few inhabitants, but of these a	GREAT	number should be endemic or peculiar; and why
0412	his conclusion was supported by looking at the	GREAT	diversity of the forms of life which, in any
0415	us. To give an example amongst insects, in one	GREAT	division of the Hymenoptera, the antennae, as
0418	nd a character nearly uniform, and common to a	GREAT	number of forms, and not common to others, th
0418	ature. Yet it has been strongly urged by those	GREAT	naturalists, Milne Edwards and Agassiz, that
0423	stant, is used in classing varieties: thus the	GREAT	agriculturist Marshall says the horns are ver
0428	r these cetaceans agree in so many characters,	GREAT	and small, that we cannot doubt that they hav
0428	recent and extinct, are included under a few	GREAT	orders, under still fewer classes, and all in
0429	ers, under still fewer classes, and all in one	GREAT	natural system. As showing how few the higher
0431	he many living and extinct members of the same	GREAT	natural class. Extinction, as we have seen in
0433	rom one dominant parent species, explains that	GREAT	and universal feature in the affinities of al
0433	d extinct forms can be grouped together in one	GREAT	system; and how the several members of each c
0434	s in widely different animals. We see the same	GREAT	law in the construction of the mouths of inse
0434	he curious folded one of a bee or bug, and the	GREAT	jaws of a beetle? yet all these organs, servi
0435	ent, so as to serve as a wing: yet in all this	GREAT	amount of modification there will be no tende
0438	by the strong principle of inheritance. In the	GREAT	class of molluscs, though we can homologise t
0438	ition of any one part, as we find in the other	GREAT	classes of the animal and vegetable kingdoms.
0448	ish and spiders, and with a few members of the	GREAT	class of insects, as with Aphis. With respect
0449	recognised by their larvae as belonging to the	GREAT	class of crustaceans. As the embryonic state
0450	ss obscured, of the common parent form of each	GREAT	class of animals. Rudimentary, atrophied, or
0456	ame view of descent with modification, all the	GREAT	facts in Morphology become intelligible, whet
0457	a corresponding period, we can understand the	GREAT	leading facts in Embryology; namely, the rese
0459	hough appearing to our imagination insuperably	GREAT,	cannot be considered real if we admit the fo
0463	several geological stages? Why do we not find	GREAT	piles of strata beneath the Silurian system,
0464	anic change; for the lapse of time has been so	GREAT	as to be utterly inappreciable by the human i
0464	eserved in a fossil condition, at least in any	GREAT	number. Widely ranging species vary most, and
0467	ed eye. This process of selection has been the	GREAT	agency in the production of the most distinct
0468	plainly proclaims that each land has undergone	GREAT	physical changes, we might have expected that
0468	riciously, can produce within a short period a	GREAT	result by adding up mere individual differenc
0471	groups subordinate to groups, all within a few	GREAT	classes, which we now see everywhere around u
0471	sive, favourable variations, it can produce no	GREAT	or sudden modification; it can act only by ve
0474	, and therefore, as we may naturally infer, of	GREAT	importance to the species; should be eminentl
0476	eory of descent with modification, most of the	GREAT	leading facts in Distribution. We can see why
0477	d marshes, most of the inhabitants within each	GREAT	class are plainly related; for they will gene
0481	s, is that we are always slow in admitting any	GREAT	change of which we do not see the intermediat
0481	ng lines of inland cliffs had been formed, and	GREAT	valleys excavated, by the slow action of the
0486	nearly in the same way as when we look at any	GREAT	mechanical invention as the summing up of the
0486	ome degree obscured, or the prototypes of each	GREAT	class. When we can feel assured that all the
0487	nd at rare intervals. The accumulation of each	GREAT	fossiliferous formation will be recognised as
0007	, i think we are driven to conclude that this	GREATER	variability is simply due to our domestic pr
0013	transmitted either exclusively, or in a much	GREATER	degree, to males alone. A much more importan
0014	a short horned cow by a long horned bull, the	GREATER	length of horn, though appearing late in lif
0026	er reverts to some one ancestor, removed by a	GREATER	number of generations. In a breed which has
0049	rked race of a Norwegian species, whereas the	GREATER	number rank it as an undoubted species pecul
0051	cases of difficulty; for he will encounter a	GREATER	number of closely allied forms. But if his o
0057	ch a number of varieties or incipient species	GREATER	than the average are now manufacturing, many
0058	ween varieties will tend to increase into the	GREATER	differences between species. There is one ot
0062	hall be treated, as it well deserves, at much	GREATER	length. The elder De Candolle and Lyell have
0067	one wedge being struck, and then another with	GREATER	force. What checks the natural tendency of a
0071	he effect on the insects must have been still	GREATER,	for six insectivorous birds were very commo
0072	rengger have shown that this is caused by the	GREATER	number in Paraguay of a certain fly, which l
0086	widely disseminated by the wind, I can see no	GREATER	difficulty in this being effected through na
0111	e between varieties become augmented into the	GREATER	difference between species? That this does h
0111	ts parent in the very same character and in a	GREATER	degree; but this alone would never account f
0112	ones by others, the differences would become	GREATER,	and would be noted as forming two sub breed
0112	inct breeds. As the differences slowly become	GREATER,	the inferior animals with intermediate char
0113	wn with several distinct genera of grasses, a	GREATER	number of plants and a greater weight of dry
0113	of grasses, a greater number of plants and a	GREATER	weight of dry herbage can thus be raised. Th
0113	d genera of grasses differ from each other, a	GREATER	number of individual plants of this species
0116	ified for different habits of life, so will a	GREATER	number of individuals be capable of there su
0117	varying species of the large genera present a	GREATER	number of varieties. We have, also, seen tha
0120	rocess of modification to be more numerous or	GREATER	in amount, to convert these three forms into
0120	species. By continuing the same process for a	GREATER	number of generations (as shown in the diagr
0123	cter between species a14 and z14 will be much	GREATER	than that between the most different of the
0128	tend to increase till they come to equal the	GREATER	differences between species of the same genu
0131	make the child like its parents. But the much	GREATER	variability, as well as the greater frequenc
0131	the much greater variability, as well as the	GREATER	frequency of monstrosities, under domesticat
0132	ose of the same species further north or from	GREATER	depths. Gould believes that birds of the sam
0134	. as Professor Owen has remarked, there is no	GREATER	anomaly in nature than a bird that cannot fl
0136	o reduce the wings, would depend on whether a	GREATER	number of individuals were saved by successf
0142	en cited for a similar purpose, and with much	GREATER	weight; but until some one will sow, during
0157	in giving to the species of the same group a	GREATER	amount of difference in their sexual charact
0158	e females. Finally, then, I conclude that the	GREATER	variability of specific characters, or those
0167	etween varieties of the same species, and the	GREATER	differences between species of the same genu
0171	aggered; but, to the best of my judgment, the	GREATER	number are only apparent, and those that are
0176	ser numbers would, as already remarked, run a	GREATER	chance of being exterminated than one existi
0177	he second chapter, presenting on an average a	GREATER	number of well marked varieties than do the
0177	s the two breeds, which originally existed in	GREATER	numbers, will come into close contact with a
0179	hich they connect; for these from existing in	GREATER	numbers will, in the aggregate, present more
0204	rse of further modification, from existing in	GREATER	numbers, will have a great advantage over th
0209	be the most serious error to suppose that the	GREATER	number of instincts have been acquired by ha
0212	e an instance of this even in England, in the	GREATER	wildness of all our large birds than of our
0212	ersecuted by man. We may safely attribute the	GREATER	wildness of our large birds to this cause; f
0221	epends merely on the slaves being captured in	GREATER	numbers in Switzerland than in England. One

Page **(Key Word)**

greater

Page ***(Key Word)***

```
0236  endered sterile is a difficulty: but not much  GREATER  than that of any other striking modification
0241  ful to the community, having been produced in  GREATER  and greater numbers through the natural sele
0241  ommunity, having been produced in greater and  GREATER  numbers through the natural selection of the
0251  at slight and mysterious causes the lesser or  GREATER  fertility of species when crossed, in compar
0256  y we have self fertilised hybrids producing a  GREATER  and greater number of seeds up to perfect fe
0256  lf fertilised hybrids producing a greater and  GREATER  number of seeds up to perfect fertility. Hyb
0263  s by no means seem to me to indicate that the  GREATER  or lesser difficulty of either grafting or c
0263  hey are imperfect. Even in first crosses, the  GREATER  or lesser difficulty, in effecting a union ap
0267  ility to the offspring. But we have seen that  GREATER  changes, or changes of a particular nature,
0267  ganic beings in some degree sterile: and that  GREATER  crosses, that is crosses between males and f
0269  between two species does not determine their  GREATER  or lesser degree of sterility when crossed:
0270  o much the less easy as their differences are  GREATER. How far these experiments may be trusted, I
0273  ccessive generations of mongrels is, perhaps,  GREATER  than in hybrids. This greater variability of
0273  ls is, perhaps, greater than in hybrids. This  GREATER  variability of mongrels than of hybrids does
0276  their vegetative systems, so in crossing, the  GREATER  or less facility of one species to unite wit
0277  tate of nature: and when we remember that the  GREATER  number of varieties have been produced under
0281  ent species of each genus, by differences not  GREATER  than we see between the varieties of the sam
0295  ial period, instead of having been really far  GREATER, that is extending from before the glacial e
0297  h species, but when they meet with a somewhat  GREATER  amount of difference between any two forms,
0314  nd whether the variations be accumulated to a  GREATER  or lesser amount, thus causing a greater or
0314  to a greater or lesser amount, thus causing a  GREATER  or lesser amount of modification in the vary
0316  cies, changing more or less quickly, and in a  GREATER  or lesser degree. A group does not reappear
0320  been produced within a given time is probably  GREATER  than that of the old forms which have been e
0327  e given my reasons for believing that all our  GREATER  fossiliferous formations were deposited duri
0353  y sub families, very many genera, and a still  GREATER  number of sections of genera are confined to
0358  nation: but for transport across the sea, the  GREATER  or less facilities may be said to be almost
0379  t migration from north to south is due to the  GREATER  extent of land in the north, and to the nort
0379  rn forms having existed in their own homes in  GREATER  numbers, and having consequently been advanc
0382  ossed at the equator: but to have flowed with  GREATER  force from the north so as to have freely in
0405  ange: but, if the variation had been a little  GREATER, the two varieties would have been ranked as
0408  tain forms and not others to enter; either in  GREATER  or lesser numbers: according or not, as thos
0415  r classification, I believe, depends on their  GREATER  constancy throughout large groups of species
0417  he latter, as A. de Jussieu has remarked, the  GREATER  number of the characters proper to the speci
0425  ough in these cases the modification has been  GREATER  in degree, and has taken a longer time to co
0429  nt groups. The more aberrant any form is, the  GREATER  must be the number of connecting forms which
0453  rule, that a rudimentary part or organ is of  GREATER  size relatively to the adjoining parts in th
0455  it in the embryo. Thus we can understand the  GREATER  relative size of rudimentary organs in the e
0463  either hand: and the latter, from existing in  GREATER  numbers, will generally be modified and impr
0470  pecies of the larger genera, which afford the  GREATER  number of varieties or incipient species, re
0470  e same species, tend to be augmented into the  GREATER  differences characteristic of species of the
0474  tincts, marvellous as some are, they offer no  GREATER  difficulty than does corporeal structure on
0489  ll organic beings are grouped, shows that the  GREATER  number of species of each genus, and all the
0001  steries, as it has been called by one of our  GREATEST  philosophers. On my return home, it occurred
0050  in the best known countries that we find the  GREATEST  number of forms of doubtful value. I have be
0110  at it is the common species which afford the  GREATEST  number of recorded varieties, or incipient s
0114  pecies. The truth of the principle, that the  GREATEST  amount of life can be supported by great div
0121  haracter, will generally tend to produce the  GREATEST  number of modified descendants: for these wi
0127  the most vigorous and best adapted males the  GREATEST  number of offspring. Sexual selection will a
0222  iduals of F. fusca were rushing about in the  GREATEST  agitation, and one was perched motionless wi
0224  their cells of the proper shape to hold the  GREATEST  possible amount of honey, with the least pos
0227  and we must further suppose, and this is the  GREATEST  difficulty, that she can somehow judge accur
0310  ande, Falconer, E. Forbes, etc., and all our  GREATEST  geologists, as Lyell, Murchison, Sedgwick, e
0325  re most widely diffused, having produced the  GREATEST  number of new varieties. It is also natural
0356  cts, which I have selected as presenting the  GREATEST  amount of difficulty on the theory of single
0386  feet. Birds of this order I can show are the  GREATEST  wanderers, and are occasionally found on the
0389  cts, which I have selected as presenting the  GREATEST  amount of difficulty, on the view that all t
0415  arisen, that almost all naturalists lay the  GREATEST  stress on resemblances in organs of high vit
0423  the forms together which were allied in the  GREATEST  number of points. In tumbler pigeons, though
0483  all away in force. But some arguments of the  GREATEST  weight extend very far. All the members of w
0020  of making distinct races by crossing has been  GREATLY  exaggerated. There can be no doubt that a ra
0021  in about the head, and this is accompanied by  GREATLY  elongated eyelids, very large external orifi
0035  ocess, only carried on more methodically, did  GREATLY  modify, even during their own lifetimes, the
0035  it is known that the English pointer has been  GREATLY  changed within the last century, and in this
0036  have insensibly passed, and come to differ so  GREATLY  from the rock pigeon. Youatt gives an excell
0043  he crossing of varieties has, I believe, been  GREATLY  exaggerated, both in regard to animals and t
0056  era have in the lapse of time often increased  GREATLY  in size: and that large genera have often co
0072  would become feral, and this would certainly  GREATLY  alter (as indeed I have observed in parts of
0095  nd by experiment that the fertility of clover  GREATLY  depends on bees visiting and moving parts of
0097  published elsewhere, that their fertility is  GREATLY  diminished if these visits be prevented. Now
0105  e very small: and fewness of individuals will  GREATLY  retard the production of new species through
0105  he production of new species. But we may thus  GREATLY  deceive ourselves, for to ascertain whether
0108  very slow process. The process will often be  GREATLY  retarded by free intercrossing. Many will ex
0157  , as Westwood has remarked, the number varies  GREATLY: and the number likewise differs in the two
0181  rs and fore arm of the Galeopithecus might be  GREATLY  lengthened by natural selection: and this, a
0237  h the working ant we have an insect differing  GREATLY  from its parents, yet absolutely sterile: so
0248  heir fertility never increased, but generally  GREATLY  decreased. I do not doubt that this is usual
0254  quite fertile together: and analogy makes me  GREATLY  doubt, whether the several aboriginal specie
0256  degree of fertility is often found to differ  GREATLY  in the several individuals raised from seed
0286  : but this source of doubt probably would not  GREATLY  affect the estimate as applied to the wester
0298  timately to new species: and this again would  GREATLY  lessen the chance of our being able to trace
0306  the Silurian age, and which probably differed  GREATLY  from any known animal. Some of the most anci
0311  is view, the difficulties above discussed are  GREATLY  diminished, or even disappear. Chapter X. On
0313  molluscs and all the Crustaceans have changed  GREATLY. The productions of the land seem to change
0320  at the rarity of a species, and yet to marvel  GREATLY  when it ceases to exist, is much the same as
0324  rge sense, at distant parts of the world, has  GREATLY  struck those admirable observers, MM. de Ver
0340  nhabitants of one continent formerly differed  GREATLY  from those of another continent, so will the
```

Page ***(Key Word)***

greatly

Page ***(Key Word)***

```
0345  ctions to the theory of natural selection are GREATLY diminished or disappear. On the other hand,
0351  igrated over vast spaces, and have not become GREATLY modified. On these views, it is obvious, tha
0366  th which their valleys were lately filled. So GREATLY has the climate of Europe changed, that in N
0377  tor. The invasion would, of course, have been GREATLY favoured by high land, and perhaps by a dry
0380  ctually exterminated, their numbers have been GREATLY reduced, and this is the first stage towards
0394  er quadrupeds have not become naturalised and GREATLY multiplied. It cannot be said, on the ordina
0403  from these considerations I think we need not GREATLY marvel at the endemic and representative spe
0405  species, and the common range would have been GREATLY reduced. Still less is it meant, that a spec
0408  ld find, as we do find, some groups of beings GREATLY, and some only slightly modified, some devel
0410  a different family of the same order, differ GREATLY. In both time and space the lower members of
0420  in classification are explained, if I do not GREATLY deceive myself, on the view that the natural
0420  blood to their common progenitor, may differ GREATLY, being due to the different degrees of modif
0445  the full grown animals; and this surprised me GREATLY, as I think it probable that the difference
0447  species, the embryonic fore limbs will differ GREATLY from the fore limbs in the mature animal; th
0448  the larvae, in the first stage, might differ GREATLY from the larvae in the second stage, as we h
0450  responding not early period. Embryology rises GREATLY in interest, when we thus look at the embryo
0461  in the conditions of life and crosses between GREATLY modified forms, lessen fertility; and on the
0462  periods; and such changes will obviously have GREATLY facilitated migration. As an example, I have
0485  eral departments of natural history will rise GREATLY in interest. The terms used by naturalists o
0185  ht, would be mistaken by any one for an auk or GREBE; nevertheless, it is essentially a petrel, bu
0471  structure fitting it for the life of an auk or GREBE! and so on in endless other cases. But on the
0185  on the surface of the sea. On the other hand, GREBES and coots are eminently aquatic, although the
0092  el. This juice, though small in quantity, is GREEDILY sought by insects. Let us now suppose a litt
0084  s be acted on. When we see leaf eating insects GREEN, and bark feeders mottled grey; the alpine pt
0129  lieve this simile largely speaks the truth. The GREEN and budding twigs may represent, existing spec
0197  stances to illustrate these latter remarks. If GREEN woodpeckers alone had existed, and we did not
0197  dare say that we should have thought that the GREEN colour was a beautiful adaptation to hide thi
0359  what a difference there is in the buoyancy of GREEN and seasoned timber; and it occurred to me th
0359  e majority sank quickly, but some which whilst GREEN floated for a very short time, when dried flo
0304  ence of the existence of whales in the upper GREENSAND, some time before the close of the secondar
0296  ssilised trees, still standing upright as they GREW, of many long intervals of time and changes of
0362  different birds of prey; and two seeds of beet GREW after having been thus retained for two days a
0386  nths, pulling up and counting each plant as it GREW; the plants were of many kinds, and were altog
0084  eating insects green, and bark feeders mottled GREY; the alpine ptarmigan white in winter, the red
0016  if, for instance, it could be shown that the GREYHOUND, bloodhound, terrier, spaniel, and bull dog
0019  that animals closely resembling the Italian GREYHOUND, the bloodhound, the bull dog, or Blenheim
0020  ce of the most extreme forms, as the Italian GREYHOUND, bloodhound, bull dog, etc., in the wild st
0029  the differences of a dray and race horse, a GREYHOUND and bloodhound, a carrier and tumbler pigeo
0091  tains in the United States, one with a light GREYHOUND like form, which pursues deer, and the othe
0214  obstinacy of greyhounds; and a cross with a GREYHOUND has given to a whole family of shepherd dog
0427  n and swedish turnip. The resemblance of the GREYHOUND and racehorse is hardly more fanciful than
0444  who have written on Dogs, maintain that the GREYHOUND and bull dog, though appearing so different
0090  n that man can improve the fleetness of his GREYHOUNDS by careful and methodical selection, or by
0214  ny generations the courage and obstinacy of GREYHOUNDS; and a cross with a greyhound has given to
0282  perimposed strata, and watch the sea at work GRINDING down old rocks and making fresh sediment, be
0290  ow and gradual rising of the land within the GRINDING action of the coast waves. We may, I think,
0226  yramid, as Huber has remarked, is manifestly a GROSS imitation of the three sided pyramidal basis
0063  same and other kinds which already clothe the GROUND. The missletoe is dependent on the apple and
0067  edlings which suffer most from germinating in GROUND already thickly stocked with other plants. Se
0067  various enemies; for instance, on a piece of GROUND three feet long and two wide, dug and cleared
0075  r on the other plants which first clothed the GROUND and thus checked the growth of the trees! Thr
0075  handful of feathers, and all must fall to the GROUND according to definite laws; but how simple is
0077  be widely distributed and fall on unoccupied GROUND. In the water beetle, the structure of its le
0091  res or rabbits, and another hunting on marshy GROUND and almost nightly catching woodcocks or snip
0113  been experimentally proved, that if a plot of GROUND be sown with one species of grass, and a simi
0113  es of wheat have been sown on equal spaces of GROUND. Hence, if any one species of grass were to g
0113  would succeed in living on the same piece of GROUND. And we well know that each species and each
0114  nts which live close round any small piece of GROUND, could live on it (supposing it not to be in
0134  as the domestic Aylesbury duck. As the larger GROUND feeding birds seldom take flight except to es
0136  stroyed. The insects in Madeira which are not GROUND feeders, and which, as the flower feeding col
0204  that there are upland geese with webbed feet, GROUND woodpeckers, diving thrushes, and petrels wit
0216  nctive purpose of allowing, as we see in wild GROUND birds, their mother to fly away. But this ins
0231  utting it away equally on both sides near the GROUND, till a smooth, very thin wall is left in the
0283  about by the waves, and then are more quickly GROUND into pebbles, sand, or mud. But how often do
0367  c forms would seize on the cleared and thawed GROUND, always ascending higher and higher, as the w
0380  that very many European productions cover the GROUND in La Plata, and in a lesser degree in Austra
0397  d adhere to the feet of birds roosting on the GROUND, and thus get transported? It occurred to me
0401  ant, for instance, would find the best fitted GROUND more perfectly occupied by distinct plants in
0471  d have been created to prey on insects on the GROUND; that upland geese, which never or rarely swi
0002  eferences, on which my conclusions have been GROUNDED; and I hope in a future work to do this. For
0320  violence. The theory of natural selection is GROUNDED on the belief that each new variety, and ult
0368  arctic regions of that country. These views, GROUNDED as they are on the perfectly well ascertaine
0319  e apparently so favourable. But how utterly GROUNDLESS was my astonishment! Professor Owen soon pe
0068  destroyed four fifths of the birds in my own GROUNDS; and this is a tremendous destruction, when
0020  g that it is always best to study some special GROUP, I have, after deliberation, taken up domesti
0048  varieties. On the islets of the little Madeira GROUP there are many insects which are characterize
0050  en a young naturalist commences the study of a GROUP of organisms quite unknown to him, he is at f
0050  the amount and kind of variation to which the GROUP is subject: and this shows, at least, how ver
0055  nt reason why more varieties should occur in a GROUP having many species, than in one having few.
0125  mount of change represented by each successive GROUP of diverging dotted lines to be very great, t
0125  have some advantage; and the largeness of any GROUP shows that its species have inherited from a
0125  re all trying to increase in number. One large GROUP will slowly conquer another large group, redu
0125  large group will slowly conquer another large GROUP, reduce its numbers, and thus lessen its chan
0126  riation and improvement. Within the same large GROUP, the later and more highly perfected sub grou
0128  e and space should be related to each other in GROUP subordinate to group, in the manner which we
0128  related to each other in group subordinate to GROUP, in the manner which we everywhere behold, na
```

Page ***(Key Word)***

Page **(Key Word)***

```
0146 ifications, having been transmitted to a whole GROUP of descendants with diverse habits, would nat
0150 would not here apply, because there is a whole GROUP of bats having wings; it would apply only if
0155 y differs in the different species of the same GROUP, the more subject it is to individual anomali
0156 also will be admitted that species of the same GROUP differ from each other more widely in their s
0157 succeeded in giving to the species of the same GROUP a greater amount of difference in their sexua
0158 t may be developed, if it be common to a whole GROUP of species; that the great variability of sec
0158 ll being mainly due to the species of the same GROUP having descended from a common progenitor, fr
0161 he exact character of the common ancestor of a GROUP, we could not distinguish these two cases: if
0162 lready occur in some other members of the same GROUP. And this undoubtedly is the case in nature.
0169 racters differ much in the species of the same GROUP. Variability in the same parts of the organis
0179 nking most closely all the species of the same GROUP together, must assuredly have existed; but th
0179 it would be easy to show that within the same GROUP carnivorous animals exist having every interm
0187 ed in each case to look to species of the same GROUP, that is to the collateral descendants from t
0210 wing facts show. I removed all the ants from a GROUP of about a dozen aphides on a dockplant, and
0218 he origin of a singular instinct in the allied GROUP of ostriches. For several hen ostriches, at l
0265 e hybrids. On the other hand, one species in a GROUP will sometimes resist great changes of condit
0265 unimpaired fertility; and certain species in a GROUP will produce unusually fertile hybrids. No on
0301 d all the past and present species of the same GROUP into one long and branching chain of life. We
0302 gh natural selection. For the development of a GROUP of forms, all of which have descended from so
0304 beds of this age, I concluded that this great GROUP had been suddenly developed at the commenceme
0304 e instance of the abrupt appearance of a great GROUP of species. But my work had hardly been publi
0305 of the apparently sudden appearance of a whole GROUP of species, is that of the teleostean fishes,
0305 ean fishes, low down in the Chalk period. This GROUP includes the large majority of existing speci
0305 uld likewise be shown that the species of this GROUP appeared suddenly and simultaneously througho
0305 d perfectly enclosed basin, in which any great GROUP of marine animals might be multiplied; and he
0306 manner in which numbers of the species of the same GROUP, suddenly appear in the lowest known fossilif
0306 d me that all the existing species of the same GROUP have descended from one progenitor, apply wit
0316 quickly, and in a greater or lesser degree. A GROUP does not reappear after it has once disappear
0316 my theory. For as all the species of the same GROUP have descended from one species, it is clear
0316 it is clear that as long as any species of the GROUP have appeared in the long succession of ages,
0316 seen in the last chapter that the species of a GROUP sometimes falsely appear to have come in abru
0316 e being a gradual increase in number, till the GROUP reaches its maximum, and then, sooner or late
0317 gradual increase in number of the species of a GROUP is strictly conformable with my theory; as th
0317 g of a great tree from a single stem, till the GROUP becomes large. On Extinction. We have as yet
0318 at the complete extinction of the species of a GROUP is generally a slower process than their prod
0318 tion: if the appearance and disappearance of a GROUP of species be represented, as before, by a ve
0321 pened that a new species belonging to some one GROUP will have seized on the place occupied by a s
0321 occupied by a species belonging to a distinct GROUP, and thus caused its extermination; and if ma
0321 a few members of the great and almost extinct GROUP of Ganoid fishes still inhabit our fresh wate
0321 sh waters. Therefore the utter extinction of a GROUP is generally, as we have seen, a slower proce
0322 ually rapid development, many species of a new GROUP have taken possession of a new area, they wil
0322 r the extinction of this particular species or GROUP of species. On the Forms of Life changing alm
0330 riters have objected to any extinct species or GROUP of species being considered as intermediate b
0344 the earth. But the utter extinction of a whole GROUP of species may often be a very slow process,
0344 g in protected and isolated situations. When a GROUP has once wholly disappeared, it does not reap
0346 require: for it is a most rare case to find a GROUP of organisms confined to any small spot, havi
0384 ding to Valenciennes, there is hardly a single GROUP of fishes confined exclusively to fresh water
0384 imagine that a marine member of a fresh water GROUP might travel far along the shores of the sea,
0393 ke fox, come nearest to an exception; but this GROUP cannot be considered as oceanic, as it lies o
0401 d yet retain the same character throughout the GROUP, just as we see on continents some species sp
0406 n regard to plants, namely, that the lower any GROUP of organisms is, the more widely it is apt to
0409 hy, in relation to the means of migration, one GROUP of beings, even within the same class, should
0409 ould have all its species endemic, and another GROUP should have all its species common to other q
0409 many facts. The endurance of each species and GROUP of species is continuous in time; for the exc
0410 he area inhabited by a single species, or by a GROUP of species, is continuous; and the exceptions
0411 ould have been of simple signification, if one GROUP had been exclusively fitted to inhabit the la
0411 ious how commonly members of even the same sub GROUP have different habits. In our second and four
0413 act in natural history of the subordination of GROUP under group, which, from its familiarity, doe
0413 al history of the subordination of group under GROUP, which, from its familiarity, does not always
0415 ifferent. No naturalist can have worked at any GROUP without being struck with this fact; and it h
0416 on of the same important organ within the same GROUP of beings. Again, no one will say that rudime
0418 of the characters which they use in defining a GROUP, or in allocating any particular species. If
0419 ould be given amongst plants and insects, of a GROUP of forms, first ranked by practised naturalis
0422 cover in nature amongst the beings of the same GROUP. Thus, on the view which I hold, the natural
0424 e subject, these tumblers are kept in the same GROUP, because allied in blood and alike in some ot
0424 o, i think he would be classed under the Negro GROUP, however much he might differ in colour and o
0426 so trifling, occur together throughout a large GROUP of beings having different habits, we may fee
0426 fication. We can understand why a species or a GROUP of species may depart, in several of its most
0426 to them; but if these same organs, in another GROUP or section of a group, are found to differ mu
0426 same organs, in another group or section of a GROUP, are found to differ much, we at once value t
0429 i attempted to show, on the principle of each GROUP having generally diverged much in character d
0430 remarked that, when a member belonging to one GROUP of animals exhibits an affinity to a quite di
0430 imals exhibits an affinity to a quite distinct GROUP, this affinity in most cases is general and n
0430 ommon progenitor, or of an early member of the GROUP. On the other hand, of all Marsupials, as Mr.
0431 h all the members of the same family or higher GROUP are connected together. For the common parent
0432 e impossible to give definitions by which each GROUP could be distinguished from other groups, as
0432 s, representing most of the characters of each GROUP, whether large or small, and thus give a gene
0433 organic beings, namely, their subordination in GROUP under group. We use the element of descent in
0433 gs, namely, their subordination in group under GROUP. We use the element of descent in classing th
0433 hysiological importance; why, in comparing one GROUP with a distinct group, we summarily reject an
0433 e; why, in comparing one group with a distinct GROUP, we summarily reject analogical or adaptive c
0433 same characters within the limits of the same GROUP. We can clearly see how it is that all living
0439 s we see in the spotted feathers in the thrush GROUP. In the cat tribe, most of the species are st
0449 ns. As the embryonic state of each species and GROUP partially shows us the structure o
0456 e attempted to show, that the subordination of GROUP to group in all organisms throughout all time
0456 ed to show, that the subordination of group to GROUP in all organisms throughout all time; that th
```

Page **(Key Word)***

Page **************************************(Key Word)**************************************

```
0458  ve all descended, each within its own class or GROUP, from common parents, and have all been modif
0461  the species of the same genus, or even higher GROUP, must have descended from common parents; and
0462  sted, linking together all the species in each GROUP by gradations as fine as our present varietie
0471  to new and dominant forms; so that each large GROUP tends to become still larger, and at the same
0478  ings constitute one grand natural system, with GROUP subordinate to group, and with extinct groups
0478  rand natural system, with group subordinate to GROUP, and with extinct groups often falling in bet
0241  act for us is, that though the workers can be GROUPED into castes of different sizes, yet they gra
0413  ny species descended from a single progenitor GROUPED into genera; and the genera are included in
0423  to have descended from one species. These are GROUPED under species, with sub varieties under var
0433  t is that all living and extinct forms can be GROUPED together in one great system; and how the se
0489  or the manner in which all organic beings are GROUPED, shows that the greater number of species of
0080  scendants from a common parent. Explains the GROUPING of all organic beings. How will the struggle
0411  fication is evidently not arbitrary like the GROUPING of the stars in constellations. The existenc
0425  t of descent have been unconsciously used in GROUPING species under genera, and genera under highe
0471  d throughout all time. The grand fact of the GROUPING of all organic beings seems to me utterly in
0028  o the many species of finches, or other large GROUPS of birds, in nature. One circumstance has str
0051  al knowledge of analogical variation in other GROUPS and in other countries, by which to correct h
0057  vided into sub genera, or sections, or lesser GROUPS. As Fries has well remarked, little groups of
0057  er groups. As Fries has well remarked, little GROUPS of species are generally clustered like satel
0057  ain other species. And what are varieties but GROUPS of forms, unequally related to each other, an
0059  e throughout the universe become divided into GROUPS subordinate to groups. Chapter III. Struggle
0059  rse become divided into groups subordinate to GROUPS. Chapter III. Struggle For Existence. Bears o
0061  e varieties of the same species? How do those GROUPS of species, which constitute what are called
0096  te animals, all insects, and some other large GROUPS of animals, pair for each birth. Modern resea
0115  turalised plants would have belonged to a few GROUPS more especially adapted to certain stations i
0116  australian marsupials, which are divided into GROUPS differing but little from each other, and fee
0123  from the eight descendants from (A); the two GROUPS, moreover, are supposed to have gone on diver
0124  ree intermediate in character between the two GROUPS descended from these species. But as these tw
0124  escended from these species. But as these two GROUPS have gone on diverging in character from the
0124  een them, but rather between types of the two GROUPS; and every naturalist will be able to bring s
0124  e, intermediate in character between existing GROUPS; and we can understand this fact, for the ext
0125  ree genera descended from (A), the two little GROUPS of genera will form two distinct families, or
0125  scendants, will mainly lie between the larger GROUPS, which are all trying to increase in number.
0126  roup, the later and more highly perfected sub GROUPS, from branching out and seizing on many new p
0126  and destroy the earlier and less improved sub GROUPS. Small and broken groups and sub groups will
0126  nd less improved sub groups. Small and broken GROUPS and sub groups will finally tend to disappear
0126  d sub groups. Small and broken groups and sub GROUPS will finally tend to disappear. Looking to th
0126  ooking to the future, we can predict that the GROUPS of organic beings which are now large and tri
0126  a long period continue to increase. But which GROUPS will ultimately prevail, no man can predict;
0126  o man can predict; for we well know that many GROUPS, formerly most extensively developed, have no
0126  e continued and steady increase of the larger GROUPS, a multitude of smaller groups will become ut
0126  of the larger groups, a multitude of smaller GROUPS will become utterly extinct, and leave no mod
0128  classes, and classes. The several subordinate GROUPS in any class cannot be ranked in a single fil
0129  d branches, in the same manner as species and GROUPS of species have tried to overmaster other spe
0129  fication of all extinct and living species in GROUPS subordinate to groups. Of the many twigs whic
0129  t and living species in groups subordinate to GROUPS. Of the many twigs which flourished when the
0136  of the almost entire absence of certain large GROUPS of beetles; elsewhere excessively numerous, a
0136  es, elsewhere excessively numerous, and which GROUPS have habits of life almost necessitating freq
0146  growth, structures which are common to whole GROUPS of species, and which in truth are simply due
0152  can see no explanation. But on the view that GROUPS of species have descended from other species,
0152  articular purpose, and perhaps in polymorphic GROUPS, we see a nearly parallel natural case; for i
0155  h is generally very constant throughout large GROUPS of species, has differed considerably in clos
0157  is a character generally common to very large GROUPS of beetles; but in the Engidae, as Westwood h
0157  e highest importance, because common to large GROUPS; but in certain genera the neuration differs
0265  in extent with systematic affinity, for whole GROUPS of animals and plants are rendered impotent b
0265  t by the same unnatural conditions; and whole GROUPS of species tend to produce sterile hybrids. O
0279  ny one formation. On the sudden appearance of GROUPS of species. On their sudden appearance in the
0299  , and has made the intervals between some few GROUPS less wide than they otherwise would have been
0302  my theory. On the sudden appearance of whole GROUPS of Allied Species. The abrupt manner in which
0302  ied Species. The abrupt manner in which whole GROUPS of species suddenly appear in certain formati
0302  have been carefully examined; we forget that GROUPS of species may elsewhere have long existed an
0303  iable we are to error in supposing that whole GROUPS of species have suddenly been produced. I may
0306  its productions. On the sudden appearance of GROUPS of Allied Species in the lowest known fossili
0310  ave existed; the sudden manner in which whole GROUPS of species appear in our European formations;
0312  of change. Species once lost do not reappear. GROUPS of species follow the same general rules in t
0316  nitor some slight characteristic differences. GROUPS of species, that is, genera and families, fol
0317  ntally of the disappearance of species and of GROUPS of species. On the theory of natural selectio
0317  of the tertiary formations, that species and GROUPS of species gradually disappear, one after ano
0318  from the world. Both single species and whole GROUPS of species last for very unequal periods; som
0318  f species last for very unequal periods; some GROUPS, as we have seen, having endured from the ear
0318  me cases, however, the extermination of whole GROUPS of beings, as of ammonites towards the close
0320  l, are bound together. In certain flourishing GROUPS, the number of new specific forms which have
0322  the manner in which single species and whole GROUPS of species become extinct, accords well with
0327  nd as these inferior forms would be allied in GROUPS by inheritance, whole groups would tend slowl
0327  uld be allied in groups by inheritance, whole GROUPS would tend slowly to disappear; though here a
0327  victorious forms, will generally be allied in GROUPS, from inheriting some inferiority in common;
0327  in common; and therefore as new and improved GROUPS spread throughout the world; old groups will
0327  roved groups spread throughout the world, old GROUPS will disappear from the world; and the succes
0329  ssils can be classed either in still existing GROUPS, or between them. That the extinct forms of l
0329  how extinct animals fall in between existing GROUPS. Cuvier ranked the Ruminants and Pachyderms,
0330  at this early epoch limited in such distinct GROUPS as they now are. Some writers have objected t
0330  red as intermediate between living species or GROUPS. If by this term it is meant that an extinct
0330  ase, especially with respect to very distinct GROUPS, such as fish and reptiles, seems to be, that
0330  aracters, the ancient members of the same two GROUPS would be distinguished by a somewhat lesser n
0330  lesser number of characters, so that the two GROUPS, though formerly quite distinct, at that peri
0330  it tends to connect by some of its characters GROUPS now widely separated from each other. This re
```

Page **************************************(Key Word)**************************************

Page **************************************(Key Word)**************************************

```
0330  s remark no doubt must be restricted to those   GROUPS which have undergone much change in the cours
0330  ing affinities directed towards very distinct    GROUPS. Yet if we compare the older Reptiles and Bat
0333  d than is represented in the diagram: for the    GROUPS will have been more numerous, they will have
0333  that we have a right to expect, is that those    GROUPS, which have within known geological periods u
0333  ters than do the existing members of the same    GROUPS; and this by the concurrent evidence of our b
0338  confirmed, at least in regard to subordinate     GROUPS, which have branched off from each other with
0338  ther than are the typical members of the same    GROUPS at the present day, it would be vain to look
0343  n falsely apparent, sudden coming in of whole     GROUPS of species. He may ask where are the remains
0343  cies has once disappeared it never reappears.    GROUPS of species increase in numbers slowly, and en
0343  . the dominant species of the larger dominant    GROUPS tend to leave many modified descendants, and
0344  e many modified descendants, and thus new sub     GROUPS and groups are formed. As these are formed, t
0344  fied descendants, and thus new sub groups and    GROUPS are formed. As these are formed, the species
0344  are formed, the species of the less vigorous     GROUPS, from their inferiority inherited from a comm
0344  nerally succeed in taking the places of those    GROUPS of species which are their inferiors in the s
0344  etween existing forms, sometimes blending two    GROUPS previously classed as distinct into one: but
0344  haracters in some degree intermediate between    GROUPS now distinct: for the more ancient a form is,
0344  nsequently resemble, the common progenitor of    GROUPS, since become widely divergent. Extinct forms
0349  o be struck by the manner in which successive    GROUPS of beings, specifically distinct, yet clearly
0350  me still further victorious, and will produce    GROUPS of modified descendants. On this principle of
0381  have been modified and have given rise to new    GROUPS of forms, and others have remained unaltered.
0386  n, as remarked by Alph. de Candolle, in large    GROUPS of terrestrial plants, which have only a very
0390  iable to modification, and will often produce    GROUPS, as of batrachians, and of terrestrial mammal
0396  or sections of classes, the absence of whole     GROUPS, as of batrachians, and of terrestrial mammal
0408  tion, and we should find, as we do find, some    GROUPS of beings greatly, and some only slightly mod
0409  r quarters of the world. We can see why whole    GROUPS of organisms, as batrachians and terrestrial
0410  e tracts. Both in time and space, species and    GROUPS of species have their points of maximum devel
0410  ies have their points of maximum development.    GROUPS of species, belonging either to a certain per
0411  bryology. Rudimentary Organs. Classification,    GROUPS subordinate to groups. Natural system. Rules
0411  organs. Classification, groups subordinate to    GROUPS. Natural system. Rules and difficulties in cl
0411  d radiating. Extinction separates and defines    GROUPS. Morphology, between members of the same clas
0411  nding degrees, so that they can be classed in    GROUPS under groups. This classification is evidentl
0411  , so that they can be classed in groups under    GROUPS. This classification is evidently not arbitra
0411  the stars in constellations. The existence of    GROUPS would have been of simple signification, if o
0412  er new and dominant species. Consequently the    GROUPS which are now large, and which generally incl
0412  ing from one progenitor become broken up into    GROUPS subordinate to groups. In the diagram each le
0412  r become broken up into groups subordinate to    GROUPS. In the diagram each letter on the uppermost
0415  s on their greater constancy throughout large    GROUPS of species; and this constancy depends on suc
0415  almost shown by the one fact, that in allied     GROUPS, in which the same organ, as we have every re
0416  highly serviceable in the definition of whole    GROUPS. For instance, whether or not there is an ope
0418  em, especial value is set on them. As in most    GROUPS of animals, important organs, such as those f
0418  ly serviceable in classification: but in some    GROUPS of animals all these, the most important vita
0419  classification, more especially in very large    GROUPS of closely allied forms. Temminck insists on
0419  or even necessity of this practice in certain    GROUPS of birds; and it has been followed by several
0419  spect to the comparative value of the various    GROUPS of species, such as orders, sub orders, famil
0420  fully. I believe that the arrangement of the     GROUPS within each class, in due subordination and r
0420  n due subordination and relation to the other    GROUPS, must be strictly genealogical in order to be
0420  ount of difference in the several branches or    GROUPS, though allied in the same degree in blood to
0422  had not been used, and only the names of the     GROUPS had been written in a linear series; it would
0422  e degrees of modification which the different    GROUPS have undergone, have to be expressed by ranki
0422  the same stock, would have to be expressed by    GROUPS subordinate to groups; but the proper or even
0422  have to be expressed by groups subordinate to    GROUPS; but the proper or even only possible arrange
0423  with pigeons. The origin of the existence of     GROUPS subordinate to groups, is the same with varie
0423  gin of the existence of groups subordinate to    GROUPS, is the same with varieties as with species,
0425  species under genera, and genera under higher    GROUPS, though in these cases the modification has b
0426  connected together by a chain of intermediate    GROUPS, we may at once infer their community of desc
0428  m has sometimes been observed between the sub    GROUPS in distinct classes. A naturalist, struck by
0428  bitrarily raising or sinking the value of the    GROUPS to which they belong large and their parents
0428  end to inherit the advantages, which made the    GROUPS thus tend to go on increasing in size; and th
0428  onomy of nature. The larger and more dominant    GROUPS. Thus we can account for the fact that all or
0428  onsequently supplant many smaller and feebler    GROUPS are in number, and how widely spread they are
0429  natural system. As showing how few the higher    GROUPS. A few old and intermediate parent forms havi
0429  e slight degree intermediate between existing    GROUPS. The more aberrant any form is, the greater m
0429  give to us our so called osculant or aberrant    GROUPS conquered by more successful competitors, wit
0429  only by looking at aberrant forms as failing     GROUPS have since undergone much modification in div
0430  d off from a common progenitor, and that both    GROUPS and sub groups, will have transmitted some of
0431  es, now broken up by extinction into distinct    GROUPS, will have transmitted some of its characters
0431  up by extinction into distinct groups and sub    GROUPS in each class. We may thus account even for t
0431  nd widening the intervals between the several    GROUPS: it has by no means made them: for if every f
0432  of affinities. Extinction has only separated     GROUPS, as all would blend together by steps as fine
0432  each group could be distinguished from other     GROUPS of modified descendants. Every intermediate l
0432  ian genera, some of which have produced large    GROUPS could be distinguished from their more immedi
0432  n by which the several members of the several    GROUPS; but we could pick out types, or forms, repre
0432  could not, as I have said, define the several    GROUPS to which such types belong. Finally, we have
0433  whether or not we can separate and define the    GROUPS of species, differing quite as much, or even
0440  given of the larvae of two species, or of two    GROUPS of animals and in certain members of other gr
0442  came visible in the embryo. And in some whole    GROUPS, the embryo does not at any period differ wid
0442  ps of animals and in certain members of other    GROUPS of animals, as with cuttle fish and spiders,
0448  s is the rule of development in certain whole    GROUPS of animal, however much they may at present d
0449  veals the structure of its progenitor. In two    GROUPS. Rudimentary organs may be utterly aborted; a
0452  e of the wings of the female moths in certain    GROUPS of organic beings; but we see so many strange
0460  d, more especially amongst broken and failing    GROUPS of allied species appear, though certainly th
0463  urged against my theory. Why, again, do whole    GROUPS of fossils? For certainly on my theory such s
0463  he remains of the progenitors of the Silurian    GROUPS round other species, in which respects they r
0470  cted ranges; and they are clustered in little    GROUPS tend to give birth to new and dominant forms;
0470  cts. Dominant species belonging to the larger    GROUPS cannot thus succeed in increasing in size, fo
0471  time more divergent in character. But as all     GROUPS
```

Page **************************************(Key Word)**************************************

```
0471  world would not hold them, the more dominant  GROUPS  beat the less dominant. This tendency in the
0471  the less dominant. This tendency in the large  GROUPS  to go on increasing in size and diverging in
0471  the arrangement of all the forms of life, in   GROUPS  subordinate to groups, all within a few great
0471  l the forms of life, in groups subordinate to  GROUPS, all within a few great classes, which we now
0475  als of time, is widely different in different  GROUPS. The extinction of species and of whole group
0475  roups. The extinction of species and of whole  GROUPS  of species, which has played so conspicuous a
0475  nd improved forms. Neither single species nor  GROUPS  of species reappear when the chain of ordinar
0476  ing either into the same or into intermediate  GROUPS, follows from the living and the extinct bein
0476  being the offspring of common parents. As the  GROUPS  which have descended from an ancient progenit
0476  gree intermediate between existing and allied  GROUPS. Recent forms are generally looked at as bein
0478  group subordinate to group, and with extinct   GROUPS  often falling in between recent groups, is in
0478  xtinct groups often falling in between recent  GROUPS, is intelligible on the theory of natural sel
0483  l can be classified on the same principle, in  GROUPS  sub ordinate to groups. Fossil remains somet
0483  the same principle, in groups sub ordinate to  GROUPS. Fossil remains sometimes tend to fill up ver
0486  e nature of long lost structures. Species and  GROUPS  of species, which are called aberrant, and wh
0489  species, belonging to the larger and dominant  GROUPS, which will ultimately prevail and procreate
0049  enced ornithologists consider our British red  GROUSE  as only a strongly marked race of a Norwegian
0068  be little doubt that the stock of partridges,  GROUSE, and hares on any large estate depends chiefl
0084  the alpine ptarmigan white in winter, the red  GROUSE  the colour of heather, and the black grouse t
0084  d grouse the colour of heather, and the black  GROUSE  that of peaty earth, we must believe that the
0084  s and insects in preserving them from danger.  GROUSE, if not destroyed at some period of their liv
0085  e in giving the proper colour to each kind of  GROUSE, and in keeping that colour, when once acquir
0049  hey flower at slightly different periods; they  GROW  in somewhat different stations: they ascend mo
0055  ule, to be now forming. Where many large trees  GROW, we expect to find saplings. Where many specie
0063  hese trees, for if too many of these parasites  GROW  on the same tree, it will languish and die. Bu
0067  turf closely browsed by quadrupeds, be let to  GROW, the more vigorous plants gradually kill the l
0068  rished from the other species being allowed to  GROW  up freely. The amount of food for each species
0249  nt species, or other allied hybrids, generally  GROW  in the same garden, the visits of insects must
0250  ries of the three first flowers soon ceased to  GROW, and after a few days perished entirely, where
0037  though he might succeed from a poor seedling   GROWING  wild, if it had come from a garden stock. Th
0063  ish and die. But several seedling missletoes,  GROWING  close together on the same branch, may more
0068  lly grown, plants: thus out of twenty species  GROWING  on a little plot of turf (three feet by four
0072  ed, it became thickly clothed with vigorously  GROWING  young firs. Yet the heath was so extremely b
0074  ; but it has been observed that the trees now  GROWING  on the ancient Indian mounds, in the Souther
0075  e proportional numbers and kinds of trees now  GROWING  on the old Indian ruins! The dependency of a
0077  seedling, whilst struggling with other plants  GROWING  vigorously all around. Look at a plant in th
0098  dlings: and whilst another species of Lobelia  GROWING  close by, which is visited by bees, seeds fr
0099  bages from some plants of different varieties  GROWING  near each other, and of these only 78 were t
0129  nct species. At each period of growth all the  GROWING  twigs have tried to branch out on all sides,
0132  cuin Tandon gives a list of plants which when  GROWING  near the sea shore have their leaves in some
0140  from seed collected by Dr. Hooker from trees  GROWING  at different heights on the Himalaya, were f
0230  ay be clearly seen by examining the edge of a  GROWING  comb, do make a rough, circumferential wall
0230  one rhombic plate which stands on the extreme  GROWING  margin, or the two plates, as the case may b
0231  de circumferential rim or wall of wax round a  GROWING  comb, flexures may sometimes be observed, co
0231  idge. We shall thus have a thin wall steadily  GROWING  upward; but always crowned by a gigantic cop
0232  xtreme margin of the circumferential rim of a  GROWING  comb, with an extremely thin layer of melted
0232  which it had been placed, and worked into the  GROWING  edges of the cells all round. The work of co
0232  d, placed directly under the middle of a comb  GROWING  downwards so that the comb has to be built o
0270  low seeds, and a tall variety with red seeds,  GROWING  near each other in his garden; and although
0373  descent; and in Sikkim, Dr. Hooker saw maize  GROWING  on gigantic ancient moraines. South of the e
0375  alian forms are clearly represented by plants  GROWING  on the summits of the mountains of Borneo. S
0375  we see that throughout the world, the plants  GROWING  on the more lofty mountains, and on the temp
0378  pical and temperate vegetation, like that now  GROWING  with strange luxuriance at the base of the H
0392  nts alone, might readily gain an advantage by  GROWING  taller and taller and overtopping the other
0392  add to the stature of herbaceous plants when  GROWING  on an island, to whatever order they belonge
0453  bsorbed, can be of any service to the rapidly  GROWING  embryonic calf by the excretion of precious
0068  gradually kill the less vigorous, though fully  GROWN  plants: thus out of twenty species growing o
0087  now, if nature had to make the beak of a full  GROWN  pigeon very short for the bird's own advantag
0129  e tree was a mere bush, only two or three, now  GROWN  into great branches, yet survive and bear all
0164  is much commoner in the foal than in the full  GROWN  animal. Without here entering on further deta
0444  k moth; or, again, in the horns of almost full  GROWN  cattle. But further than this, variations in
0445  t and race horses differed as much as the full  GROWN  animals; and this surprised me greatly, as I
0445  points were incomparably less than in the full  GROWN  birds. Some characteristic points of differen
0446  nd pigeons, for breeding, when they are nearly  GROWN  up: they are indifferent whether the desired
0446  acquired earlier or later in life, if the full  GROWN  animal possesses them. And the cases just giv
0450  presence of teeth in foetal whales, which when  GROWN  up have not a tooth in their heads; and the p
0451  se organs having become well developed in full  GROWN  males; and having secreted milk. So again the
0483  and plants created as eggs or seed, or as full  GROWN? and in the case of mammals, were they create
0184  nd on the plains of La Plata, where not a tree  GROWS, there is a woodpecker, which in every essent
0186  t there should be woodpeckers where not a tree  GROWS; that there should be diving thrushes, and so
0005  known laws of variation and of correlation of  GROWTH. In the four succeeding chapters, the most ap
0007  variability. Effects of Habit. Correlation of  GROWTH. Inheritance. Character of Domestic Varieties
0008  re or less water at some particular period of  GROWTH, will determine whether or not the plant sets
0010  parison with the laws of reproduction, and of  GROWTH, and of inheritance: for had the action of th
0011  y allude to what may be called correlation of  GROWTH. Any change in the embryo or larva will almos
0012  to the mysterious laws of the correlation of  GROWTH. The result of the various, quite unknown, or
0033  s never, the case. The laws of correlation of  GROWTH, the importance of which should never be over
0043  ws, more especially by that of correlation of  GROWTH. Something may be attributed to the direct ac
0072  si and one of them, judging from the rings of  GROWTH, had during twenty six years tried to raise i
0075  first clothed the ground and thus checked the  GROWTH  of the trees! Throw up a handful of feathers,
0077  relation to other plants. But from the strong  GROWTH  of young plants produced from such seeds (as
0077  of the nutriment in the seed is to favour the  GROWTH  of the young seedling, whilst struggling with
0085  there are many unknown laws of correlation of  GROWTH, which, when one part of the organisation is
0129  cession of extinct species. At each period of  GROWTH  all the growing twigs have tried to branch ou
0129  ving and modified descendants. From the first  GROWTH  of the tree, many a limb and branch has decay
0130  ted a protected station. As buds give rise by  GROWTH  to fresh buds, and these, if vigorous, branch
```

```
0131 nd of vision. Acclimatisation. Correlation of  GROWTH. Compensation and economy of growth. False co
0131 lation of growth. Compensation and economy of  GROWTH. False correlations. Multiple, rudimentary, a
0137 eir size with the adhesion of the eyelids and  GROWTH of fur over them, might in such case be an ad
0143 lection of innate differences. Correlation of  GROWTH. I mean by this expression that the whole orq
0143 e organisation is so tied together during its  GROWTH and development, that when slight variations
0146 y be wholly due to unknown laws of correlated  GROWTH, and without being, as far as we can see, of
0146 may often falsely attribute to correlation of  GROWTH, structures which are common to whole groups
0147 , their law of compensation or balancement of  GROWTH; or, as Goethe expressed it, in order to spen
0147 utriment from one part owing to the excess of  GROWTH in another and adjoining part. I suspect, als
0150 left to the free play of the various laws of   GROWTH, to the effects of long continued disuse, and
0168 and there are very many other correlations of  GROWTH, the nature of which we are utterly unable to
0196 rom the law of reversion; that correlation of  GROWTH will have had a most important influence in m
0197 e bamboo may have arisen from unknown laws of  GROWTH, and have been subsequently taken advantage o
0197 at this structure has arisen from the laws of  GROWTH, and has been taken advantage of in the partu
0198 are convinced that a damp climate affects the  GROWTH of the hair, and that with the hair the horns
0199 ently of any good thus gained. Correlation of  GROWTH has no doubt played a most important part, an
0199 which formerly had arisen from correlation of  GROWTH, or from other unknown cause, may reappear fr
0200 aws of inheritance, reversion, correlation of  GROWTH, etc. Hence every detail of structure in ever
0200 ly, or indirectly through the complex laws of  GROWTH. Natural selection cannot possibly produce an
0205 many modifications, wholly due to the laws of  GROWTH, and at first in no way advantageous to a spe
0206 in all cases subjected to the several laws of  GROWTH. Hence, in fact, the law of the Conditions of
0251 ted by the pollen of the hybrid made vigorous  GROWTH and rapid progress to maturity, and bore good
0261 t is incidental on differences in the laws of  GROWTH of the two plants. We can sometimes see the r
0261 on another, from differences in their rate of  GROWTH, in the hardness of their wood, in the period
0435 d form, but often affecting by correlation of  GROWTH other parts of the organisation. In changes o
0437 extremely different, are at an early stage of  GROWTH exactly alike. How inexplicable are these fac
0442 s as necessarily contingent in some manner on  GROWTH. But there is no obvious reason why, for inst
0442 verse purposes, being at this early period of  GROWTH alike: of embryos of different species within
0454 nails have appeared, not from unknown laws of  GROWTH, but in order to excrete horny matter, as tha
0466 erned by many complex laws, by correlation of  GROWTH, by use and disuse, and by the direct action
0473 in both varieties and species correlation of  GROWTH seems to have played a most important part, s
0484 , their cellular structure, and their laws of  GROWTH and reproduction. We see this even in so trif
0486 uses and laws of variation, on correlation of  GROWTH, on the effects of use and disuse, on the dir
0489 these laws, taken in the largest sense, being  GROWTH with Reproduction; Inheritance which is almos
0484 n secreted by the gall fly produces monstrous  GROWTHS on the wild rose or oak tree. Therefore I sh
0239 as they may be called, which our European ants GUARD or imprison. It will indeed be thought that I
0292 ccumulated. Nature may almost be said to have  GUARDED against the frequent discovery of her transi
0248 was enabled to rear some hybrids, carefully   GUARDING them from a cross with either pure parent, f
0089 ct by singing the females. The rock thrush of GUIANA, birds of Paradise, and some others, congrega
0047 nd judgment and wide experience seems the only GUIDE to follow. We must, however, in many cases, d
0227 derful, hardly more wonderful than those which GUIDE a bird to make its nest, I believe that the h
0484 one prototype. But analogy may be a deceitful GUIDE. Nevertheless all living things have much in
0053 d arbitrarily, and for mere convenience sake.  GUIDED by theoretical considerations, I thought that
0084 fer largely from birds of prey; and hawks are  GUIDED by their eyesight to their prey, so much so,
0425 y thus can I understand the several rules and  GUIDES which have been followed by our best systemat
0017 tes? Has the little variability of the ass or  GUINEA fowl, or the small power of endurance of warm
0450 presence of teeth, which never cut through the GUMS, in the upper jaws of our unborn calves. It ha
0480 s inherited teeth, which never cut through the GUMS of the upper jaw, from an early progenitor hav
0003 , by the effects of external conditions, or of HABIT, or of the volition of the plant itself. The
0007 n chapter I. Causes of Variability. Effects of HABIT. Correlation of Growth. Inheritance. Characte
0011 and perhaps the thickness of fur from climate. HABIT also has a decided influence, as in the perio
0021 umbler has the singular and strictly inherited HABIT of flying at a great height in a compact floc
0021 ersed feathers down the breast; and it has the HABIT of continually expanding slightly the upper p
0029 xternal conditions of life, and some little to HABIT; but he would be a bold man who would account
0039 t now does the upper part of its oesophagus, a HABIT which is disregarded by all fanciers, as it i
0091 inherited. Now, if any slight innate change of HABIT or of structure benefited an individual wolf,
0094 ous bees are to save time; for instance, their HABIT of cutting holes and sucking the nectar at th
0114 tructure, with the accompanying differences of HABIT and constitution, determine that the inhabita
0136 ttle less perfectly developed or from indolent HABIT, will have had the best chance of surviving f
0138 pe. We have some evidence of this gradation of HABIT; for, as Schiodte remarks, animals not far re
0139 l probably have been exposed. Acclimatisation. HABIT is hereditary with plants, as in the period o
0141 species to any peculiar climate is due to mere HABIT, and how much to the natural selection of var
0141 ans combined, is a very obscure question. That HABIT or custom has some influence I must believe,
0142 districts: the result must, I think, be due to HABIT. On the other hand, I can see no reason to do
0142 cannot owe their constitutional differences to HABIT. The case of the Jerusalem artichoke, which i
0142 rs. On the whole, I think we may conclude that HABIT, use, and disuse, have, in some cases, played
0167 eem to have induced some slight modifications. HABIT in producing constitutional differences, and
0182 structure highly perfected for any particular HABIT, as the wings of a bird for flight, we should
0208 der metaphysicians have compared instinct with HABIT. This comparison gives, I think, a remarkably
0209 the resemblance between what originally was a HABIT and an instinct becomes so close as not to be
0209 ater number of instincts have been acquired by HABIT in one generation, and then transmitted by in
0209 cture arise from, and are increased by, use or HABIT, and are diminished or lost by disuse, so I d
0209 h instincts. But I believe that the effects of HABIT are of quite subordinate importance to the ef
0213 o be enabled to see the respective parts which HABIT and the selection of so called accidental var
0214 ited solely from long continued and compulsory HABIT, but this, I think, is not true. No one would
0214 igeon showed a slight tendency to this strange HABIT, and that the long continued selection of the
0215 h will stand and hunt best. On the other hand, HABIT alone in some cases has sufficed: no animal i
0215 xtreme wildness to extreme tameness, simply by HABIT and long continued close confinement. Natural
0215 and if not cured, they are destroyed; so that HABIT, with some degree of selection, has probably
0215 ther hand, young chickens have lost, wholly by HABIT, that fear of the dog and cat which no doubt
0216 and natural instincts have been lost partly by HABIT, and by man selecting and accumulating during
0216 nce call an accident. In some cases compulsory HABIT alone has sufficed to produce such inherited
0216 ited mental changes; in other cases compulsory HABIT has done nothing, and all has been the result
0216 nd unconsciously; but in most cases, probably, HABIT and selection have acted together. We shall,
0217 st.if the old bird profited by this occasional HABIT, or if the young were made more vigorous by a
0217 low by inheritance the occasional and aberrant HABIT of their mother and in their turn would be ap
```

Page **********************************(Key Word)**********************************

0218 and care for her own offspring. The occasional HABIT of birds laying their eggs in other birds' ne
0219 ulty in natural selection making an occasional HABIT permanent, if of advantage to the species, an
0224 to capture workers than to procreate them, the HABIT of collecting pupae originally for food might
0242 in any manner profitable, without exercise or HABIT having come into play. For no amount of exerc
0242 come into play. For no amount of exercise, or HABIT, or volition, in the utterly sterile members
0243 extent, in any useful direction. In some cases HABIT or use and disuse have probably come into pla
0243 like the males of our distinct Kitty wrens, a HABIT wholly unlike that of any other known bird. F
0251 nea, species most widely dissimilar in general HABIT, reproduced itself as perfectly as if it had
0257 be shown that plants most widely different in HABIT and general appearance, and having strongly m
0424 t all are kept together from having the common HABIT of tumbling; but the short faced breed has ne
0424 hort faced breed has nearly or quite lost this HABIT: nevertheless, without any reasoning or think
0474 dmirable architectural powers of the hive bee. HABIT no doubt sometimes comes into play in modifyi
0474 ogeny to inherit the effects of long continued HABIT. On the view of all the species of the same g
0133 e species, when they range into the zone of HABITATION of other species, often acquiring in a very
0018 facts communicated to me by Mr. Blyth, on the HABITS, voice, and constitution, etc., of the humped
0023 thologists; and this, considering their size, HABITS, and remarkable characters, seems very improb
0023 nd the common rock pigeon, which has the same HABITS with the domestic breeds, has not been exterm
0024 termination of so many species having similar HABITS with the rock pigeon seems to me a very rash
0024 s, though agreeing generally in constitution, HABITS, voice, colouring, and in most parts of their
0027 in europe and in India; and that it agrees in HABITS and in a great number of points of structure
0038 how adaptation in their structure or in their HABITS to man's wants or fancies. We can, I think, f
0042 her hand, cats, from their nocturnal rambling HABITS, cannot be matched, and, although so much val
0074 d mr. H. Newman, who has long attended to the HABITS of humble bees, believes that more than two t
0076 or animals have so exactly the same strength, HABITS, and constitution, that the original proporti
0076 gh by no means invariably, some similarity in HABITS and constitution, and always in structure, th
0082 mely slight modifications in the structure or HABITS of one inhabitant would often give it an adva
0087 other sex, or in relation to wholly different HABITS of life in the two sexes, as is sometimes the
0089 d females of any animal have the same general HABITS of life, but differ in structure, colour, or
0091 of its young would probably inherit the same HABITS or structure, and by the repetition of this p
0092 d pistils placed, in relation to the size and HABITS of the particular insects which visited them,
0110 nearly the same structure, constitution, and HABITS, generally come into the severest competition
0112 pecies become in structure, constitution, and HABITS, by so much will they be better enabled to se
0113 y see this in the case of animals with simple HABITS. Take the case of a carnivorous quadruped, of
0113 ing less carnivorous. The more diversified in HABITS and structure the descendants of our carnivor
0116 mals and plants are diversified for different HABITS of life, so will a greater number of individu
0121 hich are most nearly related to each other in HABITS, constitution, and structure. Hence all the i
0127 ite diversity in structure, constitution, and HABITS, to be advantageous to them, I think it woul
0128 same area the more they diverge in structure, HABITS, and constitution, of which we see proof by l
0135 that the early progenitor of the ostrich had HABITS like those of a bustard, and that as natural
0136 e excessively numerous, and which groups have HABITS of life almost necessitating frequent flight;
0137 or ctenomys, is even more subterranean in its HABITS than the mole; and I was assured by a Spaniar
0137 ot indispensable to animals with subterranean HABITS, a reduction in their size with the adhesion
0141 are now all tropical or sub tropical in their HABITS, ought not to be looked at as anomalies, but
0146 to a whole group of descendants with diverse HABITS, would naturally be thought to be correlated
0148 e, when rendered superfluous by the parasitic HABITS of the Proteolepas, though effected by slow s
0156 cies, fitted to more or less widely different HABITS, in exactly the same manner: and as these so
0158 or to fit the males and females to different HABITS of life, or the males to struggle with other
0161 ral selection, in accordance with the diverse HABITS of the species, and will not be left to the m
0171 ity of transitional varieties. Transitions in HABITS of life. Diversified habits in the same speci
0171 s. transitions in habits of life. Diversified HABITS in the same species. Species with habits wide
0171 fied habits in the same species. Species with HABITS widely different from those of their allies.
0171 nimal having, for instance, the structure and HABITS of a bat, could have been formed by the modif
0171 fication of some animal with wholly different HABITS? Can we believe that natural selection could
0179 d transitions of organic beings with peculiar HABITS and structure. It has been asked by the oppon
0179 uld have been converted into one with aquatic HABITS; for how could the animal in its transitional
0179 etween truly aquatic and strictly terrestrial HABITS; and as each exists by a struggle for life, i
0179 it is clear that each is well adapted in its HABITS to its place in nature. Look at the Mustela v
0180 ive only one or two instances of transitional HABITS and structures in closely allied species of t
0180 species of the same genus; and of diversified HABITS, either constant or occasional, in the same s
0183 between structures fitted for very different HABITS of life will rarely have been developed at an
0183 three instances of diversified and of changed HABITS in the individuals of the same species. When
0183 odification of its structure, for its changed HABITS, or exclusively for one of its several differ
0183 exclusively for one of its several different HABITS. But it is difficult to tell, and immaterial
0183 icult to tell, and immaterial for us, whether HABITS generally change first and structure afterwar
0183 ht modifications of structure lead to changed HABITS; both probably often change almost simultaneo
0183 ge almost simultaneously. Of cases of changed HABITS it will suffice merely to allude to that of t
0183 vely on artificial substances. Of diversified HABITS innumerable instances could be given: I have
0184 more and more aquatic in their structure and HABITS, with larger and larger mouths, till a creatu
0184 etimes see individuals of a species following HABITS widely different from those both of their own
0184 e given rise to new species, having anomalous HABITS, and with their structure either slightly or
0185 fuego, the Puffinuria berardi, in its general HABITS, in its astonishing power of diving, its mann
0185 el would never have suspected its sub aquatic HABITS; yet this anomalous member of the strictly te
0185 urprise when he has met with an animal having HABITS and structure not at all in agreement. What c
0185 n such cases, and many others could be given, HABITS have changed without a corresponding change o
0186 any one being vary ever so little, either in HABITS or structure, and thus gain an advantage over
0186 ould be diving thrushes, and petrels with the HABITS of auks. Organs of extreme perfection and com
0193 specially if in members having very different HABITS of life, we may attribute its presence to inh
0196 ew conditions of life and with newly acquired HABITS. To give a few instances to illustrate these
0199 structures now have no direct relation to the HABITS of life of each species. Thus, we can hardly
0200 these animals having such widely diversified HABITS. Therefore we may infer that these several bo
0204 ould be in concluding that the most different HABITS of life could not graduate into each other: t
0204 s may under new conditions of life change its HABITS, or have diversified habits, with some habits
0204 f life change its habits, or have diversified HABITS, with some habits very unlike those of its ne
0204 habits, or have diversified habits, with some HABITS very unlike those of its nearest congeners. H
0204 eckers, diving thrushes, and petrels with the HABITS of auks. Although the belief that an organ so
0206 lass, and which is quite independent of their HABITS of life. On my theory, unity of type is expla

Page **********************************(Key Word)**********************************

Page ***(Key Word)***

```
0207 pter VII. Instinct. Instincts comparable with HABITS, but different in their origin. Instincts gra
0208 t they may be modified by the will or reason. HABITS easily become associated with other habits, a
0208 n. habits easily become associated with other HABITS, and with certain periods of time and states
0208 r points of resemblance between instincts and HABITS could be pointed out. As in repeating a well
0212 so, could be given, of occasional and strange HABITS in certain species, which might, if advantage
0213 f life. How stongly these domestic instincts, HABITS, and dispositions are inherited, and how curi
0216 uring successive generations, peculiar mental HABITS and actions, which at first appeared from wha
0217 ent progenitor of our European cuckoo had the HABITS of the American cuckoo! but that occasionally
0218 e modified in accordance with their parasitic HABITS: for they do not possess the pollen collectin
0219 und in the southern parts of England, and its HABITS have been attended to by Mr. F. Smith, of the
0221 rch for aphides. This difference in the usual HABITS of the masters and slaves in the two countrie
0223 t be observed what a contrast the instinctive HABITS of F. sanguinea present with those of the F.
0267 efit is derived from almost any change in the HABITS of life. Again, both with plants and animals,
0269 ner, and does not wish to alter their general HABITS of life. Nature acts uniformly and slowly dur
0349 and bizcacha, animals having nearly the same HABITS as our hares and rabbits and belonging to the
0379 at selected modifications in their structure, HABITS, and constitutions will have profited them. T
0391 kept by the others to their proper places and HABITS, and will consequently have been little liabl
0398 american species in every character, in their HABITS, gestures, and tones of voice, was manifest.
0411 ers of even the same sub group have different HABITS. In our second and fourth chapters, on Variat
0414 e parts of the structure which determined the HABITS of life, and the general place of each being
0414 of the organisation is concerned with special HABITS, the more important it becomes for classifica
0414 those which are most remotely related to the HABITS and food of an animal, I have always regarded
0426 ecies, especially those having very different HABITS of life, it assumes high value: for we can ac
0426 presence in so many forms with such different HABITS, only by its inheritance from a common parent
0426 hout a large group of beings having different HABITS, we may feel almost sure, on the theory of de
0430 g to the phascolomys having become adapted to HABITS like those of a Rodent. The elder De Candolle
0434 ers of the same class, independently of their HABITS of life, resemble each other in the general p
0442 hether adapted to the most diverse and active HABITS, or quite inactive, being fed by their parent
0446 al selection in accordance with their diverse HABITS. Then, from the many slight successive steps
0448 econdly from their following exactly the same HABITS of life with their parents: for in this case,
0448 its parents, in accordance with their similar HABITS. Some further explanation, however, of the em
0448 e other hand, it profited the young to follow HABITS of life in any degree different from those of
0448 s. the adult might become fitted for sites or HABITS, in which organs of locomotion or of the sens
0449 esent differ from each other in structure and HABITS, if they pass through the same or similar emb
0454 on: so that an organ rendered, during changed HABITS of life, useless or injurious for one purpose
0457 ecome specially modified in relation to their HABITS of life, through the principle of modificatio
0469 nising the whole constitution, structure, and HABITS of each creature, favouring the good and reje
0470 h the more as they become more diversified in HABITS and structure, so as to be enabled to seize o
0471 d that a petrel should have been created with HABITS and structure fitting it for the life of an a
0479 organ, when it has become useless by changed HABITS or under changed conditions of life: and we c
0111 e: but this alone would never account for so HABITUAL and large an amount of difference as that be
0208 ut not of its origin. How unconsciously many HABITUAL actions are performed, indeed not rarely in
0208 s generally forced to go back to recover the HABITUAL train of thought: so P. Huber found it was w
0209 the already finished work. If we suppose any HABITUAL action to become inherited, and I think it c
0011 cows and goats in countries where they are HABITUALLY milked, in comparison with the state of the
0031 bject, and to inspect the animals. Breeders HABITUALLY speak of an animal's organisation as someth
0072 these flies, numerous as they are, must be HABITUALLY checked by some means, probably by birds. H
0092 ere carried, at first occasionally and then HABITUALLY, by the pollen devouring insects from flowe
0096 tes two individuals, either occasionally or HABITUALLY, concur for the reproduction of their kind.
0096 ermaphrodite animals which certainly do not HABITUALLY pair, and a vast majority of plants are her
0099 ants have in fact separated sexes, and must HABITUALLY be crossed. How strange are these facts! Ho
0111 difference between species? That this does HABITUALLY happen, we must infer from most of the innu
0135 eft. In the Cnites apelles the tarsi are so HABITUALLY lost, that the insect has been described as
0136 er feeding coleoptera and lepidoptera, must HABITUALLY use their wings to gain their subsistence,
0142 ited States, in which certain varieties are HABITUALLY recommended for the northern, and others fo
0221 for observation, in Switzerland the slaves HABITUALLY work with their masters in making the nest,
0222 stinguish the pupae of F. fusca, which they HABITUALLY make into slaves, from those of the little
0356 of cases, namely, with all organisms which HABITUALLY unite for each birth, or which often interc
0473 blind, and then at certain moles, which are HABITUALLY blind and have their eyes covered with skin
0140 ir becoming, to a certain extent, naturally HABITUATED to different temperatures, or becoming accl
0077 and claws of the parasite which clings to the HAIR on the tiger's body. But in the beautifully pl
0090 cases under nature, for instance, the tuft of HAIR on the breast of the turkey cock, which can ha
0097 ieve, of the plant. Bees will act like a camel HAIR pencil, and it is quite sufficient just to tou
0133 ear that climate has some direct action on the HAIR of our domestic quadrupeds. Instances could be
0144 r plumage: or, again, the relation between the HAIR and teeth in the naked Turkish dog, though her
0198 that a damp climate affects the growth of the HAIR, and that with the hair the horns are correlat
0198 ects the growth of the hair, and that with the HAIR the horns are correlated. Mountain breeds alwa
0211 creted: I then tickled and stroked them with a HAIR in the same manner, as well as I could, as the
0416 grasses, the nature of the dermal covering, as HAIR or feathers, in the Vertebrata. If the Ornitho
0416 chus had been covered with feathers instead of HAIR, this external and trifling character would, I
0426 wing is folded, whether the skin be covered by HAIR or feathers, if it prevail throughout many and
0012 ons. Hairless dogs have imperfect teeth: long HAIRED and coarse haired animals are apt to have, as
0012 have imperfect teeth: long haired and coarse HAIRED animals are apt to have, as is asserted, long
0033 seberries differ in size, colour, shape, and HAIRINESS, and yet the flowers present very slight di
0012 ed individuals by certain vegetable poisons. HAIRLESS dogs have imperfect teeth: long haired and c
0061 y in the humblest parasite which clings to the HAIRS of a quadruped or feathers of a bird: in the
0452 yle remains well developed and is clothed with HAIRS as in other compositae, for the purpose of br
0013 have heard of cases of albinism, prickly skin, HAIRY bodies, etc., appearing in several members of
0024 so thoroughly domesticated in ancient times by HALF civilized man, as to be quite prolific under c
0024 antail. Hence it must be assumed not only that HALF civilized man succeeded in thoroughly domestic
0076 portions of a mixed stock could be kept up for HALF a dozen generations, if they were allowed to s
0083 uctions. He often begins his selection by some HALF monstrous form: or at least by some modificati
0220 sanguinea. The slaves are black and not above HALF the size of their red masters, so that the con
0229 te from the opposite sides: for I have noticed HALF completed rhombs at the base of a just commenc
0247 n acknowledged difficulty in the manipulation) HALF of these twenty plants had their fertility in
0264 ditions of life. But a hybrid partakes of only HALF of the nature and constitution of its mother,
```

Page ***(Key Word)***

Page ***(Key Word)***

```
0340  oncly of the present character of the southern HALF of the continent; and the southern half was fo
0340  uthern half of the continent; and the southern HALF was formerly more closely allied, than it is a
0340  allied, than it is at present, to the northern HALF. In a similar manner we know from Falconer and
0373  australia. Looking to America; in the northern HALF, ice borne fragments of rock have been observe
0389  land of Ascension aboriginally possessed under HALF a dozen flowering plants; yet many have become
0348  f open ocean extends, with not an island as a  HALTING place for emigrants; here we have a barrier
0348  barriers; and we have innumerable islands as   HALTING places, until after travelling over a hemisp
0357  ied beneath the sea, which may have served as  HALTING places for plants and for many animals durin
0382  sional means of transport, and by the aid, as  HALTING places, of existing and now sunken islands,
0396  probability of many islands having existed as  HALTING places, of which not a wreck now remains, mu
0164  ach to cream colour. I am aware that Colonel   HAMILTON Smith, who has written on this subject, beli
0184  ead; and I have many times seen and heard it   HAMMERING the seeds of the yew on a branch, and thus
0208  a caterpillar, which makes a very complicated  HAMMOCK; for if he took a caterpillar which had comp
0208  he took a caterpillar which had completed its  HAMMOCK up to, say, the sixth stage of construction,
0208  ixth stage of construction, and put it into a  HAMMOCK completed up only to the third stage, the ca
0208  if however, a caterpillar were taken out of a  HAMMOCK made up, for instance, to the third stage, a
0208  ch embarrassed; and, in order to complete its  HAMMOCK, seemed forced to start from the third stage
0220  ing May, June and August, both in Surrey and   HAMPSHIRE, and has never seen the slaves, though pres
0009  in the most sterile hybrids. When, on the one  HAND, we see domesticated animals and plants, thoug
0009  eely under confinement; and when, on the other HAND, we see individuals, though taken young from a
0041  lots, they never can be improved. On the other HAND, nurserymen, from raising large stocks of the
0042  when killed they serve for food. On the other  HAND, cats, from their nocturnal rambling habits, c
0050  orms being specifically distinct. On the other HAND, they are united by many intermediate links, a
0055  be still favourable to variation. On the other HAND, if we look at each species as a special act o
0061  ght but useful variations, given to him by the HAND of Nature. But Natural Selection, as we shall
0068  animals are now annually killed. On the other  HAND, in some cases, as with the elephant and rhino
0070  etween the parasite and its prey. On the other HAND, in many cases, a large stock of individuals o
0071  ren heath, which had never been touched by the HAND of man; but several hundred acres of exactly t
0081  ng and of procreating their kind? On the other HAND, we may feel sure that any variation in the le
0084  g of those slow changes in progress, until the HAND of time has marked the long lapse of ages, and
0092  d; and so in the long run would gain the upper HAND. Those flowers, also, which had their stamens
0095  ifferently constructed proboscis. On the other HAND, I have found by experiment that the fertility
0096  d fertility to the offspring; and on the other HAND, that close interbreeding diminishes vigour an
0097  almost inevitable. Many flowers, on the other  HAND, have their organs of fructification closely e
0100  the result was as I anticipated. On the other  HAND, Dr. Hooker has recently informed me that he f
0108  l selection. I do not believe so. On the other HAND, I do believe that natural selection will alwa
0133  nt as can well be conceived; and, on the other HAND, of different varieties being produced from th
0136  not being blown out to sea; and, on the other  HAND, those beetles which most readily took to flig
0142  t must, I think, be due to habit. On the other HAND, I can see no reason to doubt that natural sel
0147  distinguishing between the effects, on the one HAND, of a part being largely developed through nat
0147  s same process or by disuse, and, on the other HAND, the actual withdrawal of nutriment from one p
0152  constant struggle going on between, on the one HAND, the tendency to reversion to a less modified
0153  ther variability of all kinds and on the other HAND, the power of steady selection to keep the bre
0153  struggle between natural selection on the one  HAND, and the tendency to reversion and variability
0153  ency to reversion and variability on the other HAND, will in the course of time cease; and that th
0156  y should vary at the present day. On the other HAND, the points in which species differ from other
0171  at natural selection could produce, on the one HAND, organs of trifling importance, such as the ta
0171  hich serves as a fly flapper, and on the other HAND, organs of such wonderful structure, as the ey
0172  nd natural selection will, as we have seen, go HAND in hand. Hence, if we look at each species as
0172  al selection will, as we have seen, go hand in HAND. Hence, if we look at each species as descende
0185  organisation profoundly modified. On the other HAND, the acutest observer by examining the dead bo
0185  alight on the surface of the sea. On the other HAND, grebes and coots are eminently aquatic, altho
0192  balanidae or sessile cirripedes; on the other  HAND, have no ovigerous frena, the eggs lying loose
0215  s which will stand and hunt best. On the other HAND, habit alone in some cases has sufficed; no an
0215  ese domestic animals. How rarely, on the other HAND, do our civilised dogs, even when quite young,
0215  vilising by inheritance our dogs. On the other HAND, young chickens have lost, wholly by habit, th
0220  ly household slaves. The masters, on the other HAND, may be constantly seen bringing in materials
0222  th. The nest, however, must have been close at HAND, for two or three individuals of F. fusca were
0223  mercus slaves. Formica sanguinea, on the other HAND, possesses much fewer slaves; and in the early
0243  st of my judgment, annihilate it. On the other HAND, the fact that instincts are not always absolu
0246  her few or no offspring. Hybrids, on the other HAND, have their reproductive organs functionally i
0248  as gartner believes. It is certain, on the one HAND, that the sterility of various species when cr
0248  raduates away so insensibly, and, on the other HAND, that the fertility of pure species is so easi
0249  rbreeding lessens fertility, and, on the other HAND, that an occasional cross with a distinct indi
0256  e hybrids are remarkably sterile. On the other HAND, there are species which can be crossed very r
0257  matists in distinct families; and on the other HAND, by very closely allied species generally unit
0257  only with extreme difficulty; and on the other HAND of very distinct species which unite with the
0258  ence in their whole organisation. On the other HAND, these cases clearly show that the capacity fo
0260  t. the foregoing rules and facts, on the other HAND, appear to me clearly to indicate that the ste
0261  to quite distinct families; and, on the other  HAND, closely allied species, and varieties of the
0262  hus grafted were rendered barren. On the other HAND, certain species of Sorbus, when grafted on ot
0265  tend to produce sterile hybrids. On the other  HAND, one species in a group will sometimes resist
0267  n the progeny. Hence it seems that, on the one HAND, slight changes in the conditions of life bene
0267  e benefit all organic beings, and on the other HAND, that slight crosses, that is crosses between
0272  grel offspring of varieties. And, on the other HAND, they agree most closely in very many importan
0274  etween hybrid and mongrel plants. On the other HAND, the resemblance in mongrels and in hybrids to
0286  breakage of the fallen fragments. On the other HAND, I do not believe that any line of coast, ten
0287  mewhat reduce the above estimate. On the other HAND, during oscillations of level, which we know t
0292  ns before they had time to decay. On the other HAND, as long as the bed of the sea remained statio
0292  a blank in the geological record. On the other HAND, during subsidence, the inhabited area and num
0309  of which we have any record; and on the other  HAND, that where continents now exist, large tracts
0332  s line, then only the two families on the left HAND (namely, a14, etc., and b14, etc.) would have
0337  exterminate many of the natives. On the other  HAND, from what we see now occurring in New Zealand
0339  latitude, would attempt to account, on the one HAND, by dissimilar physical conditions for the dis
0340  nts of these two continents; and, on the other HAND, by similarity of conditions, for the uniformi
0345  greatly diminished or disappear. On the other  HAND, all the chief laws of palaeontology plainly p
```

Page ***(Key Word)***

Page **(Key Word)**

0348 en sea, they are wholly distinct. On the other HAND, proceeding still further westward from the ea
0360 jurious action of the salt water. On the other HAND he did not previously dry the plants or branch
0375 malacca, and are thinly scattered, on the one HAND over India and on the other as far north as Ja
0377 ave suffered to a certain extent. On the other HAND, the temperate productions, after migrating ne
0378 plants from a temperate climate. On the other HAND, the the most humid and hottest districts will
0391 easily than land birds. Bermuda, on the other HAND, which lies at about the same distance from No
0398 similarity in all these respects. On the other HAND, there is a considerable degree of resemblance
0412 ng in common. But the three genera on the left HAND have, on this same principle, much in common,
0412 hat including the next two genera on the right HAND, which diverged from a common parent at the fi
0412 ng the three genera still further to the right HAND, which diverged at a still earlier period. And
0430 of an early member of the group. On the other HAND, of all Marsupials, as Mr. Waterhouse has rema
0434 y soul. What can be more curious than that the HAND of a man, formed for grasping, that of a mole
0440 we have to believe that the same bones in the HAND of a man, wing of a bat, and fin of a porpoise
0447 ence long continued exercise or use on the one HAND, and disuse on the other, may have in modifyin
0448 rphosis is perhaps requisite. If, on the other HAND, it profited the young to follow habits of lif
0461 ased vigour and fertility. So that, on the one HAND, considerable changes in the conditions of lif
0461 fied forms, lessen fertility; and on the other HAND, lesser changes in the conditions of life and
0463 to be supplanted by the allied forms on either HAND; and the latter, from existing in greater numb
0466 t infinite number of generations. On the other HAND we have evidence that variability, when it has
0475 spring of acknowledged varieties. On the other HAND, these would be strange facts if species have
0477 inhabit oceanic islands; and why, on the other HAND, new and peculiar species of bats, which can t
0479 . the framework of bones being the same in the HAND of a man, wing of bat, fin of the porpoise, an
0075 s checked the growth of the trees! Throw up a HANDFUL of feathers, and all must fall to the ground
0080 ection, which we have seen is so potent in the HANDS of man, apply to nature? I think we shall see
0447 ification, adapted in one descendant to act as HANDS, in another as paddles, in another as wings;
0447 d of life, and having thus been converted into HANDS, or paddles, or wings. Whatever influence lon
0111 ce between species? That this does habitually HAPPEN, we must infer from most of the innumerable s
0157 n illustration two instances, the first which HAPPEN to stand on my list; and as the differences i
0209 hink it can be shown that this does sometimes HAPPEN, then the resemblance between what originally
0214 this line; and this is known occasionally to HAPPEN, as I once saw in a pure terrier. When the fi
0228 es. The bees, however, did not suffer this to HAPPEN, and they stopped their excavation in due tim
0256 pon the constitution of the individuals which HAPPEN to have been chosen for the experiment. So it
0309 ve been formerly deposited; it might well HAPPEN that strata which had subsided some miles nea
0156 a common progenitor; for it can rarely have HAPPENED that natural selection will have modified se
0292 vourable to life. Still less could this have HAPPENED during the alternate periods of elevation; o
0321 o to the same family. But it must often have HAPPENED that a new species belonging to some one gro
0385 ttle plants adhering to its back; and it has HAPPENED to me, in removing a little cuck weed from o
0400 st my views; for it may be asked, how has it HAPPENED in the several islands situated within sight
0408 rs; according or not, as those which entered HAPPENED to come in more or less direct competition w
0239 red perfect gradations of this kind. It often HAPPENS that the larger or the smaller sized workers
0394 merly transported foxes, as so frequently now HAPPENS in the arctic regions. Yet it cannot be said
0423 heir fruit, though the most important part, HAPPENS to be nearly identical; no one puts the swed
0079 t, and that the vigorous, the healthy, and the HAPPY survive and multiply. Chapter IV. Natural Sel
0195 ases) by the flies, but they are incessantly HARASSED and their strength reduced, so that they are
0391 r blown there, as I am informed by Mr. E. V. HARCOURT. So that these two islands of Bermuda and Ma
0087 sed exclusively for opening the cocoon, or the HARD tip to the beak of nestling birds, used for br
0144 nion of the petals of the corolla into a tube. HARD parts seem to affect the form of adjoining sof
0168 ologous parts tend to cohere. Modifications in HARD parts and in external parts sometimes affect s
0233 not difficult: it is known that bees are often HARD pressed to get sufficient nectar; and I am inf
0282 lines of sea coast, when formed of moderately HARD rocks, and mark the process of degradation. Th
0285 y liquid state, showed at a glance how far the HARD, rocky beds had once extended into the open oc
0361 passing through the intestines of a bird; but HARD seeds of fruit will pass uninjured through eve
0286 must remember that almost all strata contain HARDER layers or nodules, which from long resisting
0087 thin the egg, which had the most powerful and HARDEST beaks, for all with weak beaks would inevita
0090 ring that season of the year when the wolf is HARDEST pressed for food. I can under such circumsta
0110 ogress of its formation, will generally press HARDEST on its nearest kindred, and tend to extermin
019 d exclusively for respiration. I can, indeed, HARDLY doubt that all vertebrate animals having true
0009 arnivorous birds, with the rarest exceptions, HARDLY ever lay fertile eggs. Many exotic plants hav
0009 cultivation, and vary very slightly, perhaps HARDLY more than in a state of nature. A long list c
0016 must be admitted, when we find that there are HARDLY any domestic races, either amongst animals or
0016 hink it could be shown that this statement is HARDLY correct: but naturalists differ most widely i
0019 it! When we bear in mind that Britain has now HARDLY one peculiar mammal, and France but few disti
0020 o extremely different races or species; I can HARDLY believe. Sir J. Sebright expressly experiment
0020 sed one with another for several generations, HARDLY two of them will be alike, and then the extre
0032 from it, the principle would be so obvious as HARDLY to be worth notice; but its importance consis
0033 of selection is, in fact, also followed: for HARDLY any one is so careless as to allow his worst
0033 do not differ at all in other points; this is HARDLY ever, perhaps never, the case. The laws of co
0038 o slight in internal parts or organs. Man can HARDLY select, or only with much difficulty, any dev
0040 t, a breed, like a dialect of a language, can HARDLY be said to have had a definite origin. A man
0040 immediate neighbourhood. But as yet they will HARDLY have a distinct name, and from being only sli
0042 ough so much valued by women and children, we HARDLY ever see a distinct breed kept up; such breed
0046 resent an inordinate amount of variation; and HARDLY two naturalists can agree which forms to rank
0050 e crossed only with much difficulty. We could HARDLY wish for better evidence of the two forms bei
0051 ries not now continuous, in which case he can HARDLY hope to find the intermediate links between h
0090 r on the breast of the turkey cock, which can HARDLY be either useful or ornamental to this bird;
0098 varieties are planted near each other, it is HARDLY possible to raise pure seedlings, so largely
0102 in this case the effects of intercrossing can HARDLY be counterbalanced by natural selection alway
0116 ir organisation but little diversified, could HARDLY compete with a set more perfectly diversified
0144 is latter case of correlation, I think it can HARDLY be accidental, that if we pick out the two or
0147 les. With species in a state of nature it can HARDLY be maintained that the law is of universal ap
0147 , however, here give any instances, for I see HARDLY any way of distinguishing between the effects
0157 re of a very unusual nature, the relation can HARDLY be accidental. The same number of joints in t
0172 wonderful structure, as the eye, of which we HARDLY as yet fully understand the inimitable perfec
0187 n, though insuperable by our imagination, can HARDLY be considered real. How a nerve comes to be s
0187 . how a nerve comes to be sensitive to light, HARDLY concerns us more than how life itself first o
0196 gh the aid must be slight, for the hare, with HARDLY any tail, can double quickly enough. In the s

Page **(Key Word)**

```
0199  habits of life of each species. Thus, we can HARDLY believe that the webbed feet of the upland g
0217  heir own mother's care, encumbered as she can HARDLY fail to be by having eggs and young of diffe
0227  f instincts in themselves not very wonderful, HARDLY more wonderful than those which guide a bird
0235  tly such trifling importance, that they could HARDLY have been acted on by natural selection; case
0245  le, for species within the same country could HARDLY have kept distinct had they been capable of
0253  rations of the more fertile hybrid animals, I  HARDLY know of an instance in which two families of
0282  very slowly through natural selection. It is  HARDLY possible for me even to recall to the reader
0287  clined Wealden district, when upraised, could HARDLY have been great, but it would somewhat reduce
0289  cumulated. And if in each separate territory,  HARDLY any idea can be formed of the length of time
0291  ion as it has as yet suffered, but which will  HARDLY last to a distant geological age, was certai
0302  ement and close of each formation, pressed so  HARDLY on my theory. On the sudden appearance of wh
0304  of a great group of species. But my work had  HARDLY been published, when a skilful palaeontolog
0305  rioc. It is almost superfluous to remark that  HARDLY any fossil fish are known from south of the
0312  and in the waters. Lyell has shown that it is  HARDLY possible to resist the evidence on this head
0315  s for the same object, might make a new breed  HARDLY distinguishable from our present fantail; bu
0319  s of life; but what that something is, we can  HARDLY ever tell. On the supposition of the fossil
0324  australia, the most skilful naturalist would  HARDLY be able to say whether the existing or the p
0334  dified at the seventh stage; hence they could  HARDLY fail to be nearly intermediate in character
0337  we see now occurring in New Zealand, and from  HARDLY a single inhabitant of the southern hemisphe
0346  ers, under almost every temperature. There is  HARDLY a climate or condition in the Old World whic
0348  no two marine faunas are more distinct, with  HARDLY a fish, shell, or crab in common, than those
0348  defined and distinct marine faunas. Although  HARDLY one shell, crab or fish is common to the abo
0360  ng; as plants with large seeds or fruit could  HARDLY be transported by any other means; and, Alph.
0361  prise nearly all germinated. Living birds can  HARDLY fail to be highly effective agents in the tr
0363  od, bones, and the nest of a land bird, I can  HARDLY doubt that they must occasionally have trans
0384  ter; and, according to Valenciennes, there is  HARDLY a single group of fishes confined exclusivel
0394  y small islands, if close to a continent: and  HARDLY an island can be named on which our smaller
0404  y of the species have very wide ranges. I can  HARDLY doubt that this rule is generally true, thoug
0407  of occasional transport, a subject which has  HARDLY ever been properly experimented on; if we
0419  t the opposite ends of the series, which have  HARDLY a character in common; yet the species at bo
0427  resemblance of the greyhound and racehorse is  HARDLY more fanciful than the analogies which have
0438  nature having occurred, that naturalists can  HARDLY avoid employing language having this plain s
0441  this expression, though I am aware that it is  HARDLY possible to define clearly what is meant by
0445  r irstance, that of the width of mouth, could  HARDLY be detected in the young. But there was one
0461  e degree sterile, for their constitutions can  HARDLY fail to have been disturbed from being compo
0464  present states; and these many links we could  HARDLY ever expect to discover, owing to the imperf
0478  of paramount importance to the being, are of  HARDLY any importance in classification; why charac
0261  differences in their rate of growth, in the  HARDNESS of their wood, in the period of the flow or
0142  or an account has been published how much more  HARDY some seedlings appeared to be than others. On
0196  he dog, though the aid must be slight, for the  HARE, with hardly any tail, can double quickly eno
0068  oubt that the stock of partridges, grouse, and  HARES on any large estate depends chiefly on the d
0091  . st. John, bringing home winged game, another  HARES or rabbits, and another hunting on marshy gr
0214  ole family of shepherd dogs a tendency to hunt  HARES. These domestic instincts, when thus tested
0349  , animals having nearly the same habits as our  HARES and rabbits and belonging to the same order
0454  lowly to reduce the organ, until it rendered  HARMLESS and rudimentary. Any change in function, wh
0169  modifications will add to the beautiful and  HARMONIOUS diversity of nature. Whatever the cause ma
0276  larity would be an astonishing fact. But it  HARMONISES perfectly with the view that there is no e
0184  ts organisation, even in its colouring, in the  HARSH tone of its voice, and undulatory flight, to
0363  epoch. At my request Sir C. Lyell wrote to M.  HARTUNG to inquire whether he had observed erratic
0075  coloured sweet peas, they must be each year  HARVESTED separately, and the seed then mixed in due
0001  s; as I give them to show that I have not been  HASTY in coming to a decision. My work is now near
0094  trifolium pratense and incarnatum) do not on a  HASTY glance appear to differ in length; yet the h
0022  ith which the nestling birds are clothed when  HATCHED. The shape and size of the eggs vary. The ma
0144  re or less down on the young birds when first  HATCHED, with the future colour of their plumage; o
0192  secretion, to retain the eggs until they are  HATCHED within the sack. These cirripedes have no b
0217  migrate at a very early period; and the first  HATCHED young would probably have to be fed by the
0217  own nest and has eggs and young successively  HATCHED, all at the same time. It has been asserted
0218  n one nest and then in another; and these are  HATCHED by the males. This instinct may probably be
0225  of cylindrical cells, in which the young are  HATCHED, and, in addition, some large cells of wax
0385  that numbers of the extremely minute and just  HATCHED shells crawled on the feet, and clung to th
0385  e they would voluntarily drop off. These just  HATCHED molluscs, though aquatic in their nature, s
0397  t means for their transportal. Would the just  HATCHED young occasionally crawl on and adhere to t
0440  f its mother, in the egg of the bird which is  HATCHED in a nest, and in the spawn of a frog under
0445  rious breeds, within twelve hours after being  HATCHED; I carefully measured the proportions (but
0253  rids from the same parents but from different  HATCHES; and from these two birds he raised no less
0087  of it; so that fanciers assist in the act of  HATCHING. Now, if nature had to make the beak of a f
0217  his were the case, the process of laying and  HATCHING might be inconveniently long, more especial
0385  m, where many ova of fresh water shells were  HATCHING; and I found that numbers of the extremely
0202  but we ought to admire the savage instinctive  HATRED of the queen bee, which urges her instantly
0203  the community; and maternal love or maternal  HATRED, though the latter fortunately is most rare,
0472  f pollen by our fir trees; at the instinctive  HATRED of the queen bee for her own fertile daughte
0484  at present; but they will not be incessantly  HAUNTED by the shadowy doubt whether this or that f
0103  e same animal can long remain distinct, from  HAUNTING different stations, from breeding at slight
0221  struck by about a score of the slave makers  HAUNTING the same spot, and evidently not in search
0386  d be sure to fly to their natural fresh water  HAUNTS. I do not believe that botanists are aware h
0072  birds (whose numbers are probably regulated by  HAWKS or beasts of prey) were to increase in Parag
0084  nown to suffer largely from birds of prey; and  HAWKS are guided by their eyesight to their prey,
0362  ily be blown to the distance of 500 miles, and  HAWKS are known to look out for tired birds, and t
0362  carrier pigeons from France to England, as the  HAWKS on the English coast destroyed so many on th
0362  coast destroyed so many on their arrival. Some  HAWKS and owls bolt their prey whole, and after an
0315  but only an occasional scene, taken almost at  HAZARD, in a slowly changing drama. We can clearly
0487  lled museum, but as a poor collection made at  HAZARD and at rare intervals. The accumulation of e
0359  dried floated much longer; for instance, ripe  HAZEL nuts sank immediately, but when dried, they
0011  are almost always accompanied by an elongated  HEAD. Some instances of correlation are quite whim
0021  development of the carunculated skin about the  HEAD, and this is accompanied by greatly elongated
0021  ht in a compact flock, and tumbling in the air  HEAD over heels. The runt is a bird of great size,
```

Page ***************************************(Key Word)***************************************

```
0021  nd are carried so erect that in good birds the HEAD and tail touch; the oil gland is quite aborted
0031  ars, but it would take him six years to obtain HEAD and beak. In Saxony the importance of the prin
0067  y one who reflects how ignorant we are on this HEAD, even in regard to mankind, so incomparably be
0068  iefly on the destruction of vermin. If not one HEAD of game were shot during the next twenty years
0072  had during twenty six years tried to raise its HEAD above the stems of the heath, and had failed.
0112  ys been my practice, let us seek light on this HEAD from our domestic productions. We shall here f
0144  mother influences by pressure the shape of the HEAD of the child. In snakes, according to Schlegel
0145  rolls of the central and exterior flowers of a HEAD or umbel, I do not feel at all sure that C. C.
0147  n our poultry, a large tuft of feathers on the HEAD is generally accompanied by a diminished comb;
0148  hree highly important anterior segments of the HEAD enormously developed, and furnished with great
0148  ed proteolepas, the whole anterior part of the HEAD is reduced to the merest rudiment attached to
0159  nt sub varieties with reversed feathers on the HEAD and feathers on the feet, characters not posse
0184  ke a shrike, kills small birds by blows on the HEAD; and I have many times seen and heard it hamme
0187  om fossil species we can learn nothing on this HEAD. In this great class we should probably have t
0195  ulty, though of a very different kind, on this HEAD, as in the case of an organ as perfect and com
0197  and becoming a climber. The naked skin on the HEAD of a vulture is generally looked at as a direc
0197  ch inference, when we see that the skin on the HEAD of the clean feeding male turkey is likewise n
0198  logous variation, the front limbs and even the HEAD would probably be affected. The shape, also, o
0198  lvis might affect by pressure the shape of the HEAD of the young in the womb. The laborious breath
0214  cannot fly eighteen inches high without going HEAD over heels. It may be doubted whether any one
0270  ith the pollen of the other; but only a single HEAD produced any seed, and this one head produced
0270  a single head produced any seed, and this one HEAD produced only five grains. Manipulation in thi
0283  coasts are worn away. The observations on this HEAD by Hugh Miller, and by that excellent observer
0293  event us coming to any just conclusion on this HEAD. When we see a species first appearing in the
0307  ent geologists, with Sir R. Murchison at their HEAD, are convinced that we see in the organic rema
0312  hardly possible to resist the evidence on this HEAD in the case of the several tertiary stages; an
0325  to survive. We have distinct evidence on this HEAD, in the plants which are dominant, that is, wh
0351  ecent times, there is great difficulty on this HEAD. It is also obvious that the individuals of th
0395  rl has made some striking observations on this HEAD in regard to the great Malay Archipelago, whic
0397  away: but more experiments are wanted on this HEAD. The most striking and important fact for us i
0443  ly period. But we have little evidence on this HEAD, indeed the evidence rather points the other w
0182  eir wings solely as flappers, like the logger HEADED duck (Micropterus of Eyton): as fins in the w
0473  ion when we look, for instance, at the logger HEADED duck, which has wings incapable of flight, in
0145  hooker informs me, in species with the densest HEADS that the inner and outer flowers most frequen
0171  objections may be classed under the following HEADS: Firstly, why if species have descended from
0172  d their fertility is unimpaired? The two first HEADS shall be here discussed, Instinct and Hybridi
0238  lone carry a wonderful sort of shield on their HEADS, the use of which is quite unknown: in the Me
0241  ut we must suppose that the larger workmen had HEADS four instead of three times as big as those o
0450  which when grown up have not a tooth in their HEADS; and the presence of teeth, which never cut t
0001  or three more years to complete it, and as my HEALTH is far from strong, I have been urged to publ
0140  t from warmer countries which here enjoy good HEALTH. We have reason to believe that species in a
0265  cases the sterility is independent of general HEALTH, and is often accompanied by excess of size o
0265  system, independently of the general state of HEALTH, is affected by sterility in a very similar m
0009  e of nature, perfectly tamed, long lived, and HEALTHY (of which I could give numerous instances),
0079  generally prompt, and that the vigorous, the HEALTHY, and the happy survive and multiply. Chapter
0251  lants in these experiments appeared perfectly HEALTHY, and although both the ovules and pollen of
0264  ew; as hybrids, when once born, are generally HEALTHY and long lived, as we see in the case of the
0028  ious dynasty. In the time of the Romans, as we HEAR from Pliny, immense prices were given for pige
0065  here are plants which now range in India, as I HEAR from Dr. Falconer, from Cape Comorin to the Hi
0073  o high our presumption, that we marvel when we HEAR of the extinction of an organic being: and as
0076  decrease of the song thrush. How frequently we HEAR of one species of rat taking the place of anot
0085  acters of the most trifling importance: yet we HEAR from an excellent horticulturist, Downing, tha
0095  llustrative of geology; but we now very seldom HEAR the action, for instance, of the coast waves,
0138  e with some of the American cave animals, as I HEAR from Professor Dana; and some of the European
0164  of horses is so generally striped, that, as I HEAR from Colonel Poole, who examined the breed for
0214  nd near Glasgow there are house tumblers, as I HEAR from Mr. Brent, which cannot fly eighteen inch
0217  lly lays her eggs in other birds' nests; but I HEAR on the high authority of Dr. Brewer, that this
0224  o its end, without enthusiastic admiration. We HEAR from mathematicians that bees have practically
0365  se of Labrador, and nearly all the same, as we HEAR from Asa Gray, with those on the loftiest moun
0375  f borneo. Some of these Australian forms, as I HEAR from Dr. Hooker, extend along the heights of t
0394  the distance of 600 miles from the mainland. I HEAR from Mr. Tomes, who has specially studied this
0013  appearance to inheritance. Every one must have HEARD of cases of albinism, prickly skin, hairy bod
0074  e. but how false a view is this! Every one has HEARD that when an American forest is cut down, a v
0184  ws on the head; and I have many times seen and HEARD it hammering the seeds of the yew on a branch
0184  . in North America the black bear was seen by HEARNE swimming for hours with widely open mouth, th
0349  t, yet clearly related, replace each other. He HEARS from closely allied, yet distinct kinds of bi
0222  little yellow ants had crawled away, they took HEART and carried off the pupae. One evening I visi
0033  the flowers, how unlike the flowers of the HEARTSEASE are, and how alike the leaves; how much the
0037  ty which we now see in the varieties of the HEARTSEASE, rose, pelargonium, dahlia, and other plant
0037  o one would ever expect to get a first rate HEARTSEASE or dahlia from the seed of a wild plant. No
0073  d indispensable to the fertilisation of the HEARTSEASE (Viola tricolor), for other bees do not vis
0074  became extinct or very rare in England, the HEARTSEASE and red clover would become very rare, or w
0387  a heron flying to another pond and getting a HEARTY meal of fish, would probably reject from its
0010  n, we should attribute to the direct action of HEAT, moisture, light, food, etc., is most difficul
0077  it can perfectly well withstand a little more HEAT or cold, dampness or dryness, for elsewhere it
0279  ife do not graduate away quite insensibly like HEAT or moisture. I endeavoured, also, to show that
0378  lconer informs me that it is the damp with the HEAT of the tropics which is so destructive to pere
0477  nent, under the most diverse conditions, under HEAT and cold, on mountain and lowland, on deserts
0310  bare metamorphic rocks, which must have been HEATED under great pressure, have always seemed to m
0071  gation, there was a large and extremely barren HEATH, which had never been touched by the hand of
0071  e native vegetation of the planted part of the HEATH was most remarkable, more than is generally s
0071  ther: not only the proportional numbers of the HEATH plants were wholly changed, but twelve specie
0071  e plantations, which could not be found on the HEATH. The effect on the insects must have been sti
0071  plantations, which were not to be seen on the HEATH; and the heath was frequented by two or three
0071  hich were not to be seen on the heath; and the HEATH was frequented by two or three distinct insec
0072  ld examine hundreds of acres of the unenclosed HEATH, and literally I could not see a single Scotc
```

Page ***************************************(Key Word)***************************************

Page **************************************(Key Word)**************************************

```
0072  ut on looking closely between the stems of the HEATH, I found a multitude of seedlings and little
0072  tried to raise its head above the stems of the HEATH, and had failed. No wonder that, as soon as
0072  ed with vigorously growing young firs. Yet the HEATH was so extremely barren and so extensive tha
0222  or about forty yards, to a very thick clump of HEATH, whence I saw the last individual of F. sang
0222  t able to find the desolated nest in the thick HEATH. The nest, however, must have been close at
0223  its own pupa in its mouth on top of a spray of HEATH over its ravaged home. Such are the facts, t
0084  white in winter, the red grouse the colour of HEATHER, and the black grouse that of peaty earth,
0071  farnham, in Surrey. Here there are extensive HEATHS, with a few clumps of old Scotch firs on the
0465  them. I have felt these difficulties far too HEAVILY during many years to doubt their weight. Bu
0066  by a struggle at some period of its life; that HEAVY destruction inevitably falls either on the y
0165  loth to apply it to breeds so distinct as the HEAVY Belgian cart horse, Welch ponies, cobs, the
0180  ht. Here, as on other occasions, I lie under a HEAVY disadvantage, for out of the many striking c
0445  dam and of a three days old colt of a race and HEAVY cart horse, I find that the colts have by no
0021  mpact flock, and tumbling in the air head over HEELS. The runt is a bird of great size, with long
0214  y eighteen inches high without going head over HEELS. It may be doubted whether any one would hav
0107  that the flora of Madeira, according to Oswald HEER, resembles the extinct tertiary flora of Euro
0021  strictly inherited habit of flying at a great HEIGHT in a compact flock, and tumbling in the air
0175  facts ought to cause surprise, as climate and HEIGHT or depth graduate away insensibly. But when
0285  dicular cliffs of one or two thousand feet in HEIGHT; for the gentle slope of the lava streams, d
0285  one side, or thrown down on the other, to the HEIGHT or depth of thousands of feet; for since the
0286  monly wears away a line of cliff of any given HEIGHT, we could measure the time requisite to have
0286  hat the sea would eat into cliffs 500 feet in HEIGHT at the rate of one inch in a century. This w
0286  e as if we were to assume a cliff one yard in HEIGHT to be eaten back along a whole line of coast
0286  nces, I conclude that for a cliff 500 feet in HEIGHT, a denudation of one inch per century for th
0373  f a vast mound of detritus, about 800 feet in HEIGHT, crossing a valley of the Andes: and this I
0378  bout the same with that now felt there at the HEIGHT of six or seven thousand feet. During this t
0382  ly rising from the arctic lowlands to a great HEIGHT under the equator. The various beings thus le
0398  he geological nature of the islands, in their HEIGHT or climate, or in the proportions in which t
0398  the volcanic nature of the soil, in climate, HEIGHT, and size of the islands, between the Galapa
0400  , having the same geological nature, the same HEIGHT, climate, etc., that many of the immigrants
0049  stations; they ascend mountains to different HEIGHTS; they have different geographical ranges; a
0140  by dr. Hooker from trees growing at different HEIGHTS on the Himalaya, were found in this country
0367  mmits (having been exterminated on all lesser HEIGHTS) and in the artic regions of both hemisphere
0375  tain ranges of the peninsula of India, on the HEIGHTS of Ceylon, and on the volcanic cones of Java
0375  , as I hear from Dr. Hooker, extend along the HEIGHTS of the peninsula of Malacca, and are thinly
0191  for I do not know which view is now generally HELD, a part of the auditory apparatus has been wor
0390  her oceanic island which can be named. In St. HELENA there is reason to believe that the naturalis
0397  during seven days: one of these shells was the HELIX pomatia, and after it had again hybernated I
0359  s afterwards germinated: the ripe seeds of HELOSCIADIUM sank in two days, when dried they floated
0329  r between them. That the extinct forms of life HELP to fill up the wide intervals between existing
0219  arry their masters in their jaws. So utterly HELPLESS are the masters, that when Huber shut up th
0060  ough necessary as the foundation for the work, HELPS us but little in understanding how species a
0163  been seen with a double shoulder stripe. The HEMIONUS has no shoulder stripe: but traces of it, as
0165  knows of a second case) from the ass and the HEMIONUS; and this hybrid, though the ass seldom has
0165  e ass seldom has stripes on its legs and the HEMIONUS has none and has not even a shoulder stripe,
0165  face stripes on this hybrid from the ass and HEMIONUS, to ask Colonel Poole whether such face str
0167  rom one or more wild stocks, or the ass, the HEMIONUS, quagga, and zebra. He who believes that eac
0324  esembled most closely those of the southern HEMISPHERE. So, again, several highly competent observ
0337  hardly a single inhabitant of the southern HEMISPHERE having become wild in any part of Europe, w
0347  e their living productions! In the southern HEMISPHERE, if we compare large tracts of land in Aust
0348  lting places, until after travelling over a HEMISPHERE we come to the shores of Africa: and over t
0372  ow separated by a continent and by nearly a HEMISPHERE of equatorial ocean. These cases of relatio
0376  found in the southern parts of the southern HEMISPHERE, and on the mountain ranges of the intertro
0376  er regions of the earth and in the southern HEMISPHERE are of doubtful value, being ranked by some
0379  he most temperate latitudes of the opposite HEMISPHERE. Although we have reason to believe from ge
0379  ntertropical mountains, and in the southern HEMISPHERE. These being surrounded by strangers will h
0381  se and other distant points of the southern HEMISPHERE, is, on my theory of descent with modificat
0381  to look in the southern, as in the northern HEMISPHERE, to a former and warmer period, before the
0382  dispersed to various points of the southern HEMISPHERE by occasional means of transport, and by th
0353  distant points of the northern and southern HEMISPHERES? The answer, as I believe, is, that mammal
0367  r heights) and in the artic regions of both HEMISPHERES. Thus we can understand the identity of ma
0375  erate lowlands of the northern and southern HEMISPHERES, are sometimes identically the same; but t
0379  their brethren of the northern or southern HEMISPHERES, now exist in their new homes as well mark
0362  some seeds of the oat, wheat, millet, canary, HEMP, clover, and beet germinated after having been
0089  ied peacock was eminently attractive to all his HEN birds. It may appear childish to attribute any
0185  over swamps and floating plants, yet the water HEN is nearly as aquatic as the coot; and the land
0216  ctive in young pheasants, though reared under a HEN. It is not that chickens have lost all fear, b
0216  ear, but fear only of dogs and cats, for if the HEN gives the danger chuckle, they will run (more
0216  ome useless under domestication, for the mother HEN has almost lost by disuse the power of flight.
0218  t in the allied group of ostriches. For several HEN ostriches, at least in the case of the America
0012  ll feet; and those with long beaks large feet. HENCE, if man goes on selecting, and thus augmentin
0014  evert in character to their aboriginal stocks. HENCE it has been argued that no deductions can be
0023  any of the characters of the domestic breeds. HENCE the supposed aboriginal stocks must either st
0024  islets, or on the shores of the Mediterranian. HENCE the supposed extermination of so many species
0024  ; for tail feathers like those of the fantail. HENCE it must be assumed not only that half civiliz
0039  se has not given rise to any marked varieties; HENCE the Thoulouse and the common breed, which dif
0041  a large number of individuals being kept; and HENCE this comes to be of the highest importance to
0047  r the entry of doubt and conjecture is opened. HENCE, in determining whether a form should be rank
0051  s the mind with the idea of an actual passage. HENCE I look at individual differences, though of s
0052  s of structure in certain definite directions. HENCE I believe a well marked variety may be justly
0053  rked to have been recorded in botanical works. HENCE it is the most flourishing, or, as they may b
0055  tances have been favourable for variation; and HENCE we might expect that the circumstances would
0056  s to raise one or both to the rank of species. HENCE the amount of difference is one very importan
0063  eat that no country could support the product. HENCE, as more individuals are produced than can po
0065  there are very few which do not annually pair. HENCE we may confidently assert, that all plants an
0069  herefore of competitors, decreases northwards; HENCE in going northward, or in ascending a mountai
```

Page **************************************(Key Word)**************************************

age ***(Key Word)***

```
072  ally checked by some means, probably by birds.  HENCE, if certain insectivorous birds (whose number
073  tense), as other bees cannot reach the nectar.  HENCE I have very little doubt, that if the whole g
074  e to the number of cats that destroy the mice.  HENCE it is quite credible that the presence of a f
078  ame species, for the warmest or dampest spots.  HENCE, also, we can see that when a plant or animal
085  eons, as being the most liable to destruction.  HENCE I can see no reason to doubt that natural sel
093  n called the physiological division of labour:  HENCE we may believe that it would be advantageous
095  o push the pollen on to the stigmatic surface.  HENCE, again, if humble bees were to become rare in
103  , and which do not breed at a very quick rate.  HENCE in animals of this nature, for instance in bi
105  s, have been produced there, and nowhere else.  HENCE an oceanic island at first sight seems to hav
106  l thus come into competition with many others.  HENCE more new places will be formed, and the compe
107  heen less modification and less extermination.  HENCE, perhaps, it comes that the flora of Madeira,
110  r of recorded varieties, or incipient species.  HENCE, rare species will be less quickly modified o
113  heat have been sown on equal spaces of ground.  HENCE, if any one species of grass were to go on va
121  dely different places in the polity of nature:  HENCE in the diagram I have chosen the extreme spec
121  other in habits, constitution, and structure.  HENCE all the intermediate forms between the earlie
122  ich were most nearly related to their parents.  HENCE very few of the original species will have tr
123  ng (F), extinct, and have left no descendants.  HENCE the six new species descended from (I), and t
125  om a common ancestor some advantage in common.  HENCE, the struggle for the production of new and m
141  outh, and on many islands in the torrid zones.  HENCE I am inclined to look at adaptation to any sp
146  visions of the order on analogous differences.  HENCE we see that modifications of structure, viewe
154  e made constant, I can see no reason to doubt.  HENCE when an organ, however abnormal it may be, ha
162  istinct breeds of diverse colours are crossed.  HENCE, though under nature it must generally be lef
168  will be disregarded by natural selection, and  HENCE probably are variable. Specific characters, t
172  ection will, as we have seen, go hand in hand.  HENCE, if we look at each species as descended from
173  varieties between its past and present states.  HENCE we ought not to expect at the present time to
174  rer on the confines, and finally disappearing.  HENCE the neutral territory between two representat
177  the rarer forms which exist in lesser numbers.  HENCE, the more common forms, in the race for life,
183  age over other animals in the battle for life.  HENCE the chance of discovering species with transi
186  owever different it may be from its own place.  HENCE it will cause him no surprise that there shou
191  th the lungs of the higher vertebrate animals:  HENCE there seems to me to be no great difficulty i
200  itance, reversion, correlation of growth, etc.  HENCE every detail of structure in every living cre
204  ts very unlike those of its nearest congeners.  HENCE we can understand, bearing in mind that each
205  ng according to the standard of that country.  HENCE the inhabitants of one country, generally the
206  cases subjected to the several laws of growth.  HENCE, in fact, the law of the Conditions of Existe
210  numerous, slight, yet profitable, variations.  HENCE, as in the case of corporeal structures, we o
216  has almost lost by disuse the power of flight.  HENCE we may conclude, that domestic instincts have
220  d odious an instinct as that of making slaves.  HENCE I will give the observations which I have mys
220  rs in carrying them away to a place of safety.  HENCE, it is clear, that the slaves feel quite at h
220  ers in August, either leave or enter the nest.  HENCE he considers them as strictly household slave
226  ween the spheres which thus tend to intersect.  HENCE each cell consists of an outer spherical port
227  h have been made of the cells of the hive bee.  HENCE we may safely conclude that if we could sligh
234  end on a large number of bees being supported.  HENCE the saving of wax by largely saving honey mus
234  o adjoining cells, would save some little wax.  HENCE it would continually be more and more advanta
237  rns of the bulls or cows of these same breeds.  HENCE I can see no real difficulty in any character
249  refully prevented during the flowering season:  HENCE hybrids will generally be fertilised during e
266  neration the same compounded organisation, and  HENCE we need not be surprised that their sterility
267  induces weakness and sterility in the progeny.  HENCE it seems that, on the one hand, slight change
271  ther varieties when crossed with N. glutinosa.  HENCE the reproductive system of this one variety m
281  perhaps more than they differ from each other.  HENCE in all such cases, we should be unable to rec
286  sting attrition form a breakwater at the base.  HENCE, under ordinary circumstances, I conclude tha
295  icient time for the slow process of variation;  HENCE the deposit will generally have to be a very
296  at no less than sixty eight different levels.  HENCE, when the same species occur at the bottom, m
299  bjections which may be urged against my views.  HENCE it will be worth while to sum up the foregoin
305  s yet been found even in any tertiary stratum.  HENCE we now positively know that sessile cirripede
308  nant of any palaeozoic or secondary formation.  HENCE we may perhaps infer, that during the palaeoz
314  ch the varying species comes into competition.  HENCE it is by no means surprising that one species
315  d improved, will be liable to be exterminated.  HENCE we can see why all the species in the same re
321  hich are most like each other in all respects.  HENCE the improved and modified descendants of a sp
331  generally differ from its ancient progenitor.  HENCE we can understand the rule that the most anci
334  came still more modified at the seventh stage;  HENCE they could hardly fail to be nearly intermedi
344  their inferiors in the struggle for existence.  HENCE, after long intervals of time, the production
353  d been produced in two or more distinct areas!  HENCE it seems to me, as it has to many other natur
359  esh rise in the stream be washed into the sea.  HENCE I was led to dry stems in branches of 94 plan
363  rocks, which do not occur in the archipelago.  HENCE we may safely infer that icebergs formerly la
367  brethren were pursuing their northern journey.  HENCE, when the warmth had fully returned, the same
370  he climate was warmer than at the present day.  HENCE we may suppose that the organisms now living
371  ges and on the arctic lands of the two Worlds.  HENCE it has come, that when we compare the now liv
375  in regard to the plants of that large island.  HENCE we see that throughout the world, the plants
377  ithstand a much warmer climate than their own.  HENCE, it seems to me possible, bearing in mind tha
392  e, whatever the cause may be, confined ranges.  HENCE trees would be little likely to reach distant
394  ound on continents and on far distant islands.  HENCE we have only to suppose that such wandering s
401  , and that of another plant to another island.  HENCE when in former times an immigrant settled on
417  characters is very evident in natural history.  HENCE, as has often been remarked, a species may de
417  eave us in no doubt where it should be ranked.  HENCE, also, it has been found, that a classificati
434  forearm, or of the thigh and leg, transposed.  HENCE the same names can be given to the homologous
444  length, as long as it was fed by its parents.  HENCE, I conclude, that it is quite possible, that
445  e probably descended from the same wild stock;  HENCE I was curious to see how far their puppies di
453  annot be said to be in any degree rudimentary.  HENCE, also, a rudimentary organ in the adult, is o
470  e most divergent offspring of any one species.  HENCE during a long continued course of modificatio
471  it can act only by very short and slow steps.  HENCE the canon of Natura non facit saltum, which e
480  power of acting on an organ during early life;  HENCE the organ will not be much reduced or rendere
485  whereas species were formerly thus connected.  HENCE, without quite rejecting the consideration of
489  at no cataclysm has desolated the whole world.  HENCE we may look with some confidence to a secure
218  y probably be accounted for by the fact of the  HENS laying a large number of eggs; but as in the c
261  f two plants, one being woody and the other  HERBACEOUS, one being evergreen and the other deciduou
392  cing to orders which elsewhere include only  HERBACEOUS species; now trees, as Alph. de Candolle ha
```

Page *******************************(Key Word)*******************************

Page *********************************(Key Word)*********************************

0392 ly to reach distant oceanic islands; and an HERBACEOUS plant, though it would have no chance of st
0392 hed on an island and having to compete with HERBACEOUS plants alone, might readily gain an advanta
0392 n would often tend to add to the stature of HERBACEOUS plants when growing on an island, to whatev
0396 ar proportions of certain orders of.plants, HERBACEOUS forms having been developed into trees, etc
0113 number of plants and a greater weight of dry HERBAGE can thus be raised. The same has been found
0062 subject with more spirit and ability than W. HERBERT, Dean of Manchester, evidently the result of
0250 nced hybridiser, namely, the Hon. and Rev. W. HERBERT. He is as emphatic in his conclusion that so
0250 lts may, I think, be in part accounted for by HERBERT's great horticultural skill, and by his hav
0250 oduced four flowers; three were fertilised by HERBERT with their own pollen, and the fourth was su
0251 tated freely. In a letter to me, in 1839, Mr. HERBERT told me that he had then tried the experimen
0251 y of these hybrids seed freely. For instance, HERBERT asserts that a hybrid from Calceolaria integ
0213 r kind of wolf rushing round, instead of at, a HERD of deer, and driving them to a distant point,
0002 more sensible than I do of the necessity of HEREAFTER publishing in detail all the facts, with re
0003 tion. In one very limited sense, as we shall HEREAFTER see, this may be true: but it is preposterc
0011 few of which can be dimly seen, and will be HEREAFTER briefly mentioned. I will here only allude
0015 bly do occur; but natural selection, as will HEREAFTER be explained, will determine how far the ne
0016 rtility of varieties when crossed, a subject HEREAFTER to be discussed), domestic races of the sam
0038 s by a process of natural selection, as will HEREAFTER be more fully explained, two sub breeds mic
0046 d rendered definite by natural selection, as HEREAFTER will be explained. Those forms which posses
0052 f natural selection in accumulating (as will HEREAFTER be more fully explained) differences of str
0052 th rank as independent species. But we shall HEREAFTER have to return to this subject. From these
0053 ivergence of character, and other questions, HEREAFTER to be discussed. Alph. de Candolle and othe
0053 ey come into competition (which, as we shall HEREAFTER see, is a far more important circumstance)
0059 verage vary most; and varieties, as we shall HEREAFTER see, tend to become converted into new and
0059 ified and dominant descendants. But by steps HEREAFTER to be explained, the larger genera also ter
0061 f nature. But Natural Selection, as we shall HEREAFTER see, is a power incessantly ready for actic
0073 ether by a web of complex relations. I shall HEREAFTER have occasion to show that the exotic Lobel
0092 ing, we have good reason to believe (as will HEREAFTER be more fully alluded to), would produce ve
0308 views here entertained. To show that it may HEREAFTER receive some explanation, I will give the f
0324 ent day on the shores of North America would HEREAFTER be liable to be classed with somewhat older
0338 ar from proved. Yet I fully expect to see it HEREAFTER confirmed, at least in regard to subordinat
0345 on on the whole has progressed. If it should HEREAFTER be proved that ancient animals resemble to
0355 this nature are common, and are, as we shall HEREAFTER more fully see, inexplicable on the theory
0358 s, a certain degree of relation (as we shall HEREAFTER see) between the distribution of mammals an
0426 ue them less in our classification. We shall HEREAFTER, I think, clearly see why embryological cha
0449 d to confess that I only hope to see the law HEREAFTER proved true. It can be proved true in those
0485 to raise both forms to the rank of species. HEREAFTER we shall be compelled to acknowledge that t
0485 ally acknowledged to be merely varieties may HEREAFTER be thought worthy of specific names, as wit
0488 a length quite incomprehensible by us, will HEREAFTER be recognised as a mere fragment of time, c
0087 under domestication in one sex and become HEREDITARILY attached to that sex, the same fact probab
0004 n. we shall thus see that a large amount of HEREDITARY modification is at least possible; and, wha
0014 responding caterpillar or cocoon stage. But HEREDITARY diseases and some other facts make me belie
0015 ing shall be preserved. When we look to the HEREDITARY varieties or races of our domestic animals
0080 hose under nature, vary; and how strong the HEREDITARY tendency is. Under domestication, it may be
0118 nd the tendency to variability is in itself HEREDITARY, consequently they will tend to vary, and o
0139 ave been exposed. Acclimatisation. Habit is HEREDITARY with plants, as in the period of flowering,
0443 embryo, may appear late in life; as when an HEREDITARY disease, which appears in old age alone, ha
0029 ask, as I have asked, a celebrated raiser of HEREFORD cattle, whether his cattle might not have de
0096 that concerns us. But still there are many HERMAPHRODITE animals which certainly do not habitually
0101 or huxley, to discover a single case of an HERMAPHRODITE animal with the organs of reproduction so
0103 ies; and this I believe to be the case. In HERMAPHRODITE organisms which cross only occasionally,
0151 characters is clearly shown in the case of HERMAPHRODITE cirripedes; and I may here add, that I pa
0355 ve descended from a single pair, or single HERMAPHRODITE, or whether, as some authors suppose, fro
0451 that by crossing such male plants with an HERMAPHRODITE species, the rudiment of the pistil in th
0096 unite for each birth; but in the case of HERMAPHRODITES this is far from obvious. Nevertheless I
0096 troncly inclined to believe that with all HERMAPHRODITES two individuals, either occasionally or h
0096 as much diminished the number of supposed HERMAPHRODITES, and of real hermaphrodites a large numbe
0096 r of supposed hermaphrodites, and of real HERMAPHRODITES a large number pair; that is, two individ
0096 y pair, and a vast majority of plants are HERMAPHRODITES. What reason, it may be asked, is there f
0100 ce to animals: on the land there are some HERMAPHRODITES, as land mollusca and earth worms; but th
0100 animals, there are many self fertilising HERMAPHRODITES; but here currents in the water offer an
0101 viduals, though both are self fertilising HERMAPHRODITES, do sometimes cross. It must have struck
0101 isation, yet are not rarely, some of them HERMAPHRODITES, and some of them unisexual. But if, in f
0101 e of them unisexual. But if, in fact, all HERMAPHRODITES do occasionally intercross with other ind
0101 other individuals, the difference between HERMAPHRODITES and unisexual species, as far as function
0267 nt of crossing is indispensable even with HERMAPHRODITES; and that close interbreeding continued d
0424 be predicated in common of the males and HERMAPHRODITES of certain cirripedes, when adult, and ye
0441 a the larvae become developed either into HERMAPHRODITES having the ordinary structure, or into wh
0089 dividual preferences and dislikes; thus Sir R. HERON has described how one pied peacock was eminen
0385 ty hours; and in this length of time a duck or HERON might fly at least six or seven hundred miles
0387 ding to Dr. Hooker, the Nelumbium luteum) in a HERON's stomach; although I do not know the fact, y
0387 the fact, yet analogy makes me believe that a HERON flying to another pond and getting a hearty m
0387 as of the yellow water lily and Potamogeton. HERONS and other birds, century after century, have
0203 s her daughters as soon as born, or to perish HERSELF in the combat; for undoubtedly this is for t
0188 ained by the theory of descent, ought not to HESITATE to go further, and to admit that a structure
0188 too keenly to be surprised at any degree of HESITATION in extending the principle of natural selec
0458 he course of descent, that I should without HESITATION adopt this view, even if it were unsupporte
0012 mals and plants. From the facts collected by HEUSINGER, it appears that white sheep and pigs are d
0264 from observations communicated to me by Mr. HEWITT, who has had great experience in hybridising
0227 pheres in the same layer, she can prolong the HEXAGON to any length requisite to hold the stock of
0232 can lay the foundations of one wall of a new HEXAGON, in its strictly proper place, projecting be
0233 anes of intersection, and so make an isolated HEXAGON: but I am not aware that any such case has b
0233 ; nor would any good be derived from a single HEXAGON being built, as in its construction more mat
0225 le layer: each cell, as is well known, is an HEXAGONAL prism, with the basal edges of its six side
0227 formed, there will result a double layer of HEXAGONAL prisms united together by pyramidal bases f
0227 rhombs; and the rhombs and the sides of the HEXAGONAL prisms will have every angle identically th

Page *********************************(Key Word)*********************************

hexagonal

0227 ppose, but this is no difficulty, that after HEXAGONAL prisms have been formed by the intersection
0228 ntersection between the basins, so that each HEXAGONAL prism was built upon the festooned edge of
0230 upper edges of the rhombic plates, until the HEXAGONAL walls are commenced. Some of these statemen
0231 over the comb without injuring the celicate HEXAGONAL walls, which are only about one four hundre
0231 show this fact, by covering the edges of the HEXAGONAL walls of a single cell, or the extreme marg
0233 margin of wasp combs are sometimes strictly HEXAGONAL; but I have not space here to enter on this
0233 sect (as in the case of a queen wasp) making HEXAGONAL cells, if she work alternately on the insid
0235 they know what are the several angles of the HEXAGONAL prisms and of the basal rhombic plates. The
0077 g is related, in the most essential yet often HIDDEN manner, to that of all other organic beings,
0413 he similarity of organic beings, is the bond, HIDDEN as it is by various degrees of modification,
0420 enealogical: that community of descent is the HIDDEN bond which naturalists have been unconsciousl
0426 let them be ever so unimportant, betrays the HIDDEN bond of community of descent. Let two forms h
0433 and I believe this element of descent is the HIDDEN bond of connexion which naturalists have soug
0449 be genealogical. Descent being on my view the HIDDEN bond of connexion which naturalists have soug
0197 the green colour was a beautiful adaptation to HIDE this tree frequenting bird from its enemies: a
0482 ew cirectly opposite to mine. It is so easy to HIDE our ignorance under such expressions as the pl
0380 come naturalised in any part of Europe, though HIDES, wool, and other objects likely to carry seed
0046 varieties. We may instance Rubus, Rosa, and HIERACIUM amongst plants, several genera of insects,
0052 e from one stage of difference to another and HIGHER stage may be, in some cases, cue merely to th
0055 nerally much more widely diffused than plants HIGHER in the scale; and here again there is no clos
0084 ife, and should plainly bear the stamp of far HIGHER workmanship? It may be said that natural sele
0149 nature are more variable than those which are HIGHER. I presume that lowness in this case means th
0168 whole organisation more specialised, and are HIGHER in the scale. Rudimentary organs, from being
0169 y it should often still be variable in a much HIGHER degree than other parts: for variation is a l
0191 position and structure with the lungs of the HIGHER vertebrate animals: hence there seems to me t
0191 ivance by which the glottis is closed. In the HIGHER Vertebrata the branchiae have wholly disappea
0197 taken advantage of in the parturition of the HIGHER animals. We are profoundly ignorant of the ca
0206 tandard of perfection will have been rendered HIGHER. Natural selection will not necessarily produ
0206 the law of the Conditions of Existence is the HIGHER law: as it includes, through the inheritance
0258 n fertility in a small, and occasionally in a HIGHER degree. Several other singular rules could be
0314 uctions, by the more complex relations of the HIGHER beings to their organic and inorganic conditi
0325 or these latter must be victorious in a still HIGHER degree in order to be preserved and to surviv
0330 n regard to the Invertebrata, Barrande, and a HIGHER authority could not be named, asserts that he
0337 the more recent forms must, on my theory, be HIGHER than the more ancient: for each new species i
0337 roductions of Great Britain may be said to be HIGHER than those of New Zealand. Yet the most skilf
0345 ors in the race for life, and are, in so far, HIGHER in the scale of nature: and this may account
0361 es an hour: and some authors have given a far HIGHER estimate. I have never seen an instance of nu
0367 e cleared and thawed ground, always ascending HIGHER and higher, as the warmth increased, whilst t
0367 nd thawed ground, always ascending higher and HIGHER, as the warmth increased, whilst their brethe
0378 se temperate forms would naturally ascend the HIGHER mountains, being exterminated on the lowlands
0379 hrough natural selection and competition to a HIGHER stage of perfection or dominating power, than
0382 its drift in horizontal lines, though rising HIGHER on the shores where the tide rises highest, s
0406 s, generally change at a slower rate than the HIGHER forms: and consequently the lower forms will
0410 of each class generally change less than the HIGHER; but there are in both cases marked exception
0425 ouping species under genera, and genera under HIGHER groups, though in these cases the modificatio
0429 great natural system. As showing how few the HIGHER groups are in number, and how widely spread t
0431 y which all the members of the same family or HIGHER group are connected together. For the common
0441 early what is meant by the organisation being HIGHER or lower. But no one probably will dispute th
0441 e probably will cispute that the butterfly is HIGHER than the caterpillar. In some cases, however,
0443 i of the embryo apparently having sometimes a HIGHER organisation than the mature animal, into whi
0461 nd all the species of the same genus, or even HIGHER group, must have descended from common parent
0476 ally looked at as being, in some vague sense, HIGHER than ancient and extinct forms: and they are
0476 ent and extinct forms: and they are in so far HIGHER as the later and more improved forms have con
0485 l be led to weigh more carefully and to value HIGHER the actual amount of difference between them.
0490 of conceiving, namely, the production of the HIGHER animals, directly follows. There is grandeur
0004 hed and unexplained. It is, therefore, of the HIGHEST importance to gain a clear insight into the
0006 e and is rare? Yet these relations are of the HIGHEST importance, for they determine the present w
0014 the parent. I believe this rule to be of the HIGHEST importance in explaining the laws of embryol
0024 ge contingencies seem to me improbable in the HIGHEST degree. Some facts in regard to the colourin
0041 being kept: and hence this comes to be of the HIGHEST importance to success. On this principle Mar
0043 on the reproductive system, are so far of the HIGHEST importance as causing variability. I do not
0050 sidered as varieties, and in this country the HIGHEST botanical authorities and practical men can
0077 the great battle of life. A corollary of the HIGHEST importance may be deduced from the foregoing
0101 et failed, after consultation with one of the HIGHEST authorities, namely, Professor Huxley, to di
0157 neuration of the wings is a character of the HIGHEST importance, because common to large groups:
0186 ction, seems, I freely confess, absurd in the HIGHEST possible degree. Yet reason tells me, that i
0188 erfected by the long continued efforts of the HIGHEST human intellects: and we naturally infer tha
0197 es, and this contrivance, no doubt, is of the HIGHEST service to the plant: but as we see nearly a
0243 no one will cispute that instincts are of the HIGHEST importance to each animal. Therefore I can s
0299 nd such success seems to me improbable in the HIGHEST degree. Geological research, though it has a
0327 the course of time, the forms dominant in the HIGHEST degree, wherever produced, would tend everyw
0337 progress. Crustaceans, for instance, not the HIGHEST in their own class, may have beaten the high
0337 ghest in their own class, may have beaten the HIGHEST molluscs. From the extraordinary manner in w
0374 es belonging to European genera occur. On the HIGHEST mountains of Brazil, some few European gener
0376 ri as an example, I may quote a remark by the HIGHEST authority, Prof. Dana, that it is certainly
0382 ing higher on the shores where the tide rises HIGHEST, so have the living waters left their living
0408 relations of organism to organism are of the HIGHEST importance, we can see why two areas having
0415 st every author. It will suffice to quote the HIGHEST authority, Robert Brown, who in speaking of
0416 fact that the rudimentary florets are of the HIGHEST importance in the classification of the Gras
0488 origin of man and his history. Authors of the HIGHEST eminence seem to be fully satisfied with the
0005 ple organ can be changed and perfected into a HIGHLY developed being or elaborately constructed or
0018 raphical and other considerations, I think it HIGHLY probable that our domestic dogs have descende
0022 th of the ramus of the lower jaw, varies in a HIGHLY remarkable manner. The number of the caudal a
0022 and shape of the apertures in the sternum are HIGHLY variable: so is the degree of divergence and
0024 with the wild rock pigeon, yet are certainly HIGHLY abnormal in other parts of their structure, w
0025 y this, we must make one of the two following HIGHLY improbable suppositions. Either, firstly, tha

Page **************************************(Key Word)**************************************

```
0031 d quote numerous passages to this effect from HIGHLY competent authorities. Youatt, who was proba
0035 e extent since the time of that monarch. Some HIGHLY competent authorities are convinced that the
0041 ll, is, that the animal or plant should be so HIGHLY useful to man, or so much valued by him, tha
0045 same mould. These individual differences are HIGHLY important for us, as they afford materials f
0048 species, and he has entirely omitted several HIGHLY polymorphic genera. Under genera, including
0048 als which unite for each birth, and which are HIGHLY locomotive, doubtful forms, ranked by one zo
0049 st be admitted that many forms, considered by HIGHLY competent judges as varieties, have so perfe
0049 cter of species that they are ranked by other HIGHLY competent judges as good and true species. B
0050 f any animal or plant in a state of nature be HIGHLY useful to man, or from any cause closely att
0070 our native animals. When a species, owing to HIGHLY favourable circumstances, increases inordina
0073 s of bees, if not indispensable, are at least HIGHLY beneficial to the fertilisation of our clove
0093 nd more attractive flowers, had been rendered HIGHLY attractive to insects, they would, unintenti
0093 se: as soon as the plant had been rendered so HIGHLY attractive to insects that pollen was regula
0105 us see that at first sight seems to have been HIGHLY favourable for the production of new species
0106 ted areas probably have been in some respects HIGHLY favourable for the production of new species
0115 us see that these naturalised plants are of a HIGHLY diversified nature. They differ, moreover, t
0119 he modified offspring from the later and more HIGHLY improved branches in the lines of descent, w
0126 thin the same large group, the later and more HIGHLY perfected sub groups, from branching out and
0131 ble. Parts developed in an unusual manner are HIGHLY variable: specific characters more variable
0145 ets serve to attract insects, whose agency is HIGHLY advantageous in the fertilisation of plants
0148 in all other cirripedes consists of the three HIGHLY important anterior segments of the head enor
0149 hors, and I believe with truth, are apt to be HIGHLY variable. We shall have to recur to the gene
0150 the same part in allied species, tends to be HIGHLY variable. Several years ago I was much struc
0157 y of secondary sexual characters, as they are HIGHLY variable, sexual selection will have had a w
0157 ariable: variations of this part would, it is HIGHLY probable, be taken advantage of by natural a
0169 ient species. Secondary sexual characters are HIGHLY variable, and such characters differ much in
0182 ured by other fish? When we see any structure HIGHLY perfected for any particular habit, as the w
0190 umaticus for its supply, and being divided by HIGHLY vascular partitions. In these cases, one of
0190 s a good one, because it shows us clearly the HIGHLY important fact that an organ originally cons
0205 ay well produce parts, organs, and excretions HIGHLY useful or even indispensable, or highly inju
0205 tions highly useful or even indispensable, or HIGHLY injurious to another species, but in all case
0253 parent species exists, they must certainly be HIGHLY fertile. A doctrine which originated with Pa
0258 of making reciprocal crosses. Such cases are HIGHLY important, for they prove that the capacity
0273 seriously affected, and their descendants are HIGHLY variable. But to return to our comparison of
0287 . I have made these few remarks because it is HIGHLY important for us to gain some notion, howeve
0298 plants that can propagate rapidly and are not HIGHLY locomotive, there is reason to suspect, as we
0305 chalk formation, the fact would certainly be HIGHLY remarkable; but I cannot see that it would be
0307 tratum the dawn of life on this planet. Other HIGHLY competent judges, as Lyell and the late E. F
0314 ker rate of change in terrestrial and in more HIGHLY organised productions compared with marine an
0324 f the southern hemisphere. So, again, several HIGHLY competent observers believe that the existin
0326 on with many already existing forms, would be HIGHLY favourable, as would be the power of spreadin
0336 much discussion whether recent forms are more HIGHLY developed than ancient. I will not here ente
0356 nges of level in the land must also have been HIGHLY influential: a narrow isthmus now separates
0361 erminated. Living birds can hardly fail to be HIGHLY effective agents in the transportation of se
0381 ctic lands, now covered with ice, supported a HIGHLY peculiar and isolated flora. I suspect that t
0383 ed by their having become fitted, in a manner HIGHLY useful to them, for short and frequent migra
0397 there must be, on my view, some unknown, but HIGHLY efficient means for their transportal. Would
0416 and certain rudimentary bones of the leg, are HIGHLY serviceable in exhibiting the close affinity
0416 rtance, but which are universally admitted as HIGHLY serviceable in the definition of whole groups
0418 found nearly uniform, they are considered as HIGHLY serviceable in classification: but in some gr
0441 , cirripedes may be considered as either more HIGHLY or more lowly organised then they were in the
0008 or at the instant of conception. Geoffroy St. HILAIRE's experiments show that unnatural treatment
0011 y instances are given in Isidore Geoffroy St. HILAIRE's great work on this subject. Breeders belie
0144 frequently quite obscure. M. Is. Geoffroy St. HILAIRE has forcibly remarked, that certain malconfo
0149 to be a rule, as remarked by Is. Geoffroy St. HILAIRE, both in varieties and in species, that wher
0155 s to monstrosities: at least Is. Geoffroy St. HILAIRE seems to entertain no doubt, that the more a
0418 rly than by that excellent botanist, Aug. St. HILAIRE. If certain characters are always found corr
0434 in the same relative positions? Geoffroy St. HILAIRE has insisted strongly on the high importance
0177 will soon take the place of the less improved HILL breed; and thus the two breeds, which original
0177 interposition of the supplanted, intermediate HILL variety. To sum up, I believe that species com
0283 by that excellent observer Mr. Smith of Jordan HILL, are most impressive. With the mind thus impre
0375 ava raises a picture of a collection made on a HILL in Europe! Still more striking is the fact tha
0071 few clumps of old Scotch firs on the distant HILLTOPS: within the last ten years large spaces have
0177 us region; a second to a comparatively narrow, HILLY tract; and a third to wide plains at the base
0177 the small holders on the intermediate narrow, HILLY tract; and consequently the improved mountain
0065 from Dr. Falconer, from Cape Comorin to the HIMALAYA, which have been imported from America since
0140 om trees growing at different heights on the HIMALAYA, were found in this country to possess diffe
0373 at siberia was similarly affected. Along the HIMALAYA, at points 900 miles apart, glaciers have le
0375 in the intertropical parts of Africa. On the HIMALAYA, and on the isolated mountain ranges of the
0378 tives. The mountain ranges north west of the HIMALAYA, and the long line of the Cordillera, seem t
0378 o with strange luxuriance at the base of the HIMALAYA, as graphically described by Hooker. Thus, a
0313 nge and lost mammals and reptiles in the sub HIMALAYAN deposits. The Silurian Lingula differs but
0030 him. In this sense he may be said to make for HIMSELF useful breeds. The great power of this princ
0031 than almost any other individual, and who was HIMSELF a very good judge of an animal, speaks of th
0141 f enduring the most different climates by man HIMSELF and by his domestic animals, and such facts
0252 thus prevented. Any one may readily convince HIMSELF of the efficiency of insect agency by examin
0282 ach stratum. A man must for years examine for HIMSELF great piles of superimposed strata, and watc
0304 unmistakeable sessile cirripede, which he had HIMSELF extracted from the chalk of Belgium. And, as
0469 . by patience select variations most useful to HIMSELF, should nature fail in selecting variations
0143 y varying in the same manner: in the front and HIND legs, and even in the jaws and limbs, varying
0181 top of the shoulder to the tail, including the HIND legs, we perhaps see traces of an apparatus or
0198 mountainous country would probably affect the HIND limbs from exercising them more, and possibly
0450 r snakes there are rudiments of the pelvis and HIND limbs. Some of the cases of rudimentary organs
0274 over the horse, so that both the mule and the HINNY more resemble the ass than the horse: but tha
0275 ass and mare, is more like an ass, than is the HINNY, which is the offspring of the female ass and
0250 elia, and with all the species of the genus HIPPEASTRUM, which can be far more easily fertilised b
```

Page **************************************(Key Word)**************************************

Page ***(Key Word)**

```
0250  be self fertilised! For instance, a bulb of HIPPEASTRUM aulicum produced four flowers; three were
0251  confirmed by other observers in the case of HIPPEASTRUM with its sub genera; and in the case of so
0262  is latter fact of the extraordinary case of HIPPEASTRUM, Lobelia, etc., which seeded much more fre
0066  its hundreds of eggs, and another, like the HIPPOBOSCA, a single one; but this difference does not
0028  very rare birds; and, continues the courtly HISTORIAN, His Majesty by crossing the breeds, which
0282  the Principles of Geology, which the future HISTORIAN will recognise as having produced a revolut
0140  ntic cases could be given of species within HISTORICAL times having largely extended their range f
0280  become so much modified, that if we had no HISTORICAL or incirect evidence regarding their origin
0289  mammiferous remains, a single glance at the HISTORICAL table published in the Supplement to Lyell'
0111  er and inferior kinds. In Yorkshire, it is HISTORICALLY known that the ancient black cattle were d
0002  mr. Wallace, who is now studying the natural HISTORY of the Malay archipelago, has arrived at alm
0006  during the many past geological epochs in its HISTORY. Although much remains obscure, and will lon
0026  s this strong tendency to sterility: from the HISTORY of the dog I think there is some probability
0030  l cases, we know that this has not been the HISTORY. The key is man's power of accumulative sele
0034  ity. In rude and barbarous periods of English HISTORY choice animals were often imported, and laws
0040  ely, that we know nothing about the origin or HISTORY of any of our domestic breeds. But, in fact,
0040  e, and from being only slightly valued, their HISTORY will be disregarded. When further improved b
0051  y thought worth recording in works on natural HISTORY. And I look at varieties which are in any de
0087  this effect may be found in works of natural HISTORY, I cannot find one case which will bear inve
0106  l thus play an important part in the changing HISTORY of the organic world. We can, perhaps, on th
0127  w largely extinction has acted in the world's HISTORY, geology plainly declares. Natural selection
0155  i have repeatedly noticed in works on natural HISTORY, that when an author has remarked with surpr
0173  sited; there will be blanks in our geological HISTORY. The crust of the earth is a vast museum; bu
0194  is indeed shown by that old canon in natural HISTORY of Natura non facit saltum. We meet with thi
0201  y statements may be found in works on natural HISTORY to this effect, I cannot find even one which
0206  the full meaning of that old canon in natural HISTORY, Natura non facit saltum. This canon if we l
0243  nstincts of others; that the canon in natural HISTORY, of natura non facit saltum is applicable to
0300  imperfectly would they represent the natural HISTORY of the world! But we have every reason to be
0310  l look at the natural geological record, as a HISTORY of the world imperfectly kept, and written i
0310  t, and written in a changing dialect: of this HISTORY we possess the last volume alone, relating o
0311  of the slowly changing language, in which the HISTORY is supposed to be written, being more or les
0319  n of the slow breeding elephant, and from the HISTORY of the naturalisation of the domestic horse
0333  the fauna of any great period in the earth's HISTORY will be intermediate in general character be
0345  ants of each successive period in the world's HISTORY have beaten their predecessors in the race f
0395  ll soon have much light thrown on the natural HISTORY of this archipelago by the admirable zeal an
0413  to one class. Thus, the grand fact in natural HISTORY of the subordination of group under group, w
0417  gate of characters is very evident in natural HISTORY. Hence, as has often been remarked, a specie
0434  is the most interesting department of natural HISTORY, and may be said to be its very soul. What c
0450  h are second in importance to none in natural HISTORY, are explained on the principle of slight mo
0453  re imperfect and useless. In works on natural HISTORY rudimentary organs are generally said to hav
0459  fects of its adoption on the study of Natural HISTORY. Concluding remarks. As this whole volume is
0464  ent and utterly unknown epochs in the world's HISTORY. I can answer these questions and grave obje
0475  which has played so conspicuous a part in the HISTORY of the organic world, almost inevitably foll
0481  uctions was almost unavoidable as long as the HISTORY of the world was thought to be of short dura
0483  ve that at innumerable periods in the earth's HISTORY certain elemental atoms have been commanded
0484  will be a considerable revolution in natural HISTORY. Systematists will be able to pursue their l
0485  other and more general departments of natural HISTORY will rise greatly in interest. The terms use
0485  y production of nature as one which has had a HISTORY; when we contemplate every complex structure
0486  ak from experience, will the study of natural HISTORY become! A grand and almost untrodden field o
0488  of time. During early periods of the earth's HISTORY, when the forms of life were probably fewer
0488  ave been slow in an extreme degree. The whole HISTORY of the world, as at present known, although
0488  t will be thrown on the origin of man and his HISTORY. Authors of the highest eminence seem to be
0193  ame way as two men have sometimes independently HIT on the very same invention, so natural selecti
0124  nd. In the diagram, each horizontal line has HITHERTO been supposed to represent a thousand genera
0131  ons to long lost characters. Summary. I have HITHERTO sometimes spoken as if the variations, so co
0268  nvolved in hopeless difficulties; for if two HITHERTO reputed varieties be found in any degree ste
0288  in sicily, whereas not one other species has HITHERTO been found in any tertiary formation: yet it
0419  the case of crustaceans, such definition has HITHERTO been found impossible. There are crustaceans
0428  our experience shows that this valuation has HITHERTO been arbitrary), could easily extend the par
0094  sty glance appear to differ in length; yet the HIVE bee can easily suck the nectar out of the inca
0095  n ar abundant supply of precious nectar to the HIVE bee. Thus it might be a great advantage to the
0095  bee. Thus it might be a great advantage to the HIVE bee to have a slightly longer or differently c
0095  eely divided tube to its corolla, so that the HIVE bee could visit its flowers. Thus I can unders
0207  strich, and parasitic bees. Slave making ants. HIVE bee, its cell making instinct. Difficulties on
0207  lly as so wonderful an instinct as that of the HIVE bee making its cells will probably have occurr
0209  which we are acquainted, namely, those of the HIVE bee and of many ants, could not possibly have
0216  certain ants; and the comb making power of the HIVE bee: these two latter instincts have generally
0224  formica rufescens. Cell making instinct of the HIVE Bee. I will not here enter on minute details o
0224  effected by a crowd of bees working in a dark HIVE. Grant whatever instincts you please, and it s
0225  her end of the series we have the cells of the HIVE bee, placed in a double layer: each cell, as i
0225  een the extreme perfection of the cells of the HIVE bee and the simplicity of those of the humble
0225  tself is intermediate in structure between the HIVE and humble bee, but more nearly related to the
0226  three sided pyramidal basis of the cell of the HIVE bee. As in the cells of the hive bee, so here,
0226  e cell of the hive bee. As in the cells of the HIVE bee, so here, the three plane surfaces in any
0226  obably have been as perfect as the comb of the HIVE bee. Accordingly I wrote to Professor Miller,
0227  ments which have been made of the cells of the HIVE bee. Hence we may safely conclude that if we c
0227  tructure as wonderfully perfect as that of the HIVE bee. We must suppose the Melipona to make her
0227  de a bird to make its nest, I believe that the HIVE bee has acquired, through natural selection, h
0228  he case of ordinary cells. I then put into the HIVE, instead of a thick, square piece of wax, a th
0229  marked instance, I put the comb back into the HIVE, and allowed the bees to go on working for a s
0233  on, could have profited the progenitors of the HIVE bee? I think the answer is not difficult: it i
0233  fifteen pounds of dry sugar are consumed by a HIVE of bees for the secretion of each pound of wax
0233  ust be collected and consumed by the bees in a HIVE for the secretion of the wax necessary for the
0234  ees during the winter; and the security of the HIVE is known mainly to depend on a large number of
0235  na would make a comb as perfect as that of the HIVE bee. Beyond this stage of perfection in archit
0235  selection could not lead: for the comb of the HIVE bee, as far as we can see, is absolutely perfe
```

Page ***(Key Word)***

Page ***(Key Word)***

```
0235  wonderful of all known instincts, that of the HIVE bee, can be explained by natural selection hav
0236  would have been far better exemplified by the HIVE bee. If a working ant or other neuter insect h
0474  s on the admirable architectural powers of the HIVE bee. Habit no doubt sometimes comes into play
0163  imen with very distinct zebra like bars on the HOCKS. With respect to the horse, I have collected
0060  ubtful forms of British plants are entitled to HOLD, if the existence of any well marked varieties
0064  ers, all cannot do so, for the world would not HOLD them. There is no exception to the rule that e
0100  formed me that he finds that the rule does not HOLD in Australia; and I have made these few remark
0113  can thus be raised. The same has been found to HOLD good when first one variety and then several m
0151  to them, and the rule seems to me certainly to HOLD good in this class. I cannot make out that it
0179  been asked by the opponents of such views as I HOLD, how, for instance, a land carnivorous animal
0224  d have made their cells of the proper shape to HOLD the greatest possible amount of honey, with th
0225  ve humble bees, which use their old cocoons to HOLD honey, sometimes adding to them short tubes of
0227  prolong the hexagon to any length requisite to HOLD the stock of honey; in the same way as the rud
0402  their own places in nature, both probably will HOLD their own places and keep separate for almost
0422  s of the same group. Thus, on the view which I HOLD, the natural system is genealogical in its arr
0432  natural arrangement in the diagram would still HOLD good; and, on the principle of inheritance, al
0471  in increasing in size, for the world would not HOLD them, the more dominant groups beat the less c
0177  case will be strongly in favour of the great HOLDERS on the mountains or on the plains improving
0177  ving their breeds more quickly than the small HOLDERS on the intermediate narrow, hilly tract; and
0225  and, in addition, some large cells of wax for HOLDING honey. These latter cells are nearly spheric
0056  on an average many are still forming; and this HOLDS good. There are other relations between the s
0147  d to economise on the other side. I think this HOLDS true to a certain extent with our domestic pr
0151  ully convinced that the rule almost invariably HOLDS good with cirripedes. I shall, in my future w
0176  actically, as far as I can make out, this rule HOLDS good with varieties in a state of nature. I h
0339  azil. Mr. Woodward has shown that the same law HOLDS good with sea shells, but from the wide distr
0395  as far as I have gone, the relation generally HOLDS good. We see Britain separated by a shallow c
0418  generally been admitted as true. The same fact HOLDS good with flowering plants, of which the two
0094  ave time; for instance, their habit of cutting HOLES and sucking the nectar at the bases of certai
0230  , and by endeavouring to make equal spherical HOLLOWS, but never allowing the spheres to break int
0093  es of plants, presently to be alluded to. Some HOLLY trees bear only male flowers, which have four
0093  ity of pollen, and a rudimentary pistil; other HOLLY trees bear only female flowers; these have a
0271  ns which I have made on certain varieties of HOLLYHOCK, I am inclined to suspect that they present
0001  one of our greatest philosophers. On my return HOME, it occurred to me, in 1837, that something mi
0015  ngle variety should be turned loose in its new HOME. Nevertheless, as our varieties certainly do o
0078  imate may be exactly the same as in its former HOME, yet the conditions of its life will generally
0078  hed to increase its average numbers in its new HOME, we should have to modify it in a different wa
0091  i one cat, according to Mr. St. John, bringing HOME winged game, another hares or rabbits, and ano
0139  h species is adapted to the climate of its own HOME: species from an arctic or even from a tempera
0215  ound incurable in dogs which have been brought HOME as puppies from countries, such as Tierra del
0220  ce, it is clear, that the slaves feel quite at HOME. During the months of June and July, on three
0223  th on top of a spray of heath over its ravaged HOME. Such are the facts, though they did not need
0289  n the Supplement to Lyell's Manual, will bring HOME the truth, how accidental and rare is their pr
0365  re than one would be so well fitted to its new HOME, as to become naturalised. But this, as it see
0402  rles Island is at least as well fitted for its HOME as is the species peculiar to Chatham Island.
0049  to Great Britain. A wide distance between the HOMES of two doubtful forms leads many naturalists
0065  fusion of naturalised productions in their new HOMES. In a state of nature almost every plant prod
0115  ially adapted to certain stations in their new HOMES. But the case is very different; and Alph. de
0140  subsequently become acclimatised to their new HOMES. As I believe that our domestic animals were
0301  when such varieties returned to their ancient HOMES, as they would differ from their former state
0325  ant, that is, which are commonest in their own HOMES, and are most widely diffused, having produce
0350  many competitors in their own widely extended HOMES will have the best chance of seizing on new p
0350  n they spread into new countries. In their new HOMES they will be exposed to new conditions, and w
0368  d subsequently have retreated to their present HOMES: but I have met with no satisfactory evidence
0379  te northward or southward towards their former HOMES: but the forms, chiefly northern, which had c
0379  equator, would travel still further from their HOMES into the most temperate latitudes of the oppo
0379  r southern hemispheres, now exist in their new HOMES as well marked varieties or as distinct speci
0379  the northern forms having existed in their own HOMES in greater numbers, and having consequently b
0381  as common, as another species within their own HOMES. I have said that many difficulties remain to
0391  g ages have struggled together in their former HOMES, and have become mutually adapted to each oth
0391  d to each other; and when settled in their new HOMES, each kind will have been kept by the others
0395  odified through natural selection in their new HOMES in relation to their new position, and we can
0396  heir arrival, could have reached their present HOMES. But the probability of many islands having e
0403  uently modified and better fitted to their new HOMES, is of the widest application throughout natu
0406  etter adaptation of the colonists to their new HOMES. Summary of last and present Chapters. In the
0425  rous; and I might answer by the argumentum ad HOMINEM, and ask what should be done if a perfect ka
0204  uding that none could have existed, for the HOMOLOGIES of many organs and their intermediate state
0438  nct species, we can indicate but few serial HOMOLOGIES; that is, we are seldom enabled to say that
0452  this may sometimes be seen. In tracing the HOMOLOGIES of the same part in different members of a
0438  the great class of molluscs, though we can HOMOLOGISE the parts of one species with those of anot
0143  lt. The several parts of the body which are HOMOLOGOUS, and which, at an early embryonic period, a
0143  gether, for the lower jaw is believed to be HOMOLOGOUS with the limbs. These tendencies, I do not
0143  en rendered permanent by natural selection. HOMOLOGOUS parts, as has been remarked by some authors
0143  nd nothing is more common than the union of HOMOLOGOUS parts in normal structures, as the union of
0151  a marvellous amount of diversification: the HOMOLOGOUS valves in the different species being somet
0168  to have been more potent in their effects. HOMOLOGOUS parts tend to vary in the same way, and hom
0168  ous parts tend to vary in the same way, and HOMOLOGOUS parts tend to cohere. Modifications in hard
0191  physiologists admit that the swimbladder is HOMOLOGOUS, or ideally similar, in position and struct
0191  branchiae and dorsal scales of Annelids are HOMOLOGOUS with the wings and wingcovers of insects, i
0192  gerous frena in the one family are strictly HOMOLOGOUS with the branchiae of the other family; ind
0198  form of the pelvis; and then by the law of HOMOLOGOUS variation, the front limbs and even the hea
0434  s in the different species of the class are HOMOLOGOUS. The whole subject is included under the ge
0434  he high importance of relative connexion in HOMOLOGOUS organs: the parts may change to almost any
0434  d. hence the same names can be given to the HOMOLOGOUS bones in widely different animals. We see t
0435  nce perceive the plain signification of the HOMOLOGOUS construction of the limbs throughout the wh
0436  sts believe that the bones of the skull are HOMOLOGOUS with, that is correspond in number and in r
0436  rtebrate and articulate classes are plainly HOMOLOGOUS. We see the same law in comparing the wonde
```

Page ***(Key Word)***

homologous

Page **(Key Word)**

```
0438  om enabled to say that one part or organ is HOMOLOGOUS with another in the same individual. And we
0457  e look to the same pattern displayed in the HOMOLOGOUS organs, to whatever purpose applied, of re
0457  the different species of a class; or to the HOMOLOGOUS parts constructed on the same pattern in ea
0457  resemblance in an individual embryo of the HOMOLOGOUS parts, which when matured will become widel
0457  ance in different species of a class of the HOMOLOGOUS parts or organs, though fitted in the adult
0480  ins to reveal, by rudimentary organs and by HOMOLOGOUS structures, her scheme of modification, whi
0144  the naked Turkish dog, though here probably HOMOLOGY comes into play? With respect to this latter
0427  y external appearances, actually classed an HOMOPTEROUS insect as a moth. We see something of the
0020  l quarters of the world, more especially by the HON. W. Elliot from India, and by the Hon. C. Murr
0020  ly by the Hon. W. Elliot from India, and by the HON. C. Murray from Persia. Many treatises in diff
0249  third most experienced hybridiser, namely, the HON. and Rev. W. Herbert. He is as emphatic in his
0046  r as important (as some few naturalists have HONESTLY confessed) which does not vary; and, under t
0389  e globe. I have already stated that I cannot HONESTLY admit Forbes's view on continental extension
0224  shape to hold the greatest possible amount of HONEY, with the least possible consumption of preci
0225  mble bees, which use their old cocoons to hold HONEY, sometimes adding to them short tubes of wax,
0225  addition, some large cells of wax for holding HONEY. These latter cells are nearly spherical and
0227  n to any length requisite to hold the stock of HONEY; in the same way as the rude humble bee adds
0234  ing the process of secretion. A large store of HONEY is indispensable to support a large stock of
0234  ted. Hence the saving of wax by largely saving HONEY must be a most important element of success i
0234  o be altogether independent of the quantity of HONEY which the bees could collect. But let us supp
0234  e winter, and consequently required a store of HONEY: there can in this case be no doubt that it w
0235  wax; that individual swarm which wasted least HONEY in the secretion of wax, having succeeded bes
0239  sly ceveloped abdomen which secretes a sort of HONEY, supplying the place of that excreted by the
0002  k, the latter having read my sketch of 1844, HONOURED me by thinking it advisable to publish, with
0021  ed along the back of the neck that they form a HOOD, and it has, proportionally to its size, much
0230  first commencement having always been a little HOOD of wax; but I will not here enter on these det
0212  wary in England, is tame in Norway, as is the HOODED crow in Egypt. That the general disposition o
0088  e lion, the shoulder pad to the boar, and the HOOKED jaw to the male salmon; for the shield may be
0237  the nuptial plumage of many birds, and in the HOOKED jaws of the male salmon. We have even slight
0392  . some of the endemic plants have beautifully HOOKED seeds; yet few relations are more striking th
0392  ions are more striking than the adaptation of HOOKED seeds for transportal by the wool and fur of
0392  case presents no difficulty on my view, for a HOOKED seed might be transported to an island by som
0392  ng slightly modified, but still retaining its HOOKED seeds, would form an endemic species, having
0002  journal of that Society. Sir C. Lyell and Dr. HOOKER, who both knew of my work, the latter having
0003  without expressing my deep obligations to Dr. HOOKER, who for the last fifteen years has aided me
0053  re many difficulties, as did subsequently Dr. HOOKER, even in stronger terms. I shall reserve for
0053  portional numbers of the varying species. Dr. HOOKER permits me to add, that after having carefull
0100  e case in this country: and at my request Dr. HOOKER tabulated the trees of New Zealand, and Dr. A
0100  was as I anticipated. On the other hand, Dr. HOOKER has recently informed me that he finds that t
0140  dodendrons, raised from seed collected by Dr. HOOKER from trees growing at different heights on th
0145  f the Umbelliferae, it is by no means, as Dr. HOOKER informs me, in species with the densest heads
0373  their former low descent; and in Sikkim, Dr. HOOKER saw maize growing on gigantic ancient moraine
0374  identical and allied species. In America, Dr. HOOKER has shown that between forty and fifty of the
0375  of these Australian forms, as I hear from Dr. HOOKER, extend along the heights of the peninsula of
0375  ng list can be given, as I am informed by Dr. HOOKER, of European genera, found in Australia, but
0375  roduction to the Flora of New Zealand, by Dr. HOOKER, analogous and striking facts are given in re
0376  asmania, etc., of northern forms of fish. Dr. HOOKER informs me that twenty five species of Algae
0378  riking fact, lately communicated to me by Dr. HOOKER, that all the flowering plants, about forty s
0378  of the Himalaya, as graphically described by HOOKER. Thus, as I believe, a considerable number of
0379  is a remarkable fact, strongly insisted on by HOOKER in regard to America, and by Alph. de Candoll
0381  le are stated with admirable clearness by Dr. HOOKER in his botanical works on the antarctic regio
0387  uthern water lily (probably, according to Dr. HOOKER, the Nelumbium luteum) in a heron's stomach;
0391  . in the plants of the Galapagos Islands, Dr. HOOKER has shown that the proportional numbers of th
0398  d with nearly all the plants, as shown by Dr. HOOKER in his admirable memoir on the Flora of this
0399  d, and that very closely, as we know from Dr. HOOKER's account, to those of America: but on the vi
0399  ty, which, though feeble, I am assured by Dr. HOOKER is real, between the flora of the south weste
0429  in the vegetable kingdom, as I learn from Dr. HOOKER, it has added only two or three orders of sma
0030  ce, believe that the fuller's teazle, with its HOOKS, which cannot be rivalled by any mechanical c
0197  st trees by the aid of exquisitely constructed HOOKS clustered around the ends of the branches, an
0197  ice to the plant; but as we see nearly similar HOOKS on many trees which are not climbers, the hoo
0197  ooks on many trees which are not climbers, the HOOKS on the bamboo may have arisen from unknown la
0001  day I have steadily pursued the same object. I HOPE that I may be excused for entering on these pe
0002  . no doubt errors will have crept in, though I HOPE I have always been cautious in trusting to goo
0002  with a few facts in illustration, but which, I HOPE, in most cases will suffice. No one can feel m
0002  which my conclusions have been grounced: and I HOPE in a future work to do this. For I am well awa
0038  it is that neither Australia, the Cape of Good HOPE, nor any other region inhabited by quite unciv
0051  ot now continuous, in which case he can hardly HOPE to find the intermediate links between his dou
0110  as yet fully stocked, for at the Cape of Good HOPE, where more species of plants are crowded toge
0150  . i am aware of several causes of error, but I HOPE that I have made due allowance for them. It sh
0245  successive profitable degrees of sterility. I HOPE, however, to be able to show that sterility is
0282  rocks and making fresh sediment, before he can HOPE to comprehend anything of the lapse of time, t
0297  s from the reasons just assigned we can seldom HOPE to effect in any one geological section. Suppo
0323  rica, in Tierra del Fuego, at the Cape of Good HOPE, and in the peninsula of India. For at these d
0375  ives of the peculiar flora of the Cape of Good HOPE occur. At the Cape of Good Hope a very few Eur
0375  e cape of Good Hope occur. At the Cape of Good HOPE a very few European species, believed not to h
0377  esent day crowded together at the Cape of Good HOPE, and in parts of temperate Australia. As we kn
0381  and others have remained unaltered. We cannot HOPE to explain such facts, until we can say why on
0389  ith those on an equal area at the Cape of Good HOPE or in Australia, we must, I think, admit that
0399  rn corner of Australia and of the Cape of Good HOPE, is a far more remarkable case, and is at pres
0434  look to some unknown plan of creation, we may HOPE to make sure but slow progress. Morphology. We
0449  nature: but I am bound to confess that I only HOPE to see the law hereafter proved true. It can b
0111  the character of species, as is shown by the HOPELESS doubts in many cases how to rank them, yet c
0150  s come to a nearly similar conclusion. It is HOPELESS to attempt to convince any one of the truth
0219  y other slave making ant, it would have been HOPELESS to have speculated how so wonderful an insti
0268  under nature, we are immediately involved in HOPELESS difficulties; for if two hitherto reputed va
0435  h the flowers of plants. Nothing can be more HOPELESS than to attempt to explain this similarity o
```

Page **(Key Word)**

Page ***(Key Word)**

```
0354  ated thence as far as it could. It would be HOPELESSLY tedious to discuss all the exceptional case
0020  en the extreme difficulty, or rather utter HOPELESSNESS, of the task becomes apparent. Certainly,
0435  ty or by the doctrine of final causes. The HOPELESSNESS of the attempt has been expressly admitted
0117  tion. When a dotted line reaches one of the HORIZONTAL lines, and is there marked by a small numbe
0117  systematic work. The intervals between the HORIZONTAL lines in the diagram, may represent each a
0119  he lower branches not reaching to the upper HORIZONTAL lines. In some cases I do not doubt that th
0120  e suppose the amount of change between each HORIZONTAL line in our diagram to be excessively small
0121  ange supposed to be represented between the HORIZONTAL lines. After fourteen thousand generations,
0124  case before his mind. In the diagram, each HORIZONTAL line has hitherto been supposed to represen
0331  given, but this is unimportant for us. The HORIZONTAL lines may represent successive geological f
0732  re to be discovered above one of the middle HORIZONTAL lines or geological formations, for instanc
0382  the south. As the tide leaves its drift in HORIZONTAL lines, though rising higher on the shores w
0420  ifteen genera (a14 to z14) on the uppermost HORIZONTAL line. Now all these modified descendants fr
0014  w by a long horned bull, the greater length of HORN, though appearing late in life, is clearly due
0090  estic animals (as the wattle in male carriers, HORN like protuberances in the cocks of certain fow
0320  it is transported far and near, like our short HORN cattle, and takes the place of other breeds in
0014  nner as in the crossed offspring from a short HORNED cow by a long horned bull, the greater length
0014  d offspring from a short horned cow by a long HORNED bull, the greater length of horn, though appe
0015  reed our cart and race horses, long and short HORNED cattle, and poultry of various breeds, and es
0018  ere, four or five thousand years ago? But Mr. HORNER's researches have rendered it in some degree
0088  special weapons, confined to the male sex. A HORNLESS stag or spurless cock would have a poor chan
0454  the reappearance of minute dangling horns in HORNLESS breeds of cattle, more especially, according
0012  are apt to have, as is asserted, long or many HORNS: pigeons with feathered feet have skin betwee
0014  rwise: thus the inherited peculiarities in the HORNS of cattle could appear only in the offspring
0029  his cattle might not have descended from long HORNS, and he will laugh you to scorn. I have never
0086  e colour of the down of their chickens: in the HORNS of our sheep and cattle when nearly adult: so
0111  ncient black cattle were displaced by the long HORNS, and that these were swept away by the short
0111  s, and that these were swept away by the short HORNS (I quote the words of an agricultural writer)
0198  growth of the hair, and that with the hair the HORNS are correlated. Mountain breeds always differ
0237  salmon. We have even slight differences in the HORNS of different breeds of cattle in relation to
0237  le sex: for oxen of certain breeds have longer HORNS than in other breeds, in comparison with the
0237  s than in other breeds, in comparison with the HORNS of the bulls or cows of these same breeds. He
0238  always yielding oxen with extraordinarily long HORNS, could be slowly formed by carefully watching
0238  , when matched, produced oxen with the longest HORNS; and yet no one ox could ever have propagated
0275  h as albinism, melanism, deficiency of tail or HORNS, or additional fingers and toes; and do not r
0423  thus the great agriculturist Marshall says the HORNS are very useful for this purpose with cattle,
0423  lour of the body, etc.; whereas with sheep the HORNS are much less serviceable, because less const
0443  e element of one parent. Or again, as when the HORNS of cross bred cattle have been affected by th
0443  cattle have been affected by the shape of the HORNS of either parent. For the welfare of a very y
0444  ago states of the silk moth; or, again, in the HORNS of almost full grown cattle. But further than
0454  ss breeds, the reappearance of minute dangling HORNS in hornless breeds of cattle, more especially
0454  nknown laws of growth, but in order to excrete HORNY matter, as that the rudimentary nails on the
0029  gencies for the differences of a dray and race HORSE, a greyhound and bloodhound, a carrier and tu
0030  the ancon sheep. But when we compare the dray HORSE and race horse, the dromedary and camel, the
0030  p. but when we compare the dray horse and race HORSE, the dromedary and camel, the various breeds
0120  n the same way, for instance, the English race HORSE and English pointer have apparently both gone
0163  ra like bars on the hocks. With respect to the HORSE, I have collected cases in England of the spi
0164  seen in duns, and I have see a trace in a bay HORSE. My son made a careful examination and sketch
0164  nation and sketch for me of a dun Belgian cart HORSE with a double stripe on each shoulder and wit
0164  xamined the breed for the Indian Government, a HORSE without stripes is not considered as purely b
0164  mr. W. W. Edwards, that with the English race HORSE the spinal stripe is much commoner in the foa
0164  bject, believes that the several breeds of the HORSE have descended from several aboriginal specie
0165  o breeds so distinct as the heavy Belgian cart HORSE, Welch ponies, cobs, the lanky Kattywar race,
0165  effects of crossing the several species of the HORSE genus. Rollin asserts, that the common mule f
0165  asserts, that the common mule from the ass and HORSE is particularly apt to have bars on its legs.
0165  W. C. Martin, in his excellent treatise on the HORSE, has given a figure of a similar mule. In fou
0166  s? we see several very distinct species of the HORSE genus becoming, by simple variation, striped
0166  r striped on the shoulders like an ass. In the HORSE we see this tendency strong whenever a dun ti
0166  have just seen that in several species of the HORSE genus the stripes are either plainer or appea
0166  w exactly parallel is the case with that of the HORSE genus! For myself, I venture confidently to l
0167  constructed, the common parent of our domestic HORSE, whether or not it be descended from one or m
0200  the arm of the monkey, in the fore leg of the HORSE, in the wing of the bat, and in the flipper o
0200  the several bones in the limbs of the monkey, HORSE, and bat, which have been inherited from a co
0258  , i mean the case, for instance, of a stallion HORSE being first crossed with a female ass, and th
0274  in that the ass has a prepotent power over the HORSE, so that both the mule and the hinny more res
0274  e and the hinny more resemble the ass than the HORSE: but that the prepotency runs more strongly i
0281  ok to forms very distinct, for instance to the HORSE and tapir, we have no reason to suppose that
0281  much general resemblance to the tapir and the HORSE; but in some points of structure may have dif
0281  have descended from the other; for instance, a HORSE from a tapir; and in this case direct interme
0318  done. When I found in La Plata the tooth of a HORSE embedded with the remains of Mastodon, Megath
0318  filled with astonishment: for seeing that the HORSE, since its introduction by the Spaniards into
0318  could so recently have exterminated the former HORSE under conditions of life apparently so favour
0319  the tooth, though so like that of the existing HORSE, belonged to an extinct species. Had this hor
0319  orse, belonged to an extinct species. Had this HORSE been still living, but in some degree rare, n
0319  ly ever tell. On the supposition of the fossil HORSE still existing as a rare species, we might ha
0319  history of the naturalisation of the domestic HORSE in South America, that under more favourable
0319  veral contingencies, and at what period of the HORSE's life, and in what degree, they severally ac
0319  ld not have perceived the fact, yet the fossil HORSE would certainly have become rarer and rarer,
0324  alous monsters coexisted with the Mastodon and HORSE, it might at least have been inferred that th
0434  ng, that of a mole for digging, the leg of the HORSE, the paddle of the porpoise, and the wing of
0445  a three days old colt of a race and heavy cart HORSE, I find that the colts have by no means acqui
0452  s given by Owen of the bones of the leg of the HORSE, ox, and rhinoceros. It is an important fact
0473  houlder and legs of the several species of the HORSE genus and in their hybrids! How simply is thi
0479  ng of bat, fin of the porpoise, and leg of the HORSE, the same number of vertebrae forming the nec
0015  ert that we could not breed our cart and race HORSES, long and short horned cattle, and poultry of
0018  ad more than one wild parent. With respect to HORSES, from reasons which I cannot give here, I am
```

Page ***(Key Word)**

Page **(Key Word)**

```
0034  prevent their exportation: the destruction of HORSES under a certain size was ordered, and this ma
0064  rate of increase of slow breeding cattle and  HORSES in South America, and latterly in Australia,
0072  instance of this: for here neither cattle nor HORSES nor dogs have ever run wild, though they swar
0072  ay, the flies would decrease, then cattle and HORSES would become feral, and this would certainly
0112  at an early period one man preferred swifter  HORSES: another stronger and more bulky horses. The
0112  ifter horses: another stronger and more bulky HORSES. The early differences would be very slight:
0112  time, from the continued selection of swifter HORSES by some breeders, and of stronger ones by oth
0163  cted cases in England of the spinal stripe in HORSES of the most distinct breeds, and of all colou
0164  orth west part of India the Kattywar breed of HORSES is so generally striped, that, as I hear from
0164  he foal; and sometimes cuite disappear in old HORSES. Colonel Poole has seen both gray and bay Kat
0164  nel Poole has seen both gray and bay Kattywar  HORSES striped when first foaled. I have, also, reas
0164  ollected cases of leg and shoulder stripes in HORSES of very different breeds, in various countrie
0166  ur in the eminently striped Kattywar breed of HORSES, and was, as we have seen, answered in the af
0299  , that our different breeds of cattle, sheep,  HORSES, and dogs have descended from a single stock
0356  . to illustrate what I mean: our English race HORSES differ slightly from the horses of every othe
0356  english race horses differ slightly from the  HORSES of every other breed; but they do not owe the
0443  for it is notorious that breeders of cattle,  HORSES, and various fancy animals, cannot positively
0445  n, i was told that the foals of cart and race HORSES differed as much as the full grown animals; a
0446  our domestic varieties. Fanciers select their HORSES, dogs, and pigeons, for breeding, when they a
0037  y. i have seen great surprise expressed in   HORTICULTURAL works at the wonderful skill of gardeners
0062  chester, evidently the result of his great   HORTICULTURAL knowledge. Nothing is easier than to admi
0250  e in part accounted for by Herbert's great   HORTICULTURAL skill, and by his having hothouses at his
0009  sterility has been said to be the bane of   HORTICULTURAL; but on this view we owe variability to th
0085  importance: yet we hear from an excellent   HORTICULTURIST, Downing, that in the United States smoot
0237  and the individual is destroyed; but the   HORTICULTURIST sows seeds of the same stock, and confide
0032  cier. The same principles are followed by   HORTICULTURISTS; but the variations are here often more
0251  mes depends. The practical experiments of   HORTICULTURISTS, though not made with scientific precisi
0252  would have been notorious to nurserymen.   HORTICULTURISTS raise large beds of the same hybrids, an
0030  ntam so small and elegant; when we compare the HOST of agricultural, culinary, orchard, and flower
0085  d have to struggle with other trees and with a HOST of enemies, such differences would effectually
0372  l races, and others as distinct species: and a HOST of closely allied or representative forms whic
0374  on the lofty mountains of equatorial America a HOST of peculiar species belonging to European gene
0270  species. The evidence is, also, derived from HOSTILE witnesses, who in all other cases consider f
0270  s of Verbascum, by so good an observer and so HOSTILE a witness, as Gartner: namely, that yellow a
0287  ld, the land and the water has been peopled by HOSTS of living forms. What an infinite number of g
0139  n for species of the same genus to inhabit very HOT and very cold countries, and as I believe that
0374  ner, which do not exist in the wide intervening HOT countries. So on the Silla of Caraccas the ill
0375  plants of Europe, not found in the intervening HOT lowlands. A list of the genera collected on th
0250  great horticultural skill, and by his having HOTHOUSES at his command. Of his many important state
0424  her respects. If it could be proved that the HOTTENTOT had descended from the Negro, I think he wo
0078  ryness, for elsewhere it ranges into slightly HOTTER or colder, damper or drier districts. In this
0347  l areas in the Old World could be pointed out HOTTER than any in the New World, yet these are not
0378  te. On the other hand, the the most humid and HOTTEST districts will have afforded an asylum to th
0035  been chiefly effected by crosses with the fox HOUND: but what concerns us is, that the change has
0222  quickly ran away; but in about a quarter of an HOUR, shortly after all the little yellow ants had
0361  heir rate of flight would often be 35 miles an HOUR; and some authors have given a far higher esti
0084  y be said that natural selection is daily and HOURLY scrutinising, throughout the world, every var
0086  ase of those insects which live only for a few HOURS, and which never feed, a large part of their
0184  the black bear was seen by Hearne swimming for HOURS with widely open mouth, thus catching, like a
0211  and prevented their attendance during several HOURS. After this interval, I felt sure that the ap
0220  hree successive years, I have watched for many HOURS several nests in Surrey and Sussex, and never
0220  ms me that he has watched the nests at various HOURS during May, June and August, both in Surrey a
0362  do not pass into the gizzard for 12 or even 18 HOURS. A bird in this interval might easily be blow
0362  and after an interval of from twelve to twenty HOURS, disgorge pellets, which, as I know from expe
0362  ed after having been from twelve to twenty one HOURS in the stomachs of different birds of prey; a
0362  g been thus retained for two days and fourteen HOURS. Fresh water fish, I find, eat seeds of many
0362  elicans: these birds after an interval of many HOURS, either rejected the seeds in pellets or pass
0385  uck's feet, in damp air, from twelve to twenty HOURS; and in this length of time a duck or heron m
0387  when rejected in pellets or in excrement, many HOURS afterwards. When I saw the great size of the
0445  young pigeons of various breeds, within twelve HOURS after being hatched; I carefully measured the
0446  of the short faced tumbler, which when twelve HOURS old had acquired its proper proportions, prov
0214  what they now are: and near Glasgow there are HOUSE tumblers, as I hear from Mr. Brent, which can
0241  if we were to see a set of workmen building a HOUSE of whom many were five feet four inches high,
0247  , and apparently were kept in a chamber in his HOUSE. That these processes are often injurious to
0366  ffered under an Arctic climate. The ruins of a HOUSE burnt by fire do not tell their tale more pla
0220  he nest. Hence he considers them as strictly HOUSEHOLD slaves. The masters, on the other hand, may
0183  (saurophagus sulphuratus) in South America, HOVERING over one spot and then proceeding to another
0003  ould be unsatisfactory, until it could be shown HOW the innumerable species inhabiting this world
0004  what is equally or more important, we shall see HOW great is the power of man in accumulating by h
0005  th in the fourth chapter: and we shall then see HOW Natural Selection almost inevitably causes muc
0005  ifficulties of transitions, or in understanding HOW a simple being or a simple organ can be change
0008  the many cases when the male and female unite. HOW many animals there are which will not breed, t
0008  generally attributed to vitiated instincts: but HOW many cultivated plants display the utmost vigo
0009  collected on this curious subject; but to show HOW singular the laws are which determine the repr
0010  tly the same conditions of life; and this shows HOW unimportant the direct effects of the conditio
0010  obably have varied in the same manner. To judge HOW much, in the case of any variation, we should
0012  and the best on this subject. No breeder doubts HOW strong is the tendency to inheritance: like pr
0015  as will hereafter be explained, will determine HOW far the new characters thus arising shall be p
0017  ue of most of our domesticated productions: but HOW could a savage possibly know, when he first ta
0018  thousand years ago, and who will pretend to say HOW long before these ancient periods, savages, li
0023  ic breeds by the crossing of any lesser number: HOW, for instance, could a pouter be produced by c
0028  we treat of Selection. We shall then, also, see HOW it is that the breeds so often have a somewhat
0028  ons and watched the several kinds, knowing well HOW true they bred, I felt fully as much difficult
0029  ons, in his treatise on pears and apples, shows HOW utterly he disbelieves that the several sorts,
0033  s and flowers of the same set of varieties. See HOW different the leaves of the cabbage are, and h
0033  ow different the leaves of the cabbage are, and HOW extremely alike the flowers, how unlike the fl
```

Page **(Key Word)**

Page ***(Key Word)***

```
0033  bbage are, and how extremely alike the flowers,  HOW  unlike the flowers of the heartsease are, and
0033  w unlike the flowers of the heartsease are, and  HOW  alike the leaves: how much the fruit of the di
0033  f the heartsease are, and how alike the leaves:  HOW  much the fruit of the different kinds of goose
0034  squimaux their teams of dogs. Livingstone shows  HOW  much good domestic breeds are valued by the ne
0035  they carry. Lord Spencer and others have shown  HOW  the cattle of England have increased in weight
0038  tandard of usefulness to man, we can understand  HOW  it is that neither Australia, the Cape of Good
0038  by man has played, it becomes at once obvious,  HOW  it is that our domestic races show adaptation
0048  ame country, but are common in separated areas.  HOW  many of those birds and insects in North Ameri
0048  e from the American mainland, I was much struck  HOW  entirely vague and arbitrary is the distinctio
0050  t cases, will bring naturalists to an agreement  HOW  to rank doubtful forms. Yet it must be confess
0050  ome authors as species. Look at the common oak,  HOW  closely it has been studied; yet a German auth
0050  the group is subject; and this shows, at least,  HOW  very generally there is some variation. But if
0050  thin one country, he will soon make up his mind  HOW  to rank most of the doubtful forms. His genera
0058  ll it, of Divergence of Character, we shall see  HOW  this may be explained, and how the lesser diff
0058  er, we shall see how this may be explained, and  HOW  the lesser differences between varieties will
0060  i must make a few preliminary remarks, to show  HOW  the struggle for existence bears on Natural Se
0060  the work, helps us but little in understanding  HOW  species arise in nature. How have all those ex
0060  e in understanding how species arise in nature.  HOW  have all those exquisite adaptations of one pa
0061  t of the organic world. Again, it may be asked,  HOW  is it that varieties, which I have called inci
0061  more than do the varieties of the same species?  HOW  co those groups of species, which constitute w
0062  e thus constantly destroying life; or we forget  HOW  largely these songsters, or their eggs, or the
0065  dly stock every station in which they could any  HOW  exist, and that the geometrical tendency to in
0066  gle one: but this difference does not determine  HOW  many individuals of the two species can be sup
0067  ce. Nor will this surprise any one who reflects  HOW  ignorant we are on this head, even in regard t
0070  during winter: but any one who has tried, knows  HOW  troublesome it is to get seed from a few wheat
0071  here enlarge. Many cases are on record showing  HOW  complex and unexpected are the checks and rela
0071  three distinct insectivorous birds. Here we see  HOW  potent has been the effect of the introduction
0071  n enclosed, so that cattle could not enter. But  HOW  important an element enclosure is, I plainly s
0073  i am tempted to give one more instance showing  HOW  plants and animals, most remote in the scale o
0074  l numbers and kinds to what we call chance. But  HOW  false a view is this! Every one has heard that
0075  l to the ground according to definite laws; but  HOW  simple is this problem compared to the action
0076  and has caused the decrease of the song thrush.  HOW  frequently we hear of one species of rat takin
0080  t. explains the grouping of all organic beings.  HOW  will the struggle for existence, discussed too
0080  a lesser degree, those under nature, vary; and  HOW  strong the hereditary tendency is. Under domes
0080  in some degree plastic. Let it be borne in mind  HOW  infinitely complex and close fitting are the m
0081  of the former inhabitants. Let it be remembered  HOW  powerful the influence of a single introduced
0084  in the struggle for life, and so be preserved.  HOW  fleeting are the wishes and efforts of man! ho
0084  how fleeting are the wishes and efforts of man!  HOW  short his time! and consequently how poor will
0084  ts of man! how short his time! and consequently  HOW  poor will his products be, compared with those
0085  would produce little effect: we should remember  HOW  essential it is in a flock of white sheep to d
0088  s breed by careful selection of the best cocks.  HOW  low in the scale of nature this law of battle
0089  s and dislikes: thus Sir R. Heron has described  HOW  one pied peacock was eminently attractive to a
0090  of natural Selection. In order to make it clear  HOW, as I believe, natural selection acts, I must
0094  ctar for food. I could give many facts, showing  HOW  anxious bees are to save time; for instance, t
0095  could visit its flowers. Thus I can understand  HOW  a flower and a bee might slowly become, either
0097  r view are irexplicable. Every hybridizer knows  HOW  unfavourable exposure to wet is to the fertili
0099  eparated sexes, and must habitually be crossed.  HOW  strange are these facts! How strange that the
0099  tually be crossed. How strange are these facts!  HOW  strange that the pollen and stigmatic surface
0099  o many cases be mutually useless to each other!  HOW  simply are these facts explained on the view o
0099  se of the many other flowers on the same plant.  HOW, then, comes it that such a vast number of the
0111  . many curious instances could be given showing  HOW  quickly new breeds of cattle, sheep, and other
0111  s is shown by the hopeless doubts in many cases  HOW  to rank them, yet certainly differ from each o
0111  are, as I have called them, incipient species.  HOW, then, does the lesser difference between vari
0112  om each other and from their common parent. But  HOW, it may be asked, can any analogous principle
0114  ighteen genera and to eight orders, which shows  HOW  much these plants differed from each other. So
0116  places occupied by other beings. Now let us see  HOW  this principle of great benefit being derived
0126  same species making a class, we can understand  HOW  it is that there exist but very few classes ir
0127  n in the following chapters. But we already see  HOW  it entails extinction; and how largely extinct
0127  t we already see how it entails extinction; and  HOW  largely extinction has acted in the world's hi
0132  or each deviation of structure, however slight.  HOW  much direct effect difference of climate, fooc
0133  of the slightest use to a being, we cannot tell  HOW  much of it to attribute to the accumulative ac
0133  e accumulative action of natural selection, and  HOW  much to the conditions of life. Thus, it is we
0133  s under which they have lived: but who can tell  HOW  much of this difference may be due to the warm
0133  ured and preserved during many generations, and  HOW  much to the direct action of the severe climat
0133  cies under the same conditions. Such facts show  HOW  indirectly the conditions of life must act. Ac
0141  ought, under peculiar circumstances, into play.  HOW  much of the acclimatisation of species to any
0141  any peculiar climate is due to mere habit, and  HOW  much to the natural selection of varieties hav
0141  ties having different innate constitutions, and  HOW  much to both means combined, is a very obscure
0142  ever appear, for an account has been published  HOW  much more hardy some seedlings appeared to be
0166  hich have bred true for centuries, species: and  HOW  exactly parallel is the case with that of the h
0172  scoveries of profound mathematicians? Fourthly,  HOW  can we account for species, when crossed, bein
0175  mountains, and sometimes it is quite remarkable  HOW  abruptly, as Alph. de Candelle has observed, a
0179  asked by the opponents of such views as I hold,  HOW, for instance, a land carnivorous animal could
0179  een converted into one with aquatic habits: for  HOW  could the animal in its transitional state hav
0180  rent case had been taken, and it had been asked  HOW  an insectivorous quadruped could possibly have
0187  our imagination, can hardly be considered real.  HOW  a nerve comes to be sensitive to light, hardly
0187  ensitive to light, hardly concerns us more than  HOW  life itself first originated: but I may remark
0188  eyes of living crustaceans, and bearing in mind  HOW  small the number of living animals is in propo
0194  d unknown is very small, I have been astonished  HOW  rarely an organ can be named, towards which no
0196  have been checked by natural selection. Seeing  HOW  important an organ of locomotion the tail is i
0202  equently intensified, we can perhaps understand  HOW  it is that the use of the sting should so ofte
0204  rminating it. We have have seen in this chapter  HOW  cautious we should be in concluding that the m
0204  c being is trying to live wherever it can live,  HOW  it has arisen that there are upland geese with
0208  ive action is performed, but not of its origin.  HOW  unconsciously many habitual actions are perfor
0210  o instinct being known amongst extinct species,  HOW  very generally gradations, leading to the most
0213  er period, under less fixed conditions of life.  HOW  stongly these domestic instincts, habits, and
```

Page ***(Key Word)***

Page **(Key Word)**

0214	ts, habits, and dispositions are inherited, and	HOW	curiously they become mingled, is well shown w
0215	eir eggs. Familiarity alone prevents our seeing	HOW	universally and largely the minds of our domes
0215	the savages do not keep these domestic animals.	HOW	rarely, on the other hand, do our civilised do
0216	ed together. We shall, perhaps, best understand	HOW	instincts in a state of nature have become mod
0219	it would have been hopeless to have speculated	HOW	so wonderful an instinct could have been perfe
0224	ease, and it seems at first quite inconceivable	HOW	they can make all the necessary angles and pla
0229	e planes or planes of intersection. Considering	HOW	flexible thin wax is, I do not see that there
0230	i will not here enter on these details. We see	HOW	important a part excavation plays in the const
0231	ch will ultimately be left. We shall understand	HOW	they work, by supposing masons first to pile u
0231	first to add to the difficulty of understanding	HOW	the cells are made, that a multitude of bees a
0232	ty, as when two pieces of comb met at an angle,	HOW	often the bees would entirely pull down and re
0233	conditions of life, it may reasonably be asked,	HOW	a long and graduated succession of modified ar
0235	atural selection, cases, in which we cannot see	HOW	an instinct could possibly have originated: ca
0236	a single case, that of working or sterile ants.	HOW	the workers have been rendered sterile is a di
0237	r instinct to its progeny. It may well be asked	HOW	is it possible to reconcile this case with the
0237	mmunities: the difficulty lies in understanding	HOW	such correlated modifications of structure cou
0239	ei and this we do find, even often, considering	HOW	few neuter insects out of Europe have been car
0239	been carefully examined. Mr. F. Smith has shown	HOW	surprisingly the neuters of several British in
0241	from their parents, has originated. We can see	HOW	useful their production may have been to a soc
0243	can understand on the principle of inheritance,	HOW	it is that the thrush of South America lines i
0243	ame peculiar manner as does our British thrush:	HOW	it is that the male wrens (Troglodytes) of Nor
0251	cision, deserve some notice. It is notorious in	HOW	complicated a manner the species of Pelargoniu
0255	of plants. I have taken much pains to ascertain	HOW	far the rules apply to animals, and considerin
0255	far the rules apply to animals, and considering	HOW	scanty our knowledge is in regard to hybrid an
0255	o hybrid animals, I have been surprised to find	HOW	generally the same rules apply to both kingdom
0255	zero to perfect fertility. It is surprising how	HOW	many curious ways this gradation can be shown
0259	siderable degree of fertility. These facts show	HOW	completely fertility in the hybrid is independ
0268	emarkable fact; more especially when we reflect	HOW	many species there are, which, though resembli
0270	the less easy as their differences are greater.	HOW	far these experiments may be trusted, I know n
0277	perfect fertility surprising, when we remember	HOW	liable we are to argue in a circle with respec
0282	volution in natural science, yet does not admit	HOW	imcomprehensibly vast have been the past perio
0282	t observers on separate formations, and to mark	HOW	each author attempts to give an inadequate ide
0283	quickly ground into pebbles, sand, or mud. But	HOW	often do we see along the bases of retreating
0283	thickly clothed by marine productions, showing	HOW	little they are abraded and how seldom they ar
0283	ctions, showing how little they are abraded and	HOW	seldom they are rolled about! Moreover, if we
0283	which bears the stamp of time, are good to show	HOW	slowly the mass has been accumulated. Let him
0285	their formerly liquid state, showed at a glance	HOW	far the hard, rocky beds had once extended int
0289	t to Lyell's Manual, will bring home the truth,	HOW	accidental and rare is their preservation, far
0289	or is their rarity surprising, when we remember	HOW	large a proportion of the bones of tertiary ma
0291	ess of Geology, and have been surprised to note	HOW	author after author, in treating of this or th
0293	e that it then became wholly extinct. We forget	HOW	small the area of Europe is compared with the
0300	es were to be collected which ever lived there,	HOW	imperfectly would they represent the natural h
0302	o not pretend that I should ever have suspected	HOW	poor a record of the mutations of life, the be
0302	exist before that stage. We continually forget	HOW	large the world is, compared with the area ove
0303	amples to illustrate these remarks; and to show	HOW	liable we are to error in supposing that whole
0310	entertains grave doubts on this subject. I feel	HOW	rash it is to differ from these great authorit
0319	onditions of life apparently so favourable. But	HOW	utterly groundless was my astonishment! Profes
0325	resent distribution of organic beings, and find	HOW	slight is the relation between the physical co
0329	s from our great palaeontologist, Owen, showing	HOW	extinct animals fall in between existing group
0331	t there is some truth in the remark. Let us see	HOW	far these several facts and inferences accord
0336	ich includes the whole glacial period, and note	HOW	little the specific forms of the inhabitants o
0343	een our consecutive formations; he may overlook	HOW	important a part migration must have played, w
0343	rough natural selection. We can thus understand	HOW	it is that new species come in slowly and succ
0343	at new species come in slowly and successively:	HOW	species of different classes do not necessaril
0344	f generation has been broken. We can understand	HOW	the spreading of the dominant forms of life, w
0344	have changed simultaneously. We can understand	HOW	it is that all the forms of life, ancient and
0347	sm in the conditions of the Old and New Worlds,	HOW	widely different are their living productions!
0350	nheritance with modification, we can understand	HOW	it is that sections of genera, whole genera, a
0352	y cases of extreme difficulty, in understanding	HOW	the same species could possibly have migrated
0353	ly many cases occur, in which we cannot explain	HOW	the same species could have passed from one po
0358	s aid, a few experiments, it was not even known	HOW	far seeds could resist the injurious action of
0361	ation of seeds. I could give many facts showing	HOW	frequently birds of many kinds are blown by ga
0364	s in itself a rare accident. Even in this case,	HOW	small would the chance be of a seed falling on
0369	rica and Europe: and it may be reasonably asked	HOW	I account for the necessary degree of uniformi
0377	pical plants probably suffered much extinction:	HOW	much no one can say; perhaps formerly the trop
0385	as are the adults. I could not even understand	HOW	some naturalised species have rapidly spread t
0386	forty five miles distant from the nearest land:	HOW	much farther it might have flown with a favour
0386	unts. I do not believe that botanists are aware	HOW	charged the mud of ponds is with seeds! I have
0391	fined to its shores: now, though we do not know	HOW	sea shells are dispersed, yet we can see that
0396	re many and grave difficulties in understanding	HOW	several of the inhabitants of the more remote
0400	argument against my views; for it may be asked,	HOW	has it happened in the several islands situate
0400	be here fairly included, as we are considering	HOW	they have come to be modified since their arri
0407	occurred within the same period: if we remember	HOW	profoundly ignorant we are with respect to the
0407	properly experimentised on: if we bear in mind	HOW	often a species may have ranged continuously o
0407	anges on distribution, I have attempted to show	HOW	important has been the influence of the modern
0407	or at least great meridional belts. As showing	HOW	diversified are the means of occasional transp
0408	isation of sub genera, genera and families: and	HOW	it is that under different latitudes, for inst
0411	widely different in nature; for it is notorious	HOW	commonly members of even the same sub group ha
0414	ive for an essential character. So with plants,	HOW	remarkable it is that the organs of vegetation
0424	grade, or that of a species, the two sexes: and	HOW	enormously these sometimes differ in the most
0426	n, other parts of the organisation. We care not	HOW	trifling a character may be, let it be the mer
0428	land, air, and water, we can perhaps understand	HOW	it is that a numerical parallelism has sometim
0429	and all in one great natural system. As showing	HOW	few the higher groups are in number, and how w
0429	no how few the higher groups are in number, and	HOW	widely spread they are throughout the world, t
0429	ing the long continued process of modification,	HOW	it is that the more ancient forms of life ofte
0433	he limits of the same group. We can clearly see	HOW	it is that all living and extinct forms can be

Page **(Key Word)**

Page ************************************(Key Word)***

0433 an be grouped together in one great system; and HOW the several members of each class are connect
0437 are at an early stage of growth exactly alike. HOW inexplicable are these facts on the ordinary
0442 insect, we see no trace of the vermiform stage. HOW, then, can we explain these several facts in
0445 the same wild stock; hence I was curious to see HOW far their puppies differed from each other: I
0449 natural system. On this view we can understand HOW it is that, in the eyes of most naturalists,
0451 r than that wings are formed for flight, yet in HOW many insects co we see wings so reduced in si
0455 nd, on the genealogical view of classification, HOW it is that systematists have found rudimentar
0457 provide for its own wants, and bearing in mind HOW strong is the principle of inheritance, the o
0459 general belief in the immutability of species. HOW far the theory of natural selection may be ex
0460 community of ants; but I have attempted to show HOW this difficulty can be mastered. With respect
0462 . we are often wholly unable even to conjecture HOW this could have been effected. Yet, as we have
0462 ration. As an example, I have attempted to show HOW potent has been the influence of the Glacial
0466 ich we are confessedly ignorant; nor do we know HOW ignorant we are. We do not know all the possib
0466 during the long lapse of years, or that we know HOW imperfect the Geological Record is. Grave as
0466 life. There is much difficulty in ascertaining HOW much modification our domestic productions hav
0469 nd sub species, and species. Let it be observed HOW naturalists differ in the rank which they ass
0469 ndary laws. On this same view we can understand HOW it is that in each region where many species
0471 , as it seems to me, explicable on this theory. HOW strange it is that a bird, under the form of w
0473 ecies reversions to long lost characters occur. HOW inexplicable on the theory of creation is the
0473 pecies of the horse genus and in their hybrids! HOW simply is this fact explained if we believe th
0474 eir several instincts. I have attempted to show HOW much light the principle of gradation throws e
0474 ing inherited much in common, we can understand HOW it is that allied species, when placed under e
0478 e of character. On these same principles we see HOW it is, that the mutual affinities of the speci
0480 h separate organ having been specially created, HOW utterly inexplicable it is that parts, like th
0483 they consider reverent silence. It may be asked HOW far I extend the doctrine of the modification
0486 workmen; when we thus view each organic being, HOW far more interesting, I speak from experience e
0002 e of them personally unknown to me. I cannot, HOWEVER, let this opportunity pass without expressir
0004 by giving long catalogues of facts. We shall, HOWEVER, be enabled to discuss what circumstances ar
0005 stence, it follows that any being, if it vary HOWEVER slightly in any manner profitable to itself,
0015 he cabbage, in very poor soil (in which case, HOWEVER, some effect would have to be attributed to
0035 ch might serve for comparison. In some cases, HOWEVER, unchanged or but little charged individuals
0039 d it is in human nature to value any novelty, HOWEVER slight, in one's own possession. Nor must th
0041 slight varieties had been neglected. As soon, HOWEVER, as gardeners picked out individual plants w
0047 ys remove the difficulty. In very many cases, HOWEVER, one form is ranked as a variety of another,
0047 ence seems the only guide to follow. We must, HOWEVER, in many cases, decide by a majority of natu
0053 e fairly well established. The whole subject, HOWEVER, treated as it necessarily here is with much
0061 ing to this struggle for life, any variation, HOWEVER slight and from whatever cause proceeding, i
0081 rvive) that individuals having any advantage, HOWEVER slight, over others, would have the best cha
0085 produce some slight and direct effect. It is, HOWEVER, far more necessary to bear in mind that the
0105 ortance in the production of new species. If, HOWEVER, an isolated area be very small, either from
0121 d by later and improved lines of descent. If, HOWEVER, the modified offspring of a species get int
0132 e some cause for each deviation of structure, HOWEVER slight. How much direct effect difference of
0134 ll then accumulate all profitable variations, HOWEVER slight, until they become plainly developed
0141 bear widely different climates. We must not, HOWEVER, push the foregoing argument too far, on acc
0147 botanists, believe in its truth. I will not, HOWEVER, here give any instances, for I see hardly a
0148 e useful becomes less useful, any diminution, HOWEVER slight, in its development, will be seized o
0150 hat the rule by no means applies to any part, HOWEVER unusually developed, unless it be unusually
0154 see no reason to doubt. Hence when an organ, HOWEVER abnormal it may be, has been transmitted in
0154 s partly, yet only indirectly, true; I shall, HOWEVER, have to return to this subject in our chapt
0158 he not great degree of variability in a part, HOWEVER extraordinarily it may be developed, if it b
0159 osely related acts of creation. With pigeons, HOWEVER, we have another case, namely, the occasiona
0161 reversions to lost ancestral characters. As, HOWEVER, we never know the exact character of the co
0163 ccur, and seem to me very remarkable. I will, HOWEVER, give one curious and complex case, not inde
0169 in giving a fixed character to the organ, in HOWEVER extraordinary a a manner it may be developed
0175 as these species are already defined objects (HOWEVER they may have become so), not blending one i
0186 t will seize on the place of that inhabitant, HOWEVER different it may be from its own place. Henc
0208 , fifth, and sixth stages of construction. If HOWEVER, a caterpillar were taken out of a hammock m
0218 ntervals of two or three days. This instinct, HOWEVER, of the American ostrich has not as yet been
0220 d food of all kinds. During the present year, HOWEVER, in the month of July, I came across a commu
0222 desolated nest in the thick heath. The nest, HOWEVER, must have been close at hand, for two or th
0228 each other from the opposite sides. The bees, HOWEVER, did not suffer this to happen, and they sto
0245 sive profitable degrees of sterility. I hope, HOWEVER, to be able to show that sterility is not a
0252 as thoroughly well authenticated. It should, HOWEVER, be borne in mind that, owing to few animals
0253 n of the pure geese) from one nest. In India, HOWEVER, these cross bred geese must be far more fer
0264 see in the case of the common mule. Hybrids, HOWEVER, are differently circumstanced before and af
0266 degree variable, rarely diminishes. It must, HOWEVER, be confessed that we cannot understand, exc
0268 the foregoing remarks, inasmuch as varieties, HOWEVER much they may differ from each other in exte
0268 le when intercrossed. Several considerations, HOWEVER, render the fertility of domestic varieties
0273 of character could be given. The variability, HOWEVER, in the successive generations of mongrels i
0279 modification and improvement. The main cause, HOWEVER, of innumerable intermediate links not now o
0287 highly important for us to gain some notion, HOWEVER imperfect, of the lapse of years. During eac
0303 as the eocene stage. The most striking case, HOWEVER, is that of the Whale family; as these anima
0305 ectly known, are really teleostean. Assuming, HOWEVER, that the whole of them did appear, as Agass
0318 ase in numbers of the species. In some cases, HOWEVER, the extermination of whole groups of beings
0319 verally acted. If the conditions had gone on, HOWEVER slowly, becoming less and less favourable, w
0323 uld be easily correlated. These observations, HOWEVER, relate to the marine inhabitants of distant
0331 differ most from existing forms. We must not, HOWEVER, assume that divergence of character is a ne
0332 ould yet remain distinct. These two families, HOWEVER, would be less distinct from each other than
0334 n the forms of life above and below. We must, HOWEVER, allow for the entire extinction of some pre
0349 e look to the islands off the American shore, HOWEVER much they may differ in geological structure
0356 or migration, but now be impassable; I shall, HOWEVER, presently have to discuss this branch of th
0362 ed their power of germination. Certain seeds, HOWEVER, were always killed by this process. Althoug
0364 he crops or intestines of birds. These means, HOWEVER, would suffice for occasional transport acro
0379 south, than in a reversed direction. We see, HOWEVER, a few southern vegetable forms on the mount
0388 life between the individuals of the species, HOWEVER few, already occupying any pond, yet as the
0393 n, and I have found it strictly true. I have, HOWEVER, been assured that a frog exists on the moun

Page ************************************(Key Word)***

Page **(Key Word)***

```
0401  eties in the different islands. Some species,  HOWEVER, might spread and yet retain the same charac
0409  robably derived. We can see why in two areas,  HOWEVER distant from each other, there should be a c
0414  to resemblances in parts of the organisation,  HOWEVER important they may be for the welfare of the
0417  assification founded on any single character,  HOWEVER important that may be, has always failed; fo
0421  f descendants, at each successive period. If,  HOWEVER, we choose to suppose that any of the descen
0424  nk he would be classed under the Negro group,  HOWEVER much he might differ in colour and other imp
0424  several larval stages of the same individual,  HOWEVER much they may differ from each other and fro
0430  t the general order of Rodents. In this case,  HOWEVER, it may be strongly suspected that the resem
0433  e descent in classing acknowledged varieties,  HOWEVER different they may be from their parent; and
0438  r, but from some common element. Naturalists,  HOWEVER, use such language only in a metaphorical se
0440  nditions to which they are exposed. The case,  HOWEVER, is different when an animal during any part
0440  r than do their adult parents. In most cases,  HOWEVER, the larvae, though active, still obey more
0441  s higher than the caterpillar. In some cases,  HOWEVER, the mature animal is generally considered a
0448  eir similar habits. Some further explanation,  HOWEVER, of the embryo not undergoing any metamorpho
0449  e of its progenitor. In two groups of animal,  HOWEVER much they may at present differ from each ot
0449  nt. It will reveal this community of descent,  HOWEVER much the structure of the adult may have bee
0456  d acknowledged varieties of the same species,  HOWEVER different they may be in structure. If we ex
0461  cended from common parents; and therefore, in  HOWEVER distant and isolated parts of the world they
0468  mes into competition, or better adaptation in  HOWEVER slight a degree to the surrounding physical
0479  of descent by the most permanent characters,  HOWEVER slight their vital importance may be. The fr
0485  ation than it is at present; for differences,  HOWEVER slight, between any two forms, if not blende
0488  he lapse of actual time. A number of species,  HOWEVER, keeping in a body might remain for a long p
0208  stinct are universal. A little dose, as Pierre HUBER expresses it, of judgment or reason, often co
0208  o recover the habitual train of thought: so P. HUBER found it was with a caterpillar, which makes
0219  in the Formica (Polyerges) rufescens by Pierre HUBER, a better observer even than his celebrated f
0219  so utterly helpless are the masters, that when HUBER shut up thirty of them without a slave, but w
0219  feed themselves, and many perished of hunger.  HUBER then introduced a single slave (F. fusca), an
0219  sanguinea was likewise first discovered by P.  HUBER to be a slave making ant. This species is fou
0220  . although fully trusting to the statements of HUBER and Mr. Smith, I tried to approach the subjec
0221  ly in search of aphides or cocci. According to HUBER, who had ample opportunities for observation,
0221  e the doors in the morning and evening; and as HUBER expressly states, their principal office is t
0221  e to behold the masters carefully carrying, as HUBER has described, their slaves in their jaws. An
0225  ica, carefully described and figured by Pierre HUBER. The Melipona itself is intermediate in struc
0226  re united into a pyramid; and this pyramid, as HUBER has remarked, is manifestly a gross imitation
0230  from those made by the justly celebrated elder HUBER, but I am convinced of their accuracy; and if
0230  show that they are conformable with my theory. HUBER's statement that the very first cell is excav
0231  time at one cell going to another, so that, as HUBER has stated, a score of individuals work even
0088  ; male stag beetles often bear wounds from the HUGE mandibles of other males. The war is, perhaps,
0283  . at last the base of the cliff is undermined, HUGE fragments fall down, and these remaining fixed
0303  hat of the Whale family; as these animals have HUGE bones, are marine, and range over the world, t
0341  suppose that the megatherium and other allied  HUGE monsters have left behind them in South Americ
0341  this cannot for an instant be admitted. These  HUGE animals have become wholly extinct, and have l
0373  clearest evidence of former glacial action, in HUGE boulders transported far from their parent sou
0283  re worn away. The observations on this head by HUGH Miller, and by that excellent observer Mr. Smi
0039  ives extremely small differences, and it is in HUMAN nature to value any novelty, however slight,
0144  rs believe that the shape of the pelvis in the HUMAN mother influences by pressure the shape of th
0188  d by the long continued efforts of the highest HUMAN intellects; and we naturally infer that the a
0459  t by means superior to, though analogous with, HUMAN reason, but by the accumulation of innumerabl
0464  so great as to be utterly inappreciable by the HUMAN intellect. The number of specimens in all our
0073  se them. I have, also, reason to believe that  HUMBLE bees are indispensable to the fertilisation o
0073  cial to the fertilisation of our clovers; but  HUMBLE bees alone visit the common red clover (Trifo
0073  very little doubt, that if the whole genus of  HUMBLE bees became extinct or very rare in England,
0074  very rare, or wholly disappear. The number of  HUMBLE bees in any district depends in a great degre
0074  ewman, who has long attended to the habits of  HUMBLE bees, believes that more than two thirds of t
0074  ges and small towns I have found the nests of  HUMBLE bees more numerous than elsewhere, which I at
0095  of the common rec clover, which is visited by  HUMBLE bees alone: so that whole fields of the red c
0095  on to the stigmatic surface. Hence, again, if  HUMBLE bees were to become rare in any country, it m
0225  of work. At one end of a short series we have  HUMBLE bees, which use their old cocoons to hold hon
0225  e hive bee and the simplicity of those of the  HUMBLE bee, we have the cells of the Mexican Melipon
0225  ntermediate in structure between the hive and  HUMBLE bee, but more nearly related to the latter: i
0227  e stock of honey: in the same way as the rude  HUMBLE bee adds cylinders of wax to the circular mou
0234  obably often does determine, the numbers of a  HUMBLE bee which could exist in a country; and let u
0234  no doubt that it would be an advantage to our  HUMBLE bee, if a slight modification of her instinct
0234  tinually be more and more advantageous to our  HUMBLE bee, if she were to make her cells more and m
0060  letoe; and only a little less plainly in the   HUMBLEST parasite which clings to the hairs of a quad
0374  so on the Silla of Caraccas the illustrious   HUMBOLDT long ago found species belonging to genera c
0346  with the most diversified conditions: the most HUMID districts, arid deserts, lofty mountains, gra
0378  erate climate. On the other hand, the the most HUMID and hottest districts will have afforded an an
0403  wlands; thus we have in South America, Alpine  HUMMING birds, Alpine rodents, Alpine plants, etc.,
0018  habits, voice, and constitution, etc., of the HUMPED Indian cattle, that these had descended from
0254  s reason to believe that our European and the HUMPED Indian cattle are quite fertile together: but
0060  es; what rank, for instance, the two or three HUNDRED doubtful forms of British plants are entitle
0071  been touched by the hand of man; but several  HUNDRED acres of exactly the same nature had been en
0072  e cattle. In one square yard, at a point some HUNDRED yards distant from one of the old clumps, I
0124  ations, but each may represent a million or a HUNDRED million generations, and likewise a section
0160  fspring suddenly takes after an ancestor some HUNDRED generations distant, but that in each succes
0167  riation is profound. Not in one case out of a HUNDRED can we pretend to assign any reason why this
0258  ly fertile; but Kolreuter tried more than two HUNDRED times, during eight following years, to fert
0284  sippi river at the rate of only 600 feet in a HUNDRED thousand years. This estimate may be quite a
0287  have required 306,662,400 years; or say three HUNDRED million years. The action of fresh water on
0290  y any fact struck me more than when examining many HUNDRED miles of the South American coasts, which ha
0290  ican coasts, which have been upraised several HUNDRED feet within the recent period, than the abse
0324  expression relates to the same thousandth or HUNDRED thousandth year, or even that it has a very
0364  ccasional transport across tracts of sea some HUNDRED miles in breadth, or from island to island,
0385  duck or heron might fly at least six or seven HUNDRED miles, and would be sure to alight on a pool
0391  easily than land shells, across three or four HUNDRED miles of open sea. The different orders of i
```

Page **(Key Word)***

Page ***(Key Word)***

```
0398 canic islands in the Pacific, distant several HUNDRED miles from the continent, yet feels that he
0481 sibly grasp the full meaning of the term of a HUNDRED million years; it cannot add up and perceive
0066 numerous bird in the world. One fly ceposits HUNDREDS of eggs, and another, like the hippobosca, a
0068 lity, he less game than at present, although HUNDREDS of thousands of game animals are now annuall
0072 veral points of view, whence I could examine HUNDREDS of acres of the unenclosed heath, and litera
0160 after having been lost for many, perhaps for HUNDREDS of generations. But when a breed has been cr
0354 nds and on the mainland, though separated by HUNDREDS of miles of open sea. If the existence of th
0355 upheaved and formed at the distance of a few HUNDREDS of miles from a continent, would probably re
0365 untain summits, separated from each other by HUNDREDS of miles of low lands, where the Alpine spec
0231 xagonal walls, which are only about one four HUNDREDTH of an inch in thickness; the plates of the
0019 those of Germany and conversely, and so with HUNGARY, Spain, etc., but that each of these kingdom
0219 ot even feed themselves, and many perished of HUNGER. Huber then introduced a single slave (F. fus
0077 s it to compete with other aauatic irsects, to HUNT for its own prey, and to escape serving as pre
0091 ing the lowlands, would naturally be forced to HUNT different prey; and from the continued preserv
0214 a whole family of shepherd dogs a tendency to HUNT hares. These domestic instincts, when thus tes
0215 o improve the breed, dogs which will stand and HUNT best. On the other hand, habit alone in some c
0150 he term, secondary sexual characters, used by HUNTER, applies to characters which are attached to
0091 d game, another hares or rabbits, and another HUNTING on marshy ground and almost nightly catching
0218 strewed over the plains, so that in one day's HUNTING I picked up no less than twenty lost and was
0069 r species being favoured, as in this one being HURT. So it is when we travel northward, but in a s
0009 (for instance, the rabbit and ferret kept in HUTCHES), showing that their reproductive system has
0253 ly capable judges, namely Mr. Blyth and Capt. HUTTON, that whole flocks of these crossed geese are
0101 of the highest authorities, namely, Professor HUXLEY, to discover a single case of an hermaphrodit
0338 ent of recent forms. I must follow Pictet and HUXLEY in thinking that the truth of this doctrine i
0438 cases probably be more correct, as Professor HUXLEY has remarked, to speak of both skull and vert
0442 e look to the admirable drawings by Professor HUXLEY of the development of this insect, we see no
0012 some of our old cultivated plants, as on the HYACINTH, potato, even the dahlia, etc.; and it is re
0397 s the Helix pomatia, and after it had again HYBERNATED I put it in sea water for twenty days, and
0397 d? it occurred to me that land shells, when HYBERNATING and having a membranous diaphragm over the
0026 impossible, to bring forward one case of the HYBRID offspring of two animals clearly distinct bei
0047 rmediate links; nor will the commonly assumed HYBRID nature of the intermediate links always remov
0165 ble shoulder stripe. In Lord Moreton's famous HYBRID from a chestnut mare and male quagga, the hyb
0165 rid from a chestnut mare and male quagga, the HYBRID, and even the pure offspring subsequently pro
0165 , and this is another most remarkable case, a HYBRID has been figured by Dr. Gray (and he informs
0165 case) from the ass and the hemionus; and this HYBRID, though the ass seldom has stripes on its leg
0165 rom the occurence of the face stripes on this HYBRID from the ass and hemionus, to ask Colonel Poo
0246 terility of species when crossed and of their HYBRID offspring. It is impossible to study the seve
0247 uced by two species when crossed and by their HYBRID offspring, with the average number produced b
0249 to their fertility, already lessened by their HYBRID origin. I am strengthened in this conviction
0249 rtile hybrids be artificially fertilised with HYBRID pollen of the same kind, their fertility, not
0249 same plant cr from another plant of the same HYBRID nature. And thus, the strange fact of the inc
0250 uently fertilised by the pollen of a compound HYBRID descended from three other and distinct speci
0251 reas the pod impregnated by the pollen of the HYBRID made vigorous growth and rapic progress to ma
0251 freely. For instance, Herbert asserts that a HYBRID from Calceolaria integrifolia and plantaginea
0251 me that he raises stocks for grafting from a HYBRID between Rhod. Ponticum and Catawbiense, and t
0252 rhod. Ponticum and Catawbiense, and that this HYBRID seeds as freely as it is possible to imagine.
0252 ct agency the several individuals of the same HYBRID variety are allowed to freely cross with each
0252 ning the flowers of the more sterile kinds of HYBRID rhododendrons, which produce no pollen, for h
0252 doubt whether any case of a perfectly fertile HYBRID animal can be considered as thoroughly well a
0253 in successive generations of the more fertile HYBRID animals, I hardly know of an instance in whic
0253 an instance in which two families of the same HYBRID have been raised at the same time from differ
0253 well authenticated cases of perfectly fertile HYBRID animals, I have some reason to believe that t
0255 ring how scanty our knowledge is in regard to HYBRID animals, I have been surprised to find how ge
0255 pure parent species causing the flower of the HYBRID to wither earlier than it otherwise would hav
0256 d with unusual facility, and produce numerous HYBRID offspring, yet these hybrids are remarkably s
0257 duce between extremely close species a single HYBRID. Even within the limits of the same genus, we
0259 e power of impressing their likeness on their HYBRID offspring; but these two powers do not at all
0259 se facts show how completely fertility in the HYBRID is independent of its external resemblance to
0264 aced under suitable conditions of life. But a HYBRID partakes of only half of the nature and const
0270 ecies; and it is important to notice that the HYBRID plants thus raised were themselves perfectly
0272 unimportant cifferences between the so called HYBRID offspring of species, and the so called mongr
0274 , which Gartner is able to point out, between HYBRID and mongrel plants. On the other hand, the re
0274 otent power of impressing its likeness on the HYBRID; and so I believe it to be with varieties of
0274 as this prepotent power over another variety. HYBRID plants produced from a reciprocal cross, gene
0276 s ecual in cegree in a first cross and in the HYBRID produced from this cross. In the same manner
0277 species, and the degree of sterility of their HYBRID offspring should generally correspond, though
0451 te species, the rudiment of the pistil in the HYBRID offspring was much increased in size; and thi
0255 wn up from Gartner's admirable work on the HYBRIDISATION of plants. I have taken much pains to ac
0261 s prevent the two grafting together. As in HYBRIDISATION, so with grafting, the capacity is limite
0261 rafted with ease. But this capacity, as in HYBRIDISATION, is by no means absolutely governed by sy
0247 to me to be here introduced: a plant to be HYBRIDISED must be castrated, and, what is often more
0250 ividuals of certain species can actually be HYBRIDISED much more readily than they can be self fer
0249 ts arrived at by the third most experienced HYBRIDISER, namely, the Hon. and Rev. W. Herbert. He i
0248 vidence from fertility adduced by different HYBRIDISERS, or by the same author, from experiments m
0264 mr. Hewitt, who has had great experience in HYBRIDISING gallinaceous birds, that the early death o
0005 t, or the mental powers of animals; thirdly, HYBRIDISM, or the infertility of species and the fert
0049 raphical distribution, analogical variation, HYBRIDISM, etc., have been brought to bear on the att
0172 heads shall be here discussed, Instinct and HYBRIDISM in separate chapters. On the absence or rar
0245 gest live and the weakest die. Chapter VIII. HYBRIDISM. Distinction between the sterility of first
0097 h on any other view are inexplicable. Every HYBRIDIZER knows how unfavourable exposure to wet is t
0009 e same exact condition as in the most sterile HYBRINS. When, on the one hand, we see domesticated
0026 nded in treatises on inheritance. Lastly, the HYBRIDS or mongrels from between all the domestic br
0043 te disregards the extreme variability both of HYBRIDS and mongrels, and the frequent sterility of
0043 s and mongrels, and the frequent sterility of HYBRIDS; but the cases of plants not propagated by s
0050 d it is very doubtful whether these links are HYBRIDS; and there is, as it seems to me, an overwhe
0098 d that bees would thus produce a multitude of HYBRIDS between distinct species; for if you bring o
```

Page ***(Key Word)***

Page ***(Key Word)***

```
0165  four coloured drawings, which I have seen, of  HYBRIDS between the ass and zebra, the legs were muc
0166   to become striped most strongly displayed in  HYBRIDS from between several of the most distinct sp
0167  ing distant cuarters of the world, to produce  HYBRIDS resembling in their stripes, not their own p
0245  between the sterility of first crosses and of  HYBRIDS. Sterility various in degree, not universal,
0245  omestication. Laws governing the sterility of  HYBRIDS. Sterility not a special endowment, but inci
0245  uses of the sterility of first crosses and of  HYBRIDS. Parallelism between the effects of changed
0245  and of their mongrel offspring not universal.  HYBRIDS and mongrels compared independently of their
0245  ssirc freely. The importance of the fact that  HYBRIDS are very generally sterile, has, I think, be
0246  ially important, inasmuch as the sterility of  HYBRIDS could not possibly be of any advantage to th
0246   when first crossed, and the sterility of the  HYBRIDS produced from them. Pure species have of cou
0246  ssec they produce either few or no offspring.  HYBRIDS, on the other hand, have their reproductive
0248  al differences. In regard to the sterility of  HYBRIDS in successive generations: though Gartner wa
0248  ions; though Gartner was enabled to rear some  HYBRIDS, carefully guarding them from a cross with e
0249  his almost universal belief amongst breeders.  HYBRIDS are seldom raised by experimentalists in gre
0249  s; and as the parent species, or other allied  HYBRIDS, generally grow in the same garden, the visi
0249   prevented during the flowering season: hence  HYBRIDS will generally be fertilised during each gen
0249  artner, namely, that if even the less fertile  HYBRIDS be artificially fertilised with hybrid polle
0249   observer as Gartner would have castrated his  HYBRIDS, and this would have insured in each generat
0249  essive generations of artificially fertilised  HYBRIDS may, I believe, be accounted for by close in
0250  he is as emphatic in his conclusion that some  HYBRIDS are perfectly fertile, as fertile as the pur
0251  n, etc., have been crossed, yet many of these  HYBRIDS seed freely. For instance, Herbert asserts t
0252  s as freely as it is possible to imagine. Had  HYBRIDS, when fairly treated, gone on decreasing in
0252  horticulturists raise large beds of the same  HYBRIDS, and such alone are fairly treated, for by i
0252  y crossed than in the case of plants; but the  HYBRIDS themselves are, I think, more sterile. I dou
0252  es between them and the canary, or that their  HYBRIDS, should be perfectly fertile. Again, with re
0253  surprising that the inherent sterility in the  HYBRIDS should have gone on increasing. If we were t
0253  imals, I have some reason to believe that the  HYBRIDS from Cervulus vaginalis and Reevesii, and fr
0253  with P. versicolor are perfectly fertile. The  HYBRIDS from the common and Chinese geese (A. cygnoi
0253  his was effected by Mr. Eyton, who raised two  HYBRIDS from the same parents but from different hat
0253   these two birds he raised no less than eight  HYBRIDS (grandchildren of the pure geese) from one n
0254  t either at first have produced cuite fertile  HYBRIDS, or the hybrids must have become in subseque
0254  t have produced cuite fertile hybrids, or the  HYBRIDS must have become in subsequent generations q
0254  bred together and have produced cuite fertile  HYBRIDS. So again there is reason to believe that ou
0254  ee of sterility, both in first crosses and in  HYBRIDS, is an extremely general result: but that it
0254  verning the Sterility of first Crosses and of  HYBRIDS. We will now consider a little more in detai
0255  verning the sterility of first crosses and of  HYBRIDS. Our chief object will be to see whether or
0255  ee of fertility, both of first crosses and of  HYBRIDS, graduates from zero to perfect fertility. I
0255  ch the plant's own pollen will produce. So in  HYBRIDS themselves, there are some which never have
0256  e degree of sterility we have self fertilised  HYBRIDS producing a greater and greater number of se
0256  ater number of seeds up to perfect fertility.  HYBRIDS from two species which are very difficult to
0256  aking a first cross, and the sterility of the  HYBRIDS thus produced, two classes of facts which ar
0256   procuce numerous hybrid offspring, yet these  HYBRIDS are remarkably sterile. On the other hand, t
0256  y rarely, or with extreme difficulty, but the  HYBRIDS, when at last produced, are very fertile. Ev
0256   . the fertility, both of first crosses and of  HYBRIDS, is more easily affected by unfavourable con
0256  been chosen for the experiment. So it is with  HYBRIDS, for their degree of fertility is often foun
0257  of first crcsses between species, and of the  HYBRIDS produced from them, is largely governed by t
0257  systematic affinity. This is clearly shown by  HYBRIDS never having been raised between species ran
0258  lised by the pollen of M. Longiflora, and the  HYBRIDS thus produced are sufficiently fertile; but
0258  varieties. It is also a remarkable fact, that  HYBRIDS raised from reciprocal crosses, though of co
0259  ll necessarily go together. There are certain  HYBRIDS which instead of having, as is usual, an int
0259  always closely resemble one of them; and such  HYBRIDS, though externally so like one of their pure
0259  xceptions extremely sterile. So again amongst  HYBRIDS which are usually intermediate in structure
0259  resemble one of their pure parents; and these  HYBRIDS are almost always utterly sterile, even when
0259  t always utterly sterile, even when the other  HYBRIDS raised from seed from the same capsule have
0259  govern the fertility of first crosses and of  HYBRIDS, we see that when forms, which must be consi
0259  same in degree in the first cross and in the  HYBRIDS produced from this cross. That the fertility
0259  oduced from this cross. That the fertility of  HYBRIDS is not related to the degree in which they r
0260  e, in the facility of effecting an union. The  HYBRIDS, moreover, produced from reciprocal crosses
0260  s with facility, and yet produce very sterile  HYBRIDS; and other species cross with extreme diffic
0260  me difficulty, and yet produce fairly fertile  HYBRIDS? Why should their often be so great a differ
0260  , it may even be asked, has the production of  HYBRIDS been permitted? to grant to species the spec
0260  ant to species the special power of producing  HYBRIDS, and then to stop their further propagation
0260  at the sterility both of first crosses and of  HYBRIDS is simply incidental or dependent on unknown
0262  ooseberry. We have seen that the sterility of  HYBRIDS, which have their reproductive organs in an
0263  uses of the Sterility of first Crosses and of  HYBRIDS. We may now look a little closer at the prob
0263  uses of the sterility of first crosses and of  HYBRIDS. These two cases are fundamentally different
0263  emale sexual elements are perfect; whereas in  HYBRIDS they are imperfect. Even in first crosses, t
0264  st very unwilling to believe in this view; as  HYBRIDS, when once born, are generally healthy and l
0264  ed, as we see in the case of the common mule.  HYBRIDS, however, are differently circumstanced befo
0264  itions of life. In regard to the sterility of  HYBRIDS, in which the sexual elements are imperfectl
0265  n the sterility thus superinduced and that of  HYBRIDS, there are many points of similarity. In bot
0265  ole groups of species tend to produce sterile  HYBRIDS. On the other hand, one species in a group w
0265  ies in a group will produce unusually fertile  HYBRIDS. No one can tell, till he tries, whether any
0265  of a genus will produce more or less sterile  HYBRIDS. Lastly, when organic beings are placed duri
0265  ree than when sterility ensues. So it is with  HYBRIDS, for hybrids in successive generations are e
0265  sterility ensues. So it is with hybrids, for  HYBRIDS in successive generations are eminently liab
0265  under new and unnatural conditions, and when  HYBRIDS are produced by the unnatural crossing of tw
0266  reciable by us; in the other case, or that of  HYBRIDS, the external conditions have remained the s
0266  o another, or to the conditions of life. When  HYBRIDS are able to breed inter se, they transmit to
0266  everal facts with respect to the sterility of  HYBRIDS: for instance, the unequal fertility of hybr
0266  brids; for instance, the unecual fertility of  HYBRIDS produced from reciprocal crosses; or the inc
0266  crosses; or the increased sterility in those  HYBRIDS which occasionally and exceptionally resembl
0267  ome widely or specifically different, produce  HYBRIDS which are generally sterile in some degree.
0269  te sterility in the successive generations of  HYBRIDS, which were at first only slightly sterile;
0271  with the Nicotiana glutinosa, always yielded  HYBRIDS not so sterile as those which were produced
0272  invariable, sterility of first crosses and of  HYBRIDS, namely, that it is not a special endowment,
```

Page ***(Key Word)***

Page *************************************(Key Word)*************************************

```
0272  ctive systems of the forms which are crossed.   HYBRIDS and Mongrels compared, independently of thei
0272  st generation mongrels are more variable than   HYBRIDS; but Gartner admits that hybrids from specie
0272  ariable than hybrids; but Gartner admits that   HYBRIDS from species which have long been cultivated
0272  ces of this fact. Gartner further admits that   HYBRIDS between very closely allied species are more
0272  ates away. When mongrels and the more fertile   HYBRIDS are propagated for several generations an ex
0273  ring is notorious; but some few cases both of   HYBRIDS and mongrels long retaining uniformity of ch
0273  ions of mongrels is, perhaps, greater than in   HYBRIDS. This greater variability of mongrels than o
0273  this greater variability of mongrels than of   HYBRIDS does not seem to me at all surprising. For t
0273  crossing. The slight degree of variability in   HYBRIDS from the first cross or in the first generat
0273  offspring identical with the parent form. Now   HYBRIDS in the first generation are descended from s
0273  way affected, and they are not variable; but   HYBRIDS themselves have their reproductive systems s
0273  t to return to our comparison of mongrels and   HYBRIDS: Gartner states that mongrels are more liabl
0274  ner states that mongrels are more liable than   HYBRIDS to revert to either parent form; but this, i
0274  other, are crossed with a third species, the   HYBRIDS are widely different from each other: wherea
0274  species are crossed with another species, the   HYBRIDS do not differ much. But this conclusion, as
0274  ther hand, the resemblance in mongrels and in   HYBRIDS to their respective parents, more especially
0274  their respective parents, more especially in   HYBRIDS produced from nearly related species, follow
0274  s with mongrels from a reciprocal cross. Both   HYBRIDS and mongrels can be reduced to either pure p
0275  be shown that this does sometimes occur with   HYBRIDS; yet I grant much less frequently with hybri
0275  ybrics; yet I grant much less frequently with   HYBRIDS than with mongrels. Looking to the cases whi
0275  ed and semi monstrous in character, than with   HYBRIDS, which are descended from species slowly and
0276  y distinct to be ranked as species, and their   HYBRIDS, are very generally, but not universally, st
0277  e early death of the embryo. The sterility of   HYBRIDS, which have their reproductive systems imper
0277  effecting a first cross, the fertility of the   HYBRIDS produced, and the capacity of being grafted
0278  there is a close general resemblance between   HYBRIDS and mongrels. Finally, then, the facts brief
0461  it also to produce sterility. The sterility of  HYBRIDS is a very different case from that of first
0461  ditions of life, we need not feel surprise at   HYBRIDS being in some degree sterile, for their cons
0473  veral species of the horse genus and in their   HYBRIDS! How simply is this fact explained if we bel
0190  the dragonfly and in the fish Cobites. In the   HYDRA, the animal may be turned inside out, and the
0157  xes of the same species: again in fossorial   HYMENOPTERA, the manner of neuration of the wings is a
0415  ongst insects, in one great division of the   HYMENOPTERA, the antennae, as Westwood has remarked, a
0266  at we cannot understand, excepting on vague   HYPOTHESES, several facts with respect to the sterilit
0024  reed freely under domestication; yet on the   HYPOTHESIS of the multiple origin of our pigeons, it m
0026  g i think there is some probability in this   HYPOTHESIS, if applied to species closely related toge
0026  d by a single experiment. But to extend the   HYPOTHESIS so far as to suppose that species, aborigin
0160  at number of generations, the most probable   HYPOTHESIS is, not that the offspring suddenly takes a
0166  grels. I have stated that the most probable   HYPOTHESIS to account for the reappearance of very anc
0308  some explanation, I will give the following   HYPOTHESIS. From the nature of the organic remains, wh
0465  st silurian strata, I can only recur to the   HYPOTHESIS given in the ninth chapter. That the geolog
0030  ower of this principle of selection is not   HYPOTHETICAL. It is certain that several of our eminent
0161  lumage to assume this colour. This view is   HYPOTHETICAL, but could be supported by some facts; and
0343  d: i can answer this latter question only   HYPOTHETICALLY, by saying that as far as we can see, whe
0357  ise with America. Other authors have thus   HYPOTHETICALLY bridged over every ocean, and have united
019   lung, or organ used exclusively for respiration.  I can, indeed, hardly doubt that all vertebrate a
0001  tion When on board H.M.S. Beagle, as naturalist,  I was much struck with certain facts in the distr
0001  y have any bearing on it. After five years' work  I allowed myself to speculate on the subject, and
0001  the subject, and drew up some short notes; these  I enlarged in 1844 into a sketch of the conclusio
0001  me probable: from that period to the present day  I have steadily pursued the same object. I hope t
0001  ent day I have steadily pursued the same object.  I hope that I may be excused for entering on thes
0001  ve steadily pursued the same object. I hope that  I may be excused for entering on these personal d
0001  cused for entering on these personal details, as  I give them to show that I have not been hasty in
0001  se personal details, as I give them to show that  I have not been hasty in coming to a decision. My
0001  omplete it, and as my health is far from strong,  I have been urged to publish this Abstract. I hav
0001  ong, I have been urged to publish this Abstract.  I have more especially been induced to do this, a
0002  almost exactly the same general conclusions that  I have on the origin of species. Last year he sen
0002  me a memoir on this subject, with a request that  I would forward it to Sir Charles Lyell, who sent
0002  tracts from my manuscripts. This Abstract, which  I now publish, must necessarily be imperfect. I c
0002  ch i now publish, must necessarily be imperfect.  I cannot here give references and authorities for
0002  s and authorities for my several statements; and  I must trust to the reader reposing some confiden
0002  racy. No doubt errors will have crept in; though  I hope I have always been cautious in trusting to
0002  o doubt errors will have crept in, though I hope  I have always been cautious in trusting to good a
0002  cautious in trusting to good authorities alone.  I can here give only the general conclusions at w
0002  here give only the general conclusions at which  I have arrived, with a few facts in illustration,
0002  ed, with a few facts in illustration, but which,  I hope, in most cases will suffice. No one can fe
0002  will suffice. No one can feel more sensible than  I do of the necessity of hereafter publishing in
0002  on which my conclusions have been grounded; and  I hope in a future work to do this. For I am well
0002  ded; and I hope in a future work to do this. For  I am well aware that scarcely a single point is d
0002  conclusions directly opposite to those at which  I have arrived. A fair result can be obtained onl
0002  question and this cannot possibly be here done.  I much regret that want of space prevents my havi
0002  n of acknowledging the generous assistance which  I have received from very many naturalists, some
0002  uralists, some of them personally unknown to me.  I cannot, however, let this opportunity pass with
0003  f. the author of the Vestiges of Creation would,  I presume, say that after a certain unknown numbe
0004  nce of making out this obscure problem. Nor have  I been disappointed: in this and in all other per
0004  intec: in this and in all perplexing cases  I have invariably found that our knowledge, imper
0004  omestication, afforded the best and safest clue.  I may venture to express my conviction of the hig
0004  ected by naturalists. From these considerations,  I shall devote the first chapter of this Abstract
0004  o by his Selection successive slight variations.  I will then pass on to the variability of species
0004  variability of species in a state of nature; but  I shall, unfortunately, be compelled to treat thi
0005  he less improved forms of life, and induces what  I have called Divergence of Character. In the nex
0005  led Divergence of Character. In the next chapter  I shall discuss the complex and little known laws
0005  on of the Geological Record. In the next chapter  I shall consider the geological succession of org
0005  d in an embryonic condition. In the last chapter  I shall give a brief recapitulation of the whole
0006  for they determine the present welfare, and, as  I believe, the future success and modification of
0006  h remains obscure, and will long remain obscure,  I can entertain no doubt, after the most delibera
0006  berate study and dispassionate judgment of which  I am capable, that the view which most naturalist
0006  view which most naturalists entertain, and which  I formerly entertained, namely, that each species
```

Page *************************************(Key Word)*************************************

```
0006  es has been independently created, is erroneous.   I am fully convinced that species are not immutab
0006  re the descendants of that species. Furthermore,   I am convinced that Natural Selection has been th
0007  ification. Variation Under Domestication Chapter   I. Causes of Variability. Effects of Habit. Corre
0007  under the most different climates and treatment,   I think we are driven to conclude that this great
0007  have been exposed under nature. There is, also,    I think, some probability in the view propounded
0008  ar line of distinction from mere variations. But   I am strongly inclined to suspect that the most f
0008  determine whether or not the plant sets a seed.    I cannot here enter on the copious details which
0008  i cannot here enter on the copious details which   have collected on this curious subject; but to
0009  e the reproduction of animals under confinement,   may just mention that carnivorous animals, even
0009  rfectly tamed, long lived, and healthy (of which   could give numerous instances), yet having thei
0009  e of all the choicest productions of the garden.   may add, that as some organisms will breed most
0011  , nevertheless some slight amount of change may,   think, be attributed to the direct action of th
0011  imals it has a more marked effect; for instance,   find in the domestic duck that the bones of the
0011  on, than do the same bones in the wild duck; and   presume that this change may be safely attribut
0011  y seen, and will be hereafter briefly mentioned.   will here only allude to what may be called cor
0013  o males alone. A much more important rule, which   think may be trusted, is that, at whatever peri
0014  period at which it first appeared in the parent.   believe this rule to be of the highest importan
0014  ent. Having alluded to the subject of reversion,   may here refer to a statement often made by nat
0014  domestic races to species in a state of nature.   have in vain endeavoured to discover on what de
0015  ny slight deviations of structure, in such case,   grant that we could deduce nothing from domesti
0015  generations, would be opposed to all experience.   may add, that when under nature the conditions
0016  en have a somewhat monstrous character; by which   mean, that, although differing from each other,
0016  species of the same genus in a state of nature.   think this must be admitted, when we find that
0016  from each other in characters of generic value.   think it could be shown that this statement is
0016  eover, on the view of the origin of genera which   shall presently give, we have no right to expec
0017  xes, inhabiting different quarters of the world.   do not believe, as we shall presently see, that
0017  ary, and likewise to withstand diverse climates.   do not dispute that these capacities have added
0017  the common camel, prevented their domestication?   cannot doubt that if other animals and plants,
0017  f our anciently domesticated animals and plants,   do not think it is possible to come to any defi
0018  t have existed in Egypt? The whole subject must,   think, remain vague; nevertheless, I may, witho
0018  bject must, I think, remain vague; nevertheless,   may, without here entering on any details, stat
0018  hat, from geographical and other considerations,   think it highly probable that our domestic dogs
0018  veral wild species. In regard to sheep and goats   can form no opinion. I should think, from facts
0018  regard to sheep and goats I can form no opinion.   should think, from facts communicated to me by
0018  rent. With respect to horses, from reasons which   cannot give here, I am doubtfully inclined to b
0018  o horses, from reasons which I cannot give here,   am doubtfully inclined to believe, in oppositio
0018  , from his large and varied stores of knowledge,   should value more than that of almost any one,
0019  iffer considerably from each other in structure,   do not doubt that they all have descended from
0019  e of the domestic dogs of the whole world, which   fully admit have probably descended from severa
0019  ve probably descended from several wild species,   cannot doubt that there has been an immense amo
0020  etween two extremely different races or species;   can hardly believe. Sir J. Sebright expressly e
0020  n two pure breeds is tolerably and sometimes (as   have found with pigeons) extremely uniform, and
0020  treme care and long continued selection; nor can   find a single case on record of a permanent rac
0020  t it is always best to study some special group,   have, after deliberation, taken up domestic pig
0020  , after deliberation, taken up domestic pigeons.   have kept every breed which I could purchase or
0020  domestic pigeons. I have kept every breed which   could purchase or obtain, and have been most ki
0020  y important, as being of considerable antiquity,   have associated with several eminent fanciers,
0022  told that they were wild birds, would certainly,   think, be ranked by him as well defined species
0022  ranked by him as well defined species. Moreover,   do not believe that any ornithologist would pla
0023  e differences are between the breeds of pigeons,   am fully convinced that the common opinion of n
0023  ef are in some degree applicable in other cases,   will here briefly give them. If the several bre
0025  denly to acquire these characters; for instance,   crossed some uniformly white fantails with some
0025  ey produced mottled brown and black birds; these   again crossed together, and one grandchild of t
0026  of generations, been crossed by the rock pigeon:   say within a dozen or twenty generations, for w
0026  omestic breeds of pigeons are perfectly fertile.   can state this from my own observations, purpos
0026  ndency to sterility: from the history of the dog   think there is some probability in this hypothe
0027  ile; from these several reasons, taken together,   can feel no doubt that all our domestic breeds
0027  eographical sub species. In favour of this view,   may add, firstly, that C. livia, or the rock pi
0028  breeds can be kept together in the same aviary,   have discussed the probable origin of domestic
0028  me, yet quite insufficient, length: because when   first kept pigeons and watched the several kind
0028  several kinds, knowing well how true they bred,   felt fully as much difficulty in believing that
0028  animals and the cultivators of plants, with whom   have ever conversed, or whose treatises I have
0028  h whom I have ever conversed, or whose treatises   have read, are firmly convinced that the severa
0029  m so many aboriginally distinct species. Ask, as   have asked, a celebrated raiser of Hereford cat
0029  from long horns, and he will laugh you to scorn.   have never met a pigeon, or poultry, or duck, o
0029  other examples could be given. The explanation,   think, is simple: from long continued study the
0030  purposes, or so beautiful in his eyes, we must,   think, look further than to mere variability. W
0031  , which they can model almost as they please. If   had space I could quote numerous passages to th
0031  can model almost as they please. If I had space   could quote numerous passages to this effect fr
0032  reciable by an uneducated eye, differences which   for one have vainly attempted to appreciate. No
0033  ensure some differences; but, as a general rule,   cannot doubt that the continued selection of sl
0033  e been published on the subject; and the result,   may add, has been, in a corresponding degree, r
0033  m true that the principle is a modern discovery.   could give several references to the full ackno
0034  plants by nurserymen. The principle of selection   find distinctly given in an ancient Chinese enc
0034  of permanently altering the breed. Nevertheless   cannot doubt that this process, continued durin
0035  nly came from Spain, Mr. Borrow has not seen, as   am informed by him, any native dog in Spain lik
0036  existing in Britain, India, and Persia, we can,   think, clearly trace the stages through which t
0037  , to have been a fruit of very inferior quality,   have seen great surprise expressed in horticult
0037  d results from such poor materials; but the art,   cannot doubt, has been simple, and, as far as t
0037  owly and unconsciously accumulated, explains, as   believe, the well known fact, that in a vast nu
0038  their habits to man's wants or fancies. We can,   think, further understand the frequently abnorm
0039  h an expression as trying to make a fantail, is,   have no doubt, in most cases, utterly incorrect
0040  been exhibited as distinct at our poultry shows.   think these views further explain what has some
0040  d are once fully acknowledged, the principle, as   have called it, of unconscious selection will a
0040  d of such slow, varying, and insensible changes.   must now say a few words on the circumstances,
0041  such attention be paid nothing can be effected.   I have seen it gravely remarked, that it was most
```

0042	mprovement and formation of new breeds. Pigeons,	I may add, can be propagated in great numbers and
0042	some other country, often from islands. Although	I do not doubt that some domestic animals vary le
0043	gin of our Domestic Races of animals and plants.	I believe that the conditions of life, from their
0043	f the highest importance as causing variability,	I do not believe that variability is an inherent
0043	thus rendered infinitely complex. In some cases,	I do not doubt that the intercrossing of species,
0043	the importance of the crossing of varieties has,	I believe, been greatly exaggerated, both in rega
0044	only temporary. Over all these causes of Change	I am convinced that the accumulative action of Se
0044	atalogue of dry facts should be given, but these	I shall reserve for my future work. Nor shall I h
0044	se i shall reserve for my future work. Nor shall	I here discuss the various definitions which have
0044	t they graduate into varieties. By a monstrosity	I presume is meant some considerable deviation of
0045	at least some few generations? and in this case	I presume that the form would be called a variety
0045	what naturalists consider unimportant parts; but	I could show by a long catalogue of facts, that p
0045	mes vary in the individuals of the same species.	I am convinced that the most experienced naturali
0045	re, which he could collect on good authority, as	I have collected, during a course of years. It sh
0045	pare them in many specimens of the same species.	I should never have expected that the branching o
0045	ct would have been variable in the same species;	I should have expected that changes of this natur
0046	e stem of a tree. This philosophical naturalist,	I may add, has also quite recently shown that the
0046	erences, which seems to me extremely perplexing:	I refer to those genera which have sometimes been
0046	bility is independent of the conditions of life.	I am inclined to suspect that we see in these pol
0047	he variety. But cases of great difficulty, which	I will not here enumerate, sometimes occur in dec
0048	her as mere varieties. Mr. H. C. Watson, to whom	I lie under deep obligation for assistance of all
0048	ther, and with those from the American mainland,	I was much struck how entirely vague and arbitrar
0049	to bear on the attempt to determine their rank.	I will here give only a single instance, the well
0050	the greatest number of forms of doubtful value.	I have been struck with the fact, that if any ani
0051	e mind with the idea of an actual passage. Hence	I look at individual differences, though of small
0051	worth recording in works on natural history. And	I look at varieties which are in any degree more
0052	hysical conditions in two different regions; but	I have not much faith in this view; and I attribu
0052	ons; but I have not much faith in this view; and	I attribute the passage of a variety, from a stat
0052	structure in certain definite directions. Hence	I believe a well marked variety may be justly cal
0052	subject. From these remarks it will be seen that	I look at the term species, as one arbitrarily gi
0053	ence sake. Guided by theoretical considerations,	I thought that some interesting results might be
0053	med a simple task; but Mr. H. C. Watson, to whom	I am much indebted for valuable advice and assist
0053	subsequently Dr. Hooker, even in stronger terms.	I shall reserve for my future work the discussion
0054	h oftenest produce well marked varieties, or, as	I consider them, incipient species. And this, per
0054	so many causes tend to obscure this result, that	I am surprised that my tables show even a small m
0054	small majority on the side of the larger genera.	I will here allude to only two causes of obscurit
0055	only strongly marked and well defined varieties,	I was led to anticipate that the species of the l
0055	nera; for wherever many closely related species (I.e. species of the same genus) have been formed,
0055	ving few. To test the truth of this anticipation	I have arranged the plants of twelve countries, a
0057	between the species is often exceedingly small.	I have endeavoured to test this numerically by av
0057	erfect results go, they always confirm the view.	I have also consulted some sagacious and most exp
0057	s. but when we come to discuss the principle, as	I call it, of Divergence of Character, we shall s
0060	before entering on the subject of this chapter,	I must make a few preliminary remarks, to show ho
0060	ure there is some individual variability: indeed	I am not aware that this has ever been disputed.
0061	it may be asked, how is it that varieties, which	I have called incipient species, become ultimatel
0061	riodically born, but a small number can survive.	I have called this principle, by which each sligh
0062	l struggle for life, or more difficult, at least	I have found it so, than constantly to bear this
0062	t unless it be thoroughly engrained in the mind,	I am convinced that the whole economy of nature,
0062	is not so at all seasons of each recurring year.	I should premise that I use the term Struggle for
0062	ns of each recurring year. I should premise that	I use the term Struggle for Existence in a large
0063	hese several senses, which pass into each other,	I use for convenience sake the general term of st
0064	be the slowest breeder of all known animals, and	I have taken some pains to estimate its probable
0065	nd there are plants which now range in India, as	I hear from Dr. Falconer, from Cape Comorin to th
0065	iliarity with the larger domestic animals tends,	I think, to mislead us: we see no great destructi
0067	ct has been ably treated by several authors, and	I shall, in my future work, discuss some of the c
0067	gard to the feral animals of South America. Here	I will make only a few remarks, just to recall to
0067	tion of seeds, but, from some observations which	I have made, I believe that it is the seedlings w
0067	, but, from some observations which I have made,	I believe that it is the seedlings which suffer m
0067	ere there could be no choking from other plants,	I marked all the seedlings of our native weeds as
0068	d periodical seasons of extreme cold or drought,	I believe to be the most effective of all checks.
0068	believe to be the most effective of all checks.	I estimated that the winter of 1854 55 destroyed
0070	om a few wheat or other such plants in a garden;	I have in this case lost every single seed. This
0070	the same species for its preservation, explains,	I believe, some singular facts in nature, such as
0070	and thus save each other from utter destruction.	I should add that the good effects of frequent in
0071	me of these cases: but on this intricate subject	I will not here enlarge. Many cases are on record
0071	l have to struggle together in the same country.	I will give only a single instance, which, though
0071	staffordshire, on the estate of a relation where	I had ample means of investigation, there was a l
0071	nter. But how important an element enclosure is,	I plainly saw near Farnham, in Surrey. Here there
0072	es, so close together that all cannot live. When	I ascertained that these young trees had not been
0072	these young trees had not been sown or planted,	I was so much surprised at their numbers that I w
0072	d, i was so much surprised at their numbers that	I went to several points of view, whence I could
0072	rs that I went to several points of view, whence	I could examine hundreds of acres of the unenclos
0072	of acres of the unenclosed heath, and literally	I could not see a single Scotch fir, except the o
0072	looking closely between the stems of the heath,	I found a multitude of seedlings and little trees
0072	uncred yards distant from one of the old clumps,	I counted thirty two little trees; and one of the
0073	nd this would certainly greatly alter (as indeed	I have observed in parts of South America) the ve
0073	nvent laws on the duration of the forms of life!	I am tempted to give one more instance showing ho
0073	re bound together by a web of complex relations.	I shall hereafter have occasion to show that the
0073	their pollen masses and thus to fertilise them.	I have, also, reason to believe that humble bees
0073	do not visit this flower. From experiments which	I have tried, I have found that the visits of bee
0073	his flower. From experiments which I have tried,	I have found that the visits of bees, if not indi
0073	e), as other bees cannot reach the nectar. Hence	I have very little doubt, that if the whole genus
0074	d mr. Newman says, near villages and small towns	I have found the nests of humble bees more numero
0074	humble bees more numerous than elsewhere, which	I attribute to the number of cats that destroy th
0077	nd beans; when sown in the midst of long grass,	I suspect that the chief use of the nutriment in
0080	so potent in the hands of man, apply to nature?	I think we shall see that it can act most effectu

0081	tions and the rejection of injurious variations,	I	call Natural Selection. Variations neither usef
0082	, natural selection can do nothing. Not that, as	I	believe, any extreme amount of variability is n
0082	incomparably longer time at her disposal. Nor do	I	believe that any great physical change, as of c
0085	, as being the most liable to destruction. Hence	I	can see no reason to doubt that natural selecti
0086	s more and more widely disseminated by the wind,	I	can see no greater difficulty in this being eff
0087	effect may be found in works of natural history.	I	cannot find one case which will bear investigat
0088	to. And this leads me to say a few words on what	I	call Sexual Selection. This depends, not on a s
0088	the scale of nature this law of battle descends,	I	know not: male alligators have been described a
0089	ribute any effect to such apparently weak means:	I	cannot here enter on the details necessary to s
0089	is bantams, according to his standard of beauty,	I	can see no good reason to doubt that female bir
0089	andard of beauty, might produce a marked effect.	I	strongly suspect that some well known laws with
0089	he males alone, or by the males and females: but	I	have not space here to enter on this subject. T
0089	ce here to enter on this subject. Thus it is, as	I	believe, that when the males and females of any
0090	d these advantages to their male offspring. Yet,	I	would not wish to attribute all such sexual dif
0090	ral Selection. In order to make it clear how, as	I	believe, natural selection acts, I must beg per
0090	clear how, as I believe, natural selection acts,	I	must beg permission to give one or two imaginar
0090	year when the wolf is hardest pressed for food.	I	can under such circumstances see no reason to d
0090	hey might be compelled to prey on other animals.	I	can see no more reason to doubt this, than that
0091	t of intercrossing we shall soon have to return.	I	may add, that, according to Mr. Pierce, there a
0093	wer; and that they can most effectually do this,	I	could easily show by many striking instances. I
0093	I could easily show by many striking instances.	I	will give only one, not as a very striking case
0093	emale tree exactly sixty yards from a male tree,	I	put the stigmas of twenty flowers, taken from d
0093	to bees, nevertheless every female flower which	I	examined had been effectually fertilised by the
0094	ts cepended in main part on its nectar for food.	I	could give many facts, showing how anxious bees
0094	enter by the mouth. Bearing such facts in mind,	I	can see no reason to doubt that an accidental d
0095	rently constructed proboscis. On the other hand,	I	have found by experiment that the fertility of
0095	that the hive bee could visit its flowers. Thus	I	can understand how a flower and a bee might slo
0095	and slightly favourable deviations of structure.	I	am well aware that this doctrine of natural sel
0096	structure. On the Intercrossing of Individuals.	I	must here introduce a short digression. In the
0096	phrodites this is far from obvious. Nevertheless	I	am strongly inclined to believe that with all h
0096	r for the reproduction of their kind. This view,	I	may add, was first suggested by Andrew Knight.
0096	ight. We shall presently see its importance: but	I	must here treat the subject with extreme brevit
0096	e treat the subject with extreme brevity, though	I	have the materials prepared for an ample discus
0096	n? as it is impossible here to enter on details,	I	must trust to some general considerations alone
0096	eneral considerations alone. In the first place,	I	have collected so large a body of facts, showin
0097	the belief that this is a law of nature, we can,	I	think, understand several large classes of fact
0097	e visits of bees to papilionaceous flowers, that	I	have found, by experiments published elsewhere,
0097	len from one to the other, to the great good, as	I	believe, of the plant. Bees will act like a cam
0098	ertilisation, there are special contrivances, as	I	could show from the writings of C. C. Sprengel
0098	pollen from one flower on the stigma of another,	I	raised plenty of seedlings; and whilst another
0099	pollen, yet, as C. C. Sprengel has shown, and as	I	can confirm, either the anthers burst before th
0099	ed to seed near each other, a large majority, as	I	have found, of the seedlings thus raised will t
0099	hus raised will turn out mongrels: for instance,	I	raised 233 seedling cabbages from some plants o
0099	a vast number of the seedlings are mongrelized?	I	suspect that it must arise from the pollen of a
0100	as distinct individuals only in a limited sense.	I	believe this objection to be valid, but that na
0100	ir sexes more often separated than other plants,	I	find to be the case in this country; and at my
0100	hose of the United States; and the result was as	I	anticipated. On the other hand, Dr. Hooker has
0100	os that the rule does not hold in Australia; and	I	have made these few remarks on the sexes of tre
0100	usca and earth worms; but these all pair. As yet	I	have not found a single case of a terrestrial a
0100	ccasional cross. And, as in the case of flowers,	I	have as yet failed, after consultation with one
0101	y great difficulty under this point of view: but	I	have been enabled, by a fortunate chance, elsew
0101	iderations and from the many special facts which	I	have collected, but which I am not here able to
0101	special facts which I have collected, but which	I	am not here able to give, I am strongly incline
0101	collected, but which I am not here able to give,	I	am strongly inclined to suspect that, both in t
0101	s with a distinct individual is a law of nature.	I	am well aware that there are, on this view, man
0101	is view, many cases of difficulty, some of which	I	am trying to investigate. Finally then, we may
0101	perhaps only at long intervals; but in none, as	I	suspect, can self fertilisation go on for perpe
0102	e and diversified variability is favourable, but	I	believe mere individual differences suffice for
0102	mount of variability in each individual, and is,	I	believe, an extremely important element of succ
0103	lly be confined to separated countries; and this	I	believe to be the case. In hermaphrodite organi
0103	intercrosses in retarding natural selection; for	I	can bring a considerable catalogue of facts, sh
0104	th those animals which unite for each birth; but	I	have already attempted to show that we have rea
0104	even if these take place only at long intervals,	I	am convinced that the young thus produced will
0105	es; and this we are incapable of doing. Although	I	do not doubt that isolation is of considerable
0105	e in the production of new species, on the whole	I	am inclined to believe that largeness of area i
0106	y, to a certain extent, have concurred. Finally,	I	conclude that, although small isolated areas pr
0107	as the extreme intricacy of the subject permits.	I	conclude, looking to the future, that for terre
0108	selection will always act with extreme slowness,	I	fully admit. Its action depends on there being
0108	wholly to stop the action of natural selection.	I	do not believe so. On the other hand, I do beli
0108	lection. I do not believe so. On the other hand,	I	do believe that natural selection will always a
0108	inhabitants of the same region at the same time.	I	further believe, that this very slow, intermitt
0109	n do much by his powers of artificial selection,	I	can see no limit to the amount of change, to th
0110	moner species. From these several considerations	I	think it inevitably follows, that as new specie
0111	d that these were swept away by the short horns (I	I	quote the words of an agricultural writer) as i
0111	e. divergence of Character. The principle, which	I	have designated by this term, is of high import
0111	f high importance on my theory, and explains, as	I	believe, several important facts. In the first
0111	species in the process of formation, or are, as	I	have called them, incipient species. How, then,
0112	ed, can any analogous principle apply in nature?	I	believe it can and does apply most efficiently,
0114	ts utmost to increase its numbers. Consequently,	I	cannot doubt that in the course of many thousan
0114	reat diversity in its inhabitants. For instance,	I	found that a piece of turf, three feet by four
0115	an advantage over the other natives; and we may,	I	think, at least safely infer that diversificati
0116	which ought to have been much amplified, we may,	I	think, assume that the modified descendants of
0116	am by the letters standing at unequal distances.	I	have said a large genus, because we have seen i
0118	rm up to the fourteen thousandth generation. But	I	must here remark that I do not suppose that the
0118	ousandth generation. But I must here remark that	I	do not suppose that the process ever goes on so
0119	agram, though in itself made somewhat irregular.	I	am far from thinking that the most divergent va

0119 ing to the upper horizontal lines. In some cases I do not doubt that the process of modification w
0120 en a14 and m14, all descended from (A). Thus, as I believe, species are multiplied and genera are
0120 more than one species would vary. In the diagram I have assumed that a second species (I) has prod
0120 he diagram I have assumed that a second species (I) has produced, by analogous steps, after ten th
0121 es in the polity of nature: hence in the diagram I have chosen the extreme species (A), and the ne
0121 eme species (A), and the nearly extreme species (I), as those which have largely varied, and have
0122 replaced by eight new species (a14 to m14); and (I) will have been replaced by six (n14 to z14) ne
0122 , and D, than to the other species; and species (I) more to G, H, K, L, than to the others. These
0122 , than to the others. These two species (A) and (I), were also supposed to be very common and wide
0122 us exterminated, not only their parents (A) and (I), but likewise some of the original species whi
0123 even a distinct genus. The six descendants from (I) will form two sub genera or even genera. But a
0123 era or even genera. But as the original species (I) differed largely from (A), standing nearly at
0123 of the original genus, the six descendants from (I) will, owing to inheritance, differ considerabl
0123), which connected the original species (A) and (I), have all become, excepting (F), extinct, and
0123 dants. Hence the six new species descended from (I), and the eight descended from (A), will have t
0123 or even as distinct sub families. Thus it is, as I believe, that two or more genera are produced b
0124 ch stood between the two parent species (A) and (I), now supposed to be extinct and unknown, it wi
0124 eology, have to refer again to this subject, and I think we shall then see that the diagram throws
0125 he branching lines of descent had diverged less. I see no reason to limit the process of modificat
0125 so have two very distinct genera descended from (I); and as these latter two genera, both from con
0126 will transmit descendants to a remote futurity. I shall have to return to this subject in the cha
0126 is subject in the chapter on Classification, but I may add that on this view of extremely few of t
0127 in the several parts of their organisation, and I think this cannot be disputed: if there be, owi
0127 itution, and habits, to be advantageous to them, I think it would be a most extraordinary fact if
0127 y characterised. This principle of preservation, I have called, for the sake of brevity, Natural S
0128 intermediate forms of life. On these principles, I believe, the nature of the affinities of all or
0129 hat each species has been independently created, I can see no explanation of this great fact in th
0129 have sometimes been represented by a great tree. I believe this simile largely speaks the truth. T
0130 ll sides many a feebler branch; so by generation I believe it has been with the great Tree of Life
0131 er. Peversions to long lost characters. Summary. I have hitherto sometimes spoken as if the variat
0131 rs have been exposed during several generations, I have remarked in the first chapter, but a long
0132 tem being functionally disturbed in the parents, I chiefly attribute the varying or plastic condit
0134 from the facts alluded to in the first chapter, I think there can be little doubt that use in our
0134 irds seldom take flight except to escape danger, I believe that the nearly wingless condition of s
0135 ame incapable of flight. Kirby has remarked (and I have observed the same fact) that the anterior
0135 believe that mutilations are ever inherited; and I should prefer explaining the entire absence of
0137 re subterranean in its habits than the mole: and I was assured by a Spaniard, who had often caught
0137 them, that they were frequently blind; one which I kept alive was certainly in this condition, the
0137 any way injurious to animals living in darkness, I attribute their loss wholly to disuse. In one o
0138 case with some of the American cave animals, as I hear from Professor Dana: and some of the Europ
0139 roteus with reference to the reptiles of Europe, I am only surprised that more wrecks of ancient l
0139 inhabit very hot and very cold countries, and as I believe that all the species of the same genus
0140 ently become acclimatised to their new homes. As I believe that our domestic animals were original
0140 ly found capable of far extended transportation, I think the common and extraordinary capacity in
0141 , and on many islands in the torrid zones. Hence I am inclined to look at adaptation to any specia
0141 uestion. That habit or custom has some influence I must believe, both from analogy, and from the i
0142 fitted for their own districts: the result must, I think, be due to habit. On the other hand, I ca
0142 st, I think, be due to habit. On the other hand, I can see no reason to doubt that natural selecti
0142 dlings appeared to be than others. On the whole, I think we may conclude that habit, use, and disu
0143 on of innate differences. Correlation of Growth. I mean by this expression that the whole organisa
0143 be homologous with the limbs. These tendencies, I do not doubt, may be mastered more or less comp
0144 with respect to this latter case of correlation, I think it can hardly be accidental, that if we p
0144 e are likewise the most abnormal in their teeth. I know of no case better adapted to show the impo
0145 oftenest subject to peloria, and become regular. I may add, as an instance of this, and of a strik
0145 his, and of a striking case of correlation, that I have recently observed in some garden pelargoni
0145 central and exterior flowers of a head or umbel, I do not feel at all sure that C. C. Sprendel's i
0146 correlated in some necessary manner. So, again, I do not doubt that some apparent correlations, o
0146 eds are never found in fruits which do not open: I should explain the rule by the fact that seeds
0147 nature is forced to economise on the other side. I think this holds true to a certain extent with
0147 more especially botanists, believe in its truth. I will not, however, here give any instances, for
0147 will not, however, here give any instances, for I see hardly any way of distinguishing between th
0147 excess of growth in another and adjoining part. I suspect, also, that some of the cases of compen
0148 ment wasted in building up a useless structure. I can thus only understand a fact with which I wa
0148 re. I can thus only understand a fact with which I was much struck when examining cirripedes, and
0148 eloping a structure now become useless. Thus, as I believe, natural selection will always succeed
0149 e are more variable than those which are higher. I presume that lowness in this case means that th
0149 y parts, it has been stated by some authors, and I believe with truth, are apt to be highly variab
0149 l subject of rudimentary and aborted organs: and I will here only add that their variability seems
0150 , tends to be highly variable. Several years ago I was much struck with a remark, nearly to the ab
0150 o the above effect, published by Mr. Waterhouse. I infer also from an observation made by Professo
0150 ion without giving the long array of facts which I have collected; and which cannot possibly be he
0150 d, and which cannot possibly be here introduced. I can only state my conviction that it is a rule
0150 conviction that it is a rule of high generality. I am aware of several causes of error, but I hope
0150 lity. I am aware of several causes of error, but I hope that I have made due allowance for them. I
0150 ware of several causes of error, but I hope that I have made due allowance for them. It should be
0151 t displayed in any unusual manner, of which fact I think there can be little doubt. But that our r
0151 own in the case of hermaphrodite cirripedes; and I may here add, that I particularly attended to M
0151 rmaphrodite cirripedes: and I may here add, that I particularly attended to Mr. Waterhouse's remar
0151 e's remark, whilst investigating this Order, and I am fully convinced that the rule almost invaria
0151 le almost invariably holds good with cirripedes. I shall, in my future work, give a list of the mo
0151 work, give a list of the more remarkable cases; I will here only briefly give one, as it illustra
0151 same country vary in a remarkably small degree, I have particularly attended to them, and the rul
0151 eems to me certainly to hold good in this class. I cannot make out that it applies to plants; and
0152 created, with all its parts as we now see them, I can see no explanation. But on the view that gr
0152 nd have been modified through natural selection, I think we can obtain some light. In our domestic
0153 a much longer period nearly constant. And this, I am convinced, is the case. That the struggle be

0154 bnormally developed organs may be made constant,   I can see no reason to doubt. Hence when an organ
0154 variation would be a more unusual circumstance.   I have chosen this example because an explanation
0154 ce than those commonly used for classing genera.   I believe this explanation is partly, yet only in
0154 xplanation is partly, yet only indirectly, true;   I shall, however, have to return to this subject
0155 c characters are more variable than generic; but   I have repeatedly noticed in works on natural his
0155 which are closely alike in the several species?   I do not see that any explanation can be given. B
0155 neric characters; and these characters in common   I attribute to inheritance from a common progenit
0156 constant. In connexion with the present subject,   I will make only two other remarks. I think it wi
0156 ent subject, I will make only two other remarks.   I think it will be admitted, without my entering
0156 t secondary sexual characters are very variable;   I think it also will be admitted that species of
0157 same genus differ from each other. Of this fact   I will give in illustration two instances, the fi
0157 n has a clear meaning on my view of the subject: I look at all the species of the same genus as ha
0158 or the possession of the females. Finally, then,   I conclude that the greater variability of specif
0159 e normal structure of another race, the fantail.   I presume that no one will doubt that all such an
0160 ks are characteristic of the parent rock pigeon,   I presume that no one will doubt that this is a c
0160 ariation appearing in the several breeds. We may   I think confidently come to this conclusion, beca
0161 tical, but could be supported by some facts; and   I can see no more abstract improbability in a ten
0162 of the same part or organ in an allied species.   I have collected a long list of such cases; but h
0163 a long list of such cases; but here, as before,   I lie under a great disadvantage in not being abl
0163 eat disadvantage in not being able to give them.   I can only repeat that such cases certainly do oc
0163 tainly do occur, and seem to me very remarkable.   I will, however, give one curious and complex cas
0163 e plainest in the foal, and from inquiries which   I have made, I believe this to be true. It has al
0163 the foal, and from inquiries which I have made,   I believe this to be true. It has also been asser
0163 y mr. Blyth and others, occasionally appear: and   I have been informed by Colonel Poole that the fo
0163 ke bars on the hocks. With respect to the horse,   I have collected cases in England of the spinal s
0163 oulder stripe may sometimes be seen in duns, and   I have see a trace in a bay horse. My son made a
0164 h shoulder and with leg stripes; and a man, whom   I can implicitly trust, has examined for me a sma
0164 reed of horses is so generally striped, that, as   I hear from Colonel Poole, who examined the breed
0164 d bay Kattywar horses striped when first foaled.   I have, also, reason to suspect, from information
0164 nimal. Without here entering on further details,   I may state that I have collected cases of leg an
0164 re entering on further details, I may state that   I have collected cases of leg and shoulder stripe
0164 n and black to a close approach to cream colour.   I am aware that Colonel Hamilton Smith, who has w
0165 l due to ancient crosses with the dun stock. But   I am not at all satisfied with this theory, and s
0165 se is particularly apt to have bars on its legs.   I once saw a mule with its legs so much striped t
0165 a similar mule. In four coloured drawings, which   I have seen, of hybrids between the ass and zebra
0165 des of its face. With respect to this last fact,   I was so convinced that not even a stripe of colo
0165 what would commonly be called an accident, that   I was led solely from the occurence of the face s
0166 anc bars and marks to reappear in the mongrels.   I have stated that the most probable hypothesis t
0167 he case with that of the horse genus! For myself,   I venture confidently to look back thousands on t
0167 back thousands on thousands of generations, and   I see an animal striped like a zebra, but perhaps
0167 equine species was independently created, will,   I presume, assert that each species has been crea
0167 s the works of God a mere mockery and deception!   I would almost as soon believe with the old and i
0171 ader. Some of them are so grave that to this day   I can never reflect on them without being stagger
0171 only apparent, and those that are real are not,   I think, fatal to my theory. These difficulties a
0172 n the Imperfection of the geological record; and   I will here only state that I believe the answer
0172 ological record; and I will here only state that   I believe the answer mainly lies in the record be
0174 ficulty for a long time quite confounded me. But   I think it can be in large part explained. In the
0174 nuous and uniform condition than at present. But   I will pass over this way of escaping from the di
0174 er this way of escaping from the difficulty: for   I believe that many perfectly defined species hav
0174 been formed on strictly continuous areas; though   I do not doubt that the formerly broken condition
0175 e will come to be still more sharply defined. If   I am right in believing that allied or representa
0176 rrow and lesser area: and practically, as far as   I can make out, this rule holds good with varieti
0176 holds good with varieties in a state of nature.   I have met with striking instances of the rule in
0176 numbers than the forms which they connect, then,   I think, we can understand why intermediate varie
0176 f it. But a far more important consideration, as   I believe, is that, during the process of further
0177 and improved. It is the same principle which, as   I believe, accounts for the common species in eac
0177 well marked varieties than do the rarer species.   I may illustrate what I mean by supposing three v
0177 than do the rarer species. I may illustrate what   I mean by supposing three varieties of sheep to b
0177 upplanted, intermediate hill variety. To sum up,   I believe that species come to be tolerably well
0179 has been asked by the opponents of such views as   I hold, how, for instance, a land carnivorous ani
0180 question would have been far more difficult, and   I could have given no answer. Yet I think such di
0180 difficult, and I could have given no answer. Yet   I think such difficulties have very little weigh
0180 very little weight. Here, as on other occasions,   I lie under a heavy disadvantage, for out of the
0180 antage, for out of the many striking cases which   I have collected, I can give only one or two inst
0180 the many striking cases which I have collected,   I can give only one or two instances of transiti
0181 structure in a corresponding manner. Therefore,   I can see no difficulty, more especially under ch
0181 the Galeopithecus with the other Lemuridae, yet   I can see no difficulty in supposing that such li
0181 ucture had been useful to its possessor. Nor can   I see any insuperable difficulty in further belie
0183 case of species with fully developed structures.   I will now give two or three instances of diversi
0183 ied habits innumerable instances could be given: I have often watched a tyrant flycatcher (Sauroph
0184 ike, kills small birds by blows on the head; and   I have many times seen and heard it hammering the
0184 ompetitors did not already exist in the country,   I can see no difficulty in a race of bears being
0186 ld have been formed by natural selection, seems,   I freely confess, absurd in the highest possible
0187 more than how life itself first originated; but   I may remark that several facts make me suspect t
0188 n proportion to those which have become extinct,   I can see no very great difficulty (not more than
0188 reason ought to conquer his imagination; though   I have felt the difficulty far too keenly to be s
0189 ions, my theory would absolutely break down. But   I can find out no such case. No doubt many organs
0190 to the auditory organs of certain fish, or, for   I do not know which view is now generally held, a
0191 oating apparatus or swimbladder. We can thus, as   I infer from Professor Owen's interesting descrip
0191 of conversion from one function to another, that   I will give one more instance. Pedunculated cirri
0192 shell; but they have large folded branchiae. Now   I think no one will dispute that the ovigerous fr
0192 indeed, they graduate into each other. Therefore   I do not doubt that little folds of skin, which o
0193 undamental difference can generally be detected.   I am inclined to believe that in nearly the same
0194 forms to the extinct and unknown is very small,   I have been astonished how rarely an organ can be
0194 se with any unfavourable deviation of structure,   I have sometimes felt much difficulty in understa
0195 reservation of successively varying individuals.   I have sometimes felt as much difficulty, though

0195	uld be of importance or not. In a former chapter	I have given instances of most trifling character
0197	know that there were many black and pied kinds,	I dare say that we should have thought that the g
0197	en acquired through natural selection; as it is,	I have no doubt that the colour is due to some qu
0198	several known and unknown laws of variation; and	I have here alluded to them only to show that, if
0199	he slight analogous differences between species.	I might have adduced for this same purpose the di
0199	the races of man, which are so strongly marked;	I may add that some little light can apparently b
0199	rue, would be absolutely fatal to my theory. Yet	I fully admit that many structures are of no dire
0201	ound in works on natural history to this effect,	I cannot find even one which seems to me of any w
0201	own injury, namely, to warn its prey to escape.	I would almost as soon believe that the cat curls
0201	o spring, in order to warn the doomed mouse. But	I have not space here to enter on this and other
0203	inst my theory. Many of them are very grave; but	I think that in the discussion light has been thr
0207	have been worked into the previous chapters; but	I have thought that it would be more convenient t
0207	ficulty sufficient to overthrow my whole theory.	I must premise, that I have nothing to do with th
0207	overthrow my whole theory. I must premise, that	I have nothing to do with the origin of the prima
0207	igin of the primary mental powers, any more than	I have with that of life itself. We are concerned
0207	ntal qualities of animals within the same class.	I will not attempt any definition of instinct. It
0208	erformed, is usually said to be instinctive. But	I could show that none of these characters of ins
0208	ared instinct with habit. This comparison gives,	I think, a remarkably accurate notion of the fram
0209	ose any habitual action to become inherited, and	I think it can be shown that this does sometimes
0209	hown that instincts do vary ever so little, then	I can see no difficulty in natural selection pres
0209	ny extent that may be profitable. It is thus, as	I believe, that all the most complex and wonderfu
0209	habit, and are diminished or lost by disuse; so	I do not doubt it has been with instincts. But I
0209	o i do not doubt it has been with instincts. But	I believe that the effects of habit are of quite
0210	kind are possible; and this we certainly can do.	I have been surprised to find, making allowance f
0210	action for the sole good of another, with which	I am acquainted, is that of aphides voluntarily y
0210	hey do so voluntarily, the following facts show.	I removed all the ants from a group of about a do
0211	dance during several hours. After this interval,	I felt sure that the aphides would want to excret
0211	elt sure that the aphides would want to excrete.	I watched them for some time through a lens, but
0211	some time through a lens, but not one excreted;	I then tickled and stroked them with a hair in th
0211	them with a hair in the same manner, as well as	I could, as the ants do with their antennae; but
0211	their antennae; but not one excreted. Afterwards	I allowed an ant to visit them, and it immediatel
0211	excrete for the sole good of the ants. Although	I do not believe that any animal in the world per
0211	been here given; but want of space prevents me.	I can only assert, that instincts certainly do va
0212	animals. But fear of man is slowly acquired, as	I have elsewhere shown, by various animals inhabi
0212	h natural selection, to quite new instincts. But	I am well aware that these general statements, wi
0212	roduce but a feeble effect on the reader's mind.	I can only repeat my assurance, that I do not spe
0212	der's mind. I can only repeat my assurance, that	I do not speak without good evidence. The possibi
0213	dogs: it cannot be doubted that young pointers (I	have myself seen a striking instance) will some
0213	stead of at, a flock of sheep, by shepherd dogs.	I cannot see that these actions, performed withou
0213	hy she lays her eggs on the leaf of the cabbage,	I cannot see that these actions differ essentiall
0214	m long continued and compulsory habit, but this,	I think, is not true. No one would ever have thou
0214	he tumbler pigeon to tumble, an action which, as	I have witnessed, is performed by young birds, th
0214	e: and near Glasgow there are house tumblers, as	I hear from Mr. Brent, which cannot fly eighteen
0214	ne: and this is known occasionally to happen, as	I once saw in a pure terrier. When the first tend
0215	is tamer than the young of the tame rabbit; but	I do not suppose that domestic rabbits have ever
0215	abbits have ever been selected for tameness; and	I presume that we must attribute the whole of the
0216	cified by selection, by considering a few cases.	I will select only three, out of the several whic
0216	will select only three, out of the several which	I shall have to discuss in my future work, namely
0217	ionally lays her eggs in other birds' nests; but	I hear on the high authority of Dr. Brewer, that
0217	r. Brewer, that this is a mistake. Nevertheless,	I could give several instances of various birds w
0217	ir young. By a continued process of this nature,	I believe that the strange instinct of our cuckoo
0218	of our cuckoo could be, and has been, generated.	I may add that, according to Dr. Gray and to some
0218	ed over the plains, so that in one day's hunting	I picked up no less than twenty lost and wasted e
0218	s case, as with the supposed case of the cuckoo,	I can see no difficulty in natural selection maki
0220	by Mr. F. Smith, of the British Museum, to whom	I am much indebted for information on this and ot
0220	usting to the statements of Huber and Mr. Smith,	I tried to approach the subject in a sceptical fr
0220	ious an instinct as that of making slaves. Hence	I will give the observations which I have myself
0220	slaves. Hence I will give the observations which	I have myself made, in some little detail. I open
0220	which I have myself made, in some little detail,	I opened fourteen nests of F. sanguinea, and foun
0220	ths of June and July, on three successive years,	I have watched for many hours several nests in Su
0220	these months, the slaves are very few in number,	I thought that they might behave differently when
0221	the present year, however, in the month of July,	I came across a community with an unusually large
0221	ity with an unusually large stock of slaves, and	I observed a few slaves mingled with their master
0221	numbers in Switzerland than in England. One day	I fortunately chanced to witness a migration from
0221	vented from getting any pupae to rear as slaves.	I then dug up a small parcel of the pupae of F. f
0222	ictorious in their late combat. At the same time	I laid on the same place a small parcel of the pu
0222	h so small a species, it is very courageous, and	I have seen it ferociously attack other ants. In
0222	t ferociously attack other ants. In one instance	I found to my surprise an independent community o
0222	nest of the slave making F. sanguinea; and when	I had accidentally disturbed both nests, the litt
0222	heir big neighbours with surprising courage. Now	I was curious to ascertain whether F. sanguinea c
0222	ook heart and carried off the pupae. One evening	I visited another community of F. sanguinea, and
0222	that it was not a migration) and numerous pupae.	I traced the returning file burthened with booty,
0222	ty yards, to a very thick clump of heath, whence	I saw the last individual of F. sanguinea emerge,
0222	ual of F. sanguinea emerge, carrying a pupa; but	I was not able to find the desolated nest in the
0223	at steps the instinct of F. sanguinea originated	I will not pretend to conjecture. But as ants, wh
0223	but as ants, which are not slave makers, will as	I have seen, carry off pupae of other species, if
0224	its slaves than the same species in Switzerland,	I can see no difficulty in natural selection incr
0224	rufescens. Cell making instinct of the Hive Bee.	I will not here enter on minute details on this s
0224	rely give an outline of the conclusions at which	I have arrived. He must be a dull man who can exa
0224	t appears: all this beautiful work can be shown,	I think, to follow from a few very simple instinc
0225	ink, to follow from a few very simple instincts.	I was led to investigate this subject by Mr. Wate
0226	perfect as the comb of the hive bee. Accordingly	I wrote to Professor Miller, of Cambridge, and th
0227	than those which guide a bird to make its nest,	I believe that the hive bee has acquired, through
0228	riment. Following the example of Mr. Tegetmeier,	I separated two combs, and put between them a lon
0228	the basins had acquired the above stated width (i.e. about the width of an ordinary cell), and we
0228	siced pyramid as in the case of ordinary cells.	I then put into the hive, instead of a thick, squ

0229 ersection. Considering how flexible thin wax is,	I do not see that there is any difficulty in the
0229 actly the same rate from the opposite sides: for	I have noticed half completed rhombs at the base
0229 , which were slightly concave on one side, where	I suppose that the bees had excavated too quickly
0229 orked less quickly. In one well marked instance,	I put the comb back into the hive, and allowed th
0229 r a short time, and again examined the cell, and	I found that the rhombic plate had been completed
0230 fected this by gnawing away the convex side; and	I suspect that the bees in such cases stand in th
0230 push and bend the ductile and worm wax (which as	I have tried is easily done) into its proper inte
0230 e made by the justly celebrated elder Huber, but	I am convinced of their accuracy; and if I had so
0230 er, but I am convinced of their accuracy; and if	I had space, I could show that they are conformab
0230 convinced of their accuracy; and if I had space,	I could show that they are conformable with my th
0230 le parallel sided wall of wax, is not, as far as	I have seen, strictly correct: the first commence
0230 ent having always been a little hood of wax: but	I will not here enter on these details. We see ho
0231 e of intersection between two adjoining spheres.	I have several specimens showing clearly that the
0231 work even at the commencement of the first cell.	I was able practically to show this fact, by cove
0232 xtremely thin layer of melted vermilion wax: and	I invariably found that the colour was most delic
0232 te between two adjoining spheres: but, as far as	I have seen, they never gnaw away and finish off
0233 wasp combs are sometimes strictly hexagonal: but	I have not space here to enter on this subject. N
0233 tersection, and so make an isolated hexagon: but	I am not aware that any such case has been observ
0233 d have profited the progenitors of the hive bee?	I think the answer is not difficult: it is known
0233 often hard pressed to get sufficient nectar: and	I am informed by Mr. Tegetmeier that it has been
0235 absolutely perfect in economising wax. Thus, as	I believe, the most wonderful of all known instin
0236 cuired by independent acts of natural selection.	I will not here enter on these several cases, but
0236 uperable, and actually fatal to my whole theory.	I allude to the neuters or sterile females in the
0236 ll ceserves to be discussed at great length, but	I will here take only a single case, that of work
0236 n capable of work, but incapable of procreation,	I can see no very great difficulty in this being
0236 in this being effected by natural selection. But	I must pass over this preliminary difficulty. The
0236 insect had been an animal in the ordinary state,	I should have unhesitatingly assumed that all its
0237 of the bulls or cows of these same breeds. Hence	I can see no real difficulty in any character hav
0237 hough appearing insuperable, is lessened, or, as	I believe, disappears, when it is remembered that
0238 breeder goes with confidence to the same family.	I have such faith in the powers of selection, tha
0238 have such faith in the powers of selection, that	I do not doubt that a breed of cattle, always yie
0238 one ox could ever have propagated its kind. Thus	I believe it has been with social insects: a slic
0238 terile members having the same modification. And	I believe that this process has been repeated, un
0239 uard or imprison. It will indeed be thought that	I have an overweening confidence in the principle
0239 ence in the principle of natural selection, when	I do not admit that such wonderful and well estab
0239 hich have been rendered by natural selection, as	I believe to be quite possible, different from th
0239 ether by individuals taken out of the same nest!	I have myself compared perfect gradations of this
0240 ly cissected several specimens of these workers,	I can affirm that the eyes are far more rudimenta
0240 merely by their proportionally lesser size: and	I fully believe, though I dare not assert so posi
0240 ionally lesser size: and I fully believe, though	I dare not assert so positively, that the workers
0240 y some few members in an intermediate condition.	I may digress by adding, that if the smaller work
0240 e ants of this genus have well developed ocelli.	I may give one other case: so confidently did I
0240 i. i may give one other case: so confidently did	I expect to find gradations in important points o
0240 rent castes of neuters in the same species, that	I gladly availed myself of Mr. F. Smith's offer o
0241 es the widely different structure of their jaws.	I speak confidently on this latter point, as Mr.
0241 for me with the camera lucida of the jaws which	I had dissected from the workers of the several s
0241 f the several sizes. With these facts before me,	I believe that natural selection, by acting on th
0241 n intermediate structure were produced. Thus, as	I believe, the wonderful fact of two distinctly d
0242 re would have become blended. And nature has, as	I believe, effected this admirable division of la
0242 of ants, by the means of natural selection. But	I am bound to confess, that, with all my faith in
0242 fess, that, with all my faith in this principle,	I should never have anticipated that natural sele
0242 f these neuter insects convinced me of the fact.	I have, therefore, discussed this case, at some l
0242 fertile members, which alone leave descendants.	I am surprised that no one has advanced this demo
0242 nst the well known doctrine of Lamarck. Summary.	I have endeavoured briefly in this chapter to sho
0242 the variations are inherited. Still more briefly	I have attempted to show that instincts vary slig
0243 the highest importance to each animal. Therefore	I can see no difficulty, under changing condition
0243 or use and disuse have probably come into play.	I do not pretend that the facts given in this cha
0245 ct that hybrids are very generally sterile, has,	I think, been much underrated by some late writer
0245 n of successive profitable degrees of sterility.	I hope, however, to be able to show that sterilit
0248 ere perfect fertility ends and sterility begins.	I think no better evidence of this can be require
0248 ies. It is also most instructive to compare, but	I have not space here to enter on details; the ev
0248 ever increased, but generally greatly decreased.	I do not doubt that this is usually the case, and
0248 eases in the first few generations. Nevertheless	I believe that in all these experiments the ferti
0248 pendent cause, namely, from close interbreeding.	I have collected so large a body of facts, showin
0249 individual or variety increases fertility, that	I cannot doubt the correctness of this almost uni
0249 h generation by their own individual pollen: and	I am convinced that this would be injurious to th
0249 tility, already lessened by their hybrid origin.	I am strengthened in this conviction by a remarka
0249 ilisation pollen is as often taken by chance (as	I know from my own experience) from the anthers o
0249 erations of artificially fertilised hybrids may,	I believe, be accounted for by close interbreedin
0250 id gartner. The difference in their results may,	I think, be in part accounted for by Herbert's gr
0250 at his command. Of his many important statements	I will here give only a single one as an example,
0250 y c. revolutum produced a plant, which (he says)	I never saw to occur in a case of its natural fec
0251 n a natural species from the mountains of Chile.	I have taken some pains to ascertain the degree o
0251 ome of the complex crosses of Rhododendrons, and	I am assured that many of them are perfectly fert
0252 case of plants! but the hybrids themselves are,	I think, more sterile. I doubt whether any case o
0252 e hybrids themselves are, I think, more sterile.	I doubt whether any case of a perfectly fertile h
0253 generations of the more fertile hybrid animals.	I hardly know of an instance in which two familie
0253 edly be lost in a very few generations. Although	I do not know of any thoroughly well authenticate
0253 cated cases of perfectly fertile hybrid animals.	I have some reason to believe that the hybrids fr
0253 e cross bred geese must be far more fertile: for	I am assured by two eminently capable judges, nam
0254 r alternative seems to me the most probable, and	I am inclined to believe in its truth, although i
0254 truth, although it rests on no direct evidence.	I believe, for instance, that our dogs have desce
0254 but from facts communicated to me by Mr. Plyth,	I think they must be considered as distinct speci
0255 s admirable work on the hybridisation of plants.	I have taken much pains to ascertain how far the
0255 ty our knowledge is in regard to hybrid animals.	I have been surprised to find how generally the s
0258 ease. By a reciprocal cross between two species,	I mean the case, for instance, of a stallion hors

0260 to prevent their becoming confounded in nature?	I think not. For why should the sterility be so
0261 o explain a little more fully by an example what	I mean by sterility being incidental on other di
0261 nimportant for its welfare in a state of nature,	I presume that no one will suppose that this cap
0263 nown differences in their vegetative systems, so	I believe that the still more complex laws gover
0264 ative has not been sufficiently attended to: but	I believe, from observations communicated to me
0264 ry frequent cause of sterility in first crosses,	I was at first very unwilling to believe in this
0264 perfectly developed, the case is very different.	I have more than once alluded to a large body of
0264 han once alluded to a large body of facts, which	I have collected, showing that when animals and
0265 y are extremely liable to vary, which is due, as	I believe, to their reproductive systems having
0266 ally resemble closely either pure parent. Nor do	I pretend that the foregoing remarks go to the r
0266 atural conditions, is rendered sterile. All that	I have attempted to show, is that in two cases,
0266 n compounded into one. It may seem fanciful, but	I suspect that a similar parallelism extends to
0266 is an old and almost universal belief, founded,	I think, on a considerable body of evidence, tha
0267 ds, gives vigour and fertility to the offspring.	I believe, indeed, from the facts alluded to in
0267 rids which are generally sterile in some degree.	I cannot persuade myself that this parallelism i
0268 facility, and yield perfectly fertile offspring.	I fully admit that this is almost invariably the
0269 t be surprised at some difference in the result.	I have as yet spoken as if the varieties of the
0269 t of sterility in the few following cases, which	I will briefly abstract. The evidence is at leas
0270 ous, as the plants have separated sexes. No one,	I believe, has suspected that these varieties of
0270 eater. How far these experiments may be trusted,	I know not: but the forms experimentised on, are
0271 m the seed of the other. From observations which	I have made on certain varieties of hollyhock, I
0271 h I have made on certain varieties of hollyhock;	I am inclined to suspect that they present analo
0271 em: from these several considerations and facts,	I do not think that the very general fertility o
0272 eem to me sufficient to overthrow the view which	I have taken with respect to the very general, b
0272 ee most closely in very many important respects.	I shall here discuss this subject with extreme b
0272 are often variable in the first generation: and	I have myself seen striking instances of this fa
0273 for it bears on and corroborates the view which	I have taken on the cause of ordinary variability
0274 not differ much. But this conclusion, as far as	I can make out, is founded on a single experiment
0274 of impressing its likeness on the hybrid: and so	I believe it to be with varieties of plants. With
0274 y is crossed with another variety. For instance,	I think those authors are right, who maintain tha
0275 that this does sometimes occur with hybrids; yet	I grant much less frequently with hybrids than wi
0275 s than with mongrels. Looking to the cases which	I have collected of cross bred animals closely re
0275 cies slowly and naturally procuced. On the whole	I entirely agree with Dr. Prosper Lucas, who, aft
0279 known fossiliferous strata. In the sixth chapter	I enumerated the chief objections which might be
0279 ransitional links, is a very obvious difficulty.	I assigned reasons why such links do not commonly
0279 ntinous area with graduated physical conditions.	I endeavoured to show, that the life of each spec
0279 ite away quite insensibly like heat or moisture.	I endeavoured, also, to show that intermediate va
0280 rged against my theory. The explanation lies, as	I believe, in the extreme imperfection of the geo
0280 forms must, on my theory, have formerly existed.	I have found it difficult, when looking at any tw
0283 ucies the action of the sea on our shores, will,	I believe, be most deeply impressed with the slow
0284 y offers the best evidence of the lapse of time.	I remember having been much struck with the evide
0285 endeavour to grapple with the idea of eternity.	I am tempted to give one other case, the well kno
0286 formations is on an average about 1100 feet, as	I am informed by Prof. Ramsay. But if, as some ge
0286 te on one yard in nearly every twenty two years.	I doubt whether any rock, even as soft as chalk,
0286 kage of the fallen fragments. On the other hand,	I do not believe that any line of coast, ten or t
0286 t the base. Hence, under ordinary circumstances,	I conclude that for a cliff 500 feet in height, a
0287 d since the latter part of the Secondary period.	I have made these few remarks because it is highl
0288 of the sea, where sediment is not accumulating.	I believe we are continually taking a most errone
0288 y be dissolved by the percolation of rain water.	I suspect that but few of the very many animals w
0290 ving elapsed between each formation. But we can,	I think, see why the geological formations of eac
0290 the grinding action of the coast waves. We may,	I think, safely conclude that sediment must be ac
0291 resist any amount of degradation, may be formed.	I am convinced that all our ancient formations, w
0291 nce publishing my views on this subject in 1845,	I have watched the progress of Geology, and have
0291 usion that it was accumulated during subsidence.	I may add, that the only ancient tertiary formati
0292 ill almost necessarily be rendered intermittent.	I feel much confidence in the truth of these view
0293 uired a vast number of years for its deposition,	I can see several reasons why each should not inc
0293 links between the species which then lived: but	I can by no means pretend to assign due proportio
0293 od requisite to change one species into another.	I am aware that two palaeontologists, whose opini
0298 widely separated formations; but to this subject	I shall have to return in the following chapter.
0299 excepting those of the United States of America.	I fully agree with Mr. Godwin Austen, that the pr
0300 tions which we suppose to be there accumulating.	I suspect that not many of the strictly littoral
0300 remains could be preserved. In our archipelago,	I believe that fossiliferous formations could be
0302 eontologists, be ranked as distinct species. But	I do not pretend that I should ever have suspecte
0302 d as distinct species. But I do not pretend that	I should ever have suspected how poor a record of
0303 such species will appear as if suddenly created.	I may here recall a remark formerly made, namely
0303 spread rapidly and widely throughout the world.	I will now give a few examples to illustrate thes
0303 e groups of species have suddenly been produced.	I may recall the well known fact that in geologic
0304 e time before the close of the secondary period.	I may give another instance, which from having pa
0304 ck me. In a memoir on Fossil Sessile Cirripedes,	I have stated that, from the number of existing a
0304 can be recognised: from all these circumstances,	I inferred that had sessile cirripedes existed du
0304 species had been discovered in beds of this age,	I concluded that this great group had been sudden
0304 series. This was a sore trouble to me, adding as	I thought one more instance of the abrupt appeara
0305 a fact would certainly be highly remarkable; but	I cannot see that it would be an insuperable diff
0306 her and allied difficulty, which is much graver.	I allude to the manner in which numbers of specie
0306 rce to the earliest known species. For instance,	I cannot doubt that all the Silurian trilobites h
0307 t find records of these vast primordial periods,	I can give no satisfactory answer. Several of the
0308 that it may hereafter receive some explanation,	I will give the following hypothesis. From the na
0310 tly, maintained the immutability of species. But	I have reason to believe that one great authority
0310 flexion entertains grave doubts on this subject.	I feel how rash it is to differ from these great
0310 ry. For my part, following out Lyell's metaphor,	I look at the natural geological record, as a his
0314 these several facts accord well with my theory.	I believe in no fixed law of development, causing
0316 rd (though all strongly opposed to such views as	I maintain) admit its truth; and the rule strictl
0316 s existence, as long as it lasts, is continuous.	I am aware that there are some apparent exception
0316 mes falsely appear to have come in abruptly; and	I have attempted to give an explanation of this f
0318 ife, so have species a definite duration. No one	I think can have marvelled more at the extinction
0318 arvelled more at the extinction of species, than	I have done. When I found in La Plata the tooth o

18	he extinction of species, than I have done. When I	found in La Plata the tooth of a horse embedded
18	living shells at a very late geological period, I	was filled with astonishment; for seeing that t
18	as increased in numbers at an unparalleled rate,	asked myself what could so recently have exterm
20	either locally or wholly, through man's agency. I	may repeat what I published in 1845, namely, th
20	wholly, through man's agency. I may repeat what I	published in 1845, namely, that to admit that s
21	one species, the nearest allies of that species,	i.e. the species of the same genus, will be the m
21	ll be the most liable to extermination. Thus, as	believe, a number of new species descended from
24	, looking to a remotely future epoch, there can,	think, be little doubt that all the more modern
26	ication of new and dominant species; but we can,	think, clearly see that a number of individuals
27	remark connected with this subject worth making.	have given my reasons for believing that all ou
27	c remains. During these long and blank intervals	suppose that the inhabitants of each region unc
28	r for modification, extinction, and immigration.	suspect that cases of this nature have occurred
30	ving forms, the objection is probably valid. But	apprehend that in a perfectly natural classific
31	odification. As the subject is somewhat complex,	must request the reader to turn to the diagram
34	er, between the preceding and succeeding faunas.	need give only one instance, namely, the manner
36	e formations, we ought not to expect to find, as	attempted to show in the last chapter, in any o
36	nt forms are more highly developed than ancient.	will not here enter on this subject, for natura
37	ne, and a palaeozoic fauna by a secondary fauna.	do not doubt that this process of improvement h
37	omparison with the ancient and beaten forms; but	can see no way of testing this sort of progress
38	o the embryological development of recent forms.	must follow Pictet and Huxley in thinking that
38	th of this doctrine is very far from proved. Yet	fully expect to see it hereafter confirmed, fr
39	theory of natural selection. In a future chapter	shall attempt to show that the adult differs fr
39	by MM. Lund and Clausen in the caves of Brazil.	was so much impressed with these facts that I s
39	l. i was so much impressed with these facts that I	strongly insisted, in 1839 and 1845, on this La
40	nt times was peopled by numerous marsupials; and	have shown in the publications above alluded to
41	stribution. It may be asked in ridicule, whether	suppose that the megatherium and other allied h
41	. summary of the preceding and present Chapters.	have attempted to show that the geological reco
43	first bed of the Silurian system was deposited;	can answer this latter question only hypothetic
45	if then the geological record be as imperfect as	believe it to be, and it may at least be assert
50	; the relation of organism to organism being, as	have already often remarked, the most important
51	eas, as is so commonly and notoriously the case.	believe, as was remarked in the last chapter, i
53	orthern and southern hemispheres? The answer, as	believe, is; that mammals have not been able to
54	w living at distant and separated points; nor do	for a moment pretend that any explanation could
54	such cases. But after some preliminary remarks,	will discuss a few of the most striking classes
55	with a pre existing closely allied species. And	now know from correspondence, that this coincid
55	unite for each birth, or which often intercross,	believe that during the slow process of modific
56	descent from a single parent. To illustrate what	mean: our English race horses differ slightly f
56	ore discussing the three classes of facts, which	have selected as presenting the greatest amount
56	lty on the theory of single centres of creation,	must say a few words on the means of dispersal.
56	nd other authors have ably treated this subject.	can give here only the briefest abstract of the
56	high road for migration, but now be impassable;	shall, however, presently have to discuss this
57	and to the several intervening oceanic islands.	freely admit the former existence of many islan
57	ng oceans such sunken islands are now marked, as	believe, by rings of coral or atolls standing o
57	ing over them. Whenever it is fully admitted, as	believe it will some day be, that each species
58	ecurity on the former extension of the land. But	do not believe that it will ever be proved that
58	of consisting of mere piles of volcanic matter.	must now say a few words on what are called acc
58	ight be called occasional means of distribution.	shall here confine myself to plants. In botanic
58	s may be said to be almost wholly unknown. Until	tried, with Mr. Berkeley's aid, a few experimen
58	he injurious action of sea water. To my surprise	found that out of 87 kinds, 64 germinated after
59	d an immersion of 137 days. For convenience sake	chiefly tried small seeds, without the capsule
59	they were injured by the salt water. Afterwards	tried some larger fruits, capsules, etc., and s
59	rise in the stream was washed into the sea. Hence	was led to dry stems in branches of 94 plants w
60	days, and were then capable of germination. But	do not doubt that plants exposed to the waves w
60	trees, these stones being a valuable royal tax.	find on examination, that when irregularly shap
61	ars old, three dicotyledonous plants germinated:	am certain of the accuracy of this observation.
61	tain of the accuracy of this observation. Again,	can show that the carcasses of birds, when floa
61	effective agents in the transportation of seeds.	could give many facts showing how frequently bi
61	gales to vast distances across the ocean. We may	think safely assume that under such circumstanc
61	d some authors have given a far higher estimate.	have never seen an instance of nutritious seeds
61	organs of a turkey. In the course of two months,	picked up in my garden 12 kinds of seeds, out o
62	nd these seemed perfect, and some of them, which	tried, germinated. But the following fact is mo
62	astric juice, and do not in the least injure, as	know by trial, the germination of seeds; now af
62	lve to twenty hours, disgorge pellets, which, as	know from experiments made in the Zoological Ga
62	r two days and fourteen hours. Fresh water fish,	find, eat seeds of many land and water plants;
62	seeds might be transported from place to place.	forced many kinds of seeds into the stomachs of
62	aks and feet of birds are generally quite clean,	can show that earth sometimes adheres to them:
62	earth sometimes adheres to them: in one instance	removed twenty two grains of dry argillaceous e
63	would sometimes include a few minute seeds? But	shall presently have to recur to this subject.
63	d brushwood, bones, and the nest of a land bird,	can hardly doubt that they must occasionally ha
63	er of the flora in comparison with the latitude,	suspected that these islands had been partly st
64	nturies and tens of thousands of years, it would	think be a marvellous fact if many plants had n
65	d, would not receive colonists by similar means.	do not doubt that out of twenty seeds or animal
69	ently have retreated to their present homes; but	have met with no satisfactory evidence with res
69	representative species. In illustrating what, as	believe, actually took place during the Glacial
69	, actually took place during the Glacial period,	assumed that at its commencement the arctic pro
69	a and Europe; and it may be reasonably asked how	account for the necessary degree of uniformity
70	e completely separated by wider spaces of ocean.	believe the above difficulty may be surmounted
70	intermigration under a more favourable climate,	attribute the necessary amount of uniformity in
70	ted to large, but partial oscillations of level,	am strongly inclined to extend the above view,
70	mmencement of the Glacial period. We now see, as	believe, their descendants, mostly in a modifie
72	s now living in areas completely sundered. Thus,	think, we can understand the presence of many e
72	our more immediate subject, the Glacial period.	am convinced that Forbes's view may be largely
73	far below their present level. In central Chile	was astonished at the structure of a vast mound
73	height, crossing a valley of the Andes: and this	now feel convinced was a gigantic moraine, left
75	ns of Borneo. Some of these Australian forms, as	hear from Dr. Hooker, extend along the heights
75	n the lowlands; and a long list can be given, as	am informed by Dr. Hooker, of European genera,

>age **(Key Word)***

⊤65

0376 productions, similar cases occur; as an example, I may quote a remark by the highest authority, P
0378 , which must have lain or the line of march. But I do not doubt that some temperate productions wi
0378 of the Pyrenees. At this period of extreme cold, I believe that the climate under the equator at
0378 n thousand feet. During this the coldest period, I suppose that large spaces of the tropical lowl
0378 ya, as graphically described by Hooker. Thus, as I believe, a considerable number of plants, a fe
0379 forms on the mountains of Borneo and Abyssinia. I suspect that this preponderant migration from
0380 l period must have been completely isolated; and I believe that the productions of these islands
0380 continental forms, naturalised by man's agency. I am far from supposing that all difficulties ar
0380 ons. Very many difficulties remain to be solved. I do not pretend to indicate the exact lines and
0381 mmon, as another species within their own homes. I have said that many difficulties remain to be
0381 tarctic regions. These cannot be here discussed. I will only say that as far as regards the occurr
0381 mote as Kerguelen Land, New Zealand, and Fuegia, I believe that towards the close of the Glacial
0381 in radiating lines from some common centre; and I am inclined to look in the southern, as in the
0381 supported a highly peculiar and isolated flora. I suspect that before this flora was exterminate
0382 glacial period, by icebergs. By these means, as I believe, the southern shores of America, Austra
0382 nations of climate on geographical distribution. I believe that the world has recently felt one o
0383 ail in a remarkable manner throughout the world. I well remember, when first collecting in the fre
0383 ns of ranging widely, though so unexpected, can, I think, in most cases be explained by their hav
0384 re consider only a few cases. In regard to fish, I believe that the same species never occur in th
0384 ty of their ova when removed from the water. But I am inclined to attribute the dispersal of fresh
0385 ediately killed by sea water, as are the adults. I could not even understand how some naturalised
0385 hroughout the same country. But two facts, which I have observed, and no doubt many others remain
0385 enly emerges from a pond covered with duck weed, I have twice seen these little plants adhering to
0385 tle duck weed from one aquarium to another, that I have quite unintentionally stocked the one with
0385 r. but another agency is perhaps more effectual: I suspended a duck's feet, which might represent
0385 any ova of fresh water shells were hatching; and I found that numbers of the extremely minute and
0386 cquire, as if in consequence, a very wide range. I think favourable means of dispersal explain th
0386 favourable means of dispersal explain this fact. I have before mentioned that earth occasionally,
0386 t likely to have muddy feet. Birds of this order I can show are the greatest wanderers, and are oc
0386 sure to fly to their natural fresh water haunts. I do not believe that botanists are aware how cha
0386 ware how charged the mud of ponds is with seeds: I have tried several little experiments, but will
0386 but will here give only the most striking case; I took in February three tablespoons of mud from
0386 nd; this mud when dry weighed only 6 3/4 ounces! I kept it covered up in my study for six months,
0387 ned in a breakfast cup! Considering these facts, I think it would be an inexplicable circumstance
0387 known agencies probably have also played a part. I have stated that fresh water fish eat some kind
0387 ets or in excrement, many hours afterwards. When I saw the great size of the seeds of that fine wa
0387 bered Alph. de Candolle's remarks on this plant, I thought that its distribution must remain quite
0387 nelumbium luteum) in a heron's stomach; although I do not know the fact, yet analogy makes me beli
0388 same identical form or in some degree modified, I believe mainly depends on the wide dispersal of
0388 to the last of the three classes of facts, which I have selected as presenting the greatest amount
0389 ave come to inhabit distant points of the globe. I have already stated that I cannot honestly admi
0389 points of the globe. I have already stated that I cannot honestly admit Forbes's view on continen
0389 ould remove many difficulties, but it would not, I think, explain all the facts in regard to insul
0389 to insular productions. In the following remarks I shall not confine myself to the mere question o
0389 the Cape of Good Hope or in Australia, we must, I think, admit that something quite independently
0390 ts is scanty, the proportion of endemic species (i.e. those found nowhere else in the world) is of
0391 ican birds are almost every year blown there, as I am informed by Mr. E. V. Harcourt. So that they
0392 not a little doubtful. Facility of immigration, I believe, has been at least as important as the
0393 islands with which the great oceans are studded. I have taken pains to verify this assertion, and
0393 i have taken pains to verify this assertion, and I have found it strictly true. I have, however, b
0393 is assertion, and I have found it strictly true. I have, however, been assured that a frog exists
0393 ountains of the great island of New Zealand; but I suspect that this exception (if the information
0393 explain. Mammals offer another and similar case. I have carefully searched the oldest voyages; but
0393 voyages, but have not finished my search; as yet I have not found a single instance, free from dou
0394 at the distance of 600 miles from the mainland. I hear from Mr. Tomes, who has specially studied
0395 he admirable zeal and researches of Mr. Wallace. I have not as yet had time to follow up this subj
0395 n all other quarters of the world; but as far as I have gone, the relation generally holds good. W
0396 ortance of the relation of organism to organism. I do not deny that there are many and grave diffi
0397 not a wreck now remains, must not be overlooked. I will here give a single instance of one of the
0397 ily killed by salt; their eggs, at least such as I have tried, sink in sea water and are killed by
0397 mber across moderately wide arms of the sea. And I found that several species die in this state wi
0397 helix pomatia, and after it had again hybernated I put it in sea water for twenty days, and it per
0397 s this species has a thick calcareous operculum, I removed it, and when it had formed a new membra
0397 it, and when it had formed a new membranous one, I immersed it for fourteen days in sea water, and
0397 numerous instances could be given of this fact. I will give only one, that of the Galapagos Archi
0398 africa, like those of the Galapagos to America. I believe this grand fact can receive no sort of
0399 cial period. The affinity, which, though feeble, I am assured by Dr. Hooker is real, between the f
0399 is affinity is confined to the plants, and will, I do not doubt, be some day explained. The law wh
0400 ds of the Galapagos Archipelago are tenanted, as I have elsewhere shown, in a quite marvellous man
0400 mportant for its inhabitants; whereas it cannot, I think, be disputed that the nature of the other
0402 ome one island to the others. But we often take, I think, an erroneous view of the probability of
0403 he indigenous species. From these considerations I think we need not greatly marvel at the endemic
0404 her analogous facts could be given. And it will, I believe, be universally found to be true, that
0404 orld, many of the species have very wide ranges. I can hardly doubt that this rule is generally tr
0405 emote points of the world, we ought to find, and I believe as a general rule we co find, that some
0406 or island to those of the nearest mainland, are I think, utterly inexplicable on the ordinary vie
0406 of last and present Chapters. In these chapters I have endeavoured to show, that if we make due a
0407 have become extinct in the intermediate tracts, I think the difficulties in believing that all th
0407 of time being thus granted for their migration, I do not think that the difficulties are insupera
0407 the effects of climatal changes on distribution, I have attempted to show how important has been t
0407 he influence of the modern Glacial period, which I am fully convinced simultaneously affected the
0407 versified are the means of occasional transport, I have discussed at some little length the means
0408 ecies, have proceeded from some one source; then I think all the grand leading facts of geographic
0409 on these same principles, we can understand, as I have endeavoured to show, why oceanic islands s
0410 and the exceptions, which are not rare, may, as I have attempted to show, be accounted for by mig
0411 chapters, on Variation and on Natural Selection, I have attempted to show that it is the widely ra

0411 s, thus produced ultimately become converted, as I believe, into new and distinct species: and the
0412 , tend to go on increasing indefinitely in size. I further attempted to show that from the varying
0412 d by looking to certain facts in naturalisation. I attempted also to show that there is a constant
0412 vergent, the less improved, and preceding forms. I request the reader to turn to the diagram illus
0413 n order distinct from the genera descended from (I). So that we here have many species descended f
0413 ed in our classification, than mere resemblance. I believe that something more is included: and th
0414 ely related to the habits and food of an animal, I have always regarded as affording very clear in
0415 , true. But their importance for classification, I believe, depends on their greater constancy thr
0415 hat of all their parts, not only in this but, as I apprehend, in every natural family, is very une
0416 air, this external and trifling character would, I think, have been considered by naturalists as i
0417 s, even when none are important, alone explains, I think, that saying of Linnaeus, that the charac
0420 difficulties in classification are explained, if I do not greatly deceive myself, on the view that
0420 r and separating objects more or less alike. But I must explain my meaning more fully. I believe t
0420 alike. But I must explain my meaning more fully. I believe that the arrangement of the groups with
0420 iod. Species of three of these genera (A, F, and I) have transmitted modified descendants to the p
0421 itute a distinct order from those descended from I, also broken up into two families. Nor can the
0421 the same gerus with the parent A: or those from I, with the parent I. But the existing genus F14
0421 h the parent A: or those from I, with the parent I. But the existing genus F14 may be supposed to
0421 common parent, as will all the descendants from I: so will it be with each subordinate branch of
0421 o suppose that any of the descendants of A or of I have been so much modified as to have more or l
0421 ally was intermediate in character between A and I, and the several genera descended from these tw
0422 eings of the same group. Thus, on the view which I hold, the natural system is genealogical in its
0422 , had to be included, such an arrangement would, I think, be the only possible one. Yet it might b
0423 e, because less constant. In classing varieties, I apprehend if we had a real pedigree, a genealog
0424 that the Hottentot had descended from the Negro, I think he would be classed under the Negro group
0425 ? the supposition is of course preposterous: and I might answer by the argumentum ad hominem, and
0425 degree, and has taken a longer time to complete? I believe it has thus been unconsciously used: an
0425 thus been unconsciously used: and only thus can I understand the several rules and guides which h
0426 less in our classification. We shall hereafter, I think, clearly see why embryological characters
0429 new order: and that in the vegetable kingdom, as I learn from Dr. Hooker, it has added only two or
0429 ll size. In the chapter on geological succession I attempted to show, on the principle of each gro
0429 f a single one: but such richness in species, as I find after some investigation, does not commonl
0429 only fall to the lot of aberrant genera. We can, I think, account for this fact only by looking at
0432 ritance, all the forms descended from A, or from I, would have something in common. In a tree we c
0432 e two unite and blend together. We could not, as I have said, define the several groups: but we co
0433 ver different they may be from their parent: and I believe this element of descent is the hidden b
0441 of development generally rises in organisation: I use this expression, though I am aware that it
0441 s in organisation: I use this expression, though I am aware that it is hardly possible to define c
0441 ites having the ordinary structure, or into what I have called complemental males: and in the latt
0443 n the mature animal, into which it is developed. I believe that all these facts can be explained,
0443 s fully displayed. The cause may have acted, and I believe generally has acted, even before the em
0444 th, as long as it was fed by its parents. Hence, I conclude, that it is quite possible, that each
0444 may have appeared at an extremely early period. I have stated in the first chapter, that there is
0444 a corresponding age in the offspring and parent. I am far from meaning that this is invariably the
0444 om meaning that this is invariably the case: and I could give a good many cases of variations (tak
0444 wo principles, if their truth be admitted, will, I believe, explain all the above specified leadin
0445 obably descended from the same wild stock: hence I was curious to see how far their puppies differ
0445 how far their puppies differed from each other: I was told by breeders that they differed just as
0445 ing the old dogs and their six days old puppies, I found that the puppies had not nearly acquired
0445 ll amount of proportional difference. So, again, I was told that the foals of cart and race horses
0445 grown animals: and this surprised me greatly, as I think it probable that the difference between t
0445 ee days old colt of a race and heavy cart horse, I find that the colts have by no means acquired t
0445 of Figeon have descended from one wild species, I compared young pigeons of various breeds, withi
0445 breeds, within twelve hours after being hatched: I carefully measured the proportions (but will no
0445 ily in length and form of beak, that they would, I cannot doubt, be ranked in distinct genera, had
0449 agassiz believes this to be a law of nature: but I am bound to confess that I only hope to see the
0449 a law of nature: but I am bound to confess that I only hope to see the law hereafter proved true.
0450 mammae are very general in the males of mammals: I presume that the bastard wing in birds may be s
0452 yo, but afterwards wholly disappear. It is also, I believe, a universal rule, that a rudimentary p
0453 n said to have retained its embryonic condition. I have now given the leading facts with respect t
0453 imperfect nails sometimes appear on the stumps: I could as soon believe that these vestiges of na
0454 see rudiments of various parts in monsters. But I doubt whether any of these cases throw light on
0454 n by showing that rudiments can be produced: for I doubt whether species under nature ever undergo
0454 pecies under nature ever undergo abrupt changes. I believe that disuse has been the main agency: t
0456 he laws of inheritance. Summary. In this chapter I have attempted to show, that the subordination
0458 all been modified in the course of descent, that I should without hesitation adopt this view, even
0459 ent with modification through natural selection, I do not deny. I have endeavoured to give to them
0459 cation through natural selection, I do not deny. I have endeavoured to give to them their full for
0459 nstinct. The truth of these propositions cannot, I think, be disputed. It is, no doubt, extremely
0460 erile females in the same community of ants: but I have attempted to show how this difficulty can
0460 t universal fertility of varieties when crossed, I must refer the reader to the recapitulation of
0462 ve greatly facilitated migration. As an example, I have attempted to show how potent has been the
0463 one period: and all changes are slowly effected. I have also shown that the intermediate varieties
0464 d utterly unknown epochs in the world's history. I can answer these questions and grave objections
0465 ttent in their accumulation: and their duration, I am inclined to believe, has been shorter than t
0465 s formations beneath the lowest Silurian strata, I can only recur to the hypothesis given in the n
0465 it: but that it is imperfect to the degree which I require, few will be inclined to admit. If we l
0465 which may justly be urged against my theory: and I have now briefly recapitulated the answers and
0465 ers and explanations which can be given to them. I have felt these difficulties far too heavily du
0469 ature, favouring the good and rejecting the bad? I can see no limit to this power, in slowly and b
0469 than this, seems to me to be in itself probable. I have already recapitulated, as fairly as I coul
0469 able. I have already recapitulated, as fairly as I could, the opposed difficulties and objections:
0474 of the same class with their several instincts. I have attempted to show how much light the princ
0480 h it seems that we wilfully will not understand. I have now recapitulated the chief facts and cons
0481 almost infinite number of generations. Although I am fully convinced of the truth of the views gi
0481 en in this volume under the form of an abstract, I by no means expect to convince experienced natu

```
0482 f species, may be influenced by this volume; but I look with confidence to the future, to young an
0483 nsider reverent silence. It may be asked how far I extend the doctrine of the modification of spec
0483 e species closely resemble each other. Therefore I cannot doubt that the theory of descent with mo
0484 tion embraces all the members of the same class. I believe that animals have descended from at mos
0484 growths on the wild rose or oak tree. Therefore I should infer from analogy that probably all the
0484 this or that form be in essence a species. This I feel sure, and I speak after experience, will b
0484 m be in essence a species. This I feel sure, and I speak after experience, will be no slight relie
0486 ew each organic being, how far more interesting, I speak from experience, will the study of natura
0488 descendants, was created. In the distant future I see open fields for far more important research
0488 ning the birth and death of the individual. When I view all beings not as special creations, but a
0148 ll or carapace. This is the case with the male IBLA, and in a truly extraordinary manner with the
0363 d that these islands had been partly stocked by ICE borne seeds, during the glacial epoch. At my r
0366 and rocks scored by drifted icebergs and coast ICE, plainly reveal a former cold period. The form
0366 he mountains would become covered with snow and ICE, and their former Alpine inhabitants would des
0373 alia. Looking to America; in the northern half, ICE borne fragments of rock have been observed on
0381 iod, when the antarctic lands, now covered with ICE, supported a highly peculiar and isolated flor
0363 presently have to recur to this subject. As ICEBERGS are known to so sometimes loaded with earth
0363 archipelago. Hence we may safely infer that ICEBERGS formerly landed their rocky burthens on the
0366 rratic boulders, and rocks scored by drifted ICEBERGS and coast ice, plainly reveal a former cold
0381 hat towards the close of the Glacial period, ICEBERGS, as suggested by Lyell, have been largely co
0382 t the commencement of the Glacial period, by ICEBERGS. By these means, as I believe, the southern
0393 bank connected with the mainland; moreover, ICEBERGS formerly brought boulders to its western sho
0399 ed by seeds brought with earth and stones on ICEBERGS, drifted by the prevailing currents, this an
0200 g of the adder, and in the ovipositor of the ICHNEUMON, by which its eggs are deposited in the liv
0244 rothers, ants making slaves, the larvae of ICHNEUMONIDAE feeding within the live bodies of caterpi
0472 ueen bee for her own fertile daughters; at ICHNEUMONIDAE feeding within the live bodies of caterpi
0366 polished surfaces, and perched boulders, of the ICY streams with which their valleys were lately f
0029 earn a lesson of caution, when they deride the IDEA of species in a state of nature being lineal d
0051 ries; and a series impresses the mind with the IDEA of an actual passage. Hence I look at individu
0115 ere become naturalised, we can gain some crude IDEA in what manner some of the natives would have
0145 y florets in some Compositae countenances this IDEA; but, in the case of the corolla of the Umbell
0145 do not feel at all sure that C. C. Sprengel's IDEA that the ray florets serve to attract insects,
0282 how each author attempts to give an inadequate IDEA of the duration of each formation or even each
0284 tary rocks in Britain, gives but an inadequate IDEA of the time which has elapsed during their acc
0285 as does the vain endeavour to grapple with the IDEA of eternity. I am tempted to give one other ca
0289 and if in each separate territory, hardly any IDEA can be formed of the length of time which has
0296 the closest inspection of a formation give any IDEA of the time which its deposition has consumed.
0419 e view of classification tacitly including the IDEA of descent. Our classifications are often plai
0432 hether large or small, and thus give a general IDEA of the value of the differences between them.
0433 under the term of the Natural System. On this IDEA of the natural system being, in so far as it h
0481 t duration; and now that we have acquired some IDEA of the lapse of time, we are too apt to assume
0191 admit that the swimbladder is homologous, or IDEALLY similar, in position and structure with the
0306 nd from the revolution in our palaeontological IDEAS on many points, which the discoveries of even
0472 rfect; and if some of them be abhorrent to our IDEAS of fitness. We need not marvel at the sting o
0018 of the breeds closely resemble, perhaps are IDENTICAL with, those still existing. Even if this la
0273 f its proper function of producing offspring IDENTICAL with the parent form. Now hybrids in the fi
0313 rth, we have reason to believe that the same IDENTICAL form never reappears. The strongest apparen
0314 sing that one species should retain the same IDENTICAL form much longer than others; or, if changi
0316 ring, it is quite incredible that a fantail, IDENTICAL with the existing breed, could be raised fr
0371 of the New and Old Worlds, we find very few IDENTICAL species (though Asa Gray has lately shown t
0372 a gray has lately shown that more plants are IDENTICAL than was formerly supposed), but we find in
0374 can be thrown on the present distribution of IDENTICAL and allied species. In America, Dr. Hooker
0376 others as varieties; but some are certainly IDENTICAL, and many, though closely related to northe
0379 dolle in regard to Australia, that many more IDENTICAL plants and allied forms have apparently mig
0381 say that as far as regards the occurrence of IDENTICAL species at points so enormously remote as K
0382 g passage has speculated, in language almost IDENTICAL with mine, on the effects of great alternat
0388 he lower animals, whether retaining the same IDENTICAL form or in some degree modified, I believe
0395 and they are inhabited by closely allied or IDENTICAL quadrupeds. No doubt some few anomalies occ
0404 ies occur, there will likewise be found some IDENTICAL species, showing, in accordance with the fo
0409 should be a correlation, in the presence of IDENTICAL species, of varieties, of doubtful species,
0423 he most important part, happens to be nearly IDENTICAL; no one puts the swedish and common turnips
0478 osely allied species inhabit two areas, some IDENTICAL species common to both still exist. Whereve
0487 mporaneous two formations, which include few IDENTICAL species, by the general succession of their
0227 the hexagonal prisms will have every angle IDENTICALLY the same with the best measurements which
0235 atural selection; cases of instincts almost IDENTICALLY the same in animals so remote in the scale
0315 wo forms, the old and the new, would not be IDENTICALLY the same; for both would almost certainly
0323 with; for in some cases not one species is IDENTICALLY the same, but they belong to the same fami
0352 chapter, it is incredible that individuals IDENTICALLY the same should ever have been produced th
0353 alia; and some of the aboriginal plants are IDENTICALLY the same at these distant points of the no
0369 ranges, though very many of the species are IDENTICALLY the same, some present varieties, some are
0375 ic cones of Java, many plants occur, either IDENTICALLY the same or representing each other, and a
0376 ern and southern hemispheres, are sometimes IDENTICALLY the same; but they are much oftener specif
0403 nds, namely, that the inhabitants, when not IDENTICALLY the same, yet are plainly related to the i
0365 ve. Dispersal during the Glacial period. The IDENTITY of many plants and animals, on mountain summ
0367 both hemispheres. Thus we can understand the IDENTITY of many plants at points so immensely remote
0371 nderstand the relationship, with very little IDENTITY, between the productions of North America an
0372 ocean. These cases of relationship, without IDENTITY, of the inhabitants of seas now disjoined, a
0477 stand, by the aid of the Glacial period, the IDENTITY of some few plants, and the close alliance o
0234 heir combs. Moreover, many bees have to remain IDLE for many days during the process of secretion.
0062 ot see, or we forget, that the birds which are IDLY singing round us mostly live on insects or see
0003 species. Nevertheless, such a conclusion, even IF well founded, would be unsatisfactory, until I
0005 uggle for existence, it follows that any being, IF it vary however slightly in any manner profitab
0006 regard to the origin of species and varieties, IF he makes due allowance for our profound ignoran
0010 r had the action of the conditions been direct, IF any of the young had varied, all would probably
0012 ti and those with long beaks large feet. Hence, IF men goes on selecting, and thus augmenting, any
0013 ppearing in several members of the same family. IF strange and rare deviations of structure are tr
```

Page **************************************(Key Word)**************************************

0015	tral forms, it seems to me not improbable, that	IF we could succeed in naturalising, or were to cu
0015	ment itself the conditions of life are changed.	IF it could be shown that our domestic varieties m
0016	e descendants of aboriginally distinct species.	IF any marked distinction existec between domestic
0016	from one or several parent species. This point,	IF it could be cleared up, would be interesting; i
0016	f it could be cleared up, would be interesting:	IF, for instance, it could be shown that the greyh
0017	vented their domestication? I cannot doubt that	IF other animals and plants, equal in number to ou
0018	are identical with, those still existing. Even	IF this latter fact were found more strictly and c
0019	degree intermediate between their parents! and	IF we account for our several domestic races by th
0020	t a race may be modified by occasional crosses,	IF aided by the careful selection of those individ
0022	least a score of pigeons might be chosen, which	IF shown to an ornithologist, anc he were told tha
0023	in other cases, I will here briefly give them.	IF the several breeds are not varieties, and have
0025	principle of reversion to ancestral characters,	IF all the domestic breeds have descended from the
0025	breeds have descended from the rock pigeon. But	IF we deny this, we must make one of the two follo
0026	k there is some probability in this hypothesis.	IF applied to species closely related together, th
0031	ic, which they can model almost as they please.	IF I had space I could quote numerous passages to
0031	ers have done for sheep, says: It would seem as	IF they had chalked out upon a wall a form perfect
0032	more indispensable even than in ordinary cases.	IF selection consisted merely in separating some v
0032	dgment sufficient to become an eminent breeder.	IF gifted with these qualities, and he studies his
0032	will succeed, and may make great improvements!	IF he wants any of these qualities, he will assure
0036	appearance cf being quite different varieties.	IF there exist savages so barbarous as never to th
0037	ight succeed from a poor seedling growing wild,	IF it had come from a garden stock. The pear, thou
0037	t cultivated in our flower and kitchen gardens.	IF it has taken centuries or thousands of years to
0049	distance, it has been well asked, will suffice?	IF that between America and Europe is ample, will
0050	l value. I have been struck with the fact, that	IF any animal or plant in a state of nature be hig
0050	how very generally there is some variation. But	IF he confine his attention to one class within on
0051	a greater number of closely allied forms. But	IF his observations be widely extended, he will in
0052	eties of certain fossil land shells in Madeira.	IF a variety were to flourish so as to exceed in n
0054	ents to become dominant over their compatriots.	IF the plants inhabiting a country and described i
0055	ill favourable to variation. On the other hand,	IF we look at each species as a special act of cre
0056	s of the same cerus have been formed, or where,	IF we may use the expression, the manufactory of s
0056	be a slow one. And this certainly is the case,	IF varieties be looked at as incipient species: fo
0056	mall genera are now varying and increasing: for	IF this had been so, it would have been fatal to m
0058	ment is indeed scarcely more than a truism, for	IF a variety were found to have a wider range than
0059	a certain amount of difference, for two forms,	IF differing very little, are generally ranked as
0059	and we can clearly understand these analogies.	IF species have once existed as varieties, and hav
0059	ereas, these analogies are utterly inexplicable	IF each species has been independently created. We
0060	l forms of British plants are entitled to hold,	IF the existence of any well marked varieties be a
0061	ever slight and from whatever cause proceeding,	IF it be in any degree profitable to an individual
0061	this principle, by which each slight variation,	IF useful, is preserved, by the term of Natural Se
0063	sense be said to struggle with these trees, for	IF too many of these parasites grow on the same tr
0064	ing naturally increases at so high a rate, that	IF not destroyed, the earth woulc soon be covered
0064	m for his progeny. Linnaeus has calculated that	IF an annual plant produced only two seeds, and th
0064	inc forth three pair of young in this interval:	IF this be so, at the end of the fifth century the
0064	ch have run wild in several parts of the world:	IF the statements of the rate of increase of slow
0066	in the great majority of cases is an early one.	IF an animal can in any way protect its own eggs o
0066	and yet the average stock be fully kept up; but	IF many eggs or young are destroyed, many must be
0066	which lived on an average for a thousand years,	IF a single seed were produced once in a thousand
0067	5 were destroyed, chiefly by slugs and insects.	IF turf which has long been mown, and the case wou
0068	e depends chiefly on the destruction of vermin.	IF not one head of game were shot during the next
0068	twenty years in England, and at the same time,	IF no vermin were destroyed, there would, in all p
0069	om competitors for the same place and food; and	IF these enemies cr competitors be in the least de
0072	hecked by some means, probably by birds. Hence,	IF certain insectivorous birds (whose numbers are
0073	ve tried, I have found that the visits of bees,	IF not indispensable, are at least highly benefici
0073	he nectar. Hence I have very little doubt, that	IF the whole genus of humble bees became extinct i
0075	mes see the contest soon decided: for instance,	IF several varieties of wheat be sown together, an
0076	could be kept up for half a dozen generations,	IF they were allowed to struggle together, like be
0076	together, like beings in a state of nature, and	IF the seed or young were not annually sorted. As
0078	districts. In this case we can clearly see that	IF we wished in imagination to give the plant the
0078	ll generally be changed in an essential manner.	IF we wished to increase its average numbers in it
0080	ccur in the course of thousands of generations?	IF such do occur, can we doubt (remembering that m
0081	would most seriously affect many of the others.	IF the country were open on its borders, new forms
0081	ture which would assuredly be better filled up,	IF some of the original inhabitants were in some m
0084	insects in preserving them from danger. Grouse,	IF not destroyed at some period of their lives, wo
0085	far more than those with other coloured flesh.	IF, with all the aids of art, these slight differe
0086	nd by their inheritance at a corresponding age.	IF it profit a plant to have its seecs more and mo
0086	shall not be in the least degree injurious: for	IF they became so, they would cause the extinction
0087	ch individual for the benefit of the community;	IF each in consequence profits by the selected cha
0087	cture used orly once in an animal's whole life,	IF of high importance to it, might be modified to
0087	at fanciers assist in the act of hatching. Now,	IF nature had to make the beak of a full grown pig
0087	the same fact probably occurs under nature, and	IF so, natural selection will be able to modify on
0089	the details necessary to support this view; but	IF man can in a short time give elegant carriage a
0091	rather than mice is shown to be inherited. Now,	IF any slight innate change of habit or of structu
0092	ruction appears a simple loss to the plant; yet	IF a little pollen were carried, at first occasion
0094	female organs become more or less impotent: now	IF we suppose this to occur in ever so slight a de
0095	llen on to the stigmatic surface. Hence, again,	IF humble bees were to become rare in any country,
0095	ingle diluvial wave, so will natural selection,	IF it be a true principle, banish the belief of th
0097	s and stigmas fully exposed to the weather! but	IF an occasional cross be indispensable, the fulle
0097	ere, that their fertility is greatly diminished	IF these visits be prevented. Now, it is scarcely
0098	titude of hybrids between distinct species: for	IF you bring on the same brush a plant's own polle
0098	e for self fertilisation, it is well known that	IF very closely allied forms or varieties are plan
0099	ame flower, though placed so close together, as	IF for the very purpose of self fertilisation, sho
0099	individual being advantageous or indispensable!	IF several varieties of the cabbage, radish, onion
0101	hermaphrodites, and some of them unisexual. But	IF, in fact, all hermaphrodites do occasionally in
0102	o seize on each place in the economy of nature,	IF any one species does not become modified and im
0102	as better to fill up the unoccupied place. But	IF the area be large, its several districts will b
0102	present different conditions of life; and then	IF natural selection be modifying and improving a

Page **************************************(Key Word)**************************************

Page **(Key Word)**

0104	lace with all animals and with all plants. Even	IF	these take place only at long intervals, I am
0104	rcrosses even at rare intervals, will be great.	IF	there exist organic beings which never intercro
0104	ying any which depart from the proper type; but	IF	their conditions of life change and they underg
0104	ural selection. In a confined or isolated area,	IF	not very large, the organic and inorganic cond
0105	of importance in the production of new species.	IF,	however, an isolated area be very small, eithe
0105	nce of the appearance of favourable variations.	IF	we turn to nature to test the truth of these re
0106	e large number of already existing species; and	IF	some of these many species become modified and
0109	d. slow though the process of selection may be,	IF	feeble man can do much by his powers of artifi
0111	i quote the words of an agricultural writer) as	IF	by some murderous pestilence. Divergence of Cha
0113	untry has long ago arrived at its full average.	IF	its natural powers of increase be allowed to ae
0113	ly throughout all time to all animals, that is,	IF	they vary, for otherwise natural selection can
0113	plants. It has been experimentally proved, that	IF	a plot of ground be sown with one species of gr
0113	ave been sown on equal spaces of ground. Hence,	IF	any one species of grass were to go on varying
0114	stances. In an extremely small area, especially	IF	freely open to immigration, and where the conte
0117	sand generations; but it would have been better	IF	each had represented ten thousand generations.
0118	favourable to the production of new varieties.	IF,	then, these two varieties be variable, the mos
0120	this case would be represented in the diagram,	IF	all the lines proceeding from (A) were removed
0120	, from each other and from their common parent.	IF	we suppose the amount of change between each he
0121	ncured by later and improved lines of descent.	IF,	however, the modified offspring of a species g
0122	e into competition, both may continue to exist.	IF	then our diagram be assumed to represent a cons
0125	ow explained, to the formation of genera alone.	IF,	in our diagram, we suppose the amount of chang
0126	ses, as at the present day. Summary of Chapter.	IF	during the long course of ages and under varyin
0127	anisation, and I think this cannot be disputed:	IF	there be, owing to the high geometrical powers
0127	, i think it would be a most extraordinary fact	IF	no variation ever had occurred useful to each b
0127	any variations have occurred useful to man. But	IF	variations useful to any organic being do occur
0130	s give rise by growth to fresh buds, and these,	IF	vigorous, branch out and overtop on all sides n
0131	s. summary. I have hitherto sometimes spoken as	IF	the variations, so common and multiform in orga
0136	it would have been better for the good swimmers	IF	they had been able to swim still further, where
0136	it would have been better for the bad swimmers	IF	they had not been able to swim at all and had s
0137	r them, might in such case be an advantage: and	IF	so, natural selection would constantly aid the
0139	same genus have descended from a single parent,	IF	this view be correct, acclimatisation must be r
0143	ce existed with an antler only on one side; and	IF	this had been of any great use to the breed it
0144	tion, I think it can hardly be accidental, that	IF	we pick out the two orders of mammalia which ar
0146	so far fetched, as it may at first appear: and	IF	it be advantageous, natural selection may have
0147	a certain extent with our domestic productions.	IF	nourishment flows to one part or organ in exces
0147	to economise in every part of the organisation.	IF	under changed conditions of life a structure be
0150	group of bats having wings; it would apply only	IF	some one species of bat had its wings developed
0152	can obtain some light. In our domestic animals,	IF	any part, or the whole animal, be neglected and
0154	. tc explain by a simple example what is meant.	IF	some species in a large genus of plants had blu
0154	ue species varying into red, or conversely; but	IF	all the species had blue flowers, the colour wo
0158	t, however extraordinarily it may be developed,	IF	it be common to a whole group of species; that
0159	s produced by cultivation from a common parent:	IF	this be not so, the case will then be one of an
0162	roup, we could not distinguish these two cases:	IF,	for instance, we did not know that the rock pi
0171	classed under the following heads: Firstly, why	IF	species have descended from other species by in
0172	will, as we have seen, go hand in hand. Hence,	IF	we look at each species as descended from some
0173	frequent, till the one replaces the other. But	IF	we compare these species where they intermingle
0175	nge will come to be still more sharply defined.	IF	I am right in believing that allied or represen
0176	the same rule will probably apply to both; and	IF	we in imagination adapt a varying species to a
0176	rically than the forms which they connect. Now,	IF	we may trust these facts and inferences, and th
0179	, looking not to any one time, but to all time,	IF	my theory be true, numberless intermediate vari
0180	s like other polecats on mice and land animals.	IF	a different case had been taken, and it had bee
0181	gliding through the air rather than for flight.	IF	about a dozen genera of birds had become extinc
0182	ve been modified into perfectly winged animals.	IF	this had been effected, who would have ever ima
0184	n the water. Even in so extreme a case as this,	IF	the supply of insects were constant, and if bet
0184	is, if the supply of insects were constant, and	IF	better adapted competitors did not already exis
0186	y endeavouring to increase in numbers; and that	IF	any one being vary ever so little, either in ha
0186	hest possible degree. Yet reason tells me, that	IF	numerous gradations from a perfect and complex
0186	useful to its possessor, can be shown to exist;	IF	further, the eye does vary ever so slightly, an
0186	be inherited, which is certainly the case; and	IF	any variation or modification in the organ be e
0188	reat Articulate class. He who will cc thus far,	IF	he find on finishing this treatise that large b
0188	works by intellectual powers like those of man?	IF	we must compare the eye to an optical instrumen
0189	s the works of the Creator are to those of man?	IF	it could be demonstrated that any complex organ
0189	t know the transitional grades, more especially	IF	we look to much isolated species, round which,
0189	eory, there has been much extinction. Or again,	IF	we look to an organ common to all the members o
0190	ases natural selection might easily specialise,	IF	any advantage were thus gained, a part or organ
0192	and the obliteration of their adhesive glands.	IF	all pedunculated cirripedes had become extinct,
0193	n several members of the same class, especially	IF	ir members having very different habits of life
0193	s loss through disuse or natural selection. But	IF	the electric organs had been inherited from som
0197	w instances to illustrate these latter remarks.	IF	green woodpeckers alone had existed, and we did
0198	i have here alluded to them only to show that,	IF	we are unable to account for the characteristic
0199	yes of man, or for mere variety. This doctrine,	IF	true, would be absolutely fatal to my theory. Y
0201	eposited in the living bodies of other insects.	IF	it could be proved that any part of the structu
0201	g pain or for doing an injury to its possessor.	IF	a fair balance be struck between the good and e
0201	pse of time, under changing conditions of life,	IF	any part comes to be injurious, it will be modi
0201	comes to be injurious, it will be modified; or	IF	it be not so, the being will become extinct, as
0202	rfect even in that most perfect organ, the eye.	IF	our reason leads us to admire with enthusiasm a
0202	death of the insect by tearing out its viscera?	IF	we look at the sting of the bee, as having orig
0202	ould so often cause the insect's own death: for	IF	on the whole the power of stinging be useful to
0202	ugh it may cause the death of some few members.	IF	we admire the truly wonderful power of scent by
0203	the inexorable principle of natural selection.	IF	we admire the several ingenious contrivances, b
0204	stagger any one: yet in the case of any organ,	IF	we know of a long series of gradations in compl
0206	al history, Natura non facit saltum. This canon	IF	we look only to the present inhabitants of the
0206	ants of the world, is not strictly correct, but	IF	we include all those of past times, it must by
0208	one action follows another by a sort of rhythm;	IF	a person be interrupted in a song, or in repeat
0208	ar, which makes a very complicated hammock: for	IF	he took a caterpillar which had completed its h
0208	ourth, fifth, and sixth stages of construction.	IF	however, a caterpillar were taken out of a hamm

Page **(Key Word)**

age **************************************(Key Word)**************************************

209	us tried to complete the already finished work.	IF	we suppose any habitual action to become inheri
209	ct becomes so close as not to be distinguished.	IF	Mozart, instead of playing the pianoforte at th
209	instinct might be profitable to a species; and	IF	it can be shown that instincts do vary ever so
212	strange habits in certain species, which might,	IF	advantageous to the species, give rise, through
213	actions differ essentially from true instincts.	IF	we were to see one kind of wolf, when young and
215	lly do make an attack, and are then beaten; and	IF	not cured, they are destroyed; so that habit, w
216	t all fear, but fear only of dogs and cats, for	IF	the hen gives the danger chuckle, they will run
217	but at intervals of two or three days; so that,	IF	she were to make her own nest and sit on her ow
217	young birds of different ages in the same nest.	IF	this were the case, the process of laying and h
217	sionally she laid an egg in another bird's nest.	IF	the old bird profited by this occasional habit,
217	old bird profited by this occasional habit, or	IF	the young were made more vigorous by advantage
218	n collecting apparatus which would be necessary	IF	they had to store food for their own young. Som
219	selection making an occasional habit permanent,	IF	of advantage to the species, and if the insect
219	permanent, if of advantage to the species, and	IF	the insect whose nest and stored food are thus
219	xtraordinary than these well ascertained facts?	IF	we had not known of any other slave making ant,
223	i have seen, carry off pupae of other species,	IF	scattered near their nests, it is possible that
223	proper instincts, and do what work they could.	IF	their presence proved useful to the species whi
223	ed useful to the species which had seized them,	IF	it were more advantageous to this species to ca
224	ng slaves. When the instinct was once acquired,	IF	carried out to a much less extent even than in
225	uld have intersected or broken into each other,	IF	the spheres had been completed; but this is nev
226	reflecting on this case, it occurred to me that	IF	the Melipona had made its spheres at some given
226	tion, and tells me that it is strictly correct:	IF	a number of equal spheres be described with the
226	spheres in the other and parallel layer; then,	IF	the planes of intersection between the several sphe
227	the hive bee. Hence we may safely conclude that	IF	we could slightly modify the instincts already
228	ax was so thin, that the bottoms of the basins,	IF	they had been excavated to the same depth as in
230	ridge of vermilion wax, we can clearly see that	IF	the bees were to build for themselves a thin wa
230	uber, but I am convinced of their accuracy; and	IF	I had space, I could show that they are conform
233	e case of a queen wasp) making hexagonal cells,	IF	she work alternately on the inside and outside
234	hat it would be an advantage to our humble bee,	IF	a slight modification of her instinct led her t
234	e more and more advantageous to our humble bee,	IF	she were to make her cells more and more regula
234	ause, it would be advantageous to the Melipona,	IF	she were to make her cells closer together, and
236	tate of nature occasionally become sterile; and	IF	such insects had been social, and it had been p
236	ve been far better exemplified by the hive bee.	IF	a working ant or other neuter insect had been a
240	ediate condition. I may digress by adding, that	IF	the smaller workers had been the most useful to
240	te illustration: the difference was the same as	IF	we were to see a set of workmen building a hous
249	tement repeatedly made by Gartner, namely, that	IF	even the less fertile hybrids be artificially f
251	eneral habit, reproduced itself as perfectly as	IF	it had been a natural species from the mountain
252	nts have been carefully tried than with plants.	IF	our systematic arrangements can be trusted, tha
252	systematic arrangements can be trusted, that is	IF	the genera of animals are as distinct from each
253	in the hybrids should have gone on increasing.	IF	we were to act thus, and pair brothers and sist
267	tions between the nearest relations, especially	IF	these be kept under the same conditions of life
268	it that this is almost invariably the case. But	IF	we look to varieties produced under nature, we
268	ediately involved in hopeless difficulties; for	IF	two hitherto reputed varieties be found in any
268	e consequently ranks them as undoubted species.	IF	we thus argue in a circle, the fertility of all
268	under nature will assuredly have to be granted.	IF	we turn to varieties, produced, or supposed to
269	which were at first only slightly sterile; and	IF	this be so, we surely ought not to expect to fi
269	fference in the result. I have as yet spoken as	IF	the varieties of the same species were invariab
271	es in a state of nature, for a supposed variety	IF	infertile in any degree would generally be rank
273	rids to revert to either parent form; but this,	IF	it be true, is certainly only a difference in d
274	s are widely different from each other; whereas	IF	two very distinct varieties of one species are
275	g of crossed species, and of crossed varieties.	IF	we look at species as having been specially cre
280	geons have both descended from the rock pigeon;	IF	we possessed all the intermediate varieties whi
280	s, moreover, have become so much modified, that	IF	we had no historical or indirect evidence regar
281	ies, such as C. oenas. So with natural species,	IF	we look to forms very distinct, for instance to
281	he parent form of any two or more species, even	IF	we closely compared the structure of the parent
282	t have been inconceivably great. But assuredly,	IF	this theory be true, such have lived upon this
283	and how seldom they are rolled about! Moreover,	IF	we follow for a few miles any line of rocky cli
286	100 feet, as I am informed by Prof. Ramsay. But	IF,	as some geologists suppose, a range of older r
286	plied to the western extremity of the district.	IF,	then, we knew the rate at which the sea common
286	h too small an allowance; but it is the same as	IF	we were to assume a cliff one yard in height to
288	condition. The remains which do become embedded,	IF	in sand or gravel, will when the beds are uprai
289	parts of the world. The most skilful geologist,	IF	his attention had been exclusively confined to
289	ms of life, had elsewhere been accumulated. And	IF	in each separate territory, hardly any idea can
291	any thickness and extent over a shallow bottom,	IF	it continue slowly to subside. In this latter c
292	viewed as a whole, is extremely imperfect; but	IF	we confine our attention to any one formation,
296	mes, during the same geological period. So that	IF	such species were to undergo a considerable amo
297	hird, A, to be found in an underlying bed; even	IF	A were strictly intermediate between B and C, i
297	heir species; and they do this the more readily	IF	the specimens come from different sub stages of
297	which on my theory we ought to find. Moreover,	IF	we look to rather wider intervals, namely, to d
300	gions of the whole world in organic beings; yet	IF	all the species were to be collected which ever
301	nal gradations between any two or more species.	IF	such gradations were not fully preserved, trans
301	enerally be local or confined to one place, but	IF	possessed of any decided advantage, or when fur
301	logists, be ranked as new and distinct species.	IF	then, there be some degree of truth in these re
302	eri and these links, let them be ever so close,	IF	found in different stages of the same formation
302	to the belief in the transmutation of species.	IF	numerous species, belonging to the same genera
303	ucceeding formation such species will appear as	IF	suddenly created. I may here recall a remark fo
304	if extracted from the chalk of Belgium. And, as	IF	to make the case as striking as possible, this
305	north as they are at present. Even at this day,	IF	the Malay Archipelago were converted into land,
307	acters in any degree intermediate between them.	IF,	moreover, they had been the progenitors of the
307	umerous and improved descendants. Consequently,	IF	my theory be true, it is indisputable that befo
307	lated before the Silurian epoch, is very great.	IF	these most ancient beds had been wholly worn aw
309	ntervened during these enormously long periods.	IF	then we may infer anything from these facts, we
309	nd. Nor should we be justified in assuming that	IF,	for instance, the bed of the Pacific Ocean wer
312	tions, though undoubtedly of high antiquity	IF	measured by years, only one or two species are
312	r, as far as we know, on the face of the earth.	IF	we may trust the observations of Philippi in Si
313	seldom changed in exactly the same degree. Yet	IF	we compare any but the most closely related for

Page **************************************(Key Word)**************************************

Page ***************************************(Key Word)*************************************

```
0314  ame identical form much longer than others; or,  IF  changing, that it should change less. We see t
0315  all the species in the same region do at last,   IF  we look to wide enough intervals of time, beco
0315  cies when once lost should never reappear, even  IF  the very same conditions of life, organic and
0315  progenitors. For instance, it is just possible,  IF  our fantail pigeons were all destroyed, that f
0315  y distinguishable from our present fantail; but  IF  the parent rock pigeon were also destroyed, an
0316  pted to give an explanation of this fact, which  IF  true would have been fatal to my views. But su
0316  then, sooner or later, it gradually decreases.   IF  the number of the species of a genus, or the n
0318  nerally a slower process than their production:  IF  the appearance and disappearance of a group of
0319  er of species of all classes, in all countries.  IF  we ask ourselves why this or that species is r
0319  life, and in what degree, they severally acted.  IF  the conditions had gone on, however slowly, we
0321  se the extermination of the parent species; and  IF  many new forms have been developed from any on
0321  t group, and thus caused its extermination; and  IF  many allied forms be developed from the succes
0322  al selection. We need not marvel at extinction;  IF  we must marvel, let it be at our presumption i
0322  on which the existence of each species depends.  IF  we forget for an instant, that each species te
0323  pean and North American tertiary deposits. Even  IF  the new fossil species which are common to the
0323  r. we may doubt whether they have thus changed:  IF  the Megatherium, Mylodon, Macrauchenia, and To
0324  that it has a very strict geological sense: for  IF  all the marine animals which live at the prese
0324  stages, than to those which now live here; and  IF  this be so, it is evident that fossiliferous b
0325  s of life in various parts of Europe, they add,  IF  struck by this strange sequence, we turn our a
0326  and another for those in the waters of the sea.  IF  two great regions had been for a long period f
0328  surprising amount of difference in the species.  IF  the several formations in these regions have n
0329  ponding with a blank interval in the other, and  IF  in both regions the species have gone on slowl
0329  , families, and orders, cannot be disputed. For  IF  we confine our attention either to the living
0329  inct alone, the series is far less perfect than  IF  we combine both into one general system. With
0330  intermediate between living species or groups.   IF  by this term it is meant that an extinct form
0330  ties directed towards very distinct groups. Yet  IF  we compare the older Reptiles and Batrachians,
0332  day. By looking at the diagram we can see that   IF  many of the extinct forms, supposed to be embe
0332  ould be rendered less distinct from each other.  IF, for instance, the genera a1, a5, a10, f8, m3,
0332  ous course through many widely different forms.  IF  many extinct forms were to be discovered above
0333  they were before the discovery of the fossils.  IF, for instance, we suppose the existing genera
0335  districts. To compare small things with great:  IF  the principal living and extinct races of the
0337  ruggle for life over other and preceding forms.  IF  under a nearly similar climate, the eocene inh
0337  have been previously occupied, we may believe,  IF  all the animals and plants of Great Britain we
0337  ecome wild in any part of Europe, we may doubt,  IF  all the productions of New Zealand were set fr
0340  led though in some degree modified descendants.  IF  the inhabitants of one continent formerly diff
0341  s have descended from some one species; so that  IF  six genera, each having eight species, be foun
0345  that organisation on the whole has progressed.   IF  it should hereafter be proved that ancient ani
0345  erious, and is simply explained by inheritance.  IF  then the geological record be as imperfect as
0346  ne would almost suffice to prove its truth: for  IF  we exclude the northern parts where the circum
0346  ion is that between the New and Old Worlds; yet  IF  we travel over the vast American continent, fr
0347  living productions! In the southern hemisphere,  IF  we compare large tracts of land in Australia,
0349  pe. Innumerable other instances could be given.  IF  we look to the islands off the American shore,
0351  different species will be no uniform quantity.  IF, for instance, a number of species, which stan
0352  onsequently possessing the same quadrupeds. But  IF  the same species can be produced at two separa
0353  o one area. What a strange anomaly it would be,  IF, when coming one step lower in the series, to
0354  ugh separated by hundreds of miles of open sea.  IF  the existence of the same species at distant a
0354  n) from the area inhabited by their conquerors.  IF  it can be shown to be almost invariably the ca
0355  th those organic beings which never intercross (IF  such exist), the species, on my theory, must h
0357  ve united almost every island to some mainland.  IF  indeed the arguments used by Forbes are to be
0358  that they are the wrecks of sunken continents;  IF  they had originally existed as mountain ranges
0360  s of sea to another country; and when stranded,  IF  blown to a favourable spot by an inland gale,
0364  of years, it would I think be a marvellous fact  IF  many plants had not thus become widely transpo
0364  m the West Indies to our western shores, where,  IF  not killed by so long an immersion in salt wat
0365  seeds or animals transported to an island, even  IF  far less well stocked than Britain, scarcely n
0368  cts, since become too warm for their existence.  IF  the climate, since the Glacial period, has eve
0369  cation: and this we find has been the case: for  IF  we compare the present Alpine plants and animal
0370  n the broken land still nearer to the pole. Now  IF  we look at a globe, we shall see that under the
0372  y similar physical conditions of the areas: for  IF  we compare, for instance, certain parts of Sout
0373  mountains in this island, tell the same story.  IF  one account which has been published can be tr
0374  des of the southern extremity of the continent.  IF  this be admitted, it is difficult to avoid bel
0374  ly cooler. But it would suffice for my purpose,  IF  the temperature was at the same time lower alo
0377  . and it is certain that many temperate plants,  IF  protected from the inroads of competitors, can
0380  led or even outnumbered by the naturalised; and  IF  the natives have not been actually exterminate
0385  d would be sure to alight on a pool or rivulet,  IF  blown across sea to an oceanic island or to an
0386  or these latter seem immediately to acquire, as  IF  in consequence, a very wide range. I think fav
0386  birds, which frequent the muddy edges of ponds,  IF  suddenly flushed, would be the most likely to
0387  think it would be an inexplicable circumstance  IF  water birds did not transport the seeds of fre
0387  ds of fresh water plants to vast distances, and  IF  consequently the range of these plants was not
0389  forbes's view on continental extensions, which,  IF  legitimately followed out, would lead to the be
0389  its this for plants, and Wollaston for insects.  IF  we look to the large size and varied stations o
0390  re else in the world) is often extremely large.  IF  we compare, for instance, the number of the en
0392  er and taller and overtopping the other plants.  IF  so, natural selection would often tend to add
0393  new Zealand; but I suspect that this exception (IF  the information be correct) may be explained th
0394  many parts of the world on very small islands,  IF  close to a continent; and hardly an island can
0396  ion would probably have been more complete; and  IF  modification be admitted, all the forms of life
0400  ly a far more important element of success. Now  IF  we look to those inhabitants of the Galapagos A
0401  d to the attacks of somewhat different enemies.  IF  then it varied, natural selection would probab
0402  n put into free intercommunication. Undoubtedly  IF  one species has any advantage whatever over an
0402  y brief time wholly or in part supplant it; but  IF  both are equally well fitted for their own pla
0404  r degree in the Felidae and Canidae. We see it,  IF  we compare the distribution of butterflies and
0405  and the species thus has an immense range; but,  IF  the variation had been a little greater, the t
0406  these chapters I have endeavoured to show, that  IF  we make due allowance for our ignorance of the
0407  which may have occurred within the same period;  IF  we remember how profoundly ignorant we are wit
0407  as hardly ever been properly experimentised on;  IF  we bear in mind how often a species may have ra
0407  theory must have spread from one parent source!  IF  we make the same allowances as before for our
0408  means of dispersal of fresh water productions.  IF  the difficulties be not insuperable in admittin
```

Page ***************************************(Key Word)*************************************

if

```
411 groups would have been of simple signification,  IF one group had been exclusively fitted to inhabi
416 vering, as hair or feathers, in the Vertebrata.   IF the Ornithorhynchus had been covered with feath
418 croup, or in allocating any particular species.   IF they find a character nearly uniform, and commo
418 on to others, they use it as one of high value;   IF common to some lesser number, they use it as of
418 n by that excellent botanist, Aug. St. Hilaire.   IF certain characters are always found correlated
420 c difficulties in classification are explained,   IF I do not greatly deceive myself, on the view th
420  the reader will best understand what is meant,   IF he will take the trouble of referring to the di
421 anch of descendants, at each successive period.   IF, however, we choose to suppose that any of the
422 n the diagram, but in much too simple a manner.    IF a branching diagram had not been used, and only
422 lassification, by taking the case of languages.    IF we possessed a perfect pedigree of mankind, a g
422  languages now spoken throughout the world; and    IF all extinct languages, and all intermediate and
423 ss constant. In classing varieties, I apprehend    IF we had a real pedigree, a genealogical classifi
425 lied in blood and alike in some other respects.    IF it could be proved that the Hottentot had desce
425 cies. But it may be asked, what ought we to do;    IF it could be proved that one species of kangaroo
425 umentum ad hominem, and ask what should be done    IF a perfect kangaroo were seen to come out of the
425 hether the skin be covered by hair or feathers,    IF it prevail throughout many and different specie
426 orms have not a single character in common, yet    IF these extreme forms are connected together by a
426 constant, we attach especial value to them; but    IF these same organs, in another group or section
432 rated groups: it has by no means made them; for    IF every form which has ever lived on this earth w
432 ween them. This is what we should be driven to,    IF we were ever to succeed in collecting all the f
435 or the relative connexion of the several parts.    IF we suppose that the ancient progenitor, the arc
439 uld probably have retained through inheritance,    IF they had really been metamorphosed during a lon
442 nt; but in some few cases, as in that of Aphis,    IF we look to the admirable drawings by Professor
444 child than in the parent. These two principles,    IF their truth be admitted, will, I believe, expla
446 es have been acquired earlier or later in life,    IF the full grown animal possesses them. And the c
446 e appeared at an earlier period than usual, or,    IF not so, the differences must have been inherite
448 ergoing any metamorphosis is perhaps requisite.    IF, on the other hand, it profited the young to fo
449  by the finest gradations, the best, or indeed,    IF our collections were nearly perfect, the only p
449 differ from each other in structure and habits,    IF they pass through the same or similar embryonic
455 nd their lesser relative size in the adult. But    IF each step of the process of reduction were to b
455 ch the materials forming any part or structure,    IF not useful to the possessor, will be saved as f
456 classifications; the value set upon characters,    IF constant and prevalent, whether of high vital i
456 es, however different they may be in structure.    IF we extend the use of this element of descent, t
458 should without hesitation adopt this view, even    IF it were unsupported by other facts or arguments
459 on insuperably great, cannot be considered real   IF we admit the following propositions, namely, th
464 pecies as the parent of any one or more species   IF we were to examine them ever so closely, unless
464  the links between any two species are unknown,   IF any one link or intermediate variety be discove
465 modified and improved; and when they do spread,   IF discovered in a geological formation, they will
465  in a geological formation, they will appear as   IF suddenly created there, and will be simply clas
465 which I require, few will be inclined to admit.   IF we look to long enough intervals of time, geolo
468 er the changed conditions of domestication. And   IF there be any variability under nature, it would
468 under nature, it would be an unaccountable fact   IF natural selection had not come into play. It ha
469 presentative forms in Europe and North America.   IF then we have under nature variability and a pow
469  be preserved, accumulated, and inherited? Why,   IF man can by patience select variations most usef
469  of life. The theory of natural selection, even   IF we looked no further than this, seems to me to
470 o find it still in action; and this is the case   IF varieties be incipient species. Moreover, the s
470 een independently created, but are intelligible   IF all species first existed as varieties. As each
471 ovation. But why this should be a law of nature   IF each species has been independently created, no
472 tions from another land. Nor ought we to marvel  IF all the contrivances in nature be not, as far a
472 as far as we can judge, absolutely perfect; and  IF some of them be abhorrent to our ideas of fitne
473 heir hybrids! How simply is this fact explained  IF we believe that these species have descended fr
473 e likely to vary in any one species of a genus,  IF the other species, supposed to have been create
473 dently, have differently coloured flowers, than  IF all the species of the genus have the same colo
473 es of the genus have the same coloured flowers?  IF species are only well marked varieties, of whic
474  not be more variable than any other structure,  IF the part be common to many subordinate forms, t
474 t be common to many subordinate forms, that is,  IF it has been inherited for a very long period; f
475 many instincts causing other animals to suffer.  IF species be only well marked and permanent varie
475 on the other hand, these would be strange facts  IF species have been independently created, and va
475 varieties have been produced by secondary laws.  IF we admit that the geological record is imperfec
475 ife, after long intervals of time, to appear as  IF they had changed simultaneously throughout this
476 descent. Looking to geographical distribution,  IF we admit that there has been during the long co
477 se at their inhabitants being widely different,  IF they have been for a long period completely sep
481 d us plain evidence of the mutation of species;  IF they had undergone mutation. But the chief caus
484 m other forms, to be capable of definition; and  IF definable, whether the differences be sufficien
485 erences, however slight, between any two forms,  IF not blended by intermediate gradations, are loo
284  the result: Palaeozoic strata (not including IGNEOUS beds).....57,154 Feet. Secondary strata.....
006  , if he makes due allowance for our profound IGNORANCE in regard to the mutual relations of all th
073 ver another. Nevertheless so profound is our IGNORANCE, and so high our presumption, that we marve
078 so as to succeed. It will convince us of our IGNORANCE on the mutual relations of all organic bein
085 erence between species, which, as far as our IGNORANCE permits us to judge, seem to be quite unimp
131 on, but it serves to acknowledge plainly our IGNORANCE of the cause of each particular variation.
167 ls now living on the sea shore. Summary. Our IGNORANCE of the laws of variation is profound. Not i
199 , we ought not to lay too much stress on our IGNORANCE of the precise cause of the slight analogou
216 h at first appeared from what we must in our IGNORANCE call an accident. In some cases compulsory
306 similar considerations; but chiefly from our IGNORANCE of the geology of other countries beyond th
354 m a single birthplace; then, considering our IGNORANCE with respect to former climatal and geograp
406 show, that if we make due allowance for our IGNORANCE of the full effects of all the changes of c
407 e make the same allowances as before for our IGNORANCE, and remember that some forms of life chang
482 opposite to mine. It is so easy to hide our IGNORANCE under such expressions as the plan of creat
067 will this surprise any one who reflects how IGNORANT we are on this head, even in regard to manki
097 that it is a general law of nature (utterly IGNORANT though we be of the meaning of the law) that
132 should vary more or less, we are profoundly IGNORANT. Nevertheless, we can here and there dimly c
167 ould almost as soon believe with the old and IGNORANT cosmogonists, that fossil shells had never l
193 electricity, we must own that we are far too IGNORANT to argue that no transition of any kind is p
195 the eye. In the first place, we are much too IGNORANT in regard to the whole economy of any one or
197 ion of the higher animals. We are profoundly IGNORANT of the causes producing slight and unimporta
```

ignorant

Page ***************************************(Key Word)***************************************

```
0198  ion of natural selection. But we are far too IGNORANT to speculate on the relative importance of
0205  gely facilitated transitions. We are far too IGNORANT, in almost every case, to be enabled to ass
0296  us period for their accumulation; yet no one IGNORANT of this fact would have suspected the vast
0407  e same period; if we remember how profoundly IGNORANT we are with respect to the many and curious
0447  upervene, from causes of which we are wholly IGNORANT, at a very early period of life, or each st
0462  it cannot be denied that we are as yet very IGNORANT of the full extent of the various climatal
0462  roughout the world. We are as yet profoundly IGNORANT of the many occasional means of transport.
0466  ate to questions on which we are confessedly IGNORANT; nor do we know how ignorant we are. We do
0466  are confessedly ignorant; nor do we know how IGNORANT we are. We do not know all the possible tra
0029  y selecting such slight differences, yet they IGNORE all general arguments, and refuse to sum up
0483  mutability of species, on their own side they IGNORE the whole subject of the first appearance of
0070  good effects of frequent intercrossing, and the ILL effects of close interbreeding, probably come
0111  future well marked species, present slight and ILL defined differences. Mere chance, as we may c
0249  , their fertility, notwithstanding the frequent ILL effects of manipulation, sometimes decidedly
0253  time from different parents, so as to avoid the ILL effects of close interbreeding. On the contra
0345  nature; and this may account for that vague yet ILL defined sentiment, felt by many palaeontologi
0358  nical works, this or that plant is stated to be ILL adapted for wide dissemination; but for trans
0366  uthern zone became fitted for arctic beings and ILL fitted for their former more temperate inhabi
0472  arying descendants of each to any unoccupied or ILL occupied place in nature, these facts cease t
0267  f that this parallelism is an accident or an ILLUSION. Both series of facts seem to be connected
0177  varieties than do the rarer species. I may ILLUSTRATE what I mean by supposing three varieties o
0196  acquired habits. To give a few instances to ILLUSTRATE these latter remarks. If green woodpeckers
0303  he world. I will now give a few examples to ILLUSTRATE these remarks; and to show how liable we a
0356  stage, to descent from a single parent. To ILLUSTRATE what I mean: our English race horses diffe
0417  alpighiaceae. This case seems to me well to ILLUSTRATE the spirit with which our classifications
0422  ders, and classes. It may be worth while to ILLUSTRATE this view of classification, by taking the
0129  nd divergence of character, as we have seen ILLUSTRATED in the diagram. The affinities of all the
0320  y be most severe, as formerly explained and ILLUSTRATED by examples, between the forms which are
0331  divergence of character, which was formerly ILLUSTRATED by this diagram, the more recent any form
0120  into well defined species: thus the diagram ILLUSTRATES the steps by which the small differences
0151  si 1 will here only briefly give one, as it ILLUSTRATES the rule in its largest application. The
0093  t as a very striking case, but as likewise ILLUSTRATING one step in the separation of the sexes o
0369  osely allied or representative species. In ILLUSTRATING what, as I believe, actually took place d
0412  request the reader to turn to the diagram ILLUSTRATING the action, as formerly explained, of the
0002  which I have arrived, with a few facts in ILLUSTRATION, but which, I hope, in most cases will su
0036  the rock pigeon. Youatt gives an excellent ILLUSTRATION of the effects of a course of selection,
0157  om each other. Of this fact I will give in ILLUSTRATION two instances, the first which happen to
0183  te forms. Thus, to return to our imaginary ILLUSTRATION of the flying fish, it does not seem prob
0190  inct purpose, or be quite obliterated. The ILLUSTRATION of the swimbladder in fishes is a good on
0240  tual measurements, but a strictly accurate ILLUSTRATION: the difference was the same as if we wer
0280  its modified descendants. To give a simple ILLUSTRATION: the fantail and pouter pigeons have both
0299  the foregoing remarks, under an imaginary ILLUSTRATION. The Malay Archipelago is of about the si
0482  come when this will be given as a curious ILLUSTRATION of the blindness of preconceived opinion.
0090  , it would have been called a monstrosity. ILLUSTRATIONS of the action of Natural Selection. In o
0090  eg permission to give one or two imaginary ILLUSTRATIONS. Let us take the case of a wolf, which p
0329  whole pages could be filled with striking ILLUSTRATIONS from our great palaeontologist, Owen, sh
0095  ews on the modern changes of the earth, as ILLUSTRATIVE of geology; but we now very seldom hear t
0206  s of existence, so often insisted on by the ILLUSTRIOUS Cuvier, is fully embraced by the principl
0374  countries. So on the Silla of Caraccas the ILLUSTRIOUS Humboldt long ago found species belonging
0440  es afford a good instance of this: even the ILLUSTRIOUS Cuvier did not perceive that a barnacle w
0189  or in any degree, tend to produce a distincter IMAGE. We must suppose each new state of the instr
0090  ts, I must beg permission to give one or two IMAGINARY illustrations. Let us take the case of a w
0093  ee in search of nectar. But to return to our IMAGINARY case: as soon as the plant had been render
0094  ow turn to the nectar feeding insects in our IMAGINARY case: we may suppose the plant of which we
0095  natural selection, exemplified in the above IMAGINARY instances, is open to the same objections w
0119  recorded as varieties. But these breaks are IMAGINARY, and might have been inserted anywhere, aft
0183  ny subordinate forms. Thus, to return to our IMAGINARY illustration of the flying fish, it does n
0229  eye could judge, exactly along the planes of IMAGINARY intersection between the basins on the opp
0232  last completed cells, and then, by striking IMAGINARY spheres, they can build up a wall intermed
0299  le to sum up the foregoing remarks, under an IMAGINARY illustration. The Malay Archipelago io of w
0078  ase we can clearly see that if we wished in IMAGINATION to give the plant the power of increasing
0078  s or enemies. It is good thus to try in our IMAGINATION to give any form some advantage over anot
0176  e will probably apply to both; and if we in IMAGINATION adapt a varying species to a very large a
0187  atural selection, though insuperable by our IMAGINATION, can hardly be considered real. How a ner
0188  nal grades. His reason ought to conquer his IMAGINATION; though I have felt the difficulty far to
0188  e eye to an optical instrument, we ought in IMAGINATION to take a thick layer of transparent tissu
0243  t may not be a logical deduction, but to my IMAGINATION it is far more satisfactory to look at su
0459  s, this difficulty, though appearing to our IMAGINATION insuperably great, cannot be considered re
0135  well as any of the smaller quadrupeds. We may IMAGINE that the early progenitor of the ostrich ha
0137  glasses has been lost. As it is difficult to IMAGINE that eyes, though useless, could be in any w
0138  ems to have done its work. It is difficult to IMAGINE conditions of life more similar than deep li
0252  s hybrid seeds as freely as it is possible to IMAGINE. Had hybrids, when fairly treated, gone on e
0384  ed exclusively to fresh water, so that we may IMAGINE that a marine member of a fresh water group
0025  tions. Either, firstly, that all the several IMAGINED aboriginal stocks were coloured and marked w
0072  and so extensive that no one would ever have IMAGINED that cattle would have so closely and effect
0182  this had been effected, who would have ever IMAGINED that in an early transitional state they had
0192  have sessile cirripedes, who would ever have IMAGINED that the branchiae in this latter family had
0322  must marvel, let it be at our presumption in IMAGINING for a moment that we understand the many co
0444  , peculiarities in the caterpillar, cocoon, or IMAGO states of the silk moth; or, again, in the ho
0415  the existence or absence of albumen, in the IMBRICATE or valvular aestivation. Any one of these c
0282  natural science, yet does not admit how IMCOMPREHENSIBLY vast have been the past periods of time,
0226  as huber has remarked, is manifestly a gross IMITATION of the three sided pyramidal basis of the e
0060  are that this has ever been disputed. It is IMMATERIAL for us whether a multitude of doubtful form
0183  nt habits. But it is difficult to tell, and IMMATERIAL for us, whether habits generally change fir
0061  er incessantly ready for action; and is as IMMEASURABLY superior to man's feeble efforts, as the w
0309  changed in the lapse of ages? At a period IMMEASURABLY antecedent to the silurian epoch, contine
```

Page ***************************************(Key Word)***************************************

age **(Key Word)**

```
040 he improved individuals slowly spread in the IMMEDIATE neighbourhood. But as yet they will hardly
216 s. it is now commonly admitted that the more IMMEDIATE and final cause of the cuckoo's instinct is
372 y dissimilar. But we must return to our more IMMEDIATE subject, the Glacial period. I am convinced
432 roups could be distinguished from their more IMMEDIATE parents; or those parents from their ancien
081 nal numbers of its inhabitants would almost IMMEDIATELY undergo a change, and some species might b
197 ight and unimportant variations; and we are IMMEDIATELY made conscious of this by reflecting on th
211 ards I allowed an ant to visit them, and it IMMEDIATELY seemed, by its eager way of running about,
211 ach aphis, as soon as it felt the antennae, IMMEDIATELY lifted up its abdomen and excreted a limpi
268 to varieties produced under nature; we are IMMEDIATELY involved in hopeless difficulties; for if
294 the deposit, but have become extinct in the IMMEDIATELY surrounding sea; or, conversely, that some
359 longer; for instance, ripe hazel nuts sank IMMEDIATELY, but when dried, they floated for 90 days
361 floating on the sea, sometimes escape being IMMEDIATELY devoured; and seeds of many kinds in the c
366 n to the Glacial Period, which, as we shall IMMEDIATELY see, affords a simple explanation of these
385 ly to be transported by birds, and they are IMMEDIATELY killed by sea water, as are the adults. Th
386 few aquatic members; for these latter seem IMMEDIATELY to acquire, as if in consequence, a very w
393 ese animals and their spawn are known to be IMMEDIATELY killed by sea water, on my view we can see
424 times produced on the same spike, they were IMMEDIATELY included as a single species. But it may b
019 pecies, I cannot doubt that there has been an IMMENSE amount of inherited variation. Who can belie
028 he time of the Romans, as we hear from Pliny, IMMENSE prices were given for pigeons; nay, they are
028 nce of these considerations in explaining the IMMENSE amount of variation which pigeons have under
043 both of distinct species and of varieties is IMMENSE; for the cultivator here quite disregards th
137 nimals, namely, the cave rat, the eyes are of IMMENSE size; and Professor Silliman thought that it
154 have existed, according to my theory, for an IMMENSE period in nearly the same state; and thus it
308 we now possess of the Silurian deposits over IMMENSE territories in Russia and in North America,
309 ve always remained nearer to the surface. The IMMENSE areas in some parts of the world, for instan
388 fresh water productions ever can range, over IMMENSE areas, and having subsequently become extinc
405 erica and Europe, and the species thus has an IMMENSE range; but, if the variation had been a litt
027 glish carrier or short faced tumbler differs IMMENSELY in certain characters from the rock pigeon,
173 ons have been made only at intervals of time IMMENSELY remote. But it may be urged that when sever
175 cies, even in its metropolis, would increase IMMENSELY in numbers, were it not for other competing
367 and the identity of many plants at points so IMMENSELY remote as on the mountains of the United St
434 insects: what can be more different than the IMMENSELY long spiral proboscis of a sphinx moth; the
486 the study of domestic productions will rise IMMENSELY in value. A new variety raised by man will
397 d when it had formed a new membranous one, I IMMERSED it for fourteen days in sea water, and it re
358 that out of 87 kinds, 64 germinated after an IMMERSION of 28 days, and a few survived an immersion
358 immersion of 28 days, and a few survived an IMMERSION of 137 days. For convenience sake I chiefly
359 . so that as 64/87 seeds germinated after an IMMERSION of 28 days; and as 18/94 plants with ripe f
361 for instance, are killed by even a few days IMMERSION in sea water; but some taken out of the cro
364 n shores, where, if not killed by so long an IMMERSION in salt water, they could not endure our cl
397 ies did in this state withstand uninjured an IMMERSION in sea water during seven days: one of thes
401 nother island. Hence when in former times an IMMIGRANT settled on any one or more of the islands,
350 ote; on the nature, and number of the former IMMIGRANTS; and on their action and reaction, in their
355 has probably received at some former period IMMIGRANTS from this other region, my theory will be s
365 ies, through occasional means of transport, IMMIGRANTS from Europe or any other continent, that a
400 er that they would almost certainly receive IMMIGRANTS from the same original source, or from each
400 ame height, climate, etc., that many of the. IMMIGRANTS should have been differently modified, thou
408 d with the aborigines; and according as the IMMIGRANTS were capable of varying more or less rapidl
409 he nearest continent or other source whence IMMIGRANTS were probably derived. We can see why in tw
478 he inhabitants of the nearest source whence IMMIGRANTS might have been derived. We see this in nea
081 en on its borders, new forms would certainly IMMIGRATE, and this also would seriously disturb the
180 ther competing rodents or new beasts of prey IMMIGRATE, or old ones become modified, and all analo
294 the probability is that it only then first IMMIGRATED into that area. It is well known, for insta
390 species which do not become modified having IMMIGRATED with facility and in a body, so that their
081 or modified; for, had the area been open to IMMIGRATION, these same places would have been seized
082 or any unusual degree of isolation to check IMMIGRATION, is actually necessary to produce new and
104 bably acts more efficiently in checking the IMMIGRATION of better adapted organisms, after any phy
105 onstitution. Lastly, isolation, by checking IMMIGRATION and consequently competition, will give ti
107 hecked: after physical changes of any kind, IMMIGRATION will be prevented, so that new places in t
108 , which are generally very slow, and on the IMMIGRATION of better adapted forms having been checke
114 ly small area, especially if freely open to IMMIGRATION, and where the contest between individual
178 ow changes of climate, or on the occasional IMMIGRATION of new inhabitants, and, probably, in a st
322 low extermination. Moreover, when by sudden IMMIGRATION or by unusually rapid development, many sp
328 the other for modification, extinction, and IMMIGRATION. I suspect that cases of this nature have
334 and for the coming in of quite new forms by IMMIGRATION, and for a large amount of modification, d
392 ms to me not a little doubtful. Facility of IMMIGRATION, I believe, has been at least as important
487 nent in relation to their apparent means of IMMIGRATION, some light can be thrown on ancient geogr
017 great weight in making us doubt about the IMMUTABILITY of the many very closely allied and natura
310 nimously, often vehemently, maintained the IMMUTABILITY of species. But I have reason to believe t
312 better accord with the common view of the IMMUTABILITY of species, or with that of their slow and
335 professor Pictet in his firm belief in the IMMUTABILITY of species. He who is acquainted with the
459 avour. Causes of the general belief in the IMMUTABILITY of species. How far the theory of natural
482 and who have already begun to doubt on the IMMUTABILITY of species, may be influenced by this volu
006 s. i am fully convinced that species are not IMMUTABLE; but that those belonging to what are calle
340 riods. Nor can it be pretended that it is an IMMUTABLE law that marsupials should have been chiefl
340 the great law of the long enduring, but not IMMUTABLE, succession of the same types within the sa
340 re dominant forms, and there will be nothing IMMUTABLE in the laws of past and present distributio
481 gn of creation. The belief that species were IMMUTABLE productions was almost unavoidable as long
247 ty plants had their fertility in some degree IMPAIRED. Moreover, as Gartner during several years r
482 le to view both sides of the question with IMPARTIALITY. Whoever is led to believe that species ar
348 mountain chains, deserts, etc., are not as IMPASSABLE, or likely to have endured so long as the o
348 aunas are separated only by the narrow, but IMPASSABLE, isthmus of Panama. Westward of the shores
348 but from being separated from each other by IMPASSABLE barriers, either of land or open sea, they
348 pical parts of the Pacific, we encounter no IMPASSABLE barriers, and we have innumerable islands a
356 been a high road for migration, but now be IMPASSABLE; I shall, however, presently have to discus
371 y land, serving as a bridge, since rendered IMPASSABLE by cold, for the inter migration of their i
383 , and as the sea is apparently a still more IMPASSABLE barrier, that they never would have extende
```

Page **(Key Word)**

Page ***(Key Word)***

```
0207  what is meant, when it is said that instinct IMPELS the cuckoo to migrate and to lay her eggs in
0258  connected with constitutional differences IMPERCEPTIBLE by us, and confined to the reproductive
0002  ct, which I now publish, must necessarily be IMPERFECT. I cannot here give references and author
0004  i have invariably found that our knowledge, IMPERFECT though it be, of variation under domestica
0012  ertain vegetable poisons. Hairless dogs have IMPERFECT teeth; long haired and coarse haired anima
0057  s numerically by averages, and, as far as my IMPERFECT results go, they always confirm the view.
0084  s marked the long lapse of ages, and then so IMPERFECT is our view into long past geological ages
0179  ure chapter attempt to show, in an extremely IMPERFECT and intermittent record. Cn the origin and
0186  s from a perfect and complex eye to one very IMPERFECT and simple, each grade being useful to its
0188  and at their lower ends there seems to be an IMPERFECT vitreous substance. With these facts, here
0237  eds of cattle in relation to an artificially IMPERFECT state of the male sex: for oxen of certain
0251  other species, yet as they were functionally IMPERFECT in their mutual self action, we must infer
0262  , which have their reproductive organs in an IMPERFECT condition, is a very different case from t
0263  nts are perfect, whereas in hybrids they are IMPERFECT. Even in first crosses, the greater or les
0277  brids, which have their reproductive systems IMPERFECT, and which have had this system and their
0287  mportant for us to gain some notion, however IMPERFECT, of the lapse of years. During each of the
0287  at our palaeontological collections are very IMPERFECT, is admitted by every one. The remark of t
0291  and the mass when upraised will give a most IMPERFECT record of the forms of life which then exi
0292  ical record, viewed as a whole, is extremely IMPERFECT; but if we confine our attention to any on
0300  ipelago would be preserved in an excessively IMPERFECT manner in the formations which we suppose
0341  show that the geological record is extremely IMPERFECT; that only a small portion of the globe ha
0342  nded to make the geoloical record extremely IMPERFECT, and will to a large extent explain why we
0345  ritance. If then the geological record be as IMPERFECT as I believe it to be, and it may at least
0441  complex antennae; but they have a closed and IMPERFECT mouth, and cannot feed: their function an
0453  t these rudimentary or atrophied organs, are IMPERFECT and useless. In works on natural history r
0453  e? when a man's fingers have been amputated, IMPERFECT nails sometimes appear on the stumps: I co
0455  at the existence of organs in a rudimentary, IMPERFECT, and useless condition, or cuite aborted,
0464  ith that the geological record is far more IMPERFECT than most geologists believe. It cannot be
0465  ninth chapter. That the geological record is IMPERFECT all will admit; but that it is imperfect t
0465  is imperfect all will admit; but that it is IMPERFECT to the degree which I require,,few will be
0466  the long lapse of years, or that we know how IMPERFECT the Geological Record is. Grave as these s
0475  s. if we admit that the geological record is IMPERFECT in an extreme degree, then such facts as t
0005  eties when intercrossed; and fourthly, the IMPERFECTION of the Geological Record. In the next cha
0172  iscuss this cuestion in the chapter on the IMPERFECTION of the geological record; and I will here
0172  ss perfect than is generally supposed; the IMPERFECTION of the record being chiefly due to organi
0279  species and varieties. Chapter IX. On the IMPERFECTION of the Geological Record. Cn the absence
0280  anation lies, as I believe, in the extreme IMPERFECTION of the geological record. In the first pl
0289  econdary or palaeozoic formations. But the IMPERFECTION in the geological record mainly results f
0464  rdly ever expect to discover, owing to the IMPERFECTION of the geological record. Numerous existi
0487  ce of Geology loses glory from the extreme IMPERFECTION of the record. The crust of the earth wit
0143  ied. This is a very important subject, most IMPERFECTLY understood. The most obvious case is, tha
0188  with these facts; here far too briefly and IMPERFECTLY given, which show that there is much grade
0246  hey are either not at all developed, or are IMPERFECTLY developed. This distinction is important,
0264  f hybrids, in which the sexual elements are IMPERFECTLY developed, the case is very different. I
0300  be collected which ever lived there, how IMPERFECTLY would they represent the natural history
0305  fishes, of which the affinities are as yet IMPERFECTLY known, are really teleostean. Assuming, he
0310  eological record, as a history of the world IMPERFECTLY kept, and written in a changing dialect;
0164  and with leg stripes; and a man, whom I can IMPLICITLY trust, has examined for me a small dun Wel
0044  re community of descent is almost universally IMPLIED, though it can rarely be proved. We have al
0284  uffered. And what an amount of degradation is IMPLIED by the sedimentary deposits of many countri
0489  ith Reproduction; Inheritance which is almost IMPLIED by reproduction; Variability from the indir
0153  rioc. An extraordinary amount of modification IMPLIES an unusually large and long continued amoun
0203  the very process of natural selection almost IMPLIES the continual supplanting and extinction of
0273  ng been tried on natural varieties), and this IMPLIES in most cases that there has been recent va
0405  r we should never forget that to range widely IMPLIES not only the power of crossing barriers, bu
0429  ly very distinct from each cther, which again IMPLIES extinction. The genera Ornithorhynchus and
0452  ntary organs may be utterly aborted; and this IMPLIES, that we find in an animal or plant no trace
0478  d or representative species in any two areas, IMPLIES, on the theory of descent with modification
0483  state; and this in some instances necessarily IMPLIES an enormous amount of modification in the de
0281  ve existed between them. But such a case would IMPLY that one form had remained for a very long pe
0413  t that the genus gives the characters, ceem to IMPLY that something more is included in our class
0044  the term variation in a technical sense, as IMPLYING a modification directly due to the physical
0290  osition of consecutive formations, generally IMPLYING great changes in the geography of the surrou
0004  explained. It is, therefore, of the highest IMPORTANCE to gain a clear insight into the means of
0006  are? Yet these relations are of the highest IMPORTANCE, for they determine the present welfare, an
0012  ght and those of considerable physiological IMPORTANCE is endless. Dr. Prosper Lucas's treatise,
0013  o the like sex. It is a fact of some little IMPORTANCE to us, that peculiarities appearing in the
0014  t. i believe this rule to be of the highest IMPORTANCE in explaining the laws of embryology. These
0015  e experiment would succeed, is not of great IMPORTANCE for our line of argument; for by the exper
0028  geons as were the old Romans. The paramount IMPORTANCE of these considerations in explaining the
0031  ears to obtain head and beak. In Saxony the IMPORTANCE of the principle of selection in regard to
0032  vious as hardly to be worth notice; but its IMPORTANCE consists in the great effect produced by th
0033  ase. The laws of correlation of growth, the IMPORTANCE of which should never be overlooked, will e
0033  eferences to the full acknowledgment of the IMPORTANCE of the principle in works of high antiouity
0041  ; and hence this comes to be of the highest IMPORTANCE to success. On this principle Marshall has
0043  roductive system, are so far of the highest IMPORTANCE as causing variability. I do not believe th
0043  in the formation of new sub breeds; but the IMPORTANCE of the crossing of varieties has, I believe
0043  ily propagated by cuttings, buds, etc., the IMPORTANCE of the crossing both of distinct species ar
0043  plants not propagated by seed are of little IMPORTANCE to us, for their endurance is only temporar
0051  all interest to the systematist, as of high IMPORTANCE for us, as being the first step towards suc
0066  district. A large number of eggs is of some IMPORTANCE to those species, which depend on a rapidly
0066  rapidly to increase in number. But the real IMPORTANCE of a large number of eggs or seeds is to ma
0077  battle of life. A corollary of the highest IMPORTANCE may be deduced from the foregoing remarks,
0080  ection, its power on characters of trifling IMPORTANCE, its power at all ages and on both sexes. S
0084  ich we are apt to consider as very trifling IMPORTANCE, may thus be acted on. When we see leaf eat
0085  otanists as characters of the most trifling IMPORTANCE: yet we hear from an excellent horticulturi
```

Page ***************************************(Key Word)***

```
1087  once in an animal's whole life, if of high  IMPORTANCE to it, might be modified to any extent by n
1096  y andrew Knight. We shall presently see its  IMPORTANCE; but I must here treat the subject with ext
1105  owly improved; and this may sometimes be of  IMPORTANCE in the production of new species. If, howev
1105  not doubt that isolation is of considerable  IMPORTANCE in the production of new species, on the wh
1105  o believe that largeness of area is of more  IMPORTANCE, more especially in the production of speci
1111  i have designated by this term, is of high   IMPORTANCE on my theory, and explains, as I believe, s
1117  eserved or naturally selected. And here the  IMPORTANCE of the principle of benefit being derived f
1144  know of no case better adapted to show the   IMPORTANCE of the laws of correlation in modifying imp
1146  erae these differences are of such apparent  IMPORTANCE, the seeds being in some cases, according t
1152  the fair presumption is that it is of high   IMPORTANCE to that species; nevertheless the part in t
1154  are taken from parts of less physiological   IMPORTANCE than those commonly used for classing gener
1155  becomes variable, though its physiological   IMPORTANCE may remain the same. Something of the same
1157  of the wings is a character of the highest   IMPORTANCE, because common to large groups; but in cer
1171  y. natura non facit saltum. Organs of small  IMPORTANCE. Organs not in all cases absolutely perfect
1171  roduce, on the one hand, organs of trifling  IMPORTANCE, such as the tail of a giraffe, which serve
1194  nd slowest steps. Organs of little apparent  IMPORTANCE. As natural Selection acts by life and deat
1195  ng the origin of simple parts, of which the  IMPORTANCE does not seem sufficient to cause the prese
1195  o say what slight modifications would be of  IMPORTANCE or not. In a former chapter I have given in
1195  from beasts of prey. Organs now of trifling  IMPORTANCE have probably in some cases been of high im
1195  ce have probably in some cases been of high  IMPORTANCE to an early progenitor, and, after having b
1196  he second place, we may sometimes attribute  IMPORTANCE to characters which are really of very litt
1196  characters which are really of very little   IMPORTANCE, and which have originated from quite secon
1197  and consequently that it was a character of  IMPORTANCE and might have been acquired through natura
1198  r too ignorant to speculate on the relative  IMPORTANCE of the several known and unknown laws of va
1205  also, believe that a part formerly of high   IMPORTANCE has often been retained (as the tail of an
1205  ndants), though it has become of such small  IMPORTANCE that it could not, in its present state, ha
1209  e effects of habit are of quite subordinate  IMPORTANCE to the effects of the natural selection of
1235  ses of instinct of apparently such trifling  IMPORTANCE, that they could hardly have been acted on
1243  l dispute that instincts are of the highest  IMPORTANCE to each animal. Therefore I can see no diff
1245  d they been capable of crossing freely. The  IMPORTANCE of the fact that hybrids are very generally
1246  ngrel offspring, is, on my theory, of equal  IMPORTANCE with the sterility of species; for it seems
1256  re of parts which are of high physiological  IMPORTANCE and which differ little in the allied speci
1346  for by differences in physical conditions.  IMPORTANCE of barriers. Affinity of the productions of
1350  t important of all relations. Thus the high  IMPORTANCE of barriers comes into play by checking mig
1396  modified, in accordance with the paramount  IMPORTANCE of the relation of organism to organism. I
1407  al considerations, more especially from the  IMPORTANCE of barriers and from the analogical distrib
1408  new forms. We can thus understand the high  IMPORTANCE of barriers, whether of land or water, whic
1408  of organism to organism are of the highest  IMPORTANCE, we can see why two areas having nearly the
1414  he economy of nature, would be of very high  IMPORTANCE in classification. Nothing can be more fals
1414  to a whale, of a whale to a fish, as of any  IMPORTANCE. These resemblances, though so intimately c
1414  th their product the seed, are of paramount  IMPORTANCE! We must not, therefore, in classifying, tr
1415  es in organs of high vital or physiological  IMPORTANCE. No doubt this view of the classificatory i
1415  e. no doubt this view of the classificatory  IMPORTANCE of organs which are important is generally,
1415  ly, but by no means always, true. But their  IMPORTANCE for classification, I believe, depends on t
1415  itions of life. That the mere physiological  IMPORTANCE of an organ does not determine its classifi
1415  rgans in the Proteaceae, says their generic  IMPORTANCE, like that of all their parts, not only in
1415  s singly is frequently of more than generic  IMPORTANCE, though here even when all taken together t
1416  the same order are of unequal physiological  IMPORTANCE. Any number of instances could be given of
1416  of instances could be given of the varying  IMPORTANCE for the classification of the same importan
1416  d organs are of high physiological or vital  IMPORTANCE; yet, undoubtedly, organs in this condition
1416  the rudimentary florets are of the highest  IMPORTANCE in the classification of the Grasses. Numer
1416  e considered of very trifling physiological  IMPORTANCE, but which are universally admitted as high
1417  y one internal and and important organ. The  IMPORTANCE, for classification, of trifling characters
1417  th several other characters of more or less  IMPORTANCE. The value indeed of an aggregate of charac
1417  eral characters, both of high physiological  IMPORTANCE and of almost universal prevalence, and yet
1417  e organisation is universally constant. The  IMPORTANCE of an aggregate of characters, even when no
1418  derived from the embryo should be of equal  IMPORTANCE with those derived from the adult, for our
1426  ss. As we find organs of high physiological  IMPORTANCE, those which serve to preserve life under t
1426  characters are of such high classificatory  IMPORTANCE. Geographical distribution may sometimes be
1427  als. On my view of characters being of real  IMPORTANCE for classification, only in so far as they
1427  adaptive character, although of the utmost  IMPORTANCE to the welfare of the being, are almost val
1433  ely insisted, in an able paper, on the high  IMPORTANCE of looking to types, whether or not we can
1433  organs, or others of trifling physiological  IMPORTANCE; why, in comparing one group with a distinc
1434  . hilaire has insisted strongly on the high  IMPORTANCE of relative connexion in homological organs:
1441  itute of mouth, stomach, or other organs of  IMPORTANCE, excepting for reproduction. We are so much
1450  ng facts in embryology, which are second in  IMPORTANCE to none in natural history, are explained o
1455  re useful than, parts of high physiological  IMPORTANCE. Rudimentary organs may be compared with th
1456  nstant and prevalent, whether of high vital  IMPORTANCE, or of the most trifling importance, or, as
1456  h vital importance, or of the most trifling  IMPORTANCE, or, as in rudimentary organs, of no import
1456  rtance, or, as in rudimentary organs, of no  IMPORTANCE; the wide opposition in value between analo
1457  sence might have been even anticipated. The  IMPORTANCE of embryological characters and of rudiment
1474  refore, as we may naturally infer, of great  IMPORTANCE to the species, should be eminently liable
1478  hy adaptive characters; though of paramount  IMPORTANCE to the being, are of hardly any importance
1479  importance to the being, are of hardly any  IMPORTANCE in classification; why characters derived f
1479  nent characters, however slight their vital  IMPORTANCE may be. The framework of bones being the sa
0004  least possible; and, what is equally or more  IMPORTANT, we shall see how great is the power of man
0013  greater degree, to males alone. A much more  IMPORTANT rule, which I think may be trusted, is that
0020  lished on pigeons, and some of them are very  IMPORTANT, as being of considerable antiquity. I have
0033  s been, in a corresponding degree, rapid and  IMPORTANT. But it is very far from true that the prin
0034  ed from the best individual animals, is more  IMPORTANT. Thus, a man who intends keeping pointers n
0038  countries. On the view here given of the all  IMPORTANT part which selection by man has played, it
0041  lly prevent selection. But probably the most  IMPORTANT point of all, is, that the animal or plant
0042  sexes, facility in preventing crosses is an  IMPORTANT element of success in the formation of new
0043  pecies, aboriginally distinct, has played an  IMPORTANT part in the origin of our domestic producti
0045  uld. These individual differences are highly  IMPORTANT for us, as they afford materials for natura
0045  ue of facts, that parts which must be called  IMPORTANT, whether viewed under a physiological or cl
```

Page **************************************(Key Word)***

```
0045  number of the cases of variability, even in  IMPORTANT  parts of structure, which he could collect
0045  e far from pleased at finding variability in  IMPORTANT  characters, and that there are not many mer
0045  en who will laboriously examine internal and  IMPORTANT  organs, and compare them in many specimens
0046  times argue in a circle when they state that  IMPORTANT  organs never vary; for these same authors p
0046  e authors practically rank that character as  IMPORTANT  (as some few naturalists have honestly con†
0046  under this point of view, no instance of an  IMPORTANT  part varying will ever be found: but under
0047  ct species, are in several respects the most  IMPORTANT  for us. We have every reason to believe tha
0053  ch, as we shall hereafter see, is a far more  IMPORTANT  circumstance) with different sets of organ†
0056  . hence the amount of difference is one very  IMPORTANT  criterion in settling whether two forms sho
0057  arent species? Undoubtedly there is one most  IMPORTANT  point of difference between varieties and s
0060  he relation of organism to organism the most  IMPORTANT  of all relations. Before entering on the su
0062  ing on another, and including (which is more  IMPORTANT) not only the life of the individual, but s
0068  phant protected by its dam. Climate plays an  IMPORTANT  part in determining the average numbers of
0071  sed, so that cattle could not enter. But how  IMPORTANT  an element enclosure is, I plainly saw near
0088  to the male salmon; for the shield may be as  IMPORTANT  for victory, as the sword or spear. Amongst
0102  individual, and is, I believe, an extremely  IMPORTANT  element of success. Though nature grants va
0103  to pair together. Intercrossing plays a very  IMPORTANT  part in nature in keeping the individuals c
0104  avourable variations. Isolation, also, is an  IMPORTANT  element in the process of natural selectior
0106  more rapid on large areas; and what is more  IMPORTANT, that the new forms produced on large areas
0106  varieties and species, and will thus play an  IMPORTANT  part in the changing history of the organic
0111  theory, and explains, as I believe, several  IMPORTANT  facts. In the first place, varieties, even
0121  mely that of extinction, will have played an  IMPORTANT  part. As in each fully stocked country natu
0123  terrediate species; also (and this is a very  IMPORTANT  consideration), which connected the origina
0134  y, as already remarked, they seem to play an  IMPORTANT  part in affecting the reproductive system,
0143  other parts become modified. This is a very  IMPORTANT  subject, most imperfectly understood. The m
0144  etermine the position of several of the most  IMPORTANT  viscera. The nature of the bond of correlat
0144  ance of the laws of correlation in modifying  IMPORTANT  structures, independently of utility and, t
0148  ther cirripedes consists of the three highly  IMPORTANT  anterior segments of the head enormously de
0151  acles) are, in every sense of the word, very  IMPORTANT  structures, and they differ extremely littl
0151  e from each other in the characters of these  IMPORTANT  valves than do other species of distinct ge
0155  author has remarked with surprise that some  IMPORTANT  organ or part, which is generally very cons
0161  aracters thus gained would probably be of an  IMPORTANT  nature, for the presence of all important c
0161  an important nature, for the presence of all  IMPORTANT  characters will be governed by natural sele
0162  vidence is afforded by parts or organs of an  IMPORTANT  and uniform nature occasionally varying so
0163  nd complex case, not indeed as affecting any  IMPORTANT  character, but from occurring in several sp
0169  their ancient progenitors. Although new and  IMPORTANT  modifications may not arise from reversion
0170  individual, that gives rise to all the more  IMPORTANT  modifications of structure, by which the in
0174  ndition of areas now continous has played an  IMPORTANT  part in the formation of new species, more
0175  d the physical conditions of life as the all  IMPORTANT  elements of distribution, these facts ought
0175  r directly or indirectly related in the most  IMPORTANT  manner to other organic beings, we must see
0176  existing on both sides of it. But a far more  IMPORTANT  consideration, as I believe, is that, durin
0178  inhabitants; and, probably, in a still more  IMPORTANT  degree, on some of the old inhabitants beco
0190  one, because it shows us clearly the highly  IMPORTANT  fact that an organ originally constructed f
0191  considering transitions of organs, it is so  IMPORTANT  to bear in mind the probability of conversi
0196  een checked by natural selection. Seeing how  IMPORTANT  an organ of locomotion the tail is in most
0196  t correlation of growth will have had a most  IMPORTANT  influence in modifying various structures;
0199  elation of growth has no doubt played a most  IMPORTANT  part, and a useful modification of one part
0199  n rather a forced sense. But by far the most  IMPORTANT  consideration is that the chief part of the
0209  e universally admitted that instincts are as  IMPORTANT  as corporeal structure for the welfare of e
0225  e aggregated into an irregular mass. But the  IMPORTANT  point to notice, is that these cells are al
0230  not here enter on these details. We see how  IMPORTANT  a part excavation plays in the construction
0233  r place between two just commenced cells, is  IMPORTANT, as it bears on a fact, which seems at firs
0234  f wax by largely saving honey must be a most  IMPORTANT  element of success in any family of bees. O
0240  nficiently did I expect to find gradations in  IMPORTANT  points of structure between the different c
0241  in the form and number of the teeth. But the  IMPORTANT  fact for us is, that though the workers can
0245  of natural selection the case is especially  IMPORTANT, inasmuch as the sterility of hybrids could
0246  e imperfectly developed. This distinction is  IMPORTANT, when the cause of the sterility, which is
0247  d must be castrated, and, what is often more  IMPORTANT, must be secluded in order to prevent polle
0250  having hothouses at his command. Of his many  IMPORTANT  statements I will here give only a single o
0258  ng reciprocal crosses. Such cases are highly  IMPORTANT, for they prove that the capacity in any tw
0260  of which we must suppose it would be equally  IMPORTANT  to keep from blending together? Why should
0263  n the case of crossing, the difficulty is as  IMPORTANT  for the endurance and stability of specific
0269  lastly, and this seems to me by far the most  IMPORTANT  consideration, new races of animals and pla
0270  ies of maize are distinct species; and it is  IMPORTANT  to notice that the hybrid plants thus raise
0272  r hand, they agree most closely in very many  IMPORTANT  respects. I shall here discuss this subject
0272  this subject with extreme brevity. The most  IMPORTANT  distinction is, that in the first generatio
0279  t the life of each species depends in a more  IMPORTANT  manner on the presence of other already def
0287  made these few remarks because it is highly  IMPORTANT  for us to cain some notion, however imperfe
0288  ed, and no part with sufficient care, as the  IMPORTANT  discoveries made every year in Europe prove
0289  record mainly results from another and more  IMPORTANT  cause than any of the foregoing; namely, th
0296  haps very slight, changes of form. It is all  IMPORTANT  to remember that naturalists have no golden
0314  of competition, and on that of the many all  IMPORTANT  relations of organism to organism, that any
0335  tincti and carriers which are extreme in the  IMPORTANT  character of length of beak originated earl
0342  specific forms; that migration has played an  IMPORTANT  part in the first appearance of new forms i
0343  consecutive formations; he may overlook how  IMPORTANT  a part migration must have played, when the
0347  o free migration, are related in a close and  IMPORTANT  manner to the differences between the produ
0350  , as I have already often remarked, the most  IMPORTANT  of all relations. Thus the high importance
0354  at the same time to consider a point equally  IMPORTANT  for us, namely, whether the several distinc
0356  here only the briefest abstract of the more  IMPORTANT  facts. Charge of climate must have had a po
0361  . germinated. But the following fact is more  IMPORTANT: the crops of birds do not secrete gastric
0392  immigration, I believe, has been at least as  IMPORTANT  as the nature of the conditions. Many remar
0397  e wanted on this head. The most striking and  IMPORTANT  fact for us in regard to the inhabitants of
0400  physical conditions of a country the most  IMPORTANT  for its inhabitants; whereas it cannot, I t
0400  th which each has to compete, is at least as  IMPORTANT, and generally a far more important element
0400  least as important, and generally a far more  IMPORTANT  element of success. Now if we look to those
0403  inent, pre occupation has probably played an  IMPORTANT  part in checking the commingling of species
```

Page **************************************(Key Word)***

Page **(Key Word)**

```
0405 the power of crossing barriers, but the more IMPORTANT power of being victorious in distant lands
0407 n distribution, I have attempted to show how IMPORTANT has been the influence of the modern Glacia
0414 n is concerned with special habits, the more IMPORTANT it becomes for classification. As an instan
0414 lances in parts of the organisation, however IMPORTANT they may be for the welfare of the being in
0415 lassificatory importance of organs which are IMPORTANT is generally, but by no means always, true.
0416 mportance for the classification of the same IMPORTANT organ within the same group of beings. Agai
0416 hink, have been considered by naturalists as IMPORTANT an aid in determining the degree of affinit
0417 ach in structure in any one internal and and IMPORTANT organ. The importance, for classification,
0417 ion founded on any single character, however IMPORTANT that may be, has always failed; for no part
0417 aggregate of characters, even when none are IMPORTANT, alone explains, I think, that saying of Li
0417 rtinc so wonderfully in a number of the most IMPORTANT points of structure from the proper type of
0418 s set on them. As in most groups of animals, IMPORTANT organs, such as those for propelling the bl
0418 n some groups of animals all these, the most IMPORTANT vital organs, are found to offer characters
0418 y the structure of the embryo should be more IMPORTANT for this purpose than that of the adult, wh
0418 ssiz, that embryonic characters are the most IMPORTANT of any in the classification of animals; an
0419 e, not because further research has detected IMPORTANT structural differences, at first overlooked
0423 merely because their fruit, though the most IMPORTANT part, happens to be nearly identical; no on
0424 sub varieties differ from the others in the IMPORTANT character of having a longer beak, yet all
0424 ver much he might differ in colour and other IMPORTANT characters from negroes. With species in a
0424 normously these sometimes differ in the most IMPORTANT characteristics, is known to every naturalist: s
0426 f species may depart, in several of its most IMPORTANT characteristics; from its allies, and yet b
0427 we can understand, on these views, the very IMPORTANT distinction between real affinities and ana
0431 ve seen in the fourth chapter, has played an IMPORTANT part in defining and widening the intervals
0449 ts, the structure of the embryo is even more IMPORTANT for classification than that of the adult.
0451 ry or utterly aborted for one, even the more IMPORTANT purpose; and remain perfectly efficient for
0452 g of the horse, ox, and rhinoceros. It is an IMPORTANT fact that rudimentary organs, such as teeth
0466 ut it deserves especial notice that the more IMPORTANT objections relate to questions on which we
0473 lation of growth seems to have played a most IMPORTANT part, so that when one part has been modifi
0477 relation of organism to organism is the most IMPORTANT of all relations, and as the two areas will
0484 ble, whether the differences be sufficiently IMPORTANT to deserve a specific name. This latter poi
0486 new variety raised by man will be a far more IMPORTANT and interesting subject for study than one
0487 reat on and by catastrophes; and as the most IMPORTANT of all causes of organic change is one whic
0488 istant future I see open fields for far more IMPORTANT researches. Psychology will be based on a n
0034 of english history choice animals were often IMPORTED, and laws were passed to prevent their expor
0042 eds as we do sometimes see are almost always IMPORTED from some other country, often from islands.
0065 ape Comorin to the Himalaya, which have been IMPORTED from America since its discovery. In such ca
0140 quent inability to predict whether or not an IMPORTED plant will endure our climate, and from the
0380 ects likely to carry seeds have been largely IMPORTED into Europe during the last two or three cen
0204 ng conditions of life, there is no logical IMPOSSIBILITY in the acquirement of any conceivable deg
0263 causes. There must sometimes be a physical IMPOSSIBILITY in the male element reaching the ovule, a
0023 seven or eight aboriginal stocks; for it is IMPOSSIBLE to make the present domestic breeds by the
0026 tinct breeds. Now, it is difficult, perhaps IMPOSSIBLE, to bring forward one case of the hybrid of
0096 duals ever concur in reproduction? As it is IMPOSSIBLE here to enter on details, I must trust to s
0101 ct individual can be shown to be physically IMPOSSIBLE. Cirripedes long appeared to me to present
0146 th any differences in the flowers; it seems IMPOSSIBLE that they can be in any way advantageous to
0192 r another case of special difficulty; it is IMPOSSIBLE to conceive by what steps these wondrous or
0229 ad become perfectly flat: it was absolutely IMPOSSIBLE, from the extreme thinness of the little rh
0246 rossed and of their hybrid offspring. It is IMPOSSIBLE to study the several memoirs and works of t
0269 rtile when intercrossed. But it seems to me IMPOSSIBLE to resist the evidence of the existence of
0419 ns, such definition has hitherto been found IMPOSSIBLE. There are crustaceans at the opposite ends
0431 the aid of a genealogical tree, and almost IMPOSSIBLE to do this without this aid; we can underst
0432 denly to reappear, though it would be quite IMPOSSIBLE to give definitions by which each group cou
0432 t varieties. In this case it would be quite IMPOSSIBLE to give any definition by which the several
0451 mere rudiments of membrane; and here it is IMPOSSIBLE to doubt, that the rudiments represent wing
0094 etimes the female organs become more or less IMPOTENT; now if we suppose this to occur in ever so
0246 have their reproductive organs functionally IMPOTENT, as may be clearly seen in the state of the
0265 le groups of animals and plants are rendered IMPOTENT by the same unnatural conditions; and whole
0273 ns of life, being thus often rendered either IMPOTENT or at least incapable of its proper function
0461 ductive organs are more or less functionally IMPOTENT; whereas in first crosses the organs on both
0466 life; so that this system, when not rendered IMPOTENT, fails to reproduce offspring exactly like t
0251 few days perished entirely, whereas the pod IMPREGNATED by the pollen of the hybrid made vigorous
0029 from long continued study they are strongly IMPRESSED with the differences between the several ra
0051 be to make many species, for he will become IMPRESSED, just like the pigeon or poultry fancier be
0246 lives to this subject, without being deeply IMPRESSED with the high generality of some degree of
0283 our shores, will, I believe, be most deeply IMPRESSED with the slowness with which rocky coasts a
0283 ill, are most impressive. With the mind thus IMPRESSED, let any one examine beds of conglomerate m
0339 lausen in the caves of Brazil. I was so much IMPRESSED with these facts that I strongly insisted,
0488 accords better with what we know of the laws IMPRESSED on matter by the Creator, that the producti
0051 other in an insensible series; and a series IMPRESSES the mind with the idea of an actual passage
0285 swept away. The consideration of these facts IMPRESSES my mind almost in the same manner as does t
0259 f the same genus have a remarkable power of IMPRESSING their likeness on their hybrid offspring; b
0274 sed, one has sometimes a prepotent power of IMPRESSING its likeness on the hybrid; and so I believ
0010 e, light, food, etc., is most difficult: my IMPRESSION is, that with animals such agencies have pr
0132 uces on any being is extremely doubtful. My IMPRESSION is, that the effect is extremely small in t
0051 er countries; by which to correct his first IMPRESSIONS. As he extends the range of his observatio
0283 observer Mr. Smith of Jordan Hill, are most IMPRESSIVE. With the mind thus impressed, let any one
0239 be called; which our European ants guard or IMPRISON. It will indeed be thought that I have an ov
0026 e. from these several reasons, namely, the IMPROBABILITY of man having formerly got seven or eight
0161 some facts; and I can see no more abstract IMPROBABILITY in a tendency to produce any character be
0298 ists. We shall, perhaps, best perceive the IMPROBABILITY of our being enabled to connect species b
0015 ters to ancestral forms, it seems to me not IMPROBABLE, that if we could succeed in naturalising,
0023 bits, and remarkable characters, seems very IMPROBABLE; or they must have become extinct in the wi
0024 n. so many strange contingencies seem to me IMPROBABLE in the highest degree. Some facts in regard
0025 e must make one of the two following highly IMPROBABLE suppositions. Either, firstly, that all the
0080 onditions of life. Can it, then, be thought IMPROBABLE, seeing that variations useful to man have
0091 kinds of prey. Nor can this be thought very IMPROBABLE; for we often observe great differences in
```

Page **(Key Word)**

Page ***(Key Word)***

```
0299  te gradations: and such success seems to me IMPROBABLE in the highest degree. Geological research,
0034  cross their dogs with wild canine animals, to IMPROVE the breed, and they formerly did so, as is a
0035  is process, continued during centuries, would IMPROVE and modify any breed, in the same way as Bak
0037  has taken centuries or thousands of years to IMPROVE or modify most of our plants up to their pre
0088  rutal cockfighter, who knows well that he can IMPROVE his breed by careful selection of the best c
0090  more reason to doubt this, than that man can IMPROVE the fleetness of his greyhounds by careful a
0108  will be a fair field for natural selection to IMPROVE still further the inhabitants, and thus prod
0177  all trying with equal steadiness and skill to IMPROVE their stocks by selection: the chances in th
0215  ch man tries to procure, without intending to IMPROVE the breed, dogs which will stand and hunt be
0005  nevitably causes much Extinction of the less IMPROVED forms of life, and induces what I have calle
0028  which method was never practised before, has IMPROVED them astonishingly. About this same period t
0035  sed districts, where the breed has been less IMPROVED. There is reason to believe that King Charle
0038  ts, but that the native plants have not been IMPROVED by continued selection up to a standard of p
0040  best animals and thus improves them, and the IMPROVED individuals slowly spread in the immediate n
0040  ir history will be disregarded. When further IMPROVED by the same slow and gradual process, they w
0041  are mostly in small lots, they never can be IMPROVED: for in all countries, the natives have been
0083  they live, that none of them could anyhow be IMPROVED in a corresponding degree with its competito
0102  any one species does not become modified and IMPROVED variety might be quickly formed on any one s
0103  can increase at a very rapid rate, a new and IMPROVED: and this may sometimes be of importance in
0105  l give time for any new variety to be slowly IMPROVED, others will have to be improved in a corres
0106  me of these many species become modified and IMPROVED, will be able to spread over the open and co
0106  odified and improved, others will have to be IMPROVED varieties will be enabled to spread: there w
0106  new form, also, as soon as it has been much IMPROVED forms, and the relative proportional numbers
0108  be severe competition: the most favoured or IMPROVED within any given period, and they will conse
0108  d: there will be much extinction of the less IMPROVED forms by man. Many curious instances could b
0110  are species will be less quickly modified or IMPROVED branches in the lines of descent, will, it i
0110  icated productions, through the selection of IMPROVED branches: this is represented in the diagram
0119  ied offspring from the later and more highly IMPROVED descendants of any one species to supplant a
0119  ace of, and so destroy, the earlier and less IMPROVED state of a species, as well as the original
0121  ms; there will be a constant tendency in the IMPROVED lines of descent. If, however, the modified
0121  er states, that is between the less and more IMPROVED in a diversified manner at each stage of des
0121  escent, which will be conquered by later and IMPROVED sub groups. Small and broken groups and sub
0122  advantages: they have also been modified and IMPROVED and intermediate forms of life. On these pri
0126  to supplant and destroy the earlier and less IMPROVED parent or other less favoured forms with whi
0128  character and to much extinction of the less IMPROVED. It is the same principle which, as I believ
0172  of, and finally to exterminate, its own less IMPROVED mountain or plain breed will soon take the p
0177  , for these will be more slowly modified and IMPROVED hill breed; and thus the two breeds, which o
0177  te narrow, hilly tract; and consequently the IMPROVED through natural selection and gain further a
0177  n breed will soon take the place of the less IMPROVED in structure in a corresponding manner. Ther
0179  present more variation, and thus be further IMPROVED forms of life will tend to supplant the old
0181  inated, unless they also became modified and IMPROVED, they would slowly spread and supplant their
0281  ery rare event; for in all cases the new and IMPROVED descendants. Consequently, if my theory be t
0301  ided advantage, or when further modified and IMPROVED, we can understand, on the principle of comp
0307  anted and exterminated by their numerous and IMPROVED, will be liable to be exterminated. Hence we
0314  itants of a country have become modified and IMPROVED offspring, it is quite incredible that a fan
0314  does not become in some degree modified and IMPROVED forms are intimately connected together. Tan
0316  erally be supplanted and exterminated by its IMPROVED variety has been raised, it at first supplan
0317  n of old forms and the production of new and IMPROVED varieties in the same neighbourhood; when mu
0320  omestic productions: when a new and slightly IMPROVED it is transported far and near, like our sho
0320  been raised, it at first supplants the less IMPROVED and modified descendants of a species will g
0320  rieties in the same neighbourhood; when much IMPROVED, a few of the sufferers may often long be or
0321  t like each other in all respects. Hence the IMPROVED groups spread throughout the world, old grou
0321  o other species which have been modified and IMPROVED forms of life, produced by the laws of varia
0327  eriority in common: and therefore as new and IMPROVED varieties, which will never have blended wit
0345  old forms having been supplanted by new and IMPROVED, and preceding forms. I request the reader t
0355  ry, must have descended from a succession of IMPROVED at a quicker rate than the intermediate vari
0412  and exterminate the less divergent, the less IMPROVED; and when they do spread, if discovered in a
0463  ater numbers, will generally be modified and IMPROVED varieties will inevitably supplant and exter
0464  ons until they are considerably modified and IMPROVED and intermediate varieties; and thus species
0470  ristic of species of the same genus. New and IMPROVED forms. Neither single species nor groups of
0470  bly supplant and exterminate the older, less IMPROVED forms have conquered the older and less impr
0475  for old forms will be supplanted by new and IMPROVED organic beings in the struggle for life. Las
0476  y are in so far higher as the later and more IMPROVED forms. Thus, from the war of nature, from fa
0476  oved forms have conquered the older and less IMPROVEMENT or modification. It has been disputed at w
0490  ence of Character and the Extinction of less IMPROVEMENT is by no means generally due to crossing d
0008  sticated animals are still capable of rapid IMPROVEMENT in many florists' flowers, when the flower
0031  d to almost every quarter of the world. The IMPROVEMENT, through the occasional preservation of th
0032  eberry may be quoted. We see an astonishing IMPROVEMENT and formation of new breeds. Pigeons, I ma
0036  dogs. In plants the same gradual process of IMPROVEMENT. We have reason to believe, as stated in t
0042  circumstance must have, largely favoured the IMPROVEMENT of each organic being in relation to its o
0082  would thus have free scope for the work of IMPROVEMENT and modification surely but slowly follow
0084  ver and wherever opportunity offers, at the IMPROVEMENT, will naturally suffer most. And we have s
0102  o get and breed from the best animals, much IMPROVEMENT. Within the same large group, the later an
0110  tion with those undergoing modification and IMPROVEMENT. Let this process go on for millions on mi
0125  lessen its chance of further variation and IMPROVEMENT. The main cause, however, of innumerable f
0189  tion will pick out with unerring skill each IMPROVEMENT has affected in a marked and sensible mann
0279  ring the course of further modification and IMPROVEMENT: and thus they will become still further v
0337  fauna. I do not doubt that this process of IMPROVEMENT, all the individuals of each variety will
0350  frequently undergo further modification and IMPROVEMENT of one being entailing the improvement or
0355  t each successive stage of modification and IMPROVEMENT or the extermination of others: it foll
0487  utual relation of organism to organism, the IMPROVEMENTS: if he wants any of these qualities, he wi
0487  the improvement of one being entailing the IMPROVEMENT them, and the improved individuals slowly sp
0032  rance, he will succeed, and may make great IMPROVING some of the varying inhabitants. For as all
0040  usual in matching his best animals and thus IMPROVING by selection the down in the pods on his co
0082  atural selection to fill up by modifying and
0086  n, than in the cotton planter increasing and
```

Page ***(Key Word)***

Page ***************************************(Key Word)***

```
0102 d then if natural selection be modifying and IMPROVING a species in the several districts, there w
0177 at holders on the mountains or on the plains IMPROVING their breeds more quickly than the small ho
0140 errated. We may infer this from our frequent INABILITY to predict whether or not an imported plant
0442 the most diverse and active habits, or quite INACTIVE, being fed by their parents or placed in the
0282 to mark how each author attempts to give an INADEQUATE idea of the duration of each formation or e
0284 sedimentary rocks in Britain, gives but an INADEQUATE idea of the time which has elapsed during t
0032 ive generations, of differences absolutely INAPPRECIABLE by an uneducated eye, differences which I
0266 hough often in so slight a degree as to be INAPPRECIABLE by us; in the other case, or that of hybr
0464 of time has been so great as to be utterly INAPPRECIABLE by the human intellect. The number of spe
0467 idual differences so slight as to be quite INAPPRECIABLE by an uneducated eye. This process of sel
0489 e confidence to a secure future of equally INAPPRECIABLE length. And as natural selection works so
0276 d together in order to prevent them becoming INARCHED in our forests. The sterility of first cross
0056 n so, it would have been fatal to my theory; INASMUCH as geology plainly tells us that small gener
0087 ike every other structure. Sexual Selection. INASMUCH as peculiarities often appear under domestic
0149 also very liable to variation in structure, INASMUCH as this vegetative repetition, to use Prof.
0245 selection the case is especially important, INASMUCH as the sterility of hybrids could not possib
0268 be some error in all the foregoing remarks, INASMUCH as varieties, however much they may differ f
0342 is absolutely as nothing compared with the INCALCULABLE number of generations which must have pass
0105 mparison within equal times; and this we are INCAPABLE of doing. Although I do not doubt that isol
0135 more, and its wings less, until they became INCAPABLE of flight. Kirby has remarked (and I have o
0219 capturing slaves, do no other work. They are INCAPABLE of making their own nests, or of feeding th
0236 have been annually born capable of work, but INCAPABLE of procreation. I can see no very great dif
0264 element may reach the female element, but be INCAPABLE of causing an embryo to be developed, as se
0273 s often rendered either impotent or at least INCAPABLE of its proper function of producing offspri
0450 , may remain for a long period, or for ever, INCAPABLE of demonstration. Thus, as it seems to me,
0451 ee wings so reduced in size as to be utterly INCAPABLE of flight, and not rarely lying under wing
0468 en been asserted, but the assertion is quite INCAPABLE of proof, that the amount of variation unde
0473 , at the logger headed duck, which has wings INCAPABLE of flight, in nearly the same condition as
0460 is no more a special endowment than is the INCAPACITY of two trees to be grafted together; but th
0094 tubes of the corollas of the common red and INCARNATE clovers (Trifolium pratense and incarnatum)
0094 ve bee can easily suck the nectar out of the INCARNATE clover, but not out of the common red clove
0094 d incarnate clovers (Trifolium pratense and INCARNATUM) do not on a hasty glance appear to differ
0067 packed close together and driven inwards by INCESSANT blows, sometimes one wedge being struck, an
0079 e full belief, that the war of nature is not INCESSANT, that no fear is felt; that death is genera
0128 scendants of any one species, and during the INCESSANT struggle of all species to increase in numb
0141 ust believe, both from analogy, and from the INCESSANT advice given in agricultural works, even in
0290 extensive masses, in order to withstand the INCESSANT action of the waves, when first upraised an
0300 royed, almost as soon as accumulated, by the INCESSANT coast action, as we now see on the shores o
0061 tion, as we shall hereafter see, is a power INCESSANTLY ready for action, and is as immeasurably s
0195 some rare cases) by the flies, but they are INCESSANTLY harassed and their strength reduced, so th
0200 ecies; though throughout nature one species INCESSANTLY takes advantage of, and profits by, the st
0484 labours as at present; but they will not be INCESSANTLY haunted by the shadowy doubt whether this
0231 which are only about one four hundredth of an INCH in thickness; the plates of the pyramidal basi
0286 o cliffs 500 feet in height at the rate of one INCH in a century. This will at first appear much t
0287 cliff 500 feet in height, a denudation of one INCH per century for the whole length would be an a
0214 ear from Mr. Brent, which cannot fly eighteen INCHES high without going head over heels. It may be
0241 ding a house of whom many were five feet four INCHES high, and many sixteen feet high; but we must
0245 ids. Sterility not a special endowment, but INCIDENTAL on other differences. Causes of the sterili
0245 ecially acquired or endowed quality, but is INCIDENTAL on other acquired differences. In treating
0260 h of first crosses and of hybrids is simply INCIDENTAL or dependent on unknown differences, chiefl
0261 y an example what I mean by sterility being INCIDENTAL on other differences, and not a specially e
0261 endowed quality, but will admit that it is INCIDENTAL on differences in the laws of growth of the
0263 which trees can be grafted on each other as INCIDENTAL on unknown differences in their vegetative
0263 overning the facility of first crosses, are INCIDENTAL on unknown differences, chiefly in their re
0272 that it is not a special endowment, but is INCIDENTAL on slowly acquired modifications, more espe
0276 e species or variety to take on another, is INCIDENTAL on generally unknown differences in their v
0276 ty of one species to unite with another, is INCIDENTAL on unknown differences in their reproductiv
0460 rees to be grafted together; but that it is INCIDENTAL on constitutional differences in the reprod
0317 on Extinction. We have as yet spoken only INCIDENTALLY of the disappearance of species and of gro
0052 well marked variety may be justly called an INCIPIENT species; but whether this belief be justifi
0052 t need not be supposed that all varieties or INCIPIENT species necessarily attain the rank of spec
0052 the rank of species. They may whilst in this INCIPIENT state become extinct, or they may endure as
0054 ll marked varieties, or, as I consider them, INCIPIENT species. And this, perhaps, might have been
0055 e genus) have been formed, many varieties or INCIPIENT species ought, as a general rule, to be now
0056 ly is the case, if varieties be looked at as INCIPIENT species; for my tables clearly show as a ge
0056 us present a number of varieties, that is of INCIPIENT species, beyond the average. It is not that
0057 er genera; in which a number of varieties or INCIPIENT species greater than the average are now ma
0061 ow is it that varieties, which I have called INCIPIENT species, become ultimately converted into o
0110 he greatest number of recorded varieties, or INCIPIENT species. Hence, rare species will be less o
0111 of formation, or are, as I have called them, INCIPIENT species. How, then, does the lesser differe
0125 r genera which oftenest present varieties or INCIPIENT species. This, indeed, might have been expe
0169 on an average, we now find most varieties or INCIPIENT species. Secondary sexual characters are hi
0182 itants of the open ocean, and had used their INCIPIENT organs of flight exclusively, as far as we
0256 of the flower is well known to be a sign of INCIPIENT fertilisation. From this extreme degree of
0325 rally oftenest give rise to new varieties or INCIPIENT species; for these latter must be victoriou
0411 r genera, which vary most. The varieties, or INCIPIENT species; thus produced ultimately become co
0470 action; and this is the case if varieties be INCIPIENT species. Moreover, the species of the large
0470 ch afford the greater number of varieties or INCIPIENT species, retain to a certain degree the cha
0097 vigour and fertility; that these facts alone INCLINE me to believe that it is a general law of na
0134 posite climates. Such considerations as these INCLINE me to lay very little weight on the direct a
0008 tion from mere variations. Put I am strongly INCLINED to suspect that the most frequent cause of v
0018 ns which I cannot give here, I am doubtfully INCLINED to believe, in opposition to several authors
0046 independent of the conditions of life. I am INCLINED to suspect that we see in these polymorphic
0096 far from obvious. Nevertheless I am strongly INCLINED to believe that with all hermaphrodites two
0101 ch I am not here able to give, I am strongly INCLINED to suspect that, both in the vegetable and a
0105 production of new species, on the whole I am INCLINED to believe that largeness of area is of more
```

Page ***************************************(Key Word)***

Page ***(Key Word)***

```
0141 many islands in the torrid zones. Hence I am INCLINED to look at adaptation to any special climate
0193 l difference can generally be detected. I am INCLINED to believe that in nearly the same way as tw
0254 tive seems to me the most probable, and I am INCLINED to believe in its truth, although it rests o
0271 made on certain varieties of hollyhock, I am INCLINED to suspect that they present analogous facts
0287 ars. The action of fresh water on the gently INCLINED Wealden district, when upraised, could hardl
0370 partial oscillations of level, I am strongly INCLINED to extend the above view, and to infer that
0381 ting lines from some common centre; and I am INCLINED to look in the southern, as in the northern
0384 ir ova when removed from the water. But I am INCLINED to attribute the dispersal of fresh water fi
0465 their accumulation; and their duration, I am INCLINED to believe, has been shorter than the averag
0465 t to the degree which I require, few will be INCLINED to admit. If we look to long enough interval
0206 the world, is not strictly correct, but if we INCLUDE all those of past times, it must by my theor
0293 i can see several reasons why each should not INCLUDE a graduated series of links between the spec
0296 ological period, a section would not probably INCLUDE all the fine intermediate gradations which m
0362 m experiments made in the Zoological Gardens, INCLUDE seeds capable of germination. Some seeds of
0363 earth adhering to their feet would sometimes INCLUDE a few minute seeds? But I shall presently ha
0392 or bushes belonging to orders which elsewhere INCLUDE only herbaceous species; now trees, as Alph.
0412 oups which are now large, and which generally INCLUDE many dominant species, tend to go on increas
0418 the adult, for our classifications of course INCLUDE all ages of each species. But it is by no me
0434 e constructed on the same pattern, and should INCLUDE the same bones, in the same relative positio
0487 trictly contemporaneous two formations, which INCLUDE few identical species; by the general succes
0154 r and less modified condition. The principle INCLUDED in these remarks may be extended. It is noto
0164 ; by the term dun a large range of colour is INCLUDED, from one between brown and black to a close
0294 limate, on the prodigious lapse of time, all INCLUDED within this same glacial period. Yet it may
0349 ans of longitude. A third great fact, partly INCLUDED in the foregoing statements, is the affinity
0373 nce in almost every case, that the epoch was INCLUDED within the latest geological period. We have
0389 a few ferns and a few introduced plants are INCLUDED in these numbers, and the comparison in some
0400 endemic species, which cannot be here fairly INCLUDED, as we are considering how they have come to
0413 itor grouped into genera; and the genera are INCLUDED in, or subordinate to, sub families, familie
0413 acters, seem to imply that something more is INCLUDED in our classification, than mere resemblance
0413 esemblance. I believe that something more is INCLUDED; and that propinquity of descent, the only k
0422 iate and slowly changing dialects, had to be INCLUDED, such an arrangement would, I think, be the
0424 ced on the same spike, they were immediately INCLUDED as a single species. But it may be asked, wh
0428 that all organisms, recent and extinct, are INCLUDED under a few great orders, under still fewer
0434 e class are homologous. The whole subject is INCLUDED under the general name of Morphology. This i
0044 n he speaks of a species. Generally the term INCLUDES the unknown element of a distinct act of cre
0206 itions of Existence is the higher law: as it INCLUDES, through the inheritance of former adaptatio
0305 es, low down in the Chalk period. This group INCLUDES the large majority of existing species. Late
0336 climate during the pleistocene period, which INCLUDES the whole glacial period, and note how littl
0424 ught descent into his classification; for he INCLUDES in his lowest grade, or that of a species, t
0424 ne dreams of separating them. The naturalist INCLUDES as one species the several larval stages of
0424 ach other and from the adult; as he likewise INCLUDES the so called alternate generations of Steen
0424 nse be considered as the same individual. he INCLUDES monsters; he includes varieties, not solely
0424 he same individual. He includes monsters; he INCLUDES varieties, not solely because they closely r
0023 cended from the rock pigeon (Columba livia), INCLUDING under this term several geographical races
0048 ral highly polymorphic genera. Under genera, INCLUDING the most polymorphic forms, Mr. Babington g
0054 to have found in the larger genera, or those INCLUDING many species, a large proportional number o
0062 existence in a large and metaphorical sense, INCLUDING dependence of one being on another, and inc
0062 ding dependence of one being on another, and INCLUDING (which is more important) not only the life
0113 individual plants of this species of grass, INCLUDING its modified descendants, would succeed in
0124 f the successive strata of the earth's crust INCLUDING extinct remains. We shall, when we come to
0166 f pigeons: they are descended from a pigeon (INCLUDING two or three sub species or geographical ra
0181 from the corners of the jaw to the tail, and INCLUDING the limbs and the elongated fingers: the fl
0181 ed from the top of the shoulder to the tail, INCLUDING the hind legs, we perhaps see traces of an
0192 iae, the whole surface of the body and sack, INCLUDING the small frena, serving for respiration. T
0284 d this is the result: Palaeozoic strata (not INCLUDING igneous beds).....57,154 Feet. Secondary st
0294 quarter of the world, sedimentary deposits, INCLUDING fossil remains, have gone on accumulating w
0324 ormously remote period as measured by years, INCLUDING the whole glacial epoch), were to be compar
0324 remains in some degree allied, and from not INCLUDING those forms which are only found in the old
0332 e two other families (namely, a14 to f14 now INCLUDING five genera, and o14 to m14) would yet rema
0412 on the uppermost line may represent a genus INCLUDING several species; and all the genera on this
0412 n, and form a sub family, distinct from that INCLUDING the next two genera on the right hand, whic
0412 ni and they form a family distinct from that INCLUDING the three genera still further to the right
0419 uable, on the view of classification tacitly INCLUDING the idea of descent. Our classifications ar
0430 ore we must suppose either that all Rodents, INCLUDING the bizcacha, branched off from some very a
0067 n this head, even in regard to mankind, so INCOMPARABLY better known than any other animal. This s
0082 d nature, but far more easily, from having INCOMPARABLY longer time at her disposal. Nor do I beli
0172 the answer mainly lies in the record being INCOMPARABLY less perfect than is generally supposed; t
0213 election, and have been transmitted for an INCOMPARABLY shorter period, under less fixed condition
0347 different climate; and they will be found INCOMPARABLY more closely related to each other, than t
0354 is has been the universal law, seems to me INCOMPARABLY the safest. In discussing this subject, we
0400 though mostly distinct, are related in an INCOMPARABLY closer degree to each other than to the in
0445 in the above specified several points were INCOMPARABLY less than in the full grown birds. Some ch
0116 process of diversification in an early and INCOMPLETE stage of development. After the foregoing d
0488 resent known, although of a length quite INCOMPREHENSIBLE by us, will hereafter be recognised as a
0224 ts you please, and it seems at first quite INCONCEIVABLE how they can make all the necessary angle
0282 living and extinct species, must have been INCONCEIVABLY great. But assuredly, if this theory be t
0374 ng plants of Tierra del Fuego, forming no INCONSIDERABLE part of its scanty flora, are common to E
0219 eir own larvae. When the old nest is found INCONVENIENT, and they have to migrate, it is the slave
0217 e process of laying and hatching might be INCONVENIENTLY long, more especially as she has to migra
0039 is, I have no doubt, in most cases, utterly INCORRECT. The man who first selected a pigeon with a
0131 due to chance. This, of course, is a wholly INCORRECT expression, but it serves to acknowledge pl
0005 ollows from their high geometrical powers of INCREASE, will be treated of. This is the doctrine of
0058 r differences between varieties will tend to INCREASE into the greater differences between species
0060 used in a wide sense. Geometrical powers of INCREASE. Rapid increase of naturalised animals and p
0060 sense. Geometrical powers of increase. Rapid INCREASE of naturalised animals and plants. Nature of
0060 animals and plants. Nature of the checks to INCREASE. Competition universal. Effects of climate.
```

Page ***(Key Word)***

age ∗∗∗(Key Word)∗∗∗

```
0063 igh rate at which all organic beings tend to  INCREASE. Every being, which during its natural lifet
0063 , otherwise, on the principle of geometrical   INCREASE, its numbers would quickly become so inordin
0063 for in this case there can be no artificial    INCREASE of food, and no prudential restraint from ma
0064 stimate its probable minimum rate of natural   INCREASE: it will be under the mark to assume that it
0064 us recorded cases of the astonishingly rapid   INCREASE of various animals in a state of nature, whe
0064 the world: if the statements of the rate of    INCREASE of slow breeding cattle and horses in South
0065 reed. In such cases the geometrical ratio of   INCREASE, the result of which never fails to be surpr
0065 o, simply explains the extraordinarily rapid   INCREASE and wide diffusion of naturalised production
0065 , that all plants and animals are tending to   INCREASE at a geometrical ratio, that all would most
0065 exist, and that the geometrical tendency to    INCREASE must be checked by destruction at some perio
0066 mount of food, for it allows them rapidly to   INCREASE in number. But the real importance of a larg
0066 may be said to be striving to the utmost to    INCREASE in numbers; that each lives by a struggle at
0067 r of the species will almost instantaneously   INCREASE to any amount. The face of Nature may be com
0067 ecks the natural tendency of each species to   INCREASE in number is most obscure. Look at the most
0067 in numbers, by so much will its tendency to    INCREASE be still further increased. We know not exac
0068 se gives the extreme limit to which each can   INCREASE: but very frequently it is not the obtaining
0069 d by any slight change of climate, they will   INCREASE in numbers, and, as each area is already ful
0070 super abundance of food at this one season,   INCREASE in number proportionally to the supply of se
0072 navels of these animals when first born. The   INCREASE of these flies, numerous as they are, must b
0072 egulated by hawks or beasts of prey) were to   INCREASE in Paraguay, the flies would decrease, then
0075 th birds and beasts of prey, all striving to   INCREASE, and all feeding on each other or on the tre
0076 the decrease of another species. The recent   INCREASE of the missel thrush in parts of Scotland ha
0078 nged in an essential manner. If we wished to   INCREASE its average numbers in its new home, we shou
0078 mind that each organic being is striving to   INCREASE at a geometrical ratio: that each at some pe
0082 s of the same kind would often still further  INCREASE the advantage. No country can be named in wh
0084 stroyed at some period of their lives, would  INCREASE in countless numbers; they are known to suff
0103 birth, but which wander little and which can  INCREASE at a very rapid rate, a new and improved var
0109 . but as from the high geometrical powers of  INCREASE of all organic beings, each area is already
0112 es, at first barely appreciable, steadily to  INCREASE, and the breeds to diverge in character both
0112 n the polity of nature, and so be enabled to  INCREASE in numbers. We can clearly see this in the c
0113 t its full average. If its natural powers of  INCREASE be allowed to act, it can succeed in increas
0113 as it may be said, is striving its utmost to  INCREASE its numbers. Consequently, I cannot doubt th
0125 n the larger groups, which are all trying to  INCREASE in number. One large group will slowly conqu
0126 tinction, will for a long period continue to  INCREASE. But which groups will ultimately prevail, n
0126 dict that, owing to the continued and steady  INCREASE of the larger groups, a multitude of smaller
0127 be, owing to the high geometrical powers of   INCREASE of each species, at some age, season, or yea
0128 inc the incessant struggle of all species to  INCREASE in numbers, the more diversified these desce
0128 s of the same species, will steadily tend to  INCREASE till they come to equal the greater differen
0138 ften have effected other changes, such as an  INCREASE in the length of the antennae or palpi, as a
0175 every species, even in its metropolis, would  INCREASE immensely in numbers, were it not for other
0186 organic being is constantly endeavouring to  INCREASE in numbers; and that if any cne being vary e
0192 selection into branchiae, simply through an   INCREASE in their size and the obliteration of their
0198 gions would, we have some reason to believe,  INCREASE the size of the chest; and again correlation
0249 id nature. And thus, the strange fact of the  INCREASE of fertility in the successive generations o
0316 xceptional, the general rule being a gradual  INCREASE in number, till the group reaches its maximu
0317 inal extinction of the species. This gradual  INCREASE in number of the species of a group is stric
0317 enus, and the genera of the same family, can  INCREASE only slowly and progressively; for the proce
0318 er end, which marks the first appearance and  INCREASE in numbers of the species. In some cases, ho
0319 favourable conditions were which checked its  INCREASE, whether some one or several contingencies,
0319 s most difficult always to remember that the  INCREASE of every living being is constantly being ch
0322 t for an instant, that each species tends to  INCREASE inordinately, and that some check is always
0343 peared it never reappears. Groups of species  INCREASE in numbers slowly, and endure for unequal pe
0441 outh, with which they feed largely, for they  INCREASE much in size. In the second stage, answering
0461 ife and crosses between less modified forms,  INCREASE fertility. Turning to geographical distribut
0467 y follows from the high geometrical ratio of  INCREASE which is common to all organic beings. This
0467 won so all organic beings. This high rate of  INCREASE is proved by calculation, by the effects of
0467 ch shall die, which variety or species shall  INCREASE in number, and which shall decrease, or fina
0470 by its geometrical ratio of reproduction to  INCREASE inordinately in number; and as the modified
0470 scendants of each species will be enabled to  INCREASE by so much the more as they become more dive
0472 he view of each species constantly trying to  INCREASE in number, with natural selection always rea
0490 of life, and from use and disuse; a Ratio of  INCREASE so high as to lead to a Struggle for Life, a
0011 f the conditions of life, as, in some cases,  INCREASED size from amount of food, colour from parti
0035 rs have shown how the cattle of England have  INCREASED in weight and in early maturity, compared w
0037 y crossing, may plainly be recognised in the  INCREASED size and beauty which we now see in the var
0041 the chance of their appearance will be much   INCREASED by a large number of individuals being kept
0056 small genera have in the lapse of time often  INCREASED greatly in size; and that large genera have
0065 or plants has been suddenly and temporarily  INCREASED in any sensible degree. The obvious explana
0067 ll its tendency to increase be still further  INCREASED. We know not exactly what the checks are in
0090 instance, had from any change in the country  INCREASED in numbers, or that other prey had decrease
0094 individuals with this tendency more and more  INCREASED, would be continually favoured or selected,
0109 umber of specific forms has not indefinitely  INCREASED; geology shows us plainly: and indeed we ca
0109 can see reason why they should not have thus  INCREASED, for the number of places in the polity of
0119 and the more their modified progeny will be  INCREASED. In our diagram the line of succession is b
0119 nd the number of the descendants will not be  INCREASED; although the amount of divergent modificat
0120 ount of divergent modification may have been  INCREASED in the successive generations. This case wo
0120 all differences distinguishing varieties are  INCREASED into the larger differences distinguishing
0135 of a bustard, and that as natural selection  INCREASED in successive generations the size and weig
0138 struggled with the loss of light and to have  INCREASED the size of the eyes; whereas with all the
0209 s of corporeal structure arise from, and are  INCREASED by, use or habit, and are diminished or los
0248 sserts positively that their fertility never  INCREASED, but generally greatly decreased. I do not
0266 ids produced from reciprocal crosses; or the  INCREASED sterility in those hybrids which occasional
0292 the adjoining shoal parts of the sea will be  INCREASED, and new stations will often be formed; all
0318 has run wild over the whole country and has   INCREASED in numbers at an unparalleled rate, I asked
0367 s ascending higher and higher, as the warmth  INCREASED, whilst their brethren were pursuing their
0435 any extent, and the membrane connecting them  INCREASED to any extent, so as to serve as a wing: ye
0451 the pistil in the hybrid offspring was much   INCREASED in size; and this shows that the rudiment a
```

Page ∗∗(Key Word)∗∗

Page ************************************(Key Word)************************************

```
0461  gour and fertility of all organic beings are  INCREASED by slight changes in their conditions of l
0461  orms or varieties acquire from being crossed  INCREASED vigour and fertility. So that, on the one
0064  the rule that every organic being naturally   INCREASES at so high a rate, that if not destroyed,
0070  s, owing to highly favourable circumstances,   INCREASES inordinately in numbers in a small tract,
0082  acting on the reproductive system, causes or   INCREASES variability; and in the foregoing case the
0109  lows that as each selected and favoured form   INCREASES in number, so will the less favoured forms
0249  cross with a distinct individual or variety   INCREASES fertility, that I cannot doubt the correct
0249  effects of manipulation, sometimes decidedly   INCREASES, and goes on increasing. Now, in artificia
0032  give a very trifling instance, the steadily    INCREASING size of the common gooseberry may be quote
0056  e genera are now varying much, and are thus    INCREASING in the number of their species, or that no
0056  or that no small genera are now varying and    INCREASING; for if this had been so, it would have be
0064  marriage. Although some species may be now     INCREASING, more or less rapidly, in numbers, all car
0073  insectivorous birds, and so onwards in ever    INCREASING circles of complexity. We began this serie
0078  imagination to give the plant the power of     INCREASING in number, we should have to give it some
0086  tural selection, than in the cotton planter    INCREASING and improving by selection the down in the
0094  pose the plant of which we have been slowly    INCREASING the nectar by continued selection, to be a
0109  goes on perpetually and almost indefinitely    INCREASING, numbers inevitably must become extinct. T
0113  crease be allowed to act, it can succeed in    INCREASING (the country not undergoing any change in
0114  s have the best chance of succeeding and of    INCREASING in numbers, and thus of supplanting the le
0118  the common parent (A), will generally go on    INCREASING in number and diverging in character. In t
0224  can see no difficulty in natural selection    INCREASING and modifying the instinct, always supposi
0249  sometimes decidedly increases, and goes on    INCREASING. Now, in artificial fertilisation pollen i
0253  terility in the hybrids should have gone on    INCREASING. If we were to act thus, and pair brothers
0320  ber of species has not gone on indefinitely   INCREASING, at least during the later geological peri
0412  nclude many dominant species, tend to go on   INCREASING indefinitely in size. I further attempted
0412  a constant tendency in the forms which are   INCREASING in number and diverging in character, to s
0428  and more dominant groups thus tend to go on   INCREASING in size; and they consequently supplant ma
0471  r. but as all groups cannot thus succeed in   INCREASING in size, for the world would not hold them
0471  this tendency in the large groups to go on    INCREASING in size and diverging in character, togeth
0064  l authenticated, they would have been quite   INCREDIBLE. So it is with plants: cases could be give
0195  structed fly flapper; and it seems at first  INCREDIBLE that this could have been adapted for its
0238  but from each other, sometimes to an almost  INCREDIBLE degree, and are thus divided into two or e
0270  r more remarkable, and seems at first quite  INCREDIBLE; but it is the result of an astonishing num
0316  ated by its improved offspring, it is quite  INCREDIBLE that a fantail, identical with the existin
0352  or, as explained in the last chapter, it is  INCREDIBLE that individuals identically the same shou
0292  rict accordance with the general principles  INCULCATED by Sir C. Lyell; and E. Forbes independent
0368  bed; and, in accordance with the principles  INCULCATED in this volume, they will not have been li
0215  p and pigs; and this tendency has been found  INCURABLE in dogs which have been brought home as pup
0053  ask; but Mr. H. C. Watson, to whom I am much  INDEBTED for valuable advice and assistance on this s
0220  th, of the British Museum, to whom I am much  INDEBTED for information on this and other subjects.
0019  rgan used exclusively for respiration. I can,  INDEED, hardly doubt that all vertebrate animals hav
0030  races is that we see in them adaptation, not  INDEED to the animal's or plant's own good, but to r
0030  as perfect and as useful as we now see them;  INDEED, in several cases, we know that this has not
0034  attended to by the lowest savages. It would,  INDEED, have been a strange fact, had attention not
0038  excepting such as is externally visible; and  INDEED he rarely cares for what is internal. He can
0039  tablished. Many slight differences might, and  INDEED do now, arise amongst pigeons, which are refe
0058  ave much restricted ranges: this statement is  INDEED scarcely more than a truism, for if a variety
0060  nature there is some individual variability;  INDEED I am not aware that this has ever been disput
0073  l, and this would certainly greatly alter (as  INDEED I have observed in parts of South America),
0090  be either useful or ornamental to this bird;  INDEED, had the tuft appeared under domestication,
0109  tely increased, geology shows us plainly; and  INDEED we can see reason why they should not have th
0125  present varieties or incipient species. This,  INDEED, might have been expected; for as natural sel
0134  prey, has been caused by disuse. The ostrich  INDEED inhabits continents and is exposed to danger
0161  g, as we all know them to be, thus inherited  INDEED, we may sometimes observe a mere tendency to
0163  wever, give one curious and complex case, not  INDEED as affecting any important character, but fro
0192  ocous with the branchiae of the other family;  INDEED, they graduate into each other. Therefore I c
0194  is known to lead. The truth of this remark is  INDEED shown by that old canon in natural history of
0208  sciously many habitual actions are performed,  INDEED not rarely in direct opposition to our consci
0239  our European ants guard or imprison. It will  INDEED be thought that I have an overweening confide
0267  ur and fertility to the offspring. I believe,  INDEED, from the facts alluded to in our fourth chap
0325  remarks to precisely the same effect. It is,  INDEED, quite futile to look to changes of currents,
0328  cring the proximity of the two areas, unless,  INDEED, it be assumed that an isthmus separated two
0357  ited almost every island to some mainland. If  INDEED the arguments used by Forbes are to be truste
0365  having migrated from one to the other. It is  INDEED a remarkable fact to see so many of the same
0393  e accounted for by their physical conditions;  INDEED it seems that islands are peculiarly well fit
0399  thplace. Many analogous facts could be given;  INDEED it is an almost universal rule that the endem
0401  in the several islands. This difference might  INDEED have been expected on the view of the islands
0417  racters of more or less importance. The value  INDEED of an aggregate of characters is very evident
0443  od. But we have little evidence on this head,  INDEED the evidence rather points the other way; for
0449  nected by the finest gradations, the best, or  INDEED, if our collections were nearly perfect, the
0472  rpillars; and at other such cases. The wonder  INDEED is, on the theory of natural selection, that
0026  ary, may be transmitted undiminished for an    INDEFINITE number of generations. These two distinct i
0059  e to two forms the rank of species is quite    INDEFINITE. In genera having more than the average num
0102  of natural selection, she does not grant an    INDEFINITE period; for as all organic beings are striv
0437  s of successive spiral whorls of leaves. An    INDEFINITE repetition of the same part or organ is the
0438  of the class, we do not find nearly so much    INDEFINITE repetition of any one part, as we find in t
0109  cific forms goes on perpetually and almost    INDEFINITELY increasing, numbers inevitably must become
0109  that the number of specific forms has not    INDEFINITELY increased, geology shows us plainly; and i
0109  r of places in the polity of nature is not    INDEFINITELY increased, not that we have any means of knowi
0203  een that species at any one period are not    INDEFINITELY variable, and are not linked together by a
0320  that the number of species has not gone on    INDEFINITELY increasing, at least during the later geol
0412  dominant species, tend to go on increasing    INDEFINITELY in size. I further attempted to show that
0254  sed; or we must look at sterility, not as an    INDELIBLE characteristic, but as one capable of being
0286  degradation at the same time along its whole    INDENTED length; and we must remember that almost all
0046  em to show that this kind of variability is    INDEPENDENT of the conditions of life. I am inclined t
0052  s; or both might co exist, and both rank as    INDEPENDENT species. But we shall hereafter have to re
```

Page ************************************(Key Word)************************************

age ***(Key Word)***

```
068 of climate seems at first sight to be quite   INDEPENDENT of the struggle for existence; but in so f
070 en ensue: and here we have a limiting check     INDEPENDENT of the struggle for life. But even some of
139 wo continents on the ordinary view of their     INDEPENDENT creation. That several of the inhabitants
146 er thousands of generations, some other and     INDEPENDENT modification; and these two modifications,
194 why should all the parts and organs of many     INDEPENDENT beings, each supposed to have been separat
203 wn on several facts, which on the theory of     INDEPENDENT acts of creation are utterly obscure. We h
206 eings of the same class, and which is quite     INDEPENDENT of their habits of life. On my theory, uni
221 proached and were vigorously repulsed by an     INDEPENDENT community of the slave species (F. fusca)
222 . in one instance I found to my surprise an     INDEPENDENT community of F. flava under a stone beneat
234 quite distinct causes, and so be altogether     INDEPENDENT of the quantity of honey which the bees co
236 ore believe that they have been acquired by     INDEPENDENT acts of natural selection. I will not here
248 re the fertility has been diminished by an      INDEPENDENT cause, namely, from close interbreeding. I
258 ny two species to cross is often completely     INDEPENDENT of their systematic affinity, or of any re
259 w how completely fertility in the hybrid is     INDEPENDENT of its external resemblance to either pure
265 similarity. In both cases the sterility is     INDEPENDENT of general health, and is often accompanie
314 w. the variability of each species is quite     INDEPENDENT of that of all others. Whether such variab
350 over the same areas of land and water, and     INDEPENDENT of their physical conditions. The naturali
351 t. as the variability of each species is an     INDEPENDENT property, and will be taken advantage of b
355 re fully see, inexplicable on the theory of     INDEPENDENT creation. This view of the relation of spe
389 ch bear on the truth of the two theories of     INDEPENDENT creation and of descent with modification.
395 ere is also a relation, to a certain extent     INDEPENDENT of distance, between the depth of the sea
396 nt, an inexplicable relation on the view of     INDEPENDENT acts of creation. All the foregoing remark
398 sort of explanation on the ordinary view of     INDEPENDENT creation; whereas on the view here maintai
406 ly inexplicable on the ordinary view of the     INDEPENDENT creation of each species, but are explicab
435 ature of Limbs. On the ordinary view of the     INDEPENDENT creation of each being, we can only say th
478 , are utterly inexplicable on the theory of     INDEPENDENT acts of creation. The existence of closely
487 es of organic change is one which is almost     INDEPENDENT of altered and perhaps suddenly altered ph
003 conclusion that each species had not been      INDEPENDENTLY created, but had descended, like varietie
006 tained, namely, that each species has been     INDEPENDENTLY created, is erroneous. I am fully convinc
059 erly inexplicable if each species has been     INDEPENDENTLY created. We have, also, seen that it is t
081 al proportions of some of the inhabitants,     INDEPENDENTLY of the change of climate itself, would mo
129 es. On the view that each species has been     INDEPENDENTLY created, I can see no explanation of this
144 elation in modifying important structures,     INDEPENDENTLY of utility and, therefore, of natural sel
152 so? On the view that each species has been     INDEPENDENTLY created, with all its parts as we now see
155 ordinary view of each species having been     INDEPENDENTLY created, why should that part of the stru
155 which differs from the same part in other     INDEPENDENTLY created species of the same genus, be mor
159 ordinary view of each species having been     INDEPENDENTLY created, we should have to attribute this
162 s, unless all these forms be considered as     INDEPENDENTLY created species, that the one in varying
167 who believes that each equine species was     INDEPENDENTLY created, will, I presume, assert that eac
193 rly the same way as two men have sometimes     INDEPENDENTLY hit on the very same invention, so natura
196 ve originated from quite secondary causes,     INDEPENDENTLY of natural selection. We should remember
199 had some little effect on structure, quite     INDEPENDENTLY of any good thus gained. Correlation of g
245 t universal. Hybrids and mongrels compared     INDEPENDENTLY of their fertility. Summary. The view gen
265 g of two species, the reproductive system,     INDEPENDENTLY of the general state of health, is affect
272 re crossed. Hybrids and Mongrels compared.     INDEPENDENTLY of their fertility. Independently of the
272 ompared, independently of their fertility.     INDEPENDENTLY of the question of fertility. the offspri
282 ved upon this earth. On the lapse of Time.     INDEPENDENTLY of our not finding fossil remains of such
284 h the strata have in many places suffered,     INDEPENDENTLY of the rate of accumulation of the degrad
292 inculcated by Sir C. Lyell; and E. Forbes     INDEPENDENTLY arrived at a similar conclusion. One rema
365 clude that the same species must have been     INDEPENDENTLY created at several distinct points; and w
389 must, I think, admit that something quite     INDEPENDENTLY of any difference in physical conditions
408 y, there would ensue in different regions,     INDEPENDENTLY of their physical conditions, infinitely
434 e seen that the members of the same class,     INDEPENDENTLY of their habits of life, resemble each ot
470 ns on the view of each species having been     INDEPENDENTLY created, but are intelligible if all spec
471 e a law of nature if each species has been     INDEPENDENTLY created, no man can explain. Many other f
473 ordinary view of each species having been     INDEPENDENTLY created, why should the specific characte
473 her species, supposed to have been created     INDEPENDENTLY, have differently coloured flowers, than
475 ould be strange facts if species have been     INDEPENDENTLY created, and varieties have been produced
482 other species are real, that is, have been     INDEPENDENTLY created. This seems to me a strange concl
488 d with the view that each species has been     INDEPENDENTLY created. To my mind it accords better wit
019 species as distinct parent stocks? So it is in   INDIA. Even in the case of the domestic dogs of the
020 ld, more especially by the Hon. W. Elliot from   INDIA, and by the Hon. C. Murray from Persia. Many
027 ound capable of domestication in Europe and in   INDIA; and that it agrees in habits and in a great
028 ace. Pigeons were much valued by Akber Khan in   INDIA, about the year 1600; never less than 20,000
036 with these breeds as now existing in Britain,   INDIA, and Persia, we can, I think, clearly trace t
065 urope; and there are plants which now range in   INDIA, as I hear from Dr. Falconer, from Cape Comor
068 destroyed by beasts of prey: even the tiger in   INDIA most rarely dares to attack a young elephant
164 es on each shoulder. In the north west part of   INDIA the Kattywar breed of horses is so generally
253 dchildren of the pure geese) from one nest. In   INDIA, however, these cross bred geese must be far
303 ut now extinct species have been discovered in   INDIA, South America, and in Europe even as far bac
323 the Cape of Good Hope, and in the peninsula of   INDIA. For at these distant points, the organic rem
340 coner and Cautley's discoveries, that Northern   INDIA was formerly more closely related in its mamm
375 re isolated mountain ranges of the peninsula of   INDIA, on the heights of Ceylon, and on the volcani
375 and are thinly scattered, on the one hand over   INDIA and on the other as far north as Japan. On th
384 live fish not rarely dropped by whirlwinds in   INDIA, and the vitality of their ova when removed f
018 voice, and constitution, etc., of the humped   INDIAN cattle, that these had descended from a diffe
019 f poultry have proceeded from the common wild   INDIAN fowl (Gallus bankiva). In regard to ducks and
025 is of a slaty blue, and has a white rump (the   INDIAN sub species, C. intermedia of Strickland, hav
074 ved that the trees now growing on the ancient   INDIAN mounds, in the Southern United States, displa
075 ers and kinds of trees now growing on the old   INDIAN ruins! The dependency of one organic being on
164 colonel Poole, who examined the breed for the   INDIAN Government, a horse without stripes is not co
254 n to believe that our European and the humped   INDIAN cattle are quite fertile together: but from f
305 onverted into land, the tropical parts of the   INDIAN Ocean would form a large and perfectly enclos
348 yet many fish range from the Pacific into the   INDIAN Ocean, and many shells are common to the east
395 by similar channels from Australia. The West   INDIAN Islands stand on a deeply submerged bank, nea
088 fighting, bellowing, and whirling round, like   INDIANS in a war dance, for the possession of the fe
```

Page ***(Key Word)***

Page **(Key Word)**

0182 h perhaps may all have resulted from disuse, INDICATE the natural steps by which birds have acqui
0255 ject will be to see whether or not the rules INDICATE that species have specially been endowed wit
0260 ity. Now do these complex and singular rules INDICATE that species have been endowed with sterilit
0260 , on the other hand, appear to me clearly to INDICATE that the sterility both of first crosses and
0263 pressed. The facts by no means seem to me to INDICATE that the greater or lesser difficulty of eit
0380 ies remain to be solved. I do not pretend to INDICATE the exact lines and means of migration, or t
0381 he necessary degree. The facts seem to me to INDICATE that peculiar and very distinct species have
0438 hose of another and distinct species, we can INDICATE but few serial homologies; that is, we are s
0124 of an earlier genus. In our diagram, this is INDICATED by the broken lines, beneath the capital le
0307 in some of the lowest azoic rocks, probably INDICATES the former existence of life at these perio
0414 ave always regarded as affording very clear INDICATIONS of its true affinities. We are least likel
0364 h they might and do bring seeds from the West INDIES to our western shores, where, if not killed b
0446 ng, when they are nearly grown up: they are INDIFFERENT whether the desired qualities and structur
0115 ld generally have been closely allied to the INDIGENES; for these are commonly looked at as specia
0115 differ, moreover, to a large extent from the INDIGENES, for out of the 162 genera, no less than 10
0115 s which have struggled successfully with the INDIGENES of any country, and have there become natur
0115 nera, no less than 100 genera are not there. INDIGENOUS, and thus a large proportional addition is
0254 yet, with perhaps the exception of certain INDIGENOUS domestic dogs of South America, all are qui
0268 with foxes, or that certain South American INDIGENOUS domestic dogs do not readily cross with Eur
0403 which no doubt had some advantage over the INDIGENOUS species. From these considerations I think
0280 ch modified, that if we had no historical or INDIRECT evidence regarding their origin, it would no
0489 mplied by reproduction! Variability from the INDIRECT and direct action of the external conditions
0066 number of any animal or plant depends only INDIRECTLY on the number of its eggs or seeds. In look
0069 he elements. That climate acts in main part INDIRECTLY by favouring other species, we may clearly
0133 er the same conditions. Such facts show how INDIRECTLY the conditions of life must act. Again, inn
0134 he direct action of the conditions of life. INDIRECTLY, as already remarked, they seem to play an
0154 elieve this explanation is partly, yet only INDIRECTLY, true; I shall, however, have to return to
0175 at each organic being is either directly or INDIRECTLY related in the most important manner to oth
0200 scendants of this form, either directly, or INDIRECTLY through the complex laws of growth. Natural
0269 she may, either directly, or more probably INDIRECTLY, through correlation, modify the reproducti
0430 related to any one existing Marsupial, but INDIRECTLY to all or nearly all Marsupials, from havin
0032 en made, the closest selection is far more INDISPENSABLE even than in ordinary cases. If selection
0073 so, reason to believe that humble bees are INDISPENSABLE to the fertilisation of the heartsease (V
0073 have found that the visits of bees, if not INDISPENSABLE, are at least highly beneficial to the fe
0097 asionally, perhaps at very long intervals, INDISPENSABLE. On the belief that this is a law of natu
0097 the weather! but if an occasional cross be INDISPENSABLE, the fullest freedom for the entrance of
0099 distinct individual being advantageous or INDISPENSABLE! If several varieties of the cabbage, rad
0100 , on the view of an occasional cross being INDISPENSABLE, by considering the medium in which terre
0137 any animal, and as eyes are certainly not INDISPENSABLE to animals with subterranean habits, a re
0197 n, and no doubt they facilitate, or may be INDISPENSABLE for this act; but as sutures occur in the
0205 gans, and excretions highly useful or even INDISPENSABLE, or highly injurious to another species,
0211 ails on this and other such points are not INDISPENSABLE, they may be here passed over. As some de
0211 nd the inheritance of such variations, are INDISPENSABLE for the action of natural selection, as m
0234 ss of secretion. A large store of honey is INDISPENSABLE to support a large stock of bees during t
0267 pter, that a certain amount of crossing is INDISPENSABLE even with hermaphrodites: and that close
0301 pecific forms; and these contingencies are INDISPENSABLE for the preservation of all the transitio
0048 eir parents; for in this case, it would be INDISPENSABLE for the existence of the species, that th
0474 difying instincts! but it certainly is not INDISPENSABLE, as we see, in the case of neuter insects
0307 consequently, if my theory be true, it is INDISPUTABLE that before the lowest Silurian stratum wa
0413 e ingenuity and utility of this system are INDISPUTABLE. But many naturalists think that something
0293 e passed over. Although each formation has INDISPUTABLY required a vast number of years for its de
0020 if aided by the careful selection of those INDIVIDUAL mongrels, which present any desired charact
0031 rks of agriculturists than almost any other INDIVIDUAL, and who was himself a very good judge of a
0034 e trying to possess and breed from the best INDIVIDUAL animals, is more important. Thus, a man who
0040 origin. A man preserves and breeds from an INDIVIDUAL with some slight deviation of structure, or
0040 als for selection to work on: not that mere INDIVIDUAL differences are not amply sufficient, with
0041 ation in the qualities or structure of each INDIVIDUAL. Unless such attention be paid nothing can
0041 . as soon, however, as gardeners picked out INDIVIDUAL plants with slightly larger, earlier, or be
0044 er II. Variation Under Nature. Variability. INDIVIDUAL differences. Doubtful species. Wide ranging
0045 many slight differences which may be called INDIVIDUAL differences, such as are known frequently t
0045 cies are cast in the very same mould. These INDIVIDUAL differences are highly important for us, as
0045 s man can accumulate in any given direction INDIVIDUAL differences in his domesticated productions
0045 nces in his domesticated productions. These INDIVIDUAL differences generally affect what naturalis
0046 be given. There is one point connected with INDIVIDUAL differences, which seems to me extremely pe
0051 varieties, or between lesser varieties and INDIVIDUAL differences. These differences blend into e
0051 idea of an actual passage. Hence I look at INDIVIDUAL differences, though of small interest to th
0052 erm variety, again, in comparison with mere INDIVIDUAL differences, is also applied arbitrarily, a
0060 c beings in a state of nature there is some INDIVIDUAL variability; indeed I am not aware that thi
0060 ties be admitted. But the mere existence of INDIVIDUAL variability and of some few well marked var
0061 no, if it be in any degree profitable to an INDIVIDUAL of any species, in its infinitely complex r
0061 ture, will tend to the preservation of that INDIVIDUAL, and will generally be inherited by its off
0062 is more important) not only the life of the INDIVIDUAL, but success in leaving progeny. Two canine
0063 ase be a struggle for existence, either one INDIVIDUAL with another of the same species, or with t
0082 ts by adding up in any given direction mere INDIVIDUAL differences, so could Nature, but far more
0087 animals it will adapt the structure of each INDIVIDUAL for the benefit of the community; if each i
0089 confinement well know that they often take INDIVIDUAL preferences and dislikes: thus Sir R. Heron
0089 mainly caused by sexual selection; that is, INDIVIDUAL males have had, in successive generations,
0091 hange of habit or of structure benefited an INDIVIDUAL wolf, it would have the best chance of surv
0092 y inherit the nectar excreting power. Those INDIVIDUAL flowers which had the largest glands or nec
0094 ht profit a bee or other insect, so that an INDIVIDUAL so characterised would be able to obtain it
0097 generations; but that a cross with another INDIVIDUAL is occasionally, perhaps at very long inter
0097 dom for the entrance of pollen from another INDIVIDUAL will explain this state of exposure, more e
0098 s of each flower, before the stigma of that INDIVIDUAL flower is ready to receive them; and as thi
0099 view of an occasional cross with a distinct INDIVIDUAL being advantageous or indispensable! If sev
0101 and the occasional influence of a distinct INDIVIDUAL can be shown to be physically impossible. C
0101 s, ar occasional intercross with a distinct INDIVIDUAL is a law of nature. I am well aware that th

Page **(Key Word)**

Page **(Key Word)**

```
0102  riability is favourable, but I believe mere  INDIVIDUAL differences suffice for the work. A large n
0102  for a lesser amount of variability in each    INDIVIDUAL, and is, I believe, an extremely important
0113  differ from each other, a greater number of   INDIVIDUAL plants of this species of grass, including
0114  immigration, and where the contest between    INDIVIDUAL and individual must be severe, we always fi
0114  nd where the contest between individual and   INDIVIDUAL must be severe, we always find great divers
0115  ivision of labour in the organs of the same   INDIVIDUAL body, a subject so well elucidated by Milne
0131  ction of the reproductive system to produce   INDIVIDUAL differences, or very slight deviations of s
0136  ng thousands of successive generations each   INDIVIDUAL beetle which flew least, either from its wi
0146  except in fruits which opened; so that the    INDIVIDUAL plants producing seeds which were a little
0148  y natural selection, for it will profit the   INDIVIDUAL not to have its nutriment wasted in buildin
0148  d be a decided advantage to each successive   INDIVIDUAL of the species; for in the struggle for lif
0148  life to which every animal is exposed, each   INDIVIDUAL Proteolepas would have a better chance of s
0149  ted many times in the structure of the same   INDIVIDUAL (as the vertebrae in snakes, and the stamen
0155  f the same group, the more subject it is to   INDIVIDUAL anomalies. On the ordinary view of each spe
0168  which can be saved without detriment to the   INDIVIDUAL, will be saved. Changes of structure at an
0168  hat the same principle applies to the whole   INDIVIDUAL; for in a district where many species of an
0170  of such differences, when beneficial to the  INDIVIDUAL, that gives rise to all the more important
0190  imultaneously the same function in the same   INDIVIDUAL; to give one instance, there are fish with
0213  oung, and in nearly the same manner by each   INDIVIDUAL, performed with eager delight by each breed
0222  thick clump of heath, whence I saw the last   INDIVIDUAL of F. sanguinea emerge, carrying a pupa bu
0233  ructure or instinct, each profitable to the  INDIVIDUAL under its conditions of life, it may reason
0235  selection having been economy of wax: that   INDIVIDUAL swarm which wasted least honey in the secre
0236  ed through natural selection: namely, by an  INDIVIDUAL having been born with some slight profitabl
0237  be applied to the family, as well as to the  INDIVIDUAL, and may thus gain the desired end. Thus, a
0237  well flavoured vegetable is cooked, and the  INDIVIDUAL is destroyed: but the horticulturist sows s
0238  e slowly formed by carefully watching which  INDIVIDUAL bulls and cows, when matched, produced oxen
0239  did not probably at first appear in all the  INDIVIDUAL neuters in the same nest, but in a few alon
0249  d, that an occasional cross with a distinct  INDIVIDUAL or variety increases fertility, that I cann
0249  tilised during each generation by their own  INDIVIDUAL pollen; and I am convinced that this would
0250  most singular fact, namely, that there are   INDIVIDUAL plants, as with certain species of Lobelia,
0250  ertilised distinct species. So that certain  INDIVIDUAL plants and all the individuals of certain s
0318  some authors have even supposed that as the  INDIVIDUAL has a definite length of life, so have spec
0320  h the same as to admit that sickness in the  INDIVIDUAL is the forerunner of death, to feel no surp
0351  al selection, only so far as it profits the  INDIVIDUAL in its complex struggle for life, so the de
0404  many genera range over the world, and many  INDIVIDUAL species have enormous ranges. It is not mea
0411  f the same class, between parts of the same  INDIVIDUAL. Embryology, laws of, explained by variatio
0424  ecies the several larval stages of the same  INDIVIDUAL, however much they may differ from each oth
0424  a technical sense be considered as the same  INDIVIDUAL. He includes monsters: he includes varietie
0436  f the different parts or organs in the same  INDIVIDUAL. Most physiologists believe that the bones
0437  sepals, petals, stamens, and pistils in any  INDIVIDUAL flower, though fitted for such widely diffe
0438  rgan is homologous with another in the same  INDIVIDUAL. And we can understand this fact; for in mo
0439  asually remarked that certain organs in the  INDIVIDUAL, which when mature become widely different
0447  ey will not have been modified. But in each  INDIVIDUAL new species, the embryonic fore limbs will
0451  ometimes become developed and give milk. In  INDIVIDUAL plants of the same species the petals somet
0457  rts constructed on the same pattern in each  INDIVIDUAL animal and plant. On the principle of succe
0457  n embryology; namely, the resemblance in an  INDIVIDUAL embryo of the homologous parts, which when
0459  erable slight variations, each good for the  INDIVIDUAL possessor. Nevertheless, this difficulty, t
0467  y selecting, in each successive generation,  INDIVIDUAL differences so slight as to be quite inappr
0467  a grain in the balance will determine which  INDIVIDUAL shall live and which shall die, which varie
0468  ort period a great result by adding up mere  INDIVIDUAL differences in his domestic productions: an
0468  nd every one admits that there are at least  INDIVIDUAL differences in species under nature. But, b
0469  one can draw any clear distinction between   INDIVIDUAL differences and slight varieties: or betwe
0483  e that at each supposed act of creation one  INDIVIDUAL or many were produced? Were all the infinit
0488  hose determining the birth and death of the  INDIVIDUAL. When I view all beings not as special crea
0005  animal and vegetable kingdoms. As many more  INDIVIDUALS of each species are born than can possibly
0007  n's power of Selection. When we look to the  INDIVIDUALS of the same variety or sub variety of our
0007  ffer much more from each other, than do the  INDIVIDUALS of any one species or variety in a state o
0009  nement; and when, on the other hand, we see  INDIVIDUALS, though taken young from a state of nature
0010  remely valuable. When all or nearly all the  INDIVIDUALS exposed to certain conditions are affected
0012  pigs are differently affected from coloured  INDIVIDUALS by certain vegetable poisons. Hairless dog
0013  inal cause acting on both; but when amongst  INDIVIDUALS, apparently exposed to the same conditions
0013  e parent, say, once amongst several million  INDIVIDUALS, and it reappears in the child, the mere d
0013  n say why the same peculiarity in different  INDIVIDUALS of the same species, and in individuals of
0013  ent individuals of the same species, and in  INDIVIDUALS of different species, is sometimes inherit
0035  s, however, unchanged or but little changed  INDIVIDUALS of the same breed may be found in less civ
0036  uch the occasional preservation of the best  INDIVIDUALS, whether or not sufficiently distinct to b
0038  o countries very differently circumstanced,  INDIVIDUALS of the same species, having slightly diffe
0039  ded, like the present Java fantail, or like  INDIVIDUALS of other and distinct breeds, in which as
0039  rly be set on any slight differences in the  INDIVIDUALS of the same species, be judged of by the v
0040  ls and thus improves them, and the improved  INDIVIDUALS slowly spread in the immediate neighbourho
0041  will be much increased by a large number of  INDIVIDUALS being kept; and hence this comes to be of
0041  varieties. The keeping of a large number of  INDIVIDUALS of a species in any country requires that
0041  s to breed freely in that country. When the  INDIVIDUALS of any species are scanty, all the individ
0045  us arisen, being frequently observed in the  INDIVIDUALS, whatever their quality may be, will gener
0045  ined locality. No one supposes that all the  INDIVIDUALS of the same species inhabiting the same co
0045  catory point of view, sometimes vary in the  INDIVIDUALS of the same species are cast in the very s
0052  ven for the sake of convenience to a set of  INDIVIDUALS of the same species. I am convinced that t
0053  ich are most common, that is abound most in  INDIVIDUALS closely resembling each other, and that it
0054  r own country, and are the most numerous in  INDIVIDUALS, and the species which are most widely dif
0060  s of climate. Protection from the number of  INDIVIDUALS, which oftenest produce well marked variet
0060  ture. Struggle for life most severe between  INDIVIDUALS. Complex relations of all animals and plan
0061  etter chance of surviving, for, of the many  INDIVIDUALS and varieties of the same species: often s
0063  y could support the product. Hence, as more  INDIVIDUALS of any species which are periodically born
0063  h another of the same species, or with the  INDIVIDUALS are produced than can possibly survive, th
0066  this difference does not determine how many  INDIVIDUALS of distinct species, or with the physical
                                                  INDIVIDUALS of the two species can be supported in a d
```

Page **(Key Word)**

individuals

Page **(Key Word)**

```
0068  ngs on the most severe struggle between the   INDIVIDUALS, whether of the same or of distinct specie
0070  other hand, in many cases, a large stock of   INDIVIDUALS of the same species, relatively to the num
0070  plants being social, that is, abounding in    INDIVIDUALS, even on the extreme confines of their ran
0075   invariably will be most severe between the    INDIVIDUALS of the same species, for they frequent the
0078  on between some few species, or between the    INDIVIDUALS of the same species, for the warmest or da
0080   . on the generality of intercrosses between    INDIVIDUALS of the same species. Circumstances favoura
0080  namely, intercrossing, isolation, number of   INDIVIDUALS. Slow action. Extinction caused by Natural
0081  r, can we doubt (remembering that many more    INDIVIDUALS are born than can possibly survive) that i
0081  ls are born than can possibly survive) that    INDIVIDUALS having any advantage, however slight, over
0082  to arise, and which in any way favoured the    INDIVIDUALS of any of the species, by better adapting
0091   and from the continued preservation of the    INDIVIDUALS best fitted for the two sites, two varieti
0092  another flower. The flowers of two distinct    INDIVIDUALS of the same species would thus get crossed
0092  ill be a great gain to the plant: and those    INDIVIDUALS which produced more and more pollen, and h
0094  on the principle of the division of labour,    INDIVIDUALS with this tendency more and more increased
0095  ach other, by the continued preservation of   INDIVIDUALS presenting mutual and slightly favourable
0096  in their structure. On the Intercrossing of   INDIVIDUALS. I must here introduce a short digression.
0096  ate sexes, it is of course obvious that two    INDIVIDUALS must always unite for each birth; but in t
0096  to believe that with all hermaphrodites two    INDIVIDUALS, either occasionally or habitually, concur
0096  phrodites a large number pair; that is, two    INDIVIDUALS regularly unite for reproduction, which is
0096  there for supposing in these cases that two    INDIVIDUALS ever concur in reproduction? As it is impo
0096  oss between different varieties, or between    INDIVIDUALS of the same variety but of another strain,
0099   derived from the intercrossing of distinct    INDIVIDUALS of the same species. When distinct species
0100  the same tree can be considered as distinct    INDIVIDUALS only in a limited sense. I believe this ob
0100  rial animals without the concurrence of two    INDIVIDUALS. Of aquatic animals, there are many self f
0101  rtunate chance, elsewhere to prove that two    INDIVIDUALS, though both are self fertilising hermaphr
0101  dites do occasionally intercross with other    INDIVIDUALS, the difference between hermaphrodites and
0101  in many organic beings, a cross between two    INDIVIDUALS is an obvious necessity for each birth: in
0102  ces suffice for the work. A large number of    INDIVIDUALS, by giving a better chance for the appeara
0102  ection will always tend to preserve all the    INDIVIDUALS varying in the right direction, though in
0102   there will be intercrossing with the other    INDIVIDUALS of the same species on the confines of eac
0103  selection always tending to modify all the    INDIVIDUALS in each district in exactly the same manne
0103  ing took place would be chiefly between the    INDIVIDUALS of the same new variety. A local variety w
0103  ery important part in nature in keeping the    INDIVIDUALS of the same species, or of the same variet
0104  tural selection will tend to modify all the    INDIVIDUALS of a varying species throughout the area i
0104  me conditions. Intercrosses, also, with the    INDIVIDUALS of the same species, which otherwise would
0105  hysical conditions, the total number of the    INDIVIDUALS supported on it will necessarily be very s
0105  l necessarily be very small: and fewness of    INDIVIDUALS will greatly retard the production of new
0105  variations arising from the large number of   INDIVIDUALS of the same species there supported, but t
0107  the inhabitants, at this period numerous in    INDIVIDUALS and kinds, will have been subjected to ver
0107  parate islands, there will still exist many    INDIVIDUALS of the same species on each island: interc
0109   also, see that any form represented by few    INDIVIDUALS will, during fluctuations in the seasons o
0110  ore, the species which are most numerous in    INDIVIDUALS will have the best chance of producing wit
0116  habits of life, so will a greater number of    INDIVIDUALS be capable of there supporting themselves.
0127  ul to any organic being do occur, assuredly    INDIVIDUALS thus characterised will have the best chan
0133  s difference may be due to the warmest clad    INDIVIDUALS having been favoured and preserved during
0136  would depend on whether a greater number of   INDIVIDUALS were saved by successfully battling with t
0142  ion will continually tend to preserve those    INDIVIDUALS which are born with constitutions best ada
0151  n shape: and the amount of variation in the    INDIVIDUALS of several of the species is so great, tha
0152  d nearly to perfection, and frequently the    INDIVIDUALS are born which depart widely from the stan
0154  een fixed by the continued selection of the    INDIVIDUALS varying in the required manner and degree,
0155  ies, that it has also, been variable in the    INDIVIDUALS of some of the species. And this fact show
0181  s of life, in the continued preservation of   INDIVIDUALS with fuller and fuller flank membranes, ea
0183  of diversified and of changed habits in the    INDIVIDUALS of the same species. When either case occu
0184  s monstrous as a whale. As we sometimes see    INDIVIDUALS of a species following habits widely diffe
0184  s, we might expect, on my theory, that such    INDIVIDUALS would occasionally have given rise to new
0189  years: and during each year on millions of    INDIVIDUALS of many kinds: and may we not believe that
0194  s by life and death, by the preservation of   INDIVIDUALS with any favourable variation, and by the
0195  se the preservation of successively varying    INDIVIDUALS. I have sometimes felt as much difficulty,
0195  f resisting the attacks of insects: so that    INDIVIDUALS which could by any means defend themselves
0198   a certain extent to natural selection, and    INDIVIDUALS with slightly different constitutions woul
0206  larger country there will have existed more    INDIVIDUALS, and more diversified forms, and the compe
0207  any experience, and when performed by many    INDIVIDUALS in the same way, without their knowing for
0212  w in Egypt. That the general disposition of    INDIVIDUALS of the same species, born in a state of na
0214  at the long continued selection of the best    INDIVIDUALS in successive generations made tumblers wh
0222  t have been close at hand, for two or three    INDIVIDUALS of F. fusca were rushing about in the grea
0231  r, so that, as Huber has stated, a score of    INDIVIDUALS work even at the commencement of the first
0239  n sometimes be perfectly linked together by    INDIVIDUALS taken out of the same nest: I have myself
0250  that certain individual plants and all the    INDIVIDUALS of certain species can actually be hybridi
0252  y treated, for by insect agency the several    INDIVIDUALS of the same hybrid variety are allowed to
0256  epends in part upon the constitution of the    INDIVIDUALS which happen to have been chosen for the e
0256  ften found to differ greatly in the several    INDIVIDUALS raised from seed out of the same capsule a
0259  een their parents, exceptional and abnormal   INDIVIDUALS sometimes are born, which closely resemble
0260  ee of sterility be innately variable in the    INDIVIDUALS of the same species? Why should some speci
0262  sometimes an innate difference in different    INDIVIDUALS of the same two species in crossing: so Sa
0262  believes this to be the case with different   INDIVIDUALS of the same two species in being grafted t
0267  vidence, that a cross between very distinct    INDIVIDUALS of the same species, that is between membe
0275  tle from each other, namely in the union of    INDIVIDUALS of the same variety, or of different varie
0276  test. The sterility is innately variable in    INDIVIDUALS of the same species, and is eminently susc
0304  es: from the extraordinary abundance of the    INDIVIDUALS of many species all over the world, from t
0322  ly say why this species is more abundant in    INDIVIDUALS than that: why this species and not anothe
0326  can, I think, clearly see that a number of    INDIVIDUALS, from giving a better chance for the appear
0350  ction. Widely ranging species, abounding in    INDIVIDUALS, which have already triumphed over many co
0352  y on this head. It is also obvious that the    INDIVIDUALS of the same species, though now inhabiting
0352  in the last chapter, it is incredible that    INDIVIDUALS identically the same should ever have been
0353  coming one step lower in the series, to the    INDIVIDUALS of the same species, a directly opposite r
0355  her allied question, namely whether all the    INDIVIDUALS of the same species have descended from a
```

Page **(Key Word)**

Page ***(Key Word)***

```
0355  whether, as some authors suppose, from many  INDIVIDUALS  simultaneously created. With those organic
0355  s, which will never have blended with other  INDIVIDUALS  or varieties, but will have supplanted eac
0355  ge of modification and improvement, all the  INDIVIDUALS  of each variety will have descended from a
0356  during the slow process of modification the  INDIVIDUALS  of the species will have been kept nearly
0356  arly uniform by intercrossing; so that many  INDIVIDUALS  will have gone on simultaneously changing,
0356  ntinued care in selecting and training many  INDIVIDUALS  during many generations. Before discussing
0388  l always be a struggle for life between the  INDIVIDUALS  of the species, however few, already occup
0389  unt of difficulty, on the view that all the  INDIVIDUALS  both of the same and of allied species hav
0407  the difficulties in believing that all the  INDIVIDUALS  of the same species, wherever located, hav
0407  often are in this case, and in that of the  INDIVIDUALS  of the same species, extremely grave. As e
0408  mitting that in the long course of time the  INDIVIDUALS  of the same species, and likewise of allie
0425  versally been used in classing together the  INDIVIDUALS  of the same species, though the males and
0433  use the element of descent in classing the  INDIVIDUALS  of both sexes and of all ages, although ha
0452  s could be given. Rudimentary organs in the  INDIVIDUALS  of the same species are very liable to var
0452  nd which is occasionally found in monstrous  INDIVIDUALS  of the species. Thus in the snapdragon (an
0461  with modification are grave enough. All the  INDIVIDUALS  of the same species, and all the species o
0467  e may do it unconsciously by preserving the  INDIVIDUALS  most useful to him at the time, without an
0467  der nature. In the preservation of favoured  INDIVIDUALS  and races, during the constantly recurrent
0467  on, as explained in the third chapter. More  INDIVIDUALS  are born than can possibly survive. A grai
0467  decrease, or finally become extinct. As the  INDIVIDUALS  of the same species come in all respects i
0468  ossession of the females. The most vigorous  INDIVIDUALS, or those which have most successfully str
0486  lass. When we can feel assured that all the  INDIVIDUALS  of the same species, and all the closely a
0442  embryo and the adult; of parts in the same  INDIVIVIDUAL  embryo, which ultimately become very unlik
0136  r so little less perfectly developed or from  INDOLENT  habit, will have had the best chance of surv
0032  years, and devotes his lifetime to it with  INDOMITABLE  perseverance, he will succeed, and may mak
0088  owing the victor to breed might surely give  INDOMITABLE  courage, length to the spur, and strength
0135  efficient. There is not sufficient evidence to  INDUCE  us to believe that mutilations are ever inher
0001  sh this Abstract. I have more especially been  INDUCED  to do this, as Mr. Wallace, who is now study
0167  life, as climate and food, etc., seem to have  INDUCED  some slight modifications. Habit in producin
0005  ction of the less improved forms of life, and  INDUCES  what I have called Divergence of Character.
0267  ept under the same conditions of life, always  INDUCES  weakness and sterility in the progeny. Hence
0433  le for existence, and which almost inevitably  INDUCES  extinction and divergence of character in th
0134  fecting the reproductive system, and in thus  INDUCING  variability; and natural selection will then
0202  d which are ultimately slaughtered by their  INDUSTRIOUS  and sterile sisters? It may be difficult,
0097  gether that self fertilisation seems almost  INEVITABLE. Many flowers, on the other hand, have thei
0343  . the extinction of old forms is the almost  INEVITABLE  consequence of the production of new forms.
0412  everal principles; and he will see that the  INEVITABLE  result is that the modified descendants pro
0471  ging in character, together with the almost  INEVITABLE  contingency of much extinction, explains th
0004  organic beings throughout the world, which  INEVITABLY  follows from their high geometrical powers
0005  shall then see how Natural Selection almost  INEVITABLY  causes much Extinction of the less improved
0061  more fully see in the next chapter, follow  INEVITABLY  from the struggle for life. Owing to this s
0063  gle for existence. A struggle for existence  INEVITABLY  follows from the high rate at which all org
0066  period of its life; that heavy destruction  INEVITABLY  falls either on the young or old, during ea
0087  ardest beaks, for all with weak beaks would  INEVITABLY  perish: or, more delicate and more easily b
0109  and almost indefinitely increasing, numbers  INEVITABLY  must become extinct. That the number of spe
0110  rom these several considerations I think it  INEVITABLY  follows, that as new species in the course
0202  n, owing to the backward serratures, and so  INEVITABLY  causes the death of the insect by tearing o
0320  nt extinction of less favoured forms almost  INEVITABLY  follows. It is the same with our domestic p
0433  he struggle for existence, and which almost  INEVITABLY  induces extinction and divergence of charac
0467  ns of selection. The struggle for existence  INEVITABLY  follows from the high geometrical ratio of
0470  same genus. New and improved varieties will  INEVITABLY  supplant and exterminate the older, less im
0475  in the history of the organic world, almost  INEVITABLY  follows on the principle of natural selecti
0477  ourse of modification in the two areas will  INEVITABLY  be different. On this view of migration, wi
0203  nately is most rare, is all the same to the  INEXORABLE  principle of natural selection. If we admir
0059  ated: whereas, these analogies are utterly  INEXPLICABLE  if each species has been independently cre
0097  the following, which on any other view are  INEXPLICABLE. Every hybridizer knows how unfavourable e
0188  tise that large bodies of facts, otherwise  INEXPLICABLE, can be explained by the theory of descent
0243  e on the foregoing views, but is otherwise  INEXPLICABLE, all tend to corroborate the theory of nat
0308  amorphism. The case at present must remain  INEXPLICABLE; and may be truly urged as a valid argumen
0333  a satisfactory manner. And they are wholly  INEXPLICABLE  on any other view. On this same theory, it
0352  ganic beings; and, accordingly, we find no  INEXPLICABLE  cases of the same mammal inhabiting distan
0355  are, as we shall hereafter more fully see,  INEXPLICABLE  on the theory of independent creation. Thi
0372  ate lands of North America and Europe, are  INEXPLICABLE  on the theory of creation. We cannot say t
0387  dering these facts, I think it would be an  INEXPLICABLE  circumstance if water birds did not transp
0387  ht that its distribution must remain quite  INEXPLICABLE; but Audubon states that he found the seed
0396  with those of a neighbouring continent, an  INEXPLICABLE  relation on the view of independent acts o
0399  ar more remarkable case, and is at present  INEXPLICABLE: but this affinity is confined to the plan
0406  the nearest mainland, are I think, utterly  INEXPLICABLE  on the ordinary view of the independent cr
0437  n early stage of growth exactly alike. How  INEXPLICABLE  are these facts on the ordinary view of cr
0457  and their final abortion, present to us no  INEXPLICABLE  difficulties; on the contrary, their prese
0471  of all organic beings seems to me utterly  INEXPLICABLE  on the theory of creation. As natural sele
0473  ersions to long lost characters occur. How  INEXPLICABLE  on the theory of creation is the occasiona
0474  thout change for an enormous period. It is  INEXPLICABLE  on the theory of creation why a part devel
0478  r mammals, on oceanic islands, are utterly  INEXPLICABLE  on the theory of independent acts of creat
0480  having been specially created, how utterly  INEXPLICABLE  it is that parts, like the teeth in the em
0177  s, and do not at any one period present an  INEXTRICABLE  chaos of varying and intermediate links: f
0434  we shall never, probably, disentangle the  INEXTRICABLE  web of affinities between the members of a
0462  all organic beings blended together in an  INEXTRICABLE  chaos? With respect to existing forms, we
0467  racter of natural species; is shown by the  INEXTRICABLE  doubts whether very many of them are varie
0056  serve notice. We have seen that there is no  INFALLIBLE  criterion by which to distinguish species a
0486  en inherited. Rudimentary organs will speak  INFALLIBLY  with respect to the nature of long lost str
0111  ies? That this does habitually happen, we must  INFER  from most of the innumerable species througho
0115  natives; and we may, I think, at least safely  INFER  that diversification of structure, amounting
0140  der which they live is often overrated. We may  INFER  this from our frequent inability to predict w
0150  e above effect, published by Mr. Waterhouse. I  INFER  also from an observation made by Professor Ow
0188  the highest human intellects: and we naturally  INFER  that the eye has been formed by a somewhat an
```

Page ***(Key Word)***

infer

0191 ng apparatus or swimbladder. We can thus, as I INFER from Professor Owen's interesting description
0197 have only to escape from a broken egg, we may INFER that this structure has arisen from the laws
0200 ch widely diversified habits. Therefore we may INFER that these several bones might have been acqu
0251 imperfect in their mutual self action, we must INFER that the plants were in an unnatural state. N
0252 ther, as are the genera of plants, then we may INFER that animals more widely separated in the sca
0289 sed between the consecutive formations, we may INFER that this could nowhere be ascertained. The f
0293 formation, it would be rash in the extreme to INFER that it had not elsewhere previously existed.
0293 ith marine animals of all kinds, we may safely INFER a large amount of migration during climatal a
0302 rfection of the geological record, and falsely INFER, because certain genera or families have not
0308 , of which the formations are composed, we may INFER that from first to last large islands or trac
0308 c or secondary formation. Hence we may perhaps INFER, that during the palaeozoic and secondary per
0309 these enormously long periods. If then we may INFER anything from these facts, we may infer that
0309 we may infer anything from these facts, we may INFER that where our oceans now extend, oceans have
0359 ing dried, for above 28 days, as far as we may INFER anything from these scanty facts, we may conc
0363 occur in the archipelago. Hence we may safely INFER that icebergs formerly landed their rocky bur
0370 ngly inclined to extend the above view, and to INFER that during some earlier and still warmer per
0373 l range, and southward to the Pyrenees. We may INFER, from the frozen mammals and nature of the mo
0402 are common to the several islands, and we may INFER from certain facts that these have probably a
0402 ing rapidity over new countries, we are apt to INFER that most species would thus spread; but we s
0402 ed in establishing itself there? We may safely INFER that Charles Island is well stocked with its
0402 there than can possibly be reared; and we may INFER that the mocking thrush peculiar to Charles I
0426 a chain of intermediate groups, we may at once INFER their community of descent; and we put them a
0466 productions have undergone; but we may safely INFER that the amount has been large, and that modi
0474 of a genus; and therefore, as we may naturally INFER, of great importance to the species, should b
0484 the wild rose or oak tree. Therefore I should INFER from analogy that probably all the organic be
0489 ennobled. Judging from the past, we may safely INFER that not one living species will transmit its
0188 somewhat analogous process. But may not this INFERENCE be presumptuous? Have we any right to assum
0197 should be very cautious in drawing any such INFERENCE, when we see that the skin on the head of t
0176 nnect. Now, if we may trust these facts and INFERENCES, and therefore conclude that varieties link
0331 let us see how far these several facts and INFERENCES accord with the theory of descent with modi
0459 to the reader to have the leading facts and INFERENCES briefly recapitulated. That many and grave
0036 thus generally leave more offspring than the INFERIOR ones; so that in this case there would be a
0037 's description, to have been a fruit of very INFERIOR quality. I have seen great surprise expresse
0042 great numbers and at a very quick rate, and INFERIOR birds may be freely rejected, as when killed
0083 the females. He does not rigidly destroy all INFERIOR animals, but protects during each varying se
0102 withstanding a large amount of crossing with INFERIOR animals. Thus it will be in nature; for with
0111 ties of flowers, take the place of older and INFERIOR kinds. In Yorkshire, it is historically know
0112 s the differences slowly become greater, the INFERIOR animals with intermediate characters, being
0327 they would cause the extinction of other and INFERIOR forms; and as these inferior forms would be
0327 on of other and inferior forms; and as these INFERIOR forms would be allied in groups by inheritan
0348 arating continents, the differences are very INFERIOR in degree to those characteristic of distinc
0321 orms, which will suffer from some inherited INFERIORITY in common. But whether it be species belon
0322 ly be allied, for they will partake of some INFERIORITY in common. Thus, as it seems to me, the ma
0327 y be allied in groups, from inheriting some INFERIORITY in common; and therefore as new and improv
0344 ies of the less vigorous groups, from their INFERIORITY inherited from a common progenitor, tend t
0344 s of those groups of species which are their INFERIORS in the struggle for existence. Hence, after
0162 only analogous variations; but we might have INFERRED that the blueness was a case of reversion, f
0162 ple variation. More especially we might have INFERRED this, from the blue colour and marks so ofte
0182 nder all possible conditions. It must not be INFERRED from these remarks that any of the grades of
0279 their number. On the vast lapse of time, as INFERRED from the rate of deposition and of denudatio
0304 recognised: from all these circumstances, I INFERRED that had sessile cirripedes existed during t
0324 todon and Horse, it might at least have been INFERRED that they had lived during one of the later
0174 rst place we should be extremely cautious in INFERRING, because an area is now continous, that it
0271 a state of nature, for a supposed variety if INFERTILE in any degree would generally be ranked as
0005 wers of animals; thirdly, Hybridism, or the INFERTILITY of species and the fertility of varieties
0270 ly founds his classification by the test of INFERTILITY, as varieties. The following case is far m
0271 om the great difficulty of ascertaining the INFERTILITY of varieties in a state of nature, for a s
0015 eeds, and esculent vegetables, for an almost INFINITE number of generations, would be opposed to a
0109 t to the amount of change, to the beauty and INFINITE complexity of the coadaptations between all
0127 ly cannot be disputed; then, considering the INFINITE complexity of the relations of all organic b
0127 to their conditions of existence, causing an INFINITE diversity in structure, constitution, and ha
0287 en peopled by hosts of living forms. What an INFINITE number of generations, which the mind cannot
0288 ipedes) coat the rocks all over the world in INFINITE numbers: they are all strictly littoral, wit
0301 ect to find in our geological formations, an INFINITE number of those fine transitional forms, whi
0436 then natural selection will account for the INFINITE diversity in structure and function of the m
0466 , may continue to be inherited for an almost INFINITE number of generations. On the other hand we
0481 ght variations, accumulated during an almost INFINITE number of generations. Although I am fully c
0012 unknown, or dimly seen laws of variation is INFINITELY complex and diversified. It is well worth w
0040 hatever they may be. But the chance will be INFINITELY small of any record having been preserved o
0043 d disuse. The final result is thus rendered INFINITELY complex. In some cases, I do not doubt that
0061 ble to an individual of any species, in its INFINITELY complex relations to other organic beings a
0080 degree plastic. Let it be borne in mind how INFINITELY complex and close fitting are the mutual re
0084 than man's productions; that they should be INFINITELY better adapted to the most complex conditio
0098 orate contrivance by which every one of the INFINITELY numerous pollen granules are swept out of t
0106 e supported, but the conditions of life are INFINITELY complex from the large number of already ex
0119 ed by other beings; and this will depend on INFINITELY complex relations. But as a general rule, t
0189 tions, generation will multiply them almost INFINITELY, and natural selection will pick out with u
0282 y of our not finding fossil remains of such INFINITELY numerous connecting links, it may be object
0310 ur not finding in the successive formations INFINITELY numerous transitional links between the man
0343 . he may ask where are the remains of those INFINITELY numerous organisms which must have existed
0408 independently of their physical conditions, INFINITELY diversified conditions of life, there would
0435 for such different purposes, are formed by INFINITELY numerous modifications of an upper lip, man
0483 ividual or many were produced? Were all the INFINITELY numerous kinds of animals and plants create
0095 y by the preservation and accumulation of INFINITESIMALLY small inherited modifications, each pro
0463 on this doctrine of the extermination of an INFINITUDE of connecting links, between the living and
0486 or study than one more species added to the INFINITUDE of already recorded species. Our classifica

Page **************************************(Key Word)**************************************

```
0137 se, as appeared on dissection, having been INFLAMMATION of the nictitating membrane. As frequent i
0137 n of the nictitating membrane. As frequent INFLAMMATION of the eyes must be injurious to any anima
0039 nted. Perhaps the first pouter pigeon did not INFLATE its crop much more than the turbit now does
0021 ormously developed crop, which it glories in INFLATING, may well excite astonishment and even laug
0416 tely distinguishes fishes and reptiles, the INFLECTION of the angle of the jaws of the Marsupials,
0426 ling a character may be, let it be the mere INFLECTION of the angle of the jaw, the manner in whic
0011 t fur from climate. Habit also has a decided INFLUENCE, as in the period of flowering with plants
0081 tants. Let it be remembered how powerful the INFLUENCE of a single introduced tree or mammal has b
0098 y destroy, as has been shown by Gartner, any INFLUENCE from the foreign pollen. When the stamens o
0101 that access from without and the occasional INFLUENCE of a distinct individual can be shown to be
0104 c their kind: and thus, in the long run, the INFLUENCE of intercrosses even at rare intervals, wil
0132 ee everywhere throughout nature. Some little INFLUENCE may be attributed to climate, food, etc.: t
0141 cure question. That habit or custom has some INFLUENCE I must believe, both from analogy, and from
0160 aty blue, with the several marks, beyond the INFLUENCE of the mere act of crossing on the laws of
0196 food, etc., probably have some little direct INFLUENCE on the organisation: that characters reappe
0196 ion of growth will have had a most important INFLUENCE in modifying various structures: and finall
0252 ely cross with each other, and the injurious INFLUENCE of close interbreeding is thus prevented. A
0255 lant of a distinct family, it exerts no more INFLUENCE than so much inorganic dust. From this abso
0350 in a quite subordinate degree to the direct INFLUENCE of different physical conditions. The degre
0353 nd broken interspace. The great and striking INFLUENCE which barriers of every kind have had on di
0356 . change of climate must have had a powerful INFLUENCE on migration: a region when its climate was
0366 inly reveal a former cold period. The former INFLUENCE of the glacial climate on the distribution
0407 attempted to show how important has been the INFLUENCE of the modern Glacial period, which I am fu
0447 d into hands, or paddles, or wings. Whatever INFLUENCE long continued exercise or use on the one h
0447 other, may have in modifying an organ, such INFLUENCE will mainly affect the mature animal, which
0462 ve attempted to show how potent has been the INFLUENCE of the Glacial period on the distribution b
0467 the breed. It is certain that he can largely INFLUENCE the character of a breed by selecting, in e
0419 cent. Our classifications are often plainly INFLUENCED by chains of affinities. Nothing can be eas
0482 oubt on the immutability of species, may be INFLUENCED by this volume: but I look with confidence
0132 we may, at least, safely conclude that such INFLUENCES cannot have produced the many striking and
0144 the shape of the pelvis in the human mother INFLUENCES by pressure the shape of the head of the ch
0159 variation, when acted on by similar unknown INFLUENCES. In the vegetable kingdom we have a case of
0169 from a common parent and exposed to similar INFLUENCES will naturally tend to present analogous va
0369 been exposed to somewhat different climatal INFLUENCES. Their mutual relations will thus have been
0356 evel in the land must also have been highly INFLUENTIAL: a narrow isthmus now separates two marine
0164 aled. I have, also, reason to suspect, from INFORMATION given me by Mr. W. W. Edwards, that with t
0176 the genus Balanus. And it would appear from INFORMATION given me by Mr. Watson, Dr. Asa Gray, and
0220 tish Museum, to whom I am much indebted for INFORMATION on this and other subjects. Although fully
0226 the following statement, drawn up from his INFORMATION, and tells me that it is strictly correct:
0323 rought to Europe from La Plata, without any INFORMATION in regard to their geological position, no
0393 ; but I suspect that this exception (if the INFORMATION be correct) may be explained through glaci
0035 from Spain, Mr. Borrow has not seen, as I am INFORMED by him, any native dog in Spain like our poi
0100 . on the other hand, Dr. Hooker has recently INFORMED me that he finds that the rule does not hold
0163 others, occasionally appear: and I have been INFORMED by Colonel Poole that the foals of this spec
0233 d pressed to get sufficient nectar: and I am INFORMED by Mr. Tegetmeier that it has been experimen
0286 ns is on an average about 1100 feet, as I am INFORMED by Prof. Ramsay. But if, as some geologists
0375 lands: and a long list can be given, as I am INFORMED by Dr. Hooker, of European genera, found in
0391 s are almost every year blown there, as I am INFORMED by Mr. E. V. Harcourt. So that these two isl
0027 out to me by Professor Lepsius: but Mr. Birch INFORMS me that pigeons are given in a bill of fare
0140 tional powers of resisting cold. Mr. Thwaites INFORMS me that he has observed similar facts in Cey
0145 mbelliferae, it is by no means, as Dr. Hooker INFORMS me, in species with the densest heads that t
0165 a hybrid has been figured by Dr. Gray (and he INFORMS me that he knows of a second case) from the
0220 differently when more numerous: but Mr. Smith INFORMS me that he has watched the nests at various
0251 erfectly fertile. Mr. C. Noble, for instance, INFORMS me that he raises stocks for grafting from a
0285 a downthrow in Anglesea of 2300 feet; and he INFORMS me that he fully believes there is one in Me
0362 s might thus readily get scattered. Mr. Brent INFORMS me that a friend of his had to give up flyin
0376 , etc., of northern forms of fish. Dr. Hooker INFORMS me that twenty five species of Algae are com
0378 nd perhaps by a dry climate: for Dr. Falconer INFORMS me that it is the damp with the heat of the
0386 y other distant point. Sir Charles Lyell also INFORMS me that a Dyticus has been caught with an An
0203 natural selection. If we admire the several INGENIOUS contrivances, by which the flowers of the o
0355 for species) from that lately advanced in an INGENIOUS paper by Mr. Wallace, in which he concludes
0413 escription is given of each kind of dog. The INGENUITY and utility of this system are indisputable
0134 ingless condition of several birds, which now INHABIT or have lately inhabited several oceanic isl
0137 elonging to the most different classes, which INHABIT the caves of Styria and of Kentucky, are bli
0139 emely common for species of the same genus to INHABIT very hot and very cold countries; and as I b
0173 rged that when several closely allied species INHABIT the same territory we surely ought to find a
0321 d almost extinct group of Ganoid fishes still INHABIT our fresh waters. Therefore the utter extinc
0347 those north of 25 degrees, which consequently INHABIT a considerably different climate, and they w
0389 that in the course of time they have come to INHABIT distant points of the globe. I have already
0389 modification. The species of all kinds which INHABIT oceanic islands are few in number compared w
0403 the endemic and representative species, which INHABIT the several islands of the Galapagos Archipe
0405 r instance, two varieties of the same species INHABIT America and Europe, and the species thus has
0406 close relation of the distinct species which INHABIT the islets of the same archipelago, and espe
0411 , if one group had been exclusively fitted to INHABIT the land, and another the water: one to feed
0428 o live under nearly similar circumstances, to INHABIT for instance the three elements of land, air
0477 as frogs and terrestrial mammals, should not INHABIT oceanic islands: and why, on the other hand,
0478 ind that wherever many closely allied species INHABIT two areas, some identical species common to
0006 he future success and modification of every INHABITANT of this world. Still less do we know of the
0082 fications in the structure or habits of one INHABITANT would often give it an advantage over other
0186 and thus gain an advantage over some other INHABITANT of the country, it will seize on the place
0186 country, it will seize on the place of that INHABITANT, however different it may be from its own p
0337 ng in New Zealand, and from hardly a single INHABITANT of the southern hemisphere having become wi
0001 th certain facts in the distribution of the INHABITANTS of South America and in the geological rel
0001 ogical relations of the present to the past INHABITANTS of that continent. These facts seemed to m
0006 of the mutual relations of the innumerable INHABITANTS of the world during the many past geologic
0040 cording to the state of civilisation of the INHABITANTS, slowly to add to the characteristic featu
```

Page **************************************(Key Word)**************************************

Page **(Key Word)**

```
0042 land plays a part. Wandering savages or the INHABITANTS of open plains rarely possess more than on
0054 necessarily have to struggle with the other INHABITANTS of the country, the species which are alre
0069  as each area is already fully stocked with INHABITANTS, the other species will decrease. When we
0080 e of Character, related to the diverstiy of INHABITANTS of any small area, and to naturalisation.
0081 of climate. The proportional numbers of its INHABITANTS would almost immediately undergo a change,
0081 he intimate and complex manner in which the INHABITANTS of each country are bound together, that a
0081 in the numerical proportions of some of the INHABITANTS, independently of the change of climate it
0081 disturb the relations of some of the former INHABITANTS. Let it be remembered how powerful the inf
0081 e better filled up, if some of the original INHABITANTS were in some manner modified; for, had the
0082 modifying and improving some of the varying INHABITANTS. For as all the inhabitants of each countr
0082   of the varying inhabitants. For as all the INHABITANTS of each country are struggling together wi
0082 ountry can be named in which all the native INHABITANTS are now so perfectly adapted to each other
0104 my of the country are left open for the old INHABITANTS to struggle for, and become adapted to, th
0107  first have existed as a continent, and the INHABITANTS, at this period numerous in individuals an
0108 to be filled up by modifications of the old INHABITANTS; and time will be allowed for the varietie
0108 elative proportional numbers of the various INHABITANTS of the renewed continent will again be cha
0108 ural selection to improve still further the INHABITANTS, and thus produce new species. That natura
0108 which can be better occupied by some of the INHABITANTS of the country undergoing modification of
0108 robably still oftener depend on some of the INHABITANTS becoming slowly modified; the mutual relat
0108 ; the mutual relations of many of the other INHABITANTS being thus disturbed. Nothing can be effec
0108 me, and generally on only a very few of the INHABITANTS of the same region at the same time. I fur
0109 ells us of the rate and manner at which the INHABITANTS of this world have changed. Slow though th
0109 gs, each area is already fully stocked with INHABITANTS, it follows that as each selected and favo
0114 vere, we always find great diversity in its INHABITANTS. For instance, I found that a piece of tur
0114  habit and constitution, determine that the INHABITANTS, which thus jostle each other most closely
0115 em. The advantage of diversification in the INHABITANTS of the same region is, in fact, the same a
0118 nt (A) more numerous than most of the other INHABITANTS of the same country: they will likewise pa
0128 on, of which we see proof by looking at the INHABITANTS of any small spot or at naturalised produc
0138 ize of the eyes: whereas with all the other INHABITANTS of the caves, disuse by itself seems to ha
0138 d from the general resemblance of the other INHABITANTS of North America and Europe..On my view we
0138 animals of America, affinities to the other INHABITANTS of that continent, and in those of Europe,
0138 t continent, and in those of Europe, to the INHABITANTS of the European continent. And this is the
0139 ties of the blind cave animals to the other INHABITANTS of the two continents on the ordinary view
0139 r independent creation. That several of the INHABITANTS of the caves of the Old and New Worlds sho
0139 to the less severe competition to which the INHABITANTS of these dark abodes will probably have be
0175 c beings, we must see that the range of the INHABITANTS of any country by no means exclusively 'dep
0177 rd to wide plains at the base; and that the INHABITANTS are all trying with equal steadiness and s
0178 ome modification of some one or more of its INHABITANTS. And such new places will depend on slow c
0178 te, or on the occasional immigration of new INHABITANTS, and, probably, in a still more important
0178 l more important degree, on some of the old INHABITANTS becoming slowly modified, with the new for
0182 n an early transitional state they had been INHABITANTS of the open ocean, and had used their inci
0201 s, or slightly more perfect than, the other INHABITANTS of the same country with which it has to s
0205 act chiefly through the competition of the INHABITANTS one with another, and consequently will pr
0205  to the standard of that country. Hence the INHABITANTS of one country, generally the smaller one,
0205 ften yield, as we see they do yield, to the INHABITANTS of another and generally larger country. F
0206 . this canon if we look only to the present INHABITANTS of the world, is not strictly correct, but
0292 ubsidence, the inhabited area and number of INHABITANTS will decrease (excepting the productions o
0294  the ascertained amount of migration of the INHABITANTS of Europe during the Glacial period, which
0301 uch lengthy periods; and in these cases the INHABITANTS of the archipelago would have to migrate,
0301 any one formation. Very many of the marine INHABITANTS of the archipelago now range thousands of
0312 icily, the successive changes in the marine INHABITANTS of that island have been many and most ora
0314 o fixed law of development, causing all the INHABITANTS of a country to change abruptly, or simult
0314  more especially on the nature of the other INHABITANTS with which the varying species comes into
0314 ained in a former chapter. When many of the INHABITANTS of a country have become modified and impr
0317 nnected together. The old notion of all the INHABITANTS of the earth having been swept away at suc
0322 orrespondingly rapid manner many of the old INHABITANTS; and the forms which thus yield their plac
0323 observations, however, relate to the marine INHABITANTS of distant parts of the world: we have not
0324 say whether the existing or the pleistocene INHABITANTS of Europe resembled most closely those of
0325 various countries, and the nature of their INHABITANTS. This great fact of the parallel successio
0326 is probable, be slower with the terrestrial INHABITANTS of distinct continents than with the marin
0326 of distinct continents than with the marine INHABITANTS of the continuous sea. We might therefore
0326 mstanced in an equal degree, whenever their INHABITANTS met, the battle would be prolonged and sev
0327 long and blank intervals I suppose that the INHABITANTS of each region underwent a considerable am
0336 d note how little the specific forms of the INHABITANTS of the sea have been affected. On the theo
0337  under a nearly similar climate, the eocene INHABITANTS of one quarter of the world were put into
0337 were put into competition with the existing INHABITANTS of the same or some other quarter, the eoc
0339 cal conditions for the dissimilarity of the INHABITANTS of these two continents, and, on the other
0340 e same areas, is at once explained: for the INHABITANTS of each cuarter of the world will obviousl
0340 in some degree modified descendants. If the INHABITANTS of one continent formerly differed greatly
0345 ormation are intermediate in character. The INHABITANTS of each successive period in the world's h
0346 the similarity nor the dissimilarity of the INHABITANTS of various regions can be accounted for by
0347 us facts could be given with respect to the INHABITANTS of the sea. A second great fact which stri
0347 me fact in the great difference between the INHABITANTS of Australia, Africa, and South America un
0349 hey may differ in geological structure, the INHABITANTS, though they may be all peculiar species,
0350 y like each other. The dissimilarity of the INHABITANTS of different regions may be attributed to
0354 e case, that a region, of which most of its INHABITANTS are closely related to, or belong to the s
0355 , on the principle of modification, why the INHABITANTS of a region should be related to those of
0355 ll be plainly related by inheritance to the INHABITANTS of the continent. Cases of this nature are
0358 ntinent, the close relation of the tertiary INHABITANTS of several lands and even seas to their pr
0358 everal lands and even seas to their present INHABITANTS, a certain degree of relation (as we shall
0358 the nature and relative proportions of the INHABITANTS of oceanic islands likewise seem to me opp
0365 and before it had become fully stocked with INHABITANTS. On almost bare land, with few or no destr
0366 glacial climate on the distribution of the INHABITANTS of Europe, as explained with remarkable cl
0366 ill fitted for their former more temperate INHABITANTS, the latter would be supplanted and arctic
0366 ic productions would take their places. The INHABITANTS of the more temperate regions would at the
0366 with snow and ice, and their former Alpine INHABITANTS would descend to the plains. By the time t
```

Page **(Key Word)**

inhabitants

Page ***(Key Word)***

0367	hose of Europe; for the present circumpolar	INHABITANTS, which we suppose to have everywhere trave
0369	acific. During the Glacial period, when the	INHABITANTS of the Old and New Worlds lived further so
0370	acial epoch, and whilst the majority of their	INHABITANTS of the world were specifically the same as
0371	e by cold, for the inter migration of their	INHABITANTS. During the slowly decreasing warmth of th
0372	s of relationship, without identity, of the	INHABITANTS of seas now disjoined, and likewise of the
0372	oined, and likewise of the past and present	INHABITANTS of the temperate lands of North America an
0372	l their physical conditions, but with their	INHABITANTS utterly dissimilar. But we must return to t
0382	cord, full of interest to us, of the former	INHABITANTS of the surrounding lowlands. Chapter XII.
0383	ribution of fresh water productions. On the	INHABITANTS of oceanic islands. Absence of the Batrach
0383	terrestrial Mammals. On the relation of the	INHABITANTS of islands to those of the nearest mainlan
0388	nother equally well fitted for them. On the	INHABITANTS of Oceanic Islands. We now come to the las
0390	h in oceanic islands the number of kinds of	INHABITANTS is scanty, the proportion of endemic speci
0391	places are apparently occupied by the other	INHABITANTS; in the Galapagos Islands reptiles, and in
0392	le facts could be given with respect to the	INHABITANTS of remote islands. For instance, in certai
0396	the degree of affinity of the the mammalian	INHABITANTS of islands with those of a neighbouring co
0396	creation. All the foregoing remarks on the	INHABITANTS of oceanic islands, namely, the scarcity o
0396	culties in understanding how several of the	INHABITANTS of the more remote islands, whether still
0397	and important fact for us in regard to the	INHABITANTS of islands, is their affinity to those of
0398	archipelago. The naturalist, looking at the	INHABITANTS of these volcanic islands in the Pacific,
0398	an entire and absolute difference in their	INHABITANTS! The inhabitants of the Cape de Verde Isla
0398	solute difference in their inhabitants! The	INHABITANTS of the Cape de Verde Islands are related t
0399	ome day explained. The law which causes the	INHABITANTS of an archipelago, though specifically dis
0400	y very closely related species; so that the	INHABITANTS of each separate island, though mostly dis
0400	bly closer degree to each other than to the	INHABITANTS of any other part of the world. And this i
0400	but this dissimilarity between the endemic	INHABITANTS of the islands may be used as an argument
0400	of a country as the most important for its	INHABITANTS; whereas it cannot, I think, be disputed t
0400	k, be disputed that the nature of the other	INHABITANTS, with which each has to compete, is at lea
0400	element of success. Now if we look to those	INHABITANTS of the Galapagos Archipelago which are fou
0402	generally closely allied to the aboriginal	INHABITANTS, but are very distinct species, belonging
0403	flora of oceanic islands, namely, that the	INHABITANTS, when not identically the same, yet are pl
0403	ly the same, yet are plainly related to the	INHABITANTS of that region whence colonists could most
0404	the surrounding lowlands. So it is with the	INHABITANTS of lakes and marshes, excepting in so far
0406	and especially the striking relation of the	INHABITANTS of each whole archipelago or island to tho
0408	titudes, for instance in South America, the	INHABITANTS of the plains and mountains, of the forest
0408	length of time which has elapsed since new	INHABITANTS entered one region: according to the natur
0409	o show, why oceanic islands should have few	INHABITANTS, but of these a great number should be end
0409	he mainland. We can clearly see why all the	INHABITANTS of an archipelago, though specifically dis
0463	cting links, between the living and extinct	INHABITANTS of the world, and at each successive perio
0472	election acts by competition, it adapts the	INHABITANTS of each country only in relation to the de
0472	es; so that we need feel no surprise at the	INHABITANTS of any one country, although on the ordina
0477	owland, on deserts and marshes, most of the	INHABITANTS within each great class are plainly relate
0477	likewise the close alliance of some of the	INHABITANTS of the sea in the northern and southern te
0477	of life, we need feel no surprise at their	INHABITANTS being widely different, if they have been
0478	r. It is a rule of high generality that the	INHABITANTS of each area are related to the inhabitant
0478	inhabitants of each area are related to the	INHABITANTS of the nearest source whence immigrants mi
0487	mirable manner the former migrations of the	INHABITANTS of the whole world. Even at present, by co
0487	resent, by comparing the differences of the	INHABITANTS of the sea on the opposite sides of a cont
0487	a continent, and the nature of the various	INHABITANTS of that continent in relation to their app
0488	tion and extinction of the past and present	INHABITANTS of the world should have been due to secon
0038	the Cape of Good Hope, nor any other region	INHABITED by quite uncivilised man, has afforded us a
0054	be connected with the nature of the stations	INHABITED by them, and has little or no relation to t
0104	the same species, which otherwise would have	INHABITED the surrounding and differently circumstanc
0107	have endured to the present day, from having	INHABITED a confined area, and from having thus been
0130	been saved from fatal competition by having	INHABITED a protected station. As buds give rise by g
0134	eral birds, which now inhabit or have lately	INHABITED several oceanic islands, tenanted by no bea
0173	e as are specimens taken from the metropolis	INHABITED by each. By my theory these allied species
0212	on the nature and temperature of the country	INHABITED, but often from causes wholly unknown to us
0290	period. Along the whole west coast, which is	INHABITED by a peculiar marine fauna, tertiary beds a
0291	bes, we may conclude that the bottom will be	INHABITED by extremely few animals, and the mass when
0292	d. on the other hand, during subsidence, the	INHABITED area and number of inhabitants will decreas
0308	organic remains, which do not appear to have	INHABITED profound depths, in the several formations
0328	e assumed that an isthmus separated two seas	INHABITED by distinct, but contemporaneous, faunas. L
0347	than any in the New World, yet these are not	INHABITED by a peculiar fauna or flora. Notwithstandi
0349	the plains near the Straits of Magellan are	INHABITED by one species of Rhea (American ostrich),
0352	rsally admitted, that in most cases the area	INHABITED by a species is continuous: and when a plan
0354	some part of their migration) from the area	INHABITED by their progenitor. If it can be shown to
0370	large number of the same plants and animals	INHABITED the almost continuous circumpolar land: and
0371	iod, as soon as the species in common, which	INHABITED the New and Old Worlds, migrated south of t
0391	e liable to modification. Madeira, again, is	INHABITED by a wonderful number of peculiar land shel
0393	ally barren. The Falkland Islands, which are	INHABITED by a wolf like fox, come nearest to an exce
0397	ds, even the most isolated and smallest, are	INHABITED by land shells, generally by endemic specie
0403	are united by continuous land, yet they are	INHABITED by a vast number of distinct mammals, birds
0408	linked to the extinct beings which formerly	INHABITED the same continent. Bearing in mind that th
0408	the same physical conditions should often be	INHABITED by very different forms of life: for accord
0410	certainly is the general rule that the area	INHABITED by a single species, or by a group of speci
0477	on, we can see why oceanic islands should be	INHABITED by few species, but of these, that many sho
0478	modification, that the same parents formerly	INHABITED both areas: and we almost invariably find t
0395	derately deep submarine banks, and they are	INHABITIED by closely allied or identical quadrupeds.
0003	could be shown how the innumerable species	INHABITING this world have been modified, so as to acq
0017	l species, for instance, of the many foxes,	INHABITING different quarters of the world. I do not b
0045	rved in the individuals of the same species	INHABITING the same confined locality. No one supposes
0054	inant over their compatriots. If the plants	INHABITING a country and described in any Flora be div
0054	mere fact of many species of the same genus	INHABITING any country, shows that there is something
0091	parent form of wolf. Or, again, the wolves	INHABITING a mountainous district, and those frequenti
0091	pierce, there are two varieties of the wolf	INHABITING the Catskill Mountains in the United States
0105	d, although the total number of the species	INHABITING it, will be found to be small, as we shall

Page ***(Key Word)***

Page **(Key Word)**

```
0113  w kinds of prey, either dead or alive; some  INHABITING  new stations, climbing trees, frequenting w
0135  ct that 200 beetles, out of the 550 species  INHABITING  Madeira, are so far deficient in wings that
0165  onies, cobs, the lanky Kattywar race, etc.,  INHABITING  the most distant parts of the world. Now Le
0167   strong tendency, when crossed with species  INHABITING  distant quarters of the world, to produce h
0172  ord being chiefly due to organic beings not  INHABITING  profound depths of the sea, and to their re
0175  that allied or representative species, when  INHABITING  a continuous area, are generally so distrib
0176  equently, will exist in lesser numbers from  INHABITING  a narrow and lesser area; and practically,
0176  the two which exist in larger numbers from   INHABITING  larger areas, will have a great advantage o
0212   i have elsewhere shown, by various animals  INHABITING  desert islands; and we may see an instance
0243  lied, but certainly distinct, species, when  INHABITING  distant parts of the world and living under
0257  ants, deciduous and evergreen trees, plants  INHABITING  different stations and fitted for extremely
0299  ocks; or, again, whether certain sea shells  INHABITING  the shores of North America, which are rank
0304  ld, from the Arctic regions to the ecuator,  INHABITING  various zones of depths from the upper tida
0321  tted to some peculiar line of life, or from  INHABITING  some distant and isolated station, where th
0335  ally in the case of terrestrial productions  INHABITING  separated districts. To compare small thing
0336  ered that the forms of life, at least those  INHABITING  the sea, have changed almost simultaneously
0351  e several species of the same genus, though  INHABITING  the most distant quarters of the world, mus
0352  individuals of the same species, though now  INHABITING  distant and isolated regions, must have pro
0352  nd no inexplicable cases of the same mammal  INHABITING  distant points of the world. No geologist w
0393  g domesticated animals kept by the natives)  INHABITING  an island situated above 300 miles from a c
0404  ee this same principle in the blind animals  INHABITING  the caves of America and of Europe. Other a
0427   because all the species of the same genus,  INHABITING  any distinct and isolated region, have in a
0454  tary, as in the case of the eyes of animals  INHABITING  dark caverns, and of the wings of birds inh
0454  ing dark caverns, and of the wings of birds  INHABITING  oceanic islands, which have seldom been for
0462  spect to distinct species of the same genus  INHABITING  very distant and isolated regions, as the p
0473   skin; or when we look at the blind animals  INHABITING  the dark caves of America and Europe. In bo
0134  as been caused by disuse. The ostrich indeed  INHABITS  continents and is exposed to danger from whi
0288  ion of a single Mediterranean species, which  INHABITS  deep water and has been found fossil in Sici
0352  es is continuous; and when a plant or animal  INHABITS  two points so distant from each other, or wi
0017  n animals and plants having an extraordinary  INHERENT  tendency to vary, and likewise to withstand
0043  ity. I do not believe that variability is an  INHERENT  and necessary contingency, under all circums
0253  s case, it is not at all surprising that the  INHERENT  sterility in the hybrids should have gone on
0054  ch in some slight degree modified, will still  INHERIT  those advantages that enabled their parents
0091  g offspring. Some of its young would probably  INHERIT  the same habits or structure, and by the rep
0092  iving. Some of these seedlings would probably  INHERIT  the nectar excreting power. Those individual
0094  g descendants. Its descendants would probably  INHERIT  a tendency to a similar slight deviation of
0118  ng only slightly modified forms, will tend to  INHERIT  those advantages which made their common par
0315  lly the same; for both would almost certainly  INHERIT  different characters from their distinct pro
0316  newly formed fantail would be almost sure to  INHERIT  from its new progenitor some slight characte
0428  cies, belonging to the larger genera, tend to  INHERIT  the advantages, which made the groups to whi
0474  of neuter insects, which leave no progeny to  INHERIT  the effects of long continued habit. On the
0012  ant for us. But the number and diversity of  INHERITABLE  deviations of structure, both those of sli
0013  ner deviations may be freely admitted to be  INHERITABLE.  Perhaps the correct way of viewing the wh
0102  remely intricate subject. A large amount of  INHERITABLE  and diversified variability is favourable,
0005  ally selected. From the strong principle of  INHERITANCE,  any selected variety will tend to propaga
0007  y. effects of Habit. Correlation of Growth.  INHERITANCE.  Character of Domestic Varieties. Difficul
0010  laws of reproduction, and of growth, and of  INHERITANCE:  for had the action of the conditions been
0012  reeder doubts how strong is the tendency to  INHERITANCE:  like produces like is his fundamental bel
0013  compels us to attribute its reappearance to  INHERITANCE.  Every one must have heard of cases of alb
0013  the whole subject, would be, to look at the  INHERITANCE  of every character whatever as the rule, a
0013  ery character whatever as the rule, and non  INHERITANCE  as the anomaly. The laws governing inherit
0013  eritance as the anomaly. The laws governing  INHERITANCE  are quite unknown; no one can say why the
0026  cases are often confounded in treatises on  INHERITANCE.  Lastly, the hybrids or mongrels from betw
0029  alists who, knowing far less of the laws of  INHERITANCE  than does the breeder, and knowing no more
0034  ttention not been paid to breeding, for the  INHERITANCE  of good and bad qualities is so obvious. A
0043  liability are modified by various degrees of  INHERITANCE  and of reversion. Variability is governed
0086  itable variations at that age, and by their  INHERITANCE  at a corresponding age. If it profit a pla
0104  ain the same, only through the principle of  INHERITANCE,  and through natural selection destroying
0123  the six descendants from (I) will, owing to  INHERITANCE,  differ considerably from the eight descen
0125  continued divergence of character and from  INHERITANCE  from a different parent, will differ widel
0127  for life; and from the strong principle of  INHERITANCE  they will tend to produce offspring simila
0129  est of my judgment, it is explained through  INHERITANCE  and the complex action of natural selectio
0146  ecies; and which in truth are simply due to  INHERITANCE;  for an ancient progenitor may have acouir
0155  d these characters in common I attribute to  INHERITANCE  from a common progenitor, for it can rarel
0160  of the mere act of crossing on the laws of  INHERITANCE.  No doubt it is a very surprising fact tha
0193  s of life, we may attribute its presence to  INHERITANCE  from a common ancestor; and its absence in
0194  but little of their structure in common to  INHERITANCE  from the same ancestor. Although in many c
0199  rganisation of every being is simply due to  INHERITANCE;  and consequently, though each being assur
0200  we may safely attribute these structures to  INHERITANCE.  But to the progenitor of the upland goose
0200  ed formerly; as now, to the several laws of  INHERITANCE,  reversion, correlation of growth, etc. He
0206  the higher law; as it includes, through the  INHERITANCE  of former adaptations, that of Unity of Ty
0209  in one generation, and then transmitted by  INHERITANCE  to succeeding generations. It can be clear
0211  instincts under a state of nature, and the  INHERITANCE  of such variations, are incispensable for
0213  d authentic instances could be given of the  INHERITANCE  of all shades of disposition and tastes, a
0215  on, has probably concurred in civilising by  INHERITANCE  our dogs. On the other hand, young chicken
0217  young thus reared would be apt to follow by  INHERITANCE  the occasional and aberrant habit of their
0235  g succeeded best, and having transmitted by  INHERITANCE  its newly acouired economical instinct to
0236  t we cannot account for their similarity by  INHERITANCE  from a common parent, and must therefore b
0243  ance, we can understand on the principle of  INHERITANCE,  how it is that the thrush of South Americ
0327  inferior forms would be allied in groups by  INHERITANCE,  whole groups would tend slowly to disappe
0327  procuced being themselves dominant owing to  INHERITANCE,  and to having already had some advantage
0345  o be mysterious, and is simply explained by  INHERITANCE.  If then the geological record be as imper
0350  bond is. This bond, on my theory, is simply  INHERITANCE,  that cause which alone, as far as we posi
0350  modified descendants. On this principle of  INHERITANCE  with modification, we can understand how i
0355  modified, would still be plainly related by  INHERITANCE  to the inhabitants of the continent. Cases
0379  wanderers, though still plainly related by  INHERITANCE  to their brethren of the northern or south
```

Page **(Key Word)**

inheritance

```
399 be liable to modification; the principle of INHERITANCE still betraying their original birthplace.
411 nct species; and these, on the principle of INHERITANCE, tend to produce other new and dominant sp
423 more or less modification, the principle of INHERITANCE would keep the forms together which were a
426 rms with such different habits, only by its INHERITANCE from a common parent. We may err in this r
430 rely adaptive, they are due on my theory to INHERITANCE in common. Therefore we must suppose eithe
430 suppose that the bizcacha has retained, by INHERITANCE, more of the character of its ancient prog
431 on parent, together with their retention by INHERITANCE of some characters in common, we can under
432 d still hold good; and, on the principle of INHERITANCE, all the forms descended from A, or from I
438 blance, retained by the strong principle of INHERITANCE. In the great class of molluscs, though we
439 h they would probably have retained through INHERITANCE, if they had really been metamorphosed dur
448 different manner, then, on the principle of INHERITANCE at corresponding ages, the active young or
455 its full powers of action, on the principle of INHERITANCE at corresponding ages will reproduce the o
456 ed, and can be accounted for by the laws of INHERITANCE. Summary. In this chapter I have attempted
457 ring in mind how strong is the principle of INHERITANCE, the occurrence of rudimentary organs and
479 affinities of all organic beings are due to INHERITANCE or community of descent. The natural syste
480 election or disuse, and on the principle of INHERITANCE at corresponding ages have been inherited
489 gest sense, being Growth with Reproduction; INHERITANCE which is almost implied by reproduction: V
011 ng more, than its wild parent. The great and INHERITED development of the udders in cows and goats
012 he parental type. Any variation which is not INHERITED is unimportant for us. But the number and d
013 e and rare deviations of structure are truly INHERITED, less strange and commoner deviations may b
013 dividuals of different species, is sometimes INHERITED and sometimes not; why the child often r
014 cases this could not be otherwise: thus the INHERITED peculiarities in the horns of cattle could
019 ubt that there has been an immense amount of INHERITED variation. Who can believe that animals clo
021 common tumbler has the singular and strictly INHERITED habit of flying at a great height in a comp
023 lly as in each of these breeds several truly INHERITED sub breeds, or species as he might have cal
036 avages so barbarous as never to think of the INHERITED character of the offspring of their domesti
044 iations in this sense are supposed not to be INHERITED; but who can say that the dwarfed condition
045 m far northwards, would not in some cases be INHERITED in at least some few generations? and in t
061 on of that individual, and will generally be INHERITED by its offspring. The offspring, also, will
089 eason: the modifications thus produced being INHERITED at corresponding ages or seasons, either by
091 o catch rats rather than mice is shown to be INHERITED. Now, if any slight innate change of habit
095 on and accumulation of infinitesimally small INHERITED modifications, each profitable to the prese
122 en thousandth generation, will probably have INHERITED from a common ancestor some advantage in co
125 ess of any croup shows that its species have INHERITED at corresponding ages, can modify the egg,
127 lection, on the principle of qualities being INHERITED. Under free nature, we can have no standard
134 nishes them; and that such modifications are INHERITED; and I should prefer explaining the entire
135 duce us to believe that mutilations are ever INHERITED from a remote period, since that period whe
156 these so called generic characters have been INHERITED much in common, to parts which have recentl
156 rom a common progenitor, from whom they have INHERITED and have not varied, to natural selection h
158 o on varying than parts which have long been INHERITED from a common parent the same constitution
159 ue to the several races of the pigeon having INHERITED for an endless number of generations, than
161 in a tendency to produce any character being INHERITED. Indeed, we may sometimes observe a mere te
161 rgans being, as we all know them to be, thus INHERITED: for instance, in the common snapdragon (An
161 bserve a mere tendency to produce a rudiment INHERITED tendency to produce it. As all the species
161 often appears, that this plant must have an INHERITED constitution. It might further be expected
161 n of the conditions of life and of a similar INHERITED, and have not differed within this same per
168 ic characters, or those which have long been INHERITED, which is certainly the case: and if any va
186 vary ever so slightly, and the variations be INHERITED from one ancient progenitor thus provided,
193 lection. But if the electric organs had been INHERITED from a common progenitor, were formerly of
200 the monkey, horse, and bat, which have been INHERITED, and I think it can be shown that this does
209 if we suppose any habitual action to become INHERITED variations of instinct in a state of nature
212 ce. The possibility, or even probability, of INHERITED by retrievers: and a tendency to run round,
213 out; retrieving is certainly in some degree INHERITED, and how curiously they become mingled, is
214 stic instincts, habits, and dispositions are INHERITED solely from long continued and compulsory h
214 times spoken of as actions which have become INHERITED effects of compulsory training in each succ
214 once displayed, methodical selection and the INHERITED change from extreme wildness to extreme tam
215 sume that we must attribute the whole of the INHERITED mental changes: in other cases compulsory h
216 ory habit alone has sufficed to produce such INHERITED by its offspring, which again varied and we
237 itable modification of structure, this being INHERITED instincts and by inherited tools or weapons
242 is useful to civilised man. As ants work by INHERITED tools or weapons, and not by acquired knowl
242 as ants work by inherited instincts and by INHERITED. Still more briefly I have attempted to dev
242 ic animals vary, and that the variations are INHERITED inferiority in common. But whether it be sp
321 be allied forms, which will suffer from some INHERITED something in common from their ancient and
331 orm A, will form an order; for all will have INHERITED at a corresponding age. This process, whils
338 ns supervening at a not early age, and being INHERITED from a common progenitor, tend to become ex
344 less vigorous groups, from their inferiority INHERITED at a corresponding age. Rudimentary Organs;
411 s not supervening at an early age, and being INHERITED something in common. But the three genera o
412 t but unseen parent, and, consequently, have INHERITED from a common parent, and, in so far, all t
420 o or more species, are those which have been INHERITED something in common from their common paren
421 ll the modified descendants from A will have INHERITED to a certain extent their characters. This
422 ra descended from these two genera will have INHERITED from a common ancestor. And we know that su
426 d of descent, that these characters have been INHERITED their general shape of body and structure o
428 d small, that we cannot doubt that they have INHERITED by the offspring at a corresponding not ear
446 ed at an early period of life, and have been INHERITED, not at the corresponding, but at an earlie
446 r, if not so, the differences must have been INHERITED at a corresponding age, the young of the ne
446 rvened at a rather late age, and having been INHERITED at a corresponding late age, the fore limbs
447 supervening at a rather late age, and being INHERITED at a corresponding mature age. Whereas the
447 iving: and the effects thus produced will be INHERITED at an earlier period than that at which it
447 early period of life, or each step might be INHERITED at an earlier period than that at which the
450 early age, or by the variations having been INHERITED at a corresponding not early period. Embryo
450 gh perhaps caused at the earliest; and being INHERITED, not at the corresponding age, but at an ex
455 step of the process of reduction were to be INHERITED, we can understand, on the genealogical vie
455 organisation, which has long existed, to be INHERITED at a corresponding period. we can understan
457 ng at a very early period of life, and being INHERITED at corresponding ages. On this same princip
457 through the principle of modifications being INHERITED at corresponding ages. On this same princip
466 as been large, and that modifications can be INHERITED for long periods. As long as the conditions
```

```
0466   that a modification, which has already been INHERITED for many generations, may continue to be *
0466   ted for many generations, may continue to be INHERITED for an almost infinite number of generatic
0469   f life, would be preserved, accumulated, and INHERITED? Why, if man can by patience select variat
0474   than the generic characters which have been INHERITED without change for an enormous period. It
0474   y subordinate forms, that is, if it has been INHERITED for a very long period; for in this case *
0474   g descended from a common parent, and having INHERITED much in common, we can understand how it *
0479   lways supervening at an early age, and being INHERITED at a corresponding not early period of lit
0480   this early age. The calf, for instance, has INHERITED teeth, which never cut through the gums ot
0480   inheritance at corresponding ages have been INHERITED from a remote period to the present day. (
0486   characters of any kind which have long been INHRITED. Rudimentary organs will speak infallibly
0169   ary a a manner it may be developed. Species INHERITING nearly the same constitution from a commor
0327   s, will generally be allied in groups, from INHERITING some inferiority in common; and therefore
0172   which we hardly as yet fully understand the INIMITABLE perfection? Thirdly, can instincts be acqu
0186   tion. To suppose that the eye, with all its INIMITABLE contrivances for adjusting the focus to dt
0202   us to admire with enthusiasm a multitude of INIMITABLE contrivances in nature, this same reason t
0228   as acquired, through natural selection, her INIMITABLE architectural powers. But this theory can
0362   ecrete gastric juice, and do not in the least INJURE, as I know by trial, the germination of seed
0359   e spaces of the sea, whether or not they were INJURED by the salt water. Afterwards I tried some
0231   can cluster and crawl over the comb without INJURING the delicate hexagonal walls, which are onl
0044   e deviation of structure in one part, either INJURING to or not useful to the species, and not g
0069   meet with stunted forms, due to the directly INJURIOUS action of climate, than we do in proceedir
0081   sure that any variation in the least degree INJURIOUS would be rigidly destroyed. This preservat
0081   f favourable variations and the rejection of INJURIOUS variations, I call Natural Selection. Vari
0081   ral Selection. Variations neither useful nor INJURIOUS would not be affected by natural selection
0086   od of life, shall not be in the least degree INJURIOUS: for if they became so, they would cause t
0091   rently for the sake of eliminating something INJURIOUS from their sap: this is effected by glands
0137   as frequent inflammation of the eyes must be INJURIOUS to any animal, and as eyes are certainly n
0137   at eyes, though useless, could be in any way INJURIOUS to animals living in darkness; I attribute
0196   become of very slight use; and any actually INJURIOUS deviations in their structure will always
0201   ction will never produce in a being anything INJURIOUS to itself, for natural selection acts sole
0201   conditions of life, if any part comes to be INJURIOUS, it will be modified; or if it be not so,
0205   ghly useful or even indispensable, or highly INJURIOUS to another species, but in all cases at th
0247   in his house. That these processes are often INJURIOUS to the fertility of a plant cannot be doub
0249   ollen; and I am convinced that this would be INJURIOUS to their fertility, already lessened by th
0252   wed to freely cross with each other, and the INJURIOUS influence of close interbreeding is thus p
0264   ery young beings seem eminently sensitive to INJURIOUS or unnatural conditions of life. In regard
0270   anipulation in this case could not have been INJURIOUS, as the plants have separated sexes. No on
0319   g is constantly being checked by unperceived INJURIOUS agencies; and that these same unperceived
0358   ot even known how far seeds could resist the INJURIOUS action of sea water. To my surprise I foun
0360   eir flotation and of their resistance to the INJURIOUS action of the salt water. On the other han
0453   t they serve to excrete matter in excess, or INJURIOUS to the system; but can we suppose that the
0454   seful under certain conditions, might become INJURIOUS under others, as with the wings of beetles
0454   d, during changed habits of life, useless or INJURIOUS for one purpose, might easily be modified
0200   does often produce structures for the direct INJURY of other species, as we see in the fang of t
0201   snake is furnished with a rattle for its own INJURY, namely, to warn its prey to escape. I would
0201   r the purpose of causing pain or for doing an INJURY to its possessor. If a fair balance be struc
0205   hing in one species for the exclusive good or INJURY of another; though it may well produce parts
0095   s or to the formation of the longest lines of INLAND cliffs. Natural selection can act only by th
0360   stranded, if blown to a favourable spot by an INLAND gale, they would germinate. Subsequently to
0481   when Lyell first insisted that long lines of INLAND cliffs had been formed, and great valleys ex
0091   our wolf preyed, a cub might be born with an INNATE tendency to pursue certain kinds of prey. No
0091   is shown to be inherited. Now, if any slight INNATE change of habit or of structure benefited an
0141   al climate as a quality readily grafted on an INNATE wide flexibility of constitution, which is c
0141   tural selection of varieties having different INNATE constitutions, and how much to both means cor
0143   mes overmastered by, the natural selection of INNATE differences. Correlation of Growth. I mean b
0152   rsion to a less modified state, as well as an INNATE tendency to further variability of all kinds
0262   as Gartner found that there was sometimes an INNATE difference in different individuals of the sa
0256   ies. But the degree of fertility is likewise INNATELY variable; for it is not always the same whet
0259   o favourable and unfavourable conditions, is INNATELY variable. That it is by no means always the
0260   ether? Why should the degree of sterility be INNATELY variable in the individuals of the came spec
0276   ranking forms by this test. The sterility is INNATELY variable in individuals of the same species
0092   le owcet juice or nectar to be excreted by the INNER bases of the petals of a flower. In this case
0144   n that of the difference between the outer and INNER flowers in some Compositous and Umbelliferous
0145   me, in species with the densest heads that the INNER and outer flowers most frequently differ. It
0145   is a difference in the seeds of the outer and INNER florets without any difference in the corolla
0187   s, for instance, there is a double cornea, the INNER one divided into facets, within each of whict
0194   ture is prodigal in variety, but niggard in INNOVATION. Why, on the theory of Creation, should th
0471   e is prodigal in variety, though niggard in INNOVATION. But why this should be a law of nature if
0003   tisfactory, until it could be shown how the INNUMERABLE species inhabiting this world have been mo
0006   s do we know of the mutual relations of the INNUMERABLE inhabitants of the world during the many p
0029   proceeded from the seeds of the same tree. INNUMERABLE other examples could be given. The explana
0075   compared to the action and reaction of the INNUMERABLE plants and animals which have determined,
0099   in the case of a gigantic tree covered with INNUMERABLE flowers, it may be objected that pollen co
0111   ally happen, we must infer from most of the INNUMERABLE species throughout nature presenting well
0134   tly the conditions of life must act. Again, INNUMERABLE instances are known to every naturalist of
0170   nt modifications of structure, by which the INNUMERABLE beings on the face of this earth are enabl
0171   y fine gradations, do we not everywhere see INNUMERABLE transitional forms? Why is not all nature
0172   ion of the new form. Put, as by this theory INNUMERABLE transitional forms must have existed, why
0183   rtificial substances. Of diversified habits INNUMERABLE instances could be given: I have often wat
0185   to change. He who believes in separate and INNUMERABLE acts of creation will say, that in these c
0237   n? first, let it be remembered that we have INNUMERABLE instances, both in our domestic production
0279   ms, and their not being blended together by INNUMERABLE transitional links, is a very obvious diff
0279   nd improvement. The main cause, however, of INNUMERABLE intermediate links not now occurring everyw
0302   d not the difficulty of our not discovering INNUMERABLE transitional links between the species wh
0315   adapted (and no doubt this has occurred in INNUMERABLE instances) to fill the exact place of anot
0348   counter no impassable barriers, and we have INNUMERABLE islands as halting places, until after tra
```

ge **(Key Word)**

49 dest generality, and every continent offers INNUMERABLE instances. Nevertheless the naturalist in
49 and capybara, rodents of the American type. INNUMERABLE other instances could be given. If we look
59 h, human reason, but by the accumulation of INNUMERABLE slight variations, each good for the indiv
79 eck of the giraffe and of the elephant, and INNUMERABLE other such facts, at once explain themselv
83 y birth. But do they really believe that at INNUMERABLE periods in the earth's history certain ele
88 since the first creature, the progenitor of INNUMERABLE extinct and living descendants, was create
27 , is analogical. Amongst insects there are INNUMERABLE instances: thus Linnaeus, misled by extern
46 olymorphic, in which the species present an INORDINATE amount of variation; and hardly two natural
63 rease, its numbers would quickly become so INORDINATELY great that no country could support the pr
70 highly favourable circumstances, increases INORDINATELY in numbers in a small tract, epidemics, at
94 lect on the great changes of level, on the INORDINATELY great change of climate, on the prodigious
22 stent, that each species tends to increase INORDINATELY, and that some check is always in action,
70 metrical ratio of reproduction to increase INORDINATELY in number: and as the modified descendants
54 ws that there is something in the organic or INORGANIC conditions of that country favourable to th
84 organic being in relation to its organic and INORGANIC conditions of life. We see nothing of those
04 ted area, if not very large, the organic and INORGANIC conditions of life will generally be in a g
06 rying parts of each being to its organic and INORGANIC conditions of life; or by having adapted th
55 ly, it exerts no more influence than so much INORGANIC dust. From this absolute zero of fertility,
14 ns of the higher beings to their organic and INORGANIC conditions of life, as explained in a forme
66 he very same conditions of life, organic and INORGANIC, should recur. For though the offspring of
66 f almost every conceivable kind, organic and INORGANIC, that within a very recent geological perio
84 ver systems and completely prevented their INOSCULATION, seems to lead to this same conclusion. Wi
50 must feel little curiosity, who is not led to INQUIRE what this bond is. This bond, on my theory,
63 y request Sir C. Lyell wrote to M. Hartung to INQUIRE whether he had observed erratic boulders on
63 hat these are plainest in the foal, and from INQUIRIES which I have made, I believe this to be tru
86 become! A grand and almost untrodden field of INQUIRY will be opened, on the causes and laws of va
77 mediate form would be eminently liable to the INROADS of closely allied forms existing on both sid
77 many temperate plants, if protected from the INROADS of competitors, can withstand a much warmer
45 ves close to the great central ganglion of an INSECT would have been variable in the same species;
75 c its seeds by the thousand: what war between INSECT and insect, between insects, snails, and othe
75 by the thousand: what war between insect and INSECT, between insects, snails, and other animals w
86 election may modify and adapt the larva of an INSECT to a score of contingencies, wholly different
86 different from those which concern the mature INSECT. These modifications will no doubt affect, th
94 ppreciated by us, might profit a bee or other INSECT, so that an individual so characterised would
33 brighter coloured than those of islands. The INSECT species confined to sea coasts, as every coll
35 es the tarsi are so habitually lost, that the INSECT has been described as not having them. In som
36 e action of natural selection. For when a new INSECT first arrived on the island, the tendency of
02 es, and so inevitably causes the death of the INSECT by tearing out its viscera? If we look at the
02 he use of the sting should so often cause the INSECT's own death: for if on the whole the power of
03 d of many other plants are fertilised through INSECT agency, can we consider as equally perfect th
18 its own larvae to feed on, yet that when this INSECT finds a burrow already made and stored by ano
19 t, if of advantage to the species, and if the INSECT whose nest and stored food are thus felonious
33 e seem to me any great difficulty in a single INSECT (as in the case of a queen wasp) making hexag
33 ediate planes. It is even conceivable that an INSECT might, by fixing on a point at which to comme
36 i allude to the neuters or sterile females in INSECT communities: for these neuters often differ w
36 he hive bee. If a working ant or other neuter INSECT had been an animal in the ordinary state, I s
37 onwards. But with the working ant we have an INSECT differing greatly from its parents, yet absol
37 h the sterile condition of certain members of INSECT communities: the difficulty lies in understan
52 ds, and such alone are fairly treated, for by INSECT agency the several individuals of the same hy
52 readily convince himself of the efficiency of INSECT agency by examining the flowers of the more s
26 the angle of the jaw, the manner in which an INSECT's wing is folded, whether the skin be covered
27 appearances, actually classed as a homopterous INSECT as a moth. We see something of the same kind
29 ciscovery of Australia has not added a single INSECT belonging to a new order: and that in the veg
42 y professor Huxley of the development of this INSECT, we see no trace of the vermiform stage. How,
71 ects must have been still greater, for six INSECTIVOROUS birds were very common in the plantations
71 th was frequented by two or three distinct INSECTIVOROUS birds. Here we see how potent has been th
72 eans, probably by birds. Hence, if certain INSECTIVOROUS birds (whose numbers are probably regulat
73 as we just have seen in Staffordshire, the INSECTIVOROUS birds, and so onwards in ever increasing
73 les of complexity. We began this series by INSECTIVOROUS birds, and we have ended with them. Not t
80 d been taken, and it had been asked how an INSECTIVOROUS quadruped could possibly have been conver
03 ak, and tongue, so admirably adapted to catch INSECTS under the bark of trees. In the case of the
03 es absolutely requiring the agency of certain INSECTS to bring pollen from one flower to the other
46 own that the muscles in the larvae of certain INSECTS are very far from uniform. Authors sometimes
46 d hieracium amongst plants, several genera of INSECTS, and several genera of Brachiopod shells. In
48 separated areas. How many of those birds and INSECTS in North America and Europe, which differ ve
48 ts of the little Madeira group there are many INSECTS which are characterized as varieties in Mr.
55 nts of twelve countries, and the coleopterous INSECTS of two districts, into two nearly equal mass
57 n regard to plants, and Westwood in regard to INSECTS, that in large genera the amount of differen
62 hich are idly singing round us mostly live on INSECTS or seeds, and are thus constantly destroying
67 than 295 were destroyed, chiefly by slugs and INSECTS. If turf which has long been mown, and the c
71 not be found on the heath. The effect on the INSECTS must have been still greater, for six insect
72 scotch fir; but in several parts of the world INSECTS determine the existence of cattle. Perhaps P
73 getation: this again would largely affect the INSECTS; and this, as we just have seen in Staffords
73 in this part of England, is never visited by INSECTS; and consequently, from its peculiar structu
75 ; what war between insect and insect, between INSECTS, snails, and other animals with birds and be
77 ving, allows it to compete with other aquatic INSECTS, to hunt for its own prey, and to escape ser
84 may thus be acted on. When we see leaf eating INSECTS green, and bark feeders mottled grey; the al
84 these tints are of service to these birds and INSECTS in preserving them from danger. Grouse, if n
86 the adult; and probably in the case of those INSECTS which live only for a few hours, and which n
87 instance, the great jaws possessed by certain INSECTS; and used exclusively for opening the cocoon
88 the two sexes, as is sometimes the case with INSECTS. And this leads me to say a few words on wha
92 ough small in quantity, is greedily sought by INSECTS. Let us now suppose a little sweet juice or
92 bases of the petals of a flower. In this case INSECTS in seeking the nectar would get dusted with
92 ted most nectar, would be oftenest visited by INSECTS, and would be oftenest crossed; and so in th
92 tion to the size and habits of the particular INSECTS which visited them, so as to favour in any d
92 or selected. We might have taken the case of INSECTS visiting flowers for the sake of collecting

age **(Key Word)**

insects

Page ***************************************(Key Word)***************************************

0092   and then habitually, by the pollen devouring INSECTS from flower to flower, and a cross thus ef
0093   owers, had been rendered highly attractive to INSECTS, they would, unintentionally on their part
0093   ant had been rendered so highly attractive to INSECTS that pollen was regularly carried from flo
0094   fected. Let us now turn to the nectar feeding INSECTS in our imaginary case: we may suppose the
0094   ction, to be a common plant; and that certain INSECTS depended in main part on its nectar for foo
0096   ample discussion. All vertebrate animals, all INSECTS, and some other large groups of animals, pa
0098   it is useful for this end: but, the agency of INSECTS is often required to cause the stamens to s
0098   r is never visited, at least in my garden, by INSECTS, it never sets a seed; though by placing pe
0100   know of no means, analogous to the action of INSECTS and of the wind in the case of plants, by w
0114   from each other. So it is with the plants and INSECTS on small and uniform islets; and so in smal
0132   living on islands or near the coast. So with INSECTS, Wollaston is convinced that residence nea
0135   e, and therefore cannot be much used by these INSECTS. In some cases we might easily put down to
0136   lown to sea and thus have been destroyed. The INSECTS in Madeira which are not ground feeders, ar
0137   manner as in Madeira the wings of some of the INSECTS have been enlarged, and the wings of others
0138   remarked, this is not the case, and the cave INSECTS of the two continents are not more closely
0139   professor Dana; and some of the European cave INSECTS are very closely allied to those of the sur
0145   's idea that the ray florets serve to attract INSECTS, whose agency is highly advantageous in the
0182   that we have flying birds and mammals, flying INSECTS of the most diversified types, and formerly
0183   merely to allude to that of the many British INSECTS which now feed on exotic plants, or exclus
0184   dely open mouth, thus catching, like a whale, INSECTS in the water. Even in so extreme a case as
0184   n so extreme a case as this, if the supply of INSECTS were constant, and if better adapted compet
0184   woodpecker for climbing trees and for seizing INSECTS in the chinks of the bark? Yet in North Ame
0184   , and others with elongated wings which chase INSECTS on the wing: and on the plains of La Plata,
0191   e homologous with the wings and wingcovers of INSECTS, it is probable that organs which at a very
0192   re work. One of the gravest is that of neuter INSECTS, which are often very differently construct
0193   ost. The presence of luminous organs in a few INSECTS, belonging to different families and orders
0195   flesh, which, from determining the attacks of INSECTS or from being correlated with constitutiona
0195   ds on their power of resisting the attacks of INSECTS: so that individuals which could by any mea
0201   s are deposited in the living bodies of other INSECTS. If it could be proved that any part of the
0202   ful power of scent by which the males of many INSECTS find their females, can we admire the produ
0207   ral Selection of instincts. Neuter or sterile INSECTS. Summary. The subject of instinct might hav
0218   me species, likewise, of Sphegidae (wasp like INSECTS) are parasitic on other species; and M. Fab
0227   at perfectly cylindrical burrows in wood many INSECTS can make, apparently by turning round on a
0236   n of structure; for it can be shown that some INSECTS and other articulate animals in a state of
0236   ture occasionally become sterile; and if such INSECTS had been social, and it had been profitable
0238   kind. Thus I believe it has been with social INSECTS: a slight modification of structure, or ins
0238   as been produced, which we see in many social INSECTS. But we have not as yet touched on the clim
0239   late my theory. In the simpler case of neuter INSECTS all of one caste or of the same kind, which
0239   his view we ought occasionally to find neuter INSECTS of the same species, in the same nest, pres
0239   find, even often, considering how few neuter INSECTS out of Europe have been carefully examined.
0241   uction may have been to a social community of INSECTS, on the same principle that the division of
0242   gh a degree, had not the case of these neuter INSECTS convinced me of the fact. I have, therefore
0242   as advanced this demonstrative case of neuter INSECTS, against the well known doctrine of Lamarck
0247   rder to prevent pollen being brought to it by INSECTS from other plants. Nearly all the plants ex
0249   erally grow in the same garden, the visits of INSECTS must be carefully prevented during the flow
0314   instance, in the land shells and coleopterous INSECTS of Madeira having come to differ considerab
0365   almost bare land, with few or no destructive INSECTS or birds living there, nearly every seed, w
0383   surprise at the similarity of the fresh water INSECTS, shells, etc., and at the dissimilarity of
0389   lle admits this for plants; and Wollaston for INSECTS. If we look to the large size and varied st
0391   ed miles of open sea. The different orders of INSECTS in Madeira apparently present analogous fac
0415   tis from Connarus. To give an example amongst INSECTS, in one great division of the Hymenoptera,
0416   marsupials, the manner in which the wings of INSECTS are folded, mere colour in certain Algae, m
0419   . instances could be given amongst plants and INSECTS, of a group of forms, first ranked by pract
0427   se mammals and fishes, is analogical. Amongst INSECTS there are innummerable instances: thus Linn
0434   reat law in the construction of the mouths of INSECTS: what can be more different than the immens
0435   ughout the whole class. So with the mouths of INSECTS, we have only to suppose that their common
0436   ty in structure and function of the mouths of INSECTS. Nevertheless, it is conceivable that the o
0439   ch other much more closely than do the mature INSECTS; but in the case of larvae, the embryos are
0442   y to be called a metamorphosis. The larvae of INSECTS, whether adapted to the most diverse and ac
0448   and with a few members of the great class of INSECTS, as with Aphis. With respect to the final c
0451   wings are formed for flight, yet in how many INSECTS do we see wings so reduced in size as to be
0471   odpecker, should have been created to prey on INSECTS on the ground: that upland geese, which nev
0471   been created to dive and feed on sub aquatic INSECTS; and that a petrel should have been created
0474   dispensable, as we see, in the case of neuter INSECTS, which leave no progeny to inherit the effe
0489   ith birds singing on the bushes, with various INSECTS flitting about, and with worms crawling thr
0035   rms and qualities of their cattle. Slow and INSENSIBLE changes of this kind could never be recogn
0040   g been preserved of such slow, varying, and INSENSIBLE changes. I must now say a few words on the
0051   ese differences blend into each other in an INSENSIBLE series; and a series impresses the mind wi
0175   ecome so), not blending one into another by INSENSIBLE gradations, the range of any one species,
0190   alone, and thus wholly change its nature by INSENSIBLE steps. Two distinct organs sometimes perfo
0036   ly trace the stages through which they have INSENSIBLY passed, and come to differ so greatly from
0084   nd adding up all that is good; silently and INSENSIBLY working, whenever and wherever opportunity
0103   the conditions will generally graduate away INSENSIBLY from one district to another. The intercro
0171   pecies have descended from other species by INSENSIBLY fine gradations, do we not everywhere see
0175   s climate and height or depth graduate away INSENSIBLY. But when we bear in mind that almost ever
0175   country by no means exclusively depends on INSENSIBLY changing physical conditions; but in large
0203   us, and when the conditions of life did not INSENSIBLY graduate away from one part to another. Wh
0241   astes of different sizes, yet they graduate INSENSIBLY into each other, as does the widely differ
0248   o different in degree and graduates away so INSENSIBLY, and, on the other hand, that the fertilit
0279   nditions of life do not graduate away quite INSENSIBLY like heat or moisture. I endeavoured, also
0454   hange in function, which can be effected by INSENSIBLY small steps, is within the power of natura
0463   climate and other conditions of life change INSENSIBLY in going from a district occupied by one s
0119   se breaks are imaginary, and might have been INSERTED anywhere, after intervals long enough to ha
0004   re, of the highest importance to gain a clear INSIGHT into the means of modification and coadapta
0095   of the coast waves, called a trifling and INSIGNIFICANT cause, when applied to the excavation of
0136   pecially the extraordinary fact, so strongly INSISTED on by Mr. Wollaston, of the almost entire a

Page ***************************************(Key Word)***************************************

age **********************************(Key Word)**

206 | ression of conditions of existence, so often | INSISTED on by the illustrious Cuvier, is fully embra
305 | d existing species. The case most frequently | INSISTED on by palaeontologists of the apparently sud
335 | gree intermediate in character, is the fact, | INSISTED on by all palaeontologists, that fossils fro
339 | h impressed with these facts that I strongly | INSISTED, in 1839 and 1845, on this law of the succes
357 | period of existing organisms. Edward Forbes | INSISTED that all the islands in the Atlantic must re
379 | t species. It is a remarkable fact, strongly | INSISTED on by Hooker in regard to America, and by Al
409 | ive species. As the late Edward Forbes often | INSISTED, there is a striking parellelism in the laws
416 | ts and Pachyderms. Robert Brown has strongly | INSISTED on the fact that the rudimentary florets are
419 | uch as Mr. Bentham and others, have strongly | INSISTED on their arbitrary value. Instances could be
423 | ing varieties, as with species. Authors have | INSISTED on the necessity of classing varieties on a
433 | this direction; and Milne Edwards has lately | INSISTED, in an able paper, on the high importance of
434 | relative positions? Geoffroy St. Hilaire has | INSISTED strongly on the high importance of relative
481 | felt by so many geologists, when Lyell first | INSISTED that long lines of inland cliffs had been fo
273 | only a difference in degree. Gartner further | INSISTS that when any two species, although most clo
338 | could not have foreseen this result. Agassiz | INSISTS that ancient animals resemble to a certain e
419 | arge groups of closely allied forms. Temminck | INSISTS on the utility or even necessity of this pra
031 | any treatises devoted to this subject, and to | INSPECT the animals. Breeders habitually speak of an
296 | s requiring much time. Nor will the closest | INSPECTION of a formation give any idea of the time wh
003 | mere external conditions, the structure, for | INSTANCE, of the woodpecker, with its feet, tail, bea
009 | ely under the most unnatural conditions (for | INSTANCE, the rabbit and ferret kept in hutches), sho
011 | in animals it has a more marked effect; for | INSTANCE, I find in the domestic duck that the bones
011 | these organs in other countries, is another | INSTANCE of the effect of use. Not a single domestic
015 | ing many generations, the several races, for | INSTANCE, of the cabbage, in very poor soil (in which
016 | be cleared up, would be interesting: if, for | INSTANCE, it could be shown that the greyhound, blood
017 | very closely allied and natural species, for | INSTANCE, of the many foxes, inhabiting different qua
023 | the crossing of any lesser number: how, for | INSTANCE, could a pouter be produced by crossing two
025 | pt suddenly to acquire these characters; for | INSTANCE, I crossed some uniformly white fantails wit
027 | ch are mainly distinctive of each breed, for | INSTANCE the wattle and length of beak of the carrier
029 | y he disbelieves that the several sorts, for | INSTANCE a Ribston pippin or Codlin apple, could ever
030 | uddenly, or by one step; many botanists, for | INSTANCE, believe that the fuller's teazle, with its
032 | ave been kept; thus, to give a very trifling | INSTANCE, the steadily increasing size of the common
036 | er of either of them has deviated in any one | INSTANCE from the pure blood of Mr. Bakewell's flock,
046 | not vary; and, under this point of view, no | INSTANCE of an important part varying will ever be fo
046 | nk as species and which as varieties. We may | INSTANCE Rubus, Rosa, and Hieracium amongst plants, s
049 | e their rank. I will here give only a single | INSTANCE, the well known one of the primrose and cows
058 | ties, often have much restricted ranges. For | INSTANCE, Mr. H. C. Watson has marked for me in the w
060 | or sub species or varieties; what rank, for | INSTANCE, the two or three hundred doubtful forms of
067 | actly what the checks are in even one single | INSTANCE. Nor will this surprise any one who reflects
067 | oyed in vast numbers by various enemies; for | INSTANCE, on a piece of ground three feet long and tw
068 | he same kind of food. Even when climate, for | INSTANCE extreme cold, acts directly, it will be the
071 | the same country. I will give only a single | INSTANCE, which, though a simple one, has interested
072 | le. Perhaps Paraguay offers the most curious | INSTANCE of this; for here neither cattle nor horses
073 | forms of life! I am tempted to give one more | INSTANCE showing how plants and animals, most remote
075 | sometimes see the contest soon decided: for | INSTANCE, if several varieties of wheat be sown toget
078 | dvantage over another. Probably in no single | INSTANCE should we know what to do, so as to succeed.
081 | country undergoing some physical change, for | INSTANCE, of climate. The proportional numbers of its
086 | ear in the offspring at the same period; for | INSTANCE, in the seeds of the many varieties of our c
087 | fied to any extent by natural selection; for | INSTANCE, the great jaws possessed by certain insects
090 | es. We see analogous cases under nature, for | INSTANCE, the tuft of hair on the breast of the turke
090 | s suppose that the fleetest prey, a deer for | INSTANCE, had from any change in the country increase
091 | encies of our domestic animals; one cat, for | INSTANCE, taking to catch rats, another mice; one cat
095 | owing how anxious bees are to save time; for | INSTANCE, their habit of cutting holes and sucking th
095 | but we now very seldom hear the action, for | INSTANCE, of the coast waves, called a trifling and i
098 | ma receiving pollen from its own flower: for | INSTANCE, in Lobelia fulgens, there is a really beaut
099 | ings thus raised will turn out mongrels: for | INSTANCE, I raised 233 seedling cabbages from some pl
103 | k rate. Hence in animals of this nature, for | INSTANCE in birds, varieties will generally be confin
106 | ur chapter on geographical distribution; for | INSTANCE, that the productions of the smaller contine
113 | occupied by other animals: some of them, for | INSTANCE, being enabled to feed on new kinds of prey,
115 | find great diversity in its inhabitants. For | INSTANCE, I found that a piece of turf, three feet by
115 | genera than in new species. To give a single | INSTANCE: in the last edition of Dr. Asa Gray's Manua
116 | rsified in structure. It may be doubted, for | INSTANCE, whether the Australian marsupials, which ar
116 | ng that from a1 to a10. In the same way, for | INSTANCE, the English race horse and English pointer
133 | ecies, will have to say that this shell, for | INSTANCE, was created with bright colours for a warm
141 | als from several wild stocks: the blood, for | INSTANCE, of a tropical and arctic wolf or wild dog m
145 | rence in the ray and central florets of, for | INSTANCE, the daisy, and this difference is often acc
145 | eloria, and become regular. I may add, as an | INSTANCE of this, and of a striking case of correlati
146 | lone in which natural selection can act. For | INSTANCE, Alph. de Candolle has remarked that winged
152 | and no selection be applied, that part (for | INSTANCE, the comb in the Dorking fowl) or the whole
156 | er parts of their organisation; compare, for | INSTANCE, the amount of difference between the males
161 | vorable conditions, gains an ascendancy. For | INSTANCE, it is probable that in each generation of t
161 | endency to produce a rudiment inherited: for | INSTANCE, in the common snapdragon (Antirrhinum) a ru
162 | uld not distinguish these two cases: if, for | INSTANCE, we did not know that the rock pigeon was no
163 | are not rare in duns, mouse duns, and in one | INSTANCE in a chestnut: a faint shoulder stripe may s
171 | y, is it possible that an animal having, for | INSTANCE, the structure and habits of a bat, could ha
179 | opponents of such views as I hold; how, for | INSTANCE, a land carnivorous animal could have been c
184 | nces do occur in nature. Can a more striking | INSTANCE of adaptation be given than that of a woodpe
187 | e of perfection. In certain crustaceans, for | INSTANCE, there is a double cornea, the inner one div
190 | function in the same individual; to give one | INSTANCE, there are fish with gills or branchiae that
191 | nction to another; that I will give one more | INSTANCE. Pedunculated cirripedes have two minute fol
193 | difficulty. Other cases could be given; for | INSTANCE in plants, the very curious contrivance of a
201 | the endemic productions of New Zealand, for | INSTANCE, are perfect one compared with another: but
204 | at graduate into each other; that a bat, for | INSTANCE, could not have been formed by natural selec
204 | hoses in function are at least possible. For | INSTANCE, a swim bladder has apparently been converte
208 | lar were taken out of a hammock made up, for | INSTANCE, to the third stage, and were put into one f
211 | ssert, that instincts certainly do vary, for | INSTANCE, the migratory instinct, both in extent and
212 | inhabiting desert islands; and we may see an | INSTANCE of this even in England, in the greater wild

Page **********************************(Key Word)**

instance

Page ********************************(Key Word)**********************************

```
0213  oung pointers (I have myself seen a striking INSTANCE) will sometimes point and even back other do
0215  s are lost under domestication: a remarkable INSTANCE of this is seen in those breeds of fowls whi
0222  een it ferociously attack other ants. In one INSTANCE I found to my surprise an independent commun
0229  had worked less quickly. In one well marked  INSTANCE, I put the comb back into the hive, and allo
0232  d in their proper positions for working, for INSTANCE, on a slip of wood, placed directly under th
0243  ten retaining nearly the same instincts. For INSTANCE, we can understand on the principle of inher
0250  eadily than they can be self fertilised! For INSTANCE, a bulb of Hippeastrum aulicum produced four
0251  , yet many of these hybrids seed freely. For INSTANCE, Herbert asserts that a hybrid from Calceola
0251  hem are perfectly fertile. Mr. C. Noble, for INSTANCE, informs me that he raises stocks for graft
0252  few experiments have been fairly tried: for  INSTANCE, the canary bird has been crossed with nine
0253  fertile hybrid animals, I hardly know of an  INSTANCE in which two families of the same hybrid hav
0253  y with either pure parent, and in one single INSTANCE they have bred inter se. This was effected b
0254  rests on no direct evidence. I believe, for  INSTANCE, that our dogs have descended from several w
0256  ven within the limits of the same genus, for INSTANCE in Dianthus, these two opposite cases occur.
0257  enus, we meet with this same difference: for INSTANCE, the many species of Nicotiana have been mor
0258  ss between two species, I mean the case, for INSTANCE, of a stallion horse being first crossed wit
0258  s long ago observed by Kolreuter. To give an INSTANCE: Mirabilis jalappa can easily be fertilised
0259  gular rules could be given from Gartner: for INSTANCE, some species have a remarkable power of cro
0262  s is in grafting; the common gooseberry, for INSTANCE, cannot be grafted on the currant, whereas t
0266  ith respect to the sterility of hybrids; for INSTANCE, the unequal fertility of hybrids produced f
0268  e ranked by most naturalists as species. For INSTANCE, the blue and red pimpernel, the primrose an
0268  nvolved in doubt. For when it is stated, for INSTANCE, that the German Spitz dog unites more easil
0268  ng widely from each other in appearance, for INSTANCE of the pigeon or of the cabbage, is a remark
0274  variety is crossed with another variety. For INSTANCE, I think those authors are right, who mainta
0280  te between the fantail and pouter; none, for INSTANCE, combining a tail somewhat expanded with a c
0281  cies, if we look to forms very distinct, for INSTANCE to the horse and tapir, we have no reason to
0281  rms might have descended from the other; for INSTANCE, a horse from a tapir; and in this case dire
0285  is externally visible. The Craven fault, for INSTANCE, extends for upwards of 30 miles, and along
0288  en high and low watermark are preserved. For INSTANCE, the several species of the Chthamalinae (a
0289  ins is fragmentary in an extreme degree. For INSTANCE, not a land shell is known belonging to eith
0289  ey are closely consecutive. But we know, for INSTANCE, from Sir R. Murchison's great work in Russi
0294  grated into that area. It is well known, for INSTANCE, that several species appeared somewhat earl
0294  ing the whole of this period. It is not, for INSTANCE, probable that sediment was deposited during
0299  ssil links, by asking ourselves whether, for INSTANCE, geologists at some future period will be ab
0302  been urged by several palaeontologists, for INSTANCE, by Agassiz, Pictet, and by none more forcib
0303  m to some new and peculiar line of life, for INSTANCE to fly through the air; but that when this h
0304  of the secondary period. I may give another INSTANCE, which from having passed under my own eyes
0304  trouble to me, adding as I thought one more INSTANCE of the abrupt appearance of a great group of
0306  ual force to the earliest known species. For INSTANCE, I cannot doubt that all the Silurian trilob
0309  uld we be justified in assuming that if, for INSTANCE, the bed of the Pacific Ocean were now conve
0309  mmense areas in some parts of the world, for INSTANCE in South America, of bare metamorphic rocks,
0313  extinct forms. Falconer has given a striking INSTANCE of a similar fact, in an existing crocodile
0313  e than those of the sea; of which a striking INSTANCE has lately been observed in Switzerland. The
0314  same fact in geographical distribution; for INSTANCE, in the land shells and coleopterous insects
0315  racters from their distinct progenitors. For INSTANCE, it is just possible, if our fantail pigeons
0316  ied forms. Species of the genus Lingula, for INSTANCE, must have continuously existed by an unbrok
0321  re they have escaped severe competition. For INSTANCE, a single species of Trigonia, a great genus
0332  dered less distinct from each other. If, for INSTANCE, the genera a1, a5, a10, f8, m3, m6, m9, wer
0332  rizontal lines or geological formations, for INSTANCE, above No. VI., but none from beneath this l
0333  before the discovery of the fossils. If, for INSTANCE, we suppose the existing genera of the two f
0334  and succeeding faunas. I need give only one INSTANCE, namely, the manner in which the fossils of
0334  ain genera offer exceptions to the rule. For INSTANCE, mastodons and elephants, when arranged by D
0335  ote formations. Pictet gives as a well known INSTANCE, the general resemblance of the organic rema
0337  tinc this sort of progress. Crustaceans, for INSTANCE, not the highest in their own class, may hav
0338  never be capable of full proof. Seeing, for INSTANCE, that the oldest known mammals, reptiles, an
0347  itions peculiar in only a slight degree; for INSTANCE, small areas in the Old World could be point
0349  vertheless the naturalist in travelling, for INSTANCE, from north to south never fails to be struc
0351  species will be no uniform quantity. If, for INSTANCE, a number of species, which stand in direct
0355  it has been stocked. A volcanic island, for INSTANCE, upheaved and formed at the distance of a fe
0359  rt time, when dried floated much longer; for INSTANCE, ripe hazel nuts sank immediately, but when
0361  retain their vitality: peas and vetches, for INSTANCE, are killed by even a few days immersion in
0361  a far higher estimate. I have never seen an INSTANCE of nutritious seeds passing through the inte
0362  that earth sometimes adheres to them: in one INSTANCE I removed twenty two grains of dry argillace
0372  ditions of the areas; for if we compare, for INSTANCE, certain parts of South America with the sou
0388  t when a pond or stream is first formed, for INSTANCE, on a rising islet, it will be unoccupied: a
0390  is often extremely large. If we compare, for INSTANCE, the number of the endemic land shells in Ma
0392  ct to the inhabitants of remote islands. For INSTANCE, in certain islands not tenanted by mammals,
0392  s an appendage as any rudimentary organ, for INSTANCE, as the shrivelled wings under the soldered
0393  my search; as yet I have not found a single INSTANCE, free from doubt, of a terrestrial mammal (e
0397  not be overlooked. I will here give a single INSTANCE of one of the cases of difficulty. Almost al
0401  y occasional means of transport, a seed, for INSTANCE, of one plant having been brought to one isl
0401  th different sets of organisms: a plant, for INSTANCE, would find the best fitted ground more perf
0405  l largely determine their average range. For INSTANCE, two varieties of the same species inhabit A
0408  ow it is that under different latitudes, for INSTANCE in South America, the inhabitants of the pla
0413  sentence to give the characters common, for INSTANCE, to all mammals, by another those common to
0414  portant it becomes for classification. As an INSTANCE: Owen, in speaking of the dugong, says; The
0416  eable in the definition of whole groups. For INSTANCE, whether or not there is an open passage fro
0423  an artificial system; we are cautioned, for INSTANCE, not to class two varieties of the pine appl
0428  nearly similar circumstances, to inhabit for INSTANCE the three elements of land, air, and water,
0431  ctness of whole classes from each other, for INSTANCE, of birds from all other vertebrate animals,
0434  gether in the same order. We never find, for INSTANCE, the bones of the arm and forearm, or of the
0439  lly; and the wonderful fact of the jaws, for INSTANCE, of a crab retaining numerous characters, wh
0440  heir conditions of existence. We cannot, for INSTANCE, suppose that in the embryos of the vertebra
0440  ryonic resemblance. Cirripedes afford a good INSTANCE of this: even the illustrious Cuvier did not
0442  wth. But there is no obvious reason why, for INSTANCE, the wing of a bat, or the fin of a porpoise
0444  or later in life. It would not signify, for INSTANCE, to a bird which obtained its food best by h
```

Page ********************************(Key Word)**********************************

Page ***(Key Word)***

```
0444  s can only appear at corresponding ages, for  INSTANCE, peculiarities in the caterpillar, cocoon, o
0445  ome characteristic points of difference, for  INSTANCE, that of the width of mouth, could hardly be
0447  amilies or even classes. The fore limbs, for  INSTANCE, which served as legs in the parent species,
0449  een modified and obscured; we have seen, for  INSTANCE, that cirripedes can at once be recognised b
0450  are extremely common throughout nature. For  INSTANCE, rudimentary mammae are very general in the
0450  udimentary organs are extremely curious; for  INSTANCE, the presence of teeth in foetal whales, whi
0451  ary organs is often quite unmistakeable: for  INSTANCE there are beetles of the same genus (and eve
0473  to resist this conclusion when we look, for  INSTANCE, at the logger headed duck, which has wings
0473  characters in which they all agree? Why, for  INSTANCE, should the colour of a flower be more likel
0475  tincts; why the thrush of South America, for  INSTANCE, lines her nest with mud like our British sp
0480  rudimentary at this early age. The calf, for  INSTANCE, has inherited teeth, which never cut throug
0009  and healthy (of which I could give numerous  INSTANCES), yet having their reproductive system so s
0011  te distinct parts are very curious; and many  INSTANCES are given in Isidore Geoffroy St. Hilaire's
0011  lways accompanied by an elongated head. Some  INSTANCES of correlation are quite whimsical: thus ca
0046  ounc: but under any other point of view many  INSTANCES assuredly can be given. There is one point
0065  ce its discovery. In such cases, and endless  INSTANCES could be given, no one supposes that the fe
0093  o this, I could easily show by many striking  INSTANCES. I will give only one, not as a very striki
0095  election, exemplified in the above imaginary  INSTANCES, is open to the same objections which were
0111  ction of improved forms by man. Many curious  INSTANCES could be given showing how quickly new bree
0133  tion on the hair of our domestic quadrupeds.  INSTANCES could be given of the same variety being pr
0134  ditions of life must act. Again, innumerable  INSTANCES are known to every naturalist of species ke
0147  ts truth. I will not, however, here give any  INSTANCES, for I see hardly any way of distinguishing
0148  xamining cirripedes, and of which many other  INSTANCES could be given: namely, that when a cirripe
0157  of this fact I will give in illustration two  INSTANCES, the first which happen to stand on my list
0176  a state of nature. I have met with striking  INSTANCES of the rule in the case of varieties interm
0180  i have collected, I can give only one or two  INSTANCES of transitional habits and structures in cl
0183  ped structures. I will now give two or three  INSTANCES of diversified and of changed habits in the
0183  ubstances. Of diversified habits innumerable  INSTANCES could be given: I have often watched a tyra
0184  ied from that of their proper type. And such  INSTANCES do occur in nature. Can a more striking ins
0195  nce or not. In a former chapter I have given  INSTANCES of most trifling characters, such as the do
0196  nd with newly acquired habits. To give a few  INSTANCES to illustrate these latter remarks. If gree
0210  be preserved by natural selection. And such  INSTANCES of diversity of instinct in the same specie
0210  clusive good of others. One of the strongest  INSTANCES of an animal apparently performing an actio
0211  for the action of natural selection, as many  INSTANCES as possible ought to have been here given;
0213  c animals. A number of curious and authentic  INSTANCES could be given of the inheritance of all sh
0217  mistake. Nevertheless, I could give several  INSTANCES of various birds which have been known occa
0237  et it be remembered that we have innumerable  INSTANCES, both in our domestic productions and in th
0272  generation; and I have myself seen striking  INSTANCES of this fact. Gartner further admits that h
0296  time which its deposition has consumed. Many  INSTANCES could be given of beds only a few feet in t
0315  nd no doubt this has occurred in innumerable  INSTANCES) to fill the exact place of another species
0349  lity, and every continent offers innumerable  INSTANCES. Nevertheless the naturalist in travelling,
0349  ents of the American type. Innumerable other  INSTANCES could be given. If we look to the islands o
0354  d points of the earth's surface, can in many  INSTANCES be explained on the view of each species ha
0384  aving caused rivers to flow into each other.  INSTANCES, also, could be given of this having occurr
0397  ut being actually the same species. Numerous  INSTANCES could be given of this fact. I will give on
0401  go, and in a lesser degree in some analogous  INSTANCES, is that the new species formed in the sepa
0403  spread from island to island. In many other  INSTANCES, as in the several districts of the same co
0416  qual physiological importance. Any number of  INSTANCES could be given of the varying importance fo
0416  the classification of the Grasses. Numerous  INSTANCES could be given of characters derived from p
0419  strongly insisted on their arbitrary value.  INSTANCES could be given amongst plants and insects,
0427  ical. Amongst insects there are innumerable  INSTANCES: thus Linnaeus, misled by external appearan
0451  se with the mammae of male mammals, for many  INSTANCES are on record of these organs having become
0452  scent breathing organ or lung. Other similar  INSTANCES could be given. Rudimentary organs in the i
0483  in a fully developed state; and this in some  INSTANCES necessarily implies an enormous amount of m
0008  eriod of development of the embryo, or at the  INSTANT of conception. Geoffroy St. Hilaire's experi
0322  of each species depends. If we forget for an  INSTANT, that each species tends to increase inordin
0335  er, intermediate in age. But supposing for an  INSTANT, in this and other such cases, that the reco
0341  ir degenerate descendants. This cannot for an  INSTANT be admitted. These huge animals have become
0067  and the number of the species will almost  INSTANTANEOUSLY increase to any amount. The face of Natu
0202  ive hatred of the queen bee, which urges her  INSTANTLY to destroy the young queens her daughters a
0219  ntroduced a single slave (F. fusca), and she  INSTANTLY set to work, fed and saved the survivors; m
0222  them: for we have seen that they eagerly and  INSTANTLY seized the pupae of F. fusca, whereas they
0228  a long thick, square strip of wax: the bees  INSTANTLY began to excavate minute circular pits in i
0228  ged ridge, coloured with vermilion. The bees  INSTANTLY began on both sides to excavate little basi
0021  ails. The barb is allied to the carrier, but,  INSTEAD of a very long beak, has a very short and ve
0021  ntail has thirty or even forty tail feathers,  INSTEAD of twelve or fourteen, the normal number in
0092  ing flowers for the sake of collecting pollen  INSTEAD of nectar; and as pollen is formed for the s
0171  nal forms? Why is not all nature in confusion  INSTEAD of the species being, as we see them, well d
0186  uld be long toed corncrakes living in meadows  INSTEAD of in swamps; that there should be woodpecke
0209  close as not to be distinguished. If Mozart,  INSTEAD of playing the pianoforte at three years old
0213  d by retrievers; and a tendency to run round,  INSTEAD of at, a flock of sheep, by shepherd dogs. I
0213  gait; and another kind of wolf rushing round,  INSTEAD of at, a herd of deer, and driving them to a
0228  lt upon the festooned edge of a smooth basin,  INSTEAD of on the straight edges of a three sided py
0228  of ordinary cells. I then put into the hive,  INSTEAD of a thick, square piece of wax, a thin and
0241  uppose that the larger workmen had heads four  INSTEAD of three times as big as those of the smalle
0259  go together. There are certain hybrids which  INSTEAD of having, as is usual, an intermediate char
0295  ad been less than that of the glacial period,  INSTEAD of having been really far greater, that is e
0358  hists, old fossiliferous or other such rocks,  INSTEAD of consisting of mere piles of volcanic matt
0416  rnithorhynchus had been covered with feathers  INSTEAD of hair, this external and trifling characte
0423  necessity of classing varieties on a natural  INSTEAD of an artificial system; we are cautioned, f
0429  had each been represented by a dozen species  INSTEAD of a single one; but such richness in specie
0005  constructed organ; secondly, the subject of  INSTINCT, or the mental powers of animals; thirdly, F
0172  ction? What shall we say to so marvellous an  INSTINCT as that which leads the bee to make cells, w
0172  the two first heads shall be here discussed,  INSTINCT and Hybridism in separate chapters. On the a
0207  tations, that of Unity of Type. Chapter VII.  INSTINCT. Instincts comparable with habits; but diffe
0207  slave making ants. Hive bee, its cell making  INSTINCT. Difficulties on the theory of the Natural S
```

Page ***(Key Word)***

```
0207  or sterile insects. Summary. The subject of  INSTINCT  might have been worked into the previous cha
0207  ct separately, especially as so wonderful an  INSTINCT  as that of the hive bee making its cells wil
0207  e are concerned only with the diversities of  INSTINCT  and of the other mental qualities of animals
0207  class. I will not attempt any definition of  INSTINCT.  It would be easy to show that several dist
0207  erstands what is meant, when it is said that  INSTINCT  impels the cuckoo to migrate and to lay her
0208  could show that none of these characters of  INSTINCT  are universal. A little dose, as Pierre Hube
0208  al of the older metaphysicians have compared  INSTINCT  with habit. This comparison gives, I think,
0209  e between what originally was a habit and an  INSTINCT  becomes so close as not to be distinguished.
0209  least possible that slight modifications of  INSTINCT  might be profitable to a species; and if it
0209  g and continually accumulating variations of  INSTINCT  to any extent that may be profitable. It is
0209  t deviations of bodily structure. No complex  INSTINCT  can possibly be produced through natural sel
0210  ransitional gradations by which each complex  INSTINCT  has been acquired, for these could be found
0210  cept in Europe and North America, and for no  INSTINCT  being known amongst extinct species, how ver
0210  to instincts as to bodily organs. Changes of  INSTINCT  may sometimes be facilitated by the same spe
0210  etc.; in which case either one or the other  INSTINCT  might be preserved by natural selection. And
0210  election. And such instances of diversity of  INSTINCT  in the same species can be shown to occur in
0210  ructure, and conformably with my theory, the  INSTINCT  of each species is good for itself, but has
0212  rtainly do vary, for instance, the migratory  INSTINCT,  both in extent and direction, and in its to
0212  even probability, of inherited variations of  INSTINCT  in a state of nature will be strengthened by
0216  nd birds, their mother to fly away. But this  INSTINCT  retained by our chickens has become useless
0216  ve to discuss in my future work, namely, the  INSTINCT  which leads the cuckoo to lay her eggs in ot
0216  eggs in other birds' nests; the slave making  INSTINCT  of certain ants; and the comb making power o
0216  re immediate and final cause of the cuckoo's  INSTINCT  is, that she lays her eggs, not daily, but a
0217  e having been taken of the mistaken maternal  INSTINCT  of another bird, than by their own mother's
0217  s of this nature, I believe that the strange  INSTINCT  of our cuckoo could be, and has been, genera
0218  is perhaps explains the origin of a singular  INSTINCT  in the allied group of ostriches. For severa
0218  er; and these are hatched by the males. This  INSTINCT  may probably be accounted for by the fact of
0218  koo, at intervals of two or three days. This  INSTINCT,  however, of the American ostrich has not as
0219  ated, be not thus exterminated. Slave making  INSTINCT.  This remarkable instinct was first discover
0219  ated. Slave making instinct. This remarkable  INSTINCT  was first discovered in the Formica (Polyerg
0219  eless to have speculated how so wonderful an  INSTINCT  could have been perfected. Formica sanguinea
0220  the truth of so extraordinary and odious an  INSTINCT  as that of making slaves. Hence I will give
0223  nfirmation by me, in regard to the wonderful  INSTINCT  of making slaves. Let it be observed what a
0223  an they do in Switzerland. By what steps the  INSTINCT  of F. sanguinea originated I will not preten
0224  ifferent purpose of raising slaves. When the  INSTINCT  was once acquired, if carried out to a much
0224  tural selection increasing and modifying the  INSTINCT,  always supposing each modification to be of
0224  ves as is the Formica rufescens. Cell making  INSTINCT  of the Hive Bee. I will not enter on mi
0233  tion of slight modifications of structure or  INSTINCT,  each profitable to the individual under its
0234  humble bee, if a slight modification of her  INSTINCT  led her to make her waxen cells near togethe
0235  by inheritance its newly acquired economical  INSTINCT  to new swarms, which in their turn will have
0235  ection, cases, in which we cannot see how an  INSTINCT  could possibly have originated: cases, in wh
0235  late gradations are known to exist: cases of  INSTINCT  of apparently such trifling importance, that
0236  es: for these neuters often differ widely in  INSTINCT  and in structure from both the males and fer
0236  itute of wings and sometimes of eyes, and in  INSTINCT.  As far as instinct alone is concerned, the
0236  ometimes of eyes, and in instinct. As far as  INSTINCT  alone is concerned, the prodigious differenc
0237  ively acquired modifications of structure or  INSTINCT  to its progeny. It may well be asked how is
0238  ects: a slight modification of structure, or  INSTINCT,  correlated with the sterile condition of ce
0243  lection accumulating slight modifications of  INSTINCT  to any extent, in any useful direction. In s
0243  perfect and are liable to mistakes; that no  INSTINCT  has been produced for the exclusive good of
0459  gradations in the perfection of any organ or  INSTINCT,  which we may consider, either do now exist
0459  of each profitable deviation of structure or  INSTINCT.  The truth of these propositions cannot, I t
0460  tremely cautious in saying that any organ or  INSTINCT,  or any whole being, could not have arrived
0486  n we contemplate every complex structure and  INSTINCT  as the summing up of many contrivances, each
0202  ifficult, but we ought to admire the savage  INSTINCTIVE  hatred of the queen bee, which urges her t
0207  pose it is performed, is usually said to be  INSTINCTIVE.  But I could show that none of these chara
0208  notion of the frame of mind under which an  INSTINCTIVE  action is performed, but not of its origin
0211  in this manner, showing that the action was  INSTINCTIVE,  and not the result of experience. But as
0212  ear of any particular enemy is certainly an  INSTINCTIVE  quality, as may be seen in nestling birds,
0213  int, we should assuredly call these actions  INSTINCTIVE.  Domestic instincts, as they may be called
0215  le to doubt that the love of man has become  INSTINCTIVE  in the dog. All wolves, foxes, jackals, an
0215  e dog and cat which no doubt was originally  INSTINCTIVE  in them, in the same way as it is so plain
0215  n them, in the same way as it is so plainly  INSTINCTIVE  in young pheasants, though reared under a
0216  hickets: and this is evidently done for the  INSTINCTIVE  purpose of allowing, as we see in wild gro
0223  ves. Let it be observed what a contrast the  INSTINCTIVE  habits of F. sanguinea present with those
0472  no waste of pollen by our fir trees: at the  INSTINCTIVE  hatred of the queen bee for her own fertil
0209  ll, he might truly be said to have done so  INSTINCTIVELY.  But it would be the most serious error t
0211  and therefore probably the aphides do not  INSTINCTIVELY  excrete for the sole good of the ants. Al
0232  t of balance struck between many bees, all  INSTINCTIVELY  standing at the same relative distance fr
0008  ry! This is generally attributed to vitiated  INSTINCTS;  but how many cultivated plants display the
0172  tand the inimitable perfection. Thirdly, can  INSTINCTS  be acquired and modified through natural se
0207  hat of Unity of Type. Chapter VII. Instinct.  INSTINCTS  comparable with habits, but different in th
0207  with habits, but different in their origin.  INSTINCTS  graduated. Aphides and ants. Instincts vari
0207  igin. Instincts graduated. Aphides and ants.  INSTINCTS  variable. Domestic instincts, their origin.
0207  hides and ants. Instincts variable. Domestic  INSTINCTS,  their origin. Natural instincts of the cuc
0207  e. domestic instincts, their origin. Natural  INSTINCTS  of the cuckoo, ostrich, and parasitic bees.
0207  es on the theory of the Natural Selection of  INSTINCTS.  Neuter or sterile insects. Summary. The su
0208  several other points of resemblance between  INSTINCTS  and habits could be pointed out. As in repe
0208  ut. As in repeating a well known song, so in  INSTINCTS,  one action follows another by a sort of rh
0209  error to suppose that the greater number of  INSTINCTS  have been acquired by habit in one generati
0209  can be clearly shown that the most wonderful  INSTINCTS  with which we are acquainted, namely, those
0209  quired. It will be universally admitted that  INSTINCTS  are as important as corporeal structure for
0209  le to a species; and if it can be shown that  INSTINCTS  do vary ever so little, then I can see no d
0209  eve, that all the most complex and wonderful  INSTINCTS  have originated. As modifications of corpor
0209  y disuse, so I do not doubt it has been with  INSTINCTS.  But I believe that the effects of habit ar
0209  what may be called accidental variations of  INSTINCTS;  that is of variations produced by the same
0210  surprised to find, making allowance for the  INSTINCTS  of animals having been but little observed
```

instincts

age ***************************************(Key Word)***

```
210  ally gradations, leading to the most complex  INSTINCTS, can be discovered. The canon of Natura non
210  it saltum applies with almost equal force to   INSTINCTS as to bodily organs. Changes of instinct ma
210  litated by the same species having different   INSTINCTS at different periods of life, or at differe
211  each species tries to take advantage of the    INSTINCTS of others, as each takes advantage of the w
211  others. So again, in some few cases, certain    INSTINCTS cannot be considered as absolutely perfect;
211  passed over. As some degree of variation in     INSTINCTS under a state of nature, and the inheritanc
211  f space prevents me. I can only assert, that    INSTINCTS certainly do vary, for instance, the migrat
212  ise, through natural selection, to quite new    INSTINCTS. But I am well aware that these general sta
213  t these actions differ essentially from true    INSTINCTS. If we were to see one kind of wolf, when y
213  dly call these actions instinctive. Domestic    INSTINCTS, as they may be called, are certainly far l
213  ly far less fixed or invariable than natural    INSTINCTS; but they have been acted on by far less ri
213  nditions of life. How stongly these domestic    INSTINCTS, habits, and dispositions are inherited, an
214  ogs a tendency to hunt hares. These domestic    INSTINCTS, when thus tested by crossing, resemble nat
214  en thus tested by crossing, resemble natural    INSTINCTS, which in a like manner become curiously bl
214  and for a long period exhibit traces of the    INSTINCTS of either parent: for example, Le Roy descr
214  ght line to his master when called. Domestic    INSTINCTS are sometimes spoken of as actions which ha
215  nd long continued close confinement. Natural    INSTINCTS are lost under domestication: a remarkable
216  flight. Hence we may conclude, that domestic    INSTINCTS have been acquired and natural instincts ha
216  tic instincts have been acquired and natural    INSTINCTS have been lost partly by habit, and by man
216  ther. We shall, perhaps, best understand how    INSTINCTS in a state of nature have become modified b
216  king power of the hive bee: these two latter    INSTINCTS have generally, and most justly, been ranke
216  turalists as the most wonderful of all known    INSTINCTS. It is now commonly admitted that the more
218  e cuckoo; for these bees have not only their    INSTINCTS but their structure modified in accordance
223  onally reared would then follow their proper    INSTINCTS, and do what work they could. If their pres
224  bees working in a dark hive. Grant whatever    INSTINCTS you please, and it seems at first quite inc
224  n, i think, to follow from a few very simple    INSTINCTS. I was led to investigate this subject by M
227  onclude that if we could slightly modify the    INSTINCTS already possessed by the Melipona, and in t
227  of her old cocoons. By such modifications of    INSTINCTS in themselves not very wonderful, hardly mo
233  aduated succession of modified architectural    INSTINCTS, all tending towards the present perfect pl
235  s i believe, the most wonderful of all known    INSTINCTS, that of the hive bee, can be explained by
235  successive, slight modifications of simpler    INSTINCTS; natural selection having by slow degrees:
235  in the struggle for existence. No doubt many    INSTINCTS of very difficult explanation could be oppo
235  been acted on by natural selection: cases of    INSTINCTS almost identically the same in animals so r
238  e working and soldier neuters, with jaws and    INSTINCTS extraordinarily different: in Cryptocerus,
242  to civilised man. As ants work by inherited    INSTINCTS and by inherited tools or weapons, and not
242  ile, they would have intercrossed, and their    INSTINCTS and structure would have become blended. An
242  ould possibly have affected the structure or    INSTINCTS of the fertile members, which alone leave d
242  l more briefly I have attempted to show that    INSTINCTS vary slightly in a state of nature. No one
243  a state of nature. No one will dispute that    INSTINCTS are of the highest importance to each anima
243  ihilate it. On the other hand, the fact that    INSTINCTS are not always absolutely perfect and are l
243  but that each animal takes advantage of the    INSTINCTS of others; that the canon in natural histor
243  of natura non facit saltum is applicable to    INSTINCTS as well as to corporeal structure, and is p
243  gthened by some few other facts in regard to    INSTINCTS; as by that common case of closely allied,
243  of life, yet often retaining nearly the same    INSTINCTS. For instance, we can understand on the pri
244  it is far more satisfactory to look at such    INSTINCTS as the young cuckoo ejecting its foster bro
244  pillars, not as specially endowed or created    INSTINCTS, but as small consequences of one general l
459  elieve than that the more complex organs and    INSTINCTS should have been perfected, not by means su
459  , each good of its kind, that all organs and    INSTINCTS are, in ever so slight a degree, variable,
474  ong continued natural selection. Glancing at    INSTINCTS, marvellous as some are, they offer no grea
474  animals of the same class with their several    INSTINCTS. I have attempted to show how much light th
474  doubt sometimes comes into play in modifying    INSTINCTS; but it certainly is not indispensable, as
475  s of life, yet should follow nearly the same    INSTINCTS; why the thrush of South America, for insta
475  mud like our British species. On the view of    INSTINCTS having been slowly acquired through natural
475  natural selection we need not marvel at some    INSTINCTS being apparently not perfect and liable to
475  perfect and liable to mistakes, and at many    INSTINCTS causing other animals to suffer. If species
167  parents. But whenever we have the means of    INSTITUTING a comparison, the same laws appear to have
248  d to the very same species. It is also most    INSTRUCTIVE to compare, but I have not space here to e
188  ed by transparent membrane, into an optical    INSTRUMENT as perfect as is possessed by any member of
188  g the eye to a telescope. We know that this    INSTRUMENT has been perfected by the long continued ef
188  n? if we must compare the eye to an optical    INSTRUMENT, we ought in imagination to take a thick la
189  mage. We must suppose each new state of the    INSTRUMENT to be multiplied by the million; and each t
189  nd may we not believe that a living optical    INSTRUMENT might thus be formed as superior to one of
202  remote progenitor as a boring and serrated    INSTRUMENT, like that in so many members of the same g
242  not by acquired knowledge and manufactured    INSTRUMENTS, a perfect division of labour could be eff
028  gin of domestic pigeons at some, yet quite    INSUFFICIENT, length; because when I first kept pigeons
242  ussed this case, at some little but wholly    INSUFFICIENT length, in order to show the power of natu
415  e even when all taken together they appear    INSUFFICIENT to separate Cnestis from Connarus. To give
389  , i think, explain all the facts in regard to    INSULAR productions. In the following remarks I shal
392  elled wings under the soldered elytra of many    INSULAR beetles. Again, islands often possess trees
181  useful to its possessor. Nor can I see any    INSUPERABLE difficulty in further believing it impossibl
187  ould be formed by natural selection, though    INSUPERABLE by our imagination, can hardly be consider
236  l difficulty, which at first appeared to me    INSUPERABLE, and actually fatal to my whole theory. I
237  election. This difficulty, though appearing    INSUPERABLE, is lessened, or, as I believe, disappears
293  the average duration of specific forms. But    INSUPERABLE difficulties, as it seems to me, prevent u
305  kable: but I cannot see that it would be an    INSUPERABLE difficulty on my theory, unless it could l
407  ve descended from the same parents, are not    INSUPERABLE. And we are led to this conclusion, which
407  n, i do not think that the difficulties are    INSUPERABLE; though they often are in this case, and i
408  ter productions. If the difficulties be not    INSUPERABLE in admitting that in the long course of ti
459  iculty, though appearing to our imagination    INSUPERABLY great, cannot be considered real if we adm
249  ve castrated his hybrids, and this would have    INSURED in each generation a cross with the pollen f
251  ert asserts that a hybrid from Calceolaria    INTEGRIFOLIA and plantaginea, species most widely dissi
464  as to be utterly inappreciable by the human    INTELLECT. The number of specimens in all our museums
188  long continued efforts of the highest human    INTELLECTS; and we naturally infer that the eye has be
188  right to assume that the Creator works by    INTELLECTUAL powers like those of man? If we must compa
345  nimals of the same class, the fact will be    INTELLIGIBLE. The succession of the same types of struc
353  of every kind have had on distribution, is    INTELLIGIBLE only on the view that the great majority o
```

age ***************************************(Key Word)***

```
0410  al relations throughout time and space are INTELLIGIBLE; for whether we look to the forms of li*
0436  , as well as their intimate structure, are INTELLIGIBLE on the view that they consist of metamor
0456  , all the great facts in Morphology become INTELLIGIBLE, whether we look to the same pattern dis
0457  of rudimentary organs in classification is INTELLIGIBLE, on the view that an arrangement is only
0470  having been independently created, but are INTELLIGIBLE if all species first existed as varietie
0471  strictly correct, is on this theory simply INTELLIGIBLE. We can plainly see why nature is prodig
0476  ntata in America, and other such cases, is  INTELLIGIBLE, for within a confined country, the rece
0478  often falling in between recent groups, is  INTELLIGIBLE on the theory of natural selection with
0479  mens, and pistils of a flower, is likewise  INTELLIGIBLE on the view of the gradual modification
0102  ly stop his work. But when many men, without INTENDING to alter the breed, have a nearly common
0215  work, as each man tries to procure, without INTENDING to improve the breed, dogs which will sta
0034  l animals, is more important. Thus, a man who INTENDS keeping pointers naturally tries to get as
0378  tropics at the period when the cold was most INTENSE, when arctic forms had migrated some twent
0202  ginally adapted to cause galls subsequently  INTENSIFIED, we can perhaps understand how it is tha
0024  domesticating several species, but that he  INTENTIONALLY or by chance picked out extraordinarily
0189  we must suppose that there is a power always INTENTLY watching each slight accidental alteration
0026  are, should yield offspring perfectly fertile, INTER se, seems to me rash in the extreme. From t
0253  ent, and in one single instance they have bred INTER se. This was effected by Mr. Eyton, who ra*
0266  itions of life. When hybrids are able to breed INTER se, they transmit to their offspring from g
0340  er great geographical changes, permitting much INTER migration, the feebler will yield to the mo
0371  ge, since rendered impassable by cold, for the INTER migration of their inhabitants. During the
0071  ntercrossing, and the ill effects of close    INTERBREEDING, probably come into play in some of the
0096  fspring; and on the other hand, that close    INTERBREEDING diminishes vigour and fertility; that t
0245  n degree, not universal, affected by close    INTERBREEDING, removed by domestication. Laws governi
0248  y an independent cause, namely, from close    INTERBREEDING. I have collected so large a body of fa
0249  large a body of facts, showing that close    INTERBREEDING lessens fertility, and, on the other ha
0249  may, I believe, be accounted for by close    INTERBREEDING having been avoided. Now let us turn to
0252  ther, and the injurious influence of close    INTERBREEDING is thus prevented. Any one may readily
0253  s, so as to avoid the ill effects of close    INTERBREEDING. On the contrary, brothers and sisters
0267  e even with hermaphrodites; and that close    INTERBREEDING continued during several generations the
0368  satisfactory evidence with respect to this    INTERCALATED slightly warmer period, since the Glacia
0402  h other's territory, when put into free       INTERCOMMUNICATION. Undoubtedly if one species has any a*
0404  at at some former period there has been       INTERCOMMUNICATION or migration between the two regions.
0101  in fact, all hermaphrodites do occasionally   INTERCROSS with other individuals, the difference be
0101  egetable and animal kingdoms, an occasional   INTERCROSS with a distinct individual is a law of na*
0104  . if there exist organic beings which never   INTERCROSS, uniformity of character can be retained
0355  ated. With those organic beings which never   INTERCROSS (if such exist), the species, on my theory
0356  tually unite for each birth, or which often   INTERCROSS, I believe that during the slow process o*
0005  pecies and the fertility of varieties when    INTERCROSSED; and fourthly, the imperfection of the Ge
0242  for had they been fertile, they would have    INTERCROSSED, and their instincts and structure would
0245  ained by naturalists is that species, when    INTERCROSSED, have been specially endowed with the qu*
0246  roduction in a perfect condition, yet when    INTERCROSSED they produce either few or no offspring.*
0246  o have descended from common parents, when    INTERCROSSED, and likewise the fertility of their mon*
0247  other species are really so sterile, when    INTERCROSSED, as Gartner believes. It is certain, on *
0268  her most closely, are utterly sterile when    INTERCROSSED. Several considerations, however, render
0269  same species were invariably fertile when    INTERCROSSED. But it seems to me impossible to resist
0270  ties of the same species of Verbascum when    INTERCROSSED produce less seed, than do either colour*
0460  erences in the reproductive systems of the    INTERCROSSED species. We see the truth of this conclu*
0460  he mother. The fertility of varieties when    INTERCROSSED and of their mongrel offspring cannot be
0481  it cannot be maintained that species when    INTERCROSSED are invariably sterile, and varieties inv
0080  es. Sexual Selection. On the generality of    INTERCROSSES between individuals of the same species.
0103  birth, we must not overrate the effects of    INTERCROSSES in retarding natural selection; for I ca*
0104  we have reason to believe that occasional    INTERCROSSES take place with all animals and with all
0104  nd thus, in the long run, the influence of    INTERCROSSES even at rare intervals, will be great. I*
0104  manner in relation to the same conditions.   INTERCROSSES, also, with the individuals of the same s
0014  essary, in order to prevent the effects of    INTERCROSSING, that only a single variety should be tu
0015  kept in a considerable body, so that free    INTERCROSSING might check, by blending together, any s
0043  ex. In some cases, I do not doubt that the    INTERCROSSING of species, aboriginally distinct, has p
0043  ve once been established, their occasional    INTERCROSSING, with the aid of selection, has, no doub
0070  ould add that the good effects of frequent    INTERCROSSING, and the ill effects of close interbreed
0080  unfavourable to Natural Selection, namely,   INTERCROSSING, isolation, number of individuals. Slow
0091  end where they met; but to this subject of   INTERCROSSING we shall soon have to return. I may add*
0096  en modification in their structure. On the   INTERCROSSING of individuals. I must here introduce a
0099  general law of good being derived from the   INTERCROSSING of distinct individuals of the same spe*
0102  selects for some definite object, and free   INTERCROSSING will wholly stop his work. But when many
0102  es in the several districts, there will be   INTERCROSSING with the other individuals of the same s
0102  s of each. And in this case the effects of   INTERCROSSING can hardly be counterbalanced by natural
0103  sensibly from one district to another. The   INTERCROSSING will most affect those animals which uni
0103  aintain itself in a body, so that whatever   INTERCROSSING took place would be chiefly between the
0103  ants of the same variety, as the chance of   INTERCROSSING with other varieties is thus lessened. E
0103  the same kind preferring to pair together.   INTERCROSSING plays a very important part in nature in
0107  iduals of the same species on each island:   INTERCROSSING on the confines of the range of each spe
0108  ess will often be greatly retarded by free   INTERCROSSING. Many will exclaim that these several ca
0254  re aboriginal species, since commingled by   INTERCROSSING. On this view, the aboriginal species mu
0254  ooking to all the ascertained facts on the   INTERCROSSING of plants and animals, it may be conclu*
0314  ng of a beneficial nature, on the power of   INTERCROSSING, on the rate of breeding, on the slowly
0356  cies will have been kept nearly uniform by   INTERCROSSING; so that many individuals will have gone
0051  k at individual differences, though of small INTEREST to the systematist, as of high importance f
0382  every land, which serve as a record, full of INTEREST to us, of the former inhabitants of the sur
0450  ot early period. Embryology rises greatly in INTEREST, when we thus look at the embryo as a pictu
0485  ents of natural history will rise greatly in INTEREST. The terms used by naturalists of affinity,
0071  e instance, which, though a simple one, has  INTERESTED me. In Staffordshire, on the estate of a *
0016  point, if it could be cleared up, would be   INTERESTING; if, for instance, it could be shown that
0049  ies well deserve consideration; for several  INTERESTING lines of argument, from geographical dist
0053  retical considerations, I thought that some  INTERESTING results might be obtained in regard to th
0191  can thus, as I infer from Professor Owen's   INTERESTING description of these parts, understand th
```

```
221  from one nest to another, and it was a most  INTERESTING  spectacle to behold the masters carefully
228  f about the diameter of a cell. It was most   INTERESTING  to me to observe that wherever several bee
242  ry has encountered. The case, also, is very   INTERESTING, as it proves that with animals, as with p
360  ts often floating longer than the small, is   INTERESTING; as plants with large seeds or fruit could
397  ewhere. Dr. Aug. A. Gould has given several   INTERESTING  cases in regard to the land shells of the
400  e displayed on a small scale, yet in a most   INTERESTING  manner, within the limits of the same arch
434  eneral name of Morphology. This is the most   INTERESTING  department of natural history, and may be
435  been expressly admitted by Owen in his most   INTERESTING  work on the Nature of Limbs. On the ordina
486  thus view each organic being, how far more   INTERESTING, I speak from experience, will the study o
486  sed by man will be a far more important and   INTERESTING  subject for study than one more species ad
489  tend to progress towards perfection. It is   INTERESTING  to contemplate an entangled bank, clothed
034  stic breeds are valued by the negroes of the  INTERIOR  of Africa who have not associated with Europ
173  these representative species often meet and   INTERLOCK; and as the one becomes rarer and rarer, th
025  as a white rump (the Indian sub species, C.   INTERMEDIA  of Strickland, having it bluish): the tail
019  ssing we can get only forms in some degree   INTERMEDIATE  between their parents; and if we account f
020  ; but that a race could be obtained nearly   INTERMEDIATE  between two extremely different races or s
020  task becomes apparent. Certainly, a breed   INTERMEDIATE  between two very distinct breeds could not
029  r, and knowing no more than he does of the   INTERMEDIATE  links in the long lines of descent, yet ad
047  forms, or are so closely linked to them by   INTERMEDIATE  gradations, that naturalists do not like t
047  unite two forms together by others having   INTERMEDIATE  characters, he treats the one as a variety
047  r, even when they are closely connected by   INTERMEDIATE  links; nor will the commonly assumed hybri
047  the commonly assumed hybrid nature of the   INTERMEDIATE  links always remove the difficulty. In ver
050  d as a variety of another, not because the   INTERMEDIATE  links have actually been found, but becaus
050  on the other hand, they are united by many   INTERMEDIATE  links, and it is very doubtful whether the
051  which case he can hardly hope to find the   INTERMEDIATE  links between his doubtful forms, he will
056  ked varieties; and in those cases in which   INTERMEDIATE  links have not been found between doubtful
058  cies, except, firstly, by the discovery of   INTERMEDIATE  linking forms, and the occurrence of such
059  ranked as varieties, notwithstanding that   INTERMEDIATE  linking forms have not been discovered; bu
112  become greater, the inferior animals with   INTERMEDIATE  characters, being neither very swift nor v
121  constitution, and structure. Hence all the   INTERMEDIATE  forms between the earlier and later states
123  on diverging in different directions. The   INTERMEDIATE  species, also (and this is a very importan
124  nct and unknown, it will be in some degree   INTERMEDIATE  in character between the two groups descen
124  the new species (F14) will not be directly   INTERMEDIATE  between them, but rather between types of
124  now living, yet are often, in some degree,   INTERMEDIATE  in character between existing groups; and
128  o much extinction of the less improved and   INTERMEDIATE  forms of life. On these principles, I beli
162  e catalogue, also, could be given of forms   INTERMEDIATE  between two other forms, which themselves
162  racters of the other, so as to produce the   INTERMEDIATE  form. But the best evidence is afforded by
174  ed there in a fossil condition. But in the   INTERMEDIATE  region, having intermediate conditions of
174  on. But in the intermediate region, having   INTERMEDIATE  conditions of life, why do we not now find
174  life, why do we not now find close linking   INTERMEDIATE  varieties? This difficulty for a long time
174  parately formed without the possibility of   INTERMEDIATE  varieties existing in the intermediate zon
174  of intermediate varieties existing in the   INTERMEDIATE  zones. By changes in the form of the land
176  rge area, and a third variety to a narrow   INTERMEDIATE  zone. The intermediate variety, consequent
176  variety to a narrow intermediate zone. The   INTERMEDIATE  variety, consequently, will exist in lesse
176  ances of the rule in the case of varieties   INTERMEDIATE  between well marked varieties in the genus
176  . Wollaston, that generally when varieties   INTERMEDIATE  between two other forms occur, they are mu
176  nect, then, I think, we can understand why   INTERMEDIATE  varieties should not endure for very long
176  e numbers; and in this particular case the   INTERMEDIATE  form would be eminently liable to the inro
176  reas, will have a great advantage over the   INTERMEDIATE  variety, which exists in smaller numbers i
177  exists in smaller numbers in a narrow and   INTERMEDIATE  zone. For forms existing in larger numbers
177  more quickly than the small holders on the   INTERMEDIATE  narrow, hilly tract; and consequently the
177  thout the interposition of the supplanted,   INTERMEDIATE  hill variety. To sum up, I believe that sp
177  esent an inextricable chaos of varying and   INTERMEDIATE  links: firstly, because new varieties are
178  k as representative species. In this case,   INTERMEDIATE  varieties between the several representati
178  nt portions of a strictly continuous area,   INTERMEDIATE  varieties will, it is probable, at first h
178  probable, at first have been formed in the   INTERMEDIATE  zones, but they will generally have had a
178  rally have had a short duration. For these   INTERMEDIATE  varieties will, from reasons already assig
178  e of acknowledged varieties), exist in the   INTERMEDIATE  zones in lesser numbers than the varieties
178  tend to connect. From this cause alone the   INTERMEDIATE  varieties will be liable to accidental ext
179  all time, if my theory be true, numberless   INTERMEDIATE  varieties, linking most closely all the sp
179  d, to exterminate the parent forms and the   INTERMEDIATE  links. Consequently evidence of their form
179  oup carnivorous animals exist having every   INTERMEDIATE  grade between truly aquatic and strictly t
203  are not linked together by a multitude of   INTERMEDIATE  gradations, partly because the process of
203  upplanting and extinction of preceding and   INTERMEDIATE  gradations. Closely allied species, now li
203  in two districts of a continuous area, an   INTERMEDIATE  variety will often be formed, fitted for a
203  ariety will often be formed, fitted for an   INTERMEDIATE  zone; but from reasons assigned, the inter
203  diate zone; but from reasons assigned, the   INTERMEDIATE  variety will usually exist in lesser numbe
204  e a great advantage over the less numerous   INTERMEDIATE  variety, and will thus generally succeed i
204  ction. In the cases in which we know of no   INTERMEDIATE  or transitional states, we should be very
204  or the homologies of many organs and their   INTERMEDIATE  states show that wonderful metamorphoses i
225  ed by Pierre Huber. The Melipona itself is   INTERMEDIATE  in structure between the hive and humble b
229  een the basins, by stopping work along the   INTERMEDIATE  planes or planes of intersection. Consider
230  have tried is easily done) into its proper   INTERMEDIATE  plane, and thus flatten it. From the exper
232  maginary spheres, they can build up a wall   INTERMEDIATE  between two adjoining spheres; but, as far
233  pinc spheres or cylinders, and building up   INTERMEDIATE  planes. It is even conceivable that an ins
235  ssibly have originated; cases, in which no   INTERMEDIATE  gradations are known to exist; cases of in
239  e and small are numerous, with those of an   INTERMEDIATE  size scanty in numbers; Formica flava has
240  s larger and smaller workers, with some of   INTERMEDIATE  size; and, in this species, as Mr. F. Smit
240  assert so positively, that the workers of   INTERMEDIATE  size have their ocelli in an exactly inter
240  diate size have their ocelli in an exactly   INTERMEDIATE  condition. So that we here have two bodies
240  n, yet connected by some few members in an   INTERMEDIATE  condition. I may digress by adding, that i
241  s which generated them; until none with an   INTERMEDIATE  structure were produced. Thus, as I believ
259  s which instead of having, as is usual, an   INTERMEDIATE  character between their two parents, alway
259  so again amongst hybrids which are usually   INTERMEDIATE  in structure between their parents, except
279  f the Geological Record. On the absence of   INTERMEDIATE  varieties at the present day. On the natur
279  the present day. On the nature of extinct   INTERMEDIATE  varieties; on their number. On the vast la
```

Page ***************************************(Key Word)**

```
0279  f geological formations. On the absence of  INTERMEDIATE  varieteies in any one formation. On the s
0279  oisture. I endeavoured, also, to show that   INTERMEDIATE  varieties, from existing in lesser number
0279  t. the main cause, however, of innumerable   INTERMEDIATE  links not now occuring everywhere through
0280  n an enormous scale, so must the number of   INTERMEDIATE  varieties, which have formerly existed on
0280  l formation and every stratum full of such   INTERMEDIATE  links? Geology assuredly does not reveal
0280  hould always be borne in mind what sort of   INTERMEDIATE  forms must, on my theory, have formerly e
0280  avoid picturing to myself, forms directly    INTERMEDIATE  between them. But this is a wholly false
0280  alse view; we should always look for forms   INTERMEDIATE  between each species and a common but unk
0280  m the rock pigeon; if we possessed all the   INTERMEDIATE  varieties which have ever existed, we sho
0280  ; but we should have no varieties directly   INTERMEDIATE  between the fantail and pouter; none, for
0281  o suppose that links ever existed directly   INTERMEDIATE  between them, but between each and an unk
0281  time we had a nearly perfect chain of the   INTERMEDIATE  links. It is just possible by my theory,
0281  orse from a tapir; and in this case direct   INTERMEDIATE  links will have existed between them. But
0282  of each great class. So that the number of   INTERMEDIATE  and transitional links, between all livin
0296  on would not probably include all the fine   INTERMEDIATE  gradations which must on my theory have e
0297  enabled to connect them together by close    INTERMEDIATE  gradations. And this from the reasons jus
0297  an underlying bed; even if A were strictly   INTERMEDIATE  between B and C, it would simply be ranke
0297  connected with either one or both forms by   INTERMEDIATE  varieties. Nor should it be forgotten, as
0297  t might not at all necessarily be strictly   INTERMEDIATE  between them in all points of structure.
0298  tion, two forms can seldom be connected by   INTERMEDIATE  varieties and thus proved to be the same
0298  bled to connect species by numerous, fine,   INTERMEDIATE, fossil links, by asking ourselves whethe
0299  ist discovering in a fossil state numerous   INTERMEDIATE  gradations; and such success seems to me
0299  onnecting them together by numerous, fine,   INTERMEDIATE  varieties; and this not having been effec
0306  ey do not present characters in any degree   INTERMEDIATE  between them. If, moreover, they had been
0330  es or group of species being considered as   INTERMEDIATE  between living species or groups. If by t
0330  is meant that an extinct form is directly    INTERMEDIATE  in all its characters between two living
0332  living genera of three families together,    INTERMEDIATE  in character, would be justified, as they
0332  character, would be justified, as they are   INTERMEDIATE, not directly, but only by a long and cir
0333  nct genera are often in some slight degree   INTERMEDIATE  in character between their modified desce
0333  reat period in the earth's history will be   INTERMEDIATE  in general character between that which p
0334  hence they could hardly fail to be nearly    INTERMEDIATE  in character between the forms of life ab
0334  a of each geological period undoubtedly is   INTERMEDIATE  in character, between the preceding and s
0334  at once recognised by palaeontologists as    INTERMEDIATE  in character between those of the overlyi
0334  but each fauna is not necessarily exactly   INTERMEDIATE, as unequal intervals of time have elapse
0334  fauna of each period as a whole is nearly    INTERMEDIATE  in character between the preceding and su
0334  r the most recent; nor are those which are   INTERMEDIATE  in character, intermediate in age. But su
0334  those which are intermediate in character,   INTERMEDIATE  in age. But supposing for an instant, in
0335  tatement, that the organic remains from an   INTERMEDIATE  formation are in some degree intermediate
0335  intermediate formation are in some degree   INTERMEDIATE  in character, is the fact, insisted on by
0336  pter, in any one or two formations all the   INTERMEDIATE  varieties between the species which appea
0344  tly, it displays characters in some degree   INTERMEDIATE  between groups now distinct; for the more
0345  vergent. Extinct forms are seldom directly   INTERMEDIATE  between existing forms; but are intermedi
0345  termediate between existing forms; but are   INTERMEDIATE  only by a long and circuitous course throu
0345  : we can clearly see why the remains of an   INTERMEDIATE  formation are intermediate in character.
0345  e remains of an intermediate formation are   INTERMEDIATE  in character. The inhabitants of each suc
0375  genera, found in Australia, but not in the   INTERMEDIATE  torrid regions. In the admirable Introduc
0376  to Europe, but have not been found in the    INTERMEDIATE  tropical seas. It should be observed that
0388  and having subsequently become extinct in   INTERMEDIATE  regions. But the wide distribution of fres
0399  e long ago partially stocked from a nearly   INTERMEDIATE  though distant point, namely from the ant
0407  area, and then have become extinct in the   INTERMEDIATE  tracts, I think the difficulties in belie
0410  to our not having as yet discovered in an   INTERMEDIATE  deposit the forms which are therein absen
0410  y the species having become extinct in the  INTERMEDIATE  tracts. Both in time and space, species ar
0421  uch isolated, will still occupy its proper   INTERMEDIATE  position; for F originally was intermediat
0421  ntermediate position; for F originally was   INTERMEDIATE  in character between A and I, and the seve
0422  rld; and if all extinct languages, and all   INTERMEDIATE  and slowly changing dialects, had to be ir
0426  forms are connected together by a chain of   INTERMEDIATE  groups, we may at once infer their commun
0429  n present characters in some slight degree   INTERMEDIATE  between existing groups. A few old and int
0429  ate between existing groups. A few old and   INTERMEDIATE  parent forms having occasionally transmitt
0430  h will have had a character in some degree   INTERMEDIATE  with respect to all existing Marsupials; o
0432  arge groups of modified descendants. Every   INTERMEDIATE  link between these eleven genera and thers
0432  era and there primordial parent, and every  INTERMEDIATE  link in each branch and sub branch of the
0462  or by the extinction of the species in the  INTERMEDIATE  regions. It cannot be denied that we are a
0462  atural selection an interminable number of   INTERMEDIATE  forms must have existed, linking together
0463  have no just right to expect often to find  INTERMEDIATE  varieties in the intermediate zone. For we
0463  ften to find intermediate varieties in the  INTERMEDIATE  zone. For we have reason to believe that o
0463  lowly effected. I have also shown that the   INTERMEDIATE  varieties which will at first probably ex
0463  which will at first probably exist in the   INTERMEDIATE  zones, will be liable to be supplanted by
0463  ed and improved at a quicker rate than the  INTERMEDIATE  varieties, which exist in lesser numbers;
0463  which exist in lesser numbers; so that the  INTERMEDIATE  varieties will, in the long run, be suppla
0464  , unless we likewise possessed many of the  INTERMEDIATE  links between their past or parent and pre
0464  wo species are unknown, if any one link or  INTERMEDIATE  variety be discovered, it will simply be c
0464  al, both causes rendering the discovery of  INTERMEDIATE  links less likely. Local varieties will no
0470  d exterminate the older, less improved and  INTERMEDIATE  varieties; and thus species are rendered t
0475  ins of each formation being in some degree   INTERMEDIATE  in character between the fossils in the fo
0476  ve and below, is simply explained by their  INTERMEDIATE  position in the chain of descent. The gran
0476  ings, falling either into the same or into  INTERMEDIATE  groups, follows from the living and the ex
0476  r with its early descendants will often be  INTERMEDIATE  in character in comparison with its later
0476  l is, the oftener it stands in some degree  INTERMEDIATE  between existing and allied groups. Recent
0481  ny great change of which we do not see the  INTERMEDIATE  steps. The difficulty is the same as that
0485  , between any two forms, if not blended by   INTERMEDIATE  gradations, are looked at by most naturali
0485  ved, to be connected at the present day by   INTERMEDIATE  gradations, whereas species were formerly
0485  consideration of the present existence of   INTERMEDIATE  gradations between any two forms, we shall
0370  r land, and to the consequent freedom for   INTERMIGRATION  under a more favourable climate, I attrib
0342  a large extent explain why we do not find   INTERMINABLE  varieties, connecting together all the ext
0462  . as on the theory of natural selection an   INTERMINABLE  number of intermediate forms must have exi
0173  but if we compare these species where they  INTERMINGLE, they are generally as absolutely distinct
```

Page ***************************************(Key Word)**

intermingle

age **********************************(Key Word)**********************************

```
279 f our palaeontological collections. On the INTERMITTENCE of geological formations. On the absence
108 e. 1 further believe, that this very slow, INTERMITTENT action of natural selection accords perfec
179 mpt to show, in an extremely imperfect and INTERMITTENT record. On the origin and transitions of o
290 tions of each region are almost invariably INTERMITTENT; that is, have not followed each other in
292 record will almost necessarily be rendered INTERMITTENT. I feel much confidence in the truth of th
295 mations in any country, has generally been INTERMITTENT in its accumulation. When we see, as is so
315 sarily accumulated at wide and irregularly INTERMITTENT intervals; consequently the amount of orga
465 as new species. Most formations have been INTERMITTENT in their accumulation; and their duration,
038 ernal characters and relatively so slight in INTERNAL parts or organs. Man can harcly select, or o
038 ible; and indeed he rarely cares for what is INTERNAL. He can never act by selection, excepting on
045 re not many men who will laboriously examine INTERNAL and important organs, and compare them in ma
083 be useful to any being. She can act on every INTERNAL organ, on every shade of constitutional diff
146 but in regard to the differences both in the INTERNAL and external structure of the seeds, which a
168 n external parts sometimes affect softer and INTERNAL parts. When one part is largely developed, p
417 iles, as an approach in structure in any one INTERNAL and and important organ. The importance, for
437 tions. In the vertebrata, we see a series of INTERNAL vertebrae bearing certain processes and mean
177 close contact with each other, without the INTERPOSITION of the supplanted, intermediate hill vari
208 another by a sort of rhythm; if a person be INTERRUPTED in a song, or in repeating anything by rot
295 hat the process of deposition has been much INTERRUPTED, as a change in the currents of the sea an
301 at each great period of subsidence would be INTERRUPTED by oscillations of level, and that slight
311 ritten, being more or less different in the INTERRUPTED succession of chapters, may represent the
336 cumulation of each formation has often been INTERRUPTED, and as long blank intervals have interven
353 d within recent geological times, must have INTERRUPTED or rendered discontinuous the formerly con
462 r wide migration by many means. A broken or INTERRUPTED range may often be accounted for by the ex
226 f wax between the spheres which thus tend to INTERSECT. Hence each cell consists of an outer spher
227 at she always describes her spheres so as to INTERSECT largely; and then she unites the points of
234 make her waxen cells near together, so as to INTERSECT a little; for a wall in common even to two
225 earness to each other, that they would have INTERSECTED or broken into each other, if the spheres
228 they formed a part, the rims of the basins INTERSECTED or broke into each other. As soon as this
226 her and parallel layer; then, if planes of INTERSECTION between the several spheres in both layers
227 largely; and then she unites the points of INTERSECTION by perfectly flat surfaces. We have furthe
227 r hexagonal prisms have been formed by the INTERSECTION of adjoining spheres in the same layer, sh
228 build up flat walls of wax on the lines of INTERSECTION between the basins, so that each hexagonal
229 dge, exactly along the planes of imaginary INTERSECTION between the basins on the opposite sides o
229 along the intermediate planes or planes of INTERSECTION. Considering how flexible thin wax is, I d
231 oper position, that is, along the plane of INTERSECTION between two adjoining spheres. I have seve
232 ing up, or leaving ungnawed, the planes of INTERSECTION between these spheres. It was really curio
233 and from each other, strike the planes of INTERSECTION, and so make an isolated hexagon: but I am
235 p and excavate the wax along the planes of INTERSECTION. The bees, of course, no more knowing that
235 l, have migrated across the vast and broken INTERSPACE. The great and striking influence which bar
361 earth are very frequently enclosed in their INTERSTICES and behind them, so perfectly that not a p
375 und, which have not been discovered in the INTERTROPICAL parts of Africa. On the Himalaya, and on
376 isphere, and on the mountain ranges of the INTERTROPICAL regions, are not arctic, but belong to th
378 hern and southern temperate zones into the INTERTROPICAL regions, and some even crossed the equato
379 ding forms which settled themselves on the INTERTROPICAL mountains, and in the southern hemisphere
380 of the same kind must have occurred on the INTERTROPICAL mountains: no doubt before the Glacial pe
380 mountain is an island on the land; and the INTERTROPICAL mountains before the Glacial period must
380 emperate zones and on the mountains of the INTERTROPICAL regions. Very many difficulties remain to
477 erate zones, though separated by the whole INTERTROPICAL ocean. Although two areas may present the
064 , bringing forth three pair of young in this INTERVAL; if this be so, at the end of the fifth cent
118 he next thousand generations. And after this INTERVAL, variety al is supposed in the diagram to ha
211 attendance during several hours. After this INTERVAL, I felt sure that the aphides would want to
288 ation conformably covered, after an enormous INTERVAL of time, by another and later formation, wit
288 ut the underlying bed having suffered in the INTERVAL any wear and tear, seem explicable only on t
304 inct order had been suddenly produced in the INTERVAL between the latest secondary and earliest te
307 g as, or probably far longer than, the whole INTERVAL from the Silurian age to the present day; an
329 one region often corresponding with a blank INTERVAL in the other, and if in both regions the spe
352 oints so distant from each other, or with an INTERVAL of such a nature, that the space could not b
362 zard for 12 or even 18 hours. A bird in this INTERVAL might easily be blown to the distance of 500
362 and owls bolt their prey whole, and after an INTERVAL of from twelve to twenty hours, disgorge pel
362 , storks, and pelicans; these birds after an INTERVAL of many hours, either rejected the seeds in
031 y a connoisseur; this is done three times at INTERVALS of months, and the sheep are each time mark
066 old, during each generation or at recurrent INTERVALS. Lighten any check, mitigate the destructio
079 on of the year, during each generation or at INTERVALS, has to struggle for life, and to suffer or
097 vidual is occasionally, perhaps at very long INTERVALS, indispensable. On the belief that this is
101 n many others it occurs perhaps only at long INTERVALS; but in none, as I suspect, can self fertil
104 lants. Even if these take place only at long INTERVALS, I am convinced that the young thus produce
104 , the influence of intercrosses even at rare INTERVALS, will be great. If there exist organic bein
108 l always act very slowly, often only at long INTERVALS of time, and generally on only a very few o
117 appear simultaneously; but often after long INTERVALS of time; nor are they all supposed to endur
117 t worthy of record in a systematic work. The INTERVALS between the horizontal lines in the diagram
119 the line of succession is broken at regular INTERVALS by small numbered letters marking the succe
119 and might have been inserted anywhere, after INTERVALS long enough to have allowed the accumulatio
173 oncur only rarely, and after enormously long INTERVALS. Whilst the bed of the sea is stationary or
173 e natural collections have been made only at INTERVALS of time immensely remote. But it may be urg
173 a continent, we generally meet at successive INTERVALS with closely allied or representative speci
217 s, that she lays her eggs, not daily, but at INTERVALS of two or three days; so that, if she were
218 f eggs: but as in the case of the cuckoo, at INTERVALS of two or three days. This instinct, howeve
289 ions being separated from each other by wide INTERVALS of time. When we see the formations tabulat
290 een derived, accords with the belief of vast INTERVALS of time having elapsed between each formati
296 s, showing what wide, yet easily overlooked, INTERVALS have occurred in its accumulation. In other
296 standing upright as they grew, of many long INTERVALS of time and changes of level during the pro
297 o find. Moreover, if we look to rather wider INTERVALS, namely, to distinct but consecutive stages
299 xisting and extinct genera, and has made the INTERVALS between some few groups less wide than they
300 uld be separated from each other by enormous INTERVALS, during which the area would be either stat
302 e do not make due allowance for the enormous INTERVALS of time, which have probably elapsed betwee
```

Page **********************************(Key Word)**********************************

Page ************************************(Key Word)************************************

0303 or the accumulation of each formation. These INTERVALS will have given time for the multiplicatic
0308 not know what was the state of things in the INTERVALS between the successive formations; whether
0308 er europe and the United States during these INTERVALS existed as dry land, or as a submarine sur
0315 region do at last, if we look to wide enough INTERVALS of time, become modified; for those which
0315 mulated at wide and irregularly intermittent INTERVALS; consequently the amount of organic change
0321 t has been already said on the probable wide INTERVALS of time between our consecutive formations
0322 een our consecutive formations; and in these INTERVALS there may have been much slow exterminatic
0326 rtain amount of isolation, recurring at long INTERVALS of time, would probably be also favourable
0327 during periods of subsidence; and that blank INTERVALS of vast duration occurred during the peric
0327 organic remains. During these long and blank INTERVALS I suppose that the inhabitants of each reg
0329 f the several formations and during the long INTERVALS of time between them; in this case, the se
0329 tinct forms of life help to fill up the wide INTERVALS between existing genera, families, and orc
0333 , except in very rare cases, to fill up wide INTERVALS in the natural system, and thus unite dist
0334 t of modification, during the long and blank INTERVALS between the successive formations. Subject
0334 necessarily exactly intermediate, as unequal INTERVALS of time have elapsed between consecutive f
0336 as often been interrupted, and as long blank INTERVALS have intervened between successive format
0336 of these periods; but we ought to find after INTERVALS, very long as measured by years, but only
0340 same manner and degree. But after very long INTERVALS of time and after great geographical chang
0342 nough to resist future degradation, enormous INTERVALS of time have elapsed between the successiv
0342 formation. We may disbelieve in the enormous INTERVALS of time which have elapsed between our cor
0344 he struggle for existence. Hence, after long INTERVALS of time, the productions of the world will
0390 ed, species occasionally arriving after long INTERVALS in a new and isolated district, and having
0431 important part in defining and widening the INTERVALS between the several groups in each class.
0465 separated from each other by enormous blank INTERVALS of time; for fossiliferous formations, th
0465 inclined to admit. If we look to long enough INTERVALS of time, geology plainly declares that all
0475 e come on the stage slowly and at successive INTERVALS; and the amount of change, after equal int
0475 rvals; and the amount of change, after equal INTERVALS of time, is widely different in different
0475 ndants, causes the forms of life, after long INTERVALS of time, to appear as if they had changed
0483 remains sometimes tend to fill up very wide INTERVALS between existing orders. Crgans in a rudim
0487 a poor collection made at hazard and at rare INTERVALS. The accumulation of each great fossilifer
0487 concurrence of circumstances; and the blank INTERVALS between the successive stages as having be
0487 uge with some security the duration of these INTERVALS by a comparison of the preceding and succe
0301 evel, and that slight climatal changes would INTERVENE during such lengthy periods; and in these
0309 vel, which we may fairly conclude must have INTERVENED during these enormously long periods. If t
0336 terrupted, and as long blank intervals have INTERVENED between successive formations, we ought 'no
0357 ent period to each other and to the several INTERVENING oceanic islands. I freely admit the forme
0368 itted their former migration across the low INTERVENING tracts; since become too warm for their e
0374 by Gardner, which do not exist in the wide INTERVENING hot countries. So on the Silla of Caracca
0375 esenting plants of Europe, not found in the INTERVENING hot lowlands. A list of the genera collec
0074 in a district might determine, through the INTERVENTION first of mice and then of bees; the frequ
0361 nce of nutritious seeds passing through the INTESTINES of a bird; but hard seeds of fruit will pa
0764 could they be long carried in the crops or INTESTINES of birds. These means, however, would suff
0081 may conclude, from what we have seen of the INTIMATE and complex manner in which the inhabitants
0193 but, as Owen and others have remarked, their INTIMATE structure closely resembles that of common
0436 tals, stamens, and pistils, as well as their INTIMATE structure, are intelligible on' the view tha
0109 ; but it must be here alluded to from being INTIMATELY connected with natural selection. Natural
0317 he production of new and improved forms are INTIMATELY connected together. The old notion of al
0414 y importance. These resemblances, though so INTIMATELY connected with the whole life of the being
0107 to natural selection, as far as the extreme INTRICACY of the subject permits. I conclude, lookin
0071 nto play in some of these cases; but on this INTRICATE subject I will not here enlarge. Many case
0102 e to Natural Selection. This is an extremely INTRICATE subject. A large amount of inheritable anc
0096 he intercrossing of Individuals. I must here INTRODUCE a short digression. In the case of animals
0064 it is with plants: cases could be given of INTRODUCED plants which have become common throughout
0065 he exclusion of all other plants, have been INTRODUCED from Europe; and there are plants which no
0081 ered how powerful the influence of a single INTRODUCED tree or mammal has been shown to be. But i
0150 ollected, and which cannot possibly be here INTRODUCED. I can only state my conviction that it is
0202 the advancing legions of plants and animals INTRODUCED from Europe. Natural selection will not pr
0219 es, and many perished of hunger. Huber then INTRODUCED a single slave (F. fusca), and she instant
0247 rious cause of error seems to me to be here INTRODUCED: a plant to be hybridised must be castrate
0375 european species, believed not to have been INTRODUCED by man, and on the mountains, some few rep
0375 everal European species; other species, not INTRODUCED by man, occur on the lowlands; and a long
0389 of Anglesea 764, but a few ferns and a few INTRODUCED plants are included in these numbers, and
0393 tted for these animals; for frogs have been INTRODUCED into Madeira, the Azores, and Mauritius, a
0001   INTRODUCTION When on board H.M.S. Beagle, as naturalis
0071 see how potent has been the effect of the INTRODUCTION of a single tree, nothing whatever else h
0318 ment; for seeing that the horse, since its INTRODUCTION by the Spaniards into South America, has
0325 ions of species, their extinction, and the INTRODUCTION of new ones, cannot be owing to mere chan
0375 ermediate torrid regions. In the admirable INTRODUCTION to the Flora of New Zealand, by Dr. Hooke
0313 the so called colonies of M. Barrande, which INTRUDE for a period in the midst of an older forma
0321 llied forms be developed through the successful INTRUDER, many will have to yield their places; and
0388 between terrestrial 'species; consequently an INTRUDER from the waters of a foreign country, would
0082 ese same places would have been seized on by INTRUDERS. In such case, every slight modification,
0083 dvantage, so as to have better resisted such INTRUDERS. As man can produce and certainly has prod
0377 ould not have presented a firm front against INTRUDERS; that a certain number of the more vigorou
0379 se may have been wholly different with those INTRUDING forms which settled themselves on the inte
0457 seem to me to proclaim so plainly, that the INUMERABLE species, genera, and families of organic b
0382 er force from the north so as to have freely INUNDATED the south. As the tide leaves its drift in
0450 this strange condition, bearing the stamp of INUTILITY, are extremely common throughout nature. F
0480 d thus so frequently bear the plain stamp of INUTILITY! Nature may be said to have taken pains to
0302 xisted and have slowly multiplied before they INVADED the ancient archipelagoes of Europe and of
0326 and far spreading species, which already have INVADED to a certain extent the territories of othe
0402 of the probability of closely allied species INVADING each other's territory, when put into free
0213 be called, are certainly far less fixed or INVARIABLE than natural instincts; but they have been
0272 n with respect to the very general, but not INVARIABLE, sterility of first crosses and of hybrids
0004 is and in all other perplexing cases I have INVARIABLY found that our knowledge, imperfect though
0012 ite whimsical: thus cats with blue eyes are INVARIABLY deaf: colour and constitutional peculiarit

Page ************************************(Key Word)************************************

age ***(Key Word)***

```
155  maller genera on the other side, and it has  INVARIABLY  proved to be the case that a larger proport
155  e large genera which present any varieties,   INVARIABLY  present a larger average number of varietie
167  m generally to suffer most, but this is not   INVARIABLY  the case. With plants there is a vast destr
169  o north, or from a damp region to a dry, we    INVARIABLY  see some species gradually getting rarer an
175  feeding quadrupeds. But the struggle almost    INVARIABLY  will be most severe between the individuals
176  same genus have usually, though by no means    INVARIABLY, some similarity in habits and constitution
198  have such a prepotent effect, that it will     INVARIABLY  and completely destroy, as has been shown b
19   king that the most divergent varieties will    INVARIABLY  prevail and multiply: a medium form may oft
151  d i am fully convinced that the rule almost    INVARIABLY  holds good with cirripedes. I shall, in my
166  n a bluish tint, these bars and other marks    INVARIABLY  reappear; but without any other change of f
94   eated for its proper place in nature, be so    INVARIABLY  linked together by graduated steps? Why sho
132  y thin layer of melted vermilion wax; and I    INVARIABLY  found that the colour was most delicately d
161  s of the same species, can usually, but not    INVARIABLY, be grafted with ease. But this capacity, a
168  ffspring. I fully admit that this is almost    INVARIABLY  the case. But if we look to varieties produ
169  s if the varieties of the same species were    INVARIABLY  fertile when intercrossed. But it seems to
290  ogical formations of each region are almost    INVARIABLY  intermittent; that is, have not followed ea
328  having any right to conclude that this has     INVARIABLY  been the case, and that large areas have in
328  ly been the case, and that large areas have    INVARIABLY  been affected by the same movements. When t
354  progenitor. If it can be shown to be almost    INVARIABLY  the case, that a region, of which most of i
144  parent. I am far from meaning that this is     INVARIABLY  the case; and I could give a good many case
165  fossil remains from consecutive formations     INVARIABLY  being much more closely related to each oth
178  ormerly inhabited both areas; and we almost    INVARIABLY  find that wherever many closely allied spec
181  intained that species when intercrossed are    INVARIABLY  sterile, and varieties invariably fertile;
181  ossed are invariably sterile, and varieties    INVARIABLY  fertile; or that sterility is a special end
577  ave reached or even crossed the equator. The   INVASION  would, of course, have been greatly favoured
577  ra, seem to have afforded two great lines of   INVASION: and it is a striking fact, lately communica
173  e invoke cataclysms to desolate the world, or  INVENT  laws on the duration of the forms of life! I
     sometimes independently hit on the very same   INVENTION, so natural selection, working for the good
186  way as when we look at any great mechanical    INVENTION  as the summing up of the labour, the experi
530  en the pig and the camel. In regard to the     INVERTEBRATA, Barrande, and a higher authority could no
188  n optic nerve merely coated with pigment and   INVESTED  by transparent membrane, into an optical ins
101  of difficulty, some of which I am trying to    INVESTIGATE. Finally then, we may conclude that in man
225  m a few very simple instincts. I was led to    INVESTIGATE  this subject by Mr. Waterhouse, who has sh
151  ttended to Mr. Waterhouse's remark, whilst     INVESTIGATING  this Order, and I am fully convinced that
050  quently must be ranked as varieties. Close     INVESTIGATION, in most cases, will bring naturalists to
071  e of a relation where I had ample means of     INVESTIGATION, there was a large and extremely barren h
087  ry, I cannot find one case which will bear     INVESTIGATION. A structure used only once in an animal'
429  richness in species, as I find after some     INVESTIGATION, does not commonly fall to the lot of abe
073  nic being; and as we do not see the cause, we   INVOKE  cataclysms to desolate the world, or invent l
016  estic races of the same species, we are soon   INVOLVED  in doubt, from not knowing whether they have
268  es produced under nature, we are immediately   INVOLVED  in hopeless difficulties; for if two hithert
268  produced, under domestication, we are still   INVOLVED  in doubt. For when it is stated, for instanc
318  ubject of the extinction of species has been   INVOLVED  in the most gratuitous mystery. Some authors
067  sharp wedges packed close together and driven  INWARDS  by incessant blows, sometimes one wedge bein
028  ons were taken with the court. The monarchs of  IRAN and Turan sent him some very rare birds: and,
049  distinct species by many entomologists. Even   IRELAND  has a few animals, now generally regarded as
049  d the Azores, or Madeira, or the Canaries, or  IRELAND, be sufficient? It must be admitted that man
364  , from North America to the western shores of  IRELAND  and England; but seeds could be transported
046  coccus, which may almost be compared to the    IRREGULAR  branching of the stem of a tree. This philo
118  the diagram, though in itself made somewhat    IRREGULAR. I am far from thinking that the most diver
145  external flowers: we know, at least, that in   IRREGULAR  flowers, those nearest to the axis are ofte
225  f wax, and likewise making separate and very   IRREGULAR  rounded cells of wax. At the other end of t
225  arly equal sizes, and are aggregated into an   IRREGULAR  mass. But the important point to notice, is
315  almost necessarily accumulated at wide and     IRREGULARLY  intermittent intervals; consequently the a
361  royal tax. I find on examination, that when    IRREGULARLY  shaped stones are embedded in the roots of
011  very curious: and many instances are given in  ISIDORE  Geoffroy St. Hilaire's great work on this su
081  l has been shown to be. But in the case of an  ISLAND, or of a country partly surrounded by barrier
105  t any small isolated area, such as an oceanic  ISLAND, although the total number of the species inh
105  ced there, and nowhere else. Hence an oceanic  ISLAND  at first sight seems to have been highly favo
106  so largely naturalised on islands. On a small  ISLAND, the race for life will have been less severe
107  many individuals of the same species on each  ISLAND: intercrossing on the confines of the range o
108  ted, so that new places in the polity of each  ISLAND  will have to be filled up by modifications of
n.   for when a new insect first arrived on the    ISLAND, the tendency of natural selection to enlarge
308  tudded with many islands; but not one oceanic  ISLAND  is as yet known to afford even a remnant of a
312  ive changes in the marine inhabitants of that  ISLAND  have been many and most gradual. The secondar
348  wide space of open ocean extends, with not an  ISLAND  as a halting place for emigrants: here we hav
355  egion, whence it has been stocked. A volcanic  ISLAND, for instance, upheaved and formed at the dis
357  ver every ocean, and have united almost every  ISLAND  to some mainland. If indeed the arguments use
357  d, it must be admitted that scarcely a single  ISLAND  exists which has not recently been united to
364  of sea some hundred miles in breadth, or from  ISLAND  to island, or from a continent to a neighbour
364  e hundred miles in breadth, or from island to  ISLAND, or from a continent to a neighbouring island
364  island, or from a continent to a neighbouring  ISLAND, but not from one distant continent to anothe
364  at error to argue that because a well stocked  ISLAND, like Great Britain, has not, as far as is kn
365  or any other continent; that a poorly stocked  ISLAND, though standing more remote from the mainlan
365  of twenty seeds or animals transported to an   ISLAND, even if far less well stocked than Britain,
365  the long lapse of geological time, whilst an   ISLAND  was being upheaved and formed, and before it
373  , found on widely separated mountains in this  ISLAND, tell the same story. If one account which ha
375  e given in regard to the plants of that large  ISLAND. Hence we see that throughout the world, the
380  st stage towards extinction. A mountain is an  ISLAND  on the land; and the intertropical mountains
385  or rivulet, if blown across sea to an oceanic  ISLAND  or to any other distant point. Sir Charles Ly
389  y of Cambridge has 847 plants, and the little  ISLAND  of Anglesea 764, but a few ferns and a few in
     quite fair. We have evidence that the barren   ISLAND  of Ascension aboriginally possessed under hal
390  ave on New Zealand and on every other oceanic  ISLAND  which can be named. In St. Helena there is re
390  t it by no means follows, that, because in an  ISLAND  nearly all the species of one class are pecul
391  isit either periodically or occasionally this  ISLAND. Madeira does not possess one peculiar bird,
392  for a hooked seed might be transported to an   ISLAND  by some other means: and the plant then becom
```

age ***(Key Word)***

```
0392  fully developed tree, when established on an  ISLAND and having to compete with herbaceous plants
0392  ature of herbaceous plants when growing on an  ISLAND, to whatever order they belonged, and thus c
0393  t a frog exists on the mountains of the great  ISLAND of New Zealand; but I suspect that this exce
0393  herefore why they do not exist on any oceanic  ISLAND. But why, on the theory of creation, they sh
0393  ed animals kept by the natives) inhabiting an  ISLAND situated above 300 miles from a continent or
0393  0 miles from a continent or great continental  ISLAND; and many islands situated at a much less di
0394  lands, if close to a continent; and hardly an  ISLAND can be named on which our smaller quadrupeds
0394  ands, aerial mammals do occur on almost every  ISLAND. New Zealand possesses two bats found nowher
0394  bats found nowhere else in the world: Norfolk  ISLAND, the Viti Archipelago, the Bonin Islands, th
0395  e, between the depth of the sea separating an  ISLAND from the neighbouring mainland, and the pres
0399  o those of America: but on the view that this  ISLAND has been mainly stocked by seeds brought wit
0400  ies; so that the inhabitants of each separate  ISLAND, though mostly distinct, are related in an i
0401  ance, of one plant having been brought to one  ISLAND, and that of another plant to another island
0401  island, and that of another plant to another  ISLAND. Hence when in former times an immigrant set
0401  ands, or when it subsequently spread from one  ISLAND to another, it would undoubtedly be exposed
0401  perfectly occupied by distinct plants in one  ISLAND than in another, and it would be exposed to
0402  that these have probably spread from some one  ISLAND to the others. But we often take, I think, a
0402  birds, though so well adapted for flying from  ISLAND to island, are distinct on each; thus there
0402  uch so well adapted for flying from island to  ISLAND, are distinct on each; thus there are three
0402  s of mocking thrush, each confined to its own  ISLAND. Now let us suppose the mocking thrush of Ch
0402  let us suppose the mocking thrush of Chatham  ISLAND to be blown to Charles Island, which has its
0402  rush of Chatham Island to be blown to Charles  ISLAND, which has its own mocking thrush: why shoul
0402  tself there? We may safely infer that Charles  ISLAND is well stocked with its own species, for an
0402  r that the mocking thrush peculiar to Charles  ISLAND is at least as well fitted for its home as i
0402  ts home as is the species peculiar to Chatham  ISLAND. Sir C. Lyell and Mr. Wollaston have communi
0403  from Porto Santo to Madeira, yet this latter  ISLAND has not become colonised by the Porto Santo
0403  chipelago, not having universally spread from  ISLAND to island. In many other instances, as in th
0403  not having universally spread from island to  ISLAND. In many other instances, as in the several
0406  the inhabitants of each whole archipelago or  ISLAND to those of the nearest mainland, are I thin
0409  ondition, and the depth of the sea between an  ISLAND and the mainland. We can clearly see why all
0042  imported from some other country, often from  ISLANDS. Although I do not doubt that some domestic
0048  g others compare, the birds from the separate  ISLANDS of the Galapagos Archipelago, both one with
0064  nts which have become common throughout whole  ISLANDS in a period of less than ten years. Several
0106  e everywhere become so largely naturalised on  ISLANDS. On a small island, the race for life will
0107  n converted by subsidence into large separate  ISLANDS, there will still exist many individuals of
0108  d perfected. When, by renewed elevation, the  ISLANDS shall be re converted into a continental ar
0132  under a clear atmosphere, than when living on  ISLANDS or near the coast. So with insects, Wollast
0133  to mr. Gould, brighter coloured than those of  ISLANDS. The insect species confined to sea coasts,
0134  abit or have lately inhabited several oceanic  ISLANDS, tenanted by no beast of prey, has been cau
0141  nd of the Falklands in the south, and on many  ISLANDS in the torrid zones. Hence I am inclined to
0174  lmost every continent has been broken up into  ISLANDS even during the later tertiary periods; and
0174  uring the later tertiary periods; and in such  ISLANDS distinct species might have been separately
0212  e shown, by various animals inhabiting desert  ISLANDS; and we may see an instance of this even in
0212  large birds to this cause: for in uninhabited  ISLANDS large birds are not more fearful than small
0284  evidence of denudation, when viewing volcanic  ISLANDS, which have been worn by the waves and pare
0299  he malay Archipelago, with its numerous large  ISLANDS separated by wide and shallow seas, probabl
0308  d, we may infer that from first to last large  ISLANDS or tracts of land, whence the sediment was
0308  ve as the land, we see them studded with many  ISLANDS; but not one oceanic island is as yet known
0308  y periods, neither continents nor continental  ISLANDS existed where our oceans now extend; for ha
0348  soon as this is passed we meet in the eastern  ISLANDS of the Pacific, with another and totally di
0348  eding still further westward from the eastern  ISLANDS of the tropical parts of the Pacific, we en
0348  impassable barriers, and we have innumerable  ISLANDS as halting places, until after travelling ov
0348  n and Western America and the eastern Pacific  ISLANDS, yet many fish range from the Pacific into
0348  an, and many shells are common to the eastern  ISLANDS of the Pacific and the eastern shores of Af
0349  r instances could be given. If we look to the  ISLANDS off the American shore, however much they m
0354  occurrence of the same terrestrial species on  ISLANDS and on the mainland, though separated by hur
0356  s, land may at a former period have connected  ISLANDS or possibly even continents together, and t
0357  rganisms. Edward Forbes insisted that all the  ISLANDS in the Atlantic must recently have been conn
0357  other and to the several intervening oceanic  ISLANDS. I freely admit the former existence of many
0357  . I freely admit the former existence of many  ISLANDS, now buried beneath the sea, which may have
0357  on. In the coral producing oceans such sunken  ISLANDS are now marked, as I believe, by rings of co
0358  ach other, and with the many existing oceanic  ISLANDS. Several facts in distribution, such as the
0358  ive proportions of the inhabitants of oceanic  ISLANDS likewise seem to me opposed to the belief of
0358  tain ranges on the land, some at least of the  ISLANDS would have been formed, like other mountain
0360  her manner. Drift timber is thrown up on most  ISLANDS, even on those in the midst of the widest oc
0360  e widest oceans; and the natives of the coral  ISLANDS in the Pacific, procure stones for their too
0363  n comparison with the plants of other oceanic  ISLANDS nearer to the mainland, and (as remarked by
0363  son with the latitude, I suspected that these  ISLANDS had been partly stocked by ice borne seeds,
0363  her he had observed erratic boulders on these  ISLANDS, and he answered that he had found large fra
0363  cky burthens on the shores of these mid ocean  ISLANDS, and it is at least possible that they may h
0380  ore efficient workshops of the north. In many  ISLANDS the native productions are nearly equalled o
0380  i and I believe that the productions of these  ISLANDS on the land yielded to those produced within
0380  st in the same way as the productions of real  ISLANDS have everywhere lately yielded to continenta
0382  as halting places, of existing and now sunken  ISLANDS, and perhaps at the commencement of the Glac
0383  er productions. On the inhabitants of oceanic  ISLANDS. Absence of the Batrachians and of terrestr
0383  ammals. On the relation of the inhabitants of  ISLANDS to those of the nearest mainland. On coloni
0386  ver continents and to the most remote oceanic  ISLANDS. This is strikingly shown, as remarked by A
0386  asionally found on the most remote and barren  ISLANDS in the open ocean; they would not be likely
0388  itted for them. On the Inhabitants of Oceanic  ISLANDS. We now come to the last of the three classe
0389  ef that within the recent period all existing  ISLANDS have been nearly or quite joined to some con
0389  he species of all kinds which inhabit oceanic  ISLANDS are few in number compared with those on equ
0390  and animals have not been created on oceanic  ISLANDS; for man has unintentionally stocked them fr
0390  erfectly than has nature. Although in oceanic  ISLANDS the number of kinds of inhabitants is scanty
0390  y continent, and then compare the area of the  ISLANDS with that of the continent, we shall see tha
0390  ot been much disturbed. Thus in the Galapagos  ISLANDS nearly every land bird, but only two out of
0391  vious that marine birds could arrive at these  ISLANDS more easily than land birds. Bermuda, on the
```

Page **(Key Word)**

```
0391   distance from North America as the Galapagos ISLANDS do from South America, and which has a very
0391   rmed by Mr. E. V. Harcourt. So that these two ISLANDS of Bermuda and Madeira have been stocked by
0391   a apparently present analogous facts. Oceanic ISLANDS are sometimes deficient in certain classes,
0391   ed by the other inhabitants; in the Galapagos ISLANDS reptiles, and in New Zealand gigantic wingle
0391   ce of mammals. In the plants of the Galapagos ISLANDS, Dr. Hooker has shown that the proportional
0392   counted for by the physical conditions of the ISLANDS! but this explanation seems to me not a litt
0392   ven with respect to the inhabitants of remote ISLANDS. For instance, in certain islands not tenant
0392   s of remote islands. For instance, in certain ISLANDS not tenanted by mammals, some of the endemic
0392   ldered elytra of many insular beetles. Again, ISLANDS often possess trees or bushes belonging to o
0392   uld be little likely to reach distant oceanic ISLANDS; and an herbaceous plant, though it would ha
0393   ect to the absence of whole orders on oceanic ISLANDS, Bory St. Vincent long ago remarked that Bat
0393   wts) have never been found on any of the many ISLANDS with which the great oceans are studded. I h
0393   of frogs, toads, and newts on so many oceanic ISLANDS cannot be accounted for by their physical co
0393   eir physical conditions; indeed it seems that ISLANDS are peculiarly well fitted for these animals
0393   ntinent or great continental island; and many ISLANDS situated at a much less distance are equally
0393   ess distance are equally barren. The Falkland ISLANDS, which are inhabited by a wolf like fox, com
0394   tic regions. Yet it cannot be said that small ISLANDS will not support small mammals, for they occ
0394   ccur in many parts of the world on very small ISLANDS, if close to a continent; and hardly an isla
0394   me for the creation of mammals; many volcanic ISLANDS are sufficiently ancient, as shown by the st
0394   h terrestrial mammals do not occur on oceanic ISLANDS, aerial mammals do occur on almost every isl
0394   rfolk Island, the Viti Archipelago, the Bonin ISLANDS, the Caroline and Marianne Archipelagoes, an
0394   produced bats and no other mammals on remote ISLANDS? On my view this question can easily be answ
0394   nd are found on continents and on far distant ISLANDS. Hence we have only to suppose that such wan
0395   an understand the presence of endemic bats on ISLANDS, with the absence of all terrestrial mammals
0395   rial mammals in relation to the remoteness of ISLANDS from continents, there is also a relation, t
0395   distinct mammalian faunas. On either side the ISLANDS are situated on moderately deep submarine ba
0395   h sides; we meet with analogous facts on many ISLANDS separated by similar channels from Australia
0395   ilar channels from Australia. The West Indian ISLANDS stand on a deeply submerged bank, nearly 100
0396   as during changes of level it is obvious that ISLANDS separated by shallow channels are more likel
0396   d within a recent period to the mainland than ISLANDS separated by deeper channels, we can underst
0396   affinity of the the mammalian inhabitants of ISLANDS with those of a neighbouring continent, an i
0396   regoing remarks on the inhabitants of oceanic ISLANDS, namely, the scarcity of kinds, the richness
0396   f time, than with the view of all our oceanic ISLANDS having been formerly connected by continuous
0396   several of the inhabitants of the more remote ISLANDS, whether still retaining the same specific f
0396   ir present homes. But the probability of many ISLANDS having existed as halting places, of which n
0397   f the cases of difficulty. Almost all oceanic ISLANDS, even the most isolated and smallest, are in
0397   ing cases in regard to the land shells of the ISLANDS of the Pacific. Now it is notorious that lan
0397   t fact for us in regard to the inhabitants of ISLANDS, is their affinity to those of the nearest m
0398   looking at the inhabitants of these volcanic ISLANDS in the Pacific, distant several hundred mile
0398   ions of life, in the geological nature of the ISLANDS, in their height or climate, or in the propo
0398   the soil, in climate, height, and size of the ISLANDS, between the Galapagos and Cape de Verde Arc
0398   bitants! The inhabitants of the Cape de Verde ISLANDS are related to those of Africa, like those o
0398   maintained, it is obvious that the Galapagos ISLANDS would be likely to receive colonists, whethe
0399   ous land, from America! and the Cape de Verde ISLANDS from Africa; and that such colonists would b
0399   niversal rule that the endemic productions of ISLANDS are related to those of the nearest continen
0399   se of the nearest continent, or of other near ISLANDS. The exceptions are few, and most of them ca
0399   ough distant point, namely from the antarctic ISLANDS, when they were clothed with vegetation, bef
0400   its of the same archipelago. Thus the several ISLANDS of the Galapagos Archipelago are tenanted, a
0400   might have been expected on my view, for the ISLANDS are situated so near each other that they wo
0400   larity between the endemic inhabitants of the ISLANDS may be used as an argument against my views;
0400   be asked, how has it happened in the several ISLANDS situated within sight of each other, having
0401   siderable amount of difference in the several ISLANDS. This difference might indeed have been expe
0401   indeed have been expected on the view of the ISLANDS having been stocked by occasional means of t
0401   n immigrant settled on any one or more of the ISLANDS, or when it subsequently spread from one isl
0401   different conditions of life in the different ISLANDS; for it would have to compete with different
0401   y favour different varieties in the different ISLANDS. Some species, however, might spread and yet
0401   s that the new species formed in the separate ISLANDS have not quickly spread to the other islands
0401   islands have not quickly spread to the other ISLANDS. But the islands, though in sight of each ot
0401   quickly spread to the other islands. But the ISLANDS, though in sight of each other, are separate
0401   of wind are extraordinarily rare; so that the ISLANDS are far more effectually separated from each
0402   to the archipelago, are common to the several ISLANDS, and we may infer from certain facts that th
0403   by the Porto Santo species: nevertheless both ISLANDS have been colonised by some European land sh
0403   esentative species, which inhabit the several ISLANDS of the Galapagos Archipelago, not having uni
0403   l character of the fauna and flora of oceanic ISLANDS, namely, that the inhabitants, when not iden
0409   d, as I have endeavoured to show, why oceanic ISLANDS should have few inhabitants, but of these a
0409   strial mammals, should be absent from oceanic ISLANDS, whilst the most isolated islands possess th
0409   rom oceanic islands, whilst the most isolated ISLANDS possess their own peculiar species of aerial
0454   and of the wings of birds inhabiting oceanic ISLANDS, which have seldom been forced to take fligh
0454   wings of beetles living on small and exposed ISLANDS; and in this case natural selection would co
0477   bsequent modification, we can see why oceanic ISLANDS should be inhabited by few species, but of t
0477   rrestrial mammals, should not inhabit oceanic ISLANDS; and why, on the other hand, new and peculia
0477   averse the ocean, should so often be found on ISLANDS far distant from any continent. Such facts a
0478   the absence of all other mammals, on oceanic ISLANDS, are utterly inexplicable on the theory of i
0478   of Juan Fernandez, and of the other American ISLANDS being related in the most striking manner to
0478   e cape de Verde archipelago and other African ISLANDS to the African mainland. It must be admitted
0388   eam is first formed, for instance, on a rising ISLET, it will be unoccupied; and a single seed or
0402   ubject; namely, that Madeira and the adjoining ISLET of Porto Santo possess many distinct but repr
0024   inated even on several of the smaller British ISLETS, or on the shores of the Mediterranian. Hence
0048   nction between species and varieties. On the ISLETS of the little Madeira group there are many in
0114   h the plants and insects on small and uniform ISLETS; and so in small ponds of fresh water. Farmer
0406   ion of the distinct species which inhabit the ISLETS of the same archipelago, and especially the s
0409   , though specifically distinct on the several ISLETS, should be closely related to each other, and
0104   ocess of natural selection. In a confined or ISOLATED area, if not very large, the organic and ino
0105   e production of new species. If, however, an ISOLATED area be very small, either from being surrou
0105   ruth of these remarks, and look at any small ISOLATED area, such as an oceanic island, although th
0105   ourselves, for to ascertain whether a small ISOLATED area, or a large open area like a continent,
```

Page **(Key Word)**

Page **(Key Word)**

```
0106  more severe, on a large than on a small and ISOLATED area. Moreover, great areas, though now cont
0106  ed. Finally, I conclude that, although small ISOLATED areas probably have been in some respects hi
0178  ten have existed within the recent period in ISOLATED portions, in which many forms, more especial
0189  l grades, more especially if we look to much ISOLATED species, round which, according to my theory
0233  e the planes of intersection, and so make an ISOLATED hexagon: but I am not aware that any such ca
0321  of life, or from inhabiting some distant and ISOLATED station, where they have escaped severe comp
0344  few descendants, lingering in protected and ISOLATED situations. When a group has once wholly dis
0347  tude: for these countries are almost as much ISOLATED from each other as is possible. On each cont
0351  migrate in a body into a new and afterwards ISOLATED country, they will be little liable to modif
0352  e species, though now inhabiting distant and ISOLATED regions, must have proceeded from one spot,
0352  om some one point to the several distant and ISOLATED points, where now found. Nevertheless the si
0354  existence of the same species at distant and ISOLATED points of the earth's surface, can in many i
0367  nds of the Old and New Worlds, would be left ISOLATED on distant mountain summits (having been ext
0368  ation. But with our Alpine productions, left ISOLATED from the moment of the returning warmth, fir
0371  ation than with the Alpine productions, left ISOLATED, within a much more recent period, on the se
0375  parts of Africa. On the Himalaya, and on the ISOLATED mountain ranges of the peninsula of India, o
0380  the Glacial period must have been completely ISOLATED; and I believe that the productions of these
0381  ed with ice, supported a highly peculiar and ISOLATED flora. I suspect that before this flora was
0390  y arriving after long intervals in a new and ISOLATED district, and having to compete with new ass
0397  y. almost all oceanic islands, even the most ISOLATED and smallest, are inhabited by land shells,
0409  absent from oceanic islands, whilst the most ISOLATED islands possess their own peculiar species o
0421  a single genus. But this genus, though much ISOLATED, will still occupy its proper intermediate p
0427  the same genus, inhabiting any distinct and ISOLATED region, have in all probability descended fr
0461  rents; and therefore, in however distant and ISOLATED parts of the world they are now found, they
0462  f the same genus inhabiting very distant and ISOLATED regions, as the process of modification has
0080  to natural Selection, namely, intercrossing, ISOLATION, number of individuals. Slow action. Extinc
0082  nge, as of climate, or any unusual degree of ISOLATION to check immigration, is actually necessary
0104  n preserving the same favourable variations. ISOLATION, also, is an important element in the proce
0104  cumstanced districts, will be prevented. But ISOLATION probably acts more efficiently in checking
0105  in their structure and constitution. Lastly, ISOLATION, by checking immigration and consequently c
0105  pable of doing. Although I do not doubt that ISOLATION is of considerable importance in the produc
0106  roken condition, so that the good effects of ISOLATION will generally, to a certain extent, have c
0326  ng into new territories. A certain amount of ISOLATION, recurring at long intervals of time, would
0351  e to modification; for neither migration nor ISOLATION in themselves can do anything. These princi
0422  thers (owing to the spreading and subsequent ISOLATION and states of civilisation of the several r
0328  areas, unless, indeed, it be assumed that an ISTHMUS separated two seas inhabited by distinct, bu
0348  separated only by the narrow, but impassable, ISTHMUS of Panama. Westward of the shores of America
0356  t also have been highly influential: a narrow ISTHMUS now separates two marine faunas: submerge it
0019  n believe that animals closely resembling the ITALIAN greyhound, the bloodhound, the bull dog, or
0020  r existence of the most extreme forms, as the ITALIAN greyhound, bloodhound, bull dog, etc., in th
0366  he climate of Europe changed, that in Northern ITALY, gigantic moraines, left by old glaciers, are
0215  me instinctive in the dog. All wolves, foxes, JACKALS, and species of the cat genus, when kept tam
0021  lightly the upper part of the oesophagus. The JACOBIN has the feathers so much reversed along the
0024  barb; for reversed feathers like those of the JACOBIN; for a crop like that of the pouter; for tai
0258  by Kolreuter. To give an instance: Mirabilis JALAPPA can easily be fertilised by the pollen of M.
0258  iprocally M. longiflora with the pollen of M. JALAPPA, and utterly failed. Several other equally s
0372  imals, in the Mediterranean and in the seas of JAPAN, areas now separated by a continent and by ne
0375  nd over India and on the other as far north as JAPAN. On the southern mountains of Australia, Dr.
0385  when taken out of the water they could not be JARRED off, though at a somewhat more advanced age t
0039  ilfeathers somewhat expanded, like the present JAVA fantail, or like individuals of other and dist
0375  eights of Ceylon; and on the volcanic cones of JAVA, many plants occur, either identically the sam
0375  f the genera collected on the loftier peaks of JAVA raises a picture of a collection made on a hil
0022  he breadth and length of the ramus of the lower JAW, varies in a highly remarkable manner. The num
0088  n, the shoulder pad to the boar, and the hooked JAW to the male salmon; for the shield may be as i
0143  jaws and limbs, varying together, for the lower JAW is believed to be homologous with the limbs. T
0181  nk membrane, stretching from the corners of the JAW to the tail, and including the limbs and the e
0241  ters, either all of large size with one form of JAW, or all of small size with jaws having a widel
0426  t it be the mere inflection of the angle of the JAW, the manner in which an insect's wing is folde
0480  , which never cut through the gums of the upper JAW, from an early progenitor having well develope
0087  by natural selection; for instance, the great JAWS possessed by certain insects, and used exclusi
0143  ri in the front and hind legs, and even in the JAWS and limbs, varying together, for the lower jaw
0219  ion, and actually carry their masters in their JAWS. So utterly helpless are the masters, that whe
0221  as Huber has described, their slaves in their JAWS. Another day my attention was struck by anoth
0237  ptial plumage of many birds, and in the hooked JAWS of the male salmon. We have even slight differ
0238  n, there are working and soldier neuters, with JAWS and instincts extraordinarily different: in Cr
0241  times as big as those of the smaller men, and JAWS nearly five times as big. The jaws, moreover,
0241  er men, and jaws nearly five times as big. The JAWS, moreover, of the working ants of the several
0241  s does the widely different structure of their JAWS. I speak confidently on this latter point, as
0241  drawings for me with the camera lucida of the JAWS which I had dissected from the workers of the
0241  ith one form of jaw, or all of small size with JAWS having a widely different structure; or lastly
0416  ispute that the rudimentary teeth in the upper JAWS of young ruminants, and certain rudimentary bo
0416  d reptiles, the inflection of the angle of the JAWS of the Marsupials, the manner in which the win
0434  ious folded one of a bee or bug, and the great JAWS of a beetle? yet all these organs, serving for
0436  same law in comparing the wonderfully complex JAWS and legs in crustaceans. It is familiar to alm
0438  kull as formed of metamorphosed vertebrae: the JAWS of crabs as metamorphosed legs; the stamens an
0438  ed, to speak of both skull and vertebrae, both JAWS and legs, etc., as having been metamorphosed,
0438  er, have actually been modified into skulls or JAWS. Yet so strong is the appearance of a modifica
0439  used literally; and the wonderful fact of the JAWS, for instance, of a crab retaining numerous ch
0450  which never cut through the gums, in the upper JAWS of our unborn calves. It has even been stated
0452  rudimentary organs, such as teeth in the upper JAWS of whales and ruminants, can often be detected
0479  though used for such different purpose, in the JAWS and legs of a crab, in the petals, stamens, an
0142  tional differences to habit. The case of the JERUSALEM artichoke, which is never propagated by see
0031  n it existence. That most skilful breeder, Sir JOHN Sebright, used to say, with respect to pigeons
0091  s, another mice; one cat, according to Mr. St. JOHN, bringing home winged game, another hares or r
0359  would retain their power of germination. In JOHNSTON's Physical Atlas, the average rate of the se
0020  l eminent fanciers, and have been permitted to JOIN two of the London Pigeon Clubs. The diversity
```

Page **(Key Word)**

Page **(Key Word)**

```
0225 basal edges of its six sides bevelled so as to JOIN on to a pyramid, formed of three rhombs. These
0389 ll existing islands have been nearly or quite JOINED to some continent. This view would remove man
0347 g in the northern parts, where the land almost JOINS, and where, under a slightly different climat
0157 can hardly be accidental. The same number of JOINTS in the tarsi is a character generally common
0391 endemic land bird: and we know from Mr. J. M. JONES's admirable account of Permuda, that very man
0283 , and by that excellent observer Mr. Smith of JORDAN Hill, are most impressive. With the mind thus
0114 n, determine that the inhabitants, which thus JOSTLE each other most closely, shall, as a general
0002 nd it is published in the third volume of the JOURNAL of that Society. Sir C. Lyell and Dr. Hooker
0367 t their brethren were pursuing their northern JOURNEY. Hence, when the warmth had fully returned,
0478 s and animals of the Galapagos archipelago, of JUAN Fernandez, and of the other American islands b
0010 ld probably have varied in the same manner. To JUDGE how much, in the case of any variation, we sh
0031 er individual, and who was himself a very good JUDGE of an animal, speaks of the principle of sele
0085 , which, as far as our ignorance permits us to JUDGE, seem to be quite unimportant, we must not fo
0134 an have no standard of comparison, by which to JUDGE of the effects of long continued use or disus
0202 ction, nor do we always meet, as far as we can JUDGE, with this high standard under nature. The co
0206 uce absolute perfection: nor, as far as we can JUDGE by our limited faculties, can absolute perfec
0210 od for itself, but has never, as far as we can JUDGE, been produced for the exclusive good of othe
0227 the greatest difficulty, that she can somehow JUDGE accurately at what distance to stand from her
0227 spheres: but she is already so far enabled to JUDGE of distance, that she always describes her sp
0229 gnawed, were situated, as far as the eye could JUDGE, exactly along the planes of imaginary inters
0323 s of the world: we have not sufficient data to JUDGE whether the productions of the land and of fr
0425 hoose those characters which, as far as we can JUDGE, are the least likely to have been modified f
0472 ontrivances in nature be not, as far as we can JUDGE, absolutely perfect: and if some of them be a
0039 es in the individuals of the same species, be JUDGED of by the value which would now be set on the
0052 ut whether this belief be justifiable must be JUDGED of by the general weight of the several facts
0127 heir several conditions and stations, must be JUDGED of by the general tenour and balance of evide
0016 which have not been ranked by some competent JUDGES as mere varieties, and by other competent jud
0016 ges as mere varieties, and by other competent JUDGES as the descendants of aboriginally distinct s
0018 om our European cattle: and several competent JUDGES believe that these latter have had more than
0047 ranked as species by at least some competent JUDGES. That varieties of this doubtful nature are f
0049 at many forms, considered by highly competent JUDGES as varieties, have so perfectly the character
0049 hat they are ranked by other highly competent JUDGES as good and true species. But to discuss whet
0253 lei for I am assured by two eminently capable JUDGES, namely Mr. Blyth and Capt. Hutton, that whol
0307 f life on this planet. Other highly competent JUDGES, as Lyell and the late E. Forbes, dispute thi
0046 polymorphic in other countries, and likewise, JUDGING from Brachiopod shells, at former periods of
0056 ion by the amount of difference between them, JUDGING by analogy whether or not the amount suffice
0072 ted thirty two little trees: and one of them, JUDGING from the rings of growth, had curing twenty
0291 in profound depths of the sea, in which case, JUDGING from the researches of E. Forbes, we may con
0445 ered just as much as their parents; and this, JUDGING by the eye, seemed almost to be the case: bu
0489 eposited, they seem to me to become ennobled. JUDGING from the past: we may safely infer that not
0003 large stores of knowledge and his excellent JUDGMENT. In considering the Origin of Species, it is
0006 the most deliberate study and dispassionate JUDGMENT of which I am capable, that the view which m
0032 ne man in a thousand has accuracy of eye and JUDGMENT sufficient to become an eminent breeder. If
0047 ety, the opinion of naturalists having sound JUDGMENT and wide experience seems the only guide to
0129 f all organic beings: but, to the best of my JUDGMENT, it is explained through inheritance and the
0171 hout being staggered: but, to the best of my JUDGMENT, the greater number are only apparent, and t
0208 ittle dose, as Pierre Huber expresses it, of JUDGMENT or reason, often comes into play, even in an
0243 f the cases of difficulty, to the best of my JUDGMENT, annihilate it. On the other hand, the fact
0357 ves many a difficulty: but to the best of my JUDGMENT we are not authorized in admitting such enor
0395 o, and there is much difficulty in forming a JUDGMENT in some cases owing to the probable naturali
0413 not always sufficiently strike us, is in my JUDGMENT fully explained. Naturalists try to arrange
0466 ave as these several difficulties are, in my JUDGMENT they do not overthrow the theory of descent
0091 e complex case. Certain plants excrete a sweet JUICE, apparently for the sake of eliminating somet
0092 he back of the leaf of the common laurel. This JUICE, though small in quantity, is greedily sought
0092 by insects. Let us now suppose a little sweet JUICE or nectar to be excreted by the inner bases o
0211 ts abdomen and excreted a limpid drop of sweet JUICE, which was eagerly devoured by the ant. Even
0361 ant: the crops of birds do not secrete gastric JUICE, and do not in the least injure, as I know by
0220 l quite at home. During the months of June and JULY, on three successive years, I have watched for
0221 ing the present year, however, in the month of JULY, I came across a community with an unusually l
0220 laves feel quite at home. During the months of JUNE and July, on three successive years, I have wa
0220 watched the nests at various hours during May, JUNE and August, both in Surrey and Hampshire, and
0417 and degraded flowers: in the latter, as A. de JUSSIEU has remarked, the greater number of the char
0417 the order, yet M. Richard sagaciously saw, as JUSSIEU observes, that this genus should still be re
0052 cipient species: but whether this belief be JUSTIFIABLE must be judged of by the general weight of
0309 e our continents now stand. Nor should we be JUSTIFIED in assuming that if, for instance, the bed
0332 ogether, intermediate in character, would be JUSTIFIED, as they are intermediate, not directly, bu
0304 d in any secondary formation, seemed fully to JUSTIFY the belief that this great and distinct orde
0003 tion of structure and coadaptation which most JUSTLY excites our admiration. Naturalists continual
0052 hence I believe a well marked variety may be JUSTLY called an incipient species: but whether this
0216 two latter instincts have generally, and most JUSTLY, been ranked by naturalists as the most wonde
0230 hese statements differ from those made by the JUSTLY celebrated elder Huber, but I am convinced of
0279 numerated the chief objections which might be JUSTLY urged against the views maintained in this vo
0322 iven country: then, and not till then, we may JUSTLY feel surprise why we cannot account for the e
0465 l chief objections and difficulties which may JUSTLY be urged against my theory; and I have now br
0425 o, if it could be proved that one species of KANGAROO had been produced, by a long course of modif
0425 em, and ask what should be done if a perfect KANGAROO were seen to come out of the womb of a bear?
0425 then assuredly all the other species of the KANGAROO family would have to be classed under the be
0164 houlder. In the north west part of India the KATTYWAR breed of horses is so generally striped, tha
0164 es. Colonel Poole has seen both gray and bay KATTYWAR horses striped when first foaled. I have, al
0165 an cart horse, Welch ponies, cobs, the lanky KATTYWAR race, etc., inhabiting the most distant part
0166 stripes ever occur in the eminently striped KATTYWAR breed of horses, and was, as we have seen, a
0188 on: though I have felt the difficulty far too KEENLY to be surprised at any degree of hesitation i
0034 s, is more important. Thus, a man who intends KEEPING pointers naturally tries to get as good dogs
0041 rs in getting new and valuable varieties. The KEEPING of a large number of individuals of a specie
0076 t together. The same result has followed from KEEPING together different varieties of the medicina
0085 proper colour to each kind of grouse, and in KEEPING that colour, when once acquired, true and co
```

Page **(Key Word)**

Page **(Key Word)**

```
0103  sing plays a very important part in nature in KEEPING the individuals of the same species, or of t
0134  nces are known to every naturalist of species KEEPING true, or not varying at all, although living
0317  it then gradually thickens upwards, sometimes KEEPING for a space of equal thickness, and ultimate
0488  of actual time. A number of species, however, KEEPING in a body might remain for a long period unc
0137  es, which inhabit the caves of Styria and of  KENTUCKY, are blind. In some of the crabs the foot st
0138  d into the deeper and deeper recesses of the  KENTUCKY caves, as did European animals into the cave
0381  al species at points so enormously remote as  KERGUELEN Land, New Zealand, and Fuegia, I believe th
0399  of them can be explained. Thus the plants of  KERGUELEN Land, though standing nearer to Africa than
0183  e spot and then proceeding to another, like a KESTREL, and at other times standing stationary on t
0030  know that this has not been their history. The KEY is man's power of accumulative selection: natu
0028  ee and race. Pigeons were much valued by Akber KHAN in India, about the year 1600; never less than
0134  from which it cannot escape by flight, but by KICKING it can defend itself from enemies, as well a
0142  on cannot be effected! The case, also, of the KIDNEY bean has been often cited for a similar purpo
0142  will sow, during a score of generations, his  KIDNEY beans so early that a very large proportion a
0142  o differences in the constitution of seedling KIDNEY beans ever appear, for an account has been pu
0144  he remarkable diversity in the shape of their KIDNEYS. Others believe that the shape of the pelvis
0068  et to grow, the more vigorous plants gradually KILL the less vigorous, though fully grown, plants:
0129  to branch out on all sides, and to overtop and KILL the surrounding twigs and branches, in the sam
0042  nferior birds may be freely rejected, as when  KILLED they serve for food. On the other hand, cats,
0068  of thousands of game animals are now annually  KILLED. On the other hand, in some cases, as with th
0221  ve making F. sanguinea. The latter ruthlessly  KILLED their small opponents, and carried their dead
0361  vitality: peas and vetches, for instance, are  KILLED by even a few days immersion in sea water: bu
0362  mination. Certain seeds, however, were always  KILLED by this process. Although the beaks and feet
0364  t indies to our western shores, where, if not  KILLED by so long an immersion in salt water, they c
0385  ransported by birds, and they are immediately  KILLED by sea water, as are the adults. I could not
0393  s and their spawn are known to be immediately  KILLED by sea water, on my view we can see that ther
0397  is notorious that land shells are very easily  KILLED by salt: their eggs, at least such as I have
0397  ch as I have tried, sink in sea water and are  KILLED by it. Yet there must be, on my view, some un
0036  the barbarians of Tierra del Fuego, by their  KILLING and devouring their old women, in times of d
0184  lmost like a creeper. It often, like a shrike, KILLS small birds by blows on the head; and I have
0017  nd bull dog, which we all know propagate their KIND so truly, were the offspring of any single spe
0033  te from the proper standard. With animals this KIND of selection is, in fact, also followed; for h
0034  isting in the country. But, for our purpose, a KIND of Selection, which may be called Unconscious,
0035  ir cattle. Slow and insensible changes of this KIND could never be recognised unless actual measur
0036  or ones; so that in this case there would be a KIND of unconscious selection going on. We see the
0046  ry perplexing, for they seem to show that this KIND of variability is independent of the condition
0050  ieties; for he knows nothing of the amount and KIND of variation to which the group is subject; an
0068  of distinct species, which subsist on the same KIND of food. Even when climate, for instance extre
0081  t chance of surviving and of procreating their KIND? On the other hand, we may feel sure that any
0082  s; and still further modifications of the same KIND would often still further increase the advanta
0085  effective in giving the proper colour to each  KIND of grouse, and in keeping that colour, when on
0096  bitually, concur for the reproduction of their KIND. This view, I may add, was first suggested by
0099  other, and of these only 78 were true to their KIND, and some even of these were not perfectly tru
0103  fferent seasons, or from varieties of the same KIND preferring to pair together. Intercrossing pla
0104  tter chance of surviving and propagating their KIND; and thus, in the long run, the influence of i
0107  thus be checked: after physical changes of any KIND, immigration will be prevented, so that new pl
0108  of the country undergoing modification of some KIND. The existence of such places will often depen
0155  nce may remain the same. Something of the same KIND applies to monstrosities: at least Is. Geoffro
0180  ot doubt that each structure is of use to each KIND of squirrel in its own country, by enabling it
0190  been formed by transitional gradations of some KIND. Numerous cases could be given amongst the low
0193  oo ignorant to argue that no transition of any KIND is possible. The electric organs offer another
0195  as much difficulty, though of a very different KIND, on this head, as in the case of an organ as p
0199  iefly through sexual selection of a particular KIND, but without here entering on copious details
0210  ast to be able to show that gradations of some KIND are possible; and this we certainly can do. I
0213  lly from true instincts. If we were to see one KIND of wolf, when young and without any training,
0213  rawl forward with a peculiar gait: and another KIND of wolf rushing round, instead of at, a herd o
0236  rom being sterile, they cannot propagate their KIND. The subject well deserves to be discussed at
0238  d yet no one ox could ever have propagated its KIND. Thus I believe it has been with social insect
0239  neuter insects all of one caste or of the same KIND, which have been rendered by natural selection
0239  ave myself compared perfect gradations of this KIND. It often happens that the larger or the small
0249  ally fertilised with hybrid pollen of the same KIND, their fertility, notwithstanding the frequent
0257  given. No one has been able to point out what  KIND, or what amount, of difference in any recognis
0263  expected, systematic affinity, by which every  KIND of resemblance and dissimilarity between organ
0270  ion. Gartner kept during several years a dwarf KIND of maize with yellow seeds; and a tall variety
0277  view is supported by a parallelism of another KIND; namely, that the crossing of forms only sligh
0277  oth depend on the amount of difference of some KIND between the species which are crossed. Nor is
0297  of varieties; and on this view do to find the KIND of evidence of change which on my theory we ou
0347  in our general review is, that barriers of any KIND, or obstacles to free migration, are related i
0348  r emigrants; here we have a barrier of another KIND, and as soon as this is passed we meet in the
0353  and striking influence which barriers of every KIND have had on distribution, is intelligible only
0366  . we have evidence of almost every conceivable KIND, organic and inorganic, that within a very rec
0380  ty years from Australia. Something of the same KIND must have occurred on the intertropical mounta
0391  heri and when settled in their new homes, each KIND will have been kept by the others to their pro
0413  sentence, a full description is given of each  KIND of dog. The ingenuity and utility of this syst
0425  ut community of descent by resemblances of any KIND. Therefore we choose those characters which, a
0427  insect as a moth. We see something of the same KIND even in our domestic varieties, as in the thic
0438  ng course of descent, primordial organs of any KIND, vertebrae in the one case and legs in the oth
0439  casionally though rarely see something of this KIND in plants: thus the embryonic leaves of the ul
0459  exist or could have existed, each good of its  KIND, that all organs and instincts are, in ever so
0486  our natural genealogies, by characters of any KIND which have long been inherited. Rudimentary ch
0489  w living very few will transmit progeny of any KIND to a far distant futurity; for the manner in w
0020  could purchase or obtain, and have been most  KINDLY favoured with skins from several quarters of
0226  r miller, of Cambridge, and this geometer has KINDLY read over the following statement, drawn up f
0110  , will generally press hardest on its nearest KINDRED, and tend to exterminate them. We see the sa
0431  w the blood relationship between the numerous KINDRED of any ancient and noble family, even by the
0011  ze from amount of food, colour from particular KINDS of food and from light, and perhaps the thick
```

Page **(Key Word)**

Page **(Key Word)**

```
0028 n 4 first kept pigeons and watched the several KINDS, knowing well how true they bred, I felt full
0033 he leaves; how much the fruit of the different KINDS of gooseberries differ in size, colour, shape
0048 ie under deep obligation for assistance of all KINDS, has marked for me 182 British plants, which
0063 struggle with the plants of the same and other KINDS which already clothe the ground. The missleto
0064 the evidence from our domestic animals of many KINDS which have run wild in several parts of the w
0069 esser degree, for the number of species of all KINDS, and therefore of competitors, decreases nort
0074 ed to attribute their proportional numbers and KINDS to what we call chance. But how false a view
0074 the same beautiful diversity and proportion of KINDS as in the surrounding virgin forests. What a
0075 n forests. What a struggle between the several KINDS of trees must here have gone on during long c
0075 rse of centuries, the proportional numbers and KINDS of trees now growing on the old Indian ruins!
0076 mixed in due proportion, otherwise the weaker KINDS will steadily decrease in numbers and disappe
0091 born with an innate tendency to pursue certain KINDS of prey. Nor can this be thought very improba
0107 ts, at this period numerous in individuals and KINDS, will have been subjected to very severe comp
0111 flowers, take the place of older and inferior KINDS. In Yorkshire, it is historically known that
0113 em, for instance, being enabled to feed on new KINDS of prey, either dead or alive; some inhabitin
0133 ies, accords with our view that species of all KINDS are only well marked and permanent varieties.
0142 o their native countries. In treatises on many KINDS of cultivated plants, certain varieties are s
0153 innate tendency to further variability of all KINDS and on the other hand, the power of steady se
0183 any subordinate forms, for taking prey of many KINDS in many ways, on the land and in the water, u
0189 g each year on millions of individuals of many KINDS: and may we not believe that a living optical
0197 d not know that there were many black and pied KINDS, I dare say that we should have thought that
0218 s lay their eggs in the nests of bees of other KINDS. This case is more remarkable than that of th
0220 ing in materials for the nest, and food of all KINDS. During the present year, however, in the mon
0252 y by examining the flowers of the more sterile KINDS of hybrid rhododendrons, which produce no pol
0277 or systematic affinity attempts to express all KINDS of resemblance between all species. First cro
0293 h perfect accuracy. With marine animals of all KINDS, we may safely infer a large amount of migrat
0310 uch weight to the facts and arguments of other KINDS given in this volume, will undoubtedly at onc
0349 er. He hears from closely allied, yet distinct KINDS of birds, notes nearly similar, and sees thei
0358 a water. To my surprise I found that out of 87 KINDS, 64 germinated after an immersion of 28 days,
0361 being immediately devoured: and seeds of many KINDS in the crops of floating birds long retain th
0361 any facts showing how frequently birds of many KINDS are blown by gales to vast distances across t
0361 rse of two months, I picked up in my garden 12 KINDS of seeds, out of the excrement of small birds
0362 transported from place to place. I forced many KINDS of seeds into the stomachs of dead fish, and
0387 each plant as it grew; the plants were of many KINDS, and were altogether 537 in number: and yet t
0387 . i have stated that fresh water fish eat some KINDS of seeds, though they reject many other kinds
0387 kinds of seeds, though they reject many other KINDS after having swallowed them: even small fish
0388 ready occupying any pond, yet as the number of KINDS is small, compared with those on the land, th
0389 descent with modification. The species of all KINDS which inhabit oceanic islands are few in numb
0390 ure. Although in oceanic islands the number of KINDS of inhabitants is scanty, the proportion of e
0396 ts of oceanic islands, namely, the scarcity of KINDS, the richness in endemic forms in particular
0461 n. as we continually see that organisms of all KINDS are rendered in some degree sterile from thei
0475 low the same complex laws in their degrees and KINDS of resemblance to their parents, in being abs
0483 ere produced? Were all the infinitely numerous KINDS of animals and plants created as eggs or seed
0489 tangled bank, clothed with many plants of many KINDS, with birds singing on the bushes, with vario
0035 less improved. There is reason to believe that KING Charles's spaniel has been unconsciously modif
0159 similar unknown influences. In the vegetable KINGDOM we have a case of analogous variation, in th
0325 on general laws which govern the whole animal KINGDOM. M. Barrande has made forcible remarks to pr
0429 ing to a new order; and that in the vegetable KINGDOM, as I learn from Dr. Hooker, it has added on
0005 s, applied to the whole animal and vegetable KINGDOMS possesses several peculiar breeds of cattle,
0019 hungary, Spain, etc., but that each of these KINGDOMS: for in this case there can be no artificial
0063 fold force to the whole animal and vegetable KINGDOMS; for in this case there can be no artificial
0101 spect that, both in the vegetable and animal KINGDOMS, an occasional intercross with a distinct in
0126 ch main division of the animal and vegetable KINGDOMS. Although extremely few of the most ancient
0255 d how generally the same rules apply to both KINGDOMS. It has been already remarked, that the deor
0438 er great classes of the animal and vegetable KINGDOMS. Naturalists frequently speak of the skull a
0183 he margin of water, and then dashing like a KINGFISHER at a fish. In our own country the larger ti
0135 s less, until they became incapable of flight. KIRBY has remarked (and I have observed the same fa
0033 or tubers, or whatever part is valued, in the KITCHEN garden, in comparison with the flowers of th
0037 ave been longest cultivated in our flower and KITCHEN gardens. If it has taken centuries or thousa
0243 s, to roost in, like the males of our distinct KITTY wrens, a habit wholly unlike that of any othe
0002 society. Sir C. Lyell and Dr. Hooker, who both KNEW of my work, the latter having read my sketch o
0286 estern extremity of the district. If, then, we KNEW the rate at which the sea commonly wears away
0149 special purpose alone. In the same way that a KNIFE which has to cut all sorts of things may be o
0228 thick, square piece of wax, a thin and narrow, KNIFE edged ridge, coloured with vermilion. The bee
0007 probability in the view propounded by Andrew KNIGHT, that this variability may be partly connecte
0096 iew, I may add, was first suggested by Andrew KNIGHT. We shall presently see its importance; but I
0246 makes the rule universal; but then he cuts the KNOT, for in ten cases in which he found two forms,
0357 to some continent. This view cuts the Gordian KNOT of the dispersal of the same species to the mo
0006 ery inhabitant of this world. Still less do we KNOW of the mutual relations of the innumerable inh
0014 live in a wild state. In many cases we do not KNOW what the aboriginal stock was, and so could no
0017 , terrier, spaniel, and bull dog, which we all KNOW propagate their kind so truly, were the offspr
0017 d productions; but how could a savage possibly KNOW, when he first tamed an animal, whether it wou
0026 y within a dozen or twenty generations, for we KNOW of no fact countenancing the belief that the c
0029 etween the several races; and though they well KNOW that each race varies slightly, for they win t
0030 we now see them; indeed, in several cases, we KNOW that this has not been their history. The key
0037 ases we cannot recognise, and therefore do not KNOW, the wild parent stocks of the plants which ha
0040 at has sometimes been noticed, namely, that we KNOW nothing about the origin or history of any of
0047 try for a long time; for as long, as far as we KNOW, as have good and true species. Practically, w
0067 ncy to increase be still further increased. We KNOW not exactly what the checks are in even one si
0077 oes it not double or quadruple its numbers? We KNOW that it can perfectly well withstand a little
0078 ther. Probably in no single instance should we KNOW what to do, so as to succeed. It will convince
0088 scale of nature this law of battle descends, I KNOW not: male alligators have been described as fi
0089 closely attended to birds in confinement well KNOW that they often take individual preferences an
0100 the nature of the fertilising element; for we KNOW of no means, analogous to the action of insect
0110 ome naturalised, without causing, as far as we KNOW, the extinction of any natives. Furthermore, t
0113 iving on the same piece of ground. And we well KNOW that each species and each variety of grass is
```

Page **(Key Word)**

Page ***************************************(Key Word)***

```
0118  in its own country. And these circumstances we KNOW to be favourable to the production of new var
0126  ately prevail, no man can predict; for we well KNOW that many groups, formerly most extensively o
0134  ffects of long continued use or disuse, for we KNOW not the parent forms; but many animals have s
0140  udes, and conversely; but we do not positively KNOW that these animals were strictly adapted to t
0140  cases we assume such to be the case; nor do we KNOW that they have subsequently become acclimatis
0144  e likewise the most abnormal in their teeth. I KNOW of no case better adapted to show the importa
0145  t towards the central and external flowers: we KNOW, at least, that in irregular flowers, these n
0161  useless or rudimentary organs being, as we all KNOW them to be, thus inherited. Indeed, we may so
0161  st ancestral characters. As, however, we never KNOW the exact character of the common ancestor of
0162  these two cases: if, for instance, we did not KNOW that the rock pigeon was not feather footed o
0178  reasons already assigned (namely from what we KNOW of the actual distribution of closely allied
0182  ent organs of flight exclusively, as far as we KNOW, to escape being devoured by other fish? When
0188  l selection, although in this case he does not KNOW any of the transitional grades. His reason ou
0188  to avoid comparing the eye to a telescope. We KNOW that this instrument has been perfected by th
0189  no doubt many organs exist of which we do not KNOW the transitional grades, more especially if w
0190  itory organs of certain fish, or, for I do not KNOW which view is now generally held, a part of t
0191  eration from an ancient prototype, of which we KNOW nothing, furnished with a floating apparatus
0195  e being too positive even in this case, for we KNOW that the distribution and existence of cattle
0197  woodpeckers alone had existed, and we did not KNOW that there were many black and pied kinds, I
0204  r any one; yet in the case of any organ, if we KNOW of a long series of gradations in complexity
0204  gh natural selection. In the cases in which we KNOW of no intermediate or transitional states, we
0213  being known, for the young pointer can no more KNOW that he points to aid his master, than the wh
0235  particular distance from each other, than they KNOW what are the several angles of the hexagonal
0249  ation pollen is as often taken by chance (as I KNOW from my own experience) from the anthers of a
0253  s of the more fertile hybrid animals, I hardly KNOW of an instance in which two families of the s
0253  t in a very few generations. Although I do not KNOW of any thoroughly well authenticated cases of
0270  r. how far these experiments may be trusted, I KNOW not; but the forms experimentised on, are ran
0287  r hand, during oscillations of level, which we KNOW this area has undergone, the surface may have
0289  ving that they are closely consecutive. But we KNOW, for instance, from Sir R. Murchison's great
0294  h at which marine animals can flourish; for we KNOW what vast geographical changes occurred in ot
0305  any tertiary stratum. Hence we now positively KNOW that sessile cirripedes existed during the se
0308  nts of Europe and North America. But we do not KNOW what was the state of things in the intervals
0312  e first time, either locally, or, as far as we KNOW, on the face of the earth. If we may trust the
0319  tions, that rarity precedes extinction; and we KNOW that this has been the progress of events with
0320  old forms which have been exterminated; but we KNOW that the number of species has not gone on inc
0326  urse, or even their existence, would cease. We KNOW not at all precisely what are all the conditi
0340  been solely produced in South America. For we KNOW that Europe in ancient times was peopled by nu
0340  , to the northern half. In a similar manner we KNOW from Falconer and Cautley's discoveries, that
0350  hat cause which alone, as far as we positively KNOW, produces organisms quite like, or, as we see
0355  pre existing closely allied species. And I now KNOW from correspondence, that this coincidence he
0357  birthplace, and when in the course of time we KNOW something definite about the means of distribu
0362  ic juice, and do not in the least injure, as I KNOW by trial, the germination of seeds: now after
0362  to twenty hours, disgorge pellets, which, as I KNOW from experiments made in the Zoological Garder
0373  ported far from their parent source. We do not KNOW that the Glacial epoch was strictly simultaned
0377  pe, and in parts of temperate Australia. As we KNOW that many tropical plants and animals can with
0387  uteum) in a heron's stomach; although I do not KNOW the fact, yet analogy makes me believe that a
0391  does not possess one endemic land bird; and we KNOW from Mr. J. M. Jones's admirable account of Be
0391  confined to its shores: now, though we do not KNOW how sea shells are dispersed, yet we can see t
0399  ica, are related, and that very closely, as we KNOW from Dr. Hooker's account, to those of America
0426  been inherited from a common ancestor. And we KNOW that such correlated or aggregated characters
0436  multiplication of others, variations which we KNOW to be within the limits of possibility. In the
0466  n which we are confessedly ignorant; nor do we KNOW how ignorant we are. We do not know all the po
0466  nor do we know how ignorant we are. We do not KNOW all the possible transitional gradations betwe
0466  perfect organs; it cannot be pretended that we KNOW all the varied means of Distribution during th
0466  ion during the long lapse of years; or that we KNOW how imperfect the Geological Record is. Grave
0487  d from some one birthplace; and when we better KNOW the many means of migration, then, by the ligh
0488  ted. To my mind it accords better with what we KNOW of the laws impressed on matter by the Creator
0016  cies, we are soon involved in doubt, from not KNOWING whether they have descended from one or seve
0028  t kept pigeons and watched the several kinds, KNOWING well how true they bred, I felt fully as muc
0029  e generations. May not those naturalists who, KNOWING far less of the laws of inheritance than doe
0029  aws of inheritance than does the breeder, and KNOWING no more than he does of the intermediate lin
0110  finitely great, not that we have any means of KNOWING that any one region has as yet got its maxim
0207  ny individuals in the same way, without their KNOWING for what purpose it is performed, is usually
0235  of intersection. The bees, of course, no more KNOWING that they swept their spheres at one particu
0003  in every possible way by his large stores of KNOWLEDGE and his excellent judgment. In considering
0004  exinc cases I have invariably found that our KNOWLEDGE, imperfect though it be, of variation under
0018  opinion, from his large and varied stores of KNOWLEDGE, I should value more than that of almost an
0040  little free communication, the spreading and KNOWLEDGE of any new sub breed will be a slow process
0051  tinually studying; and he has little general KNOWLEDGE of analogical variation in other groups and
0062  dently the result of his great horticultural KNOWLEDGE. Nothing is easier than to admit in words t
0242  erited tools or weapons, and not by acquired KNOWLEDGE and manufactured instruments, a perfect div
0254  t that it cannot, under our present state of KNOWLEDGE, be considered as absolutely universal. Law
0255  y to animals, and considering how scanty our KNOWLEDGE is in regard to hybrid animals, I have been
0310  rities, to whom, with others, we owe all our KNOWLEDGE. Those who think the natural geological rec
0413  eems to me that nothing is thus added to our KNOWLEDGE. Such expressions as that famous one of Lin
0471  it saltum, which every fresh addition to our KNOWLEDGE tends to make more strictly correct, is on
0005  chapter I shall discuss the complex and little KNOWN laws of variation and of correlation of growt
0014  arly mature; peculiarities in the silkworm are KNOWN to appear at the corresponding caterpillar or
0023  two or three other species of rock pigeons are KNOWN; and these have not any of the characters of
0025  on! We can understand these facts, on the well KNOWN principle of reversion to ancestral character
0027  in several quarters of the world; the earliest KNOWN record of pigeons is in the fifth Aegyptian d
0030  obably been with the turnspit dog; and this is KNOWN to have been the case with the ancon sheep. B
0035  as probably been slowly altered from it. It is KNOWN that the English pointer has been greatly cha
0037  t has consisted in always cultivating the best KNOWN variety, sowing its seeds; and, when a slight
0037  accumulated, explains, as I believe, the well KNOWN fact, that in a vast number of cases we canno
0045  be called individual differences, such as are KNOWN frequently to appear in the offspring from th
```

Page ***************************************(Key Word)***

Page ***************************************(Key Word)***************************************

```
)47  y of naturalists, for few well marked and well  KNOWN  varieties can be named which have not been ra
)49  ill here give only a single instance, the well   KNOWN  one of the primrose and cowslip, or Primula v
)50  t it must be confessed, that it is in the best   KNOWN  countries that we find the greatest number of
)64  t is reckoned to be the slowest breeder of all   KNOWN  animals, and I have taken some pains to estim
)67  n in regard to mankind, so incomparably better  KNOWN  than any other animal. This subject has been
)84  would increase in countless numbers: they are   KNOWN  to suffer largely from birds of prey: and haw
)87  be selected, the thickness of the shell being   KNOWN  to vary like every other structure. Sexual Se
)89  rked effect. I strongly suspect that some well  KNOWN  laws with respect to the plumage of male and
)98  contrivance for self fertilisation, it is well  KNOWN  that if very closely allied forms or varietie
)07  r we find some of the most anomalous forms now  KNOWN  in the world, as the Ornithorhynchus and Lepi
111  ferior kinds. In Yorkshire, it is historically  KNOWN  that the ancient black cattle were displaced
130  e now no living representatives, and which are  KNOWN  to us only from having been found in a fossil
133  ch to the conditions of life. Thus, it is well  KNOWN  to furriers that animals of the same species
134  ife must act. Again, innumerable instances are  KNOWN  to every naturalist of species keeping true,
137  nstantly aid the effects of disuse. It is well  KNOWN  that several animals, belonging to the most d
139  closely related, we might expect from the well  KNOWN  relationship of most of their other productio
187  robably have to descend far beneath the lowest  KNOWN  fossiliferous stratum to discover the earlier
194  considering that the proportion of living and  KNOWN  forms to the extinct and unknown is very smal
194  named, towards which no transitional grade is  KNOWN  to lead. The truth of this remark is indeed s
198  late on the relative importance of the several  KNOWN  and unknown laws of variation: and I have her
208  s could be pointed out. As in repeating a well  KNOWN  song, so in instincts, one action follows ano
210  e and North America, and for no instinct being  KNOWN  amongst extinct species, how very generally g
213  light by each breed, and without the end being  KNOWN, for the young pointer can no more know that
214  fferent breeds of dogs are crossed. Thus it is  KNOWN  that a cross with a bulldog has affected for
214  lly shown a tendency in this line: and this is  KNOWN  occasionally to happen, as I once saw in a pu
216  ed by naturalists as the most wonderful of all  KNOWN  instincts. It is now commonly admitted that t
217  ral instances of various birds which have been  KNOWN  occasionally to lay their eggs in other bird'
219  an these well ascertained facts? If we had not  KNOWN  of any other slave making ant, it would have
225  laced in a double layer: each cell, as is well  KNOWN, is an hexagonal prism, with the basal edges
233  ee? I think the answer is not difficult: it is  KNOWN  that bees are often hard pressed to get suffi
234  ng the winter: and the security of the hive is  KNOWN  mainly to depend on a large number of bees be
235  thus, as I believe, the most wonderful of all  KNOWN  instincts, that of the hive bee, can be expla
235  cases, in which no intermediate gradations are  KNOWN  to exist: cases of instinct of apparently suc
242  ative case of neuter insects, against the well  KNOWN  doctrine of Lamarck. Summary. I have endeavou
243  wrens, a habit wholly unlike that of any other  KNOWN  bird. Finally, it may not be a logical deduct
246  e fertility of varieties, that is of the forms  KNOWN  or believed to have descended from common par
255  and the early withering of the flower is well  KNOWN  to be a sign of incipient fertilisation. From
277  tween all species. First crosses between forms  KNOWN  to be varieties, or sufficiently alike to con
279  cies. On their sudden appearance in the lowest  KNOWN  fossiliferous strata. In the sixth chapter I
285  1 am tempted to give one other case, the well  KNOWN  one of the denudation of the Weald. Though it
287  namely, that numbers of our fossil species are  KNOWN  and named from single and often broken specim
288  found in any tertiary formation: yet it is now  KNOWN  that the genus Chthamalus existed during the
289  reme degree. For instance, not a land shell is  KNOWN  belonging to either of these vast periods, wi
289  and that not a cave or true lacustrine bed is  KNOWN  belonging to the age of our secondary or pala
294  en first immigrated into that area. It is well  KNOWN, for instance, that several species appeared
298  widest range, far exceeding the limits of the  KNOWN  geological formations of Europe, which have o
303  suddenly been produced. I may recall the well  KNOWN  fact that in geological treatises, published
303  he tertiary series. And now one of the richest  KNOWN  accumulations of fossil mammals belongs to th
305  of which the affinities are as yet imperfectly  KNOWN, are really teleostean. Assuming, however, th
305  uous to remark that hardly any fossil fish are  KNOWN  from south of the equator: and by running thr
305  logy it will be seen that very few species are  KNOWN  from several formations in Europe. Some few f
306  ance of groups of Allied Species in the lowest  KNOWN  fossiliferous strata. There is another and al
306  the same group, suddenly appear in the lowest  KNOWN  fossiliferous rocks. Most of the arguments wh
306  apply with nearly equal force to the earliest  KNOWN  species. For instance, I cannot doubt that al
306  , and which probably differed greatly from any  KNOWN  animal. Some of the most ancient Silurian ani
307  rget that only a small portion of the world is  KNOWN  with accuracy. M. Barrande has lately added a
308  islands: but not one oceanic island is as yet  KNOWN  to afford even a remnant of any palaeozoic or
310  ions: the almost entire absence, as at present  KNOWN, of fossiliferous formations beneath the Silu
318  we have seen, having endured from the earliest  KNOWN  dawn of life to the present day: some having
333  xpect, is that those groups, which have within  KNOWN  geological periods undergone much modificatio
335  two remote formations. Pictet gives as a well  KNOWN  instance, the general resemblance of the orga
338  l proof. Seeing, for instance, that the oldest  KNOWN  mammals, reptiles, and fish strictly belong t
343  ontinents, formed of formations older than any  KNOWN  to us, may now all be in a metamorphosed cond
358  eley's aid, a few experiments, it was not even  KNOWN  how far seeds could resist the injurious acti
359  e of these floated for a long time. It is well  KNOWN  what a difference there is in the buoyancy of
362  wn to the distance of 500 miles, and hawks are  KNOWN  to look out for tired birds, and the contents
363  have to recur to this subject. As icebergs are  KNOWN  to be sometimes loaded with earth and stones,
364  and, like Great Britain, has not, as far as is  KNOWN  (and it would be very difficult to prove this
365  sibly exist, is one of the most striking cases  KNOWN  of the same species living at distant points,
386  tell. With respect to plants, it has long been  KNOWN  what enormous ranges many fresh water and eve
387  feeding its young, in the same way as fish are  KNOWN  sometimes to be dropped. In considering these
393  ance. But as these animals and their spawn are  KNOWN  to be immediately killed by sea water, on my
413  ded: and that propinquity of descent, the only  KNOWN  cause of the similarity of organic beings, is
423  sification of varieties, which are believed or  KNOWN  to have descended from one species. These are
424  es differ in the most important characters, is  KNOWN  to every naturalist: scarcely a single fact c
424  sly been ranked as three distinct genera, were  KNOWN  to be sometimes produced on the same spike, t
456  of this element of descent, the only certainly  KNOWN  cause of similarity in organic beings, we sha
472  have not been observed. The complex and little  KNOWN  laws governing variation are the same, as far
485  well marked varieties is, that the latter are  KNOWN, or believed, to be connected at the present
488  the whole history of the world, as at present  KNOWN, although of a length quite incomprehensible
0044  atisfied all naturalists, yet every naturalist  KNOWS  vaguely what he means when he speaks of a spe
0050  der as specific, and what as varieties: for he  KNOWS  nothing of the amount and kind of variation t
0070  cked during winter: but any one who has tried,  KNOWS  how troublesome it is to get seed from a few
0074  ber of mice is largely dependent, as every one  KNOWS, on the number of cats: and Mr. Newman says,
0088  ed leg, as well as the brutal cockfighter, who  KNOWS  well that he can improve his breed by careful
0097  other view are inexplicable. Every hybridizer  KNOWS  how unfavourable exposure to wet is to the fe
```

Page ***************************************(Key Word)***************************************

knows

Page **(Key Word)***

```
0133  ies confined to sea coasts, as every collector KNOWS, are often brassy or lurid. Plants which liv
0144  ompositous and Umbelliferous plants. Every one KNOWS the difference in the ray and central floret
0165  figured by Dr. Gray (and he informs me that he KNOWS of a second case) from the ass and the hemio
0213  ts to aid his master, than the white butterfly KNOWS why she lays her eggs on the leaf of the cab
0098  d to cause the stamens to spring forward, as KOLREUTER has shown to be the case with the barberry
0246  e two conscientious and admirable observers, KOLREUTER and Gartner, who almost devoted their live
0246  high generality of some degree of sterility. KOLREUTER makes the rule universal; but then he cuts
0247  sal; and he disputes the entire fertility of KOLREUTER's ten cases. But in these and in many othe
0248  enced observers who have ever lived, namely, KOLREUTER and Gartner, should have arrived at diamet
0250  s fertile as the pure parent species, as are KOLREUTER and Gartner that some degree of sterility
0258  he same two species was long ago observed by KOLREUTER. To give an instance: Mirabilis jalappa ca
0258  thus produced are sufficiently fertile; but KOLREUTER tried more than two hundred times, during
0271  o suspect that they present analogous facts. KOLREUTER, whose accuracy has been confirmed by ever
0274  o the results of several experiments made by KOLREUTER. These alone are the unimportant differenc
0451  owers often have a rudiment of a pistil; and KOLREUTER found that by crossing such male plants wi
0163  ually quite lost, in dark coloured asses. The KOULAN of Pallas is said to have been seen with a d
0065  lants now most numerous over the wide plains of LA Plata, clothing square leagues of surface almos
0184  chase insects on the wing; and on the plains of LA Plata, where not a tree grows, there is a wood
0318  n of species, than I have done. When I found in LA Plata the tooth of a horse embedded with the re
0323  ia, and Toxodon had been brought to Europe from LA Plata, without any information in regard to the
0339  ose of the armadillo, found in several parts of LA Plata; and Professor Owen has shown in the most
0349  (american ostrich), and northward the plains of LA Plata by another species of the same genus; and
0349  nder the same latitude. On these same plains of LA Plata, we see the agouti and bizcacha, animals
0380  y many European productions cover the ground in LA Plata, and in a lesser degree in Australia, and
0380  ope during the last two or three centuries from LA Plata, and during the last thirty or forty yea
0198  pe of the head of the young in the womb. The LABORIOUS breathing necessary in high regions would,
0045  s, and that there are not many men who will LABORIOUSLY examine internal and important organs, and
0093  has been called the physiological division of LABOUR; hence we may believe that it would be advant
0094  antageous on the principle of the division of LABOUR, individuals with this tendency more and more
0115  same as that of the physiological division of LABOUR in the organs of the same individual body, a
0242  s, on the same principle that the division of LABOUR is useful to civilised man. As ants work by
0242  nufactured instruments, a perfect division of LABOUR could be effected with them only by the worke
0242  believe, effected this admirable division of LABOUR, in the communities of ants, by the means of
0486  mechanical invention as the summing up of the LABOUR, the experience, the reason, and even the blu
0227  ly at what distance to stand from her fellow LABOURERS when several are making their spheres; but
0484  ry. Systematists will be able to pursue their LABOURS as at present; but they will not be incessan
0365  s of America, are all the same with those of LABRADOR, and nearly all the same, as we hear from As
0368  n scandinavia; those of the United States to LABRADOR; those of the mountains of Siberia to the ar
0289  have been discovered either in caves or in LACUSTRINE deposits; and that not a cave or true lacus
0289  trine deposits; and that not a cave or true LACUSTRINE bed is known belonging to the age of our se
0406  ives ranging widely, such facts, as alpine, LACUSTRINE, and marsh productions being related (with
0034  ient Chinese encyclopaedia. Explicit rules are LAID down by some of the Roman classical writers. F
0077  prey to other animals. The store of nutriment LAID up within the seeds of many plants seems at fi
0217  own nest and sit on her own eggs, those first LAID would have to be left for some time unincubate
0217  the American cuckoo; but that occasionally she LAID an egg in another bird's nest. If the old bird
0222  rious in their late combat. At the same time I LAID on the same place a small parcel of the pupae
0275  female ass and stallion. Much stress has been LAID by some authors on the supposed fact, that mon
0402  th its own species, for annually more eggs are LAID there than can possibly be reared; and we may
0462  ured by years, too much stress ought not to be LAID on the occasional wide diffusion of the same s
0378  still exist in North America, which must have LAIN on the line of march. But I do not doubt that
0403  ature. We see this on every mountain, in every LAKE and marsh. For Alpine species, excepting in so
0346  ty mountains, grassy plains, forests, marshes, LAKES, and great rivers, under almost every tempera
0383  . summary of the last and present chapters. As LAKES and river systems are separated from each oth
0404  ing lowlands. So it is with the inhabitants of LAKES and marshes, excepting in so far as great fac
0242  r insects, against the well known doctrine of LAMARCK. Summary. I have endeavoured briefly in this
0427  ties and analogical or adaptive resemblances. LAMARCK first called attention to this distinction.
0085  is in a flock of white sheep to destroy every LAMB with the faintest trace of black. In plants th
0030  s breeds of sheep fitted either for cultivated LAND or mountain pasture, with the wool of one bree
0042  other races. In this respect enclosure of the LAND plays a part. Wandering savages or the inhabit
0052  wollaston with the varieties of certain fossil LAND shells in Madeira. If a variety were to flouri
0071  having been done, with the exception that the LAND had been enclosed, so that cattle could not en
0072  and had failed. No wonder that, as soon as the LAND was enclosed, it became thickly clothed with v
0077  no doubt stands in the closest relation to the LAND being already thickly clothed by other plants;
0078  f an utter desert, will competition cease. The LAND may be extremely cold or dry, yet there will b
0083  owed foreigners to take firm possession of the LAND. And as foreigners have everywhere been
0100  ning for a very brief space to animals: on the LAND there are some hermaphrodites, as land mollusc
0100  on the land there are some hermaphrodites, as LAND mollusca and earth worms; but these all pair.
0104  change, such as of climate or elevation of the LAND, etc.; and thus new places in the natural econ
0107  l area compared with that of the sea or of the LAND; and, consequently, the competition between fr
0115  have succeeded in becoming naturalised in any LAND would generally have been closely allied to th
0116  e substances. So in the general economy of any LAND, the more widely and perfectly the animals and
0173  y the same place in the natural economy of the LAND. These representative species often meet and i
0174  ermediate zones. By changes in the form of the LAND and of climate, marine areas now continuous mu
0178  rly have existed in each broken portion of the LAND, but these links will have been supplanted and
0179  of such views as I hold, how, for instance, a LAND carnivorous animal could have been converted i
0180  ers, and preys like other polecats on mice and LAND animals. If a different case had been taken, a
0182  n); as fins in the water and front legs on the LAND, like the penguin; as sails, like the ostrich;
0182  stacea and Mollusca are adapted to live on the LAND, and seeing that we have flying birds and mamm
0183  taking prey of many kinds in many ways, on the LAND and in the water, until their organs of flight
0186  rds with webbed feet, either living on the dry LAND or most rarely alighting on the water; that th
0196  presence and use for many purposes in so many LAND animals, which in their lungs or modified swim
0285  or since the crust cracked, the surface of the LAND has been so completely planed down by the acti
0287  face may have existed for millions of years as LAND, and thus have escaped the action of the sea:
0287  each of these years, over the whole world, the LAND and the water has been peopled by hosts of liv
0289  tary in an extreme degree. For instance, not a LAND shell is known belonging to either of these va
0290  ought up by the slow and gradual rising of the LAND within the grinding action of the coast waves.
```

Page **(Key Word)***

Page ***************************************(Key Word)***************************************

```
0292  e. curing periods of elevation the area of the  LAND and of the adjoining shoal parts of the sea wi
0305  , if the Malay Archipelago were converted into  LAND, the tropical parts of the Indian Ocean would
0306  the world, as it would be for a naturalist to  LAND for five minutes on some one barren point in A
0308  from first to last large islands or tracts of  LAND, whence the sediment was derived, occurred in
0308  d states during these intervals existed as dry  LAND, or as a submarine surface near land, on which
0308  ed as dry land, or as a submarine surface near  LAND, on which sediment was not deposited, or again
0308  g oceans, which are thrice as extensive as the  LAND, we see them studded with many islands; but no
0309  at where continents now exist, large tracts of  LAND have existed, subjected no doubt to great osci
0312  ed very slowly, one after another, both on the  LAND and in the waters. Lyell has shown that it is
0313  s have changed greatly. The productions of the  LAND seem to change at a quicker rate than those of
0314  eographical distribution; for instance, in the  LAND shells and coleopterous insects of Madeira hav
0323  t data to judge whether the productions of the  LAND and of fresh water change at distant points in
0326  parallel succession in the productions of the  LAND than of the sea. Dominant species spreading fr
0326  production of new and dominant species on the  LAND, and another for those in the waters of the se
0339  as the relation between the extinct and living  LAND shells of Madeira; and between the extinct and
0346  by changes of climate and of the level of the  LAND, and by occasional means. Dispersal during the
0346  clude the northern parts where the circumpolar  LAND is almost continuous, all authors agree that o
0347  hern hemisphere, if we compare large tracts of  LAND in Australia, South Africa, and western South
0347  ds, excepting in the northern parts, where the  LAND almost joins, and where, under a slightly diff
0348  m each other by impassable barriers, either of  LAND or open sea, they are wholly distinct. Or the
0350  ouchout space and time, over the same areas of  LAND and water, and independent of their physical c
0356  ubject in some detail. Changes of level in the  LAND must also have been highly influential: a narr
0356  merly have blended: where the sea now extends,  LAND may at a former period have connected islands
0357  e with security on the former extension of the  LAND. But I do not believe that it will ever be pro
0358  d originally. existed as mountain ranges on the  LAND, some at least of the islands would have been
0362  s. fresh water fish, I find, eat seeds of many  LAND and water plants: fish are frequently devoured
0363  en carried brushwood, bones, and the nest of a  LAND bird, I can hardly doubt that they must occasi
0364  ure our climate. Almost every year, one or two  LAND birds are blown across the whole Atlantic Ocea
0365  fully stocked with inhabitants. On almost bare  LAND, with few or no destructive insects or birds l
0370  ly arctic productions then lived on the broken  LAND still nearer to the pole. Now if we look at a
0370  er the Polar Circle there is almost continuous  LAND from western Europe, through Siberia, to easte
0370  ica. And to this continuity of the circumpolar  LAND, and to the consequent freedom for intermigrat
0370  ls inhabited the almost continuous circumpolar  LAND; and that these plants and animals, both in th
0371  s will have been almost continuously united by  LAND, serving as a bridge, since rendered impassabl
0372  aturalists as specifically distinct. As on the  LAND, so in the waters of the sea, a slow southern
0377  of course, have been greatly favoured by high  LAND, and perhaps by a dry climate; for Dr. Falcone
0378  tude from their native country and covered the  LAND at the foot of the Pyrenees. At this period of
0379  north to south is due to the greater extent of  LAND in the north, and to the northern forms having
0380  rds extinction. A mountain is an island on the  LAND; and the intertropical mountains before the Gl
0380  e that the productions of these islands on the  LAND yielded to those produced within the larger ar
0381  comes naturalised by man's agency in a foreign  LAND; why one ranges twice or thrice as far, and is
0381  es at points so enormously remote as Kerguelen  LAND, New Zealand, and Fuegia, I believe that towar
0382  ing in the mountian fastnesses of almost every  LAND, which serve as a record, full of interest to
0383  s are separated from each other by barriers of  LAND, it might have been thought that fresh water p
0384  s within the recent period in the level of the  LAND, having caused rivers to flow into each other.
0384  rhine of considerable changes of level in the  LAND within a very recent geological period, and wh
0384  , and when the surface was peopled by existing  LAND and fresh water shells. The wide difference of
0385  d and adapted to the fresh waters of a distant  LAND. Some species of fresh water shells have a ver
0386  when forty five miles distant from the nearest  LAND: how much farther it might have flown with a f
0386  ould not be washed off their feet; when making  LAND, they would be sure to fly to their natural fr
0388  of kinds is small, compared with those on the  LAND, the competition will probably be less severe
0390  mpare, for instance, the number of the endemic  LAND shells in Madeira, or of the endemic birds in
0390  ed. Thus in the Galapagos Islands nearly every  LAND bird, but only two out of the eleven marine bi
0391  could arrive at these islands more easily than  LAND birds. Bermuda, on the other hand, which lies
0391  ry peculiar soil, does not possess one endemic  LAND bird; and we know from Mr. J. M. Jones's admir
0391  is inhabited by a wonderful number of peculiar  LAND shells, whereas not one species of sea shell i
0391  rds, might be transported far more easily than  LAND shells, across three or four hundred miles of
0396  s having been formerly connected by continuous  LAND with the nearest continent; for on this latter
0397  e most isolated and smallest, are inhabited by  LAND shells, generally by endemic species; but some
0397  ven several interesting cases in regard to the  LAND shells of the islands of the Pacific. Now it i
0397  lands of the Pacific. Now it is notorious that  LAND shells are very easily killed by salt; their e
0397  d thus get transported? It occurred to me that  LAND shells, when hybernating and having a membrano
0398  outh America. Here almost every product of the  LAND and water bears the unmistakeable stamp of the
0398  f the American continent. There are twenty six  LAND birds, and twenty five of these are ranked by
0398  ent, yet feels that he is standing on American  LAND. Why should this be so? Why should the species
0399  l means of transport or by formerly continuous  LAND, from America; and the Cape de Verde islands f
0399  can be explained. Thus the plants of Kerguelen  LAND, though standing nearer to Africa than to Amer
0403  santo possess many distinct but representative  LAND shells, some of which live in crevices of ston
0403  h islands have been colonised by some European  LAND shells which no doubt had some advantage over
0403  sical conditions, and are united by continuous  LAND, yet they are inhabited by a vast number of di
0407  the changes of climate and of the level of the  LAND, which have certainly occurred within the rece
0408  nd the high importance of barriers, whether of  LAND or water, which separate our several zoologica
0411  oup had been exclusively fitted to inhabit the  LAND, and another the water; one to feed on flesh,
0428  to inhabit for instance the three elements of  LAND, air, and water, we can perhaps understand how
0468  ictory. As geology plainly proclaims that each  LAND has undergone great physical changes, we might
0472  ed by the naturalised productions from another  LAND. Nor ought we to marvel if all the contrivance
0487  mer changes of climate and of the level of the  LAND, we shall surely be enabled to trace in an adm
0363  ce we may safely infer that icebergs formerly  LANDED their rocky burthens on the shores of these m
0185  en is nearly as aquatic as the coot; and the  LANDRAIL nearly as terrestrial as the quail or partri
0115  tion of plants through man's agency in foreign  LANDS. It might have been expected that the plants
0290  at changes in the geography of the surrounding  LANDS, whence the sediment has been derived, accord
0358  elation of the tertiary inhabitants of several  LANDS and even seas to their present inhabitants, a
0365  ed from each other by hundreds of miles of low  LANDS, where the Alpine species could not possibly
0371  the several mountain ranges and on the arctic  LANDS of North America and Europe, are inexplicable
0372  past and present inhabitants of the temperate  LANDS of the two Worlds. Hence it has come, that wh
0381  ment of the Glacial period, when the antarctic  LANDS, now covered with ice, supported a highly pec
```

Page ***************************************(Key Word)***************************************

Lands

Page ***(Key Word)***

```
0399  new Zealand, South America, and other southern  LANDS  were long ago partially stocked from a nearly
0405  important power of being victorious in distant  LANDS  in the struggle for life with foreign associa
0406  ore specified) to those on the surrounding low  LANDS  and dry lands, though these stations are so d
0406  to those on the surrounding low lands and dry   LANDS, though these stations are so different, the
0040  . but, in fact, a breed, like a dialect of a    LANGUAGE, can hardly be said to have had a definite o
0186  s to me only restating the fact in dignified    LANGUAGE. He who believes in the struggle for existen
0311  few lines. Each word of the slowly changing     LANGUAGE, in which the history is supposed to be writ
0382  ell in a striking passage has speculated, in    LANGUAGE almost identical with mine, on the effects o
0422  one. Yet it might be that some very ancient     LANGUAGE had altered little, and had given rise to fe
0438  mmon element. Naturalists, however, use such    LANGUAGE only in a metaphorical sense: they are far f
0438  that naturalists can hardly avoid employing     LANGUAGE having this plain signification. On my view
0485  slip; and in this case scientific and common    LANGUAGE will come into accordance. In short, we shal
0020  ray from Persia. Many treatises in different    LANGUAGES have been published on pigeons, and some of
0422  iew of classification, by taking the case of    LANGUAGES. If we possessed a perfect pedigree of mank
0422  fford the best classification of the various    LANGUAGES now spoken throughout the world; and if all
0422  ken throughout the world; and if all extinct    LANGUAGES, and all intermediate and slowly changing d
0422  ltered little, and had given rise to few new    LANGUAGES, whilst others (owing to the spreading and
0422  altered much, and had given rise to many new    LANGUAGES and dialects. The various degrees of differ
0422  ts. The various degrees of difference in the    LANGUAGES from the same stock, would have to be expre
0423  ly natural, as it would connect together all    LANGUAGES, extinct and modern, by the closest affinit
0063  ese parasites grow on the same tree, it will    LANGUISH and die. But several seedling missletoes, gr
0165  vy belgian cart horse, Welch ponies, cobs, the LANKY  Kattywar race, etc., inhabiting the most dist
0056  plainly tells us that small genera have in the  LAPSE  of time often increased greatly in size; and
0084  ss, until the hand of time has marked the long  LAPSE  of ages, and then so imperfect is our view in
0112  as forming two sub breeds; finally, after the   LAPSE  of centuries, the sub breeds would become con
0158  vinc more or less completely, according to the  LAPSE  of time, overmastered the tendency to reversi
0169  must be a very slow process, requiring a long   LAPSE  of time, in this case, natural selection may
0201  be found on the whole advantageous. After the   LAPSE  of time, under changing conditions of life, i
0279  ediate varieties; on their number. On the vast  LAPSE  of time, as inferred from the rate of deposit
0282  true, such have lived upon this earth. On the   LAPSE  of time. Independently of our not finding fos
0282  acts leading the mind feebly to comprehend any of the LAPSE  of time. He who can read Sir Charles Lyell's
0282  fore he can hope to comprehend anything of the  LAPSE  of time, the monuments of which we see around
0284  tter, probably offers the best evidence of the  LAPSE  of time. I remember having been much struck w
0287  to gain some notion, however imperfect, of the  LAPSE  of years. During each of these years, over th
0293  . although each formation may mark a very long  LAPSE  of years, each perhaps is short compared with
0294  ely great change of climate; on the prodigious  LAPSE  of time, all included within this same glacia
0296  ant of this fact would have suspected the vast  LAPSE  of time represented by the thinner formation.
0309  s of preponderant movement have changed in the  LAPSE  of ages? At a period immeasurably antecendent
0365  occasional means of transport, during the long  LAPSE  of geological time, whilst an island was bein
0396  n all cases depends to a certain degree on the  LAPSE  of time, and as during changes of level it is
0464  ient for any amount of organic change; for the  LAPSE  of time has been so great as to be utterly in
0466  e varied means of Distribution during the long  LAPSE  of years, or that we know how imperfect the G
0481  and now that we have acquired some idea of the  LAPSE  of time, we are too apt to assume, without pr
0488  tions probably serves as a fair measure of the  LAPSE  of actual time. A number of species, however,
0003  ears has aided me in every possible way by his  LARGE  stores of knowledge and his excellent judgmen
0004  under Domestication. We shall thus see that a   LARGE  amount of hereditary modification is at least
0012  aks have small feet; and those with long beaks  LARGE  feet. Hence, if man goes on selecting, and th
0012  endless. Dr. Prosper Lucas's treatise, in two   LARGE  volumes, is the fullest and the best on this
0015  action of the poor soil), that they would to a  LARGE  extent, or even wholly, revert to the wild ab
0018  wild stock. Mr. Blyth, whose opinion, from his  LARGE  and varied stores of knowledge, I should valu
0021  accompanied by greatly elongated eyelids, very  LARGE  external orifices to the nostrils, and a wide
0021  ird of great size, with long, massive beak and  LARGE  feet; some of the sub breeds of runts have ve
0028  egard to the many species of finches, or other  LARGE  groups of birds, in nature. One circumstance
0031  , even within a single lifetime, modified to a  LARGE  extent some breeds of cattle and sheep. In or
0035  s spaniel has been unconsciously modified to a  LARGE  extent since the time of that monarch. Some h
0037  the best varieties they could anywhere find. A  LARGE  amount of change in our cultivated plants, th
0041  xtreme care, to allow of the accumulation of a  LARGE  amount of modification in almost any desired
0041  f their appearance will be much increased by a  LARGE  number of individuals being kept; and hence t
0041  d. on the other hand, nurserymen, from raising  LARGE  stocks of the same plants, are generally far
0041  g new and valuable varieties. The keeping of a  LARGE  number of individuals of a species in any cou
0042  cocks, from not being very easily reared and a  LARGE  stock not kept; in geese, from being valuable
0054  ger genera, or those including many species, a  LARGE  proportional number of dominant species. But
0055  a general rule, to be now forming. Where many   LARGE  genera which present any varieties, invariabl
0055  e smaller genera. Moreover, the species of the  LARGE  genera are now varying much, and are thus inc
0056  pecies, beyond the average. It is not that all  LARGE  genera have often come to their maxima, decli
0056  time often increased greatly in size; and that LARGE  genera and their recorded varieties which des
0056  ere are other relations between the species of LARGE  genera the amount of difference between the s
0057  ts, and Westwood in regard to insects, that in LARGE  genera are related to each other, in the same
0057  nt of difference. Moreover, the species of the LARGE  genera are apt to be closely, but
0059  more than the average number of varieties. In  LARGE  genera the species are apt to be closely, but
0059  . in all these several respects the species of LARGE  genera present a strong analogy with varietie
0062  hat I use the term Struggle for Existence in a  LARGE  and metaphorical sense, including dependence
0066  onditions, a whole district, let it be ever so LARGE. The condor lays a couple of eggs and the ost
0066  two species can be supported in a district. A  LARGE  number of eggs is of some importance to those
0066  crease in number. But the real importance of a LARGE  number of eggs or seeds is to make up for muc
0068  stock of partridges, grouse, and hares on any  LARGE  estate depends chiefly on the destruction of
0070  its prey. On the other hand, in many cases, a  LARGE  stock of individuals of the same species, rel
0070  y single seed. This view of the necessity of a LARGE  stock of the same species for its preservatio
0071  had ample means of investigation, there was a  LARGE  and extremely barren heath, which had never b
0071  he distant hilltops: within the last ten years LARGE  spaces have been enclosed, and self sown firs
0074  edible that the presence of a feline animal in LARGE  numbers in a district might determine, throug
0086  only for a few hours, and which never feed, a  LARGE  part of their structure is merely the correla
0096  ertebrate animals, all insects, and some other LARGE  groups of animals: pair for each birth. Moder
0096  dermaphrodites; and of real hermaphrodites a   LARGE  number pair; that is, two individuals regular
0096  alone. In the first place, I have collected so LARGE  a body of facts, showing, in accordance with
0097  of nature, we can, I think, understand several LARGE  classes of facts, such as the following, whic
```

Page ***(Key Word)***

Page **********************************(Key Word)**********************************

Page		LARGE	
0099	plants, be allowed to seed near each other, a	LARGE	majority, as I have found, of the seedlings t
0102	ion. This is an extremely intricate subject. A	LARGE	amount of inheritable and diversified variabi
0102	individual differences suffice for the work. A	LARGE	number of individuals, by giving a better cha
0102	scious process of selection, notwithstanding A	LARGE	amount of crossing with inferior animals. Thu
0102	ll up the unoccupied place. But if the area be	LARGE	, its several districts will almost certainly
0103	, nurserymen always prefer getting seed from a	LARGE	body of plants of the same variety, as the ch
0104	n. in a confined or isolated area, if not very	LARGE	, the organic and inorganic conditions of life
0105	ical distribution; yet of these species a very	LARGE	proportion are endemic, that is, have been pr
0105	ascertain whether a small isolated area, or a	LARGE	open area like a continent, has been most fav
0105	ance of favourable variations arising from the	LARGE	number of incividuals of the same species the
0106	itions of life are infinitely complex from the	LARGE	number of already existing species; and if so
0106	etition to fill them will be more severe, on a	LARGE	than on a small and isolated area. Moreover,
0106	ication will generally have been more rapid on	LARGE	areas; and what is more important, that the n
0106	more important, that the new forms produced on	LARGE	areas, which already have been victorious ove
0107	the future, that for terrestrial productions a	LARGE	continental area, which will probably undergo
0107	competition. When converted by subsidence into	LARGE	separate islands, there will still exist many
0111	alone would never account for so habitual and	LARGE	an amount of difference as that between varie
0115	iversified nature. They differ, moreover, to a	LARGE	extent from the indigenes, for out of the 162
0115	00 genera are not there indigenous, and thus a	LARGE	proportional addition is made to the genera o
0116	t. let A to L represent the species cf a genus	LARGE	in its own country: these species are suppose
0116	s standing at unequal distances. I have said a	LARGE	genus, because we have seen in the second cha
0117	ter, that on an average more of the species of	LARGE	genera vary than of small genera: and the var
0117	f small genera: and the varying species of the	LARGE	genera present a greater number of varieties.
0117	sed, and varying species, belonging to a genus	LARGE	in its own country. The little fan of divergi
0118	genus to which the parent species belonged, a	LARGE	genus in its own country. And these circumsta
0119	on and widely diffused species, belonging to a	LARGE	genus, will tend to partake of the same advan
0120	ies are multiplied and genera are formed. In a	LARGE	genus it is probable that more than one speci
0125	hich are all trying to increase in number. One	LARGE	group will slowly conquer another large group
0125	r. one large group will slowly conquer another	LARGE	group, reduce its numbers, and thus lessen it
0125	her variation and improvement. Within the same	LARGE	group, the later and more highly perfected su
0126	hat the groups of organic beings which are now	LARGE	and triumphant, and which are least broken up
0130	me small degree connects by its affinities two	LARGE	branches of life, and which has apparently be
0136	aston, of the almost entire absence of certain	LARGE	groups of beetles, elsewhere excessively nume
0141	uncer them, may be used as an argument that a	LARGE	proportion of other animals, now in a state o
0142	rations, his kidney beans so early that a very	LARGE	proportion are destroyed by frost, and then c
0147	largely in size and quality. In our poultry, a	LARGE	tuft of feathers on the head is generally acc
0147	erally accompanied by a diminished comb, and a	LARGE	beard by diminished wattles. With species in
0148	f the prehensile antennae. Now the saving of a	LARGE	and complex structure, when rendered superflu
0153	ry amount of modification implies an unusually	LARGE	and long continued amount of variability, whi
0154	le example what is meant. If some species in a	LARGE	genus of plants had blue flowers and some had
0155	t, which is generally very constant throughout	LARGE	groups of species, has differed considerably
0157	tarsi is a character generally common to very	LARGE	groups of beetles, but in the Engidae, as Wes
0157	r of the highest importance, because common to	LARGE	groups; but in certain genera the neuration d
0164	nest in duns and mouse duns: by the term dun a	LARGE	range of colour is included, from one between
0174	quite confounded me. But I think it can be in	LARGE	part explained. In the first place we should
0174	generally find them tolerably numerous over a	LARGE	territory, then becoming somewhat abruptly ra
0175	nsensibly changing physical conditions, but in	LARGE	part on the presence of other species, on whi
0176	imagination adapt a varying species to a very	LARGE	area, we shall have to adapt two varieties to
0176	a, we shall have to adapt two varieties to two	LARGE	areas, and a third variety to a narrow interm
0176	nce of being exterminated than one existing in	LARGE	numbers; and in this particular case the inte
0188	ar, if he fird on finishing this treatise that	LARGE	bodies of facts, otherwise inexplicable, can
0189	ook to an organ common to all the members of a	LARGE	class, for in this latter case the organ must
0192	ack, in the well enclosed shell; but they have	LARGE	folded branchiae. Now I think no one will dis
0212	in england, in the greater wildness of all our	LARGE	birds than of our small birds; for the large
0212	r large birds than of our small birds; for the	LARGE	birds have been most persecuted by man. We ma
0212	y safely attribute the greater wildness of our	LARGE	birds to this cause; for in uninhabited islan
0212	irds to this cause; for in uninhabited islands	LARGE	birds are not more fearful than small; and th
0218	accounted for by the fact of the hens laying a	LARGE	number of eggs; but as in the case of the cuc
0220	d has never seen the slaves, though present in	LARGE	numbers in August, either leave or enter the
0221	y, i came across a community with an unusually	LARGE	stock of slaves, and I observed a few slaves
0225	the young are hatched, and, in addition, some	LARGE	cells of wax for holding honey. These latter
0229	x. in parts, only little bits, in other parts,	LARGE	portions of a rhombic plate had been left bet
0232	way and finish off the angles of a cell till a	LARGE	part both of that cell and of the adjoining c
0234	r many days during the process of secretion. A	LARGE	store of honey is indispensable to support a
0234	e store of honey is indispensable to support a	LARGE	stock of honey during the winter: and the secu
0234	ity of the hive is known mainly to depend on a	LARGE	number cf bees being supported. Hence the sav
0234	the cells of the Melipona: for in this case a	LARGE	part of the bounding surface of each cell wou
0239	ed workers are the most numerous: or that both	LARGE	and small are numerous, with those of an inte
0241	hould regularly produce neuters, either all of	LARGE	size with one form of jaw, or all of small si
0245	ating this subject, two classes of facts, to a	LARGE	extent fundamentally different, have generall
0248	from close interbreeding. I have collected so	LARGE	a body of facts, showing that close interbree
0252	notorious to nurserymen. Horticulturists raise	LARGE	beds of the same hybrids, and such alone are
0264	different. I have more than once alluded to a	LARGE	body of facts, which I have collected, showin
0288	serve fossil remains. Throughout an enormously	LARGE	proportion of the ocean, the bright blue tint
0289	their rarity surprising, when we remember how	LARGE	a proportion of the bones of tertiary mammals
0289	tention had been exclusively confined to these	LARGE	territories, would never have suspected that
0293	ne animals of all kinds, we may safely infer a	LARGE	amount of migration during climatal and other
0299	on of the Malay Archipelago, with its numerous	LARGE	islands separated by wide and shallow seas, p
0302	t before that stage. We continually forget how	LARGE	the world is, compared with the area over whi
0304	ile cirripede was a Chthamalus, a very commom,	LARGE	, and ubiquitous genus, of which not one speci
0305	n in the Chalk period. This group includes the	LARGE	majority of existing species. Lately, Profess
0305	ropical parts of the Indian Ocean would form a	LARGE	and perfectly enclosed basin, in which any gr
0308	composed, we may infer that from first to last	LARGE	islands or tracts of land, whence the sedimen
0309	e other hand, that where continents now exist,	LARGE	tracts of land have existed, subjected no dou
0310	nd we may perhaps believe that we see in these	LARGE	areas, the many formations long anterior to t
0317	ree from a single stem, till the group becomes	LARGE	. On Extinction. We have as yet spoken only in

Page **********************************(Key Word)**********************************

Page **(Key Word)**

0324	of life changing simultaneously, in the above	LARGE	sense, at distant parts of the world, has gre
0327	it seems to me, the parallel, and, taken in a	LARGE	sense, simultaneous, succession of the same f
0328	f the world. As we have reason to believe that	LARGE	areas are affected by the same movement, it i
0328	at this has invariably been the case, and that	LARGE	areas have invariably been affected by the sa
0334	n of quite new forms by immigration, and for a	LARGE	amount of modification, during the long and b
0342	ical record extremely imperfect, and will to a	LARGE	extent explain why we do not find interminabl
0347	ons! In the southern hemisphere, if we compare	LARGE	tracts of land in Australia, South Africa, an
0348	s, and of great deserts, and sometimes even of	LARGE	rivers, we find different productions; though
0360	mostly different from mine; but he chose many	LARGE	fruits and likewise seeds from plants which l
0360	than the small, is interesting; as plants with	LARGE	seeds or fruit could hardly be transported by
0362	eds; now after a bird has found and devoured a	LARGE	supply of food, it is positively asserted tha
0362	and in this earth there was a pebble quite as	LARGE	as the seed of a vetch. Thus seeds might occa
0363	te regions to another. In the Azores, from the	LARGE	number of the species of plants common to Eur
0363	ese islands, and he answered that he had found	LARGE	fragments of granite and other rocks, which d
0366	ow clothed by the vine and maize. Throughout a	LARGE	part of the United States, erratic boulders,
0370	he same relative position, though subjected to	LARGE,	but partial oscillations of level, I am stro
0370	r period, such as the older Pliocene period, a	LARGE	number of the same plants and animals inhabit
0375	acts are given in regard to the plants of that	LARGE	island. Hence we see that throughout the worl
0376	facts, on the belief, supported as it is by a	LARGE	body of geological evidence, that the whole w
0376	eological evidence, that the whole world, or a	LARGE	part of it, was during the glacial period sim
0378	during this the coldest period, I suppose that	LARGE	spaces of the tropical lowlands were clothed
0386	ly shown, as remarked by Alph. de Candolle, in	LARGE	groups of terrestrial plants, which have only
0388	re especially by fresh water birds, which have	LARGE	powers of flight, and naturally travel from o
0389	, and Wollaston for insects. If we look to the	LARGE	size and varied stations of New Zealand, exte
0390	nowhere else in the world) is often extremely	LARGE.	If we compare, for instance, the number of t
0402	but are very distinct species, belonging in a	LARGE	proportion of cases, as shown by Alph. de Can
0403	which live in crevices of stone; and although	LARGE	quantities of stone are annually transported
0412	species. Consequently the groups which are now	LARGE,	and which generally include many dominant sp
0415	depends on their greater constancy throughout	LARGE	groups of species; and this constancy depends
0419	ly, in classification, more especially in very	LARGE	groups of closely allied forms. Temminck insi
0426	ever so trifling, occur together throughout a	LARGE	group of beings having different habits, we m
0427	imes be brought usefully into play in classing	LARGE	and widely distributed genera, because all th
0428	es, which made the groups to which they belong	LARGE	and their parents dominant, they are almost s
0432	n silurian genera, some of which have produced	LARGE	groups of modified descendants. Every interme
0432	most of the characters of each group, whether	LARGE	or small, and thus give a general idea of the
0466	t we may safely infer that the amount has been	LARGE,	and that modifications can be inherited for
0467	t many of the breeds produced by man have to a	LARGE	extent the character of natural species, is s
0470	varieties; and thus species are rendered to a	LARGE	extent defined and distinct objects. Dominant
0471	birth to new and dominant forms; so that each	LARGE	group tends to become still larger, and at th
0471	s beat the less dominant. This tendency in the	LARGE	groups to go on increasing in size and diverg
0010	orts support my view, that variability may be	LARGELY	attributed to the ovules or pollen, or to bo
0017	not dispute that these capacities have added	LARGELY	to the value of most of our domesticated pro
0017	mestication, they would vary on an average as	LARGELY	as the parent species of our existing domest
0023	es. It is not that the varieties which differ	LARGELY	in some one point do not differ at all in ot
0042	same aviary, and this circumstance must have	LARGELY	favoured the improvement and formation of ne
0043	ne, with the aid of selection, has, no doubt,	LARGELY	aided in the formation of new sub breeds; bu
0062	length. The elder De Candolle and Lyell have	LARGELY	and philosophically shown that all organic b
0062	constantly destroying life; or we forget how	LARGELY	these songsters, or their eggs, or their nes
0073	uth America) the vegetation: this again would	LARGELY	affect the insects; and this, as we just hav
0074	d all over England. Now the number of mice is	LARGELY	dependent, as every one knows, on the number
0084	n countless numbers: they are known to suffer	LARGELY	from birds of prey; and hawks are guided by
0098	s hardly possible to raise pure seedlings, so	LARGELY	do they naturally cross. In many other cases
0100	is objection to be valid, but that nature has	LARGELY	provided against it by giving to trees a str
0106	inental productions have everywhere become so	LARGELY	naturalised on islands. On a small island, w
0120	cessive generations, will have come to differ	LARGELY,	but perhaps unequally, from each other and
0121	arly extreme species (I), as those which have	LARGELY	varied, and have given rise to new varieties
0123	era. But as the original species (I) differed	LARGELY	from (A), standing nearly at the extreme poi
0127	lready see how it entails extinction; and how	LARGELY	extinction has acted in the world's history,
0129	sented by a great tree. I believe this simile	LARGELY	speaks the truth. The green and budding twig
0140	ven of species within historical times having	LARGELY	extended their range from warmer to cooler l
0143	the effects of use and disuse have often been	LARGELY	combined with, and sometimes overmastered by
0147	uite become atrophied, the fruit itself gains	LARGELY	in size and quality. In our poultry, a large
0147	the effects, on the one hand, of a part being	LARGELY	developed through natural selection and anot
0148	ut by any means causing some other part to be	LARGELY	developed in a corresponding degree. And, co
0148	tural selection may perfectly well succeed in	LARGELY	developing any organ, without requiring as a
0158	h in common, to parts which have recently and	LARGELY	varied being more likely still to go on vary
0168	t softer and internal parts. When one part is	LARGELY	developed, perhaps it tends to draw nourishm
0184	orth America there are woodpeckers which feed	LARGELY	on fruit, and others with elongated wings wh
0196	inally, that sexual selection will often have	LARGELY	modified the external characters of animals
0205	ed whilst aided by the other, must often have	LARGELY	facilitated transitions. We are far too igno
0215	alone prevents our seeing how universally and	LARGELY	the minds of our domestic animals have been
0227	ways describes her spheres so as to intersect	LARGELY;	and then she unites the points of intersect
0231	as in every case to be finished off, by being	LARGELY	gnawed away on both sides. The manner in whi
0234	s being supported. Hence the saving of wax is	LARGELY	saving honey must be a most important elemen
0254	ctrine which originated with Pallas, has been	LARGELY	accepted by modern naturalists; namely, that
0257	es, and of the hybrids produced from them, is	LARGELY	governed by their systematic affinity. This
0257	the many species of Nicotiana have been more	LARGELY	crossed than the species of almost any other
0277	epend on several circumstances; in some cases	LARGELY	on the early death of the embryo. The steril
0305	milarly confined range, and after having been	LARGELY	developed in some one sea, might have spread
0342	y certain classes of organic beings have been	LARGELY	preserved in a fossil state; that the number
0352	e thus brought to the question which has been	LARGELY	discussed by naturalists, namely, whether sp
0373	iod. I am convinced that Forbes's view may be	LARGELY	extended. In Europe we have the plainest evi
0380	other objects likely to carry seeds have been	LARGELY	imported into Europe during the last two or
0380	lpine forms; but these have almost everywhere	LARGELY	yielded to the more dominant forms, generate
0381	d, icebergs, as suggested by Lyell, have been	LARGELY	concerned in their dispersal. But the existe
0396	of occasional means of transport having been	LARGELY	efficient in the long course of time, than w

Page **(Key Word)**

Largely

Page **(Key Word)**

```
0405   species vary and give rise to new forms will LARGELY determine their average range. For instance,
0441   a prosbosciformed mouth, with which they feed LARGELY, for they increase much in size. In the seco
0467   altering the breed. It is certain that he can LARGELY influence the character of a breed by select
0105   , or the whole I am inclined to believe that LARGENESS of area is of more importance, more especia
0125   e which already have some advantage: and the LARGENESS of any group shows that its species have in
0039   n who first selected a pigeon with a slightly LARGER tail, never dreamed what the descendants of t
0041   rs picked out individual plants with slightly LARGER, earlier, or better fruit, and raised seedlin
0044   and common species vary most. Species of the LARGER genera in any country vary more than the spec
0044   he smaller genera. Many of the species of the LARGER genera resemble varieties in being very close
0054   vided into two equal masses, all those in the LARGER genera being placed on one side, and all thos
0054   smaller genera on the other side, a somewhat LARGER number of the very common and much diffused o
0054   nant species will be found on the side of the LARGER genera. This, again, might have been anticipa
0054   , we might have expected to have found in the LARGER genera, or those including many species, a la
0054   show even a small majority on the side of the LARGER genera. I will here allude to only two causes
0055   was led to anticipate that the species of the LARGER genera in each country would oftener present
0055   o two nearly equal masses, the species of the LARGER genera on one side, and those of the smaller
0055   t has invariably proved to be the case that a LARGER proportion of the species on the side of the
0055   proportion of the species on the side of the LARGER genera present varieties, than on the side of
0055   h present any varieties, invariably present a LARGER average number of varieties than do the speci
0057   n this respect, therefore, the species of the LARGER genera resemble varieties, more than do the s
0057   another way, and it may be said, that in the LARGER genera, in which a number of varieties or inc
0059   most flourishing and dominant species of the LARGER genera which on an average vary most; and var
0059   converted into new and distinct species. The LARGER genera thus tend to become larger; and throug
0059   pecies. The larger genera thus tend to become LARGER; and throughout nature the forms of life whic
0059   . but by steps hereafter to be explained, the LARGER genera also tend to break up into smaller gen
0065   some period of life. Our familiarity with the LARGER domestic animals tends, I think, to mislead u
0092   which produced more and more pollen, and had LARGER and larger anthers, would be selected. When o
0092   uced more and more pollen, and had larger and LARGER anthers, would be selected. When our plant, b
0106   arently are now yielding, before those of the LARGER European Asiatic area. Thus, also, it is that
0120   stincuishing varieties are increased into the LARGER differences distinguishing species. By contin
0125   that in each country it is the species of the LARGER genera which oftenest present varieties or in
0125   fied descendants, will mainly lie between the LARGER groups, which are all trying to increase in n
0126   g to the continued and steady increase of the LARGER groups, a multitude of smaller groups will be
0128   and widely ranging species, belonging to the LARGER genera, which vary most; and these will tend
0134   dition as the domestic Aylesbury duck. As the LARGER ground feeding birds seldom take flight excep
0136   s: that the proportion of wingless beetles is LARGER on the exposed Dezertas than in Madeira itsel
0176   two distinct species, the two which exist in LARGER numbers from inhabiting larger areas, will ha
0176   which exist in larger numbers from inhabiting LARGER areas, will have a great advantage over the i
0177   and intermediate zone. For forms existing in LARGER numbers will always have a better chance, wit
0183   kingfisher at a fish. In our own country the LARGER titmouse (Parus major) may be seen climbing b
0184   e aquatic in their structure and habits, with LARGER and larger mouths, till a creature was produc
0184   n their structure and habits, with larger and LARGER mouths, till a creature was produced as monst
0195   us gain a great advantage. It is not that the LARGER quadrupeds are actually destroyed (except in
0205   , to the inhabitants of another and generally LARGER country. For in the larger country there will
0206   ther and generally larger country. For in the LARGER country there will have existed more individu
0239   tions of this kind. It often happens that the LARGER or the smaller sized workers are the most num
0239   ate size scanty in numbers, Formica flava has LARGER and smaller workers, with some of intermediat
0240   is species, as Mr. F. Smith has observed, the LARGER workers have simple eyes (ocelli), which thou
0241   xteen feet high: but we must suppose that the LARGER workmen had heads four instead of three times
0343   ex contingencies. The dominant species of the LARGER dominant groups tend to leave many modified d
0359   ed by the salt water. Afterwards I tried some LARGER fruits, capsules, etc., and some of these flo
0360   th, and would then germinate. The fact of the LARGER fruits often floating longer than the small,
0380   to the more dominant forms, generated in the LARGER areas and more efficient workshops of the nor
0380   the land yielded to those produced within the LARGER areas of the north, just in the same way as t
0411   that is the dominant species belonging to the LARGER genera, which vary most. The varieties, or in
0428   endants of dominant species, belonging to the LARGER genera, tend to inherit the advantages, which
0428   and more places in the economy of nature. The LARGER and more dominant groups thus tend to go on i
0470   cipient species. Moreover, the species of the LARGER genera, which afford the greater number of va
0470   enera. The closely allied species also of the LARGER genera apparently have restricted ranges, and
0470   ct objects. Dominant species belonging to the LARGER groups tend to give birth to new and dominant
0471   o that each large group tends to become still LARGER, and at the same time more divergent in chara
0489   n and widely spread species, belonging to the LARGER and dominant groups, which will ultimately pr
0092   power. Those individual flowers which had the LARGEST glands or nectaries, and which excreted most
0151   y give one, as it illustrates the rule in its LARGEST application. The opercular valves of sessile
0444   y cases of variations (taking the word in the LARGEST sense) which have supervened at an earlier a
0489   ws acting around us. These laws, taken in its LARGEST sense, being Growth with Reproduction: Inher
0011   elation of growth. Any change in the embryo or LARVA will almost certainly entail changes in the m
0086   es. Natural Selection may modify and adapt the LARVA of an insect to a score of contingencies, who
0086   ill probably often affect the structure of the LARVA; but in all cases natural selection will ensu
0143   ccumulated solely for the good of the young or LARVA, will, it may safely be concluded, affect the
0190   y canal respires, digests, and excretes in the LARVA of the dragonfly and in the fish Cobites. In
0440   ut whenever it comes on, the adaptation of the LARVA to its conditions of life is just as perfect
0440   ertainly is, a crustacean: but a glance at the LARVA shows this to be the case in an unmistakeable
0441   ally considered as lower in the scale than the LARVA, as with certain parasitic crustaceans. To re
0046   quite recently shown that the muscles in the LARVAE of certain insects are very far from uniform.
0086   successive changes in the structure of their LARVAE. So, conversely, modifications in the adult w
0218   and stores it with paralysed prey for its own LARVAE to feed on, yet that when this insect finds a
0219   kind their own nests, or of feeding their own LARVAE. When the old nest is found inconvenient, and
0219   the food which they like best, and with their LARVAE and pupae to stimulate them to work, they did
0219   the survivors: made some cells and tended the LARVAE, and put all to rights. What can be more extr
0220   nest: when the nest is much disturbed and the LARVAE and pupae are exposed, the slaves work energe
0223   slaves seem to have the exclusive care of the LARVAE, and the masters alone go on slave making exp
0223   als and food for themselves, their slaves and LARVAE. So that the masters in this country receive
0244   its foster brothers, ants making slaves, the LARVAE of ichneumonidae feeding within the live bodi
0391   dispersed, yet we can see that their eggs or LARVAE, perhaps attached to seaweed or floating timb
0425   ame species, though the males and females and LARVAE are sometimes extremely different: and as it
```

Page **(Key Word)**

Page ***(Key Word)**

```
0439  of a mammal, bird, or reptile. The vermiform LARVAE of moths, flies, beetles, etc., resemble each
0439  han do the mature insects: but in the case of LARVAE, the embryos are active, and have been adapte
0440  ch special adaptations, the similarity of the LARVAE or active embryos of allied animals is someti
0440  uch obscured: and cases could be given of the LARVAE of two species, or of two groups of species,
0440  ir adult parents. In most cases, however, the LARVAE, though active, still obey more or less close
0440  ch differ widely in external appearance, have LARVAE in all their several stages barely distinguis
0441  ceans. To refer once again to cirripedes: the LARVAE in the first stage have three pairs of legs,
0441  the larval condition. But in some genera the LARVAE become developed either into hermaphrodites h
0442  hing worthy to be called a metamorphosis. The LARVAE of insects, whether adapted to the most diver
0448  ce at corresponding ages, the active young or LARVAE might easily be rendered by natural selection
0448  successive stages of development: so that the LARVAE, in the first stage, might differ greatly fro
0448  he first stage, might differ greatly from the LARVAE in the second stage, as we have seen to be th
0449  cirripedes can at once be recognised by their LARVAE as belonging to the great class of crustacea
0457  embers for purposes as different as possible. LARVAE are active embryos, which have become special
0424  aturalist includes as one species the several LARVAL stages of the same individual, however much t
0441  or more lowly organised then they were in the LARVAL condition. But in some genera the larvae beco
0002  clusions that I have on the origin of species. LAST year he sent to me a memoir on this subject, w
0003  my deep obligations to Dr. Hooker, who for the LAST fifteen years has aided me in every possible w
0005  n mature and in an embryonic condition. In the LAST chapter I shall give a brief recapitulation of
0035  sh pointer has been greatly changed within the LAST century, and in this case the change has, it i
0042  e strawberry which have been raised during the LAST thirty or forty years. In the case of animals
0044  fore applying the principles arrived at in the LAST chapter to organic beings in a state of nature
0060  on Natural Selection. It has been seen in the LAST chapter that amongst organic beings in a state
0071  cotch firs on the distant hilltops: within the LAST ten years large spaces have been enclosed, and
0080  le for existence, discussed too briefly in the LAST chapter, act in regard to variation? Can the p
0089  e females, which standing by as spectators, at LAST choose the most attractive partner? Those who
0094  be continually favoured or selected, until at LAST a complete separation of the sexes would be af
0115  new species. To give a single instance: in the LAST edition of Dr. Asa Gray's Manual of the Flora
0138  those that are constructed for twilight: and, LAST of all, those destined for total darkness. By
0161  reproduce the character in question, which at LAST, under unknown favorable conditions, gains an
0165  on the sides of its face. With respect to this LAST fact, I was so convinced that not even a strip
0222  a very thick clump of heath, whence I saw the LAST individual of F. sanguinea emerge, carrying a
0232  nces from each other and from the walls of the LAST completed cells, and then, by striking imagina
0256  h extreme difficulty, but the hybrids, when at LAST produced, are very fertile. Even within the li
0283  ect little of nothing in wearing away rock. At LAST the base of the cliff is undermined, huge frag
0290  any recent deposits sufficiently extensive to LAST for even a short geological period. Along the
0291  it has as yet suffered, but which will hardly LAST to a distant geological age, was certainly dep
0300  ons could be formed of sufficient thickness to LAST to an age, as distant in futurity as the secon
0306  many points, which the discoveries of even the LAST dozen years have effected, it seems to me to b
0308  are composec, we may infer that from first to LAST large islands or tracts of land, whence the se
0310  anging dialect: of this history we possess the LAST volume alone, relating only to two or three co
0315  e why all the species in the same region do at LAST, if we look to wide enough intervals of time,
0316  tratum to the present day. We have seen in the LAST chapter that the species of a group sometimes
0318  oth single species and whole groups of species LAST for very unequal periods: some groups, as we h
0333  ied in various degrees. As we possess only the LAST volume of the geological record, and that in a
0335  f time: a very ancient form might occasionally LAST much longer than a form elsewhere subsequently
0336  expect to find, as I attempted to show in the LAST chapter, in any one or two formations all the
0349  we may look back to past ages, as shown in the LAST chapter, and we find American types then preva
0351  ly the case. I believe, as was remarked in the LAST chapter, in no law of necessary development. A
0351  ng physical conditions. As we have seen in the LAST chapter that some forms have retained nearly t
0352  were first produced: for, as explained in the LAST chapter, it is incredible that individuals ide
0364  difficult to prove this), received within the LAST few centuries, through occasional means of tra
0380  e been largely imported into Europe during the LAST two or three centuries from La Plata, and duri
0380  three centuries from La Plata, and during the LAST thirty or forty years from Australia. Somethin
0383  e with subsequent modification. Summary of the LAST and present chapters. As lakes and river syste
0388  bitants of Oceanic Islands. We now come to the LAST of the three classes of facts, which I have se
0406  f the colonists to their new homes. Summary of LAST and present Chapters. In these chapters I have
0441  ute, single, and very simple eye spot. In this LAST and completed state, cirripedes may be conside
0022  oes in some breeds the voice and disposition. LASTLY, in certain breeds, the males and females hav
0026  often confounded in treatises on inheritance. LASTLY, the hybrids or mongrels from between all the
0049  they have different geographical ranges; and LASTLY, according to very numerous experiments made
0105  ications in their structure and constitution. LASTLY, isolation, by checking immigration and conse
0123  tinct from the three first named species: and LASTLY, o14, e14, and m14, will be nearly related on
0165  across the legs than is even the pure quagga. LASTLY, and this is another most remarkable case, a
0179  atural selection and gain further advantages. LASTLY, looking not to any one time, but to all time
0241  jaws having a widely different structure; or LASTLY, and this is our climax of difficulty, one se
0259  ble in external appearance either parent. And LASTLY, that the facility of making a first cross be
0264  hy certain trees cannot be grafted on others. LASTLY, an embryo may be developed, and then perish
0265  us will produce more or less sterile hybrids. LASTLY, when organic beings are placed during severa
0269  ing under nearly the same conditions of life. LASTLY, and this seems to me by far the most importa
0459  e, in ever so slight a degree, variable, and, LASTLY, that there is a struggle for existence leadi
0476  oved organic beings in the struggle for life. LASTLY, the law of the long endurance of allied form
0316  e disappeared: or its existence, as long as it LASTS, is continuous. I am aware that there are som
0439  of the law of embryonic resemblance, sometimes LASTS till a rather late age: thus birds of the sam
0347  mpare the productions of South America south of LAT. 35 degrees with those north of 25 degrees, wh
0373  en observed on the eastern side as far south as LAT. 36 degrees to 37 degrees, and on the shores o
0373  he climate is now so different, as far south as LAT. 46 degrees: erratic boulders have, also, been
0373  ther south on both sides of the continent, from LAT. 41 degrees to the southernmost extremity, we
0008  be, generally act: whether during the early or LATE period of development of the embryo, or at the
0014  , the greater length of horn, though appearing LATE in life, is clearly due to the male element. H
0033  ury: it has certainly been more attended to of LATE years, and many treatises have been published
0122  l species, has transmitted descendants to the LATE stage of descent. The new species in our diagr
0221  , after all, they had been victorious in their LATE combat. At the same time I laid on the same pl
0245  le, has, I think, been much underrated by some LATE writers. On the theory of natural selection th
0287  remark of that admirable palaeontologist, the LATE Edward Forbes, should not be forgotten, namely
0307  ther highly competent judges, as Lyell and the LATE E. Forbes, dispute this conclusion. We should
```

Page ***(Key Word)***

Page ***(Key Word)***

```
0318   co existed with still living shells at a very LATE geological period, I was filled with astonishm
0346   eir climatal and other physical conditions. Of LATE, almost every author who has studied the subje
0368   rctic and temperate productions will at a very LATE period have marched a little further north, an
0409   of distinct but representative species. As the LATE Edward Forbes often insisted, there is a strik
0439   nic resemblance, sometimes lasts till a rather LATE age: thus birds of the same genus, and of clos
0443   before the formation of the embryo, may appear LATE in life: as when an hereditary disease, which
0446   eps of variation having supervened at a rather LATE age, and having been inherited at a correspond
0447   uccessive modification supervening at a rather LATE age, and being inherited at a corresponding la
0447   te age, and being inherited at a corresponding LATE age, the fore limbs in the embryos of the seve
0447   having undergone much modification at a rather LATE period of life, and having thus been converted
0482   e removed. Several eminent naturalists have of LATE published their belief that a multitude of rep
0040   olour, that most fleeting characters, have LATELY been exhibited as distinct at our poultry sho
0134   n of several birds, which now inhabit or have LATELY inhabited several oceanic islands, tenanted b
0193   esembles that of common muscle: and as it has LATELY been shown that Rays have an organ closely an
0199   rks lead me to say a few words on the protest LATELY made by some naturalists, against the utilita
0218   parasitic on other species: and M. Fabre has LATELY shown good reason for believing that although
0305   ludes the large majority of existing species. LATELY, Professor Pictet has carried their existence
0307   world is known with accuracy. M. Barrande has LATELY added another and lower stage to the Silurian
0313   of the sea, of which a striking instance has LATELY been observed in Switzerland. There is some r
0355   uting the word variety for species) from that LATELY advanced in an ingenious paper by Mr. Wallace
0366   the icy streams with which their valleys were LATELY filled. So greatly has the climate of Europe
0367   returned, the same arctic species, which had LATELY lived in a body together on the lowlands of t
0372   ry few identical species (though Asa Gray has LATELY shown that more plants are identical than was
0378   lines of invasion: and it is a striking fact, LATELY communicated to me by Dr. Hooker, that all th
0380   e productions of real islands have everywhere LATELY yielded to continental forms, naturalised by
0406   w which has long been observed, and which has LATELY been admirably discussed by Alph. de Candolle
0433   ding in this direction: and Milne Edwards has LATELY insisted, in an able paper, on the high impor
0482   y admit that a multitude of forms, which till LATELY they themselves thought were special creation
0119   ding from (A). The modified offspring from the LATER and more highly improved branches in the line
0121   the intermediate forms between the earlier and LATER states, that is between the less and more imp
0121   l lines of descent, which will be conquered by LATER and improved lines of descent. If, however, t
0126   improvement. Within the same large group, the LATER and more highly perfected sub groups, from br
0174   as been broken up into islands even during the LATER tertiary periods: and in such islands distinc
0288   r an enormous interval of time, by another and LATER formation, without the underlying bed having
0316   group reaches its maximum, and then, sooner or LATER, it gradually decreases. If the number of the
0320   n incefinitely increasing, at least during the LATER geological periods, so that looking to later
0320   e later geological periods, so that looking to LATER times we may believe that the production of n
0324   inferred that they had lived during one of the LATER tertiary stages. When the marine forms of lif
0324   to those which lived in Europe during certain LATER tertiary stages, than to those which now live
0328   l has made similar observations on some of the LATER tertiary formations. Barrande, also, shows th
0339   e same types within the same areas, during the LATER tertiary periods. Mr. Clift many years ago sh
0340   niformity of the same types in each during the LATER tertiary periods. Nor can it be pretended tha
0345   of structure within the same areas during the LATER geological periods ceases to be mysterious, a
0367   the Glacial period came on a little earlier or LATER in North America than in Europe, so will the
0367   migration there have been a little earlier or LATER: but this will make no difference in the fina
0371   e productions of Europe and America during the LATER tertiary stages were more closely related to
0440   the period of activity may come on earlier or LATER in life: but whenever it comes on, the adapta
0444   racters are fully acquired a little earlier or LATER in life. It would not signify, for instance,
0444   hat we can see, might have appeared earlier or LATER in life, tend to appear at a corresponding ag
0446   to me to explain these facts in regard to the LATER embryonic stages of our domestic varieties. F
0446   s and structures have been acquired earlier or LATER in life, if the full grown animal possesses t
0476   termediate in character in comparison with its LATER descendants: and thus we can see why the more
0476   ct forms: and they are in so far higher as the LATER and more improved forms have conquered the ol
0187   ent, and which properly act only by excluding LATERAL pencils of light, are convex at their upper
0294   erican to the European seas. In examining the LATEST deposits of various quarters of the world, it
0304   suddenly produced in the interval between the LATEST secondary and earliest tertiary formation. Bu
0374   case, that the epoch was included within the LATEST geological period. We have, also, excellent e
0339   and of parts of South America under the same LATITUDE, would attempt to account, on the one hand,
0347   ia, Africa, and South America under the same LATITUDE: for these countries are almost as much isol
0349   found in Africa and Australia under the same LATITUDE. On these same plains of La Plata, we see th
0363   haracter of the flora in comparison with the LATITUDE, I suspected that these islands had been par
0370   he organisms now living under the climate of LATITUDE 60 degrees, during the Pliocene period lived
0370   ved further north under the polar circle, in LATITUDE 66 degrees: and that the strictly
0378   rms had migrated some twenty five degrees of LATITUDE from their native country and covered the la
0389   of New Zealand, extending over 780 miles of LATITUDE, and compare its flowering plants, only 750
0140   y extended their range from warmer to cooler LATITUDES, and conversely: but we do not positively k
0347   h africa, and western South America, between LATITUDES 25 degrees and 35 degrees, we shall find pa
0376   d, in receding from polar towards equatorial LATITUDES, the Alpine or mountain floras really becom
0379   her from their homes into the most temperate LATITUDES of the opposite hemisphere. Although we hav
0408   families: and how it is that under different LATITUDES, for instance in South America, the inhabit
0002   and Dr. Hooker, who both knew of my work, the LATTER having read my sketch of 1844, honoured me by
0018   ical with, those still existing. Even if this LATTER fact were found more strictly and generally t
0018   d several competent judges believe that these LATTER have had more than one wild parent. With resp
0035   s and size the parent Arab stock, so that the LATTER, by the regulations for the Goodwood Races, a
0044   nature, we must briefly discuss whether these LATTER are subject to any variation. To treat this s
0052   ed and more permanent varieties: and at these LATTER, as leading to sub species, and to species. T
0125   tinct genera descended from (I): and as these LATTER two genera, both from continued divergence of
0144   omology comes into play? With respect to this LATTER case of correlation, I think it can hardly be
0189   all the members of a large class, for in this LATTER case the organ must have been first formed at
0190   breathe free air in their swimbladders, this LATTER organ having a ductus pneumaticus for its sup
0192   ever have imagined that the branchiae in this LATTER family had originally existed as organs for p
0196   . to give a few instances to illustrate these LATTER remarks. If green woodpeckers alone had exist
0203   maternal love or maternal hatred, though the LATTER fortunately is most rare, is all the same to
0204   forms which it connects: consequently the two LATTER, during the course of further modification, f
0216   comb making power of the hive bee: these two LATTER instincts have generally, and most justly, be
0221   he legs of the slave making F. sanguinea. The LATTER ruthlessly killed their small opponents, and
```

Page ***(Key Word)***

latter

Page **(Key Word)**

```
0223 a present with those of the F. rufescens. The LATTER does not build its own nest, does not determi
0225 nd humble bee, but more nearly related to the LATTER: it forms a nearly regular waxen comb of cyli
0225 e large cells of wax for holding honey. These LATTER cells are nearly spherical and of nearly equa
0234 s could collect. But let us suppose that this LATTER circumstance determined, as it probably often
0241 re of their jaws. I speak confidently on this LATTER point, as Mr. Lubbock made drawings for me wi
0254 tions quite fertile under domestication. This LATTER alternative seems to me the most probable, an
0260 or degree of resemblance to each other. This LATTER statement is clearly proved by reciprocal cro
0262 n on their own roots. We are reminded by this LATTER fact of the extraordinary case of Hippeastrum
0264 ped, and then perish at an early period. This LATTER alternative has not been sufficiently attende
0277 pacity of being grafted together, though this LATTER capacity evidently depends on widely differen
0286 weald within so limited a period as since the LATTER part of the Chalk formation. The distance fro
0287 than 300 million years has elapsed since the LATTER part of the Secondary period. I have made the
0291 om, if it continue slowly to subside. In this LATTER case, as long as the rate of subsidence and s
0313 ars. The strongest apparent exception to this LATTER rule, is that of the so called colonies of M.
0325 new varieties or incipient species; for these LATTER must be victorious in a still higher degree t
0342 ring the periods of elevation, and during the LATTER the record will have been least perfectly kep
0343 urian system was deposited: I can answer this LATTER question only hypothetically, by saying that
0366 their former more temperate inhabitants, the LATTER would be supplanted and arctic productions wo
0377 ctions, and these by the arctic; but with the LATTER we are not now concerned. The tropical plants
0386 ve only a very few aquatic members; for these LATTER seem immediately to acquire, as if in consequ
0396 land with the nearest continent; for on this LATTER view the migration would probably have been m
0403 sported from Porto Santo to Madeira, yet this LATTER island has not become colonised by the Porto
0417 ae, bear perfect and degraded flowers; in the LATTER, as A. de Jussieu has remarked, the greater n
0441 i have called complemental males: and in the LATTER, the development has assuredly been retrograd
0446 ese facts and the above two principles, which LATTER, though not proved true, can be shown to be i
0447 limbs in the mature animal; the limbs in the LATTER having undergone much modification at a rathe
0452 d rudimentary occasionally differs much. This LATTER fact is well exemplified in the state of the
0463 d by the allied forms on either hand; and the LATTER, from existing in greater numbers, will gener
0465 level the record will be blank. During these LATTER periods there will probably be more variabili
0484 ly important to deserve a specific name. This LATTER point will become a far more essential consid
0485 pecies and well marked varieties is, that the LATTER are known, or believed, to be connected at th
0064 ding cattle and horses in South America; and LATTERLY in Australia, had not been well authenticate
0029 ot have descended from long horns, and he will LAUGH you to scorn. I have never met a pigeon, or p
0417 the family, to the class, disappear, and thus LAUGH at our classification. But when Aspicarpa pro
0021 ted wing and tail feathers. The trumpeter and LAUGHER, as their names express, utter a very differ
0021 ating, may well excite astonishment and even LAUGHTER. The turbit has a very short and conical bea
0092 ae, and at the back of the leaf of the common LAUREL. This juice, though small in quantity, is gre
0285 nd feet in height; for the gentle slope of the LAVA streams, due to their formerly liquid state, s
0088 best cocks. How low in the scale of nature this LAW of battle descends, I know not; male alligator
0097 lone incline me to believe that it is a general LAW of nature (utterly ignorant though we be of th
0097 rly ignorant though we be of the meaning of the LAW) that no organic being self fertilises itself
0097 ls, indispensable. On the belief that this is a LAW of nature, we can, I think, understand several
0099 wn pollen; and that this is part of the general LAW of good being derived from the intercrossing o
0101 onal intercross with a distinct individual is a LAW of nature. I am well aware that there are, on
0147 the propounded, at about the same period, their LAW of compensation or balancement of growth; or,
0147 of nature it can hardly be maintained that the LAW is of universal application; but many good obs
0171 organs not in all cases absolutely perfect. The LAW of Unity of Type and of the Conditions of Exis
0196 organisation; that characters reappear from the LAW of reversion; that correlation of growth will
0198 ly even the form of the pelvis; and then by the LAW of homologous variation, the front limbs and e
0199 from other unknown cause, may reappear from the LAW of reversion, though now of no direct use. The
0206 the several laws of growth. Hence, in fact, the LAW of the Conditions of Existence is the higher l
0206 aw of the Conditions of Existence is the higher LAW; as it includes, through the inheritance of fo
0244 incts; but as small consequences of one general LAW, leading to the advancement of all organic bei
0250 erility between distinct species is a universal LAW of nature. He experimentised on some of the ve
0314 cord well with my theory. I believe in no fixed LAW of development, causing all the inhabitants of
0318 re the close of the palaeozoic period. No fixed LAW seems to determine the length of time during w
0325 as Barrande has remarked, look to some special LAW. We shall see this more clearly when we treat
0339 i strongly insisted, in 1839 and 1845, on this LAW of the succession of types, on this wonderful
0339 o the mammals of the Old World. We see the same LAW in this author's restorations of the extinct a
0339 of brazil. Mr. Woodward has shown that the same LAW holds good with sea shells, but from the wide
0339 ralo Caspian Sea. Now what does this remarkable LAW of the succession of the same types within the
0340 nor can it be pretended that it is an immutable LAW that marsupials should have been chiefly or so
0340 lications above alluded to; that in America the LAW of distribution of terrestrial mammals was for
0340 theory of descent with modification, the great LAW of the long enduring, but not immutable, succe
0348 ontinents. Turning to the sea, we find the same LAW. No two marine faunas are more distinct, with
0349 tinct at different points and stations. It is a LAW of the widest generality, and every continent
0351 eve, as was remarked in the last chapter, in no LAW of necessary development. As the variability o
0354 rt, the belief that this has been the universal LAW, seems to me incomparably the safest. In discu
0399 ill, I do not doubt, be some day explained. The LAW which causes the inhabitants of an archipelago
0406 distant transportation, probably accounts for a LAW which has long been observed, and which has la
0434 widely different animals. We see the same great LAW in the construction of the mouths of insects:
0436 classes are plainly homologous. We see the same LAW in comparing the wonderfully complex jaws and
0439 apted for special lines of life. A trace of the LAW of embryonic resemblance, sometimes lasts till
0440 ugh active, still obey more or less closely the LAW of common embryonic resemblance. Cirripedes af
0449 existing species. Agassiz believes this to be a LAW of nature; but I am bound to confess that I on
0449 am bound to confess that I only hope to see the LAW hereafter proved true. It can be proved true i
0450 should also be borne in mind, that the supposed LAW of resemblance of ancient forms of life to the
0471 niggard in innovation. But why this should be a LAW of nature if each species has been independent
0476 ic beings in the struggle for life. Lastly, the LAW of the long endurance of allied forms on the s
0490 anet has gone cycling on according to the fixed LAW of gravity, from so simple a beginning endless
0005 r I shall discuss the complex and little known LAWS of variation and correlation of growth. In
0009 curious subject; but to show how singular the LAWS are which determine the reproduction of animal
0010 conditions of life are in comparison with the LAWS of reproduction, and of growth, and of inherit
0011 rmed by danger, seems probable. There are many LAWS regulating variation, some few of which can be
0012 arts of the structure, owing to the mysterious LAWS of the correlation of growth. The result of th
0012 t of the various, quite unknown, or dimly seen LAWS of variation is infinitely complex and diversi
```

Page **(Key Word)**

```
age ****************************************( Key Word )****************************************
```

```
013  rule, and non inheritance as the anomaly. The  LAWS  governing inheritance are quite unknown; no on
014  be of the highest importance in explaining the  LAWS  of embryology. These remarks are of course con
029  those naturalists who, knowing far less of the  LAWS  of inheritance than does the breeder, and know
033  s is hardly ever, perhaps never, the case. The  LAWS  of correlation of growth, the importance of wh
034  istory choice animals were often imported, and  LAWS  were passed to prevent their exportation: the
043  rsion. Variability is governed by many unknown  LAWS, more especially by that of correlation of gro
073  kc cataclysms to desolate the world, or invent  LAWS  on the duration of the forms of life! I am tem
075  must fall to the ground according to definite  LAWS; but how simple is this problem compared to th
085  ry to bear in mind that there are many unknown  LAWS  of correlation of growth, which, when one part
086  odifications will no doubt affect, through the  LAWS  of correlation, the structure of the adult; an
089  ffect. I strongly suspect that some well known  LAWS  with respect to the plumage of male and female
131  nching and beautiful ramifications. Chapter V.  LAWS  of Variation. Effects of external conditions.
144  e better adapted to show the importance of the  LAWS  of correlation in modifying important structur
146  as of high value, may be wholly due to unknown  LAWS  of correlated growth, and without being, as fa
150  parts are left to the free play of the various  LAWS  of growth, to the effects of long continued di
160  e influence of the mere act of crossing on the  LAWS  of inheritance. No doubt it is a very surprisi
167  n the sea shore. Summary. Our ignorance of the  LAWS  of variation is profound. Not in one case out
167  he means of instituting a comparison, the same  LAWS  appear to have acted in producing the lesser d
197  oks on the bamboo may have arisen from unknown  LAWS  of growth, and have been subsequently taken ad
197  infer that this structure has arisen from the  LAWS  of growth, and has been taken advantage of in
198  ve importance of the several known and unknown  LAWS  of variation; and I have here alluded to them
200  on, subjected formerly, as now, to the several  LAWS  of inheritance, reversion, correlation of grow
200  er directly, or indirectly through the complex  LAWS  of growth. Natural selection cannot possibly p
205  eve that many modifications, wholly due to the  LAWS  of growth, and at first in no way advantageous
206  l organic beings have been formed on two great  LAWS, Unity of Type, and the Conditions of Existenc
206  nd being in all cases subjected to the several  LAWS  of growth. Hence, in fact, the law of the Cond
245  close interbreeding, removed by domestication.  LAWS  governing the sterility of hybrids. Sterility
254  wledge, be considered as absolutely universal.  LAWS  governing the Sterility of first Crosses and o
261  it that it is incidental on differences in the  LAWS  of growth of the two plants. We can sometimes
263  and as we must look at the curious and complex  LAWS  governing the facility with which trees can be
263  tems, so I believe that the still more complex  LAWS  governing the facility of first crosses, are i
274  species, follows according to Gartner the same  LAWS. When two species are crossed, one has sometim
275  to animals, comes to the conclusion, that the  LAWS  of resemblance of the child to its parents are
275  varieties as having been produced by secondary  LAWS, this similarity would be an astonishing fact.
276  but is governed by several curious and complex  LAWS. It is generally different, and sometimes wide
325  ess local and temporary, but depend on general  LAWS  which govern the whole animal kingdom. M. Barr
340  ms, and there will be nothing immutable in the  LAWS  of past and present distribution. It may be as
345  or disappear. On the other hand, all the chief  LAWS  of palaeontology plainly proclaim, as it seems
345  ew and improved forms of life, produced by the  LAWS  of variation still acting round us, and preser
409  sisted, there is a striking parellelism in the  LAWS  of life throughout time and space; the laws go
409  he laws of life throughout time and space; the  LAWS  governing the succession of forms in past time
410  ach other in time and space; in both cases the  LAWS  of variation have been the same, and modificat
411  ween parts of the same individual. Embryology,  LAWS  of, explained by variations not supervening at
435  andibles, and two pairs of maxillae. Analogous  LAWS  govern the construction of the mouths and limb
453  tiges of nails have appeared, not from unknown  LAWS  of growth, but in order to excrete horny matte
456  n anticipated, and can be accounted for by the  LAWS  of inheritance. Summary. In this chapter I hav
466  form. Variability is governed by many complex  LAWS, by correlation of growth, by use and disuse,
469  cknowledged to have been produced by secondary  LAWS. On this same view we can understand how it is
472  ot been observed. The complex and little known  LAWS  governing variation are the same, as far as we
472  n are the same, as far as we can see, with the  LAWS  which have governed the production of so calle
475  ossed offspring should follow the same complex  LAWS  in their degrees and kinds of resemblance to t
475  and varieties have been produced by secondary  LAWS. If we admit that the geological record is imp
482  ife, and which are those produced by secondary  LAWS. They admit variation as a vera causa in one c
484  vesicles; their cellular structure, and their  LAWS  of growth and reproduction. We see this even i
486  d of inquiry will be opened, on the causes and  LAWS  of variation, on correlation of growth, on the
488  ind it accords better with what we know of the  LAWS  impressed on matter by the Creator, that the p
489  so complex a manner, have all been produced by  LAWS  acting around us. These Laws, taken in the lar
489  been produced by laws acting around us. These  LAWS, taken in the largest sense, being Growth with
009  birds, with the rarest exceptions, hardly ever  LAY  fertile eggs. Many exotic plants have pollen u
134  tes. Such considerations as these incline me to  LAY  very little weight on the direct action of the
198  en through ordinary generation, we ought not to  LAY  too much stress on our ignorance of the precis
207  at instinct impels the cuckoo to migrate and to  LAY  her eggs in other birds' nests. An action, whi
216  namely, the instinct which leads the cuckoo to  LAY  her eggs in other birds' nests; the slave maki
217  ous birds which have been known occasionally to  LAY  their eggs in other bird's nests. Now let us s
217  their mother and in their turn would be apt to  LAY  their eggs in other birds' nests, and thus be
218  in the case of the American species, unite and  LAY  first a few eggs in one nest and then in anoth
218  asted eggs. Many bees are parasitic, and always  LAY  their eggs in the nests of bees of other kinds
232  one face of the slip, in this case the bees can  LAY  the foundations of one wall of a new hexagon,
415  has partly arisen, that almost all naturalists  LAY  the greatest stress on resemblances in organs
188  ument, we ought in imagination to take a thick  LAYER  of transparent tissue, with a nerve sensitive
189  t beneath, and then suppose every part of this  LAYER  to be continually changing slowly in density,
189  from each other, and with the surfaces of each  LAYER  slowly changing in form. Further we must supp
225  the cells of the hive bee, placed in a double  LAYER: each cell, as is well known, is an hexagonal
226  nd had arranged them symmetrically in a double  LAYER, the resulting structure would probably have
226  res of the six surrounding spheres in the same  LAYER; and at the same distance from the centres of
226  he adjoining spheres in the other and parallel  LAYER: then, if planes of intersection between the
227  h layers be formed, there will result a double  LAYER  of hexagonal prisms united together by pyrimi
227  intersection of adjoining spheres in the same  LAYER, she can prolong the hexagon to any length re
232  rim of a growing comb, with an extremely thin  LAYER  of melted vermilion wax; and I invariably fou
235  t a given distance from each other in a double  LAYER, and to build up and excavate the wax along t
030  reeds so little quarrelsome, with everlasting  LAYERS  which never desire to sit; and with the banta
189  ing slowly in density, so as to separate into  LAYERS  of different densities and thicknesses, place
189  ight accidental alteration in the transparent  LAYERS; and carefully selecting each alteration whic
226  bed with their centres placed in two parallel  LAYERS; with the centre of each sphere at the distan
227  ersection between the several spheres in both  LAYERS  be formed, there will result a double layer o
227  se the Melipona to arrange her cells in level  LAYERS, as she already does her cylindrical cells; a
```

```
age ****************************************( Key Word )****************************************
```

Page **(Key Word)**

```
0286  emember that almost all strata contain harder LAYERS or nodules, which from long resisting attri
0293  d a species disappearing before the uppermost LAYERS have been deposited, it would be equally ra
0217  e nest. If this were the case, the process of LAYING and hatching might be inconveniently long,
0218  own offspring. The occasional habit of birds LAYING their eggs in other birds' nests either of
0218  ably be accounted for by the fact of the hens LAYING a large number of eggs; but as in the case
0232  ells has been built. This capacity in bees of LAYING down under certain circumstances a rough wa
0275  different varieties, or of distinct species. LAYING aside the question of fertility and sterili
0400  which are found in other parts of the world (LAYING on one side for the moment the endemic spec
0066  district, let it be ever so large. The condor LAYS a couple of eggs and the ostrich a score, an
0066  he more numerous of the two: the Fulmar petrel LAYS but one egg, yet it is believed to be the mo
0072  ter number in Paraguay of a certain fly, which LAYS its eggs in the navels of these animals when
0213  master, than the white butterfly knows why she LAYS her eggs on the leaf of the cabbage, I canno
0217  al cause of the cuckoo's instinct is, that she LAYS her eggs, not daily, but at intervals of two
0217  asserted that the American cuckoo occasionally LAYS her eggs in other birds' nests: but I hear o
0214  of the instincts of either parent: for example, LE Roy describes a dog, whose great grandfather
0117  of character comes in: for this will generally LEAD to the most different or divergent variation
0174  continuous during a long period. Geology would LEAD us to believe that almost every continent ha
0181  ld ones become modified, and all analogy would, LEAD us to believe that some at least of the sout
0183  ; or whether slight modifications of structure LEAD to changed habits: both probably often chang
0193  related to each other. Nor does geology at all LEAD to the belief that formerly most fishes had
0194  owards which no transitional grade is known to LEAD. The truth of this remark is indeed shown by
0199  would appear frivolous. The foregoing remarks LEAD me to say a few words on the protest lately
0217  ung would gain an advantage. And analogy would LEAD me to believe, that the young thus reared wo
0235  n in architecture, natural selection could not LEAD: for the comb of the hive bee, as far as we
0317  nde, etc., whose general views would naturally LEAD them to this conclusion. On the contrary, we
0384  pletely prevented their inosculation, seems to LEAD to this same conclusion. With respect to all
0389  ns, which, if legitimately followed out, would LEAD to the belief that within the recent period
0452  lant no trace of an organ, which analogy would LEAD us to expect to find, and which is occasiona
0468  of the males; and the slightest advantage will LEAD to victory. As geology plainly proclaims tha
0484  from an equal or lesser number. Analogy would LEAD me one step further, namely, to the belief t
0490  and disuse: a Ratio of Increase so high as to LEAD to a Struggle for Life, and as a consequence
0002  ich facts cannot be adduced, often apparently LEADING to conclusions directly opposite to those
0051  degree more distinct and permanent, as steps LEADING to more strongly marked and more permanent
0052  permanent varieties; and at these latter, as LEADING to sub species, and to species. The passag
0210  tinct species, how very generally gradations, LEADING to the most complex instincts, can be disc
0244  but as small consequences of one general law, LEADING to the advancement of all organic beings,
0282  o may not be a practical geologist, the facts LEADING the mind feebly to comprehend the lapse of
0343  from these difficulties, all the other great LEADING facts in palaeontology seem to me simply t
0408  m some one source; then I think all the grand LEADING facts of geographical distribution are exp
0444  l, i believe, explain all the above specified LEADING facts in embryology. But first let us look
0450  f demonstration. Thus, as it seems to me, the LEADING facts in embryology, which are second in i
0453  its embryonic condition. I have now given the LEADING facts with respect to rudimentary organs.
0457  esponding period, we can understand the great LEADING facts in Embryology; namely, the resemblan
0459  t may be convenient to the reader to have the LEADING facts and inferences briefly recapitulated
0459  astly, that there is a struggle for existence LEADING to the preservation of each profitable dev
0476  descent with modification, most of the great LEADING facts in Distribution. We can see why there
0047  have actually been found, but because analogy LEADS the observer to suppose either that they do
0049  stance between the homes of two doubtful forms LEADS many naturalists to rank both as distinct s
0098  s is sometimes the case with insects. And this LEADS me to say a few words on what I call Sexual
0127  ogy plainly declares. Natural selection, also, LEADS to divergence of character; for more living
0128  natural selection, as has just been remarked, LEADS to divergence of character and to much exti
0131  mestication or cultivation, than under nature, LEADS me to believe that deviations of structure
0139  erminate, in the time of sleep, etc., and this LEADS me to say a few words on acclimatisation. A
0172  say to so marvellous an instinct as that which LEADS the bee to make cells, which have practical
0202  hat most perfect organ, the eye. If our reason LEADS us to admire with enthusiasm a multitude of
0216  in my future work, namely, the instinct which LEADS the cuckoo to lay her eggs in other birds'
0250  two distinct species. This case of the Crinum LEADS me to refer to a most singular fact, namely
0301  ands of miles beyond its confines; and analogy LEADS me to believe that it would be chiefly these
0482  only restate a fact. Any one whose disposition LEADS him to attach more weight to unexplained di
0084  es in some Leguminosae, and at the back of the LEAF eating insects green, and bark feeders mottl
0092  es in some Leguminosae, and at the back of the LEAF of the common laurel. This juice, though sma
0213  e butterfly knows why she lays her eggs on the LEAF of the cabbage, I cannot see that these acti
0065  the wide plains of La Plata, clothing square LEAGUES of surface almost to the exclusion of all o
0194  ated steps? Why should not Nature have taken a LEAP from structure to structure? On the theory o
0194  ht successive variations; she can never take a LEAP, but must advance by the shortest and slowest
0029  descended from the same parents, may they not LEARN a lesson of caution, when they deride the i
0187  ure of the eye, and from fossil species we can LEARN nothing on this head. In this great class we
0429  order; and that in the vegetable kingdom, as I LEARN from Dr. Hooker, it has added only two or th
0004  large amount of hereditary modification is at LEAST possible; and, what is equally or more impor
0019  otype. At this rate there must have existed at LEAST a score of species of wild cattle, as many s
0022  a slight degree from each other. Altogether at LEAST a score of pigeons might be chosen, which it
0023  rock pigeon, they must have descended from at LEAST seven or eight aboriginal stocks; for it is
0024  gin of our pigeons, it must be assumed that at LEAST seven or eight species were so thoroughly do
0038  always have to struggle for their own food, at LEAST during certain seasons. And in two countries
0042  t of success in the formation of new races, at LEAST, in a country which is already stocked with
0045  s, would not in some cases be inherited for at LEAST some few generations? and in this case I pre
0047  ed which have not been ranked as species by at LEAST some competent judges. That varieties of th
0050  which the group is subject; and this shows, at LEAST, how very generally there is some variation.
0062  ersal struggle for life, or more difficult, at LEAST I have found it so, than constantly to bear
0069  ce extreme cold, acts directly, it will be the LEAST vigorous, or those which have got least food
0069  be the least vigorous, or those which have not LEAST food through the advancing winter, which wil
0069  and if these enemies or competitors be in the LEAST degree favoured by any slight change of clim
0070  ely in numbers in a small tract, epidemics, at LEAST, this seems generally to occur with our game
0073  e visits of bees, if not indispensable, are at LEAST highly beneficial to the fertilisation of ou
0081  nd, we may feel sure that any variation in the LEAST degree injurious would be rigidly destroyed.
0083  s selection by some half monstrous form; or at LEAST by some modification prominent enough to cat
```

Page **(Key Word)**

ge **********************************(Key Word)***

86	different period of life, shall not be in the	LEAST degree injurious: for if they became so, they
98	them; and as this flower is never visited, at	LEAST in my garden, by insects, it never sets a see
15	ver the other natives; and we may, I think, at	LEAST safely infer that diversification of structur
22	at only one (F), of the two species which were	LEAST closely related to the other nine original sp
26	ch are now large and triumphant, and which are	LEAST broken up, that is, which as yet have suffere
26	broken up, that is, which as yet have suffered	LEAST extinction, will for a long period continue t
32	haps rather more in that of plants. We may, at	LEAST, safely conclude that such influences cannot
45	generations each individual beetle which flew	LEAST, either from its wings having been ever so li
47	the central and external flowers: we know, at	LEAST, that in irregular flowers, those nearest to
55	e part or organ in excess, it rarely flows, at	LEAST in excess, to another part; thus it is diffic
56	of the same kind applies to monstrosities: at	LEAST is. Geoffroy St. Hilaire seems to entertain n
67	uld still often be in some degree variable, at	LEAST more variable than those parts of the organis
81	s to me, to reject a real for an unreal, or at	LEAST for an unknown, cause. It makes the works of
82	analogy would lead us to believe that some at	LEAST of the squirrels would decrease in numbers or
04	ir perfect power of flight; but they serve, at	LEAST, to show what diversified means of transition
09	hat wonderful metamorphoses in function are at	LEAST possible. For instance, a swim bladder has ap
10	fe. Under changed conditions of life, it is at	LEAST possible that slight modifications of instinc
18	me evidence of such gradations; or we ought at	LEAST to be able to show that gradations of some ki
24	up of ostriches. For several hen ostriches, at	LEAST in the case of the American species, unite an
53	he greatest possible amount of honey, with the	LEAST possible consumption of precious wax in their
69	omy of wax; that individual swarm which wasted	LEAST honey in the secretion of wax, having succeed
73	any pure animal, which from any cause had the	LEAST tendency to sterility, the breed would assure
00	ch i will briefly abstract. The evidence is at	LEAST as good as that from which we believe in the
09	eing thus often rendered either impotent or at	LEAST incapable of its proper function of producing
19	ation, but the geological record would then be	LEAST perfect. It may be doubted whether the durati
20	om their wear and tear; and would have been at	LEAST partially upheaved by the oscillations of lev
36	degree rare, no naturalist would have felt the	LEAST surprise at its rarity; for rarity is the att
38	es has not gone on indefinitely increasing, at	LEAST during the later geological periods, so that
42	isted with the Mastodon and Horse, it might at	LEAST have been inferred that they had lived during
45	et it be remembered that the forms of life, at	LEAST those inhabiting the sea, have changed almost
46	fully expect to see it hereafter confirmed, at	LEAST in regard to subordinate groups, which have b
58	nd during the latter the record will have been	LEAST perfectly kept; that each single formation ha
62	imperfect as I believe it to be, and it may at	LEAST be asserted that the record cannot be proved
63	orld which cannot be paralleled in the New, at	LEAST as closely as the same species generally requ
74	xisted as mountain ranges on the land, some at	LEAST of the islands would have been formed, like o
74	o not secrete gastric juice, and do not in the	LEAST injure, as I know by trial, the germination o
85	hores of these mid ocean islands, and it is at	LEAST possible that they may have brought thither t
92	s to me probable that it was, during a part at	LEAST of the period, actually simultaneous througho
97	e distinct evidence to the contrary, we may at	LEAST admit as probable that the glacial action was
00	gitude. On this view of the whole world, or at	LEAST of broad longitudinal belts, having been simu
02	is length of time a duck or heron might fly at	LEAST six or seven hundred miles, and would be sure
05	acility of immigration, I believe, has been at	LEAST as important as the nature of the conditions.
07	are very easily killed by salt; their eggs, at	LEAST such as I have tried, sink in sea water and a
14	bitants, with which each has to compete, is at	LEAST as important, and generally a far more import
19	ocking thrush peculiar to Charles Island is at	LEAST as well fitted for its home as is the species
25	eve as a general rule we do find, that some at	LEAST of the species range very widely; for it is n
32	simultaneously affected the whole world, or at	LEAST great meridional belts. As showing how divers
65	ear indications of its true affinities. We are	LEAST likely in the modifications of these organs t
68	sub families, and genera, they seem to be, at	LEAST at present, almost arbitrary. Several of the
85	racters which, as far as we can judge, are the	LEAST likely to have been modified in relation to t
36	, nevertheless a natural classification, or at	LEAST a natural arrangement, would be possible. We
88	ses can be preserved in a fossil condition; at	LEAST in any great number. Widely ranging species v
26	ctions; and every one admits that there are at	LEAST individual differences in species under natur
20	t be a cheering prospect: but we shall have at	LEAST be freed from the vain search for the undisco
20	, and such choice animals would thus generally	LEAVE more offspring than the inferior ones; so tha
23	e best fitted for their places in nature, will	LEAVE most progeny. But in many cases, victory will
39	maller groups will become utterly extinct, and	LEAVE no modified descendants; and consequently tha
42	urrey and Sussex, and never saw a slave either	LEAVE or enter a nest. As, during these months, the
40	ugh present in large numbers in August, either	LEAVE or enter the nest. Hence he considers them as
43	ommunity. In England the masters alone usually	LEAVE the nest to collect building materials and fo
44	myrmecocystus, the workers of one caste never	LEAVE the nest; they are fed by the workers of anot
17	instincts of the fertile members; which alone	LEAVE descendants. I am surprised that no one has a
68	ch quarter of the world will obviously tend to	LEAVE in that quarter, during the next succeeding p
74	species of the larger dominant groups tend to	LEAVE many modified descendants, and thus new sub g
04	nitor, tend to become extinct together, and to	LEAVE no modified offspring on the face of the eart
33	ce and of almost universal prevalence, and yet	LEAVE us in no doubt where it should be ranked. Hen
33	with their conditions of life, will generally	LEAVE most progeny. But success will often depend o
33	s we see, in the case of neuter insects, which	LEAVE no progeny to inherit the effects of long con
04	tion seems to me to be no explanation, for it	LEAVES the case of the coadaptations of organic bein
33	pecies in the flower garden: the diversity of	LEAVES, pods, or tubers, or whatever part is valued,
33	pecies in the orchard, in comparison with the	LEAVES and flowers of the same set of varieties. See
33	same set of varieties. See how different the	LEAVES of the cabbage are, and how extremely alike t
32	wers of the heartsease are, and how alike the	LEAVES; how much the fruit of the different kinds of
33	selection of slight variations, either in the	LEAVES, the flowers, or the fruit, will produce race
80	ch when growing near the sea shore have their	LEAVES in some degree fleshy, though not elsewhere f
338	y on the sea side are very apt to have fleshy	LEAVES. He who believes in the creation of each spec
382	preys on fish; but during the long winter it	LEAVES the frozen waters, and preys like other polec
419	a corresponding age. This process, whilst it	LEAVES the embryo almost unaltered, continually adds
436	have freely inundated the south. As the tide	LEAVES its drift in horizontal lines, though rising
437	, on the number and position of the embryonic	LEAVES or cotyledons, and on the mode of development
437	n the view that they consist of metamorphosed	LEAVES, arranged in a spire. In monstrous plants, we
438	e see a series of successive spiral whorls of	LEAVES. An indefinite repetition of the same part or
439	or of flowering plants; many spiral whorls of	LEAVES. We have formerly seen that parts many times
439	amens and pistils of flowers as metamorphosed	LEAVES; but it would in these cases probably be more
	ng of this kind in plants: thus the embryonic	LEAVES of the ulex or furze, and the first leaves of
	ic leaves of the ulex or furze, and the first	LEAVES of the phyllodineous acaceas, are pinnate or

age **********************************(Key Word)***

Page ***************************************(Key Word)***************************************

0439 eas, are pinnate or divided like the ordinary LEAVES of the Leguminosae. The points of structure
0059 ominant tend to become still more dominant by LEAVING many modified and dominant descendants. But
0062 ly the life of the individual, but success in LEAVING progeny. Two canine animals in a time of de
0088 or spurless cock would have a poor chance of LEAVING offspring. Sexual selection by always allow
0091 ould have the best chance of surviving and of LEAVING offspring. Some of its young would probably
0094 ly, and so have a better chance of living and LEAVING descendants. Its descendants would probably
0221 erved a few slaves mingled with their masters LEAVING the nest, and marching along the same road
0229 oth sides, in order to have succeeded in thus LEAVING flat plates between the basins, by stopping
0232 sweep equal spheres, and then building up, or LEAVING ungnawed, the planes of intersection betwee
0023 respects. As several of the reasons which have LED me to this belief are in some degree applicab
0055 rongly marked and well defined varieties, I was LED to anticipate that the species of the larger
0165 ould commonly be called an accident, that I was LED solely from the occurence of the face stripes
0225 follow from a few very simple instincts. I was LED to investigate this subject by Mr. Waterhouse
0234 e bee, if a slight modification of her instinct LED her to make her waxen cells near together, so
0235 aving by slow degrees, more and more perfectly, LED the bees to sweep equal spheres at a given di
0309 oured map appended to my volume on Coral Reefs, LED me to conclude that the great oceans are stil
0350 turalist must feel little curiosity, who is not LED to inquire what this bond is. This bond, on m
0359 the stream be washed into the sea. Hence I was LED to dry stems in branches of 94 plants with ri
0365 of europe. Even as long ago as 1747, such facts LED Gmelin to conclude that the same species must
0407 e same parents, are not insuperable. And we are LED to this conclusion, which has been arrived at
0442 animals within the same class, that we might be LED to look at these facts as necessarily conting
0454 at disuse has been the main agency; that it has LED in successive generations to the gradual redu
0482 s of the question with impartiality. Whoever is LED to believe that species are mutable will do g
0485 e gradations between any two forms, we shall be LED to weigh more carefully and to value higher t
0076 together different varieties of the medicinal LEECH. It may even be doubted whether the varietie
0011 nes of the wing weigh less and the bones of the LEG more, in proportion to the whole skeleton, th
0022 h other and to the body: the relative length of LEG and of the feet; the number of scutellae on t
0088 d strength to the wing to strike in the spurred LEG, as well as the brutal cockfighter, who knows
0164 with a double stripe on each shoulder and with LEG stripes; and a man, whom I can implicitly tru
0164 ils, I may state that I have collected cases of LEG and shoulder stripes in horses of very differ
0200 ame bones in the arm of the monkey, in the fore LEG of the horse, in the wing of the bat, and in
0416 ruminants, and certain rudimentary bones of the LEG, are highly serviceable in exhibiting the clo
0434 d for grasping, that of a mole for digging, the LEG of the horse, the paddle of the porpoise, and
0434 nes of the arm and forearm, or of the thigh and LEG, transposed. Hence the same names can be give
0437 e been created in the formation of the wing and LEG of a bat, used as they are for such totally d
0445 stril and of eyelid, size of feet and length of LEG, in the wild stock, in pouters, fantails, run
0452 the drawings given by Owen of the bones of the LEG of the horse, ox, and rhinoceros. It is an im
0479 of a man, wing of bat, fin of the porpoise, and LEG of the horse, the same number of vertebrae fo
0479 ions. The similarity of pattern in the wing and LEG of a bat, though used for such different purp
0083 d; he does not exercise a long backed or long LEGGED quadruped in any peculiar manner; he exposes
0201 are now rapidly yielding before the advancing LEGIONS of plants and animals introduced from Europ
0389 view on continental extensions, which, if LEGITIMATELY followed out, would lead to the belief th
0021 e pouter has a much elongated body, wings, and LEGS; and its enormously developed crop, which it
0077 th ard talons of the tiger; and in that of the LEGS and claws of the parasite which clings to the
0077 he dandelion, and in the flattened and fringed LEGS of the water beetle, the relation seems at fi
0077 und. In the water beetle, the structure of its LEGS, so well adapted for diving, allows it to com
0091 s deer, and the other more bulky, with shorter LEGS, which more frequently attacks the shepherd's
0135 nerations the size and weight of its body, its LEGS were used more, and its wings less, until the
0143 ying in the same manner: in the front and hind LEGS, and even in the jaws and limbs, varying toge
0163 arely has very distinct transverse bars on its LEGS, like those on the legs of the zebra: it has
0163 transverse bars on the LEGS of the zebra: it has been asserted that these
0163 s of this species are generally striped on the LEGS and faintly on the shoulder. The quagga, thou
0163 a zebra over the body, is without bars on the LEGS; but Dr. Gray has figured one specimen with v
0163 ds, and of all colours: transverse bars on the LEGS are not rare in duns, mouse duns, and in one
0164 purely bred. The spine is always striped: the LEGS are generally barred; and the shoulder stripe
0165 horse is particularly apt to have bars on its LEGS. I once saw a mule with its legs so much stri
0165 e bars on its legs. I once saw a mule with its LEGS so much striped that any one at first would h
0165 een, of hybrids between the ass and zebra, the LEGS were much more plainly barred than the rest o
0165 sire, were much more plainly barred across the LEGS than is even the pure quagga. Lastly, and thi
0165 brid, though the ass seldom has stripes on its LEGS and the hemionus has none and has not even a
0165 a shoulder stripe, nevertheless, had all four LEGS barred, and had three short shoulder stripes,
0166 becoming, by simple variation, striped on the LEGS like a zebra, or striped on the shoulders lik
0179 and which resembles an otter in its fur, short LEGS, and form of tail; during summer this animal
0181 f the shoulder to the tail, including the hind LEGS, we perhaps see traces of an apparatus origin
0182 erus of Eyton); as fins in the water and front LEGS on the land, like the penguin; as sails, like
0221 as many as three of these ants clinging to the LEGS of the slave making F. sanguinea. The latter
0436 in comparing the wonderfully complex jaws and LEGS in crustaceans. It is familiar to almost ever
0437 of many parts, consequently always have fewer LEGS; or conversely, those with many legs have sim
0437 ave fewer legs; or conversely, those with many LEGS have simpler mouths? Why should the sepals, p
0438 vertebrae: the jaws of crabs as metamorphosed LEGS; the stamens and pistils of flowers as metamo
0438 eak of both skull and vertebrae, both jaws and LEGS, etc., as having been metamorphosed, not one
0438 ans of any kind, vertebrae in the one case and LEGS in the other, have actually been modified int
0439 osed during a long course of descent from true LEGS, or from some simple appendage, is explained.
0441 larvae in the first stage have three pairs of LEGS, a very simple single eye, and a probosciforme
0441 six pairs of beautifully constructed natatory LEGS, a pair of magnificent compound eyes, and extr
0441 is is completed they are fixed for life: their LEGS are now converted into prehensile organs: they
0447 the fore limbs, for instance, which served as LEGS in the parent species, may become, by a long
0473 onal appearance of stripes on the shoulder and LEGS of the several species of the horse genus and
0479 ed for such different purpose, in the jaws and LEGS of a crab, in the petals, stamens, and pistil
0092 glands at the base of the stipules in some LEGUMINOSAE, and at the back of the leaf of the commor
0247 ollen, and (excluding all cases such as the LEGUMINOSAE, in which there is an acknowledged difficu
0439 or divided like the ordinary leaves of the LEGUMINOSAE. The points of structure, in which the emb
0036 n of two distinct strains. The two flocks of LEICESTER sheep kept by Mr. Buckley and Mr. Burgess,
0181 duced. Now look at the Galeopithecus or flying LEMUR, which formerly was falsely ranked amongst ba
0181 now connect the Galeopithecus with the other LEMURIDAE, yet I can see no difficulty in supposing t
0435 might have all its bones, or certain bones, LENGHTENED to any extent, and the membrane connecting

Page ***************************************(Key Word)***************************************

age **(Key Word)**

```
0005  of Natural Selection will be treated at some   LENGTH in the fourth chapter; and we shall then see
0014  horned cow by a long horned bull, the greater   LENGTH of horn, though appearing late in life, is cl
0022  , the development of the bones of the face in   LENGTH and breadth and curvature differs enormously.
0022  mously. The shape, as well as the breadth and   LENGTH of the ramus of the lower jaw, varies in a hi
0022  wicth of the gape of mouth, the proportional   LENGTH of the eyelids, of the orifice of the nostril
0022  ue (not always in strict correlation with the   LENGTH of beak), the size of the crop and of the upp
0022  rimary wing and caudal feathers; the relative   LENGTH of wing and tail to each other and to the bod
0022  l to each other and to the body; the relative   LENGTH of leg and of the feet; the number of scutell
0027  ve of each breed, for instance the wattle and   LENGTH of beak of the carrier, the shortness of that
0028  stic pigeons at some, yet quite insufficient,   LENGTH; because when I first kept pigeons and watche
0062  treated, as it well deserves, at much greater   LENGTH. The elder De Candolle and Lyell have largely
0067  k, discuss some of the checks at considerable   LENGTH, more especially in regard to the feral anima
0088  breed might surely give indomitable courage,   LENGTH to the spur, and strength to the wing to stri
0094  and form of the body, or in the curvature and   LENGTH of the proboscis, etc., far too slight to be
0094  do not on a hasty glance appear to differ in   LENGTH; yet the hive bee can easily suck the nectar
0118  continue the process by similar steps for any   LENGTH of time; some of the varieties, after each th
0138  ted other changes, such as an increase in the   LENGTH of the antennae or palpi, as a compensation f
0150  n made by Professor Owen, with respect to the   LENGTH of the arms of the ourang outang, that he has
0163  shoulder stripe is certainly very variable in   LENGTH and outline. A white ass; but not an albino,
0227  ame layer, she can prolong the hexagon to any   LENGTH requisite to hold the stock of honey; in the
0236  ubject well deserves to be discussed at great   LENGTH, but I will here take only a single case, tha
0242  case, at some little but wholly insufficient   LENGTH, in order to show the power of natural select
0283  that it is only here and there, along a short   LENGTH or round a promontory, that the cliffs are at
0286  hat any line of coast, ten or twenty miles in   LENGTH, ever suffers degradation at the same time al
0286  ion at the same time along its whole indented   LENGTH; and we must remember that almost all strata
0287  udation of one inch per century for the whole   LENGTH would be an ample allowance. At this rate, on
0289  rritory, hardly any idea can be formed of the   LENGTH of time which has elapsed between the consecu
0318  c period. No fixed law seems to determine the   LENGTH of time during which any single species or an
0318  upposed that as the individual has a definite   LENGTH of life, so have species a definite duration.
0335  ich are extreme in the important character of   LENGTH of beak originated earlier than short beaked
0360  sea; and this would have favoured the average   LENGTH of their flotation and of their resistance to
0364  etain their vitality when exposed for a great   LENGTH of time to the action of sea water; nor could
0385  air, from twelve to twenty hours; and in this   LENGTH of time a duck or heron might fly at least si
0402  r own places and keep separate for almost any   LENGTH of time. Being familiar with the fact that ma
0407  al transport, I have discussed at some little   LENGTH the means of dispersal of fresh water product
0408  different forms of life; for according to the   LENGTH of time which has elapsed since new inhabitan
0444  r or not it assumed a beak of this particular   LENGTH, as long as it was fed by its parents. Hence,
0445  re give details) of the beak, width of mouth,   LENGTH of nostril and of eyelid, size of feet and le
0445  th of nostril and of eyelid, size of feet and   LENGTH of leg, in the wild stock, in pouters, fantai
0445  ds, when mature, differ so extraordinarily in   LENGTH and form of beak, that they would, I cannot d
0488  the world, as at present known, although of a   LENGTH quite incomprehensible by us, will hereafter
0489  e to a secure future of equally inappreciable   LENGTH. And as natural selection works solely by and
0181  e arm of the Galeopithecus might be greatly   LENGTHENED by natural selection; and this, as far as t
0117  ttle fan of diverging dotted lines of unequal   LENGTHS proceeding from (A), may represent its varyi
0188  nciple of natural selection to such startling   LENGTHS. It is scarcely possible to avoid comparing
0333  they will have endured for extremely unequal   LENGTHS of time, and will have been modified in vari
0335  produced necessarily endure for corresponding   LENGTHS of time; a very ancient form might occasiona
0431  er by circuitous lines of affinity of various   LENGTHS (as may be seen in the diagram so often refe
0301  climatal changes would intervene during such   LENGTHY periods; and in these cases the inhabitants
0187  d into facets, within each of which there is a   LENS shaped swelling. In other crustaceans the tran
0211  xcrete. I watched them for some time through a   LENS, but not one excreted; I then tickled and stro
0136  which, as the flower feeding coleoptera and   LEPIDOPTERA, must habitually use their wings to gain t
0107  wn in the world, as the Ornithorhynchus and   LEPIDOSIREN, which, like fossils, connect to a certain
0130  y see an animal like the Ornithorhynchus or   LEPIDOSIREN, which in some small degree connects by it
0330  y now and then even a living animal, as the   LEPIDOSIREN, is discovered having affinities directed
0429  The genera Ornithorhynchus and LEPIDOSIREN, for example, would not have been less abe
0027  O b.C., as was pointed out to me by Professor   LEPSIUS; but Mr. Birch informs me that pigeons are g
0125  her large group, reduce its numbers, and thus   LESSEN its chance of further variation and improveme
0180  on a long list of such cases is sufficient to   LESSEN the difficulty in any particular case like th
0298  to new species; and this again would greatly   LESSEN the chance of our being able to trace the sta
0461  e and crosses between greatly modified forms,   LESSEN fertility; and on the other hand, lesser chan
0103  f intercrossing with other varieties is thus   LESSENED. Even in the case of slow breeding animals,
0175  he confines of its range, where it exists in   LESSENED numbers, will, during fluctuations in the nu
0237  difficulty, though appearing insuperable, is   LESSENED, or, as I believe, disappears, when it is re
0249  uld be injurious to their fertility, already   LESSENED by their hybrid origin. I am strengthened in
0462  species of the same genus is in some degree   LESSENED. As on the theory of natural selection an in
0180  uickly, or as there is reason to believe, by   LESSENING the danger from occasional falls. But it do
0249  dy of facts, showing that close interbreeding   LESSENS fertility; and, on the other hand, that an o
0016  n the same manner as, only in most cases in a   LESSER degree than, do closely allied species of the
0023  resent domestic breeds by the crossing of any   LESSER number: how, for instance, could a pouter be
0051  species and well marked varieties, or between   LESSER varieties and individual differences. These d
0057  y be divided into sub genera, or sections, or   LESSER groups. As Fries has well remarked, little gr
0058  ll see how this may be explained, and how the   LESSER differences between varieties will tend to in
0069  s when we travel northward, but in a somewhat   LESSER degree, for the number of species of all kind
0080  liarities our domestic productions, and, in a   LESSER degree, those under nature, vary; and how str
0102  prcfitable variations, will compensate for a   LESSER amount of variability in each individual, and
0111  them, incipient species. How, then, does the   LESSER difference between varieties become augmented
0129  s divided into great branches, and these into   LESSER and lesser branches, were themselves once, wh
0129  nto great branches, and these into lesser and   LESSER branches, were themselves once, when the tree
0131  organic beings under domestication, and in a   LESSER degree in those in a state of nature, had bee
0149  of the same part or organ, when it occurs in   LESSER numbers, is constant. The same author and som
0167  me laws appear to have acted in producing the   LESSER differences between varieties of the same spe
0176  rmediate variety, consequently, will exist in   LESSER numbers from inhabiting a narrow and lesser a
0176  n lesser numbers from inhabiting a narrow and   LESSER area; and practically, as far as I can make o
0176  varieties together have generally existed in   LESSER numbers than the forms which they connect, th
0176  lly linked together. For any form existing in   LESSER numbers would, as already remarked, run a gre
```

Page **(Key Word)**

Lesser

Page ***(Key Word)***

```
0177  on, than will the rarer forms which exist in  LESSER  numbers. Hence, the more common forms, in th
0178  arieties), exist in the intermediate zones in  LESSER  numbers than the varieties which they tend t
0183  always be less, from their having existed in  LESSER  numbers, than in the case of species with fu
0203  he intermediate variety will usually exist in  LESSER  numbers than the two forms which it connects
0226  x sc. Rt. 2, or radius x 1.41421 (or at some   LESSER  distance) from the centres of the six surrou
0240  accounted for merely by their proportionally   LESSER  size; and I fully believe, though I dare not
0251  show on what slight and mysterious causes the  LESSER  or greater fertility of species when crossed
0258  g reciprocal crosses is extremely common in a  LESSER  degree. He has observed it even between form
0263  rs seem to me to indicate that the greater or  LESSER  difficulty of either grafting or crossing to
0263  erfect. Even in first crosses, the greater or  LESSER  difficulty in effecting a union apparently c
0265  s having been specially affected, though in a  LESSER  degree than when sterility ensues. So it is
0269  o species does not determine their greater or  LESSER  degree of sterility when crossed; and we may
0279  that intermediate varieties, from existing in  LESSER  numbers than the forms which they connect, w
0314  the variations be accumulated to a greater or  LESSER  amount, thus causing a greater or lesser amo
0314  r or lesser amount, thus causing a greater or  LESSER  amount of modification in the varying specie
0316  ing more or less quickly, and in a greater or  LESSER  degree. A group does not reappear after it h
0330  o groups would be distinguished by a somewhat  LESSER  number of characters; so that the two groups
0333  he early period marked VI., would differ by a  LESSER  number of characters; for at this early stag
0351  into new relations with each other, and in a  LESSER  degree with the surrounding physical conditi
0367  tain summits (having been exterminated on all  LESSER  heights) and in the artic regions of both he
0380  ctions cover the ground in La Plata, and in a  LESSER  degree in Australia, and have to a certain e
0401  s case of the Galapagos Archipelago, and in a  LESSER  degree in some analogous instances, is that
0404  see it strikingly displayed in Bats, and in a  LESSER  degree in the Felidae and Canidae. We see it
0408  and not others to enter; either in greater or  LESSER  numbers; according or not, as those which en
0418  se it as one of high value; if common to some  LESSER  number, they use it as of subordinated value
0447  a will remain unmodified, or be modified in a  LESSER  degree, by the effects of use and disuse. In
0455  f rudimentary organs in the embryo, and their  LESSER  relative size in the adult. But if each step
0461  rms, lessen fertility; and on the other hand,  LESSER  changes in the conditions of life and crosse
0463  an the intermediate varieties, which exist in  LESSER  numbers; so that the intermediate varieties
0484  five progenitors, and plants from an equal or  LESSER  number. Analogy would lead me one step furth
0029  d from the same parents, may they not learn a  LESSON  of caution; when they deride the idea of spe
0285  emoir on this subject. Yet it is an admirable  LESSON  to stand on the North Downs and to look at t
0294  n this particular deposit. It is an excellent  LESSON  to reflect on the ascertained amount of migr
0117  ines, and is there marked by a small numbered  LETTER, a sufficient amount of variation is suppose
0251  bore good seed, which vegetated freely. In a   LETTER to me, in 1839, Mr. Herbert told me that he
0332  cs. This is represented in the diagram by the  LETTER F14. All the many forms, extinct and recent,
0412  ps subordinate to groups. In the diagram each  LETTER on the uppermost line may represent a genus
0116  ure, and is represented in the diagram by the  LETTERS standing at unequal distances. I have said
0119  broken at regular intervals by small numbered  LETTERS marking the successive forms which have bec
0120  manner), we get eight species, marked by the   LETTERS between a14 and m14, all descended from (A)
0121  d generations, six new species, marked by the  LETTERS n14 to z14, are supposed to have been produ
0121  es. The other nine species (marked by capital  LETTERS) of our original genus, may for a long peri
0124  ated by the broken lines, beneath the capital  LETTERS, converging in sub branches downwards towar
0331  rth chapter. We may suppose that the numbered  LETTERS represent genera, and the dotted lines dive
0420  am in the fourth chapter. We will suppose the  LETTERS A to L to represent allied genera, which li
0432  shall see this by turning to the diagram: the  LETTERS, A to L, may represent eleven Silurian gene
0455  . rudimentary organs may be compared with the  LETTERS in a word, still retained in the spelling,
0106  hough now continuous, owing to oscillations of  LEVEL, will often have recently existed in a broke
0107  ich will probably undergo many oscillations of  LEVEL, and which consequently will exist for long
0227  t suppose the Melipona to arrange her cells in  LEVEL layers, as she already does her cylindrical
0287  ate. On the other hand, during oscillations of  LEVEL, which we know this area has undergone, the
0290  upraised and during subsequent oscillations of  LEVEL. Such thick and extensive accumulations of s
0291  nly deposited during a downward oscillation of  LEVEL, and thus gained considerable thickness. All
0291  ea has undergone numerous slow oscillations of  LEVEL, and apparently these oscillations have affe
0294  nd likewise to reflect on the great changes of  LEVEL, on the inordinately great change of climate
0296  of many long intervals of time and changes of  LEVEL during the process of deposition, which woul
0301  idence would be interrupted by oscillations of  LEVEL, and that slight climatal changes would inte
0309  east partially upheaved by the oscillations of  LEVEL, which we may fairly conclude must have inte
0309  d, subjected no doubt to great oscillations of  LEVEL, since the earliest silurian period. The col
0309  t archipelagoes still areas of oscillations of  LEVEL, and the continents areas of elevation. But
0309  y a preponderance, during many oscillations of  LEVEL, of the force of elevation; but may not the
0346  of dispersal, by changes of climate and of the  LEVEL of the land, and by occasional means. Disper
0356  anch of the subject in some detail. Changes of  LEVEL in the land must also have been highly influ
0357  geologist will dispute that great mutations of  LEVEL, have occurred within the period of existing
0357  ave abundant evidence of great oscillations of  LEVEL in our continents; but not of such vast chan
0370  ubjected to large, but partial oscillations of  LEVEL, I am strongly inclined to extend the above
0373  glaciers once extended far below their present  LEVEL. In central Chile I was astonished at the st
0378  ieve that the climate under the equator at the  LEVEL of the sea was about the same with that now
0384  slight changes within the recent period in the  LEVEL of the land, having caused rivers to flow in
0384  occurred during floods, without any change of  LEVEL. We have evidence in the loess of the Rhine
0384  loess of the Rhine of considerable changes of  LEVEL in the land within a very recent geological
0396  on the lapse of time, and as during changes of  LEVEL it is obvious that islands separated by shal
0407  fects of all the changes of climate and of the  LEVEL of the land, which have certainly occurred w
0465  ternate periods of elevation and of stationary  LEVEL the record will be blank. During these latte
0487  throw, on former changes of climate and of the  LEVEL of the land, we shall surely be enabled to t
0294  bably first appear and disappear at different   LEVELS, owing to the migration of species and to ge
0296  other, at no less than sixty eight different    LEVELS. Hence, when the same species occur at the b
0198  f flies is correlated with colour, as is the   LIABILITY to be poisoned by certain plants; so that
0383  pond to pond, or from stream to stream; and    LIABILITY to wide dispersal would follow from this c
0036  and other accidents, to which savages are so   LIABLE, and such choice animals would thus generall
0085  not to keep white pigeons, as being the most   LIABLE to destruction. Hence I can see no reason to
0143  at an early embryonic period, are alike, seem   LIABLE to vary in an allied manner: we see this in
0149  er remarked that multiple parts are also very   LIABLE to variation in structure. Inasmuch as this
0152  vertheless the part in this case is eminently   LIABLE to variation. Why should this be so? On the
0152  ge by continued selection, are also eminently   LIABLE to variation. Look at the breeds of the pige
0160  have seen, these coloured marks are eminently   LIABLE to appear in the crossed offspring of two di
```

Page ***(Key Word)***

age **************************************(Key Word)**************************************

```
175  of its prey, or in the seasons, be extremely LIABLE to utter extermination; and thus its geograph
176  case the intermediate form would be eminently LIABLE to the inroads of closely allied forms existi
179  ause alone the intermediate varieties will be LIABLE to accidental extermination; and during the p
243  cts are not always absolutely perfect and are LIABLE to mistakes; that no instinct has been produc
264  n some degree unsuitable, and consequently be LIABLE to perish at an early period; more especially
264  their natural conditions, they are extremely LIABLE to have their reproductive systems seriously
265  egrees; in both, the male element is the most LIABLE to be affected; but sometimes the female more
265  tions not natural to them, they are extremely LIABLE to vary, which is due, as I believe, to their
265  brids in successive generations are eminently LIABLE to vary, as every experimentalist has observe
273  ybrids; Gartner states that mongrels are more LIABLE than hybrids to revert to either parent form;
277  ct fertility surprising, when we remember how LIABLE we are to argue in a circle with respect to v
303  to illustrate these remarks; and to show how LIABLE we are to error in supposing that whole group
315  in some degree modified and improved, will be LIABLE to be exterminated. Hence we can see why all
321  e species of the same genus, will be the most LIABLE to extermination. Thus, as I believe, a numbe
324  he shores of North America would hereafter be LIABLE to be classed with somewhat older European an
351  erwards isolated country, they will be little LIABLE to modification; for neither migration nor is
368  cated in this volume, they will not have been LIABLE to much modification. But with our Alpine pro
369  e disturbed; consequently they will have been LIABLE to modification; and this we find has been th
390  ompete with new associates, will be eminently LIABLE to modification, and will often produce group
391  abits, and will consequently have been little LIABLE to modification. Madeira, again, is inhabited
399  from Africa; and that such colonists would be LIABLE to modification; the principle of inheritance
437  that parts many times repeated are eminently LIABLE to vary in number and structure; consequently
452  the individuals of the same species are very LIABLE to vary in degree of development and in other
463  ably exist in the intermediate zones, will be LIABLE to be supplanted by the allied forms on eithe
474  mportance to the species, should be eminently LIABLE to variation; but, on my view, this part has
475  me instincts being apparently not perfect and LIABLE to mistakes, and at many instincts causing ot
048  as mere varieties. Mr. H. C. Watson, to whom I LIE under deep obligation for assistance of all ki
125  on of new and modified descendants, will mainly LIE between the larger groups, which are all tryin
135  etles in Madeira, as observed by Mr. Wollaston, LIE much concealed, until the wind lulls and the s
163  long list of such cases; but here, as before, I LIE under a great disadvantage in not being able t
180  y little weight. Here, as on other occasions, I LIE under a heavy disadvantage, for out of the man
218  beer perfected; for a surprising number of eggs LIE strewed over the plains, so that in one day's
300  distant in futurity as the secondary formations LIE in the past, only during periods of subsidence
343  now all be in a metamorphosed condition, or may LIE buried under the ocean. Passing from these dif
069  nc in numbers, we may feel sure that the cause LIES quite as much in other species being favoured,
075  eing on another, as of a parasite on its prey, LIES generally between beings remote in the scale o
083  protects during each varying season, as far as LIES in his power, all his productions. He often be
172  re only state that I believe the answer mainly LIES in the record being incomparably less perfect
236  s preliminary difficulty. The great difficulty LIES in the working ants differing widely from both
237  members of insect communities; the difficulty LIES in understanding how such correlated modificat
280  an be urged against my theory. The explanation LIES, as I believe, in the extreme imperfection of
391  land birds. Bermuda, on the other hand, which LIES at about the same distance from North America
393  s group cannot be considered as oceanic, as it LIES on a bank connected with the mainland; moreove
004  each other and to their physical conditions of LIFE, untouched and unexplained. It is, therefore,
005  he complex and sometimes varying conditions of LIFE, will have a better chance of surviving, and t
005  much Extinction of the less improved forms of LIFE, and induces what I have called Divergence of
007  uctions having been raised under conditions of LIFE not so uniform as, and somewhat different from
007  g several generations to the new conditions of LIFE to cause any appreciable amount of variation;
008  cation. It has been disputed at what period of LIFE the causes of variability, whatever they may b
008  the action of any change in the conditions of LIFE. Nothing is more easy than to tame an animal,
010  been exposed to exactly the same conditions of LIFE; and this shows how unimportant the direct eff
010  ortant the direct effects of the comparison of LIFE are in comparison with the laws of reproductio
011  utec to the direct action of the conditions of LIFE, as, in some cases, increased size from amount
013  may be trusted, is that, at whatever period of LIFE a peculiarity first appears, it tends to appea
014  eater length of horn, though appearing late in LIFE, is clearly due to the male element. Having al
015  for by the experiment itself the conditions of LIFE are changed. If it could be shown that our dom
015  add, that when under nature the conditions of LIFE do change, variations and reversions of charac
028  ale and female pigeons can be easily mated for LIFE; and thus different breeds can be kept togethe
029  he direct action of the external conditions of LIFE, and some little to habit; but he would be a b
031  n's wand, by means of which he may summon into LIFE whatever form and mould he pleases. Lord Somer
041  hould be placed under favourable conditions of LIFE, so as to breed freely in that country. When t
042  of the same species. Pigeons can be mated for LIFE, and this is a great convenience to the fancie
043  s and plants. I believe that the conditions of LIFE, from their action on the reproductive system,
043  uted to the direct action of the conditions of LIFE. Something must be attributed to use and disus
044  ion directly due to the physical conditions of LIFE, and variations in this sense are supposed not
046  ariability is independent of the conditions of LIFE. I am inclined to suspect that we see in these
059  ome larger; and throughout nature the forms of LIFE which are now dominant tend to become still mo
059  up into smaller genera. And thus, the forms of LIFE throughout the universe become divided into gr
060  ls and plants throughout nature. Struggle for LIFE most severe between individuals and varieties
060  tion to another part, and to the conditions of LIFE, and of one distinct organic being to another
061  apter, follow inevitably from the struggle for LIFE. Owing to this struggle for life, any variatio
061  struggle for life. Owing to this struggle for LIFE, any variation, however slight and from whatev
062  words the truth of the universal struggle for LIFE, or more difficult, at least I have found it s
062  s or seeds, and are thus constantly destroying LIFE; or we forget how largely these songsters, or
062  cluding (which is more important) not only the LIFE of the individual, but success in leaving prog
062  n the edge of a desert is said to struggle for LIFE against the drought, though more properly it s
063  t suffer destruction during some period of its LIFE, and during some season or occasional year, ot
063  ct species, or with the physical conditions of LIFE. It is the doctrine of Malthus applied with ma
065  obvious explanation is that the conditions of LIFE have been very favourable, and that there has
065  st be checked by destruction at some period of LIFE. Our familiarity with the larger domestic anim
066  make up for much destruction at some period of LIFE; and this period in the great majority of case
066  each lives by a struggle at some period of its LIFE; that heavy destruction inevitably falls eithe
069  ing enormous destruction at some period of its LIFE, from enemies or from competitors for the same
069  summits, or absolute deserts, the struggle for LIFE is almost exclusively with the elements. That
070  limiting check independent of the struggle for LIFE. But even some of these so called epidemics ap
070  t could exist only where the conditions of its LIFE were so favourable that many could exist toget
```

Page **(Key Word)**

```
0073  or invent laws on the duration of the forms of  LIFE! I am tempted to give one more instance show
0074  fferent checks, acting at different periods of  LIFE, and during different seasons or years, proba
0076  victorious over another in the great battle of  LIFE. A corollary of the highest importance may be
0078  ne. Not until we reach the extreme confines of  LIFE, in the arctic regions or on the borders of a
0078  in its former home, yet the conditions of its  LIFE will generally be changed in an essential mar
0079  etrical ratio; that each at some period of its  LIFF, during some season of the year, during each
0079  eneration or at intervals, has to struggle for  LIFE, and to suffer great destruction. When we re
0080  each other and to their physical conditions of  LIFE. Can it, then, be thought improbable, seeing
0080  each being in the great and complex battle of  LIFF, should sometimes occur in the course of thou
0082  st chapter, that a change in the conditions of  LIFE, by specially acting on the reproductive sys
0082  y; and in the foregoing case the conditions of  LIFE are supposed to have undergone a change, and
0083  tutional difference, on the whole machinery of  LIFF. Man selects only for his own good; Nature or
0083  eing is placed under well suited conditions of  LIFE. Man keeps the natives of many climates in th
0084  the nicely balanced scale in the struggle for  LIFE, and so be preserved. How fleeting are the w
0084  tter adapted to the most complex conditions of  LIFE, and should plainly bear the stamp of far hig
0084  ion to its organic and inorganic conditions of  LIFE. We see nothing of those slow changes in prog
0086  mestication appear at any particular period of  LIFE, tend to reappear in the offspring at the sam
0086  n other modifications at a different period of  LIFE, shall not be in the least degree injurious:
0087  structure used only once in an animal's whole  LIFE, if of high importance to it, might be modif
0087  , or in relation to wholly different habits of  LIFE in the two sexes; as is sometimes the case w
0089  of any animal have the same general habits of  LIFE, but differ in structure, colour, or ornament
0094  der culture and placed under new conditions of  LIFE, sometimes the male organs and sometimes the
0102  most certainly present different conditions of  LIFE; and then if natural selection be modifying a
0104  d amongst them, as long as their conditions of  LIFE remain the same, only through the principle c
0104  om the proper type; but if their conditions of  LIFE change and they undergo modification,uniform
0104  large, the organic and inorganic conditions of  LIFE will generally be in a great degree uniform:
0106  species there supported, but the conditions of  LIFE are infinitely complex from the large number
0106  ed on islands. On a small island, the race for  LIFE will have been less severe, and there will ha
0107  urable for the production of many new forms of  LIFE, likely to endure long and to spread widely.
0109  another and with their physical conditions of  LIFE, which may be effected in the long course of
0110  ey will consequently be beaten in the race for  LIFE by the modified descendants of the commoner s
0114  of the principle, that the greatest amount of  LIFE can be supported by great diversification of
0116  plants are diversified for different habits of  LIFE, so will a greater number of individuals be c
0119  vantages which made their parent successful in  LIFE, they will generally go on multiplying in num
0121  form having some advantage in the struggle for  LIFE over other forms, there will be a constant te
0126  course of ages and under varying conditions of  LIFE, organic beings vary at all in the several pa
0127  me age, season, or year, a severe struggle for  LIFE, and this certainly cannot be disputed; then,
0127  chance of being preserved in the struggle for  LIFE; and from the strong principle of inheritance
0127  in modifying and adapting the various forms of  LIFE to their several conditions and stations, mus
0128  be their chance of succeeding in the battle of  LIFE. Thus the small differences distinguishing va
0128  of the less improved and intermediate forms of  LIFE. On these principles, I believe, the nature c
0129  ermaster other species in the great battle for  LIFE. The limbs divided into great branches, and t
0130  nnects by its affinities two large branches of  LIFE, and which has apparently been saved from fat
0130  n i believe it has been with the great Tree of  LIFE, which fills with its dead and broken branche
0131  ome way due to the nature of the conditions of  LIFE, to which the parents and their more remote a
0131  ly susceptible to changes in the conditions of  LIFE; and to this system being functionally distur
0133  l selection, and how much to the conditions of  LIFE. Thus, it is well known to furriers that anim
0133  ame variety being produced under conditions of  LIFE as different as can well be conceived; and, c
0134  ch facts show how indirectly the conditions of  LIFE must act. Again, innumerable instances are kr
0134  ight on the direct action of the conditions of  LIFE. Indirectly, as already remarked, they seem t
0135  no feeding beetles, they must be lost early in  LIFE, and therefore cannot be much used by these 1
0136  vely numerous, and which groups have habits of  LIFE almost necessitating frequent flight: these s
0138  work. It is difficult to imagine conditions of  LIFF more similar than deep limestone caverns unde
0139  am only surprised that more wrecks of ancient  LIFF have not been preserved, owing to the less se
0148  e organisation. If under changed conditions of  LIFE a structure before useful becomes less useful
0148  vidual of the species; for in the struggle for  LIFE to which every animal is exposed, each indivi
0158  t the males and females to different habits of  LIFF, or the males to struggle with other males fo
0160  there is nothing in the external conditions of  LIFE to cause the reappearance of the slaty blue,
0161  left to the mutual action of the conditions of  LIFE and of a similar inherited constitution. It m
0167  of the same genus. The external conditions of  LIFE, as climate and food, etc., seem to have indu
0171  anditional varieties. Transitions in habits of  LIFE. Diversified habits in the same species. Spec
0173  , each has become adapted to the conditions of  LIFE of its own region, and has supplanted and ext
0174  iate region, having intermediate conditions of  LIFE, why do we not now find close linking interme
0175  look at climate and the physical conditions of  LIFF as the all important elements of distribution
0177  hence, the more common forms, in the race for  LIFE, will tend to beat and supplant the less comm
0179  l habits; and as each exists by a struggle for  LIFE, it is clear that each is well adapted in its
0181  , more especially under changing conditions of  LIFE, in the continued preservation of individuals
0182  birds is good for it, under the conditions of  LIFF to which it is exposed, for each has to live
0183  structures fitted for very different habits of  LIFE will rarely have been developed at an early p
0183  advantage over other animals in the battle for  LIFE. Hence the chance of discovering species with
0186  eful to an animal under changing conditions of  LIFE, then the difficulty of believing that a perf
0187  ive to light, hardly concerns us more than how  LIFE itself first originated: but I may remark tha
0193  if in members having very different habits of  LIFE, we may attribute its presence to inheritance
0194  arent importance. As natural Selection acts by  LIFE and death, by the preservation of individuals
0196  endants of the species under new conditions of  LIFE and newly acquired habits. To give a few
0199  s now have no direct relation to the habits of  LIFE of each species. Thus, we can hardly believe
0201  he lapse of time, under changing conditions of  LIFE, if any part comes to be injurious, it will b
0203  was not continuous, and when the conditions of  LIFF did not insensibly graduate away from one par
0204  n concluding that the most different habits of  LIFE could not graduate into each other: that a ba
0204  een that a species may under new conditions of  LIFE change its habits, or have diversified habits
0204  possessor, then, under changing conditions of  LIFE, there is no logical impossibility in the acq
0205  n of profitable variations in the struggle for  LIFE. Natural selection will produce nothing in on
0205  duce perfection, or strength in the battle for  LIFE, only according to the standard of that count
0206  which is quite independent of their habits of  LIFF. On my theory, unity of type is explained by
0206  ing to its organic and inorganic conditions of  LIFE; or by having adapted them during long past p
0206  he direct action of the external conditions of  LIFE, and being in all cases subjected to the seve
```

age **(Key Word)**

```
207  ntal powers, any more than I have with that of LIFE itself. We are concerned only with the diversi
208  cquired, they often remain constant throughout LIFE. Several other points of resemblance between i
209  each species, under its present conditions of LIFE. Under changed conditions of life, it is at le
209  onditions of life. Under changed conditions of LIFE, it is at least possible that slight modificat
210  ng different instincts at different periods of LIFE, or at different seasons of the year, or when
213  shorter period, under less fixed conditions of LIFE. How stongly these domestic instincts, habits,
233  able to the individual under its conditions of LIFE, it may reasonably be asked, how a long and gr
243  ee no difficulty, under changing conditions of LIFE, in natural selection accumulating slight modi
243  ing under considerably different conditions of LIFE, yet often retaining nearly the same instincts
245  m between the effects of changed conditions of LIFE and crossing. Fertility of varieties when cros
264  generally placed under suitable conditions of LIFE. But a hybrid partakes of only half of the nat
264  sitive to injurious or unnatural conditions of LIFE. In regard to the sterility of hybrids, in whi
265  lar manner. In the one case, the conditions of LIFE have been disturbed, though often in so slight
266  organs one to another, or to the conditions of LIFE. When hybrids are able to breed inter se, they
266  result, in the one case from the conditions of LIFE having been disturbed, in the other case from
266  ence, that slight changes in the conditions of LIFE are beneficial to all living things. We see th
267  erived from almost any change in the habits of LIFE. Again, both with plants and animals, there is
267  if these be kept under the same conditions of LIFE, always induces weakness and sterility in the
267  one hand, slight changes in the conditions of LIFE benefit all organic beings, and on the other h
267  ich is essentially related to the principle of LIFE. Fertility of varieties when crossed, and of t
269  sappearing under nearly the same conditions of LIFE. Lastly, and this seems to me by far the most
269  does not wish to alter their general habits of LIFE. Nature acts uniformly and slowly during vast
273  y sensitive to any change in the conditions of LIFE, being thus often rendered either impotent or
277  pure species, when their natural conditions of LIFE have been disturbed. This view is supported by
277  ; and that slight changes in the conditions of LIFE are apparently favourable to the vigour and fe
279  al conditions. I endeavoured to show, that the LIFE of each species depends in a more important ma
279  efore, that the really governing conditions of LIFE do not graduate away quite insensibly like hea
281  for in all cases the new and improved forms of LIFE will tend to supplant the old and unimproved f
289  diment, charged with new and peculiar forms of LIFE, had elsewhere been accumulated. And if in eac
291  l give a most imperfect record of the forms of LIFE which then existed; or, sediment may be accumu
291  the sea will remain shallow and favourable for LIFE, and thus a fossiliferous formation thick enou
292  hallow parts, which are the most favourable to LIFE. Still less could this have happened during th
294  mpted to conclude that the average duration of LIFE of the embedded fossils had been less than tha
300  nce there would probably be much extinction of LIFE; during the periods of elevation, there would
301  ame group into one long and branching chain of LIFE. We ought only to look for a few links; some m
302  uspected how poor a record of the mutations of LIFE, the best preserved geological section present
302  e genera or families, have really started into LIFE all at once, the fact would be fatal to the th
303  t an organism to some new and peculiar line of LIFE, for instance to fly through the air: but that
307  ins of the lowest Silurian stratum the dawn of LIFE on this planet. Other highly competent judges,
307  nding with new and peculiar species. Traces of LIFE have been detected in the Longmynd beds beneat
307  ks, probably indicates the former existence of LIFE at these periods. But the difficulty of unders
311  esent the apparently abruptly changed forms of LIFE, entombed in our consecutive, but widely separ
312  ction. On simultaneous changes in the forms of LIFE throughout the world. On the affinities of ext
313  each two consecutive formations, the forms of LIFE have seldom changed in exactly the same degree
314  s to their organic and inorganic conditions of LIFE, as explained in a former chapter. When many o
315  reappear, even if the very same conditions of LIFE, organic and inorganic, should recur. For thou
318  having endured from the earliest known dawn of LIFE to the present day; some having disappeared in
318  hat as the individual has a definite length of LIFE, so have species a definite duration. No one I
318  erminated the former horse under conditions of LIFE apparently so favourable. But how utterly grou
319  something is unfavourable in its conditions of LIFE; but what that something is, we can hardly eve
319  ntingencies, and at what period of the horse's LIFE, and in what degree, they severally acted. If
321  ed, from being fitted to some peculiar line of LIFE, or from inhabiting some distant and isolated
322  r species or group of species. On the Forms of LIFE changing almost simultaneously throughout the
322  more striking than the fact, that the forms of LIFE change almost simultaneously throughout the wo
323  america, a similar parallelism in the forms of LIFE has been observed by several authors: so it is
323  general parallelism in the successive forms of LIFE, in the stages of the widely separated palaeoz
324  ater tertiary stages. When the marine forms of LIFE are spoken of as having changed simultaneously
324  n a geological sense. The fact of the forms of LIFE changing simultaneously, in the above large se
325  to the parallelism of the palaeozoic forms of LIFE in various parts of Europe, they add, if struc
325  cause of these great mutations in the forms of LIFE throughout the world, under the most different
325  act of the parallel succession of the forms of LIFE throughout the world, is explicable on the the
327  simultaneous, succession of the same forms of LIFE throughout the world, accords well with the pr
328  s, the same general succession in the forms of LIFE; but the species would not exactly correspond;
329  nce with the general succession of the form of LIFE, and the order would falsely appear to be stri
329  ps, or between them. That the extinct forms of LIFE help to fill up the wide intervals between exi
332  relation to its slightly altered conditions of LIFE, and yet retain throughout a vast period the s
333  the mutual affinities of the extinct forms of LIFE to each other and to living forms, seem to me
334  intermediate in character between the forms of LIFE above and below. We must, however, allow for t
336  e same. Let it be remembered that the forms of LIFE, at least those inhabiting the sea, have chang
337  having had some advantage in the struggle for LIFE over other and preceding forms. If under a nea
337  ion of the more recent and victorious forms of LIFE, in comparison with the ancient and beaten for
342  together all the extinct and existing forms of LIFE by the finest graduated steps. He who rejects
344  and how the spreading of the dominant forms of LIFE, which are those that oftenest vary, will in t
344  can understand how it is that all the forms of LIFE, ancient and recent, make together one grand s
345  have beaten their predecessors in the race for LIFE, and are, in so far, higher in the scale of na
345  g been supplanted by new and improved forms of LIFE, produced by the laws of variation still actin
350  on the migration of the more dominant forms of LIFE from one region into another having been effec
350  on and reaction, in their mutual struggles for LIFE; the relation of organism to organism being, a
351  its the individual in its complex struggle for LIFE, so the degree of modification in different sp
352  australia or South America? The conditions of LIFE are nearly the same, so that a multitude of Eu
379  ill have had to compete with many new forms of LIFE; and it is probable that selected modification
382  tinted by the same peculiar forms of vegetable LIFE. Sir C. Lyell in a striking passage has specul
382  bution both of the same and of allied forms of LIFE can be explained. The living waters may be sai
388  . although there will always be a struggle for LIFE between the individuals of the species, howeve
396  if modification be admitted, all the forms of LIFE would have been more equally modified, in acco
398  america? There is nothing in the conditions of LIFE, in the geological nature of the islands, in t
```

Page **(Key Word)**

Page ***********************************(Key Word)***********************************

```
0401 oubtedly be exposed to different conditions of LIFE in the different islands, for it would have t
0403 ncling of species under the same conditions of LIFE. Thus, the south east and south west corners
0405 ictorious in distant lands in the struggle for LIFE with foreign associates. But on the view of a
0407 our ignorance, and remember that some forms of LIFE change most slowly, enormous periods of time
0408 ation (generally of the more dominant forms of LIFE), together with subsequent modification and t
0408 often be inhabited by very different forms of LIFE; for according to the length of time which ha
0408 nditions, infinitely diversified conditions of LIFE, there would be an almost endless amount of o
0409 there is a striking parellelism in the laws of LIFE throughout time and space; the laws governing
0410 elligible; for whether we look to the forms of LIFE which have changed during successive ages wit
0411 gin explained. Summary. From the first dawn of LIFE, all organic beings are found to resemble eac
0412 looking at the great diversity of the forms of LIFE which, in any small area, come into the close
0414 f the structure which determined the habits of LIFE, and the general place of each being in the e
0414 though so intimately connected with the whole LIFE of the being, are ranked as merely adaptive o
0414 the organs of vegetation, on which their whole LIFE depends, are of little signification, excepti
0415 aptation of the species to their conditions of LIFE. That the mere physiological importance of an
0425 been modified in relation to the conditions of LIFE to which each species has been recently expos
0426 pecially those having very different habits of LIFE, it assumes high value; for we can account fo
0426 ical importance, those which serve to preserve LIFE under the most diverse conditions of existenc
0429 tion, how it is that the more ancient forms of LIFE often present characters in some slight degre
0431 mals, by the belief that many ancient forms of LIFE have been utterly lost, through which the ear
0431 as been less entire extinction of the forms of LIFE which once connected fishes with batrachians.
0434 e same class, independently of their habits of LIFE, resemble each other in the general plan of t
0439 ve, and have been adapted for special lines of LIFE. A trace of the law of embryonic resemblance,
0440 porpoise, are related to similar conditions of LIFE. No one will suppose that the stripes on the
0440 od of activity may come on earlier or later in LIFE; but whenever it comes on, the adaptation of
0440 e adaptation of the larva to its conditions of LIFE is just as perfect and as beautiful as in the
0441 sis. When this is completed they are fixed for LIFE: their legs are now converted into prehensile
0443 xcept when the embryo becomes at any period of LIFE active and has to provide for itself; of the
0443 ill be. The question is not, at what period of LIFE any variation has been caused, but at what pe
0443 he formation of the embryo, may appear late in LIFE; as when an hereditary disease, which appears
0444 re fully acquired a little earlier or later in LIFE. It would not signify, for instance, to a bir
0444 have supervened at a not very early period of LIFE; and some direct evidence from our domestic a
0444 n see, might have appeared earlier or later in LIFE, tend to appear at a corresponding age in the
0446 uctures have been acquired earlier or later in LIFE, if the full grown animal possesses them. And
0446 generally first appeared at an early period of LIFE, and have been inherited by the offspring at
0447 e much modification at a rather late period of LIFE, and having thus been converted into hands, o
0447 are wholly ignorant, at a very early period of LIFE, or each step might be inherited at an earlie
0448 rom their following exactly the same habits of LIFE with their parents; for in this case, it woul
0448 and, it profited the young to follow habits of LIFE in any degree different from those of their p
0449 n clearly see why ancient and extinct forms of LIFE should resemble the embryos of their descenda
0450 upposed law of resemblance of ancient forms of LIFE to the embryonic stages of recent forms, may
0450 ient progenitor, at a very early period in the LIFE of each, though perhaps caused at the earlies
0454 at an organ rendered, during changed habits of LIFE, useless or injurious for one purpose, might
0455 ed by natural selection. At whatever period of LIFE disuse or selection reduces an organ, and thi
0455 nding age, but at an extremely early period of LIFE (as we have good reason to believe to be poss
0457 enerally supervening at a very early period of LIFE, and being inherited at a corresponding perio
0457 cially modified in relation to their habits of LIFE, through the principle of modifications being
0457 ection, it will generally be at that period of LIFE when the being has to provide for its own wan
0461 ed by slightly different and new conditions of LIFE, we need not feel surprise at hybrids being i
0461 eased by slight changes in their conditions of LIFE, and that the offspring of slightly modified
0461 and, considerable changes in the conditions of LIFE and crosses between greatly modified forms, l
0461 ther hand, lesser changes in the conditions of LIFE and crosses between less modified forms, incr
0463 d of which the climate and other conditions of LIFE change insensibly in going from a district oc
0463 of the gradation and mutation of the forms of LIFE? We meet with no such evidence, and this is t
0465 l probably be more variability in the forms of LIFE; during periods of subsidence, more extinctio
0466 ly susceptible to changes in the conditions of LIFE; so that this system, when not rendered impot
0466 he direct action of the physical conditions of LIFE. There is much difficulty in ascertaining how
0466 for long periods. As long as the conditions of LIFE remain the same, we have reason to believe th
0467 ly exposes organic beings to new conditions of LIFE, and then nature acts on the organisation, an
0468 uccessfully struggled with their conditions of LIFE, will generally leave most progeny. But succe
0469 , under their excessively complex relations of LIFE, would be preserved, accumulated, and inherit
0469 ariations useful, under changing conditions of LIFE, to her living products? What limit can be pu
0469 ing each form to the most complex relations of LIFE. The theory of natural selection, even if we
0471 , explains the arrangement of all the forms of LIFE, in groups subordinate to groups, all within
0471 d with habits and structure fitting it for the LIFE of an auk or grebe! and so on in endless othe
0474 ced under considerably different conditions of LIFE, yet should follow nearly the same instincts!
0475 tion of their descendants, causes the forms of LIFE, after long intervals of time, to appear as i
0476 ss improved organic beings in the struggle for LIFE. Lastly, the law of the long endurance of all
0477 as may present the same physical conditions of LIFE, we need feel no surprise at their inhabitant
0479 herited at a corresponding not early period of LIFE, we can clearly see why the embryos of mammal
0480 changed habits or under changed conditions of LIFE! and we can clearly understand on this view t
0480 ittle power of acting on an organ during early LIFE! hence the organ will not be much reduced or
0482 ven conjecture, which are the created forms of LIFE, and which are those produced by secondary la
0484 nced from some one primordial form, into which LIFE was first breathed. When the views entertaine
0486 s in forming a picture of the ancient forms of LIFE. Embryology will reveal to us the structure,
0487 s, by the general succession of their forms of LIFE. As species are produced and exterminated by
0488 iods of the earth's history, when the forms of LIFE were probably fewer and simpler, the rate of
0488 was probably slower; and at the first dawn of LIFE, when very few forms of the simplest structur
0489 d dominant species. As all the living forms of LIFE are the lineal descendants of those which liv
0490 nd direct action of the external conditions of LIFE, and from use and disuse; a Ratio of Increáse
0490 increase so high as to lead to a Struggle for LIFE, and as a consequence to Natural Selection, e
0490 tly follows. There is grandeur in this view of LIFE, with its several powers, having been origina
0030 eminent breeders have, even within a single LIFETIME, modified to a large extent some breeds of
0032 udies his subject for years, and devotes his LIFETIME to it with indomitable perseverance, he wil
0063 rease. Every being, which during its natural LIFETIME produces several eggs or seeds, must suffer
0035 y, did greatly modify, even during their own LIFETIMES, the forms and qualities of their cattle.
```

Page ***********************************(Key Word)***********************************

Lifetimes

```
211  as soon as it felt the antennae, immediately LIFTED up its abdomen and excreted a limpid drop of
001  tinent. These facts seemed to me to throw some LIGHT on the origin of species, that mystery of mys
010  ribute to the direct action of heat, moisture, LIGHT, food, etc., is most difficult: my impression
011  colour from particular kinds of food and from LIGHT, and perhaps the thickness of fur from climat
091  ill Mountains in the United States, one with a LIGHT greyhound like form, which pursues deer, and
111  s. as has always been my practice, let us seek LIGHT on this head from our domestic productions. W
124  hink we shall then see that the diagram throws LIGHT on the affinities of extinct beings, which, t
132  can here and there dimly catch a faint ray of LIGHT, and we may feel sure that there must be some
137  hat it regained, after living some days in the LIGHT, some slight power of vision. In the same man
137  ction seems to have struggled with the loss of LIGHT and to have increased the size of the eyes; w
138  om ordinary forms, prepare the transition from LIGHT to darkness. Next follow those that are const
152  natural selection, I think we can obtain some LIGHT. In our domestic animals, if any part, or the
186  distances, for admitting different amounts of LIGHT, and for the correction of spherical and chro
187  red real. How a nerve comes to be sensitive to LIGHT, hardly concerns us more than how life itself
187  y sensitive nerve may be rendered sensitive to LIGHT, and likewise to those coarser vibrations of
187  perly act only by excluding lateral pencils of LIGHT, are convex at their upper ends and must act
188  transparent tissue, with a nerve sensitive to LIGHT beneath, and then suppose every part of the
199  so strongly marked: I may add that some little LIGHT can apparently be thrown on the origin of the
202  r nature. The correction for the aberration of LIGHT is said, on high authority, not to be perfect
203  very grave; but I think that in the discussion LIGHT has been thrown on several facts, which on th
374  simultaneously colder from pole to pole, much LIGHT can be thrown on the present distribution of
376  anked as distinct species. Now let us see what LIGHT can be thrown on the foregoing facts, on the
385  many others remain to be observed, throw some LIGHT on this subject. When a duck suddenly emerges
395  ough man's agency; but we shall soon have much LIGHT thrown on the natural history of this archipe
454  . but I doubt whether any of these cases throw LIGHT on the origin of rudimentary organs in a stat
474  l instincts. I have attempted to show how much LIGHT the principle of gradation throws on the admi
487  know the many means of migration, then, by the LIGHT which geology now throws, and will continue t
487  n to their apparent means of immigration, some LIGHT can be thrown on ancient geography. The noble
488  f each mental power and capacity by gradation. LIGHT will be thrown on the origin of man and his h
066  ng each generation or at recurrent intervals. LIGHTEN any check, mitigate the destruction ever so
003  been independently created, but had descended, LIKE varieties, from other species. Nevertheless, s
009  gularly, and producing offspring not perfectly LIKE their parents or variable. Sterility has been
012  bts how strong is the tendency to inheritance: LIKE produces like is his fundamental belief: doubt
012  is the tendency to inheritance: like produces LIKF is his fundamental belief: doubts have been th
013  lone, more commonly but not exclusively to the LIKE sex. It is a fact of some little importance to
018  ow long before these ancient periods, savages, LIKE those of Tierra del Fuego or Australia, who po
021  ort faced tumbler has a beak in outline almost LIKF that of a finch; and the common tumbler has th
024  he whole great family of Columbidae for a beak LIKE that of the English carrier, or that of the sh
024  faced tumbler, or barb; for reversed feathers LIKE those of the jacobin; for a crop like that of
024  feathers like those of the jacobin; for a crop LIKE that of the pouter; for tail feathers like tho
025  rop like that of the pouter; for tail feathers LIKE those of the fantail. Hence it must be assumed
027  ned aboriginal stocks were coloured and marked LIKE the rock pigeon, although no other existing sp
027  compared with all other Columbidae, though so LIKE in most other respects to the rock pigeon; the
031  e sheep are placed on a table and are studied, LIKE a picture by a connoisseur; this is done three
035  i am informed by him, any native dog in Spain LIKE our pointer. By a similar process of selection
039  only fourteen tailfeathers somewhat expanded, LIKE the present Java fantail, or like individuals
039  at expanded, like the present Java fantail, or LIKE individuals of other and distinct breeds; in w
040  of our domestic breeds. But, in fact, a breed, LIKE a dialect of a language, can hardly be said to
047  termediate gradations, that naturalists do not LIKE to rank them as distinct species, are in sever
051  ny species, for he will become impressed, just LIKE the pigeon or poultry fancier before alluded t
057  ttle groups of species are generally clustered LIKE satellites around certain other species. And w
066  ne fly deposits hundreds of eggs, and another, LIKE the hippobosca, a single one: but this differe
076  ns, if they were allowed to struggle together, LIKE beings in a state of nature, and if the seed o
087  the thickness of the shell being known to vary LIKF every other structure. Sexual Selection. Inasm
088  ed as fighting, bellowing, and whirling round, LIKF Indians in a war dance, for the possession of
090  animals (as the wattle in male carriers, horn LIKE protuberances in the cocks of certain fowls, e
091  the United States, one with a light greyhound LIKE form, which pursues deer, and the other more b
097  ood, as I believe, of the plant. Bees will act LIKE a camel hair pencil, and it is quite sufficien
105  er a small isolated area, or a large open area LIKE a continent, has been most favourable for the
107  as the Ornithorhynchus and Lepidosiren, which, LIKE fossils, connect to a certain extent orders no
112  not and will not admire a medium standard, but LIKE extremes, they both go on (as has actually occ
131  n its summit, so we occasionally see an animal LIKE the Ornithorhynchus or Lepidosiren, which in s
135  deviations of structure, as to make the child LIKE its parents. But the much greater variability,
159  the early progenitor of the ostrich had habits LIKE those of a bustard, and that as natural select
163  escent, and a consequent tendency to vary in a LIKE manner, but to three separate yet closely rela
163  has very distinct transverse bars on its legs, LIKE those on the legs of the zebra: it has been as
163  shoulder. The quagga, though so plainly barred LIKE a zebra over the body, is without bars on the
163  figured one specimen with very distinct zebra LIKE bars on the hocks. With respect to the horse,
165  barred, and had three short shoulder stripes, LIKE those on the dun Welch pony, and even had some
165  on the dun Welch pony, and even had some zebra LIKE stripes on the sides of its face. With respect
166  ming, by simple variation, striped on the legs LIKE a zebra, or striped on the shoulders like an a
166  legs like a zebra, or striped on the shoulders LIKE an ass. In the horse we see this tendency stro
166  ds of generations, and I see an animal striped LIKF a zebra, but perhaps otherwise very differentl
167  rticular manner, so as often to become striped LIKF other species of the genus; and that each has
180  winter it leaves the frozen waters, and preys LIKE other polecats on mice and land animals. If a
180  o lessen the difficulty in any particular case LIKE that of the bat. Look at the family of squirre
182  ted which used their wings solely as flappers, LIKE the logger headed duck (Micropterus of Eyton);
182  fins in the water and front legs on the land, LIKE the penguin; as sails, like the ostrich; and f
182  legs on the land, like the penguin; as sails, LIKE the ostrich; and functionally for no purpose l
182  e the ostrich; and functionally for no purpose LIKE the Apteryx. Yet the structure of each of thes
183  over one spot and then proceeding to another, LIKE a kestrel, and at other times standing station
184  onary on the margin of water, and then dashing LIKE a kingfisher at a fish. In our own country the
184  s major) may be seen climbing branches, almost LIKE a creeper. It often, like a shrike, kills smal
184  ing branches, almost like a creeper. It often, LIKE a shrike, kills small birds by blows on the he
184  of the yew on a branch, and thus breaking them LIKE a nuthatch. In North America the black bear wa
184  r hours with widely open mouth, thus catching, LIKE a whale, insects in the water. Even in so extr
```

Page **(Key Word)**

```
0188  that the Creator works by intellectual powers LIKE those of man? If we must compare the eye to a
0191  he slits on the sides of the neck and the loop LIKE course of the arteries still marking in the
0195  tural selection. The tail of the giraffe looks LIKE an artificially constructed fly flapper; and
0202  rogenitor as a boring and serrated instrument, LIKE that in so many members of the same great or
0213  soon as it scented its prey, stand motionless LIKE a statue, and then slowly crawl forward with
0214  ossing, resemble natural instincts, which in a LIKE manner become curiously blended together; and
0218  ng. Some species, likewise, of Sphegidae (wasp LIKE insects) are parasitic on other species; and
0219  slave, but with plenty of the food which they LIKE best, and with their larvae and pupae to stir
0220  sturbed, the slaves occasionally come out, and LIKE their masters are much agitated and defend th
0234  , nearer together, and aggregated into a mass, LIKE the cells of the Melipona; for in this case a
0243  north America, build cock nests, to roost in, LIKE the males of our distinct Kitty wrens, a hab
0259  f them; and such hybrids, though externally so LIKE one of their pure parent species, are with ra
0270  ingues crossed three varieties of gourd, which LIKE the maize has separated sexes, and he asserts
0275  he offspring of the male ass and mare, is more LIKE an ass, than is the hinny, which is the offsp
0275  t, that mongrel animals alone are born closely LIKE one of their parents; but it can be shown tha
0279  of life do not graduate away quite insensibly LIKE heat or moisture. I endeavoured, also, to sho
0295  s. It would seem that each separate formation, LIKE the whole pile of formations in any country,
0317  y equally slow steps other species, and so on, LIKE the branching of a great tree from a single s
0319  owen soon perceived that the tooth, though so LIKE that of the existing horse, belonged to an ex
0320  much improved it is transported far and near, LIKE our short horn cattle, and takes the place of
0320  by examples, between the forms which are most LIKE each other in all respects. Hence the improve
0339  educated eye, in the gigantic pieces of armour LIKE those of the armadillo, found in several part
0349  same genus; and not by a true ostrich or emeu, LIKE those found in Africa and Australia under the
0350  s we positively know, produces organisms quite LIKE, or, as we see in the case of varieties nearl
0350  or, as we see in the case of varieties nearly LIKE each other. The dissimilarity of the inhabita
0358  t least of the islands would have been formed, LIKE other mountain summits, of granite, metamorph
0360  ey were alternately wet and exposed to the air LIKE really floating plants. He tried 98 seeds, mo
0364  r to argue that because a well stocked island, LIKE Great Britain, has not, as far as is known (a
0378  h a mingled tropical and temperate vegetation, LIKE that now growing with strange luxuriance at t
0384  heir occasional transport by accidental means; LIKE that of the live fish not rarely cropped by w
0386  en caught with an Ancylus (a fresh water shell LIKE a limpet) firmly adhering to it; and a water
0388  ther and often distant piece of water. Nature, LIKE a careful gardener, thus takes her seeds from
0393  alkland Islands, which are inhabited by a wolf LIKE fox, come nearest to an exception; but this c
0398  verde Islands are related to those of Africa, LIKE those of the Galapagos to America. I believe
0411  this classification is evidently not arbitrary LIKE the grouping of the stars in constellations.
0414  sitions and of placing together the forms most LIKE each other. It might have been thought (and w
0415  the Proteaceae, says their generic importance, LIKE that of all their parts, not only in this but
0422  ral system is genealogical in its arrangement, LIKE a pedigree; but the degrees of modification w
0427  lance, in the shape of the body and in the fin LIKE anterior limbs, between the dugong, which is
0428  th another: thus the shape of the body and fin LIKE limbs are only analogical when whales are com
0428  h the water; but the shape of the body and fin LIKE limbs serve as characters exhibiting true aff
0430  he phascolomys having become adapted to habits LIKE those of a Rodent. The elder De Candolle has
0439  phylloclneous acaceas, are pinnate or divided LIKE the ordinary leaves of the Leguminosae. The p
0440  he embryos of the vertebrata the peculiar loop LIKE course of the arteries near the branchial sli
0442  nt, yet nearly all pass through a similar worm LIKE stage of development; but in some few cases,
0466  impotent, fails to reproduce offspring exactly LIKE the parent form. Variability is governed by m
0474  t may be developed in the most unusual manner, LIKE the wing of a bat, and yet not be more variab
0475  america, for instance, lines her nest with mud LIKE our British species. On the view of instincts
0479  branchial slits and arteries running in loops, LIKE those in a fish which has to breathe the air
0480  e, how utterly inexplicable it is that parts, LIKE the teeth in the embryonic calf or like the s
0480  parts, like the teeth in the embryonic calf or LIKE the shrivelled wings under the soldered wing
0488  orld should have been due to secondary causes, LIKE those determining the birth and death of the
0039  haracter was when it first appeared, the more LIKELY it would be to catch his attention. But to u
0054  s which are already dominant will be the most LIKELY to yield offspring which, though in some sli
0107  for the production of many new forms of life, LIKELY to endure long and to spread widely. For the
0142  s from one district to another; for it is not LIKELY that man should have succeeded in selecting
0158  h have recently and largely varied being more LIKELY still to go on varying than parts which have
0275  fect character of either parent would be more LIKELY to occur with mongrels, which are descended
0348  ins, deserts, etc., are not so impassable, or LIKELY to have endured so long as the oceans separa
0369  l have been somewhat different; for it is not LIKELY that all the same arctic species will have b
0380  europe, though hides, wool, and other objects LIKELY to carry seeds have been largely imported in
0385  first perplexed me much, as their ova are not LIKELY to be transported by birds, and they are imm
0386  ponds, if suddenly flushed, would be the most LIKELY to have muddy feet. Birds of this order I ca
0386  islands in the open ocean; they would not be LIKELY to alight on the surface of the sea; so that
0392  confined ranges. Hence trees would be little LIKELY to reach distant oceanic islands; and an her
0396  slands separated by shallow channels are more LIKELY to have been continuously united within a re
0398  s obvious that the Galapagos Islands would be LIKELY to receive colonists, whether by occasional
0414  ications of its true affinities. We are least LIKELY in the modifications of these organs to mist
0425  which, as far as we can judge, are the least LIKELY to have been modified in relation to the con
0460  ty surprising when we remember that it is not LIKELY that either their constitutions or their rep
0464  rinc the discovery of intermediate links less LIKELY. Local varieties will not spread into other
0473  stance, should the colour of a flower be more LIKELY to vary in any one species of a genus, if in
0474  therefore these same characters would be more LIKELY still to be variable than the generic charac
0259  have a remarkable power of impressing their LIKENESS on their hybrid offspring; but these two po
0274  ometimes a prepotent power of impressing its LIKENESS on the hybrid; and so I believe it to be wi
0274  pecially owing to prepotency in transmitting LIKENESS running more strongly in one sex than in th
0489  e living species will transmit its unaltered LIKENESS to a distant futurity. And of the species n
0017  extraordinary inherent tendency to vary, and LIKEWISE to withstand diverse climates. I do not dis
0038  bnormal character of our domestic races, and LIKEWISE their differences being so great in externa
0046  eptions, polymorphic in other countries, and LIKEWISE, judging from Brachiopod shells, at former
0092  of their pollen from flower to flower, would LIKEWISE be favoured or selected. We might have take
0093  nly one, not as a very striking case, but as LIKEWISE illustrating one step in the separation of
0103  organisms which cross only occasionally, and LIKEWISE in animals which unite for each birth, but
0118  r inhabitants of the same country; they will LIKEWISE partake of those more general advantages wh
0122  ted, not only their parents (A) and (I), but LIKEWISE some of the original species which were mos
0124  illion or a hundred million generations, and LIKEWISE a section of the successive strata of the e
```

Page **(Key Word)**

age **************************************(Key Word)**

144 | loes, scaly anteaters, etc.), that these are | LIKEWISE | the most abnormal in their teeth. I know of
147 | f compensation which have been advanced, and | LIKEWISE | some other facts, may be merged under a more
157 | d, the number varies greatly: and the number | LIKEWISE | differs in the two sexes of the same species
157 | ration differs in the different species, and | LIKEWISE | in the two sexes of the same species. This r
158 | several places in the economy of nature, and | LIKEWISE | to fit the two sexes of the same species to
178 | losely allied or representative species, and | LIKEWISE | of acknowledged varieties), exist in the int
187 | erve may be rendered sensitive to light, and | LIKEWISE | to those coarser vibrations of the air which
192 | inally served as ovigerous frena, but which, | LIKEWISE | very slightly aided the act of respiration,
197 | the head of the clean feeding male turkey is | LIKEWISE | naked. The sutures in the skulls of young ma
213 | of all shades of disposition and tastes, and | LIKEWISE | of the oddest tricks, associated with certai
218 | tore food for their own young. Some species, | LIKEWISE | of Sphegidae (wasp like insects) are parasi
219 | d have been perfected. Formica sanguinea was | LIKEWISE | first discovered by P. Huber to be a slave m
225 | times adding to them short tubes of wax, and | LIKEWISE | making separate and very irregular rounded c
242 | to show the power of natural selection, and | LIKEWISE | because this is by far the most serious spec
246 | from common parents, when intercrossed, and | LIKEWISE | the fertility of their mongrel offspring, is
256 | pure species. But the degree of fertility is | LIKEWISE | innately variable; for it is not always the
287 | for perhaps equally long periods, it would, | LIKEWISE, | have escaped the action of the coast waves.
294 | y a part of one whole geological period; and | LIKEWISE | to reflect on the great changes of level, on
305 | ble difficulty on my theory, unless it could | LIKEWISE | be shown that the species of this group appe
327 | the sea was either stationary or rising, and | LIKEWISE | when sediment was not thrown down quickly en
357 | connected with Europe or Africa, and Europe | LIKEWISE | with America. Other authors have thus hypoth
358 | rtions of the inhabitants of oceanic islands | LIKEWISE | seem to me opposed to the belief of their fo
360 | rom mine; but he chose many large fruits and | LIKEWISE | seeds from plants which live near the sea; a
363 | temperate regions of the United States would | LIKEWISE | be covered by arctic plants and animals, and
372 | f the inhabitants of seas now disjoined, and | LIKEWISE | of the past and present inhabitants of the t
404 | or representative species occur, there will | LIKEWISE | be found some identical species, showing, in
408 | ime the individuals of the same species, and | LIKEWISE | of allied species, have proceeded from some
408 | manner linked together by affinity, and are | LIKEWISE | linked to the extinct beings which formerly
409 | should be closely related to each other, and | LIKEWISE | be related, but less closely, to those of th
424 | er from each other and from the adult; as he | LIKEWISE | includes the so called alternate generations
442 | ucture between the embryo and the adult, and | LIKEWISE | a close similarity in the embryos of widely
464 | e to examine them ever so closely, unless we | LIKEWISE | possessed many of the intermediate links bet
477 | ains, under the most different climates; and | LIKEWISE | the close alliance of some of the inhabitant
478 | forms and and varieties of the same species | LIKEWISE | occur. It is a rule of high generality that
479 | petals, stamens, and pistils of a flower, is | LIKEWISE | intelligible on the view of the gradual modi
387 | seeds of moderate size, as of the yellow water | LILY | and Potamogeton. Herons and other birds, centu
387 | the great size of the seeds of that fine water | LILY | the Nelumbium, and remembered Alph. de Candoll
387 | he found the seeds of the great southern water | LILY | (probably, according to Dr. Hocker, the Nelumb
129 | nts. From the first growth of the tree, many a | LIMB | and branch has decayed and dropped off; and th
435 | pattern, or to transpose parts. The bones of a | LIMB | might be shortened or widened to any extent, a
011 | rk on this subject. Breeders believe that long | LIMBS | are almost always accompanied by an elongated
129 | ther species in the great battle for life. The | LIMBS | divided into great branches, and these into l
143 | front and hind legs, and even in the jaws and | LIMBS, | varying together, for the lower jaw is belie
143 | ower jaw is believed to be homologous with the | LIMBS. | These tendencies, I do not doubt, may be mas
180 | ing squirrels; and flying squirrels have their | LIMBS | and even the base of the tail united by a bro
181 | ners of the jaw to the tail, and including the | LIMBS | and the elongated fingers: the flank membrane
198 | tainous country would probably affect the hind | LIMBS | from exercising them more, and possibly even
198 | by the law of homologous variation, the front | LIMBS | and even the head would probably be affected.
200 | nture to believe that the several bones in the | LIMBS | of the monkey, horse, and bat, which have bee
427 | shape of the body and in the fin like anterior | LIMBS, | between the dugong, which is a pachydermatou
428 | other: thus the shape of the body and fin like | LIMBS | are only analogical when whales are compared
428 | water: but the shape of the body and fin like | LIMBS | serve as characters exhibiting true affinity
435 | d their general shape of body and structure of | LIMBS | from a common ancestor. So it is with fishes.
435 | laws govern the construction of the mouths and | LIMBS | of crustaceans. So it is with the flowers of
435 | in his most interesting work on the Nature of | LIMBS. | On the ordinary view of the independent crea
435 | e as it may be called, of all mammals, had its | LIMBS | constructed on the existing general pattern,
435 | fication of the homologous construction of the | LIMBS | throughout the whole class. So with the mouth
436 | umber of vertebrae. The anterior and posterior | LIMBS | in each member of the vertebrate and articula
447 | ew to whole families or ever classes. The fore | LIMBS, | for instance, which served as legs in the pa
447 | nherited at a corresponding late age, the fore | LIMBS | in the embryos of the several descendants of
447 | ach individual new species, the embryonic fore | LIMBS | will differ greatly from the fore limbs in th
447 | c fore limbs will differ greatly from the fore | LIMBS | in the mature animal; the limbs in the latter
447 | from the fore limbs in the mature animal; the | LIMBS | in the latter having undergone much modificat
450 | kes there are rudiments of the pelvis and hind | LIMBS. | Some of the cases of rudimentary organs are
453 | calf by the excretion of precious phosphate of | LIME? When a man's fingers have been amputated, imp
138 | ne conditions of life more similar than deep | LIMESTONE | caverns under a nearly similar climate: so
068 | d for each species of course gives the extreme | LIMIT | to which each can increase; but very frequent
109 | s powers of artificial selection. I can see no | LIMIT | to the amount of change, to the beauty and in
125 | descent had diverged less. I see no reason to | LIMIT | the process of modification, as now explained
132 | eaks confidently that shells at their southern | LIMIT, | and when living in shallow water, are more b
294 | near the mouth of the Mississippi, within that | LIMIT | of depth at which marine animals can flourish
469 | nditions of life, to her living products? What | LIMIT | can be put to this power, acting during long
469 | g the good and rejecting the bad? I can see no | LIMIT | to this power, in slowly and beautifully adap
003 | only possible cause of variation. In one very | LIMITED | sense, as we shall hereafter see, this may b
053 | ings. But my tables further show that, in any | LIMITED | country, the species which are most common,
100 | considered as distinct individuals only in a | LIMITED | sense. I believe this objection to be valid,
140 | believe that species in a state of nature are | LIMITED | in their ranges by the competition of other
206 | erfection: nor, as far as we can judge by our | LIMITED | faculties, can absolute perfection be everyw
260 | sed. The differences being of so peculiar and | LIMITED | a nature, that, in reciprocal crosses betwee
261 | ridisation, so with grafting, the capacity is | LIMITED | by systematic affinity, for no one has been
286 | hich must have covered up the Weald within so | LIMITED | a period as since the latter part of the Cha
330 | the present day, were not at this early epoch | LIMITED | in such distinct groups as they now are. Som
352 | f migrating across the sea is more distinctly | LIMITED | in terrestrial mammals, than perhaps in any
468 | mount of variation under nature is a strictly | LIMITED | quantity. Man, though acting on external cha
481 | of variation in the course of long ages is a | LIMITED | quantity; no clear distinction has been, or
070 | ame animals, often ensue: and here we have a | LIMITING | check independent of the struggle for life.

Page **************************************(Key Word)**

Limiting

```
0256 t produced, are very fertile. Even within the LIMITS of the same genus, for instance in Dianthus,
0257 lose species a single hybrid. Even within the LIMITS of the same genus, we meet with this same di
0292 oyed by being upraised and brought within the LIMITS of the coast action. Thus the geological rec
0295 nic remains, except near their upper or lower LIMITS. It would seem that each separate formation,
0298 . have had the widest range, far exceeding the LIMITS of the known geological formations of Europe
0304 various zones of depths from the upper tidal LIMITS to 50 fathoms; from the perfect manner in wh
0400 yet in a most interesting manner, within the LIMITS of the same archipelago. Thus the several is
0433 and yet use these same characters within the LIMITS of the same group. We can clearly see how it
0436 rs, variations which we know to be within the LIMITS of possibility. In the paddles of the extinc
0386 t with an Ancylus (a fresh water shell like a LIMPET) firmly adhering to it; and a water beetle o
0211 ediately lifted up its abdomen and excreted a LIMPID drop of sweet juice, which was eagerly devou
0008 monstrosities cannot be separated by any clear LINE of distinction from mere variations. But I am
0015 ld succeed, is not of great importance for our LINE of argument; for by the experiment itself the
0021 rbit has a very short and conical beak, with a LINE of reversed feathers down the breast; and it
0051 ties will rise to a climax. Certainly no clear LINE of demarcation has as yet been drawn between
0117 ccumulated by natural selection. When a dotted LINE reaches one of the horizontal lines, and is t
0119 progeny will be increased. In our diagram the LINE of succession is broken at regular intervals
0119 s of modification will be confined to a single LINE of descent, and the number of the descendants
0120 e the amount of change between each horizontal LINE in our diagram to be excessively small, these
0124 fore his mind. In the diagram, each horizontal LINE has hitherto been supposed to represent a tho
0214 e only in one way, by not coming in a straight LINE to his master when called. Domestic instincts
0214 ome one dog naturally shown a tendency in this LINE; and this is known occasionally to happen, as
0272 artner, whose strong wish was to draw a marked LINE of distinction between species and varieties,
0283 ut! Moreover, if we follow for a few miles any LINE of rocky cliff, which is undergoing degradati
0285 xtends for upwards of 30 miles, and along this LINE the vertical displacement of the strata has v
0286 he rate at which the sea commonly wears away a LINE of cliff of any given height, we could measur
0286 yard in height to be eaten back along a whole LINE of coast at the rate on one yard in nearly ev
0286 . on the other hand, I do not believe that any LINE of coast, ten or twenty miles in length, ever
0303 to adapt an organism to some new and peculiar LINE of life, for instance to fly through the air;
0317 nera of a family, be represented by a vertical LINE of varying thickness, crossing the successive
0317 formations in which the species are found, the LINE will sometimes falsely appear to begin at its
0318 ecies be represented, as before, by a vertical LINE of varying thickness, the line is found to ta
0318 , by a vertical line of varying thickness, the LINE is found to taper more gradually at its upper
0321 preserved, from being fitted to some peculiar LINE of life, or from inhabiting some distant and
0331 tions, and all the forms beneath the uppermost LINE may be considered as extinct. The three exist
0332 , the three existing families on the uppermost LINE would be rendered less distinct from each othe
0332 nce, above No. VI., but none from beneath this LINE, then only the two families on the left hand
0378 dagas north west of the Himalaya, and the long LINE of the Cordillera, seem to have afforded two
0378 in North America, which must have lain on the LINE cf march. But I do not doubt that some tempera
0382 eir living drift on our mountain summits, in a LINE gently rising from the arctic lowlands to a g
0412 s. in the diagram each letter on the uppermost LINE may represent a genus including several speci
0412 ng several species; and all the genera on this LINE form together one class, for all have descende
0420 enera (a14 to z14) on the uppermost horizontal LINE. Now all these modified descendants from a sir
0421 he descendants of the genus F, along its whole LINE of descent, are supposed to have been but lit
0469 ted as a variety, we can see why it is that no LINE of demarcation can be drawn between species, 
0006 oncing to what are called the same genera are LINEAL descendants cf some other and generally extir
0029 he idea of species in a state of nature being LINEAL descendants of other species? Selection. Let
0187 erfected, we ought to look exclusively to its LINEAL ancestors; but this is scarcely ever possible
0210 cauired, for these could be found only in the LINEAL ancestors of each species, but we ought to f
0488 l beings not as special creations, but as the LINEAL descendants of some few beings which lived lc
0489 cies. As all the living forms of life are the LINEAL descendants of those which lived long before
0422 the names of the groups had been written in a LINEAR series, it would have been still less possibl
0029 he does of the intermediate links in the long LINES of descent, yet admit that many of our domest
0049 deserve consideration; for several interesting LINES of argument, from geographical distribution,
0095 tic valleys or to the formation of the longest LINES of inland cliffs. Natural selection can act c
0117 wn country. The little fan of diverging dotted LINES of unequal lengths proceeding from (A), may r
0117 nt variations (represented by the outer dotted LINES) being preserved and accumulated by natural s
0117 en a dotted line reaches one of the horizontal LINES, and is there marked by a small numbered lett
0117 tic work. The intervals between the horizontal LINES in the diagram, may represent each a thousanc
0119 later and more highly improved branches in the LINES of descent, will, it is probable, often take
0119 branches not reaching to the upper horizontal LINES. In some cases I do not doubt that the proces
0120 ould be represented in the diagram, if all the LINES proceeding from (A) were removed, excepting t
0121 posed to be represented between the horizontal LINES. After fourteen thousand generations, six new
0121 and this is shown in the diagram by the dotted LINES not prolonged far upwards from want of space.
0121 it probably will be with many whole collateral LINES of descent, which will be conquered by later
0121 which will be conquered by later and improved LINES of descent. If, however, the modified offspri
0124 n our diagram, this is indicated by the broken LINES, beneath the capital letters, converging in s
0125 ived at very ancient epochs when the branching LINES of descent had diverged less. I see no reason
0125 d by each successive group of diverging dotted LINES to be very great, the forms marked a14 to p14
0187 , branching off in two fundamentally different LINES, can be shown to exist, until we reach a mode
0210 pecies, but we ought to find in the collateral LINES of descent some evidence of such gradations;
0228 and began to build up flat walls of wax on the LINES of intersection between the basins, so that e
0243 ce, how it is that the thrush of South America LINES its nest with mud, in the same peculiar manne
0282 h we see around us. It is good to wander along LINES of sea coast, when formed of moderately hard
0311 di end of each page, only here and there a few LINES. Each word of the slowly changing language, i
0331 bered letters represent genera, and the dotted LINES diverging from them the species in each genus
0331 but this is unimportant for us. The horizontal LINES may represent successive geological formation
0331 er with the many extinct genera on the several LINES of descent diverging from the parent form A,
0332 discovered above one of the middle horizontal LINES or geological formations, for instance, above
0348 range far northward and southward, in parallel LINES not far from each other, under corresponding
0378 he cordillera, seem to have afforded two great LINES of invasion: and it is a striking fact, latel
0380 solved. I do not pretend to indicate the exact LINES and means of migration, or the reason why cer
0381 ry distinct species have migrated in radiating LINES from some common centre; and I am inclined to
0382 th. As the tide leaves its drift in horizontal LINES, though rising higher on the shores where the
0427 t. for animals, belonging to two most distinct LINES of descent, may readily become adapted to sim
0427 nceal their blood relationship to their proper LINES of descent. We can also understand the appare
```

Page **(Key Word)**

```
0431  quently be related to each other by circuitous LINES of affinity of various lengths (as may be see
0434  ted together by the most complex and radiating LINES of affinities. We shall never, probably, dise
0439   are active, and have been adapted for special LINES of life. A trace of the law of embryonic rese
0439  most of the species are striped or spotted in LINES; and stripes can be plainly distinguished in t
0456  e united by complex, radiating, and circuitous LINES of affinities into one grand system; the rule
0475  why the thrush of South America, for instance, LINES her nest with mud like our British species. O
0479  arrangement, in which we have to discover the  LINES of descent by the most permanent characters,
0481  eologists, when Lyell first insisted that long LINES of inland cliffs had been formed, and great v
0486  have to discover and trace the many diverging  LINES of descent in our natural genealogies, by cha
0344  ess, from the survival of a few descendants,   LINGERING in protected and isolated situations. When
0306  st ancient Silurian animals, as the Nautilus,  LINGULA, etc., do not differ much from living specie
0313  s in the sub Himalayan deposits. The Silurian  LINGULA differs but little from the living species o
0316  ld and unmodified forms. Species of the genus  LINGULA, for instance, must have continuously existe
0344  lly disappeared, it does not reappear; for the LINK of generation has been broken. We can understa
0432  ps of modified descendants. Every intermediate LINK between these eleven genera and there primordi
0432  here primordial parent, and every intermediate LINK in each branch and sub branch of their descend
0464  etween any two species are unknown, if any one LINK or intermediate variety be discovered, it will
0047  imilar to some other forms, or are so closely  LINKED to them by intermediate gradations, that natu
0176  , sooner than the forms which they originally  LINKED together. For any form existing in lesser num
0194  its proper place in nature, be so invariably   LINKED together by graduated steps? Why should not N
0203  od are not indefinitely variable, and are not  LINKED together by a multitude of intermediate grada
0239  the extreme forms can sometimes be perfectly   LINKED together by individuals taken out of the same
0332  red, these three families would be so closely  LINKED together that they probably would have to be
0332  jected to call the extinct genera, which thus  LINKED the living genera of three families together,
0345  te formations; for the forms are more closely  LINKED together by generation: we can clearly see wh
0408  s, and deserts, are in so mysterious a manner  LINKED together by affinity, and are likewise linked
0408  linked together by affinity, and are likewise  LINKED to the extinct beings which formerly inhabite
0058  pt, firstly, by the discovery of intermediate  LINKING forms, and the occurrence of such links cann
0059  varieties, notwithstanding that intermediate   LINKING forms have not been discovered; but the amou
0174  ditions of life, why do we not now find close  LINKING intermediate varieties? This difficulty for
0176  rences, and therefore conclude that varieties  LINKING two other varieties together have generally
0179  y be true, numberless intermediate varieties,  LINKING most closely all the species of the same gro
0292  the frequent discovery of her transitional or LINKING forms. From the foregoing considerations it
0462  mber of intermediate forms must have existed,  LINKING together all the species in each group by gr
0462  ies, it may be asked, Why do we not see these  LINKING forms all around us? Why are not all organic
0029  owing no more than he does of the intermediate LINKS in the long lines of descent, yet admit that
0047  hen they are closely connected by intermediate LINKS; nor will the commonly assumed hybrid nature
0047  only assumed hybrid nature of the intermediate LINKS always remove the difficulty. In very many ca
0047  riety of another, not because the intermediate LINKS have actually been found, but because analogy
0050  her hand, they are united by many intermediate LINKS, and it is very doubtful whether these links
0050  e links, and it is very doubtful whether these LINKS are hybrids; and there is, as it seems to me,
0051  se he can hardly hope to find the intermediate LINKS between his doubtful forms, he will have to t
0056  ties; and in those cases in which intermediate LINKS have not been found between doubtful forms, n
0058  iate linking forms, and the occurrence of such LINKS cannot affect the actual characters of the fo
0177  inextricable chaos of varying and intermediate LINKS: firstly, because new varieties are very slow
0178   in each broken portion of the land, but these LINKS will have been supplanted and exterminated du
0179  erminate the parent forms and the intermediate LINKS. Consequently evidence of their former existe
0181  with an extensor muscle. Although no graduated LINKS of structure, fitted for gliding through the
0181  I can see no difficulty in supposing that such LINKS formerly existed, and that each had been form
0279  g blended together by innumerable transitional LINKS, is a very obvious difficulty. I assigned rea
0279  bvious difficulty. I assigned reasons why such LINKS do not commonly occur at the present day, und
0279  in cause, however, of innumerable intermediate LINKS not now occuring everywhere throughout nature
0280  on and every stratum full of such intermediate LINKS? Geology assuredly does not reveal any such f
0281  e and tapir, we have no reason to suppose that LINKS ever existed directly intermediate between th
0281  had a nearly perfect chain of the intermediate LINKS. It is just possible by my theory, that one o
0281  a tapir; and in this case direct intermediate  LINKS will have existed between them. But such a ca
0282  at the number of intermediate and transitional LINKS, between all living and extinct species, must
0282  remains of such infinitely numerous connecting LINKS, it may be objected, that time will not have
0293  each should not include a graduated series of  LINKS between the species which then lived; but I c
0298  pecies by numerous, fine, intermediate, fossil LINKS, by asking ourselves whether, for instance, g
0301  chain of life. We ought only to look for a few LINKS, some more closely, some more distantly relat
0301  ore distantly related to each other; and these LINKS, let them be ever so close, if found in diffe
0302  f our not discovering innumerable transitional LINKS between the species which appeared at the com
0310  ve formations infinitely numerous transitional LINKS between the many species which now exist or h
0329  ammals; but Owen has discovered so many fossil LINKS, that he has had to alter the whole classific
0342  in vain where are the numberless transitional  LINKS which must formerly have connected the closel
0432  cendants, may be supposed to be alive; and the LINKS to be as fine as those between the finest var
0463  in rare cases) to discover directly connecting LINKS between them, but only between each and some
0463  e extermination of an infinitude of connecting LINKS, between the living and extinct inhabitants o
0463  t every geological formation charged with such LINKS? Why does not every collection of fossil rema
0464  we likewise possessed many of the intermediate LINKS between their past or parent and present stat
0464  t or parent and present states: and these many LINKS we could hardly ever expect to discover, owin
0464  ill pretend that in future ages so many fossil LINKS will be discovered, that naturalists will be
0464  ul forms are varieties? As long as most of the LINKS between any two species are unknown, if any o
0464  causes rendering the discovery of intermediate LINKS less likely. Local varieties will not spread
0064  erally be standing room for his progeny.       LINNAEUS has calculated that if an annual plant produ
0413  edge. Such expressions as that famous one of   LINNAEUS, and which we often meet with in a more or l
0417  ant, alone explains, I think, that saying of   LINNAEUS, that the characters do not give the genus,
0427  sects there are innumerable instances: thus    LINNAEUS, misled by external appearances, actually cl
0002  d it to Sir Charles Lyell, who sent it to the  LINNEAN Society, and it is published in the third vo
0088  means of sexual selection, as the mane to the  LION, the shoulder pad to the boar, and the hooked
0439  an be plainly distinguished in the whelp of the LION. We occasionally though rarely see something o
0440  ill suppose that the stripes on the whelp of a LION, or the spots on the young blackbird, are of a
0435  y infinitely numerous modifications of an upper LIP, mandibles, and two pairs of maxillae. Analogo
0436  pose that their common progenitor had an upper LIP, mandibles, and two pair of maxillae, these pa
0285  pe of the lava streams, due to their formerly  LIQUID state, showed at a glance how far the hard, r
```

Page **(Key Word)**

Page **********************************(Key Word)**********************************

```
0009  hardly more than in a state of nature. A long LIST could easily be given of sporting plants; by t
0048  ed by botanists as species; and in making this LIST he has omitted many trifling varieties, but wh
0132  a affects their colours. Moquin Tandon gives a LIST of plants which when growing near the sea shor
0151  cirripedes. I shall, in my future work, give a LIST of the more remarkable cases; I will here only
0157  stances, the first which happen to stand on my LIST; and as the differences in these cases are of
0162  in an allied species. I have collected a long LIST of such cases; but here, as before, I lie unde
0180  d it seems to me that nothing less than a long LIST of such cases is sufficient to lessen the diff
0375  , not found in the intervening hot lowlands. A LIST of the genera collected on the loftier peaks o
0375  uced by man, occur on the lowlands; and a long LIST can be given, as I am informed by Dr. Hooker,
0064  s rate, in a few thousand years, there would LITERALLY not be standing room for his progeny. Linna
0072  ndreds of acres of the unenclosed heath, and LITERALLY I could not see a single Scotch fir, except
0439  fication. On my view these terms may be used LITERALLY; and the wonderful fact of the jaws, for ir
0010  rom the same fruit, and the young of the same LITTER, sometimes differ considerably from each othe
0005  next chapter I shall discuss the complex and LITTLE known laws of variation and of correlation of
0008  und out that very trifling changes, such as a LITTLE more or less water at some particular period
0010  with animals such agencies have produced very LITTLE direct effect, though apparently more in the
0013  usively to the like sex. It is a fact of some LITTLE importance to us, that peculiarities appearin
0017  ether it would endure other climates? Has the LITTLE variability of the ass or guinea fowl, or the
0029  from one or from several allied species. Some LITTLE effect may, perhaps, be attributed to the dir
0029  of the external conditions of life, and some LITTLE to habit; but he would be a bold man who woul
0030  pertinacious in battle, with other breeds so LITTLE quarrelsome, with everlasting layers which ne
0035  son. In some cases, however, unchanged or but LITTLE changed individuals of the same breed may be
0040  ncial name. In semi civilised countries, with LITTLE free communication, the spreading and knowlec
0042  rom only a few being kept by poor people, and LITTLE attention paid to their breeding; in peacocks
0043  cases of plants not propagated by seed are of LITTLE importance to us, for their endurance is only
0048  n species and varieties. On the islets of the LITTLE Madeira group there are many insects which ar
0051  which he is continually studying; and he has LITTLE general knowledge of analogical variation in
0054  re of the stations inhabited by them, and has LITTLE or no relation to the size of the genera to w
0057  or lesser groups. As Fries has well remarked, LITTLE groups of species are generally clustered lik
0059  difference, for two forms, if differing very LITTLE, are generally ranked as varieties, notwithsta
0059  sely, but unequally, allied together, forming LITTLE clusters round certain species. Species very
0060  as the foundation for the work, helps us but LITTLE in understanding how species arise in nature.
0060  y in the woodpecker and missletoe; and only a LITTLE less plainly in the humblest parasite which c
0062  are to those of Art. We will now discuss in a LITTLE more detail the struggle for existence. In my
0067  n any check, mitigate the destruction ever so LITTLE, and the number of the species will almost in
0068  ants: thus out of twenty species growing on a LITTLE plot of turf (three feet by four) nine specie
0068  numbers of a species. Thus, there seems to be LITTLE doubt that the stock of partridges, grouse, a
0072  e heath, I found a multitude of seedlings and LITTLE trees, which had been perpetually browsed dow
0072  m one of the old clumps, I counted thirty two LITTLE trees; and one of them, judging from the ring
0073  es cannot reach the nectar. Hence I have very LITTLE doubt, that if the whole genus of humble bees
0077  e know that it can perfectly well withstand a LITTLE more heat or cold, dampness or dryness, for e
0085  animal of any particular colour would produce LITTLE effect: we should remember how essential it i
0092  edily sought by insects. Let us now suppose a LITTLE sweet juice or nectar to be excreted by the i
0092  appears a simple loss to the plant; yet if a LITTLE pollen were carried, at first occasionally an
0094  certain flowers, which they can, with a very LITTLE more trouble, enter by the mouth. Bearing suc
0103  which unite for each birth; but which wander LITTLE and which can increase at a very rapid rate;
0116  a set of animals, with their organisation but LITTLE diversified, could hardly compete with a set
0116  , which are divided into groups differing but LITTLE from each other, and feebly representing, as
0117  cing to a genus large in its own country. The LITTLE fan of diverging dotted lines of unequal leng
0125  the three genera descended from (A), the two LITTLE groups of genera will form two distinct famil
0132  ich we see everywhere throughout nature. Some LITTLE influence may be attributed to climate, food,
0134  onsiderations as these incline me to lay very LITTLE weight on the direct action of the conditions
0134  to in the first chapter, I think there can be LITTLE doubt that use in our domestic animals streng
0136  st, either from its wings having been ever so LITTLE less perfectly developed or from indolent hab
0147  ndividual plants producing seeds which were a LITTLE better fitted to be wafted further, might get
0149  veral parts of the organisation have been but LITTLE specialised for particular functions; and as
0149  ection should have preserved or rejected each LITTLE deviation of form less carefully than when th
0151  al manner, of which fact I think there can be LITTLE doubt. But that our rule is not confined to s
0151  portant structures, and they differ extremely LITTLE even in different genera; but in the several
0152  tary organs, and in those which have been but LITTLE specialised for any particular purpose, and p
0173  sea is stationary or is rising, or when very LITTLE sediment is being deposited, there will be bl
0180  wer. Yet I think such difficulties have very LITTLE weight. Here, as on other occasions, I lie un
0186  mbers; and that if any one being vary ever so LITTLE, either in habits or structure, and thus gain
0187  earlier stages of descent, in an unaltered or LITTLE altered condition. Amongst existing Vertebrat
0192  nto each other. Therefore I do not doubt that LITTLE folds of skin, which originally served as ovi
0194  wo parts in two organic beings, which owe but LITTLE of their structure in common to inheritance f
0194  by the shortest and slowest steps. Organs of LITTLE apparent importance. As natural Selection act
0196  rtance to characters which are really of very LITTLE importance, and which have originated from qu
0196  that climate, food, etc., probably have some LITTLE direct influence on the organisation; that ch
0198  civilized countries there has been but LITTLE artificial selection. Careful observers are v
0199  h are so strongly marked; I may add that some LITTLE light can apparently be thrown on the origin
0199  s. physical conditions probably have some LITTLE effect on structure, quite independently of a
0200   structure in every living creature (making some LITTLE allowance for the direct action of physical c
0208  these characters of instinct are universal. A LITTLE dose, as Pierre Huber expresses it, of judgme
0209  ianoforte at three years old with wonderfully LITTLE practice, had played a tune with no practice
0209  t can be shown that instincts do vary ever so LITTLE, then I can see no difficulty in natural sele
0210  for the instincts of animals having been but LITTLE observed except in Europe and North America,
0220  bservations which I have myself made, in some LITTLE detail. I opened fourteen nests of F. sanguin
0222  nother species, F. flava, with a few of these LITTLE yellow ants still clinging to the fragments o
0222  i had accidentally disturbed both nests, the LITTLE ants attacked their big neighbours with surpr
0222  abitually make into slaves, from those of the LITTLE and furious F.flava, which they rarely captur
0222  t a quarter of an hour, shortly after all the LITTLE yellow ants had crawled away; they took heart
0228  rcular pits in it; and as they deepened these LITTLE pits, they make them wider and wider until th
0228  ees instantly began on both sides to excavate LITTLE basins near to each other, in the same way as
0229  o that the basins, as soon as they had been a LITTLE deepened, came to have flat bottoms; and thes
0229  ttoms; and these flat bottoms, formed by thin LITTLE plates of the vermilion wax having been left
```

Page **********************************(Key Word)**********************************

little

age **(Key Word)**

```
229  ite sides of the ridge of wax. In parts, only LITTLE bits, in other parts, large portions of a rho
229  impossible, from the extreme thinness of the LITTLE rhombic plate, that they could have effected
230  hat the very first cell is excavated out of a LITTLE parallel sided wall of wax, is not, as far as
230  : the first commencement having always been a LITTLE hood of wax: but I will not here enter on the
234  xen cells near together, so as to intersect a LITTLE: for a wall in common even to two adjoining c
234  even to two adjoining cells, would save some LITTLE wax. Hence it would continually be more and m
242  have, therefore, discussed this case, at some LITTLE but wholly insufficient length, in order to s
256  rosses and of Hybrids. We will now consider a LITTLE more in detail the circumstances and rules go
256  igh physiological importance and which differ LITTLE in the allied species. Now the fertility of f
261  direction. It will be advisable to explain a LITTLE more fully by an example what I mean by steri
263  rst Crosses and of Hybrids. We may now look a LITTLE closer at the probable causes of the sterilit
275  same, whether the two parents differ much or LITTLE from each other, namely in the union of indiv
283  reason to believe that pure water can effect LITTLE of nothing in wearing away rock. At last the
283  ly clothed by marine productions, showing how LITTLE they are abraded and how seldom they are roll
290  ill probably be preserved to a distant age. A LITTLE reflection will explain why along the rising
297  nguish species and varieties: they grant some LITTLE variability to each species, but when they me
313  an deposits. The Silurian Lingula differs but LITTLE from the living species of this genus; wherea
324  remotely future epoch, there can, I think, be LITTLE doubt that all the more modern marine formati
328  xactly correspond: for there will have been a LITTLE more time in the one region than in the other
336  cludes the whole glacial period, and note how LITTLE the specific forms of the inhabitants of the
344  o one: but more commonly only bringing them a LITTLE closer together. The more ancient a form is,
350  physical conditions. The naturalist must feel LITTLE curiosity, who is not led to inquire what thi
351  and afterwards isolated country, they will be LITTLE liable to modification: for neither migration
351  undergone during whole geological periods but LITTLE modification, there is not much difficulty in
367  may suppose that the Glacial period came on a LITTLE earlier or later in North America than in Eur
367  will the southern migration there have been a LITTLE earlier or later: but this will make no diffe
368  ons will at a very late period have marched a LITTLE further north, and subsequently have retreate
371  we can understand the relationship, with very LITTLE identity, between the productions of North Am
385  vered with duck weed, I have twice seen these LITTLE plants adhering to its back: and it has happe
385  ack; and it has happened to me, in removing a LITTLE duck weed from one aquarium to another, that
386  of ponds is with seeds: I have tried several LITTLE experiments, but will here give only the most
386  erent points, beneath water, on the edge of a LITTLE pond: this mud when dry weighed only 6 3/4 ou
389  m county of Cambridge has 847 plants, and the LITTLE island of Anglesea 764, but a few ferns and a
391  s and habits, and will consequently have been LITTLE liable to modification. Madeira, again, is in
392  lands: but this explanation seems to me not a LITTLE doubtful. Facility of immigration, I believe,
392  the nature of the conditions. Many remarkable LITTLE facts could be given with respect to the inha
392  may be, confined ranges. Hence trees would be LITTLE likely to reach distant oceanic islands: and
405  mense range: but, if the variation had been a LITTLE greater, the two varieties would have been ra
407  ccasional transport, I have discussed at some LITTLE length the means of dispersal of fresh water
410  on, we find that some organisms differ LITTLE, whilst others belonging to a different class
414  on, on which their whole life depends, are of LITTLE signification, excepting in the first main di
421  ine of descent, are supposed to have been but LITTLE modified, and they yet form a single genus. B
422  e that some very ancient language had altered LITTLE, and had given rise to few new languages, whi
429  ransmitted to the present day descendants but LITTLE modified, will give to us our so called oscul
435  ion. In changes of this nature, there will be LITTLE or no tendency to modify the original pattern
437  teristic (as Owen has observed) of all low or LITTLE modified forms: therefore we may readily beli
443  ppear at an equally early period. But we have LITTLE evidence on this head, indeed the evidence ra
444  r most of its characters are fully acquired a LITTLE earlier or later in life. It would not signif
470  restricted ranges, and they are clustered in LITTLE groups round other species, in which respects
472  ction have not been observed. The complex and LITTLE known laws governing variation are the same,
472  physical conditions seem to have produced but LITTLE direct effect: yet when varieties enter any z
480  he struggle for existence, and will thus have LITTLE power of acting on an organ during early life
288  d in infinite numbers: they are all strictly LITTORAL, with the exception of a single Mediterranea
290  sea. The explanation, no doubt, is, that the LITTORAL and sub littoral deposits are continually wo
290  ion, no doubt, is, that the littoral and sub LITTORAL deposits are continually worn away, as soon
300  ing. I suspect that not many of the strictly LITTORAL animals, or of those which lived on naked su
006  o the mutual relations of all the beings which LIVE around us. Who can explain why one species ran
014  y marked domestic varieties could not possibly LIVE in a wild state. In many cases we do not know
062  e birds which are idly singing round us mostly LIVE on insects or seeds, and are thus constantly d
062  uggle with each other which shall get food and LIVE. But a plant on the edge of a desert is said t
071  multitudes, so close together that all cannot LIVE. When I ascertained that these young trees had
082  nd to the physical conditions under which they LIVE, that none of them could anyhow be improved: f
086  nd probably in the case of those insects which LIVE only for a few hours, and which never feed, a
100  dering the medium in which terrestrial animals LIVE, and the nature of the fertilising element: fo
114  rotation. Most of the animals and plants which LIVE close round any small piece of ground, could l
114  e close round any small piece of ground, could LIVE on it (supposing it not to be in any way pecul
114  nd may be said to be striving to the utmost to LIVE there: but, it is seen, that where they come i
133  knows, are often brassy or lurid. Plants which LIVE exclusively on the sea side are very apt to ha
139  on of species to the climates under which they LIVE is often overrated. We may infer this from our
182  f life to which it is exposed, for each has to LIVE by a struggle: but it is not necessarily the b
182  s as the Crustacea and Mollusca are adapted to LIVE on the land, and seeing that we have flying bi
204  g in mind that each organic being is trying to LIVE wherever it can live, how it has arisen that t
204  rganic being is trying to live wherever it can LIVE, how it has arisen that there are upland geese
244  the larvae of ichneumonidae feeding within the LIVE bodies of caterpillars, not as specially endow
244  ngs, namely, multiply, vary, let the strongest LIVE and the weakest die. Chapter VIII. Hybridism.
264  iving in a country where their two parents can LIVE, they are generally placed under suitable cond
288  ct that but few of the very many animals which LIVE on the beach between high and low watermark ar
295  ecies undergoing modification will have had to LIVE on the same area throughout this whole time. B
295  cessary in order to enable the same species to LIVE on the same space, the supply of sediment must
324  cal sense: for if all the marine animals which LIVE at the present day in Europe, and all those th
324  later tertiary stages, than to those which now LIVE here: and if this be so, it is evident that fo
360  ge fruits and likewise seeds from plants which LIVE near the sea: and this would have favoured the
380  nge and affinities of the allied species which LIVE in the northern and southern temperate zones a
384  ransport by accidental means: like that of the LIVE fish not rarely dropped by whirlwinds in India
384  ter fish can with care be slowly accustomed to LIVE in fresh water: and, according to Valenciennes
403  but representative land shells, some of which LIVE in crevices of stone: and although large quant
```

Page **(Key Word)***

Page **(Key Word)**

```
0428  adapted by successive slight modifications to  LIVE  under nearly similar circumstances, to inhabi
0467  balance will determine which individual shall  LIVE  and which shall die, which variety or species
0472  daughters; at ichneumonidae feeding within the  LIVE  bodies of caterpillars; and at other such case
0009  from a state of nature, perfectly tamed, long  LIVED, and healthy (of which I could give numerous
0066  ce to keep up the full number of a tree, which  LIVED  on an average for a thousand years, if a sing
0125  understand this fact, for the extinct species  LIVED  at very ancient epochs when the branching lir
0129  the other branches; so with the species which  LIVED  during long past geological periods, very few
0133  re severe the climate is under which they have  LIVED; but who can tell how much of this difference
0167  ant cosmogonists, that fossil shells had never  LIVED, but had been created in stone so as to mock
0234  and let us further suppose that the community  LIVED  throughout the winter, and consequently requ
0248  e two most experienced observers who have ever  LIVED, namely, Kolreuter and Gartner, should have a
0264  when once born, are generally healthy and long  LIVED, as we see in the case of the common mule. Hy
0276  wo most careful experimentalists who have ever  LIVED, have come to diametrically opposite conclus
0282  t assuredly, if this theory be true, such have  LIVED  upon this earth. On the Lapse of Time. Indepe
0288  h respect to the terrestrial productions which  LIVED  during the Secondary and Palaeozoic periods,
0293  ted varieties between the allied species which  LIVED  at its commencement and at its close. Some ca
0293  series of links between the species which then  LIVED; but I can by no means pretend to assign due
0296  rmation, the probability is that they have not  LIVED  on the same spot during the whole period of c
0300  ll the species were to be collected which ever  LIVED  there, how imperfectly would they represent t
0300  e strictly littoral animals, or of those which  LIVED  on naked submarine rocks, would be embedded;
0302  ly slow process; and the progenitors must have  LIVED  long ages before their modified descendants.
0306  nded from some one crustacean, which must have  LIVED  long before the Silurian age, and which proba
0324  ight at least have been inferred that they had  LIVED  during one of the later tertiary stages. Wher
0324  the present day in Europe, and all those that  LIVED  in Europe during the pleistocene period (an e
0324  states are more closely related to those which  LIVED  in Europe during certain later tertiary stage
0334  at which succeeded it. Thus, the species which  LIVED  at the sixth great stage of descent in the da
0334  gram are the modified offspring of those which  LIVED  at the fifth stage, and are the parents of th
0367  ned, the same arctic species, which had lately  LIVED  in a body together on the lowlands of the Old
0370  when the inhabitants of the Old and New Worlds  LIVED  further southwards than at present, they must
0370  atitude 60 degrees, during the Pliocene period  LIVED  further north under the polar circle, in lati
0370  and that the strictly arctic productions then  LIVED  on the broken land still nearer to the pole.
0420  tters A to L to represent allied genera, which  LIVED  during the Silurian epoch, and these have desi
0432  ns made them; for if every form which has ever  LIVED  on this earth were suddenly to reappear, thou
0432  llecting all the forms in any class which have  LIVED  throughout all time and space. We shall certa
0449  ic beings, extinct and recent, which have ever  LIVED  on this earth have to be classed together, and
0484  robably all the organic beings which have ever  LIVED  on this earth have descended from some one pr
0488  he lineal descendants of some few beings which  LIVED  long before the first bed of the Silurian sys
0489  life are the lineal descendants of those which  LIVED  long before the Silurian epoch, we may feel c
0066  o the utmost to increase in numbers; that each  LIVES  by a struggle at some period of its life; tha
0084  ouse, if not destroyed at some period of their  LIVES, would increase in countless numbers; they ar
0246  olreuter and Gartner, who almost devoted their  LIVES  to this subject, without being deeply impress
0335  disappearance; for the parent rock pigeon now  LIVES; and many varieties between the rock pigeon a
0441  retrograde; for the male is a mere sack, which  LIVES  for a short time, and is destitute of mouth,
0023  l have descended from the rock pigeon (Columba  LIVIA), including under this term several geographi
0023  or willingly perching on trees. But besides C.  LIVIA, with its geographical sub species, only two
0027  omestic breeds have descended from the Columba  LIVIA  with its geographical sub species. In favour
0027  vour of this view, I may add, firstly, that C.  LIVIA, or the rock pigeon, has been found capable o
0008  nimals there are which will not breed, though  LIVING  long under not very close confinement in thei
0094  more quickly, and so have a better chance of  LIVING  and leaving descendants. Its descendants woul
0107  e. these anomalous forms may almost be called  LIVING  fossils; they have endured to the present day
0113  no its modified descendants, would succeed in  LIVING  on the same piece of ground. And we well know
0124  rders, or families, or genera, with those now  LIVING, yet are often, in some degree, intermediate
0126  endants; and consequently that of the species  LIVING  at any one period, extremely few will transmi
0126  few of the most ancient species may now have  LIVING  and modified descendants, yet at the most rem
0128  o, leads to divergence of character: for more  LIVING  beings can be supported on the same area the
0129  present the classification of all extinct and  LIVING  species in groups subordinate to groups. Of t
0129  ng past geological periods, very few now have  LIVING  and modified descendants. From the first grow
0129  rders, families, and genera which have now no  LIVING  representatives, and which are known to us on
0132  that shells at their southern limit, and when  LIVING  in shallow water, are more brightly coloured
0132  coloured under a clear atmosphere, than when  LIVING  on islands or near the coast. So with insects
0134  keeping true, or not varying at all, although  LIVING  under the most opposite climates. Such consid
0137  ess, could be in any way injurious to animals  LIVING  in darkness, I attribute their loss wholly to
0137  ssor Silliman thought that it regained, after  LIVING  some days in the light, some slight power of
0141  have a far wider range than any other rodent,  LIVING  free under the cold climate of Faroe in the n
0141  le of enduring a glacial climate, whereas the  LIVING  species are now all tropical or sub tropical
0167  created in stone so as to mock the shells now  LIVING  on the sea shore. Summary. Our ignorance of t
0178  ction, so that they will no longer exist in a  LIVING  state. Thirdly, when two or more varieties ha
0186  se and frigate birds with webbed feet, either  LIVING  on the dry land or most rarely alighting on t
0186  er; that there should be long toed corncrakes  LIVING  in meadows instead of in swamps; that there s
0188  re is much graduated diversity in the eyes of  LIVING  crustaceans, and bearing in mind how small th
0188  , and bearing in mind how small the number of  LIVING  animals is in proportion to those which have
0189  ed, and then the old ones to be destroyed. In  LIVING  bodies, variation will cause the slight alter
0189  of many kinds; and may we not believe that a  LIVING  optical instrument might thus be formed as su
0194  tate; yet, considering that the proportion of  LIVING  and known forms to the extinct and unknown is
0200  etc. Hence every detail of structure in every  LIVING  creature (making some little allowance for th
0201  eumon, by which its eggs are deposited in the  LIVING  bodies of other insects. If it could be prove
0203  diate gradations. Closely allied species, now  LIVING  on a continuous area, must often have been fo
0243  hen inhabiting distant parts of the world and  LIVING  under considerably different conditions of li
0264  stanced before and after birth: when born and  LIVING  in a country where their two parents can live
0266  the conditions of life are beneficial to all  LIVING  things. We see this acted on by farmers and g
0281  s just possible by my theory, that one of two  LIVING  forms might have descended from the other; fo
0281  forms. By the theory of natural selection all  LIVING  species have been connected with the parent s
0282  ermediate and transitional links, between all  LIVING  and extinct species, must have been inconceiv
0287  nd and the water has been peopled by hosts of  LIVING  forms. What an infinite number of generations
0306  tilus, Lingula, etc., do not differ much from  LIVING  species; and it cannot on my theory be suppos
0307  nown, periods of time, the world swarmed with  LIVING  creatures. To the question why we do not find
```

Page **(Key Word)**

Page **(Key Word)**

0312 ities of extinct species to each other and to LIVING species. On the state of development of ancie
0313 ame degree. In the oldest tertiary beds a few LIVING shells may still be found in the midst of a m
0313 silurian Lingula differs but little from the LIVING species of this genus; whereas most of the ot
0318 nct monsters, which all co existed with still LIVING shells at a very late geological period, I wa
0319 an extinct species. Had this horse been still LIVING, but in some degree rare, no naturalist would
0319 always to remember that the increase of every LIVING being is constantly being checked by unpercei
0323 suspected that they had coexisted with still LIVING sea shells; but as these anomalous monsters c
0324 al epoch), were to be compared with those now LIVING in South America or in Australia, the most sk
0329 ties of extinct Species to each other, and to LIVING forms. Let us now look to the mutual affiniti
0329 look to the mutual affinities of extinct and LIVING species. They all fall into one grand natural
0329 the more, as a general rule, it differs from LIVING forms. But, as Buckland long ago remarked, al
0329 for if we confine our attention either to the LIVING or to the extinct alone, the series is far le
0330 e same orders, families, or genera with those LIVING at the present day, were not at this early ep
0330 cies being considered as intermediate between LIVING species or groups. If by this term it is mean
0330 ntermediate in all its characters between two LIVING forms, the objection is probably valid. But I
0330 ny fossil species would have to stand between LIVING species, and some extinct genera between livi
0330 ving species, and some extinct genera between LIVING genera, even between genera belonging to dist
0330 he proposition, for every now and then even a LIVING animal, as the lepidosiren, is discovered hav
0332 all the extinct genera, which thus linked the LIVING genera of three families together, intermedia
0333 he extinct forms of life to each other and to LIVING forms, seem to me explained in a satisfactory
0335 are small things with great: if the principal LIVING and extinct races of the domestic pigeon were
0339 e australian caves were closely allied to the LIVING marsupials of that continent. In South Americ
0339 n the same continent between the dead and the LIVING. Professor Owen has subsequently extended the
0339 dded, as the relation between the extinct and LIVING land shells of Madeira; and between the extin
0339 hells of Madeira; and between the extinct and LIVING brackish water shells of the Aralo Caspian Se
0341 and in other characters to the species still LIVING in South America; and some of these fossils m
0341 hese fossils may be the actual progenitors of LIVING species. It must not be forgotten that, on my
0344 the more it generally differs from those now LIVING. Why ancient and extinct forms often tend to
0347 nd new Worlds; how widely different are their LIVING productions! In the southern hemisphere, if w
0354 he exceptional cases of the same species, now LIVING at distant and separated points; nor do I for
0361 0 days, to my surprise nearly all germinated. LIVING birds can hardly fail to be highly effective
0365 , with few or no destructive insects or birds LIVING there, nearly every seed, which chanced to ar
0365 most striking cases known of the same species LIVING at distant points, without the apparent possi
0365 rkable fact to see so many of the same plants LIVING on the snowy regions of the Alps or Pyrenees,
0367 e more especially related to the arctic forms LIVING due north or nearly due north of them: for th
0370 . hence we may suppose that the organisms now LIVING under the climate of latitude 60 degrees, dur
0371 nce it has come, that when we compare the now LIVING productions of the temperate regions of the N
0372 dification, for many closely allied forms now LIVING in areas completely sundered. Thus, I think,
0376 ecome less and less arctic. Many of the forms LIVING on the mountians of the warmer regions of the
0382 of allied forms of life can be explained. The LIVING waters may be said to have flowed during one
0382 res where the tide rises highest, so have the LIVING waters left their living drift on our mountai
0382 highest, so have the living waters left their LIVING drift on our mountain summits, in a line gent
0413 rely as a scheme for arranging together those LIVING objects which are most alike, and for separat
0421 th the parent genus F; just as some few still LIVING organic beings belong to Silurian genera. So
0431 finities which they perceive between the many LIVING and extinct members of the same great natural
0433 group. We can clearly see how it is that all LIVING and extinct forms can be grouped together in
0447 ll powers of activity and has to gain its own LIVING; and the effects thus produced will be inheri
0454 us under others, as with the wings of beetles LIVING on small and exposed islands; and in this cas
0456 the nature of the relationship, by which all LIVING and extinct beings are united by complex, rad
0463 n infinitude of connecting links, between the LIVING and extinct inhabitants of the world, and at
0469 ul, under changing conditions of life, to her LIVING products? What limit can be put to this power
0476 or into intermediate groups, follows from the LIVING and the extinct being the offspring of common
0480 y, it may be asked, have all the most eminent LIVING naturalists and geologists rejected this view
0483 ms have been commanded suddenly to flash into LIVING tissues? Do they believe that at each suppose
0484 gy may be a deceitful guide. Nevertheless all LIVING things have much in common, in their chemical
0486 aberrant, and which may fancifully be called LIVING fossils, will aid us in forming a picture of
0488 re, the progenitor of innumerable extinct and LIVING descendants, was created. In the distant futu
0489 om the past, we may safely infer that not one LIVING species will transmit its unaltered likeness
0489 to a distant futurity. And of the species now LIVING very few will transmit progeny of any kind to
0489 rocreate new and dominant species. As all the LIVING forms of life are the lineal descendants of t
0034 some of the Esquimaux their teams of dogs. LIVINGSTONE shows how much good domestic breeds are va
0436 y. in the paddles of the extinct gigantic sea LIZARDS, and in the mouths of certain suctorial crus
0482 pressing his conviction; for only thus can the LOAD of prejudice by which this subject is overwhel
0363 ubject. As icebergs are known to be sometimes LOADED with earth and stones, and have even carried
0450 n a rudimentary state: in very many snakes one LOBE of the lungs is rudimentary; in other snakes t
0073 reafter have occasion to show that the exotic LOBELIA fulgens, in this part of England, is never v
0098 pollen from its own flower: for instance, in LOBELIA fulgens, there is a really beautiful and ela
0098 y of seedlings; and whilst another species of LOBELIA growing close by, which is visited by bees,
0250 individual plants; as with certain species of LOBELIA, and with all the species of the genus Hippe
0251 era, and in the case of some other genera, as LOBELIA, Passiflora and Verbascum. Although the plan
0262 act of the extraordinary case of Hippeastrum, LOBELIA, etc., which seeded much more freely when fe
0103 een the individuals of the same new variety. A LOCAL variety when once thus formed might subsequen
0298 n, that their varieties are generally at first LOCAL: and that such local varieties do not spread
0298 es are generally at first local; and that such LOCAL varieties do not spread widely and supplant t
0298 e successive changes are supposed to have been LOCAL or confined to some one spot. Most marine ani
0298 rope, which have oftenest given rise, first to LOCAL varieties and ultimately to new species; and
0301 and the varieties would at first generally be LOCAL or confined to one place, but if possessed of
0325 n marine currents or other causes more or less LOCAL and temporary, but depend on general laws whi
0342 s; and that varieties have at first often been LOCAL. All these causes taken conjointly, must have
0353 t closely related to each other, are generally LOCAL, or confined to one area. What a strange anom
0353 opposite rule prevailed; and species were not LOCAL, but had been produced in two or more distinc
0464 es vary most, and varieties are often at first LOCAL, both causes rendering the discovery of inter
0464 e discovery of intermediate links less likely. LOCAL varieties will not spread into other and dist
0408 ical provinces. We can thus understand the LOCALISATION of sub genera, genera and families; and ho
0045 he same species inhabiting the same confined LOCALITY. No one supposes that all the individuals of
0312 ving here appeared for the first time, either LOCALLY, or, as far as we know, on the face of the e

Page **(Key Word)**

```
0320  animals which have been exterminated, either LOCALLY or wholly, through man's agency. I may repea
0407  the individuals of the same species, wherever LOCATED, have descended from the same parents, are n
0196  selection. Seeing how important an organ of LOCOMOTION the tail is in most aquatic animals, its ge
0448  ted for sites or habits, in which organs of LOCOMOTION or of the senses, etc., would be useless; a
0048  unite for each birth, and which are highly LOCOMOTIVE, doubtful forms, ranked by one zoologist as
0298  at can propagate rapidly and are not highly LOCOMOTIVE, there is reason to suspect, as we have for
0075  h each other for existence, as in the case of LOCUSTS and grass feeding quadrupeds. But the strugg
0384  t any change of level. We have evidence in the LOESS of the Rhine of considerable changes of level
0375  wlands. A list of the genera collected on the LOFTIER peaks of Java raises a picture of a collecti
0197  g bamboo in the Malay Archipelago climbs the LOFTIEST trees by the aid of exquisitely constructed
0365  as we hear from Asa Gray, with those on the LOFTIEST mountains of Europe. Even as long ago as 174
0284  ts, enormously long blank periods. So that the LOFTY pile of sedimentary rocks in Britain, gives b
0286  d coasts; though no doubt the degradation of a LOFTY cliff would be more rapid from the breakage o
0346  tions; the most humid districts, arid deserts, LOFTY mountains, grassy plains, forests, marshes, l
0348  ee the same fact; for on the opposite sides of LOFTY and continuous mountain ranges, and of great
0349  y an American type of structure. We ascend the LOFTY peaks of the Cordillera and we find an alpine
0374  there are many closely allied species. On the LOFTY mountains of equatorial America a host of pec
0375  hout the world, the plants growing on the more LOFTY mountains, and on the temperate lowlands of t
0182  used their wings solely as flappers, like the LOGGER headed duck (Micropterus of Eyton); as fins i
0473  conclusion when we look, for instance, at the LOGGER headed duck, which has wings incapable of fli
0134  ; yet there are several in this state. The LOGGERHEADED duck of South America can only flap along
0204  nder changing conditions of life, there is no LOGICAL impossibility in the acquirement of any conc
0243  ny other known bird. Finally, it may not be a LOGICAL deduction, but to my imagination it is far m
0419  as often been used, though perhaps not quite LOGICALLY, in classification, more especially in very
0021  s, and have been permitted to join two of the LONDON Pigeon Clubs. The diversity of the breeds is
0058  . watson has marked for me in the well sifted LONDON Catalogue of plants (4th edition) 63 plants w
0004  , as it can be treated properly only by giving LONG catalogues of facts. We shall, however, be ena
0006  story. Although much remains obscure, and will LONG remain obscure, I can entertain no doubt; afte
0008  there are which will not breed, though living LONG under not very close confinement in their nati
0009  young from a state of nature, perfectly tamed, LONG lived, and healthy (of which I could give nume
0009  rhaps hardly more than in a state of nature. A LONG list could easily be given of sporting plants;
0011  at work on this subject. Breeders believe that LONG limbs are almost always accompanied by an elon
0012  e poisons. Hairless dogs have imperfect teeth; LONG haired and coarse haired animals are apt to ha
0012  aired animals are apt to have, as is asserted, LONG or many horns; pigeons with feathered feet hav
0012  th short beaks have small feet; and those with LONG beaks large feet. Hence, if man goes on select
0014  crossed offspring from a short horned cow by a LONG horned bull, the greater length of horn, thoug
0015  t we could not breed our cart and race horses, LONG and short horned cattle, and poultry of variou
0018  and years ago, and who will pretend to say how LONG before these ancient periods, savages, like th
0020  eeds could not be got without extreme care and LONG continued selection; nor can I find a single c
0021  heels. The runt is a bird of great size, with LONG, massive beak and large feet: some of the sub
0021  eeti some of the sub breeds of runts have very LONG necks, others very long wings and tails, other
0021  eds of runts have very long necks, others very LONG wings and tails, others singularly short tails
0021  allied to the carrier, but, instead of a very LONG beak, has a very short and very broad one. The
0026  s perfectly fertile. Some authors believe that LONG continued domestication eliminates this strong
0029  ether his cattle might not have descended from LONG horns, and he will laugh you to scorn. I have
0029  ven. The explanation, I think, is simple: from LONG continued study they are strongly impressed wi
0029  than he does of the intermediate links in the LONG lines of descent, yet admit that many of our d
0035  awings of the breeds in question had been made LONG ago, which might serve for comparison. In some
0039  escendants of that pigeon would become through LONG continued, partly unconscious and partly metho
0044  tion. To treat this subject at all properly, a LONG catalogue of dry facts should be given, but th
0045  sider unimportant parts; but I could show by a LONG catalogue of facts, that parts which must be c
0047  ed their characters in their own country for a LONG time; for as long, as far as we know, as have
0047  s in their own country for a long time; for as LONG, as far as we know, as have good and true spec
0052  stage may be, in some cases, due merely to the LONG continued action of different physical conditi
0052  inct, or they may endure as varieties for very LONG periods, as has been shown to be the case by M
0067  for instance, on a piece of ground three feet LONG and two wide, dug and cleared, and where there
0067  hiefly by slugs and insects. If turf which has LONG been mown, and the case would be the same with
0073  recurring with varying success; and yet in the LONG run the forces are so nicely balanced, that th
0073  d, that the face of nature remains uniform for LONG periods of time, though assuredly the merest t
0074  ir combs and nests; and Mr. H. Newman, who has LONG attended to the habits of humble bees, believe
0075  l kinds of trees must here have gone on during LONG centuries, each annually scattering its seeds
0077  (as peas and beans), when sown in the midst of LONG grass, I suspect that the chief use of the nut
0083  n some peculiar and fitting manner; he feeds a LONG and a short beaked pigeon on the same food; he
0083  igeon on the same food; he does not exercise a LONG backed or long legged quadruped in any peculia
0083  me food; he does not exercise a long backed or LONG legged quadruped in any peculiar manner; he ex
0083  in any peculiar manner; he exposes sheep with LONG and short wool to the same climate. He does no
0084  rogress, until the hand of time has marked the LONG lapse of ages, and then so imperfect is our vi
0084  f ages, and then so imperfect is our view into LONG past geological ages, that we only see that th
0088  ; male salmons have been seen fighting all day LONG; male stag beetles often bear wounds from the
0092  , and would be oftenest crossed; and so in the LONG run would gain the upper hand. Those flowers,
0097  er individual is occasionally; perhaps at very LONG intervals, indispensable. On the belief that t
0101  shown to be physically impossible. Cirripedes LONG appeared to me to present a case of very great
0101  irth; in many others it occurs perhaps only at LONG intervals; but in none, as I suspect, can self
0103  he same area, varieties of the same animal can LONG remain distinct, from haunting different stati
0104  h all plants. Even if these take place only at LONG intervals, I am convinced that the young thus
0104  n vigour and fertility over the offspring from LONG continued self fertilisation, that they will h
0104  g and propagating their kind; and thus, in the LONG run, the influence of intercrosses even at rar
0104  of character can be retained amongst them, as LONG as their conditions of life remain the same, o
0105  es, which will prove capable of enduring for a LONG period, and of spreading widely. Throughout a
0107  f level, and which consequently will exist for LONG periods in a broken condition, will be the mos
0107  on of many new forms of life, likely to endure LONG and to spread widely. For the area will first
0108  ion will always act very slowly, often only at LONG intervals of time, and generally on only a ver
0109  nditions of life, which may be effected in the LONG course of time by nature's power of selection.
0111  the ancient black cattle were displaced by the LONG horns, and that these were swept away by the s
0113  umber that can be supported in any country has LONG ago arrived at its full average. If its natura
0117  all to appear simultaneously, but often after LONG intervals of time; nor are they all supposed t
```

age ***************************************(Key Word)***************************************

```
119  prevail and multiply: a medium form may often  LONG  endure, and may or may not produce more than o
119  t have been inserted anywhere, after intervals  LONG  enough to have allowed the accumulation of a c
121  ital letters) of our original genus, may for a  LONG  period continue transmitting unaltered descend
126  yet have suffered least extinction, will for a  LONG  period continue to increase. But which groups
126  present day. Summary of Chapter. If during the  LONG  course of ages and under varying conditions of
129  uced during each former year may represent the  LONG  succession of extinct species. At each period
129  anches; so with the species which lived during  LONG  past geological periods, very few now have liv
131  nus vary in an analogous manner. Reversions to  LONG  lost characters. Summary. I have hitherto some
131  s. i have remarked in the first chapter, but a  LONG  catalogue of facts which cannot be here given
134  omparison, by which to judge of the effects of  LONG  continued use or disuse, for we know not the p
135  mentary condition in some other genera, by the  LONG  continued effects of disuse in their progenito
139  cclimatisation must be readily effected during  LONG  continued descent. It is notorious that each s
148  , natural selection will always succeed in the  LONG  run in reducing and saving every part of the o
149  e specialised for particular functions; and as  LONG  as the same part has to perform diversified wo
150  the various laws of growth, to the effects of  LONG  continued disuse, and to the tendency to rever
150  e truth of this proposition without giving the  LONG  array of facts which I have collected, and whi
153  teady selection to keep the breed true. In the  LONG  run selection gains the day, and we do not exp
153  tumbler from a good short faced strain. But as  LONG  as selection is rapidly going on, there may al
153  of modification implies an unusually large and  LONG  continued amount of variability, which has con
153  developed part or organ has been so great and  LONG  continued within a period not excessively remo
156  arts of the organisation which have for a very  LONG  period remained constant. In connexion with th
158  y still to go on varying than parts which have  LONG  been inherited and have not varied, to natural
162  organ in an allied species. I have collected a  LONG  list of such cases; but here, as before, I lie
166  g of each successive generation to produce the  LONG  lost character, and that this tendency, from u
168  e than generic characters, or those which have  LONG  been inherited, and have not differed within t
169  er degree than other parts; for variation is a  LONG  continued and slow process, and natural select
169  view must be a very slow process, requiring a  LONG  lapse of time, in this case, natural selection
171  e embraced by the theory of Natural Selection.  LONG  before having arrived at this part of my work,
173  will concur only rarely, and after enormously  LONG  intervals. Whilst the bed of the sea is statio
173  intermediate varieties? This difficulty for a  LONG  time quite confounded me. But I think it can b
174  ontinous, that it has been continuous during a  LONG  period. Geology would lead us to believe that
176  ermediate varieties should not endure for very  LONG  periods; why as a general rule they should be
179  al dives for and preys on fish, but during the  LONG  winter it leaves the frozen waters, and preys
180  s. and it seems to me that nothing less than a  LONG  list of such cases is sufficient to lessen the
185  by membrane. What seems plainer than that the  LONG  toes of grallatores are formed for walking ove
186  y alighting on the water; that there should be  LONG  toed corncrakes living in meadows instead of i
188  that this instrument has been perfected by the  LONG  continued efforts of the highest human intelle
190  have to look to very ancient ancestral forms,  LONG  since become extinct. We should be extremely c
204  yet in the case of any organ, if we know of a  LONG  series of gradations in complexity each good f
206  ions of life; or by having adapted them during  LONG  past periods of time: the adaptations being ai
214  r become curiously blended together, and for a  LONG  period exhibit traces of the instincts of eith
214  ctions which have become inherited solely from  LONG  continued and compulsory habit, but this, I th
214  t tendency to this strange habit, and that the  LONG  continued selection of the best individuals in
215  dness to extreme tameness, simply to habit and  LONG  continued close confinement. Natural instincts
217  of laying and hatching might be inconveniently  LONG, more especially as she has to migrate at a ve
228  i separated two combs, and put between them a  LONG  thick, square strips of wax: the bees instantly
233  ons of life, it may reasonably be asked, how a  LONG  and graduated succession of modified architect
238  tle, always yielding oxen with extraordinarily  LONG  horns, could be slowly formed by carefully wat
239  same nest, but in a few alone; and that by the  LONG  continued selection of the fertile parents whi
258  rocal crosses between the same two species was  LONG  ago observed by Kolreuter. To give an instance
263  d be the case with a plant having a pistil too  LONG  for the pollen tubes to reach the ovarium. It
264  ids, when once born, are generally healthy and  LONG  lived, as we see in the case of the common mul
264  of its mother, and therefore before birth, as  LONG  as it is nourished within its mother's womb or
269  place, some eminent naturalists believe that a  LONG  course of domestication tends to eliminate ste
272  er admits that hybrids from species which have  LONG  been cultivated are often variable in the firs
273  ut some few cases both of hybrids and mongrels  LONG  retaining uniformity of character could be giv
273  on are descended from species (excluding those  LONG  cultivated) which have not had their reproduct
281  ld imply that one form has remained for a very  LONG  period unaltered, whilst its descendants had u
284  the opinion of most geologists, enormously  LONG  blank periods. So that the lofty pile of sedim
286  a contain harder layers or nodules, which from  LONG  resisting attrition form a breakwater at the b
287  sea: when deeply submerged for perhaps equally  LONG  periods, it would, likewise, have escaped the
287  t grasp, must have succeeded each other in the  LONG  roll of years! Now turn to our richest geologi
291  nue slowly to subside. In this latter case, as  LONG  as the rate of subsidence and supply of sedime
292  they had time to decay. On the other hand, as  LONG  as the bed of the sea remained stationary, thi
293  tions. Although each formation may mark a very  LONG  lapse of years, each perhaps is short compared
293  ration of each formation is twice or thrice as  LONG  as the average duration of specific forms. But
295  osit must have gone on accumulating for a very  LONG  period, in order to have given sufficient time
296  , still standing upright as they grew, of many  LONG  intervals of time and changes of level during
301  and present species of the same group into one  LONG  and branching chain of life. We ought only to
302  w process; and the progenitors must have lived  LONG  ages before their modified descendants. But we
302  rget that groups of species may elsewhere have  LONG  existed and have slowly multiplied before they
303  formerly made, namely that it might require a  LONG  succession of ages to adapt an organism to som
306  rom some one crustacean, which might have lived  LONG  before the Silurian age, and which probably di
307  orders, they would almost certainly have been  LONG  ago supplanted and exterminated by their numer
307  ore the lowest Silurian stratum was deposited,  LONG  periods elapsed, as long as, or probably far l
307  tratum was deposited, long periods elapsed, as  LONG  as, or probably far longer than, the whole int
309  e must have intervened during these enormously  LONG  periods. If then we may infer anything from th
310  see in these large areas, the many formations  LONG  anterior to the silurian epoch in a completely
315  ame class the average amount of change, during  LONG  and equal periods of time, may, perhaps, be ne
315  be nearly the same; but as the accumulation of  LONG  enduring fossiliferous formations depends on g
315  l destroyed, that fanciers, by striving during  LONG  ages for the same object, might make a new bre
316  it has once disappeared; or its existence, as  LONG  as it lasts, is continuous. I am aware that th
316  escended from one species, it is clear that as  LONG  as any species of the group have appeared in t
316  any species of the group have appeared in the  LONG  succession of ages, so long must its members h
316  ve appeared in the long succession of ages, so  LONG  must its members have continuously existed, in
321  and improved, a few of the sufferers may often  LONG  be preserved, from being fitted to some peculi
```

Page ***************************************(Key Word)***************************************

Page **(Key Word)**

```
0326  l changes, or on strange accidents, but in the  LONG  run the dominant forms will generally succeed
0326  s. a certain amount of isolation, recurring at   LONG  intervals of time, would probably be also fav
0326  f the sea. If two great regions had been for a   LONG  period favourably circumstanced in an equal d
0327  r; though here and there a single member might   LONG  be enabled to survive. Thus, as it seems to m
0327  bed and preserve organic remains. During these   LONG  and blank intervals I suppose that the inhabi
0329  ation of the several formations and during the   LONG  intervals of time between them; in this case,
0329  it differs from living forms. But, as Buckland   LONG  ago remarked, all fossils can be classed eith
0332  are intermediate, not directly, but only by a    LONG  and circuitous course through many widely dif
0334  for a large amount of modification, during the   LONG  and blank intervals between the successive fo
0336  h formation has often been interrupted, and as   LONG  blank intervals have intervened between succe
0336  ds; but we ought to find after intervals, very   LONG  as measured by years, but only moderately lon
0336  long as measured by years, but only moderately   LONG  as measured geologically, closely allied form
0340  escent with modification, the great law of the   LONG  enduring, but not immutable, succession of th
0340  rly the same manner and degree. But after very   LONG  intervals of time and after great geographica
0343  ely numerous organisms which must have existed   LONG  before the first bed of the Silurian system w
0343  stood ever since the Silurian epoch; but that    LONG  before that period, the world may have presen
0343  e same rate, or in the same degree; yet in the   LONG  run that all undergo modification to some ext
0344  hich are those that oftenest vary, will in the   LONG  run tend to people the world with allied, but
0344  rs in the struggle for existence. Hence, after   LONG  intervals of time, the productions of the wor
0345  existing forms; but are intermediate only by a   LONG  and circuitous course through many extinct an
0348  ot as impassable, or likely to have endured so   LONG  as the oceans separating continents, the diffe
0359  apsules, etc., and some of these floated for a   LONG  time. It is well known what a difference ther
0361  s of many kinds in the crops of floating birds   LONG  retain their vitality: peas and vetches, for
0364  to the action of sea water; nor could they be    LONG  carried in the crops or intestines of birds.
0364  our western shores, where, if not killed by so   LONG  an immersion in salt water, they could not enc
0365  d by occasional means of transport, during the   LONG  lapse of geological time, whilst an island was
0365  e on the loftiest mountains of Europe. Even as   LONG  ago as 1747, such facts led Gmelin to conclude
0368  glacial period. The arctic forms, during their   LONG  southern migration and re migration northward
0370  ns before alluded to, that our continents have   LONG  remained in nearly the same relative position
0370  te southwards as the climate became less warm,   LONG  before the commencement of the Glacial period.
0371  emperate productions are concerned, took place   LONG  ages ago. And as the plants and animals migrat
0374  han at another, but seeing that it endured for   LONG  at each, and that it was contemporaneous in a
0374  the Silla of Caraccas the illustrious Humboldt   LONG  ago found species belonging to genera characte
0375  ntroduced by man, occur on the lowlands; and a   LONG  list can be given, as I am informed by Dr. Hoc
0377  iod, as measured by years, must have been very   LONG; and when we remember over what vast spaces so
0378  ain ranges north west of the Himalaya, and the   LONG  line of the Cordillera, seem to have afforded
0379  derwent scarcely any modification during their   LONG  southern migration and re migration northward
0386  o one can tell. With respect to plants, it has   LONG  been known what enormous ranges many fresh wat
0390  explained, species occasionally arriving after   LONG  intervals in a new and isolated district, and
0391  madeira have been stocked by birds, which for   LONG  ages have struggled together in their former h
0393  le orders on oceanic islands, Bory St. Vincent   LONG  ago remarked that Batrachians (frogs, toads, n
0396  transport having been largely efficient in the   LONG  course of time, than with the view of all our
0399  , south America, and other southern lands were   LONG  ago partially stocked from a nearly intermedia
0400  modified, though only in a small degree. This    LONG  appeared to me a great difficulty: but it aris
0404  and more general way. Mr. Gould remarked to me   LONG  ago, that in those genera of birds which range
0406  rtation, probably accounts for a law which has   LONG  been observed, and which has lately been admir
0408  es be not insuperable in admitting that in the   LONG  course of time the individuals of the same spe
0410  , as of sculpture or colour. In looking to the   LONG  succession of ages, as in now looking to dista
0425  ne species of kangaroo had been produced, by a   LONG  course of modification, from a bear? Ought we
0426  this may be safely done, and is often done, as   LONG  as a sufficient number of characters, let them
0429  enerally diverged much in character during the   LONG  continued process of modification, how it is t
0432  lly diverse forms are still tied together by a,  LONG, but broken, chain of affinities. Extinction h
0434  what can be more different than the immensely    LONG  spiral proboscis of a sphinx moth, the curious
0438  uite probable that natural selection, during a   LONG  continued course of modification, should have
0438  sense: they are far from meaning that during a   LONG  course of descent, primordial organs of any ki
0439  if they had really been metamorphosed during a   LONG  course of descent from true legs, or from some
0442  osis: the cephalopodic character is manifested   LONG  before the parts of the embryo are completed:
0443  nt. For the welfare of a very young animal, as   LONG  as it remains in its mother's womb, or in the
0443  ins in its mother's womb, or in the egg, or as   LONG  as it is nourished and protected by its parent
0444  bird which obtained its food best by having a   LONG  beak, whether or not it assumed a beak of this
0444  t assumed a beak of this particular length, as   LONG  as it was fed by its parents. Hence, I conclud
0447  s legs in the parent species, may become, by a   LONG  course of modification, adapted in one descend
0447  ands, or paddles, or wings. Whatever influence   LONG  continued exercise or use on the one hand, and
0449  ated, either by the successive variations in a   LONG  course of modification having supervened at a
0450  ding far enough back in time, may remain for a   LONG  period, or for ever, incapable of demonstratio
0455  y in every part of the organisation, which has   LONG  existed, to be inherited, we can understand, o
0459  oncluding remarks. As this whole volume is one   LONG  argument, it may be convenient to the reader t
0462  have retained the same specific form for very   LONG  periods, enormously long as measured by years,
0462  pecific form for very long periods, enormously   LONG  as measured by years, too much stress ought no
0462  diffusion of the same species: for during very   LONG  periods of time there will always be a good ch
0462  igration will have been possible during a very   LONG  period: and consequently the difficulty of the
0463  form. Even on a wide area, which has during a   LONG  period remained continuous, and of which the c
0463  o that the intermediate varieties will, in the   LONG  run, be supplanted and exterminated. On this d
0464  or not these doubtful forms are varieties? As    LONG  as most of the links between any two species a
0465  , few will be inclined to admit. If we look to   LONG  enough intervals of time, geology plainly decl
0466  ll the varied means of Distribution during the   LONG  lapse of years, or that we know how imperfect
0466  e, and that modifications can be inherited for   LONG  periods. As long as the conditions of life rem
0466  ications can be inherited for long periods. As   LONG  as the conditions of life remain the same, we
0469  limit can be put to this power, acting during   LONG  ages and rigidly scrutinising the whole consti
0470  t offspring of any one species. Hence during a   LONG  continued course of modification, the slight d
0473  d. In both varieties and species reversions to   LONG  lost characters occur. How inexplicable on the
0474  , that is, if it has been inherited for a very   LONG  period: for in this case it will have been ren
0474  is case it will have been rendered constant by   LONG  continued natural selection. Glancing at insti
0474  ich leave no progeny to inherit the effects of   LONG  continued habit. On the view of all the specie
0475  r descendants, causes the forms of life, after   LONG  intervals of time, to appear as if they had ch
0476  the struggle for life. Lastly, the law of the   LONG  endurance of allied forms on the same continen
```

Page **(Key Word)**

age **************************************(Key Word)**************************************

```
0476 on, if we admit that there has been during the LONG course of ages much migration from one part of
0477 eing widely different, if they have been for a LONG period completely separated from each other; f
0481 that the amount of variation in the course of LONG ages is a limited quantity; no clear distincti
0481 mmutable productions was almost unavoidable as LONG as the history of the world was thought to be
0481 any geologists, when Lyell first insisted that LONG lines of inland cliffs had been formed, and or
0481 with a multitude of facts all viewed, during a LONG course of years, from a point of view directly
0486 ealogies, by characters of any kind which have LONG been inherited. Rudimentary organs will speak
0486 speak infallibly with respect to the nature of LONG lost structures. Species and groups of species
0488 however, keeping in a body might remain for a LONG period unchanged, whilst within this same peri
0488 eal descendants of some few beings which lived LONG before the first bed of the Silurian system wa
0489 re the lineal descendants of those which lived LONG before the Silurian epoch, we may feel certain
0082 but far more easily, from having incomparably LONGER time at her disposal. Nor do I believe that a
0095 advantage to the hive bee to have a slightly LONGER or differently constructed proboscis. On the
0112 fancier is struck by a pigeon having a rather LONGER beak; and on the acknowledged principle that
0112 igeons; choosing and breeding from birds with LONGER and longer beaks, or with shorter and shorter
0112 osing and breeding from birds with longer and LONGER beaks, or with shorter and shorter beaks. Aga
0153 organisation, which have remained for a much LONGER period nearly constant. And this, I am convin
0178 ss of natural selection, so that they will no LONGER exist in a living state. Thirdly, when two or
0237 the male sex; for oxen of certain breeds have LONGER horns than in other breeds, in comparison wit
0287 coast waves. So that in all probability a far LONGER period than 300 million years has elapsed sin
0303 y elapsed between our consecutive formations, LONGER perhaps in some cases than the time required
0307 periods elapsed, as long as, or probably far LONGER than, the whole interval from the Silurian ag
0314 es should retain the same identical form much LONGER than others; or, if changing, that it should
0335 ery ancient form might occasionally last much LONGER than a form elsewhere subsequently produced,
0359 or a very short time, when dried floated much LONGER; for instance, ripe hazel nuts sank immediate
0359 s, and some of the 18 floated for a very much LONGER period. So that as 64/87 seeds germinated aft
0360 have caused some of them to have floated much LONGER. The result was that 18/98 of his seeds float
0360 the fact of the larger fruits often floating LONGER than the small, is interesting: as plants wit
0388 ess quickly than the high; and this will give LONGER time than the average for the migration of th
0424 others in the important character of having a LONGER beak, yet all are kept together from having t
0425 n has been greater in degree, and has taken a LONGER time to complete? I believe it has thus been
0485 d will have a plain signification. When we no LONGER look at an organic being as a savage looks at
0037 d parent stocks of the plants which have been LONGEST cultivated in our flower and kitchen gardens
0095 f gigantic valleys or to the formation of the LONGEST lines of inland cliffs. Natural selection ca
0238 nd cows, when matched, produced oxen with the LONGEST horns; and yet no one ox could ever have pro
0361 at not a particle could be washed away in the LONGEST transport: out of one small portion of earth
0258 an easily be fertilised by the pollen of M. LONGIFLORA, and the hybrids thus produced are suffice
0258 llowing years, to fertilise reciprocally M. LONGIFLORA with the pollen of M. jalappa, and utterly
0349 ica, on almost exactly opposite meridians of LONGITUDE. A third great fact, partly included in the
0374 same time lower along certain broad belts of LONGITUDE. On this view of the whole world, or at lea
0374 w of the whole world, or at least of broad LONGITUDINAL belts, having been simultaneously colder f
0307 es. Traces of life have been detected in the LONGMYND beds beneath Barrande's so called primordial
0007 avourable to Man's power of Selection. When we LOOK to the individuals of the same variety or sub
0013 way of viewing the whole subject, would be, to LOOK at the inheritance of every character whatever
0015 cters thus arising shall be preserved. When we LOOK to the hereditary varieties or races of our do
0024 rmal in other parts of their structure, we may LOOK in vain throughout the whole great family of C
0030 or so beautiful in his eyes, we must, I think, LOOK further than to mere variability. We cannot su
0050 ll be often ranked by some authors as species. LOOK at the common oak, how closely it has been stu
0051 nd with the idea of an actual passage. Hence I LOOK at individual differences, though of small int
0051 h recording in works on natural history. And I LOOK at varieties which are in any degree more dist
0052 ect. From these remarks it will be seen that I LOOK at the term species, as one arbitrarily given
0055 ourable to variation. On the other hand, if we LOOK at each species as a special act of creation,
0067 species to increase in number is most obscure. LOOK at the most vigorous species; by as much as it
0074 e same species in different districts. When we LOOK at the plants and bushes clothing an entangled
0077 th other plants growing vigorously all around. LOOK at a plant in the midst of its range, why does
0105 nature to test the truth of these remarks, and LOOK at any small isolated area, such as an oceanic
0141 ds in the torrid zones. Hence I am inclined to LOOK at adaptation to any special climate as a qual
0152 ction, are also eminently liable to variation. LOOK at the breeds of the pigeon; see what a prodig
0157 s a clear meaning on my view of the subject: I LOOK at all the species of the same genus as having
0167 se genus! For myself, I venture confidently to LOOK back thousands on thousands of generations, an
0172 as we have seen, go hand in hand. Hence, if we LOOK at each species as descended from some other u
0175 epths of the sea with the dredge. To those who LOOK at climate and the physical conditions of life
0179 adapted in its habits to its place in nature. LOOK at the Mustela vison of North America, which h
0180 y in any particular case like that of the bat. LOOK at the family of squirrels; here we have the f
0181 st so called flying squirrel was produced. Now LOOK at the Galeopithecus or flying Lemur, which fo
0187 in any species has been perfected, we ought to LOOK exclusively to its lineal ancestors: but this
0187 er possible, and we are forced in each case to LOOK to species of the same group, that is to the c
0189 the transitional grades, more especially if we LOOK to much isolated species, round which, accordi
0189 here has been much extinction. Or again, if we LOOK to an organ common to all the members of a lar
0190 which the organ has passed, we should have to LOOK to very ancient ancestral forms, long since be
0202 f the insect by tearing out its viscera? If we LOOK at the sting of the bee, as having originally
0206 ory, Natura non facit saltum. This canon if we LOOK only to the present inhabitants of the world,
0213 frames of mind or periods of time. But let us LOOK to the familiar case of the several breeds of
0225 d only as a modification of his theory. Let us LOOK to the great principle of gradation, and see w
0243 my imagination it is far more satisfactory to LOOK at such instincts as the young cuckoo ejecting
0254 ct species of animals when crossed: or we must LOOK at sterility, not as an indelible characterist
0263 d of crossing distinct species. And as we must LOOK at the curious and complex laws governing the
0263 ty of first Crosses and of Hybrids. We may now LOOK a little closer at the probable causes of the
0268 this is almost invariably the case. But if we LOOK to varieties produced under nature, we are imm
0275 ossed species, and of crossed varieties. If we LOOK at species as having been specially created, a
0280 this is a wholly false view; we should always LOOK for forms intermediate between each species an
0281 ch as C. oenas. So with natural species, if we LOOK to forms very distinct, for instance to the ho
0285 able lesson to stand on the North Downs and to LOOK at the distant South Downs; for, remembering t
0297 on my theory we ought to find. Moreover, if we LOOK to rather wider intervals, namely, to distinct
0301 and branching chain of life. We ought only to LOOK for a few links, some more closely, some more
0310 for my part, following out Lyell's metaphor, I LOOK at the natural geological record, as a history
```

Page **************************************(Key Word)**************************************

Page ************************************(Key Word)************************************

```
0315  e species in the same region do at last, if we  LOOK  to wide enough intervals of time, become modif
0325  he same effect. It is, indeed, quite futile to  LOOK  to changes of currents, climate, or other phys
0325  t climates. We must, as Barrande has remarked,  LOOK  to some special law. We shall see this more cl
0329  to each other, and to living forms. Let us now  LOOK  to the mutual affinities of extinct and living
0338  groups at the present day, it would be vain to  LOOK  for animals having the common embryological ch
0349  and we find an alpine species of bizcacha; we  LOOK  to the waters, and we do not find the beaver o
0349  umerable other instances could be given. If we  LOOK  to the islands off the American shore, howeve
0349  liar species, are essentially American. We may  LOOK  back to past ages, as shown in the last chapte
0362  distance of 500 miles, and hawks are known to  LOOK  out for tired birds, and the contents of thei
0370  roken land still nearer to the pole. Now if we  LOOK  at a globe, we shall see that under the Polar
0381  from some common centre; and I am inclined to  LOOK  in the southern, as in the northern hemisphere
0389  s for plants, and Wollaston for insects. If we  LOOK  to the large size and varied stations of New Z
0400  r more important element of success. Now if we  LOOK  to those inhabitants of the Galapagos Archipel
0410  ime and space are intelligible; for whether we  LOOK  to the forms of life which have changed during
0413  but what is meant by this system? Some authors  LOOK  at it merely as a scheme for arranging togethe
0434  we have a distinct object in view, and do not  LOOK  to some unknown plan of creation, we may hope
0442  within the same class, that we might be led to  LOOK  at these facts as necessarily contingent in so
0442  in some few cases, as in that of Aphis, if we  LOOK  to the admirable drawings by Professor Huxley
0444  leading facts in embryology. But first let us  LOOK  at a few analogous cases in domestic varieties
0450  yology rises greatly in interest, when we thus  LOOK  at the embryo as a picture, more or less obscu
0457  in Morphology become intelligible, whether we  LOOK  to the same pattern displayed in the homologou
0465  require, few will be inclined to admit. If we  LOOK  to long enough intervals of time, geology plai
0473  is difficult to resist this conclusion when we  LOOK,  for instance, at the logger headed duck, whic
0473  condition as in the domestic duck; or when we  LOOK  at the burrowing tucutucu, which is occasional
0473  have their eyes covered with skin; or when we  LOOK  at the blind animals inhabiting the dark caves
0482  ecies, may be influenced by this volume; but I  LOOK  with confidence to the future, to young and ri
0485  have a plain signification. When we no longer  LOOK  at an organic being as a savage looks at a shi
0486  e possessor, nearly in the same way as when we  LOOK  at any great mechanical invention as the summ
0489  sm has desolated the whole world. Hence we may  LOOK  with some confidence to a secure future of equ
0056  d this certainly is the case, if varieties be  LOOKED  at as incipient species; for my tables clearl
0115  lied to the indigenes; for these are commonly  LOOKED  at as specially created and adapted for their
0141  sub tropical in their habits, ought not to be  LOOKED  at as anomalies, but merely as examples of a
0197  ed skin on the head of a vulture is generally  LOOKED  at as a direct adaptation for wallowing in pu
0246  r, owing to the sterility in both cases being  LOOKED  on as a special endowment, beyond the provinc
0469  . the theory of natural selection, even if we  LOOKED  no further than this, seems to me to be in it
0476  and allied groups. Recent forms are generally  LOOKED  at as being, in some vague sense, higher than
0482  e special creations, and which are still thus  LOOKED  at by the majority of naturalists, and which
0485  f not blended by intermediate gradations, are  LOOKED  at by most naturalists as sufficient to raise
0487  e earth with its embedded remains must not be  LOOKED  at as a well filled museum, but as a poor col
0055  ur chapter on geographical distribution. From  LOOKING  at species as only strongly marked and well
0066  rectly on the number of its eggs or seeds. In  LOOKING  at Nature, it is most necessary to keep the
0072  ch fir, except the old planted clumps. But on  LOOKING  closely between the stems of the heath, I fo
0085  w or purple fleshed fruit, should succeed. In  LOOKING  at many small points of difference between s
0107  intricacy of the subject permits. I conclude,  LOOKING  to the future, that for terrestrial producti
0126  nd sub groups will finally tend to disappear.  LOOKING  to the future, we can predict that the group
0126  tensively developed, have now become extinct.  LOOKING  still more remotely to the future, we may pr
0128  s, and constitution, of which we see proof by  LOOKING  at the inhabitants of any small spot or at n
0159  opositions will be most readily understood by  LOOKING  to our domestic races. The most distinct bre
0174  ith freely crossing and wandering animals. In  LOOKING  at species as they are now distributed over
0179  election and gain further advantages. Lastly,  LOOKING  not to any one time, but to all time, if my
0187  vibrations of the air which produce sound. In  LOOKING  for the gradations by which an organ in any
0254  e of being removed by domestication. Finally,  LOOKING  to all the ascertained facts on the intercro
0275  s frequently with hybrids than with mongrels.  LOOKING  to the cases which I have collected of cross
0280  erly existed. I have found it difficult, when  LOOKING  at any two species, to avoid picturing to my
0308  n as the bed of an open and unfathomable sea.  LOOKING  to the existing oceans, which are thrice as
0320  during the later geological periods, so that  LOOKING  to later times we may believe that the produ
0324  h somewhat older European beds. Nevertheless,  LOOKING  to a remotely future epoch, there can, I thi
0332  d some to have endured to the present day. By  LOOKING  at the diagram we can see that if many of th
0370  eve the above difficulty may be surmounted by  LOOKING  to still earlier changes of climate of an op
0373  ion in the south eastern corner of Australia.  LOOKING  to America; in the northern half, ice borne
0398  he flora of this archipelago. The naturalist,  LOOKING  at the inhabitants of these volcanic islands
0410  ters in common, as of sculpture or colour. In  LOOKING  to the long succession of ages, as in now Lo
0410  ing to the long succession of ages, as in now  LOOKING  to distant provinces throughout the world, w
0412  to diverge. This conclusion was supported by  LOOKING  at the great diversity of the forms of life
0412  ea, come into the closest competition, and by  LOOKING  to certain facts in naturalisation. I attemp
0429  e can, I think, account for this fact only by  LOOKING  at aberrant forms as failing groups conquere
0433  , in an able paper, on the high importance of  LOOKING  to types, whether or not we can separate and
0476  extinct will naturally be allied by descent.  LOOKING  to geographical distribution, if we admit th
0195  by natural selection. The tail of the giraffe  LOOKS  like an artificially constructed fly flapper;
0485  no longer look at an organic being as a savage  LOOKS  at a ship, as at something wholly beyond his
0191  ed, the slits on the sides of the neck and the  LOOP  like course of the arteries still marking in t
0440  in the embryos of the vertebrata the peculiar  LOOP  like course of the arteries near the branchial
0479  having branchial slits and arteries running in  LOOPS,  like those in a fish which has to breathe th
0015  g, that only a single variety should be turned  LOOSE  in its new home. Nevertheless, as our varieti
0192  hand, have no ovigerous frena, the eggs lying  LOOSE  at the bottom of the sack, in the well enclos
0019  reely in a state of nature? It has often been  LOOSELY  said that all our races of dogs have been pr
0031  into life whatever form and mould he pleases.  LORD  Somerville, speaking of what breeders have don
0035  races, are favoured in the weights they carry.  LORD  Spencer and others have shown how the cattle o
0165  of them there was a double shoulder stripe. In  LORD  Moreton's famous hybrid from a chestnut mare a
0015  ed a strong tendency to reversion, that is, to  LOSE  their acquired characters, whilst kept under u
0145  ms, that the central flower of the truss often  LOSES  the patches of darker colour in the two upper
0148  sitic within another and is thus protected, it  LOSES  more or less completely its own shell or cara
0487  ncient geography. The noble science of Geology  LOSES  glory from the extreme imperfection of the re
0092  ertilisation, its destruction appears a simple  LOSS  to the plant; yet if a little pollen were carr
0137  animals living in darkness, I attribute their  LOSS  wholly to disuse. In one of the blind animals,
0137  ral selection seems to have struggled with the  LOSS  of light and to have increased the size of the
```

Page ************************************(Key Word)************************************

age **(Key Word)**

```
193  and its absence in some of the members to its       LOSS  through disuse or natural selection. But if th
212  both in extent and direction, and in its total       LOSS. So it is with the nests of birds, which vary
026  rents to revert to a character, which has been        LOST  during some former generation, this tendency,
070  r such plants in a garden: I have in this case        LOST  every single seed. This view of the necessity
129  branch has decayed and dropped off: and these         LOST  branches of various sizes may represent those
131  ary in an analogous manner. Reversions to long        LOST  characters. Summary. I have hitherto sometimes
135  the Onites apelles the tarsi are so habitually        LOST, that the insect has been described as not hav
135  rogenitors; for as the tarsi are almost always        LOST  in many dung feeding beetles, they must be los
135  ost in many dung feeding beetles, they must be        LOST  early in life, and therefore cannot be much us
137  though the telescope with its glasses has been        LOST. As it is difficult to imagine that eyes, thou
160  t characters should reappear after having been        LOST  for many, perhaps for hundreds of generations.
160  t been crossed, but in which both parents have        LOST  some character which their progenitor possesse
160  ency, whether strong or weak, to reproduce the        LOST  character might be, as was formerly remarked,
160  f generations. When a character which has been        LOST  in a breed, reappears after a great number of
161  genus would occasionally exhibit reversions to        LOST  ancestral characters. As, however, we never kn
163  are sometimes very obscure, or actually quite         LOST, in dark coloured asses. The koulan of Pallas
166  each successive generation to produce the long        LOST  character, and that this tendency, from unknow
191  on. But it is conceivable that the now utterly        LOST  branchiae might have been gradually worked in
193  which most of their modified descendants have         LOST. The presence of luminous organs in a few inse
209  reased by, use or habit, and are diminished or        LOST  by disuse, so I do not doubt it has been with
215  inued close confinement. Natural instincts are        LOST  under domestication: a remarkable instance of
215  r dogs. On the other hand, young chickens have        LOST, wholly by habit, that fear of the dog and cat
216  ared under a hen. It is not that chickens have        LOST  all fear, but fear only of dogs and cats, for
216  r domestication, for the mother hen has almost        LOST  by disuse the power of flight. Hence we may co
216  been acquired and natural instincts have been         LOST  partly by habit, and by man selecting and accu
218  observers, the European cuckoo has not utterly        LOST  all maternal love and care for her own offspri
218  day's hunting I picked up no less than twenty         LOST  and wasted eggs. Many bees are parasitic, and
253  ncy to sterility, the breed would assuredly be        LOST  in a very few generations. Although I do not k
312  their different rates of change. Species once         LOST  do not reappear. Groups of species follow the
312  een them, and to make the percentage system of        LOST  and new forms more gradual. In some of the mos
312  measured by years, only one or two species are        LOST  forms, and only one or two are new forms, havi
313  ing crocodile associated with many strange and        LOST  mammals and reptiles in the sub Himalayan depo
315  can clearly understand why a species when once        LOST  should never reappear, even if the very same c
415  necual, and in some cases seems to be entirely        LOST. Again in another work he says, the genera of
421  ch modified as to have more or less completely        LOST  traces of their parentage, in this case, their
421  ication will have been more or less completely        LOST, as sometimes seems to have occurred with exis
424  but the short faced breed has nearly or quite         LOST  this habit; nevertheless, without any reasonin
429  n my theory have been exterminated and utterly        LOST. And we have some evidence of aberrant forms h
431  t many ancient forms of life have been utterly        LOST, through which the early progenitors of birds
436  might become so much obscured as to be finally        LOST, by the atrophy and ultimately by the complete
454  een forced to take flight, and have ultimately        LOST  the power of flying. Again, an organ useful un
455  ) the rudimentary part would tend to be wholly       LOST, and we should have a case of complete abortio
473  both varieties and species reversions to long        LOST  characters occur. How inexplicable on the theo
486  infallibly with respect to the nature of long        LOST  structures. Species and groups of species, whi
429  me investigation, does not commonly fall to the       LOT  of aberrant genera. We can, I think, account f
165  all satisfied with this theory, and should be         LOTH  to apply it to breeds so distinct as the heavy
041  belong to poor people, and are mostly in small        LOTS, they never can be improved. On the other hand
203  is for the good of the community: and maternal        LOVE  or maternal hatred, though the latter fortunat
215  ion. It is scarcely possible to doubt that the        LOVE  of man has become instinctive in the dog. All
218  opean cuckoo has not utterly lost all maternal        LOVE  and care for her own offspring. The occasional
027  watched, and terded with the utmost care, and        LOVED  by many people. They have been domesticated f
054  two causes of obscurity. Fresh water and salt        LOVING  plants have generally very wide ranges and ar
054  nera to which the species belong. Again, plants       LOW  in the scale of organisation are generally muc
088  eed by careful selection of the best cocks. How       LOW  in the scale of nature this law of battle desc
130  a thin straggling branch springing from a fork       LOW  down in a tree, and which by some chance has b
149  prof. Owen's expression, seems to be a sign of        LOW  organisation; the foregoing remark seems conne
149  ery general opinion of naturalists, that beings       LOW  in the scale of nature are more variable than
168  obably from this same cause that organic beings       LOW  in the scale of nature are more variable than
187  and without any other mechanism; and from this       LOW  stage, numerous gradations of structure, branc
208  on, often comes into play, even in animals very       LOW  in the scale of nature. Frederick Cuvier and s
288  nimals which live on the beach between high and       LOW  watermark are preserved. For instance, the sev
305  p of species, is that of the teleostean fishes,       LOW  down in the Chalk period. This group includes
313  nature, change more quickly than those that are       LOW: though there are exceptions to this rule. The
332  e formations, were discovered at several points       LOW  down in the series, the three existing familie
336  other's satisfaction what is meant by high and        LOW  forms. But in one particular sense the more re
365  parated from each other by hundreds of miles of       LOW  lands, where the Alpine species could not poss
368  ate permitted their former migration across the       LOW  intervening tracts, since become too warm for
373  t, glaciers have left the marks of their former       LOW  descent; and in Sikkim, Dr. Hooker saw maize g
388  some, perhaps many, fresh water productions are       LOW  in the scale of nature, and that we have reaso
388  e, and that we have reason to believe that such       LOW  beings change or become modified less quickly
406  believe from geological evidence that organisms       LOW  in the scale within each great class; generall
406  fact; together with the seeds and eggs of many        LOW  forms being very minute and better fitted for
406  to range. The relations just discussed, namely,       LOW  and slowly changing organisms ranging more wid
406  s before specified) to those on the surrounding       LOW  lands and dry lands, though these stations are
437  on characteristic (as Owen has observed) of all       LOW  or little modified forms; therefore we may rea
022  as the breadth and length of the ramus of the        LOWER  jaw, varies in a highly remarkable manner. Th
119  s is represented in the diagram by some of the        LOWER  branches not reaching to the upper horizontal
143  the jaws and limbs, varying together, for the         LOWER  jaw is believed to be homologous with the lim
188  ends and must act by convergence; and at their        LOWER  ends there seems to be an imperfect vitreous
190  ind. Numerous cases could be given amongst the        LOWER  animals of the same organ performing at the s
293  presenting distinct varieties in the upper and        LOWER  parts of the same formation; but, as they are
295  t gradation between two forms in the upper and        LOWER  parts of the same formation, the deposit must
295  of organic remains, except near their upper or        LOWER  limits. It would seem that each separate form
296  er formation. Many cases could be given of the        LOWER  beds of a formation having been upraised, den
297  and its several modified descendants from the        LOWER  and upper beds of a formation, and unless we
307  racy. M. Barrande has lately added another and        LOWER  stage to the Silurian system, abounding with
```

Page **(Key Word)**

Page ************************************(Key Word)************************************

```
0314 organised productions compared with marine and LOWER productions, by the more complex relations o
0317 will sometimes falsely appear to begin at its LOWER end, not in a sharp point, but abruptly; it 
0318 rks the progress of extermination, than at its LOWER end, which marks the first appearance and in
0353 anomaly it would be, if, when coming one step LOWER in the series, to the individuals of the same
0369 e forms, for some of these are the same on the LOWER mountains and on the plains of North America
0374 rpose, if the temperature was at the same time LOWER along certain broad belts of longitude. On th
0388 distribution of fresh water plants and of the LOWER animals, whether retaining the same identica
0394 and disappear at a quicker rate than other and LOWER animals. Though terrestrial mammals do not o
0406 te than the higher forms; and consequently the LOWER forms will have had a better chance of rangir
0406 candolle in regard to plants; namely, that the LOWER any group of organisms is, the more widely i
0410 er, differ greatly. In both time and space the LOWER members of each class generally change less 
0441 t is meant by the organisation being higher or LOWER. But no one probably will dispute that the b
0441 , the mature animal is generally considered as LOWER in the scale than the larva, as with certain
0034 ancient times, and is now attended to by the LOWEST savages. It would, indeed, have been a stran
0187 ould probably have to descend far beneath the LOWEST known fossiliferous stratum to discover the 
0279 of species. On their sudden appearance in the LOWEST known fossiliferous strata. In the sixth cha
0306 appearance of groups of Allied Species in the LOWEST known fossiliferous strata. There is another
0306 ies of the same group, suddenly appear in the LOWEST known fossiliferous rocks. Most of the argum
0307 y be true, it is indisputable that before the LOWEST Silurian stratum was deposited, long periods
0307 ced that we see in the organic remains of the LOWEST Silurian stratum the dawn of life on this pla
0307 nocules and bituminous matter in some of the LOWEST azoic rocks, probably indicates the former ex
0316 unbroken succession of generations, from the LOWEST Silurian stratum to the present day. We have
0338 of the Vertebrata, until beds far beneath the LOWEST Silurian strata are discovered, a discovery o
0424 to his classification; for he includes in his LOWEST grade, or that of a species, the two sexes; a
0438 stand this fact: for in molluscs, even in the LOWEST members of the class, we do not find nearly s
0465 sence of fossiliferous formations beneath the LOWEST Silurian strata, I can only recur to the hypo
0198 orrelated. Mountain breeds always differ from LOWLAND breeds; and a mountainous country would prob
0477 ditions, under heat and cold, on mountain and LOWLAND, on deserts and marshes, most of the inhabit
0091 ntainous district, and those frequenting the LOWLANDS, would naturally be forced to hunt different
0367 h had lately lived in a body together on the LOWLANDS of the Old and New Worlds, would be left iso
0375 of Europe, not found in the intervening hot LOWLANDS. A list of the genera collected on the lofti
0375 species, not introduced by man, occur on the LOWLANDS; and a long list can be given, as I am infor
0375 e more lofty mountains, and on the temperate LOWLANDS of the northern and southern hemispheres, ar
0378 ate productions entered and crossed even the LOWLANDS of the tropics at the period when the cold w
0378 i suppose that large spaces of the tropical LOWLANDS were clothed with a mingled tropical and tem
0378 higher mountains, being exterminated on the LOWLANDS; those which had not reached the equator, wo
0382 its, in a line gently rising from the arctic LOWLANDS to a great height under the equator. The var
0382 of the former inhabitants of the surrounding LOWLANDS. Chapter XII. Geographical Distribution, con
0403 och, are related to those of the surrounding LOWLANDS; thus we have in South America, Alpine hummi
0404 naturally be colonised from the surrounding LOWLANDS. So it is with the inhabitants of lakes and
0055 lation to the size of the genera. The cause of LOWLY organised plants ranging widely will be discu
0131 false correlations. Multiple, rudimentary, and LOWLY organised structures variable. Parts develope
0441 ay be considered as either more highly or more LOWLY organised then they were in the larval condit
0149 e than those which are higher. I presume that LOWNESS in this case means that the several parts of
0046 only by slow degrees: yet quite recently Mr. LUBBOCK has shown a degree of variability in these m
0241 peak confidently on this latter point, as Mr. LUBBOCK made drawings for me with the camera lucida
0012 ysiological importance is endless. Dr. Prosper LUCAS's treatise, in two large volumes, is the full
0275 on the whole I entirely agree with Dr. Prosper LUCAS, who, after arranging an enormous body of fac
0241 Lubbock made drawings for me with the camera LUCIDA of the jaws which I had dissected from the wo
0136 wollaston, lie much concealed, until the wind LULLS and the sun shines; that the proportion of wi
0193 ified descendants have lost. The presence of LUMINOUS organs in a few insects, belonging to differ
0339 nderful collection of fossil bones made by MM. LUND and Clausen in the caves of Brazil. I was so m
0191 on has actually converted a swimbladder into a LUNG, or organ used exclusively for respiration. I
0204 pparently been converted into an air breathing LUNG. The same organ having performed simultaneous
0452 me converted into a nascent breathing organ or LUNG. Other similar instances could be given. Rudim
0191 ly similar, in position and structure with the LUNGS of the higher vertebrate animals: hence there
0191 doubt that all vertebrate animals having true LUNGS have descended by ordinary generation from an
0191 he trachea, with some risk of falling into the LUNGS, notwithstanding the beautiful contrivance by
0196 rposes in so many land animals, which in their LUNGS or modified swimbladders betray their aquatic
0450 ary state: in very many snakes one lobe of the LUNGS is rudimentary; in other snakes there are rud
0133 as every collector knows, are often brassy or LURID. Plants which live exclusively on the sea sid
0307 bably, according to Dr. Hooker, the Nelumbium LUTEUM) in a heron's stomach; although I do not know
0265 ften accompanied by excess of size or great LUXURIANCE. In both cases, the sterility occurs in var
0378 etation, like that now growing with strange LUXURIANCE at the base of the Himalaya, as graphically
0002 request that I would forward it to Sir Charles LYELL, who sent it to the Linnean Society, and it i
0002 volume of the Journal of that Society. Sir C. LYELL and Dr. Hooker, who both knew of my work, the
0062 much greater length. The elder De Candolle and LYELL have largely and philosophically shown that a
0095 which were at first urged against Sir Charles LYELL's noble views on the modern changes of the ea
0282 the lapse of time. He who can read Sir Charles LYELL's grand work on the Principles of Geology, wh
0283 he mass has been accumulated. Let him remember LYELL's profound remark, that the thickness and ext
0289 riods, with one exception discovered by Sir C. LYELL in the carboniferous strata of North America.
0289 istorical table published in the Supplement to LYELL's Manual, will bring home the truth, how acci
0292 th the general principles inculcated by Sir C. LYELL; and E. Forbes independently arrived at a sim
0296 chanced to have been preserved: thus, Messrs. LYELL and Dawson found carboniferous beds 1400 feet
0304 tion. But now we may read in the Supplement to LYELL's Manual, published in 1858, clear evidence o
0307 this planet. Other highly competent judges, as LYELL and the late E. Forbes, dispute this conclusi
0310 bes, etc., and all our greatest geologists, as LYELL, Murchison, Sedgwick, etc., have unanimously,
0310 believe that one great authority, Sir Charles LYELL, from further reflexion entertains grave doub
0310 e reject my theory. For my part, following out LYELL's metaphor, I look at the natural geological
0312 r another, both on the land and in the waters. LYELL has shown that it is hardly possible to resis
0313 allow the pre existing fauna to reappear; but LYELL's explanation, namely, that it is a case of te
0323 ved by several authors: so it is, according to LYELL, with the several European and North American
0328 ited by distinct, but contemporaneous, faunas. LYELL has made similar observations on some of the
0356 means of dispersal. Means of Dispersal. Sir C. LYELL and other authors have ably treated this subj
0363 arctic and antarctic regions, as suggested by LYELL; and during the Glacial period from one part
0363 during the glacial epoch. At my request Sir C. LYELL wrote to M. Hartung to inquire whether he had
```

Page ************************************(Key Word)************************************

```
4381  the Glacial period, icebergs, as suggested by LYELL, have been largely concerned in their dispers
4382  same peculiar forms of vegetable life. Sir C. LYELL in a striking passage has speculated, in lang
4385  and or to any other distant point. Sir Charles LYELL also informs me that a Dyticus has been caugh
4402  the species peculiar to Chatham Island. Sir C. LYELL and Mr. Wollaston have communicated to me a r
4481  same as that felt by so many geologists, when LYELL first insisted that long lines of inland clif
4192  other hand, have no ovigerous frena, the eggs LYING loose at the bottom of the sack, in the well
4288  n the view of the bottom of the sea not rarely LYING for ages in an unaltered condition. The remai
4451  be utterly incapable of flight, and not rarely LYING under wing cases, firmly soldered together! T
3083  e of constitutional difference, on the whole MACHINERY of life. Man selects only for his own good;
0427  distinction, and he has been ably followed by MACLEAY and others. The resemblance, in the shape o
3323  thus changed: if the Megatherium, Mylodon, MACRAUCHENIA, and Toxodon had been brought to Europe fr
0001  o me, in 1837, that something might perhaps be MADE out on this question by patiently accumulating
0014  version, I may here refer to a statement often MADE by naturalists, namely, that our domestic vari
0014  bove statement has so often and so boldly been MADE. There would be great difficulty in proving it
0017  ere taken from a state of nature, and could be MADE to breed for an equal number of generations un
0026  state this from my own observations, purposely MADE on the most distinct breeds. Now, it is diffic
0032  y allied sub breeds. And when a cross has been MADE, the closest selection is far more indispensab
0032  of the present day are compared with drawings MADE only twenty or thirty years ago. When a race o
0035  ul drawings of the breeds in question had been MADE long ago, which might serve for comparison. In
0049  lastly, according to very numerous experiments MADE during several years by that most careful obse
0055  these results follow when another division is MADE, and when all the smallest genera, with from o
0067  eeds, but; from some observations which I have MADE, I believe that it is the seedlings which suff
0100  he rule does not hold in Australia; and I have MADE these few remarks on the sexes of trees simply
0115  ous, and thus a large proportional addition is MADE to the genera of these States. By considering
0117  nue to be exposed to the same conditions which MADE their parents variable, and the tendency to va
0118  s, will tend to inherit those advantages which MADE their common parent (A) more numerous than mos
0118  partake of those more general advantages which MADE the genus to which the parent species belonged
0118  s represented in the diagram, though in itself MADE somewhat irregular. I am far from thinking tha
0119  l tend to partake of the same advantages which MADE their parent successful in life, they will gen
0136  uent flight: these several considerations have MADE me believe that the wingless condition of so m
0140  n ceylon, and analogous observations have been MADE by Mr. H. C. Watson on European species of pla
0150  . waterhouse. I infer also from an observation MADE by Professor Owen, with respect to the length
0150  everal causes of error, but I hope that I have MADE due allowance for them. It should be understoo
0151  truth, had not the great variability in plants MADE it particularly difficult to compare their rel
0154  at the most abnormally developed organs may be MADE constant, I can see no reason to doubt. Hence
0163  t in the foal, and from inquiries which I have MADE, I believe this to be true. It has also been a
0164  and I have see a trace in a bay horse. My son MADE a careful examination and sketch for me of a d
0173  museum; but the natural collections have been MADE only at intervals of time immensely remote. Bu
0198  unimportant variations; and we are immediately MADE conscious of this by reflecting on the differe
0199  ad me to say a few words on the protest lately MADE by some naturalists, against the utilitarian d
0208  ver, a caterpillar were taken out of a hammock MADE up, for instance, to the third stage, and were
0214  the best individuals in successive generations MADE tumblers what they now are: and near Glasgow a
0217  by this occasional habit, or if the young were MADE more vigorous by advantage having been taken o
0218  t that when this insect finds a burrow already MADE and stored by another sphex it takes advantage
0219  ntly set to work, fed and saved the survivors! MADE some cells and tended the larvae, and put all
0220  will give the observations which I have myself MADE, in some little detail. I opened fourteen nest
0222  est. This species is sometimes, though rarely, MADE into slaves, as has been described by Mr. Smit
0224  actically solved a recondite problem, and have MADE their cells of the proper shape to hold the gr
0224  anes, or even perceive when they are correctly MADE. But the difficulty is not nearly so great as
0225  oint to notice, is that these cells are always MADE at that degree of nearness to each other, that
0226  se, it occurred to me that if the Melipona had MADE its spheres at some given distance from each o
0226  t some given distance from each other, and had MADE them of equal sizes and had arranged them symm
0227  ame with the best measurements which have been MADE of the cells of the hive bee. Hence we may saf
0230  ed. Some of these statements differ from those MADE by the justly celebrated elder Huber, but I am
0231  difficulty of understanding how the cells are MADE, that a multitude of bees all work together; o
0241  nfidently on this latter point, as Mr. Lubbock MADE crawings for me with the camera lucida of the
0248  isers, or by the same author, from experiments MADE during different years. It can thus be shown t
0249  onviction by a remarkable statement repeatedly MADE by Gartner, namely, that if even the less fert
0251  he pod impregnated by the pollen of the hybrid MADE vigorous growth and rapid progress to maturity
0251  cal experiments of horticulturists, though not MADE with scientific precision, deserve some notice
0270  result of an astonishing number of experiments MADE during many years on nine species of Verbascum
0271  d of the other. From observations which I have MADE on certain varieties of hollyhock, I am inclin
0274  opposed to the results of several experiments MADE by Kolreuter. These alone are the unimportant
0287  he latter part of the Secondary period. I have MADE these few remarks because it is highly importa
0288  sufficient care, as the important discoveries MADE every year in Europe prove. No organism wholly
0299  pecies to existing and extinct genera, and has MADE the intervals between some few groups less wid
0303  y created. I may here recall a remark formerly MADE, namely that it might require a long successio
0325  nthe whole animal kingdom. M. Barrande has MADE forcible remarks to precisely the same effect.
0328  stinct, but contemporaneous, faunas. Lyell has MADE similar observations on some of the later tert
0330  though formerly quite distinct, at that period MADE some small approach to each other. It is a com
0339  en in the wonderful collection of fossil bones MADE by MM. Lund and Clausen in the caves of Brazil
0362  rge pellets, which, as I know from experiments MADE in the Zoological Gardens, include seeds capab
0375  peaks of Java raises a picture of a collection MADE on a hill in Europe! Still more striking is th
0395  less modified condition. Mr. Windsor Earl has MADE some striking observations on this head in reg
0428  genera, tend to inherit the advantages, which MADE the groups to which they belong large and thei
0430  e those of a Rodent. The elder De Candolle has MADE nearly similar observations on the general nat
0432  has only separated groups: it has by no means MADE them: for if every form which has ever lived o
0445  stication: but having had careful measurements MADE of the dam and of a three days old colt of a r
0485  that genera are merely artificial combinations MADE for convienience. This may not be a cheering p
0486  ions will come to be, as far as they can be so MADE, genealogies; and will then truly give what ma
0487  a well filled museum, but as a poor collection MADE at hazard and at rare intervals. The accumulat
0048  es and varieties. On the islets of the little MADEIRA group there are many insects which are chara
0049  that between the Continent and the Azores, or MADEIRA, or the Canaries, or Ireland; be sufficient?
0052  he varieties of certain fossil land shells in MADEIRA. If a variety were to flourish so as to exce
0107  n. hence, perhaps, it comes that the flora of MADEIRA, according to Oswald Heer, resembles the ext
0135  00 beetles, out of the 550 species inhabiting MADEIRA, are so far deficient in wings that they can
```

Page ***(Key Word)***

```
0135  blown to sea and perish; that the beetles in MADEIRA, as observed by Mr. Wollaston, lie much conc
0136  les is larger on the exposed Dezertas than in MADEIRA itself; and especially the extraordinary fac
0136  elieve that the wingless condition of so many MADEIRA beetles is mainly due to the action of natur
0136  and thus have been destroyed. The insects in MADEIRA which are not ground feeders, and which, as
0137  ght power of vision. In the same manner as in MADEIRA the wings of some of the insects have been e
0314  n the land shells and coleopterous insects of MADEIRA having come to differ considerably from thei
0339  between the extinct and living land shells of MADEIRA; and between the extinct and living brackish
0390  nce, the number of the endemic land shells in MADEIRA, or of the endemic birds in the Galapagos Ar
0391  her periodically or occasionally this island. MADEIRA does not possess one peculiar bird, and many
0391  urt. So that these two islands of Bermuda and MADEIRA have been stocked by birds, which for long a
0391  ntly have been little liable to modification. MADEIRA, again, is inhabited by a wonderful number o
0391  open sea. The different orders of insects in MADEIRA apparently present analogous facts. Oceanic
0393  animals; for frogs have been introduced into MADEIRA, the Azores, and Mauritius, and have multipl
0402  le fact bearing on this subject; namely, that MADEIRA and the adjoining islet of Porto Santo posse
0403  are annually transported from Porto Santo to MADEIRA, yet this latter island has not become colon
0349  same manner. The plains near the Straits of MAGELLAN are inhabited by one species of Rhea (Americ
0031  lock, but to change it altogether. It is the MAGICIAN's wand, by means of which he may summon into
0441  ifully constructed natatory legs, a pair of MAGNIFICENT compound eyes, and extremely complex anten
0212  irds are not more fearful than small; and the MAGPIE, so wary in England, is tame in Norway, as is
0006  convinced that Natural Selection has been the MAIN but not exclusive means of modification. Varia
0029  fancier, who was not fully convinced that each MAIN breed was descended from a distinct species. V
0042  ey, peacock, goose, etc., may be attributed in MAIN part to selection not having been brought into
0045  never have expected that the branching of the MAIN nerves close to the great central ganglion of
0046  ock has shown a degree of variability in these MAIN nerves in Coccus, which may almost be compared
0069  sively with the elements. That climate acts in MAIN part indirectly by favouring other species, we
0094  on plant; and that certain insects depended in MAIN part on its nectar for food. I could give many
0126  that there exist but very few classes in each MAIN division of the animal and vegetable kingdoms.
0146  lowers, that the elder De Candolle founded his MAIN divisions of the order on analogous difference
0279  e of further modification and improvement. The MAIN cause, however, of innumerable intermediate li
0333  n the theory of descent with modification, the MAIN facts with respect to the mutual affinities of
0345  cannot be proved to be much more perfect, the MAIN objections to the theory of natural selection
0414  f little signification, excepting in the first MAIN divisions; whereas the organs of reproduction,
0418  s good with flowering plants, of which the two MAIN divisions have been founded on characters deri
0440  e in an unmistakeable manner. So again the two MAIN divisions of cirripedes, the pedunculated and
0454  pt changes. I believe that disuse has been the MAIN agency; that it has led in successive generati
0048  th another, and with those from the American MAINLAND; I was much struck how entirely vague and ar
0354  me terrestrial species on islands and on the MAINLAND, though separated by hundreds of miles of op
0357  and have united almost every island to some MAINLAND. If indeed the arguments used by Forbes and
0363  lants of other oceanic islands nearer to the MAINLAND, and (as remarked by Mr. H. C. Watson) from
0365  island, though standing more remote from the MAINLAND, would not receive colonists by similar mean
0383  habitants of islands to those of the nearest MAINLAND. On colonisation from the nearest source wit
0393  nic, as it lies on a bank connected with the MAINLAND: moreover, icebergs formerly brought boulder
0394  rmuda, at the distance of 600 miles from the MAINLAND. I hear from Mr. Tomes, who has specially st
0395  a separating an island from the neighbouring MAINLAND, and the presence in both of the same mammif
0396  nuously united within a recent period to the MAINLAND than islands separated by deeper channels, w
0397  s, is their affinity to those of the nearest MAINLAND, without being actually the same species. Nu
0399  re closely related to Australia, the nearest MAINLAND, than to any other region: and this is what
0406  rchipelago or island to those of the nearest MAINLAND, are I think, utterly inexplicable on the or
0409  e depth of the sea between an island and the MAINLAND. We can clearly see why all the inhabitants
0478  nts and animals of the neighbouring American MAINLAND; and those of the Cape de Verde archipelago
0478  ago and other African islands to the nearest MAINLAND. It must be admitted that these facts receiv
0017  ded from one or several species. The argument MAINLY relied on by those who believe in the multipl
0027  tructure. Thirdly, those characters which are MAINLY distinctive of each breed, for instance the w
0089  lour, or ornament, such differences have been MAINLY caused by sexual selection; that is, individu
0125  duction of new and modified descendants, will MAINLY lie between the larger groups, which are all
0135  difications of structure which are wholly, or MAINLY, due to natural selection. Mr. Wollaston has
0136  gless condition of so many Madeira beetles is MAINLY due to the action of natural selection, but c
0152  ur fantails, etc., these being the points now MAINLY attended to by English fanciers. Even in the
0158  nciples closely connected together. All being MAINLY due to the species of the same group having d
0172  ill here only state that I believe the answer MAINLY lies in the record being incomparably less pe
0234  winter; and the security of the hive is known MAINLY to depend on a large number of bees being sup
0270  experimentised on, are ranked by Sageret, who MAINLY founds his classification by the test of infe
0289  but the imperfection in the geological record MAINLY results from another and more important cause
0309  e to conclude that the great oceans are still MAINLY areas of subsidence, the great archipelagoes
0384  o attribute the dispersal of fresh water fish MAINLY to slight changes within the recent period in
0388  al form or in some degree modified. I believe MAINLY depends on the wide dispersal of their seeds
0399  ca: but on the view that this island has been MAINLY stocked by seeds brought with earth and stone
0417  , for classification, of trifling characters, MAINLY depends on their being correlated with severa
0447  ve in modifying an organ, such influence will MAINLY affect the mature animal, which has come to i
0466  ion we see much variability. This seems to be MAINLY due to the reproductive system being eminentl
0103  ckly formed on any one spot, and might there MAINTAIN itself in a body, so that whatever intercros
0274  stance, I think those authors are right, who MAINTAIN that the ass has a prepotent power over the
0316  ough all strongly opposed to such views as I MAINTAIN) admit its truth; and the rule strictly acco
0444  ties. Some authors who have written on Dogs, MAINTAIN that the greyhound and bull dog, though appe
0147  ecies in a state of nature it can hardly be MAINTAINED that the law is of universal application; b
0279  ich might be justly urged against the views MAINTAINED in this volume. Most of them have now been
0310  , etc., have unanimously, often vehemently, MAINTAINED the immutability of species. But I have rea
0320  ltimately each new species, is produced and MAINTAINED by having some advantage over those with wh
0398  ependent creation; whereas on the view here MAINTAINED, it is obvious that the Galapagos Islands w
0481  ies and well marked varieties. It cannot be MAINTAINED that species when intercrossed are invariab
0270  tner kept during several years a dwarf kind of MAIZE with yellow seeds, and a tall variety with re
0270  believe, has suspected that these varieties of MAIZE are distinct species; and it is important to
0270  ossed three varieties of gourd, which like the MAIZE has separated sexes, and he asserts that thei
0366  old glaciers, are now clothed by the vine and MAIZE. Throughout a large part of the United States
0373  mer low descent; and in Sikkim, Dr. Hooker saw MAIZE growing on gigantic ancient moraines. South o
0028  ds; and, continues the courtly historian, His MAJESTY by crossing the breeds, which method was nev
```

Page ***(Key Word)***

Page **(Key Word)**

0183 in our own country the larger titmouse (Parus MAJOR) may be seen climbing branches, almost like a
0047 we must, however, in many cases, decide by a MAJORITY of naturalists, for few well marked and well
0054 m surprised that my tables show even a small MAJORITY on the side of the larger genera. I will her
0066 period of life; and this period in the great MAJORITY of cases is an early one. If an animal can i
0096 certainly do not habitually pair, and a vast MAJORITY of plants are hermaphrodites. What reason, i
0099 be allowed to seed near each other, a large MAJORITY, as I have found, of the seedlings thus rais
0305 chalk period. This group includes the large MAJORITY of existing species. Lately, Professor Picte
0353 intelligible only on the view that the great MAJORITY of species have been produced on one side al
0356 e descended from a single parent. But in the MAJORITY of cases, namely, with all organisms which h
0359 e fruit, and to place them on sea water. The MAJORITY sank quickly, but some which whilst green fl
0370 od, before the Glacial epoch, and whilst the MAJORITY of the inhabitants of the world were specifi
0482 s, and which are still thus looked at by the MAJORITY of naturalists, and which consequently have
0008 rior to the act of conception. Several reasons MAKE me believe in this; but the chief one is the r
0014 . but hereditary diseases and some other facts MAKE me believe that the rule has a wider extension
0023 ght aboriginal stocks: for it is impossible to MAKE the present domestic breeds by the crossing of
0025 the rock pigeon. But if we deny this, we must MAKE one of the two following highly improbable sup
0027 y those brought from distant countries, we can MAKE an almost perfect series between the extremes
0030 useful to him. In this sense he may be said to MAKE for himself useful breeds. The great power of
0032 mitable perseverance, he will succeed, and may MAKE great improvements: if he wants any of these q
0034 selection, with a distinct object in view, to MAKE a new strain or sub breed, superior to anythin
0039 ght degree by nature. No man would ever try to MAKE a fantail, till he saw a pigeon with a tail de
0039 on. But to use such an expression as trying to MAKE a fantail, is, I have no doubt, in most cases,
0050 to one class within one country, he will soon MAKE up his mind how to rank most of the doubtful f
0051 oubtful forms. His general tendency will be to MAKE many species, for he will become impressed, ju
0051 ed, he will in the end generally be enabled to MAKE up his own mind which to call varieties and wh
0060 ntering on the subject of this chapter, I must MAKE a few preliminary remarks, to show how the str
0066 tance of a large number of eggs or seeds is to MAKE up for much destruction at some period of life
0067 he feral animals of South America. Here I will MAKE only a few remarks, just to recall to the read
0085 all the aids of art, these slight differences MAKE a great difference in cultivating the several
0087 in the act of hatching. Now, if nature had to MAKE the beak of a full grown pigeon very short for
0090 f the action of Natural Selection. In order to MAKE it clear how, as I believe, natural selection
0105 e production of new organic forms, we ought to MAKE the comparison within equal times; and this we
0107 urope. All fresh water basins, taken together, MAKE a small area compared with that of the sea or
0131 or very slight deviations of structure, as to MAKE the child like its parents. But the much great
0151 certainly to hold good in this class. I cannot MAKE out that it applies to plants, and this would
0156 in connexion with the present subject, I will MAKE only two other remarks. I think it will be adm
0172 ous an instinct as that which leads the bee to MAKE cells, which have practically anticipated the
0176 lesser area; and practically, as far as I can MAKE out, this rule holds good with varieties in a
0187 riginated; but I may remark that several facts MAKE me suspect that any sensitive nerve may be ren
0201 ecome extinct. Natural selection tends only to MAKE each organic being as perfect as, or slightly
0215 sheep, and pigs! No doubt they occasionally do MAKE an attack, and are then beaten; and if not cur
0217 of two or three days: so that, if she were to MAKE her own nest and sit on her own eggs, those fi
0222 h the pupae of F. fusca, which they habitually MAKE into slaves, from those of the little and furi
0224 and measures, would find it very difficult to MAKE cells of wax of the true form, though this is
0224 eems at first quite inconceivable how they can MAKE all the necessary angles and planes, or even p
0227 themselves not very wonderful, this bee would MAKE a structure as wonderfully perfect as that of
0227 the hive bee. We must suppose the Melipona to MAKE her cells truly spherical, and of equal sizes:
0227 y cylindrical burrows in wood many insects can MAKE, apparently by turning round on a fixed point.
0228 ; and as they deepened these little pits, they MAKE its nest, I believe that the hive bee has acqu
0230 for themselves a thin wall of wax, they could MAKE them wider and wider until they were converted
0230 ating at the same rate, and by endeavouring to MAKE their cells of the proper shape, by standing a
0230 en by examining the edge of a growing comb, do MAKE a rough, circumferential wall or rim all round
0230 rcularly as they deepen each cell. They do not MAKE the whole three sided pyramical base of any on
0231 n which the bees build is curious: they always MAKE the first rough wall from ten to twenty times
0233 her, strike the planes of intersection, and so MAKE an isolated hexagon: but I am not aware that a
0234 slight modification of her instinct led her to MAKE her waxen cells near together, so as to inters
0234 advantageous to our humble bee, if she were to MAKE her cells more and more regular, nearer togeth
0234 e advantageous to the Melipona, if she were to MAKE her cells closer together, and more regular in
0235 aced by plane surfaces; and the Melipona would MAKE a comb as perfect as that of the hive bee. Bey
0246 with the sterility of species; for it seems to MAKE a broad and clear distinction between varietie
0274 fer much. But this conclusion, as far as I can MAKE out, is founded on a single experiment; and se
0302 of Europe and of the United States. We do not MAKE due allowance for the enormous intervals of ti
0304 acted from the chalk of Belgium. And, as if to MAKE the case as striking as possible, this sessile
0312 nds to fill up the blanks between them, and to MAKE the percentage system of lost and new forms mo
0315 ng during long ages for the same object, might MAKE a new breed hardly distinguishable from our pr
0332 y forms, extinct and recent, descended from A, MAKE, as before remarked, one order: and this order
0333 h modification, should in the older formations MAKE some slight approach to each other: so that th
0342 e causes taken conjointly, must have tended to MAKE the geological record extremely imperfect, and
0344 hat all the forms of life, ancient and recent, MAKE together one grand system: for all are connect
0367 been a little earlier or later: but this will MAKE no difference in the final result. As the warm
0406 hapters I have endeavoured to show, that if we MAKE due allowance for our ignorance of the full ef
0407 must have spread from one parent source: if we MAKE the same allowances as before for our ignoranc
0413 ess concealed form, that the characters do not MAKE the genus, but that the genus gives the charac
0425 ists. We have no written pedigrees: we have to MAKE out community of descent by resemblances of an
0434 some unknown plan of creation, we may hope to MAKE sure but slow progress. Morphology. We have se
0471 every fresh addition to our knowledge tends to MAKE more strictly correct, is on this theory simpl
0221 tion was struck by about a score of the slave MAKERS haunting the same spot, and evidently not in
0223 conjecture. But as ants, which are not slave MAKERS, will as I have seen, carry off pupae of othe
0006 to the origin of species and varieties, if he MAKES due allowance for our profound ignorance in r
0050 osely it has been studied: yet a German author MAKES more than a dozen species out of forms, which
0128 modified offspring that superiority which now MAKES them dominant in their own countries. Natural
0167 unreal, or at least for an unknown, cause. It MAKES the works of God a mere mockery and deception
0208 . huber found it was with a caterpillar, which MAKES a very complicated hammock: for if he took a
0217 merican cuckoo is in this predicament: for she MAKES her own nest and has eggs and young successiv
0218 ing that although the Tachytes nigra generally MAKES its own burrow and stores it with paralysed p

Page **(Key Word)**

makes

Page ✳✳✳(Key Word)✳✳

```
0246 erality of some degree of sterility. Kolreuter MAKES the rule universal; but then he cuts the knot
0247 tingly ranks them as varieties. Gartner, also, MAKES the rule equally universal; and he disputes t
0254 a, all are quite fertile together; and analogy MAKES me greatly doubt, whether the several aborig†
0387 ; although I do not know the fact, yet analogy MAKES me believe that a heron flying to another por
0004 tivated plants would offer the best chance of MAKING out this obscure problem. Nor have I been dis
0017 s, then such facts would have great weight in MAKING us doubt about the immutability of the many v
0020 the wild state. Moreover, the possibility of MAKING distinct races by crossing has been greatly e
0048 l been ranked by botanists as species; and in MAKING this list he has omitted many trifling variet
0126 ew of all the descendants of the same species MAKING a class, we can understand how it is that the
0200 detail of structure in every living creature (MAKING some little allowance for the direct action o
0207 he cuckoo, ostrich, and parasitic bees. Slave MAKING ants. Hive bee, its cell making instinct. Dif
0207 c bees. Slave making ants. Hive bee, its cell MAKING instinct. Difficulties on the theory of the N
0207 wonderful an instinct as that of the hive bee MAKING its cells will probably have occurred to many
0210 tainly can do. I have been surprised to find, MAKING allowance for the instincts of animals having
0216 lay her eggs in other birds' nests; the slave MAKING instinct of certain ants; and the comb making
0216 making instinct of certain ants; and the comb MAKING power of the hive bee: these two latter insti
0219 i can see no difficulty in natural selection MAKING an occasional habit permanent, if of advantag
0219 appropriated, be not thus exterminated. Slave MAKING instinct. This remarkable instinct was first
0219 aves, do no other work. They are incapable of MAKING their own nests, or of feeding their own larv
0219 facts? If we had not known of any other slave MAKING ant, it would have been hopeless to have spec
0219 se first discovered by P. Huber to be a slave MAKING ant. This species is found in the southern pa
0220 traordinary and odious an instinct as that of MAKING slaves. Hence I will give the observations wh
0221 slaves habitually work with their masters in MAKING the nest, and they alone open and close the d
0221 these ants clinging to the legs of the slave MAKING F. sanguinea. The latter ruthlessly killed th
0222 ava under a stone beneath a nest of the slave MAKING F. sanguinea; and when I had accidentally dis
0223 by me, in regard to the wonderful instinct of MAKING slaves. Let it be observed what a contrast th
0223 the larvae, and the masters alone go on slave MAKING expeditions. In Switzerland the slaves and ma
0223 zerland the slaves and masters work together, MAKING and bringing materials for the nest: both, bu
0224 its slaves as is the Formica rufescens. Cell MAKING instinct of the Hive Bee. I will not here ent
0225 ding to them short tubes of wax, and likewise MAKING separate and very irregular rounded cells of
0227 nd from her fellow labourers when several are MAKING their spheres; but she is already so far enab
0233 ingle insect (as in the case of a queen wasp) MAKING hexagonal cells, if she work alternately on t
0244 ung cuckoo ejecting its foster brothers, ants MAKING slaves, the larvae of ichneumonidae feeding w
0256 but the parallelism between the difficulty of MAKING a first cross, and the sterility of the hybri
0258 widest possible difference in the facility of MAKING reciprocal crosses. Such cases are highly imp
0258 er, found that this difference of facility in MAKING reciprocal crosses is extremely common in a l
0259 ther parent. And lastly, that the facility of MAKING a first cross between any two species is not
0282 h the sea at work grinding down old rocks and MAKING fresh sediment, before he can hope to compreh
0284 ······································· 2,240 Feet. MAKING altogether······························7
0327 ther remark connected with this subject worth MAKING. I have given my reasons for believing that a
0386 dirt would not be washed off their feet; when MAKING land, they would be sure to fly to their natu
0432 nd space. We shall certainly never succeed in MAKING so perfect a collection: nevertheless, in cer
0375 extend along the heights of the peninsula of MALACCA, and are thinly scattered, on the one hand o
0002 who is now studying the natural history of the MALAY archipelago, has arrived at almost exactly th
0164 ern China; and from Norway in the north to the MALAY Archipelago in the south. In all parts of the
0197 to sexual selection. A trailing bamboo in the MALAY Archipelago climbs the loftiest trees by the
0299 remarks, under an imaginary illustration. The MALAY Archipelago is of about the size of Europe fr
0299 cwin Austen, that the present condition of the MALAY Archipelago, with its numerous large islands
0299 most of our formations were accumulating. The MALAY Archipelago is one of the richest regions of
0305 they are at present. Even at this day, if the MALAY Archipelago were converted into land, the tro
0395 servations on this head in regard to the great MALAY Archipelago, which is traversed near Celebes
0143 e of the adult; in the same manner as any MALCONFORMATION affecting the early embryo, seriously af
0144 aire has forcibly remarked, that certain MALCONFORMATIONS very frequently, and that others rarely
0008 cause of variability may be attributed to the MALE reproductive elements having been a
0008 r confinement, even in the many cases when the MALE and female unite. How many animals there are w
0014 y cause, which may have acted on the ovules or MALE element; in nearly the same manner as in the c
0014 appearing late in life, is clearly due to the MALE element. Having alluded to the subject of reve
0021 their skulls. The carrier, more especially the MALE bird, is also remarkable from the wonderful de
0028 ce for the production of distinct breeds, that MALE and female pigeons can be easily mated for lif
0088 but on having special weapons, confined to the MALE sex. A hornless stag or spurless cock would ha
0088 ature this law of battle descends, I know not; MALE alligators have been described as fighting; be
0088 war dance, for the possession of the females; MALE salmons have been seen fighting all day long;
0088 salmons have been seen fighting all day long; MALE stag beetles often bear wounds from the huge m
0088 der pad to the boar, and the hooked jaw to the MALE salmon; for the shield may be as important for
0089 well known laws with respect to the plumage of MALE and female birds, in comparison with the pluma
0090 and have transmitted these advantages to their MALE offspring. Yet, I would not wish to attribute
0090 liarities arising and becoming attached to the MALE sex in our domestic animals (as the wattle in
0090 sex in our domestic animals (as the wattle in MALE carriers, horn like protuberances in the cocks
0093 y to be alluded to. Some holly trees bear only MALE flowers, which have four stamens producing rat
0093 found a female tree exactly sixty yards from a MALE tree, I put the stigmas of twenty flowers, tak
0093 ad set for several days from the female to the MALE tree, the pollen could not thus have been carr
0094 ed under new conditions of life, sometimes the MALE organs and sometimes the female organs become
0100 es. When the sexes are separated, although the MALE and female flowers may be produced on the same
0132 ing or plastic condition of the offspring. The MALE and female sexual elements seem to be affected
0135 act) that the anterior tarsi, or feet, of many MALE dung feeding beetles are very often broken off
0148 n shell or carapace. This is the case with the MALE Ibla, and in a truly extraordinary manner with
0153 to one sex than to the other, generally to the MALE sex, as with the wattle of carriers and the en
0165 reton's famous hybrid from a chestnut mare and MALE cuagga, the hybrid, and even the pure offsprin
0196 aracters of animals having a will, to give one MALE an advantage in fighting with another or in ch
0197 that the skin on the head of the clean feeding MALE turkey is likewise naked. The sutures in the s
0217 hed young would probably have to be fed by the MALE alone. But the American cuckoo is in this pred
0237 e of many birds, and in the hooked jaws of the MALE salmon. We have even slight differences in the
0237 tion to an artificially imperfect state of the MALE sex; for oxen of certain breeds have longer ho
0240 have not even rudiments of ocelli, though the MALE and female ants of this genus have well develo
0243 as does our British thrush; how it is that the MALE wrens (Troglodytes) of North America, build co
0246 nt, as may be clearly seen in the state of the MALE element in both plants and animals; though the
```

Page ✳✳✳(Key Word)✳✳

```
)258  ng first crossed with a female ass, and then a  MALE  ass with a mare: these two species may then be
)261  in reciprocal crosses between two species the  MALE  sexual element of the one will often freely ac
)262  hesion of grafted stocks, and the union of the  MALE  and female elements in the act of reproduction
)263  remarked, in the union of two pure species the  MALE  and female sexual elements are perfect, wherea
)263  t sometimes be a physical impossibility in the  MALE  element reaching the ovule, as would be the ca
)264  ot penetrate the stigmatic surface. Again, the  MALE  element may reach the female element, but be i
)265  rility occurs in various degrees; in both, the  MALE  element is the most liable to be affected; but
)265  fected; but sometimes the female more than the  MALE. In both, the tendency goes to a certain exten
)274  that the prepotency runs more strongly in the  MALE  ass than in the female, so that the mule, whic
)275  o that the mule, which is the offspring of the  MALE  ass and mare, is more like an ass, than is the
)441  lopment has assuredly been retrograde; for the  MALE  is a mere sack, which lives for a short time,
)443  is formed; and the variation may be due to the  MALE  and female sexual elements having been affecte
)451  : this seems to be the case with the mammae of  MALE  mammals, for many instances are on record of t
)451  ped state. In plants with separated sexes, the  MALE  flowers often have a rudiment of a pistil; and
)451  til; and Kolreuter found that by crossing such  MALE  plants with an hermaphrodite species, the rudi
)452  pported on the style; but some Compositae, the  MALE  florets, which of course cannot be fecundated,
)453  papilla, which often represents the pistil in  MALE  flowers, and which is formed merely of cellula
0013  nce to us, that peculiarities appearing in the  MALES  of our domestic breeds are often transmitted
0013  r exclusively, or in a much greater degree, to  MALES  alone. A much more important rule, which I th
0022  nd disposition. Lastly, in certain breeds, the  MALES  and females have come to differ to a slight d
0083  e climate. He does not allow the most vigorous  MALES  to struggle for the females. He does not rigi
0088  e for existence, but on a struggle between the  MALES  for possession of the females; the result is
0088  atural selection. Generally, the most vigorous  MALES,  those which are best fitted for their places
0088  n bear wounds from the huge mandibles of other  MALES.  The war is, perhaps, severest between males
0088  r males. The war is, perhaps, severest between  MALES  of polygamous animals, and these seem oftenes
0088  em oftenest provided with special weapons. The  MALES  of carnivorous animals are already well armed
0089  that there is the severest rivalry between the  MALES  of many species to attract by singing the fem
0089  e, and some others, congregate; and successive  MALES  display their gorgeous plumage and perform st
0089  f generations, the most melodious or beautiful  MALES,  according to their standard of beauty, might
0089  t corresponding ages or seasons, either by the  MALES  alone, or by the males and females; but I hav
0089  seasons, either by the males alone, or by the  MALES  and females; but I have not space here to ent
0089  bject. Thus it is, as I believe, that when the  MALES  and females of any animal have the same gener
0089  aused by sexual selection; that is, individual  MALES  have had, in successive generations, some sli
0090  generations, some slight advantage over other  MALES,  in their weapons, means of defence, or charm
0090  h we cannot believe to be either useful to the  MALES  in battle, or attractive to the females. We s
0127  assuring to the most vigorous and best adapted  MALES  the greatest number of offspring. Sexual sele
0127  ection will also give characters useful to the  MALES  alone, in their struggles with other males. W
0127  the males alone, in their struggles with other  MALES.  Whether natural selection has really thus ac
0150  h the act of reproduction. The rule applies to  MALES  and females; but as females more rarely offer
0156  instance, the amount of difference between the  MALES  of gallinaceous birds, in which secondary sex
0157  nly gives fewer offspring to the less favoured  MALES.  Whatever the cause may be of the variability
0158  the same species to each other, or to fit the  MALES  and females to different habits of life, or t
0158  nd females to different habits of life, or the  MALES  to struggle with other males for the possessi
0158  s of life, or the males to struggle with other  MALES  for the possession of the females. Finally, t
0192  n very differently constructed from either the  MALES  or fertile females; but this case will be tre
0202  he truly wonderful power of scent by which the  MALES  of many insects find their females, can we ad
0218  then in another; and these are hatched by the  MALES.  This instinct may probably be accounted for
0219  certainly become extinct in a single year. The  MALES  and fertile females do no work. The workers o
0220  f F. sanguinea, and found a few slaves in all.  MALES  and fertile females of the slave species are
0236  ely in instinct and in structure from both the  MALES  and fertile females, and yet, from being ster
0236  he working ants differing widely from both the  MALES  and the fertile females in structure, as in t
0238  ous to the community: consequently the fertile  MALES  and females of the same community flourished,
0238  differ, not only from the fertile females and  MALES,  but from each other, sometimes to an almost
0239  be quite possible, different from the fertile  MALES  and females, in this case, we may safely conc
0240  en the most useful to the community, and those  MALES  and females had been continually selected, wh
0243  erica, build cock nests, to roost in, like the  MALES  of our distinct Kitty wrens, a habit wholly u
0267  at slight crosses, that is crosses between the  MALES  and females of the same species which have va
0267  that greater crosses, that is crosses between  MALES  and females which have become widely or speci
0424  single fact can be predicated in common of the  MALES  and hermaphrodites of certain cirripedes, whe
0425  he individuals of the same species, though the  MALES  and females and larvae are sometimes extremel
0441  cture, or into what I have called complemental  MALES: and in the latter, the development has assur
0450  ce, rudimentary mammae are very general in the  MALES  of mammals: I presume that the bastard wing i
0451  ans having become well developed in full grown  MALES,  and having secreted milk. So again there are
0468  there will in most cases be a struggle between  MALES  for possession of the females. The most vigor
0468  s or means of defence, or on the charms of the  MALES; and the slightest advantage will lead to vic
0417  defined. Certain plants, belonging to the  MALPIGHIACEAE,  bear perfect and degraded flowers; in th
0417  genus should still be retained amongst the  MALPIGHIACEAE.  This case seems to me well to illustrate
0005  , will be treated of. This is the doctrine of  MALTHUS,  applied to the whole animal and vegetable k
0063  cal conditions of life. It is the doctrine of  MALTHUS  applied with manifold force to the whole ani
0450  throughout nature. For instance, rudimentary  MAMMAE  are very general in the males of mammals: I p
0451  developed: this seems to be the case with the  MAMMAE  of male mammals, for many instances are on re
0019  mind that Britain has now hardly one peculiar  MAMMAL,  and France but few distinct from those of Ge
0081  the influence of a single introduced tree or  MAMMAL  has been shown to be. But in the case of an i
0303  middle of the secondary series: and one true  MAMMAL  has been discovered in the new red sandstone
0352  ly, we find no inexplicable cases of the same  MAMMAL  inhabiting distant points of the world. No ge
0352  separate points, why do we not find a single  MAMMAL  common to Europe and Australia or South Ameri
0393  e instance, free from doubt, of a terrestrial  MAMMAL  (excluding domesticated animals kept by the n
0394  on can easily be answered: for no terrestrial  MAMMAL  can be transported across a wide space of sea
0439  l, he cannot now tell whether it be that of a  MAMMAL,  bird, or reptile. The vermiform larvae of mo
0440  e related to similar conditions, in the young  MAMMAL  which is nourished in the womb of its mother,
0479  marvelling at the embryo of an air breathing  MAMMAL  or bird having branchial slits and arteries r
0144  ental, that if we pick out the two orders of  MAMMALIA  which are most abnormal in their dermal cove
0150  ng is a most abnormal structure in the class  MAMMALIA; but the rule would not here apply, because
0395  and this space separates two widely distinct  MAMMALIAN  faunas. On either side the islands are situ
0396  he sea and the degree of affinity of the the  MAMMALIAN  inhabitants of islands with those of a neig
0116  marked, our carnivorous, ruminant, and rodent  MAMMALS,  could successfully compete with these well
```

Page **(Key Word)**

```
0116 ese well pronounced orders. In the Australian MAMMALS, we see the process of diversification in a
0182 and, and seeing that we have flying birds and  MAMMALS, flying insects of the most diversified typ
0197 ise naked. The sutures in the skulls of young  MAMMALS have been advanced as a beautiful adaptatio
0289 w large a proportion of the bones of tertiary  MAMMALS have been discovered either in caves or in
0303 lished not many years ago, the great class of  MAMMALS was always spoken of as having abruptly com
0303 of the richest known accumulations of fossil   MAMMALS belongs to the middle of the secondary seri
0313 ocodile associated with many strange and lost  MAMMALS and reptiles in the sub Himalayan deposits.
0319 ve felt certain from the analogy of all other  MAMMALS, even of the slow breeding elephant, and fr
0329 achyderms, as the two most distinct orders of  MAMMALS; but Owen has discovered so many fossil lin
0331 r fish, the older Cephalopods, and the Eocene  MAMMALS, with the more recent members of the same c
0338 . seeing, for instance, that the oldest known  MAMMALS, reptiles, and fish strictly belong to thei
0339 . clift many years ago showed that the fossil  MAMMALS from the Australian caves were closely alli
0339 most striking manner that most of the fossil   MAMMALS, buried there in such numbers, are related
0339 ently extended the same generalisation to the  MAMMALS of the Old World. We see the same law in th
0340 merica the law of distribution of terrestrial  MAMMALS was formerly different from what it now is.
0340 ndia was formerly more closely related in its  MAMMALS to Africa than it is at the present time. A
0352 sea is more distinctly limited in terrestrial  MAMMALS, than perhaps in any other organic beings:
0353 mispheres? The answer, as I believe, is, that  MAMMALS have not been able to migrate, whereas some
0358 ll hereafter see) between the distribution of  MAMMALS and the depth of the sea; these and other s
0373 o the Pyrennes. We may infer, from the frozen  MAMMALS and nature of the mountain vegetation, that
0383 absence of the Batrachians and of terrestrial  MAMMALS. On the relation of the inhabitants of isla
0391 nd gigantic wingless birds, take the place of  MAMMALS. In the plants of the Galapagos Islands, Dr
0392 instance, in certain islands not tenanted by   MAMMALS, some of the endemic plants have beautifull
0393 there, it would be very difficult to explain.  MAMMALS offer another and similar case. I have care
0394 aid that small islands will not support small  MAMMALS, for they occur in many parts of the world
0394 t there has not been time for the creation of  MAMMALS; many volcanic islands are sufficiently anc
0394 classes; and on continents it is thought that  MAMMALS appear and disappear at a quicker rate than
0394 n other and lower animals. Though terrestrial  MAMMALS do not occur on oceanic islands, aerial mam
0394 mmals do not occur on oceanic islands; aerial  MAMMALS do occur on almost every island. New Zealan
0394 sed creative force produced bats and no other  MAMMALS on remote islands? On my view this question
0395 islands, with the absence of all terrestrial   MAMMALS. Besides the absence of terrestrial mammals
0395 l mammals. Besides the absence of terrestrial  MAMMALS in relation to the remoteness of islands fr
0395 ing to the probable naturalisation of certain  MAMMALS through man's agency; but we shall soon hav
0395 ted by a shallow channel from Europe, and the  MAMMALS are the same on both sides: we meet with an
0396 groups, as of batrachians, and of terrestrial  MAMMALS notwithstanding the presence of aerial bats
0403 ey are inhabited by a vast number of distinct  MAMMALS, birds, and plants. The principle which det
0404 ch it would be difficult to prove it. Amongst  MAMMALS, we see it strikingly displayed in Bats, an
0409 of organisms, as batrachians and terrestrial   MAMMALS, should be absent from oceanic islands, whi
0409 possess their own peculiar species of aerial  MAMMALS or bats. We can see why there should be som
0409 ould be some relation between the presence of  MAMMALS, in a more or less modified condition, and
0413 e the characters common, for instance, to all  MAMMALS, by another those common to all carnivora,
0427 animal, and the whale, and between both these  MAMMALS and fishes, is analogical. Amongst insects
0435 or, the archetype as it may be called, of all  MAMMALS, had its limbs constructed on the existing
0437 separate pieces in the act of parturition of  MAMMALS, will by no means explain the same construc
0450 ntary mammae are very general in the males of  MAMMALS: I presume that the bastard wing in birds ma
0451 seems to be the case with the mammae of male  MAMMALS, for many instances are on record of these
0477 ide spaces of ocean, as frogs and terrestrial  MAMMALS, should not inhabit oceanic islands: and why
0478 species of bats; and the absence of all other  MAMMALS, on oceanic islands, are utterly inexplicab
0479 f life, we can clearly see why the embryos of  MAMMALS, birds, reptiles, and fishes should be so cl
0483 or seed, or as full grown? and in the case of  MAMMALS, were they created bearing the false marks o
0289 erous strata of North America. In regard to    MAMMIFEROUS remains, a single glance at the historica
0395 nland, and the presence in both of the same    MAMMIFEROUS species or of allied species in a more or
0004 portant, we shall see how great is the power of MAN in accumulating by his Selection successive sl
0007 mestic Productions. Circumstances favourable to MAN's power of Selection. When we look to the ind
0012 thus doing with those with long beaks large feet. Hence, if MAN goes on selecting, and thus augmenting, any pe
0017 ur of this view. It has often been assumed that MAN has chosen for domestication animals and plant
0018 s have rendered it in some degree probable that MAN sufficiently civilized to have manufactured po
0024 domesticated in ancient times by half civilized MAN, as to be quite prolific under confinement. Ar
0024 it must be assumed not only that half civilized MAN succeeded in thoroughly domesticating several
0026 e several reasons, namely, the improbability of MAN having formerly got seven or eight supposed sp
0029 nd some little to habit: but he would be a bold MAN who would account by such agencies for the dif
0030 eed to the animal's or plant's own good, but to MAN's use or fancy. Some variations useful to him
0030 compare the many breeds of dogs, each good for MAN in very different ways: when we compare the ga
0030 nd flower garden races of plants most useful to MAN at different seasons and for different purpose
0030 hat this has not been their history. The key is MAN's power of accumulative selection: nature give
0030 selection: nature gives successive variations; MAN adds them up in certain directions useful to h
0032 ne have vainly attempted to appreciate. Not one MAN in a thousand has accuracy of eye and judgment
0034 individual animals, is more important. Thus, a  MAN who intends keeping pointers naturally tries t
0038 s up to their present standard of usefulness to MAN, we can understand how it is that neither Aust
0038 any other region inhabited by quite uncivilised MAN, has afforded us a single plant worth culture.
0038 ard to the domestic animals kept by uncivilised MAN, it should not be overlooked that they almost
0038 en of the all important part which selection by MAN has played, it becomes at once obvious, how it
0038 tation in their structure or in their habits to MAN's wants or fancies. We can, I think, further u
0038 latively so slight in internal parts or organs. MAN can hardly select, or only with much difficult
0039 iven to him in some slight degree by nature. No MAN would ever try to make a fantail, till he saw
0039 no doubt, in most cases, utterly incorrect. The MAN who first selected a pigeon with a slightly la
0040 hardly be said to have had a definite origin. A MAN preserves and breeds from an individual with s
0040 e circumstances, favourable, or the reverse, to MAN's power of selection. A high degree of variab
0041 as variations manifestly useful or pleasing to  MAN appear only occasionally, the chance of their
0041 e animal or plant should be so highly useful to MAN, or so much valued by him, that the closest at
0045 selection to accumulate, in the same manner as  MAN can accumulate in any given direction individu
0050 plant in a state of nature be highly useful to  MAN, or from any cause closely attract his attenti
0061 ral Selection, in order to mark its relation to MAN's power of selection. We have seen that man by
0061 to man's power of selection. We have seen that  MAN by selection can certainly produce great resul
0061 for action; and is as immeasurably superior to MAN's feeble efforts; as the works of Nature are t
0064 he progeny of a single pair. Even slow breeding MAN has doubled in twenty five years, and at this
```

Page **(Key Word)**

age **(Key Word)**

```
068  ordinarily severe mortality from epidemics with   MAN. The action of climate seems at first sight to
071  th, which had never been touched by the hand of    MAN; but several hundred acres of exactly the same
080  ion. Natural Selection, its power compared with    MAN's selection, its power on characters of trifli
080  which we have seen is so potent in the hands of    MAN, apply to nature? I think we shall see that it
080  ht improbable, seeing that variations useful to    MAN have undoubtedly occurred, that other variatio
082  extreme amount of variability is necessary; as     MAN can certainly produce great results by adding
083  o as to have better resisted such intruders. As    MAN can produce and certainly has produced a great
083  means of selection, what may not nature effect?    MAN can act only on external and visible character
083  nal difference, on the whole machinery of life.    MAN selects only for his own good; Nature only for
083  is placed under well suited conditions of life.    MAN keeps the natives of many climates in the same
084  ved. How fleeting are the wishes and efforts of    MAN! how short his time! and consequently how poor
084  oductions should be far truer in character than    MAN's productions; that they should be infinitely
089  details necessary to support this view; but if     MAN can in a short time give elegant carriage and
090  can see no more reason to doubt this, than that    MAN can improve the fleetness of his greyhounds by
090  t unconscious selection which results from each    MAN trying to keep the best dogs without any thoug
102  s competitors, it will soon be exterminated. In    MAN's methodical selection, a breeder selects for
109  ough the process of selection may be, if feeble    MAN can do much by his powers of artificial select
110  ons, through the selection of improved forms by    MAN. Many curious instances could be given showing
112  ain, we may suppose that at an early period one    MAN preferred swifter horses; another stronger and
112  have tended to disappear. Here, then, we see in    MAN's productions the action of what may be called
115  is seen in the naturalisation of plants through    MAN's agency in foreign lands. It might have been
126  e. but which groups will ultimately prevail, no    MAN can predict; for we well know that many groups
127  y as so many variations have occurred useful to    MAN. But if variations useful to any organic being
140  c animals were originally chosen by uncivilised    MAN because they were useful and bred readily unde
141  stic animals, but they have been transported by    MAN to many parts of the world, and now have a far
141  city of enduring the most different climates by    MAN himself and by his domestic animals, and such
142  district to another; for it is not likely that    MAN should have succeeded in selecting so many bre
153  ice that these variable characters, produced by    MAN's selection, sometimes become attached, from c
164  pe on each shoulder and with leg stripes; and a    MAN, whom I can implicitly trust, has examined for
188  ator works by intellectual powers like those of    MAN? If we must compare the eye to an optical inst
189  ss, as the works of the Creator are to those of    MAN? If it could be demonstrated that any complex
199  me purpose the differences between the races of    MAN, which are so strongly marked; I may add that
199  res have been created for beauty in the eyes of    MAN, or for mere variety. This doctrine, if true,
212  of the same enemy in other animals. But fear of    MAN is slowly acquired, as I have elsewhere shown,
212  or the large birds have been most persecuted by    MAN. We may safely attribute the greater wildness
215  unconscious selection is still at work, as each    MAN tries to procure, without intending to improve
215  is scarcely possible to doubt that the love of    MAN has become instinctive in the dog. All wolves,
216  nstincts have been lost partly by habit, and by    MAN selecting and accumulating during successive g
224  ions at which I have arrived. He must be a dull    MAN who can examine the exquisite structure of a c
242  t the division of labour is useful to civilised    MAN. As ants work by inherited instincts and by in
269  and plants are produced under domestication by    MAN's methodical and unconscious power of selectio
269  e in the process of selection, as carried on by    MAN and nature, we need not be surprised at some d
271  gree would generally be ranked as species; from    MAN selecting only external characters in the prod
282  ation of each formation or even each stratum. A    MAN must for years examine for himself great piles
320  exterminated, either locally or wholly, through    MAN's agency. I may repeat what I published in 184
320  feel no surprise at sickness, but when the sick    MAN dies, to wonder and to suspect 'that he died by
339  within the same areas mean? He would be a bold    MAN, who after comparing the present climate of Au
375  pecies, believed not to have been introduced by    MAN, and on the mountains, some few representative
375  opean species; other species, not introduced by    MAN, occur on the lowlands; and a long list can be
380  ly yielded to continental forms, naturalised by    MAN's agency. I am far from supposing that all dif
381  species and not another becomes naturalised by    MAN's agency in a foreign land; why one ranges twi
382  t stranded may be compared with savage races of    MAN, driven up and surviving in the mountian fastn
390  s have not been created on oceanic islands! for    MAN has unintentionally stocked them from various
395  bable naturalisation of certain mammals through    MAN's agency; but we shall soon have much light th
402  the fact that many species, naturalised through    MAN's agency, have spread with astonishing rapidit
422  ind, a genealogical arrangement of the races of    MAN would afford the best classification of the va
434  hat can be more curious than that the hand of a    MAN, formed for grasping, that of a mole for diggi
440  to believe that the same bones in the hand of a    MAN, wing of a bat, and fin of a porpoise, are rel
446  each breed, and which have been accumulated by    MAN's selection, have not generally first appeared
453  excretion of precious phosphate of lime? When a    MAN's fingers have been amputated, imperfect nails
466  by our most anciently domesticated productions.    MAN does not actually produce variability; he only
467  n the organisation, and causes variability. But    MAN can and does select the variations given to hi
467  tic breeds. That many of the breeds produced by    MAN have to a large extent the character of natura
468  on under nature is a strictly limited quantity.    MAN, though acting on external characters alone an
469  preserved, accumulated, and inherited? Why, if    MAN can by patience select variations most useful
471  each species has been independently created, no    MAN can explain. Many other facts are, as it seems
479  mework of bones being the same in the hand of a    MAN, wing of bat, fin of the porpoise, and leg of
486  ise immensely in value. A new variety raised by    MAN will be a far more important and interesting s
488  radation. Light will be thrown on the origin of    MAN and his history. Authors of the highest eminen
454  that the rudimentary nails on the fin of the     MANATEE were formed for this purpose. On my view of
062  spirit and ability than W. Herbert, Dean of      MANCHESTER, evidently the result of his great horticul
088  stag beetles often bear wounds from the huge     MANDIBLES of other males. The war is, perhaps, severe
435  tely numerous modifications of an upper lip,     MANDIBLES, and two pairs of maxillae. Analogous laws
436  at their common progenitor had an upper lip,     MANDIBLES, and two pair of maxillae, these parts bein
088  iven through means of sexual selection, as the   MANE to the lion, the shoulder pad to the boar, and
156  bility of secondary sexual characters is not     MANIFEST; but we can see why these characters should
323  aeozoic and tertiary periods, would still be     MANIFEST, and the several formations could be easily
339  in South America, a similar relationship is      MANIFEST, even to an uneducated eye, in the gigantic
398  ir habits, gestures, and tones of voice, was     MANIFEST. So it is with the other animals, and with n
435  ct each animal and plant. The explanation is     MANIFEST on the theory of the natural selection of su
015  could be shown that our domestic varieties       MANIFESTED a strong tendency to reversion, that is, to
442  etamorphosis; the cephalopodic character is      MANIFESTED long before the parts of the embryo are com
041  st any desired direction. But as variations      MANIFESTLY useful or pleasing to man appear only occas
082  to have undergone a change, and this would       MANIFESTLY be favourable to natural selection, by givi
226  and this pyramid, as Huber has remarked, is      MANIFESTLY a gross imitation of the three sided pyrami
447  the new species of our supposed genus will       MANIFESTLY tend to resemble each other much more close
```

Page **(Key Word)**

Page **************************************(Key Word)**************************************

```
0063  . it is the doctrine of Malthus applied with MANIFOLD force to the whole animal and vegetable kin
0247  there is an acknowledged difficulty in the MANIPULATION) half of these twenty plants had their fe
0249  otwithstanding the frequent ill effects of MANIPULATION, sometimes decidedly increases, and goes
0270  d this one head produced only five grains. MANIPULATION in this case could not have been injuriou
0067  norant we are on this head, even in regard to MANKIND, so incomparably better knowr than any othe
0422  guages. If we possessed a perfect pedigree of MANKIND, a genealogical arrangement of the races of
0005  any being, if it vary however slightly in any MANNER profitable to itself, under the complex and
0006  er and generally extinct species, in the same MANNER as the acknowledged varieties of any one spe
0010  d, all would probably have varied in the same MANNER. To judge how much, in the case of any varia
0014  he ovules or male elementi in nearly the same MANNER as in the crossed offspring from a short hor
0016  me species differ from each other in the same MANNER as, only in most cases in a lesser degree th
0022  the lower jaw, varies in a highly remarkable MANNER. The number of the caudal and sacral vertebr
0022  hed. The shape and size of the eggs vary. The MANNER of flight differs remarkably; as does in som
0039  developed in some slight degree in an unusual MANNER, or a pouter till he saw a pigeon with a cro
0045  natural selection to accumulate, in the same MANNER as man can accumulate in any given direction
0057  genera are related to each other, in the same MANNER as the varieties of any one species are rela
0077  lated, in the most essential yet often hidden MANNER, to that of all other organic beings, with w
0078  ife will generally be changed in an essential MANNER. If we wished to increase its average number
0081  what we have seen of the intimate and complex MANNER in which the inhabitants of each country are
0081  some of the original inhabitants were in some MANNER modified; for, had the area been open to imm
0083  lected character in some peculiar and fitting MANNER; he feeds a long and a short beaked pigeon o
0083  cked or long legged quadruped in any peculiar MANNER; he exposes sheep with long and short wool t
0095  her, modified and adapted in the most perfect MANNER to each other, by the continued preservation
0097  n between the structure of the flower and the MANNER in which bees suck the nectar: for, in doing
0103  ividuals in each district in exactly the same MANNER to the conditions of each; for in a continuo
0104  rying species throughout the area in the same MANNER in relation to the same conditions. Intercro
0109  ll with what geology tells us of the rate and MANNER at which the inhabitants of this world have
0113  h differed from each other in at all the same MANNER as distinct species and genera of grasses di
0115  uralised, we can gain some crude idea in what MANNER some of the natives would have had to be mod
0118  ary, and generally to vary in nearly the same MANNER as their parents varied. Moreover, these two
0120  in the diagram in a condensed and simplified MANNER), we get eight species, marked by the letter
0122  o been modified and improved in a diversified MANNER at each stage of descent, so as to have beco
0123  be allied to each other in a widely different MANNER. Of the eight descendants from (A) the three
0128  h other in group subordinate to group, in the MANNER which we everywhere behold, namely, varietie
0129  e surrounding twigs and branches, in the same MANNER as species and groups of species have tried
0131  tures variable. Parts developed in an unusual MANNER are highly variable: specific characters mor
0131  pecies of the same genus vary in an analogous MANNER. Reversions to long lost characters. Summary
0137  ght, some slight power of vision. In the same MANNER as in Madeira the wings of some of the insec
0143  ffect the structure of the adult; in the same MANNER as any malconformation affecting the early e
0143  , are alike, seem liable to vary in an allied MANNER: we see this in the right and left sides of
0143  nd left sides of the body varying in the same MANNER; in the front and hind legs, and even in the
0144  ng to Schlegel, the shape of the body and the MANNER of swallowing determine the position of seve
0146  be thought to be correlated in some necessary MANNER. So, again, I do not doubt that some apparen
0146  oughout whole orders, are entirely due to the MANNER alone in which natural selection can act. Fo
0148  h the male Ibla, and in a truly extraordinary MANNER with the carapace in al
0150  in any species in an extraordinary degree or MANNER, in comparison with the same part in allied
0150  at had its wings developed in some remarkable MANNER in comparison with the other species of the
0150  ual characters, when displayed in any unusual MANNER. The term, secondary sexual characters, used
0151  ters, whether or not displayed in any unusual MANNER, of which fact I think there can be little d
0151  or organ developed in a remarkable degree or MANNER in any species, the fair presumption is that
0153  a part has been developed in an extraordinary MANNER in any one species, compared with the other
0154  on of the individuals varying in the required MANNER and degree, and by the continued rejection o
0155  me to differ. Or to state the case in another MANNER: the points in which all the species of a ge
0156  widely different habits, in exactly the same MANNER: and as these so called generic characters h
0157  species: again in fossorial hymenoptera, the MANNER of neuration of the wings is a character of
0158  is developed in a species in an extraordinary MANNER in comparison with the same part in its cong
0159  , and a consequent tendency to vary in a like MANNER, but to three separate yet closely related a
0161  they would occasionally vary in an analogous MANNER; so that a variety of one species would rese
0167  e and under comestication, in this particular MANNER, so as often to become striped like other sp
0169  an extraordinary size or in an extraordinary MANNER, in comparison with the same part or organ i
0169  or to the organ, in however extraordinary a a MANNER it may be developed. Species inheriting near
0175  y or indirectly related in the most important MANNER to other organic beings, we must see that th
0181  and improved in structure in a corresponding MANNER. Therefore, I can see no difficulty, more es
0185  bits, in its astonishing power of diving, its MANNER of swimming, and of flying when unwillingly
0191  for some quite distinct purpose: in the same MANNER as, on the view entertained by some naturali
0194  as sometimes modified in very nearly the same MANNER two parts in two organic beings, which owe b
0211  kled and stroked them with a hair in the same MANNER, as well as I could, as the ants do with the
0211  even the quite young aphides behaved in this MANNER, showing that the action was instinctive, an
0213  perience by the young, and in nearly the same MANNER by each individual, performed with eager del
0214  , resemble natural instincts, which in a like MANNER become curiously blended together, and for a
0226  s obvious that the Melipona saves wax by this MANNER of building: for the flat walls between the
0231  being largely gnawed away on both sides. The MANNER in which the bees build is curious; they alw
0231  being about twice as thick. By this singular MANNER of building, strength is continually given t
0242  them accidental, variations, which are in any MANNER profitable, without exercise or habit having
0243  lines its nest with mud, in the same peculiar MANNER as does our British thrush: how it is that t
0251  notice. It is notorious in how complicated a MANNER the species of Pelargonium, Fuchsia, Calceol
0265  h, is affected by sterility in a very similar MANNER. In the one case, the conditions of life hav
0269  the same food; treats them in nearly the same MANNER, and does not wish to alter their general ha
0271  em of this one variety must have been in some MANNER and in some degree modified. From these fact
0276  hybrid produced from this cross. In the same MANNER as in grafting trees, the capacity of one sp
0279  e of each species depends in a more important MANNER on the presence of other already defined org
0285  se facts impresses my mind almost in the same MANNER as does the vain endeavour to grapple with t
0300  ould be preserved in an excessively imperfect MANNER in the formations which we suppose to be the
0302  of whole groups of Allied Species. The abrupt MANNER in which whole groups of species suddenly app
0304  tidal limits to 50 fathoms; from the perfect MANNER in which specimens are preserved in the olde
0306  iculty, which is much graver. I allude to the MANNER in which numbers of species of the same grou
```

Page **************************************(Key Word)**************************************

```
age ****************************************( Key Word )****************************************
```

```
810  s which now exist or have existed: the sudden MANNER in which whole groups of species appear in ou
810  e gravest nature. We see this in the plainest  MANNER by the fact that all the most eminent palaeon
322  have exterminated in a correspondingly rapid   MANNER many of the old inhabitants: and the forms wh
322  ority in common. Thus, as it seems to me, the  MANNER in which single species and whole groups of s
323  change at distant points in the same parallel  MANNER. We may doubt whether they have thus changed:
328  enera, yet the species themselves differ in a  MANNER very difficult to account for, considering th
332  ted into one great family, in nearly the same  MANNER as has occurred with ruminants and pachyderms
333  forms, seem to me explained in a satisfactory  MANNER. And they are wholly inexplicable on any othe
334  s. i need give only one instance, namely, the  MANNER in which the fossils of the Devonian system,
337  ovement has affected in a marked and sensible  MANNER the organisation of the more recent and victo
337  the highest molluscs. From the extraordinary  MANNER in which European productions have recently s
339  professor Owen has shown in the most striking  MANNER that most of the fossil mammals, buried there
340  t present, to the northern half. In a similar  MANNER we know from Falconer and Cautley's discoveri
340  d descendants still differ in nearly the same  MANNER and degree. But after very long intervals of
347  gration, are related in a close and important  MANNER to the differences between the productions of
349  orth to south never fails to be struck by the  MANNER in which successive groups of beings, specifi
349  alike, with eggs coloured in nearly the same   MANNER. The plains near the Straits of Magellan are
360  tens tried similar ones, but in a much better  MANNER, for he placed the seeds in a box in the actu
360  ds may be occasionally transported in another  MANNER. Drift timber is thrown up on most islands, e
368  d, seem to me to explain in so satisfactory a  MANNER the present distribution of the Alpine and Ar
376  gh related to each other in a most remarkable  MANNER. This brief abstract applies to plants alone:
379  ess powerful southern forms. Just in the same  MANNER as we see at the present day, that very many
383  e, but allied species prevail in a remarkable  MANNER throughout the world. I well remember, when f
383  explained by their having become fitted, in a  MANNER highly useful to them, for short and frequent
400  d on a small scale, yet in a most interesting  MANNER, within the limits of the same archipelago. T
400  i have elsewhere shown, in a quite marvellous  MANNER, by very closely related species: so that the
408  marshes, and deserts, are in so mysterious a   MANNER linked together by affinity, and are likewise
416  the angle of the jaws of the Marsupials, the   MANNER in which the wings of insects are folded, mer
422  per, in the diagram, but in much too simple a  MANNER. If a branching diagram had not been used, an
426  mere inflection of the angle of the jaw, the   MANNER in which an insect's wing is folded, whether
440  shows this to be the case in an unmistakeable  MANNER. So again the two main divisions of cirripede
442  these facts as necessarily contingent in some  MANNER on growth. But there is no obvious reason why
448  d be modified at a very early age in the same  MANNER with its parents, in accordance with their si
448  tly to be constructed in a slightly different  MANNER, then, on the principle of inheritance at cor
465  es have changed; and they have changed in the  MANNER which my theory requires, for they have chang
465  r they have changed slowly and in a graduated  MANNER. We clearly see this in the fossil remains fr
467  ture, and thus accumulate them in any desired  MANNER. He thus adapts animals and plants for his ow
473  cended from a striped progenitor, in the same  MANNER as the several domestic breeds of pigeon have
474  eation why a part developed in a very unusual  MANNER in any one species of a genus, and therefore,
474  t a part may be developed in the most unusual  MANNER, like the wing of a bat, and yet not be more
478  an islands being related in the most striking  MANNER to the plants and animals of the neighbouring
485  t, we shall have to treat species in the same  MANNER as those naturalists treat genera, who admit
487  ll surely be enabled to trace in an admirable  MANNER the former migrations of the inhabitants of t
489  f any kind to a far distant futurity: for the  MANNER in which all organic beings are grouped, show
489  , and dependent on each other in so complex a  MANNER, have all been produced by laws acting around
115  stance: in the last edition of Dr. Asa Gray's  MANUAL of the Flora of the Northern United States, 2
289  table published in the Supplement to Lyell's   MANUAL, will bring home the truth, how accidental an
304  now we may read in the Supplement to Lyell's   MANUAL, published in 1858, clear evidence of the exi
056  or where, if we may use the expression, the    MANUFACTORY of species has been active, we ought gener
056  been active, we ought generally to find the    MANUFACTORY still in action, more especially as we hav
169  variation and differentiation; or where the    MANUFACTORY of new specific forms has been actively at
470  hould present many varieties: for where the    MANUFACTORY of species has been active, we might expec
018  le that man sufficiently civilized to have     MANUFACTURED pottery existed in the valley of the Nile
057  manufacturing, many of the species already     MANUFACTURED still to a certain extent resemble varieti
242  weapons, and not by acquired knowledge and     MANUFACTURED instruments, a perfect division of labour
056  ave every reason to believe the process of     MANUFACTURING new species to be a slow one. And this ce
057  t species greater than the average are now     MANUFACTURING, many of the species already manufactured
053  to add, that after having carefully read my    MANUSCRIPT, and examined the tables, he thinks that th
002  cellent memoir, some brief extracts from my    MANUSCRIPTS. This Abstract, which I now publish, must
1309 ince the earliest silurian period. The coloured MAP appended to my volume on Coral Reefs, led me t
0401 ted from each other than they appear to be on a MAP. Nevertheless a good many species, both those
1238 s of cattle wish the flesh and fat to be well   MARBLED together; the animal has been slaughtered, b
0378 h america, which must have lain on the line of  MARCH. But I do not doubt that some temperate produ
1368 e productions will at a very late period have   MARCHED a little further north, and subsequently hav
0221 led with their masters leaving the nest, and   MARCHING along the same road to a tall Scotch fir tre
0165 n lord Moreton's famous hybrid from a chestnut  MARE and male quagga, the hybrid, and even the pure
0165 pure offspring subsequently produced from the   MARE by a black Arabian sire, were much more plainl
0258 with a female ass, and then a male ass with a   MARE: these two species may then be said to have be
0275 le, which is the offspring of the male ass and  MARE, is more like an ass, than is the hinny, which
0183 and at other times standing stationary on the   MARGIN of water, and then dashing like a kingfisher
0230 bic plate which stands on the extreme growing   MARGIN, or the two plates, as the case may be: and t
0232 agonal walls of a single cell, or the extreme   MARGIN of the circumferential rim of a growing comb,
0233 theory; namely, that the cells on the extreme   MARGIN of wasp combs are sometimes strictly hexagona
0394 ipelago, the Bonin Islands, the Caroline and    MARIANNE Archipelagoes, and Mauritius, all possess th
0174 anges in the form of the land and of climate,   MARINE areas now continuous must often have existed
0283 iffs rounded boulders, all thickly clothed by   MARINE productions, showing how little they are abra
0290 west coast, which is inhabited by a peculiar    MARINE fauna, tertiary beds are so scantily develope
0290 no record of several successive and peculiar    MARINE faunas will probably be preserved to a distan
0293 e been correlated with perfect accuracy. With   MARINE animals of all kinds, we may safely infer a l
0294 sissippi, within that limit of depth at which   MARINE animals can flourish; for we know what vast g
0298 been local or confined to some one spot. Most   MARINE animals have a wide range: and we have seen t
0298 sent varieties: so that with shells and other  MARINE animals, it is probably those which have had
0301 served in any one formation. Very many of the  MARINE inhabitants of the archipelago now range thou
0303 family: as these animals have huge bones, are  MARINE, and range over the world; the fact of not a
0305 y enclosed basin, in which any great group of  MARINE animals might be multiplied: and here they wo
0312 ippi in Sicily, the successive changes in the  MARINE inhabitants of that island have been many and
```

```
Page ****************************************( Key Word )****************************************
```

Page ***(Key Word)***

```
0314  llies on the continent of Europe, whereas the  MARINE  shells and birds have remained unaltered. We
0314  re highly organised productions compared with    MARINE  and lower productions, by the more complex m
0323  d. these observations, however, relate to the    MARINE  inhabitants of distant parts of the world: w
0324  ng one of the later tertiary stages. When the    MARINE  forms of life are spoken of as having change
0324  very strict geological sense; for if all the     MARINE  animals which live at the present day in Eur
0324  ink, be little doubt that all the more modern     MARINE  formations, namely, the upper pliocene, the
0325  new ones, cannot be owing to mere changes in     MARINE  currents or other causes more or less local
0326  abitants of distinct continents than with the    MARINE  inhabitants of the continuous sea. We might
0340  d be given in relation to the distribution of    MARINE  animals. On the theory of descent with modif
0348  ninc to the sea, we find the same law. No two    MARINE  faunas are more distinct, with hardly a fish
0348  nd totally distinct fauna. So that here three    MARINE  faunas range far northward and southward, ir
0348  ace we meet with no well defined and distinct    MARINE  faunas. Although hardly one shell, crab or f
0356  fluential: a narrow isthmus now separates two    MARINE  faunas; submerge it, or let it formerly have
0358  ribution, such as the great difference in the    MARINE  faunas on the opposite sides of almost every
0372  rs of the sea, a slow southern migration of a    MARINE  fauna, which during the Pliocene or even a s
0372  ana's admirable work), of some fish and other    MARINE  animals, in the Mediterranean and in the sea
0376  n the distribution of terrestrial animals. In    MARINE  productions, similar cases occur; as an exam
0378  f plants, a few terrestrial animals, and some    MARINE  productions, migrated during the glacial per
0384  to fresh water, so that we may imagine that a    MARINE  member of a fresh water group might travel f
0390  ery land bird, but only two out of the eleven    MARINE  birds, are peculiar; and it is obvious that
0391  e birds, are peculiar; and it is obvious that    MARINE  birds could arrive at these islands more eas
0136  attempt and rarely or never flying. As with     MARINERS shipwrecked near a coast, it would have bee
0061  by the term of Natural Selection, in order to    MARK   its relation to man's power of selection. We
0064  rate of natural increase: it will be under the   MARK   to assume that it breeds when thirty years ol
0282  erent observers on separate formations, and to   MARK   how each author attempts to give an inadequat
0282  ast, when formed of moderately hard rocks, and   MARK   the process of degradation. The tides in most
0293  ng considerations. Although each formation may   MARK   a very long lapse of years, each perhaps is a
0315  equal. Each formation, on this view, does not    MARK   a new and complete act of creation, but only
0011  climate to another. In animals it has a more     MARKED  effect; for instance, I find in the domestic
0014  conclude that very many of the most strongly     MARKED  domestic varieties could not possibly live t
0016  ants of aboriginally distinct species. If any    MARKED  distinction existed between domestic races a
0025  imagined aboriginal stocks were coloured and    MARKED  like the rock pigeon, although no other exis
0025  o other existing species is thus coloured and    MARKED, so that in each separate breed there might
0031  ervals of months, and the sheep are each time    MARKED  and classed, so that the very best may ultim
0039  d. the common goose has not given rise to any    MARKED  varieties; hence the Thoulouse and the commo
0047  be by a majority of naturalists, for few well    MARKED  and well known varieties can be named which
0048  p obligation for assistance of all kinds, has    MARKED  for me 182 British plants, which are general
0049  der our British red grouse as only a strongly    MARKED  race of a Norwegian species, whereas the gre
0049  o beat the air. Many of the cases of strongly    MARKED  varieties or doubtful species well deserve c
0051  cies; or, again, between sub species and well    MARKED  varieties, or between lesser varieties and t
0052  permanent, as steps leading to more strongly    MARKED  and more permanent varieties: and at these l
0052  n definite directions. Hence I believe a well    MARKED  variety may be justly called an incipient sp
0053  ften give rise to varieties sufficiently well    MARKED  to have been recorded in botanical works. He
0054  s in individuals, which oftenest produce well    MARKED  varieties, or, as I consider them, incipient
0055  ion. From looking at species as only strongly    MARKED  and well defined varieties, I was led to ant
0056  on on the view that species are only strongly    MARKED  and permanent varieties: for wherever many s
0056  rion by which to distinguish species and well    MARKED  varieties; and in those cases in which inter
0058  ed ranges. For instance, Mr. H. C. Watson has    MARKED  for me in the well sifted London Catalogue o
0058  nge, as have those very closely allied forms,    MARKED  for me by Mr. Watson as doubtful species, bu
0060  ntitled to hold, if the existence of any well    MARKED  varieties be admitted. But the mere existenc
0060  f individual variability and of some few well    MARKED  varieties, though necessary as the foundatio
0067  here could be no choking from other plants, I    MARKED  all the seedlings of our native weeds as the
0084  anges in progress, until the hand of time has    MARKED  the long lapse of ages, and then so imperfec
0089  to their standard of beauty, might produce a     MARKED  effect. I strongly suspect that some well kn
0111  in the first place, varieties, even strongly    MARKED  ones, though having somewhat of the characte
0111  ble species throughout nature presenting well    MARKED  differences; whereas varieties, the supposed
0111  upposed prototypes and parents of future well    MARKED  species, present slight and ill defined diff
0117  hes one of the horizontal lines, and is there    MARKED  by a small numbered letter, a sufficient amo
0117  been accumulated to have formed a fairly well    MARKED  variety, such as would be thought worthy of
0117  is supposed to have produced two fairly well    MARKED  varieties, namely a1 and m1. These two varie
0120  all, these three forms may still be only well    MARKED  varieties or they may have arrived at the d
0120  and simplified manner), we get eight species,    MARKED  by the letters between a14 and m14, all desc
0120  ter ten thousand generations, either two well    MARKED  varieties (w10 and z10) or two species, acco
0121  urteen thousand generations, six new species    MARKED  by the letters n14 to z14, are supposed to h
0121  arieties and species. The other nine species   (MARKED  by capital letters) of our original genus, m
0123  . of the eight descendants from (A) the three   MARKED  a14, q14, p14, will be nearly related from h
0125  ging dotted lines to be very great, the forms    MARKED  a14 to p14, those marked b14 and f14, and th
0125  ery great, the forms marked a14 to p14, those    MARKED  b14 and f14, and those marked o14 to m14, wi
0125  4 to p14, those marked b14 and f14, and those    MARKED  o14 to m14, will form three very distinct gr
0133  view that species of all kinds are only well    MARKED  and permanent varieties. Thus the species of
0155  ut on the view of species being only strongly   MARKED  and fixed varieties, we might surely expect
0161  is other species being on my view only a well    MARKED  and permanent variety. But characters thus g
0176  e case of varieties intermediate between well    MARKED  varieties in the genus Balanus. And it would
0177  enting on an average a greater number of well    MARKED  varieties than do the rarer species. I may f
0199  tween the races of man, which are so strongly    MARKED; I may add that some little light can appare
0229  the bees had worked less quickly. In one well    MARKED  instance, I put the comb back into the hive,
0257  t and general appearance, and having strongly    MARKED  differences in every part of the flower, eve
0272  cts. Gartner, whose strong wish was to draw a    MARKED  line of distinction between species and vari
0333  in this case the genera, at the early period    MARKED  VI., would differ by a lesser number of char
0337  this process of improvement has affected in a    MARKED  and sensible manner the organisation of the
0357  producing oceans such sunken islands are now    MARKED, as I believe, by rings of coral or atolls s
0379  spheres, now exist in their new homes as well    MARKED  varieties or as distinct species. It is a re
0410  than the higher; but there are in both cases    MARKED  exceptions to the rule. On my theory these s
0456  ement, with the grades of acquired difference    MARKED  by the terms varieties, species, genera, fam
0469  and slight varieties; or between more plainly    MARKED  varieties and sub species, and species. Let
0469  y. on the view that species are only strongly    MARKED  and permanent varieties, and that each speci
```

Page ***(Key Word)***

```
age ****************************************( Key Word )****************************************
```

```
473  me coloured flowers? If species are only well MARKED varieties, of which the characters have becom
475  er animals to suffer. If species be only well MARKED and permanent varieties, we can at once see w
481  en, or can be, drawn between species and well MARKED varieties. It cannot be maintained that speci
485  the only distinction between species and well MARKED varieties is, that the latter are known, or b
119  t regular intervals by small numbered letters MARKING the successive forms which have become suffi
191  nd the loop like course of the arteries still MARKING in the embryo their former position. But it
317  , and ultimately thins out in the upper beds, MARKING the decrease and final extinction of the spe
025  dency to revert to the very same colours and MARKINGS. Or, secondly, that each breed, even the pur
162  a case of reversion, from the number of the MARKINGS, which are correlated with the blue tint, an
025  the wings chequered with black. These several MARKS do not occur together in any other species of
025  king thoroughly well bred birds, all the above MARKS, even to the white edging of the outer tail f
025  hich is blue or has any of the above specified MARKS, the mongrel offspring are very apt suddenly
027  o the rock pigeon; the blue colour and various MARKS occasionally appearing in all the breeds, bot
160  dded near their bases with white. As all these MARKS are characteristic of the parent rock pigeon,
160  sion, because, as we have seen, these coloured MARKS are eminently liable to appear in the crossed
160  appearance of the slaty blue, with the several MARKS, beyond the influence of the mere act of cros
162  t have inferred this, from the blue colour and MARKS so often appearing when distinct breeds of di
166  f a bluish colour, with certain bars and other MARKS; and when any breed assumes by simple variati
166  variation a bluish tint, these bars and other MARKS invariably reappear; but without any other ch
166  strong tendency for the blue tint and bars and MARKS to reappear in the mongrels. I have stated th
318  o taper more gradually at its upper end, which MARKS the progress of extermination, than at its lo
318  of extermination, than at its lower end, which MARKS the first appearance and increase in numbers
373  points 900 miles apart, glaciers have left the MARKS of their former low descent; and in Sikkim, D
483  f mammals, were they created bearing the false MARKS of nourishment from the mother's womb? Althou
063  se of food, and no prudential restraint from MARRIAGE. Although some species may be now increasing
386  what enormous ranges many fresh water and even MARSH species have, both over continents and in the
025  see this on every mountain, in every lake and MARSH. For Alpine species, excepting in so far as t
406  widely, such facts, as alpine, lacustrine, and MARSH productions being related (with the exception
041  est importance to success. On this principle MARSHALL has remarked, with respect to the sheep of p
423  sing varieties: thus the great agriculturist MARSHALL says the horns are very useful for this purp
346  rts, lofty mountains, grassy plains, forests, MARSHES, lakes, and great rivers, under almost every
404  s. so it is with the inhabitants of lakes and MARSHES, excepting in so far as great facility of tr
408  of the plains and mountains, of the forests, MARSHES, and deserts, are in so mysterious a manner
477  cold, on mountain and lowland, on deserts and MARSHES, most of the inhabitants within each great c
091  ther hares or rabbits, and another hunting on MARSHY ground and almost nightly catching woodcocks
430  ts relations are general, and not to any one MARSUPIAL species more than to another. As the points
430  izcacha, branched off from some very ancient MARSUPIAL, which will have had a character in some de
430  not be specially related to any one existing MARSUPIAL, but indirectly to all or nearly all Marsup
116  ubted, for instance, whether the Australian MARSUPIALS, which are divided into groups differing bu
339  ian caves were closely allied to the living MARSUPIALS of that continent. In South America, a simi
340  pretended that it is an immutable law that MARSUPIALS should have been chiefly or solely produced
340  pe in ancient times was peopled by numerous MARSUPIALS; and I have shown in the publications above
416  inflection of the angle of the jaws of the MARSUPIALS, the manner in which the wings of insects a
430  nts, the bizcacha is most nearly related to MARSUPIALS; but in the points in which it approaches t
430  s the points of affinity of the bizcacha to MARSUPIALS are believed to be real and not merely adap
430  e intermediate with respect to all existing MARSUPIALS; or that both Rodents and Marsupials branch
430  isting Marsupials; or that both Rodents and MARSUPIALS branched off from a common progenitor, and
430  supial, but indirectly to all or nearly all MARSUPIALS, from having partially retained the charact
430  ber of the group. On the other hand, of all MARSUPIALS, as Mr. Waterhouse has remarked, the phasco
360  e of allied forms on the same continent, of MARSUPIALS in Australia, of edentata in America, and
360  germinate. Subsequently to my experiments, M. MARTENS tried similar ones; but in a much better man
165  ve been the product of a zebra; and Mr. W. C. MARTIN, in his excellent treatise on the horse, has
073  norance, and so high our presumption, that we MARVEL when we hear of the extinction of an organic
320  rprise at the rarity of a species, and yet to MARVEL at extinction; if we must marvel, let it be a
322  the theory of natural selection. We need not MARVEL at extinction: if we must marvel, let it be a
322  we need not marvel at extinction; if we must MARVEL, let it be at our presumption in imagining fo
403  se considerations I think we need not greatly MARVEL at the endemic and representative species, wh
472  roductions from another land. Nor ought we to MARVEL if all the contrivances in nature be not, as
472  bhorrent to our ideas of fitness. We need not MARVEL at the sting of the bee causing the bee's own
475  cquired through natural selection we need not MARVEL at some instincts being apparently not perfec
318  a definite duration. No one I think can have MARVELLED more at the extinction of species, than I h
479  be so unlike the adult forms. We may cease MARVELLING at the embryo of an air breathing mammal or
151  one genus, Pyrgoma; these valves present a MARVELLOUS amount of diversification: the homologous v
172  natural selection? What shall we say to so MARVELLOUS an instinct as that which leads the bee to
364  f thousands of years, it would I think be a MARVELLOUS fact if many plants had not thus become wid
400  nted, as I have elsewhere shown, in a quite MARVELLOUS manner, by very closely related species; so
474  d natural selection. Glancing at instincts, MARVELLOUS as some are, they offer no greater difficul
231  shall understand how they work, by supposing MASONS first to pile up a broad ridge of cement, and
231  th, very thin wall is left in the middle; the MASONS always piling up the cut away cement, and add
193  e in plants, the very curious contrivance of a MASS of pollen grains, borne on a foot stalk with a
225  al sizes, and are aggregated into an irregular MASS. But the important point to notice, is that th
234  egular, nearer together, and aggregated into a MASS, like the cells of the Melipona: for in this c
283  stamp of time, are good to show how slowly the MASS has been accumulated. Let him remember Lyell's
291  be inhabited by extremely few animals, and the MASS when upraised will give a most imperfect recor
054  cribed in any Flora be divided into two nearly equal MASSES, all those in the larger genera being placed
055  sects of two districts, into two nearly equal MASSES, the species of the larger genera on one side
073  re the visits of moths to remove their pollen MASSES and thus to fertilise them. I have, also, rea
173  mbedded and preserved to a future age only in MASSES of sediment sufficiently thick and extensive
173  of future degradation; and such fossiliferous MASSES can be accumulated only where much sediment i
285  le, in comparison with that which has removed MASSES of our palaeozoic strata, in parts ten thousa
286  ry deposits might have accumulated in thinner MASSES than elsewhere, the above estimate would be e
290  lated in extremely thick, solid, or extensive MASSES, in order to withstand the incessant action o
315  inc fossiliferous formations depends on great MASSES of sediment having been deposited on areas wh
021  the runt is a bird of great size, with long, MASSIVE beak and large feet; some of the sub breeds
090  rovided always that they retained strength to MASTER their prey at this or at some other period of
213  er can no more know that he points to aid his MASTER, than the white butterfly knows why she lays
```

```
age ****************************************( Key Word )****************************************
```

0214 way, by not coming in a straight line to his MASTER when called. Domestic instincts are sometimes
0143 bs. These tendencies, I do not doubt, may be MASTERED more or less completely by natural selection
0460 attempted to show how this difficulty can be MASTERED. With respect to the almost universal steril
0285 eet in thickness, as shown in Prof. Ramsay's MASTERLY memoir on this subject. Yet it is an admirab
0219 rmine the migration, and actually carry their MASTERS in their jaws. So utterly helpless are the s
0219 rs in their jaws. So utterly helpless are the MASTERS, that when Huber shut up thirty of them with
0220 lack and not above half the size of their red MASTERS, so that the contrast in their appearance is
0220 slaves occasionally come out, and like their MASTERS are much agitated and defend the nest: when
0220 sed, the slaves work energetically with their MASTERS in carrying them away to a place of safety.
0220 siders them as strictly household slaves. The MASTERS, on the other hand, may be constantly seen b
0221 nd i observed a few slaves mingled with their MASTERS leaving the nest, and marching along the sam
0221 zerland the slaves habitually work with their MASTERS in making the nest, and they alone open and
0221 s. this difference in the usual habits of the MASTERS and slaves in the two countries, probably de
0221 as a most interesting spectacle to behold the MASTERS carefully carrying, as Huber has described,
0223 e early part of the summer extremely few. The MASTERS determine when and where a new nest shall be
0223 t shall be formed, and when they migrate, the MASTERS carry the slaves. Both in Switzerland and En
0223 ave the exclusive care of the larvae, and the MASTERS alone go on slave making expeditions. In Sw
0223 ng expeditions. In Switzerland the slaves and MASTERS work together, making and bringing materials
0223 ollect food for the community. In England the MASTERS alone usually leave the nest to collect buil
0223 mselves, their slaves and larvae. So that the MASTERS in this country receive much less service fr
0318 ooth of a horse embedded with the remains of MASTODON, Megatherium, Toxodon, and other extinct mor
0323 these anomalous monsters coexisted with the MASTODON and Horse, it might at least have been infer
0334 offer exceptions to the rule. For instance, MASTODONS and elephants, when arranged by Dr. Falcone
0034 passages in Pliny. The savages in South Africa MATCH their draught cattle by colour, as do some of
0042 om their nocturnal rambling habits, cannot be MATCHED, and, although so much valued by women and y
0238 atching which individual bulls and cows, when MATCHED, produced oxen with the longest horns; and y
0040 structure, or takes more care than usual in MATCHING his best animals and thus improves them, and
0028 ds, that male and female pigeons can be easily MATED for life; and thus different breeds can be ke
0042 one breed of the same species. Pigeons can be MATED for life, and this is a great convenience to
0037 roduced such splendid results from such poor MATERIALS! but the art, I cannot doubt, has been sim
0040 s obviously favourable, as freely giving the MATERIALS for selection to work on; not that mere in
0045 are highly important for us, as they afford MATERIALS for natural selection to accumulate, in the
0096 ject with extreme brevity, though I have the MATERIALS prepared for an ample discussion. All verte
0220 her hand, may be constantly seen bringing in MATERIALS for the nest, and food of all kinds. During
0223 d masters work together, making and bringing MATERIALS for the nest: both, but chiefly the slaves
0223 e usually leave the nest to collect building MATERIALS and food for themselves, their slaves and b
0233 gon being built, as in its construction more MATERIALS would be required than for a cylinder. As r
0455 explained in a former chapter, by which the MATERIALS forming any part or structure, if not usefu
0203 y this is for the good of the community; and MATERNAL love or maternal hatred, though the latter f
0203 good of the community; and maternal love or MATERNAL hatred, though the latter fortunately is mos
0217 advantage having been taken of the mistaken MATERNAL instinct of another bird, than by their own
0218 the European cuckoo has not utterly lost all MATERNAL love and care for her own offspring. The occ
0172 y anticipated the discoveries of profound MATHEMATICIANS? Fourthly, how can we account for species
0224 out enthusiastic admiration. We hear from MATHEMATICIANS that bees have practically solved a recor
0116 stomach by being adapted to digest vegetable MATTER alone, or flesh alone, draws most nutriment f
0197 ossibly be due to the direct action of putrid MATTER: but we should be very cautious in drawing an
0266 t the foregoing remarks go to the root of the MATTER: no explanation is offered why an organism, w
0284 y of the rate of accumulation of the degraded MATTER, probably offers the best evidence of the lap
0307 presence of phosphatic nodules and bituminous MATTER in some of the lowest azoic rocks, probably f
0358 stead of consisting of mere piles of volcanic MATTER. I must now say a few words on what are calle
0411 ri one to feed on flesh, another on vegetable MATTER, and so on; but the case is widely different
0453 gans, by supposing that they serve to excrete MATTER in excess, or injurious to the system; but ca
0454 laws of growth, but in order to excrete horny MATTER, as that the rudimentary nails on the fin of
0488 er with what we know of the laws impressed on MATTER by the Creator, that the production and exti
0193 o the electric apparatus, and yet do not, as MATTEUCHI asserts, discharge any electricity, we must
0258 it even between forms so closely related (as MATTHIOLA annua and glabra) that many botanists rank
0005 lassification or mutual affinities, both when MATURE and in an embryonic condition. In the last ch
0011 a will almost certainly entail changes in the MATURE animal. In monstrosities, the correlations be
0014 ould appear only in the offspring when nearly MATURE: peculiarities in the silkworm are known to a
0086 wholly different from those which concern the MATURE insect. These modifications will no doubt aff
0436 male; and in flowers, that organs, which when MATURE become extremely different, are at an early s
0439 certain organs in the individual, which when MATURE become widely different and serve for differe
0439 mble each other much more closely than do the MATURE insects! but in the case of larvae, the embry
0441 the caterpillar. In some cases, however, the MATURE animal is generally considered as lower in th
0443 ving sometimes a higher organisation than the MATURE animal, into which it is developed. I believe
0445 , and tumblers. Now some of these birds, when MATURE, differ so extraordinarily in length and form
0447 ill differ greatly from the fore limbs in the MATURE animal; the limbs in the latter having underg
0447 organ, such influence will mainly affect the MATURE animal, which has come to its full powers of
0447 produced will be inherited at a corresponding MATURE age. Whereas the young will remain unmodified
0448 he young or embryo would closely resemble the MATURE parent form. We have seen that this is the ru
0480 th; and we may believe, that the teeth in the MATURE animal were reduced, during successive genera
0457 al embryo of the homologous parts, which when MATURED will become widely different from each other
0035 ngland have increased in weight and in early MATURITY, compared with the stock formerly kept in th
0063 ds, of which on an average only one comes to MATURITY, may be more truly said to struggle with the
0251 d made vigorous growth and rapid progress to MATURITY, and bore good seed, which vegetated freely,
0364 ed falling on favourable soil, and coming to MATURITY! But it would be a great error to argue that
0455 will generally be when the being has come to MATURITY and to its full powers of action, the princ
0480 ly act on each creature, when it has come to MATURITY and has to play its full part in the struggl
0393 een introduced into Madeira, the Azores, and MAURITIUS, and have multiplied so as to become a nuis
0394 the Caroline and Marianne Archipelagoes, and MAURITIUS, all possess their peculiar bats. Why, it m
0435 of an upper lip, mandibles, and two pairs of MAXILLAE. Analogous laws govern the construction of a
0436 had an upper lip, mandibles, and two pair of MAXILLAE, these parts being perhaps very simple in fo
0056 nd that large genera have often come to their MAXIMA, declined, and disappeared. All that we want
0110 nowing that any one region has as yet got its MAXIMUM of species. Probably no region is as yet ful
0247 y degree of sterility. He always compares the MAXIMUM number of seeds produced by two species wher
0284 countries! Professor Ramsay has given me the MAXIMUM thickness, in most cases from actual measure

Page ***(Key Word)***

```
0316  ncrease in number, till the group reaches its  MAXIMUM, and then, sooner or later, it gradually dec
0366  ns. By the time that the cold had reached its  MAXIMUM, we should have a uniform arctic fauna and f
0410  es and groups of species have their points of  MAXIMUM development. Groups of species, belonging ei
0001  steadily pursued the same object. I hope that I  MAY be excused for entering on these personal deta
0003  limited sense, as we shall hereafter see, this  MAY be true; but it is preposterous to attribute t
0004  stication, afforded the best and safest clue. I  MAY venture to express my conviction of the high v
0007  pounded by Andrew Knight, that this variability  MAY be partly connected with excess of food. It se
0008  f life the causes of variability, whatever they  MAY be, generally act; whether during the early or
0008  ect that the most frequent cause of variability  MAY be attributed to the male and female reproduct
0009  he reproduction of animals under confinement. I  MAY just mention that carnivorous animals, even fr
0009  f all the choicest productions of the garden. I  MAY add, that as some organisms will breed most fr
0010  fact, sports support my view, that variability  MAY be largely attributed to the ovules or pollen,
0011  ture. Nevertheless some slight amount of change  MAY, I think, be attributed to the direct action o
0011  n the wild duck; and I presume that this change  MAY be safely attributed to the domestic duck flyi
0011  efly mentioned. I will here only allude to what  MAY be called correlation of growth. Any change in
0013  the father and child, we cannot tell whether it  MAY not be due to the same original cause acting o
0013  inherited, less strange and commoner deviations  MAY be freely admitted to be inheritable. Perhaps
0013  lone. A much more important rule, which I think  MAY be trusted, is that, at whatever period of lif
0014  ecularity, and not to its primary cause, which  MAY have acted on the ovules or male element; in n
0014  . having alluded to the subject of reversion, I  MAY here refer to a statement often made by natura
0014  ld be great difficulty in proving its truth: we  MAY safely conclude that very many of the most str
0015  erations, would be opposed to all experience. I  MAY add, that when under nature the conditions of
0018  or Australia, who possess a semi domestic dog,  MAY not have existed in Egypt? The whole subject m
0018  ct must, I think, remain vague: nevertheless, I  MAY, without here entering on any details, state t
0020  exaggerated. There can be no doubt that a race  MAY be modified by occasional crosses, if aided by
0021  developed crop, which it glories in inflating,  MAY well excite astonishment and even laughter. Th
0024  abnormal in other parts of their structure, we  MAY look in vain throughout the whole great family
0026  dency, for all that we can see to the contrary,  MAY be transmitted undiminished for an indefinite
0027  raphical sub species. In favour of this view, I  MAY add, firstly, that C. livia, or the rock pigeo
0029  accumulated during many successive generations.  MAY not those naturalists who, knowing far less of
0029  tic races have descended from the same parents,  MAY they not learn a lesson of caution, when they
0029  from several allied species. Some little effect  MAY, perhaps, be attributed to the direct action o
0030  of the wild Dipsacus; and this amount of change  MAY have suddenly arisen in a seedling. So it has
0030  tain directions useful to him. In this sense he  MAY be said to make for himself useful breeds. The
0031  it is the magician's wand, by means of which he  MAY summon into life whatever form and mould he pl
0031  time marked and classed, so that the very best  MAY ultimately be selected for breeding. What Engl
0032  incomitable perseverance, he will succeed, and  MAY make great improvements; if he wants any of th
0032  eadily increasing size of the common gooseberry  MAY be quoted. We see an astonishing improvement i
0033  from each other chiefly in these characters. It  MAY be objected that the principle of selection ha
0033  een published on the subject; and the result, I  MAY add, has been, in a corresponding degree, rapi
0034  rses under a certain size was ordered, and this  MAY be compared to the roguing of plants by nurser
0034  ut, for our purpose, a kind of Selection, which  MAY be called Unconscious, and which results from
0035  ut little changed individuals of the same breed  MAY be found in less civilised districts, where th
0036  of the effects of a course of selection, which  MAY be considered as unconsciously followed, in so
0037  races have become blended together by crossing,  MAY plainly be recognised in the increased size an
0040  acteristic features of the breed, whatever they  MAY be. But the chance will be infinitely small of
0041  ty, all the individuals, whatever their quality  MAY be, will generally be allowed to breed, and th
0042  convenience to the fancier, for thus many races  MAY be kept true, though mingled in the same aviar
0042  ovement and formation of new breeds. Pigeons, I  MAY add, can be propagated in great numbers and at
0042  rs and at a very quick rate, and inferior birds  MAY be freely rejected, as when killed they serve
0042  s of the cat, the donkey, peacock, goose, etc.,  MAY be attributed in main part to selection not ha
0043  lly by that of correlation of growth. Something  MAY be attributed to the direct action of the cond
0045  y. again, we have many slight differences which  MAY be called individual differences, such as are
0045  n the offspring from the same parents, or which  MAY be presumed to have thus arisen, being frequen
0046  riability in these main nerves in Coccus, which  MAY almost be compared to the irregular branching
0046  tem of a tree. This philosophical naturalist, I  MAY add, has also quite recently shown that the mu
0046  s to rank as species and others as varieties. We  MAY instance Rubus, Rosa, and Hieracium amongst pl
0047  ose either that they do now somewhere exist, or  MAY formerly have existed; and here a wide door fo
0052  stage of difference to another and higher stage  MAY be, in some cases, due merely to the long cont
0052  rections. Hence I believe a well marked variety  MAY be justly called an incipient species; but whe
0052  es necessarily attain the rank of species. They  MAY whilst in this incipient state become extinct,
0052  in this incipient state become extinct, or they  MAY endure as varieties for very long periods, as
0053  . hence it is the most flourishing, or, as they  MAY be called, the dominant species; those which r
0056  he same genus have been formed, or where, if we  MAY use the expression, the manufactory of species
0057  the species of the smaller genera. Or the case  MAY be put in another way, and it may be said, tha
0057  . or the case may be put in another way, and it  MAY be said, that in the larger genera, in which a
0057  enus are equally distinct from each other; they  MAY generally be divided into sub genera, or secti
0058  divergence of Character, we shall see how this  MAY be explained; and how the lesser differences b
0061  d in every part of the organic world. Again, it  MAY be asked, how is it that varieties, which I ha
0062  we do not always bear in mind, that though food  MAY be now superabundant, it is not so at all seas
0062  rogeny. Two canine animals in a time of dearth,  MAY be truly said to struggle with each other whic
0063  which on an average only one comes to maturity,  MAY be more truly said to struggle with the plants
0063  oes, growing close together on the same branch,  MAY more truly be said to struggle with each other
0063  y birds, its existence depends on birds; and it  MAY metaphorically be said to struggle with other
0063  restraint from marriage. Although some species  MAY be now increasing, more or less rapidly, in nu
0065  e very few which do not annually pair. Hence we  MAY confidently assert, that all plants and animal
0066  a score, and yet in the same country the condor  MAY be the more numerous of the two: the Fulmar pe
0066  y protect its own eggs or young, a small number  MAY be produced, and yet the average stock be full
0066  rget that every single organic being around us  MAY be said to be striving to the utmost to increa
0067  usly increase to any amount. The face of Nature  MAY be compared to a yielding surface, with ten th
0069  ard and see a species decreasing in numbers, we  MAY feel sure that the cause lies quite as much in
0069  part indirectly by favouring other species, we  MAY clearly see in the prodigious number of plants
0070  confines of their range. For in such cases, we  MAY believe, that a plant could exist only where t
0075  nature. This is often the case with those which  MAY strictly be said to struggle with each other f
0076  different varieties of the medicinal leech. It  MAY even be doubted whether the varieties of any o
0077  of life. A corollary of the highest importance  MAY be deduced from the foregoing remarks, namely,
```

Page ***(Key Word)***

Page **(Key Word)**

0077 ckly clothed by other plants; so that the seeds MAY be widely distributed and fall on unoccupied
0078 utter desert, will competition cease. The land MAY be extremely cold or dry, yet there will be ce
0078 try amongst new competitors, though the climate MAY be exactly the same as in its former home, ye'
0079 struction. When we reflect on this struggle, we MAY console ourselves with the full belief, that
0080 hereditary tendency is. Under domestication, it MAY be truly said that the whole organisation bece
0081 f procreating their kind? On the other hand, we MAY feel sure that any variation in the least deg
0081 ange, and some species might become extinct. We MAY conclude, from what we have seen of the intime
0083 thus everywhere beaten some of the natives, we MAY safely conclude that the natives might have be
0083 odical and unconscious means of selection, what MAY not nature effect? Man can act only on externe
0083 thing for appearances, except in so far as they MAY be useful to any being. She can act on every
0083 ightest difference of structure or constitution MAY well turn the nicely balanced scale in the st
0084 ly bear the stamp of far higher workmanship? It MAY be said that natural selection is daily and he
0084 re apt to consider as very trifling importance, MAY thus be acted on. When we see leaf eating inse
0086 the pods on his cotton trees. Natural Selection MAY modify and adapt the larva of an insect to a s
0087 r species; and though statements to this effect MAY be found in works of natural history, I canno
0088 to them and to others, special means of defence MAY be given through means of sexual selection, ad
0088 e hooked jaw to the male salmon; for the shield MAY be as important for victory, as the sword or s
0089 s eminently attractive to all his hen birds. It MAY appear childish to attribute any effect to sue
0091 f intercrossing we shall soon have to return. I MAY add, that, according to Mr. Fierce, there are
0093 the physiological division of labour; hence we MAY believe that it would be advantageous to a pla
0094 ector feeding insects in our imaginary case: we MAY suppose the plant of which we have been slowly
0096 or the reproduction of their kind. This view, I MAY add, was first suggested by Andrew Knight. We
0096 y of plants are hermaphrodites. What reason, it MAY be asked, is there for supposing in these case
0099 antic tree covered with innumerable flowers, it MAY be objected that pollen could seldom be carrie
0100 separated, although the male and female flowers MAY be produced on the same tree, we can see that
0101 ch I am trying to investigate. Finally then, we MAY conclude that in many organic beings, a cross
0102 iod; for as all organic beings are striving, it MAY be said, to seize on each place in the economy
0105 any new variety to be slowly improved; and this MAY sometimes be of importance in the production w
0105 rable for the production of new species. But we MAY thus greatly deceive ourselves, for to ascerta
0107 ted in the natural scale. These anomalous forms MAY almost be called living fossils; they have ene
0109 e changed. Slow though the process of selection MAY be, if feeble man can do much by his powers of
0109 d with their physical conditions of life, which MAY be effected in the long course of time by nate
0109 , run a good chance of utter extinction. But we MAY go further than this: for as new forms are con
0111 and ill defined differences. Mere chance, as we MAY call it, might cause one variety to differ in
0112 s, or with shorter and shorter beaks. Again, we MAY suppose that at an early period one man prefee
0112 we see in man's productions the action of what MAY be called the principle of divergence, causing
0112 other and from their common parent. But how, it MAY be asked, can any analogous principle apply in
0113 sowing almost countless seeds; and thus, as it MAY be said, is striving its utmost to increase in
0114 the most different orders: nature follows what MAY be called a simultaneous rotation. Most of the
0114 t to be in any way peculiar in its nature), and MAY be said to be striving to the utmost to live w
0115 ned an advantage over the other natives; and we MAY, I think, at least safely infer that diversifi
0116 set more perfectly diversified in structure. It MAY be doubted, for instance, whether the Austral'
0116 on, which ought to have been much amplified, we MAY, I think, assume that the modified descendants
0117 d lines of unequal lengths proceeding from (A), MAY represent its varying offspring. The variatior
0117 s between the horizontal lines in the diagram, MAY represent each a thousand generations; but it
0118 e considerably from their common parent (A). We MAY continue the process by similar steps for any
0119 invariably prevail and multiply: a medium form MAY often long endure, and may or may not produce
0119 tiply: a medium form may often long endure, and MAY or may not produce more than one modified desc
0119 a medium form may often long endure, and may or MAY not produce more than one modified descendant s
0120 ; although the amount of divergent modification MAY have been increased in the successive generat'
0120 gram to be excessively small, these three forms MAY still be only well marked varieties; or they n
0120 ay still be only well marked varieties; or they MAY have arrived at the doubtful category of sub s
0121 rked by capital letters) of our original genus, MAY for a long period continue transmitting unalte
0122 d and parent do not come into competition, both MAY continue to exist. If then our diagram be assu
0122 eplaced by six (n14 to z14) new species. But we MAY go further than this. The original species of
0122 pring to the fourteen thousandth generation. We MAY suppose that only one (F), of the two species
0123 dely different from the other five species, and MAY constitute a sub genus or even a distinct genu
0124 d to represent a thousand generations, but each MAY represent a million or a hundred million gener
0126 . Looking still more remotely to the future, we MAY predict that, owing to the continued and stead
0126 subject in the chapter on Classification, but I MAY add that on this view of extremely few of the
0126 hough extremely few of the most ancient species MAY now have living and modified descendants, yet
0126 at the most remote geological period, the earth MAY have been as well peopled with many species of
0128 nature of the affinities of all organic beings MAY be explained. It is a truly wonderful fact, th
0129 y speaks the truth. The green and budding twigs MAY represent existing species; and those produced
0129 ies; and those produced during each former year MAY represent the long succession of extinct speci
0129 e former and present buds by ramifying branches MAY well represent the classification of all extir
0129 d off; and these lost branches of various sizes MAY represent those whole orders, families, and ge
0132 there dimly catch a faint ray of light, and we MAY feel sure that there must be some cause for ea
0132 , but perhaps rather more in that of plants. We MAY, at least, safely conclude that such influence
0132 ywhere throughout nature. Some little influence MAY be attributed to climate, food, etc.: thus, Ea
0133 d; but who can tell how much of this difference MAY be due to the warmest clad individuals having
0135 s, as well as any of the smaller quadrupeds. We MAY imagine that the early progenitor of the ostr'
0140 es under which they live is often overrated. We MAY infer this from our frequent inability to prec
0141 fectly fertile (a far severer test) under them, MAY be used as an argument that a large proportior
0141 ance, of a tropical and arctic wolf or wild dog MAY perhaps be mingled in our domestic breeds. The
0142 red to be than others. On the whole, I think we MAY conclude that habit, use, and disuse, have, in
0143 ly for the good of the young or larva, will, it MAY safely be concluded, affect the structure of t
0143 th the limbs. These tendencies, I do not doubt, MAY be mastered more or less completely by natural
0145 he corolla. Possibly, these several differences MAY be connected with some difference in the flow
0145 enest subject to peloria, and become regular. I MAY add, as an instance of this, and of a striking
0146 s of these two orders, is so far fetched, as it MAY at first appear: and if it be advantageous, na
0146 r: and if it be advantageous, natural selection MAY have come into play. But in regard to the diff
0146 cture, viewed by systematists as of high value, MAY be wholly due to unknown laws of correlated gr
0146 ee, of the slightest service to the species. We MAY often falsely attribute to correlation of grow
0146 y due to inheritance; for an ancient progenitor MAY have acquired through natural selection some q
0147 e been advanced, and likewise some other facts, MAY be merged under a more general principle, name

Page **(Key Word)**

```
148  degree. And, conversely, that natural selection MAY perfectly well succeed in largely developing a
149  at a knife which has to cut all sorts of things MAY be of almost any shape; whilst a tool for some
151  ble in the case of secondary sexual characters,  MAY be due to the great variability of these chara
151  in the case of hermaphrodite cirripedes; and I  MAY here add, that I particularly attended to Mr.
152  rn which depart widely from the standard. There  MAY be truly said to be a constant struggle going
153  as long as selection is rapidly going on, there  MAY always be expected to be much variability in t
153  ed with the other species of the same genus, we  MAY conclude that this part has undergone an extra
154  ; and that the most abnormally developed organs  MAY be made constant, I can see no reason to doubl
154  doubt. Hence when an organ, however abnormal it  MAY be, has been transmitted in approximately the
154  ought to find the generative variability, as it  MAY be called, still present in a high degree. For
154  dition. The principle included in these remarks  MAY be extended. It is notorious that specific cha
155  s variable, though its physiological importance  MAY remain the same. Something of the same kind ap
157  to the less favoured males. Whatever the cause  MAY be of the variability of secondary sexual char
157  tion will have had a wide scope for action, and  MAY thus readily have succeeded in giving to the s
158  riability in a part, however extraordinarily it  MAY be developed, if it be common to a whole group
159  en or even sixteen tail feathers in the pouter,  MAY be considered as a variation representing the
159  o called distinct species; and to these a third  MAY be added, namely, the common turnip. According
160  s variation appearing in the several breeds. We  MAY I think confidently come to this conclusion, b
161  all know them to be, thus inherited. Indeed, we  MAY sometimes observe a mere tendency to produce a
163  instance in a chestnut: a faint shoulder stripe  MAY sometimes be seen in duns, and I have see a tr
164  al. Without here entering on further details, I  MAY state that I have collected cases of leg and s
169  lapse of time, in this case, natural selection  MAY readily have succeeded in giving a fixed chara
169  e organ, in however extraordinary a a manner it  MAY be developed. Species inheriting nearly the sa
169  nt analogous variations, and these same species  MAY occasionally revert to some of the characters
169  itors. Although new and important modifications  MAY not arise from reversion and analogous variati
170  monious diversity of nature. Whatever the cause  MAY be of each slight difference in the offspring
171  to my theory. These difficulties and objections  MAY be classed under the following heads: Firstly,
173  y at intervals of time immensely remote. But it  MAY be urged that when several closely allied spec
174  egion, though they must have existed there, and  MAY be embedded there in a fossil condition. But i
175  ecies are already defined objects (however they  MAY have become so), not blending one into another
176  y than the forms which they connect. Now, if we  MAY trust these facts and inferences, and therefor
177  l marked varieties than do the rarer species. I  MAY illustrate what I mean by supposing three vari
178  ses which unite for each birth and wander much,  MAY have separately been rendered sufficiently dis
182  f wing structure here alluded to, which perhaps  MAY all have resulted from disuse, indicate the na
182  tion through natural selection. Furthermore, we  MAY conclude that transitional grades between stru
183  r own country the larger titmouse (Parus major)  MAY be seen climbing branches, almost like a creep
185  structure. The webbed feet of the upland goose  MAY be said to have become rudimentary in function
186  place of that inhabitant, however different it  MAY be from its own place. Hence it will cause him
187  re than how life itself first originated; but I  MAY remark that several facts make me suspect that
187  facts make me suspect that any sensitive nerve  MAY be rendered sensitive to light, and likewise t
188  een formed by a somewhat analogous process. But  MAY not this inference be presumptuous? Have we an
189  ch alteration which under varied circumstances,  MAY in any way, or in any degree, tend to produce
189  r on millions of individuals of many kinds; and  MAY we not believe that a living optical instrumen
190  d in the fish Cobites. In the Hydra, the animal  MAY be turned inside out, and the exterior surface
190  constructed for one purpose, namely flotation,  MAY be converted into one for a wholly different p
193  embers having very different habits of life, we  MAY attribute its presence to inheritance from a c
193  he general appearance and function of the organ  MAY be the same, yet some fundamental difference c
196  ified swimbladders betray their aquatic origin,  MAY perhaps be thus accounted for. A well develope
196  double quickly enough. In the second place, we  MAY sometimes attribute importance to characters w
196  isen from the above or other unknown causes, it  MAY at first have been of no advantage to the spec
196  t have been of no advantage to the species; but  MAY subsequently have been taken advantage of by t
197  which are not climbers, the hooks on the bamboo  MAY have arisen from unknown laws of growth, and h
197  daptation for wallowing in putridity; and so it  MAY be, or it may possibly be due to the direct ac
197  wallowing in putridity; and so it may be, or it  MAY possibly be due to the direct action of putrid
197  g parturition, and no doubt they facilitate, or  MAY be indispensable for this act; but as sutures
197  which have only to escape from a broken egg, we  MAY infer that this structure has arisen from the
199  e races of man, which are so strongly marked; I  MAY add that some little light can apparently be t
199  elation of growth, or from other unknown cause,  MAY reappear from the law of reversion, though now
200  e seal, are of special use to these animals. We  MAY safely attribute these structures to inheritan
200  re to the most aquatic of existing birds. So we  MAY believe that the progenitor of the seal had no
200  ive toes fitted for walking or grasping; and we  MAY further venture to believe that the several bo
200  ng such widely diversified habits. Therefore we  MAY infer that these several bones might have been
200  e for the direct action of physical conditions)  MAY be viewed, either as having been of special us
201  ugh natural selection. Although many statements  MAY be found in works on natural history to this e
202  in nature, this same reason tells us, though we  MAY easily err on both sides, that some other cont
202  he requirements of natural selection, though it  MAY cause the death of some few members. If we adm
202  ed by their industrious and sterile sisters? It  MAY be difficult, but we ought to admire the savag
203  clouds of pollen, in order that a few granules  MAY be wafted by a chance breeze on to the ovules?
203  d some of the difficulties and objections which  MAY be urged against my theory. Many of them are v
204  de through the air. We have seen that a species  MAY under new conditions of life change its habits
205  cumulated by means of natural selection. But we  MAY confidently believe that many modifications, w
205  urther modified descendants of this species. We  MAY, also, believe that a part formerly of high im
205  exclusive good or injury of another; though it  MAY well produce parts, organs, and excretions hig
208  rect opposition to our conscious will! yet they  MAY be modified by the will or reason. Habits easi
209  ating variations of instinct to any extent that  MAY be profitable. It is thus, as I believe, that
209  to the effects of the natural selection of what  MAY be called accidental variations of instincts;
210  tincts as to bodily organs. Changes of instinct  MAY sometimes be facilitated by the same species h
211  d other such points are not indispensable, they  MAY be here passed over. As some degree of variati
212  r enemy is certainly an instinctive quality, as  MAY be seen in nestling birds, though it is streng
212  rious animals inhabiting desert islands; and we  MAY see an instance of this even in England, in th
212  arge birds have been most persecuted by man. We  MAY safely attribute the greater wildness of our l
213  ctions instinctive. Domestic instincts, as they  MAY be called, are certainly far less fixed or inv
214  birds, that have never seen a pigeon tumble. We  MAY believe that some one pigeon showed a slight t
214  n inches high without using head over heels. It  MAY be doubted whether any one would have thought
216  st lost by disuse the power of flight. Hence we  MAY conclude, that domestic instincts have been ac
218  our cuckoo could be, and has been, generated. I  MAY add that, according to Dr. Gray and to some ot
```

Page **(Key Word)**

```
0218  d these are hatched by the males. This instinct MAY probably be accounted for by the fact of the
0220  ubject in a sceptical frame of mind, as any one MAY well be excused for doubting the truth of so
0220  e has watched the nests at various hours during MAY, June and August, both in Surrey and Hampshi
0220  usehold slaves. The masters, on the other hand, MAY be constantly seen bringing in materials for
0223  h, but chiefly the slaves, tend, and milk as it MAY be called, their aphides; and thus both coll
0225  ence of adjoining cells; and the following view MAY, perhaps, be considered only as a modificati
0227  een made of the cells of the hive bee. Hence we MAY safely conclude that if we could slightly mo
0230  spheres to break into each other. Now bees, as MAY be clearly seen by examining the edge of a g
0230  growing margin, or the two plates, as the case MAY be; and they never complete the upper edges
0231  m or wall of wax round a growing comb, flexures MAY sometimes be observed, corresponding in posi
0233  the individual under its conditions of life, it MAY reasonably be asked, how a long and graduate
0234  es. Of course the success of any species of bee MAY be dependent on the number of its parasites
0237  ons of structure or instinct to its progeny. It MAY well be asked how is it possible to reconcil
0237  isappears, when it is remembered that selection MAY be applied to the family, as well as to the
0237  o the family, as well as to the individual, and MAY thus gain the desired end. Thus, a well flav
0239  by the aphides, or the domestic cattle as they MAY be called, which our European ants guard or
0239  the fertile males and females, in this case, we MAY safely conclude from the analogy of ordinary
0240  ome few members in an intermediate condition. I MAY digress by adding, that if the smaller worke
0240  nts of this genus have well developed ocelli. I MAY give one other case: so confidently did I ex
0241  ginated. We can see how useful their production MAY have been to a social community of insects,
0243  nlike that of any other known bird. Finally, it MAY not be a logical deduction, but to my imagin
0246  r reproductive organs functionally impotent, as MAY be clearly seen in the state of the male ele
0247  l other analogous cases; it seems to me that we MAY well be permitted to doubt whether many othe
0249  generations of artificially fertilised hybrids MAY, I believe, be accounted for by close interb
0250  as did Gartner. The difference in their results MAY, I think, be in part accounted for by Herber
0252  close interbreeding is thus prevented. Any one MAY readily convince himself of the efficiency o
0252  ach other, as are the genera of plants, then we MAY infer that animals more widely separated in
0254  on the intercrossing of plants and animals, it MAY be concluded that some degree of sterility,
0255  some of these cases a first trace of fertility MAY be detected, by the pollen of one of the pur
0257  h the utmost facility. In the same family there MAY be a genus, as Dianthus, in which very many
0258  then a male ass with a mare: these two species MAY then be said, to have been reciprocally cross
0260  cal cross between the same two species? Why, it MAY even be asked, has the production of hybrids
0263  e sterility of first Crosses and of Hybrids. We MAY now look a little closer at the probable cau
0264  the stigmatic surface. Again, the male element MAY reach the female element, but be incapable o
0264  cannot be grafted on others. Lastly, an embryo MAY be developed, and then perish at an early pe
0264  thin the egg or seed produced by the mother, it MAY be exposed to conditions in some degree unsu
0266  ganisations having been compounded into one. It MAY seem fanciful, but I suspect that a similar
0267  hen crossed, and of their Mongrel offspring. It MAY be urged, as a most forcible argument, that
0268  marks, inasmuch as varieties, however much they MAY differ from each other in external appearanc
0269  lesser degree of sterility when crossed; and we MAY apply the same rule to domestic varieties. I
0269  ime on the whole organisation, in any way which MAY be for each creature's own good: and thus sh
0269  y be for each creature's own good; and thus she MAY, either directly, or more probably indirectl
0270  ferences are greater. How far these experiments MAY be trusted, I know not; but the forms experi
0272  cies when crossed and of varieties when crossed MAY be compared in several other respects. Gartn
0281  and the horse; but in some points of structure MAY have differed considerably from both, even p
0282  f such infinitely numerous connecting links, it MAY be objected, that time will not have suffice
0282  ssible for me even to recall to the reader, who MAY not be a practical geologist, the facts lead
0282  nsibly vast have been the past periods of time, MAY at once close this volume. Not that it suffi
0284  feet in a hundred thousand years. This estimate MAY be quite erroneous; yet, considering over wha
0286  weald. This, of course, cannot be done: but we MAY, in order to form some crude notion on the su
0287  ch we know this area has undergone, the surface MAY have existed for millions of years as land,
0289  elapsed between the consecutive formations, we MAY infer that this could nowhere be ascertained
0290  thin the grinding action of the coast waves. We MAY, I think, safely conclude that sediment must
0290  h thick and extensive accumulations of sediment MAY be formed in two ways; either, in profound de
0291  e, judging from the researches of E. Forbes, we MAY conclude that the bottom will be inhabited by
0291  forms of life which then existed; or, sediment MAY be accumulated to any thickness and extent o
0291  upraised, to resist any amount of degradation, MAY be formed. I am convinced that all our ancien
0291  on that it was accumulated during subsidence. I MAY add, that the only ancient tertiary formatio
0291  and extensive to resist subsequent degradation, MAY have been formed over wide spaces during per
0292  s rich in fossils have been accumulated. Nature MAY almost be said to have guarded against the f
0293  the same formation, but, as they are rare, they MAY be here passed over. Although each formation
0293  llowing considerations. Although each formation MAY mark a very long lapse of years, each perhaps
0293  accuracy. With marine animals of all kinds, we MAY safely infer a large amount of migration dur
0294  ncluded within this same glacial period. Yet it MAY be doubted whether in any quarter of the worl
0295  beds of different mineralogical composition, we MAY reasonably suspect that the process of depos
0299  d most obvious of all the many objections which MAY be urged against my views. Hence it will be u
0300  ological record would then be least perfect. It MAY be doubted whether the duration of any one gr
0302  ully examined; we forget that groups of species MAY elsewhere have long existed and have slowly m
0303  h species will appear as if suddenly created. I MAY here recall a remark formerly made, namely th
0303  roups of species have suddenly been produced. I MAY recall the well known fact that in geological
0304  ary and earliest tertiary formation. But now we MAY read in the Supplement to Lyell's Manual, pub
0304  ime before the close of the secondary period. I MAY give another instance, which from having pass
0308  e case at present must remain inexplicable; and MAY be truly urged as a valid argument against th
0308  nst the views here entertained. To show that it MAY hereafter receive some explanation, I will gi
0308  kness, of which the formations are composed, we MAY infer that from first to last large islands o
0308  any palaeozoic or secondary formation. Hence we MAY perhaps infer, that during the palaeozoic and
0309  upheaved by the oscillations of level, which we MAY fairly conclude must have intervened during t
0309  urinc these enormously long periods. If then we MAY infer anything from these facts; we may infer
0309  then we may infer anything from these facts, we MAY infer that where our oceans now extend, ocean
0309  ations of level, of the force of elevation; but MAY not the areas of preponderant movement have c
0309  y antecedent to the silurian epoch, continents MAY have existed where oceans are now spread out;
0309  s are now spread out; and clear and open oceans MAY have existed where our continents now stand.
0310  me to require some special explanation; and we MAY perhaps believe that we see in these large ar
0311  rent in the interrupted succession of chapters, MAY represent the apparently abruptly changed for
0312  far as we know, on the face of the earth. If we MAY trust the observations of Philippi in Sicily,
0313  in the oldest tertiary beds a few living shells MAY still be found in the midst of a multitude of
```

Page **(Key Word)**

age **(Key Word)**

315	change, during long and equal periods of time,	MAY, perhaps, be nearly the same: but as the accum
320	ther locally or wholly, through man's agency. I	MAY repeat what I published in 1845, namely, that
320	ical periods, so that looking to later times we	MAY believe that the production of new forms has c
321	n modified and improved, a few of the sufferers	MAY often long be preserved, from being fitted to
322	cutive formations; and in these intervals there	MAY have been much slow extermination. Moreover, w
322	in a given country; then, and not till then, we	MAY justly feel surprise why we cannot account for
323	distant points in the same parallel manner. We	MAY doubt whether they have thus changed: if the M
326	varieties and species. The process of diffusion	MAY often be very slow, being dependent on climata
326	, as before explained. One quarter of the world	MAY have been most favourable for the production o
331	o turn to the diagram in the fourth chapter. We	MAY suppose that the numbered letters represent ge
331	his is unimportant for us. The horizontal lines	MAY represent successive geological formations, an
331	s, and all the forms beneath the uppermost line	MAY be considered as extinct. The three existing g
337	r instance, not the highest in their own class,	MAY have beaten the highest molluscs. From the ext
337	es which must have been previously occupied, we	MAY believe, if all the animals and plants of Grea
337	re having become wild in any part of Europe, we	MAY doubt, if all the productions of New Zealand w
337	point of view, the productions of Great Britain	MAY be said to be higher than those of New Zealand
338	ss modified condition of each animal. This view	MAY be true, and yet it may never be capable of fu
338	each animal. This view may be true, and yet it	MAY never be capable of full proof. Seeing, for in
341	n the laws of past and present distribution. It	MAY be asked in ridicule, whether I suppose that t
341	ing in South America; and some of these fossils	MAY be the actual progenitors of living species. I
341	genera with the same number of species, then we	MAY conclude that only one species of each of the
342	rd, will rightly reject my whole theory. For he	MAY ask in vain where are the numberless transitio
342	several stages of the same great formation. he	MAY disbelieve in the enormous intervals of time w
343	elapsed between our consecutive formations; he	MAY overlook how important a part migration must h
343	on alone, as that of Europe, are considered; he	MAY urge the apparent, but often falsely apparent,
343	sudden coming in of whole groups of species. He	MAY ask where are the remains of those infinitely
343	chi but that long before that period, the world	MAY have presented a wholly different aspect; and
343	ormed of formations older than any known to us,	MAY now all be in a metamorphosed condition, or ma
343	may now all be in a metamorphosed condition, or	MAY lie buried under the ocean. Passing from these
344	he utter extinction of a whole group of species	MAY often be a very slow process, from the surviva
345	so far, higher in the scale of nature; and this	MAY account for that vague yet ill defined sentime
345	d be as imperfect as I believe it to be, and it	MAY at least be asserted that the record cannot be
347	and floras more utterly dissimilar. Or again we	MAY compare the productions of South America south
349	lands off the American shore, however much they	MAY differ in geological structure, the inhabitant
349	logical structure, the inhabitants, though they	MAY be all peculiar species, are essentially Ameri
349	peculiar species, are essentially American. We	MAY look back to past ages, as shown in the last c
350	ilarity of the inhabitants of different regions	MAY be attributed to modification through natural
351	e is not much difficulty in believing that they	MAY have migrated from the same region; for during
356	ration: a region when its climate was different	MAY have been a high road for migration; but now b
356	submerged, and the two faunas will now blend or	MAY formerly have blended: where the sea now exten
356	y have blended: where the sea now extends, land	MAY at a former period have connected islands or p
357	many islands, now buried beneath the sea, which	MAY have served as halting places for plants and f
358	across the sea, the greater or less facilities	MAY be said to be almost wholly unknown. Until I t
359	er being dried, for above 28 days, as far as we	MAY infer anything from these scanty facts, we may
359	may infer anything from these scanty facts, we	MAY conclude that the seeds of 14/100 plants of an
360	nts generally have restricted ranges. But seeds	MAY be occasionally transported in another manner.
361	by gales to vast distances across the ocean. We	MAY I think safely assume that under such circumst
363	which do not occur in the archipelago. Hence we	MAY safely infer that icebergs formerly landed the
363	islands, and it is at least possible that they	MAY have brought thither the seeds of northern pla
367	ard, are remarkably uniform round the world. We	MAY suppose that the Glacial period came on a litt
368	he same species on distant mountain summits, we	MAY almost conclude without other evidence, that a
369	the plains of North America and Europe; and it	MAY be reasonably asked how I account for the nece
370	spaces of ocean. I believe the above difficulty	MAY be surmounted by looking to still earlier chan
370	te was warmer than at the present day. Hence we	MAY suppose that the organisms now living under th
373	acial period. I am convinced that Forbes's view	MAY be largely extended. In Europe we have the pla
373	oural range, and southward to the Pyrenees. We	MAY infer, from the frozen mammals and nature of t
374	, as measured by years, at each point. The cold	MAY have come on, or have ceased, earlier at one p
374	hout some distinct evidence to the contrary, we	MAY at least admit as probable that the glacial ac
376	ductions, similar cases occur: as an example, I	MAY quote a remark by the highest authority, Prof.
379	migration and re migration northward, the case	MAY have been wholly different with those intrudin
382	rms of life can be explained. The living waters	MAY be said to have flowed during one short period
382	equator. The various beings thus left stranded	MAY be compared with savage races of man, driven u
384	confined exclusively to fresh water, so that we	MAY imagine that a marine member of a fresh water
387	hese plants was not very great. The same agency	MAY have come into play with the eggs of some of t
392	e has shown, generally have, whatever the cause	MAY be, confined ranges. Hence trees would be litt
393	this exception (if the information be correct)	MAY be explained through glacial agency. This gene
393	rought boulders to its western shores, and they	MAY have formerly transported foxes, as so frequen
394	itius, all possess their peculiar bats. Why, it	MAY be asked, has the supposed creative force prod
400	between the endemic inhabitants of the islands	MAY be used as an argument against my views; for i
400	be used as an argument against my views; for it	MAY be asked, how has it happened in the several i
402	lago, are common to the several islands, and we	MAY infer from certain facts that these have proba
402	uld it succeed in establishing itself there? We	MAY safely infer that Charles Island is well stock
402	laid there than can possibly be reared: and we	MAY infer that the mocking thrush peculiar to Char
406	cies into all quarters of the world, where they	MAY have become slightly modified in relation to t
407	cent period, and of other similar changes which	MAY have occurred within the same period; if we re
407	ised on: if we bear in mind how often a species	MAY have ranged continuously over a wide area, and
409	he exceptions to the rule are so few, that they	MAY fairly be attributed to our not having as yet
410	inuous; and the exceptions, which are not rare,	MAY, as I have attempted to show, be accounted for
412	n the diagram each letter on the uppermost line	MAY represent a genus including several species: a
414	f these resemblances we shall have to recur. It	MAY even be given as a general rule, that the less
415	rts of the organisation, however important they	MAY be for the welfare of the being in relation to
417	y, hence, as has often been remarked, a species	MAY depart from its allies in several characters,
417	on any single character, however important that	MAY be, has always failed: for no part of the orga
420	ame degree in blood to their common progenitor,	MAY differ greatly, being due to the different deg
421	ed in blood or descent to the same degree; they	MAY metaphorically be called cousins to the same m
421	, with the parent I. But the existing genus F14	MAY be supposed to have been but slightly modified

Page **(Key Word)**

0422 es, families, sections, orders, and classes. It MAY be worth while to illustrate this view of cla
0424 tages of the same individual, however much they MAY differ from each other and from the adult; as
0425 mmediately included as a single species. But it MAY be asked, what ought we to do, if it could be
0425 ometimes a considerable amount of modification, MAY not this same element of descent have been un
0426 anisation. We care not how trifling a character MAY be, let it be the mere inflection of the angl
0426 nly by its inheritance from a common parent. We MAY err in this respect in regard to single point
0426 rge group of beings having different habits, we MAY feel almost sure, on the theory of descent, t
0426 understand why a species or a group of species MAY depart, in several of its most important char
0426 lies, and yet be safely classed with them. This MAY be safely done, and is often done, as long as
0426 together by a chain of intermediate groups, we MAY at once infer their community of descent, and
0427 fficatory importance. Geographical distribution MAY sometimes be brought usefully into play in cl
0427 elonging to two most distinct lines of descent, MAY readily become adapted to similar conditions,
0430 tion in divergent directions. On either view we MAY suppose that the bizcacha has retained, by in
0430 ral order of Rodents. In this case, however, it MAY be strongly suspected that the resemblance is
0431 uitous lines of affinity of various lengths (as MAY be seen in the diagram so often referred to),
0431 ls between the several groups in each class. We MAY thus account even for the distinctness of who
0432 by turning to the diagram: the letters, A to L, MAY represent eleven Silurian genera, some of whi
0432 ach branch and sub branch of their descendants, MAY be supposed to be alive; and the links so wel
0433 acknowledged varieties, however different they MAY be from their parent; and I believe this elem
0434 o not look to some unknown plan of creation, we MAY hope to make sure but slow progress. Morpholo
0434 interesting department of natural history, and MAY be said to be its very soul. What can be more
0434 ative connexion in homologous organs: the parts MAY change to almost any extent in form and size,
0435 hat the ancient progenitor, the archetype as it MAY be called, of all mammals, had its limbs cons
0437 all low or little modified forms; therefore we MAY readily believe that the unknown progenitor o
0439 his plain signification. On my view these terms MAY be used literally; and the wonderful fact of
0440 s to provide for itself. The period of activity MAY come on earlier or later in life; but wheneve
0441 t. in this last and completed state, cirripedes MAY be considered as either more highly or more l
0443 at what period it is fully displayed. The cause MAY have acted, and I believe generally has acted
0443 before the embryo is formed; and the variation MAY be due to the male and female sexual elements
0443 eriod, even before the formation of the embryo, MAY appear late in life; as when an hereditary di
0444 ach species has acquired its present structure, MAY have supervened at a not very early period of
0444 each successive modification, or most of them, MAY have appeared at an extremely early period. I
0447 just as we have seen in the case of pigeons. We MAY extend this view to whole families or even cl
0447 ce, which served as legs in the parent species, MAY become, by a long course of modification, ada
0447 r use on the one hand, and disuse on the other, MAY have in modifying an organ, such influence wi
0449 tor. In two groups of animal, however much they MAY at present differ from each other in structur
0449 hrough the same or similar embryonic stages, we MAY feel assured that they have both descended fr
0449 escent, however much the structure of the adult MAY have been modified and obscured; we have seen
0450 f life to the embryonic stages of recent forms, MAY be true, but yet, owing to the geological rec
0450 l record not extending far enough back in time, MAY remain for a long period, or for ever, incapa
0450 mmals: I presume that the bastard wing in birds MAY be safely considered as a digit in a rudiment
0452 out of the surrounding anthers. Again, an organ MAY become rudimentary for its proper purpose, an
0452 ale moths in certain groups. Rudimentary organs MAY be utterly aborted; and this implies, that we
0452 not find a rudiment of a fifth stamen; but this MAY sometimes be seen. In tracing the homologies
0455 mer functions. An organ, when rendered useless, MAY well be variable, for its variations cannot b
0455 gh physiological importance. Rudimentary organs MAY be compared with the letters in a word, still
0455 n. on the view of descent with modification, we MAY conclude that the existence of organs in a ru
0456 ies of the same species, however different they MAY be in structure. If we extend the use of this
0459 pecies. How far the theory of natural selection MAY be extended. Effects of its adoption on the s
0459 . as this whole volume is one long argument, it MAY be convenient to the reader to have the leadi
0459 y recapitulated. That many and grave objections MAY be advanced against the theory of descent wit
0459 e perfection of any organ or instinct, which we MAY consider, either do now exist or could have e
0462 on by many means. A broken or interrupted range MAY often be accounted for by the extinction of t
0462 gradations as fine as our present varieties, it MAY be asked, Why do we not see these linking for
0463 vious and forcible of the many objections which MAY be urged against my theory. Why, again, do wh
0465 several chief objections and difficulties which MAY justly be urged against my theory; and I have
0466 our domestic productions have undergone; but we MAY safely infer that the amount has been large, a
0466 as already been inherited for many generations, MAY continue to be inherited for an almost infini
0467 and plants for its own benefit or pleasure. He MAY do this methodically, or he may do it unconsr
0467 or pleasure. He may do this methodically, or he MAY do it unconsciously by preserving the individu
0474 ny one species of a genus, and therefore, as we MAY naturally infer, of great importance to the sp
0474 part generally to be still variable. But a part MAY be developed in the most unusual manner, like
0477 e whole intertropical ocean. Although two areas MAY present the same physical conditions of life,
0479 aracters, however slight their vital importance MAY be. The framework of bones being the same in t
0479 ke, and should be so unlike the adult forms. We MAY cease marvelling at the embryo of an air breat
0480 progenitor having well developed teeth; and we MAY believe, that the teeth in the mature animal w
0480 ently bear the plain stamp of inutility! Nature MAY be said to have taken pains to reveal, by rudi
0480 uccessive slight favourable variations. Why, it MAY be asked, have all the most eminent living nat
0482 begun to doubt on the immutability of species, MAY be influenced by this volume; but I look with
0483 cies in what they consider reverent silence. It MAY be asked how far I extend the doctrine of the
0483 ecause the more distinct the forms are which we MAY consider, by so much the arguments fall away w
0484 descended from some one prototype. But analogy MAY be a deceitful guide. Nevertheless all living
0485 w generally acknowledged to be merely varieties MAY hereafter be thought worthy of specific names,
0485 ficial combinations made for convenience. This MAY not be a cheering prospect; but we shall have
0486 ade, genealogies; and will then truly give what MAY be called the plan of creation. The rules for
0486 f species, which are called aberrant, and which MAY fancifully be called living fossils, will aid
0488 simplest structure existed, the rate of change MAY have been slow in an extreme degree. The whole
0489 e to become ennobled. Judging from the past, we MAY safely infer that not one living species will
0489 which lived long before the Silurian epoch, we MAY feel certain that the ordinary succession by g
0489 taclysm has desolated the whole world. Hence we MAY look with some confidence to a secure future o
0186 here should be long toed corncrakes living in MEADOWS instead of in swamps; that there should be w
0387 on 'flying to another pond and getting a hearty MEAL of fish, would probably reject from its stomac
0009 ven of sporting plants; by this term gardeners MEAN a single bud or offset, which suddenly assumes
0016 ave a somewhat monstrous character; by which I MEAN, that, although differing from each other, and
0143 f innate differences. Correlation of Growth. I MEAN by this expression that the whole organisation
0177 do the rarer species. I may illustrate what I MEAN by supposing three varieties of sheep to be ke

```
258  . by a reciprocal cross between two species, I MEAN the case, for instance, of a stallion horse be
261  plain a little more fully by an example what I MEAN by sterility being incidental on other differe
339  ession of the same types within the same areas MEAN? He would be a bold man, who after comparing t
356  ent from a single parent. To illustrate what I MEAN: our English race horses differ slightly from
097  nature (utterly ignorant though we be of the MEANING of the law) that no organic being self ferti
157  f the same species. This relation has a clear MEANING on my view of the subject: I look at all the
206  selection we can clearly understand the full MEANING of that old canon in natural history, Natura
336  affected. On the theory of descent, the full MEANING of the fact of fossil remains from closely c
420  cts more or less alike. But I must explain my MEANING more fully. I believe that the arrangement o
438  ly in a metaphorical sense: they are far from MEANING that during a long course of descent, primor
444  ge in the offspring and parent. I am far from MEANING that this is invariably the case: and I coul
451  der wing cases, firmly soldered together! The MEANING of rudimentary organs is often quite unmista
477  ification have been the same. We see the full MEANING of the wonderful fact, which must have struc
480  nd we can clearly understand on this view the MEANING of rudimentary organs. But disuse and select
481  aves. The mind cannot possibly grasp the full MEANING of the term of a hundred million years: it c
004  st importance to gain a clear insight into the MEANS of modification and coadaptation. At the comm
006  selection has been the main but not exclusive MEANS of modification. Variation Under Domesticatio
031  e it altogether. It is the magician's wand, by MEANS of which he may summon into life whatever for
031  quarter of the world. The improvement is by no MEANS generally due to crossing different breeds; a
033  o breed. In regard to plants, there is another MEANS of observing the accumulated effects of selec
044  ts, yet every naturalist knows vaguely what he MEANS when he speaks of a species. Generally the te
071  on the estate of a relation where I had ample MEANS of investigation, there was a large and extre
072  s they are, must be habitually checked by some MEANS, probably by birds. Hence, if certain insecti
076  s of the same genus have usually, though by no MEANS invariably, some similarity in habits and con
083  great result by his methodical and unconscious MEANS of selection; what may not nature effect? Man
088  l armed: though to them and to others, special MEANS of defence may be given through means of sexu
088  special means of defence may be given through MEANS of sexual selection, as the mane to the lion,
090  o attribute any effect to such apparently weak MEANS: I cannot here enter on the details necessary
090  advantage over other males, in their weapons, MEANS of defence, or charms; and have transmitted t
100  of the fertilising element: for we know of no MEANS, analogous to the action of insects and of th
100  ut here currents in the water offer an obvious MEANS for an occasional cross. And, as in the case
110  s not indefinitely great, not that we have any MEANS of knowing that any one region has as yet got
141  ent innate constitutions, and how much to both MEANS combined, is a very obscure question. That ha
145  f the corolla of the Umbelliferae, it is by no MEANS, as Dr. Hooker informs me, in species with th
148  as it is rendered superfluous, without by any MEANS causing some other part to be largely develop
149  re higher. I presume that lowness in this case MEANS that the several parts of the organisation ha
150  m. it should be understood that the rule by no MEANS applies to any part, however unusually develo
167  part in the parents. But whenever we have the MEANS of instituting a comparison, the same laws ap
171  of their allies. Organs of extreme perfection. MEANS of transition. Cases of difficulty. Natura no
175  range of the inhabitants of any country by no MEANS exclusively depends on insensibly changing ph
182  they serve, at least, to show what diversified MEANS of transition are possible. Seeing that a few
192  the ovigerous frena, which serve, through the MEANS of a sticky secretion, to retain the eggs unt
195  nsects: so that individuals which could by any MEANS defend themselves from these small enemies, w
205  ture could not have been slowly accumulated by MEANS of natural selection. But we may confidently
242  n of labour in the communities of ants, by the MEANS of natural selection. But I am bound to confe
256  ch are generally confounded together, is by no MEANS strict. There are many cases, in which two pu
257  affinity and the facility of crossing is by no MEANS strict. A multitude of cases could be given o
259  itions, is innately variable. That it is by no MEANS always the same in degree in the first cross
261  t this capacity, as in hybridisation, is by no MEANS absolutely governed by systematic affinity. A
263  is attempted to be expressed. The facts by no MEANS seem to me to indicate that the greater or le
293  the species which then lived: but I can by no MEANS pretend to assign due proportional weight to
314  cies comes into competition. Hence it is by no MEANS surprising that one species should retain the
346  ns of the same continent. Centres of creation. MEANS of dispersal, by changes of climate and of th
346  nd of the level of the land, and by occasional MEANS. Dispersal during the Glacial period co exten
353  igrate, whereas some plants, from their varied MEANS of dispersal, have migrated across the vast a
354  nd geographical changes and various occasional MEANS of transport, the belief that this has been t
355  res of creation, I must say a few words on the MEANS of dispersal. Means of Dispersal. Sir C. Lyel
356  ust say a few words on the means of dispersal. MEANS of Dispersal. Sir C. Lyell and other authors
357  e of time we know something definite about the MEANS of distribution, we shall be enabled to specu
358  say a few words on what are called accidental MEANS, but which more properly might be called occa
358  which more properly might be called occasional MEANS of distribution. I shall here confine myself
360  fruit could hardly be transported by any other MEANS: and Alph. de Candolle has shown that such pl
363  ern plants. Considering that the several above MEANS of transport, and that several other means, w
363  ove means of transport, and that several other MEANS, which without doubt remain to be discovered,
364  had not thus become widely transported. These MEANS of transport are sometimes called accidental,
364  wind. It should be observed that scarcely any MEANS of transport would carry seeds for very great
364  ied in the crops or intestines of birds. These MEANS, however, would suffice for occasional transp
364  floras of distant continents would not by such MEANS become mingled in any great degree: but would
364  be transported by these wanderers only by one MEANS, namely, in dirt sticking to their feet, whic
364  hin the last few centuries; through occasional MEANS of transport, immigrants from Europe or any o
365  inland, would not receive colonists by similar MEANS. I do not doubt that out of twenty seeds or a
365  t against what would be effected by occasional MEANS of transport, during the long lapse of geolog
380  do not pretend to indicate the exact lines and MEANS of migration, or the reason why certain speci
382  oints of the southern hemisphere by occasional MEANS of transport, and by the aid, as halting plac
382  t of the Glacial period, by icebergs. By these MEANS, as I believe, the southern shores of America
384  ty of their occasional transport by accidental MEANS; like that of the live fish not rarely droppe
384  eographical changes, and consequently time and MEANS for much migration. In the second place, salt
386  equence, a very wide range. I think favourable MEANS of dispersal explain this fact. I have before
387  es to be dropped. In considering these several MEANS of distribution, it should be remembered that
390  e groups of modified descendants. But it by no MEANS follows, that, because in an island nearly al
392  ight be transported to an island by some other MEANS; and the plant then becoming slightly modifie
396  e to accord better with the view of occasional MEANS of transport having been largely efficient in
396  on my view, some unknown, but highly efficient MEANS for their transportal. Would the just hatched
398  ly to receive colonists, whether by occasional MEANS of transport or by formerly continuous land,
401  the islands having been stocked by occasional MEANS of transport, a seed, for instance, of one pl
407  nt we are with respect to the many and curious MEANS of occasional transport, a subject which has
```

Page ***(Key Word)**

```
0407  onal belts. As showing how diversified are the MEANS of occasional transport, I have discussed at
0407  rt, I have discussed at some little length the  MEANS of dispersal of fresh water productions. If
0409  demic or peculiar; and why, in relation to the  MEANS of migration, one group of beings, even with
0410  od under different conditions or by occasional  MEANS of transport, and by the species having beco
0413  ose which are most unlike; or as an artificial  MEANS for enunciating, as briefly as possible, gen
0415  ns which are important is generally, but by no   MEANS always, true. But their importance for class
0418  lude all ages of each species. But it is by no   MEANS obvious, on the ordinary view, why the struc
0432  nction has only separated groups: it has by no   MEANS made them; for if every form which has ever
0437  the act of parturition of mammals, will by no   MEANS explain the same construction in the skulls
0445  y cart horse, I find that the colts have by no   MEANS acquired their full amount of proportional d
0459  d instincts should have been perfected, not by  MEANS superior to, though analogous with, human re
0462  ys be a good chance for wide migration by many  MEANS. A broken or interrupted range may often be
0462  yet profoundly ignorant of the many occasional  MEANS of transport. With respect to distinct speci
0462  odification has necessarily been slow, all the  MEANS of migration will have been possible during
0466  annot be pretended that we know all the varied  MEANS of Distribution during the long lapse of yea
0467  ence, we see the most powerful and ever acting  MEANS of selection. The struggle for existence ine
0468  will often depend on having special weapons or  MEANS of defence, or on the charms of the males; a
0476  changes and to the many occasional and unknown  MEANS of dispersal, then we can understand, on the
0476  ed by the bond of ordinary generation, and the  MEANS of modification have been the same. We see t
0481  volume under the form of an abstract, I by no   MEANS expect to convince experienced naturalists w
0487  e birthplace; and when we better know the many  MEANS of migration, then, by the light which geolo
0487  f that continent in relation to their apparent  MEANS of immigration, some light can be thrown on
0044  into varieties. By a monstrosity I presume is   MEANT some considerable deviation of structure in
0154  eneric. To explain by a simple example what is  MEANT. If some species in a large genus of plants
0206  e conditions of Existence. By unity of type is  MEANT that fundamental agreement in structure, whi
0207  y this term; but every one understands what is  MEANT, when it is said that instinct impels the cu
0256  conditions. By the term systematic affinity is  MEANT, the resemblance between species in structur
0330  iving species or groups. If by this term it is  MEANT that an extinct form is directly intermediat
0336  t defined to each other's satisfaction what is  MEANT by high and low forms. But in one particular
0404  vidual species have enormous ranges. It is not  MEANT that in world ranging genera all the species
0405  ld have been greatly reduced. Still less is it  MEANT, that a species which apparently has the cap
0413  what is called the Natural System. But what is  MEANT by this system? Some authors look at it mere
0413  many naturalists think that something more is   MEANT by the Natural System: they believe that it
0413  hether order in time or space, or what else is  MEANT by the plan of the Creator, it seems to me t
0420  rders. The reader will best understand what is  MEANT, if he will take the trouble of referring to
0441  t is hardly possible to define clearly what is  MEANT by the organisation being higher or lower, B
0456  in organic beings, we shall understand what is  MEANT by the natural system: it is genealogical in
0284  of sedimentary formations are the result and   MEASURE of the degradation which the earth's crust
0286  a line of cliff of any given height, we could   MEASURE the time requisite to have denuded the Weal
0488  secutive formations probably serves as a fair   MEASURE of the lapse of actual time. A number of sp
0488  overrate the accuracy of organic change as a   MEASURE of time. During early periods of the earth'
0312  eds, though undoubtedly of high antiquity if   MEASURED by years, only one or two species are lost
0324  ocene period (an enormously remote period as   MEASURED by years, including the whole glacial epoch
0336  ought to find after intervals, very long as   MEASURED by years, but only moderately long as measu
0336  asured by years, but only moderately long as   MEASURED geologically, closely allied forms, or, as
0374  ce, that it endured for an enormous time, as   MEASURED by years, at each point. The cold may have
0377  lder than at present. The Glacial period, as   MEASURED by years, must have been very long: and whe
0445  welve hours after being hatched; I carefully   MEASURED the proportions (but will not here give det
0462  rm for very long periods, enormously long as   MEASURED by years, too much stress ought not to be L
0284  aximum thickness, in most cases from actual   MEASUREMENT, in a few cases from estimate, of each fo
0035  nd could never be recognised unless actual   MEASUREMENTS or careful drawings of the breeds in ques
0227  y angle identically the same with the best   MEASUREMENTS which have been made of the cells of the
0240  these workers, by my giving not the actual   MEASUREMENTS, but a strictly accurate illustration: th
0445  nder domestication; but having had careful   MEASUREMENTS made of the dam and of a three days old c
0224  at a skilful workman, with fitting tools and   MEASURES, would find it very difficult to make cells
0445  eemed almost to be the case; but on actually   MEASURING the old dogs and their six days old puppie
0030  its hooks, which cannot be rivalled by any   MECHANICAL contrivance, is only a variety of the wild
0098  any other cases, though there be no special   MECHANICAL contrivance to prevent the stigma of a flo
0486  n the same way as when we look at any great   MECHANICAL invention as the summing up of the labour,
0187  y coated with pigment, and without any other MECHANISM; and from this low stage, numerous gradati
0076  keeping together different varieties of the   MEDICINAL leech. It may even be doubted whether the
0288  y littoral, with the exception of a single   MEDITERRANEAN species, which inhabits deep water and h
0299  size of Europe from the North Cape to the   MEDITERRANEAN, and from Britain to Russia; and therefo
0363  illions of quails which annually cross the   MEDITERRANEAN; and can we doubt that the earth adherin
0372  some fish and other marine animals, in the   MEDITERRANEAN and in the seas of Japan, areas now sepa
0024  er british islets, or on the shores of the   MEDITERRANIAN. Hence the supposed extermination of so
0100  cross being indispensable, by considering the MEDIUM in which terrestrial animals live, and the m
0112  le that fanciers do not and will not admire a  MEDIUM standard, but like extremes, they both go on
0119  eties will invariably prevail and multiply: a  MEDIUM form may often long endure, and may or may n
0016  ntly give, we have no right to expect often to MEET with generic differences in our domesticated
0051  extends the range of his observations, he will MEET with more cases of difficulty; for he will en
0069  rd, or in ascending a mountain, we far oftener MEET with stunted forms, due to the directly injur
0173  north to south over a continent, we generally MEET at successive intervals with closely allied s
0173  f the land. These representative species often MEET and interlock; and as the one becomes rarer a
0174  we ought not to expect at the present time to  MEET with numerous transitional varieties in each
0194  natural history of Natura non facit saltum. We MEET with this admission in the writings of almost
0202  produce absolute perfection, nor do we always MEET, as far as we can judge, with this high stand
0257  . even within the limits of the same genus, we MEET with this same difference; for instance, the
0285  the west the northern and southern escarpments MEET and close, one can safely picture to oneself
0297  tle variability to each species; but when they MEET with a somewhat greater amount of difference
0346  nited States to its extreme southern point, we MEET with the most diversified conditions: the mos
0348  another kind, and as soon as this is passed we MEET in the eastern islands of the Pacific, with a
0348  shores of Africa; and over this vast space we MEET with no well defined and distinct marine faun
0395  and the mammals are the same on both sides; we MEET with analogous facts on many islands separate
0413  hat famous one of Linnaeus, and which we often MEET with in a more or less concealed form, that t
0463  radation and mutation of the forms of life? We MEET with no such evidence, and this is the most o
```

Page ***(Key Word)**************************************

```
518  orse embedded with the remains of Mastodon, MEGATHERIUM, Toxodon, and other extinct monsters, whic
523  oubt whether they have thus changed: if the MEGATHERIUM, Mylodon, Macrauchenia, and Toxodon had be
541  ked in ridicule, whether I suppose that the MEGATHERIUM and other allied huge monsters have left b
275  ch have suddenly appeared, such as albinism, MELANISM, deficiency of tail or horns, or additional
225  humble bee, we have the cells of the Mexican MELIPONA domestica, carefully described and figured b
225  y described and figured by Pierre Huber. The MELIPONA itself is intermediate in structure between
226  hree adjoining cells. It is obvious that the MELIPONA saves wax by this manner of building: for th
226  on this case, it occurred to me that if the  MELIPONA had made its spheres at some given distance
227  odify the instincts already possessed by the MELIPONA, and in themselves not very wonderful, this
227  as that of the hive bee. We must suppose the  MELIPONA to make her cells truly spherical, and of eq
227  round on a fixed point. We must suppose the  MELIPONA to arrange her cells in level layers, as she
234  ggregated into a mass, like the cells of the MELIPONA; for in this case a large part of the boundi
234  same cause, it would be advantageous to the  MELIPONA, if she were to make her cells closer togeth
234  d all be replaced by plane surfaces; and the MELIPONA would make a comb as perfect as that of the
189  g, during thousands of generations, the most MELODIOUS or beautiful males, according to their stan
232  growing comb, with an extremely thin layer of MELTED vermilion wax; and I invariably found that th
367  f the more temperate regions. And as the snow MELTED from the bases of the mountains, the arctic f
037  nt. No one would expect to raise a first rate MELTING pear from the seed of the wild pear, though
185  ed its sub aquatic habits; yet this anomalous MEMBER of the strictly terrestrial thrush family who
188  instrument as perfect as is possessed by any MEMBER of the great Articulate class. He who will go
261  distinct genus, than on the apple, which is a MEMBER of the same genus. Even different varieties o
327  to disappear; though here and there a single MEMBER might long be enabled to survive. Thus, as it
384  h water, so that we may imagine that a marine MEMBER of a fresh water group might travel far along
429  ces. Mr. Waterhouse has remarked that, when a MEMBER belonging to one group of animals exhibits an
430  er of their common progenitor, or of an early MEMBER of the group. On the other hand, of all Marsu
436  rae. The anterior and posterior limbs in each MEMBER of the vertebrate and articulate classes are
013  kin, hairy bodies, etc., appearing in several MEMBERS of the same family. If strange and rare devi
021  twelve or fourteen, the normal number in all  MEMBERS of the great pigeon family: and these feathe
162  variation) which already occur in some other  MEMBERS of the same group. And this undoubtedly is t
182  of transition are possible. Seeing that a few MEMBERS of such water breathing classes as the Crust
189  ain, if we look to an organ common to all the MEMBERS of a large class, for in this latter case th
189  emely remote period, since which all the many MEMBERS of the class have been developed; and in ord
193  erally when the same organ appears in several MEMBERS of the same class, especially if in members
193  l members of the same class, especially if in MEMBERS having very different habits of life, we may
193  mmon ancestor; and its absence in some of the MEMBERS to its loss through disuse or natural select
202  and serrated instrument, like that in so many MEMBERS of the same great order, and which has been
202  on, though it may cause the death of some few MEMBERS. If we admire the truly wonderful power of s
237  related with the sterile condition of certain MEMBERS of insect communities: the difficulty lies i
238  related with the sterile condition of certain MEMBERS of the community, has been advantageous to t
238  rtile offspring a tendency to produce sterile MEMBERS having the same modification. And I believe
240  r organs of vision, yet connected by some few MEMBERS in an intermediate condition. I may digress
242  or habit, or volition, in the utterly sterile MEMBERS of a community could possibly have affected
242  ted the structure or instincts of the fertile MEMBERS, which alone leave descendants. I am surpris
267  ividuals of the same species, that is between MEMBERS of different strains or sub breeds, gives vi
315  e which do not change will become extinct. In MEMBERS of the same class the average amount of chan
316  the long succession of ages, so long must its MEMBERS have continuously existed, in order to have
321  s, survives in the Australian seas; and a few MEMBERS of the great and almost extinct group of Gan
330  each other by a dozen characters, the ancient MEMBERS of the same two groups would be distinguishe
331  and the Eocene mammals, with the more recent  MEMBERS of the same classes, we must admit that ther
333  ght approach to each other; so that the older MEMBERS should differ less from each other in some o
333  some of their characters than do the existing MEMBERS of the same groups; and this by the concurre
338  distinct from each other than are the typical MEMBERS of the same groups at the present day, it wo
386  al plants, which have only a very few aquatic MEMBERS; for these latter seem immediately to acquir
410  fer greatly. In both time and space the lower MEMBERS of each class generally change less than the
411  rates and defines groups. Morphology, between MEMBERS of the same class, between parts of the same
411  t in nature; for it is notorious how commonly MEMBERS of even the same sub group have different ha
427  th another, but give true affinities when the MEMBERS of the same class or order are compared one
428  exhibiting true affinity between the several  MEMBERS of the whale family: for these cetaceans agr
428  m a common ancestor. So it is with fishes. As MEMBERS of distinct classes have often been adapted
429  ed by more successful competitors, with a few MEMBERS preserved by some unusual coincidence of fav
431  lex and radiating affinities by which all the MEMBERS of the same family or higher group are conne
431  perceive between the many living and extinct  MEMBERS of the same great natural class. Extinction,
432  e to give any definition by which the several MEMBERS of the several groups could be distinguished
433  ther in one great system; and how the several MEMBERS of each class are connected together by the
434  he inextricable web of affinities between the MEMBERS of any one class; but when we have a distinc
434  w progress. Morphology. We have seen that the MEMBERS of the same class, independently of their ha
435  empt to explain this similarity of pattern in MEMBERS of the same class, by utility or by the doct
436  comparison not of the same part in different  MEMBERS of a class; but of the different parts or or
438  his fact; for in molluscs, even in the lowest MEMBERS of the class, we do not find nearly so much
442  n some whole groups of animals and in certain MEMBERS of other groups, the embryo does not at any
448  with cuttle fish and spiders, and with a few  MEMBERS of the great class of insects, as with Aphis
452  the homologies of the same part in different  MEMBERS of a class, nothing is more common, or more
457  s parts or organs, though fitted in the adult MEMBERS for purposes as different as possible. Larva
483  the greatest weight extend very far. All the  MEMBERS of whole classes can be connected together b
484  of descent with modification embraces all the MEMBERS of the same class. I believe that animals ha
137  having been inflammation of the nictitating  MEMBRANE. As frequent inflammation of the eyes must b
181  amongst bats. It has an extremely wide flank  MEMBRANE, stretching from the corners of the jaw to t
181  e limbs and the elongated fingers: the flank  MEMBRANE is, also, furnished with an extensor muscle.
181  ty in further believing it possible that the  MEMBRANE connected fingers and fore arm of the Galeop
181  t it into a bat. In bats which have the wing  MEMBRANE extended from the top of the shoulder to the
185  ic, although their toes are only bordered by  MEMBRANE. What seems plainer than that the long toes
185  ure. In the frigate bird, the deeply scooped  MEMBRANE between the toes shows that structure has be
188  ted with pigment and invested by transparent  MEMBRANE, into an optical instrument as perfect as is
435  ent, and become gradually enveloped in thick  MEMBRANE, so as to serve as a fin: or a webbed foot m
435  ain bones, lengthened to any extent, and the  MEMBRANE connecting them increased to any extent, so
451  l sized wings, and another mere rudiments of  MEMBRANE: and here it is impossible to doubt, that th
```

Page **(Key Word)**

```
0181  of individuals with fuller and fuller flank MEMBRANES, each modification being useful, each bein
0397  land shells, when hybernating and having a MEMBRANOUS diaphragm over the mouth of the shell, mig
0397  i removed it, and when it had formed a new MEMBRANOUS one, I immersed it for fourteen days in se
0002  origin of species. Last year he sent to me a MEMOIR on this subject, with a request that I would
0002  able to publish, with Mr. Wallace's excellent MEMOIR, some brief extracts from my manuscripts. Th
0285  hickness, as shown in Prof. Ramsay's masterly MEMOIR on this subject. Yet it is an admirable less
0304  ed under my own eyes has much struck me. In a MEMOIR on Fossil Sessile Cirripedes, I have stated
0398  ants, as shown by Dr. Hooker in his admirable MEMOIR on the Flora of this archipelago. The natura
0246  spring. It is impossible to study the several MEMOIRS and works of those two conscientious and ad
0328  ed in Europe. Mr. Prestwich, in his admirable MEMOIRS on the eocene deposits of England and Franc
0031  rd to merino sheep is so fully recognised, that MEN follow it as a trade: the sheep are placed on
0045  portant characters, and that there are not many MEN who will laboriously examine internal and imp
0050  the highest botanical authorities and practical MEN can be quoted to show that the sessile and pe
0102  ossing will wholly stop his work. But when many MEN, without intending to alter the breed, have a
0193  d to believe that in nearly the same way as two MEN have sometimes independently hit on the very
0241  d of three times as big as those of the smaller MEN, and jaws nearly five times as big. The jaws,
0005  an; secondly, the subject of Instinct, or the MENTAL powers of animals; thirdly, Hybridism, or th
0207  nothing to do with the origin of the primary MENTAL powers, any more than I have with that of li
0207  the diversities of instinct and of the other MENTAL qualities of animals within the same class.
0207  t would be easy to show that several distinct MENTAL actions are commonly embraced by this term;
0213  ental variations have played in modifying the MENTAL qualities of our domestic animals. A number
0216  ating during successive generations, peculiar MENTAL habits and actions, which at first appeared
0216  alone has sufficed to produce such inherited MENTAL changes; in other cases compulsory habit has
0242  ured briefly in this chapter to show that the MENTAL qualities of our domestic animals vary, and
0488  on, that of the necessary acquirement of each MENTAL power and capacity by gradation. Light will
0489  for the good of each being, all corporeal and MENTAL endowments will tend to progress towards per
0009  tion of animals under confinement, I may just MENTION that carnivorous animals, even from the tro
0011  be dimly seer, and will be hereafter briefly MENTIONED. I will here only allude to what may be ca
0386  f dispersal explain this fact. I have before MENTIONED that earth occasionally, though rarely, ad
0439  of this cannot be given, than a circumstance MENTIONED by Agassiz, namely, that having forgotten
0003  e true: but it is preposterous to attribute to MERE external conditions, the structure, for insta
0008  eparated by any clear line of distinction from MERE variations. But I am strongly inclined to sus
0013  ndividuals, and it reappears in the child, the MERE doctrine of chances almost compels us to attr
0016  ve not been ranked by some competent judges as MERE varieties; and by other competent judges as t
0030  s eyes, we must, I think, look further than to MERE variability. We cannot suppose that all the b
0040  e materials for selection to work on; not that MERE individual differences are not amply sufficie
0048  ne botanist as good species, and by another as MERE varieties. Mr. H. C. Watson, to whom I lie un
0050  d oaks are either good and distinct species or MERE varieties. When a young naturalist commences
0052  s. the term variety, again, in comparison with MERE individual differences, is also applied arbit
0052  ferences, is also applied arbitrarily, and for MERE convenience sake. Guided by theoretical consi
0054  s, again, might have been anticipated; for the MERE fact of many species of the same genus inhabi
0060  any well marked varieties be admitted. But the MERE existence of individual variability and of so
0064  t we have better evidence on this subject than MERE theoretical calculations, namely, the numerou
0082  at results by adding up in any given direction MERE individual differences, so could Nature, but
0102  ified variability is favourable, but I believe MERE individual differences suffice for the work.
0111  s, present slight and ill defined differences. MERE chance, as we may call it, might cause one va
0129  any twigs flourished when the tree was a MERE bush, only two or three, now grown into great
0141  n of species to any peculiar climate is due to MERE habit, and how much to the natural selection
0160  the several marks, beyond the influence of the MERE act of crossing on the laws of inheritance. N
0161  inherited. Indeed, we may sometimes observe a MERE tendency to produce a rudiment inherited: for
0167  an unknown, cause. It makes the works of God a MERE mockery and deception; I would almost as soon
0199  created for beauty in the eyes of man, or for MERE variety. This doctrine, if true, would be abs
0262  a clear and fundamental difference between the MERE adhesion of grafted stocks, and the union of
0269  can, in the first place, be clearly shown that MERE external dissimilarity between two species do
0271  bascum present no other difference besides the MERE colour of the flower; and one variety can som
0273  ue and be super added to that arising from the MERE act of crossing. The slight degree of variabi
0278  oduced under domestication by the selection of MERE external differences, and not of differences
0281  t have been possible to have determined from a MERE comparison of their structure with that of th
0285  ed that the denudation of the Weald has been a MERE trifle, in comparison with that which has rem
0323  larly characterised in such trifling points as MERE superficial sculpture. Moreover other forms,
0325  e introduction of new ones; cannot be owing to MERE changes in marine currents or other causes mo
0358  or other such rocks, instead of consisting of MERE piles of volcanic matter. I must now say a fe
0389  wing remarks I shall not confine myself to the MERE question of dispersal; but shall consider som
0413  g more is included in our classification, than MERE resemblance. I believe that something more is
0415  species to their conditions of life. That the MERE physiological importance of an organ does not
0416  nner in which the wings of insects are folded, MERE colour in certain Algae, mere pubescence on p
0416  ects are folded, mere colour in certain Algae, MERE pubescence on parts of the flower in grasses,
0420  e enunciation of general propositions, and the MERE putting together and separating objects more
0426  how trifling a character may be, let it be the MERE inflection of the angle of the jaw, the manne
0441  s assuredly been retrograde: for the male is a MERE sack, which lives for a short time, and is de
0451  which will have full sized wings, and another MERE rudiments of membrane; and here it is impossi
0451  the same species the petals sometimes occur as MERE rudiments, and sometimes in a well developed
0468  hin a short period a great result by adding up MERE individual differences in his domestic produc
0488  sible by us, will hereafter be recognised as a MERE fragment of time, compared with the ages whic
0032  han in ordinary cases. If selection consisted MERELY in separating some very distinct variety, an
0032  raisers do not pick out the best plants, but MERELY go over their seed beds, and pull up the rog
0052  r and higher stage may be, in some cases, due MERELY to the long continued action of different ph
0086  ever feed, a large part of their structure is MERELY the correlated result of successive changes
0141  ought not to be looked at as anomalies, but MERELY as examples of a very common flexibility of
0183  y. of cases of changed habits it will suffice MERELY to allude to that of the many British insect
0187  we can commence a series with an optic nerve MERELY coated with pigment, and without any other me
0188  verted the simple apparatus of an optic nerve MERELY coated with pigment and invested by transpar
0221  slaves in the two countries, probably depends MERELY on the slaves being captured in greater numb
0224  r on minute details on this subject, but will MERELY give an outline of the conclusions at which
0240  the smaller workers than can be accounted for MERELY by their proportionally lesser size; and I f
0301  fully preserved, transitional varieties would MERELY appear as so many distinct species. It is, a
```

Page **(Key Word)**

merely

Page **(Key Word)**

```
0413  meant by this system? Some authors look at it MERELY as a scheme for arranging together those livi
0414  th the whole life of the being, are ranked as MERELY adaptive or analogical characters; but to the
0414  he modifications of these organs to mistake a MERELY adaptive for an essential character. So with
0423  ass two varieties of the pine apple together, MERELY because their fruit, though the most importan
0430  to marsupials are believed to be real and not MERELY adaptive, they are due on my theory to inheri
0451  sometimes retain their potentiality, and are MERELY not developed: this seems to be the case with
0453  nature; but this seems to me no explanation, MERELY a restatement of the fact. Would it be though
0453  e pistil in male flowers, and which is formed MERELY of cellular tissue, can thus act? Can we supp
0485  e that forms now generally acknowledged to be MERELY varieties may hereafter be thought worthy of
0485  lists treat genera, who admit that genera are MERELY artificial combinations made for convienience
0073  or long periods of time, though assuredly the MEREST trifle would often give the victory to one or
0148  e anterior part of the head is reduced to the MEREST rudiment attached to the bases of the prehens
0147  vanced, and likewise some other facts, may be MERGED under a more general principle, namely, that
0349  shores of Africa, on almost exactly opposite MERIDIANS of longitude. A third great fact, partly in
0407  affected the whole world, or at least great MERIDIONAL belts. As showing how diversified are the m
0031  ce of the principle of selection in regard to MERINO sheep is so fully recognised, that men follow
0285  me that he fully believes there is one in MERIONETHSHIRE of 12,000 feet; yet in these cases there
0443  time after the animal has been born, what its MERITS or form will ultimately turn out. We see this
0296  e trees chanced to have been preserved: thus, MESSRS. Lyell and Dawson found carboniferous beds 14
0029  s, and he will laugh you to scorn. I have never MET a pigeon, or poultry, or duck, or rabbit fanci
0091  hese varieties would cross and blend where they MET; but to this subject of intercrossing we shall
0176  ood with varieties in a state of nature. I have MET with striking instances of the rule in the cas
0185  ust occasionally have felt surprise when he has MET with an animal having habits and structure not
0232  cases of difficulty, as when two pieces of comb MET at an angle, how often the bees would entirely
0323  the Chalk. It is not that the same species are MET with; for in some cases not one species is ide
0326  in an equal degree, whenever their inhabitants MET, the battle would be prolonged and severe: and
0368  ve retreated to their present homes; but I have MET with no satisfactory evidence with respect to
0307  worn away by denudation, or obliterated by METAMORPHIC action, we ought to find only small remnan
0309  umbent water, might have undergone far more METAMORPHIC action than strata which have always remai
0309  rld, for instance in South America, of bare METAMORPHIC rocks, which must have been heated under g
0358  d, like other mountain summits, of granite, METAMORPHIC schists, old fossiliferous or other such r
0308  s suffered the extremity of denudation and METAMORPHISM. The case at present must remain inexplica
0308  and these ought to be very generally in a METAMORPHOSED condition. But the descriptions which we
0310  rior to the silurian epoch in a completely METAMORPHOSED condition. The several difficulties here
0343  than any known to us, may now all be in a METAMORPHOSED condition, or may lie buried under the oc
0436  elligible on the view that they consist of METAMORPHOSED leaves, arranged in a spire. In monstrous
0438  frequently speak of the skull as formed of METAMORPHOSED vertebrae: the jaws of crabs as metamorph
0438  amorphosed vertebrae: the jaws of crabs as METAMORPHOSED legs; the stamens and pistils of flowers
0438  egs; the stamens and pistils of flowers as METAMORPHOSED leaves; but it would in these cases proba
0438  , both jaws and legs, etc., as having been METAMORPHOSED, not one from the other, but from some co
0439  rough inheritance, if they had really been METAMORPHOSED during a long course of descent from true
0204  ir intermediate states show that wonderful METAMORPHOSES in function are at least possible. For in
0441  become attached and to undergo their final METAMORPHOSIS. When this is completed they are fixed fo
0442  rked in regard to cuttle fish, there is no METAMORPHOSIS; the cephalopodic character is manifested
0442  rs, there is nothing worthy to be called a METAMORPHOSIS. The larvae of insects, whether adapted t
0448  he young in these cases not undergoing any METAMORPHOSIS, or closely resembling their parents from
0448  however, of the embryo not undergoing any METAMORPHOSIS is perhaps requisite. If, on the other ha
0448  uld be useless; and in this case the final METAMORPHOSIS would be said to be retrograde. As all th
0310  y theory. For my part, following on Lyell's METAPHOR, I look at the natural geological record, as
0062  term Struggle for Existence in a large and METAPHORICAL sense, including dependence of one being o
0438  ists, however, use such language only in a METAPHORICAL sense: they are far from meaning that duri
0485  and aborted organs, etc., will cease to be METAPHORICAL, and will have a plain signification. When
0063  ts existence depends on birds; and it may METAPHORICALLY be said to struggle with other fruit bear
0421  d or descent to the same degree; they may METAPHORICALLY be called cousins to the same millionth d
0208  frederick Cuvier and several of the older METAPHYSICIANS have compared instinct with habit. This c
0028  an, His Majesty by crossing the breeds, which METHOD was never practised before, has improved them
0225  see whether Nature does not reveal to us her METHOD of work. At one end of a short series we have
0007  selection anciently followed, its Effects. METHODICAL and Unconscious Selection. Unknown Origin o
0033  principle of selection has been reduced to METHODICAL practice for scarcely more than three quart
0034  t the present time, eminent breeders try by METHODICAL selection, with a distinct object in view,
0039  ng continued, partly unconscious and partly METHODICAL selection. Perhaps the parent bird of all f
0083  ertainly has produced a great result by his METHODICAL and unconscious means of selection, what ma
0090  fleetness of his greyhounds by careful and METHODICAL selection, or by that unconscious selection
0102  ors, it will soon be exterminated. In man's METHODICAL selection, a breeder selects for some defin
0214  when the first tendency was once displayed, METHODICAL selection and the inherited effects of comp
0269  s are produced under domestication by man's METHODICAL and unconscious power of selection, for his
0035  is very same process, only carried on more METHODICALLY, did greatly modify, even during their own
0043  ative action of Selection, whether applied METHODICALLY and more quickly, or unconsciously and mor
0216  been the result of selection, pursued both METHODICALLY and unconsciously; but in most cases, prob
0467  is own benefit or pleasure. He may do this METHODICALLY, or he may do it unconsciously by preservi
0173  f structure as are specimens taken from the METROPOLIS inhabited by each. By my theory these allie
0175  mind that almost every species, even in its METROPOLIS, would increase immensely in numbers, were
0225  e of the humble bee, we have the cells of the MEXICAN Melipona domestica, carefully described and
0238  ds, the use of which is quite unknown: in the MEXICAN Myrmecocystus, the workers of one caste neve
0074  pends in a great degree on the number of field MICE, which destroy their combs and nests; and Mr.
0074  destroyed all over England. Now the number of MICE is largely dependent, as every one knows, on t
0074  tribute to the number of cats that destroy the MICE. Hence it is quite credible that the presence
0074  t determine, through the intervention first of MICE and then of bees, the frequency of certain flo
0091  t, for instance, taking to catch rats, another MICE; one cat, according to Mr. St. John, bringing
0091  snipes. The tendency to catch rats rather than MICE is shown to be inherited. Now, if any slight i
0180  rozen waters, and preys like other polecats on MICE and land animals. If a different case had been
0182  y as flappers, like the logger headed duck (MICROPTERUS of Eyton); as fins in the water and front
0093  s, taken from different branches, under the MICROSCOPE, and on all, without exception, there were
0246  ves are perfect in structure, as far as the MICROSCOPE reveals. In the first case the two sexual e
0363  ded their rocky burthens on the shores of these MID ocean islands; and it is at least possible tha
0231  till a smooth, very thin wall is left in the MIDDLE; the masons always piling up the cut away cem
```

Page **(Key Word)**

Page ***(Key Word)***

```
0232  on a slip of wood, placed directly under the  MIDDLE  of a comb growing downwards so that the comb
0293  when we see a species first appearing in the   MIDDLE  of any formation, it would be rash in the ext
0296  e, when the same species occur at the bottom,  MIDDLE  and top of a formation, the probability is t
0303  ccumulations of fossil mammals belongs to the  MIDDLE  of the secondary series; and one true mammal
0332  forms were to be discovered above one of the   MIDDLE  horizontal lines or geological formations, fo
0077  ch seeds (as peas and beans), when sown in the MIDST   of long grass, I suspect that the chief use o
0077  vigorously all around. Look at a plant in the  MIDST   of its range, why does it not double or quadr
0313  a few living shells may still be found in the  MIDST   of a multitude of extinct forms. Falconer has
0313  m. barrande, which intrude for a period in the MIDST   of an older formation, and then allow the pre
0360  hrown up on most islands, even on those in the MIDST   of the widest oceans; and the natives of the
0442  e, being fed by their parents or placed in the MIDST   of proper nutriment, yet nearly all pass thro
0001  me, it occurred to me, in 1837, that something MIGHT   perhaps be made out on this question by patie
0003  , geological succession, and other such facts, MIGHT   come to the conclusion that each species had
0015  considerable body, so that free intercrossing  MIGHT   check, by blending together, any slight devia
0022  te aborted. Several other less distinct breeds MIGHT   have been specified. In the skeletons of the
0022  other. Altogether at least a score of pigeons  MIGHT   be chosen, which if shown to an ornithologist
0023  l truly inherited sub breeds, or species as he MIGHT   have called them, could be shown him. Great a
0025  d marked, so that in each separate breed there MIGHT   be a tendency to revert to the very same colo
0029  raiser of Hereford cattle, whether his cattle  MIGHT   not have descended from long horns, and he wi
0035  eeds in question had been made long ago, which MIGHT   serve for comparison. In some cases, however,
0037  pear from the seed of the wild pear, though he MIGHT   succeed from a poor seedling growing wild, if
0038  eafter be more fully explained, two sub breeds MIGHT   be formed. This, perhaps, partly explains wha
0039  irly been established. Many slight differences MIGHT,  and indeed do now, arise amongst pigeons, wh
0052  species, and the species as the variety; or it MIGHT   come to supplant and exterminate the parent s
0052  nt and exterminate the parent species; or both MIGHT   co exist; and both rank as independent specie
0053  tions, I thought that some interesting results MIGHT   be obtained in regard to the nature and relat
0053  e ranges generally present varieties; and this MIGHT   have been expected, as they become exposed to
0054  er them, incipient species. And this, perhaps, MIGHT   have been anticipated; for, as varieties, in
0054  on the side of the larger genera. This, again, MIGHT   have been anticipated; for the mere fact of m
0054  favourable to the genus; and, consequently, we MIGHT   have expected to have found in the larger gen
0055  ve been favourable for variation; and hence we MIGHT   expect that the circumstances would generally
0074  a feline animal in large numbers in a district MIGHT   determine, through the intervention first of
0081  immediately undergo a change, and some species MIGHT   become extinct. We may conclude, from what we
0083  tives, we may safely conclude that the natives MIGHT   have been modified with advantage, so as to h
0085  see no reason to doubt that natural selection   MIGHT   be most effective in giving the proper colour
0087  mal's whole life, if of high importance to it, MIGHT   be modified to any extent by natural selectio
0087  r, more delicate and more easily broken shells MIGHT   be selected, the thickness of the shell being
0088  lection by always allowing the victor to breed MIGHT   surely give indomitable courage, length to th
0089  males, according to their standard of beauty,  MIGHT   produce a marked effect. I strongly suspect t
0090  or at some other period of the year, when they MIGHT   be compelled to prey on other animals. I can
0091  of the animals on which our wolf preyed, a cub MIGHT   be born with an innate tendency to pursue cer
0091  the repetition of this process, a new variety  MIGHT   be formed which would either supplant or coex
0091  s best fitted for the two sites, two varieties MIGHT   slowly be formed. These varieties would cross
0092  er, would likewise be favoured or selected. We MIGHT   have taken the case of insects visiting flowe
0092  h nine tenths of the pollen were destroyed, it MIGHT   still be a great gain to the plant; and those
0093  carried from flower to flower, another process MIGHT   commence. No naturalist doubts the advantage
0094  etc., far too slight to be appreciated by us,  MIGHT   profit a bee or other insect; so that an indi
0095  ly of precious nectar to the hive bee. Thus it MIGHT   be a great advantage to the hive bee to have
0095  le bees were to become rare in any country, it MIGHT   be a great advantage to the red clover to hav
0095  . thus I can understand how a flower and a bee MIGHT   slowly become, either simultaneously or one a
0102  ace in its polity not so perfectly occupied as MIGHT   be, natural selection will always tend to sel
0103  a very rapid rate, a new and improved variety  MIGHT   be quickly formed on any one spot, and might
0103  y might be quickly formed on any one spot, and MIGHT   there maintain itself in a body, so that what
0103  variety. A local variety when once thus formed MIGHT   subsequently slowly spread to other districts
0111  d differences. Mere chance, as we may call it, MIGHT   cause one variety to differ in some character
0115  ants through man's agency in foreign lands. It MIGHT   have been expected that the plants which have
0115  created and adapted for their own country. It  MIGHT,  also, perhaps have been expected that natura
0119  varieties. But these breaks are imaginary, and MIGHT   have been inserted anywhere, after intervals
0125  varieties or incipient species. This, indeed,  MIGHT   have been expected; for as natural selection
0135  e much used by these insects. In some cases we MIGHT   easily put down to disuse modifications of st
0137  on of the eyelids and growth of fur over them, MIGHT   in such case be an advantage; and if so, natu
0138  imilarity in their organisation and affinities MIGHT   have been expected; but, as Schiodte and othe
0138  wo continents are not more closely allied than MIGHT   have been anticipated from the general resemb
0138  ndness. Notwithstanding such modifications, we MIGHT   expect still to see in the cave animals of Am
0139  d and New Worlds should be closely related, we MIGHT   expect from the well known relationship of mo
0143  this had been of any great use to the breed it MIGHT   probably have been rendered permanent by natu
0145  r and outer flowers most frequently differ. It MIGHT   have been thought that the development of the
0147  e a little better fitted to be wafted further, MIGHT,  get an advantage over those producing seed le
0153  ued within a period not excessively remote, we MIGHT,  as a general rule, expect still to find more
0155  g only strongly marked and fixed varieties, we MIGHT   surely expect to find them still often contin
0160  trong or weak, to reproduce the lost character MIGHT   be, as was formerly remarked, for all that we
0161  ry, to have descended from a common parent, it MIGHT   be expected that they would occasionally vary
0161  fe and of a similar inherited constitution. It MIGHT   further be expected that the species of the s
0162  eversions or only analogous variations; but we MIGHT   have inferred that the blueness was a case of
0162  ther from simple variation. More especially we MIGHT   have inferred this, from the blue colour and
0174  periods; and in such islands distinct species  MIGHT   have been separately formed without the possi
0181  cted fingers and fore arm of the Galeopithecus MIGHT   be greatly lengthened by natural selection; a
0182  ould have ventured to have surmised that birds MIGHT   have existed which used their wings solely as
0182  d turning by the aid of their fluttering fins, MIGHT   have been modified into perfectly winged anim
0184  and of the other species of the same genus, we MIGHT   expect, on my theory, that such individuals w
0188  ructure even as perfect as the eye of an eagle MIGHT   be formed by natural selection, although in t
0189  e not believe that a living optical instrument MIGHT   thus be formed as superior to one of glass, a
0190  omach respire. In such cases natural selection MIGHT   easily specialise, if any advantage were thus
0190  titions. In these cases, one of the two organs MIGHT   with ease be modified and perfected so as to
0190  by the other organ; and then this other organ MIGHT   be modified for some other and quite distinct
0191  onceivable that the now utterly lost branchiae MIGHT   have been gradually worked in by natural sele
```

Page ***(Key Word)***

Page **************************************(Key Word)**************************************

```
0193  from one ancient progenitor thus provided, we  MIGHT  have expected that all electric fishes would
0195  ng correlated with constitutional differences,  MIGHT  assuredly be acted on by natural selection. T
0196  il having been formed in an aquatic animal, it  MIGHT  subsequently come to be worked in for all sor
0197  ntly that it was a character of importance and  MIGHT  have been acquired through natural selection:
0198  ly be affected. The shape, also, of the pelvis  MIGHT  affect by pressure the shape cf the head of t
0199  light analogous differences between species. I  MIGHT  have adduced for this same purpose the differ
0200  herefore we may infer that these several bones  MIGHT  have been acquired through natural selection,
0207  rile insects. Summary. The subject of instinct  MIGHT  have been worked into the previous chapters:
0209   had played a tune with no practice at all, he  MIGHT  truly be said to have done so instinctively.
0209  possible that slight modifications of instinct  MIGHT  be profitable to a species: and if it can be
0210  in which case either one or the other instinct  MIGHT  be preserved by natural selection. And such i
0212  l and strange habits in certain species, which  MIGHT, if advantageous to the species, give rise, t
0217  e the case, the process of laying and hatching  MIGHT  be inconveniently long, more especially as sh
0220  es are very few in number, I thought that they  MIGHT  behave differently when more numerous; but Mr
0223  possible that pupae originally stored as food  MIGHT  become developed; and the ants thus unintenti
0224  habit of collecting pupae originally for food  MIGHT  by natural selection be strengthened and rend
0233  planes. It is even conceivable that an insect  MIGHT, by fixing on a point at which to commence a
0263  in both cases, follow to a certain extent, as  MIGHT  have been expected, systematic affinity, by w
0273  has been recent variability: and therefore we  MIGHT  expect that such variability would often cont
0279  hapter I enumerated the chief objections which  MIGHT  be justly urged against the views maintained
0281  ble by my theory, that one of two living forms  MIGHT  have descended from the other: for instance,
0286  ks of which the overlying sedimentary deposits  MIGHT  have accumulated in thinner masses than elsew
0294  tant future, a geologist examining these beds,  MIGHT  be tempted to conclude that the average durat
0297  d it be forgotten, as before explained, that A  MIGHT  be the actual progenitor of B and C, and yet
0297  t be the actual progenitor of B and C, and yet  MIGHT  not at all necessarily be strictly intermedia
0297  en them in all points of structure. So that we  MIGHT  obtain the parent species and its several mod
0303  recall a remark formerly made, namely that it  MIGHT  require a long succession of ages to adapt an
0305  inc the secondary period: and these cirripedes  MIGHT  have been the progenitors of our many tertiar
0305  now have a confined range: the telecstean fish  MIGHT  formerly have had a similarly confined range,
0305  having been largely developed in some one sea,  MIGHT  have spread widely. Nor have we any right to
0305  in, in which any great group of marine animals  MIGHT  be multiplied; and here they would remain con
0309  g such to have been formerly deposited; for it  MIGHT  well happen that strata which had subsided so
0309  by an enormous weight of superincumbent water,  MIGHT  have undergone far more metamorphic action th
0315  recur. For though the offspring of one species  MIGHT  be adapted (and no doubt this has occurred in
0315  striving during long ages for the same object,  MIGHT  make a new breed hardly distinguishable from
0319  sil horse still existing as a rare species, we  MIGHT  have felt certain from the analogy of all oth
0324  ters coexisted with the Mastodon and Horse, it  MIGHT  at least have been inferred that they had liv
0326  e marine inhabitants of the continuous sea. We  MIGHT  therefore expect to find, as we apparently do
0326  ea. Dominant species spreading from any region  MIGHT  encounter still more dominant species; and th
0326  me from one birthplace and some from the other  MIGHT  be victorious. But in the course of time, the
0327  sappear: though here and there a single member  MIGHT  long be enabled to survive. Thus, as it seems
0331  he case of some Silurian forms, that a species  MIGHT  go on being slightly modified in relation to
0335  esponding lengths of time: a very ancient form  MIGHT  occasionally last much longer than a form els
0347  ere, under a slightly different climate, there  MIGHT  have been free migration for the northern tem
0358  lled accidental means, but which more properly  MIGHT  be called occasional means of distribution. I
0359  oned timber: and it occurred to me that floods  MIGHT  wash down plants or branches; and that these
0359  t wash down plants or branches, and that these  MIGHT  be dried on the banks, and then by a fresh ri
0359  that the seeds of 14/100 plants of any country  MIGHT  be floated by sea currents during 28 days, an
0360  eeds of 14/100 plants belonging to one country  MIGHT  be floated across 924 miles of sea to another
0362  r 12 or even 18 hours. A bird in this interval  MIGHT  easily be blown to the distance of 500 miles,
0362  ed birds, and the contents of their torn crops  MIGHT  thus readily get scattered. Mr. Brent informs
0362  equently devoured by birds, and thus the seeds  MIGHT  be transported from place to place. I forced
0363  te as large as the seed of a vetch. Thus seeds  MIGHT  occasionally be transported to great distance
0364  eds from North America to Britain; though they  MIGHT  and do bring seeds from the West Indies to ou
0365  tly created at several distinct points; and we  MIGHT  have remained in this same belief, had not Ag
0377  withstand a considerable amount of cold, many  MIGHT  have escaped extermination curing a moderate
0377  the more vigorous and dominant temperate forms  MIGHT  have penetrated the native ranks and have rea
0383  arated from each other by barriers of land, it  MIGHT  have been thought that fresh water production
0384  ne that a marine member of a fresh water group  MIGHT  travel far along the shores of the sea, and s
0385  re effectual: I suspended a duck's feet, which  MIGHT  represent those of a bird sleeping in a natur
0385  rs: and in this length of time a duck or heron  MIGHT  fly at least six or seven hundred miles, and
0386  ant from the nearest land: how much farther it  MIGHT  have flown with a favouring gale no one can t
0387  eeds of the Nelumbium undigested: or the seeds  MIGHT  be dropped by the bird whilst feeding its you
0390  ent, we shall see that this is true. This fact  MIGHT  have been expected on my theory, for, as alre
0391  oating timber, or to the feet of wading birds,  MIGHT  be transported far more easily than land shel
0392  s no difficulty on my view, for a hooked seed  MIGHT  be transported to an island by some other mea
0392  aving to compete with herbaceous plants alone,  MIGHT  readily gain an advantage by growing taller a
0397  branous diaphragm over the mouth of the shell,  MIGHT  be floated in chinks of drifted timber across
0399  nd, than to any other region: and this is what  MIGHT  have been expected: but it is also plainly re
0400  other part of the world. And this is just what  MIGHT  have been expected on my view, for the island
0401  erence in the several islands. This difference  MIGHT  indeed have been expected on the view of the
0401   the different islands. Some species, however,  MIGHT  spread and yet retain the same character thro
0414  nc together the forms most like each other. It  MIGHT  have been thought (and was in ancient times t
0422  uld, I think, be the only possible one. Yet it  MIGHT  be that some very ancient language had altere
0423   it has been attempted by some authors. For we  MIGHT  feel sure, whether there had been more or les
0424  classed under the Negro group, however much he  MIGHT  differ in colour and other important characte
0425  e supposition is of course preposterous: and I  MIGHT  answer by the argumentum ad hominem, and ask
0435  rn, or to transpose parts. The bones of a limb  MIGHT  be shortened or widened to any extent, and be
0435  ane, so as to serve as a fin: or a webbed foot  MIGHT  have all its bones, or certain bones, lenghte
0436  nceivable that the general pattern of an organ  MIGHT  become so much obscured as to be finally lost
0442  fferent animals within the same class, that we  MIGHT  be led to look at these facts as necessarily
0444  is, variations which, for all that we can see,  MIGHT  have appeared earlier or later in life, tend
0447  ertain cases the successive steps of variation  MIGHT  supervene, from causes of which we are wholly
0447  , at a very early period of life, or each step  MIGHT  be inherited at an earlier period than that a
0448  corresponding ages, the active young or larvae  MIGHT  easily be rendered by natural selection diffe
0448  le extent from their parents. Such differences  MIGHT, also, become correlated with successive stag
```

Page **************************************(Key Word)**************************************

Page ***************************************(Key Word)***************************************

```
0448  pment; so that the larvae, in the first stage, MIGHT differ greatly from the larvae in the second
0448  seen to be the case with cirripedes. The adult MIGHT become fitted for sites or habits, in which o
0454  ain, an organ useful under certain conditions, MIGHT become injurious under others, as with the wi
0454  of life, useless or injurious for one purpose, MIGHT easily be modified and used for another purpo
0454  fied and used for another purpose. Or an organ MIGHT be retained for one alone of its former funct
0456  redly do on the ordinary doctrine of creation, MIGHT even have been anticipated, and can be accoun
0457  difficulties; on the contrary, their presence MIGHT have been even anticipated. The importance of
0468  land has undergone great physical changes, we MIGHT have expected that organic beings would have
0470  the manufactory of species has been active, we MIGHT expect, as a general rule, to find it still i
0472  e, these facts cease to be strange, or perhaps MIGHT even have been anticipated. As natural select
0474  variability and modification, and therefore we MIGHT expect this part generally to be still variab
0478  itants of the nearest source whence immigrants MIGHT have been derived. We see this in nearly all
0488  number of species, however, keeping in a body MIGHT remain for a long period unchanged, whilst wi
0488  ming into competition with foreign associates, MIGHT become modified; so that we must not overrate
0207  it is said that instinct impels the cuckoo to MIGRATE and to lay her eggs in other birds' nests. A
0217  veniently long, more especially as she has to MIGRATE at a very early period; and the first hatche
0219  nest is found inconvenient, and they have to MIGRATE, it is the slaves which determine the migrat
0223  ere a new nest shall be formed, and when they MIGRATE, the masters carry the slaves. Both in Switz
0301  inhabitants of the archipelago would have to MIGRATE, and no closely consecutive record of their
0351  stand in direct competition with each other, MIGRATE in a body into a new and afterwards isolated
0353  lieve, is, that mammals have not been able to MIGRATE, whereas some plants, from their varied mean
0353  on one side alone, and have not been able to MIGRATE to the other side. Some few families, many s
0370  th in the Old and New Worlds, began slowly to MIGRATE southwards as the climate became less warm,
0378  e which had not reached the equator, would re MIGRATE northward or southward towards their former
0138  ls, having ordinary powers of vision, slowly MIGRATED by successive generations from the outer wor
0351  e geological period, so certain species have MIGRATED over vast spaces, and have not become greatl
0351  h difficulty in believing that they may have MIGRATED from the same region; for curing the vast ge
0352  ing how the same species could possibly have MIGRATED from some one point to the several distant a
0353  , from their varied means of dispersal, have MIGRATED across the vast and broken interspace. The g
0353  d in one area alone, and having subsequently MIGRATED from that area as far as its powers of migra
0354  s has been produced within one area, and has MIGRATED thence as far as it could. It would be hopel
0354  explained on the view of each species having MIGRATED from a single birthplace; then, considering
0354  descended from a common progenitor, can have MIGRATED (undergoing modification during some part of
0365  out the apparent possibility of their having MIGRATED from one to the other. It is indeed a remark
0371  mon, which inhabited the New and Old Worlds, MIGRATED south of the Polar Circle, they must have be
0371  long ages ago. And as the plants and animals MIGRATED southward, they will have become mingled in
0378  cold was most intense, when arctic forms had MIGRATED some twenty five degrees of latitude from th
0378  strial animals, and some marine productions, MIGRATED during the glacial period from the northern
0379  ical plants and allied forms have apparently MIGRATED from the north to the south, than in a rever
0380  reason why certain speciesand not other have MIGRATED; why certain species have been modified and
0381  that peculiar and very distinct species have MIGRATED in radiating lines from some common centre;
0410  or to those which have changed after having MIGRATED into distant quarters, in both cases the for
0487  e period descended from one parent, and have MIGRATED from some one birthplace; and when we better
0352  remarkable and exceptional. The capacity of MIGRATING across the sea is more distinctly limited i
0377  other hand, the temperate productions, after MIGRATING nearer to the equator, though they will hav
0488  is same period, several of these species, by MIGRATING into new countries and coming into competit
0219  igrate, it is the slaves which determine the MIGRATION, and actually carry their masters in their
0221  . one day I fortunately chanced to witness a MIGRATION from one nest to another, and it was a most
0222  odies of F. fusca (showing that it was not a MIGRATION) and numerous pupae. I traced the returning
0293  kinds, we may safely infer a large amount of MIGRATION during climatal and other changes; and when
0294  me having apparently been required for their MIGRATION from the American to the European seas. I t
0294  sson to reflect on the ascertained amount of MIGRATION of the inhabitants of Europe during the Gla
0294  disappear at different levels, owing to the MIGRATION of species and to geographical changes. And
0313  ation, namely,that it is a case of temporary MIGRATION from a distinct geographical province, seem
0327  tion and extinction, and that there was much MIGRATION from other parts of the world. As we have r
0340  geographical changes, permitting much inter MIGRATION, the feebler will yield to the more dominan
0342  the average duration of specific forms; that MIGRATION has played an important part in the first a
0343  ations; he may overlook how important a part MIGRATION must have played, when the formations of an
0347  t barriers of any kind, or obstacles to free MIGRATION, are related in a close and important manne
0347  ifferent climate, there might have been free MIGRATION for the northern temperate forms, as there
0350  e degree of dissimilarity will depend on the MIGRATION of the more dominant forms of life from one
0350  ance of barriers comes into play by checking MIGRATION; as does time for the slow process of modif
0351  e little liable to modification; for neither MIGRATION nor isolation in themselves can do anything
0351  ed since ancient times, almost any amount of MIGRATION is possible. But in many other cases, in wh
0352  causa of ordinary generation with subsequent MIGRATION, and calls in the agency of a miracle. It i
0352  the space could not be easily passed over by MIGRATION, the fact is given as something remarkable
0353  rated from that area as far as its powers of MIGRATION and subsistence under past and present cond
0354  going modification during some part of their MIGRATION) from the area inhabited by their progenito
0356  limate must have had a powerful influence on MIGRATION: a region when its climate was different ma
0356  was different may have been a high road for MIGRATION, but now be impassable; I shall, however, p
0357  for plants and for many animals during their MIGRATION. In the coral producing oceans such sunken
0367  america than in Europe, so will the southern MIGRATION there have been a little earlier or later;
0367  e north or nearly due north of them: for the MIGRATION as the cold came on, and the re migration o
0367  he migration as the cold came on, and the re MIGRATION on the returning warmth, will generally hav
0368  that a colder climate permitted their former MIGRATION across the low intervening tracts, since be
0368  the arctic forms, during their long southern MIGRATION and re migration northward, will have been
0368  during their long southern migration and re MIGRATION northward, will have been exposed to nearly
0371  e rendered impassable by cold, for the inter MIGRATION of their inhabitants. During the slowly dec
0372  so in the waters of the sea, a slow southern MIGRATION of a marine fauna, which during the Pliocen
0377  eriod will have been ample for any amount of MIGRATION. As the cold came slowly on, all the tropic
0379  any modification during their long southern MIGRATION and re migration northward, the case may ha
0379  during their long southern migration and re MIGRATION northward, the case may have been wholly di
0379  abyssinia. I suspect that this preponderant MIGRATION from north to south is due to the greater e
0380  end to indicate the exact lines and means of MIGRATION, or the reason why certain speciesand not o
0381  commencement of the Glacial period for their MIGRATION, and for their subsequent modification to t
0384  es, and consequently time and means for much MIGRATION. In the second place, salt water fish can w
```

Page ***************************************(Key Word)***************************************

```
0388  ll give longer time than the average for the  MIGRATION  of the same aquatic species, we should not
0396  arest continent; for on this latter view the   MIGRATION  would probably have been more complete; and
0404  period there has been intercommunication or    MIGRATION  between the two regions. And wherever many
0404  his relation between the power and extent of   MIGRATION  of a species, either at the present time or
0405  dents of transport; and consequently for the   MIGRATION  of some of the species into all quarters of
0407  periods of time being thus granted for their   MIGRATION, I do not think that the difficulties are i
0408  distribution are explicable on the theory of   MIGRATION (generally of the more dominant forms of li
0409  culiar; and why, in relation to the means of   MIGRATION, one group of beings, even within the same
0410  have attempted to show, be accounted for by    MIGRATION at some former period under different condi
0462  there will always be a good chance for wide    MIGRATION by many means. A broken or interrupted rang
0462  nges will obviously have greatly facilitated   MIGRATION. As an example, I have attempted to show ho
0462  has necessarily been slow, all the means of    MIGRATION will have been possible during a very long
0476  has been during the long course of ages much   MIGRATION from one part of the world to another, owin
0477  colonists. On this same principle of former    MIGRATION, combined in most cases with modification,
0477  ill inevitably be different. On this view of   MIGRATION, with subsequent modification, we can see w
0487  e; and when we better know the many means of   MIGRATION, then, by the light which geology now throw
0223  ld its own nest, does not determine its own    MIGRATIONS, does not collect food for itself or its yo
0383  ghly useful to them, for short and frequent    MIGRATIONS from pond to pond, or from stream to stream
0391  h american birds, during their great annual    MIGRATIONS, visit either periodically or occasionally
0487  to trace in an admirable manner the former     MIGRATIONS of the inhabitants of the whole world. Even
0212  stincts certainly do vary, for instance, the   MIGRATORY instinct, both in extent and direction, and
0283  rolled about! Moreover, if we follow for a few  MILES any line of rocky cliff, which is undergoing
0284  ery nearly thirteen and three quarters British MILES. Some of these formations, which are represen
0285  fault, for.instance, extends for upwards of 30 MILES, and along this line the vertical displacemen
0286  the northern to the southern Downs is about 22 MILES, and the thickness of the several formations
0286  believe that any line of coast, ten or twenty  MILES in length, ever suffers degradation at the sa
0290  act struck me more when examining many hundred MILES of the South American coasts, which have been
0301  ants of the archipelago now range thousands of MILES beyond its confines; and analogy leads me to
0308  nited States; and from the amount of sediment, MILES in thickness, of which the formations are com
0309  ell happen that strata which had subsided some  MILES nearer to the centre of the earth, and which
0354  the mainland, though separated by hundreds of  MILES of open sea. If the existence of the same spe
0355  nd formed at the distance of a few hundreds of MILES from a continent, would probably receive from
0359  ge rate of the several Atlantic currents is 33 MILES per diem (some currents running at the rate o
0359  diem (some currents running at the rate of 60 MILES per diem); on this average, the seeds of 14/1
0360  ing to one country might be floated across 924 MILES of sea to another country; and. when stranded,
0360  ed, could be floated across a space of sea 900 MILES in width, and would then germinate. The fact
0361  stances their rate of flight would often be 35 MILES an hour; and some authors have given a far hi
0362  l might easily be blown to the distance of 500 MILES, and hawks are known to look out for tired bi
0364  al transport across tracts of sea some hundred MILES in breadth, or from island to island, or from
0365  mits, separated from each other by hundreds of MILES of low lands, where the Alpine species could
0373  ly affected. Along the Himalaya, at points 900 MILES apart, glaciers have left the marks of their
0385  heron might fly at least six or seven hundred  MILES, and would be sure to alight on a pool or riv
0386  once flew on board the Beagle, when forty five MILES distant from the nearest land: how much farth
0389  ed stations of New Zealand, extending over 780 MILES of latitude, and compare its flowering plants
0391  than land shells, across three or four hundred MILES of open sea. The different orders of insects
0393  tives) inhabiting an island situated above 300 MILES from a continent or great continental island;
0394  sionally visit Bermuda, at the distance of 600 MILES from the mainland. I hear from Mr. Tomes, who
0397  ituated under the equator, between 500 and 600 MILES from the shores of South America. Here almost
0398  slands in the Pacific, distant several hundred MILES from the continent, yet feels that he is stan
0147  art; thus it is difficult to get a cow to give MILK and to fatten readily. The same varieties of t
0223  nest: both, but chiefly the slaves, tend, and  MILK as it may be called, their aphides; and thus b
0451  loped in full grown males, and having secreted MILK. So again there are normally four developed an
0451  ws the two sometimes become developed and give MILK. In individual plants of the same specie
0011  goats in countries where they are habitually   MILKED, in comparison with the state of these organs
0226  he hive bee. Accordingly I wrote to Professor  MILLER, of Cambridge, and this geometer has kindly r
0283  n away. The observations on this head by Hugh  MILLER, and by that excellent observer Mr. Smith of
0362  of germination. Some seeds of the oat, wheat,  MILLET, canary, hemp, clover, and beet germinated af
0013  ears in the parent, say, once amongst several  MILLION individuals, and it reappears in the child,
0064  so on, then in twenty years there would be a   MILLION plants. The elephant is reckoned to be the s
0064  he fifth century there would be alive fifteen  MILLION elephants, descended from the first pair. Bu
0124  housand generations, but each may represent a  MILLION or a hundred million generations; and likewi
0124  but each may represent a million or a hundred  MILLION generations, and likewise a section of the s
0189  ate of the instrument to be multiplied by the  MILLION; and each to be preserved till a better be p
0287  uired 306,662,400 years; or say three hundred  MILLION years. The action of fresh water on the gent
0287  all probability a far longer period than 300   MILLION years has elapsed since the latter part of t
0481  asp the full meaning of the term of a hundred  MILLION years; it cannot add up and perceive the ful
0189  each improvement. Let this process go on for   MILLIONS on millions of years; and curing each year o
0189  ment. Let this process go on for millions on   MILLIONS of years; and during each year on millions o
0189  n millions of years; and during each year on  MILLIONS of individuals of many kinds; and may we not
0287  undergone, the surface may have existed for    MILLIONS of years as land, and thus have escaped the
0363  rged with seeds. Reflect for a moment on the   MILLIONS of quails which annually cross the Mediterra
0421  metaphorically be called cousins to the same   MILLIONTH degree; yet they differ widely and in diffe
0116  dividual body, a subject so well elucidated by MILNE Edwards. No physiologist doubts that a stomac
0194  of almost every experienced naturalist; or, as MILNE Edwards has well expressed it, nature is prod
0418  een strongly urged by those great naturalists, MILNE Edwards and Agassiz, that embryonic character
0433  classes, we are tending in this direction; and MILNE Edwards has lately insisted, in an able paper
0019  ecies of sheep peculiar to it! When we bear in MIND that Britain has now hardly one peculiar mamma
0036  ears. There is not a suspicion existing in the MIND of any one at all acquainted with the subject
0050  s within one country, he will soon make up his MIND how to rank most of the doubtful forms. His ge
0051  he end generally be enabled to make up his own MIND which to call varieties and which species; but
0051  insensible series; and a series impresses the  MIND with the idea of an actual passage. Hence I lo
0062  so, than constantly to bear this conclusion in MIND. Yet unless it be thoroughly engrained in the
0062  . yet unless it be thoroughly engrained in the MIND, I am convinced that the whole economy of natu
0062  s and beasts of prey: we do not always bear in MIND, that though food may be now superabundant, it
0066  to keep the foregoing considerations always in MIND, never to forget that every single organic bei
0067  a few remarks, just to recall to the reader's  MIND some of the chief points. Eggs or very young a
```

Page ★★★(Key Word)★★

```
0078 re. ALL that we can do, is to keep steadily in MIND that each organic being is striving to increa
0080 t can act most effectually. Let it be borne in MIND in what an endless number of strange peculiar
0080 mes in some degree plastic. Let it be borne in MIND how infinitely complex and close fitting are
0085 it is, however, far more necessary to bear in MIND that there are many unknown laws of correlati
0094 ble, enter by the mouth. Bearing such facts in MIND, I can see no reason to doubt that an acciden
0124 ill be able to bring some such case before his MIND. In the diagram, each horizontal line has hit
0175 graduate away insensibly. But when we bear in MIND that almost every species, even in its metrop
0182 wings of a bird for flight, we should bear in MIND that animals displaying early transitional gr
0188 the eyes of living crustaceans, and bearing in MIND how small the number of living animals is in
0191 tions of organs, it is so important to bear in MIND the probability of conversion from one funct★
0204 congeners. Hence we can understand, bearing in MIND that each organic being is trying to live whe
0208 , a remarkably accurate notion of the frame of MIND under which an instinctive action is performe
0212 an produce but a feeble effect on the reader's MIND. I can only repeat my assurance, that I do no
0213 dest tricks, associated with certain frames of MIND or periods of time. But let us look to the fa
0220 o approach the subject in a sceptical frame of MIND, as any one may well be excused for doubting
0252 authenticated. It should, however, be borne in MIND that, owing to few animals breeding freely un
0280 n the first place it should always be borne in MIND what sort of intermediate forms must, on my t
0282 e a practical geologist, the facts leading the MIND feebly to comprehend the lapse of time. He wh
0283 of Jordan Hill, are most impressive. With the MIND thus impressed, let any one examine beds of c
0285 the consideration of these facts impresses my MIND almost in the same manner as does the vain en
0287 t an infinite number of generations, which the MIND cannot grasp, must have succeeded each other
0352 procued within a single region captivates the MIND. He who rejects it, rejects the vera causa of
0377 e warmest spots. But the great fact to bear in MIND is, that all tropical productions will have s
0377 wn. Hence, it seems to me possible, bearing in MIND that the tropical productions were in a suffe
0405 tribution of certain genera, we should bear in MIND that some are extremely ancient, and must hav
0407 been properly experimentised on: if we bear in MIND how often a species may have ranged continuou
0408 merly inhabited the same continent. Bearing in MIND that the mutual relations of organism to orga
0450 hey first appeared. It should also be borne in MIND, that the supposed law of resemblance of anci
0456 view of classification, it should be borne in MIND that the element of descent has been universa
0457 g ages. On this same principle, and bearing in MIND, that when organs are reduced in size, either
0457 s to provide for its own wants, and bearing in MIND how strong is the principle of inheritance, t
0481 ed, by the slow action of the coast waves. The MIND cannot possibly grasp the full meaning of the
0482 naturalists, endowed with much flexibility of MIND, and who have already begun to doubt on the f
0488 species has been independently created. To my MIND it accords better with what we know of the la
0029 neral arguments, and refuse to sum up in their MINDS slight differences accumulated during many s
0215 nts our seeing how universally and largely the MINDS of our domestic animals have been modified b
0481 pect to convince experienced naturalists whose MINDS are stocked with a multitude of facts all vi
0360 ants. He tried 98 seeds, mostly different from MINE; but he chose many large fruits and likewise
0382 speculated, in language almost identical with MINE, on the effects of great alternations of clim
0481 ars, from a point of view directly opposite to MINE. It is so easy to hide our ignorance under su
0322 fferent climates, where not a fragment of the MINERAL chalk itself can be found; namely, in North
0290 ned. The frequent and great changes in the MINERALOGICAL composition of consecutive formations, g
0295 a formation composed of beds of different MINERALOGICAL composition, we may reasonably suspect t
0042 for thus many races may be kept true, though MINGLED in the same aviary, and this circumstance m
0141 al and arctic wolf or wild dog may perhaps be MINGLED in our domestic breeds. The rat and mouse ca
0214 are inherited, and how curiously they become MINGLED, is well shown when different breeds of dogs
0221 stock of slaves, and I observed a few slaves MINGLED with their masters leaving the nest, and ma
0364 ant continents would not by such means become MINGLED in any great degree; but would remain as dis
0369 ey will, also, in all probability have become MINGLED with ancient Alpine species, which must have
0371 als migrated southward, they will have become MINGLED in the one great region with the native Ame
0378 of the tropical lowlands were clothed with a MINGLED tropical and temperate vegetation, like that
0064 ave taken some pains to estimate its probable MINIMUM rate of natural increase: it will be under t
0191 re instance. Pedunculated cirripedes have two MINUTE folds of skin, called by me the ovigerous fre
0224 nct of the Hive Bee. I will not here enter on MINUTE details on this subject; but will merely give
0228 of wax: the bees instantly began to excavate MINUTE circular pits in it; and as they deepened the
0363 g to their feet would sometimes include a few MINUTE seeds? But I shall presently have to recur to
0385 ng; and I found that numbers of the extremely MINUTE and just hatched shells crawled on the feet,
0406 e seeds and eggs of many low forms being very MINUTE and better fitted for distant transportation
0441 and their two eyes are now reconverted into a MINUTE, single, and very simple eye spot. In this la
0453 us to the system; but can we suppose that the MINUTE papilla, which often represents the pistil in
0454 an ear in earless breeds, the reappearance of MINUTE dangling horns in hornless breeds of cattle,
0306 it would be for a naturalist to land for five MINUTES on some one barren point in Australia, and t
0258 observed by Kolreuter. To give an instance: MIRABILIS jalappa can easily be fertilised by the pol
0352 quent migration, and calls in the agency of a MIRACLE. It is universally admitted, that in most ca
0483 n. these authors seem no more startled at a MIRACULOUS act of creation than at an ordinary birth.
0487 cting and still existing causes, and not by MIRACULOUS acts of creation and by catastrophes; and a
0065 he larger domestic animals tends, I think, to MISLEAD us: we see no great destruction falling on t
0427 re are innumerable instances: thus Linnaeus, MISLED by external appearances, actually classed an
0076 f another species. The recent increase of the MISSEL thrush in parts of Scotland has caused the de
0003 under the bark of trees. In the case of the MISSELTOE, which draws its nourishment from certain t
0004 birth to a woodpecker, and some plant to the MISSELTOE, and that these had been produced perfect a
0284 ted that sediment is deposited by the great MISSISSIPPI river at the rate of only 600 feet in a hu
0294 of the glacial period near the mouth of the MISSISSIPPI, within that limit of depth at which marin
0294 ited in shallow water near the mouth of the MISSISSIPPI during some part of the glacial period sha
0060 aptations most plainly in the woodpecker and MISSELTOE; and only a little less plainly in the humb
0063 r kinds which already clothe the ground. The MISSELTOE is dependent on the apple and a few other t
0063 be said to struggle with each other. As the MISSELTOE is disseminated by birds, its existence dep
0063 will languish and die. But several seedling MISSELTOES, growing close together on the same branch,
0217 high authority of Dr. Brewer, that this is a MISTAKE. Nevertheless, I could give several instance
0414 ikely in the modifications of these organs to MISTAKE a merely adaptive for an essential character
0185 g when unwillingly it takes flight, would be MISTAKEN by any one for an auk or grebe: nevertheless
0217 gorous by advantage having been taken of the MISTAKEN maternal instinct of another bird, than by t
0243 always absolutely perfect and are liable to MISTAKES; that no instinct has been produced for the
0475 s being apparently not perfect and liable to MISTAKES, and at many instincts causing other animals
0062 and variation, will be dimly seen or quite MISUNDERSTOOD. We behold the face of nature bright with
0066 r at recurrent intervals. Lighten any check, MITIGATE the destruction ever so little, and the numb
```

Page ★★(Key Word)★★★

```
age ******************************************( Key Word )******************************************
```

```
)75 l varieties of wheat be sown together, and the MIXED seed be resown, some of the varieties which b
)75 ite supplant the other varieties. To keep up a MIXED stock of even such extremely close varieties
)75 h year harvested separately, and the seed then MIXED in due proportion, otherwise the weaker kinds
)76 nstitution, that the original proportions of a MIXED stock could be kept up for half a dozen gener
113 d good when first one variety and then several MIXED varieties of wheat have been sown on equal sp
167 lived, but had been created in stone so as to MOCK the shells now living on the sea shore. Summar
167 nown, cause. It makes the works of God a mere MOCKERY and deception: I would almost as soon believ
162 our systematic works, is due to its varieties MOCKING, as it were, some of the other species of th
402 hus there are three closely allied species of MOCKING thrush, each confined to its own island. Now
402 ned to its own island. Now let us suppose the MOCKING thrush of Chatham Island to be blown to Char
402 be blown to Charles Island, which has its own MOCKING thrush: why should it succeed in establishin
402 possibly be reared: and we may infer that the MOCKING thrush peculiar to Charles Island is at leas
419 the embryonic leaves or cotyledons, and on the MODE of development of the plumule and radicle. In
031 ion as something quite plastic, which they can MODEL almost as they please. If I had space I could
377 ny might have escaped extermination during a MODERATE fall of temperature, more especially by esca
387 lowed them: even small fish swallow seeds of MODERATE size, as of the yellow water lily and Potamo
155 their structure which have varied within a MODERATELY recent period, and which have thus come to
185 es, can be shown to exist, until we reach a MODERATELY high stage of perfection. In certain crusta
282 er along lines of sea coast, when formed of MODERATELY hard rocks, and mark the process of degrada
336 s, very long as measured by years, but only MODERATELY long as measured geologically, closely alli
395 on either side the islands are situated on MODERATELY deep submarine banks, and they are inhabiti
397 floated in chinks of drifted timber across MODERATELY wide arms of the sea. And I found that seve
033 is very far from true that the principle is a MODERN discovery. I could give several references to
095 gainst Sir Charles Lyell's noble views on the MODERN changes of the earth, as illustrative of geol
095 ach profitable to the preserved being: and as MODERN geology has almost banished such views as the
086 large groups of animals, pair for each birth. MODERN research has much diminished the number of su
254 ted with Pallas, has been largely accepted by MODERN naturalists: namely, that most of our domesti
324 n, i think, be little doubt that all the more MODERN marine formations, namely, the upper pliocene
324 upper pliocene, the pleistocene and strictly MODERN beds, of Europe, North and South America, and
407 w how important has been the influence of the MODERN Glacial period, which I am fully convinced si
423 d connect together all languages, extinct and MODERN, by the closest affinities, and would give th
462 changes which have affected the earth during MODERN periods: and such changes will obviously have
004 to gain a clear insight into the means of MODIFICATION and coadaptation. At the commencement of m
004 thus see that a large amount of hereditary MODIFICATION is at least possible: and, what is equally
006 and, as I believe, the future success and MODIFICATION of every inhabitant of this world. Still l
006 s been the main but not exclusive means of MODIFICATION. Variation Under Domestication Chapter I.
008 are still capable of rapid improvement or MODIFICATION. It has been disputed at what period of li
041 w of the accumulation of a large amount of MODIFICATION in almost any desired direction. But as va
044 iation in a technical sense, as implying a MODIFICATION directly due to the physical conditions of
082 n by intruders. In such case, every slight MODIFICATION, which in the course of ages chanced to ar
083 e half monstrous form: or at least by some MODIFICATION prominent enough to catch his eye, or to b
087 r the bird's own advantage, the process of MODIFICATION would be very slow, and there would be sim
096 organic beings, or of any great and sudden MODIFICATION in their structure. On the Intercrossing o
102 rom the best animals, much improvement and MODIFICATION surely but slowly follow from this unconsc
104 conditions of life change and they undergo MODIFICATION.uniformity of character can be given to th
106 ion of new species, yet that the course of MODIFICATION will generally have been more rapid on lar
106 less severe, and there will have been less MODIFICATION and less extermination. Hence, perhaps, it
108 the inhabitants of the country undergoing MODIFICATION of some kind. The existence of such places
110 closest competition with those undergoing MODIFICATION and improvement, will naturally suffer mos
119 e cases I do not doubt that the process of MODIFICATION will be confined to a single line of desce
120 ncreased: although the amount of divergent MODIFICATION may have been increased in the successive
120 nly to suppose the steps in the process of MODIFICATION to be more numerous or greater in amount,
121 m want of space. But during the process of MODIFICATION, represented in the diagram, another of ou
122 umec to represent a considerable amount of MODIFICATION, species (A) and all the earlier varieties
123 t the first commencement of the process of MODIFICATION, will be widely different from the other f
123 more genera are produced by descent, with MODIFICATION, from two or more species of the same genu
125 s. i see no reason to limit the process of MODIFICATION, as now explained, to the formation of gen
125 ders, according to the amount of divergent MODIFICATION supposed to be represented in the diagram.
128 uralised productions. Therefore during the MODIFICATION of the descendants of any one species, and
143 e cases, played a considerable part in the MODIFICATION of the constitution, and of the structure
146 ccuired through natural selection some one MODIFICATION in structure, and, after thousands of gene
146 of generations, some other and independent MODIFICATION: and these two modifications, having been
153 ch variability in the structure undergoing MODIFICATION. It further deserves notice that these var
153 t has undergone an extraordinary amount of MODIFICATION, since the period when the species branche
153 logical period. An extraordinary amount of MODIFICATION implies an unusually large and long contin
154 re. It is only in those cases in which the MODIFICATION has been comparatively recent and extraord
169 ve gone through an extraordinary amount of MODIFICATION since the genus arose: and thus we can und
171 difficulties on the theory of descent with MODIFICATION. Transitions. Absence or rarity of transit
171 ts of a bat, could have been formed by the MODIFICATION of some animal with wholly different habit
173 a common parent: and during the process of MODIFICATION, each has become adapted to the conditions
176 ve, is that, during the process of further MODIFICATION, by which two varieties are supposed on my
179 f the country can be better filled by some MODIFICATION of some one or more of its inhabitants. An
179 ination: and during the process of further MODIFICATION through natural selection, they will almos
181 th fuller and fuller flank membranes, each MODIFICATION being useful, each being propagated, until
183 tural selection to fit the animal, by some MODIFICATION of its structure, for its changed habits,
186 ertainly the case: and if any variation or MODIFICATION in the organ be ever useful to an animal u
190 itself, being aided during the process of MODIFICATION by the other organ: and then this other or
196 r in charming the females. Moreover when a MODIFICATION of structure has primarily arisen from the
197 vantage of by the plant undergoing further MODIFICATION and becoming a climber. The naked skin on
199 played a most important part, and a useful MODIFICATION of one part will often have entailed on ot
200 ural selection cannot possibly produce any MODIFICATION in any one species exclusively for the goo
204 e two latter, during the course of further MODIFICATION, from existing in greater numbers, will ha
224 ifying the instinct, always supposing each MODIFICATION to be of use to the species, until an ant
225 view may, perhaps, be considered only as a MODIFICATION of his theory. Let us look to the great pr
234 n advantage to our humble bee, if a slight MODIFICATION of her instinct led her to make her waxen
236 ch greater than that of any other striking MODIFICATION of structure: for it can be shown that som
```

```
age ******************************************( Key Word )******************************************
```

Page **(Key Word)**

0237	ving been born with some slight profitable	MODIFICATION of structure, this being inherited by its
0238	it has been with social insects: a slight	MODIFICATION of structure, or instinct, correlated wit
0238	to produce sterile members having the same	MODIFICATION. And I believe that this process has been
0239	, that each successive, slight, profitable	MODIFICATION did not probably at first appear in all t
0239	produced most neuters with the profitable	MODIFICATION, all the neuters ultimately came to have
0242	ith animals, as with plants, any amount of	MODIFICATION in structure can be effected by the accum
0279	exterminated during the course of further	MODIFICATION and improvement. The main cause, however,
0295	very thick one; and the species undergoing	MODIFICATION will have had to live on the same area th
0296	s were to undergo a considerable amount of	MODIFICATION during any one geological period, a secti
0302	e fatal to the theory of descent with slow	MODIFICATION through natural selection. For the develo
0312	es, or with that of their slow and gradual	MODIFICATION, through descent and natural selection. N
0314	sly, or to an equal degree. The process of	MODIFICATION must be extremely slow. The variability o
0314	thus causing a greater or lesser amount of	MODIFICATION in the varying species, depends on many c
0317	owly and progressively; for the process of	MODIFICATION and the production of a number of allied
0327	region underwent a considerable amount of	MODIFICATION and extinction, and that there was much m
0328	me in the one region than in the other for	MODIFICATION, extinction, and immigration. I suspect t
0331	ces accord with the theory of descent with	MODIFICATION. As the subject is somewhat complex, I mu
0333	in known geological periods undergone much	MODIFICATION, should in the older formations make some
0333	case. Thus, on the theory of descent with	MODIFICATION, the main facts with respect to the mutua
0334	by immigration, and for a large amount of	MODIFICATION, during the long and blank intervals betw
0340	ine animals. On the theory of descent with	MODIFICATION, the great law of the long enduring, but r
0343	ly to follow on the theory of descent with	MODIFICATION through natural selection. We can thus unc
0343	gree; yet in the long run that all undergo	MODIFICATION to some extent. The extinction of old form
0343	nequal periods of time; for the process of	MODIFICATION is necessarily slow, and depends on many c
0350	of different regions may be attributed to	MODIFICATION through natural selection, and in a quite
0350	tion; as does time for the slow process of	MODIFICATION through natural selection. Widely ranging
0350	tions, and will frequently undergo further	MODIFICATION and improvement; and thus they will become
0350	nts. On this principle of inheritance with	MODIFICATION, we can understand how it is that sections
0351	omplex struggle for life, so the degree of	MODIFICATION in different species will be no uniform qu
0351	ted country, they will be little liable to	MODIFICATION; for neither migration nor isolation in th
0351	during whole geological periods but little	MODIFICATION, there is not much difficulty in believing
0354	progenitor, can have migrated (undergoing	MODIFICATION during some part of their migration) from
0355	an clearly understand, on the principle of	MODIFICATION, why the inhabitants of a region should be
0355	incidence he attributes to generation with	MODIFICATION. The previous remarks on single and multip
0355	theri so that, at each successive stage of	MODIFICATION and improvement, all the individuals of ea
0356	i believe that during the slow process of	MODIFICATION the individuals of the species will have b
0356	aneously changing, and the whole amount of	MODIFICATION will not have been due, at each stage, to
0368	me, they will not have been liable to much	MODIFICATION. But with our Alpine productions, left iso
0369	consequently they will have been liable to	MODIFICATION; and this we find has been the case; for i
0371	e have here everything favourable for much	MODIFICATION, for far more modification than with the A
0371	urable for much modification, for far more	MODIFICATION than with the Alpine procuctions, left iso
0372	lar Circle, will account, on the theory of	MODIFICATION, for many closely allied forms now living
0379	dy of arctic shells underwent scarcely any	MODIFICATION during their long southern migration and r
0381	misphere, is, on my theory of descent with	MODIFICATION, a far more remakable case of difficulty.
0381	their migration, and for their subsequent	MODIFICATION to the necessary degree. The facts seem to
0382	ange; and that on this view, combined with	MODIFICATION through natural selection, a multitude of
0383	on from the nearest source with subsequent	MODIFICATION. Summary of the last and present chapters.
0389	f independent creation and of descent with	MODIFICATION. The species of all kinds which inhabit oc
0390	ew associates, will be eminently liable to	MODIFICATION, and will often produce groups of modified
0391	ll consequently have been little liable to	MODIFICATION. Madeira, again, is inhabited by a wonderf
0395	the genera are distinct. As the amount of	MODIFICATION in all cases depends to a certain degree o
0396	d probably have been more complete; and if	MODIFICATION be admitted, all the forms of life would h
0399	and that such colonists would be liable to	MODIFICATION; the principle of inheritance still betray
0404	rms showing us the steps in the process of	MODIFICATION. This relation between the power and exten
0405	ied parent should range widely, undergoing	MODIFICATION during its diffusion, and should place its
0406	diest source, together with the subsequent	MODIFICATION and better adaptation of the colonists to
0408	t forms of life), together with subsequent	MODIFICATION and the multiplication of new forms. We ca
0411	n, explained on the theory of descent with	MODIFICATION. Classification of varieties. Descent alwa
0413	ond, hidden as it is by various degrees of	MODIFICATION, which is partially revealed to us by our
0420	natural system is founded on descent with	MODIFICATION; that the characters which naturalists con
0420	tly, being due to the different degrees of	MODIFICATION which they have undergone; and this is exp
0422	ement, like a pedigree; but the degrees of	MODIFICATION which the different groups have undergone,
0423	oseness of descent with various degrees of	MODIFICATION. Nearly the same rules are followed in cla
0423	sure, whether there had been more or less	MODIFICATION, the principle of inheritance would keep t
0425	roo had been produced, by a long course of	MODIFICATION, from a bear? Ought we to rank this one sp
0425	in, and sometimes a considerable amount of	MODIFICATION, may not this same element of descent have
0425	r higher groups, though in these cases the	MODIFICATION has been greater in degree, and has taken
0429	acter during the long continued process of	MODIFICATION, how it is that the more ancient forms of
0430	that both groups have since undergone much	MODIFICATION in divergent directions. On either view we
0435	n of successive slight modifications, each	MODIFICATION being profitable in some way to the modifi
0435	as a wing; yet in all this great amount of	MODIFICATION there will be no tendency to alter the fra
0438	lection, during a long continued course of	MODIFICATION, should have seized on a certain number of
0438	verse purposes. And as the whole amount of	MODIFICATION will have been effected by slight successi
0438	jaws. Yet so strong is the appearance of a	MODIFICATION of this nature having occurred, that natur
0443	d, as follows, on the view of descent with	MODIFICATION. It is commonly assumed, perhaps from mons
0444	it is quite possible that each successive	MODIFICATION, or most of them, may have appeared at an
0447	t species, may become, by a long course of	MODIFICATION, adapted in one descendant to act as hands
0447	two principles, namely of each successive	MODIFICATION supervening at a rather late age, and bein
0447	limbs in the latter having undergone much	MODIFICATION at a rather late period of life, and havin
0448	irstly, from the young, during a course of	MODIFICATION carried on for many generations, having to
0449	successive variations in a long course of	MODIFICATION having supervened at a very early age, or
0454	r this purpose. On my view of descent with	MODIFICATION, the origin of rudimentary organs is simpl
0455	ts derivation. On the view of descent with	MODIFICATION, we may conclude that the existence of org
0456	naturalists as allied, together with their	MODIFICATION through natural selection, with its contin
0456	classes. On this same view of descent with	MODIFICATION, all the great facts in Morphology become
0459	dvanced against the theory of descent with	MODIFICATION through natural selection, I do not deny.

Page **(Key Word)**

Page **(Key Word)***

```
0461   encountered on the theory of descent with   MODIFICATION are grave enough. All the individuals of t
0462   nt and isolated regions, as the process of   MODIFICATION has necessarily been slow, all the means o
0466   o not overthrow the theory of descent with   MODIFICATION. Now let us turn to the other side of the
0466   s much difficulty in ascertaining how much   MODIFICATION our domestic productions have undergone; b
0466   the same, we have reason to believe that a   MODIFICATION, which has already been inherited for many
0470   s. hence during a long continued course of   MODIFICATION, the slight differences, characteristic of
0471   iations, it can produce no great or sudden   MODIFICATION; it can act only by very short and slow st
0474   itor, an unusual amount of variability and   MODIFICATION, and therefore we might expect this part c
0475   gives, support the theory of descent with   MODIFICATION. New species have come on the stage slowly
0475   diffusion of dominant forms, with the slow   MODIFICATION of their descendants, causes the forms of
0476   understand, on the theory of descent with   MODIFICATION, most of the great leading facts in Distri
0477   d of ordinary generation, and the means of   MODIFICATION have been the same. We see the full meanin
0477   mer migration, combined in most cases with   MODIFICATION, we can understand, by the aid of the Glac
0477   nd in different proportions, the course of   MODIFICATION in the two areas will inevitably be differ
0477   on this view of migration, with subsequent   MODIFICATION, we can see why oceanic islands should be
0478   as, implies, on the theory of descent with   MODIFICATION, that the same parents formerly inhabited
0479   se intelligible on the view of the gradual   MODIFICATION of parts or organs, which were alike in th
0480   nd by homologous structures, her scheme of   MODIFICATION, which it seems that we wilfully will not
0483   asked how far I extend the doctrine of the   MODIFICATION of species. The question is difficult to a
0483   necessarily implies an enormous amount of   MODIFICATION in the descendants. Throughout whole class
0483   nnot doubt that the theory of descent with   MODIFICATION embraces all the members of the same class
0082   h nicely balanced forces, extremely slight   MODIFICATIONS in the structure or habits of one inhabit
0082   n advantage over others; and still further   MODIFICATIONS of the same kind would often still furthe
0085   ion is modified through variation, and the   MODIFICATIONS are accumulated by natural selection for
0086   or the 'good of the being, will cause other   MODIFICATIONS, often of the most expected nature. As we
0086   ose which concern the mature insect. These   MODIFICATIONS will no doubt affect, through the laws of
0086   structure of their larvae. So, conversely,   MODIFICATIONS in the adult will probably often affect t
0086   l cases natural selection will ensure that   MODIFICATIONS consequent on other modifications at a di
0086   ure that modifications consequent on other   MODIFICATIONS at a different period of life, shall not
0089   ing age or during the breeding season: the   MODIFICATIONS thus produced being inherited at correspo
0095   ulation of infinitesimally small inherited   MODIFICATIONS, each profitable to the preserved being;
0104   ruggle for, and become adapted to, through   MODIFICATIONS in their structure and constitution. Last
0108   f each island will have to be filled up by   MODIFICATIONS of the old inhabitants; and time will be
0134   and disuse diminishes them; and that such   MODIFICATIONS are inherited. Under free nature, we can
0135   e cases we might easily put down to disuse   MODIFICATIONS of structure which are wholly, or mainly,
0138   sation for blindness. Notwithstanding such   MODIFICATIONS, we might expect still to see in the cave
0143   understood. The most obvious case is, that   MODIFICATIONS accumulated solely for the good of the yo
0146   n analogous differences. Hence we see that   MODIFICATIONS of structure, viewed by systematists as o
0146   nd independent modification; and these two   MODIFICATIONS, having been transmitted to a whole group
0167   od, etc., seem to have induced some slight   MODIFICATIONS. Habit in producing constitutional differ
0168   way, and homologous parts tend to cohere.   MODIFICATIONS in hard parts and in external parts somet
0168   to any particular function, so that their   MODIFICATIONS have not been closely checked by natural
0169   nt progenitors. Although new and important   MODIFICATIONS may not arise from reversion and analogou
0169   om reversion and analogous variation, such   MODIFICATIONS will add to the beautiful and harmonious
0170   that gives rise to all the more important   MODIFICATIONS of structure, by which the.innumerable be
0172   s solely by the preservation of profitable   MODIFICATIONS, each new form will tend in a fully stock
0178   nly to see a few species presenting slight   MODIFICATIONS of structure in some degree permanent; an
0183   nd structure afterwards; or whether slight   MODIFICATIONS of structure lead to changed habits; both
0189   een formed by numerous, successive, slight   MODIFICATIONS, my theory would absolutely break down. B
0195   any one organic being, to say what slight   MODIFICATIONS would be of importance or not. In a forme
0195   r its present purpose by successive slight   MODIFICATIONS, each better and better, for so trifling
0205   portant for the welfare of a species; that   MODIFICATIONS in its structure could not have been slow
0205   . but we may confidently believe that many   MODIFICATIONS, wholly due to the laws of growth, and at
0209   life, it is at least possible that slight   MODIFICATIONS of instinct might be profitable to a spec
0209   nd wonderful instincts have originated. As   MODIFICATIONS of corporeal structure arise from, and ar
0227   ircular mouths of her old cocoons. By such   MODIFICATIONS of instincts in themselves not very wonde
0233   on acts only by the accumulation of slight   MODIFICATIONS of structure or instinct, each profitable
0235   advantage of numerous, successive, slight   MODIFICATIONS of simpler instincts; natural selection h
0237   ver have transmitted successively acquired   MODIFICATIONS of structure or instinct to its progeny.
0237   lies in understanding how such correlated   MODIFICATIONS of structure could have been slowly accum
0243   , in natural selection accumulating slight   MODIFICATIONS of instinct to any extent, in any useful
0272   ment, but is incidental on slowly acquired   MODIFICATIONS, more especially in the reproductive syst
0301   and no closely consecutive record of their   MODIFICATIONS could be preserved in any one formation.
0325   ena, it will appear certain that all these   MODIFICATIONS of species, their extinction, and the int
0379   of life; and it is probable that selected   MODIFICATIONS in their structure, habits, and constitut
0410   laws of variation have been the same, and   MODIFICATIONS have been accumulated by the same power o
0414   rue affinities. We are least likely in the   MODIFICATIONS of these organs to mistake a merely adapt
0428   ve often been adapted by successive slight   MODIFICATIONS to live under nearly similar circumstance
0435   urposes, are formed by infinitely numerous   MODIFICATIONS of an upper lip, mandibles, and two pairs
0435   the natural selection of successive slight   MODIFICATIONS, each modification being profitable in so
0444   possible, that each of the many successive   MODIFICATIONS, by which each species has acquired its p
0450   , are explained on the principle of slight   MODIFICATIONS not appearing, in the many descendants ty
0457   r habits of life, through the principle of   MODIFICATIONS being inherited at corresponding ages. On
0466   r that the amount has been large, and that   MODIFICATIONS can be inherited for long periods. As lon
0474   tion of successive, slight, but profitable   MODIFICATIONS. The similarity of pattern in the wing an
0479   f descent with slow and slight successive   MODIFICATIONS. We can thus understand why nature moves
0003   able species inhabiting this world have been   MODIFIED. This fundamental subject of Natural Se
0005   d variety will tend to propagate its new and   MODIFIED form. This fundamental subject of Natural Se
0020   ed. There can be no doubt that a race may be   MODIFIED by occasional crosses, if aided by the caref
0030   reeders have, even within a single lifetime,   MODIFIED to a large extent some breeds of cattle and
0035   ing Charles's spaniel has been unconsciously   MODIFIED to a large extent since the time of that mon
0043   have thought. The effects of variability are   MODIFIED by various degrees of inheritance and of rev
0054   ffspring which, though in some slight degree   MODIFIED, will still inherit those advantages that en
0059   o become still more dominant by leaving many   MODIFIED and dominant descendants. But by steps herea
0081   the original inhabitants were in some manner   MODIFIED; for, had the area been open to immigration,
0083   ly conclude that the natives might have been   MODIFIED with advantage, so as to have better resiste
```

Page **(Key Word)***

Page ***(Key Word)***

0085	which, when one part of the organisation is	MODIFIED through variation, and the modifications are
0087	life, if of high importance to it, might be	MODIFIED to any extent by natural selection; for inst
0089	d on the view of plumage having been chiefly	MODIFIED by sexual selection, acting when the birds h
0095	ither simultaneously or one after the other,	MODIFIED and adapted in the most perfect manner to ea
0102	f nature, if any one species does not become	MODIFIED and improved in a corresponding degree with
0104	niformity of character can be given to their	MODIFIED offspring, solely by natural selection prese
0106	es; and if some of these many species become	MODIFIED and improved, others will have to be improve
0108	wed for the varieties in each to become well	MODIFIED and perfected. When, by renewed elevation, t
0108	d on some of the inhabitants becoming slowly	MODIFIED; the mutual relations of many of the other i
0110	es. Hence, rare species will be less quickly	MODIFIED or improved within any given period, and the
0110	uently be beaten in the race for life by the	MODIFIED descendants of the commoner species. From th
0113	ants of this species of grass, including its	MODIFIED descendants, would succeed in living on the
0115	ner some of the natives would have had to be	MODIFIED, in order to have gained an advantage over t
0116	amplified, we may, I think, assume that the	MODIFIED descendants of any one species will succeed
0118	er, these two varieties, being only slightly	MODIFIED forms, will tend to inherit those advantages
0118	nly a single variety, but in a more and more	MODIFIED condition, some producing two or three varie
0118	ailing to produce any. Thus the varieties or	MODIFIED descendants, proceeding from the common pare
0119	re, and may or may not produce more than one	MODIFIED descendant; for natural selection will alway
0119	l be enabled to seize on, and the more their	MODIFIED progeny will be increased. In our diagram th
0119	le amount of divergent variation. As all the	MODIFIED descendants from a common and widely diffuse
0119	divergent branches proceeding from (A). The	MODIFIED offspring from the later and more highly imp
0121	rally tend to produce the greatest number of	MODIFIED descendants; for these will have the best ch
0122	improved lines of descent. If, however, the	MODIFIED offspring of a species get into some distinc
0122	ost of the other species of the genus. Their	MODIFIED descendants, fourteen in number at the fourt
0122	of the same advantages: they have also been	MODIFIED and improved in a diversified manner at each
0125	, the struggle for the production of new and	MODIFIED descendants, will mainly lie between the lar
0126	ps will become utterly extinct, and leave no	MODIFIED descendants; and consequently that of the sp
0126	most ancient species may now have living and	MODIFIED descendants, yet at the most remote geologic
0128	st; and these will tend to transmit to their	MODIFIED offspring that superiority which now makes t
0129	ogical periods, very few now have living and	MODIFIED descendants. From the first growth of the tr
0143	hrough natural selection, other parts become	MODIFIED. This is a very important subject, most impe
0152	descended from other species, and have been	MODIFIED through natural selection, I think we can ob
0152	ne hand, the tendency to reversion to a less	MODIFIED state, as well as an innate tendency to furt
0154	in approximately the same condition to many	MODIFIED descendants, as in the case of the wing of t
0154	those tending to revert to a former and less	MODIFIED condition. The principle included in these r
0156	ve happened that natural selection will have	MODIFIED several species, fitted to more or less wide
0169	rther variability and to reversion to a less	MODIFIED state. But when a species with any extraordi
0169	eveloped organ has become the parent of many	MODIFIED descendants, which on my view must be a very
0172	tion? Thirdly, can instincts be acquired and	MODIFIED through natural selection? What shall we say
0177	common forms; for these will be more slowly	MODIFIED and improved. It is the same principle which
0178	some of the old inhabitants becoming slowly	MODIFIED, with the new forms thus produced and the ol
0181	beasts of prey immigrate, or old ones become	MODIFIED, and all analogy would lead us to believe th
0181	become exterminated, unless they also became	MODIFIED and improved in structure in a corresponding
0182	id of their fluttering fins, might have been	MODIFIED into perfectly winged animals. If this had b
0184	ir structure either slightly or considerably	MODIFIED from that of their proper type. And such ins
0185	th many parts of its organisation profoundly	MODIFIED. On the other hand, the acutest observer wo
0190	es, one of the two organs might with ease be	MODIFIED and perfected so as to perform all the work
0190	er organ; and then this other organ might be	MODIFIED for some other and quite distinct purpose, o
0193	hes had electric organs, which most of their	MODIFIED descendants have lost. The presence of lumin
0194	ntage of analogous variations, has sometimes	MODIFIED in very nearly the same manner two parts in
0196	o many land animals, which in their lungs or	MODIFIED swimbladders betray their aquatic origin, ma
0196	hat sexual selection will often have largely	MODIFIED the external characters of animals having a
0201	f any part comes to be injurious, it will be	MODIFIED; or if it be not so, the being will become e
0202	of the same great order, and which has been	MODIFIED but not perfected for its present purpose, w
0205	ntly taken advantage of by the still further	MODIFIED descendants of this species. We may, also, b
0208	ition to our conscious will! yet they may be	MODIFIED by the will or reason. Habits easily become
0215	the minds of our domestic animals have been	MODIFIED by domestication. It is scarcely possible to
0216	w instincts in a state of nature have become	MODIFIED by selection, by considering a few cases. I
0218	not only their instincts but their structure	MODIFIED in accordance with their parasitic habits; f
0233	sked, how a long and graduated succession of	MODIFIED architectural instincts, all tending towards
0271	have been in some manner and in some degree	MODIFIED. From these facts; from the great difficulty
0280	have differed in some respects from all its	MODIFIED descendants. To give a simple illustration:
0280	se two breeds, moreover, have become so much	MODIFIED, that if we had no historical or indirect ev
0281	the structure of the parent with that of its	MODIFIED descendants, unless at the same time we had
0297	ht obtain the parent species and its several	MODIFIED descendants from the lower and upper beds of
0298	lant their parent forms until they have been	MODIFIED and perfected in some considerable degree. A
0301	ed of any decided advantage, or when further	MODIFIED and improved, they would slowly spread and s
0302	itors must have lived long ages before their	MODIFIED descendants. But we continually over rate th
0314	of the inhabitants of a country have become	MODIFIED and improved, we can understand, on the prin
0314	ny form which does not become in some degree	MODIFIED and improved, will be liable to be extermina
0315	ook to wide enough intervals of time, become	MODIFIED; for those which do not change will become e
0316	d, in order to have generated either new and	MODIFIED or the same old and unmodified forms. Specie
0321	ther in all respects. Hence the improved and	MODIFIED descendants of a species will generally caus
0321	heir places to other species which have been	MODIFIED and improved, a few of the sufferers may of
0332	s, that a species might go on being slightly	MODIFIED in relation to its slightly altered conditio
0333	gree intermediate in character between their	MODIFIED descendants, or between their collateral rel
0333	unequal lengths of time, and will have been	MODIFIED in various degrees. As we possess only the l
0334	reat stage of descent in the diagram are the	MODIFIED offspring of those which lived at the fifth
0334	the parents of those which became still more	MODIFIED at the seventh stage: hence they could hardl
0338	preserved by nature, of the ancient and less	MODIFIED condition of each animal. This view may be t
0340	f time, closely allied though in some degree	MODIFIED descendants. If the inhabitants of one conti
0340	om those of another continent, so will their	MODIFIED descendants still differ in nearly the same
0341	ies of each of the six older genera have left	MODIFIED descendants, constituting the six new genera
0341	till fewer genera and species will have left	MODIFIED blood descendants. Summary of the preceding
0343	he larger dominant groups tend to leave many	MODIFIED descendants, and thus new sub groups and gro
0344	to become extinct together, and to leave no	MODIFIED offspring on the face of the earth. But the

Page ***(Key Word)***

Page **(Key Word)**

```
0344  un tend to people the world with allied, but  MODIFIED, descendants; and these will generally succe
0350  rther victorious, and will produce groups of  MODIFIED descendants. On this principle of inheritanc
0351  ver vast spaces, and have not become greatly  MODIFIED. On these views, it is obvious, that the sev
0355  few colonists, and their descendants, though  MODIFIED, would still be plainly related by inheritan
0371  as i believe, their descendants, mostly in a  MODIFIED condition, in the central parts of Europe an
0381  have migrated; why certain species have been  MODIFIED and have given rise to new groups of forms,
0385  e chores of the sea, and subsequently become  MODIFIED and adapted to the fresh waters of a distant
0388  elieve that such low beings change or become  MODIFIED less quickly than the high; and this will gi
0388  ng the same identical form or in some degree  MODIFIED, I believe mainly depends on the wide disper
0390  dification, and will often produce groups of  MODIFIED descendants. But it by no means follows, tha
0390  to depend on the species which do not become  MODIFIED having immigrated with facility and in a bod
0392  means; and the plant then becoming slightly   MODIFIED, but still retaining its hooked seeds, would
0394  uppose that such wandering species have been  MODIFIED through natural selection in their new homes
0395  ecies or of allied species in a more or less  MODIFIED condition. Mr. Windsor Earl has made some st
0396  e forms of life would have been more equally  MODIFIED, in accordance with the paramount importance
0396  er still retaining the same specific form or  MODIFIED since their arrival, could have reached thei
0400  the immigrants should have been differently   MODIFIED since their arrival), we find a considerable
0400  we are considering how they have come to be   MODIFIED since their arrival, we find a considerable
0403  ived, the colonists having been subsequently  MODIFIED and better fitted to their new homes, is of
0406  e world, where they may have become slightly  MODIFIED in relation to their new conditions. There i
0408  ps of beings greatly, and some only slightly  MODIFIED, some developed in great force, some existin
0409  n the presence of mammals, in a more or less  MODIFIED condition, and the depth of the sea between
0412  l see that the inevitable result is that the  MODIFIED descendants proceeding from one progenitor b
0420  these genera (A, F, and I) have transmitted   MODIFIED descendants to the present day, represented
0420  the uppermost horizontal line. Now all these  MODIFIED descendants from a single species, are repre
0421  14 may be supposed to have been but slightly  MODIFIED; and it will then rank with the parent genus
0421  t each successive period of descent. All the  MODIFIED descendants from A will have inherited somet
0421  e descendants of A or of I have been so much  MODIFIED as to have more or less completely lost trac
0421  escent, are supposed to have been but little  MODIFIED, and they yet form a single genus. But this
0425  can judge, are the least likely to have been  MODIFIED in relation to the conditions of life to whi
0428  classifications have probably arisen. As the  MODIFIED descendants of dominant species, belonging t
0429  ed to the present day descendants but little  MODIFIED, will give to us our so called osculant or a
0431  ill have transmitted some of its characters,  MODIFIED in various ways and degrees, to all; and the
0432  some of which have produced large groups of  MODIFIED descendants. Every intermediate link between
0435  fication being profitable in some way to the  MODIFIED form, but often affecting by correlation of
0437  (as Owen has observed) of all low or little  MODIFIED forms; therefore we may readily believe that
0438  se and legs in the other, have actually been  MODIFIED into skulls or jaws. Yet so strong is the ap
0446  of which the several new species have become  MODIFIED through natural selection in accordance with
0447  h other closely, for they will not have been  MODIFIED. But in each individual new species, the emb
0447  reas the young will remain unmodified, or be  MODIFIED in a lesser degree, by the effects of use an
0448  nce of the species, that the child should be  MODIFIED at a very early age in the same manner with
0449  lt. For the embryo is the animal in its less  MODIFIED state; and in so far it reveals the structur
0449  uch the structure of the adult may have been  MODIFIED and obscured; we have seen, for instance, th
0449  rtially shows us the structure of their less  MODIFIED ancient progenitors, we can clearly see why
0454  r injurious for one purpose, might easily be  MODIFIED and used for another purpose. Or an organ mi
0457  active embryos, which have become specially  MODIFIED in relation to their habits of life, through
0458  roup, from common parents, and have all been  MODIFIED in the course of descent, that I should with
0460  oductive systems should have been profoundly  MODIFIED. Moreover, most of the varieties which have
0461  of life, and that the offspring of slightly  MODIFIED forms or varieties acquire from being crosse
0461  nditions of life and crosses between greatly  MODIFIED forms, lessen fertility; and on the other ha
0461  conditions of life and crosses between less  MODIFIED forms, increase fertility. Turning to geogra
0463  isting in greater numbers, will generally be  MODIFIED and improved at a quicker rate than the inte
0464  distant regions until they are considerably  MODIFIED and improved; and when they do spread, if di
0470  increase inordinately in number: and as the  MODIFIED descendants of each species will be enabled
0473  portant part, so that when one part has been  MODIFIED other parts are necessarily modified. In bot
0473  as been modified other parts are necessarily  MODIFIED. In both varieties and species reversions to
0488  tition with foreign associates, might become  MODIFIED; so that we must not overrate the accuracy o
0012  arity; he will almost certainly unconsciously  MODIFY other parts of the structure, owing to the my
0031  which enables the agriculturist, not only to  MODIFY the character of his flock, but to change it
0035  continued during centuries, would improve and  MODIFY any breed, in the same way as Bakewell, Colli
0035  nly carried on more methodically, did greatly  MODIFY, even during their own lifetimes, the forms a
0038  centuries or thousands of years to improve or  MODIFY most of our plants up to their present standa
0078  ge numbers in its new home, we should have to  MODIFY it in a different way to what we should have
0086  tural selection will be enabled to act on and  MODIFY organic beings at any age, by the accumulatio
0086  ds on his cotton trees. Natural Selection may  MODIFY and adapt the larva of an insect to a score o
0086  nction of the species. Natural selection will  MODIFY the structure of the young in relation to the
0087  ange. What natural selection cannot do, is to  MODIFY the structure of one species, without giving
0087  and if so, natural selection will be able to  MODIFY one sex in its functional relations to the ot
0103  lanced by natural selection always tending to  MODIFY all the individuals in each district in exact
0104  iform; so that natural selection will tend to  MODIFY all the individuals of a varying species thro
0127  es being inherited at corresponding ages, can  MODIFY the egg, seed, or young, as easily as the adu
0227  may safely conclude that if we could slightly  MODIFY the instincts already possessed by the Melipo
0269  ore probably indirectly, through correlation,  MODIFY the reproductive system in the several descen
0435  ature, there will be little or no tendency to  MODIFY the original pattern, or to transpose parts.
0082  d places for natural selection to fill up by  MODIFYING and improving some of the varying inhabitan
0091  to keep the best dogs without any thought to  MODIFYING the breed. Even without any change in the p
0102  ns of life; and then if natural selection be  MODIFYING and improving a species in the several dist
0127  election has really thus acted in nature, in  MODIFYING and adapting the various forms of life to t
0144  the importance of the laws of correlation in  MODIFYING important structures, independently of util
0196  will have had a most important influence in  MODIFYING various structures; and finally, that sexua
0213  called accidental variations have played in  MODIFYING the mental qualities of our domestic animal
0224  fficulty in natural selection increasing and  MODIFYING the instinct, always supposing each modific
0447  e hand, and disuse on the other, may have in  MODIFYING an organ, such influence will mainly affect
0474  habit ro doubt sometimes comes into play in  MODIFYING instincts; but it certainly is not indispen
0010  ould attribute to the direct action of heat,  MOISTURE, light, food, etc., is most difficult: my im
0062  rly it should be said to be dependent on the  MOISTURE. A plant which annually produces a thousand
```

Page **(Key Word)**

Page **(Key Word)**

```
0279  graduate away quite insensibly like heat or MOISTURE. I endeavoured, also, to show that intermedi
0137  even more subterranean in its habits than the MOLE; and I was assured by a Spaniard, who had ofte
0434  hard of a man, formed for grasping, that of a MOLE for digging, the leg of the horse, the paddle
0137  at all and had stuck to the wreck. The eyes of MOLES and of some burrowing rodents are rudimentary
0473  ich is occasionally blind, and then at certain MOLES, which are habitually blind and have their ey
0100  land there are some hermaphrodites, as land MOLLUSCA and earth worms; but these all pair. As yet
0182  water breathing classes as the Crustacea and MOLLUSCA are adapted to live on the land, and seeing
0288  hamalus existed during the chalk period. The MOLLUSCAN genus Chiton offers a partially analogous c
0313  is genus; whereas most of the other Silurian MOLLUSCS and all the Crustaceans have changed greatly
0337  their own class, may have beaten the highest MOLLUSCS. From the extraordinary manner in which Euro
0339  from the wide distribution of most genera of MOLLUSCS, it is not well displayed by them. Other cas
0385  uld voluntarily drop off. These just hatched MOLLUSCS, though aquatic in their nature, survived on
0438  nciple of inheritance. In the great class of MOLLUSCS, though we can homologise the parts of one s
0438  ual. And we can understand this fact; for in MOLLUSCS, even in the lowest members of the class, we
0124  nd genera. It is worth while to reflect for a MOMENT on the character of the new species f14, whic
0322  t it be at our presumption in imagining for a MOMENT that we understand the many complex contingen
0354  distant and separated points; nor do I for a MOMENT pretend that any explanation could be offered
0363  erywhere is charged with seeds. Reflect for a MOMENT on the millions of quails which annually cros
0368  ur alpine productions, left isolated from the MOMENT of the returning warmth, first at the bases a
0400  arts of the world (laying on one side for the MOMENT the endemic species, which cannot be here fai
0035  fied to a large extent since the time of that MONARCH. Some highly competent authorities are convi
0028  0,000 pigeons were taken with the court. The MONARCHS of Iran and Turan sent him some very rare bi
0025  or has any of the above specified marks, the MONGREL offspring are very apt suddenly to acquire t
0027  ds, both when kept pure and when crossed; the MONGREL offspring being perfectly fertile; from thes
0245  tility of varieties when crossed and of their MONGREL offspring not universal. Hybrids and mongrel
0246  rcrossed, and likewise the fertility of their MONGREL offspring, is, on my theory, of equal import
0267  ility of varieties when crossed, and of their MONGREL offspring. It may be urged, as a most forcib
0271  ly, by reciprocal crosses, and he found their MONGREL offspring perfectly fertile. But one of thes
0272  ybrid offspring of species, and the so called MONGREL offspring of varieties. And, on the other ha
0274  tner is able to point out, between hybrid and MONGREL plants. On the other hand, the resemblance i
0275  id by some authors on the supposed fact, that MONGREL animals alone are born closely like one of t
0277  y alike to considered as varieties, and their MONGREL offspring, are very generally, but not quite
0460  y of varieties when intercrossed and of their MONGREL offspring cannot be considered as universal;
0099  hat such a vast number of the seedlings are MONGRELIZED? I suspect that it must arise from the pol
0020  by the careful selection of those individual MONGRELS, which present any desired character; but th
0020  erything seems simple enough; but when these MONGRELS are crossed one with another for several gen
0026  tises on inheritance. Lastly, the hybrids or MONGRELS from between all the domestic breeds of pige
0043  the extreme variability both of hybrids and MONGRELS, and the frequent sterility of hybrids: but
0099  , of the seedlings thus raised will turn out MONGRELS: for instance, I raised 233 seedling cabbage
0166  e tint and bars and marks to reappear in the MONGRELS. I have stated that the most probable hypoth
0245  mongrel offspring not universal. Hybrids and MONGRELS compared independently of their fertility. S
0272  of the forms which are crossed. Hybrids and MONGRELS compared, independently of their fertility.
0272  distinction is, that in the first generation MONGRELS are more variable than hybrids; but Gartner
0272  e degree of variability graduates away. When MONGRELS and the more fertile hybrics are propagated
0273  ious; but some few cases both of hybrids and MONGRELS long retaining uniformity of character could
0273  y, however, in the successive generations of MONGRELS: is, perhaps, greater than in hybrids. This g
0273  than in hybrids. This greater variability of MONGRELS than of hybrids does not seem to me at all s
0273  to me at all surprising. For the parents of MONGRELS are varieties, and mostly domestic varieties
0273  variable. But to return to our comparison of MONGRELS and hybrids: Gartner states that mongrels ar
0273  of mongrels and hybrids: Gartner states that MONGRELS are more liable than hybrids to revert to ei
0274  lants. On the other hand, the resemblance in MONGRELS and in hybrids to their respective parents,
0274  semble each other closely; and so it is with MONGRELS from a reciprocal cross. Both hybrids and mo
0274  ls from a reciprocal cross. Both hybrids and MONGRELS can be reduced to either pure parent form, b
0275  much less frequently with hybrids than with MONGRELS. Looking to the cases which I have collected
0275  er parent would be more likely to occur with MONGRELS, which are descended from varieties often su
0278  lose general resemblance between hybrids and MONGRELS. Finally, then, the facts briefly given in t
0200  believe that the same bones in the arm of the MONKEY, in the fore leg of the horse, in the wing of
0200  ve that the several bones in the limbs of the MONKEY, horse, and bat, which have been inherited fr
0303  his great series. Cuvier used to urge that no MONKEY occurred in any tertiary stratum; but now ext
0424  inition. As soon as three Orchidean forms (MONOCHANTHUS, Myanthus, and Catasetum), which had previ
0029  eed was descended from a distinct species. Von MONS, in his treatise on pears and apples, shows ho
0318  don, Megatherium, Toxodon, and other extinct MONSTERS, which all co existed with still living shel
0323  ll living sea shells; but as these anomalous MONSTERS coexisted with the Mastodon and Horse, it mi
0341  e that the megatherium and other allied huge MONSTERS have left behind them in South America the s
0424  nsidered as the same individual. He includes MONSTERS; he includes varieties, not solely because t
0454  . we often see rudiments of various parts in MONSTERS. But I doubt whether any of these cases thro
0008  t unnatural treatment of the embryo causes MONSTROSITIES; and monstrosities cannot be separated by
0008  nt of the embryo causes monstrosities; and MONSTROSITIES cannot be separated by any clear line of
0011  ly entail changes in the mature animal. In MONSTROSITIES, the correlations between quite distinct
0044  ly be proved. We have also what are called MONSTROSITIES; but they graduate into varieties. By a m
0131  ility, as well as the greater frequency of MONSTROSITIES, under domestication or cultivation, than
0155  ame. Something of the same kind applies to MONSTROSITIES: at least Is. Geoffroy St. Hilaire seems
0443  tion. It is commonly assumed, perhaps from MONSTROSITIES often affecting the embryo at a very earl
0044  ies; but they graduate into varieties. By a MONSTROSITY I presume is meant some considerable devia
0090  domestication, it would have been called a MONSTROSITY. Illustrations of the action of Natural Se
0016  he same species, also, often have a somewhat MONSTROUS character; by which I mean, that, although
0028  is that the breeds so often have a somewhat MONSTROUS character. It is also a most favourable cir
0083  . he often begins his selection by some half MONSTROUS form; or at least by some modification prom
0143  thors, tend to cohere; this is often seen in MONSTROUS plants; and nothing is more common than the
0184  rger mouths, till a creature was produced as MONSTROUS as a whale. As we sometimes see individuals
0275  s seem chiefly confined to characters almost MONSTROUS in their nature, and which have suddenly ap
0275  m varieties often suddenly produced and semi MONSTROUS in character, than with hybrids, which are
0436  etamorphosed leaves, arranged in a spire. In MONSTROUS plants, we often get direct evidence of the
0452  to find, and which is occasionally found in MONSTROUS individuals of the species. Thus in the sna
0484  the poison secreted by the gall fly produces MONSTROUS growths on the wild rose or oak tree. There
0220  inds. During the present year, however, in the MONTH of July, I came across a community with an un
```

Page **(Key Word)**

Page ***(Key Word)***

```
0031  euri this is done three times at intervals of MONTHS, and the sheep are each time marked and class
0220  hat the slaves feel quite at home. During the MONTHS of June and July, on three successive years,
0220  ither leave or enter a nest. As, during these MONTHS, the slaves are very few in number, I thought
0361  tive organs of a turkey. In the course of two MONTHS, I picked up in my garden 12 kinds of seeds,
0386  cesi I kept it covered up in my study for six MONTHS, pulling up and counting each plant as it gre
0018  most ancient records, more especially on the MONUMENTS of Egypt, much diversity in the breeds; and
0282  omprehend anything of the lapse of time, the MONUMENTS of which we see around us. It is good to wa
0132  residence near the sea affects their colours. MOQUIN Tandon gives a list of plants which when grow
0373  and this I now feel convinced was a gigantic MORAINE, left far below any existing glacier. Furthe
0366  pe changed, that in Northern Italy, gigantic MORAINES, left by old glaciers, are now clothed by th
0373  hooker saw maize growing on gigantic ancient MORAINES. South of the equator, we have some direct e
0016  such valuations being at present empirical. MOREOVER, on the view of the origin of genera which I
0020  oodhound, bull dog, etc., in the wild state. MOREOVER, the possibility of making distinct races by
0022  k, be ranked by him as well defined species. MOREOVER, I do not believe that any ornithologist wou
0024  k pigeon seems to me a very rash assumption. MOREOVER, the several above named domesticated breeds
0025  thers, sometimes concur perfectly developed. MOREOVER, when two birds belonging to two distinct br
0050  versally be found recorded. These varieties, MOREOVER, will be often ranked by some authors as spe
0051  ften be disputed by other naturalists. When, MOREOVER, he comes to study allied forms brought from
0055  ies, than on the side of the smaller genera. MOREOVER, the species of the large genera which prese
0057  r by a less than usual amount of difference. MOREOVER, the species of the large genera are related
0106  n a large than on a small and isolated area. MOREOVER, great areas, though now continuous, owing t
0115  of a highly diversified nature. They differ, MOREOVER, to a large extent from the indigenes, for o
0118  rly the same manner as their parents varied. MOREOVER, these two varieties, being only slightly mo
0123  he original eleven species. The new species, MOREOVER, will be allied to each other in a widely di
0123  eight descendants from (A); the two groups, MOREOVER, are supposed to have gone on diverging in d
0164  mes treble, is common; the side of the face, MOREOVER, is sometimes striped. The stripes are plain
0175  of others, will tend to be sharply defined. MOREOVER, each species on the confines of its range,
0196  ing with another or in charming the females. MOREOVER when a modification of structure has primari
0234  cessary for the construction of their combs. MOREOVER, many bees have to remain idle for many days
0238  d into two or even three castes. The castes, MOREOVER, do not generally graduate into each other,
0241  and jaws nearly five times as big. The jaws, MOREOVER, of the working ants of the several sizes di
0247  had their fertility in some degree impaired. MOREOVER, as Gartner during several years repeatedly
0249  y on the same plant, would be thus effected. MOREOVER, whenever complicated experiments are in pro
0258  act with certain sea weeds or Fuci. Gartner, MOREOVER, found that this difference of facility in m
0260  fac*lity of effecting an union. The hybrids, MOREOVER, produced from reciprocal crosses often diff
0270  with pollen from their own coloured flowers. MOREOVER, he asserts that when yellow and white varie
0280  tures of these two breeds. These two breeds, MOREOVER, have become so much modified, that if we ha
0283  braded and how seldom they are rolled about! MOREOVER, if we follow for a few miles any line of ro
0284  sands of feet in thickness on the Continent. MOREOVER, between each successive formation, we have,
0297  change which on my theory we ought to find. MOREOVER, if we look to rather wider intervals, namel
0307  in any degree intermediate between them. If, MOREOVER, they had been the progenitors of these orde
0322  there may have been much slow extermination. MOREOVER, when by sudden immigration or by unusually
0323  ifling points as mere superficial sculpture. MOREOVER other forms, which are not found in the Chal
0393  lies on a bank connected with the mainland; MOREOVER, icebergs formerly brought boulders to its w
0452  degree of development and in other respects. MOREOVER, in closely allied species, the degree to wh
0460  ystems should have been profoundly modified. MOREOVER, most of the varieties which have been exper
0470  the case if varieties be incipient species. MOREOVER, the species of the larger genera, which aff
0165  m there was a double shoulder stripe. In lord MORETON's famous hybrid from a chestnut mare and mal
0221  nd they alone open and close the doors in the MORNING and evening; and as Huber expressly states,
0411  xiiI. Mutual Affinities of Organic Beings. MORPHOLOGY. Embryology. Rudimentary Organs. Classifica
0411  g. extinction separates and defines groups. MORPHOLOGY, between members of the same class, between
0434  we may hope to make sure but slow progress. MORPHOLOGY. We have seen that the members of the same
0434  bject is included under the general name of MORPHOLOGY. This is the most interesting department of
0456  t with modification, all the great facts in MORPHOLOGY become intelligible, whether we look to the
0485  relationship, community of type, paternity, MORPHOLOGY, adaptive characters, rudimentary and abort
0068  t ten per cent. is an extraordinarily severe MORTALITY from epidemics with man. The action of clim
0041  they generally belong to poor people, and are MOSTLY in small lots, they never can be improved. On
0062  hat the birds which are idly singing round us MOSTLY live on insects or seeds, and are thus consta
0273  or the parents of mongrels are varieties, and MOSTLY domestic varieties (very few experiments havi
0360  ke really floating plants. He tried 98 seeds, MOSTLY different from mine; but he chose many large
0371  we now see, as I believe, their descendants, MOSTLY in a modified condition, in the central parts
0400  e inhabitants of each separate island, though MOSTLY distinct, are related in an incomparably clos
0427  s, actually classed an homopterous insect as a MOTH. We see something of the same kind even in our
0434  he immensely long spiral proboscis of a sphinx MOTH, the curious folded one of a bee or bug, and t
0444  terpillar, cocoon, or imago states of the silk MOTH; or, again, in the horns of almost full grown
0144  eve that the shape of the pelvis in the human MOTHER influences by pressure the shape of the head
0216  lowing, as we see in wild ground birds, their MOTHER to fly away. But this instinct retained by ou
0216  s become useless under domestication, for the MOTHER hen has almost lost by disuse the power of fl
0217  l instinct of another bird, than by their own MOTHER's care, encumbered as she can hardly fail to
0217  ce the occasional and aberrant habit of their MOTHER and in their turn would be apt to lay their e
0258  first been used as the father and then as the MOTHER, generally differ in fertility in a small, an
0260  ies or the other is used as the father or the MOTHER, there is generally some difference, and occa
0264  ly half of the nature and constitution of its MOTHER, and therefore before birth, as long as it is
0264  birth, as long as it is nourished within the MOTHER's womb or within the egg or seed produced by
0264  omb or within the egg or seed produced by the MOTHER, it may be exposed to conditions in some degr
0271  five varieties, when used either as father or MOTHER, and crossed with the Nicotiana glutinosa, al
0440  mammal which is nourished in the womb of its MOTHER, in the egg of the bird which is hatched in a
0443  ry young animal, as long as it remains in its MOTHER's womb, or in the egg, or as long as it is no
0460  s is first used as the father and then as the MOTHER. The fertility of varieties when intercrossed
0483  aring the false marks of nourishment from the MOTHER's womb? Although naturalists very properly de
0073  aceous plants absolutely require the visits of MOTHS to remove their pollen masses and thus to fer
0439  mal, bird, or reptile. The vermiform larvae of MOTHS, flies, beetles, etc., resemble each other mu
0452  lified in the state of the wings of the female MOTHS in certain groups. Rudimentary organs may be
0213  ning, as soon as it scented its prey, stand MOTIONLESS like a statue, and then slowly crawl forwar
0223  the greatest agitation, and one was perched MOTIONLESS with its own pupa in its mouth on top of a
0235  l prisms and of the basal rhombic plates. The MOTIVE power of the process of natural selection hav
```

Page ***(Key Word)***

Page ************************************(Key Word)************************************

```
0025 some uniformly black barbs, and they produced MOTTLED brown and black birds; these I again crossed
0084 e leaf eating insects green, and bark feeders MOTTLED grey; the alpine ptarmigan white in winter,
0031 hich he may summon into life whatever form and MOULD he pleases. Lord Somerville, speaking of what
0045 of the same species are cast in the very same MOULD. These individual differences are highly impo
0373 le i was astonished at the structure of a vast MOUND of detritus, about 800 feet in height, crossi
0074 t the trees now growing on the ancient Indian MOUNDS, in the Southern United States, display the s
0030 f sheep fitted either for cultivated land or MOUNTAIN pasture, with the wool of one breed good for
0069 hence in going northward, or in ascending a MOUNTAIN, we far oftener meet with stunted forms, due
0069 in proceeding southwards or in descending a MOUNTAIN. When we reach the Arctic regions, or snow c
0076 of sheep: it has been asserted that certain MOUNTAIN varieties will starve out other mountain var
0076 ain mountain varieties will starve out other MOUNTAIN varieties, so that they cannot be kept toget
0177 , hilly tract and consequently the improved MOUNTAIN or plain breed will soon take the place of t
0198 that with the hair the horns are correlated. MOUNTAIN breeds always differ from lowland breeds; an
0348 n the opposite sides of lofty and continuous MOUNTAIN ranges, and of great deserts, and sometimes
0348 rs, we find different productions: though as MOUNTAIN chains, deserts, etc., are not as impassable
0354 f the same species on the summits of distant MOUNTAIN ranges, and at distant points in the arctic
0358 ontinents: if they had originally existed as MOUNTAIN ranges on the land, some at least of the isl
0358 e islands would have been formed, like other MOUNTAIN summits, of granite, metamorphic schists, ol
0365 the identity of many plants and animals, on MOUNTAIN summits, separated from each other by hundre
0367 ew worlds, would be left isolated on distant MOUNTAIN summits (having been exterminated on all les
0367 tand the fact that the Alpine plants of each MOUNTAIN range are more especially related to the arc
0368 regions we find the same species on distant MOUNTAIN summits, we may almost conclude without othe
0369 e same arctic species will have been left on MOUNTAIN ranges distant from each other, and have sur
0369 ts and animals of the several great European MOUNTAIN ranges, though very many of the species are
0371 in a much more recent period, on the several MOUNTAIN ranges and on the arctic lands of the two Wo
0373 r, from the frozen mammals and nature of the MOUNTAIN vegetation, that Siberia was similarly affec
0375 africa. On the Himalaya, and on the isolated MOUNTAIN ranges of the peninsula of India, on the hei
0376 parts of the southern hemisphere, and on the MOUNTAIN ranges of the intertropical regions, are not
0376 towards equatorial latitudes, the Alpine or MOUNTAIN floras really become less and less arctic. M
0378 orded an asylum to the tropical natives. The MOUNTAIN ranges north west of the Himalaya, and the l
0380 his is the first stage towards extinction. A MOUNTAIN is an island on the land; and the intertropi
0382 living waters left their living drift on our MOUNTAIN summits, in a line gently rising from the ar
0384 of the fish on opposite sides of continuous MOUNTAIN ranges, which from an early period must have
0403 tion throughout nature. We see this on every MOUNTAIN, in every lake and marsh. For Alpine species
0404 tly American forms, and it is obvious that a MOUNTAIN, as it became slowly upheaved, would natural
0477 diverse conditions, under heat and cold, on MOUNTAIN and lowland, on deserts and marshes, most of
0091 of wolf. Or, again, the wolves inhabiting a MOUNTAINOUS district, and those frequenting the lowlan
0177 eep to be kept, one adapted to an extensive MOUNTAINOUS region; a second to a comparatively narrow
0198 ds always differ from lowland breeds; and a MOUNTAINOUS country would probably affect the hind lim
0049 in somewhat different stations; they ascend MOUNTAINS to different heights; they have different g
0091 arieties of the wolf inhabiting the Catskill MOUNTAINS in the United States, one with a light grey
0174 r to each. We see the same fact in ascending MOUNTAINS, and sometimes it is quite remarkable how a
0177 rongly in favour of the great holders on the MOUNTAINS or on the plains improving their breeds mor
0251 as if it had been a natural species from the MOUNTAINS of Chile. I have taken some pains to ascert
0346 he most humid districts, arid deserts, lofty MOUNTAINS, grassy plains, forests, marshes, lakes, an
0365 ore remarkable, that the plants on the White MOUNTAINS, in the United States of America, are all t
0365 ar from Asa Gray, with those on the loftiest MOUNTAINS of Europe. Even as long ago as 1747, such f
0366 ot tell their tale more plainly, than do the MOUNTAINS of Scotland and Wales, with their scored fl
0366 rriers, in which case they would perish. The MOUNTAINS would become covered with snow and ice, and
0367 and as the snow melted from the bases of the MOUNTAINS, the arctic forms would seize on the cleare
0367 ants at points so immensely remote as on the MOUNTAINS of the United States and of Europe. We can
0368 the United States to Labrador; those of the MOUNTAINS of Siberia to the arctic regions of that co
0368 e bases and ultimately on the summits of the MOUNTAINS, the case will have been somewhat different
0369 pine species, which must have existed on the MOUNTAINS before the commencement of the Glacial epoc
0369 for some of these are the same on the lower MOUNTAINS and on the plains of North America and Euro
0373 d the same plants, found on widely separated MOUNTAINS in this island, tell the same story. If one
0373 ulders have, also, been noticed on the Rocky MOUNTAINS. In the Cordillera of Equatorial South Amer
0374 re many closely allied species. On the lofty MOUNTAINS of equatorial America a host of peculiar sp
0374 ing to European genera occur. On the highest MOUNTAINS of Brazil, some few European genera were fo
0375 era characteristic of the Cordillera. On the MOUNTAINS of Abyssinia, several European forms and so
0375 t to have been introduced by man, and on the MOUNTAINS, some few representative European forms are
0375 nted by plants growing on the summits of the MOUNTAINS of Borneo. Some of these Australian forms,
0375 other as far north as Japan. On the southern MOUNTAINS of Australia, Dr. F. Muller has discovered
0375 world, the plants growing on the more lofty MOUNTAINS, and on the temperate lowlands of the north
0378 rate forms would naturally ascend the higher MOUNTAINS, being exterminated on the lowlands; those
0379 hich settled themselves on the intertropical MOUNTAINS, and in the southern hemisphere. These bein
0379 wever, a few southern vegetable forms on the MOUNTAINS of Borneo and Abyssinia. I suspect that thi
0380 kind must have occurred on the intertropical MOUNTAINS: no doubt before the Glacial period they we
0380 an island on the land; and the intertropical MOUNTAINS before the Glacial period must have been co
0380 hern and southern temperate zones and on the MOUNTAINS of the intertropical regions. Very many dif
0393 ever, been assured that a frog exists on the MOUNTAINS of the great island of New Zealand; but I s
0408 h america, the inhabitants of the plains and MOUNTAINS, of the forests, marshes, and deserts, are
0477 alliance of many others, on the most distant MOUNTAINS, under the most different climates: and lik
0382 races of man, driven up and surviving in the MOUNTIAN fastnesses of almost every land, which serve
0376 less arctic. Many of the forms living on the MOUNTIANS of the warmer regions of the earth and in t
0431 e seen in the diagram so often referred to), MOUNTING up through many predecessors. As it is diffi
0141 be mingled in our domestic breeds. The rat and MOUSE cannot be considered as domestic animals, but
0163 nsverse bars on the legs are not rare in duns, MOUSE duns, and in one instance in a chestnut: a fa
0164 d these stripes occur far oftenest in duns and MOUSE duns; by the term dun a large range of colour
0201 eparing to spring, in order to warn the doomed MOUSE. But I have not space here to enter on this a
0414 e. no one regards the external similarity of a MOUSE to a shrew, of a dugong to a whale, of a whal
0021 l orifices to the nostrils, and a wide gape of MOUTH. The short faced tumbler has a beak in outlin
0022 furcula. The proportional width of the gape of MOUTH, the proportional length of the eyelids, of t
0094 with a very little more trouble, enter by the MOUTH. Bearing such facts in mind, I can see no rea
0184 by Hearne swimming for hours with widely open MOUTH, thus catching, like a whale, insects in the
0223 as perched motionless with its own pupa in its MOUTH on top of a spray of heath over its ravaged h
```

Page ************************************(Key Word)************************************

Page **(Key Word)**

0294 uring the whole of the glacial period near the MOUTH of the Mississippi, within that limit of dept
0294 ds as were deposited in shallow water near the MOUTH of the Mississippi during some part of the gl
0397 ing and having a membranous diaphragm over the MOUTH of the shell, might be floated in chinks of d
0416 re is an open passage from the nostrils to the MOUTH, the only character, according to Owen, which
0437 one crustacean, which has an extremely complex MOUTH formed of many parts, consequently always hav
0441 a very simple single eye, and a probosciformed MOUTH, with which they feed largely, for they incre
0441 antennae; but they have a closed and imperfect MOUTH, and cannot feed: their function at this stag
0441 e organs: they again obtain a well constructed MOUTH; but they have no antennae, and their two eye
0441 ch lives for a short time, and is destitute of MOUTH, stomach, or other organs of importance, exce
0445 l not here give details) of the beak, width of MOUTH, length of nostril and of eyelid, size of fee
0445 difference, for instance, that of the width of MOUTH, could hardly be detected in the young. But t
0184 structure and habits, with larger and larger MOUTHS, till a creature was produced as monstrous as
0227 ble bee adds cylinders of wax to the circular MOUTHS of her old cocoons. By such modifications of
0434 the same great law in the construction of the MOUTHS of insects: what can be more different than t
0435 analogous laws govern the construction of the MOUTHS and limbs of crustaceans. So it is with the f
0435 limbs throughout the whole class. So with the MOUTHS of insects, we have only to suppose that thei
0436 te diversity in structure and function of the MOUTHS of insects. Nevertheless, it is conceivable t
0436 the extinct gigantic sea lizards, and in the MOUTHS of certain suctorial crustaceans, the general
0437 conversely, those with many legs have simpler MOUTHS? Why should the sepals, petals, stamens, and
0098 suddenly spring towards the pistil, or slowly MOVE one after the other towards it, the contrivanc
0295 nced the amount of subsidence. But this same MOVEMENT of subsidence will often tend to sink the ar
0295 thus diminish the supply whilst the downward MOVEMENT continues. In fact, this nearly exact balanc
0309 ation; but may not the areas of preponderant MOVEMENT have changed in the lapse of ages? At a peri
0328 ve that large areas are affected by the same MOVEMENT, it is probable that strictly contemporaneou
0360 less time than those protected from violent MOVEMENT as in our experiments. Therefore it would pe
0285 thing on the surface to show such prodigious MOVEMENT: the pile of rocks on the one or other side
0328 as have invariably been affected by the same MOVEMENTS. When two formations have been deposited in
0474 difications. We can thus understand why nature MOVES by graduated steps in endowing different anim
0095 f clover greatly depends on bees visiting and MOVING parts of the corolla, so as to push the polle
0233 a point at which to commence a cell, and then MOVING outside, first to one point, and then to five
0067 slugs and insects. If turf which has long been MOWN, and the case would be the same with turf clos
0209 comes so close as not to be distinguished. If MOZART, instead of playing the pianoforte at three y
0243 the thrush of South America lines its nest with MUD, in the same peculiar manner as does our Briti
0283 are more quickly ground into pebbles, sand, or MUD. But how often do we see along the bases of re
0386 elieve that botanists are aware how charged the MUD of ponds is with seeds: I have tried several l
0386 g case: I took in February three tablespoons of MUD from three different points, beneath water, on
0386 neath water, on the edge of a little pond: this MUD when dry weighed only 6 3/4 ounces; I kept it
0387 re altogether 537 in number; and yet the viscid MUD was all contained in a breakfast cup! Consider
0475 outh America, for instance, lines her nest with MUD like our British species. On the view of insti
0290 ormous degradation of the coast rocks and from MUDDY streams entering the sea. The explanation, no
0386 aks of birds. Wading birds, which frequent the MUDDY edges of ponds, if suddenly flushed, would be
0386 enly flushed, would be the most likely to have MUDDY feet. Birds of this order I can show are the
0165 e horse genus. Rollin asserts, that the common MULE from the ass and horse is particularly apt to
0165 rly apt to have bars on its legs. I once saw a MULE with its legs so much striped that any one at
0165 on the horse, has given a figure of a similar MULE. In four coloured drawings, which I have seen,
0264 ong lived, as we see in the case of the common MULE. Hybrids, however, are differently circumstanc
0274 epotent power over the horse, so that both the MULE and the hinny more resemble the ass than the h
0275 n the male ass than in the female, so that the MULE, which is the offspring of the male ass and ma
0010 er, though both the young and the parents, as MULLER has remarked, have apparently been exposed to
0375 n the southern mountains of Australia, Dr. F. MULLER has discovered several European species: othe
0131 s spoken as if the variations, so common and MULTIFORM in organic beings under domestication, and
0017 mainly relied on by those who believe in the MULTIPLE origin of our domestic animals is, that we f
0024 domestication: yet on the hypothesis of the MULTIPLE origin of our pigeons, it must be assumed th
0131 n and economy of growth. False correlations. MULTIPLE, rudimentary, and lowly organised structures
0149 nd some botanists have further remarked that MULTIPLE parts are also very liable to variation in s
0168 f which we are utterly unable to understand. MULTIPLE parts are variable in number and in structur
0355 fication. The previous remarks on single and MULTIPLE centres of creation do not directly bear on
0303 se intervals will have given time for the MULTIPLICATION of species from some one or some few pare
0326 ll the conditions most favourable for the MULTIPLICATION of new and dominant species; but we can,
0408 ther with subsequent modification and the MULTIPLICATION of new forms. We can thus understand the
0430 orders of plants. On the principle of the MULTIPLICATION and gradual divergence in character of th
0436 er of other parts, and by the doubling or MULTIPLICATION of others, variations which we know to be
0120 d from (A). Thus, as I believe, species are MULTIPLIED and genera are formed. In a large genus th
0189 pose each new state of the instrument to be MULTIPLIED by the million; and each to be preserved ti
0302 elsewhere have long existed and have slowly MULTIPLIED before they invaded the ancient archipelago
0305 any great group of marine animals might be MULTIPLIED; and here they would remain confined, until
0393 adeira, the Azores, and Mauritius; and have MULTIPLIED so as to become a nuisance. But as these an
0394 eds have not become naturalised and greatly MULTIPLIED. It cannot be said, on the ordinary view of
0079 rous, the healthy, and the happy survive and MULTIPLY. Chapter IV. Natural Selection. Natural Sele
0119 ergent varieties will invariably prevail and MULTIPLY: a medium form may often long endure, and ma
0189 ause the slight alterations, generation will MULTIPLY them almost infinitely, and natural selectio
0244 e advancement of all organic beings, namely, MULTIPLY, vary, let the strongest live and the weakes
0119 ccessful in life, they will generally go on MULTIPLYING in number as well as diverging in characte
0060 disputed. It is immaterial for us whether a MULTITUDE of doubtful forms be called species or sub
0072 ly between the stems of the heath, I found a MULTITUDE of seedlings and little trees, which had be
0097 to the fertilisation of a flower, yet what a MULTITUDE of flowers have their anthers and stigmas f
0098 t be supposed that bees would thus produce a MULTITUDE of hybrids between distinct species: for if
0126 and steady increase of the larger groups, a MULTITUDE of smaller groups will become utterly extin
0202 reason leads us to admire with enthusiasm a MULTITUDE of inimitable contrivances in nature, this
0203 y variable; and are not linked together by a MULTITUDE of intermediate gradations, partly because
0212 is extremely diversified, can be shown by a MULTITUDE of facts. Several cases also, could be give
0231 understanding how the cells are made, that a MULTITUDE of bees all work together; one bee after wo
0257 acility of crossing is by no means strict. A MULTITUDE of cases could be given of very closely all
0261 flow or nature of their sap, etc., but in a MULTITUDE of cases we can assign no reason whatever.
0270 from which we believe in the sterility of a MULTITUDE of species. The evidence is, also, derived
0313 shells may still be found in the midst of a MULTITUDE of extinct forms. Falconer has given a stri

Page **(Key Word)**

Page ***(Key Word)***

0337 in new Zealand, that in the course of time a MULTITUDE of British forms would become thoroughly na
0353 tions of life are nearly the same, so that a MULTITUDE of European animals and plants have become
0382 th modification through natural selection, a MULTITUDE of facts in the present distribution both o
0481 d naturalists whose minds are stocked with a MULTITUDE of facts all viewed, during a long course o
0482 s have of late published their belief that a MULTITUDE of reputed species in each genus are not re
0482 e conclusion to arrive at. They admit that a MULTITUDE of forms, which till lately they themselves
0071 and self sown firs are now springing up in MULTITUDES, so close together that all cannot live. Wh
0289 tive. But we know, for instance, from Sir R. MURCHISON's great work in Russia, what wide gaps ther
0307 of the most eminent geologists, with Sir R. MURCHISON at their head, are convinced that we see in
0310 , and all our greatest geologists, as Lyell, MURCHISON, Sedgwick, etc., have unanimously, often ve
0317 en by those geologists, as Elie de Beaumont, MURCHISON, Barrande, etc., whose general views would
0111 rds of an agricultural writer) as if by some MURDEROUS pestilence. Divergence of Character. The pr
0020 hon. W. Elliot from India, and by the Hon. C. MURRAY from Persia. Many treatises in different Lang
0181 membrane is, also, furnished with an extensor MUSCLE. Although no graduated links of structure, fi
0193 te structure closely resembles that of common MUSCLE; and as it has lately been shown that Rays ha
0011 that the drooping is due to the disuse of the MUSCLES of the ear, from the animals not being much
0046 y add, has also quite recently shown that the MUSCLES in the larvae of certain insects are very fa
0148 eveloped, and furnished with great nerves and MUSCLES; but in the parasitic and protected Proteole
0173 cal history. The crust of the earth is a vast MUSEUM; but the natural collections have been made o
0220 n attended to by Mr. F. Smith, of the British MUSEUM, to whom I am much indebted for information o
0487 emains must not be looked at as a well filled MUSEUM, but as a poor collection made at hazard and
0287 of years! Now turn to our richest geological MUSEUMS, and what a paltry display we behold! On the
0342 of specimens and of species, preserved in our MUSEUMS, is absolutely as nothing compared with the
0464 intellect. The number of specimens in all our MUSEUMS is absolutely as nothing compared with the c
0349 o the waters, and we do not find the beaver or MUSK rat, but the coypu and capybara, rodents of th
0002 nuscripts. This Abstract, which I now publish, MUST necessarily be imperfect. I cannot here give r
0002 d authorities for my several statements; and I MUST trust to the reader reposing some confidence i
0003 hment from certain trees, which has seeds that MUST be transported by certain birds, and which has
0007 ood. It seems pretty clear that organic beings MUST be exposed during several generations to the n
0013 ute its reappearance to inheritance. Every one MUST have heard of cases of albinism, prickly skin,
0016 same genus in a state of nature. I think this MUST be admitted, when we find that there are hardl
0018 y not have existed in Egypt? The whole subject MUST, I think, remain vague; nevertheless, I may, w
0019 has had its wild prototype. At this rate there MUST have existed at least a score of species of wi
0019 ral peculiar breeds of cattle, sheep, etc., we MUST admit that many domestic breeds have originate
0020 our several domestic races by this process, we MUST admit the former existence of the most extreme
0023 have not proceeded from the rock pigeon, they MUST have descended from at least seven or eight ab
0023 enormous crop? The supposed aboriginal stocks MUST all have been rock pigeons, that is, not breed
0023 c breeds. Hence the supposed aboriginal stocks MUST either still exist in the countries where they
0023 ble characters, seems very improbable; or they MUST have become extinct in the wild state. But bir
0024 rts of the world, and, therefore, some of them MUST have been carried back again into their native
0024 esis of the multiple origin of our pigeons, it MUST be assumed that at least seven or eight specie
0024 l feathers like those of the fantail. Hence it MUST be assumed not only that half civilized man su
0025 from the rock pigeon. But if we deny this, we MUST make one of the two following highly improbabl
0030 rent purposes, or so beautiful in his eyes, we MUST, I think, look further than to mere variabilit
0039 , however slight, in one's own possession. Nor MUST the value which would formerly be set on any s
0040 such slow, varying, and insensible changes. I MUST now say a few words on the circumstances, favo
0042 glec in the same aviary, and this circumstance MUST have largely favoured the improvement and form
0043 ct action of the conditions of life. Something MUST be attributed to use and disuse. The final res
0044 ter to organic beings in a state of nature, we MUST briefly discuss whether these latter are subje
0045 by a long catalogue of facts, that parts which MUST be called important, whether viewed under a ph
0047 experience seems the only guide to follow. We MUST, however, in many cases, decide by a majority
0049 or the Canaries, or Ireland, be sufficient? It MUST be admitted that many forms, considered by hig
0050 descend from common parents, and consequently MUST be ranked as varieties. Close investigation, i
0050 n agreement how to rank doubtful forms. Yet it MUST be confessed, that it is in the best known cou
0052 pecies; but whether this belief be justifiable MUST be judged of by the general weight of the seve
0060 ore entering on the subject of this chapter, I MUST make a few preliminary remarks, to show how th
0063 tural lifetime produces several eggs or seeds, MUST suffer destruction during some period of its l
0063 are produced than can possibly survive, there MUST in every case be a struggle for existence, eit
0065 and that the geometrical tendency to increase MUST be checked by destruction at some period of li
0066 but if many eggs or young are destroyed, many MUST be produced, or the species will become extinc
0071 found on the heath. The effect on the insects MUST have been still greater, for six insectivorous
0072 increase of these flies, numerous as they are, MUST be habitually checked by some means, probably
0073 ver be as simple as this. Battle within battle MUST ever be recurring with varying success; and ye
0075 a struggle between the several kinds of trees MUST here have gone on during long centuries; each
0075 trees! Throw up a handful of feathers, and all MUST fall to the ground according to definite laws;
0075 ies as the variously coloured sweet peas, they MUST be each year harvested separately, and the see
0084 , and the black grouse that of peaty earth, we MUST believe that these tints are of service to the
0085 us to judge, seem to be quite unimportant, we MUST not forget that climate, food, etc., probably
0090 r how, as I believe, natural selection acts, I MUST beg permission to give one or two imaginary il
0096 ucture. On the Intercrossing of Individuals. I MUST here introduce a short digression. In the case
0096 , it is of course obvious that two individuals MUST always unite for each birth; but in the case o
0096 . we shall presently see its importance; but I MUST here treat the subject with extreme brevity, t
0096 s it is impossible here to enter on details, I MUST trust to some general considerations alone. In
0097 the same brush to ensure fertilisation; but it MUST not be supposed that bees would thus produce a
0099 these plants have in fact separated sexes, and MUST habitually be crossed. How strange are these f
0099 e seedlings are mongrelized? I suspect that it MUST arise from the pollen of a distinct variety ha
0100 duced on the same tree, we can see that pollen MUST be regularly carried from flower to flower; an
0101 ilising hermaphrodites, do sometimes cross. It MUST have struck most naturalists as a strange anom
0103 eeding animals, which unite for each birth, we MUST not overrate the effects of intercrosses in re
0109 ly discussed in our chapter on Geology; but it MUST be here alluded to from being intimately conne
0109 st indefinitely increasing, numbers inevitably MUST become extinct. That the number of specific fo
0111 species? That this does habitually happen, we MUST infer from most of the innumerable species thr
0114 the contest between individual and individual MUST be severe, we always find great diversity in i
0118 p to the fourteen thousandth generation. But I MUST here remark that I do not suppose that the pro
0122 mmon and widely diffused species, so that they MUST originally have had some advantage over most o
0127 life to their several conditions and stations, MUST be judged of by the general tenour and balance

Page ***(Key Word)***

age ***(Key Word)***

```
132   ray of light, and we may feel sure that there    MUST  be some cause for each deviation of structure,
134   cts show how indirectly the conditions of life    MUST  act. Again, innumerable instances are known to
135   always lost in many dung feeding beetles, they    MUST  be lost early in life, and therefore cannot be
136   the flower feeding coleoptera and lepidoptera,    MUST  habitually use their wings to gain their subsi
137   membrane. As frequent inflammation of the eyes    MUST  be injurious to any animal, and as eyes are ce
138   nts of North America and Europe. On my view we    MUST  suppose that American animals, having ordinary
139   rent, if this view be correct, acclimatication    MUST  be readily effected during long continued desc
141   brought to bear widely different climates. We     MUST  not, however, push the foregoing argument too
141   ion. That habit or custom has some influence I    MUST  believe, both from analogy, and from the inces
142   lly fitted for their own districts: the result    MUST, I think, be due to habit. On the other hand,
154   nts, as in the case of the wing of the bat, it    MUST  have existed, according to my theory, for an i
161   fifth stamen so often appears, that this plant    MUST  have an inherited tendency to produce it. As a
162   urs are crossed. Hence, though under nature it    MUST  generally be left doubtful, what cases are rev
162   iate between two other forms, which themselves    MUST  be doubtfully ranked as either varieties or sp
165   at any one at first would have thought that it    MUST  have been the product of a zebra; and Mr. W. C
167   the same part or organ in the allied species,    MUST  have gone through an extraordinary amount of m
169   of many modified descendants, which on my view   MUST  be a very slow process, requiring a long lapse
170   princ from their parents, and a cause for each    MUST  exist, it is the steady accumulation, through
172   by this theory innumerable transitional forms    MUST  have existed, why do we not find them embedded
174   sitional varieties in each region, though they    MUST  have existed there, and may be embedded there
174   nd and of climate, marine areas now continuous    MUST  often have existed within recent times in a fa
175   t important manner to other organic beings, we    MUST  see that the range of the inhabitants of any c
178   edly we do see. Secondly, areas now continuous    MUST  often have existed within the recent period in
178   epresentative species and their common parent,    MUST  formerly have existed in each broken portion o
179   ly all the species of the same group together,    MUST  assuredly have existed; but the very process o
182   est possible under all possible conditions. It    MUST  not be inferred from these remarks that any of
185   each being has been created as we now see it,    MUST  occasionally have felt surprise when he has me
188   s of light, are convex at their upper ends and    MUST  act by convergence; and at their lower ends th
189   y intellectual powers like those of man? If we    MUST  compare the eye to an optical instrument, we o
189   each layer slowly changing in form. Further we    MUST  suppose that there is a power always intently
189   degree, tend to produce a distincter image. We    MUST  suppose each new state of the instrument to be
192   large class, for in this latter case the organ    MUST  have been first formed at an extremely remote
192   from being washed out of the sack? Although we    MUST  be extremely cautious in concluding that any o
193   tteuchi asserts, discharge any electricity, we    MUST  own that we are far too ignorant to argue that
194   ive variations; she can never take a leap, but    MUST  advance by the shortest and slowest steps. Org
196   id in turning, as with the dog, though the aid    MUST  be slight, for the hare, with hardly any tail,
203   lied species, now living on a continuous area,    MUST  often have been formed when the area was not c
205   ving been perfected whilst aided by the other,    MUST  often have largely facilitated transitions. We
205   atural selection in each well stocked country,    MUST  act chiefly through the competition of the inh
206   but if we include all those of past times, it    MUST  by my theory be strictly true. It is generally
207   lly sufficient to overthrow my whole theory. I    MUST  premise, that I have nothing to do with the or
215   n selected for tameness: and I presume that we    MUST  attribute the whole of the inherited change fr
216   actions, which at first appeared from what we    MUST  in our ignorance call an accident. In some cas
222   ed nest in the thick heath. The nest, however,    MUST  have been close at hand, for two or three indi
224   of the conclusions at which I have arrived. We    MUST  be a dull man who can examine the exquisite st
227   onderfully perfect as that of the hive bee. We    MUST  suppose the Melipona to make her cells truly s
227   parently by turning round on a fixed point. We    MUST  suppose the Melipona to arrange her cells in l
227   she already does her cylindrical cells: and we    MUST  further suppose, and this is the greatest diff
229   hings, had not been neatly performed. The bees    MUST  have worked at very nearly the same rate on th
233   so that a prodigious quantity of fluid nectar    MUST  be collected and consumed by the bees in a hiv
234   ence the saving of wax by largely saving honey    MUST  be a most important element of success in any
236   arity by inheritance from a common parent, and   MUST  therefore believe that they have been acquired
236   his being effected by natural selection. But I   MUST  pass over this preliminary difficulty. The gre
241   nches high, and many sixteen feet high: but we   MUST  suppose that the larger workmen had heads four
242   he accumulation of numerous, slight, and as we   MUST  call them accidental, variations, which are in
247   o be here introduced: a plant to be hybridised   MUST  be castrated, and, what is often more importan
247   castrated; and, what is often more important,   MUST  be secluded in order to prevent pollen being b
249   grow in the same garden, the visits of insects   MUST  be carefully prevented during the flowering se
251   ally imperfect in their mutual self action, we   MUST  infer that the plants were in an unnatural sta
253   est. In India, however, these cross bred geese   MUST  be far more fertile; for I am assured by two e
253   where neither pure parent species exists, they   MUST  certainly be highly fertile. A doctrine which
254   crossing. On this view, the aboriginal species   MUST  either at first have produced quite fertile hy
254   produced quite fertile hybrids, or the hybrids   MUST  have become in subsequent generations quite fe
254   communicated to me by Mr. Blyth, I think they   MUST  be considered as distinct species. On this vie
254   the origin of many of our domestic animals, we   MUST  either give up the belief of the almost univer
254   istinct species of animals when crossed: or we   MUST  look at sterility, not as an indelible charact
259   and of hybrids, we see that when forms, which    MUST  be considered as good and distinct species, ar
260   n various degrees are crossed, all of which we   MUST  suppose it would be equally important to keep
263   ng and of crossing distinct species. And as we   MUST  look at the curious and complex laws governing
263   ntly depends on several distinct causes. There   MUST  sometimes be a physical impossibility in the m
266   in some degree variable, rarely diminishes. It   MUST, however, be confessed that we cannot understa
268   urged, as a most forcible argument, that there   MUST  be some essential distinction between species
268   between species and varieties; and that there   MUST  be some error in all the foregoing remarks, in
271   ce the reproductive system of this one variety   MUST  have been in some manner and in some degree mo
280   termination has acted on an enormous scale; so   MUST  the number of intermediate varieties, which ha
280   borne in mind what sort of intermediate forms    MUST, on my theory, have formerly existed. I have f
282   links, between all living and extinct species,   MUST  have been inconceivably great. But assuredly,
282   of each formation or even each stratum. A man    MUST  for years examine for himself great piles of s
284   during their accumulation; yet what time this    MUST  have consumed! Good observers have estimated t
284   a, the process of accumulation in any one area   MUST  be extremely slow. But the amount of denudatio
285   one of the denudation of the Weald. Though it    MUST  be admitted that the denudation of the Weald h
286   cture to oneself the great dome of rocks which    MUST  have covered up the Weald within so limited a
286   e time along its whole indented length; and we   MUST  remember that almost all strata contain harder
287   on the above data, the denudation of the Weald   MUST  have required 306,662,400 years; or say three
287   r of generations, which the mind cannot grasp,   MUST  have succeeded each other in the long roll of
290   ywhere be found, though the supply of sediment   MUST  for ages have been great, from the enormous de
```

Page ***(Key Word)***

Page **(Key Word)**

```
0290  we may, I think, safely conclude that sediment  MUST  be accumulated in extremely thick, solid, or
0295  lower parts of the same formation, the deposit  MUST  have gone on accumulating for a very long pe
0295  live on the same space, the supply of sediment  MUST  nearly have counterbalanced the amount of su
0296  here thousands of feet in thickness, and which  MUST  have required an enormous period for their a
0296  ude all the fine intermediate gradations which  MUST  on my theory have existed between them, but
0302  which have descended from some one progenitor,  MUST  have been an extremely slow process; and the
0302  an extremely slow process; and the progenitors  MUST  have lived long ages before their modified d
0306  have descended from some one crustacean, which  MUST  have lived long before the Silurian age, and
0308  nudation and metamorphism. The case at present  MUST  remain inexplicable; and may be truly urged
0309  lations of level, which we may fairly conclude  MUST  have intervened during these enormously long
0310  outh America, of bare metamorphic rocks, which  MUST  have been heated under great pressure, have
0314  o an equal degree. The process of modification  MUST  be extremely slow. The variability of each s
0316  peared in the long succession of ages, so long  MUST  its members have continuously existed, in or
0316  s. species of the genus Lingula, for instance,  MUST  have continuously existed by an unbroken suc
0317  and the production of a number of allied forms  MUST  be slow and gradual, one species giving rise
0321  ld genus, belonging to the same family. But it  MUST  often have happened that a new species belon
0321  nites at the close of the secondary period, we  MUST  remember what has been already said on the p
0322  ction. We need not marvel at extinction; if we  MUST  marvel, let it be at our presumption in imag
0324  hanged simultaneously throughout the world, it  MUST  not be supposed that this expression relates
0325  e world, under the most different climates. We  MUST, as Barrande has remarked, look to some spec
0325  rieties or incipient species; for these latter  MUST  be victorious in a still higher degree in or
0330  eparated from each other. This remark no doubt  MUST  be restricted to those groups which have und
0331  he more recent members of the same classes, we  MUST  admit that there is some truth in the remark
0331  ication. As the subject is somewhat complex, we  MUST  request the reader to turn to the diagram in
0331  nt fossils differ most from existing forms. We  MUST  not, however, assume that divergence of char
0334  between the forms of life above and below. We  MUST, however, allow for the entire extinction of
0337  in one particular sense the more recent forms  MUST, on my theory, be higher than the more ancie
0337  r new Zealand, and have seized on places which  MUST  have been previously occupied, we may believ
0338  e embryological development of recent forms. I  MUST  follow Pictet and Huxley in thinking that th
0341  h the actual progenitors of living species. It  MUST  not be forgotten that, on my theory, all the
0342  h the incalculable number of generations which  MUST  have passed away even during a single format
0342  been local. All these causes taken conjointly,  MUST  have tended to make the geological record ex
0342  e are the numberless transitional links which  MUST  formerly have connected the closely allied o
0343  he may overlook how important a part migration  MUST  have played, when the formations of any one
0343  s of those infinitely numerous organisms which  MUST  have existed long before the first bed of th
0350  t of their physical conditions. The naturalist  MUST  feel little curiosity, who is not led to inqu
0351  biting the most distant quarters of the world,  MUST  originally have proceeded from the same sour
0352  h now inhabiting distant and isolated regions,  MUST  have proceeded from one spot, where their pa
0353  ainly occurred within recent geological times,  MUST  have interrupted or rendered discontinuous th
0355  ss (if such exist), the species, on my theory,  MUST  have descended from a succession of improved
0356  on the theory of single centres of creation, I  MUST  say a few words on the means of dispersal. M
0356  of the more important facts. Change of climate  MUST  have had a powerful influence on migration: a
0356  t in some detail. Changes of level in the land  MUST  also have been highly influential: a narrow
0357  insisted that all the islands in the Atlantic  MUST  recently have been connected with Europe or A
0357  arguments used by Forbes are to be trusted, it  MUST  be admitted that scarcely a single island ex
0358  consisting of mere piles of volcanic matter. I  MUST  now say a few words on what are called accide
0363  t of a land bird, I can hardly doubt that they  MUST  occasionally have transported seeds from one,
0365  s led Gmelin to conclude that the same species  MUST  have been independently created at several di
0369  ome mingled with ancient Alpine species, which  MUST  have existed on the mountains before the comm
0370  lived further southwards than at present, they  MUST  have been still more completely separated by
0371  rlds, migrated south of the Polar Circle, they  MUST  have been completely cut off from each other.
0372  h their inhabitants utterly dissimilar. But we  MUST  return to our more immediate subject, the Gla
0376  any, though closely related to northern forms,  MUST  be ranked as distinct species. Now let us see
0377  ent. The Glacial period, as measured by years,  MUST  have been very long; and when we remember ove
0378  to Europe still exist in North America, which  MUST  have lain on the line of march. But I do not
0380  ars from Australia. Something of the same kind  MUST  have occurred on the intertropical mountains:
0380  ertropical mountains before the Glacial period  MUST  have been completely isolated; and I believe
0384  us mountain ranges, which from an early period  MUST  have parted river systems and completely prev
0385  theory, are descended from a common parent and  MUST  have proceeded from a single source, prevail
0387  on this plant, I thought that its distribution  MUST  remain quite inexplicable; but Audubon states
0389  e at the Cape of Good Hope or in Australia, we  MUST, I think, admit that something quite independ
0396  ting places, of which not a wreck now remains,  MUST  not be overlooked. I will here give a single
0397  k in sea water and are killed by it. Yet there  MUST  be, on my view, some unknown, but highly eff
0405  r in mind that some are extremely ancient, and  MUST  have branched off from a common parent at a r
0407  species of the same genus, which on my theory  MUST  have spread from one parent source; if we mak
0414  duct the seed, are of paramount importance! We  MUST  not, therefore, in classifying, trust to rese
0416  e given of characters derived from parts which  MUST  be considered of very trifling physiological
0420  d separating objects more or less alike. But I  MUST  explain my meaning more fully. I believe that
0420  ubordination and relation to the other groups,  MUST  be strictly genealogical in order to be natur
0429  ps. The more aberrant any form is, the greater  MUST  be the number of connecting forms which on my
0430  theory to inheritance in common. Therefore we  MUST  suppose either that all Rodents, including th
0443  t is nourished and protected by its parent, it  MUST  be quite unimportant whether most of its char
0446  rule; for here the characteristic differences  MUST  either have appeared at an earlier period tha
0446  iod than usual, or, if not so, the differences  MUST  have been inherited, not at the corresponding
0453  ntary organs. In reflecting on them, every one  MUST  be struck with astonishment: for the same rea
0460  i state by many graduated steps. There are, it  MUST  be admitted, cases of special difficulty on t
0460  iversal fertility of varieties when crossed, I  MUST  refer the reader to the recapitulation of the
0461  ecies of the same genus, or even higher group,  MUST  have descended from common parents; and there
0461  ed parts of the world they are now found, they  MUST  in the course of successive generations have
0462  n an interminable number of intermediate forms  MUST  have existed, linking together all the specie
0464  ossils? For certainly on my theory such strata  MUST  somewhere have been deposited at these ancien
0477  the full meaning of the wonderful fact, which  MUST  have struck every traveller, namely, that on
0478  er african islands to the African mainland. It  MUST  be admitted that these facts receive no expla
0487  e crust of the earth with its embedded remains  MUST  not be looked at as a well filled museum; but
0487  the preceding and succeeding organic forms. We  MUST  be cautious in attempting to correlate as str
0488  associates, might become modified; so that we  MUST  not overrate the accuracy of organic change a
```

Page **(Key Word)**

ge **(Key Word)***

```
79  ts habits to its place in nature. Look at the MUSTELA vison of North America, which has webbed fee
80  ts and geologists rejected this view of the MUTABILITY of species? It cannot be asserted that orga
82  y. whoever is led to believe that species are MUTABLE will do good service by conscientiously expr
83  ry difficulty from those who believe in the MUTABLIITY of species, on their own side they ignore t
36  h evidence of the slow and scarcely sensible MUTATION of specific forms, as we have a just right t
63  s afford plain evidence of the gradation and MUTATION of the forms of life? We meet with no such e
95  would have afforded us plain evidence of the MUTATION of species, if they had undergone mutation.
81  e mutation of species, if they had undergone MUTATION. But the chief cause of our natural unwillin
02  ever have suspected how poor a record of the MUTATIONS of life, the best preserved geological sect
25  ical conditions, as the cause of these great MUTATIONS in the forms of life throughout the world,
57  . no other geologist will dispute that great MUTATIONS of level, have occurred within the period o
35  cient evidence to induce us to believe that MUTILATIONS are ever inherited; and I should prefer ex
03  ceivable that a naturalist, reflecting on the MUTUAL affinities of organic beings, on their embryo
05  e; in the thirteenth, their classification or MUTUAL affinities, both when mature and in an embryo
06  e for our profound ignorance in regard to the MUTUAL relations of all the beings which live around
t   t of this world. Still less do we know of the MUTUAL relations of the innumerable inhabitants of t
78  . it will convince us of our ignorance on the MUTUAL relations of all organic beings; a conviction
80  infinitely complex and close fitting are the MUTUAL relations of all organic beings to each other
t   tinued preservation of individuals presenting MUTUAL and slightly favourable deviations of structu
08  the inhabitants becoming slowly modified; the MUTUAL relations of many of the other inhabitants be
61  s of the species, and will not be left to the MUTUAL action of the conditions of life and of a sim
51  as they were functionally imperfect in their MUTUAL self action, we must infer that the plants we
66  in the development, or periodical action, or MUTUAL relation of the different parts and organs on
70  as separated sexes; and he asserts that their MUTUAL fertilisation is by so much the less easy as
29  , and to living forms. Let us now look to the MUTUAL affinities of extinct and living species. The
33  ification, the main facts with respect to the MUTUAL affinities of the extinct forms of life to ea
34  coner in two series, first according to their MUTUAL affinities and then according to their period
t   s; and on their action and reaction, in their MUTUAL struggles for life; the relation of organism
68  e kept in a body together; consequently their MUTUAL relations will not have been much disturbed,
69  somewhat different climatal influences. Their MUTUAL relations will thus have been in some degree
90  ed with facility and in a body, so that their MUTUAL relations have not been much disturbed. Thus
08  the same continent. Bearing in mind that the MUTUAL relations of organism to organism are of the
11  ame power of natural selection. Chapter XIII. MUTUAL Affinities of Organic Beings. Morphology. Emb
78  se same principles we see how it is, that the MUTUAL affinities of the species and genera within e
87  enly altered physical conditions, namely, the MUTUAL relation of organism to organism, the improve
91  lf fertilisation, should in so many cases be MUTUALLY useless to each other! How simply are these
99  ether in their former homes, and have become MUTUALLY adapted to each other; and when settled in t
01  called by one of our greatest philosophers. On MY return home, it occurred to me, in 1837, that s
01  i have not been hasty in coming to a decision. MY work is now nearly finished; but as it will tak
01  two or three more years to complete it, and as MY health is far from strong, I have been urged to
02  . sir C. Lyell and Dr. Hooker, who both knew of MY work, the latter having read my sketch of 1844,
02  ho both knew of my work, the latter having read MY sketch of 1844, honoured me by thinking it advi
02  ce's excellent memoir, some brief extracts from MY manuscripts. This Abstract, which I now publish
02  cannot here give references and authorities for MY several statements; and I must trust to the rea
02  trust to the reader reposing some confidence in MY accuracy. No doubt errors will have crept in, t
02  detail all the facts, with references, on which MY conclusions have been grounded; and I hope in a
02  done. I much regret that want of space prevents MY having the satisfaction of acknowledging the ge
03  r, let this opportunity pass without expressing MY deep obligations to Dr. Hooker, who for the las
04  cation and coadaptation. At the commencement of MY observations it seemed to me probable that a ca
04  best and safest clue. I may venture to express MY conviction of the high value of such studies, a
10  of formation; so that, in fact, sports support MY view, that variability may be largely attribute
10  moisture, light, food, etc., is most difficult: MY impression is, that with animals such agencies
26  ns are perfectly fertile. I can state this from MY own observations, purposely made on the most di
44  should be given, but these I shall reserve for MY future work. Nor shall I here discuss the vario
53  er, even in stronger terms. I shall reserve for MY future work the discussion of these difficultie
53  its me to add, that after having carefully read MY manuscript, and examined the tables, he thinks
53  nce) with different sets of organic beings. But MY tables further show that, in any limited countr
54  o obscure this result, that I am surprised that MY tables show even a small majority on the side o
54  arieties be looked at as incipient species; for MY tables clearly show as a general rule that, whe
56  f this had been so, it would have been fatal to MY theory; inasmuch as geology plainly tells us th
57  st this numerically by averages, and, as far as MY imperfect results go, they always confirm the v
62  ttle more detail the struggle for existence. In MY future work this subject shall be treated, as i
67  bly treated by several authors, and I shall, in MY future work, discuss some of the checks at cons
68  f 1854 55 destroyed four fifths of the birds in MY own grounds; and this is a tremendous destructi
98  ow from the writings of C. C. Sprengel and from MY own observations, which effectually prevent the
t   nd as this flower is never visited, at least in MY garden, by insects, it never sets a seed, thoug
00  , i find to be the case in this country; and at MY request Dr. Hooker tabulated the trees of New Z
11  signated by this term, is of high importance on MY theory, and explains, as I believe, several imp
11  nd distinct species. Nevertheless, according to MY view, varieties are species in the process of f
11  d species of the same genus. As has always been MY practice, let us seek light on this head from o
29  tion of all organic beings; but, to the best of MY judgment, it is explained through inheritance a
32  ., produces on any being is extremely doubtful. MY impression is, that the effect is extremely sma
38  her inhabitants of North America and Europe. On MY view we must suppose that American animals, hav
50  t possibly be here introduced. I can only state MY conviction that it is a rule of high generality
51  ariably holds good with cirripedes. I shall, in MY future work, give a list of the more remarkable
51  to plants, and this would seriously have shaken MY belief in its truth, had not the great variabil
54  of the bat, it must have existed, according to MY theory, for an immense period in nearly the sam
56  r remarks. I think it will be admitted, without MY entering on details, that secondary sexual char
57  o instances; the first which happen to stand on MY list; and as the differences in these cases are
57  e species. This relation has a clear meaning on MY view of the subject; I look at all the species
61  the species of the same genus are supposed, on MY theory, to have descended from a common parent,
62  rs another species; this other species being on MY view only a well marked and permanent variety.
62  new but analogous variations, yet we ought, on MY theory, sometimes to find the varying offspring
64  in cuns, and I have see a trace in a bay horse. MY son made a careful examination and sketch for m
69  e parent of many modified descendants, which on MY view must be a very slow process, requiring a l
71  ion. Long before having arrived at this part of MY work, a crowd of difficulties will have occurre
```

age **(Key Word)***

0171 em without being staggered; but, to the best of MY judgment, the greater number are only apparent
0171 those that are real are not, I think, fatal to MY theory. These difficulties and objections may
0173 taken from the metropolis inhabited by each. By MY theory these allied species have descended fro
0176 ication, by which two varieties are supposed on MY theory to be converted and perfected into two
0179 ooking not to any one time, but to all time, if MY theory be true, numberless intermediate variet
0184 species of the same genus, we might expect, on MY theory, that such individuals would occasional
0189 by numerous, successive, slight modifications, MY theory would absolutely break down. But I can
0189 uch isolated species, round which, according to MY theory, there has been much extinction. Or aga
0192 culty occur, some of which will be discussed in MY future work. One of the gravest is that of neu
0199 d, but without here entering on copious details MY reasoning would appear frivolous. The foregoin
0199 doctrine, if true, would be absolutely fatal to MY theory. Yet I fully admit that many structures
0201 ve good of another species, it would annihilate MY theory, for such could not have been produced
0203 lties and objections which may be urged against MY theory. Many of them are very grave; but I thi
0206 we include all those of past times, it must by MY theory be strictly true. It is generally ackno
0206 s quite independent of their habits of life. On MY theory, unity of type is explained by unity of
0207 eaders, as a difficulty sufficient to overthrow MY whole theory. I must premise, that I have noth
0210 se of corporeal structure, and conformably with MY theory, the instinct of each species is good f
0212 effect on the reader's mind. I can only repeat MY assurance, that I do not speak without good ev
0216 of the several which I shall have to discuss in MY future work, namely, the instinct which leads
0221 cribed, their slaves in their jaws. Another day MY attention was struck by about a score of the s
0222 y attack other ants. In one instance I found to MY surprise an independent community of F. flava
0230 ce, I could show that they are conformable with MY theory. Huber's statement that the very first
0236 peared to me insuperable, and actually fatal to MY whole theory. I allude to the neuters or steri
0239 l and well established facts at once annihilate MY theory. In the simpler case of neuter insects
0240 e the amount of difference in these workers, by MY giving not the actual measurements, but a stri
0242 tion. But I am bound to confess, that, with all MY faith in this principle, I should never have a
0242 far the most serious special difficulty, which MY theory has encountered. The case, also, is ver
0243 in this chapter strengthen in any great degree MY theory; but none of the cases of difficulty, to
0243 none of the cases of difficulty, to the best of MY judgment, annihilate it. On the other hand, th
0243 ally, it may not be a logical deduction, but to MY imagination it is far more satisfactory to loo
0246 he fertility of their mongrel offspring, is, on MY theory, of equal importance with the sterility
0249 len is as often taken by chance (as I know from MY own experience) from the anthers of another fl
0280 nd gravest objection which can be urged against MY theory. The explanation lies, as I believe, in
0280 n mind what sort of intermediate forms must, on MY theory, have formerly existed. I have found it
0281 the intermediate links. It is just possible by MY theory, that one of two living forms might hav
0285 way. The consideration of these facts impresses MY mind almost in the same manner as does the vai
0291 been formed during subsidence. Since publishing MY views on this subject in 1845, I have watched
0296 the fine intermediate gradations which must on MY theory have existed between them, but abrupt,
0297 do find the kind of evidence of change which on MY theory we ought to find. Moreover, if we look
0299 the many objections which may be urged against MY views. Hence it will be worth while to sum up
0301 mber of those fine transitional forms, which on MY theory assuredly have connected all the past an
0302 d close of each formation, pressed so hardly on MY theory. On the sudden appearance of whole grou
0304 nother instance, which from having passed under MY own eyes has much struck me. In a memoir on Fo
0304 upt appearance of a great group of species. But MY work had hardly been published, when a skilful
0305 e that it would be an insuperable difficulty on MY theory, unless it could likewise be shown that
0306 ffer much from living species; and it cannot on MY theory be supposed, that these old species were
0307 rous and improved descendants. Consequently, if MY theory be true, it is indisputable that before
0307 of vast piles of fossiliferous strata, which on MY theory no doubt were somewhere accumulated bef
0309 t silurian period. The coloured map appended to MY volume on Coral Reefs, led me to conclude that
0310 in this volume, will undoubtedly at once reject MY theory. For my part, following out Lyell's met
0310 will undoubtedly at once reject my theory. For MY part, following out Lyell's metaphor, I look a
0314 isfactory. These several facts accord well with MY theory. I believe in no fixed law of developme
0316 t its truth; and the rule strictly accords with MY theory. For as all the species of the same gro
0316 is fact, which if true would have been fatal to MY views. But such cases are certainly exceptiona
0317 species of a group is strictly conformable with MY theory; as the species of the same genus, and
0319 y so favourable. But how utterly groundless was MY astonishment! Professor Owen soon perceived th
0327 ed with this subject worth making. I have given MY reasons for believing that all our greater foss
0337 particular sense the more recent forms must, on MY theory, be higher than the more ancient; for ea
0341 ving species. It must not be forgotten that, on MY theory, all the species of the same genus have
0342 e of the geological record, will rightly reject MY whole theory. For he may ask in vain where are
0350 led to inquire what this bond is. This bond, on MY theory, is simply inheritance, that cause which
0354 e several distinct species of a genus, which on MY theory have all descended from a common progen
0355 ormer period immigrants from this other region, MY theory will be strengthened; for we can clearly
0355 ver intercross (if such exist), the species, on MY theory, must have descended from a succession o
0357 d removes many a difficulty; but to the best of MY judgment we are not authorized in admitting suc
0358 ld resist the injurious action of sea water. To MY surprise I found that out of 87 kinds, 64 germi
0360 and gale, they would germinate. Subsequently to MY experiments, M. Martens tried similar ones, bu
0361 loated on artificial salt water for 30 days, to MY surprise nearly all germinated. Living birds ca
0361 ey. In the course of two months, I picked up in MY garden 12 kinds of seeds, out of the excrement
0363 y ice borne seeds, during the glacial epoch. At MY request Sir C. Lyell wrote to M. Hartung to inc
0374 simultaneously cooler. But it would suffice for MY purpose, if the temperature was at the same tim
0381 stant points of the southern hemisphere, is, on MY theory of descent with modification, a far more
0385 very wide range, and allied species, which, on MY theory, are descended from a common parent and
0386 ghed only 6 3/4 ounces; I kept it covered up in MY study for six months, pulling up and counting e
0390 is true. This fact might have been expected on MY theory, for, as already explained, species occa
0392 quadrupeds. This case presents no difficulty on MY view, for a hooked seed might be transported to
0393 known to be immediately killed by sea water, on MY view we can see that there would be great diffi
0393 rched the oldest voyages, but have not finished MY search; as yet I have not found a single instar
0394 bats and no other mammals on remote islands? On MY view this question can easily be answered; for
0397 ter and are killed by it. Yet there must be, on MY view, some unknown, but highly efficient means
0400 d this is just what might have been expected on MY view, for the islands are situated so near each
0400 the islands may be used as an argument against MY views; for it may be asked, how has it happened
0407 he distinct species of the same genus, which on MY theory must have spread from one parent source;
0410 in both cases marked exceptions to the rule. On MY theory these several relations throughout time
0413 , does not always sufficiently strike us, is in MY judgment fully explained. Naturalists try to ar
0420 objects more or less alike. But I must explain MY meaning more fully. I believe that the arrangem

```
age *****************************************( Key Word )*****************************************
*27  some authors between very distinct animals. On MY view of characters being of real importance for
*29  must be the number of connecting forms which on MY theory have been exterminated and utterly lost.
*30  e real and not merely adaptive, they are due on MY theory to inheritance in common. Therefore we m
*38  ng language having this plain signification. On MY view these terms may be used literally; and the
*46  ure. Let us take a genus of birds, descended on MY theory from some one parent species, and of whi
*49  gement, would be genealogical. Descent being on MY view the hidden bond of connexion which natural
*54  of the manatee were formed for this purpose. On MY view of descent with modification, the origin o
*63  the many objections which may be urged against MY theory. Why, again, do whole groups of allied s
*63  he silurian groups of fossils? For certainly on MY theory such strata must somewhere have been dep
*65  nged; and they have changed in the manner which MY theory requires, for they have changed slowly a
*65  difficulties which may justly be urged against MY theory; and I have now briefly recapitulated th
*66  is. Grave as these several difficulties are, on MY judgment they do not overthrow the theory of de
*74  hould be eminently liable to variation; but, on MY view, this part has undergone, since the severa
*82  a certain number of facts will certainly reject MY theory. A few naturalists, endowed with much fl
*88  each species has been independently created. To MY mind it accords better with what we know of the
*24  soon as three Orchidean forms (Monochanthus, MYANTHUS, and Catasetum), which had previously been r
*23  r they have thus changed: if the Megatherium, MYLODON, Macrauchenia, and Toxodon had been brought
*01  be not so, the being will become extinct, as MYRIADS have become extinct. Natural selection tends
*38  of which is quite unknown: in the Mexican MYRMECOCYSTUS, the workers of one caste never leave the
*40  ry nearly in the same condition with those of MYRMICA. For the workers of Myrmica have not even ru
*40  ion with those of Myrmica. For the workers of MYRMICA have not even rudiments of ocelli, though th
*01  aring on it. After five years' work I allowed MYSELF to speculate on the subject, and drew up some
*67  is the case with that of the horse genus! For MYSELF, I venture confidently to look back thousands
*13  cannot be doubted that young pointers (I have MYSELF seen a striking instance) will sometimes poin
*20  nce I will give the observations which I have MYSELF made, in some little detail. I opened fourtee
*36  nter on these several cases, but will confine MYSELF to one special difficulty, which at first app
*39  ndividuals taken out of the same nest: I have MYSELF compared perfect gradations of this kind. It
*80  rs in the same species, that I gladly availed MYSELF of Mr. F. Smith's offer of numerous specimens
*67  lly sterile in some degree. I cannot persuade MYSELF that this parallelism is an accident or an il
*72  variable in the first generation; and I have MYSELF seen striking instances of this fact. Gartner
*80  ing at any two species, to avoid picturing to MYSELF, forms directly intermediate between them. Bu
*18  d in numbers at an unparalleled rate, I asked MYSELF what could so recently have exterminated the
*58  l means of distribution. I shall here confine MYSELF to plants. In botanical works, this or that p
*89  in the following remarks I shall not confine MYSELF to the mere question of dispersal; but shall
*20  on are explained, if I do not greatly deceive MYSELF, on the view that the natural system is found
*01  ht on the origin of species, that mystery of MYSTERIES, as it has been called by one of our greate
*12  other parts of the structure, owing to the MYSTERIOUS laws of the correlation of growth. The resu
*51  theless these facts show on what slight and MYSTERIOUS causes the lesser or greater fertility of s
*45  g the later geological periods ceases to be MYSTERIOUS, and is simply explained by inheritance. If
*08  he forests, marshes, and deserts, are in so MYSTERIOUS a manner linked together by affinity, and a
*01  row some light on the origin of species, that MYSTERY of mysteries, as it has been called by one o
*18  cies has been involved in the most gratuitous MYSTERY. Some authors have even supposed that as the
*22  o m14); and (I) will have been replaced by six (N14 to z14) new species. But we may go further tha
*53  a man's fingers have been amputated, imperfect NAILS sometimes appear on the stumps: I could as so
*53  i could as soon believe that these vestiges of NAILS have appeared, not from unknown laws of growt
*54  excrete horny matter, as that the rudimentary NAILS on the fin of the manatee were formed for thi
*44  the relation between the hair and teeth in the NAKED Turkish dog, though here probably homology co
*97  rther modification and becoming a climber. The NAKED skin on the head of a vulture is generally lo
*00  d of the clean feeding male turkey is likewise NAKED. The sutures in the skulls of young mammals h
*00  y littoral animals, or of those which lived on NAKED submarine rocks, would be embedded; and those
*40  d. but as yet they will hardly have a distinct NAME, and from being only slightly valued, their hi
*40  will then probably first receive a provincial NAME. In semi civilised countries, with little free
*34  he whole subject is included under the general NAME of Morphology. This is the most interesting de
*84  e sufficiently important to deserve a specific NAME. This latter point will become a far more esse
*11  ct of use. Not a single domestic animal can be NAMED which has not in some country drooping ears;
*24  y rash assumption. Moreover, the several above NAMED domesticated breeds have been transported to
*47  ew well marked and well known varieties can be NAMED which have not been ranked as species by at l
*82  ther increase the advantage. No country can be NAMED in which all the native inhabitants are now s
*23  e in some degree distinct from the three first NAMED species; and lastly, o14, e14, and m14, will
*94  ave been astonished how rarely an organ can be NAMED, towards which no transitional grade is known
*87  at numbers of our fossil species are known and NAMED from single and often broken specimens, or fr
*30  barrande, and a higher authority could not be NAMED, asserts that he is every day taught that pal
*48  one shell, crab or fish is common to the above NAMED three approximate faunas of Eastern and Weste
*90  and on every other oceanic island which can be NAMED. In St. Helena there is reason to believe tha
*94  se to a continent; and hardly an island can be NAMED on which our smaller quadrupeds have not beco
*64  ord. Numerous existing doubtful forms could be NAMED which are probably varieties; but who will or
*05  est difficulties on the theory will be given: NAMELY, first, the difficulties of transitions; or i
*06  entertain, and which I formerly entertained, NAMELY, that each species has been independently cre
*14  fer to a statement often made by naturalists, NAMELY, that our domestic varieties, when run wild,
*23  the common opinion of naturalists is correct, NAMELY, that all have descended from the rock pigeon
*26  h in the extreme. From these several reasons, NAMELY, the improbability of man having formerly got
*31  nature. One circumstance has struck me much; NAMELY, that all the breeders of the various domesti
*33  bserving the accumulated effects of selection NAMELY, by comparing the diversity of flowers in the
*36  hed to have produced the result which ensued, NAMELY, the production of two distinct strains. The
*38  lains what has been remarked by some authors, NAMELY, that the varieties kept by savages have more
*40  ther explain what has sometimes been noticed, NAMELY, that we know nothing about the origin or his
*57  of difference between varieties and species, NAMELY, that the amount of difference between variet
*64  s subject than mere theoretical calculations, NAMELY, the numerous recorded cases of the astonishi
*77  ce may be deduced from the foregoing remarks, NAMELY, that the structure of every organic being is
*80  urable and unfavourable to Natural Selection, NAMELY, intercrossing, isolation, number of individu
*17  ultation with one of the highest authorities, NAMELY, Professor Huxley, to discover a single case
*17  ve produced two fairly well marked varieties, NAMELY a1 and m1. These two varieties will generally
*18  1 is supposed to have produced two varieties, NAMELY m2 and s2, differing from each other, and mor
*21  ed in the diagram, another of our principles, NAMELY that of extinction, will have played an impor
*28  up, in the manner which we everywhere behold, NAMELY, varieties of the same species most closely r
age *****************************************( Key Word )*****************************************
```

Page **(Key Word)**

```
0135  eir species in this condition! Several facts,  NAMELY, that beetles in many parts of the world are
0137  holly to disuse. In one of the blind animals,  NAMELY, the cave rat, the eyes are of immense size;
0147  may be merged under a more general principle,  NAMELY, that natural selection is continually tryin
0148  of which many other instances could be given:  NAMELY, that when a cirripede is parasitic within a
0154  icable, which most naturalists would advance,  NAMELY, that specific characters are more variable
0159  t species; and to these a third may be added,  NAMELY, the common turnip. According to the ordinar
0159  with pigeons, however, we have another case,  NAMELY, the occasional appearance in all the breeds
0178  arieties will, from reasons already assigned  (NAMELY from what we know of the actual distribution
0190  organ originally constructed for one purpose,  NAMELY flotation, may be converted into one for a w
0190  rted into one for a wholly different purpose,  NAMELY respiration. The swimbladder has, also, been
0201  s furnished with a rattle for its own injury,  NAMELY, to warn its prey to escape. I would almost
0209  erful instincts with which we are acquainted,  NAMELY, those of the hive bee and of many ants, cou
0216  ch i shall have to discuss in my future work,  NAMELY, the instinct which leads the cuckoo to lay
0233  rst quite subversive of the foregoing theory!  NAMELY, that the cells on the extreme margin of was
0236  en slowly acquired through natural selection;  NAMELY, by an individual having been born with some
0238  yet touched on the climax of the difficulty;  NAMELY, the fact that the neuters of several ants d
0244  ing to the advancement of all organic beings,  NAMELY, multiply, vary, let the strongest live and
0245  ent, have generally been confounded together;  NAMELY, the sterility of two species when first cro
0248  st experienced observers who have ever lived,  NAMELY, Kolreuter and Gartner, should have arrived
0248  has been diminished by an independent cause,  NAMELY, from close interbreeding. I have collected
0249  arkable statement repeatedly made by Gartner,  NAMELY, that if even the less fertile hybrids be ar
0249  at by the third most experienced hybridiser,  NAMELY, the Hon. and Rev. W. Herbert. He is as emph
0250  ll here give only a single one as an example,  NAMELY, that every ovule in a pod of Crinum capense
0250  um leads me to refer to a most singular fact,  NAMELY, that there are individual plants, as with c
0253  i am assured by two eminently capable judges,  NAMELY Mr. Blyth and Capt. Hutton, that whole flock
0254  been largely accepted by modern naturalists;  NAMELY, that most of our domestic animals have desc
0270  bserver and so hostile a witness, as Gartner:  NAMELY, that yellow and white varieties of the same
0271  s, and which he tested by the severest trial,  NAMELY, by reciprocal crosses, and he found their m
0272  e, sterility of first crosses and of hybrids,  NAMELY, that it is not a special endowment, but is
0273  e taken on the cause of ordinary variability;  NAMELY, that it is due to the reproductive system b
0275  arents differ much or little from each other,  NAMELY in the union of individuals of the same vari
0277  s supported by a parallelism of another kind;  NAMELY, that the crossing of forms only slightly di
0279  e. most of them have now been discussed. One,  NAMELY the distinctness of specific forms, and thei
0279  pparently most favourable for their presence,  NAMELY on an extensive and continous area with grad
0287  late Edward Forbes, should not be forgotten,  NAMELY, that numbers of our fossil species are know
0289  re important cause than any of the foregoing;  NAMELY, from the several formations being separated
0293  whose opinions are worthy of much deference,  NAMELY Bronn and Woodward, have concluded that the
0297  reover, if we look to rather wider intervals,  NAMELY, to distinct but consecutive stages of the s
0303  ed. I may here recall a remark formerly made,  NAMELY that it might require a long succession of a
0310  ion. The several difficulties here discussed,  NAMELY our not finding in the successive formations
0310  t that all the most eminent palaeontologists,  NAMELY Cuvier, Owen, Agassiz, Barrande, Falconer, E
0313  g fauna to reappear: but Lyell's explanation,  NAMELY,that it is a case of temporary migration fro
0320  gency. I may repeat what I published in 1845,  NAMELY, that to admit that species generally become
0322  ent of the mineral chalk itself can be found;  NAMELY, in North America, in equatorial South Ameri
0324  t that all the more modern marine formations,  NAMELY, the upper pliocene, the pleistocene and str
0332  then only the two families on the left hand  (NAMELY, a14, etc., and b14, etc.) would have to be
0332  into one family; and the two other families  (NAMELY, a14 to f14 now including five genera, and o
0334  eeding faunas. I need give only one instance,  NAMELY, the manner in which the fossils of the Devo
0352  ch has been largely discussed by naturalists,  NAMELY, whether species have been created at one or
0354  a few of the most striking classes of facts;  NAMELY, the existence of the same species on the su
0354  to consider a point equally important for us,  NAMELY, whether the several distinct species of a g
0355  not directly bear on another allied question,  NAMELY whether all the individuals of the same spec
0356  single parent. But in the majority of cases,  NAMELY, with all organisms which habitually unite f
0364  sported by these wanderers only by one means,  NAMELY, in dirt sticking to their feet, which is in
0396  emarks on the inhabitants of oceanic islands,  NAMELY, the scarcity of kinds, the richness in ende
0399  m a nearly intermediate though distant point,  NAMELY from the antarctic islands, when they were c
0402  me a remarkable fact bearing on this subject;  NAMELY, that Madeira and the adjoining islet of Por
0403  er of the fauna and flora of oceanic islands,  NAMELY, that the inhabitants, when not identically
0406  sed by Alph. de Candolle in regard to plants,  NAMELY, that the lower any group of organisms is, t
0406  s apt to range. The relations just discussed,  NAMELY, low and slowly changing organisms ranging m
0423  , is the same with varieties as with species,  NAMELY, closeness of descent with various degrees o
0433  ture in the affinities of all organic beings;  NAMELY, their subordination in group under group. W
0436  qually curious branch of the present subject;  NAMELY, the comparison not of the same part in diffe
0439  en, than a circumstance mentioned by Agassiz,  NAMELY, that having forgotten to ticket the embryo
0442  we explain these several facts in embryology,  NAMELY the very general, but not universal differenc
0447  er as wings; and on the above two principles,  NAMELY of each successive modification supervening a
0457  rstand the great leading facts in Embryology;  NAMELY, the resemblance in an individual embryo of
0459  real if we admit the following propositions,  NAMELY, that gradations in the perfection of any org
0461  allel, but directly opposite, class of facts;  NAMELY, that the vigour and fertility of all organic
0477  fact, which must have struck every traveller,  NAMELY, that on the same continent, under the most c
0484  mber. Analogy would lead me one step further,  NAMELY, to the belief that all animals and plants ha
0487  perhaps suddenly altered physical conditions,  NAMELY, the mutual relation of organism to organism,
0490  ed object which we are capable of conceiving,  NAMELY, the production of the higher animals, direct
0021  feathers. The trumpeter and laugher, as their  NAMES express, utter a very different coo from the
0422  nching diagram had not been used, and only the  NAMES of the groups had been written in a linear se
0434  the thigh and leg, transposed. Hence the same  NAMES can be given to the homologous bones in widel
0485  es may hereafter be thought worthy of specific  NAMES, as with the primrose and cowslip; and in th
0006  umerous; and why another allied species has a  NARROW range and is rare? Yet these relations are o
0174  tween two representative species is generally  NARROW in comparison with the territory proper to ea
0175  t each has a wide range, with a comparatively  NARROW neutral territory between them, in which they
0176  to two large areas, and a third variety to a  NARROW intermediate zone. The intermediate variety,
0176  ill exist in lesser numbers from inhabiting a  NARROW and lesser area; and practically, as far as
0177  variety, which exists in smaller numbers in a  NARROW and intermediate zone. For forms existing in
0177  untainous region; a second to a comparatively  NARROW, hilly tract; and a third to wide plains at t
0177  ly than the small holders on the intermediate  NARROW, hilly tract; and consequently the improved m
0228  d of a thick, square piece of wax, a thin and  NARROW, knife edged ridge, coloured with vermilion.
```

Page **(Key Word)**

age **(Key Word)**

```
348  these great faunas are separated only by the   NARROW, but impassable, isthmus of Panama. Westward
356  and must also have been highly influential: a    NARROW isthmus now separates two marine faunas; subm
452  ing buoyancy, but has become converted into a    NASCENT breathing organ or lung. Other similar insta
441  ey have six pairs of beautifully constructed     NATATORY legs, a pair of magnificent compound eyes, a
008  ong under not very close confinement in their    NATIVE country! This is generally attributed to viti
024  must have been carried back again into their     NATIVE country; but not one has ever become wild or
035  ow has not seen, as I am informed by him, any    NATIVE dog in Spain like our pointer. By a similar p
038  nal stocks of any useful plants, but that the    NATIVE plants have not been improved by continued se
067  her plants, I marked all the seedlings of our    NATIVE weeds as they came up, and out of the 357 no
069  naturalised, for they cannot compete with our    NATIVE plants, nor resist destruction by our native
069  native plants, nor resist destruction by our     NATIVE animals. When a species, owing to highly favo
071  nd planted with Scotch fir. The change in the    NATIVE vegetation of the planted part of the heath w
078  ferent way to what we should have done in its    NATIVE country; for we should have to give it some a
082  age. No country can be named in which all the    NATIVE inhabitants are now so perfectly adapted to e
115  sation, proportionally with the number of the    NATIVE genera and species, far more in new genera th
140  these animals were strictly adapted to their    NATIVE climate, but in all ordinary cases we assume
142  born with constitutions best adapted to their   NATIVE countries. In treatises on many kinds of cult
337  nabled to seize on places now occupied by our    NATIVE plants and animals. Under this point of view,
371  come mingled in the one great region with the    NATIVE American productions, and have had to compete
377  ant temperate forms might have penetrated the    NATIVE ranks and have reached or even crossed the eq
378  me twenty five degrees of latitude from their   NATIVE country and covered the land at the foot of t
380  t workshops of the north. In many islands the    NATIVE productions are nearly equalled or even outnu
390  nimals have nearly or quite exterminated many    NATIVE productions. He who admits the doctrine of th
083  anyhow be improved: for in all countries, the    NATIVES have been so far conquered by naturalised pr
083  gners have thus everywhere beaten some of the    NATIVES, we may safely conclude that the natives mig
083  the natives, we may safely conclude that the     NATIVES might have been modified with advantage, so
083  well suited conditions of life. Man keeps the    NATIVES of many climates in the same country; he sel
110  ing, as far as we know, the extinction of any    NATIVES. Furthermore, the species which are most num
115  in some crude idea in what manner some of the    NATIVES would have had to be modified, in order to h
115  er to have gained an advantage over the other    NATIVES; and we may, I think, at least safely infer
337  ized there, and would exterminate many of the    NATIVES. On the other hand, from what we see now occ
360  se in the midst of the widest oceans; and the    NATIVES of the coral islands in the Pacific, procure
378  will have afforded an asylum to the tropical     NATIVES. The mountain ranges north west of the Himal
380  alia, and have to a certain extent beaten the    NATIVES; whereas extremely few southern forms have b
380  en outnumbered by the naturalised; and if the    NATIVES have not been actually exterminated, their n
393  l (excluding domesticated animals kept by the    NATIVES) inhabiting an island situated above 300 mil
171  on. Means of transition. Cases of difficulty.    NATURA non facit saltum. Organs of small importance.
194  shown by that old canon in natural history of    NATURA non facit saltum. We meet with this admission
206  meaning of that old canon in natural history.    NATURA non facit saltum. This canon if we look only
210  ex instincts, can be discovered. The canon of    NATURA non facit saltum applies with almost equal fo
243  others; that the canon in natural history, of    NATURA non facit saltum is applicable to instincts a
460  ons in nature, as is proclaimed by the canon,    NATURA non facit saltum, that we ought to be extreme
471  very short and slow steps. Hence the canon of    NATURA non facit saltum, which every fresh addition
008  d the happy survive and multiply. Chapter IV.    NATURAL Selection. Natural Selection, its power comp
002  this, as Mr. Wallace, who is now studying the    NATURAL history of the Malay archipelago, has arrive
005  nd modified form. This fundamental subject of    NATURAL Selection will be treated at some length in
005  the fourth chapter: and we shall then see how    NATURAL Selection almost inevitably causes much Exti
006  hat species. Furthermore, I am convinced that   NATURAL Selection has been the main but not exclusiv
015  eversions of character probably do occur: but    NATURAL selection, as will hereafter be explained, w
017  utability of the many very closely allied and    NATURAL species; for instance, of the many foxes, o
032  uredly fail. Few would readily believe in the    NATURAL capacity and years of practice requisite to
038  y than in the other, and thus by a process of    NATURAL selection, as will hereafter be more fully e
045  mportant for us, as they afford materials for    NATURAL selection to accumulate, in the same manner
046  e not been seized on and rendered definite by    NATURAL selection, as hereafter will be explained.
051  re barely thought worth recording in works on    NATURAL history. And I look at varieties which are i
052  ne in which it differs more, to the action of    NATURAL selection in accumulating (as will hereafter
060  chapter III. The Struggle For Existence. Bears on NATURAL selection. The term used in a wide sense. Ge
060  show how the struggle for existence bears on     NATURAL Selection. It has been seen in the last chap
061  tion, if useful, is preserved, by the term of    NATURAL Selection, in order to mark its relation to
061  ions, given to him by the hand of Nature. But    NATURAL Selection, as we shall hereafter see, is a p
063  nd to increase. Every being, which during its    NATURAL lifetime produces several eggs or seeds, mus
064  ains to estimate its probable minimum rate of    NATURAL increase: it will be under the mark to assum
067  n another with greater force. What checks the    NATURAL tendency of each species to increase in numb
080  and multiply. Chapter IV. Natural Selection.     NATURAL Selection, its power compared with man's sel
080  circumstances favourable and unfavourable to     NATURAL Selection, namely, intercrossing, isolation,
080  ndividuals. Slow action. Extinction caused by    NATURAL Selection. Divergence of Character, related
080  small area, and to naturalisation. Action of    NATURAL Selection, through Divergence of Character a
081  the rejection of injurious variations, I call   NATURAL Selection. Variations neither useful nor inj
081  useful nor injurious would not be affected by    NATURAL selection, and would be left a fluctuating e
081  shall best understand the probable course of     NATURAL selection by taking the case of a country un
082  d conditions, would tend to be preserved; and    NATURAL selection would thus have free scope for the
082  e, and this would manifestly be favourable to    NATURAL selection, by giving a better chance of prof
082  gi and unless profitable variations do occur,    NATURAL selection can do nothing. Not that, as I hel
082  sary to produce new and unoccupied places for    NATURAL selection to fill up by modifying and improv
084  f far higher workmanship? It may be said that    NATURAL selection is daily and hourly scrutinising,
085  ferent from what they formerly were. Although    NATURAL selection can act only through and for the g
085  tion. Hence I can see no reason to doubt that    NATURAL selection might be most effective in giving
085  ion, and the modifications are accumulated by    NATURAL selection for the good of the being, will ca
086  e when nearly adult; so in a state of nature,    NATURAL selection will be enabled to act on and modi
086  ter difficulty in this being effected through   NATURAL selection, than in the cotton planter increa
086  ion the down in the pods on his cotton trees.    NATURAL Selection may modify and adapt the larva of
086  the structure of the larva; but in all cases    NATURAL selection will ensure that modifications con
086  ey would cause the extinction of the species.    NATURAL selection will modify the structure of the y
087  sequence profits by the selected change. What    NATURAL selection cannot do, is to modify the struct
087  ments to this effect may be found in works of    NATURAL history, I cannot find one case which will b
087  nce to it, might be modified to any extent by    NATURAL selection; for instance, the great jaws poss
```

Page **(Key Word)**

Page ***(Key Word)**

```
0087 fact probably occurs under nature, and if so, NATURAL selection will be able to modify one sex in
0088 l selection is, therefore, less rigorous than NATURAL selection. Generally, the most vigorous mal
0090 a monstrosity. Illustrations of the action of NATURAL Selection. In order to make it clear how, a
0090 in order to make it clear how, as I believe, NATURAL selection acts, I must beg permission to gi
0091 for we often observe great differences in the NATURAL tendencies of our domestic animals: one cat
0093 this process of the continued preservation or NATURAL selection of more and more attractive flowe
0095 ucture. I am well aware that this doctrine of NATURAL selection, exemplified in the above imagina
0095 mation of the longest lines of inland cliffs. NATURAL selection can act only by the preservation
0095 eat valley by a single diluvial wave, so will NATURAL selection, if it be a true principle, banis
0101 n for perpetuity. Circumstances favourable to NATURAL Selection. This is an extremely intricate s
0102 e grants vast periods of time for the work of NATURAL selection, she does not grant an indefinite
0102 polity not so perfectly occupied as might be, NATURAL selection will always tend to preserve all
0102 ent different conditions of life; and then if NATURAL selection be modifying and improving a spec
0103 ntercrossing can hardly be counterbalanced by NATURAL selection always tending to modify all the
0103 rate the effects of intercrosses in retarding NATURAL selection; for I can bring a considerable c
0104 ugh the principle of inheritance, and through NATURAL selection destroying any which depart from
0104 given to their modified offspring, solely by NATURAL selection preserving the same favourable va
0104 so, is an important element in the process of NATURAL selection. In a confined or isolated area,
0104 nerally be in a great degree uniform; so that NATURAL selection will tend to modify all the indiv
0104 of the land, etc.; and thus new places in the NATURAL economy of the country are left open for th
0105 retard the production of new species through NATURAL selection, by decreasing the chance of the
0107 ain extent orders now widely separated in the NATURAL scale. These anomalous forms may almost be
0107 circumstances favourable and unfavourable to NATURAL selection, as far as the extreme intricacy
0108 ged; and again there will be a fair field for NATURAL selection to improve still further the inha
0108 habitants, and thus produce new species. That NATURAL selection will always act with extreme slow
0108 forms having been checked. But the action of NATURAL selection will probably still oftener depen
0108 amply sufficient wholly to stop the action of NATURAL selection. I do not believe so. On the othe
0108 ieve so. On the other hand, I do believe that NATURAL selection will always act very slowly, ofte
0109 , that this very slow, intermittent action of NATURAL selection accords perfectly well with what
0109 luded to from being intimately connected with NATURAL selection. Natural selection acts solely th
0109 intimately connected with natural selection. NATURAL selection acts solely through the preservat
0110 cies in the course of time are formed through NATURAL selection, others will become rarer and rar
0113 long ago arrived at its full average. If its NATURAL powers of increase be allowed to act, it ca
0113 animals, that is, if they vary, for otherwise NATURAL selection can do nothing. So it will be wit
0114 ersification of structure, is seen under many NATURAL circumstances. In an extremely small area,
0116 of character, combined with the principles of NATURAL selection and of extinction, will tend to a
0117 ted lines) being preserved and accumulated by NATURAL selection. When a dotted line reaches one o
0119 roduce more than one modified descendant; for NATURAL selection will always act according to the
0121 ortant part. As in each fully stocked country NATURAL selection necessarily acts by the selected
0122 become adapted to many related places in the NATURAL economy of their country. It seems, therefo
0123 in number. Owing to the divergent tendency of NATURAL selection, the extreme amount of difference
0125 his, indeed, might have been expected: for as NATURAL selection acts through one form having some
0127 tion, I have called, for the sake of brevity, NATURAL Selection. Natural selection, on the princi
0127 , for the sake of brevity, Natural Selection. NATURAL selection, on the principle of qualities be
0127 in their struggles with other males. Whether NATURAL selection has really thus acted in nature,
0127 he world's history, geology plainly declares. NATURAL selection, also, leads to divergence of cha
0128 w makes them dominant in their own countries. NATURAL selection, as has just been remarked, leads
0129 through inheritance and the complex action of NATURAL selection, entailing extinction and diverge
0131 nal conditions. Use and disuse, combined with NATURAL selection; organs of flight and of vision.
0133 it to attribute to the accumulative action of NATURAL selection, and how much to the conditions o
0134 system, and in thus inducing variability; and NATURAL selection will then accumulate all profitab
0135 d habits like those of a bustard, and that as NATURAL selection increased in successive generatio
0135 structure which are wholly, or mainly, due to NATURAL selection. Mr. Wollaston has discovered the
0136 adeira beetles is mainly due to the action of NATURAL selection, but combined probably with disus
0136 , this is quite compatible with the action of NATURAL selection. For when a new insect first arri
0136 first arrived on the island, the tendency of NATURAL selection to enlarge or to reduce the wings
0137 l reduction from disuse, but aided perhaps by NATURAL selection. In South America, a burrowing ro
0137 ight in such case be an advantage; and if so, NATURAL selection would constantly aid the effects
0137 and the wings of others have been reduced by NATURAL selection aided by use and disuse, so in th
0137 se and disuse, so in the case of the cave rat NATURAL selection seems to have struggled with the
0138 e or less perfectly obliterated its eyes, and NATURAL selection will often have effected other ch
0141 ate is due to mere habit, and how much to the NATURAL selection of varieties having different inn
0142 other hand, I can see no reason to doubt that NATURAL selection will continually tend to preserve
0143 ined with, and sometimes overmastered by, the NATURAL selection of innate differences. Correlatio
0143 y one part occur, and are accumulated through NATURAL selection, other parts become modified. Thi
0143 t, may be mastered more or less completely by NATURAL selection: thus a family of stags once exis
0143 ight probably have been rendered permanent by NATURAL selection. Homologous parts, as has been re
0144 , independently of utility and, therefore, of NATURAL selection, than that of the difference betw
0146 y at first appear: and if it be advantageous, NATURAL selection may have come into play. But in r
0146 ancient progenitor may have acquired through NATURAL selection some one modification in structur
0146 are entirely due to the manner alone in which NATURAL selection can act. For instance, Alph. de C
0146 eds could not gradually become winged through NATURAL selection, except in fruits which opened: s
0147 nd, of a part being largely developed through NATURAL selection and another and adjoining part be
0147 under a more general principle, namely, that NATURAL selection is continually trying to economis
0148 ght, in its development, will be seized on by NATURAL selection, for it will profit the individua
0148 cture now become useless. Thus, as I believe, NATURAL selection will always succeed in the long r
0148 a corresponding degree. And, conversely, that NATURAL selection may perfectly well succeed in lar
0149 e why it should remain variable, that is, why NATURAL selection should have preserved or rejected
0149 bject had better be of some particular shape. NATURAL selection, it should never be forgotten, ca
0149 owing to their uselessness, and therefore to NATURAL selection having no power to check deviatio
0152 other species, and have been modified through NATURAL selection, I think we can obtain some light
0152 polymorphic groups, we see a nearly parallel NATURAL case: for in such cases natural selection e
0152 arly parallel natural case; for in such cases NATURAL selection either has not or cannot come int
0153 ty, which has continually been accumulated by NATURAL selection for the benefit of the species. B
0153 inced, is the case. That the struggle between NATURAL selection on the one hand, and the tendency
0155 ic; but I have repeatedly noticed in works on NATURAL history, that when an author has remarked w
```

Page ***(Key Word)**

156	genitor, for it can rarely have happened that	NATURAL selection will have modified several species
157	is highly probable, be taken advantage of by	NATURAL and sexual selection, in order to fit the se
158	e long been inherited and have not varied, to	NATURAL selection having more or less completely, ac
158	in the same parts having been accumulated by	NATURAL and sexual selection, and thus adapted for s
161	all important characters will be governed by	NATURAL selection, in accordance with the diverse ha
168	odifications have not been closely checked by	NATURAL selection. It is probably from this same cau
168	s, from being useless, will be disregarded by	NATURAL selection, and hence probably are variable.
169	ion is a long continued and slow process, and	NATURAL selection will in such cases not as yet have
169	requiring a long lapse of time, in this case,	NATURAL selection may readily have succeeded in givi
170	exist, it is the steady accumulation, through	NATURAL selection, of such differences, when benefic
171	itions of Existence embraced by the theory of	NATURAL Selection. Long before having arrived at thi
171	wholly different habits? Can we believe that	NATURAL selection could produce, on the one hand, or
172	an instincts be acquired and modified through	NATURAL selection? What shall we say to so marvellou
172	sence or rarity of transitional varieties. As	NATURAL selection acts solely by the preservation of
172	t comes into competition. Thus extinction and	NATURAL selection will, as we have seen, go hand in
173	crust of the earth is a vast museum; but the	NATURAL collections have been made only at intervals
173	vidently filling nearly the same place in the	NATURAL economy of the land. These representative sp
177	presenting further favourable variations for	NATURAL selection to seize on, than will the rarer f
177	ed, for variation is a very slow process, and	NATURAL selection can do nothing until favourable va
177	ons chance to occur, and until a place in the	NATURAL polity of the country can be better filled b
178	lanted and exterminated during the process of	NATURAL selection, so that they will no longer exist
179	g the process of further modification through	NATURAL selection, they will almost certainly be bea
179	riation, and thus be further improved through	NATURAL selection and gain further advantages. Lastl
179	suredly have existed; but the very process of	NATURAL selection constantly tends, as has been so o
180	est that it is possible to conceive under all	NATURAL conditions. Let the climate and vegetation c
181	by the accumulated effects of this process of	NATURAL selection, a perfect so called flying squirr
181	galeopithecus might be greatly lengthened by	NATURAL selection; and this, as far as the organs of
182	y all have resulted from disuse, indicate the	NATURAL steps by which birds have acquired their per
182	ted by the very process of perfection through	NATURAL selection. Furthermore, we may conclude that
183	when either case occurs, it would be easy for	NATURAL selection to fit the animal, by some modific
184	ficulty in a race of bears being rendered, by	NATURAL selection, more and more aquatic in their st
186	truggle for existence and in the principle of	NATURAL selection, will acknowledge that every organ
186	romatic aberration, could have been formed by	NATURAL selection, seems, I freely confess, absurd i
186	a perfect and complex eye could be formed by	NATURAL selection, though insuperable by our imagina
188	e of many other structures) in believing that	NATURAL selection has converted the simple apparatus
188	ect as the eye of an eagle might be formed by	NATURAL selection, although in this case he does not
188	e of hesitation in extending the principle of	NATURAL selection, to such startling lengths. It is s
189	ion will multiply them almost infinitely, and	NATURAL selection will pick out with unerring skill
190	digest and the stomach respire. In such cases	NATURAL selection might easily specialise, if any ad
191	e to be no great difficulty in believing that	NATURAL selection has actually converted a swimbladd
191	nchiae might have been gradually worked in by	NATURAL selection for some quite distinct purpose: i
192	respiration, have been gradually converted by	NATURAL selection into branchiae, simply through an
193	of the members to its loss through disuse or	NATURAL selection. But if the electric organs had be
194	ependently hit on the very same invention, so	NATURAL selection, working for the good of each bein
194	s remark is indeed shown by that old canon in	NATURAL history of Natura non facit saltum. We meet
194	from structure to structure? On the theory of	NATURAL selection, we can clearly understand why she
194	an clearly understand why she should not; for	NATURAL selection can act only by taking advantage o
194	eps. Organs of little apparent importance. As	NATURAL Selection acts by life and death, by the pre
195	l differences, might assuredly be acted on by	NATURAL selection. The tail of the giraffe looks lik
196	ir structure will always have been checked by	NATURAL selection. Seeing how important an organ of
196	from quite secondary causes, independently of	NATURAL selection. We should remember that climate,
197	portance and might have been acquired through	NATURAL selection; as it is, I have no doubt that th
198	, and would be exposed to a certain extent to	NATURAL selection, and individuals with slightly dif
198	lour would be thus subjected to the action of	NATURAL selection. But we are far too ignorant to sp
200	everal bones might have been acquired through	NATURAL selection, subjected formerly, as now, to th
200	ndirectly through the complex laws of growth.	NATURAL selection cannot possibly produce any modifi
200	and profits by, the structure of another. But	NATURAL selection can and does often produce structu
201	for such could not have been produced through	NATURAL selection. Although many statements may be f
201	ough many statements may be found in works on	NATURAL history to this effect, I cannot find even o
201	e here to enter on this and other such cases.	NATURAL selection will never produce in a being anyt
201	in a being anything injurious to itself, for	NATURAL selection acts solely by and for the good of
201	come extinct, as myriads have become extinct.	NATURAL selection tends only to make each organic be
202	of plants and animals introduced from Europe.	NATURAL selection will not produce absolute perfecti
202	unity, it will fulfil all the requirements of	NATURAL selection, though it may cause the death of
203	s all the same to the inexorable principle of	NATURAL selection. If we admire the several ingeniou
203	ate gradations, partly because the process of	NATURAL selection will always be very slow, and will
203	forms; and partly because the very process of	NATURAL selection almost implies the continual suppl
204	, for instance, could not have been formed by	NATURAL selection from an animal which at first coul
204	perfect as the eye could have been formed by	NATURAL selection, is more than enough to stagger an
204	any conceivable degree of perfection through	NATURAL selection. In the cases in which we know of
205	not have been slowly accumulated by means of	NATURAL selection. But we may confidently believe th
205	, in its present state, have been acquired by	NATURAL selection, a power which acts solely by the
205	ofitable variations in the struggle for life.	NATURAL selection will produce nothing in one specie
206	l cases at the same time useful to the owner.	NATURAL selection in each well stocked country, must
206	of perfection will have been rendered higher.	NATURAL selection will not necessarily produce absol
206	fection be everywhere found. On the theory of	NATURAL selection we can clearly understand the full
206	erstand the full meaning of that old canon in	NATURAL history, Natura non facit saltum. This canon
206	cuvier, is fully embraced by the principle of	NATURAL selection. For natural selection acts by eit
206	ed by the principle of natural selection. For	NATURAL selection acts by either now adapting the va
207	s variable. Domestic instincts; their origin.	NATURAL instincts of the cuckoo, ostrich, and parasi
207	g instinct. Difficulties on the theory of the	NATURAL Selection of instincts. Neuter or sterile in
209	er so little, then I can see no difficulty in	NATURAL selection preserving and continually accumul
209	subordinate importance to the effects of the	NATURAL selection of what may be called accidental v
210	lex instinct can possibly be produced through	NATURAL selection, except by the slow and gradual ac
210	e or the other instinct might be preserved by	NATURAL selection. And such instances of diversity o
211	riations, are indispensable for the action of	NATURAL selection, as many instances as possible oug

Page *********************************(Key Word)*********************************

```
0212 vantageous to the species, give rise, through NATURAL selection, to quite new instincts. But I am
0213 e certainly far less fixed or invariable than NATURAL instincts: but they have been acted on by f
0214 incts, when thus tested by crossing, resemble NATURAL instincts, which in a like manner become cu
0215 o habit and long continued close confinement. NATURAL instincts are lost under domestication: a r
0216 hat domestic instincts have been acquired and NATURAL instincts have been lost partly by habit, a
0219 ase of the cuckoo, I can see no difficulty in NATURAL selection making an occasional habit perman
0224 collecting pupae originally for food might by NATURAL selection be strengthened and rendered perm
0224 es in Switzerland, I can see no difficulty in NATURAL selection increasing and modifying the inst
0228 lieve that the hive bee has acquired, through NATURAL selection, her inimitable architectural pow
0233 als would be required than for a cylinder. As NATURAL selection acts only by the accumulation of
0235 ond this stage of perfection in architecture, NATURAL selection could not lead; for the comb of t
0235 ts, that of the hive bee, can be explained by NATURAL selection having taken advantage of numerou
0235 e, slight modifications of simpler instincts; NATURAL selection having by slow degrees, more and
0235 ic plates. The motive power of the process of NATURAL selection having been economy of wax: that
0235 explanation could be opposed to the theory of NATURAL selection, cases, in which we cannot see ho
0235 that they could hardly have been acted on by NATURAL selection: cases of instincts almost identi
0236 hey have been acquired by independent acts of NATURAL selection. I will not here enter on these s
0236 ry great difficulty in this being effected by NATURAL selection. But I must pass over this prelim
0236 s characters had been slowly acquired through NATURAL selection; namely, by an individual having
0237 ble to reconcile this case with the theory of NATURAL selection? First, let it be remembered that
0237 ructure could have been slowly accumulated by NATURAL selection. This difficulty, though appearin
0239 an overweening confidence in the principle of NATURAL selection, when I do not admit that such wo
0239 of the same kind, which have been rendered by NATURAL selection, as I believe to be quite possibl
0241 s. with these facts before me, I believe that NATURAL selection, by acting on the fertile parents
0241 ed in greater and greater numbers through the NATURAL selection of the parents which generated th
0242 r in the communities of ants, by the means of NATURAL selection. But I am bound to confess, that,
0242 inciple, I should never have anticipated that NATURAL selection could have been efficient in so h
0242 ficient length, in order to show the power of NATURAL selection, and likewise because this is by
0243 iculty, under changing conditions of life, in NATURAL selection accumulating slight modifications
0243 of the instincts of others; that the canon in NATURAL history, of natura non facit saltum is appl
0243 icable, all tend to corroborate the theory of NATURAL selection. This theory is, also, strengthen
0245 rrated by some late writers. On the theory of NATURAL selection the case is especially important,
0250 e says) I never saw to occur in a case of its NATURAL fecundation. So that we here have perfect,
0251 duced itself as perfectly as if it had been a NATURAL species from the mountains of Chile. I have
0264 hen animals and plants are removed from their NATURAL conditions, they are extremely liable to ha
0265 ring several generations under conditions not NATURAL to them, they are extremely liable to vary,
0273 es (very few experiments having been tried on NATURAL varieties), and this implies in most cases
0277 o frequently affects pure species, when their NATURAL conditions of life have been disturbed. Thi
0280 oughout nature depends on the very process of NATURAL selection, through which new varieties cont
0281 her allied species, such as C. oenas. So with NATURAL species, if we look to forms very distinct,
0281 he old and unimproved forms. By the theory of NATURAL selection all living species have been conn
0282 nges having been effected very slowly through NATURAL selection. It is hardly possible for me eve
0282 recognise as having produced a revolution in NATURAL science, yet does not admit how imcomprehen
0300 ere, how imperfectly would they represent the NATURAL history of the world! But we have every rea
0302 ory of descent with slow modification through NATURAL selection. For the development of a group o
0310 we owe all our knowledge. Those who think the NATURAL geological record in any degree perfect, an
0310 following out Lyell's metaphor, I look at the NATURAL geological record, as a history of the worl
0312 and gradual modification, through descent and NATURAL selection. New species have appeared very s
0314 her such variability be taken advantage of by NATURAL selection, and whether the variations be ac
0317 es and of groups of species. On the theory of NATURAL selection the extinction of old forms and t
0320 some unknown deed of violence. The theory of NATURAL selection is grounded on the belief that ea
0320 orms and the disappearance of old forms, both NATURAL and artificial, are bound together. In cert
0322 come extinct, accords well with the theory of NATURAL selection. We need not marvel at extinction
0325 out the world, is explicable on the theory of NATURAL selection. New species are formed by new va
0325 greatest number of new varieties. It is also NATURAL that the dominant, varying, and far spreadi
0329 living species. They all fall into one grand NATURAL system; and this fact is at once explained
0330 ly valid. But I apprehend that in a perfectly NATURAL classification many fossil species would ha
0333 rare cases, to fill up wide intervals in the NATURAL system, and thus unite distinct families or
0338 ne of Agassiz accords well with the theory of NATURAL selection. In a future chapter I shall atte
0343 e theory of descent with modification through NATURAL selection. We can thus understand how it is
0345 perfect, the main objections to the theory of NATURAL selection are greatly diminished or disappe
0345 ation still acting round us, and preserved by NATURAL Selection. Chapter XI. Geographical Distribu
0350 ons may be attributed to modification through NATURAL selection, and in a quite subordinate degre
0350 for the slow process of modification through NATURAL selection. Widely ranging species, abounding
0351 t property, and will be taken advantage of by NATURAL selection, only so far as it profits the in
0352 e same should ever have been produced through NATURAL selection from parents specifically distinc
0353 bserved by several naturalists, that the most NATURAL genera, or those genera in which the specie
0379 and having consequently been advanced through NATURAL selection and competition to a higher stage
0382 this view, combined with modification through NATURAL selection, a multitude of facts in the pres
0385 might represent those of a bird sleeping in a NATURAL pond, in an aquarium, where many ova of fre
0386 king land, they would be sure to fly to their NATURAL fresh water haunts. I do not believe that b
0392 ller and overtopping the other plants. If so, NATURAL selection would often tend to add to the st
0395 wandering species have been modified through NATURAL selection in their new homes in relation to
0395 t we shall soon have much light thrown on the NATURAL history of this archipelago by the admirabl
0401 omewhat different enemies. If then it varied, NATURAL selection would probably favour different va
0410 ns have been accumulated by the same power of NATURAL selection. Chapter XIII. Mutual Affinities
0411 classification, groups subordinate to groups. NATURAL system. Rules and difficulties in classific
0411 cond and fourth chapters, on Variation and on NATURAL Selection, I have attempted to show that it
0413 nited into one class. Thus, the grand fact in NATURAL history of the subordination of group under
0413 families in each class, on what is called the NATURAL System. But what is meant by this system? S
0413 sts think that something more is meant by the NATURAL System; they believe that it reveals the pl
0415 ot only in this but, as I apprehend, in every NATURAL family, is very unequal, and in some cases
0417 an aggregate of characters is very evident in NATURAL history. Hence, as has often been remarked,
0420 greatly deceive myself, on the view that the NATURAL system is founded on descent with modificat
0420 must be strictly genealogical in order to be NATURAL: but that the amount of difference in the se
0421 ir parentage, in this case, their places in a NATURAL classification will have been more or less
```

Page *********************************(Key Word)*********************************

age **************************************(Key Word)**************************************

422	ed to a certain extent their characters. This	NATURAL arrangement is shown, as far as is possible
422	have been still less possible to have given a	NATURAL arrangement: and it is notoriously not possi
422	me group. Thus, on the view which I hold, the	NATURAL system is genealogical in its arrangement, l
422	l be genealogical: and this would be strictly	NATURAL, as it would connect together all languages;
423	d on the necessity of classing varieties on a	NATURAL instead of an artificial system: we are caut
429	der still fewer classes, and all in one great	NATURAL system. As showing how few the higher groups
431	living and extinct members of the same great	NATURAL class. Extinction, as we have seen in the fo
432	the finest existing varieties, nevertheless a	NATURAL classification, or at least a natural arrang
432	eless a natural classification, or at least a	NATURAL arrangement, would be possible. We shall see
432	their ancient and unknown progenitor. Yet the	NATURAL arrangement in the diagram would still hold
433	such types belong. Finally, we have seen that	NATURAL selection, which results from the struggle f
433	naturalists have sought under the term of the	NATURAL System. On this idea of the natural system b
433	rm of the Natural System. On this idea of the	NATURAL system being, in so far as it is perfe
434	y, this is the most interesting department of	NATURAL history, and may be said to be its very soul
435	explanation is manifest on the theory of the	NATURAL selection of successive slight modifications
436	s being perhaps very simple in form; and then	NATURAL selection will account for the infinite dive
437	tructed on the same pattern? On the theory of	NATURAL selection, we can satisfactorily answer thes
438	cture: consequently it is quite probable that	NATURAL selection, during a long continued course of
445	, be ranked in distinct genera, had they been	NATURAL productions. But when the nestling birds of
446	eral new species have become modified through	NATURAL selection in accordance with their diverse h
448	e young or larvae might easily be rendered by	NATURAL selection different to any conceivable exten
449	lists have been seeking under the term of the	NATURAL system. On this view we can understand how i
450	gy, which are second in importance to none in	NATURAL history, are explained on the principle of s
454	rgans, are imperfect and useless. In works on	NATURAL history rudimentary organs are generally sai
454	n small and exposed islands: and in this case	NATURAL selection would continue slowly to reduce th
455	nsensibly small steps, is within the power of	NATURAL selection: so that an organ rendered, during
455	able, for its variations cannot be checked by	NATURAL selection. At whatever period of life disuse
456	ied, together with their modification through	NATURAL selection, with its contingencies of extinct
456	ngs, we shall understand what is meant by the	NATURAL system: it is genealogical in its attempted
457	n the view that an arrangement is only so far	NATURAL as it is genealogical. Finally, the several
459	tutation of the difficulties on the theory of	NATURAL Selection. Recapitulation of the general and
459	mmutability of species. How far the theory of	NATURAL selection may be extended. Effects of its ad
459	nded. Effects of its adoption on the study of	NATURAL history. Concluding remarks. As this whole v
460	e theory of descent with modification through	NATURAL selection; and one of the most curious of th
462	in some degree lessened. As on the theory of	NATURAL selection an interminable number of intermed
467	y man have to a large extent the character of	NATURAL species, is shown by the inextricable doubts
468	nature, it would be an unaccountable fact if	NATURAL selection had not come into play. It has oft
469	most complex relations of life. The theory of	NATURAL selection, even if we looked no further than
470	nature, there will be a constant tendency in	NATURAL selection to preserve the most divergent off
471	ly inexplicable on the theory of creation. As	NATURAL selection acts solely by accumulating slight
472	constantly trying to increase in number, with	NATURAL selection always ready to adapt the slowly v
472	perhaps might even have been anticipated. As	NATURAL selection acts by competition, it adapts the
472	cases. The wonder indeed is, on the theory of	NATURAL selection, that more cases of the want of ab
474	have been rendered constant by long continued	NATURAL selection. Glancing at instincts, marvellous
474	does corporeal structure on the thoery of the	NATURAL selection of successive, slight, but profita
475	instincts having been slowly acquired through	NATURAL selection we need not marvel at some instinc
475	almost inevitably follows on the principle of	NATURAL selection: for old forms will be supplanted
478	d present organic beings constitute one grand	NATURAL system, with group subordinate to group, and
478	cent groups, is intelligible on the theory of	NATURAL selection with its contingencies of extincti
479	e to inheritance or community of descent. The	NATURAL system is a genealogical arrangement, in whi
479	veloped branchiae. Disuse, aided sometimes by	NATURAL selection, will often tend to reduce an orga
480	y the tongue and palate having been fitted by	NATURAL selection to browse without their aid: where
481	ndergone mutation. But the chief cause of our	NATURAL unwillingness to admit that one species has
484	at there will be a considerable revolution in	NATURAL history. Systematists will be able to pursue
485	es. The other and more general departments of	NATURAL history will rise greatly in interest. The t
486	o, i speak from experience, will the study of	NATURAL history become! A grand and almost untrodden
486	ce the many diverging lines of descent in our	NATURAL genealogies, by characters of any kind which
489	uture of equally inappreciable length. And as	NATURAL Selection works solely by and for the good o
490	a struggle for Life, and as a consequence to	NATURAL Selection, entailing Divergence of Character
080	of inhabitants of any small area, and to	NATURALISATION. Action of Natural Selection, through Div
114	orders. The same principle is seen in the	NATURALISATION of plants through man's agency in foreign
115	t and admirable work, that floras gain by	NATURALISATION, proportionally with the number of the na
319	ing elephant, and from the history of the	NATURALISATION of the domestic horse in South America, t
395	gment in some cases owing to the probable	NATURALISATION of certain mammals through man's agency:
412	ition, and by looking to certain facts in	NATURALISATION. I attempted also to show that there is a
467	f peculiar seasons, and by the results of	NATURALISATION, as explained in the third chapter. More
060	rical powers of increase. Rapid increase of	NATURALISED animals and plants. Nature of the checks t
065	narily rapid increase and wide diffusion of	NATURALISED productions in their new homes. In a state
069	endure our climate, but which never become	NATURALISED, for they cannot compete with our native p
083	, the natives have been so far conquered by	NATURALISED productions, that they have allowed foreig
106	oductions have everywhere become so largely	NATURALISED on islands. On a small island, the race fo
110	the world, some foreign plants have become	NATURALISED, without causing, as far as we know, the e
115	the plants which have succeeded in becoming	NATURALISED in any land would generally have been clos
115	ight, also, perhaps have been expected that	NATURALISED plants would have belonged to a few groups
115	he flora of the Northern United States, 260	NATURALISED plants are enumerated, and these belong to
115	elong to 162 genera. We thus see that these	NATURALISED plants are of a highly diversified nature.
115	genes of any country, and have there become	NATURALISED, we can gain some crude idea in what manne
128	at the inhabitants of any small spot or at	NATURALISED productions. Therefore during the modifica
322	at; why this species and not another can be	NATURALISED in a given country: then, and not till the
353	of European animals and plants have become	NATURALISED in America and Australia: and some of the
365	o well fitted to its new home, as to become	NATURALISED. But this, as it seems to me, is no valid
377	when we remember over what vast spaces some	NATURALISED plants and animals have spread within a fe
380	as extremely few southern forms have become	NATURALISED in any part of Europe, though hides, wool,
380	nearly equalled or even outnumbered by the	NATURALISED; and if the natives have not been actually
380	ywhere lately yielded to continental forms,	NATURALISED by man's agency. I am far from supposing t

Page **************************************(Key Word)**************************************

Page ***(Key Word)***

```
0381   say why one species and not another becomes NATURALISED by man's agency in a foreign land: why on
0385   dults. I could not even understand how some NATURALISED species have rapidly spread throughout th
0390   ozen flowering plants: yet many have become NATURALISED on it, as they have on New Zealand and on
0390   helena there is reason to believe that the NATURALISED plants and animals have nearly or quite e
0394   hich our smaller quadrupeds have not become NATURALISED and greatly multiplied. It cannot be said
0402   g familiar with the fact that many species, NATURALISED through man's agency, have spread with as
0402   should remember that the forms which become NATURALISED in new countries are not generally closel
0472   country, being beaten and supplanted by the NATURALISED productions from another land. Nor ought
0015   ot improbable, that if we could succeed in NATURALISING, or were to cultivate, during many genera
0001   ntroduction when on board H.M.S. Beagle, as NATURALIST, I was much struck with certain facts in t
0003   of Species, it is quite conceivable that a NATURALIST, reflecting on the mutual affinities of or
0028   have descended from a common parent, as any NATURALIST could in coming to a similar conclusion in
0044   as yet satisfied all naturalists, yet every NATURALIST knows vaguely what he means when he speaks
0045   s. I am convinced that the most experienced NATURALIST would be surprised at the number of the ca
0046   g of the stem of a tree. This philosophical NATURALIST, I may add, has also quite recently shown
0047   good and true species. Practically, when a NATURALIST can unite two forms together by others hav
0048   each other, have been ranked by one eminent NATURALIST as undoubted species, and by another as va
0050   nct species or mere varieties. When a young NATURALIST commences the study of a group of organism
0057   y one species are related to each other. No NATURALIST pretends that all the species of a genus a
0093   flower, another process might commence. No NATURALIST doubts the advantage of what has been call
0124   between types of the two groups: and every NATURALIST will be able to bring some such case befor
0134   n, innumerable instances are known to every NATURALIST of species keeping true, or not varying at
0194   in the writings of almost every experienced NATURALIST; or, as Milne Edwards has well expressed t
0306   throughout the world, as it would be for a NATURALIST to land for five minutes on some one barre
0319   n still living, but in some degree rare, no NATURALIST would have felt the least surprise at its
0324   h america or in Australia, the most skilful NATURALIST would hardly be able to say whether the ex
0337   those of New Zealand. Yet the most skilful NATURALIST from an examination of the species of the
0349   ers innumerable instances. Nevertheless the NATURALIST in travelling, for instance, from north to
0350   dependent of their physical conditions. The NATURALIST must feel little curiosity, who is not led
0398   emoir on the Flora of this archipelago. The NATURALIST, looking at the inhabitants of these volca
0415   lassificatory value is widely different. No NATURALIST can have worked at any group without being
0424   s. with species in a state of nature, every NATURALIST has in fact brought descent into his class
0424   ost important characters, is known to every NATURALIST: scarcely a single fact can be predicated
0424   d yet no one dreams of separating them. The NATURALIST includes as one species the several larval
0428   tween the sub groups in distinct classes. A NATURALIST, struck by a parallelism of this nature in
0002   stance which I have received from very many NATURALISTS, some of them personally unknown to me. I
0003   n which most justly excites our admiration. NATURALISTS continually refer to external conditions,
0004   h they have been very commonly neglected by NATURALISTS. From these considerations, I shall devot
0006   hich I am capable, that the view which most NATURALISTS entertain, and which I formerly entertain
0014   may here refer to a statement often made by NATURALISTS, namely, that our domestic varieties, whe
0016   that this statement is hardly correct; but NATURALISTS differ most widely in determining what ch
0023   fully convinced that the common opinion of NATURALISTS is correct, namely, that all have descend
0029   many successive generations. May not those NATURALISTS who, knowing far less of the laws of inhe
0044   no one definition has as yet satisfied all NATURALISTS, yet every naturalist knows vaguely what
0045   ndividual differences generally affect what NATURALISTS consider unimportant parts: but I could s
0046   nk that character as important (as some few NATURALISTS have honestly confessed) which does not v
0046   rdinate amount of variation; and hardly two NATURALISTS can agree which forms to rank as species
0047   ed to them by intermediate gradations, that NATURALISTS do not like to rank them as distinct spec
0047   d as a species or a variety, the opinion of NATURALISTS having sound judgment and wide experience
0047   ver, in many cases, decide by a majority of NATURALISTS, for few well marked and well known varie
0049   the homes of two doubtful forms leads many NATURALISTS to rank both as distinct species; but wha
0050   se investigation, in most cases, will bring NATURALISTS to an agreement how to rank doubtful form
0051   s admission will often be disputed by other NATURALISTS. When, moreover, he comes to study allied
0051   is, the forms which in the opinion of some NATURALISTS come very near to, but do not quite arriv
0056   have not been found between doubtful forms, NATURALISTS are compelled to come to a determination
0101   o sometimes cross. It must have struck most NATURALISTS as a strange anomaly that, in the case of
0149   connected with the very general opinion of NATURALISTS, that beings low in the scale of nature a
0154   is not in this case applicable, which most NATURALISTS would advance, namely, that specific chara
0191   manner as, on the view entertained by some NATURALISTS that the branchiae and dorsal scales of An
0199   ew words on the protest lately made by some NATURALISTS, against the utilitarian doctrine that eve
0216   generally, and most justly, been ranked by NATURALISTS as the most wonderful of all known instin
0245   summary. The view generally entertained by NATURALISTS is that species, when intercrossed, have b
0254   pallas, has been largely accepted by modern NATURALISTS; namely, that most of our domestic animals
0268   e together, they are at once ranked by most NATURALISTS as species. For instance, the blue and ree
0269   arieties. In the second place, some eminent NATURALISTS believe that a long course of domesticatio
0296   form. It is all important to remember that NATURALISTS have no golden rule by which to distingui
0336   i will not here enter on this subject, for NATURALISTS have not as yet defined to each other's sa
0352   uestion which has been largely discussed by NATURALISTS, namely, whether species have been created
0353   region; and it has been observed by several NATURALISTS, that the most natural genera, or those ge
0353   nce it seems to me, as it has to many other NATURALISTS, that the view of each species having been
0372   in every great class many forms, which some NATURALISTS rank as geographical races, and others as
0372   epresentative forms which are ranked by all NATURALISTS as specifically distinct. As on the land,
0376   are of doubtful value, being ranked by some NATURALISTS as specifically distinct, by others as var
0404   , there will be found many forms which some NATURALISTS rank as distinct species, and some as var
0407   nclusion, which has been arrived at by many NATURALISTS under the designation of single centres of
0413   rike us, is in my judgment fully explained. NATURALISTS try to arrange the species, genera, and fa
0413   y of this system are indisputable. But many NATURALISTS think that something more is meant by the
0415   cause it has partly arisen, that almost all NATURALISTS lay the greatest stress on resemblances in
0416   ter would, I think, have been considered by NATURALISTS as important an aid in determining the deg
0417   times necessarily founded. Practically when NATURALISTS are at work, they do not trouble themselve
0418   rinciple has been broadly confessed by some NATURALISTS to be the true one; and by none more clear
0418   t it has been strongly urged by those great NATURALISTS, Milne Edwards and Agassiz, that embryonic
0419   a group of forms, first ranked by practised NATURALISTS as only a genus, and then raised to the ra
0420   ith modification; that the characters which NATURALISTS consider as showing true affinity between
0420   mmunity of descent is the hidden bond which NATURALISTS have been unconsciously seeking, and not s
0431   derstand the extraordinary difficulty which NATURALISTS have experienced in describing, without th
```

Page ***(Key Word)***

```
Page ****************************************( Key Word )****************************************
```

```
433  scent is the hidden bond of connexion which   NATURALISTS have sought under the term of the Natural
438  asses of the animal and vegetable kingdoms.    NATURALISTS frequently speak of the skull as formed of
438  om the other, but from some common element.    NATURALISTS, however, use such language only in a meta
438  cation of this nature having occurred, that    NATURALISTS can hardly avoid employing language having
449   my view the hidden bond of connexion which   NATURALISTS have been seeking under the term of the na
449  erstand how it is that, in the eyes of most    NATURALISTS, the structure of the embryo is even more
456  ollowed and the difficulties encountered by    NATURALISTS in their classifications; the value set up
456  tage of those forms which are considered by    NATURALISTS as allied, together with their modificatio
464  many fossil links will be discovered, that     NATURALISTS will be able to decide, on the common view
468  nature. But, besides such differences, all     NATURALISTS have admitted the existence of varieties,
469  pecies, and species. Let it be observed how    NATURALISTS differ in the rank which they assign to th
480  be asked, have all the most eminent living     NATURALISTS and geologists rejected this view of the m
481  by no means expect to convince experienced     NATURALISTS whose minds are stocked with a multitude o
482  acts will certainly reject my theory. A few    NATURALISTS, endowed with much flexibility of mind, an
482  nfidence to the future; to young and rising    NATURALISTS, who will be able to view both sides of th
482  is overwhelmed be removed. Several eminent     NATURALISTS have of late published their belief that a
482  are still thus looked at by the majority of    NATURALISTS, and which consequently have every externa
483  ourishment from the mother's womb? Although    NATURALISTS very properly demand a full explanation of
485  ermediate gradations, are looked at by most    NATURALISTS as sufficient to raise both forms to the r
485  o treat species in the same manner as those    NATURALISTS treat genera, who admit that genera are me
485  rise greatly in interest. The terms used by    NATURALISTS of affinity, relationship, community of ty
337  de of British forms would become thoroughly    NATURALIZED there, and would exterminate many of the n
005  ve a better chance of surviving, and thus be   NATURALLY selected. From the strong principle of inhe
026  o any character derived from such cross will    NATURALLY become less and less, as in each succeeding
034  nt. Thus, a man who intends keeping pointers    NATURALLY tries to get as good dogs as he can, and af
037  fruit, in some small degree, to their having   NATURALLY chosen and preserved the best varieties the
064  ception to the rule that every organic being    NATURALLY increases at so high a rate, that if not de
075  which best suit the soil or climate, or are    NATURALLY the most fertile, will beat the others and
091  t, and those frequenting the lowlands, would    NATURALLY be forced to hunt different prey; and from
098  to raise pure seedlings, so largely do they    NATURALLY cross. In many other cases, far from there
110  ndergoing modification and improvement, will    NATURALLY suffer most. And we have seen in the chapte
117  in some way profitable will be preserved or    NATURALLY selected. And here the importance of the pr
140  nts, of their becoming, to a certain extent,    NATURALLY habituated to different temperatures, or be
146  up of descendants with diverse habits, would   NATURALLY be thought to be correlated in some necessa
169  arent and exposed to similar influences will   NATURALLY tend to present analogous variations, and t
188  orts of the highest human intellects; and we   NATURALLY infer that the eye has been formed by a som
214  raining a dog to point, had not some one dog   NATURALLY shown a tendency in this line; and this is
270  hese plants have separated sexes, they never   NATURALLY crossed. He then fertilised thirteen flower
275  which are descended from species slowly and    NATURALLY produced. On the whole I entirely agree wit
317  n, barrande, etc.; whose general views would   NATURALLY lead them to this conclusion. On the contra
325  the other forms in their own country, would    NATURALLY oftenest give rise to new varieties or inci
378  warmth returned, these temperate forms would   NATURALLY ascend the higher mountains, being extermin
388  irds, which have large powers of flight, and   NATURALLY travel from one to another and often distan
404  ountain, as it became slowly upheaved, would   NATURALLY be colonised from the surrounding lowlands.
456  of true affinity; and other such rules; all   NATURALLY follow on the view of the common parentage
474  species of a genus, and therefore, as we may   NATURALLY infer, of great importance to the species,
476  ned country, the recent and the extinct will   NATURALLY be allied by descent. Looking to geographic
004  n to the variability of species in a state of  NATURE; but I shall, unfortunately, be compelled to
007  s of any one species or variety in a state of  NATURE. When we reflect on the vast diversity of the
007  ch the parent species have been exposed under  NATURE. There is, also, I think, some probability in
009  dividuals, though taken young from a state of  NATURE, perfectly tamed, long lived, and healthy (of
009  ghtly, perhaps hardly more than in a state of  NATURE. A long list could easily be given of sportin
010  y seed. These sports are extremely rare under  NATURE, but far from rare under cultivation; and in
014  from domestic races to species in a state of  NATURE. I have in vain endeavoured to discover on wh
015  to all experience. I may add, that when under  NATURE the conditions of life do change, variations
016  ecially when compared with all the species in  NATURE to which they are nearest allied. With these
016  llied species of the same genus in a state of  NATURE. I think this must be admitted, when we find
017  ses and countries, were taken from a state of  NATURE, and could be made to breed for an equal numb
019  ld canidae, ever existed freely in a state of  NATURE? It has often been loosely said that all our
028  f finches, or other large groups of birds, in  NATURE. One circumstance has struck me much; namely,
029  they deride the idea of species in a state of  NATURE being lineal descendants of other species? Se
030  key is man's power of accumulative selection:  NATURE gives successive variations; man adds them up
039  e first given to him in some slight degree by  NATURE. No man would ever try to make a fantail, til
039  tremely small differences, and it is in human  NATURE to value any novelty, however slight, in one'
044  redominant Power. Chapter II. Variation Under  NATURE. Variability. Individual differences. Doubtfu
044  last chapter to organic beings in a state of  NATURE, we must briefly discuss whether these latter
045  ; i should have expected that changes of this  NATURE could have been effected only by slow degrees
047  e links; nor will the commonly assumed hybrid  NATURE of the intermediate links always remove the c
048  etent judges. That varieties of this doubtful  NATURE are far from uncommon cannot be disputed. Com
050  ct, that if any animal or plant in a state of  NATURE be highly useful to man, or from any cause cl
053  nc results might be obtained in regard to the  NATURE and relations of the species which vary most,
054  used, but this seems to be connected with the  NATURE of the stations inhabited by them, and has li
059  ra thus tend to become larger; and throughout  NATURE the forms of life which are now dominant tend
060  d increase of naturalised animals and plants.  NATURE of the checks to increase. Competition univer
060  lations of all animals and plants throughtout  NATURE. Struggle for Life most severe between indivi
060  ter that amongst organic beings in a state of  NATURE there is some individual variability; indeed
060  little in understanding how species arise in  NATURE. How have all those exquisite adaptations of
061  tions to other organic beings and to external  NATURE, will tend to the preservation of that indivi
061  seful variations, given to him by the hand of  NATURE. But Natural Selection, as we shall hereafter
061  rior to man's feeble efforts, as the works of  NATURE are to those of Art. We will now discuss in a
062  ind, I am convinced that the whole economy of  NATURE, with every fact on distribution, rarity, abu
062  or quite misunderstood. We behold the face of  NATURE bright with gladness, we often see superabund
064  pid increase of various animals in a state of  NATURE, when circumstances have been favourable to t
065  productions in their new homes. In a state of  NATURE almost every plant produces seed, and amongst
065  slaughtered for food, and that in a state of  NATURE an equal number would have somehow to be disp
066  he number of its eggs or seeds. In looking at  NATURE, it is most necessary to keep the foregoing c
```

```
Page ****************************************( Key Word )****************************************
```

Page ***(Key Word)***

```
0067  taneously increase to any amount. The face of  NATURE may be compared to a yielding surface, with
0070  , explains, I believe, some singular facts in   NATURE, such as that of very rare plants being some
0071  but several hundred acres of exactly the same   NATURE had been enclosed twenty five years previous
0073  rds, and we have ended with them. Not that in   NATURE the relations can ever be as simple as this.
0073  rces are so nicely balanced, that the face of   NATURE remains uniform for long periods of time, the
0073  ants and animals, most remote in the scale of   NATURE, are bound together by a web of complex rela
0075  nerally between beings remote in the scale of   NATURE. This is often the case with those which may
0076  struggle together, like beings in a state of   NATURE, and if the seed or young were not annually
0076  fill nearly the same place in the economy of   NATURE; but probably in no one case could we precis
0079  rselves with the full belief, that the war of   NATURE is not incessant, that no fear is felt, that
0080  en is so potent in the hands of man, apply to   NATURE? I think we shall see that it can act most e
0080  uctions, and, in a lesser degree, those under   NATURE, vary; and how strong the hereditary tendenc
0081  we should then have places in the economy of   NATURE which would assuredly be better filled up, i
0082  rection mere individual differences, so could   NATURE, but far more easily, from having incomparab
0083  unconscious means of selection; what may not   NATURE effect? Man can act only on external and vis
0083  act only on external and visible characters:   NATURE cares nothing for appearances, except in so
0083  y of life. Man selects only for his own good;   NATURE only for that of the being which she tends.
0083  is eye, or to be plainly useful to him. Under   NATURE, the slightest difference of structure or co
0084  oducts be, compared with those accumulated by   NATURE during whole geological periods. Can we wond
0084  geological periods. Can we wonder, then, that   NATURE's productions should be far truer in charact
0085  e several varieties, assuredly, in a state of   NATURE, where the trees would have to struggle with
0086  her modifications, often of the most expected   NATURE. As we see that those variations which under
0086  nd cattle when nearly adult; so in a state of   NATURE, natural selection will be enabled to act on
0087  nciers assist in the act of hatching. Now, if   NATURE had to make the beak of a full grown pigeon
0087  that sex, the same fact probably occurs under   NATURE, and if so, natural selection will be able t
0088  ose which are best fitted for their places in   NATURE, will leave most progeny. But in many cases,
0088  on of the best cocks. How low in the scale of   NATURE this law of battle descends, I know not: mal
0090  to the females. We see analogous cases under   NATURE, for instance, the tuft of hair on the breas
0094  his to occur in ever so slight a degree under   NATURE, then as pollen is already carried regularly
0097  ine me to believe that it is a general law of   NATURE (utterly ignorant though we be of the meanin
0097  pensable. On the belief that this is a law of   NATURE, we can, I think, understand several large c
0100  believe this objection to be valid, but that   NATURE has largely provided against it by giving to
0100  um in which terrestrial animals live, and the   NATURE of the fertilising element; for we know of n
0101  rcross with a distinct individual is a law of   NATURE. I am well aware that there are, on this view
0102  xtremely important element of success. Though   NATURE grants vast periods of time for the work of n
0102  aid, to seize on each place in the economy of   NATURE, if any one species does not become modified
0102  inc with inferior animals. Thus it will be in   NATURE: for within a confined area, with some place
0103  t a very quick rate. Hence in animals of this   NATURE, for instance in birds, varieties will genera
0103  intercrossing plays a very important part in   NATURE in keeping the individuals of the same specie
0105  rance of favourable variations. If we turn to   NATURE to test the truth of these remarks, and look
0108  epends on there being places in the polity of   NATURE, which can be better occupied by some of the
0109  may be effected in the long course of time by   NATURE's power of selection. Extinction. This subje
0109  ed, for the number of places in the polity of   NATURE is not indefinitely great, not that we have a
0111  om most of the innumerable species throughout   NATURE presenting well marked differences; whereas v
0112  e asked, can any analogous principle apply in   NATURE? I believe it can and does apply most efficie
0112  nd widely diversified places in the polity of   NATURE, and so be enabled to increase in numbers. We
0114  lants belonging to the most different orders:   NATURE follows what may be called a simultaneous rot
0114  osing it not to be in any way peculiar in its   NATURE), and may be said to be striving to the utmos
0115  aturalised plants are of a highly diversified   NATURE. They differ, moreover, to a large extent fro
0115  he genera of these States. By considering the   NATURE of the plants or animals which have struggled
0116  equal degrees, as is so generally the case in   NATURE, and is represented in the diagram by the let
0117  extremely slight, but of the most diversified   NATURE; they are not supposed all to appear simultar
0119  at selection will always act according to the   NATURE of the places which are either unoccupied or
0121  and widely different places in the polity of   NATURE: hence in the diagram I have chosen the extre
0122  equal degrees, as is so generally the case in   NATURE: species (A) being more nearly related to B,
0124  w species will be of a curious and circuitous   NATURE. Having descended from a form which stood bet
0126  d seizing on many new places in the polity of   NATURE, will constantly tend to supplant and destroy
0127  er natural selection has really thus acted in   NATURE, in modifying and adapting the various forms
0128  of life. On these principles, I believe, the   NATURE of the affinities of all organic beings may b
0131  and in a lesser degree in those in a state of   NATURE, had been due to chance. This, of course, is
0131  nder domestication or cultivation, than under   NATURE, leads me to believe that deviations of struc
0131  tions of structure are in some way due to the   NATURE of the conditions of life, to which the parer
0132  d another, which we see everywhere throughout   NATURE. Some little influence may be attributed to t
0134  such modifications are inherited. Under free   NATURE, we can have no standard of comparison, by wh
0134  has remarked, there is no greater anomaly in   NATURE than a bird that cannot fly: yet there are se
0140  reason to believe that species in a state of   NATURE are limited in their ranges by the competitic
0141  roportion of other animals, now in a state of   NATURE, could easily be brought to bear widely diffe
0144  of several of the most important viscera. The   NATURE of the bond of correlation is very frequently
0147  expressed it, in order to spend on one side,   NATURE is forced to economise on the other side. I t
0147  iminished wattles. With species in a state of   NATURE it can hardly be maintained that the law is o
0149  naturalists, that beings low in the scale of   NATURE are more variable than those which are higher
0153  enlarged crop of pouters. Now let us turn to   NATURE. When a part has been developed in an extraor
0157  ferences in these cases are of a very unusual   NATURE, the relation can hardly be accidental. The s
0158  ies to their several places in the economy of   NATURE, and likewise to fit the two sexes of the sam
0161  thus gained would probably be of an important   NATURE, for the presence of all important characters
0162  erse colours are crossed. Hence, though under   NATURE it must generally be left doubtful, what case
0162  me group. And this undoubtedly is the case in   NATURE. A considerable part of the difficulty in rec
0162  y parts or organs of an important and uniform   NATURE occasionally varying so as to acquire; in som
0163  , partly under domestication and partly under   NATURE. It is a case apparently of reversion. The as
0167  n created with a tendency to vary, both under   NATURE and under domestication, in this particular m
0168  e very many other correlations of growth, the   NATURE of which we are utterly unable to understand.
0168  cause that organic beings low in the scale of   NATURE are more variable than those which have their
0169  to the beautiful and harmonious diversity of   NATURE. Whatever the cause may be of each slight dif
0171  nnumerable transitional forms? Why is not all   NATURE in confusion instead of the species being, as
0176  rule holds good with varieties in a state of   NATURE. I have met with striking instances of the ru
0179  is well adapted in its habits to its place in   NATURE. Look at the Mustela vison of North America,
```

Page ***(Key Word)***

age **************************************(Key Word)**************************************

```
0184  r proper type. And such instances do occur in  NATURE. Can a more striking instance of adaptation b
0190  ne function alone, and thus wholly change its  NATURE by insensible steps. Two distinct organs some
0194  ; or, as Milne Edwards has well expressed it,  NATURE is prodigal in variety, but niggard in innova
0194  en separately created for its proper place in  NATURE, be so invariably linked together by graduate
0194  d together by graduated steps? Why should not  NATURE have taken a leap from structure to structure
0199  ing assuredly is well fitted for its place in  NATURE, many structures now have no direct relation
0200  he good of another species; though throughout  NATURE one species incessantly takes advantage of, a
0201  is is the degree of perfection attained under  NATURE. The endemic productions of New Zealand, for
0202  s we can judge, with this high standard under  NATURE. The correction for the aberration of light i
0202  asm a multitude of inimitable contrivances in  NATURE, this same reason tells us, though we may eas
0208  lay, even in animals very low in the scale of  NATURE. Frederick Cuvier and several of the older me
0210  of corporeal structures, we ought to find in  NATURE, not the actual transitional gradations by wh
0210  in the same species can be shown to occur in  NATURE. Again as in the case of corporeal structure,
0211  ee of variation in instincts under a state of  NATURE, and the inheritance of such variations, are
0212  ependence on the situations chosen and on the  NATURE and temperature of the country inhabited, but
0212  duals of the same species, born in a state of  NATURE, is extremely diversified, can be shown by a
0212  nherited variations of instinct in a state of  NATURE will be strengthened by briefly considering a
0216  , best understand how instincts in a state of  NATURE have become modified by selection, by conside
0217  g their young. By a continued process of this  NATURE, I believe that the strange instinct of our c
0225  great principle of gradation, and see whether  NATURE does not reveal to us her method of work. At
0235  the same in animals so remote in the scale of  NATURE, that we cannot account for their similarity
0236  ts and other articulate animals in the state of NATURE occasionally become sterile: and if such inse
0237  mestic productions and in those in a state of  NATURE, of all sorts of differences of structure whi
0242  and structure would have become blended. And  NATURE has, as I believe, effected this admirable di
0243  ow that instincts vary slightly in a state of  NATURE. No one will dispute that instincts are of th
0247  ced by both pure parent species in a state of  NATURE. But a serious cause of error seems to me to
0249  lant or from another plant of the same hybrid  NATURE. And thus, the strange fact of the increase o
0250  etween distinct species is a universal law of  NATURE. He experimentised on some of the very same s
0252  animals more widely separated in the scale of  NATURE can be more easily crossed than in the case o
0260  imply to prevent their becoming confounded in  NATURE? I think not. For why should the sterility of
0260  ifferences being of so peculiar and limited a  NATURE, that, in reciprocal crosses between two spec
0261  ely unimportant for its welfare in a state of  NATURE, I presume that no one will suppose that this
0261  s of their wood, in the period of the flow or  NATURE of their sap, etc., but in a multitude of cas
0264  fe. But a hybrid partakes of only half of the  NATURE and constitution of its mother, and therefore
0267  t greater changes, or changes of a particular  NATURE, often render organic beings in some degree s
0268  e. but if we look to varieties produced under  NATURE, we are immediately involved in hopeless diff
0268  the fertility of all varieties produced under  NATURE will assuredly have to be granted. If we turn
0269  t wish to alter their general habits of life.  NATURE acts uniformly and slowly during vast periods
0271  rocess of selection, as carried on by man and  NATURE, we need not be surprised at some difference
0271  ng the infertility of varieties in a state of  NATURE, for a supposed variety if infertile in any d
0275  fined to characters almost monstrous in their  NATURE, and which have suddenly appeared, such as al
0276  lity to prevent them crossing and blending in  NATURE, than to think that trees have been specially
0277  ircle with respect to varieties in a state of  NATURE; and when we remember that the greater number
0279  rmediate varieties at the present day. On the  NATURE of extinct intermediate varieties; on their n
0279  links not now occurring everywhere throughout  NATURE depends on the very process of natural select
0289  d in written works, or when we follow them in  NATURE, it is difficult to avoid believing that they
0292  posits rich in fossils have been accumulated.  NATURE may almost be said to have guarded against th
0295  e sea and a supply of sediment of a different  NATURE will generally have been due to geographical
0308  will give the following hypothesis. From the  NATURE of the organic remains, which do not appear t
0310  an strata, are all undoubtedly of the gravest  NATURE. We see this in the plainest manner by the fa
0313  at organisms, considered high in the scale of  NATURE, change more quickly than those that are low:
0314  ies, on the variability being of a beneficial  NATURE, on the power of intercrossing, on the rate o
0314  ns of the country, and more especially on the  NATURE of the other inhabitants with which the varyi
0315  ct place of another species in the economy of  NATURE, and thus supplant it; yet the two forms, the
0315  arent rock pigeon were also destroyed, and in  NATURE we have every reason to believe that the pare
0322  seldom perceived by us, the whole economy of  NATURE will be utterly obscured. Whenever we can pre
0325  ical conditions of various countries; and the  NATURE of their inhabitants. This great fact of the
0328  and immigration. I suspect that cases of this  NATURE have occurred in Europe. Mr. Prestwich, in hi
0331  n many and different places in the economy of  NATURE. Therefore it is quite possible, as we have s
0333  ts, or between their collateral relations. In  NATURE the case will be far more complicated than is
0338  to be left as a sort of picture, preserved by  NATURE, of the ancient and less modified condition o
0342  ated steps. He who rejects these views on the  NATURE of the geological record, will rightly reject
0345  e, and are, in so far, higher in the scale of  NATURE: and this may account for that vague yet ill
0350  ease, at periods more or less remote; on the  NATURE and number of the former immigrants; and on t
0352  rom each other, or with an interval of so grave a NATURE, that the space could not be easily passed ov
0354  ty of range are so numerous and of so grave a  NATURE, that we ought to give up the belief, rendere
0355  e inhabitants of the continent. Cases of this  NATURE are common, and are, as we shall hereafter mo
0358  orbes and admitted by his many followers. The  NATURE and relative proportions of the inhabitants o
0370  ill earlier changes of climate of an opposite  NATURE. We have good reason to believe that during t
0373  es. We may infer, from the frozen mammals and  NATURE of the mountain vegetation, that Siberia was
0385  ust hatched molluscs, though aquatic in their  NATURE, survived on the duck's feet, in damp air, fr
0388  esh water productions are low in the scale of  NATURE, and that we have reason to believe that such
0388  to another and often distant piece of water.  NATURE, like a careful gardener, thus takes her seed
0388  us takes her seeds from a bed of a particular  NATURE, and drops them in another equally well fitte
0390  sources far more fully and perfectly than has  NATURE. Although in oceanic islands the number of ki
0392  elieve, has been at least as important as the  NATURE of the conditions. Many remarkable little fac
0398  in the conditions of life, in the geological  NATURE of the islands, in their height or climate, o
0398  derable degree of resemblance in the volcanic  NATURE of the soil, in climate, height, and size of
0400  oht of each other, having the same geological  NATURE, the same height, climate, etc., that many of
0400  reas it cannot, I think, be disputed that the  NATURE of the other inhabitants, with which each has
0402  e equally well fitted for their own places in  NATURE, both probably will hold their own places and
0403  omes, is of the widest application throughout  NATURE. We see this on every mountain, in every lake
0408  abitants entered one region: according to the  NATURE of the communication which allowed certain fo
0411  nd so on; but the case is widely different in  NATURE: for it is notorious how commonly members of
0412  ifferent places as possible in the economy of  NATURE, there is a constant tendency in their charac
0414  general place of each being in the economy of  NATURE, would be of very high importance in classifi
```

Page **************************************(Key Word)**************************************

Page ++(Key Word)++

0416 scence on parts of the flower in grasses, the NATURE of the dermal covering, as hair or feathers,
0418 h alone plays its full part in the economy of NATURE. Yet it has been strongly urged by those grea
0422 surface, the affinities which we discover in NATURE amongst the beings of the same group. Thus, c
0424 ters from negroes. With species in a state of NATURE, every naturalist has in fact brought descent
0428 a naturalist, struck by a parallelism of this NATURE in any one class, by arbitrarily raising or s
0428 ize on more and more places in the economy of NATURE. The larger and more dominant groups thus ter
0430 de nearly similar observations on the general NATURE of the affinities of distinct orders of plant
0435 d by Owen in his most interesting work on the NATURE of Limbs. On the ordinary view of the indeper
0435 parts of the organisation. In changes of this NATURE, there will be little or no tendency to modif
0438 g is the appearance of a modification of this NATURE having occurred, that naturalists can hardly
0446 ome degree probable, to species in a state of NATURE. Let us take a genus of birds, descended on m
0449 species. Agassiz believes this to be a law of NATURE: but I am bound to confess that I only hope t
0450 of inutility, are extremely common throughout NATURE. For instance, rudimentary mammae are very ge
0451 d the perfect pistil are essentially alike in NATURE. An organ serving for two purposes, many becc
0453 mmetry, or in order to complete the scheme of NATURE: but this seems to me no explanation, merely
0453 ke of symmetry, and to complete the scheme of NATURE? An eminent physiologist accounts for the pre
0454 he origin of rudimentary organs in a state of NATURE, further than by showing that rudiments can t
0454 e produced: for I doubt whether species under NATURE ever undergo abrupt changes. I believe that c
0456 n all organisms throughout all time: that the NATURE of the relationship, by which all living and
0460 ngs: but we see so many strange gradations in NATURE, as is proclaimed by the canon, Natura non fa
0467 ic beings to new conditions of life, and then NATURE acts on the organisation, and causes variabil
0467 nd does select the variations given to him by NATURE, and thus accumulate them in any desired manr
0467 der domestication should not have acted under NATURE. In the preservation of favoured individuals
0468 re between beings most remote in the scale of NATURE. The slightest advantage in one being, at any
0468 d that organic beings would have varied under NATURE, in the same way as they generally have varie
0468 cation. And if there be any variability under NATURE, it would be an unaccountable fact if natural
0468 of proof, that the amount of variation under NATURE is a strictly limited quantity. Man, though a
0468 least individual differences in species under NATURE. But, besides such differences, all naturalis
0469 rope and North America. If then we have under NATURE variability and a powerful agent always ready
0469 ect variations most useful to himself, should NATURE fail in selecting variations useful, under ch
0470 and widely different places in the economy of NATURE, there will be a constant tendency in natural
0471 y simply intelligible. We can plainly see why NATURE is prodigal in variety, though niggard in inr
0471 n innovation. But why this should be a law of NATURE if each species has been independently create
0472 ch to any unoccupied or ill occupied place in NATURE, these facts cease to be strange, or perhaps
0472 ought we to marvel if all the contrivances in NATURE be not, as far as we can judge, absolutely pe
0474 ble modifications. We can thus understand why NATURE moves by graduated steps in endowing differer
0480 frequently bear the plain stamp of inutility! NATURE may be said to have taken pains to reveal, by
0481 be asserted that organic beings in a state of NATURE are subject to no variation: it cannot be pro
0485 ion: when we regard every production of NATURE as one which has had a history: when we conte
0486 ans will speak infallibly with respect to the NATURE of long lost structures. Species and groups o
0487 on the opposite sides of a continent, and the NATURE of the various inhabitants of that continent
0490 of less improved forms. Thus, from the war of NATURE, from famine and death, the most exalted obje
0306 of the most ancient Silurian animals, as the NAUTILUS, Lingula, etc., do not differ much from livi
0072 of a certain fly, which lays its eggs in the NAVELS of these animals when first born. The increas
0028 m pliny, immense prices were given for pigeons: NAY, they are come to this pass, that they can rec
0234 were to make her cells more and more regular, NEARER together, and aggregated into a mass, like th
0309 pen that strata which had subsided some miles NEARER to the centre of the earth, and which had bee
0309 action than strata which have always remained NEARER to the surface. The immense areas in some par
0363 ison with the plants of cther oceanic islands NEARER to the mainland, and (as remarked by Mr. H. C
0370 oductions then lived on the broken land still NEARER to the pole. Now if we look at a globe, we sh
0377 d, the temperate productions, after migrating NEARER to the equator, though they will have been pl
0399 the plants of Kerguelen Land, though standing NEARER to Africa than to America, are related, and t
0410 early any two forms are related in blood, the NEARER they will generally stand to each other in ti
0016 h all the species in nature to which they are NEAREST allied. With these exceptions (and with that
0110 ormation, will generally press hardest on its NEAREST kindred, and tend to exterminate them. We se
0145 w, at least, that in irregular flowers, those NEAREST to the axis are oftenest subject to peloria,
0204 ts, with some habits very unlike those of its NEAREST congeners. Hence we can understand, bearing
0267 tinued during several generations between the NEAREST relations, especially if these be kept under
0314 having come to differ considerably from their NEAREST allies on the continent of Europe, whereas t
0321 have been developed from any one species, the NEAREST allies of that species, i.e. the species of
0383 of the inhabitante of islands to those of the NEAREST mainland. On colonisation from the nearest s
0383 he nearest mainland. Cn colonisation from the NEAREST source with subsequent modification. Summary
0386 eagle, when forty five miles distant from the NEAREST land: how much farther it might have flown w
0393 which are inhabited by a wolf like fox, come NEAREST to an exception: but this group cannot be co
0396 ormerly connected by continuous land with the NEAREST continent: for on this latter view the migra
0397 of islands, is their affinity to those of the NEAREST mainland, without being actually the same sp
0399 ctions of islands are related to those of the NEAREST continent, or of other near islands. The exc
0399 s much more closely related to Australia, the NEAREST mainland, than to any other region: and this
0399 ed to South America, which, although the next NEAREST continent, is so enormously remote, that the
0400 istinct, to be closely allied to those of the NEAREST continent, we sometimes see displayed on a sm
0406 h whole archipelago or island to those of the NEAREST mainland, are I think, utterly inexplicable
0406 plicable on the view of colonisation from the NEAREST and readiest source, together with the subse
0409 be related, but less closely, to those of the NEAREST continent or other source whence immigrants
0478 ch area are related to the inhabitants of the NEAREST source whence immigrants might have been der
0001 hasty in coming to a decision. My work is now NEARLY finished: but as it will take me two or three
0010 n plants seem extremely valuable. When all or NEARLY all the individuals exposed to certain condit
0014 attle could appear only in the offspring when NEARLY mature: peculiarities in the silkworm are kno
0014 have acted on the ovules or male element: in NEARLY the same manner as in the crossed offspring f
0014 ock was, and so could not tell whether or not NEARLY perfect reversion had ensued. It would be qui
0020 character: but that a race could be obtained NEARLY intermediate between two extremely different
0055 leopterous insects of two districts, into two NEARLY equal masses, the species of the larger gener
0058 so that the acknowledged varieties have very NEARLY the same restricted average range, as have th
0065 ss cestruction of the old and young, and that NEARLY all the young have been enabled to breed. In
0076 most severe between allied forms, which fill NEARLY the same place in the economy of nature: but
0086 nsi in the horns of our sheep and cattle when NEARLY adult: so in a state of nature, natural selec
0102 without intending to alter the breed, have a NEARLY common standard of perfection, and all try to

Page ++(Key Word)++

Page **(Key Word)**

0110	enus or of related genera, which, from having	NEARLY the same structure, constitution, and habits,
0118	y will tend to vary, and generally to vary in	NEARLY the same manner as their parents varied. More
0121	have chosen the extreme species (A), and the	NEARLY extreme species (I), as those which are larg
0121	ost severe between those forms which are most	NEARLY related to each other in habits, constitution
0122	ly the case in nature; species (A) being more	NEARLY related to B, C, and D, than to the other spe
0122	some of the original species which were most	NEARLY related to their parents. Hence very few of t
0123	m (a) the three marked a14, o14, p14, will be	NEARLY related from having recently branched off fro
0123	ecies; and lastly, o14, e14, and m14, will be	NEARLY related one to the other, but from having div
0123	ecies (I) differed largely from (A), standing	NEARLY at the extreme points of the original genus,
0134	he surface of the water, and has its wings in	NEARLY the same condition as the domestic Aylesbury
0134	t except to escape danger, I believe that the	NEARLY wingless condition of several birds, which no
0138	e similar than deep limestone caverns under a	NEARLY similar climate; so that on the common view a
0150	al years ago I was much struck with a remark,	NEARLY to the above effect, published by Mr. Waterho
0150	s of the ourang outang, that he has come to a	NEARLY similar conclusion. It is hopeless to attempt
0152	fowl) or the whole breed will cease to have a	NEARLY uniform character. The breed will then be sai
0152	, and perhaps in polymorphic groups, we see a	NEARLY parallel natural case; for in such cases natu
0152	er, it is notoriously difficult to breed them	NEARLY to perfection, and frequently individuals are
0153	which have remained for a much longer period	NEARLY constant. And this, I am convinced, is the ca
0154	ording to my theory, for an immense period in	NEARLY the same state; and thus it comes to be no mo
0169	anner it may be developed. Species inheriting	NEARLY the same constitution from a common parent an
0173	or representative species, evidently filling	NEARLY the same place in the natural economy of the
0175	were it not for other competing species; that	NEARLY all either prey on or serve as prey for other
0185	mps and floating plants, yet the water hen is	NEARLY as aquatic as the coot; and the landrail near
0185	arly as aquatic as the coot; and the landrail	NEARLY as terrestrial as the quail or partridge. In
0193	be detected. I am inclined to believe that in	NEARLY the same way as two men have sometimes indepe
0194	us variations, has sometimes modified in very	NEARLY the same manner two parts in two organic bein
0196	at a former period, have been transmitted in	NEARLY the same state, although now become of very s
0197	e highest service to the plant; but as we see	NEARLY similar hooks on many trees which are not cli
0213	ormed without experience by the young, and in	NEARLY the same manner by each individual, performed
0224	are correctly made. But the difficulty is not	NEARLY so great as it at first appears: all this bea
0225	ure between the hive and humble bee, but more	NEARLY related to the latter: it forms a nearly regu
0225	more nearly related to the latter; it forms a	NEARLY regular waxen comb of cylindrical cells, in w
0225	wax for holding honey. These latter cells are	NEARLY spherical and of nearly equal sizes, and are
0225	hese latter cells are nearly spherical and of	NEARLY equal sizes, and are aggregated into an irreg
0226	ee other cells, which, from the spheres being	NEARLY of the same size, is very frequently and nece
0229	performed. The bees must have worked at very	NEARLY the same rate on the opposite sides of the ri
0238	he same stock, and confidently expects to get	NEARLY the same variety: breeders of cattle wish the
0240	n have had a species of ant with neuters very	NEARLY in the same condition with those of Myrmica.
0241	as big as those of the smaller men, and jaws	NEARLY five times as big. The jaws, moreover, of the
0243	erent conditions of life, yet often retaining	NEARLY the same instincts. For instance, we can unde
0247	g brought to it by insects from other plants.	NEARLY all the plants experimentised on by Gartner w
0255	dation in the number of seeds produced, up to	NEARLY complete or even quite complete fertility; an
0269	erility both appearing and disappearing under	NEARLY the same conditions of life. Lastly, and this
0269	varieties with the same food; treats them in	NEARLY the same manner, and does not wish to alter t
0274	nts, more especially in hybrids produced from	NEARLY related species, follows according to Gartner
0277	t not quite universally, fertile. Nor is this	NEARLY general and perfect fertility surprising, whe
0281	descendants, unless at the same time we had a	NEARLY perfect chain of the intermediate links. It i
0284	................72,584 feet; that is, very	NEARLY thirteen and three quarters British miles. So
0286	hole line of coast at the rate on one yard in	NEARLY every twenty two years. I doubt whether any r
0288	rselves that sediment is being deposited over	NEARLY the whole bed of the sea, at a rate sufficien
0291	the rate of subsidence and supply of sediment	NEARLY balance each other, the sea will remain shall
0295	n the same space, the supply of sediment must	NEARLY have counterbalanced the amount of subsidence
0295	he downward movement continues. In fact, this	NEARLY exact balancing between the supply of sedimen
0301	ey would differ from their former state, in a	NEARLY uniform, though perhaps extremely slight degr
0303	s been discovered in the new red sandstone at	NEARLY the commencement of this great series. Cuvier
0306	ave descended from one progenitor, apply with	NEARLY equal force to the earliest known species. Fo
0315	g and equal periods of time, may, perhaps, be	NEARLY the same; but as the accumulation of long end
0328	ons have been deposited in two regions during	NEARLY, but not exactly the same period, we should f
0332	d have to be united into one great family, in	NEARLY the same manner as has occurred with ruminant
0333	cter from the common progenitor of the order,	NEARLY so much as they subsequently diverged. Thus i
0334	nth stage; hence they could hardly fail to be	NEARLY intermediate in character between the forms o
0334	, that the fauna of each period as a whole is	NEARLY intermediate in character between the precedi
0336	nditions of the ancient areas having remained	NEARLY the same. Let it be remembered that the forms
0337	fe over other and preceding forms. If under a	NEARLY similar climate, the eocene inhabitants of on
0340	ll their modified descendants still differ in	NEARLY the same manner and degree. But after very lo
0344	nct; for the more ancient a form is, the more	NEARLY it will be related to, and consequently resem
0347	the productions of Australia or Africa under	NEARLY the same climate. Analogous facts could be gi
0347	gions. We see this in the great difference of	NEARLY all the terrestrial productions of the New an
0349	ly allied, yet distinct kinds of birds, notes	NEARLY similar, and sees their nests similarly const
0349	d, but not quite alike, with eggs coloured in	NEARLY the same manner. The plains near the Straits
0349	e see the agouti and bizcacha, animals having	NEARLY the same habits as our hares and rabbits and
0350	like, or, as we see in the case of varieties	NEARLY like each other. The dissimilarity of the inh
0351	he last chapter that some forms have retained	NEARLY the same character from an enormously remote
0353	or South America? The conditions of life are	NEARLY the same, so that a multitude of European ani
0356	ndividuals of the species will have been kept	NEARLY uniform by intercrossing: so that many indivi
0361	ficial salt water for 30 days, to my surprise	NEARLY all germinated. Living birds can hardly fail
0365	no destructive insects or birds living there,	NEARLY every seed, which chanced to arrive, would be
0365	are all the same with those of Labrador, and	NEARLY all the same, as we hear from Asa Gray, with
0367	arctic plants and animals, and these would be	NEARLY the same with those of Europe; for the presen
0367	lated to the arctic forms living due north or	NEARLY due north of them: for the migration as the c
0368	igration northward, will have been exposed to	NEARLY the same climate; and, as is especially to be
0370	on, that our continents have long remained in	NEARLY the same relative position, though subjected
0372	iocene or even a somewhat earlier period, was	NEARLY uniform along the continuous shores of the Po
0372	an, areas now separated by a continent and by	NEARLY a hemisphere of equatorial ocean. These cases
0372	een created alike, in correspondence with the	NEARLY similar physical conditions of the areas: for
0380	h. in many islands the native productions are	NEARLY equalled or even outnumbered by the naturalis

Page **(Key Word)**

Page **************************************(Key Word)**************************************

```
0389  recent period all existing islands have been NEARLY or quite joined to some continent. This view
0390  that the naturalised plants and animals have NEARLY or quite exterminated many native productions
0390  no means follows, that, because in an island NEARLY all the species of one class are peculiar, th
0390  much disturbed. Thus in the Galapagos Islands NEARLY every land bird, but only two out of the elev
0395  ian Islands stand on a deeply submerged bank, NEARLY 1000 fathoms in depth, and here we find Amer*
0398  st. So it is with the other animals, and with NEARLY all the plants, as shown by Dr. Hooker in his
0399  lands were long ago partially stocked from a  NEARLY intermediate though distant point, namely fre
0403  east and south west corners of Australia have NEARLY the same physical conditions, and are united
0408  t importance, we can see why two areas having NEARLY the same physical conditions should often be
0409  c the succession of forms in past times being NEARLY the same with those governing at the present
0410  ame bond of ordinary generation: and the more NEARLY any two forms are related in blood, the neare
0415  rgan, as we have every reason to suppose, has NEARLY the same physiological value, its classificat
0418  particular species. If they find a character NEARLY uniform, and common to a great number of form
0418  or those for propagating the race, are found NEARLY uniform, they are considerec as highly servic
0423  descent with various degrees of modification. NEARLY the same rules are followed in classifying va
0423  though the most important part, happens to be NEARLY identical; no one puts the swedish and commor
0424  it of tumbling; but the short faced breed has NEARLY or quite lost this habit; nevertheless, withc
0428  successive slight modifications to live under NEARLY similar circumstances, to inhabit for instanc
0430  erhouse, of all Rodents, the bizcacha is most NEARLY related to Marsupials; but in the points in w
0430  existing Marsupial, but indirectly to all or  NEARLY all Marsupials, from having partially retaine
0430  has remarkec, the phascolomys resembles most NEARLY, not any one species, but the general order c
0430  e of a Rodent. The elder De Candolle has made NEARLY similar observations on the general nature of
0438  e lowest members of the class, we do not find NEARLY so much indefinite repetition of any one part
0442  placed in the midst of proper nutriment, yet  NEARLY all pass through a similar worm like stage of
0445  old puppies, I found that the puppies had not NEARLY acquired their full amount of proportional d*
0446  ogs, and pigeons, for breeding, when they are NEARLY grown up: they are indifferent whether the de
0449  the best, or indeed, if our collections were  NEARLY perfect, the only possible arrangement, woulc
0449  hat they have both descended from the same or NEARLY similar parents, and are therefore in that de
0473  duck, which has wings incapable of flight, in NEARLY the same condition as in the domestic duck; o
0475  fferent conditions of life, yet should follow-NEARLY the same instincts; why the thrush of South A
0478  rants might have been derived. We see this in NEARLY all the plants and animals of the Galapagos a
0486  y contrivances, each useful to the possessor, NEARLY in the same way as when we look at any great
0225  hese cells are always made at that degree of  NEARNESS to each other, that they would have intersec
0229  m the unnatural state of things, had not been NEATLY performed. The bees must have worked at very
0002  s. this Abstract, which I now publish, must   NECESSARILY be imperfect. I cannot here give reference
0010  ese cases anyhow show that variation is not   NECESSARILY connected, as some authors have supposed,
0052  sed that all varieties or incipient species   NECESSARILY attain the rank of species. They may whilsi
0053  . the whole subject, however, treated as it   NECESSARILY here is with much brevity, is rather perpl
0054  in order to become in any degree permanent,   NECESSARILY have to struggle with the other inhabitant
0105  ber of the individuals supported on it will   NECESSARILY be very small; and fewness of individuals
0121  ach fully stocked country natural selection   NECESSARILY acts by the selected form having some adva
0182  ch has to live by a struggle; but it is not   NECESSARILY the best possible under all possible condi
0206  rendered higher. Natural selection will not   NECESSARILY produce absolute perfection; nor, as far a
0226  ly of the same size, is very frequently and   NECESSARILY the case, the three flat surfaces are unit
0226  e, the three plane surfaces in any one cell   NECESSARILY enter into the construction of three adjoi
0259  fspring; but these two powers do not at all   NECESSARILY go together. There are certain hybrids whi
0292  ion. Thus the geological record will almost   NECESSARILY be rendered intermittent. I feel much conf
0297  enitor of B anc C, and yet might not at all   NECESSARILY be strictly intermediate between them in a
0315  subsiding, our formations have been almost    NECESSARILY accumulated at wide and irregularly inter
0334  ying Silurian system. But each fauna is not   NECESSARILY exactly intermediate, as unecual intervals
0335  to believe that forms successively produced  NECESSARILY endure for corresponding lengths of time:
0343  ly; how species of different classes do not  NECESSARILY change together, or at the same rate, or i
0343  of time: for the process of modification is  NECESSARILY slow, and depends on many complex continge
0405  se of certain powerfully winged birds, will  NECESSARILY range widely; for we should never forget t
0417  ith which our classifications are sometimes  NECESSARILY founded. Fractically when naturalists are
0442  t we might be led to look at these facts as  NECESSARILY contingent in some manner on growth. But t
0443  a very early period, that slight variations  NECESSARILY appear at an equally early period. But we
0457  nciple of successive slight variations, not  NECESSARILY or generally supervening at a very early p
0462  regions, as the process of mocification has NECESSARILY been slow, all the means of migration will
0473  one part has been modified other parts are  NECESSARILY modified. In both varieties and species re
0483  developed state; and this in some instances NECESSARILY implies an enormous amount of modification
0014  fect reversion had ensued. It would be quite NECESSARY, in order to prevent the effects of intercr
0031  to realise what they have done, it is almost NECESSARY to read several of the many treatises devot
0039  t some great deviation of structure would be NECESSARY to catch the fancier's eye: he perceives ex
0043  believe that variability is an inherent and  NECESSARY contingency, under all circumstances, with
0059  red: but the amount of difference corsidered NECESSARY to give to two forms the rank of species is
0060  nd of some few well marked varieties, though NECESSARY as the foundation for the work, helps us bu
0066  s or seeds. In looking at Nature, it is most NECESSARY to keep the foregoing considerations always
0070  to the numbers of its enemies, is absolutely NECESSARY for its preservation. Thus we can easily ra
0078  tions of all organic beings; a conviction as NECESSARY, as it seers to be difficult to acquire. Al
0082  elieve, any extreme amount of variability is NECESSARY; as man can certainly procuce great results
0082  isolation to check immigration, is actually  NECESSARY to produce new and unoccupied places for na
0085  and direct effect. It is, however, far more  NECESSARY to bear in mind that there are many unknown
0089  ak means: I cannot here enter on the details NECESSARY to support this view; but if man can in a s
0097  gma, or bring pollen from another flower. So NECESSARY are the visits of bees to papilionaceous fl
0131  of facts which cannot be here given would be NECESSARY to show the truth of the remark, that the r
0146  aturally be thought to be correlated in some NECESSARY manner. So, again, I do nct doubt that some
0148  developing any organ, without requiring as a NECESSARY compensation the reduction of some adjoinin
0198  e young in the womb. The laborious breathing NECESSARY in high regions would, we have some reason
0218  e pollen collecting apparatus which would be NECESSARY if they had to store food for their own you
0224  uite inconceivable how they can make all the NECESSARY angles and planes, or even perceive when th
0234  bees in a hive for the secretion of the wax  NECESSARY for the construction of their combs. Moreov
0295  p the depth approximately the same, which is NECESSARY in order to enable the same species to live
0303  ganisms, a comparatively short time would be NECESSARY to produce many divergent forms, which woul
0331  er, assume that divergence of character is a NECESSARY contingency; it depends solely on the desce
0342  e formation: that, owing to subsidence being NECESSARY for the accumulation of fossiliferous depos
```

Page **************************************(Key Word)**************************************

Page **(Key Word)**

0351 s remarked in the last chapter, in no law of NECESSARY development. As the variability of each spe
0369 ay be reasonably asked how I account for the NECESSARY degree of uniformity of the sub arctic and
0370 r a more favourable climate, I attribute the NECESSARY amount of uniformity in the sub arctic and
0381 and for their subsequent modification to the NECESSARY degree. The facts seem to me to indicate th
0383 would follow from this capacity as an almost NECESSARY consequence. We can more consider only a fe
0405 of the species range very widely; for it is NECESSARY that the unmodified parent should range wid
0452 of a class, nothing is more common, or more NECESSARY, than the use and discovery of rudiments. T
0480 ll be based on a new foundation, that of the NECESSARY acquirement of each mental power and capaci
0358 volutions within the recent period, as are NECESSITATED on the view advanced by Forbes and admitte
0136 nd which groups have habits of life almost NECESSITATING frequent flight; these several considerat
0002 one can feel more sensible than I do of the NECESSITY of hereafter publishing in detail all the f
0070 ase lost every single seed. This view of the NECESSITY of a large stock of the same species for it
0101 cross between two individuals is an obvious NECESSITY for each birth; in many others it occurs pe
0419 rms. Temminck insists on the utility or even NECESSITY of this practice in certain groups of birds
0423 s with species. Authors have insisted on the NECESSITY of classing varieties on a natural instead
0021 eathers so much reversed along the back of the NECK that they form a hood, and it has, proportiona
0191 lly disappeared, the slits on the sides of the NECK and the loop like course of the arteries still
0479 orse, the same number of vertebrae forming the NECK of the giraffe and of the elephant, and innume
0021 some of the sub breeds of runts have very long NECKS, others very long wings and tails, others sin
0073 ium pratense), as other bees cannot reach the NECTAR. Hence I have very little doubt, that if the
0092 s. Let us now suppose a little sweet juice or NECTAR to be excreted by the inner bases of the peta
0092 a flower. In this case insects in seeking the NECTAR would get dusted with pollen, and would certa
0092 of these seedlings would probably inherit the NECTAR excreting power. Those individual flowers whi
0092 glands or nectaries, and which excreted most NECTAR, would be oftenest visited by insects, and wo
0092 for the sake of collecting pollen instead of NECTAR; and as pollen is formed for the sole object
0093 , having flown from tree to tree in search of NECTAR. But to return to our imaginary case: as soon
0094 xes would be affected. Let us now turn to the NECTAR feeding insects in our imaginary case: we may
0094 t of which we have been slowly increasing the NECTAR by continued selection, to be a common plant;
0094 certain insects depended in main part on its NECTAR for food. I could give many facts, showing ho
0094 their habit of cutting holes and sucking the NECTAR at the bases of certain flowers, which they c
0094 length; yet the hive bee can easily suck the NECTAR out of the incarnate clover, but not out of t
0095 offer in vain an abundant supply of precious NECTAR to the hive bee. Thus it might be a great adv
0097 flower and the manner in which bees suck the NECTAR; for, in doing this, they either push the flo
0233 bees are often hard pressed to get sufficient NECTAR; and I am informed by Mr. Tegetmeier that it
0233 f wax; so that a prodigious quantity of fluid NECTAR must be collected and consumed by the bees in
0092 dual flowers which had the largest glands or NECTARIES, and which excreted most nectar, would be o
0145 tals; and that when this occurs, the adherent NECTARY is quite aborted; when the colour is absent
0145 nt from only one of the two upper petals, the NECTARY is only much shortened. With respect to the
0009 by unperceived causes as to fail in acting, we NEED not be surprised at this system, when it does
0052 facts and views given throughout this work. It NEED not be supposed that all varieties or incipien
0223 home. Such are the facts, though they did not NEED confirmation by me, in regard to the wonderful
0266 the same compounded organisation, and hence we NEED not be surprised that their sterility, though
0269 selection, as carried on by man and nature, we NEED not be surprised at some difference in the res
0322 well with the theory of natural selection. We NEED not marvel at extinction; if we must marvel, l
0334 between the preceding and succeeding faunas. I NEED give only one instance; namely, the manner in
0403 species. From these considerations I think we NEED not greatly marvel at the endemic and represen
0438 e been effected by slight successive steps, we NEED not wonder at discovering in such parts or org
0461 ghtly different and new conditions of life, we NEED not feel surprise at hybrids being in some deg
0472 of perfection of their associates; so that we NEED feel no surprise at the inhabitants of any one
0472 them be abhorrent to our ideas of fitness. We NEED not marvel at the sting of the bee causing the
0475 n slowly acquired through natural selection we NEED not marvel at some instincts being apparently
0477 esent the same physical conditions of life, we NEED feel no surprise at their inhabitants being wi
0004 udies, although they have been very commonly NEGLECTED by naturalists. From these considerations,
0041 ultivated, but the slight varieties had been NEGLECTED. As soon, however, as gardeners picked out
0112 r very swift nor very strong, will have been NEGLECTED, and will have tended to disappear. Here, t
0152 nimals, if any part, or the whole animal, be NEGLECTED and no selection be applied, that part (for
0424 oved that the Hottentot had descended from the NEGRO, I think he would be classed under the Negro
0424 e negro, I think he would be classed under the NEGRO group, however much he might differ in colour
0034 w much good domestic breeds are valued by the NEGROES of the interior of Africa who have not assoc
0424 in colour and other important characters from NEGROES. With species in a state of nature, every na
0040 individuals slowly spread in the immediate NEIGHBOURHOOD. But as yet they will hardly have a disti
0308 the sediment was derived, occurred in the NEIGHBOURHOOD of the existing continents of Europe and
0320 ts the less improved varieties in the same NEIGHBOURHOOD; when much improved it is transported far
0294 versely, that some are now abundant in the NEIGHBOURING sea, but are rare or absent in this partic
0364 island to island, or from a continent to a NEIGHBOURING island, but not from one distant continent
0395 h of the sea separating an island from the NEIGHBOURING mainland, and the presence in both of the
0396 ian inhabitants of islands with those of a NEIGHBOURING continent, an inexplicable relation on the
0478 ng manner to the plants and animals of the NEIGHBOURING American mainland; and those of the Cape d
0222 h nests, the little ants attacked their big NEIGHBOURS with surprising courage. Now I was curious
0025 belonging to two distinct breeds are crossed, NEITHER of which is blue or has any of the above spe
0038 ness to man, we can understand how it is that NEITHER Australia, the Cape of Good Hope, nor any ot
0072 s the most curious instance of this; for here NEITHER cattle nor horses nor dogs have ever run wil
0081 iations, I call Natural Selection. Variations NEITHER useful nor injurious would not be affected b
0112 r animals with intermediate characters, being NEITHER very swift nor very strong, will have been n
0248 ng different years. It can thus be shown that NEITHER sterility nor fertility affords any clear di
0253 untry; and as they are kept for profit, where NEITHER pure parent species exists, they must certai
0269 f selection, for his own use and pleasure; he NEITHER wishes to select, nor could select, slight d
0308 during the palaeozoic and secondary periods, NEITHER continents nor continental islands existed w
0312 are more broken; but, as Bronn has remarked, NEITHER the appearance nor disappearance of their ma
0346 he first great fact which strikes us is, that NEITHER the similarity nor the dissimilarity of the
0351 ey will be little liable to modification; for NEITHER migration nor isolation in themselves can do
0475 will be supplanted by new and improved forms. NEITHER single species nor groups of species reappea
0387 ize of the seeds of that fine water lily the NELUMBIUM, and remembered Alph. de Candolle's remarks
0387 lily (probably, according to Dr. Hooker, the NELUMBIUM luteum) in a heron's stomach: although I do
0387 stomach a pellet containing the seeds of the NELUMBIUM undigested; or the seeds might be dropped b
0187 gination, can hardly be considered real. How a NERVE comes to be sensitive to light, hardly concer

Page **(Key Word)**

Page ***(Key Word)***

0187 veral facts make me suspect that any sensitive NERVE may be rendered sensitive to light, and likew
0187 iculata we can commence a series with an optic NERVE merely coated with pigment, and without any o
0188 has converted the simple apparatus of an optic NERVE merely coated with pigment and invested by tr
0188 ke a thick layer of transparent tissue, with a NERVE sensitive to light beneath, and then suppose
0045 have expected that the branching of the main NERVES close to the great central ganglion of an ins
0046 s shown a degree of variability in these main NERVES in Coccus, which may almost be compared to th
0148 normously developed, and furnished with great NERVES and muscles; but in the parasitic and protect
0217 ree days; so that, if she were to make her own NEST and sit on her own eggs, those first laid woul
0217 and young birds of different ages in the same NEST. If this were the case, the process of laying
0217 is in this predicament; for she makes her own NEST and has eggs and young successively hatched, a
0217 occasionally she laid an egg in another bird's NEST.If the old bird profited by this occasional ha
0218 species, unite and lay first a few eggs in one NEST and then in another; and these are hatched by
0219 antage to the species, and if the insect whose NEST and stored food are thus feloniously appropria
0219 , or of feeding their own larvae. When the old NEST is found inconvenient, and they have to migrat
0220 st in their appearance is very great. When the NEST is slightly disturbed, the slaves occasionally
0220 their masters are much agitated and defend the NEST: when the nest is much disturbed and the larva
0220 re much agitated and defend the nest: when the NEST is much disturbed and the larvae and pupae are
0220 and never saw a slave either leave or enter a NEST. As, during these months, the slaves are very
0220 e numbers in August; either leave or enter the NEST. Hence he considers them as strictly household
0220 constantly seen bringing in materials for the NEST, and food of all kinds. During the present yea
0221 slaves mingled with their masters leaving the NEST, and marching along the same road to a tall Sc
0221 bitually work with their masters in making the NEST, and they alone open and close the doors in th
0221 nately chanced to witness a migration from one NEST to another, and it was a most interesting spec
0221 and carried their dead bodies as food to their NEST, twenty nine yards distant, but they were prev
0221 l parcel of the pupae of F. fusca from another NEST, and put them down on a bare spot near the pla
0222 ow ants still clinging to the fragments of the NEST. This species is sometimes, though rarely, mad
0222 community of F. flava under a stone beneath a NEST of the slave making F. sanguinea; and when I h
0222 e across the pupae, or even the earth from the NEST of F. flava, and quickly ran away; but in abou
0222 nd found a number of these ants entering their NEST, carrying the dead bodies of F. fusca (showing
0222 pupa; but I was not able to find the desolated NEST in the thick heath. The nest, however, must ha
0222 ind the desolated nest in the thick heath. The NEST, however, must have been close at hand, for tw
0223 . rufescens. The latter does not build its own NEST, does not determine its own migrations, does n
0223 ew. The masters determine when and where a new NEST shall be formed, and when they migrate; the ma
0223 ogether, making and bringing materials for the NEST: both, but chiefly the slaves, tend, and milk
0223 in england the masters alone usually leave the NEST to collect building materials and food for the
0227 rful than those which guide a bird to make its NEST, I believe that the hive bee has acquired, thr
0239 stus, the workers of one caste never leave the NEST: they are fed by the workers of another caste,
0239 pear in all the individual neuters in the same NEST, but in a few alone; and that by the long cont
0239 euter insects of the same species, in the same NEST, presenting gradations of structure; and this
0239 together by individuals taken out of the same NEST: I have myself compared perfect gradations of
0240 have two bodies of sterile workers in the same NEST, differing not only in size, but in their orga
0240 th's offer of numerous specimens from the same NEST of the driver ant (Anomma) of West Africa. The
0241 castes of sterile workers existing in the same NEST, both widely different from each other and fro
0243 is that the thrush of South America lines its NEST with mud, in the same peculiar manner as does
0253 ids (grandchildren of the pure geese) from one NEST. In India, however, these cross bred geese mus
0363 nd have even carried brushwood, bones, and the NEST of a land bird, I can hardly doubt that they m
0440 , in the egg of the bird which is hatched in a NEST, and in the spawn of a frog under water. We ha
0475 rush of South America, for instance, lines her NEST with mud like our British species. On the view
0022 as does the state of the down with which the NESTLING birds are clothed when hatched. The shape an
0087 g the cocoon, or the hard tip to the beak of NESTLING birds, used for breaking the egg. It has bee
0212 ly an instinctive quality, as may be seen in NESTLING birds, though it is strengthened by experie
0445 they been natural productions. But when the NESTLING birds of these several breeds were placed in
0062 ely these songsters, or their eggs, or their NESTLINGS, are destroyed by birds and beasts of prey;
0074 r of field mice, which destroy their combs and NESTS; and Mr. H. Newman, has long attended to
0074 near villages and small towns I have found the NESTS of humble bees more numerous than elsewhere,
0207 to migrate and to lay her eggs in other birds' NESTS. An action, which we ourselves should require
0212 tion, and in its total loss. So it is with the NESTS of birds, which vary partly in dependence on
0212 ven several remarkable cases of differences in NESTS of the same species in the northern and south
0216 ads the cuckoo to lay her eggs in other birds' NESTS; the slave making instinct of certain ants; a
0217 koo occasionally lays her eggs in other birds' NESTS; but I hear on the high authority of Dr. Brew
0217 occasionally to lay their eggs in other bird's NESTS. Now let us suppose that the ancient progenit
0217 would be apt to lay their eggs in other birds' NESTS, and thus be successful in rearing their youn
0218 bit of birds laying their eggs in other birds' NESTS either of the same or of a distinct species,
0218 re parasitic, and always lay their eggs in the NESTS of bees of other kinds. This case is more rem
0219 r work. They are incapable of making their own NESTS, or of feeding their own larvae. When the old
0220 made, in some little detail. I opened fourteen NESTS of F. sanguinea, and found a few slaves in al
0220 mmunities, and have never been observed in the NESTS of F. sanguinea. The slaves are black and not
0220 e years, I have watched for many hours several NESTS in Surrey and Sussex, and never saw a slave e
0220 t mr. Smith informs me that he has watched the NESTS at various hours during May, June and August,
0222 ea; and when I had accidentally disturbed both NESTS, the little ants attacked their big neighbour
0223 upae of other species, if scattered near their NESTS, it is possible that pupae originally stored
0243 ens (Troglodytes) of North America, build cock NESTS, to roost in, like the males of our distinct
0349 of birds, notes nearly similar, and sees their NESTS similarly constructed, but not quite alike, w
0157 gain in fossorial hymenoptera, the manner of NEURATION of the wings is a character of the highest
0157 n to large groups; but in certain genera the NEURATION differs in the different species, and likew
0192 my future work. One of the gravest is that of NEUTER insects, which are often very differently con
0207 theory of the Natural Selection of instincts. NEUTER or sterile insects. Summary. The subject of i
0236 ed by the hive bee. If a working ant or other NEUTER insect had been an animal in the ordinary sta
0239 annihilate my theory. In the simpler case of NEUTER insects all of one caste or of the same kind,
0239 r. on this view we ought occasionally to find NEUTER insects of the same species, in the same nest
0239 s we do find, even often, considering how few NEUTER insects out of Europe have been carefully exa
0242 n so high a degree, had not the case of these NEUTER insects convinced me of the fact. I have, the
0242 o one has advanced this demonstrative case of NEUTER insects, against the well known doctrine of L
0474 not indispensable, as we see, in the case of NEUTER insects, which leave no progeny to inherit th
0236 lly fatal to my whole theory. I allude to the NEUTERS or sterile females in insect communities: fo
0236 rile females in insect communities: for these NEUTERS often differ widely in instinct and in struc

Page ***(Key Word)***

0238	of the difficulty; namely, the fact that the	NEUTERS of several ants differ, not only from the fe
0238	thus in Eciton, there are working and soldier	NEUTERS, with jaws and instincts extraordinarily dif
0239	robably at first appear in all the individual	NEUTERS in the same nest, but in a few alone; and th
0239	on of the fertile parents which produced most	NEUTERS with the profitable modification, all the ne
0239	ers with the profitable modification, all the	NEUTERS ultimately came to have the desired characte
0239	. mr. F. Smith has shown how surprisingly the	NEUTERS of several British ants differ from each oth
0240	we should then have had a species of ant with	NEUTERS very nearly in the same condition with those
0240	of structure between the different castes of	NEUTERS in the same species, that I gladly availed m
0241	form a species which should regularly produce	NEUTERS, either all of large size with one form of j
0174	confines, and finally disappearing. Hence the	NEUTRAL territory between two representative species
0175	has a wide range, with a comparatively narrow	NEUTRAL territory between them, in which they become
0008	s display the utmost vigour, and yet rarely or	NEVER seed! In some few such cases it has been foun
0028	d by Akber Khan in India, about the year 1600;	NEVER less than 20,000 pigeons were taken with the
0028	jesty by crossing the breeds, which method was	NEVER practised before, has improved them astonishi
0029	horns, and he will laugh you to scorn. I have	NEVER met a pigeon, or poultry, or cuck, or rabbit
0030	tle quarrelsome, with everlasting layers which	NEVER desire to sit; and with the bantam so small a
0033	in other points; this is hardly ever, perhaps	NEVER, the case. The laws of correlation of growth,
0033	tion of growth, the importance of which should	NEVER be overlooked; will ensure some differences;
0035	slow and insensible changes of this kind could	NEVER be recognised unless actual measurements or c
0036	ly followed, in so far that the breeders could	NEVER have expected or even have wished to have pro
0036	ieties. If there exist savages so barbarous as	NEVER to think of the inherited character of the of
0037	o cultivated the best pear they could procure,	NEVER thought what splendid fruit we should eat; th
0038	d he rarely cares for what is internal. He can	NEVER act by selection, excepting on variations whi
0039	selected a pigeon with a slightly larger tail,	NEVER dreamed what the descendants of that pigeon w
0041	oor people, and are mostly in small lots, they	NEVER can be improved. On the other hand, nurseryme
0045	n many specimens of the same species. I should	NEVER have expected that the branching of the main
0046	a circle when they state that important organs	NEVER vary; for these same authors practically rank
0065	etrical ratio of increase, the result of which	NEVER fails to be surprising, simply explains the e
0066	thousand years, supposing that this seed were	NEVER destroyed, and could be ensured to germinate
0066	p the foregoing considerations always in mind,	NEVER to forget that every single organic being are
0069	n perfectly well endure our climate, but which	NEVER become naturalised, for they cannot compete w
0071	a large and extremely barren heath, which had	NEVER been touched by the hand of man: but several
0073	c Lobelia fulgens, in this part of England, is	NEVER visited by insects, and consequently, from it
0073	and consequently, from its peculiar structure,	NEVER can set a seed. Many of our orchidaceous plan
0086	cts which live only for a few hours, and which	NEVER feed, a large part of their structure is mere
0098	s ready to receive them: and as this flower is	NEVER visited, at least in my garcen, by insects, i
0098	visited, at least in my garden, by insects, it	NEVER sets a seed, though by placing pollen from on
0104	be great. If there exist organic beings which	NEVER intercross, uniformity of character can be re
0111	anc in a greater degree: but this alone would	NEVER account for so habitual and large an amount o
0136	nds, or by giving up the attempt and rarely or	NEVER flying. As with mariners shipwrecked near a c
0142	the case of the Jerusalem artichoke, which is	NEVER propagated by seed, and of which consequently
0146	de candolle has remarked that winged seeds are	NEVER found in fruits which do not open: I should e
0149	particular shape. Natural selection, it should	NEVER be forgotten, can act on each part of each be
0161	to lost ancestral characters. As, however, we	NEVER know the exact character of the common ancest
0167	ignorant cosmogonists; that fossil shells had	NEVER lived, but had been created in stone so as to
0171	me of them are so grave that to this day I can	NEVER reflect on them without being staggered; but,
0184	r common species: yet it is a woodpecker which	NEVER climbs a tree! Petrels are the most aerial an
0185	amining the dead body of the water ouzel would	NEVER have suspected its sub aquatic habits; yet th
0185	upland geese with webbed feet which rarely or	NEVER go near the water! and no one except Audubon
0194	ntage of slight successive variations; she can	NEVER take a leap, but must advance by the shortest
0201	s and other such cases. Natural selection will	NEVER produce in a being anything injurious to itse
0210	ct of each species is good for itself, but has	NEVER, as far as we can judge, been produced for th
0214	nessed, is performed by young birds, that have	NEVER seen a pigeon tumble. We may believe that som
0215	in those breeds of fowls which very rarely or	NEVER become broody, that is, never wish to sit on
0215	h very rarely or never become broody, that is,	NEVER wish to sit on their eggs. Familiarity alone
0220	only in their own proper communities, and have	NEVER been observed in the nests of F. sanguinea. T
0220	hours several nests in Surrey and Sussex, and	NEVER saw a slave either leave or enter a nest. As,
0220	august, both in Surrey and Hampshire, and has	NEVER seen the slaves, though present in large numb
0225	if the spheres had been completed; but this is	NEVER permitted, the bees building perfectly flat w
0230	eavouring to make equal spherical hollows, but	NEVER allowing the spheres to break into each other
0230	r the two plates, as the case may be; and they	NEVER complete the upper edges of the rhombic plate
0232	ning spheres: but, as far as I have seen, they	NEVER gnaw away and finish off the angles of a cell
0237	ents, yet absolutely sterile: so that it could	NEVER have transmitted successively acquired modifi
0239	exican Myrmecocystus; the workers of one caste	NEVER leave the nest: they are fed by the workers o
0242	with all my faith in this principle, I should	NEVER have anticipated that natural selection could
0248	yet he asserts positively that their fertility	NEVER increased, but generally greatly decreased. I
0250	revolutum produced a plant, which (he says) I	NEVER saw to occur in a case of its natural fecunda
0255	so in hybrids themselves, there are some which	NEVER have produced, and probably never would produ
0255	e some which never have produced, and probably	NEVER would produce, even with the pollen of either
0257	tic affinity. This is clearly shown by hybrids	NEVER having been raised between species ranked by
0270	though these plants have separated sexes, they	NEVER naturally crossed. He then fertilised thirtee
0289	ely confined to these large territories, would	NEVER have suspected that during the periods which
0296	during the process of deposition, which would	NEVER even have been suspected, had not the trees c
0313	reason to believe that the same identical form	NEVER reappears. The strongest apparent exception t
0315	understand why a species when once lost should	NEVER reappear, even if the very same conditions of
0338	animal. This view may be true, and yet it may	NEVER be capable of full proof. Seeing, for instanc
0343	and why when a species has once disappeared it	NEVER reappears. Groups of species increase in numb
0349	travelling, for instance, from north to south	NEVER fails to be struck by the manner in which suc
0355	ously created. With those organic beings which	NEVER intercross (if such exist), the species, on m
0355	a succession of improved varieties, which will	NEVER have blended with other individuals or variet
0361	thors have given a far higher estimate. I have	NEVER seen an instance of nutritious seeds passing
0364	to be. The currents, from their course, would	NEVER bring seeds from North America to Britain, th
0383	tly a still more impassable barrier, that they	NEVER would have extended to distant countries. But
0384	egard to fish, I believe that the same species	NEVER occur in the fresh waters of distant continen
0393	ed that Batrachians (frogs, toads, newts) have	NEVER been found on any of the many islands with wh
0405	, will necessarily range widely; for we should	NEVER forget that to range widely implies not only

Page **(Key Word)**

```
0432 oughout all time and space. We shall certainly NEVER succeed in making so perfect a collection: n
0434 ex and radiating lines of affinities. We shall NEVER, probably, disentangle the inextricable web
0434 emain connected together in the same order. We NEVER find, for instance, the bones of the arm and
0450 their heads; and the presence of teeth, which NEVER cut through the gums, in the upper jaws of o
0471 nsects on the ground: that upland geese, which NEVER or rarely swim should have been created with
0480 calf, for instance, has inherited teeth, which NEVER cut through the gums of the upper jaw, from
0489 that the ordinary succession by generation has NEVER once been broken, and that no cataclysm has
0003 ended, like varieties, from other species. NEVERTHELESS, such a conclusion, even if well founded,
0011 ions produce similar changes of structure. NEVERTHELESS some slight amount of change may, I think
0015 ty should be turned loose in its new home. NEVERTHELESS, as our varieties certainly do occasional
0018 whole subject must, I think, remain vague; NEVERTHELESS, I may, without here entering on any deta
0034 ctation of permanently altering the breed. NEVERTHELESS I cannot doubt that this process, continu
0048 omitted many trifling varieties, but which NEVERTHELESS have been ranked by some botanists as spe
0073 victory to one organic being over another. NEVERTHELESS so profound is our ignorance, and so high
0093 ous, and therefore not favourable to bees. NEVERTHELESS every female flower which I examined had
0096 f hermaphrodites this is far from obvious. NEVERTHELESS I am strongly inclined to believe that wi
0111 ar less than do good and distinct species. NEVERTHELESS, according to my view, varieties are spec
0132 more or less, we are profoundly ignorant. NEVERTHELESS, we can here and there dimly catch a fain
0152 it is of high importance to that species; NEVERTHELESS the part in this case is eminently liable
0165 s none and has not even a shoulder stripe, NEVERTHELESS, had all four legs barred, and had three
0185 e mistaken by any one for an auk or grebe; NEVERTHELESS, it is essentially a petrel, but with many
0198 differences of our domestic breeds, which NEVERTHELESS we generally admit to have arisen through
0217 ity of Dr. Brewer, that this is a mistake. NEVERTHELESS, I could give several instances of variou
0248 ly decreases in the first few generations. NEVERTHELESS I believe that in all these experiments th
0251 hat the plants were in an unnatural state. NEVERTHELESS these facts show on what slight and myste
0268 rom several aboriginally distinct species. NEVERTHELESS the perfect fertility of so many domestic
0324 classed with somewhat older European beds. NEVERTHELESS, looking to a remotely future epoch, there
0328 urian deposits of Bohemia and Scandinavia: NEVERTHELESS he finds a surprising amount of difference
0329 ld falsely appear to be strictly parallel; NEVERTHELESS the species would not all be the same in t
0349 ry continent offers innumerable instances. NEVERTHELESS the naturalist in travelling, for instance
0352 tant and isolated points, where now found. NEVERTHELESS the simplicity of the view that each spec
0401 ach other than they appear to be on a map. NEVERTHELESS a good many species, both those found in c
0403 come colonised by the Porto Santo species. NEVERTHELESS both islands have been colonised by some E
0421 in blood, has come to be widely different. NEVERTHELESS their genealogical arrangement remains str
0424 breed has nearly or quite lost this habit; NEVERTHELESS, without any reasoning or thinking on the
0432 ose between the finest existing varieties. NEVERTHELESS a natural classification, or at least a na
0433 succeed in making so perfect a collection: NEVERTHELESS, in certain classes, we are tending in th
0436 ure and function of the mouths of insects. NEVERTHELESS, it is conceivable that the general patter
0443 nt, or their ancestors, have been exposed. NEVERTHELESS an effect thus caused at a very early peri
0459 s, each good for the individual possessor. NEVERTHELESS, this difficulty, though appearing to our
0482 o other and very slightly different forms. NEVERTHELESS they do not pretend that they can define,
0484 ype. But analogy may be a deceitful guide. NEVERTHELESS all living things have much in common, in
0005 any selected variety will tend to propagate its NEW and modified form. This fundamental subject of
0007 st be exposed during several generations to the NEW conditions of life to cause any appreciable am
0008 ivated plants, such as wheat, still often yield NEW varieties: our oldest domesticated animals are
0009 single bud or offset, which suddenly assumes a NEW and sometimes very different character from th
0015 a single variety should be turned loose in its NEW home. Nevertheless, as our varieties certainly
0015 eafter be explained, will determine how far the NEW characters thus arising shall be preserved. Wh
0034 tion, with a distinct object in view, to make a NEW strain or sub breed, superior to anything exis
0040 mmunication, the spreading and knowledge of any NEW sub breed will be a slow process. As soon as t
0040 process. As soon as the points of value of the NEW sub breed are once fully acknowledged, the pri
0041 ly far more successful than amateurs in getting NEW and valuable varieties. The keeping of a large
0042 mportant element of success in the formation of NEW races, at least, in a country which is already
0042 rgely favoured the improvement and formation of NEW breeds. Pigeons, I may add, can be propagated
0043 as, no doubt, largely aided in the formation of NEW sub breeds; but the importance of the crossing
0056 reason to believe the process of manufacturing NEW species to be a slow one. And this certainly i
0059 ll hereafter see, tend to become converted into NEW and distinct species. The larger genera thus t
0065 e diffusion of naturalised productions in their NEW homes. In a state of nature almost every plant
0078 see that when a plant or animal is placed in a NEW country amongst new competitors, though the cl
0078 nt or animal is placed in a new country amongst NEW competitors, though the climate may be exactly
0078 e wished to increase its average numbers in its NEW home, we should have to modify it in a differe
0081 thers. If the country were open on its borders, NEW forms would certainly immigrate, and this also
0081 untry partly surrounded by barriers, into which NEW and better adapted forms could not freely ente
0082 k immigration, is actually necessary to produce NEW and unoccupied places for natural selection to
0091 cture, and by the repetition of this process, a NEW variety might be formed which would either sup
0094 plant. In plants under culture and placed under NEW conditions of life, sometimes the male organs
0095 banish the belief of the continued creation of NEW organic beings, or of any great and sudden mod
0100 at my request Dr. Hooker tabulated the trees of NEW Zealand, and Dr. Asa Gray those of the United
0103 and which can increase at a very rapid rate, a NEW and improved variety might be quickly formed o
0103 be chiefly between the individuals of the same NEW variety. A local variety when once thus formed
0104 limate or elevation of the land, etc.; and thus NEW places in the natural economy of the country a
0105 onsecuently competition, will give time for any NEW variety to be slowly improved: and this may so
0105 sometimes be of importance in the production of NEW species. If, however, an isolated area be very
0105 dividuals will greatly retard the production of NEW species through natural selection, by decreasi
0105 ve been highly favourable for the production of NEW species. But we may thus greatly deceive ourse
0105 has been most favourable for the production of NEW organic forms, we ought to make the comparison
0105 of considerable importance in the production of NEW species, on the whole I am inclined to believe
0106 nding degree or they will be exterminated. Each NEW form, also, as soon as it has been much improv
0106 e into competition with many others. Hence more NEW places will be formed, and the competition to
0106 espects highly favourable for the production of NEW species, yet that the course of modification w
0106 rge areas; and what is more important, that the NEW forms produced on large areas, which already h
0106 will spread most widely, will give rise to most NEW varieties and species, and will thus play an i
0107 ions will have been less severe than elsewhere; NEW forms will have been more slowly formed, and o
0107 the most favourable for the production of many NEW forms of life, likely to endure long and to sp
0108 ny kind, immigration will be prevented, so that NEW places in the polity of each island will have
0108 still further the inhabitants, and thus produce NEW species. That natural selection will always ac
```

Page **(Key Word)**

Page **************************************(Key Word)***************************************

0109	nction. But we may go further than this: for as	NEW
0110	erations I think it inevitably follows, that as	NEW
0110	competition with each other. Consequently, each	NEW
0111	us instances could be given showing how quickly	NEW
0113	of them, for instance, being enabled to feed on	NEW
0113	of prey, either dead or alive; some inhabiting	NEW
0115	especially adapted to certain stations in their	NEW
0115	r of the native genera and species, far more in	NEW
0115	era and species, far more in new genera than in	NEW
0115	that diversification of structure, amounting to	NEW
0118	s we know to be favourable to the production of	NEW
0121	lines. After fourteen thousand generations, six	NEW
0121	for these will have the best chance of filling	NEW
0121	ich have largely varied, and have given rise to	NEW
0122	ountry, or become quickly adapted to some quite	NEW
0122	e become extinct, having been replaced by eight	NEW
0122	(i) will have been replaced by six (n14 to z14)	NEW
0123	descendants to this late stage of descent. The	NEW
0123	t different of the original eleven species. The	NEW
0123	ct, and have left no descendants. Hence the six	NEW
0124	cies, the supposed single parent of our several	NEW
0124	to reflect for a moment on the character of the	NEW
0124	this case, its affinities to the other fourteen	NEW
0124	n character from the type of their parents, the	NEW
0125	d to be represented in the diagram. And the two	NEW
0125	mmon. Hence, the struggle for the production of	NEW
0126	groups, from branching out and seizing on many	NEW
0132	efore that union takes place which is to form a	NEW
0136	ith the action of natural selection. For when a	NEW
0139	of the inhabitants of the caves of the Old and	NEW
0140	have subsequently become acclimatised to their	NEW
0142	r propagated by seed, and of which consequently	NEW
0160	that this is a case of reversion, and not of a	NEW
0162	to an anciently existing character and what are	NEW
0166	companied by any change of form or by any other	NEW
0169	nd differentiation, or where the manufactory of	NEW
0169	aracters of their ancient progenitors. Although	NEW
0172	preservation of profitable modifications, each	NEW
0172	very process of formation and perfection of the	NEW
0174	as played an important part in the formation of	NEW
0177	arying and intermediate links: firstly, because	NEW
0178	f some one or more of its inhabitants. And such	NEW
0178	of climate, or on the occasional immigration of	NEW
0178	inhabitants becoming slowly modified, with the	NEW
0180	getation change, let other competing rodents or	NEW
0184	dividuals would occasionally have given rise to	NEW
0189	roduce a distincter image. We must suppose each	NEW
0195	hese small enemies, would be able to range into	NEW
0196	tage of by the descendants of the species under	NEW
0201	tained under nature. The endemic productions of	NEW
0204	the air. We have seen that a species may under	NEW
0212	give rise, through natural selection, to quite	NEW
0223	ely few. The masters determine when and where a	NEW
0232	e bees can lay the foundations of one wall of a	NEW
0235	tance its newly acquired economical instinct to	NEW
0265	e see that when organic beings are placed under	NEW
0269	to me by far the most important consideration,	NEW
0280	ery process of natural selection, through which	NEW
0281	er this a very rare event; for in all cases the	NEW
0289	country, great piles of sediment, charged with	NEW
0292	g shoal parts of the sea will be increased, and	NEW
0292	, as previously explained, for the formation of	NEW
0292	ce, though there will be much extinction, fewer	NEW
0298	ise, first to local varieties and ultimately to	NEW
0301	ar ranging species which would oftenest produce	NEW
0301	followed by many palaeontologists, be ranked as	NEW
0303	succession of ages to adapt an organism to some	NEW
0303	and one true mammal has been discovered in the	NEW
0307	er stage to the Silurian system, abounding with	NEW
0312	eings. On the slow and successive appearance of	NEW
0312	ication, through descent and natural selection.	NEW
0312	, and to make the percentage system of lost and	NEW
0312	species are lost forms, and only one or two are	NEW
0315	. each formation, on this view, does not mark a	NEW
0315	supplant it; yet the two forms, the old and the	NEW
0315	ing long ages for the same object, might make a	NEW
0316	antail would be almost sure to inherit from its	NEW
0316	usly existed, in order to have generated either	NEW
0317	e extinction of old forms and the production of	NEW
0320	l selection is grounded on the belief that each	NEW
0320	lief that each new variety, and ultimately each	NEW
0320	the same with our domestic productions: when a	NEW
0320	eeds in other countries. Thus the appearance of	NEW
0320	r. in certain flourishing groups, the number of	NEW
0320	ter times we may believe that the production of	NEW
0321	xtermination of the parent species; and if many	NEW

NEW	forms are continually and slowly being produce
NEW	species in the course of time are formed throu
NEW	variety or species, during the progress of its
NEW	breeds of cattle, sheep, and other animals, an
NEW	kinds of prey, either dead or alive; some inha
NEW	stations, climbing trees, frequenting water, a
NEW	homes. But the case is very different; and Alp
NEW	genera than in new species. To give a single i
NEW	species. To give a single instance: in the las
NEW	generic differences, would have been profitabl
NEW	varieties. If, then, these two varieties be va
NEW	species, marked by the letters n14 to z14, are
NEW	and widely different places in the polity of n
NEW	varieties and species. The other nine species
NEW	station, in which child and parent do not come
NEW	species (a14 to m14); and (I) will have been r
NEW	species. But we may go further than this. The
NEW	species in our diagram descended from the orig
NEW	species, moreover, will be allied to each othe
NEW	species descended from (I), and the eight desc
NEW	sub genera and genera. It is worth while to re
NEW	species f14, which is supposed not to have div
NEW	species will be of a curious and circuitous na
NEW	species (F14) will not be directly intermediat
NEW	families, or orders, will have descended from
NEW	and modified descendants, will mainly lie betw
NEW	places in the polity of Nature, will constantl
NEW	being. In the case of sporting plants, the bud
NEW	insect first arrived on the island, the tenden
NEW	Worlds should be closely related, we might exp
NEW	homes. As I believe that our domestic animals
NEW	varieties have not been produced, has even bee
NEW	yet analogous variation appearing in the sever
NEW	but analogous variations, yet we ought, on my
NEW	character. We see this tendency to become stri
NEW	specific forms has been actively at work, ther
NEW	and important modifications may not arise from
NEW	form will tend in a fully stocked country to t
NEW	form. But, as by this theory innumerable trans
NEW	species, more especially with freely crossing
NEW	varieties are very slowly formed, for variatio
NEW	places will depend on slow changes of climate,
NEW	inhabitants, and, probably, in a still more im
NEW	forms thus produced and the old ones acting an
NEW	beasts of prey immigrate, or old ones become m
NEW	species, having anomalous habits, and with the
NEW	state of the instrument to be multiplied by th
NEW	pastures and thus gain a great advantage. It i
NEW	conditions of life and with newly acquired hab
NEW	Zealand, for instance, are perfect one compare
NEW	conditions of life change its habits, or have
NEW	instincts. But I am well aware that these gene
NEW	nest shall be formed, and when they migrate, t
NEW	hexagon, in its strictly proper place, project
NEW	swarms, which in their turn will have had the
NEW	and unnatural conditions, and when hybrids are
NEW	races of animals and plants are produced under
NEW	varieties continually take the places of and e
NEW	and improved forms of life will tend to suppla
NEW	and peculiar forms of life, had elsewhere been
NEW	stations will often be formed; all circumstanc
NEW	varieties and species; but during such periods
NEW	varieties or species will be formed; and it is
NEW	species: and this again would greatly lessen t
NEW	varieties; and the varieties would at first ge
NEW	and distinct species. If then, there be some d
NEW	and peculiar line of life, for instance to fly
NEW	red sandstone at nearly the commencement of th
NEW	and peculiar species. Traces of life have been
NEW	species. On their different rates of change. S
NEW	species have appeared very slowly, one after a
NEW	forms more gradual. In some of the most recent
NEW	forms, having here appeared for the first time
NEW	and complete act of creation, but only an occa
NEW	, would not be identically the same; for both w
NEW	breed hardly distinguishable from our present
NEW	progenitor some slight characteristic differen
NEW	and modified or the same old and unmodified fo
NEW	and improved forms are intimately connected to
NEW	variety, and ultimately each new species, is p
NEW	species, is produced and maintained by having
NEW	and slightly improved variety has been raised,
NEW	forms and the disappearance of old forms, both
NEW	specific forms which have been produced within
NEW	forms has caused the extinction of about the s
NEW	forms have been developed from any one species

Page **************************************(Key Word)***************************************

Page **(Key Word)***************************************

```
0321  extermination. Thus, as I believe, a number of NEW species descended from one species, that is a
0321  w species descended from one species, that is a NEW genus, comes to supplant an old genus, belong
0321  family. But it must often have happened that a NEW species belonging to some one group will have
0322  unusually rapid development, many species of a NEW group have taken possession of a new area, the
0322  ecies of a new group have taken possession of a NEW area, they will have exterminated in a corresp
0323  d north American tertiary deposits. Even if the NEW fossil species which are common to the Old and
0323  fossil species which are common to the Old and NEW Worlds be kept wholly out of view, the general
0325  cies, their extinction, and the introduction of NEW ones, cannot be owing to mere changes in marin
0325  explicable on the theory of natural selection. NEW species are formed by new varieties arising, w
0325  of natural selection. New species are formed by NEW varieties arising, which have some advantage o
0325  country, would naturally oftenest give rise to NEW varieties or incipient species; for these latt
0325  iffused, having produced the greatest number of NEW varieties. It is also natural that the dominan
0326  spreading still further, and of giving rise to NEW countries to new varieties and species. The pr
0326  further, and of giving rise in new countries to NEW varieties and species. The process of diffusio
0326  tions most favourable for the multiplication of NEW and dominant species; but we can, I think, cle
0326  urable, as would be the power of spreading into NEW territories. A certain amount of isolation, re
0326  have been most favourable for the production of NEW and dominant species on the land, and another
0327  t the world, accords well with the principle of NEW species having been formed by dominant species
0327  inant species spreading widely and varying; the NEW species thus produced being themselves dominan
0327  ; these again spreading, varying, and producing NEW species. The forms which are beaten and which
0327  are beaten and which yield their places to the NEW and victorious forms, will generally be allied
0327  ng some inferiority in common; and therefore as NEW and improved groups spread throughout the worl
0334  preceding forms, and for the coming in of quite NEW forms by immigration, and for a large amount o
0337  eory, be higher than the more ancient; for each NEW species is formed by having had some advantage
0337  european productions have recently spread over NEW Zealand, and have seized on places which must
0337  ls and plants of Great Britain were set free in NEW Zealand, that in the course of time a multitud
0337  e other hand, from what we see now occurring in NEW Zealand, and from hardly a single inhabitant o
0337  europe, we may doubt, if all the productions of NEW Zealand were set free in Great Britain, whethe
0337  britain may be said to be higher than those of NEW Zealand. Yet the most skilful naturalist from
0339  storations of the extinct and gigantic birds of NEW Zealand. We see it also in the birds of the ca
0341  left modified descendants, constituting the six NEW genera. The other seven species of the old gen
0341  er genera will have been the parents of the six NEW genera; the other old species and the other wh
0342  ed an important part in the first appearance of NEW forms in any one area and formation; that wide
0342  ve varied most, and have oftenest given rise to NEW species; and that varieties have at first ofte
0343  election. We can thus understand how it is that NEW species come in slowly and successively; how s
0343  ost inevitable consequence of the production of NEW forms. We can understand why when a species ha
0344  nd to leave many modified descendants, and thus NEW sub groups and groups are formed. As these are
0345  generation: old forms having been supplanted by NEW and improved forms of life, produced by the la
0346  n geographical distribution is that between the NEW and Old Worlds; yet if we travel over the vast
0346  the Old World which cannot be paralleled in the NEW, at least as closely as the same species gener
0347  rld could be pointed out hotter than any in the NEW World, yet these are not inhabited by a peculi
0347  is parallelism in the conditions of the Old and NEW Worlds, how widely different are their living
0347  f nearly all the terrestrial productions of the NEW and Old Worlds, excepting in the northern part
0350  d homes will have the best chance of seizing on NEW places, when they spread into new countries. I
0350  of seizing on new places, when they spread into NEW countries. In their new homes they will be exp
0350  , when they spread into new countries. In their NEW homes they will be exposed to new conditions,
0350  ies. In their new homes they will be exposed to NEW conditions, and will frequently undergo furthe
0351  ition with each other, migrate in a body into a NEW and afterwards isolated country, they will be
0351  come into play only by bringing organisms into NEW relations with each other, and in a lesser deg
0365  ly more than one would be so well fitted to its NEW home, as to become naturalised. But this, as I
0366  follow the changes more readily, by supposing a NEW glacial period to come slowly on, and then pas
0367  a body together on the lowlands of the Old and NEW Worlds, would be left isolated on distant moun
0369  d northern temperate productions of the Old and NEW Worlds are separated from each other by the At
0370  ial period, when the inhabitants of the Old and NEW Worlds lived further southwards than at presen
0370  d northern temperate productions of the Old and NEW Worlds, at a period anterior to the Glacial ep
0370  t these plants and animals, both in the Old and NEW Worlds, began slowly to migrate southwards as
0371  armer periods than the northern parts of the Old and NEW Worlds will have been almost continuously unit
0371  n as the species in common, which inhabited the NEW and Old Worlds, migrated south of the Polar Ci
0371  ing productions of the temperate regions of the NEW and Old Worlds, we find very few identical spe
0373  ome direct evidence of former glacial action in NEW Zealand; and the same plants, found on widely
0375  . in the admirable Introduction to the Flora of NEW Zealand, by Dr. Hooker, analogous and striking
0376  ana, that it is certainly a wonderful fact that NEW Zealand should have a closer resemblance in it
0376  o, speaks of the re appearance on the shores of NEW Zealand, Tasmania, etc., of northern forms of
0376  that twenty five species of Algae are common to NEW Zealand and to Europe, but have not been found
0377  hough they will have been placed under somewhat NEW conditions, will have suffered less. And it is
0379  ern or southern hemispheres, now exist in their NEW forms of life; and it is probable that selecte
0379  ecies have been modified and have given rise to NEW homes as well marked varieties or as distinct
0381  points so enormously remote as Kerguelen Land, NEW Zealand, and Fuegia, I believe that towards th
0382  eve, the southern shores of America, Australia, NEW Zealand have become slightly tinted by the sam
0389  e look to the large size and varied stations of NEW Zealand, extending over 780 miles of latitude,
0390  have become naturalised on it, as they have on NEW Zealand and on every other oceanic island whic
0390  occasionally arriving after long intervals in a NEW and isolated district, and having to compete w
0390  d isolated district, and having to compete with NEW associates, will be eminently liable to modifi
0391  dapted to each other; and when settled in their NEW homes, each kind will have been kept by the ot
0391  ants; in the Galapagos Islands reptiles, and in NEW Zealand gigantic wingless birds, take the plac
0393  exists on the mountains of the great island of NEW Zealand; but I suspect that this exception (if
0394  aerial mammals do occur on almost every island. NEW Zealand possesses two bats found nowhere else
0395  een modified through natural selection in their NEW homes in relation to their new position, and w
0395  lection in their new homes in relation to their NEW position, and we can understand the presence o
0397  erculum, I removed it, and when it had formed a NEW membranous one, I immersed it for fourteen day
0399  e prevailing currents, this anomaly disappears. NEW Zealand in its endemic plants is much more clo
0399  ficulty almost disappears on the view that both NEW Zealand, South America, and other southern lan
0401  degree in some analogous instances, is that the NEW species formed in the separate islands have no
0402  ncy, have spread with astonishing rapidity over NEW countries, we are apt to infer that most speci
0402  mber that the forms which become naturalised in NEW countries are not generally closely allied to
```

Page ***************************************(Key Word)***************************************

Page **(Key Word)**

0403	ubsecuently modified and better fitted to their	NEW	homes, is of the widest application throughout
0405	ch widely ranging species vary and give rise to	NEW	forms will largely determine their average ran
0405	r the conversion of its offspring, firstly into	NEW	varieties and ultimately into new species. In
0405	firstly into new varieties and ultimately into	NEW	species. In considering the wide distribution
0406	e become slightly modified in relation to their	NEW	conditions. There is, also, some reason to bel
0406	and better adaptation of the colonists to their	NEW	homes. Summary of last and present Chapters. I
0408	hsecuent mocification and the multiplication of	NEW	forms. We can thus understand the high importa
0408	g to the length of time which has elapsed since	NEW	inhabitants entered one region; according to t
0411	ultimately become converted, as I believe, into	NEW	and distinct species; and these, on the princi
0411	principle of inheritance, tend to produce other	NEW	and dominant species. Consequently the groups
0422	e had altered little, and had given rise to few	NEW	languages, whilst others (owing to the spreadi
0422	e) had altered much, and had given rise to many	NEW	languages and dialects. The various degrees of
0429	la has not added a single insect belonging to a	NEW	order; and that in the vegetable kingdom, as I
0446	me one parent species, and of which the several	NEW	epecies have become modified through natural s
0447	erited at a corresponding age, the young of the	NEW	species of our supposed genus will manifestly
0447	not have been modified. But in each individual	NEW	species, the embryonic fore limbs will differ
0461	having been disturbed by slightly different and	NEW	conditions of life, we need not feel surprise
0465	ly created there, and will be simply classed as	NEW	species. Most formations have been intermitten
0466	once come into play, does not wholly ceasei for	NEW	varieties are still occasionally produced by o
0467	only unintentionally exposes organic beings to	NEW	conditions of life, and then nature acts on th
0470	es characteristic of species of the same genus.	NEW	and improved varieties will inevitably supplan
0470	cing to the larger groups tend to give birth to	NEW	and dominant forms; so that each large group t
0475	upport the theory of descent with modification.	NEW	species have come on the stage slowly and at s
0475	selection: for old forms will be supplanted by	NEW	and improved forms. Neither single species nor
0477	it oceanic islands; and why, on the other hand,	NEW	and peculiar species of bats, which can traver
0486	tic productions will rise immensely in value. A	NEW	variety raised by man will be a far more impor
0488	od, several of these species, by migrating into	NEW	countries and coming into competition with for
0488	rtant researches. Psychology will be based on a	NEW	foundation, that of the necessary acquirement
0489	ps, which will ultimately prevail and procreate	NEW	and dominant species. As all the living forms
0370	we have good reason to believe that during the	NEWER	Pliocene period, before the Glacial epoch, an
0196	species under new conditions of life and with	NEWLY	acquired habits. To give a few instances to i
0235	est, and having transmitted by inheritance its	NEWLY	acquired economical instinct to new swarms, w
0316	ablished races of the domestic pigeon, for the	NEWLY	formed fantail would be almost sure to inheri
0074	ich destroy their combs and nests; and Mr. H.	NEWMAN,	who has long attended to the habits of humbl
0074	ery one knows, on the number of cats; and Mr.	NEWMAN	says, near villages and small towns I have fo
0393	g ago remarked that Batrachians (frogs, toads, | NEWTS)	have never been found on any of the many isl
0393	ncy. This general absence of frogs, toads, and	NEWTS	on so many oceanic islands cannot be accounte
0073	ssi and yet in the long run the forces are so	NICELY	balanced, that the face of nature remains uni
0082	of each country are struggling together with	NICELY	balanced forces, extremely slight modificatio
0083	f structure or constitution may well turn the	NICELY	balanced scale in the struggle for life, and
0257	ifference: for instance, the many species of	NICOTIANA	have been more largely crossed than the spe
0257	ised by, no less than eight other species of	NICOTIANA.	Very many analogous facts could be given.
0271	er as father or mother, and crossed with the	NICOTIANA	glutinosa, always yielded hybrids not so st
0137	dissection; having been inflammation of the	NICTITATING	membrane. As frequent inflammation of the
0194	ressed it, nature is prodigal in variety, but	NIGGARD	in innovation. Why, on the theory of Creatio
0471	see why nature is prodigal in variety, though	NIGGARD	in innovation. But why this should be a law
0091	d another hunting on marshy ground and almost	NIGHTLY	catching woodcocks or snipes. The tendency t
0218	eason for believing that although the Tachytes	NIGRA	generally makes its own burrow and stores it
0018	ufactured pottery existed in the valley of the	NILE	thirteen or fourteen thousand years ago, and w
0068	on a little plot of turf (three feet by four)	NINE	species perished from the other species being
0092	to flower, and a cross thus effected, although	NINE	tenths of the pollen were destroyed, it might
0121	n rise to new varieties and species. The other	NINE	species (marked by capital letters) of our ori
0122	which were least closely related to the other	NINE	original species, has transmitted descendants
0135	s that they cannot fly; and that of the twenty	NINE	endemic genera, no less than twenty three gene
0221	heir dead bodies as food to their nest, twenty	NINE	yards distant; but they were prevented from ge
0252	nstance, the canary bird has been crossed with	NINE	other finches; but as not one of these nine sp
0252	th nine other finches; but as not one of these	NINE	species breeds freely in confinement, we have
0270	umber of experiments made during many years on	NINE	species of Verbascum, by so good an observer a
0064	n thirty years old, and goes on breeding till	NINETY	years old, bringing forth three pair of young
0465	can only recur to the hypothesis given in the	NINTH	chapter. That the geological record is imperf
0095	ere at first urged against Sir Charles Lyell's	NOBLE	views on the modern changes of the earth, as
0251	hat many of them are perfectly fertile. Mr. C.	NOBLE,	for instance, informs me that he raises stoc
0431	etween the numerous kindred of any ancient and	NOBLE	family, even by the aid of a genealogical tre
0487	light can be thrown on ancient geography. The	NOBLE	science of Geology loses glory from the extre
0042	or food. On the other hand, cats, from their	NOCTURNAL	rambling habits, cannot be matched, and, al
0286	at almost all strata contain harder layers or	NODULES,	which from long resisting attrition form a
0307	d primordial zone. The presence of phosphatic	NODULES	and bituminous matter in some of the lowest
0013	ce of every character whatever as the rule, and	NON	inheritance as the anomaly. The laws governing
0171	eans of transition. Cases of difficulty. Natura	NON	facit saltum. Organs of small importance. Orga
0194	g of that old canon in natural history of Natura	NON	facit saltum. We meet with this admission in t
0206	ng of that old canon in natural history, Natura	NON	facit saltum. This canon if we look only to th
0210	stincts, can be discovered. The canon of Natura	NON	facit saltum applies with almost equal force t
0243	si that the canon in natural history, of natura	NON	facit saltum is applicable tc instincts as wel
0460	n nature, as is proclaimed by the canon, Natura	NON	facit saltum, that we ought to be extremely ca
0471	short and slow steps. Hence the canon of Natura	NON	facit saltum, which every fresh addition to ou
0068	me cases, as with the elephant and rhinoceros,	NONE	are destroyed by beasts of prey: even the tige
0082	hysical conditions under which they live, that	NONE	of them could anyhow be improved: for in all c
0101	occurs perhaps only at long intervals: but in	NONE,	as I suspect, can self fertilisation go on fo
0165	m has stripes on its legs and the hemionus has	NONE	and has not even a shoulder stripe, neverthele
0204	we should be very cautious in concluding that	NONE	could have existed, for the homologies of many
0208	said to be instinctive. But I could show that	NONE	of these characters of instinct are universal.
0241	ion of the parents which generated them; until	NONE	with an intermediate structure were produced.
0243	strengthen in any great degree my theory; but	NONE	of the cases of difficulty, to the best of my
0280	y intermediate between the fantail and pouteri	NONE,	for instance, combining a tail somewhat expan
0302	ists, for instance, by Agassiz, Pictet, and by	NONE	more forcibly than by Professor Sedgwick, as a
0332	l formations, for instance, above No. VI., but	NONE	from beneath this line, then only the two fami

Page **(Key Word)**

Page **************************************(Key Word)**

0417	tance of an aggregate of characters, even when	NONE are important, alone explains, I think, that si
0418	by some naturalists to be the true one; and by	NONE more clearly than by that excellent botanist,
0450	embryology, which are second in importance to	NONE in natural history, are explained on the princ
0004	best chance of making out this obscure problem.	NOR have I been disappointed; in this and in all o
0020	hout extreme care and long continued selection;	NOR can I find a single case on record of a perman
0038	that neither Australia, the Cape of Good Hope,	NOR any other region inhabited by quite uncivilise
0039	s, as it is not one of the points of the breed.	NOR let it be thought that some great deviation of
0039	velty, however slight, in one's own possession.	NOR must the value which would formerly be set on
0044	, but these I shall reserve for my future work.	NOR shall I here discuss the various definitions w
0047	ey are closely connected by intermediate links;	NOR will the commonly assumed hybrid nature of the
0067	hat the checks are in even one single instance.	NOR will this surprise any one who reflects how ig
0069	for they cannot compete with our native plants,	NOR resist destruction by our native animals. When
0070	ed with the number of birds which feed on them;	NOR can the birds, though having a super abundance
0072	rious instance of this; for here neither cattle	NOR dogs have ever run wild, though they swarm sou
0072	nce of this; for here neither cattle nor horses	NOR injurious would not be affected by natural sel
0081	ll natural Selection. Variations neither useful	NOR do I believe that any great physical change, a
0082	aving incomparably longer time at her disposal.	NOR ought we to think that the occasional destruct
0085	colour, when once acquired, true and constant,	NOR can this be thought very improbable: for we of
0091	nnate tendency to pursue certain kinds of prey.	NOR very strong, will have been neglected, and wil
0112	termediate characters, being neither very swift	NOR are they all supposed to endure for equal peri
0117	eously, but often after long intervals of time;	NOR do we know that they have subsequently become
0140	l ordinary cases we assume such to be the case;	NOR let it be supposed that no differences in the
0142	eriment cannot be said to have been even tried.	NOR can I see any insuperable difficulty in furthe
0181	of structure had been useful to its possessor.	NOR does geology at all lead to the belief that fo
0193	ould have been specially related to each other.	NOR do we always meet, as far as we can judge, wit
0202	selection will not produce absolute perfection;	NOR, as far as we can judge by our limited faculti
0206	ll not necessarily produce absolute perfection;	NOR does there seem to me any great difficulty in
0233	i have not space here to enter on this subject.	NOR would any good be derived from a single hexago
0233	not aware that any such case has been observed;	NOR fertility affords any clear distinction betwee
0248	rs. It can thus be shown that neither sterility	NOR can he tell, till he tries, whether any two sp
0265	inement or any plant seed freely under culture;	NOR do I pretend that the foregoing remarks go to
0266	eptionally resemble closely either pure parent.	NOR could select, slight differences in the reprod
0269	use and pleasure: he neither wishes to select,	NOR is it surprising that the facility of effectin
0277	ome kind between the species which are crossed.	NOR is this nearly general and perfect fertility s
0277	generally, but not quite universally, fertile.	NOR is their rarity surprising, when we remember h
0289	preservation, far better than pages of detail.	NOR have the several stages of the same formation
0293	europe is compared with the rest of the world;	NOR will the closest inspection of a formation giv
0296	ue to geographical changes requiring much time.	NOR should it be forgotten, as before explained, t
0297	er one or both forms by intermediate varieties.	NOR have we any right to suppose that the seas of
0305	oped in some one sea, might have spread widely,	NOR continental islands existed where our oceans n
0308	ozoic and secondary periods, neither continents	NOR should we be justified in assuming that if, fo
0309	ay have existed where our continents now stand.	NOR disappearance of their many now extinct specie
0313	, as Bronn has remarked, neither the appearance	NOR are those which are intermediate in character,
0334	aracter are not the oldest, or the most recent;	NOR can it be pretended that it is an immutable la
0340	ypes in each during the later tertiary periods.	NOR the dissimilarity of the inhabitants of variou
0346	hich strikes us is, that neither the similarity	NOR isolation in themselves can do anything. These
0351	e liable to modification: for neither migration	NOR do I for a moment pretend that any explanation
0354	es, now living at distant and separated points;	NOR does their almost universally volcanic composi
0358	ief of their former continuity with continents.	NOR is the direction of prevalent gales of wind. I
0364	ct: the currents of the sea are not accidental,	NOR could they be long carried in the crops or int
0364	reat length of time to the action of sea water:	NOR can the existing species, descended from A, be
0421	ended from I, also broken up into two families.	NOR is their very general fertility surprising whe
0460	el offspring cannot be considered as universal:	NOR do we know how ignorant we are. We do not know
0466	questions on which we are confessedly ignorant;	NOR ought we to marvel if all the contrivances in
0472	the naturalised productions from another land.	NOR groups of species reappear when the chain of o
0475	new and improved forms. Neither single species	NORFOLK Island, the Viti Archipelago, the Bonin Isla
0394	ses two bats found nowhere else in the world:	NORFOLK Island, the Viti Archipelago, the Bonin Isla
0021	feathers, instead of twelve or fourteen, the	NORMAL number in all members of the great pigeon fam
0143	common than the union of homologous parts in	NORMAL structures, as the union of the petals of the
0159	be considered as a variation representing the	NORMAL structure of another race, the fantail. I pre
0155	o entertain no doubt, that the more an organ	NORMALLY differs in the different species of the same
0451	and having secreted milk. So again there are	NORMALLY four developed and two rudimentary teats in
0048	areas. How many of those birds and insects in	NORTH America and Europe, which differ very slightl
0069	will suffer most. When we travel from south to	NORTH, or from a damp region to a dry, we invariabl
0132	oloured than those of the same species further	NORTH or from greater depths. Gould believes that b
0138	eneral resemblance of the other inhabitants of	NORTH America and Europe. On my view we must suppos
0141	ng free under the cold climate of Faroe in the	NORTH and of the Falklands in the south, and on man
0164	hort parellel stripes on each shoulder. In the	NORTH west part of India the Kattywar breed of hors
0164	itain to Eastern China; and from Norway in the	NORTH to the Malay Archipelago in the south. In all
0173	let us take a simple case: in travelling from	NORTH to south over a continent, we generally meet
0179	place in nature. Look at the Mustela vison of	NORTH America, which has webbed feet and which rese
0184	ch, and thus breaking them like a nuthatch. In	NORTH America the black bear was seen by Hearne swi
0184	zinc insects in the chinks of the bark? Yet in	NORTH America there are woodpeckers which feed larg
0210	been but little observed except in Europe and	NORTH America, and for no instinct being known amon
0243	how it is that the male wrens (Troglodytes) of	NORTH America, build cock nests, to roost in, like
0285	yet it is an admirable lesson to stand on the	NORTH Downs and to look at the distant South Downs;
0289	by sir C. Lyell in the carboniferous strata of	NORTH America. In regard to mammiferous remains, a
0289	tween the superimposed formations; so it is in	NORTH America, and in many other parts of the world
0294	red somewhat earlier in the palaeozoic beds of	NORTH America than in those of Europe; time having
0299	er certain sea shells inhabiting the shores of	NORTH America, which are ranked by some conchologis
0299	pelago is of about the size of Europe from the	NORTH Cape to the Mediterranean, and from Britain t
0305	have always been so freely open from south to	NORTH as they are at present. Even at this day, if
0308	sits over immense territories in Russia and in	NORTH America, do not support the view, that the ol
0308	rhood of the existing continents of Europe and	NORTH America. But we do not know what was the stat
0322	mineral chalk itself can be found; namely, in	NORTH America, in equatorial South America, in Tier
0323	ozoic formations of Russia, Western Europe and	NORTH America, a similar parallelism in the forms o

Page **************************************(Key Word)**

north

Page **(Key Word)**

```
0323 ording to Lyell, with the several European and NORTH American tertiary deposits. Even if the new f
0324 deposited at the present day on the shores of NORTH America would hereafter be liable to be class
0324 eistocene and strictly modern beds, cf Europe, NORTH and South America, and Australia, from contai
0325 his strange sequence, we turn our attention to NORTH America, and there discover a series of analo
0340 ls was formerly different from what it now is. NORTH America formerly partook strongly of the pres
0347 th america south of Lat. 35 degrees with those NORTH of 25 degrees, which consequently inhabit a c
0349 e naturalist in travelling, for instance, from NORTH to south never fails to be struck by the mann
0364 rom their course, would never bring seeds from NORTH America to Britain, though they might and do
0364 re blown across the whole Atlantic Ocean, from NORTH America to the western shores of Ireland and
0366 y recent geological period, central Europe and NORTH America suffered under an Arctic climate. The
0367 al period came on a little earlier or later in NORTH America than in Europe, so will the southern
0367 ecially related to the arctic forms living due NORTH or nearly due north of them: for the migratio
0367 he arctic forms living due north or nearly due NORTH of them: for the migration as the cold came o
0367 warmth, will generally have been due south and NORTH. The Alpine plants, for example, of Scotland,
0368 very late period have marched a little further NORTH, and subsequently have retreated to their pre
0369 me on the lower mountains and on the plains of NORTH America and Europe; and it may be reasonably
0370 rees, during the Pliocene period lived further NORTH under the polar circle, in latitude 66 degree
0371 ry little identity, between the productions of NORTH America and Europe, a relationship which is m
0372 on the eastern and western shores of temperate NORTH America; and the still more striking case of
0372 present inhabitants of the temperate lands of NORTH America and Europe, are inexplicable on the t
0374 multaneous on the eastern and western sides of NORTH America, in the Cordillera under the equator
0375 he one hand over India and on the other as far NORTH as Japan. On the southern mountains of Austra
0378 m to the tropical natives. The mountain ranges NORTH west of the Himalaya, and the long line of th
0378 tierra del Fuego and to Europe still exist in NORTH America, which must have lain on the line of
0379 allied forms have apparently migrated from the NORTH to the south, than in a reversed direction. W
0379 suspect that this preponderant migration from NORTH to south is due to the greater extent of land
0379 th is due to the greater extent of land in the NORTH, and to the northern forms having existed in
0380 rger areas and more efficient workshops of the NORTH. In many islands the native productions are n
0380 those produced within the larger areas of the NORTH; just in the same way as the productions of r
0382 o have flowed during one short period from the NORTH and from the south, and to have crossed at th
0382 but to have flowed with greater force from the NORTH so as to have freely inundated the south. As
0391 nd, which lies at about the same distance from NORTH America as the Galapagos Islands do from Sout
0391 s admirable account of Bermuda, that very many NORTH American birds; during their great annual mig
0394 ng by day far over the Atlantic Ocean; and two NORTH American species either regularly or occasion
0469 to the many representative forms in Europe and NORTH America. If then we have under nature variabi
0115 of dr. Asa Gray's Manual of the Flora of the NORTHERN United States, 260 naturalised plants are en
0142 varieties are habitually recommended for the NORTHERN, and others for the southern States; and as
0212 ferences in nests of the same species in the NORTHERN and southern United States. Fear of any part
0285 ng that at no great distance to the west the NORTHERN and southern escarpments meet and close, one
0286 f the Chalk formation. The distance from the NORTHERN to the southern Downs is about 22 miles, and
0340 losely allied, than it is at present, to the NORTHERN half. In a similar manner we know from Falco
0340 rom Falconer and Cautley's discoveries, that NORTHERN India was formerly more closely related to i
0346 ce to prove its truth: for if we exclude the NORTHERN parts where the circumpolar land is almost c
0347 of the New and Old Worlds, excepting in the NORTHERN parts, where the land almost joins, and wher
0347 there might have been free migration for the NORTHERN temperate forms, as there now is for the str
0353 ally the same at these distant points of the NORTHERN and southern hemispheres? The answer, as I b
0363 arked by Mr. H. C. Watson) from the somewhat NORTHERN character of the flora in comparison with th
0363 t they may have brought thither the seeds of NORTHERN plants. Considering that the several above m
0365 of the Alps or Pyrenees, and in the extreme NORTHERN parts of Europe; but it is far more remarkab
0366 y has the climate of Europe changed, that in NORTHERN Italy, gigantic moraines, left by old glacie
0367 d, whilst their brethren were pursuing their NORTHERN journey. Hence, when the warmth had fully re
0368 are more especially allied to the plants of NORTHERN Scandinavia; those of the United States to L
0369 but also to many sub arctic and to some few NORTHERN temperate forms, for some of these are the s
0369 y degree of uniformity of the sub arctic and NORTHERN temperate forms round the world, at the comm
0369 riod. At the present day, the sub arctic and NORTHERN temperate productions of the Old and New Wor
0369 her by the Atlantic Ocean and by the extreme NORTHERN part of the Pacific. During the Glacial peri
0370 y amount of uniformity in the sub arctic and NORTHERN temperate productions of the Old and New Wor
0371 nt time; for during these warmer periods the NORTHERN parts of the Old and New Worlds will have be
0373 ner of Australia. Looking to America; in the NORTHERN half, ice borne fragments of rock have been
0375 ntains, and on the temperate lowlands of the NORTHERN and southern hemispheres, are sometimes iden
0376 he shores of New Zealand, Tasmania, etc., of NORTHERN forms of fish. Dr. Hooker informs me that tw
0376 ropical seas. It should be observed that the NORTHERN species and forms found in the southern part
0376 l regions, are not arctic, but belong to the NORTHERN temperate zones. As Mr. H. C. Watson has rec
0376 entical, and many, though closely related to NORTHERN forms, must be ranked as distinct species. N
0378 migrated during the glacial period from the NORTHERN and southern temperate zones into the intert
0379 s their former homes; but the forms, chiefly NORTHERN, which had crossed the equator, would travel
0379 ated by inheritance to their brethren of the NORTHERN or southern hemispheres, now exist in their
0379 ater extent of land in the north, and to the NORTHERN forms having existed in their own homes in g
0379 me commingled during the Glacial period, the NORTHERN forms were enabled to beat the less powerful
0380 ties of the allied species which live in the NORTHERN and southern temperate zones and on the moun
0381 inclined to look in the southern, as in the NORTHERN hemisphere, to a former and warmer period, b
0477 of some of the inhabitants of the sea in the NORTHERN and southern temperate zones, though separat
0069 this one being hurt. So it is when we travel NORTHWARD, but in a somewhat lesser degree, for the n
0069 titors, decreases northwards; hence in going NORTHWARD, or in ascending a mountain, we far oftener
0072 er run wild, though they swarm southward and NORTHWARD in a feral state; and Azara and Rengger hav
0348 . so that here three marine faunas range far NORTHWARD and southward, in parallel lines not far fr
0349 one species of Rhea (American ostrich), and NORTHWARD the plains of La Plata by another species o
0367 mth returned, the arctic forms would retreat NORTHWARD, closely followed up in their retreat by th
0368 eir long southern migration and re migration NORTHWARD, will have been exposed to nearly the same
0378 ad not reached the equator, would re migrate NORTHWARD or southward towards their former homes; bu
0379 eir long southern migration and re migration NORTHWARD, the case may have been wholly different wi
0045 s; or the thicker fur of an animal from far NORTHWARDS, would not in some cases be inherited for a
0069 ds, and therefore of competitors, decreases NORTHWARDS; hence in going northward, or in ascending
0164 tries from Britain to Eastern China; and from NORWAY in the north to the Malay Archipelago in the
0212 nd the magpie, so wary in England, is tame in NORWAY, as is the hooded crow in Egypt. That the gen
0049 d grouse as only a strongly marked race of a NORWEGIAN species, whereas the greater number rank it
```

Page **(Key Word)**

norwegian

Page **(Key Word)**

```
0445  tails) of the beak, width of mouth, length of NOSTRIL and of eyelid, size of feet and length of le
0021  eyelids, very large external orifices to the NOSTRILS, and a wide gape of mouth. The short faced t
0022  length of the eyelids, of the orifice of the NOSTRILS, of the tongue (not always in strict correla
0416  her or not there is an open passage from the NOSTRILS to the mouth, the only character, according
0012  e dahlia, etc.; and it is really surprising to NOTE the endless points in structure and constituti
0232  etween these spheres. It was really curious to NOTE in cases of difficulty, as when two pieces of
0291  rogress of Geology, and have been surprised to NOTE how author after author, in treating of this o
0336  , which includes the whole glacial period, and NOTE how little the specific forms of the inhabitan
0112  differences would become greater, and would be NOTED as forming two sub breeds; finally, after the
0294  quarters of the world, it has everywhere been NOTED, that some few still existing species are com
0001  eculate on the subject, and drew up some short NOTES; these I enlarged in 1844 into a sketch of th
0349  m closely allied, yet distinct kinds of birds, NOTES nearly similar, and sees their nests similarl
0008  tion of any change in the conditions of life. NOTHING is more easy than to tame an animal, and few
0015  e, in such case, I grant that we could deduce NOTHING from domestic varieties in regard to species
0040  sometimes been noticed, namely, that we know NOTHING about the origin or history of any of our do
0041  ach individual. Unless such attention be paid NOTHING can be effected. I have seen it gravely rema
0050  specific, and what as varieties; for he knows NOTHING of the amount and kind of variation to which
0062  result of his great horticultural knowledge. NOTHING is easier than to admit in words the truth o
0071  effect of the introduction of a single tree, NOTHING whatever else having been cone, with the exc
0082  variations do occur, natural selection can do NOTHING. Not that, as I believe, any extreme amount
0083  external and visible characters: nature cares NOTHING for appearances, except in so far as they ma
0084  anic and inorganic conditions of life. We see NOTHING of those slow changes in progress, until the
0108  f the other inhabitants being thus disturbed. NOTHING can be effected, unless favourable variation
0113  vary, for otherwise natural selection can do NOTHING. So it will be with plants. It has been expe
0143  ; this is often seen in monstrous plants; and NOTHING is more common than the union of homologous
0160  ly coloured breeds; and in this case there is NOTHING in the external conditions of life to cause
0177  ry slow process, and natural selection can do NOTHING until favourable variations chance to occur,
0180  in the same species. And it seems to me that NOTHING less than a long list of such cases is suffi
0187  the eye, and from fossil species we can learn NOTHING on this head. In this great class we should
0191  n from an ancient prototype, of which we know NOTHING, furnished with a floating apparatus or swim
0205  ggle for life. Natural selection will produce NOTHING in one species for the exclusive good or inj
0207  my whole theory. I must premise, that I have NOTHING to do with the origin of the primary mental
0216  ges; in other cases compulsory habit has done NOTHING, and all has been the result of selection, p
0219  and pupae to stimulate them to work, they did NOTHING; they could not even feed themselves, and ma
0283  believe that pure water can effect little of NOTHING in wearing away rock. At last the base of th
0285  e of 12,000 feet; yet in these cases there is NOTHING on the surface to show such prodigious movem
0340  to the more dominant forms, and there will be NOTHING immutable in the laws of past and present di
0342  s, preserved in our museums, is absolutely as NOTHING compared with the incalculable number of gen
0398  ffinity to those created in America? There is NOTHING in the conditions of life, in the geological
0413  the plan of the Creator, it seems to me that NOTHING is thus added to our knowledge. Such express
0414  be of very high importance in classification. NOTHING can be more false. No one regards the extern
0419  n plainly influenced by chains of affinities. NOTHING can be easier than to define a number of cha
0435  taceans. So it is with the flowers of plants. NOTHING can be more hopeless than to attempt to expl
0442  are completed; and again in spiders, there is NOTHING worthy to be called a metamorphosis. The lar
0451  ctec in the beaks of certain embryonic birds. NOTHING can be plainer than that wings are formed fo
0452  he same part in different members of a class, NOTHING is more common, or more necessary, than the
0459  endeavoured to give to them their full force. NOTHING at first can appear more difficult to believ
0464  specimens in all our museums is absolutely as NOTHING compared with the countless generations of c
0032  ple would be so obvious as hardly to be worth NOTICE; but its importance consists in the great eff
0056  ra and their recorded varieties which deserve NOTICE. We have seen that there is no infallible cri
0058  re is one other point which seems to me worth NOTICE. Varieties generally have much restricted ran
0153  undergoing modification. It further deserves NOTICE that these variable characters, produced by m
0225  an irregular mass. But the important point to NOTICE, is that these cells are always made at that
0251  made with scientific precision, deserve some NOTICE. It is notorious in how complicated a manner
0270  are distinct species; and it is important to NOTICE that the hybrid plants thus raised were thems
0292  onclusion. One remark is here worth a passing NOTICE. During periods of elevation the area of the
0298  ing chapter. One other consideration is worth NOTICE: with animals and plants that can propagate r
0466  doubt their weight. But it deserves especial NOTICE that the more important objections relate to
0040  views further explain what has sometimes been NOTICED, namely, that we know nothing about the orig
0155  variable than generic; but I have repeatedly NOTICED in works on natural history, that when an au
0175  ne species disappears. The same fact has been NOTICED by Forbes in sounding the depths of the sea
0229  same rate from the opposite sides; for I have NOTICED half completed rhombs at the base of a just
0368  the same climate, and, as is especially to be NOTICED, they will have kept in a body together; con
0373  46 degrees; erratic boulders have, also, been NOTICED on the Rocky Mountains. In the Cordillera of
0208  parison gives, I think, a remarkably accurate NOTION of the frame of mind under which an instincti
0286  done; but we may, in order to form some crude NOTION on the subject, assume that the sea would eat
0287  se it is highly important for us to gain some NOTION, however imperfect, of the lapse of years. Du
0317  ms are intimately connected together. The old NOTION of all the inhabitants of the earth having be
0139  ffected during long continued descent. It is NOTORIOUS that each species is adapted to the climate
0154  uded in these remarks may be extended. It is NOTORIOUS that specific characters are more variable
0251  ntific precision, deserve some notice. It is NOTORIOUS in how complicated a manner the species of
0252  ves to be the case, the fact would have been NOTORIOUS to nurserymen. Horticulturists raise large
0272  amount of variability in their offspring is NOTORIOUS; but some few cases both of hybrids and mon
0297  to rank them all as distinct species. It is NOTORIOUS on what excessively slight differences many
0397  lls of the islands of the Pacific. Now it is NOTORIOUS that land shells are very easily killed by
0411  ase is widely different in nature; for it is NOTORIOUS how commonly members of even the same sub g
0443  dence rather points the other way; for it is NOTORIOUS that breeders of cattle, horses, and variou
0152  reeds, as in the short faced tumbler, it is NOTORIOUSLY difficult to breed them nearly to perfecti
0351  ed to the same areas, as is so commonly and NOTORIOUSLY the case. I believe, as was remarked in th
0422  have given a natural arrangement; and it is NOTORIOUSLY not possible to represent in a series, on
0059  ittle, are generally ranked as varieties, NOTWITHSTANDING that intermediate linking forms have not
0102  om this unconscious process of selection, NOTWITHSTANDING a large amount of crossing with inferior
0138  r palpi, as a compensation for blindness, NOTWITHSTANDING such modifications, we might expect stil
0191  with some risk of falling into the lungs, NOTWITHSTANDING the beautiful contrivance by which the g
0249  pollen of the same kind, their fertility, NOTWITHSTANDING the frequent ill effects of manipulation
0250  ouch quite sterile with their own pollen, NOTWITHSTANDING that their own pollen was found to be pe
```

Page **(Key Word)**

Page **************************************(Key Word)**************************************

```
0347  t inhabited by a peculiar fauna or flora. NOTWITHSTANDING this parallelism in the conditions of th
0389  e all proceeded from a common birthplace, NOTWITHSTANDING that in the course of time they have com
0396  f batrachians, and of terrestrial mammals NOTWITHSTANDING the presence of aerial bats, the singula
0264  and therefore before birth, as long as it is NOURISHED within its mother's womb or within the egg
0440  lar conditions, in the young mammal which is NOURISHED in the womb of its mother, in the egg of th
0443  r's womb, or in the egg, or as long as it is NOURISHED and protected by its parent, it must be qui
0003  the case of the misseltoe, which draws its NOURISHMENT from certain trees, which has seeds that m
0145  he development of the ray petals by drawing NOURISHMENT from certain other parts of the flower had
0147  in extent with our domestic productions: if NOURISHMENT flows to one part or organ in excess, it r
0168  largely developed, perhaps it tends to draw NOURISHMENT from the adjoining parts; and every part o
0483  ere they created bearing the false marks of NOURISHMENT from the mother's womb? Although naturalis
0296  on found carboniferous beds 1400 feet thick in NOVA Scotia, with ancient root bearing strata, one
0039  ences, and it is in human nature to value any NOVELTY, however slight, in one's own possession. No
0027  e unknown in a wild state, and their becoming NOWHERE feral: these species having very abnormal ch
0105  demic, that is, have been produced there, and NOWHERE else. Hence an oceanic island at first sight
0289  tive formations, we may infer that this could NOWHERE be ascertained. The frequent and great chang
0390  oportion of endemic species (i.e. those found NOWHERE else in the world) is often extremely large.
0394  island. New Zealand possesses two bats found NOWHERE else in the world: Norfolk Island, the Viti
0398  een created in the Galapagos Archipelago, and NOWHERE else, bear so plain a stamp of affinity to t
0393  itius, and have multiplied so as to become a NUISANCE. But as these animals and their spawn are kn
0003  , i presume, say that after a certain unknown NUMBER of generations, some bird had given birth to
0012  not inherited is unimportant for us. But the NUMBER and diversity of inheritable deviations of st
0015  d esculent vegetables, for an almost infinite NUMBER of generations, would be opposed to all exper
0017  bt that if other animals and plants, equal in NUMBER to our domesticated productions, and belongin
0017  ture, and could be made to breed for an equal NUMBER of generations under domestication, they woul
0019  , as these several countries do not possess a NUMBER of peculiar species as distinct parent stocks
0021  rs, instead of twelve or fourteen, the normal NUMBER in all members of the great pigeon family; an
0022  aw, varies in a highly remarkable manner. The NUMBER of the caudal and sacral vertebrae vary; as d
0022  caudal and sacral vertebrae vary; as does the NUMBER of the ribs, together with their relative bre
0022  evelopment and abortion of the oil gland; the NUMBER of the primary wing and caudal feathers; the
0022  e relative length of leg and of the feet; the NUMBER of scutellae on the toes, the development of
0023  domestic breeds by the crossing of any lesser NUMBER: how, for instance, could a pouter be produce
0026  ts to some one ancestor, removed by a greater NUMBER of generations. In a breed which has been cro
0026  be transmitted undiminished for an indefinite NUMBER of generations. These two distinct cases are
0027  ; and that it agrees in habits and in a great NUMBER of points of structure with all the domestic
0027  the shortness of that of the tumbler, and the NUMBER of tail feathers in the fantail, are in each
0037  believe, the well known fact, that in a vast NUMBER of cases we cannot recognise, and therefore d
0041  appearance will be much increased by a large NUMBER of individuals being kept; and hence this com
0041  nd valuable varieties. The keeping of a large NUMBER of individuals of a species in any country re
0045  erienced naturalist would be surprised at the NUMBER of the cases of variability, even in importan
0048  ifferent botanists, and see what a surprising NUMBER of forms have been ranked by one botanist as
0049  e of a Norwegian species; whereas the greater NUMBER rank it as an undoubted species peculiar to G
0050  est known countries that we find the greatest NUMBER of forms of doubtful value. I have been struc
0051  f difficulty; for he will encounter a greater NUMBER of closely allied forms. But if his observati
0054  r genera on the other side, a somewhat larger NUMBER of the very common and much diffused or domin
0054  including many species, a large proportional NUMBER of dominant species. But so many causes tend
0055  arieties, invariably present a larger average NUMBER of varieties than do the species of the small
0056  n formed, the species of that genus present a NUMBER of varieties, that is of incipient species, b
0056  varying much, and are thus increasing in the NUMBER of their species, or that no small genera are
0057  e said, that in the larger genera, in which a NUMBER of varieties or incipient species greater tha
0059  inite. In genera having more than the average NUMBER of species in any country, the species of the
0059  es of these genera have more than the average NUMBER of varieties. In large genera the species are
0060  rsal. Effects of climate. Protection from the NUMBER of individuals. Complex relations of all anim
0061  cies which are periodically born; but a small NUMBER can survive. I have called this principle, by
0065  food, and that in a state of nature an equal NUMBER would have somehow to be disposed of. The onl
0066  ecies can be supported in a district. A large NUMBER of eggs is of some importance to those specie
0066  od, for it allows them rapidly to increase in NUMBER. But the real importance of a large number of
0066  in number. But the real importance of a large NUMBER of eggs or seeds is to make up for much destr
0066  ny way protect its own eggs or young, a small NUMBER may be produced, and yet the average stock be
0066  extinct. It would suffice to keep up the full NUMBER of a tree, which lived on an average for a th
0066  ting place. So that in all cases, the average NUMBER of any animal or plant depends only indirectl
0066  nimal or plant depends only indirectly on the NUMBER of its eggs or seeds. In looking at Nature, i
0067  igate the destruction ever so little, and the NUMBER of the species will almost instantaneously in
0067  tural tendency of each species to increase in NUMBER is most obscure. Look at the most vigorous sp
0069  ard; but in a somewhat lesser degree, for the NUMBER of species of all kinds, and therefore of com
0069  species, we may clearly see in the prodigious NUMBER of plants in our gardens which can perfectly
0070  e seeds are in great excess compared with the NUMBER of birds which feed on them: nor can the bird
0070  dance of food at this one season, increase in NUMBER proportionally to the supply of seed, as thei
0072  have shown that this is caused by the greater NUMBER in Paraguay of a certain fly, which lays its
0074  ld become very rare, or wholly disappear. The NUMBER of humble bees in any district depends in a g
0074  any district depends in a great degree on the NUMBER of field mice, which destroy their combs and
0074  are thus destroyed all over England. Now the NUMBER of mice is largely dependent, as every one kn
0074  largely dependent, as every one knows, on the NUMBER of cats; and Mr. Newman says, near villages a
0074  rous than elsewhere, which I attribute to the NUMBER of cats that destroy the mice. Hence it is qu
0074  but all concurring in determining the average NUMBER or even the existence of the species. In some
0078  to give the plant the power of increasing in NUMBER, we should have to give it some advantage ove
0080  selection, namely, intercrossing, isolation, NUMBER of individuals. Slow action. Extinction cause
0080  y. Let it be borne in mind in what an endless NUMBER of strange peculiarities our domestic product
0096  irth. Modern research has much diminished the NUMBER of supposed hermaphrodites, and of real herma
0096  phrodites; ard of real hermaphrodites a large NUMBER pair; that is, two individuals regularly unit
0099  e plant. How, then, comes it that such a vast NUMBER of the seedlings are mongrelized? I suspect t
0102  ual differences suffice for the work. A large NUMBER of individuals, by giving a better chance for
0105  very peculiar physical conditions, the total NUMBER of the individuals supported on it will neces
0105  such as an oceanic island, although the total NUMBER of the species inhabiting it, will be found t
0105  favourable variations arising from the large NUMBER of individuals of the same species there supp
0106  of life are infinitely complex from the large NUMBER of already existing species; and if some of t
```

Page **************************************(Key Word)**************************************

Page ***************************************(Key Word)**

```
0109   each selected and favoured form increases in NUMBER, so will the less favoured forms decrease and
0109   during fluctuations in the seasons or in the NUMBER of its enemies, run a good chance of utter ex
0109   ly being produced, unless we believe that the NUMBER of specific forms goes on perpetually and alm
0109   bers inevitably must become extinct. That the NUMBER of specific forms has not indefinitely increa
0109   they should not have thus increased, for the  NUMBER of places in the polity of nature is not inde
0110   the common species which afford the greatest  NUMBER of recorded varieties, or incipient species.
0113   case of a carnivorous quadruped, of which the NUMBER that can be supported in any country has long
0113   several distinct genera of grasses, a greater NUMBER of plants and a greater weight of dry herbage
0113   of grasses differ from each other, a greater  NUMBER of individual plants of this species of grass
0115   in by naturalisation, proportionally with the NUMBER of the native genera and species, far more in
0116   r different habits of life, so will a greater  NUMBER of individuals be capable of there supporting
0117   species of the large genera present a greater NUMBER of varieties. We have, also, seen that the sp
0118   arent (A), will generally go on increasing in NUMBER and diverging in character. In the diagram th
0119   ife, they will generally go on multiplying in NUMBER as well as diverging in character: this is re
0119   confined to a single line of descent, and the NUMBER of the descendants will not be increased; alt
0120   by continuing the same process for a greater  NUMBER of generations (as shown in the diagram in a
0121   , will generally tend to produce the greatest NUMBER of modified descendants; for these will have
0122   enus. Their modified descendants, fourteen in NUMBER at the fourteen thousandth generation, will p
0123   iginal eleven species, will now be fifteen in NUMBER. Owing to the divergent tendency of natural s
0125   r groups, which are all trying to increase in NUMBER. One large group will slowly conquer another
0127   vigorous and best adapted males the greatest  NUMBER of offspring. Sexual selection will also give
0136   the wings, would depend on whether a greater  NUMBER of individuals were saved by successfully bat
0140   d plant will endure our climate, and from the NUMBER of plants and animals brought from warmer cou
0149   , and the stamens in polyandrous flowers) the NUMBER is variable; whereas the number of the same p
0149   flowers) the number is variable; whereas the NUMBER of the same part or organ, when it occurs in
0157   e relation can hardly be accidental. The same NUMBER of joints in the tarsi is a character general
0157   in the Engidae, as Westwood has remarked, the NUMBER varies greatly; and the number likewise diffe
0157   remarked, the number varies greatly; and the NUMBER likewise differs in the two sexes of the same
0160   e to the contrary, transmitted for almost any NUMBER of generations. When a character which has be
0160   been lost in a breed, reappears after a great NUMBER of generations, the most probable hypothesis
0161   any character being inherited for an endless  NUMBER of generations, than in quite useless or rudi
0162   he blueness was a case of reversion, from the NUMBER of the markings, which are correlated with th
0168   to understand. Multiple parts are variable in NUMBER and in structure, perhaps arising from such p
0171   but, to the best of my judgment, the greater  NUMBER are only apparent, and those that are real ar
0175   ned numbers, will, during fluctuations in the NUMBER of its enemies or of its prey, or in the seas
0177   d chapter, presenting on an average a greater NUMBER of well marked varieties than do the rarer sp
0188   rustaceans, and bearing in mind how small the NUMBER of living animals is in proportion to those w
0209   ost serious error to suppose that the greater NUMBER of instincts have been acquired by habit in o
0213   e mental qualities of our domestic animals. A NUMBER of curious and authentic instances could be g
0218   ed for by the fact of the hens laying a large NUMBER of eggs; but as in the case of the cuckoo, at
0218   s not as yet been perfected: for a surprising NUMBER of eggs lie strewed over the plains, so that
0220   ring these months, the slaves are very few in NUMBER, I thought that they might behave differently
0222   nother community of F. sanguinea, and found a NUMBER of these ants entering their nest, carrying t
0226   nd tells me that it is strictly correct: If a NUMBER of equal spheres be described with their cent
0234   the hive is known mainly to depend on a large NUMBER of bees being supported. Hence the saving of
0234   of any species of bee may be dependent on the NUMBER of its parasites or other enemies, or on quit
0236   t had been profitable to the community that a NUMBER should have been annually born capable of wor
0241   red wonderfully in shape, and in the form and NUMBER of the teeth. But the important fact for us i
0247   of sterility. He always compares the maximum  NUMBER of seeds produced by two species when crossed
0247   d by their hybrid offspring, with the average NUMBER produced by both pure parent species in a sta
0255   ne species, yields a perfect gradation in the NUMBER of seeds produced, up to nearly complete or e
0256   lised hybrids producing a greater and greater NUMBER of seeds up to perfect fertility. Hybrids fro
0270   dible; but it is the result of an astonishing NUMBER of experiments made during many years on nine
0277   nature: and when we remember that the greater NUMBER of varieties have been produced under domesti
0279   e of extinct intermediate varieties; on their NUMBER. On the vast lapse of time, as inferred from
0280   n has acted on an enormous scale, so must the NUMBER of intermediate varieties, which have formerl
0282   mon ancestor of each great class. So that the NUMBER of intermediate and transitional links, betwe
0287   ed by hosts of living forms. What an infinite NUMBER of generations, which the mind cannot grasp,
0292   nd, during subsidence, the inhabited area and NUMBER of inhabitants will decrease (excepting the p
0293   ch formation has indisputably required a vast NUMBER of years for its deposition. I can see severa
0301   ind in our geological formations, an infinite NUMBER of those fine transitional forms, which on my
0304   sile Cirripedes, I have stated that, from the NUMBER of existing and extinct tertiary species; fro
0306   n point in Australia, and then to discuss the NUMBER and range of its productions. On the sudden a
0316   the general rule being a gradual increase in  NUMBER, till the group reaches its maximum, and then
0317   oner or later, it gradually decreases. If the NUMBER of the species of a genus, or the number of t
0317   the number of the species of a genus, or the NUMBER of the genera of a family, be represented by
0317   tion of the species. This gradual increase in NUMBER of the species of a group is strictly conform
0317   ocess of modification and the production of a NUMBER of allied forms must be slow and gradual, one
0319   rarity; for rarity is the attribute of a vast NUMBER of species of all classes, in all countries.
0320   together. In certain flourishing groups, the NUMBER of new specific forms have been produce
0320   have been exterminated; but we know that the  NUMBER of species has not gone on indefinitely incre
0320   s has caused the extinction of about the same NUMBER of old forms. The competition will generally
0321   iable to extermination. Thus, as I believe, a NUMBER of new species descended from one species, th
0325   widely diffused, having produced the greatest NUMBER of new varieties. It is also natural that the
0326   cies; but we can, I think, clearly see that a NUMBER of individuals, from giving a better chance o
0330   s would be distinguished by a somewhat lesser NUMBER of characters, so that the two groups, though
0333   y period marked VI., would differ by a lesser NUMBER of characters; for at this early stage of, des
0337   ee in Great Britain, whether any considerable NUMBER would be enabled to seize or places now occup
0341   allied or representative genera with the same NUMBER of species, then we may conclude that only on
0342   largely preserved in a fossil state; that the NUMBER of both of specimens and of species, preserved i
0342   ely as nothing compared with the incalculable NUMBER of generations which must have passed away ev
0350   eriods more or less remote; on the nature and NUMBER of the former immigrants; and on their action
0351   l be no uniform quantity. If, for instance, a NUMBER of species, which stand in direct competition
0353   milies, very many genera, and a still greater NUMBER of sections of genera are confined to a singl
0363   ons to another. In the Azores, from the large NUMBER of the species of plants common to Europe, as
0370   d, such as the older Pliocene period, a large NUMBER of the same plants and animals inhabited the
```

Page ***************************************(Key Word)**

Page **(Key Word)**

```
0377  firm front against intruders, that a certain  NUMBER of the more vigorous and dominant temperate f
0378  all the flowering plants, about forty six in   NUMBER, common to Tierra del Fuego and to Europe sti
0378  by hooker. Thus, as I believe, a considerable  NUMBER of plants, a few terrestrial animals, and som
0387  ere of many kinds, and were altogether 537 in  NUMBER; and yet the viscid mud was all contained in
0388  r few, already occupying any pond, yet as the  NUMBER of kinds is small, compared with those on the
0389  inds which inhabit oceanic islands are few in  NUMBER compared with those on equal continental area
0389  and compare its flowering plants, only 750 in  NUMBER, with those on an equal area at the Cape of G
0389  onditions has caused so great a difference in  NUMBER. Even the uniform county of Cambridge has 847
0390  species, will have to admit that a sufficient  NUMBER of the best adapted plants and animals have n
0390  n has nature. Although in oceanic islands the  NUMBER of kinds of inhabitants is scanty, the propor
0390  emely large. If we compare, for instance, the  NUMBER of the endemic land shells in Madeira, or of
0390  birds in the Galapagos Archipelago, with the  NUMBER found on any continent, and then compare the
0391  . madeira, again, is inhabited by a wonderful  NUMBER of peculiar land shells, whereas not one spec
0403  inuous land, yet they are inhabited by a vast  NUMBER of distinct mammals, birds, and plants. The p
0409  ld have few inhabitants, but of these a great  NUMBER should be endemic or peculiar; and why, in re
0412  tendency in the forms which are increasing in  NUMBER and diverging in character; to supplant and e
0416  are of unequal physiological importance. Any   NUMBER of instances could be given of the varying im
0417  r, as A. de Jussieu has remarked, the greater  NUMBER of the characters proper to the species, to t
0417  graded flowers, departing so wonderfully in a  NUMBER of the most important points of structure fro
0418  aracter nearly uniform, and common to a great  NUMBER of forms, and not common to others, they use
0418  s one of high value; if common to some lesser  NUMBER, they use it as of subordinated value. This p
0419  ities. Nothing can be easier than to define a  NUMBER and position of the embryonic leaves or cotyl
0423  ms together which were allied in the greatest  NUMBER of characters common to all birds; but in the
0426  e, and is often done, as long as a sufficient  NUMBER of points. In tumbler pigeons, though some su
0429  . as showing how few the higher groups are in  NUMBER, and how widely spread they are throughout th
0429  aberrant any form is, the greater must be the  NUMBER of connecting forms which on my theory have b
0436  ll are homologous with, that is correspond in  NUMBER and in relative connexion with, the elemental
0436  nexion with, the elemental parts of a certain  NUMBER of vertebrae. The anterior and posterior limb
0437  imes repeated are eminently liable to vary in  NUMBER and structure; consequently it is quite proba
0438  modification, should have seized on a certain  NUMBER of the primordially similar elements, many ti
0462  e theory of natural selection an interminable  NUMBER of intermediate forms must have existed, link
0464  rly inappreciable by the human intellect. The  NUMBER of specimens in all our museums is absolutely
0464  in a fossil condition, at least in any great   NUMBER. Widely ranging species vary most, and variet
0466  ntinue to be inherited for an almost infinite  NUMBER of generations. On the other hand we have evi
0467  e, which variety or species shall increase in  NUMBER, and which shall decrease, or finally become
0470  f the larger genera, which afford the greater  NUMBER of varieties or incipient species, retain to
0470  o of reproduction to increase inordinately in  NUMBER; and as the modified descendants of each spec
0472  each species constantly trying to increase in  NUMBER, with natural selection always ready to adapt
0479  the porpoise, and leg of the horse, the same  NUMBER of vertebrae forming the neck of the giraffe
0481  ations, accumulated during an almost infinite  NUMBER of generations. Although I am fully convinced
0482  iculties than to the explanation of a certain  NUMBER of facts will certainly reject my theory. A f
0484  ogenitors, and plants from an equal or lesser  NUMBER. Analogy would lead me one step further, name
0488  a fair measure of the lapse of actual time. A  NUMBER of species, however, keeping in a body might
0489  ic beings are grouped, shows that the greater  NUMBER of species of each genus, and all the species
0117  zontal lines, and is there marked by a small   NUMBERED letter, a sufficient amount of variation is
0119  sion is broken at regular intervals by small   NUMBERED letters marking the successive forms which h
0331  the fourth chapter. We may suppose that the   NUMBERED letters represent genera, and the dotted lin
0138  the time that an animal had reached, after   NUMBERLESS generations, the deepest recesses, disuse w
0179  ime, but to all time, if my theory be true,   NUMBERLESS intermediate varieties, linking most closel
0342  heory. For he may ask in vain where are the   NUMBERLESS transitional links which must formerly have
0042  igeons, I may add, can be propagated in great  NUMBERS and at a very quick rate, and inferior birds
0052  a variety were to flourish so as to exceed in  NUMBERS the parent species, it would then rank as th
0053  and the tables themselves of the proportional  NUMBERS of the varying species. Dr. hooker permits m
0063  on the principle of geometrical increase, its  NUMBERS would quickly become so inordinately great t
0064  y be now increasing, more or less rapidly, in  NUMBERS, all cannot do so, for the world would not h
0066  d to be striving to the utmost to increase in  NUMBERS; that each lives by a struggle at some perio
0067  vigorous species; by as much as it swarms in  NUMBERS, by so much will its tendency to increase be
0067  lants. Seedlings, also, are destroyed in vast  NUMBERS by various enemies; for instance, on a piece
0068  o other animals, which determines the average  NUMBERS of a species. Thus, there seems to be little
0068  an important part in determining the average  NUMBERS of a species, and periodical seasons of extr
0069  ight change of climate, they will increase in  NUMBERS, and, as each area is already fully stocked
0069  vel southward and see a species decreasing in  NUMBERS, we may feel sure that the cause lies quite
0070  able circumstances, increases inordinately in  NUMBERS in a small tract, epidemics, at least, this
0070  iduals of the same species, relatively to the  NUMBERS of its enemies, is absolutely necessary for
0070  roportionally to the supply of seed, as their  NUMBERS are checked during winter; but any one who h
0071  nt soil to another: not only the proportional  NUMBERS of the heath plants were wholly changed, but
0072  or planted, I was so much surprised at their  NUMBERS that I went to several points of view, whenc
0072  hence, if certain insectivorous birds (whose  NUMBERS are probably regulated by hawks or beasts of
0074  that the presence of a feline animal in large  NUMBERS in a district might determine, through the i
0074  e are tempted to attribute their proportional  NUMBERS and kinds to what we call chance. But how fa
0075  in the course of centuries, the proportional  NUMBERS and kinds of trees now growing on the old In
0076  se the weaker kinds will steadily decrease in  NUMBERS and disappear. So again with the varieties o
0077  ange, why does it not double or quadruple its  NUMBERS? We know that it can perfectly well withstan
0078  manner. If we wished to increase its average  NUMBERS in its new home, we should have to modify it
0081  e, for instance, of climate. The proportional  NUMBERS of its inhabitants would almost immediately
0084  d of their lives, would increase in countless  NUMBERS; they are known to suffer largely from birds
0090  d from any change in the country increased in  NUMBERS, or that other prey had decreased in numbers
0090  numbers, or that other prey had decreased in  NUMBERS, during that season of the year when the wol
0091  . even without any change in the proportional  NUMBERS of the animals on which our wolf preyed, a c
0108  improved forms, and the relative proportional  NUMBERS of the various inhabitants of the renewed co
0109  rpetually and almost indefinitely increasing,  NUMBERS inevitably must become extinct. That the num
0112  y of nature, and so be enabled to increase in  NUMBERS. We can clearly see this in the case of anim
0113  said, is striving its utmost to increase its  NUMBERS. Consequently, I cannot doubt that in the co
0114  est chance of succeeding and of increasing in  NUMBERS, and thus of supplanting the less distinct v
0125  lowly conquer another large group, reduce its  NUMBERS, and thus lessen its chance of further varia
```

Page **(Key Word)**

Page **(Key Word)**

```
0128  essant struggle of all species to increase in NUMBERS, the more diversified these descendants bec
0149  same part or organ, when it occurs in lesser NUMBERS, is constant. The same author and some botan
0172  why do we not find them embedded in countless NUMBERS in the crust of the earth? It will be much r
0175  n its metropolis, would increase immensely in NUMBERS, were it not for other competing species; th
0175  nes of its range, where it exists in lessened NUMBERS, will, during fluctuations in the number of
0176  e variety, consequently, will exist in lesser NUMBERS from inhabiting a narrow and lesser area; an
0176  ies together have generally existed in lesser NUMBERS than the forms which they connect, then, I t
0176  ked together. For any form existing in lesser NUMBERS would, as already remarked, run a greater ch
0176  being exterminated than one existing in large NUMBERS; and in this particular case the intermediat
0176  stinct species, the two which exist in larger NUMBERS from inhabiting larger areas, will have a gr
0177  intermediate variety, which exists in smaller NUMBERS in a narrow and intermediate zone. For forms
0177  termediate zone. For forms existing in larger NUMBERS will always have a better chance, within any
0177  an will the rarer forms which exist in lesser NUMBERS. Hence, the more common forms, in the race f
0177  o breeds, which originally existed in greater NUMBERS, will come into close contact with each othe
0178  s), exist in the intermediate zones in lesser NUMBERS than the varieties which they tend to connec
0179  y connect; for these from existing in greater NUMBERS will, in the aggregate, present more variati
0181  e at least of the squirrels would decrease in NUMBERS or become exterminated, unless they also bec
0183  ve been developed at an early period in great NUMBERS and under many subordinate forms. Thus, to m
0183  be less, from their having existed in lesser NUMBERS, than in the case of species with fully deve
0186  ing is constantly endeavouring to increase in NUMBERS; and that if any one being vary ever so litt
0203  rmediate variety will usually exist in lesser NUMBERS than the two forms which it connects; consec
0204  urther modification, from existing in greater NUMBERS, will have a great advantage over the less r
0220  ever seen the slaves, though present in large NUMBERS in August, either leave or enter the nest. M
0221  erely on the slaves being captured in greater NUMBERS in Switzerland than in England. One day I fo
0234  ned, as it probably often does determine, the NUMBERS of a humble bee which could exist in a count
0239  with those of an intermediate size scanty in NUMBERS, Formica flava has larger and smaller worker
0241  , having been produced in greater and greater NUMBERS through the natural selection of the parents
0249  re seldom raised by experimentalists in great NUMBERS; and as the parent species, or other allied
0279  termediate varieties, from existing in lesser NUMBERS than the forms which they connect, will gene
0287  forbes, should not be forgotten, namely, that NUMBERS of our fossil species are known and named fr
0288  coat the rocks all over the world in infinite NUMBERS: they are all strictly littoral, with the ex
0306  much graver. I allude to the manner in which NUMBERS of species of the same group, suddenly appea
0318  ch marks the first appearance and increase in NUMBERS of the species. In some cases, however, the
0318  d over the whole country and has increased in NUMBERS at an unparalleled rate, I asked myself what
0328  he finds in both a curious accordance in the NUMBERS of the species belonging to the same genera,
0339  t of the fossil mammals, buried there in such NUMBERS, are related to South American types. This r
0341  rs, with the genera and species decreasing in NUMBERS, as apparently is the case of the Edentata o
0343  ever reappears. Groups of species increase in NUMBERS slowly, and endure for unequal periods of ti
0379  having existed in their own homes in greater NUMBERS, and having consequently been advanced throu
0380  es have not been actually exterminated, their NUMBERS have been greatly reduced, and this is the f
0385  water shells were hatching; and I found that NUMBERS of the extremely minute and just hatched she
0389  a few introduced plants are included in these NUMBERS, and the comparison in some other respects t
0391  s, dr. Hooker has shown that the proportional NUMBERS of the different orders are very different f
0408  others to enter, either in greater or lesser NUMBERS; according or not, as those which entered ha
0409  loped in great force, some existing in scanty NUMBERS, in the different great geographical provinc
0463  and; and the latter, from existing in greater NUMBERS, will generally be modified and improved at
0463  intermediate varieties, which exist in lesser NUMBERS; so that the intermediate varieties will, in
0472  death; at crones being produced in such vast NUMBERS for one single act, and being then slaughter
0081  y are bound together, that any change in the NUMERICAL proportions of some of the inhabitants, ind
0428  , we can perhaps understand how it is that a NUMERICAL parallelism has sometimes been observed bet
0057  ngly small. I have endeavoured to test this NUMERICALLY by averages; and, as far as my imperfect r
0176  two other forms occur, they are much rarer NUMERICALLY than the forms which they connect. Now, if
0006  in why one species ranges widely and is very NUMEROUS, and why another allied species has a narrow
0009  ng lived, and healthy (of which I could give NUMEROUS instances), yet having their reproductive sy
0031  as they please. If I had space I could quote NUMEROUS passages to this effect from highly competen
0049  phical ranges; and lastly, according to very NUMEROUS experiments made during several years by the
0054  fused in their own country, and are the most NUMEROUS in individuals, which oftenest produce well
0064  n mere theoretical calculations, namely, the NUMEROUS recorded cases of the astonishingly rapid in
0065  an ten years. Several of the plants now most NUMEROUS over the wide plains of La Plata, clothing s
0066  the same country the condor may be the more NUMEROUS of the two: the Fulmar petrel lays but one e
0066  t one egg, yet it is believed to be the most NUMEROUS bird in the world. One fly deposits hundreds
0072  hen first born. The increase of these flies, NUMEROUS as they are, must be habitually checked by s
0074  s i have found the nests of humble bees more NUMEROUS than elsewhere, which I attribute to the num
0098  rivance by which every one of the infinitely NUMEROUS pollen granules are swept out of the conjoin
0107  ntirent, and the inhabitants, at this period NUMEROUS in individuals and kinds, will have been sub
0110  ves. Furthermore, the species which are most NUMEROUS in individuals will have the best chance of
0118  ages which made their common parent (A) more NUMEROUS than most of the other inhabitants of the sa
0120  ps in the process of modification to become NUMEROUS or greater in amount, to convert these three
0136  rge groups of beetles, elsewhere excessively NUMEROUS, and which groups have habits of life almost
0174  t to expect at the present time to meet with NUMEROUS transitional varieties in each region, thoug
0174  wide area, we generally find them tolerably NUMEROUS over a large territory, then becoming somewh
0186  ossible degree. Yet reason tells me, that if NUMEROUS gradations from a perfect and complex eye to
0187  ny other mechanism; and from this low stage, NUMEROUS gradations of structure, branching off in tw
0189  which could not possibly have been formed by NUMEROUS, successive, slight modifications, my theory
0190  med by transitional gradations of some kind. NUMEROUS cases could be given amongst the lower anima
0204  s, will have a great advantage over the less NUMEROUS intermediate variety, and will thus generall
0210  cept by the slow and gradual accumulation of NUMEROUS, slight, yet profitable, variations. Hence,
0220  that they might behave differently when more NUMEROUS; but Mr. Smith informs me that he has watche
0222  ca (showing that it was not a migration) and NUMEROUS pupae. I traced the returning file burthened
0223  ed itself; it is absolutely dependent on its NUMEROUS slaves. Formica sanguinea, on the other hand
0235  natural selection having taken advantage of NUMEROUS, successive, slight modifications of simpler
0239  er or the smaller sized workers are the most NUMEROUS; or that both large and small are numerous,
0239  t numerous; or that both large and small are NUMEROUS, with those of an intermediate size scanty i
0240  ly availed myself of Mr. F. Smith's offer of NUMEROUS specimens from the same nest of the driver a
0242  cture can be effected by the accumulation of NUMEROUS, slight, and as we must call them accidental
0256  be united with unusual facility, and produce NUMEROUS hybrid offspring, yet these hybrids are rema
```

Page **(Key Word)**

age **(Key Word)***

```
282 ot finding fossil remains of such infinitely NUMEROUS connecting links, it may be objected, that t
291 tell us plainly that each area has undergone NUMEROUS slow oscillations of level, and apparently t
297 beds of a formation, and unless we obtained NUMEROUS transitional gradations, we should not recog
298 y of our being enabled to connect species by NUMEROUS, fine, intermediate, fossil links, by asking
299 ture geologist discovering in a fossil state NUMEROUS intermediate gradations: and such success se
299 ee. Geological research, though it has added NUMEROUS species to existing and extinct genera, and
299 ween species, by connecting them together hy NUMEROUS, fine, intermediate varieties: and this not
299 condition of the Malay Archipelago, with its NUMEROUS large islands separated by wide and shallow
302 e belief in the transmutation of species. If NUMEROUS species, belonging to the same genera or fam
307 ong ago supplanted and exterminated by their NUMEROUS and improved descendants. Consequently, if m
310 ding in the successive formations infinitely NUMEROUS transitional links between the many species
333 diagram: for the groups will have been more NUMEROUS, they will have endured for extremely unequa
340 that Europe in ancient times was peopled by NUMEROUS marsupials: and I have shown in the publicat
343 sk where are the remains of those infinitely NUMEROUS organisms which must have existed long befor
354 the exceptions to continuity of range are so NUMEROUS and of so grave a nature, that we ought to g
397 nd, without being actually the same species. NUMEROUS instances could be given of this fact. I wil
416 rtance in the classification of the Grasses. NUMEROUS instances could be given of characters deriv
419 ifferences, at first overlooked, but because NUMEROUS allied species, with slightly different grad
431 t to show the blood relationship between the NUMEROUS kindred of any ancient and noble family, eve
435 different purposes, are formed by infinitely NUMEROUS modifications of an upper lip, mandibles, an
437 brain be enclosed in a box composed of such NUMEROUS and such extraordinarily shaped pieces of bo
439 the jaws, for instance, of a crab retaining NUMEROUS characters, which they would probably have r
464 o the imperfection of the geological record. NUMEROUS existing doubtful forms could be named which
483 many were produced? Were all the infinitely NUMEROUS kinds of animals and plants created as eggs
486 rience, the reason, and even the blunders of NUMEROUS workmen; when we thus view each organic bein
1237 the reproductive system is active, as in the NUPTIAL plumage of many birds, and in the hooked jaw
1034 may be compared to the roguing of plants by NURSERYMEN. The principle of selection I find distinc
1041 y never can be improved. On the other hand, NURSERYMEN, from raising large stocks of the same plan
1103 to other districts. On the above principle, NURSERYMEN always prefer getting seed from a large bod
1252 case, the fact would have been notorious to NURSERYMEN. Horticulturists raise large beds of the sa
1184 w on a branch, and thus breaking them like a NUTHATCH. In North America the black bear was seen by
1077 rving as prey to other animals. The store of NUTRIMENT laid up within the seeds of many plants see
1077 g grass, I suspect that the chief use of the NUTRIMENT in the seed is to favour the growth of the
1116 ble matter alone, or flesh alone, draws most NUTRIMENT from these substances. So in the general ec
1145 onnected with some difference in the flow of NUTRIMENT towards the central and external flowers: w
1147 on the other hand, the actual withdrawal of NUTRIMENT from one part owing to the excess of growth
1148 t will profit the individual not to have its NUTRIMENT wasted in building up an useless structure.
1148 better chance of supporting itself, by less NUTRIMENT being wasted in developing a structure now
1442 eir parents or placed in the midst of proper NUTRIMENT, yet nearly all pass through a similar worm
1147 es of the cabbage do not yield abundant and NUTRITIOUS foliage and a copious supply of oil bearing
1361 estimate. I have never seen an instance of NUTRITIOUS seeds passing through the intestines of a b
1359 floated much longer: for instance, ripe hazel NUTS sank immediately, but when dried, they floated
1050 by some authors as species. Look at the common OAK, how closely it has been studied; yet a German
1361 of earth thus completely enclosed by wood in an OAK about 50 years old, three dicotyledonous plant
1484 produces monstrous growths on the wild rose or OAK tree. Therefore I should infer from analogy th
1050 oted to show that the sessile and pedunculated OAKS are either good and distinct species or mere v
0362 seeds capable of germination. Some seeds of the OAT, wheat, millet, canary, hemp, clover, and beet
0001 ses, however, the larvae, though active, still OBEY more or less closely the law of common embryon
0020 present day I have steadily pursued the same OBJECT. I hope that I may be excused for entering in
0020 j. sebright expressly experimentised for this OBJECT, and failed. The offspring from the first cro
0034 try by methodical selection, with a distinct OBJECT in view: to make a new strain or sub breed, s
0092 nectar: and as pollen is formed for the sole OBJECT of fertilisation, its destruction appears a s
0102 election, a breeder selects for some definite OBJECT, and free intercrossing will wholly stop his
0149 any shape: whilst a tool for some particular OBJECT had better be of some particular shape. Natur
0195 s, each better and better, for so trifling an OBJECT as driving away flies; yet we should pause be
0255 ty of first crosses and of hybrids. Our chief OBJECT will be to see whether or not the rules hold
0315 rs, by striving during long ages for the same OBJECT, might make a new breed hardly distinguishabl
0434 of any one class; but when we have a distinct OBJECT in view, and do not look to some unknown plan
0452 ts proper purpose, and be used for a distinct OBJECT: in certain fish the swim bladder seems to be
0486 doubt become simpler when we have a definite OBJECT in view. We possess no pedigrees or armorial
0490 ture, from famine and death, the most exalted OBJECT which we are capable of conceiving, namely, t
0033 other chiefly in these characters. It may be OBJECTED that the principle of selection has been red
0099 covered with innumerable flowers, it may be OBJECTED that pollen could seldom be carried from tre
0282 initely numerous connecting links, it may be OBJECTED, that time will not have sufficed for so gre
0330 ct groups as they now are. Some writers have OBJECTED to any extinct species or group of species b
0332 ed with ruminants and pachycerms. Yet he who OBJECTED to call the extinct genera, which thus linke
0464 t than most geologists believe. It cannot be OBJECTED that there has not been time sufficient for
0100 uals only in a limited sense. I believe this OBJECTION to be valid; but that nature has largely pr
0280 is, perhaps, is the most obvious and gravest OBJECTION which can be urged against my theory. The e
0302 cibly than by Professor Sedgwick, as a fatal OBJECTION to the belief in the transmutation of speci
0330 its characters between two living forms, the OBJECTION is probably valid. But I apprehend that in
0334 etween consecutive formations. It is no real OBJECTION to the truth of the statement, that the fau
0095 ve imaginary instances, is open to the same OBJECTIONS which were at first urged against Sir Charl
0171 fatal to my theory. These difficulties and OBJECTIONS may be classed under the following heads: F
0203 pter discussed some of the difficulties and OBJECTIONS which may be urged against my theory. Many
0279 in the sixth chapter I enumerated the chief OBJECTIONS which might be justly urged against the vie
0299 he gravest and most obvious of all the many OBJECTIONS which may be urged against my views. Hence
0345 be proved to be much more perfect, the main OBJECTIONS to the theory of natural selection are grea
0459 briefly recapitulated. That many and grave OBJECTIONS may be advanced against the theory of desce
0463 s the most obvious and forcible of the many OBJECTIONS which may be urged against my theory. Why,
0464 ory. I can answer these questions and grave OBJECTIONS only on the supposition that the geological
0465 time. Such is the sum of the several chief OBJECTIONS and difficulties which may justly be urged
0466 ves especial notice that the more important OBJECTIONS relate to questions on which we are confess
0469 ly as I could, the opposed difficulties and OBJECTIONS: now let us turn to the special facts and a
0175 ioni and as these species are already defined OBJECTS (however they may have become so), not blend
0177 hat species come to be tolerably well defined OBJECTS, and do not at any one period present an ine
```

Page **(Key Word)***

Page **************************************(Key Word)**

```
0380 part of Europe, though hides, wool, and other OBJECTS likely to carry seeds have been largely imp
0413 a scheme for arranging together those living OBJECTS which are most alike, and for separating th
0420 and the mere putting together and separating OBJECTS more or less alike. But I must explain my m
0470 ndered to a large extent defined and distinct OBJECTS. Dominant species belonging to the larger g
0048 mr. H. C. Watson, to whom I lie under deep OBLIGATION for assistance of all kinds, has marked fo
0003 opportunity pass without expressing my deep OBLIGATIONS to Dr. Hooker, who for the last fifteen y
0247 in these and in many other cases, Gartner is OBLIGED carefully to count the seeds, in order to s
0138 ll on this view have more or less perfectly OBLITERATED its eyes, and natural selection will ofte
0190 her and quite distinct purpose, or be quite OBLITERATED. The illustration of the swimbladder in f
0307 had been wholly worn away by denudation, or OBLITERATED by metamorphic action, we ought to find o
0449 e represented in many embryos, has not been OBLITERATED, either by the successive variations in a
0192 through an increase in their size and the OBLITERATION of their adhesive glands. If all peduncul
0455 ay; and this will tend to cause the entire OBLITERATION of a rudimentary organ. As the presence o
0004 ould offer the best chance of making out this OBSCURE problem. Nor have I been disappointed: in t
0006 epochs in its history. Although much remains OBSCURE, and will long remain obscure, I can entert
0006 gh much remains obscure, and will long remain OBSCURE, I can entertain no doubt, after the most d
0054 dominant species. But so many causes tend to OBSCURE this result, that I am surprised that my ta
0067 of each species to increase in number is most OBSCURE. Look at the most vigorous species; by as m
0141 nd how much to both means combined, is a very OBSCURE question. That habit or custom has some inf
0144 bond of correlation is very frequently quite OBSCURE. M. Is. Geoffroy St. Hilaire has forcibly r
0163 stripe; and these stripes are sometimes very OBSCURE, or actually quite lost, in dark coloured a
0203 y of independent acts of creation are utterly OBSCURE. We have seen that species at any one perio
0322 the whole economy of nature will be utterly OBSCURED. Whenever we can precisely say why this spe
0436 ral pattern of an organ might become so much OBSCURED as to be finally lost, by the atrophy and u
0436 seems to have been thus to a certain extent OBSCURED. There is another and equally curious branc
0440 embryos of allied animals is sometimes much OBSCURED; and cases could be given of the larvae of
0449 ture of the adult may have been modified and OBSCURED; we have seen, for instance, that cirripede
0450 ook at the embryo as a picture, more or less OBSCURED, of the common parent form of each great cl
0486 l reveal to us the structure, in some degree OBSCURED, or the prototypes of each great class. Whe
0054 ra. I will here allude to only two causes of OBSCURITY. Fresh water and salt loving plants have g
0150 hed by Mr. Waterhouse. I infer also from an OBSERVATION made by Professor Owen, with respect to t
0221 g to Huber, who had ample opportunities for OBSERVATION, in Switzerland the slaves habitually wor
0361 nated: I am certain of the accuracy of this OBSERVATION. Again, I can show that the carcasses of
0004 nd coadaptation. At the commencement of my OBSERVATIONS it seemed to me probable that a careful s
0026 ctly fertile. I can state this from my own OBSERVATIONS, purposely made on the most distinct bree
0051 mpressions. As he extends the range of his OBSERVATIONS, he will meet with more cases of difficul
0051 number of closely allied forms. But if his OBSERVATIONS be widely extended, he will in the end ge
0067 vast destruction of seeds, but, from some OBSERVATIONS which I have made, I believe that it is t
0098 writings of C. C. Sprengel and from my own OBSERVATIONS, which effectually prevent the stigma rec
0140 ved similar facts in Ceylon, and analogous OBSERVATIONS have been made by Mr. H. C. Watson on Eur
0220 at of making slaves. Hence I will give the OBSERVATIONS which I have myself made, in some little
0264 ficiently attended to: but I believe, from OBSERVATIONS communicated to me by Mr. Hewitt, who has
0271 be raised from the seed of the other. From OBSERVATIONS which I have made on certain varieties of
0283 with which rocky coasts are worn away. The OBSERVATIONS on this head by Hugh Miller, and by that
0312 the face of the earth. If we may trust the OBSERVATIONS of Philippi in Sicily, the successive cha
0323 rmations could be easily correlated. These OBSERVATIONS, however, relate to the marine inhabitants
0328 mporaneous; faunas. Lyell has made similar OBSERVATIONS on some of the later tertiary formations.
0395 n. mr. Windsor Earl has made some striking OBSERVATIONS on this head in regard to the great Malay
0430 elder De Candolle has made nearly similar OBSERVATIONS on the general nature of the affinities o
0091 this be thought very improbable; for we often OBSERVE great differences in the natural tendencies
0161 be, thus inherited. Indeed, we may sometimes OBSERVE a mere tendency to produce a rudiment inher
0166 een several of the most distinct species. Now OBSERVE the case of the several breeds of pigeons:
0228 r of a cell. It was most interesting to me to OBSERVE that wherever several bees had begun to exca
0045 esumed to have thus arisen, being frequently OBSERVED in the individuals of the same species inhab
0073 ld certainly greatly alter (as indeed I have OBSERVED in parts of South America: the vegetation:
0074 erent vegetation springs up: but it has been OBSERVED that the trees now growing on the ancient Ir
0135 le of flight. Kirby has remarked (and I have OBSERVED the same fact) that the anterior tarsi, or
0135 and perish; that the beetles in Madeira, as OBSERVED by Mr. Wollaston, lie much concealed, until
0140 ng cold. Mr. Thwaites informs me that he has OBSERVED similar facts in Ceylon, and analogous obse
0145 ng case of correlation, that I have recently OBSERVED in some garden pelargoniums, that the centr
0175 kable how abruptly, as Alph. de Candolle has OBSERVED, a common alpine species disappears. The sam
0193 ently the same anomalous organ, it should be OBSERVED that, although the general appearance and fu
0210 instincts of animals having been but little OBSERVED except in Europe and North America, and that
0220 own proper communities, and have never been OBSERVED in the nests of F. sanguinea. The slaves are
0221 th an unusually large stock of slaves, and I OBSERVED a few slaves mingled with their masters leav
0223 nderful instinct of making slaves. Let it be OBSERVED what a contrast the instinctive habits of F.
0231 nd a growing comb, flexures may sometimes be OBSERVED, corresponding in position to the planes of
0233 t i am not aware that any such case has been OBSERVED: nor would any good be derived from a single
0240 e; and, in this species, as Mr. F. Smith has OBSERVED, the larger workers have simple eyes (ocell
0258 es between the same two species was long ago OBSERVED by Kolreuter. To give an instance: Mirabilis
0258 ly striking cases could be given. Thuret has OBSERVED the same fact with certain sea weeds or Fuc
0258 extremely common in a lesser degree. He has OBSERVED it even between forms so closely related (as
0263 tubes to reach the ovarium. It has also been OBSERVED that when pollen of one species is placed or
0265 liable to vary, as every experimentalist has OBSERVED. Thus we see that when organic beings are pl
0295 probably a rare contingency: for it has been OBSERVED by more than one palaeontologist, that very
0313 of which a striking instance has lately been OBSERVED in Switzerland. There is some reason to beli
0323 ar parallelism in the forms of life has been OBSERVED by several authors: so it is, according to L
0353 confined to a single region: and it has been OBSERVED by several naturalists, that the most natura
0363 rote to M. Hartung to inquire whether he had OBSERVED erratic boulders on these islands, and he ar
0364 ion of prevalent gales of wind. It should be OBSERVED that scarcely any means of transport would c
0373 half, ice borne fragments of rock have been OBSERVED on the eastern side as far south as lat. 36
0376 the intermediate tropical seas. It should be OBSERVED that the northern species and forms found in
0385 he same country. But two facts, which I have OBSERVED, and no doubt many others remain to be obser
0385 erved, and no doubt many others remain to be OBSERVED, throw some light on this subject. When a du
0406 bably accounts for a law which has long been OBSERVED, and which has lately been admirably discuss
0428 t a numerical parallelism has sometimes been OBSERVED between the sub groups in distinct classes.
```

Page *************************************(Key Word)************************************

```
age ******************************************( Key Word )******************************************
```

37	an is the common characteristic (as Owen has	OBSERVED) of all low or little modified forms: theref
69	ties and sub species, and species. Let it be	OBSERVED how naturalists differ in the rank which the
72	he want of absolute perfection have not been	OBSERVED. The complex and little known laws governing
47	ly been found, but because analogy leads the	OBSERVER to suppose either that they co now somewhere
50	de during several years by that most careful	OBSERVER Gartner, they can be crossed only with much
85	dly modified. On the other hand, the acutest	OBSERVER by examining the dead body of the water ouze
98	nstitution and colour are correlated. A good	OBSERVER, also, states that in cattle suscept¹b³lity
19	lyerges) rufescens by Pierre Huber, a better	OBSERVER even than his celebrated father. This ant is
49	d experiments are in progress, so careful an	OBSERVER as Gartner would have castrated his hybrids,
70	on nine species of Verbascum, by so good an	OBSERVER and so hostile a witness, as Gartner: namely
71	uracy has been confirmed by every subsequent	OBSERVER, has proved the remarkable fact, that one va
83	s head by Hugh Miller, and by that excellent	OBSERVER Mr. Smith of Jordan Hill, are most impressiv
57	onsulted some sagacious and most experienced	OBSERVERS, and, after deliberation, they concur in th
47	w is of universal application; but many good	OBSERVERS, more especially botanists, believe in its
98	een but little artificial selection. Careful	OBSERVERS are convinced that a damp climate affects t
18	hat, according to Dr. Gray and to some other	OBSERVERS, the European cuckoo has not utterly lost a
46	rks of those two conscientious and admirable	OBSERVERS, Kolreuter and Gartner, who almost devoted
48	required than that the two most experienced	OBSERVERS who have ever lived, namely, Kolreuter and
51	is result has, also, been confirmed by other	OBSERVERS in the case of Hippeastrum with its sub gen
82	y, or to read special treatises by different	OBSERVERS on separate formations, and to mark how eac
84	yet what time this must have consumed! Good	OBSERVERS have estimated that sediment is deposited b
24	isphere. So, again, several highly competent	OBSERVERS believe that the existing productions of th
24	he world, has greatly struck those admirable	OBSERVERS, MM. de Verneuil and d'Archiac. After refer
71	and the singular fact remarked on by several	OBSERVERS, that the productions of Europe and America
17	, yet M. Richard sagaciously saw, as Jussieu	OBSERVES, that this genus should still be retained am
33	regard to plants, there is another means of	OBSERVING the accumulated effects of selection namely
47	ral review is, that barriers of any kind, or	OBSTACLES to free migration, are related in a close a
14	ffected for many generations the courage and	OBSTINACY of greyhounds: and a cross with a greyhound
57	ich is not a particularly distinct species,	OBSTINATELY failed to fertilise, or to be fertilised b
20	ve kept every breed which I could purchase or	OBTAIN, and have been most kindly favoured with skin
31	ree years, but it would take him six years to	OBTAIN head and beak. In Saxony the importance of th
94	individual so characterised would be able to	OBTAIN its food more quickly, and so have a better c
52	ied through natural selection, I think we can	OBTAIN some light. In our domestic animals, if any p
97	in all points of structure. So that we might	OBTAIN the parent species and its several modified d
41	converted into prehensile organs: they again	OBTAIN a well constructed mouth; but they have no an
02	t which I have arrived. A fair result can be	OBTAINED only by fully stating and balancing the fact
20	desired character; but that a race could be	OBTAINED nearly intermediate between two extremely di
53	ought that some interesting results might be	OBTAINED in regard to the nature and relations of the
97	and upper beds of a formation, and unless we	OBTAINED numerous transitional gradations, we should
44	d not signify, for instance, to a bird which	OBTAINED its food best by having a long beak, whether
68	increase: but very frequently it is not the	OBTAINING food, but the serving as prey to other anim
27	ble: and the explanation of this fact will be	OBVIOUS when we come to treat of selection. Fourthly
24	riation which pigeons have undergone, will be	OBVIOUS when we treat of Selection. We shall then, a
32	d breeding from it, the principle would be so	OBVIOUS as hardly to be worth notice: but its import
34	e inheritance of good and bad qualities is so	OBVIOUS. At the present time, eminent breeders try b
38	lection by man has played, it becomes at once	OBVIOUS, how it is that our domestic races show adap
65	orarily increased in any sensible degree. The	OBVIOUS explanation is that the conditions of life h
77	has to escape, or on which it preys. This is	OBVIOUS in the structure of the teeth and talons of
96	d plants with separate sexes, it is of course	OBVIOUS that two individuals must always unite for e
96	n the case of hermaphrodites this is far from	OBVIOUS. Nevertheless I am strongly inclined to beli
00	ites: but here currents in the water offer an	OBVIOUS means for an occasional cross. And, as in th
01	beings, a cross between two individuals is an	OBVIOUS necessity for each birth: in many others it
43	ubject, most imperfectly understood. The most	OBVIOUS case is, that modifications accumulated sole
22	construction of three adjoining cells. It is	OBVIOUS that the Melipona saves wax by this manner o
79	by innumerable transitional links, is a very	OBVIOUS difficulty. I assigned reasons why such link
80	organic chain: and this, perhaps, is the most	OBVIOUS and gravest objection which can be urged aga
99	en effected, is probably the gravest and most	OBVIOUS of all the many objections which may be urge
36	s distinct species, being closely related, is	OBVIOUS. As the accumulation of each formation has o
51	ecome greatly modified. On these views, it is	OBVIOUS, that the several species of the same genus,
52	is great difficulty on this head. It is also	OBVIOUS, that the individuals of the same species, th
90	eleven marine birds, are peculiar: and it is	OBVIOUS that marine birds could arrive at these isla
96	of time, and as during changes of level it is	OBVIOUS that islands separated by shallow channels a
98	ni whereas on the view here maintained, it is	OBVIOUS that the Galapagos Islands would be likely t
03	c., all of strictly American forms, and it is	OBVIOUS that a mountain, as it became slowly upheave
18	l ages of each species. But it is by no means	OBVIOUS, on the ordinary view, why the structure of
42	ent in some manner on growth. But there is no	OBVIOUS reason why, for instance, the wing of a bat,
63	t with no such evidence, and this is the most	OBVIOUS and forcible of the many objections which ma
63	varieties or aboriginal species. There is no	OBVIOUS reason why the principles which have acted s
40	f selection. A high degree of variability is	OBVIOUSLY favourable, as freely giving the materials
61	od and distinct species, which in most cases	OBVIOUSLY differ from each other far more than do the
03	iety, true and uniform in character. It will	OBVIOUSLY thus act far more efficiently with those an
40	nhabitants of each quarter of the world will	OBVIOUSLY tend to leave in that quarter, during the n
62	during modern periods: and such changes will	OBVIOUSLY have greatly facilitated migration. As an e
73	of complex relations. I shall hereafter have	OCCASION to show that the exotic Lobelia fulgens, in
18	advantage of the prize, and becomes for the	OCCASION parasitic. In this case, as with the suppose
20	be no doubt that a race may be modified by	OCCASIONAL crosses, if aided by the careful selection
36	gradual process of improvement, through the	OCCASIONAL preservation of the best individuals, wheth
43	ic breeds have once been established, their	OCCASIONAL intercrossing, with the aid of selection, h
63	riod of its life, and during some season or	OCCASIONAL year, otherwise, on the principle of geomet
85	nd constant. Nor ought we to think that the	OCCASIONAL destruction of an animal of any particular
97	mas fully exposed to the weather: but if an	OCCASIONAL cross be indispensable, the fullest freedom
99	are these facts explained on the view of an	OCCASIONAL cross with a distinct individual being adva
00	with terrestrial plants, on the view of an	OCCASIONAL cross being indispensable, by considering t
00	the wind in the case of plants, by which an	OCCASIONAL cross could be effected with terrestrial an
00	in the water offer an obvious means for an	OCCASIONAL cross. And, as in the case of flowers, I ha
01	the body, that access from without and the	OCCASIONAL influence of a distinct individual can be s

```
age ******************************************( Key Word )******************************************
```

Page *********************************(Key Word)*********************************

```
0101 th in the vegetable and animal kingdoms, an OCCASIONAL intercross with a distinct individual is a
0104 to show that we have reason to believe that OCCASIONAL intercrosses take place with all animals a
0159 however, we have another case, namely, the OCCASIONAL appearance in all the breeds, of slaty blu
0178 epend on slow changes of climate, or on the OCCASIONAL immigration of new inhabitants, and, proba
0180 d of diversified habits, either constant or OCCASIONAL, in the same species. And it seems to me t
0180 on to believe, by lessening the danger from OCCASIONAL falls. But it does not follow from this fa
0212 cts. Several cases also, could be given, of OCCASIONAL and strange habits in certain species, whi
0217 ird's nest.If the old bird profited by this OCCASIONAL habit, or if the young were made more vigo
0217 d would be apt to follow by inheritance the OCCASIONAL and aberrant habit of their mother and in
0218 al love and care for her own offspring. The OCCASIONAL habit of birds laying their eggs in other
0219 o difficulty in natural selection making an OCCASIONAL habit permanent, if of advantage to the sp
0249 fertility, and, on the other hand, that an OCCASIONAL cross with a distinct individual or variet
0315 w and complete act of creation, but only an OCCASIONAL scene, taken almost at hazard, in a slowly
0346 limate and of the level of the land, and by OCCASIONAL means. Dispersal during the Glacial period
0354 imatal and geographical changes and various OCCASIONAL means of transport, the belief that this h
0358 ns, but which more properly might be called OCCASIONAL means of distribution. I shall here confin
0364 ds. These means, however, would suffice for OCCASIONAL transport across tracts of sea some hundre
0364 ived within the last few centuries, through OCCASIONAL means of transport, immigrants from Europe
0365 argument against what would be effected by OCCASIONAL means of transport, during the long lapse
0382 arious points of the southern hemisphere by OCCASIONAL means of transport, and by the aid, as hal
0384 cts seem to favour the possibility of their OCCASIONAL transport by accidental means: like that o
0396 eem to me to accord better with the view of OCCASIONAL means of transport having been largely eff
0398 be likely to receive colonists, whether by OCCASIONAL means of transport or by formerly continuo
0401 view of the islands having been stocked by OCCASIONAL means of transport, a seed, for instance,
0407 th respect to the many and curious means of OCCASIONAL transport, a subject which has hardly ever
0407 as showing how diversified are the means of OCCASIONAL transport, I have discussed at some little
0410 mer period under different conditions or by OCCASIONAL means of transport, and by the species hav
0462 too much stress ought not to be laid on the OCCASIONAL wide diffusion of the same species: for du
0462 are as yet profoundly ignorant of the many OCCASIONAL means of transport. With respect to distin
0473 explicable on the theory of creation is the OCCASIONAL appearance of stripes on the shoulder and
0476 al and geographical changes and to the many OCCASIONAL and unknown means of dispersal, then we ca
0015 evertheless, as our varieties certainly do OCCASIONALLY revert in some of their characters to anc
0027 pigeon; the blue colour and various marks OCCASIONALLY appearing in all the breeds, both when ke
0041 stly useful or pleasing to man appear only OCCASIONALLY, the chance of their appearance will be me
0092 if a little pollen were carried, at first OCCASIONALLY and then habitually, by the pollen devour
0096 all hermaphrodites two individuals, either OCCASIONALLY or habitually, concur for the reproductio
0097 ut that a cross with another individual is OCCASIONALLY, perhaps at very long intervals, indispen
0100 will give a better chance of pollen being OCCASIONALLY carried from tree to tree. That trees bel
0101 al. But if, in fact, all hermaphrodites do OCCASIONALLY intercross with other individuals, the di
0103 n hermaphrodite organisms which cross only OCCASIONALLY, and likewise in animals which unite for
0130 ed and is still alive on its summit, so we OCCASIONALLY see an animal like the Ornithorhynchus or
0160 ly once by some other breed, the offspring OCCASIONALLY show a tendency to revert in character to
0161 rent, it might be expected that they would OCCASIONALLY vary in an analogous manner: so that a va
0161 d that the species of the same genus would OCCASIONALLY exhibit reversions to lost ancestral chara
0162 organs of an important and uniform nature OCCASIONALLY varying so as to acquire, in some degree,
0163 of it, as stated by Mr. Blyth and others, OCCASIONALLY appear: and I have been informed by Colone
0169 ous variations, and these same species may OCCASIONALLY revert to some of the characters of their
0184 on my theory, that such individuals would OCCASIONALLY have given rise to new species, having and
0185 ng has been created as we now see it, must OCCASIONALLY have felt surprise when he has met with ar
0214 a tendency in this line: and this is known OCCASIONALLY to happen, as I once saw in a pure terrier
0215 ck poultry, sheep, and pigs! No doubt they OCCASIONALLY do make an attack, and are then beaten: ar
0217 has been asserted that the American cuckoo OCCASIONALLY lays her eggs in other birds' nests: but I
0217 ces of various birds which have been known OCCASIONALLY to lay their eggs in other bird's nests. M
0217 he habits of the American cuckoo: but that OCCASIONALLY she laid an egg in another bird's nest.If
0220 the nest is slightly disturbed, the slaves OCCASIONALLY come out, and like their masters are much
0236 er articulate animals in a state of nature OCCASIONALLY become sterile; and if such insects had be
0239 e desired character. On this view we ought OCCASIONALLY to find neuter insects of the same species
0258 erally differ in fertility in a small, and OCCASIONALLY in a higher degree. Several other singular
0260 r, there is generally some difference, and OCCASIONALLY the widest possible difference, in the fac
0266 increased sterility in those hybrids which OCCASIONALLY and exceptionally resemble closely either
0335 lengths of time: a very ancient form might OCCASIONALLY last much longer than a form elsewhere sub
0360 y have restricted ranges. But seeds may be OCCASIONALLY transported in another manner. Drift timbe
0363 e as the seed of a vetch. Thus seeds might OCCASIONALLY be transported to great distances: for mar
0363 nd bird, I can hardly doubt that they must OCCASIONALLY have transported seeds from one part to ar
0386 s fact. I have before mentioned that earth OCCASIONALLY, though rarely, adheres in some quantity t
0386 n show are the greatest wanderers, and are OCCASIONALLY found on the most remote and barren island
0390 theory, for, as already explained, species OCCASIONALLY arriving after long intervals in a new and
0391 l migrations, visit either periodically or OCCASIONALLY this island. Madeira does not possess one
0394 north American species either regularly or OCCASIONALLY visit Bermuda, at the distance of 600 mile
0397 transportal. Would the just hatched young OCCASIONALLY crawl on and adhere to the feet of birds r
0429 w old and intermediate parent forms having OCCASIONALLY transmitted to the present day descendants
0439 distinguised in the whelp of the lion. We OCCASIONALLY though rarely see something of this kind i
0452 e same organ has been rendered rudimentary OCCASIONALLY differs much. This latter fact is well exe
0452 ld lead us to expect to find, and which is OCCASIONALLY found in monstrous individuals of the spec
0466 wholly cease: for new varieties are still OCCASIONALLY produced by our most anciently domesticate
0472 ti yet when varieties enter any zone, they OCCASIONALLY assume some of the characters of the speci
0473 e look at the burrowing tucutucu, which is OCCASIONALLY blind, and then at certain moles, which ar
0180 s have very little weight. Here, as on other OCCASIONS, I lie under a heavy disadvantage, for out
0403 everal districts of the same continent, pre OCCUPATION has probably played an important part in ch
0102 th some place in its polity not so perfectly OCCUPIED as might be, natural selection will always t
0108 in the polity of nature, which can be better OCCUPIED by some of the inhabitants of the country un
0113 ing descendants seizing on places at present OCCUPIED by other animals: some of them, for instance
0116 , and are thus enabled to encroach on places OCCUPIED by other beings. Now let us see how this pri
0119 which are either unoccupied or not perfectly OCCUPIED by other beings: and this will depend on inf
0321 some one group will have seized on the place OCCUPIED by a species belonging to a distinct group.
0337 ed on places which must have been previously OCCUPIED, we may believe, if all the animals and plan
```

Page *********************************(Key Word)*********************************

```
 337  mber would be enabled to seize on places now OCCUPIED by our native plants and animals. Under this
 391  ain classes, and their places are apparently OCCUPIED by the other inhabitants; in the Galapagos I
 401  d find the best fitted ground more perfectly OCCUPIED by distinct plants in one island than in ano
 463  e change insensibly in going from a district OCCUPIED by one species into another district occupie
 463  ccupied by one species into another district OCCUPIED by a closely allied species, we have no just
 472  descendants of each to any unoccupied or ill OCCUPIED place in nature, these facts cease to be str
 113  ame, the more places they would be enabled to OCCUPY. What applies to one animal will apply throug
 412  varying descendants of each species trying to OCCUPY as many and as different places as possible i
 421  this genus, though much isolated, will still OCCUPY its proper intermediate position; for F origi
 388  viduals of the species, however few, already OCCUPYING any pond, yet as the number of kinds is sma
 015  ations and reversions of character probably do OCCUR; but natural selection, as will hereafter be
 025  equered with black. These several marks do not OCCUR together in any other species of the whole fa
 047  ty, which I will not here enumerate, sometimes OCCUR in deciding whether or not to rank one form as
 055  s no apparent reason why more varieties should OCCUR in a group having many species, than in one h
 070  , epidemics, at least, this seems generally to OCCUR with our game animals, often ensue: and here
 070  remely abundant in the few spots where they do OCCUR: and that of some social plants being social,
 080  t and complex battle of life, should sometimes OCCUR in the course of thousands of generations? If
 080  course of thousands of generations? If such do OCCUR, can we doubt (remembering that many more ind
 082  occurring; and unless profitable variations do OCCUR, natural selection can do nothing. Not that,
 094  re or less impotent; now if we suppose this to OCCUR in ever so slight a degree under nature, then
 108  can be effected, unless favourable variations OCCUR, and variation itself is apparently always a
 127  t if variations useful to any organic being do OCCUR, assuredly individuals thus characterised wil
 143  t, that when slight variations in any one part OCCUR, and are accumulated through natural selectio
 162  ion or from analogous variation) which already OCCUR in some other members of the same group. And
 163  i can only repeat that such cases certainly do OCCUR, and seem to me very remarkable. I will, howe
 164  south. In all parts of the world these stripes OCCUR far oftenest in duns and mouse duns; by the t
 166  k colonel Poole whether such face stripes ever OCCUR in the eminently striped Kattywar breed of ho
 176  varieties intermediate between two other forms OCCUR, they are much rarer numerically than the for
 177  nothing until favourable variations chance to OCCUR, and until a place in the natural polity of t
 184  at of their proper type. And such instances do OCCUR in nature. Can a more striking instance of ad
 192  s, yet, undoubtedly, grave cases of difficulty OCCUR, some of which will be discussed in my future
 193  her and even more serious difficulty; for they OCCUR in only about a dozen fishes, of which severa
 197  be indispensable for this act; but as sutures OCCUR in the skulls of young birds and reptiles, wh
 210  f instinct in the same species can be shown to OCCUR in nature. Again as in the case of corporeal
 250  oduced a plant, which (he says) I never saw to OCCUR in a case of its natural fecundation. So that
 256  instance in Dianthus, these two opposite cases OCCUR. The fertility, both of first crosses and of
 268  with European dogs, the explanation which will OCCUR to everyone, and probably the true one, is th
 275  ; but it can be shown that this does sometimes OCCUR with hybrids; yet I grant much less frequentl
 275  acter of either parent would be more likely to OCCUR with mongrels, which are descended from varie
 279  ssigned reasons why such links do not commonly OCCUR at the present day, under the circumstances a
 296  different levels. Hence, when the same species OCCUR at the bottom, middle, and top of a formation
 323  re not found in the Chalk of Europe, but which OCCUR in the formations either above or below, are
 353  , is the most probable. Undoubtedly many cases OCCUR, in which we cannot explain how the same spec
 363  ments of granite and other rocks, which do not OCCUR in the archipelago. Hence we may safely infer
 374  peculiar species belonging to European genera OCCUR. On the highest mountains of Brazil, some few
 375  of the peculiar flora of the Cape of Good Hope OCCUR. At the Cape of Good Hope a very few European
 375  and on the volcanic cones of Java, many plants OCCUR, either identically the same or representing
 375  species; other species, not introduced by man, OCCUR on the lowlands: and a long list can be given
 376  animals. In marine productions, similar cases OCCUR; as an example, I may quote a remark by the h
 384  to fish, I believe that the same species never OCCUR in the fresh waters of distant continents. Bu
 394  lands will not support small mammals, for they OCCUR in many parts of the world on very small isla
 394  wer animals. Though terrestrial mammals do not OCCUR on oceanic islands, aerial mammals do occur o
 394  ot occur on oceanic islands, aerial mammals do OCCUP on almost every island. New Zealand possesses
 395  ntical quadrupeds. No doubt some few anomalies OCCUR in this great archipelago, and there is much
 404  many closely allied or representative species OCCUR, there will likewise be found some identical
 404  ions. And wherever many closely allied species OCCUR, there will be found many forms which some na
 410  the forms which are therein absent, but which OCCUR above and below: so in space, it certainly is
 426  eral characters, let them be ever so trifling, OCCUR together throughout a large group of beings h
 429  extremely few species; and such species as do OCCUR are generally very distinct from each other,
 451  lants of the same species the petals sometimes OCCUR as mere rudiments, and sometimes in a well de
 473  and species reversions to long lost characters OCCUR. How inexplicable on the theory of creation i
 478  rever many closely allied yet distinct species OCCUR, many doubtful forms and varieties of the
 478  and and varieties of the same species likewise OCCUR. It is a rule of high generality that the inh
 165  an accident, that I was led solely from the OCCURENCE of the face stripes on this hybrid from the
 272  f varieties can be proved to be of universal OCCURENCE, or to form a fundamental distinction betwe
 266  ompounded into one, without some disturbance OCCURING in the development, or periodical action, or
 279  r, of innumerable intermediate links not now OCCURING everywhere throughout nature depends on the
 001  greatest philosophers. On my return home, it OCCURRED to me, in 1837, that something might perhaps
 080  at variations useful to man have undoubtedly OCCURRED, that other variations useful in some way to
 112  e extremes, they both go on (as has actually OCCURRED with tumbler pigeons) choosing and breeding
 127  extraordinary fact if no variation ever had OCCURRED useful to each being's own welfare, in the s
 127  , in the same way as so many variations have OCCURRED useful to man. But if variations useful to a
 171  f my work, a crowd of difficulties will have OCCURRED to the reader. Some of them are so grave tha
 226  hive bee making its cells will probably have OCCURRED to many readers, as a difficulty sufficient
 226  rt of two cells. Reflecting on this case, it OCCURRED to me that if the Melipona had made its sphe
 228  ed or broke into each other. As soon as this OCCURRED, the bees ceased to excavate, and began to b
 294  ; for we know what vast geographical changes OCCURRED in other parts of America during this space
 296  wide, yet easily overlooked, intervals have OCCURRED in its accumulation. In other cases we have
 303  t series. Cuvier used to urge that no monkey OCCURRED in any tertiary stratum: but now extinct spe
 308  ts of land, whence the sediment was derived, OCCURRED in the neighbourhood of the existing contine
 315  cies might be adapted (and no doubt this has OCCURRED in innumerable instances) to fill the exact
 327  ei and that blank intervals of vast duration OCCURRED during the periods when the bed of the sea w
 328  on. I suspect that cases of this nature have OCCURRED in Europe. Mr. Prestwich, in his admirable M
 332  eat family, in nearly the same manner as has OCCURRED with ruminants and pachyderms. Yet he who ob
 353  l and climatal changes, which have certainly OCCURRED within recent geological times, must have in
 357  dispute that great mutations of level, have OCCURRED within the period of existing organisms. Edw
```

Page ***********************************(Key Word)***********************************

```
0359  uoyancy of green and seasoned timber; and it OCCURRED to me that floods might wash down plants or
0366  e slowly on, and then pass away, as formerly OCCURRED. As the cold came on, and as each more south
0380  tralia. Something of the same kind must have OCCURRED on the intertropical mountains: no doubt bef
0384  stances, also, could be given of this having OCCURRED during floods, without any change of level.
0397  on the ground, and thus get transported? It OCCURRED to me that land shells, when hybernating and
0407  the level of the land, which have certainly OCCURRED within the recent period, and of other simil
0407  and of other similar changes which may have OCCURRED within the same period: if we remember how p
0421  completely lost, as sometimes seems to have OCCURRED with existing organisms. All the descendants
0438  ance of a modification of this nature having OCCURRED, that naturalists can hardly avoid employing
0058  very of intermediate linking forms, and the OCCURRENCE of such links cannot affect the actual char
0354  f fresh water productions; and thirdly, the OCCURRENCE of the same terrestrial species on islands
0368  they are on the perfectly well ascertained OCCURRENCE of a former Glacial period, seem to me to e
0381  i will only say that as far as regards the OCCURRENCE of identical species at points so enormousl
0457  strong is the principle of inheritance, the OCCURRENCE of rudimentary organs and their final abort
0082  ing a better chance of profitable variations OCCURRING; and unless profitable variations do occur
0146  o not doubt that some apparent correlations, OCCURRING throughout whole orders, are entirely due t
0163  affecting any important character, but from OCCURRING in several species of the same genus, partl
0337  ves. On the other hand, from what we see now OCCURRING in New Zealand, and from hardly a single ir
0384  ion. With respect to allied fresh water fish OCCURRING at very distant points of the world, no dou
0087  attached to that sex, the same fact probably OCCURS under nature, and if so, natural selection wi
0101  s necessity for each birth; in many others it OCCURS perhaps only at long intervals; but in none,
0145  r in the two upper petals; and that when this OCCURS, the adherent nectary is quite aborted; when
0149  the number of the same part or organ, when it OCCURS in lesser numbers, is constant. The same auth
0183  viduals of the same species. When either case OCCURS, it would be easy for natural selection to fu
0262  certain extent parallel. Something analogous OCCURS in grafting; for Thouin found that three spec
0265  reat luxuriance. In both cases, the sterility OCCURS in various degrees; in both, the male element
0182  al state they had been inhabitants of the open OCEAN, and had used their incipient organs of fligh
0285  rd, rocky beds had once extended into the open OCEAN. The same story is still more plainly told by
0288  roughout an enormously large proportion of the OCEAN, the bright blue tint of the water bespeaks i
0305  ed into land, the tropical parts of the Indian OCEAN would form a large and perfectly enclosed bas
0309  that if, for instance, the bed of the Pacific OCEAN were now converted into a continent, we shoul
0343  rphosed condition, or may lie buried under the OCEAN. Passing from these difficulties, all the oth
0348  of the shores of America, a wide space of open OCEAN extends, with not an island as a halting plac
0348  ny fish range from the Pacific into the Indian OCEAN, and many shells are common to the eastern is
0357  rs have thus hypothetically bridged over every OCEAN, and have united almost every island to some
0361  re blown by gales to vast distances across the OCEAN. We may I think safely assume that under such
0363  heir rocky burthens on the shores of these mid OCEAN islands, and it is at least possible that the
0364  land birds are blown across the whole Atlantic OCEAN, from North America to the western shores of
0369  are separated from each other by the Atlantic OCEAN and by the extreme northern part of the Pacif
0370  l more completely separated by wider spaces of OCEAN. I believe the above difficulty may be surmou
0371  wo areas, anc their separation by the Atlantic OCEAN. We can further understand the singular fact
0372  inent and by nearly a hemisphere of equatorial OCEAN. These cases of relationship, without identif
0386  the most remote and barren islands in the open OCEAN; they would not be likely to alight on the su
0394  en seen wandering by day far over the Atlantic OCEAN; and two North American species either regula
0395  h is traversed near Celebes by a space of deep OCEAN; and this space separates two widely distinct
0477  s, though separated by the whole intertropical OCEAN. Although two areas may present the same phys
0477  hose animals which cannot cross wide spaces of OCEAN, as frogs and terrestrial mammals, should not
0477  cular species of bats, which can traverse the OCEAN, should so often be found on islands far dist
0105  d look at any small isolated area, such as an OCEANIC island, although the total number of the spe
0105  en produced there, and nowhere else. Hence an OCEANIC island at first sight seems to have been hig
0134  now inhabit or have lately inhabited several OCEANIC islands, tenanted by no beast of prey, has b
0184  limbs a tree! Petrels are the most aerial and OCEANIC of birds, yet in the quiet Sounds of Tierra
0308  e them studded with many islands; but not one OCEANIC island is as yet known to afford even a remn
0357  to each other and to the several intervening OCEANIC islands. I freely admit the former existence
0358  d with each other, and with the many existing OCEANIC islands. Several facts in distribution, such
0358  nd relative proportions of the inhabitants of OCEANIC islands likewise seem to me opposed to the b
0363  urope, in comparison with the plants of other OCEANIC islands nearer to the mainland, and (as rema
0383  resh water productions. On the inhabitants of OCEANIC islands. Absence of the Batrachians and of t
0385  a pool or rivulet, if blown across sea to an OCEANIC island or to any other distant point. Sir Ch
0386  ; both over continents and to the most remote OCEANIC islands. This is strikingly shown, as remark
0388  y well fitted for them. On the Inhabitants of OCEANIC islands. We now come to the last of the thre
0389  ation. The species of all kinds which inhabit OCEANIC islands are few in number compared with thos
0390  s they have on New Zealand and on every other OCEANIC island which can be named. In St. Helena hav
0390  d plants and animals have not been created on OCEANIC islands; for man has unintentionally stocked
0390  ly and perfectly than has nature. Although in OCEANIC islands the number of kinds of inhabitants i
0391  n madeira apparently present analogous facts. OCEANIC islands are sometimes deficient in certain c
0392  trees would be little likely to reach distant OCEANIC islands; and an herbaceous plant, though it
0393  ith respect to the absence of whole orders on OCEANIC islands, Bory St. Vincent long ago remarked
0393  absence of frogs, toads, and newts on so many OCEANIC islands cannot be accounted for by their phy
0393  a, and therefore why they do not exist on any OCEANIC island. But why, on the theory of creation,
0393  ption? but this group cannot be considered as OCEANIC, as it lies on a bank connected with the mai
0394  s. though terrestrial mammals do not occur on OCEANIC islands, aerial mammals do occur on almost e
0396  l the foregoing remarks on the inhabitants of OCEANIC islands, namely, the scarcity of kinds, the
0396  course of time, than with the view of all our OCEANIC islands having been formerly connected by co
0397  of one of the cases of difficulty. Almost all OCEANIC islands, even the most isolated and smallest
0403  e general character of the fauna and flora of OCEANIC islands, namely, that the inhabitants, when
0409  nderstand, as I have endeavoured to show, why OCEANIC islands should have few inhabitants, but of
0409  nd terrestrial mammals, should be absent from OCEANIC islands, whilst the most isolated islands po
0454  caverns, and of the wings of birds inhabiting OCEANIC islands, which have seldom been forced to ta
0477  with subsequent modification, we can see why OCEANIC islands should be inhabited by few species,
0477  s and terrestrial mammals, should not inhabit OCEANIC islands; and why, on the other hand, new and
0478  ats, and the absence of all other mammals, on OCEANIC islands, are utterly inexplicable on the the
0308  and unfathomable sea. Looking to the existing OCEANS, which are thrice as extensive as the land, w
0308  nts nor continental islands existed where our OCEANS now extend; for had they existed there, palae
0309  from these facts, we may infer that where our OCEANS now extend, oceans have extended from the rem
0309  e may infer that where our oceans now extend, OCEANS have extended from the remotest period of whi
```

Page ***********************************(Key Word)***********************************

Page **(Key Word)**

0309 oral Reefs, led me to conclude that the great OCEANS are still mainly areas of subsidence, the gre
0309 rian epoch, continents may have existed where OCEANS are now spread out; and clear and open oceans
0309 oceans are now spread out; and clear and open OCEANS may have existed where our continents now sta
0343 y saying that as far as we can see, where our OCEANS now extend they have for an enormous period e
0348 ble, or likely to have endured so long as the OCEANS separating continents, the differences are ve
0357 uring their migration. In the coral producing OCEANS such sunken islands are now marked, as I beli
0360 nds, even on those in the midst of the widest OCEANS; and the natives of the coral islands in the
0393 any of the many islands with which the great OCEANS are studded. I have taken pains to verify thi
0240 bserved, the larger workers have simple eyes (OCELLI), which though small can be plainly distingui
0240 ished, whereas the smaller workers have their OCELLI rudimentary. Having carefully dissected sever
0240 t the workers of intermediate size have their OCELLI in an exactly intermediate condition. So that
0240 workers of Myrmica have not even rudiments of OCELLI, though the male and female ants of this genu
0240 female ants of this genus have well developed OCELLI. I may give one other case: so confidently di
0213 f disposition and tastes, and likewise of the ODDEST tricks, associated with certain frames of min
0220 or doubting the truth of so extraordinary an ODIOUS an instinct as that of making slaves. Hence I
0049 have a different flavour and emit a different ODOUR; they flower at slightly different periods: t
0281 or from some other allied species, such as C. OENAS. So with natural species, if we look to forms
0021 ly expanding slightly the upper part of the OESOPHAGUS. The Jacobin has the feathers so much rever
0022 ze of the crop and of the upper part of the OESOPHAGUS, the development and abortion of the oil gl
0039 n the turbit now does the upper part of its OESOPHAGUS, a habit which is disregarded by all fancie
0004 ticated animals and of cultivated plants would OFFER the best chance of making out this obscure pr
0095 alone; so that whole fields of the red clover OFFER in vain an abundant supply of precious nectar
0100 hermaphrodites; but here currents in the water OFFER an obvious means for an occasional cross. And
0150 males and females; but as females more rarely OFFER remarkable secondary sexual character, it app
0192 he next chapter. The electric organs of fishes OFFER another case of special difficulty; it is imp
0193 n of any kind is possible. The electric organs OFFER another and even more serious difficulty; for
0240 that I gladly availed myself of Mr. F. Smith's OFFER of numerous specimens from the same nest of t
0334 ing and succeeding faunas, that certain genera OFFER exceptions to the rule. For instance, mastodo
0393 it would be very difficult to explain. Mammals OFFER another and similar case. I have carefully se
0418 the most important vital organs, are found to OFFER characters of quite subordinate value. We can
0474 ing at instincts, marvellous as some are, they OFFER no greater difficulty than does corporeal str
0266 to the root of the matter: no explanation is OFFERED why an organism, when placed under unnatural
0354 moment pretend that any explanation could be OFFERED of many such cases. But after some prelimina
0072 ine the existence of cattle. Perhaps Paraguay OFFERS the most curious instance of this; for here n
0084 ly working, whenever and wherever opportunity OFFERS, at the improvement of each organic being in
0100 we can understand this remarkable fact, which OFFERS so strong a contrast with terrestrial plants,
0193 , belonging to different families and orders, OFFERS a parallel case of difficulty. Other cases co
0284 accumulation of the degraded matter, probably OFFERS the best evidence of the lapse of time. I rem
0288 the chalk period. The molluscan genus Chiton OFFERS a partially analogous case. With respect to t
0349 of the widest generality, and every continent OFFERS innumerable instances. Nevertheless the natur
0221 nd as Huber expressly states, their principal OFFICE is to search for aphides. This difference in
0451 efficient for the other. Thus in plants, the OFFICE of the pistil is to allow the pollen tubes to
0009 ; by this term gardeners mean a single bud or OFFSET, which suddenly assumes a new and sometimes v
0010 treatment of the parent has affected a bud or OFFSET, and not the ovules or pollen. But it is the
0009 t, acting not quite regularly, and producing OFFSPRING not perfectly like their parents or variabl
0013 ity first appears, it tends to appear in the OFFSPRING at a corresponding age, though sometimes ea
0014 the horns of cattle could appear only in the OFFSPRING when nearly mature; peculiarities in the si
0014 age, yet that it does tend to appear in the OFFSPRING at the same period at which it first appear
0014 in nearly the same manner as in the crossed OFFSPRING from a short horned cow by a long horned bu
0017 know propagate their kind so truly, were the OFFSPRING of any single species; then such facts woul
0020 erimentised for this object, and failed. The OFFSPRING from the first cross between two pure breed
0025 ny of the above specified marks, the mongrel OFFSPRING are very apt suddenly to acquire these char
0026 ble, to bring forward one case of the hybrid OFFSPRING of two animals clearly distinct being thems
0026 pouters, and fantails now are; should yield OFFSPRING perfectly fertile, inter se, seems to me ra
0027 when kept pure and when crossed; the mongrel OFFSPRING being perfectly fertile; from these several
0036 r to think of the inherited character of the OFFSPRING of their domestic animals, yet any one anim
0036 oice animals would thus generally leave more OFFSPRING than the inferior ones; so that in this cas
0045 uch as are known frequently to appear in the OFFSPRING from the same parents, or which may be pres
0054 dy dominant will be the most likely to yield OFFSPRING which, though in some slight degree modifie
0061 dual, and will generally be inherited by its OFFSPRING. The offspring, also, will thus have a bett
0061 generally be inherited by its offspring. The OFFSPRING, also, will thus have a better chance of su
0086 ular period of life, tend to reappear in the OFFSPRING at the same period; for instance, in the se
0088 o the unsuccessful competitor, but few or no OFFSPRING. Sexual selection is, therefore, less rigor
0088 ess cock would have a poor chance of leaving OFFSPRING. Sexual selection by always allowing the vi
0090 e transmitted these advantages to their male OFFSPRING. Yet, I would not wish to attribute all suc
0091 the best chance of surviving and of leaving OFFSPRING. Some of its young would probably inherit t
0096 er strain, gives vigour and fertility to the OFFSPRING; and on the other hand, that close interbre
0104 ain so much in vigour and fertility over the OFFSPRING from long continued self fertilisation, tha
0104 of character can be given to their modified OFFSPRING, solely by natural selection preserving the
0111 in some character from its parents, and the OFFSPRING of this variety again to differ from its pa
0117 oceeding from (A), may represent its varying OFFSPRING. The variations are supposed to be extremel
0119 t branches proceeding from (A). The modified OFFSPRING from the later and more highly improved bra
0122 lines of descent. If, however, the modified OFFSPRING of a species get into some distinct country
0122 f the original species will have transmitted OFFSPRING to the fourteen thousandth generation. We m
0127 ple of inheritance they will tend to produce OFFSPRING similarly characterised. This principle of
0127 d best adapted males the greatest number of OFFSPRING. Sexual selection will also give characters
0128 hese will tend to transmit to their modified OFFSPRING that superiority which now makes them domin
0132 bute the varying or plastic condition of the OFFSPRING. The male and female sexual elements seem t
0157 does not entail death, but only gives fewer OFFSPRING to the less favoured males. Whatever the ca
0160 re eminently liable to appear in the crossed OFFSPRING of two distinct and differently coloured br
0160 n crossed only once by some other breed, the OFFSPRING occasionally show a tendency to revert in c
0160 he most probable hypothesis is, not that the OFFSPRING suddenly takes after an ancestor some hundr
0162 on my theory, sometimes to find the varying OFFSPRING of a species assuming characters (either fr
0165 d male quagga, the hybrid, and even the pure OFFSPRING subsequently produced from the mare by a bl
0170 ause may be of each slight difference in the OFFSPRING from their parents, and a cause for each mu
0172 crossed, being sterile and producing sterile OFFSPRING, whereas, when varieties are crossed their

Page **(Key Word)**

Page **(Key Word)**

```
0218  lost all maternal love and care for her own OFFSPRING. The occasional habit of birds laying thei
0237  on of structure, this being inherited by its OFFSPRING, which again varied and were again selecte
0238  flourished, and transmitted to their fertile OFFSPRING a tendency to produce sterile members havi
0245  varieties when crossed and of their mongrel  OFFSPRING not universal. Hybrids and mongrels compare
0246  n intercrossed they produce either few or no OFFSPRING. Hybrids, on the other hand, have their re
0246  and likewise the fertility of their mongrel  OFFSPRING, is, on my theory, of equal importance wit
0246  of species when crossed and of their hybrid  OFFSPRING. It is impossible to study the several mem
0247  two species when crossed and by their hybrid OFFSPRING, with the average number produced by both
0256  icult to cross, and which rarely produce any OFFSPRING, are generally very sterile; but the paral
0256  nusual facility, and produce numerous hybrid OFFSPRING, yet these hybrids are remarkably sterile.
0259  of impressing their likeness on their hybrid OFFSPRING; but these two powers do not at all necess
0266  le to breed inter se, they transmit to their OFFSPRING from generation to generation the same com
0267  ub breeds, gives vigour and fertility to the OFFSPRING. I believe, indeed, from the facts alluded
0267  different, give vigour and fertility to the  OFFSPRING. But we have seen that greater changes, or
0267  varieties when crossed, and of their Mongrel OFFSPRING. It may be urged, as a most forcible argume
0268  erfect facility, and yield perfectly fertile OFFSPRING. I fully admit that this is almost invariab
0271  ciprocal crosses, he found their mongrel     OFFSPRING perfectly fertile. But one of these five va
0272  ependently of the question of fertility, the OFFSPRING of species when crossed and of varieties wh
0272  ant differences between the so called mongrel OFFSPRING of species, and the so called mongrel offsp
0272  spring of species, and the so called mongrel OFFSPRING of varieties. And, on the other hand, they
0272  ns an extreme amount of variability in their OFFSPRING is notorious; but some few cases both of hy
0273  ncapable of its proper function of producing OFFSPRING identical with the parent form. Now hybrids
0275  n the female, so that the mule, which is the OFFSPRING of the male ass and mare, is more like an a
0275  like an ass, than is the hinny, which is the OFFSPRING of the female ass and stallion. Much stress
0275  to be a general and close similarity in the OFFSPRING of crossed species; and of crossed varietie
0277  ourable to the vigour and fertility of their OFFSPRING; and that slight changes in the conditions
0277  and the degree of sterility of their hybrid OFFSPRING should generally correspond, though due to
0277  o considered as varieties, and their mongrel OFFSPRING, are very generally, but not quite universa
0315  and inorganic, should recur. For though the OFFSPRING of one species might be adapted (and no dou
0316  supplanted and exterminated by its improved OFFSPRING, it is quite incredible that a fantail, ide
0334  e of descent in the diagram are the modified OFFSPRING of those which lived at the fifth stage, ar
0344  e extinct together, and to leave no modified OFFSPRING on the face of the earth. But the utter ex
0405  ditions favourable for the conversion of its OFFSPRING, firstly into new varieties and ultimately
0443  old age alone, has been communicated to the OFFSPRING from the reproductive element of one parent
0444  ds to reappear at a corresponding age in the OFFSPRING. Certain variations can only appear at corr
0444  tend to appear at a corresponding age in the OFFSPRING and parent. I am far from meaning that this
0446  riod of life, and have been inherited by the OFFSPRING at a corresponding not early period. But th
0451  es, the rudiment of the pistil in their male OFFSPRING was much increased in size; and this shows
0460  eties when intercrossed and of their mongrel OFFSPRING cannot be considered as universal; nor is t
0461  es in their conditions of life, and that the OFFSPRING of slightly modified forms or varieties acc
0466  en not rendered impotent, fails to reproduce OFFSPRING exactly like the parent form. Variability
0470  ral selection to preserve the most divergent OFFSPRING of any one species. Hence during a long cor
0475  ieties, we can at once see why their crossed OFFSPRING should follow the same complex laws in the
0475  and in other such points, as do the crossed OFFSPRING of acknowledged varieties. On the other har
0476  ws from the living and the extinct being the OFFSPRING of common parents. As the groups which have
0002  this volume on which facts cannot be adduced, OFTEN apparently leading to conclusions directly op
0008  oldest cultivated plants, such as wheat, still OFTEN yield new varieties: our oldest domesticated
0009  we see domesticated animals and plants, though OFTEN weak and sickly, yet breeding quite freely ur
0013  inherited and sometimes not so: why the child OFTEN reverts in certain characters to its grandfat
0013  uch more remote ancestor: why a peculiarity is OFTEN transmitted from one sex to both sexes, or to
0013  earing in the males of our domestic breeds are OFTEN transmitted either exclusively, or in a much
0014  of reversion, I may here refer to a statement OFTEN made by naturalists, namely, that our domesti
0014  what decisive facts the above statement has so OFTEN and so boldly been made. There would be great
0016  ies. Domestic races of the same species, also, OFTEN have a somewhat monstrous character: by which
0016  same genus, in several trifling respects, they OFTEN differ in an extreme degree in some one part
0016  f doubt could not so perpetually recur. It has OFTEN been stated that domestic races do not differ
0016  all presently give, we have no right to expect OFTEN to meet with generic differences in our domes
0017  trong, evidence in favour of this view. It has OFTEN been assumed that man has chosen for domestic
0019  er existed freely in a state of nature. It has OFTEN been loosely said that all our races of dogs
0026  r of generations. These two distinct cases are OFTEN confounded in treatises on inheritance. Lastl
0028  l then, also, see how it is that the breeds so OFTEN have a somewhat monstrous character. It is al
0032  y horticulturists; but the variations are here OFTEN more abrupt. No one supposes that our choices
0034  periods of English history choice animals were OFTEN imported, and laws were passed to prevent the
0038  ly different constitutions or structure, would OFTEN succeed better in the one country than in the
0042  lmost always imported from some other country, OFTEN from islands. Although I do not doubt that so
0048  , and by another as varieties, or, as they are OFTEN called, as geographical races? Many years ago
0050  d recorded. These varieties, moreover, will be OFTEN ranked by some authors as species. Look at th
0051  ariation, and the truth of this admission will OFTEN be disputed by other naturalists. When, moreo
0053  nge, and to a certain extent from commonness), OFTEN give rise to varieties sufficiently well mark
0056  us that small genera have in the lapse of time OFTEN increased greatly in size: and that large gen
0056  ed greatly in size: and that large genera have OFTEN come to their maxima, declined, and disappear
0057  he amount of difference between the species is OFTEN exceedingly small. I have endeavoured to test
0058  her species, and in so far resemble varieties, OFTEN have much restricted ranges. For instance, Mr
0060  individuals and varieties of the same species: OFTEN severe between species of the same genus. The
0062  lc the face of nature bright with gladness, we OFTEN see superabundance of food: we do not see, or
0070  eems generally to occur with our game animals, OFTEN ensue: and here we have a limiting check inde
0073  time, though assuredly the merest trifle would OFTEN give the victory to one organic being over an
0075  beings remote in the scale of nature. This is OFTEN the case with those which may strictly be sai
0077  ic being is related, in the most essential yet OFTEN hidden manner, to that of all other organic b
0082  he structure or habits of one inhabitant would OFTEN give it an advantage over others; and still f
0082  l further modifications of the same kind would OFTEN still further increase the advantage. No coun
0083  as lies in his power, all his productions. He OFTEN begins his selection by some half monstrous f
0086  of the being, will cause other modifications, OFTEN of the most expected nature. As we see that t
0086  sely, modifications in the adult will probably OFTEN affect the structure of the larva; but in all
0087  e. sexual Selection. Inasmuch as peculiarities OFTEN appear under domestication in one sex and bec
0088  seen fighting all day long: male stag beetles OFTEN bear wounds from the huge mandibles of other
```

Page **(Key Word)**

```
088  sword or spear. Amongst birds, the contest is  OFTEN of a more peaceful character. All those who h
089  ed to birds in confinement well know that they  OFTEN take individual preferences and dislikes: thu
091  or can this be thought very improbable; for we  OFTEN observe great differences in the natural tend
092  ld get dusted with pollen, and would certainly  OFTEN transport the pollen from one flower to the s
098  ul for this end: but, the agency of insects is  OFTEN required to cause the stamens to spring forwa
100  belonging to all Orders have their sexes more   OFTEN separated than other plants, I find to be the
106  ntinuous, owing to oscillations of level, will  OFTEN have recently existed in a broken condition,
108  f some kind. The existence of such places will  OFTEN depend on physical changes, which are general
108  y always a very slow process. The process will  OFTEN be greatly retarded by free intercrossing. Ma
108  natural selection will always act very slowly,  OFTEN only at long intervals of time, and generally
117  not supposed to all appear simultaneously, but  OFTEN after long intervals of time; nor are they al
119  riably prevail and multiply: a medium form may  OFTEN long endure, and may or may not produce more
119  in the lines of descent, will, it is probable,  OFTEN take the place of, and so destroy, the earlie
124  ies, or genera, with those now living, yet are  OFTEN, in some degree, intermediate in character be
133  into the zone of habitation of other species,  OFTEN acquiring in a very slight degree some of the
133  d to sea coasts, as every collector knows, are  OFTEN brassy or lurid. Plants which live exclusivel
135  et, of many male dung feeding beetles are very  OFTEN broken off: he examined seventeen specimens f
137  mole: and I was assured by a Spaniard, who had  OFTEN caught them, that they were frequently blind:
138  literated its eyes, and natural selection will  OFTEN have effected other changes, such as an incre
139  ecies to the climates under which they live is  OFTEN overrated. We may infer this from our frequen
142  d! the case, also, of the kidney bean has been  OFTEN cited for a similar purpose, and with much gr
143  s: but that the effects of use and disuse have  OFTEN been largely combined with, and sometimes ove
143  arked by some authors, tend to cohere; this is  OFTEN seen in monstrous plants; and nothing is more
145  or instance, the daisy, and this difference is  OFTEN accompanied with the abortion of parts of the
145  rgoniums, that the central flower of the truss  OFTEN loses the patches of darker colour in the two
146  f the slightest service to the species. We may  OFTEN falsely attribute to correlation of growth, s
155  s in value and becomes only of specific value,  OFTEN becomes variable, though its physiological im
155  ies, we might surely expect to find them still  OFTEN continuing to vary in those parts of their st
156  genitor, it is probable that they should still  OFTEN be in some degree variable, at least more var
159  ogous variations; and a variety of one species  OFTEN assumes some of the characters of an allied s
161  (antirrhinum) a rudiment of a fifth stamen so   OFTEN appears, that this plant must have an inherit
162  ferred this, from the blue colour and marks so  OFTEN appearing when distinct breeds of diverse col
167  omestication, in this particular manner, so as  OFTEN to become striped like other species of the g
169  rose; and thus we can understand why it should  OFTEN still be variable in a much higher degree tha
173  nomy of the land. These representative species  OFTEN meet and interlock; and as the one becomes ra
174  d of climate, marine areas now continuous must  OFTEN have existed within recent times in a far les
178  we do see. Secondly, areas now continuous must  OFTEN have existed within the recent period in isol
179  ral selection constantly tends, as has been so  OFTEN remarked, to exterminate the parent forms and
182  tructure lead to changed habits: both probably  OFTEN change almost simultaneously. Of cases of cha
183  s innumerable instances could be given: I have  OFTEN watched a tyrant flycatcher (Saurophagus sulp
184  n climbing branches, almost like a creeper. It  OFTEN, like a shrike, kills small birds by blows on
192  e gravest is that of neuter insects, which are  OFTEN very differently constructed from either the
196  tures: and finally, that sexual selection will  OFTEN have largely modified the external characters
198  animals kept by savages in different countries  OFTEN have to struggle for their own subsistence, a
199  rt, and a useful modification of one part will  OFTEN have entailed on other parts diversified chan
200  of another. But natural selection can and does  OFTEN produce structures for the direct injury of o
202  how it is that the use of the sting should so  OFTEN cause the insect's own death: for if on the w
203  species, now living on a continuous area, must  OFTEN have been formed when the area was not contin
203  continuous area, an intermediate variety will  OFTEN be formed, fitted for an intermediate zone: b
205  been perfected whilst aided by the other, must  OFTEN have largely facilitated transitions. We are
205  ve that a part formerly of high importance has  OFTEN been retained (as the tail of an aquatic anim
205  f one country, generally the smaller one, will  OFTEN yield, as we see they do yield, to the inhabi
206  the expression of conditions of existence, so  OFTEN insisted on by the illustrious Cuvier, is ful
208  rre tuber expresses it, of judgment or reason,  OFTEN comes into play, even in animals very low in
208  d states of the body. When once acquired, they  OFTEN remain constant throughout life. Several othe
212  and temperature of the country inhabited, but  OFTEN from causes wholly unknown to us: Audubon has
232  s when two pieces of comb met at an angle, how  OFTEN the bees would entirely pull down and rebuild
233  er is not difficult: it is known that bees are  OFTEN hard pressed to get sufficient nectar; and I
234  latter circumstance determined, as it probably  OFTEN does determine, the numbers of a humble bee w
236  males in insect communities: for these neuters  OFTEN differ widely in instinct and in structure fr
239  ations of structure: and this we do find, even  OFTEN, considering how few neuter insects out of Eu
239  f compared perfect gradations of this kind. It  OFTEN happens that the larger or the smaller sized
243  considerably different conditions of life, yet  OFTEN retaining nearly the same instincts. For inst
247  be hybridised must be castrated, and, what is  OFTEN more important, must be secluded in order to
247  chamber in his house. That these processes are  OFTEN injurious to the fertility of a plant cannot
248  is is usually the case, and that the fertility  OFTEN suddenly decreases in the first few generatio
248  now, in artificial fertilisation pollen is as  OFTEN taken by chance (as I know from my own experi
253  are generally ranked in distinct species, have  OFTEN bred in this country with either pure parent,
256  with hybrids, for their degree of fertility is  OFTEN found to differ greatly in the several indivi
257  d fitted for extremely different climates, can  OFTEN be crossed with ease. By a reciprocal cross b
258  id to have been reciprocally crossed. There is  OFTEN the widest possible difference in the facilit
258  at the capacity in any two species to cross is  OFTEN completely independent of their systematic af
260  ds, moreover, produced from reciprocal crosses  OFTEN differ in fertility. Now do these complex and
260  oduce fairly fertile hybrids? Why should there  OFTEN be so great a difference in the result of a r
261  pecies the male sexual element of the one will  OFTEN freely act on the female sexual element of th
262  crosses, the facility of effecting an union is  OFTEN very far from equal, so it sometimes is in gr
265  ility is independent of general health, and is  OFTEN accompanied by excess of size or great luxuri
265  conditions of life have been disturbed, though  OFTEN in so slight a degree as to be inappreciable
267  er changes, or changes of a particular nature,  OFTEN render organic beings in some degree sterile;
272  om species which have long been cultivated are  OFTEN variable in the first generation; and I have
273  re we might expect that such variability would  OFTEN continue and be super added to that arising f
273  y change in the conditions of life, being thus  OFTEN rendered either impotent or at least incapabl
274  of plants. With animals one variety certainly  OFTEN has this prepotent power over another variety
275  h mongrels, which are descended from varieties  OFTEN suddenly produced and semi monstrous in chara
276  erile. The sterility is of all degrees, and is  OFTEN so slight that the two most careful experimen
283  kly ground into pebbles, sand, or mud. But how  OFTEN do we see along the bases of retreating cliff
```

Page **(Key Word)***************************************

```
0287 il species are known and named from single and OFTEN broken specimens, or from a few specimens co
0292 e sea will be increased, and new stations will OFTEN be formed; all circumstances most favourable
0295 nce. But this same movement of subsidence will OFTEN tend to sink the area whence the sediment is
0295 ent in its accumulation. When we see, as is so OFTEN the case, a formation composed of beds of di
0310 , murchison, Sedgwick, etc., have unanimously, OFTEN vehemently, maintained the immutability of s
0321 nus, belonging to the same family. But it must OFTEN have happened that a new species belonging t
0321 ified and improved, a few of the sufferers may OFTEN long be preserved, from being fitted to some
0326 ties and species. The process of diffusion may OFTEN be very slow, being dependent on climatal an
0328 that strictly contemporaneous formations have OFTEN been accumulated over very wide spaces in th
0329 same exact periods, a formation in one region OFTEN corresponding with a blank interval in the o
0333 s it comes that ancient and extinct genera may OFTEN in some slight degree intermediate in charac
0336 ous. As the accumulation of each formation has OFTEN been interrupted, and as long blank interval
0342 new species; and that varieties have at first OFTEN been local. All these causes taken conjointl
0343 are considered; he may urge the apparent, but OFTEN falsely apparent, sudden coming in of whole
0344 ter extinction of a whole group of species may OFTEN be a very slow process, from the survival of
0344 hose now living. Why ancient and extinct forms OFTEN tend to fill up gaps between existing forms,
0344 together. The more ancient a form is, the more OFTEN, apparently, it displays characters in some
0350 organism to organism being, as I have already OFTEN remarked, the most important of all relation
0356 hich habitually unite for each birth, or which OFTEN intercross, I believe that during the slow p
0360 then germinate. The fact of the larger fruits OFTEN floating longer than the small, is interest*
0361 such circumstances their rate of flight would OFTEN be 35 miles an hour; and some authors have g
0384 tinents. But on the same continent the species OFTEN range widely and almost capriciously; for tw
0388 , and naturally travel from one to another and OFTEN distant piece of water. Nature, like a caref
0390 i.e. those found nowhere else in the world) is OFTEN extremely large. If we compare, for instance
0390 be eminently liable to modification, and will OFTEN produce groups of modified descendants. But
0392 elytra of many insular beetles. Again, islands OFTEN possess trees or bushes belonging to orders
0392 e other plants. If so, natural selection would OFTEN tend to add to the stature of herbaceous pla
0402 ead from some one island to the others. But we OFTEN take, I think, an erroneous view of the prob
0407 erly experimentised on; if we bear in mind how OFTEN a species may have ranged continuously over
0407 the difficulties are insuperable; though they OFTEN are in this case, and in that of the individ
0408 ing nearly the same physical conditions should OFTEN be inhabited by very different forms of life
0409 resentative species. As the late Edward Forbes OFTEN insisted, there is a striking parellelism in
0410 tain period of time, or to a certain area, are OFTEN characterised by trifling characters in comm
0413 s as that famous one of Linnaeus, and which we OFTEN meet with in a more or less concealed form.
0416 yet, undoubtedly, organs in this condition are OFTEN of high value in classification. No one will
0417 very evident in natural history. Hence, as has OFTEN been remarked, a species may depart from its
0419 g the idea of descent. Our classifications are OFTEN plainly influenced by chains of affinities.
0419 s of Articulata. Geographical distribution has OFTEN been used, though perhaps not quite logicall
0426 sed with them. This may be safely done, and is OFTEN done, as long as a sufficient number of char
0428 th fishes. As members of distinct classes have OFTEN been adapted by successive slight modificat*
0429 how it is that the more ancient forms of life OFTEN present characters in some slight degree int
0431 ious lengths (as may be seen in the diagram so OFTEN referred to), mounting up through many prede
0434 lan of their organisation. This resemblance is OFTEN expressed by the term unity of type; or by s
0435 ofitable in some way to the modified form, but OFTEN affecting by correlation of growth other par
0436 , arranged in a spire. In monstrous plants, we OFTEN get direct evidence of the possibility of on
0439 of distinct animals within the same class are OFTEN strikingly similar: a better proof of this c
0439 the same genus, and of closely allied genera, OFTEN resemble each other in their first and secon
0439 animals of the same class resemble each other, OFTEN have no direct relation to their conditions
0443 s commonly assumed, perhaps from monstrosities OFTEN affecting the embryo at a very early period,
0451 together! The meaning of rudimentary organs is OFTEN quite unmistakeable: for instance there are
0451 plants with separated sexes, the male flowers OFTEN have a rudiment of a pistil: and Kolreuter f
0452 in the upper jaws of whales and ruminants, can OFTEN be detected in the embryo, but afterwards wh
0453 ce, also, a rudimentary organ in the adult, is OFTEN said to have retained its embryonic conditio
0453 can we suppose that the minute papilla, which OFTEN represents the pistil in male flowers, and w
0454 ate of the whole flower in the cauliflower. We OFTEN see rudiments of various parts in monsters.
0455 be saved as far as is possible, will probably OFTEN come into play; and this will tend to cause
0461 assed from some one part to the others. We are OFTEN wholly unable even to conjecture how this co
0462 many means. A broken or interrupted range may OFTEN be accounted for by the extinction of the sp
0463 llied species, we have no just right to expect OFTEN to find intermediate varieties in the interm
0463 f allied species appear, though certainly they OFTEN falsely appear, to have come in suddenly on
0464 y ranging species vary most, and varieties are OFTEN at first local, both causes rendering the di
0468 ecies of the same genus. But the struggle will OFTEN be very severe between beings most remote in
0468 generally leave most progeny. But success will OFTEN depend on having special weapons or means of
0468 tural selection had not come into play. It has OFTEN been asserted, but the assertion is quite in
0468 though acting on external characters alone and OFTEN capriciously, can produce within a short per
0476 the progenitor with its early descendants will OFTEN be intermediate in character in comparison w
0477 bats, which can traverse the ocean, should so OFTEN be found on islands far distant from any con
0478 subordinate to group, and with extinct groups OFTEN falling in between recent groups, is intelli
0479 parts, though of no service to the being, are OFTEN of high classificatory value; and why embryo
0479 se, aided sometimes by natural selection, will OFTEN tend to reduce an organ, when it has become
0484 rifling a circumstance as that the same poison OFTEN similarly affects plants and animals; or tha'
0055 es of the larger genera in each country would OFTENER present varieties, than the species of the
0069 northward, or in ascending a mountain, we far OFTENER meet with stunted forms, due to the directl
0108 tion of natural selection will probably still OFTENER depend on some of the inhabitants becoming
0376 times identically the same; but they are much OFTENER specifically distinct, though related to ea
0476 can see why the more ancient a fossil is, the OFTENER it stands in some degree intermediate betwe
0054 are the most numerous in individuals, which OFTENEST produce well marked varieties, or, as I con
0088 males of polygamous animals, and these seem OFTENEST provided with special weapons. The males of
0092 es, and which excreted most nectar, would be OFTENEST visited by insects, and would be oftenest c
0092 be oftenest visited by insects, and would be OFTENEST crossed; and so in the long run would gain
0125 it is the species of the larger genera which OFTENEST present varieties or incipient species. This
0136 etles which most readily took to flight will OFTENEST have been blown to sea and thus have been de
0145 gular flowers, those nearest to the axis are OFTENEST subject to peloria, and become regular. I ma
0164 l parts of the world these stripes occur far OFTENEST in duns and mouse duns; by the term dun a la
0298 t is those which have the widest range, that OFTENEST present varieties; so that with shells and c
0298 geological formations of Europe, which have OFTENEST given rise, first to local varieties and ult
```

Page **(Key Word)***************************************

age **(Key Word)**

```
301  hiefly these far ranging species which would OFTENEST produce new varieties; and the varieties wou
325  forms in their own country, would naturally OFTENEST give rise to new varieties or incipient spec
342  s are those which have varied most, and have OFTENEST given rise to new species; and that varietie
344  dominant forms of life, which are those that OFTENEST vary, will in the long run tend to people th
022  that in good birds the head and tail touch;  the OIL gland is quite aborted. Several other less dis
022  oesophagus, the development and abortion of the OIL gland; the number of the primary wing and caud
147  and nutritious foliage and a copious supply of OIL bearing seeds. When the seeds in our fruits be
012  the several treatises published on some of our OLD cultivated plants, as on the hyacinth, potato,
028  e dutch were as eager about pigeons as were the OLD Romans. The paramount importance of these cons
035  ually, and yet so effectually, that, though the OLD Spanish pointer certainly came from Spain, Mr.
035  his country. By comparing the accounts given in OLD pigeon treatises of carriers and tumblers with
036  del Fuego, by their killing and devouring their OLD women, in times of dearth, as of less value th
035  mark to assume that it breeds when thirty years OLD, and goes on breeding till ninety years old, b
064  ars old, and goes on breeding till ninety years OLD, bringing forth three pair of young in this in
065  e has consequently been less destruction of the OLD and young, and that nearly all the young have
066  ruction inevitably falls either on the young or OLD, during each generation or at recurrent interv
071  here are extensive heaths, with a few clumps of OLD Scotch firs on the distant hilltops: within th
072  i could not see a single Scotch fir, except the OLD planted clumps. But on looking closely between
072  oint some hundred yards distant from one of the OLD clumps, I counted thirty two little trees; and
075  l numbers and kinds of trees now growing on the OLD Indian ruins! The dependency of one organic be
104  al economy of the country are left open for the OLD inhabitants to struggle for, and become adapte
107  ew forms will have been more slowly formed, and OLD forms more slowly exterminated. And it is in f
108  ll have to be filled up by modifications of the OLD inhabitants; and time will be allowed for the
139  several of the inhabitants of the caves of the OLD and New Worlds should be closely related, we m
164  t in the foal; and sometimes quite disappear in OLD horses. Colonel Poole has seen both gray and b
166  r appear more commonly in the young than in the OLD. Call the breeds of pigeons, some of which hav
167  eption! I would almost as soon believe with the OLD and ignorant cosmogonists, that fossil shells
178  n a still more important degree, on some of the OLD inhabitants becoming slowly modified, with the
178  ified, with the new forms thus produced and the OLD ones acting and reacting on each other. So tha
180  ing rodents or new beasts of prey immigrate, or OLD ones become modified, and all analogy would le
189  eserved till a better be produced, and then the OLD ones to be destroyed. In living bodies, variat
194  he truth of this remark is indeed shown by that OLD canon in natural history of Natura non facit s
206  can clearly understand the full meaning of that OLD canon in natural history, Natura non facit sal
209  nstead of playing the pianoforte at three years OLD with wonderfully little practice, had played a
217  y she laid an egg in another bird's nest.If the OLD bird profited by this occasional habit, or if
217  ng of different ages at the same time; then the OLD birds or the fostered young would gain an adva
219  nests, or of feeding their own larvae. When the OLD nest is found inconvenient, and they have to m
225  ort series we have humble bees, which use their OLD cocoons to hold honey, sometimes adding to the
227  cylinders of wax to the circular mouths of her OLD cocoons. By such modifications of instincts in
266  ied yet very different class of facts. It is an OLD and almost universal belief, founded, I think,
281  mproved forms of life will tend to supplant the OLD and unimproved forms. By the theory of natural
282  strata, and watch the sea at work grinding down OLD rocks and making fresh sediment, before he can
306  it carries on my theory be supposed, that these OLD species were the progenitors of all the specie
315  e, and thus supplant it; yet the two forms, the OLD and the new, would not be identically the same
316  e generated either new and modified or the same OLD and unmodified forms. Species of the genus Lin
317  e theory of natural selection the extinction of OLD forms and the production of new and improved f
317  ed forms are intimately connected together. The OLD notion of all the inhabitants of the earth hav
320  ppearance of new forms and the disappearance of OLD forms, both natural and artificial, are bound
320  given time is probably greater than that of the OLD forms which have been exterminated; but we kno
320  used the extinction of about the same number of OLD forms. The competition will generally be most
321  cies, that is a new genus, comes to supplant an OLD genus, belonging to the same family. But it mu
322  d in a correspondingly rapid manner many of the OLD inhabitants; and the forms which thus yield th
323  the new fossil species which are common to the OLD and New Worlds be kept wholly out of view, the
327  nd improved groups spread throughout the world, OLD groups will disappear from the world; and the
338  their own proper classes, though some of these OLD forms are in a slight degree less distinct fro
339  d the same generalisation to the mammals of the OLD World. We see the same law ir this author's re
341  six new genera. The other seven species of the OLD genera have all died out and have left no prog
341  en the parents of the six new genera; the other OLD species and the other whole genera having beco
343  modification to some extent. The extinction of OLD forms is the almost inevitable consequence of
345  cies have been produced by ordinary generation: OLD forms having been supplanted by new and improv
346  phical distribution is that between the New and OLD Worlds; yet if we travel over the vast America
346  . there is hardly a climate or condition in the OLD World which cannot be paralleled in the New, a
347  slight degree; for instance, small areas in the OLD World could be pointed out hotter than any in
347  nding this parallelism in the conditions of the OLD and New Worlds, how widely different are their
347  all the terrestrial productions of the New and OLD Worlds, excepting in the northern parts, where
358  ntain summits, of granite, metamorphic schists, OLD fossiliferous or other such rocks, instead of
361  etely enclosed by wood in an oak about 50 years OLD, three dicotyledonous plants germinated: I am
366  t in Northern Italy, gigantic moraines, left by OLD glaciers, are now clothed by the vine and maiz
367  lived in a body together on the lowlands of the OLD and New Worlds, would be left isolated on dist
369  rctic and northern temperate productions of the OLD and New Worlds are separated from each other b
370  the Glacial period, when the inhabitants of the OLD and New Worlds lived further southwards than a
370  rctic and northern temperate productions of the OLD and New Worlds, at a period anterior to the Gl
370  and that these plants and animals, both in the OLD and New Worlds, began slowly to migrate southw
371  these warmer periods the northern parts of the OLD and New Worlds will have been almost continuou
371  species in common, which inhabited the New and OLD Worlds, migrated south of the Polar Circle, th
371  nd in the other great region, with those of the OLD World. Consequently we have here everything fa
371  uctions of the temperate regions of the New and OLD Worlds, we find very few identical species (th
372  uth America with the southern continents of the OLD World, we see countries closely corresponding
429  ree intermediate between existing groups. A few OLD and intermediate parent forms having occasiona
443  as when an hereditary disease, which appears in OLD age alone, has been communicated to the offspr
445  t to be the case; but on actually measuring the OLD dogs and their six days old puppies, I found t
445  ually measuring the old dogs and their six days OLD puppies, I found that the puppies had not near
445  easurements made of the dam and of a three days OLD colt of a race and heavy cart horse, I find th
446  he short faced tumbler, which when twelve hours OLD had acquired its proper proportions, proves th
475  lows on the principle of natural selection; for OLD forms will be supplanted by new and improved f
007  uals of the same variety or sub variety of our OLDER cultivated plants and animals, one of the fir
```

Page **(Key Word)**

```
0037  hlia, and other plants, when compared with the OLDER varieties or with their parent stocks. No on
0111  s, and varieties of flowers, take the place of OLDER and inferior kinds. In Yorkshire, it is hist
0208  of nature. Frederick Cuvier and several of the OLDER metaphysicians have compared instinct with h
0286  but if, as some geologists suppose, a range of OLDER rocks underlies the Weald, on the flanks of
0305  ome palaeontologists believe that certain much OLDER fishes, of which the affinities are as yet i
0308  rth America, do not support the view, that the OLDER a formation is, the more it has suffered the
0309  o a continent, we should there find formations OLDER than the Silurian strata, supposing such to
0313  which intrude for a period in the midst of an OLDER formation, and then allow the pre existing f
0324  ereafter be liable to be classed with somewhat OLDER European beds. Nevertheless, looking to a re
0324  luding those forms which are only found in the OLDER underlying deposits, would be correctly rank
0325  ieties arising, which have some advantage over OLDER forms; and those forms, which are already do
0330  ds very distinct groups. Yet if we compare the OLDER Reptiles and Batrachians, the older Fish, th
0331  ompare the older Reptiles and Batrachians, the OLDER Fish, the older Cephalopods, and the Eocene
0331  reptiles and Batrachians, the older Fish, the OLDER Cephalopods, and the Eocene mammals, with th
0333  ods undergone much modification, should in the OLDER formations make some slight approach to each
0333  ome slight approach to each other; so that the OLDER members should differ less from each other i
0341  clude that only one species of each of the six OLDER genera has left modified descendants, consti
0341  three species of two or three alone of the six OLDER genera will have been the parents of the six
0343  sented a wholly different aspect; and that the OLDER continents, formed of formations older than
0343  hat the older continents, formed of formations OLDER than any known to us, may now all be in a me
0370  e earlier and still warmer period, such as the OLDER Pliocene period, a large number of the same
0463  uccessive period between the extinct and still OLDER species, why is not every geological formati
0470  s will inevitably supplant and exterminate the OLDER, less improved and intermediate varieties; a
0476  ter and more improved forms have conquered the OLDER and less improved organic beings in the stru
0008  ceasing to be variable under cultivation. Our OLDEST cultivated plants, such as wheat, still ofte
0008  s wheat, still often yield new varieties; our OLDEST domesticated animals are still capable of ra
0166  y other change of form or character. When the OLDEST and truest breeds of various colours are cro
0304  anner in which specimens are preserved in the OLDEST tertiary beds; from the ease with which even
0313  the same rate, or in the same degree. In the OLDEST tertiary beds a few living shells may still
0334  the species extreme in character are not the OLDEST, or the most recent; nor are those which are
0338  of full proof. Seeing, for instance, that the OLDEST known mammals, reptiles, and fish strictly b
0393  d similar case. I have carefully searched the OLDEST voyages, but have not finished my search; as
0048  ts as species; and in making this list he has OMITTED many trifling varieties, but which neverthe
0048  ome botanists as species, and he has entirely OMITTED several highly polymorphic genera. Under ge
0286  nts meet and close, one can safely picture to ONESELF the great dome of rocks which must have cov
0099  ! if several varieties of the cabbage, radish, ONION, and some other plants, be allowed to seed n
0135  on, and not one had even a relic left. In the ONITES apelles the tarsi are so habitually lost, th
0037  y has chanced to appear, selecting it, and so ONWARDS. But the gardeners of the classical period,
0073  taffordshire, the insectivorous birds, and so ONWARDS in ever increasing circles of complexity. W
0237  again varied and were again selected, and so ONWARDS. But with the working ant we have an insect
0419  allied to others, and these to others, and so ONWARDS, can be recognised as unequivocally belongi
0042  part. Wandering savages or the inhabitants of OPEN plains rarely possess more than one breed of
0081  affect many of the others. If the country were OPEN on its borders, new forms would certainly imm
0081  n some manner modified; for, had the area been OPEN to immigration, these same places would have
0095  emplified in the above imaginary instances, is OPEN to the same objections which were at first ur
0104  in the natural economy of the country are left OPEN for the old inhabitants to struggle for, and
0105  tain whether a small isolated area, or a large OPEN area like a continent, has been most favourab
0105  nd of spreading widely. Throughout a great and OPEN area, not only will there be a better chance
0106  much improved, will be able to spread over the OPEN and continuous area, and will thus come into
0114  an extremely small area, especially if freely OPEN to immigration, and where the contest between
0146  d seeds are never found in fruits which do not OPEN; I should explain the rule by the fact that s
0147  ould not possibly go on in fruit which did not OPEN. The elder Geoffroy and Goethe propounded, at
0182  itional state they had been inhabitants of the OPEN ocean, and had used their incipient organs of
0184  seen by Hearne swimming for hours with widely OPEN mouth, thus catching, like a whale, insects i
0221  eir masters in making the nest, and they alone OPEN and close the doors in the morning and evening
0285  he hard, rocky beds had once extended into the OPEN ocean. The same story is still more plainly t
0305  e seas of the world have always been so freely OPEN from south to north as they are at present. E
0308  t was not deposited, or again as the bed of an OPEN and unfathomable sea. Looking to the existing
0309  where oceans are now spread out; and clear and OPEN oceans may have existed where our continents r
0348  ward of the shores of America, a wide space of OPEN ocean extends, with not an island as a halting
0348  ther by impassable barriers, either of land or OPEN sea, they are wholly distinct. On the other ha
0354  land, though separated by hundreds of miles of OPEN sea. If the existence of the same species at
0386  d on the most remote and barren islands in the OPEN ocean; they would not be likely to alight on
0391  shells, across three or four hundred miles of OPEN sea. The different orders of insects in Madei
0416  oups. For instance, whether or not there is an OPEN passage from the nostrils to the mouth, the or
0488  ants, was created. In the distant future I see OPEN fields for far more important researches. Psy
0047  door for the entry of doubt and conjecture is OPENED. Hence, in determining whether a form should
0146  ugh natural selection, except in fruits which OPENED; so that the individual plants producing seed
0220  i have myself made, in some little detail. I OPENED fourteen nests of F. sanguinea, and found a
0486  and almost untrodden field of inquiry will be OPENED, on the causes and laws of variation, on cor
0087  by certain insects, and used exclusively for OPENING the cocoon, or the hard tip to the beak of r
0151  tes the rule in its largest application. The OPERCULAR valves of sessile cirripedes (rock barnacle
0397  erec. As this species has a thick calcareous OPERCULUM, I removed it, and when it had formed a ne
0010  , and not the ovules or pollen. But it is the OPINION of most physiologists that there is no esser
0018  s. in regard to sheep and goats I can form no OPINION. I should think, from facts communicated to
0018  scended from one wild stock. Mr. Blyth, whose OPINION, from his large and varied stores of knowlec
0023  pigeons, I am fully convinced that the common OPINION of naturalists is correct, namely, that all
0047  ould be ranked as a species or a variety, the OPINION of naturalists having sound judgment and wic
0051  sub species; that is, the forms which in the OPINION of some naturalists come very near to, but c
0149  remark seems connected with the very general OPINION of naturalists, that beings low in the scale
0284  en each successive formation, we have, in the OPINION of most geologists, enormously long blank pe
0483  illustration of the blindness of preconceived OPINION. These authors seem no more startled at a m
0293  i am aware that two palaeontologists, whose OPINIONS are worthy of much deference, namely Bronn a
0179  bits and structure. It has been asked by the OPPONENTS of such views as I hold, how, for instance,
0221  ea. The latter ruthlessly killed their small OPPONENTS, and carried their dead bodies as food to t
0221  r cocci. According to Huber, who had ample OPPORTUNITIES for observation, in Switzerland the slave
```

age **(Key Word)**

```
003  unknown to me. I cannot, however, let this OPPORTUNITY pass without expressing my deep obligation
084  d insensibly working, whenever and wherever OPPORTUNITY offers, at the improvement of each organic
015  most infinite number of generations, would be OPPOSED to all experience. I may add, that when unde
032  nt breeds; all the best breeders are strongly OPPOSED to this practice, except sometimes amongst c
229  of a rhombic plate had been left between the OPPOSED basins, but the work, from the unnatural sta
229  had excavated too quickly, and convex on the OPPOSED side, where the bees had worked less quickly
230  pect that the bees in such cases stand in the OPPOSED cells and push and bend the ductile and warm
235  tincts of very difficult explanation could be OPPOSED to the theory of natural selection, cases, i
274  ed on a single experiment: and seems directly OPPOSED to the results of several experiments made b
278  iefly given in this chapter do not seem to me OPPOSED to, but even rather to support the view, tha
316  es, Pictet, and Woodward (though all strongly OPPOSED to such views as I maintain) admit its truth
358  he sea, these and other such facts seem to me OPPOSED to the admission of such prodigious geograph
358  itants of oceanic islands likewise seem to me OPPOSED to the belief of their former continuity wit
469  eady recapitulated, as fairly as I could, the OPPOSED difficulties and objections: now let us turn
002  n apparently leading to conclusions directly OPPOSITE to those at which I have arrived. A fair res
010  but in some cases it can be shown that quite OPPOSITE conditions produce similar changes of struct
134  rying at all, although living under the most OPPOSITE climates. Such considerations as these incli
225  of the bases of three adjoining cells on the OPPOSITE side. In the series between the extreme perf
228  , would have broken into each other from the OPPOSITE sides. The bees, however, cic not suffer thi
229  inary intersection between the basins on the OPPOSITE sides of the ridge of wax. In parts, only li
229  e worked at very nearly the same rate on the OPPOSITE sides of the ridge of vermilion wax, as they
229  in working at exactly the same rate from the OPPOSITE sides: for I have noticed half completed rho
230  d the comb: and they gnaw into this from the OPPOSITE sides, always working circularly as they dee
248  artner, should have arrived at diametrically OPPOSITE conclusions in regard to the very same speci
256  e genus, for instance in Dianthus, these two OPPOSITE cases occur. The fertility, both of first cr
276  have ever lived, have come to diametrically OPPOSITE conclusions in ranking forms by this test. T
335  than short beaked tumblers, which are at the OPPOSITE end of the series in this same respect. Clos
347  nent, also, we see the same fact: for on the OPPOSITE sides of lofty and continuous mountain range
349  eastern shores of Africa, on almost exactly OPPOSITE meridians of longitude. A third great fact,
353  individuals of the same species, a directly OPPOSITE rule prevailed: and species were not local,
358  great difference in the marine faunas on the OPPOSITE sides of almost every continent, the close r
370  ng to still earlier changes of climate of an OPPOSITE nature. We have good reason to believe that
373  neous at these several far distant points on OPPOSITE sides of the world. But we have good evidenc
379  mes into the most temperate latitudes of the OPPOSITE hemisphere. Although we have reason to belie
384  r shells. The wide difference of the fish on OPPOSITE sides of continuous mountain ranges, which f
419  und impossible. There are crustaceans at the OPPOSITE ends of the series, which have hardly a char
461  supported by another parallel, but directly OPPOSITE, class of facts: namely, that the vigour and
481  urse of years, from a point of view directly OPPOSITE to mine. It is so easy to hide our ignorance
487  erences of the inhabitants of the sea on the OPPOSITE sides of a continent, and the nature of the
018  re, I am doubtfully inclined to believe, in OPPOSITION to several authors, that all the races have
208  are performed, indeed not rarely in direct OPPOSITION to our conscious will! yet they may be modi
253  n crossed in each successive generation, in OPPOSITION to the constantly repeated admonition of ev
456  imentary organs, of no importance: the wide OPPOSITION in value between analogical or adaptive cha
187  he articulata we can commence a series with an OPTIC nerve merely coated with pigment, and without
188  ction has converted the simple apparatus of an OPTIC nerve merely coated with pigment and invested
188  and invested by transparent membrane, into an OPTICAL instrument as perfect as is possessed by any
188  hose of man? If we must compare the eye to an OPTICAL instrument, we ought in imagination to take
189  y kinds; and may we not believe that a living OPTICAL instrument might thus be formed as superior
297  ow sinking many of the very fine species of D'ORBIGNY and others into the rank of varieties; and o
033  e compare the host of agricultural, culinary, ORCHARD, and flower garden races of plants most usef
073  diversity of fruit of the same species in the ORCHARD, in comparison with the leaves and flowers o
073  ructure, never can set a seed. Many of our ORCHIDACEOUS plants absolutely require the visits of mo
424  gives a single definition. As soon as three ORCHIDEAN forms (Monochanthus, Myanthus, and Catasetu
193  in a sticky gland at the end, is the same in ORCHIS and Asclepias, genera almost as remote as pos
203  ous contrivances, by which the flowers of the ORCHIS and of many other plants are fertilised throu
014  on had ensued. It would be quite necessary, in ORDER to prevent the effects of intercrossing, that
031  rge extent some breeds of cattle and sheep. In ORDER fully to realise what they have done, it is a
054  t have been anticipated: for, as varieties, in ORDER to become in any degree permanent, necessaril
061  reserved, by the term of Natural Selection, in ORDER to mark its relation to man's power of select
063  o struggle with other fruit bearing plants, in ORDER to tempt birds to devour and thus disseminate
090  rations of the action of Natural Selection. In ORDER to make it clear how, as I believe, natural s
107  ganoid fishes, remnants of a once proponderant ORDER: and in fresh water we find some of the most
115  the natives would have had to be modified, in ORDER to have gained an advantage over the other na
146  de Candolle founded his main divisions of the ORDER on analogous differences. Hence we see that m
147  ment of growth: or, as Goethe expressed it, in ORDER to spend on one side, nature is forced to eco
151  waterhouse's remark, whilst investigating this ORDER, and I am fully convinced that the rule almos
158  vantage of by natural and sexual selection, in ORDER to fit the several species to their several p
187  endants from the same original parent form, in ORDER to see what gradations are possible, and for
189  mbers of the class have been developed; and in ORDER to discover the early transitional grades thr
201  e end of its tail when preparing to spring, in ORDER to warn the doomed mouse. But I have not spac
202  like that in so many members of the same great ORDER, and which has been modified but not perfecte
203  by our fir trees of dense clouds of pollen, in ORDER that a few granules may be wafted by a chance
208  efit of this, it was much embarrassed, and, in ORDER to complete its hammock, seemed forced to sta
229  away and deepenec the basins on both sides, in ORDER to have succeeded in thus leaving flat plates
242  some little but wholly insufficient length, in ORDER to show the power of natural selection, and l
245  ally endowed with the quality of sterility, in ORDER to prevent the confusion of all organic forms
247  er is obliged carefully to count the seeds, in ORDER to show that there is any degree of sterility
247  t is often more important, must be secluded in ORDER to prevent pollen being brought to it by inse
255  e specially been endowed with this quality, in ORDER to prevent their crossing and blending togeth
276  ees of difficulty in being grafted together in ORDER to prevent them becoming inarched in our fore
286  his, of course, cannot be done: but we may, in ORDER to form some crude notion on the subject, ass
290  xtremely thick, solid, or extensive masses, in ORDER to withstand the incessant action of the wave
295  efore the glacial epoch to the present day. In ORDER to get a perfect gradation between two forms
295  one on accumulating for a very long period, in ORDER to have given sufficient time for the slow fr
295  approximately the same, which is necessary in ORDER to enable the same species to live on the sam
304  ustify the belief that this great and distinct ORDER had been suddenly produced in the interval be
```

Page **(Key Word)**

Page **(Key Word)**

```
0316 must its members have continuously existed, in ORDER to have generated either new and modified or
0325 must be victorious in a still higher degree in ORDER to be preserved and to survive. We have dist
0329  the two regions could be arranged in the same ORDER, in accordance with the general succession o
0329 eneral succession of the form of life, and the ORDER would falsely appear to be strictly parallel
0329 has placed certain pachyderms in the same sub   ORDER with ruminants: for example, he dissolves hp
0331 diverging from the parent form A, will form an  ORDER; for all will have inherited something in co
0332 escended from A, make, as before remarked, one  ORDER; and this order, from the continued effects
0332 make, as before remarked, one order; and this   ORDER, from the continued effects of extinction an
0333 in character from the common progenitor of the  ORDER, nearly so much as they subsequently diverge
0335 arrangement would not closely accord with the   ORDER in time of their production, and still less
0335 e of their production, and still less with the  ORDER of their disappearance; for the parent rock
0349 ur hares and rabbits and belonging to the same  ORDER of Rodents, but they plainly display an Amer
0386  most likely to have muddy feet. Birds of this  ORDER I can show are the greatest wanderers, and a
0392 plants when growing on an island, to whatever   ORDER they belonged, and thus convert them first i
0410 onging to a different class, or to a different  ORDER, or even only to a different family of the s
0410 or even only to a different family of the same  ORDER, differ greatly. In both time and space the
0412  all these genera, descended from (A), form an  ORDER distinct from the genera descended from (I).
0413 he creator; but unless it be specified whether  ORDER in time or space, or what else is meant by t
0416 he antennae in these two divisions of the same  ORDER are of unequal physiological importance. Any
0417 oints of structure from the proper type of the  ORDER, yet M. Richard sagaciously saw, as Jussieu
0420 other groups, must be strictly genealogical in  ORDER to be natural; but that the amount of differ
0421 o two or three families, constitute a distinct  ORDER from those descended from I, also broken up
0427 me characters are analogical when one class or  ORDER is compared with another, but give true affi
0428 finities when the members of the same class or  ORDER are compared one with another: thus the shap
0429 s not added a single insect belonging to a new  ORDER; and that in the vegetable kingdom, as I lea
0430  but in the points in which it approaches this  ORDER, its relations are general, and not to any o
0430 t nearly, not any one species, but the general  ORDER of Rodents. In this case, however, it may be
0434 y always remain connected together in the same  ORDER. We never find, for instance, the bones of t
0453 e been created for the sake of symmetry, or in  ORDER to complete the scheme of nature; but this s
0454 eared, not from unknown laws of growth, but in  ORDER to excrete horny matter, as that the rudimen
0034 estruction of horses under a certain size was   ORDERED, and this may be compared to the roguing of
0100 rom tree to tree. That trees belonging to all   ORDERS have their sexes more often separated than o
0107 ch, like fossils, connect to a certain extent   ORDERS now widely separated in the natural scale. T
0114 hese belonged to eighteen genera and to eight   ORDERS, which shows how much these plants differed
0114 ion of plants belonging to the most different   ORDERS: nature follows what may be called a simulta
0114 , belong to what we call different genera and   ORDERS. The same principle is seen in the naturalis
0116 ccessfully compete with these well pronounced   ORDERS. In the Australian mammals, we see the proce
0124 which, though generally belonging to the same   ORDERS, or families, or genera, with those now livi
0125 nera will form two distinct families, or even   ORDERS, according to the amount of divergent modifi
0125  in the diagram. And the two new families, or   ORDERS, will have descended from two species of the
0126 d with many species of many genera, families,  ORDERS, and classes, as at the present day. Summary
0128 rent degrees, forming sub families, families,  ORDERS, sub classes, and classes. The several subor
0129 es of various sizes may represent those whole   ORDERS, families, and genera which have now no livi
0144 ly be accidental, that if we pick out the two   ORDERS of mammalia which are most abnormal in their
0146 s in the fertilisation of plants of these two   ORDERS, is so far fetched, as it may at first appea
0146 rent correlations, occurring throughout whole  ORDERS, are entirely due to the manner alone in whi
0193 insects, belonging to different families and   ORDERS, offers a parallel case of difficulty. Other
0306 ere the progenitors of all the species of the  ORDERS to which they belong, for they do not presen
0307 eover, they had been the progenitors of these  ORDERS, they would almost certainly have been long
0321 tly sudden extermination of whole families or  ORDERS, as of Trilobites at the close of the palaeo
0329 ervals between existing genera, families, and  ORDERS, cannot be disputed. For if we confine our a
0329 ants and Pachyderms, as the two most distinct  ORDERS of mammals; but Owen has discovered so many
0329 o alter the whole classification of these two  ORDERS; and has placed certain pachyderms in the sa
0330 aeozoic animals, though belonging to the same  ORDERS, families, or genera with those living at th
0333 l system, and thus unite distinct families or  ORDERS. All that we have a right to expect, is that
0341 era having become utterly extinct. In failing  ORDERS, with the genera and species decreasing in n
0391 four hundred miles of open sea. The different  ORDERS of insects in Madeira apparently present ana
0391 hat the proportional numbers of the different  ORDERS are very different from what they are elsewhe
0392 ds often possess trees or bushes belonging to  ORDERS which elsewhere include only herbaceous spec
0392 o trees. With respect to the absence of whole  ORDERS on oceanic islands, Bory St. Vincent long ag
0396 ial bats, the singular proportions of certain  ORDERS of plants, herbaceous forms having been deve
0413 r subordinate to, sub families, families, and  ORDERS, all united into one class. Thus, the grand
0419 lue of the various groups of species, such as  ORDERS, sub orders, families, sub families, and gene
0419 arious groups of species, such as orders, sub  ORDERS, families, sub families, and genera, they se
0420 nder different genera, families, sections, or  ORDERS. The reader will best understand what is mean
0422 led genera, sub families, families, sections,  ORDERS, and classes. It may be worth while to illus
0429 t and extinct, are included under a few great  ORDERS, under still fewer classes, and all in one g
0429 om dr. Hooker, it has added only two or three  ORDERS of small size. In the chapter on geological
0430  general nature of the affinities of distinct  ORDERS of plants. On the principle of the multiplica
0433 ent, expressed by the terms genera, families,  ORDERS, etc., we can understand the rules which we a
0456 e terms varieties, species, genera, families,  ORDERS, and classes. On this same view of descent w
0483 fill up very wide intervals between existing   ORDERS. Organs in a rudimentary condition plainly sh
0032 ction is far more indispensable even than in   ORDINARY cases. If selection consisted merely in sepa
0127 imals, sexual selection will give its aid to   ORDINARY selection, by assuring to the most vigorous
0138 e must suppose that American animals, having   ORDINARY powers of vision, slowly migrated by success
0138 chiodte remarks, animals not far remote from   ORDINARY forms, prepare the transition from light to
0139 her inhabitants of the two continents on the   ORDINARY view of their independent creation. That se
0140 adapted to their native climate, but in all   ORDINARY cases we assume such to be the case; nor do
0155 ubject it is to individual anomalies. On the   ORDINARY view of each species having been independent
0157 tion, which is less rigid in its action than  ORDINARY selection, as it does not entail death, but
0158 ly allied species; that secondary sexual and  ORDINARY specific differences are generally displayed
0158 y, to sexual selection being less rigid than  ORDINARY selection, or to variations in the same par
0158 d thus adapted for secondary sexual, and for  ORDINARY specific purposes. Distinct species present
0159  namely, the common turnip. According to the  ORDINARY view of each species having been independent
0191 animals having true lungs have descended by   ORDINARY generation from an ancient prototype, of wh
0198 ss we generally admit to have arisen through  ORDINARY generation, we ought not to lay too much str
```

Page **(Key Word)**

age ***(Key Word)***

```
228  ove stated width (i.e. about the width of an ORDINARY cell), and were in depth about one sixth of
228  s of a three sided pyramid as in the case of ORDINARY cells. I then put into the hive, instead of
229  r thinness, and then stopping their work. In ORDINARY combs it has appeared to me that the bees do
236  ther neuter insect had been an animal in the ORDINARY state, I should have unhesitatingly assumed
239  , we may safely conclude from the analogy of ORDINARY variations, that each successive, slight, pr
273  the view which I have taken on the cause of  ORDINARY variability; namely, that it is due to the r
286  form a breakwater at the base. Hence, under  ORDINARY circumstances, I conclude that for a cliff 5
345  ms to me, that species have been produced by ORDINARY generation: old forms having been supplanted
352  he who rejects it, rejects the vera causa of ORDINARY generation with subsequent migration, and ca
394  reatly multiplied. It cannot be said, on the ORDINARY view of creation, that there has not been ti
398  ct can receive no sort of explanation on the ORDINARY view of independent creation: whereas on the
406  nd, are I think, utterly inexplicable on the ORDINARY view of the independent creation of each spe
410  lass have been connected by the same bond of ORDINARY generation: and the more nearly any two form
418  ecies. But it is by no means obvious, on the ORDINARY view, why the structure of the embryo should
435  eresting work on the Nature of Limbs. On the ORDINARY view of the independent creation of each bei
437  ike. How inexplicable are these facts on the ORDINARY view of creation! Why should the brain be en
439  ous acaceas, are pinnate or divided like the ORDINARY leaves of the Leguminosae. The points of str
441  eloped either into hermaphrodites having the  ORDINARY structure, or into what I have called comple
456  ange difficulty, as they assuredly do on the ORDINARY doctrine of creation, might even have been a
472  abitants of any one country, although on the ORDINARY view supposed to have been specially created
473  from the blue and barred rock pigeon! On the ORDINARY view of each species having been independent
475  groups of species reappear when the chain of ORDINARY generation has once been broken. The gradual
476  he beings have been connected by the bond of ORDINARY generation, and the means of modification ha
483  d at a miraculous act of creation than at an ORDINARY birth. But do they really believe that at in
489  silurian epoch, we may feel certain that the ORDINARY succession by generation has never once been
483  ssified on the same principle, in groups sub ORDINATE to groups. Fossil remains sometimes tend to
005  n understanding how a simple being or a simple ORGAN can be changed and perfected into a highly de
005  hly developed being or elaborately constructed ORGAN; secondly, the subject of Instinct, or the me
083  ul to any being. She can act on every internal ORGAN, on every shade of constitutional difference,
147  oductions: if nourishment flows to one part or ORGAN in excess, it rarely flows, at least in exces
148  rfectly well succeed in largely developing any ORGAN, without requiring as a necessary compensatio
149  arieties and in species, that when any part or ORGAN is repeated many times in the structure of th
149  riable: whereas the number of the same part or ORGAN, when it occurs in lesser numbers, is constan
151  egrees of variability. When we see any part or ORGAN developed in a remarkable degree or manner in
153  ility of the extraordinarily developed part or ORGAN has been so great and long continued within a
154  t, i can see no reason to doubt. Hence when an  ORGAN, however abnormal it may be, has been transmi
155  has remarked with surprise that some important ORGAN or part, which is generally very constant thr
155  seems to entertain no doubt, that the more an  ORGAN normally differs in the different species of
162  some degree, the character of the same part or ORGAN in an allied species. I have collected a long
169  several species of the same genus. Any part or ORGAN developed to an extraordinary size or in an e
169  ry manner, in comparison with the same part or ORGAN in the allied species, must have gone through
169  n a species with any extraordinarily developed ORGAN has become the parent of many modified descen
169  e succeeded in giving a fixed character to the ORGAN, in however extraordinary a a manner it may b
186  ei: and if any variation or modification in the ORGAN be ever useful to an animal under changing co
187  und. In looking for the gradations by which an  ORGAN in any species has been perfected, we ought t
189  ? if it could be demonstrated that any complex  ORGAN existed, which could not possibly have been f
189  en much extinction. Or again, if we look to an  ORGAN common to all the members of a large class, f
189  of a large class, for in this latter case the  ORGAN must have been first formed at an extremely r
189  he early transitional grades through which the ORGAN has passed, we should have to look to very an
190  ld be extremely cautious in concluding that an  ORGAN could not have been formed by transitional gr
190  be given amongst the lower animals of the same ORGAN performing at the same time wholly distinct f
190  , if any advantage were thus gained, a part or ORGAN, which had performed two functions, for one f
190  he free air in their swimbladders; this latter ORGAN having a ductus pneumaticus for its supply, a
190  uring the process of modification by the other ORGAN; and then this other organ might be modified
190  cation by the other organ: and then this other ORGAN might be modified for some other and quite di
190  s us clearly the highly important fact that an  ORGAN originally constructed for one purpose, namel
192  tually converted a swimbladder into a lung, or  ORGAN used exclusively for respiration. I can, inde
192  t be extremely cautious in concluding that any  ORGAN could not possibly have been produced by succ
193  as it has lately been shown that Rays have an   ORGAN closely analogous to the electric apparatus,
193  e in their affinities. Generally when the same  ORGAN appears in several members of the same class,
193  s furnished with apparently the same anomalous  ORGAN, it should be observed that, although the gen
193  ugh the general appearance and function of the ORGAN may be the same, yet some funcamental differe
194  difficult to conjecture by what transitions an ORGAN could have arrived at its present state: yet,
194  ry small, I have been astonished how rarely an  ORGAN can be named, towards which no transitional g
195  erent kind, on this head, as in the case of an ORGAN as perfect and complex as the eye. In the fir
196  by natural selection. Seeing how important an  ORGAN of locomotion the tail is in most aquatic ani
196  or all sorts of purposes, as a fly flapper, an ORGAN of prehension, or as an aid in turning, as wh
201  on acts solely by and for the good of each. No ORGAN will be formed, as Paley has remarked, for th
202  y, not to be perfect even in that most perfect ORGAN, the eye. If our reason leads us to admire wi
204  he habits of auks. Although the belief that an ORGAN so perfect as the eye could have been formed
204  ugh to stagger any one: yet in the case of any ORGAN, if we know of a long series of gradations in
204  converted into an air breathing lung. The same ORGAN having performed simultaneously very differen
205  case, to be enabled to assert that any part or ORGAN is so unimportant to the welfare of a specie
392  ing as useless an appendage as any rudimentary ORGAN, for instance, as the shrivelled wings under
415  . that the mere physiological importance of an ORGAN does not determine its classificatory value,
415  fact, that in allied groups, in which the same ORGAN, as we have every reason to suppose, has near
416  e for the classification of the same important ORGAN within the same group of beings. Again, no on
417  tructure in any one internal and and important ORGAN. The importance, for classification, of trifl
436  is conceivable that the general pattern of an  ORGAN might become so much obscured as to be finall
436  get direct evidence of the possibility of one  ORGAN being transformed into another: and we can ac
437  . an indefinite repetition of the same part or ORGAN is the common characteristic (as Owen has obs
438  we are seldom enabled to say that one part or  ORGAN is homologous with another in the same indivi
447  disuse on the other, may have in modifying an  ORGAN, such influence will mainly affect the mature
451  ect pistil are essentially alike in nature. An ORGAN serving for two purposes, many become rudimen
452  llen out of the surrounding anthers. Again, an ORGAN may become rudimentary for its proper purpose
452  has become converted into a nascent breathing  ORGAN or lung. Other similar instances could be giv
```

Page ***************************************(Key Word)***

Page ***(Key Word)***

```
0452  y allied species, the degree to which the same  ORGAN  has been rendered rudimentary occasionally d
0452  t we find in an animal or plant no trace of an   ORGAN, which analogy would lead us to expect to fir
0453  , a universal rule, that a rudimentary part or   ORGAN  is of greater size relatively to the adjoinir
0453  in the embryo, than in the adult; so that the    ORGAN  at this early age is less rudimentary, or eve
0453  degree rudimentary. Hence, also, a rudimentary   ORGAN  in the adult, is often said to have retained
0454  ultimately lost the power of flying. Again, an   ORGAN  useful under certain conditions, might become
0454  selection would continue slowly to reduce the    ORGAN, until it rendered harmless and rudimentary.
0454  him the power of natural selection; so that an   ORGAN  rendered, during changed habits of life, usel
0454  e modified and used for another purpose. Or an   ORGAN  might be retained for one alone of its former
0455  ined for one alone of its former functions. An   ORGAN, when rendered useless, may well be variable,
0455  period of life disuse or selection reduces an   ORGAN, and this will generally be when the being ha
0455  tance at corresponding ages will reproduce the   ORGAN  in its reduced state at the same age, and cor
0455  cause the entire obliteration of a rudimentary   ORGAN. As the presence of rudimentary organs is thu
0459  mely, that gradations in the perfection of any   ORGAN  or instinct, which we may consider, either dc
0460  ht to be extremely cautious in saying that any   ORGAN  or instinct, or any whole being, could not ha
0479  atural selection, will often tend to reduce an   ORGAN, when it has become useless by changed habits
0480  nd will thus have little power of acting on an   ORGAN  during early life; hence the organ will not b
0480  cting on an organ during early life; hence the   ORGAN  will not be much reduced or rendered rudiment
0480  e view of each organic being and each separate   ORGAN  having been specially created, how utterly ir
0483  plainly show that an early progenitor had the    ORGAN  in a fully developed state; and this in some
0003  alist, reflecting on the mutual affinities of    ORGANIC  beings, on their embryological relations, th
0003  asite, with its relations to several distinct    ORGANIC  beings, by the effects of external conditior
0004  or it leaves the case of the coadaptations of    ORGANIC  beings to each other and to their physical c
0004  hapter the Struggle for Existence amongst all    ORGANIC  beings throughout the world, which inevitabl
0005  I shall consider the geological succession of    ORGANIC  beings throughout time; in the eleventh and
0007  th excess of food. It seems pretty clear that    ORGANIC  beings must be exposed during several genera
0043  ontingency, under all circumstances, with all    ORGANIC  beings, as some authors have thought. The ef
0044  principles arrived at in the last chapter to    ORGANIC  beings in a state of nature, we must briefly
0053  mportant circumstance) with different sets of    ORGANIC  beings. But my tables further show that, in
0054  country, shows that there is something in the    ORGANIC  or inorganic conditions of that country favc
0060  as been seen in the last chapter that amongst    ORGANIC  beings in a state of nature there is some in
0060  o the conditions of life, and of one distinct    ORGANIC  being to another being, been perfected? We s
0061  aptations everywhere and in every part of the    ORGANIC  world. Again, it may be asked, how is it tha
0061  in its infinitely complex relations to other    ORGANIC  beings and to external nature, will tend to
0061  ertainly produce great results, and can adapt    ORGANIC  beings to his own uses, through the accumula
0062  ve largely and philosophically shown that all    ORGANIC  beings are exposed to severe competition. Ir
0063  tably follows from the high rate at which all    ORGANIC  beings tend to increase. Every being, which
0064  there is no exception to the rule that every    ORGANIC  being naturally increases at so high a rate,
0066  ys in mind, never to forget that every single    ORGANIC  being arounc us may be said to be striving t
0071  expected are the checks and relations between    ORGANIC  beings, which have to struggle together in t
0073  st trifle would often give the victory to one    ORGANIC  being over another. Nevertheless so profounc
0073  e marvel when we hear of the extinction of an    ORGANIC  being; and as we do not see the cause, we in
0075  n the old Indian ruins! The dependency of one    ORGANIC  being on another, as of a parasite on its pr
0077  remarks, namely, that the structure of every    ORGANIC  being is related, in the most essential yet
0077  yet often hidden manner, to that of all other    ORGANIC  beings, with which it comes into competition
0078  our ignorance on the mutual relations of all    ORGANIC  beings; a conviction as necessary, as it see
0078  can do, is to keep steadily in mind that each    ORGANIC  being is striving to increase at a geometric
0080  a common parent. Explains the grouping of all    ORGANIC  beings. How will the struggle for existence,
0080  close fitting are the mutual relations of all    ORGANIC  beings to each other and to their physical c
0084  pportunity offers, at the improvement of each    ORGANIC  being in relation to its organic and inorgan
0084  ment of each organic being in relation to its    ORGANIC  and inorganic conditions of life. We see not
0086  election will be enabled to act on and modify    ORGANIC  beings at any age, by the accumulation of pr
0095  h the belief of the continued creation of new    ORGANIC  beings, or of any great and sudden modificat
0097  ough we be of the meaning of the law) that no    ORGANIC  being self fertilises itself for an eternity
0101  e. finally then, we may conclude that in many    ORGANIC  beings, a cross between two individuals is a
0102  es not grant an indefinite period: for as all    ORGANIC  beings are striving, it may be said, to seiz
0104  rare intervals, will be great. If there exist    ORGANIC  beings which never intercross, uniformity of
0104  ined or isolated area, if not very large, the    ORGANIC  and inorganic conditions of life will genera
0105  een most favourable for the production of new    ORGANIC  forms, we ought to make the comparison withi
0106  important part in the changing history of the    ORGANIC  world. We can, perhaps, on these views, unde
0109  e complexity of the coadaptations between all    ORGANIC  beings, one with another and with their phys
0109  he high geometrical powers of increase of all    ORGANIC  beings, each area is already fully stocked w
0126  the future, we can predict that the croups of    ORGANIC  beings which are now large and triumphant, a
0126  of ages and under varying conditions of life,    ORGANIC  beings vary at all in the several parts of t
0127  e infinite complexity of the relations of all    ORGANIC  beings to each other and to their conditions
0127  seful to man. But if variations useful to any    ORGANIC  being do occur, assuredly individuals thus c
0128  believe, the nature of the affinities of all    ORGANIC  beings may be explained. It is a truly wonde
0129  this great fact in the classification of all    ORGANIC  beings; but, to the best of my judgment, all
0131  if the variations, so common and multiform in    ORGANIC  beings under domestication, and in a lesser
0132  mplex cc adaptations of structure between one    ORGANIC  being and another, which we see everywhere t
0140  d in their ranges by the competition of other    ORGANIC  beings quite as much as, or more than, by th
0168  ion. It is probably from this same cause that    ORGANIC  beings low in the scale of nature are more v
0172  perfection of the record being chiefly due to    ORGANIC  beings not inhabiting profound depths of the
0175  serve as prey for others: in short, that each    ORGANIC  being is either directly or indirectly relat
0175  related in the most important manner to other    ORGANIC  beings, we must see that the range of the in
0179  tent record. On the origin and transitions of    ORGANIC  beings with peculiar habits and structure. I
0186  atural selection, will acknowledge that every    ORGANIC  being is constantly endeavouring to increase
0194  very nearly the same manner two parts in two    ORGANIC  beings, which owe but little of their struct
0195  ant in regard to the whole economy of any one    ORGANIC  being, to say what slight modifications woul
0201  ct. Natural selection tends only to make each    ORGANIC  being as perfect as, or slightly more perfec
0204  we can understand; bearing in mind that each    ORGANIC  being is trying to live wherever it can live
0206  y true. It is generally acknowledged that all    ORGANIC  beings have been formed on two great laws, U
0206  ental agreement in structure, which we see in    ORGANIC  beings of the same class, and which is quite
0206  apting the varying parts of each being to its    ORGANIC  and inorganic conditions of life: or by havi
0244  eneral law, leading to the advancement of all    ORGANIC  beings, namely, multiply, vary, let the stro
0245  ity, in order to prevent the confusion of all    ORGANIC  forms. This view certainly seems at first pr
```

Page ***(Key Word)***

Page ***(Key Word)***

0263	kind of resemblance and dissimilarity between	ORGANIC beings is attempted to be expressed. The fac
0265	ce more or less sterile hybrids. Lastly, when	ORGANIC beings are placed during several generations
0265	mentalist has observed. Thus we see that when	ORGANIC beings are placed under new and unnatural co
0267	changes in the conditions of life benefit all	ORGANIC beings, and on the other hand, that slight c
0267	changes of a particular nature, often render	ORGANIC beings in some degree sterile; and that grea
0277	favourable to the vigour and fertility of all	ORGANIC beings. It is not surprising that the degree
0279	nner on the presence of other already defined	ORGANIC forms, than on climate; and, therefore, that
0280	dly does not reveal any such finely graduated	ORGANIC chain; and this, perhaps, is the most obviou
0282	l not have sufficed for so great an amount of	ORGANIC change, all changes having been effected ver
0294	the glacial period shall have been upraised,	ORGANIC remains will probably first appear and disap
0295	hat very thick deposits are usually barren of	ORGANIC remains, except near their upper or lower li
0300	of the richest regions of the whole world in	ORGANIC beings; yet if all the species were to be co
0300	ot accumulate at a sufficient rate to protect	ORGANIC bodies from decay, no remains could be prese
0306	rash in us to dogmatize on the succession of	ORGANIC beings throughout the world, as it would be
0307	their head, are convinced that we see in the	ORGANIC remains of the lowest Silurian stratum the d
0308	following hypothesis. From the nature of the	ORGANIC remains, which do not appear to have inhabit
0312	r. chapter X. On the Geological Succession of	ORGANIC Beings. On the slow and successive appearanc
0312	ules relating to the geological succession of	ORGANIC beings, better accord with the common view o
0313	re are exceptions to this rule. The amount of	ORGANIC change, as Pictet has remarked, does not str
0314	mplex relations of the higher beings to their	ORGANIC and inorganic conditions of life, as explain
0315	mittent intervals; consequently the amount of	ORGANIC change exhibited by the fossils embedded in
0315	ar, even if the very same conditions of life,	ORGANIC and inorganic, should recur. For though the
0323	la of India. For at these distant points, the	ORGANIC remains in certain beds present an unmistake
0325	when we treat of the present distribution of	ORGANIC beings, and find how slight is the relation
0327	own down quickly enough to embed and preserve	ORGANIC remains. During these long and blank interva
0335	losely connected with the statement, that the	ORGANIC remains from an intermediate formation are i
0335	nown instance, the general resemblance of the	ORGANIC remains from the several stages of the chalk
0342	lorec with care; that only certain classes of	ORGANIC beings have been largely preserved in a foss
0345	y different forms. We can clearly see why the	ORGANIC remains of closely consecutive formations ar
0346	the world. In considering the distribution of	ORGANIC beings over the face of the globe, the first
0350	merican seas. We see in these facts some deep	ORGANIC bond, prevailing throughout space and time,
0352	errestrial mammals, than perhaps in any other	ORGANIC beings; and, accordingly, we find no inexpli
0355	ndividuals simultaneously created. With those	ORGANIC beings which never intercross (if such exist
0366	ve evidence of almost every conceivable kind,	ORGANIC and inorganic, that within a very recent geo
0408	e, there would be an almost endless amount of	ORGANIC action and reaction, and we should find, as
0411	selection. Chapter XIII. Mutual Affinities of	ORGANIC Beings. Morphology. Embryology. Rudimentary
0411	ed. Summary. From the first dawn of life, all	ORGANIC beings are found to resemble each other in d
0413	nt, the only known cause of the similarity of	ORGANIC beings, is the bond, hidden as it is by vari
0421	parent genus F; just as some few still living	ORGANIC beings belong to Silurian genera. So that th
0421	he amount or value of the differences between	ORGANIC beings all related to each other in the same
0433	nd universal feature in the affinities of all	ORGANIC beings, namely, their subordination in group
0448	is would be said to be retrograde. As all the	ORGANIC beings, extinct and recent, which have ever
0456	e only certainly known cause of similarity in	ORGANIC beings, we shall understand what is meant by
0457	e inumerable species, genera, and families of	ORGANIC beings, with which this world is peopled, ha
0460	pecially amongst broken and failing groups of	ORGANIC beings; but we see so many strange gradation
0461	namely, that the vigour and fertility of all	ORGANIC beings are increased by slight changes in th
0462	linking forms all around us? Why are not all	ORGANIC beings blended together in an inextricable c
0464	as not been time sufficient for any amount of	ORGANIC change; for the lapse of time has been so or
0464	he world has been geologically explored. Only	ORGANIC beings of certain classes can be preserved i
0467	variability; he only unintentionally exposes	ORGANIC beings to new conditions of life, and then n
0467	ical ratio of increase which is common to all	ORGANIC beings. This high rate of increase is proved
0468	physical changes, we might have expected that	ORGANIC beings would have varied under nature, in th
0471	l time. The grand fact of the grouping of all	ORGANIC beings seems to me utterly inexplicable on t
0475	d so conspicuous a part in the history of the	ORGANIC world, almost inevitably follows on the prin
0476	n of descent. The grand fact that all extinct	ORGANIC beings belong to the same system with recent
0476	ms have conquered the older and less improved	ORGANIC beings in the struggle for life. Lastly, the
0476	striking a parallelism in the distribution of	ORGANIC beings throughout space, and in their geolog
0478	t, as we have seen, that all past and present	ORGANIC beings constitute one grand natural system,
0479	t valuable of all. The real affinities of all	ORGANIC beings are due to inheritance or community o
0480	eriod to the present day. On the view of each	ORGANIC being and each separate organ having been sp
0481	bility of species? It cannot be asserted that	ORGANIC beings in a state of nature are subject to n
0484	ould infer from analogy that probably all the	ORGANIC beings which have ever lived on this earth h
0485	n signification. When we no longer look at an	ORGANIC being as a savage looks at a ship, as at som
0486	s of numerous workmen; when we thus view each	ORGANIC being, how far more interesting, I speak fro
0487	a comparison of the preceding and succeeding	ORGANIC forms. We must be cautious in attempting to
0487	si and as the most important of all causes of	ORGANIC change is one which is almost independent of
0488	of of others; it follows, that the amount of	ORGANIC change in the fossils of consecutive formati
0488	so that we must not overrate the accuracy of	ORGANIC change as a measure of time. During early pe
0489	distant futurity; for the manner in which all	ORGANIC beings are grouped, shows that the greater n
0007	ble amount of variation; and that when the	ORGANISATION has once begun to vary, it generally conti
0008	ore susceptible than any other part of the	ORGANISATION, to the action of any change in the condit
0012	differ slightly from each other. The whole	ORGANISATION seems to have become plastic, and tends to
0031	breeders habitually speak of an animal's	ORGANISATION as something quite plastic, which they can
0054	belong. Again, plants low in the scale of	ORGANISATION are generally much more widely diffused th
0060	e exquisite adaptations of one part of the	ORGANISATION to another part, and to the conditions of
0080	ation, it may be truly said that the whole	ORGANISATION becomes in some degree plastic. Let it be
0085	ion of growth, which, when one part of the	ORGANISATION is modified through variation, and the mod
0101	sely with each other in almost their whole	ORGANISATION, yet are not rarely, some of them hermaphr
0116	g themselves. A set of animals, with their	ORGANISATION but little diversified, could hardly compe
0127	vary at all in the several parts of their	ORGANISATION, and I think this cannot be disputed; if t
0138	uropean caverns; close similarity in their	ORGANISATION and affinities might have been expected; b
0143	i mean by this expression that the whole	ORGANISATION is so tied together during its growth and
0143	early embryo, seriously affects the whole	ORGANISATION of the adult. The several parts of the bod
0147	g trying to economise in every part of the	ORGANISATION. If under changed conditions of life a str
0148	n in reducing and saving every part of the	ORGANISATION, as soon as it is rendered superfluous, wi
0149	en's expression, seems to be a sign of low	ORGANISATION; the foregoing remark seems connected with

Page ***(Key Word)***

Page *********************************(Key Word)*********************************#*****

```
0149  s case means that the several parts of the  ORGANISATION  have been but little specialised for parti
0152  r cannot come into full play, and thus the  ORGANISATION  is left in a fluctuatuing condition. But w
0153  y in such parts than in other parts of the  ORGANISATION, which have remained for a much longer per
0156  east more variable than those parts of the  ORGANISATION  which have for a very long period remained
0156  l characters, than in other parts of their  ORGANISATION; compare, for instance, the amount of diff
0156  constant and uniform as other parts of the  ORGANISATION; for secondary sexual characters have been
0157  ly displayed in the very same parts of the  ORGANISATION  in which the different species of the same
0158  nerally displayed in the same parts of the  ORGANISATION, are all principles closely connected toge
0168  variable than those which have their whole  ORGANISATION  more specialised, and are higher in the sc
0169  roup. Variability in the same parts of the  ORGANISATION  has generally been taken advantage of in g
0184  cker, which in every essential part of its  ORGANISATION, even in its colouring, in the harsh tone
0185  ially a petrel, but with many parts of its  ORGANISATION  profoundly modified. On the other hand, th
0196  y have some little direct influence on the  ORGANISATION; that characters reappear from the law of
0199  onsideration is that the chief part of the  ORGANISATION  of every being is simply due to inheritanc
0258  any recognisable difference in their whole  ORGANISATION. On the other hand, these cases clearly sh
0266  conditions have remained the same, but the  ORGANISATION  has been disturbed by two different struct
0266  neration to generation the same compounded  ORGANISATION, and hence we need not be surprised that t
0266  been disturbed, in the other case from the  ORGANISATION  having been disturbed by two organisations
0269  y during vast periods of time on the whole  ORGANISATION, in any way which may be for each creature
0277  which have had this system and their whole  ORGANISATION  disturbed by being compounded of two disti
0281  e common parent will have had in its whole  ORGANISATION  much general resemblance to the tapir and
0337  fected in a marked and sensible manner the  ORGANISATION  of the more recent and victorious forms of
0345  iment, felt by many palaeontologists, that  ORGANISATION  on the whole has progressed. If it should
0414  eneral rule, that the less any part of the  ORGANISATION  is concerned with special habits, the more
0414  inc, trust to resemblances in parts of the  ORGANISATION, however important they may be for the wel
0417  be, has always failed; for no part of the  ORGANISATION  is universally constant. The importance of
0425  sometimes better than, other parts of the  ORGANISATION. We care not how trifling a character may
0434  le each other in the general plan of their  ORGANISATION. This resemblance is often expressed by th
0435  y correlation of growth other parts of the  ORGANISATION. In changes of this nature, there will be
0441  e course of development generally rises in  ORGANISATION: I use this expression, though I am aware
0441  ble to define clearly what is meant by the  ORGANISATION  being higher or lower. But no one probably
0443  mbryo apparently having sometimes a higher  ORGANISATION  than the mature animal, into which it is d
0455  s due to the tendency in every part of the  ORGANISATION, which has long existed, to be inherited,
0467  tions of life, and then nature acts on the  ORGANISATION, and causes variability. But man can and d
0266  one. For it is scarcely possible that two  ORGANISATIONS  should be compounded into one, without so
0266  organisation having been disturbed by two  ORGANISATIONS  having been compounded into one. It may s
0461  rbed from being compounded of two distinct  ORGANISATIONS. This parallelism is supported by another
0055  o the size of the genera. The cause of lowly  ORGANISED  plants ranging widely will be discussed in
0131  rrelations. Multiple, rudimentary, and lowly  ORGANISED  structures variable. Parts developed in an
0314  of change in terrestrial and in more highly  ORGANISED  productions compared with marine and lower
0441  nsidered as either more highly or more lowly  ORGANISED  then they were in the larval condition. But
0060  n species of the same genus. The relation of  ORGANISM  to organism the most important of all relati
0060  the same genus. The relation of organism to  ORGANISM  the most important of all relations. Before
0266  the matter: no explanation is offered why an  ORGANISM, when placed under unnatural conditions, is
0281  ge; and the principle of competition between  ORGANISM  and organism, between child and parent, will
0281  rinciple of competition between organism and  ORGANISM, between child and parent, will render this
0288  coveries made every year in Europe prove. No  ORGANISM  wholly soft can be preserved. Shells and bon
0303  equire a long succession of ages to adapt an  ORGANISM  to some new and peculiar line of life, for i
0314  that of the many all important relations of  ORGANISM  to organism, that any form which does not be
0314  many all important relations of organism to  ORGANISM, that any form which does not become in some
0350  r mutual struggles for life; the relation of  ORGANISM  to organism being, as I have already often r
0350  uggles for life; the relation of organism to  ORGANISM  being, as I have already often remarked, the
0396  the paramount importance of the relation of  ORGANISM  to organism. I do not deny that there are ma
0396  nt importance of the relation of organism to  ORGANISM. I do not deny that there are many and grave
0408  bearing in mind that the mutual relations of  ORGANISM  to organism are of the highest importance, w
0408  ind that the mutual relations of organism to  ORGANISM  are of the highest importance, we can see wh
0477  ated from each other; for as the relation of  ORGANISM  to organism is the most important of all rel
0477  ch other; for as the relation of organism to  ORGANISM  is the most important of all relations, and
0487  l conditions, namely, the mutual relation of  ORGANISM  to organism, the improvement of one being en
0487  , namely, the mutual relation of organism to  ORGANISM, the improvement of one being entailing the
0009  tions of the garden, I may add, that as some  ORGANISMS  will breed most freely under the most unnat
0050  naturalist commences the study of a group of  ORGANISMS  quite unknown to him, he is at first much p
0065  be disposed of. The only difference between  ORGANISMS  which annually produce eggs or seeds by the
0103  s i believe to be the case. In hermaphrodite  ORGANISMS  which cross only occasionally, and likewise
0104  n checking the immigration of better adapted  ORGANISMS, after any physical change, such as of clim
0303  d thus acquired a great advantage over other  ORGANISMS, a comparatively short time would be necess
0313  erland. There is some reason to believe that  ORGANISMS, considered high in the scale of nature, ch
0343  are the remains of those infinitely numerous  ORGANISMS  which must have existed long before the fir
0346  or it is a most rare case to find a group of  ORGANISMS  confined to any small spot, having conditio
0350  lone, as far as we positively know, produces  ORGANISMS  quite like, or, as we see in the case of va
0351  e principles come into play only by bringing  ORGANISMS  into new relations with each other, and in
0356  t in the majority of cases, namely, with all  ORGANISMS  which habitually unite for each birth, or w
0357  have occurred within the period of existing  ORGANISMS. Edward Forbes insisted that all the island
0370  e present day. Hence we may suppose that the  ORGANISMS  now living under the climate of latitude 60
0401  would have to compete with different sets of  ORGANISMS: a plant, for instance, would find the best
0406  son to believe from geological evidence that  ORGANISMS  low in the scale within each great class, g
0406  plants, namely, that the lower any group of  ORGANISMS  is, the more widely it is apt to range. The
0406  t discussed, namely, low and slowly changing  ORGANISMS  ranging more widely than the high, some of
0409  of the world. We can see why whole groups of  ORGANISMS, as batrachians and terrestrial mammals, sh
0410  nces throughout the world, we find that some  ORGANISMS  differ little, whilst others belonging to a
0421  metimes seems to have occurred with existing  ORGANISMS. All the descendants of the genus F, along
0428  s. thus we can account for the fact that all  ORGANISMS, recent and extinct, are included under a f
0456  t the subordination of group to group in all  ORGANISMS  throughout all time; that the nature of the
0461  erfect condition. As we continually see that  ORGANISMS  of all kinds are rendered in some degree st
0011  milked, in comparison with the state of these  ORGANS  in other countries, is another instance of th
0038  and relatively so slight in internal parts or  ORGANS. Man can hardly select, or only with much dif
```

Page *********************************(Key Word)*********************************#*****

organs

Page **************************************(Key Word)**************************************

```
0045  ll laboriously examine internal and important ORGANS, and compare them in many specimens of the sa
0046  ue in a circle when they state that important ORGANS never vary; for these same authors practicall
0094  er new conditions of life, sometimes the male ORGANS and sometimes the female organs become more o
0094  imes the male organs and sometimes the female ORGANS become more or less impotent; now if we suppo
0097  . many flowers, on the other hand, have their ORGANS of fructification closely enclosed, as in the
0101  ngle case of an hermaphrodite animal with the ORGANS of reproduction so perfectly enclosed within
0115  f the physiological division of labour in the ORGANS of the same individual body, a subject so wel
0131  and disuse, combined with natural selection; ORGANS of flight and of vision. Acclimatisation. Cor
0143  constitution, and of the structure of various ORGANS; but that the effects of use and disuse have
0149  he general subject of rudimentary and aborted ORGANS: and I will here only add that their variabil
0152  n be said to have degenerated. In rudimentary  ORGANS, and in those which have been but little spec
0154  cease; and that the most abnormally developed  ORGANS may be made constant, I can see no reason to
0161  rations, thar in quite useless or rudimentary  ORGANS being, as we all know them to be, thus inheri
0162  but the best evidence is afforded by parts or  ORGANS of an important and uniform nature occasional
0168  ning, and disuse in weakening and diminishing  ORGANS, seem to have been more potent in their effec
0168  sed, and are higher in the scale. Rudimentary  ORGANS, from being useless, will be disregarded by n
0168  remarks we have referred to special parts or   ORGANS being still variable, because they have recen
0171  widely different from those of their allies.   ORGANS of extreme perfection. Means of transition. C
0171  cases of difficulty. Natura non facit saltum.  ORGANS of small importance. Organs not in all cases
0171  non facit saltum. Organs of small importance.  ORGANS not in all cases absolutely perfect. The law
0171  ral selection could produce, on the one hand,  ORGANS of trifling importance, such as the tail of a
0171  rves as a fly flapper, and on the other hand,  ORGANS of such wonderful structure, as the eye, of w
0181  by natural selection? and this, as far as the  ORGANS of flight are concerned, would convert it int
0182  the open ocean, and had used their incipient   ORGANS of flight exclusively, as far as we know, to
0183  ys, on the land and in the water, until their  ORGANS of flight had come to a high stage of perfect
0186  hrushes, and petrels with the habits of auks.  ORGANS of extreme perfection and complication. To su
0189  ut i can find out no such case. No doubt many  ORGANS exist of which we do not know the transitiona
0190  its nature by insensible steps. Two distinct   ORGANS sometimes perform simultaneously the same fun
0190  ar partitions. In these cases, one of the two  ORGANS might with ease be modified and perfected so
0190  een worked in as an accessory to the auditory  ORGANS of certain fish, or, for I do not know which
0191  nd wingcovers of insects, it is probable that  ORGANS which at a very ancient period served for res
0191  respiration have been actually converted into  ORGANS of flight. In considering transitions of orga
0191  gans of flight. In considering transitions of  ORGANS, it is so important to bear in mind the proba
0192  this latter family had originally existed as   ORGANS for preventing the ova from being washed out
0192  treated of in the next chapter. The electric   ORGANS of fishes offer another case of special diffi
0192  ible to conceive by what steps these wondrous  ORGANS have been produced; but, as Owen and others h
0193  nsition of any kind is possible. The electric  ORGANS offer another and even more serious difficult
0193  use or natural selection. But if the electric  ORGANS had been inherited from one ancient progenito
0193  belief that formerly most fishes had electric  ORGANS, which most of their modified descendants hav
0193  scendants have lost. The presence of luminous  ORGANS in a few insects, belonging to different fami
0194  ould this be so? Why should all the parts and  ORGANS of many independent beings, each supposed to
0194  st advance by the shortest and slowest steps.  ORGANS of little apparent importance. As natural Sel
0195  h for food, or to escape from beasts of prey.  ORGANS now of trifling importance have probably in s
0204  ould have existed, for the homologies of many  ORGANS and their intermediate states show that wonde
0205  lised for one function; and two very distinct  ORGANS having performed at the same time the same fu
0205  of another; though it may well produce parts,  ORGANS, and excretions highly useful or even indispe
0210  almost equal force to instincts as to bodily   ORGANS. Changes of instinct may sometimes be facilit
0240  est, differing not only in size, but in their  ORGANS of vision, yet connected by some few members
0246  from them. Pure species have of course their   ORGANS of reproduction in a perfect condition, yet w
0246  s, on the other hand, have their reproductive  ORGANS functionally impotent, as may be clearly seen
0246  lement in both plants and animals; though the  ORGANS themselves are perfect in structure, as far a
0262  ity of hybrids, which have their reproductive  ORGANS in an imperfect condition, is a very differen
0262  o pure species, which have their reproductive  ORGANS perfect; yet these two distinct cases run to
0266  or mutual relation of the different parts and  ORGANS one to another, or to the conditions of life.
0361  ill pass uninjured through even the digestive  ORGANS of a turkey. In the course of two months, T p
0411  c beings. Morphology. Embryology. Rudimentary  ORGANS. Classification, groups subordinate to groups
0411  inherited at a corresponding age. Rudimentary  ORGANS; their origin explained. Summary. From the fi
0414  speaking of the dugong, says, The generative   ORGANS being those which are most remotely related t
0414  re least likely in the modifications of these  ORGANS to mistake a merely adaptive for an essential
0414  so with plants, how remarkable it is that the  ORGANS of vegetation, on which their whole life depe
0414  ting in the first main divisions! whereas the  ORGANS of reproduction, with their product the seed,
0415  ts lay the greatest stress on resemblances in  ORGANS of high vital or physiological importance. No
0415  this view of the classificatory importance of  ORGANS which are important is generally, but by no m
0415  f species; and this constancy depends on such  ORGANS having generally been subjected to less chang
0415  ity. Robert Brown, who in speaking of certain  ORGANS in the Proteaceae, says their generic importa
0416  no one will say that rudimentary or atrophied  ORGANS are of high physiological or vital importance
0416  ogical or vital importance; yet, undoubtedly,  ORGANS in this condition are often of high value in
0418  them. As in most groups of animals, important  ORGANS, such as those for propelling the blood, or f
0418  f animals all these, the most important vital  ORGANS, are found to offer characters of quite subor
0426  put them all into the same class. As we find   ORGANS of high physiological importance, those which
0426  ach especial value to them; but if these same  ORGANS, in another group or section of a group, are
0433  are permitted to use rudimentary and useless  ORGANS, or others of trifling physiological importan
0434  type; or by saying that the several parts and  ORGANS in the different species of the class are hom
0434  mportance of relative connexion in homologous  ORGANS: the parts may change to almost any extent in
0434  and the great jaws of a beetle? yet all these  ORGANS, serving for such different purposes, are for
0436  ers of a class, but of the different parts or  ORGANS in the same individual. Most physiologists be
0436  d in many other animals, and in flowers, that  ORGANS, which when mature become extremely different
0438  ec rot wonder at discovering in such parts or  ORGANS, a certain degree of fundamental resemblance,
0438  t during a long course of descent, primordial  ORGANS of any kind, vertebrae in the one case and le
0439  s already been casually remarked that certain  ORGANS in the individual, which when mature become w
0441  s stage is, to search by their well developed  ORGANS of sense, and to reach by their active powers
0441  their legs are now converted into prehensile  ORGANS; they again obtain a well constructed mouth;
0441  and is destitute of mouth, stomach, or other  ORGANS of importance, excepting for reproduction. We
0448  t become fitted for sites or habits, in which  ORGANS of locomotion or of the senses, etc., would b
0450  f animals. Rudimentary, atrophied, or aborted  ORGANS. Organs or parts in this strange condition, b
0450  s. rudimentary, atrophied, or aborted organs.  ORGANS or parts in this strange condition, bearing t
```

Page **************************************(Key Word)**************************************

Page ***(Key Word)**

```
0450  hind limbs. Some of the cases of rudimentary ORGANS are extremely curious; for instance, the pres
0451  soldered together! The meaning of rudimentary ORGANS is often quite unmistakeable: for instance th
0451  at the rudiments represent wings. Rudimentary ORGANS sometimes retain their potentiality, and are
0451  ls, for many instances are on record of these ORGANS having become well developed in full grown ma
0452  similar instances could be given. Rudimentary ORGANS in the individuals of the same species are ve
0452  e female moths in certain groups. Rudimentary ORGANS may be utterly aborted; and this implies, tha
0452  ros. It is an important fact that rudimentary ORGANS, such as teeth in the upper jaws of whales an
0453  the leading facts with respect to rudimentary ORGANS. In reflecting on them, every one must be str
0453  er which tells us plainly that most parts and ORGANS are exquisitely adapted for certain purposes,
0453  plainness that these rudimentary or atrophied ORGANS, are imperfect and useless. In works on natur
0453  less. In works on natural history rudimentary ORGANS are generally said to have been created for t
0453  gist accounts for the presence of rudimentary ORGANS, by supposing that they serve to excrete matt
0454  with modification, the origin of rudimentary ORGANS is simple. We have plenty of cases of rudimen
0454  imple. We have plenty of cases of rudimentary ORGANS in our domestic productions, as the stump of
0454  ases throw light on the origin of rudimentary ORGANS in a state of nature, further than by showing
0454  nerations to the gradual reduction of various ORGANS, until they become rudimentary, as in the cas
0455  tanc the greater relative size of rudimentary ORGANS in the embryo, and their lesser relative size
0455  mentary organ. As the presence of rudimentary ORGANS is thus due to the tendency in every part of
0455  of high physiological importance. Rudimentary ORGANS may be compared with the letters in a word, s
0455  cation, we may conclude that the existence of ORGANS in a rudimentary, imperfect, and useless cond
0456  st trifling importance, or, as in rudimentary ORGANS, of no importance; the wide opposition in val
0457  the same pattern displayed in the homologous ORGANS, to whatever purpose applied, of the differen
0457  species of a class of the homologous parts or ORGANS, though fitted in the adult members for purpo
0457  ame principle, and bearing in mind, that when ORGANS are reduced in size, either from disuse or se
0457  of inheritance, the occurrence of rudimentary ORGANS and their final abortion, present to us no in
0457  f embryological characters and of rudimentary ORGANS in classification is intelligible, on the vie
0459  fficult to believe than that the more complex ORGANS and instincts should have been perfected, not
0459  have existed, each good of its kind, that all ORGANS and instincts are, in ever so slight a degree
0461  that of first crosses, for their reproductive ORGANS are more or less functionally impotent; where
0461  onally impotent; whereas in first crosses the ORGANS on both sides are in a perfect condition. As
0466  ons between the simplest and the most perfect ORGANS; it cannot be pretended that we know all the
0479  view of the gradual modification of parts or ORGANS, which were alike in the early progenitor of
0480  stand on this view the meaning of rudimentary ORGANS. But disuse and selection will generally act
0480  to have taken pains to reveal, by rudimentary ORGANS and by homologous structures, her scheme of m
0483  very wide intervals between existing orders. ORGANS in a rudimentary condition plainly show that
0485  adaptive characters, rudimentary and aborted ORGANS, etc., will cease to be metaphorical, and wil
0486  d which have long been inherited. Rudimentary ORGANS will speak infallibly with respect to the nat
0022  he proportional length of the eyelids, of the ORIFICE of the nostrils, of the tongue (not always i
0191  d drink which we swallow has to pass over the ORIFICE of the trachea, with some risk of falling in
0021  eatly elongated eyelids, very large external ORIFICES to the nostrils, and a wide gape of mouth. T
0001  facts seemed to me to throw some light on the ORIGIN of species, that mystery of mysteries, as it
0002  e same general conclusions that I have on the ORIGIN of species. Last year he sent to me a memoir
0003  nd his excellent judgment. In considering the ORIGIN of Species, it is quite conceivable that a na
0006  remaining as yet unexplained in regard to the ORIGIN of species and varieties, if he makes due all
0007  distinguishing between Varieties and Species. ORIGIN of Domestic Varieties from one or more Specie
0007  cies. Domestic Pigeons, their Differences and ORIGIN. Principle of Selection anciently followed, i
0007  methodical and Unconscious Selection. Unknown ORIGIN of our Domestic Productions. Circumstances fa
0016  esert empirical. Moreover, on the view of the ORIGIN of genera which I shall presently give, we ha
0017  elied on by those who believe in the multiple ORIGIN of our domestic animals is, that we find in t
0019  mon wild duck and rabbit. The doctrine of the ORIGIN of our several domestic races from several ab
0024  cation! yet on the hypothesis of the multiple ORIGIN of our pigeons, it must be assumed that at le
0028  he same aviary. I have discussed the probable ORIGIN of domestic pigeons at some, yet quite insuff
0040  ticed, namely, that we know nothing about the ORIGIN or history of any of our domestic breeds. But
0040  ge, can hardly be said to have had a definite ORIGIN. A man preserves and breeds from an individua
0043  display of distinct breeds. To sum up on the ORIGIN of our Domestic Races of animals and plants.
0043  distinct, has played an important part in the ORIGIN of our domestic productions. When in any coun
0141  argument too far, on account of the probable ORIGIN of some of our domestic animals from several
0142  and as most of these varieties are of recent ORIGIN, they cannot owe their constitutional differe
0179  ely imperfect and intermittent record. On the ORIGIN and transitions of organic beings with peculi
0195  mes felt much difficulty in understanding the ORIGIN of simple parts, of which the importance does
0196  or modified swimbladders betray their aquatic ORIGIN, may perhaps be thus accounted for. A well de
0199  little light can apparently be thrown on the ORIGIN of these differences, chiefly through sexual
0207  omparable with habits, but different in their ORIGIN. Instincts graduated. Aphides and ants. Insti
0207  instincts variable. Domestic instincts, their ORIGIN. Natural instincts of the cuckoo, ostrich, an
0207  t premise, that I have nothing to do with the ORIGIN of the primary mental powers, any more than I
0208  stinctive action is performed, but not of its ORIGIN. How unconsciously many habitual actions are
0218  he gallinaceae; and this perhaps explains the ORIGIN of a singular instinct in the allied group of
0249  r fertility, already lessened by their hybrid ORIGIN. I am strengthened in this conviction by a re
0254  ered as distinct species. On this view of the ORIGIN of many of our domestic animals, we must eith
0280  storical or indirect evidence regarding their ORIGIN, it would not have been possible to have dete
0411  corresponding age. Rudimentary Organs; their ORIGIN explained. Summary. From the first dawn of li
0423  affinities, and would give the filiation and ORIGIN of each tongue. In confirmation of this view,
0423  requisite, as we have seen with pigeons. The ORIGIN of the existence of groups subordinate to gro
0454  on my view of descent with modification, the ORIGIN of rudimentary organs is simple. We have plen
0454  whether any of these cases throw light on the ORIGIN of rudimentary organs in a state of nature, f
0484  n the views entertained in this volume on the ORIGIN of species, or when analogous views are gener
0488  ity by gradation. Light will be thrown on the ORIGIN of man and his history. Authors of the highes
0013  t tell whether it may not be due to the same ORIGINAL cause acting on both; but when amongst indiv
0036  uatt remarks, have been purely bred from the ORIGINAL stock of Mr. Bakewell for upwards of fifty y
0076  strength, habits, and constitution, that the ORIGINAL proportions of a mixed stock could be kept u
0081  ssuredly be better filled up, if some of the ORIGINAL inhabitants were in some manner modified; fo
0120  on slowly diverging in character from their ORIGINAL stocks, without either having given off any
0121  e species (marked by capital letters) of our ORIGINAL genus, may for a long period continue transm
0121  tage of descent their predecessors and their ORIGINAL parent. For it should be remembered that the
0121  improved state of a species, as well as the ORIGINAL parent species itself, will generally tend t
0122  pecies. But we may go further than this. The ORIGINAL species of our genus were supposed to resemb
```

Page ***************************************(Key Word)***************************************

Page **(Key Word)**

```
0122  arents (A) and (I), but likewise some of the  ORIGINAL  species which were most nearly related to th
0122  ated to their parents. Hence very few of the  ORIGINAL  species will have transmitted offspring to t
0122  were least closely related to the other nine  ORIGINAL  species, has transmitted descendants to this
0123  ew species in our diagram descended from the  ORIGINAL  eleven species, will now be fifteen in numbe
0123  than that between the most different of the   ORIGINAL  eleven species. The new species, moreover, w
0123  rm two sub genera or even genera. But as the  ORIGINAL  species (I) differed largely from (A), stand
0123  standing nearly at the extreme points of the  ORIGINAL  genus, the six descendants from (I) will, ow
0123  mportant consideration), which connected the  ORIGINAL  species (A) and (I), have all become, except
0125  will have descended from two species of the   ORIGINAL  genus: and these two species are supposed to
0156  roposition will be granted. The cause of the  ORIGINAL  variability of secondary sexual characters i
0173  ion, and has supplanted and exterminated its  ORIGINAL  parent and all the transitional varieties be
0187  to the collateral descendants from the same   ORIGINAL  parent form, in order to see what gradations
0399  inciple of inheritance still betraying their  ORIGINAL  birthplace. Many analogous facts could be gi
0400  t certainly receive immigrants from the same  ORIGINAL  source, or from each other. But this dissimi
0435  will be little or no tendency to modify the   ORIGINAL  pattern, or to transpose parts. The bones of
0023  till exist in the countries where they were   ORIGINALLY domesticated, and yet be unknown to ornitho
0122  widely diffused species, so that they must    ORIGINALLY have had some advantage over most of the ot
0140  as i believe that our domestic animals were   ORIGINALLY chosen by uncivilised man because they were
0176  disappear, sooner than the forms which they   ORIGINALLY linked together. For any form existing in l
0177  hill breed; and thus the two breeds, which    ORIGINALLY existed in greater numbers, will come into
0181  legs, we perhaps see traces of an apparatus   ORIGINALLY constructed for gliding through the air rat
0190  rly the highly important fact that an organ   ORIGINALLY constructed for one purpose, namely flotati
0192  not doubt that little folds of skin, which    ORIGINALLY served as ovigerous frena, but which, likew
0192  hat the branchiae in this latter family had   ORIGINALLY existed as organs for preventing the ova fr
0202  we look at the sting of the bee, as having   ORIGINALLY existed in a remote progenitor as a boring
0202  ed for its present purpose, with the poison   ORIGINALLY adapted to cause galls subsequently intensi
0209  s happen, then the resemblance between what   ORIGINALLY was a habit and an instinct becomes so clos
0215  fear of the dog and cat which no doubt was   ORIGINALLY instinctive in them, in the same way as it
0223  near their rests, it is possible that pupae   ORIGINALLY stored as food might become developed; and
0224  ocreate them, the habit of collecting pupae  ORIGINALLY for food might by natural selection be stre
0351  he most distant quarters of the world, must   ORIGINALLY have proceeded from the same source, as the
0358  he wrecks of sunken continents: if they had   ORIGINALLY existed as mountain ranges on the land, som
0421  upy its proper intermediate position: for F  ORIGINALLY was intermediate in character between A and
0490  life, with its several powers, having been   ORIGINALLY breathed into a few forms or into one: and
0018  t does it show, but that some of our breeds   ORIGINATED there, four or five thousand years ago? But
0019  e must admit that many domestic breeds have  ORIGINATED in Europe: for whence could they have been
0059  ve once existed as varieties, and have thus  ORIGINATED: whereas, these analogies are utterly inexp
0187  concerns us more than how life itself first  ORIGINATED: but I may remark that several facts make m
0196  y of very little importance, and which have  ORIGINATED from quite secondary causes, independently
0209  e most complex and wonderful instincts have  ORIGINATED. As modifications of corporeal structure ar
0223  by what steps the instinct of F. sanguinea  ORIGINATED I will not pretend to conjecture. But as an
0235  not see how an instinct could possibly have  ORIGINATED; cases, in which no intermediate gradations
0241  from each other and from their parents, has  ORIGINATED. We can see how useful their production may
0253  rtainly be highly fertile. A doctrine which  ORIGINATED with Pallas, has been largely accepted by m
0335  n the important character of length of beak  ORIGINATED earlier than short beaked tumblers, which a
0089  of life, but differ in structure, colour, or ORNAMENT, such differences have been mainly caused by
0090  cock, which can hardly be either useful or   ORNAMENTAL to this bird: indeed, had the tuft appeared
0022  eons might be chosen, which if shown to an   ORNITHOLOGIST, and he were told that they were wild bir
0022  ecies. Moreover, I do not believe that any  ORNITHOLOGIST would place the English carrier, the shor
0023  nally domesticated, and yet be unknown to   ORNITHOLOGISTS: and this, considering their size, habits
0049  some zoologists. Several most experienced   ORNITHOLOGISTS consider our British red grouse as only a
0107  lous forms now known in the world, as the   ORNITHORHYNCHUS and Lepidosiren, which, like fossils, co
0130  so we occasionally see an animal like the   ORNITHORHYNCHUS or Lepidosiren, which in some small degr
0416  ir or feathers, in the Vertebrata. If the   ORNITHORHYNCHUS had been covered with feathers instead o
0429  hich again implies extinction. The genera   ORNITHORHYNCHUS and Lepidosiren, for example, would not
0146  being in some cases, according to Tausch,   ORTHOSPERMOUS in the exterior flowers and coelospermous
0343  an enormous period extended, and where our  OSCILLATING continents now stand they have stood ever
0291  , was certainly deposited during a downward  OSCILLATION of level, and thus gained considerable thi
0106  eat areas, though now continuous, owing to  OSCILLATIONS of level, will often have recently existed
0107  tal area, which will probably undergo many  OSCILLATIONS of level, and which consequently will exis
0287  above estimate. On the other hand, during  OSCILLATIONS of level, which we know this area has unde
0290  when first upraised and during subsequent  OSCILLATIONS of level. Such thick and extensive accumul
0291  that each area has undergone numerous slow  OSCILLATIONS of level, and apparently these oscillation
0291  scillations of level, and apparently these  OSCILLATIONS have affected wide spaces. Consequently fo
0301  riod of subsidence would be interrupted by  OSCILLATIONS of level, and that slight climatal changes
0309  ve been at least partially upheaved by the  OSCILLATIONS of level, which we may fairly conclude mus
0309  have existed, subjected no doubt to great   OSCILLATIONS of level, since the earliest silurian peri
0309  ce, the great archipelagoes still areas of  OSCILLATIONS of level, and the continents areas of elev
0309  een formed by a preponderance, during many  OSCILLATIONS of level, of the force of elevation: but m
0357  me that we have abundant evidence of great  OSCILLATIONS of level in our continents: but not of suc
0370  on, though subjected to large, but partial  OSCILLATIONS of level, I am strongly inclined to extend
0429  ttle modified, will give to us our so called OSCULANT or aberrant groups. The more aberrant any fo
0066  rge. The condor lays a couple of eggs and the OSTRICH a score, and yet in the same country the con
0134  beast of prey, has been caused by disuse. The OSTRICH indeed inhabits continents and is exposed to
0135  may imagine that the early progenitor of the  OSTRICH had habits like those of a bustard, and that
0182  he land, like the penguin: as sails, like the OSTRICH; and functionally for no purpose like the Ap
0207  heir origin. Natural instincts of the cuckoo, OSTRICH, and parasitic bees. Slave making ants. Hive
0218  days. This instinct, however, of the American OSTRICH has not as yet been perfected: for a surpris
0349  re inhabited by one species of Rhea (American OSTRICH), and northward the plains of La Plata by an
0349  species of the same genus: and not by a true  OSTRICH or emeu, like those found in Africa and Aust
0218  f a singular instinct in the allied group of OSTRICHES. For several hen ostriches, at least in the
0218  e allied group of ostriches. For several hen OSTRICHES, at least in the case of the American speci
0107  comes that the flora of Madeira, according to OSWALD Heer, resembles the extinct tertiary flora of
0014  mes earlier. In many cases this could not be OTHERWISE: thus the inherited peculiarities in the ho
0063  , and during some season or occasional year, OTHERWISE, on the principle of geometrical increase,
0076  , and the seed then mixed in due proportion, OTHERWISE the weaker kinds will steadily decrease in
```

Page **(Key Word)**

```
0104  h the individuals of the same species, which OTHERWISE would have inhabited the surrounding and d
0113  e to all animals, that is, if they vary, for OTHERWISE natural selection can do nothing. So it wi
0167  an animal striped like a zebra, but perhaps OTHERWISE very differently constructed, the common p
0188  ng this treatise that large bodies of facts, OTHERWISE inexplicable, can be explained by the theo
0243  ly explicable on the foregoing views, but is OTHERWISE inexplicable, all tend to corroborate the
0255  ower of the hybrid to wither earlier than it OTHERWISE would have done; and the early withering o
0299  between some few groups less wide than they OTHERWISE would have been, yet has done scarcely any
0179  , which has webbed feet and which resembles an OTTER in its fur, short legs, and form of tail; du
0006  ole work, and a few concluding remarks. No one OUGHT to feel surprise at much remaining as yet un
0055  en formed, many varieties or incipient species OUGHT, as a general rule, to be now forming. Where
0056  the manufactory of species has been active, we OUGHT generally to find the manufactory still in a
0058  s supposed parent species, their denominations OUGHT to be reversed. But there is also reason to
0085  ur, when once acquired, true and constant. Nor OUGHT we to think that the occasional destruction
0105  le for the production of new organic forms, we OUGHT to make the comparison within equal times; a
0116  lopment. After the foregoing discussion, which OUGHT to have been much amplified, we may, I think
0141  all tropical or sub tropical in their habits, OUGHT not to be looked at as anomalies, but merely
0154  ively recent and extraordinarily great that we OUGHT to find the generative variability, as it ma
0162  what are new but analogous variations, yet we OUGHT, on my theory, sometimes to find the varying
0173  d species inhabit the same territory we surely OUGHT to find at the present time many transitional
0173  between its past and present states. Hence we OUGHT not to expect at the present time to meet wit
0175  mportant elements of distribution, these facts OUGHT to cause surprise, as climate and height or c
0178  hat, in any one region and at any one time, we OUGHT only to see a few species presenting slight r
0187  an organ in any species has been perfected, we OUGHT to look exclusively to its lineal ancestors;
0188  le, can be explained by the theory of descent, OUGHT not to hesitate to go further, and to admit t
0188  now any of the transitional grades. His reason OUGHT to conquer his imagination; though I have fel
0188  t compare the eye to an optical instrument, we OUGHT in imagination to take a thick layer of trans
0198  to have arisen through ordinary generation, we OUGHT not to lay too much stress on our ignorance o
0202  d sterile sisters? It may be difficult, but we OUGHT to admire the savage instinctive hatred of th
0210  ce, as in the case of corporeal structures, we OUGHT to find in nature, not the actual transitiona
0210  n the lineal ancestors of each species, but we OUGHT to find in the collateral lines of descent so
0210  escent some evidence of such gradations; or we OUGHT at least to be able to show that gradations c
0211  tural selection, as many instances as possible OUGHT to have been here given; but want of space pr
0239  to have the desired character. On this view we OUGHT occasionally to find neuter insects of the sa
0269  slightly sterile; and if this be so, we surely OUGHT not to expect to find sterility both appearin
0297  nd of evidence of change which on my theory we OUGHT to find. Moreover, if we look to rather wider
0301  into one long and branching chain of life. We OUGHT only to look for a few links, some more close
0307  tion, or obliterated by metamorphic action, we OUGHT to find only small remnants of the formations
0307  mations next succeeding them in age, and these OUGHT to be very generally in a metamorphosed condi
0336  e intervened between successive formations, we OUGHT not to expect to find, as I attempted to show
0336  ommencement and close of these periods; but we OUGHT to find after intervals, very long as measure
0354  so numerous and of so grave a nature, that we OUGHT to give up the belief, rendered probable by g
0405  ted to the most remote points of the world, we OUGHT to find, and I believe as a general rule we d
0425  as a single species. But it may be asked, what OUGHT we to do, if it could be proved that one spec
0425  by a long course of modification, from a bear? OUGHT we to rank this one species with bears, and w
0460  by the canon, Natura non facit saltum, that we OUGHT to be extremely cautious in saying that any o
0461  on apparently tends to eliminate sterility, we OUGHT not to expect it also to produce sterility. T
0462  sly long as measured by years, too much stress OUGHT not to be laid on the occasional wide diffusi
0472  naturalised productions from another land. Nor OUGHT we to marvel if all the contrivances in natur
0386  le pond; this mud when dry weighed only 6 3/4 OUNCES; I kept it covered up in my study for six mon
0373  iod, from the western whores of Britain to the OURAL range, and southward to the Pyrennes. We may
0150  with respect to the length of the arms of the OURANG outang, that he has come to a nearly similar
0079  we reflect on this struggle, we may console OURSELVES with the full belief, that the war of natur
0105  new species. But we may thus greatly deceive OURSELVES, for to ascertain whether a small isolated
0207  s in other birds' nests. An action, which we OURSELVES should require experience to enable us to p
0288  ost erroneous view, when we tacitly admit to OURSELVES that sediment is being deposited over nearl
0299  fine, intermediate, fossil links, by asking OURSELVES whether, for instance, geologists at some f
0319  of all classes, in all countries. If we ask OURSELVES why this or that species is rare, we answer
0150  spect to the length of the arms of the ourang OUTANG, that he has come to a nearly similar conclus
0021  mouth. The short faced tumbler has a beak in OUTLINE almost like that of a finch; and the common
0163  ripe is certainly very variable in length and OUTLINE. A white ass, but not an albino, has been de
0224  ails on this subject; but will merely give an OUTLINE of the conclusions at which I have arrived.
0255  on can be shown to exist; but only the barest OUTLINE of the facts can here be given. When pollen
0380  ive productions are nearly equalled or even OUTNUMBERED by the naturalised; and if the natives hav
0185  server by examining the dead body of the water OUZEL would never have suspected its sub aquatic ha
0192  originally existed as organs for preventing the OVA from being washed out of the sack? Although we
0384  whirlwinds in India, and the vitality of their OVA when removed from the water. But I am inclined
0385  stribution at first perplexed me much; as their OVA are not likely to be transported by birds, and
0385  g in a natural pond, in an aquarium, where many OVA of fresh water shells were hatching; and I fou
0415  the Connaraceae differ in having one or more OVARIA, in the existence or absence of albumen, in t
0250  and distinct species: the result was that the OVARIES of the three first flowers soon ceased to gr
0263  il too long for the pollen tubes to reach the OVARIUM. It has also been observed that when pollen
0451  en tubes to reach the ovules protected in the OVARIUM at its base. The pistil consists of a stigma
0145  so differ in shape and sculpture; and even the OVARY itself, with its accessory parts, differs, as
0169  ll in such cases not as yet have had time to OVERCOME the tendency to further variability and to r
0128  rful fact, the wonder of which we are apt to OVERLOOK from familiarity, that all animals and all p
0343  d between our consecutive formations; he may OVERLOOK how important a part migration must have pla
0033  th, the importance of which should never be OVERLOOKED, will ensure some differences; but, as a ge
0038  s kept by uncivilised man, it should not be OVERLOOKED that they almost always have to struggle fo
0296  ation; facts, showing what wide, yet easily OVERLOOKED, intervals have occurred in its accumulatio
0396  which not a wreck now remains, must not be OVERLOOKED. I will here give a single instance of one
0419  important structural differences, at first OVERLOOKED, but because numerous allied species, with
0286  erlies the Weald, on the flanks of which the OVERLYING sedimentary deposits might have accumulated
0334  termediate in character between those of the OVERLYING carboniferous, and underlying Silurian syst
0129  species and groups of species have tried to OVERMASTER other species in the great battle for life.
0143  been largely combined with, and sometimes OVERMASTERED by, the natural selection of innate differ
0158  ompletely, according to the lapse of time, OVERMASTERED the tendency to reversion and to further v
```

Page **(Key Word)**

Page ********************************(Key Word)********************************

```
0103  als, which unite for each birth, we must not  OVERRATE  the effects of intercrosses in retarding nat
0488  , might become modified; so that we must not  OVERRATE  the accuracy of organic change as a measure
0139  the climates under which they live is often  OVERRATED. We may infer this from our frequent inabil
0207  many readers, as a difficulty sufficient to  OVERTHROW  my whole theory. I must premise, that I hav
0272  varieties does not seem to me sufficient to  OVERTHROW  the view which I have taken with respect to
0466  difficulties are, in my judgment they do not  OVERTHROW  the theory of descent with modification. No
0129  have tried to branch cut on all sides, and to  OVERTOP  and kill the surrounding twigs and branches,
0130  buds, and these, if vigorous, branch out and  OVERTOP  on all sides many a feebler branch, so by ge
0392  advantage by growing taller and taller and  OVERTOPPING  the other plants. If so, natural selection
0239  n. it will indeed be thought that I have an  OVERWEENING  confidence in the principle of natural sel
0482  load of prejudice by which this subject is  OVERWHELMED be removed. Several eminent naturalists ha
0050  brids; and there is, as it seems to me, an  OVERWHELMING  amount of experimental evidence, showing t
0192  e two minute folds of skin, called by me the  OVIGEROUS  frena, which serve, through the means of a
0192  ssile cirripedes, on the other hand, have no  OVIGEROUS  frena, the eggs lying loose at the bottom o
0192  ae. Now I think no one will dispute that the  OVIGEROUS  frena in the one family are strictly homolo
0192  le folds of skin, which originally served as  OVIGEROUS  frena, but which, likewise, very slightly a
0200  we see in the fang of the adder, and in the  OVIPOSITOR  of the ichneumon, by which its eggs are dep
0010  s no essential difference between a bud and an  OVULE  in their earliest stages of formation: so tha
0132  does not apparently differ essentially from an  OVULE, is alone affected. But why, because the repr
0250  a single one as an example, namely, that every  OVULE  in a pod of Crinum capense fertilised by C. r
0263  impossibility in the male element reaching the  OVULE, as would be the case with a plant having a p
0010  ent has affected a bud or offset, and not the  OVULES  or pollen. But it is the opinion of most phys
0010  variability may be largely attributed to the  OVULES  or pollen, or to both, having been affected b
0014  ts primary cause, which may have acted on the  OVULES  or male element: in nearly the same manner as
0203  es may be wafted by a chance breeze on to the  OVULES? Summary of Chapter. We have in this chapter
0251  ared perfectly healthy, and although both the  OVULES  and pollen of the same flower were perfectly
0451  til is to allow the pollen tubes to reach the  OVULES  protected in the ovarium at its base. The pis
0009  e the bane of horticulture; but on this view we  OWE  variability to the same cause which produces s
0037  ht what splendid fruit we should eat; though we  OWE  our excellent fruit, in some small degree, to
0142  ese varieties are of recent origin, they cannot  OWE  their constitutional differences to habit. The
0194  e manner two parts in two organic beings, which  OWE  but little of their structure in common to inh
0310  ese great authorities, to whom, with others, we  OWE  all our knowledge. Those who think the natural
0356  he horses of every other breed; but they do not  OWE  their difference and superiority to descent fr
0134  plained by the effects of disuse. As Professor  OWEN  has remarked, there is no greater anomaly in n
0149  ch as this vegetative repetition, to use Prof.  OWEN's  expression, seems to be a sign of low organi
0150  fer also from an observation made by Professor  OWEN, with respect to the length of the arms of the
0191  ladder. We can thus, as I infer from Professor  OWEN's  interesting description of these parts, unde
0192  se wondrous organs have been produced; but, as  OWEN  and others have remarked, their intimate struc
0310  most eminent palaeontologists, namely Cuvier,  OWEN, Agassiz, Barrande, Falconer, E. Forbes, etc.,
0319  erly groundless was my astonishment! Professor  OWEN  soon perceived that the tooth, though so like
0329  illustrations from our great palaeontologist,  OWEN, showing how extinct animals fall in between e
0329  s the two most distinct orders of mammals; but  OWEN  has discovered so many fossil links, that he h
0339  nd in several parts of La Plata; and Professor  OWEN  has shown in the most striking manner that mos
0339  ent between the dead and the living. Professor  OWEN  has subsequently extended the same generalisat
0414  it becomes for classification. As an instance:  OWEN, in speaking of the dugong, says, The generati
0416  to the mouth, the only character, according to  OWEN, which absolutely distinguishes fishes and rep
0435  of the attempt has been expressly admitted by  OWEN  in his most interesting work on the Nature of
0437  such extraordinarily shaped pieces of bone? As  OWEN  has remarked, the benefit derived from the yie
0437  part or organ is the common characteristic (as  OWEN  has observed) of all low or little modified fo
0442  any period differ widely from the adult: thus  OWEN  has remarked in regard to cuttle fish, there i
0452  s. this is well shown in the drawings given by  OWEN  of the bones of the leg of the horse, ox, and
0012  nsciously modify other parts of the structure,  OWING  to the mysterious laws of the correlation of
0061  follow inevitably from the struggle for life.  OWING  to this struggle for life, any variation, how
0070  ruction by our native animals. When a species,  OWING  to highly favourable circumstances, increases
0106  moreover, great areas, though now continuous,  OWING  to oscillations of level, will often have rec
0118  agram to have produced variety a2, which will,  OWING  to the principle of divergence, differ more f
0123  eleven species, will now be fifteen in number.  OWING  to the divergent tendency of natural selectio
0123  inal genus, the six descendants from (I) will,  OWING  to inheritance, differ considerably from the
0126  e remotely to the future, we may predict that,  OWING  to the continued and steady increase of the l
0127  i think this cannot be disputed; if there be,  OWING  to the high geometrical powers of increase of
0139  recks of ancient life have not been preserved,  OWING  to the less severe competition to which the i
0147  e actual withdrawal of nutriment from one part  OWING  to the excess of growth in another and adjoin
0149  re only add that their variability seems to be  OWING  to their uselessness, and therefore to natura
0202  t many attacking animals, cannot be withdrawn,  OWING  to the backward serratures, and so inevitably
0246  he distinction has probably been slurred over,  OWING  to the sterility in both cases being looked o
0252  ed. It should, however, be borne in mind that,  OWING  to few animals breeding freely under confinem
0274  ubject is here excessively complicated, partly  OWING  to the existence of secondary sexual characte
0274  condary sexual characters: but more especially  OWING  to prepotency in transmitting likeness runnin
0294  irst appear and disappear at different levels,  OWING  to the migration of species and to geographic
0325  n, and the introduction of new ones, cannot be  OWING  to mere changes in marine currents or other c
0327  pecies thus produced being themselves dominant  OWING  to inheritance, and to having already had som
0338  o show that the adult differs from its embryo,  OWING  to variations supervening at a not early age,
0342  sed away even during a single formation; that,  OWING  to subsidence being necessary for the accumul
0395  difficulty in forming a judgment in some cases  OWING  to the probable naturalisation of certain mam
0422  iven rise to few new languages, whilst others (OWING  to the spreading and subsequent isolation and
0430  ected that the resemblance is only analogical,  OWING  to the phascolomys having become adapted to h
0450  stages of recent forms, may be true, but yet,  OWING  to the geological record not extending far en
0464  links we could hardly ever expect to discover,  OWING  to the imperfection of the geological record.
0476  gration from one part of the world to another,  OWING  to former climatal and geographical changes a
0362  royed so many on their arrival. Some hawks and  OWLS  bolt their prey whole, and after an interval o
0036  ne at all acquainted with the subject that the  OWNER  of either of them has deviated in any one ins
0205  ut in all cases at the same time useful to the  OWNER. Natural selection in each well stocked count
0238  ced oxen with the longest horns; and yet no one  OX  could ever have propagated its kind. Thus I bel
0452  n by Owen of the bones of the leg of the horse,  OX, and rhinoceros. It is an important fact that r
0237  ifficially imperfect state of the male sex; for  OXEN  of certain breeds have longer horns than in ot
0238  doubt that a breed of cattle, always yielding  OXEN  with extraordinarily long horns, could be slow
```

Page ********************************(Key Word)********************************

```
0238 ividual bulls and cows, when matched, produced OXEN with the longest horns; and yet no one ox coul
0427 ior limbs, between the dugong, which is a PACHYDERMATOUS animal, and the whale, and between both t
0329 ing groups. Cuvier ranked the Ruminants and PACHYDERMS, as the two most distinct orders of mammals
0329 of these two orders; and has placed certain PACHYDERMS in the same sub order with ruminants: for e
0332 e manner as has occurred with ruminants and PACHYDERMS. Yet he who objected to call the extinct ge
0416 ng the close affinity between Ruminants and PACHYDERMS. Robert Brown has strongly insisted on the
0309 ssuming that if, for instance, the bed of the PACIFIC Ocean were now converted into a continent, w
0348 passed we meet in the eastern islands of the PACIFIC, with another and totally distinct fauna. So
0348 eastern islands of the tropical parts of the PACIFIC, we encounter no impassable barriers, and we
0348 f eastern and Western America and the eastern PACIFIC Islands, yet many fish range from the Pacifi
0348 pacific Islands, yet many fish range from the PACIFIC into the Indian Ocean, and many shells are c
0348 ells are common to the eastern islands of the PACIFIC and the eastern shores of Africa, on almost
0360 ; and the natives of the coral islands in the PACIFIC, procure stones for their tools, solely from
0369 ocean and by the extreme northern part of the PACIFIC. During the Glacial period, when the inhabit
0373 grees to 37 degrees, and on the shores of the PACIFIC, where the climate is now so different, as f
0397 gard to the land shells of the islands of the PACIFIC. Now it is notorious that land shells are ve
0398 inhabitants of these volcanic islands in the PACIFIC, distant several hundred miles from the cont
0067 lding surface, with ten thousand sharp wedges PACKED close together and driven inwards by incessan
0088 election, as the mane to the lion, the shoulder PAD to the boar, and the hooked jaw to the male sa
0434 a mole for digging, the leg of the horse, the PADDLE of the porpoise, and the wing of the bat, sho
0436 o be within the limits of possibility. In the PADDLES of the extinct gigantic sea lizards, and in
0447 one descendant to act as hands, in another as PADDLES, in another as wings; and on the above two p
0447 and having thus been converted into hands, or PADDLES, or wings. Whatever influence long continued
0311 short chapter has been preserved; and of each PAGE, only here and there a few lines. Each word of
0289 nd rare is their preservation, far better than PAGES of detail. Nor is their rarity surprising, wh
0329 system. With respect to the Vertebrata, whole PAGES could be filled with striking illustrations f
0034 ve been a strange fact, had attention not been PAID to breeding, for the inheritance of good and b
0041 d by him, that the closest attention should be PAID to even the slightest deviation in the qualiti
0041 e of each individual. Unless such attention be PAID nothing can be effected. I have seen it gravel
0042 eing kept by poor people, and little attention PAID to their breeding; in peacocks, from not being
0201 paley has remarked, for the purpose of causing PAIN or for doing an injury to its possessor. If a
0064 er of all known animals, and I have taken some PAINS to estimate its probable minimum rate of natu
0251 from the mountains of Chile. I have taken some PAINS to ascertain the degree of fertility of some
0255 the hybridisation of plants. I have taken much PAINS to ascertain how far the rules apply to anima
0393 ich the great oceans are studded. I have taken PAINS to verify this assertion, and I have found it
0480 of inutility! Nature may be said to have taken PAINS to reveal, by rudimentary organs and by homol
0232 tely diffused by the bees, as delicately as a PAINTER could have done with his brush, by atoms of
0064 uld soon be covered by the progeny of a single PAIR. Even slow breeding man has doubled in twenty
0064 ng till ninety years old, bringing forth three PAIR of young in this interval; if this be so, at t
0064 en million elephants, descended from the first PAIR. But we have better evidence on this subject t
0065 imals there are very few which do not annually PAIR. Hence we may confidently assert, that all pla
0096 sects, and some other large groups of animals, PAIR for each birth. Modern research has much dimin
0096 tes, and of real hermaphrodites a large number PAIR: that is, two individuals regularly unite for
0096 dite animals which certainly do not habitually PAIR, and a vast majority of plants are hermaphrodi
0100 s land mollusca and earth worms; but these all PAIR. As yet I have not found a single case of a te
0103 from varieties of the same kind preferring to PAIR together. Intercrossing plays a very important
0253 one on increasing. If we were to act thus, and PAIR brothers and sisters in the case of any pure a
0355 the same species have descended from a single PAIR, or single hermaphrodite, or whether, as some
0356 nce and superiority to descent from any single PAIR, but to continued care in selecting and traini
0436 rogenitor had an upper lip, mandibles, and two PAIR of maxillae, these parts being perhaps very si
0441 rs of beautifully constructed natatory legs, a PAIR of magnificent compound eyes, and extremely co
0042 ht into play: in cats, from the difficulty in PAIRING them; in donkeys, from only a few being kept
0435 ifications of an upper lip, mandibles, and two PAIRS of maxillae. Analogous laws govern the constr
0441 edes: the larvae in the first stage have three PAIRS of legs, a very simple single eye, and a prob
0441 chrysalis stage of butterflies, they have six PAIRS of beautifully constructed natatory legs, a p
0279 nd of denudation. On the poorness of our PALAEONTOLOGICAL collections. On the intermittence of geo
0287 isplay we behold! On the poorness of our PALAEONTOLOGICAL collections. That our palaeontological c
0287 r palaeontological collections. That our PALAEONTOLOGICAL collections are very imperfect, is admit
0306 d states; and from the revolution in our PALAEONTOLOGICAL ideas on many points, which the discover
0322 ously throughout the World. Scarcely any PALAEONTOLOGICAL discovery is more striking than the fact
0287 y every one. The remark of that admirable PALAEONTOLOGIST, the late Edward Forbes, should not be f
0295 for it has been observed by more than one PALAEONTOLOGIST, that very thick deposits are usually ba
0304 had hardly been published, when a skilful PALAEONTOLOGIST, M. Bosquet, sent me a drawing of a perf
0329 ith striking illustrations from our great PALAEONTOLOGIST, Owen, showing how extinct animals fall
0293 pecies into another. I am aware that two PALAEONTOLOGISTS, whose opinions are worthy of much defer
0297 what excessively slight differences many PALAEONTOLOGISTS have founded their species; and they do
0298 species this could rarely be effected by PALAEONTOLOGISTS. We shall, perhaps, best perceive the im
0301 rding to the principles followed by many PALAEONTOLOGISTS, be ranked as new and distinct species.
0301 es of the same formation, would, by most PALAEONTOLOGISTS, be ranked as distinct species. But I do
0302 in formations, has been urged by several PALAEONTOLOGISTS, for instance, by Agassiz, Pictet, and b
0305 the case most frequently insisted on by PALAEONTOLOGISTS of the apparently sudden appearance of a
0305 nce one sub stage further back; and some PALAEONTOLOGISTS believe that certain much older fishes,
0310 er by the fact that all the most eminent PALAEONTOLOGISTS, namely Cuvier, Owen, Agassiz, Barrande,
0333 s by the concurrent evidence of our best PALAEONTOLOGISTS seems frequently to be the case. Thus, o
0334 t discovered, were at once recognised by PALAEONTOLOGISTS as intermediate in character between the
0335 aracter, is the fact, insisted on by all PALAEONTOLOGISTS, that fossils from two consecutive forma
0345 yet ill defined sentiment, felt by many PALAEONTOLOGISTS, that organisation on the whole has prog
0305 e equator; and by running through Pictet's PALAEONTOLOGY it will be seen that very few species are
0343 ties, all the other great leading facts in PALAEONTOLOGY seem to me simply to follow on the theory
0345 . on the other hand, all the chief laws of PALAEONTOLOGY plainly proclaim, as it seems to me, that
0284 s of Great Britain; and this is the result: PALAEOZOIC strata (not including igneous beds).....57,
0285 n with that which has removed masses of our PALAEOZOIC strata, in parts ten thousand feet in thick
0288 ctions which lived during the Secondary and PALAEOZOIC periods, it is superfluous to state that ou
0289 wn belonging to the age of our secondary or PALAEOZOIC formations. But the imperfection in the geo
0294 al species appeared somewhat earlier in the PALAEOZOIC beds of North America than in those of Euro
0308 s yet known to afford even a remnant of any PALAEOZOIC or secondary formation. Hence we may perhap
```

Page **(Key Word)**

0308 hence we may perhaps infer, that during the PALAEOZOIC and secondary periods, neither continents n
0308 ans now extend; for had they existed there, PALAEOZOIC and secondary formations would in all proba
0318 having disappeared before the close of the PALAEOZOIC period. No fixed law seems to determine the
0321 rders, as of Trilobites at the close of the PALAEOZOIC period and of Ammonites at the close of the
0323 nts of the world. In the several successive PALAEOZOIC formations of Russia, Western Europe and No
0323 life, in the stages of the widely separated PALAEOZOIC and tertiary periods, would still be manife
0325 . after referring to the parallelism of the PALAEOZOIC forms of life in various parts of Europe, t
0330 d, asserts that he is every day taught that PALAEOZOIC animals, though belonging to the same order
0337 would a secondary fauna by an eocene, and a PALAEOZOIC fauna by a secondary fauna. I do not doubt
0480 e generations, by disuse or the tongue and PALATE having been fitted by natural selection to br
0201 the good of each. No organ will be formed, as PALEY has remarked, for the purpose of causing pain
0163 e lost, in dark coloured asses. The koulan is PALLAS is said to have been seen with a double shoul
0253 hly fertile. A doctrine which originated with PALLAS, has been largely accepted by modern naturali
0138 s an increase in the length of the antennae or PALPI, as a compensation for blindness. Notwithstan
0287 to our richest geological museums, and what a PALTRY display we behold! On the poorness of our Pal
0348 nly by the narrow, but impassable, isthmus of PANAMA. Westward of the shores of America, a wide sp
0355 ies) from that lately advanced in an ingenious PAPER by Mr. Wallace, in which he concludes, that e
0422 arrangement is shown, as far as is possible on PAPER, in the diagram, but in much too simple a man
0433 milne Edwards has lately insisted, in an able PAPER, on the high importance of looking to types,
0097 ication closely enclosed, as in the great PAPILIONACEOUS or pea family; but in several, perhaps in
0097 r. so necessary are the visits of bees to PAPILIONACEOUS flowers, that I have found, by experiment
0453 he system; but can we suppose that the minute PAPILLA, which often represents the pistil in male f
0180 y a broad expanse of skin, which serves as a PARACHUTE and allows them to glide through the air to
0089 females. The rock thrush of Guiana, birds of PARADISE, and some others, congregate; and successive
0427 descent. We can also understand the apparent PARADOX, that the very same characters are analogica
0328 from the causes explained in the foregoing PARAGRAPHS, the same general succession in the forms o
0072 s determine the existence of cattle. Perhaps PARAGUAY offers the most curious instance of this; fo
0072 that this is caused by the greater number in PARAGUAY of a certain fly, which lays its eggs in the
0072 hawks or beasts of prey) were to increase in PARAGUAY, the flies would decrease, then cattle and h
0152 rhaps in polymorphic groups, we see a nearly PARALLEL natural case; for in such cases natural sele
0166 true for centuries, species; and how exactly PARALLEL is the case with that ofthe horse genus! For
0193 g to different families and orders, offers a PARALLEL case of difficulty. Other cases could be giv
0226 e described with their centres placed in two PARALLEL layers; with the centre of each sphere at th
0226 es of the adjoining spheres in the other and PARALLEL layer; then, if planes of intersection betwe
0230 very first cell is excavated out of a little PARALLEL sided wall of wax, is not, as far as I have
0262 e two distinct cases run to a certain extent PARALLEL. Something analogous occurs in grafting; for
0277 tances, should all run, to a certain extent, PARALLEL with the systematic affinity of the forms wh
0323 h water change at distant points in the same PARALLEL manner. We may doubt whether they have thus
0325 of their inhabitants. This great fact of the PARALLEL succession of the forms of life throughout t
0326 apparently do find, a less strict degree of PARALLEL succession in the productions of the land th
0327 led to survive. Thus, as it seems to me, the PARALLEL, and, taken in a large sense, simultaneous,
0329 he order would falsely appear to be strictly PARALLEL; nevertheless the species would not all be t
0338 uccession of extinct forms is in some degree PARALLEL to the embryological development of recent f
0348 faunas range far northward and southward, in PARALLEL lines not far from each other, under corresp
0461 ns. This parallelism is supported by another PARALLEL, but directly opposite, class of facts; name
0346 condition in the Old World which cannot be PARALLELED in the New, at least as closely as the same
0245 sterility of first crosses and of hybrids. PARALLELISM between the effects of changed conditions
0256 spring, are generally very sterile; but the PARALLELISM between the difficulty of making a first c
0262 duction, yet that there is a rude degree of PARALLELISM in the results of grafting and of crossing
0266 seem fanciful, but I suspect that a similar PARALLELISM extends to an allied yet very different cl
0267 degree. I cannot persuade myself that this PARALLELISM is an accident or an illusion. Both series
0277 been disturbed. This view is supported by a PARALLELISM of another kind; namely, that the crossing
0323 western Europe and North America, a similar PARALLELISM in the forms of life has been observed by
0323 lds be kept wholly out of view, the general PARALLELISM in the successive forms of life, in the st
0325 neuil and d'Archiac. After referring to the PARALLELISM of the palaeozoic forms of life in various
0328 and France, is able to draw a close general PARALLELISM between the successive stages in the two c
0328 lso, shows that there is a striking general PARALLELISM in the successive Silurian deposits of Boh
0347 culiar fauna or flora. Notwithstanding this PARALLELISM in the conditions of the Old and New World
0428 rhaps understand how it is that a numerical PARALLELISM has sometimes been observed between the su
0428 distinct classes. A naturalist, struck by a PARALLELISM of this nature in any one class, by arbitr
0428 to been arbitrary), could easily extend the PARALLELISM over a wide range; and thus the septenary,
0461 pounded of two distinct organisations. This PARALLELISM is supported by another parallel, but dire
0476 e can see why there should be so striking a PARALLELISM in the distribution of organic beings thro
0218 ally makes its own burrow and stores it with PARALYSED prey for its own larvae to feed on, yet tha
0028 er about pigeons as were the old Romans. The PARAMOUNT importance of these considerations in expla
0396 ore equally modified, in accordance with the PARAMOUNT importance of the relation of organism to o
0414 duction, with their product the seed, are of PARAMOUNT importance! We must not, therefore, in clas
0478 fication; why adaptive characters, though of PARAMOUNT importance to the being, are of hardly any
0003 sterous to account for the structure of this PARASITE, with its relations to several distinct orga
0060 d only a little less plainly in the humblest PARASITE which clings to the hairs of a quadruped or
0070 here comes in a sort of struggle between the PARASITE and its prey. On the other hand, in many cas
0075 ncy of one organic being on another, as of a PARASITE on its prey, lies generally between beings r
0077 eri and in that of the legs and claws of the PARASITE which clings to the hair on the tiger's body
0063 e with these trees, for if too many of these PARASITES grow on the same tree, it will languish and
0234 of bee may be dependent on the number of its PARASITES or other enemies, or on quite distinct caus
0070 hese so called epidemics appear to be due to PARASITIC worms, which have from some cause, possibly
0148 d be given: namely, that when a cirripede is PARASITIC within another and is thus protected, it lo
0148 ed with great nerves and muscles; but in the PARASITIC and protected Proteolepas, the whole anteri
0148 structure, when rendered superfluous by the PARASITIC habits of the Proteolepas, though effected
0207 atural instincts of the cuckoo, ostrich, and PARASITIC bees. Slave making ants. Hive bee, its cell
0218 n twenty lost and wasted eggs. Many bees are PARASITIC, and always lay their eggs in the nests of
0218 structure modified in accordance with their PARASITIC habits: for they do not possess the pollen
0218 kewise, of Sphegidae (wasp like insects) are PARASITIC on other species; and M. Fabre has lately s
0218 e of the prize, and becomes for the occasion PARASITIC. In this case, as with the supposed case of
0441 in the scale than the larva, as with certain PARASITIC crustaceans. To refer once again to cirripe
0221 upae to rear as slaves. I then dug up a small PARCEL of the pupae of F. fusca from another nest, a

Page **(Key Word)**

Page ***(Key Word)**

```
0222  he same time I laid on the same place a small  PARCEL  of the pupae of another species, F. flava, wi
0361  nes are embedded in the roots of trees, small  PARCELS  of earth are very frequently enclosed in the
0285  islands, which have been worn by the waves and  PARED  all round into perpendicular cliffs of one or
0164  r me a small dun Welch pony with three short  PARELLEL  stripes on each shoulder. In the north west
0409  forbes often insisted, there is a striking  PARELLELISM  in the laws of life throughout time and sp
0007  d somewhat different from, those to which the  PARENT  species have been exposed under nature. There
0010  in this case we see that the treatment of the  PARENT  has affected a bud or offset, and not the ovu
0010  having been affected by the treatment of the  PARENT  prior to the act of conception. These cases a
0011  ng much less, and walking more, than its wild  PARENT. The great and inherited development of the u
0013  combination of circumstances, appears in the  PARENT, say, once amongst several million individual
0014  same period at which it first appeared in the  PARENT. I believe this rule to be of the highest imp
0016  ether they have descended from one or several  PARENT  species. This point, if it could be cleared u
0017  ey would vary on an average as largely as the  PARENT  species of our existing domesticated producti
0018  that these latter have had more than one wild  PARENT. With respect to horses, from reasons which I
0019  sess a number of peculiar species as distinct  PARENT  stocks? So it is in India. Even in the case o
0023  uced by crossing two breeds unless one of the  PARENT  stocks possessed the characteristic enormous
0028  they could ever have descended from a common  PARENT, as any naturalist could in coming to a simil
0035  ave come to surpass in fleetness and size the  PARENT  Arab stock, so that the latter, by the regula
0037  mpared with the older varieties or with their  PARENT  stocks. No one would ever expect to get a fir
0037  ecognise, and therefore do not know, the wild  PARENT  stocks of the plants which have been longest
0039  and partly methodical selection. Perhaps the  PARENT  bird of all fantails had only fourteen tailfe
0052  te in which it differs very slightly from its  PARENT  to one in which it differs more, to the actio
0052  re to flourish so as to exceed in numbers the  PARENT  species, it would then rank as the species, a
0052  it might come to supplant and exterminate the  PARENT  species; or both might co exist, and both ran
0057  red round certain forms, that is, round their  PARENT  species? Undoubtedly there is one most import
0057  , when compared with each other or with their  PARENT  species, is much less than that between the s
0058  have a wider range than that of its supposed  PARENT  species, their denominations ought to be reve
0080  extinction, on the descendants from a common  PARENT. Explains the grouping of all organic beings.
0087  the structure of the young in relation to the  PARENT, and of the parent in relation to the young.
0087  e young in relation to the parent, and of the  PARENT  in relation to the young. In social animals i
0091  ich would either supplant or coexist with the  PARENT  form of wolf. Or, again, the wolves inhabitin
0111  ring of this variety again to differ from its  PARENT  in the very same character and in a greater d
0112  er both from each other and from their common  PARENT. But how, it may be asked, can any analogous
0118  erit those advantages which made their common  PARENT  (A) more numerous than most of the other inha
0118  advantages which made the genus to which the  PARENT  species belonged, a large genus in its own co
0118  ther, and more considerably from their common  PARENT  (A). We may continue the process by similar s
0118  ified descendants, proceeding from the common  PARENT  (A), will generally go on increasing in numbe
0119  rtake of the same advantages which made their  PARENT  successful in life, they will generally go on
0120  qually, from each other and from their common  PARENT. If we suppose the amount of change between e
0121  descent their predecessors and their original  PARENT. For it should be remembered that the competi
0121  d state of a species, as well as the original  PARENT  species itself, will generally tend to become
0122  to some quite new station, in which child and  PARENT  do not come into competition, both may contin
0124  pecies of the same genus. And the two or more  PARENT  species are supposed to have descended from s
0124  senting a single species, the supposed single  PARENT  of our several new sub genera and genera. It
0124  ended from a form which stood between the two  PARENT  species (A) and (I), now supposed to be extin
0125  aracter and from inheritance from a different  PARENT, will differ widely from the three genera des
0134  continued use or disuse, for we know not the  PARENT  forms; but many animals have structures which
0139  f the same genus have descended from a single  PARENT, if this view be correct, acclimatisation mus
0159  of the pigeon having inherited from a common  PARENT  the same constitution and tendency to variati
0159  rieties produced by cultivation from a common  PARENT: if this be not so, the case will then be one
0160  as all these marks are characteristic of the  PARENT  rock pigeon, I presume that no one will doubt
0161  on my theory, to have descended from a common  PARENT, it might be expected that they would occasio
0167  wise very differently constructed, the common  PARENT  of our domestic horse, whether or not it be d
0168  of the same genus branched off from a common  PARENT, are more variable than generic characters, o
0169  xtraordinarily developed organ has become the  PARENT  of many modified descendants, which on my vie
0169  ng nearly the same constitution from a common  PARENT  and exposed to similar influences will natura
0172  finally to exterminate, its own less improved  PARENT  or other less favoured forms with which it co
0172  cended from some other unknown form, both the  PARENT  and all the transitional varieties will gener
0173  e allied species have descended from a common  PARENT: and during the process of modification, each
0173  has supplanted and exterminated its original  PARENT  and all the transitional varieties between it
0178  veral representative species and their common  PARENT, must formerly have existed in each broken po
0179  as been so often remarked, to exterminate the  PARENT  forms and the intermediate links. Consequentl
0187  collateral descendants from the same original  PARENT  form, in order to see what gradations are pos
0214  iod exhibit traces of the instincts of either  PARENT: for example, Le Roy describes a dog, whose p
0236  their similarity by inheritance from a common  PARENT, and must therefore believe that they have be
0247  with the average number produced by both pure  PARENT  species in a state of nature. But a serious c
0248  y guarding them from a cross with either pure  PARENT, for six or seven, and in one case for ten ge
0249  experimentalists in great numbers: and as the  PARENT  species; or other allied hybrids, generally g
0250  are perfectly fertile, as fertile as the pure  PARENT  species, as Kolreuter and Gartner that so
0253  e often bred in this country with either pure  PARENT, and in one single instance they have bred in
0253  they are kept for profit, where neither pure  PARENT  species exists, they must certainly be highly
0255  produce, even with the pollen of either pure  PARENT, a single fertile seed: but in some of these
0255  be detected, by the pollen of one of the pure  PARENT  species causing the flower of the hybrid to w
0259  , though externally so like one of their pure  PARENT  species, are with rare exceptions extremely s
0259  nt of its external resemblance to either pure  PARENT. Considering the several rules now given, whi
0259  h they resemble in external appearance either  PARENT. And lastly, that the facility of making a fi
0266  nd exceptionally resemble closely either pure  PARENT. Nor do I pretend that the foregoing remarks
0273  ion of producing offspring identical with the  PARENT  form. Now hybrids in the first generation are
0273  more liable than hybrids to revert to either  PARENT  form; but this, if it be true, is certainly c
0274  ds and mongrels can be reduced to either pure  PARENT  form, by repeated crosses in successive gener
0274  crosses in successive generations with either  PARENT. These several remarks are apparently applica
0275  of cross bred animals closely resembling one  PARENT, the resemblances seem chiefly confined to ch
0275  reversions to the perfect character of either  PARENT  would be more likely to occur with mongrels,
0280  ally take the places of and exterminate their  PARENT  forms. But just in proportion as this process
0281  them, but between each and an unknown common  PARENT. The common parent will have had in its whole
0281  each and an unknown common parent. The common  PARENT  will have had in its whole organisation much
```

Page ***(Key Word)**************************************

Page ***(Key Word)***

```
0281 h cases, we should be unable to recognise the  PARENT form of any two or more species, even if we c
0281 n if we closely compared the structure of the  PARENT with that of its modified descendants, unless
0281 ween organism and organism, between child and  PARENT, will render this a very rare event; for in a
0281 l living species have been connected with the  PARENT species of each genus, by differences not gre
0282 he same species at the present day; and these  PARENT species, now generally extinct, have in their
0297 nts of structure. So that we might obtain the  PARENT species and its several modified descendants
0298 eties do not spread widely and supplant their  PARENT forms until they have been modified and perfe
0301 , they would slowly spread and supplant their  PARENT forms. When such varieties returned to their
0303 lication of species from some one or some few  PARENT forms; and in the succeeding formation such s
0315 uishable from our present fantail; but if the  PARENT rock pigeon were also destroyed, and in natur
0315 ture we have every reason to believe that the  PARENT form will generally be supplanted and extermi
0321 will generally cause the extermination of the  PARENT species; and if many new forms have been deve
0331 e several lines of descent diverging from the  PARENT form A, will form an order; for all will have
0335 ith the order of their disappearance; for the  PARENT rock pigeon now lives; and many varieties bet
0356 ach variety will have descended from a single  PARENT. But in the majority of cases, namely, with a
0356 due, at each stage, to descent from a single  PARENT. To illustrate what I mean: our English race
0373 , in huge boulders transported far from their  PARENT source. We do not know that the glacial epoch
0385 ch, on my theory, are descended from a common  PARENT and must have proceeded from a single source,
0389 f allied species have descended from a single  PARENT; and therefore have all proceeded from a comm
0405 ies of a genus having descended from a single  PARENT, though now distributed to the most remote po
0405 dely; for it is necessary that the unmodified  PARENT should range widely, undergoing modification
0405 ent, and must have branched off from a common  PARENT at a remote epoch; so that in such cases ther
0407 which on my theory must have spread from one  PARENT source; if we make the same allowances as bef
0412 ll have descended from one ancient but unseen  PARENT, and, consequently, have inherited something
0412 the right hand, which diverged from a common  PARENT at the fifth stage of descent. These five gen
0420 those which have been inherited from a common  PARENT, and, in so far, all true classification is g
0421 from A, be ranked in the same genus with the  PARENT A; or those from I, with the parent I. But th
0421 with the parent A; or those from I, with the  PARENT I. But the existing genus F14 may be supposed
0421 htly modified; and it will then rank with the  PARENT genus F; just as some few still living organi
0421 herited something in common from their common  PARENT, as will all the descendants from I; so will
0424 not solely because they closely resemble the  PARENT form, but because they are descended from it.
0426 habits, only by its inheritance from a common  PARENT. We may err in this respect in regard to sing
0429 n existing groups. A few old and intermediate  PARENT forms having occasionally transmitted to the
0431 racter of the species descended from a common  PARENT, together with their retention by inheritance
0431 group are connected together. For the common  PARENT of a whole family of species, now broken up b
0432 ween these eleven genera and there primordial  PARENT, and every intermediate link in each branch a
0433 ter in the many descendants from one dominant  PARENT species, explains that great and universal fe
0433 ies, however different they may be from their  PARENT; and I believe this element of descent is the
0433 ference between the descendants from a common  PARENT, expressed by the terms genera, families, ord
0443 en affected by the conditions to which either  PARENT, or their ancestors, have been exposed. Never
0443 ffspring from the reproductive element of one  PARENT. Or again, as when the horns of cross bred ca
0443 affected by the shape of the horns of either  PARENT. For the welfare of a very young animal, as l
0443 long as it is nourished and protected by its  PARENT, it must be quite unimportant whether most of
0444 atever age any variation first appears in the  PARENT, it tends to reappear at a corresponding age
0444 r at a corresponding age in the offspring and  PARENT. I am far from meaning that this is invariabl
0444 ed at an earlier age in the child than in the  PARENT. These two principles, if their truth be admi
0446 f birds, descended on my theory from some one  PARENT species, and of which the several new species
0447 bs, for instance, which served as legs in the  PARENT species, may become, by a long course of modi
0447 the embryos of the several descendants of the  PARENT species will still resemble each other closel
0448 o or embryo would closely resemble the mature  PARENT form. We have seen that this is the rule of d
0448 e in any degree different from those of their  PARENT, and consequently to be constructed in a slig
0450 picture, more or less obscured, of the common  PARENT form of each great class of animals. Rudiment
0464 uld not be able to recognise a species as the  PARENT of any one or more species if we were to exam
0464 the intermediate links between their past or  PARENT and present states; and these many links we c
0466 fails to reproduce offspring exactly like the  PARENT form. Variability is governed by many complex
0474 the same genus having descended from a common  PARENT, and having inherited much in common, we can
0487 n a not very remote period descended from one  PARENT, and have migrated from some one birthplace;
0214 olf, and this dog showed a trace of its wild  PARENTAGE only in one way, by not coming in a straigh
0421 more or less completely lost traces of their  PARENTAGE of those forms which are considered by natu
0456 l naturally follow on the view of the common  PARENTAGE of those forms which are considered by natu
0012 depart in some small degree from that of the  PARENTAL type. Any variation which is not inherited i
0009 producing offspring not perfectly like their  PARENTS or variable. Sterility has been said to be t
0010 rom each other, though both the young and the  PARENTS, as Muller has remarked, have apparently bee
0019 rms in some degree intermediate between their  PARENTS; and if we account for our several domestic
0026 stinct breed, and there is a tendency in both  PARENTS to revert to a character, which has been los
0029 r domestic races have descended from the same  PARENTS, may they not learn a lesson of caution, whe
0045 ntly to appear in the offspring from the same  PARENTS, or which may be presumed to have thus arise
0050 idence, showing that they descend from common  PARENTS, and consequently must be ranked as varietie
0054 l inherit those advantages that enabled their  PARENTS to become dominant over their compatriots. I
0111 hereas varieties, the supposed prototypes and  PARENTS of future well marked species, present sligh
0111 variety to differ in some character from its  PARENTS, and the offspring of this variety again to
0117 posed to the same conditions which made their  PARENTS variable, and the tendency to variability is
0118 ly to vary in nearly the same manner as their  PARENTS varied. Moreover, these two varieties, being
0122 ces of, and thus exterminated, not only their  PARENTS (A) and (I), but likewise some of the origin
0122 ecies which were most nearly related to their  PARENTS. Hence very few of the original species will
0124 diverging in character from the type of their  PARENTS, the new species (F14) will not be directly
0131 s of structure, as to make the child like its  PARENTS. But the much greater variability, as well a
0131 ature of the conditions of life, to which the  PARENTS and their more remote ancestors have been ex
0132 is system being functionally disturbed in the  PARENTS, I chiefly attribute the varying or plastic
0160 which has not been crossed, but in which both  PARENTS have lost some character which their progeni
0167 ds resembling in their stripes, not their own  PARENTS, but other species of the genus. To admit th
0167 fers, more or less, from the same part in the  PARENTS. But whenever we have the means of instituti
0170 slight difference in the offspring from their  PARENTS, and a cause for each must exist, it is the
0237 we have an insect differing greatly from its  PARENTS, yet absolutely sterile; so that it could ne
0239 y the long continued selection of the fertile  PARENTS which produced most neuters with the profita
0241 t natural selection, by acting on the fertile  PARENTS, could form a species which should regularly
```

Page ***(Key Word)***

```
0241  numbers through the natural selection of the  PARENTS  which generated them; until none with an int
0241  dely different from each other and from their  PARENTS,  has originated. We can see how useful their
0246  own or believed to have descended from common  PARENTS,  when intercrossed, and likewise the fertili
0253  e been raised at the same time from different  PARENTS,  so as to avoid the ill effects of close int
0253  . eyton, who raised two hybrics from the same  PARENTS  but from different hatches; and from these t
0259  , an intermediate character between their two  PARENTS,  always closely resemble one of them; and su
0259  ually intermediate in structure between their  PARENTS,  exceptional and abnormal individuals someti
0259  orn, which closely resemble one of their pure  PARENTS;  and these hybrids are almost always utterly
0260  the facility of the first union between their  PARENTS,  seems to be a strange arrangement. The fore
0264  born and living in a country where their two  PARENTS  can live, they are generally placed under su
0273  oes not seem to me at all surprising. For the  PARENTS  of mongrels are varieties, and mostly domest
0274  n mongrels and in hybrids to their respective  PARENTS,  more especially in hybrids produced from ne
0275  mals alone are born closely like one of their  PARENTS;  but it can be shown that this does sometime
0275  t the laws of resemblance of the child to its  PARENTS  are the same, whether the two parents differ
0275  to its parents are the same, whether the two  PARENTS  differ much or little from each other, namel
0327  having already had some advantage over their  PARENTS  or over other species; these again spreading
0334  e which lived at the fifth stage, and are the  PARENTS  of those which became still more modified at
0341  ne of the six older genera will have been the  PARENTS  of the six new genera; the other old species
0352  ust have proceeded from one spot, where their  PARENTS  were first produced: for, as explained in th
0352  been produced through natural selection from  PARENTS  specifically distinct. We are thus brought t
0407  herever located, have descended from the same  PARENTS,  are not insuperable. And we are led to this
0427  ve in all probability descended from the same  PARENTS.  We can understand, on these views, the very
0428  e groups to which they belong large and their  PARENTS  dominant, they are almost sure to spread wid
0432  ld be distinguished from their more immediate  PARENTS;  or those parents from their ancient and unk
0432  d from their more immediate parents; or those  PARENTS  from their ancient and unknown progenitor. Y
0440  ven more, from each other than do their adult  PARENTS.  In most cases, however, the larvae, though
0442  habits, or quite inactive, being fed by their  PARENTS  or placed in the midst of proper nutriment,
0444  rticular length, as long as it was fed by its  PARENTS.  Hence, I conclude, that it is quite possibl
0445  ders that they differed just as much as their  PARENTS,  and this, judging by the eye, seemed almost
0448  ny metamorphosis; or closely resembling their  PARENTS  from their earliest age, we can see that thi
0448  ng exactly the same habits of life with their  PARENTS;  for in this case, it would be indispensable
0448  a very early age in the same manner with its  PARENTS,  in accordance with their similar habits. So
0448  ifferent to any conceivable extent from their  PARENTS.  Such differences might, also, become correl
0449  oth descended from the same or nearly similar  PARENTS,  and are therefore in that degree closely re
0458  ch within its own class or group, from common  PARENTS,  and have all been modified in the course of
0461  higher group, must have descended from common  PARENTS;  and therefore, in however distant and isola
0475  eir degrees and kinds of resemblance to their  PARENTS,  in being absorbed into each other by success
0476  and the extinct being the offspring of common  PARENTS.  As the groups which have descended from an
0478  y of descent with modification, that the same  PARENTS  formerly inhabited both areas; and we almost
0118  tants of the same country; they will likewise  PARTAKE  of those more general advantages which made
0119  ies, belonging to a large genus, will tend to  PARTAKE  of the same advantages which made their pare
0322  places will commonly be allied, for they will  PARTAKE  of some inferiority in common. Thus, as it s
0264  er suitable conditions of life. But a hybrid  PARTAKES  of only half of the nature and constitution
0384  ranges, which from an early period must have  PARTED  river systems and completely prevented their
0370  tive position, though subjected to large, but  PARTIAL  oscillations of level, I am strongly incline
0288  period. The molluscan genus Chiton offers a  PARTIALLY  analogous case. With respect to the terrest
0309  wear and tear; and would have been at least  PARTIALLY  upheaved by the oscillations of level, whic
0399  rica, and other southern lands were long ago  PARTIALLY  stocked from a nearly intermediate though d
0414  by various degrees of modification, which is  PARTIALLY  revealed to us by our classifications. Let
0430  to all or nearly all Marsupials, from having  PARTIALLY  retained the character of their common prog
0449  c state of each species and group of species  PARTIALLY  shows us the structure of their less modifi
0191  arts, understand the strange fact that every  PARTICLE  of food and drink which we swallow has to pa
0361  ces and behind them, so perfectly that not a  PARTICLE  could be washed away in the longest transpor
0008  such as a little more or less water at some  PARTICULAR  period of growth, will determine whether or
0011  eased size from amount of food, colour from  PARTICULAR  kinds of food and from light, and perhaps t
0014  ason why a peculiarity should appear at any  PARTICULAR  age, yet that it does tend to appear in the
0085  occasional destruction of an animal of any  PARTICULAR  colour would produce little effect: we shou
0086  ons which under domestication appear at any  PARTICULAR  period of life, tend to reappear in the off
0092  , in relation to the size and habits of the  PARTICULAR  insects which visited them, so as to favour
0131  plainly our ignorance of the cause of each  PARTICULAR  variation. Some authors believe it to be as
0140  as much as, or more than, by adaptation to  PARTICULAR  climates. But whether or not the adaptation
0149  sation have been but little specialised for  PARTICULAR  functions; and as long as the same part has
0149  of almost any shape; whilst a tool for some  PARTICULAR  object had better be of some particular sha
0149  ome particular object had better be of some  PARTICULAR  shape. Natural selection, it should never b
0152  ch have been but little specialised for any  PARTICULAR  purpose, and perhaps in polymorphic groups,
0167  der nature and under domestication, in this  PARTICULAR  manner, so as often to become striped like
0168  not having been closely specialised to any  PARTICULAR  function, so that their modifications have
0176  one existing in large numbers; and in this  PARTICULAR  case the intermediate form would be eminent
0180  sufficient to lessen the difficulty in any  PARTICULAR  case like that of the bat. Look at the fami
0182  see any structure highly perfected for any  PARTICULAR  habit, as the wings of a bird for flight, w
0199  nces, chiefly through sexual selection of a  PARTICULAR  kind, but without here entering on copious
0212  ern and southern United States. Fear of any  PARTICULAR  enemy is certainly an instinctive quality,
0235  nowing that they swept their spheres at one  PARTICULAR  distance from each other, than they know wh
0265  no one can tell, till he tries, whether any  PARTICULAR  animal will breed under confinement or any
0267  seen that greater changes, or changes of a  PARTICULAR  nature, often render organic beings in some
0294  bouring sea, but are rare or absent in this  PARTICULAR  deposit. It is an excellent lesson to refle
0322  e cannot account for the extinction of this  PARTICULAR  species or group of species. On the Forms o
0336  is meant by high and low forms. But in one  PARTICULAR  sense the more recent forms must, on my the
0388  dener, thus takes her seeds from a bed of a  PARTICULAR  nature, and drops them in another equally w
0396  of kinds, the richness in endemic forms in  PARTICULAR  classes or sections of classes, the absence
0418  e in defining a group, or in allocating any  PARTICULAR  species. If they find a character nearly un
0444  k, whether or not it assumed a beak of this  PARTICULAR  length, as long as it was fed by its parent
0036  their domestic animals, yet any one animal  PARTICULARLY  useful to them, for any special purpose, w
0151  ite cirripedes; and I may here add, that I  PARTICULARLY  attended to Mr. Waterhouse's remark, whils
0151  vary in a remarkably small degree, I have  PARTICULARLY  attended to them, and the rule seems to me
0151  ot the great variability in plants made it  PARTICULARLY  difficult to compare their relative degree
```

Page **(Key Word)**

0165 the common mule from the ass and horse is PARTICULARLY apt to have bars on its legs. I once saw a
0257 er found that N. acuminata, which is not a PARTICULARLY distinct species, obstinately failed to fe
0190 upply, and being divided by highly vascular PARTITIONS. In these cases, one of the two organs migh
0007 y andrew Knight, that this variability may be PARTLY connected with excess of food. It seems prett
0038 wo sub breeds might be formed. This, perhaps, PARTLY explains what has been remarked by some autho
0039 t pigeon would become through long continued, PARTLY unconscious and partly methodical selection.
0039 hrough long continued, partly unconscious and PARTLY methodical selection. Perhaps the parent bird
0081 but in the case of an island, or of a country PARTLY surrounded by barriers, into which new and be
0154 lassing genera. I believe this explanation is PARTLY, yet only indirectly, true; I shall, however,
0163 curring in several species of the same genus, PARTLY under domestication and partly under nature.
0163 he same genus, partly under domestication and PARTLY under nature. It is a case apparently of reve
0203 er by a multitude of intermediate gradations, PARTLY because the process of natural selection will
0203 t any one time, only on a very few forms; and PARTLY because the very process of natural selection
0212 so it is with the nests of birds, which vary PARTLY in dependence on the situations chosen and on
0216 acquired and natural instincts have been lost PARTLY by habit, and by man selecting and accumulati
0274 the subject is here excessively complicated, PARTLY owing to the existence of secondary sexual ch
0349 e meridians of longitude. A third great fact, PARTLY included in the foregoing statements, is the
0363 tude, I suspected that these islands had been PARTLY stocked by ice borne seeds, during the glacia
0415 e outer world. Perhaps from this cause it has PARTLY arisen, that almost all naturalists lay the g
0089 pectators, at last choose the most attractive PARTNER. Those who have closely attended to birds in
0340 t from what it now is. North America formerly PARTOOK strongly of the present character of the sou
0185 ndrail nearly as terrestrial as the quail or PARTRIDGE. In such cases, and many others could be gi
0362 of dry argillaceous earth from one foot of a PARTRIDGE, and in this earth there was a pebble quite
0068 seems to be little doubt that the stock of PARTRIDGES, grouse, and hares on any large estate depe
0197 vanced as a beautiful adaptation for aiding PARTURITION, and no doubt they facilitate, or may be i
0197 wth, and has been taken advantage of in the PARTURITION of the higher animals. We are profoundly i
0437 elding of the separate pieces in the act of PARTURITION of mammals, will by no means explain the s
0183 fish. In our own country the larger titmouse (PARUS major) may be seen climbing branches, almost
0003 to me. I cannot, however, let this opportunity PASS without expressing my deep obligations to Dr.
0004 tion successive slight variations. I will then PASS on to the variability of species in a state of
0028 given for pigeons; nay, they are come to this PASS, that they can reckon up their pedigree and ra
0063 f other plants. In these several senses, which PASS into each other, I use for convenience sake th
0174 uniform condition than at present. But I will PASS over this way of escaping from the difficulty;
0191 icle of food and drink which we swallow has to PASS over the orifice of the trachea, with some ris
0236 eing effected by natural selection. But I must PASS over this preliminary difficulty. The great di
0356 d thus have allowed terrestrial productions to PASS from one to the other. No other geologist will
0361 stines of a bird; but hard seeds of fruit will PASS uninjured through even the digestive organs of
0362 positively asserted that all the grains do not PASS into the gizzard for 12 or even 18 hours. A bi
0366 new glacial period to come slowly on, and then PASS away, as formerly occurred. As the cold came o
0442 the midst of proper nutriment, yet nearly all PASS through a similar worm like stage of developme
0449 on each other in structure and habits, if they PASS through the same or similar embryonic stages,
0051 impresses the mind with the idea of an actual PASSAGE. Hence I look at individual differences, tho
0052 s leading to sub species, and to species. The PASSAGE from one stage of difference to another and
0052 much faith in this view; and I attribute the PASSAGE of a variety, from a state in which it diffe
0382 of vegetable life. Sir C. Lyell in a striking PASSAGE has speculated, in language almost identical
0416 for instance, whether or not there is an open PASSAGE from the nostrils to the mouth, the only cha
0031 lease. If I had space I could quote numerous PASSAGES to this effect from highly competent authori
0034 by some of the Roman classical writers. From PASSAGES in Genesis, it is clear that the colour of d
0034 and they formerly did so, as is attested by PASSAGES in Pliny. The savages in South Africa match
0034 ce animals were often imported, and laws were PASSED to prevent their exportation: the destruction
0036 the stages through which they have insensibly PASSED, and come to differ so greatly from the rock
0190 ansitional grades through which the organ has PASSED, we should have to look to very ancient ances
0211 oints are not indispensable, they may be here PASSED over. As some degree of variation in instinct
0293 tior, but, as they are rare, they may be here PASSED over. Although each formation has indisputabl
0304 may give another instance, which from having PASSED under my own eyes has much struck me. In a me
0342 culable number of generations which must have PASSED away even during a single formation; that, ow
0348 rrier of another kind, and as soon as this is PASSED we meet in the eastern islands of the Pacific
0352 a nature, that the space could not be easily PASSED over by migration, the fact is given as somet
0353 annot explain how the same species could have PASSED from one point to the other. But the geograph
0362 ours, either rejected the seeds in pellets or PASSED them in their excrement; and several of these
0461 in the course of successive generations have PASSED from some one part to the others. We are ofte
0251 the case of some other genera, as Lobelia, PASSIFLORA and Verbascum. Although the plants in these
0071 st remarkable, more than is generally seen in PASSING from one quite different soil to another: no
0292 imilar conclusion. One remark is here worth a PASSING notice. During periods of elevation the area
0343 condition, or may lie buried under the ocean. PASSING from these difficulties, all the other great
0361 ve never seen an instance of nutritious seeds PASSING through the intestines of a bird; but hard s
0001 the geological relations of the present to the PAST inhabitants of that continent. These facts see
0006 rable inhabitants of the world during the many PAST geological epochs in its history. Although muc
0084 s, and then so imperfect is our view into long PAST geological ages, that we only see that the for
0129 s; so with the species which lived during long PAST geological periods, very few now have living a
0173 and all the transitional varieties between its PAST and present states. Hence we ought not to expe
0206 rictly correct, but if we include all those of PAST times, it must by my theory be strictly true.
0206 of life; or by having adapted them during long PAST periods of time: the adaptations being aided i
0282 admit how imcomprehensibly vast have been the PAST periods of time; may at once close this volume
0300 uturity as the secondary formations lie in the PAST, only during periods of subsidence. These peri
0301 on my theory assuredly have connected all the PAST and present species of the same group into one
0340 there will be nothing immutable in the laws of PAST and present distribution. It may be asked in r
0349 are essentially American. We may look back to PAST ages, as shown in the last chapter, and we fin
0353 its powers of migration and subsistence under PAST and present conditions permitted, is the most
0372 nts of seas now disjoined, and likewise of the PAST and present inhabitants of the temperate lands
0409 the laws governing the succession of forms in PAST times being nearly the same with those governi
0464 d many of the intermediate links between their PAST or parent and present states; and these many l
0478 creation. The fact, as we have seen, that all the PAST and present organic beings constitute one gran
0488 tor, that the production and extinction of the PAST and present inhabitants of the world should ha
0489 eem to me to become ennobled. Judging from the PAST, we may safely infer that not one living speci
0030 fitted either for cultivated land or mountain PASTURE, with the wool of one breed good for one pur

Page **(Key Word)**

Page **(Key Word)**

```
0195  all enemies, would be able to range into new PASTURES and thus gain a great advantage. It is not t
0145  e central flower of the truss often loses the PATCHES of darker colour in the two upper petals; an
0485  f affinity, relationship, community of type, PATERNITY, morphology, adaptive characters, rudimenta
0469  cumulated, and inherited? Why, if man can by PATIENCE select variations most useful to himself, sh
0001  ight perhaps be made out on this question by PATIENTLY accumulating and reflecting on all sorts of
0434  he bat, should all be constructed on the same PATTERN, and should include the same bones, in the s
0435  than to attempt to explain this similarity of PATTERN in members of the same class, by utility or
0435  little or no tendency to modify the original PATTERN, or to transpose parts. The bones of a limb
0435  its limbs constructed on the existing general PATTERN, for whatever purpose they served, we can at
0436  ertheless, it is conceivable that the general PATTERN of an organ might become so much obscured by
0436  of certain suctorial crustaceans, the general PATTERN seems to have been thus to a certain extent
0437  rent purposes, be all constructed on the same PATTERN? On the theory of natural selection, we can
0457  ome intelligible, whether we look to the same PATTERN displayed in the homologous organs, to whate
0457  the homologous parts constructed on the same PATTERN in each individual animal and plant. On the
0479  t successive modifications. The similarity of PATTERN in the wing and leg of a bat, though used fo
0483  ses various structures are formed on the same PATTERN, and at an embryonic age the species closely
0195  an object as driving away flies; yet we should PAUSE before driving too positive even in this case,
0097  ely enclosed, as in the great papilionaceous or PEA family; but in several, perhaps in all, such f
0088  mongst birds, the contest is often of a more PEACEFUL character. All those who have attended to th
0262  i so do different varieties of the apricot and PEACH on certain varieties of the plum. As Gartner
0085  hereas another disease attacks yellow fleshed PEACHES far more than those with other coloured fles
0042  ce of distinct breeds of the cat, the donkey, PEACOCK, goose, etc., may be attributed in main part
0089  thus Sir R. Heron has described how one pied PEACOCK was eminently attractive to all his hen bird
0042  little attention paid to their breeding; in PEACOCKS, from not being very easily reared and a lar
0349  merican type of structure. We ascend the lofty PEAKS of the Cordillera and we find an alpine speci
0375  a list of the genera collected on the loftier PEAKS of Java raises a picture of a collection made
0037  one would expect to raise a first rate melting PEAR from the seed of the wild pear, though he migh
0037  st rate melting pear from the seed of the wild PEAR, though he might succeed from a poor seedling
0037  wild, if it had come from a garden stock. The PEAR, though cultivated in classical times, appears
0037  the classical period, who cultivated the best PEAR they could procure, never thought what splendi
0261  he same genus will not take or each other. The PEAR can be grafted far more readily on the quince,
0261  he same genus. Even different varieties of the PEAR take with different degrees of facility on the
0029  distinct species. Van Mons, in his treatise on PEARS and apples, shows how utterly he disbelieves
0075  lose varieties as the variously coloured sweet PEAS, they must be each year harvested separately,
0077  h of young plants produced from such seeds (as PEAS and beans), when sown in the midst of long gra
0361  of floating birds long retain their vitality; PEAS and vetches, for instance, are killed by even
0084  olour of heather, and the black grouse that of PEATY earth, we must believe that these tints are o
0362  of a partridge, and in this earth there was a PEBBLE quite as large as the seed of a vetch. Thus s
0283  them only when they are charged with sand or PEBBLES! for there is reason to believe that pure wa
0283  waves, and then are more quickly ground into PEBBLES, sand, or mud. But how often do we see along
0283  s, yet, from being formed of worn and rounded PEBBLES, each of which bears the stamp of time, are
0019  n great Britain eleven wild species of sheep PECULIAR to it! When we bear in mind that Britain has
0019  bear in mind that Britain has now hardly one PECULIAR mammal, and France but few distinct from tho
0019  hat each of these kingdoms possesses several PECULIAR breeds of cattle, sheep, etc., we must admit
0019  several countries do not possess a number of PECULIAR species as distinct parent stocks? So it is
0049  eater number rank it as an undoubted species PECULIAR to Great Britain. A wide distance between th
0073  sited by insects, and consequently, from its PECULIAR structure, never can set a seed. Many of our
0083  om exercises each selected character in some PECULIAR and fitting manner: he feeds a long and a sh
0083  long backed or long legged quadruped in any PECULIAR manner: he exposes sheep with long and short
0105  surrounded by barriers, or from having very PECULIAR physical conditions, the total number of the
0114  ive on it (supposing it not to be in any way PECULIAR in its nature), and may be said to be strivi
0141  flexibility of constitution, brought, under PECULIAR circumstances, into play. How much of the ac
0141  uch of the acclimatisation of species to any PECULIAR climate is due to mere habit, and how much t
0179  rigin and transitions of organic beings with PECULIAR habits and structure. It has been asked by t
0213  statue, and then slowly crawl forward with a PECULIAR gait; and another kind of wolf rushing round
0216  accumulating during successive generations, PECULIAR mental habits and actions, which at first ap
0243  america lines its nest with mud, in the same PECULIAR manner as does our British thrush: how it is
0260  ich are crossed. The differences being of so PECULIAR and limited a nature, that, in reciprocal cr
0289  reat piles of sediment, charged with new and PECULIAR forms of life, had elsewhere been accumulate
0290  he whole west coast, which is inhabited by a PECULIAR marine fauna, tertiary beds are so scantily
0290  ed, that no record of several successive and PECULIAR marine faunas will probably be preserved to
0303  of ages to adapt an organism to some new and PECULIAR line of life, for instance to fly through th
0307  the Silurian system, abounding with new and PECULIAR species. Traces of life have been detected i
0321  long be preserved, from being fitted to some PECULIAR line of life, or from inhabiting some distan
0346  onfined to any small spot, having conditions PECULIAR in only a slight degree; for instance, small
0347  new World, yet these are not inhabited by a PECULIAR fauna or flora. Notwithstanding this paralle
0349  ure, the inhabitants, though they may be all PECULIAR species, are essentially American. We may lo
0374  ty mountains of equatorial America a host of PECULIAR species belonging to European genera occur.
0375  an forms and some few representatives of the PECULIAR flora of the Cape of Good Hope occur. At the
0381  egree. The facts seem to me to indicate that PECULIAR and very distinct species have migrated in r
0381  ds, now covered with ice, supported a highly PECULIAR and isolated flora. I suspect that before th
0382  land have become slightly tinted by the same PECULIAR forms of vegetable life. Sir C. Lyell in a s
0390  land nearly all the species of one class are PECULIAR, those of another class, or of another secti
0390  or of another section of the same class, are PECULIAR; and this difference seems to depend on the
0390  only two out of the eleven marine birds, are PECULIAR; and it is obvious that marine birds could a
0391  do from South America, and which has a very PECULIAR soil, does not possess one endemic land bird
0391  ly this island. Madeira does not possess one PECULIAR bird, and many European and African birds ar
0391  again, is inhabited by a wonderful number of PECULIAR land shells, whereas not one species of sea
0394  hipelagoes, and Mauritius, all possess their PECULIAR bats. Why, it may be asked, has the supposed
0402  ed; and we may infer that the mocking thrush PECULIAR to Charles Island is at least as well fitted
0402  s well fitted for its home as is the species PECULIAR to Chatham Island. Sir C. Lyell and Mr. Woll
0409  of these a great number should be endemic or PECULIAR; and why, in relation to the means of migrat
0409  the most isolated islands possess their own PECULIAR species of aerial mammals or bats. We can se
0440  se that in the embryos of the vertebrata the PECULIAR loop like course of the arteries near the br
0467  lculation; by the effects of a succession of PECULIAR seasons; and by the results of naturalisatio
0477  W species; but of these, that many should be PECULIAR. We can clearly see why those animals which
```

Page **(Key Word)**

```
0477  islands; and why, on the other hand, new and PECULIAR species of bats, which can traverse the ocea
0478  any continent. Such facts as the presence of PECULIAR species of bats, and the absence of all othe
0012  invariably deaf; colour and constitutional PECULIARITIES go together, of which many remarkable cas
0013  fact of some little importance to us, that PECULIARITIES appearing in the males of our domestic br
0014  could not be otherwise: thus the inherited PECULIARITIES in the horns of cattle could appear only
0014  only in the offspring when nearly mature; PECULIARITIES in the silkworm are known to appear at th
0080  mind in what an endless number of strange PECULIARITIES our domestic productions, and, in a lesse
0087  r structure. Sexual Selection. Inasmuch as PECULIARITIES often appear under domestication in one s
0090  ual differences to this agency: for we see PECULIARITIES arising and becoming attached to the male
0444  ppear at corresponding ages; for instance, PECULIARITIES in the caterpillar, coccon, or imago stat
0012  goes on selecting, and thus augmenting, any PECULIARITY, he will almost certainly unconsciously mo
0013  quite unknown; no one can say why the same PECULIARITY in different individuals of the same speci
0013  r or other much more remote ancestor; why a PECULIARITY is often transmitted from one sex to both
0013  sted, is that, at whatever period of life a PECULIARITY first appears, it tends to appear in the o
0014  that when there is no apparent reason why a PECULIARITY should appear at any particular age, yet t
0014  rse confined to the first appearance of the PECULIARITY, and not to its primary cause, which may h
0393  onditions; indeed it seems that islands are PECULIARLY well fitted for these animals; for frogs ha
0028  to this pass, that they can reckon up their PEDIGREE and race. Pigeons were much valued by Akber
0031  normous prices given for animals with a good PEDIGREE; and these have now been exported to almost
0422  m is genealogical in its arrangement, like a PEDIGREE; but the degrees of modification which the d
0422  case of languages. If we possessed a perfect PEDIGREE of mankind, a genealogical arrangement of th
0423  sing varieties, I apprehend if we had a real PEDIGREE, a genealogical classification would be univ
0425  by our best systematists. We have no written PEDIGREES; we have to make out community of descent b
0486  ave a definite object in view. We possess no PEDIGREES or armorial bearings; and we have to discov
0191  other, that I will give one more instance. PEDUNCULATED cirripedes have two minute folds of skin,
0192  iteration of their adhesive glands. If all PEDUNCULATED cirripedes had become extinct, and they ha
0440  the two main divisions of cirripedes, the PEDUNCULATED and sessile, which differ widely in extern
0037  e in the varieties of the heartsease, rose, PELARGONIUM, dahlia, and other plants, when compared w
0251  in how complicated a manner the species of PELARGONIUM, Fuchsia, Calceolaria, Petunia, Rhododendr
0145  at i have recently observed in some garden PELARGONIUMS, that the central flower of the truss ofte
0362  their bodies to fishing eagles, storks, and PELICANS; these birds after an interval of many hours
0387  ish, would probably reject from its stomach a PELLET containing the seeds of the Nelumbium undiges
0362  rval of from twelve to twenty hours, disgorge PELLETS, which, as I know from experiments made in t
0362  l of many hours, either rejected the seeds in PELLETS or passed them in their excrement; and sever
0387  their power of germination, when rejected in PELLETS or in excrement, many hours afterwards. When
0145  e nearest to the axis are oftenest subject to PELORIA, and become regular. I may add, as an instan
0144  uthors that the diversity in the shape of the PELVIS in birds causes the remarkable diversity in t
0144  kidneys. Others believe that the shape of the PELVIS in the human mother influences by pressure th
0198  them more, and possibly even the form of the PELVIS; and then by the law of homologous variation,
0198  probably be affected. The shape, also, of the PELVIS might affect by pressure the shape of the hea
0450  yi in other snakes there are rudiments of the PELVIS and hind limbs. Some of the cases of rudiment
0097  of the plant. Bees will act like a camel hair PENCIL, and it is quite sufficient just to touch the
0187  which properly act only by excluding lateral PENCILS of light, are convex at their upper ends and
0263  hough the pollen tubes protrude, they do not PENETRATE the stigmatic surface. Again, the male elem
0377  ous and dominant temperate forms might have PENETRATED the native ranks and have reached or even c
0323  fuego, at the Cape of Good Hope, and in the PENINSULA of India. For at these distant points, the
0375  , and on the isolated mountain ranges of the PENINSULA of India, on the heights of Ceylon, and on
0375  dr. Hooker, extend along the heights of the PENINSULA of Malacca, and are thinly scattered, on th
0182  he water and front legs on the land, like the PENGUIN; as sails, like the ostrich; and functionall
0027  ended with the utmost care, and loved by many PEOPLE. They have been domesticated for thousands of
0041  kshire, that as they generally belong to poor PEOPLE, and are mostly in small lots, they never can
0042  n donkeys, from only a few being kept by poor PEOPLE, and little attention paid to their breeding:
0065  ow breeders would require a few more years to PEOPLE, under favourable conditions, a whole distric
0344  t oftenest vary, will in the long run tend to PEOPLE the world with allied, but modified, descenda
0126  gical period, the earth may have been as well PEOPLED with many species of many genera, families,
0287  whole world, the land and the water has been PEOPLED by hosts of living forms. What an infinite n
0340  for we know that Europe in ancient times was PEOPLED by numerous marsupials; and I have shown in
0384  t geological period, and when the surface was PEOPLED by existing land and fresh water shells. The
0458  s of organic beings, with which this world is PEOPLED, have all descended, each within its own cla
0068  emendous destruction, when we remember that ten PER cent. is an extraordinarily severe mortality f
0287  ff 500 feet in height, a denudation of one inch PER century for the whole length would be an ample
0359  te of the several Atlantic currents is 33 miles PER diem (some currents running at the rate of 60
0360  (some currents running at the rate of 60 miles PER diem); on this average, the seeds of 14/100 pl
0015  pecies closely allied together, we generally PERCEIVE in each domestic race, as already remarked,
0224  all the necessary angles and planes, or even PERCEIVE when they are correctly made. But the diffic
0298  by palaeontologists. We shall, perhaps, best PERCEIVE the improbability of our being enabled to co
0431  a diagram, the various affinities which they PERCEIVE between the many living and extinct members
0435  whatever purpose they served, we can at once PERCEIVE the plain signification of the homologous co
0440  of this: even the illustrious Cuvier did not PERCEIVE that a barnacle was, as it certainly is, a c
0481  hundred million years; it cannot add up and PERCEIVE the full effects of many slight variations,
0319  ess was my astonishment! Professor Owen soon PERCEIVED that the tooth, though so like that of the
0319  ess favourable, we assuredly should not have PERCEIVED the fact, yet the fossil horse would certai
0322  t some check is always in action, yet seldom PERCEIVED by us, the whole economy of nature will be
0039  be necessary to catch the fancier's eye: he PERCEIVES extremely small differences, and it is in h
0229  at work on the two sides of a strip of wax, PERCEIVING when they have gnawed the wax away to the p
0312  up the blanks between them, and to make the PERCENTAGE system of lost and new forms more gradual.
0223  about in the greatest agitation, and one was PERCHED motionless with its own pupa in its mouth on
0366  h their scored flanks, polished surfaces, and PERCHED boulders, of the icy streams with which thei
0023  pigeons, that is, not breeding or willingly PERCHING on trees. But besides C. livia, with its geo
0288  are upraised generally be dissolved by the PERCOLATION of rain water. I suspect that but few of t
0257  n the cotyledons, can be crossed. Annual and PERENNIAL plants, deciduous and evergreen trees, plan
0378  at of the tropics which is so destructive to PERENNIAL plants from a temperate climate. On the oth
0004  e misseltoe, and that these had been produced PERFECT as we now see them; but this assumption seem
0014  , and so could not tell whether or not nearly PERFECT reversion had ensued. It would be quite nece
0016  . with these exceptions (and with that of the PERFECT fertility of varieties when crossed, a subje
```

0022 e which are variable. The period at which the PERFECT plumage is acquired varies, as does the stat
0027 from distant countries, we can make an almost PERFECT series between the extremes of structure. Th
0030 that all the breeds were suddenly produced as PERFECT and as useful as we now see them: indeed, in
0031 as if they had chalked out upon a wall a form PERFECT in itself, and then had given it existence.
0095 r the other, modified and adapted in the most PERFECT manner to each other, by the continued prese
0171 mportance. Organs not in all cases absolutely PERFECT. The Law of Unity of Type and of the Conditi
0172 ly lies in the record being incomparably less PERFECT than is generally supposed: the imperfection
0181 fects of this process of natural selection, a PERFECT so called flying squirrel was produced. Now
0182 ural steps by which birds have acquired their PERFECT power of flight: but they serve, at least, t
0186 tells me, that if numerous gradations from a PERFECT and complex eye to one very imperfect and si
0186 life, then the difficulty of believing that a PERFECT and complex eye could be formed by natural s
0188 arent membrane, into an optical instrument as PERFECT as is possessed by any member of the great A
0188 urther, and to admit that a structure even as PERFECT as the eye of an eagle might be formed by na
0195 , on this head, as in the case of an organ as PERFECT and complex as the eye. In the first place,
0201 tior tends only to make each organic being as PERFECT as, or slightly more perfect than, the other
0201 organic being as perfect as, or slightly more PERFECT than, the other inhabitants of the same coun
0201 productions of New Zealand, for instance, are PERFECT one compared with another: but they are now
0202 f light is said, on high authority, not to be PERFECT even in that most perfect organ, the eye. If
0202 uthority, not to be perfect even in that most PERFECT organ, the eye. If our reason leads us to ac
0202 sices, that some other contrivances are less PERFECT. Can we consider the sting of the wasp or of
0202 nsider the sting of the wasp or of the bee as PERFECT, which, when used against many attacking ani
0203 uch insect agency, can we consider as equally PERFECT the elaboration by our fir trees of dense cl
0204 of auks. Although the belief that an organ so PERFECT as the eye could have been formed by natural
0211 instincts cannot be considered as absolutely PERFECT; but as details on this and other such point
0226 sulting structure would probably have been as PERFECT as the comb of the hive bee. Accordingly I w
0227 his bee would make a structure as wonderfully PERFECT as that of the hive bee. We must suppose the
0233 al instincts, all tending towards the present PERFECT plan of construction, could have profited th
0235 rfaces; and the Melipona would make a comb as PERFECT as that of the hive bee. Beyond this stage o
0235 hive bee, as far as we can see, is absolutely PERFECT in economising wax. Thus, as I believe, the
0236 e in this respect between the workers and the PERFECT females, would have been far better exemplif
0239 out of the same nest: I have myself compared PERFECT gradations of this kind. It often happens th
0242 red knowledge and manufactured instruments, a PERFECT division of labour could be effected with th
0243 fact that instincts are not always absolutely PERFECT and are liable to mistakes: that no instinct
0246 e of course their organs of reproduction in a PERFECT condition, yet when intercrossed they produc
0246 and animals; though the organs themselves are PERFECT in structure, as far as the microscope revea
0246 xual elements which go to form the embryo are PERFECT: in the second case they are either not at a
0248 al purposes it is most difficult to say where PERFECT fertility ends and sterility begins. I think
0250 its natural fecundation. So that we here have PERFECT, or even more than commonly perfect, fertili
0250 here have perfect, or even more than commonly PERFECT, fertility in a first cross between two dist
0255 rosses and of hybrids, graduates from zero to PERFECT fertility. It is surprising in how many curi
0255 d to the stigma of some one species, yields a PERFECT gradation in the number of seeds produced, u
0256 g a greater and greater number of seeds up to PERFECT fertility. Hybrids from two species which ar
0259 nited, their fertility graduates from zero to PERFECT fertility, or even to fertility under certai
0262 species, which have their reproductive organs PERFECT; yet these two distinct cases run to a certa
0263 ecies the male and female sexual elements are PERFECT, whereas in hybrids they are imperfect. Even
0268 each other in external appearance, cross with PERFECT facility, and yield perfectly fertile offspr
0268 originally distinct species. Nevertheless the PERFECT fertility of so many domestic varieties, dif
0275 ction. Consequently, sudden reversions to the PERFECT character of either parent would be more lik
0276 pecies, which have their reproductive systems PERFECT, seems to depend on several circumstances; i
0277 ally, fertile. Nor is this nearly general and PERFECT fertility surprising, when we remember how l
0281 ants, unless at the same time we had a nearly PERFECT chain of the intermediate links. It is just
0293 mation throughout Europe been correlated with PERFECT accuracy. With marine animals of all kinds,
0295 l epoch to the present day. In order to get a PERFECT gradation between two forms in the upper and
0298 t be forgotten, that at the present day, with PERFECT specimens for examination, two forms can sel
0300 but the geological record would then be least PERFECT. It may be doubted whether the duration of a
0304 he upper tidal limits to 50 fathoms; from the PERFECT manner in which specimens are preserved in t
0304 ntologist, M. Bosquet, sent me a drawing of a PERFECT specimen of an unmistakeable sessile cirripe
0310 k the natural geological record in any degree PERFECT, and who do not attach much weight to the fa
0329 to the extinct alone, the series is far less PERFECT than if we combine both into one general sys
0335 pearance and disappearance of the species was PERFECT, we have no reason to believe that forms suc
0345 t the record cannot be proved to be much more PERFECT, the main objections to the theory of natura
0361 he excrement of small birds, and these seemed PERFECT, and some of them, which I tried, germinated
0417 plants, belonging to the Malpighiaceae, bear PERFECT and degraded flowers: in the latter, as A. d
0422 king the case of languages. If we possessed a PERFECT pedigree of mankind, a genealogical arrangem
0425 ad hominem, and ask what should be done if a PERFECT kangaroo were seen to come out of the womb o
0433 we shall certainly never succeed in making so PERFECT a collection: nevertheless, in certain class
0440 he larva to its conditions of life is just as PERFECT and as beautiful as in the adult animal. Fro
0449 st, or indeed, if our collections were nearly PERFECT, the only possible arrangement, would be gen
0451 ize; and this shows that the rudiment and the PERFECT pistil are essentially alike in nature. An o
0461 rst crosses the organs on both sides are in a PERFECT condition. As we continually see that organi
0466 gradations between the simplest and the most PERFECT organs: it cannot be pretended that we know
0472 re be not, as far as we can judge, absolutely PERFECT; and if some of them be abhorrent to our ide
0475 marvel at some instincts being apparently not PERFECT and liable to mistakes; and at many instinct
0481 thout proof, that the geological record is so PERFECT that it would have afforded us plain evidenc
0005 e being or a simple organ can be changed and PERFECTED into a highly developed being or elaboratel
0060 istinct organic being to another being, been PERFECTED? We see these beautiful co adaptations most
0108 arieties in each to become well modified and PERFECTED. When, by renewed elevation, the islands sh
0126 same large group, the later and more highly PERFECTED sub groups, from branching out and seizing
0176 re supposed on my theory to be converted and PERFECTED into two distinct species, the two which ex
0182 other fish? When we see any structure highly PERFECTED for any particular habit, as the wings of a
0187 ns by which an organ in any species has been PERFECTED, we ought to look exclusively to its lineal
0187 he earlier stages, by which the eye has been PERFECTED. In the Articulata we can commence a series
0188 scope. We know that this instrument has been PERFECTED by the long continued efforts of the highes
0190 e two organs might with ease be modified and PERFECTED so as to perform all the work by itself, to
0195 ly progenitor; and, after having been slowly PERFECTED at a former period, have been transmitted i
0202 t order, and which has been modified but not PERFECTED for its present purpose, with the poison or

Page ***(Key Word)***

```
205   time the same function, the one having been  PERFECTED whilst aided by the other, must often have
218   of the American ostrich has not as yet been  PERFECTED; for a surprising number of eggs lie strewe
219   how so wonderful an instinct could have been  PERFECTED. Formica sanguinea was likewise first disco
298   rent forms until they have been modified and  PERFECTED in some considerable degree. According to t
433   tural system being, in so far as it has been  PERFECTED, genealogical in its arrangement, with the
459   omplex organs and instincts should have been  PERFECTED, not by means superior to, though analogous
460   by what gradations many structures have been  PERFECTED, more especially amongst broken and failing
003   d have been modified, so as to acquire that  PERFECTION of structure and coadaptation which most ju
038   by continued selection up to a standard of  PERFECTION comparable with that given to the plants in
039   s faults or deviations from the standard of  PERFECTION of each breed. The common goose has not giv
102   the breed, have a nearly common standard of  PERFECTION, and all try to get and breed from the best
152   toriously difficult to breed them nearly to  PERFECTION, and frequently individuals are born which
171   om those of their allies. Organs of extreme  PERFECTION. Means of transition. Cases of difficulty.
172   rdly as yet fully understand the inimitable  PERFECTION? Thirdly, can instincts be acquired and mod
172   inated by the very process of formation and  PERFECTION of the new form. But, as by this theory inn
182   have been supplanted by the very process of  PERFECTION through natural selection. Furthermore, we
183   rgans of flight had come to a high stage of  PERFECTION, so as to have given them a decided advanta
186   with the habits of auks. Organs of extreme  PERFECTION and complication. To suppose that the eye,
187   , until we reach a moderately high stage of  PERFECTION. In certain crustaceans, for instance, ther
201   ence. And we see that this is the degree of  PERFECTION attained under nature. The endemic producti
202   natural selection will not produce absolute  PERFECTION, nor do we always meet, as far as we can ju
204   he acquirement of any conceivable degree of  PERFECTION through natural selection. In the cases in
205   with another, and consequently will produce  PERFECTION, or strength in the battle for life, only a
206   have been severer, and thus the standard of  PERFECTION will have been rendered higher. Natural sel
206   ction will not necessarily produce absolute  PERFECTION; nor, as far as we can judge by our limited
206   udge by our limited faculties, can absolute  PERFECTION be everywhere found. On the theory of natur
235   ite side. In the series between the extreme  PERFECTION of the cells of the hive bee and the simpli
235   that of the hive bee. Beyond this stage of  PERFECTION in architecture, natural selection could no
302   scendants. But we continually over rate the  PERFECTION of the geological record, and falsely infer
379   ection and competition to a higher stage of  PERFECTION or dominating power, than the southern form
459   ropositions, namely, that gradations in the  PERFECTION of any organ or instinct, which we may cons
472   h country only in relation to the degree of  PERFECTION of their associates; so that we need feel n
472   on, that more cases of the want of absolute  PERFECTION have not been observed. The complex and lit
489   al endowments will tend to progress towards  PERFECTION. It is interesting to contemplate an entang
009   , though taken young from a state of nature,  PERFECTLY tamed, long lived, and healthy (of which I
009   quite regularly, and producing offspring not  PERFECTLY like their parents or variable. Sterility h
025   of the outer tail feathers, sometimes concur  PERFECTLY developed. Moreover, when two birds belongi
026   tween all the domestic breeds of pigeons are  PERFECTLY fertile. I can state this from my own obser
026   wo animals clearly distinct being themselves  PERFECTLY fertile. Some authors believe that long con
026   and fantails now are, should yield offspring  PERFECTLY fertile, inter se, seems to me rash in the
027   nd when crossed; the mongrel offspring being  PERFECTLY fertile: from these several reasons, taken
049   ighly competent judges as varieties, have so  PERFECTLY the character of species that they are rank
069   us number of plants in our gardens which can  PERFECTLY well endure our climate; but which never be
077   r quadruple its numbers? We know that it can  PERFECTLY well withstand a little more. heat or cold,
082   which all the native inhabitants are now so  PERFECTLY adapted to each other and to the physical c
099   their kind, and some even of these were not  PERFECTLY true. Yet the pistil of each cabbage flower
101   te animal with the organs of reproduction so  PERFECTLY enclosed within the body, that access from
102   d area, with some place in its polity not so  PERFECTLY occupied as might be, natural selection wil
109   rmittent action of natural selection accords  PERFECTLY well with what geology tells us of the rate
116   ral economy of any land, the more widely and  PERFECTLY the animals and plants are diversified for
116   sified, could hardly compete with a set more  PERFECTLY diversified in structure. It may be doubted
119   he places which are either unoccupied or not  PERFECTLY occupied by other beings: and this will dep
136   om its wings having been ever so little less  PERFECTLY developed or from indolent habit, will have
138   , disuse will on this view have more or less  PERFECTLY obliterated its eyes, and natural selection
140   ing the most different climates but of being  PERFECTLY fertile (a far severer test) under them, ma
148   and, conversely, that natural selection may  PERFECTLY well succeed in largely developing any orga
174   from the difficulty: for I believe that many  PERFECTLY defined species have been formed on strictl
181   by the same steps as in the case of the less  PERFECTLY gliding squirrels: and that each grade of s
182   uttering fins, might have been modified into  PERFECTLY winged animals. If this had been effected,
224   ells of wax of the true form, though this is  PERFECTLY effected by a crowd of bees working in a da
226   t this is never permitted, the bees building  PERFECTLY flat walls of wax between the spheres which
226   spherical portion and of two, three, or more  PERFECTLY flat surfaces, according as the cell adjoin
227   does so to a certain extent, and seeing what  PERFECTLY cylindrical burrows in wood many insects ca
228   hen she unites the points of intersection by  PERFECTLY flat surfaces. We have further to suppose,
228   ed into shallow basins, appearing to the eye  PERFECTLY true or parts of a sphere, and of about the
229   bic plate had been completed, and had become  PERFECTLY flat: it was absolutely impossible, from th
235   ection having by slow degrees, more and more  PERFECTLY, led the bees to sweep equal spheres at a g
238   generally graduate into each other, but are  PERFECTLY well defined; being as distinct from each o
239   and that the extreme forms can sometimes be  PERFECTLY linked together by individuals taken out of
250   atic in his conclusion that some hybrids are  PERFECTLY fertile, as fertile as the pure parent spec
250   anding that their own pollen was found to be  PERFECTLY good, for it fertilised distinct species. S
251   uch the plants in these experiments appeared  PERFECTLY healthy, and although both the ovules and p
251   he ovules and pollen of the same flower were  PERFECTLY good with respect to other species, yet as
251   milar in general habit, reproduced itself as  PERFECTLY as if it had been a natural species from th
251   rons; and I am assured that many of them are  PERFECTLY fertile. Mr. C. Noble, for instance, inform
252   more sterile. I doubt whether any case of a  PERFECTLY fertile hybrid animal can be considered as
252   the canary, or that their hybrids, should be  PERFECTLY fertile. Again, with respect to the fertili
253   f any thoroughly well authenticated cases of  PERFECTLY fertile hybrid animals, I have some reason
253   with P. torquatus and with P. versicolor are  PERFECTLY fertile. The hybrids from the common and Ch
268   ance, cross with perfect facility, and yield  PERFECTLY fertile offspring. I fully admit that this
270   he hybrid plants thus raised were themselves  PERFECTLY fertile; so that even Gartner did not ventu
271   rosses; and he found their mongrel offspring  PERFECTLY fertile. But one of these five varieties, w
276   be an astonishing fact. But it harmonises  PERFECTLY with the view that there is no essential di
305   s of the Indian Ocean would form a large and  PERFECTLY enclosed basin, in which any great group of
330   is probably valid. But I apprehend that in a  PERFECTLY natural classification many fossil species
342   g the latter the record will have been least  PERFECTLY kept; that each single formation has not be
```

Page ***(Key Word)***

Page ***(Key Word)***

```
0361 sed in their interstices and behind them, so PERFECTLY that not a particle could be washed away +
0368 ry. These views, grounded as they are on the PERFECTLY well ascertained occurrence of a former GL
0390 them from various sources far more fully and PERFECTLY than has nature. Although in oceanic islan
0397 put it in sea water for twenty days, and it PERFECTLY recovered. As this species has a thick cal
0401 ance, would find the best fitted ground more PERFECTLY occupied by distinct plants in one island
0451 even the more important purpose; and remain PERFECTLY efficient for the other. Thus in plants, t
0089 sive males display their gorgeous plumage and PERFORM strange antics before the females, which st
0149 unctions; and as long as the same part has to PERFORM diversified work, we can perhaps see why it
0190 sensible steps. Two distinct organs sometimes PERFORM simultaneously the same function in the sam
0190 with ease be modified and perfected so as to PERFORM all the work by itself, being aided during
0207 ves should require experience to enable us to PERFORM, when performed by an animal, more especial
0190 were thus gained, a part or organ, which had PERFORMED two functions, for one function alone, and
0204 an air breathing lung. The same organ having PERFORMED simultaneously very different functions, a
0205 unction; and two very distinct organs having PERFORMED at the same time the same function, the on
0207 ire experience to enable us to perform, when PERFORMED by an animal, more especially by a very yo
0207 young one, without any experience, and when PERFORMED by many individuals in the same way, witho
0207 without their knowing for what purpose it is PERFORMED, is usually said to be instinctive. But I
0208 of mind under which an instinctive action is PERFORMED, but not of its origin. How unconsciously
0208 how unconsciously many habitual actions are PERFORMED, indeed not rarely in direct opposition to
0208 o the third stage, the caterpillar simply re PERFORMED the fourth, fifth, and sixth stages of con
0213 pherd dogs. I cannot see that these actions, PERFORMED without experience by the young, and in ne
0213 n nearly the same manner by each individual, PERFORMED with eager delight by each breed, and with
0214 le, an action which, as I have witnessed, is PERFORMED by young birds, that have never seen a pig
0229 natural state of things, had not been neatly PERFORMED. The bees must have worked at very nearly
0190 amongst the lower animals of the same organ PERFORMING at the same time wholly distinct functions
0210 strongest instances of an animal apparently PERFORMING an action for the sole good of another, wh
0211 do not believe that any animal in the world PERFORMS an action for the exclusive good of another
0001 occurred to me, in 1837, that something might PERHAPS be made out on this question by patiently a
0009 ation or cultivation, and vary very slightly, PERHAPS hardly more than in a state of nature. A lo
0011 particular kinds of food and from light, and PERHAPS the thickness of fur from climate. Habit al
0013 ons may be freely admitted to be inheritable. PERHAPS the correct way of viewing the whole subjec
0018 and that some of the breeds closely resemble, PERHAPS are identical with, those still existing. F
0026 e most distinct breeds. Now, it is difficult, PERHAPS impossible, to bring forward one case of th
0029 veral allied species. Some little effect may, PERHAPS, be attributed to the direct action of the
0033 at all in other points: this is hardly ever, PERHAPS never, the case. The laws of correlation of
0038 lained, two sub breeds might be formed. This, PERHAPS, partly explains what has been remarked by
0039 unconscious and partly methodical selection. PERHAPS the parent bird of all fantails had only fo
0039 as seventeen tail feathers have been counted. PERHAPS the first pouter pigeon did not inflate its
0040 t, of unconscious selection will always tend, PERHAPS more at one period than at another, as the
0040 ther, as the breed rises or falls in fashion, PERHAPS more in one district than in another, accor
0054 i consider them, incipient species. And this, PERHAPS, might have been anticipated; for, as varie
0072 ld insects determine the existence of cattle. PERHAPS Paraguay offers the most curious instance o
0081 , and would be left a fluctuating element, as PERHAPS we see in the species called polymorphic.' W
0088 he huge mandibles of other males. The war is, PERHAPS, severest between males of polygamous anima
0097 ross with another individual is occasionally, PERHAPS at very long intervals, indispensable. On t
0097 papilionaceous or pea family; but in several, PERHAPS in all, such flowers, there is a very curio
0101 sity for each birth; in many others it occurs PERHAPS only at long intervals: but in none, as I s
0106 hanging history of the organic world. We can, PERHAPS, on these views, understand some facts whic
0107 s modification and less extermination. Hence, 'PERHAPS, it comes that the flora of Madeira, accord
0113 , climbing trees, frequenting water, and some PERHAPS becoming less carnivorous. The more diversi
0115 dapted for their own country. It might, also, PERHAPS have been expected that naturalised plants
0120 ations, will have come to differ largely, but PERHAPS unequally, from each other and from their c
0132 s extremely small in the case of animals, but PERHAPS rather more in that of plants. We may, at l
0137 e to gradual reduction from disuse, but aided PERHAPS by natural selection. In South America, a b
0141 of a tropical and arctic wolf or wild dog may PERHAPS be mingled in our domestic breeds. The rat
0149 part has to perform diversified work, we can PERHAPS see why it should remain variable, that is,
0152 e specialised for any particular purpose, and PERHAPS in polymorphic groups, we see a nearly para
0160 uld reappear after having been lost for many, PERHAPS for hundreds of generations. But when a bree
0167 and I see an animal striped like a zebra, but PERHAPS otherwise very differently constructed, the
0168 al parts. When one part is largely developed, PERHAPS it tends to draw nourishment from the adjoi
0168 arts are variable in number and in structure, PERHAPS arising from such parts not having been clos
0181 lder to the tail, including the hind legs, we PERHAPS see traces of an apparatus originally const
0182 ades of wing structure here alluded to, which PERHAPS may all have resulted from disuse, indicate
0196 swimbladders betray their aquatic origin, may PERHAPS be thus accounted for. A well developed tai
0202 cause galls subsequently intensified, we can PERHAPS understand how it is that the use of the st
0216 and selection have acted together. We shall, PERHAPS, best understand how instincts in a state o
0218 very uncommon with the Gallinaceae: and this PERHAPS explains the origin of a singular instinct
0221 y seized, and carried off by the tyrants, who PERHAPS fancied that, after all, they had been victe
0225 adjoining cells; and the following view may, PERHAPS, be considered only as a modification of his
0240 ant (Anomma) of West Africa. The reader will PERHAPS best appreciate the amount of difference in
0254 descended from several wild stocks; yet, with PERHAPS the exception of certain indigenous domestic
0273 in the successive generations of mongrels is, PERHAPS, greater than in hybrids. This greater varia
0280 uch finely graduated organic chains; and this, PERHAPS, is the most obvious and gravest objection w
0281 ay have differed considerably from both, even PERHAPS more than they differ from each other. Hence
0287 action of the sea: when deeply submerged for PERHAPS equally long periods, it would, likewise, ha
0293 ion may mark a very long lapse of years, each PERHAPS is short compared with the period requisite
0296 osition, but have disappeared and reappeared, PERHAPS many times, during the same geological perio
0296 have existed between them, but abrupt, though PERHAPS very slight, changes of form. It is all impo
0298 ly be effected by palaeontologists. We shall, PERHAPS, best perceive the improbability of our bein
0301 eir former state, in a nearly uniform, though PERHAPS extremely slight degree, they would, accordi
0303 ed between our consecutive formations, longer PERHAPS in some cases than the time required for the
0308 laeozoic or secondary formation. Hence we may PERHAPS infer, that during the palaeozoic and second
0310 require some special explanation; and we may PERHAPS believe that we see in these large areas, th
0314 lls and birds have remained unaltered. We can PERHAPS understand the apparently quicker rate of ch
0315 , during long and equal periods of time, may, PERHAPS, be nearly the same; but as the accumulation
0342 ited; that the duration of each formation is, PERHAPS, short compared with the average duration of
```

Page ***(Key Word)***

age **(Key Word)**

552 stinctly limited in terrestrial mammals, than PERHAPS in any other organic beings; and, accordingl
560 ent as in our experiments. Therefore it would PERHAPS be safer to assume that the seeds of about 1
577 red much extinction; how much no one can say; PERHAPS formerly the tropics supported as many speci
577 have been greatly favoured by high land, and PERHAPS by a dry climate; for Dr. Falconer informs m
582 aces, of existing and now sunken islands, and PERHAPS at the commencement of the Glacial period, b
585 shells from the other. But another agency is PERHAPS more effectual: I suspended a duck's feet, w
588 lonists. We should, also, remember that some, PERHAPS many, fresh water productions are low in the
591 ed, yet we can see that their eggs or larvae, PERHAPS attached to seaweed or floating timber, or t
615 of the being in relation to the outer world. PERHAPS from this cause it has partly arisen, that a
619 ical distribution has often been used, though PERHAPS not quite logically, in classification, more
628 hree elements of land, air, and water, we can PERHAPS understand how it is that a numerical parall
636 , and two pair of maxillae, these parts being PERHAPS very simple in form; and then natural select
643 nt with modification. It is commonly assumed, PERHAPS from monstrosities often affecting the embry
648 he embryo not undergoing any metamorphosis is PERHAPS requisite. If, on the other hand, it profite
650 very early period in the life of each, though PERHAPS caused at the earliest, anc being inherited
672 n nature, these facts cease to be strange, or PERHAPS might even have been anticipated. As natural
687 ne which is almost independent of altered and PERHAPS suddenly altered physical conditions, namely
001 , which then seemed to me probable: from that PERIOD to the present day I have steadily pursued th
008 or modification. It has been disputed at what PERIOD of life the causes of variability, whatever i
008 nerally act; whether during the early or late PERIOD of development of the embryo, or at the insta
008 little more or less water at some particular PERIOD of growth, will determine whether or not the
011 habit also has a decided influence, as in the PERIOD of flowering with plants when transported fro
013 I think may be trusted, is that, at whatever PERIOD of life a peculiarity first appears, it tends
014 s tend to appear in the offspring at the same PERIOD at which it first appeared in the parent. I b
022 l points of structure which are variable. The PERIOD at which the perfect plumage is acquired vari
028 improved then astonishingly. About this same PERIOD the Dutch were as eager about pigeons as were
034 colour of domestic animals was at that early PERIOD attended to. Savages now sometimes cross thei
037 o onwards. But the gardeners of the classical PERIOD, who cultivated the best pear they could proc
040 lection will always tend, perhaps more at one PERIOD than at another, as the breed rises or falls
063 or seeds, must suffer destruction during some PERIOD of its life, and during some season or occasi
064 e become common throughout whole islands in a PERIOD of less than ten years. Several of the plants
065 crease must be checked by destruction at some PERIOD of life. Our familiarity with the larger dome
065 ds is to make up for much destruction at some PERIOD of life; and this period in the great majorit
066 destruction at some period of life; and this PERIOD in the great majority of cases is an early on
066 umbers! that each lives by a struggle at some PERIOD of its life; that heavy destruction inevitabl
069 tantly suffering enormous destruction at some PERIOD of its life, from enemies or from competitors
079 ase at a geometrical ratio; that each at some PERIOD of its life, during some season of the year,
084 from danger. Grouse, if not destroyed at some PERIOD of their lives, would increase in countless n
086 under domestication appear at any particular PERIOD of life, tend to reappear in the offspring at
086 tend to reappear in the offspring at the same PERIOD; for instance, in the seeds of the many varie
086 sequent on other modifications at a different PERIOD of life, shall not be in the least degree inj
090 to master their prey at this or at some other PERIOD of the year, when they might be compelled to
102 er chance for the appearance within any given PERIOD of profitable variations, will compensate for
102 l selection, she does not grant an indefinite PERIOD! for as all organic beings are striving, it m
105 ich will prove capable of enduring for a long PERIOD, and of spreading widely. Throughout a great
107 as a continent, and the inhabitants, at this PERIOD numerous in individuals and kinds, will have
110 the best chance of producing within any given PERIOD favourable variations. We have evidence of th
110 quickly modified or improved within any given PERIOD, and they will consequently be beaten in the
112 beaks. Again, we may suppose that at an early PERIOD one man preferred swifter horses; another str
121 etters) of our original genus, may for a long PERIOD continue transmitting unaltered descendants;
123 4 and f14, from having diverged at an earlier PERIOD from a5, will be in some degree distinct from
126 ve suffered least extinction, will for a long PERIOD continue to increase. But which groups will u
126 quently that of the species living at any one PERIOD, extremely few will transmit descendants to a
126 escendants; yet at the most remote geological PERIOD, the earth may have been as well peopled with
129 e long succession of extinct species. At each PERIOD of growth all the growing twigs have tried to
139 n. habit is hereditary with plants, as in the PERIOD of flowering, in the amount of rain requisite
147 froy and Goethe propounded, at about the same PERIOD, are alike, seem liable to vary in an allied
153 traordinary amount of modification, since the PERIOD, their law of compensation or balancement of
153 from the common progenitor of the genus. This PERIOD when the species branched off from the common
153 ry rarely endure for more than one geological PERIOD will seldom be remote in any extreme degree.
153 has been so great and long continued within a PERIOD. An extraordinary amount of modification impl
153 sation, which have remained for a much longer PERIOD not excessively remote, we might, as a genera
155 isted, according to my theory, for an immense PERIOD nearly constant. And this, I am convinced, is
155 which have varied within a moderately recent PERIOD in nearly the same state; and thus it comes t
156 characters have been inherited from a remote PERIOD, and which have thus come to differ. Or to st
156 en inherited from a remote period, since that PERIOD, since that period when the species first bra
156 ers have varied and come to differ within the PERIOD when the species first branched off from thei
156 f the organisation which have for a very long PERIOD of the branching off of the species from a co
168 rited, and have not differed within this same PERIOD remained constant. In connexion with the pres
174 us, that it has been continuous during a long PERIOD. In these remarks we have referred to special
177 always have a better chance, within any given PERIOD. Geology would lead us to believe that almost
177 y well defined objects, and do not at any one PERIOD, of presenting further favourable variations
178 ous must often have existed within the recent PERIOD present an inextricable chaos of varying and
183 e will rarely have been developed at an early PERIOD in isolated portions, in which many forms, mo
189 have been first formed at an extremely remote PERIOD in great numbers and under many subordinate f
191 probable that organs which at a very ancient PERIOD, since which all the many members of the clas
196 fter having been slowly perfected at a former PERIOD served for respiration have been actually con
203 obscure. We have seen that species at any one PERIOD, have been transmitted in nearly the same sta
213 been transmitted for an incomparably shorter PERIOD are not indefinitely variable, and are not li
214 me curiously blended together, and for a long PERIOD, under less fixed conditions of life. How sto
217 ecially as she has to migrate at a very early PERIOD exhibit traces of the instincts of either par
237 elated not only to one sex, but to that short PERIOD! and the first hatched young would probably h
261 growth, in the hardness of their wood, in the PERIOD alone when the reproductive system is active,
264 may be developed, and then perish at an early PERIOD of the flow or nature of their sap, etc., but
264 consequently be liable to perish at an early PERIOD! This latter alternative has not been suffici
264 PERIOD; more especially as all very young beings see

age **(Key Word)**

Page **(Key Word)***

```
0281 ly that one form had remained for a very long   PERIOD unaltered, whilst its descendants had underg
0286 have covered up the Weald within so limited a   PERIOD as since the latter part of the Chalk format
0287 aves. So that in all probability a far longer   PERIOD than 300 million years has elapsed since the
0287 lapsed since the latter part of the Secondary   PERIOD. I have made these few remarks because it is
0288 the genus Chthamalus existed during the chalk   PERIOD. The molluscan genus Chiton offers a partial
0290 raised several hundred feet within the recent   PERIOD, than the absence of any recent deposits suf
0290 extensive to last for even a short geological   PERIOD. Along the whole west coast, which is inhabi
0293 ears, each perhaps is short compared with the   PERIOD requisite to change one species into another
0294  the inhabitants of Europe during the Glacial   PERIOD, which forms only a part of one whole geolog
0294 ich forms only a part of one whole geological   PERIOD; and likewise to reflect on the great change
0294 f time, all included within this same glacial   PERIOD. Yet it may be doubted whether in any quarte
0294 within the same area during the whole of this   PERIOD. It is not, for instance, probable that sedi
0294 was deposited during the whole of the glacial   PERIOD near the mouth of the Mississippi, within th
0294 e mississippi during some part of the glacial   PERIOD shall have been upraised, organic remains wil
0295 ossils had been less than that of the glacial   PERIOD, instead of having been really far greater,
0295 ust have gone on accumulating for a very long   PERIOD, in order to have given sufficient time for
0295 ss formation can only be accumulated during a   PERIOD of subsidence: and to keep the depth approxi
0296 ess, and which must have required an enormous   PERIOD for their accumulation; yet no one ignorant
0296 e not lived on the same spot during the whole   PERIOD of deposition, but have disappeared and reap
0296 erhaps many times, during the same geological   PERIOD. So that if such species were to undergo a c
0296 unt of modification during any one geological   PERIOD, a section would not probably include all the
0299 ther, for instance, geologists at some future   PERIOD will be able to prove, that our different bre
0300 doubted whether the duration of any one great   PERIOD of subsidence over the whole or part of the a
0301 pecies. It is, also, probable that each great   PERIOD of subsidence would be interrupted by oscilla
0304 , some time before the close of the secondary   PERIOD. I may give another instance, which from hav
0305 ssile cirripedes existed during the secondary   PERIOD; and these cirripedes might have been the pre
0305  the teleostean fishes, low down in the Chalk   PERIOD. This group includes the large majority of ex
0305 ultaneously throughout the world at this same   PERIOD. It is almost superfluous to remark that harc
0309 xtend, oceans have extended from the remotest   PERIOD of which we have any record; and on the other
0309 lations of level, since the earliest silurian   PERIOD. The coloured map appended to my volume on Co
0309 ement have changed in the lapse of ages? At a   PERIOD immeasurably antecedent to the silurian epoc
0313 colonies of M. Barrande, which intrude for a   PERIOD in the midst of an older formation, and then
0318 isappeared before the close of the palaeozoic   PERIOD. No fixed law seems to determine the length c
0318 ammonites towards the close of the secondary   PERIOD, has been wonderfully sudder. The whole subje
0318 still living shells at a very late geological   PERIOD, I was filled with astonishment: for seeing t
0319 ome one or several contingencies, and at what   PERIOD of the horse's life, and in what degree, they
0321 of Trilobites at the close of the palaeozoic   PERIOD and of Ammonites at the close of the secondar
0321 nd of Ammonites at the close of the secondary   PERIOD, we must remember what has been already said
0324 e that lived in Europe during the pleistocene   PERIOD (an enormously remote period as measured by y
0324  the pleistocene period (an enormously remote   PERIOD as measured by years, including the whole gla
0326 sea. If two great regions had been for a long   PERIOD favourably circumstanced in an equal degree,
0328 gions during nearly, but not exactly the same   PERIOD, we should find in both, from the causes expl
0330 oups, though formerly quite distinct, at that   PERIOD made some small approach to each other. It is
0332 ons of life, and yet retain throughout a vast   PERIOD the same general characteristics. This is rec
0333 acters, in this case the genera, at the early   PERIOD marked VI., would differ by a lesser number c
0333 ry, it is evident that the fauna of any great   PERIOD in the earth's history will be intermediate i
0334 hese allowances, the fauna of each geological   PERIOD undoubtedly is intermediate in character, bet
0334 ruth of the statement, that the fauna of each   PERIOD as a whole is nearly intermediate in characte
0336 icissitudes of climate during the pleistocene   PERIOD, which includes the whole glacial period, and
0336 cene period, which includes the whole glacial   PERIOD, and note how little the specific forms of th
0340 e in that quarter, during the next succeeding   PERIOD of time, closely allied though in some degree
0343 r oceans now extend they have for an enormous   PERIOD extended; and where our oscillating continent
0343 the Silurian epoch; but that long before that   PERIOD, the world may have presented a wholly differ
0345 character. The inhabitants of each successive   PERIOD in the world's history have beaten their pred
0346 ccasional means. Dispersal during the Glacial   PERIOD co extensive with the world. In considering t
0351 haracter from an enormously remote geological   PERIOD, so certain species have migrated over vast s
0355 region, has probably received at some former   PERIOD immigrants from this other region, my theory
0356 ere the sea row extends, land may at a former   PERIOD have connected islands or possibly even conti
0357 mutations of level, have occurred within the   PERIOD of existing organisms. Edward Forbes insisted
0357 such enormous geographical changes within the   PERIOD of existing species. It seems to me that we h
0357 ion, as to have united them within the recent   PERIOD to each other and to the several intervening
0357 it will ever be proved that within the recent   PERIOD continents which are now quite separate, have
0358 us geographical revolutions within the recent   PERIOD, as are necessitated on the view advanced by
0359 some of the 18 floated for a very much longer   PERIOD. So that as 64/87 seeds germinated after an
0363 as suggested by Lyell; and during the Glacial   PERIOD from one part of the now temperate regions to
0365 ate and survive. Dispersal during the Glacial   PERIOD. The identity of many plants and animals, on
0366  others called vivid attention to the Glacial   PERIOD, which, as we shall immediately see, affords
0366 organic, that within a very recent geological   PERIOD, central Europe and North America suffered un
0366 s and coast ice, plainly reveal a former cold   PERIOD. The former influence of the glacial climate
0366 nges more readily, by supposing a new glacial   PERIOD to come slowly on, and then pass away, as for
0367 nd the world. We may suppose that the Glacial   PERIOD came on a little earlier or later in North Am
0368 ll ascertained occurrence of a former Glacial   PERIOD, seem to me to explain in so satisfactory a m
0368  existence. If the climate, since the Glacial   PERIOD, has ever been in any degree warmer than at p
0368 and temperate productions will at a very late   PERIOD have marched a little further north, and subs
0368 respect to this intercalated slightly warmer   PERIOD, since the Glacial period. The arctic forms,
0368 ted slightly warmer period, since the Glacial   PERIOD. The arctic forms, during their long southern
0369 e glacial epoch, and which during its coldest   PERIOD will have been temporarily driven down to the
0369 lieve, actually took place during the Glacial   PERIOD, I assumed that at its commencement the arcti
0369 the world, at the commencement of the Glacial   PERIOD. At the present day, the sub arctic and north
0369 thern part of the Pacific. During the Glacial   PERIOD, when the inhabitants of the Old and New Worl
0370 son to believe that during the newer Pliocene   PERIOD, before the Glacial epoch, and whilst the maj
0370 e of latitude 60 degrees, during the Pliocene   PERIOD lived further north under the polar circle,
0370 e productions of the Old and New Worlds, at a   PERIOD anterior to the Glacial epoch. Believing, fro
0370 fer that during some earlier and still warmer   PERIOD, such as the older Pliocene period, a large n
0370 ill warmer period, such as the older Pliocene   PERIOD, a large number of the same plants and animal
0371 , long before the commencement of the Glacial   PERIOD. We now see, as I believe, their descendants,
```

Page **(Key Word)***

```
age ****************************************( Key Word )****************************************
```

```
571  the slowly decreasing warmth of the Pliocene PERIOD, as soon as the species in common, which inha
571  ons, left isolated, within a much more recent PERIOD, on the several mountain ranges and on the ar
572  uring the Pliocene or even a somewhat earlier PERIOD, was nearly uniform along the continuous shor
572  rn to our more immediate subject, the Glacial PERIOD. I am convinced that Forbes's view may be lar
573  one we have the plainest evidence of the cold PERIOD, from the western whores of Britain to the Ou
574  och was included within the latest geological PERIOD. We have, also, excellent evidence, that it e
574  le that it was, during a part at least of the PERIOD, actually simultaneous throughout the world.
574  he temperature of the whole world was at this PERIOD simultaneously cooler. But it would suffice f
576  or a large part of it, was during the Glacial PERIOD simultaneously much colder than at present. T
577  usly much colder than at present. The Glacial PERIOD, as measured by years, must have been very lo
577  mals have spread within a few centuries, this PERIOD will have been ample for any amount of migrat
578  ossed even the lowlands of the tropics at the PERIOD when the cold was most intense, when arctic f
578  the land at the foot of the Pyrenees. At this PERIOD of extreme cold, I believe that the climate u
578  seven thousand feet. During this the coldest  PERIOD, I suppose that large spaces of the tropical
578  rine productions, migrated during the glacial PERIOD from the northern and southern temperate zone
379  hen they became commingled during the Glacial PERIOD the northern forms were enabled to beat the
380  opical mountains: no doubt before the Glacial PERIOD they were stocked with endemic Alpine forms:
380  he intertropical mountains before the Glacial PERIOD must have been completely isolated; and I bel
381  believe that towards the close of the Glacial PERIOD, icebergs, as suggested by Lyell, have been l
381  en time since the commencement of the Glacial PERIOD for their migration, and for their subsequent
381  e northern hemisphere, to a former and warmer PERIOD, before the commencement of the Glacial perio
381  eriod, before the commencement of the Glacial PERIOD, when the antarctic lands, now covered with i
382  nd perhaps at the commencement of the Glacial PERIOD, by icebergs. By these means, as I believe, t
382  s may be said to have flowed during one short PERIOD from the north and from the south, and to hav
384  sh mainly to slight changes within the recent PERIOD in the level of the land, having caused river
384  l in the land within a very recent geological PERIOD, and when the surface was peopled by existing
384  ntinuous mountain ranges, which from an early PERIOD must have parted river systems and completely
389  uld lead to the belief that within the recent PERIOD all existing islands have been nearly or quit
396  have been continuously united within a recent PERIOD to the mainland than islands separated by dee
399  ation, before the commencement of the Glacial PERIOD. The affinity, which, though feeble, I am ass
401  eason to suppose that they have at any former  PERIOD been continuously united. The currents of the
404  with the foregoing view, that at some former  PERIOD there has been intercommunication or migratio
404  either at the present time or at some former  PERIOD under different physical conditions, and the
407  ich have certainly occurred within the recent PERIOD, and of other similar changes which may have
407  anges which may have occurred within the same PERIOD; if we remember how profoundly ignorant we ar
407  has been the influence of the modern Glacial  PERIOD, which I am fully convinced simultaneously af
410  be accounted for by migration at some former  PERIOD under different conditions or by occasional m
410  ups of species, belonging either to a certain PERIOD of time, or to a certain area, are often char
412  right hand, which diverged at a still earlier PERIOD. And all these genera, descended from (A), fo
420  species which existed at an unknown anterior  PERIOD. Species of three of these genera (A, F, and
421  y at the present time, but at each successive PERIOD of descent. All the modified descendants from
421  ate branch of descendants, at each successive PERIOD. If, however, we choose to suppose that any o
443  is active, and has to provide for itself. The PERIOD of activity may come on earlier or later in l
442  s of other groups, the embryo does not at any PERIOD differ widely from the adult: thus Owen has r
442  rve for diverse purposes, being at this early PERIOD of growth alike: of embryos of different spec
443  sterce, except when the embryo becomes at any PERIOD of life active and has to provide for itself;
443  es often affecting the embryo at a very early PERIOD, that slight variations necessarily appear at
443  ations necessarily appear at an equally early PERIOD. But we have little evidence on this head, in
443  eatures will be. The question is not, at what PERIOD of life any variation has been caused, but at
443  fe any variation has been caused, but at what PERIOD it is fully displayed. The cause may have act
443  theless an effect thus caused at a very early PERIOD, even before the formation of the embryo, may
444  ture, may have supervened at a not very early PERIOD of life: and some direct evidence from our do
444  them, may have appeared at an extremely early PERIOD. I have stated in the first chapter, that the
446  have not generally first appeared at an early PERIOD of life, and have been inherited by the offsp
446  by the offspring at a corresponding not early PERIOD. But the case of the short faced tumbler, whi
446  ences must either have appeared at an earlier PERIOD than usual, or, if not so, the differences mu
447  undergone much modification at a rather late  PERIOD of life, and having thus been converted into
447  which we are wholly ignorant, at a very early PERIOD of life, or each step might be inherited at a
447  or each step might be inherited at an earlier PERIOD than that at which it first appeared. In eith
450  ariations having been inherited at an earlier PERIOD than that at which they first appeared. It sh
450  ar enough back in time, may remain for a long PERIOD, or for ever, incapable of demonstration. Thu
450  some one ancient progenitor, at a very early  PERIOD in the life of each, though perhaps caused at
450  being inherited at a corresponding not early  PERIOD. Embryology rises greatly in interest, when w
455  be checked by natural selection. At whatever  PERIOD of life disuse or selection reduces an organ,
455  corresponding age, but at an extremely early  PERIOD of life (as we have good reason to believe to
457  rily or generally supervening at a very early PERIOD of life, and being inherited at a correspondi
457  life, and being inherited at a corresponding  PERIOD, we can understand the great leading facts in
457  se or selection, it will generally be at that PERIOD of life when the being has to provide for its
462  potent has been the influence of the Glacial  PERIOD on the distribution both of the same and of r
462  on will have been possible during a very long PERIOD; and consequently the difficulty of the wide
463  even on a wide area, which has during a long  PERIOD remained continuous, and of which the climate
463  few species are undergoing change at any one  PERIOD; and all changes are slowly effected. I have
463  abitants of the world, and at each successive PERIOD between the extinct and still older species,
468  ften capriciously, can produce within a short PERIOD a great result by adding up mere individual d
474  been inherited without change for an enormous PERIOD. It is inexplicable on the theory of creation
474  is, if it has been inherited for a very long  PERIOD; for in this case it will have been rendered
477  we can understand, by the aid of the Glacial  PERIOD, the identity of some few plants, and the clo
477  idely different, if they have been for a long PERIOD completely separated from each other: for as
479  being inherited at a corresponding not early  PERIOD of life, we can clearly see why the embryos o
480  onding ages have been inherited from a remote PERIOD to the present day. On the view of each organ
486  of most genera, have within a not very remote PERIOD descended from one parent, and have migrated
488  er, keeping in a body might remain for a long PERIOD unchanged, whilst within this same period, se
488  ong period unchanged, whilst within this same PERIOD, several of these species, by migrating into
068  ining the average numbers of a species, and  PERIODICAL seasons of extreme cold or drought, I belie
061  disturbance occuring in the development, or   PERIODICAL action, or mutual relation of the different
266  many individuals of any species which are PERIODICALLY born, but a small number can survive. I ha
```

```
age ****************************************( Key Word )****************************************
```

Page **(Key Word)**

```
0391 heir great annual migrations, visit either PERIODICALLY or occasionally this island. Madeira does
0018 pretend to say how long before these ancient PERIODS, savages, like those of Tierra del Fuego or
0034 orks of high antiquity. In rude and barbarous PERIODS of English history choice animals were ofte
0046 se, judging from Brachiopod shells, at former PERIODS of time. These facts seem to be very perple
0049 rent odour; they flower at slightly different PERIODS; they grow in somewhat different stations:
0052 or they may endure as varieties for very long PERIODS, as has been shown to be the case by Mr. Wo
0073 t the face of nature remains uniform for long PERIODS of time, though assuredly the merest trifle
0074 s, many different checks, acting at different PERIODS of life, and during different seasons or ye
0084 accumulated by nature during whole geological PERIODS. Can we wonder, then, that nature's product
0102 element of success. Though nature grants vast PERIODS of time for the work of natural selection,
0107 l, and which consequently will exist for long PERIODS in a broken condition, will be the most fav
0117 nor are they all supposed to endure for equal PERIODS. Only those variations which are in some wa
0129 ecies which lived during long past geological PERIODS, very few now have living and modified desc
0174 p into islands even during the later tertiary PERIODS: and in such islands distinct species might
0176 ate varieties should not endure for very long PERIODS; why as a general rule they should be exter
0206 ei or by having adapted them during long past PERIODS of time: the adaptations being aided in som
0208 ssociated with other habits, and with certain PERIODS of time and states of the body. When once a
0210 ecies having different instincts at different PERIODS of life, or at different seasons of the yea
0213 ks, associated with certain frames of mind or PERIODS of time. But let us look to the familiar ca
0269 nature acts uniformly and slowly during vast PERIODS of time on the whole organisation, in any w
0282 how imcomprehensibly vast have been the past PERIODS of time, may at once close this volume. Not
0284 ion of most geologists, enormously long blank PERIODS. So that the lofty pile of sedimentary rock
0287 hen deeply submerged for perhaps equally long PERIODS, it would, likewise, have escaped the actio
0288 ich lived during the Secondary and Palaeozoic PERIODS, it is superfluous to state that our eviden
0289 ll is known belonging to either of these vast PERIODS, with one exception discovered by Sir C. Ly
0289 s, would never have suspected that during the PERIODS which were blank and barren in his own coun
0291 may have been formed over wide spaces during PERIODS of subsidence, but only where the supply of
0292 could this have happened during the alternate PERIODS of elevation; or, to speak more accurately,
0292 remark is here worth a passing notice. During PERIODS of elevation the area of the land and of th
0292 of new varieties and species: but during such PERIODS there will generally be a blank in the geol
0292 s will be formed: and it is during these very PERIODS of subsidence, that our great deposits rich
0300 ndary formations lie in the past, only during PERIODS of subsidence. These periods of subsidence
0300 ast, only during periods of subsidence. These PERIODS of subsidence would be separated from each
0300 ee on the shores of South America. During the PERIODS of subsidence there would probably be much
0300 obably be much extinction of life: during the PERIODS of elevation, there would be much variation
0301 l changes would intervene during such lengthy PERIODS: and in these cases the inhabitants of the
0304 ssile cirripedes existed during the secondary PERIODS, they would certainly have been preserved in
0307 e lowest Silurian stratum was deposited, long PERIODS elapsed, as long as, or probably far longer
0307 nd that during these vast, yet quite unknown, PERIODS of time, the world swarmed with living crea
0307 do not find records of these vast primordial PERIODS, I can give no satisfactory answer. Several
0307 dicates the former existence of life at these PERIODS. But the difficulty of understanding the ab
0308 fer, that during the palaeozoic and secondary PERIODS, neither continents nor continental islands
0309 have intervened during these enormously long PERIODS. If then we may infer anything from these f
0315 erage amount of change, during long and equal PERIODS of time, may, perhaps, be nearly the same:
0317 he earth having been swept away at successive PERIODS by catastrophes, is very generally given up
0318 whole groups of species last for very unequal PERIODS: some groups, as we have seen, having endur
0320 reasing, at least during the later geological PERIODS, so that looking to later times we may beli
0323 the widely separated palaeozoic and tertiary PERIODS, would still be manifest, and the several f
0327 ossiliferous formations were deposited during PERIODS of subsidence: and that blank intervals of
0327 ntervals of vast duration occurred during the PERIODS when the bed of the sea was either stationa
0329 have not been deposited during the same exact PERIODS, a formation in one region often correspond
0332 ch are supposed to have perished at different PERIODS, and some to have endured to the present da
0333 se groups, which have within known geological PERIODS undergone much modification, should in the
0334 mutual affinities and then according to their PERIODS of existence, do not accord in arrangement.
0336 peared at the commencement and close of these PERIODS: but we ought to find after intervals, very
0339 hin the same areas, during the later tertiary PERIODS. Mr. Clift many years ago showed that the f
0340 same types in each during the later tertiary PERIODS. Nor can it be pretended that it is an immu
0342 has probably been more extinction during the PERIODS of subsidence, and more variation during th
0342 of subsidence, and more variation during the PERIODS of elevation, and during the latter the rec
0343 ase in numbers slowly, and endure for unequal PERIODS of time: for the process of modification is
0345 in the same areas during the later geological PERIODS ceases to be mysterious, and is simply expl
0350 ving been effected with more or less ease, at PERIODS more or less remote: on the nature and numb
0351 which have undergone during whole geological PERIODS but little modification, there is not much
0371 at the present time: for during these warmer PERIODS the northern parts of the Old and New World
0407 me forms of life change most slowly, enormous PERIODS of time being thus granted for their migrat
0462 retained the same specific form for very long PERIODS, enormously long as measured by years, too
0462 ion of the same species: for during very long PERIODS of time there will always be a good chance
0462 s which have affected the earth during modern PERIODS: and such changes will obviously have great
0465 ubsiding bed of the sea. During the alternate PERIODS of elevation and of stationary level the re
0465 the record will be blank. During these latter PERIODS there will probably be more variability in
0465 more variability in the forms of life: during PERIODS of subsidence, more extinction. With respec
0466 that modifications can be inherited for long PERIODS. As long as the conditions of life remain t
0477 e third source or from each other, at various PERIODS and in different proportions, the course of
0483 ut do they really believe that at innumerable PERIODS in the earth's history certain elemental at
0488 nic change as a measure of time. During early PERIODS of the earth's history, when the forms of l
0087 of the best short beaked tumbler pigeons more PERISH in the egg than are able to get out of it: s
0087 aks, for all with weak beaks would inevitably PERISH: or, more delicate and more easily broken sh
0135 he world are very frequently blown to sea and PERISH; that the beetles in Madeira, as observed by
0203 c queens her daughters as soon as born, or to PERISH herself in the combat: for undoubtedly this
0264 lastly, an embryo may be developed, and then PERISH at an early period. This latter alternative
0264 ree unsuitable, and consequently be liable to PERISH at an early period: more especially as all v
0366 stopped by barriers, in which case they would PERISH. The mountains would become covered with sno
0068 ot of turf (three feet by four) nine species PERISHED from the other species being allowed to gro
0219 hey could not even feed themselves, and many PERISHED of hunger. Huber then introduced a single s
0251 rs soon ceased to grow, and after a few days PERISHED entirely, whereas the pod impregnated by th
0332 families, some of which are supposed to have PERISHED at different periods, and some to have endu
```

Page **(Key Word)**

age ***(Key Word)***

```
020  nor can I find a single case on record of a PERMANENT race having been thus formed. On the Breeds
051  es which are in any degree more distinct and PERMANENT, as steps leading to more strongly marked a
052  eps leading to more strongly marked and more PERMANENT varieties; and at these latter, as leading
054  varieties, in order to become in any degree PERMANENT, necessarily have to struggle with the othe
056  ew that species are only strongly marked and PERMANENT varieties; for wherever many species of the
133  pecies of all kinds are only well marked and PERMANENT varieties. Thus the species of shells which
143  e breed it might probably have been rendered PERMANENT by natural selection. Homologous parts, as
161  cies being on my view only a well marked and PERMANENT variety. But characters thus gained would p
178  ht modifications of structure in some degree PERMANENT; and this assuredly we do see. Secondly, ar
219  natural selection making an occasional habit PERMANENT, if of advantage to the species; and if the
224  tural selection be strengthened and rendered PERMANENT for the very different purpose of raising s
469  ew that species are only strongly marked and PERMANENT varieties, and that each species first exis
473  the characters have become in a high degree PERMANENT, we can understand this fact; for they have
475  o suffer. If species be only well marked and PERMANENT varieties, we can at once see why their cro
479  to discover the lines of descent by the most PERMANENT characters, however slight their vital impo
034  dogs, but he has no wish or expectation of PERMANENTLY altering the breed. Nevertheless I cannot
047  hese doubtful and closely allied forms have PERMANENTLY retained their characters in their own cou
090  believe, natural selection acts, I must beg PERMISSION to give one or two imaginary illustrations.
053  al numbers of the varying species. Dr. Hooker PERMITS me to add, that after having carefully read
085  tween species, which, as far as our ignorance PERMITS us to judge, seem to be quite unimportant, w
107  s far as the extreme intricacy of the subject PERMITS. I conclude, looking to the future, that for
020  with several eminent fanciers, and have been PERMITTED to join two of the London Pigeon Clubs. The
225  pheres had been completed; but this is never PERMITTED, the bees building perfectly flat walls of
247  us cases; it seems to me that we may well be PERMITTED to doubt whether many other species are rea
260  be asked, has the production of hybrids been PERMITTED? to grant to species the special power of p
353  ubsistence under past and present conditions PERMITTED, is the most probable. Undoubtedly many cas
368  ithout other evidence, that a colder climate PERMITTED their former migration across the low inter
433  esemblances far more than others; why we are PERMITTED to use rudimentary and useless organs; or o
340  time and after great geographical changes, PERMITTING much inter migration, the feebler will yiel
285  worn by the waves and pared all round into PERPENDICULAR cliffs of one or two thousand feet in hei
016  species, this source of doubt could not so PERPETUALLY recur. It has often been stated that domes
072  seedlings and little trees, which had been PERPETUALLY browsed down by the cattle. In one square
109  e that the number of specific forms goes on PERPETUALLY and almost indefinitely increasing, number
101  i suspect, can self fertilisation go on for PERPETUITY. Circumstances favourable to Natural Select
050  ms quite unknown to him, he is at first much PERPLEXED to determine what differences to consider a
385  ghout the world. Their distribution at first PERPLEXED me much, as their ova are not likely to be
004  been disappointed: in this and in all other PERPLEXING cases I have invariably found that our know
046  al differences, which seems to me extremely PERPLEXING: I refer to those genera which have sometim
046  eriods of time. These facts seem to be very PERPLEXING, for they seem to show that this kind of va
053  sarily here is with much brevity, is rather PERPLEXING, and allusions cannot be avoided to the str
116  am will aid us in understanding this rather PERPLEXING subject. Let A to L represent the species o
212  l birds; for the large birds have been most PERSECUTED by man. We may safely attribute the greater
257  nother genus, as Silene, in which the most PERSERVERING efforts have failed to produce between ext
032  evotes his lifetime to it with indomitable PERSEVERANCE, he will succeed, and may make great impro
020  ot from India, and by the Hon. C. Murray from PERSIA. Many treatises in different languages have b
036  breeds as now existing in Britain, India, and PERSIA, we can, I think, clearly trace the stages th
208  ion follows another by a sort of rhythm; if a PERSON be interrupted in a song, or in repeating any
001  that I may be excused for entering on these PERSONAL details, as I give them to show that I have
002  ed from very many naturalists, some of them PERSONALLY unknown to me. I cannot, however, let this
085  y, so much so, that on parts of the Continent PERSONS are warned not to keep white pigeons, as bei
267  e generally sterile in some degree. I cannot PERSUADE myself that this parallelism is an accident
030  nt ways: when we compare the game cock, so PERTINACIOUS in battle, with other breeds so little qua
111  gricultural writer) as if by some murderous PESTILENCE. Divergence of Character. The principle, wh
092  ctar to be excreted by the inner bases of the PETALS of a flower. In this case insects in seeking
144  rts in normal structures, as the union of the PETALS of the corolla into a tube. Hard parts seem t
145  been thought that the development of the ray PETALS by drawing nourishment from certain other par
145  the patches of darker colour in the two upper PETALS; and that when this occurs, the adherent nect
145  lour is absent from only one of the two upper PETALS, the nectary is only much shortened. With res
436  a flower the relative position of the sepals, PETALS, stamens, and pistils, as well as their intim
437  s have simpler mouths? Why should the sepals, PETALS, stamens, and pistils in any individual flowe
451  in individual plants of the same species the PETALS sometimes occur as mere rudiments, and someti
479  rpose, in the jaws and legs of a crab, in the PETALS, stamens, and pistils of a flower, is likewis
066  y be the more numerous of the two: the Fulmar PETREL lays but one egg, yet it is believed to be th
185  k or grebe; nevertheless, it is essentially a PETREL, but with many parts of its organisation prof
471  e and feed on sub aquatic insects; and that a PETREL should have been created with habits and stru
184  it is a woodpecker which never climbs a tree! PETRELS are the most aerial and oceanic of birds; ye
186  ws: that there should be diving thrushes, and PETRELS with the habits of auks. Organs of extreme p
204  eet, ground woodpeckers, diving thrushes, and PETRELS with the habits of auks. Although the belief
251  species of Pelargonium, Fuchsia, Calceolaria, PETUNIA, Rhododendron, etc., have been crossed, yet
430  upials, as Mr. Waterhouse has remarked, the PHASCOLOMYS resembles most nearly, not any one species
430  esemblance is only analogical, owing to the PHASCOLOMYS having become adapted to habits like those
253  om cervulus vaginalis and Reevesii, and from PHASIANUS colchicus with P. torquatus and with P. ver
216  way as it is so plainly instinctive in young PHEASANTS, though reared under a hen. It is not that
325  ca, and there discover a series of analogous PHENOMENA, it will appear certain that all these modi
312  e earth. If we may trust the observations of PHILIPPI in Sicily, the successive changes in the mar
001  it has been called by one of our greatest PHILOSOPHERS. On my return home, it occurred to me, in
046  ular branching of the stem of a tree. This PHILOSOPHICAL naturalist, I may add, has also quite rec
062  er de Candolle and Lyell have largely and PHILOSOPHICALLY shown that all organic beings are expose
453  embryonic calf by the excretion of precious PHOSPHATE of lime? When a man's fingers have been amp
307  so called primordial zone. The presence of PHOSPHATIC nodules and bituminous matter in some of th
439  ulex or furze, and the first leaves of the PHYLLODINEOUS acaceas, are pinnate or divided like the
004  of organic beings to each other and to their PHYSICAL conditions of life, untouched and unexplane
044  implying a modification directly due to the PHYSICAL conditions of life, and variations in this s
052  ly to the long continued action of different PHYSICAL conditions in two different regions; but I h
053  expected, as they become exposed to diverse PHYSICAL conditions, and as they come into competitio
063  individuals of distinct species, or with the PHYSICAL conditions of life. It is the doctrine of Ma
```

Page ***(Key Word)***

```
0080  ll organic beings to each other and to their  PHYSICAL  conditions of life. Can it, then, be though
0081  taking the case of a country undergoing some   PHYSICAL  change, for instance, of climate. The propo
0082  er disposal. Nor do I believe that any great   PHYSICAL  change, as of climate, or any unusual degre
0082  o perfectly adapted to each other and to the   PHYSICAL  conditions under which they live, that none
0104  ation of better adapted organisms, after any   PHYSICAL  change, such as of climate or elevation of
0105  ed by barriers, or from having very peculiar   PHYSICAL  conditions, the total number of the individ
0107  of each species will thus be checked: after    PHYSICAL  changes of any kind, immigration will be pr
0108  xistence of such places will often depend on   PHYSICAL  changes, which are generally very slow, and
0109  anic beings, one with another and with their   PHYSICAL  conditions of life, which may be effected
0175  dredge. To those who look at climate and the   PHYSICAL  conditions of life as the all important ele
0175  s exclusively depends on insensibly changing   PHYSICAL  conditions, but in large part on the presen
0199  es are of no direct use to their possessors.   PHYSICAL  conditions probably have had some little ef
0200  me little allowance for the direct action of   PHYSICAL  conditions) may be viewed, either as having
0263  l distinct causes. There must sometimes be a   PHYSICAL  impossibility in the male element reaching
0279  extensive and continous area with graduated   PHYSICAL  conditions. I endeavoured to show, that the
0314  the rate of breeding, on the slowly changing   PHYSICAL  conditions of the country, and more especia
0325  ok to changes of currents, climate, or other   PHYSICAL  conditions, as the cause of these great mut
0325  find how slight is the relation between the    PHYSICAL  conditions of various countries, and the na
0336  es in closely consecutive formations, by the   PHYSICAL  conditions of the ancient areas having rema
0339  t to account, on the one hand, by dissimilar   PHYSICAL  conditions for the dissimilarity of the inh
0346  on cannot be accounted for by differences in   PHYSICAL  conditions. Importance of barriers. Affinit
0346  be accounted for by their climatal and other   PHYSICAL  conditions. Of late, almost every author wh
0350  of land and water, and independent of their   PHYSICAL  conditions. The naturalist must feel little
0350  degree to the direct influence of different   PHYSICAL  conditions. The degree of dissimilarity wil
0351  and in a lesser degree with the surrounding   PHYSICAL  conditions. As we have seen in the last cha
0359  in their power of germination. In Johnston's   PHYSICAL  Atlas, the average rate of the several Atla
0372  e, in correspondence with the nearly similar   PHYSICAL  conditions of the areas; for if we compare,
0372  countries closely corresponding in all their   PHYSICAL  conditions, but with their inhabitants utte
0389  ing quite independently of any difference in   PHYSICAL  conditions has caused so great a difference
0392  uch cases are generally accounted for by the   PHYSICAL  conditions of the islands; but this explana
0393  nic islands cannot be accounted for by their   PHYSICAL  conditions; indeed it seems that islands ar
0400  m the deeply seated error of considering the   PHYSICAL  conditions of a country as the most importa
0403  st corners of Australia have nearly the same   PHYSICAL  conditions, and are united by continuous la
0404  ime or at some former period under different   PHYSICAL  conditions, and the existence at remote poi
0408  can see why two areas having nearly the same   PHYSICAL  conditions should often be inhabited by ver
0408  in different regions, independently of their   PHYSICAL  conditions, infinitely diversified conditio
0466  and disuse, and by the direct action of the   PHYSICAL  conditions of life. There is much difficult
0468  n however slight a degree to the surrounding   PHYSICAL  conditions, will turn the balance. With ani
0468  proclaims that each land has undergone great   PHYSICAL  changes, we might have expected that organi
0472  n of so called specific forms. In both cases   PHYSICAL  conditions seem to have produced but little
0477  ean. Although two areas may present the same   PHYSICAL  conditions of life, we need feel no surpris
0487  dent of altered and perhaps suddenly altered   PHYSICAL  conditions, namely, the mutual relation of
0101  of a distinct individual can be shown to be   PHYSICALLY impossible. Cirripedes long appeared to me
0012  those of slight and those of considerable     PHYSIOLOGICAL  importance is endless. Dr. Prosper Lucas
0045  e called important, whether viewed under a     PHYSIOLOGICAL  or classificatory point of view, sometim
0093  the advantage of what has been called the      PHYSIOLOGICAL  division of labour: hence we may believe
0115  egion is, in fact, the same as that of the      PHYSIOLOGICAL  division of labour in the organs of the
0154  because they are taken from parts of less       PHYSIOLOGICAL  importance than those commonly used for
0155  value, often becomes variable, though its       PHYSIOLOGICAL  importance may remain the same. Somethin
0256  n the structure of parts which are of high      PHYSIOLOGICAL  importance and which differ little in th
0415  on resemblances in organs of high vital or      PHYSIOLOGICAL  importance. No doubt this view of the cl
0415  to their conditions of life. That the mere      PHYSIOLOGICAL  importance of an organ does not determin
0415  ery reason to suppose, has nearly the same      PHYSIOLOGICAL  value, its classificatory value is widel
0416  divisions of the same order are of unequal      PHYSIOLOGICAL  importance. Any number of instances coul
0416  udimentary or atrophied organs are of high      PHYSIOLOGICAL  or vital importance; yet, undoubtedly, o
0416  which must be considered of very trifling      PHYSIOLOGICAL  importance, but which are universally ad
0417  allies in several characters, both of high      PHYSIOLOGICAL  importance and of almost universal preva
0418  , they do not trouble themselves about the      PHYSIOLOGICAL  value of the characters which they use i
0426  the same class. As we find organs of high      PHYSIOLOGICAL  importance, those which serve to preserv
0433  and useless organs, or others of trifling      PHYSIOLOGICAL  importance; why, in comparing one group
0455  sometimes more useful than, parts of high      PHYSIOLOGICAL  importance. Rudimentary organs may be co
0116  ct so well elucidated by Milne Edwards. No      PHYSIOLOGIST  doubts that a stomach by being adapted to
0453  complete the scheme of nature? An eminent      PHYSIOLOGIST  accounts for the presence of rudimentary
0010  s or pollen. But it is the opinion of most      PHYSIOLOGISTS  that there is no essential difference be
0191  in as a complement to the swimbladder. All      PHYSIOLOGISTS  admit that the swimbladder is homologous
0436  rts or organs in the same individual. Most      PHYSIOLOGISTS  believe that the bones of the skull are
0209  nguished. If Mozart, instead of playing the   PIANOFORTE  at three years old with wonderfully little
0032  etty well established, the seed raisers do not  PICK  out the best plants; but merely go over their
0144  think it can hardly be accidental, that if we   PICK  out with unerring skill each improvement. Let
0189  almost infinitely, and natural selection will   PICK  out with unerring skill each improvement. Let
0432  said, define the several groups; but we could   PICK  out types, or forms, representing most of the
0024  ecies; but that he intentionally or by chance   PICKED  out extraordinarily abnormal species; and fu
0041  een neglected. As soon, however, as gardeners   PICKED  out individual plants with slightly larger,
0041  it, and raised seedlings from them, and again   PICKED  out the best seedlings and bred from them, t
0218  er the plains, so that in one day's hunting I   PICKED  up no less than twenty lost and wasted eggs.
0361  s of a turkey. In the course of two months, I   PICKED  up in my garden 12 kinds of seeds, out of th
0302  l palaeontologists, for instance, by Agassiz,  PICTET, and by none more forcibly than by Professor
0305  jority of existing species. Lately, Professor   PICTET  has carried their existence one sub stage fu
0305  south of the equator; and by running through   PICTET's Palaeontology it will be seen that very fe
0313  o this rule. The amount of organic change, as   PICTET  has remarked, does not strictly correspond w
0316  are surprisingly few, so few, that E. Forbes,   PICTET, and Woodward (though all strongly opposed t
0335  n are the fossils from two remote formations.   PICTET  gives as a well known instance, the general
0335  ts generality, seems to have shaken Professor   PICTET  in his firm belief in the immutability of sp
0338  al development of recent forms. I must follow   PICTET  and Huxley in thinking that the truth of thi
0031  are placed on a table and are studied, like a   PICTURE  by a connoisseur; this is cone three times
0285  rn escarpments meet and close, one can safely   PICTURE  to oneself the great dome of rocks which mu
```

Page **************************************(Key Word)**************************************

```
0338 thus the embryo comes to be left as a sort of PICTURE, preserved by nature, of the ancient and les
0375 llected on the loftier peaks of Java raises a PICTURE of a collection made on a hill in Europe! St
0450 nterest, when we thus look at the embryo as a PICTURE, more or less obscured, of the common parent
0486 lled living fossils, will aid us in forming a PICTURE of the ancient forms of life. Embryology wil
0280 t, when looking at any two species, to avoid PICTURING to myself, forms directly intermediate betw
0067 numbers by various enemies: for instance, on a PIECE of ground three feet long and two wide, dug a
0113 scendants, would succeed in living on the same PIECE of ground. And we well know that each species
0114 its inhabitants. For instance, I found that a PIECE of turf, three feet by four in size, which ha
0114 ls and plants which live close round any small PIECE of ground, could live on it (supposing it not
0228 put into the hive, instead of a thick, square PIECE of wax, a thin and narrow, knife edged ridge,
0388 y travel from one to another and often distant PIECE of water. Nature, like a careful gardener, th
0232 s to note in cases of difficulty, as when two PIECES of comb met at an angle, how often the bees w
0339 t, even to an uneducated eye, in the gigantic PIECES of armour like those of the armadillo, found
0437 such numerous and such extraordinarily shaped PIECES of bone? As Owen has remarked, the benefit de
0437 fit derived from the yielding of the separate PIECES in the act of parturition of mammals, will by
0089 likes: thus Sir R. Heron has described how one PIED peacock was eminently attractive to all his he
0197 we did not know that there were many black and PIED kinds, I dare say that we should have thought
0091 to return. I may add, that, according to Mr. PIERCE, there are two varieties of the wolf inhabiti
0208 of instinct are universal. A little dose, as PIERRE Huber expresses it, of judgment or reason, of
0219 vered in the Formica (Polyerges) rufescens by PIERRE Huber, a better observer even than his celebr
0225 domestica, carefully described and figured by PIERRE Huber. The Melipona itself is intermediate in
0330 ions the apparently wide difference between the PIG and the camel. In regard to the Invertebrata,
0020 en thus formed. On the Breeds of the Domestic PIGEON. Believing that it is always best to study so
0021 have been permitted to join two of the London PIGEON Clubs. The diversity of the breeds is somethi
0021 the normal number in all members of the great PIGEON family; and these feathers are kept expanded,
0023 namely, that all have descended from the rock PIGEON (Columba livia), including under this term se
0023 rieties, and have not proceeded from the rock PIGEON, they must have descended from at least seven
0023 ikely to be exterminated, and the common rock PIGEON, which has the same habits with the domestic
0024 y species having similar habits with the rock PIGEON seems to me a very rash assumption. Moreover,
0024 ever become wild or feral, though the dovecot PIGEON, which is the rock pigeon in a very slightly
0024 though the dovecot pigeon, which is the rock PIGEON in a very slightly altered state, has become
0024 parts of their structure, with the wild rock PIGEON, yet are certainly highly abnormal in other p
0025 pigeons well deserve consideration. The rock PIGEON is of a slaty blue, and has a white rump (the
0025 d white edged tail feathers, as any wild rock PIGEON! We can understand these facts, on the well k
0025 domestic breeds have descended from the rock PIGEON. But if we deny this, we must make one of the
0026 stocks were coloured and marked like the rock PIGEON, although no other existing species is thus c
0026 core of generations, been crossed by the rock PIGEON: I say within a dozen or twenty generations,
0027 ch so like in most other respects to the rock PIGEON; the blue colour and various marks occasional
0027 may add, firstly, that C. livia, or the rock PIGEON, has been found capable of domestication in E
0027 immensely in certain characters from the rock PIGEON, yet by comparing the several sub breeds of t
0029 eyhound and bloodhound, a carrier and tumbler PIGEON. I have never met a pigeon, or poultry, or duck, or rabbit fancier, who
0032 f practice requisite to become even a skilful PIGEON fancier. The same principles are followed by
0035 untry. By comparing the accounts given in old PIGEON treatises of carriers and tumblers with these
0036 , and come to differ so greatly from the rock PIGEON. Youatt gives an excellent illustration of th
0039 uld ever try to make a fantail, till he saw a PIGEON with a tail developed in some slight degree i
0039 an unusual manner, or a pouter till he saw a PIGEON with a crop of somewhat unusual size; and the
0039 terly incorrect. The man who first selected a PIGEON with a slightly larger tail, never dreamed wh
0039 l, never dreamed what the descendants of that PIGEON would become through long continued, partly u
0051 s have been counted. Perhaps the first pouter PIGEON did not inflate its crop much more than the t
0083 , for he will become impressed, just like the PIGEON or poultry fancier before alluded to, with th
0087 ng manner: he feeds a long and a short beaked PIGEON on the same food; he does not exercise a long
0112 f nature had to make the beak of a full grown PIGEON very short for the bird's own advantage, the
0112 something analogous. A fancier is struck by a PIGEON having a slightly shorter beak: another fanci
0112 shorter beak: another fancier is struck by a PIGEON having a rather longer beak: and on the ackno
0152 iable to variation. Look at the breeds of the PIGEON; see what a prodigious amount of difference t
0159 aracters not possessed by the aboriginal rock PIGEON: these then are analogous variations in two o
0159 ariations are due to the several races of the PIGEON having inherited from a common parent the sam
0160 e marks are characteristic of the parent rock PIGEON, I presume that no one will doubt that this i
0161 probable that in each generation of the barb PIGEON, which produces most rarely a blue and black
0162 , for instance, we did not know that the rock PIGEON was not feather footed or turn crowned, we co
0166 breeds of pigeons: they are descended from a PIGEON (including two or three sub species or geogra
0214 ng or probably could have taught, the tumbler PIGEON to tumble, an action which, as I have witness
0214 formed by young birds, that have never seen a PIGEON tumble. We may believe that some one pigeon s
0214 a pigeon tumble. We may believe that some one PIGEON showed a slight tendency to this strange habi
0268 each other in appearance, for instance of the PIGEON or of the cabbage, is a remarkable fact; more
0280 ter pigeons have both descended from the rock PIGEON: if we possessed all the intermediate varieti
0280 remely close series between both and the rock PIGEON; but we should have no varieties directly int
0281 ison of their structure with that of the rock PIGEON, whether they had descended from this species
0315 m our present fancil; but if the parent rock PIGEON were also destroyed, and in nature we have ev
0316 ed, could be raised from any other species of PIGEON, or even from the other well established race
0316 other well established races of the domestic PIGEON, for the newly formed fantail would be almost
0335 ipal living and extinct races of the domestic PIGEON were arranged as well as they could be in ser
0335 r of their disappearance: for the parent rock PIGEON now lives; and many varieties between the roc
0335 ow lives; and many varieties between the rock PIGEON and the carrier have become extinct: and carr
0361 ea water; but some taken out of the crop of a PIGEON, which had floated on artificial salt water f
0445 nclusive, that the several domestic breeds of PIGEON have descended from one wild species, I compa
0446 bler differed from the young of the wild rock PIGEON and of the other breeds, in all its proportio
0473 same manner as the several domestic breeds of PIGEON have descended from the blue and barred rock
0473 have descended from the blue and barred rock PIGEON! On the ordinary view of each species having
0007 varieties from one or more Species. Domestic PIGEONS, their Differences and Origin. Principle of
0012 to have, as is asserted, long or many horns; PIGEONS with feathered feet have skin between their
0012 ered feet have skin between their outer toes; PIGEONS with short beaks have small feet; and those
0020 tolerably and sometimes (as I have found with PIGEONS) extremely uniform, and everything seems sim
0020 i have, after deliberation, taken up domestic PIGEONS. I have kept every breed which I could purch
0020 in different languages have been published on PIGEONS, and some of them are very important, as bei
```

Page **************************************(Key Word)**************************************

Page **(Key Word)**

```
0022  om each other. Altogether at least a score of PIGEONS might be chosen, which if shown to an ornitl
0023  as the differences are between the breeds of PIGEONS, I am fully convinced that the common opini«
0023  sed aboriginal stocks must all have been rock PIGEONS, that is, not breeding or willingly perchin«
0023  cies, only two or three other species of rock PIGEONS are known; and these have not any of the cha
0024  the hypothesis of the multiple origin of our PIGEONS, it must be assumed that at least seven or «
0025  ree. Some facts in regard to the colouring of PIGEONS well deserve consideration. The rock pigeon
0026  grels from between all the domestic breeds of PIGEONS are perfectly fertile. I can state this fror
0027  rmerly got seven or eight supposed species of PIGEONS to breed freely under domestication; these s
0027  when we come to treat of selection. Fourthly, PIGEONS have been watched, and tended with the utmos
0027  rs of the world; the earliest known record of PIGEONS is in the fifth Aegyptian dynasty, about 30«
0027  fessor Lepsius; but Mr. Birch informs me that PIGEONS are given in a bill of fare in the previous
0028  ear from Pliny, immense prices were given for PIGEONS; nay, they are come to this pass, that they
0028  t they can reckon up their pedigree and race. PIGEONS were much valued by Akber Khan in India, abc
0028  , about the year 1600; never less than 20,000 PIGEONS were taken with the court. The monarchs of `
0028  his same period the Dutch were as eager about PIGEONS as were the old Romans. The paramount import
0028  laining the immense amount of variation which PIGEONS have undergone, will be obvious when we trea
0028  tion of distinct breeds, that male and female PIGEONS can be easily mated for life; and thus diffe
0028  ave discussed the probable origin of domestic PIGEONS at some, yet quite insufficient, length; bec
0028  sufficient, length; because when I first kept PIGEONS and watched the several kinds, knowing well
0031  r John Sebright, used to say, with respect to PIGEONS, that he would produce any given feather in
0039  ences might, and indeed do now, arise amongst PIGEONS, which are rejected as faults or deviations
0042  sess more than one breed of the same species. PIGEONS can be mated for life, and this is a great «
0042  the improvement and formation of new breeds. PIGEONS, I may add, can be propagated in great numbe
0085  ontinent persons are warned not to keep white PIGEONS, as being the most liable to destruction. He
0087  serted, that of the best short beaked tumbler PIGEONS more perish in the egg than are able to get
0112  go on (as has actually occurred with tumbler PIGEONS) choosing and breeding from birds with longe
0144  hered feet and skin between the outer toes in PIGEONS, and the presence of more or less down on tl
0159  r domestic races. The most distinct breeds of PIGEONS, in countries most widely set apart, presen«
0159  te yet closely related acts of creation. With PIGEONS, however, we have another case, namely, the
0166  now observe the case of the several breeds of PIGEONS: they are descended from a pigeon (includin«
0166  the young than in the old. Call the breeds of PIGEONS, some of which have bred true for centuries,
0280  a simple illustration: the fantail and pouter PIGEONS have both descended from the rock pigeon; i-
0315  instance, it is just possible, if our fantail PIGEONS were all destroyed, that fanciers, by striv-
0362  a friend of his had to give up flying carrier PIGEONS from France to England, as the hawks on the
0423  ifference are requisite, as we have seen with PIGEONS. The origin of the existence of groups subo«
0423  in the greatest number of points. In tumbler PIGEONS, though some sub varieties differ from the «
0445  ended from one wild species, I compared young PIGEONS of various breeds, within twelve hours afte«
0446  ties. Fanciers select their horses, dogs, and PIGEONS, for breeding, when they are nearly grown up
0446  the cases just given, more especially that of PIGEONS, seem to show that the characteristic diffe«
0447  e adults, just as we have seen in the case of PIGEONS. We may extend this view to whole families «
0187  series with an optic nerve merely coated with PIGMENT, and without any other mechanism; and from «
0187  ans the transparent cones which are coated by PIGMENT, and which properly act only by excluding la
0188  pparatus of an optic nerve merely coated with PIGMENT and invested by transparent membrane, into a
0012  by Heusinger, it appears that white sheep and PIGS are differently affected from coloured indivi«
0215  e, are most eager to attack poultry, sheep and PIGS! and this tendency has been found incurable ir
0215  to be taught not to attack poultry, sheep, and PIGS! No doubt they occasionally do make an attack«
0231  nd how they work, by supposing masons first to PILE up a broad ridge of cement, and then to begin
0284  ormously long blank periods. So that the lofty PILE of sedimentary rocks in Britain, gives but an
0285  surface to show such prodigious movements; the PILE of rocks on the one or other side having been
0295  m that each separate formation, like the whole PILE of formations in any country, has generally be
0282  a man must for years examine for himself great PILES of superimposed strata, and watch the sea at
0289  ere blank and barren in his own country, great PILES of sediment, charged with new and peculiar f«
0307  ifficulty of understanding the absence of vast PILES of fossiliferous strata, which on my theory w
0358  ther such rocks, instead of consisting of mere PILES of volcanic matter. I must now say a few word
0463  al geological stages? Why do we not find great PILES of strata beneath the Silurian system, stored
0231  wall is left in the middle; the masons always PILING up the cut away cement, and adding fresh ceme
0268  s as species. For instance, the blue and red PIMPERNEL, the primrose and cowslip, which are consi«
0247  e seed; as he found the common red and blue PIMPERNELS (Anagallis arvensis and coerulea), which tl
0423  or instance, not to class two varieties of the PINE apple together, merely because their fruit, tl
0140  mperatures, or becoming acclimatised: thus the PINES and rhododendrons, raised from seed collectec
0439  irst leaves of the phyllodineous acaceas, are PINNATE or divided like the ordinary leaves of the l
0029  hat the several sorts, for instance a Ribston PIPPIN or Codlin apple, could ever have proceeded fr
0093  a small quantity of pollen, and a rudimentary PISTIL; other holly trees bear only female flowers;
0093  only female flowers; these have a full sized PISTIL, and four stamens with shrivelled anthers, ir
0097  ore especially as the plant's own anthers and PISTIL generally stand so close together that self «
0098  amens of a flower suddenly spring towards the PISTIL, or slowly move one after the other towards `
0099  ven of these were not perfectly true. Yet the PISTIL of each cabbage flower is surrounded not only
0263  e, as would be the case with a plant having a PISTIL too long for the pollen tubes to reach the o«
0451  , the male flowers often have a rudiment of a PISTIL; and Kolreuter found that by crossing such ma
0451  an hermaphrodite species, the rudiment of the PISTIL in the hybrid offspring was much increased ir
0451  this shows that the rudiment and the perfect PISTIL are essentially alike in nature. An organ ser
0451  the other. Thus in plants, the office of the PISTIL is to allow the pollen tubes to reach the ovu
0451  les protected in the ovarium at its base. The PISTIL consists of a stigma supported on the style;
0452  which of course cannot be fecundated, have a PISTIL, which is in a rudimentary state, for it is r
0453  he minute papilla, which often represents the PISTIL in male flowers, and which is formed merely «
0092  se flowers; also, which had their stamens and PISTILS placed, in relation to the size and habits «
0093  lone in one flower or on one whole plant, and PISTILS alone in another flower or on another plant«
0436  position of the sepals, petals, stamens, and PISTILS, as well as their intimate structure, are ir
0437  ? why should the sepals, petals, stamens, and PISTILS in any individual flower, though fitted for
0438  crabs as metamorphosed legs: the stamens and PISTILS of flowers as metamorphosed leaves; but it w
0479  d legs of a crab, in the petals, stamens, and PISTILS of a flower, is likewise intelligible on the
0228  es instantly began to excavate minute circular PITS in it; and as they deepened these little pits«
0228  pits in it; and as they deepened these little PITS, they make them wider and wider until they wer
0022  i do not believe that any ornithologist would PLACE the English carrier, the short faced tumbler «
0066  and could be ensured to germinate in a fitting PLACE. So that in all cases, the average number of
0069  from enemies or from competitors for the same PLACE and food; and if these enemies or competitors
```

Page **(Key Word)**

Page **(Key Word)**

```
0076 ently we hear of one species of rat taking the  PLACE of another species under the most different c
0076 tween allied forms, which fill nearly the same  PLACE in the economy of nature; but probably in no
0096 ome general considerations alone. In the first  PLACE, I have collected so large a body of facts, s
0102 are striving, it may be said, to seize on each   PLACE in the economy of nature, if any one species
0102 nature; for within a confined area, with some   PLACE in its polity not so perfectly occupied as mi
0102 egrees, so as better to fill up the unoccupied   PLACE. But if the area be large, its several distri
0103 in a body, so that whatever intercrossing took   PLACE would be chiefly between the individuals of t
0104 n to believe that occasional intercrosses take   PLACE with all animals and with all plants. Even if
0104 nimals and with all plants. Even if these take   PLACE only at long intervals, I am convinced that t
0111 er animals, and varieties of flowers, take the   PLACE of older and inferior kinds. In Yorkshire, it
0111 believe, several important facts. In the first   PLACE, varieties, even strongly marked ones, though
0119 descent, will, it is probable, often take the    PLACE of, and so destroy, the earlier and less impr
0132 ts seem to be affected before that union takes   PLACE which is to form a new being. In the case of
0172 ll tend in a fully stocked country to take the   PLACE of, and finally to exterminate, its own less
0173 ive species, evidently filling nearly the same   PLACE in the natural economy of the land. These rep
0174 t can be in large part explained. In the first   PLACE we should be extremely cautious in inferring,
0177 ved mountain or plain breed will soon take the   PLACE of the less improved hill breed; and thus the
0177 urable variations chance to occur, and until a   PLACE in the natural polity of the country can be b
0179 that each is well adapted in its habits to its   PLACE in nature. Look at the Mustela vison of North
0186 eator to cause a being of one type to take the   PLACE of one of another type; but this seems to me
0186 nhabitant of the country, it will seize on the   PLACE of that inhabitant, however different it may
0186 tant, however different it may be from its own   PLACE. Hence it will cause him no surprise that the
0194 to have been separately created for its proper   PLACE in nature, be so invariably linked together b
0195 s perfect and complex as the eye. In the first  PLACE, we are much too ignorant in regard to the wh
0199 tail, can double quickly enough. In the second  PLACE, we may sometimes attribute importance to cha
0199 gh each being assuredly is well fitted for its  PLACE in nature, many structures now have no direct
0220 with their masters in carrying them away to a   PLACE of safety. Hence, it is clear, that the slave
0221 est, and put them down on a bare spot near the  PLACE of combat: they were eagerly seized, and carr
0222 te combat. At the same time I laid on the same  PLACE a small parcel of the pupae of another specie
0232 h they had at first rejected. When bees have a  PLACE on which they can stand in their proper posit
0232 wall of a new hexagon, in its strictly proper   PLACE, projecting beyond the other completed cells.
0232 rtain circumstances a rough wall in its proper  PLACE between two just commenced cells, is importan
0239 which secretes a sort of honey, supplying the   PLACE of that excreted by the aphides, or the domes
0269 le than at first appears. It can, in the first  PLACE, be clearly shown that mere external dissimil
0269 same rule to domestic varieties. In the second PLACE, some eminent naturalists believe that a long
0280 fection of the geological record. In the first  PLACE it should always be borne in mind what sort o
0301 at first generally be local or confined to one  PLACE, but if possessed of any decided advantage, o
0315 ed in innumerable instances) to fill the exact  PLACE of another species in the economy of nature,
0319 come rarer and rarer, and finally extinct; its  PLACE being seized on by some more successful compe
0320 ear, like our short horn cattle, and takes the  PLACE of other breeds in other countries. Thus the
0321 ging to some one group will have seized on the  PLACE occupied by a species belonging to a distinct
0348 ocean extends, with not an island as a halting  PLACE for emigrants: here we have a barrier of anot
0359 branches of 94 plants with ripe fruit, and to   PLACE them on sea water. The majority sank quickly,
0362 , and thus the seeds might be transported from  PLACE to place. I forced many kinds of seeds into t
0362 s the seeds might be transported from place to  PLACE. I forced many kinds of seeds into the stomac
0369 illustrating what, as I believe, actually took  PLACE during the Glacial period, I assumed that at
0371 more temperate productions are concerned, took  PLACE long ages ago. And as the plants and animals
0384 me and means for much migration. In the second PLACE, salt water fish can with care be slowly accu
0388 ry, would have a better chance of seizing on a  PLACE, than in the case of terrestrial colonists. W
0391 new Zealand gigantic wingless birds, take the   PLACE of mammals. In the plants of the Galapagos Is
0405 modification during its diffusion, and should   PLACE itself under diverse conditions favourable fo
0414 determined the habits of life, and the general  PLACE of each being in the economy of nature, would
0441 h by their active powers of swimming, a proper  PLACE on which to become attached and to undergo th
0472 ants of each to any unoccupied or ill occupied  PLACE in nature, these facts cease to be strange, o
0031 that men follow it as a trade: the sheep are    PLACED on a table and are studied, like a picture by
0041 y country requires that the species should be   PLACED under favourable conditions of life, so as to
0054 masses, all those in the larger genera being    PLACED on one side, and all those in the smaller gen
0078 so, we can see that when a plant or animal is   PLACED in a new country amongst new competitors, tho
0083 r is fully exercised by her; and the being is   PLACED under well suited conditions of life. Man kee
0092 rs, also, which had their stamens and pistils   PLACED, in relation to the size and habits of the pa
0094 on another plant. In plants under culture and   PLACED under new conditions of life, sometimes the m
0099 stigmatic surface of the same flower, though    PLACED so close together, as if for the very purpose
0189 ayers of different densities and thicknesses,   PLACED at different distances from each other, and w
0210 or at different seasons of the year, or when    PLACED under different circumstances, etc.: in which
0225 the series we have the cells of the hive bee,   PLACED in a double layer: each cell, as is well know
0226 equal spheres be described with their centres   PLACED in two parallel layers; with the centre of ea
0232 been taken from the spot on which it had been   PLACED, and worked into the growing edges of the cel
0232 for working, for instance, on a slip of wood,   PLACED directly under the middle of a comb growing d
0255 en. When pollen from a plant of one family is   PLACED on the stigma of a plant of a distinct family
0263 n observed that when pollen of one species is   PLACED on the stigma of a distantly allied species,
0264 heir two parents can live, they are generally   PLACED under suitable conditions of life. But a hybr
0265 rile hybrids. Lastly, when organic beings are   PLACED during several generations under conditions n
0265 ved. Thus we see that when organic beings are   PLACED under new and unnatural conditions; and when
0266 explanation is offered why an organism, when   PLACED under unnatural conditions, is rendered steri
0329 e classification of these two orders: and has   PLACED certain pachyderms in the same sub order with
0360 lar ones; but in a much better manner, for he   PLACED the seeds in a box in the actual sea, so that
0377 er to the equator, though they will have been  PLACED under somewhat new conditions, will have suff
0442 quite inactive, being fed by their parents or  PLACED in the midst of proper nutriment, yet nearly
0445 e nestling birds of these several breeds were  PLACED in a row, though most of them could be distin
0474 nderstand how it is that allied species, when  PLACED under considerably different conditions of li
0024 ly altered state, has become feral in several  PLACES. Again, all recent experience shows that it i
0081 s should not freely enter, we should then have PLACES in the economy of nature which would assuredl
0082 the area been open to immigration, these same  PLACES would have been seized on by intruders. In su
0082 ually necessary to produce new and unoccupied  PLACES for natural selection to fill up by modifying
0088 males, those which are best fitted for their   PLACES in nature, will leave most progeny. But in ma
0104 or elevation of the land, etc.; and thus new   PLACES in the natural economy of the country are lef
```

Page **(Key Word)**

Page ***************************************(Key Word)*************************************

```
0106  competition with many others. Hence more new PLACES will be formed, and the competition to fill t
0108  d, immigration will be prevented, so that new PLACES in the polity of each island will have to be
0108  ully admit. Its action depends on there being PLACES in the polity of nature, which can be better
0108  ification of some kind. The existence of such PLACES will often depend on physical changes, which
0109  ld not have thus increased, for the number of PLACES in the polity of nature is not indefinitely g
0112  abled to seize on many and widely diversified PLACES in the polity of nature, and so be enabled to
0113  s) only by its varying descendants seizing on PLACES at present occupied by other animals: some of
0113  ts of our carnivorous animal became, the more PLACES they would be enabled to occupy. What applies
0116  tructure, and are thus enabled to encroach on PLACES occupied by other beings. Now let us see how
0119  ill always act according to the nature of the PLACES which are either unoccupied or not perfectly
0119  rom any one species can be rendered, the more PLACES they will be enabled to seize on, and the mor
0121  st chance of filling new and widely different PLACES in the polity of nature: hence in the diagram
0122  so as to have become adapted to many related PLACES in the natural economy of their country. It s
0122  remely probable that they will have taken the PLACES of, and thus exterminated, not only their par
0126  s, from branching out and seizing on many new PLACES in the polity of Nature, will constantly tend
0158  r to fit the several species to their several PLACES in the economy of nature, and likewise to fit
0178  one or more of its inhabitants. And such new PLACES will depend on slow changes of climate, or on
0280  ough which new varieties continually take the PLACES of and exterminate their parent forms. But ju
0284  t of denudation which the strata have in many PLACES suffered, independently of the rate of accumu
0298  many specimens have been collected from many PLACES: and in the case of fossil species this could
0321  ssful intruder, many will have to yield their PLACES: and it will generally be allied forms, which
0321  ame or to a distinct class, which yield their PLACES to other species which have been modified and
0322  bitants: and the forms which thus yield their PLACES will commonly be allied, for they will partak
0327  forms which are beaten and which yield their PLACES to the new and victorious forms, will general
0331  g thus enabled to seize on many and cifferent PLACES in the economy of nature. Therefore it is qui
0337  y spread over New Zealand, and have seized on PLACES which must have been previously occupied, we
0337  siderable number would be enabled to seize on PLACES now occupied by our native plants and animals
0344  nd these will generally succeed in taking the PLACES of those groups of species which are their li
0348  s, and we have innumerable islands as halting PLACES, until after travelling over a hemisphere we
0350  s will have the best chance of seizing on new PLACES, when they spread into new countries. In thei
0357  ath the sea, which may have served as halting PLACES for plants and for many animals during their
0366  anted and arctic productions would take their PLACES. The inhabitants of the more temperate region
0382  eans of transport, and by the aid, as halting PLACES, of existing and now sunken islands, and perh
0391  have been kept by the others to their proper PLACES and habits, and will consequently have been l
0391  times deficient in certain classes, and their PLACES are apparently occupied by the other inhabita
0396  ity of many islands having existed as halting PLACES, of which not a wreck now remains, must not b
0402  if both are equally well fitted for their own PLACES in nature, both probably will hold their own
0402  in nature, both probably will hold their own PLACES and keep separate for almost any length of ti
0412  ies trying to occupy as many and as different PLACES as possible in the economy of nature, there
0421  races of their parentage, in this case, their PLACES in a natural classification will have been mo
0428  spread widely, and to seize on more and more PLACES in the economy of nature. The larger and more
0470  enabled to seize on many and widely different PLACES in the economy of nature, there will be a cor
0098  , by insects, it never sets a seed, though by PLACING pollen from one flower on the stigma of anot
0414  e for enunciating general propositions and of PLACING together the forms most like each other. It
0056  y excluded from the tables. These facts are of PLAIN signification on the view that species are or
0177  acti and consequently the improved mountain or PLAIN breed will soon take the place of the less im
0398  Lapagos Archipelago, and nowhere else, bear so PLAIN a stamp of affinity to those created in Ameri
0435  rpose they served, we can at once perceive the PLAIN signification of the homologous construction
0438  an hardly avoid employing language having this PLAIN signification. On my view these terms may be
0463  not every collection of fossil remains afford PLAIN evidence of the gradation and mutation of the
0480  me beetles, should thus so frequently bear the PLAIN stamp of inutility! Nature may be said to hav
0481  d is so perfect that it would have afforded us PLAIN evidence of the mutation of species, if they
0485  will cease to be metaphorical, and will have a PLAIN signification. When we no longer look at an o
0166  ies of the horse genus the stripes are either PLAINER or appear more commonly in the young than in
0185  tructure not at all in agreement. What can be PLAINER than that the webbed feet of ducks and geese
0185  oes are only bordered by membrane. What seems PLAINER than that the long toes of grallatores are f
0451  ks of certain embryonic birds. Nothing can be PLAINER than that wings are formed for flight, yet t
0163  e zebra: it has been asserted that these are PLAINEST in the foal, and from inquiries which I have
0164  eover, is sometimes striped. The stripes are PLAINEST in the foal; and sometimes quite disappear t
0296  its accumulation. In other cases we have the PLAINEST evidence in great fossilised trees, still st
0310  ly of the gravest nature. We see this in the PLAINEST manner by the fact that all the most eminent
0373  y be largely extended. In Europe we have the PLAINEST evidence of the cold period, from the wester
0037  have been blended together by crossing, may PLAINLY be recognised in the increased size and beau
0056  been fatal to my theory! inasmuch as geology PLAINLY tells us that small genera have in the lapse
0060  d? we see these beautiful co adaptations most PLAINLY in the woodpecker and missletoe; and only a
0060  pecker and missletoe: and only a little less PLAINLY in the humblest parasite which clings to the
0071  but how important an element enclosure is, I PLAINLY saw near Farnham, in Surrey. Here there are
0083  n prominent enough to catch his eye, or to be PLAINLY useful to him. Under nature, the slightest o
0084  e most complex conditions of life, and should PLAINLY bear the stamp of far higher workmanship?
0109  not incefinitely increased, geology shows us PLAINLY: and indeed we can see reason why they shoul
0127  ion has acted in the world's history, geology PLAINLY declares. Natural selection, also, leads to
0131  rect expression; but it serves to acknowledge PLAINLY our ignorance of the cause of each particula
0134  variations, however slight, until they become PLAINLY developed and appreciable by us. Effects of
0151  pplies more rarely to them. The rule being so PLAINLY applicable in the case of secondary sexual c
0163  aintly on the shoulder. The quagga, though so PLAINLY barred like a zebra over the body, is withou
0165  en the ass and zebra, the legs were much more PLAINLY barred than the rest of the body: and in one
0165  mare by a black Arabian sire, were much more PLAINLY barred across the legs than is even the pure
0184  of its voice, and undulatory flight, told me PLAINLY of its close blood relationship to our commo
0215  tinctive in them, in the same way as it is so PLAINLY instinctive in young pheasants, though reare
0240  mple eyes (ocelli), which though small can be PLAINLY distinguished, whereas the smaller workers h
0243  cts as well as to corporeal structure, and is PLAINLY explicable on the foregoing views, but is ot
0267  gain. During the convalescence of animals, we PLAINLY see that great benefit is cerived from almos
0285  the open ocean. The same story is still more PLAINLY told by faults, those great cracks along whi
0291  rable thickness. All geological facts tell us PLAINLY that each area has undergone numerous slow o
0345  her hand, all the chief laws of palaeontology PLAINLY proclaim, as it seems to me, that species ha
0349  onging to the same order of Rodents, but they PLAINLY display an American type of structure. We as
```

Page ***************************************(Key Word)*************************************

Page **(Key Word)**

```
0355   descendants, though modified, would still be PLAINLY related by inheritance to the inhabitants of
0366   use burnt by fire do not tell their tale more PLAINLY, than do the mountains of Scotland and Wales
0366   cks scored by drifted icebergs and coast ice, PLAINLY reveal a former cold period. The former infl
0379   m. thus many of these wanderers, though still PLAINLY related by inheritance to their brethren of
0399   what might have been expected: but it is also  PLAINLY related to South America, which, although th
0403   tants, when not identically the same, yet are  PLAINLY related to the inhabitants of that region wh
0419   dea of descent. Our classifications are often  PLAINLY influenced by chains of affinities. Nothing
0419   mon; yet the species at both ends, from being  PLAINLY allied to others, and these to others, and s
0436   of the vertebrate and articulate classes are  PLAINLY homologous. We see the same law in comparing
0439   riped or spotted in lines; and stripes can be  PLAINLY distinguished in the whelp of the lion. We oc
0443   or form will ultimately turn out. We see this  PLAINLY in our own children; we cannot always tell w
0453   : for the same reasoning power which tells us  PLAINLY that most parts and organs are exquisitely a
0457   ed in this chapter, seem to me to proclaim so  PLAINLY, that the inumerable species, genera, and fa
0465   ook to long enough intervals of time, geology  PLAINLY declares that all species have changed: and
0468   st advantage will lead to victory. As geology  PLAINLY proclaims that each land has undergone great
0469   erences and slight varieties; or between more  PLAINLY marked varieties and sub species, and specie
0471   is on this theory simply intelligible. We can  PLAINLY see why nature is prodigal in variety, thoug
0477   f the inhabitants within each great class are  PLAINLY related; for they will generally be descenda
0483   ing orders. Organs in a rudimentary condition  PLAINLY show that an early progenitor had the organ
0453   ed for certain purposes, tells us with equal   PLAINNESS that these rudimentary or atrophied organs,
0042   wandering savages or the inhabitants of open   PLAINS rarely possess more than one breed of the sam
0065   of the plants now most numerous over the wide  PLAINS of La Plata, clothing square leagues of surfa
0177   vely narrow, hilly tract; and a third to wide  PLAINS at the base; and that the inhabitants are all
0177   the great holders on the mountains or on the   PLAINS of La Plata, where not a tree grows, there is
0184   s which chase insects on the wing; and on the  PLAINS, so that in one day's hunting I picked up no
0218   uprising number of eggs lie strewed over the   PLAINS, forests, marshes, lakes, and great rivers, u
0346   tricts, arid deserts, lofty mountains, grassy  PLAINS near the Straits of Magellar are inhabited by
0349   eggs coloured in nearly the same manner. The   PLAINS of La Plata by another species of the same ge
0349   of rhea (American ostrich), and northward the  PLAINS of La Plata, we see the agouti and bizcacha,
0349   tralia under the same latitude. On these same  PLAINS. By the time that the cold had reached its ma
0366   ormer Alpine inhabitants would descend to the  PLAINS: they will, also, have been exposed to somewh
0369   will have been temporarily driven down to the  PLAINS of North America and Europe; and it may be re
0369   re the same on the lower mountains and on the  PLAINS and mountains, of the forests, marshes, and d
0408   ance in South America, the inhabitants of the  PLAN of construction, could have profited the proge
0233   incts, all tending towards the present perfect PLAN of the Creator; but unless it be specified whe
0413   tural System; they believe that it reveals the PLAN of the Creator, it seems to me that nothing is
0413   in time or space, or what else is meant by the PLAN of creation, or is simply a scheme for enuncia
0414   that classification either gives some unknown  PLAN of creation, or the enunciation of general pro
0420   en unconsciously seeking, and not some unknown PLAN of creation, we may hope to make sure but slow
0434   bject in view, and do not look to some unknown PLAN of their organisation. This resemblance is oft
0434   ts of life, resemble each other in the general PLAN of creation, unity of design, etc., and to thi
0482   de our ignorance under such expressions as the PLAN of creation. The rules for classifying will no
0486   nd will then truly give what may be called the PLANE surfaces in any one cell necessarily enter in
0226   the cells of the hive bee, so here, the three  PLANE, and thus flatten it. From the experiment of
0230   c is easily done) into its proper intermediate PLANE of intersection between two adjoining spheres
0231   wax in the proper position, that is, along the PLANE surfaces; and the Melipona would make a comb
0234   wholly disappear, and would all be replaced by PLANED down by the action of the sea, that no trace
0285   he surface of the land has been so completely  PLANES, or even perceive when they are correctly mad
0224   ow they can make all the necessary angles and  PLANES of intersection between the several spheres i
0226   res in the other and parallel layer; then, if  PLANES of imaginary intersection between the basins
0229   far as the eye could judge, exactly along the  PLANES or planes of intersection. Considering how fl
0229   sins, by stopping work along the intermediate  PLANES of intersection. Considering how flexible thi
0229   topping work along the intermediate planes or  PLANES of the rhombic basal plates of future cells.
0231   be observed, corresponding in position to the  PLANES of intersection between these spheres. It was
0232   nd then building up, or leaving ungnawed, the  PLANES. It is even conceivable that an insect might,
0233   es or cylinders, and building up intermediate  PLANES of intersection, and so make an isolated hexa
0233   central point and from each other, strike the  PLANES of intersection. The bees, of course, no more
0235   nd to build up and excavate the wax along the  PLANES of intersection. Other highly competent judges, as Lyell and
0307   est Silurian stratum the dawn of life on this  PLANET has gone cycling on according to the fixed la
0490   few forms or into one; and that, whilst this   PLANETS revolve in elliptic courses round the sun, s
0453   it be thought sufficient to say that because   PLANETS, for the sake of symmetry, and to complete t
0453   , satellites follow the same course round the  PLANT itself. The author of the Vestiges of Creatio
0003   itions, or of habit, or of the volition of the PLANT to the misseltoe, and that these had been pro
0004   bird had given birth to a woodpecker, and some PLANT sets a seed. I cannot then enter on the copio
0008   d of growth, will determine whether or not the PLANT. Such buds can be propagated by grafting, etc
0009   fferent character from that of the rest of the PLANT's own good, but to man's use or fancy. Some v
0030   them adaptation, not indeed to the animal's or PLANT. No one would expect to raise a first rate me
0037   e heartsease or dahlia from the seed of a wild PLANT worth culture. It is not that these countries
0038   uite uncivilised man, has afforded us a single PLANT should be so highly useful to man, or so much
0041   important point of all, is, that the animal or PLANT. No doubt the strawberry had always varied si
0041   when gardeners began to attend closely to this PLANT in a state of nature highly useful to man,
0050   en struck with the fact, that if any animal or PLANT on the edge of a desert is said to struggle f
0062   ach other which shall get food and live. But a PLANT which annually produces a thousand seeds, of
0063   uld be said to be dependent on the moisture. A  PLANT produced only two seeds, and there is no plan
0064   eny. Linnaeus has calculated that if an annual PLANT so unproductive as this, and their seedlings
0064   plant produced only two seeds, and there is no PLANT produces seed, and amongst animals there are
0065   r new homes. In a state of nature almost every PLANT depends only indirectly on the number of its
0066   all cases, the average number of any animal or PLANT could exist only where the conditions of its
0070   nge. For in such cases, we may believe, that a PLANT in the midst of its range, why does it not do
0077   lants growing vigorously all around. Look at a PLANT the power of increasing in number, we should
0078   e that if we wished in imagination to give the PLANT; but we have reason to believe that only a fe
0078   o climate would clearly be an advantage to our PLANT or animal is placed in a new country amongst
0078   est spots. Hence, also, we can see that when a PLANT to have its seeds more and more widely dissem
0086   rítance at a corresponding age. If it profit a PLANT; yet if a little pollen were carried, at firs
0092   , its destruction appears a simple loss to the PLANT: and those individuals which produced more an
0092   stroyed, it might still be a great gain to the
```

Page **(Key Word)**

Page **************************************(Key Word)***************************************

```
0093 nd larger anthers, would be selected. When our PLANT, by this process of the continued preservatic
0093 o return to our imaginary case: as soon as the PLANT had been rendered so highly attractive to ins
0093 may believe that it would be advantageous to a PLANT to produce stamens alone in one flower or on a
0093 ce stamens alone in one flower or on one whole PLANT, and pistils alone in another flower or on ar
0094 pistils alone in another flower or on another PLANT. In plants under culture and placed under new
0094 a more complete separation of the sexes of our PLANT would be advantageous on the principle of the
0094 ects in our imaginary case: we may suppose the PLANT of which we have been slowly increasing the r
0094 nectar by continued selection, to be a common PLANT: and that certain insects depended in main pa
0097 this state of exposure, more especially as the PLANT's own anthers and pistil generally stand so c
0097 other, to the great good, as I believe, of the PLANT. Bees will act like a camel hair pencil, and
0098 species: for if you bring on the same brush a PLANT's own pollen and pollen from another species,
0099 by those of the many other flowers on the same PLANT. How, then, comes it that such a vast number
0099 rossed the case is directly the reverse, for a PLANT's own pollen is always prepotent over foreign
0140 nability to predict whether or not an imported PLANT will endure our climate, and from the number
0146 hat they can be in any way advantageous to the PLANT: yet in the Umbelliferae these differences ar
0161 of a fifth stamen so often appears, that this PLANT must have an inherited tendency to produce it
0197 ce, no doubt, is of the highest service to the PLANT: but as we have nearly similar hooks on many t
0197 ve been subsequently taken advantage of by the PLANT undergoing further modification and becoming
0247 of error seems to me to be here introduced: a PLANT to be hybridised must be castrated, and, what
0247 sses are often injurious to the fertility of a PLANT cannot be doubted; for Gartner gives in his t
0249 tween two flowers, though probably on the same PLANT, would be thus effected. Moreover, whenever c
0249 n from a distinct flower, either from the same PLANT or from another plant of the same hybrid natu
0249 er, either from the same plant or from another PLANT of the same hybrid nature. And thus, the stra
0250 capense fertilised by C. revolutum produced a PLANT, which (he says) I never saw to occur in a ca
0255 he facts can here be given. When pollen from a PLANT of one family is placed on the stigma of a pl
0255 ant of one family is placed on the stigma of a PLANT of a distinct family, it exerts no more influ
0255 an excess of fertility, beyond that which the PLANT's own pollen will produce. So in hybrids them
0261 cially endowed quality. As the capacity of one PLANT to be grafted on another is so enti
0263 eaching the ovule, as would be the case with a PLANT having a pistil too long for the pollen tubes
0265 lar animal will breed under confinement or any PLANT seed freely under culture: nor can he tell, t
0352 habited by a species is continuous; and when a PLANT or animal inhabits two points so distant from
0358 if to plants. In botanical works, this or that PLANT is stated to be ill adapted for wide dissémin
0359 rds when planted they germinated: an asparagus PLANT with ripe berries floated for 23 days; when d
0386 y for six months, pulling up and counting each PLANT as it grew; the plants were of many kinds, an
0387 remembered Alph. de Candolle's remarks on this PLANT, I thought that its distribution must remain
0392 rted to an island by some other means: and the PLANT then becoming slightly modified, but still re
0392 ach distant oceanic islands: and an herbaceous PLANT, though it would have no chance of successful
0401 ans of transport, a seed, for instance, of one PLANT having been brought to one island, and that o
0401 een brought to one island, and that of another PLANT to another island. Hence when in former times
0401 to compete with different sets of organisms: a PLANT, for instance, would find the best fitted gro
0435 eased the creator to construct each animal and PLANT. The explanation is manifest on the theory of
0452 and this implies, that we find in an animal or PLANT no trace of an organ, which analogy would lea
0457 the same pattern in each individual animal and PLANT. On the principle of successive slight variat
0251 a hybrid from Calceolaria integrifolia and PLANTAGINEA, species most widely dissimilar in general
0071 ting grasses and carices) flourished in the PLANTATIONS, which could not be found on the heath: a
0071 insectivorous birds were very common in the PLANTATIONS, which were not to be seen on the heath: a
0071 een enclosed twenty five years previously and PLANTED with Scotch fir. The change in the native ve
0071 r. the change in the native vegetation of the PLANTED part of the heath was most remarkable, more
0072 d that these young trees had not been sown or PLANTED, I was so much surprised at their numbers th
0072 d not see a single Scotch fir, except the old PLANTED clumps. But on looking closely between the s
0098 if very closely allied forms or varieties are PLANTED near each other, it is hardly possible to ra
0359 they floated for 90 days and afterwards when PLANTED they germinated; an asparagus plant with rip
0086 through natural selection, than in the cotton PLANTER increasing and improving by selection the qu
0009 der confinement, with the exception of the PLANTIGRADES or bear family; whereas, carnivorous birds
0004 udy of domesticated animals and of cultivated PLANTS would offer the best chance of making out thi
0007 ariety or sub variety of our older cultivated PLANTS and animals, one of the first points which st
0007 when we reflect on the vast diversity of the PLANTS and animals which have been cultivated, and w
0008 able under cultivation. Our oldest cultivated PLANTS, such as wheat, still often yield new varieti
0008 o vitiated instincts: but how many cultivated PLANTS display the utmost vigour, and yet rarely or
0009 ns, hardly ever lay fertile eggs. Many exotic PLANTS have pollen utterly worthless, in the same ex
0009 the one hand, we see domesticated animals and PLANTS, though often weak and sickly, yet breeding o
0009 been thus affected; so will some animals and PLANTS withstand domestication or cultivation, and v
0009 a long list could easily be given of sporting PLANTS; by this term gardeners mean a single bud or
0010 effect, though apparently more in the case of PLANTS. Under this point of view, Mr. Buckman's rece
0010 of view, Mr. Buckman's recent experiments on PLANTS seem extremely valuable. When all or nearly a
0011 influence, as in the period of flowering with PLANTS when transported from one climate to another.
0012 able cases could be given amongst animals and PLANTS. From the facts collected by Heusinger, it ap
0012 tises published on some of our old cultivated PLANTS, as on the hyacinth, potato, even the dahlia,
0015 arieties or races of our domestic animals and PLANTS, and compare them with species closely allied
0016 any domestic races, either amongst animals or PLANTS, which have not been ranked by some competent
0017 man has chosen for domestication animals and PLANTS having an extraordinary inherent tendency to
0017 ion? I cannot doubt that if other animals and PLANTS, equal in number to our domesticated producti
0017 ost of our anciently domesticated animals and PLANTS, I do not think it is possible to come to any
0028 rious domestic animals and the cultivators of PLANTS, with whom I have ever conversed, or whose tr
0030 culinary, orchard, and flower garden races of PLANTS most useful to man at different seasons and f
0032 ly twenty or thirty years ago. When a race of PLANTS is once pretty well established, the seed rai
0032 ed, the seed raisers do not pick out the best PLANTS, but merely go over their seed beds, and pull
0032 eds, and pull up the rogues, as they call the PLANTS that deviate from the proper standard. With a
0033 llow his worst animals to breed. In regard to PLANTS, there is another means of observing the accu
0034 d, and this may be compared to the roguing of PLANTS by nurserymen. The principle of selection I f
0036 dearth, as of less value than their dogs. In PLANTS the same gradual process of improvement, thro
0037 rtsease, rose, pelargonium, dahlia, and other PLANTS, when compared with the older varieties or wi
0037 d. a large amount of change in our cultivated PLANTS, thus slowly and unconsciously accumulated, e
0037 re do not know,'the wild parent stocks of the PLANTS which have been longest cultivated in our flo
0038 nds of years to improve or modify most of our PLANTS up to their present standard of usefulness to
0038 e possess the aboriginal stocks of any useful PLANTS, but that the native plants have not been imp
```

Page **************************************(Key Word)***************************************

Page **(Key Word)**

0038 cks of any useful plants, but that the native	PLANTS have not been improved by continued selection
0038 perfection comparable with that given to the	PLANTS in countries anciently civilised. In regard t
0041 erymen, from raising large stocks of the same	PLANTS, are generally far more successful than amate
0041 , however, as gardeners picked out individual	PLANTS with slightly larger, earlier, or better frui
0043 e origin of our Domestic Races of animals and	PLANTS. I believe that the conditions of life, from
0043 rated, both in regard to animals and to those	PLANTS which are propagated by seec. In plants which
0043 those plants which are propagated by seed. In	PLANTS which are temporarily propagated by cuttings,
0043 equent sterility of hybrids: but the cases of	PLANTS not propagated by seed are of little importan
0045 the brackish waters of the Baltic, or dwarfed	PLANTS on Alpine summits, or the thicker fur of an a
0046 y instance Rubus, Rosa, and Hieracium amongst	PLANTS, several genera of insects, and several gener
0048 e of all kinds, has marked for me 182 British	PLANTS, which are generally considered as varieties,
0049 cowslip, or Primula veris and elaticr. These	PLANTS differ considerably in appearance; they have
0053 alph. de Candolle and others have shown that	PLANTS which have very wide ranges generally present
0054 ecome dominant over their compatriots. If the	PLANTS inhabiting a country and described in any Flo
0054 ses of obscurity. Fresh water and salt loving	PLANTS have generally very wide ranges and are much
0054 he genera to which the species belong. Again,	PLANTS low in the scale of organisation are generall
0055 are generally much more widely diffused than	PLANTS higher in the scale; and here again there is
0055 e of the genera. The cause of lowly organised	PLANTS ranging widely will be discussed in our chapt
0055 ruth of this anticipation I have arranged the	PLANTS of twelve countries, and the coleopterous ins
0057 arieties. Now Fries has remarked in regard to	PLANTS, and Westwood in regard to insects, that in l
0058 for me in the well sifted London Catalogue of	PLANTS (4th edition) 63 plants which are therein ran
0058 d london Catalogue of plants (4th edition) 63	PLANTS which are therein ranked as species, but whic
0060 se. Rapid increase of naturalised animals and	PLANTS. Nature of the checks to increase. Competitio
0060 viduals. Complex relations of all animals and	PLANTS throughout nature. Struggle for life most se
0060 wo or three hundred doubtful forms of British	PLANTS are entitled to hold, if the existence of any
0062 e exposed to severe competition. In regard to	PLANTS, no one has treated this subject with more sp
0063 , may be more truly said to struggle with the	PLANTS of the same and other kinds which already clo
0063 be said to struggle with other fruit bearing	PLANTS, in order to tempt birds to devour and thus d
0063 seminate its seeds rather than those of other	PLANTS. In these several senses, which pass into eac
0064 then in twenty years there would be a million	PLANTS. The elephant is reckoned to be the slowest b
0064 uld have been quite incredible. So it is with	PLANTS: cases could be given of introduced plants wh
0065 th plants: cases could be given of introduced	PLANTS which have become common throughout whole isl
0065 period of less than ten years. Several of the	PLANTS now most numerous over the wide plains of La
0065 surface almost to the exclusion of all other	PLANTS, have been introduced from Europe; and there
0065 ve been introduced from Europe; and there are	PLANTS which have now range in India, as I hear from Dr.
0065 pposes that the fertility of these animals or	PLANTS has been suddenly and temporarily increased i
0065 ir. Hence we may confidently assert, that all	PLANTS and animals are tending to increase at a geom
0067 st, but this is not invariably the case. With	PLANTS there is a vast destruction of seeds, but, fr
0067 in ground already thickly stocked with other	PLANTS. Seedlings, also, are destroyed in vast numbe
0067 nd where there could be no choking from other	PLANTS, I marked all the seedlings of our native wee
0068 quadrupeds, be let to grow, the more vigorous	PLANTS gradually kill the less vigorous, though full
0068 y kill the less vigorous, though fully grown,	PLANTS: thus out of twenty species growing on a litt
0069 e may clearly see in the prodigious number of	PLANTS in our gardens which can perfectly well endur
0069 isec: for they cannot compete with our native	PLANTS, nor resist destruction by our native animals
0070 is to get seed from a few wheat or other such	PLANTS in a garden; I have in this case lost every s
0070 ar facts in nature, such as that of very rare	PLANTS being sometimes extremely abundant in the few
0070 where they do occur; and that of some social	PLANTS being social, that is, abounding in individua
0071 ot only the proportional numbers of the heath	PLANTS were wholly changed, but twelve species of pl
0071 ts were wholly changed, but twelve species of	PLANTS (not counting grasses and carices) flourished
0073 tempted to give one more instance showing how	PLANTS and animals, most remote in the scale of natu
0073 ever can set a seed. Many of our orchidaceous	PLANTS absolutely require the visits of moths to rem
0074 s in different districts. When we look at the	PLANTS and bushes clothing an entangled bank, we are
0075 or their seeds and seedlings, or on the other	PLANTS which first clothed the ground and thus check
0075 to the action and reaction of the innumerable	PLANTS and animals which have determined, in the cou
0076 ther the varieties of any one of our domestic	PLANTS or animals have so exactly the same strength,
0077 e land being already thickly clothed by other	PLANTS: so that the seeds may be widely distributed
0077 of nutriment laid up within the seeds of many	PLANTS seems at first sight to have no sort of relat
0077 st sight to have no sort of relation to other	PLANTS. But from the strong growth of young plants p
0077 r plants. But from the strong growth of young	PLANTS produced from such seeds (as peas and beans),
0077 young seedling, whilst struggling with other	PLANTS growing vigorously all around. Look at a plan
0078 but we have reason to believe that only a few	PLANTS or animals range so far, that they are destro
0085 ery lamb with the faintest trace of black. In	PLANTS the down of the fruit and the colour of the f
0086 ny varieties of our culinary and agricultural	PLANTS! in the caterpillar and cocoon stages of the
0091 let us now take a more complex case. Certain	PLANTS excrete a sweet juice, apparently for the sak
0093 no one step in the separation of the sexes of	PLANTS, presently to be alluded to. Some holly trees
0094 one in another flower or on another plant. In	PLANTS under culture and placed under new conditions
0096 short digression. In the case of animals and	PLANTS with separate sexes, it is of course obvious
0096 o not habitually pair, and a vast majority of	PLANTS are hermaphrodites. What reason, it may be as
0096 sal belief of breeders, that with animals and	PLANTS a cross between different varieties, or betwe
0099 pollen of that flower is ready, so that these	PLANTS have in fact separated sexes, and must habitu
0099 of the cabbage, radish, onion, and some other	PLANTS, be allowed to seed near each other, a large
0099 nce, I raised 233 seedling cabbages from some	PLANTS of different varieties growing near each othe
0100 e their sexes more often separated than other	PLANTS, I find to be the case in this country; and a
0100 offers so strong a contrast with terrestrial	PLANTS, on the view of an occasional cross being ind
0100 ion of insects and of the wind in the case of	PLANTS, by which an occasional cross could be effect
0101 anomaly that, in the case of both animals and	PLANTS, species of the same family and even of the s
0103 ways prefer getting seed from a large body of	PLANTS of the same variety, as the chance of intercr
0104 sses take place with all animals and with all	PLANTS. Even if these take place only at long interv
0110 the Cape of Good Hope, where more species of	PLANTS are crowded together than in any other quarte
0110 any other quarter of the world; some foreign	PLANTS have become naturalised, without causing, as
0113 selection can do nothing. So it will be with	PLANTS. It has been experimentally proved, that if a
0113 stinct genera of grasses, a greater number of	PLANTS and a greater weight of dry herbage can thus
0113 om each other, a greater number of individual	PLANTS of this species of grass, including its modif
0114 same conditions, supported twenty species of	PLANTS, and these belonged to eighteen genera and to
0114 d to eight orders, which shows how much these	PLANTS differed from each other. So it is with the p
0114 s differed from each other. So it is with the	PLANTS and insects on small and uniform islets; and

Page **(Key Word)**

Page ***************************************(Key Word)***

```
0114  hat they can raise most food by a rotation of  PLANTS  belonging to the most different orders: natur
0114  imultaneous rotation. Most of the animals and  PLANTS  which live close round any small piece of gro
0115  me principle is seen in the naturalisation of  PLANTS  through man's agency in foreign lands. It mig
0115  n lands. It might have been expected that the  PLANTS  which have succeeded in becoming naturalised
0115  , perhaps have been expected that naturalised  PLANTS  would have belonged to a few groups more espe
0115  f the Northern United States, 260 naturalised  PLANTS  are enumerated, and these belong to 162 gener
0115  62 genera. We thus see that these naturalised  PLANTS  are of a highly diversified nature. They diff
0115  hese States. By considering the nature of the  PLANTS  or animals which have struggled successfully
0116  the more widely and perfectly the animals and  PLANTS  are diversified for different habits of life,
0128  ok from familiarity, that all animals and all  PLANTS  throughout all time and space should be relat
0132  to form a new being. In the case of sporting   PLANTS, the bud, which in its earliest condition doe
0132  f animals, but perhaps rather more in that of  PLANTS. We may, at least, safely conclude that such
0132  their colours. Moquin Tandon gives a list of  PLANTS  which when growing near the sea shore have th
0133  y collector knows, are often brassy or lurid.  PLANTS  which live exclusively on the sea side are ve
0139  ed. Acclimatisation. Habit is hereditary with  PLANTS, as in the period of flowering, in the amount
0139  mate, or conversely. So again, many succulent  PLANTS  cannot endure a damp climate. But the degree
0140  ll endure our climate, and from the number of  PLANTS  and animals brought from warmer countries whi
0140  se, we have evidence, in the case of some few  PLANTS, of their becoming, to a certain extent, natu
0140  de by Mr. H. C. Watson on European species of  PLANTS  brought from the Azores to England. In regard
0142  ies. In treatises on many kinds of cultivated  PLANTS, certain varieties are said to withstand cert
0143  nd to cohere: this is often seen in monstrous  PLANTS: and nothing is more common than the union of
0144  flowers in some Compositous and Umbelliferous  PLANTS. Every one knows the difference in the ray an
0145  parts of the flower. But, in some Compositous  PLANTS, the seeds also differ in shape and sculpture
0145  s highly advantageous in the fertilisation of  PLANTS  of these two orders, is so far fetched, as it
0146  n fruits which opened: so that the individual  PLANTS  producing seeds which were a little better fi
0151  s class. I cannot make out that it applies to  PLANTS, and this would seriously have shaken my beli
0151  n its truth, had not the great variability in  PLANTS  made it particularly difficult to compare the
0154  is meant. If some species in a large genus of  PLANTS  had blue flowers and some had red, the colour
0159  called, of the Swedish turnip and Ruta baga,   PLANTS  which several botanists rank as varieties pro
0159  milarity in the enlarged stems of these three  PLANTS, not to the vera causa of community of descen
0183  many British insects which now feed on exotic  PLANTS, or exclusively on artificial substances. Of
0185  e formed for walking over swamps and floating  PLANTS, yet the water hen is nearly as aquatic as th
0193  . other cases could be given: for instance in  PLANTS, the very curious contrivance of a mass of po
0193  lmost as remote as possible amongst flowering  PLANTS. In all these cases of two very distinct spec
0198  as is the liability to be poisoned by certain  PLANTS: so that colour would be thus subjected to th
0201  idly yielding before the advancing legions of  PLANTS  and animals introduced from Europe. Natural s
0203  h the flowers of the orchis and of many other  PLANTS  are fertilised through insect agency, can we
0242  ting, as it proves that with animals, as with  PLANTS, any amount of modification in structure can
0246  seen in the state of the male element in both  PLANTS  and animals: though the organs themselves are
0247  len being brought to it by insects from other  PLANTS. Nearly all the plants experimentised on by G
0247  by insects from other plants. Nearly all the  PLANTS  experimentised on by Gartner were potted, and
0247  gives in his table about a score of cases of  PLANTS  which he castrated, and artificially fertilis
0247  lty in the manipulation) half of these twenty  PLANTS  had their fertility in some degree impaired.
0250  gular fact, namely, that there are individual  PLANTS, as with certain species of Lobelia, and with
0250  species, than by their own pollen. For these  PLANTS  have been found to yield seed to the pollen o
0250  distinct species. So that certain individual  PLANTS  and all the individuals of certain species ca
0251  belia, Passiflora and Verbascum. Although the  PLANTS  in these experiments appeared perfectly healt
0251  ir mutual self action, we must infer that the  PLANTS  were in an unnatural state. Nevertheless thes
0252  periments have been carefully tried than with  PLANTS. If our systematic arrangements can be truste
0252  istinct from each other, as are the genera of  PLANTS, then we may infer that animals more widely s
0252  an be more easily crossed than in the case of  PLANTS: but the hybrids themselves are, I think, mor
0254  the ascertained facts on the intercrossing of  PLANTS  and animals, it may be concluded that some de
0255  tner's admirable work on the hybridisation of  PLANTS. I have taken much pains to ascertain how far
0257  nt two species crossing. It can be shown that  PLANTS  most widely different in habit and general ap
0257  yledons, can be crossed. Annual and perennial  PLANTS, deciduous and evergreen trees, plants inhabi
0257  ennial plants, deciduous and evergreen trees,  PLANTS  inhabiting different stations and fitted for
0261  differences in the laws of growth of the two  PLANTS. We can sometimes see the reason why one tree
0261  whatever. Great diversity in the size of two  PLANTS, one being woody and the other herbaceous, on
0264  have collected, showing that when animals and  PLANTS  are removed from their natural conditions, th
0265  tic affinity, for whole groups of animals and  PLANTS  are rendered impotent by the same unnatural c
0267  hange in the habits of life. Again, both with  PLANTS  and animals, there is abundant evidence, that
0269  rtant consideration, new races of animals and  PLANTS  are produced under domestication by man's met
0270  each other in his garden: and although these  PLANTS  have separated sexes, they never naturally cr
0270  is case could not have been injurious, as the  PLANTS  thus raised were themselves perfectly fertile
0270  and it is important to notice that the hybrid  PLANTS. On the other hand, the resemblance in mongre
0274  able to point out, between hybrid and mongrel  PLANTS. With animals one variety certainly often has
0274  ; and so I believe it to be with varieties of  PLANTS  produced from a reciprocal cross, generally r
0274  prepotent power over another variety. Hybrid  PLANTS  that can propagate rapidly and are not highly
0298  nsideration is worth notice: with animals and  PLANTS  it is those which have the widest range, that
0298  have a wide range: and we have seen that with  PLANTS  which are dominant, that is, which are common
0325  e have distinct evidence on this head, in the  PLANTS  of Great Britain were set free in New Zealand
0337  upied, we may believe, if all the animals and  PLANTS  and animals. Under this point of view, the pr
0337  to seize on places now occupied by our native  PLANTS  have become naturalised in America and Austra
0353  , so that a multitude of European animals and  PLANTS  are identically the same at these distant poi
0353  ica and Australia: and some of the aboriginal  PLANTS, from their varied means of dispersal, have m
0353  s have not been able to migrate, whereas some  PLANTS  and for many animals during their migration.
0357  , which may have served as halting places for  PLANTS. In botanical works, this or that plant is st
0358  distribution. I shall here confine myself to  PLANTS  or branches, and that these might be dried on
0359  it occurred to me that floods might wash down  PLANTS  with ripe fruit, and to place them on sea wat
0359  ence I was led to dry stems in branches of 94  PLANTS, 18 floated for above 28 days, and some of th
0359  ds germinated. Altogether out of the 94 dried  PLANTS  with ripe fruit (but not all the same species
0359  d after an immersion of 28 days: and as 18/94  PLANTS  of any country might be floated by sea curren
0359  cts, we may conclude that the seeds of 14/100  PLANTS  belonging to one country might be floated acr
0360  r diem): or this average, the seeds of 14/100  PLANTS  which live near the sea: and this would have
0360  t and exposed to the air like really floating  PLANTS. He tried 98 seeds, mostly different from min
0360  ose many large fruits and likewise seeds from  PLANTS  which live near the sea: and this would have
```

Page **********************************(Key Word)**

Page **************************************(Key Word)**************************************

0360	the other hand he did not previously dry the	PLANTS or branches with the fruit: and this, as we h
0360	pable of germination. But I do not doubt that	PLANTS exposed to the waves would float for a less t
0360	afer to assume that the seeds of about 10/100	PLANTS of a flora, after having been dried, could be
0360	inc longer than the small, is interesting: as	PLANTS with large seeds or fruit could hardly be tra
0360	nsi and Alph. de Candolle has shown that such	PLANTS generally have restricted ranges. But seeds m
0361	oak about 50 years old, three dicotyledonous	PLANTS germinated: I am certain of the accuracy of t
0362	ish, I find, eat seeds of many land and water	PLANTS: fish are frequently devoured by birds, and t
0363	ores, from the large number of the species of	PLANTS common to Europe, in comparison with the plan
0363	ants common to Europe, in comparison with the	PLANTS of other oceanic islands nearer to the mainla
0363	ay have brought thither the seeds of northern	PLANTS. Considering that the several above means of
0364	it would I think be a marvellous fact if many	PLANTS had not thus become widely transported. These
0365	ring the Glacial period. The identity of many	PLANTS and animals, on mountain summits, separated f
0365	a remarkable fact to see so many of the same	PLANTS living on the snowy regions of the Alps or Py
0365	rope: but it is far more remarkable, that the	PLANTS on the White Mountains, in the United States
0367	ed states would likewise be covered by arctic	PLANTS and animals, and these would be nearly the sa
0367	. thus we can understand the identity of many	PLANTS at points so immensely remote as on the mount
0367	thus also unerstand the fact that the Alpine	PLANTS of each mountain range are more especially re
0367	lly have been due south and north. The Alpine	PLANTS, for example, of Scotland, as remarked by Mr.
0368	by Ramond, are more especially allied to the	PLANTS of northern Scandinavia: those of the United
0370	he case: for if we compare the present Alpine	PLANTS and animals of the several great European mou
0370	r pliocene period, a large number of the same	PLANTS and animals inhabited the almost continuous c
0370	t continuous circumpolar land: and that these	PLANTS and animals, both in the Old and New Worlds,
0371	ncerned, took place long ages ago. And as the	PLANTS and animals migrated southward, they will hav
0372	s (though Asa Gray has lately shown that more	PLANTS are identical than was formerly supposed), bu
0373	r glacial action in New Zealand: and the same	PLANTS, found on widely separated mountains in this
0374	that between forty and fifty of the flowering	PLANTS of Tierra del Fuego, forming no inconsiderabl
0375	ylon, and on the volcanic cones of Java, many	PLANTS occur, either identically the same or represe
0375	each other, and at the same time representing	PLANTS of Europe, not found in the intervening hot l
0375	n australian forms are clearly represented by	PLANTS growing on the summits of the mountains of Bo
0375	and striking facts are given in regard to the	PLANTS of that large island. Hence we see that throu
0375	. hence we see that throughout the world, the	PLANTS growing on the more lofty mountains, and on t
0376	rkable manner. This brief abstract applies to	PLANTS alone: some strictly analogous facts could be
0377	member over what vast spaces some naturalised	PLANTS and animals have spread within a few centurie
0377	as the cold came slowly on, all the tropical	PLANTS and other productions will have retreated fro
0377	latter we are not now concerned. The tropical	PLANTS probably suffered much extinction: how much n
0377	rate Australia. As we know that many tropical	PLANTS and animals can withstand a considerable amou
0377	d less. And it is certain that many temperate	PLANTS, if protected from the inroads of competitors
0378	tropics which is so destructive to perennial	PLANTS from a temperate climate. On the other hand,
0378	d to me by Dr. Hooker, that all the flowering	PLANTS, about forty six in number, common to Tierra
0378	thus, as I believe, a considerable number of	PLANTS, a few terrestrial animals, and some marine p
0379	regard to Australia, that many more identical	PLANTS and allied forms have apparently migrated fro
0385	ith duck weed, I have twice seen these little	PLANTS adhering to its back: and it has happened to
0386	vouring gale no one can tell. With respect to	PLANTS, it has long been known what enormous ranges
0386	. de Candolle, in large groups of terrestrial	PLANTS, which have only a very few aquatic members:
0386	ng up and counting each plant as it grew: the	PLANTS were of many kinds, and were altogether 537 i
0387	ds did not transport the seeds of fresh water	PLANTS to vast distances, and if consequently the ra
0387	ances, and if consequently the range of these	PLANTS was not very great. The same agency may have
0388	ons. But the wide distribution of fresh water	PLANTS and of the lower animals, whether retaining t
0389	ntal areas: Alph. de Candolle admits this for	PLANTS, and Wollaston for insects. If we look to the
0389	miles of latitude, and compare its flowering	PLANTS, only 750 in number, with those on an equal a
0389	even the uniform county of Cambridge has 847	PLANTS, and the little island of Anglesea 764, but a
0389	sea 764, but a few ferns and a few introduced	PLANTS are included in these numbers; and the compar
0390	inally possessed under half a dozen flowering	PLANTS: yet many have become naturalised on it, as t
0390	ere is reason to believe that the naturalised	PLANTS and animals have nearly or quite exterminated
0390	that a sufficient number of the best adapted	PLANTS and animals have not been created on oceanic
0391	less birds, take the place of mammals. In the	PLANTS of the Galapagos Islands, Dr. Hooker has show
0392	not tenanted by mammals, some of the endemic	PLANTS have beautifully hooked seeds: yet few relati
0392	island and having to compete with herbaceous	PLANTS alone, might readily gain an advantage by gro
0392	g taller and taller and overtopping the other	PLANTS. If so, natural selection would often tend to
0392	ften tend to add to the stature of herbaceous	PLANTS when growing on an island, to whatever order
0396	the singular proportions of certain orders of	PLANTS, herbaceous forms having been developed into
0398	th the other animals, and with nearly all the	PLANTS, as shown by Dr. Hooker in his admirable memo
0399	, and most of them can be explained. Thus the	PLANTS of Kerguelen Land, though standing nearer to
0399	nomaly disappears. New Zealand in its endemic	PLANTS is much more closely related to Australia, th
0399	licable: but this affinity is confined to the	PLANTS, and will, I do not doubt, be some day explai
0401	ed ground more perfectly occupied by distinct	PLANTS in one island than in another, and it would b
0403	a vast number of distinct mammals, birds, and	PLANTS. The principle which determines the general c
0403	pting in so far as the same forms, chiefly of	PLANTS, have spread widely throughout the world duri
0403	alpine humming birds, Alpine rodents, Alpine	PLANTS, etc., all of strictly American forms, and it
0406	y discussed by Alph. de Candolle in regard to	PLANTS, namely, that the lower any group of organism
0414	adaptive for an essential character. So with	PLANTS, how remarkable it is that the organs of vege
0417	esemblance, too slight to be defined. Certain	PLANTS, belonging to the Malpighiaceae, bear perfect
0418	true. The same fact holds good with flowering	PLANTS, of which the two main divisions have been fo
0419	trary value. Instances could be given amongst	PLANTS and insects, of a group of forms, first ranke
0430	ature of the affinities of distinct orders of	PLANTS. On the principle of the multiplication and g
0435	of crustaceans. So it is with the flowers of	PLANTS. Nothing can be more hopeless than to attempt
0436	sed leaves, arranged in a spire. In monstrous	PLANTS, we often get direct evidence of the possibil
0437	bearing external appendages: and in flowering	PLANTS, we see a series of successive spiral whorls
0437	ents: and the unknown progenitor of flowering	PLANTS, many spiral whorls of leaves. We have former
0439	y though rarely see something of this kind in	PLANTS: thus the embryonic leaves of the ulex or fur
0451	become developed and give milk. In individual	PLANTS of the same species the petals sometimes occu
0451	, and sometimes in a well developed state. In	PLANTS with separated sexes, the male flowers often
0451	nd kolreuter found that by crossing such male	PLANTS with an hermaphrodite species, the rudiment o
0451	in perfectly efficient for the other. Thus in	PLANTS, the office of the pistil is to allow the pol
0467	ny desired manner. He thus adapts animals and	PLANTS for his own benefit or pleasure. He may do th
0477	the Glacial period, the identity of some few	PLANTS, and the close alliance of many others, on th

Page **************************************(Key Word)**************************************

Page **(Key Word)***

```
0478  e been derived. We see this in nearly all the  PLANTS  and animals of the Galapagos archipelago, of
0478  ng related in the most striking manner to the  PLANTS  and animals of the neighbouring American main
0483  the infinitely numerous kinds of animals and  PLANTS  created as eggs or seed, or as full grown? an
0484  om at most only four or five progenitors, and  PLANTS  from an equal or lesser number. Analogy would
0484  r, namely, to the belief that all animals and  PLANTS  have descended from some one prototype. But a
0484  that the same poison often similarly affects   PLANTS  and animals; or that the poison secreted by t
0489  template an entangled bank, clothed with many  PLANTS  of many kinds, with birds singing on the bush
0012  . the whole organisation seems to have become  PLASTIC, and tends to depart in some small degree fr
0031  f an animal's organisation as something quite  PLASTIC, which they can model almost as they please.
0080  the whole organisation becomes in some degree  PLASTIC. Let it be borne in mind how infinitely comp
0132  e parents, I chiefly attribute the varying or  PLASTIC  condition of the offspring. The male and fem
0065  s now most numerous over the wide plains of La  PLATA, clothing square leagues of surface almost to
0184  e insects on the wing; and on the plains of La  PLATA, where not a tree grows, there is a woodpecke
0318  species, than I have done. When I found in La  PLATA  the tooth of a horse embedded with the remain
0323  and Toxodon had been brought to Europe from La  PLATA, without any information in regard to their g
0339  of the armadillo, found in several parts of La  PLATA; and Professor Owen has shown in the most str
0349  rican ostrich), and northward the plains of La  PLATA  by another species of the same genus; and not
0349  the same latitude. On these same plains of La  PLATA, we see the agouti and bizcacha, animals havi
0380  ny european productions cover the ground in La  PLATA, and in a lesser degree in Australia, and hav
0380  during the last two or three centuries from La  PLATA, and during the last thirty or forty years fr
0229  s, in other parts, large portions of a rhombic  PLATE  had been left between the opposed basins; but
0229  xamined the cell, and I found that the rhombic  PLATE  had been completed, and had become perfectly
0229  rom the extreme thinness of the little rhombic  PLATE, that they could have effected this by gnawin
0230  ell at the same time, but only the one rhombic  PLATE  which stands on the extreme growing margin, o
0229  and these flat bottoms, formed by thin little  PLATES  of the vermilion wax having been left ungnawe
0229  order to have succeeded in thus leaving flat  PLATES  between the basins, by stopping work along th
0230  nds on the extreme growing margin, or the two  PLATES, as the case may be; and they never complete
0230  never complete the upper edges of the rhombic  PLATES, until the hexagonal walls are commenced. Som
0231  n position to the planes of the rhombic basal  PLATES  of future cells. But the rough wall of wax ha
0231  e four hundredth of an inch in thickness; the  PLATES  of the pyramidal basis being about twice as t
0235  the hexagonal prisms and of the basal rhombic  PLATES. The motive power of the process of natural s
0042  part to selection not having been brought into  PLAY: in cats, from the difficulty in pairing them;
0071  cts of close interbreeding, probably come into  PLAY; some one check or some few being generally th
0074  different seasons or years, probably come into  PLAY; some one check or some few being generally th
0106  most new varieties and species, and will thus  PLAY an important part in the changing history of t
0134  indirectly, as already remarked, they seem to  PLAY an important part in affecting the reproductiv
0141  n, brought, under peculiar circumstances, into  PLAY. How much of the acclimatisation of species to
0144  dog, though here probably homology comes into  PLAY? With respect to this latter case of correlati
0146  ntageous, natural selection may have come into  PLAY. But in regard to the differences both in the
0150  e. thus rudimentary parts are left to the free  PLAY of the various laws of growth, to the effects
0152  ection either has not or cannot come into full  PLAY, and thus the organisation is left in a fluctu
0198  e chest; and again correlation would come into  PLAY. Animals kept by savages in different countrie
0208  es it, of judgment or reason, often comes into  PLAY, even in animals very low in the scale of natu
0211  rich flock it had discovered; it then began to  PLAY with its antennae on the abdomen first of one
0242  le, without exercise or habit having come into  PLAY. For no amount of exercise, or habit, or volit
0243  abit or use and disuse have probably come into  PLAY. I do not pretend that the facts given in this
0350  hus the high importance of barriers comes into  PLAY by checking migration; as does time for the sl
0351  es can do anything. These principles come into  PLAY only by bringing organisms into new relations
0387  very great. The same agency may have come into  PLAY with the eggs of some of the smaller fresh wat
0427  ibution may sometimes be brought usefully into  PLAY in classing large and widely distributed gener
0455  as is possible, will probably often come into  PLAY; and this will tend to cause the entire oblite
0466  e that variability, when it has once come into  PLAY, does not wholly cease; for new varieties are
0468  le fact if natural selection had not come into  PLAY. It has often been asserted, but the assertion
0474  hive bee. Habit no doubt sometimes comes into  PLAY in modifying instincts; but it certainly is no
0480  ature, when it has come to maturity and has to  PLAY its full part in the struggle for existence,
0038  all important part which selection by man has  PLAYED, it becomes at once obvious, how it is that o
0043  ossing of species, aboriginally distinct, has  PLAYED an important part in the origin of our domest
0121  nciples, namely that of extinction, will have  PLAYED an important part. As in each fully stocked c
0143  habit, use, and disuse, have, in some cases,  PLAYED a considerable part in the modification of th
0174  y broken condition of areas now continuous has  PLAYED an important part in the formation of new spe
0199  us gained. Correlation of growth has no doubt  PLAYED a most important part, and a useful modificat
0209  ars old with wonderfully little practice, had  PLAYED a tune with no practice at all, he might trul
0213  ction of so called accidental variations have  PLAYED in modifying the mental qualities of our dome
0342  uration of specific forms; that migration has  PLAYED an important part in the first appearance of
0343  look how important a part migration must have  PLAYED, when the formations of any one great region
0387  other and unknown agencies probably have also  PLAYED a part. I have stated that fresh water fish e
0403  e same continent, pre occupation has probably  PLAYED an important part in checking the commingling
0431  n, as we have seen in the fourth chapter, has  PLAYED an important part in defining and widening th
0473  d species correlation of growth seems to have  PLAYED a most important part, so that when one part
0475  ies and of whole groups of species, which has  PLAYED so conspicuous a part in the history of the o
0209  ot to be distinguished. If Mozart, instead of  PLAYING the pianoforte at three years old with wonde
0042  r races. In this respect enclosure of the land  PLAYS a part. Wandering savages or the inhabitants
0068  a young elephant protected by its dam. Climate  PLAYS an important part in determining the average
0103  ind preferring to pair together. Intercrossing  PLAYS a very important part in nature in keeping th
0230  etails. We see how important a part excavation  PLAYS in the construction of the cells; but the hi
0418  is purpose than that of the adult, which alone  PLAYS its full part in the economy of nature. Yet i
0031  plastic, which they can model almost as they  PLEASE. If I had space I could quote numerous passag
0224  in a dark hive. Grant whatever instincts you  PLEASE, and it seems at first quite inconceivable ho
0045  be remembered that systematists are far from  PLEASED at finding variability in important characte
0185  creation will say, that in these cases it has  PLEASED the Creator to cause a being of one type to
0435  we can only say that so it is; that it has so  PLEASED the creator to construct each animal and pla
0031  y summon into life whatever form and mould he  PLEASES. Lord Somerville, speaking of what breeders
0041  tion. But as variations manifestly useful or  PLEASING to man appear only occasionally, the chance
0042  od and feathers, and more especially from no  PLEASURE having been felt in the display of distinct
0269  ious power of selection, for his own use and  PLEASURE; he neither wishes to select, nor could sele
0467  ts animals and plants for his own benefit or  PLEASURE. He may do this methodically, or he may do i
```

Page **(Key Word)***

Page **********************************(Key Word)***********************************

0324 d all those that lived in Europe during the PLEISTOCENE period (an enormously remote period as mea
0324 be able to say whether the existing or the PLEISTOCENE inhabitants of Europe resembled most close
0324 formations, namely, the upper pliocene, the PLEISTOCENE and strictly modern beds, of Europe, North
0336 odigious vicissitudes of climate during the PLEISTOCENE period, which includes the whole glacial p
0070 or its preservation. Thus we can easily raise PLENTY of corn and rape seed, etc., in our fields, b
0098 one flower on the stigma of another, I raised PLENTY of seedlings: and whilst another species of L
0219 t up thirty of them without a slave, but with PLENTY of the food which they like best, and with th
0252 no pollen, for he will find on their stigmas PLENTY of pollen brought from other flowers. In regu
0454 igir of rudimentary organs is simple. We have PLENTY of cases of rudimentary organs in our domesti
0028 ty. In the time of the Romans, as we hear from PLINY, immense prices were given for pigeons: nay,
0034 formerly did so, as is attested by passages in PLINY. The savages in South Africa match their drau
0037 h cultivated in classical times, appears, from PLINY's description, to have been a fruit of very i
0324 modern marine formations, namely, the upper PLIOCENE, the pleistocene and strictly modern beds, o
0370 good reason to believe that during the newer PLIOCENE period, before the Glacial epoch, and whilst
0370 e climate of latitude 60 degrees, during the PLIOCENE period lived further north under the polar c
0370 r and still warmer period, such as the older PLIOCENE period, a large number of the same plants an
0371 . during the slowly decreasing warmth of the PLIOCENE period, as soon as the species in common, wh
0372 igration of a marine fauna, which during the PLIOCENE or even a somewhat earlier period, was nearl
0068 thus out of twenty species growing on a little PLOT of turf (three feet by four) nine species peri
0113 . it has been experimentally proved, that if a PLOT of ground be sown with one species of grass, a
0113 sown with one species of grass, and a similar PLOT be sown with several distinct genera of grasse
0262 apricot and peach on certain varieties of the PLUM. As Gartner found that there was sometimes an
0022 are variable. The period at which the perfect PLUMAGE is acquired varies, as does the state of the
0089 ; and successive males display their gorgeous PLUMAGE and perform strange antics before the female
0089 that some well known laws with respect to the PLUMAGE of male and female birds, in comparison with
0089 male and female birds, in comparison with the PLUMAGE of the young, can be explained on the view o
0089 of the young, can be explained on the view of PLUMAGE having been chiefly modified by sexual selec
0144 irst hatched, with the future colour of their PLUMAGE; or, again, the relation between the hair an
0161 has been a tendency in each generation in the PLUMAGE to assume this colour. This view is hypothet
0237 roductive system is active, as in the nuptial PLUMAGE of many birds, and in the hooked jaws of the
0439 resemble each other in their first and second PLUMAGE; as we see in the spotted feathers in the th
0061 beetle which dives through the water; in the PLUMED seed which is wafted by the gentlest breeze;
0077 r on the tiger's body. But in the beautifully PLUMED seed of the dandelion, and in the flattened a
0077 ements of air and water. Yet the advantage of PLUMED seeds no doubt stands in the closest relation
0085 a curculio, than those with down; that purple PLUMS suffer far more from a certain disease than y
0085 er far more from a certain disease than yellow PLUMS: whereas another disease attacks yellow flesh
0419 ledons, and on the mode of development of the PLUMULE and radicle. In our discussion on embryology
0190 bladders, this latter organ having a ductus PNEUMATICUS for its supply, and being divided by highl
0250 ne as an example, namely, that every ovule in a POD of Crinum capense fertilised by C. revolutum p
0251 after a few days perished entirely, whereas the POD impregnated by the pollen of the hybrid made v
0033 in the flower garden; the diversity of leaves, PODS, or tubers, or whatever part is valued, in the
0086 ing and improving by selection the down in the PODS on his cotton trees. Natural Selection may mod
0002 is. For I am well aware that scarcely a single POINT is discussed in this volume on which facts ca
0010 arently more in the case of plants. Under this POINT of view, Mr. Buckman's recent experiments on
0016 ended from one or several parent species. This POINT, if it could be cleared up, would be interest
0033 the varieties which differ largely in some one POINT do not differ at all in other points; this is
0041 ent selection. But probably the most important POINT of all, is, that the animal or plant should b
0045 viewed under a physiological or classificatory POINT of view, sometimes vary in the individuals of
0046 onfessed) which does not vary; and, under this POINT of view, no instance of an important part var
0046 arying will ever be found: but under any other POINT of view many instances assuredly can be given
0046 instances assuredly can be given. There is one POINT connected with individual differences, which
0057 ecies? Undoubtedly there is one most important POINT of difference between varieties and species;
0058 ifferences between species. There is one other POINT which seems to me worth notice. Varieties gen
0072 d down by the cattle. In one square yard, at a POINT some hundred yards distant from one of the ol
0101 ent a case of very great difficulty under this POINT of view: but I have been enabled, by a fortun
0124 ing in sub branches downwards towards a single POINT; this point representing a single species, th
0124 ranches downwards towards a single point; this POINT representing a single species, the supposed s
0213 yself seen a striking instance) will sometimes POINT and even back other dogs the very first time
0213 a herd of deer, and driving them to a distant POINT, we should assuredly call these actions insti
0214 ny one would have thought of training a dog to POINT, had not some one dog naturally shown a tende
0225 ated into an irregular mass. But the important POINT to notice, is that these cells are always mad
0227 n make, apparently by turning round on a fixed POINT. We must suppose the Melipona to arrange her
0233 nceivable that an insect might, by fixing on a POINT at which to commence a cell, and then moving
0233 a cell, and then moving outside, first to one POINT, and then to five other points, at the proper
0233 the proper relative distances from the central POINT and from each other, strike the planes of int
0241 their jaws. I speak confidently on this latter POINT, as Mr. Lubbock made drawings for me with the
0257 facts could be given. No one has been able to POINT out what kind, or what amount, of difference
0274 mportant differences, which Gartner is able to POINT out, between hybrid and mongrel plants. On th
0306 st to land for five minutes on some one barren POINT in Australia, and then to discuss the number
0317 pear to begin at its lower end, not in a sharp POINT, but abruptly: it then gradually thickens upw
0337 d by our native plants and animals. Under this POINT of view, the productions of Great Britain may
0346 s of the United States to its extreme southern POINT, we meet with the most diversified conditions
0347 ir conditions, yet it would not be possible to POINT out three faunas and floras more utterly diss
0352 ies could possibly have migrated from some one POINT to the several distant and isolated points, w
0353 ow the same species could have passed from one POINT to the other. But the geographical and climat
0354 hall be enabled at the same time to consider a POINT equally important for us, namely, whether the
0374 n enormous time, as measured by years, at each POINT. The cold may have come on, or have ceased, e
0374 y have come on, or have ceased, earlier at one POINT of the globe than at another; but seeing that
0385 a to an oceanic island or to any other distant POINT. Sir Charles Lyell also informs me that a Dyt
0399 cked from a nearly intermediate though distant POINT, namely from the antarctic islands, when they
0481 viewed, during a long course of years, from a POINT of view directly opposite to mine. It is so e
0484 ortant to deserve a specific name. This latter POINT will become a far more essential consideratio
0027 th aegyptian dynasty, about 3000 B.C., as was POINTED out to me by Professor Lepsius; but Mr. Birc
0208 mblance between instincts and habits could be POINTED out. As in repeating a well known song, so i
0347 stance, small areas in the Old World could be POINTED out hotter than any in the New World, yet th
0035 altered from it. It is known that the English POINTER has been greatly changed within the last cen

Page **********************************(Key Word)***********************************

Page **(Key Word)**

```
0035  so effectually, that, though the old Spanish  POINTER  certainly came from Spain, Mr. Borrow has no
0035  rmed by him, any native dog in Spain like our  POINTER. By a similar process of selection, and by c
0120  instance, the English race horse and English  POINTER  have apparently both gone on slowly divergin
0213  nd without the end being known, for the young  POINTER  can no more know that he points to aid his m
0034  e important. Thus, a man who intends keeping   POINTERS  naturally tries to get as good dogs as he ca
0213  eds of dogs: it cannot be doubted that young   POINTERS  (I have myself seen a striking instance) wil
0007  ltivated plants and animals, one of the first  POINTS  which strikes us, is, that they generally dif
0012  d it is really surprising to note the endless  POINTS  in structure and constitution in which the va
0022  development of skin between the toes, are all  POINTS  of structure which are variable. The period a
0027  it agrees in habits and in a great number of  POINTS  of structure with all the domestic breeds. Se
0033  some one point do not differ at all in other   POINTS; this is hardly ever, perhaps never, the case
0039  rded by all fanciers, as it is not one of the  POINTS  of the breed. Nor let it be thought that some
0040  breed will be a slow process. As soon as the   POINTS  of value of the new sub breed are once fully
0046  see in these polymorphic genera variations in  POINTS  of structure which are of no service or disse
0067  recall to the reader's mind some of the chief  POINTS. Eggs or very young animals seem generally to
0072  rised at their numbers that I went to several  POINTS  of view, whence I could examine hundreds of a
0085  uit, should succeed. In looking at many small  POINTS  of difference between species, which, as far
0123  gely from (A), standing nearly at the extreme  POINTS  of the original genus, the six descendants fr
0129  e file, but seem rather to be clustered round  POINTS, and these round other points, and so on in a
0129  clustered round points, and these round other  POINTS, and so on in almost endless cycles. On the v
0152  rns us is, that in our domestic animals those  POINTS, which at the present time are undergoing rap
0152  d tail of our fantails, etc., these being the  POINTS  now mainly attended to by English fanciers. E
0155  . or to state the case in another manner: the  POINTS  in which all the species of a genus resemble
0156  ry at the present day. On the other hand, the  POINTS  in which species differ from other species of
0208  emain constant throughout life. Several other  POINTS  of resemblance between instincts and habits c
0211  erfect; but as details on this and other such  POINTS  are not indispensable, they may be here passe
0213  or the young pointer can no more know that he  POINTS  to aid his master, than the white butterfly k
0227  to intersect largely; and then she unites the  POINTS  of intersection by perfectly flat surfaces. W
0233  e, first to one point, and then to five other  POINTS, at the proper relative distances from the ce
0240  did I expect to find gradations in important   POINTS  of structure between the different castes of
0265  erinduced and that of hybrids, there are many  POINTS  of similarity. In both cases the sterility is
0281  lance to the tapir and the horse; but in some  POINTS  of structure may have differed considerably f
0297  be strictly intermediate between them in all   POINTS  of structure. So that we might obtain the par
0306  olution in our palaeontological ideas on many  POINTS, which the discoveries of even the last dozen
0323  the peninsula of India. For at these distant   POINTS, the organic remains in certain beds present
0323  are similarly characterised in such trifling  POINTS  as mere superficial sculpture. Moreover other
0323  below, are similarly distant at these distant  POINTS  of the world. In the several successive palae
0323  the land and of fresh water change at distant  POINTS  in the same parallel manner. We may doubt whe
0332  essive formations, were discovered at several  POINTS  low down in the series, the three existing fa
0349  species themselves are distinct at different   POINTS  and stations. It is a law of the widest gener
0352  ther species have been created at one or more  POINTS  of the earth's surface. Undoubtedly there are
0352  one point to the several distant and isolated  POINTS, where now found. Nevertheless the simplicity
0352  uous; and when a plant or animal inhabits two  POINTS  so distant from each other, or with an interv
0352  e cases of the same mammal inhabiting distant  POINTS  of the world. No geologist will feel any diff
0352  same species can be produced at two separate  POINTS, why do we not find a single mammal common to
0353  nts are identically the same at these distant  POINTS  of the northern and southern hemispheres? The
0354  species, now living at distant and separated  POINTS; nor do I for a moment pretend that any expla
0354  ts of distant mountain ranges, and at distant  POINTS  in the arctic and antarctic regions; and seco
0354  e of the same species at distant and isolated  POINTS  of the earth's surface, can in many instances
0357  ersal of the same species to the most distant  POINTS, and removes many a difficulty: but to the be
0365  s known of the same species living at distant  POINTS, without the apparent possibility of their ha
0365  een independently created at several distinct  POINTS: and we might have remained in this same beli
0367  can understand the identity of many plants at  POINTS  so immensely remote as on the mountains of th
0373  as similarly affected. Along the Himalaya, at  POINTS  900 miles apart, glaciers have left the marks
0373  tly simultaneous at these several far distant  POINTS  on opposite sides of the world. But we have g
0374  mon to Europe, enormously remote as these two  POINTS  are; and there are many closely allied specie
0381  egards the occurrence of identical species at  POINTS  so enormously remote as Kerguelen Land, New Z
0381  ined to the south, at these and other distant  POINTS  of the southern hemisphere, is, on my theory
0382  a few forms were widely dispersed to various   POINTS  of the southern hemisphere by occasional mean
0384  ed fresh water fish occurring at very distant  POINTS  of the world, no doubt there are many cases w
0386  three tablespoons of mud from three different  POINTS, beneath water, on the edge of a little pond;
0389  rse of time they have come to inhabit distant  POINTS  of the globe. I have already stated that I ca
0404  sical conditions, and the existence at remote  POINTS  of the world of other species allied to fit, i
0405  nt, though now distributed to the most remote  POINTS  of the world, we ought to find, and I believe
0410  ace, species and groups of species have their  POINTS  of maximum development. Groups of species, be
0417  s founded on an appreciation of many trifling  POINTS  of resemblance, too slight to be defined. Cer
0417  wonderfully in a number of the most important  POINTS  of structure from the proper type of the orde
0423  r which were allied in the greatest number of  POINTS. In tumbler pigeons, though some sub varietie
0426  e may err in this respect in regard to single  POINTS  of structure, but when several characters, le
0430  most nearly related to Marsupials: but in the  POINTS  in which it approaches this order, its relati
0430  arsupial species more than to another. As the  POINTS  of affinity of the bizcacha to Marsupials are
0439  e the ordinary leaves of the Leguminosae. The  POINTS  of structure, in which the embryos of widely
0443  ence on this head; indeed the evidence rather  POINTS  the other way: for it is notorious that breed
0445  al differences in the above specified several  POINTS  were incomparably less than in the full grown
0445  in the full grown birds. Some characteristic  POINTS  of difference, for instance, that of the widt
0475  ther by successive crosses, and in other such  POINTS, as do the crossed offspring of acknowledged
0201  ht. It is admitted that the rattlesnake has a  POISON  fang for its own defence and for the destruct
0202  t perfected for its present purpose, with the  POISON  originally adapted to cause galls subsequentl
0484  n so trifling a circumstance as that the same  POISON  often similarly affects plants and animals: o
0484  larly affects plants and animals: or that the  POISON  secreted by the gall fly produces monstrous g
0198  lated with colour, as is the liability to be   POISONED  by certain plants: so that colour would be t
0012  rom coloured individuals by certain vegetable  POISONS. Hairless dogs have imperfect teeth: long ha
0369  e arctic productions were as uniform round the  POLAR  regions as they are at the present day. But t
0370  pliocene period lived further north under the  POLAR  circle, in latitude 66 degrees 67 degrees; an
0370  e look at a globe, we shall see that under the  POLAR  Circle there is almost continuous land from w
0371  the New and Old Worlds, migrated south of the  POLAR  Circle, they must have been completely cut of
```

Page **(Key Word)**

Page ***************************************(Key Word)***************************************

```
0372  rly uniform along the continuous shores of the  POLAR Circle, will account, on the theory of modifi
0376  watson has recently remarked, In receding from   POLAR towards equatorial latitudes, the Alpine or m
0370  n lived on the broken land still nearer to the   POLE. Now if we look at a globe, we shall see that
0374  belts, having been simultaneously colder from    POLE to pole, much light can be thrown on the prese
0374  having been simultaneously colder from pole to   POLE, much light can be thrown on the present distr
0180  aves the frozen waters, and preys like other     POLECATS on mice and land animals. If a different cas
0366  cotland and Wales, with their scored flanks,     POLISHED surfaces, and perched boulders, of the icy s
0102  ithin a confined area, with some place in its    POLITY not so perfectly occupied as might be, natura
0108  will be prevented, so that new places in the     POLITY of each island will have to be filled up by m
0108  s action depends on there being places in the    POLITY of nature, which can be better occupied by so
0109  us increased, for the number of places in the    POLITY of nature is not indefinitely great, not that
0112  on many and widely diversified places in the     POLITY of nature, and so be enabled to increase in n
0121  illing new and widely different places in the    POLITY of nature: hence in the diagram I have chosen
0126  ing out and seizing on many new places in the    POLITY of Nature, will constantly tend to supplant a
0178  ce to occur, and until a place in the natural    POLITY of the country can be better filled by some m
0003  uiring the agency of certain insects to bring    POLLEN from one flower to the other, it is equally p
0009  ver lay fertile eggs. Many exotic plants have    POLLEN utterly worthless, in the same exact conditio
0010  fected a bud or offset, and not the ovules or    POLLEN. But it is the opinion of most physiologists
0010  ty may be largely attributed to the ovules or    POLLEN, or to both, having been affected by the trea
0073  y require the visits of moths to remove their    POLLEN masses and thus to fertilise them. I have, al
0092  s in seeking the nectar would get dusted with    POLLEN, and would certainly often transport the poll
0092  llen, and would certainly often transport their  POLLEN from one flower to the stigma of another flow
0092  favour in any degree the transportal of their    POLLEN from flower to flower, would likewise be favo
0092  s visiting flowers for the sake of collecting    POLLEN instead of nectar; and as pollen is formed fo
0092  f collecting pollen instead of nectar; and as    POLLEN is formed for the sole object of fertilisatio
0092  s a simple loss to the plant; yet if a little    POLLEN were carried, at first occasionally and then
0092  irst occasionally and then habitually, by the    POLLEN devouring insects from flower to flower, and
0092  ss thus effected, although nine tenths of the    POLLEN were destroyed, it might still be a great gai
0092  hose individuals which produced more and more    POLLEN, and had larger and larger anthers, would be
0093  nintentionally on their part, regularly carry    POLLEN from flower to flower! and that they can most
0093  stamens producing rather a small quantity of    POLLEN, and a rudimentary pistil: other holly trees
0093  h shrivelled anthers, in which not a grain of    POLLEN can be detected. Having found a female tree e
0093  pe, and on all, without exception, there were    POLLEN grains, and on some a profusion of pollen. As
0093  ere pollen grains, and on some a profusion of    POLLEN. As the wind had set for several days from th
0093  al days from the female to the male tree, the    POLLEN could not thus have been carried. The weather
0093  tilised by the bees, accidentally dusted with    POLLEN, having flown from tree to tree in search of
0093  rendered so highly attractive to insects that    POLLEN was regularly carried from flower to flower,
0094  ever so slight a degree under nature, then as    POLLEN is already carried regularly from flower to f
0095  oving parts of the corolla, so as to push the    POLLEN on to the stigmatic surface. Hence, again, if
0097  able, the fullest freedom for the entrance of    POLLEN from another individual will explain this sta
0097  doing this, they either push the flowers own    POLLEN on the stigma, or bring pollen from another f
0097  he flowers own pollen on the stigma, or bring    POLLEN from another flower. So necessary are the vis
0097  ould fly from flower to flower, and not carry    POLLEN from one to the other, to the great good, as
0098  if you bring on the same brush a plant's own    POLLEN and pollen from another species, the former w
0098  ng on the same brush a plant's own pollen and    POLLEN from another species, the former will have su
0098  wn by Gartner, any influence from the foreign    POLLEN. When the stamens of a flower suddenly spring
0098  hich effectually prevent the stigma receiving    POLLEN from its own flower: for instance, in Lobelia
0098  by which every one of the infinitely numerous    POLLEN granules are swept out of the conjoined anthe
0098  ects, it never sets a seed; though by placing    POLLEN from one flower on the stigma of another, I r
0098  vent the stigma of a flower receiving its own    POLLEN, yet, as C. C. Sprengel has shown, and as I c
0099  tilisation, or the stigma is ready before the    POLLEN of that flower is ready, so that these plants
0099  strange are these facts! How strange that the    POLLEN and stigmatic surface of the same flower, tho
0099  elized? I suspect that it must arise from the    POLLEN of a distinct variety having a prepotent effe
0099  having a prepotent effect over a flower's own    POLLEN: and that this is part of the general law of
0099  se is directly the reverse, for a plant's own    POLLEN is always prepotent over foreign pollen: but
0099  s own pollen is always prepotent over foreign    POLLEN: but to this subject we shall return in a fut
0099  innumerable flowers, it may be objected that    POLLEN could seldom be carried from tree to tree, an
0100  be produced on the same tree, we can see that    POLLEN must be regularly carried from flower to flow
0100  flower; and this will give a better chance of    POLLEN being occasionally carried from tree to tree.
0193  ts, the very curious contrivance of a mass of    POLLEN grains, borne on a foot stalk with a sticky g
0203  aboration by our fir trees of dense clouds of    POLLEN, in order that a few granules may be wafted b
0218  parasitic habits; for they do not possess the    POLLEN collecting apparatus which would be necessary
0247  portant, must be secluded in order to prevent    POLLEN being brought to it by insects from other pla
0247  d, and artificially fertilised with their own    POLLEN, and (excluding all cases such as the Legumin
0249  uring each generation by their own individual    POLLEN: and I am convinced that this would be injuri
0249  ybrids be artificially fertilised with hybrid    POLLEN of the same kind, their fertility, notwithsta
0249  increasing. Now, in artificial fertilisation    POLLEN is as often taken by chance (as I know from m
0249  e insured in each generation a cross with the    POLLEN from a distinct flower, either from the same
0250  hich can be far more easily fertilised by the    POLLEN of another and distinct species, than by thei
0250  other and distinct species, than by their own    POLLEN. For these plants have been found to yield se
0250  e plants have been found to yield seed to the    POLLEN of a distinct species, though quite sterile w
0250  species, though quite sterile with their own    POLLEN, notwithstanding that their own pollen was fo
0250  ir own pollen, notwithstanding that their own    POLLEN was found to be perfectly good, for it fertil
0250  ree were fertilised by Herbert with their own    POLLEN, and the fourth was subsequently fertilised b
0250  the fourth was subsequently fertilised by the    POLLEN of a compound hybrid descended from three oth
0251  entirely, whereas the pod impregnated by the    POLLEN of the hybrid made vigorous growth and rapid
0251  tly healthy, and although both the ovules and    POLLEN of the same flower were perfectly good with r
0252  nds of hybrid rhododendrons, which produce no    POLLEN, for he will find on their stigmas plenty of
0252  , for he will find on their stigmas plenty of    POLLEN brought from other flowers. In regard to anim
0255  outline of the facts can here be given. When    POLLEN from a plant of one family is placed on the s
0255  st. From this absolute zero of fertility, the    POLLEN of different species of the same genus applie
0255  fertility, beyond that which the plant's own    POLLEN will produce. So in hybrids themselves, there
0255  d probably never would produce, even with the    POLLEN of either pure parent, a single fertile seed:
0255  st trace of fertility may be detected, by the    POLLEN of one of the pure parent species causing the
0257  nces in every part of the flower, even in the    POLLEN, in the fruit, and in the cotyledons, can be
0258  bilis jalappa can easily be fertilised by the    POLLEN of M. longiflora, and the hybrids thus produc
```

Page ***************************************(Key Word)***************************************

Page ***(Key Word)***

```
0258 fertilise reciprocally M. longiflora with the POLLEN of M. jalappa, and utterly failed. Several ot
0262 ded much more freely when fertilized with the POLLEN of distinct species, than when self fertilise
0262 ies, than when self fertilised with their own POLLEN. We thus see, that although there is a clear
0263 with a plant having a pistil too long for the POLLEN tubes to reach the ovarium. It has also been
0263 ovarium. It has also been observed that when POLLEN of one species is placed on the stigma of a d
0263 gma of a distantly allied species, though the POLLEN tubes protrude, they do not penetrate the sti
0270 rtilised thirteen flowers of the one with the POLLEN of the other; but only a single head produced
0270 ither coloured varieties when fertilised with POLLEN from their own coloured flowers. Moreover, he
0451 nts, the office of the pistil is to allow the POLLEN tubes to reach the ovules protected in the ov
0452 r compositae, for the purpose of brushing the POLLEN out of the surrounding anthers. Again, an org
0472 sterile sisters! at the astonishing waste of POLLEN by our fir trees! at the instinctive hatred o
0149 the vertebrae in snakes, and the stamens in POLYANDROUS flowers) the number is variable; whereas t
0219 nstinct was first discovered in the Formica (POLYERGES) rufescens by Pierre Huber, a better observ
0088 war is, perhaps, severest between males of POLYGAMOUS animals; and these seem oftenest provided w
0046 which have sometimes been called protean or POLYMORPHIC, in which the species present an inordinat
0046 everal genera of Brachiopod shells. In most POLYMORPHIC genera some of the species have fixed and
0046 d and definite characters. Genera which are POLYMORPHIC in one country seem to be, with some few e
0046 untry seem to be, with some few exceptions, POLYMORPHIC in other countries; and likewise, judging
0046 am inclined to suspect that we see in these POLYMORPHIC genera variations in points of structure w
0048 and he has entirely omitted several highly POLYMORPHIC genera. Under genera, including the most p
0048 ic genera. Under genera, including the most POLYMORPHIC forms, Mr. Babington gives 251 species, wh
0081 nt, as perhaps we see in the species called POLYMORPHIC. We shall best understand the probable cou
0152 for any particular purpose, and perhaps in POLYMORPHIC groups, we see a nearly parallel natural c
0397 seven days: one of these shells was the Helix POMATIA, and after it had again hybernated I put it
0383 o them, for short and frequent migrations from POND to pond, or from stream to stream; and liabili
0383 for short and frequent migrations from pond to POND, or from stream to stream; and liability to wi
0385 s subject. When a duck suddenly emerges from a POND covered with duck weed, I have twice seen thes
0385 epresent those of a bird sleeping in a natural POND, in an aquarium, where many ova of fresh water
0386 points, beneath water, on the edge of a little POND; this mud when dry weighed only 6 3/4 ounces!
0387 akes me believe that a heron flying to another POND and getting a hearty meal of fish, would proba
0388 tribution, it should be remembered that when a POND or stream is first formed, for instance, on a
0388 he species, however few, already occupying any POND, yet as the number of kinds is small, compared
0114 s on small and uniform islets; and so in small PONDS of fresh water. Farmers find that they can ra
0386 ading birds, which frequent the muddy edges of PONDS, if suddenly flushed, would be the most likel
0386 hat botanists are aware how charged the mud of PONDS is with seeds: I have tried several little ex
0165 stinct as the heavy Belgian cart horse, Welch PONIES, cobs, the lanky Kattywar race, etc., inhabit
0252 cks for grafting from a hybrid between Rhod. PONTICUM and Catawbiense, and that this hybrid seeds
0164 y trust, has examined for me a small dun Welch PONY with three short parellel stripes on each shou
0165 shoulder stripes, like those on the dun Welch PONY, and even had some zebra like stripes on the s
0385 undred miles, and would be sure to alight on a POOL or rivulet, if blown across sea to an oceanic
0163 ly appear: and I have been informed by Colonel POOLE that the foals of this species are generally
0164 enerally striped, that, as I hear from Colonel POOLE, who examined the breed for the Indian Govern
0164 metimes quite disappear in old horses. Colonel POOLE has seen both gray and bay Kattywar horses st
0166 brid from the ass and hemionus, to ask Colonel POOLE whether such face stripes ever occur in the e
0015 l races, for instance, of the cabbage, in very POOR soil (in which case, however, some effect woul
0015 e to be attributed to the direct action of the POOR soil), that they would to a large extent, or e
0037 the wild pear, though he might succeed from a POOR seedling growing wild, if it had come from a g
0037 aving produced such splendid results from such POOR materials; but the art, I cannot doubt, has be
0041 of yorkshire, that as they generally belong to POOR people, and are mostly in small lots, they nev
0042 hem; in donkeys, from only a few being kept by POOR people, and little attention paid to their bre
0084 man! how short his time! and consequently how POOR will his products be, compared with those accu
0088 a hornless stag or spurless cock would have a POOR chance of leaving offspring. Sexual selection
0302 pretend that I should ever have suspected how POOR a record of the mutations of life, the best pr
0487 be looked at as a well filled museum, but as a POOR collection made at hazard and at rare interval
0365 ts from Europe or any other continent, that a POORLY stocked island; though standing more remote f
0279 rate of deposition and of denudation. On the POORNESS of our palaeontological collections. On the
0287 and what a paltry display we behold! On the POORNESS of our Palaeontological collections. That ou
0434 ing, the leg of the horse, the paddle of the PORPOISE, and the wing of the bat, should all be cons
0440 e hand of a man, wing of a bat, and fin of a PORPOISE, are related to similar conditions of life.
0442 instance, the wing of a bat, or the fin of a PORPOISE, should not have been sketched out with all
0479 n the hand of a man, wing of bat, fin of the PORPOISE, and leg of the horse, the same number of we
0178 nt, must formerly have existed in each broken PORTION of the land, but these links will have been
0226 ence each cell consists of an outer spherical PORTION and of two, three, or more perfectly flat su
0226 e outer spherical portions, and yet each flat PORTION forms a part of two cells. Reflecting on thi
0287 mens collected on some one spot. Only a small PORTION of the surface of the earth has been geologi
0307 usion. We should not forget that only a small PORTION of the world is known with accuracy. M. Barr
0341 ord is extremely imperfect; that only a small PORTION of the globe has been geologically explored
0361 ay in the longest transport: out of one small PORTION of earth thus completely enclosed by wood in
0464 as another and distinct species. Only a small PORTION of the world has been geologically explored.
0178 existed within the recent period in isolated PORTIONS, in which many forms, more especially among
0178 more varieties have been formed in different PORTIONS of a strictly continuous area, intermediate
0226 of the same thickness as the outer spherical PORTIONS, and yet each flat portion forms a part of t
0229 rts, only little bits, in other parts, large PORTIONS of a rhombic plate had been left between the
0403 amely, that Madeira and the adjoining islet of PORTO Santo possess many distinct but representativ
0403 ntities of stone are annually transported from PORTO Santo to Madeira, yet this latter island has
0403 latter island has not become colonised by the PORTO Santo species: nevertheless both islands have
0144 y and the manner of swallowing determine the POSITION of several of the most important viscera. Th
0191 ladder is homologous, or ideally similar, in POSITION and structure with the lungs of the higher v
0191 ies still marking in the embryo their former POSITION. But it is conceivable that the now utterly
0231 t build up a rough wall of wax in the proper POSITION, that is, along the plane of intersection be
0231 may sometimes be observed, corresponding in POSITION to the planes of the rhombic basal plates of
0323 ny information in regard to their geological POSITION, no one would have suspected that they had c
0357 nents; but not of such vast changes in their POSITION and extension, as to have united them within
0370 ve long remained in nearly the same relative POSITION, though subjected to large, but partial osci
0395 in their new homes in relation to their new POSITION, and we can understand the presence of endem
0418 s derived from the embryo, on the number and POSITION of the embryonic leaves or cotyledons, and o
```

Page ***(Key Word)***

Page **(Key Word)***************************************

```
0421  d, will still occupy its proper intermediate POSITION; for F originally was intermediate in charac
0436  ost every one, that in a flower the relative POSITION of the sepals, petals, stamens, and pistils,
0476  w, is simply explained by their intermediate POSITION in the chain of descent. The grand fact that
0232  lace on which they can stand in their proper POSITIONS for working, for instance, on a slip of woo
0434  include the same bones, in the same relative POSITIONS? Geoffroy St. Hilaire has insisted strongly
0195  flies; yet we should pause before being too POSITIVE even in this case, for we know that the dist
0140  er latitudes, and conversely; but we do not POSITIVELY know that these animals were strictly adapt
0240  fully believe, though I dare not assert so POSITIVELY, that the workers of intermediate size have
0248  ne case for ten generations, yet he asserts POSITIVELY that their fertility never increased, but g
0305  even in any tertiary stratum. Hence we now POSITIVELY know that sessile cirripedes existed during
0350  tance, that cause which alone, as far as we POSITIVELY know, produces organisms quite like, or, as
0362  and devoured a large supply of food, it is POSITIVELY asserted that all the grains do not pass in
0443  , horses, and various fancy animals, cannot POSITIVELY tell, until some time after the animal has
0018  e those of Tierra del Fuego or Australia, who POSSESS a semi domestic dog, may not have existed in
0019  en derived, as these several countries do not POSSESS a number of peculiar species as distinct par
0034  s, and which results from every one trying to POSSESS and breed from the best individual animals,
0038  o rich in species, do not by a strange chance POSSESS the aboriginal stocks of any useful plants,
0042  ages or the inhabitants of open plains rarely POSSESS more than one breed of the same species. Pig
0047  ereafter will be explained. Those forms which POSSESS in some considerable degree the character of
0140  n the Himalaya, were found in this country to POSSESS different constitutional powers of resisting
0158  eneric characters, or those which the species POSSESS in common: that the frequent extreme variabi
0218  with their parasitic habits: for they do not POSSESS the pollen collecting apparatus which would
0308  condition. But the descriptions which we now POSSESS of the Silurian deposits over immense territ
0310  ten in a changing dialect: of this history we POSSESS the last volume alone, relating only to two
0333  have been modified in various degrees. As we POSSESS only the last volume of the geological recor
0391  and which has a very peculiar soil, does not POSSESS one endemic land bird; and we know from Mr.
0391  or occasionally this island. Madeira does not POSSESS one peculiar bird, and many European and Afr
0392  of many insular beetles. Again, islands often POSSESS trees or bushes belonging to orders which el
0394  nd marianne Archipelagoes, and Mauritius, all POSSESS their peculiar bats. Why, it may be asked, h
0403  adeira and the adjoining islet of Porto Santo POSSESS many distinct but representative land shells
0409  nic islands, whilst the most isolated islands POSSESS their own peculiar species of aerial mammals
0486  er when we have a definite object in view. We POSSESS no pedigrees or armorial bearings; and we ha
0023  g two breeds unless one of the parent stocks POSSESSED the characteristic enormous crop? The suppo
0036  ck, and yet the difference between the sheep POSSESSED by these two gentlemen is so great that the
0087  ural selection: for instance, the great jaws POSSESSED by certain insects, and used exclusively fo
0159  ead and feathers on the feet, characters not POSSESSED by the aboriginal rock pigeon: these then a
0188  into an optical instrument as perfect as is POSSESSED by any member of the great Articulate class
0227  could slightly modify the instincts already POSSESSED by the Melipona, and in themselves not very
0280  e both descended from the rock pigeon: if we POSSESSED all the intermediate varieties which have e
0301  ly be local or confined to one place, but if POSSESSED of any decided advantage, or when further m
0389  the barren island of Ascension aboriginally POSSESSED under half a dozen flowering plants: yet ma
0422  tion, by taking the case of languages. If we POSSESSED a perfect pedigree of mankind, a genealogic
0437  hat the unknown progenitor of the vertebrata POSSESSED many vertebrae: the unknown progenitor of t
0464  ine them ever so closely, unless we likewise POSSESSED many of the intermediate links between thei
0019  spain, etc., but that each of these kingdoms POSSESSES several peculiar breeds of cattle, sheep, e
0223  laves. Formica sanguinea, on the other hand, POSSESSES much fewer slaves, and in the early part of
0394  do occur on almost every island. New Zealand POSSESSES two bats found nowhere else in the world: N
0446  r or later in life, if the full grown animal POSSESSES them. And the cases just given, more especi
0352  formerly united to Europe, and consequently POSSESSING the same quadrupeds. But if the same specie
0039  e any novelty, however slight, in one's own POSSESSION. Nor must the value which would formerly be
0083  t they have allowed foreigners to take firm POSSESSION of the land. And as foreigners have thus ev
0088  ce, but on a struggle between the males for POSSESSION of the females: the result is not death to
0088  round, like Indians in a war dance, for the POSSESSION of the females: male salmons have been seen
0158  males to struggle with other males for the POSSESSION of the females. Finally, then, I conclude t
0322  ent, many species of a new group have taken POSSESSION of a new area, they will have exterminated
0468  most cases be a struggle between males for POSSESSION of the females. The most vigorous individua
0181  ch grade of structure had been useful to its POSSESSOR. Nor can I see any insuperable difficulty i
0186  t and simple, each grade being useful to its POSSESSOR, can be shown to exist; if further, the eye
0199  ucture has been produced for the good of its POSSESSOR. They believe that very many structures hav
0201  f causing pain or for doing an injury to its POSSESSOR. If a fair balance be struck between the go
0204  f gradations in complexity each good for its POSSESSOR, then, under changing conditions of life, t
0455  any part or structure, if not useful to the POSSESSOR, will be saved as far as is possible, will
0459  ght variations, each good for the individual POSSESSOR. Nevertheless, this difficulty, though appe
0486  up of many contrivances, each useful to the POSSESSOR, nearly in the same way as when we look at
0199  ny structures are of no direct use to their POSSESSORS. Physical conditions probably have had some
0020  dog, etc., in the wild state. Moreover, the POSSIBILITY of making distinct races by crossing has b
0174  hat have been separately formed without the POSSIBILITY of intermediate varieties existing in the
0212  t i do not speak without good evidence. The POSSIBILITY, or even probability, of inherited variati
0365  ing at distant points, without the apparent POSSIBILITY of their having migrated from one to the o
0384  e different. A few facts seem to favour the POSSIBILITY of their occasional transport by accidenta
0436  ns which we know to be within the limits of POSSIBILITY. In the paddles of the extinct gigantic se
0436  plants, we often get direct evidence of the POSSIBILITY of one organ being transformed into anothe
0003  the last fifteen years has aided me in every POSSIBLE way by his large stores of knowledge and his
0003  ns, such as climate, food, etc., as the only POSSIBLE cause of variation. In one very limited sens
0004  mount of hereditary modification is at least POSSIBLE; and, what is equally or more important, we
0017  ted animals and plants, I do not think it is POSSIBLE to come to any definite conclusion, whether
0097  ese visits be prevented. Now, it is scarcely POSSIBLE that bees should fly from flower to flower,
0098  es are planted near each other, it is hardly POSSIBLE to raise pure seedlings, so largely do they
0171  s we see them, well defined? Secondly, is it POSSIBLE to conceive under all natural conditions. Le
0180  ture of each squirrel is the best that it is POSSIBLE to conceive under all natural conditions. Le
0181  superable difficulty in further believing it POSSIBLE that the membrane connected fingers and fore
0182  struggle: but it is not necessarily the best POSSIBLE under all possible conditions. It must not b
0182  not necessarily the best possible under all POSSIBLE conditions. It must not be inferred from the
0182  how what diversified means of transition are POSSIBLE. Seeing that a few members of such water bre
0186  ems, I freely confess, absurd in the highest POSSIBLE degree. Yet reason tells me, that if numerou
```

Page **(Key Word)***************************************

Page **************************************(Key Word)***************************************

```
0187  lineal ancestors; but this is scarcely ever  POSSIBLE, and we are forced in each case to look to s
0187  nt form, in order to see what gradations are  POSSIBLE, and for the chance of some gradations havin
0188  on to such startling lengths. It is scarcely  POSSIBLE to avoid comparing the eye to a telescope. W
0193  t to argue that no transition of any kind is  POSSIBLE. The electric organs offer another and even
0193  is and Asclepias, genera almost as remote as  POSSIBLE amongst flowering plants. In all these cases
0204  erful metamorphoses in function are at least  POSSIBLE. For instance, a swim bladder has apparently
0209  r changed conditions of life, it is at least  POSSIBLE that slight modifications of instinct might
0210  ble to show that gradations of some kind are  POSSIBLE; and this we certainly can do. I have been s
0211  n of natural selection, as many instances as  POSSIBLE ought to have been here given; but want of s
0215  en modified by domestication. It is scarcely  POSSIBLE to doubt that the love of man has become ins
0223  pecies, if scattered near their nests, it is  POSSIBLE that pupae originally stored as food might b
0224  lls of the proper shape to hold the greatest  POSSIBLE amount of honey, with the least possible con
0224  est possible amount of honey, with the least  POSSIBLE consumption of precious wax in their constru
0237  its progeny. It may well be asked how is it  POSSIBLE to reconcile this case with the theory of na
0239  natural selection, as I believe to be quite  POSSIBLE, different from the fertile males and female
0252  nd that this hybrid seeds as freely as it is  POSSIBLE to imagine. Had hybrids, when fairly treated
0258  iprocally crossed. There is often the widest  POSSIBLE difference in the facility of making recipro
0260  some difference, and occasionally the widest  POSSIBLE difference, in the facility of effecting an
0266  ng been blended into one. For it is scarcely  POSSIBLE that two organisations should be compounded
0280  garding their origin, it would not have been  POSSIBLE to have determined from a mere comparison of
0281  chain of the intermediate links. It is just  POSSIBLE by my theory, that one of two living forms m
0282  owly through natural selection. It is hardly  POSSIBLE for me even to recall to the reader, who may
0304  . and, as if to make the case as striking as  POSSIPLE, this sessile cirripede was a Chthamalus, a
0312  he waters. Lyell has shown that it is hardly  POSSIBLE to resist the evidence on this head in the c
0315  stinct progenitors. For instance, it is just  POSSIBLE, if our fantail pigeons were all destroyed,
0331  the economy of nature. Therefore it is quite  POSSIBLE, as we have seen in the case of some Siluria
0347  in all their conditions; yet it would not be  POSSIBLE to point out three faunas and floras more ut
0347  lmost as much isolated from each other as is  POSSIBLE. On each continent, also, we see the same fa
0351  ent times, almost any amount of migration is  POSSIBLE. But in many other cases, in which we have r
0363  these mid ocean islands, and it is at least  POSSIBLE that they may have brought thither the seeds
0377  limate than their own. Hence, it seems to me  POSSIBLE, bearing in mind that the tropical productio
0412  to occupy as many and as different places as  POSSIBLE in the economy of nature, there is a constan
0413  ificial means for enunciating, as briefly as  POSSIBLE, general propositions, that is, by one sente
0422  s natural arrangement is shown, as far as is  POSSIBLE on paper, in the diagram, but in much too si
0422  linear series, it would have been still less  POSSIBLE to have given a natural arrangement; and it
0422  tural arrangement; and it is notoriously not  POSSIBLE to represent in a series, on a flat surface,
0422  h an arrangement would, I think, be the only  POSSIBLE one. Yet it might be that some very ancient
0422  inate to groups; but the proper or even only  POSSIBLE arrangement would still be genealogical; and
0432  or at least a natural arrangement, would be  POSSIBLE. We shall see this by turning to the diagram
0441  ression, though I am aware that it is hardly  POSSIBLE to define clearly what is meant by the organ
0444  parents. Hence, I conclude, that it is quite  POSSIBLE, that each of the many successive modificati
0444  ts this view. But in other cases it is quite  POSSIBLE that each successive modification, or most o
0449  ur collections were nearly perfect, the only  POSSIBLE arrangement, would be genealogical. Descent
0455  ife (as we have good reason to believe to be  POSSIBLE) the rudimentary part would tend to be wholl
0455  to the possessor, will be saved as far as is  POSSIBLE, will probably often come into play; and thi
0457  e adult members for purposes as different as  POSSIBLE. Larvae are active embryos, which have becom
0462  w, all the means of migration will have been  POSSIBLE during a very long period; and consequently
0466  how ignorant we are. We do not know all the  POSSIBLE transitional gradations between the simplest
0485  ount of difference between them. It is quite  POSSIBLE that forms now generally acknowledged to be
0001  reflecting on all sorts of facts which could  POSSIBLY have any bearing on it. After five years' wo
0002  both sides of each question; and this cannot  POSSIBLY be here done. I much regret that want of spa
0005  ndividuals of each species are born than can  POSSIBLY survive; and as, consequently, there is a fr
0014  strongly marked domestic varieties could not  POSSIBLY live in a wild state. In many cases we do no
0017  sticated productions; but how could a savage  POSSIBLY know, when he first tamed an animal, whether
0063  e, as more individuals are produced than can  POSSIBLY survive, there must in every case be a strug
0070  parasitic worms, which have from some cause,  POSSIBLY in part through facility of diffusion amongs
0081  that many more individuals are born than can  POSSIBLY survive) that individuals having any advanta
0145  orets without any difference in the corolla.  POSSIBLY, these several differences may be connected
0147  ed for dispersal; and this process could not  POSSIBLY go on in fruit which did not open. The elder
0150  cts which I have collected, and which cannot  POSSIBLY be here introduced. I can only state my conv
0180  n asked how an insertivorous quadruped could  POSSIBLY have been converted into a flying bat; the u
0189  t any complex organ existed, which could not  POSSIBLY have been formed by numerous, successive, sl
0192  tious in concluding that any organ could not  POSSIBLY have been produced by successive transitiona
0197  ng in putridity; and so it may be, or it may  POSSIBLY be due to the direct action of putrid matter
0198  he hind limbs from exercising them more, and  POSSIBLY even the form of the pelvis; and then by the
0200  lex laws of growth. Natural selection cannot  POSSIBLY produce any modification in any one species
0209  of the hive bee and of many ants, could not  POSSIBLY have been thus acquired. It will be universa
0209  of bodily structure. No complex instinct can  POSSIBLY be produced through natural selection, excep
0235  in which we cannot see how an instinct could  POSSIBLY have originated; cases, in which no intermed
0242  utterly sterile members of a community could  POSSIBLY have affected the structure or instincts of
0245  asmuch as the sterility of hybrids could not  POSSIBLY be of any advantage to them, and therefore c
0352  in understanding how the same species could  POSSIBLY have migrated from some one point to the sev
0356  at a former period have connected islands or  POSSIBLY even continents together, and thus have allo
0365  ow lands, where the Alpine species could not  POSSIBLY exist, is one of the most striking cases kno
0402  r annually more eggs are laid there than can  POSSIBLY be reared; and we may infer that the mocking
0467  chapter. More individuals are born than can  POSSIBLY survive. A grain in the balance will determi
0481  w action of the coast waves. The mind cannot  POSSIBLY grasp the full meaning of the term of a hund
0180  as Sir J. Richardson has remarked, with the  POSTERIOR part of their bodies rather wide and with t
0436  ertain number of vertebrae. The anterior and  POSTERIOR limbs in each member of the vertebrate and
0387  erate size, as of the yellow water lily and  POTAMOGETON. Herons and other birds, century after cen
0012  ur old cultivated plants, as on the hyacinth,  POTATO, even the dahlia, etc.; and it is really surp
0071  distinct insectivorous birds. Here we see how  POTENT has been the effect of the introduction of a
0074  ne check or some few being generally the most  POTENT, but all concurring in determining the averag
0080  nciple of selection, which we have seen is so  POTENT in the hands of man, apply to nature? I think
0168  nd diminishing organs, seem to have been more  POTENT in their effects. Homologous parts tend to va
0462  . as an example, I have attempted to show how  POTENT has been the influence of the Glacial period
```

Page **************************************(Key Word)***************************************

Page ***********************************(Key Word)***

0451 rudimentary organs sometimes retain their POTENTIALITY, and are merely not developed: this seems
0247 the plants experimentised on by Gartner were POTTED, and apparently were kept in a chamber in his
0018 n sufficiently civilized to have manufactured POTTERY existed in the valley of the Nile thirteen o
0015 ace horses, long and short horned cattle, and POULTRY of various breeds, and esculent vegetables,
0018 almost any one, thinks that all the breeds of POULTRY have proceeded from the common wild Indian f
0029 h you to scorn. I have never met a pigeon, or a POULTRY, or duck, or rabbit fancier, who was not ful
0040 have lately been exhibited as distinct at our POULTRY shows. I think these views further explain w
0051 ill become impressed, just like the pigeon or POULTRY fancier before alluded to, with the amount o
0086 the varieties of the silkworm; in the eggs of POULTRY, and in the colour of the down of their chic
0147 elf gains largely in size and quality. In our POULTRY, a large tuft of feathers on the head is gen
0215 nus, when kept tame, are most eager to attack POULTRY, sheep and pigs; and this tendency has been
0215 ite young, require to be taught not to attack POULTRY, sheep, and pigs! No doubt they occasionally
0233 ed by a hive of bees for the secretion of each POUND of wax; so that a prodigious quantity of flui
0233 ound that no less than from twelve to fifteen POUNDS of dry sugar are consumed by a hive of bees f
0021 eak, has a very short and very broad one. The POUTER has a much elongated body, wings, and legs; a
0023 the short faced tumbler, the runt, the barb, POUTER, and fantail in the same genus; more especial
0023 any lesser number: how, for instance, could a POUTER be produced by crossing two breeds unless one
0024 e of the jacobin: for a crop like that of the POUTER; for tail feathers like those of the fantail.
0039 some slight degree in an unusual manner, or a POUTER till he saw a pigeon with a crop of somewhat
0039 feathers have been counted. Perhaps the first POUTER, may be considered as a variation representin
0159 fourteen or even sixteen tail feathers in the POUTER pigeons have both descended from the rock pig
0280 o give a simple illustration: the fantail and POUTER pigeons; none, for instance, combining a tail somewha
0280 directly intermediate between the fantail and POUTER; none, for instance, combining a tail somewha
0026 originally as distinct as carriers, tumblers, POUTERS, and fantails now are, should yield offsprin
0153 e wattle of.carriers and the enlarged crop of POUTERS. Now let us turn to nature. When a part has
0445 feet and length of leg, in the wild stock, in POUTERS, fantails, runts, barbs, dragons, carriers,
0004 more important, we shall see how great is the POWER of man in accumulating by his Selection succe
0007 productions. Circumstances favourable to Man's POWER of Selection. When we look to the individuals
0017 bility of the ass or guinea fowl, or the small POWER of endurance of warmth by the rein deer, or o
0030 s has not been their history. The key is man's POWER of accumulative selection: nature gives succe
0030 d to make for himself useful breeds. The great POWER of this principle of selection is not hypothe
0040 mstances, favourable, or the reverse, to man's POWER of selection. A high degree of variability is
0043 ut more efficiently, is by far the predominant POWER. Chapter II. Variation Under Nature. Variabil
0061 ection, in order to mark its relation to man's POWER of selection. We have seen that man by select
0061 ral Selection, as we shall hereafter see, is a POWER incessantly ready for action, and is as immea
0078 we wished in imagination to give the plant the POWER of increasing in number, we should have to gi
0080 iv. Natural Selection. Natural Selection, its POWER compared with man's selection, its power on c
0080 , its power compared with man's selection, its POWER on characters of trifling importance, its pow
0080 ower on characters of trifling importance, its POWER at all ages and on both sexes. Sexual Selecti
0083 ing each varying season, as far as lies in his POWER, all his productions. He often begins his sel
0092 as would probably inherit the nectar excreting POWER. Those individual flowers which had the large
0109 ffected in the long course of time by nature's POWER of selection. Extinction. This subject will b
0137 ter living some days in the light, some slight POWER of vision. In the same manner as in Madeira t
0149 , and therefore to natural selection having no POWER to check deviations in their structure. Thus
0153 bility of all kinds and on the other hand, the POWER of steady selection to keep the breed true. I
0182 eps by which birds have acquired their perfect POWER of flight; but they serve, at least, to show
0185 rdi, in its general habits, in its astonishing POWER of diving, its manner of swimming, and of fly
0189 form. Further we must suppose that there is a POWER always intently watching each slight accident
0195 s in South America absolutely depends on their POWER of resisting the attacks of insects: so that
0202 he insect's own death: for if on the whole the POWER of stinging be useful to the community, it wi
0202 few members. If we admire the truly wonderful POWER of scent by which the males of many insects f
0205 te, have been acquired by natural selection, a POWER which acts solely by the preservation of prof
0216 r the mother hen has almost lost by disuse the POWER of flight. Hence we may conclude, that domest
0216 instinct of certain ants: and the comb making POWER of the hive bee: these two latter instincts h
0235 ms and of the basal rhombic plates. The motive POWER of the process of natural selection having be
0242 olly insufficient length, in order to show the POWER of natural selection, and likewise because th
0259 : for instance, some species have a remarkable POWER of crossing with other species: other species
0259 er species of the same genus have a remarkable POWER of impressing their likeness on their hybrid
0260 een permitted? to grant to species the special POWER of producing hybrids, and then to stop their
0269 estication by man's methodical and unconscious POWER of selection, for his own use and pleasure: h
0274 ies are crossed, one has sometimes a prepotent POWER of impressing its likeness on the hybrid: and
0274 one variety certainly often has this prepotent POWER over another variety. Hybrid plants produced
0274 cht, who maintain that the ass has a prepotent POWER over the horse, so that both the mule and the
0314 riability being of a beneficial nature, on the POWER of intercrossing, on the rate of breeding, on
0326 s, would be highly favourable, as would be the POWER of spreading into new territories. A certain
0359 urrents during 28 days, and would retain their POWER of germination. In Johnston's Physical Atlas,
0362 ent: and several of these seeds retained their POWER of germination. Certain seecs, however, were
0379 to a higher stage of perfection or dominating POWER, than the southern forms. And thus, when they
0383 ings, compared with those of Britain. But this POWER in fresh water productions of ranging widely,
0387 seal and we have seen that seeds retain their POWER of germination, when reiected in pellets or i
0404 ess of modification. This relation between the POWER and extent of migration of a species, either
0405 rget that to range widely implies not only the POWER of crossing barriers, but the more important
0405 r of crossing barriers, but the more important POWER of being victorious in distant lands in the s
0410 odifications have been accumulated by the same POWER of natural selection. Chapter XIII. Mutual Af
0453 ruck with astonishment: for the same reasoning POWER which tells us plainly that most parts and or
0454 d to take flight, and have ultimately lost the POWER of flying. Again, an organ useful under certa
0454 ected by insensibly small steps, is within the POWER of natural selection: so that an organ render
0469 living products? What limit can be put to this POWER, acting during long ages anc rigidly scrutini
0469 rejecting the bad? I can see no limit to this POWER, in slowly and beautifully adapting each form
0480 uggle for existence, and will thus have little POWER of acting on an organ during early life: henc
0488 at of the necessary acquirement of each mental POWER and capacity by gradation. Light will be thro
0081 former inhabitants. Let it be remembered how POWERFUL the influence of a single introduced tree or
0087 ung birds within the egg, which had the most POWERFUL and hardest beaks, for all with weak beaks w
0356 ant facts. Change of climate must have had a POWERFUL influence on migration: a region when its cl
0379 northern forms were enabled to beat the less POWERFUL southern forms. Just in the same manner as w
0467 rent Struggle for Existence, we see the most POWERFUL and ever acting means of selection. The stru

Page ***********************************(Key Word)***

Page **(Key Word)**

```
0469    then we have under nature variability and a POWERFUL agent always ready to act and select, why sh
0405  d ranging widely, as in the case of certain POWERFULLY winged birds, will necessarily range widely
0004  nevitably follows from their high geometrical POWERS of increase, will be treatec of. This is the
0005  ondly, the subject of Instinct, or the mental POWERS of animals; thirdly, Hybridism, or the infert
0060  n. the term used in a wide sense. Geometrical POWERS of increase. Rapid increase of naturalised an
0109  tion may be, if feeble man can do much by his POWERS of artificial selection, I can see no limit t
0109  ntly endure. But as from the high geometrical POWERS of increase of all organic beings, each area
0113  o arrived at its full average. If its natural POWERS of increase be allowed to act, it can succeed
0127  d; if there be, owing to the high geometrical POWERS of increase of each species, at some age, sea
0138  uppose that American animals, having ordinary POWERS of vision, slowly migrated by successive gene
0140  s country to possess different constitutional POWERS of resisting cold. Mr. Thwaites informs me th
0188  assume that the Creator works by intellectual POWERS like those of man? If we must compare the eye
0207  c to do with the origin of the primary mental POWERS, any more than I have with that of life itsel
0228  tural selection, her inimitable architectural POWERS. But this theory can be tested by experiment.
0238    to the same family. I have such faith in the POWERS of selection, that I do not doubt that a bree
0246  cowment, beyond the province of our reasoning POWERS. The fertility of varieties, that is of the f
0259  ness on their hybrid offspring; but these two POWERS do not at all necessarily go together. There
0353  quently migrated from that area as far as its POWERS of migration and subsistence under past and p
0388  cially by fresh water birds, which have large POWERS of flight, and naturally travel from one to a
0441  organs of sense, and to reach by their active POWERS of swimming, a proper place on which to becom
0447  the mature animal, which has come to its full POWERS of activity and has to gain its own living: a
0455  he being has come to maturity and to its full POWERS of action, the principle of inheritance at co
0474  adation throws on the admirable architectural POWERS of the hive bee. Habit no doubt sometimes com
0490  andeur in this view of life, with its several POWERS, having been originally breathed into a few f
0050  ountry the highest botanical authorities and PRACTICAL men can be quoted to show that the sessile
0248  ected by various circumstances, that for all PRACTICAL purposes it is most difficult to say where
0251  when self fertilised, sometimes depends. The PRACTICAL experiments of horticulturists, though not
0282  en to recall to the reader, who may not be a PRACTICAL geologist, the facts leading the mind feebl
0046  t organs never vary; for these same authors PRACTICALLY rank that character as important (as some
0047  as we know, as have good and true species. PRACTICALLY, when a naturalist can unite two forms tog
0172  ich leads the bee to make cells, which have PRACTICALLY anticipated the discoveries of profound ma
0176  om inhabiting a narrow and lesser area; and PRACTICALLY, as far as I can make out, this rule holds
0224  we hear from mathematicians that bees have PRACTICALLY solved a recondite problem, and have made
0231  commencement of the first cell. I was able PRACTICALLY to show this fact, by covering the edges o
0417  ications are sometimes necessarily founded. PRACTICALLY when naturalists are at work, they do not
0032  e best breeders are strongly opposed to this PRACTICE, except sometimes amongst closely allied sub
0032  believe in the natural capacity and years of PRACTICE requisite to become even a skilful pigeon fa
0033  of selection has been reduced to methodical PRACTICE for scarcely more than three quarters of a c
0111  ies of the same genus. As has always been my PRACTICE, let us seek light on this head from our dom
0209  e at three years old with wonderfully little PRACTICE, had played a tune with no practice at all,
0209  y little practice, had played a tune with no PRACTICE at all, he might truly be said to have done
0419  sts on the utility or even necessity of this PRACTICE in certain groups of birds; and it has been
0028  crossing the breeds, which method was never PRACTISED before, has improved them astonishingly. Ab
0419  nsects, of a group of forms, first ranked by PRACTISED naturalists as only a genus, and then raise
0073  alone visit the common red clover (Trifolium PRATENSE), as other bees cannot reach the nectar. Hen
0094  common red and incarnate clovers (Trifolium PRATENSE and incarnatum) do not on a hasty glance app
0313  midst of an older formation, and then allow the PRE existing fauna to reappear? But Lyell's explan
0355  stence coincident both in space and time with a PRE existing closely allied species. And I now kno
0403  in the several districts of the same continent, PRE occupation has probably played an important pa
0142  et seed from these seedlings, with the same PRECAUTIONS, the experiment cannot be said to have bee
0334  iate in general character between that which PRECEDED and that which succeeded it. Thus, the speci
0319  more recent tertiary formations, that rarity PRECEDES extinction: and we know that this has been t
0203    the continual supplanting and extinction of PRECEDING and intermediate gradations. Closely allied
0312  same types within the same areas. Summary of PRECEDING and present chapters. Let us now see whethe
0334  ver, allow for the entire extinction of some PRECEDING forms, and for the coming in of quite new f
0334  ly is intermediate in character, between the PRECEDING and succeeding faunas. I need give only one
0334  nearly intermediate in character between the PRECEDING and succeeding faunas, that certain genera
0337  tage in the struggle for life over other and PRECEDING forms. If under a nearly similar climate, t
0341  t modified blood descendants. Summary of the PRECEDING and present Chapters. I have attempted to s
0412  e the less divergent, the less improved, and PRECEDING forms. I request the reader to turn to the
0487  on of these intervals by a comparison of the PRECEDING and succeeding organic forms. We must be ca
0095  d clover offer in vain an abundant supply of PRECIOUS nectar to the hive bee. Thus it might be a g
0224  oney, with the least possible consumption of PRECIOUS wax in their construction. It has been remar
0453  y growing embryonic calf by the excretion of PRECIOUS phosphate of lime? When a man's fingers have
0023  ct in the wild state. But birds breeding on PRECIPICES, and good fliers, are unlikely to be exterm
0199  o lay too much stress on our ignorance of the PRECISE cause of the slight analogous differences be
0443  the child will be tall or short, or what its PRECISE features will be. The question is not, at wh
0076  nature: but probably in no one case could we PRECISELY say why one species has been victorious ove
0322  re will be utterly obscured. Whenever we can PRECISELY say why this species is more abundant in in
0325  om. M. Barrande has made forcible remarks to PRECISELY the same effect. It is, indeed, quite futil
0326  r existence, would cease. We know not at all PRECISELY what are all the conditions most favourable
0251  iculturists, though not made with scientific PRECISION, deserve some notice. It is notorious in ho
0483  a curious illustration of the blindness of PRECONCEIVED opinion. These authors seem no more startl
0109  me rare. Rarity, as geology tells us, is the PRECURSOR to extinction. We can, also, see that any f
0121  exterminate in each stage of descent their PREDECESSORS and their original parent. For it should b
0345  d in the world's history have beaten their PREDECESSORS in the race for life, and are, in so far,
0431  ten referred to), mounting up through many PREDECESSORS. As it is difficult to show the blood rela
0217  e alone. But the American cuckoo is in this PREDICAMENT; for she makes her own nest and has eggs a
0424  y naturalist: scarcely a single fact can be PREDICATED in common of the males and hermaphrodites o
0126  d to disappear. Looking to the future, we can PREDICT that the groups of organic beings which are
0126  ch groups will ultimately prevail, no man can PREDICT; for we well know that many groups, formerly
0126  inc still more remotely to the future, we may PREDICT that, owing to the continued and steady incr
0140  may infer this from our frequent inability to PREDICT whether or not an imported plant will endure
0043  slowly, but more efficiently, is by far the PREDOMINANT Power. Chapter II. Variation Under Nature.
0103  ts. On the above principle, nurserymen always PREFER getting seed from a large body of plants of t
0135    mutilations are ever inherited; and I should PREFER explaining the entire absence of the anterior
```

Page **(Key Word)**

Page **(Key Word)**

```
0089 t well know that they often take individual PREFERENCES and dislikes: thus Sir R. Heron has descri
0112  may suppose that at an early period one man PREFERRED swifter horses: another stronger and more b
0423 alogical classification would be universally PREFERRED: and it has been attempted by some authors.
0103 seasons, or from varieties of the same kind PREFERRING to pair together. Intercrossing plays a ver
0148 erest rudiment attached to the bases of the PREHENSILE antennae. Now the saving of a large and com
0441 for life: their legs are now converted into PREHENSILE organs: they again obtain a well constructe
0196 of purposes, as a fly flapper, an organ of PREHENSION, or as an aid in turning, as with the dog,
0482 is conviction: for only thus can the load of PREJUDICE by which this subject is overwhelmed be rem
0060 subject of this chapter, I must make a few PRELIMINARY remarks, to show how the struggle for exis
0236 atural selection. But I must pass over this PRELIMINARY difficulty. The great difficulty lies in t
0354 offered of many such cases. But after some PRELIMINARY remarks, I will discuss a few of the most
0062 all seasons of each recurring year. I should PREMISE that I use the term Struggle for Existence i
0207 fficient to overthrow my whole theory. I must PREMISE, that I have nothing to do with the origin o
0138 , animals not far remote from ordinary forms, PREPARE the transition from light to darkness. Next
0096 extreme brevity, though I have the materials PREPARED for an ample discussion. All vertebrate anim
0201 that the cat curls the end of its tail when PREPARING to spring, in order to warn the doomed mous
0309 ontinents seem to to have been formed by a PREPONDERANCE, during many oscillations of level, of th
0309 rce of elevation: but may not the areas of PREPONDERANT movement have changed in the lapse of ages
0379 borneo and Abyssinia. I suspect that this PREPONDERANT migration from north to south is due to th
0003 hereafter see, this may be true; but it is PREPOSTEROUS to attribute to mere external conditions,
0003 rom one flower to the other, it is equally PREPOSTEROUS to account for the structure of this paras
0425 ther species? The supposition is of course PREPOSTEROUS; and I might answer by the argumentum ad h
0425 ed under the bear genus. The whole case is PREPOSTEROUS; for where there has been close descent in
0274 al characters; but more especially owing to PREPOTENCY in transmitting likeness running more stron
0274 semble the ass than the horse; but that the PREPOTENCY runs more strongly in the male ass than in
0098 another species, the former will have such a PREPOTENT effect, that it will invariably and complet
0099 om the pollen of a distinct variety having a PREPOTENT effect over a flower's own pollen; and that
0099 reverse, for a plant's own pollen is always PREPOTENT over foreign pollen; but to this subject we
0274 two species are crossed, one has sometimes a PREPOTENT power of impressing its likeness on the hyb
0274 animals one variety certainly often has this PREPOTENT power over another variety. Hybrid plants p
0274 s are right, who maintain that the ass has a PREPOTENT power over the horse, so that both the mule
0022 together with their relative breadth and the PRESENCE of processes. The size and shape of the aper
0074 he mice. Hence it is quite credible that the PRESENCE of a feline animal in large numbers in a dis
0144 in between the outer toes in pigeons, and the PRESENCE of more or less down on the young birds when
0159 in two or more distinct races. The frequent PRESENCE of fourteen or even sixteen tail feathers in
0161 probably be of an important nature, for the PRESENCE of all important characters will be governed
0175 hysical conditions, but in large part on the PRESENCE of other species, on which it depends, or by
0193 fferent habits of life, we may attribute its PRESENCE to inheritance from a common ancestor: and i
0193 of their modified descendants have lost. The PRESENCE of luminous organs in a few insects, belongi
0196 tail is in most aquatic animals, its general PRESENCE and use for many purposes in so many land an
0223 incts, and do what work they could. If their PRESENCE proved useful to the species which had seize
0225 of the cell stands in close relation to the PRESENCE of adjoining cells: and the following view m
0279 stances apparently most favourable for their PRESENCE, namely on an extensive and continous area w
0279 es depends in a more important manner on the PRESENCE of other already defined organic forms, than
0307 th barrande's so called primordial zone. The PRESENCE of phosphatic nodules and bituminous matter
0372 ndered. Thus, I think, we can understand the PRESENCE of many existing and tertiary representative
0395 heir new position, and we can understand the PRESENCE of endemic bats on islands, with the absence
0395 land from the neighbouring mainland, and the PRESENCE in both of the same mammiferous species or o
0396 d of terrestrial mammals notwithstanding the PRESENCE of aerial bats, the singular proportions of
0409 hy there should be some relation between the PRESENCE of mammals, in a more or less modified condi
0409 other, there should be a correlation, in the PRESENCE of identical species, of varieties, of doubt
0426 sumes high value: we can account for its PRESENCE in so many forms with such different habits,
0450 ans are extremely curious; for instance, the PRESENCE of teeth in foetal whales, which when grown
0450 up have not a tooth in their heads: and the PRESENCE of teeth, which never cut through the gums,
0453 re? An eminent physiologist accounts for the PRESENCE of rudimentary organs, by supposing that the
0455 obliteration of a rudimentary organ. As the PRESENCE of rudimentary organs is thus due to the ten
0457 licable difficulties; on the contrary, their PRESENCE might have been even anticipated. The import
0478 istant from any continent. Such facts as the PRESENCE of peculiar species of bats, and the absence
0001 merica and in the geological relations of the PRESENT to the past inhabitants of that continent. T
0001 eemed to me probable: from that period to the PRESENT day I have steadily pursued the same object.
0006 he highest importance, for they determine the PRESENT welfare, and, as I believe, the future succe
0016 f generic value; all such valuations being at PRESENT empirical. Moreover, on the view of the orig
0020 selection of those individual mongrels, which PRESENT any desired character; but that a race could
0023 inal stocks: for it is impossible to make the PRESENT domestic breeds by the crossing of any lesse
0032 ny florists' flowers, when the flowers of the PRESENT day are compared with drawings made only twe
0033 ur, shape, and hairiness, and yet the flowers PRESENT very slight differences. It is not that the
0034 good and bad qualities is so obvious. At the PRESENT time, eminent breeders try by methodical sel
0038 rove or modify most of our plants up to their PRESENT standard of usefulness to man; we can unders
0039 teen tailfeathers somewhat expanded, like the PRESENT Java fantail, or like individuals of other a
0046 protean or polymorphic, in which the species PRESENT an inordinate amount of variation; and hardl
0053 plants which have very wide ranges generally PRESENT varieties; and this might have been expected
0055 e larger genera in each country would oftener PRESENT varieties, than the species of the smaller g
0055 the species on the side of the larger genera PRESENT varieties, than on the side of the smaller g
0055 reover, the species of the large genera which PRESENT any varieties, invariably present a larger a
0055 enera which present any varieties, invariably PRESENT a larger average number of varieties than do
0056 s have been formed, the species of that genus PRESENT a number of varieties, that is of incipient
0059 several respects the species of large genera PRESENT a strong analogy with varieties. And we can
0068 uld, in all probability, be less game than at PRESENT, although hundreds of thousands of game anim
0101 impossible. Cirripedes long appeared to me to PRESENT a case of very great difficulty under this p
0102 , its several districts will almost certainly PRESENT different conditions of life: and then if na
0107 lled living fossils; they have endured to the PRESENT day, from having inhabited a confined area,
0111 es and parents of future well marked species; PRESENT slight and ill defined differences. Mere cha
0113 its varying descendants seizing on places at PRESENT occupied by other animals: some of them, for
0117 ; and the varying species of the large genera PRESENT a greater number of varieties. We have, also
0125 e species of the larger genera which oftenest PRESENT varieties or incipient species. This, indeed
0126 era, families, orders, and classes, as at the PRESENT day. Summary of Chapter. If during the long
```

Page **(Key Word)**

```
0129  g twigst and this connexion of the former and PRESENT buds by ramifying branches may well represen
0135  ot having them. In some other genera they are PRESENT, but in a rudimentary condition. In the Ateu
0151  l species of one genus, Pyrgoma, these valves PRESENT a marvellous amount of diversification: the
0152  r domestic animals those points, which at the PRESENT time are undergoing rapid change by continue
0154  ative variability, as it may be called, still PRESENT in a high degree. For in this case the varia
0156  is not probable that they should vary at the PRESENT day. On the other hand, the points in which
0156  riod remained constant. In connexion with the PRESENT subject, I will make only two other remarks.
0159  ordinary specific purposes. Distinct species  PRESENT analogous variations; and a variety of one s
0159  pigeons, in countries most widely set apart,  PRESENT sub varieties with reversed feathers on the
0169  to similar influences will naturally tend to  PRESENT analogous variations, and these same species
0173  same territory we surely ought to find at the PRESENT time many transitional forms. Let us take a
0173  e transitional varieties between its past and PRESENT states. Hence we ought not to expect at the
0174  t states. Hence we ought not to expect at the PRESENT time to meet with numerous transitional vari
0174  less continuous and uniform condition than at PRESENT. But I will pass over this way of escaping f
0177  defined objects, and do not at any one period PRESENT an inextricable chaos of varying and interme
0179  ng in greater numbers will, in the aggregate, PRESENT more variation, and thus be further improved
0182  tructure will seldom continue to exist to the PRESENT day, for they will have been supplanted by t
0194  ransitions an organ could have arrived at its PRESENT state; yet, considering that the proportion
0195  ble that this could have been adapted for its PRESENT purpose by successive slight modifications,
0202  h has been modified but not perfected for its PRESENT purpose, with the poison originally adapted
0205  ch small importance that it could not, in its PRESENT state, have been acquired by natural selecti
0206  cit saltum. This canon if we look only to the PRESENT inhabitants of the world, is not strictly co
0209  re for the welfare of each species, under its PRESENT conditions of life. Under changed conditions
0220  pshire, and has never seen the slaves, though PRESENT in large numbers in August, either leave or
0220  r the nest, and food of all kinds. During the PRESENT year, however, in the month of July, I came
0223  ntrast the instinctive habits of F. sanguinea PRESENT with those of the F. rufescens. The latter d
0233  hitectural instincts, all tending towards the PRESENT perfect plan of construction, could have pro
0234  gether, and more regular in every way than at PRESENT; for then, as we have seen, the spherical su
0254  general result: but that it cannot, under our PRESENT state of knowledge, be considered as absolut
0271  ly coloured. Yet these varieties of Verbascum  PRESENT no other difference besides the mere colour
0271  hollyhock, I am inclined to suspect that they PRESENT analogous facts. Kolreuter, whose accuracy h
0279  the absence of intermediate varieties at the PRESENT day. On the nature of extinct intermediate v
0279  s why such links do not commonly occur at the PRESENT day, under the circumstances apparently most
0281  ween the varieties of the same species at the PRESENT day; and these parent species, now generally
0283  ound a promontory, that the cliffs are at the PRESENT time suffering. The appearance of the surfac
0295  xtending from before the glacial epoch to the PRESENT day. In order to get a perfect gradation bet
0298  se which have the widest range, that oftenest PRESENT varieties; so that with shells and other mar
0298  tion. It should not be forgotten, that at the PRESENT day, with perfect specimens for examination,
0299  fully agree with Mr. Godwin Austen, that the PRESENT condition of the Malay Archipelago, with its
0301  ory assuredly have connected all the past and PRESENT species of the same group into one long and
0305  reely open from south to north as they are at PRESENT. Even at this day, if the Malay Archipelago
0306  orders to which they belong, for they do not PRESENT characters in any degree intermediate betwee
0307  e whole interval from the Silurian age to the PRESENT day; and that during these vast, yet quite u
0308  y of denudation and metamorphism. The case at PRESENT must remain inexplicable; and may be truly u
0310  formations; the almost entire absence, as at PRESENT known, of fossiliferous formations beneath t
0312  thin the same areas. Summary of preceding and PRESENT chapters. Let us now see whether the several
0315  e a new breed hardly distinguishable from our PRESENT fantail: but if the parent rock pigeon were
0316  ions, from the lowest Silurian stratum to the PRESENT day. We have seen in the last chapter that t
0318  d from the earliest known dawn of life to the PRESENT day; some having disappeared before the clos
0323  t points, the organic remains in certain beds PRESENT an unmistakeable degree of resemblance to th
0324  r if all the marine animals which live at the PRESENT day in Europe, and all those that lived in E
0324  dent that fossiliferous beds deposited at the PRESENT day on the shores of North America would her
0325  ll see this more clearly when we treat of the PRESENT distribution of organic beings, and find how
0330  families, or genera with those living at the PRESENT day, were not at this early epoch limited in
0330  hat supposing them to be distinguished at the PRESENT day from each other by a dozen characters, t
0332  rent periods, and some to have endured to the PRESENT day. By looking at the diagram we can see th
0338  the typical members of the same groups at the PRESENT day, it would be vain to look for animals ha
0339  would be a bold man, who after comparing the PRESENT climate of Australia and of parts of South A
0340  orth America formerly partook strongly of the PRESENT character of the southern half of the contin
0340  s formerly more closely allied, than it is at PRESENT, to the northern half. In a similar manner w
0340  ed in its mammals to Africa than it is at the PRESENT time. Analogous facts could be given in rela
0340  be nothing immutable in the laws of past and PRESENT distribution. It may be asked in ridicule, w
0341  ood descendants. Summary of the preceding and PRESENT Chapters. I have attempted to show that the
0346          Chapter XI. Geographical Distribution. PRESENT distribution cannot be accounted for by diff
0353  s of migration and subsistence under past and PRESENT conditions permitted, is the most probable.
0358  tants of several lands and even seas to their PRESENT inhabitants, a certain degree of relation (a
0367  nearly the same with those of Europe: for the PRESENT circumpolar inhabitants, which we suppose to
0368  me to explain in so satisfactory a manner the PRESENT distribution of the Alpine and Arctic produc
0368  d, has ever been in any degree warmer than at PRESENT (as some geologists in the United States bel
0368  rth, and subsequently have retreated to their PRESENT homes; but I have met with no satisfactory e
0369  find has been the case: for if we compare the PRESENT Alpine plants and animals of the several gre
0369  of the species are identically the same, some PRESENT varieties, some are ranked as doubtful forms
0369  rm round the polar regions as they are at the PRESENT day. But the foregoing remarks on distributi
0369  he commencement of the Glacial period. At the PRESENT day, the sub arctic and northern temperate p
0370  d new Worlds lived further southwards than at PRESENT, they must have been still more completely s
0370  me as now, the climate was warmer than at the PRESENT day. Hence we may suppose that the organisms
0371  ly related to each other than they are at the PRESENT time: for during these warmer periods the no
0372  s now disjoined, and likewise of the past and PRESENT inhabitants of the temperate lands of North
0373  erica, glaciers once extended far below their PRESENT level. In central Chile I was astonished at
0374  pole to pole, much light can be thrown on the PRESENT distribution of identical and allied species
0377  ial period simultaneously much colder than at PRESENT. The Glacial period, as measured by years, m
0377  cs supported as many species as we see at the PRESENT day crowded together at the Cape of Good Hop
0379  rms. Just in the same manner as we see at the PRESENT day, that very many European productions cov
0382  atural selection, a multitude of facts in the PRESENT distribution both of the same and of allied
0383  sequent modification. Summary of the last and PRESENT chapters. As lakes and river systems are sep
0384  no doubt there are many cases which cannot at PRESENT be explained: but some fresh water fish belo
```

Page **(Key Word)**

0391 erent orders of insects in Madeira apparently PRESENT analogous facts. Oceanic islands are sometim
0396 since their arrival, could have reached their PRESENT homes. But the probability of many islands h
0399 one, is a far more remarkable case, and is at PRESENT inexplicable: but this affinity is confined
0404 tent of migration of a species, either at the PRESENT time or at some former period under differen
0406 nists to their new homes. Summary of last and PRESENT Chapters. In these chapters I have endeavour
0409 g nearly the same with those governing at the PRESENT time the differences in different areas. We
0419 ies, and genera, they seem to be, at least at PRESENT, almost arbitrary. Several of the best botan
0420 have transmitted modified descendants to the PRESENT day, represented by the fifteen genera (a14
0421 gement remains strictly true, not only at the PRESENT time, but at each successive period of desce
0429 is that the more ancient forms of life often PRESENT characters in some slight degree intermediat
0429 forms having occasionally transmitted to the PRESENT day descendants but little modified, will gi
0436 is another and equally curious branch of the PRESENT subject; namely, the comparison not of the s
0444 tions, by which each species has acquired its PRESENT structure, may have supervened at a not very
0449 wo groups of animal, however much they may at PRESENT differ from each other in structure and habi
0457 rudimentary organs and their final abortion, PRESENT to us no inexplicable difficulties: on the c
0460 ny whole being, could not have arrived at its PRESENT state by many graduated steps. There are, it
0462 es in each group by gradations as fine as our PRESENT varieties, it may be asked, Why do we not se
0464 ediate links between their past or parent and PRESENT states; and these many links we could hardly
0470 they now flourish, these same species should PRESENT many varieties; for where the manufactory of
0477 e intertropical ocean. Although two areas may PRESENT the same physical conditions of life, we nee
0478 the fact, as we have seen, that all past and PRESENT organic beings constitute one grand natural
0480 ve been inherited from a remote period to the PRESENT day. On the view of each organic being and e
0484 ts will be able to pursue their labours as at PRESENT; but they will not be incessantly haunted by
0485 ar more essential consideration than it is at PRESENT; for differences, however slight, between an
0485 re known, or believed, to be connected at the PRESENT day by intermediate gradations, whereas spec
0485 hout quite rejecting the consideration of the PRESENT existence of intermediate gradations between
0487 f the inhabitants of the whole world. Even at PRESENT, by comparing the differences of the inhabit
0488 degree. The whole history of the world, as at PRESENT known, although of a length quite incomprehe
0488 the production and extinction of the past and PRESENT inhabitants of the world should have been du
0302 life, the best preserved geological section PRESENTED, had not the difficulty of our not discover
0343 long before that period, the world may have PRESENTED a wholly different aspect; and that the old
0377 were in a suffering state and could not have PRESENTED a firm front against intruders, that a cert
0095 y the continued preservation of individuals PRESENTING mutual and slightly favourable deviations o
0111 f the innumerable species throughout nature PRESENTING well marked differences; whereas varieties,
0177 better chance, within any given period, of PRESENTING further favourable variations for natural s
0177 ch country, as shown in the second chapter, PRESENTING on an average a greater number of well mark
0178 ne time, we ought only to see a few species PRESENTING slight modifications of structure in some d
0239 ects of the same species, in the same nest, PRESENTING gradations of structure; and this we do fin
0293 ome cases are on record of the same species PRESENTING distinct varieties in the upper and lower p
0356 classes of facts; which I have selected as PRESENTING the greatest amount of difficulty on the th
0389 classes of facts, which I have selected as PRESENTING the greatest amount of difficulty, on the v
0456 eless condition, or quite aborted, far from PRESENTING a strange difficulty, as they assuredly do
0016 e view of the origin of genera which I shall PRESENTLY give, we have no right to expect often to m
0017 of the world. I do not believe, as we shall PRESENTLY see, that all our dogs have descended from
0093 ep in the separation of the sexes of plants, PRESENTLY to be alluded to. Some holly trees bear onl
0096 s first suggested by Andrew Knight. We shall PRESENTLY see its importance; but I must here treat t
0356 on, but now be impassable; I shall, however, PRESENTLY have to discuss this branch of the subject
0363 imes include a few minute seeds? But I shall PRESENTLY have to recur to this subject. As icebergs
0392 by the wool and fur of quadrupeds. This case PRESENTS no difficulty on my view, for a hooked seed
0036 ess of improvement, through the occasional PRESERVATION of the best individuals, whether or not su
0061 s and to external nature, will tend to the PRESERVATION of that individual, and will generally be
0070 s enemies, is absolutely necessary for its PRESERVATION. Thus we can easily raise plenty of corn a
0070 a large stock of the same species for its PRESERVATION, explains, I believe, some singular facts
0081 injurious would be rigidly destroyed. This PRESERVATION of favourable variations and the rejection
0091 unt different prey; and from the continued PRESERVATION of the individuals best fitted for the two
0093 ur plant, by this process of the continued PRESERVATION or natural selection of more and more attr
0095 ect manner to each other, by the continued PRESERVATION of individuals presenting mutual and sligh
0095 ffs. Natural selection can act only by the PRESERVATION and accumulation of infinitesimally small
0109 natural selection acts solely through the PRESERVATION of variations in some way advantageous, wh
0127 similarly characterised. This principle of PRESERVATION, I have called, for the sake of brevity, N
0172 s. as natural selection acts solely by the PRESERVATION of profitable modifications, each new form
0181 nging conditions of life, in the continued PRESERVATION of individuals with fuller and fuller flan
0194 l selection acts by life and death, by the PRESERVATION of individuals with any favourable variati
0195 ance does not seem sufficient to cause the PRESERVATION of successively varying individuals. I hav
0205 election, a power which acts solely by the PRESERVATION of profitable variations in the struggle f
0245 ld not have been acquired by the continued PRESERVATION of successive profitable degrees of steril
0289 he truth, how accidental and rare is their PRESERVATION, far better than pages of detail. Nor is t
0301 se contingencies are indispensable for the PRESERVATION of all the transitional gradations between
0459 is a struggle for existence leading to the PRESERVATION of each profitable deviation of structure
0467 should not have acted under nature. In the PRESERVATION of favoured individuals and races, during
0480 nged, and are still slowly changing by the PRESERVATION and accumulation of successive slight favo
0102 ht be, natural selection will always tend to PRESERVE all the individuals varying in the right dir
0142 t natural selection will continually tend to PRESERVE those individuals which are born with consti
0288 a, at a rate sufficiently quick to embed and PRESERVE fossil remains. Throughout an enormously lar
0292 ent to keep the sea shallow and to embed and PRESERVE the remains before they had time to decay. O
0327 not thrown down quickly enough to embed and PRESERVE organic remains. During these long and blank
0426 ysiological importance, those which serve to PRESERVE life under the most diverse conditions of ex
0470 a constant tendency in natural selection to PRESERVE the most divergent offspring of any one spec
0015 far the new characters thus arising shall be PRESERVED. When we look to the hereditary varieties o
0036 for any special purpose, would be carefully PRESERVED during famines and other accidents, to whic
0037 degree, to their having naturally chosen and PRESERVED the best varieties they could anywhere find
0040 e infinitely small of any record having been PRESERVED of such slow, varying, and insensible chang
0061 y which each slight variation, if useful, is PRESERVED, by the term of Natural Selection, in order
0082 o their altered conditions, would tend to be PRESERVED; and natural selection would thus have free
0084 ed scale in the struggle for life, and so be PRESERVED. How fleeting are the wishes and efforts of
0090 have the best chance of surviving, and so be PRESERVED or selected, provided always that they reta

Page **(Key Word)**

Page ***(Key Word)**

```
0095  erited modifications, each profitable to the  PRESERVED  being; and as modern geology has almost ban
0117  ons which are in some way profitable will be   PRESERVED  or naturally selected. And here the importa
0117  represented by the outer dotted lines) being   PRESERVED  and accumulated by natural selection. When
0118  ergent of their variations will generally be   PRESERVED  during the next thousand generations. And a
0127  acterised will have the best chance of being   PRESERVED  in the struggle for life; and from the stro
0133  st clad individuals having been favoured and   PRESERVED  during many generations; and how much to th
0139  at more wrecks of ancient life have not been   PRESERVED, owing to the less severe competition to wh
0149  , that is, why natural selection should have   PRESERVED  or rejected each little deviation of form l
0173  sea, and to their remains being embedded and   PRESERVED  to a future age only in masses of sediment
0179  found only amongst fossil remains, which are   PRESERVED, as we shall in a future chapter attempt to
0189  be multiplied by the million; and each to be   PRESERVED  till a better be produced, and then the old
0210  se either one or the other instinct might be   PRESERVED  by natural selection. And such instances of
0288  europe prove. No organism wholly soft can be   PRESERVED. Shells and bones will decay and disappear
0288  the beach between high and low watermark are   PRESERVED. For instance, the several species of the C
0290  and peculiar marine faunas will probably be    PRESERVED  to a distant age. A little reflection will
0296  cted, had not the trees chanced to have been   PRESERVED: thus, Messrs. Lyell and Dawson found carbo
0300  rial productions of the archipelago would be   PRESERVED  in an excessively imperfect manner in the f
0300  ganic bodies from decay, no remains could be   PRESERVED. In our archipelago, I believe that fossili
0301  e species. If such gradations were not fully   PRESERVED, transitional varieties would merely appear
0301  utive record of their modifications could be   PRESERVED  in any one formation. Very many of the mari
0302  a record of the mutations of life, the best   PRESERVED  geological section presented, had not the d
0304  om the perfect manner in which specimens are   PRESERVED  in the oldest tertiary beds; from the ease
0304  dary periods, they would certainly have been   PRESERVED  and discovered; and as not one species had
0311  only here and there a short chapter has been   PRESERVED; and of each page, only here and there a fe
0321  ed, a few of the sufferers may often long be   PRESERVED, from being fitted to some peculiar line of
0325  ious in a still higher degree in order to be   PRESERVED  and to survive. We have distinct evidence o
0338  mbryo comes to be left as a sort of picture,   PRESERVED  by nature, of the ancient and less modified
0342  classes of organic beings have been largely   PRESERVED  in a fossil state; that the number both of
0342  the number both of specimens and of species,  PRESERVED  in our museums, is absolutely as nothing co
0345  laws of variation still acting round us, and   PRESERVED  by Natural Selection. Chapter XI. Geographi
0429  e successful competitors, with a few members   PRESERVED  by some unusual coincidence of favourable c
0464  nly organic beings of certain classes can be   PRESERVED  in a fossil condition, at least in any grea
0469  essively complex relations of life, would be   PRESERVED, accumulated, and inherited? Why, if man ca
0040  be said to have had a definite origin. A man   PRESERVES  and breeds from an individual with some sli
0084  the slightest; rejecting that which is bad,   PRESERVING  and adding up all that is good; silently an
0084  re of service to these birds and insects in   PRESERVING  them from danger. Grouse, if not destroyed
0104  fied offspring, solely by natural selection  PRESERVING  the same favourable variations. Isolation,
0209  can see no difficulty in natural selection   PRESERVING  and continually accumulating variations of
0467  hodically, or he may do it unconsciously by  PRESERVING  the individuals most useful to him at the t
0110  the progress of its formation, will generally  PRESS  hardest on its nearest kindred, and tend to e
0090  t season of the year when the wolf is hardest  PRESSED  for food. I can under such circumstances see
0233  fficult: it is known that bees are often hard  PRESSED  to get sufficient nectar; and I am informed
0302  the commencement and close of each formation,  PRESSED  so hardly on my theory. On the sudden appear
0309  o the centre of the earth, and which had been  PRESSED  on by an enormous weight of superincumbent w
0144  the pelvis in the human mother influences by   PRESSURE  the shape of the head of the child. In snake
0145  nces have been attributed by some authors to   PRESSURE, and the shape of the seeds in the ray flore
0198  e shape, also, of the pelvis might affect by   PRESSURE  the shape of the head of the young in the wo
0310  cks, which must have been heated under great   PRESSURE, have always seemed to me to require some sp
0328  of this nature have occurred in Europe. Mr.    PRESTWICH, in his admirable Memoirs on the eocene dep
0003  e author of the Vestiges of Creation would,  I PRESUME, say that after a certain unknown number of
0011  han do the same bones in the wild duck; and  I PRESUME  that this change may be safely attributed to
0044  y graduate into varieties. By a monstrosity  I PRESUME  is meant some considerable deviation of stru
0045  east some few generations? and in this case  I PRESUME  that the form would be called a variety. Aga
0149  more variable than those which are higher.   I PRESUME  that lowness in this case means that the sev
0159  mal structure of another race, the fantail.  I PRESUME  that no one will doubt that all such analogo
0160  e characteristic of the parent rock pigeon,  I PRESUME  that no one will doubt that this is a case o
0167  ne species was independently created, will,  I PRESUME, assert that each species has been created w
0215  s have ever been selected for tameness; and  I PRESUME  that we must attribute the whole of the inhe
0261  rtant for its welfare in a state of nature,  I PRESUME  that no one will suppose that this capacity
0450  e are very general in the males of mammals:  I PRESUME  that the bastard wing in birds may be safely
0045  pring from the same parents, or which may be  PRESUMED  to have thus arisen, being frequently observ
0073  profound is our ignorance, and so high our   PRESUMPTION, that we marvel when we hear of the extinc
0152  e degree or manner in any species, the fair   PRESUMPTION  is that it is of high importance to that s
0322  nction; if we must marvel, let it be at our   PRESUMPTION  in imagining for a moment that we understa
0017  case of some other domestic races, there is   PRESUMPTIVE, or even strong, evidence in favour of thi
0188  ous process. But may not this inference be    PRESUMPTUOUS? Have we any right to assume that the Crea
0018  or fourteen thousand years ago, and who will   PRETEND  to say how long before these ancient periods
0167  ound. Not in one case out of a hundred can we  PRETEND  to assign any reason why this or that part o
0223  nstinct of F. sanguinea originated I will not  PRETEND  to conjecture. But as ants, which are not sl
0243  disuse have probably come into play. I do not  PRETEND  that the facts given in this chapter strengt
0266  resemble closely either pure parent. Nor do I  PRETEND  that the foregoing remarks go to the root of
0293  ecies which then lived; but I can by no means  PRETEND  to assign due proportional weight to the fol
0302  , be ranked as distinct species. But I do not  PRETEND  that I should ever have suspected how poor a
0354  t and separated points; nor do I for a moment  PRETEND  that any explanation could be offered of man
0380  ny difficulties remain to be solved. I do not  PRETEND  to indicate the exact lines and means of mig
0464  ed which are probably varieties; but who will  PRETEND  that in future ages so many fossil links wil
0482  tly different forms. Nevertheless they do not  PRETEND  that they can define, or even conjecture, wh
0340  ng the later tertiary periods. Nor can it be   PRETENDED  that it is an immutable law that marsupials
0466  st and the most perfect organs; it cannot be   PRETENDED  that we know all the varied means of Distri
0057  ies are related to each other. No naturalist   PRETENDS  that all the species of a genus are equally
0007  artly connected with excess of food. It seems  PRETTY  clear that organic beings must be exposed dur
0009  even from the tropics, breed in this country   PRETTY  freely under confinement, with the exception
0032  irty years ago. When a race of plants is once  PRETTY  well established, the seed raisers do not pic
0119  the most divergent varieties will invariably  PREVAIL  and multiply: a medium form may often long e
0126  to increase. But which groups will ultimately  PREVAIL, no man can predict; for we well know that m
0327  , wherever produced, would tend everywhere to  PREVAIL. As they prevailed, they would cause the ext
```

Page ***(Key Word)**

Page **(Key Word)**

```
0383 lasses, an enormous range, but allied species PREVAIL in a remarkable manner throughout the world.
0385 and must have proceeded from a single source, PREVAIL throughout the world. Their distribution at
0426 he skin be covered by hair or feathers, if it PREVAIL throughout many and different species, espec
0489 er and dominant groups, which will ultimately PREVAIL and procreate new and dominant species. As a
0327 d, would tend everywhere to prevail. As they PREVAILED, they would cause the extinction of other a
0353 f the same species, a directly opposite rule PREVAILED; and species were not local, but had been p
0471 now see everywhere around us, and which has PREVAILED throughout all time. The grand fact of the
0350 see in these facts some deep organic bond, PREVAILING throughout space and time, over the same ar
0399 arth and stones on icebergs, drifted by the PREVAILING currents, this anomaly disappears. New Zeal
0166 his tendency, from unknown causes, sometimes PREVAILS. And we have just seen that in several speci
0417 ological importance and of almost universal PREVALENCE, and yet leave us in no doubt where it shou
0349 ast chapter, and we find American types then PREVALENT on the American continent and in the Americ
0364 are not accidental, nor is the direction of PREVALENT gales of wind. It should be observed that s
0456 e value set upon characters, if constant and PREVALENT, whether of high vital importance, or of th
0014 ued. It would be quite necessary, in order to PREVENT the effects of intercrossing, that only a si
0034 were often imported, and laws were passed to PREVENT their exportation: the destruction of horses
0041 e allowed to breed, and this will effectually PREVENT selection. But probably the most important p
0098 d from my own observations, which effectually PREVENT the stigma receiving pollen from its own flo
0098 there be no special mechanical contrivance to PREVENT the stigma of a flower receiving its own pol
0142 ect seed from the few survivors, with care to PREVENT accidental crosses, and then again get seed
0245 ed with the quality of sterility, in order to PREVENT the confusion of all organic forms. This vie
0247 more important, must be secluded in order to PREVENT pollen being brought to it by insects from o
0255 y been endowed with this quality, in order to PREVENT their crossing and blending together in utte
0257 n any recognisable character is sufficient to PREVENT two species crossing. It can be shown that p
0260 es have been endowed with sterility simply to PREVENT their becoming confounded in nature? I think
0261 to widely different climates, does not always PREVENT the two grafting together. As in hybridisati
0276 endowed with various degrees of sterility to PREVENT them crossing and blending in nature, than t
0276 ficulty in being grafted together in order to PREVENT them becoming inarched in our forests. The s
0293 insuperable difficulties, as it seems to me, PREVENT us coming to any just conclusion on this hea
0017 e rein deer, or of cold by the common camel, PREVENTED their domestication? I cannot doubt that if
0097 ity is greatly diminished if these visits be PREVENTED. Now, it is scarcely possible that bees sho
0104 differently circumstanced districts, will be PREVENTED. But isolation probably acts more efficient
0107 cal changes of any kind, immigration will be PREVENTED, so that new places in the polity of each i
0211 of about a dozen aphides on a dockplant, and PREVENTED their attendance during several hours. Afte
0221 st, twenty nine yards distant, but they were PREVENTED from getting any pupae to rear as slaves. I
0249 den, the visits of insects must be carefully PREVENTED during the flowering season: hence hybrids
0252 ous influence of close interbreeding is thus PREVENTED. Any one may readily convince himself of th
0384 ust have parted river systems and completely PREVENTED their inosculation, seems to lead to this s
0042 of animals with separate sexes, facility in PREVENTING crosses is an important element of success
0192 family had originally existed as organs for PREVENTING the ova from being washed out of the sack?
0002 here done. I much regret that want of space PREVENTS my having the satisfaction of acknowledging
0211 t to have been here given; but want of space PREVENTS me. I can only assert, that instincts certai
0215 wish to sit on their eggs. Familiarity alone PREVENTS our seeing how universally and largely the m
0028 t pigeons are given in a bill of fare in the PREVIOUS dynasty. In the time of the Romans, as we he
0207 of instinct might have been worked into the PREVIOUS chapters; but I have thought that it would b
0355 ributes to generation with modification. The PREVIOUS remarks on single and multiple centres of cr
0071 nature had been enclosed twenty five years PREVIOUSLY and planted with Scotch fir. The change in
0292 rmed; all circumstances most favourable, as PREVIOUSLY explained, for the formation of new varieti
0293 extreme to infer that it had not elsewhere PREVIOUSLY existed. So again when we find a species di
0337 have seized on places which must have been PREVIOUSLY occupied, we may believe, if all the animal
0344 isting forms, sometimes blending two groups PREVIOUSLY classed as distinct forms one; but more comm
0360 he salt water. On the other hand he did not PREVIOUSLY dry the plants or branches with the fruit:
0424 anthus, Myanthus, and Catasetum, which had PREVIOUSLY been ranked as three distinct genera, were
0062 estlings, are destroyed by birds and beasts of PREY; we do not always bear in mind, that though fo
0068 is not the obtaining food, but the serving as PREY to other animals, which determines the average
0068 nd rhinoceros, none are destroyed by beasts of PREY: even the tiger in India most rarely dares to
0070 sort of struggle between the parasite and its PREY. On the other hand, in many cases, a large sto
0072 s are probably regulated by hawks or beasts of PREY) were to increase in Paraguay, the flies would
0075 ls, and other animals with birds and beasts of PREY, all striving to increase, and all feeding on
0075 anic being on another, as of a parasite on its PREY, lies generally between beings remote in the s
0077 ith their aquatic insects, to hunt for its own PREY, and to escape serving as prey to other animal
0077 unt for its own prey; and to escape serving as PREY to other animals. The store of nutriment laid
0084 they are known to suffer largely from birds of PREY; and hawks are guided by their eyesight to the
0084 nd hawks are guided by their eyesight to their PREY, so much so, that on parts of the Continent pe
0090 leetness; and let us suppose that the fleetest PREY, a deer for instance, had from any change in t
0090 he country increased in numbers, or that other PREY had decreased in numbers, during that season o
0090 ys that they retained strength to master their PREY at this or at some other period of the year, w
0090 d of the year, when they might be compelled to PREY on other animals. I can see no more reason to
0091 an innate tendency to pursue certain kinds of PREY. Nor can this be thought very improbable; for
0091 s, would naturally be forced to hunt different PREY; and from the continued preservation of the in
0113 nstance, being enabled to feed on new kinds of PREY, either dead or alive; some inhabiting new sta
0134 veral oceanic islands, tenanted by no beast of PREY, has been caused by disuse. The ostrich indeed
0175 ther competing species; that nearly all either PREY on or serve as prey for others; in short, that
0175 es; that nearly all either prey on or serve as PREY for others; in short, that each organic being
0175 uations in the number of its enemies or of its PREY, or in the seasons, be extremely liable to utt
0180 y, by enabling it to escape birds or beasts of PREY, or to collect food more quickly, or as there
0180 , let other competing rodents or new beasts of PREY immigrate, or old ones become modified, and al
0183 loped under many subordinate forms, for taking PREY of many kinds in many ways, on the land and in
0195 o search for food, or to escape from beasts of PREY. Organs now of trifling importance have probab
0201 its own defence and for the destruction of its PREY; but some authors suppose that at the same tim
0201 rattle for its own injury, namely, to warn its PREY to escape. I would almost as soon believe that
0213 ithout any training, as soon as it scented its PREY, stand motionless like a statue, and then slow
0218 es its own burrow and stores it with paralysed PREY for its own larvae to feed on, yet that when t
0362 their arrival. Some hawks and owls bolt their PREY whole, and after an interval of from twelve to
0362 ne hours in the stomachs of different birds of PREY; and two seeds of beet grew after having been
0471 orm of woodpecker, should have been created to PREY on insects on the ground: that upland geese, w
```

Page **(Key Word)**

0078 er its competitors, or over the animals which PREYED on it. On the confines of its geographical ra
0091 onal numbers of the animals on which our wolf PREYED, a cub might be born with an innate tendency
0077 or from which it has to escape, or on which it PREYS. This is obvious in the structure of the teet
0090 rations. Let us take the case of a wolf, which PREYS on various animals, securing some by craft, s
0179 tail; during summer this animal dives for and PREYS on fish, but during the long winter it leaves
0180 e long winter it leaves the frozen waters, and PREYS like other polecats on mice and land animals.
0028 of the Romans, as we hear from Pliny, immense PRICES were given for pigeons; nay, they are come to
0031 e actually effected is proved by the enormous PRICES given for animals with a good pedigree; and t
0013 ery one must have heard of cases of albinism, PRICKLY skin, hairy bodies, etc., appearing in sever
0196 oreover when a modification of structure has PRIMARILY arisen from the above or other unknown caus
0014 appearance of the peculiarity, and not to its PRIMARY cause, which may have acted on the ovules or
0022 abortion of the oil gland; the number of the PRIMARY wing and caudal feathers; the relative lengt
0207 t i have nothing to do with the origin of the PRIMARY mental powers, any more than I have with tha
0307 on why we do not find records of these vast PRIMORDIALLY similar periods, I can give no satisfactory answer.
0307 longmynd beds beneath Barrande's so called PRIMORDIAL zone. The presence of phosphatic nodules an
0432 link between these eleven genera and there PRIMORDIAL parent, and every intermediate link in each
0438 aning that during a long course of descent, PRIMORDIAL organs of any kind, vertebrae in the one ca
0484 on this earth have descended from some one PRIMORDIAL form, into which life was first breathed. W
0438 uld have seized on a certain number of the PRIMORDIALLY similar elements, many times repeated, and
0049 a single instance, the well known one of the PRIMROSE and cowslip, or Primula veris and elatior. T
0247 during several years repeatedly crossed the PRIMROSE and cowslip, which we have such good reason
0268 or instance, the blue and red pimpernel, the PRIMROSE and cowslip, which are considered by many of
0424 ieves that the cowslip is descended from the PRIMROSE, or conversely, ranks them together as a sin
0485 hought worthy of specific names, as with the PRIMROSE and cowslip; and in this case scientific and
0049 ell known one of the primrose and cowslip, or PRIMULA veris and elatior. These plants differ consi
0221 vening; and as Huber expressly states, their PRINCIPAL office is to search for aphides. This diffe
0335 . to compare small things with great: if the PRINCIPAL living and extinct races of the domestic pi
0005 thus be naturally selected. From the strong PRINCIPLE of inheritance, any selected variety will t
0007 estic Pigeons, their Differences and Origin. PRINCIPLE of Selection anciently followed, its Effect
0012 ntal belief: doubts have been thrown on this PRINCIPLE by theoretical writers alone. When a deviat
0025 an understand these facts, on the well known PRINCIPLE of reversion to ancestral characters, if al
0030 mself useful breeds. The great power of this PRINCIPLE of selection is not hypothetical. It is cer
0031 very good judge of an animal, speaks of the PRINCIPLE of selection as that which enables the agri
0031 ad and beak. In Saxony the importance of the PRINCIPLE of selection in regard to merino sheep is s
0032 distinct variety, and breeding from it, the PRINCIPLE would be so obvious as hardly to be worth n
0033 hese characters. It may be objected that the PRINCIPLE of selection has been reduced to methodical
0033 rtant. But it is very far from true that the PRINCIPLE is a modern discovery. I could give several
0033 full acknowledgment of the importance of the PRINCIPLE in works of high antiquity. In rude and bar
0034 to the roguing of plants by nurserymen. The PRINCIPLE of selection I find distinctly given in an
0040 w sub breed are once fully acknowledged, the PRINCIPLE, as I have called it, of unconscious select
0041 f the highest importance to success. On this PRINCIPLE Marshall has remarked, with respect to the
0057 same genus. But when we come to discuss the PRINCIPLE, as I call it, of Divergence of Character,
0061 small number can survive. I have called this PRINCIPLE, by which each slight variation, if useful,
0063 season or occasional year, otherwise, on the PRINCIPLE of geometrical increase, its numbers would
0080 chapter, act in regard to variation? Can the PRINCIPLE of selection, which we have seen is so pote
0094 es of our plant would be advantageous on the PRINCIPLE of the division of labour, individuals with
0095 , so will natural selection, if it be a true PRINCIPLE, banish the belief of the continued creatio
0103 owly spread to other districts. On the above PRINCIPLE, nurserymen always prefer getting seed from
0104 ns of life remain the same, only through the PRINCIPLE of inheritance, and through natural selecti
0111 ous pestilence. Divergence of Character. The PRINCIPLE, which I have designated by this term, is o
0112 rather longer beak; and on the acknowledged PRINCIPLE that fanciers do not and will not admire a
0112 uctions the action of what may be called the PRINCIPLE of divergence, causing differences, at firs
0112 but how, it may be asked, can any analogous PRINCIPLE apply in nature? I believe it can and does
0114 , take the rank of species. The truth of the PRINCIPLE, that the greatest amount of life can be su
0114 e call different genera and orders. The same PRINCIPLE is seen in the naturalisation of plants thr
0116 ied by other beings. Now let us see how this PRINCIPLE of great benefit being derived from diverge
0117 lly selected. And here the importance of the PRINCIPLE of benefit being derived from divergence of
0118 roduced variety a2, which will, owing to the PRINCIPLE of divergence, differ more from (A) than di
0127 n the struggle for life; and from the strong PRINCIPLE of inheritance they will tend to produce of
0127 duce offspring similarly characterised. This PRINCIPLE of preservation, I have called, for the sak
0127 natural Selection. Natural selection, on the PRINCIPLE of qualities being inherited at correspondi
0147 er facts, may be merged under a more general PRINCIPLE, namely, that natural selection is continua
0154 to a former and less modified condition. The PRINCIPLE included in these remarks may be extended.
0168 lso seen in the second Chapter that the same PRINCIPLE applies to the whole individual: for in a d
0177 slowly modified and improved. It is the same PRINCIPLE which, as I believe, accounts for the commo
0186 ves in the struggle for existence and in the PRINCIPLE of natural selection, will acknowledge that
0188 at any degree of hesitation in extending the PRINCIPLE of natural selection to such startling leng
0203 most rare, is all the same to the inexorable PRINCIPLE of natural selection. If we admire the seve
0206 illustrious Cuvier, is fully embraced by the PRINCIPLE of natural selection. For natural selection
0225 tion of his theory. Let us look to the great PRINCIPLE of gradation, and see whether Nature does n
0239 that I have an overweening confidence in the PRINCIPLE of natural selection, when I do not admit t
0241 o a social community of insects, on the same PRINCIPLE that the division of labour is useful to ci
0242 to confess; that, with all my faith in this PRINCIPLE, I should never have anticipated that natur
0243 ncts. For instance, we can understand on the PRINCIPLE of inheritance, how it is that the thrush o
0267 wn bond, which is essentially related to the PRINCIPLE of life. Fertility of varieties when crosse
0281 d undergone a vast amount of change; and the PRINCIPLE of competition between organism and organis
0314 fied and improved, we can understand, on the PRINCIPLE of competition, and on that of the many all
0327 throughout the world, accords well with the PRINCIPLE of new species having been formed by domina
0329 mi and this fact is at once explained on the PRINCIPLE of descent. The more ancient any form is, t
0331 their ancient and common progenitor. On the PRINCIPLE of the continued tendency to divergence of
0350 duce groups of modified descendants. On this PRINCIPLE of inheritance with modification, we can un
0355 hened: for we can clearly understand, on the PRINCIPLE of modification, why the inhabitants of a r
0399 lonists would be liable to modification; the PRINCIPLE of inheritance still betraying their origin
0403 of distinct mammals, birds, and plants. The PRINCIPLE which determines the general character of t
0404 l forms to the whole world. We see this same PRINCIPLE in the blind animals inhabiting the caves o
0411 new and distinct species; and these, on the PRINCIPLE of inheritance, tend to produce other new a

Page **************************************(Key Word)**************************************

```
0412 e genera on the left hand have, on this same PRINCIPLE, much in common, and form a sub family, dis
0418 , they use it as of subordinated value. This PRINCIPLE has been broadly confessed by some naturali
0423 here had been more or less modification, the PRINCIPLE of inheritance would keep the forms togethe
0429 gical succession I attempted to show, on the PRINCIPLE of each group having generally diverged muc
0430 inities of distinct orders of plants. On the PRINCIPLE of the multiplication and gradual divergenc
0432 e diagram would still hold good; and, on the PRINCIPLE of inheritance, all the forms descended fro
0438 damental resemblance, retained by the strong PRINCIPLE of inheritance. In the great class of mollu
0448 in a slightly different manner, then, on the PRINCIPLE of inheritance at corresponding ages, the a
0450 one in natural history, are explained on the PRINCIPLE of slight modifications not appearing, in t
0455 turity and to its full powers of action, the PRINCIPLE of inheritance at corresponding ages will r
0455 should have a case of complete abortion. The PRINCIPLE, also, of economy, explained in a former ch
0457 in each individual animal and plant. On the PRINCIPLE of successive slight variations, not necess
0457 elation to their habits of life, through the PRINCIPLE of modifications being inherited at corresp
0457 nherited at corresponding ages. On this same PRINCIPLE, and bearing in mind; that when organs are
0457 wants, and bearing in mind how strong is the PRINCIPLE of inheritance, the occurrence of rudimenta
0474 i have attempted to show how much light the PRINCIPLE of gradation throws on the admirable archit
0475 anic world, almost inevitably follows on the PRINCIPLE of natural selection: for old forms will be
0477 rogenitors and early colonists. On this same PRINCIPLE of former migration, combined in most cases
0479 n the early progenitor of each class. On the PRINCIPLE of successive variations not always superve
0480 untouched by selection or disuse, and on the PRINCIPLE of inheritance at corresponding ages have b
0483 ities, and all can be classified on the same PRINCIPLE, in groups sub ordinate to groups. Fossil r
0032 ome even a skilful pigeon fancier. The same PRINCIPLES are followed by horticulturists: but the va
0044 ving restricted ranges. Before applying the PRINCIPLES arrived at in the last chapter to organic b
0116 divergence of character, combined with the PRINCIPLES of natural selection and of extinction, wil
0121 represented in the diagram, another of our PRINCIPLES, namely that of extinction, will have playe
0128 ed and intermediate forms of life. On these PRINCIPLES, I believe, the nature of the affinities of
0158 the same parts of the organisation, are all PRINCIPLES closely connected together. All being mainl
0282 read Sir Charles Lyell's grand work on the PRINCIPLES of Geology, which the future historian will
0282 s volume. Not that it suffices to study the PRINCIPLES of geology, or to read special treatises by
0292 y are in strict accordance with the general PRINCIPLES inculcated by Sir C. Lyell; and E. Forbes i
0301 slight degree, they would, according to the PRINCIPLES followed by many palaeontologists, be ranke
0351 lation in themselves can do anything. These PRINCIPLES come into play only by bringing organisms i
0368 much disturbed, and, in accordance with the PRINCIPLES inculcated in this volume, they will not ha
0409 hical provinces of the world. On these same PRINCIPLES, we can understand, as I have endeavoured t
0412 on, as formerly explained, of these several PRINCIPLES: and he will see that the inevitable result
0444 in the child than in the parent. These two PRINCIPLES, if their truth be admitted, will, I believ
0446 ctly as much as in the adult state. The two PRINCIPLES above given seem to me to explain these fac
0446 let us apply these facts and the above two PRINCIPLES, which latter, though not proved true, can
0447 , in another as wings; and on the above two PRINCIPLES, namely of each successive modification sup
0467 species. There is no obvious reason why the PRINCIPLES which have acted so efficiently under domes
0478 and divergence of character. On these same PRINCIPLES we see how it is, that the mutual affinitie
0008 ale reproductive elements having been affected PRIOR to the act of conception. Several reasons mak
0010 g been affected by the treatment of the parent PRIOR to the act of conception. These cases anyhow
0225 : each cell, as is well known, is an hexagonal PRISM, with the basal edges of its six sides bevell
0228 ion between the basins; so that each hexagonal PRISM was built upon the festooned edge of a smooth
0227 there will result a double layer of hexagonal PRISMS united together by pyrimidal bases formed of
0227 and the rhombs and the sides of the hexagonal PRISMS will have every angle identically the same wi
0227 t this is no difficulty, that after hexagonal PRISMS have been formed by the intersection of adjoi
0235 what are the several angles of the hexagonal PRISMS and of the basal rhombic plates. The motive p
0218 red by another sphex it takes advantage of the PRIZE, and becomes for the occasion parasitic. In t
0029 each race varies slightly, for they win their PRIZES by selecting such slight differences, yet the
0007 under nature. There is, also, I think, some PROBABILITY in the view propounded by Andrew Knight, t
0026 he history of the dog I think there is some PROBABILITY in this hypothesis, if applied to species
0068 vermin were destroyed, there would, in all PROBABILITY, be less game than at present, although hu
0191 ars, it is so important to bear in mind the PROBABILITY of conversion from one function to another
0212 out good evidence. The possibility, or even PROBABILITY, of inherited variations of instinct in a
0287 e action of the coast waves. So that in all PROBABILITY a far longer period than 300 million years
0293 ecies first appearing in any formation, the PROBABILITY is that it only then first immigrated into
0296 bottom, middle, and top of a formation, the PROBABILITY is that they have not lived on the same sp
0308 ozoic and secondary formations would in all PROBABILITY have been accumulated from sediment derive
0369 d there ever since; they will, also, in all PROBABILITY have become mingled with ancient Alpine sp
0388 e aquatic species; we should not forget the PROBABILITY of many species having formerly ranged as
0396 have reached their present homes. But the PROBABILITY of many islands having existed as halting
0402 ten take, I think, an erroneous view of the PROBABILITY of closely allied species invading each ot
0427 y distinct and isolated region, have in all PROBABILITY descended from the same parents. We can un
0001 of the conclusions, which then seemed to me PROBABLE: from that period to the present day I have
0004 mencement of my observations it seemed to me PROBABLE that a careful study of domesticated animals
0011 mals not being much alarmed by danger, seems PROBABLE. There are many laws regulating variation, s
0018 s researches have rendered it in some degree PROBABLE that man sufficiently civilized to have manu
0018 and other considerations, I think it highly PROBABLE that our domestic dogs have descended from s
0028 her in the same aviary. I have discussed the PROBABLE origin of domestic pigeons at some, yet quit
0064 and I have taken some pains to estimate its PROBABLE minimum rate of natural increase: it will be
0081 ed polymorphic. We shall best understand the PROBABLE course of natural selection by taking the ca
0119 ranches in the lines of descent, will, it is PROBABLE, often take the place of, and so destroy, th
0120 nd genera are formed. In a large genus it is PROBABLE that more than one species would vary. In th
0122 ountry. It seems, therefore, to me extremely PROBABLE that they will have taken the places of, and
0141 oregoing argument too far, on account of the PROBABLE origin of some of our domestic animals from
0156 egree, or only in a slight degree, it is not PROBABLE that they should vary at the present day. On
0156 the species from a common progenitor, it is PROBABLE that they should still often be in some degr
0157 variations of this part would, it is highly PROBABLE, be taken advantage of by natural and sexual
0160 fter a great number of generations, the most PROBABLE hypothesis is, not that the offspring sudden
0161 ns, gains an ascendancy. For instance, it is PROBABLE that in each generation of the barb pigeon,
0162 the blue tint, and which it does not appear PROBABLE would all appear together from simple variat
0166 in the mongrels. I have stated that the most PROBABLE hypothesis to account for the reappearance o
0178 ous area, intermediate varieties will, it is PROBABLE, at first have been formed in the intermedia
0183 tration of the flying fish, it does not seem PROBABLE that fishes capable of true flight would hav
```

Page *************************************(Key Word)**************************************

Page ***************************************(Key Word)***************************************

```
0191  h, the wings and wingcovers of insects, it is  PROBABLE  that organs which at a very ancient period
0245  ic forms. This view certainly seems at first  PROBABLE,  for species within the same country could
0254  this latter alternative seems to me the most  PROBABLE,  and I am inclined to believe in its truth,
0263  rics. We may now look a little closer at the  PROBABLE  causes of the sterility of first crosses an
0294  ole of this period. It is not, for instance,  PROBABLE  that sediment was deposited during the whol
0301  ar as so many distinct species. It is, also,  PROBABLE  that each great period of subsidence would
0321  t remember what has been already said on the  PROBABLE  wide intervals of time between our consecut
0326  eed in spreading. The diffusion would, it is  PROBABLE,  be slower with the terrestrial inhabitants
0328  eas are affected by the same movement, it is  PROBABLE  that strictly contemporaneous formations ha
0353  nd present conditions permitted, is the most  PROBABLE.  Undoubtedly many cases occur, in which we
0354  hat we ought to give up the belief, rendered  PROBABLE  by general considerations, that each specie
0374  aneous in a geological sense, it seems to me  PROBABLE  that it was, during a part at least of the
0374  ce to the contrary, we may at least admit as  PROBABLE  that the glacial action was simultaneous on
0379  mpete with many new forms of life; and it is  PROBABLE  that selected modifications in their structu
0395  orming a judgment in some cases owing to the  PROBABLE  naturalisation of certain mammals through ma
0437  mber and structure; consequently it is quite  PROBABLE  that natural selection, during a long contir
0444  er, that there is some evidence to render it  PROBABLE,  that at whatever age any variation first ap
0445  and this surprised me greatly, as I think it  PROBABLE  that the difference between these two breeds
0446  oved true, can be shown to be in some degree  PROBABLE,  to species in a state of nature. Let us tak
0469  rther than this, seems to me to be in itself  PROBABLE.  I have already recapitulated, as fairly as
0010  t, if any of the young had varied, all would  PROBABLY  have varied in the same manner. To judge how
0015  ange, variations and reversions of character  PROBABLY  do occur: but natural selection, as will her
0019  of the whole world, which I fully admit have  PROBABLY  descended from several wild species, I canno
0030  or fancy. Some variations useful to him have  PROBABLY  arisen suddenly, or by one step; many botani
0030  ave suddenly arisen in a seedling. So it has  PROBABLY  been with the turnspit dog: and this is know
0031  ighly competent authorities. Youatt, who was  PROBABLY  better acquainted with the works of agricult
0035  s directly derived from the spaniel, and has  PROBABLY  been slowly altered from it. It is known tha
0040  mething distinct and valuable, and will then  PROBABLY  first receive a provincial name. In semi civ
0041  this will effectually prevent selection. But  PROBABLY  the most important point of all, is, that th
0071  and the ill effects of close interbreeding,  PROBABLY  come into play in some of these cases: but o
0072  e, must be habitually checked by some means,  PROBABLY  by birds. Hence, if certain insectivorous bi
0072  rtain insectivorous birds (whose numbers are  PROBABLY  regulated by hawks or beasts of prey) were t
0074  life, and during different seasons or years,  PROBABLY  come into play: some one check or some few b
0076  the same place in the economy of nature; but  PROBABLY  in no one case could we precisely say why on
0078  o give any form some advantage over another.  PROBABLY  in no single instance should we know what to
0085  we must not forget that climate, food, etc.,  PROBABLY  produce some slight and direct effect. It is
0086  correlation, the structure of the adult: and  PROBABLY  in the case of those insects which live only
0086  conversely, modifications in the adult will  PROBABLY  often affect the structure of the larva; but
0087  ditarily attached to that sex, the same fact  PROBABLY  occurs under nature, and if so, natural sele
0091  f leaving offspring. Some of its young would  PROBABLY  inherit the same habits or structure, and by
0092  and surviving. Some of these seedlings would  PROBABLY  inherit the nectar excreting power. Those in
0094  d leaving descendants. Its descendants would  PROBABLY  inherit a tendency to a similar slight devía
0104  districts, will be prevented. But isolation  PROBABLY  acts more efficiently in checking the immigr
0106  conclude that, although small isolated areas  PROBABLY  have been in some respects highly favourable
0107  uctions a large continental area, which will  PROBABLY  undergo many oscillations of level, and whic
0108  ed. But the action of natural selection will  PROBABLY  still oftener depend on some of the inhabita
0110  egion has as yet got its maximum of species.  PROBABLY  no region is as yet fully stocked, for at th
0121  will generally tend to become extinct. So it  PROBABLY  will be with many whole collateral lines of
0122  at the fourteen thousandth generation, will  PROBABLY  have inherited some of the same advantages:
0136  he action of natural selection, but combined  PROBABLY  with disuse. For during thousands of success
0137  p by skin and fur. This state of the eyes is  PROBABLY  due to gradual reduction from disuse, but ai
0139  ch the inhabitants of these dark abodes will  PROBABLY  have been exposed. Acclimatisation. Habit is
0143  been of any great use to the breed it might  PROBABLY  have been rendered permanent by natural sele
0144  teeth in the naked Turkish dog, though here  PROBABLY  homology comes into play? With respect to th
0161  nt variety. But characters thus gained would  PROBABLY  be of an important nature, for the presence
0168  closely checked by natural selection. It is  PROBABLY  from this same cause that organic beings low
0168  disregarded by natural selection, and hence  PROBABLY  are variable. Specific characters, that is,
0176  ally differ from species, the same rule will  PROBABLY  apply to both: and if we in imagination adap
0178  asional immigration of new inhabitants, and,  PROBABLY,  in a still more important degree, on some o
0183  ns of structure lead to changed habits: both  PROBABLY  often change almost simultaneously. Of cases
0187  on this head. In this great class we should  PROBABLY  have to descend far beneath the lowest known
0195  prey. Organs now of trifling importance have  PROBABLY  in some cases been of high importance to an
0196  we should remember that climate, food, etc.,  PROBABLY  have some little direct influence on the org
0197  colour is due to some quite distinct cause,  PROBABLY  to sexual selection. A trailing bamboo in th
0198  land breeds; and a mountainous country would  PROBABLY  affect the hind limbs from exercising them m
0198  ion, the front limbs and even the head would  PROBABLY  be affected. The shape, also, of the pelvis
0199  use to their possessors. Physical conditions  PROBABLY  have had some little effect on structure, qu
0207  s that of the hive bee making its cells will  PROBABLY  have occurred to many readers, as a difficul
0211  as the excretion is extremely viscid, it is  PROBABLY  a convenience to the aphides to have it remo
0211  he aphides to have it removed: and therefore  PROBABLY  the aphides do not instinctively excrete for
0214  o one would ever have thought of teaching or  PROBABLY  could have taught, the tumbler pigeon to tum
0215  at habit, with some degree of selection, has  PROBABLY  concurred in civilising by inheritance our d
0216  ically and unconsciously: but in most cases,  PROBABLY,  habit and selection have acted together. We
0217  ly period: and the first hatched young would  PROBABLY  have to be fed by the male alone. Put the Am
0218  are hatched by the males. This instinct may  PROBABLY  be accounted for by the fact of the hens lay
0221  yards distant, which they ascended together  PROBABLY  in search of aphides or cocci. According to
0221  the masters and slaves in the two countries,  PROBABLY  depends merely on the slaves being captured
0226  double layer, the resulting structure would  PROBABLY  have been as perfect as the comb of the hive
0234  t this latter circumstance determined, as it  PROBABLY  often does determine, the numbers of a humbl
0239  ive, slight, profitable modification did not  PROBABLY  at first appear in all the individual reuter
0243  . in some cases habit or use and disuse have  PROBABLY  come into play. I do not pretend that the fa
0246  s, has to be considered. The distinction has  PROBABLY  been slurred over, owing to the sterility of
0249  so that a cross between two flowers, though  PROBABLY  on the same plant, would be thus effected. M
0255  here are some which never have produced, and  PROBABLY  never would produce, even with the pollen of
0268  xplanation which will occur to everyone, and  PROBABLY  the true one, is that these dogs have descen
0269  i and thus she may, either directly, or more  PROBABLY  indirectly, through correlation, modify the
```

Page ***************************************(Key Word)***************************************

probably

age ***************************************(Key Word)***************************************

```
283  ny thousand feet in thickness, which, though  PROBABLY  formed at a quicker rate than many other dep
284  rate of accumulation of the degraded matter,  PROBABLY  offers the best evidence of the lapse of tim
286  would be erroneous: but this source of doubt  PROBABLY  would not greatly affect the estimate as app
290  l successive and peculiar marine faunas will  PROBABLY  be preserved to a distant age. A little refl
294  all have been upraised, organic remains will  PROBABLY  first appear and disappear at different leve
295  of sediment and the amount of subsidence is   PROBABLY  a rare contingency; for it has been observed
296  y one geological period, a section would not  PROBABLY  include all the fine intermediate gradations
298  with shells and other marine animals, it is   PROBABLY  those which have had the widest range, far e
299  eties; and this not having been effected, is  PROBABLY  the gravest and most obvious of all the many
300  during the periods of subsidence there would  PROBABLY  represents the former state of Europe, when
303  r the enormous intervals of time, which have  PROBABLY  be much extinction of life: during the perio
306  ived long before the Silurian age, and which  PROBABLY  elapsed between our consecutive formations,
307  osited, long periods elapsed, as long as, or  PROBABLY  differed greatly from any known animal. Some
307  us matter in some of the lowest azoic rocks,  PROBABLY  far longer than, the whole interval from the
320  ch have been produced within a given time is  PROBABLY  indicates the former existence of life at th
326  , recurring at long intervals of time, would  PROBABLY  greater than that of the old forms which hav
330  s between two living forms, the objection is  PROBABLY  be also favourable, as before explained. One
332  ould be so closely linked together that they  PROBABLY  valid. But I apprehend that in a perfectly n
341  ut and have left no progeny. Or, which would  PROBABLY  would have to be united into one great famil
342  en the successive formations; that there has  PROBABLY  be a far commoner case, two or three species
355  era with the species of a second region, has  PROBABLY  been more extinction during the periods of s
355  ew hundreds of miles from a continent, would  PROBABLY  received at some former period immigrants fr
377  e are not now concerned. The tropical plants  PROBABLY  receive from it in the course of time a few
387  sh water animals. Other and unknown agencies  PROBABLY  suffered much extinction; how much no one ca
387  the seeds of the great southern water lily  (PROBABLY, have also played a part. I have stated that
387  ond and getting a hearty meal of fish, would  PROBABLY  according to Dr. Hooker, the Nelumbium lute
388  with those on the land, the competition will  PROBABLY  reject from its stomach a pellet containing
396  for on this latter view the migration would  PROBABLY  be less severe between aquatic than between
401  . if then it varied, natural selection would  PROBABLY  have been been more complete; and if modification
402  may infer from certain facts that these have  PROBABLY  favour different varieties in the different
402  fitted for their own places in nature, both  PROBABLY  spread from some one island to the others. B
403  ts of the same continent, pre occupation has  PROBABLY  will hold their own places and keep separate
406  nd better fitted for distant transportation,  PROBABLY  played an important part in checking the com
409  inent or other source whence immigrants were  PROBABLY  accounts for a law which has long been obser
416  ordinate value in classification: yet no one  PROBABLY  derived. We can see why in two areas, howeve
428  quaternary, and ternary classifications have  PROBABLY  will say that the antennae in these two divi
434  diating lines of affinities. We shall never,  PROBABLY, arisen. As the modified descendants of domin
438  orphosed leaves: but it would in these cases  PROBABLY  be more correct, as Professor Huxley has rem
439  aining numerous characters, which they would  PROBABLY  have retained through inheritance, if they h
441  ganisation being higher or lower. But no one  PROBABLY  will dispute that the butterfly is higher th
444  ally varieties most closely allied, and have  PROBABLY  descended from the same wild stock; hence I
455  r, will be saved as far as is possible, will  PROBABLY  often come into play; and this will tend to
463  e intermediate varieties which will at first  PROBABLY  exist in the intermediate zones, will be lia
464  ting doubtful forms could be named which are  PROBABLY  varieties: but who will pretend that in futu
484  lank. During these latter periods there will  PROBABLY  be more variability in the forms of life: du
484  . therefore I should infer from analogy that  PROBABLY  all the organic beings which have ever lived
488  nge in the fossils of consecutive formations  PROBABLY  serves as a fair measure of the lapse of act
488  earth's history, when the forms of life were  PROBABLY  fewer and simpler, the rate of change was pr
488  ly fewer and simpler, the rate of change was  PROBABLY  slower; and at the first dawn of life, when
0004 er the best chance of making out this obscure  PROBLEM. Nor have I been disappointed: in this and i
0075 ding to definite laws: but how simple is this  PROBLEM compared to the action and reaction of the i
0224 that bees have practically solved a recondite  PPORLEM, and have made their cells of the proper sha
0441 of legs, a very simple single eye, and a      PROBOSCIFORMED mouth, with which they feed largely, for
0094 body, or in the curvature and length of the   PROBOSCIS, etc., far too slight to be appreciated by
0095 a slightly longer or differently constructed  PROBOSCIS. On the other hand, I have found by experim
0434 ore different than the immensely long spiral  PROBOSCIS of a sphinx moth, the curious folded one of
0018 , thinks that all the breeds of poultry have  PROCEEDED from the common wild Indian fowl (Gallus ba
0023 veral breeds are not varieties, and have not  PROCEEDED from the rock pigeon, they must have descen
0029 ston pippin or Codlin apple, could ever have  PROCEEDED from the seeds of the same tree. Innumerabl
0351 quarters of the world, must originally have  PROCEEDED from the same source, as they have descende
0352 ting distant and isolated regions, must have  PROCEEDED from one spot, where their parents were fir
0357 e it will some day be, that each species has  PROCEEDED from a single birthplace; and when in the c
0385 descended from a common parent and must have  PROCEEDED from a single source, prevail throughout th
0389 from a single parent; and therefore have all  PROCEEDED from a common birthplace, notwithstanding t
0408 pecies, and likewise of allied species, have  PROCEEDED from some one source; then I think all the
0061 ion, however slight and from whatever cause   PROCEEDING, if it be in any degree profitable to an in
0069 injurious action of climate, than we do in   PROCEEDING southwards or in descending a mountain. Whe
0117 f diverging dotted lines of unequal lengths   PROCEEDING from (A), may represent its varying offspri
0118 thus the varieties or modified descendants,   PROCEEDING from the common parent (A), will generally
0119 e diagram by the several divergent branches   PROCEEDING from (A). The modified offspring from the l
0120 epresented in the diagram, if all the lines   PROCEEDING from (A) were removed, excepting that from
0183 th america, hovering over one spot and then   PROCEEDING to another, like a kestrel, and at other ti
0348 hey are wholly distinct. On the other hand,   PROCEEDING still further westward from the eastern isl
0412 ble result is that the modified descendants   PROCEEDING from one progenitor become broken up into g
0020 ccount for our several domestic races by this  PROCESS, we must admit the former existence of the m
0035 breed. Nevertheless I cannot doubt that this  PROCESS, continued during centuries, would improve a
0035 as bakewell, Collins, etc., by this very same PROCESS, only carried on more methodically, did grea
0035 e dog in Spain like our pointer. By a similar PROCESS of selection, and by careful training, the w
0036 e than their dogs. In plants the same gradual PROCESS of improvement, through the occasional prese
0038 one country than in the other, and thus by a  PROCESS of natural selection, as will hereafter be m
0040 further improved by the same slow and gradual PROCESS, they will spread more widely, and will get
0040 knowledge of any new sub breed will be a slow PROCESS. As soon as the points of value of the new s
0056 cially as we have every reason to believe the PROCESS of manufacturing new species to be a slow on
0087 very short for the bird's own advantage, the  PROCESS of modification would be very slow, and ther
0091 s or structure, and by the repetition of this PROCESS, a new variety might be formed which would e
```

Page ***(Key Word)***************************************

Page ***(Key Word)***

```
0093  s, would be selected. When our plant, by this  PROCESS of the continued preservation or natural se
0093  ularly carried from flower to flower, another  PROCESS might commence. No naturalist doubts the adv
0102  urely but slowly follow from this unconscious  PROCESS of selection, notwithstanding a large amount
0104  olation, also, is an important element in the  PROCESS of natural selection. In a confined or isola
0108  ation itself is apparently always a very slow  PROCESS. The process will often be greatly retarded
0108  is apparently always a very slow process. The  PROCESS will often be greatly retarded by free inter
0109  s of this world have changed. Slow though the  PROCESS of selection may be, if feeble man can do mu
0110  and tend to exterminate them. We see the same  PROCESS of extermination amongst our domesticated pr
0111  ding to my view, varieties are species in the  PROCESS of formation, or are, as I have called them,
0116  orders. In the Australian mammals, we see the  PROCESS of diversification in an early and incomplet
0118  their common parent (A). We may continue the  PROCESS by similar steps for any length of time: som
0118  nd diverging in character. In the diagram the  PROCESS is represented up to the ten thousandth gene
0118  st here remark that I do not suppose that the  PROCESS ever goes on so regularly as is represented
0119  lines. In some cases I do not doubt that the  PROCESS of modification will be confined to a single
0120  but we have only to suppose the steps in the  PROCESS of modification to be more numerous or great
0120  istinguishing species. By continuing the same  PROCESS for a greater number of generations (as show
0121  ar upwards from want of space. But during the  PROCESS of modification, represented in the diagram
0123  ing diverged at the first commencement of the  PROCESS of modification, will be widely different fr
0125  d diverged less. I see no reason to limit the  PROCESS of modification, as now explained, to the fo
0147  cing seed less fitted for dispersal; and this  PROCESS could not possibly go on in fruit which did
0147  and adjoining part being reduced by this same  PROCESS or by disuse, and, on the other hand, the ac
0169  si for variation is a long continued and slow  PROCESS, and natural selection will in such cases no
0169  endants, which on my view must be a very slow  PROCESS, requiring a long lapse of time, in this cas
0172  generally have been exterminated by the very  PROCESS of formation and perfection of the new form.
0173  escended from a common parent; and during the  PROCESS of modification, each has become adapted to
0176  sideration, as I believe, is that, during the  PROCESS of further modification, by which two variet
0177  y slowly formed, for variation is a very slow  PROCESS, and natural selection can do nothing until
0178  e been supplanted and exterminated during the  PROCESS of natural selection, so that they will no l
0179  e to accidental extermination; and during the  PROCESS of further modification through natural sele
0179  er, must assuredly have existed; but the very  PROCESS of natural selection constantly tends, as ha
0181  ted, until by the accumulated effects of this  PROCESS of natural selection, a perfect so called fl
0182  or they will have been supplanted by the very  PROCESS of perfection through natural selection. Fur
0188  e eye has been formed by a somewhat analogous  PROCESS. But may not this inference be presumptuous?
0189  ith unerring skill each improvement. Let this  PROCESS go on for millions on millions of years; and
0190  ll the work by itself, being aided during the  PROCESS of modification by the other organi and then
0203  f intermediate gradations, partly because the  PROCESS of natural selection will always be very slo
0203  a very few forms; and partly because the very  PROCESS of natural selection almost implies the cont
0217  in the same nest. If this were the case, the  PROCESS of laying and hatching might be inconvenient
0217  essful in rearing their young. By a continued  PROCESS of this nature, I believe that the strange i
0234  have to remain idle for many days during the  PROCESS of secretion. A large store of honey is indi
0235  basal rhombic plates. The motive power of the  PROCESS of natural selection having been economy of
0238  he same modification. And I believe that this  PROCESS has been repeated, until that prodigious amo
0269  ny one species. Seeing this difference in the  PROCESS of selection, as carried on by man and natur
0280  rywhere throughout nature depends on the very  PROCESS of natural selection, through which new vari
0280  parent forms. But just in proportion as this  PROCESS of extermination has acted on an enormous sc
0283  formed of moderately hard rocks, and mark the  PROCESS of degradation. The tides in most cases reac
0284  s transported by the currents of the sea, the  PROCESS of accumulation in any one area must be extr
0295  er to have given sufficient time for the slow  PROCESS of variation; hence the deposit will general
0295  mposition, we may reasonably suspect that the  PROCESS of deposition has been much interrupted, as
0296  rvals of time and changes of level during the  PROCESS of deposition, which would never even have b
0302  progenitor, must have been an extremely slow  PROCESS; and the progenitors must have lived long ag
0314  or simultaneously, or to an equal degree. The  PROCESS of modification must be extremely slow. The
0317  crease only slowly and progressively: for the  PROCESS of modification and the production of a numb
0318  the species of a group is generally a slower  PROCESS than their production: if the appearance and
0321  group is generally, as we have seen, a slower  PROCESS than its production. With respect to the app
0326  w countries to new varieties and species. The  PROCESS of diffusion may often be very slow, being d
0337  y a secondary fauna. I do not doubt that this  PROCESS of improvement has affected in a marked and
0338  being inherited at a corresponding age. This  PROCESS, whilst it leaves the embryo almost unaltere
0343  d endure for unequal periods of time: for the  PROCESS of modification is necessarily slow, and dep
0344  ole group of species may often be a very slow  PROCESS, from the survival of a few descendants, lin
0350  checking migration; as does time for the slow  PROCESS of modification through natural selection. W
0356  en intercross; I believe that during the slow  PROCESS of modification the individuals of the speci
0362  in seeds, however, were always killed by this  PROCESS. Although the beaks and feet of birds are ge
0404  se doubtful forms showing us the steps in the  PROCESS of modification. This relation between the p
0429  d much in character during the long continued  PROCESS of modification, how it is that the more anc
0455  ve size in the adult. But if each step of the  PROCESS of reduction were to be inherited, not at th
0462  ing very distant and isolated regions, as the  PROCESS of modification has necessarily been slow, a
0467  uite inappreciable by an uneducated eye. This  PROCESS of selection has been the great agency in th
0022  h their relative breadth and the presence of  PROCESSES. The size and shape of the apertures in the
0247  e kept in a chamber in his house. That these  PROCESSES are often injurious to the fertility of a p
0437  series of internal vertebrae bearing certain  PROCESSES and appendages; in the articulata, we see t
0345  all the chief laws of palaeontology plainly  PROCLAIM, as it seems to me, that species have been p
0457  en considered in this chapter, seem to me to  PROCLAIM so plainly, that the inumerable species, gen
0460  so many strange gradations in nature, as is  PROCLAIMED by the canon, Natura non facit saltum, that
0468  age will lead to victory. As geology plainly  PROCLAIMS that each land has undergone great physical
0224  s to this species to capture workers than to  PROCREATE them, the habit of collecting pupae origina
0489  nt groups, which will ultimately prevail and  PROCREATE new and dominant species. As all the living
0081  ld have the best chance of surviving and of  PROCREATING their kind? On the other hand, we may feel
0236  ally born capable of work, but incapable of  PROCREATION, I can see no very great difficulty in thi
0037  riod, who cultivated the best pear they could  PROCURE, never thought what splendid fruit we should
0215  ection is still at work, as each man tries to  PROCURE, without intending to improve the breed, dog
0360  natives of the coral islands in the Pacific,  PROCURE stones for their tools, solely from the root
0194  lne Edwards has well expressed it, nature is  PRODIGAL in variety, but niggard in innovation. Why,
0471  telligible. We can plainly see why nature is  PRODIGAL in variety, though niggard in innovation. Bu
0069  ng other species, we may clearly see in the  PRODIGIOUS number of plants in our gardens which can p
0152  ook at the breeds of the pigeon: see what a  PRODIGIOUS amount of difference there is in the beak o
```

Page ***(Key Word)***

prodigious

```
233  e secretion of each pound of wax; so that a  PRODIGIOUS  quantity of fluid nectar must be collected
236  as far as instinct alone is concerned, the   PRODIGIOUS  difference in this respect between the work
238  this process has been repeated, until that   PRODIGIOUS  amount of difference between the fertile an
285  here is nothing on the surface to show such  PRODIGIOUS  movements; the pile of rocks on the one or
294  nordinately great change of climate, on the  PRODIGIOUS  lapse of time, all included within this sam
336  erent climates and conditions. Consider the  PRODIGIOUS  vicissitudes of climate during the pleistoc
358  seem to me opposed to the admission of such  PRODIGIOUS  geographical revolutions within the recent
010  t can be shown that quite opposite conditions PRODUCE  similar changes of structure. Nevertheless s
031  o say, with respect to pigeons, that he would PRODUCE  any given feather in three years; but it wou
033  n the leaves, the flowers, or the fruit, will PRODUCE  races differing from each other chiefly in t
054  most numerous in individuals, which oftenest  PRODUCE  well marked varieties, or, as I consider the
061  have seen that man by selection can certainly PRODUCE  great results, and can adapt organic beings
065  y difference between organisms which annually PRODUCE  eggs or seeds by the thousand, and those whi
065  ggs or seeds by the thousand, and those which PRODUCE  extremely few, is, that the slow breeders wo
082  ariability is necessary: as man can certainly PRODUCE  great results by adding up in any given dire
082  o check immigration, is actually necessary to PRODUCE  new and unoccupied places for natural select
083  ve better resisted such intruders. As man can PRODUCE  and certainly has produced a great result by
085  n of an animal of any particular colour would PRODUCE  little effect: we should remember how essent
085  not forget that climate, food, etc., probably PRODUCE  some slight and direct effect. It is, howeve
089  according to their standard of beauty, might  PRODUCE  a marked effect. I strongly suspect that som
092  ll hereafter be more fully alluded to), would PRODUCE  very vigorous seedlings, which consequently
093  e that it would be advantageous to a plant to PRODUCE  stamens alone in one flower or on one whole
098  it must not be supposed that bees would thus  PRODUCE  a multitude of hybrids between distinct spec
108  prove still further the inhabitants, and thus PRODUCE  new species. That natural selection will alw
118  o two or three varieties, and some failing to PRODUCE  any. Thus the varieties or modified descenda
119  orm may often long endure, and may or may not PRODUCE  more than one modified descendant: for natur
121  ifferent in character, will generally tend to PRODUCE  the greatest number of modified descendants;
127  nc principle of inheritance they will tend to PRODUCE  offspring similarly characterised. This prin
131  ch the function of the reproductive system to PRODUCE  individual differences, or very slight devia
161  more abstract improbability in a tendency to  PRODUCE  any character being inherited for an endless
161  , we may sometimes observe a mere tendency to PRODUCE  a rudiment inherited: for instance, in the c
161  this plant must have an inherited tendency to PRODUCE  it. As all the species of the same genus are
162  some of the characters of the other, so as to PRODUCE  the intermediate form. But the best evidence
166  in the young of each successive generation to PRODUCE  the long lost character, and that this tende
171  inhabiting distant quarters of the world, to  PRODUCE  hybrids resembling in their stripes, not the
171  ? can we believe that natural selection could PRODUCE, on the one hand, organs of trifling importa
187  to those coarser vibrations of the air which  PRODUCE  sound. In looking for the gradations by whic
189  es, may in any way, or in any degree, tend to PRODUCE  a distincter image. We must suppose each new
200  of growth. Natural selection cannot possibly  PRODUCE  any modification in any one species exclusiv
200  her. But natural selection can and does often PRODUCE  structures for the direct injury of other sp
201  ther such cases. Natural selection will never PRODUCE  in a being anything injurious to itself, for
202  duced from Europe. Natural selection will not PRODUCE  absolute perfection, nor do we always meet,
205  the struggle for life. Natural selection will PRODUCE  nothing in one species for the exclusive goo
205  gooc or injury of another: though it may well PRODUCE  parts, organs, and excretions highly useful
205  tants one with another, and consequently will PRODUCE  perfection, or strength in the battle for li
206  igher. Natural selection will not necessarily PRODUCE  absolute perfection; nor, as far as we can j
209  ons produced by the same unknown causes which PRODUCE  slight deviations of bodily structure. No co
212  tatements, without facts given in detail, can PRODUCE  but a feeble effect on the reader's mind. I
216  cases compulsory habit alone has sufficed to  PRODUCE  such inherited mental changes: in other case
238  tted to their fertile offspring a tendency to PRODUCE  sterile members having the same modification
241  , could form a species which should regularly PRODUCE  neuters, either all of large size with one f
246  perfect condition, yet when intercrossed they PRODUCE  either few or no offspring. Hybrids, on the
252  sterile kinds of hybrid rhododendrons, which  PRODUCE  no pollen, for he will find on their stigmas
255  beyond that which the plant's own pollen will PRODUCE. So in hybrids themselves, there are some wh
255  never have produced, and probably never would PRODUCE, even with the pollen of either pure parent,
256  are very difficult to cross, and which rarely PRODUCE  any offspring, are generally very sterile: b
256  cies can be united with unusual facility, and PRODUCE  numerous hybrid offspring, yet these hybrids
257  the most perservering efforts have failed to  PRODUCE  between extremely close species a single hyb
260  uld some species cross with facility, and yet PRODUCE  very sterile hybrids: and other species cros
260  pecies cross with extreme difficulty, and yet PRODUCE  fairly fertile hybrids? Why should there oft
265  nditions: and whole groups of species tend to PRODUCE  sterile hybrids. On the other hand, one spec
265  ertility: and certain species in a group will PRODUCE  unusually fertile hybrids. No one can tell,
265  ries, whether any two species of a genus will PRODUCE  more or less sterile hybrics. Lastly, when o
267  have become widely or specifically different, PRODUCE  hybrids which are generally sterile in some
270  e same species of Verbascum when intercrossed PRODUCE  less seed, than do either coloured varieties
271  ieties; and from not wishing or being able to PRODUCE  recondite and functional differences in the
301  hese far ranging species which would oftenest PRODUCE  new varieties: and the varieties would at fi
303  omparatively short time would be necessary to PRODUCE  many divergent forms, which would be able to
317  y converted into species, which in their turn PRODUCE  by equally slow steps other species, and so
350  ill become still further victorious, and will PRODUCE  groups of modified descendants. On this prin
390  nently liable to modification, and will often PRODUCE  groups of modified descendants. But it by no
411  ce, on the principle of inheritance, tend to  PRODUCE  other new and dominant species. Consequently
461  sterility, we ought not to expect it also to  PRODUCE  sterility. The sterility of hybrids is a ver
466  mesticated productions. Man does not actually PRODUCE  variability; he only unintentionally exposes
468  characters alone and often capriciously, can  PRODUCE  within a short period a great result by addi
471  ht, successive, favourable variations, it can PRODUCE  no great or sudden modification: it can act
004  nt to the misseltoe, and that these had been  PRODUCED  perfect as we now see them: but this assump
010  ion is, that with animals such agencies have  PRODUCED  very little direct effect, though apparently
019  ly said that all our races of dogs have been  PRODUCED  by the crossing of a few aboriginal species:
023  number: how, for instance, could a pouter be  PRODUCED  by crossing two breeds unless one of the par
025  ls with some uniformly black barbs, and they  PRODUCED  mottled brown and black birds: these I again
029  the steps by which domestic races have been  PRODUCED, either from one or from several allied spec
030  ot suppose that all the breeds were suddenly  PRODUCED  as perfect and as useful as we now see them:
032  its importance consists in the great effect  PRODUCED  by the accumulation in one direction, during
032  oses that our choicest productions have been  PRODUCED  by a single variation from the aboriginal st
036  er have expected or even have wished to have  PRODUCED  the result which ensued, namely, the product
```

Page **********************************(Key Word)**********************************

```
0037  the wonderful skill of gardeners, in having PRODUCED such splendid results from such poor materia
0063  the product. Hence, as more individuals are PRODUCED than can possibly survive, there must in eve
0064  naeus has calculated that if an annual plant PRODUCED only two seeds, and there is no plant so unp
0064  ctive as this, and their seedlings next year PRODUCED two, and so on, then in twenty years there w
0066  its own eggs or young, a small number may be PRODUCED, and yet the average stock be fully kept up;
0066  ny eggs or young are destroyed, many must be PRODUCED, or the species will become extinct. It woul
0066  for a thousand years, if a single seed were PRODUCED once in a thousand years, supposing that thi
0077  . but from the strong growth of young plants PRODUCED from such seeds (as peas and beans), when so
0083  ruders. As man can produce and certainly has PRODUCED a great result by his methodical and unconsc
0089  the breeding season; the modifications thus PRODUCED being inherited at corresponding ages or sea
0092  in to the plant; and those individuals which PRODUCED more and more pollen, and had larger and lar
0100  although the male and female flowers may be PRODUCED on the same tree, we can see that pollen mus
0104  ntervals, I am convinced that the young thus PRODUCED will gain so much in vigour and fertility ov
0105  e proportion are endemic, that is, have been PRODUCED there, and nowhere else. Hence an oceanic is
0106  d what is more important, that the new forms PRODUCED on large areas, which already have been vict
0109  s new forms are continually and slowly being PRODUCED, unless we believe that the number of specif
0117  generations, species (A) is supposed to have PRODUCED two fairly well marked varieties, namely a1
0118  ariety a1 is supposed in the diagram to have PRODUCED variety a2, which will, owing to the princip
0118  d variety a1. Variety m1 is supposed to have PRODUCED two varieties, namely m2 and s2, differing f
0120  generations, species (A) is supposed to have PRODUCED three forms, a10, f10, and m10, which, from
0120  i have assumed that a second species (I) has PRODUCED, by analogous steps, after ten thousand gene
0121  etters n14 to z14, are supposed to have been PRODUCED. In each genus, the species, which are alrea
0123  s, as I believe, that two or more genera were PRODUCED by descent, with modification, from two or m
0129  gs may represent existing species; and those PRODUCED during each former year may represent the lo
0132  ly conclude that such influences cannot have PRODUCED the many striking and complex co adaptations
0133  ces could be given of the same variety being PRODUCED under conditions of life as different as can
0133  the other hand, of different varieties being PRODUCED from the same species under the same conditi
0142  ich consequently new varieties have not been PRODUCED, has even been advanced, for it is now as te
0153  erves notice that these variable characters, PRODUCED by man's selection, sometimes become attache
0159  ts which several botanists rank as varieties PRODUCED by cultivation from a common parent: if this
0165  id, and even the pure offspring subsequently PRODUCED from the mare by a black Arabian sire, were
0178  ing slowly modified, with the new forms thus PRODUCED and the old ones acting and reacting on each
0181  ion, a perfect so called flying squirrel was PRODUCED. Now look at the Galeopithecus or flying Lem
0184  arger and larger mouths, till a creature was PRODUCED as monstrous as a whale. As we sometimes see
0189  ni and each to be preserved till a better be PRODUCED, and then the old ones to be destroyed. In l
0192  that any organ could not possibly have been PRODUCED by successive transitional gradations, yet,
0192  y what steps these wondrous organs have been PRODUCED; but, as Owen and others have remarked, the
0199  rine that every detail of structure has been PRODUCED for the good of its possessor. They believe
0201  late my theory, for such could not have been PRODUCED through natural selection. Although many sta
0209  riations of instincts; that is of variations PRODUCED by the same unknown causes which produce sli
0209  ructure. No complex instinct can possibly be PRODUCED through natural selection, except by the slo
0210  but has never, as far as we can judge, been PRODUCED for the exclusive good of others. One of the
0238  ich individual bulls and cows, when matched, PRODUCED oxen with the longest horns; and yet no one
0238  sterile females of the same species has been PRODUCED, which we see in many social insects. But we
0239  inued selection of the fertile parents which PRODUCED most neuters with the profitable modificatio
0240  females had been continually selected, which PRODUCED more and more of the smaller workers, until
0241  he most useful to the community, having been PRODUCED in greater and greater numbers through the n
0241  til none with an intermediate structure were PRODUCED. Thus, as I believe, the wonderful fact of t
0243  iable to mistakes; that no instinct has been PRODUCED for the exclusive good of other animals, but
0246  st crossed, and the sterility of the hybrids PRODUCED from them. Pure species have of course their
0247  always compares the maximum number of seeds PRODUCED by two species when crossed and by their hyb
0247  ir hybrid offspring, with the average number PRODUCED by both pure parent species in a state of na
0250  of crinum capense fertilised by C. revolutum PRODUCED a plant, which (he says) I never saw to occu
0250  for instance, a bulb of Hippeastrum aulicum PRODUCED four flowers; three were fertilised by Herbe
0254  aboriginal species must either at first have PRODUCED quite fertile hybrids, or the hybrids must h
0254  at first have freely bred together and have PRODUCED quite fertile hybrids. So again there is rea
0255  s a perfect gradation in the number of seeds PRODUCED, up to nearly complete or even quite complet
0255  themselves, there are some which never have PRODUCED, and probably never would produce, even with
0256  cross, and the sterility of the hybrids thus PRODUCED, two classes of facts which are generally co
0256  me difficulty, but the hybrids, when at last PRODUCED, are very fertile. Even within the limits of
0257  crosses between species, and of the hybrids PRODUCED from them, is largely governed by their syst
0258  ollen of M. longiflora, and the hybrids thus PRODUCED are sufficiently fertile; but Kolreuter trie
0259  degree in the first cross and in the hybrids PRODUCED from this cross. That the fertility of hybri
0260  f effecting an union. The hybrids, moreover, PRODUCED from reciprocal crosses often differ in fert
0264  its mother's womb or within the egg or seed PRODUCED by the mother, it may be exposed to conditio
0265  d unnatural conditions, and when hybrids are PRODUCED by the unnatural crossing of two species, th
0266  r instance, the unequal fertility of hybrids PRODUCED from reciprocal crosses; or the increased st
0268  riably the case. But if we look to varieties PRODUCED under nature, we are immediately involved in
0268  in a circle, the fertility of all varieties PRODUCED under nature will assuredly have to be grant
0268  have to be granted. If we turn to varieties, PRODUCED, or supposed to have been produced, under do
0268  arieties, produced, or supposed to have been PRODUCED, under domestication, we are still involved
0269  eration, new races of animals and plants are PRODUCED under domestication by man's methodical and
0270  pollen of the other; but only a single head PRODUCED any seed, and this one head produced only fi
0270  le head produced any seed, and this one head PRODUCED only five grains. Manipulation in this case
0271  arieties of a distinct species, more seed is PRODUCED by the crosses between the same coloured flo
0271  d hybrids not so sterile as those which were PRODUCED from the four other varieties when crossed w
0274  spective parents, more especially in hybrids PRODUCED from nearly related species, follows accordi
0274  nt power over another variety. Hybrid plants PRODUCED from a reciprocal cross, generally resemble
0275  are descended from varieties often suddenly PRODUCED and semi monstrous in character, than with h
0275  descended from species slowly and naturally PRODUCED. On the whole I entirely agree with Dr. Pros
0275  lly created, and at varieties as having been PRODUCED by secondary laws, this similarity would be
0276  in degree in a first cross and in the hybrid PRODUCED from this cross. In the same manner as in gr
0277  a first cross, the fertility of the hybrids PRODUCED, and the capacity of being grafted together,
0277  at the greater number of varieties have been PRODUCED under domestication by the selection of mere
0282  he future historian will recognise as having PRODUCED a revolution in natural science, yet does no
0303  t whole groups of species have suddenly been PRODUCED. I may recall the well known fact that in ge
```

Page **********************************(Key Word)**********************************

Page **(Key Word)***

```
0304  s great and distinct order had been suddenly  PRODUCED  in the interval between the latest secondary
0320  variety, and ultimately each new species, is   PRODUCED  and maintained by having some advantage over
0320  number of new specific forms which have been   PRODUCED  within a given time is probably greater than
0325  homes, and are most widely diffused, having    PRODUCED  the greatest number of new varieties. It is
0327  rms dominant in the highest degree, wherever   PRODUCED,  would tend everywhere to prevail. As they p
0327  ing widely and varying: the new species thus   PRODUCED  being themselves dominant owing to inheritan
0335  no reason to believe that forms successively   PRODUCED  necessarily endure for corresponding lengths
0335  ch longer than a form elsewhere subsequently   PRODUCED,  especially in the case of terrestrial produ
0340  arsupials should have been chiefly or solely   PRODUCED  in Australia; or that Edentata and other Ame
0340  other American types should have been solely   PRODUCED  in South America. For we know that Europe in
0345  m, as it seems to me, that species have been   PRODUCED  by ordinary generation: old forms having bee
0345  upplanted by new and improved forms of life,   PRODUCED  by the laws of variation still acting round
0351  elieve that the species of a genus have been   PRODUCED  within comparatively recent times, there is
0352  rom one spot, where their parents were first   PRODUCED:  for, as explained in the last chapter, it i
0352  s identically the same should ever have been   PRODUCED  through natural selection from parents speci
0352  city of the view that each species was first   PRODUCED  within a single region captivates the mind.
0352  e quadrupeds. But if the same species can be   PRODUCED  at two separate points, why do we not find a
0353  that the great majority of species have been   PRODUCED  on one side alone, and have not been able to
0353  ed; and species were not local, but had been   PRODUCED  in two or more distinct areas! Hence it seem
0353  s, that the view of each species having been   PRODUCED  in one area alone, and having subsequently m
0354  l considerations, that each species has been   PRODUCED  within one area, and has migrated thence as
0380  f these islands on the land yielded to those   PRODUCED  within the larger areas of the north, just i
0394  ay be asked, has the supposed creative force   PRODUCED  bats and no other mammals on remote islands?
0411  t. the varieties, or incipient species, thus   PRODUCED  ultimately become converted, as I believe, i
0417  gh at our classification. But when Aspicarpa   PRODUCED  in France, during several years, only degrad
0424  distinct genera, were known to be sometimes   PRODUCED  on the same spike, they were immediately inc
0425  proved that one species of kangaroo had been   PRODUCED,  by a long course of modification, from a be
0432  t eleven Silurian genera, some of which have   PRODUCED  large groups of modified descendants. Every
0447  to gain its own living; and the effects thus   PRODUCED  will be inherited at a corresponding mature
0454  urther than by showing that rudiments can be   PRODUCED:  for I doubt whether species under nature ev
0461  which have been experimentised on have been   PRODUCED  under domestication; and as domestication ap
0466  se; for new varieties are still occasionally   PRODUCED  by our most anciently domesticated productio
0467  ful domestic breeds. That many of the breeds   PRODUCED  by man have to a large extent the character
0469  ween species, commonly supposed to have been   PRODUCED  by special acts of creation, and varieties w
0469  arieties which are acknowledged to have been   PRODUCED  by secondary laws. On this same view we can
0470  gion where many species of a genus have been   PRODUCED,  and where they now flourish, these same spe
0472  causing the bee's own death: at drones being   PRODUCED  in such vast numbers for one single act, and
0472  both cases physical conditions seem to have   PRODUCED  but little direct effect: yet when varieties
0472  ies and species, use and disuse seem to have   PRODUCED  some effect: for it is difficult to resist t
0475  dependently created, and varieties have been   PRODUCED  by secondary laws. If we admit that the geol
0482  rue species, they admit that these have been   PRODUCED  by variation, but they refuse to extend the
0482  e created forms of life, and which are those   PRODUCED  by secondary laws. They admit variation as a
0483  act of creation one individual or many were   PRODUCED?  Were all the infinitely numerous kinds of a
0487  ssion of their forms of life. As species are   PRODUCED  and exterminated by slowly acting and still
0489  other in so complex a manner, have all been   PRODUCED  by laws acting around us. These laws, taken
0009  w we owe variability to the same cause which   PRODUCES  sterility; and variability is the source of
0012  strong is the tendency to inheritance: like   PRODUCES  like is his fundamental belief: doubts have
0063  dent on the moisture. A plant which annually   PRODUCES  a thousand seeds, of which on an average onl
0063  ery being, which during its natural lifetime   PRODUCES  several eggs or seeds, must suffer destructi
0065  mes. In a state of nature almost every plant   PRODUCES  seed, and amongst animals there are very few
0132  ct effect difference of climate, food, etc.,   PRODUCES  on any being is extremely doubtful. My impre
0161  in each generation of the barb pigeon, which   PRODUCES  most rarely a blue and black barred bird, th
0350  e which alone, as far as we positively know,   PRODUCES  organisms quite like, or, as we see in the c
0484  or that the poison secreted by the gall fly   PRODUCES  monstrous growths on the wild rose or oak tr
0009  confinement, acting not quite regularly, and   PRODUCING  offspring not perfectly like their parents
0093  r only male flowers, which have four stamens   PRODUCING  rather a small quantity of pollen, and a ru
0110  in individuals will have the best chance of   PRODUCING  within any given period favourable variatio
0118  varieties, after each thousand generations,   PRODUCING  only a single variety, but in a more and mo
0118  in a more and more modified condition, some   PRODUCING  two or three varieties, and some failing to
0146  which opened: so that the individual plants   PRODUCING  seeds which were a little better fitted to
0147  d further, might get an advantage over those   PRODUCING  seed less fitted for dispersal: and this pr
0167  rison, the same laws appear to have acted in   PRODUCING  the lesser differences between varieties of
0167  induced some slight modifications. Habit in   PRODUCING  constitutional differences, and use in stre
0172  for species, when crossed, being sterile and   PRODUCING  sterile offspring, whereas, when varieties
0197  ls. We are profoundly ignorant of the causes   PRODUCING  slight and unimportant variations; and we a
0256  of sterility we have self fertilised hybrids   PRODUCING  a greater and greater number of seeds up to
0260  ed? to grant to species the special power of   PRODUCING  hybrids, and then to stop their further pro
0273  at least incapable of its proper function of   PRODUCING  offspring identical with the parent form. N
0327  species: these again spreading, varying, and   PRODUCING  new species. The forms which are beaten and
0357  animals during their migration. In the coral   PRODUCING  oceans such sunken islands are now marked,
0063  ately great that no country could support the  PRODUCT.  Hence, as more individuals are produced tha
0165  would have thought that it must have been the  PRODUCT  of a zebra; and Mr. W. C. Martin, in his exc
0398  he shores of South America. Here almost every  PRODUCT  of the land and water bears the unmistakeabl
0414  hereas the organs of reproduction, with their  PRODUCT  the seed, are of paramount importance! We mu
0028  also a most favourable circumstance for the    PRODUCTION  of distinct breeds, that male and female pi
0036  oduced the result which ensued, namely, the    PRODUCTION  of two distinct strains. The two flocks of
0105  this may sometimes be of importance in the     PRODUCTION  of new species. If, however, an isolated ar
0105  ness of individuals will greatly retard the    PRODUCTION  of new species through natural selection, b
0105  eems to have been highly favourable for the    PRODUCTION  of new species. But we may thus greatly dec
0105  continent, has been most favourable for the    PRODUCTION  of new organic forms, we ought to make the
0105  lation is of considerable importance in the    PRODUCTION  of new species, on the whole I am inclined
0105  of more importance, more especially in the     PRODUCTION  of species, which will prove capable of end
0106  in some respects highly favourable for the     PRODUCTION  of new species, yet that the course of modi
0107  dition, will be the most favourable for the    PRODUCTION  of many new forms of life, likely to endure
0118  rcumstances we know to be favourable to the    PRODUCTION  of new varieties. If, then, these two varie
0125  tage in common. Hence, the struggle for the    PRODUCTION  of new and modified descendants, will mainl
```

Page **(Key Word)***

Page **************************************(Key Word)**************************************

```
0202 sects find their females, can we admire the PRODUCTION for this single purpose of thousands of dr
0241 has originated. We can see how useful their PRODUCTION may have been to a social community of ins
0260 species? Why, it may even be asked, has the PRODUCTION of hybrids been permitted? to grant to spe
0271 n selecting only external characters in the PRODUCTION of the most distinct domestic varieties, a
0317 ly; for the process of modification and the PRODUCTION of a number of allied forms must be slow a
0317 lection the extinction of old forms and the PRODUCTION of new and improved forms are intimately c
0318 up is generally a slower process than their PRODUCTION: if the appearance and disappearance of a
0320 king to later times we may believe that the PRODUCTION of new forms has caused the extinction of
0321 as we have seen, a slower process than its PRODUCTION. With respect to the apparently sudden ext
0326 world may have been most favourable for the PRODUCTION of new and dominant species on the land, a
0335 sely accord with the order in time of their PRODUCTION, and still less with the order of their di
0343 is the almost inevitable consequence of the PRODUCTION of new forms. We can understand why when a
0394 ry strata: there has also been time for the PRODUCTION of endemic species belonging to other clas
0467 selection has been the great agency in the PRODUCTION of the most distinct and useful domestic b
0472 see, with the laws which have governed the PRODUCTION of so called specific forms. In both cases
0485 ond his comprehension; when we regard every PRODUCTION of nature as one which has had a history;
0488 mpressed on matter by the Creator, that the PRODUCTION and extinction of the past and present inh
0490 h we are capable of conceiving, namely, the PRODUCTION of the higher animals, directly follows. T
0007 s selection. Unknown Origin of our Domestic PRODUCTIONS. Circumstances favourable to Man's power
0007 r variability is simply due to our domestic PRODUCTIONS having been raised under conditions of li
0009 riability is the source of all the choicest PRODUCTIONS of the garden. I may add, that as some or
0016 ith generic differences in our domesticated PRODUCTIONS. When we attempt to estimate the amount o
0017 ly to the value of most of our domesticated PRODUCTIONS; but how could a savage possibly know, wh
0017 plants, equal in number to our domesticated PRODUCTIONS, and belonging to equally diverse classes
0017 parent species of our existing domesticated PRODUCTIONS have varied. In the case of most of our a
0032 e abrupt. No one supposes that our choicest PRODUCTIONS have been produced by a single variation
0043 mportant part in the origin of our domestic PRODUCTIONS. When in any country several domestic bre
0045 individual differences in his descendant PRODUCTIONS. These individual differences generally a
0065 increase and wide diffusion of naturalised PRODUCTIONS in their new homes. In a state of nature
0080 umber of strange peculiarities our domestic PRODUCTIONS, and, in a lesser degree, those under nat
0083 s have been so far conquered by naturalised PRODUCTIONS, that they have allowed foreigners to tak
0083 eason, as far as lies in his power, all his PRODUCTIONS. He often begins his selection by some ha
0084 periods. Can we wonder, then, that nature's PRODUCTIONS should be far truer in character than man
0084 should be far truer in character than man's PRODUCTIONS; that they should be infinitely better ad
0106 phical distribution; for instance, that the PRODUCTIONS of the smaller continent of Australia hav
0106 ic area. Thus, also, it is that continental PRODUCTIONS have everywhere become so largely natural
0107 uently, the competition between fresh water PRODUCTIONS will have been less severe than elsewhere
0107 looking to the future, that for terrestrial PRODUCTIONS a large continental area, which will prob
0110 s of extermination amongst our domesticated PRODUCTIONS, through the selection of improved forms
0112 s seek light on this head from our domestic PRODUCTIONS. We shall here find something analogous.
0112 d to disappear. Here, then, we see in man's PRODUCTIONS the action of what may be called the prin
0128 bitants of any small spot or at naturalised PRODUCTIONS. Therefore during the modification of the
0139 l known relationship of most of their other PRODUCTIONS. Far from feeling any surprise that some
0147 true to a certain extent with our domestic PRODUCTIONS: if nourishment flows to one part or orga
0201 rfection attained under nature. The endemic PRODUCTIONS of New Zealand, for instance, are perfect
0237 innumerable instances, both in our domestic PRODUCTIONS and in those in a state of nature, of all
0283 ded boulders, all thickly clothed by marine PRODUCTIONS, showing how little they are abraded and
0288 ogous case. With respect to the terrestrial PRODUCTIONS which lived during the Secondary and Pala
0292 of inhabitants will decrease (excepting the PRODUCTIONS on the shores of a continent when first b
0300 very reason to believe that the terrestrial PRODUCTIONS of the archipelago would be preserved in
0306 then to discuss the number and range of its PRODUCTIONS. On the sudden appearance of groups of Al
0313 l the Crustaceans have changed greatly. The PRODUCTIONS of the land seem to change at a quicker r
0314 in terrestrial and in more highly organised PRODUCTIONS compared with marine and lower production
0314 productions compared with marine and lower PRODUCTIONS, by the more complex relations of the hig
0320 y follows. It is the same with our domestic PRODUCTIONS: when a new and slightly improved variety
0323 ve not sufficient data to judge whether the PRODUCTIONS of the land and of fresh water change at
0324 mpetent observers believe that the existing PRODUCTIONS of the United States are more closely rel
0326 strict degree of parallel succession in the PRODUCTIONS of the land than of the sea. Dominant spe
0335 uced, especially in the case of terrestrial PRODUCTIONS inhabiting separated districts. To compar
0337 the extraordinary manner in which European PRODUCTIONS have recently spread over New Zealand, an
0337 ny part of Europe, we may doubt, if all the PRODUCTIONS of New Zealand were set free in Great Bri
0337 and animals. Under this point of view, the PRODUCTIONS of Great Britain may be said to be higher
0344 e. hence, after long intervals of time, the PRODUCTIONS of the world will appear to have changed
0346 ns. Importance of barriers. Affinity of the PRODUCTIONS of the same continent. Centres of creatio
0347 rlds, how widely different are their living PRODUCTIONS! In the southern hemisphere, if we compar
0347 rly dissimilar. Or again we may compare the PRODUCTIONS of South America south of lat. 35 degrees
0347 related to each other, than they are to the PRODUCTIONS of Australia or Africa under nearly the s
0347 rtant manner to the differences between the PRODUCTIONS of various regions. We see this in the gr
0347 at difference of nearly all the terrestrial PRODUCTIONS of the New and Old Worlds, excepting in t
0347 ms, as there now is for the strictly arctic PRODUCTIONS. We see the same fact in the great differe
0348 mes even of large rivers, we find different PRODUCTIONS; though as mountain chains, deserts, etc.
0349 oregoing statements, is the affinity of the PRODUCTIONS of the same continent or sea, though the
0354 pter), the wide distribution of fresh water PRODUCTIONS: and thirdly, the occurrence of the same
0356 together, and thus have allowed terrestrial PRODUCTIONS to pass from one to the other. No other g
0366 , the latter would be supplanted and arctic PRODUCTIONS would take their places. The inhabitants
0367 closely followed up in their retreat by the PRODUCTIONS of the more temperate regions. And as the
0368 esent distribution of the Alpine and Arctic PRODUCTIONS of Europe and America, that when in other
0368 l gnathodon), then the arctic and temperate PRODUCTIONS will at a very late period have marched a
0368 e to much modification. But with our Alpine PRODUCTIONS, left isolated from the moment of the ret
0369 assumed that at its commencement the arctic PRODUCTIONS were as uniform round the polar regions a
0369 day, the sub arctic and northern temperate PRODUCTIONS of the Old and New Worlds are separated f
0370 es 67 degrees; and that the strictly arctic PRODUCTIONS then lived on the broken land still neare
0370 ty in the sub arctic and northern temperate PRODUCTIONS of the Old and New Worlds, at a period an
0371 hip, with very little identity, between the PRODUCTIONS of North America and Europe, a relationsh
0371 remarked on by several observers, that the PRODUCTIONS of Europe and America during the later te
0371 is separation, as far as the more temperate PRODUCTIONS are concerned, took place long ages ago.
```

Page **************************************(Key Word)**************************************

Page ***(Key Word)***

```
0371  e one great region with the native American  PRODUCTIONS, and have had to compete with them; and in
0371   far more mocification than with the Alpine  PRODUCTIONS, left isolated, within a much more recent
0371  s come, that when we compare the now living  PRODUCTIONS of the temperate regions of the New and Ol
0376  tribution of terrestrial animals. In marine  PRODUCTIONS, similar cases occur; as an example, I may
0377  lowly on, all the tropical plants and other  PRODUCTIONS will have retreated from both sides toward
0377  ator, followed in the rear by the temperate  PRODUCTIONS, and these by the arctic: but with the lat
0377   fact to bear in mind is, that all tropical  PRODUCTIONS will have suffered to a certain extent. On
0377  in extent. On the other hand, the temperate  PRODUCTIONS, after migrating nearer to the equator, th
0377  possible, bearing in mind that the tropical  PRODUCTIONS were in a suffering state and could not ha
0378  rch. But I do not doubt that some temperate  PRODUCTIONS entered and crossed even the lowlands of t
0378   a few terrestrial animals, and some marine  PRODUCTIONS, migrated during the glacial period from t
0380  at the present day, that very many European  PRODUCTIONS cover the ground in La Plata, and in a les
0380  ps of the north. In many islands the native  PRODUCTIONS are nearly equalled or even outnumbered by
0380  completely isolated; and I believe that the  PRODUCTIONS of these islands on the land yielded to th
0380  s of the north, in the same way as the naturalised  PRODUCTIONS. On the inhabitants of oceanic islands. Ab
0383  ion, continued. Distribution of fresh water  PRODUCTIONS would not have ranged widely within the sa
0383   it might have been thought that fresh water  PRODUCTIONS of ranging widely, though so unexpected, c
0383  e of Britain. But this power in fresh water  PRODUCTIONS are low in the scale of nature, and that w
0388  member that some, perhaps many, fresh water  PRODUCTIONS ever can range, over immense areas, and ha
0388  merly ranged as continuously as fresh water  PRODUCTIONS. In the following remarks I shall not conf
0389   explain all the facts in regard to insular  PRODUCTIONS. He who admits the doctrine of the creatio
0390  ve nearly or quite exterminated many native  PRODUCTIONS of islands are related to those of the nea
0399  s an almost universal rule that the endemic  PRODUCTIONS, in which so many genera range over the wo
0404  and beetles. So it is with most fresh water  PRODUCTIONS being related (with the exceptions before
0406  uch facts, as alpine, lacustrine, and marsh  PRODUCTIONS. If the difficulties be not insuperable in
0407  eath the means of dispersal of fresh water  PRODUCTIONS, several other grades of difference are re
0423  ties under varieties; and with our domestic  PRODUCTIONS. But when the nestling birds of these seve
0445  d in distinct genera, had they been natural  PRODUCTIONS, as the stump of a tail in tailless breeds
0454  cases of rudimentary organs in our domestic  PRODUCTIONS have undergone; but we may safely infer th
0466  rtaining how much modification our domestic  PRODUCTIONS. Man does not actually produce variability
0466  produced by our most anciently domesticated  PRODUCTIONS; and every one admits that there are at le
0468  mere individual differences in his domestic  PRODUCTIONS from another land. Nor ought we to marvel
0472  ng beaten and supplanted by the naturalised  PRODUCTIONS was almost unavoidable as long as the hist
0481  ion. The belief that species were immutable  PRODUCTIONS will rise immensely in value. A new variet
0486  itions, and so forth. The study of domestic  PRODUCTS be, compared with those accumulated by natur
0084  his time! and consequently how poor will his  PRODUCTS? What limit can be put to this power, acting
0469  r changing conditions of life, to her living  PROF. Owen's expression, seems to be a sign of low
0149  inasmuch as this vegetative repetition, to use  PROF. Ramsay has published an account of a downthro
0285  f the strata has varied from 600 to 3000 feet.  PROF. Ramsay's masterly memoir on this subject. Yet
0285  ts ten thousand feet in thickness, as shown in  PROF. Ramsay. But if, as some geologists suppose, a
0286  n average about 1100 feet, as I am informed by  PROF. Dana, that it is certainly a wonderful fact t
0376  I may quote a remark by the highest authority,  PROFESSOR Lepsius; but Mr. Birch informs me that pige
0027  about 3000 B.C., as was pointed out to me by  PROFESSOR Huxley, to discover a single case of an her
0101  with one of the highest authorities, namely,  PROFESSOR Owen has remarked, there is no greater anom
0134  an be explained by the effects of disuse. As  PROFESSOR Silliman thought that it regained, after li
0137   cave rat, the eyes are of immense size; and  PROFESSOR Dana; and some of the European cave insects
0139  of the American cave animals, as I hear from  PROFESSOR Owen, with respect to the length of the arm
0150  se. I think also from an observation made by  PROFESSOR Owen's interesting description of these par
0191  or swimbladder. We can thus, as I infer from  PROFESSOR Miller, of Cambridge, and this geometer has
0226  comb of the hive bee. Accordingly I wrote to  PROFESSOR Ramsay has given me the maximum thickness,
0284   the sedimentary deposits of many countries!  PROFESSOR Sedgwick, as a fatal objection to the belie
0302  z, pictet, and by none more forcibly than by  PROFESSOR Pictet has carried their existence one sub
0305  large majority of existing species. Lately,  PROFESSOR Owen soon perceived that the tooth, though
0319  how utterly groundless was my astonishment!  PROFESSOR Pictet in his firm belief in the immutabili
0335  e, from its generality, seems to have shaken  PROFESSOR Owen has shown in the most striking manner
0339  llo, found in several parts of La Plata; and  PROFESSOR Owen has subsequently extended the same gen
0339  e continent between the dead and the living.  PROFESSOR Huxley has remarked, to speak of both skull
0438   in these cases probably be more correct, as  PROFESSOR Huxley of the development of this insect, w
0442  his, if we look to the admirable drawings by  PROFIT a plant to have its seeds more and more widel
0086  eir inheritance at a corresponding age. If it  PROFIT a bee or other insect, so that an individual
0094  far too slight to be appreciated by us, might  PROFIT the individual not to have its nutriment wast
0148  e seized on by natural selection, for it will  PROFIT, where neither pure parent species exists, th
0253  arts of the country; and as they are kept for  PROFITABLE to itself, under the complex and sometimes
0005  , if it vary however slightly in any manner  PROFITABLE to an individual of any species, in its inf
0061  er cause proceeding. if it be in any degree  PROFITABLE variations occurring: and unless profitable
0082  ral selection, by giving a better chance of  PROFITABLE variations do occur, natural selection can
0082  profitable variations occurring: and unless  PROFITABLE variations at that age, and by their inheri
0086  c beings at any age, by the accumulation of  PROFITABLE to the preserved being: and as modern geolo
0095  similally small inherited modifications, each  PROFITABLE variations, will compensate for a lesser am
0102  r the appearance within any given period of  PROFITABLE to them. The advantage of diversification i
0115  to new generic differences, would have been  PROFITABLE will be preserved or naturally selected. An
0117  only those variations which are in some way  PROFITABLE variations, however slight, until they beco
0134   natural selection will then accumulate all  PROFITABLE modifications, each new form will tend in a
0172  election acts solely by the preservation of  PROFITABLE variations in the struggle for life. Natura
0205  er which acts solely by the preservation of  PROFITABLE to a species; and if it can be shown that i
0209  t slight modifications of instinct might be  PROFITABLE. It is thus, as I believe, that all the mos
0209  tions of instinct to any extent that may be  PROFITABLE, variations. Hence, as in the case of corpo
0210  adual accumulation of numerous, slight, yet  PROFITABLE to the individual under its conditions of l
0233  odifications of structure or instinct, each  PROFITABLE to the community that a number should have
0236  ch insects had been social, and it had been  PROFITABLE modification of structure, this being inher
0237  ndividual having been born with some slight  PROFITABLE modification did not probably at first appe
0239  y variations, that each successive, slight,  PROFITABLE modification, all the neuters ultimately ca
0239  arents which produced most neuters with the  PROFITABLE, without exercise or habit having come into
0242  dental, variations, which are in any manner  PROFITABLE degrees of sterility. I hope, however, to b
0245  by the continued preservation of successive  PROFITABLE in some way to the modified form, but often
0435  ight modifications, each modification being  PROFITABLE in some way to the modified form, but often
```

Page ***(Key Word)***

```
0459  istence leading to the preservation of each PROFITABLE deviation of structure or instinct. The tr
0474  atural selection of successive, slight, but PROFITABLE modifications. We can thus understand why
0217  n egg in another bird's nest.If the old bird PROFITED by this occasional habit, cr if the young w
0233  ent perfect plan of construction, could have PROFITED the progenitors of the hive bee? I think th
0379  ructure, habits, and constitutions will have PROFITED them. Thus many of these wanderers, though
0448  perhaps requisite. If, on the other hand, it PROFITED the young to follow habits of life in any d
0087  efit of the community; if each in consequence PROFITS by the selected change. What natural select
0200  e species incessantly takes advantage of, and PROFITS by, the structure of another. But natural s
0351  ge of by natural selection, only so far as it PROFITS the individual in its complex struggle for
0006  varieties, if he makes due allowance for our PROFOUND ignorance in regard to the mutual relations
0073  organic being over another. Nevertheless so PROFOUND is our ignorance, and so high our presumpti
0167  y. our ignorance of the laws of variation is PROFOUND. Not in one case out of a hundred can we pr
0172  e practically anticipated the discoveries of PROFOUND mathematicians? Fourthly, how can we accoun
0173  chiefly due to organic beings not inhabiting PROFOUND depths of the sea, and to their remains bei
0283  s been accumulated. Let him remember Lyell's PROFOUND remark, that the thickness and extent of se
0291  diment may be formed in two ways: either, in PROFOUND depths of the sea, in which case, judging f
0308  mains, which do not appear to have inhabited PROFOUND depths, in the several formations of Europe
0132  that part should vary more or less, we are PROFOUNDLY ignorant. Nevertheless, we can here and the
0185  el, but with many parts of its organisation PROFOUNDLY modified. On the other hand, the acutest o
0197  e parturition of the higher animals. We are PROFOUNDLY ignorant of the causes producing slight an
0407  within the same period; if we remember how PROFOUNDLY ignorant we are with respect to the many a
0460  their reproductive systems should have been PROFOUNDLY modified. Moreover, most of the varieties
0462  species throughout the world. We are as yet PROFOUNDLY ignorant of the many occasional means of t
0093  ion, there were pollen grains, and on some a PROFUSION of pollen. As the wind hac set for several
0135  r quadrupeds. We may imagine that the early PROGENITOR of the ostrich had habits like those of a
0146  e simply due to inheritance; for an ancient PROGENITOR may have acquired through natural selectio
0153  en the species branched off from the common PROGENITOR of the genus. This period will seldom be r
0156  on i attribute to inheritance from a common PROGENITOR, for it can rarely have happened that natu
0156  pecies first branched off from their common PROGENITOR, and subsequently have not varied or come
0156  branching off of the species from a common PROGENITOR, it is probable that they should still oft
0157  having as certainly descended from the same PROGENITOR, as have the two sexes of any one of the s
0157  hatever part of the structure of the common PROGENITOR, or of its early descendants, became varia
0158  e same group having descended from a common PROGENITOR, from whom they have inherited much in com
0159  verts to some of the characters of an early PROGENITOR. These propositions will be most readily u
0160  arents have lost some character which their PROGENITOR possessed, the tendency, whether strong or
0193  organs had been inherited from one ancient PROGENITOR thus provided, we might have expected that
0195  e cases been of high importance to an early PROGENITOR, and, after having been slowly perfected a
0200  these structures to inheritance. But to the PROGENITOR of the upland goose and of the frigate bir
0200  existing birds. So we may believe that the PROGENITOR of the seal had not a flipper, but a foot
0200  at, which have been inherited from a common PROGENITOR, were formerly of more special use to that
0200  , were formerly of more special use to that PROGENITOR, or its progenitors, than they now are to
0202  e, as having originally existed in a remote PROGENITOR as a boring and serrated instrument, like
0217  nests. Now let us suppose that the ancient PROGENITOR of our European cuckoo hac the habits of t
0280  tween each species and a common but unknown PROGENITOR; and the progenitor will generally have di
0280  nd a common but unknown progenitor; and the PROGENITOR will generally have differed in some respe
0297  efore explained, that A might be the actual PROGENITOR of B and C, and yet might not at all neces
0302  , all of which have descended from some one PROGENITOR, must have been an extremely slow process:
0306  s of the same group have descended from one PROGENITOR, apply with nearly equal force to the earl
0316  ould be almost sure to inherit from its new PROGENITOR some slight characteristic differences. Gr
0331  ing in common from their ancient and common PROGENITOR. On the principle of the continued tendenc
0331  e it will generally differ from its ancient PROGENITOR. Hence we can understand the rule that the
0333  e not diverged in character from the common PROGENITOR of the order, nearly so much as they subse
0344  m their inferiority inherited from a common PROGENITOR, tend to become extinct together, and to l
0344  d to, and consequently resemble, the common PROGENITOR of groups, since become widely divergent.
0351  ource, as they have descended from the same PROGENITOR. In the case of those species, which have
0354  my theory have all descended from a common PROGENITOR, can have migrated (undergoing modificatio
0354  migration) from the area inhabited by their PROGENITOR. If it can be shown to be almost invariabl
0412  he modified descendants proceeding from one PROGENITOR become broken up into groups subordinate t
0413  e have many species descended from a single PROGENITOR grouped into genera; and the genera are in
0420  in the same degree in blood to their common PROGENITOR, may differ greatly, being due to the diff
0430  s and Marsupials branched off from a common PROGENITOR, and that both groups have since undergone
0430  tance, more of the character of its ancient PROGENITOR than have other Rodents; and therefore it
0430  ally retained the character of their common PROGENITOR, or of an early member of the group. On th
0432  hose parents from their ancient and unknown PROGENITOR. Yet the natural arrangement in the diagram
0435  veral parts. If we suppose that the ancient PROGENITOR, the archetype as it may be called, of all
0436  , we have only to suppose that their common PROGENITOR had an upper lip, mandibles, and two pair c
0437  ore we may readily believe that the unknown PROGENITOR of the vertebrata possessed many vertebrae
0437  brata possessed many vertebrae; the unknown PROGENITOR of the articulata, many segments; and the u
0437  articulata, many segments; and the unknown PROGENITOR of flowering plants, many spiral whorls of
0449  d in so far it reveals the structure of its PROGENITOR. In two groups of animal, however much they
0450  the many descendants from some one ancient PROGENITOR, at a very early period in the life of each
0473  these species have descended from a striped PROGENITOR, in the same manner as the several domesti
0473  aried since they branched off from a common PROGENITOR in certain characters, by which they have c
0474  several species branched off from a common PROGENITOR, an unusual amount of variability and modi
0476  croups which have descended from an ancient PROGENITOR have generally diverged in character, the p
0476  r have generally diverged in character, the PROGENITOR with its early descendants will often be ir
0479  ts or organs, which were alike in the early PROGENITOR of each class. On the principle of success
0480  oh the gums of the upper jaw, from an early PROGENITOR having well developed teeth; and we may bel
0483  entary condition plainly show that an early PROGENITOR had the organ in a fully developed state; a
0488  have elapsed since the first creature, the PROGENITOR of innumerable extinct and living descendar
0135  e long continued effects of disuse in their PROGENITORS; for as the tarsi are almost always lost
0169  to some of the characters of their ancient PROGENITORS. Although new and important modifications
0200  more special use to that progenitor, or its PROGENITORS, than they now are to these animals having
0233  an of construction, could have profited the PROGENITORS of the hive bee? I think the answer is not
0302  ave been an extremely slow process: and the PROGENITORS must have lived long ages before their moc
0305  d; and these cirripedes might have been the PROGENITORS of our many tertiary and existing species.
```

Page ***(Key Word)***

```
9306  e supposed, that these old species were the  PROGENITORS of all the species of the orders to which
9307  tween them. If, moreover, they had been the  PROGENITORS of these orders, they would almost certain
9315  it cifferent characters from their distinct  PROGENITORS. For instance, it is just possible, if our
9341  and some of these fossils may be the actual  PROGENITORS of living species. It must not be forgotte
9431  been utterly lost, through which the early   PROGENITORS of birds were formerly connected with the
9431  irds were formerly connected with the early  PROGENITORS of the other vertebrate classes. There has
9449  he structure of their less modified ancient  PROGENITORS, we can clearly see why ancient and extinc
9463  rian system, stored with the remains of the  PROGENITORS of the Silurian groups of fossils? For cer
9477  y will generally be descendants of the same  PROGENITORS and early colonists. On this same principl
9484  ve descended from at most only four or five  PROGENITORS, and plants from an equal or lesser number
9062  ife of the individual, but success in leaving PROGENY. Two canine animals in a time of dearth, may
9064  royed, the earth would soon be covered by the PROGENY of a single pair. Even slow breeding man has
9064  would literally not be standing room for his PROGENY. Linnaeus has calculated that if an annual p
9088  d for their places in nature, will leave most PROGENY. But in many cases, victory will depend not
9119  bled to seize on, and the more their modified PROGENY. It may well be asked how is it possible to
9237  modifications of structure or instinct to its PROGENY. It may well be asked how is it possible to
9267  always induces weakness and sterility in the  PROGENY. Hence it seems that, on the one hand, sligh
9341  have become wholly extinct, and have left no  PROGENY. But in the caves of Brazil, there are many
9341  old genera have all died out and have left no PROGENY. Or, which would probably be a far commoner
9468  conditions of life, will generally leave most PROGENY. But success will often depend on having spe
9474  in the case of neuter insects, which leave no PROGENY to inherit the effects of long continued hab
9489  the species now living very few will transmit PROGENY of any kind to a far distant futurity; for t
9084  ife. We see nothing of those slow changes in  PROGRESS, until the hand of time has marked the long
9110  tly, each new variety or species, during the  PROGRESS of its formation, will generally press harde
9249  ver, whenever complicated experiments are in  PROGRESS, so careful an observer as Gartner would hav
9251  of the hybrid made vigorous growth and rapid  PROGRESS to maturity, and bore good seed, which veget
9291  on this subject in 1845, I have watched the   PROGRESS of Geology, and have been surprised to note
9318  gradually at its upper end, which marks the   PROGRESS of extermination, than at its lower end, whi
9319  tinction, and we know that this has been the  PROGRESS of events with those animals which have been
9337  but I can see no way of testing this sort of  PROGRESS. Crustaceans, for instance, not the highest
9434  creation, we may hope to make sure but slow   PROGRESS. Morphology. We have seen that the members o
9489  corporeal and mental endowments will tend to  PROGRESS towards perfection. It is interesting to con
9345  logists, that organisation on the whole has   PROGRESSED. If it should hereafter be proved that anci
9317  same family, can increase only slowly and     PROGRESSIVELY: for the process of modification and the
9232  new hexagon, in its strictly proper place,    PROJECTING beyond the other completec cells. It suffic
9024  times by half civilized man, as to be quite   PROLIFIC under confinement. An argument, as it seems
9227  adjoining spheres in the same layer, she can  PROLONG the hexagon to any length requisite to hold
9121  shown in the diagram by the dotted lines not  PROLONGED far upwards from want of space. But during
9326  r their inhabitants met, the battle would be  PROLONGED and severe; and some from one birthplace an
9083  trous form; or at least by some modification  PROMINENT enough to catch his eye, or to be plainly u
9283  and there, along a short length or round a    PROMONTORY, that the cliffs are at the present time su
9079  that no fear is felt, that death is generally PROMPT, and that the vigorous, the healthy, and the
9116  could successfully compete with these well    PRONOUNCED orders. In the Australian mammals, we see t
9455  in the spelling, but become useless in the    PRONUNCIATION, but which serve as a clue in seeking for
9128  ure, habits, and constitution, of which we see PROOF by looking at the inhabitants of any small sp
9338  true, and yet it may never be capable of full PROOF. Seeing, for instance, that the oldest known
9439  e class are often strikingly similar: a better PROOF of this cannot be given, than a circumstance
9468  erted, but the assertion is quite incapable of PROOF, that the amount of variation under nature is
9481  ose of time, we are too apt to assume, without PROOF, that the geological record is so perfect tha
9032  variation from the aboriginal stock. We have  PROOFS that this is not so in some cases, in which e
9005  heritance, any selected variety will tend to  PROPAGATE its new and modified form. This fundamental
9017  er, spaniel, and bull dog, which we all know  PROPAGATE their kind so truly, were the offspring of
9236  es, and yet, from being sterile, they cannot  PROPAGATE their kind. The subject well deserves to be
9298  rth notice: with animals and plants that can  PROPAGATE rapidly and are not highly locomotive, ther
9010  of the rest of the plant. Such buds can be    PROPAGATED by grafting, etc., and sometimes by seed. T
9042  n of new breeds. Pigeons, I may add, can be   PROPAGATED in great numbers and at a very quick rate,
9043  rd to animals and to those plants which are   PROPAGATED by seed. In plants which are temporarily pr
9043  ed by seed. In plants which are temporarily   PROPAGATED by cuttings, buds, etc., the importance of
9043  ity of hybrics; but the cases of plants not   PROPAGATED by seed are of little importance to us, for
9044  ot useful to the species, and not generally   PROPAGATED. Some authors use the term variation in a t
9142  of the Jerusalem artichoke, which is never    PROPAGATED by seed, and of which consequently new vari
9181  each modification being useful, each being    PROPAGATED, until by the accumulated effects of this p
9238  st horns; and yet no one ox could ever have   PROPAGATED its kind. Thus I believe it has been with s
9272  n mongrels and the more fertile hybrids are   PROPAGATED for several generations an extreme amount o
9104  will have a better chance of surviving and    PROPAGATING their kind; and thus, in the long run, the
9418  the blood, or for aerating it, or those for   PROPAGATING the race, are found nearly uniform, they a
9260  ing hybrids, and then to stop their further   PROPAGATION by different degrees of sterility, not str
9418  nimals, important organs, such as those for   PROPELLING the blood, or for aerating it, or those for
9032  as they call the plants that deviate from the PROPER standard. With animals this kind of selection
9085  lection might be most effective in giving the PROPER colour to each kind of grouse, and in keeping
9104  election destroying any which depart from the PROPER type; but if their conditions of life change
9174  rally narrow in comparison with the territory PROPER to each. We see the same fact in ascending mo
9184  y or considerably modified from that of their PROPER type. And such instances do occur in nature.
9194  posec to have been separately created for its PROPER place in nature, be so invariably linked toge
9220  the slave species are found only in their own PROPER communities, and have never been observed in
9223  nintentionally reared would then follow their PROPER instincts, and do what work they could. If th
9224  ite problem, and have made their cells of the PROPER shape to hold the greatest possible amount of
9229  ing when they have gnawed the wax away to the PROPER thinness, and then stopping their work. In or
9230  hich as I have tried is easily done) into its PROPER intermediate plane, and thus flatten it. From
9230  ll of wax, they could make their cells of the PROPER shape, by standing at the proper distance fro
9230  cells of the proper shape, by standing at the PROPER distance from each other, by excavating at th
9230  es cannot build up a rough wall of wax in the PROPER position, that is, along the plane of interse
9232  have a place on which they can stand in their PROPER positions for working, for instance, on a sli
9232  of one wall of a new hexagon, in its strictly PROPER place, projecting beyond the other completed
9232  the bees should be enabled to stand at their  PROPER relative distances from each other and from t
9232  der certain circumstances a rough wall in its PROPER place between two just commenced cells, is im
```

Page ***(Key Word)***

```
0233  nced at the same time, always standing at the  PROPER  relative distance from the parts of the cells
0233  point, and then to five other points, at the  PROPER  relative distances from the central point and
0273  either impotent or at least incapable of its  PROPER  function of producing offspring identical wit
0338  ptiles, and fish strictly belong to their own  PROPER  classes, though some of these old forms are
0391  nd will have been kept by the others to their  PROPER  places and habits, and will consequently have
0417  emarked, the greater number of the characters  PROPER  to the species, to the genus, to the family,
0417  e most important points of structure from the  PROPER  type of the order, yet M. Richard sagaciously
0421  , though much isolated, will still occupy its  PROPER  intermediate position: for F originally was
0422  ssed by groups subordinate to groups; but the  PROPER  or even only possible arrangement would still
0427  to conceal their blood relationship to their  PROPER  lines of descent. We can also understand the
0441  o reach by their active powers of swimming, a  PROPER  place on which to become attached and to unde
0442  have been sketched out with all the parts in  PROPER  proportion, as soon as any structure became
0442  ed by their parents or placed in the midst of  PROPER  nutriment, yet nearly all pass through a simi
0446  which when twelve hours old had acquired its  PROPER  proportions, proves that this is not the univ
0452  gain, an organ may become rudimentary for its  PROPER  purpose, and be used for a distinct object:
0452  swim bladder seems to be rudimentary for its  PROPER  function of giving buoyancy, but has become
0472  assume some of the characters of the species  PROPER  to that zone. In both varieties and species,
0004  ubject far too briefly, as it can be treated  PROPERLY  only by giving long catalogues of facts. We
0044  any variation. To treat this subject at all  PROPERLY,  a long catalogue of dry facts should be giv
0062  le for life against the drought, though more  PROPERLY  it should be said to be dependent on the mo
0187  cones which are coated by pigment, and which  PROPERLY  act only by excluding lateral pencils of lig
0358  are called accidental means, but which more  PROPERLY  might be called occasional means of distribu
0407  nsport, a subject which has hardly ever been  PROPERLY  experimentised on; if we bear in mind how of
0483  the mother's womb? Although naturalists very  PROPERLY  demand a full explanation of every difficult
0351  ariability of each species is an independent  PROPERTY,  and will be taken advantage of by natural s
0489  become utterly extinct. We can so far take a  PROPHETIC  glance into futurity as to foretel that it
0413  e that something more is included; and that  PROPINQUITY  of descent, the only known cause of the s
0107  enera of Ganoid fishes, remnants of a once  PROPONDERANT  order: and in fresh water we find some of
0011  eigh less and the bones of the leg more, in  PROPORTION  to the whole skeleton, than do the same bon
0055  ariably proved to be the case that a larger  PROPORTION  of the species on the side of the larger ge
0074  s, display the same beautiful diversity and  PROPORTION  of kinds as in the surrounding virgin fores
0075  separately, and the seed then mixed in due  PROPORTION,  otherwise the weaker kinds will steadily d
0105  ribution; yet of these species a very large  PROPORTION  are endemic, that is, have been produced th
0136  the wind lulls and the sun shines; that the  PROPORTION  of wingless beetles is larger on the expose
0141  em, may be used as an argument that a large  PROPORTION  of other animals, now in a state of nature,
0142  his kidney beans so early that a very large  PROPORTION  are destroyed by frost, and then collect se
0160  generations. After twelve generations, the  PROPORTION  of blood, to use a common expression, of an
0160  to reversion is retained by this very small  PROPORTION  of foreign blood. In a breed which has not
0188  ow small the number of living animals is in  PROPORTION  to those which have become extinct, I can s
0194  ts present state: yet, considering that the  PROPORTION  of living and known forms to the extinct an
0280  exterminate their parent forms. But just in  PROPORTION  as this process of extermination has acted
0288  sil remains. Throughout an enormously large  PROPORTION  of the ocean, the bright blue tint of the w
0289  ty surprising, when we remember how large a  PROPORTION  of the bones of tertiary mammals have been
0390  mber of kinds of inhabitants is scanty, the  PROPORTION  of endemic species (i.e. those found nowher
0402  very distinct species, belonging in a large  PROPORTION  of cases, as shown by Alph. de Candolle, to
0442  n sketched out with all the parts in proper  PROPORTION,  as soon as any structure became visible in
0022  e size of the two arms of the furcula. The  PROPORTIONAL  width of the gape of mouth, the proportion
0022  oportional width of the gape of mouth, the  PROPORTIONAL  length of the eyelids, of the orifice of t
0053  iculties, and the tables themselves of the  PROPORTIONAL  numbers of the varying species. Dr. Hooker
0054  , or those including many species, a large  PROPORTIONAL  number of dominant species. But so many ca
0071  te different soil to another: not only the  PROPORTIONAL  numbers of the heath plants were wholly ch
0074  ed bank, we are tempted to attribute their  PROPORTIONAL  numbers and kinds to what we call chance.
0075  etermined, in the course of centuries, the  PROPORTIONAL  numbers and kinds of trees now growing on
0081  ical change, for instance, of climate. The  PROPORTIONAL  numbers of its inhabitants would almost im
0091  the breed. Even without any change in the  PROPORTIONAL  numbers of the animals on which our wolf p
0108  the less improved forms, and the relative  PROPORTIONAL  numbers of the various inhabitants of the
0115  are not there indigenous, and thus a large  PROPORTIONAL  addition is made to the genera of these St
0293  ut I can by no means pretend to assign due  PROPORTIONAL  weight to the following considerations. Al
0391  gos Islands, Dr. Hooker has shown that the  PROPORTIONAL  numbers of the different orders are very d
0445  d not nearly acquired their full amount of  PROPORTIONAL  difference. So, again, I was told that the
0445  by no means acquired their full amount of  PROPORTIONAL  difference. As the evidence appears to me
0445  e distinguished from each other, yet their  PROPORTIONAL  differences in the above specified several
0021  e neck that they form a 'hood, and it has,  PROPORTIONALLY  to its size, much elongated wing and tail
0070  od at this one season, increase in number  PROPORTIONALLY  to the supply of seed, as their numbers a
0115  work, that floras gain by naturalisation,  PROPORTIONALLY  with the number of the native genera and
0240  than can be accounted for merely by their  PROPORTIONALLY  lesser size; and I fully believe, though
0076  habits, and constitution, that the original  PROPORTIONS  of a mixed stock could be kept up for half
0081  together, that any change in the numerical  PROPORTIONS  of some of the inhabitants, independently
0358  his many followers. The nature and relative  PROPORTIONS  of the inhabitants of oceanic islands like
0396  o the presence of aerial bats, the singular  PROPORTIONS  of certain orders of plants, herbaceous fo
0398  ands, in their height or climate, or in the  PROPORTIONS  in which the several classes are associate
0445  ter being hatched: I carefully measured the  PROPORTIONS  (but will not here give details) of the be
0446  pigeon and of the other breeds, in all its  PROPORTIONS,  almost exactly as much as in the adult st
0446  en twelve hours old had acquired its proper  PROPORTIONS,  proves that this is not the universal rul
0477  other, at various periods and in different  PROPORTIONS,  the course of modification in the two are
0150  pt to convince any one of the truth of this  PROPOSITION  without giving the long array of facts whi
0156  etween their females: and the truth of this  PROPOSITION  will be granted. The cause of the original
0330  ould be difficult to prove the truth of the  PROPOSITION,  for every now and then even a living anim
0159  e characters of an early progenitor. These  PROPOSITIONS  will be most readily understood by looking
0413  unciating, as briefly as possible, general  PROPOSITIONS,  that is, by one sentence to give the char
0414  is simply a scheme for enunciating general  PROPOSITIONS  and of placing together the forms most lik
0420  of creation, or the enunciation of general  PROPOSITIONS,  and the mere putting together and separat
0459  considered real if we admit the following  PROPOSITIONS,  namely, that gradations in the perfection
0459  structure or instinct. The truth of these  PROPOSITIONS  cannot, I think, be disputed. It is, no do
0007  also, I think, some probability in the view  PROPOUNDED  by Andrew Knight, that this variability may
0147  did not open. The elder Geoffroy and Goethe  PROPOUNDED,  at about the same period, their law of com
```

```
0485 for convienience. This may not be a cheering PROSPECT; but we shall have at least be freed from th
0012 able physiological importance is endless. Dr. PROSPER Lucas's treatise, in two large volumes, is t
0275 duced. On the whole I entirely agree with Dr. PROSPER Lucas, who, after arranging an enormous body
0415 n, who in speaking of certain organs in the PROTEACEAE, says their generic importance, like that o
0046 those genera which have sometimes been called PROTEAN or polymorphic, in which the species present
0066 is an early one. If an animal can in any way PROTECT its own eggs or young, a small number may be
0300 it did not accumulate at a sufficient rate to PROTECT organic bodies from decay, no remains could
0068 most rarely dares to attack a young elephant PROTECTED by its dam. Climate plays an important part
0130 from fatal competition by having inhabited a PROTECTED station. As buds give rise by growth to fre
0148 pede is parasitic within another and is thus PROTECTED, it loses more or less completely its own s
0148 nerves and muscles; but in the parasitic and PROTECTED Proteolepas, the whole anterior part of the
0344 survival of a few descendants, lingering in PROTECTED and isolated situations. When a group has o
0360 waves would float for a less time than those PROTECTED from violent movement as in our experiments
0377 it is certain that many temperate plants, if PROTECTED from the inroads of competitors, can withst
0443 n the egg, or as long as it is nourished and PROTECTED by its parent, it must be quite unimportant
0451 o allow the pollen tubes to reach the ovules PROTECTED in the ovarium at its base. The pistil cons
0060 competition universal. Effects of climate. PROTECTION from the number of individuals. Complex rel
0083 ot rigidly destroy all inferior animals, but PROTECTS during each varying season, as far as lies i
0148 nd in a truly extraordinary manner with the PROTEOLEPAS: for the carapace in all other cirripedes
0148 muscles; but in the parasitic and protected PROTEOLEPAS, the whole anterior part of the head is re
0148 superfluous by the parasitic habits of the PROTEOLEPAS, though effected by slow steps, would be a
0148 ch every animal is exposed, each individual PROTEOLEPAS would have a better chance of supporting i
0199 ing remarks lead me to say a few words on the PROTEST lately made by some naturalists, against the
0139 amblyopsis, and as is the case with the blind PROTEUS with reference to the reptiles of Europe, I
0019 aracters be ever so slight, has had its wild PROTOTYPE. At this rate there must have existed at le
0191 ended by ordinary generation from an ancient PROTOTYPE, of which we know nothing, furnished with a
0484 mals and plants have descended from some one PROTOTYPE. But analogy may be a deceitful guide. Neve
0111 ifferences: whereas varieties, the supposed PROTOTYPES and parents of future well marked species,
0486 structure, in some degree obscured, or the PROTOTYPES of each great class. When we can feel assur
0263 ntly allied species, though the pollen tubes PROTRUDE, they do not penetrate the stigmatic surface
0090 (as the wattle in male carriers, horn like PROTUBERANCES in the cocks of certain fowls, etc.) whic
0101 n enabled, by a fortunate chance, elsewhere to PROVE that two individuals, though both are self fe
0105 ially in the production of species, which will PROVE capable of enduring for a long period, and of
0258 ses. Such cases are highly important, for they PROVE that the capacity in any two species to cross
0288 mportant discoveries made every year in Europe PROVE. No organism wholly soft can be preserved. Sh
0299 ologists at some future period will be able to PROVE, that our different breeds of cattle, sheep,
0330 geological ages; and it would be difficult to PROVE the truth of the proposition, for every now a
0346 case of America alone would almost suffice to PROVE its truth: for if we exclude the northern par
0364 as is known (and it would be very difficult to PROVE this), received within the last few centuries
0404 enerally true, though it would be difficult to PROVE it. Amongst mammals, we see it strikingly dis
0031 at english breeders have actually effected is PROVED by the enormous prices given for animals with
0044 universally implied, though it can rarely be PROVED. We have also what are called monstrosities!
0055 nera on the other side, and it has invariably PROVED to be the case that a larger proportion of th
0113 ll be with plants. It has been experimentally PROVED, that if a plot of ground be sown with one sp
0201 iving bodies of other insects. If it could be PROVED that any part of the structure of any one spe
0223 nd do what work they could. If their presence PROVED useful to the species which had seized them,
0260 each other. This latter statement is clearly PROVED by reciprocal crosses between the same two sp
0271 n confirmed by every subsequent observer, has PROVED the remarkable fact, that one variety of the
0272 he very general fertility of varieties can be PROVED to be of universal occurence, or to form a fu
0298 connected by intermediate varieties and thus PROVED to be the same species, until many specimens
0338 t the truth of this doctrine is very far from PROVED. Yet I fully expect to see it hereafter confi
0345 ole has progressed. If it should hereafter be PROVED that ancient animals resemble to a certain ex
0345 t least be asserted that the record cannot be PROVED to be much more perfect, the main objections
0357 nd. But I do not believe that it will ever be PROVED that within the recent period continents whic
0424 alike in some other respects. If it could be PROVED that the Hottentot had descended from the Neg
0425 be asked, what ought we to do, if it could be PROVED that one species of kangaroo had been produce
0446 bove two principles, which latter, though not PROVED true, can be shown to be in some degree proba
0449 ess that I only hope to see the law hereafter PROVED true. It can be proved true in those cases al
0449 see the law hereafter proved true. It can be PROVED true in those cases alone in which the ancien
0467 organic beings. This high rate of increase is PROVED by calculation, by the effects of a successio
0481 ure are subject to no variation; it cannot be PROVED that the amount of variation in the course of
0242 d. the case, also, is very interesting, as it PROVES that with animals, as with plants, any amount
0446 ours old had acquired its proper proportions, PROVES that this is not the universal rule: for here
0440 of its embryonic career is active, and has to PROVIDE for itself. The period of activity may come
0443 comes at any period of life active, and has to PROVIDE for itself; of the embryo apparently having
0448 on carried on for many generations, having to PROVIDE for their own wants at a very early stage of
0457 at that period of life when the being has to PROVIDE for its own wants, and bearing in mind how s
0088 polygamous animals, and these seem oftenest PROVIDED with special weapons. The males of carnivoro
0090 surviving, and so be preserved or selected, PROVIDED always that they retained strength to master
0100 ion to be valid, but that nature has largely PROVIDED against it by giving to trees a strong tende
0193 n inherited from one ancient progenitor thus PROVIDED, we might have expected that all electric fi
0246 looked on as a special endowment, beyond the PROVINCE of our reasoning powers. The fertility of va
0313 orary migration from a distinct geographical PROVINCE, seems to me satisfactory. These several fac
0058 species range on an average over 6.9 of the PROVINCES into which Mr. Watson has divided Great Bri
0058 eties are recorded, and these range over 7.7 PROVINCES: whereas, the species to which these variet
0058 which these varieties belong range over 14.3 PROVINCES. So that the acknowledged varieties have ve
0408 eparate our several zoological and botanical PROVINCES. We can thus understand the localisation of
0409 numbers, in the different great geographical PROVINCES of the world. On these same principles, we
0410 ession of ages, as in now looking to distant PROVINCES throughout the world, we find that some org
0040 ble, and will then probably first receive a PROVINCIAL name. In semi civilised countries, with lit
0014 been made. There would be great difficulty in PROVING its truth: we may safely conclude that very
0142 d, for it is now as tender as ever it was, as PROVING that acclimatisation cannot be effected! The
0328 ry difficult to account for, considering the PROXIMITY of the two areas, unless, indeed, it be ass
0063 n be no artificial increase of food, and no PRUDENTIAL restraint from marriage. Although some spec
0488 n fields for far more important researches. PSYCHOLOGY will be based on a new foundation, that of
0084 n, and bark feeders mottled grey: the alpine PTARMIGAN white in winter, the red grouse the colour
```

```
0416   folded, mere colour in certain Algae, mere PUBESCENCE on parts of the flower in grasses, the natu
0340   merous marsupials; and I have shown in the PUBLICATIONS above alluded to, that in America the law
0001   alth is far from strong, I have been urged to PUBLISH this Abstract. I have more especially been
0002   1844, honoured me by thinking it advisable to PUBLISH, with Mr. Wallace's excellent memoir, some b
0002   om my manuscripts. This Abstract, which I now PUBLISH, must necessarily be imperfect. I cannot her
0002   ho sent it to the Linnean Society, and it is PUBLISHED in the third volume of the Journal of that
0012   ile carefully to study the several treatises PUBLISHED on some of our old cultivated plants, as or
0020   y treatises in different languages have been PUBLISHED on pigeons, and some of them are very impor
0033   of late years, and many treatises have been PUBLISHED on the subject; and the result, I may add,
0097   s flowers, that I have found, by experiments PUBLISHED elsewhere, that their fertility is greatly
0142   ery strikingly shown in works on fruit trees PUBLISHED in the United States, in which certain vari
0142   y beans ever appear, for an account has been PUBLISHED how much more hardy some seedlings appeared
0150   k with a remark, nearly to the above effect, PUBLISHED by Mr. Waterhouse. I infer also from an obs
0285   ried from 600 to 3000 feet. Prof. Ramsay has PUBLISHED an account of a downthrow in Anglesea of 23
0289   ins, a single glance at the historical table PUBLISHED in the Supplement to Lyell's Manual, will b
0303   ell known fact that in geological treatises, PUBLISHED not many years ago, the great class of mamm
0304   ay read in the Supplement to Lyell's Manual, PUBLISHED in 1858, clear evidence of the existence of
0304   roup of species. But my work had hardly been PUBLISHED, when a skilful palaeontologist, M. Bosquet
0320   y, through man's agency. I may repeat what I PUBLISHED in 1845, namely, that to admit that species
0373   he same story. If one account which has been PUBLISHED can be trusted, we have direct evidence of
0482   s. Several eminent naturalists have of late PUBLISHED their belief that a multitude of reputed sp
0002   ble than I do of the necessity of hereafter PUBLISHING in detail all the facts, with references, o
0291   e thus been formed during subsidence. Since PUBLISHING my views on this subject in 1845, I have wa
0184   n the quiet Sounds of Tierra del Fuego, the PUFFINURIA berardi, in its general habits, in its asto
0032   lants, but merely go over their seed beds, and PULL up the rogues, as they call the plants that de
0232   at an angle, how often the bees would entirely PULL down and rebuild in different ways the same ce
0386   ept it covered up in my study for six months, PULLING up and counting each plant as it grew; the p
0222   individual of F. sanguinea emerge, carrying a PUPA; but I was not able to find the desolated nest
0223   n, and one was perched motionless with its own PUPA in its mouth on top of a spray of heath over i
0219   hich they like best, and with their larvae and PUPAE to stimulate them to work, they did nothing;
0220   the nest is much disturbed and the larvae and PUPAE are exposed, the slaves work energetically wi
0221   tant, but they were prevented from getting any PUPAE to rear as slaves. I then dug up a small parc
0221   as slaves. I then dug up a small parcel of the PUPAE of F. fusca from another nest, and put them d
0222   i laid on the same place a small parcel of the PUPAE of another species, F. flava, with a few of t
0222   ain when F. sanguinea could distinguish the PUPAE of F. fusca, which they habitually make into
0222   een that they eagerly and instantly seized the PUPAE of F. fusca, whereas they were much terrified
0222   were much terrified when they came across the PUPAE, or even the earth from the nest of F. flava,
0222   wled away, they took heart and carried off the PUPAE. One evening I visited another community of F
0222   wing that it was not a migration) and numerous PUPAE. I traced the returning file burthened with b
0223   t slave makers, will as I have seen, carry off PUPAE of other species, if scattered near their nes
0223   cattered near their nests, it is possible that PUPAE originally stored as food might become develo
0224   han to procreate them, the habit of collecting PUPAE originally for food might by natural selectio
0215   rable in dogs which have been brought home as PUPPIES from countries, such as Tierra del Fuego and
0445   ock; hence I was curious to see how far their PUPPIES differed from each other: I was told by bree
0445   measuring the old dogs and their six days old PUPPIES, I found that the puppies had not nearly acq
0445   their six days old puppies, I found that the PUPPIES had not nearly acquired their full amount of
0020   geons. I have kept every breed which I could PURCHASE or obtain, and have been most kindly favoure
0020   the offspring from the first cross between two PURE breeds is tolerably and sometimes (as I have f
0025   in crossed together, and one grandchild of the PURE white fantail and pure black barb was of as be
0025   d one grandchild of the pure white fantail and PURE black barb was of as beautiful a blue colour,
0027   ly appearing in all the breeds, both when kept PURE and when crossed; the mongrel offspring being
0036   them has deviated in any one instance from the PURE blood of Mr. Bakewell's flock, and yet the dif
0098   ear each other, it is hardly possible to raise PURE seedlings, so largely do they naturally cross.
0165   mare and male quagga, the hybrid, and even the PURE offspring subsequently produced from the mare
0165   lainly barred across the legs than is even the PURE quagga. Lastly, and this is another most remar
0214   own occasionally to happen, as I once saw in a PURE terrier. When the first tendency was once disp
0246   e sterility of the hybrids produced from them. PURE species have of course their organs of reprodu
0247   ring, with the average number produced by both PURE parent species in a state of nature. But a ser
0248   and, on the other hand, that the fertility of PURE species is so easily affected by various circu
0248   refully guarding them from a cross with either PURE parent, for six or seven, and in one case for
0250   brids are perfectly fertile, as fertile as the PURE parent species, as are Kolreuter and Gartner t
0253   d pair brothers and sisters in the case of any PURE animal, which from any cause had the least ten
0253   a, have often bred in this country with either PURE parent, and in one single instance they have b
0253   less than eight hybrids (grandchildren of the PURE geese) from one nest. In India, however, these
0253   and as they are kept for profit, where neither PURE parent species exists, they must certainly be
0255   would produce, even with the pollen of one of PURE parent, a single fertile seed: but in some of
0255   y may be detected, by the pollen of one of the PURE parent species causing the flower of the hybri
0256   ans strict. There are many cases, in which two PURE species can be united with unusual facility, a
0256   avourable conditions, than is the fertility of PURE species. But the degree of fertility is likewi
0259   ybrids, though externally so like one of their PURE parent species, are with rare exceptions extre
0259   are born, which closely resemble one of their PURE parents; and these hybrids are almost always u
0259   ependent of its external resemblance to either PURE parent. Considering the several rules now give
0262   ferent case from the difficulty of uniting two PURE species, which have their reproductive organs
0263   nt, for, as just remarked, in the union of two PURE species the male and female sexual elements ar
0266   ally and exceptionally resemble closely either PURE parent. Nor do I pretend that the foregoing re
0274   hybrids and mongrels can be reduced to either PURE parent form, by repeated crosses in successive
0276   orests. The sterility of first crosses between PURE species, which have their reproductive systems
0277   to that sterility which so frequently affects PURE species, when their natural conditions of life
0283   r pebbles: for there is reason to believe that PURE water can effect little of nothing in wearing
0036   mr. Burgess, as Mr. Youatt remarks, have been PURELY bred from the original stock of Mr. Bakewell
0164   a horse without stripes is not considered as PURELY bred. The spine is always striped; the legs a
0026   ings. Or, secondly, that each breed, even the PUREST, has within a dozen or, at most, within a sco
0288   he bright blue tint of the water bespeaks its PUPITY. The many cases on record of a formation conf
0085   eetle, a curculio, than those with down; that PURPLE plums suffer far more from a certain disease
0085   riety, whether a smooth or downy, a yellow or PURPLE fleshed fruit, should succeed. In looking at
0030   ture, with the wool of one breed good for one PURPOSE, and that of another breed for another purpo
```

Page **************************************(Key Word)**

0030	urpose, and that of another breed for another	PURPOSE: when we compare the many breeds of dogs, ea
0034	nything existing in the country. But, for our	PURPOSE, a kind of Selection, which may be called Un
0036	particularly useful to them, for any special	PURPOSE, would be carefully preserved during famines
0099	placed so close together, as if for the very	PURPOSE of self fertilisation, should in so many cas
0142	idney bean has been often cited for a similar	PURPOSE, and with much greater weight: but until som
0149	an when the part has to serve for one special	PURPOSE alone. In the same way that a knife which ha
0152	een but little specialised for any particular	PURPOSE, and perhaps in polymorphic groups, we see a
0182	ls, like the ostrich: and functionally for no	PURPOSE like the Apteryx. Yet the structure of each
0190	be modified for some other and quite distinct	PURPOSE, or be quite obliterated. The illustration o
0190	that an organ originally constructed for one	PURPOSE, namely flotation, may be converted into one
0190	be converted into one for a wholly different	PURPOSE, namely respiration. The swimbladder has, al
0191	by natural selection for some quite distinct	PURPOSE: in the same manner as, on the view entertai
0195	this could have been adapted for its present	PURPOSE by successive slight modifications, each bet
0199	n species. I might have adduced for this same	PURPOSE the differences between the races of man, wh
0201	ill have formed, as Paley has remarked, for the	PURPOSE of causing pain or for doing an injury to it
0202	en modified but not perfected for its present	PURPOSE, with the poison originally adapted to cause
0202	can we admire the production for this single	PURPOSE of thousands of drones, which are utterly us
0207	the same way, without their knowing for what	PURPOSE it is performed, is usually said to be insti
0216	nd this is evidently done for the instinctive	PURPOSE of allowing, as we see in wild ground birds,
0224	and rendered permanent for the very different	PURPOSE of raising slaves. When the instinct was onc
0374	taneously cooler. But it would suffice for my	PURPOSE, if the temperature was at the same time low
0418	the embryo should be more important for this	PURPOSE than that of the adult, which alone plays it
0423	shall says the horns are very useful for this	PURPOSE with cattle, because they are less variable
0435	on the existing general pattern, for whatever	PURPOSE they served, we can at once perceive the pla
0451	erly aborted for one, even the more important	PURPOSE; and remain perfectly efficient for the othe
0452	ed with hairs as in other compositae, for the	PURPOSE of brushing the pollen out of the surroundin
0452	n organ may become rudimentary for its proper	PURPOSE, and be used for a distinct object: in certa
0454	n the fin of the manatee have formed for this	PURPOSE. On my view of descent with modification, th
0454	habits of life, useless or injurious for one	PURPOSE, might easily be modified and used for anoth
0454	might easily be modified and used for another	PURPOSE. Or an organ might be retained for one alone
0457	splayed in the homologous organs, to whatever	PURPOSE applied, of the different species of a class
0479	leg of a bat, though used for such different	PURPOSE, in the jaws and legs of a crab, in the peta
0026	. i can state this from my own observations,	PURPOSELY made on the most distinct breeds. Now, it i
0030	o man at different seasons and for different	PURPOSES, or so beautiful in his eyes, we must, I thi
0042	: in geese, from being valuable only for two	PURPOSES, food and feathers, and more especially from
0158	secondary sexual, and for ordinary specific	PURPOSES. Distinct species present analogous variatio
0196	imals, its general presence and use for many	PURPOSES in so many land animals, which in their lung
0196	uently come to be worked in for all sorts of	PURPOSES, as a fly flapper, an organ of prehension, o
0248	arious circumstances, that for all practical	PURPOSES it is most difficult to say where perfect fe
0435	all these organs, serving for such different	PURPOSES, are formed by infinitely numerous modificat
0437	used as they are for such totally different	PURPOSES? Why should one crustacean, which has an ext
0437	wer, though fitted for such widely different	PURPOSES, be all constructed on the same pattern? On
0438	d, and have adapted them to the most diverse	PURPOSES. And as the whole amount of modification wil
0439	ome widely different and serve for different	PURPOSES, are in the embryo exactly alike. The embryo
0442	ely become very unlike and serve for diverse	PURPOSES, being at this early period of growth alike:
0451	ly alike in nature. An organ serving for two	PURPOSES, many become rudimentary or utterly aborted
0453	d organs are exquisitely adapted for certain	PURPOSES, tells us with equal plainness that these ru
0457	gans, though fitted in the adult members for	PURPOSES as different as possible. Larvae are active
0091	cub might be born with an innate tendency to	PURSUE certain kinds of prey. Nor can this be though
0484	natural history. Systematists will be able to	PURSUE their labours as at present: but they will no
0001	hat period to the present day I have steadily	PURSUED the same object. I hope that I may be excuse
0216	ng, and all has been the result of selection,	PURSUED both methodically and unconsciously: but in
0091	, one with a light greyhound like form, which	PURSUES deer, and the other more bulky, with shorter
0367	warmth increased, whilst their brethern were	PURSUING their northern journey. Hence, when the warm
0095	ting and moving parts of the corolla, so as to	PUSH the pollen on to the stigmatic surface. Hence,
0097	ck the nectar: for, in doing this, they either	PUSH the flowers own pollen on the stigma, or bring
0141	dely different climates. We must not, however,	PUSH the foregoing argument too far, on account of
0230	s in such cases stand in the opposed cells and	PUSH and bend the ductile and warm wax (which as I
0197	t may possibly be due to the direct action of	PUTRID matter: but we should be very cautious in dra
0197	d at as a direct adaptation for wallowing in	PUTRIDITY: and so it may be, or it may possibly be du
0225	its six sides bevelled so as to join on to a	PYRAMID, formed of three rhombs. These rhombs have c
0226	se, the three flat surfaces are united into a	PYRAMID: and this pyramid, as Huber has remarked, is
0226	surfaces are united into a pyramid: and this	PYRAMID, as Huber has remarked, is manifestly a gros
0228	ead of on the straight edges of a three sided	PYRAMID as in the case of ordinary cells. I then put
0225	certain angles, and the three which form the	PYRAMIDAL base of a single cell on one side of the co
0226	ifestly a gross imitation of the three sided	PYRAMIDAL basis of the cell of the hive bee. As in th
0230	cell. They do not make the whole three sided	PYRAMIDAL base of any one cell at the same time, but
0231	h of an inch in thickness: the plates of the	PYRAMIDAL basis being about twice as thick. By this s
0365	s living on the snowy regions of the Alps or	PYRENEES, and in the extreme northern parts of Europe
0367	arts of Europe, as far south as the Alps and	PYRENEES, and even stretching into Spain. The now tem
0368	marked by Mr. H. C. Watson, and those of the	PYRENEES, as remarked by Ramond, are more especially
0378	ntry and covered the land at the foot of the	PYRENEES. At this period of extreme cold, I believe t
0373	ain to the Oural range, and southward to the	PYRENNES. We may infer, from the frozen mammals and n
0151	era: but in the several species of ore genus,	PYRGOMA, these valves present a marvellous amount of
0227	layer of hexagonal prisms united together by	PYRIMIDAL bases formed of three rhombs: and the rhomb
0061	lest parasite which clings to the hairs of a	QUADRUPED or feathers of a bird: in the structure of
0083	es not exercise a long backed or long legged	QUADRUPED in any peculiar manner: he exposes sheep wi
0113	imple habits. Take the case of a carnivorous	QUADRUPED, of which the number that can be supported
0180	, and it had been asked how an insectivorous	QUADRUPED could possibly have been converted into a f
0067	ld be the same with turf closely browsed by	QUADRUPEDS, be let to grow, the more vigorous plants g
0075	as in the case of locusts and grass feeding	QUADRUPEDS. But the struggle almost invariably will be
0133	e direct action on the hair of our domestic	QUADRUPEDS. Instances could be given of the same varie
0135	from enemies, as well as any of the smaller	QUADRUPEDS. We may imagine that the early progenitor o
0195	great advantage. It is not that the larger	QUADRUPEDS are actually destroyed (except in some rare
0352	urope, and consequently possessing the same	QUADRUPEDS. But if the same species can be produced at
0392	eeds for transportal by the wool and fur of	QUADRUPEDS. This case presents no difficulty on my vie

Page **************************************(Key Word)**

quadrupeds

Page ***(Key Word)***

```
0394  an island can be named on which our smaller QUADRUPEDS have not become naturalised and greatly mul
0395  e inhabitied by closely allied or identical QUADRUPEDS. No doubt some few anomalies occur in this
0077  idst of its range, why does it not double or QUADRUPLE its numbers? We know that it can perfectly
0163  on the legs and faintly on the shoulder. The QUAGGA, though so plainly barred like a zebra over t
0165  s famous hybrid from a chestnut mare and male QUAGGA, the hybrid, and even the pure offspring subs
0165  barred across the legs than is even the pure QUAGGA. Lastly, and this is another most remarkable
0167  r more wild stocks; or the ass, the hemionus, QUAGGA, and zebra. He who believes that each equine
0185  and the landrail nearly as terrestrial as the QUAIL or partridge. In such cases, and many others
0363  eeds. Reflect for a moment on the millions of QUAILS which annually cross the Mediterranean; and c
0032  ome an eminent breeder. If gifted with these QUALITIES, and he studies his subject for years, and
0032  great improvements; if he wants any of these QUALITIES, he will assuredly fail. Few would readily
0034  reeding, for the inheritance of good and bad QUALITIES is so obvious. At the present time, eminent
0035  en during their own lifetimes, the forms and QUALITIES of their cattle. Slow and insensible change
0041  paid to even the slightest deviation in the QUALITIES or structure of each individual. Unless suc
0127  tion. Natural selection, on the principle of QUALITIES being inherited at corresponding ages, can
0207  ersities of instinct and of the other mental QUALITIES of animals within the same class. I will no
0213  riations of instinct in modifying the mental QUALITIES of our domestic animals. A number of curiou
0242  efly in this chapter to show that the mental QUALITIES of our domestic animals vary, and that the
0446  up: they are indifferent whether the desired QUALITIES and structures have been acquired earlier o
0037  iption, to have been a fruit of very inferior QUALITY. I have seen great surprise expressed in hor
0041  e scanty, all the individuals, whatever their QUALITY may be, will generally be allowed to breed,
0141  ook at adaptation to any special climate as a QUALITY readily grafted on an innate wide flexibilit
0147  d, the fruit itself gains largely in size and QUALITY. In our poultry, a large tuft of feathers on
0212  particular enemy is certainly an instinctive QUALITY, as may be seen in nestling birds, though it
0245  crossed, have been specially endowed with the QUALITY of sterility, in order to prevent the confus
0245  rility is not a specially acquired or endowed QUALITY, but is incidental on other acquired differe
0255  species have specially been endowed with this QUALITY, in order to prevent their crossing and blen
0261  ther differences, and not a specially endowed QUALITY. As the capacity of one plant to be grafted
0261  ose that this capacity is a specially endowed QUALITY, but will admit that it is incidental on dif
0403  ve in crevices of stone: and although large QUANTITIES of stone are annually transported from Port
0092  e common laurel. This juice, though small in QUANTITY, is greedily sought by insects. Let us now s
0093  h have four stamens producing rather a small QUANTITY of pollen, and a rudimentary pistil; other h
0233  n of each pound of wax; so that a prodigious QUANTITY of fluid nectar must be collected and consum
0234  ses, and so be altogether independent of the QUANTITY of honey which the bees could collect. But l
0351  tion in different species will be no uniform QUANTITY. If, for instance, a number of species, whic
0386  occasionally, though rarely, adheres in some QUANTITY to the feet and beaks of birds. Wading birds
0468  variation under nature is a strictly limited QUANTITY. Man, though acting on external characters a
0481  tion in the course of long ages is a limited QUANTITY; no clear distinction has been, or can be, d
0030  ious in battle, with other breeds so little QUARRELSOME, with everlasting layers which never desir
0031  these have now been exported to almost every QUARTER of the world. The improvement is by no means
0110  plants are crowded together than in any other QUARTER of the world, some foreign plants have becom
0222  . flava, and quickly ran away; but in about a QUARTER of an hour, shortly after all the little yel
0294  period. Yet it may be doubted whether in any QUARTER of the world, sedimentary deposits, includin
0326  be also favourable, as before explained. One QUARTER of the world may have been most favourable f
0328  accumulated over very wide spaces in the same QUARTER of the world; but we are far from having any
0337  imilar climate, the eocene inhabitants of one QUARTER of the world were put into competition with
0337  xisting inhabitants of the same or some other QUARTER of the world, the eocene fauna or flora would certainly b
0340  t once explained; for the inhabitants of each QUARTER of the world will obviously tend to leave in
0340  he world will obviously tend to leave in that QUARTER, during the next succeeding period of time,
0410  hanged during successive ages within the same QUARTER of the world, or to those which have changed
0017  nce, of the many foxes, inhabiting different QUARTERS of the world. I do not believe, as we shall
0020  most kindly favoured with skins from several QUARTERS of the world, more especially by the Hon. W.
0027  mesticated for thousands of years in several QUARTERS of the world; the earliest known record of p
0033  odical practice for scarcely more than three QUARTERS of a century; it has certainly been more att
0167  when crossed with species inhabiting distant QUARTERS of the world, to produce hybrids resembling
0284  eet; that is, very nearly thirteen and three QUARTERS British miles. Some of these formations, whi
0294  in examining the latest deposits of various QUARTERS of the world, it has everywhere been noted,
0351  me genus, though inhabiting the most distant QUARTERS of the world, must originally have proceeded
0395  time to follow up this subject in all other QUARTERS of the world; but as far as I have gone, the
0406  he migration of some of the species into all QUARTERS of the world, where they may have become sli
0409  should have all its species common to other QUARTERS of the world. We can see why whole groups of
0410  e changed after having migrated into distant QUARTERS, in both cases the forms within each class h
0428  ide range: and thus the septenary, quinary, QUATERNARY, and ternary classifications have probably
0202  to admire the savage instinctive hatred of the QUEEN bee, which urges her instantly to destroy the
0233  iculty in a single insect (as in the case of a QUEEN wasp) making hexagonal cells, if she work alt
0472  ur fir trees: at the instinctive hatred of the QUEEN bee for her own fertile daughters; at ichneum
0203  hich urges her instantly to destroy the young QUEENS her daughters as soon as born, or to perish h
0001  something might perhaps be made out on this QUESTION by patiently accumulating and reflecting on
0002  he facts and arguments on both sides of each QUESTION; and this cannot possibly be here done. I mu
0035  rements or careful drawings of the breeds in QUESTION had been made long ago, which might serve fo
0141  ch to both means combined, is a very obscure QUESTION. That habit or custom has some influence I m
0161  een a tendency to reproduce the character in QUESTION, which at last, under unknown favorable cond
0172  will be much more convenient to discuss this QUESTION in the chapter on the Imperfection of the ge
0180  y have been converted into a flying bat, the QUESTION would have been far more difficult, and I co
0248  idence advanced by our best botanists on the QUESTION whether certain doubtful forms should be ran
0272  tly of their fertility. Independently of the QUESTION of fertility, the offspring of species when
0275  es, or of distinct species. Laying aside the QUESTION of fertility and sterility, in all other res
0307  world swarmed with living creatures. To the QUESTION why we do not find records of these vast pri
0343  stem was deposited: I can answer this latter QUESTION only hypothetically, by saying that as far a
0352  fically distinct. We are thus brought to the QUESTION which has been largely discussed by naturali
0355  ation do not directly bear on another allied QUESTION, namely whether all the individuals of the s
0389  marks I shall not confine myself to the mere QUESTION of dispersal: but shall consider some other
0394  r mammals on remote islands? On my view this QUESTION can easily be answered: for no terrestrial m
0443  t, or what its precise features will be. The QUESTION is not, at what period of life any variation
0482  , who will be able to view both sides of the QUESTION with impartiality. Whoever is led to believe
0483  doctrine of the modification of species. The QUESTION is difficult to answer, because the more dis
```

Page ***(Key Word)***

Page **********************************(Key Word)**********************************

```
0053 xistence, divergence of character, and other QUESTIONS, hereafter to be discussed. Alph. de Candol
0437 election, we can satisfactorily answer these QUESTIONS. In the vertebrata, we see a series of inte
0464 s in the world's history. I can answer these QUESTIONS and grave objections only on the suppositio
0466 that the more important objections relate to QUESTIONS on which we are confessedly ignorant; nor d
0042 n be propagated in great numbers and at a very QUICK rate, and inferior birds may be freely reject
0103 wander much, and which do not breed at a very QUICK rate. Hence in animals of this nature, for in
0288 e whole bed of the sea, at a rate sufficiently QUICK to embed and preserve fossil remains. Through
0283 thickness, which, though probably formed at a QUICKER rate than many other deposits, yet, from bei
0313 e productions of the land seem to change at a QUICKER rate than those of the sea, of which a strik
0314 red. We can perhaps understand the apparently QUICKER rate of change in terrestrial and in more hi
0394 hought that mammals appear and disappear at a QUICKER rate than other and lower animals. Though te
0463 will generally be modified and improved at a QUICKER rate than the intermediate varieties, which
0043 ection, whether applied methodically and more QUICKLY, or unconsciously and more slowly, but more
0063 le of geometrical increase, its numbers would QUICKLY become so inordinately great that no country
0094 terised would be able to obtain its food more QUICKLY, and so have a better chance of living and l
0103 pid rate, a new and improved variety might be QUICKLY formed on any one spot, and might there main
0110 ent species. Hence, rare species will be less QUICKLY modified or improved within any given period
0111 curious instances could be given showing how QUICKLY new breeds of cattle, sheep, and other anima
0122 ies get into some distinct country, or become QUICKLY adapted to some quite new station, in which
0177 or on the plains improving their breeds more QUICKLY than the small holders on the intermediate n
0180 ds or beasts of prey, or to collect food more QUICKLY, or as there is reason to believe, by lessen
0196 or the hare, with hardly any tail, can double QUICKLY enough. In the second place, we may sometime
0222 even the earth from the nest of F. flava, and QUICKLY ran away; but in about a quarter of an hour,
0229 ere I suppose that the bees had excavated too QUICKLY, and convex on the opposed side, where the b
0229 opposed side, where the bees had worked less QUICKLY. In one well marked instance, I put the comb
0283 rolled about by the waves, and then are more QUICKLY ground into pebbles, sand, or mud. But how o
0313 ered high in the scale of nature, change more QUICKLY than those that are low: though there are ex
0316 e as do single species, changing more or less QUICKLY, and in a greater or lesser degree. A group
0327 nd likewise when sediment was not thrown down QUICKLY enough to embed and preserve organic remains
0359 to place them on sea water. The majority sank QUICKLY, but some which whilst green floated for a v
0388 uch low beings change or become modified less QUICKLY than the high; and this will give longer tim
0401 ecies formed in the separate islands have not QUICKLY spread to the other islands. But the islands
0184 most aerial and oceanic of birds, yet in the QUIET Sounds of Tierra del Fuego, the Puffinuria be
0428 sm over a wide range; and thus the septenary, QUINARY, quaternary, and ternary classifications hav
0261 e pear can be grafted far more readily on the QUINCE, which is ranked as a distinct genus, than on
0262 ake with different degrees of facility on the QUINCE; so do different varieties of the apricot and
0003 t. In considering the Origin of Species, it is QUITE conceivable that a naturalist, reflecting on
0009 s, though often weak and sickly, yet breeding QUITE freely under confinement; and when, on the ot
0009 when it does act under confinement, acting not QUITE regularly, and producing offspring not perfec
0010 itions; but in some cases it can be shown that QUITE opposite conditions produce similar changes o
0011 al. In monstrosities, the correlations between QUITE distinct parts are very curious; and many ins
0011 ngated head. Some instances of correlation are QUITE whimsical: thus cats with blue eyes are invar
0012 relation of growth. The result of the various, QUITE unknown, or dimly seen laws of variation is i
0013 he anomaly. The laws governing inheritance are QUITE unknown; no one can say why the same peculiar
0014 arly perfect reversion had ensued. It would be QUITE necessary, in order to prevent the effects of
0022 irds the head and tail touch; the oil gland is QUITE aborted. Several other less distinct breeds m
0024 ancient times by half civilized man, as to be QUITE prolific under confinement. An argument, as i
0027 er domestication: these supposed species being QUITE unknown in a wild state, and their becoming n
0028 obable origin of domestic pigeons at some, yet QUITE insufficient, length: because when I first ke
0031 speak of an animal's organisation as something QUITE plastic, which they can model almost as they
0036 o great that they have the appearance of being QUITE different varieties. If there exist savages o
0038 f good hope, nor any other region inhabited by QUITE uncivilised man, has afforded us a single pla
0043 varieties is immense; for the cultivator here QUITE disregards the extreme variability both of hy
0046 d have been effected only by slow degrees; yet QUITE recently Mr. Lubbock has shown a degree of va
0046 philosophical naturalist, I may add, has also QUITE recently shown that the muscles in the larvae
0050 st commences the study of a group of organisms QUITE unknown to him, he is at first much perplexed
0051 some naturalists come very near to, but do not QUITE arrive at the rank of species; or, again, bet
0059 ry to give to two forms the rank of species is QUITE indefinite. In genera having more than the av
0062 tinction, and variation, will be dimly seen or QUITE misunderstood. We behold the face of nature b
0064 been well authenticated, they would have been QUITE incredible. So it is with plants: cases could
0068 e action of climate seems at first sight to be QUITE independent of the struggle for existence; bu
0069 numbers, we may feel sure that the cause lies QUITE as much in other species being favoured, as i
0071 ore than is generally seen in passing from one QUITE different soil to another: not only the propo
0074 ber of cats that destroy the mice. Hence it is QUITE credible that the presence of a feline animal
0075 ore seed, and will consequently in a few years QUITE supplant the other varieties. To keep up a mi
0085 our ignorance permits us to judge, seem to be QUITE unimportant, we must not forget that climate,
0097 s will act like a camel hair pencil, and it is QUITE sufficient just to touch the anthers of one f
0122 nct country, or become quickly adapted to some QUITE new station, in which child and parent do not
0136 not at all reduced, but even enlarged. This is QUITE compatible with the action of natural selecti
0137 are rudimentary in size, and in some cases are QUITE covered up by skin and fur. This state of the
0140 ges by the competition of other organic beings QUITE as much as, or more than, by adaptation to pa
0144 of the bond of correlation is very frequently QUITE obscure. M. Is. Geoffroy St. Hilaire has forc
0145 that when this occurs, the adherent nectary is QUITE aborted; when the colour is absent from only
0153 ection, sometimes become attached, from causes QUITE unknown to us, more to one sex than to the ot
0161 for an endless number of generations, than in QUITE useless or rudimentary organs being, as we al
0163 tripes are sometimes very obscure, or actually QUITE lost, in dark coloured asses. The koulan of P
0164 tripes are plainest in the foal; and sometimes QUITE disappear in old horses. Colonel Poole has se
0174 ate varieties? This difficulty for a long time QUITE confounded me. But I think it can be in large
0175 ct in ascending mountains, and sometimes it is QUITE remarkable how abruptly, as Alph. de Candolle
0190 her organ might be modified for some other and QUITE distinct purpose, or be quite obliterated. Th
0190 r some other and quite distinct purpose, or be QUITE obliterated. The illustration of the swimblad
0191 dually worked in by natural selection for some QUITE distinct purpose: in the same manner as, on t
0196 tle importance, and which have originated from QUITE secondary causes, independently of natural se
0197 i have no doubt that the colour is due to some QUITE distinct cause, probably to sexual selection.
0199 ably have had some little effect on structure, QUITE independently of any good thus gained. Correl
0206 organic beings of the same class, and which is QUITE independent of their habits of life. On my th
```

Page **********************************(Key Word)**********************************

Page **(Key Word)**

0209	but I believe that the effects of habit are of	QUITE	subordinate importance to the effects of the
0211	hich was eagerly devoured by the ant. Even the	QUITE	young aphides behaved in this manner, showing
0212	cies, give rise, through natural selection, to	QUITE	new instincts. But I am well aware that these
0215	e other hand, do our civilised dogs, even when	QUITE	young, require to be taught not to attack pou
0220	fety. Hence, it is clear, that the slaves feel	QUITE	at home. During the months of June and July,
0224	er instincts you please, and it seems at first	QUITE	inconceivable how they can make all the neces
0233	t, as it bears on a fact, which seems at first	QUITE	subversive of the foregoing theory; namely, t
0234	umber of its parasites or other enemies, or on	QUITE	distinct causes, and so be altogether indepen
0238	of shield or their heads, the use of which is	QUITE	unknown: in the Mexican Myrmecocystus, the wo
0239	dered by natural selection, as I believe to be	QUITE	possible, different from the fertile males on
0246	onsidered by most authors as distinct species,	QUITE	fertile together, he unhesitatingly ranks the
0250	ed to the pollen of a distinct species, though	QUITE	sterile with their own pollen, notwithstandin
0254	nal species must either at first have produced	QUITE	fertile hybrids, or the hybrids must have bec
0254	ids must have become in subsequent generations	QUITE	fertile under domestication. This latter alte
0254	genous domestic dogs of South America, all are	QUITE	fertile together! and analogy makes me greatl
0254	st have freely bred together and have produced	QUITE	fertile hybrids. So again there is reason to
0254	our European and the humped Indian cattle are	QUITE	fertile together; but from facts communicated
0255	seeds produced, up to nearly complete or even	QUITE	complete fertility; and, as we have seen, in
0261	been able to graft trees together belonging to	QUITE	distinct families; and, on the other hand, cl
0268	ts as varieties, are said by Gartner not to be	QUITE	fertile when crossed, and he consequently ran
0270	ase is far more remarkable, and seems at first	QUITE	incredible; but it is the result of an astoni
0272	s, could find very few and, as it seems to me,	QUITE	unimportant differences between the so called
0277	mongrel offspring, are very generally, but not	QUITE	universally, fertile. Nor is this nearly gene
0279	erning conditions of life do not graduate away	QUITE	insensibly like heat or moisture. I endeavour
0284	a hundred thousand years. This estimate may be	QUITE	erroneous; yet, considering over what wide sp
0307	e present day; and that during these vast, yet	QUITE	unknown, periods of time, the world swarmed w
0314	emely slow. The variability of each species is	QUITE	independent of that of all others. Whether su
0316	exterminated by its improved offspring, it is	QUITE	incredible that a fantail, identical with the
0325	s to precisely the same effect. It is, indeed,	QUITE	futile to look to changes of currents, climat
0330	cters, so that the two groups, though formerly	QUITE	distinct, at that period made some small appr
0331	aces in the economy of nature. Therefore it is	QUITE	possible, as we have seen in the case of some
0334	some preceding forms, and for the coming in of	QUITE	new forms by immigration, and for a large amo
0349	ees their nests similarly constructed, but not	QUITE	alike, with eggs coloured in nearly the same
0350	far as we positively know, produces organisms	QUITE	like, or, as we see in the case of varieties
0350	dification through natural selection, and in a	QUITE	subordinate degree to the direct influence of
0357	hin the recent period continents which are now	QUITE	separate, have been continuously, or almost c
0362	ough the beaks and feet of birds are generally	QUITE	clean, I can show that earth sometimes adhere
0362	artridge, and in this earth there was a pebble	QUITE	as large as the seed of a vetch. Thus seeds a
0381	their dispersal. But the existence of several	QUITE	distinct species, belonging to genera exclusi
0383	ly have many fresh water species, belonging to	QUITE	different classes, an enormous range, but all
0385	weed from one aquarium to another, that I have	QUITE	unintentionally stocked the one with fresh wa
0387	t, I thought that its distribution must remain	QUITE	inexplicable; but Audubon states that he foun
0389	eriod all existing islands have been nearly or	QUITE	joined to some continent. This view would rem
0389	tralia, we must, I think, admit that something	QUITE	independently of any difference in physical c
0389	d the comparison in some other respects is not	QUITE	fair. We have evidence that the barren island
0390	naturalised plants and animals have nearly or	QUITE	exterminated many native productions. He who
0400	are tenanted, as I have elsewhere shown, in a	QUITE	marvellous manner, by very closely related sp
0416	n they differ much, and the differences are of	QUITE	subordinate value in classification; yet no o
0418	vital organs, are found to offer characters of	QUITE	subordinate value. We can see why characters
0419	bution has often been used, though perhaps not	QUITE	logically, in classification, more especially
0424	bling; but the short faced breed has nearly or	QUITE	lost this habit; nevertheless, without any re
0430	one group of animals exhibits an affinity to a	QUITE	distinct group, this affinity in most cases i
0432	were suddenly to reappear, though it would be	QUITE	impossible to give definitions by which each
0432	the finest varieties. In this case it would be	QUITE	impossible to give any definition by which th
0437	ry in number and structure; consequently it is	QUITE	probable that natural selection, during a lon
0440	pecies, or of two groups of species, differing	QUITE	as much, or even more, from each other than d
0442	pted to the most diverse and active habits, or	QUITE	inactive, being fed by their parents or place
0443	rished and protected by its parent, it must be	QUITE	unimportant whether most of its characters ar
0444	by its parents. Hence, I conclude, that it is	QUITE	possible, that each of the many successive mo
0444	s supports this view. But in other cases it is	QUITE	possible that each successive modification, o
0451	er! The meaning of rudimentary organs is often	QUITE	unmistakeable: for instance there are beetles
0455	imentary, imperfect, and useless condition, or	QUITE	aborted, far from presenting a strange diffic
0467	ion, individual differences so slight as to be	QUITE	inappreciable by an uneducated eye. This proc
0468	has often been asserted, but the assertion is	QUITE	incapable of proof, that the amount of variat
0485	s were formerly thus connected. Hence, without	QUITE	rejecting the consideration of the present ex
0485	ctual amount of difference between them. It is	QUITE	possible that forms now generally acknowledge
0488	rld, as at present known, although of a length	QUITE	incomprehensible by us, will hereafter be rec
0031	almost as they please. If I had space I could	QUOTE	numerous passages to this effect from highly
0111	at these were swept away by the short horns (I	QUOTE	the words of an agricultural writer) as if by
0376	ons, similar cases occur; as an example, I may	QUOTE	a remark by the highest authority, Prof. Dana
0415	ngs of almost every author. It will suffice to	QUOTE	the highest authority, Robert Brown, who in s
0032	creasing size of the common gooseberry may be	QUOTED.	We see an astonishing improvement in many fl
0050	otanical authorities and practical men can be	QUOTED	to show that the sessile and pedunculated oak
0009	most unnatural conditions (for instance, the	RABBIT	and ferret kept in hutches), showing that the
0019	have descended from the common wild duck and	RABBIT.	The doctrine of the origin of our several do
0029	e never met a pigeon, or poultry, or duck, or	RABBIT	fancier, who was not fully convinced that eac
0215	difficult to tame than the young of the wild	RABBIT:	scarcely any animal is tamer than the young
0215	ny animal is tamer than the young of the tame	RABBIT:	but I do not suppose that domestic rabbits h
0019	fowl (Gallus bankiva). In regard to ducks and	RABBITS,	the breeds of which differ considerably fro
0091	, bringing home winged game; another hares or	RABBITS,	and another hunting on marshy ground and al
0215	me rabbit; but I do not suppose that domestic	RABBITS	have ever been selected for tameness; and I
0349	aving nearly the same habits as our hares and	RABBITS	and belonging to the same order of Rodents,
0015	to assert that we could not breed our cart and	RACE	horses, long and short horned cattle, and poul
0015	gether, we generally perceive in each domestic	RACE,	as already remarked, less uniformity of chara
0019	treme by some authors. They believe that every	RACE	which breeds true, let the distinctive charact
0020	atly exaggerated. There can be no doubt that a	RACE	may be modified by occasional crosses, if aide

Page **(Key Word)**

Page **(Key Word)**

0020 hich present any desired character; but that a RACE could be obtained nearly intermediate between
0020 i find a single case on record of a permanent RACE having been thus formed. On the Breeds of the
0028 ss, that they can reckon up their pedigree and RACE. Pigeons were much valued by Akber Khan in Ind
0029 ral races; and though they well know that each RACE varies slightly, for they win their prizes by
0029 uch agencies for the differences of a dray and RACE horse, a greyhound and bloodhound, a carrier a
0030 sheep. But when we compare the dray horse and RACE horse, the dromedary and camel, the various br
0032 s made only twenty or thirty years ago. When a RACE of plants is once pretty well established, the
0049 r british red grouse as only a strongly marked RACE of a Norwegian species, whereas the greater nu
0106 naturalised on islands. On a small island, the RACE for life will have been less severe, and there
0110 d, and they will consequently be beaten in the RACE for life by the modified descendants of the co
0120 10. In the same way, for instance, the English RACE horse and English pointer have apparently both
0159 n representing the normal structure of another RACE, the fantail. I presume that no one will doubt
0164 me by Mr. W. W. Edwards, that with the English RACE horse the spinal stripe is much commoner in th
0165 horse, Welch ponies, cobs, the lanky Kattywar RACE, etc., inhabiting the most distant parts of th
0177 numbers. Hence, the more common forms, in the RACE for life, will tend to beat and supplant the l
0184 t in the country, I can see no difficulty in a RACE of bears being rendered, by natural selection,
0345 history have beaten their predecessors in the RACE for life, and are, in so far, higher in the sc
0356 parent. To illustrate what I mean: our English RACE horses differ slightly from the horses of ever
0418 for aerating it, or those for propagating the RACE, are found nearly uniform, they are considered
0422 of the several races, descended from a common RACE) had altered much, and had given rise to many
0445 , again, I was told that the foals of cart and RACE horses differed as much as the full grown anim
0445 e of the dam and of a three days old colt of a RACE and heavy cart horse, I find that the colts ha
0427 turnip. The resemblance of the greyhound and RACEHORSE is hardly more fanciful than the analogies
0035 careful training, the whole body of English RACEHORSES have come to surpass in fleetness and size
0014 that no deductions can be drawn from domestic RACES to species in a state of nature. I have in va
0015 ultivate, during many generations, the several RACES, for instance, of the cabbage, in very poor s
0015 d. when we look to the hereditary varieties or RACES of our domestic animals and plants, and compa
0015 ty of character than in true species. Domestic RACES of the same species, also, often have a somew
0016 a subject hereafter to be discussed), domestic RACES of the same species differ from each other in
0016 hen we find that there are hardly any domestic RACES, either amongst animals or plants, which have
0016 ny marked distinction existed between domestic RACES and species, this source of doubt could not s
0016 recur. It has often been stated that domestic RACES do not differ from each other in characters o
0016 of structural difference between the domestic RACES of the same species, we are soon involved in
0017 ecies; but, in the case of some other domestic RACES, there is presumptive, or even strong, eviden
0018 in opposition to several authors, that all the RACES have descended from one wild stock. Mr. Blyth
0019 doctrine of the origin of our several domestic RACES from several aboriginal stocks, has been carr
0019 e? it has often been loosely said that all our RACES of dogs have been produced by the crossing of
0020 ts; and if we account for our several domestic RACES by this process, we must admit the former exi
0020 , moreover, the possibility of making distinct RACES by crossing has been greatly exaggerated. The
0020 y intermediate between two extremely different RACES or species, I can hardly believe. Sir J. Sebr
0023 including under this term several geographical RACES or sub species, which differ from each other
0029 essed with the differences between the several RACES; and though they well know that each race var
0029 f descent, yet admit that many of our domestic RACES have descended from the same parents, may the
0029 w briefly consider the steps by which domestic RACES have been produced, either from one or from s
0029 e most remarkable features in our domesticated RACES is that we see in them adaptation, not indeed
0030 cultural, culinary, orchard, and flower garden RACES of plants most useful to man at different sea
0033 eaves, the flowers, or the fruit, will produce RACES differing from each other chiefly in these ch
0035 he latter, by the regulations for the Goodwood RACES, are favoured in the weights they carry. Lord
0037 ies, and whether or not two or more species or RACES have become blended together by crossing, may
0038 s at once obvious, how it is that our domestic RACES show adaptation in their structure or in thei
0038 frequently abnormal character of our domestic RACES, and likewise their differences being so grea
0042 ant element of success in the formation of new RACES, at least, in a country which is already stoc
0042 a country which is already stocked with other RACES. In this respect enclosure of the land plays
0042 reat convenience to the fancier, for thus many RACES may be kept true, though mingled in the same
0043 reeds. To sum up on the origin of our Domestic RACES! Many years ago, when comparing, and seeing o
0048 or, as they are often called, as geographical RACES. After ten thousand generations, species (A)
0120 either having given off any fresh branches or RACES. The most distinct breeds of pigeons, in coun
0159 readily understood by looking to our domestic RACES. The frequent presence of fourteen or even si
0159 e analogous variations in two or more distinct RACES. The frequent presence of fourteen or even si
0159 ch analogous variations are due to the several RACES of the pigeon having inherited from a common
0166 uding two or three sub species or geographical RACES) of a bluish colour, with certain bars and ot
0199 this same purpose the differences between the RACES of man, which are so strongly marked; I may a
0269 e by far the most important consideration, new RACES of animals and plants are procued under dome
0316 igeon, or even from the other well established RACES of the domestic pigeon, for the newly formed
0335 ith great: if the principal living and extinct RACES of the domestic pigeon were arranged as well
0372 s, which some naturalists rank as geographical RACES, and others as distinct species: and a host o
0382 thus left stranded may be compared with savage RACES of man, driven up and surviving in the mounti
0422 of mankind, a genealogical arrangement of the RACES of man would afford the best classification o
0422 tion and states of civilisation of the several RACES, descended from a common race) had altered mu
0467 n the preservation of favoured individuals and RACES, during the constantly recurrent Struggle for
0381 r and very distinct species have migrated in RADIATING lines from some common centre; and I am inc
0411 characters. Affinities, general, complex and RADIATING. Extinction separates and defines groups. M
0431 e can understand the excessively complex and RADIATING affinities by which all the members of the
0433 e connected together by the most complex and RADIATING lines of affinities. We shall never, probab
0456 ng and extinct beings are united by complex, RADIATING, and circuitous lines of affinities into o
0419 on the mode of development of the plumule and RADICLE. In our discussion on embryology, we shall s
0099 ensable! If several varieties of the cabbage, RADISH, onion, and some other plants, be allowed to
0226 the centre of each sphere at the distance of RADIUS x sq. Rt. 2, or radius x 1.41421 (or at some
0226 ere at the distance of radius x sq. Rt. 2, or RADIUS x 1.41421 (or at some lesser distance) from t
0139 s in the period of flowering, in the amount of RAIN requisite for seeds to germinate, in the time
0288 d generally be dissolved by the percolation of RAIN water. I suspect that but few of the very many
0037 e seed of a wild plant. No one would expect to RAISE a first rate melting pear from the seed of th
0056 analogy whether or not the amount suffices to RAISE one or both to the rank of species. Hence the
0070 ssary for its preservation. Thus we can easily RAISE plenty of corn and rape seed, etc., in our fi
0072 f growth, had during twenty six years tried to RAISE its head above the stems of the heath, and ha
0098 nted near each other, it is hardly possible to RAISE pure seedlings, so largely do they naturally

Page **(Key Word)**

Page ************************************(Key Word)************************************

```
0114  nds of fresh water. Farmers find that they can RAISE most food by a rotation of plants belonging t
0252  been notorious to nurserymen. Horticulturists RAISE large beds of the same hybrids, and such alon
0485  looked at by most naturalists as sufficient to RAISE both forms to the rank of species. Hereafter
0007  y due to our domestic productions having been RAISED under conditions of life not so uniform as, a
0041  lightly larger, earlier, or better fruit, and RAISED seedlings from them, and again picked out the
0042  e varieties of the strawberry which have been RAISED during the last thirty or forty years. In the
0098  n from one flower on the stigma of another, I RAISED plenty of seedlings: and whilst another speci
0099  ority, as I have found, of the seedlings thus RAISED will turn out mongrels: for instance, I raise
0099  aised will turn out mongrels: for instance, I RAISED 233 seedling cabbages from some plants of dif
0113  d a greater weight of dry herbage can thus be RAISED. The same has been found to hold good when fi
0140  climatised: thus the pines and rhododendrons, RAISED from seed collected by Dr. Hooker from trees
0249  l belief amongst breeders. Hybrids are seldom RAISED by experimentalists in great numbers; and as
0253  ich two families of the same hybrid have been RAISED at the same time from different parents, so a
0253  inter se. This was effected by Mr. Eyton, who RAISED two hybrids from the same parents but from di
0253  ifferent hatches; and from these two birds he RAISED no less than eight hybrids (grandchildren of
0256  to differ greatly in the several individuals RAISED from seed out of the same capsule and exposed
0257  is clearly shown by hybrids never having been RAISED between species ranked by systematists in dis
0258  s. it is also a remarkable fact, that hybrids RAISED from reciprocal crosses, though of course com
0259  utterly sterile, even when the other hybrids RAISED from seed from the same capsule have a consid
0270  portant to notice that the hybrid plants thus RAISED were themselves perfectly fertile; so that ev
0271  the flower; and one variety can sometimes be RAISED from the seed of the other. From observations
0316  , identical with the existing breed, could be RAISED from any other species of pigeon, or even fro
0320  a new and slightly improved variety has been RAISED, it at first supplants the less improved vari
0419  actised naturalists as only a genus; and then RAISED to the rank of a sub family or family; and th
0486  s will rise immensely in value. A new variety RAISED by man will be a far more important and inter
0029  t species. Ask, as I have asked, a celebrated RAISER of Hereford cattle, whether his cattle might
0251  r. c. Noble, for instance, informs me that he RAISES stocks for grafting from a hybrid between Rho
0375  genera collected on the loftier peaks of Java RAISES a picture of a collection made on a hill in E
0041  improved. On the other hand, nurserymen, from RAISING large stocks of the same plants, are general
0224  d permanent for the very different purpose of RAISING slaves. When the instinct was once acquired,
0428  this nature in any one class, by arbitrarily RAISING or sinking the value of the groups in other
0042  n the other hand, cats, from their nocturnal RAMBLING habits, cannot be matched, and, although so
0130  face with its ever branching and beautiful RAMIFICATIONS. Chapter V. Laws of Variation.
0129  connexion of the former and present buds by RAMIFYING branches well represent the classificat
0368  on, and those of the Pyrenees, as remarked by RAMOND, are more especially allied to the plants of
0284  mentary deposits of many countries! Professor RAMSAY has given me the maximum thickness, in most c
0285  trata has varied from 600 to 3000 feet. Prof. RAMSAY has published an account of a downthrow in An
0285  thousand feet in thickness, as shown in Prof. RAMSAY's masterly memoir on this subject. Yet it is
0286  ge about 1100 feet, as I am informed by Prof. RAMSAY. But if, as some geologists suppose, a range
0022  hape, as well as the breadth and length of the RAMUS of the lower jaw, varies in a highly remarkab
0222  he earth from the nest of F. flava, and quickly RAN away; but in about a quarter of an hour, short
0006  s, and why another allied species has a narrow RANGE and is rare? Yet these relations are of the h
0051  rrect his first impressions. As he extends the RANGE of his observations, he will meet with more c
0053  nd this is a different consideration from wide RANGE, and to a certain extent from commonness), of
0054  y be called, the dominant species, those which RANGE widely over the world, are the most diffused
0058  m, for if a variety were found to have a wider RANGE than that of its supposed parent species, the
0058  be of doubtful value: these 63 reputed species RANGE on an average over 6.9 of the provinces into
0058  acknowledged varieties are recorded, and these RANGE over 7.7 provinces; whereas, the species to w
0058  s, the species to which these varieties belong RANGE over 14.3 provinces. So that the acknowledged
0058  s have very nearly the same restricted average RANGE, as have those very closely allied forms, mar
0065  ed from Europe; and there are plants which now RANGE in India, as I hear from Dr. Falconer, from C
0070  viduals, even on the extreme confines of their RANGE. For in such cases, we may believe, that a ch
0077  ll around. Look at a plant in the midst of its RANGE, why does it not double or quadruple its numb
0078  yed on it. On the confines of its geographical RANGE, a change of constitution with respect to cli
0078  n to believe that only a few plants or animals RANGE so far, that they are destroyed by the rigour
0107  h island: intercrossing on the confines of the RANGE of each species will thus be checked: after p
0132  he fact of varieties of one species, when they RANGE into the zone of habitation of other species,
0140  historical times having largely extended their RANGE from warmer to cooler latitudes, and converse
0141  y parts of the world, and now have a far wider RANGE than any other rodent, living free under the
0164  n duns and mouse duns: by the term dun a large RANGE of colour is included, from one between brown
0175  to other organic beings, we must see that the RANGE of the inhabitants of any country by no means
0175  one into another by insensible gradations, the RANGE of any one species, depending as it does on t
0175  f any one species, depending as it does on the RANGE of others, will tend to be sharply defined. M
0175  moreover, each species on the confines of its RANGE, where it exists in lessened numbers, will, d
0175  utter extermination: and thus its geographical RANGE will come to be still more sharply defined. I
0175  generally so distributed that each has a wide RANGE, with a comparatively narrow neutral territor
0195  ves from these small enemies, would be able to RANGE into new pastures and thus gain a great advan
0286  ramsay. But if, as some geologists suppose, a RANGE of older rocks underlies the Weald, on the fl
0298  some one spot. Most marine animals have a wide RANGE! and we have seen that with plants it is thos
0298  with plants it is those which have the widest RANGE, that oftenest present varieties; so that wit
0298  it is probably those which have had the widest RANGE, far exceeding the limits of the known geolog
0301  the marine inhabitants of the archipelago now RANGE thousands of miles beyond its confines; and a
0303  these animals have huge bones, are marine, and RANGE over the world, the fact of not a single bone
0305  some few families of fish now have a confined RANGE; the teleostean fish might formerly have had
0305  h might formerly have had a similarly confined RANGE, and after having been largely developed in s
0306  australia, and then to discuss the number and RANGE of its productions. On the sudden appearance
0348  stinct fauna. So that here three marine faunas RANGE far northward and southward, in parallel line
0348  and the eastern Pacific Islands, yet many fish RANGE from the Pacific into the Indian Ocean, and m
0353  rendered discontinuous the formerly continuous RANGE of many species. So that we are reduced to co
0354  nsider whether the exceptions to continuity of RANGE are so numerous and of so grave a nature, tha
0367  e fact that the Alpine plants of each mountain RANGE are more especially related to the arctic for
0373  rom the western whores of Britain to the Oural RANGE, and southward to the Pyrennes. We may infer,
0380  emoved on the view here given in regard to the RANGE and affinities of the allied species which li
0383  onging to quite different classes, an enormous RANGE, but allied species prevail in a remarkable m
0384  s. but on the same continent the species often RANGE widely and almost capriciously: for two river
```

Page ************************************(Key Word)************************************

Page ***(Key Word)***

```
0385  species of fresh water shells have a very wide  RANGE, and allied species, which, on my theory, are
0386  to acquire, as if in consequence, a very wide   RANGE. I think favourable means of dispersal explai
0387  nts to vast distances, and if consequently the  RANGE of these plants was not very great. The same
0388  ntinuously as fresh water productions ever can  RANGE, over immense areas, and having subsequently
0404  long ago, that in those genera of birds which   RANGE over the world, many of the species have very
0404  esh water productions, in which so many genera  RANGE over the world, and many individual species h
0405  rld ranging genera all the species have a wide  RANGE, or even that they have on an average a wide
0405  e, or even that they have on an average a wide  RANGE: but only that some of the species range very
0405  wide range: but only that some of the species  RANGE very widely; for the facility with which wide
0405  new forms will largely determine their average  RANGE. For instance, two varieties of the same spec
0405  nd europe, and the species thus has an immense  RANGE: but, if the variation had been a little grea
0405  een ranked as distinct species, and the common  RANGE would have been greatly reduced. Still less i
0405  tain powerfully winged birds, will necessarily  RANGE widely; for we should never forget that to ra
0405  nge widely; for we should never forget that to  RANGE widely implies not only the power of crossing
0405  we do find, that some at least of the species   RANGE very widely; for it is necessary that the unm
0405  is necessary that the unmodified parent should  RANGE widely, undergoing modification during its di
0406  of organisms is, the more widely it is apt to   RANGE. The relations just discussed, namely, low an
0428  ould easily extend the parallelism over a wide  RANGE: and thus the septenary, quinary, quaternary,
0462  gration by many means. A broken or interrupted  RANGE may often be accounted for by the extinction
0133  l became bright coloured by variation when it   RANGED into warmer or shallower waters. When a varia
0383  t that fresh water productions would not have   RANGED widely within the same country, and as the se
0398  e probability of many species having formerly   RANGED as continuously as fresh water productions ev
0407  we bear in mind how often a species may have    RANGED continuously over a wide area, and then have
0006  ve around us. Who can explain why one species   RANGES widely and is very numerous, and why another
0044  lated to each other, and in having restricted   RANGES. Before applying the principles arrived at in
0049  ent heights: they have different geographical   RANGES; and lastly, according to very numerous exper
0053  s have shown that plants which have very wide   RANGES generally present varieties; and this might h
0054  d salt loving plants have generally very wide   RANGES and are much diffused, but this seems to be c
0058  ice. Varieties generally have much restricted   RANGES: this statement is indeed scarcely more than
0058  esemble varieties, often have much restricted   RANGES. For instance, Mr. H. C. Watson has marked fo
0059  d to other species apparently have restricted   RANGES. In all these several respects the species of
0077  r cold, dampness or dryness, for elsewhere it   RANGES into slightly hotter or colder, damper or dri
0117  , vary more than rare species with restricted   RANGES. Let (A) be a common, widely diffused, and va
0140  ies in a state of nature are limited in their   RANGES by the competition of other organic beings qu
0348  posite sides of lofty and continuous mountain   RANGES, and of great deserts, and sometimes even of
0354  me species on the summits of distant mountain   RANGES, and at distant points in the arctic and anta
0358  s; if they had originally existed as mountain   RANGES on the land, some at least of the islands wou
0360  wn that such plants generally have restricted   RANGES. But seeds can be occasionally transported th
0369  rctic species will have been left on mountain   RANGES distant from each other, and have survived th
0369  nimals of the several great European mountain   RANGES, though very many of the species are identica
0371  h more recent period, on the several mountain   RANGES and on the arctic lands of the two Worlds. He
0375  on the Himalaya, and on the isolated mountain   RANGES of the peninsula of India, on the heights of
0376  the southern hemisphere, and on the mountain   RANGES of the intertropical regions, are not arctic,
0378  asylum to the tropical natives. The mountain   RANGES north west of the Himalaya, and the long line
0381  ed by man's agency in a foreign land; why one  RANGES twice or thrice as far, and is twice or thric
0384  fish on opposite sides of continuous mountain  RANGES, which from an early period must have parted
0386  plants, it has long been known what enormous   RANGES many fresh water and even marsh species have,
0392  lly have, whatever the cause may be, confined  RANGES. Hence trees would be little likely to reach
0394  , that many of the same species have enormous  RANGES, and are found on continents and on far dista
0404  the world, many of the species have very wide  RANGES. I can hardly doubt that this rule is general
0404  ld, and many individual species have enormous  RANGES. It is not meant that in world ranging genera
0470  the larger genera apparently have restricted   RANGES, and they are clustered in little groups roun
0044  ndividual differences. Doubtful species. Wide  RANGING, much diffused, and common species vary most
0055  e genera. The cause of lowly organised plants  RANGING widely will be discussed in our chapter on g
0128  s the common, the widely diffused, and widely  RANGING species, belonging to the larger genera, whi
0301  to believe that it would be chiefly these far  RANGING species which would oftenest produce new var
0342  ms in any one area and formation: that widely  RANGING species are those which have varied most, an
0350  odification through natural selection. Widely  RANGING species, abounding in individuals, which hav
0383  but this power in fresh water productions of   RANGING widely, though so unexpected, can, I think,
0404  normous ranges. It is not meant that in world  RANGING genera all the species have a wide range, or
0405  ry widely; for the facility with which widely  RANGING species vary and give rise to new forms will
0405  tly has the capacity of crossing barriers and  RANGING widely, as in the case of certain powerfully
0406  lower forms will have had a better chance of   RANGING widely and of still retaining the same speci
0406  ed, namely, low and slowly changing organisms  RANGING more widely than the high, some of the speci
0406  than the high, some of the species of widely   RANGING genera themselves ranging widely, such facts
0406  e species of widely ranging genera themselves  RANGING widely, such facts, as alpine, lacustrine, a
0411  have attempted to show that it is the widely   RANGING, the much diffused and common, that is the d
0464  ndition; at least in any great number. Widely  RANGING species vary most, and varieties are often a
0046  never vary; for these same authors practically  RANK that character as important (as some few natur
0046  ardly two naturalists can agree which forms to  RANK as species and which as varieties. We may inst
0047  te gradations, that naturalists do not like to  RANK them as distinct species, are in several respe
0047  sometimes occur in deciding whether or not to   RANK one form as a variety of another, even when th
0049  norwegian species; whereas the greater number  RANK it as an undoubted species peculiar to Great B
0049  f two doubtful forms leads many naturalists to  RANK both as distinct species: but what distance, i
0049  ught to bear on the attempt to determine their  RANK. I will here give only a single instance, the
0050  will bring naturalists to an agreement how to   RANK doubtful forms. Yet it must be confessed, that
0050  country, he will soon make up his mind how to  RANK most of the doubtful forms. His general tenden
0051  e very near to, but do not arrive at the       RANK of species; or, again, between sub species and
0052  es or incipient species necessarily attain the  RANK of species. They may whilst in this incipient
0052  d in numbers the parent species, it would then  RANK as the species, and the species as the variety
0052  rent species; or both might co exist, and both  RANK as independent species. But we shall hereafter
0056  he amount suffices to raise one or both to the  RANK of species. Hence the amount of difference is
0059  considered necessary to give to two forms the  RANK of species is quite indefinite. In genera havi
0060  lled species or sub species or varieties; what  RANK, for instance, the two or three hundred doubtf
0111  wn by the hopeless doubts in many cases how to  RANK them, yet certainly differ from each other far
0114  ndered very distinct from each other, take the  RANK of species. The truth of the principle, that t
```

Page ***(Key Word)***

```
0159   and Ruta baga, plants which several botanists RANK as varieties produced by cultivation from a co
0178   arately been rendered sufficiently distinct to RANK as representative species. In this case, inter
0247   versis and coerulea), which the best botanists RANK as varieties, absolutely sterile together; and
0258   atthiola annua and glabra) that many botanists RANK them only as varieties. It is also a remarkabl
0297   ount of difference between any two forms, they RANK both as species, unless they are enabled to co
0297   nship, and should consequently be compelled to RANK them all as distinct species. It is notorious
0297   fine species of D'Orbigny and others into the  RANK of varieties; and on this view we do find the
0372   great class many forms, which some naturalists RANK as geographical races, and others as distinct
0404   ill be found many forms which some naturalists RANK as distinct species, and some as varieties; th
0419   alists as only a genus, and then raised to the RANK of a sub family or family; and this has been o
0421   e been but slightly modified; and it will then RANK with the parent genus F; just as some few stil
0425   urse of modification, from a bear? Ought we to RANK this one species with bears, and what should w
0469   t it be observed how naturalists differ in the RANK which they assign to the many representative f
0485   lists as sufficient to raise both forms to the RANK of species. Hereafter we shall be compelled to
0016   mongst animals or plants, which have not been RANKED by some competent judges as mere varieties, a
0022   wild birds, would certainly, I think, be RANKED by him as well defined species. Moreover, I d
0036   s, whether or not sufficiently distinct to be RANKED at their first appearance as distinct varieti
0047   lty. In very many cases, however, one form is RANKED as a variety of another, not because the inte
0047   ence, in determining whether a form should be RANKED as a species or a variety, the opinion of nat
0047   wn varieties can be named which have not been RANKED as species by at least some competent judges.
0048   e what a surprising number of forms have been RANKED by one botanist as good species, and by anoth
0048   sidered as varieties, but which have all been RANKED by botanists as species; and in making this l
0048   g varieties, but which nevertheless have been RANKED by some botanists as species, and he has enti
0048   which are highly locomotive, doubtful forms, RANKED by one zoologist as a species and by another
0048   ffer very slightly from each other, have been RANKED by one eminent naturalist as undoubted specie
0049   work, but which it cannot be doubted would be RANKED as distinct species by many entomologists. Ev
0049   ly regarded as varieties, but which have been RANKED as species by some zoologists. Several most e
0049   fectly the character of species that they are RANKED by other highly competent judges as good and
0050   from common parents, and consequently must be RANKED as varieties. Close investigation, in most ca
0050   ded. These varieties, moreover, will be often RANKED by some authors as species. Look at the commo
0057   erion in settling whether two forms should be RANKED as species or varieties. Now Fries has remark
0058   nts (4th edition) 63 plants which are therein RANKED as species, but which he considers as so clos
0058   ful species, but which are almost universally RANKED by British botanists as good and true species
0059   orms, if differing very little, are generally RANKED as varieties, notwithstanding that intermedia
0123   the eight descended from (A), will have to be RANKED as very distinct genera, or even as distinct
0129   ral subordinate groups in any class cannot be RANKED in a single file, but seem rather to be clust
0162   er forms, which themselves must be doubtfully RANKED as either varieties or species; and this show
0181   s or flying Lemur, which formerly was falsely RANKED amongst bats. It has an extremely wide flank
0216   stincts have generally, and most justly, been RANKED by naturalists as the most wonderful of all k
0248   tion whether certain doubtful forms should be RANKED as species or varieties, with the evidence fr
0253   hich are so different that they are generally RANKED in distinct genera, have often bred in this c
0257   rids never having been raised between species RANKED by systematists in distinct families; and on
0261   fted far more readily on the quince, which is RANKED as a distinct genus, than on the apple, which
0268   any degree sterile together, they are at once RANKED by most naturalists as species. For instance,
0270   now not; but the forms experimentised on, are RANKED by Sageret, who mainly founds his classificat
0271   if infertile in any degree would generally be RANKED as species; from man selecting only external
0276   ses between forms sufficiently distinct to be RANKED as species, and their hybrids, are very gener
0297   ermediate between B and C, it would simply be RANKED as a third and distinct species, unless at th
0297   e embedded fossils, though almost universally RANKED as specifically different, yet are far more c
0299   biting the shores of North America, which are RANKED by some conchologists as distinct species fro
0301   nciples followed by many palaeontologists, be RANKED as new and distinct species. If then, there b
0302   ormation, would, by most palaeontologists, be RANKED as distinct species. But I do not pretend tha
0324   older underlying deposits, would be correctly RANKED as simultaneous in a geological sense. The fa
0329   imals fall in between existing groups. Cuvier RANKED the Ruminants and Pachyderms, as the two most
0336   s from closely consecutive formations, though RANKED as distinct species, being closely related, i
0369   ly the same, some present varieties, some are RANKED as doubtful forms, and some few are distinct
0372   sely allied or representative forms which are RANKED by all naturalists as specifically distinct.
0376   thern hemisphere are of doubtful value, being RANKED by some naturalists as specifically distinct,
0376   gh closely related to northern forms, must be RANKED as distinct species. Now let us see what ligh
0398   six land birds, and twenty five of these are RANKED by Mr. Gould as distinct species, supposed to
0405   le greater, the two varieties would have been RANKED as distinct species, and the common range wou
0414   nnected with the whole life of the being, are RANKED as merely adaptive or analogical characters;
0417   d yet leave us in no doubt where it should be RANKED. Hence, also, it has been found, that a class
0419   lants and insects, of a group of forms, first RANKED by practised naturalists as only a genus, and
0420   one; and this is expressed by the forms being RANKED under different genera, families, sections, o
0421   an the existing species, descended from A, be RANKED in the same genus with the parent A; or those
0424   us, and Catasetum), which had previously been RANKED as three distinct genera, were known to be so
0425   a bear? According to all analogy, it would be RANKED with bears; but then assuredly all the other
0445   of beak, that they would, I cannot doubt, be RANKED in distinct genera, had they been natural pro
0047   he treats the one as a variety of the other, RANKING the most common; but sometimes the one first
0276   come to diametrically opposite conclusions in RANKING forms by this test. The sterility is innate
0422   roups have undergone, have to be expressed by RANKING them under different so called genera, sub f
0456   ement of descent has been universally used in RANKING together the sexes, ages, and acknowledged v
0247   ies, quite fertile together, he unhesitatingly RANKS them as varieties. Gartner, also, makes the r
0268   uite fertile when crossed, and he consequently RANKS them as undoubted species. If we thus argue i
0377   mperate forms might have penetrated the native RANKS and have reached or even crossed the equator.
0424   is descended from the primrose, or conversely, RANKS them together as a single species, and gives
0070   n. thus we can easily raise plenty of corn and RAPE seed, etc., in our fields, because the seeds a
0008   dest domesticated animals are still capable of RAPID improvement or modification. It has been disp
0033   may add, has been, in a corresponding degree, RAPID and important. But it is very far from true t
0060   a wide sense. Geometrical powers of increase. RAPID increase of naturalised animals and plants. N
0064   e numerous recorded cases of the astonishingly RAPID increase of various animals in a state of nat
0065   urprising, simply explains the extraordinarily RAPID increase and wide diffusion of naturalised pr
0103   wander little and which can increase at a very RAPID rate, a new and improved variety might be qui
0106   of modification will generally have been more RAPID on large areas; and what is more important, t
0152   ints, which at the present time are undergoing RAPID change by continued selection, are also emine
```

Page **(Key Word)**

```
0251  pollen of the hybrid made vigorous growth and RAPID progress to maturity, and bore good seed, whi
0286  the degradation of a lofty cliff would be more RAPID from the breakage of the fallen fragments. On
0322  er, when by sudden immigration or by unusually RAPID development, many species of a new group have
0322  ey will have exterminated in a correspondingly RAPID manner many of the old inhabitants; and the f
0401  ntinuously united. The currents of the sea are RAPID and sweep across the archipelago, and gales o
0402  h man's agency, have spread with astonishing RAPIDITY over new countries, we are apt to infer that
0064  e species may be now increasing, more or less RAPIDLY, in numbers, all cannot do so, for the world
0065  e at a geometrical ratio, that all would must RAPIDLY stock every station in which they could any
0066  mportance to those species, which depend on a RAPIDLY fluctuating amount of food, for it allows th
0066  luctuating amount of food, for it allows them RAPIDLY to increase in number. But the real importan
0153  ort faced strain. But as long as selection is RAPIDLY going on, there may always be expected to be
0201  t one compared with another; but they are now RAPIDLY yielding before the advancing legions of pla
0298  e: with animals and plants that can propagate RAPIDLY and are not highly locomotive, there is reas
0303  ivergent forms, which would be able to spread RAPIDLY and widely throughout the world. I will now
0385  understand how some naturalised species have RAPIDLY spread throughout the same country. But two
0408  migrants were capable of varying more or less RAPIDLY, there would ensue in different regions, inc
0453  uently absorbed, can be of any service to the RAPIDLY growing embryonic calf by the excretion of p
0006  other allied species has a narrow range and is RARE? Yet these relations are of the highest import
0010  sometimes by seed. These sports are extremely RARE under nature, but far from rare under cultivat
0010  are extremely rare under nature, but far from RARE under cultivation; and in this case we see tha
0013  ently exposed to the same conditions, any very RARE deviation, due to some extraordinary combinati
0013  ral members of the same family. If strange and RARE deviations of structure are truly inherited, l
0028  monarchs of Iran and Turan sent him some very RARE birds; and, continues the courtly historian, H
0070  singular facts in nature, such as that of very RARE plants being sometimes extremely abundant in t
0074  le genus of humble bees became extinct or very RARE in England, the heartsease and red clover woul
0074  he heartsease and red clover would become very RARE, or wholly disappear. The number of humble bee
0095  e. hence, again, if humble bees were to become RARE in any country, it might be a great advantage
0104  ong run, the influence of intercrosses even at RARE intervals, will be great. If there exist organ
0109  ll the less favoured forms decrease and become RARE. Rarity, as geology tells us, is the precursor
0110  corded varieties, or incipient species. Hence, RARE species will be less quickly modified or impro
0117  t and the most widely diffused, vary more than RARE species with restricted ranges. Let (A) be a c
0163  l colours; transverse bars on the legs are not RARE in duns, mouse duns, and in one instance in a
0195  drupeds are actually destroyed (except in some RARE cases) by the flies, but they are incessantly
0203  hatred, though the latter fortunately is most RARE, is all the same to the inexorable principle o
0259  ike one of their pure parent species, are with RARE exceptions extremely sterile. So again amongst
0281  ween child and parent, will render this a very RARE event; for in all cases the new and improved f
0289  will bring home the truth, how accidental and RARE is their preservation, far better than pages o
0293  parts of the same formation, but, as they are RARE, they may be here passed over. Although each f
0294  now abundant in the neighbouring sea, but are RARE or absent in this particular deposit. It is an
0295  ent and the amount of subsidence is probably a RARE contingency; for it has been observed by more
0319  is horse been still living, but in some degree RARE, no naturalist would have felt the least surpr
0319  f we ask ourselves why this or that species is RARE, we answer that something is unfavourable in i
0319  sition of the fossil horse still existing as a RARE species, we might have felt certain from the a
0320  y, that to admit that species generally become RARE before they become extinct, to feel no surpris
0333  on, we have no right to expect, except in very RARE cases, to fill up wide intervals in the natura
0346  me species generally require; for it is a most RARE case to find a group of organisms confined to
0364  t sticking to their feet, which is in itself a RARE accident. Even in this case, how small would t
0401  ipelago, and gales of wind are extraordinarily RARE; so that the islands are far more effectually
0410  continuous; and the exceptions, which are not RARE, may, as I have attempted to show, be accounte
0462  that we have no right to expect (excepting in RARE cases) to discover directly connecting links b
0487  but as a poor collection made at hazard and at RARE intervals. The accumulation of each great foss
0008  ted plants display the utmost vigour, and yet RARELY or never seed! In some few such cases it has
0038  such as is externally visible; and indeed he RARELY cares for what is internal. He can never act
0042  ing savages or the inhabitants of open plains RARELY possess more than one breed of the same speci
0044  is almost universally implied, though it can RARELY be proved. We have also what are called monst
0048  as a species and by another as a variety, can RARELY be found within the same country, but are com
0068  beasts of prey: even the tiger in India most RARELY dares to attack a young elephant protected by
0101  almost their whole organisation, yet are not RARELY, some of them hermaphrodites, and some of the
0136  th the winds, or by giving up the attempt and RARELY or never flying. As with mariners shipwrecked
0144  onformations very frequently, and that others RARELY coexist, without our being able to assign any
0147  ment flows to one part or organ in excess, it RARELY flows; at least in excess, to another part; t
0150  ies to males and females; but as females more RARELY offer remarkable secondary sexual character,
0151  e secondary sexual character, it applies more RARELY to them. The rule being so plainly applicable
0153  remote in any extreme degree, as species very RARELY endure for more than one geological period. A
0156  eritance from a common progenitor, for it can RARELY have happened that natural selection will hav
0161  ation of the barb pigeon, which produces most RARELY a blue and black barred bird, there has been
0163  s a case apparently of reversion. The ass ver RARELY has very distinct transverse bars on its legs
0173  ubsides. These contingencies will concur only RARELY, and after enormously long intervals. Whilst
0183  fitted for very different habits of life will RARELY have been developed at an early period in gre
0185  there are upland geese with webbed feet which RARELY or never go near the water; and no one except
0186  d feet, either living on the dry land or most RARELY alighting on the water; that there should be
0194  own is very small, I have been astonished how RARELY an organ can be named, towards which no trans
0208  ny habitual actions are performed, indeed most RARELY in direct opposition to our conscious will! y
0215  s is seen in those breeds of fowls which very RARELY or never become broody, that is, never wish t
0215  vages do not keep these domestic animals. How RARELY, on the other hand, do our civilised dogs, ev
0222  f the nest. This species is sometimes, though RARELY, made into slaves, as has been described by M
0222  of the little and furious F.flava, which they RARELY capture, and it was evident that they did at
0256  which are very difficult to cross, and which RARELY produce any offspring; are generally very ste
0256  , there are species which can be crossed very RARELY, or with extreme difficulty, but the hybrids,
0266  ir sterility, though in some degree variable, RARELY diminishes. It must, however, be confessed th
0288  only on the view of the bottom of the sea not RARELY lying for ages in an unaltered condition. The
0298  and in the case of fossil species this could RARELY be effected by palaeontologists. We shall, pe
0384  idental means; like that of the live fish not RARELY dropped by whirlwinds in India, and the vital
0386  ore mentioned that earth occasionally, though RARELY, adheres in some quantity to the feet and bea
0439  the whelp of the lion. We occasionally though RARELY see something of this kind in plants: thus th
0451  as to be utterly incapable of flight, and not RARELY lying under wing cases, firmly soldered toget
```

Page **(Key Word)**

rarely

Page ***(Key Word)**

```
0471  the ground; that upland geese, which never or RARELY swim should have been created with webbed fee
0069  invariably see some species gradually getting RARER and rarer, and finally disappearing; and the
0069  y see some species gradually getting rarer and RARER, and finally disappearing; and the change of
0110  through natural selection, others will become RARER, and rarer, and finally extinct. The forms whi
0110  atural selection, others will become rarer and RARER, and finally extinct. The forms which stand i
0173  ten meet and interlock; and as the one becomes RARER and rarer, the other becomes more and more fr
0173  nd interlock; and as the one becomes rarer and RARER, the other becomes more and more frequent, ti
0174  rge territory, then becoming somewhat abruptly RARER and rarer on the confines, and finally disapp
0174  ory, then becoming somewhat abruptly rarer and RARER on the confines, and finally cisappearing. He
0175  een them, in which they become rather suddenly RARER and rarer; then, as varieties do not essentia
0175  in which they become rather suddenly rarer and RARER; then, as varieties do not essentia
0176  e between two other forms occur, they are much RARER numerically than the forms which they connect
0177  r natural selection to seize on, than will the RARER forms which exist in lesser numbers. Hence, t
0177  nt number of well marked varieties than do the RARER species. I may illustrate what I mean by supp
0319  t the fossil horse would certainly have become RARER and rarer, and finally extinct; its place bei
0319  il horse would certainly have become rarer and RARER, and finally extinct; its place being seized
0009  family; whereas, carnivorous birds, with the RAREST exceptions, hardly ever lay fertile eggs. Man
0042  mestic animals vary less than others, yet the RARITY or absence of distinct breeds of the cat, the
0062  y of nature, with every fact on distribution, RARITY, abundance, extinction, and variation, will b
0109  less favoured forms decrease and become rare. RARITY, as geology tells us, is the precursor to ext
0171  nt with modification. Transitions. Absence or RARITY of transitional varieties. Transitions in hab
0172  idism in separate chapters. On the absence or RARITY of transitional varieties. As natural selecti
0289  far better than pages of detail. Nor is their RARITY surprising, when we remember how large a prop
0319  1st would have felt the least surprise at its RARITY; for rarity is the attribute of a vast number
0319  ve felt the least surprise at its rarity; for RARITY is the attribute of a vast number of species
0319  ceived agencies are amply sufficient to cause RARITY, and finally extinction. We see in many cases
0319  in the more recent tertiary formations, that RARITY precedes extinction; and we know that this ha
0320  ey become extinct, to feel no surprise at the RARITY of a species, and yet to marvel greatly when
0024  habits with the rock pigeon seems to me a very RASH assumption. Moreover, the several above named
0026  princ perfectly fertile, inter se, seems to me RASH in the extreme. From these several reasons, na
0293  nc in the middle of any formation, it would be RASH in the extreme to infer that it had not elsewh
0293  ayers have been deposited, it would be equally RASH to suppose that it then became wholly extinct.
0306  s have effected, it seems to me to be about as RASH in us to dogmatize on the succession of organi
0310  tains grave doubts on this subject. I feel how RASH it is to differ from these great authorities,
0076  hrush. How frequently we hear of one species of RAT taking the place of another species under the
0137  . in one of the blind animals, namely, the cave RAT, the eyes are of immense size; and Professor S
0137  d by use and disuse, so in the case of the cave RAT natural selection seems to have struggled with
0141  perhaps be mingled in our domestic breeds. The RAT and mouse cannot be considered as domestic ani
0349  e waters, and we do not find the beaver or musk RAT, but the coypu and capybara, rodents of the Am
0019  so slight, has had its wild prototype. At this RATE there must have existed at least a score of sp
0037  tocks. No one would ever expect to get a first RATE heartsease or dahlia from the seed of a wild p
0037  ld plant. No one would expect to raise a first RATE melting pear from the seed of the wild pear, t
0042  ropagated in great numbers and at a very quick RATE, and inferior birds may be freely rejected, as
0063  for existence inevitably follows from the high RATE at which all organic beings tend to increase.
0064  organic being naturally increases at so high a RATE, that if not destroyed, the earth would soon b
0064  has doubled in twenty five years, and at this RATE, in a few thousand years, there would literall
0064  en some pains to estimate its probable minimum RATE of natural increase: it will be under the mark
0064  l parts of the world: if the statements of the RATE of increase of slow breeding cattle and horses
0103  r much, and which do not breed at a very quick RATE. Hence in animals of this nature, for instance
0103  little and which can increase at a very rapid RATE, a new and improved variety might be quickly f
0109  rfectly well with what geology tells us of the RATE and manner at which the inhabitants of this wo
0229  bees must have worked at very nearly the same RATE on the opposite sides of the ridge of vermilio
0229  always succeed in working at exactly the same RATE from the opposite sides: for I have noticed ha
0230  nce from each other, by excavating at the same RATE, and by endeavouring to make ecual spherical h
0261  not take on another, from differences in their RATE of growth, in the hardness of their wood, in t
0279  n the vast lapse of time, as inferred from the RATE of deposition and of denudation. On the poorne
0283  ss, which, though probably formed at a quicker RATE than many other deposits, yet, from being form
0284  eposited by the great Mississippi river at the RATE of only 600 feet in a hundred thousand years.
0284  in many places suffered, independently of the RATE of accumulation of the degraced matter, probab
0286  tremity of the district. If, then, we knew the RATE at which the sea commonly wears away a line of
0286  ould eat into cliffs 500 feet in height at the RATE of one inch in a century. This will at first a
0286  eaten back along a whole line of coast at the RATE on one yard in nearly every twenty two years.
0286  ck, even as soft as chalk, would yield at this RATE excepting on the most exposed coasts; though n
0287  le length woulc be an ample allowance. At this RATE, on the above data, the denucation of the Weal
0288  ted over nearly the whole bed of the sea, at a RATE sufficiently quick to embed and preserve fossi
0291  o subside. In this latter case, as long as the RATE of subsidence and supply of seciment nearly ba
0300  or where it cid not accumulate at a sufficient RATE to protect organic bodies from decay, no remai
0302  modified descendants. But we continually over RATE the perfection of the geological record, and f
0313  enera and classes have not changed at the same RATE, or in the same degree. In the oldest tertiary
0313  ctions of the land seem to change at a quicker RATE than those of the sea, of which a striking ins
0314  nature, on the power of intercrossing, on the RATE of breeding, on the slowly charging physical c
0314  can perhaps understand the apparently quicker RATE of change in terrestrial and in more highly or
0318  nd has increased in numbers at an unparalleled RATE, I asked myself what could so recently have ex
0343  ot necessarily change together, or at the same RATE, or in the same degree; yet in the long run th
0359  ion. In Johnston's Physical Atlas, the average RATE of the several Atlantic currents is 33 miles p
0359  3 miles per diem (some currents running at the RATE of 60 miles per diem); on this average, the se
0361  ely assume that under such circumstances their RATE of flight would often be 35 miles an houri and
0394  that mammals appear and disappear at a quicker RATE than other and lower animals. Though terrestri
0406  each great class, generally change at a slower RATE than the higher forms; and consequently the lo
0463  enerally be modified and improved at a quicker RATE than the intermediate varieties, which exist i
0467  ich is common to all organic beings. This high RATE of increase is proved by calculation, by the e
0488  s of life were probably fewer and simpler, the RATE of change was probably slower; and at the firs
0488  w forms of the simplest structure existed, the RATE of change may have been slow in an extreme deg
0312  appearance of new species. On their different RATES of change. Species once lost do not reappear.
0020  be alike, and then the extreme difficulty, or RATHER utter hopelessness, of the task becomes appar
0053  it necessarily here is with much brevity, is RATHER perplexing, and allusions cannot be avoided t
```

Page ***(Key Word)**

610   rather

Page **(Key Word)**

```
0063  irds to devour and thus disseminate its seeds RATHER than those of other plants. In these several
0091  odcocks or snipes. The tendency to catch rats RATHER than mice is shown to be inherited. Now, if a
0093  le flowers, which have four stamens producing RATHER a small quantity of pollen, and a rudimentary
0112  nother fancier is struck by a pigeon having a RATHER longer beak; and on the acknowledged principl
0116  ing diagram will aid us in understanding this RATHER perplexing subject. Let A to L represent the
0124  ot be directly intermediate between them, but RATHER between types of the two groups; and every na
0129  e cannot be ranked in a single file, but seem RATHER to be clustered round points, and these round
0132  ely small in the case of animals, but perhaps RATHER more in that of plants. We may, at least, saf
0175  territory between them, in which they become RATHER suddenly rarer and rarer; then, as varieties
0180  rked, with the posterior part of their bodies RATHER wide and with the skin on their flanks rather
0180  rather wide and with the skin on their flanks RATHER full, to the so called flying squirrels; and
0181  nally constructed for gliding through the air RATHER than for flight. If about a dozen genera of b
0199  arm the females, can be called useful only in RATHER a forced sense. But by far the most important
0238  as are any two species of the same genus, or RATHER as any two genera of the same family. Thus in
0278  hapter do not seem to be opposed to, but even RATHER to support the view, that there is no fundame
0297  ory we ought to find. Moreover, if we look to RATHER wider intervals, namely, to distinct but cons
0427  ; but such resemblances will not reveal, will RATHER tend to conceal their blood relationship to t
0439  embryonic resemblance, sometimes lasts till a RATHER late age: thus birds of the same genus, and o
0443  le evidence on this head, indeed the evidence RATHER points the other way; for it is notorious tha
0446  ive steps of variation having supervened at a RATHER late age, and having been inherited at a corr
0447  each successive modification supervening at a RATHER late age, and being inherited at a correspond
0447  after having undergone much modification at a RATHER late period of life, and having thus been con
0065  nabled to breed. In such cases the geometrical RATIO of increase, the result of which never fails
0065  imals are tending to increase at a geometrical RATIO, that all would most rapidly stock every stat
0079  being is striving to increase at a geometrical RATIO; that each at some period of its life, during
0467  e inevitably follows from the high geometrical RATIO of increase which is common to all organic be
0470  ties. As each species tends by its geometrical RATIO of reproduction to increase inordinately in n
0490  conditions of life, and from use and disuse; a RATIO of Increase so high as to Lead to a Struggle
0139  ntry. It would be most difficult to give any RATIONAL explanation of the affinities of the blind c
0091  nimals: one cat, for instance, taking to catch RATS, another mice; one cat, according to Mr. St. J
0091  ing woodcocks or snipes. The tendency to catch RATS rather than mice is shown to be inherited. Now
0201  the same time this snake is furnished with a RATTLE for its own injury, namely, to warn its prey
0201  o me of any weight. It is admitted that the RATTLESNAKE has a poison fang for its own defence and
0223  its mouth on top of a spray of heath over its RAVAGED home. Such are the facts, though they did no
0132  less, we can here and there dimly catch a faint RAY of light, and we may feel sure that there must
0145  s plants. Every one knows the difference in the RAY and central florets of, for instance, the dais
0145  to pressure, and the shape of the seeds in the RAY florets in some Compositae countenances this i
0145  t have been thought that the development of the RAY petals by drawing nourishment from certain oth
0145  at all sure that C. C. Sprengel's idea that the RAY florets serve to attract insects, whose agency
0193  n muscle; and as it has lately been shown that RAYS have an organ closely analogous to the electri
0069  outhwards or in descending a mountain. When we REACH the Arctic regions, or snow capped summits, o
0073  ver (Trifolium pratense), as other bees cannot REACH the nectar. Hence I have very little doubt, t
0078  the rigour of the climate alone. Not until we REACH the extreme confines of life, in the arctic r
0187  fferent lines, can be shown to exist, until we REACH a moderately high stage of perfection. In cer
0263  ving a pistil too long for the pollen tubes to REACH the ovarium. It has also been observed that w
0264  stigmatic surface. Again, the male element may REACH the female element, but be incapable of causi
0283  rocess of degradation. The tides in most cases REACH the cliffs only for a short time twice a day,
0306  outhern capes of Africa or Australia, and thus REACH other and distant seas. From these and simila
0392  ranges. Hence trees would be little likely to REACH distant oceanic islands; and an herbaceous pl
0441  y their well developed organs of sense, and to REACH by their active powers of swimming, a proper
0451  of the pistil is to allow the pollen tubes to REACH the ovules protected in the ovarium at its ba
0138  otal darkness. By the time that an animal had REACHED, after numberless generations, the deepest r
0366  to the plains. By the time that the cold had REACHED its maximum, we should have a uniform arctic
0377  ght have penetrated the native ranks and have REACHED or even crossed the equator. The invasion wo
0378  rminated on the lowlands: those which had not REACHED the equator, would re migrate northward or s
0396  m or modified since their arrival, could have REACHED their present homes. But the probability of
0117  ated by natural selection. When a dotted line REACHES one of the horizontal lines, and is there ma
0316  a gradual increase in number, till the group REACHES its maximum, and then, sooner or later, it g
0119  he diagram by some of the lower branches not REACHING to the upper horizontal lines. In some cases
0263  a physical impossibility in the male element REACHING the ovule, as would be the case with a plant
0178  ms thus produced and the old ones acting and REACTING on each other. So that, in any one region an
0075  e is this problem compared to the action and REACTION of the innumerable plants and animals which
0350  e former immigrants; and on their action and REACTION, in their mutual struggles for life; the rel
0408  almost endless amount of organic action and REACTION, and we should find, as we do find, some gro
0002  r, who both knew of my work, the latter having READ my sketch of 1844, honoured me by thinking it
0028  have ever conversed, or whose treatises I have READ, are firmly convinced that the several breeds
0031  what they have done, it is almost necessary to READ several of the many treatises devoted to this
0053  permits me to add, that after having carefully READ my manuscript, and examined the tables, he thi
0226  er, of Cambridge, and this geometer has kindly READ over the following statement, drawn up from hi
0282  ly to comprehend the lapse of time. He who can READ Sir Charles Lyell's grand work on the Principl
0282  ices to study the principles of geology, or to READ special treatises by different observers on se
0304  nd earliest tertiary formation. But now we may READ in the Supplement to Lyell's Manual, published
0002  y several statements; and I must trust to the READER reposing some confidence in my accuracy. No d
0067  ake only a few remarks, just to recall to the READER's mind some of the chief points. Eggs or very
0171  owd of difficulties will have occurred to the READER. Some of them are so grave that to this day I
0212  etail, can produce but a feeble effect on the READER's mind. I can only repeat my assurance, that
0240  f the driver ant (Anomma) of West Africa. The READER will perhaps best appreciate the amount of di
0282  hardly possible for me even to recall to the READER, who may not be a practical geologist, the fa
0331  bject is somewhat complex, I must request the READER to turn to the diagram in the fourth chapter.
0412  improved, and preceding forms. I request the READER to turn to the diagram illustrating the actio
0420  nt genera, families, sections, or orders. The READER will best understand what is meant, if he wil
0459  ne long argument, it may be convenient to the READER to have the leading facts and inferences brie
0460  of varieties when crossed, I must refer the READER to the recapitulation of the facts given at t
0207  its cells will probably have occurred to many READERS, as a difficulty sufficient to overthrow my
0406  he view of colonisation from the nearest and READIEST source, together with the subsequent modific
0032  qualities, he will assuredly fail. Few would READILY believe in the natural capacity and years of
```

Page **(Key Word)**

Page **************************************(Key Word)**************************************

```
0136 , on the other hand, those beetles which most READILY took to flight will oftenest have been blown
0139 this view be correct, acclimatisation must be READILY effected during long continued descent. It i
0140 vilised man because they were useful and bred READILY under confinement, and not because they were
0141 daptation to any special climate as a quality READILY grafted on an innate wide flexibility of con
0147 icult to get a cow to give milk and to fatten READILY. The same varieties of the cabbage do not yi
0157 ave had a wide scope for action, and may thus READILY have succeeded in giving to the species of t
0159 y progenitor. These propositions will be most READILY understood by looking to our domestic races.
0169 of time, in this case, natural selection may READILY have succeeded in giving a fixed character t
0250 species can actually be hybridised much more READILY than they can be self fertilised! For instan
0252 interbreeding is thus prevented. Any one may READILY convince himself of the efficiency of insect
0257 dianthus, in which very many species can most READILY be crossed: and another genus, as Silene, in
0261 each other. The pear can be grafted far more READILY on the quince, which is ranked as a distinct
0268 outh American indigenous domestic dogs do not READILY cross with European dogs, the explanation wh
0297 nded their species: and they do this the more READILY if the specimens come from different sub sta
0362 d the contents of their torn crops might thus READILY get scattered. Mr. Brent informs me that a f
0366 follows. But we shall follow the changes more READILY, by supposing a new glacial period to come s
0392 o compete with herbaceous plants alone, might READILY gain an advantage by growing taller and tall
0403 ts of that region whence colonists could most READILY have been derived, the colonists having been
0427 ng to two most distinct lines of descent, may READILY become adapted to similar conditions, and th
0437 ow or little modified forms: therefore we may READILY believe that the unknown progenitor of the v
0061 we shall hereafter see, is a power incessantly READY for action, and is as immeasurably superior t
0098 before the stigma of that individual flower is READY to receive them: and as this flower is never
0099 either the anthers burst before the stigma is READY for fertilisation, or the stigma is ready bef
0099 a is ready for fertilisation, or the stigma is READY before the pollen of that flower is ready, so
0099 a is ready before the pollen of that flower is READY, so that these plants have in fact separated
0469 nature variability and a powerful agent always READY to act and select, why should we doubt that v
0472 rease in number, with natural selection always READY to adapt the slowly varying descendants of ea
0066 ws them rapidly to increase in number. But the REAL importance of a large number of eggs or seeds
0096 the number of supposed hermaphrodites, and of REAL hermaphrodites a large number pair: that is, t
0167 t this view is, as it seems to me, to reject a REAL for an unreal, or at least for an unknown, cau
0171 r number are only apparent, and those that are REAL are not, I think, fatal to my theory. These di
0187 e by our imagination, can hardly be considered REAL. How a nerve comes to be sensitive to light, h
0237 cows of these same breeds. Hence I can see no REAL difficulty in any character having become corr
0334 apsed between consecutive formations. It is no REAL objection to the truth of the statement, that
0380 th, just in the same way as the productions of REAL islands have everywhere lately yielded to cont
0399 , though feeble, I am assured by Dr. Hooker is REAL. between the flora of the south western corner
0423 in classing varieties, I apprehend if we had a REAL pedigree, a genealogical classification would
0427 views, the very important distinction between REAL affinities and analogical or acaptive resembla
0427 nct animals. On my view of characters being of REAL importance for classification, only in so far
0430 the bizcacha to Marsupials are believed to be REAL and not merely adaptive, they are due on my th
0459 nation insuperably great, cannot be considered REAL if we admit the following propositions, namely
0479 l characters are the most valuable of all. The REAL affinities of all organic beings are due to in
0482 itude of reputed species in each genus are not REAL species: but that other species are real, tha
0482 e not real species: but that other species are REAL, that is, have been independently created. Thi
0031 breeds of cattle and sheep. In order fully to REALISE what they have done, it is almost necessary
0012 nth, potato, even the dahlia, etc.: and it is REALLY surprising to note the endless points in stru
0098 for instance, in Lobelia fulgens, there is a REALLY beautiful and elaborate contrivance by which
0127 th other males. Whether natural selection has REALLY thus acted in nature, in modifying and adapti
0196 attribute importance to characters which are REALLY of very little importance, and which have ori
0232 of intersection between these spheres. It was REALLY curious to note in cases of difficulty, as wh
0247 itted to doubt whether many other species are REALLY so sterile, when intercrossed, as Gartner bel
0279 ms, than on climate: and, therefore, that the REALLY governing conditions of life co not graduate
0295 of the glacial period, instead of having been REALLY far greater, that is extending from before th
0299 by other conchologists as only varieties, are REALLY varieties or are, as it is called, specifical
0302 elonging to the same genera or families, have REALLY started into life all at once, the fact would
0305 affinities are as yet imperfectly known, are REALLY teleostean. Assuming, however, that the whole
0360 e alternately wet and exposed to the air like REALLY floating plants. He tried 98 seeds, mostly di
0376 rial latitudes, the Alpine or mountain floras REALLY become less and less arctic. Many of the form
0401 spreading widely and remaining the same. The REALLY surprising fact in this case of the Galapagos
0439 ave retained through inheritance, if they had REALLY been metamorphosed during a long course of de
0444 bull dog, though appearing so different, are REALLY varieties most closely allied, and have proba
0483 eation than at an ordinary birth. But do they REALLY believe that at innumerable periods in the ea
0086 ar at any particular period of life, tend to REAPPEAR in the offspring at the same period: for ins
0160 very surprising fact that characters should REAPPEAR after having been lost for many, perhaps for
0166 tint, these bars and other marks invariably REAPPEAR: but without any other change of form or cha
0166 ency for the blue tint and bars and marks to REAPPEAR in the mongrels. I have stated that the most
0196 fluence on the organisation: that characters REAPPEAR from the law of reversion: that correlation
0199 of growth, or from other unknown cause, may REAPPEAR from the law of reversion, though now of no
0312 nt rates of change. Species once lost do not REAPPEAR. Groups of species follow the same general r
0313 on, and then allow the pre existing fauna to REAPPEAR: but Lyell's explanation, namely, that it is
0315 nd why a species when once lost should never REAPPEAR, even if the very same conditions of life, o
0316 a greater or lesser degree. A group does not REAPPEAR after it has once disappeared: or its existe
0344 oup has once wholly disappeared, it does not REAPPEAR; for the link of generation has been broken.
0432 as ever lived on this earth were suddenly to REAPPEAR, though it would be quite impossible to give
0444 ion first appears in the parent, it tends to REAPPEAR at a corresponding age in the offspring. Cer
0475 neither single species nor groups of species REAPPEAR when the chain of ordinary generation has on
0013 chances almost compels us to attribute its REAPPEARANCE to inheritance. Every one must have heard
0160 e external conditions of life to cause the REAPPEARANCE of the slaty blue, with the several marks,
0166 ost probable hypothesis to account for the REAPPEARANCE of very ancient character, is, that there
0454 e vestige of an ear in earless breeds, the REAPPEARANCE of minute dangling horns in hornless bree
0296 iod of deposition, but have disappeared and REAPPEARED, perhaps many times, during the same geolog
0013 amongst several million individuals, and it REAPPEARS in the child, the mere doctrine of chances
0160 a character which has been lost in a breed, REAPPEARS after a great number of generations, the mo
0313 o believe that the same identical form never REAPPEARS. The strongest apparent exception to this l
0343 when a species has once disappeared it never REAPPEARS. Groups of species increase in numbers slow
0221 they were prevented from getting any pupae to REAR as slaves. I then dug up a small parcel of the
```

Page **************************************(Key Word)**************************************

Page **(Key Word)**

```
0248 ive generations; though Gartner was enabled to REAR some hybrids, carefully guarding them from a c
0377 oth sides towards the equator, followed in the REAR by the temperate productions, and these by the
0042 ding; in peacocks, from not being very easily REARED and a large stock not kept; in geese, from be
0216 lainly instinctive in young pheasants, though REARED under a hen. It is not that chickens have los
0217 would lead me to believe, that the young thus REARED would be apt to follow by inheritance the occ
0223 developed; and the ants thus unintentionally REARED would then follow their proper instincts, and
0402 more eggs are laid there than can possibly be REARED; and we may infer that the mocking thrush pec
0217 other birds' nests, and thus be successful in REARING their young. By a continued process of this
0014 extension, and that when there is no apparent REASON why a peculiarity should appear at any partic
0035 re the breed has been less improved. There is REASON to believe that King Charles's spaniel has be
0047 ects the most important for us. We have every REASON to believe that many of these doubtful and cl
0055 special act of creation, there is no apparent REASON why more varieties should occur in a group ha
0056 l in action, more especially as we have every REASON to believe the process of manufacturing new s
0058 tions ought to be reversed. But there is also REASON to believe, that those species which are very
0073 ses and thus to fertilise them. I have, also, REASON to believe that humble bees are indispensable
0078 rly be an advantage to our plant; but we have REASON to believe, that only a few plants or animals
0082 ee scope for the work of improvement. We have REASON to believe, as stated in the first chapter, t
0085 ost liable to destruction. Hence I can see no REASON to doubt that natural selection might be most
0089 to his standard of beauty, I can see no good REASON to doubt that female birds, by selecting, dur
0090 r food. I can under such circumstances see no REASON to doubt that the swiftest and slimmest wolve
0090 d to prey on other animals. I can see no more REASON to doubt this, than that man can improve the
0092 rossed; and the act of crossing, we have good REASON to believe (as will hereafter be more fully a
0094 uth. Bearing such facts in mind, I can see no REASON to doubt that an accidental deviation in the
0096 t majority of plants are hermaphrodites. What REASON, it may be asked, is there for supposing in t
0104 i have already attempted to show that we have REASON to believe that occasional intercrosses take
0109 ology shows us plainly; and indeed we can see REASON why they should not have thus increased, for
0125 lines of descent had diverged less. I see no REASON to limit the process of modification, as now
0140 untries which here enjoy good health. We have REASON to believe that species in a state of nature
0142 due to habit. On the other hand, I can see no REASON to doubt that natural selection will continua
0144 coexist, without our being able to assign any REASON. What can be more singular than the relation
0154 ped organs may be made constant. I can see no REASON to doubt. Hence when an organ, however abnorm
0164 rses striped when first foaled. I have, also, REASON to suspect, from information given me by Mr.
0167 out of a hundred can we pretend to assign any REASON why this or that part differs, more or less,
0180 to collect food more quickly, or as there is REASON to believe, by lessening the danger from occa
0186 s, absurd in the highest possible degree. Yet REASON tells me, that if numerous gradations from a
0188 not know any of the transitional grades. His REASON ought to conquer his imagination; though I ha
0198 necessary in high regions would, we have some REASON to believe, increase the size of the chest; a
0198 d best under different climates; and there is REASON to believe that constitution and colour are c
0202 n in that most perfect organ, the eye. If our REASON leads us to admire with enthusiasm a multitud
0202 inimitable contrivances in nature, this same REASON tells us, though we may easily err on both si
0208 as Pierre Huber expresses it, of judgment or REASON, often comes into play, even in animals very
0208 will! yet they may be modified by the will or REASON. Habits easily become associated with other h
0218 r species; and M. Fabre has lately shown good REASON for believing that although the Tachytes nigr
0247 primrose and cowslip, which we have such good REASON to believe to be varieties, and only once or
0253 perfectly fertile hybrid animals, I have some REASON to believe that the hybrids from Cervulus vag
0254 uced quite fertile hybrids. So again there is REASON to believe that our European and the humped I
0261 h of the two plants. We can sometimes see the REASON why one tree will not take on another, from d
0261 but in a multitude of cases we can assign no REASON whatever. Great diversity in the size of two
0276 their reproductive systems. There is no more REASON to think that species have been specially end
0281 r instance to the horse and tapir, we have no REASON to suppose that links ever existed directly i
0283 re charged with sand or pebbles; for there is REASON to believe that pure water can effect little
0298 pidly and are not highly locomotive, there is REASON to suspect, as we have formerly seen, that th
0300 tural history of the world! But we have every REASON to believe that the terrestrial productions o
0310 ained the immutability of species. But I have REASON to believe that one great authority, Sir Char
0313 y been observed in Switzerland. There is some REASON to believe that organisms, considered high in
0313 sappeared from the face of the earth, we have REASON to believe that the same identical form never
0315 e also destroyed, and in nature we have every REASON to believe that the parent form will generall
0317 is conclusion. On the contrary, we have every REASON to believe, from the study of the tertiary fo
0318 species or any single genus endures. There is REASON to believe that the complete extinction of th
0328 ion from other parts of the world. As we have REASON to believe that large areas are affected by t
0335 arance of the species was perfect, we have no REASON to believe that forms successively produced n
0351 le. But in many other cases, in which we have REASON to believe that the species of a genus have b
0370 f climate of an opposite nature. We have good REASON to believe that during the newer Pliocene per
0379 of the opposite hemisphere. Although we have REASON to believe from geological evidence that the
0380 he exact lines and means of migration, or the REASON why certain species and not other have migrate
0388 low in the scale of nature, and that we have REASON to believe that such low beings change or bec
0390 nd which can be named. In St. Helena there is REASON to believe that the naturalised plants and an
0401 der than the British Channel, and there is no REASON to suppose that they have at any former perio
0406 to their new conditions. There is, also, some REASON to believe from geological evidence that orga
0415 ps, in which the same organ, as we have every REASON to suppose, has nearly the same physiological
0440 spawn of a frog under water. We have no more REASON to believe in such a relation, than we have t
0442 ome manner on growth. But there is no obvious REASON why, for instance, the wing of a bat, or the
0455 tremely early period of life (as we have good REASON to believe to be possible) the rudimentary pa
0459 ans superior to, though analogous with, human REASON, but by the accumulation of innumerable sligh
0462 his could have been effected. Yet, as we have REASON to believe that some species have retained th
0463 rieties in the intermediate zone. For we have REASON to believe that only a few species are underg
0466 e conditions of life remain the same, we have REASON to believe that a modification, which has alr
0467 es or aboriginal species. There is no obvious REASON why the principles which have acted so effici
0486 summing up of the labour, the experience, the REASON, and even the blunders of numerous workmen; w
0233 vidual under its conditions of life, it may REASONABLY be asked, how a long and graduated successi
0295 different mineralogical composition, we may REASONABLY suspect that the process of deposition has
0369 of North America and Europe; and it may be REASONABLY asked how I account for the necessary degre
0199 without here entering on copious details my REASONING would appear frivolous. The foregoing remar
0246 pecial endowment, beyond the province of our REASONING powers. The fertility of varieties, that is
0424 e lost this habit; nevertheless, without any REASONING or thinking on the subject, these tumblers
0453 st be struck with astonishment; for the same REASONING power which tells us plainly that most part
```

Page **(Key Word)**

reasoning

Page *************************************(Key Word)*************************************

```
0008 ected prior to the act of conception. Several REASONS make me believe in this; but the chief one i
0018 one wild parent. With respect to horses, from REASONS which I cannot give here, I am doubtfully in
0023 the most trifling respects. As several of the REASONS which have led me to this belief are in some
0026 to me rash in the extreme. From these several REASONS, namely, the improbability of man having for
0027 g being perfectly fertile; from these several REASONS, taken together, I can feel no doubt that al
0178 . for these intermediate varieties will, from REASONS already assigned (namely from what we know o
0203 ed, fitted for an intermediate zone; but from REASONS assigned, the intermediate variety will usua
0279 nks, is a very obvious difficulty. I assigned REASONS why such links do not commonly occur at the
0293 f years for its deposition, I can see several REASONS why each should not include a graduated seri
0297 se intermediate gradations. And this from the REASONS just assigned we can seldom hope to effect i
0327 th this subject worth making. I have given my REASONS for believing that all our greater fossilife
0370 nterior to the Glacial epoch. Believing, from REASONS before alluded to, that our continents have
0232 w often the bees would entirely pull down and REBUILD in different ways the same cell, sometimes r
0067 here I will make only a few remarks, just to RECALL to the reader's mind some of the chief points
0282 lection. It is hardly possible for me even to RECALL to the reader, who may not be a practical geo
0303 ill appear as if suddenly created. I may here RECALL a remark formerly made, namely that it might
0303 of species have suddenly been produced. I may RECALL the well known fact that in geological treati
0459 e the leading facts and inferences briefly RECAPITULATED. That many and grave objections may be ad
0465 against my theory; and I have now briefly RECAPITULATED the answers and explanations which can be
0469 e to be in itself probable. I have already RECAPITULATED, as fairly as I could, the opposed diffic
0480 e wilfully will not understand. I have now RECAPITULATED the chief facts and considerations which
0006 in the last chapter I shall give a brief RECAPITULATION of the whole work, and a few concluding r
0459 by other facts or arguments. Chapter XIV. RECAPITULATION and Conclusion. Recapitulation of the dif
0459 apter XIV. Recapitulation and Conclusion. RECAPITULATION of the difficulties on the theory of Natu
0459 lties on the theory of Natural Selection. RECAPITULATION of the general and special circumstances
0460 n crossed, I must refer the reader to the RECAPITULATION of the facts given at the end of the eigh
0376 s mr. H. C. Watson has recently remarked, In RECEDING from polar towards equatorial latitudes, the
0040 ct and valuable, and will then probably first RECEIVE a provincial name. In semi civilised countri
0098 stigma of that individual flower is ready to RECEIVE them; and as this flower is never visited, a
0223 d larvae. So that the masters in this country RECEIVE much less service from their slaves than the
0308 re entertained. To show that it may hereafter RECEIVE some explanation, I will give the following
0355 eds of miles from a continent, would probably RECEIVE from it in the course of time a few colonist
0365 ding more remote from the mainland, would not RECEIVE colonists by similar means. I do not doubt t
0398 gos to America. I believe this grand fact can RECEIVE no sort of explanation on the ordinary view
0398 that the Galapagos Islands would be likely to RECEIVE colonists, whether by occasional means of tr
0400 r each other that they would almost certainly RECEIVE immigrants from the same original source, or
0478 ainland. It must be admitted that these facts RECEIVE no explanation on the theory of creation. Th
0002 ledging the generous assistance which I have RECEIVED from very many naturalists, some of them per
0355 the species of a second region, has probably RECEIVED at some former period immigrants from this o
0364 d it would be very difficult to prove this), RECEIVED within the last few centuries, through occas
0477 ll relations, and as the two areas will have RECEIVED colonists from some third source or from eac
0098 ations, which effectually prevent the stigma RECEIVING pollen from its own flower: for instance, i
0098 ontrivance to prevent the stigma of a flower RECEIVING its own pollen, yet, as C. C. Sprengel has
0010 ants. Under this point of view, Mr. Buckman's RECENT experiments on plants seem extremely valuable
0024 as become feral in several places. Again, all RECENT experience shows that it is most difficult to
0076 pecies of distinct genera. We see this in the RECENT extension over parts of the United States of
0076 g caused the decrease of another species. We RECENT increase of the missel thrush in parts of Sco
0142 states; and as most of these varieties are of RECENT origin, they cannot owe their constitutional
0154 which the modification has been comparatively RECENT and extraordinarily great that we ought to fi
0155 ructure which have varied within a moderately RECENT period, and which have thus come to differ. O
0174 now continuous must often have existed within RECENT times in a far less continuous and uniform co
0178 continuous must often have existed within the RECENT period in isolated portions, in which many fo
0273 his implies in most cases that there has been RECENT variability; and therefore we might expect th
0290 been upraised several hundred feet within the RECENT period, than the absence of any recent deposi
0290 in the recent period, than the absence of any RECENT deposits sufficiently extensive to last for e
0299 f south America, no extensive formations with RECENT or tertiary remains can anywhere be found, th
0312 d new forms more gradual. In some of the most RECENT beds, though undoubtedly of high antiquity if
0319 extinction. We see in many cases in the more RECENT tertiary formations, that rarity precedes ext
0331 lopods, and the Eocene mammals, with the more RECENT members of the same classes, we must admit th
0331 ormerly illustrated by this diagram, the more RECENT any form is, the more it will generally diffe
0332 e letter F14. All the many forms, extinct and RECENT, descended from A, make, as before remarked,
0334 in character are not the oldest, or the most RECENT; nor are those which are intermediate in char
0336 forms. There has been much discussion whether RECENT forms are more highly developed than ancient.
0337 w forms. But in one particular sense the more RECENT forms must, on my theory, be higher than the
0337 sensible manner the organisation of the more RECENT and victorious forms of life, in comparison w
0338 s resemble to a certain extent the embryos of RECENT animals of the same classes; or that the geol
0338 parallel to the embryological development of RECENT forms. I must follow Pictet and Huxley in thi
0338 ched off from each other within comparatively RECENT times. For this doctrine of Agassiz accords w
0344 it is that all the forms of life, ancient and RECENT, make together one grand system: for all are
0345 emble to a certain extent the embryos of more RECENT animals of the same class, the fact will be j
0351 genus have been produced within comparatively RECENT times; there is great difficulty on this head
0353 changes, which have certainly occurred within RECENT geological times, must have interrupted or re
0357 extension, as to have united them within the RECENT period to each other and to the several inter
0357 e that it will ever be proved that within the RECENT period continents which are now quite separat
0358 rodigious geographical revolutions within the RECENT period, as are necessitated on the view advan
0366 nd, organic and inorganic, that within a very RECENT geological period, central Europe and North A
0371 roductions, left isolated, within a much more RECENT period, on the several mountain ranges and on
0384 ater fish mainly to slight changes within the RECENT period in the level of the land, having cause
0384 le changes of level in the land within a very RECENT geological period, and when the surface was p
0389 out, would lead to the belief that within the RECENT period all existing islands have been nearly
0396 ely to have been continuously united within a RECENT period to the mainland than islands separated
0403 spread widely throughout the world during the RECENT Glacial epoch, are related to those of the su
0407 and, which have certainly occurred within the RECENT period, and of other similar changes which ma
0428 can account for the fact that all organisms, RECENT and extinct, are included under a few great o
0448 grade. As all the organic beings, extinct and RECENT, which have ever lived on this earth have to
0450 ient forms of life to the embryonic stages of RECENT forms, may be true, but yet, owing to the geo
```

Page *************************************(Key Word)*************************************

Page **(Key Word)**

```
0476  organic beings belong to the same system with RECENT beings, falling either into the same or into
0476  ermediate between existing and allied groups.  RECENT forms are generally looked at as being, in so
0476  ellig'ble, for within a confined country, the  RECENT and the extinct will naturally be allied by d
0478  with extinct groups often falling in between   RECENT groups, is intelligible on the theory of natu
0046  een effected only by slow degrees: yet quite   RECENTLY Mr. Lubbock has shown a degree of variabilit
0046  phical naturalist, I may add, has also quite   RECENTLY shown that the muscles in the larvae of cert
0100  ticipated. On the other hand, Dr. Hooker has   RECENTLY informed me that he finds that the rule does
0106  no to oscillations of level, will often have   RECENTLY existed in a broken condition, so that the g
0123  a14, p14, will be nearly related from having   RECENTLY branched off from a10; b14 and f14, from hav
0145  a striking case of correlation, that I have    RECENTLY observed in some garden pelargoniums, that t
0158  nherited much in common, to parts which have   RECENTLY and largely varied being more likely still t
0168  gans being still variable, because they have   RECENTLY varied and thus come to differ; but we have
0318  aralleled rate, I asked myself what could so   RECENTLY have exterminated the former horse under con
0337  ry manner in which European productions have   RECENTLY spread over New Zealand, and have seized on
0357  ed that all the islands in the Atlantic must   RECENTLY have been connected with Europe or Africa, a
0357  carcely a single island exists which has not   RECENTLY been united to some continent. This view cut
0376  ern temperate zones. As Mr. H. C. Watson has   RECENTLY remarked, In receding from polar towards equ
0382  l distribution. I believe that the world has   RECENTLY felt one of his great cycles of change: and
0425  tions of life to which each species has been   RECENTLY exposed. Rudimentary structures on this view
0138  m the outer world into the deeper and deeper   RECESSES of the Kentucky caves, as did European anima
0138  d, after numberless generations, the deepest   RECESSES, disuse will on this view have more or less
0258  mates, can often be crossed with ease. By a    RECIPROCAL cross between two species, I mean the case,
0258  ssible difference in the facility of making    RECIPROCAL crosses. Such cases are highly important, f
0258  ve system. This difference in the result of    RECIPROCAL crosses between the same two species was lo
0258  that this difference of facility in making    RECIPROCAL crosses is extremely common in a lesser deg
0258  a remarkable fact, that hybrids raised from    RECIPROCAL crosses, though of course compounded of the
0260  this latter statement is clearly proved by    RECIPROCAL crosses between the same two species, for a
0260  union. The hybrids, moreover, produced from    RECIPROCAL crosses often differ in fertility. Now do t
0260  be so great a difference in the result of a    RECIPROCAL cross between the same two species? Why, it
0261  so peculiar and limited a nature, that, in    RECIPROCAL crosses between two species the male sexual
0262  wo species in being grafted together. As in    RECIPROCAL crosses, the facility of effecting an union
0266  unequal fertility of hybrids produced from    RECIPROCAL crosses; or the increased sterility in thos
0271  he tested by the severest trial, namely, by   RECIPROCAL crosses; and he found their mongrel offspri
0274  ther variety. Hybrid plants produced from a   RECIPROCAL cross, generally resemble each other closel
0274  closely; and so it is with mongrels from a    RECIPROCAL cross. Both hybrids and mongrels can be red
0276  fferent, and sometimes widely different, in   RECIPROCAL crosses between the same two species. It is
0258  two species may then be said to have been    RECIPROCALLY crossed. There is often the widest possibl
0258  during eight following years, to fertilise   RECIPROCALLY M. longiflora with the pollen of M. jalapp
0460  ult, when the same two species are crossed   RECIPROCALLY; that is, when one species is first used a
0028  ay, they are come to this pass, that they can RECKON up their pedigree and race. Pigeons were much
0064  e would be a million plants. The elephant is  RECKONED to be the slowest breeder of all known anima
0257  kind, or what amount, of difference in any    RECOGNISABLE character is sufficient to prevent two spe
0258  nt of their systematic affinity, or of any    RECOGNISABLE difference in their whole organisation. On
0037  ct, that in a vast number of cases we cannot  RECOGNISE, and therefore do not know, the wild parent
0281  ce in all such cases, we should be unable to  RECOGNISE the parent form of any two or more species,
0282  of Geology, which the future historian will   RECOGNISE as having produced a revolution in natural
0297  erous transitional gradations, we should not  RECOGNISE their relationship, and should consequently
0464  ainly have existed. We should not be able to  RECOGNISE a species as the parent of any one or more
0031  ction in regard to merino sheep is so fully   RECOGNISED, that men follow it as a trade: the sheep a
0035  ensible changes of this kind could never be   RECOGNISED unless actual measurements or careful drawi
0037  lended together by crossing, may plainly be   RECOGNISED in the increased size and beauty which we n
0040  they will spread more widely, and will get    RECOGNISED as something distinct and valuable, and wil
0304  ith which even a fragment of a valve can be   RECOGNISED; from all these circumstances, I inferred t
0322  d. thus our European Chalk formation can be   RECOGNISED in many distant parts of the world, under t
0334  s system was first discovered, were at once   RECOGNISED by palaeontologists as intermediate in char
0419  and these to others, and so onwards, can be   RECOGNISED as unequivocally belonging to this, and to
0449  or instance, that cirripedes can at once be   RECOGNISED by their larvae as belonging to the great c
0487  each great fossiliferous formation will be    RECOGNISED as having depended on an unusual concurrenc
0488  e incomprehensible by us, will hereafter be   RECOGNISED as a mere fragment of time, compared with t
0162  e. a considerable part of the difficulty in   RECOGNISING a variable species in our systematic works
0142  , in which certain varieties are habitually   RECOMMENDED for the northern, and others for the south
0237  . it may well be asked how is it possible to  RECONCILE this case with the theory of natural select
0224  aticians that bees have practically solved a  RECONDITE problem, and have made their cells of the p
0271  nd from not wishing or being able to produce  RECONDITE and functional differences in the reproduct
0441  ave no antennae, and their two eyes are now   RECONVERTED into a minute, single, and very simple eye
0005  fourthly, the imperfection of the Geological  RECORD. In the next chapter I shall consider the geo
0008  s to vary for many generations. No case is on RECORD of a variable being ceasing to be variable un
0020  ed selection; nor can I find a single case on RECORD of a permanent race having been thus formed.
0027  ral cuarters of the world; the earliest known RECORD of pigeons is in the fifth Aegyptian dynasty,
0040  ut the chance will be infinitely small of any RECORD having been preserved of such slow, varying,
0071  ct i will not here enlarge. Many cases are on RECORD showing how complex and unexpected are the ch
0117  d variety, such as would be thought worthy of RECORD in a systematic work. The intervals between t
0172  chapter on the Imperfection of the geological RECORD; and I will here only state that I believe th
0172  that I believe the answer mainly lies in the  RECORD being incomparably less perfect than is gener
0172  s generally supposed; the imperfection of the RECORD being chiefly due to organic beings not inhab
0179  w, in an extremely imperfect and intermittent RECORD. On the origin and transitions of organic bei
0279  ter IX. On the Imperfection of the Geological RECORD. On the absence of intermediate varieties at
0280  in the extreme imperfection of the geological RECORD. In the first place it should always be borne
0288  water bespeaks its purity. The many cases on  RECORD of a formation conformably covered, after an
0289  tions. But the imperfection in the geological RECORD mainly results from another and more importan
0290  tiary beds are so scantily developed, that no  RECORD of several successive and peculiar marine fau
0291  mass when upraised will give a most imperfect RECORD of the forms of life which then existed; or,
0292  mits of the coast action. Thus the geological RECORD will almost necessarily be rendered intermitt
0292  e will generally be a blank in the geological RECORD. On the other hand, during subsidence, the in
0292  ions it cannot be doubted that the geological RECORD, viewed as a whole, is extremely imperfect; b
0293  mencement and at its close. Some cases are on RECORD of the same species presenting distinct varie
```

Page **(Key Word)**

615   record

Page ***(Key Word)***

```
0300  e would be much variation, but the geological RECORD would then be least perfect. It may be doubte
0301  d have to migrate, and no closely consecutive RECORD of their modifications could be preserved in
0302  that I should ever have suspected how poor a RECORD of the mutations of life, the best preserved
0302  ly over rate the perfection of the geological RECORD, and falsely infer, because certain genera or
0309  from the remotest period of which we have any RECORD; and on the other hand, that where continents
0310  ledge. Those who think the natural geological RECORD in any degree perfect, and who do not attach
0310  's metaphor, I look at the natural geological RECORD, as a history of the world imperfectly kept,
0333  ossess only the last volume of the geological RECORD, and that in a very broken condition, we have
0335  stant, in this and other such cases, that the RECORD of the first appearance and disappearance of
0341  1 have attempted to show that the geological RECORD is extremely imperfect; that only a small por
0342  riocs of elevation, and during the latter the RECORD will have been least perfectly kept; that eac
0342  ntly, must have tended to make the geological RECORD extremely imperfect, and will to a large exte
0342  s these views on the nature of the geological RECORD, will rightly reject my whole theory. For he
0345  lained by inheritance. If then the geological RECORD be as imperfect as I believe it to be, and it
0345  be, and it may at least be asserted that the RECORD cannot be proved to be much more perfect, the
0382  nesses of almost every land, which serve as a RECORD, full of interest to us, of the former inhabi
0450  may be true, but yet, owing to the geological RECORD not extending far enough back in time, may re
0451  ae of male mammals, for many instances are on RECORD of these organs having become well developed
0464  s only on the supposition that the geological RECORD is far more imperfect than most geologists be
0464  , owing to the imperfection of the geological RECORD. Numerous existing doubtful forms could be na
0465  iods of elevation and of stationary level the RECORD will be blank. During these latter periods th
0465  ven in the ninth chapter. That the geological RECORD is imperfect all will admit: but that it is i
0466  or that we know how imperfect the Geological RECORD is. Grave as these several difficulties are,
0469  y think sufficiently distinct to be worthy of RECORD in systematic works. No one can draw any clea
0475  condary laws. If we admit that the geological RECORD is imperfect in an extreme degree, then such
0475  in an extreme degree, then such facts as the RECORD gives, support the theory of descent with mod
0481  to assume, without proof, that the geological RECORD is so perfect that it would have afforded us
0487  es glory from the extreme imperfection of the RECORD. The crust of the earth with its embedded rem
0050  eties of it will almost universally be found RECORDED. These varieties, moreover, will be often ra
0053  ieties sufficiently well marked to have been RECORDED in botanical works. Hence it is the most flo
0056  etween the species of large genera and their RECORDED varieties which deserve notice. We have seen
0058  ame catalogue, 53 acknowledged varieties are RECORDED, and these range over 7.7 provinces; whereas
0064  eoretical calculations, namely, the numerous RECORDED cases of the astonishingly rapid increase of
0110  species which afford the greatest number of RECORDED varieties, or incipient species. Hence, rare
0119  hich have become sufficiently distinct to be RECORDED as varieties. But these breaks are imaginary
0486  e species added to the infinitude of already RECORDED species. Our classifications will come to be
0051  slight varieties as are barely thought worth RECORDING in works on natural history. And I look at
0018  animals is, that we find in the most ancient RECORDS, more especially on the monuments of Egypt,
0032  this is not so in some cases, in which exact RECORDS have been kept; thus, to give a very triflin
0307  creatures. To the question why we do not find RECORDS of these vast primordial periods, I can give
0208  by rote, he is generally forced to go back to RECOVER the habitual train of thought: so P. Huber f
0397  sea water for twenty days, and it perfectly RECOVERED. As this species has a thick calcareous ope
0397  ed it for fourteen days in sea water, and it RECOVERED and crawled away: but more experiments are
0016  this source of doubt could not so perpetually RECUR. It has often been stated that domestic races
0149  re apt to be highly variable. We shall have to RECUR to the general subject of rudimentary and abo
0315  ditions of life, organic and inorganic, should RECUR. For though the offspring of one species migh
0363  ew minute seeds? But I shall presently have to RECUR to this subject. As icebergs are known to be
0414  eration of these resemblances we shall have to RECUR. It may even be given as a general rule, that
0465  beneath the lowest Silurian strata, I can only RECUR to the hypothesis given in the ninth chapter.
0066  e young or old, during each generation or at RECURRENT intervals. Lighten any check, mitigate the
0467  individuals and races, during the constantly RECURRENT Struggle for Existence, we see the most pow
0005  and as, consequently, there is a frequently RECURRING struggle for existence, it follows that any
0062  bundant, it is not so at all seasons of each RECURRING year. I should premise that I use the term
0073  e as this. Battle within battle must ever be RECURRING with varying success; and yet in the long r
0232  d in different ways the same cell, sometimes RECURRING to a shape which they had at first rejected
0326  territories. A certain amount of isolation, RECURRING at long intervals of time, would probably b
0049  experienced ornithologists consider our British RED grouse as only a strongly marked race of a Nor
0073  clovers; but humble bees alone visit the common RED clover (Trifolium pratense), as other bees can
0074  nct or very rare in England, the heartsease and RED clover would become very rare, or wholly disap
0084  grey; the alpine ptarmigan white in winter, the RED grouse the colour of heather, and the black or
0094  ucture. The tubes of the corollas of the common RED and incarnate clovers (Trifolium pratense and
0094  the incarnate clover, but not out of the common RED clover, which is visited by humble bees alone;
0095  humble bees alone; so that whole fields of the RED clover offer in vain an abundant supply of pre
0095  y country, it might be a great advantage to the RED clover to have a shorter or more deeply divide
0154  e genus of plants had blue flowers and some had RED, the colour would by only a specific character
0154  rprised at one of the blue species varying into RED, or conversely; but if all the species had blu
0220  are black and not above half the size of their RED masters, so that the contrast in their appeara
0247  in getting fertile seed: as he found the common RED and blue pimpernels (Anagallis arvensis and co
0268  uralists as species. For instance, the blue and RED pimpernel, the primrose and cowslip, which are
0270  aize with yellow seeds, and a tall variety with RED seeds, growing near each other in his garden;
0303  one true mammal has been discovered in the new RED sandstone at nearly the commencement of this g
0125  roup will slowly conquer another large group, REDUCE its numbers, and thus lessen its chance of fu
0136  endency of natural selection to enlarge or to REDUCE the wings, would depend on whether a greater
0287  hardly have been great, but it would somewhat REDUCE the above estimate. On the other hand, during
0454  se natural selection would continue slowly to REDUCE the organ, until it rendered harmless and rud
0455  e age, and consequently will seldom affect or REDUCE it in the embryo. Thus we can understand the
0479  imes by natural selection, will often tend to REDUCE an organ, when it has become useless by chang
0033  cted that the principle of selection has been REDUCED to methodical practice for scarcely more tha
0136  r. Wollaston suspects, their wings not at all REDUCED, but even enlarged. This is quite compatible
0137  n enlarged, and the wings of others have been REDUCED by natural selection aided by use and disuse
0147  election and another and adjoining part being REDUCED by this same process or by disuse, and, on t
0148  lepas, the whole anterior part of the head is REDUCED to the merest rudiment attached to the bases
0195  y are incessantly harassed and their strength REDUCED, so that they are more subject to disease, o
0274  rocal cross. Both hybrids and mongrels can be REDUCED to either pure parent form, by repeated cros
0283  and, have to be worn away, atom by atom, until REDUCED in size they can be rolled about by the wave
0353  tinuous range of many species. So that we are REDUCED to consider whether the exceptions to contin
```

Page ***(Key Word)***

Page **(Key Word)**

```
0380  exterminated, their numbers have been greatly REDUCED, and this is the first stage towards extinct
0405   and the common range would have been greatly REDUCED. Still less is it meant, that a species whic
0451  t, yet in how many insects do we see wings so REDUCED in size as to be utterly incapable of flight
0455  sponging ages will reproduce the organ in its REDUCED state at the same age, and consequently will
0457  le, and bearing in mind, that when organs are REDUCED in size, either from disuse or selection, it
0480   early life; hence the organ will not be much REDUCED or rendered rudimentary at this early age. T
0480  eve, that the teeth in the mature animal were REDUCED, during successive generations, by disuse or
0455  t whatever period of life disuse or selection REDUCES an organ, and this will generally be when th
0068  ce; but in so far as climate chiefly acts in REDUCING food, it brings on the most severe struggle
0148  ction will always succeed in the long run in REDUCING and saving every part of the organisation, a
0137  state of the eyes is probably due to gradual REDUCTION from disuse, but aided perhaps by natural s
0137  sable to animals with subterranean habits, a REDUCTION in their size with the adhesion of the eyel
0148  ut requiring as a necessary compensation the REDUCTION of some adjoining part. It seems to be a ru
0454  led in successive generations to the gradual REDUCTION of various organs, until they become rudime
0455  he adult. But if each step of the process of REDUCTION were to be inherited, not at the correspond
0309  he coloured map appended to my volume on Coral REEFS, led me to conclude that the great oceans are
0253  that the hybrids from Cervulus vaginalis and REEVESII, and from Phasianus colchicus with P. torqua
0003  xcites our admiration. Naturalists continually REFER to external conditions, such as climate, food
0014  lluded to the subject of reversion, I may here REFER to a statement often made by naturalists, nam
0046  ces, which seems to me extremely perplexing: I REFER to those genera which have sometimes been cal
0124  hen we come to our chapter on Geology, have to REFER again to this subject, and I think we shall t
0250  t species. This case of the Crinum leads me to REFER to a most singular fact, namely, that there a
0441  rva, as with certain parasitic crustaceans. To REFER once again to cirripedes: the larvae in the f
0460  al fertility of varieties when crossed, I must REFER the reader to the recapitulation of the facts
0139  d as is the case with the blind Proteus with REFERENCE to the reptiles of Europe, I am only surpri
0002  ecessarily be imperfect. I cannot here give REFERENCES and authorities for my several statements;
0002  er publishing in detail all the facts, with REFERENCES, on which my conclusions have been grounded
0033  is a modern discovery. I could give several REFERENCES to the full acknowledgment of the importanc
0168  n this same period. In these remarks we have REFERRED to special parts or organs being still varia
0431  gths (as may be seen in the diagram so often REFERRED to), mounting up through many predecessors.
0325  ervers, MM. de Verneuil and d'Archiac. After REFERRING to the parallelism of the palaeozoic forms
0420  hat is meant, if he will take the trouble of REFERRING to the diagram in the fourth chapter. We wi
0007  cies or variety in a state of nature. When we REFLECT on the vast diversity of the plants and anim
0079  ife, and to suffer great destruction. When we REFLECT on this struggle, we may console ourselves w
0124  w sub genera and genera. It is worth while to REFLECT for a moment on the character of the new spe
0171  hem are so grave that to this day I can never REFLECT on them without being staggered; but, to the
0268  is a remarkable fact: more especially when we REFLECT how many species there are, which, though re
0294  ticular deposit. It is an excellent lesson to REFLECT on the ascertained amount of migration of th
0294   one whole geological period; and likewise to REFLECT on the great changes of level, on the inordi
0363  soil almost everywhere is charged with seeds. REFLECT for a moment on the millions of quails which
0489  worms crawling through the damp earth, and to REFLECT that these elaborately constructed forms, so
0001  this question by patiently accumulating and REFLECTING on all sorts of facts which could possibly
0003   it is quite conceivable that a naturalist, REFLECTING on the mutual affinities of organic beings,
0198  e are immediately made conscious of this by REFLECTING on the differences in the breeds of our dom
0226  ach flat portion forms a part of two cells. REFLECTING on this case, it occurred to me that if the
0453  acts with respect to rudimentary organs. In REFLECTING on them, every one must be struck with asto
0290  bly be preserved to a distant age. A little REFLECTION will explain why along the rising coast of
0067  instance. Nor will this surprise any one who REFLECTS how ignorant we are on this head, even in re
0310  t authority, Sir Charles Lyell, from further REFLEXION entertains grave doubts on this subject. I
0029  s, yet they ignore all general arguments, and REFUSE to sum up in their minds slight differences a
0492  ese have been produced by variation, but they REFUSE to extend the same view to other and very sli
0137  size; and Professor Silliman thought that it REGAINED, after living some days in the light, some s
0006  prise at much remaining as yet unexplained in REGARD to the origin of species and varieties, if he
0006  s due allowance for our profound ignorance in REGARD to the mutual relations of all the beings whi
0015  uld deduce nothing from domestic varieties in REGARD to species. But there is not a shadow of evid
0018   have descended from several wild species. In REGARD to sheep and goats I can form no opinion. I s
0019   common wild Indian fowl (Gallus bankiva). In REGARD to ducks and rabbits, the breeds of which dif
0025  probable in the highest degree. Some facts in REGARD to the colouring of pigeons well deserve cons
0028  st could in coming to a similar conclusion in REGARD to the many species of finches, or other larg
0031  e importance of the principle of selection in REGARD to merino sheep is so fully recognised, that
0033  ss as to allow his worst animals to breed. In REGARD to plants, there is another means of observin
0038  e plants in countries anciently civilised. In REGARD to the domestic animals kept by uncivilised m
0043  i believe, been greatly exaggerated, both in REGARD to animals and to those plants which are prop
0053  some interesting results might be obtained in REGARD to the nature and relations of the species wh
0057  ecies or varieties. Now Fries has remarked in REGARD to plants, and Westwood in regard to insects,
0057  remarked in regard to plants, and Westwood in REGARD to insects, that in large genera the amount o
0062  beings are exposed to severe competition. In REGARD to plants, no one has treated this subject wi
0067  cts how ignorant we are on this head, even in REGARD to mankind, so incomparably better known than
0067  ks at considerable length, more especially in REGARD to the feral animals of South America. Here I
0080  ussed too briefly in the last chapter, act in REGARD to variation? Can the principle of selection,
0139  be very anomalous, as Agassiz has remarked in REGARD to the blind fish, the Amblyopsis, and as is
0140  plants brought from the Azores to England. In REGARD to animals, several authentic cases could be
0146  ral selection may have come into play. But in REGARD to the differences both in the internal and e
0195   the first place, we are much too ignorant in REGARD to the whole economy of any one organic being
0223  ouch they did not need confirmation by me, in REGARD to the wonderful instinct of making slaves. L
0243  also, strengthened by some few other facts in REGARD to instincts; as by that common case of close
0248  ived at diametrically opposite conclusions in REGARD to the very same species. It is also most ins
0248  constitutional and structural differences. In REGARD to the sterility of hybrids in successive gen
0252  enty of pollen brought from other flowers. In REGARD to animals, much fewer experiments have been
0255  nd considering how scanty our knowledge is in REGARD to hybrid animals, I have been surprised to f
0264  injurious or unnatural conditions of life. In REGARD to the sterility of hybrids, in which the sex
0289  the carboniferous strata of North America. In REGARD to mammiferous remains, a single glance at th
0323  ope from La Plata, without any information in REGARD to their geological position, no one would ha
0330   difference between the pig and the camel. In REGARD to the Invertebrata, Barrande, and a higher a
0338  ct to see it hereafter confirmed, at least in REGARD to subordinate groups, which have branched of
0375  er, analogous and striking facts are given in REGARD to the plants of that large island. Hence we
```

Page **(Key Word)**

Page **(Key Word)**

```
0379 kable fact, strongly insisted on by Hooker in REGARD to America, and by Alph. de Candolle in regar
0379 egard to America, and by Alph. de Candolle in REGARD to Australia, that many more identical plants
0380 culties are removed on the view here given in REGARD to the range and affinities of the allied spe
0383 ce. We can here consider only a few cases. In REGARD to fish, I believe that the same species neve
0389 would not, I think, explain all the facts in REGARD to insular productions. In the following rema
0395 de some striking observations on this head in REGARD to the great Malay Archipelago, which is trav
0397 gould has given several interesting cases in REGARD to the land shells of the islands of the Paci
0397 he most striking and important fact for us in REGARD to the inhabitants of islands, is their affin
0406 n admirably discussed by Alph. de Candolle in REGARD to plants, namely, that the lower any group o
0426 common parent. We may err in this respect in REGARD to single points of structure, but when sever
0442 ely from the adult: thus Owen has remarked in REGARD to cuttle fish, there is no metamorphosis; th
0446 ve given seem to me to explain these facts in REGARD to the later embryonic stages of our domestic
0485 hing wholly beyond his comprehension; when we REGARD every production of nature as one which has h
0049 ven Ireland has a few animals, now generally REGARDED as varieties, but which have been ranked as
0414 habits and food of an animal, I have always REGARDED as affording very clear indications of its t
0280 if we had no historical or indirect evidence REGARDING their origin, it would not have been possib
0381 ere discussed. I will only say that as far as REGARDS the occurrence of identical species at point
0414 sification. Nothing can be more false. No one REGARDS the external similarity of a mouse to a shre
0038 stralia, the Cape of Good Hope, nor any other REGION inhabited by quite uncivilised man, has affor
0069 we travel from south to north, or from a damp REGION to a dry, we invariably see some species grad
0108 nly a very few of the inhabitants of the same REGION at the same time. I further believe, that thi
0110 hat we have any means of knowing that any one REGION has as yet got its maximum of species. Probab
0110 s yet got its maximum of species. Probably no REGION is as yet fully stocked, for at the Cape of G
0115 iversification in the inhabitants of the same REGION is, in fact, the same as that of the physiolo
0139 ecies from an arctic or even from a temperate REGION cannot endure a tropical climate, or converse
0173 adapted to the conditions of life of its own REGION, and has supplanted and exterminated its orig
0174 with numerous transitional varieties in each REGION, though they must have existed there, and may
0174 n a fossil condition. But in the intermediate REGION, having intermediate conditions of life, why
0177 kept, one adapted to an extensive mountainous REGION; a second to a comparatively narrow, hilly tr
0178 d reacting on each other. So that, in any one REGION and at any one time, we ought only to see a f
0290 nk, see why the geological formations of each REGION are almost invariably intermittent; that is,
0315 ce we can see why all the species in the same REGION do at last, if we look to wide enough interva
0326 the sea. Dominant species spreading from any REGION might encounter still more dominant species,
0327 ervals I suppose that the inhabitants of each REGION underwent a considerable amount of modificati
0328 will have been a little more time in the one REGION than in the other for modification, extinctio
0329 ng the same exact periods, a formation in one REGION often corresponding with a blank interval in
0343 played, when the formations of any one great REGION alone, as that of Europe, are considered; he
0350 n of the more dominant forms of life from one REGION into another having been effected with more o
0351 ing that they may have migrated from the same REGION; for during the vast geographical and climata
0352 ch species was first produced within a single REGION captivates the mind. He who rejects it, rejec
0353 f sections of genera are confined to a single REGION; and it has been observed by several naturali
0354 hown to be almost invariably the case, that a REGION, of which most of its inhabitants are closely
0354 the same genera with the species of a second REGION, has probably received at some former period
0355 some former period immigrants from this other REGION, my theory will be strengthened; for we can c
0355 ple of modification, why the inhabitants of a REGION should be related to those of another region,
0355 region should be related to those of another REGION, whence it has been stocked. A volcanic islan
0355 . this view of the relation of species in one REGION to those in another, does not differ much (by
0356 have had a powerful influence on migration: a REGION when its climate was different may have been
0371 hey will have become mingled in the one great REGION with the native American productions, and hav
0371 to compete with them; and in the other great REGION, with those of the Old World. Consequently we
0399 alia, the nearest mainland, than to any other REGION: and this is what might have been expected: b
0403 re plainly related to the inhabitants of that REGION whence colonists could most readily have been
0408 has elapsed since new inhabitants entered one REGION; according to the nature of the communication
0427 e genus, inhabiting any distinct and isolated REGION, have in all probability descended from the s
0469 view we can understand how it is that in each REGION where many species of a genus have been produ
0052 ifferent physical conditions in two different REGIONS; but I have not much faith in this view; and
0069 scending a mountain. When we reach the Arctic REGIONS, or snow capped summits, or absolute deserts
0078 h the extreme confines of life, in the arctic REGIONS or on the borders of an utter desert, will c
0198 mb. The laborious breathing necessary in high REGIONS would, we have some reason to believe, incre
0299 . the Malay Archipelago is one of the richest REGIONS of the whole world in organic beings; yet if
0304 y species all over the world, from the Arctic REGIONS to the equator, inhabiting various zones of
0326 those in the waters of the sea. If two great REGIONS had been for a long period favourably circum
0328 hen two formations have been deposited in two REGIONS during nearly, but not exactly the same peri
0328 e species. If the several formations in these REGIONS have not been deposited during the same exac
0329 a blank interval in the other, and if in both REGIONS the species have gone on slowly changing dur
0329 this case, the several formations in the two REGIONS could be arranged in the same order, in acco
0329 he apparently corresponding stages in the two REGIONS. On the Affinities of extinct Species to eac
0346 e dissimilarity of the inhabitants of various REGIONS can be accounted for by their climatal and o
0347 ifferences between the productions of various REGIONS. We see this in the great difference of near
0350 dissimilarity of the inhabitants of different REGIONS may be attributed to modification through na
0352 s, though now inhabiting distant and isolated REGIONS, must have proceeded from one spot, where th
0354 at distant points in the arctic and antarctic REGIONS; and secondly (in the following chapter), th
0363 e part to another of the arctic and antarctic REGIONS, as suggested by Lyell: and during the Glaci
0363 ial period from one part of the now temperate REGIONS to another. In the Azores, from the large nu
0365 o many of the same plants living on the snowy REGIONS of the Alps or Pyrenees, and in the extreme
0366 places. The inhabitants of the more temperate REGIONS would at the same time travel southward, unl
0367 even stretching into Spain. The now temperate REGIONS of the United States would likewise be cover
0367 reat by the productions of the more temperate REGIONS. And as the snow melted from the bases of th
0367 nated on all lesser heights) and in the artic REGIONS of both hemispheres. Thus we can understand
0368 ose of the mountains of Siberia to the arctic REGIONS of that country. These views, grounded as th
0368 ons of Europe and America, that when in other REGIONS we find the same species on distant mountain
0369 f productions were as uniform round the polar REGIONS as they are at the present day. But the fore
0371 e the now living productions of the temperate REGIONS of the New and Old Worlds, we find very few
0375 australia, but not in the intermediate torrid REGIONS. In the admirable Introduction to the Flora
0376 d on the mountain ranges of the intertropical REGIONS, are not arctic, but belong to the northern
0376 e forms living on the mountians of the warmer REGIONS of the earth and in the southern hemisphere
```

Page **(Key Word)**

Page **(Key Word)**

```
0378 uthern temperate zones into the intertropical REGIONS, and some even crossed the equator. As the w
0380 nes and on the mountains of the intertropical REGIONS. Very many difficulties remain to be solved.
0381 ooker in his botanical works on the antarctic REGIONS. These cannot be here discussed. I will only
0388 g subsequently become extinct in intermediate REGIONS. But the wide distribution of fresh water pl
0394 s, as so frequently now happens in the arctic REGIONS. Yet it cannot be said that small islands wi
0404 rsally found to be true, that wherever in two REGIONS, let them be ever so distant, many closely a
0404 tercommunication or migration between the two REGIONS. And wherever many closely allied species oc
0408 less rapidly, there would ensue in different REGIONS, independently of their physical conditions,
0462 extinction of the species in the intermediate REGIONS. It cannot be denied that we are as yet very
0462 me genus inhabiting very distant and isolated REGIONS, as the process of modification has necessar
0464 ieties will not spread into other and distant REGIONS until they are considerably modified and imp
0002 and this cannot possibly be here done. I much REGRET that want of space prevents my having the sat
0119 r diagram the line of succession is broken at REGULAR intervals by small numbered letters marking
0145 s are oftenest subject to peloria, and become REGULAR. I may add, as an instance of this, and of a
0225 arly related to the latter: it forms a nearly REGULAR waxen comb of cylindrical cells, in which th
0234 , if she were to make her cells more and more REGULAR, nearer together, and aggregated into a mass
0234 e to make her cells closer together, and more REGULAR in every way than at present; for then, as w
0009 does act under confinement, acting not quite REGULARLY, and producing offspring not perfectly like
0093 , they would, unintentionally on their part, REGULARLY carry pollen from flower to flower: and tha
0093 highly attractive to insects that pollen was REGULARLY carried from flower to flower, another proc
0094 er nature, then as pollen is already carried REGULARLY from flower to flower, and as a more comple
0096 large number pair: that is, two individuals REGULARLY unite for reproduction, which is all that c
0100 he same tree, we can see that pollen must be REGULARLY carried from flower to flower! and this wil
0118 not suppose that the process ever goes on so REGULARLY as is represented in the diagram, though in
0241 e parents, could form a species which should REGULARLY produce neuters, either all of large size w
0394 ocean: and two North American species either REGULARLY or occasionally visit Bermuda, at the dista
0072 ectivorous birds (whose numbers are probably REGULATED by hawks or beasts of prey) were to increas
0011 danger, seems probable. There are many laws REGULATING variation, some few of which can be dimly s
0035 rent Arab stock, so that the latter, by the REGULATIONS for the Goodwood Races, are favoured in th
0017 the small power of endurance of warmth by the REIN deer, or of cold by the common camel, prevente
0167 to admit this view is, as it seems to me, to REJECT a real for an unreal, or at least for an unkn
0310 iven in this volume, will undoubtedly at once REJECT my theory. For my part, following out Lyell's
0342 nature of the geological record, will rightly REJECT my whole theory. For he may ask in vain where
0387 ter fish eat some kinds of seeds, though they REJECT many other kinds after having swallowed them:
0387 getting a hearty meal of fish, would probably REJECT from its stomach a pellet containing the seed
0433 one group with a distinct group, we summarily REJECT analogical or adaptive characters, and yet us
0482 n of a certain number of facts will certainly REJECT my theory. A few naturalists, endowed with mu
0482 as a vera causa in one case, they arbitrarily REJECT it in another, without assigning any distinct
0039 eed do now, arise amongst pigeons, which are REJECTED as faults or deviations from the standard of
0042 quick rate, and inferior birds may be freely REJECTED, as when killed they serve for food. On the
0149 y natural selection should have preserved or REJECTED each little deviation of form less carefully
0232 recurring to a shape which they had at first REJECTED. When bees have a place on which they can st
0362 irds after an interval of many hours, either REJECTED the seeds in pellets or passed them in their
0387 eeds retain their power of germination, when REJECTED in pellets or in excrement, many hours after
0480 st eminent living naturalists and geologists REJECTED this view of the mutability of species? It c
0084 world, every variation, even the slightest! REJECTING that which is bad, preserving and adding up
0469 its of each creature, favouring the good and REJECTING the bad? I can see no limit to this power,
0485 ormerly thus connected. Hence, without quite REJECTING the consideration of the present existence
0081 reservation of favourable variations and the REJECTION of injurious variations, I call Natural Sel
0154 ired manner and degree, and by the continued REJECTION of those tending to revert to a former and
0342 of life by the finest graduated steps. He who REJECTS these views on the nature of the geological
0352 n a single region captivates the mind. He who REJECTS it, rejects the vera causa of ordinary gener
0352 egion captivates the mind. He who rejects it, REJECTS the vera causa of ordinary generation with s
0275 s, or additional fingers and toes: and do not RELATE to characters which have been slowly acquired
0323 sily correlated. These observations; however, RELATE to the marine inhabitants of distant parts of
0466 ial notice that the more important objections RELATE to questions on which we are confessedly igno
0026 his hypothesis, if applied to species closely RELATED together, though it is unsupported by a sing
0044 rieties in being very closely, but unequally, RELATED to each other, and in having restricted rang
0055 the smaller genera: for wherever many closely RELATED species (i.e. species of the same genus) hav
0057 moreover, the species of the large genera are RELATED to each other, in the same manner as the var
0057 anner as the varieties of any one species are RELATED to each other. No naturalist pretends that a
0057 are varieties but groups of forms, unequally RELATED to each other, and clustered round certain f
0077 that the structure of every organic being is RELATED, in the most essential yet often hidden mann
0080 y natural Selection. Divergence of Character, RELATED to the diverstiy of inhabitants of any small
0110 species, and species of the same genus or of RELATED genera, which, from having nearly the same s
0121 ere between those forms which are most nearly RELATED to each other in habits, constitution, and s
0122 case in nature: species (A) being more nearly RELATED to B, C, and D, than to the other species: a
0122 descent, so as to have become adapted to many RELATED places in the natural economy of their count
0122 f the original species which were most nearly RELATED to their parents. Hence very few of the orig
0122 , of the two species which were least closely RELATED to the other nine original species, has tran
0123 he three marked a14, c14, p14, will be nearly RELATED from having recently branched off from a10;
0123 and lastly, o14, e14, and m14, will be nearly RELATED one to the other, but from having diverged a
0128 lants throughout all time and space should be RELATED to each other in group subordinate to group,
0128 y, varieties of the same species most closely RELATED together, species of the same genus less clo
0128 of the same genus less closely and unequally RELATED together, forming sections and sub genera, s
0128 species of distinct genera much less closely RELATED, and genera related in different degrees, fo
0128 genera much less closely related, and genera RELATED in different degrees, forming sub families,
0139 s of the Old and New Worlds should be closely RELATED, we might expect from the well known relatio
0159 ike manner, but to three separate yet closely RELATED acts of creation. With pigeons, however, we
0175 rganic being is either directly or indirectly RELATED in the most important manner to other organi
0193 all electric fishes would have been specially RELATED to each other. Nor does geology at all lead
0225 ween the hive and humble bee, but more nearly RELATED to the latter: it forms a nearly regular wax
0258 has observed it even between forms so closely RELATED (as Matthiola annua and glabra) that many bo
0259 s cross. That the fertility of hybrids is not RELATED to the degree in which they resemble in exte
0260 different degrees of sterility, not strictly RELATED to the facility of the first union between t
0267 common but unknown bond, which is essentially RELATED to the principle of life. Fertility of varie
```

Page **(Key Word)**

Page ***(Key Word)***

```
0274  re especially in hybrids produced from nearly  RELATED  species, follows according to Gartner the s
0301  links, some more closely, some more distantly  RELATED  to each other; and these links, let them be
0313  e. yet if we compare any but the most closely  RELATED  formations, all the species will be found t
0324  uctions of the United States are more closely  RELATED  to those which lived in Europe during certa
0335  o consecutive formations are far more closely  RELATED  to each other, than are the fossils from tw
0336  ugh ranked as distinct species, being closely  RELATED, is obvious. As the accumulation of each fo
0339  il mammals, buried there in such numbers, are  RELATED  to South American types. This relationship
0340  that Northern India was formerly more closely  RELATED  in its mammals to Africa than it is at the
0344  ancient a form is, the more nearly it will be  RELATED  to, and consequently resemble, the common p
0347  they will be found incomparably more closely  RELATED  to each other, than they are to the product
0347  any kind, or obstacles to free migration, are  RELATED  in a close and important manner to the diff
0349  of beings, specifically distinct, yet clearly  RELATED, replace each other. He hears from closely
0353  genera in which the species are most closely  RELATED  to each other, are generally local, or conf
0354  of which most of its inhabitants are closely  RELATED  to, or belong to the same genera with the sp
0355  on, why the inhabitants of a region should be  RELATED  to those of another region, whence it has b
0355  ants, though modified, would still be plainly  RELATED  by inheritance to the inhabitants of the cor
0367  ts of each mountain range are more especially  RELATED  to the arctic forms living due north or near
0371  g the later tertiary stages were more closely  RELATED  to each other than they are at the present t
0376  re much oftener specifically distinct, though  RELATED  to each other in a most remarkable manner. T
0376  certainly identical, and many, though closely  RELATED  to northern forms, must be ranked as distinc
0379  many of these wanderers, though still plainly  RELATED  by inheritance to their brethren of the nort
0398  inhabitants of the Cape de Verde Islands are  RELATED  to those of Africa, like those of the Galapa
0399  e that the endemic productions of islands are  RELATED  to those of the nearest continent, or of oth
0399  tanding nearer to Africa than to America, are  RELATED, and that very closely, as we know from Dr.
0399  nd in its endemic plants is much more closely  RELATED  to Australia, the nearest mainland, than to
0399  ht have been expected; but it is also plainly  RELATED  to South America, which, although the next n
0400  in a quite marvellous manner, by very closely  RELATED  species: so that the inhabitants of each sep
0400  separate island, though mostly distinct, are  RELATED  in an incomparably closer degree to each oth
0403  hen not identically the same, yet are plainly  RELATED  to the inhabitants of that region whence col
0403  he world during the recent Glacial epoch, are  RELATED  to those of the surrounding lowlands; thus w
0406  pine, lacustrine, and marsh productions being  RELATED  (with the exceptions before specified) to th
0409  inct on the several islets, should be closely  RELATED  to each other, and likewise be related, but
0409  losely related to each other, and likewise be  RELATED, but less closely, to those of the nearest c
0410  ration; and the more nearly any two forms are  RELATED  in blood, the nearer they will generally sta
0414  ve organs being those which are most remotely  RELATED  to the habits and food of an animal, I have
0420  nts from a single species, are represented as  RELATED  in blood or descent to the same degree; they
0421  of the differences between organic beings all  RELATED  to each other in the same degree in blood, h
0430  , of all Rodents, the bizcacha is most nearly  RELATED  to Marsupials; but in the points in which it
0430  dents; and therefore it will not be specially  RELATED  to any one existing Marsupial, but indirectl
0431  and the several species will consequently be  RELATED  to each other by circuitous lines of affinit
0440  of the arteries near the branchial slits are  RELATED  to similar conditions, in the young mammal w
0440  an, wing of a bat, and fin of a porpoise, are  RELATED  to similar conditions of life. No one will s
0440  bird, are of any use to these animals, or are  RELATED  to the conditions to which they are exposed.
0442  the structure of the embryo not being closely  RELATED  to its conditions of existence, except when
0449  nts, and are therefore in that degree closely  RELATED. Thus, community in embryonic structure reve
0465  formations invariably being much more closely  RELATED  to each other, than are the fossils from for
0477  habitants within each great class are plainly  RELATED: for they will generally be descendants of t
0478  erality that the inhabitants of each area are  RELATED  to the inhabitants of the nearest source whe
0478  ndez, and of the other American islands being  RELATED  in the most striking manner to the plants an
0324  it must not be supposed that this expression  RELATES  to the same thousandth or hundred thousandth
0310  is history we possess the last volume alone,  RELATING  only to two or three countries. Of this volu
0312  now see whether the several facts and rules  RELATING  to the geological succession of organic bein
0054  ions inhabited by them, and has little or no  RELATION  to the size of the genera to which the speci
0055  the scale: and here again there is no close  RELATION  to the size of the genera. The cause of lowl
0060  evere between species of the same genus. The  RELATION  of organism to organism the most important o
0061  m of Natural Selection, in order to mark its  RELATION  to man's power of selection. We have seen th
0071  ted me. In Staffordshire, on the estate of a  RELATION  where I had ample means of investigation, th
0077  ed and fringed legs of the water beetle, the  RELATION  seems at first confined to the elements of a
0077  plumed seeds no doubt stands in the closest  RELATION  to the land being already thickly clothed by
0077  ants seems at first sight to have no sort of  RELATION  to other plants. But from the strong growth
0084  at the improvement of each organic being in  RELATION  to its organic and inorganic conditions of l
0087  on will modify the structure of the young in  RELATION  to the parent, and of the parent in relation
0087  on to the parent, and of the parent in  RELATION  to the young. In social animals it will adap
0087  functional relations to the other sex, or in  RELATION  to wholly different habits of life in the tw
0092  ich had their stamens and pistils placed, in  RELATION  to the size and habits of the particular ins
0104  es throughout the area in the same manner in  RELATION  to the same conditions. Intercrosses, also,
0144  y reason. What can be more singular than the  RELATION  between blue eyes and deafness in cats, and
0144  ture colour of their plumage; or, again, the  RELATION  between the hair and teeth in the naked Turk
0157  hese cases are of a very unusual nature, the  RELATION  can hardly be accidental. The same number of
0157  e in the two sexes of the same species. This  RELATION  has a clear meaning on my view of the subjec
0199  n nature, many structures now have no direct  RELATION  to the habits of life of each species. Thus,
0225  wn that the form of the cell stands in close  RELATION  to the presence of adjoining cells: and the
0237  n the horns of different breeds of cattle in  RELATION  to an artificially imperfect state of the ma
0266  development; or periodical action; or mutual  RELATION  of the different parts and organs one to ano
0325  f organic beings, and find how slight is the  RELATION  between the physical conditions of various c
0332  ecies might go on being slightly modified in  RELATION  to its slightly altered conditions of life,
0339  by them. Other cases could be added, as the  RELATION  between the extinct and living land shells a
0340  sent time. Analogous facts could be given in  RELATION  to the distribution of marine animals. On th
0350  ion, in their mutual struggles for life; the  RELATION  of organism to organism being, as I have alr
0355  ry of independent creation. This view of the  RELATION  of species in one region to those in another
0358  e sides of almost every continent, the close  RELATION  of the tertiary inhabitants of several lands
0358  eir present inhabitants, a certain degree of  RELATION  (as we shall hereafter see) between the dist
0383  trachians and of terrestrial Mammals. On the  RELATION  of the inhabitants of islands to those of th
0395  ough natural selection in their new homes in  RELATION  to their new position, and we can understand
0395  esies the absence of terrestrial mammals in  RELATION  to the remoteness of islands from continents
0395  of islands from continents, there is also a  RELATION, to a certain extent independent of distance
```

Page ***(Key Word)***

relation

```
395 of the world; but as far as I have gone, the  RELATION  generally holds good. We see Britain separat
396 per channels, we can understand the frequent   RELATION  between the depth of the sea and the degree
396 of a neighbouring continent, an inexplicable   RELATION  on the view of independent acts of creation.
396 ordance with the paramount importance of the   RELATION  of organism to organism. I do not deny that
404 e steps in the process of modification. This   RELATION  between the power and extent of migration of
406 re they may have become slightly modified in   RELATION  to their new conditions. There is, also, som
406 se stations are so different, the very close   RELATION  of the distinct species which inhabit the is
406 ame archipelago, and especially the striking   RELATION  of the inhabitants of each whole archipelago
409 r should be endemic or peculiar; and why, in   RELATION  to the means of migration, one group of bein
409 or bats. We can see why there should be some   RELATION  between the presence of mammals, in a more o
415 they may be for the welfare of the being in    RELATION  to the outer world. Perhaps from this cause
420 within each class, in due subordination and    RELATION  to the other groups, must be strictly geneal
425 re the least likely to have been modified in   RELATION  to the conditions of life to which each spec
439 ss resemble each other, often have no direct   RELATION  to their conditions of existence. We cannot,
440 we have no more reason to believe in such a    RELATION, than we have to believe that the same bones
457 yos, which have become specially modified in   RELATION  to their habits of life, through the princip
472 apts the inhabitants of each country only in   RELATION  to the degree of perfection of their associa
477 lately separated from each other! for as the   RELATION  of organism to organism is the most importan
487 the various inhabitants of that continent in   RELATION  to their apparent means of immigration, some
487 ered physical conditions, namely, the mutual   RELATION  of organism to organism, the improvement of
001 tants of South America and in the geological   RELATIONS  of the present to the past inhabitants of t
003 es of organic beings, on their embryological   RELATIONS, their geographical distribution, geologica
003 for the structure of this parasite, with its  RELATIONS  to several distinct organic beings, by the
006 r profound ignorance in regard to the mutual   RELATIONS  of all the beings which live around us. Who
006 es has a narrow range and is rare? Yet these   RELATIONS  are of the highest importance, for they det
006 s world. Still less do we know of the mutual   RELATIONS  of the innumerable inhabitants of the world
053 ight be obtained in regard to the nature and   RELATIONS  of the species which vary most, by tabulati
056 orming; and this holds good. There are other   RELATIONS  between the species of large genera and the
060 tion from the number of individuals. Complex   RELATIONS  of all animals and plants throughtout natur
060 ganism to organism the most important of all   RELATIONS. Before entering on the subject of this cha
061 al of any species, in its infinitely complex  RELATIONS  to other organic beings and to external nat
071 ow complex and unexpected are the checks and   RELATIONS  between organic beings, which have to struc
073 have ended with them. Not that in nature the   RELATIONS  can ever be as simple as this. Battle withi
073 ture, are bound together by a web of complex   RELATIONS. I shall hereafter have occasion to show th
078 l convince us of our ignorance on the mutual   RELATIONS  of all organic beings; a conviction as nece
080 ely complex and close fitting are the mutual   RELATIONS  of all organic beings to each other and to
081 e, and this also would seriously disturb the   RELATIONS  of some of the former inhabitants. Let it b
087 be able to modify one sex in its functional   RELATIONS  to the other sex, or in relation to wholly
108 bitants becoming slowly modified; the mutual   RELATIONS  of many of the other inhabitants being thus
119 ; and this will depend on infinitely complex  RELATIONS. But as a general rule, the more diversifie
127 , considering the infinite complexity of the  RELATIONS  of all organic beings to each other and to
267 ring several generations between the nearest  RELATIONS, especially if these be kept under the same
314 e and lower productions, by the more complex  RELATIONS  of the higher beings to their organic and i
314 ition, and on that of the many all important  RELATIONS  of organism to organism, that any form whic
333 ied descendants; or between their collateral  RELATIONS. In nature the case will be far more compli
350 dy often remarked, the most important of all  RELATIONS. Thus the high importance of barriers comes
351 nto play only by bringing organisms into new  RELATIONS  with each other, and in a lesser degree wit
368 n a body together; consequently their mutual  RELATIONS  will not have been much disturbed, and, in
369 different climatal influences. Their mutual   RELATIONS  will thus have been in some degree disturbe
390 facility and in a body, so that their mutual  RELATIONS  have not been much disturbed. Thus in the G
392 lants have beautifully hooked seeds; yet few  RELATIONS  are more striking than the adaptation of ho
406 's, the more widely it is apt to range. The   RELATIONS  just discussed, namely, low and slowly chan
408 e continent. Bearing in mind that the mutual  RELATIONS  of organism to organism are of the highest
410 ions to the rule. On my theory these several  RELATIONS  throughout time and space are intelligible;
430 oints in which it approaches this order, its  RELATIONS  are general, and not to any one marsupial s
469 l to beings, under their excessively complex  RELATIONS  of life, would be preserved, accumulated, a
469 fully adapting each form to the most complex  RELATIONS  of life. The theory of natural selection, e
470 s they resemble varieties. These are strange  RELATIONS  on the view of each species having been ind
477 ism to organism is the most important of all  RELATIONS, and as the two areas will have received co
139 lated, we might expect from the well known    RELATIONSHIP  of most of their other productions. Far fr
184 flight, told me plainly of its close blood    RELATIONSHIP  to our common species; yet it is a woodpec
297 gradations, we should not recognise their     RELATIONSHIP, and should consequently be compelled to r
339 hat continent. In South America, a similar    RELATIONSHIP  is manifest, even to an uneducated eye, in
339 are related to South American types. This     RELATIONSHIP  is even more clearly seen in the wonderful
339 the succession of types, on this wonderful    RELATIONSHIP  in the same continent between the dead and
371 states. On this view we can understand the    RELATIONSHIP, with very little identity, between the pr
371 productions of North America and Europe, a    RELATIONSHIP  which is most remarkable, considering the
372 sphere of equatorial ocean. These cases of    RELATIONSHIP, without identity, of the inhabitants of s
427 l, will rather tend to conceal their blood    RELATIONSHIP  to their proper lines of descent. We can a
431 sors. As it is difficult to show the blood    RELATIONSHIP  between the numerous kindred of any ancien
456 hroughout all time; that the nature of the    RELATIONSHIP, by which all living and extinct beings ar
485 the terms used by naturalists of affinity,   RELATIONSHIP, community of type, paternity, morphology,
022 the number of the ribs, together with their   RELATIVE  breadth and the presence of processes. The s
022 variable; so is the degree of divergence and  RELATIVE  size of the two arms of the furcula. The pro
022 of the primary wing and caudal feathers; the  RELATIVE  length of wing and tail to each other and to
022 and tail to each other and to the body; the   RELATIVE  length of leg and of the feet; the number of
108 tinction of the less improved forms, and the  RELATIVE  proportional numbers of the various inhabita
151 e it particularly difficult to compare their  RELATIVE  degrees of variability. When we see any part
198 we are far too ignorant to speculate on the   RELATIVE  importance of the several known and unknown
232 bees, all instinctively standing at the same  RELATIVE  distance from each other, all trying to swee
232 s should be enabled to stand at their proper  RELATIVE  distances from each other and from the walls
233 the same time, always standing at the proper  RELATIVE  distance from the parts of the cells just be
233 and then to five other points, at the proper  RELATIVE  distances from the central point and from ea
358 mitted by his many followers. The nature and  RELATIVE  proportions of the inhabitants of oceanic is
370 inents have long remained in nearly the same  RELATIVE  position, though subjected to large, but par
434 d should include the same bones, in the same  RELATIVE  positions? Geoffroy St. Hilaire has insisted
```

Page ************************************(Key Word)************************************

0434 insisted strongly on the high importance of RELATIVE connexion in homologous organs: the parts ma
0435 dency to alter the framework of bones or the RELATIVE connexion of the several parts. If we suppos
0436 us with, that is correspond in number and in RELATIVE connexion with, the elemental parts of a cer
0436 ar to almost every one, that in a flower the RELATIVE position of the sepals, petals, stamens, and
0455 e embryo. Thus we can understand the greater RELATIVE size of rudimentary organs in the embryo, an
0455 ntary organs in the embryo, and their lesser RELATIVE size in the adult. But if each step of the p
0038 s being so great in external characters and RELATIVELY so slight in internal parts or organs. Man
0070 e stock of individuals of the same species, RELATIVELY to the numbers of its enemies, is absolutel
0453 udimentary part or organ is of greater size RELATIVELY to the adjoining parts in the embryo; than
0135 in his own collection, and not one had even a RELIC left. In the Onites apelles the tarsi are so
0017 m one or several species. The argument mainly RELIED on by those who believe in the multiple origi
0484 d i speak after experience, will be no slight RELIEF. The endless disputes whether or not some fif
0006 although much remains obscure, and will long REMAIN obscure, I can entertain no doubt, after the
0018 ed in Egypt? The whole subject must, I think, REMAIN vague; nevertheless, I may, without here ente
0103 e area, varieties of the same animal can long REMAIN distinct, from haunting different stations, f
0104 gst them, as long as their conditions of life REMAIN the same, only through the principle of inher
0149 sified work, we can perhaps see why it should REMAIN variable, that is, why natural selection shou
0155 able, though its physiological importance may REMAIN the same. Something of the same kind applies
0208 s of the body. When once acquired, they often REMAIN constant throughout life. Several other point
0234 n of their combs. Moreover, many bees have to REMAIN idle for many days during the process of secr
0291 iment nearly balance each other, the sea will REMAIN shallow and favourable for life, and thus a f
0306 mals might be multiplied; and here they would REMAIN confined, until some of the species became ad
0308 on and metamorphism. The case at present must REMAIN inexplicable; and may be truly urged as a val
0332 luding five genera, and o14 to m14) would yet REMAIN distinct. These two families, however, would
0363 that several other means, which without doubt REMAIN to be discovered; have been in action year af
0364 become mingled in any great degree; but would REMAIN as distinct as we now see them to be. The cur
0380 intertropical regions. Very many difficulties REMAIN to be solved. I do not pretend to indicate th
0381 own homes. I have said that many difficulties REMAIN to be solved: some of the most remarkable are
0385 ich I have observed, and no doubt many others REMAIN to be observed; throw some light on this subj
0387 s plant, I thought that its distribution must REMAIN quite inexplicable; but Audubon states that h
0434 extent in form and size, and yet they always REMAIN connected together in the same order. We neve
0447 responding mature age. Whereas the young will REMAIN unmodified, or be modified in a lesser degree
0450 rd not extending far enough back in time, may REMAIN for a long period, or for ever, incapable of
0451 for one, ever the more important purpose; and REMAIN perfectly efficient for the other. Thus in pl
0466 ng periods. As long as the conditions of life REMAIN the same, we have reason to believe that a mo
0488 of species, however, keeping in a body might REMAIN for a long period unchanged, whilst within th
0153 other parts of the organisation, which have REMAINED for a much longer period nearly constant. An
0156 ganisation which have for a very long period REMAINED constant. In connexion with the present subj
0266 hat of hybrids, the external conditions have REMAINED the same, but the organisation has been dist
0281 ut such a case would imply that one form had REMAINED for a very long period unaltered, whilst its
0292 he other hand, as long as the bed of the sea REMAINED stationary, thick deposits could not have be
0309 we any right to assume that things have thus REMAINED from eternity? Our continents seem to to hav
0309 morphic action than strata which have always REMAINED nearer to the surface. The immense areas in
0314 pe, whereas the marine shells and birds have REMAINED unaltered. We can perhaps understand the app
0336 sical conditions of the ancient areas having REMAINED nearly the same. Let it be remembered that t
0365 t several distinct points; and we might have REMAINED in this same belief, had not Agassiz and oth
0370 re alluded to, that our continents have long REMAINED in nearly the same relative position, though
0381 rise to new groups of forms, and others have REMAINED unaltered. We cannot hope to explain such fa
0463 a wide area, which has during a long period REMAINED continuous, and of which the climate and oth
0006 marks. No one ought to feel surprise at much REMAINING as yet unexplained in regard to the origin
0283 ermined, huge fragments fall down, and these REMAINING fixed, have to be worn away, atom by atom,
0401 continents some species spreading widely and REMAINING the same. The really surprising fact in thi
0006 ological epochs in its history. Although much REMAINS obscure, and will long remain obscure, I can
0073 e so nicely balanced, that the face of nature REMAINS uniform for long periods of time, though ass
0124 strata of the earth's crust including extinct REMAINS. We shall, when we come to our chapter on Ge
0137 some of the crabs the foot stalk for the eye REMAINS, though the eye is gone; the stand for the t
0173 ting profound depths of the sea, and to their REMAINS being embedded and preserved to a future age
0179 existence could be found only amongst fossil REMAINS, which are preserved, as we shall in a futur
0282 time. Independently of our not finding fossil REMAINS of such infinitely numerous connecting links
0288 fficiently quick to embed and preserve fossil REMAINS. Throughout an enormously large proportion o
0288 lying for ages in an unaltered condition. The REMAINS which do become embedded, if in sand or grav
0289 fluous to state that our evidence from fossil REMAINS is fragmentary in an extreme degree. For ins
0289 ta of North America. In regard to mammiferous REMAINS, a single glance at the historical table pub
0290 extensive formations with recent or tertiary REMAINS can anywhere be found, though the supply of
0292 the sea shallow and to embed and preserve the REMAINS before they had time to decay. On the other
0294 world, sedimentary deposits, including fossil REMAINS, have gone on accumulating within the same a
0294 cial period shall have been upraised, organic REMAINS will probably first appear and disappear at
0295 thick deposits are usually barren of organic REMAINS, except near their upper or lower limits. It
0300 rate to protect organic bodies from decay, no REMAINS could be preserved. In our archipelago, I be
0307 ead, are convinced that we see in the organic REMAINS of the lowest Silurian stratum the dawn of l
0308 ng hypothesis. From the nature of the organic REMAINS, which do not appear to have inhabited profo
0318 plata the tooth of a horse embedded with the REMAINS of Mastodon, Megatherium, Toxodon, and other
0323 dia. For at these distant points, the organic REMAINS in certain beds present an unmistakeable deg
0324 merica, and Australia, from containing fossil REMAINS in some degree allied, and from not includin
0327 quickly enough to embed and preserve organic REMAINS. During these long and blank intervals I sup
0335 onnected with the statement, that the organic REMAINS from an intermediate formation are in some d
0335 tance, the general resemblance of the organic REMAINS from the several stages of the chalk formati
0336 scent; the full meaning of the fact of fossil REMAINS from closely consecutive formations, though
0343 e groups of species. He may ask where are the REMAINS of those infinitely numerous organisms which
0345 ent forms. We can clearly see why the organic REMAINS of closely consecutive formations are more c
0345 her by generation: we can clearly see why the REMAINS of an intermediate formation are intermediat
0396 d as halting places, of which not a wreck now REMAINS, must not be overlooked. I will here give a
0421 . nevertheless their genealogical arrangement REMAINS strictly true, not only at the present time,
0443 welfare of a very young animal, as long as it REMAINS in its mother's womb, or in the egg, or as l
0452 t is not crowned with a stigma; but the style REMAINS well developed and is clothed with hairs as
0463 inks? Why does not every collection of fossil REMAINS afford plain evidence of the gradation and m

Page ************************************(Key Word)************************************

remains

```
0463  beneath the Silurian system, stored with the REMAINS of the progenitors of the Silurian groups of
0465  ted manner. We clearly see this in the fossil REMAINS from consecutive formations invariably being
0475  throughout the world. The fact of the fossil REMAINS of each formation being in some degree inter
0483  ple, in groups sub ordinate to groups. Fossil REMAINS sometimes tend to fill up very wide interval
0487  ord. The crust of the earth with its embedded REMAINS must not be looked at as a well filled museu
0381  ory of descent with modification, a far more REMARKABLE case of difficulty. For some of these speci
0118  urteen thousandth generation. But I must here REMARK that I do not suppose that the process ever g
0131  n would be necessary to show the truth of the REMARK, that the reproductive system is eminently su
0149  be a sign of low organisation; the foregoing REMARK seems connected with the very general opinion
0150  e. several years ago I was much struck with a REMARK, nearly to the above effect, published by Mr.
0151  t i particularly attended to Mr. Waterhouse's REMARK, whilst investigating this order, and I am fu
0187  n how life itself first originated; but I may REMARK that several facts make me suspect that any s
0194  nal grade is known to lead. The truth of this REMARK is indeed shown by that old canon in natural
0283  ccumulated. Let him remember Lyell's profound REMARK, that the thickness and extent of sedimentary
0287  very imperfect, is admitted by every one. The REMARK of that admirable palaeontologist, the late E
0292  endently arrived at a similar conclusion. One REMARK is here worth a passing notice. During period
0303  r as if suddenly created. I may here recall a REMARK formerly made, namely that it might require a
0305  this same period. It is almost superfluous to REMARK that hardly any fossil fish are known from so
0327  ywhere tend to correspond. There is one other REMARK connected with this subject worth making. I h
0330  ps now widely separated from each other. This REMARK no doubt must be restricted to those groups w
0331  we must admit that there is some truth in the REMARK. Let us see how far these several facts and i
0376  lar cases occur; as an example, I may quote a REMARK by the highest authority, Prof. Dana, that it
0008  e believe in this; but the chief one is the REMARKABLE effect which confinement or cultivation ca
0012  al peculiarities go together, of which many REMARKABLE cases could be given amongst animals and pl
0021  ier, more especially the male bird, is also REMARKABLE from the wonderful development of the carun
0022  ramus of the lower jaw, varies in a highly REMARKABLE manner. The number of the caudal and sacral
0023  d this, considering their size, habits, and REMARKABLE characters, seems very improbable; or they
0029  carrier and tumbler pigeon. One of the most REMARKABLE features in our domesticated races is that
0071  n of the planted part of the heath was most REMARKABLE, more than is generally seen in passing fro
0100  h fertilises itself. We can understand this REMARKABLE fact, which offers so strong a contrast wit
0135  selection. Mr. Wollaston has discovered the REMARKABLE fact that 200 beetles, out of the 550 speci
0144  the shape of the pelvis in birds causes the REMARKABLE diversity in the shape of their kidneys. Ot
0150  cies of bat had its wings developed in some REMARKABLE manner in comparison with the other species
0150  d females; but as females more rarely offer REMARKABLE secondary sexual character, it applies more
0151  in my future work, give a list of the more REMARKABLE cases; I will here only briefly give one, a
0151  hen we see any part or organ developed in a REMARKABLE degree or manner in any species, the fair p
0157  n other parts of their structure. It is a REMARKABLE fact, that the secondary sexual differences
0163  ses certainly do occur, and seem to me very REMARKABLE. I will, however, give one curious and comp
0165  re cuagga. Lastly, and this is another most REMARKABLE case, a hybrid has been figured by Dr. Gray
0175  ending mountains, and sometimes it is quite REMARKABLE how abruptly, as Alph. de Candolle has obse
0212  ly unknown to us; Audubon has given several REMARKABLE cases of differences in nests of the same s
0215  l instincts are lost under domestication: a REMARKABLE instance of this is seen in those breeds of
0218  s of bees of other kinds. This case is more REMARKABLE than that of the cuckoo; for these bees hav
0219  s exterminated. Slave making instinct. This REMARKABLE instinct was first discovered in the Formic
0249  . i am strengthened in this conviction by a REMARKABLE statement repeatedly made by Gartner, namel
0258  s rank them only as varieties. It is also a REMARKABLE fact, that hybrids raised from reciprocal c
0259  gartner: for instance, some species have a REMARKABLE power of crossing with other species; other
0259  ies; other species of the same genus have a REMARKABLE power of impressing their likeness on their
0268  tance of the pigeon or of the cabbage, is a REMARKABLE fact; more especially when we reflect how m
0268  er the fertility of domestic varieties less REMARKABLE than at first appears. It can, in the first
0270  s varieties. The following case is far more REMARKABLE, and seems at first quite incredible; but i
0271  y every subsequent observer, has proved the REMARKABLE fact, that one variety of the common tobacc
0305  rmation, the fact would certainly be highly REMARKABLE; but I cannot see that it would be an insup
0339  f the Aralo Caspian Sea. Now what does this REMARKABLE law of the succession of the same types wit
0352  y migration, the fact is given as something REMARKABLE and exceptional. The capacity of migrating
0365  rated from one to the other. It is indeed a REMARKABLE fact to see so many of the same plants livi
0365  orthern parts of Europe: but it is far more REMARKABLE, that the plants on the White Mountains, in
0366  he inhabitants of Europe, as explained with REMARKABLE clearness by Edward Forbes, is substantiall
0371  ca and Europe, a relationship which is most REMARKABLE, considering the distance of the two areas,
0376  nct, though related to each other in a most REMARKABLE manner. This brief abstract applies to plan
0379  d varieties or as distinct species. It is a REMARKABLE fact, strongly insisted on by Hooker in reg
0381  lties remain to be solved: some of the most REMARKABLE are stated with admirable clearness by Dr.
0383  mous range, but allied species prevail in a REMARKABLE manner throughout the world. I well remembe
0392  rtant as the nature of the conditions. Many REMARKABLE little facts could be given with respect to
0399  and of the Cape of Good Hope, is a far more REMARKABLE case, and is at present inexplicable: but t
0402  and Mr. Wollaston have communicated to me a REMARKABLE fact bearing on this subject: namely, that
0414  an essential character. So with plants, how REMARKABLE it is that the organs of vegetation, on whi
0446  be detected in the young. But there was one REMARKABLE exception to this rule, for the young of th
0460  species when first crossed, which forms so REMARKABLE a contrast with the almost universal fertil
0022  the eggs vary. The manner of flight differs REMARKABLY; as does in some breeds the voice and dispo
0151  as birds within the same country vary in a REMARKABLY small degree, I have particularly attended
0208  th habit. This comparison gives, I think, a REMARKABLY accurate notion of the frame of mind under
0256  ous hybrid offspring, yet these hybrids are REMARKABLY sterile. On the other hand, there are speci
0367  to have everywhere travelled southward, are REMARKABLY uniform round the world. We may suppose tha
0010  oth the young and the parents, as Muller has REMARKED, have apparently been exposed to exactly the
0015  y perceive in each domestic race, as already REMARKED by some authors, namely, that the varieties
0038  this, perhaps, partly explains what has been REMARKED with respect to the sheep of parts of Yorks
0041  e to success. On this principle Marshall has REMARKED, that it was most fortunate that the strawbe
0057  anked as species or varieties. Now Fries has REMARKED in regard to plants, and Westwood in regard
0057  ections, or lesser groups. As Fries has well REMARKED, little groups of species are generally clus
0115  ry different; and Alph. de Candolle has well REMARKED in his great and admirable work, that floras
0116  resenting, as Mr. Waterhouse and others have REMARKED, our carnivorous, ruminant, and rodent mamma
0128  untries. Natural selection, as has just been REMARKED, leads to divergence of character and to muc
0131  n exposed during several generations. I have REMARKED in the first chapter, but a long catalogue o
0134  e conditions of life. Indirectly, as already REMARKED, they seem to play an important part in affe
```

Page **************************************(Key Word)***************************************

0134 the effects of disuse. As Professor Owen has REMARKED, there is no greater anomaly in nature than
0135 l they became incapable of flight. Kirby has REMARKED (and I have observed the same fact) that the
0138 n expected; but, as Schiodte and others have REMARKED, this is not the case, and the cave insects
0139 als should be very anomalous, as Agassiz has REMARKED in regard to the blind fish, the Amblyopsis,
0143 ral selection. Homologous parts, as has been REMARKED by some authors, tend to cohere; this is of
0144 re. M. Is. Geoffroy St. Hilaire has forcibly REMARKED, that certain malconformations very frequent
0146 can act. For instance, Alph. de Candolle has REMARKED that winged seeds are never found in fruits
0149 me adjoining part. It seems to be a rule, as REMARKED by Is. Geoffroy St. Hilaire, both in variet
0149 same author and some botanists have further REMARKED that multiple parts are also very liable to
0155 on natural history, that when an author has REMARKED with surprise that some important organ or p
0157 beetles, but in the Engidae, as Westwood has REMARKED, the number varies greatly; and the number l
0160 the lost character might be, as was formerly REMARKED, for all that we can see to the contrary, tr
0176 existing in lesser numbers would, as already REMARKED, run a greater chance of being exterminated
0179 ction constantly tends, as has been so often REMARKED, to exterminate the parent forms and the in
0180 d, and from others, as Sir J. Richardson has REMARKED, with the posterior part of their bodies rat
0192 been produced; but, as Owen and others have REMARKED, their intimate structure closely resembles
0201 each. No organ will be formed, as Paley has REMARKED, for the purpose of causing pain or for doir
0224 cious wax in their construction. It has been REMARKED that a skilful workman, with fitting tools a
0226 to a pyramid; and this pyramid, as Huber has REMARKED, is manifestly a gross imitation of the thre
0255 apply to both kingdoms. It has been already REMARKED, that the degree of fertility, both of first
0263 es are fundamentally different, for, as just REMARKED, in the union of two pure species the male a
0312 ormations are more broken; but, as Bronn has REMARKED, neither the appearance nor disappearance of
0313 the amount of organic change, as Pictet has REMARKED, does not strictly correspond with the succe
0325 different climates. We must, as Barrande has REMARKED, look to some special law. We shall see this
0329 from living forms. But, as Buckland long ago REMARKED, all fossils can be classed either in still
0332 nd recent, descended from A, make, as before REMARKED, one order; and this order, from the continu
0350 m to organism being, as I have already often REMARKED, the most important of all relations. Thus t
0351 and notoriously the case. I believe, as was REMARKED in the last chapter, in no law of necessary
0363 anic islands nearer to the mainland, and (as REMARKED by Mr. H. C. Watson) from the somewhat north
0367 alpine plants, for example, of Scotland, as REMARKED by Mr. H. C. Watson, and those of the Pyrene
0368 h. C. Watson, and those of the Pyrenees, as REMARKED by Ramond, are more especially allied to the
0371 we can further understand the singular fact REMARKED on by several observers, that the production
0376 rate zones. As Mr. H. C. Watson has recently REMARKED, In receding from polar towards equatorial l
0386 ceanic islands. This is strikingly shown, as REMARKED by Alph. de Candolle, in large groups of ter
0393 n oceanic islands, Bory St. Vincent long ago REMARKED that Batrachians (frogs, toads, newts) have
0404 n in another and more general way. Mr. Gould REMARKED to me long ago, that in those genera of bird
0415 e hymenoptera, the antennae, as Westwood has REMARKED, are most constant in structure: in another
0417 in natural history. Hence, as has often been REMARKED, a species may depart from its allies in sev
0417 flowers; in the latter, as A. de Jussieu has REMARKED, the greater number of the characters proper
0429 favourable circumstances. Mr. Waterhouse has REMARKED that, when a member belonging to one group o
0430 nd, of all Marsupials, as Mr. Waterhouse has REMARKED, the phascolomys resembles most nearly, not
0437 rdinarily shaped pieces of bone? As Owen has REMARKED, the benefit derived from the, yielding of th
0438 bly be more correct, as Professor Huxley has REMARKED, to speak of both skull and vertebrae, both
0439 ed. Embryology. It has already been casually REMARKED that certain organs in the individual, which
0442 differ widely from the adult: thus Owen has REMARKED in regard to cuttle fish, there is no metamo
0006 ation of the whole work, and a few concluding REMARKS. No one ought to feel surprise at much remai
0014 e in explaining the laws of embryology. These REMARKS are of course confined to the first appearan
0036 by mr. Buckley and Mr. Burgess, as Mr. Youatt REMARKS, have been purely bred from the original sto
0052 er have to return to this subject. From these REMARKS it will be seen that I look at the term spec
0060 f this chapter, I must make a few preliminary REMARKS, to show how the struggle for existence bear
0067 of south America. Here I will make only a few REMARKS, just to recall to the reader's mind some of
0077 importance may be deduced from the foregoing REMARKS, namely, that the structure of every organic
0100 hold in Australia; and I have made these few REMARKS on the sexes of trees simply to call attenti
0105 we turn to nature to test the truth of these REMARKS, and look at any small isolated area, such a
0138 of this gradation of habit; for, as Schiodte REMARKS, animals not far remote from ordinary forms,
0154 ed condition. The principle included in these REMARKS may be extended. It is notorious that specif
0156 e present subject, I will make only two other REMARKS. I think it will be admitted, without my ent
0168 ot differed within this same period. In these REMARKS we have referred to special parts or organs
0182 onditions. It must not be inferred from these REMARKS that any of the grades of wing structure her
0197 ve a few instances to illustrate these latter REMARKS. If green woodpeckers alone had existed, and
0199 asoning would appear frivolous. The foregoing REMARKS lead me to say a few words on the protest la
0266 e parent. Nor do I pretend that the foregoing REMARKS go to the root of the matter: no explanation
0268 there must be some error in all the foregoing REMARKS, inasmuch as varieties, however much they ma
0274 generations with either parent. These several REMARKS are apparently applicable to animals; but th
0287 f the Secondary period. I have made these few REMARKS because it is highly important for us to gai
0299 t will be worth while to sum up the foregoing REMARKS, under an imaginary illustration. The Malay
0301 then, there be some degree of truth in these REMARKS, we have no right to expect to find in our g
0303 l now give a few examples to illustrate these REMARKS; and to show how liable we are to error in s
0325 animal kingdom. M. Barrande has made forcible REMARKS to precisely the same effect. It is, indeed,
0354 f many such cases. But after some preliminary REMARKS, I will discuss a few of the most striking c
0355 to generation with modification. The previous REMARKS on single and multiple centres of creation o
0369 hey are at the present day. But the foregoing REMARKS on distribution apply not only to strictly a
0387 nelumbium, and remembered Alph. de Candolle's REMARKS on this plant, I thought that its distributi
0389 gard to insular productions. In the following REMARKS I shall not confine myself to the mere quest
0396 dependent acts of creation. All the foregoing REMARKS on the inhabitants of oceanic islands, namel
0459 n on the study of Natural history. Concluding REMARKS. As this whole volume is one long argument,
0068 nd this is a tremendous destruction, when we REMEMBER that ten per cent. is an extraordinarily sev
0085 olour would produce little effect: we should REMEMBER how essential it is in a flock of white shee
0196 ndependently of natural selection. We should REMEMBER that climate, food, etc., probably have some
0277 al and perfect fertility surprising, when we REMEMBER how liable we are to argue in a circle with
0277 varieties in a state of nature; and when we REMEMBER that the greater number of varieties have be
0283 lowly the mass has been accumulated. Let him REMEMBER Lyell's profound remark, that the thickness
0284 rs the best evidence of the lapse of time. I REMEMBER having been much struck with the evidence of
0286 along its whole indented length: and we must REMEMBER that almost all strata contain harder layers
0289 ail. Nor is their rarity surprising, when we REMEMBER how large a proportion of the bones of terti
0296 ght, changes of form. It is all important to REMEMBER that naturalists have no golden rule by whic

Page **************************************(Key Word)***************************************

```
0319  l competitor. It is most difficult always to REMEMBER that the increase of every living being is c
0321  t the close of the secondary period, we must REMEMBER what has been already said on the probable w
0377  years, must have been very long; and when we REMEMBER over what vast spaces some naturalised plant
0383  markable manner throughout the world. I well REMEMBER, when first collecting in the fresh waters o
0388  e of terrestrial colonists. We should, also, REMEMBER that some, perhaps many, fresh water product
0402  ost species would thus spread; but we should REMEMBER that the forms which become naturalised in n
0407  have occurred within the same period; if we  REMEMBER how profoundly ignorant we are with respect
0407  allowances as before for our ignorance, and  REMEMBER that some forms of life change most slowly,
0460  ir very general fertility surprising when we  REMEMBER that it is not likely that either their cons
0462  s? with respect to existing forms, we should REMEMBER that we have no right to expect (excepting i
0045  ted, during a course of years. It should be  REMEMBERED that systematists are far from pleased at f
0081  f some of the former inhabitants. Let it be  REMEMBERED how powerful the influence of a single intr
0121  and their original parent. For it should be  REMEMBERED that the competition will generally be most
0237  eory of natural selection? First, let it be  REMEMBERED that we have innumerable instances, both in
0237  d, or, as I believe, disappears, when it is  REMEMBERED that selection may be applied to the family
0336  having remained nearly the same. Let it be  REMEMBERED that the forms of life, at least those inha
0387  of that fine water lily the Nelumbium, and  REMEMBERED Alph. de Candolle's remarks on this plant,
0388  several means of distribution, it should be  REMEMBERED that when a pond or stream is first formed,
0080  enerations? If such do occur, can we doubt  (REMEMBERING that many more individuals are born than c
0285  nd to look at the distant South Downs for,  REMEMBERING that at no great distance to the west the
0262  uch fruit as when on their own roots. We are REMINDED by this latter fact of the extraordinary cas
0308  eanic island is as yet known to afford even a REMNANT of any palaeozoic or secondary formation. He
0107  that we find seven genera of Ganoid fishes,  REMNANTS of a once proponderant order: and in fresh w
0307  amorphic action, we ought to find only small REMNANTS of the formations next succeeding them in ag
0013  grandfather or grandmother or other much more REMOTE ancestor: why a peculiarity is often transmit
0073  instance showing how plants and animals, most REMOTE in the scale of nature, are bound together by
0075  te on its prey, lies generally between beings REMOTE in the scale of nature. This is often the cas
0126  extremely few will transmit descendants to a REMOTE futurity. I shall have to return to this subj
0131  ing and modified descendants, yet at the most REMOTE geological period, the earth may have been as
0131  of life, to which the parents and their more REMOTE ancestors have been exposed during several ge
0138  it: for, as Schiodte remarks, animals not far REMOTE from ordinary forms, prepare the transition f
0153  itor of the genus. This period will seldom be REMOTE in any extreme degree, as species very rarely
0153  ong continued within a period not excessively REMOTE, we might, as a general rule, expect still to
0156  generic characters have been inherited from a REMOTE period, since that period when the species fi
0173  been made only at intervals of time immensely REMOTE. But it may be urged that when several closel
0189  n must have been first formed at an extremely REMOTE period, since which all the many members of t
0193  t a dozen fishes, of which several are widely REMOTE in their affinities. Generally when the same
0193  ame in Orchis and Asclepias, genera almost as REMOTE as possible amongst flowering plants. In all
0202  of the bee, as having originally existed in a REMOTE progenitor as a boring and serrated instrumen
0235  cts almost identically the same in animals so REMOTE in the scale of nature, that we cannot accoun
0324  during the pleistocene period (an enormously REMOTE period as measured by years, including the wh
0335  to each other, than are the fossils from two REMOTE formations. Pictet gives as a well known inst
0345  osely allied to each other, than are those of REMOTE formations: for the forms are more closely li
0350  th more or less ease, at periods more or less REMOTE: on the nature and number of the former immig
0351  nearly the same character from an enormously REMOTE geological period, so certain species have mi
0365  a poorly stocked island, though standing more REMOTE from the mainland, would not receive colonist
0367  dentity of many plants at points so immensely REMOTE as on the mountains of the United States and
0374  canty flora, are common to Europe, enormously REMOTE as these two points are: anc there are many c
0381  of identical species at points so enormously REMOTE as Kerguelen Land, New Zealand, and Fuegia, I
0386  es have, both over continents and to the most REMOTE oceanic islands. This is strikingly shown, as
0386  erers, and are occasionally found on the most REMOTE and barren islands in the open ocean: they wo
0392  d be given with respect to the inhabitants of REMOTE islands. For instance, in certain islands not
0394  e force produced bats and no other mammals on REMOTE islands? On my view this question can easily
0396  nc how several of the inhabitants of the more REMOTE islands, whether still retaining the same spe
0399  the next nearest continent, is so enormously REMOTE, that the fact becomes an anomaly. But this c
0404  ent physical conditions, and the existence at REMOTE points of the world of other species allied t
0405  le parent, though now distributed to the most REMOTE points of the world, we ought to find, and I
0405  t have branched off from a common parent at a REMOTE epoch: so that in such cases there will have
0468  will often be very severe between beings most REMOTE in the scale of nature. The slightest advanta
0480  corresponding ages have been inherited from a REMOTE period to the present day. On the view of eac
0486  pecies of most genera, have within a not very REMOTE period descended from one parent, and have mi
0126  have now become extinct. Looking still more REMOTELY to the future, we may predict that, owing to
0324  er european beds. Nevertheless, looking to a REMOTELY future epoch, there can, I think, be little
0414  generative organs being those which are most REMOTELY related to the habits and food of an animal,
0395  e of terrestrial mammals in relation to the REMOTENESS of islands from continents, there is also a
0309  ns now extend, oceans have extended from the REMOTEST period of which we have any record: and on t
0047  ybrid nature of the intermediate links always REMOVE the difficulty. In very many cases, however,
0073  nts absolutely require the visits of moths to REMOVE their pollen masses and thus to fertilise the
0389  ite joined to some continent. This view would REMOVE many difficulties; but it would not, I think,
0026  the child ever reverts to some one ancestor, REMOVED by a greater number of generations. In a bre
0120  am, if all the lines proceeding from (A) were REMOVED, excepting that from a1 to a10. In the same
0210  o so voluntarily, the following facts show. I REMOVED all the ants from a group of about a dozen a
0211  bably a convenience to the aphides to have it REMOVED; and therefore probably the aphides do not i
0245  t universal, affected by close interbreeding, REMOVED by domestication. Laws governing the sterili
0254  e characteristic, but as one capable of being REMOVED by domestication. Finally, looking to all th
0264  ted, showing that when animals and plants are REMOVED from their natural conditions, they are extr
0285  ere trifle, in comparison with that which has REMOVED masses of our palaeozoic strata, in parts te
0362  sometimes adheres to them: in one instance I REMOVED twenty two grains of dry argillaceous earth
0380  far from supposing that all difficulties are REMOVED on the view here given in regard to the rang
0384  in India; and the vitality of their ova when REMOVED from the water. But I am inclined to attribu
0397  s species has a thick calcareous operculum, I REMOVED it, and when it had formed a new membranous
0482  uclice by which this subject is overwhelmed be REMOVED. Several eminent naturalists have of late pu
0357  same species to the most distant points, and REMOVES many a difficulty: but to the best of my jud
0385  g to its back; and it has happened to me, in REMOVING a little duck weed from one aquarium to anot
0267  ges, or changes of a particular nature, often RENDER organic beings in some degree sterile; and th
0268  ntercrossed. Several considerations, however, RENDER the fertility of domestic varieties less rema
```

Page **(Key Word)**

```
0281   and organism, between child and parent, will RENDER this a very rare event; for in all cases the
0444   first chapter, that there is some evidence to RENDER it probable, that at whatever age any variat
0018   years ago? But Mr. Horner's researches have RENDERED it in some degree probable that man suffici
0043   to use and disuse. The final result is thus RENDERED infinitely complex. In some cases, I do not
0046   ich consequently have not been seized on and RENDERED definite by natural selection, as hereafter
0093   f more and more attractive flowers, had been RENDERED highly attractive to insects, they would, u
0093   maginary case: as soon as the plant had been RENDERED so highly attractive to insects that pollen
0114   less distinct varieties: and varieties, when RENDERED very distinct from each other, take the ran
0119   the descendants from any one species can be RENDERED, the more places they will be enabled to se
0143   use to the breed it might probably have been RENDERED permanent by natural selection. Homologous
0148   aving of a large and complex structure, when RENDERED superfluous by the parasitic habits of the
0148   y part of the organisation, as soon as it is RENDERED superfluous, without by any means causing s
0156   ee why these characters should not have been RENDERED as constant and uniform as other parts of t
0178   th and wander much, may have separately been RENDERED sufficiently distinct to rank as representa
0184   n see no difficulty in a race of bears being RENDERED, by natural selection, more and more aquati
0187   e me suspect that any sensitive nerve may be RENDERED sensitive to light, and likewise to those c
0206   us the standard of perfection will have been RENDERED higher. Natural selection will not necessar
0224   ght by natural selection be strengthened and RENDERED permanent for the very different purpose of
0236   g or sterile ants. How the workers have been RENDERED sterile is a difficulty: but not much great
0239   e caste or of the same kind, which have been RENDERED by natural selection, as I believe to be qu
0262   y on another species, when thus grafted were RENDERED barren. On the other hand, certain species
0265   , for whole groups of animals and plants are RENDERED impotent by the same unnatural conditions;
0266   , when placed under unnatural conditions, is RENDERED sterile. All that I have attempted to show,
0273   in the conditions of life, being thus often RENDERED either impotent or at least incapable of it
0292   geological record will almost necessarily be RENDERED intermittent. I feel much confidence in the
0332   ting families on the uppermost line would be RENDERED less distinct from each other. If, for inst
0353   t geological times, must have interrupted or RENDERED discontinuous the formerly continuous range
0354   nature, that we ought to give up the belief, RENDERED probable by general considerations, that ea
0371   y united by land, serving as a bridge, since RENDERED impassable by cold, for the inter migration
0448   , the active young or larvae might easily be RENDERED by natural selection different to any conce
0452   the degree to which the same organ has been RENDERED rudimentary occasionally differs much. This
0454   ontinue slowly to reduce the organ, until it RENDERED harmless and rudimentary. Any change in fun
0454   power of natural selection; so that an organ RENDERED, during changed habits of life, useless or
0455   lone of its former functions. An organ, when RENDERED useless, may well be variable, for its varia
0461   tinually seen that organisms of all kinds are RENDERED in some degree sterile from their constitut
0466   tions of life; so that this system, when not RENDERED impotent, fails to reproduce offspring exac
0470   intermediate varieties; and thus species are RENDERED to a large extent defined and distinct obje
0474   g period; for in this case it will have been RENDERED constant by long continued natural selectio
0480   hence the organ will not be much reduced or RENDERED rudimentary at this early age. The calf, for
0464   ieties are often at first local, both causes RENDERING the discovery of intermediate links less l
0108   become well modified and perfected. When, by RENEWED elevation, the islands shall be re converte
0108   nal numbers of the various inhabitants of the RENEWED continent will again be changed; and again t
0072   and northward in a feral state; and Azara and RENGGER have shown that this is caused by the greate
0163   ge in not being able to give them. I can only REPEAT that such cases certainly do occur, and seem
0212   eeble effect on the reader's mind. I can only REPEAT my assurance, that I do not speak without goo
0320   ocally or wholly, through man's agency. I may REPEAT what I published in 1845, namely, that to adm
0149   d in species, that when any part or organ is REPEATED many times in the structure of the same ind
0238   on. And I believe that this process has been REPEATED, until that prodigious amount of difference
0253   generation, in opposition to the constantly REPEATED admonition of every breeder. And in this cas
0274   an be reduced to either pure parent form, by REPEATED crosses in successive generations with eithe
0437   we have formerly seen that parts many times REPEATED are eminently liable to vary in number and s
0438   he primordially similar elements, many times REPEATED, and have adapted them to the most diverse
0155   are more variable than generic; but I have REPEATEDLY noticed in works on natural history, that
0247   , moreover, as Gartner during several years REPEATEDLY crossed the primrose and cowslip, which we
0249   n this conviction by a remarkable statement REPEATEDLY made by Gartner, namely, that if even the
0208   incts and habits could be pointed out. As in REPEATING a well known song, so in instincts, one act
0208   if a person be interrupted in a song, or in REPEATING anything by rote, he is generally forced to
0091   it the same habits or structure, and by the REPETITION of this process, a new variety might be fo
0149   n in structure. Inasmuch as this vegetative REPETITION, to use Prof. Owen's expression, seems to
0437   sive spiral whorls of leaves. An indefinite REPETITION of the same part or organ is the common cha
0438   s, we do not find nearly so much indefinite REPETITION of any one part, as we find in the other or
0349   , specifically distinct, yet clearly related, REPLACE each other. He hears from closely allied, ye
0122   ieties will have become extinct, having been REPLACED by eight new species (a14 to m14); and (I) w
0122   species (a14 to m14); and (I) will have been REPLACED by six (n14 to z14) new species. But we may
0234   ces would wholly disappear, and would all be REPLACED by plane surfaces; and the Melipona would ma
0173   becomes more and more frequent, till the one REPLACES the other. But if we compare these species
0002   l statements; and I must trust to the reader REPOSING some confidence in my accuracy. No doubt er
0116   g this rather perplexing subject. Let A to L REPRESENT the species of a genus large in its own coa
0117   of unequal lengths proceeding from (A), may REPRESENT its varying offspring. The variations are s
0117   een the horizontal lines in the diagram, may REPRESENT each a thousand generations; but it would h
0122   to exist. If then our diagram be assumed to REPRESENT a considerable amount of modification, spec
0124   orizontal line has hitherto been supposed to REPRESENT a thousand generations, but each may repres
0124   present a thousand generations, but each may REPRESENT a million or a hundred million generations
0129   s the truth. The green and budding twigs may REPRESENT existing species; and those produced during
0129   d those produced during each former year may REPRESENT the long succession of extinct species. At
0129   present buds by ramifying branches may well REPRESENT the classification of all extinct and livin
0129   and these lost branches of various sizes may REPRESENT those whole orders, families, and genera wh
0300   ever lived there, how imperfectly would they REPRESENT the natural history of the world! But we ha
0311   the interrupted succession of chapters, may REPRESENT the apparently abruptly changed forms of li
0331   er. We may suppose that the numbered letters REPRESENT genera, and the dotted lines diverging from
0331   unimportant for us. The horizontal lines may REPRESENT successive geological formations, and all t
0385   tual: I suspended a duck's feet, which might REPRESENT those of a bird sleeping in a natural pond,
0412   iagram each letter on the uppermost line may REPRESENT a genus including several species; and all
0420   apter. We will suppose the letters A to L to REPRESENT allied genera, which lived during the Silur
0422   ement and it is notoriously not possible to REPRESENT in a series, on a flat surface, the affini
0432   ing to the diagram: the letters, A to L, may REPRESENT eleven Silurian genera, some of which have
```

Page **(Key Word)**

Page **(Key Word)**

```
0451  t is impossible to doubt, that the rudiments REPRESENT wings. Rudimentary organs sometimes retain
0173  ccessive intervals with closely allied or REPRESENTATIVE species, evidently filling nearly the sam
0173  in the natural economy of the land. These REPRESENTATIVE species often meet and interlock; and as
0174  . hence the neutral territory between two REPRESENTATIVE species is generally narrow in comparison
0175  if i am right in believing that allied or REPRESENTATIVE species, when inhabiting a continuous are
0178  rendered sufficiently distinct to rank as REPRESENTATIVE species. In this case, intermediate varie
0178  ntermediate varieties between the several REPRESENTATIVE species and their common parent, must for
0178  actual distribution of closely allied or REPRESENTATIVE species, and likewise of acknowledged var
0336  as they have been called by some authors, REPRESENTATIVE species; and these we assuredly do find.
0341  ng formation there be six other allied or REPRESENTATIVE genera with the same number of species, t
0342  erly have connected the closely allied or REPRESENTATIVE species, found in the several stages of t
0369  me few are distinct yet closely allied or REPRESENTATIVE species. In illustrating what, as I belie
0372  species; and a host of closely allied or REPRESENTATIVE forms which are ranked by all naturalists
0372  he presence of many existing and tertiary REPRESENTATIVE forms on the eastern and western shores o
0375  ed by man, and on the mountains, some few REPRESENTATIVE European forms are found, which have not
0403  of Porto Santo possess many distinct but REPRESENTATIVE land shells; some of which live in crevic
0403  eed not greatly marvel at the endemic and REPRESENTATIVE species, which inhabit the several island
0404  e ever so distant: many closely allied or REPRESENTATIVE species occur, there will likewise be fou
0409  of doubtful species, and of distinct but REPRESENTATIVE species. As the late Edward Forbes often
0462  the distribution both of the same and of REPRESENTATIVE species throughout the world. We are as y
0469  in the rank which they assign to the many REPRESENTATIVE forms in Europe and North America. If the
0478  ation. The existence of closely allied or REPRESENTATIVE species in any two areas, implies, on the
0129  lies, and genera which have now no living REPRESENTATIVES, and which are known to us only from hav
0299  s as distinct species from their European REPRESENTATIVES, and by other conchologists as only vari
0375  inia, several European forms and some few REPRESENTATIVES of the peculiar flora of the Cape of Goo
0109  extinction. We can, also, see that any form REPRESENTED by few individuals will, during fluctuatio
0116  is so generally the case in nature, and is REPRESENTED in the diagram by the letters standing at
0117  the most different or divergent variations REPRESENTED by the outer dotted lines) being preserved
0117  ; but it would have been better if each had REPRESENTED ten thousand generations. After a thousand
0118  in character. In the diagram the process is REPRESENTED up to the ten thousandth generation, and u
0118  the process ever goes on so regularly as is REPRESENTED in the diagram, though in itself made some
0119  as well as diverging in character: this is REPRESENTED in the diagram by the several divergent br
0119  earlier and less improved branches: this is REPRESENTED in the diagram by some of the lower branch
0120  successive generations. This case would be REPRESENTED in the diagram, if all the lines proceedin
0120  ding to the amount of change supposed to be REPRESENTED between the horizontal lines. After fourte
0121  ce. But during the process of modification, REPRESENTED in the diagram, another of our principles,
0125  ur diagram, we suppose the amount of change REPRESENTED by each successive group of diverging dott
0125  nt of divergent modification supposed to be REPRESENTED in the diagram. And the two new families,
0129  eings of the same class have sometimes been REPRESENTED by a great tree. I believe this simile lar
0284  miles. Some of these formations, which are REPRESENTED in England by thin beds, are thousands of
0296  would have suspected the vast lapse of time REPRESENTED by the thinner formation. Many cases could
0317  or the number of the genera of a family, be REPRESENTED by a vertical line of varying thickness, c
0318  and disappearance of a group of species be REPRESENTED, as before, by a vertical line of varying
0332  d the same general characteristics. This is REPRESENTED in the diagram by the letter F14. All the
0333  e case will be far more complicated than is REPRESENTED in the diagram: for the groups will have b
0375  that southern Australian forms are clearly REPRESENTED by plants growing on the summits of the mo
0420  ed modified descendants to the present day, REPRESENTED by the fifteen genera (a14 to z14) on the
0420  fied descendants from a single species, are REPRESENTED as related in blood or descent to the same
0429  ely from extinction, for they are generally REPRESENTED by extremely few species; and such species
0429  d not have been less aberrant had each been REPRESENTED by a dozen species instead of a single one
0449  which the ancient state, now supposed to be REPRESENTED in many embryos, has not been obliterated,
0116  ing but little from each other, and feebly REPRESENTING, as Mr. Waterhouse and others have remarke
0124  wnwards towards a single point; this point REPRESENTING a single species, the supposed single pare
0159  e pouter, may be considered as a variation REPRESENTING the normal structure of another race, the
0296  iven of beds only a few feet in thickness, REPRESENTING formations, elsewhere thousands of feet in
0375  ants occur, either identically the same or REPRESENTING each other, and at the same time represent
0375  resenting each other, and at the same time REPRESENTING plants of Europe, not found in the interve
0432  psi but we could pick out types, or forms, REPRESENTING most of the characters of each group, whet
0299  eparated by wide and shallow seas, probably REPRESENTS the former state of Europe, when most of ou
0453  uppose that the minute papilla, which often REPRESENTS the pistil in male flowers, and which is fo
0160  ed, the tendency, whether strong or weak, to REPRODUCE the lost character might be, as was formerl
0161  sive generation there has been a tendency to REPRODUCE the character in question, which at last, u
0455  le of inheritance at corresponding ages will REPRODUCE the organ in its reduced state at the same
0466  system, when not rendered impotent, fails to REPRODUCE offspring exactly like the parent form. Var
0251  es most widely dissimilar in general habit, REPRODUCED itself as perfectly as if it had been a nat
0009  singular the laws are which determine the REPRODUCTION of animals under confinement, I may just m
0010  of life are in comparison with the laws of REPRODUCTION, and of growth, and of inheritance; for ha
0096  occasionally or habitually, concur for the REPRODUCTION of their kind. This view, I may add, was f
0096  at is, two individuals regularly unite for REPRODUCTION, which is all that concerns us. But still
0096  cases that two individuals ever concur in REPRODUCTION? As it is impossible here to enter on deta
0101  an hermaphrodite animal with the organs of REPRODUCTION so perfectly enclosed within the body, tha
0150  are not directly connected with the act of REPRODUCTION. The rule applies to males and females; bu
0246  ure species have of course their organs of REPRODUCTION in a perfect condition, yet when intercros
0262  the male and female elements in the act of REPRODUCTION, yet that there is a rude degree of parall
0414  irst main divisions; whereas the organs of REPRODUCTION, with their product the seed, are of param
0441  other organs of importance, excepting for REPRODUCTION. We are so much accustomed to see differen
0470  species tends by its geometrical ratio of REPRODUCTION to increase inordinately in number; and as
0484  ar structure, and their laws of growth and REPRODUCTION. We see this even in so trifling a circums
0489  en in the largest sense, being Growth with REPRODUCTION: Inheritance which is almost implied by re
0489  on: Inheritance which is almost implied by REPRODUCTION: Variability from the indirect and direct
0008  y may be attributed to the male and female REPRODUCTIVE elements having been affected prior to the
0008  or cultivation has on the functions of the REPRODUCTIVE system: this system appearing to be far mo
0009  give numerous instances), yet having their REPRODUCTIVE system so seriously affected by unperceive
0009  erret kept in hutches), showing that their REPRODUCTIVE system has not been thus affected; so will
0043  nditions of life, from their action on the REPRODUCTIVE system, are so far of the highest importan
0082  itions of life, by specially acting on the REPRODUCTIVE system, causes or increases variability: a
```

Page **(Key Word)**

0131 lieve it to be as much the function of the REPRODUCTIVE system to produce individual differences,
0131 to show the truth of the remark, that the REPRODUCTIVE system is eminently susceptible to changes
0132 e, is alone affected. But why, because the REPRODUCTIVE system is disturbed, this or that part sho
0134 to play an important part in affecting the REPRODUCTIVE system, and in thus inducing variability;
0237 x, but to that short period alone when the REPRODUCTIVE system is active, as in the nuptial plumag
0246 ng. Hybrids, on the other hand, have their REPRODUCTIVE organs functionally impotent, as may be cl
0258 s imperceptible by us, and confined to the REPRODUCTIVE system. This difference in the result of r
0260 ent on unknown differences, chiefly in the REPRODUCTIVE systems, of the species which are crossed.
0262 the sterility of hybrids, which have their REPRODUCTIVE organs in an imperfect condition, is a ver
0262 uniting two pure species, which have their REPRODUCTIVE organs perfect; yet these two distinct cas
0263 l on unknown differences, chiefly in their REPRODUCTIVE systems. These differences, in both cases,
0264 s, they are extremely liable to have their REPRODUCTIVE systems seriously affected. This, in fact,
0265 vary, which is due, as I believe, to their REPRODUCTIVE systems having been specially affected, th
0265 the unnatural crossing of two species, the REPRODUCTIVE system, independently of the general state
0269 or could select, slight differences in the REPRODUCTIVE system, or other constitutional difference
0269 titutional differences correlated with the REPRODUCTIVE system. He supplies his several varieties
0269 ndirectly, through correlation, modify the REPRODUCTIVE system in the several descendants from any
0271 when crossed with N. glutinosa. Hence the REPRODUCTIVE system of this one variety must have been
0271 econdite and functional differences in the REPRODUCTIVE system; from these several considerations
0272 ired modifications, more especially in the REPRODUCTIVE systems of the forms which are crossed. Hy
0273 variability; namely, that it is due to the REPRODUCTIVE system being eminently sensitive to any ch
0273 long cultivated) which have not had their REPRODUCTIVE systems in any way affected, and they are
0273 ariable; but hybrids themselves have their REPRODUCTIVE systems seriously affected, and their desc
0276 incidental on unknown differences in their REPRODUCTIVE systems. There is no more reason to think
0276 ses between pure species, which have their REPRODUCTIVE systems perfect, seems to depend on severa
0277 the sterility of hybrids, which have their REPRODUCTIVE systems imperfect, and which have had this
0278 differences, and not of differences in the REPRODUCTIVE system. In all other respects, excluding f
0443 een communicated to the offspring from the REPRODUCTIVE element of one parent. Or again, as when t
0460 ental on constitutional differences in the REPRODUCTIVE systems of the intercrossed species. We se
0460 y that either their constitutions or their REPRODUCTIVE systems should have been profoundly modifi
0461 case from that of first crosses, for their REPRODUCTIVE organs are more or less functionally impot
0466 bility. This seems to be mainly due to the REPRODUCTIVE system being eminently susceptible to chan
0439 tell whether it be that of a mammal, bird, or REPTILE. The vermiform larvae of moths, flies, beetl
0139 with the blind Proteus with reference to the REPTILES of Europe, I am only surprised that more wre
0182 t diversified types, and formerly had flying REPTILES, it is conceivable that flying fish, which n
0197 tures occur in the skulls of young birds and REPTILES, which have only to escape from a broken egg
0313 iated with many strange and lost mammals and REPTILES in the sub Himalayan deposits. The Silurian
0330 ct to very distinct groups, such as fish and REPTILES, seems to be, that supposing them to be dist
0330 distinct groups. Yet if we compare the older REPTILES and Batrachians, the older Fish, the older C
0338 for instance, that the oldest known mammals, REPTILES, and fish strictly belong to their own prope
0391 other inhabitants; in the Galapagos Islands REPTILES, and in New Zealand gigantic wingless birds,
0416 n, which absolutely distinguishes fishes and REPTILES, the inflection of the angle of the jaws of
0417 finity of this strange creature to birds and REPTILES, as an approach in structure in any one inte
0479 early see why the embryos of mammals, birds, REPTILES, and fishes should be so closely alike, and
0221 of food; they approached and were vigorously REPULSED by an independent community of the slave spe
0058 species as to be of doubtful value: these 63 REPUTED species range on an average over 6.9 of the
0268 n hopeless difficulties; for if two hitherto REPUTED varieties be found in any degree sterile tog
0271 erimentised on five forms, which are commonly REPUTED to be varieties, and which he tested by the
0482 te published their belief that a multitude of REPUTED species in each genus are not real species;
0002 e sent to me a memoir on this subject, with a REQUEST that I would forward it to Sir Charles Lyell
0100 ind to be the case in this country; and at my REQUEST Dr. Hooker tabulated the trees of New Zealan
0331 n. as the subject is somewhat complex, I must REQUEST the reader to turn to the diagram in the fou
0363 borne seeds, during the glacial epoch. At my REQUEST Sir C. Lyell wrote to M. Hartung to inquire
0412 nt, the less improved, and preceding forms. I REQUEST the reader to turn to the diagram illustrati
0065 tremely few, is, that the slow breeders would REQUIRE a few more years to people, under favourable
0073 d. many of our orchidaceous plants absolutely REQUIRE the visits of moths to remove their pollen m
0075 pecies, for they frequent the same districts, REQUIRE the same food, and are exposed to the same d
0207 ' nests. An action, which we ourselves should REQUIRE experience to enable us to perform, when per
0215 do our civilised dogs, even when quite young, REQUIRE to be taught not to attack poultry, sheep, a
0303 a remark formerly made, namely that it might REQUIRE a long succession of ages to adapt an organi
0310 r great pressure, have always seemed to me to REQUIRE some special explanation; and we may perhaps
0346 east as closely as the same species generally REQUIRE; for it is a most rare case to find a group
0465 ut that it is imperfect to the degree which I REQUIRE, few will be inclined to admit. If we look t
0098 his end: but, the agency of insects is often REQUIRED to cause the stamens to spring forward, as K
0154 selection of the individuals varying in the REQUIRED manner and degree, and by the continued reje
0233 in its construction more materials would be REQUIRED than for a cylinder. As natural selection ac
0234 ived throughout the winter, and consequently REQUIRED a store of honey: there can in this case be
0248 s. i think no better/evidence of this can be REQUIRED than that the two most experienced observers
0287 data, the denudation of the Weald must have REQUIRED 306,662,400 years; or say three hundred mill
0293 er. Although each formation has indisputably REQUIRED a vast number of years for its deposition, I
0294 those of Europe; time having apparently been REQUIRED for their migration from the American to the
0296 ds of feet in thickness; and which must have REQUIRED an enormous period for their accumulation; I
0303 , longer perhaps in some cases than the time REQUIRED for the accumulation of each formation. Thes
0202 l to the community, it will fulfil all the REQUIREMENTS of natural selection, though it may cause
0041 r of individuals of a species in any country REQUIRES that the species should be placed under favo
0465 y have changed in the manner which my theory REQUIRES, for they have changed slowly and in a gradu
0003 h has flowers with separate sexes absolutely REQUIRING the agency of certain insects to bring poll
0148 eed in largely developing any organ, without REQUIRING as a necessary compensation the reduction o
0169 hich on my view must be a very slow process, REQUIRING a long lapse of time, in this case, natural
0296 erally have been due to geographical changes REQUIRING much time. Nor will the closest inspection
0032 n the natural capacity and years of practice REQUISITE to become even a skilful pigeon fancier. Th
0139 e period of flowering, in the amount of rain REQUISITE for seeds to germinate, in the time of slee
0227 r, she can prolong the hexagon to any length REQUISITE to hold the stock of honey; in the same way
0286 any given height, we could measure the time REQUISITE to have denuded the Weald. This, of course,
0293 ch perhaps is short compared with the period REQUISITE to change one species into another. I am aw
0423 ions, several other grades of difference are REQUISITE, as we have seen with pigeons. The origin o

Page ***(Key Word)***

0448 not undergoing any metamorphosis is perhaps REQUISITE. If, on the other hand, it profited the you
0096 oups of animals, pair for each birth. Modern RESEARCH has much diminished the number of supposed h
0299 improbable in the highest degree. Geological RESEARCH, though it has added numerous species to exi
0419 and this has been done, not because further RESEARCH has detected important structural difference
0018 r five thousand years ago? But Mr. Horner's RESEARCHES have rendered it in some degree probable th
0291 of the sea, in which case, judging from the RESEARCHES of E. Forbes, we may conclude that the bott
0395 this archipelago by the admirable zeal and RESEARCHES of Mr. Wallace. I have not as yet had time
0488 re I see open fields for far more important RESEARCHES. Psychology will be based on a new foundati
0138 ight have been anticipated from the general RESEMBLANCE of the other inhabitants of North America
0208 nt throughout life. Several other points of RESEMBLANCE between instincts and habits could be poin
0209 n that this does sometimes happen, then the RESEMBLANCE between what originally was a habit and an
0256 the term systematic affinity is meant, the RESEMBLANCE between species in structure and in consti
0259 n the hybrid is independent of its external RESEMBLANCE to either pure parent. Considering the sev
0260 d by their systematic affinity or degree of RESEMBLANCE to each other. This latter statement is cl
0263 systematic affinity, by which every kind of RESEMBLANCE and dissimilarity between organic beings i
0274 and mongrel plants. On the other hand, the RESEMBLANCE in mongrels and in hybrids to their respec
0275 , comes to the conclusion, that the laws of RESEMBLANCE of the child to its parents are the same,
0277 c affinity attempts to express all kinds of RESEMBLANCE between all species. First crosses between
0278 cluding fertility, there is a close general RESEMBLANCE between hybrids and mongrels. Finally, the
0281 had in its whole organisation much general RESEMBLANCE to the tapir and the horse; but in some po
0323 ain beds present an unmistakeable degree of RESEMBLANCE to those of the Chalk. It is not that the
0335 gives as a well known instance, the general RESEMBLANCE of the organic remains from the several st
0335 , will not attempt to account for the close RESEMBLANCE of the distinct species in closely consecu
0376 fact that New Zealand should have a closer RESEMBLANCE in its crustacea to Great Britain, its ant
0398 her hand, there is a considerable degree of RESEMBLANCE in the volcanic nature of the soil, in cli
0413 s included in our classification, than mere RESEMBLANCE. I believe that something more is included
0417 an appreciation of many trifling points of RESEMBLANCE, too slight to be defined. Certain plants,
0425 nt in common, there will certainly be close RESEMBLANCE or affinity. As descent has universally be
0427 en ably followed by Macleay and others. The RESEMBLANCE, in the shape of the body and in the fin l
0427 stems of the common and swedish turnip. The RESEMBLANCE of the greyhound and racehorse is hardly m
0427 onditions, and thus assume a close external RESEMBLANCE; but such resemblances will not reveal, wi
0430 ever, it may be strongly suspected that the RESEMBLANCE is only analogical, owing to the phascolom
0434 he general plan of their organisation. This RESEMBLANCE is often expressed by the term unity of ty
0438 or organs, a certain degree of fundamental RESEMBLANCE, retained by the strong principle of inher
0439 es of life. A trace of the law of embryonic RESEMBLANCE, sometimes lasts till a rather late age: t
0440 or less closely the law of common embryonic RESEMBLANCE. Cirripedes afford a good instance of this
0450 be borne in mind, that the supposed law of RESEMBLANCE of ancient forms of life to the embryonic
0457 at leading facts in Embryology; namely, the RESEMBLANCE in an individual embryo of the homologous
0457 ch other in structure and function; and the RESEMBLANCE in different species of a class of the hom
0475 complex laws in their degrees and kinds of RESEMBLANCE to their parents, in being absorbed into e
0275 animals closely resembling one parent, the RESEMBLANCES seem chiefly confined to characters almost
0414 ale to a fish, as of any importance. These RESEMBLANCES, though so intimately connected with the w
0414 racters; but to the consideration of these RESEMBLANCES we shall have to recur. It may even be giv
0414 t not, therefore, in classifying, trust to RESEMBLANCES in parts of the organisation, however impo
0415 all naturalists lay the greatest stress on RESEMBLANCES in organs of high vital or physiological i
0425 e have to make out community of descent by RESEMBLANCES of any kind. Therefore we choose those cha
0427 real affinities and analogical or adaptive RESEMBLANCES. Lamarck first called attention to this di
0427 ume a close external resemblance; but such RESEMBLANCES will not reveal, will rather tend to conce
0433 on. We can understand why we value certain RESEMBLANCES far more than others; why we are permitted
0018 breeds; and that some of the breeds closely RESEMBLE, perhaps are identical with, those still exi
0044 ra. Many of the species of the larger genera RESEMBLE varieties in being very closely, but unequal
0057 therefore, the species of the larger genera RESEMBLE varieties, more than do the species of the s
0057 ready manufactured still to a certain extent RESEMBLE varieties, for they differ from each other b
0058 osely allied to other species, and in so far RESEMBLE varieties, often have much restricted ranges
0116 s own country; these species are supposed to RESEMBLE each other in unequal degrees, as is so gene
0122 iginal species of our genus were supposed to RESEMBLE each other in unequal degrees, as is so gene
0155 e points in which all the species of a genus RESEMBLE each other, and in which they differ from th
0161 nner; so that a variety of one species would RESEMBLE in some of its characters another species; t
0214 tic instincts, when thus tested by crossing, RESEMBLE natural instincts, which in a like manner be
0259 er between their two parents, always closely RESEMBLE one of them; and such hybrids, though extern
0259 ndividuals sometimes are born, which closely RESEMBLE one of their pure parents; and these hybrids
0259 s is not related to the degree in which they RESEMBLE in external appearance either parent. And la
0266 hybrids which occasionally and exceptionally RESEMBLE closely either pure parent. Nor do I pretend
0274 produced from a reciprocal cross, generally RESEMBLE each other closely; and so it is with mongre
0274 se, so that both the mule and the hinny more RESEMBLE the ass than the horse; but that the prepote
0338 result. Agassiz insists that ancient animals RESEMBLE to a certain extent the embryos of recent an
0344 arly it will be related to, and consequently RESEMBLE, the common progenitor of groups, since beco
0345 uld hereafter be proved that ancient animals RESEMBLE to a certain extent the embryos of more rece
0411 awn of life, all organic beings are found to RESEMBLE each other in descending degrees, so that th
0424 s varieties, not solely because they closely RESEMBLE the parent form, but because they are descen
0434 lass, independently of their habits of life, RESEMBLE each other in the general plan of their orga
0439 iform larvae of moths, flies, beetles, etc., RESEMBLE each other much more closely than do the mat
0439 e genus, and of closely allied genera, often RESEMBLE each other in their first and second plumage
0439 f widely different animals of the same class RESEMBLE each other, often have no direct relation to
0447 f our supposed genus will manifestly tend to RESEMBLE each other much more closely than do the adu
0447 descendants of the parent species will still RESEMBLE each other closely, for they will not have b
0448 d tumbler) the young or embryo would closely RESEMBLE the mature parent form. We have seen that th
0449 why ancient and extinct forms of life should RESEMBLE the embryos of their descendants, our existi
0470 round other species, in which respects they RESEMBLE varieties. These are strange relations on th
0483 and at an embryonic age the species closely RESEMBLE each other. Therefore I cannot doubt that th
0324 ing or the pleistocene inhabitants of Europe RESEMBLED most closely those of the southern hemisphe
0107 flora of Madeira, according to Oswald Heer, RESEMBLES the extinct tertiary flora of Europe. All f
0179 rth America, which has webbed feet and which RESEMBLES an otter in its fur, short legs, and form o
0193 e remarked, their intimate structure closely RESEMBLES that of common muscle: and as it has lately
0398 veral classes are associated together, which RESEMBLES closely the conditions of the South America
0430 mr. Waterhouse has remarked, the phascolomys RESEMBLES most nearly, not any one species, but the g

Page ***(Key Word)***

Page **(Key Word)***

```
0019  ation. Who can believe that animals closely RESEMBLING the Italian greyhound, the bloodhound, the
0052  convenience to a set of individuals closely RESEMBLING each other, and that it does not essentiall
0167  t quarters of the world, to produce hybrids RESEMBLING in their stripes, not their own parents, bu
0268  t how many species there are, which, though RESEMBLING each other most closely, are utterly steril
0275  ave collected of cross bred animals closely RESEMBLING one parent, the resemblances seem chiefly c
0442  same class, generally, but not universally, RESEMBLING each other; of the structure of the embryo
0448  ot undergoing any metamorphosis, or closely RESEMBLING their parents from their earliest age, we c
0451  e same genus (and even of the same species) RESEMBLING each other most closely in all respects, on
0044  dry facts should be given, but these I shall RESERVE for my future work. Nor shall I here discuss
0053  y dr. Hooker, even in stronger terms. I shall RESERVE for my future work the discussion of these d
0077  which it comes into competition for food or RESIDENCE, or from which it has to escape, or on whic
0132  so with insects, Wollaston is convinced that RESIDENCE near the sea affects their colours. Moquin
0069  ey cannot compete with our native plants, nor RESIST destruction by our native animals. When a spe
0265  r hand, one species in a group will sometimes RESIST great changes of conditions with unimpaired f
0269  ntercrossed. But it seems to me impossible to RESIST the evidence of the existence of a certain am
0291  ous formation thick enough, when upraised, to RESIST any amount of degradation, may be formed. I a
0291  south America, which has been bulky enough to RESIST such degradation as it has as yet suffered, b
0291  ssils and sufficiently thick and extensive to RESIST subsequent degradation, may have been formed
0312  Lyell has shown that it is hardly possible to RESIST the evidence on this head in the case of the
0342  ion of fossiliferous deposits thick enough to RESIST future degradation, enormous intervals of tim
0358  ts, it was not even known how far seeds could RESIST the injurious action of sea water. To my surp
0465  for fossiliferous formations, thick enough to RESIST future degradation, can be accumulated only w
0472  produced some effect; for it is difficult to RESIST this conclusion when we look, for instance, a
0360  rage length of their flotation and of their RESISTANCE to the injurious action of the salt water.
0083  odified with advantage, so as to have better RESISTED such intruders. As man can produce and certa
0140  o possess different constitutional powers of RESISTING cold. Mr. Thwaites informs me that he has o
0195  america absolutely depends on their power of RESISTING the attacks of insects: so that individuals
0286  in harder layers or nodules, which from long RESISTING attrition form a breakwater at the base. He
0075  wheat be sown together, and the mixed seed be RESOWN, some of the varieties which best suit the so
0018  tter have had more than one wild parent. With RESPECT to horses, from reasons which I cannot give
0031  breeder, Sir John Sebright, used to say, with RESPECT to pigeons, that he would produce any given
0041  on this principle Marshall has remarked, with RESPECT to the sheep of parts of Yorkshire, that as
0042  is already stocked with other races. In this RESPECT enclosure of the land plays a part. Wanderin
0057  liberation, they concur in this view. In this RESPECT, therefore, the species of the larger genera
0078  raphical range, a change of constitution with RESPECT to climate would clearly be an advantage to
0089  roncly suspect that some well known laws with RESPECT to the plumage of male and female birds, in
0144  here probably homology comes into play? With RESPECT to this latter case of correlation, I think
0145  als, the nectary is only much shortened. With RESPECT to the difference in the corolla of the cent
0150  m an observation made by Professor Owen, with RESPECT to the length of the arms of the ourang outa
0163  y distinct zebra like bars on the hocks. With RESPECT to the horse, I have collected cases in Engl
0165  a like stripes on the sides of its face. With RESPECT to this last fact, I was so convinced that n
0236  concerned, the prodigious difference in this RESPECT between the workers and the perfect females,
0251  n of the same flower were perfectly good with RESPECT to other species, yet as they were functiona
0252  ids, should be perfectly fertile. Again, with RESPECT to the fertility in successive generations o
0266  pting on vague hypotheses, several facts with RESPECT to the sterility of hybrids: for instance, t
0272  to overthrow the view which I have taken with RESPECT to the very general, but not invariable, ste
0275  fter arranging an enormous body of facts with RESPECT to animals, comes to the conclusion, that th
0277  r how liable we are to argue in a circle with RESPECT to varieties in a state of nature: and when
0288  hiton offers a partially analogous case. With RESPECT to the terrestrial productions which lived d
0321  n, a slower process than its production. With RESPECT to the apparently sudden extermination of wh
0329  we combine both into one general system. With RESPECT to the Vertebrata, whole pages could be fill
0330  milies. The most common case, especially with RESPECT to very distinct groups, such as fish and ro
0333  escent with modification, the main facts with RESPECT to the mutual affinities of the extinct form
0335  t the opposite end of the series in this same RESPECT. Closely connected with the statement, that
0347  climate. Analogous facts could be given with RESPECT to the inhabitants of the sea. A second grea
0354  thplace; then, considering our ignorance with RESPECT to former climatal and geographical changes
0368  i have met with no satisfactory evidence with RESPECT to this intercalated slightly warmer period,
0384  , seems to lead to this same conclusion. With RESPECT to allied fresh water fish occurring at very
0386  n with a favouring gale no one can tell. With RESPECT to plants, it has long been known what enorm
0392  y remarkable little facts could be given with RESPECT to the inhabitants of remote islands. For in
0392  t into bushes and ultimately into trees. With RESPECT to the absence of whole orders on oceanic is
0407  remember how profoundly ignorant we are with RESPECT to the many and curious means of occasional
0407  ion of sub genera, genera, and families. With RESPECT to the distinct species of the same genus, w
0419  al entomologists and botanists. Finally, with RESPECT to the comparative value of the various grou
0426  ance from a common parent. We may err in this RESPECT in regard to single points of structure, but
0430  a character in some degree intermediate with RESPECT to all existing Marsupials; or that both Rod
0448  e great class of insects, as with Aphis. With RESPECT to the final cause of the young in these cas
0453  tion. I have now given the leading facts with RESPECT to rudimentary organs. In reflecting on them
0460  how how this difficulty can be mastered. With RESPECT to the almost universal sterility of species
0462  the many occasional means of transport. With RESPECT to distinct species of the same genus inhabi
0462  ended together in an inextricable chaos? With RESPECT to existing forms, we should remember that w
0465  periods of subsidence, more extinction. With RESPECT to the absence of fossiliferous formations b
0486  rudimentary organs will speak infallibly with RESPECT to the nature of long lost structures. Speci
0213  n. we shall thus also be enabled to see the RESPECTIVE parts which habit and the selection of so c
0274  mblance in mongrels and in hybrids to their RESPECTIVE parents, more especially in hybrids produce
0016  ecies of the same genus, in several trifling RESPECTS, they often differ in an extreme degree in s
0023  differ from each other in the most trifling RESPECTS. As several of the reasons which have led me
0027  s having very abnormal characters in certain RESPECTS, as compared with all other Columbidae, thou
0027  her Columbidae, though so like in most other RESPECTS to the rock pigeon: the blue colour and vari
0047  ake them as distinct species, are in several RESPECTS the most important for us. We have every rea
0059  have restricted ranges. In all these several RESPECTS the species of large genera present a strong
0106  ll isolated areas probably have been in some RESPECTS highly favourable for the production of new
0266  mpted to show, is that in two cases, in some RESPECTS allied, sterility is the common result, in t
0272  hen crossed may be compared in several other RESPECTS. Gartner, whose strong wish was to draw a ma
0272  ey agree most closely in very many important RESPECTS. I shall here discuss this subject with extr
0275  ion of fertility and sterility, in all other RESPECTS there seems to be a general and close simila
```

Page **(Key Word)***

Page **(Key Word)**

```
0278  ces in the reproductive system. In all other RESPECTS, excluding fertility, there is a close gener
0280  genitor will generally have differed in some RESPECTS from all its modified descendants. To give a
0320  forms which are most like each other in all RESPECTS. Hence the improved and modified descendants
0389  se numbers, and the comparison in some other RESPECTS is not quite fair. We have evidence that the
0398  is a considerable dissimilarity in all these RESPECTS. On the other hand, there is a considerable
0424  ause allied in blood and alike in some other RESPECTS. If it could be proved that the Hottentot ha
0451  s) resembling each other most closely in all RESPECTS, one of which will have full sized wings, an
0452  o vary in degree of development and in other RESPECTS. Moreover, in closely allied species, the de
0468  individuals of the same species come in all RESPECTS into the closest competition with each other
0470  little groups round other species, in which RESPECTS they resemble varieties. These are strange r
0190  one for a wholly different purpose, namely RESPIRATION. The swimbladder has, also, been worked in
0191  into a lung, or organ used exclusively for RESPIRATION. I can, indeed, hardly doubt that all vert
0191  s which at a very ancient period served for RESPIRATION have been actually converted into organs o
0192  ack, including the small frena, serving for RESPIRATION. The Balanidae or sessile cirripedes, on t
0192  h, likewise, very slightly aided the act of RESPIRATION, have been gradually converted by natural
0190  rior surface will then digest and the stomach RESPIRES, digests, and excretes in the larva of the d
0190  istinct functions; thus the alimentary canal RESPIRES, digests, and excretes in the larva of the d
0009  imes very different character from that of the REST of the plant. Such buds can be propagated by g
0165  he legs were much more plainly barred than the REST of the body; and in one of them there was a do
0293  small the area of Europe is compared with the REST of the world; nor have the several stages of t
0482  hink that we give an explanation when we only RESTATE a fact. Any one whose disposition leads him
0453  t this seems to me no explanation, merely a RESTATEMENT of the fact. Would it be thought sufficien
0186  e of another type; but this seems to me only RESTATING the fact in dignified language. He who beli
0339  orld. We see the same law in this author's RESTORATIONS of the extinct and gigantic birds of New Z
0063  tificial increase of food, and no prudential RESTRAINT from marriage. Although some species may be
0044  ually, related to each other, and in having RESTRICTED ranges: this statement is indeed scarcely m
0058  worth notice. Varieties generally have much RESTRICTED ranges. Before applying the principles arri
0058  so far resemble varieties, often have much RESTRICTED ranges. For instance, Mr. H. C. Watson has
0058  wledged varieties have very nearly the same RESTRICTED average range, as have those very closely a
0059  ely allied to other species apparently have RESTRICTED ranges. In all these several respects the s
0117  diffused, vary more than rare species with RESTRICTED ranges. Let (A) be a common, widely diffuse
0330  om each other. This remark no doubt must be RESTRICTED to those groups which have undergone much c
0360  e has shown that such plants generally have RESTRICTED ranges. But seeds may be occasionally trans
0470  s also of the larger genera apparently have RESTRICTED ranges, and they are clustered in little gr
0254  inclined to believe in its truth, although it RESTS on no direct evidence. I believe, for instanc
0002  site to those at which I have arrived. A fair RESULT can be obtained only by fully stating and bal
0012  erious laws of the correlation of growth. The RESULT of the various, quite unknown, or dimly seen
0033  s have been published on the subject; and the RESULT, I may add, has been, in a corresponding degr
0036  cted or even have wished to have produced the RESULT which ensued, namely, the production of two d
0037  bt, has been simple, and, as far as the final RESULT is concerned, has been followed almost uncons
0043  st be attributed to use and disuse. The final RESULT is thus rendered infinitely complex. In some
0054  cies. But so many causes tend to obscure this RESULT, that I am surprised that my tables show even
0062  W. Herbert, Dean of Manchester, evidently the RESULT of his great horticultural knowledge. Nothing
0065  cases the geometrical ratio of increase, the RESULT of which never fails to be surprising, simply
0076  o that they cannot be kept together. The same RESULT has followed from keeping together different
0083  an produce and certainly has produced a great RESULT by his methodical and unconscious means of se
0086  t of their structure is merely the correlated RESULT of successive changes in the structure of the
0088  the males for possession of the females; the RESULT is not death to the unsuccessful competitor,
0100  asa Gray those of the United States, and the RESULT was as I anticipated. On the other hand, Dr.
0142  specially fitted for their own districts: the RESULT must, I think, be due to habit. On the other
0211  that the action was instinctive, and not the RESULT of experience. But as the excretion is extrem
0216  habit has done nothing, and all has been the RESULT of selection, pursued both methodically and u
0227  spheres in both layers be formed, there will RESULT a double layer of hexagonal prisms united tog
0250  ed from three other and distinct species: the RESULT was that the ovaries of the three first flowe
0251  al subsequent years, and always with the same RESULT. This result has, also, been confirmed by oth
0251  years, and always with the same result. This RESULT has, also, been confirmed by other observers
0254  osses and in hybrids, is an extremely general RESULT; but that it cannot, under our present state
0258  e reproductive system. This difference in the RESULT of reciprocal crosses between the same two sp
0260  d there often be so great a difference in the RESULT of a reciprocal cross between the same two sp
0266  some respects allied, sterility is the common RESULT, in the one case from the conditions of life
0269  ed not be surprised at some difference in the RESULT. I have as yet spoken as if the varieties of
0270  eems at first quite incredible; but it is the RESULT of an astonishing number of experiments made
0284  and extent of sedimentary formations are the RESULT and measure of the degradation which the eart
0284  erent parts of Great Britain; and this is the RESULT: Palaeozoic strata (not including igneous bed
0338  he two countries could not have foreseen this RESULT. Agassiz insists that ancient animals resembl
0360  some of them to have floated much longer. The RESULT was that 18/98 of his seeds floated for 42 da
0367  but this will make no difference in the final RESULT. As the warmth returned, the arctic forms wou
0412  inciples; and he will see that the inevitable RESULT is that the modified descendants proceeding f
0448  heir earliest age, we can see that this would RESULT from the two following contingencies; firstly
0460  this conclusion in the vast difference in the RESULT, when the same two species are crossed recipr
0468  ly, can produce within a short period a great RESULT by adding up mere individual differences in h
0182  here alluded to, which perhaps may all have RESULTED from disuse, indicate the natural steps by w
0226  ed them symmetrically in a double layer, the RESULTING structure would probably have been as perfe
0034  n, which may be called Unconscious, and which RESULTS from every one trying to possess and breed f
0037  f gardeners, in having produced such splendid RESULTS from such poor materials; but the art, I can
0053  nsiderations, I thought that some interesting RESULTS might be obtained in regard to the nature an
0055  o the species of the small genera. Both these RESULTS follow when another division is made, and wh
0057  ally by averages, and, as far as my imperfect RESULTS go, they always confirm the view. I have als
0061  e species of the same genus, arise? All these RESULTS, as we shall more fully see in the next chap
0061  man by selection can certainly produce great RESULTS, and can adapt organic beings to his own use
0082  necessary; as man can certainly produce great RESULTS by adding up in any given direction mere ind
0090  ction, or by that unconscious selection which RESULTS from each man trying to keep the best dogs w
0249  g having been avoided. Now let us turn to the RESULTS arrived at by the third most experienced hyb
0250  ecies as did Gartner. The difference in their RESULTS may, I think, be in part accounted for by He
0262  there is a rude degree of parallelism in the RESULTS of grafting and of crossing distinct species
0274  experiment; and seems directly opposed to the RESULTS of several experiments made by Kolreuter. Th
```

Page **(Key Word)**

Page **(Key Word)**

```
0289  imperfection in the geological record mainly RESULTS from another and more important cause than a
0433  y, we have seen that natural selection, which RESULTS from the struggle for existence, and which a
0467   a succession of peculiar seasons, and by the RESULTS of naturalisation, as explained in the third
0192   , through the means of a sticky secretion, to RETAIN the eggs until they are hatched within the sa
0314   y no means surprising that one species should RETAIN the same identical form much longer than othe
0332   slightly altered conditions of life, and yet RETAIN throughout a vast period the same general cha
0359  ted by sea currents during 28 days, and would RETAIN their power of germination. In Johnston's Phy
0361   any kinds in the crops of floating birds long RETAIN their vitality: peas and vetches, for instanc
0364  ds for very great distances: for seeds do not RETAIN their vitality when exposed for a great lengt
0387   n across the sea; and we have seen that seeds RETAIN their power of germination, when rejected in
0401   . some species, however, might spread and yet RETAIN the same character throughout the group, just
0451   represent wings. Rudimentary organs sometimes RETAIN their potentiality, and are merely not develo
0470  ter number of varieties or incipient species, RETAIN to a certain degree the character of varietie
0047  ul and closely allied forms have permanently RETAINED their characters in their own country for a
0090  erved or selected, provided always that they RETAINED strength to master their prey at this or at
0104  r intercross, uniformity of character can be RETAINED amongst them, as long as their conditions of
0124  have diverged much in character, but to have RETAINED the form of (F), either unaltered or altered
0160  lly believed that a tendency to reversion is RETAINED by this very small proportion of foreign blo
0205  t formerly of high importance has often been RETAINED (as the tail of an aquatic animal by its ter
0216   their mother to fly away. But this instinct RETAINED by our chickens has become useless under dom
0351  een in the last chapter that some forms have RETAINED nearly the same character from an enormously
0362  wo seeds of beet grew after having been thus RETAINED for two days and fourteen hours. Fresh water
0362   their excrement; and several of these seeds RETAINED their power of germination. Certain seeds, h
0417  eu observes, that this genus should still be RETAINED amongst the Malpighiaceae. This case seems t
0430  er view we may suppose that the bizcacha has RETAINED, by inheritance, more of the character of it
0430  nearly all Marsupials, from having partially RETAINED the character of their common progenitor, or
0438   a certain degree of fundamental resemblance, RETAINED by the strong principle of inheritance. In t
0439  s characters, which they would probably have RETAINED through inheritance, if they had really been
0453  ry organ in the adult, is often said to have RETAINED its embryonic condition. I have now given th
0454  ed for another purpose. Or an organ might be RETAINED for one alone of its former functions. An or
0455  e compared with the letters in a word, still RETAINED in the spelling, but become useless in the p
0462  ave reason to believe that some species have RETAINED the same specific form for very long periods
0243  ably different conditions of life, yet often RETAINING nearly the same instincts. For instance, we
0273   few cases both of hybrids and mongrels long RETAINING uniformity of character could be given. The
0388  ter plants and of the lower animals, whether RETAINING the same identical form or in some degree m
0392  t then becoming slightly modified, but still RETAINING its hooked seeds, would form an endemic spe
0396  ts of the more remote islands, whether still RETAINING the same specific form or modified since th
0406  better chance of ranging widely and of still RETAINING the same specific character. This fact, too
0439  ul fact of the jaws, for instance, of a crab RETAINING numerous characters, which they would proba
0105  malli and fewness of individuals will greatly RETARD the production of new species through natural
0108  w process. The process will often be greatly RETARDED by free intercrossing. Many will exclaim tha
0103   not overrate the effects of intercrosses in RETARDING natural selection; for I can bring a consid
0431  ed from a common parent, together with their RETENTION by inheritance of some characters in common
0367  s the warmth returned, the arctic forms would RETREAT northward, closely followed up in their retr
0367  treat northward, closely followed up in their RETREAT by the productions of the more temperate reg
0368   little further north, and subsequently have RETREATED to their present homes; but I have met with
0377  pical plants and other productions will have RETREATED from both sides towards the equator, follow
0283   but how often do we see along the bases of PRETREATING cliffs rounded boulders, all thickly clothe
0213  ng is certainly in some degree inherited by RETRIEVERS; and a tendency to run round, instead of at
0213  he very first time that they are taken out: RETRIEVING is certainly in some degree inherited by re
0441   latter, the development has assuredly been RETROGRADE: for the male is a mere sack, which lives f
0448  the final metamorphosis would be said to be RETROGRADE. As all the organic beings, extinct and rec
0001  ed by one of our greatest philosophers. On my RETURN home, it occurred to me, in 1837, that someth
0052  ndent species. But we shall hereafter have to RETURN to this subject. From these remarks it will b
0091  ubject of intercrossing we shall soon have to RETURN. I may add, that, according to Mr. Pierce, is
0093  from tree to tree in search of nectar. But to RETURN to our imaginary case: as soon as the plant h
0099   foreign pollen; but to this subject we shall RETURN in a future chapter. In the case of a giganti
0126  endants to a remote futurity. I shall have to RETURN to this subject in the chapter on Classificat
0154  y indirectly, true; I shall, however, have to RETURN to this subject in our chapter on Classificat
0183  rs and under many subordinate forms. Thus, to RETURN to our imaginary illustration of the flying f
0273  their descendants are highly variable. But to RETURN to our comparison of mongrels and hybrids: Gä
0298  rmations; but to this subject I shall have to RETURN in the following chapter. One other considera
0372  r inhabitants utterly dissimilar. But we must RETURN to our more immediate subject, the Glacial pe
0301  lant their parent forms. When such varieties RETURNED to their ancient homes, as they would differ
0367  ifference in the final result. As the warmth RETURNED, the arctic forms would retreat northward, c
0367  rn journey. Hence, when the warmth had fully RETURNED, the same arctic species, which had lately l
0378  some even crossed the equator. As the warmth RETURNED, these temperate forms would naturally ascen
0222   migration) and numerous pupae. I traced the RETURNING file burthened with booty, for about forty
0367  he cold came on, and the re migration on the RETURNING warmth, will generally have been due south
0368  ctions, left isolated from the moment of the RETURNING warmth, first at the bases and ultimately o
0250  st experienced hybridiser, namely, the Hon. and REV. W. Herbert. He is as emphatic in his conclusi
0225  of gradation, and see whether Nature does not REVEAL to us her method of work. At one end of a sho
0280  ntermediate links? Geology assuredly does not REVEAL any such finely graduated organic chain; and
0366  ed by drifted icebergs and coast ice, plainly REVEAL a former cold period. The former influence of
0427  ce for classification, only in so far as they REVEAL descent, we can clearly understand why analog
0427  l resemblance; but such resemblances will not REVEAL, will rather tend to conceal their blood rela
0449  ructure reveals community of descent. It will REVEAL this community of descent, however much the s
0480  ty! Nature may be said to have taken pains to REVEAL, by rudimentary organs and by homologous stru
0486  of the ancient forms of life. Embryology will REVEAL to us the structure, in some degree obscured,
0414   degrees of modification, which is partially REVEALED to us by our classifcications. Let us now co
0246  erfect in structure, as far as the microscope REVEALS. In the first case the two sexual elements w
0413  t by the Natural System; they believe that it REVEALS the plan of the Creator; but unless it be sp
0449  in its less modified state; and in so far it REVEALS the structure of its progenitor. In two grou
0449  lated. Thus, community in embryonic structure REVEALS community of descent. It will reveal this co
0483  appearance of species in what they consider REVERENT silence. It may be asked how far I extend th
0040  ords on the circumstances, favourable, or the REVERSE, to man's power of selection. A high degree
```

Page **(Key Word)**

Page **(Key Word)**

```
0099  species are crossed the case is directly the  REVERSE, for a plant's own pollen is always prepoten
0383  istant countries. But the case is exactly the  REVERSE. Not only have many fresh water species, bel
0021  very short and conical beak, with a line of    REVERSED feathers down the breast; and it has the hab
0021  phagus. The Jacobin has the feathers so much   REVERSED along the back of the neck that they form a
0024  hat of the short faced tumbler, or barb; for   REVERSED feathers like those of the jacobin; for a cr
0058  ent species, their denominations ought to be   REVERSED. But there is also reason to believe, that t
0159  widely set apart, present sub varieties with   REVERSED feathers on the head and feathers on the fee
0261  le sexual element of the other, but not in a   REVERSED direction. It will be advisable to explain a
0379  rated from the north to the south, than in a   REVERSED direction. We see, however, a few southern v
0014  le element. Having alluded to the subject of   REVERSION, I may here refer to a statement often made
0014  could not tell whether or not nearly perfect   REVERSION had ensued. It would be quite necessary, in
0015  ic varieties manifested a strong tendency to   REVERSION, that is, to lose their acquired characters
0025  these facts, on the well known principle of    REVERSION to ancestral characters, if all the domesti
0026  ce with some distinct breed, the tendency to   REVERSION to any character derived from such cross wi
0043  ied by various degrees of inheritance and of   REVERSION. Variability is governed by many unknown la
0150  ong continued disuse, and to the tendency to   REVERSION. A part developed in any species in an extr
0152  on between, on the one hand, the tendency to   REVERSION to a less modified state, as well as an inn
0153  lection on the one hand, and the tendency to   REVERSION and variability on the other hand, will in
0158  lapse of time, overmastered the tendency to    REVERSION and to further variability, to sexual selec
0160  hat no one will doubt that this is a case of   REVERSION, and not of a new yet analogous variation a
0160  it is generally believed that a tendency to    REVERSION is retained by this very small proportion o
0162  ave inferred that the blueness was a case of   REVERSION, from the number of the markings, which are
0162  f a species assuming characters (either from   REVERSION or from analogous variation) which already
0163  tly under nature. It is a case apparently of   REVERSION. The ass not rarely has very distinct trans
0169  e the tendency to further variability and to   REVERSION to a less modified state. But when a specie
0169  d important modifications may not arise from   REVERSION and analogous variation, such modifications
0196  on; that characters reappear from the law of   REVERSION; that correlation of growth will have had a
0199  unknown cause, may reappear from the law of    REVERSION, though now of no direct use. The effects o
0200  as now, to the several laws of inheritance,    REVERSION, correlation of growth, etc. Hence every de
0015  onditions of life do change, variations and    REVERSIONS of character probably do occur; but natural
0131  the same genus vary in an analogous manner.    REVERSIONS to long lost characters. Summary. I have hi
0161  f the same genus would occasionally exhibit    REVERSIONS to lost ancestral characters. As, however,
0162  hese characters in our domestic breeds were    REVERSIONS or only analogous variations; but we might
0162  generally be left doubtful, what cases are     REVERSIONS to an anciently existing character and what
0275  acquired by selection. Consequently, sudden    REVERSIONS to the perfect character of either parent w
0473  ily modified. In both varieties and species   REVERSIONS to long lost characters occur. How inexplic
0014  eties, when run wild, gradually but certainly  REVERT in character to their aboriginal stocks. Henc
0015  s, as our varieties certainly do occasionally  REVERT in some of their characters to ancestral form
0015  they would to a large extent, or even wholly,  REVERT to the wild aboriginal stock. Whether or not
0025  h separate breed there might be a tendency to  REVERT to the very same colours and markings. Or, se
0026  d, and there is a tendency in both parents to  REVERT to a character, which has been lost during so
0154  y the continued rejection of those tending to  REVERT to a former and less modified condition. The
0160  the offspring occasionally show a tendency to  REVERT in character to the foreign breed for many ge
0169  ions, and these same species may occasionally  REVERT to some of the characters of their ancient pr
0273  that mongrels are more liable than hybrids to  REVERT to either parent form; but this, if it be tru
0013  ted and sometimes not so; why the child often  REVERTS in certain characters to its grandfather or
0026  countenancing the belief that the child ever   REVERTS to some one ancestor, removed by a greater n
0159  me of the characters of an allied species, or  REVERTS to some of the characters of an early progen
0347  nd great fact which strikes us in our general  REVIEW is, that barriers of any kind, or obstacles t
0282  storian will recognise as having produced a    REVOLUTION in natural science, yet does not admit how
0306  europe and the United States; and from the     REVOLUTION in our palaeontological ideas or many point
0484  y foresee that there will be a considerable    REVOLUTION in natural history. Systematists will be ab
0358  e admission of such prodigious geographical    REVOLUTIONS within the recent period, as are necessita
0250  in a pod of Crinum capense fertilised by C.    REVOLUTUM produced a plant, which (he says) I never s
0453  hought sufficient to say that because planets  REVOLVE in elliptic courses round the sun, satellite
0349  ts of Magellan are inhabited by one species of  RHEA (American ostrich), and northward the plains o
0384  of level. We have evidence in the loess of the  RHINE of considerable changes of level in the land
0068  nd, in some cases, as with the elephant and    RHINOCEROS, none are destroyed by beasts of prey: even
0141  as that former species of the elephant and     RHINOCEROS were capable of enduring a glacial climate,
0452  the bones of the leg of the horse, ox, and     RHINOCEROS. It is an important fact that rudimentary o
0252  ises stocks for grafting from a hybrid between  RHOD. Ponticum and Catawbiense, and that this hybri
0251  elargonium, Fuchsia, Calceolaria, Petunia,     RHODODENDRON, etc., have been crossed, yet many of thes
0140  becoming acclimatised: thus the pines and      RHODODENDRONS, raised from seed collected by Dr. Hooker
0251  ertility of some of the complex crosses of     RHODODENDRONS, and I am assured that many of them are p
0252  lowers of the more sterile kinds of hybrid     RHODODENDRONS, which produce no pollen, for he will fin
0229  tle bits, in other parts, large portions of a  RHOMBIC plate had been left between the opposed basi
0229  again examined the cell, and I found that the  RHOMBIC plate had been completed, and had become per
0229  ible, from the extreme thinness of the little  RHOMBIC plate, that they could have effected this by
0230  y one cell at the same time, but only the one  RHOMBIC plate which stands on the extreme growing ma
0230  nd they never complete the upper edges of the  RHOMBIC plates, until the hexagonal walls are commen
0231  orresponding in position to the planes of the  RHOMBIC basal plates of future cells. But the rough
0235  gles of the hexagonal prisms and of the basal  RHOMBIC plates. The motive power of the process of n
0225  o as to join on to a pyramid, formed of three  RHOMBS. These rhombs have certain angles, and the th
0225  n to a pyramid, formed of three rhombs. These  RHOMBS have certain angles, and the three which form
0227  d together by pyrimidal bases formed of three  RHOMBS; and the rhombs and the sides of the hexagona
0227  rimidal bases formed of three rhombs; and the  RHOMBS and the sides of the hexagonal prisms will ha
0229  site sides; for I have noticed half completed  RHOMBS at the base of a just commenced cell, which w
0208  ncts, one action follows another by a sort of  RHYTHM; if a person be interrupted in a song, or in
0022  cral vertebrae vary; as does the number of the  RIBS, together with their relative breadth and the
0029  lieves that the several sorts, for instance a  RIBSTON pippin or Codlin apple, could ever have proc
0038  th culture. It is not that these countries, so  RICH in species, do not by a strange chance possess
0211  way of running about, to be well aware what a  RICH flock it had discovered; it then began to play
0291  ced that all our ancient formations, which are  RICH in fossils, have thus been formed during subsi
0291  affected wide spaces. Consequently formations  RICH in fossils and sufficiently thick and extensiv
0292  periods of subsidence; that our great deposits  RICH in fossils have been accumulated. Nature may a
0417  ure from the proper type of the order, yet M.  RICHARD sagaciously saw, as Jussieu observes, that t
```

Page **(Key Word)**

0180 ghtly flattened, and from others, as Sir J. RICHARDSON has remarked, with the posterior part of th
0376 than to any other part of the world. Sir J. RICHARDSON, also, speaks of the re appearance on the s
0287 er in the long roll of years! Now turn to our RICHEST geological museums, and what a paltry displa
0299 mulating. The Malay Archipelago is one of the RICHEST regions of the whole world in organic beings
0303 nt of the tertiary series. And now one of the RICHEST known accumulations of fossil mammals belong
0396 islands, namely, the scarcity of kinds, the RICHNESS in endemic forms in particular classes or se
0429 en species instead of a single one; but such RICHNESS in species, as I find after some investigati
0228 e piece of wax, a thin and narrow, knife edged RIDGE, coloured with vermilion. The bees instantly
0228 each other, in the same way as before; but the RIDGE of wax was so thin, that the bottoms of the b
0229 etween the basins on the opposite sides of the RIDGE of wax. In parts, only little bits, in other
0229 rly the same rate on the opposite sides of the RIDGE of vermilion wax, as they circularly gnawed a
0230 nd thus flatten it. From the experiment of the RIDGE of vermilion wax, we can clearly see that if
0231 , by supposing masons first to pile up a broad RIDGE of cement, and then to begin cutting it away
0231 and adding fresh cement, on the summit of the RIDGE of wax. We shall thus have a thin wall steadily grow
0341 and present distribution. It may be asked in RIDICULE, whether I suppose that the megatherium and
0016 enera which I shall presently give, we have no RIGHT to expect often to meet with generic differen
0102 to preserve all the individuals varying in the RIGHT direction, though in different degrees, so as
0143 o vary in an allied manner: we see this in the RIGHT and left sides of the body varying in the sam
0175 come to be still more sharply defined. If I am RIGHT in believing that allied or representative sp
0188 ot this inference be presumptuous? Have we any RIGHT to assume that the Creator works by intellect
0252 ecies breeds freely. In confinement, we have no RIGHT to expect that the first crosses between them
0274 riety. For instance, I think those authors are RIGHT, who maintain that the ass has a prepotent po
0301 e degree of truth in these remarks, we have no RIGHT to expect to find in our geological formation
0305 sea, might have spread widely. Nor have we any RIGHT to suppose that the seas of the world have al
0309 continents areas of elevation. But have we any RIGHT to assume that things have thus remained from
0328 r of the world; but we are far from having any RIGHT to conclude that this has invariably been the
0333 nd that in a very broken condition, we have no RIGHT to expect, except in very rare cases, to fill
0333 istinct families or orders. All that we have a RIGHT to expect, is that those groups, which have w
0336 mutation of specific forms, as we have a just RIGHT to expect to find. On the state of Developmen
0412 from that including the next two genera on the RIGHT hand, which diverged from a common parent at
0412 ncluding the three genera still further to the RIGHT hand, which diverged at a still earlier perio
0462 ting forms, we should remember that we have no RIGHT to expect (excepting in rare cases) to discov
0463 d by a closely allied species, we have no just RIGHT to expect often to find intermediate varietie
0049 true species. But to discuss whether they are RIGHTLY called species or varieties, before any defi
0342 on the nature of the geological record, will RIGHTLY reject my whole theory. For he may ask in va
0219 e cells and tended the larvae, and put all to RIGHTS. What can be more extraordinary than these we
0157 accumulated by sexual selection, which is less RIGID in its action than ordinary selection, as it
0158 er variability, to sexual selection being less RIGID than ordinary selection, and to variations in
0081 iation in the least degree injurious would be RIGIDLY destroyed. This preservation of favourable v
0083 ales to struggle for the females. He does not RIGIDLY destroy all inferior animals, but protects d
0469 ut to this power, acting during long ages and RIGIDLY scrutinising the whole constitution, structu
0087 , and there would be simultaneously the most RIGOROUS selection of the young birds within the egg,
0088 spring. Sexual selection is, therefore, less RIGOROUS than natural selection. Generally, the most
0213 cts; but they have been acted on by far less RIGOROUS selection, and have been transmitted for an
0078 range so far, that they are destroyed by the RIGOUR of the climate alone. Not until we reach the
0230 comb, do make a rough, circumferential wall or RIM all round the comb; and they gnaw into this fr
0231 y can do this. Even in the rude circumferential RIM or wall of wax round a growing comb, flexures
0232 l, or the extreme margin of the circumferential RIM of a growing comb, with an extremely thin laye
0228 of the sphere of which they formed a part, the RIMS of the basins intersected or broke into each o
0072 ittle trees: and one of them, judging from the RINGS of growth, had during twenty six years tried
0357 unken islands are now marked, as I believe, by RINGS of coral or atolls standing over them. Whenev
0359 led to dry stems in branches of 94 plants with RIPE fruit, and to place them on sea water. The maj
0359 when dried floated much longer; for instance, RIPE hazel nuts sank immediately, but when dried, t
0359 anted they germinated: an asparagus plant with RIPE berries floated for 23 days, when dried it flo
0359 days, and the seeds afterwards germinated: the RIPE seeds of Helosciadium sank in two days, when d
0359 immersion of 28 days: and as 18/94 plants with RIPE fruit (but not all the same species as in the
0039 of each breed. The common goose has not given RISE to any marked varieties: hence the Thoulouse a
0051 entirely to analogy, and his difficulties will RISE to a climax. Certainly no clear line of demarc
0053 a certain extent from commonness), often give RISE to varieties sufficiently well marked to have
0106 those that will spread most widely, will give RISE to most new varieties and species, and will th
0121 hose which have largely varied, and have given RISE to new varieties and species. The other nine s
0130 ng inhabited a protected station. As buds give RISE by growth to fresh buds, and these, if vigorou
0170 when beneficial to the individual, that gives RISE to all the more important modifications of str
0184 such individuals would occasionally have given RISE to new species, having anomalous habits, and w
0212 ch might, if advantageous to the species, give RISE, through natural selection, to quite new insti
0298 ormations of Europe, which have oftenest given RISE, first to local varieties and ultimately to ne
0317 s must be slow and gradual, one species giving RISE first to two or three varieties, these being s
0325 eir own country, would naturally oftenest give RISE to new varieties or incipient species; for the
0326 ance of spreading still further, and of giving RISE in new countries to new varieties and species.
0342 hich have varied most, and have oftenest given RISE to new species; and that varieties have at fir
0359 ght be dried on the banks, and then by a fresh RISE in the stream be washed into the sea. Hence I
0381 tain species have been modified and have given RISE to new groups of forms, and others have remain
0405 ith which widely ranging species vary and give RISE to new forms will largely determine their aver
0422 ent language had altered little, and had given RISE to few new languages, whilst others (owing to
0422 a common race) had altered much, and had given RISE to many new languages and dialects. The variou
0485 re general departments of natural history will RISE greatly in interest. The terms used by natural
0486 forth. The study of domestic productions will RISE immensely in value. A new variety raised by ma
0040 re at one period than at another, as the breed RISES or falls in fashion, perhaps more in one dist
0382 ugh rising higher on the shores where the tide RISES highest, so have the living waters left their
0441 embryo in the course of development generally RISES in organisation: I use this expression, thoug
0450 t a corresponding not early period. Embryology RISES greatly in interest, when we thus look at the
0173 whilst the bed of the sea is stationary or is RISING, or when very little sediment is being deposi
0182 which now glide far through the air, slightly RISING and turning by the aid of their fluttering fi
0290 little reflection will explain why along the RISING coast of the western side of South America, n
0290 s they are brought up by the slow and gradual RISING of the land within the grinding action of the
0300 which the area would be either stationary or RISING: whilst rising, each fossiliferous formation

Page **********************************(Key Word)**********************************

```
0300   would be either stationary or rising; whilst RISING, each fossiliferous formation would be destro
0327   n the bed of the sea was either stationary or RISING, and likewise when sediment was not thrown do
0382   leaves its drift in horizontal lines, though RISING higher on the shores where the tide rises hig
0382   ift on our mountain summits, in a line gently RISING from the arctic lowlands to a great height un
0388   or stream is first formed; for instance, on a RISING islet, it will be unoccupied; and a single se
0482   k with confidence to the future, to young and RISING naturalists, who will be able to view both si
0191   ass over the orifice of the trachea, with some RISK of falling into the lungs, notwithstanding the
0030   er's teazle, with its hooks, which cannot be RIVALLED by any mechanical contrivance, is only a var
0089   e subject, believe that there is the severest RIVALRY between the males of many species to attract
0284   sediment is deposited by the great Mississippi RIVER at the rate of only 600 feet in a hundred tho
0383   of the last and present chapters. As lakes and RIVER systems are separated from each other by barr
0384   range widely and almost capriciously; for two RIVER systems will have some fish in common and som
0384   s, which from an early period must have parted RIVER systems and completely prevented their inoscu
0346   sy plains, forests, marshes, lakes, and great RIVERS, under almost every temperature. There is har
0348   of great deserts, and sometimes even of large RIVERS, we find different productions; though as mou
0384   eriod in the level of the land, having caused RIVERS to flow into each other. Instances, also, cou
0385   les, and would be sure to alight on a pool or RIVULET, if blown across sea to an oceanic island or
0221   leaving the nest, and marching along the same ROAD to a tall Scotch fir tree, twenty five yards d
0356   its climate was different may have been a high ROAD for migration, but now be impassable; I shall,
0415   will suffice to quote the highest authority, ROBERT Brown, who in speaking of certain organs in t
0416   se affinity between Ruminants and Pachyderms. ROBERT Brown has strongly insisted on the fact that
0262   fting; for Thouin found that three species of ROBINIA, which seeded freely on their own roots, and
0023   rect, namely, that all have descended from the ROCK pigeon (Columba livia), including under this t
0023   not varieties, and have not proceeded from the ROCK pigeon, they must have descended from at least
0023   supposed aboriginal stocks must all have been ROCK pigeons, that is, not breeding or willingly pe
0023   ub species, only two or three other species of ROCK pigeons are known; and these have not any of t
0023   re unlikely to be exterminated, and the common ROCK pigeon, which has the same habits with the dom
0024   so many species having similar habits with the ROCK pigeon seems to me a very rash assumption. Mor
0024   feral, though the dovecot pigeon, which is the ROCK pigeon in a very slightly altered state, has b
0024   n most parts of their structure, with the wild ROCK pigeon, yet are certainly highly abnormal in o
0025   inc of pigeons well deserve consideration. The ROCK pigeon is of a slaty blue, and has a white rum
0025   red and white edged tail feathers, as any wild ROCK pigeon! We can understand these facts, on the
0025   ll the domestic breeds have descended from the ROCK pigeon. But if we deny this, we must make one
0025   ginal stocks were coloured and marked like the ROCK pigeon, although no other existing species is
0026   in a score of generations, been crossed by the ROCK pigeon: I say within a dozen or twenty generat
0027   , though so like in most other respects to the ROCK pigeon; the blue colour and various marks occa
0027   iew, I may add, firstly, that C. livia, or the ROCK pigeon, has been found capable of domesticatio
0027   ffers immensely in certain characters from the ROCK pigeon, yet by comparing the several sub breed
0036   passed, and come to differ so greatly from the ROCK pigeon. Youatt gives an excellent illustration
0089   species to attract by singing the females. The ROCK thrush of Guiana, birds of Paradise, and some
0151   n. the opercular valves of sessile cirripedes (ROCK barnacles) are, in every sense of the word, ve
0159   et, characters not possessed by the aboriginal ROCK pigeon; these then are analogous variations in
0160   l these marks are characteristic of the parent ROCK pigeon, I presume that no one will doubt that
0162   es: if, for instance, we did not know that the ROCK pigeon was not feather footed or turn crowned,
0280   nd pouter pigeons have both descended from the ROCK pigeon; if we possessed all the intermediate v
0280   an extremely close series between both and the ROCK pigeon; but we should have no varieties direct
0281   comparison of their structure with that of the ROCK pigeon, whether they had descended from this s
0283   r can effect little of nothing in wearing away ROCK. At last the base of the cliff is undermined,
0286   ly every twenty two years. I doubt whether any ROCK, even as soft as chalk, would yield at this ra
0315   le from our present fantail; but if the parent ROCK pigeon were also destroyed, and in nature we h
0335   e order of their disappearance; for the parent ROCK pigeon now lives; and many varieties between t
0335   geon now lives; and many varieties between the ROCK pigeon and the carrier have become extinct; an
0373   ; in the northern half, ice borne fragments of ROCK have been observed on the eastern side as far
0446   ec tumbler differed from the young of the wild ROCK pigeon and of the other breeds, in all its pro
0473   pigeon have descended from the blue and barred ROCK pigeon! On the ordinary view of each species h
0282   a, and watch the sea at work grinding down old ROCKS and making fresh sediment, before he can hope
0282   s of sea coast, when formed of moderately hard ROCKS, and mark the process of degradation. The tid
0284   periods. So that the lofty pile of sedimentary ROCKS in Britain, gives but an inadequate idea of t
0285   to show such prodigious movements; the pile of ROCKS on the one or other side having been smoothly
0286   an safely picture to oneself the great dome of ROCKS which must have covered up the Weald within s
0286   , as some geologists suppose, a range of older ROCKS underlies the Weald, on the flanks of which t
0288   (a sub family of sessile cirripedes) coat the ROCKS all over the world in infinite numbers: they
0290   at, from the enormous degradation of the coast ROCKS and from muddy streams entering the sea. The
0300   ls, or of those which lived on naked submarine ROCKS, would be embedded; and those embedded in gra
0306   denly appear in the lowest known fossiliferous ROCKS. Most of the arguments which have convinced m
0307   bituminous matter in some of the lowest azoic ROCKS, probably indicates the former existence of l
0309   instance in South America, of bare metamorphic ROCKS, which must have been heated under great pres
0358   rphic schists, old fossiliferous or other such ROCKS, instead of consisting of mere piles of volca
0363   had found large fragments of granite and other ROCKS, which do not occur in the archipelago. Hence
0366   rt of the United States, erratic boulders, and ROCKS scored by drifted icebergs and coast ice, pla
0283   over, if we follow for a few miles any line of ROCKY cliff, which is undergoing degradation, we fi
0283   deeply impressed with the slowness with which ROCKY coasts are worn away. The observations on thi
0285   id state, showed at a glance how far the hard, ROCKY beds had once extended into the open ocean. T
0363   fely infer that icebergs formerly landed their ROCKY burthens on the shores of these mid ocean isl
0373   ratic boulders have, also, been noticed on the ROCKY Mountains. In the Cordillera of Equatorial So
0116   have remarked, our carnivorous, ruminant, and RODENT mammals, could successfully compete with thes
0137   ural selection. In South America, a burrowing RODENT, the tuco tuco, or Ctenomys, is even more sub
0141   and now have a far wider range than any other RODENT, living free under the cold climate of Faroe
0430   ving become adapted to habits like those of a RODENT. The elder De Candolle has made nearly simila
0137   reck. The eyes of moles and of some burrowing RODENTS are rudimentary in size, and in some cases a
0180   te and vegetation change, let other competing RODENTS or new beasts of prey immigrate, or old ones
0349   nd rabbits and belonging to the same order of RODENTS, but they plainly display an American type o
0349   aver or musk rat, but the coypu and capybara, RODENTS of the American type. Innumerable other inst
0403   n south America, Alpine humming birds, Alpine RODENTS, Alpine plants, etc., all of strictly Americ
0430   al; thus, according to Mr. Waterhouse, of all RODENTS, the bizcacha is most nearly related to Mars
0430   on. Therefore we must suppose either that all RODENTS, including the bizcacha, branched off from s
```

Page **********************************(Key Word)**********************************

Page ***(Key Word)***

```
0430  pect to all existing Marsupials: or that both RODENTS and Marsupials branched off from a common pr
0430  ter of its ancient progenitor than have other RODENTS; and therefore it will not be specially rela
0430  not any one species, but the general order of RODENTS. In this case, however, it may be strongly s
0032  rely go over their seed beds, and pull up the ROGUES, as they call the plants that deviate from th
0034  were ordered, and this may be compared to the ROGUING of plants by nurserymen. The principle of se
0287  sp, must have succeeded each other in the long ROLL of years! Now turn to our richest geological m
0283  om by atom, until reduced in size they can be  ROLLED about by the waves, and then are more quickly
0283  ttle they are abraded and how seldom they are  ROLLED about! Moreover, if we follow for a few miles
0165  ssing the several species of the horse genus.  ROLLIN asserts, that the common mule from the ass an
0034  a. explicit rules are laid down by some of the ROMAN classical writers. From passages in Genesis,
0028  e in the previous dynasty. In the time of the  ROMANS, as we hear from Pliny, immense prices were g
0028  h were as eager about pigeons as were the old  ROMANS. The paramount importance of these considerat
0064  d years, there would literally not be standing ROOM for his progeny. Linnaeus has calculated that
0243  odytes) of North America, build cock nests, to ROOST in, like the males of our distinct Kitty wren
0397  lly crawl on and adhere to the feet of birds   ROOSTING on the ground, and thus get transported? It
0266  i pretend that the foregoing remarks go to the ROOT of the matter: no explanation is offered why a
0296  s 1400 feet thick in Nova Scotia, with ancient ROOT bearing strata, one above the other, at no les
0159  analogous variation, in the enlarged stems, or ROOTS as commonly called, of the Swedish turnip and
0262  s of Robinia, which seeded freely on their own ROOTS, and which could be grafted with no great dif
0262  elded twice as much fruit as when on their own ROOTS. We are reminded by this latter fact of the e
0361  rocure stones for their tools, solely from the ROOTS of drifted trees, these stones being a valuab
0361  irregularly shaped stones are embedded in the  ROOTS of trees, small parcels of earth are very fre
0046  and which as varieties. We may instance Rubus, ROSA, and Hieracium amongst plants, several genera
0037  we now see in the varieties of the heartsease, ROSE, pelargonium, dahlia, and other plants, when c
0484  all fly produces monstrous growths on the wild ROSE or oak tree. Therefore I should infer from ana
0114  mers find that they can raise most food by a   ROTATION of plants belonging to the most different or
0114  re follows what may be called a simultaneous   ROTATION. Most of the animals and plants which live c
0208  rrupted in a song, or in repeating anything by ROTE, he is generally forced to go back to recover
0230  xamining the edge of a growing comb, do make a ROUGH, circumferential wall or rim all round the co
0230  ror to suppose that the bees cannot build up a ROUGH wall of wax in the proper position, that is,
0231  rhombic basal plates of future cells. But the  ROUGH wall of wax has in every case to be finished
0231  s build is curious: they always make the first ROUGH wall from ten to twenty times thicker than th
0232  s of laying down under certain circumstances a ROUGH wall in its proper place between two just com
0057  unequally related to each other, and clustered ROUND certain forms, that is, round their parent sp
0057  r, and clustered round certain forms, that is, ROUND their parent species? Undoubtedly there is on
0059  ally, allied together, forming little clusters ROUND certain species. Species very closely allied
0062  forget, that the birds which are idly singing  ROUND us mostly live on insects or seeds, and are t
0088  described as fighting, bellowing, and whirling ROUND, like Indians in a war dance, for the possess
0114  ost of the animals and plants which live close ROUND any small piece of ground, could live on it (
0129  a single file, but seem rather to be clustered ROUND points, and these round other points, and so
0129  rather to be clustered round points, and these ROUND other points, and so on in almost endless cyc
0189  specially if we look to much isolated species, ROUND which, according to my theory, there has been
0213  inherited by retrievers: and a tendency to run ROUND, instead of at, a flock of sheep, by shepherd
0213  eculiar gait; and another kind of wolf rushing ROUND, instead of at, a herd of deer, and driving t
0227  d many insects can make, apparently by turning ROUND on a fixed point. We must suppose the Melipon
0230  make a rough, circumferential wall or rim all  ROUND the comb: and they gnaw into this from the op
0231  in the rude circumferential rim or wall of wax ROUND a growing comb, flexures may sometimes be obs
0232  worked into the growing edges of the cells all ROUND. The work of construction seems to be a sort
0283  s only here and there, along a short length or ROUND a promontory, that the cliffs are at the pres
0285  hich have been worn by the waves and pared all ROUND into perpendicular cliffs of one or two thous
0345  produced by the laws of variation still acting ROUND us, and preserved by Natural Selection. Chapt
0367  re travelled southward, are remarkably uniform ROUND the world. We may suppose that the Glacial pe
0369  ncerent the arctic productions were as uniform ROUND the polar regions as they are at the present
0369  of the sub arctic and northern temperate forms ROUND the world, at the commencement of the Glacial
0453  at because planets revolve in elliptic courses ROUND the sun, satellites follow the same course ro
0453  und the sun, satellites follow the same course ROUND the planets, for the sake of symmetry, and to
0470  anges, and they are clustered in little groups ROUND other species, in which respects they resembl
0225  d likewise making separate and very irregular  ROUNDED cells of wax. At the other end of the series
0283  o we see along the bases of retreating cliffs  ROUNDED boulders, all thickly clothed by marine prod
0283  deposits, yet, from being formed of worn and   ROUNDED pebbles, each of which bears the stamp of ti
0445  birds of these several breeds were placed in a ROW, though most of them could be distinguished fr
0214  the instincts of either parent: for example,   Le ROY describes a dog, whose great grandfather was a
0361  f drifted trees, these stones being a valuable ROYAL tax. I find on examination, that when irregul
0046  pecies and which as varieties. We may instance RUBUS, Rosa, and Hieracium amongst plants, several
0033  f the principle in works of high antiquity. In RUDE and barbarous periods of English history choic
0227  old the stock of honey: in the same way as the RUDE humble bee adds cylinders of wax to the circul
0231  ing clearly that they can do this. Even in the RUDE circumferential rim or wall of wax round a gro
0262  n the act of reproduction, yet that there is a RUDE degree of parallelism in the results of grafti
0148  or part of the head is reduced to the merest   RUDIMENT attached to the bases of the prehensile ante
0161  metimes observe a mere tendency to produce a   RUDIMENT inherited: for instance, in the common snapd
0161  ce, in the common snapdragon (Antirrhinum) a   RUDIMENT of a fifth stamen so often appears, that thi
0451  parated sexes, the male flowers often have a   RUDIMENT of a pistil; and Kolreuter found that by cro
0451  le plants with an hermaphrodite species, the   RUDIMENT of the pistil in the hybrid offspring was mu
0451  h increased in size: and this shows that the   RUDIMENT and the perfect pistil are essentially alike
0452  gon (antirrhinum) we generally do not find a   RUDIMENT of a fifth stamen: but this may sometimes be
0093  ng rather a small quantity of pollen, and a    RUDIMENTARY pistil; other holly trees bear only female
0131  my of growth. False correlations. Multiple,    RUDIMENTARY, and lowly organised structures variable.
0135  ome other genera they are present, but in a    RUDIMENTARY condition. In the Ateuchus or sacred beetl
0135  f the anterior tarsi in Ateuchus, and their    RUDIMENTARY condition in some other genera, by the lon
0137  of moles and of some burrowing rodents are     RUDIMENTARY in size, and in some cases are quite cover
0149  eing, solely through and for its advantage.    RUDIMENTARY parts, it has been stated by some authors,
0149  all have to recur to the general subject of    RUDIMENTARY and aborted organs; and I will here only a
0150  o check deviations in their structure. Thus    RUDIMENTARY parts are left to the free play of the var
0152  d will then be said to have degenerated. In    RUDIMENTARY organs, and in those which have been but l
0161  er of generations, than in quite useless or    RUDIMENTARY organs being, as we all know them to be, t
0168  e specialised, and are higher in the scale.    RUDIMENTARY organs, from being useless, will be disreg
```

Page ***(Key Word)***

rudimentary

Page ***************************************(Key Word)***

```
0185  the upland goose may be said to have become  RUDIMENTARY in function, though not in structure. In t
0240  ereas the smaller workers have their ocelli  RUDIMENTARY. Having carefully dissected several specim
0240  rs, I can affirm that the eyes are far more  RUDIMENTARY in the smaller workers than can be account
0392  cies, having as useless an appendage as any  RUDIMENTARY organ, for instance, as the shrivelled win
0411  of Organic Beings. Morphology. Embryology.   RUDIMENTARY Organs. Classification, groups subordinate
0411  and being inherited at a corresponding age.  RUDIMENTARY Organs; their origin explained. Summary. F
0416  roup of beings. Again, no one will say that  RUDIMENTARY or atrophied organs are cf high physiologi
0416  lassification. No one will dispute that the  RUDIMENTARY teeth in the upper jaws of young ruminants
0416  upper jaws of young ruminants, and certain  RUDIMENTARY bones of the leg, are highly serviceable i
0416  has strongly insisted on the fact that the  RUDIMENTARY florets are of the highest importance in t
0425  ich each species has been recently exposed.  RUDIMENTARY structures on this view are as good as, or
0433  re than others! why we are permitted to use  RUDIMENTARY and useless organs, or others of trifling
0450  parent form of each great class of animals.  RUDIMENTARY, atrophied, or aborted organs. Organs or p
0450  ely common throughout nature. For instance,  RUDIMENTARY mammae are very general in the males of ma
0450  ds may be safely considered as a digit in a  RUDIMENTARY state: in very many snakes one lobe of the
0450  n very many snakes one lobe of the lungs is  RUDIMENTARY: in other snakes there are rudiments of th
0450  pelvis and hind limbs. Some of the cases of  RUDIMENTARY organs are extremely curious; for instance
0451  s, firmly soldered together! The meaning of  RUDIMENTARY organs is often quite unmistakeable: for i
0451  doubt, that the rudiments represent wings.   RUDIMENTARY organs sometimes retain their potentiality
0451  n there are normally four developed and two  RUDIMENTARY teats in the udders of the genus Bos, but
0451  organ serving for two purposes, many become  RUDIMENTARY or utterly aborted for one, even the more
0452  be fecundated, have a pistil, which is in a  RUDIMENTARY state, for it is not crowned with a stigma
0452  ounding anthers. Again, an organ may become  RUDIMENTARY for its proper purpose, and be used for a
0452  n certain fish the swim bladder seems to be  RUDIMENTARY for its proper function cf giving buoyancy
0452  ng. Cther similar instances could be given.  RUDIMENTARY organs in the individuals of the same spec
0452  e to which the same organ has been rendered  RUDIMENTARY occasionally differs much. This latter fac
0452  ings of the female moths in certain groups.  RUDIMENTARY organs may be utterly aborted; and this im
0452  nd rhinoceros. It is an important fact that  RUDIMENTARY organs, such as teeth in the upper jaws of
0453  s also, I believe, a universal rule, that a  RUDIMENTARY part or organ is of greater size relativel
0453  so that the organ at this early age is less  RUDIMENTARY, or even cannot be said to be in any degre
0453  or even cannot be said to be in any degree  RUDIMENTARY. Hence, also, a rudimentary organ in the a
0453  e in any degree rudimentary. Hence, also, a  RUDIMENTARY organ in the adult, is often said to be
0453  now given the leading facts with respect to  RUDIMENTARY organs. In reflecting on them, every one m
0453  s, tells us with equal plainness that these  RUDIMENTARY or atrophied organs, are imperfect and use
0453  ct and useless. In works on natural history  RUDIMENTARY organs are generally said to have been cre
0453  t physiologist accounts for the presence of  RUDIMENTARY organs, by supposing that they serve to ex
0453  s act? Can we suppose that the formation of  RUDIMENTARY teeth which are subsequertly absorbed, can
0454  order to excrete horny matter, as that the  RUDIMENTARY nails on the fin of the manatee were forme
0454  of descent with modification, the origin of  RUDIMENTARY organs is simple. We have plenty of cases
0454  rgans is simple. We have plenty of cases of  RUDIMENTARY organs in our domestic productions, as the
0454  of these cases throw light on the origin of  RUDIMENTARY organs in a state of nature, further than
0454  uction of various organs, until they become  RUDIMENTARY, as in the case of the eyes of animals inh
0454  e the organ, until it rendered harmless and  RUDIMENTARY. Any change in function, which can be effe
0455  can understand the greater relative size of  RUDIMENTARY organs in the embryo, anc their lesser rel
0455  good reason to believe to be possible) the  RUDIMENTARY part would tend to be wholly lost, and we
0455  tend to cause the entire obliteration of a  RUDIMENTARY organ. As the presence of rudimentary orga
0455  of a rudimentary organ. As the presence of  RUDIMENTARY organs is thus due to the tendency in ever
0455  ion, how it is that systematists have found  RUDIMENTARY parts as useful as, or even sometimes more
0455  an, parts of high physiological importance.  RUDIMENTARY organs may be compared with the letters in
0455  conclude that the existence of organs in a  RUDIMENTARY, imperfect, and useless condition, or quit
0456  of the most trifling importance, or, as in  RUDIMENTARY organs, of no importance: the wide opposit
0457  principle of inheritance, the occurrence of  RUDIMENTARY organs and their final abortion, present t
0457  portance of embryological characters and of  RUDIMENTARY organs in classification is intelligible,
0479  classification: why characters derived from  RUDIMENTARY parts, though of no service to the being,
0480  arly understand on this view the meaning of  RUDIMENTARY organs. But disuse and selection will gene
0480  organ will not be much reduced or rendered  RUDIMENTARY at this early age. The calf, for instance,
0480  y be said to have taken pains to reveal, by  RUDIMENTARY organs and by homologous structures, her s
0483  ervals between existing orders. Organs in a  RUDIMENTARY condition plainly show that an early proge
0485  paternity, morphology, adaptive characters,  RUDIMENTARY and aborted organs, etc., will cease to be
0486  of any kind which have long been inherited.  RUDIMENTARY organs will speak infallibly with respect
0240  ca. For the workers of Myrmica have not even  RUDIMENTS of ocelli, though the male and female ants
0450  gs is rudimentary; in other snakes there are  RUDIMENTS of the pelvis and hind limbs. Some of the c
0450  has even been stated on good authority that  RUDIMENTS of teeth can be detected in the beaks of ce
0451  will have full sized wings, and another mere  RUDIMENTS of membrane; and here it is impossible to d
0451  and here it is impossible to doubt, that the  RUDIMENTS represent wings. Rudimentary organs sometim
0451  e species the petals sometimes occur as mere  RUDIMENTS, and sometimes in a well developed state. I
0452  ore necessary, than the use and discovery of  RUDIMENTS. This is well shown in the drawings given b
0454  hole flower in the cauliflower. We often see  RUDIMENTS of various parts in monsters. But I doubt w
0454  tate of nature, further than by showing that  RUDIMENTS can be produced; for I doubt whether specie
0219  first discovered in the Formica (Polyerges)  RUFESCENS by Pierre Huber, a better observer even tha
0223  of f. sanguinea present with those of the F.  RUFESCENS. The latter does not build its own nest, do
0224  ly dependent on its slaves as is the Formica  RUFESCENS. Cell making instinct of the Hive Bee. I wi
0075  d kinds of trees now growing on the old Indian  RUINS! The dependency of one organic being on anoth
0366  america suffered under an Arctic climate. The  RUINS of a house burnt by fire do not tell their ta
0013  inheritance of every character whatever as the  RULE, and non inheritance as the anomaly. The laws
0013  degree, to males alone. A much more important  RULE, which I think may be trusted, is that, at wha
0014  and some other facts make me believe that the  RULE has a wider extension, and that when there is
0014  t first appeared in the parent. I believe this  RULE to be of the highest importance in explaining
0033  ill ensure some differences: but, as a general  RULE, I cannot doubt that the continued selection o
0055  eties or incipient species ought, as a general  RULE, to be now forming. Where many large trees gro
0056  ecies: for my tables clearly show as a general  RULE that, wherever many species of a genus have be
0064  ld not hold them. There is no exception to the  RULE that every organic being naturally increases a
0100  as recently informed me that he finds that the  RULE does not hold in Australia: and I have made th
0114  e each other most closely, shall, as a general  RULE, belong to what we call different genera and o
0119  infinitely complex relations. But as a general  RULE, the more diversified in structure the descend
0146  fruits which do not open: I should explain the  RULE by the fact that seeds could not gradually bec
```

Page ***************************************(Key Word)***

Page **(Key Word)**

```
0149 ction of some adjoining part. It seems to be a RULE, as remarked by Is. Geoffroy St. Hilaire, both
0150 d. I can only state my conviction that it is a RULE of high generality. I am aware of several caus
0150 nce for them. It should be understood that the RULE by no means applies to any part, however unusu
0150 ormal structure in the class mammalia; but the RULE would not here apply, because there is a whole
0150 with the other species of the same genus. The RULE applies very strongly in the case of secondary
0150 ly connected with the act of reproduction. The RULE applies to males and females; but as females m
0151 character, it applies more rarely to them. The RULE being so plainly applicable in the case of sec
0151 think there can be little doubt. But that our RULE is not confined to secondary sexual characters
0151 this Order, and I am fully convinced that the RULE almost invariably holds good with cirripedes.
0151 e only briefly give one, as it illustrates the RULE in its largest application. The opercular valv
0151 i have particularly attended to them, and the RULE seems to me certainly to hold good in this cla
0153 not excessively remote, we might, as a general RULE, expect still to find more variability in such
0176 not essentially differ from species, the same RULE will probably apply to both; and if we in imag
0176 nd practically, as far as I can make out, this RULE holds good with varieties in a state of nature
0176 ure. I have met with striking instances of the RULE in the case of varieties intermediate between
0176 endure for very long periods; why as a general RULE they should be exterminated and disappear, soo
0246 some degree of sterility. Kolreuter makes the RULE universal; but then he cuts the knot, for in t
0247 ks them as varieties. Gartner, also, makes the RULE equally universal; and he disputes the entire
0269 rility when crossed; and we may apply the same RULE to domestic varieties. In the second place, so
0297 nt to remember that naturalists have no golden RULE by which to distinguish species and varieties;
0313 t are low: though there are exceptions to this RULE. The amount of organic change, as Pictet has r
0313 he strongest apparent exception to this latter RULE, is that of the so called colonies of M. Barra
0316 hat there are some apparent exceptions to this RULE, but the exceptions are surprisingly few, so f
0316 views as I maintain) admit its truth; and the RULE strictly accords with my theory. For as all th
0316 h cases are certainly exceptional; the general RULE being a gradual increase in number, till the g
0329 re ancient any form is, the more, as a general RULE, it differs from living forms. But, as Bucklan
0331 ncient progenitor. Hence we can understand the RULE that the most ancient fossils differ most from
0334 s, that certain genera offer exceptions to the RULE. For instance, mastodons and elephants, when a
0353 duals of the same species, a directly opposite RULE prevailed; and species were not local, but had
0399 uld be given: indeed it is an almost universal RULE that the endemic productions of islands are re
0404 very wide ranges. I can hardly doubt that this RULE is generally true, though it would be difficul
0405 , we ought to find, and I believe as a general RULE we do find, that some at least of the species
0409 continuous in time; for the exceptions to the RULE are so few, that they may fairly be attributed
0410 elow: so in space, it certainly is the general RULE that the area inhabited by a single species, o
0410 ere are in both cases marked exceptions to the RULE. On my theory these several relations througho
0414 ve to recur. It may even be given as a general RULE, that the less any part of the organisation is
0446 but there was one remarkable exception to this RULE, for the young of the short faced tumbler diff
0446 ortions, proves that this is not the universal RULE; for here the characteristic differences must
0448 ure parent form. We have seen that this is the RULE of development in certain whole groups of anim
0453 disappear. It is also, I believe, a universal RULE, that a rudimentary part or organ is of greate
0470 has been active, we might expect, as a general RULE, to find it still in action; and this is the c
0478 es of the same species likewise occur. It is a RULE of high generality that the inhabitants of eac
0034 in an ancient Chinese encyclopaedia. Explicit RULES are laid down by some of the Roman classical
0255 a little more in detail the circumstances and RULES governing the sterility of first crosses and
0255 chief object will be to see whether or not the RULES indicate that species have specially been end
0255 ing together in utter confusion. The following RULES and conclusions are chiefly drawn up from Gar
0255 have taken much pains to ascertain how far the RULES apply to animals, and considering how scanty
0255 been surprised to find how generally the same RULES apply to both kingdoms. It has been already r
0258 lly in a higher degree. Several other singular RULES could be given from Gartner: for instance, so
0259 to either pure parent. Considering the several RULES now given, which govern the fertility of firs
0260 n fertility. Now do these complex and singular RULES indicate that species have been endowed with
0260 ems to be a strange arrangement. The foregoing RULES and facts, on the other hand, appear to me cl
0312 ear. Groups of species follow the same general RULES in their appearance and disappearance as do s
0312 . Let us now see whether the several facts and RULES relating to the geological succession of orga
0316 , genera and families, follow the same general RULES in their appearance and disappearance as do s
0411 groups subordinate to groups. Natural system. RULES and difficulties in classification, explained
0414 our classifications. Let us now consider the RULES followed in classification, and the difficult
0420 een subsequently discovered. All the foregoing RULES and aids and difficulties in classification a
0423 rious degrees of modification. Nearly the same RULES are followed in classifying varieties, as wit
0425 edi and only thus can I understand the several RULES and guides which have been followed by our be
0433 families, orders, etc., we can understand the RULES which we are compelled to follow in our class
0456 lines of affinities into one grand system; the RULES followed and the difficulties encountered by
0456 nd characters of true affinity; and other such RULES; all naturally follow on the view of the comm
0486 e what may be called the plan of creation. The RULES for classifying will no doubt become simpler
0116 e and others have remarked, our carnivorous, RUMINANT, and rodent mammals, could successfully comp
0329 n between existing groups. Cuvier ranked the RUMINANTS and Pachyderms, as the two most distinct or
0329 ertain pachyderms in the same sub order with RUMINANTS: for example, he dissolves by fine gradatio
0332 nearly the same manner as has occurred with RUMINANTS and pachyderms. Yet he who objected to call
0416 rudimentary teeth in the upper jaws of young RUMINANTS, and certain rudimentary bones of the leg,
0416 ble in exhibiting the close affinity between RUMINANTS and Pachyderms. Robert Brown has strongly i
0452 uch as teeth in the upper jaws of whales and RUMINANTS, can often be detected in the embryo, but a
0025 ock pigeon is of a slaty blue, and has a white RUMP (the Indian sub species, C. intermedia of Stri
0025 of as beautiful a blue colour, with the white RUMP, double black wing bar, and barred and white e
0160 irds with two black bars on the wings, a white RUMP, a bar at the end of the tail, with the outer
0014 ists, namely, that our domestic varieties; when RUN wild, gradually but certainly revert in charac
0064 m our domestic animals of many kinds which have RUN wild in several parts of the world: if the sta
0072 re neither cattle nor horses nor dogs have ever RUN wild, though they swarm southward and northwar
0073 rrinc with varying success; and yet in the long RUN the forces are so nicely balanced, that the fa
0092 d would be oftenest crossed; and so in the long RUN would gain the upper hand. Those flowers, also
0104 d propagating their kind; and thus, in the long RUN, the influence of intercrosses even at rare in
0109 in the seasons or in the number of its enemies; RUN a good chance of utter extinction. But we may
0148 tural selection will always succeed in the long RUN in reducing and saving every part of the organ
0153 y selection to keep the breed true. In the long RUN selection gains the day, and we do not expect
0176 g in lesser numbers would, as already remarked, RUN a greater chance of being exterminated than on
0213 gree inherited by retrievers; and a tendency to RUN round, instead of at, a flock of sheep, by she
0216 if the hen gives the danger chuckle, they will RUN (more especially young turkeys) from under her
```

Page **(Key Word)**

run

```
0262 ve organs perfect; yet these two distinct cases RUN to a certain extent parallel. Something analog
0277 s on widely different circumstances, should all RUN, to a certain extent, parallel with the system
0318 uction by the Spaniards into South America, has RUN wild over the whole country and has increased
0326 anges, or on strange accidents, but in the long RUN the dominant forms will generally succeed in s
0343 me rate, or in the same degree; yet in the long RUN that all undergo modification to some extent.
0344 are those that oftenest vary, will in the long RUN tend to people the world with allied, but modi
0463 at the intermediate varieties arise, will, in the long RUN, be supplanted and exterminated. On this doctr
0211 nd it immediately seemed, by its eager way of RUNNING about, to be well aware what a rich flock it
0274 owing to prepotency in transmitting likeness RUNNING more strongly in one, sex than in the other,
0305 h are known from south of the equator; and by RUNNING through Pictet's Palaeontology it will be se
0359 currents is 33 miles per diem (some currents RUNNING at the rate of 60 miles per diem); on this a
0479 l or bird having branchial slits and arteries RUNNING in loops, like those in a fish which has to
0274 he ass than the horse; but that the prepotency RUNS more strongly in the male ass than in the fema
0021 , and tumbling in the air head over heels. The RUNT is a bird of great size, with long, massive be
0023 english carrier, the short faced tumbler, the RUNT, the barb, pouter, and fantail in the same gen
0021 beak and large feet; some of the sub breeds of RUNTS have very long necks, others very long wings
0445 legs, in the wild stock; in pouters, fantails, RUNTS, barbs, dragons, carriers, and tumblers. Now
0213 ith a peculiar gait; and another kind of wolf RUSHING round, instead of at, a herd of deer, and dr
0222 for two or three individuals of F. fusca were RUSHING about in the greatest agitation, and one was
0076 species under the most different climates. In RUSSIA the small Asiatic cockroach has everywhere dr
0289 stance, from Sir R. Murchison's great work in RUSSIA, what wide gaps there are in that country bet
0299 ape to the Mediterranean, and from Britain to RUSSIA; and therefore equals all the geological form
0308 silurian deposits over immense territories in RUSSIA and in North America, do not support the view
0323 e several successive palaeozoic formations of RUSSIA, Western Europe and North America, a similar
0159 as commonly called, of the Swedish turnip and RUTA baga, plants which several botanists rank as v
0221 f the slave-making F. sanguinea. The latter RUTHLESSLY killed their small opponents, and carried t
0192 ain the eggs until they are hatched within the SACK. These cirripedes have no branchiae, the whole
0192 o branchiae, the whole surface of the body and SACK, including the small frena, serving for respir
0192 ena, the eggs lying loose at the bottom of the SACK, in the well enclosed shell; but they have lar
0192 reventing the ova from being washed out of the SACK? Although we must be extremely cautious in con
0441 uredly been retrograde; for the male is a mere SACK, which lives for a short time, and is destitut
0022 markable manner. The number of the caudal and SACRAL vertebrae vary; as does the number of the rib
0135 n a rudimentary condition. In the Ateuchus or SACRED beetle of the Egyptians, they are totally def
0270 ther cases consider fertility and sterility as SAFE criterions of specific distinction. Gartner ke
0011 d duck; and I presume that this change may be SAFELY attributed to the domestic duck flying much l
0014 great difficulty in proving its truth: we may SAFELY conclude that very many of the most strongly
0083 everywhere beaten some of the natives, we may SAFELY conclude that the natives might have been mod
0115 other natives; and we may, I think, at least SAFELY infer that diversification of structure, amou
0132 her more in that of plants. We may, at least, SAFELY conclude that such influences cannot have pro
0143 the good of the young or larva, will, it may SAFELY be concluded, affect the structure of the adu
0200 , are of special use to these animals. We may SAFELY attribute these structures to inheritance. Bu
0212 irds have been most persecuted by man. We may SAFELY attribute the greater wildness of our large b
0227 de of the cells of the hive bee. Hence we may SAFELY conclude that if we could slightly modify the
0239 rtile males and females, in this case, we may SAFELY conclude from the analogy of ordinary variati
0285 southern escarpments meet and close, one can SAFELY picture to oneself the great dome of rocks wh
0290 g action of the coast waves. We may, I think, SAFELY conclude that sediment must be accumulated in
0293 acy. With marine animals of all kinds, we may SAFELY infer a large amount of migration during clim
0361 st distances across the ocean. We may I think SAFELY assume that under such circumstances their ra
0363 do not occur in the archipelago. Hence we may SAFELY infer that icebergs formerly landed their roc
0402 succeed in establishing itself there? We may SAFELY infer that Charles Island is well stocked wit
0426 characteristics, from its allies, and yet be SAFELY classed with them. This may be safely done, a
0426 yet be safely classed with them. This may be SAFELY done, and is often done, as long as a suffici
0450 presume that the bastard wing in birds may be SAFELY considered as a digit in a rudimentary state;
0466 mestic productions have undergone; but we may SAFELY infer that the amount has been large, and tha
0489 ecome ennobled. Judging from the past, we may SAFELY infer that not one living species will transm
0360 our experiments. Therefore it would perhaps be SAFER to assume that the seeds of about 10/100 plan
0004 on under domestication, afforded the best and SAFEST clue. I may venture to express my conviction
0354 e universal law, seems to me incomparably the SAFEST. In discussing this subject, we shall be enab
0220 r masters in carrying them away to a place of SAFETY. Hence, it is clear, that the slaves fed qui
0057 confirm the view. I have also consulted some SAGACIOUS and most experienced observers, and, after
0417 he proper type of the order, yet M. Richard SAGACIOUSLY saw, as Jussieu observes, that this genus
0262 duals of the same two species in crossing; so SAGARET believes this to be the case with different
0270 ut the forms experimentised on, are ranked by SAGARET, who mainly founds his classification by the
0182 d front legs on the land, like the penguin; as SAILS, like the ostrich; and functionally for no pu
0052 term species, as one arbitrarily given for the SAKE of convenience to a set of individuals closely
0052 applied arbitrarily, and for mere convenience SAKE. Guided by theoretical considerations, I thoug
0063 ch pass into each other, I use for convenience SAKE the general term of struggle for existence. A
0091 ants excrete a sweet juice, apparently for the SAKE of eliminating something injurious from their
0092 n the case of insects visiting flowers for the SAKE of collecting pollen instead of nectar; and as
0127 nciple of preservation, I have called, for the SAKE of brevity, Natural Selection. Natural selecti
0359 ived an immersion of 137 days. For convenience SAKE I chiefly tried small seeds, without the capsu
0453 re generally said to have been created for the SAKE of symmetry, or in order to complete the schem
0453 low the same course round the planets, for the SAKE of symmetry, and to complete the scheme of nat
0088 d to the boar, and the hooked jaw to the male SALMON; for the shield may be as important for victo
0237 any birds, and in the hooked jaws of the male SALMON. We have even slight differences in the horns
0088 ance, for the possession of the females; male SALMONS have been seen fighting all day long; male s
0054 only two causes of obscurity. Fresh water and SALT loving plants have generally very wide ranges
0359 e sea, whether or not they were injured by the SALT water. Afterwards I tried some larger fruits,
0360 heir resistance to the injurious action of SALT water. On the other hand he did not previously
0361 p of a pigeon, which had floated on artificial SALT water for 30 days, to my surprise nearly all g
0364 here, if not killed by so long an immersion in SALT water, they could not endure our climate. Almo
0384 means for much migration. In the second place, SALT water fish can with care be slowly accustomed
0397 ous that land shells are very easily killed by SALT; their eggs, at least such as I have tried, si
0171 sition. Cases of difficulty. Natura non facit SALTUM. Organs of small importance. Organs not in al
0194 canon in natural history of Natura non facit SALTUM. We meet with this admission in the writings
0206 ld canon in natural history, Natura non facit SALTUM. This canon if we look only to the present in
```

```
0210  be discovered. The canon of Natura non facit SALTUM applies with almost equal force to instincts
0243  canon in natural history, of natura non facit SALTUM is applicable to instincts as well as to corp
0460  is proclaimed by the canon, Natura non facit SALTUM, that we ought to be extremely cautious in sa
0471  ow steps. Hence the canon of Natura non facit SALTUM, which every fresh addition to our knowledge
0283  eat into them only when they are charged with SAND or pebbles; for there is reason to believe tha
0283  and then are more quickly ground into pebbles, SAND, or mud. But how often do we see along the bas
0288  n. the remains which do become embedded, if in SAND or gravel, will when the beds are upraised gen
0300  d be embedded; and those embedded in gravel or SAND, would not endure to a distant epoch. Wherever
0303  ue mammal has been discovered in the new red SANDSTONE at nearly the commencement of this great se
0219  instinct could have been perfected. Formica SANGUINEA was likewise first discovered by P. Huber t
0220  little detail. I opened fourteen nests of F. SANGUINEA, and found a few slaves in all. Males and f
0220  have never been observed in the nests of F. SANGUINEA. The slaves are black and not above half th
0221  clinging to the legs of the slave making F. SANGUINEA. The latter ruthlessly killed their small o
0222  stone beneath a nest of the slave making F. SANGUINEA; and when I had accidentally disturbed both
0222  e. now I was curious to ascertain whether F. SANGUINEA could distinguish the pupae of F. fusca, wh
0222  ne evening I visited another community of F. SANGUINEA, and found a number of these ants entering
0222  eath, whence I saw the last individual of F. SANGUINEA emerge, carrying a pupa; but I was not able
0223  what a contrast the instinctive habits of F. SANGUINEA present with those of the F. rufescens. The
0223  ly dependent on its numerous slaves. Formica SANGUINEA, on the other hand, possesses much fewer sl
0223  witzerland. By what steps the instinct of F. SANGUINEA originated I will not pretend to conjecture
0224  much less extent even than in our British F. SANGUINEA, which, as we have seen, is less aided by i
0359  hout the capsule or fruit; and as all of these SANK in a few days, they could not be floated acros
0359  , and to place them on sea water. The majority SANK quickly, but some which whilst green floated f
0359  ted much longer; for instance, ripe hazel nuts SANK immediately, but when dried, they floated for
0359  rds germinated: the ripe seeds of Helosciadium SANK in two days, when dried they floated for above
0403  that Madeira and the adjoining islet of Porto SANTO possess many distinct but representative land
0403  s of stone are annually transported from Porto SANTO to Madeira, yet this latter island has not be
0403  r island has not become colonised by the Porto SANTO species: nevertheless both islands have repre
0091  e of eliminating something injurious from their SAP: this is effected by glands at the base of the
0261  d, in the period of the flow or nature of their SAP, etc., but in a multitude of cases we can assi
0055  ere many large trees grow, we expect to find SAPLINGS. Where many species of a genus have been for
0057  ups of species are generally clustered like SATELLITES around certain other species. And what are
0453  revolve in elliptic courses round the sun, SATELLITES follow the same course round the planets, f
0002  that want of space prevents my having the SATISFACTION of acknowledging the generous assistance w
0336  ts have not as yet defined to each other's SATISFACTION what is meant by high and low forms. But i
0437  n the theory of natural selection, we can SATISFACTORILY answer these questions. In the vertebrata
0243  tion, but to my imagination it is far more SATISFACTORY to look at such instincts as the young cuc
0307  ese vast primordial periods, I can give no SATISFACTORY answer. Several of the most eminent geolog
0313  istinct geographical province, seems to me SATISFACTORY. These several facts accord well with my t
0333  to living forms, seem to me explained in a SATISFACTORY manner. And they are wholly inexplicable o
0368  lacial period, seem to me to explain in so SATISFACTORY a manner the present distribution of the A
0368  heir present homes; but I have met with no SATISFACTORY evidence with respect to this intercalated
0044  e term species. No one definition has as yet SATISFIED all naturalists, yet every naturalist knows
0165  sses with the dun stock. But I am not at all SATISFIED with this theory, and should be loth to app
0488  ors of the highest eminence seem to be fully SATISFIED with the view that each species has been in
0183  : i have often watched a tyrant flycatcher (SAUROPHAGUS sulphuratus) in South America, hovering ov
0017  our domesticated productions; but how could a SAVAGE possibly know, when he first tamed an animal,
0202  may be difficult; but we ought to admire the SAVAGE instinctive hatred of the queen bee, which ur
0382  eings thus left stranded may be compared with SAVAGE races of man, driven up and surviving in the
0485  en we no longer look at an organic being as a SAVAGE looks at a ship, as at something wholly beyon
0018  to say how long before these ancient periods, SAVAGES, like those of Tierra del Fuego or Australia
0034  animals was at that early period attended to. SAVAGES now sometimes cross their dogs with wild can
0034  so, as is attested by passages in Pliny. The SAVAGES in South Africa match their draught cattle b
0034  t times, and is now attended to by the lowest SAVAGES. It would, indeed, have been a strange fact,
0036  ing quite different varieties. If there exist SAVAGES so barbarous as never to think of the inheri
0036  during famines and other accidents, to which SAVAGES are so liable, and such choice animals would
0038  e authors, namely, that the varieties kept by SAVAGES have more of the character of species than t
0042  enclosure of the land plays a part. Wandering SAVAGES or the inhabitants of open plains rarely pos
0198  elation would come into play. Animals kept by SAVAGES in different countries often have to struggl
0215  as Tierra del Fuego and Australia, where the SAVAGES do not keep these domestic animals. How rare
0070  rable that many could exist together, and thus SAVE each other from utter destruction. I should ad
0094  ve many facts, showing how anxious bees are to SAVE time; for instance, their habit of cutting hol
0234  l in common even to two adjoining cells, would SAVE some little wax. Hence it would continually be
0130  ranches of life, and which has apparently been SAVED from fatal competition by having inhabited a
0136  n whether a greater number of individuals were SAVED by successfully battling with the winds, or b
0168  ; and every part of the structure which can be SAVED without detriment to the individual, will be
0168  d without detriment to the individual, will be SAVED. Changes of structure at an early age will ge
0219  fusca), and she instantly set to work, fed and SAVED the survivors; made some cells and tended the
0234  ve to bound other cells, and much wax would be SAVED. Again, from the same cause, it would be adva
0455  cture, if not useful to the possessor, will be SAVED as far as is possible, will probably often co
0226  joining cells. It is obvious that the Melipona SAVES wax by this manner of building; for the flat
0148  the bases of the prehensile antennae. Now the SAVING of a large and complex structure, when render
0148  lways succeed in the long run in reducing and SAVING every part of the organisation, as soon as it
0234  rge number of bees being supported. Hence the SAVING of wax by largely saving honey must be a most
0234  supported. Hence the saving of wax by largely SAVING honey must be a most important element of suc
0031  ake him six years to obtain head and beak. In SAXONY the importance of the principle of selection
0054  h the species belong. Again, plants low in the SCALE of organisation are generally much more widel
0055  more widely diffused than plants higher in the SCALE; and here again there is no close relation to
0073  ing how plants and animals, most remote in the SCALE of nature, are bound together by a web of com
0075  y, lies generally between beings remote in the SCALE of nature. This is often the case with those
0083  constitution may well turn the nicely balanced SCALE in the struggle for life, and so be preserved
0088  ul selection of the best cocks. How low in the SCALE of nature this law of battle descends, I know
0107  ent orders now widely separated in the natural SCALE. These anomalous forms may almost be called l
0149  opinion of naturalists, that beings low in the SCALE of nature are more variable than those which
0168  this same cause that organic beings low in the SCALE of nature are more variable than those which
0168  sation more specialised, and are higher in the SCALE. Rudimentary organs, from being useless, will
```

Page **(Key Word)**

```
0208 mes into play, even in animals very low in the  SCALE of nature. Frederick Cuvier and several of th
0235 entically the same in animals so remote in the   SCALE of nature, that we cannot account for their s
0252 nfer that animals more widely separated in the   SCALE of nature can be more easily crossed than in
0280 cess of extermination has acted on an enormous   SCALE, so must the number of intermediate varieties
0313 believe that organisms, considered high in the   SCALE of nature, change more quickly than those tha
0345 ce for life, and are, in so far, higher in the   SCALE of nature; and this may account for that vagu
0388 s many, fresh water productions are low in the   SCALE of nature, and that we have reason to believe
0400 ntinent, we sometimes see displayed on a small   SCALE, yet in a most interesting manner, within the
0406 geological evidence that organisms low in the    SCALE within each great class, generally change at
0441 animal is generally considered as lower in the   SCALE than the larva, as with certain parasitic cru
0468 very severe between beings most remote in the    SCALE of nature. The slightest advantage in one bei
0191 ome naturalists that the branchiae and dorsal    SCALES of Annelids are homologous with the wings and
0144 z. cetacea (whales) and Edentata (armadilloes,   SCALY anteaters, etc.), that these are likewise the
0328 successive Silurian deposits of Bohemia and      SCANDINAVIA; nevertheless he finds a surprising amount
0368 especially allied to the plants of northern      SCANDINAVIA; those of the United States to Labrador; t
0290 peculiar marine fauna, tertiary beds are so      SCANTILY developed, that no record of several success
0041 ntry. When the individuals of any species are    SCANTY, all the individuals, whatever their quality
0239 numerous, with those of an intermediate size     SCANTY in numbers, Formica flava has larger and smal
0255 e rules apply to animals, and considering how    SCANTY our knowledge is in regard to hybrid animals,
0359 s, as far as we may infer anything from these    SCANTY facts, we may conclude that the seeds of 14/1
0374 fuego, forming no inconsiderable part of its     SCANTY flora, are common to Europe, enormously remot
0390 islands the number of kinds of inhabitants is    SCANTY, the proportion of endemic species (i.e. thos
0409 me developed in great force, some existing in    SCANTY numbers, in the different great geographical
0002 re work to do this. For I am well aware that     SCARCELY a single point is discussed in this volume o
0033 has been reduced to methodical practice for      SCARCELY more than three quarters of a century; it ha
0058 restricted ranges: this statement is indeed      SCARCELY more than a truism, for if a variety were fo
0097 hed if these visits be prevented. Now, it is     SCARCELY possible that bees should fly from flower to
0187 usively to its lineal ancestors; but this is     SCARCELY ever possible, and we are forced in each cas
0188 l selection to such startling lengths. It is     SCARCELY possible to avoid comparing the eye to a tel
0215 t to tame than the young of the wild rabbit;     SCARCELY any animal is tamer than the young of the ta
0215 s have been modified by domestication. It is     SCARCELY possible to doubt that the love of man has b
0266 ions having been blended into one. For it is     SCARCELY possible that two organisations should be co
0290 e not followed each other in close sequence.     SCARCELY any fact struck me more when examining many
0299 they otherwise would have been, yet has done     SCARCELY anything in breaking down the distinction be
0322 almost simultaneously throughout the World.      SCARCELY any palaeontological discovery is more strik
0336 ind, in short, such evidence of the slow and     SCARCELY sensible mutation of specific forms, as we h
0357 are to be trusted, it must be admitted that     SCARCELY a single island exists which has not recentl
0364 nt gales of wind. It should be observed that     SCARCELY any means of transport would carry seeds for
0365 even if far less well stocked than Britain,     SCARCELY more than one would be so well fitted to its
0379 at the whole body of arctic shells underwent    SCARCELY any modification during their long southern
0424 nt characters, is known to every naturalist:    SCARCELY a single fact can be predicated in common of
0396 inhabitants of oceanic islands, namely, the     SCARCITY of kinds, the richness in endemic forms in p
0223 e seen, carry off pupae of other species, if    SCATTERED near their nests, it is possible that pupae
0362 s of their torn crops might thus readily get    SCATTERED. Mr. Brent informs me that a friend of his
0375 of the peninsula of Malacca, and are thinly     SCATTERED, on the one hand over India and on the othe
0075 one on during long centuries, each annually     SCATTERING its seeds by the thousand; what war between
0315 mplete act of creation, but only an occasional  SCENE, taken almost at hazard, in a slowly changing
0202 ers. If we admire the truly wonderful power of  SCENT by which the males of many insects find their
0213 young and without any training, as soon as it   SCENTED its prey, stand motionless like a statue, an
0220 smith, I tried to approach the subject in a     SCEPTICAL frame of mind, as any one may well be excus
0413 s system? Some authors look at it merely as a   SCHEME for arranging together those living objects w
0414 some unknown plan of creation, or is simply a   SCHEME for enunciating general propositions and of p
0453 sake of symmetry, or in order to complete the   SCHEME of nature; but this seems to me no explanatio
0453 for the sake of symmetry, and to complete the   SCHEME of nature? An eminent physiologist accounts f
0480 tary organs and by homologous structures, her   SCHEME of modification, which it seems that we wilfu
0138 affinities might have been expected; but, as    SCHIODTE and others have remarked, this is not the ca
0138 evidence of this gradation of habit; for, as    SCHIODTE remarks, animals not far remote from ordinar
0358 her mountain summits, of granite, metamorphic   SCHISTS, old fossiliferous or other such rocks, inst
0144 e head of the child. In snakes, according to    SCHLEGEL, the shape of the body and the manner of swa
0282 se as having produced a revolution in natural   SCIENCE, yet does not admit how incomprehensibly vas
0487 can be thrown on ancient geography. The noble   SCIENCE of Geology loses glory from the extreme impe
0251 ts of horticulturists, though not made with     SCIENTIFIC precision, deserve some notice. It is notor
0485 the primrose and cowslip; and in this case      SCIENTIFIC and common language will come into accordan
0185 in structure. In the frigate bird, the deeply   SCOOPED membrane between the toes shows that structu
0082 ed; and natural selection would thus have free  SCOPE for the work of improvement. We have reason t
0157 ariable, sexual selection will have had a wide  SCOPE for action, and may thus readily have succeed
0019 t this rate there must have existed at least a  SCORE of species of wild cattle, as many sheep, and
0022 degree from each other. Altogether at least a   SCORE of pigeons might be chosen, which if shown to
0026 rest, has within a dozen or, at most, within a  SCORE of generations, been crossec by the rock pige
0066 condor lays a couple of eggs and the ostrich a  SCORE, and yet in the same country the condor may b
0086 y modify and adapt the larva of an insect to a  SCORE of contingencies, wholly different. from those
0142 weight; but until some one will sow, during a   SCORE of generations, his kidney beans so early tha
0160 y generations, some say, for a dozen or even a  SCORE of generations. After twelve generations, the
0221 another day my attention was struck by about a  SCORE of the slave makers haunting the same spot, a
0231 ng to another, so that, as Huber has stated, a  SCORE of individuals work even at the commencement
0247 oubted; for Gartner gives in his table about a  SCORE of cases of plants which he castrated, and ar
0366 e mountains of Scotland and Wales, with their   SCORED flanks, polished surfaces, and perched boulde
0366 he united States, erratic boulders, and rocks   SCORED by drifted icebergs and coast ice, plainly re
0029 nded from long horns, and he will laugh you to  SCORN. I have never met a pigeon, or poultry, or du
0071 twenty five years previously and planted with   SCOTCH fir. The change in the native vegetation of t
0071 re extensive heaths, with a few clumps of old   SCOTCH firs on the distant hilltops: within the last
0072 heath, and literally I could not see a single   SCOTCH fir, except the old planted clumps. But on lo
0072 tle absolutely determine the existence of the   SCOTCH fir; but in several parts of the world insect
0221 t, and marching along the same road to a tall   SCOTCH fir tree, twenty five yards distant, which th
0296 nd carboniferous beds 1400 feet thick in Nova   SCOTIA, with ancient root bearing strata, one above
0076 nt increase of the missel thrush in parts of    SCOTLAND has caused the decrease of the song thrush.
```

Page **(Key Word)**

Page **(Key Word)**

```
0366   tale more plainly, than do the mountains of SCOTLAND and Wales, with their scored flanks, polishe
0367   nd north. The Alpine plants, for example, of SCOTLAND, as remarked by Mr. H. C. Watson, and those
0084   that natural selection is daily and hourly SCRUTINISING, throughout the world, every variation, ev
0469   power, acting during long ages and rigidly SCRUTINISING the whole constitution, structure, and hab
0145   s plants, the seeds also differ in shape and SCULPTURE; and even the ovary itself, with its access
0323   in such trifling points as mere superficial SCULPTURE. Moreover other forms, which are not found
0410   ised by trifling characters in common, as of SCULPTURE or colour. In looking to the long successio
0022   length of leg and of the feet; the number of SCUTELLAE on the toes, the development of skin betwee
0026   should yield offspring perfectly fertile, inter SE, seems to me rash in the extreme. From these se
0253   and in one single instance they have bred inter SE. This was effected by Mr. Eyton, who raised two
0266   s of life. When hybrids are able to breed inter SE, they transmit to their offspring from generati
0107   er, make a small area compared with that of the SEA or of the land; and, consequently, the competi
0132   wollaston is convinced that residence near the SEA affects their colours. Moquin Tandon gives a l
0132   es a list of plants which when growing near the SEA shore have their leaves in some degree fleshy,
0133   hose of islands. The insect species confined to SEA coasts, as every collector knows, are often br
0133   or lurid. Plants which live exclusively on the SEA side are very apt to have fleshy leaves. He wh
0133   nce, was created with bright colours for a warm SEA; but that this other shell became bright colou
0135   parts of the world are very frequently blown to SEA and perish; that the beetles in Madeira, as ob
0136   chance of surviving from not being blown out to SEA; and, on the other hand, those beetles which m
0136   took to flight will oftenest have been blown to SEA and thus have been destroyed. The insects in M
0167   tone so as to mock the shells now living on the SEA shore. Summary. Our ignorance of the laws of v
0173   ic beings not inhabiting profound depths of the SEA, and to their remains being embedded and prese
0173   sediment is deposited on the shallow bed of the SEA, whilst it slowly subsides. These contingencie
0173   normously long intervals. Whilst the bed of the SEA is stationary or is rising, or when very littl
0175   noticed by Forbes in sounding the depths of the SEA with the dredge. To those who look at climate
0185   four toes webbed, alight on the surface of the SEA. On the other hand, grebes and coots are emine
0258   thuret has observed the same fact with certain SEA weeds or Fuci. Gartner, moreover, found that t
0282   eat piles of superimposed strata, and watch the SEA at work grinding down old rocks and making fre
0282   around us. It is good to wander along lines of SEA coast, when formed of moderately hard rocks, a
0283   . he who most closely studies the action of the SEA on our shores, will, I believe, be most deeply
0284   sediment is transported by the currents of the SEA, the process of accumulation in any one area m
0285   so completely planed down by the action of the SEA, that no trace of these vast dislocations is e
0286   strict. If, then, we knew the rate at which the SEA commonly wears away a line of cliff of any giv
0286   me crude notion on the subject, assume that the SEA would eat into cliffs 500 feet in height at th
0287   s land, and thus have escaped the action of the SEA: when deeply submerged for perhaps equally lon
0288   ay and disappear when left on the bottom of the SEA, where sediment is not accumulating. I believe
0288   eing deposited over nearly the whole bed of the SEA, at a rate sufficiently quick to embed and pre
0288   xplicable only on the view of the bottom of the SEA not rarely lying for ages in an unaltered cond
0290   coast rocks and from muddy streams entering the SEA. The explanation, no doubt, is, that the litto
0291   in two ways: either, in profound depths of the SEA, in which case, judging from the researches of
0291   pply of sediment nearly balance each other, the SEA will remain shallow and favourable for life, a
0291   e supply of sediment was sufficient to keep the SEA shallow and to embed and preserve the remains
0292   y. on the other hand, as long as the bed of the SEA remained stationary, thick deposits could not
0292   he land and of the adjoining shoal parts of the SEA will be increased, and new stations will often
0294   e become extinct in the immediately surrounding SEA; or, conversely, that some are now abundant in
0294   that some are now abundant in the neighbouring SEA, but are rare or absent in this particular dep
0295   interrupted, as a change in the currents of the SEA, and a supply of sediment of a different nature
0299   l aboriginal stocks; or, again, whether certain SEA shells inhabiting the shores of North America,
0300   r sediment did not accumulate on the bed of the SEA, or where it did not accumulate at a sufficien
0305   after having been largely developed in some one SEA, might have spread widely. Nor have we any rig
0308   or again as the bed of an open and unfathomable SEA. Looking to the existing oceans, which are thr
0313   m to change at a quicker rate than those of the SEA, of which a striking instance has lately been
0323   ected that they had coexisted with still living SEA shells; but as these anomalous monsters coexis
0326   n with the marine inhabitants of the continuous SEA. We might therefore expect to find, as we appa
0326   sion in the productions of the land than of the SEA. Dominant species spreading from any region mi
0326   and, and another for those in the waters of the SEA. If two great regions had been for a long peri
0327   occurred during the periods when the bed of the SEA was either stationary or rising, and likewise
0336   he forms of life, at least those inhabiting the SEA, have changed almost simultaneously throughout
0336   le the specific forms of the inhabitants of the SEA have been affected. On the theory of descent,
0339   ard has shown that the same law holds good with SEA shells, but from the wide distribution of most
0339   ving brackish water shells of the Aralo Caspian SEA. Now what does this remarkable law of the succ
0347   be given with respect to the inhabitants of the SEA. A second great fact which strikes us in our g
0348   teristic of distinct continents. Turning to the SEA, we find the same law. No two marine faunas ar
0348   by impassable barriers, either of land or open SEA, they are wholly distinct. On the other hand,
0349   ity of the productions of the same continent or SEA, though the species themselves are distinct at
0352   ceptional. The capacity of migrating across the SEA is more distinctly limited in terrestrial mamm
0354   , though separated by hundreds of miles of open SEA. If the existence of the same species at dista
0356   w blend or may formerly have blended: where the SEA now extends, land may at a former period have
0357   istence of many islands, now buried beneath the SEA, which may have served as halting places for p
0358   he distribution of mammals and the depth of the SEA, these and other such facts seem to me opposed
0358   ide dissemination; but for transport across the SEA, the greater or less facilities may be said to
0358   far seeds could resist the injurious action of SEA water. To my surprise I found that out of 87 k
0359   could not be floated across wide spaces of the SEA, whether or not they were injured by the salt
0359   y a fresh rise in the stream be washed into the SEA. Hence I was led to dry stems in branches of 9
0359   94 plants with ripe fruit, and to place them on SEA water. The majority sank quickly, but some whi
0359   4/100 plants of any country might be floated by SEA currents during 28 days, and would retain thei
0360   ne country might be floated across 924 miles of SEA to another country; and when stranded, if blow
0360   for he placed the seeds in a box in the actual SEA, so that they were alternately wet and exposed
0360   likewise seeds from plants which live near the SEA; and this would have favoured the average leng
0360   been dried, could be floated across a space of SEA 900 miles in width, and would then germinate.
0361   at the carcasses of birds, when floating on the SEA, sometimes escape being immediately devoured;
0361   nce, are killed by even a few days immersion in SEA water; but some taken out of the crop of a pig
0364   is is not strictly correct: the currents of the SEA are not accidental, nor is the direction of pr
0364   sed for a great length of time to the action of SEA water: nor could they be long carried in the c
0364   ffice for occasional transport across tracts of SEA some hundred miles in breadth, or from island
0372   stirct. As on the land, so in the waters of the SEA, a slow southern migration of a marine fauna,
```

Page **(Key Word)**

Page **(Key Word)***

```
0378  e climate under the equator at the level of the SEA was about the same with that now felt there at
0383  nged widely within the same country, and as the SEA is apparently a still more impassable barrier,
0384  group might travel far along the shores of the SEA, and subsequently become modified and adapted
0385  ed by birds, and they are immediately killed by SEA water, as are the adults. I could not even und
0385  to alight on a pool or rivulet, if blown across SEA to an oceanic island or to any other distant p
0386  d not be likely to alight on the surface of the SEA, so that the dirt would not be washed off thei
0387  and go to other waters, or are blown across the SEA; and we have seen that seeds retain their powe
0391  ecular land shells, whereas not one species of SEA shell is confined to its shores: now, though w
0391  d to its shores: now, though we do not know how SEA shells are dispersed, yet we can see that the i
0391  lls, across three or four hundred miles of open SEA. The different orders of insects in Madeira ap
0393  eir spawn are known to be immediately killed by SEA water, on my view we can see that there would
0393  reat difficulty in their transportal across the SEA, and therefore why they do not exist on any oc
0394  ammal can be transported across a wide space of SEA, but bats can fly across. Bats have been seen
0395  dependent of distance, between the depth of the SEA separating an island from the neighbouring mai
0396  the frequent relation between the depth of the SEA and the degree of affinity of the mammalia
0397  ir eggs, at least such as I have tried, sink in SEA water and are killed by it. Yet there must be,
0397  ifted timber across moderately wide arms of the SEA. And I found that several species did in this
0397  this state withstand uninjured an immersion in SEA water during seven days: one of these shells w
0397  , and after it had again hybernated I put it in SEA water for twenty days, and it perfectly recove
0397  branous one, I immersed it for fourteen days in SEA water, and it recovered and crawled away: but
0401  f each other, are separated by deep arms of the SEA, in most cases wider than the British Channel,
0401  d been continuously united. The currents of the SEA are rapid and sweep across the archipelago, an
0409  r less modified condition, and the depth of the SEA between an island and the mainland. We can cle
0436  ibility. In the paddles of the extinct gigantic SEA lizards, and in the mouths of certain suctoria
0465  diment is deposited on the subsiding bed of the SEA. During the alternate periods of elevation and
0477  lose alliance of some of the inhabitants of the SEA in the northern and southern temperate zones,
0487  aring the differences of the inhabitants of the SEA on the opposite sides of a continent, and the
0200  the wing of the bat, and in the flipper of the SEAL, are of special use to these animals. We may s
0200  . so we may believe that the progenitor of the SEAL had not a flipper, but a foot with five toes f
0093  ith pollen, having flown from tree to tree in SEARCH of nectar. But to return to our imaginary cas
0195  or not so well enabled in a coming dearth to SEARCH for food, or to escape from beasts of prey. O
0221  ant, which they ascended together probably in SEARCH of aphides or cocci. According to Huber, who
0221  xpressly states, their principal office is to SEARCH for aphides. This difference in the usual hab
0221  haunting the same spot, and evidently not in SEARCH of food; they approached and were vigorously
0393  the oldest voyages, but have not finished my SEARCH; as yet I have not found a single instance, f
0441  not feed: their function at this stage is, to SEARCH by their well developed organs of sense, and
0485  we shall have at least be freed from the vain SEARCH for the undiscovered and undiscoverable essen
0072  cattle would have so closely and effectually SEARCHED it for food. Here we see that cattle absolut
0393  r another and similar case. I have carefully SEARCHED the oldest voyages, but have not finished my
0133  lls which are confined to tropical and shallow SEAS are generally brighter coloured than those con
0133  oloured than those confined to cold and deeper SEAS. The birds which are confined to continents ar
0294  ir migration from the American to the European SEAS. In examining the latest deposits of various o
0299  us large islands separated by wide and shallow SEAS, probably represents the former state of Europ
0305  ely. Nor have we any right to suppose that the SEAS of the world have always been so freely open f
0306  or australia, and thus reach other and distant SEAS. From these and similar considerations, but ch
0321  condary formations, survives in the Australian SEAS; and a few members of the great and almost ext
0328  d, it be assumed that an isthmus separated two SEAS inhabited by distinct, but contemporaneous, fa
0350  on the American continent and in the American SEAS. We see in these facts some deep organic bond,
0358  tertiary inhabitants of several lands and even SEAS to their present inhabitants, a certain degree
0372  arine animals, in the Mediterranean and in the SEAS of Japan, areas now separated by a continent a
0372  nship, without identity, of the inhabitants of SEAS now disjoined, and likewise of the past and pr
0376  ve not been found in the intermediate tropical SEAS. It should be observed that the northern speci
0063  ring some period of its life, and during some SEASON or occasional year, otherwise, on the princip
0070  having a super abundance of food at this one SEASON, increase in number proportionally to the sup
0079  each at some period of its life, during some SEASON of the year, during each generation or at int
0083  lor animals, but protects during each varying SEASON, as far as lies in his power, all his product
0089  me to the breeding age or during the breeding SEASON; the modifications thus produced being inheri
0090  er prey had decreased in numbers, during that SEASON of the year when the wolf is hardest pressed
0127  ers of increase of each species, at some age, SEASON, or year, a severe struggle for life, and thi
0249  t be carefully prevented during the flowering SEASON: hence hybrids will generally be fertilised d
0468  antage in one being, at any age or during any SEASON, over those with which it comes into competit
0359  erence there is in the buoyancy of green and SEASONED timber; and it occurred to me that floods mi
0030  ces of plants most useful to man at different SEASONS and for different purposes, or so beautiful
0038  e for their own food, at least during certain SEASONS. And in two countries very differently circu
0062  may be now superabundant, it is not so at all SEASONS of each recurring year. I should premise tha
0064  ourable to them during two or three following SEASONS. Still more striking is the evidence from ou
0068  average numbers of a species, and periodical SEASONS of extreme cold or drought, I believe to be
0074  fferent periods of life, and curing different SEASONS or years, probably come into play: some one
0089  uced being inherited at corresponding ages or SEASONS, either by the males alone, or by the males
0103  stations, from breeding at slightly different SEASONS, or from varieties of the same kind preferri
0109  individuals will, during fluctuations in the SEASONS or in the number of its enemies, run a good
0175  mber of its enemies or of its prey, or in the SEASONS, be extremely liable to utter extermination;
0210  at different periods of life, or at different SEASONS of the year, or when placed under different
0467  n, by the effects of a succession of peculiar SEASONS, and by the results of naturalisation, as ex
0400  : but it arises in chief part from the deeply SEATED error of considering the physical conditions
0391  hat their eggs or larvae, perhaps attached to SEAWEED or floating timber, or to the feet of wading
0020  ces or species, I can hardly believe. Sir John SEBRIGHT expressly experimentised for this object, an
0031  istence. That most skilful breeder, Sir John SEBRIGHT, used to say, with respect to pigeons, that
0247  , and, what is often more important, must be SECLUDED in order to prevent pollen being brought to
0131  cific characters more variable than generic: SECONDARY sexual characters variable. Species of the
0150  he rule applies very strongly in the case of SECONDARY sexual characters, when displayed in any un
0150  n displayed in any unusual manner. The term, SECONDARY sexual characters, used by Hunter, applies
0150  but as females more rarely offer remarkable SECONDARY sexual character, it applies more rarely to
0151  e being so plainly applicable in the case of SECONDARY sexual characters, may be due to the great
0151  doubt. But that our rule is not confined to SECONDARY sexual characters is clearly shown in the c
0156  mitted, without my entering on details, that SECONDARY sexual characters are very variable: I thin
```

Page **(Key Word)***

Page ***************************************(Key Word)***

```
0156  differ from each other more widely in their  SECONDARY  sexual characters, than in other parts of t
0156  en the males of gallinaceous birds, in which  SECONDARY  sexual characters are strongly displayed, w
0156  ed. The cause of the original variability of  SECONDARY  sexual characters is not manifest; but we c
0156  form as other parts of the organisation; for  SECONDARY  sexual characters have been accumulated by
0157  tever the cause may be of the variability of  SECONDARY  sexual characters, as they are highly varia
0157  structure. It is a remarkable fact, that the  SECONDARY  sexual differences between the two sexes of
0158  up of species; that the great variability of  SECONDARY  sexual characters, and the great amount of
0158  racters between closely allied species; that  SECONDARY  sexual and ordinary specific differences ar
0158  l and sexual selection, and thus adapted for  SECONDARY  sexual, and for ordinary specific purposes.
0169  ow find most varieties or incipient species.  SECONDARY  sexual characters are highly variable, and
0169  generally been taken advantage of in giving   SECONDARY  sexual differences to the sexes of the same
0196  rtance, and which have originated from quite  SECONDARY  causes, independently of natural selection.
0274  omplicated, partly owing to the existence of  SECONDARY  sexual characters; but more especially owin
0275  and at varieties as having been produced by   SECONDARY  laws, this similarity would be an astonishi
0284  not including igneous beds).....57,154 Feet.  SECONDARY  strata........................................13
0287  ars has elapsed since the latter part of the  SECONDARY  period. I have made these few remarks becau
0288  rrestrial productions which lived during the  SECONDARY  and Palaeozoic periods, it is superfluous t
0289  ine bed is known belonging to the age of our  SECONDARY  or palaeozoic formations. But the imperfect
0300  ast to an age, as distant in futurity as the  SECONDARY  formations lie in the past, only during per
0303  fossil mammals belongs to the middle of the  SECONDARY  series; and one true mammal has been discov
0304  one of a whale having been discovered in any  SECONDARY  formation, seemed fully to justify the beli
0304  procuced in the interval between the latest   SECONDARY  and earliest tertiary formation. But now we
0304  greensand, some time before the close of the  SECONDARY  period. I may give another instance, which
0304  at had sessile cirripedes existed during the  SECONDARY  periods, they would certainly have been pre
0305  w that sessile cirripedes existed during the  SECONDARY  period; and these cirripedes might have bee
0308  o afford even a remnant of any palaeozoic or  SECONDARY  formation. Hence we may perhaps infer, that
0308  erhaps infer, that during the palaeozoic and  SECONDARY  periods, neither continents nor continental
0308  ; for had they existed there, palaeozoic and  SECONDARY  formations would in all probability have be
0312  island have been many and most gradual. The  SECONDARY  formations are more broken; but, as Bronn h
0318  gs, as of ammonites towards the close of the  SECONDARY  period, has been wonderfully sudden. The wh
0321  of Trigonia, a great genus of shells in the  SECONDARY  formations, survives in the Australian seas
0321  period and of Ammonites at the close of the  SECONDARY  period, we must remember what has been alre
0337  ainly be beaten and exterminated; as would a  SECONDARY  fauna by an eocene, and a palaeozoic fauna
0337  na by an eocene, and a palaeozoic fauna by a  SECONDARY  fauna. I do not doubt that this process of
0469  ch are acknowledged to have been produced by  SECONDARY  laws. On this same view we can understand h
0475  created, and varieties have been produced by  SECONDARY  laws. If we admit that the geological recor
0482  rms of life, and which are those produced by  SECONDARY  laws. They admit variation as a vera causa
0488  bitants of the world should have been due to  SECONDARY  causes, like those determining the birth an
0005  oped being or elaborately constructed organ;  SECONDLY,  the subject of Instinct, or the mental powe
0025  t to the very same colours and markings. Or,  SECONDLY,  that each breed, even the purest, has withi
0027  s of structure with all the domestic breeds.  SECONDLY,  although an English carrier or short faced
0058  of the forms which they connect; and except,  SECONDLY,  by a certain amount of difference, for two
0171  species being, as we see them, well defined?  SECONDLY,  is it possible that an animal having, for i
0178  ree permanent; and this assuredly we do see.  SECONDLY,  areas now continuous must often have existe
0354  nts in the arctic and antarctic regions; and  SECONDLY  (in the following chapter), the wide distrib
0448  ts at a very early stage of development, and  SECONDLY  from their following exactly the same habits
0361  is more important: the crops of birds do not  SECRETE  gastric juice, and do not in the least injur
0451  ll ceveloped in full grown males, and having  SECRETED  milk. So again there are normally four devel
0484  fects plants and animals; or that the poison  SECRETED  by the gall fly produces monstrous growths o
0239  y have an enormously developed abdomen which  SECRETES  a sort of honey, supplying the place of that
0192  , which serve, through the means of a sticky  SECRETION,  to retain the eggs until they are hatched
0233  sugar are consumed by a hive of bees for the  SECRETION  of each pound of wax: so that a prodigious
0234  d and consumed by the bees in a hive for the  SECRETION  of the wax necessary for the construction o
0234  ain idle for many days during the process of  SECRETION.  A large store of honey is indispensable to
0235  vidual swarm which wasted least honey in the  SECRETION  of wax, having succeeded best, and having t
0124  a hundred million generations, and likewise a  SECTION  of the successive strata of the earth's crus
0296  ification during any one geological period, a  SECTION  would not probably include all the fine inte
0297  n seldom hope to effect in any one geological  SECTION.  Supposing B and C to be two species, and a
0302  ations of life, the best preserved geological  SECTION  presented, had not the difficulty of our not
0390  culiar, those of another class, or of another  SECTION  of the same class, are peculiar; and this di
0426  but if these same organs, in another group or  SECTION  of a group, are found to differ much, we at
0057  may generally be divided into sub genera, or   SECTIONS,  or lesser groups. As Fries has well remarke
0128  sely and unequally related together, forming   SECTIONS  and sub genera, species of distinct genera m
0323  hey belong to the same families, genera, and   SECTIONS  of genera, and sometimes are similarly chara
0350  dification, we can understand how it is that   SECTIONS  of genera, whole genera, and even families a
0353  y many genera, and a still greater number of   SECTIONS  of genera are confined to a single region; a
0396  ss in endemic forms in particular classes or   SECTIONS  of classes, the absence of whole groups, as
0420  ing ranked under different genera, families,   SECTIONS,  or orders. The reader will best understand
0422  nt so called genera, sub families, families,   SECTIONS,  orders, and classes. It may be worth while
0489  , hence we may look with some confidence to a  SECURE  future of equally inappreciable length. And a
0090  e of a wolf, which preys on various animals,   SECURING  some by craft, some by strength, and some by
0234  rge stock of bees during the winter; and the  SECURITY  of the hive is known mainly to depend on a l
0357  ution, we shall be enabled to speculate with  SECURITY  on the former extension of the land. But I d
0487  ion. But we shall be able to gauge with some  SECURITY  the duration of these intervals by a compari
0302  and by none more forcibly than by Professor   SEDGWICK,  as a fatal objection to the belief in the t
0310  ur greatest geologists, as Lyell, Murchison,  SEDGWICK,  etc., have unanimously, often vehemently, m
0173  preserved to a future age only in masses of  SEDIMENT  sufficiently thick and extensive to withstan
0173  us masses can be accumulated only where much  SEDIMENT  is deposited on the shallow bed of the sea,
0173  stationary or is rising, or when very little  SEDIMENT  is being deposited; there will be blanks in
0282  ork grinding down old rocks and making fresh  SEDIMENT,  before he can hope to comprehend anything o
0284  consumed! Good observers have estimated that  SEDIMENT  is deposited by the great Mississippi river
0284  considering over what wide spaces very fine  SEDIMENT  is transported by the currents of the sea, t
0288  ar when left on the bottom of the sea, where  SEDIMENT  is not accumulating. I believe we are contin
0288  iew, when we tacitly admit to ourselves that  SEDIMENT  is being deposited over nearly the whole bed
0289  nd barren in his own country, great piles of  SEDIMENT,  charged with new and peculiar forms of life
0290  ography of the surrounding lands, whence the  SEDIMENT  has been derived, accords with the belief of
```

Page ***************************************(Key Word)***

Page **(Key Word)**

```
0290  can anywhere be found, though the supply of  SEDIMENT must for ages have been great, from the enor
0290  waves. We may, I think, safely conclude that  SEDIMENT must be accumulated in extremely thick, soli
0290  l. such thick and extensive accumulations of  SEDIMENT may be formed in two ways; either, in profou
0291  of the forms of life which then existed; or,  SEDIMENT may be accumulated to any thickness and exte
0291  long as the rate of subsidence and supply of  SEDIMENT nearly balance each other, the sea will rema
0291  of subsidence, but only where the supply of   SEDIMENT was sufficient to keep the sea shallow and t
0294  riod. It is not, for instance, probable that  SEDIMENT was deposited during the whole of the glacia
0295  ies to live on the same space, the supply of  SEDIMENT must nearly have counterbalanced the amount
0295  will often tend to sink the area whence the   SEDIMENT is derived, and thus diminish the supply whi
0295  nearly exact balancing between the supply of  SEDIMENT and the amount of subsidence is probably a r
0295  e in the currents of the sea and a supply of  SEDIMENT of a different nature will generally have be
0300  ould not endure to a distant epoch. Wherever  SEDIMENT did not accumulate on the bed of the sea, or
0300  ether with a contemporaneous accumulation of  SEDIMENT, would exceed the average duration of the sa
0308  of the United States; and from the amount of  SEDIMENT, miles in thickness, of which the formations
0308  large islands or tracts of land, whence the   SEDIMENT was derived, occurred in the neighbourhood o
0308  r as a submarine surface near land, on which  SEDIMENT was not deposited, or again as the bed of an
0308  n all probability have been accumulated from  SEDIMENT derived from their wear and tear; and would
0315  ferous formations depends on great masses of  SEDIMENT having been deposited on areas whilst subsid
0327  their stationary or rising, and likewise when SEDIMENT was not thrown down quickly enough to embed
0465  radation, can be accumulated only where much  SEDIMENT is deposited on the subsiding bed of the sea
0283  nd remark, that the thickness and extent of   SEDIMENTARY formations are the result and measure of t
0284  an amount of degradation is implied by the    SEDIMENTARY deposits of many countries! Professor Rams
0284  ng blank periods. So that the lofty pile of   SEDIMENTARY rocks in Britain, gives but an inadequate
0286  weald, on the flanks of which the overlying   SEDIMENTARY deposits might have accumulated in thinner
0294  oubted whether in any quarter of the world,   SEDIMENTARY deposits, including fossil remains, have g
0008  lay the utmost vigour, and yet rarely or never SEED! In some few such cases it has been found out
0008  will determine whether or not the plant sets a SEED. I cannot here enter on the copious details wh
0010  propagated by grafting, etc., and sometimes by SEED. These sports are extremely rare under nature,
0032  of plants is once pretty well established, the SEED raisers do not pick out the best plants, but m
0032  out the best plants, but merely go over their  SEED beds, and pull up the rogues, as they call the
0037  get a first rate heartsease or dahlia from the SEED of a wild plant. No one would expect to raise
0037  ct to raise a first rate melting pear from the SEED of the wild pear, though he might succeed from
0043  ls and to those plants which are propagated by SEED. In plants which are temporarily propagated by
0043  ids; but the cases of plants not propagated by SEED are of little importance to us, for their endu
0061  e which dives through the water; in the plumed SEED which is wafted by the gentlest breeze; in sho
0065  n an average for a thousand years, if a single SEED, and amongst animals there are very few which
0066  once in a thousand years, supposing that this  SEED were produced once in a thousand years, suppos
0070  us we can easily raise plenty of corn and rape SEED were never destroyed, and could be ensured to
0070  ease in number proportionally to the supply of SEED, etc., in our fields, because the seeds are in
0070  has tried, knows how troublesome it is to get  SEED, as their numbers are checked during winter: b
0070  garden! I have in this case lost every single  SEED. This view of the necessity of a large stock o
0073  , from its peculiar structure, never can set a SEED. Many of our orchidaceous plants absolutely re
0075  eties of wheat be sown together, and the mixed SEED be resown, some of the varieties which best su
0075  ertile, will beat the others and so yield more SEED, and will consequently in a few years quite su
0075  ust be each year harvested separately, and the SEED then mixed in due proportion, otherwise the we
0076  , like beings in a state of nature, and if the SEED or young were not annually sorted. As species
0077  he tiger's body. But in the beautifully plumed SEED of the dandelion, and in the flattened and fri
0077  ect that the chief use of the nutriment in the SEED is to favour the growth of the young seedling,
0098  east in my garden, by insects, it never sets a SEED, though by placing pollen from one flower on t
0099  h, onion, and some other plants, be allowed to SEED near each other, a large majority, as I have f
0103  ve principle, nurserymen always prefer getting SEED from a large body of plants of the same variet
0127  ted at corresponding ages, can modify the egg, SEED, or young, as easily as the adult. Amongst man
0140  thus the pines and rhododendrons, raised from  SEED collected by Dr. Hooker from trees growing at
0142  usalem artichoke, which is never propagated by SEED, and of which consequently new varieties have
0142  rtion are destroyed by frost, and then collect SEED from the few survivors, with care to prevent a
0142  prevent accidental crosses, and then again get SEED from these seedlings, with the same precaution
0147  r, might get an advantage over those producing SEED less fitted for dispersal: and this process co
0247  nly once or twice succeeded in getting fertile SEED: as he found the common red and blue pimpernel
0250  len. For these plants have been found to yield SEED to the pollen of a distinct species, though qu
0251  and rapid progress to maturity, and bore good  SEED, which vegetated freely. In a letter to me, in
0251  , have been crossed, yet many of these hybrids SEED freely. For instance, Herbert asserts that a h
0255  pollen of either pure parent, a single fertile SEED: but in some of these cases a first trace of f
0256  greatly in the several individuals raised from SEED out of the same capsule and exposed to exactly
0259  erile, even when the other hybrids raised from SEED from the same capsule have a considerable degr
0264  within its mother's womb or within the egg or  SEED produced by the mother, it may be exposed to c
0265  imal will breed under confinement or any plant SEED freely under culture; nor can he tell, till he
0267  s and gardeners in their frequent exchanges of SEED, tubers, etc, from one soil or climate to anot
0270  the other; but only a single head produced any SEED, and this one head produced only five grains.
0270  es of Verbascum when intercrossed produce less SEED, than do either coloured varieties when fertil
0271  nd white varieties of a distinct species, more SEED is produced by the crosses between the same co
0271  d one variety can sometimes be raised from the SEED of the other. From observations which I have m
0363  earth there was a pebble quite as large as the SEED of a vetch. Thus seeds might occasionally be f
0364  this case, how small would the chance be of a  SEED falling on favourable soil, and coming to matu
0365  ve insects or birds living there, nearly every SEED, which chanced to arrive, would be sure to ger
0388  ing islet, it will be unoccupied; and a single SEED or egg will have a good chance of succeeding.
0392  resents no difficulty on my view, for a hooked SEED might be transported to an island by some othe
0401  en stocked by occasional means of transport, a SEED, for instance, of one plant having been brough
0414  organs of reproduction, with their product the SEED, are of paramount importance! We must not, the
0483  kinds of animals and plants created as eggs or SEED, or as full grown? and in the case of mammals,
0262  in found that three species of Robinia, which  SEEDED freely on their own roots; and which could be
0262  ary case of Hippeastrum, Lobelia, etc., which  SEEDED much more freely when fertilized with the pol
0030  ount of change may have suddenly arisen in a   SEEDLING. So it has probably been with the turnspit d
0037  ld pear, though he might succeed from a poor   SEEDLING growing wild, if it had come from a garden s
0063  tree, it will languish and die. But several    SEEDLING missletoes, growing close together on the sa
0077  he seed is to favour the growth of the young   SEEDLING, whilst struggling with other plants growing
```

Page **(Key Word)**

Page ********************************.*********************(Key Word)**

```
0099  urn out mongrels: for instance, I raised 233  SEEDLING cabbages from some plants of different varie
0142  d that no differences in the constitution of   SEEDLING kidney beans ever appear, for an account has
0010  s have supposed, with the act of generation.   SEEDLINGS from the same fruit, and the young of the s
0041  larger, earlier, or better fruit, and raised   SEEDLINGS from them, and again picked out the best se
0041  ngs from them, and again picked out the best    SEEDLINGS from them, and bred from them, then, there appeared (a
0064  no plant so unproductive as this, and their    SEEDLINGS next year produced two, and so on, then in
0067  which I have made, I believe that it is the     SEEDLINGS which suffer most from germinating in groun
0067  d already thickly stocked with other plants.    SEEDLINGS, also, are destroyed in vast numbers by var
0067  choking from other plants, I marked all the     SEEDLINGS of our native weeds as they came up, and ou
0072  e stems of the heath, I found a multitude of    SEEDLINGS and little trees, which had been perpetuall
0075  ach other or on the trees or their seeds and    SEEDLINGS, or on the other plants which first clothed
0092  lly would produce very vigorous                 SEEDLINGS, which consequently would have the best cha
0092  of flourishing and surviving. Some of these     SEEDLINGS would probably inherit the nectar excreting
0098  h other, it is hardly possible to raise pure    SEEDLINGS, so largely do they naturally cross. In man
0098  on the stigma of another, I raised plenty of    SEEDLINGS: and whilst another species of Lobelia grow
0099  r, a large majority, as I have found, of the    SEEDLINGS thus raised will turn out mongrels: for ins
0099  hen, comes it that such a vast number of the    SEEDLINGS are mongrelized? I suspect that it must ari
0142  crosses, and then again get seed from these     SEEDLINGS, with the same precautions, the experiment
0142  has been published how much more hardy some     SEEDLINGS appeared to be than others. On the whole, I
0003  its nourishment from certain trees, which has   SEEDS that must be transported by certain birds, an
0029  dlin apple, could ever have proceeded from the  SEEDS of the same tree. Innumerable other examples
0037  cultivating the best known variety, sowing its  SEEDS, and, when a slightly better variety has chan
0062  dly singing round us mostly live on insects or  SEEDS, and are thus constantly destroying life; or
0063  re. A plant which annually produces a thousand  SEEDS, of which on an average only one comes to mat
0063  tempt birds to devour and thus disseminate its  SEEDS rather than those of other plants. In these s
0063  its natural lifetime produces several eggs or   SEEDS, must suffer destruction during some period o
0064  ated that if an annual plant produced only two  SEEDS, and there is no plant so unproductive as thi
0065  tween organisms which annually produce eggs or  SEEDS by the thousand, and those which produce extr
0066  e real importance of a large number of eggs or  SEEDS is to make up for much destruction at some pe
0066  s only indirectly on the number of its eggs or  SEEDS. In looking at Nature, it is most necessary t
0067  se. With plants there is a vast destruction of  SEEDS, but, from some observations which I have mad
0070  nd rape seed, etc., in our fields, because the  SEEDS are in great excess compared with the number
0075  g long centuries, each annually scattering its  SEEDS by the thousand; what war between insect and
0075  feeding on each other or on the trees or their  SEEDS and seedlings; or on the other plants which f
0077  of air and water. Yet the advantage of plumed   SEEDS no doubt stands in the closest relation to th
0077  y thickly clothed by other plants; so that the  SEEDS may be widely distributed and fall on unoccup
0077  als. The store of nutriment laid up within the  SEEDS of many plants seems at first sight to have n
0077  ronc growth of young plants produced from such  SEEDS (as peas and beans), when sown in the midst o
0086  pring at the same period; for instance, in the  SEEDS of the many varieties of our culinary and agr
0086  sponding age. If it profit a plant to have its  SEEDS more and more widely disseminated by the wind
0098  ia growing close by, which is visited by bees,  SEEDS freely. In very many other cases, though ther
0113  y of grass is annually sowing almost countless  SEEDS; and thus, as it may be said, is striving its
0139  flowering, in the amount of rain requisite for  SEEDS to germinate, in the time of sleep, etc., and
0145  e flower. But, in some Compositous plants, the  SEEDS also differ in shape and sculpture; and even
0145  some authors to pressure, and the shape of the  SEEDS in the ray florets in some Compositae counten
0145  n some Compositae there is a difference in the  SEEDS of the outer and inner florets without any di
0146  in the internal and external structure of the  SEEDS, which are not always correlated with any dif
0146  fferences are of such apparent importance, the  SEEDS being in some cases, according to Tausch, ort
0146  ce, Alph. de Candolle has remarked that winged  SEEDS are never found in fruits which do not open:
0146  en: I should explain the rule by the fact that  SEEDS could not gradually become winged through nat
0147  pened; so that the individual plants producing  SEEDS which were a little better fitted to be wafte
0147  us foliage and a copious supply of oil bearing  SEEDS. When the seeds in our fruits become atrophie
0147  copious supply of oil bearing seeds. When the   SEEDS in our fruits become atrophied, the fruit its
0184  ave many times seen and heard it hammering the  SEEDS of the yew on a branch, and thus breaking the
0237  dual is destroyed; but the horticulturist sows  SEEDS of the same stock, and confidently expects to
0247  ses, Gartner is obliged carefully to count the  SEEDS, in order to show that there is any degree of
0247  lity. He always compares the maximum number of  SEEDS produced by two species when crossed and by t
0252  ponticum and Catawbiense, and that this hybrid  SEEDS as freely as it is possible to imagine. Had h
0255  s, yields a perfect gradation in the number of  SEEDS produced, up to nearly complete or even quite
0256  rids producing a greater and greater number of  SEEDS up to perfect fertility. Hybrids from two spe
0270  everal years a dwarf kind of maize with yellow  SEEDS, and a tall variety with red seeds, growing n
0270  with yellow seeds, and a tall variety with red  SEEDS, growing near each other in his garden; and a
0358  few experiments, it was not even known how far  SEEDS could resist the injurious action of sea wate
0359  ys. For convenience sake I chiefly tried small  SEEDS, without the capsule or fruit; and as all of
0359  ys, when dried it floated for 85 days, and the  SEEDS afterwards germinated: the ripe seeds of Helo
0359  and the seeds afterwards germinated: the ripe   SEEDS of Helosciadium sank in two days, when dried
0359  or a very much longer period. So that as 64/87  SEEDS germinated after an immersion of 28 days; and
0359  m these scanty facts, we may conclude that the  SEEDS of 14/100 plants of any country might be floa
0360  te of 60 miles per diem); on this average, the  SEEDS of 14/100 plants belonging to one country mig
0360  but in a much better manner, for he placed the  SEEDS in a box in the actual sea, so that they were
0360  e air like really floating plants. He tried 98  SEEDS, mostly different from mine; but he chose man
0360  ei; but he chose many large fruits and likewise SEEDS from plants which live near the sea; and this
0360  much longer. The result was that 18/98 of his   SEEDS floated for 42 days, and were then capable of
0360  e it would perhaps be safer to assume that the  SEEDS of about 10/100 plants of a flora, after havi
0360  he small, is interesting; as plants with large  SEEDS or fruit could hardly be transported by any o
0360  h plants generally have restricted ranges. But  SEEDS may be occasionally transported in another ma
0361  metimes escape being immediately devoured; and  SEEDS of many kinds in the crops of floating birds
0361  ghly effective agents in the transportation of  SEEDS. I could give many facts showing how frequent
0361  e. I have never seen an instance of nutritious  SEEDS passing through the intestines of a bird; but
0361  ing through the intestines of a bird; but hard  SEEDS of fruit will pass uninjured through even the
0361  o months; I picked up in my garden 12 kinds of  SEEDS, out of the excrement of small birds, and the
0362  injure, as I know by trial, the germination of  SEEDS; now after a bird has found and devoured a la
0362  iments made in the Zoological Gardens, include  SEEDS capable of germination. Some seeds of the oat
0362  ns, include seeds capable of germination. Some  SEEDS of the oat, wheat, millet, canary, hemp, clov
0362  e stomachs of different birds of prey; and two  SEEDS of beet grew after having been thus retained
0362  fourteen hours. Fresh water fish, I find, eat   SEEDS of many land and water plants: fish are frequ
```

Page **(Key Word)***

seeds

Page **(Key Word)**

```
0362  are frequently devoured by birds, and thus the  SEEDS might be transported from place to place. I f
0362  ed from place to place. I forced many kinds of   SEEDS into the stomachs of dead fish, and then gave
0362  an interval of many hours, either rejected the   SEEDS in pellets or passed them in their excrement;
0362  them in their excrement; and several of these    SEEDS retained their power of germination. Certain
0362  s retained their power of germination. Certain   SEEDS, however, were always killed by this process.
0363  le quite as large as the seed of a vetch. Thus   SEEDS might occasionally be transported to great di
0363  nq that soil almost everywhere is charged with   SEEDS. Reflect for a moment on the millions of quai
0363  heir feet would sometimes include a few minute   SEEDS? But I shall presently have to recur to this
0363  t that they must occasionally have transported   SEEDS from one part to another of the arctic and an
0363  e islands had been partly stocked by ice borne   SEEDS, during the glacial epoch. At my request Sir
0363  ossible that they may have brought thither the   SEEDS of northern plants. Considering that the seve
0364  at scarcely any means of transport would carry   SEEDS for very great distances; for seeds do not re
0364  ould carry seeds for very great distances; for   SEEDS do not retain their vitality when exposed for
0364  currents, from their course, would never bring   SEEDS from North America to Britain, though they mi
0364  ica to Britain, though they might and do bring   SEEDS from the West Indies to our western shores, w
0364  the western shores of Ireland and England; but   SEEDS could be transported by these wanderers only
0365  milar means. I do not doubt that out of twenty   SEEDS or animals transported to an island, even if
0380  hides, wool, and other objects likely to carry   SEEDS have been largely imported into Europe during
0386  are aware how charged the mud of ponds is with   SEEDS: I have tried several little experiments, but
0387  cumstance if water birds did not transport the   SEEDS of fresh water plants to vast distances, and
0387  stated that fresh water fish eat some kinds of   SEEDS, though they reject many other kinds after ha
0387  having swallowed them; even small fish swallow   SEEDS of moderate size, as of the yellow water lily
0387  re blown across the sea; and we have seen that   SEEDS retain their power of germination, when rejec
0387  s afterwards. When I saw the great size of the   SEEDS of that fine water lily the Nelumbium, and re
0387  plicable; but Audubon states that he found the   SEEDS of the great southern water lily (probably, a
0387  eject from its stomach a pellet containing the   SEEDS of the Nelumbium undigested; or the seeds mig
0388  the seeds of the Nelumbium undigested; or the   SEEDS might be dropped by the bird whilst feeding i
0388  mainly depends on the wide dispersal of their   SEEDS and eggs by animals, more especially by fresh
0388  ature, like a careful gardener, thus takes her   SEEDS from a bed of a particular nature, and drops
0392  of the endemic plants have beautifully hooked   SEEDS; yet few relations are more striking than the
0392  re more striking than the adaptation of hooked   SEEDS for transportal by the wool and fur of quadru
0392  ghtly modified, but still retaining its hooked   SEEDS, would form an endemic species, having as use
0399  ew that this island has been mainly stocked by   SEEDS brought with earth and stones on icebergs, dr
0406  ecific character. This fact, together with the   SEEDS and eggs of many low forms being very minute
0111  genus. As has always been my practice, let us   SEEK light on this head from our domestic productio
0092  e petals of a flower. In this case insects in   SEEKING the nectar would get dusted with pollen, and
0420  ond which naturalists have been unconsciously   SEEKING, and not some unknown plan of creation, or t
0449  bond of connexion which naturalists have been   SEEKING under the term of the natural system. On thi
0455  e pronunciation, but which serve as a clue in   SEEKING for its derivation. On the view of descent w
0010  ew, Mr. Buckman's recent experiments on plants  SEEM extremely valuable. When all or nearly all the
0024  inct or unknown. So many strange contingencies  SEEM to me improbable in the highest degree. Some f
0031  t breeders have done for sheep, says: It would  SEEM as if they had chalked out upon a wall a form
0046  s. genera which are polymorphic in one country  SEEM to be, with some few exceptions, polymorphic i
0046  shells, at former periods of time. These facts  SEEM to be very perplexing, for they seem to show t
0046  ese facts seem to be very perplexing, for they  SEEM to show that this kind of variability is indep
0067  f the chief points. Eggs or very young animals  SEEM generally to suffer most, but this is not inva
0085  , as far as our ignorance permits us to judge,  SEEM to be quite unimportant, we must not forget th
0088  between males of polygamous animals, and these  SEEM oftenest provided with special weapons. The ma
0129  y class cannot be ranked in a single file, but  SEEM rather to be clustered round points, and these
0132  offspring. The male and female sexual elements  SEEM to be affected before that union takes place w
0134  of life. Indirectly, as already remarked, they  SEEM to play an important part in affecting the rep
0143  hich, at an early embryonic period, are alike,  SEEM liable to vary in an allied manner: we see thi
0144  petals of the corolla into a tube. Hard parts  SEEM to affect the form of adjoining soft parts; it
0163  repeat that such cases certainly do occur, and  SEEM to me very remarkable. I will, however, give o
0167  conditions of life, as climate and food, etc.,  SEEM to have induced some slight modifications. Hab
0168  nd disuse in weakening and diminishing organs,  SEEM to have been more potent in their effects. Hom
0183  y illustration of the flying fish, it does not  SEEM probable that fishes capable of true flight wo
0195  simple parts, of which the importance does not  SEEM sufficient to cause the preservation of succes
0223  es. Both in Switzerland and England the slaves  SEEM to have the exclusive care of the larvae, and
0233  here to enter on this subject. Nor does there  SEEM to me any great difficulty in a single insect
0263  tempted to be expressed. The facts by no means  SEEM to me to indicate that the greater or lesser d
0264  riod; more especially as all very young beings  SEEM eminently sensitive to injurious or unnatural
0266  ations having been compounded into one. It may  SEEM fanciful, but I suspect that a similar paralle
0267  accident or an illusion. Both series of facts  SEEM to be connected together by some common but un
0272  s. the general fertility of varieties does not  SEEM to me sufficient to overthrow the view which I
0273  riability of mongrels than of hybrids does not  SEEM to me at all surprising. For the parents of mo
0275  losely resembling one parent, the resemblances  SEEM chiefly confined to characters almost monstrou
0278  the facts briefly given in this chapter do not  SEEM to me opposed to, but even rather to support t
0288  ng suffered in the interval any wear and tear,  SEEM explicable only on the view of the bottom of t
0295  ept near their upper or lower limits. It would  SEEM that each separate formation, like the whole p
0309  ve thus remained from eternity? Our continents  SEEM to to have been formed by a preponderance, dur
0313  e changed greatly. The productions of the land  SEEM to change at a quicker rate than those of the
0333  rms of life to each other and to living forms,  SEEM to me explained in a satisfactory manner. And
0343  the other great leading facts in palaeontology  SEEM to me simply to follow on the theory of descen
0358  e depth of the sea, these and other such facts  SEEM to me opposed to the admission of such prodigi
0358  of the inhabitants of oceanic islands likewise  SEEM to me opposed to the belief of their former co
0368  rtained occurrence of a former Glacial period,  SEEM to me to explain in so satisfactory a manner t
0378  himalaya, and the long line of the Cordillera,  SEEM to have afforded two great lines of invasion:
0381  odification to the necessary degree. The facts  SEEM to me to indicate that peculiar and very disti
0384  fish in common and some different. A few facts  SEEM to favour the possibility of their occasional
0386  y a very few aquatic members: for these latter  SEEM immediately to acquire, as if in consequence,
0396  forms having been developed into trees, etc.,  SEEM to me to accord better with the view of occasi
0413  enus, but that the genus gives the characters,  SEEM to imply that something more is included in ou
0419  ders, families, sub families, and genera, they  SEEM to be, at least at present, almost arbitrary,
0446  he adult state. The two principles above given  SEEM to me to explain these facts in regard to the
0446  s just given, more especially that of pigeons,  SEEM to show that the characteristic differences wh
```

Page **(Key Word)**

0457 ts which have been considered in this chapter, SEEM to me to proclaim so plainly, that the inumera
0460 given at the end of the eighth chapter, which SEEM to me conclusively to show that this sterility
0472 cific forms. In both cases physical conditions SEEM to have produced but little direct effect; yet
0472 in both varieties and species, use and disuse SEEM to have produced some effect; for it is diffic
0483 indness of preconceived opinion. These authors SEEM no more startled at a miraculous act of creati
0488 d his history. Authors of the highest eminence SEEM to be fully satisfied with the view that each
0489 bed of the Silurian system was deposited, they SEEM to me to become ennobled. Judging from the pas
0001 st inhabitants of that continent. These facts SEEMED to me to throw some light on the origin of sp
0001 into a sketch of the conclusions, which then SEEMED to me probable: from that period to the prese
0004 on. At the commencement of my observations it SEEMED to me probable that a careful study of the expo
0053 in several well worked floras. At first this SEEMED a simple task; but Mr. H. C. Watson, to whom
0208 assed, and, in order to complete its hammock, SEEMED forced to start from the third stage, where i
0211 owed an ant to visit them, and it immediately SEEMED, by its eager way of running about, to be wel
0304 g been discovered in any secondary formation, SEEMED fully to justify the belief that this great a
0310 been heated under great pressure, have always SEEMED to me to require some special explanation; an
0361 ut of the excrement of small birds, and these SEEMED perfect, and some of them, which I tried, ger
0445 their parents, and this, judging by the eye, SEEMED almost to be the case; but on actually measur
0004 erfect as we now see them; but this assumption SEEMS to me to be no explanation, for it leaves the
0007 ay be partly connected with excess of food. It SEEMS pretty clear that organic beings must be expo
0011 the animals not being much alarmed by danger, SEEMS probable. There are many laws regulating vari
0012 ightly from each other. The whole organisation SEEMS to have become plastic, and tends to depart i
0015 ome of their characters to ancestral forms, it SEEMS to me not improbable, that if we could succee
0018 re found more strictly and generally true than SEEMS to me to be the case, what does it show, but
0020 ith pigeons) extremely uniform, and everything SEEMS simple enough; but when these mongrels are cr
0023 their size, habits, and remarkable characters, SEEMS very improbable; or they must have become ext
0024 ies having similar habits with the rock pigeon SEEMS to me a very rash assumption. Moreover, the s
0024 prolific under confinement. An argument, as it SEEMS to me, of great weight, and applicable in sev
0026 d yield offspring perfectly fertile, inter se, SEEMS to me rash in the extreme. From these several
0046 t connected with individual differences, which SEEMS to me extremely perplexing: I refer to those
0047 ists having sound judgment and wide experience SEEMS the only guide to follow. We must, however, f
0050 r these links are hybrids; and there is, as it SEEMS to me, an overwhelming amount of experimental
0054 ry wide ranges and are much diffused, but this SEEMS to be connected with the nature of the statio
0058 etween species. There is one other point which SEEMS to me worth notice. Varieties generally have
0068 the average numbers of a species. Thus, there SEEMS to be little doubt that the stock of partridg
0068 from epidemics with man. The action of climate SEEMS at first sight to be quite independent of the
0070 rs in a small tract, epidemics, at least, this SEEMS generally to occur with our game animals, oft
0077 fringed legs of the water beetle, the relation SEEMS at first confined to the elements of air and
0077 riment laid up within the seeds of many plants SEEMS at first sight to have no sort of relation to
0078 ganic beings; a conviction as necessary, as it SEEMS to be difficult to acquire. All that we can d
0097 tand so close together that self fertilisation SEEMS almost inevitable. Many flowers, on the other
0098 ne after the other towards it, the contrivance SEEMS adapted solely to ensure self fertilisation:
0098 berry; and curiously in this very genus, which SEEMS to have a special contrivance for self fertil
0105 e else. Hence an oceanic island at first sight SEEMS to have been highly favourable for the produc
0122 es in the natural economy of their country. It SEEMS, therefore, to me extremely probable that the
0137 in the case of the cave rat natural selection SEEMS to have struggled with the loss of light and
0138 her inhabitants of the caves, disuse by itself SEEMS to have done its work. It is difficult to ima
0146 elated with any differences in the flowers, it SEEMS impossible that they can be in any way advant
0149 ation the reduction of some adjoining part. It SEEMS to be a rule, as remarked by Is. Geoffroy St.
0149 ve repetition, to use Prof. Owen's expression, SEEMS to be a sign of low organisation; the foregoi
0149 sign of low organisation; the foregoing remark SEEMS connected with the very general opinion of na
0149 nd I will here only add that their variability SEEMS to be owing to their uselessness, and therefo
0151 ve particularly attended to them, and the rule SEEMS to me certainly to hold good in this class. I
0155 nstrosities: at least Is. Geoffroy St. Hilaire SEEMS to entertain no doubt, that the more an organ
0167 ies of the genus. To admit this view is, as it SEEMS to me, to reject a real for an unreal, or at
0180 ant or occasional, in the same species. And it SEEMS to me that nothing less than a long list of s
0185 their toes are only bordered by membrane. What SEEMS plainer than that the long toes of grallatore
0186 ake the place of one of another type; but this SEEMS to me only restating the fact in dignified la
0186 , could have been formed by natural selection, SEEMS, I freely confess, absurd in the highest poss
0188 by convergence; and at their lower ends there SEEMS to be an imperfect vitreous substance. With t
0191 of the higher vertebrate animals: hence there SEEMS to me to be no great difficulty in believing
0195 n artificially constructed fly flapper! and il SEEMS al first incredible that this could have been
0201 y to this effect, I cannot find even one which SEEMS to me of any weight. It is admitted that the
0224 e. grant whatever instincts you please, and it SEEMS at first quite inconceivable how they can mak
0231 b, with the utmost ultimate economy of wax. It SEEMS at first to add to the difficulty of understa
0232 the cells all round. The work of construction SEEMS to be a sort of balance struck between many b
0233 ls, is important, as it bears on a fact, which SEEMS at first quite subversive of the foregoing th
0245 sion of all organic forms. This view certainly SEEMS at first probable, for species within the sam
0246 portance with the sterility of species; for it SEEMS to make a broad and clear distinction between
0247 state of nature. But a serious cause of error SEEMS to me to be here introduced: a plant to be hy
0247 onclusion in several other analogous cases; it SEEMS to me that we may well be permitted to doubt
0254 e under domestication. This latter alternative SEEMS to me the most probable, and I am inclined to
0260 lity of the first union between their parents, SEEMS to be a strange arrangement. The foregoing ru
0264 pable of causing an embryo to be developed, as SEEMS to have been the case with some of Thuret's e
0267 eakness and sterility in the progeny. Hence it SEEMS that, on the one hand, slight changes in the
0269 the same conditions of life. Lastly, and this SEEMS to me by far the most important consideration
0269 e invariably fertile when intercrossed. But it SEEMS to me impossible to resist the evidence of th
0270 the following case is far more remarkable, and SEEMS at first quite incredible; but it is the resu
0272 and varieties, could find very few and, as it SEEMS to me, quite unimportant differences between
0274 ke out, is founded on a single experiment; and SEEMS directly opposed to the results of several ex
0275 ity and sterility, in all other respects there SEEMS to be a general and close similarity in the o
0276 which have their reproductive systems perfect, SEEMS to depend on several circumstances; in some c
0277 d by being compounded of two distinct species, SEEMS closely allied to that sterility which so fre
0293 fic forms. But insuperable difficulties, as it SEEMS to me, prevent us coming to any just conclusi
0299 rous intermediate gradations! and such success SEEMS to me improbable in the highest degree. Geolo
0306 of even the last dozen years have effected, it SEEMS to me to be about as rash in us to dogmatize
0313 gration from a distinct geographical province, SEEMS to me satisfactory. These several facts accor

Page **(Key Word)**

```
0318 e close of the palaeozoic period. No fixed law    SEEMS to determine the length of time during which
0322 ake of some inferiority in common. Thus, as it    SEEMS to me, the manner in which single species and
0327  might long be enabled to survive. Thus, as it    SFEMS to me, the parallel, and, taken in a large se
0330 ry distinct groups, such as fish and reptiles,    SEEMS to be, that supposing them to be distinguishe
0333 ncurrent evidence of our best palaeontologists    SEEMS frequently to be the case. Thus, on the theor
0335 h stage. This fact alone, from its generality,    SEEMS to have shaken Professor Pictet in his firm b
0345  laws of palaeontology plainly proclaim, as it    SFEMS to me, that species have been produced by ord
0353 oduced in two or more distinct areas! Hence it    SCEMS to me, as it has to many other naturalists, t
0354 e belief that this has been the universal law,    SEEMS to me incomparably the safest. In discussing
0357 nges within the period of existing species. It    SFEMS to me that we have abundant evidence of great
0365 ome, as to become naturalised. But this, as it    SFEMS to me, is no valid argument against what woul
0374  was contemporaneous in a geological sense, it    SEEMS to me probable that it was, during a part at
0377  much warmer climate than their own. Hence, it    SEEMS to me possible, bearing in mind that the trop
0384 s and completely prevented their inosculation,    SEEMS to lead to this same conclusion. With respect
0390  same class, are peculiar; and this difference    SEEMS to depend on the species which do not become
0392 onditions of the islands! but this explanation    SEEMS to me not a little doubtful. Facility of immi
0393 ed for by their physical conditions; indeed it    SEEMS that islands are peculiarly well fitted for t
0413 t else is meant by the plan of the Creator, it    SEEMS to me that nothing is thus added to our knowl
0415 ral family, is very unequal, and in some cases    SEEMS to be entirely lost. Again in another work he
0417 he genus gives the characters; for this saying  SFEMS founded on an appreciation of many trifling p
0417  retained amongst the Malpighiaceae. This case  SEEMS to me well to illustrate the spirit with whic
0421 een more or less completely lost, as sometimes  SEEMS to have occurred with existing organisms. All
0436 ain suctorial crustaceans, the general pattern  SEEMS to have been thus to a certain extent obscure
0450 ever, incapable of demonstration. Thus, as it   SEES to me, the leading facts in embryology, which
0451 tentiality, and are merely not developed: this  SEEMS to be the case with the mammae of male mammal
0452 tinct object: in certain fish the swim bladder  SEEMS to be rudimentary for its proper function of
0453 der to complete the scheme of nature! but this  SEEMS to me no explanation, merely a restatement of
0466 er domestication we see much variability. This  SEEMS to be mainly due to the reproductive system b
0469 ction, even if we looked no further than this,  SEEMS to me to be in itself probable. I have alread
0471 and fact of the grouping of all organic beings  SEEMS to me utterly inexplicable on the theory of c
0471 o man can explain. Many other facts are, as it  SEEMS to me, explicable on this theory. How strange
0473 th varieties and species correlation of growth  SEEMS to have played a most important part, so that
0480 ructures, her scheme of modification, which it  SEEMS that we wilfully will not understand. I have
0482 that is, have been independently created. This  SEEMS to me a strange conclusion to arrive at. They
0349 inct kinds of birds, notes nearly similar, and  SEES their nests similarly constructed, but not qui
0148 sists of the three highly important anterior    SEGMENTS of the head enormously developed, and furnis
0437 ta, we see the body divided into a series of    SEGMENTS, bearing external appendages; and in floweri
0437 e unknown progenitor of the articulata, many   SEGMENTS! and the unknown progenitor of flowering pla
0102 rganic beings are striving, it may be said, to  SEIZE on each place in the economy of nature, if an
0112 its, by so much will they be better enabled to  SEIZE on many and widely diversified places in the
0119 dered, the more places they will be enabled to  SFIZE on, and the more their modified progeny will
0177 favourable variations for natural selection to  SEIZE on, than will the rarer forms which exist in
0186  some other inhabitant of the country, it will  SEIZE on the place of that inhabitant, however diff
0331 scendants from a species being thus enabled to  SEIZE on many and different places in the economy o
0337 er any considerable number would be enabled to  SFIZE on places now occupied by our native plants a
0367 bases of the mountains, the arctic forms would  SEIZE on the cleared and thawed ground, always asce
0428  they are almost sure to spread widely, and to  SEIZE on more and more places in the economy of nat
0470 n habits and structure, so as to be enabled to  SFIZE on many and widely different places in the ec
0046 species, and which consequently have not been   SEIZED on and rendered definite by natural selection
0082 mmigration, these same places would have been   SEIZED on by intruders. In such case, every slight m
0148 , however slight, in its development, will be   SEIZED on by natural selection, for it will profit t
0221 t near the place of combat: they were eagerly   SEIZED, and carried off by the tyrants, who perhaps
0222  we have seen that they eagerly and instantly   SEIZED the pupae of F. fusca, whereas they were much
0223 esence proved useful to the species which had   SEIZED them, if it were more advantageous to this sp
031°  d rarer, and finally extinct: its place being  SEIZED on by some more successful competitor. It is
0321 species belonging to some one group will have   SEIZED on the place occupied by a species belonging
0337 ve recently spread over New Zealand, and have   SEIZED on places which must have been previously occ
0438 continued course of modification, should have   SEIZED on a certain number of the primordially simil
0113 s conditions) only by its varying descendants   SEIZING on places at present occupied by other anima
0126  perfected sub groups, from branching out and  SEIZING on many new places in the polity of Nature,
0184 at of a woodpecker for climbing trees and for  SEIZING insects in the chinks of the bark? Yet in No
0350 y extended homes will have the best chance of  SEIZING on new places, when they spread into new cou
0388  oreign country, would have a better chance of SEIZING on a place, than in the case of terrestrial
0083 ives of many climates in the same country; he  SELDOM exercises each selected character in some pec
0095 , as illustrative of geology; but we now very  SELDOM hear the action, for instance, of the coast w
0099 flowers, it may be objected that pollen could  SELDOM be carried from tree to tree, and at most onl
0134 bury duck. As the larger ground feeding birds  SELDOM take flight except to escape danger, I believ
0153 mon progenitor of the genus. This period will  SELDOM be remote in any extreme degree, as species v
0154 degree. For in this case the variability will  SELDOM as yet have been fixed by the continued selec
0165 the hemionus; and this hybrid, though the ass  SELDOM has stripes on its legs and the hemionus has
0182 rly transitional grades of the structure will  SELDOM continue to exist to the present day, for the
0249 niversal belief amongst breeders. Hybrids are  SELDOM raised by experimentalists in great numbers;
0283 , showing how little they are abraded and how  SELDOM they are rolled about! Moreover, if we follow
0297 nd this from the reasons just assigned we can  SELDOM hope to effect in any one geological section.
0298 fect specimens for examination, two forms can  SELDOM be connected by intermediate varieties and th
0313 onsecutive formations, the forms of life have  SELDOM changed in exactly the same degree. Yet if we
0322  and that some check is always in action, yet  SELDOM perceived by us, the whole economy of nature
0345 ce become widely divergent. Extinct forms are  SELDOM directly intermediate between existing forms;
0438 te but few serial homologies; that is, we are  SELDOM enabled to say that one part or organ is homo
0454 birds inhabiting oceanic islands, which have   SELDOM been forced to take flight, and have ultimate
0455  state at the same age, and consequently will  SELDOM affect or reduce it in the embryo. Thus we ca
0038 t in internal parts or organs. Man can hardly  SELECT, or only with much difficulty, any deviation
0216 selection, by considering a few cases. I will  SELECT only three, out of the several which I shall
0269 is own use and pleasure: he neither wishes to  SELECT, nor could select, slight differences in the reproductive syste
0269 asure! he neither wishes to select, nor could  SELECT, slight differences in the reproductive syste
0446 ic stages of our domestic varieties. Fanciers  SELECT their horses, dogs, and pigeons, for breeding
```

Page **(Key Word)**

select

Page **(Key Word)**

```
0467   and causes variability. But man can and does  SELECT the variations given to him by nature, and th
0469   and a powerful agent always ready to act and  SELECT, why should we doubt that variations in any w
0469   d, and inherited? Why, if man can by patience  SELECT variations most useful to himself, should nat
0005   r chance of surviving, and thus be naturally  SELECTED. From the strong principle of inheritance, a
0005   rom the strong principle of inheritance, any  SELECTED variety will tend to propagate its new and m
0031   sed, so that the very best may ultimately be  SELECTED for breeding. What English breeders have act
0039   cases, utterly incorrect. The man who first  SELECTED a pigeon with a slightly larger tail, never
0083   for that of the being which she tends. Every  SELECTED character is fully exercised by her; and the
0083   n the same country; he seldom exercises each  SELECTED character in some peculiar and fitting manne
0087   unity; if each in consequence profits by the  SELECTED change. What natural selection cannot do, is
0087   icate and more easily broken shells might be  SELECTED, the thickness of the shell being known to w
0090   chance of surviving, and so be preserved or  SELECTED, provided always that they retained strength
0092   wer to flower, would likewise be favoured or  SELECTED. We might have taken the case of insects vis
0092   and had larger and larger anthers, would be  SELECTED. When our plant, by this process of the cont
0094   increased, would be continually favoured or  SELECTED, until at last a complete separation of the
0109   ed with inhabitants, it follows that as each  SELECTED and favoured form increases in number, so wi
0113   arying, and those varieties were continually  SELECTED which differed from each other in at all the
0117   ay profitable will be preserved or naturally  SELECTED. And here the importance of the principle of
0121   ry natural selection necessarily acts by the  SELECTED form having some advantage in the struggle f
0215   suppose that domestic rabbits have ever been  SELECTED for tameness; and I presume that we must att
0237   offspring, which again varied and were again  SELECTED, and so onwards. But with the working ant we
0240   those males and females had been continually  SELECTED, which produced more and more of the smaller
0356   ing the three classes of facts, which I have  SELECTED as presenting the greatest amount of difficu
0379   y new forms of life; and it is probable that  SELECTED modifications in their structure, habits, an
0389   of the three classes of facts, which I have  SELECTED as presenting the greatest amount of difficu
0012   long beaks large feet. Hence, if man goes on  SELECTING, and thus augmenting, any peculiarity, he w
0029   aries slightly, for they win their prizes by  SELECTING such slight differences, yet they ignore al
0037   ightly better variety has chanced to appear,  SELECTING it, and so onwards. But the gardeners of th
0089   o good reason to doubt that female birds, by  SELECTING, during thousands of generations, the most
0142   not likely that man should have succeeded in  SELECTING so many breeds and sub breeds with constitu
0189   ion in the transparent layers; and carefully  SELECTING each alteration which under varied circumst
0216   s have been lost partly by habit, and by man  SELECTING and accumulating during successive generati
0271   uld generally be ranked as species; from man  SELECTING only external characters in the production
0356   om any single pair, but to continued care in  SELECTING and training many individuals during many g
0467   argely influence the character of a breed by  SELECTING, in each successive generation, individual
0469   ost useful to himself, should nature fail in  SELECTING variations useful, under changing condition
008    py survive and multiply. Chapter IV. Natural  SELECTION. Natural Selection, its power compared with
0004   t is the power of man in accumulating by his  SELECTION successive slight variations. I will then p
0005   ed form. This fundamental subject of Natural  SELECTION will be treated at some length in the fourt
0005   h chapter; and we shall then see how Natural  SELECTION almost inevitably causes much Extinction of
0006   es. Furthermore, I am convinced that Natural  SELECTION has been the main but not exclusive means o
0007   , their Differences and Origin. Principle of  SELECTION anciently followed, its Effects. Methodical
0007   wed, its Effects. Methodical and Unconscious  SELECTION. Unknown Origin of our Domestic Productions
0007   . circumstances favourable to Man's power of  SELECTION. When we look to the individuals of the sam
0015   of character probably do occur; but natural  SELECTION, as will hereafter be explained, will deter
0020   occasional crosses, if aided by the careful  SELECTION of those individual mongrels, which present
0020   got without extreme care and long continued  SELECTION: nor can I find a single case on record of
0027   act will be obvious when we come to treat of  SELECTION. Fourthly, pigeons have been watched, and t
0028   uncergone, will be obvious when we treat of  SELECTION. We shall then, also, see how it is that th
0029   e being lineal descendants of other species?  SELECTION. Let us now briefly consider the steps by w
0030   tory. The key is man's power of accumulative  SELECTION: nature gives successive variations; man ad
0030   breeds. The great power of this principle of  SELECTION is not hypothetical. It is certain that sev
0031   dge of an animal, speaks of the principle of  SELECTION as that which enables the agriculturist, no
0031   in saxony the importance of the principle of  SELECTION in regard to merino sheep is so fully recog
0032   and when a cross has been made, the closest  SELECTION is far more indispensable even than in ordi
0032   ndispensable even than in ordinary cases. If  SELECTION consisted merely in separating some very di
0033   e proper standard. With animals this kind of  SELECTION is, in fact, also followed; for hardly any
0033   eans of observing the accumulated effects of  SELECTION namely, by comparing the diversity of flowe
0033   eral rule, I cannot doubt that the continued  SELECTION of slight variations, either in the leaves,
0033   rs. It may be objected that the principle of  SELECTION has been reduced to methodical practice for
0034   ng of plants by nurserymen. The principle of  SELECTION I find distinctly given in an ancient Chine
0034   eans. Some of these facts do not show actual  SELECTION, but they show that the breeding of domesti
0034   ent time, eminent breeders try by methodical  SELECTION, with a distinct object in view, to make a
0034   the country. But, for our purpose, a kind of  SELECTION, which may be called Unconscious, and which
0035   in like our pointer. By a similar process of  SELECTION, and by careful training, the whole body of
0036   t illustration of the effects of a course of  SELECTION, which may be considered as unconsciously f
0036   is case there would be a kind of unconscious  SELECTION going on. We see the value set on animals w
0038   e plants have not been improved by continued  SELECTION up to a standard of perfection comparable w
0038   the other, and thus by a process of natural  SELECTION, as will hereafter be more fully explained,
0038   w here given of the all important part which  SELECTION by man has played, it becomes at once obvio
0038   es for what is internal. He can never act by  SELECTION, excepting on variations which are first gi
0039   ed, partly unconscious and partly methodical  SELECTION. Perhaps the parent bird of all fantails ha
0040   inciple, as I have called it, of unconscious  SELECTION will always tend, perhaps more at one perio
0040   avourable, or the reverse, to man's power of  SELECTION. A high degree of variability is obviously
0040   vourable, as freely giving the materials for  SELECTION to work on: not that mere individual differ
0041   to breed, and this will effectually prevent  SELECTION. But probably the most important point of a
0042   ose, etc., may be attributed in main part to  SELECTION not having been brought into play: in cats,
0043   ir occasional intercrossing, with the aid of  SELECTION, has, no doubt, largely aided in the format
0043   am convinced that the accumulative action of  SELECTION, whether applied methodically and more quic
0045   for us, as they afford materials for natural  SELECTION to accumulate, in the same manner as man ca
0046   n seized on and rendered definite by natural  SELECTION, as hereafter will be explained. Those form
0052   ch it differs more, to the action of natural  SELECTION in accumulating (as will hereafter be more
0060   ii. Struggle For Existence. Bears on natural  SELECTION. The term used in a wide sense. Geometrical
0060   the struggle for existence bears on Natural  SELECTION. It has been seen in the last chapter that
0061   useful, is preserved, by the term of Natural  SELECTION, in order to mark its relation to man's pow
0061   order to mark its relation to man's power of  SELECTION. We have seen that man by selection can cer
```

Page **(Key Word)**

Page **(Key Word)**************************************

```
0061  power of selection. We have seen that man by   SELECTION can certainly produce great results, and ca
0061  en to him by the hand of Nature. But Natural   SELECTION, as we shall hereafter see, is a power ince
0080  iply. Chapter IV. Natural Selection. Natural   SELECTION, its power compared with man's selection, i
0080  ral Selection, its power compared with man's   SELECTION, its power on characters of trifling import
0080  power at all ages and on both sexes. Sexual   SELECTION. On the generality of intercrosses between
0080  ances favourable and unfavourable to Natural   SELECTION, namely, intercrossing, isolation, number o
0080  s. slow action. Extinction caused by Natural   SELECTION. Divergence of Character, related to the di
0080  ea, and to naturalisation. Action of Natural   SELECTION, through Divergence of Character and Extinc
0080  in regard to variation? Can the principle of   SELECTION, which we have seen is so potent in the han
0081  tion of injurious variations, I call Natural   SELECTION. Variations neither useful nor injurious wo
0081  r injurious would not be affected by natural   SELECTION, and would be left a fluctuating element, a
0081  st understand the probable course of natural   SELECTION by taking the case of a country undergoing
0082  ons, would tend to be preserved; and natural   SELECTION would thus have free scope for the work of
0082  is would manifestly be favourable to natural   SELECTION, by giving a better chance of profitable va
0082  less profitable variations do occur, natural   SELECTION can do nothing. Not that, as I believe, any
0082  roduce new and unoccupied places for natural   SELECTION to fill up by modifying and improving some
0083  t by his methodical and unconscious means of   SELECTION, what may not nature effect? Man can act on
0083  er, all his productions. He often begins his   SELECTION by some half monstrous form; or at least by
0084  her workmanship? It may be said that natural   SELECTION is daily and hourly scrutinising, throughou
0084  om what they formerly were. Although natural   SELECTION can act only through and for the good of ea
0085  ce i can see no reason to doubt that natural   SELECTION might be most effective in giving the prope
0085  the modifications are accumulated by natural   SELECTION for the good of the being, will cause other
0086  arly adult; so in a state of nature, natural   SELECTION will be enabled to act on and modify organi
0086  culty in this being effected through natural   SELECTION, than in the cotton planter increasing and
0086  e cotton planter increasing and improving by   SELECTION the down in the pods on his cotton trees. N
0086  own in the pods on his cotton trees. Natural   SELECTION may modify and adapt the larva of an insect
0086  cture of the larva; but in all cases natural   SELECTION will ensure that modifications consequent o
0086  cause the extinction of the species. Natural   SELECTION will modify the structure of the young in r
0087  profits by the selected change. What natural   SELECTION cannot do, is to modify the structure of on
0087  , might be modified to any extent by natural   SELECTION; for instance, the great laws possessed by
0087  re would be simultaneously the most rigorous   SELECTION of the young birds within the egg, which ha
0087  n to vary like every other structure. Sexual   SELECTION. Inasmuch as peculiarities often appear und
0087  ably occurs under nature, and if so, natural   SELECTION will be able to modify one sex in its funct
0088  me to say a few words on what I call Sexual   SELECTION. This depends, not on a struggle for existe
0088  competitor, but few or no offspring. Sexual   SELECTION is, therefore, less rigorous than natural s
0088  on is, therefore, less rigorous than natural   SELECTION. Generally, the most vigorous males, those
0088  e a poor chance of leaving offspring. Sexual   SELECTION by always allowing the victor to breed migh
0088  ell that he can improve his breed by careful   SELECTION of the best cocks. How low in the scale of
0088  defence may be given through means of sexual   SELECTION, as the mane to the lion, the shoulder pad
0089  umage having been chiefly modified by sexual   SELECTION, acting when the birds have come to the bre
0089  ifferences have been mainly caused by sexual   SELECTION; that is, individual males have had, in suc
0090  sity. Illustrations of the action of Natural   SELECTION. In order to make it clear how, as I believ
0090  to make it clear how, as I believe, natural   SELECTION acts, I must beg permission to give one or
0090  of his greyhounds by careful and methodical   SELECTION, or by that unconscious selection which res
0090  methodical selection, or by that unconscious   SELECTION which results from each man trying to keep
0093  ess of the continued preservation or natural   SELECTION of more and more attractive flowers, had be
0094  en slowly increasing the nectar by continued   SELECTION, to be a common plant; and that certain ins
0095  am well aware that this doctrine of natural   SELECTION, exemplified in the above imaginary instanc
0095  the longest lines of inland cliffs. Natural   SELECTION can act only by the preservation and accumu
0095  y by a single diluvial wave, so will natural   SELECTION, if it be a true principle, banish the beli
0101  petuity. Circumstances favourable to Natural   SELECTION. This is an extremely intricate subject. A
0102  vast periods of time for the work of natural   SELECTION, she does not grant an indefinite period; f
0102  ll soon be exterminated. In man's methodical   SELECTION, a breeder selects for some definite object
0102  owly follow from this unconscious process of   SELECTION, notwithstanding a large amount of crossing
0102  t so perfectly occupied as might be, natural   SELECTION will always tend to preserve all the indivi
0102  rent conditions of life; and then if natural   SELECTION be modifying and improving a species in the
0103  ing can hardly be counterbalanced by natural   SELECTION always tending to modify all the individual
0103  effects of intercrosses in retarding natural   SELECTION; for I can bring a considerable catalogue o
0104  rinciple of inheritance, and through natural   SELECTION destroying any which depart from the proper
0104  their modified offspring, solely by natural   SELECTION preserving the same favourable variations.
0104  important element in the process of natural   SELECTION. In a confined or isolated area, if not ver
0104  e in a great degree uniform; so that natural   SELECTION will tend to modify all the individuals of
0105  he production of new species through natural   SELECTION, by decreasing the chance of the appearance
0107  ances favourable and unfavourable to natural   SELECTION, as far as the extreme intricacy of the sub
0108  again there will be a fair field for natural   SELECTION to improve still further the inhabitants, a
0108  , and thus produce new species. That natural   SELECTION will always act with extreme slowness, I fu
0108  ving been checked. But the action of natural   SELECTION will probably still oftener depend on some
0108  ficient wholly to stop the action of natural   SELECTION. I do not believe so. On the other hand, I
0108  on the other hand, I do believe that natural   SELECTION will always act very slowly, often only at
0109  is very slow, intermittent action of natural   SELECTION accords perfectly well with what geology te
0109  rld have changed. Slow though the process of   SELECTION may be, if feeble man can do much by his po
0109  man can do much by his powers of artificial   SELECTION, I can see no limit to the amount of change
0109  the long course of time by nature's power of   SELECTION. Extinction. This subject will be more full
0109  from being intimately connected with natural   SELECTION. Natural selection acts solely through the
0109  ly connected with natural selection. Natural   SELECTION acts solely through the preservation of var
0110  he course of time are formed through natural   SELECTION, others will become rarer and rarer, and fi
0110  st our domesticated productions, through the   SELECTION of improved forms by man. Many curious inst
0112  t; in the course of time, from the continued   SELECTION of swifter horses by some breeders, and of
0113  that is, if they vary, for otherwise natural   SELECTION can do nothing. So it will be with plants.
0116  ter, combined with the principles of natural   SELECTION and of extinction, will tend to act. The ac
0117  ) being preserved and accumulated by natural   SELECTION. When a dotted line reaches one of the hori
0119  re than one modified descendant; for natural   SELECTION will always act according to the nature of
0121  rt. As in each fully stocked country natural   SELECTION necessarily acts by the selected form havin
0123  . owing to the divergent tendency of natural   SELECTION acts through one form having some advantage
0125  ed, might have been expected; for as natural   SELECTION acts through one form having some advantage
0127  ave called, for the sake of brevity, Natural   SELECTION. Natural selection, on the principle of qua
```

Page **(Key Word)**************************************

Page ***(Key Word)***

```
0127  sake of brevity, Natural Selection. Natural  SELECTION, on the principle of qualities being inheri
0127  y as the adult. Amongst many animals, sexual  SELECTION will give its aid to ordinary selection, by
0127  xual selection will give its aid to ordinary  SELECTION, by assuring to the most vigorous and best
0127  les the greatest number of offspring. Sexual  SELECTION will also give characters useful to the mal
0127  struggles with other males. Whether natural  SELECTION has really thus acted in nature, in modifyi
0127  s history, geology plainly declares. Natural  SELECTION, also, leads to divergence of character; fo
0128  hem dominant in their own countries. Natural  SELECTION, as has just been remarked, leads to diverg
0129  nheritance and the complex action of natural  SELECTION, entailing extinction and divergence of cha
0131  tions. Use and disuse, combined with natural  SELECTION; organs of flight and of vision. Acclimatis
0133  ribute to the accumulative action of natural  SELECTION, and how much to the conditions of life. Th
0134  nd in thus inducing variability; and natural  SELECTION will then accumulate all profitable variati
0135  like those of a bustard, and that as natural  SELECTION increased in successive generations the siz
0135  which are wholly, or mainly, due to natural  SELECTION. Mr. Wollaston has discovered the remarkabl
0136  etles is mainly due to the action of natural  SELECTION, but combined probably with disuse. For dur
0136  cuite compatible with the action of natural  SELECTION. For when a new insect first arrived on the
0136  rived on the island, the tendency of natural  SELECTION to enlarge or to reduce the wings, would de
0137  on from disuse, but aided perhaps by natural  SELECTION. In South America, a burrowing rodent, the
0137  uch case be an advantage; and if so, natural  SELECTION would constantly aid the effects of disuse.
0137  wings of others have been reduced by natural  SELECTION aided by use and disuse, so in the case of
0137  suse, so in the case of the cave rat natural  SELECTION seems to have struggled with the loss of li
0138  perfectly obliterated its eyes, and natural  SELECTION will often have effected other changes, suc
0141  e to mere habit, and how much to the natural  SELECTION of varieties having different innate consti
0142  d, i can see no reason to doubt that natural  SELECTION will continually tend to preserve those ind
0143  , and sometimes overmastered by, the natural  SELECTION of innate differences. Correlation of Growt
0143  t occur, and are accumulated through natural  SELECTION, other parts become modified. This is a ver
0143  mastered more or less completely by natural  SELECTION: thus a family of stags once existed with a
0143  ably have been rendered permanent by natural  SELECTION. Homologous parts, as has been remarked by
0144  dently of utility and, therefore, of natural  SELECTION, than that of the difference between the ou
0146  t appear: and if it be advantageous, natural  SELECTION may have come into play. Put in regard to t
0146  progenitor may have acquired through natural  SELECTION some one modification in structure, and, af
0146  ely due to the manner alone in which natural  SELECTION can act. For instance, Alph. de Candolle ha
0146  not gradually become winged through natural  SELECTION, except in fruits which openec so that the
0147  part being largely developed through natural  SELECTION and another and adjoining part being reduce
0147  more general principle, namely, that natural  SELECTION is continually trying to economise in every
0148  ts development, will be seized on by natural  SELECTION, for it will profit the individual not to h
0148  become useless. Thus, as I believe, natural  SELECTION will always succeed in the long run in redu
0148  onding degree. And, conversely, that natural  SELECTION may perfectly well succeed in largely devel
0149  should remain variable, that is, why natural  SELECTION should have preserved or rejected each litt
0149  better be of some particular shape. Natural  SELECTION, it should never be forgotten, can act on e
0149  their uselessness, and therefore to natural  SELECTION having no power to check deviations in thei
0152  cies, and have been modified through natural  SELECTION, I think we can obtain some light. In our d
0152  rt, or the whole animal, be neglected and no  SELECTION be applied, that part (for instance, the co
0152  llel natural case; for in such cases natural  SELECTION either has not or cannot come into full pla
0152  ime are undergoing rapid change by continued  SELECTION, are also eminently liable to variation. Lo
0153  s and on the other hand, the power of steady  SELECTION to keep the breed true. In the long run sel
0153  tion to keep the breed true. In the long run  SELECTION gains the day, and we do not expect to fail
0153  om a good short faced strain. But as long as  SELECTION is rapidly going on, there may always be ex
0153  these variable characters, produced by man's  SELECTION, sometimes become attached, from causes qui
0153  has continually been accumulated by natural  SELECTION for the benefit of the species. Put as the
0153  the case. That the struggle between natural  SELECTION on the one hand, and the tendency to revers
0154  ldom as yet have been fixed by the continued  SELECTION of the individuals varying in the required
0156  for it can rarely have happened that natural  SELECTION will have modified several species, fitted
0156  l characters have been accumulated by sexual  SELECTION, which is less rigid in its action than ord
0157  ch is less rigid in its action than ordinary  SELECTION, as it does not entail death, but only give
0157  racters, as they are highly variable, sexual  SELECTION will have had a wide scope for action, and
0157  he taken advantage of by natural and sexual  SELECTION, in order to fit the several species to the
0158  en inherited and have not varied, to natural  SELECTION having more or less completely, according t
0158  ersion and to further variability, to sexual  SELECTION being less rigid than ordinary selection, a
0158  ual selection being less rigid than ordinary  SELECTION, and to variations in the same parts having
0158  aving been accumulated by natural and sexual  SELECTION, and thus adapted for secondary sexual, and
0161  rtant characters will be governed by natural  SELECTION, in accordance with the diverse habits of t
0168  ons have not been closely checked by natural  SELECTION. It is probably from this same cause that o
0168  eing useless, will be disregarded by natural  SELECTION, and hence probably are variable. Specific
0169  long continued and slow process, and natural  SELECTION will in such cases not as yet have had time
0169  a long lapse of time, in this case, natural  SELECTION may readily have succeeded in giving a fixe
0170  is the steady accumulation, through natural  SELECTION, of such differences, when beneficial to th
0171  existence embraced by the theory of Natural  SELECTION. Long before having arrived at this part of
0171  fferent habits? Can we believe that natural  SELECTION could produce, on the one hand, organs of t
0172  cts be acquired and modified through natural  SELECTION? What shall we say to so marvellous an inst
0172  rarity of transitional varieties. As natural  SELECTION acts solely by the preservation of profitab
0172  nto competition. Thus extinction and natural  SELECTION will, as we have seen, go hand in hand. Hen
0177  ng further favourable variations for natural  SELECTION to seize on, than will the rarer forms whic
0177  adiness and skill to improve their stocks by  SELECTION; the chances in this case will be strongly
0177  ariation is a very slow process, and natural  SELECTION can do nothing until favourable variations
0178  d exterminated during the process of natural  SELECTION, so that they will no longer exist in a liv
0179  cess of further modification through natural  SELECTION, they will almost certainly be beaten and s
0179  and thus be further improved through natural  SELECTION and gain further advantages. Lastly, lookin
0179  ave existed: but the very process of natural  SELECTION constantly tends, as has been so often rema
0181  cumulated effects of this process of natural  SELECTION, a perfect so called flying squirrel was pr
0181  hecus might be greatly lengthened by natural  SELECTION; and this, as far as the organs of flight a
0182  e very process of perfection through natural  SELECTION. Furthermore, we may conclude that transiti
0183  er case occurs, it would be easy for natural  SELECTION to fit the animal, by some modification of
0184  n a race of bears being rendered, by natural  SELECTION, more and more aquatic in their structure,
0186  or existence and in the principle of natural  SELECTION, will acknowledge that every organic being
0186  berration, could have been formed by natural  SELECTION, seems, I freely confess, absurd in the hig
0187  t and complex eye could be formed by natural  SELECTION, though insuperable by our imagination, can
```

Page *******************************--*****************(Key Word)***

Page **(Key Word)**

```
0188  other structures) in believing that natural  SELECTION has converted the simple apparatus of an op
0188  e eye of an eagle might be formed by natural  SELECTION, although in this case he does not know any
0188  tation in extending the principle of natural  SELECTION to such startling lengths. It is scarcely p
0189  multiply them almost infinitely, and natural  SELECTION will pick out with unerring skill each impr
0190  d the stomach respire. In such cases natural  SELECTION might easily specialise, if any advantage w
0191  o great difficulty in believing that natural  SELECTION has actually converted a swimbladder into a
0191  ght have been gradually worked in by natural  SELECTION for some quite distinct purpose: in the sam
0192  on, have been gradually converted by natural  SELECTION into branchiae, simply through an increase
0193  embers to its loss through disuse or natural  SELECTION. But if the electric organs had been inheri
0194  y hit on the very same invention, so natural  SELECTION, working for the good of each being and tak
0194  cture to structure? On the theory of natural  SELECTION, we can clearly understand why she should n
0194  y understand why she should not: for natural  SELECTION can act only by taking advantage of slight
0195  ns of little apparent importance. As natural  SELECTION acts by life and death, by the preservation
0195  nces, might assuredly be acted on by natural  SELECTION. The tail of the giraffe looks like an arti
0196  ure will always have been checked by natural  SELECTION. Seeing how important an organ of locomotio
0196  e secondary causes, independently of natural  SELECTION. We should remember that climate, food, etc
0196  various structures: and finally, that sexual  SELECTION will often have largely modified the extern
0197  and might have been acquired through natural  SELECTION; as it is, I have no doubt that the colour
0197  ome quite distinct cause, probably to sexual  SELECTION. A trailing bamboo in the Malay Archipelago
0198  s where there has been but little artificial  SELECTION. Careful observers are convinced that a dam
0198  ld be exposed to a certain extent to natural  SELECTION, and individuals with slightly different co
0198  d be thus subjected to the action of natural  SELECTION. But we are far too ignorant to speculate o
0199  of these differences, chiefly through sexual  SELECTION of a particular kind, but without here ente
0199  now of no direct use. The effects of sexual  SELECTION, when displayed in beauty to charm the fema
0200  nes might have been acquired through natural  SELECTION, subjected formerly, as now, to the several
0200  through the complex laws of growth. Natural  SELECTION cannot possibly produce any modification in
0200  ts by, the structure of another. But natural  SELECTION can and does often produce structures for t
0201  could not have been produced through natural  SELECTION. Although many statements may be found in w
0201  enter on this and other such cases. Natural  SELECTION will never produce in a being anything inju
0201  ng anything injurious to itself, for natural  SELECTION acts solely by and for the good of each. No
0201  nct, as myriads have become extinct. Natural  SELECTION tends only to make each organic being as pe
0202  and animals introduced from Europe. Natural  SELECTION will not produce absolute perfection, nor d
0202  will fulfil all the requirements of natural  SELECTION, though it may cause the death of some few
0203  same to the inexorable principle of natural  SELECTION. If we admire the several ingenious contriv
0203  tions, partly because the process of natural  SELECTION will always be very slow, and will act, at
0203  d partly because the very process of natural  SELECTION almost implies the continual supplanting an
0204  tance, could not have been formed by natural  SELECTION from an animal which at first could only gl
0204  as the eye could have been formed by natural  SELECTION, is more than enough to stagger any one: ye
0204  eivable degree of perfection through natural  SELECTION. In the cases in which we know of no interm
0205  been slowly accumulated by means of natural  SELECTION. But we may confidently believe that many m
0205  present state, have been acquired by natural  SELECTION, a power which acts solely by the preservat
0205  variations in the struggle for life. Natural  SELECTION will produce nothing in one species for the
0205  t the same time useful to the owner. Natural  SELECTION in each well stocked country, must act chie
0206  tion will have been rendered higher. Natural  SELECTION will not necessarily produce absolute perfe
0206  e everywhere found. On the theory of natural  SELECTION we can clearly understand the full meaning
0206  s fully embraced by the principle of natural  SELECTION. For natural selection acts by either now a
0206  principle of natural selection. For natural  SELECTION acts by either now adapting the varying par
0207  t. difficulties on the theory of the Natural  SELECTION of instincts. Neuter or sterile insects. Su
0209  tle, then I can see no difficulty in natural  SELECTION preserving and continually accumulating var
0209  ate importance to the effects of the natural  SELECTION of what may be called accidental variations
0210  nct can possibly be produced through natural  SELECTION, except by the slow and gradual accumulatio
0210  other instinct might be preserved by natural  SELECTION. And such instances of diversity of instinc
0211  are indispensable for the action of natural  SELECTION, as many instances as possible ought to hav
0212  s to the species, give rise, through natural  SELECTION, to quite new instincts. But I am well awar
0213  see the respective parts which habit and the  SELECTION of so called accidental variations have pla
0213  they have been acted on by far less rigorous  SELECTION, and have been transmitted for an incompara
0214  s strange habit, and that the long continued  SELECTION of the best individuals in successive gener
0214  irst tendency was once displayed, methodical  SELECTION and the inherited effects of compulsory tra
0215  ould soon complete the work; and unconscious  SELECTION is still at work, as each man tries to proc
0215  estroyed: so that habit, with some degree of  SELECTION, has probably concurred in civilising by in
0216  done nothing, and all has been the result of  SELECTION, pursued both methodically and unconsciousl
0216  usly: but in most cases, probably, habit and  SELECTION have acted together. We shall, perhaps, bes
0216  in a state of nature have become modified by  SELECTION, by considering a few cases. I will select
0219  e cuckoo. I can see no difficulty in natural  SELECTION making an occasional habit permanent, if of
0224  g pupae originally for food might by natural  SELECTION be strengthened and rendered permanent for
0224  tzerland, I can see no difficulty in natural  SELECTION increasing and modifying the instinct, alwa
0228  t the hive bee has acquired, through natural  SELECTION, her inimitable architectural powers. But I
0233  be required than for a cylinder. As natural  SELECTION acts only by the accumulation of slight mod
0235  stage of perfection in architecture, natural  SELECTION could not lead; for the comb of the hive be
0235  of the hive bee, can be explained by natural  SELECTION having taken advantage of numerous, success
0235  modifications of simpler instincts: natural  SELECTION having by slow degrees, more and more perfe
0235  . the motive power of the process of natural  SELECTION having been economy of wax: that individual
0235  on could be opposed to the theory of natural  SELECTION, cases, in which we cannot see how an insti
0235  y could hardly have been acted on by natural  SELECTION; cases of instincts almost identically the
0236  been acquired by independent acts of natural  SELECTION. I will not here enter on these several cas
0236  difficulty in this being effected by natural  SELECTION. But I must pass over this preliminary diff
0236  ers had been slowly acquired through natural  SELECTION: namely, by an individual having been born
0237  concile this case with the theory of natural  SELECTION? First, let it be remembered that we have i
0237  ould have been slowly accumulated by natural  SELECTION. This difficulty, though appearing insupera
0237  ieve, disappears, when it is remembered that  SELECTION may be applied to the family, as well as to
0238  e family. I have such faith in the powers of  SELECTION, that I do not doubt that a breed of cattle
0239  ening confidence in the principle of natural  SELECTION, when I do not admit that such wonderful an
0239  me kind, which have been rendered by natural  SELECTION, as I believe to be quite possible, differe
0239  a few alone; and that by the long continued  SELECTION of the fertile parents which produced most
0241  hese facts before me, I believe that natural  SELECTION, by acting on the fertile parents, could fo
0241  ater and greater numbers through the natural  SELECTION of the parents which generated them; until
```

Page **(Key Word)**

Page **(Key Word)**

```
0242  communities of ants, by the means of natural  SELECTION. But I am bound to confess, that, with all
0242  i should never have anticipated that natural  SELECTION could have been efficient in so high a degr
0242  ength, in order to show the power of natural  SELECTION, and likewise because this is by far the mo
0243  nder changing conditions of life, in natural  SELECTION accumulating slight modifications of instin
0243  ll tend to corroborate the theory of natural  SELECTION. This theory is, also, strengthened by some
0245  some late writers. On the theory of natural  SELECTION the case is especially important, inasmuch
0269  by man's methodical and unconscious power of  SELECTION, for his own use and pleasure: he neither w
0269  es. Seeing this difference in the process of  SELECTION, as carried on by man and nature, we need n
0275  haracters which have been slowly acquired by  SELECTION. Consequently, sudden reversions to the per
0278  ave been produced under domestication by the  SELECTION of mere external differences, and not of di
0280  ature depends on the very process of natural  SELECTION, through which new varieties continually ta
0281  d unimproved forms. By the theory of natural  SELECTION all living species have been connected with
0282  ng been effected very slowly through natural  SELECTION. It is hardly possible for me even to recal
0302  scent with slow modification through natural  SELECTION. For the development of a group of forms, a
0312  al modification, through descent and natural  SELECTION. New species have appeared very slowly, one
0314  variability be taken advantage of by natural  SELECTION, and whether the variations be accumulated
0317  groups of species. On the theory of natural  SELECTION the extinction of old forms and the product
0320  nown deed of violence. The theory of natural  SELECTION is grounded on the belief that each new var
0322  nct, accords well with the theory of natural  SELECTION. We need not marvel at extinction: if we mu
0325  orld, is explicable on the theory of natural  SELECTION. New species are formed by new varieties ar
0338  ssiz accords well with the theory of natural  SELECTION. In a future chapter I shall attempt to sho
0343  of descent with modification through natural  SELECTION. We can thus understand how it is that new
0345  the main objections to the theory of natural  SELECTION are greatly diminished or disappear. On the
0345  ll acting round us, and preserved by Natural  SELECTION. Chapter XI. Geographical Distribution.
0350  e attributed to modification through natural  SELECTION, and in a quite subordinate degree to the d
0350  slow process of modification through natural  SELECTION. Widely ranging species, abounding in indiv
0351  y, and will be taken advantage of by natural  SELECTION, only so far as it profits the individual i
0352  ould ever have been produced through natural  SELECTION from parents specifically distinct. We are
0379  g consequently been advanced through natural  SELECTION and competition to a higher stage of perfec
0382  , combined with modification through natural  SELECTION, a multitude of facts in the present distri
0392  overtopping the other plants. If so, natural  SELECTION would often tend to add to the stature of h
0395  g species have been modified through natural  SELECTION in their new homes in relation to their new
0401  ifferent enemies. If then it varied, natural  SELECTION would probably favour different varieties i
0410  een accumulated by the same power of natural  SELECTION. Chapter XIII. Mutual Affinities of Organic
0411  fourth chapters, on Variation and on Natural  SELECTION, I have attempted to show that it is the wi
0433  s belong. Finally, we have seen that natural  SELECTION, which results from the struggle for existe
0435  ion is manifest on the theory of natural  SELECTION of successive slight modifications, each mo
0436  erhaps very simple in form; and then natural  SELECTION will account for the infinite diversity in
0437  n the same pattern? On the theory of natural  SELECTION, we can satisfactorily answer these questio
0438  nsequently it is quite probable that natural  SELECTION, during a long continued course of modifica
0445  n these two breeds has been wholly caused by  SELECTION under domestication; but having had careful
0446  ed, and which have been accumulated by man's  SELECTION, have not generally first appeared at an ea
0446  species have become modified through natural  SELECTION in accordance with their diverse habits. Th
0448  r larvae might easily be rendered by natural  SELECTION different to any conceivable extent from th
0454  nd exposed islands; and in this case natural  SELECTION would continue slowly to reduce the organ,
0454  small steps, is within the power of natural  SELECTION; so that an organ rendered, during changed
0455  its variations cannot be checked by natural  SELECTION. At whatever period of life or selec
0455  ection. At whatever period of life disuse or  SELECTION reduces an organ, and this will generally b
0456  ther with their modification through natural  SELECTION, with its contingencies of extinction and d
0457  s are reduced in size, either from disuse or  SELECTION, it will generally be at that period of lif
0459  of the difficulties on the theory of Natural  SELECTION. Recapitulation of the general and special
0459  ty of species. How far the theory of natural  SELECTION may be extended. Effects of its adoption on
0459  of descent with modification through natural  SELECTION, I do not deny. I have endeavoured to give
0460  special difficulty on the theory of natural  SELECTION; and one of the most curious of these is th
0462  degree lessened. As on the theory of natural  SELECTION an interminable number of intermediate form
0467  ciable by an uneducated eye. This process of  SELECTION has been the great agency in the production
0467  e the most powerful and ever acting means of  SELECTION. The struggle for existence inevitably foll
0468  it would be an unaccountable fact if natural  SELECTION had not come into play. It has often been a
0469  lex relations of life. The theory of natural  SELECTION, even if we looked no further than this, se
0470  there will be a constant tendency in natural  SELECTION to preserve the most divergent offspring of
0471  icable on the theory of creation. As natural  SELECTION acts solely by accumulating slight, success
0472  y trying to increase in number, with natural  SELECTION always ready to adapt the slowly varying de
0472  might even have been anticipated. As natural  SELECTION acts by competition, it adapts the inhabita
0472  e wonder indeed is, on the theory of natural  SELECTION, that more cases of the want of absolute pe
0474  rendered constant by long continued natural  SELECTION. Glancing at instincts, marvellous as some
0474  oreal structure on the thoery of the natural  SELECTION of successive, slight, but profitable modif
0475  having been slowly acquired through natural  SELECTION we need not marvel at some instincts being
0475  evitably follows on the principle of natural  SELECTION; for old forms will be supplanted by new an
0478  ps, is intelligible on the theory of natural  SELECTION with its contingencies of extinction and di
0479  ranchiae. Disuse, aided sometimes by natural  SELECTION, will often tend to reduce an organ, when i
0480  earing of rudimentary organs. But disuse and  SELECTION will generally act on each creature, when i
0480  gue and palate having been fitted by natural  SELECTION to browse without their aid: whereas in the
0480  calf, the teeth have been left untouched by  SELECTION or disuse, and on the principle of inherita
0489  equally inappreciable length. And as natural  SELECTION works solely by and for the good of each be
0490  le for Life, and as a consequence to Natural  SELECTION, entailing Divergence of Character and the
0083  fference, on the whole machinery of life. Man  SELECTS only for his own good; Nature only for that
0102  ted. In man's methodical selection, a breeder  SELECTS for some definite object, and free intercros
0071  ten years large spaces have been enclosed, and  SELF sown firs are now springing up in multitudes,
0097  the meaning of the law) that no organic being  SELF fertilises itself for an eternity of generatio
0097  pistil generally stand so close together that  SELF fertilisation seems almost inevitable. Many fl
0098  the contrivance seems adapted solely to ensure  SELF fertilisation; and no doubt it is useful for t
0098  which seems to have a special contrivance for  SELF fertilisation, it is well known that if very c
0098  other cases, far from there being any aids for  SELF fertilisation, there are special contrivances,
0098  close together, as if for the very purpose of  SELF fertilisation, should in so many cases be mutu
0100  ndividuals. Of aquatic animals, there are many  SELF fertilising hermaphrodites: but here currents
0101  to prove that two individuals, though both are  SELF fertilising hermaphrodites, do sometimes cross
```

Page **(Key Word)**

Page ************************************(Key Word)************************************

```
0101 long intervals; but in none, as I suspect, can SELF fertilisation go on for perpetuity. Circumstan
0104 rtility over the offspring from long continued SELF fertilisation, that they will have a better ch
0250 hybridised much more readily than they can be SELF fertilised! For instance, a bulb of Hippeastru
0251 ey were functionally imperfect in their mutual SELF action, we must infer that the plants were in
0251 ssed, in comparison with the same species when SELF fertilised, sometimes depends. The practical e
0256 from this extreme degree of sterility we have SELF fertilised hybrids producing a greater and gre
0262 with the pollen of distinct species, than when SELF fertilised with their own pollen. We thus see,
0018 f tierra del Fuego or Australia, who possess a SEMI domestic dog, may not have existed in Egypt? T
0025 ith white; the wings have two black bars; some SEMI domestic breeds and some apparently truly wild
0040 n probably first receive a provincial name. In SEMI civilised countries, with little free communic
0275 ded from varieties often suddenly produced and SEMI monstrous in character, than with hybrids, whi
0003 ssible cause of variation. In one very limited SENSE, as we shall hereafter see, this may be true;
0030 p in certain directions useful to him. In this SENSE he may be said to make for himself useful bre
0044 authors use the term variation in a technical SENSE, as implying a modification directly due to t
0044 cal conditions of life, and variations in this SENSE are supposed not to be inherited: but who can
0060 on natural selection. The term used in a wide SENSE. Geometrical powers of increase. Rapid increa
0062 ogle for Existence in a large and metaphorical SENSE, including dependence of one being on another
0063 few other trees, but can only in a far fetched SENSE be said to struggle with these trees, for if
0100 ered as distinct individuals only in a limited SENSE. I believe this objection to be valid, but th
0151 sile cirripedes (rock barnacles) are, in every SENSE of the word, very important structures, and t
0199 , can be called useful only in rather a forced SENSE. But by far the most important consideration
0324 , or even that it has a very strict geological SENSE; for if all the marine animals which live at
0324 rrectly ranked as simultaneous in a geological SENSE. The fact of the forms of life changing simul
0324 fe changing simultaneously, in the above large SENSE, at distant parts of the world, has greatly s
0327 ems to me, the parallel, and, taken in a large SENSE, simultaneous, succession of the same forms o
0336 t by high and low forms. But in one particular SENSE the more recent forms must, on my theory, be
0374 nd that it was contemporaneous in a geological SENSE, it seems to me probable that it was, during
0424 s of Steenstrup, which can only in a technical SENSE be considered as the same individual. He incl
0438 ever, use such language only in a metaphorical SENSE: they are far from meaning that during a long
0441 s, to search by their well developed organs of SENSE, and to reach by their active powers of swimm
0444 of variations (taking the word in the largest SENSE) which have supervened at an earlier age in t
0476 re generally looked at as being, in some vague SENSE, higher than ancient and extinct forms; and t
0489 ng around us. These laws, taken in the largest SENSE, being Growth with ºeproduction; Inheritance
0063 than those of other plants. In these several SENSES, which pass into each other, I use for conven
0448 bits, in which organs of locomotion cr of the SENSES, etc., would be useless: and in this case the
0002 ost cases will suffice. No one can feel more SENSIBLE than I do of the necessity of hereafter publ
0065 en suddenly and temporarily increased in any SENSIBLE degree. The obvious explanation is that the
0336 hort, such evidence of the slow and scarcely SENSIBLE mutation of specific forms, as we have a jus
0337 of improvement has affected in a marked and SENSIBLE manner the organisation of the more recent a
0187 be considered real. How a nerve comes to be SENSITIVE to light, hardly concerns us more than how
0187 that several facts make me suspect that any SENSITIVE nerve may be rendered sensitive to light, a
0187 ect that any sensitive nerve may be rendered SENSITIVE to light, and likewise to those coarser vib
0188 ck ·layer of transparent tissue, with a nerve SENSITIVE to light beneath, and then suppose every pa
0264 ally as all very young beings seem eminently SENSITIVE to injurious or unnatural conditions of lif
0273 e to the reproductive system being eminently SENSITIVE to any change in the conditions of life, be
0002 i have on the origin of species. Last year he SENT to me a memoir on this subject, with a request
0002 t i would forward it to Sir Charles Lyell, who SENT it to the Linnean Society, and it is published
0028 with the court. The monarchs of Iran and Turan SENT him some very rare birds; and, continues the c
0304 d, when a skilful palaeontologist, M. Bosquet, SENT me a drawing of a perfect specimen of an unmis
0413 sible, general propositions, that is, by one SENTENCE to give the characters common, for instance,
0413 o the dog genus, and then by adding a single SENTENCE, a full description is given of each kind of
0345 s may account for that vague yet ill defined SENTIMENT, felt by many palaeontologists, that organi
0436 that in a flower the relative position of the SEPALS, petals, stamens, and pistils, as well as the
0437 many legs have simpler mouths? Why should the SEPALS, petals, stamens, and pistils in any individu
0003 by certain birds, and which has flowers with SEPARATE sexes absolutely requiring the agency of cer
0025 is thus coloured and marked, so that in each SEPARATE breed there might be a tendency to revert to
0042 or forty years. In the case of animals with SEPARATE sexes, facility in preventing crosses is an
0048 nd seeing others compare, the birds from the SEPARATE islands of the Galapagos Archipelago, both o
0096 sion. In the case of animals and plants with SEPARATE sexes, it is of course obvious that two indi
0107 ion. When converted by subsidence into large SEPARATE islands, there will still exist many individ
0159 dency to vary in a like manner, but to three SEPARATE yet closely related acts of creation. With p
0172 be here discussed, Instinct and Hybridism in SEPARATE chapters. On the absence or rarity of transi
0185 ture has begun to change. He who believes in SEPARATE and innumerable acts of creation will say, t
0189 inually changing slowly in density, so as to SEPARATE into layers of different densities and thick
0225 them short tubes of wax, and likewise making SEPARATE and very irregular rounded cells of wax. At
0282 special treatises by different observers on SEPARATE formations, and to mark how each author atte
0289 d elsewhere been accumulated. And if in each SEPARATE territory, hardly any idea can be formed of
0295 per or lower limits. It would seem that each SEPARATE formation, like the whole pile of formations
0313 xtinct species has been simultaneous in each SEPARATE formation. Species of different genera and c
0352 t if the same species can be produced at two SEPARATE points, why do we not find a single mammal c
0357 recent period continents which are now quite SEPARATE, have been continuously, or almost continuou
0390 admits the doctrine of the creation of each SEPARATE species, will have to admit that a sufficien
0400 ted species; so that the inhabitants of each SEPARATE island, though mostly distinct, are related
0401 ances, is that the new species formed in the SEPARATE islands have not quickly spread to the other
0402 probably will hold their own places and keep SEPARATE for almost any length of time. Being familia
0408 of barriers, whether of land or water, which SEPARATE our several zoological and botanical provinc
0415 l taken together they appear insufficient to SEPARATE Cnestis from Connarus. To give an example am
0433 e of looking to types, whether or not we can SEPARATE and define the groups to which such types be
0437 the benefit derived from the yielding of the SEPARATE pieces in the act of parturition of mammals,
0480 . on the view of each organic being and each SEPARATE organ having been specially created, how utt
0008 s monstrosities; and monstrosities cannot be SEPARATED by any clear line of distinction from mere
0048 d within the same country, but are common in SEPARATED areas. How many of those birds and insects
0099 is ready, so that these plants have in fact SEPARATED sexes, and must habitually be crossed. How
0100 trees a strong tendency to bear flowers with SEPARATED sexes. When the sexes are separated, althou
0100 ers with separated sexes. When the sexes are SEPARATED, although the male and female flowers may b
0100 ng to all Orders have their sexes more often SEPARATED than other plants, I find to be the case in
```

Page ************************************(Key Word)************************************

Page ********************************(Key Word)**

```
0103  rds, varieties will generally be confined to  SEPARATED  countries; and this I believe to be the cas
0107  onnect to a certain extent orders now widely  SEPARATED  in the natural scale. These anomalous forms
0228  . following the example of Mr. Tegetmeier, I  SEPARATED  two combs, and put between them a long thic
0252  , then we may infer that animals more widely  SEPARATED  in the scale of nature can be more easily c
0270  n his garden; and although these plants have  SEPARATED  sexes, they never naturally crossed. He the
0270  not have been injurious, as the plants have  SEPARATED  sexes. No one, I believe, has suspected tha
0270  varieties of gourd, which like the maize has  SEPARATED  sexes, and he asserts that their mutual fer
0289  gi namely, from the several formations being  SEPARATED  from each other by wide intervals of time.
0298  er than are the species found in more widely  SEPARATED  formations; but to this subject I shall hav
0299  archipelago, with its numerous large islands  SEPARATED  by wide and shallow seas, probably represen
0300  idence. These periods of subsidence would be  SEPARATED  from each other by enormous intervals, duri
0311  ife, entombed in our consecutive, but widely  SEPARATED,  formations. On this view, the difficulties
0323  e forms of life, in the stages of the widely  SEPARATED  palaeozoic and tertiary periods, would stil
0328  nless, indeed, it be assumed that an isthmus  SEPARATED  two seas inhabited by distinct, but contemp
0330  by some of its characters groups now widely  SEPARATED  from each other. This remark no doubt must
0335  e case of terrestrial productions inhabiting  SEPARATED  districts. To compare small things with gre
0348  central America; yet these great faunas are  SEPARATED  only by the narrow, but impassable, isthmus
0348  under corresponding climates; but from being  SEPARATED  from each other by impassable barriers, eit
0354  the same species, now living at distant and  SEPARATED  points; nor do I for a moment pretend that
0354  ecies on islands and on the mainland, though  SEPARATED  by hundreds of miles of open sea. If the ex
0365  any plants and animals, on mountain summits,  SEPARATED  from each other by hundreds of miles of low
0369  te productions of the Old and New Worlds are  SEPARATED  from each other by the Atlantic Ocean and b
0370  t, they must have been still more comple.  by  SEPARATED  by wider spaces of ocean. I believe the abo
0372  erranean and in the seas of Japan, areas now  SEPARATED  by a continent and by nearly a hemisphere o
0373  ealand; and the same plants, found on widely  SEPARATED  mountains in this island, tell the same sto
0383  ent chapters. As lakes and river systems are  SEPARATED  from each other by barriers of land, it mig
0395  elation generally holds good. We see Britain  SEPARATED  by _ shallow channel from Europe, and the m
0395  we meet with analogous facts on many islands  SEPARATED  by similar channels from Australia. The Wes
0396  changes of level it is obvious that islands  SEPARATED  by shallow channels are more likely to have
0396  a recent period to the mainland than islands  SEPARATED  by deeper channels, we can understand the f
0401  islands, though in sight of each other, are  SEPARATED  by deep arms of the sea, in most cases wide
0401  so that the islands are far more effectually  SEPARATED  from each other than they appear to be on a
0432  en, chain of affinities. Extinction has only  SEPARATED  groups: it has by no means made them; for i
0451  es in a well developed state. In plants with  SEPARATED  sexes, the male flowers often have a rudime
0465  of specific forms. Successive formations are  SEPARATED  from each other by enormous blank intervals
0468  , will turn the balance. With animals having  SEPARATED  sexes there will in most cases be a struggl
0477  orthern and southern temperate zones, though  SEPARATED  by the whole intertropical ocean. Although
0477  they have been for a long period completely  SEPARATED  from each other; for as the relation of org
0075  weet peas, they must be each year harvested  SEPARATELY, and the seed then mixed in due proportion,
0138  ommon view of the blind animals having been  SEPARATELY  created for the American and European caver
0174  ch islands distinct species might have been  SEPARATELY  formed without the possibility of intermedi
0178  te for each birth and wander much, may have  SEPARATELY  been rendered sufficiently distinct to rank
0194  eperdent beings, each supposec to have been  SEPARATELY  created for its proper place in nature, be
0207  uld be more convenient to treat the subject  SEPARATELY, especially as so wonderful an instinct as
0356  een highly influential: a narrow isthmus now  SEPARATES  two marine faunas; submerge it, or let it f
0395  bes by a space of deep ocean; and this space  SEPARATES  two widely distinct mammalian faunas. On ei
0411  , general, complex and radiating. Extinction  SEPARATES  and defines groups. Morphology, between mem
0032  ary cases. If selection consisted merely in  SEPARATING  some very distinct variety, and breeding fr
0348  ikely to have endured so long as the oceans  SEPARATING  continents, the differences are very inferi
0395  t of distance, between the depth of the sea  SEPARATING  an island from the neighbouring mainland, a
0413  iving objects which are most alike, and for  SEPARATING  those which are most unlike; or as an artif
0420  ositions, and the mere putting together and  SEPARATING  objects more or less alike. But I must expl
0424  pedes, when adult, and yet no one dreams of  SEPARATING  them. The naturalist includes as one specie
0093  ut as likewise illustrating one step in the  SEPARATION  of the sexes of plants, presently to be all
0094  om flower to flower, and as a more complete  SEPARATION  of the sexes of our plant would be advantag
0094  oured or selected, until at last a complete  SEPARATION  of the sexes would be affected. Let us now
0371  ng the distance of the two areas, and their  SEPARATION  by the Atlantic Ocean. We can further under
0371  en completely cut off from each other. This  SEPARATION, as far as the more temperate productions a
0428  parallelism over a wide range; and thus the  SEPTENARY, quinary, quaternary, and ternary classific
0290  at is, have not followed each other in close  SEQUENCE. Scarcely any fact struck me more when exami
0325  europe, they add, if struck by this strange  SEQUENCE, we turn our attention to North America, and
0335  eon were arranged as well as they could be in  SERIAL  affinity, this arrangement would not closely
0438  and distinct species, we can indicate but few  SERIAL  homologies; that is, we are seldom enabled to
0027  tant countries, we can make an almost perfect  SERIES  between the extremes of structure. Thirdly, t
0051  rences blend into each other in an insensible  SERIES; and a series impresses the mind with the ide
0051  nto each other in an insensible series; and a  SERIES  impresses the mind with the idea of an actual
0073  creasing circles of complexity. We began this  SERIES  by insectivorous birds, and we have ended wit
0187  erfected. In the Articulata we can commence a  SERIES  with an optic nerve merely coated with pigmen
0204  n the case of any organ, if we know of a long  SERIES  of gradations in complexity each good for its
0225  us her method of work. At one end of a short  SERIES  we have humble bees, which use their old coco
0225  rounded cells of wax. At the other end of the  SERIES  we have the cells of the hive bee, placed in
0225  adjoining cells on the opposite side. In the  SERIES  between the extreme perfection of the cells o
0241  f a different size and structure; a graduated  SERIES  having been first formed, as in the case of t
0267  rallelism is an accident or an illusion. Both  SERIES  of facts seem to be connected together by som
0280  er existed, we should have an extremely close  SERIES  between both and the rock pigeon; but we shou
0293  asons why each should not include a graduated  SERIES  of links between the species which then lived
0303  y come in at the commencement of the tertiary  SERIES. And now one of the richest known accumulatio
0303  ammals belongs to the middle of the secondary  SERIES; and one true mammal has been discovered in t
0303  tone at nearly the commencement of this great  SERIES. Cuvier used to urge that no monkey occurred
0304  developed at the commencement of the tertiary  SERIES. This was a sore trouble to me, adding as I t
0325  ention to North America, and there discover a  SERIES  of analogous phenomena, it will appear certai
0329  er to the living or to the extinct alone, the  SERIES  is far less perfect than if we combine both i
0332  discovered at several points low down in the  SERIES, the three existing families on the uppermost
0334  ephants, when arranged by Dr. Falconer in two  SERIES, first according to their mutual affinities i
0335  umblers, which are at the opposite end of the  SERIES  in this same respect. Closely connected with
0353  uld be, if, when coming one step lower in the  SERIES, to the individuals of the same species, a di
```

Page ********************************(Key Word)**

Page ***(Key Word)***

```
0419 e are crustaceans at the opposite ends of the SERIES, which have hardly a character in common; yet
0422 es of the groups had been written in a linear SERIES, it would have been still less possible to ha
0422 is notoriously not possible to represent in a SERIES, on a flat surface, the affinities which we d
0437 these questions. In the vertebrata, we see a SERIES of internal vertebrae bearing certain process
0437 he articulata, we see the body divided into a SERIES of segments, bearing external appendages; and
0437 appendages; and in flowering plants, we see a SERIES of successive spiral whorls of leaves. An ind
0193 e electric organs offer another and even more SERIOUS difficulty; for they occur in only about a d
0209 ne so instinctively. But it would be the most SERIOUS error to suppose that the greater number of
0242 and likewise because this is by far the most SERIOUS special difficulty, which my theory has enco
0247 re parent species in a state of nature. But a SERIOUS cause of error seems to me to be here introd
0009 es), yet having their reproductive system so SERIOUSLY affected by unperceived causes as to fail i
0081 of the change of climate itself, would most SERIOUSLY affect many of the others. If the country w
0081 uld certainly immigrate, and this also would SERIOUSLY disturb the relations of some of the former
0143 malconformation affecting the early embryo, SERIOUSLY affects the whole organisation of the adult
0151 ut that it applies to plants, and this would SERIOUSLY have shaken my belief in its truth, had not
0264 ly liable to have their reproductive systems SERIOUSLY affected. This, in fact, is the great bar t
0273 s themselves have their reproductive systems SERIOUSLY affected, and their descendants are highly
0202 isted in a remote progenitor as a boring and SERRATED instrument, like that in so many members of
0202 cannot be withdrawn, owing to the backward SERRATURES, and so inevitably causes the death of the
0035 n question had been made long ago, which might SERVE for comparison. In some cases, however, uncha
0042 ds may be freely rejected, as when killed they SERVE for food. On the other hand, cats, from their
0145 hat C. C. Sprengel's idea that the ray florets SERVE to attract insects, whose agency is highly ad
0149 form less carefully than when the part has to SERVE for one special purpose alone. In the same wa
0175 ing species; that nearly all either prey on or SERVE as prey for others; in short, that each organ
0182 quired their perfect power of flight; but they SERVE, at least, to show what diversified means of
0234 skin, called by me the ovigerous frena, which SERVE, through the means of a sticky secretion, to
0234 art of the bounding surface of each cell would SERVE to bound other cells, and much wax would be s
0382 ountian fastnesses of almost every land, which SERVE as a record, full of interest to us, of the f
0426 of high physiological importance, those which SERVE to preserve life under the most diverse condi
0428 : but the shape of the body and fin like limbs SERVE as characters exhibiting true affinity betwee
0435 radually enveloped in thick membrane, so as to SERVE as a fin: or a webbed foot might have all its
0435 necting them increased to any extent, so as to SERVE as a wing: yet in all this great amount of mo
0439 which when mature become widely different and SERVE for different purposes, are in the embryo exa
0442 mbryo, which ultimately become very unlike and SERVE for diverse purposes, being at this early per
0453 of rudimentary organs, by supposing that they SERVE to excrete matter in excess, or injurious to
0455 become useless in the pronunciation, but which SERVE as a clue in seeking for its derivation. On t
0191 le that organs which at a very ancient period SERVED for respiration have been actually converted
0192 t that little folds of skin, which originally SERVED as ovigerous frena, but which, likewise, very
0357 s, now buried beneath the sea, which may have SERVED as halting places for plants and for many ani
0435 ng general pattern, for whatever purpose they SERVED, we can at once perceive the plain significat
0447 classes. The fore limbs, for instance, which SERVED as legs in the parent species, may become, by
0131 rse, is a wholly incorrect expression, but it SERVES to acknowledge plainly our ignorance of the c
0171 ortance, such as the tail of a giraffe, which SERVES as a fly flapper, and on the other hand, orga
0180 tail united by a broad expanse of skin, which SERVES as a parachute and allows them to glide throu
0488 he fossils of consecutive formations probably SERVES as a fair measure of the lapse of actual time
0046 ations in points of structure which are of no SERVICE or disservice to the species; and which cons
0084 arth, we must believe that these tints are of SERVICE to these birds and insects in preserving the
0146 being, as far as we can see, of the slightest SERVICE to the species. We may often falsely attribu
0197 this contrivance, no doubt, is of the highest SERVICE to the plant; but as we see nearly similar h
0223 the masters in this country receive much less SERVICE from their slaves than they do in Switzerlan
0453 hich are subsequently absorbed, can be of any SERVICE to the rapidly growing embryonic calf by the
0479 derived from rudimentary parts, though of no SERVICE to the being, are often of high classificato
0482 believe that species are mutable will do good SERVICE by conscientiously expressing his conviction
0416 in rudimentary bones of the leg, are highly SERVICEABLE in exhibiting the close affinity between R
0416 ut which are universally admitted as highly SERVICEABLE in the definition of whole groups. For ins
0418 arly uniform, they are considered as highly SERVICEABLE in classification; but in some groups of a
0423 whereas with sheep the horns are much less SERVICEABLE, because less constant. In classing variet
0478 we see why certain characters are far more SERVICEABLE than others for classification; why adapti
0068 quently it is not the obtaining food, but the SERVING as prey to other animals, which determines t
0077 ects, to hunt for its own prey, and to escape SERVING as prey to other animals. The store of nutri
0192 the body and sack, including the small frena, SERVING for respiration. The Balanidae or sessile ci
0371 have been almost continuously united by land, SERVING as a bridge, since rendered impassable by co
0434 great jaws of a beetle? yet all these organs, SERVING for such different purposes, are formed by i
0451 til are essentially alike in nature. An organ SERVING for two purposes, many become rudimentary or
0050 practical men can be quoted to show that the SESSILE and pedunculated oaks are either good and di
0151 largest application. The opercular valves of SESSILE cirripedes (rock barnacles) are, in every se
0192 na, serving for respiration. The Balanidae or SESSILE cirripedes, on the other hand, have no ovige
0192 lready suffered far more extinction than have SESSILE cirripedes, who would ever have imagined tha
0288 species of the Chthamalinae (a sub family of SESSILE cirripedes) coat the rocks all over the worl
0304 yes has much struck me. In a memoir on Fossil SESSILE Cirripedes, I have stated that, from the num
0304 all these circumstances, I inferred that had SESSILE cirripedes existed during the secondary peri
0304 ing of a perfect specimen of an unmistakeable SESSILE cirripede, which he had himself extracted fr
0304 o make the case as striking as possible, this SESSILE cirripede was a Chthamalus, a very commom, l
0305 ry stratum. Hence we now positively know that SESSILE cirripedes existed during the secondary peri
0440 divisions of cirripedes; the pedunculated and SESSILE, which differ widely in external appearance,
0033 parison with the leaves and flowers of the same SET of varieties. See how different the leaves of
0036 nconscious selection going on. We see the value SET on animals even by the barbarians of Tierra de
0039 ion. Nor must the value which would formerly be SET on any slight differences in the individuals o
0039 s, be judged of by the value which would now be SET on them, after several breeds have once fairly
0052 itrarily given for the sake of convenience to a SET of individuals closely resembling each other,
0073 quently, from its peculiar structure, never can SET a seed. Many of our orchidaceous plants absolu
0078 have to give it some advantage over a different SET of competitors or enemies. It is good thus to
0093 on some a profusion of pollen. As the wind had SET for several days from the female to the male t
0116 ls be capable of there supporting themselves. A SET of animals, with their organisation but little
0116 little diversified, could hardly compete with a SET more perfectly diversified in structure. It ma
0159 nct breeds of pigeons, in countries most widely SET apart, present sub varieties with reversed fea
```

Page ***(Key Word)***

Page ***(Key Word)***

```
0219  ed a single slave (F. fusca), and she instantly SET to work, fed and saved the survivors; made som
0240  difference was the same as if we were to see a SET of workmen building a house of whom many were
0241  stly, and this is our climax of difficulty, one SET of workers of one size and structure, and simu
0241  size and structure, and simultaneously another SET of workers of a different size and structure;
0337  ll the animals and plants of Great Britain were SET free in New Zealand, that in the course of tim
0337  ubt, if all the productions of New Zealand were SET free in Great Britain, whether any considerabl
0418  n be discovered between them, especial value is SET on them. As in most groups of animals, importa
0456  naturalists in their classifications; the value SET upon characters, if constant and prevalent, wh
0008  rowth, will determine whether or not the plant SETS a seed. I cannot here enter on the copious det
0053  ar more important circumstance) with different SETS of organic beings. But my tables further show
0098  d, at least in my garden, by insects, it never SETS a seed, though by placing pollen from one flow
0401  s, for it would have to compete with different SETS of organisms: a plant, for instance, would fin
0035  competent authorities are convinced that the SETTER is directly derived from the spaniel, and has
0085  f enemies, such differences would effectually SETTLE which variety, whether a smooth or downy, a y
0379  ly different with those intruding forms which SETTLED themselves on the intertropical mountains, a
0391  come mutually adapted to each other; and when SETTLED in their new homes, each kind will have been
0401  land. Hence when in former times an immigrant SETTLED on any one or more of the islands, or when i
0056  ifference is one very important criterion in SETTLING whether two forms should be ranked as specie
0023  pigeon, they must have descended from at least SEVEN or eight aboriginal stocks; for it is impossi
0024  our pigeons, it must be assumed that at least SEVEN or eight species were so thoroughly domestica
0026  , the improbability of man having formerly got SEVEN or eight supposed species of pigeons to breed
0107  minated. And it is in fresh water that we find SEVEN genera of Ganoid fishes, remnants of a once p
0248  om a cross with either pure parent, for six or SEVEN, and in one case for ten generations, yet he
0341  ts, constituting the six new genera. The other SEVEN species of the old genera have all died out a
0378  th that now felt there at the height of six or SEVEN thousand feet. During this the coldest period
0385  time a duck or heron might fly at least six or SEVEN hundred miles, and would be sure to alight on
0397  and uninjured an immersion in sea water during SEVEN days: one of these shells was the Helix pomat
0039  her and distinct breeds, in which as many as SEVENTEEN tail feathers have been counted. Perhaps in
0135  etles are very often broken off; he examined SEVENTEEN specimens in his own collection, and not on
0334  those which became still more modified at the SEVENTH stage; hence they could hardly fail to be ne
0002  t here give references and authorities for my SEVERAL statements; and I must trust to the reader r
0003  cture of this parasite, with its relations to SEVERAL distinct organic beings, by the effects of e
0007  ar that organic beings must be exposed during SEVERAL generations to the new conditions of life to
0008  been affected prior to the act of conception. SEVERAL reasons make me believe in this; but the chi
0012  it is well worth while carefully to study the SEVERAL treatises published on some of our old culti
0013  ces, appears in the parent, say, once amongst SEVERAL million individuals, and it reappears in the
0013  rickly skin, hairy bodies, etc., appearing in SEVERAL members of the same family. If strange and r
0015  re to cultivate, during many generations, the SEVERAL races, for instance, of the cabbage, in very
0016  from the other species of the same genus, in SEVERAL trifling respects, they often differ in an e
0016  owing whether they have descended from one or SEVERAL parent species. This point, if it could be c
0017  sion, whether they have descended from one or SEVERAL species. The argument mainly relied on by th
0018  le that our domestic dogs have descended from SEVERAL wild species. In regard to sheep and goats I
0018  boriginal stock from our European cattle; and SEVERAL competent judges believe that these latter h
0018  btfully inclined to believe, in opposition to SEVERAL authors, that all the races have descended f
0019  and rabbit. The doctrine of the origin of our SEVERAL domestic races from several aboriginal stock
0019  the origin of our several domestic races from SEVERAL aboriginal stocks, has been carried to an ab
0019  of species of wild cattle, as many sheep, and SEVERAL goats in Europe alone, and several even with
0019  sheep, and several goats in Europe alone, and SEVERAL even within Great Britain. One author believ
0019  c., but that each of these kingdoms possesses SEVERAL peculiar breeds of cattle, sheep, etc., we m
0019  whence could they have been derived, as these SEVERAL countries do not possess a number of peculia
0019  ch I fully admit have probably descended from SEVERAL wild species, I cannot doubt that there has
0020  ween their parents; and if we account for our SEVERAL domestic races by this process, we must admi
0020  ese mongrels are crossed one with another for SEVERAL generations, hardly two of them will be alik
0020  ave been most kindly favoured with skins from SEVERAL quarters of the world, more especially by th
0020  onsiderable antiquity. I have associated with SEVERAL eminent fanciers, and have been permitted to
0022  d tail touch; the oil gland is quite aborted. SEVERAL other less distinct breeds might have been s
0022  have been specified. In the skeletons of the SEVERAL breeds, the development of the bones of the
0023  si more especially as in each of these breeds SEVERAL truly inherited sub breeds, or species as he
0023  on (Columba livia), including under this term SEVERAL geographical races or sub species, which dif
0023  each other in the most trifling respects. As SEVERAL of the reasons which have led me to this bel
0023  cases, I will here briefly give them. If the SEVERAL breeds are not varieties, and have not proce
0024  tic breeds, has not been exterminated even on SEVERAL of the smaller British islets, or on the sho
0024  s to me a very rash assumption. Moreover, the SEVERAL above named domesticated breeds have been tr
0024  y slightly altered state, has become feral in SEVERAL places. Again, all recent experience shows t
0024  ems to me, of great weight, and applicable in SEVERAL other cases, is, that the above specified br
0024  zed man succeeded in thoroughly domesticating SEVERAL species, but that he intentionally or by cha
0025  k bars, the wings chequered with black. These SEVERAL marks do not occur together in any other spe
0025  e suppositions. Either, firstly, that all the SEVERAL imagined aboriginal stocks were coloured and
0026  , seems to me rash in the extreme. From these SEVERAL reasons, namely, the improbability of man ha
0027  offspring being perfectly fertile; from these SEVERAL reasons, taken together, I can feel no doubt
0027  rs from the rock pigeon, yet by comparing the SEVERAL sub breeds of these breeds, more especially
0027  e been domesticated for thousands of years in SEVERAL quarters of the world; the earliest known re
0028  use when I first kept pigeons and watched the SEVERAL kinds, knowing well how true they bred, I fe
0028  es I have read, are firmly convinced that the SEVERAL breeds to which each has attended, are desce
0029  es, shows how utterly he disbelieves that the SEVERAL sorts, for instance a Ribston pippin or Codl
0029  ly impressed with the differences between the SEVERAL races; and though they well know that each r
0029  s have been produced, either from one or from SEVERAL allied species. Some little effect may, perh
0030  and as useful as we now see them; indeed, in SEVERAL cases, we know that this has not been their
0030  ction is not hypothetical. It is certain that SEVERAL of our eminent breeders have, even within a
0031  hey have done, it is almost necessary to read SEVERAL of the many treatises devoted to this subjec
0033  principle is a modern discovery. I could give SEVERAL references to the full acknowledgment of the
0039  e value which would now be set on them, after SEVERAL breeds have once fairly been established. Ma
0043  our domestic productions. When in any country SEVERAL domestic breeds have once been established,
0046  ce rubus, Rosa, and Hieracium amongst plants, SEVERAL genera of insects, and several genera of Bra
0046  mongst plants, several genera of insects, and SEVERAL genera of Brachiopod shells. In most polymor
0047  like to rank them as distinct species, are in SEVERAL respects the most important for us. We have
```

Page ***(Key Word)***

Page ***(Key Word)***

0048	from uncommon cannot be disputed. Compare the	SEVERAL floras of Great Britain, of France or of the
0048	nists as species, and he has entirely omitted	SEVERAL highly polymorphic genera. Under genera, inc
0049	ve been ranked as species by some zoologists.	SEVERAL most experienced ornithologists consider our
0049	btful species well deserve consideration; for	SEVERAL interesting lines of argument, from geograph
0049	cing to very numerous experiments made during	SEVERAL years by that most careful observer Gartner,
0052	ust be judged of by the general weight of the	SEVERAL facts and views given throughout this work.
0053	vary most, by tabulating all the varieties in	SEVERAL well worked floras. At first this seemed a s
0059	parently have restricted ranges. In all these	SEVERAL respects the species of large genera present
0063	the same tree, it will languish and die. But	SEVERAL seedling missletoes, growing close together
0063	s rather than those of other plants. In these	SEVERAL senses, which pass into each other, I use fo
0063	g, which during its natural lifetime produces	SEVERAL eggs or seeds, must suffer destruction durin
0064	animals of many kinds which have run wild in	SEVERAL parts of the world: if the statements of sev
0064	e islands in a period of less than ten years.	SEVERAL of the plants now most numerous over the wid
0067	animal. This subject has been ably treated by	SEVERAL authors, and I shall, in my future work, dis
0071	ad never been touched by the hand of man; but	SEVERAL hundred acres of exactly the same nature had
0072	uch surprised at their numbers that I went to	SEVERAL points of view, whence I could examine hundr
0072	rmine the existence of the Scotch fir; but in	SEVERAL parts of the world insects determine the exi
0074	g virgin forests. What a struggle between the	SEVERAL kinds of trees must here have gone on during
0075	ee the contest soon decided: for instance, if	SEVERAL varieties of wheat be sown together, and the
0085	es make a great difference in cultivating the	SEVERAL varieties, assuredly, in a state of nature,
0093	profusion of pollen. As the wind had set for	SEVERAL days from the female to the male tree, the p
0097	a law of nature, we can, I think, understand	SEVERAL large classes of facts, such as the followin
0097	he great papilionaceous or pea family; but in	SEVERAL, perhaps in all, such flowers, there is a ve
0099	idual being advantageous or indispensable! If	SEVERAL varieties of the cabbage, radish, onion, and
0101	is concerned, becomes very small. From these	SEVERAL considerations and from the many special fac
0102	occupied place. But if the area be large, its	SEVERAL districts will almost certainly present diff
0102	n be modifying and improving a species in the	SEVERAL districts, there will be intercrossing with
0108	e intercrossing. Many will exclaim that these	SEVERAL causes are amply sufficient wholly to stop t
0110	scendants of the commoner species. From these	SEVERAL considerations I think it inevitably follows
0111	nce on my theory, and explains, as I believe,	SEVERAL important facts. In the first place, varieti
0113	ies of grass, and a similar plot be sown with	SEVERAL distinct genera of grasses, a greater number
0113	to hold good when first one variety and then	SEVERAL mixed varieties of wheat have been sown on e
0119	er: this is represented in the diagram by the	SEVERAL divergent branches proceeding from (A). The
0124	le species, the supposed single parent of our	SEVERAL new sub genera and genera. It is worth while
0127	ns of life, organic beings vary at all in the	SEVERAL parts of their organisation, and I think thi
0127	d adapting the various forms of life to their	SEVERAL conditions and stations, must be judged of b
0128	milies, orders, sub classes, and classes. The	SEVERAL subordinate groups in any class cannot be ra
0131	ore remote ancestors have been exposed during	SEVERAL generations. I have remarked in the first ch
0132	e degree fleshy, though not elsewhere fleshy.	SEVERAL other such cases could be given. The fact of
0134	re than a bird that cannot fly; yet there are	SEVERAL in this state. The loggerheaded duck of Sout
0134	believe that the nearly wingless condition of	SEVERAL birds, which now inhabit or have lately inha
0134	s, which now inhabit or have lately inhabited	SEVERAL oceanic islands, tenanted by no beast of pre
0135	era have all their species in this condition!	SEVERAL facts, namely, that beetles in many parts of
0136	e almost necessitating frequent flight: these	SEVERAL considerations have made me believe that the
0137	the effects of disuse. It is well known that	SEVERAL animals, belonging to the most different cla
0139	nary view of their independent creation. That	SEVERAL of the inhabitants of the caves of the Old a
0140	the Azores to England. In regard to animals,	SEVERAL authentic cases could be given of species wi
0141	e origin of some of our domestic animals from	SEVERAL wild stocks: the blood, for instance, of a t
0143	ects the whole organisation of the adult. The	SEVERAL parts of the body which are homologous, and
0144	anner of swallowing determine the position of	SEVERAL of the most important viscera. The nature of
0145	ny difference in the corolla. Possibly, these	SEVERAL differences may be connected with some diffe
0149	sume that lowness in this case means that the	SEVERAL parts of the organisation have been but litt
0150	allied species, tends to be highly variable.	SEVERAL years ago I was much struck with a remark, n
0150	t is a rule of high generality. I am aware of	SEVERAL causes of error, but I hope that I have made
0151	y little even in different genera; but in the	SEVERAL species of one genus, Pyrgoma, these valves
0151	the amount of variation in the individuals of	SEVERAL of the species is so great, that it is no ex
0155	an those parts which are closely alike in the	SEVERAL species? I do not see that any explanation c
0156	ned that natural selection will have modified	SEVERAL species, fitted to more or less widely diffe
0158	ral and sexual selection, in order to fit the	SEVERAL species to their several places in the econo
0158	in order to fit the several species to their	SEVERAL places in the economy of nature, and likewis
0159	all such analogous variations are due to the	SEVERAL races of the pigeon having inherited from a
0159	he swedish turnip and Ruta baga, plants which	SEVERAL botanists rank as varieties produced by cult
0160	new yet analogous variation appearing in the	SEVERAL breeds. We may I think confidently come to t
0160	the reappearance of the slaty blue, with the	SEVERAL marks, beyond the influence of the mere act
0163	ny important character; but from occurring in	SEVERAL species of the same genus, partly under dome
0164	as written on this subject, believes that the	SEVERAL breeds of the horse have descended from seve
0164	veral breeds of the horse have descended from	SEVERAL aboriginal species, one of which, the dun, w
0165	ow let us turn to the effects of crossing the	SEVERAL species of the horse genus. Rollin asserts,
0166	affirmative. What now are we to say to these	SEVERAL facts? We see several very distinct species
0166	are we to say to these several facts? We see	SEVERAL very distinct species of the horse genus bec
0166	st strongly displayed in hybrids from between	SEVERAL of the most distinct species. Now observe th
0166	distinct species. Now observe the case of the	SEVERAL breeds of pigeons: they are descended from a
0166	times prevails. And we have just seen that in	SEVERAL species of the horse genus the stripes are e
0168	haracters which have come to differ since the	SEVERAL species of the same genus branched off from
0169	same species, and specific differences to the	SEVERAL species of the same genus. Any part or organ
0173	mensely remote. But it may be urged that when	SEVERAL closely allied species inhabit the same terr
0178	this case, intermediate varieties between the	SEVERAL representative species and their common pare
0183	changed habits, or exclusively for one of its	SEVERAL different habits. But it is difficult to tel
0187	tself first originated: but I may remark that	SEVERAL facts make me suspect that any sensitive ner
0193	occur in only about a dozen fishes, of which	SEVERAL are widely remote in their affinities. Gener
0193	ies. Generally when the same organ appears in	SEVERAL members of the same class, especially if in
0198	o speculate on the relative importance of the	SEVERAL known and unknown laws of variation; and I h
0200	nd we may further venture to believe that the	SEVERAL bones in the limbs of the monkey, horse, and
0200	ied habits. Therefore we may infer that these	SEVERAL bones might have been acquired through natur
0200	selection, subjected formerly, as now, to the	SEVERAL laws of inheritance, reversion, correlation
0203	nciple of natural selection. If we admire the	SEVERAL ingenious contrivances, by which the flowers

Page ***(Key Word)***

Page **************************************(Key Word)**************************************

```
0203 at in the discussion light has been thrown on SEVERAL facts, which on the theory of independent a
0206 life, and being in all cases subjected to the SEVERAL laws of growth. Hence, in fact, the law of
0207 on of instinct. It would be easy to show that SEVERAL distinct mental actions are commonly embrace
0208 in the scale of nature. Frederick Cuvier and SEVERAL of the older metaphysicians have compared ir
0208 , they often remain constant throughout life. SEVERAL other points of resemblance between instinct
0211 kplant, and prevented their attendance during SEVERAL hours. After this interval, I felt sure that
0212 auses wholly unknown to us: Audubon has given SEVERAL remarkable cases of differences in nests of
0212 sified, can be shown by a multitude of facts. SEVERAL cases also, could be given, of occasional ar
0213 . but let us look to the familiar case of the SEVERAL breeds of dogs: it cannot be doubted that yo
0216 w cases. I will select only three, out of the SEVERAL which I shall have to discuss in my future w
0217 this is a mistake. Nevertheless, I could give SEVERAL instances of various birds which have been h
0218 nstinct in the allied group of ostriches. For SEVERAL hen ostriches, at least in the case of the A
0220 ccessive years, I have watched for many hours SEVERAL nests in Surrey and Sussex, and never saw a
0226 ; then, if planes of intersection between the SEVERAL spheres in both layers be formed, there will
0227 tance to stand from her fellow labourers when SEVERAL are making their spheres: but she is already
0228 st interesting to me to observe that wherever SEVERAL bees had begun to excavate these basins near
0231 section between two adjoining spheres. I have SEVERAL specimens showing clearly that they can do t
0235 from each other, than they know what are the SEVERAL angles of the hexagonal prisms and of the ba
0236 ral selection. I will not here enter on these SEVERAL cases, but will confine myself to one specia
0238 ficulty; namely, the fact that the neuters of SEVERAL ants differ, not only from the fertile femal
0239 ith has shown how surprisingly the neuters of SEVERAL British ants differ from each other in size
0240 celli rudimentary. Having carefully dissected SEVERAL specimens of these workers, I can affirm tha
0241 he jaws, moreover, of the working ants of the SEVERAL sizes differed wonderfully in shape, and in
0241 which I had dissected from the workers of the SEVERAL sizes. With these facts before me, I believe
0246 brid offspring. It is impossible to study the SEVERAL memoirs and works of those two conscientious
0247 degree impaired. Moreover, as Gartner during SEVERAL years repeatedly crossed the primrose and co
0247 her: and as he came to the same conclusion in SEVERAL other analogous cases; it seem's to me that w
0251 five years, and he continued to try it during SEVERAL subsequent years, and always with the same r
0252 are fairly treated, for by insect agency the SEVERAL individuals of the same hybrid variety are a
0254 r instance, that our dogs have descended from SEVERAL wild stocks; yet, with perhaps the exception
0254 d analogy makes me greatly doubt, whether the SEVERAL aboriginal species would at first have freel
0256 ility is often found to differ greatly in the SEVERAL individuals raised from seed out of the same
0258 the pollen of M. jalappa, and utterly failed. SEVERAL other equally striking cases could be given.
0258 a small, and occasionally in a higher degree. SEVERAL other singular rules could be given from Gar
0259 blance to either pure parent. Considering the SEVERAL rules now given, which govern the fertility
0263 ty in effecting a union apparently depends on SEVERAL distinct causes. There must sometimes be a p
0265 lastly, when organic beings are placed during SEVERAL generations under conditions not natural to
0266 ot understand, excepting on vague hypotheses, SEVERAL facts with respect to the sterility of hybri
0267 and that close interbreeding continued during SEVERAL generations between the nearest relations, e
0268 e one, is that these dogs have descended from SEVERAL aboriginally distinct species. Nevertheless
0268 osely, are utterly sterile when intercrossed. SEVERAL considerations, however, render the fertilit
0269 with the reproductive system. He supplies his SEVERAL varieties with the same food: treats them in
0269 lation, modify the reproductive system in the SEVERAL descendants from any one species. Seeing thi
0270 of specific distinction. Gartner kept during SEVERAL years a dwarf kind of maize with yellow seed
0271 rences in the reproductive system: from these SEVERAL considerations and facts, I do not think tha
0272 of varieties when crossed may be compared in SEVERAL other respects. Gartner, whose strong wish w
0272 d the more fertile hybrids are propagated for SEVERAL generations an extreme amount of variability
0274 and seems directly opposed to the results of SEVERAL experiments made by Kolreuter. These alone a
0274 cessive generations with either parent. These SEVERAL remarks are apparently applicable to animals
0276 ollow systematic affinity, but is governed by SEVERAL curious and complex laws. It is generally di
0277 roductive systems perfect, seems to depend on SEVERAL circumstances; in some cases largely on the
0286 s is about 22 miles, and the thickness of the SEVERAL formations is on an average about 1100 feet,
0288 ow watermark are preserved. For instance, the SEVERAL species of the Chthamalinae (a sub family of
0289 e than any of the foregoing; namely, from the SEVERAL formations being separated from each other b
0290 uth American coasts, which have been upraised SEVERAL hundred feet within the recent period, than
0290 are so scantily developed, that no record of SEVERAL successive and peculiar marine faunas will p
0293 number of years for its deposition, I can see SEVERAL reasons why each should not include a gradua
0293 ared with the rest of the world: nor have the SEVERAL stages of the same formation throughout Euro
0294 at area. It is well known, for instance, that SEVERAL species appeared somewhat earlier in the pal
0297 at we might obtain the parent species and its SEVERAL modified descendants from the lower and uppe
0299 go have descended from a single stock or from SEVERAL aboriginal stocks; or, again, whether certai
0302 pear in certain formations, has been urged by SEVERAL palaeontologists; for instance, by Agassiz,
0305 be seen that very few species are known from SEVERAL formations in Europe. Some few families of f
0307 l periods, I can give no satisfactory answer. SEVERAL of the most eminent geologists, with Sir R.
0308 ear to have inhabited profound depths, in the SEVERAL formations of Europe and of the United State
0310 in a completely metamorphosed condition. The SEVERAL difficulties here discussed, namely our not
0312 present chapters. Let us now see whether the SEVERAL facts and rules relating to the geological s
0312 the evidence on this head in the case of the SEVERAL tertiary stages: and every year tends to fil
0314 cal province, seems to me satisfactory. These SEVERAL facts accord well with my theory. I believe
0319 ich checked its increase, whether some one or SEVERAL contingencies, and at what period of the wor
0323 at these distant points of the world. In the SEVERAL successive palaeozoic formations of Russia,
0323 ism in the forms of life has been observed by SEVERAL authors: so it is, according to Lyell, with
0323 thors: so it is, according to Lyell, with the SEVERAL European and North American tertiary deposit
0323 ary periods, would still be manifest, and the SEVERAL formations could be easily correlated. These
0324 those of the southern hemisphere. So, again, SEVERAL highly competent observers believe that the
0328 g amount of difference in the species. If the SEVERAL formations in these regions have not been de
0329 lowly changing during the accumulation of the SEVERAL formations and during the long intervals of
0329 rvals of time between them: in this case, the SEVERAL formations in the two regions could be arran
0331 truth in the remark. Let us see how far these SEVERAL facts and inferences accord with the theory
0331 together with the many extinct genera on the SEVERAL lines of descent diverging from the parent f
0332 ergence of character, has become divided into SEVERAL sub families and families, some of which are
0332 the successive formations, were discovered at SEVERAL points low down in the series, the three exi
0335 l resemblance of the organic remains from the SEVERAL stages of the chalk formation, though the sp
0339 armour like those of the armadillo, found in SEVERAL parts of La Plata; and Professor Owen has sh
0342 llied or representative species, found in the SEVERAL stages of the same great formation. He may d
0351 fied. On these views, it is obvious, that the SEVERAL species of the same genus, though inhabiting
```

Page **************************************(Key Word)**************************************

Page ***(Key Word)***

0352 ibly have migrated from some one point to the	SEVERAL	distant and isolated points, where now found
0353 a single region; and it has been observed by	SEVERAL	naturalists, that the most natural genera, o
0354 equally important for us, namely, whether the	SEVERAL	distinct species of a genus, which on my the
0357 in the recent period to each other and to the	SEVERAL	intervening oceanic islands. I freely admit
0358 , and with the many existing oceanic islands.	SEVERAL	facts in distribution, such as the great dif
0358 close relation of the tertiary inhabitants of	SEVERAL	lands and even seas to their present inhabit
0359 ton's Physical Atlas, the average rate of the	SEVERAL	Atlantic currents is 33 miles per diem (some
0362 ellets or passed them in their excrement; and	SEVERAL	of these seeds retained their power of germi
0363 eeds of northern plants. Considering that the	SEVERAL	above means of transport, and that several o
0363 he several above means of transport, and that	SEVERAL	other means, which without doubt remain to b
0365 ecies must have been independently created at	SEVERAL	distinct points; and we might have remained
0369 the present Alpine plants and animals of the	SEVERAL	great European mountain ranges, though very
0371 r understand the singular fact remarked on by	SEVERAL	observers, that the productions of Europe an
0371 ted, within a much more recent period, on the	SEVERAL	mountain ranges and on the arctic lands of t
0373 cial epoch was strictly simultaneous at these	SEVERAL	far distant points on opposite sides of the
0375 he cordillera. On the mountains of Abyssinia,	SEVERAL	European forms and some few representatives
0375 ns of Australia, Dr. F. Muller has discovered	SEVERAL	European species; other species, not introdu
0381 rned in their dispersal. But the existence of	SEVERAL	quite distinct species, belonging to genera
0386 the mud of ponds is with seeds: I have tried	SEVERAL	little experiments, but will here give only
0387 sometimes to be dropped. In considering these	SEVERAL	means of distribution, it should be remember
0396 y and grave difficulties in understanding how	SEVERAL	of the inhabitants of the more remote island
0397 found elsewhere. Dr. Aug. A. Gould has given	SEVERAL	interesting cases in regard to the land shel
0397 rately wide arms of the sea. And I found that	SEVERAL	species did in this state withstand uninjure
0398 hese volcanic islands in the Pacific, distant	SEVERAL	hundred miles from the continent, yet feels
0398 r climate, or in the proportions in which the	SEVERAL	classes are associated together, which resem
0400 the limits of the same archipelago. Thus the	SEVERAL	islands of the Galapagos Archipelago are ten
0400 r it may be asked, how has it happened in the	SEVERAL	islands situated within sight of each other,
0401 nd a considerable amount of difference in the	SEVERAL	islands. This difference might indeed have b
0402 onfined to the archipelago, are common to the	SEVERAL	islands, and we may infer from certain facts
0403 and representative species, which inhabit the	SEVERAL	islands of the Galapagos Archipelago, not ha
0403 to island. In many other instances, as in the	SEVERAL	districts of the same continent, pre occupat
0408 whether of land or water, which separate our	SEVERAL	zoological and botanical provinces. We can t
0409 hipelago, though specifically distinct on the	SEVERAL	islets, should be closely related to each ot
0410 ed exceptions to the rule. On my theory these	SEVERAL	relations throughout time and space are inte
0412 g the action, as formerly explained, of these	SEVERAL	principles; and he will see that the inevita
0412 ppermost line may represent a genus including	SEVERAL	species; and all the genera on this line for
0417 mainly depends on their being correlated with	SEVERAL	other characters of more or less importance.
0417 rked, a species may depart from its allies in	SEVERAL	characters, both of high physiological impor
0417 but when Aspicarpa produced in France, during	SEVERAL	years, only degraded flowers, departing so w
0419 groups of birds; and it has been followed by	SEVERAL	entomologists and botanists. Finally, with r
0419 to be, at least at present, almost arbitrary.	SEVERAL	of the best botanists, such as Mr. Bentham a
0420 ral; but that the amount of difference in the	SEVERAL	branches or groups, though allied in the sam
0421 mediate in character between A and I, and the	SEVERAL	genera descended from these two genera will
0422 t isolation and states of civilisation of the	SEVERAL	races, descended from a common race) had alt
0423 varieties; and with our domestic productions,	SEVERAL	other grades of difference are requisite, as
0424 m. the naturalist includes as one species the	SEVERAL	larval stages of the same individual, howeve
0425 usly used; and only thus can I understand the	SEVERAL	rules and guides which have been followed by
0426 egard to single points of structure, but when	SEVERAL	characters, let them be ever so trifling, oc
0426 species or a group of species may depart, in	SEVERAL	of its most important characteristics, from
0428 aracters exhibiting true affinity between the	SEVERAL	members of the whale family: for these cetac
0431 in various ways and degrees, to all; and the	SEVERAL	species will consequently be related to each
0431 fining and widening the intervals between the	SEVERAL	groups in each class. We may thus account ev
0432 mpossible to give any definition by which the	SEVERAL	members of the several groups could be disti
0432 efinition by which the several members of the	SEVERAL	groups could be distinguished from their mor
0432 her. We could not, as I have said, define the	SEVERAL	groups; but we could pick out types, or form
0433 ped together in one great system: and how the	SEVERAL	members of each class are connected together
0434 the term unity of type; or by saying that the	SEVERAL	parts and organs in the different species of
0435 ork of bones or the relative connexion of the	SEVERAL	parts. If we suppose that the ancient progen
0440 external appearance, have larvae in all their	SEVERAL	stages barely distinguishable. The embryo in
0442 miform stage. How, then, can we explain these	SEVERAL	facts in embryology, namely the very general
0445 e evidence appears to me conclusive, that the	SEVERAL	domestic breeds of Pigeon have descended fro
0445 uctions. But when the nestling birds of these	SEVERAL	breeds were placed in a row, though most of
0445 oportional differences in the above specified	SEVERAL	points were incomparably less than in the fu
0446 rom some one parent species, and of which the	SEVERAL	new species have become modified through nat
0447 ate age, the fore limbs in the embryos of the	SEVERAL	descendants of the parent species will still
0457 r natural as it is genealogical. Finally, the	SEVERAL	classes of facts which have been considered
0463 lsely appear, to have come in suddenly on the	SEVERAL	geological stages? Why do we not find great
0465 om each other in time. Such is the sum of the	SEVERAL	chief objections and difficulties which may
0466 fect the Geological Record is. Grave as these	SEVERAL	difficulties are, in my judgment they do not
0473 ce of stripes on the shoulder and legs of the	SEVERAL	species of the horse genus and in their hybr
0473 striped progenitor, in the same manner as the	SEVERAL	domestic breeds of pigeon have descended fro
0474 n my view, this part has undergone, since the	SEVERAL	species branched off from a common progenito
0474 ifferent animals of the same class with their	SEVERAL	instincts. I have attempted to show how much
0482 which this subject is overwhelmed be removed.	SEVERAL	eminent naturalists have of late published t
0488 od unchanged, whilst within this same period,	SEVERAL	of these species, by migrating into new coun
0490 re is grandeur in this view of life, with its	SEVERAL	powers, having been originally breathed into
0319 f the horse's life, and in what degree, they	SEVERALLY	acted. If the conditions had gone on, howev
0060 ts throughout nature. Struggle for life most	SEVERE	between individuals and varieties of the same
0060 uals and varieties of the same species; often	SEVERE	between species of the same genus. The relati
0062 shown that all organic beings are exposed to	SEVERE	competition. In regard to plants, no one has
0068 mber that ten per cent. is an extraordinarily	SEVERE	mortality from epidemics with man. The action
0068 acts in reducing food, it brings on the most	SEVERE	struggle between the individuals, whether of
0075 t the struggle almost invariably will be most	SEVERE	between the individuals of the same species,
0075 the struggle will generally be almost equally	SEVERE,	and we sometimes see the contest soon decide
0076 tructure, the struggle will generally be more	SEVERE	between species of the same genus, when they
0076 dimly see why the competition should be most	SEVERE	between allied forms, which fill nearly the s

Page ***(Key Word)***

```
0106  and the competition to fill them will be more SEVERE, on a large than on a small and isolated area
0106  island, the race for life will have been less SEVERE, and there will have been less modification a
0107  n fresh water productions will have been less SEVERE than elsewhere: new forms will have been more
0107  ea, and from having thus been exposed to less SEVERE competition. To sum up the circumstances favo
0107  s and kinds, will have been subjectec to very SEVERE competition. When converted by subsidence int
0108   into a continental area, there will again be SEVERE competition: the most favoured or improved va
0114  est between individual and individual must be SEVERE, we always find great diversity in its inhabi
0121  d that the competition will generally be most SEVERE between those forms which are most nearly rel
0127  each species, at some age, season, or year, a SEVERE struggle for life, and this certainly cannot
0133   species have thicker and better fur the more SEVERE the climate is under which they have lived: b
0133  ons, and how much to the direct action of the SEVERE climate? for it would appear that climate has
0139  fe have not been preserved, owing to the less SEVERE competition to which the inhabitants of these
0320  forms. The competition will generally be most SEVERE, as formerly explained and illustrated by exa
0321  and isolated station, where they have escaped SEVERE competition. For instance, a single species o
0326  appearance of favourable variations, and that SEVERE competition with many already existing forms,
0326  itants met, the battle would be prolonged and SEVERE; and some from one birthplace and some from t
0388  e land, the competition will probably be less SEVERE between aquatic than between terrestrial spec
0468  ch other, the struggle will generally be most SEVERE between them; it will be almost equally sever
0468  evere between them! it will be almost equally SEVERE between the varieties of the same species, an
0468  me genus. But the struggle will often be very SEVERE between beings most remote in the scale of na
0429  e evidence of aberrant forms having suffered SEVERELY from extinction, for they are generally repr
0141  limates but of being perfectly fertile (a far SEVERER test) under them, may be used as an argument
0206  ied forms, and the competition will have been SEVERER, and thus the standard of perfection will ha
0088  ndibles of other males. The war is, perhaps, SEVEREST between males of polygamous animals, and the
0089  ed to the subject, believe that there is the SEVEREST rivalry between the males of many species to
0110  itution, and habits, generally come into the SEVEREST competition with each other. Consequently,
0271  to be varieties, and which he tested by the SEVEREST trial, namely, by reciprocal crosses, and he
0468  e varieties of the same species, and next in SEVERITY between the species of the same genus. But t
0013  why a peculiarity is often transmitted from one SEX to both sexes, or to one sex alone, more commo
0013  ansmitted from one sex to both sexes, or to one SEX alone, more commonly but not exclusively to th
0013  , more commonly but not exclusively to the like SEX. It is a fact of some little importance to us,
0087  arities often appear under domestication in one SEX and become hereditarily attached to that sex,
0087  ne sex and become hereditarily attached to that SEX, the same fact probably occurs under nature, a
0087  o, natural selection will be able to modify one SEX in its functional relations to the other sex,
0087  ne sex in its functional relations to the other SEX, or in relation to wholly different habits of
0088  on having special weapons, confined to the male SEX. A hornless stag or spurless cock would have a
0090  ities arising and becoming attached to the male SEX in our domestic animals (as the wattle in male
0144  , and the tortoise shell colour with the female SEX; the feathered feet and skin between the outer
0150  applies to characters which are attached to one SEX, but are not directly connected with the act o
0153  d, from causes quite unknown to us, more to one SEX than to the other, generally to the male sex,
0153  ne sex than to the other, generally to the male SEX, as with the wattle of carriers and the enlarg
0237  ecome correlated to certain ages, and to either SEX. We have differences correlated not only to on
0237  we have differences correlated not only to one SEX, but to that short period alone when the repro
0237  to an artificially imperfect state of the male SEX! for oxen of certain breeds have longer horns
0274  nsmitting likeness running more strongly in one SEX than in the other; both when one species is cr
0003  ain birds, and which has flowers with separate SEXES absolutely requiring the agency of certain in
0013  rity is often transmitted from one sex to both SEXES, or to one sex alone, more commonly but not e
0042  ty years. In the case of animals with separate SEXES, facility in preventing crosses is an importa
0080  importance, its power at all ages and on both SEXES. Sexual Selection. On the generality of inter
0087  to wholly different habits of life in the two SEXES, as is sometimes the case with insects. And t
0093  illustrating one step in the separation of the SEXES of plants, presently to be alluded to. Some h
0094  ower, and as a more complete separation of the SEXES of our plant would be advantageous on the pri
0094  ec, until at last a complete separation of the SEXES would be affected. Let us now turn to the nec
0096  n the case of animals and plants with separate SEXES, it is of course obvious that two individuals
0099  y, so that these plants have in fact separated SEXES, and must habitually be crossed. How strange
0100  strong tendency to bear flowers with separated SEXES. When the sexes are separated, although the m
0100  to bear flowers with separated sexes. When the SEXES are separated, although the male and female f
0100   that trees belonging to all Orders have their SEXES more often separated than other plants, I fin
0100  alia; and I have made these few remarks on the SEXES of trees simply to call attention to the subj
0157  e secondary sexual differences between the two SEXES of the same species are generally displayed i
0157  lyi and the number likewise differs in the two SEXES of the same species: again in fossorial hymen
0157  the different species, and likewise in the two SEXES of the same species. This relation has a clea
0157  nded from the same progenitor, as have the two SEXES of any one of the species. Consequently, what
0158  ecoromy of nature, and likewise to fit the two SEXES of the same species to each other, or to fit
0169   in giving secondary sexual differences to the SEXES of the same species, and specific differences
0270  rden; and although these plants have separated SEXES, they never naturally crossed. He then fertil
0270  e been injurious, as the plants have separated SEXES. No one, I believe, has suspected that these
0270  s of gourd, which like the maize has separated SEXES, and he asserts that their mutual fertilisati
0424  is lowest grade, or that of a species, the two SEXES; and how enormously these sometimes differ in
0433  of descent in classing the individuals of both SEXES and of all ages, although having few characte
0451  well developed state. In plants with separated SEXES, the male flowers often have a rudiment of a
0456  been universally used in ranking together the SEXES, ages, and acknowledged varieties of the same
0468  urn the balance. With animals having separated SEXES there will in most cases be a struggle betwee
0080  nce, its power at all ages and on both sexes. SEXUAL Selection. On the generality of intercrosses
0087  ing known to vary like every other structure. SEXUAL Selection. Inasmuch as peculiarities often ap
0088  is leads me to say a few words on what I call SEXUAL Selection. This depends, not on a struggle fo
0088  ccessful competitor, but few or no offspring. SEXUAL selection is, therefore, less rigorous than n
0088  ould have a poor chance of leaving offspring. SEXUAL selection by always allowing the victor to br
0088  eans of defence may be given through means of SEXUAL selection, as the mane to the lion, the shoul
0089  ew of plumage having been chiefly modified by SEXUAL selection, acting when the birds have come to
0089  , such differences have been mainly caused by SEXUAL selection; that is, individual males have had
0090  . yet, I would not wish to attribute all such SEXUAL differences to this agency: for we see peculi
0127  as easily as the adult. Amongst many animals, SEXUAL selection will give its aid to ordinary selec
0127  apted males the greatest number of offspring, SEXUAL selection will also give characters useful to
0131  racters more variable than generic: secondary SEXUAL characters variable. Species of the same genu
0132  ndition of the offspring. The male and female SEXUAL elements seem to be affected before that unio
```

Page ***(Key Word)***

```
0150 pplies very strongly in the case of secondary SEXUAL characters, when displayed in any unusual man
0150 ed in any unusual manner. The term, secondary SEXUAL characters, used by Hunter, applies to charac
0150 emales more rarely offer remarkable secondary SEXUAL character, it applies more rarely to them. Th
0151 o plainly applicable in the case of secondary SEXUAL characters, may be due to the great variabili
0151 ut that our rule is not confined to secondary SEXUAL characters is clearly shown in the case of he
0156 ithout my entering on details, that secondary SEXUAL characters are very variable; I think it also
0156 rom each other more widely in their secondary SEXUAL characters, than in other parts of their orga
0156 les of gallinaceous birds, in which secondary SEXUAL characters are strongly displayed, with the o
0156 ause of the original variability of secondary SEXUAL characters is not manifest; but we can see wh
0156 ther parts of the organisation; for secondary SEXUAL characters have been accumulated by sexual se
0156 ry sexual characters have been accumulated by SEXUAL selection, which is less rigid in its action
0157 cause may be of the advantages of secondary SEXUAL characters, as they are highly variable, sexu
0157 xual characters, as they are highly variable, SEXUAL selection will have had a wide scope for acti
0157 group a greater amount of difference in their SEXUAL characters, than in other parts of their stru
0157 . it is a remarkable fact, that the secondary SEXUAL differences between the two sexes of the same
0157 robable, be taken advantage of by natural and SEXUAL selection, in order to fit the several specie
0158 cies; that the great variability of secondary SEXUAL characters, and the great amount of differenc
0158 etween closely allied species; that secondary SEXUAL and ordinary specific differences are general
0158 y to reversion and to further variability, to SEXUAL selection being less rigid than ordinary sele
0158 parts having been accumulated by natural and SEXUAL selection, and thus adapted for secondary sex
0158 ual selection, and thus adapted for secondary SEXUAL, and for ordinary specific purposes. Distinct
0169 ost varieties or incipient species. Secondary SEXUAL characters are highly variable, and such char
0169 y been taken advantage of in giving secondary SEXUAL differences to the sexes of the same species,
0196 difying various structures; and finally, that SEXUAL selection will often have largely modified th
0197 due to some quite distinct cause, probably to SEXUAL selection. A trailing bamboo in the Malay Arc
0199 origin of these differences, chiefly through SEXUAL selection of a particular kind, but without h
0199 , though now of no direct use. The effects of SEXUAL selection, when displayed in beauty to charm
0246 microscope reveals. In the first case the two SEXUAL elements which go to form the embryo are perf
0261 ciprocal crosses between two species the male SEXUAL element of the one will often freely act on t
0261 f the one will often freely act on the female SEXUAL element of the other, but not in a reversed d
0263 union of two pure species the male and female SEXUAL elements are perfect, whereas in hybrids they
0264 ard to the sterility of hybrids, in which the SEXUAL elements are imperfectly developed, the case
0274 d, partly owing to the existence of secondary SEXUAL characters; but more especially owing to prep
0443 e variation may be due to the male and female SEXUAL elements having been affected by the conditio
0083 she can act on every internal organ, on every SHADE of constitutional difference, on the whole ma
0213 nces could be given of the inheritance of all SHADES of disposition and tastes, and likewise of th
0015 ties in regard to species. But there is not a SHADOW of evidence in favour of this view: to assert
0484 t they will not be incessantly haunted by the SHADOWY doubt whether this or that form be in essenc
0151 lies to plants, and this would seriously have SHAKEN my belief in its truth, had not the great var
0335 act alone, from its generality, seems to have SHAKEN Professor Pictet in his firm belief in the im
0003 of variation. In one very limited sense, as we SHALL hereafter see, this may be true; but it is pr
0004 d by naturalists. From these considerations, I SHALL devote the first chapter of this Abstract to
0004 abstract to Variation under Domestication. We SHALL thus see that a large amount of hereditary mo
0004 lei and, what is equally or more important, we SHALL see how great is the power of man in accumula
0004 ability of species in a state of nature; but I SHALL, unfortunately, be compelled to treat this su
0004 ly only by giving long catalogues of facts. We SHALL, however, be enabled to discuss what circumst
0005 d at some length in the fourth chapter; and we SHALL then see how Natural Selection almost inevita
0005 divergence of Character. In the next chapter I SHALL discuss the complex and little known laws of
0005 f the Geological Record. In the next chapter I SHALL consider the geological succession of organic
0005 an embryonic condition. In the last chapter I SHALL give a brief recapitulation of the whole work
0015 ermine how far the new characters thus arising SHALL be preserved. When we look to the hereditary
0016 r, on the view of the origin of genera which I SHALL presently give, we have no right to expect of
0017 quarters of the world. I do not believe, as we SHALL presently see, that all our dogs have descend
0028 will be obvious when we treat of Selection. We SHALL then, also, see how it is that the breeds so
0044 ogue of dry facts should be given, but these I SHALL reserve for my future work. Nor shall I here
0044 these I shall reserve for my future work. Nor SHALL I here discuss the various definitions which
0052 , and both rank as independent species. But we SHALL hereafter have to return to this subject. Fro
0053 equently Dr. Hooker, even in stronger terms. I SHALL reserve for my future work the discussion of
0053 nd as they come into competition (which, as we SHALL hereafter see, is a far more important circum
0058 , as I call it, of Divergence of Character, we SHALL see how this may be explained, and how the le
0059 on an average vary most; and varieties, as we SHALL hereafter see, tend to become converted into
0061 he same genus, arise? All these results, as we SHALL more fully see in the next chapter, follow in
0061 e hand of Nature. But Natural Selection, as we SHALL hereafter see, is a power incessantly ready f
0062 for existence. In my future work this subject SHALL be treated, as it well deserves, at much grea
0062 e truly said to struggle with each other which SHALL get food and live. But a plant on the edge of
0067 as been ably treated by several authors, and I SHALL, in my future work, discuss some of the check
0073 ound together by a web of complex relations. I SHALL hereafter have occasion to show that the exot
0080 the hands of man, apply to nature? I think we SHALL see that it can act most effectually. Let it
0081 s we see in the species called polymorphic. We SHALL best understand the probable course of natura
0086 r modifications at a different period of life, SHALL not be in the least degree injurious: for if
0091 y met; but to this subject of intercrossing we SHALL soon have to return. I may add, that, accordi
0096 add, was first suggested by Andrew Knight. We SHALL presently see its importance; but I must here
0099 nt over foreign pollen; but to this subject we SHALL return in a future chapter. In the case of a
0105 nhabiting it, will be found to be small; as we SHALL see in our chapter on geographical distributi
0108 ected. When, by renewed elevation, the islands SHALL be re converted into a continental area, ther
0112 on this head from our domestic productions. We SHALL here find something analogous. A fancier is s
0114 ts, which thus jostle each other most closely, SHALL, as a general rule, belong to what we call di
0124 he earth's crust including extinct remains. We SHALL, when we come to our chapter on Geology, have
0124 to refer again to this subject, and I think we SHALL then see that the diagram throws light on the
0125 m14, will form three very distinct genera. We SHALL also have two very distinct genera descended
0126 l transmit descendants to a remote futurity. I SHALL have to return to this subject in the chapter
0149 with truth, are apt to be highly variable. We SHALL have to recur to the general subject of rudim
0151 lmost invariably holds good with cirripedes. I SHALL, in my future work, give a list of the more r
0154 nation is partly, yet only indirectly, true; I SHALL, however, have to return to this subject in o
0172 d and modified through natural selection? What SHALL we say to so marvellous an instinct as that w
0172 r fertility is unimpaired? The two first heads SHALL be here discussed, Instinct and Hybridism in
```

Page ***(Key Word)***

Page **(Key Word)***

```
0176  apt a varying species to a very large area, we  SHALL  have to adapt two varieties to two large area
0179  gst fossil remains, which are preserved, as we  SHALL  in a future chapter attempt to show, in an ex
0213  onsidering a few cases under domestication. We  SHALL  thus also be enabled to see the respective pa
0216  y, habit and selection have acted together. We  SHALL, perhaps, best understand how instincts in a
0216  select only three, out of the several which I  SHALL  have to discuss in my future work, namely, th
0223  he masters determine when and where a new nest  SHALL  be formed, and when they migrate, the masters
0231  of the cell, which will ultimately be left. We  SHALL  understand how they work, by supposing masons
0231  g fresh cement, on the summit of the ridge. We  SHALL  thus have a thin wall steadily growing upward
0272  ost closely in very many important respects. I  SHALL  here discuss this subject with extreme brevit
0294  issippi during some part of the glacial period  SHALL  have been upraised, organic remains will prob
0298  ly separated formations; but to this subject I  SHALL  have to return in the following chapter. One
0298  uld rarely be effected by palaeontolcgists. We  SHALL, perhaps, best perceive the improbability of
0325  nde has remarked, look to some special law. We  SHALL  see this more clearly when we treat of the pr
0338  ry of natural selection. In a future chapter I  SHALL  attempt to show that the adult differs from i
0347  etween latitudes 25 degrees and 35 degrees, we  SHALL  find parts extremely similar in all their con
0354  bly the safest. In discussing this subject, we  SHALL  be enabled at the same time to consider a poi
0355  ases of this nature are common, and are, as we  SHALL  hereafter more fully see, inexplicable on the
0356  h road for migration, but now be impassable; I  SHALL, however, presently have to discuss this bran
0357  g definite about the means of distribution, we  SHALL  be enabled to speculate with security on the
0358  habitants, a certain degree of relation (as we  SHALL  hereafter see) between the distribution of ma
0358  be called occasional means of distribution. I  SHALL  here confine myself to plants. In botanical w
0363  ld sometimes include a few minute seeds? But I  SHALL  presently have to recur to this subject. As i
0366  attention to the Glacial Period, which, as we  SHALL  immediately see, affords a simple explanation
0366  rd forbes, is substantially as follows. But we  SHALL  follow the changes more readily, by supposing
0370  rer to the pole. Now if we look at a globe, we  SHALL  see that under the Polar Circle 'there is almo
0389  nsular productions. In the following remarks- I  SHALL  not confine myself to the mere question of di
0389  myself to the mere question of dispersal; but  SHALL  consider some other facts, which bear on the
0390  of the islands with that of the continent, we  SHALL  see that this is true. This fact might have b
0395  f certain mammals through man's agency; but we  SHALL  soon have much light thrown on the natural hi
0414  to the consideration of these resemblances we  SHALL  have to recur. It may even be given as a gene
0419  d radicle. In our discussion on embryology, we  SHALL  see why such characters are so valuable, on t
0426  once value them less in our classification. We  SHALL  hereafter, I think, clearly see why embryolog
0432  t a natural arrangement, would be possible. We  SHALL  see this by turning to the diagram: the lette
0432  h have lived throughout all time and space. We  SHALL  certainly never succeed in making so perfect
0434  complex and radiating lines of affinities. We  SHALL  never, probably, disentangle the inextricable
0456  nown cause of similarity in organic beings, we  SHALL  understand what is meant by the natural syste
0467  in the balance will determine which individual  SHALL  live and which shall die, which variety or sp
0467  etermine which individual shall live and which  SHALL  die, which variety or species shall increase
0467  and which shall die, which variety or species  SHALL  increase in number, and which shall decrease,
0467  or species shall increase in number, and which  SHALL  decrease, or finally become extinct. As the i
0485  oth forms to the rank of species. Hereafter we  SHALL  be compelled to acknowledge that the only dis
0485  ermediate gradations between any two forms, we  SHALL  be led to weigh more carefully and to value h
0485  nguage will come into accordance. In short, we  SHALL  have to treat species in the same manner as t
0485  e. this may not be a cheering prospect; but we  SHALL  have at least be freed from the vain search f
0487  es of climate and of the level of the land, we  SHALL  surely be enabled to trace in an admirable ma
0487  stages as having been of vast duration. But we  SHALL  be able to gauge with some security the durat
0132  s at their southern limit, and when living in  SHALLOW  water, are more brightly coloured than those
0133  of shells which are confined to tropical and  SHALLOW  seas are generally brighter coloured than th
0173  only where much sediment is deposited on the  SHALLOW  bed of the sea; whilst it slowly subsides. T
0228  ider and wider until they were converted into  SHALLOW  basins, appearing to the eye perfectly true
0291  ccumulated to any thickness and extent over a  SHALLOW  bottom, if it continue slowly to subside. In
0291  early balance each other, the sea will remain  SHALLOW  and favourable for life, and thus a fossilif
0291  ly of sediment was sufficient to keep the sea  SHALLOW  and to embed and preserve the remains before
0292  posits could not have been accumulated in the  SHALLOW  parts, which are the most favourable to life
0294  of time. When such beds as were deposited in  SHALLOW  water near the mouth of the Mississippi duri
0299  numerous large islands separated by wide and  SHALLOW  seas, probably represents the former state o
0395  lly holds good. We see Britain separated by a  SHALLOW  channel from Europe, and the mammals are the
0396  level it is obvious that islands separated by  SHALLOW  channels are more likely to have been contin
0133  d by variation when it ranged into warmer or  SHALLOWER  waters. When a variation is of the slightes
0022  breadth and curvature differs enormously. The  SHAPE, as well as the breadth and length of the ram
0022  th and the presence of processes. The size and  SHAPE  of the apertures in the sternum are highly va
0022  e nestling birds are clothed when hatched. The  SHAPE  and size of the eggs vary. The manner of flig
0033  kinds of gooseberries differ in size, colour,  SHAPE, and hairiness, and yet the flowers present v
0144  eved by some authors that the diversity in the  SHAPE  of the pelvis in birds causes the remarkable
0144  n birds causes the remarkable diversity in the  SHAPE  of their kidneys. Others believe that the sha
0144  hape of their kidneys. Others believe that the  SHAPE  of the pelvis in the human mother influences
0144  in the human mother influences by pressure the  SHAPE  of the head of the child. In snakes, accordin
0144  e child. In snakes, according to Schlegel, the  SHAPE  of the body and the manner of swallowing dete
0145  e compositous plants, the seeds also differ in  SHAPE  and sculpture; and even the ovary itself, wit
0145  ttributed by some authors to pressure, and the  SHAPE  of the seeds in the ray florets in some Compo
0149  o cut all sorts of things may be of almost any  SHAPE; whilst a tool for some particular object had
0149  icular object had better be of some particular  SHAPE. Natural selection, it should never be forgot
0151  erent species being sometimes wholly unlike in  SHAPE; and the amount of variation in the individua
0198  even the head would probably be affected. The  SHAPE, also, of the pelvis might affect by pressure
0198  so, of the pelvis might affect by pressure the  SHAPE  of the head of the young in the womb. The lab
0224  oblem, and have made their cells of the proper  SHAPE  to hold the greatest possible amount of honey
0230  wax, they could make their cells of the proper  SHAPE, by standing at the proper distance from each
0232  t ways the same cell, sometimes recurring to a  SHAPE  which they had at first rejected. When bees h
0236  nd the fertile females in structure, as in the  SHAPE  of the thorax and in being destitute of wings
0241  s of the several sizes differed wonderfully in  SHAPE, and in the form and number of the teeth. But
0423  attle, because they are less variable than the  SHAPE  or colour of the body, etc.: whereas with she
0427  by macleay and others. The resemblance, in the  SHAPE  of the body and in the fin like anterior limb
0428  order are compared one with another: thus the  SHAPE  of the body and fin like limbs are only analo
0428  lasses for swimming through the water; but the  SHAPE  of the body and fin like limbs serve as chara
0428  I doubt that they have inherited their general  SHAPE  of body and structure of limbs from a common
0443  of cross bred cattle have been affected by the  SHAPE  of the horns of either parent. For the welfar
```

Page ***(Key Word)***

shape

Page **************************************(Key Word)**************************************

0187 facets, within each of which there is a lens SHAPED swelling. In other crustaceans the transparen
0361 i find on examination, that when irregularly SHAPED stones are embedded in the roots of trees, sm
0437 sed of such numerous and such extraordinarily SHAPED pieces of bone? As Owen has remarked, the ben
0067 pared to a yielding surface, with ten thousand SHARP wedges packed close together and driven inwar
0317 ely appear to begin at its lower end, not in a SHARP point, but abruptly; it then gradually thicke
0175 does or the range of others, will tend to be SHARPLY defined. Moreover, each species on the confi
0175 geographical range will come to be still more SHARPLY defined. If I am right in believing that all
0012 collected by Heusinger, it appears that white SHEEP and pigs are differently affected from colour
0018 cended from several wild species. In regard to SHEEP and goats I can form no opirion. I should thi
0019 ast a score of species of wild cattle, as many SHEEP, and several goats in Europe alone, and sever
0019 xisted in Great Britain eleven wild species of SHEEP peculiar to it! When we bear in mind that Bri
0019 s possesses several peculiar breeds of cattle, SHEEP, etc., we must admit that many domestic breed
0030 is known to have been the case with the ancon SHEEP. But when we compare the dray horse and race
0030 the dromedary and camel, the various breeds of SHEEP fitted either for cultivatec land or mountain
0031 ed to a large extent some breeds of cattle and SHEEP. In order fully to realise what they have don
0031 ville, speaking of what breeders have done for SHEEP, says: It would seem as if they had chalked o
0031 the principle of selection in regard to merino SHEEP is so fully recognised, that men follow it as
0031 recognised, that men follow it as a trade: the SHEEP are placed on a table and are studied, like a
0031 ne three times at intervals of months; and the SHEEP are each time marked and classed, so that the
0036 distinct strains. The two flocks of Leicester SHEEP kept by Mr. Buckley and Mr. Burgess, as Mr. Y
0036 ll's flock, and yet the difference between the SHEEP possessed by these two gentlemen is so great
0041 ple Marshall has remarked, with respect to the SHEEP of parts of Yorkshire, that as they generally
0076 and disappear. So again with the varieties of SHEEP: it has been asserted that certain mountain v
0083 d quadruped in any peculiar manner; he exposes SHEEP with long and short wool to the same climate.
0085 member how essential it is in a flock of white SHEEP to destroy every lamb with the faintest trace
0086 he down of their chickens; in the horns of our SHEEP and cattle when nearly adult; so in a state o
0111 iven showing how quickly new breeds of cattle, SHEEP, and other animals, and varieties of flowers,
0177 te what I mean by supposing three varieties of SHEEP to be kept, one adapted to an extensive mount
0213 ndency to run round, instead of at, a flock of SHEEP, by shepherd dogs. I cannot see that these ac
0215 n kept tame, are most eager to attack poultry, SHEEP and pigs; and this tendency has been found in
0215 c, require to be taught not to attack poultry, SHEEP, and pigs! No doubt they occasionally do make
0299 to prove, that our different breeds of cattle, SHEEP, horses, and dogs have descended from a sinal
0423 hape or colour of the body, etc.; whereas with SHEEP the horns are much less serviceable, because
0087 shells might be selected, the thickness of the SHELL being known to vary like every other structur
0133 on of each species, will have to say that this SHELL, for instance, was created with bright colour
0133 ht colours for a warm sea; but that this other SHELL became bright coloured by variation when it r
0144 ue eyes and deafness in cats; and the tortoise SHELL colour with the female sex; the feathered fee
0148 cted, it loses more or less completely its own SHELL or carapace. This is the case with the male I
0192 t the bottom of the sack, in the well enclosed SHELL: but they have large folded branchiae. Now I
0289 in an extreme degree. For instance, not a land SHELL is known belonging to either of these vast pe
0348 faunas are more distinct, with hardly a fish, SHELL, or crab in common, than those of the eastern
0348 nd distinct marine faunas. Although hardly one SHELL, crab or fish is common to the above named fo
0386 has been caught with an Ancylus (a fresh water SHELL like a limpet) firmly adhering to it; and a w
0391 ar land shells, whereas not one species of sea SHELL is confined to its shores: now, though we do
0397 g a membranous diaphragm over the mouth of the SHELL, might be floated in chinks of drifted timber
0044 but who can say that the dwarfed condition of SHELLS in the brackish waters of the Baltic, or dwar
0046 of insects, and several genera of Brachiopod SHELLS. In most polymorphic genera some of the speci
0046 ntries, and likewise, judging from Brachiopod SHELLS, at former periods of time. These facts seem
0052 ton with the varieties of certain fossil land SHELLS in Madeira. If a variety were to flourish so
0087 ish: or, more delicate and more easily broken SHELLS might be selected, the thickness of the shell
0132 etc.: thus, E. Forbes speaks confidently that SHELLS at their southern limit, and when living in s
0133 and permanent varieties. Thus the species of SHELLS which are confined to tropical and shallow se
0167 he old and ignorant cosmogonists, that fossil SHELLS had never lived, but had been created in ston
0167 t had been created in stone so as to mock the SHELLS now living on the sea shore. Summary. Our ign
0288 ve. No organism wholly soft can be preserved. SHELLS and bones will decay and disappear when left
0298 that oftenest present varieties; so that with SHELLS and other marine animals, it is probably thos
0299 iginal stocks: or, again, whether certain sea SHELLS inhabiting the shores of North America, which
0313 ree. In the oldest tertiary beds a few living SHELLS may still be found in the midst of a multitud
0314 hical distribution; for instance, in the land SHELLS and coleopterous insects of Madeira having co
0314 n the continent of Europe, whereas the marine SHELLS and birds have remained unaltered. We can per
0318 sters, which all co existed with still living SHELLS at a very late geological period, I was fille
0321 single species of Trigonia, a great genus of SHELLS in the secondary formations, survives in the
0323 that they had coexisted with still living sea SHELLS; but as these anomalous monsters coexisted wi
0339 s shown that the same law holds good with sea SHELLS, but from the wide distribution of most gener
0339 relation between the extinct and living land SHELLS of Madeira; and between the extinct and livin
0339 between the extinct and living brackish water SHELLS of the Aralo Caspian Sea. Now what does this
0348 m the Pacific into the Indian Ocean, and many SHELLS are common to the eastern islands of the Paci
0379 ogical evidence that the whole body of arctic SHELLS underwent scarcely any modification during th
0383 at the similarity of the fresh water insects, SHELLS, etc., and at the dissimilarity of the surrou
0384 was peopled by existing land and fresh water SHELLS. The wide difference of the fish on opposite
0385 f a distant land. Some species of fresh water SHELLS have a very wide range, and allied species, w
0385 ntentionally stocked the one with fresh water SHELLS from the other. But another agency is perhaps
0385 in an aquarium, where many ova of fresh water SHELLS were hatching; and I found that numbers of th
0385 bers of the extremely minute and just hatched SHELLS crawled on the feet, and clung to them so fir
0390 for instance, the number of the endemic land SHELLS in Madeira, or of the endemic birds in the Ga
0391 abited by a wonderful number of peculiar land SHELLS, whereas not one species of sea shell is conf
0391 ts shores: now, though we do not know how sea SHELLS are dispersed, yet we can see that their eggs
0391 ight be transported far more easily than land SHELLS, across three or four hundred miles of open s
0397 isolated and smallest, are inhabited by land SHELLS, generally by endemic species, but sometimes
0397 veral interesting cases in regard to the land SHELLS of the islands of the Pacific. Now it is noto
0397 of the Pacific. Now it is notorious that land SHELLS are very easily killed by salt; their eggs, a
0397 get transported? It occurred to me that land SHELLS, when hybernating and having a membranous dia
0397 in sea water during seven days: one of these SHELLS was the Helix pomatia, and after it had again
0403 possess many distinct but representative land SHELLS, some of which live in crevices of stone; and
0403 nds have been colonised by some European land SHELLS which no doubt had some advantage over the in
0091 rter legs, which more frequently attacks the SHEPHERD's flocks. Let us now take a more complex cas

Page **************************************(Key Word)**************************************

Page ***(Key Word)***

```
0213 n round, instead of at, a flock of sheep, by SHEPHERD dogs. I cannot see that these actions, perfo
0214 h a greyhound has given to a whole family of SHEPHERD dogs a tendency to hunt hares. These domesti
0088 nd the hooked jaw to the male salmon; for the SHIELD may be as important for victory, as the sword
0238 of one caste alone carry a wonderful sort of SHIELD on their heads, the use of which is quite unk
0136 h concealed, until the wind lulls and the sun SHINES; that the proportion of wingless beetles is l
0485 ook at an organic being as a savage looks at a SHIP, as at something wholly beyond his comprehensi
0136 nd rarely or never flying. As with mariners SHIPWRECKED near a coast, it would have been better fo
0292 tion the area of the land and of the adjoining SHOAL parts of the sea will be increased, and new s
0132 list of plants which when growing near the sea SHORE have their leaves in some degree fleshy, thou
0167 so as to mock the shells now living on the sea SHORE. Summary. Our ignorance of the laws of variat
0349 en. If we look to the islands off the American SHORE, however much they may differ in geological s
0024 eral of the smaller British islets, or on the SHORES of the Mediterranian. Hence the supposed exte
0283 closely studies the action of the sea on our SHORES, will, I believe, be most deeply impressed wi
0292 ll decrease (excepting the productions on the SHORES of a continent when first broken up into an a
0299 in, whether certain sea shells inhabiting the SHORES of North America, which are ranked by some co
0300 incessant coast action, as we now see on the SHORES of South America. During the periods of subsi
0324 rous beds deposited at the present day on the SHORES of North America would hereafter be liable to
0348 common, than those of the eastern and western SHORES of South and Central America; yet these great
0348 mpassable, isthmus of Panama. Westward of the SHORES of America, a wide space of open ocean extend
0348 r travelling over a hemisphere we come to the SHORES of Africa; and over this vast space we meet w
0349 astern islands of the Pacific and the eastern SHORES of Africa, on almost exactly opposite meridia
0363 s formerly landed their rocky burthens on the SHORES of these mid ocean islands, and it is at leas
0364 ing seeds from the West Indies to our western SHORES, where, if not killed by so long an immersion
0364 ntic Ocean, from North America to the western SHORES of Ireland and England; but seeds could be tr
0372 riod, was nearly uniform along the continuous SHORES of the Polar Circle, will account, on the the
0372 presentative forms on the eastern and western SHORES of temperate North America; and the still mor
0373 as lat. 36 degrees to 37 degrees, and on the SHORES of the Pacific, where the climate is now so d
0376 son, also, speaks of the re appearance on the SHORES of New Zealand, Tasmania, etc., of northern f
0382 s. by these means, as I believe, the southern SHORES of America, Australia, New Zealand have becom
0382 horizontal lines, though rising higher on the SHORES where the tide rises highest, so have the liv
0384 fresh water group might travel far along the SHORES of the sea, and subsequently become modified
0391 t one species of sea shell is confined to its SHORES: now, though we do not know how sea shells ar
0393 ergs formerly brought boulders to its western SHORES, and they may have formerly transported foxes
0397 e equator, between 500 and 600 miles from the SHORES of South America. Here almost every product o
0001 to speculate on the subject, and drew up some SHORT notes; these I enlarged in 1844 into a sketch
0012 ve skin between their outer toes: pigeons with SHORT beaks have small feet; and those with long be
0014 same manner as in the crossed offspring from a SHORT horned cow by a long horned bull, the greater
0015 d not breed our cart and race horses, long and SHORT horned cattle, and poultry of various breeds,
0021 tonishing. Compare the English carrier and the SHORT faced tumbler, and see the wonderful differen
0021 to the nostrils, and a wide gape of mouth. The SHORT faced tumbler has a beak in outline almost li
0021 s very long wings and tails, others singularly SHORT tails. The barb is allied to the carrier, but
0021 , but, instead of a very long beak, has a very SHORT and very broad one. The pouter has a much elo
0021 hment and even laughter. The turbit has a very SHORT and conical beak, with a line of reversed fea
0023 thologist would place the English carrier, the SHORT faced tumbler, the runt, the barb, pouter, an
0024 ke that of the English carrier, or that of the SHORT faced tumbler, or barbi for reversed feathers
0027 eeds. Secondly, although an English carrier or SHORT faced tumbler differs immensely in certain ch
0061 eed which is wafted by the gentlest breeze; in SHORT, we see beautiful adaptations everywhere and
0083 liar and fitting manner: he feeds a long and a SHORT beaked pigeon on the same food: he does not e
0083 ecuilar manner: he exposes sheep with long and SHORT wool to the same climate. He does not allow t
0084 leeting are the wishes and efforts of man! how SHORT his time! and consequently how poor will his
0087 he egg. It has been asserted that of the best SHORT beaked tumbler pigeons more perish in the egg
0087 d to make the beak of a full grown pigeon very SHORT for the bird's own advantage, the process of
0089 sary to support this view; but if man can in a SHORT time give elegant carriage and beauty to his
0096 ossing of individuals. I must here introduce a SHORT digression. In the case of animals and plants
0111 g horns, and that these were swept away by the SHORT horns (I quote the words of an agricultural w
0152 sh fanciers. Even in the sub breeds, as in the SHORT faced tumbler, it is notoriously difficult to
0153 bird as coarse as a common tumbler from a good SHORT faced strain. But as long as selection is rap
0164 mined for me a small dun Welch pony with three SHORT parellel stripes on each shoulder. In the nor
0165 eless, had all four legs barred, and had three SHORT shoulder stripes, like those on the dun Welch
0175 either prey on or serve as prey for others; in SHORT, that each organic being is either directly o
0178 late zones, but they will generally have had a SHORT duration. For these intermediate varieties wi
0179 feet and which resembles an otter in its fur, SHORT legs, and form of tail; during summer this an
0225 veal to us her method of work. At one end of a SHORT series we have humble bees, which use their o
0225 ocoons to hold honey, sometimes adding to them SHORT tubes of wax, and likewise making separate an
0229 e, and allowed the bees to go on working for a SHORT time, and again examined the cell, and I foun
0231 ees all work together: one bee after working a SHORT time at one cell going to another, so that, a
0237 es correlated not only to one sex, but to that SHORT period alone when the reproductive system is
0283 ides in most cases reach the cliffs only for a SHORT time twice a day, and the waves eat into them
0283 e find that it is only here and there, along a SHORT length or round a promontory, that the cliffs
0290 sits sufficiently extensive to last for even a SHORT geological period. Along the whole west coast
0293 rk a very long lapse of years, each perhaps is SHORT compared with the period requisite to change
0303 dvantage over other organisms, a comparatively SHORT time would be necessary to produce many diver
0310 untries. Of this volume, only here and there a SHORT chapter has been preserved; and of each page,
0320 rovec it is transported far and near, like our SHORT horn cattle, and takes the place of other bre
0335 cter of length of beak originated earlier than SHORT beaked tumblers, which are at the opposite en
0336 si and these we assuredly do find. We find, in SHORT, such evidence of the slow and scarcely sensi
0342 at the duration of each formation is, perhaps, SHORT compared with the average duration of specifi
0359 but some which whilst green floated for a very SHORT time, when dried floated much longer; for ins
0382 g waters may be said to have flowed during one SHORT period from the north and from the south, and
0383 fitted, in a manner highly useful to them, for SHORT and frequent migrations from pond to pond, or
0424 m having the common habit of tumbling; but the SHORT faced breed has nearly or quite lost this hab
0441 for the male is a mere sack, which lives for a SHORT time, and is destitute of mouth, stomach, or
0443 always tell whether the child will be tall or SHORT, or what its precise features will be. The qu
0446 e exception to this rule, for the young of the SHORT faced tumbler differed from the young of the
0446 sponding not early period. But the case of the SHORT faced tumbler, which when twelve hours old ha
0447 it first appeared. In either case (as with the SHORT faced tumbler) the young or embryo would clos
```

Page ***************************************(Key Word)**

Page **(Key Word)**

```
0468 e and often capriciously, can produce within a SHORT period a great result by adding up mere indiv
0471 r sudden modification; it can act only by very SHORT and slow steps. Hence the canon of Natura non
0481 the history of the world was thought to be of SHORT duration; and now that we have acquired some
0485 common language will come into accordance. In SHORT, we shall have to treat species in the same m
0145 e two upper petals, the nectary is only much SHORTENED. With respect to the difference in the coro
0435 ranspose parts. The bones of a limb might be SHORTENED or widened to any extent, and become gradua
0091 pursues deer; and the other more bulky, with SHORTER legs, which more frequently attacks the shep
0095 a great advantage to the red clover to have a SHORTER or more deeply divided tube to its corolla.
0112 ncier is struck by a pigeon having a slightly SHORTER beak; another fancier is struck by a pigeon
0112 m birds with longer and longer beaks, or with SHORTER and shorter beaks. Again, we may suppose tha
0112 longer and longer beaks, or with shorter and SHORTER beaks. Again, we may suppose that an earl
0213 and have been transmitted for an incomparably SHORTER period, under less fixed conditions of life.
0465 duration, I am inclined to believe, has been SHORTER than the average duration of specific forms.
0194 n never take a leap, but must advance by the SHORTEST and slowest steps. Organs of little apparent
0222 ran away; but in about a quarter of an hour, SHORTLY after all the little yellow ants had crawled
0027 attle and length of beak of the carrier, the SHORTNESS of that of the tumbler, and the number of t
0068 uction of vermin. If not one head of game were SHOT during the next twenty years in England, and a
0010 ge how much, in the case of any variation, we SHOULD attribute to the direct action of heat, moist
0014 there is no apparent reason why a peculiarity SHOULD appear at any particular age, yet that it doe
0015 of intercrossing, that only a single variety SHOULD be turned loose in its new home. Nevertheless
0018 d to sheep and goats I can form no opinion. I SHOULD think, from facts communicated to me by Mr. B
0018 m his large and varied stores of knowledge, I SHOULD value more than that of almost any one, think
0026 ers, tumblers, pouters, and fantails now are, SHOULD yield offspring perfectly fertile, inter se,
0033 orrelation of growth, the importance of which SHOULD never be overlooked, will ensure some differe
0037 procure, never thought what splendid fruit we SHOULD eat; though we owe our excellent fruit, in so
0038 domestic animals kept by uncivilised man, it SHOULD not be overlooked that they almost always hav
0041 cies in any country requires that the species SHOULD be placed under favourable conditions of life
0041 nt point of all, is, that the animal or plant SHOULD be so highly useful to man, or so much valued
0041 uch valued by him, that the closest attention SHOULD be paid to even the slightest deviation in th
0044 t all properly, a long catalogue of dry facts SHOULD be given, but these I shall reserve for my fu
0045 have collected during a course of years. It SHOULD be remembered that systematists are far from
0045 them in many specimens of the same species. I SHOULD never have expected that the branching of the
0045 uld have been variable in the same species; I SHOULD have expected that changes of this nature cou
0047 opened. Hence, in determining whether a form SHOULD be ranked as a species or a variety, the opin
0055 here is no apparent reason why more varieties SHOULD occur in a group having many species, than in
0057 rtant criterion in settling whether two forms SHOULD be ranked as species or varieties. Now Fries
0062 t so at all seasons of each recurring year. I SHOULD premise that I use the term Struggle for Exis
0062 against the drought, though more properly it SHOULD be said to be dependent on the moisture. A pl
0070 hus save each other from utter destruction. I SHOULD add that the good effects of frequent intercr
0076 r cases. We can dimly see why the competition SHOULD be most severe between allied forms, which fi
0078 e plant the power of increasing in number, we SHOULD have to give it some advantage over its compe
0078 rease its average numbers in its new home, we SHOULD have to modify it in a different way to what
0078 ve to modify it in a different way to what we SHOULD have done in its native country; for we shoul
0078 hould have done in its native country; for we SHOULD have to give it some advantage over a differe
0078 over another. Probably in no single instance SHOULD we know what to do, so as to succeed. It will
0080 eing in the great and complex battle of life, SHOULD sometimes occur in the course of thousands of
0081 tter adapted forms could not freely enter, we SHOULD then have places in the economy of nature whi
0084 an we wonder, then, that nature's productions SHOULD be far truer in character than man's producti
0084 n character than man's productions; that they SHOULD be infinitely better adapted to the most comp
0084 d to the most complex conditions of life, and SHOULD plainly bear the stamp of far higher workmans
0085 icular colour would produce little effect: we SHOULD remember how essential it is in a flock of wh
0085 h or downy, a yellow or purple fleshed fruit, SHOULD succeed. In looking at many small points of d
0097 ented. Now, it is scarcely possible that bees SHOULD fly from flower to flower, and not carry poll
0099 f for the very purpose of self fertilisation, SHOULD in so many cases be mutually useless to each
0109 lainly: and indeed we can see reason why they SHOULD not have thus increased, for the number of pl
0121 redecessors and their original parent. For it SHOULD be remembered that the competition will gener
0128 and all plants throughout all time and space SHOULD be related to each other in group subordinate
0132 uctive system is disturbed, this or that part SHOULD vary more or less, we are profoundly ignorant
0135 ve that mutilations are ever inherited; and I SHOULD prefer explaining the entire absence of the a
0139 itants of the caves of the Old and New Worlds SHOULD be closely related, we might expect from the
0139 ng any surprise that some of the cave animals SHOULD be very anomalous, as Agassiz has remarked in
0142 ict to another; for it is not likely that man SHOULD have succeeded in selecting so many breeds an
0146 re never found in fruits which do not open: I SHOULD explain the rule by the fact that seeds could
0149 m diversified work, we can perhaps see why it SHOULD remain variable, that is, why natural selecti
0149 main variable, that is, why natural selection SHOULD have preserved or rejected each little deviat
0149 some particular shape. Natural selection, it SHOULD never be forgotten, can act on each part of e
0150 e that I have made due allowance for them. It SHOULD be understood that the rule by no means appli
0152 is case is eminently liable to variation. Why SHOULD this be so? On the view that each species has
0155 pecies having been independently created, why SHOULD that part of the structure, which differs fro
0156 a slight degree, it is not probable that they SHOULD vary at the present day. On the other hand, t
0156 a common progenitor, it is probable that they SHOULD still often be in some degree variable, at le
0156 manifest; but we can see why these characters SHOULD not have been rendered as constant and unifor
0159 species having been independently created, we SHOULD have to attribute this similarity in the enla
0160 it is a very surprising fact that characters SHOULD reappear after having been lost for many, per
0165 am not at all satisfied with this theory, and SHOULD be loth to apply it to breeds so distinct as
0169 enus arose; and thus we can understand why it SHOULD often still be variable in a much higher deor
0174 n large part explained. In the first place we SHOULD be extremely cautious in inferring, because a
0176 we can understand why intermediate varieties SHOULD not endure for very long periods; why as a ge
0176 very long periods; why as a general rule they SHOULD be exterminated and disappear, sooner than th
0182 habit, as the wings of a bird for flight, we SHOULD bear in mind that animals displaying early tr
0186 ence it will cause him no surprise that there SHOULD be geese and frigate birds with webbed feet,
0186 ost rarely alighting on the water; that there SHOULD be long toed corncrakes living in meadows ins
0186 g in meadows instead of in swamps; that there SHOULD be woodpeckers where not a tree grows; that t
0186 oodpeckers where not a tree grows; that there SHOULD be diving thrushes, and petrels with the habi
0187 nothing on this head. In this great class we SHOULD probably have to descend far beneath the lowe
0190 grades through which the organ has passed, we SHOULD have to look to very ancient ancestral forms,
```

Page **(Key Word)**

should

Page **(Key Word)**

0190	ncestral forms, long since become extinct. We	SHOULD	be extremely cautious in concluding that an o
0193	with apparently the same anomalous organ, it	SHOULD	be observed that, although the general appear
0194	n innovation. Why, on the theory of Creation,	SHOULD	this be so? Why should all the parts and orga
0194	he theory of Creation, should this be so? Why	SHOULD	all the parts and organs of many independent
0194	iably linked together by graduated steps? Why	SHOULD	not Nature have taken a leap from structure t
0194	selection, we can clearly understand why she	SHOULD	not: for natural selection can act only by ta
0195	fling an object as driving away flies; yet we	SHOULD	pause before being too positive even in this
0196	auses, independently of natural selection. We	SHOULD	remember that climate, food, etc., probably h
0197	many black and pied kinds, I dare say that we	SHOULD	have thought that the green colour was a beau
0197	to the direct action of putrid matter: but we	SHOULD	be very cautious in drawing any such inferenc
0202	nderstand how it is that the use of the sting	SHOULD	so often cause the insect's own death: for if
0204	ave have seen in this chapter how cautious we	SHOULD	be in concluding that the most different habi
0204	of no intermediate or transitional states, we	SHOULD	be very cautious in concluding that nome coul
0207	r birds' nests. An action, which we ourselves	SHOULD	require experience to enable us to perform, w
0213	deer, and driving them to a distant point, we	SHOULD	assuredly call these actions instinctive. Dom
0232	er completed cells. It suffices that the bees	SHOULD	be enabled to stand at their proper relative
0236	een profitable to the community that a number	SHOULD	have been annually born capable of work, but
0236	t had been an animal in the ordinary state, I	SHOULD	have unhesitatingly assumed that all its char
0240	workers had come to be in this condition: we	SHOULD	then have had a species of ant with neuters v
0241	e fertile parents, could form a species which	SHOULD	regularly produce neuters, either all of larg
0242	that, with all my faith in this principle, I	SHOULD	never have anticipated that natural selection
0248	ve ever lived, namely, Kolreuter and Gartner,	SHOULD	have arrived at diametrically opposite conclu
0248	n the question whether certain doubtful forms	SHOULD	be ranked as species or varieties, with the e
0252	nsidered as thoroughly well authenticated. It	SHOULD,	however, be borne in mind that, owing to few
0252	n them and the canary, or that their hybrids,	SHOULD	be perfectly fertile. Again, with respect to
0253	ng that the inherent sterility in the hybrids	SHOULD	have gone on increasing. If we were to act th
0260	ng confounded in nature? I think not. For why	SHOULD	the sterility be so extremely different in de
0260	important to keep from blending together? Why	SHOULD	the degree of sterility be innately variable
0260	e in the individuals of the same species? Why	SHOULD	some species cross with facility, and yet pro
0260	, and yet produce fairly fertile hybrids? Why	SHOULD	there often be so great a difference in the r
0266	t is scarcely possible that two organisations	SHOULD	be compounded into one, without some disturba
0277	degree of sterility of their hybrid offspring	SHOULD	generally correspond, though due to distinct
0277	ly cepends on widely different circumstances,	SHOULD	all run, to a certain extent, parallel with t
0280	the geological record. In the first place it	SHOULD	always be borne in mind what sort of intermed
0280	een them. But this is a wholly false view; we	SHOULD	always look for forms intermediate between ea
0280	mediate varieties which have ever existed, we	SHOULD	have an extremely close series between both a
0280	ries between both and the rock pigeon; but we	SHOULD	have no varieties directly intermediate betwe
0281	from each other. Hence in all such cases, we	SHOULD	be unable to recognise the parent form of any
0287	able palaeontologist, the late Edward Forbes,	SHOULD	not be forgotten; namely, that numbers of our
0293	eposition, I can see several reasons why each	SHOULD	not include a graduated series of links betwe
0297	or both forms by intermediate varieties. Nor	SHOULD	it be forgotten, as before explained, that A
0297	obtained numerous transitional gradations, we	SHOULD	not recognise their relationship, and should
0297	should not recognise their relationship, and	SHOULD	consequently be compelled to rank them all as
0298	ransition in any one geological formation. It	SHOULD	not be forgotten, that at the present day, wi
0302	distinct species. But I do not pretend that I	SHOULD	ever have suspected how poor a record of the
0307	e late E. Forbes, dispute this conclusion. We	SHOULD	not forget that only a small portion of the w
0309	e existed where our continents now stand. Nor	SHOULD	we be justified in assuming that if, for inst
0309	ocean were now converted into a continent, we	SHOULD	there find formations older than the Silurian
0314	it is by no means surprising that one species	SHOULD	retain the same identical form much longer th
0314	longer than others; or, if changing, that it	SHOULD	change less. We see the same fact in geograph
0315	early understand why a species when once lost	SHOULD	never reappear, even if the very same conditi
0315	me conditions of life, organic and inorganic,	SHOULD	recur. For though the offspring of one specie
0319	coming less and less favourable, we assuredly	SHOULD	not have perceived the fact, yet the fossil h
0326	tain extent the territories of other species,	SHOULD	be those which would have the best chance of
0328	g nearly, but not exactly the same period, we	SHOULD	find in both, from the causes explained in th
0333	ological periods undergone much modification,	SHOULD	in the older formations make some slight appr
0333	oach to each other: so that the older members	SHOULD	differ less from each other in some of their
0340	d that it is an immutable law that marsupials	SHOULD	have been chiefly or solely produced in Austr
0340	ia; or that Edentata and other American types	SHOULD	have been solely produced in South America. F
0345	ganisation on the whole has progressed. If it	SHOULD	hereafter be proved that ancient animals rese
0352	redible that individuals identically the same	SHOULD	ever have been produced through natural selec
0355	modification, why the inhabitants of a region	SHOULD	be related to those of another region, whence
0364	the direction of prevalent gales of wind. It	SHOULD	be observed that scarcely any means of transp
0366	ime that the cold had reached its maximum, we	SHOULD	have a uniform arctic fauna and flora, coveri
0376	s certainly a wonderful fact that New Zealand	SHOULD	have a closer resemblance in its crustacea to
0376	d found in the intermediate tropical seas. It	SHOULD	be observed that the northern species and for
0388	ering these several means of distribution, it	SHOULD	be remembered that when a pond or stream is f
0388	than in the case of terrestrial colonists. We	SHOULD,	also, remember that some, perhaps many, fres
0388	the migration of the same aquatic species, we	SHOULD	not forget the probability of many species ha
0393	and. But why, on the theory of creation, they	SHOULD	not have been created there, it would be very
0398	els that he is standing on American land. Why	SHOULD	this be so? Why should the species which are
0398	on American land. Why should this be so? Why	SHOULD	the species which are supposed to have been c
0400	t, climate, etc., that many of the immigrants	SHOULD	have been differently modified, though only i
0402	r that most species would thus spread; but we	SHOULD	remember that the forms which become naturali
0402	island, which has its own mocking thrush: why	SHOULD	it succeed in establishing itself there? We m
0405	birds, will necessarily range widely; for we	SHOULD	never forget that to range widely implies not
0405	or it is necessary that the unmodified parent	SHOULD	range widely, undergoing modification during
0405	racing modification during its diffusion, and	SHOULD	place itself under diverse conditions favoura
0405	g the wide distribution of certain genera, we	SHOULD	bear in mind that some are extremely ancient,
0408	as having nearly the same physical conditions	SHOULD	often be inhabited by very cifferent forms of
0408	amount of organic action and reaction, and we	SHOULD	find, as we do find, some groups of beings gr
0409	have endeavoured to show, why oceanic islands	SHOULD	have few inhabitants, but of these a great nu
0409	few inhabitants, but of these a great number	SHOULD	be endemic or peculiar; and why, in relation
0409	croup of beings, even within the same class,	SHOULD	have all its species endemic, and another gro
0409	ve all its species endemic, and another group	SHOULD	have all its species common to other quarters
0409	isms, as batrachians and terrestrial mammals,	SHOULD	be absent from oceanic islands, whilst the mo

Page **(Key Word)**

should

Page **(Key Word)***

```
0409  aerial mammals or bats. We can see why there  SHOULD  be some relation between the presence of mamm
0409  specifically distinct on the several islets,  SHOULD  be closely related to each other, and likewis
0409  areas, however distant from each other, there  SHOULD  be a correlation, in the presence of identica
0417  alence, and yet leave us in no doubt where it  SHOULD  be ranked. Hence, also, it has been found, th
0417  sly saw, as Jussieu observes, that this genus  SHOULD  still be retained amongst the Malpighiaceae.
0418  an see why characters derived from the embryo  SHOULD  be of equal importance with those derived fro
0418  rdinary view, why the structure of the embryo  SHOULD  be more important for this purpose than that
0425  to rank this one species with bears, and what  SHOULD  we do with the other species? The supposition
0425  er by the argumentum ad hominem, and ask what  SHOULD  be done if a perfect kangaroo were seen to co
0432  the differences between them. This is what we  SHOULD  be driven to, if we were ever to succeed in c
0434  dle of the porpoise, and the wing of the bat,  SHOULD  all be constructed on the same pattern, and s
0434  d all be constructed on the same pattern, and  SHOULD  include the same bones, in the same relative
0437  e facts on the ordinary view of creation! Why  SHOULD  the brain be enclosed in a box composed of su
0437  same construction in the skulls of birds. Why  SHOULD  similar bones have been created in the format
0437  are for such totally different purposes? Why  SHOULD  one crustacean, which has an extremely comple
0437  those with many legs have simpler mouths? Why  SHOULD  the sepals, petals, stamens, and pistils in a
0438  ring a long continued course of modification,  SHOULD  have seized on a certain number of the primor
0442  the wing of a bat, or the fin of a porpoise,  SHOULD  not have been sketched out with all the parts
0448  the existence of the species, that the child  SHOULD  be modified at a very early age in the same m
0449  rly see why ancient and extinct forms of life  SHOULD  resemble the embryos of their descendants, ou
0450  od than that at which they first appeared. It  SHOULD  also be borne in mind, that the supposed law
0455  ary part would tend to be wholly lost, and we  SHOULD  have a case of complete abortion. The princip
0456  n considering this view of classification, it  SHOULD  be borne in mind that the element of descent
0458  een modified in the course of descent; that I  SHOULD  without hesitation adopt this view, even if i
0459  an that the more complex organs and instincts  SHOULD  have been perfected, not by means superior to
0460  r constitutions or their reproductive systems  SHOULD  have been profoundly modified. Moreover, most
0462  ble chaos? With respect to existing forms, we  SHOULD  remember that we have no right to expect (exc
0464  less species which certainly have existed. We  SHOULD  not be able to recognise a species as the par
0467  have acted so efficiently under domestication  SHOULD  not have acted under nature. In the preservat
0469  ful agent always ready to act and select, why  SHOULD  we doubt that variations in any way useful to
0469  nce select variations most useful to himself,  SHOULD  nature fail in selecting variations useful, u
0470  d where they now flourish, these same species  SHOULD  present many varieties; for where the manufac
0471  y, though niggard in innovation. But why this  SHOULD  be a law of nature if each species has been i
0471  is that a bird, under the form of woodpecker,  SHOULD  have been created to prey on insects on the g
0471  that upland geese, which never or rarely swim  SHOULD  have been created with webbed feet; that a th
0471  been created with webbed feet; that a thrush  SHOULD  have been created to dive and feed on sub aqu
0471  eed on sub aquatic insects; and that a petrel  SHOULD  have been created with habits and structure f
0473  pecies having been independently created, why  SHOULD  the specific characters, or those by which th
0473  s in which they all agree? Why, for instance,  SHOULD  the colour of a flower be more likely to vary
0474  ly infer, of great importance to the species,  SHOULD  be eminently liable to variation; but, on my
0475  onsiderably different conditions of life, yet  SHOULD  follow nearly the same instincts; why the thr
0475  e can at once see why their crossed offspring  SHOULD  follow the same complex laws in their degrees
0476  g facts in Distribution. We can see why there  SHOULD  be so striking a parallelism in the distribut
0477  modification, we can see why oceanic islands  SHOULD  be inhabited by few species; but of these, th
0477  bited by few species; but of these, that many  SHOULD  be peculiar. We can clearly see why those ani
0477  s of ocean, as frogs and terrestrial mammals,  SHOULD  not inhabit oceanic islands; and why, on the
0477  pecies of bats, which can traverse the ocean,  SHOULD  so often be found on islands far distant from
0479  bryos of mammals, birds, reptiles, and fishes  SHOULD  be so closely alike, and should be so unlike
0479  s, and fishes should be so closely alike, and  SHOULD  be so unlike the adult forms. We may cease ma
0480  der the soldered wing covers of some beetles,  SHOULD  thus so frequently bear the plain stamp of in
0484  ths on the wild rose or oak tree. Therefore I  SHOULD  infer from analogy that probably all the orga
0488  the past and present inhabitants of the world  SHOULD  have been due to secondary causes, like those
0088  xual selection, as the mane to the lion, the  SHOULDER  pad to the boar, and the hooked jaw to the m
0163  s also been asserted that the stripe on each  SHOULDER  is sometimes double. The shoulder stripe is
0163  pe or each shoulder is sometimes double. The  SHOULDER  stripe is certainly very variable in length
0163  has been described without either spinal or  SHOULDER  stripe; and these stripes are sometimes very
0163  llas is said to have been seen with a double  SHOULDER  stripe. The hemionus has no shoulder stripe;
0163  double shoulder stripe. The hemionus has no  SHOULDER  stripe; but traces of it, as stated by Mr. B
0163  rally interest on the legs and faintly on the  SHOULDER. The quagga, though so plainly barred like a
0163  , and in one instance in a chestnut: a faint  SHOULDER  stripe may sometimes be seen in duns, and I
0164  gian cart horse with a double stripe on each  SHOULDER  and with leg stripes; and a mare, whom I can
0164  ny with three short parallel stripes on each  SHOULDER. In the north west part of India the Kattywa
0164  iped; the legs are generally barred; and the  SHOULDER  stripe, which is sometimes double and someti
0164  state that I have collected cases of leg and  SHOULDER  stripes in horses of very different breeds,
0165  body; and in one of them there was a double  SHOULDER  stripe. In Lord Moreton's famous hybrid from
0165  and the hemionus has none and has not even a  SHOULDER  stripe, nevertheless, had all four legs barr
0165  ad all four legs barred, and had three short  SHOULDER  stripes, like those on the dun Welch pony, a
0181  e wing membrane extended from the top of the  SHOULDER  to the tail, including the hind legs, we per
0473  the occasional appearance of stripes on the  SHOULDER  and legs of the several species of the horse
0166  on the legs like a zebra, or striped on the  SHOULDERS  like an ass. In the horse we see this tende
0001  g on these personal details, as I give them to  SHOW  that I have not been hasty in coming to a deci
0008  conception. Geoffroy St. Hilaire's experiments  SHOW  that unnatural treatment of the embryo causes
0009  have collected on this curious subject; but to  SHOW  how singular the laws are which determine the
0010  r to the act of conception. These cases anyhow  SHOW  that variation is not necessarily connected, a
0018  than seems to me to be the case, What does it  SHOW, but that some of our breeds originated there,
0034  ted with Europeans. Some of these facts do not  SHOW  actual selection, but they show that the breed
0034  e facts do not show actual selection, but they  SHOW  that the breeding of domestic animals was care
0038  nce obvious, how it is that our domestic races  SHOW  adaptation in their structure or in their habi
0045  alists consider unimportant parts; but I could  SHOW  by a long catalogue of facts, that parts which
0046  s seem to be very perplexing, for they seem to  SHOW  that this kind of variability is independent o
0050  authorities and practical men can be quoted to  SHOW  that the sessile and pedunculated oaks are eit
0053  sets of organic beings. But my tables further  SHOW  that, in any limited country, the species whic
0054  his result, that I am surprised that my tables  SHOW  even a small majority on the side of the large
0056  at as incipient species; for my tables clearly  SHOW  as a general rule that, wherever many species
0056  declined; and disappeared. All that we want to  SHOW  is, that where many species of a genus have be
0060  ter, I must make a few preliminary remarks, to  SHOW  how the struggle for existence bears on Natura
```

Page **(Key Word)***

Page **(Key Word)***

```
0073  relations. I shall hereafter have occasion to  SHOW  that the exotic Lobelia fulgens, in this part
0093  y can most effectually do this, I could easily  SHOW  by many striking instances. I will give only o
0098  on, there are special contrivances, as I could  SHOW  from the writings of C. C. Sprengel and from m
0104  or each birth; but I have already attempted to  SHOW  that we have reason to believe that occasional
0131  ich cannot be here given would be necessary to  SHOW  the truth of the remark, that the reproductive
0133  species under the same conditions. Such facts   SHOW  how indirectly the conditions cf life must act
0144  eir teeth. I know of no case better adapted to  SHOW  the importance of the laws of correlation in m
0160  y some other breed, the offspring occasionally  SHOW  a tendency to revert in character to the forei
0179  ed, as we shall in a future chapter attempt to  SHOW,  in an extremely imperfect and intermittent re
0179  onal state have subsisted? It would be easy to  SHOW  that within the same group carnivorous animals
0182  power of flight; but they serve, at least, to   SHOW  what diversified means of transition are possi
0188  e far too briefly and imperfectly given, which  SHOW  that there is much graduated diversity in the
0198  ation; and I have here alluded to them only to  SHOW  that, if we are unable to account for the char
0204  s of many organs and their intermediate states  SHOW  that wonderful metamorphoses in function are a
0207  ny definition of instinct. It would be easy to  SHOW  that several distinct mental actions are commo
0208  is usually said to be instinctive. But I could  SHOW  that none of these characters of instinct are
0210  gradations; or we ought at least to be able to  SHOW  that gradations of some kind are possible; and
0210  at they do so voluntarily, the following facts  SHOW.  I removed all the ants from a group of about
0230  of their accuracy! and if I had space, I could  SHOW  that they are conformable with my theory. Hube
0231  t of the first cell. I was able practically to  SHOW  this fact, by covering the edges of the hexago
0242  le but wholly insufficient length, in order to  SHOW  the power of natural selection, and likewise b
0242  i have endeavoured briefly in this chapter to   SHOW  that the mental qualities of our domestic anim
0242  erited. Still more briefly I have attempted to  SHOW  that instincts vary slightly in a state of nat
0245  s of sterility. I hope, however, to be able to  SHOW  that sterility is not a specially acquired or
0247  iged carefully to count the seeds, in order to  SHOW  that there is any degree of sterility. He alwa
0251  n an unnatural state. Nevertheless these facts  SHOW  on what slight and mysterious causes the lesse
0258  sation. On the other hand, these cases clearly  SHOW  that the capacity for crossing is connected wi
0259  considerable degree of fertility. These facts  SHOW  how completely fertility in the hybrid is inde
0266  rendered sterile. All that I have attempted to  SHOW,  is that in two cases, in some respects allied
0279  raduated physical conditions. I endeavoured to  SHOW,  that the life of each species depends in a mo
0279  like heat or moisture. I endeavoured, also, to  SHOW  that intermediate varieties, from existing in
0283  e appearance of the surface and the vegetation  SHOW  that elsewhere years have elapsed since the wa
0283  of which bears the stamp of time, are good to   SHOW  how slowly the mass has been accumulated. Let
0285  these cases there is nothing on the surface to  SHOW  such prodigious movements; the pile of rocks o
0303  w examples to illustrate these remarks; and to  SHOW  how liable we are to error in supposing that w
0308  rgument against the views here entertained. To  SHOW  that it may hereafter receive some explanation
0336  ought not to expect to find, as I attempted to  SHOW  in the last chapter, in any one or two formati
0338  ection. In a future chapter I shall attempt to  SHOW  that the adult differs from its embryo, owing
0341  ding and present Chapters. I have attempted to  SHOW  that the geological record is extremely imperf
0361  the accuracy of this observation. Again, I can  SHOW  that the carcasses of birds, when floating ab
0362  feet of birds are generally quite clean, I can  SHOW  that earth sometimes adheres to them: in one i
0386  to have muccy feet. Birds of this order I can   SHOW  are the greatest wanderers, and are occasional
0406  pters. In these chapters I have endeavoured to  SHOW,  that if we make due allowance for our ignoran
0407  l changes on distribution, I have attempted to  SHOW  how important has been the influence of the mo
0409  s, we can understand, as I have endeavoured to  SHOW,  why oceanic islands should have few inhabitan
0410  hich are not rare, may, as I have attempted to  SHOW,  be accounted for by migration at some former
0411  and on Natural Selection, I have attempted to  SHOW  that it is the widely ranging, the much diffus
0412  g indefinitely in size. I further attempted to  SHOW  that from the varying descencants of each spec
0412  n facts in naturalisation. I attemptec also to  SHOW  that there is a constant tendency in the forms
0429  hapter on geological succession I attempted to  SHOW,  on the principle of each group having general
0431  rough many predecessors. As it is difficult to  SHOW  the blood relationship between the numerous ki
0446  iven, more especially that of pigeons, seem to  SHOW  that the characteristic differences which give
0456  . summary. In this chapter I have attempted to  SHOW,  that the subordination of group to group in a
0460  ame community of ants! but I have attempted to  SHOW  how this difficulty can be mastered. With resp
0460  ghth chapter, which seem to me conclusively to  SHOW  that this sterility is no more a special endow
0462  migration. As an example, I have attempted to  SHOW  how potent has been the influence of the Glaci
0474  h their several instincts. I have attempted to  SHOW  how much light the principle of gradation thro
0483  ers. Organs in a rudimentary condition plainly  SHOW  that an early progenitor had the organ in a fu
0214  se great grandfather was a wolf, and this dog   SHOWED  a trace of its wild parentage only in one way
0214  n tumble. We may believe that some one pigeon   SHOWED  a slight tendency to this strange habit, and
0285  streams, due to their formerly liquid state,  SHOWED  at a glance how far the hard, rocky beds had
0339  er tertiary periods. Mr. Clift many years ago  SHOWED  that the fossil mammals from the Australian c
0009  ance, the rabbit and ferret kept in hutches),  SHOWING  that their reproductive system has not been
0050  overwhelming amount of experimental evidence,  SHOWING  that they descend from common parents, and c
0071  ll not here enlarge. Many cases are on record  SHOWING  how complex and unexpected are the checks an
0073  life! I am tempted to give one more instance  SHOWING  how plants and animals, most remote in the s
0094  its nectar for food. I could give many facts,  SHOWING  how anxious bees are to save time: for insta
0096  e, i have collected so large a body of facts,  SHOWING,  in accordance with the almost universal bel
0103  can bring a considerable catalogue of facts,  SHOWING  that within the same area, varieties of the
0110  is, in the facts given in the second chapter,  SHOWING  that it is the common species which afford t
0111  by man. Many curious instances could be given  SHOWING  how quickly new breeds of cattle, sheep, and
0211  e quite young aphides behaved in this manner,  SHOWING  that the action was instinctive, and not the
0222  r nest, carrying the cead bodies of F. fusca (SHOWING  that it was not a migration) and numerous pu
0231  o adjoining spheres. I have several specimens  SHOWING  clearly that they can do this. Even in the r
0248  g. i have collected so large a body of facts,  SHOWING  that close interbreeding lessens fertility,
0264  large body of facts, which I have collected,  SHOWING  that when animals and plants are removed fro
0283  s, all thickly clothed by marine productions,  SHOWING  how little they are abradec and how seldom t
0296  the upper beds of the same formation! facts,  SHOWING  what wide, yet easily overlooked, intervals
0329  rations from our great palaeontologist, Owen,  SHOWING  how extinct animals fall in between existing
0361  nsportation of seeds. I could give many facts  SHOWING  how frequently birds of many kinds are blown
0363  reat distances; for many facts could be given  SHOWING  that soil almost everywhere is charged with
0404  ill likewise be found some identical species,  SHOWING,  in accordance with the foregoing view, that
0404  , and some as varieties; these doubtful forms  SHOWING  us the steps in the process of modification.
0407  worlc, or at least great meridional belts. As  SHOWING  how diversified are the means of occasional
0420  the characters which naturalists consider as  SHOWING  true affinity between any two or more specie
0429  sses, and all in one great natural system. As  SHOWING  how few the higher groups are in number, and
```

Page **(Key Word)***

Page **(Key Word)**

```
0454  organs in a state of nature, further than by  SHOWING  that rudiments can be produced; for I doubt
0003  ed, would be unsatisfactory, until it could be  SHOWN  how the innumerable species inhabiting this w
0010  o such conditions; but in some cases it can be  SHOWN  that quite opposite conditions produce simila
0015  conditions of life are changed. If it could be  SHOWN  that our domestic varieties manifested a stro
0016  aracters of generic value. I think it could be  SHOWN  that this statement is hardly correct; but na
0016  be interesting; if, for instance, it could be  SHOWN  that the greyhound, bloodhound, terrier, span
0022  t a score of pigeons might be chosen, which if  SHOWN  to an ornithologist, and he were told that th
0023  species as he might have called them, could be  SHOWN  him. Great as the differences are between the
0035  ights they carry. Lord Spencer and others have  SHOWN  how the cattle of England have increased in w
0046  ow degrees: yet quite recently Mr. Lubbock has  SHOWN  a degree of variability in these main nerves
0046  naturalist, I may add, has also quite recently  SHOWN  that the muscles in the larvae of certain ins
0052  s varieties for very long periods, as has been  SHOWN  to be the case by Mr. Wollaston with the vari
0053  e discussed. Alph. de Candolle and others have  SHOWN  that plants which have very wide ranges gener
0062  lle and Lyell have largely and philosophically  SHOWN  that all organic beings are exposed to severe
0072  d in a feral state; and Azara and Rengger have  SHOWN  that this is caused by the greater number in
0074  stence of the species. In some cases it can be  SHOWN  that widely different checks act on the same
0081  of a single introduced tree or mammal has been  SHOWN  to be. But in the case of an island, or of a
0091  the tendency to catch rats rather than mice is  SHOWN  to be inherited. Now, if any slight innate ch
0098  invariably and completely destroy, as has been  SHOWN  by Gartner, any influence from the foreign po
0098  he stamens to spring forward, as Kolreuter has  SHOWN  to be the case with the barberry; and curious
0099  ing its own pollen; yet, as C. C. Sprengel has  SHOWN, and as I can confirm, either the anthers bur
0101  onal influence of a distinct individual can be  SHOWN  to be physically impossible. Cirripedes long
0111  ng somewhat of the character of species, as is  SHOWN  by the hopeless doubts in many cases how to r
0120  rocess for a greater number of generations (as  SHOWN  in the diagram in a condensed and simplified
0121  ransmitting unaltered descendants; and this is  SHOWN  in the diagram by the dotted lines not prolon
0142  es better than others: this is very strikingly  SHOWN  in works on fruit trees published in the Unit
0151  ined to secondary sexual characters is clearly  SHOWN  in the case of hermaphrodite cirripedes; and
0177  nts for the common species in each country, as  SHOWN  in the second chapter, presenting on an avera
0186  ch grade being useful to its possessor, can be  SHOWN  to exist; if further, the eye does vary ever
0187  f in two fundamentally different lines, can be  SHOWN  to exist, until we reach a moderately high st
0193  at of common muscle: and as it has lately been  SHOWN  that Rays have an organ closely analogous to
0194  wn to lead. The truth of this remark is indeed  SHOWN  by that old canon in natural history of Natur
0209  ion to become inherited, and I think it can be  SHOWN  that this does sometimes happen, then the res
0209  e tc succeeding generations. It can be clearly  SHOWN  that the most wonderful instincts with which
0209  t be profitable to a species; and if it can be  SHOWN  that instincts do vary ever so little, then I
0210  versity of instinct in the same species can be  SHOWN  to occur in nature. Again as in the case of c
0212  of man is slowly acquired, as I have elsewhere  SHOWN, by various animals inhabiting desert islands
0212  te of nature, is extremely diversified, can be  SHOWN  by a multitude of facts. Several cases also,
0214  and how curiously they become mingled, is well  SHOWN  when different breeds of dogs are crossed. Th
0214  a dog to point, had not some one dog naturally  SHOWN  a tendency in this line; and this is known oc
0218  itic on other species; and M. Fabre has lately  SHOWN  good reason for believing that although the T
0224  first appears: all this beautiful work can be  SHOWN, I think, to follow from a few very simple in
0225  tigate this subject by Mr. Waterhouse, who has  SHOWN  that the form of the cell stands in close rel
0236  iking modification of structure; for it can be  SHOWN  that some insects and other articulate animal
0239  have been carefully examined. Mr. F. Smith has  SHOWN  how surprisingly the neuters of several Briti
0248  ts made during different years. It can thus be  SHOWN  that neither sterility nor fertility affords
0255  in how many curious ways this gradation can be  SHOWN  to exist; but only the barest outline of the
0257  by their systematic affinity. This is clearly  SHOWN  by hybrids never having been raised between s
0257  ent to prevent two species crossing. It can be  SHOWN  that plants most widely different in habit an
0269  ppears. It can, in the first place, be clearly  SHOWN  that mere external dissimilarity between two
0275  osely like one of their parents; but it can be  SHOWN  that this does sometimes occur with hybrids:
0285  a, in parts ten thousand feet in thickness, as  SHOWN  in Prof. Ramsay's masterly memoir on this sub
0305  ulty on my theory, unless it could likewise be  SHOWN  that the species of this group appeared sudde
0312  both on the land and in the waters. Lyell has  SHOWN  that it is hardly possible to resist the evid
0339  eral parts of La Plata; and Professor Owen has  SHOWN  in the most striking manner that most of the
0339  birds of the caves of Brazil. Mr. Woodward has  SHOWN  that the same law holds good with sea shells,
0340  was peopled by numerous marsupials; and I have  SHOWN  in the publications above alluded to, that in
0349  ly american. We may look back to past ages, as  SHOWN  in the last chapter, and we find American typ
0354  ea inhabited by their progenitor. It can be  SHOWN  to be almost invariably the case, that a regi
0360  by any other means; and Alph. de Candolle has  SHOWN  that such plants generally have restricted ra
0372  identical species (though Asa Gray has lately  SHOWN  that more plants are identical than was forme
0374  and allied species. In America, Dr. Hooker has  SHOWN  that between forty and fifty of the flowering
0386  ost remote oceanic islands. This is strikingly  SHOWN, as remarked by Alph. de Candolle, in large g
0391  lants of the Galapagos Islands, Dr. Hooker has  SHOWN  that the proportional numbers of the differen
0392  s species; now trees; as Alph. de Candolle has  SHOWN, generally have, whatever the cause may be, c
0394  volcanic islands are sufficiently ancient, as  SHOWN  by the stupendous degradation which they have
0398  er animals, and with nearly all the plants, as  SHOWN  by Dr. Hooker in his admirable memoir on the
0400  archipelago are tenanted; as I have elsewhere  SHOWN, in a quite marvellous manner, by very closel
0402  belonging in a large proportion of cases, as  SHOWN  by Alph. de Candolle, to distinct genera. In
0404  of the world of other species allied to it, is  SHOWN  in another and more general way. Mr. Gould re
0415  determine its classificatory value, is almost  SHOWN  by the one fact, that in allied groups, in wh
0422  their characters. This natural arrangement is  SHOWN, as far as is possible on paper, in the diagr
0446  , which latter, though not proved true, can be  SHOWN  to be in some degree probable, to species in
0452  e use and discovery of rudiments. This is well  SHOWN  in the drawings given by Owen of the bones of
0463  d all changes are slowly effected. I have also  SHOWN  that the intermediate varieties which will at
0467  ge extent the character of natural species, is  SHOWN  by the inextricable doubts whether very many
0010  exactly the same conditions of life; and this  SHOWS  how unimportant the direct effects of the con
0024  n several places. Again, all recent experience  SHOWS  that it is most difficult to get any wild ani
0029  van Mons, in his treatise on pears and apples,  SHOWS  how utterly he disbelieves that the several s
0034  the Esquimaux their teams of dogs. Livingstone  SHOWS  how much good domestic breeds are valued by t
0040  tely been exhibited as distinct at our poultry  SHOWS. I think these views further explain what has
0050  lation to which the group is subject; and this  SHOWS, at least, how very generally there is some v
0054  cies of the same genus inhabiting any country,  SHOWS  that there is something in the organic or ino
0109  forms has not indefinitely increased, geology  SHOWS  us plainly; and indeed we can see reason why
0114  to eighteen genera and to eight orders, which  SHOWS  how much these plants differed from each othe
0125  some advantage; and the largeness of any group  SHOWS  that its species have inherited from a common
```

Page **(Key Word)**

0155 ividuals of some of the species. And this fact SHOWS that a character,'which is generally of gener
0162 anked as either varieties or species; and this SHOWS, unless all these forms be considered as inde
0185 , the deeply scooped membrane between the toes SHOWS that structure has begun to change. He who be
0190 wimbladder in fishes is a good one, because it SHOWS us clearly the highly important fact that an
0272 han those from very distinct species; and this SHOWS that the difference in the degree of variabil
0328 the later tertiary formations. Barrande, also, SHOWS that there is a striking general parallelism
0428 roups in other classes (and all our experience SHOWS that this valuation has hitherto been arbitra
0440 ly is, a crustacean; but a glance at the larva SHOWS this to be the case in an unmistakeable manne
0449 of each species and group of species partially SHOWS us the structure of their less modified ancie
0451 offspring was much increased in size; and this SHOWS that the rudiment and the perfect pistil are
0489 anner in which all organic beings are grouped, SHOWS that the greater number of species of each ge
0414 egards the external similarity of a mouse to a SHREW, of a dugong to a whale, of a whale to a fish
0184 ches, almost like a creeper. It often, like a SHRIKE, k+lls small birds by blows on the head; and
0093 a full sized pistil, and four stamens with SHRIVELLED anthers, in which not a grain of pollen can
0392 any rudimentary organ, for instance, as the SHRIVELLED wings under the soldered elytra of many ins
0480 the teeth in the embryonic calf or like the SHRIVELLED wings under the soldered wing covers of som
0219 erly helpless are the masters, that when Huber SHUT up thirty of them without a slave, but with pl
0368 states to Labrador; those of the mountains of SIBERIA to the arctic regions of that country. These
0370 continuous land from western Europe, through SIBERIA, to eastern America. And to this continuity
0373 s and nature of the mountain vegetation, that SIBERIA was similarly affected. Along the Himalaya,
0288 abits deep water and has been found fossil in SICILY, whereas not one other species has hitherto b
0312 we may trust the observations of Philippi in SICILY, the successive changes in the marine inhabit
0320 to feel no surprise at sickness, but when the SICK man dies, to wonder and to suspect that he die
0009 ted animals and plants, though often weak and SICKLY, yet breeding quite freely under confinement:
0320 to exist, is much the same as to admit that SICKNESS in the individual is the forerunner of death
0320 forerunner of death, to feel no surprise at SICKNESS, but when the sick man dies, to wonder and t
0226 , is manifestly a gross imitation of the three SIDED pyramidal basis of the cell of the hive bee.
0228 n, instead of on the straight edges of a three SIDED pyramid as in the case of ordinary cells. I t
0230 en each cell. They do not make the whole three SIDED pyramidal base of any one cell at the same ti
0230 rst cell is excavated out of a little parallel SIDED wall of wax, is not, as far as I have seen, s
0002 and balancing the facts and arguments on both SIDES of each question; and this cannot possibly be
0129 growing twigs have tried to branch out on all SIDES, and to overtop and kill the surrounding twig
0130 se, if vigorous, branch out and overtop on all SIDES many a feebler branch, so by generation I bel
0143 lied manner: we see this in the right and left SIDES of the body varying in the same manner; in th
0165 y, and even had some zebra like stripes on the SIDES of its face. With respect to this last fact,
0176 roads of closely allied forms existing on both SIDES of it. But a far more important consideration
0191 hiae have wholly disappeared, the slits on the SIDES of the neck and the loop like course of the a
0202 son tells us, though we may easily err on both SIDES, that some other contrivances are less perfec
0225 xagonal prism, with the basal edges of its six SIDES bevelled so as to join on to a pyramid, forme
0227 formed of three rhombs; and the rhombs and the SIDES of the hexagonal prisms will have every angle
0228 th vermilion. The bees instantly began on both SIDES to excavate little basins near to each other,
0228 have broken into each other from the opposite SIDES. The bees, however, did not suffer this to ha
0229 ntersection between the basins on the opposite SIDES of the ridge of wax. In parts, only little bi
0229 d at very nearly the same rate on the opposite SIDES of the ridge of vermilion wax, as they circul
0229 ly gnawed away and deepened the basins on both SIDES, in order to have succeeded in thus leaving f
0229 ficulty in the bees, whilst at work on the two SIDES of a strip of wax, perceiving when they have
0229 ing at exactly the same rate from the opposite SIDES; for I have noticed half completed rhombs at
0230 ombs and they gnaw into this from the opposite SIDES, always working circularly as they deepen eac
0231 shed off, by being largely gnawed away on both SIDES. The manner in which the bees build is curiou
0231 then to begin cutting it away equally on both SIDES near the ground, till a smooth, very thin wal
0347 lso, we see the same fact; for on the opposite SIDES of lofty and continuous mountain ranges, and
0358 ifference in the marine faunas on the opposite SIDES of almost every continent, the close relation
0373 ow any existing glacier. Further south on both SIDES of the continent, from lat. 41 degrees to the
0373 t these several far distant points on opposite SIDES of the world. But we have good evidence in al
0374 on was simultaneous on the eastern and western SIDES of North America, in the Cordillera under the
0374 under the warmer temperate zones, and on both SIDES of the southern extremity of the continent. I
0377 ther productions will have retreated from both SIDES towards the equator, followed in the rear by
0384 s, the wide difference of the fish on opposite SIDES of continuous mountain ranges, which from an
0395 m europe, and the mammals are the same on both SIDES; we meet with analogous facts on many islands
0461 t; whereas in first crosses the organs on both SIDES are in a perfect condition. As we continually
0482 ing naturalists, who will be able to view both SIDES of the question with impartiality. Whoever is
0487 of the inhabitants of the sea on the opposite SIDES of a continent, and the nature of the various
0058 r. h. C. Watson has marked for me in the well SIFTED London Catalogue of plants (4th edition) 63 p
0068 with man. The action of climate seems at first SIGHT to be quite independent of the struggle for e
0077 within the seeds of many plants seems at first SIGHT to have no sort of relation to other plants.
0105 nowhere else. Hence an oceanic island at first SIGHT seems to have been highly favourable for the
0212 it is strengthened by experience, and by the SIGHT of fear of the same enemy in other animals. B
0400 appened in the several islands situated within SIGHT of each other, having the same geological nat
0401 the other islands. But the islands, though in SIGHT of each other, are separated by deep arms of
0149 to use Prof. Owen's expression, seems to be a SIGN of low organisation: the foregoing remark seem
0255 withering of the flower is well known to be a SIGN of incipient fertilisation. From this extreme
0481 ; or that sterility is a special endowment and SIGN of creation. The belief that species were immu
0056 from the tables. These facts are of plain SIGNIFICATION on the view that species are only strongl
0411 stence of groups would have been of simple SIGNIFICATION, if one group had been exclusively fitted
0414 ch their whole life depends, are of little SIGNIFICATION, excepting in the first main divisions; w
0435 served, we can at once perceive the plain SIGNIFICATION of the homologous construction of the lim
0438 avoid employing language having this plain SIGNIFICATION. On my view these terms may be used liter
0485 to be metaphorical, and will have a plain SIGNIFICATION. When we no longer look at an organic bei
0444 little earlier or later in life. It would not SIGNIFY, for instance, to a bird which obtained its
0373 the marks of their former low descent; and in SIKKIM, Dr. Hooker saw maize growing on gigantic anc
0483 nce of species in what they consider reverent SILENCE. It may be asked how far I extend the doctri
0257 ost readily be crossed; and another genus, as SILENE, in which the most perservering efforts have
0084 , preserving and adding up all that is good; SILENTLY and insensibly working, whenever and whereve
0444 he caterpillar, cocoon, or imago states of the SILK moth; or, again, in the horns of almost full g
0014 ing when nearly mature; peculiarities in the SILKWORM are known to appear at the corresponding cat
0086 ar and cocoon stages of the varieties of the SILKWORM; in the eggs of poultry, and in the colour o

Page ***(Key Word)***

```
0374  the wide intervening hot countries. So on the SILLA of Caraccas the illustrious Humboldt long ago
0137  the eyes are of immense size: and Professor SILLIMAN thought that it regained, after living some
0306  s. for instance, I cannot doubt that all the SILURIAN trilobites have descended from some one crus
0306  acean, which must have lived long before the SILURIAN age, and which probably differed greatly fro
0306  m any known animal. Some of the most ancient SILURIAN animals, as the Nautilus, Lingula, etc., do
0307  e, it is indisputable that before the lowest SILURIAN stratum was deposited, long periods elapsed,
0307  far longer than, the whole interval from the SILURIAN age to the present day: and that during thes
0307  we see in the organic remains of the lowest SILURIAN stratum the dawn of life on this planet. Oth
0307  lately added another and lower stage to the SILURIAN system, abounding with new and peculiar spec
0307  doubt were somewhere accumulated before the SILURIAN epoch, is very great. If these most ancient
0308  the descriptions which we now possess of the SILURIAN deposits over immense territories in Russia
0309  at oscillations of level, since the earliest SILURIAN period. The coloured map appended to my volu
0309  at a period immeasurably antecedent to the SILURIAN epoch, continents may have existed where oce
0309  should there find formations older than the SILURIAN strata, supposing such to have been formerly
0310  as, the many formations long anterior to the SILURIAN epoch in a completely metamorphosed conditio
0310  own, of fossiliferous formations beneath the SILURIAN strata, are all undoubtedly of the gravest n
0313  reptiles in the sub Himalayan deposits. The SILURIAN Lingula differs but little from the living s
0313  ies of this genus: whereas most of the other SILURIAN Molluscs and all the Crustaceans have change
0316  n succession of generations, from the lowest SILURIAN stratum to the present day. We have seen in
0328  riking general parallelism in the successive SILURIAN deposits of Bohemia and Scandinavia: neverth
0331  ossible, as we have seen in the case of some SILURIAN forms, that a species might go on being slig
0334  the overlying carboniferous, and underlying SILURIAN system. But each fauna is not necessarily ex
0338  ertebrata, until beds far beneath the lowest SILURIAN strata are discovered, a discovery of which
0343  ave existed long before the first bed of the SILURIAN system was deposited: I can answer this latt
0343  nts now stand they have stood ever since the SILURIAN epoch: but that long before that period, the
0420  resent alliec genera, which lived during the SILURIAN epoch, and these have descended from a speci
0421  me few still living organic beings belong to SILURIAN genera. So that the amount or value of the d
0432  m: the letters, A to L, may represent eleven SILURIAN genera, some of which have produced large gr
0463  e not find great piles of strata beneath the SILURIAN system, stored with the remains of the proge
0463  d with the remains of the progenitors of the SILURIAN groups of fossils? For certainly on my theor
0465  fossiliferous formations beneath the lowest SILURIAN strata, I can only recur to the hypothesis g
0489  which lived long before the first bed of the SILURIAN system was deposited, they seem to me to bec
0489  endants of those which lived long before the SILURIAN epoch, we may feel certain that the ordinary
0011  shown that quite opposite conditions produce SIMILAR changes of structure. Nevertheless some slig
0024  posed extermination of so many species having SIMILAR habits with the rock pigeon seems to me a ve
0028  arent, as any naturalist could in coming to a SIMILAR conclusion in regard to the many species of
0035  ny native dog in Spain like our pointer. By a SIMILAR process of selection, and ty careful trainin
0047  haracter of species, but which are so closely SIMILAR to some other forms, or are so closely linke
0094  ndants would probably inherit a tendency to a SIMILAR slight deviation of structure. The tubes of
0113  ound be sown with one species of grass, and a SIMILAR plot be sown with several distinct genera of
0118  on parent (A). We may continue the process by SIMILAR steps for any length of time: some of the va
0138  difficult to imagine conditions of life more SIMILAR than deep limestone caverns under a nearly s
0138  ar than deep limestone caverns under a nearly SIMILAR climate: so that on the common view of the b
0140  mr. Thwaites informs me that he has observed SIMILAR facts in Ceylon, and analogous observations
0142  of the kidney bean has been often cited for a SIMILAR purpose, and with much greater weight: but u
0150  e ourang outang, that he has come to a nearly SIMILAR conclusion. It is hopeless to attempt to con
0159  n and tendency to variation, when acted on by SIMILAR unknown influences. In the vegetable kingdom
0161  ual action of the conditions of life and of a SIMILAR inherited constitution. It might further be
0165  reatise on the horse, has given a figure of a SIMILAR mule. In four coloured drawings, which I hav
0169  stitution from a common parent and exposed to SIMILAR influences will naturally tend to present an
0191  hat the swimbladder is homologous, or ideally SIMILAR, in position and structure with the lungs of
0197  st service to the plant: but as we see nearly SIMILAR hooks on many trees which are not climbers,
0265  of health, is affected by sterility in a very SIMILAR manner. In the one case, the conditions of l
0266  e. it may seem fanciful, but I suspect that a SIMILAR parallelism extends to an allied yet very di
0292  elli and E. Forbes independently arrived at a SIMILAR conclusion. One remark is here worth a passi
0306  reach other and distant seas. From these and SIMILAR considerations, but chiefly from our ignoran
0313  . falconer has given a striking instance of a SIMILAR fact, in an existing crocodile associated wi
0323  f russia, Western Europe and North America, a SIMILAR parallelism in the forms of life has been ob
0328  , but contemporaneous, faunas. Lyell has made SIMILAR observations on some of the later tertiary f
0337  other and preceding forms. If under a nearly SIMILAR climate, the eocene inhabitants of one quart
0339  upials of that continent. In South America, a SIMILAR relationship is manifest, even to an uneduca
0340  it is at present, to the northern half. In a SIMILAR manner we know from Falconer and Cautley's d
0347  and 35 degrees, we shall find parts extremely SIMILAR in all their conditions, yet it would not be
0349  ed, yet distinct kinds of birds, notes nearly SIMILAR, and sees their nests similarly constructed,
0360  sequently to my experiments, M. Martens tried SIMILAR ones, but in a much better manner, for he pl
0365  the mainland, would not receive colonists by SIMILAR means. I do not doubt that out of twenty see
0372  ated alike, in correspondence with the nearly SIMILAR physical conditions of the areas: for if we
0376  f terrestrial animals. In marine productions, SIMILAR cases occur: as an example, I may quote a re
0393  fficult to explain. Mammals offer another and SIMILAR case. I have carefully searched the oldest v
0395  analogous facts on many islands separated by SIMILAR channels from Australia. The West Indian Isl
0407  curred within the recent period, and of other SIMILAR changes which may have occurred within the s
0423  hough the esculent and thickened stems are so SIMILAR. Whatever part is found to be most constant,
0427  nes of descent, may readily become adapted to SIMILAR conditions, and thus assume a close external
0428  ive slight modifications to live under nearly SIMILAR circumstances, to inhabit for instance the t
0430  rodent. The elder De Candolle has made nearly SIMILAR observations on the general nature of the af
0437  nstruction in the skulls of birds. Why should SIMILAR bones have been created in the formation of
0438  eized on a certain number of the primordially SIMILAR elements, many times repeated, and have adap
0439  ls within the same class are often strikingly SIMILAR: a better proof of this cannot be given, tha
0440  eries near the branchial slits are related to SIMILAR conditions, in the young mammal which is nou
0440  a bat, and fin of a porpoise, are related to SIMILAR conditions of life. No one will suppose that
0442  oper nutriment, yet nearly all pass through a SIMILAR worm like stage of development: but in some
0448  er with its parents, in accordance with their SIMILAR habits. Some further explanation, however, o
0449  and habits, if they pass through the same or SIMILAR embryonic stages, we may feel assured that t
0449  y have both descended from the same or nearly SIMILAR parents, and are therefore in that degree cl
0452  into a nascent breathing organ or lunc. Other SIMILAR instances could be given. Rudimentary organs
0076  sually, though by no means invariably, some SIMILARITY in habits and constitution, and always in s
```

Page ***(Key Word)***

Page ***(Key Word)***

```
0138  or the American and European caverns, close  SIMILARITY  in their organisation and affinities might
0159  y created, we should have to attribute this  SIMILARITY  in the enlarged stems of these three plants
0236  of nature, that we cannot account for their  SIMILARITY  by inheritance from a common parent, and mu
0265  d that of hybrids, there are many points of  SIMILARITY.  In both cases the sterility is independent
0275  pects there seems to be a general and close  SIMILARITY  in the offspring of crossed species, and of
0275  aving been produced by secondary laws, this  SIMILARITY  would be an astonishing fact. But it harmon
0340  two continents, and, on the other hand, by  SIMILARITY  of conditions, for the uniformity of the sa
0346  fact which strikes us is, that neither the  SIMILARITY  nor the dissimilarity of the inhabitants of
0383  ers of Brazil, feeling much surprise at the  SIMILARITY  of the fresh water insects, shells, etc., a
0413  ity of descent, the only known cause of the  SIMILARITY  of organic beings, is the bond, hidden as i
0414  be more false. No one regards the external  SIMILARITY  of a mouse to a shrew, of a dugong to a wha
0435  re hopeless than to attempt to explain this  SIMILARITY  of pattern in members of the same class, by
0440  animal. From such special adaptations, the  SIMILARITY  of the larvae or active embryos of allied a
0442  embryo and the adult, and likewise a close  SIMILARITY  in the embryos of widely different animals
0456  descent, the only certainly known cause of  SIMILARITY  in organic beings, we shall understand what
0479  ow and slight successive modifications. The  SIMILARITY  of pattern in the wing and leg of a bat, th
0127  eritance they will tend to produce offspring  SIMILARLY  characterised. This principle of preservati
0282  w generally extinct, have in their turn been  SIMILARLY  connected with more ancient species; and so
0305  he teleostean fish might formerly have had a  SIMILARLY  confined range, and after having been large
0323  a, and sections of genera, and sometimes are  SIMILARLY  characterised in such trifling points as me
0323  in the formations either above or below, are  SIMILARLY  absent at these distant points of the world
0349  , notes nearly similar, and sees their nests  SIMILARLY  constructed, but not quite alike, with eggs
0373  of the mountain vegetation, that Siberia was  SIMILARLY  affected. Along the Himalaya, at points 900
0484  a circumstance as that the same poison often  SIMILARLY  affects plants and animals; or that the poi
0129  n represented by a great tree. I believe this  SIMILE  largely speaks the truth. The green and buddi
0005  ies of transitions, or in understanding how a  SIMPLE  being or a simple organ can be changed and pe
0005  , or in understanding how a simple being or a  SIMPLE  organ can be changed and perfected into a hig
0020  eons) extremely uniform, and everything seems  SIMPLE  enough; but when these mongrels are crossed o
0029  could be given. The explanation, I think, is  SIMPLE:  from long continued study they are strongly
0037  erials: but the art, I cannot doubt, has been  SIMPLE,  and, as far as the final result is concerned
0053  al well worked floras. At first this seemed a  SIMPLE  task; but Mr. H. C. Watson, to whom I am much
0071  give only a single instance, which, though a  SIMPLE  one, has interested me. In Staffordshire, on
0073  t that in nature the relations can ever be as  SIMPLE  as this. Battle within battle must ever be re
0075  he ground according to definite laws; but how  SIMPLE  is this problem compared to the action and re
0092  t of fertilisation, its destruction appears a  SIMPLE  loss to the plant; yet if a little pollen wer
0112  can and does apply most efficiently, from the  SIMPLE  circumstance that the more diversified the de
0113  clearly see this in the case of animals with  SIMPLE  habits. Take the case of a carnivorous quadru
0154  e more variable than generic. To explain by a  SIMPLE  example what is meant. If some species in a l
0162  ppear probable would all appear together from  SIMPLE  variation. More especially we might have infe
0166  tinct species of the horse genus becoming, by  SIMPLE  variation, striped on the legs like a zebra,
0166  nd other marks; and when any breed assumes by  SIMPLE  variation a bluish tint, these bars and other
0173  t time many transitional forms. Let us take a  SIMPLE  case: in travelling from north to south over
0186  ect and complex eye to one very imperfect and  SIMPLE,  each grade being useful to its possessor, ca
0188  ving that natural selection has converted the  SIMPLE  apparatus of an optic nerve merely coated wit
0195  uch difficulty in understanding the origin of  SIMPLE  parts, of which the importance does not seem
0224  be shown, I think, to follow from a few very  SIMPLE  instincts. I was led to investigate this subj
0240  . smith has observed, the larger workers have  SIMPLE  eyes (ocelli), which though small can be plai
0280  from all its modified descendants. To give a  SIMPLE  illustration: the fantail and pouter pigeons
0331  pecies in each genus. The diagram is too much  SIMPLE,  too few genera and too few species being giv
0366  which, as we shall immediately see, affords a  SIMPLE  explanation of these facts. We have evidence
0411  s. the existence of groups would have been of  SIMPLE  signification, if one group had been exclusiv
0422  ble on paper, in the diagram, but in much too  SIMPLE  a manner. If a branching diagram had not been
0436  r of maxillae, these parts being perhaps very  SIMPLE  in form; and then natural selection will acco
0439  ourse of descent from true legs, or from some  SIMPLE  appendage, is explained. Embryology. It has a
0441  first stage have three pairs of legs, a very  SIMPLE  single eye, and a proboscifcrmed mouth, with
0441  w reconverted into a minute, single, and very  SIMPLE  eye spot. In this last and completed state, c
0454  fication, the origin of rudimentary organs is  SIMPLE.  We have plenty of cases of rudimentary organ
0490  ccording to the fixed law of gravity, from so  SIMPLE  a beginning endless forms most beautiful and
0235  numerous, successive, slight modifications of  SIMPLER  instincts; natural selection having by slow
0239  ed facts at once annihilate my theory. In the  SIMPLER  case of neuter insects all of one caste or o
0437  egs; or conversely, those with many legs have  SIMPLER  mouths? Why should the pedals, petals, stame
0486  he rules for classifying will no doubt become  SIMPLER  when we have a definite object in view. We p
0488  hen the forms of life were probably fewer and  SIMPLER,  the rate of change was probably slower; and
0466  possible transitional gradations between the  SIMPLEST  and the most perfect organs; it cannot be pr
0488  rst dawn of life, when very few forms of the  SIMPLEST  structure existed, the rate of change may ha
0225  ection of the cells of the hive bee and the  SIMPLICITY  of those of the humble bee, we have the cel
0352  d points, where now found. Nevertheless the  SIMPLICITY  of the view that each species was first pro
0118  ancth generation, and under a condensed and  SIMPLIFIED  form up to the fourteen thousandth generati
0120  (as shown in the diagram in a condensed and  SIMPLIFIED  manner), we get eight species, marked by th
0007  to conclude that this greater variability is  SIMPLY  due to our domestic productions having been r
0065  result of which never fails to be surprising.  SIMPLY  explains the extraordinarily rapid increase a
0099  cases be mutually useless to each other! How  SIMPLY  are these facts explained on the view of an a
0100  made these few remarks on the sexes of trees  SIMPLY  to call attention to the subject. Turning for
0146  ole groups of species, and which in truth are  SIMPLY  due to inheritance: for an ancient progenitor
0192  onverted by natural selection into branchiae,  SIMPLY  through an increase in the size and the obl
0199  ef part of the organisation of every being is  SIMPLY  due to inheritance; and consecuently, though
0208  d up only to the third stage, the caterpillar  SIMPLY  re performed the fourth, fifth, and sixth sta
0215  ge from extreme wildness to extreme tameness,  SIMPLY  to habit and long continued close confinement
0260  that species have been endowed with sterility  SIMPLY  to prevent their becoming confounded in natur
0260  ility both of first crosses and of hybrids is  SIMPLY  incidental or dependent on unknown difference
0297  rictly intermediate between B and C, it would  SIMPLY  be ranked as a third and distinct species, un
0343  eat leading facts in palaeontology seem to me  SIMPLY  to follow on the theory of descent with modif
0345  gical periods ceases to be mysterious, and is  SIMPLY  explained by inheritance. If then the geologi
0350  hat this bond is. This bond, on my theory, is  SIMPLY  inheritance, that cause which alone, as far a
0414  er gives some unknown plan of creation, or is  SIMPLY  a scheme for enunciating general propositions
0464  r intermediate variety be discovered, it will  SIMPLY  be classed as another and distinct species. C
```

Page ***(Key Word)***

Page **(Key Word)**

```
0465 ear as if suddenly created there, and will be SIMPLY classed as new species. Most formations have
0471 make more strictly correct, is on this theory SIMPLY intelligible. We can plainly see why nature i
0473  of the horse genus and in their hybrids! How SIMPLY is this fact explained if we believe that the
0476 fossils in the formations above and below, is SIMPLY explained by their intermediate position in t
0114 rders: nature follows what may be called a SIMULTANEOUS rotation. Most of the animals and plants w
0312 ce as do single species. On Extinction. On SIMULTANEOUS changes in the forms of life throughout th
0313 of their many now extinct species has been SIMULTANEOUS in each separate formation. Species of dif
0324 ing deposits, would be correctly ranked as SIMULTANEOUS in a geological sense. The fact of the for
0327 the parallel, and, taken in a large sense, SIMULTANEOUS, succession of the same forms of life thro
0373 t know that the Glacial epoch was strictly SIMULTANEOUS at these several far distant points on opp
0374 ng a part at least of the period, actually SIMULTANEOUS throughout the world. Without some distinc
0374 it as probable that the glacial action was SIMULTANEOUS on the eastern and western sides of North
0087 on would be very slow, and there would be SIMULTANEOUSLY the most rigorous selection of the young
0095 wer and a bee might slowly become, either SIMULTANEOUSLY or one after the other, modified and adap
0117 ture; they are not supposed all to appear SIMULTANEOUSLY, but often after long intervals of time:
0183 habits: both probably often change almost SIMULTANEOUSLY. Of cases of changed habits it will suffi
0190 ps. Two distinct organs sometimes perform SIMULTANEOUSLY the same function in the same individual:
0205 ing lung. The same organ having performed SIMULTANEOUSLY very different functions, and then having
0241 of workers of one size and structure, and SIMULTANEOUSLY another set of workers of a different siz
0305 ecies of this group appeared suddenly and SIMULTANEOUSLY throughout the world at this same period.
0314 tants of a country to change abruptly, or SIMULTANEOUSLY, or to an equal degree. The process of mo
0322 ies. On the Forms of Life changing almost SIMULTANEOUSLY throughout the World. Scarcely any palaeo
0322 act, that the forms of life change almost SIMULTANEOUSLY throughout the world. Thus our European C
0324 s of life are spoken of as having changed SIMULTANEOUSLY throughout the world, it must not be supp
0324 e. the fact of the forms of life changing SIMULTANEOUSLY, in the above large sense, at distant par
0336 e inhabiting the sea, have changed almost SIMULTANEOUSLY throughout the world, and therefore under
0344  of the world will appear to have changed SIMULTANEOUSLY. We can understand how it is that all the
0355 me authors suppose, from many individuals SIMULTANEOUSLY created. With those organic beings which
0356 o that many individuals will have gone on SIMULTANEOUSLY changing, and the whole amount of modific
0374 ure of the whole world was at this period SIMULTANEOUSLY cooler. But it would suffice for my purpo
0374  of broad longitudinal belts, having been SIMULTANEOUSLY colder from pole to pole, much light can
0376 part of it, was during the Glacial period SIMULTANEOUSLY much colder than at present. The Glacial
0407 lacial period, which I am fully convinced SIMULTANEOUSLY affected the whole world, or at least gre
0475 of time, to appear as if they had changed SIMULTANEOUSLY throughout the world. The fact of the fos
0024 ies; and further, that these very species have SINCE all become extinct or unknown. So many strang
0035  been unconsciously modified to a large extent SINCE the time of that monarch. Some highly compete
0041 ant. No doubt the strawberry had always varied SINCE it was cultivated, but the slight varieties h
0065 imalaya, which have been imported from America SINCE its discovery. In such cases, and endless ins
0153 rgone an extraordinary amount of modification, SINCE the period when the species branched off from
0156 ters have been inherited from a remote period, SINCE that period when the species first branched o
0168 t is, the characters which have come to differ SINCE the several species of the same genus branche
0169 hrough an extraordinary amount of modification SINCE the genus arose: and thus we can understand w
0189 en first formed at an extremely remote period, SINCE which all the many members of the class have
0190  to look to very ancient ancestral forms, long SINCE become extinct. We should be extremely cautio
0254 descended from two or more aboriginal species, SINCE commingled by intercrossing. On this view, th
0283 etation show that elsewhere years have elapsed SINCE the waters washed their base. He who most clo
0285  the height or depth of thousands of feet: and SINCE the crust cracked, the surface of the land ha
0286 red up the Weald within so limited a period as SINCE the latter part of the Chalk formation. The d
0287 nger period than 300 million years has elapsed SINCE the latter part of the Secondary period. I ha
0291 sils, have thus been formed during subsidence. SINCE publishing my views on this subject in 1845,
0309 ected no doubt to great oscillations of level, SINCE the earliest silurian ma. The coloured ma
0318  with astonishment: for seeing that the horse, SINCE its introduction by the Spaniards into South
0343 ting continents now stand they have stood ever SINCE the Silurian epoch: but that long before the
0344 tly resemble, the common progenitor of groups, SINCE become widely divergent. Extinct forms are se
0351 nd climatal changes which will have supervened SINCE ancient times, almost any amount of migration
0368 r migration across the low intervening tracts, SINCE become too warm for their existence. If the c
0368  too warm for their existence. If the climate, SINCE the Glacial period, has ever been in any degr
0368 t to this intercalated slightly warmer period, SINCE the Glacial period. The arctic forms, during
0369  from each other, and have survived there ever SINCE: they will, also, in all probability have bec
0371 tinuously united by land, serving as a bridge, SINCE rendered impassable by cold, for the inter mi
0381 hat we cannot suppose that there has been time SINCE the commencement of the Glacial period for th
0396 l retaining the same specific form or modified SINCE their arrival, could have reached their prese
0400  considering how they have come to be modified SINCE their arrival), we find a considerable amount
0408 ording to the length of time which has elapsed SINCE new inhabitants entered one region; according
0430 a common progenitor, and that both groups have SINCE undergone much modification in divergent dire
0473 rstand this fact: for they have already varied SINCE they branched off from a common progenitor in
0474 ion; but, on my view, this part has undergone, SINCE the several species branched off from a commo
0488 ime, compared with the ages which have elapsed SINCE the first creature, the progenitor of innumer
0062 , or we forget, that the birds which are idly SINGING round us mostly live on insects or seeds, an
0089 tween the males of many species to attract by SINGING the females. The rock thrush of Guiana, bird
0489 ed with many plants of many kinds, with birds SINGING on the bushes, with various insects flitting
0002  do this. For I am well aware that scarcely a SINGLE point is discussed in this volume on which fa
0009 porting plants: by this term gardeners mean a SINGLE bud or offset, which suddenly assumes a new a
0011  another instance of the effect of use. Not a SINGLE domestic animal can be named which has not in
0015 ent the effects of intercrossing, that only a SINGLE variety should be turned loose in its new hom
0017 heir kind so truly, were the offspring of any SINGLE species, then such facts would have great wei
0020 nd long continued selection: nor can I find a SINGLE case on record of a permanent race having bee
0026 lated together, though it is unsupported by a SINGLE experiment. But to extend the hypothesis so f
0030 l of our eminent breeders have, even within a SINGLE lifetime, modified to a large extent some bre
0032 choicest productions have been produced by a SINGLE variation from the aboriginal stock. We have
0038 d by quite uncivilised man, has afforded us a SINGLE plant worth culture. It is not that these cou
0049 determine their rank. I will here give only a SINGLE instance, the well known one of the primrose
0064 rth would soon be covered by the progeny of a SINGLE pair. Even slow breeding man has doubled in t
0066  of eggs; and another, like the hippobosca, a SINGLE one: but this difference does not determine h
0066 ived on an average for a thousand years, if a SINGLE seed were produced once in a thousand years,
0066 ns always in mind, never to forget that every SINGLE organic being around us may be said to be str
```

Page **(Key Word)**

single

Page **************************************(Key Word)**************************************

```
0067  w not exactly what the checks are in even one  SINGLE  instance. Nor will this surprise any one who
0070  s in a garden: I have in this case lost every  SINGLE  seed. This view of the necessity of a large s
0071  ether in the same country. I will give only a  SINGLE  instance, which, though a simple one, has int
0071  has been the effect of the introduction of a  SINGLE  tree, nothing whatever else having been done,
0072  closed heath, and literally I could not see a  SINGLE  Scotch fir, except the old planted clumps. Bu
0078  m some advantage over another. Probably in no  SINGLE  instance should we know what to do, so as to
0081  be remembered how powerful the influence of a  SINGLE  introduced tree or mammal has been shown to b
0095  iews as the excavation of a great valley by a  SINGLE  diluvial wave, so will natural selection, if
0100  but these all pair. As yet I have not found a  SINGLE  case of a terrestrial animal which fertilises
0101  ties, namely, Professor Huxley, to discover a  SINGLE  case of an hermaphrodite animal with the orga
0115  in new genera than in new species. To give a  SINGLE  instance: in the last edition of Dr. Asa Gray
0118  r each thousand generations, producing only a  SINGLE  variety, but in a more and more modified cond
0119  process of modification will be confined to a  SINGLE  line of descent, and the number of the descen
0124  onverging in sub branches downwards towards a  SINGLE  point: this point representing a single speci
0124  rds a single point: this point representing a  SINGLE  species, the supposed single parent of our se
0124  t representing a single species, the supposed  SINGLE  parent of our several new sub genera and gene
0129  ate groups in any class cannot be ranked in a  SINGLE  file, but seem rather to be clustered round p
0139  ecies of the same genus have descended from a  SINGLE  parent, if this view be correct, acclimatisat
0202  emales, can we admire the production for this  SINGLE  purpose of thousands of drones, which are utt
0219  e species would certainly become extinct in a  SINGLE  year. The males and fertile females do no wor
0219  y perished of hunger. Huber then introduced a  SINGLE  slave (F. fusca), and she instantly set to wo
0225  the three which form the pyramidal base of a  SINGLE  cell on one side of the comb, enter into the
0232  overing the edges of the hexagonal walls of a  SINGLE  cell, or the extreme margin of the circumfere
0233  es there seem to me any great difficulty in a  SINGLE  insect (as in the case of a queen wasp) makin
0233  bserved: nor would any good be derived from a  SINGLE  hexagon being built, as in its construction m
0236  at great length, but I will here take only a  SINGLE  case, that of working or sterile ants. How th
0250  important statements I will here give only a  SINGLE  one as an example, namely, that every ovule i
0253  s country with either pure parent, and in one  SINGLE  instance they have bred inter se. This was ef
0255  even with the pollen of either pure parent, a  SINGLE  fertile seed: but in some of these cases a fi
0257  to produce between extremely close species a  SINGLE  hybrid. Even within the limits of the same ge
0270  one with the pollen of the other: but only a  SINGLE  head produced any seed, and this one head pro
0274  on, as far as I can make out, is founded on a  SINGLE  experiment: and seems directly opposed to the
0287  f our fossil species are known and named from  SINGLE  and often broken specimens, or from a few spe
0288  ll strictly littoral, with the exception of a  SINGLE  Mediterranean species, which inhabits deep wa
0289  america. In regard to mammiferous remains, a  SINGLE  glance at the historical table published in t
0299  sheep, horses, and dogs have descended from a  SINGLE  stock or from several aboriginal stocks: or,
0303  , and range over the world, the fact of not a  SINGLE  bone of a whale having been discovered in any
0312  s in their appearance and disappearance as do  SINGLE  species. On Extinction. On simultaneous chang
0316  s in their appearance and disappearance as do  SINGLE  species, changing more or less quickly, and i
0317  on, like the branching of a great tree from a  SINGLE  stem, till the group becomes large. On Extinc
0318  rom another, and finally from the world. Both  SINGLE  species and whole groups of species last for
0318  determine the length of time during which any  SINGLE  species or any single genus endures. There is
0318  f time during which any single species or any  SINGLE  genus endures. There is reason to believe tha
0321  e escaped severe competition. For instance, a  SINGLE  species of Trigonia, a great genus of shells
0322  thus, as it seems to me, the manner in which  SINGLE  species and whole groups of species become ex
0327  slowly to disappear: though here and there a  SINGLE  member might long be enabled to survive. Thus
0337  w occurring in New Zealand, and from hardly a  SINGLE  inhabitant of the southern hemisphere having
0342  ons which must have passed away even during a  SINGLE  formation: that, owing to subsidence being ne
0342  ill have been least perfectly kept: that each  SINGLE  formation has not been continuously deposited
0352  that each species was first produced within a  SINGLE  region captivates the mind. He who rejects it
0352  at two separate points, why do we not find a  SINGLE  mammal common to Europe and Australia or Sout
0353  umber of sections of genera are confined to a  SINGLE  region: and it has been observed by several n
0354  e view of each species having migrated from a  SINGLE  birthplace: then, considering our ignorance w
0355  on with modification. The previous remarks on  SINGLE  and multiple centres of creation do not direc
0355  als of the same species have descended from a  SINGLE  pair, or single hermaphrodite, or whether, as
0355  species have descended from a single pair, or  SINGLE  hermaphrodite, or whether, as some authors su
0356  ls of each variety will have descended from a  SINGLE  parent. But in the majority of cases, namely,
0356  ve been due, at each stage, to descent from a  SINGLE  parent. To illustrate what I mean: our Englis
0356  ifference and superiority to descent from any  SINGLE  pair, but to continued care in selecting and
0356  reatest amount of difficulty on the theory of  SINGLE  centres of creation, I must say a few words o
0357  trusted, it must be admitted that scarcely a  SINGLE  island exists which has not recently been uni
0357  ay be, that each species has proceeded from a  SINGLE  birthplace, and when in the course of time we
0384  according to Valenciennes, there is hardly a  SINGLE  group of fishes confined exclusively to fresh
0385  common parent and must have proceeded from a  SINGLE  source, prevail throughout the world. Their d
0388  a rising islet, it will be unoccupied: and a  SINGLE  seed or egg will have a good chance of succee
0389  e and of allied species have descended from a  SINGLE  parent: and therefore have all proceeded from
0393  finished my search: as yet I have not found a  SINGLE  instance, free from doubt, of a terrestrial m
0397  s must not be overlooked. I will here give a  SINGLE  instance of one of the cases of difficulty. A
0405  he species of a genus having descended from a  SINGLE  parent, though now distributed to the most re
0407  by many naturalists under the designation of  SINGLE  centres of creation, by some general consider
0410  the general rule that the area inhabited by a  SINGLE  species, or by a group of species, is continu
0413  at we here have many species descended from a  SINGLE  progenitor grouped into genera: and the gener
0413  common to the dog genus, and then by adding a  SINGLE  sentence, a full description is given of each
0417  n found, that a classification founded on any  SINGLE  character, however important that may be, has
0420  ne. Now all these modified descendants from a  SINGLE  species, are represented as related in blood
0421  been but little modified, and they yet form a  SINGLE  genus. But this genus, though much isolated,
0424  ers, is known to every naturalist: scarcely a  SINGLE  fact can be predicated in common of the membe
0424  rose, or conversely, ranks them together as a  SINGLE  species, and gives a single definition. As so
0424  hem together as a single species, and gives a  SINGLE  definition. As soon as three Orchidean forms
0424  me spike, they were immediately included as a  SINGLE  species. But it may be asked, what ought we t
0426  rent. We may err in this respect in regard to  SINGLE  points of structure, but when several charact
0426  ommunity of descent. Let two forms have not a  SINGLE  character in common, yet if these extreme for
0429  at the discovery of Australia has not added a  SINGLE  insect belonging to a new order: and that in
0429  n represented by a dozen species instead of a  SINGLE  one: but such richness in species, as I find
0441  stage have three pairs of legs, a very simple  SINGLE  eye, and a probosciformed mouth, with which c
0441  r two eyes are now reconverted into a minute,  SINGLE, and very simple eye spot. In this last and c
```

Page **************************************(Key Word)**************************************

Page **(Key Word)**

0472 s being produced in such vast numbers for one SINGLE act, and being then slaughtered by their ster
0475 supplanted by new and improved forms. Neither SINGLE species nor groups of species reappear when t
0415 ular aestivation. Any one of these characters SINGLY is frequently of more than generic importance
0009 ted on this curious subject; but to show how SINGULAR the laws are which determine the reproductio
0021 t of a finch; and the common tumbler has the SINGULAR and strictly inherited habit of flying at a
0070 its preservation, explains, I believe, some SINGULAR facts in nature, such as that of very rare p
0144 able to assign any reason. What can be more SINGULAR than the relation between blue eyes and deaf
0218 e; and this perhaps explains the origin of a SINGULAR instinct in the allied group of ostriches. F
0231 al basis being about twice as thick. By this SINGULAR manner of building, strength is continually
0250 se of the Crinum leads me to refer to a most SINGULAR fact, namely, that there are individual plan
0258 casionally in a higher degree. Several other SINGULAR rules could be given from Gartner: for insta
0260 iffer in fertility. Now do these complex and SINGULAR rules indicate that species have been endowe
0371 tlantic Ocean. We can further understand the SINGULAR fact remarked on by several observers, that
0396 ithstanding the presence of aerial bats, the SINGULAR proportions of certain orders of plants, her
0021 s, others very long wings and tails, others SINGULAPLY short tails. The barb is allied to the carr
0295 same movement of subsidence will often tend to SINK the area whence the sediment is derived, and t
0397 lti their eggs, at least such as I have tried, SINK in sea water and are killed by it. Yet there m
0297 ation. Some experienced conchologists are now SINKING many of the very fine species of D'Orbigny a
0428 e in any one class, by arbitrarily raising or SINKING the value of the groups in other classes (an
0155 , which is generally of generic value, when it SINKS in value and becomes only of specific value,
0165 ntly produced from the mare by a black Arabian SIRE, were much more plainly barred across the legs
0202 slaughtered by their industrious and sterile SISTERS? It may be difficult, but we ought to admire
0253 interbreeding. On the contrary, brothers and SISTERS have usually been crossed in each successive
0253 if we were to act thus, and pair brothers and SISTERS in the case of any pure animal, which from a
0472 , and being then slaughtered by their sterile SISTERS; at the astonishing waste of pollen by our f
0030 , with everlasting layers which never desire to SIT, and with the bantam so small and elegant; whe
0215 or never become broody, that is, never wish to SIT on their eggs. Familiarity alone prevents our
0217 ; so that, if she were to make her own nest and SIT on her own eggs, those first laid would have t
0091 ion of the individuals best fitted for the two SITES, two varieties might slowly be formed. These
0448 cirripedes. The adult might become fitted for SITES or habits, in which organs of locomotion or o
0229 ermilion wax having been left ungnawed, were SITUATED, as far as the eye could judge, exactly alon
0393 ls kept by the natives) inhabiting an island SITUATED above 300 miles from a continent or great co
0393 r great continental island; and many islands SITUATED at a much less distance are equally barren.
0395 alian faunas. On either side the islands are SITUATED on moderately deep submarine banks; and they
0397 only one, that of the Galapagos Archipelago, SITUATED under the equator, between 500 and 600 miles
0400 een expected on my view, for the islands are SITUATED so near each other that they would almost ce
0400 , how has it happened in the several islands SITUATED within sight of each other, having the same
0212 rds, which vary partly in dependence on the SITUATIONS chosen and on the nature and temperature of
0344 ndants, lingering in protected and isolated SITUATIONS. When a group has once wholly disappeared,
0031 n feather in three years, but it would take him SIX years to obtain head and beak. In Saxony the i
0071 n the insects must have been still greater, for SIX insectivorous birds were very common in the. pl
0072 ing from the rings of growth; had during twenty SIX years tried to raise its head above the stems
0099 abbage flower is surrounded not only by its own SIX stamens, but by those of the many other flower
0121 tal lines. After fourteen thousand generations, SIX new species, marked by the letters n14 to z14,
0122 a14 to m14); and (I) will have been replaced by SIX (n14 to z14) new species. But we may go furthe
0123 itute a sub genus or even a distinct genus. The SIX descendants from (I) will form two sub genera
0123 t the extreme points of the original genus, the SIX descendants from (I) will, owing to inheritanc
0123 xtinct, and have left no descendants. Hence the SIX new species descended from (I), and the eight
0225 an hexagonal prism, with the basal edges of its SIX sides bevelled so as to join on to a pyramid,
0226 t some lesser distance) from the centres of the SIX surrounding spheres in the same layer; and at
0248 them from a cross with either pure parent, for SIX or seven, and in one case for ten generations,
0341 ave descended from some one species; so that if SIX genera, each having eight species, be found in
0341 , and in the next succeeding formation there be SIX other allied or representative genera with the
0341 y conclude that only one species of each of the SIX older genera has left modified descendants, co
0341 has left modified descendants, constituting the SIX new genera. The other seven species of the old
0341 o or three species of two or three alone of the SIX older genera will have been the parents of the
0341 older genera will have been the parents of the SIX new genera; the other old species and the othe
0378 ker, that all the flowering plants, about forty SIX in number, common to Tierra del Fuego and to E
0378 same with that now felt there at the height of SIX or seven thousand feet. During this the coldes
0385 ngth of time a duck or heron might fly at least SIX or seven hundred miles, and would be sure to a
0386 /4 ounces; I kept it covered up in my study for SIX months, pulling up and counting each plant as
0398 amp of the American continent. There are twenty SIX land birds, and twenty five of these are ranke
0441 o the chrysalis stage of butterflies, they have SIX pairs of beautifully constructed natatory legs
0445 ut on actually measuring the old dogs and their SIX days old puppies, I found that the puppies had
0159 es. The frequent presence of fourteen or even SIXTEEN tail feathers in the pouter, may be consider
0241 any were five feet four inches high, and many SIXTEEN feet high; but we must suppose that the larg
0208 hich had completed its hammock up to, say, the SIXTH stage of construction, and put it into a hamm
0208 lar simply re performed the fourth, fifth, and SIXTH stages of construction. If however, a caterpi
0208 tage, and were put into one finished up to the SIXTH stage, so that much of its work was already d
0228 an ordinary cell), and were to them in depth about one SIXTH of the diameter of the sphere of which they f
0279 the lowest known fossiliferous strata. In the SIXTH chapter I enumerated the chief objections whi
0334 eeded it. Thus, the species which lived at the SIXTH great stage of descent in the diagram are the
0093 e detected. Having found a female tree exactly SIXTY yards from a male tree, I put the stigmas of
0296 g strata, one above the other, at no less than SIXTY eight different levels. Hence, when the same
0011 nditions of life, as, in some cases, increased SIZE from amount of food, colour from particular ki
0021 r head over heels. The runt is a bird of great SIZE, with long, massive beak and large feet; some
0021 form a hood, and it has, proportionally to its SIZE, much elongated wing and tail feathers. The tr
0022 ive breadth and the presence of processes. The SIZE and shape of the apertures in the sternum are
0022 e; so is the degree of divergence and relative SIZE of the two arms of the furcula. The proportion
0022 rict correlation with the length of beak); the SIZE of the crop and of the upper part of the oesop
0022 birds are clothed when hatched. The shape and SIZE of the eggs vary. The manner of flight differs
0023 to ornithologists; and this, considering their SIZE, habits, and remarkable characters, seems very
0032 ery trifling instance, the steadily increasing SIZE of the common gooseberry may be quoted. We see
0033 the different kinds of gooseberries differ in SIZE, colour, shape, and hairiness; and yet the flo
0034 ion: the destruction of horses under a certain SIZE was ordered, and this may be compared to the r
0035 cehorses have come to surpass in fleetness and SIZE the parent Arab stock, so that the latter, by

Page **(Key Word)**

Page *********************************(Key Word)*********************************

```
0037 ng, may plainly be recognised in the increased   SIZE and beauty which we now see in the varieties o
0039 e saw a pigeon with a crop of somewhat unusual    SIZE; and the more abnormal or unusual any characte
0054 by them, and has little or no relation to the     SIZE of the genera to which the species belong. Aga
0055 d here again there is no close relation to the     SIZE of the genera. The cause of lowly organised pl
0056 n the lapse of time often increased greatly in     SIZE; and that large genera have often come to thei
0092 stamens and pistils placed, in relation to the    SIZE and habits of the particular insects which vis
0094 n to doubt that an accidental deviation in the     SIZE and form of the body, or in the curvature and
0114 nd that a piece of turf, three feet by four in     SIZE, which had been exposed for many years to exac
0135 ection increased in successive generations the    SIZE and weight of its body, its legs were used mor
0137 d of some burrowing rodents are rudimentary in    SIZE, and in some cases are quite covered up by ski
0137 with subterranean habits, a reduction in their    SIZE with the adhesion of the eyelids and growth of
0137 namely, the cave rat, the eyes are of immense     SIZE; and Professor Silliman thought that it regain
0138 th the loss of light and to have increased the    SIZE of the eyes; whereas with all the other inhabi
0147 e atrophied, the fruit itself gains largely in    SIZE and quality. In our poultry, a large tuft of f
0169 ny part or organ developed to an extraordinary    SIZE or in an extraordinary manner, in comparison w
0192 branchiae, simply through an increase in their    SIZE and the obliteration of their adhesive glands.
0198 , we have some reason to believe, increase the    SIZE of the chest; and again correlation would come
0220 a. the slaves are black and not above half the    SIZE of their red masters, so that the contrast in
0226 ich, from the spheres being nearly of the same    SIZE, is very frequently and necessarily the case,
0239 several British ants differ from each other in    SIZE and sometimes in colour; and that the extreme
0239 ll are numerous, with those of an intermediate    SIZE scanty in numbers, Formica flava has larger an
0240 and smaller workers, with some of intermediate    SIZE; and, in this species, as Mr. F. Smith has obs
0240 nted for merely by their proportionally lesser    SIZE; and I fully believe, though I dare not assert
0240 o positively, that the workers of intermediate    SIZE have their ocelli in an exactly intermediate c
0240 orkers in the same nest, differing not only in    SIZE, but in their organs of vision, yet connected
0241 regularly produce neuters, either all of large   SIZE with one form of jaw, or all of small size wit
0241 rge size with one form of jaw, or all of small    SIZE with jaws having a widely different structure;
0241 limax of difficulty, one set of workers of one    SIZE and structure, and simultaneously another set
0241 aneously another set of workers of a different    SIZE and structure; a graduated series having been
0261 ign no reason whatever. Great diversity in the    SIZE of two plants, one being woody and the other h
0265 health, and is often accompanied by excess of     SIZE or great luxuriance. In both cases, the steril
0283 o be worn away, atom by atom, until reduced in    SIZE they can be rolled about by the waves; and the
0299 tration. The Malay Archipelago is of about the    SIZE of Europe from the North Cape to the Mediterra
0341 ny extinct species which are closely allied in    SIZE and in other characters to the species still l
0387 hem; even small fish swallow seeds of moderate    SIZE, as of the yellow water lily and Potamogeton.
0387 t, many hours afterwards. When I saw the great    SIZE of the seeds of that fine water lily the Nelum
0389 wollaston for insects. If we look to the large    SIZE and varied stations of New Zealand, extending
0398 ic nature of the soil, in climate, height, and    SIZE of the islands, between the Galapagos and Cape
0412 cies, tend to go on increasing indefinitely in    SIZE. I further attempted to show that from the var
0428 minant groups thus tend to go on increasing in    SIZE; and they consequently supplant many smaller a
0429 it has added only two or three orders of small    SIZE. In the chapter on geological succession I att
0434 ts may change to almost any extent in form and    SIZE, and yet they always remain connected together
0441 h they feed largely, for they increase much in    SIZE. In the second stage, answering to the chrysal
0445 dth of mouth, length of nostril and of eyelid,    SIZE of feet and length of leg, in the wild stock,
0451 how many insects do we see wings so reduced in    SIZE as to be utterly incapable of flight, and not
0451 in the hybrid offspring was much increased in     SIZE; and this shows that the rudiment and the perf
0453 that a rudimentary part or organ is of greater    SIZE relatively to the adjoining parts in the embry
0455 o. thus we can understand the greater relative    SIZE of rudimentary organs in the embryo, and their
0455 rgans in the embryo, and their lesser relative    SIZE in the adult. But if each step of the process
0457 aring in mind, that when organs are reduced in    SIZE, either from disuse or selection, it will gene
0471 ll groups cannot thus succeed in increasing in    SIZE, for the world would not hold them, the more d
0471 ncy in the large groups to go on increasing in    SIZE and diverging in character, together with the
0093 es bear only female flowers; these have a full    SIZED pistil, and four stamens with shrivelled anth
0239 t often happens that the larger or the smaller    SIZED workers are the most numerous; or that both l
0451 y in all respects, one of which will have full    SIZED wings, and another mere rudiments of membrane
0129 ropped off; and these lost branches of various    SIZES may represent those whole orders, families, a
0225 cells are nearly spherical and of nearly equal    SIZES, and are aggregated into an irregular mass. B
0226 ce from each other, and had made them of equal    SIZES and had arranged them symmetrically in a doub
0227 o make her cells truly spherical, and of equal    SIZES; and this would not be very surprising, seein
0241 , moreover, of the working ants of the several    SIZES differed wonderfully in shape, and in the for
0241 orkers can be grouped into castes of different    SIZES, yet they graduate insensibly into each other
0241 had dissected from the workers of the several    SIZES. With these facts before me, I believe that n
0011 of the leg more, in proportion to the whole      SKELETON, than do the same bones in the wild duck; an
0022 nct breeds might have been specified. In the      SKELETONS of the several breeds, the development of t
0001 short notes; these I enlarged in 1844 into a      SKETCH of the conclusions, which then seemed to me p
0002 th knew of my work, the latter having read my     SKETCH of 1844, honoured me by thinking it advisable
0164 horse. My son made a careful examination and     SKETCH for me of a dun Belgian cart horse with a dou
0442 the fin of a porpoise, should not have been       SKETCHED out with all the parts in proper proportion,
0031 f, and then had given it existence. That most     SKILFUL breeder, Sir John Sebright, used to say, wit
0032 years of practice requisite to become even a     SKILFUL pigeon fancier. The same principles are foll
0224 eir construction. It has been remarked that a    SKILFUL workman, with fitting tools and measures, wo
0289 nd in many other parts of the world. The most    SKILFUL geologist, if his attention had been exclusi
0304 but my work had hardly been published, when a    SKILFUL palaeontologist, M. Bosquet, sent me a drawi
0324 ng in South America or in Australia, the most    SKILFUL naturalist would hardly be able to say wheth
0337 igher than those of New Zealand. Yet the most    SKILFUL naturalist from an examination of the specie
0037 ressed in horticultural works at the wonderful   SKILL of gardeners, in having produced such splendi
0177 tants are all trying with equal steadiness and   SKILL to improve their stocks by selection; the cha
0189 natural selection will pick out with unerring    SKILL each improvement. Let this process go on for
0250 accounted for by Herbert's great horticultural   SKILL, and by his having hothouses at his command.
0012 r many horns; pigeons with feathered feet have    SKIN between their outer toes; pigeons with short b
0013 must have heard of cases of albinism, prickly    SKIN, hairy bodies, etc., appearing in several membe
0021 the wonderful development of the carunculated    SKIN about the head, and this is accompanied by gre
0022 r of scutellae on the toes, the development of    SKIN between the toes, are all points of structure
0137 ize, and in some cases are quite covered up by    SKIN and fur. This state of the eyes is probably du
0144 ur with the female sex; the feathered feet and    SKIN between the outer toes in pigeons, and the pre
0180 part of their bodies rather wide and with the     SKIN on their flanks rather full, to the so called
```

Page *********************************(Key Word)*********************************

Page **(Key Word)**

0180 base of the tail united by a broad expanse of SKIN, which serves as a parachute and allows them t
0191 dunculated cirripedes have two minute folds of SKIN, called by me the ovigerous frena, which serve
0192 therefore I do not doubt that little folds of SKIN, which originally served as ovigerous frena, b
0197 modification and becoming a climber. The naked SKIN on the head of a vulture is generally looked a
0197 awing any such inference, when we see that the SKIN on the head of the clean feeding male turkey i
0426 which an insect's wing is folded, whether the SKIN be covered by hair or feathers, if it prevail
0473 itually blind and have their eyes covered with SKIN; or when we look at the blind animals inhabiti
0085 st, Downing, that in the United States smooth SKINNED fruits suffer far more from a beetle, a curc
0020 btain, and have been most kindly favoured with SKINS from several quarters of the world, more espe
0436 st physiologists believe that the bones of the SKULL are homologous with, that is correspond in nu
0438 kingdoms. Naturalists frequently speak of the SKULL as formed of metamorphosed vertebrae: the jaw
0021 rofessor Huxley has remarked, to speak of both SKULL and vertebrae, both jaws anc legs, etc., as h
0197 entailing corresponding differences in their SKULLS. The carrier, more especially the male bird,
0197 turkey is likewise naked. The sutures in the SKULLS of young mammals have been advanced as a beau
0197 ble for this act: but as sutures occur in the SKULLS of young birds and reptiles, which have only
0437 no means explain the same construction in the SKULLS of birds. Why should similar bones have been
0438 n the other, have actually been modified into SKULLS or jaws. Yet so strong is the appearance of a
0025 deserve consideration. The rock pigeon is of a SLATY blue, and has a white rump (the Indian sub sp
0159 he occasional appearance in all the breeds, of SLATY blue birds with two black bars on the wings,
0160 tions of life to cause the reappearance of the SLATY blue, with the several marks, beyond the infl
0065 , and we forget that thousands are annually SLAUGHTERED for food, and that in a state of nature an
0202 for any other end, and which are ultimately SLAUGHTERED by their industrious and sterile sisters?
0238 well marbled together; the animal has been SLAUGHTERED, but the breeder goes with confidence to t
0472 numbers for one single act, and being then SLAUGHTERED by their sterile sisters: at the astonishi
0207 ts of the cuckoo, ostrich, and parasitic bees. SLAVE making ants. Hive bee, its cell making instin
0216 koo to lay her eggs in other birds' nests: the SLAVE making instinct of certain ants: and the comb
0219 iously appropriated, be not thus exterminated. SLAVE making instinct. This remarkable instinct was
0219 at when Huber shut up thirty of them without a SLAVE, but with plenty of the food which they like
0219 shed of hunger. Huber then introduced a single SLAVE (F. fusca), and she instantly set to work, fe
0219 tained facts? If we had not known of any other SLAVE making ant, it would have been hopeless to ha
0219 likewise first discovered by P. Huber to be a SLAVE making ant. This species is found in the sout
0220 laves in all. Males and fertile females of the SLAVE species are found only in their own proper co
0220 al nests in Surrey and Sussex, and never saw a SLAVE either leave or enter a nest. As, during thes
0221 y attention was struck by about a score of the SLAVE makers haunting the same spot, and evidently
0221 ly repulsed by an independent community of the SLAVE species (F. fusca) sometimes as many as three
0221 hree of these ants clinging to the legs of the SLAVE making F. sanguinea. The latter ruthlessly ki
0222 f f. flava under a stone beneath a nest of the SLAVE making F. sanguinea: and when I had accidenta
0223 are of the larvae, and the masters alone go on SLAVE making expeditions. In Switzerland the slaves
0223 tend to conjecture. But as ants, which are not SLAVE makers, will as I have seen, carry off pupae
0219 ther. This ant is absolutely dependent on its SLAVES: without their aid, the species would certain
0219 gh most energetic and courageous in capturing SLAVES, do no other work. They are incapable of maki
0219 nvenient, and they have to migrate, it is the SLAVES which determine the migration, and actually c
0220 nary and odious an instinct as that of making SLAVES. Hence I will give the observations which I h
0220 urteen nests of F. sanguinea, and found a few SLAVES in all. Males and fertile females of the slav
0220 en observed in the nests of F. sanguinea. The SLAVES are black and not above half the size of thei
0220 eat. When the nest is slightly disturbed, the SLAVES occasionally come out, and like their masters
0220 bed and the larvae and pupae are exposed, the SLAVES work energetically with their masters in carr
0220 place of safety. Hence, it is clear, that the SLAVES feel quite at home. During the months of June
0220 or enter a nest. As, during these months, the SLAVES are very few in number, I thought that they m
0220 surrey and Hampshire, and has never seen the SLAVES, though present in large numbers in August, e
0220 hence he considers them as strictly household SLAVES. The masters, on the other hand, may be const
0221 a community with an unusually large stock of SLAVES, and I observed a few slaves mingled with the
0221 y large stock of slaves, and I observed a few SLAVES mingled with their masters leaving the nest,
0221 rtunities for observation, in Switzerland the SLAVES habitually work with their masters in making
0221 erence in the usual habits of the masters and SLAVES in the two countries, probably depends merely
0221 two countries, probably depends merely on the SLAVES being captured in greater numbers in Switzerl
0221 fully carrying, as Huber has described, their SLAVES in their jaws. Another day my attention was s
0221 e prevented from getting any pupae to rear as SLAVES. I then dug up a small parcel of the pupae of
0222 pecies is sometimes, though rarely, made into SLAVES, as has been described by Mr. Smith. Although
0222 of F. fusca, which they habitually make into SLAVES, from those of the little and furious F.flava
0223 in regard to the wonderful instinct of making SLAVES. Let it be observed what a contrast the insti
0223 f: it is absolutely dependent on its numerous SLAVES. Formica sanguinea, on the other hand, posses
0223 inea, on the other hand, possesses much fewer SLAVES, and in the early part of the summer extremel
0223 and when they migrate, the masters carry the SLAVES. Both in Switzerland and England the slaves s
0223 e slaves. Both in Switzerland and England the SLAVES seem to have the exclusive care of the larvae
0223 slave making expeditions. In Switzerland the SLAVES and masters work together, making and bringin
0223 materials for the nest: both, but chiefly the SLAVES, tend, and milk as it may be called, their ap
0223 ding materials and food for themselves, their SLAVES and larvae. So that the masters in this count
0223 country receive much less service from their SLAVES than they do in Switzerland. By what steps th
0224 ent for the very different purpose of raising SLAVES. When the instinct was once acquired, if carr
0224 which, as we have seen, is less aided by its SLAVES than the same species in Switzerland, I can s
0224 n ant was formed as abjectly dependent on its SLAVES as is the Formica rufescens. Cell making inst
0244 koo ejecting its foster brothers, ants making SLAVES, the larvae of Ichneumonidae feeding within t
0139 quisite for seeds to germinate, in the time of SLEEP, etc., and this leads me to say a few words o
0385 feet, which might represent those of a bird SLEEPING in a natural pond, in an aquarium, where man
0004 n in accumulating by his Selection successive SLIGHT variations. I will then pass on to the variab
0011 milar changes of structure. Nevertheless some SLIGHT amount of change may, I think, be attributed
0012 itable deviations of structure, both those of SLIGHT and those of considerable physiological impor
0015 ossing might check, by blending together, any SLIGHT deviations of structure, in such case, I gran
0019 ue, let the distinctive characters be ever so SLIGHT, has had its wild prototype. At this rate the
0022 he males and females have come to differ to a SLIGHT degree from each other. Altogether at least a
0029 , for they win their prizes by selecting such SLIGHT differences, yet they ignore all general argu
0029 rguments, and refuse to sum up in their minds SLIGHT differences accumulated during many successiv
0033 d hairiness, and yet the flowers present very SLIGHT differences. It is not that the varieties whi
0033 cannot doubt that the continued selection of SLIGHT variations, either in the leaves, the flowers
0038 reat in external characters and relatively so SLIGHT in internal parts or organs. Man can hardly s

Page **(Key Word)**

Page ********************************(Key Word)********************************

```
0039  riations which are first given to him in some  SLIGHT  degree by nature. No man would ever try to ma
0039  he saw a pigeon with a tail developed in some  SLIGHT  degree in an unusual manner, or a pouter till
0039  in human nature to value any novelty, however  SLIGHT, in one's own possession. Nor must the value
0039  the value which would formerly be set on any   SLIGHT  differences in the individuals of the same sp
0039  reeds have once fairly been established. Many  SLIGHT  differences might, and indeed do now, arise a
0040  erves and breeds from an individual with some  SLIGHT  deviation of structure, or takes more care th
0041  lways varied since it was cultivated, but the  SLIGHT  varieties had been neglected. As soon, howeve
0045  ould be called a variety. Again, we have many  SLIGHT  differences which may be called individual di
0051  for us, as being the first step towards such   SLIGHT  varieties as are barely thought worth recordi
0054  kely to yield offspring which, though in some  SLIGHT  degree modified, will still inherit those adv
0061  his struggle for life, any variation, however  SLIGHT  and from whatever cause proceeding, if it be
0061  . i have called this principle, by which each  SLIGHT  variation, if useful, is preserved, by the te
0061  to his own uses, through the accumulation of   SLIGHT  but useful variations, given to him by the ha
0069  titors be in the least degree favoured by any  SLIGHT  change of climate, they will increase in numb
0081  hat individuals having any advantage, however  SLIGHT, over others, would have the best chance of s
0082  n seized on by intruders. In such case, every  SLIGHT  modification, which in the course of ages cha
0082  gether with nicely balanced forces, extremely  SLIGHT  modifications in the structure or habits of o
0085  ed flesh. If, with all the aids of art, these  SLIGHT  differences make a great difference in cultiv
0085  at climate, food, etc., probably produce some  SLIGHT  and direct effect. It is, however, far more n
0089  les have had, in successive generations, some  SLIGHT  advantage over other males, in their weapons,
0091  an mice is shown to be inherited. Now, if any  SLIGHT  innate change of habit or of structure benefi
0094  t; now if we suppose this to occur in ever so  SLIGHT  a degree under nature, then as pollen is alre
0094  re and length of the proboscis, etc., far too  SLIGHT  to be appreciated by us, might profit a bee o
0094  ould probably inherit a tendency to a similar  SLIGHT  deviation of structure. The tubes of the coro
0111  arents of future well marked species, present  SLIGHT  and ill defined differences. Mere chance, as
0112  y horses. The early differences would be very  SLIGHT; in the course of time, from the continued se
0117  . the variations are supposed to be extremely  SLIGHT, but of the most diversified nature; they are
0124  of (F), either unaltered or altered only in a  SLIGHT  degree. In this case, its affinities to the o
0131  em to produce individual differences, or very  SLIGHT  deviations of structure, as to make the child
0132  ause for each deviation of structure, however  SLIGHT. How much direct effect difference of climate
0133  n of other species, often acquiring in a very  SLIGHT  degree some of the characters of such species
0134  accumulate all profitable variations, however  SLIGHT, until they become plainly developed and appr
0137  ed, after living some days in the light, some  SLIGHT  power of vision. In the same manner as in Mad
0143  during its growth and development, that when   SLIGHT  variations in any one part occur, and are acc
0148  becomes less useful, any diminution, however  SLIGHT, in its development, will be seized on by nat
0156  or come to differ in any degree, or only in a  SLIGHT  degree, it is not probable that they should v
0167  ate and food, etc., seem to have induced some  SLIGHT  modifications. Habit in producing constitutio
0170  of nature. Whatever the cause may be of each   SLIGHT  difference in the offspring from their parent
0178  we ought only to see a few species presenting  SLIGHT  modifications of structure in some degree per
0183  ge first and structure afterwards; or whether  SLIGHT  modifications of structure lead to changed ha
0189  here is a power always intently watching each  SLIGHT  accidental alteration in the transparent laye
0189  d. in living bodies, variation will cause the  SLIGHT  alterations, generation will multiply them al
0189  bly have been formed by numerous, successive,  SLIGHT  modifications, my theory would absolutely bre
0194  selection can act only by taking advantage of  SLIGHT  successive variations; she can never take a l
0195  economy of any one organic being, to say what  SLIGHT  modifications would be of importance or not.
0195  adapted for its present purpose by successive  SLIGHT  modifications, each better and better, for so
0196  y the same state, although now become of very  SLIGHT  use; and any actually injurious deviations in
0196  ning, as with the dog, though the aid must be  SLIGHT, for the hare, with hardly any tail, can doub
0197  e profoundly ignorant of the causes producing  SLIGHT  and unimportant variations; and we are immedi
0199  on our ignorance of the precise cause of the   SLIGHT  analogous differences between species. I migh
0209  ditions of life, it is at least possible that  SLIGHT  modifications of instinct might be profitable
0209  uced by the same unknown causes which produce  SLIGHT  deviations of bodily structure. No complex in
0210  he slow and gradual accumulation of numerous,  SLIGHT, yet profitable, variations. Hence, as in the
0214  we may believe that some one pigeon showed a   SLIGHT  tendency to this strange habit, and that the
0233  al selection acts only by the accumulation of  SLIGHT  modifications of structure or instinct, each
0234  would be an advantage to our humble bee, if a  SLIGHT  modification of her instinct led her to make
0235  ving taken advantage of numerous, successive,  SLIGHT  modifications of simpler instincts; natural s
0237  , by an individual having been born with some  SLIGHT  profitable modification of structure, this be
0237  hooked jaws of the male salmon. We have even   SLIGHT  differences in the horns of different breeds
0238  i believe it has been with social insects: a   SLIGHT  modification of structure, or instinct, corre
0239  of ordinary variations, that each successive,  SLIGHT, profitable modification did not probably ar
0242  be effected by the accumulation of numerous,   SLIGHT, and as we must call them accidental, variati
0243  ns of life, in natural selection accumulating  SLIGHT  modifications of instinct to any extent, in a
0251  state. Nevertheless these facts show on what   SLIGHT  and mysterious causes the lesser or greater f
0265  life have been disturbed, though often in so   SLIGHT  a degree as to be inappreciable by us; in the
0266  ink, on a considerable body of evidence, that  SLIGHT  changes in the conditions of life are benefic
0267  rogeny. Hence it seems that, on the one hand,  SLIGHT  changes in the conditions of life benefit all
0267  l organic beings, and on the other hand, that  SLIGHT  crosses, that is crosses between the males an
0269  e neither wishes to select, nor could select,  SLIGHT  differences in the reproductive system, or ot
0273  at arising from the mere act of crossing. The  SLIGHT  degree of variability in hybrids from the fir
0276  sterility is of all degrees, and is often so   SLIGHT  that the two most careful experimentalists wh
0277  ur and fertility of their offspring; and that  SLIGHT  changes in the conditions of life are apparen
0296  between them, but abrupt, though perhaps very  SLIGHT, changes of form. It is all important to reme
0297  species. It is notorious on what excessively   SLIGHT  differences many palaeontologists have founde
0301  interrupted by oscillations of level, and that  SLIGHT  climatal changes would intervene during such
0301  in a nearly uniform, though perhaps extremely  SLIGHT  degree, they would, according to the principl
0316  sure to inherit from its new progenitor some   SLIGHT  characteristic differences. Groups of species
0325  distribution of organic beings, and find how   SLIGHT  is the relation between the physical conditio
0333  ancient and extinct genera are often in some   SLIGHT  degree intermediate in character between thei
0333  ion, should in the older formations make some  SLIGHT  approach to each other; so that the older mem
0338  sses, though some of these old forms are in a  SLIGHT  degree less distinct from each other than are
0346  ll spot, having conditions peculiar in only a  SLIGHT  degree; for instance, small areas in the Old
0384  e the dispersal of fresh water fish mainly to  SLIGHT  changes within the recent period in the level
0417  n of many trifling points of resemblance, too  SLIGHT  to be defined. Certain plants, belonging to t
0428  classes have often been adapted by successive  SLIGHT  modifications to live under nearly similar ci
0429  orms of life often present characters in some  SLIGHT  degree intermediate between existing groups.
```

Page ********************************(Key Word)********************************

Page **(Key Word)**

```
0435  theory of the natural selection of successive  SLIGHT  modifications, each modification being profit
0438  nt of modification will have been effected by   SLIGHT  successive steps, we need not wonder at disco
0443  cting the embryo at a very early period, that   SLIGHT  variations necessarily appear at an equally e
0446  ith their diverse habits. Then, from the many   SLIGHT  successive steps of variation having superven
0450  al history, are explained on the principle of   SLIGHT  modifications not appearing, in the many desc
0457  mal and plant. On the principle of successive   SLIGHT  variations, not necessarily or generally supe
0459  eason, but by the accumulation of innumerable   SLIGHT  variations, each good for the individual poss
0459  that all organs and instincts are, in ever so   SLIGHT  a degree, variable, and, lastly, that there i
0461  tility of all organic beings are increased by   SLIGHT  changes in their conditions of life, and that
0467  cessive generation, individual differences so   SLIGHT  as to be quite inappreciable by an uneducated
0468  competition, or better adaptation in however   SLIGHT  a degree to the surrounding physical conditio
0469  istinction between individual differences and   SLIGHT  varieties; or between more plainly marked var
0470  a long continued course of modification, the   SLIGHT  differences, characteristic of varieties of t
0471  natural selection acts solely by accumulating   SLIGHT, successive, favourable variations, it can pr
0474  hoery of the natural selection of successive,   SLIGHT, but profitable modifications. We can thus un
0479  ent by the most permanent characters, however   SLIGHT  their vital importance may be. The framework
0479  selves on the theory of descent with slow and   SLIGHT  successive modifications. The similarity of p
0480  e preservation and accumulation of successive   SLIGHT  favourable variations. Why, it may be asked,
0481  add up and perceive the full effects of many   SLIGHT  variations, accumulated during an almost infi
0484  ure, and I speak after experience, will be no   SLIGHT  relief. The endless disputes whether or not s
0485  an it is at present; for differences, however   SLIGHT, between any two forms, if not blended by int
0041  closest attention should be paid to even the   SLIGHTEST  deviation in the qualities or structure of
0083  be plainly useful to him. Under nature, the   SLIGHTEST  difference of structure or constitution may
0084  oughout the world, every variation, even the   SLIGHTEST; rejecting that which is bad, preserving an
0133  shallower waters. When a variation is of the   SLIGHTEST  use to a being, we cannot tell how much of
0146  without being, as far as we can see, of the   SLIGHTEST  service to the species. We may often falsel
0468  ings most remote in the scale of nature. The   SLIGHTEST  advantage in one being, at any age or durin
0468  ence, or on the charms of the males; and the   SLIGHTEST  advantage will lead to victory. As geology
0005  t follows that any being, if it vary however   SLIGHTLY  in any manner profitable to itself, under th
0009  domestication or cultivation, and vary very   SLIGHTLY, perhaps hardly more than in a state of natu
0012  which the varieties and sub varieties differ   SLIGHTLY  from each other. The whole organisation seem
0021  nd it has the habit of continually expanding   SLIGHTLY  the upper part of the oesophagus. The Jacobi
0024  t pigeon, which is the rock pigeon in a very   SLIGHTLY  altered state, has become feral in several p
0029  though they well know that each race varies   SLIGHTLY, for they win their prizes by selecting such
0037  known variety, sowing its seeds, and, when a   SLIGHTLY  better variety has chanced to appear, select
0038  ced, individuals of the same species, having   SLIGHTLY  different constitutions or structure, would
0039  . the man who first selected a pigeon with a   SLIGHTLY  larger tail, never dreamed what the descenda
0040  ly have a distinct name, and from being only   SLIGHTLY  valued, their history will be disregarded. W
0041  gardeners picked out individual plants with   SLIGHTLY  larger, earlier, or better fruit, and raised
0048  north America and Europe, which differ very   SLIGHTLY  from each other, have been ranked by one emi
0049  r and emit a different odour; they flower at   SLIGHTLY  different periods; they grow in somewhat dif
0052  riety, from a state in which it differs very   SLIGHTLY  from its parent to one in which it differs m
0078  ess or dryness, for elsewhere it ranges into   SLIGHTLY  hotter or colder, damper or drier districts.
0095  a great advantage to the hive bee to have a   SLIGHTLY  longer or differently constructed proboscis.
0095  rvation of individuals presenting mutual and   SLIGHTLY  favourable deviations of structure. I am wel
0103  aunting different stations, from breeding at   SLIGHTLY  different seasons, or from varieties of the
0112  us. A fancier is struck by a pigeon having a   SLIGHTLY  shorter beak; another fancier is struck by a
0118  d. moreover, these two varieties, being only   SLIGHTLY  modified forms, will tend to inherit those a
0180  gradation from animals with their tails only   SLIGHTLY  flattened, and from others, as Sir J. Richar
0182  g fish, which now glide far through the air,   SLIGHTLY  rising and turning by the aid of their flutt
0184  lous habits, and with their structure either   SLIGHTLY  or considerably modified from that of their
0186  exist; if further, the eye does vary ever so   SLIGHTLY, and the variations be inherited, which is c
0192  s ovigerous frena, but which, likewise, very   SLIGHTLY  aided the act of respiration, have been grad
0198  t to natural selection, and individuals with   SLIGHTLY  different constitutions would succeed best u
0201  to make each organic being as perfect as, or   SLIGHTLY  more perfect than, the other inhabitants of
0206  aided in some cases by use and disuse, being   SLIGHTLY  affected by the direct action of the externa
0220  r appearance is very great. When the nest is   SLIGHTLY  disturbed, the slaves occasionally come out,
0227  ence we may safely conclude that if we could   SLIGHTLY  modify the instincts already possessed by th
0229  he base of a just commenced cell, which were   SLIGHTLY  concave on one side, where I suppose that th
0243  i have attempted to show that instincts vary   SLIGHTLY  in a state of nature. No one will dispute th
0267  he same species which have varied and become   SLIGHTLY  different, give vigour and fertility to the
0269  rations of hybrids, which were at first only   SLIGHTLY  sterile; and if this be so, we surely ought
0277  ind: namely, that the crossing of forms only   SLIGHTLY  different is favourable to the vigour and fe
0320  ith our domestic productions: when a new and   SLIGHTLY  improved variety has been raised, it at firs
0331  rian forms, that a species might go on being   SLIGHTLY  modified in relation to its slightly altered
0332  n being slightly modified in relation to its   SLIGHTLY  altered conditions of life, and yet retain t
0347  re the land almost joins, and where, under a   SLIGHTLY  different climate, there might have been fre
0356  what I mean: our English race horses differ   SLIGHTLY  from the horses of every other breed; but th
0368  y evidence with respect to this intercalated   SLIGHTLY  warmer period, since the Glacial period. The
0382  america, Australia, New Zealand have become   SLIGHTLY  tinted by the same peculiar forms of vegetab
0392  ome other means; and the plant then becoming   SLIGHTLY  modified, but still retaining its hooked see
0406  ers of the world, where they may have become   SLIGHTLY  modified in relation to their new conditions
0408  some groups of beings greatly, and some only   SLIGHTLY  modified, some developed in great force, som
0419  d, but because numerous allied species, with   SLIGHTLY  different grades of difference, have been su
0421  g genus F14 may be supposed to have been but   SLIGHTLY  modified; and it will then rank with the par
0448  ent, and consequently to be constructed in a   SLIGHTLY  different manner, then, on the principle of
0461  their constitutions having been disturbed by   SLIGHTLY  different and new conditions of life, we nee
0461  onditions of life, and that the offspring of   SLIGHTLY  modified forms or varieties acquire from bei
0482  se to extend the same view to other and very   SLIGHTLY  different forms. Nevertheless they do not pr
0090  see no reason to doubt that the swiftest and   SLIMMEST  wolves would have the best chance of survivi
0232  oper positions for working, for instance, on a   SLIP  of wood, placed directly under the middle of a
0232  the comb has to be built over one face of the   SLIP, in this case the bees can lay the foundations
0191  ata the branchiae have wholly disappeared, the   SLITS  on the sides of the neck and the loop like co
0440  like course of the arteries near the branchial   SLITS  are related to similar conditions, in the you
0479  air breathing mammal or bird having branchial   SLITS  and arteries running in loops, like those in
0285  or two thousand feet in height; for the gentle   SLOPE  of the lava streams, due to their formerly li
```

Page **(Key Word)**

Page *********************************(Key Word)*********************************

0341 ers have left behind them in South America the SLOTH, armadillo, and anteater, as their degenerate
0035 imes, the forms and qualities of their cattle. SLOW and insensible changes of this kind could neve
0040 disregarded. When further improved by the same SLOW and gradual process, they will spread more wid
0040 g and knowledge of any new sub breed will be a SLOW process. As soon as the points of value of the
0040 ll of any record having been preserved of such SLOW, varying, and insensible changes. I must now s
0046 f this nature could have been effected only by SLOW cegrees: yet quite recently Mr. Lubbock has sh
0056 e process of manufacturing new species to be a SLOW one. And this certainly is the case, if variet
0064 covered by the progeny of a single pair. Even SLOW breeding man has doubled in twenty five years,
0064 : if the statements of the rate of increase of SLOW breeding cattle and horses in South America, a
0065 hose which produce extremely few, is, that the SLOW breeders would require a few more years to peo
0080 tercrossing, isolation, number of individuals. SLOW action. Extinction caused by Natural Selection
0084 ic conditions of life. We see nothing of those SLOW changes in progress, until the hand of time ha
0087 age, the process of modification would be very SLOW, and there would be simultaneously the most ri
0103 arieties is thus lessened. Even in the case of SLOW breeding animals, which unite for each birth,
0108 on physical changes, which are generally very SLOW, and on the immigration of better adapted form
0108 d variation itself is apparently always a very SLOW process. The process will often be greatly ret
0108 e same time. I further believe, that this very SLOW, intermittent action of natural selection acco
0109 ch the inhabitants of this world have changed. SLOW though the process of selection may be, if fee
0148 habits of the Proteolepas, though effected by SLOW steps, would be a decided advantage to each su
0169 r parts; for variation is a long continued and SLOW process, and natural selection will in such ca
0169 d descendants, which on my view must be a very SLOW process, requiring a long lapse of time, in th
0177 re very slowly formed, for variation is a very SLOW process, and natural selection can do nothing
0178 nhabitants. And such new places will depend on SLOW changes of climate, or on the occasional immig
0203 ocess of natural selection will always be very SLOW, and will act, at any one time, only on a very
0210 duced through natural selection, except by the SLOW and gradual accumulation of numerous, slight,
0235 simpler instincts; natural selection having by SLOW degrees, more and more perfectly, led the bees
0284 accumulation in any one area must be extremely SLOW. But the amount of denudation which the strata
0290 rn away, as soon as they are brought up by the SLOW and gradual rising of the land within the grin
0291 plainly that each area has undergone numerous SLOW oscillations of level, and apparently these os
0295 in crder to have given sufficient time for the SLOW process of variation: hence the deposit will g
0302 t would be fatal to the theory of descent with SLOW modification through natural selection. For th
0302 me one progenitor, must have been an extremely SLOW process; and the progenitors must have lived l
0312 eological Succession of Organic Beings. On the SLOW and successive appearance of new species. On t
0312 immutability of species, or with that of their SLOW and gradual modification, through descent and
0314 the process of modification must be extremely SLOW. The variability of each species is quite inde
0317 production of a number of allied forms must be SLOW and gradual, one species giving rise first to
0317 pecies, which in their turn produce by equally SLOW steps other species, and so on, like the branc
0319 the analogy of all other mammals, even of the SLOW breeding elephant, and from the history of the
0322 nd in these intervals there may have been much SLOW extermination. Moreover, when by sudden immior
0326 es. The process of diffusion may often be very SLOW, being dependent on climatal and geographical
0336 find. We find, in short, such evidence of the SLOW and scarcely sensible mutation of specific for
0343 for the process of modification is necessarily SLOW, and depends on many complex contingencies. Th
0344 f a whole group of species may often be a very SLOW process, from the survival of a few descendant
0350 ay by checking migration; as does time for the SLOW process of mocification through natural select
0356 ch often intercross, I believe that during the SLOW process of modification the individuals of the
0372 as on the land, so in the waters of the sea, a SLOW southern migration of a marine fauna, which du
0434 plan of creation, we may hope to make sure but SLOW progress. Morphology. We have seen that the me
0462 e process of modification has necessarily been SLOW, all the means of migration will have been pos
0471 odification; it can act only by very short and SLOW steps. Hence the canon of Natura non facit sal
0475 gradual diffusion of dominant forms, with the SLOW modification of their descendants, causes the
0479 plain themselves on the theory of descent with SLOW and slight successive modifications. The simpl
0481 er and distinct species, is that we are always SLOW in admitting any great change of which we do n
0481 en formed, and great valleys excavated, by the SLOW action of the coast waves. The mind cannot pos
0488 ture existed, the rate of change may have been SLOW in an extreme degree. The whole history of the
0318 tion of the species of a group is generally a SLOWER process than their production: if the appeara
0321 n of a group is generally, as we have seen, a SLOWER process than its production. With respect to
0326 ding. The diffusion would, it is probable, be SLOWER with the terrestrial inhabitants of distinct
0406 ithin each great class, cenerally change at a SLOWER rate than the higher forms; and consequently
0488 and simpler, the rate of change was probably SLOWER; and at the first dawn of life, when very few
0064 on plants. The elephant is reckoned to be the SLOWEST breeder of all known animals, and I have tak
0194 a leap, but must advance by the shortest and SLOWEST steps. Organs of little apparent importance.
0035 rived from the spaniel, and has probably been SLOWLY altered from it. It is known that the English
0037 ount of change in our cultivated plants, thus SLOWLY and unconsciously accumulated, explains, as I
0040 s improves them, and the improved individuals SLOWLY spread in the immediate neighbourhood. But as
0040 the state of civilisation of the inhabitants, SLOWLY to add to the characteristic features of the
0043 y and more quickly, or unconsciously and more SLOWLY, but more efficiently, is by far the predomin
0091 fitted for the two sites, two varieties might SLOWLY be formed. These varieties would cross and bl
0094 e may suppose the plant of which we have been SLOWLY increasing the nectar by continued selection,
0095 i can understand how a flower and a bee might SLOWLY become, either simultaneously or one after th
0098 flower suddenly spring towards the pistil, or SLOWLY move one after the other towards it, the cont
0102 much improvement and modification surely but SLOWLY follow from this unconscious process of selec
0103 iety when once thus formed might subsequently SLOWLY spread to other districts. Cn the above princ
0105 ion, will give time for any new variety to be SLOWLY improved; and this may sometimes be of import
0107 than elsewhere; new forms will have been more SLOWLY formed, and old forms more slowly exterminate
0107 e been more slowly formed, and old forms more SLOWLY exterminated. And it is in fresh water that w
0108 er depend on some of the inhabitants becoming SLOWLY modified; the mutual relations of many of the
0108 e that natural selection will always act very SLOWLY, often only at long intervals of time, and ce
0109 an this; for as new forms are continually and SLOWLY being produced, unless we believe that the nu
0112 ished and distinct breeds. As the differences SLOWLY become greater, the inferior animals with int
0120 english pointer have apparently both gone on SLOWLY diverging in character from their original st
0125 g to increase in number. One large group will SLOWLY conquer another large group, reduce its numbe
0138 an animals, having ordinary powers of vision, SLOWLY migrated by successive generations from the o
0173 ited on the shallow bed of the sea, whilst it SLOWLY subsides. These contingencies will concur onl
0177 the less common forms, for these will be more SLOWLY modified and improved. It is the same princip
0177 inks: firstly, because new varieties are very SLOWLY formed, for variation is a very slow process,
0178 gree, on some of the old inhabitants becoming SLOWLY modified, with the new forms thus produced an

Page *********************************(Key Word)*********************************

Page ***(Key Word)***

```
0189 part of this layer to be continually changing SLOWLY in density, so as to separate into layers of
0189 ch other, and with the surfaces of each layer SLOWLY changing in form. Further we must suppose tha
0195 o an early progenitor, and, after having been SLOWLY perfected at a former period, have been trans
0205 ications in its structure could not have been SLOWLY accumulated by means of natural selection. Pu
0212 me enemy in other animals. But fear of man is SLOWLY acquired, as I have elsewhere shown, by vario
0213 rey, stand motionless like a statue, and then SLOWLY crawl forward with a peculiar gait; and anoth
0236 ngly assumed that all its characters had been SLOWLY acquired through natural selection; namely, b
0237 ed modifications of structure could have been SLOWLY accumulated by natural selection. This diffic
0238 xen with extraordinarily long horns, could be SLOWLY formed by carefully watching which individual
0269 ral habits of life. Nature acts uniformly and SLOWLY during vast periods of time on the whole orga
0272 not a special endowment, but is incidental on SLOWLY acquired modifications, more especially in th
0275 d do not relate to characters which have been SLOWLY acquired by selection. Consequently, sudden r
0275 ith hybrids, which are descended from species SLOWLY and naturally produced. On the whole I entire
0282 change, all changes having been effected very SLOWLY through natural selection. It is hardly possi
0283 bears the stamp of time, are good to show how SLOWLY the mass has been accumulated. Let him rememb
0291 extent over a shallow bottom, if it continue SLOWLY to subside. In this latter case, as long as t
0301 hen further modified and improved, they would SLOWLY spread and supplant their parent forms. When
0302 cies may elsewhere have long existed and have SLOWLY multiplied before they invaded the ancient ar
0311 here and there a few lines. Each word of the SLOWLY changing language, in which the history is su
0312 ral selection. New species have appeared very SLOWLY, one after another, both on the land and in t
0314 ntercrossing, on the rate of breeding, on the SLOWLY changing physical conditions of the country,
0315 ccasional scene, taken almost at hazard, in a SLOWLY changing drama. We can clearly understand why
0317 genera of the same family, can increase only SLOWLY and progressively; for the process of modific
0317 first to two or three varieties; these being SLOWLY converted into species, which in their turn p
0319 acted. If the conditions had gone on, however SLOWLY, becoming less and less favourable, we assure
0327 roups by inheritance, whole groups would tend SLOWLY to disappear; though here and there a single
0329 d if in both regions the species have gone on SLOWLY changing during the accumulation of the sever
0343 understand how it is that new species come in SLOWLY and successively; how species of different cl
0343 ppears. Groups of species increase in numbers SLOWLY, and endure for unequal periods of time; for
0366 ly, by supposing a new glacial period to come SLOWLY on, and then pass away, as formerly occurred.
0370 nimals, both in the Old and New Worlds, began SLOWLY to migrate southwards as the climate became l
0371 er migration of their inhabitants. During the SLOWLY decreasing warmth of the Pliocene period, as
0377 for any amount of migration. As the cold came SLOWLY on, all the tropical plants and other product
0384 econd place, salt water fish can with care be SLOWLY accustomed to live in fresh water; and, accor
0404 d it is obvious that a mountain, as it became SLOWLY upheaved, would naturally be colonised from t
0406 the relations just discussed, namely, low and SLOWLY changing organisms ranging more widely than t
0407 remember that some forms of life change most SLOWLY, enormous periods of time being thus granted
0422 l extinct languages, and all intermediate and SLOWLY changing dialects, had to be included, such a
0454 l in this case natural selection would continue SLOWLY to reduce the organ, until it rendered harmle
0463 change at any one period; and all changes are SLOWLY effected. I have also shown that the intermed
0465 ich my theory requires, for they have changed SLOWLY and in a graduated manner. We clearly see thi
0469 the bad? I can see no limit to this power, in SLOWLY and beautifully adapting each form to the mos
0472 h natural selection always ready to adapt the SLOWLY varying descendants of each to any unoccupied
0475 species. On the view of instincts having been SLOWLY acquired through natural selection we need no
0475 ification. New species have come on the stage SLOWLY and at successive intervals; and the amount o
0480 d me that species have changed, and are still SLOWLY changing by the preservation and accumulation
0487 . as species are produced and exterminated by SLOWLY acting and still existing causes, and not by
0108 tural selection will always act with extreme SLOWNESS, I fully admit. Its action depends on there
0283 i believe, be most deeply impressed with the SLOWNESS with which rocky coasts are worn away. The o
0067 57 no less than 295 were destroyed, chiefly by SLUGS and insects. If turf which has long been mown
0246 considered. The distinction has probably been SLURRED over, owing to the sterility in both cases b
0012 heir outer toes; pigeons with short beaks have SMALL feet; and those with long beaks large feet. H
0012 ve become plastic, and tends to depart in some SMALL degree from that of the parental type. Any va
0017 variability of the ass or guinea fowl, or the SMALL power of endurance of warmth by the rein deer
0030 ch never desire to sit, and with the bantam to SMALL and elegant; when we compare the host of agri
0037 at; though we owe our excellent fruit, in some SMALL degree, to their having naturally chosen and
0039 atch the fancier's eye: he perceives extremely SMALL differences, and it is in human nature to val
0040 they may be. But the chance will be infinitely SMALL of any record having been preserved of such s
0041 rally belong to poor people, and are mostly in SMALL lots, they never can be improved. On the othe
0051 ce i look at individual differences, though of SMALL interest to the systematist, as of high impor
0054 that I am surprised that my tables show even a SMALL majority on the side of the larger genera. I
0055 number of varieties than do the species of the SMALL genera. Both these results follow when anothe
0056 ing in the number of their species, or that no SMALL genera are now varying and increasing; for if
0056 ory; inasmuch as geology plainly tells us that SMALL genera have in the lapse of time often increa
0057 rence between the species is often exceedingly SMALL. I have endeavoured to test this numerically
0061 any species which are periodically born, but a SMALL number can survive. I have called this princi
0066 an in any way protect its own eggs or young, a SMALL number may be produced, and yet the average s
0070 tances, increases inordinately in numbers in a SMALL tract, epidemics, at least, this seems genera
0074 f cats; and Mr. Newman says, near villages and SMALL towns I have found the nests of humble bees m
0076 der the most different climates. In Russia the SMALL Asiatic cockroach has everywhere driven befor
0080 related to the diverstiy of inhabitants of any SMALL area, and to naturalisation. Action of Natura
0085 shed fruit, should succeed. In looking at many SMALL points of difference between species, which,
0092 leaf of the common laurel. This juice, though SMALL in quantity, is greedily sought by insects. L
0093 rs, which have four stamens producing rather a SMALL quantity of pollen, and a rudimentary pistil;
0095 eservation and accumulation of infinitesimally SMALL inherited modifications, each profitable to t
0101 as far as function is concerned, becomes very SMALL. From these several considerations and from t
0105 species. If, however, an isolated area be very SMALL, either from being surrounded by barriers, or
0105 duals supported on it will necessarily be very SMALL; and fewness of individuals will greatly reta
0105 st the truth of these remarks, and look at any SMALL isolated area, such as an oceanic island, alt
0105 the species inhabiting it, will be found to be SMALL, as we shall see in our chapter on geographic
0105 deceive ourselves; for to ascertain whether a SMALL isolated area, or a large open area like a co
0106 them will be more severe, on a large than on a SMALL and isolated area. Moreover, great areas, tho
0106 concurred. Finally, I conclude that, although SMALL isolated areas probably have been in some res
0106 become so largely naturalised on islands. On a SMALL island, the race for life will have been less
0107 all fresh water basins, taken together, make a SMALL area compared with that of the sea or of the
0114 er many natural circumstances. In an extremely SMALL area, especially if freely open to immigratio
```

Page ***(Key Word)***

Page ***(Key Word)***

```
0114  other. So it is with the plants and insects on   SMALL  and uniform islets; and so in small ponds of
0114  insects on small and uniform islets; and so in    SMALL  ponds of fresh water. Farmers find that they
0114  animals and plants which live close round any     SMALL  piece of ground, could live on it (supposing
0117  re of the species of large genera vary than of    SMALL  genera; and the varying species of the large
0117  the horizontal lines, and is there marked by a    SMALL  numbered letter, a sufficient amount of varia
0119  f succession is broken at regular intervals by    SMALL  numbered letters marking the successive forms
0120  rizontal line in our diagram to be excessively    SMALL, these three forms may still be only well mar
0120  the diagram illustrates the steps by which the    SMALL  differences distinguishing varieties are incr
0126  troy the earlier and less improved sub groups.    SMALL  and broken groups and sub groups will finally
0128  see proof by looking at the inhabitants of any    SMALL  spot or at naturalised productions. Therefore
0128  of succeeding in the battle of life. Thus the     SMALL  differences distinguishing varieties of the s
0129  nches, were themselves once, when the tree was    SMALL, budding twigs; and this connexion of the for
0130  ornithorhynchus or Lepidosiren, which in some     SMALL  degree connects by its affinities two large b
0132  my impression is, that the effect is extremely    SMALL  in the case of animals, but perhaps rather mo
0151  s within the same country vary in a remarkably    SMALL  degree, I have particularly attended to them,
0160  tendency to reversion is retained by this very    SMALL  proportion of foreign blood. In a breed which
0164  i can implicitly trust, has examined for me a     SMALL  dun Welch pony with three short parellel stri
0171  difficulty. Natura non facit saltum. Organs of    SMALL  importance. Organs not in all cases absolutel
0177  s improving their breeds more quickly than the    SMALL  holders on the intermediate narrow, hilly tra
0184  like a creeper. It often, like a shrike, kills    SMALL  birds by blows on the head; and I have many t
0187  on. Amongst existing Vertebrata, we find but a    SMALL  amount of gradation in the structure of the e
0188  of living crustaceans, and bearing in mind how    SMALL  the number of living animals is in proportion
0192  le surface of the body and sack, including the    SMALL  frena, serving for respiration. The Balanidae
0194  known forms to the extinct and unknown is very    SMALL, I have been astonished how rarely an organ o
0195  ould by any means defend themselves from these    SMALL  enemies, would be able to range into new past
0205  ial descendants), though it has become of such    SMALL  importance that it could not, in its present
0212  er wildness of all our large birds than of our    SMALL  birds; for the large birds have been most per
0212  islands large birds are not more fearful than     SMALL; and the magpie, so wary in England, is tame
0221  sanguinea. The latter ruthlessly killed their     SMALL  opponents, and carried their dead bodies as f
0221  g any pupae to rear as slaves. I then dug up a     SMALL  parcel of the pupae of F. fusca from another
0222  t. at the same time I laid on the same place a    SMALL  parcel of the pupae of another species, F. fl
0222  s has been described by Mr. Smith. Although so     SMALL  a species, it is very courageous, and I have
0239  are the most numerous; or that both large and     SMALL  are numerous, with those of an intermediate s
0240  orkers have simple eyes (ocelli), which though    SMALL  can be plainly distinguished; whereas the sma
0241  of large size with one form of jaw, or all of     SMALL  size with jaws having a widely different stru
0244  specially encowed or created instincts, but as    SMALL  consequences of one general law, leading to t
0258  the mother, generally differ in fertility in a    SMALL, and occasionally in a higher degree. Several
0286  a century. This will at first appear much too     SMALL  an allowance; but it is the same as if we wer
0287  w specimens collected on some one spot. Only a     SMALL  portion of the surface of the earth has been
0293  t it then became wholly extinct. We forget how    SMALL  the area of Europe is compared with the rest
0298  stages of transition between any two forms, is    SMALL, for the successive changes are supposed to h
0307  s conclusion. We should not forget that only a    SMALL  portion of the world is known with accuracy.
0307  d by metamorphic action, we ought to find only    SMALL  remnants of the formations next succeeding th
0330  merly quite distinct, at that period made some    SMALL  approach to each other. It is a common belief
0331  ee existing genera, a14, o14, p14, will form a    SMALL  family; b14 and f14 a closely allied family o
0335  ons inhabiting separated districts. To compare    SMALL  things with great: if the principal living an
0338  vered, a discovery of which the chance is very    SMALL. On the Succession of the same types within t
0341  cal record is extremely imperfect; that only a    SMALL  portion of the globe has been geologically ex
0346  e to find a group of organisms confined to any    SMALL  spot, having conditions peculiar in only a sl
0347  eculiar in only a slight degree; for instance,    SMALL  areas in the Old World could be pointed out h
0359  137 days. For convenience sake I chiefly tried    SMALL  seeds, without the capsule or fruit; and as a
0360  e larger fruits often floating longer than the    SMALL, is interesting; as plants with large seeds o
0361  ped stones are embedded in the roots of trees,    SMALL  parcels of earth are very frequently enclosed
0361  shed away in the longest transport: out of one    SMALL  portion of earth thus completely enclosed by
0361  den 12 kinds of seeds, out of the excrement of    SMALL  birds, and these seemed perfect, and some of
0364  itself a rare accident. Even in this case, how    SMALL  would the chance be of a seed falling on favo
0387  other kinds after having swallowed them; even     SMALL  fish swallow seeds of moderate size, as of th
0388  upying any pond, yet as the number of kinds is    SMALL, compared with those on the land, the competi
0394  the arctic regions. Yet it cannot be said that    SMALL  islands will not support small mammals, for t
0394  ot be said that small islands will not support    SMALL  mammals, for they occur in many parts of the
0394  they occur in many parts of the world on very    SMALL  islands, if close to a continent; and hardly
0400  est continent, we sometimes see displayed on a    SMALL  scale, yet in a most interesting manner, with
0400  ve been differently modified, though only in a    SMALL  degree. This long appeared to me a great diff
0412  t diversity of the forms of life which, in any    SMALL  area, come into the closest competition, and
0428  taceans agree in so many characters, great and    SMALL, that we cannot doubt that they have inherite
0429  oker, it has added only two or three orders of    SMALL  size. In the chapter on geological succession
0432  the characters of each group, whether large or    SMALL, and thus give a general idea of the value of
0454  others, as with the wings of beetles living on    SMALL  and exposed islands; and in this case natural
0454  function, which can be effected by insensibly     SMALL  steps, is within the power of natural selecti
0464  lassed as another and distinct species. Only a    SMALL  portion of the world has been geologically ex
0024  not been exterminated even on several of the      SMALLER  British islets; or on the shores of the Medi
0044  any country vary more than the species of the     SMALLER  genera. Many of the species of the larger ge
0054  eing placed on one side, and all those in the     SMALLER  genera on the other side, a somewhat larger
0055  er present varieties, than the species of the     SMALLER  genera; for wherever many closely related sp
0055  e larger genera on one side, and those of the     SMALLER  genera on the other side, and it has invaria
0055  ra present varieties, than on the side of the     SMALLER  genera. Moreover, the species of the large g
0057  le varieties, more than do the species of the     SMALLER  genera. Or the case may be put in another wa
0059  the larger genera also tend to break up into      SMALLER  genera. And thus, the forms of life througho
0106  on; for instance, that the productions of the     SMALLER  continent of Australia have formerly yielded
0126  increase of the larger groups, a multitude of     SMALLER  groups will become utterly extinct, and leav
0134  nd itself from enemies, as well as any of the     SMALLER  quadrupeds. We may imagine that the early pr
0177  ver the intermediate variety, which exists in     SMALLER  numbers in a narrow and intermediate zone. F
0205  the inhabitants of one country, generally the     SMALLER  one, will often yield, as we see they do yie
0239  kind. It often happens that the larger or the     SMALLER  sized workers are the most numerous; or that
0240  anty in numbers, Formica flava has larger and     SMALLER  workers, with some of intermediate size; and
0240  all can be plainly distinguished, whereas the     SMALLER  workers have their ocelli rudimentary. Havin
```

Page ***(Key Word)***

smaller

Page **(Key Word)**

```
0240  that the eyes are far more rudimentary in the SMALLER workers than can be accounted for merely by
0240  ndition. I may digress by adding, that if the SMALLER workers had been the most useful to the comm
0240  selected, which produced more and more of the SMALLER workers, until all the workers had come to b
0241  instead of three times as big as those of the SMALLER men, and jaws nearly five times as big. The
0387  e come into play with the eggs of some of the SMALLER fresh water animals. Other and unknown agenc
0394  nd hardly an island can be named on which our SMALLER quadrupeds have not become naturalised and q
0428  in size; and they consequently supplant many SMALLER and feebler groups. Thus we can account for
0470  s amount of difference than do the species of SMALLER genera. The closely allied species also of t
0055  n another division is made, and when all the SMALLEST genera, with from only one to four species,
0397  oceanic islands, even the most isolated and SMALLEST, are inhabited by land shells, generally by
0164  cream colour. I am aware that Colonel Hamilton SMITH, who has written on this subject, believes th
0219  and its habits have been attended to by Mr. F. SMITH, of the British Museum, to whom I am much ind
0220  ly trusting to the statements of Huber and Mr. SMITH, I tried to approach the subject in a sceptic
0220  behave differently when more numerous; but Mr. SMITH informs me that he has watched the nests at v
0222  made into slaves, as has been described by Mr. SMITH. Although so small a species, it is very cour
0239  of europe have been carefully examined. Mr. F. SMITH has shown how surprisingly the neuters of sev
0240  rmediate size; and, in this species, as Mr. F. SMITH has observed, the larger workers have simple
0240  pecies, that I gladly availed myself of Mr. F. SMITH's offer of numerous specimens from the same n
0283  ugh Miller, and by that excellent observer Mr. SMITH of Jordan Hill, are most impressive. With the
0085  culturist, Downing, that in the United States SMOOTH skinned fruits suffer far more from a beetle,
0085  d effectually settle which variety, whether a SMOOTH or downy, a yellow or purple fleshed fruit, s
0228  prism was built upon the festooned edge of a SMOOTH basin, instead of on the straight edges of a
0231  equally on both sides near the ground, till a SMOOTH, very thin wall is left in the middle; the ma
0285  f rocks on the one or other side having been SMOOTHLY swept away. The consideration of these facts
0075  r between insect and insect, between insects, SNAILS, and other animals with birds and beasts of p
0201  ome authors suppose that at the same time this SNAKE is furnished with a rattle for its own injury
0144  essure the shape of the head of the child. In SNAKES, according to Schlegel, the shape of the body
0149  e of the same individual (as the vertebrae in SNAKES, and the stamens in polyandrous flowers) the
0450  a digit in a rudimentary state: in very many SNAKES one lobe of the lungs is rudimentary; in othe
0450  ne lobe of the lungs is rudimentary; in other SNAKES there are rudiments of the pelvis and hind li
0161  ment inherited: for instance, in the common SNAPDRAGON (Antirrhinum) a rudiment of a fifth stamen
0452  ous individuals of the species. Thus in the SNAPDRAGON (antirrhinum) we generally do not find a ru
0091  ound and almost nightly catching woodcocks or SNIPES. The tendency to catch rats rather than mice
0069  mountain. When we reach the Arctic regions, or SNOW capped summits, or absolute deserts, the strug
0366  erish. The mountains would become covered with SNOW and ice, and their former Alpine inhabitants w
0367  ions of the more temperate regions. And as the SNOW melted from the bases of the mountains, the ar
0365  o see so many of the same plants living on the SNOWY regions of the Alps or Pyrenees, and in the e
0070  w spots where they do occur; and that of some SOCIAL plants being social, that is, abounding in in
0070  o occur; and that of some social plants being SOCIAL, that is, abounding in individuals, even on t
0087  nd of the parent in relation to the young. In SOCIAL animals it will adapt the structure of each i
0236  become sterile; and if such insects had been SOCIAL, and it had been profitable to the community
0238  ted its kind. Thus I believe it has been with SOCIAL insects: a slight modification of structure,
0238  ecies has been produced, which we see in many SOCIAL insects. But we have not as yet touched on th
0241  ow useful their production may have been to a SOCIAL community of insects, on the same principle t
0002  sir Charles Lyell, who sent it to the Linnean SOCIETY, and it is published in the third volume of
0002  ed in the third volume of the Journal of that SOCIETY. Sir C. Lyell and Dr. Hooker, who both knew
0144  ard parts seem to affect the form of adjoining SOFT parts; it is believed by some authors that the
0286  y two years. I doubt whether any rock, even as SOFT as chalk, would yield at this rate excepting o
0288  every year in Europe prove. No organism wholly SOFT can be preserved. Shells and bones will decay
0168  parts and in external parts sometimes affect SOFTER and internal parts. When one part is largely
0015  es, for instance, of the cabbage, in very poor SOIL (in which case, however, some effect would hav
0015  be attributed to the direct action of the poor SOIL), that they would to a large extent, or even w
0071  rally seen in passing from one quite different SOIL to another: not only the proportional numbers
0075  own, some of the varieties which best suit the SOIL or climate, or are naturally the most fertile,
0267  quent exchanges of seed, tubers, etc, from one SOIL or climate to another, and back again. During
0363  es: for many facts could be given showing that SOIL almost everywhere is charged with seeds. Refle
0364  the chance be of a seed falling on favourable SOIL, and coming to maturity! But it would be a gre
0391  m south America, and which has a very peculiar SOIL, does not possess one endemic land bird; and w
0398  e of resemblance in the volcanic nature of the SOIL, in climate, height, and size of the islands,
0392  instance, as the shrivelled wings under the SOLDERED elytra of many insular beetles. Again, islan
0451  nd not rarely lying under wing cases, firmly SOLDERED together! The meaning of rudimentary organs
0480  calf or like the shrivelled wings under the SOLDERING wing covers of some beetles, should thus so
0436  e complete abortion of certain parts, by the SOLDERING together of other parts, and by the doublin
0238  family. Thus in Eciton, there are working and SOLDIER neuters, with jaws and instincts extraordina
0092  ead of nectar; and as pollen is formed for the SOLE object of fertilisation, its destruction appea
0210  animal apparently performing an action for the SOLE good of another, with which I am acquainted, i
0211  e aphides do not instinctively excrete for the SOLE good of the ants. Although I do not believe th
0098  her towards it, the contrivance seems adapted SOLELY to ensure self fertilisation; and no doubt it
0104  ter can be given to their modified offspring; SOLELY by natural selection preserving the same favo
0109  ith natural selection. Natural selection acts SOLELY through the preservation of variations in som
0143  vious case is, that modifications accumulated SOLELY for the good of the young or larva, will, it
0149  orgotten, can act on each part of each being, SOLELY through and for its advantage. Rudimentary pa
0165  ommonly be called an accident, that I was led SOLELY from the occurence of the face stripes on thi
0172  sitional varieties. As natural selection acts SOLELY by the preservation of profitable modificatio
0182  rds might have existed which used their wings SOLELY as flappers, like the logger headed duck (Mic
0201  jurious to itself, for natural selection acts SOLELY by and for the good of each. No organ will be
0205  ired by natural selection, a power which acts SOLELY by the preservation of profitable variations
0214  ken of as actions which have become inherited SOLELY from long continued and compulsory habit, but
0331  racter is a necessary contingency; it depends SOLELY on the descendants from a species being thus
0340  w that marsupials should have been chiefly or SOLELY produced in Australia; or that Edentata and o
0340  ata and other American types should have been SOLELY produced in South America. For we know that E
0361  the Pacific, procure stones for their tools, SOLELY from the roots of drifted trees, these stones
0424  includes monsters; he includes varieties, not SOLELY because they closely resemble the parent form
0471  theory of creation. As natural selection acts SOLELY by accumulating slight, successive, favourabl
0489  ciable length. And as natural selection works SOLELY by and for the good of each being, all corpor
0290  diment must be accumulated in extremely thick, SOLID, or extensive masses, in order to withstand t
```

Page **(Key Word)**

solid

Page ***(Key Word)***

```
0224  rom mathematicians that bees have practically SOLVED a recondite problem, and have made their cell
0380  regions. Very many difficulties remain to be SOLVED. I do not pretend to indicate the exact lines
0381  have said that many difficulties remain to be SOLVED: some of the most remarkable are stated with
0065  a state of nature an equal number would have SOMEHOW to be disposed of. The only difference betwe
0227  this is the greatest difficulty, that she can SOMEHOW judge accurately at what distance to stand f
0307  ous strata, which on my theory no doubt were SOMEWHERE accumulated before the Silurian epoch, is v
0031  le whatever form and mould he pleases. Lord SOMERVILLE, speaking of what breeders have done for sh
0001  eturn home, it occurred to me, in 1837, that SOMETHING might perhaps be made out on this question
0021  pigeon Clubs. The diversity of the breeds is SOMETHING astonishing. Compare the English carrier an
0031  itually speak of an animal's organisation as SOMETHING quite plastic, which they can model almost
0040  read more widely, and will get recognised as SOMETHING distinct and valuable, and will then probab
0043  especially by that of correlation of growth. SOMETHING may be attributed to the direct action of
0043  the direct action of the conditions of life. SOMETHING must be attributed to use and disuse. The f
0054  inhabiting any country, shows that there is SOMETHING in the organic or inorganic conditions of t
0091  uice, apparently for the sake of eliminating SOMETHING injurious from their sap: this is effected
0112  our domestic productions. We shall here find SOMETHING analogous. A fancier is struck by a pigeon
0155  hysiological importance may remain the same. SOMETHING of the same kind applies to monstrosities:
0262  inct cases run to a certain extent parallel. SOMETHING analogous occurs in grafting; for Thouin fo
0319  this or that species is rare, we answer that SOMETHING is unfavourable in its conditions of life:
0319  ble in its conditions of life; but what that SOMETHING is, we can hardly ever tell. On the supposi
0331  l form an order; for all will have inherited SOMETHING in common from their ancient and common pro
0352  ssed over by migration, the fact is given as SOMETHING remarkable and exceptional. The capacity of
0357  lace, and when in the course of time we know SOMETHING definite about the means of distribution, w
0380  e last thirty or forty years from Australia. SOMETHING of the same kind must have occurred on the
0389  r in Australia, we must, I think, admit that SOMETHING quite independently of any difference in ph
0412  en parent, and, consequently, have inherited SOMETHING in common. Put the three genera on the left
0413  ndisputable. But many naturalists think that SOMETHING more is meant by the Natural System; they b
0413  nus gives the characters, seem to imply that SOMETHING more is included in our classification, tha
0413  ation, than mere resemblance. I believe that SOMETHING more is included; and that propinquity of d
0421  ified descendants from A will have inherited SOMETHING in common from their common parent, as will
0427  ssed an homopterous insect as a moth. We see SOMETHING of the same kind even in our domestic varie
0432  orms descended from A, or from I, would have SOMETHING in common. In a tree we can specify this or
0439  the lion. We occasionally though rarely see SOMETHING of this kind in plants: thus the embryonic
0485  nic being as a savage looks at a ship, as at SOMETHING wholly beyond his comprehension; when we re
0007  er conditions of life not so uniform as, and SOMEWHAT different from, those to which the parent sp
0016  aces of the same species, also, often have a SOMEWHAT monstrous character; by which I mean, that,
0028  ee how it is that the breeds so often have a SOMEWHAT monstrous character. It is also a most favou
0039  a pouter till he saw a pigeon with a crop of SOMEWHAT unusual size; and the more abnormal or unusu
0039  all fantails had only fourteen tailfeathers SOMEWHAT expanded, like the present Java fantail, or
0049  at slightly different periods; they grow in SOMEWHAT different stations: they ascend mountains to
0054  e in the smaller genera on the other side, a SOMEWHAT larger number of the very common and much di
0069  so it is when we travel northward, but in a SOMEWHAT lesser degree, for the number of species of
0111  es, even strongly marked ones, though having SOMEWHAT of the character of species, as is shown by
0118  sented in the diagram, though in itself made SOMEWHAT irregular. I am far from thinking that the m
0174  merous over a large territory, then becoming SOMEWHAT abruptly rarer and rarer on the confines, an
0188  ally infer that the eye has been formed by a SOMEWHAT analogous process. But may not this inferenc
0276  have been specially endowed with various and SOMEWHAT analogous degrees of difficulty in being gra
0280  pouter; none, for instance, combining a tail SOMEWHAT expanded with a crop somewhat enlarged, the
0280  mbining a tail somewhat expanded with a crop SOMEWHAT enlarged, the characteristic features of the
0287  , could hardly have been great, but it would SOMEWHAT reduce the above estimate. On the other hand
0294  for instance, that several species appeared SOMEWHAT earlier in the palaeozoic beds of North Amer
0297  y to each species, but when they meet with a SOMEWHAT greater amount of difference between any two
0324  would hereafter be liable to be classed with SOMEWHAT older European beds. Nevertheless, looking t
0330  same two groups would be distinguished by a SOMEWHAT lesser number of characters, so that the two
0331  descent with modification. As the subject is SOMEWHAT complex, I must request the reader to turn t
0363  d (as remarked by Mr. H. C. Watson) from the SOMEWHAT northern character of the flora in compariso
0368  ts of the mountains, the case will have been SOMEWHAT different; for it is not likely that all the
0369  lains; they will, also, have been exposed to SOMEWHAT different climatal influences. Their mutual
0372  e fauna, which during the Pliocene or even a SOMEWHAT earlier period, was nearly uniform along the
0377  tor, though they will have been placed under SOMEWHAT new conditions, will have suffered less. And
0385  er they could not be jarred off, though at a SOMEWHAT more advanced age they would voluntarily dro
0401  r, and it would be exposed to the attacks of SOMEWHAT different enemies. If then it varied, natura
0047  observer to suppose either that they do now SOMEWHERE exist, or may formerly have existed; and he
0464  for certainly on my theory such strata must SOMEWHERE have been deposited at these ancient and ut
0164  duns, and I have see a trace in a bay horse. My SON made a careful examination and sketch for me o
0076  rts of Scotland has caused the decrease of the SONG thrush. How frequently we hear of one species
0208  d be pointed out. As in repeating a well known SONG, so in instincts, one action follows another b
0208  ort of rhythm; if a person be interrupted in a SONG, or in repeating anything by rote, he is gener
0062  troying life; or we forget how largely these SONGSTERS, or their eggs, or their nestlings, are des
0176  le they should be exterminated and disappear, SOONER than the forms which they originally linked t
0316  till the group reaches its maximum, and then, SOONER or later, it gradually decreases. If the numb
0262  barren. On the other hand, certain species of SORBUS, when grafted on the same species, yielded twice
0304  ommencement of the tertiary series. This was a SORE trouble to me, adding as I thought one more in
0070  isproportionably favoured: and here comes in a SORT of struggle between the parasite and its prey.
0077  of many plants seems at first sight to have no SORT of relation to other plants. But from the stro
0208  in instincts, one action follows another by a SORT of rhythm; if a person be interrupted in a son
0232  round. The work of construction seems to be a SORT of balance struck between many bees, all insti
0238  e workers of one caste alone carry a wonderful SORT of shield on their heads, the use of which is
0239  enormously developed abdomen which secretes a SORT of honey, supplying the place of that excreted
0280  t place it should always be borne in mind what SORT of intermediate forms must, on my theory, have
0337  en forms: but I can see no way of testing this SORT of progress. Crustaceans, for instance, not th
0338  e adult. Thus the embryo comes to be left as a SORT of picture, preserved by nature, of the ancien
0398  rica. I believe this grand fact can receive no SORT of explanation on the ordinary view of indepen
0076  e, and if the seed or young were not annually SORTED. As species of the same genus have usually, t
0001  y patiently accumulating and reflecting on all SORTS of facts which could possibly have any bearin
0029  ws how utterly he disbelieves that the several SORTS, for instance a Ribston pippin or Codlin appl
```

Page ***(Key Word)***

Page **(Key Word)**

```
0149 the same way that a knife which has to cut all SORTS of things may be of almost any shape; whilst
0196 ight subsequently come to be worked in for all SORTS of purposes, as a fly flapper, an organ of pr
0237 ions and in those in a state of nature, of all SORTS of differences of structure which have become
0092 juice, though small in quantity, is greedily SOUGHT by insects. Let us now suppose a little sweet
0433 dden bond of connexion which naturalists have SOUGHT under the term of the Natural System. On this
0434 atural history, and may be said to be its very SOUL. What can be more curious than that the hand o
0047 r a variety, the opinion of naturalists having SOUND judgment and wide experience seems the only g
0187 se coarser vibrations of the air which produce SOUND. In looking for the gradations by which an or
0175 the same fact has been noticed by Forbes in SOUNDING the depths of the sea with the dredge. To th
0184 aerial and oceanic of birds, yet in the quiet SOUNDS of Tierra del Fuego, the Puffinuria berardi,
0009 ch produces sterility; and variability is the SOURCE of all the choicest productions of the garden
0016 sted between domestic races and species, this SOURCE of doubt could not so perpetually recur. It h
0248 nd varieties; but that the evidence from this SOURCE graduates away, and is doubtful in the same d
0286 e above estimate would be erroneous: but this SOURCE of doubt probably would not greatly affect th
0351 must originally have proceeded from the same SOURCE, as they have descended from the same progeni
0373 ge boulders transported far from their parent SOURCE. We do not know that the Glacial epoch was st
0383 st mainland. On colonisation from the nearest SOURCE with subsequent modification. Summary of the
0385 parent and must have proceeded from a single SOURCE, prevail throughout the world. Their distribu
0400 nly receive immigrants from the same original SOURCE, or from each other. But this dissimilarity b
0406 of colonisation from the nearest and readiest SOURCE, together with the subsequent modification an
0407 on my theory must have spread from one parent SOURCE: if we make the same allowances as before for
0408 allied species, have proceeded from some one SOURCE: then I think all the grand leading facts of
0409 y, to those of the nearest continent or other SOURCE whence immigrants were probably derived. We c
0477 will have received colonists from some third SOURCE or from each other, at various periods and in
0478 are related to the inhabitants of the nearest SOURCE whence immigrants might have been derived. We
0390 has unintentionally stocked them from various SOURCES far more fully and perfectly than has nature
0001 acts in the distribution of the inhabitants of SOUTH America and in the geological relations of th
0034 attested by passages in Pliny. The savages in SOUTH Africa match their draught cattle by colour,
0064 increase of slow breeding cattle and horses in SOUTH America, and latterly in Australia, had not b
0067 e especially in regard to the feral animals of SOUTH America. Here I will make only a few remarks,
0069 r, which will suffer most. When we travel from SOUTH to north, or from a damp region to a dry, we
0073 y alter (as indeed I have observed in parts of SOUTH America) the vegetation: this again would lar
0134 everal in this state. The loggerheaded duck in SOUTH America can only flap along the surface of th
0137 se, but aided perhaps by natural selection. In SOUTH America, a burrowing rodent, the tuco tuco, o
0141 faroe in the north and of the Falklands in the SOUTH, and on many islands in the torrid zones. Hen
0164 y in the north to the Malay Archipelago in the SOUTH. In all parts of the world these stripes occu
0173 ake a simple case: in travelling from north to SOUTH over a continent, we generally meet at succes
0183 tyrant flycatcher (Saurophagus sulphuratus) in SOUTH America, hovering over one spot and then proc
0195 n and existence of cattle and other animals in SOUTH America absolutely depends on their power of
0243 e of inheritance, how it is that the thrush of SOUTH America lines its nest with mud, in the same
0254 ception of certain indigenous domestic dogs of SOUTH America, all are quite fertile together; and
0268 ly than other dogs with foxes, or that certain SOUTH American indigenous domestic dogs do not read
0285 on the North Downs and to look at the distant SOUTH Downs; for, remembering that at no great dist
0290 more when examining many hundred miles of the SOUTH American coasts, which have been upraised sev
0290 along the rising coast of the western side of SOUTH America, no extensive formations with recent
0291 ncient tertiary formation on the west coast of SOUTH America, which has been bulky enough to resis
0300 t coast action, as we now see on the shores of SOUTH America. During the periods of subsidence the
0303 extinct species have been discovered in India, SOUTH America, and in Europe even as far back as th
0305 ark that hardly any fossil fish are known from SOUTH of the equator: and by running through Pictet
0305 the world have always been so freely open from SOUTH to north as they are at present. Even at this
0309 as in some parts of the world, for instance in SOUTH America, of bare metamorphic rocks, which mus
0318 , since its introduction by the Spaniards into SOUTH America, has run wild over the whole country
0319 of the naturalisation of the domestic horse in SOUTH America, that under more favourable condition
0323 found: namely, in North America, in equatorial SOUTH America, in Tierra del Fuego, at the Cape of
0324 , were to be compared with those now living in SOUTH America or in Australia, the most skilful nat
0324 and strictly modern beds, of Europe, North and SOUTH America, and Australia, from containing fossi
0339 to the living marsupials of that continent. In SOUTH America, a similar relationship is manifest,
0339 , buried there in such numbers, are related to SOUTH American types. This relationship is even mor
0339 e present climate of Australia and of parts of SOUTH America under the same latitude, would attemp
0340 ican types should have been solely produced in SOUTH America. For we know that Europe in ancient t
0341 allied huge monsters have left behind them in SOUTH America the sloth, armadillo, and anteater, a
0341 ther characters to the species still living in SOUTH America; and some of these fossils may be the
0341 , as apparently is the case of the Edentata of SOUTH America, still fewer genera and species will
0347 we compare large tracts of land in Australia, SOUTH Africa, and western South America, between la
0347 f land in Australia, South Africa, and western SOUTH America, between latitudes 25 degrees and 35
0347 ar. Or again we may compare the productions of SOUTH America south of lat. 35 degrees with those n
0347 e may compare the productions of South America SOUTH of lat. 35 degrees with those north of 25 deg
0347 ween the inhabitants of Australia, Africa, and SOUTH America under the same latitude: for these co
0348 han those of the eastern and western shores of SOUTH and Central America; yet these great faunas a
0349 ist in travelling, for instance, from north to SOUTH never fails to be struck by the manner in whi
0352 ingle mammal common to Europe and Australia or SOUTH America? The conditions of life are nearly th
0367 , covering the central parts of Europe, as far SOUTH as the Alps and Pyrenees, and even stretching
0367 returning warmth, will generally have been due SOUTH and north. The Alpine plants, for example, of
0371 ich inhabited the New and Old Worlds, migrated SOUTH of the Polar Circle, they must have been comp
0372 if we compare, for instance, certain parts of SOUTH America with the southern continents of the O
0373 aw maize growing on gigantic ancient moraines. SOUTH of the equator, we have some direct evidence
0373 have direct evidence of glacial action in the SOUTH eastern corner of Australia. Looking to Ameri
0373 have been observed on the eastern side as far SOUTH as lat. 36 degrees to 37 degrees, and on the
0373 where the climate is now so different, as far SOUTH as lat. 46 degrees: erratic boulders have, al
0373 cky Mountains. In the Cordillera of Equatorial SOUTH America, glaciers once extended far below the
0373 , left far below any existing glacier. Further SOUTH on both sides of the continent, from lat. 41
0379 have apparently migrated from the north to the SOUTH, than in a reversed direction. We see, howeve
0379 that this preponderant migration from north to SOUTH is due to the greater extent of land in the n
0381 elonging to genera exclusively confined to the SOUTH, at these and other distant points of the sou
0382 g one short period from the north and from the SOUTH, and to have crossed at the equator; but to h
0382 m the north so as to have freely inundated the SOUTH. As the tide leaves its drift in horizontal l
```

Page **(Key Word)**

Page ***(Key Word)***

```
0391  north America as the Galapagos Islands do from SOUTH America, and which has a very peculiar soil,
0397  , between 500 and 600 miles from the shores of SOUTH America. Here almost every product of the Lan
0398  which resembles closely the conditions of the SOUTH American coast: in fact there is a considerab
0399  en expected: but it is also plainly related to SOUTH America, which, although the next nearest con
0399  disappears on the view that both New Zealand,  SOUTH America, and other southern lands were long a
0399  y dr. Hooker is real, between the flora of the SOUTH western corner of Australia and of the Cape o
0403  s under the same conditions of life. Thus, the SOUTH east and south west corners of Australia have
0403  e conditions of life. Thus, the south east and SOUTH west corners of Australia have nearly the sam
0403  e of the surrounding lowlands; thus we have in  SOUTH America, Alpine humming birds, Alpine rodents
0408  hat under different latitudes, for instance in  SOUTH America, the inhabitants of the plains and mo
0475  w nearly the same instincts; why the thrush of  SOUTH America, for instance, lines her nest with mu
0074  growing on the ancient Indian mounds, in the    SOUTHERN United States, display the same beautiful di
0132  rbes speaks confidently that shells at their    SOUTHERN limit, and when living in shallow water, are
0142  mmended for the northern, and others for the    SOUTHERN States; and as most of these varieties are o
0212  ests of the same species in the northern and    SOUTHERN United States. Fear of any particular enemy
0219  ave making ant. This species is found in the    SOUTHERN parts of England, and its habits have been a
0285  great distance to the west the northern and     SOUTHERN escarpments meet and close, one can safely p
0286  ation. The distance from the northern to the    SOUTHERN Downs is about 22 miles, and the thickness o
0306  oler climate, and were enabled to double the    SOUTHERN capes of Africa or Australia, and thus reach
0324  f europe resembled most closely those of the    SOUTHERN hemisphere. So, again, several highly compet
0337  , and from hardly a single inhabitant of the    SOUTHERN hemisphere having become wild in any part of
0340  ook strongly of the present character of the    SOUTHERN half of the continent; and the southern half
0340  the southern half of the continent; and the     SOUTHERN half was formerly more closely allied, than
0346  al parts of the United States to its extreme    SOUTHERN point, we meet with the most diversified con
0347  fferent are their living productions! In the     SOUTHERN hemisphere, if we compare large tracts of la
0353  at these distant points of the northern and     SOUTHERN hemispheres? The answer, as I believe, is, t
0366  urred. As the cold came on, and as each more    SOUTHERN zone became fitted for arctic beings and ill
0367  in north America than in Europe, so will the    SOUTHERN migration there have been a little earlier o
0368  period. The arctic forms, during their long    SOUTHERN migration and re migration northward, will h
0372  he land, so in the waters of the sea, a slow    SOUTHERN migration of a marine fauna, which during th
0372  nce, certain parts of South America with the    SOUTHERN continents of the Old World, we see countrie
0374  er temperate zones, and on both sides of the    SOUTHERN extremity of the continent. If this be admit
0375  europe! Still more striking is the fact that    SOUTHERN Australian forms are clearly connected by d
0375  d on the other as far north as Japan. On the    SOUTHERN mountains of Australia, Dr. F. Muller has di
0375  n the temperate lowlands of the northern and    SOUTHERN hemispheres, are sometimes identically the s
0376  the northern species and forms found in the    SOUTHERN parts of the southern hemisphere, and on the
0376  and forms found in the southern parts of the    SOUTHERN hemisphere, and on the mountain ranges of th
0376  f the warmer regions of the earth and in the    SOUTHERN hemisphere are of doubtful value, being rank
0378  ing the glacial period from the northern and    SOUTHERN temperate zones into the intertropical regio
0379  scarcely any modification during their long    SOUTHERN migration and re migration northward, the ca
0379  s on the intertropical mountains, and in the    SOUTHERN hemisphere. These being surrounded by strang
0379  ritance to their brethren of the northern or    SOUTHERN hemispheres, now exist in their new homes as
0379  a reversed direction. We see, however, a few    SOUTHERN vegetable forms on the mountains of Borneo a
0379  of perfection or dominating power, than the    SOUTHERN forms. And thus, when they became commingled
0379  forms were enabled to beat the less powerful    SOUTHERN forms. Just in the same manner as we see at
0380  nt beaten the natives; whereas extremely few    SOUTHERN forms have become naturalised in any part of
0380  llied species which live in the northern and    SOUTHERN temperate zones and on the mountains of the
0381  th, at these and other distant points of the    SOUTHERN hemisphere, is, on my theory of descent with
0381  mon centre; and I am inclined to look in the    SOUTHERN, as in the northern hemisphere, to a former
0382  re widely dispersed to various points of the    SOUTHERN hemisphere by occasional means of transport,
0382  icebergs. By these means, as I believe, the    SOUTHERN shores of America, Australia, New Zealand ha
0387  states that he found the seeds of the great    SOUTHERN water lily (probably, according to Dr. Hooke
0399  t both New Zealand, South America, and other    SOUTHERN lands were long ago partially stocked from a
0477  e inhabitants of the sea in the northern and    SOUTHERN temperate zones, though separated by the who
0373  the continent, from lat. 41 degrees to the    SOUTHERNMOST extremity, we have the clearest evidence o
0069  other species will decrease. When we travel    SOUTHWARD and see a species decreasing in numbers, we
0072  r dogs have ever run wild, though they swarm    SOUTHWARD and northward in a feral state: and Azara a
0348  three marine faunas range far northward and    SOUTHWARD, in parallel lines not far from each other,
0366  perate regions would at the same time travel    SOUTHWARD, unless they were stopped by barriers, in w
0367  hich we suppose to have everywhere travelled    SOUTHWARD, are remarkably uniform round the world. We
0371  ago. And as the plants and animals migrated    SOUTHWARD, they will have become mingled in the one g
0373  rn whores of Britain to the Oural range, and    SOUTHWARD to the Pyrenees. We may infer, from the fro
0378  d the equator, would re migrate northward or    SOUTHWARD towards their former homes; but the forms,
0069  action of climate, than we do in proceeding    SOUTHWARDS or in descending a mountain. When we reach
0370  nts of the Old and New Worlds lived further    SOUTHWARDS than at present; they must have been still
0370  old and New Worlds, began slowly to migrate    SOUTHWARDS as the climate became less warm, long befor
0142  th much greater weight; but until some one will SOW, during a score of generations, his kidney bea
0037  in always cultivating the best known variety,   SOWING its seeds, and, when a slightly better variet
0113  species and each variety of grass is annually   SOWING almost countless seeds; and thus, as it may b
0071  ears large spaces have been enclosed, and self  SOWN firs are now springing up in multitudes, so cl
0072  scertained that these young trees had not been  SOWN or planted, I was so much surprised at their n
0075  for instance, if several varieties of wheat be  SOWN together, and the mixed seed be resown, some o
0077  uced from such seeds (as peas and beans); when  SOWN in the midst of long grass, I suspect that the
0113  rimentally proved, that if a plot of ground be  SOWN with one species of grass, and a similar plot
0113  th one species of grass, and a similar plot be  SOWN with several distinct genera of grasses, a gre
0113  hen several mixed varieties of wheat have been  SOWN on equal spaces of ground. Hence, if any one s
0237  ndividual is destroyed; but the horticulturist  SOWS seeds of the same stock, and confidently expec
0002  sibly be here done. I much regret that want of  SPACE prevents my having the satisfaction of acknow
0005  th, their geographical distribution throughout  SPACE; in the thirteenth, their classification or m
0031  they can model almost as they please. If I had  SPACE I could quote numerous passages to this effec
0089  e, or by the males and females; but I have not  SPACE here to enter on this subject. Thus it is, as
0100  ntion to the subject. Turning for a very brief  SPACE to animals: on the land there are some hermap
0121  d lines not prolonged far upwards from want of  SPACE. But during the process of modification, repr
0128  animals and all plants throughout all time and  SPACE should be related to each other in group subo
0201  order to warn the doomed mouse. But I have not  SPACE here to enter on this and other such cases. N
0211  ble ought to have been here given; but want of  SPACE prevents me. I can only assert, that instinct
```

Page ***(Key Word)***

Page ***(Key Word)***

```
0230  i am convinced of their accuracy; and if I had SPACE, I could show that they are conformable with
0233  e sometimes strictly hexagonal; but I have not SPACE here to enter on this subject. Nor does there
0248  so most instructive to compare, but I have not SPACE here to enter on details, the evidence advanc
0294  occurred in other parts of America during this SPACE of time. When such beds as were deposited in
0295  to enable the same species to live on the same SPACE, the supply of sediment must nearly have coun
0317  ally thickens upwards, sometimes keeping for a SPACE of equal thickness, and ultimately thins out
0348  ama. Westward of the shores of America, a wide SPACE of open ocean extends, with not an island as
0348  me to the shores of Africa; and over this vast SPACE we meet with no well defined and distinct mar
0350  some deep organic bond, prevailing throughout SPACE and time, over the same areas of land and wat
0352  or with an interval of such a nature, that the SPACE could not be easily passed over by migration,
0355  ies has come into existence coincident both in SPACE and time with a pre existing closely allied s
0360  r having been dried, could be floated across a SPACE of sea 900 miles in width, and would then ger
0394  strial mammal can be transported across a wide SPACE of sea, but bats can fly across. Bats have be
0395  hipelago, which is traversed near Celebes by a SPACE of deep ocean; and this space separates two w
0395  ear Celebes by a space of deep ocean; and this SPACE separates two widely distinct mammalian fauna
0409  lelism in the laws of life throughout time and SPACE: the laws governing the succession of forms i
0410  absent, but which occur above and below: so in SPACE, it certainly is the general rule that the ar
0410  t in the intermediate tracts. Both in time and SPACE, species and groups of species have their poi
0410  e same order, differ greatly. In both time and SPACE the lower members of each class generally cha
0410  ry these several relations throughout time and SPACE are intelligible; for whether we look to the
0410  will generally stand to each other in time and SPACE: in both cases the laws of variation have bee
0413  nless it be specified whether order in time or SPACE, or what else is meant by the plan of the Cre
0432  class which have lived throughout all time and SPACE. We shall certainly never succeed in making s
0476  the distribution of organic beings throughout SPACE, and in their geological succession throughou
0071  ant hilltops: within the last ten years large SPACES have been enclosed, and self sown firs are no
0113  ed varieties of wheat have been sown on equal SPACES of ground. Hence, if any one species of grass
0284  te erroneous; yet, considering over what wide SPACES very fine sediment is transported by the curr
0291  arently these oscillations have affected wide SPACES. Consequently formations rich in fossils and
0291  t degradation, may have been formed over wide SPACES during periods of subsidence, but only where
0328  ns have often been accumulated over very wide SPACES in the same quarter of the world; but we are
0351  d, so certain species have migrated over vast SPACES, and have not become greatly modified. On the
0359  w days, they could not be floated across wide SPACES of the sea, whether or not they were injured
0370  been still more completely separated by wider SPACES of ocean. I believe the above difficulty may
0377  ery long; and when we remember over what vast SPACES some naturalised plants and animals have spre
0378  this the coldest period, I suppose that large SPACES of the tropical lowlands were clothed with a
0477  see why those animals which cannot cross wide SPACES of ocean, as frogs and terrestrial mammals, s
0019  f germany and conversely, and so with Hungary, SPAIN, etc., but that each of these kingdoms posses
0035  gh the old Spanish pointer certainly came from SPAIN, Mr. Borrow has not seen, as I am informed by
0035  en, as I am informed by him, any native dog in SPAIN like our pointer. By a similar process of sel
0367  he alps and Pyrenees, and even stretching into SPAIN. The now temperate regions of the United Stat
0137  habits than the mole; and I was assured by a SPANIARD, who had often caught them, that they were f
0318  hat the horse, since its introduction by the SPANIARDS into South America, has run wild over the w
0017  hown that the greyhound, bloodhound, terrier, SPANIEL, and bull dog, which we all know propagate t
0019  nd, the bloodhound, the bull dog, or Blenheim SPANIEL, etc., so unlike all wild Canidae, ever exis
0035  here is reason to believe that King Charles's SPANIEL has been unconsciously modified to a large e
0035  that the setter is directly derived from the SPANIEL, and has probably been slowly altered from i
0035  and yet so effectually, that, though the old SPANISH pointer certainly came from Spain, Mr. Borro
0393  ome a nuisance. But as these animals and their SPAWN are known to be immediately killed by sea wat
0440  he bird which is hatched in a nest, and in the SPAWN of a frog under water. We have no more reason
0031  nd to inspect the animals. Breeders habitually SPEAK of an animal's organisation as something quit
0212  i can only repeat my assurance, that I do not SPEAK without good evidence. The possibility, or ev
0241  he widely different structure of their jaws. I SPEAK confidently on this latter point, as Mr. Lubb
0292  ing the alternate periods of elevation; or, to SPEAK more accurately, the beds which were then acc
0438  and vegetable kingdoms. Naturalists frequently SPEAK of the skull as formed of metamorphosed verte
0438  correct, as Professor Huxley has remarked, to SPEAK of both skull and vertebrae, both jaws and le
0484  in essence a species. This I feel sure, and I SPEAK after experience, will be no slight relief. T
0486  ach organic being, how far more interesting, I SPEAK from experience, will the study of natural hi
0486  e long been inherited. Rudimentary organs will SPEAK infallibly with respect to the nature of long
0031  form and mould he pleases. Lord Somerville, SPEAKING of what breeders have done for sheep, says:
0414  for classification. As an instance: Owen, in SPEAKING of the dugong, says: The generative organs b
0415  the highest authority, Robert Brown, who in SPEAKING of certain organs in the Proteaceae, says th
0031  o was himself a very good judge of an animal, SPEAKS of the principle of selection as that which e
0044  aturalist knows vaguely what he means when he SPEAKS of a species. Generally the term includes the
0129  y a great tree. I believe this simile largely SPEAKS the truth. The green and budding twigs may re
0132  buted to climate, food, etc.: thus, E. Forbes SPEAKS confidently that shells at their southern lim
0376  r part of the world. Sir J. Richardson, also, SPEAKS of the re appearance on the shores of New Zea
0088  y be as important for victory, as the sword or SPEAR. Amongst birds, the contest is often of a mor
0020  elieving that it is always best to study some SPECIAL group, I have, after deliberation, taken up
0036  e animal particularly useful to them, for any SPECIAL purpose, would be carefully preserved during
0055  e other hand, if we look at each species as a SPECIAL act of creation, there is no apparent reason
0088  l depend not on general vigour, but on having SPECIAL weapons, confined to the male sex. A hornles
0088  nimals, and these seem oftenest provided with SPECIAL weapons. The males of carnivorous animals ar
0088  ady well armed; though to them and to others, SPECIAL means of defence may be given through means
0098  sly in this very genus, which seems to have a SPECIAL contrivance for self fertilisation, it is we
0098  ng any odds for self fertilisation, there are SPECIAL contrivances, as I could show from the writi
0098  in very many other cases, though there be no SPECIAL mechanical contrivance to prevent the stigma
0101  hese several considerations and from the many SPECIAL facts which I have collected, but which I am
0141  ce i am inclined to look at adaptation to any SPECIAL climate as a quality readily grafted on an i
0149  fully than when the part has to serve for one SPECIAL purpose alone. In the same way that a knife
0168  period. In these remarks we have referred to SPECIAL parts or organs being still variable, becaus
0192  ectric organs of fishes offer another case of SPECIAL difficulty; it is impossible to conceive by
0200  he upland goose or of the frigate bird are of SPECIAL use to these birds; we cannot believe that t
0200  e bat, and in the flipper of the seal, are of SPECIAL use to these animals. We may safely attribut
0200  om a common progenitor, were formerly of more SPECIAL use to that progenitor, or its progenitors,
0200  ions) may be viewed, either as having been of SPECIAL use to some ancestral form, or as being now
0200  se to some ancestral form, or as being now of SPECIAL use to the descendants of this form, either
```

Page ***(Key Word)***

Page ***(Key Word)***

0236 several cases, but will confine myself to one SPECIAL difficulty, which at first appeared to me in
0242 ewise because this is by far the most serious SPECIAL difficulty, which my theory has encountered.
0245 ing the sterility of hybrids. Sterility not a SPECIAL endowment, but incidental on other differenc
0246 sterility in both cases being looked on as a SPECIAL endowment, beyond the province of our reason
0260 brids been permitted? to grant to species the SPECIAL power of producing hybrids, and then to stop
0263 crossing together various species has been a SPECIAL endowment; although in the case of crossing,
0272 sses and of hybrids, namely, that it is not a SPECIAL endowment, but is incidental on slowly acqui
0282 o study the principles of geology, or to read SPECIAL treatises by different observers on separate
0310 ure, have always seemed to me to require some SPECIAL explanation; and we may perhaps believe that
0325 must, as Barrande has remarked, look to some SPECIAL law. We shall see this more clearly when we
0414 ny part of the organisation is concerned with SPECIAL habits, the more important it becomes for cl
0430 his affinity in most cases is general and not SPECIAL; thus, according to Mr. Waterhouse, of all R
0439 embryos are active, and have been adapted for SPECIAL lines of life. A trace of the law of embryon
0440 s beautiful as in the adult animal. From such SPECIAL adaptations, the similarity of the larvae or
0459 selection. Recapitulation of the general and SPECIAL circumstances in its favour. Causes of the g
0460 eps. There are, it must be admitted, cases of SPECIAL difficulty on the theory of natural selectio
0460 vely to show that this sterility is no more a SPECIAL endowment than is the incapacity of two tree
0468 geny. But success will often depend on having SPECIAL weapons or means of defence, or on the charm
0469 ulties and objections: now let us turn to the SPECIAL facts and arguments in favour of the theory.
0469 s, commonly supposed to have been produced by SPECIAL acts of creation, and varieties which are ac
0481 es invariably fertile; or that sterility is a SPECIAL endowment and sign of creation. The belief t
0482 hich till lately they themselves thought were SPECIAL creations, and which are still thus looked a
0488 the individual. When I view all beings not as SPECIAL creations, but as the lineal descendants of
0190 n such cases natural selection might easily SPECIALISE, if any advantage were thus gained, a part
0149 ts of the organisation have been but little SPECIALISED for particular functions: and as long as t
0152 ns, and in those which have been but little SPECIALISED for any particular purpose, and perhaps in
0168 ing from such parts not having been closely SPECIALISED to any particular function, so that their
0168 se which have their whole organisation more SPECIALISED, and are higher in the scale. Rudimentary
0205 y different functions, and then having been SPECIALISED for one function; and two very distinct or
0082 that a change in the conditions of life, by SPECIALLY acting on the reproductive system, causes o
0115 digenes: for these are commonly looked at as SPECIALLY created and adapted for their own country.
0142 any breeds and sub breeds with constitutions SPECIALLY fitted for their own districts: the result
0193 ted that all electric fishes would have been SPECIALLY related to each other. Nor does geology at
0244 thin the live bodies of caterpillars, not as SPECIALLY endowed or created instincts, but as small
0245 s that species, when intercrossed, have been SPECIALLY endowed with the quality of sterility, in o
0245 , to be able to show that sterility is not a SPECIALLY acquired or endowed quality, but is inciden
0255 or not the rules indicate that species have SPECIALLY been endowed with this quality, in order to
0261 g incidental on other differences, and not a SPECIALLY endowed quality. As the capacity of one pla
0261 no one will suppose that this capacity is a SPECIALLY endowed quality, but will admit that it is
0265 e, to their reproductive systems having been SPECIALLY affected, though in a lesser degree than wh
0275 ieties. If we look at species as having been SPECIALLY created, and at varieties as having been pr
0276 more reason to think that species have been SPECIALLY endowed with various degrees of sterility t
0276 n nature, than to think that trees have been SPECIALLY endowed with various and somewhat analogous
0394 the mainland. I hear from Mr. Tomes, who has SPECIALLY studied this family, that many of the same
0430 other Rodents: and therefore it will not be SPECIALLY related to any one existing Marsupial, but
0457 larvae are active embryos, which have become SPECIALLY modified in relation to their habits of lif
0472 h on the ordinary view supposed to have been SPECIALLY created and adapted for that country, being
0480 ic being and each separate organ having been SPECIALLY created, how utterly inexplicable it is tha
0001 ed to me to throw some light on the origin of SPECIES, that mystery of mysteries, as it has been c
0002 eral conclusions that I have on the origin of SPECIES. Last year he sent to me a memoir on this su
0003 ellent judgment. In considering the Origin of SPECIES, it is quite conceivable that a naturalist,
0003 facts, might come to the conclusion that each SPECIES had not been independently created, but had
0003 but had descended, like varieties, from other SPECIES. Nevertheless, such a conclusion, even if we
0003 , until it could be shown how the innumerable SPECIES inhabiting this world have been modified, so
0004 ns. I will then pass on to the variability of SPECIES in a state of nature: but I shall, unfortuna
0005 le kingdoms. As many more individuals of each SPECIES are born than can possibly survive; and as,
0005 ls; thirdly, Hybridism, or the infertility of SPECIES and the fertility of varieties when intercro
0006 as yet unexplained in regard to the origin of SPECIES and varieties, if he makes due allowance for
0006 which live around us. Who can explain why one SPECIES ranges widely and is very numerous, and why
0006 and is very numerous, and why another allied SPECIES has a narrow range and is rare? Yet these ve
0006 ich I formerly entertained, namely, that each SPECIES had been independently created, is erroneous
0006 ated, is erroneous. I am fully convinced that SPECIES are not immutable; but that those belonging
0006 scendants of some other and generally extinct SPECIES, in the same manner as the acknowledged vari
0006 nner as the acknowledged varieties of any one SPECIES are the descendants of that species. Further
0006 f any one species are the descendants of that SPECIES. Furthermore, I am convinced that Natural Se
0007 culty of distinguishing between Varieties and SPECIES. Origin of Domestic Varieties from one or mo
0007 origin of Domestic Varieties from one or more SPECIES. Domestic Pigeons, their Differences and Ori
0007 ach other, than do the individuals of any one SPECIES or variety in a state of nature. When we ref
0007 hat different from, those to which the parent SPECIES have been exposed under nature. There is, al
0013 uliarity in different individuals of the same SPECIES, and in individuals of different species, is
0013 same species, and in individuals of different SPECIES, is sometimes inherited and sometimes not so
0014 eductions can be drawn from domestic races to SPECIES in a state of nature. I have in vain endeavo
0015 nothing from domestic varieties in regard to SPECIES. But there is not a shadow of evidence in fa
0015 tic animals and plants, and compare them with SPECIES closely allied together, we generally percei
0015 ed, less uniformity of character than in true SPECIES. Domestic races of the same species; also, o
0016 n in true species. Domestic races of the same SPECIES, also, often have a somewhat monstrous chara
0016 differing from each other, and from the other SPECIES of the same genus, in several trifling respe
0016 nd more especially when compared with all the SPECIES in nature to which they are nearest allied.
0016 to be discussed), domestic races of the same SPECIES differ from each other in the same manner as
0016 es in a lesser degree than, do closely allied SPECIES of the same genus in a state of nature. I th
0016 s as the descendants of aboriginally distinct SPECIES. If any marked distinction existed between d
0016 istinction existed between domestic races and SPECIES, this source of doubt could not so perpetual
0016 erence between the domestic races of the same SPECIES, we are soon involved in doubt, from not kno
0016 hey have descended from one or several parent SPECIES. This point, if it could be cleared up, woul
0017 nd so truly, were the offspring of any single SPECIES, then such facts would have great weight in
0017 y of the many very closely allied and natural SPECIES, for instance, of the many foxes, inhabiting

Page ***(Key Word)***

Page **(Key Word)**

```
0017 all our dogs have descended from any one wild SPECIES; but, in the case of some other domestic rac
0017 d vary on an average as largely as the parent SPECIES of our existing domesticated productions hav
0017 ether they have descended from one or several SPECIES. The argument mainly relied on by those who
0018 omestic dogs have descended from several wild SPECIES. In regard to sheep and goats I can form no
0019 e there must have existed at least a score of  SPECIES of wild cattle, as many sheep, and several g
0019 formerly existed in Great Britain eleven wild SPECIES of sheep peculiar to it! When we bear in min
0019 countries do not possess a number of peculiar SPECIES as distinct parent stocks? So it is in India
0019 mit have probably descended from several wild SPECIES, I cannot doubt that there has been an immen
0019 produced by the crossing of a few aboriginal  SPECIES; but by crossing we can get only forms in so
0020 iate between two extremely different races or  SPECIES, I can hardly believe. Sir J. Sebright expre
0022 ly, I think, be ranked by him as well defined  SPECIES. Moreover, I do not believe that any ornitho
0023 breeds several truly inherited sub breeds, or  SPECIES as he might have called them, could be shown
0023 r this term several geographical races or sub  SPECIES, which differ from each other in the most tr
0023 t besides C. Livia, with its geographical sub SPECIES, only two or three other species of rock pig
0023 raphical sub species, only two or three other  SPECIES of rock pigeons are known; and these have no
0024 . hence the supposed extermination of so many  SPECIES having similar habits with the rock pigeon s
0024 must be assumed that at least seven or eight  SPECIES were so thoroughly domesticated in ancient t
0024 succeeded in thoroughly domesticating several SPECIES, but that he intentionally or by chance pick
0024 by chance picked out extraordinarily abnormal SPECIES; and further, that these very species have s
0024 bnormal species; and further, that these very SPECIES have since all become extinct or unknown. So
0025 ty blue, and has a white rump (the Indian sub SPECIES, C. intermedia of Strickland, having it blui
0025 eral marks do not occur together in any other SPECIES of the whole family. Now, in every one of th
0025 e the rock pigeon, although no other existing SPECIES is thus coloured and marked, so that in each
0026 probability in this hypothesis, if applied to SPECIES closely related together, though it is unsup
0026 tend the hypothesis so far as to suppose that SPECIES, aboriginally as distinct as carriers, tumbl
0027 n having formerly got seven or eight supposed SPECIES of pigeons to breed freely under domesticati
0027 ed freely under domestication; these supposed SPECIES being quite unknown in a wild state, and the
0027 tate, and their becoming nowhere feral; these SPECIES having very abnormal characters in certain r
0027 m the Columba livia with its geographical sub SPECIES. In favour of this view, I may add, firstly,
0028 to a similar conclusion in regard to the many SPECIES of finches, or other large groups of birds,
0028 descended from so many aboriginally distinct SPECIES. Ask, as I have asked, a celebrated raiser o
0029 each main breed was descended from a distinct SPECIES. Van Mons, in his treatise on pears and appl
0029 sson of caution, when they deride the idea of SPECIES in a state of nature being lineal descendant
0029 e of nature being lineal descendants of other SPECIES? Selection. Let us now briefly consider the
0029 duced, either from one or from several allied SPECIES. Some little effect may, perhaps, be attribu
0033 lowers in the different varieties of the same SPECIES in the flower garden; the diversity of leave
0033 eties; and the diversity of fruit of the same SPECIES in the orchard, in comparison with the leave
0037 nct varieties, and whether or not two or more SPECIES or races have become blended together by cro
0038 e. it is not that these countries, so rich in SPECIES, do not by a strange chance possess the abor
0038 rently circumstanced, individuals of the same SPECIES, having slightly different constitutions or
0038 kept by savages have more of the character of SPECIES than the varieties kept in civilised countri
0039 ht cifferences in the individuals of the same SPECIES, be judged of by the value which would now b
0041 keeping of a large number of individuals of a SPECIES in any country requires that the species sho
0041 of a species in any country requires that the SPECIES should be placed under favourable conditions
0041 in that country. When the individuals of any SPECIES are scanty, all the individuals, whatever th
0042 ppeared (aided by some crossing with distinct SPECIES) those many admirable varieties of the straw
0042 arely possess more than one breed of the same SPECIES. Pigeons can be mated for life, and this is
0043 ses, I do not doubt that the intercrossing of SPECIES, aboriginally distinct, has played an import
0043 e importance of the crossing both of distinct SPECIES and of varieties is immense; for the cultiva
0044 variability. Individual differences. Doubtful SPECIES. Wide ranging, much diffused, and common spe
0044 cies. Wide ranging, much diffused, and common SPECIES vary most. Species of the larger genera in a
0044 much diffused, and common species vary most. SPECIES of the larger genera in any country vary mor
0044 rger genera in any country vary more than the SPECIES of the smaller genera. Many of the species o
0044 he species of the smaller genera. Many of the SPECIES of the larger genera resemble varieties in b
0044 definitions which have been given of the term SPECIES. No one definition has as yet satisfied all
0044 ows vaguely what he means when he speaks of a SPECIES. Generally the term includes the unknown ele
0044 art, either injurious to or not useful to the SPECIES, and not generally propagated. Some authors
0045 ently observed in the individuals of the same SPECIES inhabiting the same confined locality. No on
0045 supposes that all the individuals of the same SPECIES are cast in the very same mould. These indiv
0045 sometimes vary in the individuals of the same SPECIES. I am convinced that the most experienced na
0045 nd compare them in many specimens of the same SPECIES. I should never have expected that the branc
0045 n insect would have been variable in the same SPECIES; I should have expected that changes of this
0046 n called protean or polymorphic, in which the SPECIES present an inordinate amount of variation; a
0046 naturalists can agree which forms to rank as SPECIES and which as varieties. We may instance Rubu
0046 hells. In most polymorphic genera some of the SPECIES have fixed and definite characters. Genera w
0046 which are of no service or disservice to the SPECIES, and which consequently have not been seized
0047 in some considerable degree the character of SPECIES, but which are so closely similar to some ot
0047 uralists do not like to rank them as distinct SPECIES, are in several respects the most important
0047 ong, as far as we know, as have good and true SPECIES. Practically, when a naturalist can unite tw
0047 but sometimes the one first describec, as the SPECIES, and the other as the variety. But cases of
0047 ermining whether a form should be ranked as a SPECIES or a variety, the opinion of naturalists hav
0047 es can be named which have not been ranked as SPECIES by at least some competent judges. That vari
0048 orms have been ranked by one botanist as good SPECIES, and by another as mere varieties. Mr. H. C.
0048 ut which have all been ranked by botanists as SPECIES; and in making this list he has omitted many
0048 theless have been ranked by some botanists as SPECIES, and he has entirely omitted several highly
0048 st polymorphic forms, Mr. Babington gives 251 SPECIES, whereas Mr. Bentham gives only 112, a diffe
0048 doubtful forms, ranked by one zoologist as a SPECIES and by another as a variety, can rarely be f
0048 ranked by one eminent naturalist as undoubted SPECIES, and by another as varieties, or, as they ar
0048 ague and arbitrary is the distinction between SPECIES and varieties. On the islets of the little M
0049 cannot be doubted would be ranked as distinct SPECIES by many entomologists. Even Ireland has a fe
0049 d as varieties, but which have been ranked as SPECIES by some zoologists. Several most experienced
0049 as only a strongly marked race of a Norwegian SPECIES, whereas the greater number rank it as an un
0049 as the greater number rank it as an undoubted SPECIES peculiar to Great Britain. A wide distance b
0049 ads many naturalists to rank both as distinct SPECIES; but what distance, it has been well asked,
0049 varieties, have so perfectly the character of SPECIES that they are ranked by other highly compete
0049 ther highly competent judges as good and true SPECIES. But to discuss whether they are rightly cal
```

Page **(Key Word)**

Page **************************************(Key Word)***************************************

```
0049 ut to discuss whether they are rightly called SPECIES or varieties, before any definition of these
0049 ases of strongly marked varieties or doubtful SPECIES well deserve consideration; for several inte
0050 over, will be often ranked by some authors as SPECIES. Look at the common oak, how closely it has
0050 ; yet a German author makes more than a dozen SPECIES out of forms, which are very generally consi
0050 edcunculated oaks are either good and distinct SPECIES or mere varieties. When a young naturalist c
0051 ms. His general tendency will be to make many SPECIES, for he will become impressed, just like the
0051 is own mind which to call varieties and which SPECIES; but he will succeed in this at the expense
0051 of demarcation has as yet been drawn between SPECIES and sub species, that is, the forms which in
0051 has as yet been drawn between species and sub SPECIES, that is, the forms which in the opinion of
0051 ar to, but do not quite arrive at the rank of SPECIES; or, again, between sub species and well mar
0051 t the rank of species; or, again, between sub SPECIES and well marked varietie#, or between lesser
0052 eties; and at these latter, as leading to sub SPECIES, and to species. The passage from one stage
0052 ese latter, as leading to sub species, and to SPECIES. The passage from one stage of difference to
0052 ked variety may be justly called an incipient SPECIES; but whether this belief be justifiable must
0052 t be supposed that all varieties or incipient SPECIES necessarily attain the rank of species. They
0052 ipient species necessarily attain the rank of SPECIES. They may whilst in this incipient state bec
0052 lourish so as to exceed in numbers the parent SPECIES, it would then rank as the species, and the
0052 the parent species, it would then rank as the SPECIES, and the species as the variety; or it might
0052 s, it would then rank as the species, and the SPECIES as the variety; or it might come to supplant
0052 t come to supplant and exterminate the parent SPECIES; or both might co exist, and both rank as in
0052 might co exist, and both rank as independent SPECIES. But we shall hereafter have to return to th
0052 marks it will be seen that I look at the term SPECIES, as one arbitrarily given for the sake of co
0053 in regard to the nature and relations of the SPECIES which vary most, by tabulating all the varie
0053 es of the proportional numbers of the varying SPECIES. Dr. Hooker permits me to add, that after ha
0053 urther show that, in any limited country, the SPECIES which are most common, that is abound most i
0053 , that is abound most in individuals, and the SPECIES which are most widely diffused within their
0053 hing, or, as they may be called, the dominant SPECIES, those which range widely over the world, ar
0054 varieties, or, as I consider them, incipient SPECIES. And this, perhaps, might have been anticipa
0054 ith the other inhabitants of the country, the SPECIES which are already dominant will be the most
0054 the very common and much diffused or dominant SPECIES will be found on the side of the larger gene
0054 e been anticipated; for the mere fact of many SPECIES of the same genus inhabiting any country, sh
0054 in the larger genera, or those including many SPECIES, a large proportional number of dominant spe
0054 cies, a large proportional number of dominant SPECIES. But so many causes tend to obscure this res
0054 lation to the size of the genera to which the SPECIES belong. Again, plants low in the scale of or
0055 on geographical distribution. From looking at SPECIES as only strongly marked and well defined var
0055 d varieties, I was led to anticipate that the SPECIES of the larger genera in each country would o
0055 try would oftener present varieties, than the SPECIES of the smaller genera; for wherever many clo
0055 ler genera; for wherever many closely related SPECIES (i.e. species of the same genus) have been f
0055 r wherever many closely related species (i.e. SPECIES of the same genus) have been formed, many va
0055 have been formed, many varieties or incipient SPECIES ought, as a general rule, to be now forming.
0055 grow, we expect to find saplings. Where many SPECIES of a genus have been formed through variatio
0055 iation. On the other hand, if we look at each SPECIES as a special act of creation, there is no ap
0055 varieties should occur in a group having many SPECIES, than in one having few. To test the truth o
0055 districts, into two nearly equal masses, the SPECIES of the larger genera on one side, and those
0055 o be the case that a larger proportion of the SPECIES on the side of the larger genera present var
0055 the side of the smaller genera. Moreover, the SPECIES of the large genera which present any variet
0055 arger average number of varieties than do the SPECIES of the small genera. Both these results foll
0055 e smallest genera, with from only one to four SPECIES, are absolutely excluded from the tables. Th
0056 s are of plain signification on the view that SPECIES are only strongly marked and permanent varie
0056 ed and permanent varieties; for wherever many SPECIES of the same genus have been formed, or where
0056 we may use the expression, the manufactory of SPECIES has been active, we ought generally to find
0056 n to believe the process of manufacturing new SPECIES to be a slow one. And this certainly is the
0056 case, if varieties be looked at as incipient SPECIES; for my tables clearly show as a general rul
0056 ly show as a general rule that, wherever many SPECIES of a genus have been formed, the species of
0056 many species of a genus have been formed, the SPECIES of that genus present a number of varieties,
0056 t a number of varieties, that is of incipient SPECIES, beyond the average. It is not that all larg
0056 nd are thus increasing in the number of their SPECIES, or that no small genera are now varying and
0056 all that we want to show is, that where many SPECIES of a genus have been formed, on an average m
0056 s good. There are other relations between the SPECIES of large genera and their recorded varieties
0056 infallible criterion by which to distinguish SPECIES and well marked varieties; and in those case
0056 suffices to raise one or both to the rank of SPECIES. Hence the amount of difference is one very
0057 ettling whether two forms should be ranked as SPECIES or varieties. Now Fries has remarked in rega
0057 e genera the amount of difference between the SPECIES is often exceedingly small. I have endeavour
0057 in this view. In this respect, therefore, the SPECIES of the larger genera resemble varieties, mor
0057 r genera resemble varieties, more than do the SPECIES of the smaller genera. Or the case may be pu
0057 , in which a number of varieties or incipient SPECIES greater than the average are now manufacturi
0057 he average are now manufacturing, many of the SPECIES already manufactured still to a certain exte
0057 han usual amount of difference. Moreover, the SPECIES of the large genera are related to each othe
0057 n the same manner as the varieties of any one SPECIES are related to each other. No naturalist pre
0057 ch other. No naturalist pretends that all the SPECIES of a genus are equally distinct from each ot
0057 as Fries has well remarked, little groups of SPECIES are generally clustered like satellites arou
0057 lustered like satellites around certain other SPECIES. And what are varieties but groups of forms,
0057 nd certain forms, that is, round their parent SPECIES? Undoubtedly there is one most important poi
0057 ant point of difference between varieties and SPECIES; namely, that the amount of difference betwe
0057 compared with each other or with their parent SPECIES, is much less than that between the species
0057 t species, is much less than that between the SPECIES of the same genus. But when we come to discu
0058 increase into the greater differences between SPECIES. There is one other point which seems to me
0058 wider range than that of its supposed parent SPECIES, their denominations ought to be reversed. B
0058 t there is also reason to believe, that those SPECIES which are very closely allied to other speci
0058 pecies which are very closely allied to other SPECIES, and in so far resemble varieties, often hav
0058 dition) 63 plants which are therein ranked as SPECIES, but which he considers as so closely allied
0058 ch he considers as so closely allied to other SPECIES as to be of doubtful value: these 63 reputed
0058 as to be of doubtful value: these 63 reputed SPECIES range on an average over 6.9 of the province
0058 these range over 7.7 provinces; whereas, the SPECIES to which these varieties belong range over 1
0058 orms, marked for me by Mr. Watson as doubtful SPECIES, but which are almost universally ranked by
0058 ranked by British botanists as good and true SPECIES. Finally, then, varieties have the same gene
```

Page **************************************(Key Word)**************************************

Page ***(Key Word)***

```
0058  varieties have the same general characters as  SPECIES, for they cannot be distinguished from speci
0058  pecies, for they cannot be distinguished from  SPECIES, except, firstly, by the discovery of interm
0059  ed necessary to give to two forms the rank of  SPECIES is quite indefinite. In genera having more t
0059  genera having more than the average number of  SPECIES in any country, the species of these genera
0059  average number of species in any country, the  SPECIES of these genera have more than the average n
0059  rage number of varieties. In large genera the  SPECIES are apt to be closely, but unequally, allied
0059  gether, forming little clusters round certain  SPECIES. Species very closely allied to other specie
0059  orming little clusters round certain species.  SPECIES very closely allied to other species apparen
0059  species. Species very closely allied to other  SPECIES apparently have restricted ranges. In all th
0059  ted ranges. In all these several respects the  SPECIES of large genera present a strong analogy wit
0059  we can clearly understand these analogies, if  SPECIES have once existed as varieties, and have thu
0059  se analogies are utterly inexplicable if each  SPECIES has been independently created. We have, als
0059  that it is the most flourishing and dominant   SPECIES of the larger genera which on an average var
0059  end to become converted into new and distinct  SPECIES. The larger genera thus tend to become large
0060  between individuals and varieties of the same  SPECIES; often severe between species of the same ge
0060  ies of the same species; often severe between  SPECIES of the same genus. The relation of organism
0060  ether a multitude of doubtful forms be called  SPECIES or sub species or varieties: what rank, for
0060  de of doubtful forms be called species or sub  SPECIES or varieties: what rank, for instance, the t
0060  ork, helps us but little in understanding how  SPECIES arise in nature. How have all those exquisit
0061  that varieties, which I have called incipient  SPECIES, become ultimately converted into good and d
0061  e ultimately converted into good and distinct  SPECIES, which in most cases obviously differ from e
0061  er far more than do the varieties of the same  SPECIES? How do those groups of species, which const
0061  s of the same species? How do those groups of  SPECIES, which constitute what are called distinct g
0061  which differ from each other more than do the  SPECIES of the same genus, arise? All these results,
0061  any degree profitable to an individual of any  SPECIES, in its infinitely complex relations to othe
0061  urviving, for, of the many individuals of any  SPECIES which are periodically born, but a small num
0063  ither one individual with another of the same  SPECIES, or with the individuals of distinct species
0063  species, or with the individuals of distinct   SPECIES, or with the physical conditions of life. It
0063  ential restraint from marriage. Although some  SPECIES may be now increasing, more or less rapidly,
0066  not determine how many individuals of the two  SPECIES can be supported in a district. A large numb
0066  number of eggs is of some importance to those  SPECIES, which depend on a rapidly fluctuating amoun
0066  are destroyed, many must be produced, or the   SPECIES will become extinct. It would suffice to kee
0067  ruction ever so little, and the number of the  SPECIES will almost instantaneously increase to any
0067  rce. What checks the natural tendency of each  SPECIES to increase in number is most obscure. Look
0067  er is most obscure. Look at the most vigorous  SPECIES; by as much as it swarms in numbers, by so m
0068  hough fully grown, plants: thus out of twenty  SPECIES growing on a little plot of turf (three feet
0068  little plot of turf (three feet by four) nine  SPECIES perished from the other species being allowe
0068  by four) nine species perished from the other  SPECIES being allowed to grow up freely. The amount
0068  o grow up freely. The amount of food for each  SPECIES of course gives the extreme limit to which e
0068  ls, which determines the average numbers of a  SPECIES. Thus, there seems to be little doubt that t
0068  part in determining the average numbers of a  SPECIES, and periodical seasons of extreme cold or d
0068  dividuals, whether of the same or of distinct  SPECIES, which subsist on the same kind of food, Eve
0069  damp region to a dry, we invariably see some  SPECIES gradually getting rarer and rarer, and final
0069  his is a very false view: we forget that each  SPECIES, even where it most abounds, is constantly s
0069  ady fully stocked with inhabitants, the other  SPECIES will decrease. When we travel southward and
0069  decrease. When we travel southward and see a  SPECIES decreasing in numbers, we may feel sure that
0069  re that the cause lies quite as much in other  SPECIES being favoured, as in this one being hurt. S
0069  n a somewhat lesser degree, for the number of  SPECIES of all kinds, and therefore of competitors,
0069  ts in main part indirectly by favouring other  SPECIES, we may clearly see in the prodigious number
0070  1st destruction by our native animals. When a  SPECIES, owing to highly favourable circumstances, i
0070  ses, a large stock of individuals of the same  SPECIES, relatively to the numbers of its enemies, i
0070  of the necessity of a large stock of the same  SPECIES for its preservation, explains, I believe, s
0071  heath plants were wholly changed, but twelve   SPECIES of plants (not counting grasses and carices)
0074  flowers in that district! In the case of many  SPECIES, many different checks, acting at different
0074  e average number or even the existence of the  SPECIES. In some cases it can be shown that widely d
0074  that widely different checks act on the same   SPECIES in different districts. When we look at the
0075  st severe between the individuals of the same  SPECIES, for they frequent the same districts, requi
0075  dangers. In the case of varieties of the same  SPECIES, the struggle will generally be almost equal
0076  he seed or young were not annually sorted. As  SPECIES of the same genus have usually, though by no
0076  truggle will generally be more severe between  SPECIES of the same genus, when they come into compe
0076  nto competition with each other, than between  SPECIES of distinct genera. We see this in the recen
0076  ension over parts of the United States of one  SPECIES of swallow having caused the decrease of ano
0076  swallow having caused the decrease of another  SPECIES. The recent increase of the missel thrush in
0076  he song thrush. How frequently we hear of one  SPECIES of rat taking the place of another species u
0076  ne species of rat taking the place of another  SPECIES under the most different climates. In Russia
0076  here driven before it its great congener. One  SPECIES of charlock will supplant another, and so in
0076  in no one case could we precisely say why one  SPECIES has been victorious over another in the grea
0078  et there will be competition between some few  SPECIES, or between the individuals of the same spec
0078  ecies, or between the individuals of the same  SPECIES, for the warmest or dampest spots. Hence, al
0080  intercrosses between individuals of the same  SPECIES. Circumstances favourable and unfavourable t
0081  fluctuating element, as perhaps we see in the  SPECIES called polymorphic. We shall best understand
0081  almost immediately undergo a change, and some  SPECIES might become extinct. We may conclude, from
0082  ny way favoured the individuals of any of the  SPECIES, by better adapting them to their altered co
0085  ng at many small points of difference between  SPECIES, which, as far as our ignorance permits us t
0086  me so, they would cause the extinction of the  SPECIES. Natural selection will modify the structure
0087  cannot do, is to modify the structure of one  SPECIES, without giving it any advantage, for the go
0087  ing it any advantage, for the good of another  SPECIES; and though statements to this effect may be
0089  he severest rivalry between the males of many  SPECIES to attract by singing the females. The rock
0092  owers of two distinct individuals of the same  SPECIES would thus get crossed: and the act of cross
0098  oduce a multitude of hybrids between distinct  SPECIES; for if you bring on the same brush a plant'
0098  a plant's own pollen and pollen from another  SPECIES, the former will have such a prepotent effec
0098  aised plenty of seedlings: and whilst another  SPECIES of Lobelia growing close by, which is visite
0099  rcrossing of distinct individuals of the same  SPECIES. When distinct species are crossed the case
0099  ndividuals of the same species. When distinct  SPECIES are crossed the case is directly the reverse
0101  that, in the case of both animals and plants,  SPECIES of the same family and even of the same genu
0101  fference between hermaphrodites and unisexual  SPECIES, as far as function is concerned, becomes ve
```

Page ***(Key Word)***

```
0102 ch place in the economy of nature, if any one  SPECIES does not become modified and improved in a c
0102 atural selection be modifying and improving a  SPECIES in the several districts, there will be inte
0102 ossing with the other individuals of the same  SPECIES on the confines of each. And in this case th
0103 nature in keeping the individuals of the same  SPECIES, or of the same variety, true and uniform in
0104 nd to modify all the individuals of a varying  SPECIES throughout the area in the same manner in re
0104 osses, also, with the individuals of the same  SPECIES, which otherwise would have inhabited the su
0105 mes be of importance in the production of new  SPECIES. If, however, an isolated area be very small
0105 als will greatly retard the production of new  SPECIES through natural selection, by decreasing the
0105 anic island, although the total number of the  SPECIES inhabiting it, will be found to be small, as
0105 er on geographical distribution: yet of these  SPECIES a very large proportion are endemic, that is
0105 n highly favourable for the production of new  SPECIES. But we may thus greatly deceive ourselves,
0105 siderable importance in the production of new  SPECIES, on the whole I am inclined to believe that
0105 ortance, more especially in the production of  SPECIES, which will prove capable of enduring for a
0105 m the large number of individuals of the same  SPECIES there supported, but the conditions of life
0106 lex from the large number of already existing  SPECIES: and if some of these many species become mo
0106 y existing species: and if some of these many  SPECIES become modified and improved, others will ha
0106 s highly favourable for the production of new  SPECIES, yet that the course of modification will ge
0106 ely, will give rise to most new varieties and  SPECIES, and will thus play an important part in the
0107 will still exist many individuals of the same  SPECIES on each island: intercrossing on the confine
0107 crossing on the confines of the range of each  SPECIES will thus be checked: after physical changes
0108 further the inhabitants, and thus produce new  SPECIES. That natural selection will always act with
0110 any one region has as yet got its maximum of  SPECIES. Probably no region is as yet fully stocked,
0110 ked, for at the Cape of Good Hope, where more  SPECIES of plants are crowded together than in any o
0110 e extinction of any natives. Furthermore, the  SPECIES which are most numerous in individuals will
0110 second chapter, showing that it is the common  SPECIES which afford the greatest number of recorded
0110 st number of recorded varieties, or incipient  SPECIES. Hence, rare species will be less quickly mo
0110 varieties, or incipient species. Hence, rare  SPECIES will be less quickly modified or improved wi
0110 e by the modified descendants of the commoner  SPECIES. From these several considerations I think I
0110 ns I think it inevitably follows, that as new  SPECIES in the course of time are formed through nat
0110 t closely allied forms, varieties of the same  SPECIES, and species of the same genus or of related
0110 ied forms, varieties of the same species, and  SPECIES of the same genus or of related genera, whic
0110 each other. Consequently, each new variety or  SPECIES, during the progress of its formation, will
0111 s, though having somewhat of the character of  SPECIES, as is shown by the hopeless doubts in many
0111 each other far less than do good and distinct  SPECIES. Nevertheless, according to my view, variet
0111 rtheless, according to my view, varieties are  SPECIES in the process of formation, or are, as I ha
0111 ion, or are, as I have called them, incipient  SPECIES. How, then, does the lesser difference betwe
0111 augmented into the greater difference between  SPECIES? That this does habitually happen, we must i
0111 n, we must infer from most of the innumerable  SPECIES throughout nature presenting well marked dif
0111 prototypes and parents of future well marked  SPECIES, present slight and ill defined differences.
0111 ference as that between varieties of the same  SPECIES and species of the same genus. As has always
0111 hat between varieties of the same species and  SPECIES of the same genus. As has always been my pra
0112 more diversified the descendants of any one  SPECIES become in structure, constitution, and habit
0113 ed, that if a plot of ground be sown with one  SPECIES of grass, and a similar plot be sown with se
0113 on equal spaces of ground. Hence, if any one  SPECIES of grass were to go on varying, and those va
0113 h other in at all the same manner as distinct  SPECIES and genera of grasses differ from each other
0113 a greater number of individual plants of this  SPECIES of grass, including its modified descendants
0113 e piece of ground. And we well know that each  SPECIES and each variety of grass is annually sowing
0114 tions, the most distinct varieties of any one  SPECIES of grass would always have the best chance o
0114 ry distinct from each other, take the rank of  SPECIES. The truth of the principle, that the greate
0114 exactly the same conditions, supported twenty  SPECIES of plants, and these belonged to eighteen ge
0115 ally with the number of the native genera and  SPECIES, far more in new genera than in new species.
0115 d species, far more in new genera than in new  SPECIES. To give a single instance: in the last edit
0116 sume that the modified descendants of any one  SPECIES will succeed by so much the better as they b
0116 perplexing subject. Let A to L represent the  SPECIES of a genus large in its own country; these
0116 es of a genus large in its own country; these  SPECIES are supposed to resemble each other in unequ
0117 econd chapter, that on an average more of the  SPECIES of large genera vary than of small genera: a
0117 ra vary than of small genera: and the varying  SPECIES of the large genera present a greater number
0117 er of varieties. We have, also, seen that the  SPECIES, which are the commonest and the most widely
0117 the most widely diffused, vary more than rare  SPECIES with restricted ranges. Let (A) be a common,
0117 (a) be a common, widely diffused, and varying  SPECIES, belonging to a genus large in its own count
0117 nd generations. After a thousand generations,  SPECIES (A) is supposed to have produced two fairly
0118 ages which made the genus to which the parent  SPECIES belonged, a large genus in its own country.
0119 ied in structure the descendants from any one  SPECIES can be rendered, the more places they will b
0119 descendants from a common and widely diffused  SPECIES, belonging to a large genus, will tend to pa
0120 hes or races. After ten thousand generations,  SPECIES (A) is supposed to have produced three forms
0120 have arrived at the doubtful category of sub  SPECIES: but we have only to suppose the steps in th
0120 o convert these three forms into well defined  SPECIES: thus the diagram illustrates the steps by w
0120 ed into the larger differences distinguishing  SPECIES. By continuing the same process for a greate
0120 ondensed and simplified manner); we get eight  SPECIES, marked by the letters between a14 and m14,
0120 , all descended from (A). Thus, as I believe,  SPECIES are multiplied and genera are formed. In a l
0120 large genus it is probable that more than one  SPECIES would vary. In the diagram I have assumed th
0120 . in the diagram I have assumed that a second  SPECIES (I) has produced, by analogous steps, after
0120 wo well marked varieties (w10 and z10) or two  SPECIES, according to the amount of change supposed
0121 after fourteen thousand generations, six new  SPECIES, marked by the letters n14 to z14, are suppo
0121 sed to have been produced. In each genus, the  SPECIES, which are already extremely different in ch
0121 ence in the diagram I have chosen the extreme  SPECIES (A), and the nearly extreme species (I), as
0121 e extreme species (A), and the nearly extreme  SPECIES (I), as those which have largely varied, and
0121 ied, and have given rise to new varieties and  SPECIES. The other nine species (marked by capital l
0121 to new varieties and species. The other nine  SPECIES (marked by capital letters) of our original
0121 ndency in the improved descendants of any one  SPECIES to supplant and exterminate in each stage of
0121 between the less and more improved state of a  SPECIES, as well as the original parent species itse
0121 of a species, as well as the original parent  SPECIES itself, will generally tend to become extinc
0122 ent. If, however, the modified offspring of a  SPECIES get into some distinct country, or become qu
0122 resent a considerable amount of modification,  SPECIES (A) and all the earlier varieties will have
0122 me extinct, having been replaced by eight new  SPECIES (a14 to m14); and (I) will have been replace
0122 ll have been replaced by six (n14 to z14) new  SPECIES. But we may go further than this. The origin
```

Page **(Key Word)**

```
0122 but we may co further than this. The original SPECIES of our genus were supposed to resemble each
0122 grees, as is so generally the case in nature; SPECIES (A) being more nearly related to B, C, and D
0122 rly related to B, C, and D, than to the other SPECIES; and species (I) more to G, H, K, L, than to
0122 o B, C, and D, than to the other species; and SPECIES (I) more to G, H, K, L, than to the others.
0122 to G, H, K, L, than to the others. These two SPECIES (A) and (I), were also supposed to be very c
0122 upposed to be very common and widely diffused SPECIES, so that they must originally have had some
0122 ave had some advantage over most of the other SPECIES of the genus. Their modified descendants, fo
0122 a) and (I), but likewise some of the original SPECIES which were most nearly related to their pare
0122 their parents. Hence very few of the original SPECIES will have transmitted offspring to the fourt
0122 we may suppose that only one (F), of the two SPECIES which were least closely related to the othe
0122 st closely related to the other nine original SPECIES, has transmitted descendants to this late st
0123 ndants to this late stage of descent. The new SPECIES in our diagram descended from the original e
0123 ur diagram descended from the original eleven SPECIES, will now be fifteen in number. Owing to the
0123 eme amount of difference in character between SPECIES a14 and z14 will be much greater than that b
0123 een the most different of the original eleven SPECIES. The new species, moreover, will be allied t
0123 erent of the original eleven species. The new SPECIES, moreover, will be allied to each other in a
0123 me degree distinct from the three first named SPECIES; and lastly, o14, e14, and m14, will be near
0123 will be widely different from the other five SPECIFS, and may constitute a sub genus or even a di
0123 ub genera or even genera. But as the original SPECIES (I) differed largely from (A), standing near
0123 ing in different directions. The intermediate SPECIES, also (and this is a very important consider
0123 consideration), which connected the original SPECIES (A) and (I), have all become, excepting (F),
0123 d have left no descendants. Hence the six new SPECIES descended from (I), and the eight descended
0123 descent, with modification, from two or more SPECIES of the same genus. And the two or more paren
0124 of the same genus. And the two or more parent SPECIES are supposed to have descended from some one
0124 are supposed to have descended from some one SPECIES of an earlier genus. In our diagram, this is
0124 ingle point; this point representing a single SPECIES, the supposed single parent of our several n
0124 lect for a moment on the character of the new SPECIES f14, which is supposed not to have diverged
0124 ase, its affinities to the other fourteen new SPECIES will be of a curious and circuitous nature.
0124 rom a form which stood between the two parent SPECIES (A) and (I), now supposed to be extinct and
0124 r between the two groups descended from these SPECIES. But as these two groups have gone on diverg
0124 acter from the type of their parents, the new SPECIES (F14) will not be directly intermediate betw
0125 we can understand this fact, for the extinct SPECIES lived at very ancient epochs when the branch
0125 lies, or orders, will have descended from two SPECIES of the original genus; and these two species
0125 species of the original genus; and these two SPECIES are supposed to have descended from one spec
0125 ecies are supposed to have descended from one SPECIES of a still more ancient and unknown genus. W
0125 . we have seen that in each country it is the SPECIES of the larger genera which oftenest present
0125 which oftenest present varieties or incipient SPECIES. This, indeed, might have been expected; for
0125 and the largeness of any group shows that its SPECIES have inherited from a common ancestor some a
0126 ied descendants; and consequently that of the SPECIES living at any one period, extremely few will
0126 his view of extremely few of the more ancient SPECIES having transmitted descendants, and on the v
0126 n the view of all the descendants of the same SPECIES making a class, we can understand how it is
0126 s. although extremely few of the most ancient SPECIES may now have living and modified descendants
0126 earth may have been as well peopled with many SPECIES of many genera, families, orders, and classe
0127 e high geometrical powers of increase of each SPECIES, at some age, season, or year, a severe stru
0128 he modification of the descendants of any one SPECIES, during the incessant struggle of all sp
0128 ies, and during the incessant struggle of all SPECIES to increase in numbers, the more diversified
0128 ferences distinguishing varieties of the same SPECIES, will steadily tend to increase till they co
0128 come to equal the greater differences between SPECIES of the same genus, or even of distinct gener
0128 mmon, the widely diffused, and widely ranging SPECIES, belonging to the larger genera, which vary
0128 rywhere behold, namely, varieties of the same SPECIES most closely related together, species of th
0128 e same species most closely related together, SPECIES of the same genus less closely and unequally
0128 ed together, forming sections and sub genera, SPECIES of distinct genera much less closely related
0129 almost endless cycles. On the view that each SPECIES has been independently created, I can see no
0129 reen and budding twigs may represent existing SPECIES; and those produced during each former year
0129 may represent the long succession of extinct SPECIES. At each period of growth all the growing tw
0129 ing twigs and branches, in the same manner as SPECIES and groups of species have tried to overmast
0129 , in the same manner as species and groups of SPECIES have tried to overmaster other species in th
0129 ups of species have tried to overmaster other SPECIES in the great battle for life. The limbs divi
0129 the classification of all extinct and living SPECIES in groups subordinate to groups. Of the many
0129 and bear all the other branches; so with the SPECIES which lived during long past geological peri
0131 eneric: secondary sexual characters variable. SPECIES of the same genus vary in an analogous manne
0132 more brightly coloured than those of the same SPECIES further north or from greater depths. Gould
0132 depths. Gould believes that birds of the same SPECIES are more brightly coloured under a clear atm
0132 could be given. The fact of varieties of one SPECIES, when they range into the zone of habitation
0133 ey range into the zone of habitation of other SPECIES, often acquiring in a very slight degree som
0133 slight degree some of the characters of such SPECIES, accords with our view that species of all k
0133 s of such species, accords with our view that SPECIES of all kinds are only well marked and perman
0133 well marked and permanent varieties. Thus the SPECIES of shells which are confined to tropical and
0133 er coloured than those of islands. The insect SPECIES confined to sea coasts; as every collector k
0133 aves. He who believes in the creation of each SPECIES, will have to say that this shell, for insta
0133 ll known to furriers that animals of the same SPECIES have thicker and better fur the more severe
0133 ferent varieties being produced from the same SPECIES under the same conditions. Such facts show h
0134 le instances are known to every naturalist of SPECIES keeping true, or not varying at all, althoug
0135 arkable fact that 200 beetles, out of the 550 SPECIES inhabiting Madeira, are so far deficient in
0135 less than twenty three genera have all their SPECIES in this condition! Several facts, namely, th
0139 cclimatisation. As it is extremely common for SPECIES of the same genus to inhabit very hot and ve
0139 cold countries, and as I believe that all the SPECIES of the same genus have descended from a sing
0139 continued descent. It is notorious that each SPECIES is adapted to the climate of its own home: s
0139 es is adapted to the climate of its own home: SPECIES from an arctic or even from a temperate regi
0139 damp climate. But the degree of adaptation of SPECIES to the climates under which they live is oft
0140 y good health. We have reason to believe that SPECIES in a state of nature are limited in their ra
0140 ave been made by Mr. H. C. Watson on European SPECIES of plants brought from the Azores to England
0140 ls, several authentic cases could be given of SPECIES within historical times having largely exten
0141 mestic animals, and such facts as that former SPECIES of the elephant and rhinoceros were capable
0141 nduring a glacial climate, whereas the living SPECIES are now all tropical or sub tropical in thei
0141 into play. How much of the acclimatisation of SPECIES to any peculiar climate is due to mere habit
```

Page **(Key Word)**

Page **************************************(Key Word)**************************************

```
0145  is by no means, as Dr. Hooker informs me, in  SPECIES with the densest heads that the inner and ou
0146  s we can see, of the slightest service to the  SPECIES. We may often falsely attribute to correlati
0146  tructures which are common to whole groups of  SPECIES, and which in truth are simply due to inheri
0147  and a large beard by diminished wattles. With  SPECIES in a state of nature it can hardly be mainta
0148  dvantage to each successive individual of the  SPECIES; for in the struggle for life to which every
0149  eoffroy St. Hilaire, both in varieties and in  SPECIES, that when any part or organ is repeated man
0150  endency to reversion. A part developed in any  SPECIES in an extraordinary degree or manner, in com
0150  r, in comparison with the same part in allied  SPECIES, tends to be highly variable. Several years
0150  mparison with the same part in closely allied  SPECIES. Thus, the bat's wing is a most abnormal str
0150  having wings; it would apply only if some one  SPECIES of bat had its wings developed in some remar
0150  emarkable manner in comparison with the other  SPECIES of the same genus. The rule applies very str
0151  even in different genera; but in the several  SPECIES of one genus, Pyrgoma, these valves present
0151  ation: the homologous valves in the different  SPECIES being sometimes wholly unlike in shape; and
0151  ariation in the individuals of several of the  SPECIES is so great, that it is no exaggeration to s
0151  cters of these important valves than do other  SPECIES of distinct genera. As birds within the same
0151  loped in a remarkable degree or manner in any  SPECIES, the fair presumption is that it is of high
0152  tion is that it is of high importance to that  SPECIES; nevertheless the part in this case is emine
0152  why should this be so? On the view that each  SPECIES has been independently created, with all its
0152  o explanation. But on the view that groups of  SPECIES have descended from other species, and have
0152  t groups of species have descended from other  SPECIES, and have been modified through natural sele
0153  veloped in an extraordinary manner in any one  SPECIES, compared with the other species of the same
0153  r in any one species, compared with the other  SPECIES of the same genus, we may conclude that this
0153  nt of modification, since the period when the  SPECIES branched off from the common progenitor of t
0153  ll seldom be remote in any extreme degree, as  SPECIES very rarely endure for more than one geologi
0153  d by natural selection for the benefit of the  SPECIES. But as the variability of the extraordinari
0154  in by a simple example what is meant. If some  SPECIES in a large genus of plants had blue flowers
0154  no one would be surprised at one of the blue  SPECIES varying into red, or conversely; but if all
0154  rying into red, or conversely; but if all the  SPECIES had blue flowers, the colour would become a
0155  ally very constant throughout large groups of  SPECIES, has differed considerably in closely allied
0155  , has differed considerably in closely allied  SPECIES, that it has also, been variable in the indi
0155  en variable in the individuals of some of the  SPECIES. And this fact shows that a character, which
0155  re an organ normally differs in the different  SPECIES of the same group, the more subject it is to
0155  idual anomalies. On the ordinary view of each  SPECIES having been independently created, why shoul
0155  the same part in other independently created  SPECIES of the same genus, be more variable than tho
0155  parts which are closely alike in the several  SPECIES? I do not see that any explanation can be gi
0155  explanation can be given. But on the view of  SPECIES being only strongly marked and fixed varieti
0155  n another manner: the points in which all the  SPECIES of a genus resemble each other, and in which
0155  each other, and in which they differ from the  SPECIES of some other genus, are called generic char
0156  natural selection will have modified several  SPECIES, fitted to more or less widely different hab
0156  m a remote period, since that period when the  SPECIES first branched off from their common progeni
0156  t day. On the other hand, the points in which  SPECIES differ from other species of the same genus,
0156  the points in which species differ from other  SPECIES of the same genus, are called specific chara
0156  within the period of the branching off of the  SPECIES from a common progenitor, it is probable tha
0156  riable; I think it also will be admitted that  SPECIES of the same group differ from each other mor
0157  thus readily have succeeded in giving to the  SPECIES of the same group a greater amount of differ
0157  differences between the two sexes of the same  SPECIES are generally displayed in the very same par
0157  ts of the organisation in which the different  SPECIES of the same genus differ from each other. Of
0157  likewise differs in the two sexes of the same  SPECIES: again in fossorial hymenoptera, the manner
0157  genera the neuration differs in the different  SPECIES, and likewise in the two sexes of the same s
0157  es, and likewise in the two sexes of the same  SPECIES. This relation has a clear meaning on my vie
0157  on my view of the subject: I look at all the  SPECIES of the same genus as having as certainly des
0157  itor, as have the two sexes of any one of the  SPECIES. Consequently, whatever part of the structur
0158  sexual selection, in order to fit the several  SPECIES to their several places in the economy of na
0158  and likewise to fit the two sexes of the same  SPECIES to each other, or to fit the males and femal
0158  ecific characters, or those which distinguish  SPECIES from species, than of generic characters, or
0158  ters, or those which distinguish species from  SPECIES, than of generic characters, or those which
0158  han of generic characters, or those which the  SPECIES possess in common; that the frequent extreme
0158  riability of any part which is developed in a  SPECIES in an extraordinary manner in comparison wit
0158  eveloped, if it be common to a whole group of  SPECIES; that the great variability of secondary sex
0158  these same characters between closely allied  SPECIES; that secondary sexual and ordinary specific
0158  nnected together. All being mainly due to the  SPECIES of the same group having descended from a co
0159  and for ordinary specific purposes. Distinct  SPECIES present analogous variations; and a variety
0159  nt analogous variations; and a variety of one  SPECIES often assumes some of the characters of an a
0159  n assumes some of the characters of an allied  SPECIES, or reverts to some of the characters of an
0159  analogous variation in two so called distinct  SPECIES; and to these a third may be added, namely,
0159  urnip. According to the ordinary view of each  SPECIES having been independently created, we should
0161  inherited tendency to produce it. As all the  SPECIES of the same genus are supposed, on my theory
0161  an analogous manner; so that a variety of one  SPECIES would resemble in some of its characters ano
0161  ld resemble in some of its characters another  SPECIES; this other species being on my view only a
0161  of its characters another species; this other  SPECIES being on my view only a well marked and perm
0161  in accordance with the diverse habits of the  SPECIES, and will not be left to the mutual action o
0161  tution. It might further be expected that the  SPECIES of the same genus would occasionally exhibit
0162  sometimes to find the varying offspring of a  SPECIES assuming characters (either from reversion o
0162  t of the difficulty in recognising a variable  SPECIES in our systematic works, is due to its varie
0162  ieties mocking, as it were, some of the other  SPECIES of the same genus. A considerable catalogue,
0162  t be doubtfully ranked as either varieties or  SPECIES; and this shows, unless all these forms be c
0162  forms be considered as independently created  SPECIES, that the one in varying has assumed some of
0162  racter of the same part or organ in an allied  SPECIES. I have collected a long list of such cases;
0163  tant character, but from occurring in several  SPECIES of the same genus, partly under domesticatio
0163  ormed by Colonel Poole that the foals of this  SPECIES are generally striped on the legs and faintl
0164  horse have descended from several aboriginal  SPECIES, one of which, the dun, was striped; and tha
0165  s turn to the effects of crossing the several  SPECIES of the horse genus. Rollin asserts, that the
0166  e several facts? We see several very distinct  SPECIES of the horse genus becoming, by simple varia
0166  to that of the general colouring of the other  SPECIES of the genus. The appearance of the stripes
0166  ids from between several of the most distinct  SPECIES. Now observe the case of the several breeds
0166  ded from a pigeon (including two or three sub  SPECIES or geographical races) of a bluish colour, w
```

Page **************************************(Key Word)**************************************

Page ***(Key Word)***

```
0166  evails. And we have just seen that in several  SPECIES  of the horse genus the stripes are either pl
0166  , some of which have bred true for centuries,  SPECIES; and how exactly parallel is the case with t
0167  , and zebra. He who believes that each equine  SPECIES  was independently created, will, I presume,
0167  ly created, will, I presume, assert that each  SPECIES  has been created with a tendency to vary, bo
0167  ner, so as often to become striped like other  SPECIES  of the genus; and that each has been created
0167  ted with a strong tendency, when crossed with  SPECIES  inhabiting distant quarters of the world, to
0167  eir stripes, not their own parents, but other  SPECIES  of the genus. To admit this view is, as it s
0167  ser differences between varieties of the same  SPECIES, and the greater differences between species
0167  species, and the greater differences between  SPECIES  of the same genus. The external conditions o
0168  s which have come to differ since the several  SPECIES  of the same genus branched off from a common
0168  hole individual: for in a district where many  SPECIES  of any genus are found, that is, where there
0169  rage, we now find most varieties or incipient  SPECIES. Secondary sexual characters are highly vari
0169  iable, and such characters differ much in the  SPECIES  of the same group. Variability in the same p
0169  y sexual differences to the sexes of the same  SPECIES, and specific differences to the several spe
0169  cies, and specific differences to the several  SPECIES  of the same genus. Any part or organ develop
0169  son with the same part or organ in the allied  SPECIES, must have gone through an extraordinary amo
0169  eversion to a less modified state. But when a  SPECIES  with any extraordinarily developed organ has
0169  extraordinary a a manner it may be developed.  SPECIES  inheriting nearly the same constitution from
0169  present analogous variations, and these same  SPECIES  may occasionally revert to some of the chara
0171  abits of life. Diversified habits in the same  SPECIES. Species with habits widely different from t
0171  life. Diversified habits in the same species.  SPECIES  with habits widely different from those of t
0171  ed under the following heads: Firstly, why if  SPECIES  have descended from other species by insensi
0171  tly, why if species have descended from other  SPECIES  by insensibly fine gradations, do we not eve
0171  is not all nature in confusion instead of the  SPECIES  being, as we see them, well defined? Secondl
0172  hematicians? Fourthly, how can we account for  SPECIES, when crossed, being sterile and producing s
0172  n, go hand in hand. Hence, if we look at each  SPECIES  as descended from some other unknown form, b
0173  may be urged that when several closely allied  SPECIES  inhabit the same territory we surely ought t
0173  tervals with closely allied or representative  SPECIES, evidently filling nearly the same place in
0173  ral economy of the land. These representative  SPECIES  often meet and interlock; and as the one bec
0173  e replaces the other. But if we compare these  SPECIES  where they intermingle, they are generally a
0173  inhabited by each. By my theory these allied  SPECIES  have descended from a common parent; and dur
0174  ertiary periods; and in such islands distinct  SPECIES  might have been separately formed without th
0174  ty; for I believe that many perfectly defined  SPECIES  have been formed on strictly continuous area
0174  yed an important part in the formation of new  SPECIES, more especially with freely crossing and wa
0174  crossing and wandering animals. In looking at  SPECIES  as they are now distributed over a wide area
0174  neutral territory between two representative  SPECIES  is generally narrow in comparison with the t
0175  ph. de Candolle has observed, a common alpine  SPECIES  disappears. The same fact has been noticed b
0175  y. but when we bear in mind that almost every  SPECIES, even in its metropolis, would increase imme
0175  y in numbers, were it not for other competing  SPECIES; that nearly all either prey on or serve as
0175  s, but in large part on the presence of other  SPECIES, on which it depends, or by which it is dest
0175  with it comes into competition; and as these  SPECIES  are already defined objects (however they ma
0175  y insensible gradations, the range of any one  SPECIES, depending as it does on the range of others
0175  ll tend to be sharply defined. Moreover, each  SPECIES  on the confines of its range, where it exist
0175  ht in believing that allied or representative  SPECIES, when inhabiting a continuous area, are gene
0175  , as varieties do not essentially differ from  SPECIES, the same rule will probably apply to both;
0176  oth; and if we in imagination adapt a varying  SPECIES  to a very large area, we shall have to adapt
0176  be converted and perfected into two distinct  SPECIES, the two which exist in larger numbers from
0177  which, as I believe, accounts for the common  SPECIES  in each country, as shown in the second chap
0177  er of well marked varieties than do the rarer  SPECIES. I may illustrate what I mean by supposing t
0177  diate hill variety. To sum up, I believe that  SPECIES  come to be tolerably well defined objects, a
0178  d at any one time, we ought only to see a few  SPECIES  presenting slight modifications of structure
0178  fficiently distinct to rank as representative  SPECIES. In this case, intermediate varieties betwee
0178  varieties between the several representative  SPECIES  and their common parent, must formerly have
0178  tribution of closely allied or representative  SPECIES, and likewise of acknowledged varieties), ex
0179  diate varieties, linking most closely all the  SPECIES  of the same group together, must assuredly h
0180  ional habits and structures in closely allied  SPECIES  of the same genus; and of diversified habits
0180  s, either constant or occasional, in the same  SPECIES. And it seems to me that nothing less than a
0183  tle for life. Hence the chance of discovering  SPECIES  with transitional grades of structure in a f
0183  xisted in lesser numbers, than in the case of  SPECIES  with fully developed structures. I will now
0183  changed habits in the individuals of the same  SPECIES. When either case occurs, it would be easy f
0184  a whale. As we sometimes see individuals of a  SPECIES  following habits widely different from those
0184  widely different from those both of their own  SPECIES  and of the other species of the same genus,
0184  se both of their own species and of the other  SPECIES  of the same genus; we might expect, on my th
0184  als would occasionally have given rise to new  SPECIES, having anomalous habits, and with their str
0184  of its close blood relationship to our common  SPECIES; yet it is a woodpecker which never climbs a
0187  o for the gradations by which an organ in any  SPECIES  has been perfected, we ought to look exclusi
0187  le, and we are forced in each case to look to  SPECIES  of the same group, that is to the collateral
0187  in the structure of the eye, and from fossil  SPECIES  we can learn nothing on this head. In this g
0189  , more especially if we look to much isolated  SPECIES, round which, according to my theory, there
0193  ants. In all these cases of two very distinct  SPECIES  furnished with apparently the same anomalous
0196  may at first have been of no advantage to the  SPECIES, but may subsequently have been taken advant
0196  taken advantage of by the descendants of the  SPECIES  under new conditions of life and with newly
0199  e of the slight analogous differences between  SPECIES. I might have adduced for this same purpose
0199  direct relation to the habits of life of each  SPECIES. Thus, we can hardly believe that the webbed
0200  possibly produce any modification in any one  SPECIES  exclusively for the good of another species;
0200  e species exclusively for the good of another  SPECIES; though throughout nature one species incess
0200  another species; though throughout nature one  SPECIES  incessantly takes advantage of, and profits
0200  uce structures for the direct injury of other  SPECIES, as we see in the fang of the adder, and in
0201  ved that any part of the structure of any one  SPECIES  had been formed for the exclusive good of an
0201  been formed for the exclusive good of another  SPECIES, it would annihilate my theory, for such cou
0203  eation are utterly obscure. We have seen that  SPECIES  at any one period are not indefinitely varia
0203  g and intermediate gradations. Closely allied  SPECIES, now living on a continuous area, must often
0204  ly glide through the air. We have seen that a  SPECIES  may under new conditions of life change its
0205  organ is so unimportant for the welfare of a  SPECIES, that modifications in its structure could n
0205  wth, and at first in no way advantageous to a  SPECIES, have been subsequently taken advantage of b
0205  he still further modified descendants of this  SPECIES. We may, also, believe that a part formerly
```

Page ***(Key Word)***

Page **************************************(Key Word)**************************************

```
0205  natural selection will produce nothing in one  SPECIES  for the exclusive good or injury of another;
0205  indispensable, or highly injurious to another  SPECIES,  but in all cases at the same time useful to
0209  s corporeal structure for the welfare of each  SPECIES,  under its present conditions of life. Under
0209  ications of instinct might be profitable to a  SPECIES;  and if it can be shown that instincts do va
0210  be found only in the lineal ancestors of each  SPECIES,  but we ought to find in the collateral line
0210  d for no instinct being known amongst extinct  SPECIES,  how very generally gradations, leading to t
0210  inct may sometimes be facilitated by the same  SPECIES  having different instincts at different peri
0210  nstances of diversity of instinct in the same  SPECIES  can be shown to occur in nature. Again as in
0210  formably with my theory, the instinct of each  SPECIES  is good for itself, but has never, as far as
0211  r the exclusive good of another of a distinct  SPECIES,  yet each species tries to take advantage of
0211  od of another of a distinct species, yet each  SPECIES  tries to take advantage of the instincts of
0212  ble cases of differences in nests of the same  SPECIES  in the northern and southern United States.
0212  eneral disposition of individuals of the same  SPECIES,  born in a state of nature, is extremely div
0212  , of occasional and strange habits in certain  SPECIES,  which might, if advantageous to the species
0212  species, which might, if advantageous to the  SPECIES,  give rise, through natural selection, to au
0215  e in the dog. All wolves, foxes, jackals, and  SPECIES  of the cat genus, when kept tame, are most e
0218  ds' nests either of the same or of a distinct  SPECIES,  is not very uncommon with the Gallinaceae:
0218  triches, at least in the case of the American  SPECIES,  unite and lay first a few eggs in one nest
0218  y had to store food for their own young. Some  SPECIES,  likewise, of Sphegidae (wasp like insects)
0218  ae (wasp like insects) are parasitic on other  SPECIES:  and M. Fabre has lately shown good reason f
0219  ional habit permanent, if of advantage to the  SPECIES,  and if the insect whose nest and stored foo
0219  pendent on its slaves: without their aid, the  SPECIES  would certainly become extinct in a single y
0219  ed by P. Huber to be a slave making ant. This  SPECIES  is found in the southern parts of England, a
0220  n all. Males and fertile females of the slave  SPECIES  are found only in their own proper communiti
0221  lsed by an independent community of the slave  SPECIES  (F. fusca) sometimes as many as three of the
0222  place a small parcel of the pupae of another  SPECIES,  F. flava, with a few of these little yellow
0222  l clinging to the fragments of the nest. This  SPECIES  is sometimes, though rarely, made into slave
0222  n described by Mr. Smith. Although so small a  SPECIES,  it is very courageous, and I have seen it f
0223  will as I have seen, carry off pupae of other  SPECIES,  if scattered near their nests, it is possib
0223  could. If their presence proved useful to the  SPECIES  which had seized them, if it were more advan
0224  ed them, if it were more advantageous to this  SPECIES  to capture workers than to procreate them, t
0224  en, is less aided by its slaves than the same  SPECIES  in Switzerland, I can see no difficulty in n
0224  pposing each modification to be of use to the  SPECIES,  until an ant was formed as abjectly depende
0234  family of bees. Of course the success of any  SPECIES  of bee may be dependent on the number of its
0238  n the fertile and sterile females of the same  SPECIES  has been produced, which we see in many soci
0238  g as distinct from each other, as are any two  SPECIES  of the same genus, or rather as any two gene
0239  casionally to find neuter insects of the same  SPECIES,  in the same nest, presenting gradations of
0240  with some of intermediate size; and, in this  SPECIES,  as Mr. F. Smith observed, the larger wo
0240  in this condition; we should then have had a  SPECIES  of ant with neuters very nearly in the same
0240  n the different castes of neuters in the same  SPECIES,  that I gladly availed myself of Mr. F. Smit
0241  y acting on the fertile parents, could form a  SPECIES  which should regularly produce neuters, eith
0243  se of closely allied, but certainly distinct,  SPECIES,  when inhabiting distant parts of the world
0245  generally entertained by naturalists is that  SPECIES,  when intercrossed, have been specially endo
0245  s view certainly seems at first probable, for  SPECIES  within the same country could hardly have ke
0246  ounded together; namely, the sterility of two  SPECIES  when first crossed, and the sterility of the
0246  ility of the hybrids produced from them. Pure  SPECIES  have of course their organs of reproduction
0246  ry, of equal importance with the sterility of  SPECIES:  for it seems to make a bread and clear dist
0246  d and clear distinction between varieties and  SPECIES.  First, for the sterility of species when cr
0246  ties and species. First, for the sterility of  SPECIES  when crossed and of their hybrid offspring.
0246  forms, considered by most authors as distinct  SPECIES,  quite fertile together, he unhesitatingly r
0247  s the maximum number of seeds produced by two  SPECIES  when crossed and by their hybrid offspring,
0247  e average number produced by both pure parent  SPECIES  in a state of nature. But a serious cause of
0247  well be permitted to doubt whether many other  SPECIES  are really so sterile, when intercrossed, as
0248  n the one hand, that the sterility of various  SPECIES  when crossed is so different in degree and g
0248  on the other hand, that the fertility of pure  SPECIES  is so easily affected by various circumstanc
0248  posite conclusions in regard to the very same  SPECIES.  It is also most instructive to compare, but
0248  er certain doubtful forms should be ranked as  SPECIES  or varieties, with the evidence from fertili
0248  rtility affords any clear distinction between  SPECIES  and varieties: but that the evidence from th
0249  entalists in great numbers; and as the parent  SPECIES,  or other allied hybrids, generally grow in
0250  fectly fertile, as fertile as the pure parent  SPECIES,  as are Kolreuter and Gartner that some degr
0250  hat some degree of sterility between distinct  SPECIES  is a universal law of nature. He experimenti
0250  e. he experimentised on some of the very same  SPECIES  as did Gartner. The difference in their resu
0250  rtility in a first cross between two distinct  SPECIES.  This case of the Crinum leads me to refer t
0250  there are individual plants, as with certain  SPECIES  of Lobelia, and with all the species of the
0250  certain species of Lobelia, and with all the  SPECIES  of the genus Hippeastrum, which can be far m
0250  tilised by the pollen of another and distinct  SPECIES,  than by their own pollen. For these plants
0250  und to yield seed to the pollen of a distinct  SPECIES,  though quite sterile with their own pollen,
0250  be perfectly good, for it fertilised distinct  SPECIES.  So that certain individual plants and all t
0250  ual plants and all the individuals of certain  SPECIES  can actually be hybridised much more readily
0250  ybrid descended from three other and distinct  SPECIES:  the result was that the ovaries of the thre
0251  wer were perfectly good with respect to other  SPECIES,  yet as they were functionally imperfect in
0251  ous causes the lesser or greater fertility of  SPECIES  when crossed, in comparison with the same sp
0251  ies when crossed, in comparison with the same  SPECIES  when self fertilised, sometimes depends. He
0251  is notorious in how complicated a manner the  SPECIES  of Pelargonium, Fuchsia, Calceolaria, Petuni
0251  rom Calceolaria integrifolia and plantaginea,  SPECIES  most widely dissimilar in general habit, rep
0251  self as perfectly as if it had been a natural  SPECIES  from the mountains of Chile. I have taken so
0252  e other species, but as not one of these nine  SPECIES  breeds freely in confinement, we have no rig
0253  the common and Chinese geese (A. cygnoides),  SPECIES  which are so different that they are general
0253  re kept for profit, where neither pure parent  SPECIES  exists, they must certainly be highly fertil
0254  ls have descended from two or more aboriginal  SPECIES,  since commingled by intercrossing. On this
0254  y intercrossing. On this view, the aboriginal  SPECIES  must either at first have produced quite fer
0254  greatly doubt, whether the several aboriginal  SPECIES  would at first have freely bred together and
0254  , i think they must be considered as distinct  SPECIES.  On this view of the origin of many of our d
0254  of the almost universal sterility of distinct  SPECIES  of animals when crossed; or we must look at
0255  to see whether or not the rules indicate that  SPECIES  have specially been endowed with this qualit
0255  te zero of fertility, the pollen of different  SPECIES  of the same genus applied to the stigma of s
```

Page **************************************(Key Word)**************************************

Page ***(Key Word)***

```
0255   same genus applied to the stigma of some one SPECIES, yields a perfect gradation in the number of
0255   cted, by the pollen of one of the pure parent SPECIES causing the flower of the hybrid to wither e
0256   eds up to perfect fertility. Hybrids from two SPECIES which are very difficult to cross, and which
0256   rict. There are many cases, in which two pure SPECIES can be united with unusual facility, and pro
0256   arkably sterile. On the other hand, there are  SPECIES which can be crossed very rarely, or with ex
0256   ble conditions, than is the fertility of pure  SPECIES. But the degree of fertility is likewise inn
0256   r it is not always the same when the same two  SPECIES are crossed under the same circumstances, bu
0256   ic affinity is meant, the resemblance between  SPECIES in structure and in constitution, more espec
0256   ortance and which differ little in the allied  SPECIES. Now the fertility of first crosses between
0257   s. now the fertility of first crosses between  SPECIES, and of the hybrids produced from them, is l
0257   n by hybrids never having been raised between  SPECIES ranked by systematists in distinct families;
0257   and on the other hand, by very closely allied  SPECIES generally uniting with facility. But the cor
0257   f cases could be given of very closely allied  SPECIES which will not unite, or only with extreme d
0257   culty; and on the other hand of very distinct  SPECIES which unite with the utmost facility. In the
0257   y be a genus, as Dianthus, in which very many  SPECIES can most readily be crossed: and another gen
0257   ave failed to produce between extremely close  SPECIES a single hybrid. Even within the limits of t
0257   this same difference; for instance, the many  SPECIES of Nicotiana have been more largely crossed
0257   tiana have been more largely crossed than the  SPECIES of almost any other genus: but Gartner found
0257   uminata, which is not a particularly distinct  SPECIES, obstinately failed to fertilise, or to be f
0257   to be fertilised by, no less than eight other  SPECIES of Nicotiana. Very many analogous facts coul
0257   isable character is sufficient to prevent two  SPECIES crossing. It can be shown that plants most w
0258   with ease. By a reciprocal cross between the same two SPECIES, I mean the case, for instance, of a stallio
0258   s, and then a male ass with a mare: these two  SPECIES may then be said to have been reciprocally c
0258   , for they prove that the capacity in any two  SPECIES to cross is often completely independent of
0258   lt of reciprocal crosses between the same two  SPECIES was long ago observed by Kølreuter. To give
0258   ugh of course compounded of the very same two  SPECIES, the one species having first been used as t
0258   pounded of the very same two species, the one  SPECIES having first been used as the father and the
0259   uld be given from Gartner: for instance, some  SPECIES have a remarkable power of crossing with oth
0259   ave a remarkable power of crossing with other  SPECIES; other species of the same genus have a rema
0259   e power of crossing with other species; other  SPECIES of the same genus have a remarkable power of
0259   h externally so like one of their pure parent  SPECIES, are with rare exceptions extremely sterile.
0259   which must be considered as good and distinct  SPECIES, are united, their fertility graduates from
0259   ility of making a first cross between any two  SPECIES is not always governed by their systematic a
0260   ed by reciprocal crosses between the same two  SPECIES, for according as the one species or the oth
0260   he same two species, for according as the one  SPECIES or the other is used as the father or the mo
0260   hese complex and singular rules indicate that  SPECIES have been endowed with sterility simply to p
0260   o extremely different in degree, when various  SPECIES are crossed, all of which we must suppose it
0260   ately variable in the individuals of the same  SPECIES? Why should some species cross with facility
0260   d yet produce very sterile hybrids; and other  SPECIES cross with facility, and yet produce very st
0260   lt of a reciprocal cross between the same two  SPECIES cross with extreme difficulty, and yet produ
0260   uction of hybrids been permitted? to grant to  SPECIES? Why, it may even be asked, has the producti
0260   , chiefly in the reproductive systems, of the  SPECIES the special power of producing hybrids, and
0261   ture, that, in reciprocal crosses between two  SPECIES which are crossed. The differences being of
0261   ilies; and, on the other hand, closely allied  SPECIES the male sexual element of the one will ofte
0261   ely allied species, and varieties of the same  SPECIES, and varieties of the same species, can usua
0261   ly have been grafted together, in other cases  SPECIES, can usually, but not invariably, be grafted
0262   ence in different individuals of the same two  SPECIES of the same genus will not take on each othe
0262   se with different individuals of the same two  SPECIES in crossing; so Sagaret believes this to be
0262   case from the difficulty of uniting two pure  SPECIES in being grafted together. As in reciprocal
0262   curs in grafting; for Thouin found that three  SPECIES, which have their reproductive organs perfec
0262   e grafted with no great difficulty on another  SPECIES of Robinia, which seeded freely on their own
0262   e rendered barren. On the other hand, certain  SPECIES, when thus grafted were rendered barren. On
0262   tain species of Sorbus, when grafted on other  SPECIES of Sorbus, when grafted on other species, yi
0262   y when fertilized with the pollen of distinct  SPECIES, yielded twice as much fruit as when on thei
0263   results of grafting and of crossing distinct  SPECIES, than when self fertilised with their own po
0263   either grafting or crossing together various  SPECIES. And as we must look at the curious and comp
0263   r, as just remarked, in the union of two pure  SPECIES has been a special endowment: although in th
0263   as also been observed that when pollen of one  SPECIES the male and female sexual elements are perf
0263   is placed on the stigma of a distantly allied  SPECIES is placed on the stigma of a distantly allie
0265   ame unnatural conditions; and whole groups of  SPECIES, though the pollen tubes protrude, they do n
0265   oduce sterile hybrids. On the other hand, one  SPECIES tend to produce sterile hybrids. On the othe
0265   itions with unimpaired fertility; and certain  SPECIES in a group will sometimes resist great chang
0265   r can he tell, till he tries, whether any two  SPECIES in a group will produce unusually fertile hy
0265   are produced by the unnatural crossing of two  SPECIES of a genus will produce more or less sterile
0267   between very distinct individuals of the same  SPECIES, the reproductive system, independently of t
0267   ses between the males and females of the same  SPECIES, that is between members of different strain
0268   re must be some essential distinction between  SPECIES which have varied and become slightly differ
0268   hey are at once ranked by most naturalists as  SPECIES and varieties, and that there must be some e
0268   , and he consequently ranks them as undoubted  SPECIES. For instance, the blue and red pimpernel, t
0268   descended from several aboriginally distinct  SPECIES. If we thus argue in a circle, the fertility
0268   act; more especially when we reflect how many  SPECIES. Nevertheless the perfect fertility of so ma
0269   that mere external dissimilarity between two  SPECIES there are, which, though resembling each oth
0269   ystem in the several descendants from any one  SPECIES does not determine their greater or lesser d
0269   as yet spoken as if the varieties of the same  SPECIES. Seeing this difference in the process of se
0270   we believe in the sterility of a multitude of  SPECIES were invariably fertile when intercrossed. B
0270   ed that these varieties of maize are distinct  SPECIES. The evidence is, also, derived from hostile
0270   of experiments made during many years on nine  SPECIES; and it is important to notice that the hybr
0270   , that yellow and white varieties of the same  SPECIES of Verbascum, by so good an observer and so
0271   s that when yellow and white varieties of one  SPECIES of Verbascum when intercrossed produce less
0271   with yellow and white varieties of a distinct  SPECIES are crossed with yellow and white varieties
0271   fertile, when crossed with a widely distinct  SPECIES, more seed is produced by the crosses betwee
0271   le in any degree would generally be ranked as  SPECIES; than are the other varieties. He experiment
0272   fundamental distinction between varieties and  SPECIES; from man selecting only external characters
0272   f the question of fertility; the offspring of  SPECIES. The general fertility of varieties does not
0272   to draw a marked line of distinction between  SPECIES when crossed and of varieties when crossed m
                                                   SPECIES and varieties, could find very few and, as i
```

Page ***(Key Word)***

```
0272 ces between the so called hybrid offspring of   SPECIES, and the so called mongrel offspring of vari
0272 hybrids; but Gartner admits that hybrids from    SPECIES which have long been cultivated are often va
0272 mits that hybrids between very closely allied    SPECIES are more variable than those from very disti
0272 e more variable than those from very distinct    SPECIES; and this shows that the difference in the d
0273 ds in the first generation are descended from    SPECIES (excluding those long cultivated) which have
0273 ee. Gartner further insists that when any two    SPECIES, although most closely allied to each other,
0274 llied to each other, are crossed with a third    SPECIES, the hybrids are widely different from each
0274 whereas if two very distinct varieties of one    SPECIES are crossed with another species, the hybrid
0274 eties of one species are crossed with another    SPECIES, the hybrids do not differ much. But this co
0274 ially in hybrids produced from nearly related    SPECIES, follows according to Gartner the same laws.
0274 according to Gartner the same laws. When two     SPECIES are crossed, one has sometimes a prepotent p
0274 y in one sex than in the other, both when one    SPECIES is crossed with another, and when one variet
0275 , than with hybrids, which are descended from    SPECIES slowly and naturally produced. On the whole
0275 ty, or of different varieties, or of distinct    SPECIES. Laying aside the question of fertility and
0275 close similarity in the offspring of crossed    SPECIES, and of crossed varieties. If we look at spe
0275 cies, and of crossed varieties. If we look at    SPECIES as having been specially created, and at var
0276 hat there is no essential distinction between    SPECIES and varieties. Summary of Chapter. First cro
0276 n forms sufficiently distinct to be ranked as    SPECIES, and their hybrids, are very generally, but
0276 innately variable in individuals of the same    SPECIES, and is eminently susceptible of favourable
0276 t, in reciprocal crosses between the same two    SPECIES. It is not always equal in degree in a first
0276 ner as in grafting trees, the capacity of one    SPECIES or variety to take on another, is incidental
0276 crossing, the greater or less facility of one    SPECIES to unite with another, is incidental on unkn
0276 ystems. There is no more reason to think that    SPECIES have been specially endowed with various deg
0276 . the sterility of first crosses between pure    SPECIES, which have their reproductive' systems perfe
0277 disturbed by being compounded of two distinct    SPECIES, seems closely allied to that sterility whic
0277 at sterility which so frequently affects pure    SPECIES, when their natural conditions of life have
0277 that the degree of difficulty in uniting two    SPECIES, and the degree of sterility of their hybrid
0277 amount of difference of some kind between the    SPECIES which are crossed. Nor is it surprising that
0277 express all kinds of resemblance between all    SPECIES. First crosses between forms known to be var
0279 t there is no fundamental distinction between    SPECIES and varieties. Chapter IX. On the Imperfecti
0279 mation. On the sudden appearance of groups of    SPECIES. On their sudden appearance in the lowest kn
0279 i endeavoured to show, that the life of each    SPECIES depends in a more important manner on the pr
0280 e found it difficult, when looking at any two    SPECIES, to avoid picturing to myself, forms directl
0280 ways look for forms intermediate between each    SPECIES and a common but unknown progenitor; and the
0281 pigeon, whether they had descended from this    SPECIES or from some other allied species, such as C
0281 d from this species or from some other allied    SPECIES, such as C. oenas. So with natural species,
0281 ed species, such as C. oenas. So with natural    SPECIES, if we look to forms very distinct, for inst
0281 recognise the parent form of any two or more    SPECIES, even if we closely compared the structure o
0281 by the theory of natural selection all living    SPECIES have been connected with the parent species
0281 g. species have been connected with the parent    SPECIES of each genus, by differences not greater, th
0281 than we see between the varieties of the same    SPECIES at the present day; and these parent species
0282 species at the present day; and these parent    SPECIES, now generally extinct, have in their turn b
0282 rn been similarly connected with more ancient    SPECIES; and so on backwards, always converging to t
0282 itional links, between all living and extinct    SPECIES, must have been inconceivably great. But ass
0287 forgotten, namely, that numbers of our fossil    SPECIES are known and named from single and often br
0288 mark are preserved. For instance, the several    SPECIES of the Chthamalinae (a sub family of sessile
0288 with the exception of a single Mediterranean    SPECIES, which inhabits deep water and has been foun
0288 found fossil in Sicily, whereas not one other    SPECIES has hitherto been found in any tertiary form
0292 ained, for the formation of new varieties and    SPECIES; but during such periods there will generall
0292 ll be much extinction, fewer new varieties or    SPECIES will be formed; and it is during these very
0293 losely graduated varieties between the allied    SPECIES which lived at its commencement and at its c
0293 s close. Some cases are on record of the same    SPECIES presenting distinct varieties in the upper a
0293 clude a graduated series of links between the    SPECIES which then lived; but I can by no means pret
0293 pared with the period requisite to change one    SPECIES into another. I am aware that two palaeontol
0293 y just conclusion on this head. When we see a    SPECIFS first appearing in the middle of any formati
0293 e previously existed. So again when we find a    SPECIES disappearing before the uppermost layers hav
0293 climatal and other changes; and when we see a    SPECIES first appearing in any formation, the probab
0294 it is well known, for instance, that several    SPECIES appeared somewhat earlier in the palaeozoic
0294 here been noted, that some few still existing    SPECIES are common in the deposit, but have become e
0294 t different levels, owing to the migration of    SPECIES and to geographical changes. And in the dist
0295 encrally have to be a very thick one; and the    SPECIES undergoing modification will have had to liv
0295 hich is necessary in order to enable the same    SPECIES to live on the same space, the supply of sed
0296 eight different levels. Hence, when the same    SPECIES occur at the bottom, middle, and top of a fo
0296 g the same geological period. So that if such    SPECIES were to undergo a considerable amount of mod
0297 s have no golden rule by which to distinguish    SPECIES and varieties; they grant some little variab
0297 s; they grant some little variability to each    SPECIES, but when they meet with a somewhat greater
0297 ence between any two forms, they rank both as    SPECIES, unless they are enabled to connect them tog
0297 ological section. Supposing B and C to be two    SPECIES, and a third, A, to be found in an underlyin
0297 ould simply be ranked as a third and distinct    SPECIES, unless at the same time it could be most cl
0297 structure. So that we might obtain the parent    SPECIES and its several modified descendants from th
0297 tly be compelled to rank them all as distinct    SPECIES. It is notorious on what excessively slight
0297 nces many palaeontologists have founded their    SPECIES; and they do this the more readily if the sp
0297 logists are now sinking many of the very fine    SPECIES of D'Orbigny and others into the rank of var
0298 ore closely allied to each other than are the    SPECIES found in more widely separated formations; b
0298 irst to local varieties and ultimately to new    SPECIES; and this again would greatly lessen the cha
0298 iate varieties and thus proved to be the same    SPECIES, until many specimens have been collected fr
0298 d from many places! and in the case of fossil    SPECIES this could rarely be effected by palaeontolo
0298 improbability of our being enabled to connect    SPECIES by numerous, fine, intermediate, fossil link
0299 are ranked by some conchologists as distinct    SPECIES from their European representatives, and by
0299 ogical research, though it has added numerous    SPECIES to existing and extinct genera, and has made
0299 hing in breaking down the distinction between    SPECIES, by connecting them together by numerous, fi
0300 whole world in organic beings; yet if all the    SPECIES were to be collected which ever lived there,
0301 ansitional gradations between any two or more    SPECIES. If such gradations were not fully preserved
0301 eties would merely appear as so many distinct    SPECIES. It is, also, probable that each great perio
0301 ve that it would be chiefly these far ranging    SPECIES which would oftenest produce new varieties;
0301 laeontologists, be ranked as new and distinct    SPECIES. If then, there be some degree of truth in t
```

Page **(Key Word)**

```
0301  redly have connected all the past and present  SPECIES of the same group into one long and branchin
0302  most palaeontologists, be ranked as distinct    SPECIES. But I do not pretend that I should ever hav
0302  ng innumerable transitional links between the   SPECIES which appeared at the commencement and close
0302  e sudden appearance of whole groups of Allied   SPECIES. The abrupt manner in which whole groups of
0302  s. the abrupt manner in which whole groups of   SPECIES suddenly appear in certain formations, has b
0302  jection to the belief in the transmutation of   SPECIES. If numerous species, belonging to the same
0302  in the transmutation of species. If numerous    SPECIES, belonging to the same genera or families, h
0302  carefully examined; we forget that groups of    SPECIES may elsewhere have long existed and have slo
0303  ill have given time for the multiplication of   SPECIES from some one or some few parent forms; and
0303  t forms; and in the succeeding formation such   SPECIES will appear as if suddenly created. I may he
0303  t that when this had been effected, and a few   SPECIES had thus acquired a great advantage over oth
0303  re to error in supposing that whole groups of   SPECIES have suddenly been produced. I may recall th
0303  rred in any tertiary stratum; but now extinct   SPECIES have been discovered in India, South America
0304  m the number of existing and extinct tertiary   SPECIES; from the extraordinary abundance of the ind
0304  ordinary abundance of the individuals of many   SPECIES all over the world, from the Arctic regions
0304  been preserved and discovered; and as not one   SPECIES had been discovered in beds of this age, I c
0304  of the abrupt appearance of a great group of    SPECIES. But my work had hardly been published, when
0305  progenitors of our many tertiary and existing   SPECIES. The case most frequently insisted on by pal
0305  arently sudden appearance of a whole group of   SPECIES, is that of the teleostean fishes, low down
0305  group includes the large majority of existing   SPECIES. Lately, Professor Pictet has carried their
0305  y, unless it could likewise be shown that the   SPECIES of this group appeared suddenly and simultan
0305  s palaeontology it will be seen that very few   SPECIES are known from several formations in Europe.
0306  they would remain confined, until some of the   SPECIES became adapted to a cooler climate, and were
0306  on the sudden appearance of groups of Allied    SPECIES in the lowest known fossiliferous strata. Th
0306  r. 1 allude to the manner in which numbers of   SPECIES of the same group, suddenly appear in the lo
0306  which have convinced me that all the existing   SPECIES of the same group have descended from one pr
0306  with nearly equal force to the earliest known   SPECIES. For instance, I cannot doubt that all the S
0306  lingula, etc., do not differ much from living   SPECIES; and it cannot on my theory be supposed, tha
0306  nnot on my theory be supposed, that these old   SPECIES were the progenitors of all the species of t
0306  e old species were the progenitors of all the   SPECIES of the orders to which they belong, for they
0307  urian system, abounding with new and peculiar   SPECIES. Traces of life have been detected in the lo
0310  numerous transitional links between the many   SPECIES which now exist or have existed; the sudden
0310  di the sudden manner in which whole groups of   SPECIES appear in our European formations; the almos
0310  en vehemently, maintained the immutability of   SPECIES. But I have reason to believe that one great
0312  on the slow and successive appearance of new    SPECIES. On their different rates of change. Species
0312  species. On their different rates of change.    SPECIES once lost do not reappear. Groups of species
0312  species once lost do not reappear. Groups of    SPECIES follow the same general rules in their appea
0312  eir appearance and disappearance as do single   SPECIES. On Extinction. On simultaneous changes in t
0312  ghout the world. On the affinities of extinct   SPECIES to each other and to living species. On the
0312  f extinct species to each other and to living   SPECIES. On the state of development of ancient form
0312  d with the common view of the immutability of   SPECIES, or with that of their slow and gradual modi
0312  n, through descent and natural selection. New   SPECIES have appeared very slowly, one after another
0312  tiquity if measured by years, only one or two   SPECIES are lost forms, and only one or two are new
0313  e nor disappearance of their many now extinct   SPECIES has been simultaneous in each separate forma
0313  been simultaneous in each separate formation.   SPECIES of different genera and classes have not cha
0313  an lingula differs but little from the living   SPECIES of this genus; whereas most of the other Sil
0313  the most closely related formations, all the   SPECIES will be found to have undergone some change.
0313  e found to have undergone some change. When a   SPECIES has once disappeared from the face of the ea
0314  st be extremely slow. The variability of each   SPECIES is quite independent of that of all others.
0314  lesser amount of modification in the varying   SPECIES, depends on many complex contingencies, on t
0314  the other inhabitants with which the varying   SPECIES comes into competition. Hence it is by no me
0314  . hence it is by no means surprising that one   SPECIES should retain the same identical form much l
0315  be exterminated. Hence we can see why all the   SPECIES in the same region do at last, if we look to
0315  anging drama. We can clearly understand why a   SPECIES when once lost should never reappear, even i
0315  should recur. For though the offspring of one   SPECIES might be adapted (and no doubt this has occu
0315  instances) to fill the exact place of another   SPECIES in the economy of nature, and thus supplant
0316  xisting breed, could be raised from any other   SPECIES of pigeon, or even from the other well estab
0316  slight characteristic differences. Groups of   SPECIES, that is, genera and families, follow the sa
0316  eir appearance and disappearance as do single   SPECIES, changing more or less quickly, and in a gre
0316  rictly accords with my theory. For as all the   SPECIES of the same group have descended from one sp
0316  ies of the same group have descended from one   SPECIES, it is clear that as long as any species of
0316  one species, it is clear that as long as any   SPECIES of the group have appeared in the long succe
0316  odified or the same old and unmodified forms.   SPECIES of the genus Lingula, for instance, must hav
0317  ay. We have seen in the last chapter that the   SPECIES of a group sometimes falsely appear to have
0317  it gradually decreases. If the number of the   SPECIES of a genus, or the number of the genera of a
0317  successive geological formations in which the   SPECIES are found, the line will sometimes falsely a
0317  king the decrease and final extinction of the   SPECIES. This gradual increase in number of the spec
0317  ecies. This gradual increase in number of the   SPECIES of a group is strictly conformable with my t
0317  s strictly conformable with my theory; as the   SPECIES of the same genus, and the genera of the sam
0317  of allied forms must be slow and gradual, one   SPECIES giving rise first to two or three varieties,
0317  varieties, these being slowly converted into   SPECIES, which in their turn produce by equally slow
0317  heir turn produce by equally slow steps other   SPECIES, and so on, like the branching of a great tr
0317  ken only incidentally of the disappearance of   SPECIES and of groups of species. On the theory of n
0317  the disappearance of species and of groups of   SPECIES. On the theory of natural selection the exti
0317  om the study of the tertiary formations, that   SPECIES and groups of species gradually disappear, o
0317  rtiary formations, that species and groups of   SPECIES gradually disappear, one after another, firs
0318  ther, and finally from the world. Both single   SPECIES and whole groups of species last for very un
0318  orlc. Both single species and whole groups of   SPECIES last for very unequal periods; some groups,
0318  ne the length of time during which any single   SPECIES or any single genus endures. There is reason
0318  o believe that the complete extinction of the   SPECIES of a group is generally a slower process tha
0318  he appearance and disappearance of a group of   SPECIES be represented, as before, by a vertical lin
0318  rst appearance and increase in numbers of the   SPECIES. In some cases, however, the extermination o
0318  udden. The whole subject of the extinction of   SPECIES has been involved in the most gratuitous mys
0318  vidual has a definite length of life, so have   SPECIES a definite duration. No one I think can have
0318  can have marvelled more at the extinction of   SPECIES, than I have done. When I found in La Plata
0319  of the existing horse, belonged to an extinct   SPECIES. Had this horse been still living, but in so
```

Page **(Key Word)**

Page **(Key Word)**

```
0319  r rarity is the attribute of a vast number of  SPECIES of all classes, in all countries. If we ask
0319  untries. If we ask ourselves why this or that    SPECIES is rare, we answer that something is unfavou
0319  of the fossil horse still existing as a rare    SPECIES, we might have felt certain from the analogy
0320  published in 1845, namely, that to admit that   SPECIES generally become rare before they become ext
0320  tinct, to feel no surprise at the rarity of a   SPECIES, and yet to marvel greatly when it ceases to
0320  hat each new variety, and ultimately each new   SPECIES, is produced and maintained by having some a
0320  exterminated; but we know that the number of    SPECIES has not gone on indefinitely increasing, and
0321  ce the improved and modified descendants of a   SPECIES will generally cause the extermination of th
0321  nerally cause the extermination of the parent   SPECIES; and if many new forms have been developed f
0321  ny new forms have been developed from any one   SPECIES, the nearest allies of that species, i.e. th
0321  m any one species, the nearest allies of that   SPECIES, i.e. the species of the same genus, will be
0321  the nearest allies of that species, i.e. the    SPECIES of the same genus, will be the most liable t
0321  mination. Thus, as I believe, a number of new   SPECIES descended from one species, that is a new ge
0321  e, a number of new species descended from one   SPECIES, that is a new genus, comes to supplant an o
0321  y. but it must often have happened that a new   SPECIES belonging to some one group will have seized
0321  p will have seized on the place occupied by a   SPECIES belonging to a distinct group, and thus caus
0321  ited inferiority in common. But whether it be   SPECIES belonging to the same or to a distinct class
0321  inct class, which yield their places to other   SPECIES which have been modified and improved, a few
0321  ed severe competition. For instance, a single   SPECIES of Trigonia, a great genus of shells in the
0322  ation or by unusually rapid development, many   SPECIES of a new group have taken possession of a ne
0322  as it seems to me, the manner in which single   SPECIES and whole groups of species become extinct,
0322  r in which single species and whole groups of   SPECIES become extinct, accords well with the theory
0322  contingencies, on which the existence of that   SPECIES depends. If we forget for an instant, that e
0322  pends. If we forget for an instant, that each   SPECIES tends to increase inordinately, and that som
0322  cured. Whenever we can precisely say why this   SPECIES is more abundant in individuals than that: w
0322  e abundant in individuals than that: why this   SPECIES and not another can be naturalised in a give
0322  account for the extinction of this particular   SPECIES or group of species. On the Forms of Life ch
0322  nction of this particular species or group of   SPECIES. On the Forms of Life changing almost simult
0323  o those of the Chalk. It is not that the same   SPECIES are met with; for in some cases not one spec
0323  ecies are met with; for in some cases not one   SPECIES is identically the same; but they belong to
0323  can tertiary deposits. Even if the new fossil   SPECIES which are common to the Old and New Worlds b
0325  ppear certain that all these modifications of   SPECIES, their extinction, and the introduction of n
0325  cable on the theory of natural selection. New   SPECIES are formed by new varieties arising, which h
0325  enest give rise to new varieties or incipient   SPECIES: for these latter must be victorious in a st
0326  that the dominant, varying, and far spreading   SPECIES, which already have invaded to a certain ext
0326  to a certain extent the territories of other   SPECIES, should be those which would have the best c
0326  no rise in new countries to new varieties and   SPECIES. The process of diffusion may often be very
0326  uctions of the land than of the sea. Dominant   SPECIES spreading from any region might encounter st
0326  ny region might encounter still more dominant   SPECIES, and then their triumphant course, or even t
0326  le for the multiplication of new and dominant   SPECIES: but we can, I think, clearly see that a num
0326  urable for the production of new and dominant   SPECIES on the land, and another for those in the wa
0327  world, accords well with the principle of new   SPECIES having been formed by dominant species sprea
0327  of new species having been formed by dominant   SPECIES spreading widely and varying; the new specie
0327  species spreading widely and varying; the new   SPECIES thus produced being themselves dominant owin
0327  me advantage over their parents or over other   SPECIES: these again spreading, varying, and produci
0327  e again spreading, varying, and producing new   SPECIES. The forms which are beaten and which yield
0328  eral succession in the forms of life; but the   SPECIES would not exactly correspond; for there will
0328  th a curious accordance in the numbers of the   SPECIFS belonging to the same genera, yet the specie
0328  species belonging to the same genera, yet the   SPECIES themselves differ in a manner very difficult
0328  inds a surprising amount of difference in the   SPECIES. If the several formations in these regions
0329  rval in the other, and if in both regions the   SPECIES have gone on slowly changing during the accu
0329  ear to be strictly parallel: nevertheless the   SPECIES would not all be the same in the apparently
0329  the two regions. On the Affinities of extinct   SPECIES to each other, and to living forms. Let us n
0329  o the mutual affinities of extinct and living   SPECIES. They all fall into one grand natural system
0330  re. Some writers have objected to any extinct   SPECIES or group of species being considered as inte
0330  e objected to any extinct species or group of   SPECIES being considered as intermediate between liv
0330  ing considered as intermediate between living   SPECIES or groups. If by this term it is meant that
0330  perfectly natural classification many fossil   SPECIES would have to stand between living species,
0330  il species would have to stand between living   SPECIES, and some extinct genera between living gene
0331  and the dotted lines diverging from them the   SPECIES in each genus. The diagram is much too simpl
0331  s much too simple, too few genera and too few   SPECIES being given, but this is unimportant for us.
0331  : it depends solely on the descendants from a   SPECIES being thus enabled to seize on many and diff
0331  en in the case of some Silurian forms, that a   SPECIES might go on being slightly modified in relat
0334  eceded and that which succeeded it. Thus, the   SPECIES which lived at the sixth great stage of desc
0334  existence, do not accord in arrangement. The   SPECIES extreme in character are not the oldest, or
0335  the first appearance and disappearance of the   SPECIES was perfect, we have no reason to believe th
0335  ral stages of the chalk formation, though the   SPECIES are distinct in each stage. This fact alone,
0335  tet in his firm belief in the immutability of   SPECIES. He who is acquainted with the distribution
0335  acquainted with the distribution of existing   SPECIES over the globe, will not attempt to account
0335  unt for the close resemblance of the distinct   SPECIES in closely consecutive formations, by the ph
0336  ecutive formations, though ranked as distinct   SPECIES, being closely related, is obvious. As the a
0336  ns all the intermediate varieties between the   SPECIES which appeared at the commencement and close
0336  e been called by some authors, representative   SPECIES: and these we assuredly do find. We find, in
0337  be higher than the more ancient: for each new   SPECIES is formed by having had some advantage in th
0338  skilful naturalist from an examination of the   SPECIES of the two countries could not have foreseen
0341  n the caves of Brazil, there are many extinct   SPECIES which are closely allied in size and in othe
0341  allied in size and in other characters to the   SPECIES still living in South America: and some of t
0341  ssils may be the actual progenitors of living   SPECIES. It must not be forgotten that, on my theory
0341  not be forgotten that, on my theory, all the   SPECIES of the same genus have descended from some o
0341  f the same genus have descended from some one   SPECIES: so that if six genera, each having eight sp
0341  ies: so that if six genera, each having eight   SPECIES, be found in one geological formation, and i
0341  representative genera with the same number of   SPECIES, then we may conclude that only one species
0341  f species, then we may conclude that only one   SPECIES of each of the six older genera has left mod
0341  stituting the six new genera. The other seven   SPECIES of the old genera have all died out and have
0341  probably be a far commoner case, two or three   SPECIES of two or three alone of the six older gener
0341  parents of the six new genera: the other old   SPECIES and the other whole genera having become utt
```

Page **(Key Word)**

Page ***(Key Word)***

```
0341 tinct. In failing orders, with the genera and   SPECIES decreasing in numbers, as apparently is the
0341 tata of South America, still fewer genera and    SPECIES will have left modified blood descendants. S
0342 ate; that the number both of specimens and of    SPECIES, preserved in our museums, is absolutely as
0342 y one area and formation; that widely ranging     SPECIES are those which have varied most, and have o
0342 ied most, and have oftenest given rise to new     SPECIES; and that varieties have at first often been
0342 onnected the closely allied or representative     SPECIES, found in the several stages of the same gre
0343 apparent, sudden coming in of whole groups of     SPECIES. He may ask where are the remains of those i
0343 on. We can thus understand how it is that new     SPECIES come in slowly and successively; how species
0343 species come in slowly and successively; how      SPECIES of different classes do not necessarily chan
0343 on of new forms. We can understand why when a     SPECIES has once disappeared it never reappears. Gro
0343 nce disappeared it never reappears. Groups of     SPECIES increase in numbers slowly, and endure for u
0343 s on many complex contingencies. The dominant     SPECIES of the larger dominant groups tend to leave
0344 d groups are formed. As these are formed, the     SPECIES of the less vigorous groups, from their infe
0344 but the utter extinction of a whole group of      SPECIES may often be a very slow process, from the s
0344 cceed in taking the places of those groups of     SPECIES which are their inferiors in the struggle fo
0345 ogy plainly proclaim, as it seems to me, that     SPECIES have been produced by ordinary generation: o
0346 d in the New, at least as closely as the same     SPECIES generally require; for it is a most rare cas
0349 ions of the same continent or sea; though the     SPECIES themselves are distinct at different points
0349 the Straits of Magellan are inhabited by one      SPECIES of Rhea (American ostrich), and northward th
0349 d northward the plains of La Plata by another     SPECIES of the same genus; and not by a true ostrich
0349 peaks of the Cordillera and we find an alpine     SPECIES of bizcacha; we look to the waters, and we d
0349 inhabitants, though they may be all peculiar      SPECIES, are essentially American. We may look back
0350 ion through natural selection. Widely ranging     SPECIES, abounding in individuals, which have alread
0351 ssary development. As the variability of each     SPECIES is an independent property, and will be take
0351 e, so the degree of modification in different     SPECIES will be no uniform quantity. If, for instanc
0351 iform quantity. If, for instance, a number of     SPECIES, which stand in direct competition with each
0351 ormously remote geological period, so certain     SPECIES have migrated over vast spaces, and have not
0351 these views, it is obvious, that the several     SPECIES of the same genus, though inhabiting the mos
0351 rom the same progenitor. In the case of those     SPECIES, which have undergone during whole geologica
0351 , in which we have reason to believe that the     SPECIES of a genus have been produced within compara
0352 also obvious that the individuals of the same     SPECIES, though now inhabiting distant and isolated
0352 ely discussed by naturalists, namely, whether     SPECIES have been created at one or more points of t
0352 eme difficulty, in understanding how the same     SPECIES could possibly have migrated from some one p
0352 rtheless the simplicity of the view that each     SPECIES was first produced within a single region ca
0352 d, that in most cases the area inhabited by a     SPECIES is continuous; and when a plant or animal in
0352 ssessing the same quadrupeds. But if the same     SPECIES can be produced at two separate points, why
0353 e only on the view that the great majority of     SPECIES have been produced on one side alone, and ha
0353 natural genera, or those genera in which the     SPECIES are most closely related to each other, are
0353 in the series, to the individuals of the same     SPECIES, a directly opposite rule prevailed; and sne
0353 cies, a directly opposite rule prevailed; and     SPECIES were not local, but had been produced in two
0353 many other naturalists, that the view of each     SPECIES having been produced in one area alone, and
0353 ccur, in which we cannot explain how the same     SPECIES could have passed from one point to the othe
0353 tinuous the formerly continuous range of many     SPECIES. So that we are reduced to consider whether
0354 probable by general considerations, that each     SPECIES has been produced within one area, and has m
0354 discuss all the exceptional cases of the same     SPECIES, now living at distant and separated points;
0354 s of facts; namely, the existence of the same     SPECIES on the summits of distant mountain ranges, a
0354 irdly, the occurrence of the same terrestrial     SPECIES on islands and on the mainland, though separ
0354 les of open sea. If the existence of the same     SPECIES at distant and isolated points of the earth'
0354 ny instances be explained on the view of each     SPECIES having migrated from a single birthplace; th
0354 for us; namely, whether the several distinct      SPECIES of a genus, which on my theory have all desc
0354 ted to, or belong to the same genera with the     SPECIES of a second region, has probably received at
0355 endent creation. This view of the relation to     SPECIES in one region to those in another, does not
0355 er much (by substituting the word variety for     SPECIES) from that lately advanced in an ingenious p
0355 r. wallace, in which he concludes, that every     SPECIES has come into existence coincident both in s
0355 e and time with a pre existing closely allied     SPECIES. And I now know from correspondence, that th
0355 amely whether all the individuals of the same     SPECIES have descended from a single pair, or single
0355 s which never intercross (if such exist), the     SPECIES, on my theory, must have descended from a su
0356 rocess of modification the individuals of the     SPECIES will have been kept nearly uniform by interc
0357 the Gordian knot of the dispersal of the same     SPECIES to the most distant points, and removes many
0357 aphical changes within the period of existing     SPECIES. It seems to me that we have abundant eviden
0357 , as I believe it will some day be, that each     SPECIES has proceeded from a single birthplace; and
0359 plants with ripe fruit (but not all the same     SPECIES as in the foregoing experiment) floated, aft
0363 . in the Azores, from the large number of the     SPECIES of plants common to Europe, in comparison wi
0365 dreds of miles of low lands, where the Alpine     SPECIES could not possibly exist, is one of the most
0365 of the most striking cases known of the same     SPECIES living at distant points, without the appare
0365 ch facts led Gmelin to conclude that the same     SPECIES must have been independently created at seve
0367 he warmth had fully returned, the same arctic     SPECIES, which had lately lived in a body together o
0368 , that when in other regions we find the same     SPECIES on distant mountain summits, we may almost c
0369 for it is not likely that all the same arctic     SPECIES will have been left on mountain ranges dista
0369 ility have become mingled with ancient Alpine     SPECIES, which must have existed on the mountains be
0369 pean mountain ranges, though very many of the     SPECIES are identically the same, some present varie
0369 distinct yet closely allied or representative     SPECIES. In illustrating what, as I believe, actuall
0371 warmth of the Pliocene period, as soon as the     SPECIES in common, which inhabited the New and Old W
0372 ew and Old Worlds, we find very few identical     SPECIES (though Asa Gray has lately shown that more
0372 as geographical races, and others as distinct     SPECIES; and a host of closely allied or representat
0374 present distribution of identical and allied     SPECIES. In America, Dr. Hooker has shown that betwe
0374 points are; and there are many closely allied     SPECIES. On the lofty mountains of equatorial Americ
0374 ains of equatorial America a host of peculiar     SPECIES belonging to European genera occur. On the h
0374 accas the illustrious Humboldt long ago found     SPECIES belonging to genera characteristic of the Co
0375 at the Cape of Good Hope a very few European      SPECIES, believed not to have been introduced by man
0375 dr. F. Muller has discovered several European     SPECIES; other species, not introduced by man, occur
0375 as discovered several European species; other     SPECIES, not introduced by man, occur on the lowland
0376 fish. Dr. Hooker informs me that twenty five      SPECIES of Algae are common to New Zealand and to Eu
0376 seas. It should be observed that the northern     SPECIES and forms found in the southern parts of the
0376 to northern forms, must be ranked as distinct     SPECIES. Now let us see what light can be thrown on
0377 erhaps formerly the tropics supported as many     SPECIES as we see at the present day crowded togethe
```

Page ***(Key Word)***

Page ***(Key Word)**

```
0379 homes as well marked varieties or as distinct SPECIES. It is a remarkable fact, strongly insisted
0380 ard to the range and affinities of the allied SPECIES which live in the northern and southern temp
0380 means of migration, or the reason why certain SPECIESand not other have migrated: why certain spec
0381 eciesand not other have migrated; why certain SPECIES have been modified and have given rise to ne
0381 explain such facts, until we can say why one SPECIES and not another becomes naturalised by man's
0381 and is twice or thrice as common, as another SPECIES within their own homes. I have said that man
0381 as far as regards the occurrence of identical SPECIES at points so enormously remote as Kerguelen
0381 . but the existence of several quite distinct SPECIES, belonging to genera exclusively confined to
0381 makable case of difficulty. For some of these SPECIES are so distinct, that we cannot suppose that
0381 e to indicate that peculiar and very distinct SPECIES have migrated in radiating lines from some c
0383 y the reverse. Not only have many fresh water SPECIES, belonging to quite different classes, an en
0383 ferent classes, an enormous range, but allied SPECIES prevail in a remarkable manner throughout th
0384 s. in regard to fish, I believe that the same SPECIES never occur in the fresh waters of distant c
0384 ant continents. But on the same continent the SPECIES often range widely and almost capriciously:
0385 d to the fresh waters of a distant land. Some SPECIFS of fresh water shells have a very wide range
0385 ter shells have a very wide range, and allied SPECIES, which, on my theory, are descended from a c
0385 ould not even understand how some naturalised SPECIES have rapidly spread throughout the same coun
0386 ormous ranges many fresh water and even marsh SPECIES have, both over continents and to the most r
0388 uggle for life between the individuals of the SPECIES, however few, already occupying any pond, ye
0388 vere between aquatic than between terrestrial SPECIES: consequently an intruder from the waters of
0388 average for the migration of the same aquatic SPECIES, we should not forget the probability of man
0388 we should not forget the probability of many SPECIES having formerly ranged as continuously as fr
0389 he individuals both of the same and of allied SPECIES have descended from a single parent; and the
0389 reation and of descent with modification. The SPECIES of all kinds which inhabit oceanic islands a
0390 the doctrine of the creation of each separate SPECIES, will have to admit that a sufficient number
0390 abitants is scanty, the proportion of endemic SPECIES (i.e. those found nowhere else in the world)
0390 cted on my theory, for, as already explained, SPECIES occasionally arriving after long intervals i
0390 ws, that, because in an island nearly all the SPECIFS of one class are peculiar, those of another
0390 ri and this difference seems to depend on the SPECIES which do not become modified having immigrat
0391 mber of peculiar land shells; whereas not one SPECIES of sea shell is confined to its shores: now,
0392 ining its hooked seeds, would form an endemic SPECIES, having as useless an appendage as any rudim
0392 rders which elsewhere include only herbaceous SPECIES; now trees, as Alph. de Candolle has shown,
0394 also been time for the production of endemic SPFCIES belonging to other classes: and on continent
0394 er the Atlantic Ocean; and two North American SPECIES either regularly or occasionally visit Bermu
0394 ly studied this family, that many of the same SPECIES have enormous ranges, and are found on conti
0394 e we have only to suppose that such wandering SPECIES have been modified through natural selection
0395 the presence in both of the same mammiferous SPECIES or of allied species in a more or less modif
0395 of the same mammiferous species or of allied SPECIES in a more or less modified condition. Mr. Wi
0395 pth, and here we find American forms, but the SPECIES and even the genera are distinct. As the amo
0397 nhabited by land shells, generally by endemic SPECIES, but sometimes by species found elsewhere. D
0397 enerally by endemic species, but sometimes by SPECIES found elsewhere. Dr. Aug. A. Gould has given
0397 ide arms of the sea. And I found that several SPECIES did in this state withstand uninjured an imm
0397 nty days, and it perfectly recovered. As this SPECIFS has a thick calcareous operculum, I removed
0397  not mainland, without being actually the same SPECIES. Numerous instances could be given of this f
0398 of these are ranked by Mr. Gould as distinct SPECIES, supposed to have been created here; yet the
0398 e affinity of most of these birds to American SPECIES in every character, in their habits, gesture
0398 n land. Why should this be so? Why should the SPECIES which are supposed to have been created in t
0400 te marvellous manner, by very closely related SPECIES; so that the inhabitants of each separate is
0400 laying on one side for the moment the endemic SPECIES, which cannot be here fairly included, as we
0401 rent varieties in the different islands. Some SPECIES, however, might spread and yet retain the sa
0401 the group, just as we see on continents some SPECIES spreading widely and remaining the same. The
0401 in some analogous instances, is that the new SPECIES formed in the separate islands have not quic
0401 pear to be on a map. Nevertheless a good many SPECIES, both those found in other parts of the worl
0402 ous view of the probability of closely allied SPECIES invading each other's territory, when put in
0402 o free intercommunication. Undoubtedly if one SPECIES has any advantage whatever over another, it
0402 time. Being familiar with the fact that many SPECIES, naturalised through man's agency, have spre
0402 new countries, we are apt to infer that most SPECIES would thus spread: but we should remember th
0402 aboriginal inhabitants, but are very distinct SPECIES, belonging in a large proportion of cases, a
0402 on each; thus there are three closely allied SPECIES of mocking thrush, each confined to its own
0402 t charles Island is well stocked with its own SPECIES, for annually more eggs are laid there than
0402 t least as well fitted for its home as is the SPECIES peculiar to Chatham Island. Sir C. Lyell and
0403 d has not become colonised by the Porto Santo SPECIES: nevertheless both islands have been colonis
0403 doubt had some advantage over the indigenous SPECIES. From these considerations I think we need n
0403 atly marvel at the endemic and representative SPECIES, which inhabit the several islands of the Ga
0403 important part in checking the commingling of SPECIES under the same conditions of life. Thus, the
0403 mountain, in every lake and marsh. For Alpine SPECIES, excepting in so far as the same forms, chie
0404 istant, many closely allied or representative SPECIES occur, there will likewise be found some ide
0404 , there will likewise be found some identical SPECIES, showing, in accordance with the foregoing v
0404 two regions. And wherever many closely allied SPECIES occur, there will be found many forms which
0404 forms which some naturalists rank as distinct SPECIES, and some as varieties; these doubtful forms
0404 etween the power and extent of migration of a SPECIES, either at the present time or at some forme
0404 stence at remote points of the world of other SPECIES allied to it, is shown in another and more g
0404 birds which range over the world, many of the SPECIES have very wide ranges. I can hardly doubt th
0404 era range over the world, and many individual SPECIES have enormous ranges. It is not meant that i
0405 ot meant that in world ranging genera all the SPECIES have a wide range, or even that they have on
0405 erage a wide range; but only that some of the SPECIES range very widely: for the facility with whi
0405 y; for the facility with which widely ranging SPECIES vary and give rise to new forms will largely
0405 ange. For instance, two varieties of the same SPECIES inhabit America and Europe, and the species
0405 e species inhabit America and Europe, and the SPECIES thus has an immense range: but, if the varia
0405 varieties would have been ranked as distinct SPECIES, and the common range would have been greatl
0405 eatly reduced. Still less is it meant, that a SPECIES which apparently has the capacity of crossin
0405 oreign associates. But on the view of all the SPECIES of a genus having descended from a single pa
0405 al rule we do find, that some at least of the SPECIES range very widely: for it is necessary that
0405 ly into new varieties and ultimately into new SPECIES. In considering the wide distribution of cer
0405 consequently for the migration of some of the SPECIES into all quarters of the world, where they m
0406 anging more widely than the high, some of the SPECIES of widely ranging genera themselves ranging
```

Page ***(Key Word)**

Page **(Key Word)**

```
0406  rent, the very close relation of the distinct  SPECIES which inhabit the islets of the same archipe
0406  nary view of the independent creation of each  SPECIES, but are explicable on the view of colonisat
0407  rimentised on: if we bear in mind how often a  SPECIES may have ranged continuously over a wide are
0407  elieving that all the individuals of the same  SPECIES, wherever located, have descended from the s
0407  a, and families. With respect to the distinct  SPECIES of the same genus, which on my theory must h
0407  e, and in that of the individuals of the same  SPECIES, extremely grave. As exemplifying the effect
0408  ng course of time the individuals of the same  SPECIES, and likewise of allied species; have procee
0408  s of the same soecies, and likewise of allied  SPECIES, have proceeded from some one source; then I
0409  en within the same class, should have all its  SPECIFS endemic, and another group should have all i
0409  ndemic, and another group should have all its  SPECIES common to other quarters of the world. We ca
0409  t isolated islands possess their own peculiar  SPECIES of aerial mammals or bats. We can see why th
0409  e a correlation, in the presence of identical  SPECIES, of varieties, of doubtful species, and of d
0409  identical species, of varieties, of doubtful  SPECIES, and of distinct but representative species.
0409  l species; and of distinct but representative  SPECIES. As the Late Edward Forbes often insisted, t
0409  see this in many facts. The endurance of each  SPECIES and group of species is continuous in time;
0409  s. the endurance of each species and group of  SPECIES is continuous in time; for the exceptions to
0410  eral rule that the area inhabited by a single  SPECIES, or by a group of species, is continuous: an
0410  habited by a single species; or by a group of  SPECIES, is continuous; and the exceptions, which ar
0410  by occasional means of transport, and by the  SPECIES having become extinct in the intermediate tr
0410  intermediate tracts. Both in time and space,  SPECIES and groups of species have their points of m
0410  both in time and space, species and groups of  SPECIES have their points of maximum development. Gr
0410  heir points of maximum development. Groups of  SPECIES, belonging either to a certain period of tim
0411  uch diffused and common, that is the dominant  SPECIES belonging to the larger genera, which vary m
0411  which vary most. The varieties, or incipient  SPECIES, thus produced ultimately become converted,
0411  onverted, as I believe, into new and distinct  SPECIES; and these, on the principle of inheritance,
0412  tance, tend to produce other new and dominant  SPECIES. Consequently the groups which are now large
0412  ge, and which generally include many dominant  SPECIES, tend to go on increasing indefinitely in si
0412  how that from the varying descendants of each  SPECIES trying to occupy as many and as different pl
0412  line may represent a genus including several  SPECIES; and all the genera on this line form togeth
0413  descended from (I). So that we here have many  SPECIFS descended from a single progenitor grouped i
0413  lly explained. Naturalists try to arrange the  SPECIES, genera, and families in each class, on what
0415  greater constancy throughout large groups of  SPECIES; and this constancy depends on such organs h
0415  ected to less change in the adaptation of the  SPECIES to their conditions of life. That the mere p
0417  history. Hence, as has often been remarked, a  SPECIFS may depart from its allies in several charac
0417  reater number of the characters proper to the  SPECIES, to the genus, to the family, to the class,
0418  ning a group; or in allocating any particular  SPECIES. If they find a character nearly uniform, an
0418  ifications of course include all ages of each  SPECIES. But it is by no means obvious, on the ordin
0419  ch have hardly a character in common; yet the  SPECIES at both ends, from being plainly allied to o
0419  he comparative value of the various groups of  SPECIES, such as orders, sub orders, families, sub f
0419  first overlooked, but because numerous allied  SPECIES, with slightly different grades of differenc
0420  showing true affinity between any two or more  SPECIES, are those which have been inherited from a
0420  lurian epoch, and these have descended from a  SPECIES which existed at an unknown anterior period.
0420  which existed at an unknown anterior period.  SPECIES of three of these genera (A, F, and I) have
0421  en up into two families. Nor can the existing  SPECIES, are represented as related in blood or desc
0423  believed or known to have descended from one  SPECIES. These are grouped under species, with sub v
0423  ded from one species. These are grouped under  SPECIES, with sub varieties under varieties; and wit
0423  to groups, is the same with varieties as with  SPECIES, namely, closeness of descent with various d
0423  re followed in classifying varieties, as with  SPECIES. Authors have insisted on the necessity of c
0424  other important characters from negroes. With  SPECIFS in a state of nature, every naturalist has i
0424  he includes in his lowest grade, or that of a  SPECIES, the two sexes; and how enormously these som
0424  parating them. The naturalist includes as one  SPECIES the several larval stages of the same indivi
0424  r conversely, ranks them together as a single  SPECIES, and gives a single definition. As soon as t
0424  e, they were immediately included as a single  SPECIES. But it may be asked, what ought we to do, i
0425  ught we to do, if it could be proved that one  SPECIES of kangaroo had been produced, by a long cou
0425  ation, from a bear? Ought we to rank this one  SPECIES with bears, and what should we do with the o
0425  h bears, and what should we do with the other  SPECIES? The supposition is of course preposterous;
0425  with bears; but then assuredly all the other  SPECIES of the kangaroo family would have to be clas
0425  classing together the individuals of the same  SPECIES, though the males and females and larvae are
0425  cent have been unconsciously used in grouping  SPECIES under genera, and genera under higher groups
0425  ation to the conditions of life to which each  SPECIES has been recently exposed. Rudimentary struc
0426  , if it prevail throughout many and different  SPECIES, especially those having very cifferent habi
0426  ue in classification. We can understand why a  SPECIES or a group of species may depart, in several
0426  we can understand why a species or a group of  SPECIES may depart, in several of its most important
0427  nd widely distributed genera, because all the  SPECIES of the same genus, inhabiting any distinct a
0428  isen. As the modified descendants of dominant  SPECIES, belonging to the larger genera, tend to inh
0429  ey are generally represented by extremely few  SPECIES; and such species as do occur are generally
0429  epresented by extremely few species; and such  SPECIES as do occur are generally very distinct from
0429  aberrant had each been represented by a dozen  SPECIES instead of a single one; but such richness i
0429  instead of a single one; but such richness in  SPECIES, as I find after some investigation, does no
0430  ons are general, and not to any one marsupial  SPECIES more than to another. As the points of affin
0430  hascolomys resembles most nearly, not any one  SPECIES, but the general order of Rodents. In this c
0430  on and gradual divergence in character of the  SPECIES descended from a common parent, together wit
0431  r. for the common parent of a whole family of  SPECIES, now broken up by extinction into distinct g
0431  ous ways and degrees; to all: and the several  SPECIES will consequently be related to each other b
0433  the many descendants from one dominant parent  SPECIES, explains that great and universal feature i
0433  gh having few characters in common, under one  SPECIES; we use descent in classing acknowledged var
0434  the several parts and organs in the different  SPECIES of the class are homologous. The whole subje
0438  cs, though we can homologise the parts of one  SPECIES with those of another and cistinct species;
0438  ne species with those of another and distinct  SPECIES, we can indicate but few serial homologies;
0439  e thrush group. In the cat tribe, most of the  SPECIES are striped or spotted in lines; and stripes
0440  and cases could be given of the larvae of two  SPECIES, or of two groups of species, ciffering guit
0440  he larvae of two species, or of two groups of  SPECIFS, differing quite as much, or even more, from
0442  riod of growth alike; of embryos of different  SPECIES within the same class, generally, but not un
0444  many successive modifications, by which each  SPECIES has acquired its present structure, may have
0445  breeds of Pigeon have descended from one wild  SPECIES, I compared young pigeons cf various breeds,
```

Page **(Key Word)**

```
0446  an be shown to be in some degree probable, to  SPECIES in a state of nature. Let us take a genus of
0446  , descended on my theory from some one parent  SPECIES, and of which the several new species have b
0446  parent species, and of which the several new   SPECIES have become modified through natural selecti
0447  at a corresponding age, the young of the new   SPECIES of our supposed genus will manifestly tend t
0447  instance, which served as legs in the parent   SPECIES, may become, by a long course of modificatio
0447  ryos of the several descendants of the parent  SPECIES will still resemble each other closely, for
0447  ave been modified. But in each individual new   SPECIES, the embryonic fore limbs will differ greatl
0448  uld be indispensable for the existence of the  SPECIES, that the child should be modified at a very
0449  f crustaceans. As the embryonic state of each  SPECIES and group of species partially shows us the
0449  embryonic state of each species and group of   SPECIES partially shows us the structure of their le
0449  he embryos of their descendants, our existing  SPECIES. Agassiz believes this to be a law of nature
0451  etles of the same genus (and even of the same  SPECIES) resembling each other most closely in all r
0451  d give milk. In individual plants of the same  SPECIES the petals sometimes occur as mere rudiments
0451  ossing such male plants with an hermaphrodite  SPECIES, the rudiment of the pistil in the hybrid of
0452  mentary organs in the individuals of the same  SPECIES are very liable to vary in degree of develop
0452  n other respects. Moreover, in closely allied  SPECIES, the degree to which the same organ has been
0452  ionally found in monstrous individuals of the  SPECIES. Thus in the snapdragon (antirrhinum) we gen
0454  udiments can be produced: for I doubt whether  SPECIES under nature ever undergo abrupt changes. I
0456  ages, and acknowledged varieties of the same   SPECIES, however different they may be in structure.
0456  red difference marked by the terms varieties,  SPECIES, genera, families, orders, and classes. On t
0457  to whatever purpose applied, of the different  SPECIES of a class; or to the homologous parts const
0457  nd function; and the resemblance in different  SPECIES of a class of the homologous parts or organs
0457  e to proclaim so plainly, that the inumerable  SPECIES, genera, and families of organic beings, wit
0459  of the general belief in the immutability of  SPECIES. How far the theory of natural selection may
0460  respect to the almost universal sterility of  SPECIES when first crossed, which forms so remarkabl
0460  the reproductive systems of the intercrossed  SPECIES. We see the truth of this conclusion in the
0460  t difference in the result, when the same two  SPECIES are crossed reciprocally; that is, when one
0460  s are crossed reciprocally; that is, when one  SPECIES is first used as the father and then as the
0461  grave enough. All the individuals of the same  SPECIES, and all the species of the same genus, or e
0461  individuals of the same species, and all the  SPECIES of the same genus, or even higher group, mus
0462  . yet, as we have reason to believe that some  SPECIES have retained the same specific form for ver
0462  on the occasional wide diffusion of the same  SPECIES; for during very long periods of time there
0462  ten be accounted for by the extinction of the  SPECIES in the intermediate regions. It cannot be de
0462  bution both of the same and of representative  SPECIES throughout the world. We are as yet profound
0462  means of transport. With respect to distinct  SPECIES of the same genus inhabiting very distant an
0462  ently the difficulty of the wide diffusion of  SPECIES of the same genus is in some degree lessened
0462  s must have existed, linking together all the  SPECIES in each group by gradations as fine as our p
0463  ibly in going from a district occupied by one  SPECIES into another district occupied by a closely
0463  another district occupied by a closely allied  SPECIES, we have no just right to expect often to fi
0463  for we have reason to believe that only a few  SPECIES are undergoing change at any one period; and
0463  ve period between the extinct and still older  SPECIES, why is not every geological formation charg
0463  theory. Why, again, do whole groups of allied  SPECIES appear, though certainly they often falsely
0464  d with the countless generations of countless  SPECIES which certainly have existed. We should not
0464  existed. We should not be able to recognise a  SPECIES as the parent of any one or more species if
0464  se a species as the parent of any one or more  SPECIES if we were to examine them ever so closely,
0464  as long as most of the links between any two  SPECIES are unknown, if any one link or intermediate
0464  ill simply be classed as another and distinct  SPECIES. Only a small portion of the world has been
0464  at least in any great number. Widely ranging  SPECIES vary most, and varieties are often at first
0465  ated there, and will be simply classed as new  SPECIES. Most formations have been intermittent in t
0465  ls of time, geology plainly declares that all  SPECIES have changed; and they have changed in the m
0467  ve to a large extent the character of natural  SPECIES, is shown by the inextricable doubts whether
0467  very many of them are varieties or aboriginal  SPECIES. There is no obvious reason why the principl
0467  ll live and which shall die, which variety or  SPECIES shall increase in number, and which shall de
0468  ecome extinct. As the individuals of the same  SPECIES come in all respects into the closest compet
0468  ally severe between the varieties of the same  SPECIES, and next in severity between the species of
0468  ame species, and next in severity between the  SPECIES of the same genus. But the struggle will oft
0468  there are at least individual differences in  SPECIES under nature. But, besides such differences,
0469  between more plainly marked varieties and sub  SPECIES, and species. Let it be observed how natural
0469  plainly marked varieties and sub species, and  SPECIES. Let it be observed how naturalists differ i
0469  nts in favour of the theory. On the view that  SPECIES are only strongly marked and permanent varie
0469  marked and permanent varieties, and that each  SPECIES first existed as a variety, we can see why i
0469  t no line of demarcation can be drawn between  SPECIES, commonly supposed to have been produced by
0470  tand how it is that in each region where many  SPECIES of a genus have been produced, and where the
0470  uced, and where they now flourish, these same  SPECIES should present many varieties; for where the
0470  many varieties; for where the manufactory of  SPECIES has been active, we might expect, as a gener
0470  nd this is the case if varieties be incipient  SPECIES. Moreover, the species of the larger genera,
0470  varieties be incipient species. Moreover, the  SPECIES of the larger genera, which afford the great
0470  the greater number of varieties or incipient  SPECIES, retain to a certain degree the character of
0470  er by a less amount of difference than do the  SPECIES of smaller genera. The closely allied specie
0470  species of smaller genera. The closely allied  SPECIES also of the larger genera apparently have re
0470  ey are clustered in little groups round other  SPECIES, in which respects they resemble varieties.
0470  ese are strange relations on the view of each  SPECIES having been independently created, but are i
0470  endently created, but are intelligible if all  SPECIES first existed as varieties. As each species
0470  l species first existed as varieties. As each  SPECIES tends by its geometrical ratio of reproducti
0470  mber; and as the modified descendants of each  SPECIES will be enabled to increase by so much the m
0470  serve the most divergent offspring of any one  SPECIES. Hence during a long continued course of mod
0470  nces, characteristic of varieties of the same  SPECIES, tend to be augmented into the greater diffe
0470  nto the greater differences characteristic of  SPECIES of the same genus. New and improved varietie
0470  improved and intermediate varieties; and thus  SPECIES are rendered to a large extent defined and d
0470  extent defined and distinct objects. Dominant  SPECIES belonging to the larger groups tend to give
0471  ut why this should be a law of nature if each  SPECIES has been independently created, no man can e
0472  endless other cases. But on the view of each  SPECIES constantly trying to increase in number, wit
0472  sionally assume some of the characters of the  SPECIES proper to that zone. In both varieties and s
0472  es proper to that zone. In both varieties and  SPECIES, use and disuse seem to have produced some e
0473  of America and Europe. In both varieties and  SPECIES correlation of growth seems to have played a
0473  e necessarily modified. In both varieties and  SPECIES reversions to long lost characters occur. Ho
```

Page **************************************(Key Word)**************************************

0473 ripes on the shoulder and legs of the several SPECIES of the horse genus and in their hybrids! How
0473 this fact explained if we believe that these SPECIES have descended from a striped progenitor, in
0473 red rock pigeon! On the ordinary view of each SPECIES having been independently created, why shoul
0473 he specific characters, or those by which the SPECIES of the same genus differ from each other, be
0473 of a flower be more likely to vary in any one SPECIES of a genus, if the other species, supposed t
0473 y in any one species of a genus, if the other SPECIES, supposed to have been created independently
0473 differently coloured flowers, than if all the SPECIES of the genus have the same coloured flowers?
0473 the genus have the same coloured flowers? If SPECIES are only well marked varieties, of which the
0474 developed in a very unusual manner in any one SPECIES of a genus, and therefore, as we may natural
0474 y naturally infer, of great importance to the SPECIES, should be eminently liable to variation; bu
0474 w, this part has undergone, since the several SPECIES branched off from a common progenitor, an un
0474 long continued habit. On the view of all the SPECIES of the same genus having descended from a co
0474 mmon, we can understand how it is that allied SPECIES, when placed under considerably different co
0475 nce, lines her nest with mud like our British SPECIES. On the view of instincts having been slowly
0475 instincts causing other animals to suffer. If SPECIES be only well marked and permanent varieties,
0475 e other hand, these would be strange facts if SPECIES have been independently created, and varieti
0475 the theory of descent with modification. New SPECIES have come on the stage slowly and at success
0475 ferent in different groups. The extinction of SPECIES and of whole groups of species, which has pl
0475 extinction of species and of whole groups of SPECIES, which has played so conspicuous a part in t
0475 ted by new and improved forms. Neither single SPECIES nor groups of species reappear when the chai
0475 d forms. Neither single species nor groups of SPECIES reappear when the chain of ordinary generati
0477 hy oceanic islands should be inhabited by few SPECIES, but of these, that many should be peculiar.
0477 and why, on the other hand, new and peculiar SPECIES of bats, which can traverse the ocean, shoul
0478 inent. Such facts as the presence of peculiar SPECIES of bats, and the absence of all other mammal
0478 existence of closely allied or representative SPECIES in any two areas, implies, on the theory of
0478 riably find that wherever many closely allied SPECIES inhabit two areas, some identical species co
0478 ied species inhabit two areas, some identical SPECIES common to both still exist. Wherever many cl
0478 st. Wherever many closely allied yet distinct SPECIES occur, many doubtful forms and varieties
0478 doubtful forms and and varieties of the same SPECIES likewise occur. It is a rule of high general
0478 how it is, that the mutual affinities of the SPECIES and genera within each class are so complex
0480 tions which have thoroughly convinced me that SPECIES have changed, and are still slowly changing
0480 gists rejected this view of the mutability of SPECIES? It cannot be asserted that organic beings i
0481 istinction has been, or can be, drawn between SPECIES and well marked varieties. It cannot be main
0481 arked varieties. It cannot be maintained that SPECIES when intercrossed are invariably sterile, an
0481 dowment and sign of creation. The belief that SPECIES were immutable productions was almost unavoi
0481 afforded us plain evidence of the mutation of SPECIES, if they had undergone mutation. But the chi
0481 f our natural unwillingness to admit that one SPECIES has given birth to other and distinct specie
0481 species has given birth to other and distinct SPECIES, is that we are always slow in admitting any
0482 already begun to doubt on the immutability of SPECIES, may be influenced by this volume; but I loo
0482 impartiality. Whoever is led to believe that SPECIES are mutable will do good service to conscien
0482 shed their belief that a multitude of reputed SPECIES in each genus are not real species; but that
0482 of reputed species in each genus are not real SPECIES; but that other species are real, that is, h
0482 ch genus are not real species; but that other SPECIES are real, that is, have been independently c
0482 every external characteristic feature of true SPECIES, they admit that these have been produced by
0483 y from those who believe in the mutability of SPECIES, on their own side they ignore the whole sub
0483 the whole subject of the first appearance of SPECIES in what they consider reverent silence. It m
0483 i extend the doctrine of the modification of SPECIES. The question is difficult to answer, becaus
0483 the same pattern, and at an embryonic age the SPECIES closely resemble each other. Therefore I can
0484 s entertained in this volume on the origin of SPECIES, or when analogous views are generally admit
0484 ubt whether this or that form be in essence a SPECIES. This I feel sure, and I speak after experie
0484 he endless disputes whether or not some fifty SPECIES of British brambles are true species will ce
0484 me fifty species of British brambles are true SPECIES will cease. Systematists will have only to d
0485 sufficient to raise both forms to the rank of SPECIES. Hereafter we shall be compelled to acknowle
0485 acknowledge that the only distinction between SPECIES and well marked varieties is, that the latte
0485 esent day by intermediate gradations, whereas SPECIES were formerly thus connected. Hence, without
0485 accordance. In short, we shall have to treat SPECIES in the same manner as those naturalists trea
0485 overed and undiscoverable essence of the term SPECIES. The other and more general departments of n
0486 d interesting subject for study than one more SPECIES added to the infinitude of already recorded
0486 s added to the infinitude of already recorded SPECIES. Our classifications will come to be, as far
0486 espect to the nature of long lost structures. SPECIES and groups of species, which are called aber
0486 f long lost structures. Species and groups of SPECIES, which are called aberrant, and which may fa
0486 assured that all the individuals of the same SPECIES, and all the closely allied species of most
0486 the same species, and all the closely allied SPECIES of most genera, have within a not very remot
0487 s two formations, which include few identical SPECIES, by the general succession of their forms of
0487 general succession of their forms of life. As SPECIES are produced and exterminated by slowly acti
0488 sure of the lapse of actual time. A number of SPECIES, however, keeping in a body might remain for
0488 lst within this same period, several of these SPECIES, by migrating into new countries and coming
0488 to be fully satisfied with the view that each SPECIES has been independently created. To my mind i
0489 past, we may safely infer that not one living SPECIES will transmit its unaltered likeness to a di
0489 ed likeness to a distant futurity. And of the SPECIES now living very few will transmit progeny of
0489 are grouped, shows that the greater number of SPECIES of each genus, and all the species of many g
0489 number of species of each genus, and all the SPECIES of many genera, have left no descendants, bu
0489 that it will be the common and widely spread SPECIES, belonging to the larger and dominant groups
0489 mately prevail and procreate new and dominant SPECIES. As all the living forms of life are the lin
0050 to determine what differences to consider as SPECIFIC, and what as varieties: for he knows nothing
0109 oduced, unless we believe that the number of SPECIFIC forms goes on perpetually and almost indefin
0109 ably must become extinct. That the number of SPECIFIC forms has not indefinitely increased, geolog
0131 ed in an unusual manner are highly variable: SPECIFIC characters more variable than generic: secon
0154 emarks may be extended. It is notorious that SPECIFIC characters are more variable than generic. T
0154 and some had red, the colour would be only a SPECIFIC character, and no one would be surprised at
0154 most naturalists would advance, namely, that SPECIFIC characters are more variable than generic, b
0155 ence in support of the above statement; that SPECIFIC characters are more variable than generic; b
0155 , when it sinks in value and becomes only of SPECIFIC value, often becomes variable, though its ph
0156 other species of the same genus, are called SPECIFIC characters; and as these specific characters
0156 are called specific characters; and as these SPECIFIC characters have varied and come to differ wi
0158 , i conclude that the greater variability of SPECIFIC characters, or those which distinguish speci

Page **************************************(Key Word)**************************************

Page ***(Key Word)***

```
0158  species; that secondary sexual and ordinary SPECIFIC differences are generally displayed in the s
0158  apted for secondary sexual, and for ordinary SPECIFIC purposes. Distinct species present analogous
0168  selection, and hence probably are variable. SPECIFIC characters, that is, the characters which ha
0169  erentiation, or where the manufactory of new SPECIFIC forms has been actively at work, there, on a
0169  rences to the sexes of the same species, and SPECIFIC differences to the several species of the sa
0263  important for the endurance and stability of SPECIFIC forms, as in the case of grafting it is unim
0270  ertility and sterility as safe criterions of SPECIFIC distinction. Gartner kept during several yea
0279  n discussed. One, namely the distinctness of SPECIFIC forms, and their not being blended together
0293  or thrice as long as the average duration of SPECIFIC forms. But insuperable difficulties, as it s
0300  ould exceed the average duration of the same SPECIFIC forms; and these contingencies are indispens
0320  ertain flourishing groups, the number of new SPECIFIC forms which have been produced within a give
0336  hole glacial period, and note how little the SPECIFIC forms of the inhabitants of the sea have bee
0336  f the slow and scarcely sensible mutation of SPECIFIC forms, as we have a just right to expect to
0342  short compared with the average duration of SPECIFIC forms; that migration has played an importan
0396  te islands, whether still retaining the same SPECIFIC form or modified since their arrival, could
0406  nging widely and of still retaining the same SPECIFIC character. This fact, together with the seed
0462  eve that some species have retained the same SPECIFIC form for very long periods, enormously long
0465  as been shorter than the average duration of SPECIFIC forms. Successive formations are separated f
0472  ch have governed the production of so called SPECIFIC forms. In both cases physical conditions see
0473  g been independently created, why should the SPECIFIC characters, or those by which the species of
0484  ences be sufficiently important to deserve a SPECIFIC name. This latter point will become a far mo
0485  varieties may hereafter be thought worthy of SPECIFIC names, as with the primrose and cowslip; and
0050  for better evidence of the two forms being SPECIFICALLY distinct. On the other hand, they are unit
0267  es and females which have become widely or SPECIFICALLY different, produce hybrids which are gener
0270  t venture to consider the two varieties as SPECIFICALLY distinct. Girou de Buzareingues crossed th
0297  ssils, though almost universally ranked as SPECIFICALLY different, yet are far more closely allied
0299  really varieties or are, as it 's called, SPECIFICALLY distinct. This could be effected only by t
0349  nner in which successive groups of beings, SPECIFICALLY distinct, yet clearly related, replace eac
0352  ced through natural selection from parents SPECIFICALLY distinct. We are thus brought to the quest
0370  ority of the inhabitants of the world were SPECIFICALLY the same as now, the climate was warmer th
0372  rms which are ranked by all naturalists as SPECIFICALLY distinct. As on the land, so in the waters
0376  ically the same; but they are much oftener SPECIFICALLY distinct, though related to each other in
0376  value, being ranked by some naturalists as SPECIFICALLY distinct, by others as varieties; but some
0400  the inhabitants of an archipelago, though SPECIFICALLY distinct, to be closely allied to those of
0409  the inhabitants of an archipelago, though SPECIFICALLY distinct on the several islets, should be
0473  characters, by which they have come to be SPECIFICALLY distinct from each other; and therefore th
0022  l other less distinct breeds might have been SPECIFIED. In the skeletons of the several breeds, th
0024  e in several other cases, is, that the above SPECIFIED breeds, though agreeing generally in consti
0025  her of which is blue or has any of the above SPECIFIED marks, the mongrel offspring are very apt s
0406  ns being related (with the exceptions before SPECIFIED) to those on the surrounding low lands and
0413  ls the plan of the Creator; but unless it be SPECIFIED whether order in time or space, or what els
0444  tted, will, I believe, explain all the above SPECIFIED leading facts in embryology. But first let
0445  their proportional differences in the above SPECIFIED several points were incomparably less than
0432  ld have something in common. In a tree we can SPECIFY this or that branch, though at the actual fo
0163  rs on the legs; but Dr. Gray has figured one SPECIMEN with very distinct zebra like bars on the ho
0304  , m. Bosquet, sent me a drawing of a perfect SPECIMEN of an unmistakeable sessile cirripede, which
0304  arge, and ubiquitous genus, of which not one SPECIMEN has as yet been found even in any tertiary s
0045  d important organs, and compare them in many SPECIMENS of the same species. I should never have ex
0135  very often broken off; he examined seventeen SPECIMENS in his own collection, and not one had even
0173  ch other in every detail of structure as are SPECIMENS taken from the metropolis inhabited by each
0231  etween two adjoining spheres. I have several SPECIMENS showing clearly that they can do this. Even
0240  imentary. Having carefully dissected several SPECIMENS of these workers, I can affirm that the eye
0240  d myself of Mr. F. Smith's offer of numerous SPECIMENS from the same nest of the driver ant (Anomm
0287  known and named from single and often broken SPECIMENS, or from a few specimens collected on some
0287  le and often broken specimens, or from a few SPECIMENS collected on some one spot. Only a small po
0297  es; and they do this the more readily if the SPECIMENS come from different sub stages of the same
0298  otten, that at the present day, with perfect SPECIMENS for examination, two forms can seldom be co
0298  us proved to be the same species, until many SPECIMENS have been collected from many places; and i
0304  50 fathoms; from the perfect manner in which SPECIMENS are preserved in the oldest tertiary beds;
0342  d in a fossil state; that the number both of SPECIMENS and of species, preserved in our museums, i
0464  ciable by the human intellect. The number of SPECIMENS in all our museums is absolutely as nothing
0221  st to another, and it was a most interesting SPECTACLE to behold the masters carefully carrying, a
0089  cs before the females, which standing by as SPECTATORS, at last choose the most attractive partner
0001  . after five years' work I allowed myself to SPECULATE on the subject, and drew up some short note
0198  al selection. But we are far too ignorant to SPECULATE on the relative importance of the several k
0357  eans of distribution, we shall be enabled to SPECULATE with security on the former extension of th
0219  ng ant, it would have been hopeless to have SPECULATED how so wonderful an instinct could have bee
0382  ife. Sir C. Lyell in a striking passage has SPECULATED, in language almost identical with mine, on
0455  the letters in a word, still retained in the SPELLING, but become useless in the pronunciation, bu
0035  are favoured in the weights they carry. Lord SPENCER and others have shown how the cattle of Engl
0147  rowth; or, as Goethe expressed it, in order to SPEND on one side, nature is forced to economise on
0218  their own young. Some species, likewise, of SPHEGIDAE (wasp like insects) are parasitic on other
0226  two parallel layers; with the centre of each SPHERE at the distance of radius x so. Rt. 2, or rad
0228  aring to the eye perfectly true or parts of a SPHERE, and of about the diameter of a cell. It was
0228  depth about one sixth of the diameter of the SPHERE of which they formed a part, the rims of the
0225  intersected or broken into each other, if the SPHERES had been completed; but this is never permit
0226  lding perfectly flat walls of wax between the SPHERES which thus tend to intersect. Hence each cel
0226  ntact with three other cells, which, from the SPHERES being nearly of the same size, is very frequ
0226  urred to me that if the Melipona had made its SPHERES at some given distance from each other, and
0226  it is strictly correct: If a number of equal SPHERES be described with their centres placed in tw
0226  ance) from the centres of the six surrounding SPHERES in the same layer; and at the same distance
0226  me distance from the centres of the adjoining SPHERES in the other and parallel layer; then, if pl
0226  if planes of intersection between the several SPHERES in both layers be formed, there will result
0227  ellow labourers when several are making their SPHERES; but she is already so far enabled to judge
0227  ge of distance; that she always describes her SPHERES so as to intersect largely; and then she uni
0227  been formed by the intersection of adjoining SPHERES in the same layer, she can prolong the hexag
```

Page ***(Key Word)***

Page ***(Key Word)***

```
0230 ual spherical hollows, but never allowing the SPHERES to break into each other. Now bees, as may b
0231 e plane of intersection between two adjoining SPHERES. I have several specimens showing clearly th
0232 ce from each other, all trying to sweep equal SPHERES, and then building up, or leaving ungnawed,
0232 wed, the planes of intersection between these SPHERES. It was really curious to note in cases of d
0232 pleted cells, and then, by striking imaginary SPHERES, they can build up a wall intermediate betwe
0232 up a wall intermediate between two adjoining SPHERES; but, as far as I have seen, they never gnaw
0233 m the parts of the cells just begun, sweeping SPHERES or cylinders, and building up intermediate p
0235 d more perfectly, led the bees to sweep equal SPHERES at a given distance from each other in a dou
0235 course, no more knowing that they swept their SPHERES at one particular distance from each other,
0186 amounts of light, and for the correction of SPHERICAL and chromatic aberration, could have been f
0225 holding honey. These latter cells are nearly SPHERICAL and of nearly equal sizes, and are aggregat
0226 ersect. Hence each cell consists of an outer SPHERICAL portion and of two, three, or more perfectl
0226 , but are of the same thickness as the outer SPHERICAL portions, and yet each flat portion forms a
0227 suppose the Melipona to make her cells truly SPHERICAL, and of equal sizes; and this would not be
0230 same time, and by endeavouring to make equal SPHERICAL hollows, but never allowing the spheres to
0234 n at present; for then, as we have seen, the SPHERICAL surfaces would wholly disappear, and would
0218 ds a burrow already made and stored by another SPHEX it takes advantage of the prize, and becomes
0434 than the immensely long spiral proboscis of a SPHINX moth, the curious folded one of a bee or bug,
0442 rts of the embryo are completed; and again in SPIDERS, there is nothing worthy to be called a meta
0448 le groups of animals, as with cuttle fish and SPIDERS, and with a few members of the great class o
0424 ere known to be sometimes produced on the same SPIKE, they were immediately included as a single s
0163 an albino, has been described without either SPINAL or shoulder stripe; and these stripes are som
0163 rse, I have collected cases in England of the SPINAL stripe in horses of the most distinct breeds,
0164 edwards; that with the English race horse the SPINAL stripe is much commoner in the foal than in t
0164 stripes is not considered as purely bred. The SPINE is always striped; the legs are generally bar
0434 can be more different than the immensely long SPIRAL proboscis of a sphinx moth, the curious folde
0437 owering plants, we see a series of successive SPIRAL whorls of leaves. An indefinite repetition of
0437 unknown progenitor of flowering plants, many SPIRAL whorls of leaves. We have formerly seen that
0436 consist of metamorphosed leaves, arranged in a SPIRE. In monstrous plants, we often get direct evi
0062 ts, no one has treated this subject with more SPIRIT and ability than W. Herbert, Dean of Manchest
0417 this case seems to me well to illustrate the SPIRIT with which our classifications are sometimes
0268 en it is stated, for instance, that the German SPITZ dog unites more easily than other dogs with f
0037 skill of gardeners, in having produced such SPLENDID results from such poor materials; but the ar
0037 pear they could procure, never thought what SPLENDID fruit we should eat; though we owe our excel
0131 haracters. Summary. I have hitherto sometimes SPOKEN as if the variations, so common and multiform
0214 when called. Domestic instincts are sometimes SPOKEN of as actions which have become inherited sol
0269 some difference in the result. I have as yet SPOKEN as if the varieties of the same species were
0303 rs ago, the great class of mammals was always SPOKEN of as having abruptly come in at the commence
0317 becomes large. On Extinction. We have as yet SPOKEN only incidentally of the disappearance of spe
0324 ary stages. When the marine forms of life are SPOKEN of as having changed simultaneously throughou
0422 t classification of the various languages now SPOKEN throughout the world; and if all extinct lang
0009 nature. A long list could easily be given of SPORTING plants; by this term gardeners mean a single
0132 which is to form a new being. In the case of SPORTING plants, the bud, which in its earliest condi
0010 grafting, etc., and sometimes by seed. These SPORTS are extremely rare under nature, but far from
0010 rliest stages of formation; so that, in fact, SPORTS support my view, that variability may be larg
0103 ved variety might be quickly formed on any one SPOT, and might there maintain itself in a body, so
0128 oof by looking at the inhabitants of any small SPOT or at naturalised productions. Therefore durin
0183 (phuratus) in South America, hovering over one SPOT and then proceeding to another, like a kestrel
0221 a score of the slave makers haunting the same SPOT, and evidently not in search of food; they app
0221 from another nest, and put them down on a bare SPOT near the place of combat: they were eagerly se
0232 of the coloured wax having been taken from the SPOT on which it had been placed, and worked into t
0287 or from a few specimens collected on some one SPOT. Only a small portion of the surface of the ea
0296 bility is that they have not lived on the same SPOT during the whole period of deposition, but hav
0298 sed to have been local or confined to some one SPOT. Most marine animals have a wide range: and we
0317 y disappear, one after another, first from one SPOT, the from another, and finally from the world.
0346 ind a group of organisms confined to any small SPOT, having conditions peculiar in only a slight d
0352 isolated regions, must have proceeded from one SPOT, where their parents were first produced: for,
0360 y; and when stranded, if blown to a favourable SPOT by an inland gale, they would germinate. Subse
0441 ted into a minute, single, and very simple eye SPOT. In this last and completed state, cirripedes
0070 being sometimes extremely abundant in the few SPOTS where they do occur; and that of some social
0078 f the same species, for the warmest or dampest SPOTS. Hence, also, we can see that when a plant or
0377 , more especially by escaping into the warmest SPOTS. But the great fact to bear in mind is, that
0440 hat the stripes on the whelp of a lion, or the SPOTS on the young blackbird, are of any use to the
0439 ir first and second plumage; as we see in the SPOTTED feathers in the thrush group. In the cat tri
0439 cat tribe, most of the species are striped or SPOTTED in lines; and stripes can be plainly disting
0223 ess with its own pupa in its mouth on top of a SPRAY of heath over its ravaged home. Such are the
0040 ves them, and the improved individuals slowly SPREAD in the immediate neighbourhood. But as yet th
0040 the same slow and gradual process, they will SPREAD more widely, and will get recognised as somet
0103 en once thus formed might subsequently slowly SPREAD to other districts. On the above principle, n
0106 as it has been much improved, will be able to SPREAD over the open and continuous area, and will t
0106 ver many competitors, will be those that will SPREAD most widely, will give rise to most new varie
0107 w forms of life, likely to endure long and to SPREAD widely. For the area will first have existed
0108 ured or improved varieties will be enabled to SPREAD: there will be much extinction of the less im
0298 t local; and that such local varieties do not SPREAD widely and supplant their parent forms until
0301 ther modified and improved, they would slowly SPREAD and supplant their parent forms. When such va
0303 many divergent forms, which would be able to SPREAD rapidly and widely throughout the world. I wi
0305 largely developed in some one sea, might have SPREAD widely. Nor have we any right to suppose that
0309 tinents may have existed where oceans are now SPREAD out: and clear and open oceans may have exist
0327 mon; and therefore as new and improved groups SPREAD throughout the world, old groups will disappe
0337 r in which European productions have recently SPREAD over New Zealand, and have seized on places w
0350 st chance of seizing on new places; when they SPREAD into new countries. In their new homes they w
0377 aces some naturalised plants and animals have SPREAD within a few centuries, this period will have
0385 and how some naturalised species have rapidly SPREAD throughout the same country. But two facts, w
0401 more of the islands, or when it subsequently SPREAD from one island to another, it would undoubte
0401 fferent islands. Some species, however, might SPREAD and yet retain the same character throughout
0401 rmed in the separate islands have not quickly SPREAD to the other islands. But the islands, though
```

Page ***(Key Word)***

Page **************************************(Key Word)**************************************

0402 r from certain facts that these have probably SPREAD from some one island to the others. But we of
0402 ecies, naturalised through man's agency, have SPREAD with astonishing rapidity over new countries,
0402 are apt to infer that most species would thus SPREAD: but we should remember that the forms which
0403 galapagos Archipelago, not having universally SPREAD from island to island. In many other instance
0403 ar as the same forms, chiefly of plants, have SPREAD widely throughout the world during the recent
0407 the same genus, which on my theory must have SPREAD from one parent source; if we make the same a
0428 eir parents dominant, they are almost sure to SPREAD widely, and to seize on more and more places
0429 e higher groups are in number, and how widely SPREAD they are throughout the world, the fact is st
0464 e links less likely. Local varieties will not SPREAD into other and distant regions until they are
0465 rably modified and improved; and when they do SPREAD, if discovered in a geological formation, the
0489 foretel that it will be the common and widely SPREAD species, belonging to the larger and dominant
0040 untries, with little free communication, the SPREADING and knowledge of any new sub breed will be
0105 apable of enduring for a long period, and of SPREADING widely. Throughout a great and open area, n
0326 natural that the dominant, varying, and far SPREADING species, which already have invaded to a ce
0326 be those which would have the best chance of SPREADING still further, and of giving rise in new co
0326 the dominant forms will generally succeed in SPREADING. The diffusion would, it is probable, be sl
0326 f the land than of the sea. Dominant species SPREADING from any region might encounter still more
0326 highly favourable, as would be the power of SPREADING into new territories. A certain amount of i
0327 ecies having been formed by dominant species SPREADING widely and varying; the new species thus pr
0327 r parents or over other species; these again SPREADING, varying, and producing new species. The fo
0344 n has been broken. We can understand how the SPREADING of the dominant forms of life, which are th
0401 p, just as we see on continents some species SPREADING widely and remaining the same. The really s
0422 w new languages, whilst others (owing to the SPREADING and subsequent isolation and states of civi
0098 , as I could show from the writings of C. C. SPRENGEL and from my own observations, which effectua
0099 ower receiving its own pollen, yet, as C. C. SPRENGEL has shown, and as I can confirm, either the
0145 umbel, I do not feel at all sure that C. C. SPRENGEL's idea that the ray florets serve to attract
0098 pollen. When the stamens of a flower suddenly SPRING towards the pistil, or slowly move one after
0098 cts is often required to cause the stamens to SPRING forward, as Kolreuter has shown to be the cas
0201 t curls the end of its tail when preparing to SPRING, in order to warn the doomed mouse. But I hav
0071 ve been enclosed, and self sown firs are now SPRINGING up in multitudes, so close together that al
0130 here and there see a thin straggling branch SPRINGING from a fork low down in a tree, and which b
0074 rest is cut down, a very different vegetation SPRINGS up: but it has been observed that the trees
0088 surely give indomitable courage, length to the SPUR, and strength to the wing to strike in the spu
0088 confined to the male sex. A hornless stag or SPURLESS cock would have a poor chance of leaving off
0088 ur, and strength to the wing to strike in the SPURRED leg, as well as the brutal cockfighter, who
0065 us over the wide plains of La Plata, clothing SQUARE leagues of surface almost to the exclusion of
0072 erpetually browsed down by the cattle. In one SQUARE yard, at a point some hundred yards distant f
0228 two combs, and put between them a long thick, SQUARE strip of wax: the bees instantly began to exc
0228 i then put into the hive, instead of a thick, SQUARE piece of wax, a thin and narrow, knife edged
0180 hat each structure is of use to each kind of SQUIRREL in its own country, by enabling it to escape
0180 ow from this fact that the structure of each SQUIRREL is the best that it is possible to conceive
0181 atural selection, a perfect so called flying SQUIRREL was produced. Now look at the Galeopithecus
0180 like that of the bat. Look at the family of SQUIRRELS: here we have the finest gradation from ani
0180 flanks rather full, to the so called flying SQUIRRELS; and flying squirrels have their limbs and
0190 o the so called flying squirrels; and flying SQUIRRELS have their limbs and even the base of the t
0181 lead us to believe that some at least of the SQUIRRELS would decrease in numbers or become extermi
0181 as in the case of the less perfectly gliding SQUIRRELS; and that each grade of structure had been
0263 iculty is as important for the endurance and STABILITY of specific forms, as in the case of grafti
0071 though a simple one, has interested me. In STAFFORDSHIRE, on the estate of a relation where I had
0073 insects; and this, as we just have seen in STAFFORDSHIRE, the insectivorous birds, and so onwards
0088 weapons, confined to the male sex. A hornless STAG or spurless cock would have a poor chance of l
0088 ons have been seen fighting all day long; male STAG beetles often bear wounds from the huge mandib
0014 ear at the corresponding caterpillar or cocoon STAGE. But hereditary diseases and some other facts
0052 species, and to species. The passage from one STAGE of difference to another and higher stage may
0052 one stage of difference to another and higher STAGE may be, in some cases, due merely to the long
0116 of diversification in an early and incomplete STAGE of development. After the foregoing discussio
0121 ne species to supplant and exterminate in each STAGE of descent their predecessors and their origi
0122 d and improved in a diversified manner at each STAGE of descent, so as to have become adapted to m
0122 cies, has transmitted descendants to this late STAGE of descent. The new species in our diagram de
0183 ntil their organs of flight had come to a high STAGE of perfection, so as to have given them a dec
0187 without any other mechanism; and from this low STAGE, numerous gradations of structure, branching
0187 own to exist, until we reach a moderately high STAGE of perfection. In certain crustaceans, for in
0208 ad completed its hammock up to, say, the sixth STAGE of construction, and put it into a hammock co
0208 into a hammock completed up only to the third STAGE, the caterpillar simply re performed the four
0208 a hammock made up, for instance, to the third STAGE, and were put into one finished up to the six
0208 and were put into one finished up to the sixth STAGE, so that much of its work was already done fo
0208 hammock, seemed forced to start from the third STAGE, where it had left off, and thus tried to com
0235 s perfect as that of the hive bee. Beyond this STAGE of perfection in architecture, natural select
0302 families have not been found beneath a certain STAGE, that they did not exist before that stage. W
0302 ain stage, that they did not exist before that STAGE. We continually forget how large the world is
0303 , and in Europe even as far back as the eocene STAGE. The most striking case, however, is that of
0305 sor Pictet has carried their existence one sub STAGE further back; and some palaeontologists belie
0307 m. barrande has lately added another and lower STAGE to the Silurian system, abounding with new an
0333 lesser number of characters; for at this early STAGE of descent they have not diverged in characte
0334 us, the species which lived at the sixth great STAGE of descent in the diagram are the modified or
0334 ed offspring of those which lived at the fifth STAGE, and are the parents of those which became st
0334 hich became still more modified at the seventh STAGE; hence they could hardly fail to be nearly in
0335 ation, though the species are distinct in each STAGE. This fact alone, from its generality, seems
0355 lanted each other; so that, at each successive STAGE of modification and improvement, all the indi
0356 f modification will not have been due, at each STAGE, to descent from a single parent. To illustra
0379 natural selection and competition to a higher STAGE of perfection or dominating power, than the s
0380 ve been greatly reduced, and this is the first STAGE towards extinction. A mountain is an island o
0412 ich diverged from a common parent at the fifth STAGE of descent. These five genera have also much,
0437 re become extremely different, are at an early STAGE of growth exactly alike. How inexplicable are
0441 e again to cirripedes: the larvae in the first STAGE have three pairs of legs, a very simple singl
0441 for they increase much in size. In the second STAGE, answering to the chrysalis stage of butterfl

Page **************************************(Key Word)**************************************

Page **(Key Word)**

```
0441  n the second stage, answering to the chrysalis  STAGE  of butterflies, they have six pairs of beauti
0441  mouth, and cannot feed: their function at this   STAGE  is, to search by their well developed organs
0442  et nearly all pass through a similar worm like   STAGE  of development; but in some few cases, as in
0442  this insect, we see no trace of the vermiform    STAGE. How, then, can we explain these several fact
0448  to provide for their own wants at a very early   STAGE  of development, and secondly from their follo
0448  development; so that the larvae, in the first    STAGE, might differ greatly from the larvae in the
0448  t differ greatly from the larvae in the second   STAGE, as we have seen to be the case with cirriped
0475  ith modification. New species have come on the   STAGE  slowly and at successive intervals; and the a
0010  between a bud and an ovule in their earliest     STAGES  of formation; so that, in fact, sports suppor
0036  nd persia, we can, I think, clearly trace the    STAGES  through which they have insensibly passed, an
0086  ultural plants; in the caterpillar and cocoon    STAGES  of the varieties of the silkworm; in the eggs
0187  ions having been transmitted from the earlier    STAGES  of descent, in an unaltered or little altered
0187  fossiliferous stratum to discover the earlier    STAGES, by which the eye has been perfected. In the
0208  ply re performed the fourth, fifth, and sixth    STAGES  of construction. If however, a caterpillar we
0293  h the rest of the world; nor have the several    STAGES  of the same formation throughout Europe been
0297  dily if the specimens come from different sub    STAGES  of the same formation. Some experienced conch
0297  ntervals, namely, to distinct but consecutive    STAGES  of the same great formation, we find that the
0298  a formation in any one country all the early    STAGES  of transition between any two forms, is small
0298  sen the chance of our being able to trace the    STAGES  of transition in any one geological formation
0301  them be ever so close, if found in different    STAGES  of the same formation, would, by most palaeon
0312  this head in the case of the several tertiary   STAGES; and every year tends to fill up the blanks b
0323  elism in the successive forms of life, in the   STAGES  of the widely separated palaeozoic and tertia
0324  ey had lived during one of the later tertiary   STAGES. When the marine forms of life are spoken of
0324  lived in Europe during certain later tertiary   STAGES, than to those which now live here; and if th
0328  se general parallelism between the successive   STAGES  in the two countries; but when he compares ce
0328  e two countries; but when he compares certain   STAGES  in England with those in France, although he
0329  l be the same in the apparently corresponding   STAGES  in the two regions. On the Affinities of exti
0335  lance of the organic remains from the several   STAGES  of the chalk formation, though the species ar
0342  representative species, found in the several   STAGES  of the same great formation. He may disbeliev
0371  europe and America during the later tertiary   STAGES  were more closely related to each other than
0424  st includes as one species the several larval  STAGES  of the same individual, however much they may
0440  appearance, have larvae in all their several   STAGES  barely distinguishable. The embryo in the cou
0446  these facts in regard to the later embryonic   STAGES  of our domestic varieties. Fanciers select th
0448  ight, also, become correlated with successive  STAGES  of development; so that the larvae, in the fi
0449  ey pass through the same or similar embryonic  STAGES, we may feel assured that they have both desc
0450  nce of ancient forms of life to the embryonic  STAGES  of recent forms, may be true, but yet, owing
0463  ve come in suddenly on the several geological  STAGES? Why do we not find great piles of strata ben
0487  nd the blank intervals between the successive  STAGES  as having been of vast duration. But we shall
0204  by natural selection, is more than enough to   STAGGER  any one; yet in the case of any organ, if we
0171  ay i can never reflect on them without being   STAGGERED; but, to the best of my judgment, the great
0143  pletely by natural selection: thus a family of STAGS  once existed with an antler only on one side;
0137  ucky, are blind. In some of the crabs the foot STALK  for the eye remains, though the eye is gone;
0193  ce of a mass of pollen grains, borne on a foot STALK  with a sticky gland at the end, is the same i
0258  species, I mean the case, for instance, of a   STALLION  horse being first crossed with a female ass,
0275  which is the offspring of the female ass and   STALLION. Much stress has been laid by some authors o
0161  napdragon (Antirrhinum) a rudiment of a fifth  STAMEN  so often appears, that this plant must have a
0452  e generally do not find a rudiment of a fifth  STAMEN; but this may sometimes be seen. In tracing t
0092  er hand. Those flowers, also, which had their  STAMENS  and pistils placed, in relation to the size
0093  trees bear only male flowers, which have four  STAMENS  producing rather a small quantity of pollen,
0093  ers; these have a full sized pistil, and four STAMENS  with shrivelled anthers, in which not a grai
0093  t would be advantageous to a plant to produce  STAMENS  alone in one flower or on one whole plant, a
0098  y influence from the foreign pollen. When the  STAMENS  of a flower suddenly spring towards the pist
0098  ncy of insects is often required to cause the  STAMENS  to spring forward, as Kolreuter has shown to
0099  flower is surrounded not only by its own six   STAMENS, but by those of the many other flowers on t
0149  dividual (as the vertebrae in snakes, and the  STAMENS  in polyandrous flowers) the number is variab
0436  the relative position of the sepals, petals,  STAMENS, and pistils, as well as their intimate stru
0437  impler mouths? Why should the sepals, petals, STAMENS, and pistils in any individual flower, thoug
0438  the jaws of crabs as metamorphosed legs; the  STAMENS  and pistils of flowers as metamorphosed leav
0479  n the jaws and legs of a crab, in the petals, STAMENS, and pistils of a flower, is likewise intell
0084  onditions of life, and should plainly bear the STAMP  of far higher workmanship? It may be said tha
0283  n and rounded pebbles, each of which bears the STAMP  of time, are good to show how slowly the mass
0398  of the land and water bears the unmistakeable STAMP  of the American continent. There are twenty s
0398  archipelago, and nowhere else, bear so plain a STAMP  of affinity to those created in America? Ther
0450  r parts in this strange condition, bearing the STAMP  of inutility, are extremely common throughout
0480  tles, should thus so frequently bear the plain STAMP  of inutility! Nature may be said to have taken
0097  s the plant's own anthers and pistil generally STAND  so close together that self fertilisation see
0110  nd rarer, and finally extinct. The forms which STAND  in closest competition with those undergoing
0137  r the eye remains, though the eye is gone; the STAND  for the telescope is there, though the telesc
0157  ation two instances, the first which happen to STAND  on my list; and as the differences in these c
0213  any training, as soon as it scented its prey,  STAND  motionless like a statue, and then slowly cra
0215  ntending to improve the breed, dogs which will STAND  and hunt best. On the other hand, habit alone
0227  n somehow judge accurately at what distance to STAND  from her fellow labourers when several are ma
0230  ide; and I suspect that the bees in such cases STAND  in the opposed cells and push and bend the du
0232  cted. When bees have a place on which they can STAND  in their proper positions for working, for in
0232  it suffices that the bees should be enabled to STAND  at their proper relative distances from each
0285  this subject. Yet it is an admirable lesson to STAND  on the North Downs and to look at the distant
0309  eans may have existed where our continents now STAND. Nor should we be justified in assuming that
0330  assification many fossil species would have to STAND  between living species, and some extinct gene
0343  nded, and where our oscillating continents now STAND  they have stood ever since the Silurian epoch
0351  . if, for instance, a number of species, which STAND  in direct competition with each other, migrat
0395  annels from Australia. The West Indian Islands STAND  on a deeply submerged bank, nearly 1000 fatho
0410  lated in blood, the nearer they will generally STAND  to each other in time and space; in both case
0032  call the plants that deviate from the proper   STANDARD. With animals this kind of selection is, in
0038  odify most of our plants up to their present   STANDARD  of usefulness to man, we can understand how
0038  been improved by continued selection up to a   STANDARD  of perfection comparable with that given to
0039  re rejected as faults or deviations from the   STANDARD  of perfection of each breed. The common goos
```

Page **(Key Word)**

Page ***(Key Word)***************************************

```
0089   and beauty to his bantams, according to his   STANDARD of beauty, I can see no good reason to doubt
0089   dious or beautiful males, according to their   STANDARD of beauty, might produce a marked effect. I
0102   ing to alter the breed, have a nearly common   STANDARD of perfection, and all try to get and breed
0112   fanciers do not and will not admire a medium   STANDARD, but like extremes, they both go on (as has
0134   inherited. Under free nature, we can have no   STANDARD of comparison, by which to judge of the effe
0152   iduals are born which depart widely from the   STANDARD. There may be truly said to be a constant st
0202   meet, as far as we can judge, with this high   STANDARD under nature. The correction for the aberrat
0205   n the battle for life, only according to the   STANDARD of that country. Hence the inhabitants of on
0206   etition will have been severer, and thus the   STANDARD of perfection will have been rendered higher
0064   thousand years, there would literally not be   STANDING room for his progeny. Linnaeus has calculate
0089   orm strange antics before the females, which   STANDING by as spectators, at last choose the most at
0116   is represented in the diagram by the letters   STANDING at unequal distances. I have said a large ge
0123   ginal species (I) differed largely from (A),   STANDING nearly at the extreme points of the original
0183   another, like a kestrel, and at other times   STANDING stationary on the margin of water, and then
0230   uld make their cells of the proper shape, by   STANDING at the proper distance from each other, by e
0232   struck between many bees, all instinctively   STANDING at the same relative distance from each othe
0233   ree cells commenced at the same time, always   STANDING at the proper relative distance from the par
0296   st evidence in great fossilised trees, still   STANDING upright as they grew, of many long intervals
0357   d, as I believe, by rings of coral or atolls   STANDING over them. Whenever it is fully admitted, as
0365   tinent, that a poorly stocked island, though   STANDING more remote from the mainland, would not rec
0398   les from the continent, yet feels that he is   STANDING on American land. Why should this be so? Why
0399   d. thus the plants of Kerguelen Land, though   STANDING nearer to Africa than to America, are relate
0077   r. yet the advantage of plumed seeds no doubt   STANDS in the closest relation to the land being alr
0225   ouse, who has shown that the form of the cell   STANDS in close relation to the presence of adjoinin
0230   me time, but only the one rhombic plate which   STANDS on the extreme growing margin, or the two pla
0476   the more ancient a fossil is, the oftener it   STANDS in some degree intermediate between existing
0411   idently not arbitrary like the grouping of the   STARS in constellations. The existence of groups wo
0208   rder to complete its hammock, seemed forced to   START from the third stage, where it had left off,
0302   g to the same genera or families, have really   STARTED into life all at once, the fact would be fat
0483   onceived opinion. These authors seem no more   STARTLED at a miraculous act of creation than at an o
0188   g the principle of natural selection to such   STARTLING lengths. It is scarcely possible to avoid c
0076   asserted that certain mountain varieties will   STARVE out other mountain varieties, so that they ca
0004   hen pass on to the variability of species in a   STATE of nature; but I shall, unfortunately, be com
0007   individuals of any one species or variety in a   STATE of nature. When we reflect on the vast divers
0009   we see individuals, though taken young from a   STATE of nature, perfectly tamed, long lived, and h
0009   y very slightly, perhaps hardly more than in a   STATE of nature. A long list could easily be given
0011   are habitually milked, in comparison with the   STATE of these organs in other countries, is anothe
0014   n be drawn from domestic races to species in a   STATE of nature. I have in vain endeavoured to disc
0014   ic varieties could not possibly live in a wild   STATE. In many cases we do not know what the aborig
0016   closely allied species of the same genus in a   STATE of nature. I think this must be admitted, whe
0017   verse classes and countries, were taken from a   STATE of nature, and could be made to breed for an
0018   , I may, without here entering on any details,   STATE that, from geographical and other considerati
0019   ike all wild Canidae, ever existed freely in a   STATE of nature? It has often been loosely said tha
0020   hound, bloodhound, bull dog, etc., in the wild   STATE. Moreover, the possibility of making distinct
0022   erfect plumage is acquired varies, as does the   STATE of the down with which the nestling birds are
0023   ; or they must have become extinct in the wild   STATE. But birds breeding on precipices, and good f
0024   is the rock pigeon in a very slightly altered   STATE, has become feral in several places. Again, a
0026   breeds of pigeons are perfectly fertile. I can   STATE this from my own observations, purposely made
0027   supposed species being quite unknown in a wild   STATE, and their becoming nowhere feral; these spec
0029   ion, when they deride the idea of species in a   STATE of nature being lineal descendants of other s
0040   one district than in another, according to the   STATE of civilisation of the inhabitants, slowly to
0044   at in the last chapter to organic beings in a   STATE of nature, we must briefly discuss whether th
0046   authors sometimes argue in a circle when they   STATE that important organs never vary; for these s
0050   ith the fact, that if any animal or plant in a   STATE be highly useful to man, or from an
0052   d i attribute the passage of a variety, from a   STATE in which it differs very slightly from its pa
0052   of species. They may whilst in this incipient   STATE become extinct, or they may endure as varieti
0060   last chapter that amongst organic beings in a   STATE of nature there is some individual variabilit
0064   shingly rapid increase of various animals in a   STATE of nature, when circumstances have been favou
0065   turalised productions in their new homes. In a   STATE of nature almost every plant produces seed, a
0065   e annually slaughtered for food, and that in a   STATE of nature an equal number would have somehow
0072   they swarm southward and northward in a feral   STATE; and Azara and Rengger have shown that this i
0076   allowed to struggle together, like beings in a   STATE of nature, and if the seed or young were not
0085   ivating the several varieties, assuredly, in a   STATE of nature, where the trees would have to stru
0085   ur sheep and cattle when nearly adult; so in a   STATE of nature, natural selection will be enabled
0097   llen from another individual will explain this   STATE of exposure, more especially as the plant's o
0121   es, that is between the less and more improved   STATE of a species, as well as the original parent
0130   to us only from having been found in a fossil   STATE. As we here and there see a thin straggling b
0131   tication, and in a lesser degree in those in a   STATE of nature, had been due to chance. This, of c
0134   that cannot fly; yet there are several in this   STATE. The loggerheaded duck of South America can o
0137   ses are quite covered up by skin and fur. This   STATE of the eyes is probably due to gradual reduct
0140   h. we have reason to believe that species in a   STATE of nature are limited in their ranges by the
0141   a large proportion of other animals, now in a   STATE of nature, could easily be brought to bear wi
0147   beard by diminished wattles. With species in a   STATE of nature it can hardly be maintained that th
0150   cannot possibly be here introduced. I can only   STATE my conviction that it is a rule of high gener
0151   ies is so great, that it is no exaggeration to   STATE that the varieties differ more from each othe
0152   , the tendency to reversion to a less modified   STATE, as well as an innate tendency to further var
0154   eory, for an immense period in nearly the same   STATE; and thus it comes to be no more variable tha
0155   iod, and which have thus come to differ. Or to   STATE the case in another manner: the points in whi
0164   ithout even entering on further details, I may   STATE that I have collected cases of leg and should
0169   ariability and to reversion to a less modified   STATE. But when a species with any extraordinarily
0172   of the geological record: and I will here only   STATE that I believe the answer mainly lies in the
0176   out, this rule holds good with varieties in a   STATE of nature. I have met with striking instances
0178   so that they will no longer exist in a living   STATE. Thirdly, when two or more varieties have bee
0179   ; for how could the animal in its transitional   STATE have subsisted? It would be easy to show that
0182   ve ever imagined that in an early transitional   STATE they had been inhabitants of the open ocean,
0189   e a distincter image. We must suppose each new   STATE of the instrument to be multiplied by the mil
```

Page ***(Key Word)***************************************

state

Page **************************************(Key Word)**************************************

```
0194 ons an organ could have arrived at its present STATE; yet, considering that the proportion of livi
0196 riod, have been transmitted in nearly the same STATE, although now become of very slight use; and
0205 l importance that it could not, in its present  STATE, have been acquired by natural selection, a p
0211 some degree of variation in instincts under a  STATE of nature, and the inheritance of such variat
0212 of individuals of the same species, born in a  STATE of nature, is extremely diversified, can be s
0212 lity, of inherited variations of instinct in a STATE of nature will be strengthened by briefly con
0216 l, perhaps, best understand how instincts in a STATE of nature have become modified by selection,
0229 posed basins, but the work, from the unnatural STATE of things, had not been neatly performed. The
0236 some insects and other articulate animals in a STATE of nature occasionally become sterile; and if
0236 uter insect had been an animal in the ordinary STATE, I should have unhesitatingly assumed that al
0237 in our domestic productions and in those in a  STATE of nature, of all sorts of differences of str
0237 attle in relation to an artificially imperfect STATE of the male sex; for oxen of certain breeds h
0243 ptec to show that instincts vary slightly in a STATE of nature. No one will dispute that instincts
0246 onally impotent, as may be clearly seen in the STATE of the male element in both plants and animal
0247 mber produced by both pure parent species in a STATE of nature. But a serious cause of error seems
0251 ust infer that the plants were in an unnatural STATE. Nevertheless these facts show on what slight
0254 result; but that it cannot, under our present  STATE of knowledge, be considered as absolutely uni
0261 s so entirely unimportant for its welfare in a STATE of nature, I presume that no one will suppose
0265 roductive system, independently of the general STATE of health, is affected by sterility in a very
0271 ascertaining the infertility of varieties in a STATE of nature, for a supposed variety if infertil
0277 gue in a circle with respect to varieties in a STATE of nature; and when we remember that the grea
0285 the lava streams, due to their formerly liquid STATE, showed at a glance how far the hard, rocky b
0288 y and Palaeozoic periods, it is superfluous to STATE that our evidence from fossil remains is frag
0299 y the future geologist discovering in a fossil STATE numerous intermediate gradations; and such su
0299 d shallow seas, probably represents the former STATE of Europe, when most of our formations were a
0301 homes, as they would differ from their former  STATE, in a nearly uniform, though perhaps extremel
0308 north America. But we do not know what was the STATE of things in the intervals between the succes
0312 es to each other and to living species. On the STATE of development of ancient forms. On the succe
0336 we have a just right to expect to find. On the STATE of Development of Ancient Forms. There has be
0342 beings have been largely preserved in a fossil STATE; that the number both of specimens and of spe
0377 t the tropical productions were in a suffering STATE and could not have presented a firm front aga
0397 . and I found that several species did in this STATE withstand uninjured an immersion in sea water
0424 ant characters from negroes. With species in a STATE of nature, every naturalist has in fact broug
0441 ry simple eye spot. In this last and completed STATE, cirripedes may be considered as either more
0446 rtions, almost exactly as much as in the adult STATE. The two principles above given seem to me to
0446 to be in some degree probable, to species in a STATE of nature. Let us take a genus of birds, desc
0449 the embryo is the animal in its less modified  STATE; and in so far it reveals the structure of it
0449 e great class of crustaceans. As the embryonic STATE of each species and group of species partiall
0449 true in those cases alone in which the ancient STATE, now supposed to be represented in many embry
0450 safely considered as a digit in a rudimentary  STATE; in very many snakes one lobe of the lungs is
0451 e rudiments, and sometimes in a well developed STATE. In plants with separated sexes, the male flo
0452 ated, have a pistil, which is in a rudimentary STATE, for it is not crowned with a stigma; but the
0452 h. this latter fact is well exemplified in the STATE of the wings of the female moths in certain g
0454 according to Youatt, in young animals, and the STATE of the whole flower in the cauliflower. We of
0454 light on the origin of rudimentary organs in a STATE of nature, further than by showing that rudim
0455 g ages will reproduce the organ in its reduced STATE at the same age, and consequently will seldom
0460 e being, could not have arrived at its present STATE by many graduated steps. There are, it must b
0481 it cannot be asserted that organic beings in a STATE of nature are subject to no variation; it can
0483 progenitor had the organ in a fully developed  STATE; and this in some instances necessarily impli
0016 d not so perpetually recur. It has often been  STATED that domestic races do not differ from each o
0082 of improvement. We have reason to believe, as  STATED in the first chapter, that a change in the co
0149 its advantage. Rudimentary parts, it has been  STATED by some authors, and I believe with truth, ar
0163 has no shoulder stripe; but traces of it, as   STATED by Mr. Blyth and others, occasionally appear:
0166 and marks to reappear in the mongrels. I have  STATED that the most probable hypothesis to account
0228 by the time the basins had acquired the above  STATED width (i.e. about the width of an ordinary ce
0231 cell going to another; so that, as Huber has   STATED, a score of individuals work even at the comm
0268 e are still involved in doubt. For when it is  STATED, for instance, that the German Spitz dog unit
0304 a memoir on Fossil Sessile Cirripedes, I have  STATED that, from the number of existing and extinct
0358 ts. In botanical works, this or that plant is  STATED to be ill adapted for wide dissemination: but
0381 to be solved: some of the most remarkable are  STATED with admirable clearness by Dr. Hooker in his
0387 cies probably have also played a part. I have  STATED that fresh water fish eat some kinds of seeds
0389 t distant points of the globe. I have already  STATED that I cannot honestly admit Forbes's view on
0444 appeared at an extremely early period. I have  STATED in the first chapter, that there is some evid
0450 r jaws of our unborn calves. It has even been  STATED on good authority that rudiments of teeth can
0014 subject of reversion, I may here refer to a   STATEMENT often made by naturalists, namely, that our
0014 to discover on what decisive facts the above  STATEMENT has so often and so boldly been made. There
0016 c value. I think it could be shown that this  STATEMENT is hardly correct; but naturalists differ m
0058 generally have much restricted ranges: this   STATEMENT is indeed scarcely more than a truism, for
0155 s to adduce evidence in support of the above  STATEMENT, that specific characters are more variable
0226 geometer has kindly read over the following   STATEMENT, drawn up from his information, and tells m
0230 they are conformable with my theory. Huber's  STATEMENT that the very first cell is excavated out o
0249 engthened in this conviction by a remarkable  STATEMENT repeatedly made by Gartner, namely, that if
0260 ee of resemblance to each other. This latter  STATEMENT is clearly proved by reciprocal crosses bet
0334 it is no real objection to the truth of the   STATEMENT, that the fauna of each period as a whole i
0335 his same respect. Closely connected with the  STATEMENT, that the organic remains from an intermedi
0002 e references and authorities for my several   STATEMENTS; and I must trust to the reader reposing so
0053 ed the tables, he thinks that the following   STATEMENTS are fairly well established. The whole subj
0064 wild in several parts of the world: if the    STATEMENTS of the rate of increase of slow breeding ca
0087 for the good of another species: and though   STATEMENTS to this effect may be found in works of nat
0201 ed through natural selection. Although many   STATEMENTS may be found in works on natural history to
0212 cts. But I am well aware that these general   STATEMENTS, without facts given in detail, can produce
0220 er subjects. Although fully trusting to the   STATEMENTS of Huber and Mr. Smith, I tried to approach
0230 exagonal walls are commenced. Some of these   STATEMENTS differ from those made by the justly celebr
0250 ouses at his command. Of his many important   STATEMENTS I will here give only a single one as an ex
0349 reat fact, partly included in the foregoing   STATEMENTS, is the affinity of the productions of the
0048 of Great Britain, of France or of the United STATES, drawn up by different botanists, and see wha
```

Page **************************************(Key Word)**************************************

states

Page ***(Key Word)**

```
0074  ancient Indian mounds, in the Southern United STATES, display the same beautiful diversity and pro
0076  the recent extension over parts of the United STATES of one species of swallow having caused the d
0085  t horticulturist, Downing, that in the United STATES smooth skinned fruits suffer far more from a
0091  habiting the Catskill Mountains in the United STATES, one with a light greyhound like form, which
0100  zealand, and Dr. Asa Gray those of the United STATES, and the result was as I anticipated. On the
0115  's manual of the Flora of the Northern United STATES, 260 naturalised plants are enumerated, and t
0115  ional addition is made to the genera of these STATES. By considering the nature of the plants or a
0121  ermediate forms between the earlier and later STATES, that is between the less and more improved s
0142  works on fruit trees published in the United STATES, in which certain varieties are habitually re
0142  for the northern, and others for the southern STATES; and as most of these varieties are of recent
0173  tional varieties between its past and present STATES. Hence we ought not to expect at the present
0198  colour are correlated. A good observer, also, STATES that in cattle susceptibility to the attacks
0204  ch we know of no intermediate or transitional STATES, we should be very cautious in concluding tha
0204  ologies of many organs and their intermediate STATES show that wonderful metamorphoses in function
0208  habits, and with certain periods of time and STATES of the body. When once acquired, they often r
0212  e species in the northern and southern United STATES. Fear of any particular enemy is certainly an
0221  e morning and evening; and as Huber expressly STATES, their principal office is to search for aphi
0273  r comparison of mongrels and hybrids: Gartner STATES that mongrels are more liable than hybrids to
0299  h any accuracy, excepting those of the United STATES of America. I fully agree with Mr. Godwin Aus
0302  ent archipelagoes of Europe and of the United STATES. We do not make due allowance for the enormou
0306  beyond the confines of Europe and the United STATES; and from the revolution in our palaeontologi
0308  everal formations of Europe and of the United STATES; and from the amount of sediment, miles in th
0308  ive formations; whether Europe and the United STATES during these intervals existed as dry land, o
0324  e that the existing productions of the United STATES are more closely related to those which lived
0346  ntinent, from the central parts of the United STATES to its extreme southern point, we meet with t
0365  plants on the White Mountains, in the United STATES of America, are all the same with those of La
0366  maize. Throughout a large part of the United STATES, erratic boulders, and rocks scored by drifte
0367  pair. The now temperate regions of the United STATES would likewise be covered by arctic plants an
0367  sely remote as on the mountains of the United STATES and of Europe. We can thus also understand th
0368  of northern Scandinavia; those of the United STATES to Labrador; those of the mountains of Siberi
0368  at present (as some geologists in the United STATES believe to have been the case, chiefly from t
0371  in the central parts of Europe and the United STATES. On this view we can understand the relations
0387  n must remain quite inexplicable; but Audubon STATES that he found the seeds of the great southern
0422  to the spreading and subsequent isolation and STATES of civilisation of the several races, descend
0444  iarities in the caterpillar, cocoon, or imago STATES of the silk moth; or, again, in the horns of
0464  inks between their past or parent and present STATES; and these many links we could hardly ever ex
0002  . a fair result can be obtained only by fully STATING and balancing the facts and arguments on bot
0065  atio, that all would most rapidly stock every STATION in which they could any how exist, and that
0122  , or become quickly adapted to some quite new STATION, in which child and parent do not come into
0130  l competition by having inhabited a protected STATION. As buds give rise by growth to fresh buds,
0321  or from inhabiting some distant and isolated STATION, where they have escaped severe competition.
0173  ong intervals. Whilst the bed of the sea is STATIONARY or is rising, or when very little sediment
0183  like a kestrel, and at other times standing STATIONARY on the margin of water, and then dashing li
0292  and, as long as the bed of the sea remained STATIONARY, thick deposits could not have been accumul
0300  vals, during which the area would be either STATIONARY or rising; whilst rising, each fossiliferou
0327  periods when the bed of the sea was either STATIONARY or rising, and likewise when sediment was n
0465  g the alternate periods of elevation and of STATIONARY level the record will be blank. During thes
0049  ent periods; they grow in somewhat different STATIONS; they ascend mountains to different heights;
0054  seems to be connected with the nature of the STATIONS inhabited by them, and has little or no rela
0103  ong remain distinct, from haunting different STATIONS, from breeding at slightly different seasons
0113  y, either dead or alive; some inhabiting new STATIONS, climbing trees, frequenting water, and some
0115  ew groups more especially adapted to certain STATIONS in their new homes. But the case is very dif
0127  orms of life to their several conditions and STATIONS, must be judged of by the general tenour and
0257  evergreen trees, plants inhabiting different STATIONS and fitted for extremely different climates,
0292  parts of the sea will be increased, and new STATIONS will often be formed; all circumstances most
0349  mselves are distinct at different points and STATIONS. It is a law of the widest generality, and e
0389  cts. If we look to the large size and varied STATIONS of New Zealand, extending over 780 miles of
0406  unding low lands and dry lands, though these STATIONS are so different, the very close relation of
0213  it scented its prey, stand motionless like a STATUE, and then slowly crawl forward with a peculia
0392  d have no chance of successfully competing in STATURE with a fully developed tree, when establishe
0392  ural selection would often tend to add to the STATURE of herbaceous plants when growing on an isla
0001  : from that period to the present day I have STEADILY pursued the same object. I hope that I may b
0032  thus, to give a very trifling instance, the STEADILY increasing size of the common gooseberry may
0076  proportion, otherwise the weaker kinds will STEADILY decrease in numbers and disappear. So again
0078  t to acquire. All that we can do, is to keep STEADILY in mind that each organic being is striving
0112  ng differences, at first barely appreciable, STEADILY to increase; and the breeds to diverge in ch
0128  quishing varieties of the same species, will STEADILY tend to increase till they come to equal the
0231  of the ridge. We shall thus have a thin wall STEADILY growing upward; but always crowned by a giga
0177  t the inhabitants are all trying with equal STEADINESS and skill to improve their stocks by select
0126  may predict that, owing to the continued and STEADY increase of the larger groups, a multitude of
0153  all kinds and on the other hand, the power of STEADY selection to keep the breed true. In the long
0170  s, and a cause for each must exist, it is the STEADY accumulation, through natural selection, of s
0424  udes the so called alternate generations of STEENSTRUP, which can only in a technical sense be con
0046  be compared to the irregular branching of the STEM of a tree. This philosophical naturalist, I ma
0317  ke the branching of a great tree from a single STEM, till the group becomes large. On Extinction.
0072  ted clumps. But on looking closely between the STEMS of the heath, I found a multitude of seedling
0072  ty six years tried to raise its head above the STEMS of the heath, and had failed. No wonder that,
0159  a case of analogous variation, in the enlarged STEMS, or roots as commonly called, of the Swedish
0159  e to attribute this similarity in the enlarged STEMS of these three plants, not to the vera causa
0359  be washed into the sea. Hence I was led to dry STEMS in branches of 94 plants with ripe fruit, and
0423  ps together, though the esculent and thickened STEMS are so similar. Whatever part is found to be
0427  in our domestic varieties, as in the thickened STEMS of the common and swedish turnip. The resembl
0030  o him have probably arisen suddenly, or by one STEP; many botanists, for instance, believe that th
0051  of high importance for us, as being the first STEP towards such slight varieties as are barely th
0093  triking case, but as likewise illustrating one STEP in the separation of the sexes of plants, pres
0353  range anomaly it would be, if, when coming one STEP lower in the series, to the individuals of the
```

Page ***(Key Word)**

Page **(Key Word)**

```
0447 orant, at a very early period of life, or each   STEP might be inherited at an earlier period than t
0455 lesser relative size in the adult. But if each   STEP of the process of reduction were to be inherit
0484 al or lesser number. Analogy would lead me one   STEP further, namely, to the belief that all animal
0029 es? Selection. Let us now briefly consider the   STEPS by which domestic races have been produced, e
0051 in any degree more distinct and permanent, as   STEPS leading to more strongly marked and more perm
0059 many modified and dominant descendants. But by   STEPS hereafter to be explained, the larger genera
0118 nt (A). We may continue the process by similar   STEPS for any length of time; some of the varieties
0120 f sub species; but we have only to suppose the   STEPS in the process of modification to be more num
0120 ined species: thus the diagram illustrates the   STEPS by which the small differences distinguishing
0120 second species (I) has produced, by analogous   STEPS, after ten thousand generations, either two w
0148 ts of the Proteolepas, though effected by slow   STEPS, would be a decided advantage to each success
0181 ted, and that each had been formed by the same   STEPS as in the case of the less perfectly gliding
0182 ave resulted from disuse, indicate the natural   STEPS by which birds have acquired their perfect po
0190 nd thus wholly change its nature by insensible   STEPS. Two distinct organs sometimes perform simult
0192 fficulty; it is impossible to conceive by what   STEPS these wondrous organs have been produced; but
0194 be so invariably linked together by graduated   STEPS? Why should not Nature have taken a leap from
0194 , but must advance by the shortest and slowest   STEPS. Organs of little apparent importance. As nat
0223 ir slaves than they do in Switzerland. By what   STEPS the instinct of F. sanguinea originated I wil
0317 s, which in their turn produce by equally slow   STEPS other species; and so on, like the branching
0342 existing forms of life by the finest graduated   STEPS. He who rejects these views on the nature of
0404 varieties; these doubtful forms showing us the   STEPS in the process of modification. This relation
0432 m other groups, as all would blend together by   STEPS as fine as those between the finest existing
0438 n will have been effected by slight successive   STEPS, we need not wonder at discovering in such pa
0446 habits. Then, from the many slight successive   STEPS of variation having supervened at a rather la
0447 se and disuse. In certain cases the successive   STEPS of variation might supervene, from causes of
0454 ion, which can be effected by insensibly small   STEPS, is within the power of natural selection; so
0460 arrived at its present state by many graduated   STEPS. There are, it must be admitted, cases of spe
0471 cation; it can act only by very short and slow   STEPS. Hence the canon of Natura non facit saltum,
0474 thus understand why nature moves by graduated   STEPS in endowing different animals of the same cla
0481 change of which we do not see the intermediate   STEPS. The difficulty is the same as that felt by s
0009 s, in the same exact condition as in the most   STERILE hybrids. When, on the one hand, we see domes
0172 n we account for species, when crossed, being   STERILE and producing sterile offspring, whereas, wh
0172 es, when crossed, being sterile and producing   STERILE offspring, whereas, when varieties are cross
0202 timately slaughtered by their industrious and   STERILE sisters? It may be difficult, but we ought t
0207 the Natural Selection of instincts. Neuter or   STERILE insects. Summary. The subject of instinct mi
0219 nd fertile females do no work. The workers or   STERILE females, though most energetic and courageou
0236 o my whole theory. I allude to the neuters or   STERILE females in insect communities: for these neu
0236 ales and fertile females, and yet, from being   STERILE, they cannot propagate their kind. The subje
0236 e take only a single case, that of working or   STERILE ants. How the workers have been rendered ste
0236 rile ants. How the workers have been rendered   STERILE is a difficulty; but not much greater than t
0236 mals in a state of nature occasionally become   STERILE; and if such insects had been social, and it
0237 ring greatly from its parents, yet absolutely   STERILE; so that it could never have transmitted suc
0237 y character having become correlated with the   STERILE condition of certain members of insect commu
0238 f structure, or instinct, correlated with the   STERILE condition of certain members of the communit
0238 their fertile offspring a tendency to produce   STERILE members having the same modification. And I
0238 amount of difference between the fertile and   STERILE females of the same species has been produce
0240 condition. So that we here have two bodies of   STERILE workers in the same nest, differing not only
0241 rful fact of two distinctly defined castes of   STERILE workers existing in the same nest, both wide
0242 effected with them only by the workers being   STERILE; for had they been fertile, they would have
0242 ercise, or habit, or volition, in the utterly   STERILE members of a community could possibly have a
0245 e of the fact that hybrids are very generally   STERILE, has, I think, been much underrated by some
0247 best botanists rank as varieties, absolutely   STERILE together; and as he came to the same conclus
0247 oubt whether many other species are really so   STERILE, when intercrossed, as Gartner believes. It
0250 he pollen of a distinct species, though quite   STERILE with their own pollen, notwithstanding that
0252 t agency by examining the flowers of the more   STERILE kinds of hybrid rhododendrons, which produce
0252 but the hybrids themselves are, I think, more   STERILE. I doubt whether any case of a perfectly fer
0256 ely produce any offspring, are generally very   STERILE; but the parallelism between the difficulty
0256 d offspring, yet these hybrids are remarkably   STERILE. On the other hand, there are species which
0259 t species, are with rare exceptions extremely   STERILE, even when the other hybrids raised from see
0259 ; and these hybrids are almost always utterly   STERILE, when intercrossed, as Gartner believes. It
0260 ies cross with facility, and yet produce very   STERILE hybrids; and other species cross with extrem
0265 ; and whole groups of species tend to produce   STERILE hybrids. On the other hand, one species in a
0265 species of a genus will produce more or less   STERILE hybrids. Lastly, when organic beings are pla
0266 laced under unnatural conditions, is rendered   STERILE. All that I have attempted to show, is that
0267 e, often render organic beings in some degree   STERILE; and that greater crosses, that is crosses b
0267 ifferent, produce hybrids which are generally   STERILE in some degree. I cannot persuade myself tha
0268 erto reputed varieties be found in any degree   STERILE together, they are at once ranked by most na
0268 sembling each other most closely, are utterly   STERILE when intercrossed. Several considerations, h
0269 of hybrids, which were at first only slightly   STERILE; and if this be so, we surely ought not to e
0271 iana glutinosa, always yielded hybrids not so   STERILE as those which were produced from the four o
0276 ids, are very generally, but not universally,   STERILE. The sterility is of all degrees, and is oft
0460 of two or three defined castes of workers or   STERILE females in the same community of ants; but I
0461 isms of all kinds are rendered in some degree   STERILE from their constitutions having been disturb
0461 feel surprise at hybrids being in some degree   STERILE, for their constitutions can hardly fail to
0472 ngle act, and being then slaughtered by their   STERILE sisters; at the astonishing waste of pollen
0481 that species when intercrossed are invariably   STERILE, and varieties invariably fertile; or that s
0009 ot perfectly like their parents or variable.   STERILITY has been said to be the bane of horticultur
0009 variability to the same cause which produces   STERILITY; and variability is the source of all the c
0026 stication eliminates this strong tendency to   STERILITY: from the history of the dog I think there
0043 th of hybrids and mongrels; and the frequent   STERILITY of hybrids; but the cases of plants not pro
0245 ter VIII. Hybridism. Distinction between the   STERILITY of first crosses and of hybrids. Sterility
0245 e sterility of first crosses and of hybrids.   STERILITY various in degree, not universal, affected
0245 removed by domestication. Laws governing the   STERILITY of hybrids. Sterility not a special endowme
0245 on. Laws governing the sterility of hybrids.   STERILITY not a special endowment, but incidental on
0245 cidental on other differences. Causes of the   STERILITY of first crosses and of hybrids. Parallelis
0245 e been specially endowed with the quality of   STERILITY, in order to prevent the confusion of all o
```

Page **(Key Word)**

Page ***(Key Word)***

```
0245  ase is especially important, inasmuch as the  STERILITY of hybrids could not possibly be of any adv
0245  ervation of successive profitable degrees of   STERILITY. I hope, however, to be able to show that s
0245  ty. I hope, however, to be able to show that   STERILITY is not a specially acquired or endowed qual
0245  erally been confounded together; namely, the   STERILITY of two species when first crossed, and the
0246  y of two species when first crossed, and the   STERILITY of the hybrids produced from them. Pure spe
0246  tinction is important, when the cause of the   STERILITY, which is common to the two cases, has to b
0246  has probably been slurred over, owing to the   STERILITY in both cases being looked on as a special
0246  , on my theory, of equal importance with the   STERILITY of species; for it seems to make a broad an
0246  etween varieties and species. First, for the   STERILITY of species when crossed and of their hybrid
0246  d with the high generality of some degree of   STERILITY. Kolreuter makes the rule universal; but th
0247  in order to show that there is any degree of   STERILITY. He always compares the maximum number of s
0248  es. It is certain, on the one hand, that the   STERILITY of various species when crossed is so diffe
0248  cult to say where perfect fertility ends and   STERILITY begins. I think no better evidence of this
0248  ent years. It can thus be shown that neither   STERILITY nor fertility affords any clear distinction
0248  and structural differences. In regard to the   STERILITY of hybrids in successive generations; thoug
0250  re kolreuter and Gartner that some degree of   STERILITY between distinct species is a universal law
0253  t is not at all surprising that the inherent   STERILITY in the hybrids should have gone on increasi
0253  ich from any cause had the least tendency to   STERILITY, the breed would assuredly be lost in a ver
0254  r give up the belief of the almost universal   STERILITY of distinct species of animals when crossed
0254  of animals when crossed; or we must look at   STERILITY, not as an indelible characteristic, but as
0254  als, it may be concluded that some degree of   STERILITY, both in first crosses and in hybrids, is a
0254  as absolutely universal. Laws governing the   STERILITY of first Crosses and of Hybrids. We will no
0255  il the circumstances and rules governing the   STERILITY of first crosses and of hybrids. Our chief
0256  t fertilisation. From this extreme degree of   STERILITY we have self fertilised hybrids producing a
0256  difficulty of making a first cross, and the   STERILITY of the hybrids thus produced, two classes o
0260  indicate that species have been endowed with   STERILITY simply to prevent their becoming confounded
0260  d in nature? I think not. For why should the   STERILITY be so extremely different in degree, when v
0260  blending together? Why should the degree of   STERILITY be innately variable in the individuals of
0260  further propagation by different degrees of   STERILITY, not strictly related to the facility of th
0260  d, appear to me clearly to indicate that the   STERILITY both of first crosses and of hybrids is sim
0261  ttle more fully by an example what I mean by   STERILITY being incidental on other differences, and
0262  ty, on the gooseberry. We have seen that the   STERILITY of hybrids, which have their reproductive o
0263  unimportant for their welfare. Causes of the   STERILITY of first Crosses and of Hybrids. We may now
0263  little closer at the probable causes of the   STERILITY of first crosses and of hybrids. These two
0264  th of the embryo is a very frequent cause of   STERILITY in first crosses. I was at first very unwil
0264  natural conditions of life. In regard to the   STERILITY of hybrids, in which the sexual elements ar
0265  to the domestication of animals. Between the   STERILITY thus superinduced and that of hybrids, ther
0265  many points of similarity. In both cases the   STERILITY is independent of general health, and is of
0265  size or great luxuriance. In both cases, the   STERILITY occurs in various degrees; in both, the mal
0265  ffected, though in a lesser degree than when   STERILITY ensues. So it is with hybrids, for hybrids
0265  the general state of health, is affected by   STERILITY in a very similar manner. In the one case,
0266  nd hence we need not be surprised that their   STERILITY, though in some degree variable, rarely dim
0266  ypotheses, several facts with respect to the   STERILITY of hybrids; for instance, the unequal ferti
0266  ed from reciprocal crosses; or the increased   STERILITY in those hybrids which occasionally and exc
0266  that in two cases, in some respects allied,   STERILITY is the common result, in the one case from
0267  ditions of life, always induces weakness and   STERILITY in the progeny. Hence it seems that, on the
0269  determine their greater or lesser degree of   STERILITY when crossed; and we may apply the same rul
0269  g course of domestication tends to eliminate   STERILITY in the successive generations of hybrids, w
0269  be so, we surely ought not to expect to find   STERILITY both appearing and disappearing under nearl
0269  ence of the existence of a certain amount of   STERILITY in the few following cases, which I will br
0270  as good as that from which we believe in the   STERILITY of a multitude of species. The evidence is,
0270  ho in all other cases consider fertility and   STERILITY as safe criterions of specific distinction.
0272  ect to the very general, but not invariable,   STERILITY of first crosses and of hybrids; namely, th
0275  . Laying aside the question of fertility and   STERILITY, in all other respects there seems to be a
0276  generally, but not universally, sterile. The   STERILITY is of all degrees, and is often so slight t
0276  nclusions in ranking forms by this test. The   STERILITY is innately variable in individuals of the
0276  e and unfavourable conditions. The degree of   STERILITY does not strictly follow systematic affinit
0276  en specially endowed with various degrees of   STERILITY to prevent them crossing and blending in na
0276  t them becoming inarched in our forests. The   STERILITY of first crosses between pure species, whic
0277  argely on the early death of the embryo. The   STERILITY of hybrids, which have their reproductive s
0277  stinct species, seems closely allied to that   STERILITY which so frequently affects pure species, w
0277  ty in uniting two species, and the degree of   STERILITY of their hybrid offspring should generally
0460  stered. With respect to the almost universal   STERILITY of species when first crossed, which forms
0460  ch seem to me conclusively to show that this   STERILITY is no more a special endowment than is the
0461  domestication apparently tends to eliminate   STERILITY, we ought not to expect it also to produce
0461  y, we ought not to expect it also to produce   STERILITY. The sterility of hybrids is a very differe
0461  to expect it also to produce sterility. The   STERILITY of hybrids is a very different case from th
0481  e, and varieties invariably fertile; or that   STERILITY is a special endowment and sign of creation
0022  s. the size and shape,of the apertures in the   STERNUM are highly variable; so is the degree of div
0364  wanderers only by one means, namely, in dirt   STICKING to their feet, which is in itself a rare acc
0192  us frena, which serve, through the means of a   STICKY secretion, to retain the eggs until they are
0193  f pollen grains, borne on a foot stalk with a   STICKY gland at the end, is the same in Orchis and A
0092  n transport the pollen from one flower to the   STIGMA of another flower. The flowers of two distinc
0097  hey either push the flowers own pollen on the   STIGMA, or bring pollen from another flower. So nece
0097  touch the anthers of one flower and then the   STIGMA of another with the same brush to ensure fert
0098  n observations, which effectually prevent the   STIGMA receiving pollen from its own flower: for ins
0098  conjoined anthers of each flower, before the   STIGMA of that individual flower is ready to receive
0098  ough by placing pollen from one flower on the   STIGMA of another, I raised plenty of seedlings; and
0098  special mechanical contrivance to prevent the   STIGMA of a flower receiving its own pollen, yet, as
0099  confirm, either the anthers burst before the   STIGMA is ready for fertilisation, or the stigma is
0099  the stigma is ready for fertilisation, or the   STIGMA is ready before the pollen of that flower is
0255  n from a plant of one family is placed on the   STIGMA of a plant of a distinct family, it exerts no
0255  rent species of the same genus applied to the   STIGMA of some one species, yields a perfect gradati
0263  t when pollen of one species is placed on the   STIGMA of a distantly allied species, though the pol
0451  ovarium at its base. The pistil consists of a   STIGMA supported on the style; but some Compositae,
0452  dimentary state, for it is not crowned with a   STIGMA: but the style remains well developed and is
```

Page ***(Key Word)***

stigma

Page ***(Key Word)***

```
0093 actly sixty yards from a male tree, I put the STIGMAS of twenty flowers, taken from different bran
0097 a multitude of flowers have their anthers and STIGMAS fully exposed to the weather! but if an occa
0252 produce no pollen, for he will find on their STIGMAS plenty of pollen brought from other flowers.
0095 corolla, so as to push the pollen on to the STIGMATIC surface. Hence, again, if humble bees were
0099 these facts! How strange that the pollen and STIGMATIC surface of the same flower, though placed s
0263 en tubes protrude, they do not penetrate the STIGMATIC surface. Again, the male element may reach
0219 ike best, and their larvae and pupae to STIMULATE them to work, they did nothing! they could
0202 rivances are less perfect. Can we consider the STING of the wasp or of the bee as perfect, which,
0202 by tearing out its viscera? If we look at the STING of the bee, as having originally existed in a
0202 rhaps understand how it is that the use of the STING should so often cause the insect's own death:
0472 ur ideas of fitness. We need not marvel at the STING of the bee causing the bee's own death: at dr
0202 own death: for if on the whole the power of STINGING be useful to the community, it will fulfil a
0092 his is effected by glands at the base of the STIPULES in some Leguminosae, and at the back of the
0014 many cases we do not know what the aboriginal STOCK was, and so could not tell whether or not nea
0015 or even wholly, revert to the wild aboriginal STOCK. Whether or not the experiment would succeed,
0018 hese had descended from a different aboriginal STOCK from our European cattle; and several compete
0018 hat all the races have descended from one wild STOCK. Mr. Blyth, whose opinion, from his large and
0032 uced by a single variation from the aboriginal STOCK. We have proofs that this is not so in some c
0035 surpass in fleetness and size the parent Arab STOCK, so that the latter, by the regulations for t
0035 eight and in early maturity, compared with the STOCK formerly kept in this country. By comparing t
0036 marks, have been purely bred from the original STOCK of Mr. Bakewell for upwards of fifty years. T
0037 ing growing wild, if it had come from a garden STOCK. The pear, though cultivated in classical tim
0042 from not being very easily reared and a large STOCK not kept! in geese, from being valuable only
0065 geometrical ratio, that all would most rapidly STOCK every station in which they could any how exi
0066 ll number may be produced, and yet the average STOCK be fully kept up! but if many eggs or young a
0068 thus, there seems to be little doubt that the STOCK of partridges, grouse, and hares on any large
0070 rey. On the other hand, in many cases, a large STOCK of individuals of the same species, relativel
0070 le seed. This view of the necessity of a large STOCK of the same species for its preservation, exp
0075 pplant the other varieties. To keep up a mixed STOCK of even such extremely close varieties as the
0076 tion, that the original proportions of a mixed STOCK could be kept up for half a dozen generations
0165 es are all due to ancient crosses with the dun STOCK. But I am not at all satisfied with this theo
0221 ame across a community with an unusually large STOCK of slaves, and I observed a few slaves mingle
0227 he hexagon to any length requisite to hold the STOCK of honey! in the same way as the rude humble
0234 e of honey is indispensable to support a large STOCK of bees during the winter! and the security o
0237 but the horticulturist sows seeds of the same STOCK, and confidently expects to get nearly the sa
0299 horses, and dogs have descended from a single STOCK or from several aboriginal stocks! or, again,
0422 s of difference in the languages from the same STOCK, would have to be expressed by groups subordi
0445 and have probably descended from the same wild STOCK! hence I was curious to see how far their pup
0445 d, size of feet and length of leg, in the wild STOCK, in pouters, fantails, runts, barbs, dragons,
0042 aces, at least, in a country which is already STOCKED with other races. In this respect enclosure
0067 st from germinating in ground already thickly STOCKED with other plants. Seedlings, also, are dest
0069 n numbers, and, as each area is already fully STOCKED with inhabitants, the other species will dec
0109 ll organic beings, each area is already fully STOCKED with inhabitants, it follows that as each se
0110 f species. Probably no region is as yet fully STOCKED, for at the Cape of Good Hope, where more sp
0121 ve played an important part. As in each fully STOCKED country natural selection necessarily acts b
0172 fications, each new form will tend in a fully STOCKED country to take the place of, and finally to
0205 to the owner. Natural selection in each well STOCKED country, must act chiefly through the compet
0319 conditions it would in a very few years have STOCKED the whole continent. But we could not have t
0355 o those of another region, whence it has been STOCKED. A volcanic island, for instance, upheaved a
0363 suspected that these islands had been partly STOCKED by ice borne seeds, during the glacial epoch
0364 be a great error to argue that because a well STOCKED island, like Great Britain, has not, as far
0365 europe or any other continent, that a poorly STOCKED island, though standing more remote from the
0365 ansported to an island, even if far less well STOCKED than Britain, scarcely more than one would b
0365 ed and formed, and before it had become fully STOCKED with inhabitants. On almost bare land, with
0380 no doubt before the Glacial period they were STOCKED with endemic Alpine forms! but these have al
0385 to another, that I have quite unintentionally STOCKED the one with fresh water shells from the oth
0390 oceanic islands; for man has unintentionally STOCKED them from various sources far more fully and
0391 two islands of Bermuda and Madeira have been STOCKED by birds, which for long ages have struggled
0399 on the view that this island has been mainly STOCKED by seeds brought with earth and stones on ic
0399 other southern lands were long ago partially STOCKED from a nearly intermediate though distant po
0401 pected on the view of the islands having been STOCKED by occasional means of transport, a seed, fo
0402 may safely infer that Charles Island is well STOCKED with its own species, for annually more eggs
0481 vince experienced naturalists whose minds are STOCKED with a multitude of facts all viewed, during
0014 ainly revert in character to their aboriginal STOCKS. Hence it has been argued that no deductions
0019 everal domestic races from several aboriginal STOCKS, has been carried to an absurd extreme by som
0019 number of peculiar species as distinct parent STOCKS? So it is in India. Even in the case of the d
0023 ended from at least seven or eight aboriginal STOCKS! for it is impossible to make the present dom
0023 crossing two breeds unless one of the parent STOCKS possessed the characteristic enormous crop? T
0023 ristic enormous crop? The supposed aboriginal STOCKS must all have been rock pigeons, that is, not
0023 omestic breeds. Hence the supposed aboriginal STOCKS must either still exist in the countries wher
0025 tly, that all the several imagined aboriginal STOCKS were coloured and marked like the rock pigeon
0037 with the older varieties or with their parent STOCKS. No one would ever expect to get a first rate
0037 e, and therefore do not know, the wild parent STOCKS of the plants which have been longest cultiva
0038 ot by a strange chance possess the aboriginal STOCKS of any useful plants, but that the native pla
0041 he other hand, nurserymen, from raising large STOCKS of the same plants, are generally far more su
0120 ly diverging in character from their original STOCKS, without either having given off any fresh br
0141 ome of our domestic animals from several wild STOCKS: the blood, for instance, of a tropical and a
0167 or not it be descended from one or more wild STOCKS, or the ass, the hemionus, quagga, and zebra.
0177 h ecual steadiness and skill to improve their STOCKS by selection; the chances in this case will b
0251 oble, for instance, informs me that he raises STOCKS for grafting from a hybrid between Rhod. Pont
0254 hat our dogs have descended from several wild STOCKS; yet, with perhaps the exception of certain i
0262 fference between the mere adhesion of grafted STOCKS, and the union of the male and female element
0299 rom a single stock or from several aboriginal STOCKS! or, again, whether certain sea shells inhabi
0116 milne Edwards. No physiologist doubts that a STOMACH by being adapted to digest vegetable matter
0190 the exterior surface will then digest and the STOMACH respire. In such cases natural selection mig
0387 r. hooker, the Nelumbium luteum) in a heron's STOMACH! although I do not know the fact, yet analog
```

Page ***(Key Word)***

Page ***(Key Word)***

```
0387  meal of fish, would probably reject from its STOMACH a pellet containing the seeds of the Nelumbi
0441  for a short time, and is destitute of mouth, STOMACH, or other organs of importance, excepting fo
0362  been from twelve to twenty one hours in the STOMACHS of different birds of prey; and two seeds of
0362  place. I forced many kinds of seeds into the STOMACHS of dead fish, and then gave their bodies to
0167  hells had never lived, but had been created in STONE so as to mock the shells now living on the se
0222  e an independent community of F. flava under a STONE beneath a nest of the slave making F. sanguin
0403  land shells, some of which live in crevices of STONE; and although large quantities of stone are a
0403  ces of stone; and although large quantities of STONE are annually transported from Porto Santo to
0185  amily wholly subsists by diving, grasping the STONES with its feet and using its wings under water
0361  of the coral islands in the Pacific, procure STONES for their tools, solely from the roots of dri
0361  solely from the roots of drifted trees, these STONES being a valuable royal tax. I find on examina
0361  on examination, that when irregularly shaped STONES are embedded in the roots of trees, small par
0363  e known to be sometimes loaded with earth and STONES, and have even carried brushwood, bones, and
0399  ainly stocked by seeds brought with earth and STONES on icebergs, drifted by the prevailing curren
0213  iod, under less fixed conditions of life. How STONGLY these domestic instincts, habits, and dispos
0124  ous nature. Having descended from a form which STOOD between the two parent species (A) and (I), n
0343  our oscillating continents now stand they have STOOD ever since the Silurian epoch; but that long
0102  ite object, and free intercrossing will wholly STOP his work. But when many men, without intending
0108  several causes are amply sufficient wholly to STOP the action of natural selection. I do not beli
0260  pecial power of producing hybrids, and then to STOP their further propagation by different degrees
0228  ever, did not suffer this to happen, and they STOPPED their excavation in due time; so that the ba
0366  same time travel southward, unless they were STOPPED by barriers, in which case they would perish
0229  s leaving flat plates between the basins, by STOPPING work along the intermediate planes or planes
0229  he wax away to the proper thinness, and then STOPPING their work. In ordinary combs it has appeare
0077  o escape serving as prey to other animals. The STORE of nutriment laid up within the seeds of many
0218  aratus which would be necessary if they had to STORE food for their own young. Some species, likew
0234  days during the process of secretion. A large STORE of honey is indispensable to support a large
0234  uchout the winter, and consequently required a STORE of honey: there can in this case be no doubt
0218  n this insect finds a burrow already made and STORED by another sphex it takes advantage of the pr
0219  the species, and if the insect whose nest and STORED food are thus feloniously appropriated, be no
0223  r nests, it is possible that pupae originally STORED as food might become developed; and the ants
0463  piles of strata beneath the Silurian system, STORED with the remains of the progenitors of the Si
0003  s aided me in every possible way by his large STORES of knowledge and his excellent judgment. In c
0018  yth, whose opinion, from his large and varied STORES of knowledge, I should value more than that o
0218  ytes nigra generally makes its own burrow and STORES it with paralysed prey for its own larvae to
0362  and then gave their bodies to fishing eagles, STORKS, and pelicans; these birds after an interval
0285  ad once extended into the open ocean. The same STORY is still more plainly told by faults, those g
0373  arated mountains in this island, tell the same STORY. If one account which has been published can
0130  ssil state. As we here and there see a thin STRAGGLING branch springing from a fork low down in a
0214  arentage only in one way, by not coming in a STRAIGHT line to his master when called. Domestic ins
0228  ed edge of a smooth basin, instead of on the STRAIGHT edges of a three sided pyramid as in the cas
0034  with a distinct object in view, to make a new STRAIN or sub breed, superior to anything existing i
0096  ndividuals of the same variety but of another STRAIN, gives vigour and fertility to the offspring;
0153  e as a common tumbler from a good short faced STRAIN. But as long as selection is rapidly going on
0036  nsued, namely, the production of two distinct STRAINS. The two flocks of Leicester sheep kept by M
0267  species, that is between members of different STRAINS or sub breeds, gives vigour and fertility to
0349  n nearly the same manner. The plains near the STRAITS of Magellan are inhabited by one species of
0360  24 miles of sea to another country; and when STRANDED, if blown to a favourable spot by an inland
0382  er the equator. The various beings thus left STRANDED may be compared with savage races of man, dr
0013  ing in several members of the same family. If STRANGE and rare deviations of structure are truly i
0013  ations of structure are truly inherited, less STRANGE and commoner deviations may be freely admitt
0024  since all become extinct or unknown. So many STRANGE contingencies seem to me improbable in the h
0034  lowest savages. It would, indeed, have been a STRANGE fact, had attention not been paid to breedin
0038  se countries, so rich in species, do not by a STRANGE chance possess the aboriginal stocks of any
0080  be borne in mind in what an endless number of STRANGE peculiarities our domestic productions, and,
0089  es display their gorgeous plumage and perform STRANGE antics before the females, which standing by
0099  ed sexes, and must habitually be crossed. How STRANGE are these facts! How strange that the pollen
0099  be crossed. How strange are these facts! How STRANGE that the pollen and stigmatic surface of the
0101  ss. It must have struck most naturalists as a STRANGE anomaly that, in the case of both animals an
0191  ng description of these parts, understand the STRANGE fact that every particle of food and drink w
0212  cases also, could be given, of occasional and STRANGE habits in certain species, which might, if a
0214  e one pigeon showed a slight tendency to this STRANGE habit, and that the long continued selection
0217  ed process of this nature, I believe that the STRANGE instinct of our cuckoo could be; and has bee
0249  lant of the same hybrid nature. And thus, the STRANGE fact of the increase of fertility in the suc
0260  st union between their parents, seems to be a STRANGE arrangement. The foregoing rules and facts,
0313  in an existing crocodile associated with many STRANGE and lost mammals and reptiles in the sub Him
0325  parts of Europe, they add, if struck by this STRANGE sequence, we turn our attention to North Ame
0326  t on climatal and geographical changes, or on STRANGE accidents, but in the long run the dominant
0353  erally local, or confined to one area. What a STRANGE anomaly it would be, if, when coming one ste
0378  perate vegetation, like that now growing with STRANGE luxuriance at the base of the Himalaya, as g
0416  in determining the degree of affinity of this STRANGE creature to birds and reptiles, as an approa
0450  d, or aborted organs. Organs or parts in this STRANGE condition, bearing the stamp of inutility, a
0456  tion, or quite aborted, far from presenting a STRANGE difficulty, as they assuredly do on the ordi
0460  groups of organic beings: but we see so many STRANGE gradations in nature, as is proclaimed by th
0470  h respects they resemble varieties. These are STRANGE relations on the view of each species having
0471  t seems to me, explicable on this theory. How STRANGE it is that a bird, under the form of woodpec
0472  pied place in nature, these facts cease to be STRANGE, or perhaps might even have been anticipated
0475  varieties. On the other hand, these would be STRANGE facts if species have been independently cre
0482  een independently created. This seems to me a STRANGE conclusion to arrive at. They admit that a m
0379  uthern hemisphere. These being surrounded by STRANGERS will have had to compete with many new form
0124  ons, and likewise a section of the successive STRATA of the earth's crust including extinct remain
0279  appearance in the lowest known fossiliferous STRATA. In the sixth chapter I enumerated the chief
0282  amine for himself great piles of superimposed STRATA, and watch the sea at work grinding down old
0284  t britain: and this is the result: Palaeozoic STRATA (not including igneous beds).....57,154 Feet.
0284  ding igneous beds).....57,154 Feet. Secondary STRATA...............................13,190 Feet
0284  ...............13,190 Feet. Tertiary STRATA............................. 2,240 Fee
```

Page ***(Key Word)***

Page ***(Key Word)***

```
0284   slow. But the amount of denudation which the STRATA have in many places suffered, independently o
0285   by faults, those great cracks along which the STRATA have been upheaved on one side, or thrown dow
0285   no this line the vertical displacement of the STRATA has varied from 600 to 3000 feet. Prof. Ramsa
0285   at which has removed masses of our palaeozoic STRATA, in parts ten thousand feet in thickness, as
0286   length: and we must remember that almost all STRATA contain harder layers or nodules, which from
0289   scovered by Sir C. Lyell in the carboniferous STRATA of North America. In regard to mammiferous re
0296   ick in Nova Scotia, with ancient root bearing STRATA, one above the other, at no less than sixty e
0306   ied Species in the lowest known fossiliferous STRATA. There is another and allied difficulty, whic
0307   ng the absence of vast piles of fossiliferous STRATA, which on my theory no doubt were somewhere a
0309   there find formations older than the Silurian STRATA, supposing such to have been formerly deposit
0309   erly deposited; for it might well happen that STRATA which had subsided some miles nearer to the c
0309   ve undergone far more metamorphic action than STRATA which have always remained nearer to the surf
0310   fossiliferous formations beneath the Silurian STRATA, are all undoubtedly of the gravest nature. W
0338   a, until beds far beneath the lowest Silurian STRATA are discovered, a discovery of which the chan
0394   hich they have suffered and by their tertiary STRATA: there has also been time for the production
0463   cal stages? Why do we not find great piles of STRATA beneath the Silurian system, stored with the
0464   s of fossils? For certainly on my theory such STRATA must somewhere have been deposited at these a
0465   ferous formations beneath the lowest Silurian STRATA, I can only recur to the hypothesis given in
0187   nd far beneath the lowest known fossiliferous STRATUM to discover the earlier stages, by which the
0280   n is not every geological formation and every STRATUM full of such intermediate links? Geology ass
0282   f the duration of each formation or even each STRATUM. A man must for years examine for himself or
0303   urge that no monkey occurred in any tertiary STRATUM; but now extinct species have been discovere
0305   en has as yet been found even in any tertiary STRATUM. Hence we now positively know that sessile c
0307   indisputable that before the lowest Silurian STRATUM was deposited, long periods elapsed, as long
0307   in the organic remains of the lowest Silurian STRATUM the dawn of life on this planet. Other highl
0316   sion of generations, from the lowest Silurian STRATUM to the present day. We have seen in the last
0041   marked, that it was most fortunate that the STRAWBERRY began to vary just when gardeners began to
0041   attend closely to this plant. No doubt the STRAWBERRY had always varied since it was cultivated,
0042   cies) those many admirable varieties of the STRAWBERRY which have been raised during the last thir
0359   on the banks, and then by a fresh rise in the STREAM be washed into the sea. Hence I was led to dr
0383   requent migrations from pond to pond, or from STREAM to stream; and liability to wide dispersal wo
0383   grations from pond to pond, or from stream to STREAM; and liability to wide dispersal would follow
0388   , it should be remembered that when a pond or STREAM is first formed, for instance, on a rising is
0285   t in height; for the gentle slope of the lava STREAMS, due to their formerly liquid state, showed
0290   degradation of the coast rocks and from muddy STREAMS entering the sea. The explanation, no doubt,
0366   ed surfaces, and perched boulders, of the icy STREAMS with which their valleys were lately filled.
0212   ay be seen in nestling birds, though it is STRENGTHENED by experience, and by the sight of fear o
0076   c plants or animals have so exactly the same STRENGTH, habits, and constitution, that the original
0088   indomitable courage, length to the spur, and STRENGTH to the wing to strike in the spurred leg, as
0090   ous animals, securing some by craft, some by STRENGTH, and some by fleetness; and let us suppose t
0090   selected, provided always that they retained STRENGTH to master their prey at this or at some othe
0195   but they are incessantly harassed and their STRENGTH reduced, so that they are more subject to di
0205   and consequently will produce perfection, or STRENGTH in the battle for life, only according to th
0231   thick. By this singular manner of building, STRENGTH is continually given to the comb, with the u
0243   retend that the facts given in this chapter STRENGTHEN in any great degree my theory; but none of
0212   s of instinct in a state of nature will be STRENGTHENED by briefly considering a few cases under d
0224   lly for food might by natural selection be STRENGTHENED and rendered permanent for the very differ
0243   f natural selection. This theory is, also, STRENGTHENED by some few other facts in regard to insti
0249   eady lessened by their hybrid origin. I am STRENGTHENED in this conviction by a remarkable stateme
0355   from this other region, my theory will be STRENGTHENED; for we can clearly understand, on the pri
0168   ing constitutional differences, and use in STRENGTHENING, and disuse in weakening and diminishing
0134   ttle doubt that use in our domestic animals STRENGTHENS and enlarges certain parts, and disuse dim
0198   nary generation, we ought not to lay too much STRESS on our ignorance of the precise cause of the
0275   ffspring of the female ass and stallion. Much STRESS has been laid by some authors on the supposed
0415   that almost all naturalists lay the greatest STRESS on resemblances in organs of high vital or ph
0462   normously long as measured by years, too much STRESS ought not to be laid on the occasional wide d
0181   s. It has an extremely wide flank membrane, STRETCHING from the corners of the jaw to the tail, an
0367   ar south as the Alps and Pyrenees, and even STRETCHING into Spain. The now temperate regions of th
0218   erfected; for a surprising number of eggs lie STREWED over the plains, so that in one day's huntin
0025   p (the Indian sub species, C. intermedia of STRICKLAND, having it bluish); the tail has a terminal
0022   of the nostrils, of the tongue (not always in STRICT correlation with the length of beak), the siz
0256   generally confounded together, is by no means STRICT. There are many cases, in which two pure spec
0257   y and the facility of crossing is by no means STRICT. A multitude of cases could be given of very
0292   in the truth of these views; for they are in STRICT accordance with the general principles inculc
0324   d thousandth year, or even that it has a very STRICT geological sense: for if all the marine anima
0326   ect to find, as we apparently do find, a less STRICT degree of parallel succession in the producti
0018   ng. Even if this latter fact were found more STRICTLY and generally true than seems to me to be th
0021   and the common tumbler has the singular and STRICTLY inherited habit of flying at a great height
0075   this is often the case with those which may STRICTLY be said to struggle with each other for exis
0140   not positively know that these animals were STRICTLY adapted to their native climate, but in all
0174   erfectly defined species have been formed on STRICTLY continuous areas; though I do not doubt that
0178   have been formed in different portions of a STRICTLY continuous area, intermediate varieties will
0179   intermediate grade between truly aquatic and STRICTLY terrestrial habits; and as each exists by a
0185   tic habits; yet this anomalous member of the STRICTLY terrestrial thrush family wholly subsists by
0192   at the ovigerous frena in the one family are STRICTLY homologous with the branchiae of the other f
0206   the present inhabitants of the world, is not STRICTLY correct, but if we include all those of past
0206   those of past times, it must by my theory be STRICTLY true. It is generally acknowledged that all
0220   r enter the nest. Hence he considers them as STRICTLY household slaves. The masters, on the other
0226   rom his information, and tells me that it is STRICTLY correct: If a number of equal spheres be des
0230   wall of wax, is not, as far as I have seen, STRICTLY correct; the first commencement having alway
0232   dations of one wall of a new hexagon, in its STRICTLY proper place, projecting beyond the other co
0233   e extreme margin of wasp combs are sometimes STRICTLY hexagonal; but I have not space here to ente
0240   my giving not the actual measurements, but a STRICTLY accurate illustration: the difference was th
0260   ation by different degrees of sterility, not STRICTLY related to the facility of the first union b
0276   conditions. The degree of sterility does not STRICTLY follow systematic affinity, but is governed
0288   the world in infinite numbers: they are all STRICTLY littoral, with the exception of a single Med
```

Page ***(Key Word)***

Page ***************************************(Key Word)***************************************

```
0297  e found in an underlying bed; even if A were  STRICTLY  intermediate between B and C, it would simpl
0297  d c, and yet might not at all necessarily be  STRICTLY  intermediate between them in all points of s
0300  accumulating. I suspect that not many of the  STRICTLY  littoral animals, or of those which lived on
0313  nic change, as Pictet has remarked, does not  STRICTLY  correspond with the succession of our geolog
0316  as I maintain) admit its truth; and the rule  STRICTLY  accords with my theory. For as all the speci
0317  rease in number of the species of a group is  STRICTLY  conformable with my theory; as the species o
0324  ely, the upper pliocene, the pleistocene and  STRICTLY  modern beds, of Europe, North and South Amer
0328  ed by the same movement, it is probable that  STRICTLY  contemporaneous formations have often been a
0329  fe, and the order would falsely appear to be  STRICTLY  parallel; nevertheless the species would not
0338  the oldest known mammals, reptiles, and fish  STRICTLY  belong to their own proper classes, though s
0347  ern temperate forms, as there now is for the  STRICTLY  arctic productions. We see the same fact in
0364  sometimes called accidental, but this is not  STRICTLY  correct: the currents of the sea are not acc
0369  ng remarks on distribution apply not only to  STRICTLY  arctic forms, but also to many sub arctic an
0370  latitude 66 degrees 67 degrees; and that the  STRICTLY  arctic productions then lived on the broken
0373  e. we do not know that the Glacial epoch was  STRICTLY  simultaneous at these several far distant po
0376  brief abstract applies to plants alone: some  STRICTLY  analogous facts could be given on the distri
0393  o verify this assertion, and I have found it  STRICTLY  true. I have, however, been assured that a f
0403  alpine rodents, Alpine plants, etc., all of  STRICTLY  American forms, and it is obvious that a mou
0420  on and relation to the other groups, must be  STRICTLY  genealogical in order to be natural; but tha
0421  eless their genealogical arrangement remains  STRICTLY  true, not only at the present time, but at e
0422  uld still be genealogical; and this would be  STRICTLY  natural, as it would connect together all s
0468  at the amount of variation under nature is a  STRICTLY  limited quantity. Man, though acting on exte
0471  addition to our knowledge tends to make more  STRICTLY  correct, is on this theory simply intelligib
0487  st be cautious in attempting to correlate as  STRICTLY  contemporaneous two formations, which includ
0088  ngth to the spur, and strength to the wing to  STRIKE  in the spurred leg, as well as the brutal coc
0233  s from the central point and from each other,  STRIKE  the planes of intersection, and so make an is
0413  its familiarity, does not always sufficiently  STRIKE  us, is in my judgment fully explained. Natura
0007  ts and animals, one of the first points which  STRIKES  us, is, that they generally differ much more
0346  face of the globe, the first great fact which  STRIKES  us is, that neither the similarity nor the d
0347  bitants of the sea. A second great fact which  STRIKES  us in our general review is, that barriers o
0064  c two or three following seasons. Still more  STRIKING  is the evidence from our domestic animals of
0093  could easily do this, I could easily show by many  STRIKING  instances. I will give only one, not as a ve
0093  stances. I will give only one, not as a very  STRIKING  case, but as likewise illustrating one step
0132  uch influences cannot have produced the many  STRIKING  and complex co adaptations of structure betw
0145  i may add, as an instance of this, and of a  STRIKING  case of correlation, that I have recently ob
0176  ieties in a state of nature. I have met with  STRIKING  instances of the rule in the case of varieti
0180  er a heavy disadvantage, for out of the many  STRIKING  cases which I have collected, I can give onl
0184  uch instances do occur in nature. Can a more  STRIKING  instance of adaptation be given than that of
0213  ed that young pointers (I have myself seen a  STRIKING  instance) will sometimes point and even back
0232  ls of the last completed cells, and then, by  STRIKING  imaginary spheres, they can build up a wall
0236  but not much greater than that of any other  STRIKING  modification of structure; for it can be sho
0258  a, and utterly failed. Several other equally  STRIKING  cases could be given. Thuret has observed th
0272  the first generation; and I have myself seen  STRIKING  instances of this fact. Gartner further admi
0303  en as far back as the eocene stage. The most  STRIKING  case, however, is that of the Whale family;
0304  k of Belgium. And, as if to make the case as  STRIKING  as possible, this sessile cirripede was a Ch
0313  itude of extinct forms. Falconer has given a  STRIKING  instance of a similar fact, in an existing c
0313  icker rate than those of the sea, of which a  STRIKING  instance has lately been observed in Switzer
0322  rcely any palaeontological discovery is more  STRIKING  than the fact, that the forms of life change
0328  tions. Barrande, also, shows that there is a  STRIKING  general parallelism in the successive Siluri
0329  vertebrata, whole pages could be filled with  STRIKING  illustrations from our great palaeontologist
0339  tai and Professor Owen has shown in the most  STRIKING  manner that most of the fossil mammals, buri
0353  he vast and broken interspace. The great and  STRIKING  influence which barriers of every kind have
0354  ry remarks, I will discuss a few of the most  STRIKING  classes of facts: namely, the existence of t
0365  could not possibly exist, is one of the most  STRIKING  cases known of the same species living at di
0372  temperate North America; and the still more  STRIKING  case of many closely allied crustaceans (as
0375  lection made on a hill in Europe! Still more  STRIKING  is the fact that southern Australian forms a
0375  of new Zealand, by Dr. Hooker, analogous and  STRIKING  facts are given in regard to the plants of t
0378  ded two great lines of invasion; and it is a  STRIKING  fact, lately communicated to me by Dr. Hooke
0382  r forms of vegetable life. Sir C. Lyell in a  STRIKING  passage has speculated, in language almost i
0386  xperiments, but will here give only the most  STRIKING  case; I took in February three tablespoons o
0392  lly hooked seeds; yet few relations are more  STRIKING  than the adaptation of hooked seeds for tran
0395  ed condition. Mr. Windsor Earl has made some  STRIKING  observations on this head in regard to the g
0397  xperiments are wanted on this head. The most  STRIKING  and important fact for us in regard to the i
0406  of the same archipelago, and especially the  STRIKING  relation of the inhabitants of each whole ar
0409  ate Edward Forbes often insisted, there is a  STRIKING  parellelism in the laws of life throughout t
0429  d they are throughout the world, the fact is  STRIKING,  that the discovery of Australia has not add
0476  tribution. We can see why there should be so  STRIKING  a parallelism in the distribution of organic
0478  r american islands being related in the most  STRIKING  manner to the plants and animals of the neig
0142  n climates better than others: this is very  STRIKINGLY  shown in works on fruit trees published in
0386  to the most remote oceanic islands. This is  STRIKINGLY  shown, as remarked by Alph. de Candolle, in
0404  ult to prove it. Amongst mammals, we see it  STRIKINGLY  displayed in Bats, and in a lesser degree i
0439  nct animals within the same class are often  STRIKINGLY  similar: a better proof of this cannot be g
0228  mbs, and put between them a long thick, square  STRIP  of wax: the bees instantly began to excavate
0229  the bees, whilst at work on the two sides of a  STRIP  of wax, perceiving when they have gnawed the
0163  o be true. It has also been asserted that the  STRIPE  on each shoulder is sometimes double. The sho
0163  ch shoulder is sometimes double. The shoulder  STRIPE  is certainly very variable in length and outl
0163  n described without either spinal or shoulder  STRIPE;  and these stripes are sometimes very obscure
0163  said to have been seen with a double shoulder  STRIPE.  The hemionus has no shoulder stripe; but tra
0163  shoulder stripe. The hemionus has no shoulder  STRIPE;  but traces of it, as stated by Mr. Blyth and
0163  have collected cases in England of the spinal  STRIPE  in horses of the most distinct breeds, and of
0163  one instance in a chestnut: a faint shoulder  STRIPE  may sometimes be seen in duns, and I have see
0164  me of a dun Belgian cart horse with a double  STRIPE  on each shoulder and with leg stripes; and a
0164  e legs are generally barred; and the shoulder  STRIPE,  which is sometimes double and sometimes treb
0164  , that with the English race horse the spinal  STRIPE  is much commoner in the foal than in the full
0165  nd in one of them there was a double shoulder  STRIPE.  In lord Moreton's famous hybrid from a chest
0165  hemionus has none and has not even a shoulder  STRIPE,  nevertheless, had all four legs barred, and
```

Page ***************************************(Key Word)***************************************

Page ***(Key Word)**

```
0165 last fact, I was so convinced that not even a STRIPE of colour appears from what would commonly be
0163 that the foals of this species are generally STRIPED on the legs and faintly on the shoulder. The
0164 the Kattywar breed of horses is so generally STRIPED, that, as I hear from Colonel Poole, who exa
0164 onsidered as purely bred. The spine is always STRIPED; the legs are generally barred; and the shou
0164 the side of the face, moreover, is sometimes STRIPED. The stripes are plainest in the foal; and s
0164 le has seen both gray and bay Kattywar horses STRIPED when first foaled. I have, also, reason to s
0164 boriginal species, one of which, the dun, was STRIPED; and that the above described appearances ar
0165 legs. I once saw a mule with its legs so much STRIPED that any one at first would have thought tha
0166 such face stripes ever occur in the eminently STRIPED Kattywar breed of horses, and was, as we hav
0166 he horse genus becoming, by simple variation, STRIPED on the legs like a zebra, or striped on the
0166 riation, striped on the legs like a zebra, or STRIPED on the shoulders like an ass. In the horse w
0166 new character. We see this tendency to become STRIPED most strongly displayed in hybrids from betw
0167 thousands of generations, and I see an animal STRIPED like a zebra, but perhaps otherwise very dif
0167 this particular manner, so as often to become STRIPED like other species of the genus; and that ea
0439 up. In the cat tribe, most of the species are STRIPED or spotted in lines; and stripes can be plai
0473 ieve that these species have descended from a STRIPED progenitor, in the same manner as the severa
0163 t either spinal or shoulder stripe; and these STRIPES are sometimes very obscure, or actually quit
0164 a double stripe on each shoulder and with leg STRIPES; and a man, whom I can implicitly trust, has
0164 mall dun Welch pony with three short parellel STRIPES on each shoulder. In the north west part of
0164 ed for the Indian Government, a horse without STRIPES is not considered as purely bred. The spine
0164 the face, moreover, is sometimes striped. The STRIPES are plainest in the foal; and sometimes quit
0164 at i have collected cases of leg and shoulder STRIPES in horses of very different breeds, in vario
0164 in the south. In all parts of the world these STRIPES occur far oftenest in duns and mouse duns: b
0165 si and this hybrid, though the ass seldom has STRIPES on its legs and the hemionus has none and ha
0165 our legs barred, and had three short shoulder STRIPES, like those on the dun Welch pony, and even
0165 dun Welch pony, and even had some zebra like STRIPES on the sides of its face. With respect to th
0165 was led solely from the occurence of the face STRIPES on this hybrid from the ass and hemionus, to
0166 ionus, to ask Colonel Poole whether such face STRIPES ever occur in the eminently striped Kattywar
0166 r species of the genus. The appearance of the STRIPES is not accompanied by any change of form or
0166 hat in several species of the horse genus the STRIPES are either plainer or appear more commonly i
0167 world, to produce hybrids resembling in their STRIPES, not their own parents, but other species of
0439 species are striped or spotted in lines; and STRIPES can be plainly distinguised in the whelp of
0440 ditions of life. No one will suppose that the STRIPES on the whelp of a lion, or the spots on the
0473 y of creation is the occasional appearance of STRIPES on the shoulder and legs of the several spec
0066 le organic being around us may be said to be STRIVING to the utmost to increase in numbers; that e
0075 r animals with birds and beasts of prey, all STRIVING to increase, and all feeding on each other o
0078 steadily in mind that each organic being is STRIVING to increase at a geometrical ratio; that eac
0102 finite period; for as all organic beings are STRIVING, it may be said, to seize on each place in t
0113 tless seeds; and thus, as it may be said, is STRIVING its utmost to increase its numbers. Conseque
0114 culiar in its nature), and may be said to be STRIVING to the utmost to live there; but, it is seen
0315 igeons were all destroyed, that fanciers, by STRIVING during long ages for the same object, might
0211 ens, but not one excreted; I then tickled and STROKED them with a hair in the same manner, as well
0001 to complete it, and as my health is far from STRONG, I have been urged to publish this Abstract.
0005 ing, and thus be naturally selected. From the STRONG principle of inheritance, any selected variet
0012 e best on this subject. No breeder doubts how STRONG is the tendency to inheritance: like produces
0015 hown that our domestic varieties manifested a STRONG tendency to reversion, that is, to lose their
0017 domestic races, there is presumptive, or even STRONG, evidence in favour of this view. It has ofte
0026 long continued domestication eliminates this STRONG tendency to sterility: from the history of th
0059 espects the species of large genera present a STRONG analogy with varieties. And we can clearly un
0077 ort of relation to other plants. But from the STRONG growth of young plants produced from such see
0080 ser degree, those under nature, vary; and how STRONG the hereditary tendency is. Under domesticati
0100 gely provided against it by giving to trees a STRONG tendency to bear flowers with separated sexes
0100 erstand this remarkable fact, which offers so STRONG a contrast with terrestrial plants, on the vi
0112 characters, being neither very swift nor very STRONG, will have been neglected, and will have tend
0127 served in the struggle for life; and from the STRONG principle of inheritance they will tend to pr
0160 r progenitor possessed, the tendency, whether STRONG or weak, to reproduce the lost character migh
0166 ike an ass. In the horse we see this tendency STRONG whenever a dun tint appears, a tint which app
0166 eeds of various colours are crossed, we see a STRONG tendency for the blue tint and bars and marks
0167 genus; and that each has been created with a STRONG tendency, when crossed with species inhabitin
0231 and those completed, being thus crowned by a STRONG coping of wax, the bees can cluster and crawl
0272 red in several other respects. Gartner, whose STRONG wish was to draw a marked line of distinction
0438 e of fundamental resemblance, retained by the STRONG principle of inheritance. In the great class
0438 lly been modified into skulls or jaws. Yet so STRONG is the appearance of a modification of this n
0457 de for its own wants, and bearing in mind how STRONG is the principle of inheritance, the occurren
0053 ies, as did subsequently Dr. Hooker, even in STRONGER terms. I shall reserve for my future work th
0112 od one man preferred swifter horses; another STRONGER and more bulky horses. The early differences
0112 n of swifter horses by some breeders; and of STRONGER ones by others, the differences would become
0210 for the exclusive good of others. One of the STRONGEST instances of an animal apparently performin
0244 anic beings; namely, multiply, vary, let the STRONGEST live and the weakest die. Chapter VIII. Hyb
0313 the same identical form never reappears. The STRONGEST apparent exception to this latter rule, is
0008 f distinction from mere variations. But I am STRONGLY inclined to suspect that the most frequent c
0014 y safely conclude that very many of the most STRONGLY marked domestic varieties could not possibly
0029 s simple: from long continued study they are STRONGLY impressed with the differences between the s
0032 different breeds; all the best breeders are STRONGLY opposed to this practice, except sometimes a
0049 ts consider our British red grouse as only a STRONGLY marked race of a Norwegian species, whereas
0049 vainly to beat the air. Many of the cases of STRONGLY marked varieties or doubtful species well de
0052 inct and permanent, as steps leading to more STRONGLY marked and more permanent varieties; and at
0055 istribution. From looking at species as only STRONGLY marked and well defined varieties, I was led
0056 nification on the view that species are only STRONGLY marked and permanent varieties; for wherever
0089 of beauty, might produce a marked effect. I STRONGLY suspect that some well known laws with respe
0096 this is far from obvious. Nevertheless I am STRONGLY inclined to believe that with all hermaphrod
0101 , but which I am not here able to give, I am STRONGLY inclined to suspect that, both in the vegeta
0111 t facts. In the first place, varieties, even STRONGLY marked ones, though having somewhat of the c
0136 fi and especially the extraordinary fact, so STRONGLY insisted on by Mr. Wollaston, of the almost
0150 ies of the same genus. The rule applies very STRONGLY in the case of secondary sexual characters,
0155 given. But on the view of species being only STRONGLY marked and fixed varieties, we might surely
```

Page ***(Key Word)**

```
0156 ds, in which secondary sexual characters are STRONGLY displayed, with the amount of difference bet
0166 we see this tendency to become striped most STRONGLY displayed in hybrids from between several of
0177 selection; the chances in this case will be STRONGLY in favour of the great holders on the mounta
0199 ences between the races of man, which are so STRONGLY marked; I may add that some little light can
0257 in habit and general appearance, and having STRONGLY marked differences in every part of the flow
0274 otency in transmitting likeness running more STRONGLY in one sex than in the other, both when one
0274 the horse; but that the prepotency runs more STRONGLY in the male ass than in the female, so that
0316 e. Forbes, Pictet, and Woodward (though all STRONGLY opposed to such views as I maintain) admit i
0339 as so much impressed with these facts that I STRONGLY insisted, in 1839 and 1845, on this law of t
0340 at it now is. North America formerly partook STRONGLY of the present character of the southern hal
0370 rge, but partial oscillations of level, I am STRONGLY inclined to extend the above view, and to in
0379 s distinct species. It is a remarkable fact, STRONGLY insisted on by Hooker in regard to America,
0416 n ruminants and Pachyderms. Robert Brown has STRONGLY insisted on the fact that the rudimentary fl
0418 rt in the economy of nature. Yet it has been STRONGLY urged by those great naturalists, Milne Edwa
0419 anists, such as Mr. Bentham and others, have STRONGLY insisted on their arbitrary value. Instances
0430 of rodents. In this case, however, it may be STRONGLY suspected that the resemblance is only analo
0434 positions? Geoffroy St. Hilaire has insisted STRONGLY on the high importance of relative connexion
0469 he theory. On the view that species are only STRONGLY marked and permanent varieties, and that eac
0001 oard H.M.S. Beagle, as naturalist, I was much STRUCK with certain facts in the distribution of the
0028 ups of birds, in nature. One circumstance has STRUCK me much; namely, that all the breeders of the
0048 those from the American mainland, I was much STRUCK how entirely vague and arbitrary is the disti
0050 umber of forms of doubtful value. I have been STRUCK with the fact, that if any animal or plant in
0067 by incessant blows, sometimes one wedge being STRUCK, and then another with greater force. What ch
0101 maphrodites, do sometimes cross. It must have STRUCK most naturalists as a strange anomaly that, i
0112 l here find something analogous. A fancier is STRUCK by a pigeon having a slightly shorter beak; a
0112 g a slightly shorter beak; another fancier is STRUCK by a pigeon having a rather longer beak; and
0148 only understand a fact with which I was much STRUCK when examining cirripedes, and of which many
0150 highly variable. Several years ago I was much STRUCK with a remark, nearly to the above effect, pu
0201 injury to its possessor. If a fair balance be STRUCK between the good and evil caused by each part
0221 s in their jaws. Another day my attention was STRUCK by about a score of the slave makers haunting
0232 of construction seems to be a sort of balance STRUCK between many bees, all instinctively standing
0284 he lapse of time. I remember having been much STRUCK with the evidence of denudation, when viewing
0290 ch other in close sequence. Scarcely any fact STRUCK me more when examining many hundred miles of
0304 from having passed under my own eyes has much STRUCK me. In a memoir on Fossil Sessile Cirripedes,
0324 e, at distant parts of the world, has greatly STRUCK those admirable observers, MM. de Verneuil an
0325 life in various parts of Europe, they add, if STRUCK by this strange sequence, we turn our attenti
0349 stance, from north to south never fails to be STRUCK by the manner in which successive groups of b
0415 st can have worked at any group without being STRUCK with this fact; and it has been most fully ac
0428 sub groups in distinct classes. A naturalist, STRUCK by a parallelism of this nature in any one cl
0453 ans. In reflecting on them, every one must be STRUCK with astonishment: for the same reasoning pow
0477 eaning of the wonderful fact, which must have STRUCK every traveller, namely, that on the same con
0016 . when we attempt to estimate the amount of STRUCTURAL difference between the domestic races of th
0248 dence derived from other constitutional and STRUCTURAL differences. In regard to the sterility of
0419 use further research has detected important STRUCTURAL differences, at first overlooked, but becau
0003 odified, so as to acquire that perfection of STRUCTURE and coadaptation which most justly excites
0003 o attribute to mere external conditions, the STRUCTURE, for instance, of the woodpecker, with its
0003 t is equally preposterous to account for the STRUCTURE of this parasite, with its relations to sev
0011 posite conditions produce similar changes of STRUCTURE. Nevertheless some slight amount of change
0012 inly unconsciously modify other parts of the STRUCTURE, owing to the mysterious laws of the correl
0012 lly surprising to note the endless points in STRUCTURE and constitution in which the varieties and
0012 r and diversity of inheritable deviations of STRUCTURE, both those of slight and those of consider
0013 me family. If strange and rare deviations of STRUCTURE are truly inherited, less strange and commo
0015 blending together, any slight deviations of STRUCTURE, in such case, I grant that we could deduce
0019 which differ considerably from each other in STRUCTURE, I do not doubt that they all have descende
0022 of skin between the toes, are all points of STRUCTURE which are variable. The period at which the
0024 voice, colouring, and in most parts of their STRUCTURE, with the wild rock pigeon, yet are certain
0024 inly highly abnormal in other parts of their STRUCTURE, we may look in vain throughout the whole g
0027 in habits and in a great number of points of STRUCTURE with all the domestic breeds. Secondly, alt
0027 lmost perfect series between the extremes of STRUCTURE. Thirdly, those characters which are mainly
0038 , having slightly different constitutions or STRUCTURE, would often succeed better in the one coun
0038 our domestic races show adaptation in their STRUCTURE or in their habits to man's wants or fancie
0038 only with much difficulty, any deviation of STRUCTURE excepting such as is externally visible; an
0039 t it be thought that some great deviation of STRUCTURE would be necessary to catch the fancier's e
0040 an individual with some slight deviation of STRUCTURE, or takes more care than usual in matching
0041 the slightest deviation in the qualities or STRUCTURE of each individual. Unless such attention b
0044 sume is meant some considerable deviation of STRUCTURE in one part, either injurious to or not use
0045 s of variability, even in important parts of STRUCTURE, which he could collect on good authority,
0046 e polymorphic genera variations in points of STRUCTURE which are of no service or disservice to th
0052 fter be more fully explained) differences of STRUCTURE in certain definite directions. Hence I bel
0061 of a quadruped or feathers of a bird: in the STRUCTURE of the beetle which dives through the water
0073 insects, and consequently, from its peculiar STRUCTURE, never can set a seed. Many of our orchidac
0076 ty in habits and constitution, and always in STRUCTURE, the struggle will generally be more severe
0077 from the foregoing remarks, namely, that the STRUCTURE of every organic being is related, in the m
0077 or on which it preys. This is obvious in the STRUCTURE of the teeth and talons of the tiger; and i
0077 unoccupied ground. In the water beetle, the STRUCTURE of its legs, so well adapted for diving, al
0082 orces, extremely slight modifications in the STRUCTURE or habits of one inhabitant would often giv
0083 m. under nature, the slightest difference of STRUCTURE or constitution may well turn the nicely ba
0086 affect, through the laws of correlation, the STRUCTURE of the adult; and probably in the case of t
0086 and which never feed, a large part of their STRUCTURE is merely the correlated result of successi
0086 rrelated result of successive changes in the STRUCTURE of their larvae. So, conversely, modificati
0086 in the adult will probably often affect the STRUCTURE of the larva; but in all cases natural sele
0086 e species. Natural selection will modify the STRUCTURE of the young in relation to the parent, and
0087 e young. In social animals it will adapt the STRUCTURE of each individual for the benefit of the c
0087 atural selection cannot do, is to modify the STRUCTURE of one species, without giving it any advan
0087 nd one case which will bear investigation. A STRUCTURE used only once in an animal's whole life, i
0087 e shell being known to vary like every other STRUCTURE. Sexual Selection. Inasmuch as peculiaritie
```

Page **(Key Word)**

```
0089  e same general habits of life, but differ in  STRUCTURE, colour, or ornament, such differences have
0091  , if any slight innate change of habit or of   STRUCTURE benefited an individual wolf, it would have
0091  ng would probably inherit the same habits or   STRUCTURE, and by the repetition of this process, a n
0094  a tendency to a similar slight deviation of    STRUCTURE. The tubes of the corollas of the common re
0095  mutual and slightly favourable deviations of   STRUCTURE. I am well aware that this doctrine of natu
0096  f any great and sudden modification in their   STRUCTURE. On the Intercrossing of Incividuals. I mus
0097  ere is a very curious adaptation between the   STRUCTURE of the flower and the manner in which bees
0105  e adapted to, through modifications in their   STRUCTURE and constitution..Lastly, isolation. by che
0110  d genera, which, from having nearly the same   STRUCTURE, constitution, and habits, generally come i
0112  e descendants from any one species become in   STRUCTURE, constitution, and habits, by so much will
0113  nivorous. The more diversified in habits and   STRUCTURE the descendants of our carnivorous animal b
0114  can be supported by great diversification of   STRUCTURE, is seen under many natural circumstances.
0114  other, the advantages of diversification of   STRUCTURE, with the accompanying differences of habit
0115  t least safely infer that diversification of   STRUCTURE, amounting to new generic differences, woul
0116  ete with a set more perfectly diversified in   STRUCTURE. It may be doubted, for instance, whether t
0116  he better as they become more diversified in   STRUCTURE, and are thus enabled to encroach on places
0119  t as a general rule, the more diversified in   STRUCTURE the descendants from any one species can be
0121  d to each other in habits, constitution, and   STRUCTURE. Hence all the intermediate forms between t
0127  existence, causing an infinite diversity in   STRUCTURE, constitution, and habits, to be advantageo
0128  ed on the same area the more they diverge in   STRUCTURE, habits, and constitution, of which we see
0131  al differences, or very slight deviations of   STRUCTURE, as to make the child like its parents. But
0131  ture, leads me to believe that deviations of   STRUCTURE are in some way due to the nature of the co
0132  ere must be some cause for each deviation of   STRUCTURE, however slight. How much direct effect dif
0132  many striking and complex co adaptations of   STRUCTURE between one organic being and another, whic
0135  t easily put down to disuse modifications of   STRUCTURE which are wholly, or mainly, due to natural
0143  modification of the constitution, and of the   STRUCTURE of various organs; but that the effects of
0143  will, it may safely be concluded, affect the   STRUCTURE of the adult; in the same manner as any mal
0146  ifferences both in the internal and external   STRUCTURE of the seeds; which are not always correlat
0146  ferences. Hence we see that modifications in   STRUCTURE, viewed by systematists as of high value, m
0146  h natural selection some one modification in   STRUCTURE, and, after thousands of cenerations, some
0148  ation. If under changed conditions of life a   STRUCTURE before useful becomes less useful, any dimi
0148  s nutriment wasted in building up an useless   STRUCTURE. I can thus only understand a fact with whi
0148  ennae. Now the saving of a large and complex   STRUCTURE, when rendered superfluous by the parasitic
0148  less nutriment being wasted in developing a   STRUCTURE now become useless. Thus, as I believe, nat
0149  part or organ is repeated many times in the   STRUCTURE of the same individual (as the vertebrae in
0149  e parts are also very liable to variation in   STRUCTURE. Inasmuch as this vegetat*ve repetition, to
0149  having no power to check deviations in their   STRUCTURE. Thus rudimentary parts are left to the fre
0150  ies. Thus, the bat's wing is a most abnormal   STRUCTURE in the class mammalia; but the rule would n
0153  ys be expected to be much variability in the   STRUCTURE undergoing modification. It further deserve
0154  comes to be no more variable than any other   STRUCTURE. It is only in those cases in which the mod
0155  ndently created, why should that part of the   STRUCTURE, which differs from the same part in other
0155  n continuing to vary in those parts of their   STRUCTURE which have varied within a moderately recen
0157  ual characters, than in other parts of their   STRUCTURE. It is a remarkable fact, that the secondar
0157  species. Consequently, whatever part of the   STRUCTURE of the common progenitor, or of its early d
0159  dered as a variation representing the normal   STRUCTURE of another race, the fantail. I presume tha
0168  m the adjoining parts; and every part of the   STRUCTURE which can be saved without detriment to the
0168  to the individual, will be saved. Changes of   STRUCTURE at an early age will generally affect parts
0168  multiple parts are variable in number and in   STRUCTURE, perhaps arising from such parts not having
0170  e to all the more important modifications of   STRUCTURE, by which the innumerable beings on the fac
0171  ble that an animal having, for instance, the   STRUCTURE and habits of a bat, could have been formed
0172  on the other hand, organs of such wonderful   STRUCTURE, as the eye, of which we hardly as yet full
0173  distinct from each other in every detail of   STRUCTURE as are specimens taken from the metropolis
0178  w species presenting slight modifications of   STRUCTURE in some degree permanent; and this assuredl
0179  s of organic beings with peculiar habits and   STRUCTURE. It has been asked by the opponents of such
0180  from tree to tree. We cannot doubt that each   STRUCTURE is of use to each kind of squirrel in its o
0180  t it does not follow from this fact that the   STRUCTURE of each squirrel is the best that it is pos
0181  ss they also became modified and improved in   STRUCTURE in a corresponding manner. Therefore, I can
0181  ensor muscle. Although no graduated links of   STRUCTURE, fitted for gliding through the air, now co
0181  ly gliding squirrels; and that each grade of   STRUCTURE had been useful to its possessor. Nor can I
0182  lly for no purpose like the Apteryx. Yet the   STRUCTURE of each of these birds is good for it, unde
0182  these remarks that any of the grades of wing   STRUCTURE here alluded to, which perhaps may all have
0182  eing devoured by other fish? When we see any   STRUCTURE highly perfected for any particular habit,
0182  displaying early transitional grades of the   STRUCTURE will seldom continue to exist to the presen
0183  covering species with transitional grades of   STRUCTURE in a fossil condition will always be less,
0183  fit the animal, by some modification of its   STRUCTURE, for its changed habits, or exclusively for
0183  s, whether habits generally change first and   STRUCTURE afterwards; or whether slight modifications
0183  terwards; or whether slight modifications of   STRUCTURE lead to changed habits; both probably often
0184  al selection, more and more aquatic in their   STRUCTURE and habits, with larger and larger mouths,
0184  ies, having anomalous habits, and with their   STRUCTURE either slightly or considerably modified fr
0185  he has met with an animal having habits and   STRUCTURE not at all in agreement. What can be plaine
0185  ve changed without a corresponding change of   STRUCTURE. The webbed feet of the upland goose may be
0185  ecome rudimentary in function, though not in   STRUCTURE. In the frigate bird, the deeply scooped me
0185  scooped membrane between the toes shows that   STRUCTURE has begun to change. He who believes in sep
0186  ing vary ever so little, either in habits or   STRUCTURE, and thus gain an advantage over some other
0187  find but a small amount of gradation in the   STRUCTURE of the eye, and from fossil species we can
0187  from this low stage, numerous gradations of   STRUCTURE, branching off in two fundamentally differe
0188  hesitate to go further, and to admit that a   STRUCTURE even as perfect as the eye of an eagle migh
0191  ologous, or ideally similar, in position and   STRUCTURE with the lungs of the higher vertebrate ani
0193  wen and others have remarked, their intimate   STRUCTURE closely resembles that of common muscle; an
0194  rganic beings, which owe but little of their   STRUCTURE in common to inheritance from the same ance
0194  why should not Nature have taken a leap from   STRUCTURE to structure? On the theory of natural sele
0194  t nature have taken a leap from structure to   STRUCTURE? On the theory of natural selection, we can
0194  of those with any unfavourable deviation of   STRUCTURE, I have sometimes felt much difficulty in u
0196  d any actually injurious deviations in their   STRUCTURE will always have been checked by natural se
0196  the females. Moreover when a modification of   STRUCTURE has primarily arisen from the above or othe
0197  pe from a broken egg, we may infer that this   STRUCTURE has arisen from the laws of growth, and has
```

Page **(Key Word)**

Page ***(Key Word)***

```
0199  he utilitarian doctrine that every detail of  STRUCTURE has been produced for the good of its posse
0199  ions probably have had some little effect on   STRUCTURE, quite independently of any good thus gaine
0200  lation of growth, etc. Hence every detail of   STRUCTURE in every living creature (making some littl
0200  ntly takes advantage of, and profits by, the   STRUCTURE of another. But natural selection can and d
0201  . if it could be proved that any part of the   STRUCTURE of any one species had been formed for the
0205  fare of a species, that modifications in its   STRUCTURE could not have been slowly accumulated by m
0206  type is meant that fundamental agreement in    STRUCTURE, which we see in organic beings of the same
0209  that instincts are as important as corporeal   STRUCTURE for the welfare of each species, under its
0209  ve originated. As modifications of corporeal   STRUCTURE arise from, and are increased by, use or ha
0209  es which produce slight deviations of bodily   STRUCTURE. No complex instinct can possibly be produc
0210  in nature. Again as in the case of corporeal   STRUCTURE, and conformably with my theory, the instin
0211  as each takes advantage of the weaker bodily   STRUCTURE of others. So again, in some few cases, cer
0218  bees have not only their instincts but their   STRUCTURE modified in accordance with their parasitic
0224  be a dull man who can examine the exquisite   STRUCTURE of a comb, so beautifully adapted to its en
0225  uber. The Melipona itself is intermediate in   STRUCTURE between the hive and humble bee, but more n
0226  mmetrically in a double layer, the resulting   STRUCTURE would probably have been as perfect as the
0227  es not very wonderful, this bee would make a   STRUCTURE as wonderfully perfect as that of the hive
0233  the accumulation of slight modifications of   STRUCTURE or instinct, each profitable to the individ
0236  uters often differ widely in instinct and in   STRUCTURE from both the males and fertile females, an
0236  n that of any other striking modification of   STRUCTURE; for it can be shown that some insects and
0236  om both the males and the fertile females in   STRUCTURE, as in the shape of the thorax and in being
0237  with some slight profitable modification of   STRUCTURE, this being inherited by its offspring, whi
0237  itted successively acquired modifications of   STRUCTURE or instinct to its progeny. It may well be
0237  te of nature, of all sorts of differences of   STRUCTURE which have become correlated to certain age
0237  tanding how such correlated modifications of   STRUCTURE could have been slowly accumulated by natur
0238  ith social insects: a slight modification of   STRUCTURE, or instinct, correlated with the sterile c
0239  , in the same nest, presenting gradations of   STRUCTURE; and this we do find, even often, consideri
0240  ct to find gradations in important points of   STRUCTURE between the different castes of neuters in
0241  nto each other, as does the widely different   STRUCTURE of their jaws. I speak confidently on this
0241  all size with jaws having a widely different   STRUCTURE; or lastly, and this is our climax of diffi
0241  fficulty, one set of workers of one size and   STRUCTURE, and simultaneously another set of workers
0241  other set of workers of a different size and   STRUCTURE; a graduated series having been first forme
0241  erated them; until none with an intermediate   STRUCTURE were produced. Thus, as I believe, the wond
0242  d have intercrossed, and their instincts and   STRUCTURE would have become blended. And nature has,
0242  s with plants, any amount of modification in   STRUCTURE can be effected by the accumulation of nume
0242  a community could possibly have affected the   STRUCTURE or instincts of the fertile members, which
0243  licable to instincts as well as to corporeal   STRUCTURE, and is plainly explicable on the foregoing
0246  though the organs themselves are perfect in   STRUCTURE, as far as the microscope reveals. In the f
0256  is meant, the resemblance between species in   STRUCTURE and in constitution, more especially in the
0256  and in constitution, more especially in the   STRUCTURE of parts which are of high physiological im
0259  st hybrids which are usually intermediate in   STRUCTURE between their parents, exceptional and abno
0281  e determined from a mere comparison of their   STRUCTURE with that of the rock pigeon, whether they
0281  e tapir and the horse; but in some points of   STRUCTURE may have differed considerably from both, e
0281  ore species, even if we closely compared the   STRUCTURE of the parent with that of its modified des
0297  y intermediate between them in all points of   STRUCTURE. So that we might obtain the parent species
0345  ligible. The succession of the same types of   STRUCTURE within the same areas during the later geol
0349  but they plainly display an American type of   STRUCTURE. We ascend the lofty peaks of the Cordiller
0349  , however much they may differ in geological   STRUCTURE, the inhabitants, though they may be all pe
0373  el. In central Chile I was astonished at the   STRUCTURE of a vast mound of detritus, about 800 feet
0379  robable that selected modifications in their   STRUCTURE, habits, and constitutions will have profit
0414  cient times thought) that those parts of the   STRUCTURE which determined the habits of life, and th
0415  westwood has remarked, are most constant in   STRUCTURE; in another division they differ much, and
0417  ure to birds and reptiles, as an approach in   STRUCTURE in any one internal and and important organ
0417  in a number of the most important points of   STRUCTURE from the proper type of the order, yet M. R
0418  means obvious, on the ordinary view, why the   STRUCTURE of the embryo should be more important for
0426  n this respect in regard to single points of   STRUCTURE, but when several characters, let them be e
0428  ve inherited their general shape of body and   STRUCTURE of limbs from a common ancestor. So it is w
0436  n will account for the infinite diversity in   STRUCTURE and function of the mouths of insects. Neve
0436  mens, and pistils, as well as their intimate   STRUCTURE, are intelligible on the view that they con
0437  d are eminently liable to vary in number and   STRUCTURE; consequently it is quite probable that nat
0439  ary leaves of the Leguminosae. The points of   STRUCTURE, in which the embryos of wicely different a
0441  ther into hermaphrodites having the ordinary   STRUCTURE, or into what I have called complemental ma
0442  are so much accustomed to see differences in   STRUCTURE between the embryo and the adult, and likew
0442  e parts in proper proportion, as soon as any   STRUCTURE became visible in the embryo. And in some w
0442  ery general, but not universal difference in   STRUCTURE between the embryo and the adult; of parts
0442  t universally, resembling each other: of the   STRUCTURE of the embryo not being closely related to
0444  which each species has acquired its present   STRUCTURE, may have supervened at a not very early pe
0449  s that, in the eyes of most naturalists, the   STRUCTURE of the embryo is even more important for cl
0449  modified state; and in so far it reveals the   STRUCTURE of its progenitor. In two groups of animal,
0449  hey may at present differ from each other in   STRUCTURE and habits, if they pass through the same o
0449  losely related. Thus, community in embryonic   STRUCTURE reveals community of descent. It will revea
0449  this community of descent, however much the   STRUCTURE of the adult may have been modified and obs
0449  and group of species partially shows us the   STRUCTURE of their less modified ancient progenitors,
0455  , by which the materials forming any part or   STRUCTURE, if not useful to the possessor, will be sa
0456  me species, however different they may be in   STRUCTURE. If we extend the use of this element of de
0457  l become widely different from each other in   STRUCTURE and function; and the resemblance in differ
0459  preservation of each profitable deviation of   STRUCTURE or instinct. The truth of these proposition
0469  rigidly scrutinising the whole constitution,   STRUCTURE, and habits of each creature, favouring the
0470  s they become more diversified in habits and   STRUCTURE, so as to be enabled to seize on many and w
0471  rel should have been created with habits and   STRUCTURE fitting it for the life of an auk or grebe!
0474  and yet not be more variable than any other   STRUCTURE, if the part be common to many subordinate
0474  er no greater difficulty than does corporeal   STRUCTURE on the thoery of the natural selection of s
0484  ion, their germinal vesicles, their cellular   STRUCTURE, and their laws of growth and reproduction,
0485  a history: when we contemplate every complex   STRUCTURE and instinct as the summing up of many cont
0486  ms of life. Embryology will reveal to us the   STRUCTURE, in some degree obscured, or the prototypes
0488  of life, when very few forms of the simplest   STRUCTURE existed, the rate of change may have been s
```

Page ***(Key Word)***

```
0084  the good of each being, yet characters and STRUCTURES, which we are apt to consider as very trifl
0131  multiple, rudimentary, and lowly organised STRUCTURES variable. Parts developed in an unusual man
0134  not the parent forms; but many animals have STRUCTURES which can be explained by the effects of di
0143  han the union of homologous parts in normal STRUCTURES, as the union of the petals of the corolla
0144  laws of correlation in modifying important  STRUCTURES, independently of utility and, therefore, o
0146  falsely attribute to correlation of growth, STRUCTURES which are common to whole groups of species
0151  in every sense of the word, very important  STRUCTURES, and they differ extremely little even in d
0180  or two instances of transitional habits and STRUCTURES in closely allied species of the same genus
0183  y conclude that transitional grades between STRUCTURES fitted for very different habits of life wi
0183  in the case of species with fully developed STRUCTURES. I will now give two or three instances of
0188  ty (not more than in the case of many other STRUCTURES) in believing that natural selection has co
0196  st important influence in modifying various  STRUCTURES; and finally, that sexual selection will of
0199  its possessor. They believe that very many  STRUCTURES have been created for beauty in the eyes of
0199  l to my theory. Yet I fully admit that many  STRUCTURES are of no direct use to their possessors. c
0199  s well fitted for its place in nature, many  STRUCTURES now have no direct relation to the habits o
0200  hese animals. We may safely attribute these  STRUCTURES to inheritance. But to the progenitor of th
0200  atural selection can and does often produce  STRUCTURES for the direct injury of other species, as
0210  iations. Hence, as in the case of corporeal  STRUCTURES, we ought to find in nature, not the actual
0266  isation has been disturbed by two different  STRUCTURES and constitutions having been blended into
0425  cies has been recently exposed. Rudimentary  STRUCTURES on this view are as good as, or even someti
0446  different whether the desired qualities and  STRUCTURES have been acquired earlier or later in life
0460  even to conjecture by what gradations many  STRUCTURES have been perfected, more especially amongs
0480  al, by rudimentary organs and by homologous  STRUCTURES, her scheme of modification, which it seems
0483  scendants. Throughout whole classes various  STRUCTURES are formed on the same pattern, and at an e
0486  bly with respect to the nature of long lost  STRUCTURES. Species and groups of species, which are c
0004  urable to variation. In the next chapter the STRUGGLE for Existence amongst all organic beings thr
0005  onsequently, there is a frequently recurring STRUGGLE for existence, it follows that any being, if
0038  e overlooked that they almost always have to STRUGGLE for their own food, at least during certain
0053  xing, and allusions cannot be avoided to the STRUGGLE for existence, divergence of character, and
0054  in any degree permanent, necessarily have to STRUGGLE with the other inhabitants of the country, t
0060  groups subordinate to groups. Chapter III.   STRUGGLE For Existence. Bears on natural selection. T
0060  f all animals and plants throughout nature.  STRUGGLE for life most severe between individuals and
0060  e a few preliminary remarks, to show how the STRUGGLE for existence bears on Natural Selection. It
0061  the next chapter, follow inevitably from the STRUGGLE for life. Owing to this struggle for life, a
0061  ly from the struggle for life. Owing to this STRUGGLE for life, any variation, however slight and
0062  will now discuss in a little more detail the STRUGGLE for existence. In my future work this subjec
0062  to admit in words the truth of the universal STRUGGLE for life, or more difficult; at least I have
0062  g year. I should premise that I use the term STRUGGLE for Existence in a large and metaphorical se
0062  ls in a time of dearth, may be truly said to STRUGGLE with each other which shall get food and liv
0062  t a plant on the edge of a desert is said to STRUGGLE for life against the drought, though more pr
0063  comes to maturity, may be more truly said to STRUGGLE with the plants of the same and other kinds
0063  t can only in a far fetched sense be said to STRUGGLE with these trees, for if too many of these p
0063  n the same branch, may more truly be said to STRUGGLE with each other. As the missletoe is dissemi
0063  birds; and it may metaphorically be said to  STRUGGLE with other fruit bearing plants, in order to
0063  use for convenience sake the general term of STRUGGLE for existence. A struggle for existence inev
0063  he general term of struggle for existence. A STRUGGLE for existence inevitably follows from the hi
0063  sibly survive, there must in every case be a STRUGGLE for existence, either one individual with an
0066  to increase in numbers; that each lives by a STRUGGLE at some period of its life; that heavy destr
0068  t first sight to be quite independent of the STRUGGLE for existence; but in so far as climate chie
0068  recucing food, it brings on the most severe  STRUGGLE between the individuals, whether of the same
0069  now capped summits, or absolute deserts, the STRUGGLE for life is almost exclusively with the elem
0070  we have a limiting check independent of the  STRUGGLE for life. But even some of these so called e
0070  onably favoured; and here comes in a sort of STRUGGLE between the parasite and its prey. On the ot
0071  ations between organic beings, which have to STRUGGLE together in the same country. I will give on
0074  as in the surrounding virgin forests. What a STRUGGLE between the several kinds of trees must here
0075  ase with those which may strictly be said to STRUGGLE with each other for existence, as in the cas
0075  ocusts and grass feeding quadrupeds. But the STRUGGLE almost invariably will be most severe betwee
0075  e case of varieties of the same species, the STRUGGLE will generally be almost equally severe, and
0076  a dozen generations, if they were allowed to STRUGGLE together, like beings in a state of nature,
0076  d constitution, and always in structure, the STRUGGLE will generally be more severe between specie
0079  ring each generation or at intervals, has to STRUGGLE for life, and to suffer great destruction. W
0079  r great destruction. When we reflect on this STRUGGLE, we may console ourselves with the full beli
0080  grouping of all organic beings. How will the STRUGGLE for existence, discussed too briefly in the
0083  he does not allow the most vigorous males to STRUGGLE for the females. He does not rigidly destroy
0084  y well turn the nicely balanced scale in the STRUGGLE for life, and so be preserved. How fleeting
0085  ate of nature, where the trees would have to STRUGGLE with other trees and with a host of enemies,
0088  all Sexual Selection. This depends, not on a STRUGGLE for existence, but on a struggle between the
0088  s, not on a struggle for existence, but on a STRUGGLE between the males for possession of the fema
0104  try are left open for the old inhabitants to STRUGGLE for, and become adapted to, through modifica
0110  most. And we have seen in the chapter on the STRUGGLE for Existence that it is the most closely al
0121  e selected form having some advantage in the STRUGGLE for life over other forms, there will be a c
0125  aving some advantage over other forms in the STRUGGLE for existence, it will chiefly act on those
0125  ncestor some advantage in common. Hence, the STRUGGLE for the production of new and modified desce
0127  cies, at some age, season, or year, a severe STRUGGLE for life, and this certainly cannot be dispu
0127  ve the best chance of being preserved in the STRUGGLE for life; and from the strong principle of i
0128  of any one species, and during the incessant STRUGGLE of all species to increase in numbers, the m
0148  essive individual of the species; for in the STRUGGLE for life to which every animal is exposed, e
0152  rd. There may be truly said to be a constant STRUGGLE going on between, on the one hand, the tende
0153  this, I am convinced, is the case. That the  STRUGGLE between natural selection on the one hand, a
0158  to different habits of life, or the males to STRUGGLE with other males for the possession of the f
0170  ncs on the face of this earth are enabled to STRUGGLE with each other, and the best adapted to sur
0179  terrestrial habits; and as each exists by a  STRUGGLE for life, it is clear that each is well adap
0182  ich it is exposed; for each has to live by a STRUGGLE; but it is not necessarily the best possible
0186  n struggle language. He who believes in the  STRUGGLE for existence and in the principle of natura
0198  savages in different countries often have to STRUGGLE for their own subsistence, and would be expo
0201  nts of the same country with which it has to STRUGGLE for existence. And we see that this is the d
```

Page **(Key Word)**

```
0205 preservation of profitable variations in the STRUGGLE for life. Natural selection will produce not
0235 ave had the best chance of succeeding in the STRUGGLE for existence. No doubt many instincts of ve
0337 s formed by having had some advantage in the STRUGGLE for life over other and preceding forms. If
0344 of species which are their inferiors in the  STRUGGLE for existence. Hence, after long intervals o
0351 as it profits the individual in its complex  STRUGGLE for life, so the degree of modification in d
0388 succeeding. Although there will always be a  STRUGGLE for life between the individuals of the spec
0405 of being victorious in distant lands in the  STRUGGLE for life with foreign associates. But on the
0433 at natural selection, which results from the STRUGGLE for existence, and which almost inevitably i
0459 gree, variable, and, lastly, that there is a STRUGGLE for existence leading to the preservation of
0467 s and races, during the constantly recurrent STRUGGLE for Existence, we see the most powerful and
0467 rful and ever acting means of selection. The STRUGGLE for existence inevitably follows from the hi
0468 the closest competition with each other, the STRUGGLE will generally be most severe between them;
0468 tween the species of the same genus. But the STRUGGLE will often be very severe between beings mos
0468 eparated sexes there will in most cases be a  STRUGGLE between males for possession of the females.
0476 lder and less improved organic beings in the STRUGGLE for life. Lastly, the law of the long endura
0480 aturity and has to play its full part in the STRUGGLE for existence, and will thus have little pow
0490 a ratio of increase so high as to lead to a  STRUGGLE for Life, and as a consequence to Natural Se
0115 e nature of the plants or animals which have STRUGGLED successfully with the indigenes of any coun
0137 the cave rat natural selection seems to have STRUGGLED with the loss of light and to have increase
0391 n stocked by birds, which for long ages have STRUGGLED together in their former homes, and have be
0468 duals, or those which have most successfully STRUGGLED with their conditions of life, will general
0127 aracters useful to the males alone, in their STRUGGLES with other males. Whether natural selection
0350 n their action and reaction, in their mutual STRUGGLES for life; the relation of organism to organ
0077 ur the growth of the young seedling, whilst  STRUGGLING with other plants growing vigorously all ar
0082 as all the inhabitants of each country are   STRUGGLING together with nicely balanced forces, extre
0136 they had not been able to swim at all and had STUCK to the wreck. The eyes of moles and of some b
0308 thrice as extensive as the land, we see them STUDDED with many islands; but not one oceanic islan
0393 many islands with which the great oceans are STUDDED. I have taken pains to verify this assertion
0031 rade: the sheep are placed on a table and are STUDIED, like a picture by a connoisseur; this is do
0050 ok at the common oak, how closely it has been STUDIED; yet a German author makes more than a dozen
0346 ditions. Of late, almost every author who has STUDIED the subject has come to this conclusion. The
0394 and. I hear from Mr. Tomes, who has specially STUDIED this family, that many of the same species h
0004 press my conviction of the high value of such STUDIES, although they have been very commonly negle
0032 eeder. If gifted with these qualities, and he STUDIES his subject for years, and devotes his lifet
0283 waters washed their base. He who most closely STUDIES the action of the sea on our shores, will, I
0004 ations it seemed to me probable that a careful STUDY of domesticated animals and of cultivated pla
0006 entertain no doubt, after the most deliberate STUDY and dispassionate judgment of which I am capa
0012 versified. It is well worth while carefully to STUDY the several treatises published on some of ou
0020 ic pigeon. Believing that it is always best to STUDY some special group, I have, after deliberatio
0029 ation, I think, is simple: from long continued STUDY they are strongly impressed with the differen
0050 rieties. When a young naturalist commences the STUDY of a group of organisms quite unknown to him,
0051 other naturalists. When, moreover, he comes to STUDY allied forms brought from countries not now c
0246 of their hybrid offspring. It is impossible to STUDY the several memoirs and works of those two co
0282 nce close this volume. Not that it suffices to STUDY the principles of geology, or to read special
0317 ary, we have every reason to believe, from the STUDY of the tertiary formations, that species and
0386 only 6 3/4 ounces; I kept it covered up in my STUDY for six months, pulling up and counting each
0459 ay be extended. Effects of its adoption on the STUDY of Natural history. Concluding remarks. As th
0486 interesting, I speak from experience, will the STUDY of natural history become! A grand and almost
0486 tion of external conditions, and so forth. The STUDY of domestic productions will rise immensely i
0486 far more important and interesting subject for STUDY than one more species added to the infinitude
0001 duced to do this, as Mr. Wallace, who is now STUDYING the natural history of the Malay archipelago
0051 ference in the forms which he is continually STUDYING; and he has little general knowledge of anal
0454 ary organs in our domestic productions, as the STUMP of a tail in tailless breeds, the vestige of
0453 ated, imperfect nails sometimes appear on the STUMPS: I could as soon believe that these vestiges
0069 scending a mountain, we far oftener meet with STUNTED forms, due to the directly injurious action
0394 s are sufficiently ancient, as shown by the STUPENDOUS degradation which they have suffered and by
0452 e pistil consists of a stigma supported on the STYLE; but some Compositae, the male florets, which
0452 , for it is not crowned with a stigma; but the STYLE remains well developed and is clothed with ha
0137 different classes, which inhabit the caves of STYRIA and of Kentucky, are blind. In some of the cr
0007 look to the individuals of the same variety or SUB variety of our older cultivated plants and ani
0002 ture and constitution in which the varieties and SUB varieties differ slightly from each other. The
0021 long, massive beak and large feet; some of the SUB breeds of runts have very long necks, others v
0023 in each of these breeds several truly inherited SUB breeds, or species as he might have called the
0023 g under this term several geographical races or SUB species, which differ from each other in the m
0023 es. But besides C. livia, with its geographical SUB species, only two or three other species of ro
0025 a slaty blue, and has a white rump (the Indian SUB species, C. intermedia of Strickland, having i
0027 ed from the Columba livia with its geographical SUB species. In favour of this view, I may add, fi
0027 m the rock pigeon, yet by comparing the several SUB breeds of these breeds, more especially those
0032 actice, except sometimes amongst closely allied SUB breeds. And when a cross has been made, the cl
0034 istinct object in view, to make a new strain or SUB breed, superior to anything existing in the co
0038 as will hereafter be more fully explained, two SUB breeds might be formed. This, perhaps, partly
0040 ication, the spreading and knowledge of any new SUB breed will be a slow process. As soon as the p
0040 cess. As soon as the points of value of the new SUB breed are once fully acknowledged, the princip
0043 no doubt, largely aided in the formation of new SUB breeds; but the importance of the crossing of
0051 ation has as yet been drawn between species and SUB species, that is, the forms which in the opini
0051 rive at the rank of species; or, again, between SUB species and well marked varieties, or between
0052 t varieties; and at these latter, as leading to SUB species, and to species. The passage from one
0057 each other; they may generally be divided into SUB genera, or sections, or lesser groups. As Frie
0060 ultitude of doubtful forms be called species or SUB species or varieties; what rank, for instance,
0112 come greater, and would be noted as forming two SUB breeds; finally, after the lapse of centuries,
0112 eds; finally, after the lapse of centuries, the SUB breeds would become converted into two well es
0120 ey may have arrived at the doubtful category of SUB genus; but we have only to suppose the steps
0123 om the other five species, and may constitute a SUB genus or even a distinct genus. The six descen
0123 nus. The six descendants from (I) will form two SUB genera or even genera. But as the original spe
0123 ed as very distinct genera, or even as distinct SUB families. Thus it is, as I believe, that two o
0124 nes, beneath the capital letters, converging in SUB branches downwards towards a single point; thi
```

Page **(Key Word)**

Page ***(Key Word)***

```
0124  , the supposed single parent of our several new  SUB  genera and genera. It is worth while to reflec
0126  arge group, the later and more highly perfected  SUB  groups, from branching out and seizing on many
0126  plant and destroy the earlier and less improved  SUB  groups. Small and broken groups and sub groups
0126  mproved sub groups. Small and broken groups and  SUB  groups will finally tend to disappear. Looking
0128  nequally related together, forming sections and  SUB  genera, species of distinct genera much less c
0128  nd genera related in different degrees, forming  SUB  families, families, orders, sub classes, and c
0128  egrees, forming sub families, families, orders,  SUB  classes, and classes. The several subordinate
0141  reas the living species are now all tropical or  SUB  tropical in their habits, ought not to be look
0142  have succeeded in selecting so many breeds and  SUB  breeds with constitutions specially fitted for
0152  ly attended to by English fanciers. Even in the  SUB  breeds, as in the short faced tumbler, it is n
0159  ns, in countries most widely set apart, present  SUB  varieties with reversed feathers on the head a
0166  descended from a pigeon (including two or three  SUB  species or geographical races) of a bluish col
0185  the water ouzel would never have suspected its  SUB  aquatic habits; yet this anomalous member of t
0251  r observers in the case of Hippeastrum with its  SUB  genera, and in the case of some other genera,
0267  that is between members of different strains or  SUB  breeds, gives vigour and fertility to the offs
0288  nce, the several species of the Chthamalinae (a  SUB  family of sessile cirripedes) coat the rocks a
0290  xplanation, no doubt, is, that the littoral and  SUB  littoral deposits are continually worn away, a
0297  re readily if the specimens come from different  SUB  stages of the same formation. Some experienced
0305  rofessor Pictet has carried their existence one  SUB  stage further back; and some palaeontologists
0313  ny strange and lost mammals and reptiles in the  SUB  Himalayan deposits. The Silurian Lingula diffe
0329  ; and has placed certain pachyderms in the same  SUB  order with ruminants: for example, he dissolve
0331  family; b14 and f14 a closely allied family or  SUB  family; and o14, e14, m14, a third family. The
0332  e of character, has become divided into several  SUB  families and families, some of which are suppo
0344  o leave many modified descendants, and thus new  SUB  groups and groups are formed. As these are for
0353  rate to the other side. Some few families, many  SUB  families, very many genera, and a still greate
0369  only to strictly arctic forms, but also to many  SUB  arctic and to some few northern temperate form
0369  t for the necessary degree of uniformity of the  SUB  arctic and northern temperate forms round the
0369  of the Glacial period. At the present day, the  SUB  arctic and northern temperate productions of t
0370  ibute the necessary amount of uniformity in the  SUB  arctic and northern temperate productions of t
0407  arriers and from the analogical distribution of  SUB  genera, genera, and families. With respect to
0408  ces. We can thus understand the localisation of  SUB  genera, genera and families; and how it is tha
0411  notorious how commonly members of even the same  SUB  group have different habits. In our second and
0412  this same principle, much in common, and form a  SUB  family, distinct from that including the next
0413  the genera are included in, or subordinate to,  SUB  families, families, and orders, all united int
0419  the various groups of species, such as orders,  SUB  orders, families, sub families, and genera, th
0419  species, such as orders, sub orders, families,  SUB  families, and genera, they seem to be, at leas
0419  only a genus, and then raised to the rank of a  SUB  family or family; and this has been done, not
0422  ranking them under different so called genera,  SUB  families, families, sections, orders, and clas
0423  species. These are grouped under species, with  SUB  varieties under varieties; and with our domest
0423  mber of points. In tumbler pigeons, though some  SUB  varieties differ from the others in the import
0428  llelism has sometimes been observed between the  SUB  groups in distinct classes. A naturalist, stru
0431  roken up by extinction into distinct groups and  SUB  groups, will have transmitted some of its char
0432  and every intermediate link in each branch and  SUB  branch of their descendants, may be supposed t
0469  si or between more plainly marked varieties and  SUB  species, and species. Let it be observed how n
0471  sh should have been created to dive and feed on  SUB  aquatic insects; and that a petrel should have
0483  be classified on the same principle, in groups  SUB  ordinate to groups. Fossil remains sometimes t
0001  rs' work I allowed myself to speculate on the  SUBJECT, and drew up some short notes; these I enlar
0002  ies. Last year he sent to me a memoir on this  SUBJECT, with a request that I would forward it to S
0004  ll, unfortunately, be compelled to treat this  SUBJECT far too briefly, as it can be treated proper
0005  e its new and modified form. This fundamental  SUBJECT of Natural Selection will be treated at some
0005  elaborately constructed organ; secondly, the  SUBJECT of Instinct, or the mental powers of animals
0009  etails which I have collected on this curious  SUBJECT; but to show how singular the laws are which
0011  ore Geoffroy St. Hilaire's great work on this  SUBJECT. Breeders believe that long limbs are almost
0012  volumes, is the fullest and the best on this  SUBJECT. No breeder doubts how strong is the tendenc
0013  perhaps the correct way of viewing the whole  SUBJECT, would be, to look at the inheritance of eve
0014  ue to the male element. Having alluded to the  SUBJECT of reversion, I may here refer to a statemen
0016  erfect fertility of varieties when crossed, a  SUBJECT hereafter to be discussed), domestic races o
0018  dog, may not have existed in Egypt? The whole  SUBJECT must, I think, remain vague; nevertheless, I
0031  several of the many treatises devoted to this  SUBJECT, and to inspect the animals. Breeders habitu
0032  fted with these qualities, and he studies his  SUBJECT for years, and devotes his lifetime to it wi
0033  and many treatises have been published on the  SUBJECT; and the result, I may add, has been, in a c
0036  he mind of any one at all acquainted with the  SUBJECT that the owner of either of them has deviate
0044  must briefly discuss whether these latter are  SUBJECT to any variation. To treat this subject at a
0044  r are subject to any variation. To treat this  SUBJECT at all properly, a long catalogue of dry fac
0050  t and kind of variation to which the group is  SUBJECT; and this shows, at least, how very generall
0052  but we shall hereafter have to return to this  SUBJECT. From these remarks it will be seen that I l
0053  ed for valuable advice and assistance on this  SUBJECT, soon convinced me that there were many diff
0053  ements are fairly well established. The whole  SUBJECT, however, treated as it necessarily here is
0060  tant of all relations. Before entering on the  SUBJECT of this chapter, I must make a few prelimina
0062  truggle for existence. In my future work this  SUBJECT shall be treated, as it well deserves, at mu
0062  in regard to plants, no one has treated this  SUBJECT with more spirit and ability than W. Herbert
0064  rst pair. But we have better evidence on this  SUBJECT than mere theoretical calculations; namely,
0067  ably better known than any other animal. This  SUBJECT has been ably treated by several authors; an
0071  in some of these cases; but on this intricate  SUBJECT I will not here enlarge. Many cases are on r
0088  character. All those who have attended to the  SUBJECT, believe that there is the severest rivalry
0089  s; but I have not space here to enter on this  SUBJECT. Thus it is, as I believe, that when the mal
0091  d cross and blend where they met: but to this  SUBJECT of intercrossing we shall soon have to retur
0096  see its importance; but I must here treat the  SUBJECT with extreme brevity, though I have the mate
0099  ys prepotent over foreign pollen; but to this  SUBJECT we shall return in a future chapter. In the
0100  exes of trees simply to call attention to the  SUBJECT. Turning for a very brief space to animals:
0102  ral Selection. This is an extremely intricate  SUBJECT. A large amount of inheritable and diversifi
0107  ction, as far as the extreme intricacy of the  SUBJECT permits. I conclude, looking to the future,
0109  nature's power of selection. Extinction. This  SUBJECT will be more fully discussed in our chapter
0115  in the organs of the same individual body, a  SUBJECT so well elucidated by Milne Edwards. No phys
0116  id us in understanding this rather perplexing  SUBJECT. Let A to L represent the species of a genus
0124  apter on Geology, have to refer again to this  SUBJECT, and I think we shall then see that the diag
```

Page ***(Key Word)***

Page **(Key Word)**

```
0126 mote futurity. I shall have to return to this SUBJECT in the chapter on Classification, but I may
0143 rts become modified. This is a very important SUBJECT, most imperfectly understood. The most obvio
0145 owers, those nearest to the axis are oftenest SUBJECT to peloria, and become regular. I may add, a
0149 riable. We shall have to recur to the general SUBJECT of rudimentary and aborted organs; and I wil
0155 rue; I shall, however, have to return to this SUBJECT in our chapter on Classification. It would b
0155 different species of the same group, the more SUBJECT it is to individual anomalies. On the ordina
0156 ained constant. In connexion with the present SUBJECT, I will make only two other remarks. I think
0157 elation has a clear meaning on my view of the SUBJECT: I look at all the species of the same genus
0164 Lonel Hamilton Smith, who has written on this SUBJECT, believes that the several breeds of the hor
0195 their strength reduced, so that they are more SUBJECT to disease, or not so well enabled in a comi
0207 ncts. Neuter or sterile insects. Summary. The SUBJECT of instinct might have been worked into the
0207 that it would be more convenient to treat the SUBJECT separately, especially as so wonderful an in
0220 huber and Mr. Smith, I tried to approach the  SUBJECT in a sceptical frame of mind, as any one may
0224 will not here enter on minute details on this SUBJECT, but will merely give an outline of the conc
0225 mple instincts. I was led to investigate this SUBJECT by Mr. Waterhouse, who has shown that the fo
0233 l; but I have not space here to enter on this SUBJECT. Nor does there seem to me any great difficu
0236 terile, they cannot propagate their kind. The SUBJECT well deserves to be discussed at great lengt
0245 other acquired differences. In treating this  SUBJECT, two classes of facts, to a large extent fun
0246 rtner, who almost devoted their lives to this SUBJECT, without being deeply impressed with the hig
0272 important respects. I shall here discuss this SUBJECT with extreme brevity. The most important dis
0274 are apparently applicable to animals; but the SUBJECT is here excessively complicated, partly owin
0285 own in Prof. Ramsay's masterly memoir on this SUBJECT. Yet it is an admirable lesson to stand on t
0286 ay, in order to form some crude notion on the SUBJECT, assume that the sea would eat into cliffs 5
0291 subsidence. Since publishing my views on this SUBJECT in 1845, I have watched the progress of Geol
0298 more widely separated formations; but to this SUBJECT I shall have to return in the following chap
0310 her reflexion entertains grave doubts on this SUBJECT. I feel how rash it is to differ from these
0318 eriod, has been wonderfully sudden. The whole SUBJECT of the extinction of species has been involv
0327 there is one other remark connected with this SUBJECT worth making. I have given my reasons for be
0331 e theory of cescent with modification. As the SUBJECT is somewhat complex, I must request the read
0334 intervals between the successive formations.  SUBJECT to these allowances, the fauna of each geolo
0336 d than ancient. I will not here enter on this SUBJECT, for naturalists have not as yet defined to
0346 late, almost every author who has studied the SUBJECT has come to this conclusion. The case of Ame
0354 e incomparably the safest. In discussing this SUBJECT, we shall be enabled at the same time to con
0356 yell and other authors have ably treated this SUBJECT. I can give here only the briefest abstract
0356 presently have to discuss this branch of the  SUBJECT in some detail. Changes of level in the Land
0363 ? but I shall presently have to recur to this SUBJECT. As icebergs are known to be sometimes loade
0372 lar. But we must return to our more immediate SUBJECT, the Glacial period. I am convinced that For
0385 main to be observed, throw some light on this SUBJECT. When a duck suddenly emerges from a pond co
0395 i have not as yet had time to follow up this  SUBJECT in all other quarters of the world; but as f
0402 cated to me a remarkable fact bearing on this SUBJECT; namely, that Madeira and the adjoining isle
0407 and curious means of occasional transport, a  SUBJECT which has hardly ever been properly experime
0424 ess, without any reasoning or thinking on the SUBJECT, these tumblers are kept in the same group,
0434 pecies of the class are homologous. The whole SUBJECT is included under the general name of Morpho
0436 her and equally curious branch of the present SUBJECT; namely, the comparison not of the same part
0481 that organic beings in a state of nature are  SUBJECT to no variation; it cannot be proved that th
0482 thus can the load of prejudice by which this  SUBJECT is overwhelmed be removed. Several eminent n
0483 cies, on their own side they ignore the whole SUBJECT of the first appearance of species in what t
0486 will be a far more important and interesting  SUBJECT for study than one more species added to the
0107 ous in individuals and kinds, will have been  SUBJECTED to very severe competition. When converted
0198 certain plants; so that colour would be thus  SUBJECTED to the action of natural selection. But we
0200 ave been acquired through natural selection,  SUBJECTED formerly, as now, to the several laws of in
0206 l conditions of life, and being in all cases  SUBJECTED to the several laws of growth. Hence, in fa
0277 e systematic affinity of the forms which are  SUBJECTED to experiment; for systematic affinity atte
0309 ow exist, large tracts of land have existed,  SUBJECTED no doubt to great oscillations of level, si
0370 in nearly the same relative position, though  SUBJECTED to large, but partial oscillations of level
0415 depends on such organs having generally been  SUBJECTED to less change in the adaptation of the spe
0220 h indebted for information on this and other  SUBJECTS. Although fully trusting to the statements o
0300 al animals; or of those which lived on naked  SUBMARINE rocks, would be embedded; and those embedde
0308 these intervals existed as dry land, or as a  SUBMARINE surface near land, on which sediment was no
0395 the islands are situated on moderately deep   SUBMARINE banks, and they are inhabitied by closely a
0356 row isthmus now separates two marine faunas;  SUBMERGE it, or let it formerly have been submerged,
0287 e escaped the action of the sea: when deeply  SUBMERGED for perhaps equally long periods, it would,
0296 f a formation having been upraised, denuded,  SUBMERGED, and then re covered by the upper beds of t
0356 s: submerge it, or let it formerly have been  SUBMERGED, and the two faunas will now blend or may f
0395 a. the West Indian Islands stand on a deeply  SUBMERGED bank, nearly 1000 fathoms in depth, and her
0059 out the universe become divided into groups   SUBORDINATE to groups. Chapter III. Struggle For Exist
0128 ce should be related to each other in group   SUBORDINATE to group, in the manner which we everywher
0128 ders, sub classes, and classes. The several   SUBORDINATE groups in any class cannot be ranked in a
0129 of all extinct and living species in groups   SUBORDINATE to groups. Of the many twigs which flouris
0183 arly period in great numbers and under many   SUBORDINATE forms. Thus, to return to our imaginary il
0183 flight would have been developed under many   SUBORDINATE forms, for taking prey of many kinds in ma
0209 ieve that the effects of habit are of quite   SUBORDINATE importance to the effects of the natural s
0338 hereafter confirmed, at least in regard to    SUBORDINATE groups, which have branched off from each
0350 n through natural selection, and in a quite   SUBORDINATE degree to the direct influence of differen
0411 rudimentary Organs. Classification, groups    SUBORDINATE to groups. Natural system. Rules and diffi
0412 one progenitor become broken up into groups   SUBORDINATE to groups. In the diagram each letter on t
0413 genera: and the genera are included in, or    SUBORDINATE to, sub families, families, and orders, al
0416 ffer much, and the differences are of quite   SUBORDINATE value in classification: yet no one probab
0418 ans, are found to offer characters of quite   SUBORDINATE value. We can see why characters derived f
0421 descendants from I; so will it be with each   SUBORDINATE branch of descendants, at each successive
0422 stock, would have to be expressed by groups   SUBORDINATE to groups: but the proper or even only pos
0423 eons. The origin of the existence of groups   SUBORDINATE to groups, is the same with varieties as w
0471 ngement of all the forms of life, in groups   SUBORDINATE to groups, all within a few great classes,
0474 er structure, if the part be common to many   SUBORDINATE forms, that is, if it has been inherited f
0478 titute one grand natural system, with group   SUBORDINATE to group, and with extinct groups often fa
0418 n to some lesser number, they use it as of    SUBORDINATED value. This principle has been broadly con
```

Page **(Key Word)**

Page **(Key Word)***

```
0413  , the grand fact in natural history of the  SUBORDINATION  of group under group, which, from its fam
0420  nt of the groups within each class, in due   SUBORDINATION  and relation to the other groups, must be
0433  ities of all organic beings, namely, their  SUBORDINATION  in group under group. We use the element
0456  chapter I have attempted to show, that the  SUBORDINATION  of group to group in all organisms throug
0251  , and he continued to try it during several  SUBSEQUENT    years, and always with the same result. Thi
0254  hybrids, or the hybrids must have become in  SUBSEQUENT    generations quite fertile under domesticati
0271  whose accuracy has been confirmed by every  SUBSEQUENT    observer, has proved the remarkable fact, t
0290  f the waves, when first upraised and during  SUBSEQUENT    oscillations of level. Such thick and exten
0291  sufficiently thick and extensive to resist  SUBSEQUENT    degradation, may have been formed over wide
0352  the vera causa of ordinary generation with  SUBSEQUENT    migration, and calls in the agency of a mir
0381  l period for their migration, and for their  SUBSEQUENT    modification to the necessary degree. The f
0383  n colonisation from the nearest source with  SUBSEQUENT    modification. Summary of the last and prese
0406  rest and readiest source, together with the  SUBSEQUENT    modification and better adaptation of the c
0408  more dominant forms of life), together with  SUBSEQUENT    modification and the multiplication of new
0422  , whilst others (owing to the spreading and  SUBSEQUENT    isolation and states of civilisation of the
0477  different. On this view of migration, with   SUBSEQUENT    modification, we can see why oceanic island
0053  that there were many difficulties, as did   SUBSEQUENTLY  Dr. Hooker, even in stronger terms. I shal
0103  local variety when once thus formed might    SUBSEQUENTLY  slowly spread to other districts. On the a
0140  be the case; nor do we know that they have   SUBSEQUENTLY  become acclimatised to their new homes. As
0140  der confinement, and not because they were   SUBSEQUENTLY  found capable of far extended transportati
0156  ched off from their common progenitor, and   SUBSEQUENTLY  have not varied or come to differ in any d
0165  a, the hybrid, and even the pure offspring   SUBSEQUENTLY  produced from the mare by a black Arabian
0168  t an early age will generally affect parts   SUBSEQUENTLY  developed; and there are very many other c
0196  been formed in an aquatic animal, it might   SUBSEQUENTLY  come to be worked in for all sorts of purp
0196  en of no advantage to the species, but may   SUBSEQUENTLY  have been taken advantage of by the descen
0197  from unknown laws of growth, and have been   SUBSEQUENTLY  taken advantage of by the plant undergoing
0202  e poison originally adapted to cause galls   SUBSEQUENTLY  intensified, we can perhaps understand how
0205  o way advantageous to a species, have been   SUBSEQUENTLY  taken advantage of by the still further mo
0250  with their own pollen, and the fourth was   SUBSEQUENTLY  fertilised by the pollen of a compound hyb
0333  nitor of the order; nearly so much as they   SUBSEQUENTLY  diverged. Thus it comes that ancient and e
0335  lly last much longer than a form elsewhere   SUBSEQUENTLY  produced, especially in the case of terres
0339  he dead and the living. Professor Owen has   SUBSEQUENTLY  extended the same generalisation to the ma
0353  een produced in one area alone, and having   SUBSEQUENTLY  migrated from that area as far as its powe
0360  t by an inland gale, they would germinate.   SUBSEQUENTLY  to my experiments, M. Martens tried simila
0368  d have marched a little further north, and   SUBSEQUENTLY  have retreated to their present homes; but
0384  ravel far along the shores of the sea, and   SUBSEQUENTLY  become modified and adapted to the fresh w
0388  can range, over immense areas, and having   SUBSEQUENTLY  become extinct in intermediate regions. Bu
0401  any one or more of the islands, or when it   SUBSEQUENTLY  spread from one island to another, it woul
0403  ve been derived, the colonists having been   SUBSEQUENTLY  modified and better fitted to their new ho
0419  different grades of difference, have been   SUBSEQUENTLY  discovered. All the foregoing rules and ai
0453  e formation of rudimentary teeth which are   SUBSEQUENTLY  absorbed, can be of any service to the rap
0291  er a shallow bottom, if it continue slowly to  SUBSIDE. In this latter case, as long as the rate of
0309  r it might well happen that strata which had  SUBSIDED    some miles nearer to the centre of the earth
0107  very severe competition. When converted by   SUBSIDENCE  into large separate islands, there will sti
0291  in this latter case, as long as the rate of  SUBSIDENCE  and supply of sediment nearly balance each
0291  ch in fossils, have thus been formed during  SUBSIDENCE. Since publishing my views on this subject
0291  e conclusion that it was accumulated during  SUBSIDENCE. I may add, that the only ancient tertiary
0291  n formed over wide spaces during periods of  SUBSIDENCE, but only where the supply of sediment was
0292  eological record. On the other hand, during  SUBSIDENCE, the inhabited area and number of inhabitan
0292  to an archipelago), and consequently during  SUBSIDENCE, though there will be much extinction, fewe
0292  med; and it is during these very periods of  SUBSIDENCE, that our great deposits rich in fossils ha
0295  can only be accumulated during a period of  SUBSIDENCE; and to keep the depth approximately the sa
0295  t nearly have counterbalanced the amount of  SUBSIDENCE. But this same movement of subsidence will
0295  nt of subsidence. But this same movement of  SUBSIDENCE will often tend to sink the area whence the
0295  en the supply of sediment and the amount of  SUBSIDENCE is probably a rare contingency; for it has
0300  ons lie in the past, only during periods of  SUBSIDENCE. These periods of subsidence would be separ
0300  ing periods of subsidence. These periods of  SUBSIDENCE would be separated from each other by enorm
0300  res of South America. During the periods of  SUBSIDENCE there would probably be much extinction of
0300  her the duration of any one great period of  SUBSIDENCE over the whole or part of the archipelago,
0301  s, also, probable that each great period of  SUBSIDENCE would be interrupted by oscillations of lev
0309  the great oceans are still mainly areas of  SUBSIDENCE, the great archipelagoes still areas of osc
0327  formations were deposited during periods of  SUBSIDENCE; and that blank intervals of vast duration
0342  n during a single formation; that, owing to  SUBSIDENCE being necessary for the accumulation of fos
0342  been more extinction during the periods of  SUBSIDENCE, and more variation during the periods of e
0465  ity in the forms of life; during periods of  SUBSIDENCE, more extinction. With respect to the absen
0173  the shallow bed of the sea, whilst it slowly  SUBSIDES. These contingencies will concur only rarely
0315  diment having been deposited on areas whilst  SUBSIDING, our formations have been almost necessaril
0465  only where much sediment is deposited on the  SUBSIDING  bed of the sea. During the alternate period
0068  her of the same or of distinct species, which  SUBSIST  on the same kind of food. Even when climate,
0179  ld the animal in its transitional state have  SUBSISTED? It would be easy to show that within the s
0136  st habitually use their wings to gain their   SUBSISTENCE, have, as Mr. Wollaston suspects, their wi
0198  ntries often have to struggle for their own   SUBSISTENCE, and would be exposed to a certain extent
0353  area as far as its powers of migration and   SUBSISTENCE  under past and present conditions permitte
0185  he strictly terrestrial thrush family wholly  SUBSISTS  by diving, grasping the stones with its feet
0188  ends there seems to be an imperfect vitreous  SUBSTANCE? With these facts, here far too briefly and
0116  lesh alone, draws most nutriment from these   SUBSTANCES. So in the general economy of any land, the
0183  exotic plants, or exclusively on artificial   SUBSTANCES. Of diversified habits innumerable instance
0366  remarkable clearness by Edward Forbes, is    SUBSTANTIALLY  as follows. But we shall follow the chang
0355  those in another, does not differ much (by   SUBSTITUTING  the word variety for species) from that la
0137  , the tuco tuco, or Ctenomys, is even more   SUBTERRANEAN  in its habits than the mole; and I was ass
0137  ertainly not indispensable to animals with   SUBTERRANEAN  habits, a reduction in their size with the
0233  bears on a fact, which seems at first quite  SUBVERSIVE  of the foregoing theory; namely, that the c
0015  seems to me not improbable, that if we could  SUCCEED  in naturalising, or were to cultivate, durin
0015  al stock. Whether or not the experiment would  SUCCEED, is not of great importance for our line of
0032  to it with indomitable perseverance, he will  SUCCEED, and may make great improvements; if he want
0037  om the seed of the wild pear, though he might  SUCCEED  from a poor seedling growing wild, if it had
```

Page **(Key Word)***

Page **(Key Word)**

```
0038  erent constitutions or structure, would often  SUCCEED  better in the one country than in the other,
0051  call varieties and which species; but he will   SUCCEED  in this at the expense of admitting much var
0078  instance should we know what to do, so as to    SUCCEED.  It will convince us of our ignorance in the
0085  wny, a yellow or purple fleshed fruit, should   SUCCEED.  In looking at many small points of differen
0113  powers of increase be allowed to act, it can    SUCCEED  in increasing (the country not undergoing an
0113  ss, including its modified descendants, would   SUCCEED  in living on the same piece of ground. And w
0116  modified descendants of any one species will    SUCCEED  by so much the better as they become more di
0148  , as I believe, natural selection will always   SUCCEED  in the long run in reducing and saving every
0148  ly, that natural selection may perfectly well   SUCCEED  in largely developing any organ, without req
0198  s with slightly different constitutions would   SUCCEED  best under different climates; and there is
0204  intermediate variety, and will thus generally   SUCCEED  in supplanting and exterminating it. We have
0229  as appeared to me that the bees do not always   SUCCEED  in working at exactly the same rate from the
0326  he long run the dominant forms will generally   SUCCEED  in spreading. The diffusion would, it is pro
0344  dified, descendants; and these will generally   SUCCEED  in taking the places of those groups of spec
0402  ich has its own mocking thrush: why should it   SUCCEED  in establishing itself there? We may safely
0432  at we should be driven to, if we were ever to   SUCCEED  in collecting all the forms in any class whi
0432  all time and space. We shall certainly never   SUCCEED  in making so perfect a collection: neverthel
0471  t in character. But as all groups cannot thus   SUCCEED  in increasing in size, for the world would n
0024  be assumed not only that half civilized man    SUCCEEDED  in thoroughly domesticating several species
0115  ave been expected that the plants which have   SUCCEEDED  in becoming naturalised in any land would g
0142  ri for it is not likely that man should have   SUCCEEDED  in selecting so many breeds and sub breeds
0157  scope for action, and may thus readily have    SUCCEEDED  in giving to the species of the same group
0169  his case, natural selection may readily have   SUCCEEDED  in giving a fixed character to the organ, i
0229  d the basins on both sides, in order to have   SUCCEEDED  in thus leaving flat plates between the bas
0235  least honey in the secretion of wax, having   SUCCEEDED  best, and having transmitted by inheritance
0247  ieve to be varieties, and only once or twice   SUCCEEDED  in getting fertile seed; as he found the co
0287  ions, which the mind cannot grasp, must have   SUCCEEDED  each other in the long roll of years! Now t
0334  r between that which preceded and that which   SUCCEEDED  it. Thus, the species which lived at the si
0005  n and of correlation of growth. In the four   SUCCEEDING  chapters, the most apparent and gravest dif
0017  t tamed an animal, whether it would vary in   SUCCEEDING  generations, and whether it would endure ot
0026  naturally become less and less, as in each   SUCCEEDING  generation there will be less of the foreig
0114  grass would always have the best chance of   SUCCEEDING  and of increasing in numbers, and thus of s
0128  become, the better will be their chance of   SUCCEEDING  in the battle of life. Thus the small diffe
0209  ion, and then transmitted by inheritance to   SUCCEEDING  generations. It can be clearly shown that t
0235  their turn will have had the best chance of   SUCCEEDING  in the struggle for existence. No doubt man
0273  trast with their extreme variability in the   SUCCEEDING  generations, is a curious fact and deserves
0303  me one or some few parent forms; and in the   SUCCEEDING  formation such species will appear as if su
0307  only small remnants of the formations next   SUCCEEDING  them in age, and these ought to be very gen
0334  ate in character, between the preceding and   SUCCEEDING  faunas. I need give only one instance, name
0334  iate in character between the preceding and   SUCCEEDING  faunas, that certain genera offer exception
0340  d to leave in that quarter, during the next   SUCCEEDING  period of time, closely allied though in so
0341  n one geological formation, and in the next   SUCCEEDING  formation there be six other allied or repr
0388  ngle seed or egg will have a good chance of   SUCCEEDING. Although there will always be a struggle f
0487  ervals by a comparison of the preceding and   SUCCEEDING  organic forms. We must be cautious in attem
0006  resent welfare, and, as I believe, the future   SUCCESS  and modification of every inhabitant of this
0041  this comes to be of the highest importance to   SUCCESS.  On this principle Marshall has remarked, wi
0042  preventing crosses is an important element of   SUCCESS  in the formation of new races, at least, in
0062  ant) not only the life of the individual, but   SUCCESS  in leaving progeny. Two canine animals in a
0073  in battle must ever be recurring with varying   SUCCESS;  and yet in the long run the forces are so n
0102  i believe, an extremely important element of   SUCCESS.  Though nature grants vast periods of time f
0234  ing honey must be a most important element of   SUCCESS  in any family of bees. Of course the success
0234  success in any family of bees. Of course the   SUCCESS  of any species of bee may be dependent on th
0299  te numerous intermediate gradations; and such   SUCCESS  seems to me improbable in the highest degree
0400  and generally a far more important element of   SUCCESS.  Now if we look to those inhabitants of the
0468  life, will generally leave most progeny. But   SUCCESS  will often depend on having special weapons
0041  of the same plants, are generally far more   SUCCESSFUL  than amateurs in getting new and valuable v
0119  the same advantages which made their parent   SUCCESSFUL  in life, they will generally go on multiply
0217  eir eggs in other birds' nests, and thus be   SUCCESSFUL  in rearing their young. By a continued proc
0319  ncti its place being seized on by some more   SUCCESSFUL  competitor. It is most difficult always to
0321  if many allied forms be developed from the   SUCCESSFUL  intruder, many will have to yield their pla
0429  l forms as failing groups conquered by more   SUCCESSFUL  competitors, with a few members preserved b
0115  the plants or animals which have struggled   SUCCESSFULLY  with the indigenes of any country, and hav
0116  orous, ruminant, and rodent mammals, could   SUCCESSFULLY  compete with these well pronounced orders.
0136  reater number of individuals were saved by   SUCCESSFULLY  battling with the winds, or by giving up t
0392  s plant, though it would have no chance of   SUCCESSFULLY  competing in stature with a fully develope
0468  rous individuals, or those which have most   SUCCESSFULLY  struggled with their conditions of life, w
0003  their geographical distribution, geological   SUCCESSION, and other such facts, might come to the co
0005  ext chapter I shall consider the geological   SUCCESSION  of organic beings throughout time: in the e
0119  ll be increased. In our diagram the line of   SUCCESSION  is broken at regular intervals by small num
0129  ing each former year may represent the long   SUCCESSION  of extinct species. At each period of growt
0233  asonably be asked, how a long and graduated   SUCCESSION  of modified architectural instincts, all te
0303  y made, namely that it might require a long   SUCCESSION  of ages to adapt an organism to some new an
0306  be about as rash in us to dogmatize on the   SUCCESSION  of organic beings throughout the world, as
0311  g more or less different in the interrupted   SUCCESSION  of chapters, may represent the apparently a
0312  ven disappear. Chapter X. On the Geological   SUCCESSION  of Organic Beings. On the slow and successi
0312  ate of development of ancient forms. On the   SUCCESSION  of the same types within the same areas. Su
0312  facts and rules relating to the geological   SUCCESSION  of organic beings, better accord with the c
0313  rked, does not strictly correspond with the   SUCCESSION  of our geological formations: so that betwe
0316  cies of the group have appeared in the long   SUCCESSION  of ages, so long must its members have cont
0316  st have continuously existed by an unbroken   SUCCESSION  of generations, from the lowest Silurian st
0325  nhabitants. This great fact of the parallel   SUCCESSION  of the forms of life throughout the world,
0326  y do find, a less strict degree of parallel   SUCCESSION  in the productions of the land than of the
0327  and, taken in a large sense, simultaneous,   SUCCESSION  of the same forms of life throughout the wo
0327  oups will disappear from the world; and the   SUCCESSION  of forms in both ways will everywhere tend
0328  the foregoing paragraphs, the same general   SUCCESSION  in the forms of life; but the species would
0329  same order, in accordance with the general   SUCCESSION  of the form of life, and the order would fa
```

Page **(Key Word)**

Page ***(Key Word)***

0338	of the same classes; or that the geological	SUCCESSION of extinct forms is in some degree parallel
0338	y of which the chance is very small. On the	SUCCESSION of the same types within the same areas; du
0339	isted, in 1839 and 1845, on this law of the	SUCCESSION of types, on this wonderful relationship in
0339	a. now what does this remarkable law of the	SUCCESSION of the same types within the same areas mea
0340	aw of the long enduring, but not immutable,	SUCCESSION of the same types within the same areas, is
0345	e class, the fact will be intelligible. The	SUCCESSION of the same types of structure within the s
0355	s, on my theory, must have descended from a	SUCCESSION of improved varieties, which will never hav
0409	hout time and space; the laws governing the	SUCCESSION of forms in past times being nearly the sam
0410	sculpture or colour. In looking to the long	SUCCESSION of ages, as in now looking to distant provi
0429	of small size. In the chapter on geological	SUCCESSION I attempted to show, on the principle of ea
0467	proved by calculation, by the effects of a	SUCCESSION of peculiar seasons, and by the results of
0476	s throughout space, and in their geological	SUCCESSION throughout time; for in both cases the bein
0487	clude few identical species, by the general	SUCCESSION of their forms of life. As species are prod
0489	poch, we may feel certain that the ordinary	SUCCESSION by generation has never once been broken, a
0004	wer of man in accumulating by his Selection	SUCCESSIVE slight variations. I will then pass on to t
0029	slight differences accumulated during many	SUCCESSIVE generations. May not those naturalists who,
0030	wer of accumulative selection: nature gives	SUCCESSIVE variations; man adds them up in certain dir
0032	y the accumulation in one direction, during	SUCCESSIVE generations, of differences absolutely inap
0086	tructure is merely the correlated result of	SUCCESSIVE changes in the structure of their larvae. S
0089	paradise, and some others, congregate; and	SUCCESSIVE males display their gorgeous plumage and pe
0089	ion; that is, individual males have had, in	SUCCESSIVE generations, some slight advantage over oth
0119	rvals by small numbered letters marking the	SUCCESSIVE forms which have become sufficiently distin
0120	modification may have been increased in the	SUCCESSIVE generations. This case would be represented
0120	rom having diverged in character during the	SUCCESSIVE generations, will have come to differ large
0124	generations, and likewise a section of the	SUCCESSIVE strata of the earth's crust including extin
0125	se the amount of change represented by each	SUCCESSIVE group of diverging dotted lines to be very
0135	and that as natural selection increased in	SUCCESSIVE generations the size and weight of its body
0136	obably with disuse. For during thousands of	SUCCESSIVE generations each individual beetle which fl
0138	dinary powers of vision, slowly migrated by	SUCCESSIVE generations from the outer world into the d
0148	steps, would be a decided advantage to each	SUCCESSIVE individual of the species: for in the strug
0161	ndred generations distant, but that in each	SUCCESSIVE generation there has been a tendency to rep
0166	at there is a tendency in the young of each	SUCCESSIVE generation to produce the long lost charact
0173	outh over a continent, we generally meet at	SUCCESSIVE intervals with closely allied or representa
0189	not possibly have been formed by numerous,	SUCCESSIVE, slight modifications, my theory would abso
0192	an could not possibly have been produced by	SUCCESSIVE transitional gradations, yet, undoubtedly,
0194	can act only by taking advantage of slight	SUCCESSIVE variations; she can never take a leap, but
0195	ave been adapted for its present purpose by	SUCCESSIVE slight modifications, each better and bette
0214	tinued selection of the best individuals in	SUCCESSIVE generations made tumblers what they now are
0214	ited effects of compulsory training in each	SUCCESSIVE generation would soon complete the work; an
0216	nd by man selecting and accumulating during	SUCCESSIVE generations, peculiar mental habits and act
0220	uring the months of June and July, on three	SUCCESSIVE years, I have watched for many hours severa
0235	lection having taken advantage of numerous,	SUCCESSIVE, slight modifications of simpler instincts;
0239	e analogy of ordinary variations, that each	SUCCESSIVE, slight, profitable modification did not pr
0245	n acquired by the continued preservation of	SUCCESSIVE profitable degrees of sterility. I hope, ho
0248	s. in regard to the sterility of hybrids in	SUCCESSIVE generations; though Gartner was enabled to
0249	ge fact of the increase of fertility in the	SUCCESSIVE generations of artificially fertilised hybr
0252	ed, gone on decreasing in fertility in each	SUCCESSIVE generation, as Gartner believes to be the c
0252	le. Again, with respect to the fertility in	SUCCESSIVE generations of the more fertile hybrid anim
0253	d sisters have usually been crossed in each	SUCCESSIVE generation, in opposition to the constantly
0265	sues. So it is with hybrids, for hybrids in	SUCCESSIVE generations are eminently liable to vary, a
0269	ication tends to eliminate sterility in the	SUCCESSIVE generations of hybrids, which were at first
0273	be given. The variability, however, in the	SUCCESSIVE generations of mongrels is, perhaps, greate
0274	er pure parent form, by repeated crosses in	SUCCESSIVE generations with either parent. These sever
0284	ss on the Continent. Moreover, between each	SUCCESSIVE formation, we have, in the opinion of most
0290	antily developed, that no record of several	SUCCESSIVE and peculiar marine faunas will probably be
0298	on between any two forms, is small, for the	SUCCESSIVE changes are supposed to have been local or
0308	tate of things in the intervals between the	SUCCESSIVE formations; whether Europe and the United S
0310	re discussed, namely our not finding in the	SUCCESSIVE formations infinitely numerous transitional
0312	ccession of Organic Beings. On the slow and	SUCCESSIVE appearance of new species. On their differe
0312	the observations of Philippi in Sicily, the	SUCCESSIVE changes in the marine inhabitants of that i
0317	cal line of varying thickness, crossing the	SUCCESSIVE geological formations in which the species
0317	ants of the earth having been swept away at	SUCCESSIVE periods by catastrophes, is very generally
0323	distant points of the world. In the several	SUCCESSIVE palaeozoic formations of Russia, Western Eu
0323	out of view, the general parallelism in the	SUCCESSIVE forms of life, in the stages of the widely
0328	raw a close general parallelism between the	SUCCESSIVE stages in the two countries; but when he co
0328	re is a striking general parallelism in the	SUCCESSIVE Silurian deposits of Bohemia and Scandinavi
0331	for us. The horizontal lines may represent	SUCCESSIVE geological formations, and all the forms be
0332	tinct forms, supposed to be embedded in the	SUCCESSIVE formations, were discovered at several poin
0334	ng the long and blank intervals between the	SUCCESSIVE formations. Subject to these allowances, th
0336	ong blank intervals have intervened between	SUCCESSIVE formations, we ought not to expect to find,
0338	altered, continually adds, in the course of	SUCCESSIVE generations, more and more difference to th
0342	intervals of time have elapsed between the	SUCCESSIVE formations; that there has probably been mo
0345	diate in character. The inhabitants of each	SUCCESSIVE period in the world's history have beaten t
0349	r fails to be struck by the manner in which	SUCCESSIVE groups of beings, specifically distinct, ye
0355	ave supplanted each other; so that, at each	SUCCESSIVE stage of modification and improvement, all
0410	the forms of life which have changed during	SUCCESSIVE ages within the same quarter of the world,
0421	, not only at the present time, but at each	SUCCESSIVE period of descent. All the modified descend
0421	subordinate branch of descendants, at each	SUCCESSIVE period. If, however, we choose to suppose t
0428	distinct classes have often been adapted by	SUCCESSIVE slight modifications to live under nearly a
0435	t on the theory of the natural selection of	SUCCESSIVE slight modifications, each modification bei
0437	and in flowering plants, we see a series of	SUCCESSIVE spiral whorls of leaves. An indefinite repe
0438	ification will have been effected by slight	SUCCESSIVE steps, we need not wonder at discovering in
0444	it is quite possible, that each of the many	SUCCESSIVE modifications, by which each species has ac
0444	other cases it is quite possible that each	SUCCESSIVE modification, or most of them, may have app
0446	diverse habits. Then, from the many slight	SUCCESSIVE steps of variation having supervened at a r
0447	on the above two principles, namely of each	SUCCESSIVE modification supervening at a rather late a

Page ***(Key Word)***

Page **(Key Word)***

```
0447  cts of use and disuse. In certain cases the  SUCCESSIVE  steps of variation might supervene, from ca
0448  erences might, also, become correlated with  SUCCESSIVE  stages of development; so that the larvae,
0449  os, has not been obliterated, either by the  SUCCESSIVE  variations in a long course of modification
0454  as been the main agency; that it has led in  SUCCESSIVE  generations to the gradual reduction of var
0457  idual animal and plant. On the principle of  SUCCESSIVE  slight variations, not necessarily or gener
0461  y are now found, they must in the course of  SUCCESSIVE  generations have passed from some one part
0463  tinct inhabitants of the world, and at each  SUCCESSIVE  period between the extinct and still older
0465  han the average duration of specific forms.  SUCCESSIVE  formations are separated from each other by
0467   character of a breed by selecting, in each  SUCCESSIVE  generation, individual differences so sligh
0471  lection acts solely by accumulating slight,  SUCCESSIVE, favourable variations, it can produce no g
0474  e on the thoery of the natural selection of  SUCCESSIVE, slight, but profitable modifications. We c
0475  rents, in being absorbed into each other by  SUCCESSIVE  crosses; and in other such points, as do th
0475  pecies have come on the stage slowly and at  SUCCESSIVE  intervals; and the amount of change, after
0479   the theory of descent with slow and slight  SUCCESSIVE  modifications. The similarity of pattern in
0479  ogenitor of each class. On the principle of  SUCCESSIVE  variations not always supervening at an ear
0480  h in the mature animal were reduced, during  SUCCESSIVE  generations, by disuse or by the tongue and
0480  ing by the preservation and accumulation of  SUCCESSIVE  slight favourable variations. Why, it may b
0487  tances, and the blank intervals between the  SUCCESSIVE  stages as having been of vast duration. But
0195  em sufficient to cause the preservation of  SUCCESSIVELY  varying individuals. I have sometimes felt
0217   makes her own nest and has eggs and young  SUCCESSIVELY  hatched, all at the same time. It has been
0237  ei so that it could never have transmitted  SUCCESSIVELY  acquired modifications of structure or ins
0335  t, we have no reason to believe that forms  SUCCESSIVELY  produced necessarily endure for correspond
0343  it is that new species come in slowly and  SUCCESSIVELY; how species of different classes do not n
0139  pical climate, or conversely. So again, many  SUCCULENT  plants cannot endure a damp climate. But th
0094  differ in length; yet the hive bee can easily  SUCK  the nectar out of the incarnate clover, but no
0097  ure of the flower and the manner in which bees  SUCK  the nectar; for, in doing this, they either pu
0094  or instance, their habit of cutting holes and  SUCKING  the nectar at the bases of certain flowers,
0436  ic sea lizards, and in the mouths of certain-SUCTORIAL  crustaceans, the general pattern seems to h
0096  on of new organic beings, or of any great and  SUDDEN  modification in their structure. On the Inter
0275  n slowly acquired by selection. Consequently,  SUDDEN  reversions to the perfect character of either
0279  diate varieteies in any one formation. On the  SUDDEN  appearance of groups of species. On their sud
0279  den appearance of groups of species. On their  SUDDEN  appearance in the lowest known fossiliferous
0302  ation, pressed so hardly on my theory. On the  SUDDEN  appearance of whole groups of Allied Species.
0305  sted on by palaeontologists of the apparently  SUDDEN  appearance of a whole group of species, is th
0306  e number and range of its productions. On the  SUDDEN  appearance of groups of Allied Species in the
0310   species which now exist or have existed; the  SUDDEN  manner in which whole groups of species appea
0318  of the secondary period, has been wonderfully  SUDDEN. The whole subject of the extinction of speci
0321  ts production. With respect to the apparently  SUDDEN  extermination of whole families or orders, as
0322  en much slow extermination. Moreover, when by  SUDDEN  immigration or by unusually rapid development
0343  rge the apparent, but often falsely apparent,  SUDDEN  coming in of whole groups of species. He may
0471  urable variations, it can produce no great or  SUDDEN  modification; it can act only by very short a
0009  gardeners mean a single bud or offset, which  SUDDENLY  assumes a new and sometimes very different c
0025  ed marks, the mongrel offspring are very apt  SUDDENLY  to acquire these characters; for instance, I
0030  ariations useful to him have probably arisen  SUDDENLY, or by one step; many botanists, for instanc
0030  dipsacus; and this amount of change may have  SUDDENLY  arisen in a seedling. So it has probably bee
0030  . we cannot suppose that all the breeds were  SUDDENLY  produced as perfect and as useful as we now
0065  ertility of these animals or plants has been  SUDDENLY  and temporarily increased in any sensible de
0098  foreign pollen. When the stamens of a flower  SUDDENLY  spring towards the pistil, or slowly move on
0160  obable hypothesis is, not that the offspring  SUDDENLY  takes after an ancestor some hundred generat
0175  ry between them, in which they become rather  SUDDENLY  rarer and rarer; then, as varieties do not e
0248  ually the case, and that the fertility often  SUDDENLY  decreases in the first few generations. Neve
0275  st monstrous in their nature, and which have  SUDDENLY  appeared, such as albinism, melanism, defici
0275  ls, which are descended from varieties often  SUDDENLY  produced and semi monstrous in character, th
0302  rupt manner in which whole groups of species  SUDDENLY  appear in certain formations, has been urged
0303  ing formation such species will appear as if  SUDDENLY  created. I may here recall a remark formerly
0303  supposing that whole groups of species have  SUDDENLY  been produced. I may recall the well known f
0304  that this great and distinct order had been  SUDDENLY  produced in the interval between the latest
0304  , i concluded that this great group had been  SUDDENLY  developed at the commencement of the tertiar
0305  hown that the species of this group appeared  SUDDENLY  and simultaneously throughout the world at t
0306  which numbers of species of the same group,  SUDDENLY  appear in the lowest known fossiliferous roc
0385  hrow some light on this subject. When a duck  SUDDENLY  emerges from a pond covered with duck weed,
0386  which frequent the muddy edges of ponds, if  SUDDENLY  flushed, would be the most likely to have mu
0432  form which has ever lived on this earth were  SUDDENLY  to reappear, though it would be quite imposs
0463  y they often falsely appear, to have come in  SUDDENLY  on the several geological stages? Why do we
0465  geological formation, they will appear as if  SUDDENLY  created there, and will be simply classed as
0483   certain elemental atoms have been commanded  SUDDENLY  to flash into living tissues? Do they believ
0487  is almost independent of altered and perhaps  SUDDENLY  altered physical conditions; namely, the mut
0063  lifetime produces several eggs or seeds, must  SUFFER  destruction during some period of its life, a
0067  eggs or very young animals seem generally to  SUFFER  most, but this is not invariably the case. Wi
0067  ade, I believe that it is the seedlings which  SUFFER  most from germinating in ground already thick
0069  food through the advancing winter, which will  SUFFER  most. When we travel from south to north, or
0079  t intervals, has to struggle for life, and to  SUFFER  great destruction. When we reflect on this st
0084  rease in countless numbers; they are known to  SUFFER  largely from birds of prey; and hawks are qui
0085  at in the United States smooth skinned fruits  SUFFER  far more from a beetle, a curculio, than thos
0085  ulio, than those with down; that purple plums  SUFFER  far more from a certain disease than yellow p
0110  modification and improvement, will naturally  SUFFER  most. And we have seen in the chapter on the
0228  he opposite sides. The bees, however, did not  SUFFER  this to happen, and they stopped their excava
0321  it will generally be allied forms, which will  SUFFER  from some inherited inferiority in common. Bu
0475  nd at many instincts causing other animals to  SUFFER. If species be only well marked and permanent
0126  least broken up, that is, which as yet have  SUFFERED  least extinction, will for a long period con
0192  es had become extinct, and they have already  SUFFERED  far more extinction than have sessile cirrip
0284  dation which the earth's crust has elsewhere  SUFFERED. And what an amount of degradation is implie
0284  udation which the strata have in many places  SUFFERED, independently of the rate of accumulation o
0288  formation, without the underlying bed having  SUFFERED  in the interval any wear and tear, seem expl
0291   to resist such degradation as it has as yet  SUFFERED, but which will hardly last to a distant geo
0308  at the older a formation is, the more it has  SUFFERED  the extremity of denudation and metamorphism
```

Page **(Key Word)***

Page ***************************************(Key Word)***************************************

```
0366 cal period, central Europe and North America SUFFERED under an Arctic climate. The ruins of a hous
0377 now concerned. The tropical plants probably SUFFERED much extinction: how much no one can say: pe
0377 is, that all tropical productions will have SUFFERED to a certain extent. On the other hand, the
0377 ced under somewhat new conditions, will have SUFFERED less. And it is certain that many temperate
0394 y the stupendous degradation which they have SUFFERED and by their tertiary strata: there has also
0429 have some evidence of aberrant forms having SUFFERED severely from extinction, for they are gener
0321 ave been modified and improved, a few of the SUFFERERS may often long be preserved, from being fit
0069 s, even where it most abounds, is constantly SUFFERING enormous destruction at some period of its
0283 ory, that the cliffs are at the present time SUFFERING. The appearance of the surface and the vege
0377 mind that the tropical productions were in a SUFFERING state and could not have presented a firm f
0286 of coast, ten or twenty miles in length, ever SUFFERS degradation at the same time along its whole
0002 ration, but which, I hope, in most cases will SUFFICE. No one can feel more sensible than I do of
0049 t what distance, it has been well asked, will SUFFICE? If that between America and Europe is ample
0066 or the species will become extinct. It would SUFFICE to keep up the full number of individuals,
0102 le, but I believe mere individual differences SUFFICE for the work. A large number of individuals,
0183 taneously. Of cases of changed habits it will SUFFICE merely to allude to that of the many British
0346 usion. The case of America alone would almost SUFFICE to prove its truth: for if we exclude the no
0364 estines of birds. These means, however, would SUFFICE for occasional transport across tracts of se
0374 is period simultaneously cooler. But it would SUFFICE for my purpose, if the temperature was at th
0415 the writings of almost every author. It will SUFFICE to quote the highest authority, Robert Brown
0215 he other hand, habit alone in some cases has SUFFICED: no animal is more difficult to tame than th
0216 nt. In some cases compulsory habit alone has SUFFICED to produce such inherited mental changes: in
0282 it may be objected, that time will not have SUFFICED for so great an amount of organic change, al
0056 judging by analogy whether or not the amount SUFFICES to raise one or both to the rank of species.
0232 jecting beyond the other completed cells. It SUFFICES that the bees should be enabled to stand at
0282 , may at once close this volume. Not that it SUFFICES to study the principles of geology, or to re
0032 a thousand has accuracy of eye and judgment SUFFICIENT to become an eminent breeder. If gifted wit
0041 t mere individual differences are not amply SUFFICIENT, with extreme care, to allow of the accumul
0049 or madeira, or the Canaries, or Ireland, be SUFFICIENT? It must be admitted that many forms, consi
0097 t like a camel hair pencil, and it is quite SUFFICIENT just to touch the anthers of one flower and
0108 exclaim that these several causes are amply SUFFICIENT wholly to stop the action of natural select
0117 there marked by a small numbered letter, a SUFFICIENT amount of variation is supposed to have bee
0135 s, there is not SUFFICIENT evidence to induce us to believe that mutil
0180 hing less than a long list of such cases is SUFFICIENT to lessen the difficulty in any particular
0195 arts, of which the importance does not seem SUFFICIENT to cause the preservation of successively v
0207 e occurred to many readers, as a difficulty SUFFICIENT to overthrow my whole theory. I must premis
0233 own that bees are often hard pressed to get SUFFICIENT nectar: and I am informed by Mr. Tegetmeier
0257 difference in any recognisable character is SUFFICIENT to prevent two species crossing. It can be
0272 fertility of varieties does not seem to me SUFFICIENT to overthrow the view which I have taken wi
0288 een geologically explored, and no part with SUFFICIENT care, as the important discoveries made eve
0291 , but only where the supply of sediment was SUFFICIENT to keep the sea shallow and to embed and pr
0295 a very long period, in order to have given SUFFICIENT time for the slow process of variation: hen
0300 he sea, or where it did not accumulate at a SUFFICIENT rate to protect organic bodies from decay,
0300 fossiliferous formations could be formed of SUFFICIENT thickness to last to an age, as distant in
0319 t these same unperceived agencies are amply SUFFICIENT to cause rarity, and finally extinction. We
0323 of distant parts of the world: we have not SUFFICIENT data to judge whether the productions of th
0390 separate species, will have to admit that a SUFFICIENT number of the best adapted plants and anima
0426 afely done, and is often done, as long as a SUFFICIENT number of characters, let them be ever so u
0453 estatement of the fact. Would it be thought SUFFICIENT to say that because planets revolve in elli
0464 ot be objected that there has not been time SUFFICIENT for any amount of organic change: for the l
0485 tions, are looked at by most naturalists as SUFFICIENT to raise both forms to the rank of species.
0018 ndered it in some degree probable that man SUFFICIENTLY civilized to have manufactured pottery exi
0036 on of the best individuals, whether or not SUFFICIENTLY distinct to be ranked at their first appea
0053 commonness), often give rise to varieties SUFFICIENTLY well marked to have been recorded in botan
0119 ing the successive forms which have become SUFFICIENTLY distinct to be recorded as varieties. But
0173 to a future age only in masses of sediment SUFFICIENTLY thick and extensive to withstand an enormo
0178 er much, may have separately been rendered SUFFICIENTLY distinct to rank as representative species
0258 giflora, and the hybrids thus produced are SUFFICIENTLY fertile: but Kolreuter tried more than two
0264 rioc. This latter alternative has not been SUFFICIENTLY attended to: but I believe, from observati
0276 ry of Chapter. First crosses between forms SUFFICIENTLY distinct to be ranked as species, and thei
0277 es between forms known to be varieties, or SUFFICIENTLY alike to be considered as varieties, and thei
0288 nearly the whole bed of the sea, at a rate SUFFICIENTLY quick to embed and preserve fossil remains
0290 d, than the absence of any recent deposits SUFFICIENTLY extensive to last for even a short geologi
0291 onsequently formations rich in fossils and SUFFICIENTLY thick and extensive to resist subsequent d
0394 tion of mammals: many volcanic islands are SUFFICIENTLY ancient, as shown by the stupendous degrad
0413 ich, from its familiarity, does not always SUFFICIENTLY strike us, is in my judgment fully explain
0469 e existence of varieties, which they think SUFFICIENTLY distinct to be worthy of record in systema
0484 hat this will be easy) whether any form be SUFFICIENTLY constant and distinct from other forms, to
0484 d if definable, whether the differences be SUFFICIENTLY important to deserve a specific name. This
0233 less than from twelve to fifteen pounds of dry SUGAR are consumed by a hive of bees for the secret
0011 in some country drooping ears: and the view SUGGESTED by some authors, that the drooping is due t
0096 their kind. This view, I may add, was first SUGGESTED by Andrew Knight. We shall presently see it
0363 ther of the arctic and antarctic regions, as SUGGESTED by Lyell: and during the Glacial period fro
0381 he close of the Glacial period, icebergs, as SUGGESTED by Lyell, have been largely concerned in th
0075 ed be resown, some of the varieties which best SUIT the soil or climate, or are naturally the most
0264 ts can live, they are generally placed under SUITABLE conditions of life. But a hybrid partakes of
0083 ed by her: and the being is placed under well SUITED conditions of life. Man keeps the natives of
0183 en watched a tyrant flycatcher (Saurophagus SULPHURATUS) in South America, hovering over one spot
0029 hey ignore all general arguments; and refuse to SUM up in their minds slight differences accumulat
0043 been felt in the display of distinct breeds. To SUM up on the origin of our Domestic Races of anim
0107 hus been exposed to less severe competition. To SUM up the circumstances favourable and unfavourab
0177 f the supplanted, intermediate hill variety. To SUM up, I believe that species come to be tolerabl
0299 ainst my views. Hence it will be worth while to SUM up the foregoing remarks, under an imaginary i
0465 ns distant from each other in time. Such is the SUM of the several chief objections and difficulti
0433 omparing one group with a distinct group, we SUMMARILY reject analogical or adaptive characters, a
0126 , orders, and classes, as at the present day. SUMMARY of Chapter. If during the long course of age
```

Page ***************************************(Key Word)***************************************

Page **(Key Word)**

```
0131 s manner. Reversions to long lost characters. SUMMARY. I have hitherto sometimes spoken as if the
0167 mock the shells now living on the sea shore. SUMMARY. Our ignorance of the laws of variation is p
0203 e wafted by a chance breeze on to the ovules? SUMMARY of Chapter. We have in this chapter discusse
0207 tion of instincts. Neuter or sterile insects. SUMMARY. The subject of instinct might have been wor
0242 , against the well known doctrine of Lamarck. SUMMARY. I have endeavoured briefly in this chapter
0245 ls compared independently of their fertility. SUMMARY. The view generally entertained by naturalis
0276 al distinction between species and varieties. SUMMARY of Chapter. First crosses between forms suff
0312 sion of the same types within the same areas. SUMMARY of preceding and present chapters. Let us no
0341 es will have left modified blood descendants. SUMMARY of the preceding and present Chapters. I hav
0383 nearest source with subsequent modification. SUMMARY of the last and present chapters. As lakes a
0406 aptation of the colonists to their new homes. SUMMARY of last and present Chapters. In these chapt
0411 . rudimentary Organs: their origin explained. SUMMARY. From the first dawn of life, all organic be
0456 be accounted for by the laws of inheritance. SUMMARY. In this chapter I have attempted to show, t
0179 its fur, short legs, and form of tail: during SUMMER this animal dives for and preys on fish, but
0223 ch fewer slaves, and in the early part of the SUMMER extremely few. The masters determine when and
0486 e every complex structure and instinct as the SUMMING up of many contrivances, each useful to the
0486 look at any great mechanical invention as the SUMMING up of the labour, the experience, the reason
0130 e has been favoured and is still alive on its SUMMIT, so we occasionally see an animal like the Or
0231 away cement, and adding fresh cement, on the SUMMIT of the ridge. We shall thus have a thin wall
0045 rs of the Baltic, or dwarfed plants on Alpine SUMMITS, or the thicker fur of an animal from far no
0069 n we reach the Arctic regions, or snow capped SUMMITS, or absolute deserts, the struggle for life
0354 ely, the existence of the same species on the SUMMITS of distant mountain ranges; and at distant p
0358 s would have been formed, like other mountain SUMMITS, of granite, metamorphic schists, old fossil
0365 ntity of many plants and animals, on mountain SUMMITS, separated from each other by hundreds of mi
0367 s, would be left isolated on distant mountain SUMMITS (having been exterminated on all lesser heig
0368 we find the same species on distant mountain SUMMITS, we may almost conclude without other eviden
0368 mth, first at the bases and ultimately on the SUMMITS of the mountains, the case will have been so
0375 clearly represented by plants growing on the SUMMITS of the mountains of Borneo. Some of these ha
0382 aters left their living drift on our mountain SUMMITS, in a line gently rising from the arctic low
0031 the magician's wand, by means of which he may SUMMON into life whatever form and mould he pleases.
0136 ie much concealed, until the wind lulls and the SUN shines; that the proportion of wingless beetle
0453 e planets revolve in elliptic courses round the SUN, satellites follow the same course round the p
0372 allied forms now living in areas completely SUNDERED. Thus, I think, we can understand the presen
0357 migration. In the coral producing oceans such SUNKEN islands are now marked, as I believe, by ring
0358 our the admission that they are the wrecks of SUNKEN continents; if they had originally existed as
0382 e aid, as halting places, of existing and now SUNKEN islands; and perhaps at the commencement of t
0070 ed on them: nor can the birds, though having a SUPER abundance of food at this one season, increas
0273 t such variability would often continue and be SUPER added to that arising from the mere act of cr
0062 nature bright with gladness, we often see SUPERABUNDANCE of food: we do not see, or we forget, tha
0062 bear in mind, that though food may be now SUPERABUNDANT, it is not at all seasons of each recu
0323 aracterised in such trifling points as mere SUPERFICIAL sculpture. Moreover other forms, which are
0148 large and complex structure, when rendered SUPERFLUOUS by the parasitic habits of the Proteolepas
0148 the organisation, as soon as it is rendered SUPERFLUOUS, without by any means causing some other p
0155 apter on Classification. It would be almost SUPERFLUOUS to adduce evidence in support of the above
0288 the Secondary and Palaeozoic periods, it is SUPERFLUOUS to state that our evidence from fossil rem
0305 the world at this same period. It is almost SUPERFLUOUS to remark that hardly any fossil fish are
0282 r years examine for himself great piles of SUPERIMPOSED strata, and watch the sea at work grinding
0289 gaps there are in that country between the SUPERIMPOSED formations; so it is in North America, and
0309 been pressed on by an enormous weight of SUPERINCUMBENT water, might have undergone far more meta
0265 ion of animals. Between the sterility thus SUPERINDUCED and that of hybrids, there are many points
0034 in view, to make a new stud or sub breed, SUPERIOR to anything existing in the country. But, fo
0061 tly ready for action, and is as immeasurably SUPERIOR to man's feeble efforts, as the works of Nat
0189 g optical instrument might thus be formed as SUPERIOR to one of glass, as the works of the Creator
0459 cts should have been perfected, not by means SUPERIOR to, though analogous with, human reason, but
0128 o transmit to their modified offspring that SUPERIORITY which now makes them dominant in their own
0356 di but they do not owe their difference and SUPERIORITY to descent from any single pair, but to co
0447 ases the successive steps of variation might SUPERVENE, from causes of which we are wholly ignoran
0351 phical and climatal changes which will have SUPERVENED since ancient times, almost any amount of m
0444 as acquired its present structure, may have SUPERVENED at a not very early period of life; and som
0444 g the word in the largest sense) which have SUPERVENED at an earlier age in the child than in the
0446 slight successive steps of variation having SUPERVENED at a rather late age, and having been inher
0449 ons in a long course of modification having SUPERVENED at a very early age, or by the variations h
0338 iffers from its embryo, owing to variations SUPERVENING at a not early age, and being inherited at
0411 ology, laws of, explained by variations not SUPERVENING at an early age, and being inherited at a
0447 les, namely of each successive modification SUPERVENING at a rather late age, and being inherited
0457 ht variations, not necessarily or generally SUPERVENING at a very early period of life, and being
0479 inciple of successive variations not always SUPERVENING at an early age, and being inherited at a
0052 species as the variety: or it might come to SUPPLANT and exterminate the parent species: or both
0075 , and will consequently in a few years quite SUPPLANT the other varieties. To keep up a mixed stoc
0076 great congener. One species of charlock will SUPPLANT another, and so in other cases. We can dimly
0091 w variety might be formed which would either SUPPLANT or coexist with the parent form of wolf. Or,
0121 e improved descendants of any one species to SUPPLANT and exterminate in each stage of descent the
0126 he polity of Nature, will constantly tend to SUPPLANT and destroy the earlier and less improved su
0177 in the race for life, will tend to beat and SUPPLANT the less common forms; for these will be mor
0281 new and improved forms of life will tend to SUPPLANT the old and unimproved forms. By the theory
0298 uch local varieties do not spread widely and SUPPLANT their parent forms until they have been modi
0301 d and improved, they would slowly spread and SUPPLANT their parent forms. When such varieties retu
0315 r species in the economy of nature, and thus SUPPLANT it; yet the two forms, the old and the new,
0321 m one species, that is a new genus, comes to SUPPLANT an old genus, belonging to the same family.
0402 will in a very brief time wholly or in part SUPPLANT it; but if both are equally well fitted for
0412 ing in number and diverging in character, to SUPPLANT and exterminate the less divergent, the less
0428 on increasing in size; and they consequently SUPPLANT many smaller and feebler groups. Thus we can
0470 . new and improved varieties will inevitably SUPPLANT and exterminate the older, less improved and
0173 nditions of life of its own region, and has SUPPLANTED and exterminated its original parent and al
0177 ach other, without the interposition of the SUPPLANTED, intermediate hill variety. To sum up, I be
0178 of the land, but these links will have been SUPPLANTED and exterminated during the process of natu
```

Page **(Key Word)**

Page **(Key Word)**

```
0179  n, they will almost certainly be beaten and  SUPPLANTED by the forms which they connect; for these
0182  to the present day, for they will have been   SUPPLANTED by the very process of perfection through n
0307  y would almost certainly have been long ago   SUPPLANTED and exterminated by their numerous and impr
0315  ieve that the parent form will generally be   SUPPLANTED and exterminated by its improved offspring,
0345  ordinary generation: old forms having been    SUPPLANTED by new and improved forms of life, produced
0355  her individuals or varieties, but will have   SUPPLANTED each other; so that, at each successive sta
0366  temperate inhabitants, the latter would be    SUPPLANTED and arctic productions would take their pla
0463  but only between each and some extinct and    SUPPLANTED form. Even on a wide area, which has during
0463  he intermediate zones, will be liable to be   SUPPLANTED by the allied forms on either hand; and the
0463  mediate varieties will, in the long run, be   SUPPLANTED and exterminated. On this doctrine of the e
0472  adapted for that country, being beaten and    SUPPLANTED by the naturalised productions from another
0475  of natural selection; for old forms will be   SUPPLANTED by new and improved forms. Neither single s
0114  g and of increasing in numbers, and thus of   SUPPLANTING the less distinct varieties; and varieties
0203  ural selection almost implies the continual   SUPPLANTING and extinction of preceding and intermedia
0204  variety, and will thus generally succeed in   SUPPLANTING and exterminating it. We have have seen in
0320  mproved variety has been raised, it at first  SUPPLANTS the less improved varieties in the same nei
0289  ce at the historical table published in the   SUPPLEMENT to Lyell's Manual, will bring home the trut
0304  tiary formation. But now we may read in the   SUPPLEMENT to Lyell's Manual, published in 1858, clear
0269  correlated with the reproductive system. He   SUPPLIES his several varieties with the same food; tr
0070  son, increase in number proportionally to the SUPPLY of seed, as their numbers are checked during
0095  s of the red clover offer in vain an abundant SUPPLY of precious nectar to the hive bee. Thus it m
0147  abundant and nutritious foliage and a copious SUPPLY of oil bearing seeds. When the seeds in our f
0184  er. Even in so extreme a case as this, if the SUPPLY of insects were constant, and if better adapt
0190  ter organ having a ductus pneumaticus for its SUPPLY, and being divided by highly vascular partiti
0290  ary remains can anywhere be found, though the SUPPLY of sediment must for ages have been great, fr
0291  r case, as long as the rate of subsidence and SUPPLY of sediment nearly balance each other, the se
0291  ing periods of subsidence, but only where the SUPPLY of sediment was sufficient to keep the sea sh
0295  e same species to live on the same space, the SUPPLY of sediment must nearly have counterbalanced
0295  he sediment is derived, and thus diminish the SUPPLY whilst the downward movement continues. In fa
0295  fact, this nearly exact balancing between the SUPPLY of sediment and the amount of subsidence is p
0295  as a change in the currents of the sea and a  SUPPLY of sediment of a different nature will genera
0362  w after a bird has found and devoured a large  SUPPLY of food, it is positively asserted that all t
0239  oped abdomen which secretes a sort of honey,   SUPPLYING the place of that excreted by the aphides,
0010  stages of formation; so that, in fact, sports  SUPPORT my view, that variability may be largely att
0063  e so inordinately great that no country could  SUPPORT the product. Hence, as more individuals are
0089  cannot here enter on the details necessary to  SUPPORT this view; but if man can in a short time gi
0155  d be almost superfluous to adduce evidence in  SUPPORT of the above statement, that specific chara
0234  n. a large store of honey is indispensable to  SUPPORT a large stock of bees during the winter; and
0278  not seem to me opposed to, but even rather to  SUPPORT the view, that there is no fundamental disti
0308  tories in Russia and in North America, do not  SUPPORT the view, that the older a formation is, the
0394  it cannot be said that small islands will not  SUPPORT small mammals, for they occur in many parts
0475  degree, then such facts as the record gives,   SUPPORT the theory of descent with modification. New
0066  w many individuals of the two species can be   SUPPORTED in a district. A large number of eggs is of
0105  ditions, the total number of the individuals   SUPPORTED on it will necessarily be very small; and f
0106  ber of individuals of the same species there   SUPPORTED, but the conditions of life are infinitely
0113  s quadruped, of which the number that can be   SUPPORTED in any country has long ago arrived at its
0114  ple, that the greatest amount of life can be   SUPPORTED by great diversification of structure, is s
0114  r many years to exactly the same conditions,   SUPPORTED twenty species of plants, and these belonge
0128  of character; for more living beings can be    SUPPORTED on the same area the more they diverge in s
0161  our. This view is hypothetical, but could be   SUPPORTED by some facts; and I can see no more abstra
0234  ly to depend on a large number of bees being   SUPPORTED. Hence the saving of wax by largely saving
0277  ns of life have been disturbed. This view is   SUPPORTED by a parallelism of another kind; namely, t
0376  hrown on the foregoing facts, on the belief,   SUPPORTED as it is by a large body of geological evid
0377  no one can say; perhaps formerly the tropics   SUPPORTED as many species as we see at the present da
0381  n the antarctic lands, now covered with ice,   SUPPORTED a highly peculiar and isolated flora. I sus
0412  r characters to diverge. This conclusion was   SUPPORTED by looking at the great diversity of the fo
0452  at its base. The pistil consists of a stigma   SUPPORTED on the style; but some Compositae, the male
0461  distinct organisations. This parallelism is   SUPPORTED by another parallel, but directly opposite,
0116  r number of individuals to be capable of there SUPPORTING themselves. A set of animals, with their or
0148  l proteolepas would have a better chance of    SUPPORTING itself, by less nutriment being wasted in d
0444  me direct evidence from our domestic animals   SUPPORTS this view. But in other cases it is quite po
0026  nt. But to extend the hypothesis so far as to  SUPPOSE that species, aboriginally as distinct as ca
0030  k further than to mere variability. We cannot  SUPPOSE that all the breeds were suddenly produced a
0047  nd, but because analogy leads the observer to  SUPPOSE either that they do now somewhere exist, or
0090  y strength, and some by fleetness; and let us  SUPPOSE that the fleetest prey, a deer for instance,
0092  ty, is greedily sought by insects. Let us now  SUPPOSE a little sweet juice or nectar to be excrete
0094  rgans become more or less impotent; now if we  SUPPOSE this to occur in ever so slight a degree und
0094  feeding insects in our imaginary case: we may  SUPPOSE the plant of which we have been slowly incre
0112  with shorter and shorter beaks. Again, we may  SUPPOSE that at an early period one man preferred sw
0118  eration. But I must here remark that I do not  SUPPOSE that the process ever goes on so regularly a
0120  ach other and from their common parent. If we  SUPPOSE the amount of change between each horizontal
0120  category of sub species; but we have only to   SUPPOSE the steps in the process of modification to
0122  to the fourteen thousandth generation. We may  SUPPOSE that only one (F), of the two species which
0125  ation of genera alone. If, in our diagram, we  SUPPOSE the amount of change represented by each suc
0138  north America and Europe. On my view we must   SUPPOSE that American animals, having ordinary power
0186  ns of extreme perfection and complication. To  SUPPOSE that the eye, with all its inimitable contri
0188  a nerve sensitive to light beneath, and then   SUPPOSE every part of this layer to be continually c
0189  ayer slowly changing in form. Further we must  SUPPOSE that there is a power always intently watchi
0189  , tend to produce a distincter image. We must  SUPPOSE each new state of the instrument to be multi
0201  the destruction of its prey; but some authors  SUPPOSE that at the same time this snake is furnishe
0209  to complete the already finished work. If we   SUPPOSE any habitual action to become inherited, and
0209  ly. But it would be the most serious error to  SUPPOSE that the greater number of instincts have be
0215  an the young of the tame rabbit; but I do not  SUPPOSE that domestic rabbits have ever been selecte
0217  their eggs in other bird's nests. Now let us   SUPPOSE that the ancient progenitor of our European
0227  ully perfect as that of the hive bee. We must  SUPPOSE the Melipona to make her cells truly spheric
0227  ly by turning round on a fixed point. We must  SUPPOSE the Melipona to arrange her cells in level l
```

Page **(Key Word)**

0227 es her cylindrical cells; and we must further SUPPOSE, and this is the greatest difficulty, that s
0227 y perfectly flat surfaces. We have further to SUPPOSE, but this is no difficulty, that after hexag
0229 ch were slightly concave on one side, where I SUPPOSE that the bees had excavated too quickly, and
0230 f the cells; but it would be a great error to SUPPOSE that the bees cannot build up a rough wall o
0234 oney which the bees could collect. But let us SUPPOSE that this latter circumstance determined, as
0234 could exist in a country; and let us further SUPPOSE that the community lived throughout the wint
0241 high, and many sixteen feet high; but we must SUPPOSE that the larger workmen had heads four inste
0260 ous species are crossed, all of which we must SUPPOSE it would be equally important to keep from b
0261 a state of nature, I presume that no one will SUPPOSE that this capacity is a specially endowed qu
0281 to the horse and tapir, we have no reason to SUPPOSE that links ever existed directly intermediat
0286 d by Prof. Ramsay. But if, as some geologists SUPPOSE, a range of older rocks underlies the Weald
0293 e been deposited, it would be equally rash to SUPPOSE that it then became wholly extinct. We forge
0300 y imperfect manner in the formations which we SUPPOSE to be there accumulating. I suspect that not
0305 have spread widely. Nor have we any right to SUPPOSE that the seas of the world have always been
0327 ains. During these long and blank intervals I SUPPOSE that the inhabitants of each region underwen
0331 to the diagram in the fourth chapter. We may SUPPOSE that the numbered letters represent genera,
0333 iscovery of the fossils. If, for instance, we SUPPOSE the existing genera of the two families to d
0341 ution. It may be asked in ridicule, whether I SUPPOSE that the megatherium and other allied huge m
0355 le hermaphrodite, or whether, as some authors SUPPOSE, from many individuals simultaneously create
0367 the present circumpolar inhabitants, which we SUPPOSE to have everywhere travelled southward, are
0367 re remarkably uniform round the world. We may SUPPOSE that the Glacial period came on a little ear
0370 warmer than at the present day. Hence we may SUPPOSE that the organisms now living under the clim
0378 usand feet. During this the coldest period, I SUPPOSE that large spaces of the tropical lowlands w
0381 these species are so distinct, that we cannot SUPPOSE that there has been time since the commencem
0394 on far distant islands. Hence we have only to SUPPOSE that such wandering species have been modifi
0401 he British Channel, and there is no reason to SUPPOSE that they have at any former period been con
0402 , each confined to its own island. Now let us SUPPOSE the mocking thrush of Chatham Island to be b
0415 ch the same organ, as we have every reason to SUPPOSE, has nearly the same physiological value, it
0420 to the diagram in the fourth chapter. We will SUPPOSE the letters A to L to represent allied gener
0421 successive period. If, however, we choose to SUPPOSE that any of the descendants of A or of I hav
0430 y to inheritance in common. Therefore we must SUPPOSE either that all Rodents, including the bizca
0430 n divergent directions. On either view we may SUPPOSE that the bizcacha has retained, by inheritan
0435 elative connexion of the several parts. If we SUPPOSE that the ancient progenitor, the archetype a
0436 o with the mouths of insects, we have only to SUPPOSE that their common progenitor had an upper li
0440 itions of existence. We cannot, for instance, SUPPOSE that in the embryos of the vertebrata the pe
0440 ed to similar conditions of life. No one will SUPPOSE that the stripes on the whelp of a lion, or
0453 xcess, or injurious to the system; but can we SUPPOSE that the minute papilla, which often represe
0453 rely of cellular tissue, can thus act? Can we SUPPOSE that the formation of rudimentary teeth whic
0010 necessarily connected, as some authors have SUPPOSED, with the act of generation. Seedlings from
0023 sessed the characteristic enormous crop? The SUPPOSED aboriginal stocks must all have been rock pi
0023 characters of the domestic breeds. Hence the SUPPOSED aboriginal stocks must either still exist in
0024 n the shores of the Mediterranean. Hence the SUPPOSED extermination of so many species having simi
0026 ty of man having formerly got seven or eight SUPPOSED species of pigeons to breed freely under dom
0027 s to breed freely under domestication; these SUPPOSED species being quite unknown in a wild state,
0044 ns of life, and variations in this sense are SUPPOSED not to be inherited; but who can say that th
0052 s given throughout this work. It need not be SUPPOSED that all varieties or incipient species nece
0058 found to have a wider range than that of its SUPPOSED parent species, their denominations ought to
0082 he foregoing case the conditions of life are SUPPOSED to have undergone a change, and this would m
0096 n research has much diminished the number of SUPPOSED hermaphrodites, and of real hermaphrodites a
0098 to ensure fertilisation; but it must not be SUPPOSED that bees would thus produce a multitude of
0111 l marked differences; whereas varieties, the SUPPOSED prototypes and parents of future well marked
0116 large in its own country; these species are SUPPOSED to resemble each other in unequal degrees, a
0117 nt its varying offspring. The variations are SUPPOSED to be extremely slight, but of the most dive
0117 of the most diversified nature; they are not SUPPOSED all to appear simultaneously, but often afte
0117 ter long intervals of time; nor are they all SUPPOSED to endure for equal periods. Only those vari
0117 letter, a sufficient amount of variation is SUPPOSED to have been accumulated to have formed a fa
0117 after a thousand generations, species (A) is SUPPOSED to have produced two fairly well marked vari
0118 ions. And after this interval, variety a1 is SUPPOSED in the diagram to have produced variety a2,
0118 from (A) than did variety a1. Variety m1 is SUPPOSED to have produced two varieties, namely m2 an
0120 ter ten thousand generations, species (A) is SUPPOSED to have produced three forms, a10, f10, and
0120 o species, according to the amount of change SUPPOSED to be represented between the horizontal lin
0121 ecies, marked by the letters n14 to z14, are SUPPOSED to have been produced. In each genus, the sp
0122 this. The original species of our genus were SUPPOSED to resemble each other in unequal degrees, a
0122 rs. These two species (A) and (I), were also SUPPOSED to be very common and widely diffused specie
0123 ants from (A); the two groups, moreover, are SUPPOSED to have gone on diverging in different direc
0124 enus. And the two or more parent species are SUPPOSED to have descended from some one species of a
0124 his point representing a single species, the SUPPOSED single parent of our several new sub-genera
0124 e character of the new species f14, which is SUPPOSED not to have diverged much in character, but
0124 ween the two parent species (A) and (I), now SUPPOSED to be extinct and unknown, it will be in som
0124 gram, each horizontal line has hitherto been SUPPOSED to represent a thousand generations, but eac
0125 ding to the amount of divergent modification SUPPOSED to be represented in the diagram. And the tw
0125 he original genus; and these two species are SUPPOSED to have descended from one species of a stil
0142 said to have been even tried. Nor let it be SUPPOSED that no differences in the constitution of a
0161 it. As all the species of the same genus are SUPPOSED, on my theory, to have descended from a comm
0172 incomparably less perfect than is generally SUPPOSED: the imperfection of the record being chiefl
0176 her modification, by which two varieties are SUPPOSED on my theory to be converted and perfected i
0194 and organs of many independent beings, each SUPPOSED to have been separately created for its prop
0218 ccasion parasitic. In this case, as with the SUPPOSED case of the cuckoo, I can see no difficulty
0268 anted. If we turn to varieties, produced, or SUPPOSED to have been produced, under domestication,
0271 ity of varieties in a state of nature, for a SUPPOSED variety if infertile in any degree would gen
0275 stress has been laid by some authors on the SUPPOSED fact, that mongrel animals alone are born cl
0298 ms, is small, for the successive changes are SUPPOSED to have been local or confined to some one s
0306 iving species; and it cannot on my theory be SUPPOSED, that these old species were the progenitors
0311 y changing language, in which the history is SUPPOSED to be written, being more or less different
0318 t gratuitous mystery. Some authors have even SUPPOSED that as the individual has a definite length
0324 neously throughout the world, it must not be SUPPOSED that this expression relates to the same tho

Page **************************************(Key Word)**************************************

```
0332  sub families and families, some of which are  SUPPOSED  to have perished at different periods, and s
0332  e can see that if many of the extinct forms,  SUPPOSED  to be embedded in the successive formations,
0372  more plants are identical than was formerly   SUPPOSED), but we find in every great class many form
0394  peculiar bats. Why, it may be asked, has the  SUPPOSED  creative force produced bats and no other ma
0398  are ranked by Mr. Gould as distinct species,  SUPPOSED  to have been created here; yet the close aff
0398  this be so? Why should the species which are  SUPPOSED  to have been created in the Galapagos Archip
0421  parent I. But the existing genus F14 may be   SUPPOSED  to have been but slightly modified; and it w
0421  enus F, along its whole line of descent, are  SUPPOSED  to have been but little modified; and they y
0432  and sub branch of their descendants, may be   SUPPOSED  to be alive; and the links to be as fine as
0447  ing age, the young of the new species of our  SUPPOSED  genus will manifestly tend to resemble each
0449  cases alone in which the ancient state, now   SUPPOSED  to be represented in many embryos, has not b
0450  d. it should also be borne in mind, that the  SUPPOSED  law of resemblance of ancient forms of life
0469  ation can be drawn between species, commonly  SUPPOSED  to have been produced by special acts of cre
0472  y one country, although on the ordinary view  SUPPOSED  to have been specially created and adapted f
0473  ne species of a genus, if the other species,  SUPPOSED  to have been created independently, have dif
0483  living tissues? Do they believe that at each  SUPPOSED  act of creation one individual or many were
0032  ariations are here often more abrupt. No one  SUPPOSES  that our choicest productions have been prod
0045  nhabiting the same confined locality. No one  SUPPOSES  that all the individuals of the same species
0065  and endless instances could be given, no one  SUPPOSES  that the fertility of these animals or plant
0066  seed were produced once in a thousand years,  SUPPOSING  that this seed were never destroyed, and co
0096  . what reason, it may be asked, is there for  SUPPOSING  in these cases that two individuals ever co
0114  any small piece of ground, could live on it  (SUPPOSING  it not to be in any way peculiar in its nat
0177  rer species. I may illustrate what I mean by  SUPPOSING  three varieties of sheep to be kept; one ad
0181  er lemuridae, yet I can see no difficulty in  SUPPOSING  that such links formerly existed, and that
0224  ncreasing and modifying the instinct, always  SUPPOSING  each modification to be of use to the speci
0231  left. We shall understand how they work, by  SUPPOSING  masons first to pile up a broad ridge of ce
0297  ope to effect in any one geological section.  SUPPOSING  B and C to be two species, and a third, A,
0303  s: and to show how liable we are to error in  SUPPOSING  that whole groups of species have suddenly
0309  d formations older than the Silurian strata,  SUPPOSING  such to have been formerly deposited; for i
0330  such as fish and reptiles, seems to be, that  SUPPOSING  them to be distinguished at the present day
0335  diate in character, intermediate in age. But  SUPPOSING  for an instant, in this and other such case
0366  we shall follow the changes more readily, by  SUPPOSING  a new glacial period to come slowly on, and
0380  , naturalised by man's agency. I am far from  SUPPOSING  that all difficulties are removed on the vi
0453  s for the presence of rudimentary organs, by  SUPPOSING  that they serve to excrete matter in excess
0319  mething is, we can hardly ever tell. On the   SUPPOSITION  of the fossil horse still existing as a ra
0425  at should we do with the other species? The  SUPPOSITION  is of course preposterous; and I might ans
0464  questions and grave objections only on the   SUPPOSITION  that the geological record is far more imp
0025  one of the two following highly improbable   SUPPOSITIONS. Either, firstly, that all the several ima
0069  e a species decreasing in numbers, we may feel  SURE  that the cause lies quite as much in other spe
0081  ing their kind? On the other hand, we may feel  SURE  that any variation in the least degree injurio
0132  ly catch a faint ray of light, and we may feel  SURE  that there must be some cause for each deviati
0145  owers of a head or umbel, I do not feel at all  SURE  that C. C. Sprengel's idea that the ray floret
0211  ing several hours. After this interval, I felt  SURE  that the aphides would want to excrete. I watc
0316  , for the newly formed fantail would be almost  SURE  to inherit from its new progenitor some slight
0365  every seed, which chanced to arrive, would be  SURE  to germinate and survive. Dispersal during the
0385  least six or seven hundred miles, and would be  SURE  to alight on a pool or rivulet, if blown acros
0386  ff their feet: when making land, they would be  SURE  to fly to their natural fresh water haunts. I
0423  n attempted by some authors. For we might feel  SURE, whether there had been more or less modificat
0426  gs having different habits, we may feel almost  SURE, on the theory of descent, that these characte
0428  ge and their parents dominant, they are almost  SURE  to spread widely, and to seize on more and mor
0434  unknown plan of creation, we may hope to make  SURE  but slow progress. Morphology. We have seen th
0484  that form be in essence a species. This I feel  SURE, and I speak after experience, will be no slig
0088  by always allowing the victor to breed might  SURELY  give indomitable courage. Length to the spur,
0102  st animals, much improvement and modification  SURELY  but slowly follow from this unconscious proce
0155  strongly marked and fixed varieties, we might  SURELY  expect to find them still often continuing to
0173  allied species inhabit the same territory we  SURELY  ought to find at the present time many transi
0269  only slightly sterile; and if this be so, we  SURELY  ought not to expect to find sterility both ap
0487  limate and of the level of the land, we shall  SURELY  be enabled to trace in an admirable manner th
0065  lairs of La Plata, clothing square leagues of  SURFACE  almost to the exclusion of all other plants,
0067  face of Nature may be compared to a yielding  SURFACE, with ten thousand sharp wedges packed close
0095  so as to push the pollen on to the stigmatic  SURFACE. Hence, again, if humble bees were to become
0099  ts! How strange that the pollen and stigmatic  SURFACE  of the same flower, though placed so close t
0130  anches the crust of the earth, and covers the  SURFACE  with its ever branching and beautiful ramifi
0134  duck of South America can only flap along the  SURFACE  of the water, and has its wings in nearly th
0185  h has all its four toes webbed, alight on the  SURFACE  of the sea. On the other hand, grebes and co
0190  al may be turned inside out, and the exterior  SURFACE  will then digest and the stomach respire. In
0192  these cirripedes have no branchiae, the whole  SURFACE  of the body and sack, including the small fr
0234  for in this case a large part of the bounding  SURFACE  of each cell would serve to bound other cell
0263  protrude, they do not penetrate the stigmatic  SURFACE. Again, the male element may reach the femal
0283  present time suffering. The appearance of the  SURFACE  and the vegetation show that elsewhere years
0285  nds of feet: for since the crust cracked, the  SURFACE  of the land has been so completely planed do
0285  ti: yet in these cases there is nothing on the  SURFACE  to show such prodigious movements: the pile
0287  l, which we know this area has undergone, the  SURFACE  may have existed for millions of years as la
0287  on some one spot. Only a small portion of the  SURFACE  of the earth has been geologically explored,
0308  ervals existed as dry land, or as a submarine  SURFACE  near land, on which sediment was not deposit
0309  rata which have always remained nearer to the  SURFACE. The immense areas in some parts of the worl
0352  created at one or more points of the earth's  SURFACE. Undoubtedly there are very many cases of ex
0354  at distant and isolated points of the earth's  SURFACE, can in many instances be explained on the v
0384  a very recent geological period, and when the  SURFACE  was peopled by existing land and fresh water
0386  an: they would not be likely to alight on the  SURFACE  of the sea, so that the dirt would not be wa
0422  possible to represent in a series, on a flat  SURFACE, the affinities which we discover in nature
0189  rent distances from each other, and with the  SURFACES  of each layer slowly changing in form. Furth
0226  on and of two, three, or more perfectly flat  SURFACES, according as the cell adjoins two, three or
0226  tly and necessarily the case, the three flat  SURFACES  are united into a pyramid; and this pyramid,
0226  ls of the hive bee, so here, the three plane  SURFACES  in any one cell necessarily enter into the c
0227  the points of intersection by perfectly flat  SURFACES. We have further to suppose, but this is no
```

Page **************************************(Key Word)**************************************

surfaces

Page **********************************(Key Word)**********************************

```
0234 nt; for then, as we have seen, the spherical SURFACES would wholly disappear, and would all be rep
0234 isappear, and would all be replaced by plane SURFACES; and the Melipona would make a comb as perfe
0366 nd wales, with their scored flanks, polished SURFACES, and perched boulders, of the icy streams wi
0182 ere unknown, who would have ventured to have SURMISED that birds might have existed which used the
0370 cean. I believe the above difficulty may be SURMOUNTED by looking to still earlier changes of clim
0035 whole body of English racehorses have come to SURPASS in fleetness and size the parent Arab stock,
0006 few concluding remarks. No one ought to feel SURPRISE at much remaining as yet unexplained in rega
0037 of very inferior quality. I have seen great SURPRISE expressed in horticultural works at the wond
0067 e in even one single instance. Nor will this SURPRISE any one who reflects how ignorant we are on
0139 heir other productions. Far from feeling any SURPRISE that some of the cave animals should be very
0155 story, that when an author has remarked with SURPRISE that some important organ or part, which is
0175 of distribution, these facts ought to cause SURPRISE, as climate and height or depth graduate awa
0185 s we now see it, must occasionally have felt SURPRISE when he has met with an animal having habits
0186 om its own place. Hence it will cause him no SURPRISE that there should be geese and frigate birds
0222 ck other ants. In one instance I found to my SURPRISE an independent community of F. flava under a
0319 are, no naturalist would have felt the least SURPRISE at its rarity; for rarity is the attribute o
0320 rare before they become extinct, to feel no SURPRISE at the rarity of a species, and yet to marve
0320 idual is the forerunner of death, to feel no SURPRISE at sickness, but when the sick man dies, to
0322 then, and not till then, we may justly feel SURPRISE why we cannot account for the extinction of
0358 ist the injurious action of sea water. To my SURPRISE I found that out of 87 kinds, 64 germinated
0361 on artificial salt water for 30 days, to my SURPRISE nearly all germinated. Living birds can hard
0383 in the fresh waters of Brazil, feeling much SURPRISE at the similarity of the fresh water insects
0461 and new conditions of life, we need not feel SURPRISE at hybrids being in some degree sterile, for
0472 of their associates; so that we need feel no SURPRISE at the inhabitants of any one country, altho
0477 physical conditions of life, we need feel no SURPRISE at their inhabitants being widely different,
0009 causes as to fail in acting, we need not be SURPRISED at this system, when it does act under conf
0045 hat the most experienced naturalist would be SURPRISED at the number of the cases of variability,
0054 auses tend to obscure this result, that I am SURPRISED that my tables show even a small majority o
0072 had not been sown or planted, I was so much SURPRISED at their numbers that I went to several poi
0139 ference to the reptiles of Europe, I am only SURPRISED that more wrecks of ancient life have not b
0154 ly a specific character, and no one would be SURPRISED at one of the blue species varying into red
0188 ave felt the difficulty far too keenly to be SURPRISED at any degree of hesitation in extending th
0210 e; and this we certainly can do. I have been SURPRISED to find, making allowance for the instincts
0242 members, which alone leave descendants. I am SURPRISED that no one has advanced this demonstrative
0255 is in regard to hybrid animals, I have been SURPRISED to find how generally the same rules apply
0266 unded organisation; and hence we need not be SURPRISED that their sterility, though in some degree
0269 carried on by man and nature, we need not be SURPRISED at some difference in the result. I have as
0291 tched the progress of Geology, and have been SURPRISED to note how author after author, in treatin
0445 as much as the full grown animals; and this SURPRISED me greatly, as I think it probable that the
0012 to, even the dahlia, etc.; and it is really SURPRISING to note the endless points in structure and
0048 n up by different botanists, and see what a SURPRISING number of forms have been ranked by one bot
0065 ease, the result of which never fails to be SURPRISING, simply explains the extraordinarily rapid
0160 laws of inheritance. No doubt it is a very SURPRISING fact that characters should reappear after
0218 strich has not as yet been perfected; for a SURPRISING number of eggs lie strewed over the plains,
0222 tle ants attacked their big neighbours with SURPRISING courage. Now I was curious to ascertain whe
0227 of equal sizes; and this would not be very SURPRISING, seeing that she already does so to a certa
0253 breeder. And in this case, it is not at all SURPRISING that the inherent sterility in the hybrids
0255 uates from zero to perfect fertility. It is SURPRISING in how many curious ways this gradation can
0273 than of hybrids does not seem to me at all SURPRISING. For the parents of mongrels are varieties,
0277 fertility of all organic beings. It is not SURPRISING that the degree of difficulty in uniting tw
0277 en the species which are crossed. Nor is it SURPRISING that the facility of effecting a first cros
0277 s this nearly general and perfect fertility SURPRISING, when we remember how liable we are to argu
0289 r than pages of detail. Nor is their rarity SURPRISING, when we remember how large a proportion of
0314 s into competition. Hence it is by no means SURPRISING that one species should retain the same ide
0328 ia and Scandinavia; nevertheless he finds a SURPRISING amount of difference in the species. If the
0401 g widely and remaining the same. The really SURPRISING fact in this case of the Galapagos Archipel
0460 versal; nor is their very general fertility SURPRISING when we remember that it is not likely that
0239 fully examined. Mr. F. Smith has shown how SURPRISINGLY the neuters of several British ants differ
0316 tions to this rule, but the exceptions are SURPRISINGLY few, so few, that E. Forbes, Pictet, and W
0071 enclosure is, I plainly saw near Farnham, in SURREY. Here there are extensive heaths, with a few
0220 have watched for many hours several nests in SURREY and Sussex, and never saw a slave either leav
0220 us hours during May, June and August, both in SURREY and Hampshire, and has never seen the slaves,
0081 e case of an island, or of a country partly SURROUNDED by barriers, into which new and better adap
0099 e. yet the pistil of each cabbage flower is SURROUNDED not only by its own six stamens; but by tho
0105 lated area be very small, either from being SURROUNDED by barriers, or from having very peculiar p
0379 and in the southern hemisphere. These being SURROUNDED by strangers will have had to compete with
0074 diversity and proportion of kinds as in the SURROUNDING virgin forests. What a struggle between th
0104 s, which otherwise would have inhabited the SURROUNDING and differently circumstanced districts, w
0129 t on all sides, and to overtop and kill the SURROUNDING twigs and branches, in the same manner as
0139 cts are very closely allied to those of the SURROUNDING country. It would be most difficult to giv
0216 om under her, and conceal themselves in the SURROUNDING grass or thickets; and this is evidently d
0226 esser distance) from the centres of the six SURROUNDING spheres in the same layer; and at the same
0290 lying great changes in the geography of the SURROUNDING lands, whence the sediment has been derive
0294 but have become extinct in the immediately SURROUNDING sea; or, conversely, that some are now abu
0351 each other, and in a lesser degree with the SURROUNDING physical conditions. As we have seen in th
0382 est to us, of the former inhabitants of the SURROUNDING lowlands. Chapter XII.
0383 ells, etc., and at the dissimilarity of the SURROUNDING terrestrial beings, compared with those of
0403 glacial epoch, are related to those of the SURROUNDING lowlands; thus we have in South America, A
0404 aved, would naturally be colonised from the SURROUNDING lowlands. So it is with the inhabitants of
0406 xceptions before specified) to those on the SURROUNDING low lands and dry lands, though these stat
0452 e purpose of brushing the pollen out of the SURROUNDING anthers. Again, an organ may become rudime
0468 daptation in however slight a degree to the SURROUNDING physical conditions, will turn the balance
0344 s may often be a very slow process, from the SURVIVAL of a few descendants, lingering in protected
0005 ls of each species are born than can possibly SURVIVE; and as, consequently, there is a struggle
0061 are periodically born, but a small number can SURVIVE. I have called this principle, by which each
0063 re individuals are produced than can possibly SURVIVE, there must in every case be a struggle for
```

Page **********************************(Key Word)**********************************

survive

```
0079  that the vigorous, the healthy, and the happy  SURVIVE and multiply. Chapter IV. Natural Selection.
0081  y more individuals are born than can possibly  SURVIVE) that individuals having any advantage, howe
0129  or three, now grown into great branches, yet  SURVIVE and bear all the other branches; so with the
0170  ggle with each other, and the best adapted to  SURVIVE. Chapter VI. Difficulties on Theory. Difficu
0325  higher degree in order to be preserved and to  SURVIVE. We have distinct evidence on this head, in
0327  here a single member might long be enabled to  SURVIVE. Thus, as it seems to me, the parallel, and,
0365  ced to arrive, would be sure to germinate and  SURVIVE. Dispersal during the Glacial period. The id
0467  . more individuals are born than can possibly  SURVIVE. A grain in the balance will determine which
0358  ted after an immersion of 28 days, and a few  SURVIVED an immersion of 137 days. For convenience sa
0369  ain ranges distant from each other, and have  SURVIVED there ever since; they will, also, in all pr
0385  ed molluscs, though aquatic in their nature,  SURVIVED on the duck's feet, in damp air, from twelve
0321  genus of shells in the secondary formations,  SURVIVES in the Australian seas; and a few members of
0005  itions of life, will have a better chance of  SURVIVING, and thus be naturally selected. From the s
0061  ing, also, will thus have a better chance of  SURVIVING, for, of the many individuals of any specie
0081  , over others, would have the best chance of  SURVIVING and of procreating their kind? On the other
0090  limmest wolves would have the best chance of  SURVIVING, and so be preserved or selected, provided
0091  idual wolf, it would have the best chance of  SURVIVING and of leaving offspring. Some of its young
0092  ould have the best chance of flourishing and  SURVIVING. Some of these seedlings would probably inh
0104  tior, that they will have a better chance of  SURVIVING and propagating their kind; and thus, in th
0136  lent habit, will have had the best chance of  SURVIVING from not being blown out to sea; and, on th
0382  ared with savage races of man, driven up and  SURVIVING in the mountian fastnesses of almost every
0142  by frost, and then collect seed from the few  SURVIVORS, with care to prevent accidental crosses, a
0219  she instantly set to work, fed and saved the  SURVIVORS; made some cells and tended the larvae, and
0198  ood observer, also, states that in cattle  SUSCEPTIBILITY to the attacks of flies is correlated wit
0008  ystem; this system appearing to be far more  SUSCEPTIBLE than any other part of the organisation, t
0131  , that the reproductive system is eminently  SUSCEPTIBLE to changes in the conditions of life; and
0259  at their fertility, besides being eminently  SUSCEPTIBLE to favourable and unfavourable conditions,
0276  duals of the same species, and is eminently  SUSCEPTIBLE of favourable and unfavourable conditions.
0466  to the reproductive system being eminently  SUSCEPTIBLE to changes in the conditions of life; so t
0008  ere variations. But I am strongly inclined to  SUSPECT that the most frequent cause of variability
0046  t of the conditions of life. I am inclined to  SUSPECT that we see in these polymorphic genera vari
0077  ans), when sown in the midst of long grass, I  SUSPECT that the chief use of the nutriment in the s
0089  ty, might produce a marked effect. I strongly  SUSPECT that some well known laws with respect to th
0099  st number of the seedlings are mongrelized? I  SUSPECT that it must arise from the pollen of a dist
0101  here able to give, I am strongly inclined to  SUSPECT that, both in the vegetable and animal kingd
0101  aps only at long intervals; but in none, as I  SUSPECT, can self fertilisation go on for perpetuity
0147  ss of growth in another and adjoining part. I  SUSPECT, also, that some of the cases of compensatio
0164  ed when first foaled. I have, also, reason to  SUSPECT, from information given me by Mr. W. W. Edwa
0187  ; but I may remark that several facts make me  SUSPECT that any sensitive nerve may be rendered sen
0230  d this by gnawing away the convex side; and I  SUSPECT that the bees in such cases stand in the opp
0266  pounded into one. It may seem fanciful, but I  SUSPECT that a similar parallelism extends to an all
0271  tain varieties of hollyhock, I am inclined to  SUSPECT that they present analogous facts. Kolreuter
0288  dissolved by the percolation of rain water. I  SUSPECT that but few of the very many animals which
0295  mineralogical composition, we may reasonably  SUSPECT that the process of deposition has been much
0298  are not highly locomotive, there is reason to  SUSPECT, as we have formerly seen, that their variet
0300  which we suppose to be there accumulating. I  SUSPECT that not many of the strictly littoral anima
0320  but when the sick man dies, to wonder and to  SUSPECT that he died by some unknown deed of violenc
0328  mocification, extinction, and immigration. I  SUSPECT that cases of this nature have occurred in E
0379  s on the mountains of Borneo and Abyssinia. I  SUSPECT that this preponderant migration from north
0381  orted a highly peculiar and isolated flora. I  SUSPECT that before this flora was exterminated by t
0393  ins of the great island of New Zealand; but I  SUSPECT that this exception (if the information be c
0185  ead body of the water ouzel would never have  SUSPECTED its sub aquatic habits; yet this anomalous
0270  have separated sexes. No one, I believe, has  SUSPECTED that these varieties of maize are distinct
0289  to these large territories, would never have  SUSPECTED that during the periods which were blank an
0296  yet not one ignorant of this fact would have  SUSPECTED the vast lapse of time represented by the t
0296  deposition, which would never even have been  SUSPECTED, had not the trees chanced to have been pre
0302  but I do not pretend that I should ever have  SUSPECTED how poor a record of the mutations of life,
0323  their geological position, no one would have  SUSPECTED that they had coexisted with still living s
0363  the flora in comparison with the latitude, I  SUSPECTED that these islands had been partly stocked
0430  s. in this case, however, it may be strongly  SUSPECTED that the resemblance is only analogical, ow
0136  in their subsistence, have, as Mr. Wollaston  SUSPECTS, their wings not at all reduced, but even en
0385  another agency is perhaps more effectual: I  SUSPENDED a duck's feet, which might represent those
0036  l for upwards of fifty years. There is not a  SUSPICION existing in the mind of any one at all acqu
0220  ed for many hours several nests in Surrey and  SUSSEX, and never saw a slave either leave or enter
0197  an feeding male turkey is likewise naked. The  SUTURES in the skulls of young mammals have been adv
0197  or may be indispensable for this act; but as  SUTURES occur in the skulls of young birds and repti
0076  parts of the United States of one species of  SWALLOW having caused the decrease of another specie
0191  hat every particle of food and drink which we  SWALLOW has to pass over the orifice of the trachea,
0387  after having swallowed them; even small fish  SWALLOW seeds of moderate size, as of the yellow wat
0387  gh they reject many other kinds after having  SWALLOWED them; even small fish swallow seeds of mode
0144  el, the shape of the body and the manner of  SWALLOWING determine the position of several of the mo
0185  es of grallatores are formed for walking over  SWAMPS and floating plants, yet the water hen is nea
0186  ed corncrakes living in meadows instead of in  SWAMPS; that there should be woodpeckers where not a
0072  orses nor dogs have ever run wild, though they  SWARM southward and northward in a feral state; and
0235  on having been economy of wax; that individual  SWARM which wasted least honey in the secretion of
0307  yet quite unknown, periods of time, the world  SWARMED with living creatures. To the question why w
0067  t the most vigorous species; by as much as it  SWARMS in numbers, by so much will its tendency to i
0235  its newly acquired economical instinct to new  SWARMS, which in their turn will have had the best c
0159  ed stems, or roots as commonly called, of the  SWEDISH turnip and Ruta baga, plants which several b
0423  ppens to be nearly identical; no one puts the  SWEDISH and common turnips together, though the escu
0427  , as in the thickened stems of the common and  SWEDISH turnip. The resemblance of the greyhound and
0232  lative distance from each other, all trying to  SWEEP equal spheres, and then building up, or leavi
0235  rees, more and more perfectly, led the bees to  SWEEP equal spheres at a given distance from each o
0401  united. The currents of the sea are rapid and  SWEEP across the archipelago, and gales of wind are
0233  ance from the parts of the cells just begun,  SWEEPING spheres or cylinders, and building up interm
0075  mely close varieties as the variously coloured  SWEET peas, they must be each year harvested separa
```

Page **(Key Word)**

```
0091  a more complex case. Certain plants excrete a SWEET juice, apparently for the sake of eliminating
0092  sought by insects. Let us now suppose a little SWEET juice or nectar to be excreted by the inner b
0210  is that of aphides voluntarily yielding their SWEET excretion to ants: that they do so voluntaril
0211  d up its abdomen and excreted a limpid drop of SWEET juice, which was eagerly devoured by the ant.
0187  within each of which there is a lens shaped SWELLING. In other crustaceans the transparent cones
0098  of the infinitely numerous pollen granules are SWEPT out of the conjoined anthers of each flower,
0111  splaced by the long horns, and that these were SWEPT away by the short horns (I quote the words of
0235  the bees, of course, no more knowing that they SWEPT their spheres at one particular distance from
0285  on the one or other side having been smoothly SWEPT away. The consideration of these facts impres
0317  f all the inhabitants of the earth having been SWEPT away at successive periods by catastrophes, i
0112  th intermediate characters, being neither very SWIFT nor very strong, will have been neglected, an
0112  ose that at an early period one man preferred SWIFTER horses; another stronger and more bulky hors
0112  urse of time, from the continued selection of SWIFTER horses by some breeders, and of stronger one
0090  ircumstances see no reason to doubt that the SWIFTEST and slimmest wolves would have the best chan
0136  for the good swimmers if they had been able to SWIM still further, whereas it would have been bett
0136  the bad swimmers if they had not been able to SWIM at all and had stuck to the wreck. The eyes of
0204  unction are at least possible. For instance, a SWIM bladder has apparently been converted into an
0452  sed for a distinct object: in certain fish the SWIM bladder seems to be rudimentary for its proper
0471  ound: that upland geese, which never or rarely SWIM should have been created with webbed feet: tha
0190  quite obliterated. The illustration of the SWIMBLADDER in fishes is a good one, because it shows
0190  different purpose, namely respiration. The SWIMBLADDER has, also, been worked in as an accessory
0191  s has been worked in as a complement to the SWIMBLADDER. All physiologists admit that the swimblad
0191  imbladder. All physiologists admit that the SWIMBLADDER is homologous, or ideally similar, in posi
0191  natural selection has actually converted a SWIMBLADDER into a lung, or organ used exclusively for
0191  ing, furnished with a floating apparatus or SWIMBLADDER. We can thus, as I infer from Professor Ow
0190  e time that they breathe free air in their SWIMBLADDERS, this latter organ having a ductus pneumat
0196  animals, which in their lungs or modified SWIMBLADDERS betray their aquatic origin, may perhaps b
0136  oast, it would have been better for the good SWIMMERS if they had been able to swim still further,
0136  hereas it would have been better for the bad SWIMMERS if they had not been able to swim at all and
0184  th america the black bear was seen by Hearne SWIMMING for hours with widely open mouth, thus catch
0185  s astonishing power of diving, its manner of SWIMMING, and of flying when unwillingly it takes fli
0185  ebbed feet of cucks and geese are formed for SWIMMING? yet there are upland geese with webbed feet
0428  ishes, being adaptations in both classes for SWIMMING through the water; but the shape of the body
0441  ense, and to reach by their active powers of SWIMMING, a proper place on which to become attached
0221  had ample opportunities for observation, in SWITZERLAND the slaves habitually work with their mast
0221  slaves being captured in greater numbers in SWITZERLAND than in England. One day I fortunately cha
0223  rate, the masters carry the slaves. Both in SWITZERLAND and England the slaves seem to have the ex
0223  rs alone go on slave making expeditions. In SWITZERLAND the slaves and masters work together, maki
0223  s service from their slaves than they do in SWITZERLAND. By what steps the instinct of F. sanguine
0224  ided by its slaves than the same species in SWITZERLAND, I can see no difficulty in natural select
0313  riking instance has lately been observed in SWITZERLAND. There is some reason to believe that orga
0088  shield may be as important for victory, as the SWORD or spear. Amongst birds, the contest is often
0226  them of equal sizes and had arranged them SYMMETRICALLY in a double layer, the resulting structur
0453  ly said to have been created for the sake of SYMMETRY, or in order to complete the scheme of natur
0453  me course round the planets, for the sake of SYMMETRY, and to complete the scheme of nature? An em
0008  tion has on the functions of the reproductive SYSTEM; this system appearing to be far more suscept
0008  he functions of the reproductive system; this SYSTEM appearing to be far more susceptible than any
0009  ous instances), yet having their reproductive SYSTEM so seriously affected by unperceived causes a
0009  l in acting, we need not be surprised at this SYSTEM, when it does act under confinement; acting n
0009  in hutches), showing that their reproductive SYSTEM has not been thus affected; so will some anim
0043  f life, from their action on the reproductive SYSTEM, are so far of the highest importance as caus
0082  life, by specially acting on the reproductive SYSTEM, causes or increases variability; and in the
0131  o be as much the function of the reproductive SYSTEM to produce individual differences, or very sl
0131  he truth of the remark, that the reproductive SYSTEM is eminently susceptible to changes in the co
0132  hanges in the conditions of life; and to this SYSTEM being functionally disturbed in the parents,
0132  e affected. But why, because the reproductive SYSTEM is disturbed, this or that part should vary m
0134  important part in affecting the reproductive SYSTEM, and in thus inducing variability; and natura
0237  that short period alone when the reproductive SYSTEM is active, as in the nuptial plumage of many
0258  tible by us, and confined to the reproductive SYSTEM. This difference in the result of reciprocal
0265  ral crossing of two species, the reproductive SYSTEM, independently of the general state of health
0269  elect, slight differences in the reproductive SYSTEM, or other constitutional differences correlat
0269  differences correlated with the reproductive SYSTEM. He supplies his several varieties with the s
0269  through correlation, modify the reproductive SYSTEM in the several descendants from any one speci
0271  sed with N. glutinosa. Hence the reproductive SYSTEM of this one variety must have been in some ma
0271  nd functional differences in the reproductive SYSTEM; from these several considerations and facts,
0273  yi namely, that it is due to the reproductive SYSTEM being eminently sensitive to any change in th
0277  ve systems imperfect, and which have had this SYSTEM and their whole organisation disturbed by bei
0278  s, and not of differences in the reproductive SYSTEM. In all other respects, excluding fertility,
0307  added another and lower stage to the Silurian SYSTEM, abounding with new and peculiar species. Tra
0312  anks between them, and to make the percentage SYSTEM of lost and new forms more gradual. In some o
0329  species. They all fall into one grand natural SYSTEM; and this fact is at once explained on the pr
0329  fect than if we combine both into one general SYSTEM. With respect to the Vertebrata, whole pages
0333  ses, to fill up wide intervals in the natural SYSTEM, and thus unite distinct families or orders.
0334  e manner in which the fossils of the Devonian SYSTEM, when this system was first discovered, were
0334  the fossils of the Devonian system, when this SYSTEM was first discovered, were at once recognised
0334  rlying carboniferous, and underlying Silurian SYSTEM. But each fauna is not necessarily exactly in
0343  ted long before the first bed of the Silurian SYSTEM was deposited. I can answer this latter quest
0344  , ancient and recent, make together one grand SYSTEM; for all are connected by generation. We can
0411  cation, groups subordinate to groups. Natural SYSTEM. Rules and difficulties in classification, ex
0413  in each class, on what is called the Natural SYSTEM. But what is meant by this system? Some autho
0413  the Natural System. But what is meant by this SYSTEM? Some authors look at it merely as a scheme f
0413  ind of dog. The ingenuity and utility of this SYSTEM are indisputable. But many naturalists think
0413  k that something more is meant by the Natural SYSTEM; they believe that it reveals the plan of the
0420  deceive myself, on the view that the natural SYSTEM is founded on descent with modification; that
0422  . thus, on the view which I hold, the natural SYSTEM is genealogical in its arrangement, like a pe
0423  rieties on a natural instead of an artificial SYSTEM; we are cautioned, for instance, not to class
```

Page **(Key Word)**

Page ***(Key Word)***

```
0429 l fewer classes, and all in one great natural SYSTEM. As showing how few the higher groups are in
0433 sts have sought under the term of the Natural SYSTEM. On this idea of the natural system being, in
0433 e natural System. On this idea of the natural SYSTEM being, in so far as it has been perfected, ge
0433 ct forms can be grouped together in one great SYSTEM; and how the several members of each class ar
0449 ve been seeking under the term of the natural SYSTEM. On this view we can understand how it is tha
0453 excrete matter in excess, or injurious to the SYSTEM; but can we suppose that the minute papilla,
0456 circuitous lines of affinities into one grand SYSTEM; the rules followed and the difficulties enco
0456 shall understand what is meant by the natural SYSTEM; it is genealogical in its attempted arrangem
0463 nd great piles of strata beneath the Silurian SYSTEM, stored with the remains of the progenitors o
0466 is seems to be mainly due to the reproductive SYSTEM being eminently susceptible to changes in the
0466 anges in the conditions of life; so that this SYSTEM, when not rendered impotent, fails to reprodu
0476 all extinct organic beings belong to the same SYSTEM with recent beings, falling either into the s
0478 t organic beings constitute one grand natural SYSTEM, with group subordinate to group, and with ex
0479 eritance or community of descent. The natural SYSTEM is a genealogical arrangement, in which we ha
0489 ved long before the first bed of the Silurian SYSTEM was deposited, they seem to me to become enno
0117 h as would be thought worthy of record in a SYSTEMATIC work. The intervals between the horizontal
0162 ty in recognising a variable species in our SYSTEMATIC works, is due to its varieties mocking, as
0252 en carefully tried than with plants. If our SYSTEMATIC arrangements can be trusted, that is if the
0256 to exactly the same conditions. By the term SYSTEMATIC affinity is meant, the resemblance between
0257 ced from them, is largely governed by their SYSTEMATIC affinity. This is clearly shown by hybrids
0257 th facility. But the correspondence between SYSTEMATIC affinity and the facility of crossing is by
0258 ss is often completely independent of their SYSTEMATIC affinity, or of any recognisable difference
0259 two species is not always governed by their SYSTEMATIC affinity or degree of resemblance to each o
0261 o with grafting, the capacity is limited by SYSTEMATIC affinity, for no one has been able to graft
0261 tion, is by no means absolutely governed by SYSTEMATIC affinity. Although many distinct genera wit
0263 ertain extent, as might have been expected, SYSTEMATIC affinity, by which every kind of resemblanc
0265 the tendency goes to a certain extent with SYSTEMATIC affinity, for whole groups of animals and p
0276 egree of sterility does not strictly follow SYSTEMATIC affinity, but is governed by several curiou
0277 run, to a certain extent, parallel with the SYSTEMATIC affinity of the forms which are subjected t
0277 orms which are subjected to experiment; for SYSTEMATIC affinity attempts to express all kinds of r
0469 iciently distinct to be worthy of record in SYSTEMATIC works. No one can draw any clear distinctio
0051 ifferences, though of small interest to the SYSTEMATIST, as of high importance for us, as being th
0427 e of the being, are almost valueless to the SYSTEMATIST. For animals, belonging to two most distin
0045 rse of years. It should be remembered that SYSTEMATISTS are far from pleased at finding variabilit
0146 that modifications of structure, viewed as SYSTEMATISTS as of high value, may be wholly due to unk
0257 ving been raised between species ranked by SYSTEMATISTS in distinct families; and on the other han
0425 uides which have been followed by our best SYSTEMATISTS. We have no written pedigrees; we have to
0455 cal view of classification, how it is that SYSTEMATISTS have found rudimentary parts as useful as,
0484 onsiderable revolution in natural history. SYSTEMATISTS will be able to pursue their labours as at
0484 tish brambles are true species will cease. SYSTEMATISTS will have only to decide (not that this wi
0260 nown differences, chiefly in the reproductive SYSTEMS, of the species which are crossed. The diffe
0263 al on unknown differences in their vegetative SYSTEMS, so I believe that the still more complex la
0263 wn differences, chiefly in their reproductive SYSTEMS. These differences, in both cases, follow to
0264 e extremely liable to have their reproductive SYSTEMS seriously affected. This, in fact, is the gr
0265 h is due, as I believe, to their reproductive SYSTEMS having been specially affected, though in a
0272 ications, more especially in the reproductive SYSTEMS of the forms which are crossed. Hybrids and
0273 ivated) which have not had their reproductive SYSTEMS in any way affected, and they are not variab
0273 ut hybrids themselves have their reproductive SYSTEMS seriously affected, and their descendants ar
0276 rally unknown differences in their vegetative SYSTEMS, so in crossing, the greater or less facilit
0276 on unknown differences in their reproductive SYSTEMS. There is no more reason to think that speci
0277 ity of hybrids, which have their reproductive SYSTEMS perfect, seems to depend on several circumst
0277 ity of hybrids, which have their reproductive SYSTEMS imperfect, and which have had this system an
0383 last and present chapters. As lakes and river SYSTEMS are separated from each other by barriers of
0384 widely and almost capriciously; for two river SYSTEMS will have some fish in common and some diffe
0384 h from an early period must have parted river SYSTEMS and completely prevented their inosculation.
0460 onstitutional differences in the reproductive SYSTEMS of the intercrossed species. We see the trut
0460 her their constitutions or their reproductive SYSTEMS should have been profoundly modified. Moreov
0031 ollow it as a trade: the sheep are placed on a TABLE and are studied, like a picture by a connoiss
0247 nt cannot be doubted; for Gartner gives in his TABLE about a score of cases of plants which he cas
0289 ous remains, a single glance at the historical TABLE published in the Supplement to Lyell's Manual
0053 the discussion of these difficulties, and the TABLES themselves of the proportional numbers of the
0053 arefully read my manuscript, and examined the TABLES, he thinks that the following statements are
0053 with different sets of organic beings. But my TABLES further show that, in any limited country, th
0054 cure this result, that I am surprised that my TABLES show even a small majority on the side of the
0055 our species, are absolutely excluded from the TABLES. These facts are of plain signification on th
0056 ies be looked at as incipient species; for my TABLES clearly show as a general rule that, wherever
0386 ost striking case; I took in February three TABLESPOONS of mud from three different points, beneat
0100 n this country; and at my request Dr. Hooker TABULATED the trees of New Zealand; and Dr. Asa Gray
0289 ntervals of time. When we see the formations TABULATED in written works, or when we follow them in
0053 elations of the species which vary most, by TABULATING all the varieties in several well worked fl
0218 good reason for believing that although the TACHYTES nigra generally makes its own burrow and sto
0288 inually taking a most erroneous view, when we TACITLY admit to ourselves that sediment is being de
0419 re so valuable, on the view of classification TACITLY including the idea of descent. Our classific
0003 or instance, of the woodpecker, with its feet, TAIL, beak, and tongue, so admirably adapted to cat
0021 rtionally to its size, much elongated wing and TAIL feathers. The trumpeter and laugher, as their
0021 r breeds. The fantail has thirty or even forty TAIL feathers, instead of twelve or fourteen, the n
0021 ried so erect that in good birds the head and TAIL touch; the oil gland is quite aborted. Several
0022 udal feathers; the relative length of wing and TAIL to each other and to the body; the relative le
0024 cobin; for a crop like that of the pouter; for TAIL feathers like those of the fantail. Hence it m
0025 termedia of Strickland, having it bluish); the TAIL has a terminal dark bar, with the bases of the
0025 e marks, even to the white edging of the outer TAIL feathers, sometimes concur perfectly developed
0025 ble black wing bar, and barred and white edged TAIL feathers, as any wild rock pigeon! We can unde
0027 ness of that of the tumbler, and the number of TAIL feathers in the fantail, are in each breed emi
0039 to make a fantail, till he saw a pigeon with a TAIL developed in some slight degree in an unusual
0039 first selected a pigeon with a slightly larger TAIL, never dreamed what the descendants of that pi
0039 distinct breeds, in which as many as seventeen TAIL feathers have been counted. Perhaps the first
```

Page **(Key Word)***

Page **************************************(Key Word)**************************************

```
0152 of the different carriers, in the carriage and TAIL of our fantails, etc., these being the points
0159 frequent presence of fourteen or even sixteen TAIL feathers in the pouter, may be considered as a
0160 e wings, a white rump, a bar at the end of the TAIL, with the outer feathers externally edged near
0171 nd, organs of trifling importance, such as the TAIL of a giraffe, which serves as a fly flapper, a
0179 s an otter in its fur, short legs, and form of TAIL: during summer this animal dives for and preys
0180 rels have their limbs and even the base of the TAIL united by a broad expanse of skin, which serve
0181 stretching from the corners of the jaw to the TAIL, and including the limbs and the elongated fin
0181 e extended from the top of the shoulder to the TAIL, including the hind legs, we perhaps see trace
0195 ssuredly be acted on by natural selection. The TAIL of the giraffe looks like an artificial cons
0196 eeing how important an organ of locomotion the TAIL is in most aquatic animals, its general presen
0196 erhaps be thus accounted for. A well developed TAIL having been formed in an aquatic animal, it mi
0196 must be slight, for the hare, with hardly any TAIL, can double quickly enough. In the second plac
0201 soon believe that the cat curls the end of its TAIL when preparing to spring, in order to warn the
0205 igh importance has often been retained (as the TAIL of an aquatic animal by its terrestrial descen
0275 red, such as albinism, melanism, deficiency of TAIL or horns, or additional fingers and toes; and
0280 il and pouter; none, for instance, combining a TAIL somewhat expanded with a crop somewhat enlarge
0454 in our domestic productions, as the stump of a TAIL in tailless breeds, the vestige of an ear in e
0039 ent bird of all fantails had only fourteen TAILFEATHERS somewhat expanded, like the present Java f
0454 estic productions, as the stump of a tail in TAILLESS breeds, the vestige of an ear in earless bre
0021 ve very long necks, others very long wings and TAILS, others singularly short tails. The barb is a
0021 long wings and tails, others singularly short TAILS. The barb is allied to the carrier, but, inst
0180 e the finest gradation from animals with their TAILS only slightly flattened, and from others, as
0366 ins of a house burnt by fire do not tell their TALE more plainly, than do the mountains of Scotlan
0221 he nest, and marching along the same road to a TALL Scotch fir tree, twenty five yards distant, wh
0270 a dwarf kind of maize with yellow seeds, and a TALL variety with red seeds, growing near each othe
0443 e cannot always tell whether the child will be TALL or short, or what its precise features will be
0392 e, might readily gain an advantage by growing TALLER and taller and overtopping the other plants.
0392 adily gain an advantage by growing taller and TALLER and overtopping the other plants. If so, natu
0077 is obvious in the structure of the teeth and TALONS of the tiger; and in that of the legs and cla
0008 nditions of life. Nothing is more easy than to TAME an animal, and few things more difficult than
0212 small; and the magpie, so wary in England, is TAME in Norway, as is the hooded crow in Egypt. Tha
0215 s has sufficed: no animal is more difficult to TAME than the young of the wild rabbit; scarcely an
0215 cely any animal is tamer than the young of the TAME rabbit; but I do not suppose that domestic rab
0215 ckals, and species of the cat genus, when kept TAME, are most eager to attack poultry, sheep and p
0009 taken young from a state of nature, perfectly TAMED, long lived, and healthy (of which I could gi
0017 ow could a savage possibly know, when he first TAMED an animal, whether it would vary in succeedin
0215 domestic rabbits have ever been selected for TAMENESS; and I presume that we must attribute the wh
0215 ited change from extreme wildness to extreme TAMENESS, simply to habit and long continued close co
0215 ung of the wild rabbit: scarcely any animal is TAMER than the young of the tame rabbit; but I do n
0132 ce near the sea affects their colours. Moquin TANDON gives a list of plants which when growing nea
0318 ine of varying thickness, the line is found to TAPER more gradually at its upper end, which marks
0281 s very distinct, for instance to the horse and TAPIR, we have no reason to suppose that links ever
0281 e organisation much general resemblance to the TAPIR and the horse; but in some points of structur
0281 d from the other; for instance, a horse from a TAPIR; and in this case direct intermediate links w
0135 have observed the same fact) that the anterior TARSI, or feet, of many male dung feeding beetles a
0135 d even a relic left. In the Onites apelles the TARSI are so habitually lost, that the insect has b
0135 explaining the entire absence of the anterior TARSI in Ateuchus, and their rudimentary condition
0135 cts of disuse in their progenitors; for as the TARSI are almost always lost in many dung feeding b
0157 e accidental. The same number of joints in the TARSI is a character generally common to very large
0020 fficulty, or rather utter hopelessness, of the TASK becomes apparent. Certainly, a breed intermedi
0053 l worked floras. At first this seemed a simple TASK; but Mr. H. C. Watson, to whom I am much indeb
0376 re appearance on the shores of New Zealand, TASMANIA, etc., of northern forms of fish. Dr. Hooker
0213 inheritance of all shades of disposition and TASTES, and likewise of the oddest tricks, associate
0214 ve thought of teaching or probably could have TAUGHT, the tumbler pigeon to tumble, an action whic
0215 ed dogs, even when quite young, require to be TAUGHT not to attack poultry, sheep, and pigs! No do
0330 ld not be named, asserts that he is every day TAUGHT that palaeozoic animals, though belonging to
0146 , the seeds being in some cases, according to TAUSCH, orthospermous in the exterior flowers and co
0361 fted trees, these stones being a valuable royal TAX. I find on examination, that when irregularly
0214 not true. No one would ever have thought of TEACHING or probably could have taught, the tumbler p
0034 e by colour, as do some of the Esquimaux their TEAMS of dogs. Livingstone shows how much good dome
0288 d having suffered in the interval any wear and TEAR, ndem explicable only on the view of the botto
0309 ated from sediment derived from their wear and TEAR; and would have been at least partially upheav
0202 inevitably causes the death of the insect by TEARING out its viscera? If we look at the sting of
0451 re normally four developed and two rudimentary TEATS in the udders of the genus Bos, but in our do
0030 ists, for instance, believe that the fuller's TEAZLE, with its hooks, which cannot be rivalled by
0044 ed. Some authors use the term variation in a TECHNICAL sense, as implying a modification directly
0424 nerations of Steenstrup, which can only in a TECHNICAL sense be considered as the same individual.
0354 ce as far as it could. It would be hopelessly TEDIOUS to discuss all the exceptional cases of the
0012 egetable poisons. Hairless dogs have imperfect TEETH; long haired and coarse haired animals are ap
0077 preys. This is obvious in the structure of the TEETH and talons of the tiger; and in that of the l
0144 ; or, again, the relation between the hair and TEETH in the naked Turkish dog, though here probabl
0144 these are likewise the most abnormal in their TEETH. I know of no case better adapted to show the
0241 ly in shape, and in the form and number of the TEETH. But the important fact for us is, that thoug
0416 tion. No one will dispute that the rudimentary TEETH in the upper jaws of young ruminants, and cer
0450 tremely curious; for instance, the presence of TEETH in foetal whales, which when grown up have no
0450 ot a tooth in their heads; and the presence of TEETH, which never cut through the gums, in the upp
0450 een stated on good authority that rudiments of TEETH can be detected in the beaks of certain embry
0452 mportant fact that rudimentary organs, such as TEETH in the upper jaws of whales and ruminants, ca
0453 n we suppose that the formation of rudimentary TEETH which are subsequently absorbed, can be of an
0480 rly age. The calf, for instance, has inherited TEETH, which never cut through the gums of the uppe
0480 from an early progenitor having well developed TEETH; and we may believe, that the teeth in the ma
0480 developed teeth: and we may believe, that the TEETH in the mature animal were reduced, during suc
0480 se without their aid: whereas in the calf, the TEETH have been left untouched by selection or disu
0480 tterly inexplicable it is that parts, like the TEETH in the embryonic calf or like the shrivelled
0228 by experiment. Following the example of Mr. TEGETMEIER, I separated two combs, and put between the
0233 sufficient nectar: and I am informed by Mr. TEGETMEIER that it has been experimentally found that
```

Page **************************************(Key Word)**************************************

Page **************************************(Key Word)**************************************

```
0305  of a whole group of species, is that of the  TELEOSTEAN  fishes, low down in the Chalk period. This
0305  es are as yet imperfectly known, are really  TELEOSTEAN. Assuming, however, that the whole of them
0305  lies of fish now have a confined range; the  TELEOSTEAN  fish might formerly have had a similarly co
0137  s, though the eye is gone; the stand for the  TELESCOPE  is there, though the telescope with its gla
0137  stand for the telescope is there, though the  TELESCOPE  with its glasses has been lost. As it is di
0188  ely possible to avoid comparing the eye to a  TELESCOPE. We know that this instrument has been perf
0056  atal to my theory; inasmuch as geology plainly  TELLS  us that small genera have in the lapse of tim
0109  ction accords perfectly well with what geology  TELLS  us of the rate and manner at which the inhabi
0109  s decrease and become rare. Rarity, as geology  TELLS  us, is the precursor.to extinction. We can, a
0186  urd in the highest possible degree. Yet reason  TELLS  me, that if numerous gradations from a perfec
0202  table contrivances in nature, this same reason  TELLS  us, though we may easily err on both sides, t
0226  statement, drawn up from his information, and   TELLS  me that it is strictly correct: If a number o
0453  tonishment: for the same reasoning power which  TELLS  us plainly that most parts and organs are exc
0453  are exquisitely adapted for certain purposes,  TELLS  us with equal plainness that these rudimentar
0419  n very large groups of closely allied forms.  TEMMINCK  insists on the utility or even necessity of
0139  home: species from an arctic or even from a  TEMPERATE  region cannot endure a tropical climate, or
0347  ht have been free migration for the northern  TEMPERATE  forms, as there now is for the strictly arc
0363  the Glacial period from one part of the now   TEMPERATE  regions to another. In the Azores, from the
0366  beings and ill fitted for their former more   TEMPERATE  inhabitants, the latter would be supplanted
0366  ke their places. The inhabitants of the more  TEMPERATE  regions would at the same time travel south
0367  ees, and even stretching into Spain. The now  TEMPERATE  regions of the United States would likewise
0367  their retreat by the productions of the more  TEMPERATE  regions. And as the snow melted from the ba
0368  f the fossil Gnathodon), then the arctic and  TEMPERATE  productions will at a very late period have
0369  to many sub arctic and to some few northern  TEMPERATE  forms, for some of these are the same on th
0369  of uniformity of the sub arctic and northern  TEMPERATE  forms round the world, at the commencement
0369  the present-day, the sub arctic and northern  TEMPERATE  productions of the Old and New Worlds are s
0370  of uniformity in the sub arctic and northern  TEMPERATE  productions of the Old and New Worlds, at a
0371  h other. This separation, as far as the more  TEMPERATE  productions are concerned, took place long
0371  we compare the now living productions of the  TEMPERATE  regions of the New and Old Worlds, we find
0372  e forms on the eastern and western shores of  TEMPERATE  North America; and the still more striking
0372  e of the past and present inhabitants of the  TEMPERATE  lands of North America and Europe, are inex
0374  llera under the equator and under the warmer  TEMPERATE  zones, and on both sides of the southern ex
0375  wing on the more lofty mountains, and on the  TEMPERATE  lowlands of the northern and southern hemis
0376  , are not arctic, but belong to the northern  TEMPERATE  zones. As Mr. H. C. Watson has recently rem
0377  rds the equator, followed in the rear by the  TEMPERATE  productions, and these by the arctic: but w
0377  er at the Cape of Good Hope, and in parts of  TEMPERATE  Australia. As we know that many tropical pl
0377  to a certain extent. On the other hand, the  TEMPERATE  productions, after migrating nearer to the
0377  e suffered less. And it is certain that many  TEMPERATE  plants, if protected from the inroads of co
0378  ain number of the more vigorous and dominant  TEMPERATE  forms might have penetrated the native rank
0378  is so destructive to perennial plants from a  TEMPERATE  climate. On the other hand, the the most hu
0378  line of march. But I do not doubt that some  TEMPERATE  productions entered and crossed even the lo
0378  nds were clothed with a mingled tropical and  TEMPERATE  vegetation, like that now growing with stra
0378  lacial period from the northern and southern  TEMPERATE  zones into the intertropical regions, and s
0378  d the equator. As the warmth returned, these  TEMPERATE  forms would naturally ascend the higher mou
0379  still further from their homes into the most  TEMPERATE  latitudes of the opposite hemisphere. Altho
0380  cies which live in the northern and southern  TEMPERATE  zones and on the mountains of the intertrop
0477  ants of the sea in the northern and southern  TEMPERATE  zones, though separated by the whole intert
0212  the situations chosen and on the nature and  TEMPERATURE  of the country inhabited, but often from c
0346  lakes, and great rivers, under almost every  TEMPERATURE. There is hardly a climate or condition in
0374  it is difficult to avoid believing that the  TEMPERATURE  of the whole world was at this period simu
0374  but it would suffice for my purpose, if the  TEMPERATURE  was at the same time lower along certain b
0377  ped extermination during a moderate fall of  TEMPERATURE, more especially by escaping into the warm
0140  extent, naturally habituated to different  TEMPERATURES, or becoming acclimatised: thus the pines
0043  are propagated by seed. In plants which are  TEMPORARILY  propagated by cuttings, buds, etc., the im
0065  ese animals or plants has been suddenly and  TEMPORARILY  increased in any sensible degree. The obvi
0369  ch during its coldest period will have been  TEMPORARILY  driven down to the plains; they will, also
0043  mportance to us, for their endurance is only  TEMPORARY. Over all these causes of Change I am convi
0313  l's explanation, namely,that it is a case of  TEMPORARY  migration from a distinct geographical prov
0325  rents or other causes more or less local and  TEMPORARY, but depend on general laws which govern th
0063  e with and other fruit bearing plants, in order to  TEMPT  birds to devour and thus disseminate its seed
0069  e change of climate being conspicuous, we are  TEMPTED  to attribute the whole effect to its direct
0073  ws on the duration of the forms of life! I am  TEMPTED  to give one more instance showing how plants
0074  and bushes clothing an entangled bank, we are  TEMPTED  to attribute their proportional numbers and
0285  ur to grapple with the idea of eternity. I am  TEMPTED  to give one other case, the well known one o
0294  e, a geologist examining these beds, might be  TEMPTED  to conclude that the average duration of lif
0064  roughout whole islands in a period of less than  TEN  years. Several of the plants now most numerous
0067  ure may be compared to a yielding surface, with  TEN  thousand sharp wedges packed close together an
0068  a tremendous destruction, when we remember that  TEN  per cent. is an extraordinarily severe mortali
0071  h firs on the distant hilltops: within the last  TEN  years large spaces have been enclosed, and sel
0117  would have been better if each had represented  TEN  thousand generations. After a thousand generat
0118  he diagram the process is represented up to the  TEN  thousandth generation, and under a condensed a
0120  no given off any fresh branches or races. After  TEN  thousand generations, species (A) is supposed
0120  ies (I) has produced, by analogous steps, after  TEN  thousand generations, either two well marked v
0231  ous; they always make the first rough wall from  TEN  to twenty times thicker than the excessively t
0246  le universal: but then he cuts the knot, for in  TEN  cases in which he found two forms, considered
0247  he disputes the entire fertility of Kolreuter's  TEN  cases. But in these and in many other cases, G
0248  e parent, for six or seven, and in one case for  TEN  generations, yet he asserts positively that th
0285  moved masses of our palaeozoic strata, in parts  TEN  thousand feet in thickness, as shown in Prof.
0286  hard, I do not believe that any line of coast,  TEN  or twenty miles in length, ever suffers degrad
0134  ve lately inhabited several oceanic islands,  TENANTED  by no beast of prey, has been caused by disu
0392  slands. For instance, in certain islands not  TENANTED  by mammals, some of the endemic plants have
0400  ral islands of the Galapagos Archipelago are  TENANTED, as I have elsewhere shown, in a quite marve
0005  iple of inheritance, any selected variety will  TEND  to propagate its new and modified form. This f
0014  appear at any particular age, yet that it does  TEND  to appear in the offspring at the same period
0040  alled it, of unconscious selection will always  TEND, perhaps more at one period than at another, a
0054  number of dominant species. But so many causes  TEND  to obscure this result, that I am surprised th
```

Page **************************************(Key Word)**************************************

Page ***(Key Word)***

```
0058  the lesser differences between varieties will  TEND  to increase into the greater differences betwe
0059  ost; and varieties, as we shall hereafter see,  TEND  to become converted into new and distinct spec
0059  w and distinct species. The larger genera thus  TEND  to become larger; and throughout nature the fo
0059  ature the forms of life which are now dominant  TEND  to become still more dominant by leaving many
0059  eafter to be explained, the larger genera also  TEND  to break up into smaller genera. And thus, the
0061  er organic beings and to external nature, will  TEND  to the preservation of that individual, and wi
0063  from the high rate at which all organic beings  TEND  to increase. Every being, which during its nat
0082  apting them to their altered conditions, would  TEND  to be preserved; and natural selection would t
0086  atior appear at any particular period of life,  TEND  to reappear in the offspring at the same perio
0102  ied as might be, natural selection will always  TEND  to preserve all the individuals varying in the
0104  degree uniform; so that natural selection will  TEND  to modify all the individuals of a varying spe
0110  ally press hardest on its nearest kindred, and  TEND  to exterminate them. We see the same process o
0116  s of natural selection and of extinction, will  TEND  to act. The accompanying diagram will aid us i
0118  s in itself hereditary, consequently they will  TEND  to vary, and generally to vary in nearly the s
0118  ties, being only slightly modified forms, will  TEND  to inherit those advantages which made their c
0119  used species, belonging to a large genus, will  TEND  to partake of the same advantages which made t
0121  tremely different in character, will generally  TEND  to produce the greatest number of modified des
0121  original parent species itself, will generally  TEND  to become extinct. So it probably will be with
0126  laces in the polity of Nature, will constantly  TEND  to supplant and destroy the earlier and less i
0126  and broken groups and sub groups will finally  TEND  to disappear. Looking to the future, we can pr
0127  the strong principle of inheritance they will  TEND  to produce offspring similarly characterised.
0128  g varieties of the same species, will steadily  TEND  to increase till they come to equal the greate
0128  larger genera, which vary most; and these will  TEND  to transmit to their modified offspring that s
0142  doubt that natural selection will continually  TEND  to preserve those individuals which are born w
0143  s parts, as has been remarked by some authors,  TEND  to cohere; this is often seen in monstrous pla
0168  more potent in their effects. Homologous parts  TEND  to vary in the same way, and homologous parts
0168  to vary in the same way, and homologous parts  TEND  to cohere. Modifications in hard parts and in
0169  d exposed to similar influences will naturally  TEND  to present analogous variations, and these sam
0172  f profitable modifications, each new form will  TEND  in a fully stocked country to take the place o
0175  ending as it does on the range of others, will  TEND  to be sharply defined. Moreover, each species
0177  more common forms, in the race for life, will  TEND  to beat and supplant the less common forms, fo
0178  n lesser numbers than the varieties which they  TEND  to connect. From this cause alone the intermed
0189  rcumstances, may in any way, or in any degree,  TEND  to produce a cistincter image. We must suppose
0223  ls for the nest; both, but chiefly the slaves,  TEND, and milk as it may be called, their aphides;
0226  at walls of wax between the spheres which thus  TEND  to intersect. Hence each cell consists of an o
0243  cing views, but is otherwise inexplicable, all  TEND  to corroborate the theory of natural selection
0265  atural conditions; and whole groups of species  TEND  to produce sterile hybrids. On the other hand,
0281  cases the new and improved forms of life will  TEND  to supplant the old and unimproved forms. By t
0295  ut this same movement of subsidence will often  TEND  to sink the area whence the sediment is derive
0327  n the highest degree, wherever produced, would  TEND  everywhere to prevail. As they prevailed, they
0327  d in groups by inheritance, whole groups would  TEND  slowly to disappear; though here, and there a s
0327  ccession of forms in both ways will everywhere  TEND  to correspond. There is one other remark conne
0340  ts of each cuarter of the world will obviously  TEND  to leave in that quarter, during the next succ
0343  dominant species of the larger dominant groups  TEND  to leave many modified descendants, and thus n
0344  nferiority inherited from a common progenitor,  TEND  to become extinct together, and to leave no mo
0344  those that oftenest vary, will in the long run  TEND  to people the world with allied, but modified,
0344  ow living. Why ancient and extinct forms often  TEND  to fill up gaps between existing forms, someti
0392  r plants. If so, natural selection would often  TEND  to add to the stature of herbaceous plants whe
0411  si and these, on the principle of inheritance,  TEND  to produce other new and dominant species. Con
0412  which generally include many dominant species,  TEND  to go on increasing indefinitely in size. I fu
0427  such resemblances will not reveal, will rather  TEND  to conceal their blood relationship to their p
0428  inant species, belonging to the larger genera,  TEND  to inherit the advantages, which made the grou
0428  ture. The larger and more dominant groups thus  TEND  to go on increasing in size; and they conseque
0444  might have appeared earlier or later in life,  TEND  to appear at a corresponding age in the offspr
0447  species of our supposed genus will manifestly  TEND  to resemble each other much more closely than
0455  eve to be possible) the rudimentary part would  TEND  to be wholly lost, and we should have a case o
0455  l probably often come into play; and this will  TEND  to cause the entire obliteration of a rudiment
0470  aracteristic of varieties of the same species,  TEND  to be augmented into the greater differences c
0470  ominant species belonging to the larger groups  TEND  to give birth to new and dominant forms; so th
0479  ded sometimes by natural selection, will often  TEND  to reduce an organ, when it has become useless
0483  b ordinate to groups. Fossil remains sometimes  TEND  to fill up very wide intervals between existin
0489  eing, all corporeal and mental endowments will  TEND  to progress towards perfection. It is interest
0027  ion. Fourthly, pigeons have been watched, and  TENDED  with the utmost care, and loved by many peopl
0112  ronc, will have been neglected, and will have  TENDED  to disappear. Here, then, we see in man's pro
0219  and saved the survivors; made some cells and  TENDED  the larvae, and put all to rights. What can b
0342  all these causes taken conjointly, must have  TENDED  to make the geological record extremely imper
0091  en observe great differences in the natural  TENDENCIES  of our domestic animals; one cat, for insta
0143  eved to dispose with the limbs. These  TENDENCIES, I do not doubt, may be mastered more or le
0012  subject. No breeder doubts how strong is the  TENDENCY  to inheritance: like produces like is his fu
0015  t our domestic varieties manifested a strong  TENDENCY  to reversion, that is, to lose their acquire
0017  and plants having an extraordinary inherent  TENDENCY  to vary, and likewise to withstand diverse c
0025  that in each separate breed there might be a  TENDENCY  to revert to the very same colours and marki
0026  ssed only once with some distinct breed, the  TENDENCY  to reversion to any character derived from s
0026  cross with a distinct breed, and there is a  TENDENCY  in both parents to revert to a character, wh
0026  een lost during some former generation, the  TENDENCY, for all that we can see to the contrary, ma
0026  ntinued domestication eliminates this strong  TENDENCY  to sterility: from the history of the dog I
0051  rank most of the doubtful forms. His general  TENDENCY  will be to make many species; for he will be
0065  ould any how exist, and that the geometrical  TENDENCY  to increase must be checked by destruction a
0067  with greater force. What checks the natural  TENDENCY  of each species to increase in number is mos
0067  as it swarms in numbers, by so much will its  TENDENCY  to increase be still further increased. We k
0080  nature, vary; and how strong the hereditary  TENDENCY  is. Under domestication, it may be truly sai
0091  f preyed, a cub might be born with an innate  TENDENCY  to pursue certain kinds of prey. Nor can thi
0091  st nightly catching woodcocks or snipes. The  TENDENCY  to catch rats rather than mice is shown to b
0094  he division of labour, individuals with this  TENDENCY  more and more increased, would be continuall
0094  ts. Its descendants would probably inherit a  TENDENCY  to a similar slight deviation of structure.
0100  vided against it by giving to trees a strong  TENDENCY  to bear flowers with separated sexes. When t
```

Page ***(Key Word)***

tendency

Page **(Key Word)**

```
0118  s which made their parents variable, and the TENDENCY to variability is in itself hereditary, cons
0121  e over other forms, there will be a constant TENDENCY in the improved descendants of any one speci
0123  be fifteen in number. Owing to the divergent TENDENCY of natural selection, the extreme amount of
0136   new insect first arrived on the island, the TENDENCY of natural selection to enlarge or to reduce
0150  effects of long continued disuse, and to the TENDENCY to reversion. A part developed in any specie
0152  uggle going on between, on the one hand, the TENDENCY to reversion to a less modified state, as we
0152   a less modified state, as well as an innate TENDENCY to further variability of all kinds and on t
0153  n natural selection on the one hand, and the TENDENCY to reversion and variability on the other ha
0158  rding to the lapse of time, overmastered the TENDENCY to reversion and to further variability, to
0159  om a common parent the same constitution and TENDENCY to variation, when acted on by similar unkno
0159  sa of community of descent, and a consequent TENDENCY to vary in a like manner, but to three separ
0160  her breed, the offspring occasionally show a TENDENCY to revert in character to the foreign breed
0160  yet, we see, it is generally believed that a TENDENCY to reversion is retained by this very small
0160  racter which their progenitor possessed, the TENDENCY, whether strong or weak, to reproduce the lo
0161  each successive generation there has been a TENDENCY to reproduce the character in question, whic
0161  blue and black barred bird, there has been a TENDENCY in each generation in the plumage to assume
0161   can see no more abstract improbability in a TENDENCY to produce any character being inherited for
0161  ted. Indeed, we may sometimes observe a mere TENDENCY to produce a rudiment inherited: for instanc
0161  ears, that this plant must have an inherited TENDENCY to produce it. As all the species of the sam
0166  ulders like an ass. In the horse we see this TENDENCY strong whenever a dun tint appears, a tint w
0166  m or by any other new character. We see this TENDENCY to become striped most strongly displayed in
0166  various colours are crossed, we see a strong TENDENCY for the blue tint and bars and marks to reap
0166  very ancient character, is, that there is a TENDENCY in the young of each successive generation t
0166  oduce the long lost character, and that this TENDENCY, from unknown causes, sometimes prevails. An
0167  rt that each species has been created with a TENDENCY to vary, both under nature and under domesti
0167  and that each has been created with a strong TENDENCY, when crossed with species inhabiting distan
0169  ses not as yet have had time to overcome the TENDENCY to further variability and to reversion to a
0213  n some degree inherited by retrievers; and a TENDENCY to run round, instead of at, a flock of shee
0214  s given to a whole family of shepherd dogs a TENDENCY to hunt hares. These domestic instincts, whe
0214  believe that some one pigeon showed a slight TENDENCY to this strange habit, and that the long con
0214  oint, had not some one dog naturally shown a TENDENCY in this line: and this is known occasionally
0214  i once saw in a pure terrier. When the first TENDENCY was once displayed, methodical selection and
0215  to attack poultry, sheep and pigs; and this TENDENCY has been found incurable in dogs which have
0238  and transmitted to their fertile offspring a TENDENCY to produce sterile members having the same m
0253  e animal, which from any cause had the least TENDENCY to sterility, the breed would assuredly be l
0265   the female more than the male. In both, the TENDENCY goes to a certain extent with systematic aff
0331  rogenitor. On the principle of the continued TENDENCY to divergence of character, which was former
0344  ation. We can understand, from the continued TENDENCY to divergence of character, why the more and
0412  n the economy of nature, there is a constant TENDENCY in their characters to diverge. This conclus
0412  empted also to show that there is a constant TENDENCY in the forms which are increasing in number
0435  s of this nature, there will be little or no TENDENCY to modify the original pattern, or to transp
0435  reat amount of modification there will be no TENDENCY to alter the framework of bones or the relat
0455  nce of rudimentary organs is thus due to the TENDENCY in every part of the organisation, which has
0470   economy of nature, there will be a constant TENDENCY in natural selection to preserve the most di
0471  dominant groups beat the less dominant. This TENDENCY in the large groups to go on increasing in s
0142  ced, has even been advanced, for it is now as TENDER as ever it was, as proving that acclimatisati
0065  ently assert, that all plants and animals are TENDING to increase at a geometrical ratio, that all
0103  e counterbalanced by natural selection always TENDING to modify all the individuals in each distri
0154  gree, and by the continued rejection of those TENDING to revert to a former and less modified cond
0233  sion of modified architectural instincts, all TENDING towards the present perfect plan of construc
0433  ion: nevertheless, in certain classes, we are TENDING in this direction; and Milne Edwards has lat
0012  organisation seems to have become plastic, and TENDS to depart in some small degree from that of t
0013  period of life a peculiarity first appears, it TENDS to appear in the offspring at a corresponding
0065  r familiarity with the larger domestic animals TENDS, I think, to mislead us: we see no great dest
0083  d: nature only for that of the being which she TENDS. Every selected character is fully exercised
0150  mparison with the same part in allied species, TENDS to be highly variable. Several years ago I wa
0168  when one part is largely developed, perhaps it TENDS to draw nourishment from the adjoining parts:
0179  e very process of natural selection constantly TENDS, as has been so often remarked, to exterminat
0201  myriads have become extinct. Natural selection TENDS only to make each organic being as perfect as
0269  ts believe that a long course of domestication TENDS to eliminate sterility in the successive gene
0312  of the several tertiary stages; and every year TENDS to fill up the blanks between them, and to ma
0322  if we forget for an instant, that each species TENDS to increase inordinately, and that some check
0330  more ancient a form is, by so much the more. It TENDS to connect by some of its characters groups n
0444   any variation first appears in the parent, it TENDS to reappear at a corresponding age in the off
0461  domestication; and as domestication apparently TENDS to eliminate sterility, we ought not to expec
0470  es first existed as varieties. As each species TENDS by its geometrical ratio of reproduction to i
0471  w and dominant forms: so that each large group TENDS to become still larger, and at the same time
0471  m, which every fresh addition to our knowledge TENDS to make more strictly correct, is on this the
0127  nd stations, must be judged of by the general TENOUR and balance of evidence given in the followin
0363  n in action year after year, for centuries and TENS of thousands of years, it would I think be a m
0092  wer, and a cross thus effected, although nine TENTHS of the pollen were destroyed, it might still
0009  ld easily be given of sporting plants: by this TERM gardeners mean a single bud or offset, which s
0023  k pigeon (Columba livia), including under this TERM several geographical races or sub species, whi
0044  rious definitions which have been given of the TERM species. No one definition has as yet satisfie
0044  ans when he speaks of a species. Generally the TERM includes the unknown element of a distinct act
0044  own element of a distinct act of creation. The TERM variety is almost equally difficult to define,
0044  not generally propagated. Some authors use the TERM variation in a technical sense, as implying a
0052  ese remarks it will be seen that I look at the TERM species, as one arbitrarily given for the sake
0052  d that it does not essentially differ from the TERM variety, which is given to less distinct and m
0052   less distinct and more fluctuating forms. As TERM variety, again, in comparison with mere indivi
0060  for Existence. Bears on natural selection. The TERM used in a wide sense. Geometrical powers of in
0061  ght variation, if useful, is preserved, by the TERM of Natural Selection, in order to mark its rel
0062  ecurring year. I should premise that I use the TERM Struggle for Existence in a large and metaphor
0063  other, I use for convenience sake the general TERM of struggle for existence. A struggle for exis
0111  the principle, which I have designated by this TERM, is of high importance on my theory, and expla
0150  ers, when displayed in any unusual manner. The TERM, secondary sexual characters, used by Hunter,
```

Page **(Key Word)**

Page **************************************(Key Word)**************************************

```
0164 ur far oftenest in duns and mouse duns; by the TERM dun a large range of colour is included, from
0207 t mental actions are commonly embraced by this TERM; but every one understands what is meant, when
0256 exposed to exactly the same conditions. By the TERM systematic affinity is meant, the resemblance
0330 e between living species or groups. If by this TERM it is meant that an extinct form is directly i
0433 nexion which naturalists have sought under the TERM of the Natural System. On this idea of the nat
0434 on. This resemblance is often expressed by the TERM unity of type; or by saying that the several p
0449 which naturalists have been seeking under the TERM of the natural system. On this view we can und
0481 cannot possibly grasp the full meaning of the TERM of a hundred million years; it cannot add up a
0485 undiscovered and undiscoverable essence of the TERM species. The other and more general department
0025 trickland, having it bluish); the tail has a TERMINAL dark bar, with the bases of the outer feathe
0049 s or varieties, before any definition of these TERMS has been generally accepted, is vainly to bea
0053 did subsequently Dr. Hooker, even in stronger TERMS. I shall reserve for my future work the discu
0433 endants from a common parent, expressed by the TERMS genera, families, orders, etc., we can unders
0439 ing this plain signification. On my view these TERMS may be used literally; and the wonderful fact
0456 he grades of acquired difference marked by the TERMS varieties, species, genera, families, orders,
0485 ral history will rise greatly in interest. The TERMS used by naturalists of affinity, relationship
0428 thus the septenary, quinary, quaternary, and TERNARY classifications have probably arisen. As the
0100 as yet I have not found a single case of a TERRESTRIAL animal which fertilises itself. We can und
0100 act, which offers so strong a contrast with TERRESTRIAL plants, on the view of an occasional cross
0100 ensable, by considering the medium in which TERRESTRIAL animals live, and the nature of the fertil
0100 an occasional cross could be effected with TERRESTRIAL animals without the concurrence of two two
0107 i conclude, looking to the future, that for TERRESTRIAL productions a large continental area, whic
0179 te grade between truly aquatic and strictly TERRESTRIAL habits; and as each exists by a struggle f
0185 i yet this anomalous member of the strictly TERRESTRIAL thrush family wholly subsists by diving, g
0185 tic as the coot! and the landrail nearly as TERRESTRIAL as the quail or partridge. In such cases,
0205 ed (as the tail of an aquatic animal by its TERRESTRIAL descendants), though it has become of such
0288 rtially analogous case. With respect to the TERRESTRIAL productions which lived during the Seconda
0300 ut we have every reason to believe that the TERRESTRIAL productions of the archipelago would be pr
0314 nd the apparently quicker rate of change in TERRESTRIAL and in more highly organised productions c
0326 n would, it is probable, be slower with the TERRESTRIAL inhabitants of distinct continents than wi
0335 quently produced, especially in the case of TERRESTRIAL productions inhabiting separated districts
0340 that in America the law of distribution of TERRESTRIAL mammals was formerly different from what i
0347 s in the great difference of nearly all the TERRESTRIAL productions of the New and Old Worlds, exc
0352 cross the sea is more distinctly limited in TERRESTRIAL mammals, than perhaps in any other organic
0354 ns; and thirdly, the occurrence of the same TERRESTRIAL species on islands and on the mainland, th
0356 continents together, and thus have allowed TERRESTRIAL productions to pass from one to the other.
0376 facts could be given on the distribution of TERRESTRIAL animals. In marine productions, similar ca
0378 eve, a considerable number of plants, a few TERRESTRIAL animals, and some marine productions, migr
0383 islands. Absence of the Batrachians and of TERRESTRIAL Mammals. On the relation of the inhabitant
0383 and at the dissimilarity of the surrounding TERRESTRIAL beings, compared with those of Britain. Bu
0386 ed by Alph. de Candolle, in large groups of TERRESTRIAL plants, which have only a very few aquatic
0388 be less severe between aquatic than between TERRESTRIAL species; consequently an intruder from the
0388 of seizing on a place, than in the case of TERRESTRIAL colonists. We should, also, remember that
0393 nd a single instance, free from doubt, of a TERRESTRIAL mammal (excluding domesticated animals kep
0394 r rate than other and lower animals. Though TERRESTRIAL mammals do not occur on oceanic islands, a
0394 his question can easily be answered: for no TERRESTRIAL mammal can be transported across a wide sp
0395 ic bats on islands, with the absence of all TERRESTRIAL mammals. Besides the absence of terrestria
0395 terrestrial mammals. Besides the absence of TERRESTRIAL mammals in relation to the remoteness of i
0396 of whole groups, as of batrachians, and of TERRESTRIAL mammals notwithstanding the presence of ae
0409 ole groups of organisms, as batrachians and TERRESTRIAL mammals, should be absent from oceanic isl
0477 ot cross wide spaces of ocean, as frogs and TERRESTRIAL mammals, should not inhabit oceanic island
0017 ould be shown that the greyhound, bloodhound, TERRIER, spaniel, and bull dog, which we all know pr
0214 casionally to happen, as I once saw in a pure TERRIFIED. When the first tendency was once displayed,
0222 he pupae of F. fusca, whereas they were much TERRIFIED when they came across the pupae, or even th
0289 ad been exclusively confined to these large TERRITORIES, would never have suspected that during th
0308 ssess of the Silurian deposits over immense TERRITORIES in Russia and in North America, do not sup
0326 lready have invaded to a certain extent the TERRITORIES of other species, should be those which wo
0326 as would be the power of spreading into new TERRITORIES. A certain amount of isolation, recurring
0173 eral closely allied species inhabit the same TERRITORY we surely ought to find at the present time
0174 ly find them tolerably numerous over a large TERRITORY, then becoming somewhat abruptly rarer and
0174 and finally disappearing. Hence the neutral TERRITORY between two representative species is gener
0174 s is generally narrow in comparison with the TERRITORY proper to each. We see the same fact in asc
0175 e range, with a comparatively narrow neutral TERRITORY between them, in which they become rather s
0289 re been accumulated. And if in each separate TERRITORY, hardly any idea can be formed of the lengt
0402 closely allied species invading each other's TERRITORY, when put into free intercommunication. Und
0107 ording to Oswald Heer, resembles the extinct TERTIARY flora of Europe. All fresh water basins, tak
0174 broken up into islands even during the later TERTIARY periods; and in such islands distinct specie
0284 ..........................13,190 Feet. TERTIARY strata.......................... 2
0288 other species has hitherto been found in any TERTIARY formation: yet it is now known that the genu
0289 ember how large a proportion of the bones of TERTIARY mammals have been discovered either in caves
0290 ich is inhabited by a peculiar marine fauna, TERTIARY beds are so scantily developed, that no reco
0290 rica, no extensive formations with recent or TERTIARY remains can anywhere be found; though the su
0291 subsidence. I may add, that the only ancient TERTIARY formation on the west coast of South America
0303 abruptly come in at the commencement of the TERTIARY series. And now one of the richest known acc
0303 used to urge that no monkey occurred in any TERTIARY stratum: but now extinct species have been d
0304 al between the latest secondary and earliest TERTIARY formation. But now we may read in the Supple
0304 hat, from the number of existing and extinct TERTIARY species; from the extraordinary abundance of
0304 which specimens are preserved in the oldest TERTIARY beds; from the ease with which even a fragme
0304 uddenly developed at the commencement of the TERTIARY series. This was a sore trouble to me; addin
0304 e specimen has as yet been found even in any TERTIARY stratum. Hence we now positively know that s
0305 might have been the progenitors of our many TERTIARY and existing species. The case most frequent
0312 ence on this head in the case of the several TERTIARY stages; and every year tends to fill up the
0313 e rate, or in the same degree. In the oldest TERTIARY beds a few living shells may still be found
0317 ery reason to believe, from the study of the TERTIARY formations, that species and groups of speci
0319 ion. We see in many cases in the more recent TERTIARY formations, that rarity precedes extinction;
0323 with the several European and North American TERTIARY deposits. Even if the new fossil species whi
```

Page **************************************(Key Word)**************************************

Page ***(Key Word)**

```
0323  tages of the widely separated palaeozoic and TERTIARY periods, would still be manifest, and the se
0324  that they had lived during one of the later  TERTIARY stages. When the marine forms of life are sp
0324  e which lived in Europe during certain later TERTIARY stages, than to those which now live here: a
0328  de similar observations on some of the later TERTIARY formations. Barrande, also, shows that there
0339  ypes within the same areas, during the later TERTIARY periods. Mr. Clift many years ago showed tha
0340  y of the same types in each during the later  TERTIARY periods. Nor can it be pretended that it is
0358  t every continent, the close relation of the  TERTIARY inhabitants of several lancs and even seas t
0371  tions of Europe and America during the later  TERTIARY stages were more closely related to each oth
0372  understand the presence of many existing and  TERTIARY representative forms on the eastern and west
0394  dation which they have suffered and by their  TERTIARY strata: there has also been time for the pro
0055  aving many species, than in one having few. To TEST the truth of this anticipation I have arranged
0057  often exceedingly small. I have endeavoured to TEST this numerically by averages, and, as far as m
0105  favourable variations. If we turn to nature to TEST the truth of these remarks, and look at any sm
0141  but of being perfectly fertile (a far severer  TEST) under them, may be used as an argument that a
0270  t, who mainly founds his classification by this TEST of infertility, as varieties. The following ca
0276  opposite conclusions in ranking forms by this  TEST. The sterility is innately variable in individ
0214  nt hares. These domestic instincts, when thus  TESTED by crossing, resemble natural instincts, whic
0228  architectural powers. But this theory can be  TESTED by experiment. Following the example of Mr. T
0271  ommonly reputed to be varieties, and which he  TESTED by the severest trial, namely, by reciprocal
0337  ent and beaten forms: but I can see no way of  TESTING this sort of progress. Crustaceans, for inst
0367  e arctic forms would seize on the cleared and  THAWED ground, always ascending higher and higher, a
0026  princ of two animals clearly distinct being   THEMSELVES perfectly fertile. Some authors believe tha
0053  ssion of these difficulties, and the tables   THEMSELVES of the proportional numbers of the varying
0116  incividuals to be capable of there supporting THEMSELVES. A set of animals, with their organisation
0129  these into lesser and lesser branches, were   THEMSELVES once, when the tree was small, budding twig
0162  intermediate between two other forms, which   THEMSELVES must be doubtfully ranked as either varieti
0195  individuals which could by any means defend   THEMSELVES from these small enemies, would be able to
0216  young turkeys) from under her, and conceal    THEMSELVES in the surrounding grass or thickets: and t
0219  they did nothing: they could not even feed    THEMSELVES, and many perished of hunger. Huber then in
0223  to collect building materials and food for    THEMSELVES, their slaves and larvae. So that the maste
0227  s already possessed by the Melipona, and in   THEMSELVES not very wonderful, this bee would make a s
0227  oons. By such modifications of instincts in   THEMSELVES not very wonderful, hardly more wonderful t
0230  arly see that if the bees were to build for   THEMSELVES a thin wall of wax, they could make their c
0246  both plants and animals: though the organs    THEMSELVES are perfect in structure, as far as the mic
0252  than in the case of plants: but the hybrids   THEMSELVES are, I think, more sterile. I doubt whether
0255  nt's own pollen will produce. So in hybrids   THEMSELVES, there are some which never have produced,
0270  ice that the hybrid plants thus raised were   THEMSELVES perfectly fertile: so that even Gartner did
0273  ted, and they are not variable: but hybrids   THEMSELVES have their reproductive systems seriously a
0327  arying: the new species thus produced being  THEMSELVES dominant owing to inheritance, and to havin
0328  longing to the same genera, yet the species  THEMSELVES differ in a manner very difficult to accoun
0349  e same continent or sea, though the species  THEMSELVES are distinct at different points and static
0351  ion: for neither migration nor isolation in  THEMSELVES can do anything. These principles come into
0379  nt with those intruding forms which settled  THEMSELVES on the intertropical mountains, and in the
0406  ome of the species of widely ranging genera  THEMSELVES ranging widely, such facts, as alpine, lacu
0418  aturalists are at work, they do not trouble  THEMSELVES about the physiological value of the charac
0479  numerable other such facts, at once explain  THEMSELVES on the theory of descent with slow and slig
0482  multitude of forms, which till lately they   THEMSELVES thought were special creations, and which a
0354  en produced within one area, and has migrated THENCE as far as it could. It would be hopelessly te
0012  oubts have been thrown on this principle by  THEORETICAL writers alone. When a deviation appears no
0053  y, and for mere convenience sake. Guided by  THEORETICAL considerations, I thought that some intere
0064  e better evidence on this subject than mere  THEORETICAL calculations, namely, the numerous recorde
0389  er facts, which bear on the truth of the two THEORIES of independent creation and of descent with
0005  most apparent and gravest difficulties on the THEORY will be given: namely, first, the difficultie
0056  s had been so, it would have been fatal to my THEORY: inasmuch as geology plainly tells us that sm
0111  ted by this term, is of high importance on my THEORY, and explains, as I believe, several importan
0154  he bat, it must have existed, according to my THEORY, for an immense period in nearly the same sta
0161  species of the same genus are supposed, on my THEORY, to have descended from a common parent, it m
0162  but analogous variations, yet we ought, on my THEORY, sometimes to find the varying offspring of a
0165  tock. But I am not at all satisfied with this THEORY, and should be loth to apply it to breeds so
0171  apted to survive. Chapter VI. Difficulties on THEORY. Difficulties on the theory of descent with m
0171  . difficulties on Theory. Difficulties on the THEORY of descent with modification. Transitions. Ab
0171  f the Conditions of Existence embraced by the THEORY of Natural Selection. Long before having arri
0171  e that are real are not, I think, fatal to my THEORY. These difficulties and objections may be cla
0172  d perfection of the new form. But, as by this THEORY innumerable transitional forms must have exis
0173  from the metropolis inhabited by each. By my  THEORY these allied species have descended from a co
0176  on, by which two varieties are supposed on my THEORY to be converted and perfected into two distin
0179  c not to any one time, but to all time, if my THEORY be true, numberless intermediate varieties, l
0184  ies of the same genus, we might expect, on my THEORY, that such individuals would occasionally hav
0188  herwise inexplicable, can be explained by the THEORY of descent, ought not to hesitate to go furth
0189  umerous, successive, slight modifications, my THEORY would absolutely break down. But I can find o
0189  solated species, round which, according to my THEORY, there has been much extinction. Or again, if
0194  riety, but niggard in innovation. Why, on the THEORY of Creation, should this be so? Why should al
0194  en a leap from structure to structure? On the THEORY of natural selection, we can clearly understa
0199  ine, if true, would be absolutely fatal to my THEORY. Yet I fully admit that many structures are o
0201  od of another species, it would annihilate my THEORY, for such could not have been produced throug
0203  and objections which may be urged against my  THEORY. Many of them are very grave: but I think tha
0203  as been thrown on several facts, which on the THEORY of independent acts of creation are utterly o
0206  solute perfection be everywhere found. On the THEORY of natural selection we can clearly understan
0206  nclude all of those of past times, it must by my THEORY be strictly true. It is generally acknowledge
0206  te independent of each other in their habits of life. On my THEORY, unity of type is explained by unity of desce
0207  its cell making instinct. Difficulties on the THEORY of the Natural Selection of instincts. Neuter
0207  a difficulty sufficient to overthrow my whole THEORY. I must premise, that I have nothing to do wi
0210  corporeal structure, and conformably with my  THEORY, the instinct of each species is good for its
0225  , be considered only as a modification of his THEORY. Let us look to the great principle of gradat
0228  her inimitable architectural powers. But this THEORY can be tested by experiment. Following the ex
0230  could show that they are conformable with my  THEORY. Huber's statement that the very first cell i
```

Page ***(Key Word)**

theory

Page **********************************(Key Word)**********************************

0233 ms at first quite subversive of the foregoing THEORY; namely, that the cells on the extreme margin
0235 difficult explanation could be opposed to the THEORY of natural selection, cases, in which we cann
0236 e insuperable, and actually fatal to my whole THEORY. I allude to the neuters or sterile females i
0237 s it possible to reconcile this case with the THEORY of natural selection? First, let it be rememb
0239 well established facts at once annihilate my THEORY. In the simpler case of neuter insects all of
0242 the most serious special difficulty, which my THEORY has encountered. The case, also, is very inte
0243 his chapter strengthen in any great degree my THEORY; but none of the cases of difficulty, to the
0243 ise inexplicable, all tend to corroborate the THEORY of natural selection. This theory is, also, s
0243 oborate the theory of natural selection. This THEORY is, also, strengthened by some few other fact
0245 much underrated by some late writers. On the THEORY of natural selection the case is especially i
0246 rtility of their mongrel offspring, is, on my THEORY, of equal importance with the sterility of sp
0280 avest objection which can be urged against my THEORY. The explanation lies, as I believe, in the e
0280 d what sort of intermediate forms must, on my THEORY, have formerly existed. I have found it diffi
0281 intermediate links. It is just possible by my THEORY, that one of two living forms might have desc
0281 supplant the old and unimproved forms. By the THEORY of natural selection all living species have
0282 n inconceivably great. But assuredly, if this THEORY be true, such have lived upon this earth. On
0296 fine intermediate gradations which must on my THEORY have existed between them, but abrupt, though
0297 nd the kind of evidence of change which on my THEORY we ought to find. Moreover, if we look to rat
0301 of those fine transitional forms, which on my THEORY assuredly have connected all the past and pre
0302 se of each formation, pressed so hardly on my THEORY. On the sudden appearance of whole groups of
0302 e all at once, the fact would be fatal to the THEORY of descent with slow modification through nat
0305 t it would be an insuperable difficulty on my THEORY, unless it could likewise be shown that the s
0306 much from living species; and it cannot on my THEORY be supposed, that these old species were the
0307 and improved descendants. Consequently, if my THEORY be true, it is indisputable that before the l
0307 st piles of fossiliferous strata, which on my THEORY no doubt were somewhere accumulated before th
0310 is volume, will undoubtedly at once reject my THEORY. For my part, following out Lyell's metaphor,
0314 tory. These several facts accord well with my THEORY. I believe in no fixed law of development, ca
0316 much underrated the rule strictly accords with my THEORY. For as all the species of the same group hav
0317 es of a group is strictly conformable with my THEORY; as the species of the same genus, and the ge
0317 e of species and of groups of species. On the THEORY of natural selection the extinction of old fo
0320 he died by some unknown deed of violence. The THEORY of natural selection is grounded on the belie
0322 species become extinct, accords well with the THEORY of natural selection. We need not marvel at e
0325 fe throughout the world, is explicable on the THEORY of natural selection. New species are formed
0331 several facts and inferences accord with the THEORY of descent with modification. As the subject
0333 seems frequently to be the case. Thus, on the THEORY of descent with modification, the main facts
0333 inexplicable on any other view. On this same THEORY, it is evident that the fauna of any great pe
0336 bitants of the sea have been affected. On the THEORY of descent, the full meaning of the fact of f
0337 cular sense the more recent forms must, on my THEORY, be higher than the more ancient; for each ne
0338 his doctrine of Agassiz accords well with the THEORY of natural selection. In a future chapter I s
0340 to the distribution of marine animals. On the THEORY of descent with modification, the great law o
0341 species. It must not be forgotten that, on my THEORY, all the species of the same genus have desce
0342 ological record, will rightly reject my whole THEORY. For he may ask in vain where are the numberl
0343 aeontology seem to me simply to follow on the THEORY of descent with modification through natural
0345 much more perfect, the main objections to the THEORY of natural selection are greatly diminished o
0350 o inquire what this bond is. This bond, on my THEORY, is simply inheritance, that cause which alon
0354 eral distinct species of a genus, which on my THEORY have all descended from a common progenitor,
0355 period immigrants from this other region, my THEORY will be strengthened; for we can clearly unde
0355 hereafter more fully see, inexplicable on the THEORY of independent creation. This view of the rel
0355 ntercross (if such exist), the species, on my THEORY, must have descended from a succession of imp
0356 ting the greatest amount of difficulty on the THEORY of single centres of creation, I must say a f
0372 res of the Polar Circle, will account, on the THEORY of modification, for many closely allied form
0372 h america and Europe, are inexplicable on the THEORY of creation. We cannot say that they have bee
0381 points of the southern hemisphere, is, on my THEORY of descent with modification, a far more rema
0385 wide range, and allied species, which, on my THEORY, are descended from a common parent and must
0390 rue. This fact might have been expected on my THEORY, for, as already explained, species occasiona
0393 exist on any oceanic island. But why, on the THEORY of creation, they should not have been create
0407 stinct species of the same genus, which on my THEORY must have spread from one parent source; if w
0408 ographical distribution are explicable on the THEORY of migration (generally of the more dominant
0410 th cases marked exceptions to the rule. On my THEORY these several relations throughout time and s
0411 ficulties in classification, explained on the THEORY of descent with modification. Classification
0426 erent habits, we may feel almost sure, on the THEORY of descent, that these characters have been i
0429 be the number of connecting forms which on my THEORY have been exterminated and utterly lost. And
0430 l and not merely adaptive, they are due on my THEORY to inheritance in common. Therefore we must s
0435 and plant. The explanation is manifest on the THEORY of the natural selection of successive slight
0437 e all constructed on the same pattern? On the THEORY of natural selection, we can satisfactorily a
0446 let us take a genus of birds, descended on my THEORY from some one parent species, and of which th
0459 on. Recapitulation of the difficulties on the THEORY of Natural Selection. Recapitulation of the g
0459 f in the immutability of species. How far the THEORY of natural selection may be extended. Effects
0459 grave objections may be advanced against the THEORY of descent with modification through natural
0460 admitted, cases of special difficulty on the THEORY of natural selection; and one of the most cur
0461 ribution, the difficulties encountered on the THEORY of descent with modification are grave enough
0462 e genus is in some degree lessened. As on the THEORY of natural selection an interminable number o
0463 many objections which may be urged against my THEORY. Why, again, do whole groups of allied specie
0463 lurian groups of fossils? For certainly on my THEORY such strata must somewhere have been deposite
0465 and they have changed in the manner which my THEORY requires, for they have changed slowly and in
0465 iculties which may justly be urged against my THEORY; and I have now briefly recapitulated the ans
0466 are, in my judgment they do not overthrow the THEORY of descent with modification. Now let us turn
0469 special facts and arguments in favour of the THEORY of natural selection, even if we looked no fu
0469 eings seems to me utterly inexplicable on the THEORY of creation. As natural selection acts solely
0471 nds to make more strictly correct, is on this THEORY simply intelligible. We can plainly see why n
0471 ts are, as it seems to me, explicable on this THEORY. How strange it is that a bird, under the for
0472 ther such cases. The wonder indeed is, on the THEORY of natural selection, that more cases of the
0473 ost characters occur. How inexplicable on the THEORY of creation is the occasional appearance of s
0474 an enormous period. It is inexplicable on the THEORY of creation why a part developed in a very un
0475 n such facts as the record gives, support the THEORY of descent with modification. New species hav

Page **********************************(Key Word)**********************************

Page **(Key Word)***

```
0476  of dispersal, then we can understand, on the  THEORY of descent with modification, most of the gre
0478  anic islands, are utterly inexplicable on the  THEORY of independent acts of creation. The existenc
0478  ive species in any two areas, implies, on the  THEORY of descent with modification, that the same p
0478  hat these facts receive no explanation on the  THEORY of creation. The fact, as we have seen, that
0478  ween recent groups, is intelligible on the     THEORY of natural selection with its contingencies o
0479  such facts, at once explain themselves on the  THEORY of descent with slow and slight successive mo
0482  tain number of facts will certainly reject my  THEORY. A few naturalists, endowed with much flexibi
0483  each other. Therefore I cannot doubt that the  THEORY of descent with modification embraces all the
0004  s of life, untouched and unexplained. It is,   THEREFORE, of the highest importance to gain a clear
0024  transported to all parts of the world, and,    THEREFORE, some of them must have been carried back a
0037  ast number of cases we cannot recognise, and   THEREFORE do not know, the wild parent stocks of the
0057  , they concur in this view. In this respect,   THEREFORE, the species of the larger genera resemble
0069  for the number of species of all kinds, and   THEREFORE of competitors, decreases northwards: hence
0088  ut few or no offspring. Sexual selection is,   THEREFORE, less rigorous than natural selection. Gene
0093  he weather had been cold and boisterous, and   THEREFORE not favourable to bees, nevertheless every
0122  natural economy of their country. It seems,    THEREFORE, to me extremely probable that they will ha
0128  ny small spot or at naturalised productions.   THEREFORE during the modification of the descendants
0135  eetles, they must be lost early in life, and   THEREFORE cannot be much used by these insects. In so
0144  nt structures, independently of utility, and  THEREFORE, of natural selection, than that of the dif
0149  seems to be owing to their uselessness, and   THEREFORE to natural selection having no power to che
0176  we may trust these facts and inferences, and  THEREFORE conclude that varieties linking two other v
0181  oved in structure in a corresponding manner.  THEREFORE, I can see no difficulty, more especially u
0192  mily: indeed, they graduate into each other.  THEREFORE I do not doubt that little folds of skin, w
0200  imals having such widely diversified habits.  THEREFORE we may infer that these several bones might
0211  ience to the aphides to have it removed: and  THEREFORE probably the aphides do not instinctively e
0236  y inheritance from a common parent, and must  THEREFORE believe that they have been acquired by ind
0242  er insects convinced me of the fact. I have,  THEREFORE, discussed this case, at some little but wh
0243  re of the highest importance to each animal.  THEREFORE I can see no difficulty, under changing con
0245  ot possibly be of any advantage to them, and  THEREFORE could not have been acquired by the continu
0264  e nature and constitution of its mother, and  THEREFORE before birth, as long as it is nourished wi
0273  that there has been recent variability; and   THEREFORE we might expect that such variability would
0279  defined organic forms, than on climate: and,  THEREFORE, that the really governing conditions of li
0299  diterranean, and from Britain to Russia; and  THEREFORE equals all the geological formations which
0321  anoid fishes still inhabit our fresh waters.  THEREFORE the utter extinction of a group is generall
0326  inhabitants of the continuous seas. We might  THEREFORE expect to find, as we apparently do find, a
0327  m inheriting some inferiority in common: and  THEREFORE as new and improved groups spread throughou
0331  d different places in the economy of nature.  THEREFORE it is quite possible, as we have seen in th
0336  ost simultaneously throughout the world, and  THEREFORE under the most different climates and condi
0360  from violent movement as in our experiments.  THEREFORE it would perhaps be safer to assume that th
0389  ies have descended from a single parent; and  THEREFORE have all proceeded from a common birthplace
0393  lty in their transportal across the sea, and  THEREFORE why they do not exist on any oceanic island
0414  d, are of paramount importance! We must not,  THEREFORE, in classifying, trust to resemblances in p
0425  nity of descent by resemblances of any kind.  THEREFORE we choose those characters which, as far as
0430  e due on my theory to inheritance in common.  THEREFORE we must suppose either that all Rodents, in
0430  ient progenitor than have other Rodents: and  THEREFORE it will not be specially related to any one
0437  served) of all low or little modified forms:  THEREFORE we may readily believe that the unknown pro
0449  the same or nearly similar parents, and are   THEREFORE in that degree closely related. Thus, commu
0461  must have descended from common parents: and  THEREFORE, in however distant and isolated parts of t
0474  e specifically distinct from each other; and  THEREFORE these same characters would be more likely
0474  al manner in any one species of a genus, and  THEREFORE, as we may naturally infer, of great import
0474  amount of variability and modification, and  THEREFORE we might expect this part generally to be s
0483  age the species closely resemble each other.  THEREFORE I cannot doubt that the theory of descent w
0484  strous growths on the wild rose or oak tree.  THEREFORE I should infer from analogy that probably a
0058  of plants (4th edition) 63 plants which are   THEREIN ranked as species, but which he considers as
0293  s more difficult to understand, why we do not  THEREIN find closely graduated varieties between the
0410  n an intermediate deposit the forms which are  THEREIN absent, but which occur above and below: so
0173  re age only in masses of sediment sufficiently  THICK and extensive to withstand an enormous amount
0188  instrument, we ought in imagination to take a  THICK layer of transparent tissue, with a nerve sen
0222  d with booty, for about forty yards, to a very  THICK clump of heath, whence I saw the last individ
0222  was not able to find the desolated nest in the  THICK heath. The nest, however, must have been clos
0228  parated two combs, and put between them a long  THICK, square strip of wax: the bees instantly bega
0228  cells. I then put into the hive, instead of a  THICK, square piece of wax, a thin and narrow, knif
0231  es of the pyramidal basis being about twice as  THICK. By this singular manner of building, strengt
0290  that sediment must be accumulated in extremely  THICK, solid, or extensive masses, in order to with
0290  during subsequent oscillations of level. Such  THICK and extensive accumulations of sediment may b
0291  e for life, and thus a fossiliferous formation  THICK enough, when upraised, to resist any amount o
0291  ly formations rich in fossils and sufficiently  THICK and extensive to resist subsequent degradatio
0292  ong as the bed of the sea remained stationary.  THICK deposits could not have been accumulated in t
0295  e the deposit will generally have to be a very  THICK one: and the species undergoing modification
0295  ghout this whole time. But we have seen that a  THICK fossiliferous formation can only be accumulat
0295  ed by more than one palaeontologist, that very  THICK deposits are usually barren of organic remain
0296  and Dawson found carboniferous beds 1400 feet  THICK in Nova Scotia, with ancient root bearing str
0342  for the accumulation of fossiliferous deposits  THICK enough to resist future degradation, enormous
0397  it perfectly recovered. As this species has a  THICK calcareous operculum, I removed it, and when
0435  any extent, and become gradually enveloped in  THICK membrane, so as to serve as a fin: or a webbe
0465  tervals of time: for fossiliferous formations,  THICK enough to resist future degradation, can be a
0423  on turnips together, though the esculent and   THICKENED stems are so similar. Whatever part is foun
0427  nd even in our domestic varieties, as in the   THICKENED stems of the common and swedish turnip. The
0317  sharp point, but abruptly: it then gradually  THICKENS upwards, sometimes keeping for a space of eq
0045  , or dwarfed plants on Alpine summits, or the  THICKER fur of an animal from far northwards, would
0133  urriers that animals of the same species have  THICKER and better fur the more severe the climate i
0231  the first rough wall from ten to twenty times  THICKER than the excessively thin finished wall of t
0216  nceal themselves in the surrounding grass or   THICKETS; and this is evidently done for the instinct
0067  uffer most from germinating in ground already  THICKLY stocked with other plants. Seedlings, also,
0072  , as soon as the land was enclosed, it became  THICKLY clothed with vigorously growing young firs.
0077  he closest relation to the land being already  THICKLY clothed by other plants; so that the seeds m
```

Page **(Key Word)***

Page ***************************************(Key Word)***************************************

0283 es of retreating cliffs rounded boulders, all THICKLY clothed by marine productions, showing how l
0011 inds of food and from light, and perhaps the THICKNESS of fur from climate. Habit also has a decid
0087 easily broken shells might be selected, the THICKNESS of the shell being known to vary like every
0226 ng cells are not double, but are of the same THICKNESS as the outer spherical portions, and yet ea
0231 only about one four hundredth of an inch in THICKNESS: the plates of the pyramidal basis being ab
0283 e beds of conglomerate many thousand feet in THICKNESS, which, though probably formed at a quicker
0283 m remember Lyell's profound remark, that the THICKNESS and extent of sedimentary formations are th
0284 s! professor Ramsay has given me the maximum THICKNESS, in most cases from actual measurement, in
0284 gland by thin beds, are thousands of feet in THICKNESS on the Continent. Moreover, between each su
0285 eozoic strata, in parts ten thousand feet in THICKNESS, as shown in Prof. Ramsay's masterly memoir
0286 he southern Downs is about 22 miles, and the THICKNESS of the several formations is on an average
0291 sted; or, sediment may be accumulated to any THICKNESS and extent over a shallow bottom, if it con
0291 ation of level, and thus gained considerable THICKNESS. All geological facts tell us plainly that
0296 es could be given of beds only a few feet in THICKNESS, representing formations, elsewhere thousan
0296 g formations, elsewhere thousands of feet in THICKNESS, and which must have required an enormous p
0300 ous formations could be formed of sufficient THICKNESS to last to an age, as distant in futurity a
0308 s; and from the amount of sediment, miles in THICKNESS, of which the formations are composed, we m
0317 be represented by a vertical line of varying THICKNESS, crossing the successive geological formati
0317 ards, sometimes keeping for a space of equal THICKNESS, and ultimately thins out in the upper beds
0318 ed, as before, by a vertical line of varying THICKNESS, the line is found to taper more gradually
0189 rate into layers of different densities and THICKNESSES, placed at different distances from each o
0434 e, the bones of the arm and forearm, or of the THIGH and leg, transposed. Hence the same names can
0130 in a fossil state. As we here and there see a THIN straggling branch springing from a fork low do
0228 ve, instead of a thick, square piece of wax, a THIN and narrow, knife edged ridge, coloured with v
0228 ame way as before; but the ridge of wax was so THIN, that the bottoms of the basins, if they had b
0229 lat bottoms; and these flat bottoms, formed by THIN little plates of the vermilion wax having been
0229 anes of intersection. Considering how flexible THIN wax is, I do not see that there is any difficu
0230 hat if the bees were to build for themselves a THIN wall of wax, they could make their cells of th
0231 n to twenty times thicker than the excessively THIN finished wall of the cell, which will ultimate
0231 oth sides near the ground, till a smooth, very THIN wall is left in the middle; the masons always
0231 the summit of the ridge. We shall thus have a THIN wall steadily growing upward: but always crown
0232 ntial rim of a growing comb, with an extremely THIN layer of melted vermilion wax: and I invariabl
0284 ormations, which are represented in England by THIN beds, are thousands of feet in thickness on th
0007 r the most different climates and treatment, I THINK we are driven to conclude that this greater v
0007 e been exposed under nature. There is, also, I THINK, some probability in the view propounded by A
0011 vertheless some slight amount of change may, I THINK, be attributed to the direct action of the co
0013 les alone. A much more important rule, which I THINK may be trusted, is that, at whatever period o
0016 cies of the same genus in a state of nature. I THINK this must be admitted, when we find that ther
0016 m each other in characters of generic value. I THINK it could be shown that this statement is hard
0017 ntly domesticated animals and plants, I do not THINK it is possible to come to any definite conclu
0018 ve existed in Egypt? The whole subject must, I THINK, remain vague; nevertheless, I may, without h
0018 from geographical and other considerations, I THINK it highly probable that our domestic dogs hav
0018 heep and goats I can form no opinion. I should THINK, from facts communicated to me by Mr. Blyth,
0022 that they were wild birds, would certainly, I THINK, be ranked by him as well defined species. Mo
0026 cy to sterility: from the history of the dog I THINK there is some probability in this hypothesis,
0029 er examples could be given. The explanation, I THINK, is simple: from long continued study they ar
0030 poses, or so beautiful in his eyes, we must, I THINK, look further than to mere variability. We ca
0036 sting in Britain, India, and Persia, we can, I THINK, clearly trace the stages through which they
0036 f there exist savages so barbarous as never to THINK of the inherited character of the offspring o
0038 ir habits to man's wants or fancies. We can, I THINK, further understand the frequently abnormal c
0040 ished as distinct at our poultry shows. We can, I THINK, these views further explain what has sometime
0065 rity with the larger domestic animals tends, I THINK, to mislead us: we see no great destruction f
0080 potent in the hands of man, apply to nature? I THINK we shall see that it can act most effectually
0085 e acquired, true and constant. Nor ought we to THINK that the occasional destruction of an animal
0097 belief that this is a law of nature, we can, I THINK, understand several large classes of facts, s
0110 r species. From these several considerations, I THINK it inevitably follows, that as new species in
0115 dvantage over the other natives; and we may, I THINK, at least safely infer that diversification o
0116 h ought to have been much amplified, we may, I THINK, assume that the modified descendants of any
0124 gy, have to refer again to this subject, and I THINK we shall then see that the diagram throws lig
0127 the several parts of their organisation, and I THINK this cannot be disputed; if there be, owing t
0127 ion, and habits, to be advantageous to them, I THINK it would be a most extraordinary fact if no v
0134 m the facts alluded to in the first chapter, I THINK there can be little doubt that use in our dom
0140 ound capable of far extended transportation, I THINK the common and extraordinary capacity in our
0142 ed for their own districts: the result must, I THINK, be due to habit. On the other hand, I can se
0142 gs appeared to be than others. On the whole, I THINK we may conclude that habit, use, and disuse,
0144 respect to this latter case of correlation, I THINK it can hardly be accidental, that if we pick
0147 re is forced to economise on the other side. I THINK this holds true to a certain extent with our
0151 splayed in any unusual manner, of which fact I THINK there can be little doubt. But that our rule
0152 ave been modified through natural selection, I THINK we can obtain some light. In our domestic ani
0156 subject, I will make only two other remarks. I THINK it will be admitted, without my entering on d
0156 condary sexual characters are very variable; I THINK it also will be admitted that species of the
0160 tion appearing in the several breeds. We may I THINK confidently come to this conclusion, because,
0171 y apparent, and those that are real are not, I THINK, fatal to my theory. These difficulties and o
0174 lty for a long time quite confounded me. But I THINK it can be in large part explained. In the fir
0176 ers than the forms which they connect, then, I THINK, we can understand why intermediate varieties
0180 icult; and I could have given no answer. Yet I THINK such difficulties have very little weight. H
0192 li; but they have large folded branchiae. Now I THINK no one will dispute that the ovigerous frena
0203 my theory. Many of them are very grave; but I THINK that in the discussion light has been thrown
0208 instinct with habit. This comparison gives, I THINK, a remarkably accurate notion of the frame of
0209 any habitual action to become inherited, and I THINK it can be shown that this does sometimes happ
0214 ng continued and compulsory habit, but this, I THINK, is not true. No one would ever have thought
0224 pears: all this beautiful work can be shown, I THINK, to follow from a few very simple instincts.
0233 ve profited the progenitors of the hive bee? I THINK the answer is not difficult: it is known that
0245 hat hybrids are very generally sterile, has, I THINK, been much underrated by some late writers. O
0248 perfect fertility ends and sterility begins, I THINK no better evidence of this can be required th
0250 artner. The difference in their results may, I THINK, be in part accounted for by Herbert's great

Page ***************************************(Key Word)***************************************

```
0252  e of plants; but the hybrids themselves are, I THINK, more sterile. I doubt whether any case of a
0254  from facts communicated to me by Mr. Blyth, I THINK they must be considered as distinct species.
0260  prevent their becoming confounded in nature? I THINK not. For why should the sterility be so extre
0266  an old and almost universal belief, founded, I THINK, on a considerable body of evidence, that sli
0271  ese several considerations and facts, I do not THINK that the very general fertility of varieties
0274  crossed with another variety. For instance, I THINK those authors are right, who maintain that th
0276  productive systems. There is no more reason to THINK that species have been specially endowed with
0276  them crossing and blending in nature, than to THINK that trees have been specially endowed with v
0290  elapsed between each formation. But we can, I THINK, see why the geological formations of each re
0290  grinding action of the coast waves. We may, I THINK, safely conclude that sediment must be accumu
0310  th others, we owe all our knowledge. Those who THINK the natural geological record in any degree p
0318  so have species a definite duration. No one I THINK can have marvelled more at the extinction of
0324  oking to a remotely future epoch, there can, I THINK, be little doubt that all the more modern mar
0326  ion of new and dominant species; but we can, I THINK, clearly see that a number of individuals, fr
0361  s to vast distances across the ocean. We may I THINK safely assume that under such circumstances t
0364  ies and tens of thousands of years, it would I THINK be a marvellous fact if many plants had not t
0372  w living in areas completely sundered. Thus, I THINK, we can understand the presence of many exist
0383  f ranging widely, though so unexpected, can, I THINK, in most cases be explained by their having b
0386  re, as if in consequence, a very wide range. I THINK favourable means of dispersal explain this fa
0387  in a breakfast cup! Considering these facts, I THINK it would be an inexplicable circumstance if w
0389  remove many difficulties, but it would not, I THINK, explain all the facts in regard to insular p
0389  cape of Good Hope or in Australia, we must, I THINK, admit that something quite independently of v
0400  tant for its inhabitants; whereas it cannot, I THINK, be disputed that the nature of the other inh
0402  one island to the others. But we often take, I THINK, an erroneous view of the probability of clos
0403  ndigenous species. From these considerations I THINK we need not greatly marvel at the endemic and
0406  island to those of the nearest mainland, are I THINK, utterly inexplicable on the ordinary view of
0407  e become extinct in the intermediate tracts, I THINK the difficulties in believing that all the in
0407  ing thus granted for their migration, I do not THINK that the difficulties are insuperable; though
0408  s, have proceeded from some one source; then I THINK all the grand leading facts of geographical d
0413  system are indisputable. But many naturalists THINK that something more is meant by the Natural S
0416  this external and trifling character would, I THINK, have been considered by naturalists as impor
0417  ven when none are important, alone explains, I THINK, that saying of Linnaeus, that the characters
0422  d to be included, such an arrangement would, I THINK, be the only possible one. Yet it might be th
0424  the Hottentot had descended from the Negro, I THINK he would be classed under the Negro group, ho
0426  s in our classification. We shall hereafter, I THINK, clearly see why embryological characters are
0429  fall to the lot of aberrant genera. We can, I THINK, account for this fact only by looking at abe
0445  n animals; and this surprised me greatly, as I THINK it probable that the difference between these
0459  nct. The truth of these propositions cannot, I THINK, be disputed. It is, no doubt, extremely diff
0469  dmitted the existence of varieties, which they THINK sufficiently distinct to be worthy of record
0482  lan of creation, unity of design, etc., and to THINK that we give an explanation when we only rest
0002  aving read my sketch of 1844, honoured me by THINKING it advisable to publish, with Mr. Wallace's
0119  tself made somewhat irregular. I am far from THINKING that the most divergent varieties will invar
0338  nt forms. I must follow Pictet and Huxley in THINKING that the truth of this doctrine is very far
0424  abit; nevertheless, without any reasoning or THINKING on the subject, these tumblers are kept in t
0018  hould value more than that of almost any one, THINKS that all the breeds of poultry have proceeded
0053  ad my manuscript, and examined the tables, he THINKS that the following statements are fairly well
0375  heights of the peninsula of Malacca, and are THINLY scattered, on the one hand over India and on
0286  ecimentary deposits might have accumulated in THINNER masses than elsewhere, the above estimate wo
0296  ted the vast lapse of time represented by the THINNER formation. Many cases could be given of the
0229  they have gnawed the wax away to the proper THINNESS, and then stopping their work. In ordinary c
0229  was absolutely impossible, from the extreme THINNESS of the little rhombic plate, that they could
0317  for a space of equal thickness, and ultimately THINS out in the upper beds, marking the decrease a
0002  he Linnean Society, and it is published in the THIRD volume of the Journal of that Society. Sir C.
0159  two so called distinct species; and to these a THIRD may be added, namely, the common turnip. Acco
0176  adapt two varieties to two large areas, and a THIRD variety to a narrow intermediate zone. The in
0177  to a comparatively narrow, hilly tract; and a THIRD to wide plains at the base; and that the inha
0208  put it into a hammock completed up only to the THIRD stage, the caterpillar simply re performed th
0208  out of a hammock made up, for instance, to the THIRD stage, and were put into one finished up to t
0208  e its hammock, seemed forced to start from the THIRD stage, where it had left off, and thus tried
0249  w let us turn to the results arrived at by the THIRD most experienced hybridiser, namely, the Hon.
0274  osely allied to each other, are crossed with a THIRD species, the hybrids are widely different fro
0297  on. Supposing B and C to be two species, and a THIRD, A, to be found in an underlying bed; even if
0297  etween B and C, it would simply be ranked as a THIRD and distinct species, unless at the same time
0331  ied family or sub family; and o14, e14, m14, a THIRD family. These three families, together with t
0349  ost exactly opposite meridians of longitude. A THIRD great fact, partly included in the foregoing
0467  results of naturalisation, as explained in the THIRD chapter. More individuals are born than can p
0477  o areas will have received colonists from some THIRD source or from each other, at various periods
0005  of instinct, or the mental powers of animals; THIRDLY, Hybridism, or the infertility of species an
0027  ect series between the extremes of structure. THIRDLY, those characters which are mainly distincti
0172  t fully understand the inimitable perfection? THIRDLY, can instincts be acquired and modified thro
0178  they will no longer exist in a living state. THIRDLY, when two or more varieties have been formed
0354  distribution of fresh water productions; and THIRDLY, the occurrence of the same terrestrial spec
0074  s of humble bees, believes that more than two THIRDS of them are thus destroyed all over England.
0018  ed pottery existed in the valley of the Nile THIRTEEN or fourteen thousand years ago, and who will
0270  never naturally crossed. He then fertilised THIRTEEN flowers of the one with the pollen of the ot
0284  ...........72,584 Feet; that is, very nearly THIRTEEN and three quarters British miles. Some of th
0005  hical distribution throughout space; in the THIRTEENTH, their classification or mutual affinities
0021  nt coo from the other breeds. The fantail has THIRTY or even forty tail feathers, instead of twelv
0032  re compared with drawings made only twenty or THIRTY years ago. When a race of plants is once pret
0042  wberry which have been raised during the last THIRTY or forty years. In the case of animals with s
0064  under the mark to assume that it breeds when THIRTY years old, and goes on breeding till ninety y
0072  distant from one of the old clumps, I counted THIRTY two little trees; and one of them, judging fr
0219  less are the masters, that when Huber shut up THIRTY of them without a slave, but with plenty of t
0380  centuries from La Plata, and during the last THIRTY or forty years from Australia. Something of t
0363  at least possible that they may have brought THITHER the seeds of northern plants. Considering th
0474  fficulty than does corporeal structure on the THOERY of the natural selection of successive, sligh
```

Page **(Key Word)**

```
0236  females in structure, as in the shape of the  THORAX  and in being destitute of wings and sometimes
0024  hat at least seven or eight species were so     THOROUGHLY  domesticated in ancient times by half civil
0024  t only that half civilized man succeeded in     THOROUGHLY  domesticating several species, but that he
0025  in every one of the domestic breeds, taking     THOROUGHLY  well bred birds, all the above marks, even
0062  r this conclusion in mind. Yet unless it be     THOROUGHLY  engrained in the mind, I am convinced that
0252  fertile hybrid animal can be considered as      THOROUGHLY  well authenticated. It should, however, be
0253  generations. Although I do not know of any      THOROUGHLY  well authenticated cases of perfectly ferti
0337  e a multitude of British forms would become     THOROUGHLY  naturalized there, and would exterminate ma
0480  e chief facts and considerations which have     THOROUGHLY  convinced me that species have changed, and
0002  accuracy. No doubt errors will have crept in,   THOUGH  I hope I have always been cautious in trustin
0004  nvariably found that our knowledge, imperfect   THOUGH  it be, of variation under domestication, affo
0008  many animals there are which will not breed,    THOUGH  living long under not very close confinement
0009  hand, we see domesticated animals and plants,   THOUGH  often weak and sickly, yet breeding quite fre
0009  when, on the other hand, we see individuals,    THOUGH  taken young from a state of nature, perfectly
0010  ometimes differ considerably from each other,   THOUGH  both the young and the parents, as Muller has
0010  cies have produced very little direct effect,   THOUGH  apparently more in the case of plants. Under
0013  pear in the offspring at a corresponding age,   THOUGH  sometimes earlier. In many cases this could n
0014  long horned bull, the greater length of horn,   THOUGH  appearing late in life, is clearly due to the
0024  y; but not one has ever become wild or feral,   THOUGH  the dovecot pigeon, which is the rock pigeon
0024  r cases, is, that the above specified breeds,   THOUGH  agreeing generally in constitution, habits, v
0026  applied to species closely related together,   THOUGH  it is unsupported by a single experiment. But
0027  pects, as compared with all other Columbidae,  THOUGH  so like in most other respects to the rock pi
0029  he differences between the several races: and  THOUGH  they well know that each race varies slightly
0035  and gradually, and yet so effectually, that,   THOUGH  the old Spanish pointer certainly came from S
0037  melting pear from the seed of the wild pear,   THOUGH  he might succeed from a poor seedling growing
0037  if it had come from a garden stock. The pear,  THOUGH  cultivated in classical times, appears, from
0037  er thought what splendid fruit we should eat;  THOUGH  we owe our excellent fruit, in some small deg
0042  ancier, for thus many races may be kept true,  THOUGH  mingled in the same aviary, and this circumst
0044  ity of descent is almost universally implied,  THOUGH  it can rarely be proved. We have also what ar
0051  sage. Hence I look at individual differences,  THOUGH  of small interest to the systematist, as of h
0054  be the most likely to yield offspring which,   THOUGH  in some slight degree modified, will still in
0060  bility and of some few well marked varieties,  THOUGH  necessary as the foundation for the work, hel
0062  of prey: we do not always bear in mind, that   THOUGH  food may be now superabundant, it is not so a
0062  aid to struggle for life against the drought,  THOUGH  more properly it should be said to be depende
0068  rous plants gradually kill the less vigorous,  THOUGH  fully grown, plants: thus out of twenty speci
0070  birds which feed on them; nor can the birds,   THOUGH  having a super abundance of food at this one
0071  y. i will give only a single instance, which,  THOUGH  a simple one, has interested me. In Staffords
0072  attle nor horses nor dogs have ever run wild,  THOUGH  they swarm southward and northward in a feral
0073  ure remains uniform for long periods of time,  THOUGH  assuredly the merest trifle would often give
0076  d. as species of the same genus have usually,  THOUGH  by no means invariably, some similarity in ha
0078  ced in a new country amongst new competitors,  THOUGH  the climate may be exactly the same as in its
0087  vantage, for the good of another species: and  THOUGH  statements to this effect may be found in wor
0088  f carnivorous animals are already well armed:  THOUGH  to them and to others, special means of defen
0092  of the leaf of the common laurel. This juice,  THOUGH  small in quantity, is greedily sought by inse
0096  here treat the subject with extreme brevity,   THOUGH  I have the materials prepared for an ample di
0097  is a general law of nature (utterly ignorant   THOUGH  we be of the meaning of the law) that no orga
0098  my garden, by insects, it never sets a seed,   THOUGH  by placing pollen from one flower on the stig
0098  bees, seeds freely. In very many other cases,  THOUGH  there be no special mechanical contrivance to
0099  len and stigmatic surface of the same flower,  THOUGH  placed so close together, as if for the very
0101  nce, elsewhere to prove that two individuals,  THOUGH  both are self fertilising hermaphrodites, do
0101  f the same family and even of the same genus,  THOUGH  agreeing closely with each other in almost th
0102  e, an extremely important element of success.  THOUGH  nature grants vast periods of time for the wo
0102  e individuals varying in the right direction,  THOUGH  in different degrees, so as better to fill up
0106  all and isolated area. Moreover, great areas,  THOUGH  now continuous, owing to oscillations of leve
0109  inhabitants of this world have changed. Slow   THOUGH  the process of selection may be, if feeble ma
0111  place, varieties, even strongly marked ones,  THOUGH  having somewhat of the character of species,
0118  o regularly as is represented in the diagram,  THOUGH  in itself made somewhat irregular. I am far f
0124  t on the affinities of extinct beings, which,  THOUGH  generally belonging to the same orders, or fa
0132  hore have their leaves in some degree fleshy,  THOUGH  not elsewhere fleshy. Several other such case
0137  the crabs the foot stalk for the eye remains,  THOUGH  the eye is gone; the stand for the telescope
0137  s gone; the stand for the telescope is there,  THOUGH  the telescope with its glasses has been lost.
0137  ost. As it is difficult to imagine that eyes,  THOUGH  useless, could be in any way injurious to ani
0144  the hair and teeth in the naked Turkish dog,   THOUGH  here probably homology comes into play? With
0148  s by the parasitic habits of the Proteolepas,  THOUGH  effected by slow steps, would be a decided ad
0155  ly of specific value, often becomes variable,  THOUGH  its physiological importance may remain the s
0162  breeds of diverse colours are crossed,        THOUGH  under nature it must generally be left doubtf
0163  legs and faintly on the shoulder. The quagga,  THOUGH  so plainly barred like a zebra over the body,
0165  om the ass and the hemionus; and this hybrid,  THOUGH  the ass seldom has stripes on its legs and th
0174  merous transitional varieties in each region,  THOUGH  they must have existed there, and may be embe
0174  ave been formed on strictly continuous areas;  THOUGH  I do not doubt that the formerly broken condi
0185  said to have become rudimentary in function,  THOUGH  not in structure. In the frigate bird, the de
0187  lex eye could be formed by natural selection;  THOUGH  insuperable by our imagination, can hardly be
0188  his reason ought to conquer his imagination;  THOUGH  I have felt the difficulty far too keenly to
0195  ls. I have sometimes felt as much difficulty,  THOUGH  of a very different kind, on this head, as in
0196  on, or as an aid in turning, as with the dog,  THOUGH  the aid must be slight, for the hare, with ha
0199  ause, may reappear from the law of reversion,  THOUGH  now of no direct use. The effects of sexual s
0199  simply due to inheritance: and consequently,  THOUGH  each being assuredly is well fitted for its p
0200  exclusively for the good of another species;  THOUGH  throughout nature one species incessantly tak
0202  ivances in nature, this same reason tells us,  THOUGH  we may easily err on both sides, that some ot
0202  il all the requirements of natural selection,  THOUGH  it may cause the death of some few members. I
0203  munity; and maternal love or maternal hatred,  THOUGH  the latter fortunately is most rare, is all t
0205  uatic animal by its terrestrial descendants),  THOUGH  it has become of such small importance that i
0205  for the exclusive good or injury of another;  THOUGH  it may well produce parts, organs; and excret
0212  ve quality, as may be seen in nestling birds,  THOUGH  it is strenghthened by experience; and by the
0216  is so plainly instinctive in young pheasants,  THOUGH  reared under a hen. It is not that chickens h
0219  s do no work. The workers or sterile females,  THOUGH  most energetic and courageous in capturing sl
```

Page **(Key Word)**

Page **(Key Word)**

```
0220  and Hampshire, and has never seen the slaves,   THOUGH  present in large numbers in August, either le
0222  ments of the nest. This species is sometimes,   THOUGH  rarely, made into slaves, as has been describ
0223  th over its ravaged home. Such are the facts,   THOUGH  they did not need confirmation by me, in reca
0224  ficult to make cells of wax of the true form,   THOUGH  this is perfectly effected by a crowd of bees
0237  ulated by natural selection. This difficulty,   THOUGH  appearing insuperable, is lessened, or, as I
0240  rger workers have simple eyes (ocelli), which   THOUGH  small can be plainly distinguished, whereas t
0240  portionally lesser size; and I fully believe,   THOUGH  I dare not assert so positively, that the wor
0240  of myrmica have not even rudiments of ocelli,  THOUGH  the male and female ants of this genus have w
0241  teeth. But the important fact for us is, that  THOUGH  the workers can be grouped into castes of dif
0246  the male element in both plants and animals:   THOUGH  the organs themselves are perfect in structur
0248  erility of hybrids in successive generations:  THOUGH  Gartner was enabled to rear some hybrids, car
0249  tilised; so that a cross between two flowers,  THOUGH  probably on the same plant, would be thus eff
0250  eld seed to the pollen of a distinct species,  THOUGH  quite sterile with their own pollen, notwiths
0251  the practical experiments of horticulturists,  THOUGH  not made with scientific precision, deserve s
0258  that hybrids raised from reciprocal crosses,   THOUGH  of course compounded of the very same two spe
0259  osely resemble one of them; and such hybrids,  THOUGH  externally so like one of their pure parent s
0262  n the currant, whereas the currant will take,  THOUGH  with difficulty, on the gooseberry. We have s
0263  on the stigma of a distantly allied species,  THOUGH  the pollen tubes protrude, they do not penetr
0265  ctive systems having been specially affected,  THOUGH  in a lesser degree than when sterility ensues
0265  , the conditions of life have been disturbed,  THOUGH  often in so slight a degree as to be inapprec
0266  e need not be surprised that their sterility,  THOUGH  in some degree variable, rarely diminishes. I
0268  we reflect how many species there are, which,  THOUGH  resembling each other most closely, are utter
0277  hybrid offspring should generally correspond,  THOUGH  due to distinct causes: for both depend on th
0277  , and the capacity of being grafted together,  THOUGH  this latter capacity evidently depends on wid
0283  erate many thousand feet in thickness, which,  THOUGH  probably formed at a quicker rate than many o
0285  ell known one of the denudation of the Weald.  THOUGH  it must be admitted that the denudation of th
0286  is rate excepting on the most exposed coasts:  THOUGH  no doubt the degradation of a lofty cliff wou
0290  nt or tertiary remains can anywhere be found,  THOUGH  the supply of sediment must for ages have bee
0292  ipelago), and consequently during subsidence,  THOUGH  there will be much extinction, fewer new vari
0296  theory have existed between them, but abrupt,  THOUGH  perhaps very slight, changes of form. It is a
0297  formation, we find that the embedded fossils,  THOUGH  almost universally ranked as specifically dif
0299  e in the highest degree. Geological research,  THOUGH  it has added numerous species to existing and
0301  from their former state, in a nearly uniform,  THOUGH  perhaps extremely slight degree, they would,
0312  ore gradual. In some of the most recent beds,  THOUGH  undoubtedly of high antiquity if measured by
0313  change more quickly than those that are low:  THOUGH  there are exceptions to this rule. The amount
0315  ife, organic and inorganic, should recur. For  THOUGH  the offspring of one species might be adapted
0316  so few, that E. Forbes, Pictet, and Woodward  (THOUGH  all strongly opposed to such views as I maint
0319  professor Owen soon perceived that the tooth,  (THOUGH  so like that of the existing horse, belonged
0327  whole groups would tend slowly to disappear:  THOUGH  here and there a single member might long be
0330  is every day taught that palaeozoic animals,  THOUGH  belonging to the same orders, families, or ge
0330  number of characters, so that the two groups,  THOUGH  formerly quite distinct, at that period made
0335  om the several stages of the chalk formation,  THOUGH  the species are distinct in each stage. This
0336  remains from closely consecutive formations,  THOUGH  ranked as distinct species, being closely rel
0338  strictly belong to their own proper classes,  THOUGH  some of these old forms are in a slight degre
0340  ext succeeding period of time, closely allied  THOUGH  in some degree modified descendants. If the i
0348  large rivers, we find different productions:  THOUGH  as mountain chains, deserts, etc., are not as
0349  the productions of the same continent or sea,  THOUGH  the species themselves are distinct at differ
0349  fer in geological structure, the inhabitants,  THOUGH  they may be all peculiar species, are essenti
0351  , that the several species of the same genus,  THOUGH  inhabiting the most distant quarters of the w
0352  ous that the individuals of the same species,  THOUGH  now inhabiting distant and isolated regions,
0354  trial species on islands and on the mainland,  THOUGH  separated by hundreds of miles of open sea. I
0355  time a few colonists; and their descendants,  THOUGH  modified, would still be plainly related by i
0364  er bring seeds from North America to Britain,  THOUGH  they might and do bring seeds from the West I
0365  ther continent; that a poorly stocked island,  THOUGH  standing more remote from the mainland, would
0369  f the several great European mountain ranges,  THOUGH  very many of the species are identically the
0370  emained in nearly the same relative position,  THOUGH  subjected to large, but partial oscillations
0372  d worlds, we find very few identical species  (THOUGH  Asa Gray has lately shown that more plants ar
0376  they are much oftener specifically distinct,  THOUGH  related to each other in a most remarkable ma
0376  ; but some are certainly identical, and many,  THOUGH  closely related to northern forms, must be ra
0377  tions, after migrating nearer to the equator,  THOUGH  they will have been placed under somewhat new
0379  profited them. Thus many of these wanderers,  THOUGH  still plainly related by inheritance to their
0380  ave become naturalised in any part of Europe,  THOUGH  hides, wool, and other objects likely to carr
0382  he tide leaves its drift in horizontal lines,  THOUGH  rising higher on the shores where the tide ri
0383  in fresh water productions of ranging widely,  THOUGH  so unexpected, can, I think, in most cases be
0385  ut of the water they could not be jarred off,  THOUGH  at a somewhat more advanced age they would vo
0385  tarily drop off. These just hatched molluscs,  THOUGH  aquatic in their nature, survived on the duck
0386  ave before mentioned that earth occasionally,  THOUGH  rarely, adheres in some quantity to the feet
0387  hat fresh water fish eat some kinds of seeds,  THOUGH  they reject many other kinds after having swa
0391  of sea shell is confined to its shores: now,  THOUGH  we do not know how sea shells are dispersed,
0392  ant oceanic islands; and an herbaceous plant,  THOUGH  it would have no chance of successfully compe
0394  a quicker rate than other and lower animals.  THOUGH  terrestrial mammals do not occur on oceanic i
0399  explained. Thus the plants of Kerguelen Land,  THOUGH  standing nearer to Africa than to America, ar
0399  partially stocked from a nearly intermediate  THOUGH  distant point, namely from the antarctic isla
0399  t of the Glacial period. The affinity, which,  THOUGH  feeble, I am assured by Dr. Hooker is real, b
0400  ich causes the inhabitants of an archipelago,  THOUGH  specifically distinct, to be closely allied t
0400  that the inhabitants of each separate island,  THOUGH  mostly distinct, are related in an incomparab
0400  grants should have been differently modified,  THOUGH  only in a small degree. This long appeared to
0401  spread to the other islands. But the islands,  THOUGH  in sight of each other, are separated by deep
0402  alapagos Archipelago, many even of the birds,  THOUGH  so well adapted for flying from island to isl
0404  ardly doubt that this rule is generally true,  THOUGH  it would be difficult to prove it. Amongst ma
0405  genus having descended from a single parent,  THOUGH  now distributed to the most remote points of
0406  e on the surrounding low lands and dry lands,  THOUGH  these stations are so different, the very clo
0407  think that the difficulties are insuperable:  THOUGH  they often are in this case, and in that of t
0409  ee why all the inhabitants of an archipelago,  THOUGH  specifically distinct on the several islets,
0412  of descent. These five genera have also much,  THOUGH  less, in common; and they form a family disti
0414  sh, as of any importance. These resemblances,  THOUGH  so intimately connected with the whole life o
```

Page **(Key Word)**

Page **************************************(Key Word)**************************************

```
0415  s frequently of more than generic importance,   THOUGH  here even when all taken together they appear
0418  ters are always found correlated with others,   THOUGH  no apparent bond of connexion can be discover
0419  eographical distribution has often been used,   THOUGH  perhaps not quite logically, in classificatio
0420  difference in the several branches or groups,   THOUGH  allied in the same degree in blood to their c
0421  they yet form a single genus. But this genus,   THOUGH  much isolated, will still occupy its proper i
0423  e apple together, merely because their fruit,   THOUGH  the most important part, happens to be nearly
0423  puts the swedish and common turnips together,   THOUGH  the esculent and thickened stems are so simil
0423  reatest number of points. In tumbler pigeons,   THOUGH  some sub varieties differ from the others in
0425  together the individuals of the same species,   THOUGH  the males and females and larvae are sometime
0425  under genera, and genera under higher groups,   THOUGH  in these cases the modification has been grea
0432  ived on this earth were suddenly to reappear,   THOUGH  it would be quite impossible to give definiti
0432  in a tree we can specify this or that branch,   THOUGH  at the actual fork the two unite and blend to
0437  tamens, and pistils in any individual flower,   THOUGH  fitted for such widely different purposes, be
0438  inheritance. In the great class of molluscs,   THOUGH  we can homologise the parts of one species wi
0439  sed in the whelp of the lion. We occasionally   THOUGH  rarely see something of this kind in plants:
0440  parents. In most cases, however, the larvae,   THOUGH  active, still obey more or less closely the l
0441  rises in organisation: I use this expression,   THOUGH  I am aware that it is hardly possible to defi
0444  gs, maintain that the greyhound and bull dog,   THOUGH  appearing so different, are really varieties
0445  of these several breeds were placed in a row,   THOUGH  most of them could be distinguished from each
0446  s and the above two principles, which latter,   THOUGH  not proved true, can be shown to be in some c
0450  , at a very early period in the life of each,   THOUGH  perhaps caused at the earliest, and being inh
0457  of a class of the homologous parts or organs,   THOUGH  fitted in the adult members for purposes as d
0459  ave been perfected, not by means superior to,   THOUGH  analogous with, human reason, but by the accu
0459  ual possessor. Nevertheless, this difficulty,   THOUGH  appearing to our imagination insuperably grea
0463  in, do whole groups of allied species appear,   THOUGH  certainly they often falsely appear, to have
0468  r nature is a strictly limited quantity. Man,   THOUGH  acting on external characters alone and often
0471  lainly see why nature is prodigal in variety,   THOUGH  niggard in innovation. But why this should be
0477  in the northern and southern temperate zones,   THOUGH  separated by the whole intertropical ocean. A
0478  for classification: why adaptive characters,   THOUGH  of paramount importance to the being, are of
0479  hy characters derived from rudimentary parts,   THOUGH  of no service to the being, are often of high
0479  rity of pattern in the wing and leg of a bat,   THOUGH  used for such different purpose, in the jaws
0037  vated the best pear they could procure, never   THOUGHT  what splendid fruit we should eat; though we
0039  one of the points of the breed. Nor let it be   THOUGHT  that some great deviation of structure would
0043  with all organic beings, as some authors have   THOUGHT.  The effects of variability are modified by
0051  p towards such slight varieties as are barely   THOUGHT  worth recording in works on natural history.
0053  sake. Guided by theoretical considerations, I   THOUGHT  that some interesting results might be obtai
0080  physical conditions of life. Can it, then, be   THOUGHT  improbable, seeing that variations useful to
0091  man trying to keep the best dogs without any   THOUGHT  to modifying the breed. Even without any cha
0091  pursue certain kinds of prey. Nor can this be   THOUGHT  very improbable; for we often observe great
0117  fairly well marked variety, such as would be   THOUGHT  worthy of record in a systematic work. The i
0137  s are of immense size; and Professor Silliman  THOUGHT  that it regained, after living some days in
0145  rs most frequently differ. It might have been  THOUGHT  that the development of the ray petals by dr
0146  dants with diverse habits, would naturally be  THOUGHT  to be correlated in some necessary manner. S
0165  much striped that any one at first would have  THOUGHT  that it must have been the product of a zebr
0197  nd pied kinds, I dare say that we should have  THOUGHT  that the green colour was a beautiful adapta
0207  worked into the previous chapters; but I have  THOUGHT  that it would be more convenient to treat th
0208  d to go back to recover the habitual train of  THOUGHT:  so P. Huber found it was with a caterpillar
0214  I think, is not true. No one would ever have   THOUGHT  of teaching or probably could have taught, t
0214  it may be doubted whether any one would have   THOUGHT  of training a dog to point, had not some one
0220  months, the slaves are very few in number, I   THOUGHT  that they might behave differently when more
0239  ean ants guard or imprison. It will indeed be  THOUGHT  that I have an overweening confidence in the
0304  s. this was a sore trouble to me, adding as I  THOUGHT  one more instance of the abrupt appearance o
0383  other by barriers of land, it might have been  THOUGHT  that fresh water productions would not have
0387  alph. de Candolle's remarks on this plant, I   THOUGHT  that its distribution must remain quite inex
0394  ing to other classes; and on continents it is  THOUGHT  that mammals appear and disappear at a quick
0414  orms most like each other. It might have been  THOUGHT  (and was in ancient times thought) that thos
0414  t have been thought (and was in ancient times  THOUGHT)  that those parts of the structure which det
0453  merely a restatement of the fact. Would it be  THOUGHT  sufficient to say that because planets revol
0467  s most useful to him at the time, without any  THOUGHT  of altering the breed. It is certain that he
0481  dable as long as the history of the world was  THOUGHT  to be of short duration; and now that we hav
0482  e of forms, which till lately they themselves  THOUGHT  were special creations, and which are still
0485  edged to be merely varieties may hereafter be  THOUGHT  worthy of specific names, as with the primro
0262  . something analogous occurs in grafting; for  THOUIN  found that three species of Robinia, which se
0039  iven rise to any marked varieties; hence the   THOULOUSE  and the common breed, which differ only in
0018  of our breeds originated there, four or five   THOUSAND  years ago? But Mr. Horner's researches have
0018  the valley of the Nile thirteen or fourteen   THOUSAND  years ago, and who will pretend to say how l
0032  ly attempted to appreciate. Not one man in a   THOUSAND  has accuracy of eye and judgment sufficient
0063  moisture. A plant which annually produces a   THOUSAND  seeds, of which on an average only one comes
0064  wenty five years, and at this rate, in a few   THOUSAND  years, there would literally not be standing
0065  which annually produce eggs or seeds by the   THOUSAND,  and those which produce extremely few, is,
0066  r of a tree, which lived on an average for a   THOUSAND  years, if a single seed were produced once i
0066  rs, if a single seed were produced once in a   THOUSAND  years, supposing that this seed were never d
0067  be compared to a yielding surface, with ten   THOUSAND  sharp wedges packed close together and drive
0075  s, each annually scattering its seeds by the   THOUSAND:  what war between insect and insect, between
0117  l lines in the diagram, may represent each a   THOUSAND  generations; it would have been better i
0117  have been better if each had represented ten  THOUSAND  generations. After a thousand generations, s
0117  epresented ten thousand generations. After a  THOUSAND  generations, species (A) is supposed to have
0118  will generally be preserved during the next  THOUSAND  generations. And after this interval, variet
0118  h of time; some of the varieties, after each  THOUSAND  generations, producing only a single variety
0120  n off any fresh branches or races. After ten  THOUSAND  generations, species (A) is supposed to have
0120  has produced, by analogous steps, after ten  THOUSAND  generations, either two well marked varietie
0121  between the horizontal lines. After fourteen  THOUSAND  generations, six new species, marked by the
0124  ne has hitherto been supposed to represent a  THOUSAND  generations, but each may represent a millio
0283  et any one examine beds of conglomerate many  THOUSAND  feet in thickness; which, though probably fo
0284  er at the rate of only 600 feet in a hundred  THOUSAND  years. This estimate may be quite erroneous;
0285  ound into perpendicular cliffs of one or two  THOUSAND  feet in height; for the gentle slope of the
```

Page **************************************(Key Word)**************************************

Page ************************************(Key Word)************************************

```
0285 asses of our palaeozoic strata, in parts ten THOUSAND feet in thickness, as shown in Prof. Ramsay'
0378 now felt there at the height of six or seven THOUSAND feet. During this the coldest period, I supp
0027 many people. They have been domesticated for THOUSANDS of years in several quarters of the world:
0037 itchen gardens. If it has taken centuries or THOUSANDS of years to improve or modify most of our p
0065 truction falling on them, and we forget that THOUSANDS are annually slaughtered for food, and that
0068 s game than at present, although hundreds of THOUSANDS of game animals are now annually killed. On
0080 ife, should sometimes occur in the course of THOUSANDS of generations? If such do occur, can we do
0089 oubt that female birds, by selecting, during THOUSANDS of generations, the most melodious or beaut
0114 y, i cannot doubt that in the course of many THOUSANDS of generations, the most distinct varieties
0136 ut combined probably with disuse. For during THOUSANDS of successive generations each individual b
0146 me one modification in structure, and, after THOUSANDS of generations, some other and independent
0167 r myself, I venture confidently to look back THOUSANDS on thousands of generations, and I see an a
0167 enture confidently to look back thousands on THOUSANDS of generations, and I see an animal striped
0202 re the production for this single purpose of THOUSANDS of drones, which are utterly useless to the
0284 are represented in England by thin beds, are THOUSANDS of feet in thickness on the Continent. More
0285 down on the other, to the height or depth of THOUSANDS of feet; for since the crust cracked, the s
0296 hickness, representing formations, elsewhere THOUSANDS of feet in thickness, and which must have r
0301 ine inhabitants of the archipelago now range THOUSANDS of miles beyond its confines; and analogy l
0363 n year after year, for centuries and tens of THOUSANDS of years, it would I think be a marvellous
0118 am the process is represented up to the ten THOUSANDTH generation; and under a condensed and simpl
0118 nsed and simplified form up to the fourteen THOUSANDTH generation. But I must here remark that I d
0122 endants, fourteen in number at the fourteen THOUSANDTH generation, will probably have inherited so
0122 have transmitted offspring to the fourteen THOUSANDTH generation. We may suppose that only one (F
0324 ed that this expression relates to the same THOUSANDTH or hundred thousandth year, or even that it
0324 n relates to the same thousandth or hundred THOUSANDTH year, or even that it has a very strict geo
0293 verage duration of each formation is twice as long as the average duration of specific f
0308 ea. Looking to the existing oceans, which are THRICE as extensive as the land, we see them studded
0381 cy in a foreign land; why one ranges twice or THRICE as far, and is twice or thrice as common, as
0381 anges twice or thrice as far, and is twice or THRICE as common, as another species within their ow
0004 le for Existence amongst all organic beings THROUGHOUT the world, which inevitably follows from th
0005 the geological succession of organic beings THROUGHOUT time: in the eleventh and twelfth, their ge
0005 nd twelfth, their geographical distribution THROUGHOUT space: in the thirteenth, their classificat
0024 rts of their structure, we may look in vain THROUGHOUT the whole great family of Columbidae for a
0052 weight of the several facts and views given THROUGHOUT this work. It need not be supposed that all
0059 rger genera thus tend to become larger; and THROUGHOUT nature the forms of life which are now domi
0059 smaller genera. And thus, the forms of life THROUGHOUT the universe become divided into groups sub
0064 introduced plants which have become common THROUGHOUT whole islands in a period of less than ten
0084 selection is daily and hourly scrutinising, THROUGHOUT the world, every variation, even the slight
0104 fy all the individuals of a varying species THROUGHOUT the area in the same manner in relation to
0105 for a long period, and of spreading widely. THROUGHOUT a great and open area, not only will there
0111 infer from most of the innumerable species THROUGHOUT nature presenting well marked differences:
0113 cupy. What applies to one animal will apply THROUGHOUT all time to all animals, that is, if they v
0128 amiliarity, that all animals and all plants THROUGHOUT all time and space should be related to eac
0132 being and another, which we see everywhere THROUGHOUT nature. Some little influence may be attrib
0146 that some apparent correlations, occurring THROUGHOUT whole orders, are entirely due to the manne
0155 n or part, which is generally very constant THROUGHOUT large groups of species, has differed consi
0200 ely for the good of another species: though THROUGHOUT nature one species incessantly takes advant
0208 n once acquired, they often remain constant THROUGHOUT life. Several other points of resemblance b
0234 us further suppose that the community lived THROUGHOUT the winter, and consequently required a sto
0279 ermediate links not now occuring everywhere THROUGHOUT nature depends on the very process of natur
0288 quick to embed and preserve fossil remains. THROUGHOUT an enormously large proportion of the ocean
0293 ve the several stages of the same formation THROUGHOUT Europe been correlated with perfect accurac
0295 tion will have had to live on the same area THROUGHOUT this whole time. But we have seen that a th
0303 would be able to spread rapidly and widely THROUGHOUT the world. I will now give a few examples t
0305 group appeared suddenly and simultaneously THROUGHOUT the world at this same period. It is almost
0306 gmatize on the succession of organic beings THROUGHOUT the world, as it would be for a naturalist
0312 n simultaneous changes in the forms of life THROUGHOUT the World. On the affinities of extinct spe
0322 orms of Life changing almost simultaneously THROUGHOUT the World. Scarcely any palaeontological di
0322 forms of Life change almost simultaneously THROUGHOUT the world. Thus our European Chalk formatio
0324 spoken of as having changed simultaneously THROUGHOUT the world, it must not be supposed that thi
0325 these great mutations in the forms of life THROUGHOUT the world, under the most different climate
0325 he parallel succession of the forms of life THROUGHOUT the world, is explicable on the theory of n
0327 neous, succession of the same forms of life THROUGHOUT the world, accords well with the principle
0327 therefore as new and improved groups spread THROUGHOUT the world, old groups will disappear from t
0332 altered conditions of life, and yet retain THROUGHOUT a vast period the same general characterist
0336 the sea, have changed almost simultaneously THROUGHOUT the world, and therefore under the most dif
0350 se facts some deep organic bond, prevailing THROUGHOUT space and time, over the same areas of land
0366 ers, are now clothed by the vine and maize. THROUGHOUT a large part of the United States, erratic
0374 least of the period, actually simultaneous THROUGHOUT the world. Without some distinct evidence t
0375 nts of that large island. Hence we see that THROUGHOUT the world, the plants growing on the more l
0383 liec species prevail in a remarkable manner THROUGHOUT the world. I well remember, when first coll
0385 ave proceeded from a single source, prevail THROUGHOUT the world. Their distribution at first perp
0385 ome naturalised species have rapidly spread THROUGHOUT the same country. But two facts, which I ha
0401 ht spread and yet retain the same character THROUGHOUT the group, just as we see on continents som
0403 eir new homes, is of the widest application THROUGHOUT nature. We see this on every mountain, in e
0403 orms, chiefly of plants, have spread widely THROUGHOUT the world during the recent Glacial epoch,
0409 a striking parellelism in the laws of life THROUGHOUT time and space; the laws governing the succ
0410 ges, as in now looking to distant provinces THROUGHOUT the world, we find that some organisms diff
0410 rule. On my theory these several relations THROUGHOUT time and space are intelligible: for whethe
0415 believe, depends on their greater constancy THROUGHOUT large groups of species; and this constancy
0422 ication of the various languages now spoken THROUGHOUT the world; and if all extinct languages, an
0426 covered by hair or feathers, if it prevail THROUGHOUT many and different species, especially thos
0426 et them be ever so trifling, occur together THROUGHOUT a large group of beings having different ha
0429 e in number, and how widely spread they are THROUGHOUT the world, the fact is striking, that the d
0432 all the forms in any class which have lived THROUGHOUT all time and space. We shall certainly neve
0435 of the homologous construction of the limbs THROUGHOUT the whole class. So with the mouths of inse
```

Page ************************************(Key Word)************************************

Page **(Key Word)***************************************

```
0450  he stamp of inutility, are extremely common THROUGHOUT nature. For instance, rudimentary mammae ar
0456  dination of group to group in all organisms THROUGHOUT all time; that the nature of the relationsh
0462  h of the same and of representative species THROUGHOUT the world. We are as yet profoundly ignoran
0471  erywhere around us, and which has prevailed THROUGHOUT all time. The grand fact of the grouping of
0475  ppear as if they had changed simultaneously THROUGHOUT the world. The fact of the fossil remains o
0476  elism in the distribution of organic beings THROUGHOUT space, and in their geological succession t
0476  t space, and in their geological succession THROUGHOUT time; for in both cases the beings have bee
0483  amount of modification in the descendants. THROUGHOUT whole classes various structures are formed
0060  complex relations of all animals and plants THROUGHTOUT nature. Struggle for life most severe betw
0001  of that continent. These facts seemed to me to THROW some light on the origin of species, that mys
0075  ound and thus checked the growth of the trees! THROW up a handful of feathers, and all must fall t
0385  nd no doubt many others remain to be observed, THROW some light on this subject. When a duck sudde
0454  nsters. But I doubt whether any of these cases THROW light on the origin of rudimentary organs in
0487  which geology now throws, and will continue to THROW, on former changes of climate and of the leve
0012  e is his fundamental belief: doubts have been THROWN on this principle by theoretical writers alon
0199  add that some little light can apparently be THROWN on the origin of these differences, chiefly t
0203  i think that in the discussion light has been THROWN on several facts, which on the theory of inde
0285  the strata have been upheaved on one side, or THROWN down on the other, to the height or depth of
0327  or rising, and likewise when sediment was not THROWN down quickly enough to embed and preserve org
0360  ransported in another manner. Drift timber is THROWN up on most islands, even on those in the mids
0374  y colder from pole to pole, much light can be THROWN on the present distribution of identical and
0376  nct species. Now let us see what light can be THROWN on the foregoing facts, on the belief, suppor
0395  n's agency; but we shall soon have much light THROWN on the natural history of this archipelago by
0487  arent means of immigration, some light can be THROWN on ancient geography. The noble science of Ge
0488  ower and capacity by gradation. Light will be THROWN on the origin of man and his history. Authors
0124  nd i think we shall then see that the diagram THROWS light on the affinities of extinct beings, wh
0474  how how much light the principle of gradation THROWS on the admirable architectural powers of the
0487  gration, then, by the light which geology now THROWS, and will continue to throw, on former change
0076  er species. The recent increase of the missel THRUSH in parts of Scotland has caused the decrease
0076  scotland has caused the decrease of the song THRUSH. How frequently we hear of one species of rat
0089  s to attract by singing the females. The rock THRUSH of Guiana, birds of Paradise, and some others
0185  anomalous member of the strictly terrestrial THRUSH family wholly subsists by diving, grasping. th
0243  principle of inheritance, how it is that the THRUSH of South America lines its nest with mud, in
0243  the same peculiar manner as does our British THRUSH: how it is that the male wrens (Troglodytes)
0402  e are three closely allied species of mocking THRUSH, each confined to its own island. Now let us
0402  ts own island. Now let us suppose the mocking THRUSH of Chatham Island to be blown to Charles Isla
0402  to Charles Island, which has its own mocking THRUSH: why should it succeed in establishing itself
0402  be reared: and we may infer that the mocking THRUSH peculiar to Charles Island is at least as wel
0439  age: as we see in the spotted feathers in the THRUSH group. In the cat tribe, most of the species
0471  ld have been created with webbed feet; that a THRUSH should have been created to dive and feed on
0475  uld follow nearly the same instincts; why the THRUSH of South America, for instance, lines her nes
0186  ot a tree grows; that there should be diving THRUSHES, and petrels with the habits of auks. Organs
0204  with webbed feet, ground woodpeckers, diving THRUSHES, and petrels with the habits of auks. Althou
0258  other equally striking cases could be given. THURET has observed the same fact with certain sea w
0264  , as seems to have been the case with some of THURET's experiments on Fuci. No explanation can be
0004  act to Variation under Domestication. We shall THUS see that a large amount of hereditary modifica
0005  e, will have a better chance of surviving, and THUS be naturally selected. From the strong princip
0009  ng that their reproductive system has not been THUS affected; so will some animals and plants with
0011  instances of correlation are quite whimsical: THUS cats with blue eyes are invariably deaf; colou
0012  rge feet. Hence, if man goes on selecting, and THUS augmenting, any peculiarity, he will almost ce
0014  er. In many cases this could not be otherwise: THUS the inherited peculiarities in the horns of ca
0015  ned, will determine how far the new characters THUS arising shall be preserved. When we look to th
0020  case on record of a permanent race having been THUS formed. On the Breeds of the Domestic Pigeon.
0025  pigeon, although no other existing species is THUS coloured and marked, so that in each separate
0028  male pigeons can be easily mated for life; and THUS different breeds can be kept together in the s
0032  cases, in which exact records have been kept; THUS, to give a very trifling instance, the steadil
0034  he best individual animals, is more important. THUS, a man who intends keeping pointers naturally
0036  s are so liable, and such choice animals would THUS generally leave more offspring than the inferi
0037  rge amount of change in our cultivated plants, THUS slowly and unconsciously accumulated, explains
0038  tter in the one country than in the other, and THUS by a process of natural selection, as will her
0040  re than usual in matching his best animals and THUS improves them, and the improved individuals sl
0042  his is a great convenience to the fancier, for THUS many races may be kept true, though mingled in
0043  ributed to use and disuse. The final result is THUS rendered infinitely complex. In some cases, I
0045  same parents, or which may be presumed to have THUS arisen, being frequently observed in the indiv
0056  all large genera are now varying much, and are THUS increasing in the number of their species, or
0059  ecies have once existed as varieties, and have THUS originated: whereas, these analogies are utter
0059  to new and distinct species. The larger genera THUS tend to become larger; and throughout nature t
0059  also tend to break up into smaller genera. And THUS, the forms of life throughout the universe bec
0061  ed by its offspring. The offspring, also, will THUS have a better chance of surviving, for, of the
0062  nd us mostly live on insects or seeds, and are THUS constantly destroying life; or we forget how l
0063  plants, in order to tempt birds to devour and THUS disseminate its seeds rather than those of oth
0068  the less vigorous, though fully grown, plants: THUS out of twenty species growing on a little plot
0068  h determines the average numbers of a species. THUS, there seems to be little doubt that the stock
0070  is absolutely necessary for its preservation. THUS we can easily raise plenty of corn and rape se
0070  favourable that many could exist together, and THUS save each other from utter destruction. I shou
0073  its of moths to remove their pollen masses and THUS to fertilise them. I have, also, reason to bel
0074  believes that more than two thirds of them are THUS destroyed all over England. Now the number of
0075  ther plants which first clothed the ground and THUS checked the growth of the trees! Throw up a ha
0078  rent set of competitors or enemies. It is good THUS to try in our imagination to give any form som
0082  d to be preserved: and natural selection would THUS have free scope for the work of improvement. W
0083  possession of the land. And as foreigners have THUS everywhere beaten some of the natives, we may
0084  t to consider as very trifling importance, may THUS be acted on. When we see leaf eating insects g
0089  ften take individual preferences and dislikes: THUS Sir R. Heron has described how one pied peacoc
0089  during the breeding season: the modifications THUS produced being inherited at corresponding ages
0089  have not space here to enter on this subject. THUS it is, as I believe, that when the males and f
0092  distinct individuals of the same species would THUS get crossed: and the act of crossing, we have
```

Page **************************************(Key Word)***************************************

Page ***(Key Word)**

```
0092  ing insects from flower to flower, and a cross   THUS effected, although nine tenths of the pollen w
0093  female to the male tree, the pollen could not    THUS have been carried. The weather had been cold a
0095  ant supply of precious nectar to the hive bee.   THUS it might be a great advantage to the hive bee
0095   so that the hive bee could visit its flowers.   THUS I can understand how a flower and a bee might
0098  n; but it must not be supposed that bees would   THUS produce a multitude of hybrids between distinc
0099  ge majority, as I have found, of the seedlings   THUS raised will turn out mongrels: for instance, I
0102  arge amount of crossing with inferior animals.   THUS it will be in nature: for within a confined ar
0103  he same new variety. A local variety when once   THUS formed might subsequently slowly spread to oth
0103  hance of intercrossing with other varieties is   THUS lessened. Even in the case of slow breeding an
0103  ue and uniform in character. It will obviously   THUS act far more efficiently with those animals wh
0104  long intervals, I am convinced that the young   THUS produced will gain so much in vigour and ferti
0104  e of surviving and propagating their kind; and   THUS, in the long run, the influence of intercrosse
0104  of climate or elevation of the land, etc.; and   THUS new places in the natural economy of the count
0105  for the production of new species. But we may   THUS greatly deceive ourselves, for to ascertain wh
0106  ad over the open and continuous area, and will   THUS come into competition with many others. Hence
0106  se to most new varieties and species, and will   THUS play an important part in the changing history
0106  ore those of the larger Europaeo Asiatic area.   THUS, also, it is that continental productions have
0107  ing inhabited a confined area, and from having   THUS been exposed to less severe competition. To su
0107  the confines of the range of each species will   THUS be checked: after physical changes of any kind
0108   to improve still further the inhabitants, and   THUS produce new species. That natural selection wi
0108  lations of many of the other inhabitants being   THUS disturbed. Nothing can be effected, unless fav
0109  eed we can see reason why they should not have   THUS increased, for the number of places in the pol
0113  plants and a greater weight of dry herbage can   THUS be raised. The same has been found to hold goo
0113  is annually sowing almost countless seeds; and   THUS, as it may be said, is striving its utmost to
0114  f succeeding and of increasing in numbers, and   THUS of supplanting the less distinct varieties; an
0114  itution, determine that the inhabitants, which   THUS jostle each other most closely, shall, as a ge
0115  enumerated, and these belong to 162 genera. We   THUS see that these naturalised plants are of a hig
0115  than 100 genera are not there indigenous, and   THUS a large proportional addition is made to the g
0116  become more diversified in structure, and are   THUS enabled to encroach on places occupied by othe
0118  ee varieties, and some failing to produce any.   THUS the varieties or modified descendants, proceed
0120  t these three forms into well defined species:   THUS the diagram illustrates the steps by which the
0120  s between a14 and m14, all descended from (A).   THUS, as I believe, species are multiplied and gene
0122  e that they will have taken the places of, and   THUS exterminated, not only their parents (A) and (
0123  inct genera, or even as distinct sub families.   THUS it is, as I believe, that two or more genera a
0125  r another large group, reduce its numbers, and   THUS lessen its chance of further variation and imp
0127  organic being do occur, assuredly individuals   THUS characterised will have the best chance of bei
0127  er males. Whether natural selection has really   THUS acted in nature, in modifying and adapting the
0128  ir chance of succeeding in the battle of life.   THUS the small differences distinguishing varieties
0132  ence may be attributed to climate, food, etc.:   THUS, E. Forbes speaks confidently that shells at t
0133  are only well marked and permanent varieties.   THUS the species of shells which are confined to tr
0133  ction, and how much to the conditions of life.   THUS, it is well known to furriers that animals of
0134  t in affecting the reproductive system, and in   THUS inducing variability; and natural selection wi
0136  light will oftenest have been blown to sea and   THUS have been destroyed. The insects in Madeira wh
0140  ferent temperatures, or becoming acclimatised:   THUS the pines and rhododendrons, raised from seed
0143   more or less completely by natural selection:   THUS a family of stags once existed with an antler
0147  ly flows, at least in excess, to another part;   THUS it is difficult to get a cow to give milk and
0148  ted in building up an useless structure. I can   THUS only understand a fact with which I was much s
0148  a cirripede is parasitic within another and is   THUS protected, it loses more or less completely it
0148   in developing a structure now become useless.   THUS, as I believe, natural selection will always s
0149  power to check deviations in their structure.   THUS rudimentary parts are left to the free play of
0150  with the same part in closely allied species.   THUS, the bat's wing is a most abnormal structure i
0152  her has not or cannot come into full play, and   THUS the organisation is left in a fluctuating con
0154  n immense period in nearly the same state; and   THUS it comes to be no more variable than any other
0155  hin a moderately recent period, and which have   THUS come to differ. Or to state the case in anothe
0157  will have had a wide scope for action, and may   THUS readily have succeeded in giving to the specie
0158  cumulated by natural and sexual selection, and   THUS adapted for secondary sexual, and for ordinary
0161  ntary organs being, as we all know them to be,   THUS inherited. Indeed, we may sometimes observe a
0161  l marked and permanent variety. But characters   THUS gained would probably be of an important natur
0168  ariable, because they have recently varied and   THUS come to differ; but we have also seen in the s
0169  unt of modification since the genus arose; and   THUS we can understand why it should often still be
0172  ed forms with which it comes into competition.   THUS extinction and natural selection will, as we h
0175  e extremely liable to utter extermination; and   THUS its geographical range will come to be still m
0177  the place of the less improved hill breed; and   THUS the two breeds, which originally existed in gr
0178  s becoming slowly modified, with the new forms   THUS produced and the old ones acting and reacting
0179   in the aggregate, present more variation, and   THUS be further improved through natural selection
0183  reat numbers and under many subordinate forms.   THUS, to return to our imaginary illustration of th
0184  ammering the seeds of the yew on a branch, and   THUS breaking them like a nuthatch. In North Americ
0184  rne swimming for hours with widely open mouth,   THUS catching, like a whale, insects in the water.
0186   so little, either in habits or structure, and   THUS gain an advantage over some other inhabitant o
0188  of the great Articulate class. He who will go   THUS far, if he find on finishing this treatise tha
0189  believe that a living optical instrument might   THUS be formed as superior to one of glass, as the
0190  ng at the same time wholly distinct functions;   THUS the alimentary canal respires, digests, and ex
0190  might easily specialise, if any advantage were   THUS gained, a part or organ, which had performed t
0190  med two functions, for one function alone, and   THUS wholly change its nature by insensible steps.
0191  th a floating apparatus or swimbladder. We can   THUS, as I infer from Professor Owen's interesting
0193  had been inherited from one ancient progenitor   THUS provided, we might have expected that all elec
0195  , would be able to range into new pastures and   THUS gain a great advantage. It is not that the lar
0196  rs betray their aquatic origin, may perhaps be   THUS accounted for. A well developed tail having be
0198  ned by certain plants; so that colour would be   THUS subjected to the action of natural selection.
0199   on structure, quite independently of any good   THUS gained. Correlation of growth has no doubt pla
0199  elation to the habits of life of each species.   THUS, we can hardly believe that the webbed feet of
0204  e less numerous intermediate variety, and will   THUS generally succeed in supplanting and extermina
0206  nd the competition will have been severer, and   THUS the standard of perfection will have been rend
0208  om the third stage, where it had left off, and   THUS tried to complete the already finished work. I
0209  and of many ants, could not possibly have been   THUS acquired. It will be universally admitted that
0209  ct to any extent that may be profitable. It is   THUS, as I believe, that all the most complex and w
```

Page ***(Key Word)**

Page **(Key Word)**

```
0213  rin'g a few cases under domestication. We shall  THUS  also be enabled to see the respective parts wh
0214  own when different breeds of dogs are crossed.    THUS  it is known that a cross with a bulldog has af
0214  to hunt hares. These domestic instincts, when     THUS  tested by crossing, resemble natural instincts
0217  alogy would lead me to believe, that the young    THUS  reared would be apt to follow by inheritance t
0217  t to lay their eggs in other birds' nests, and    THUS  be successful in rearing their young. By a con
0219  d if the insect whose nest and stored food are    THUS  feloniously appropriated, be not thus extermin
0219  food are thus feloniously appropriated, be not    THUS  exterminated. Slave making instinct. This rema
0223  d milk as it may be called, their aphides; and    THUS  both collect food for the community. In Englan
0223  d as food might become developed; and the ants    THUS  unintentionally reared would then follow their
0226  ly flat walls of wax between the spheres which    THUS  tend to intersect. Hence each cell consists of
0229  s on both sides, in order to have succeeded in    THUS  leaving flat plates between the basins, by sto
0230  done) into its proper intermediate plane, and    THUS  flatten it. From the experiment of the ridge o
0231  h cement, on the summit of the ridge. We shall    THUS  have a thin wall steadily growing upward; but
0231  hose just commenced and those completed, being    THUS  crowned by a strong coping of wax, the bees ca
0235  see, is absolutely perfect in economising wax.    THUS, as I believe, the most wonderful of all known
0237  family, as well as to the individual, and may    THUS  gain the desired end. Thus, a well flavoured v
0237  individual, and may thus gain the desired end.    THUS, a well flavoured vegetable is cooked, and the
0238  no one ox could ever have propagated its kind.    THUS  I believe it has been with social insects: a s
0238  etimes to an almost incredible degree, and are    THUS  divided into two or even three castes. The cas
0238  r rather as any two genera of the same family.    THUS  in Eciton, there are working and soldier neute
0241  with an intermediate structure were produced.    THUS, as I believe, the wonderful fact of two disti
0248  xperiments made during different years. It can    THUS  be shown that neither sterility nor fertility
0249  s, though probably on the same plant, would be    THUS  effected. Moreover, whenever complicated exper
0249  m another plant of the same hybrid nature. And    THUS, the strange fact of the increase of fertility
0252  injurious influence of close interbreeding is    THUS  prevented. Any one may readily convince himsel
0253  uld have gone on increasing. If we were to act    THUS, and pair brothers and sisters in the case of
0256  first cross, and the sterility of the hybrids    THUS  produced, two classes of facts which are gener
0258  y the pollen of M. longiflora, and the hybrids    THUS  produced are sufficiently fertile; but Kolreut
0262  h no great difficulty on another species, when    THUS  crafted were rendered barren. On the other han
0262  when self fertilised with their own pollen. We    THUS  see, that although there is a clear and fundam
0265  omestication of animals. Between the sterility    THUS  superinduced and that of hybrids, there are ma
0265  o vary, as every experimentalist has observed.    THUS  we see that when organic beings are placed und
0268  quently ranks them as undoubted species. If we    THUS  argue in a circle, the fertility of all variet
0269  which may be for each creature's own good; and    THUS  she may, either directly, or more probably ind
0270  is important to notice that the hybrid plants    THUS  raised were themselves perfectly fertile; so t
0273  to any change in the conditions of life, being    THUS  often rendered either impotent or at least inc
0283  ordan Hill, are most impressive. With the mind    THUS  impressed, let any one examine beds of conglom
0287  ave existed for millions of years as land, and    THUS  have escaped the action of the sea: when deepl
0291  ll remain shallow and favourable for life, and    THUS  a fossiliferous formation thick enough, when u
0291  nt formations, which are rich in fossils, have    THUS  been formed during subsidence. Since publishin
0291  ed during a downward oscillation of level, and    THUS  gained considerable thickness. All geological
0292  brought within the limits of the coast action.    THUS  the geological record will almost necessarily
0295  k the area whence the sediment is derived, and    THUS  diminish the supply whilst the downward moveme
0296  not the trees chanced to have been preserved:    THUS, Messrs. Lyell and Dawson found carboniferous
0298  dom be connected by intermediate varieties and    THUS  proved to be the same species, until many spec
0303  this had been effected, and a few species had    THUS  acquired a great advantage over other organism
0306  the southern capes of Africa or Australia, and    THUS  reach other and distant seas. From these and s
0309  t have we any right to assume that things have    THUS  remained from eternity? Our continents seem to
0314  he accumulated to a greater or lesser amount,    THUS  causing a greater or lesser amount of modifica
0315  another species in the economy of nature, and    THUS  supplant it; yet the two forms, the old and th
0320  the place of other breeds in other countries.    THUS  the appearance of new forms and the disappeara
0321  nus, will be the most liable to extermination.    THUS, as I believe, a number of new species descend
0321  y a species belonging to a distinct group, and    THUS  caused its extermination; and if many allied f
0322  ny of the old inhabitants; and the forms which    THUS  yield their places will commonly be allied, fo
0322  ey will partake of some inferiority in common.    THUS, as it seems to me, the manner in which single
0322  ge almost simultaneously throughout the world.    THUS  our European Chalk formation can be recognised
0323  arallel manner. We may doubt whether they have    THUS  changed: if the Megatherium, Mylodon, Macrauch
0327  ingle member might long be enabled to survive.    THUS, as it seems to me, the parallel, and, taken i
0327  spreading widely and varying; the new species    THUS  produced being themselves dominant owing to in
0331  solely on the descendants from a species being    THUS  enabled to seize on many and different places
0332  who objected to call the extinct genera, which    THUS  linked the living genera of three families tog
0333  nearly so much as they subsequently diverged.    THUS  it comes that ancient and extinct genera are o
0333  l up wide intervals in the natural system, and    THUS  unite distinct families or orders. All that we
0333  aeontologists seems frequently to be the case.    THUS, on the theory of descent with modification, t
0334  at which preceded and that which succeeded it.    THUS, the species which lived at the sixth great st
0338  ations, more and more difference to the adult.    THUS  the embryo comes to be left as a sort of pictu
0343  modification through natural selection. We can    THUS  understand how it is that new species come in
0344  s tend to leave many modified descendants, and    THUS  new sub groups and groups are formed. As these
0350  remarked, the most important of all relations.    THUS  the high importance of barriers comes into pla
0350  ergo further modification and improvement; and    THUS  they will become still further victorious, and
0352  ion from parents specifically distinct. We are    THUS  brought to the question which has been largely
0356  ands or possibly even continents together, and    THUS  have allowed terrestrial productions to pass f
0357  rope likewise with America. Other authors have    THUS  hypothetically bridged over every ocean, and h
0361  t transport: out of one small portion of earth    THUS  completely enclosed by wood in an oak about 50
0362  ds, and the contents of their torn crops might    THUS  readily get scattered. Mr. Brent informs me th
0362  ; and two seeds of beet grew after having been    THUS  retained for two days and fourteen hours. Fres
0362  ts: fish are frequently devoured by birds, and    THUS  the seeds might be transported from place to p
0363  pebble quite as large as the seed of a vetch.    THUS  seeds might occasionally be transported to gre
0364  nk be a marvellous fact if many plants had not    THUS  become widely transported. These means of tran
0367  and in the artic regions of both hemispheres.    THUS  we can understand the identity of many plants
0367  ins of the United States and of Europe. We can    THUS  also understand the fact that the Alpine plant
0369  imatal influences. Their mutual relations will    THUS  have been in some degree disturbed; consequent
0372  forms now living in areas completely sundered.    THUS, I think, we can understand the presence of ma
0378  himalaya, as graphically described by Hooker.    THUS, as I believe, a considerable number of plants
0379  ts, and constitutions will have profited them.    THUS, many of these wanderers, though still plainly
0379  dominating power, than the southern forms. And    THUS, when they became commingled during the Glacia
```

Page **(Key Word)**

Page **************************************(Key Word)**************************************

0382	t height under the equator. The various beings	THUS left stranded may be compared with savage race
0388	ece of water. Nature, like a careful gardener,	THUS takes her seeds from a bed of a particular nat
0390	mutual relations have not been much disturbed.	THUS in the Galapagos Islands nearly every land bir
0392	n island, to whatever order they belonged, and	THUS convert them first into bushes and ultimately
0397	the feet of birds roosting on the ground, and	THUS get transported? It occurred to me that land s
0399	ns are few, and most of them can be explained.	THUS the plants of Kerguelen Land, though standing
0400	er, within the limits of the same archipelago.	THUS the several islands of the Galapagos Archipela
0402	s, we are apt to infer that most species would	THUS spread; but we should remember that the forms
0402	g from island to island, are distinct on each;	THUS there are three closely allied species of mock
0403	of species under the same conditions of life.	THUS the south east and south west corners of Aust
0403	related to those of the surrounding lowlands;	THUS we have in South America, Alpine humming birds
0405	es inhabit America and Europe, and the species	THUS has an immense range; but, if the variation ha
0407	ae most slowly, enormous periods of time being	THUS granted for their migration, I do not think th
0408	on and the multiplication of new forms. We can	THUS understand the high importance of barriers, wh
0408	ral zoological and botanical provinces. We can	THUS understand the localisation of sub genera, gen
0411	ary most. The varieties, or incipient species,	THUS produced ultimately become converted, as I bel
0413	milies, and orders, all united into one class.	THUS, the grand fact in natural history of the subo
0413	of the Creator, it seems to me that nothing is	THUS added to our knowledge. Such expressions as th
0417	s, to the family, to the class, disappear, and	THUS laugh at our classification. But when Aspicarp
0422	n nature amongst the beings of the same group.	THUS, on the view which I hold, the natural system
0423	most constant, is used in classing varieties:	THUS the great agriculturist Marshall says the horn
0425	en a longer time to complete? I believe it has	THUS been unconsciously used; and only thus can I u
0425	it has thus been unconsciously used; and only	THUS can I understand the several rules and guides
0427	ngst insects there are innumerable instances;	THUS Linnaeus, misled by external appearances, actu
0427	dily become adapted to similar conditions; and	THUS assume a close external resemblance; but such
0428	class or order are compared one with another:	THUS the shape of the body and fin like limbs are o
0428	extend the parallelism over a wide range; and	THUS the septenary, quinary, quaternary, and ternar
0428	of nature. The larger and more dominant groups	THUS tend to go on increasing in size; and they con
0428	ntly supplant many smaller and feebler groups.	THUS we can account for the fact that all organisms
0430	nity in most cases is general and not special:	THUS, according to Mr. Waterhouse, of all Rodents,
0431	tween the several groups in each class. We may	THUS account even for the distinctness of whole cla
0432	ers of each group, whether large or small, and	THUS give a general idea of the value of the differ
0436	oceans, the general pattern seems to have been	THUS to a certain extent obscured. There is another
0439	lance; sometimes lasts till a rather late age:	THUS birds of the same genus, and of closely allied
0439	h rarely see something of this kind in plants:	THUS the embryonic leaves of the ulex or furze, and
0442	ot at any period differ widely from the adult:	THUS Owen has remarked in regard to cuttle fish, th
0443	ors, have been exposed. Nevertheless an effect	THUS caused at a very early period, even before the
0447	on at a rather late period of life, and having	THUS been converted into hands, or paddles, or wing
0447	nd has to gain its own living; and the effects	THUS produced will be inherited at a corresponding
0449	are therefore in that degree closely related.	THUS, community in embryonic structure reveals comm
0450	rioc, or for ever, incapable of demonstration.	THUS, as it seems to me, the leading facts in embry
0450	embryology rises greatly in interest, when we	THUS look at the embryo as a picture, more or less
0451	and remain perfectly efficient for the other.	THUS in plants, the office of the pistil is to allo
0452	found in monstrous individuals of the species.	THUS in the snapdragon (antirrhinum) we generally d
0453	which is formed merely of cellular tissue, can	THUS act? Can we suppose that the formation of rudi
0455	will seldom affect or reduce it in the embryo.	THUS we can understand the greater relative size of
0455	rgan. As the presence of rudimentary organs is	THUS so due to the tendency in every part of the organ
0467	ect the variations given to him by nature, and	THUS accumulate them in any desired manner. He thus
0467	thus accumulate them in any desired manner. He	THUS adapts animals and plants for his own benefit
0470	less improved and intermediate varieties; and	THUS species are rendered to a large extent defined
0471	vergent in character. But as all groups cannot	THUS succeed in increasing in size, for the world w
0474	, slight, but profitable modifications. We can	THUS understand why nature moves by graduated steps
0476	in comparison with its later descendants; and	THUS we can see why the more ancient a fossil is, t
0480	l part in the struggle for existence, and will	THUS have little power of acting on an organ during
0480	e soldered wing covers of some beetles, should	THUS so frequently bear the plain stamp of inutilit
0482	ientiously expressing his conviction; for only	THUS can the load of prejudice by which this subjec
0482	ht were special creations, and which are still	THUS looked at by the majority of naturalists, and
0485	late gradations, whereas species were formerly	THUS connected. Hence, without quite rejecting the
0486	even the blunders of numerous workmen; when we	THUS view each organic being, how far more interest
0490	ter and the Extinction of less improved forms.	THUS, from the war of nature, from famine and death
0140	constitutional powers of resisting cold. Mr.	THWAITES informs me that he has observed similar fact
0439	by Agassiz, namely, that having forgotten to	TICKET the embryo of some vertebrate animal, he cann
0211	through a lens; but not one excreted: I then	TICKLED and stroked them with a hair in the same man
0304	abiting various zones of depths from the upper	TIDAL limits to 50 fathoms; from the perfect manner
0382	as to have freely inundated the south. As the	TIDE leaves its drift in horizontal lines, though r
0382	, though rising higher on the shores where the	TIDE rises highest, so have the living waters left
0283	ocks, and mark the process of degradation. The	TIDES in most cases reach the cliffs only for a sho
0143	s expression that the whole organisation is so	TIED together during its growth and development, th
0431	e most wonderfully diverse forms are still	TIED together by a long, but broken, chain of affin
0018	these ancient periods, savages, like those of	TIERRA del Fuego or Australia, who possess a semi do
0036	alue set on animals even by the barbarians of	TIERRA del Fuego, by their killing and devouring the
0184	oceanic of birds, yet in the quiet Sounds of	TIERRA del Fuego, the Puffinuria berardi, in its gen
0215	ought home as puppies from countries; such as	TIERRA del Fuego and Australia, where the savages do
0323	orth America, in equatorial South America, in	TIERRA del Fuego, at the Cape of Good Hope, and in t
0374	en forty and fifty of the flowering plants of	TIERRA del Fuego, forming no inconsiderable part of
0378	plants, about forty six in number, common to	TIERRA del Fuego and to Europe still exist in North
0068	none are destroyed by beasts of prey: even the	TIGER in India most rarely dares to attack a young
0077	n the structure of the teeth and talons of the	TIGER; and in that of the legs and claws of the par
0077	f the parasite which clings to the hair on the	TIGER's body. But in the beautifully plumed seed of
0039	ture. No man would ever try to make a fantail;	TILL he saw a pigeon with a tail developed in some
0039	light degree in an unusual manner, or a pouter	TILL he saw a pigeon with a crop of somewhat unusua
0064	ds when thirty years old, and goes on breeding	TILL ninety years old, bringing forth three pair of
0128	e same species, will steadily tend to increase	TILL they come to equal the greater differences bet
0173	rer, the other becomes more and more frequent,	TILL the one replaces the other. But if we compare
0184	ure and habits, with larger and larger mouths,	TILL a creature was produced as monstrous as a whal
0189	plied by the million! and each to be preserved	TILL a better be produced, and then the old ones to

Page **************************************(Key Word)**************************************

```
0231  it away equally on both sides near the ground,  TILL a smooth, very thin wall is left in the middle
0232  gnaw away and finish off the angles of a cell  TILL a large part both of that cell and of the adjo
0265  ce unusually fertile hybrids. No one can tell,  TILL he tries, whether any particular animal will b
0265  nt seed freely under culture; nor can he tell,  TILL he tries, whether any two species of a genus w
0316  neral rule being a gradual increase in number,  TILL the group reaches its maximum, and then, soone
0317  branching of a great tree from a single stem,  TILL the group becomes large. On Extinction. We hav
0322  naturalised in a given country; then, and not  TILL then, we may justly feel surprise why we canno
0439  law of embryonic resemblance, sometimes lasts  TILL a rather late age: thus birds of the same genu
0482  t. they admit that a multitude of forms, which  TILL lately they themselves thought were special cr
0359  here is in the buoyancy of green and seasoned  TIMBER: and it occurred to me that floods might wash
0360  sionally transported in another manner. Drift  TIMBER is thrown up on most islands, even on those i
0391  rvae, perhaps attached to seaweed or floating  TIMBER, or to the feet of wading birds, might be tra
0397  shell, might be floated in chinks of drifted  TIMBER across moderately wide arms of the sea. And I
0005  ogical succession of organic beings throughout  TIME: in the eleventh and twelfth, their geographic
0028  a bill of fare in the previous dynasty. In the  TIME of the Romans, as we hear from Pliny, immense
0031  at intervals of months, and the sheep are each  TIME marked and classed, so that the very best may
0034  nd bad qualities is so obvious. At the present  TIME, eminent breeders try by methodical selection,
0035  nsciously modified to a large extent since the  TIME of that monarch. Some highly competent authori
0046  g from Brachiopod shells, at former periods of  TIME. These facts seem to be very perplexing, for t
0047  eir characters in their own country for a long  TIME; for as long, as far as we know, as have good
0056  ells us that small genera have in the lapse of  TIME often increased greatly in size: and that larg
0062  ss in leaving progeny. Two canine animals in a  TIME of dearth, may be truly said to struggle with
0068  next twenty years in England, and at the same  TIME, if no vermin were destroyed, there would, in
0073  of nature remains uniform for long periods of  TIME, though assuredly the merest trifle would ofte
0082  r more easily, from having incomparably longer  TIME at her disposal. Nor do I believe that any gre
0084  e the wishes and efforts of man! how short his  TIME! and consequently how poor will his products b
0084  se slow changes in progress, until the hand of  TIME has marked the long lapse of ages, and then so
0089  o support this view; but if man can in a short  TIME give elegant carriage and beauty to his bantam
0094  ny facts, showing how anxious bees are to save  TIME; for instance, their habit of cutting holes an
0102  success. Though nature grants vast periods of  TIME for the work of natural selection, she does no
0105  ration and consequently competition, will give  TIME for any new variety to be slowly improved; and
0108  p by modifications of the old inhabitants; and  TIME will be allowed for the varieties in each to b
0108  t very slowly, often only at long intervals of  TIME, and generally on only a very few of the inhab
0108  the inhabitants of the same region at the same  TIME. I further believe, that this very slow, inter
0109  e, which may be effected in the long course of  TIME by nature's power of selection. Extinction. Th
0110  follows, that as new species in the course of  TIME are formed through natural selection, others w
0112  erences would be very slight; in the course of  TIME, from the continued selection of swifter horse
0113  pplies to one animal will apply throughout all  TIME to all animals, that is, if they vary, for oth
0117  ultaneously, but often after long intervals of  TIME; nor are they all supposed to endure for equal
0118  the process by similar steps for any length of  TIME; some of the varieties, after each thousand ge
0128  that all animals and all plants throughout all  TIME and space should be related to each other in o
0138  all, those destined for total darkness. By the  TIME that an animal had reached, after numberless g
0139  rain requisite for seeds to germinate, in the  TIME of sleep, etc., and this leads me to say a few
0152  tic animals those points, which at the present  TIME are undergoing rapid change by continued selec
0154  ility on the other hand, will in the course of  TIME cease: and that the most abnormally developed
0158  or less completely, according to the lapse of  TIME, overmastered the tendency to reversion and to
0169  lection will in such cases not as yet have had  TIME to overcome the tendency to further variabilit
0169  a very slow process, requiring a long lapse of  TIME, in this case, natural selection may readily h
0173  ollections have been made only at intervals of  TIME immensely remote. But it may be urged that whe
0173  rritory we surely ought to find at the present  TIME many transitional forms. Let us take a simple
0174  s. hence we ought not to expect at the present  TIME to meet with numerous transitional varieties i
0174  rmediate varieties? This difficulty for a long  TIME quite confounded me. But I think it can be in
0178  her. So that, in any one region and at any one  TIME, we ought only to see a few species presenting
0179  her advantages. Lastly, looking not to any one  TIME, but to all time, if my theory be true, number
0179  astly, looking not to any one time, but to all  TIME, if my theory be true, numberless intermediate
0190  imals of the same organ performing at the same  TIME wholly distinct functions; thus the alimentary
0190  he the air dissolved in the water, at the same  TIME that they breathe free air in their swimbladde
0201  rey: but some authors suppose that at the same  TIME this snake is furnished with a rattle for its
0201  on the whole advantageous. After the lapse of  TIME, under changing conditions of life, if any par
0203  always be very slow, and will act, at any one  TIME, only on a very few forms; and partly because
0205  y distinct organs having performed at the same  TIME the same function, the one having been perfect
0205  another species, but in all cases at the same  TIME useful to the owner. Natural selection in each
0206  aving adapted them during long past periods of  TIME: the adaptations being aided in some cases by
0208  with other habits, and with certain periods of  TIME and states of the body. When once acquired, th
0211  would want to excrete. I watched them for some  TIME through a lens, but not one excreted: I then t
0213  ated with certain frames of mind or periods of  TIME. But let us look to the familiar case of the s
0213  point and even back other dogs the very first  TIME that they are taken out: retrieving is certain
0217  hose first laid would have to be left for some  TIME unincubated, or there would be eggs and young
0217  nd young successively hatched, all at the same  TIME. It has been asserted that the American cuckoo
0217  g eggs and young of different ages at the same  TIME; then the old birds or the fostered young woul
0222  n victorious in their late combat. At the same  TIME I laid on the same place a small parcel of the
0228  t such a distance from each other, that by the  TIME the basins had acquired the above stated width
0229  open, and they stopped their excavation in due  TIME; so that the basins, as soon as they had been
0229  allowed the bees to go on working for a short  TIME; and again examined the cell, and I found that
0230  ced pyramidal base of any one cell at the same  TIME; but only the one rhombic plate which stands o
0231  l work together: one bee after working a short  TIME at one cell going to another, so that, as Hube
0233  de of two or three cells commenced at the same  TIME, always standing at the proper relative distan
0253  f the same hybrid have been raised at the same  TIME from different parents, so as to avoid the ill
0269  ts uniformly and slowly during vast periods of  TIME on the whole organisation, in any way which ma
0279  rieties: on their number. On the vast lapse of  TIME, as inferred from the rate of deposition and a
0281  f its modified descendants, unless at the same  TIME we had a nearly perfect chain of the intermedi
0282  ch have lived upon this earth. On the lapse of  TIME. Independently of our not finding fossil remai
0282  ous connecting links, it may be objected, that  TIME will not have sufficed for so great an amount
0282  ing the mind feebly to comprehend the lapse of  TIME. He who can read Sir Charles Lyell's grand wor
0282  prehensibly vast have been the past periods of  TIME, may at once close this volume. Not that it su
0282  an hope to comprehend anything of the lapse of  TIME, the monuments of which we see around us. It i
```

Page ***(Key Word)***

```
0283  n most cases reach the cliffs only for a short  TIME  twice a day, and the waves eat into them only
0283  promontory, that the cliffs are at the present   TIME  suffering. The appearance of the surface and t
0283  nded pebbles, each of which bears the stamp of   TIME, are good to show how slowly the mass has been
0284  n britain, gives but an inadequate idea of the   TIME  which has elapsed during their accumulation; y
0284  as elapsed during their accumulation: yet what   TIME  this must have consumed! Good observers have e
0284  bably offers the best evidence of the lapse of   TIME. I remember having been much struck with the e
0286  liff of any given height, we could measure the   TIME  requisite to have denuded the Weald. This, of
0286  n length, ever suffers degradation at the same   TIME  along its whole indented length: and we must r
0288  ormally covered, after an enormous interval of   TIME, by another and later formation, without the u
0289  separated from each other by wide intervals of   TIME. When we see the formations tabulated in writt
0289  hardly any idea can be formed of the length of   TIME  which has elapsed between the consecutive form
0290  , accords with the belief of vast intervals of   TIME  having elapsed between each formation. But we
0292  embed and preserve the remains before they had  TIME  to decay. On the other hand, as long as the be
0294  beds of North America than in those of Europe;  TIME  having apparently been required for their migr
0294  change of climate, on the prodigious lapse of   TIME, all included within this same glacial period.
0294  in other parts of America during this space of  TIME. When such beds as were deposited in shallow w
0295  long period, in order to have given sufficient  TIME  for the slow process of variation; hence the d
0295  to live on the same area throughout this whole  TIME. But we have seen that a thick fossiliferous f
0296  een due to geographical changes requiring much  TIME. Nor will the closest inspection of a formatio
0296  inspection of a formation give any idea of the  TIME  which its deposition has consumed. Many instan
0296  is fact would have suspected the vast lapse of  TIME  represented by the thinner formation. Many cas
0296  bright as they grew, of many long intervals of  TIME  and changes of level during the process of dep
0297  third and distinct species, unless at the same  TIME  it could be most closely connected with either
0302  ke due allowance for the enormous intervals of  TIME, which have probably elapsed between our conse
0303  mations, longer perhaps in some cases than the  TIME  required for the accumulation of each formatio
0303  ach formation. These intervals will have given  TIME  for the multiplication of species from some on
0303  ge over other organisms, a comparatively short  TIME  would be necessary to produce many divergent f
0304  istence of whales in the upper greensand, some  TIME  before the close of the secondary period. I ma
0307  ring these vast, yet quite unknown, periods of  TIME, the world swarmed with living creatures. To t
0312  new forms, having here appeared for the first  TIME, either locally, or, as far as we know, on the
0315  t last, if we look to wide enough intervals of  TIME, become modified; for those which do not chang
0315  nt of change, during long and equal periods of  TIME, may, perhaps, be nearly the same: but as the
0318  no fixed law seems to determine the length of   TIME  during which any single species or any single
0320  forms which have been produced within a given   TIME  is probably greater than that of the old forms
0321  already said on the probable wide intervals of  TIME  between our consecutive formations; and in the
0326  t of isolation, recurring at long intervals of  TIME, would probably be also favourable, as before
0326  ther might be victorious. But in the course of  TIME, the forms dominant in the highest degree, whe
0328  espond; for there will have been a little more  TIME  in the one region than in the other for modifi
0329  al formations and during the long intervals of  TIME  between them; in this case, the several format
0333  have endured for extremely unequal lengths of  TIME, and will have been modified in various degree
0334  exactly intermediate, as unequal intervals of  TIME  have elapsed between consecutive formations. I
0335  ecessarily endure for corresponding lengths of TIME: a very ancient form might occasionally last m
0335  ent would not closely accord with the order in  TIME  of their production, and still less with the o
0337  set free in New Zealand, that in the course of  TIME  a multitude of British forms would become thor
0340  ts mammals to Africa than it is at the present  TIME. Analogous facts could be given in relation to
0340  quarter, during the next succeeding period of  TIME, closely allied though in some degree modified
0340  r and degree. But after very long intervals of  TIME  and after great geographical changes, permitti
0342  sist future degradation, enormous intervals of  TIME  have elapsed between the successive formations
0342  he may disbelieve in the enormous intervals of  TIME  which have elapsed between our consecutive for
0343  bers slowly, and endure for unequal periods of  TIME: for the process of modification is necessaril
0344  for existence. Hence, after long intervals of  TIME, the productions of the world will appear to h
0350  organic bond, prevailing throughout space and  TIME, over the same areas of land and water, and in
0350  comes into play by checking migration: as does  TIME  for the slow process of modification through n
0354  this subject, we shall be enabled at the same  TIME  to consider a point equally important for us,
0355  ould probably receive from it in the course of  TIME  a few colonists, and their descendants, though
0355  me into existence coincident both in space and  TIME  with a pre existing closely allied species. An
0357  a single birthplace, and when in the course of  TIME  we know something definite about the means of
0359  es, etc., and some of these floated for a long  TIME. It is well known what a difference there is i
0359  me whilst green floated for a very short  TIME, when dried floated much longer; for instance,
0360  ts exposed to the waves would float for a less  TIME  than those protected from violent movement as
0364  ir vitality when exposed for a great length of  TIME  to the action of sea water; nor could they be
0365  transport, during the long lapse of geological  TIME, whilst an island was being upheaved and forme
0366  f the more temperate regions would at the same  TIME  travel southward; unless they were stopped by
0366  nhabitants would descend to the plains. By the  TIME  that the cold had reached its maximum, we shou
0371  ted to each other than they are at the present  TIME; for during these warmer periods the northern
0374  lent evidence, that it endured for an enormous  TIME, as measured by years, at each point. The cold
0374  my purpose, if the temperature was at the same  TIME  lower along certain broad belts of longitude.
0375  me or representing each other, and at the same  TIME  representing plants of Europe, not found in th
0381  ct, that we cannot suppose that there has been  TIME  since the commencement of the Glacial period f
0384  , and in such cases there will have been ample  TIME  for great geographical changes, and consequent
0384  r great geographical changes, and consequently TIME  and means for much migration. In the second pl
0385  twelve to twenty hours; and in this length of  TIME  a duck or heron might fly at least six or seve
0388  ickly than the high; and this will give longer TIME  than the average for the migration of the same
0389  thplace, notwithstanding that in the course of TIME  they have come to inhabit distant points of th
0394  nary view of creation; that there has not been TIME  for the creation of mammals; many volcanic isl
0394  by their tertiary strata: there has also been  TIME  for the production of endemic species belongin
0395  searches of Mr. Wallace. I have not as yet had TIME  to follow up this subject in all other quarter
0396  es depends to a certain degree on the lapse of TIME, and as during changes of level it is obvious
0396  g been largely efficient in the long course of TIME, than with the view of all our oceanic islands
0402  whatever over another, it will in a very brief TIME  wholly or in part supplant it; but if both are
0402  ces and keep separate for almost any length of TIME. Being familiar with the fact that many specie
0404  migration of a species, either at the present  TIME  or at some former period under different physi
0405  that in such cases there will have been ample  TIME  for great climatal and geographical changes an
0407  f life change most slowly, enormous periods of TIME  being thus granted for their migration, I do n
0408  erable in admitting that in the long course of TIME  the individuals of the same species, and likew
0408  forms of life; for according to the length of TIME  which has elapsed since new inhabitants entere
```

Page ***(Key Word)***

Page ***************************************(Key Word)***************************************

```
0409  ing parellelism in the laws of life throughout  TIME  and space; the laws governing the succession o
0409  y the same with those governing at the present  TIME  the differences in different areas. We see thi
0409   species and group of species is continuous in  TIME; for the exceptions to the rule are so few, th
0410  me extinct in the intermediate tracts. Both in  TIME  and space, species and groups of species have
0410  ecies, belonging either to a certain period of  TIME, or to a certain area, are often characterised
0410  ily of the same order, differ greatly. In both  TIME  and space the lower members of each class gene
0410  n my theory these several relations throughout  TIME  and space are intelligible; for whether we loo
0410  rer they will generally stand to each other in  TIME  and space; in both cases the laws of variation
0413  r; but unless it be specified whether order in  TIME  or space, or what else is meant by the plan of
0421  remains strictly true, not only at the present  TIME, but at each successive period of descent. All
0425  been greater in degree, and has taken a longer  TIME  to complete? I believe it has thus been uncons
0432  s in any class which have lived throughout all  TIME  and space. We shall certainly never succeed in
0441  e male is a mere sack, which lives for a short  TIME, and is destitute of mouth, stomach, or other
0443  cy animals, cannot positively tell, until some  TIME  after the animal has been born, what its merit
0450  ogical record not extending far enough back in  TIME, may remain for a long period, or for ever, in
0456  group to group in all organisms throughout all  TIME; that the nature of the relationship, by which
0462   same species; for during very long periods of  TIME  there will always be a good chance for wide mi
0464  it cannot be objected that there has not been   TIME  sufficient for any amount of organic change; f
0464  any amount of organic change; for the lapse of  TIME  has been so great as to be utterly inappreciab
0465  from each other by enormous blank intervals of  TIME; for fossiliferous formations, thick enough to
0465  admit. If we look to long enough intervals of   TIME, geology plainly declares that all species hav
0465  ils from formations distant from each other in  TIME. Such is the sum of the several chief objectio
0467  vinc the individuals most useful to him at the  TIME, without any thought of altering the breed. It
0471  tends to become still larger, and at the same   TIME  more divergent in character. But as all groups
0471  und us, and which has prevailed throughout all  TIME. The grand fact of the grouping of all organic
0475  the amount of change, after equal intervals of  TIME, is widely different in different groups. The
0475  ses the forms of life, after long intervals of  TIME, to appear as if they had changed simultaneous
0476   and in their geological succession throughout  TIME; for in both cases the beings have been connec
0481  hat we have acquired some idea of the lapse of  TIME, we are too apt to assume, without proof, that
0488  erves as a fair measure of the lapse of actual  TIME. A number of species, however, keeping in a bo
0488  the accuracy of organic change as a measure of  TIME. During early periods of the earth's history,
0488  hereafter be recognised as a mere fragment of   TIME, compared with the ages which have elapsed sin
0024  ies were so thoroughly domesticated in ancient  TIMES  by half civilized man, as to be quite prolifi
0031  a picture by a connoisseur; this is done three  TIMES  at intervals of months, and the sheep are eac
0034  c animals was carefully attended to in ancient  TIMES, and is now attended to by the lowest savages
0036  heir killing and devouring their old women, in  TIMES  of dearth, as of less value than their dogs.
0037  tock. The pear, though cultivated in classical  TIMES, appears, from Pliny's description, to have b
0105  , we ought to make the comparison within equal  TIMES! and this we are incapable of doing. Although
0140  es could be given of species within historical  TIMES  having largely extended their range from warm
0149  , that when any part or organ is repeated many  TIMES  in the structure of the same individual (as t
0174  ntinuous must often have existed within recent  TIMES  in a far less continuous and uniform conditio
0183  eding to another, like a kestrel, and at other  TIMES  standing stationary on the margin of water, a
0184  ll birds by blows on the head; and I have many  TIMES  seen and heard it hammering the seeds of the
0206  y correct, but if we include all those of past  TIMES, it must by my theory be strictly true. It is
0231  s make the first rough wall from ten to twenty  TIMES  thicker than the excessively thin finished wa
0241  larger workmen had heads four instead of three  TIMES  as big as those of the smaller men, and jaws
0241  those of the smaller men, and jaws nearly five  TIMES  as big. The jaws, moreover, of the working an
0258  ile; but Kölreuter tried more than two hundred  TIMES, during eight following years, to fertilise r
0296   have disappeared and reappeared, perhaps many  TIMES, during the same geological period. So that i
0320  r geological periods, so that looking to later  TIMES  we may believe that the procuction of new for
0338  ff from each other within comparatively recent  TIMES. For this doctrine of Agassiz accords well wi
0340  th america. For we know that Europe in ancient  TIMES  was peopled by numerous marsupials; and I hav
0351  anges which will have supervened since ancient  TIMES, almost any amount of migration is possible.
0351  have been produced within comparatively recent  TIMES, there is great difficulty on this head. It i
0353  ve certainly occurred within recent geological  TIMES, must have interrupted or rencered discontinu
0401  plant to another island. Hence when in former   TIMES  an immigrant settled on any one or more of th
0409  laws governing the succession of forms in past  TIMES  being nearly the same with those governing at
0414  it might have been thought (and was in ancient  TIMES  thought) that those parts of the structure wh
0437  leaves. We have formerly seen that parts many   TIMES  repeated are eminently liable to vary in numb
0438  ber of the primordially similar elements, many  TIMES  repeated, and have adapted them to the most d
0162  e markings, which are correlated with the blue  TINT; and which it does not appear probable would a
0166  rse we see this tendency strong whenever a dun  TINT  appears, a tint which approaches to that of th
0166  tendency strong whenever a dun tint appears, a  TINT  which approaches to that of the general colour
0166  any breed assumes by simple variation a bluish  TINT, these bars and other marks invariably reappea
0166  crossed, we see a strong tendency for the blue  TINT  and bars and marks to reappear in the mongrels
0288  large proportion of the ocean, the bright blue  TINT  of the water bespeaks its purity. The many cas
0382  , australia, New Zealand have become slightly   TINTED  by the same peculiar forms of vegetable life.
0084  hat of peaty earth, we must believe that these  TINTS  are of service to these birds and insects in
0087  exclusively for opening the cocoon, or the hard  TIP  to the beak of nestling birds, used for breaki
0362  500 miles, and hawks are known to look out for  TIRED  birds, and the contents of their torn crops m
0188  gination to take a thick layer of transparent   TISSUE, with a nerve sensitive to light beneath, and
0453  owers, and which is formed merely of cellular   TISSUE, can thus act? Can we suppose that the format
0483   been commanded suddenly to flash into living   TISSUES? Do they believe that at each supposed act o
0183  her at a fish. In our own country the larger   TITMOUSE (Parus major) may be seen climbing branches,
0393  ent long ago remarked that Batrachians (frogs,  TOADS, newts) have never been found on any of the m
0393  glacial agency. This general absence of frogs,  TOADS, and newts on so many oceanic islands cannot
0271  markable fact, that one variety of the common  TOBACCO  is more fertile, when crossed with a widely
0186  chting on the water; that there should be long  TOED  corncrakes living in meadows instead of in swa
0012  h feathered feet have skin between their outer  TOES; pigeons with short beaks have small feet; and
0022  nd of the feet; the number of scutellae on the  TOES, the development of skin between the toes, are
0022  the toes, the development of skin between the   TOES, are all points of structure which are variabl
0144  the feathered feet and skin between the outer  TOES  in pigeons, and the presence of more or less d
0185  seen the frigate bird, which has all its four   TOES  webbed, alight on the surface of the sea. On t
0185  nd coots are eminently aquatic, although their  TOES  are only bordered by membrane. What seems plai
0185  embrane. What seems plainer than that the long  TOES  of grallatores are formed for walking over swa
0185  bird, the deeply scooped membrane between the   TOES  shows that structure has begun to change. He w
```

Page ***************************************(Key Word)***************************************

Page ***(Key Word)***

```
0200  e seal had not a flipper, but a foot with five TOES fitted for walking or grasping; and we may fur
0275  cy of tail or horns, or additional fingers and TOES; and do not relate to characters which have be
0022  hich if shown to an ornithologist, and he were TOLD that they were wild birds, would certainly, I
0162  ther footed or turn crowned, we could not have TOLD, whether these characters in our domestic bree
0184  arsh tone of its voice, and undulatory flight, TOLD me plainly of its close blood relationship to
0251  reely. In a letter to me, in 1839, Mr. Herbert TOLD me that he had then tried the experiment durin
0285  en ocean. The same story is still more plainly TOLD by faults, those great cracks along which the
0319  ked the whole continent. But we could not have TOLD what the unfavourable conditions were which ch
0445  their puppies differed from each other: I was TOLD by breeders that they differed just as much as
0445  t of proportional difference. So, again, I was TOLD that the foals of cart and race horses differe
0020  m the first cross between two pure breeds is TOLERABLY and sometimes (as I have found with pigeons
0174  ted over a wide area, we generally find them TOLERABLY numerous over a large territory, then becom
0177  to sum up, I believe that species come to be TOLERABLY well defined objects, and do not at any one
0394  f 600 miles from the mainland. I hear from Mr. TOMES, who has specially studied this family, that
0184  anisation, even in its colouring, in the harsh TONE of its voice, and undulatory flight, told me p
0398  very character, in their habits, gestures, and TONES of voice, was manifest. So it is with the oth
0003  he woodpecker, with its feet, tail, beak, and TONGUE, so admirably adapted to catch insects under
0022  elids, of the orifice of the nostrils, of the TONGUE (not always in strict correlation with the le
0423  d would give the filiation and origin of each TONGUE. In confirmation of this view, let us glance
0480  g successive generations, by disuse or by the TONGUE and palate having been fitted by natural sele
0149  of things may be of almost any shape: whilst a TOOL for some particular object had better be of so
0224  remarked that a skilful workman, with fitting TOOLS and measures, would find it very difficult to
0242  s work by inherited instincts and by inherited TOOLS or weapons, and not by acquired knowledge and
0361  lands in the Pacific, procure stones for their TOOLS, solely from the roots of drifted trees, thes
0318  than I have done. When I found in La Plata the TOOTH of a horse embedded with the remains of Masto
0319  shment! Professor Owen soon perceived that the TOOTH, though so like that of the existing horse, b
0450  foetal whales, which when grown up have not a TOOTH in their heads: and the presence of teeth, wh
0181  which have the wing membrane extended from the TOP of the shoulder to the tail, including the hin
0223  ed motionless with its own pupa in its mouth on TOP of a spray of heath over its ravaged home. Suc
0296  e same species occur at the bottom, middle, and TOP of a formation, the probability is that they h
0362  out for tired birds, and the contents of their TORN crops might thus readily get scattered. Mr. Br
0253  evesii, and from Phasianus colchicus with P. TORQUATUS and with P. versicolor are perfectly fertil
0141  ands in the south, and on many islands in the TORRID zones. Hence I am inclined to look at adaptat
0375  und in Australia, but not in the intermediate TORRID regions. In the admirable Introduction to the
0144  ween blue eyes and deafness in cats; and the TORTOISE shell colour with the female sex; the feathe
0105  having very peculiar physical conditions, the TOTAL number of the individuals supported on it wil
0105  area, such as an oceanic island, although the TOTAL number of the species inhabiting it, will be
0138  twilight; and, last of all, those destined for TOTAL darkness. By the time that an animal had reac
0212  inct, both in extent and direction, and in its TOTAL loss. So it is with the nests of birds, which
0135  s or sacred beetle of the Egyptians, they are TOTALLY deficient. There is not sufficient evidence
0348  tern islands of the Pacific, with another and TOTALLY distinct fauna. So that here three marine fa
0437  g and leg of a bat, used as they are for such TOTALLY different purposes? Why should one crustacea
0022  so erect that in good birds the head and tail TOUCH; the oil gland is quite aborted. Several othe
0097  air pencil, and it is quite sufficient just to TOUCH the anthers of one flower and then the stigma
0071  extremely barren heath, which had never been TOUCHED by the hand of man; but several hundred acre
0238  n many social insects. But we have not as yet TOUCHED on the climax of the difficulty; namely, the
0051  gh importance for us, as being the first step TOWARDS such slight varieties as are barely thought
0098  when the stamens of a flower suddenly spring TOWARDS the pistil, or slowly move one after the oth
0098  he pistil, or slowly move one after the other TOWARDS it, the contrivance seems adapted solely to
0124  letters, converging in sub branches downwards TOWARDS a single point; this point representing a si
0145  with some difference in the flow of nutriment TOWARDS the central and external flowers: we know, a
0194  astonished how rarely an organ can be named, TOWARDS which no transitional grade is known to lead
0233  modified architectural instincts, all tending TOWARDS the present perfect plan of construction, co
0318  on of whole groups of beings, as of ammonites TOWARDS the close of the secondary period, has been
0330  ren, is discovered having affinities directed TOWARDS very distinct groups. Yet if we compare the
0376  has recently remarked, In receding from polar TOWARDS equatorial latitudes, the Alpine or mountain
0377  oductions will have retreated from both sides TOWARDS the equator, followed in the rear by the tem
0378  ator, would re migrate northward or southward TOWARDS their former homes; but the forms, chiefly in
0380  greatly reduced, and this is the first stage TOWARDS extinction. A mountain is an island on the l
0381  land, New Zealand, and Fuegia, I believe that TOWARDS the close of the Glacial period, icebergs, a
0489  l and mental endowments will tend to progress TOWARDS perfection. It is interesting to contemplate
0074  ; and Mr. Newman says, near villages and small TOWNS I have found the nests of humble bees more nu
0318  ed with the remains of Mastodon, Megatherium, TOXODON, and other extinct monsters, which all co ex
0323  f the Megatherium, Mylodon, Macrauchenia, and TOXODON had been brought to Europe from La Plata, wi
0036  n, India, and Persia, we can, I think, clearly TRACE the stages through which they have insensibly
0085  sheep to destroy every lamb with the faintest TRACE of black. In plants the down of the fruit and
0163  ay sometimes be seen in duns, and I have see a TRACE in a bay horse. My son made a careful examina
0214  grandfather was a wolf, and this dog showed a TRACE of its wild parentage only in one way, by not
0255  rtile seed: but in some of these cases a first TRACE of fertility may be detected, by the pollen o
0285  planed down by the action of the sea, that no TRACE of these vast dislocations is externally visi
0298  greatly lessen the chance of our being able to TRACE the stages of transition in any one geologica
0439  have been adapted for special lines of life. A TRACE of the law of embryonic resemblance, sometime
0442  y of the development of this insect, we see no TRACE of the vermiform stage. How, then, can we exp
0452  implies, that we find in an animal or plant no TRACE of an organ, which analogy would lead us to e
0486  armorial bearings! and we have to discover and TRACE the many diverging lines of descent in our na
0487  vel of the land, we shall surely be enabled to TRACE in an admirable manner the former migrations
0222  it was not a migration) and numerous pupae. I TRACED the returning file burthened with booty, for
0163  ipe. The hemionus has no shoulder stripe; but TRACES of it, as stated by Mr. Blyth and others, occ
0181  tail, including the hind legs, we perhaps see TRACES of an apparatus originally constructed for gl
0214  ended together, and for a long period exhibit TRACES of the instincts of either parent: for exampl
0307  tem, abounding with new and peculiar species. TRACES of life have been detected in the longmynd be
0421  ified as to have more or less completely lost TRACES of their parentage, in this case, their place
0191  e swallow has to pass over the orifice of the TRACHEA, with some risk of falling into the lungs, n
0452  th stamen: but this may sometimes be seen. In TRACING the homologies of the same part in different
0070  , increases inordinately in numbers in a small TRACT, epidemics, at least, this seems generally to
0177  ioni a second to a comparatively narrow, hilly TRACT; and a third to wide plains at the base; and
```

Page ***(Key Word)***

Page **(Key Word)***

```
0177  mall holders on the intermediate narrow, hilly TRACT; and consequently the improved mountain or pl
0308  nfer that from first to last large islands or TRACTS of land, whence the sediment was derived, occ
0309  hand, that where continents now exist, large TRACTS of land have existed, subjected no doubt to g
0347  the southern hemisphere, if we compare large TRACTS of land in Australia, South Africa, and weste
0364  would suffice for occasional transport across TRACTS of sea some hundred miles in breadth, or from
0368  r former migration across the low intervening TRACTS, since become too warm for their existence. I
0407  then have become extinct in the intermediate TRACTS, I think the difficulties in believing that a
0410  ies having become extinct in the intermediate TRACTS. Both in time and space, species and groups o
0031  s so fully recognised, that men follow it as a TRADE: the sheep are placed on a table and are stud
0197  tinct cause, probably to sexual selection. A TRAILING bamboo in the Malay Archipelago climbs the l
0208  ally forced to go back to recover the habitual TRAIN of thought: so P. Huber found it was with a c
0035  similar process of selection, and by careful TRAINING, the whole body of English racehorses have c
0213  one kind of wolf, when young and without any TRAINING, as soon as it scented its prey, stand motio
0214  oubted whether any one would have thought of TRAINING a dog to point, had not some one dog natural
0214  tion and the inherited effects of compulsory TRAINING in each successive generation would soon com
0356  pair, but to continued care in selecting and TRAINING many individuals during many generations. Be
0436  dence of the possibility of one organ being TRANSFORMED into another; and we can actually see in e
0138  far remote from ordinary forms, prepare the TRANSITION from light to darkness. Next follow those t
0171  ies. Organs of extreme perfection. Means of TRANSITION. Cases of difficulty. Natura non facit salt
0182  at least, to show what diversified means of TRANSITION are possible. Seeing that a few members of
0193  at we are far too ignorant to argue that no TRANSITION of any kind is possible. The electric organ
0298  in any one country all the early stages of TRANSITION between any two forms, is small, for the su
0298  ce of our being able to trace the stages of TRANSITION in any one geological formation. It should
0171  ication. Transitions. Absence or rarity of TRANSITIONAL varieties. Transitions in habits of life.
0171  ions, do we not everywhere see innumerable TRANSITIONAL forms? Why is not all nature in confusion
0172  rate chapters. On the absence or rarity of TRANSITIONAL varieties. As natural selection acts solel
0172  unknown form, both the parent and all the TRANSITIONAL varieties will generally have been extermi
0172  w form. But, as by this theory innumerable TRANSITIONAL forms must have existed, why do we not fin
0173  ely ought to find at the present time many TRANSITIONAL forms. Let us take a simple case: in trave
0173  terminated its original parent and all the TRANSITIONAL varieties between its past and present sta
0174  at the present time to meet with numerous TRANSITIONAL varieties in each region, though they must
0179  ic habits; for how could the animal in its TRANSITIONAL state have subsisted? It would be easy to
0180  d, I can give only one or two instances of TRANSITIONAL habits and structures in closely allied sp
0182  would have ever imagined that in an early TRANSITIONAL state they had been inhabitants of the ope
0182  bear in mind that animals displaying early TRANSITIONAL grades of the structure will seldom contin
0182  lection. Furthermore, we may conclude that TRANSITIONAL grades between structures fitted for very
0183  nce the chance of discovering species with TRANSITIONAL grades of structure in a fossil condition
0188  h in this case he does not know any of the TRANSITIONAL grades. His reason ought to conquer his im
0189  y organs exist of which we do not know the TRANSITIONAL grades, more especially if we look to much
0189  eloped; and in order to discover the early TRANSITIONAL grades through which the organ has passed,
0190  hat an organ could not have been formed by TRANSITIONAL gradations of some kind. Numerous cases co
0192  possibly have been produced by successive TRANSITIONAL gradations, yet, undoubtedly, grave cases
0194  ly an organ can be named, towards which no TRANSITIONAL grade is known to lead. The truth of this
0204  ses in which we know of no intermediate or TRANSITIONAL states, we should be very cautious in conc
0210  we ought to find in nature, not the actual TRANSITIONAL gradations by which each complex instinct
0279  not being blended together by innumerable TRANSITIONAL links, is a very obvious difficulty. I ass
0282  ss. So that the number of intermediate and TRANSITIONAL links, between all living and extinct spec
0292  rded against the frequent discovery of her TRANSITIONAL or linking forms. From the foregoing consi
0297  formation, and unless we obtained numerous TRANSITIONAL gradations, we should not recognise their
0301  ispensable for the preservation of all the TRANSITIONAL gradations between any two or more species
0301  such gradations were not fully preserved, TRANSITIONAL varieties would merely appear as so many d
0301  rmations, an infinite number of those fine TRANSITIONAL forms, which on my theory assuredly have c
0302  ficulty of our not discovering innumerable TRANSITIONAL links between the species which appeared a
0310  successive formations infinitely numerous TRANSITIONAL links between the many species which now e
0342  e may ask in vain where are the numberless TRANSITIONAL links which must formerly have connected t
0466  nt we are. We do not know all the possible TRANSITIONAL gradations between the simplest and the mo
0005  e given: namely, first, the difficulties of TRANSITIONS, or in understanding how a simple being or
0171  on the theory of descent with modification. TRANSITIONS. Absence or rarity of transitional varieti
0171  bsence or rarity of transitional varieties. TRANSITIONS in habits of life. Diversified habits in t
0179  and intermittent record. On the origin and TRANSITIONS of organic beings with peculiar habits and
0191  erted into organs of flight. In considering TRANSITIONS of organs, it is so important to bear in m
0194  it is most difficult to conjecture by what TRANSITIONS an organ could have arrived at its present
0205  other, must often have largely facilitated TRANSITIONS. We are far too ignorant, in almost every
0126  living at any one period, extremely few will TRANSMIT descendants to a remote futurity. I shall ha
0128  era, which vary most; and these will tend to TRANSMIT to their modified offspring that superiority
0266  hen hybrids are able to breed inter se, they TRANSMIT to their offspring from generation to genera
0489  afely infer that not one living species will TRANSMIT its unaltered likeness to a distant futurity
0489  and of the species now living very few will TRANSMIT progeny of any kind to a far distant futurit
0013  remote ancestor; why a peculiarity is often TRANSMITTED from one sex to both sexes, or to one sex
0013  the males of our domestic breeds are often TRANSMITTED either exclusively, or in a much greater d
0026  all that we can see to the contrary, may be TRANSMITTED undiminished for an indefinite number of g
0090  pons, means of defence, or charms; and have TRANSMITTED these advantages to their male offspring.
0122  very few of the original species will have TRANSMITTED offspring to the fourteen thousandth gener
0122  ted to the other nine original species, has TRANSMITTED descendants to this late stage of descent.
0126  mely few of the more ancient species having TRANSMITTED descendants, and on the view of all the de
0146  n; and these two modifications, having been TRANSMITTED to a whole group of descendants with diver
0154  organ, however abnormal it may be, has been TRANSMITTED in approximately the same condition to man
0160  d, for all that we can see to the contrary, TRANSMITTED for almost any number of generations. When
0187  r the chance of some gradations having been TRANSMITTED from the earlier stages of descent, in an
0196  wly perfected at a former period, have been TRANSMITTED in nearly the same state, although now bec
0209  quired by habit in one generation, and then TRANSMITTED by inheritance to succeeding generations.
0213  far less rigorous selection, and have been TRANSMITTED for an incomparably shorter period, under
0235  n of wax, having succeeded best, and having TRANSMITTED by inheritance its newly acquired economic
0237  lutely sterile; so that it could never have TRANSMITTED successively acquired modifications of str
0238  males of the same community flourished, and TRANSMITTED to their fertile offspring a tendency to p
0420  of three of these genera (A, F, and I) have TRANSMITTED modified descendants to the present day, r
```

Page **(Key Word)***

Page ********`**********************************(Key Word)**************************************

0429 termediate parent forms having occasionally TRANSMITTED to the present day descendants but little
0431 o distinct groups and sub groups, will have TRANSMITTED some of its characters, modified in variou
0121 inal genus, may for a long period continue TRANSMITTING unaltered descendants; and this is shown i
0274 but more especially owing to prepotency in TRANSMITTING likeness running more strongly in one sex
0302 as a fatal objection to the belief in the TRANSMUTATION of species. If numerous species, belongin
0187 s shaped swelling. In other crustaceans the TRANSPARENT cones which are coated by pigment, and whi
0188 merely coated with pigment and invested by TRANSPARENT membrane, into an optical instrument as pe
0188 ght in imagination to take a thick layer of TRANSPARENT tissue, with a nerve sensitive to light be
0189 ng each slight accidental alteration in the TRANSPARENT layers; and carefully selecting each alter
0092 usted with pollen, and would certainly often TRANSPORT the pollen from one flower to the stigma of
0354 ical changes and various occasional means of TRANSPORT, the belief that this has been the universa
0358 ill adapted for wide dissemination; but for TRANSPORT across the sea, the greater or less facilit
0361 particle could be washed away in the longest TRANSPORT: out of one small portion of earth thus com
0363 considering that the several above means of TRANSPORT, and that several other means, which withou
0364 is become widely transported. These means of TRANSPORT are sometimes called accidental, but this i
0364 hould be observed that scarcely any means of TRANSPORT would carry seeds for very great distances;
0364 means, however, would suffice for occasional TRANSPORT across tracts of sea some hundred miles in
0365 t few centuries, through occasional means of TRANSPORT, immigrants from Europe or any other contin
0365 hat would be effected by occasional means of TRANSPORT, during the long lapse of geological time,
0382 e southern hemisphere by occasional means of TRANSPORT, and by the aid, as halting places, of exis
0384 o favour the possibility of their occasional TRANSPORT by accidental means: like that of the live
0387 plicable circumstance if water birds did not TRANSPORT the seeds of fresh water plants to vast dis
0396 better with the view of occasional means of TRANSPORT having been largely efficient in the long c
0398 ve colonists, whether by occasional means of TRANSPORT or by formerly continuous land, from Americ
0401 s having been stocked by occasional means of TRANSPORT, a seed, for instance, of one plant having
0404 es, excepting in so far as great facility of TRANSPORT has given the same general forms to the who
0405 nd geographical changes and for accidents of TRANSPORT; and consequently for the migration of some
0407 to the many and curious means of occasional TRANSPORT, a subject which has hardly ever been prope
0407 how diversified are the means of occasional TRANSPORT, I have discussed at some little length the
0410 fferent conditions or by occasional means of TRANSPORT, and by the species having become extinct i
0462 dly ignorant of the many occasional means of TRANSPORT. With respect to distinct species of the sa
0092 ted them, so as to favour in any degree the TRANSPORTAL of their pollen from flower to flower, wou
0392 ing than the adaptation of hooked seeds for TRANSPORTAL by the wool and fur of quadrupeds. This ca
0393 at there would be great difficulty in their TRANSPORTAL across the sea, and therefore why they do
0397 known, but highly efficient means for their TRANSPORTAL. Would the just hatched young occasionally
0140 ubsequently found capable of far extended TRANSPORTATION, I think the common and extraordinary cap
0361 fail to be highly effective agents in the TRANSPORTATION of seeds. I could give many facts showing
0406 very minute and better fitted for distant TRANSPORTATION, probably accounts for a law which has lo
0003 certain trees, which has seeds that must be TRANSPORTED by certain birds, and which has flowers wi
0011 in the period of flowering with plants when TRANSPORTED from one climate to another. In animals it
0024 l above named domesticated breeds have been TRANSPORTED to all parts of the world, and, therefore,
0141 red as domestic animals, but they have been TRANSPORTED by man to many parts of the world, and now
0284 over what wide spaces very fine sediment is TRANSPORTED by the currents of the sea, the process of
0320 ame neighbourhood: when much improved it is TRANSPORTED far and near, like our short horn cattle,
0360 s with large seeds or fruit could hardly be TRANSPORTED by any other means; and Alph. de Candolle
0360 icted ranges. But seeds may be occasionally TRANSPORTED in another manner. Drift timber is thrown
0362 oured by birds, and thus the seeds might be TRANSPORTED from place to place. I forced many kinds o
0363 f a vetch. Thus seeds might occasionally be TRANSPORTED to great distances; for many facts could b
0363 rdly doubt that they must occasionally have TRANSPORTED seeds from one part to another of the arct
0364 t if many plants had not thus become widely TRANSPORTED. These means of transport are sometimes ca
0364 of Ireland and England: but seeds could be TRANSPORTED by these wanderers only by one means, name
0365 I doubt that out of twenty seeds or animals TRANSPORTED to an island, even if far less well stocke
0373 of former glacial action, in huge boulders TRANSPORTED far from their parent source. We do not kn
0385 me much, as their ova are not likely to be TRANSPORTED by birds, and they are immediately killed
0391 r, or to the feet of wading birds, might be TRANSPORTED far more easily than land shells, across t
0392 ulty on my view, for a hooked seed might be TRANSPORTED to an island by some other means: and the
0394 western shores, and they may have formerly TRANSPORTED foxes, as so frequently now happens in the
0394 answered; for no terrestrial mammal can be TRANSPORTED across a wide space of sea, but bats can f
0397 birds roosting on the ground, and thus get TRANSPORTED? It occurred to me that land shells, when
0403 ough large quantities of stone are annually TRANSPORTED from Porto Santo to Madeira, yet this latt
0435 ndency to modify the original pattern, or to TRANSPOSE parts. The bones of a limb might be shorten
0434 e arm and forearm, or of the thigh and leg, TRANSPOSED. Hence the same names can be given to the h
0142 clopaedias of China, to be very cautious in TRANSPOSING animals from one district to another; for
0163 rsion. The ass not rarely has very distinct TRANSVERSE bars on its legs, like those on the legs of
0163 e most distinct breeds; and of all colours: TRANSVERSE bars on the legs are not rare in duns, mous
0069 ncing winter, which will suffer most. When we TRAVEL from south to north, or from a damp region to
0069 nts, the other species will decrease. When we TRAVEL southward and see a species decreasing in num
0069 , as in this one being hurt. So it is when we TRAVEL northward, but in a somewhat lesser degree, f
0346 hat between the New and Old Worlds: yet if we TRAVEL over the vast American continent, from the ce
0366 more temperate regions would at the same time TRAVEL southward, unless they were stopped by barrie
0379 orthern, which had crossed the equator, would TRAVEL still further from their homes into the most
0384 a marine member of a fresh water group might TRAVEL far along the shores of the sea, and subseque
0388 ch have large powers of flight, and naturally TRAVEL from one to another and often distant piece o
0367 bitants, which we suppose to have everywhere TRAVELLED southward, are remarkably uniform round the
0477 woncerful fact, which must have struck every TRAVELLER, namely, that on the same continent, under
0173 tional forms. Let us take a simple case: in TRAVELLING from north to south over a continent, we ge
0348 able islands as halting places, until after TRAVELLING over a hemisphere we come to the shores of
0349 e instances. Nevertheless the naturalist in TRAVELLING, for instance, from north to south never fa
0477 new and peculiar species of bats, which can TRAVERSE the ocean, should so often be found on islan
0395 ard to the great Malay Archipelago, which is TRAVERSED near Celebes by a space of deep ocean; and
0004 e; but I shall, unfortunately, be compelled to TREAT this subject far too briefly, as it can be tr
0027 n of this fact will be obvious when we come to TREAT of selection. Fourthly, pigeons have been wat
0028 igeons have undergone, will be obvious when we TREAT of Selection. We shall then, also, see how it
0044 these latter are subject to any variation. To TREAT this subject at all properly, a long catalogu
0096 presently see its importance; but I must here TREAT the subject with extreme brevity, though I ha
0207 ve thought that it would be more convenient to TREAT the subject separately, especially as so wond

Page **(Key Word)**************************************

```
0325 al law. We shall see this more clearly when we TREAT of the present distribution of organic beings
0485 me into accordance. In short, we shall have to TREAT species in the same manner as those naturalis
0485 pecies in the same manner as those naturalists TREAT genera, who admit that genera are merely arti
0004 at this subject far too briefly, as it can be TREATED properly only by giving long catalogues of f
0005 high geometrical powers of increase, will be TREATED of. This is the doctrine of Malthus, applied
0005 damental subject of Natural Selection will be TREATED at some length in the fourth chapter; and we
0053 well established. The whole subject, however, TREATED as it necessarily here is with much brevity,
0062 ence. In my future work this subject shall be TREATED, as it well deserves, at much greater length
0062 competition. In regard to plants, no one has TREATED this subject with more spirit and ability th
0067 any other animal. This subject has been ably TREATED by several authors, and I shall, in my futur
0192 les or fertile females; but this case will be TREATED of in the next chapter. The electric organs
0252 possible to imagine. Had hybrids, when fairly TREATED, gone on decreasing in fertility in each suc
0252 f the same hybrids, and such alone are fairly TREATED, for by insect agency the several individual
0356 sal. Sir C. Lyell and other authors have ably TREATED this subject. I can give here only the brief
0245 incidental on other acquired differences. In TREATING this subject, two classes of facts, to a lar
0291 urprised to note how author after author, in TREATING of this or that great formation, has come to
0012 l importance is endless. Dr. Prosper Lucas's TREATISE, in two large volumes, is the fullest and th
0029 ed from a distinct species. Van Mons, in his TREATISE on pears and apples, shows how utterly he di
0165 ebra; and Mr. W. C. Martin, in his excellent TREATISE on the horse, has given a figure of a simila
0188 ll go thus far, if he find on finishing this TREATISE that large bodies of facts, otherwise inexpl
0012 l worth while carefully to study the several TREATISES published on some of our old cultivated pla
0020 and by the Hon. C. Murray from Persia. Many TREATISES in different languages have been published
0026 e two distinct cases are often confounded in TREATISES on inheritance. Lastly, the hybrids or mong
0028 s, with whom I have ever conversed, or whose TREATISES I have read, are firmly convinced that the
0031 almost necessary to read several of the many TREATISES devoted to this subject, and to inspect the
0033 een more attended to of late years, and many TREATISES have been published on the subject; and the
0035 y comparing the accounts given in old pigeon TREATISES of carriers and tumblers with these breeds
0142 s best adapted to their native countries. In TREATISES on many kinds of cultivated plants, certain
0282 he principles of geology, or to read special TREATISES by different observers on separate formatio
0303 ecall the well known fact that in geological TREATISES, published not many years ago, the great cl
0007 l ages under the most different climates and TREATMENT, I think we are driven to conclude that thi
0008 t. Hilaire's experiments show that unnatural TREATMENT of the embryo causes monstrosities; and mon
0010 ultivation; and in this case we see that the TREATMENT of the parent has affected a bud or offset,
0010 len, or to both, having been affected by the TREATMENT of the parent prior to the act of conceptio
0047 by others having intermediate characters, he TREATS the one as a variety of the other, ranking th
0269 ies his several varieties with the same food; TREATS them in nearly the same manner, and does not
0164 ripe, which is sometimes double and sometimes TREBLE, is common; the side of the face, moreover, i
0029 ever have proceeded from the seeds of the same TREE. Innumerable other examples could be given. Th
0046 ed to the irregular branching of the stem of a TREE. This philosophical naturalist, I may add, has
0063 f too many of these parasites grow on the same TREE, it will languish and die. But several seedlin
0066 would suffice to keep up the full number of a TREE, which lived on an average for a thousand year
0071 een the effect of the introduction of a single TREE, nothing whatever else having been done, with
0081 powerful the influence of a single introduced TREE or mammal has been shown to be. But in the cas
0093 pollen can be detected. Having found a female TREE exactly sixty yards from a male tree, I put th
0093 a female tree exactly sixty yards from a male TREE, I put the stigmas of twenty flowers, taken fr
0093 t for several days from the female to the male TREE, the pollen could not thus have been carried.
0093 dentally dusted with pollen, having flown from TREE to tree in search of nectar. But to return to
0093 dusted with pollen, having flown from tree to TREE in search of nectar. But to return to our imag
0099 in a future chapter. In the case of a gigantic TREE covered with innumerable flowers, it may be ob
0099 ected that pollen could seldom be carried from TREE to tree, and at most only from flower to flowe
0099 at pollen could seldom be carried from tree to TREE, and at most only from flower to flower on the
0100 at most only from flower to flower on the same TREE, and that flowers on the same tree can be cons
0100 on the same tree, and that flowers on the same TREE can be considered as distinct individuals only
0100 and female flowers may be produced on the same TREE, we can see that pollen must be regularly carr
0100 ance of pollen being occasionally carried from TREE to tree. That trees belonging to all Orders ha
0100 pollen being occasionally carried from tree to TREE. That trees belonging to all Orders have their
0129 ass have sometimes been represented by a great TREE. I believe this simile largely speaks the trut
0129 esser branches, were themselves once, when the TREE was small, budding twigs; and this connexion o
0129 s. of the many twigs which flourished when the TREE was a mere bush, only two or three, now grown
0129 fied descendants. From the first growth of the TREE, many a limb and branch has decayed and droppe
0130 ing branch springing from a fork low down in a TREE, and which by some chance has been favoured an
0130 eneration I believe it has been with the great TREE of Life, which fills with its dead and broken
0180 hrough the air to an astonishing distance from TREE to tree. We cannot doubt that each structure i
0180 he air to an astonishing distance from tree to TREE. We cannot doubt that each structure is of use
0184 ng; and on the plains of La Plata, where not a TREE grows, there is a woodpecker, which in every e
0184 s; yet it is a woodpecker which never climbs a TREE! Petrels are the most aerial and oceanic of bi
0186 ; that there should be woodpeckers where not a TREE grows; that there should be diving thrushes, a
0197 colour was a beautiful adaptation to hide this TREE frequenting bird from its enemies; and consequ
0221 ching along the same road to a tall Scotch fir TREE, twenty five yards distant, which they ascende
0261 lants. We can sometimes see the reason why one TREE will not take on another; from differences in
0317 cies, and so on, like the branching of a great TREE from a single stem, till the group becomes lar
0392 ly competing in stature with a fully developed TREE, when established on an island and having to c
0431 oble family, even by the aid of a genealogical TREE, and almost impossible to do this without this
0432 r from I, would have something in common. In a TREE we can specify this or that branch, though at
0484 uces monstrous growths on the wild rose or oak TREE. Therefore I should infer from analogy that pr
0003 bly adapted to catch insects under the bark of TREES. In the case of the misseltoe, which draws it
0003 ltoe, which draws its nourishment from certain TREES, which has seeds that must be transported by
0023 that is, not breeding or willingly perching on TREES. But besides C. Livia, with its geographical
0055 eral rule, to be now forming. Where many large TREES grow, we expect to find saplings. Where many
0063 etoe is dependent on the apple and a few other TREES, but can only in a far fetched sense be said
0063 r fetched sense be said to struggle with these TREES, for if too many of these parasites grow on t
0072 nnot live. When I ascertained that these young TREES had not been sown or planted, I was so much s
0072 h, i found a multitude of seedlings and little TREES, which had been perpetually browsed down by t
0072 of the old clumps, I counted thirty two little TREES; and one of them, judging from the rings of g
0074 springs up; but it has been observed that the TREES now growing on the ancient Indian mounds, in
0074 . what a struggle between the several kinds of TREES must here have gone on during long centuries,
```

Page ***(Key Word)***

```
0075  rease, and all feeding on each other or on the  TREES or their seeds and seedlings, or on the other
0075   the ground and thus checked the growth of the  TREES! Throw up a handful of feathers, and all must
0075  nturies, the proportional numbers and kinds of  TREES now growing on the old Indian ruins! The depe
0085  es, assuredly, in a state of nature, where the  TREES would have to struggle with other trees and w
0085  re the trees would have to struggle with other  TREES and with a host of enemies, such differences
0086  y selection the down in the pods on his cotton  TREES. Natural Selection may modify and adapt the l
0093  plants, presently to be alluded to. Some holly  TREES bear only male flowers, which have four stame
0093  pollen, and a rudimentary pistil; other holly  TREES bear only female flowers; these have a full s
0100  e has largely provided against it by giving to  TREES a strong tendency to bear flowers with separa
0100  g occasionally carried from tree to tree. That  TREES belonging to all Orders have their sexes more
0100  ry; and at my request Dr. Hooker tabulated the  TREES of New Zealand, and Dr. Asa Gray those of the
0100   i have made these few remarks on the sexes of  TREES simply to call attention to the subject. Turn
0113  alive; some inhabiting new stations, climbing  TREES, frequenting water, and some perhaps becoming
0140  raised from seed collected by Dr. Hooker from  TREES growing at different heights on the Himalaya,
0142  his is very strikingly shown in works on fruit  TREES published in the United States, in which cert
0184  e given that of a woodpecker for climbing  TREES and for seizing insects in the chinks of the
0197  o in the Malay Archipelago climbs the loftiest  TREES by the aid of exquisitely constructed hooks c
0197  nti but as we see nearly similar hooks on many  TREES which are not climbers, the hooks on the bamb
0203  as equally perfect the elaboration by our fir  TREES of dense clouds of pollen, in order that a fe
0257  and perennial plants, deciduous and evergreen  TREES, plants inhabiting different stations and fit
0261  ic affinity, for no one has been able to graft  TREES together belonging to quite distinct families
0263  complex laws governing the facility with which  TREES can be grafted on each other as incidental on
0264  iven of these facts, any more than why certain  TREES cannot be grafted on others. Lastly, an embry
0276  this cross. In the same manner as in grafting  TREES, the capacity of one species or variety to ta
0276  ing and blending in nature, than to think that  TREES have been specially endowed with various and
0296  have the plainest evidence in great fossilised  TREES, still standing upright as they grew, of many
0296  ld never even have been suspected, had not the  TREES chanced to have been preserved: thus, Messrs.
0361  their tools, solely from the roots of drifted  TREES, these stones being a valuable royal tax. I f
0361  rly shaped stones are embedded in the roots of  TREES, small parcels of earth are very frequently e
0392  insular beetles. Again, islands often possess  TREES or bushes belonging to orders which elsewhere
0392  elsewhere include only herbaceous species; now  TREES, as Alph. de Candolle has shown, generally ha
0392  tever the cause may be, confined ranges. Hence  TREES would be little likely to reach distant ocean
0392  ert them first into bushes and ultimately into  TREES. With respect to the absence of whole orders
0396  s, herbaceous forms having been developed into  TREES, etc., seem to me to accord better with the v
0460  pecial endowment than is the incapacity of two  TREES to be grafted together; but that it is incide
0472  at the astonishing waste of pollen by our fir  TREES; at the instinctive hatred of the queen bee f
0068  the birds in my own grounds; and this is a  TREMENDOUS destruction, when we remember that ten per
0271  varieties, and which he tested by the severest  TRIAL, namely, by reciprocal crosses, and he found
0362  , and do not in the least injure, as I know by  TRIAL, the germination of seeds; now after a bird h
0439  otted feathers in the thrush group. In the cat  TRIBE, most of the species are striped or spotted i
0213  sition and tastes, and likewise of the oddest  TRICKS, associated with certain frames of mind or pe
0073  o the fertilisation of the heartsease (Viola  TRICOLOR), for other bees do not visit this flower. F
0070  are checked during winter: but any one who has  TRIED, knows how troublesome it is to get seed from
0072  e rings of growth, had during twenty six years  TRIED to raise its head above the stems of the heat
0073  sit this flower. From experiments which I have  TRIED, I have found that the visits of bees, if not
0129  ch period of growth all the growing twigs have  TRIED to branch out on all sides, and to overtop an
0129  e manner as species and groups of species have  TRIED to overmaster other species in the great batt
0142  he experiment cannot be said to have been even  TRIED. Nor let it be supposed that no differences i
0208  e third stage, where it had left off, and thus  TRIED to complete the already finished work. If we
0220  ng to the statements of Huber and Mr. Smith; I  TRIED to approach the subject in a sceptical frame
0230  bend the ductile and warm wax (which as I have  TRIED is easily done) into its proper intermediate
0251  in 1839, Mr. Herbert told me that he had then  TRIED the experiment during five years, and he cont
0252  ls, much fewer experiments have been carefully  TRIED than with plants. If our systematic arrangeme
0252  confinement, few experiments have been fairly  TRIED: for instance, the canary bird has been cross
0258  oduced are sufficiently fertile; but Kolreuter  TRIED more than two hundred times. during eight fol
0273  ic varieties (very few experiments having been  TRIED on natural varieties), and this implies in mo
0358  y be said to be almost wholly unknown. Until I  TRIED, with Mr. Berkeley's aid, a few experiments,
0359  on of 137 days. For convenience sake I chiefly  TRIED small seeds, without the capsule or fruit; an
0359  y were injured by the salt water. Afterwards I  TRIED some larger fruits, capsules, etc., and some
0360  te. Subsequently to my experiments, M. Martens  TRIED similar ones, but in a much better manner, fo
0360  sed to the air like really floating plants. He  TRIED 98 seeds, mostly different from mine; but he
0361  hese seemed perfect, and some of them, which I  TRIED, germinated. But the following fact is more i
0386  charged the mud of ponds is with seeds: I have  TRIED several little experiments, but will here giv
0397  d by salt; their eggs, at least such as I have  TRIED, sink in sea water and are killed by it. Yet
0034  , a man who intends keeping pointers naturally  TRIES to get as good dogs as he can, and afterwards
0211  nother of a distinct species, yet each species  TRIES to take advantage of the instincts of others,
0215  scious selection is still at work, as each man  TRIES to procure, without intending to improve the
0265  ally fertile hybrids. No one can tell, till he  TRIES, whether any particular animal will breed und
0265  freely under culture: nor can he tell, till he  TRIES, whether any two species of a genus will prod
0073  periods of time, though assuredly the merest  TRIFLE would often give the victory to one organic b
0285  t the denudation of the Weald has been a mere  TRIFLE, in comparison with that which has removed ma
0008  w such cases it has been found out that very  TRIFLING changes, such as a little more or less water
0016  other species of the same genus, in several  TRIFLING respects, they often differ in an extreme de
0023  es, which differ from each other in the most  TRIFLING respects. As several of the reasons which ha
0032  records have been kept: thus, to give a very  TRIFLING instance, the steadily increasing size of th
0048  and in making this list he has omitted many  TRIFLING varieties, but which nevertheless have been
0080  man's selection, its power on characters of  TRIFLING importance, its power at all ages and on bot
0084  ctures, which we are apt to consider as very  TRIFLING importance, may thus be acted on. When we se
0085  dered by botanists as characters of the most  TRIFLING importance: yet we hear from an excellent ho
0095  , for instance, of the coast waves, called a  TRIFLING and insignificant cause, when applied to the
0171  on could produce; on the one hand, organs of  TRIFLING importance, such as the tail of a giraffe, w
0195  ormer chapter I have given instances of most  TRIFLING characters, such as the down on fruit and th
0195  odifications, each better and better, for so  TRIFLING an object as driving away flies; yet we shou
0195  to escape from beasts of prey. Organs now of  TRIFLING importance have probably in some cases been
0235  exist; cases of instinct of apparently such  TRIFLING importance, that they could hardly have been
0323  ometimes are similarly characterised in such  TRIFLING points as mere superficial sculpture. Moreov
```

Page ***(Key Word)***

Page **(Key Word)**

```
0410  o a certain area, are often characterised by TRIFLING characters in common, as of sculpture or col
0416  from parts which must be considered of very TRIFLING physiological importance, but which are univ
0416  feathers instead of hair, this external and TRIFLING character would, I think, have been consider
0417  rgan. The importance, for classification, of TRIFLING characters, mainly depends on their being co
0417  ing seems founded on an appreciation of many TRIFLING points of resemblance, too slight to be defi
0426  r parts of the organisation. We care not how TRIFLING a character may be, let it be the mere infle
0426  when several characters, let them be ever so TRIFLING, occur together throughout a large group of
0433  rudimentary and useless organs, or others of TRIFLING physiological importance: why, in comparing
0456  her of high vital importance, or of the most TRIFLING importance, or, as in rudimentary organs, of
0484  wth and reproduction. We see this even in so TRIFLING a circumstance as that the same poison often
0073  mble bees alone visit the common red clover (TRIFOLIUM pratense), as other bees cannot reach the n
0094  las of the common red and incarnate clovers (TRIFOLIUM pratense and incarnatum) do not on a hasty
0321  mpetition. For instance, a single species of TRIGONIA, a great genus of shells in the secondary fo
0306  tance, I cannot doubt that all the Silurian TRILOBITES have descended from some one crustacean, wh
0321  mination of whole families or orders, as of TRILOBITES at the close of the palaeozoic period and o
0126  s of organic beings which are now large and TRIUMPHANT, and which are least broken up, that is, wh
0326  still more dominant species, and then their TRIUMPHANT course, or even their existence, would ceas
0350  abounding in individuals, which have already TRIUMPHED over many competitors in their own widely e
0243  tish thrush: how it is that the male wrens (TROGLODYTES) of North America, build cock nests, to ro
0133  the species of shells which are confined to TROPICAL and shallow seas are generally brighter colo
0139  even from a temperate region cannot endure a TROPICAL climate, or conversely. So again, many succu
0141  l wild stocks: the blood, for instance, of a TROPICAL and arctic wolf or wild dog may perhaps be m
0141  mate, whereas the living species are now all TROPICAL or sub tropical in their habits, ought not t
0141  e living species are now all tropical or sub TROPICAL in their habits, ought not to be looked at a
0305  ay archipelago were converted into land, the TROPICAL parts of the Indian Ocean would form a large
0348  her westward from the eastern islands of the TROPICAL parts of the Pacific, we encounter no impass
0376  but have not been found in the intermediate TROPICAL seas. It should be observed that the norther
0377  gration. As the cold came slowly on, all the TROPICAL plants and other productions will have retre
0377  ith the latter we are not now concerned. The TROPICAL plants probably suffered much extinction: ho
0377  of temperate Australia. As we know that many TROPICAL plants and animals can withstand a considera
0377  the great fact to bear in mind is, that all TROPICAL productions will have suffered to a certain
0377  ems to me possible, bearing in mind that the TROPICAL productions were in a suffering state and co
0378  istricts will have afforded an asylum to the TROPICAL natives. The mountain ranges north west of t
0378  t period, I suppose that large spaces of the TROPICAL lowlands were clothed with a mingled tropica
0378  ropical lowlands were clothed with a mingled TROPICAL and temperate vegetation, like that now grow
0009  ntion that carnivorous animals, even from the TROPICS, breed in this country pretty freely under c
0377  how much no one can say: perhaps formerly the TROPICS supported as many species as we see at the p
0378  s me that it is the damp with the heat of the TROPICS which is so destructive to perennial plants
0378  entered and crossed even the lowlands of the TROPICS at the period when the cold was most intense
0094  wers, which they can, with a very little more TROUBLE, enter by the mouth. Bearing such facts in m
0304  ement of the tertiary series. This was a sore TROUBLE to me, adding as I thought one more instance
0418  lly when naturalists are at work, they do not TROUBLE themselves about the physiological value of
0420  understand what is meant, if he will take the TROUBLE of referring to the diagram in the fourth ch
0070  inter: but any one who has tried, knows how TROUBLESOME it is to get seed from a few wheat or othe
0003  sense, as we shall hereafter see, this may be TRUE; but it is preposterous to attribute to mere e
0015  remarked, less uniformity of character than in TRUE species. Domestic races of the same species, a
0018  er fact were found more strictly and generally TRUE than seems to me to be the case, what does it
0019  ors. They believe that every race which breeds TRUE. Let the distinctive characters be ever so sli
0028  nd watched the several kinds, knowing well how TRUE they bred, I felt fully as much difficulty in
0033  , rapid and important. But it is very far from TRUE that the principle is a modern discovery. I co
0042  o the fancier, for thus many races may be kept TRUE, though mingled in the same aviary, and this c
0047  r as long, as far as we know, as have good and TRUE species. Practically, when a naturalist can un
0049  d by other highly competent judges as good and TRUE species. But to discuss whether they are right
0058  rsally ranked by British botanists as good and TRUE species. Finally, then, varieties have the sam
0085  nd in keeping that colour, when once acquired, TRUE and constant. Nor ought we to think that the o
0095  al wave, so will natural selection, if it be a TRUE principle, banish the belief of the continued
0099  ing near each other, and of these only 78 were TRUE to their kind, and some even of these were not
0099  ind, and some even of these were not perfectly TRUE. Yet the pistil of each cabbage flower is surr
0103  s of the same species, or of the same variety, TRUE and uniform in character. It will obviously th
0134  e known to every naturalist of species keeping TRUE, or not varying at all, although living under
0147  conomise on the other side. I think this holds TRUE to a certain extent with our domestic producti
0153  he power of steady selection to keep the breed TRUE. In the long run selection gains the day, and
0154  is explanation is partly, yet only indirectly, TRUE; I shall, however, have to return to this subj
0163  uiries which I have made, I believe this to be TRUE. It has also been asserted that the stripe on
0166  the breeds of pigeons, some of which have bred TRUE for centuries, species; and how exactly parall
0179  any one time, but to all time, if my theory be TRUE, numberless intermediate varieties, linking mo
0183  does not seem probable that fishes capable of TRUE flight would have been developed under many su
0191  ardly doubt that all vertebrate animals having TRUE lungs have descended by ordinary generation fr
0199  of man, or for mere variety. This doctrine, if TRUE, would be absolutely fatal to my theory. Yet I
0206  f past times, it must by my theory be strictly TRUE. It is generally acknowledged that all organic
0213  see that these actions differ essentially from TRUE instincts. If we were to see one kind of wolf,
0214  nd compulsory habit, but this, I think, is not TRUE. No one would ever have thought of teaching or
0224  it very difficult to make cells of wax of the TRUE form, though this is perfectly effected by a c
0228  shallow basins, appearing to the eye perfectly TRUE or parts of a sphere, and of about the diamete
0268  which will occur to everyone, and probably the TRUE one, is that these dogs have descended from se
0273  vert to either parent form: but this, if it be TRUE, is certainly only a difference in degree. Gar
0282  ivably great. But assuredly, if this theory be TRUE, such have lived upon this earth. On the lapse
0289  in lacustrine deposits; and that not a cave or TRUE lacustrine bed is known belonging to the age o
0303  to the middle of the secondary series: and one TRUE mammal has been discovered in the new red sand
0307  ved descendants. Consequently, if my theory be TRUE, it is indisputable that before the lowest Sil
0316  to give an explanation of this fact, which if TRUE would have been fatal to my views. But such ca
0338  ied condition of each animal. This view may be TRUE, and yet it may never be capable of full proof
0349  nother species of the same genus: and not by a TRUE ostrich or emeu, like those found in Africa an
0390  at of the continent, we shall see that this is TRUE. This fact might have been expected on my theo
0393  y this assertion, and I have found it strictly TRUE. I have, however, been assured that a frog exi
0404  it will, I believe, be universally found to be TRUE, that wherever in two regions, let them be eve
```

Page **(Key Word)**

Page **(Key Word)**

```
0404 i can hardly doubt that this rule is generally TRUE, though it would be difficult to prove it. Amo
0414 ded as affording very clear indications of its TRUE affinities. We are least likely in the modific
0415 mportant is generally, but by no means always, TRUE. But their importance for classification, I be
0418 roadly confessed by some naturalists to be the TRUE one; and by none more clearly than by that exc
0418 s doctrine has very generally been admitted as TRUE. The same fact holds good with flowering plant
0420 aracters which naturalists consider as showing TRUE affinity between any two or more species, are
0420 ited from a common parent, and, in so far, all TRUE classification is genealogical; that community
0421 heir genealogical arrangement remains strictly TRUE, not only at the present time, but at each suc
0427 ss or order is compared with another, but give TRUE affinities when the members of the same class
0428 fin like limbs serve as characters exhibiting TRUE affinity between the several members of the wh
0439 morphosed during a long course of descent from TRUE legs, or from some simple appendage, is explai
0446 wo principles, which latter, though not proved TRUE, can be shown to be in some degree probable, t
0449 at i only hope to see the law hereafter proved TRUE. It can be proved true in those cases alone in
0449 he law hereafter proved true. It can be proved TRUE in those cases alone in which the ancient stat
0450 o the embryonic stages of recent forms, may be TRUE, but yet, owing to the geological record not e
0456 ical or adaptive characters, and characters of TRUE affinity; and other such rules; all naturally
0482 have every external characteristic feature of TRUE species; they admit that these have been produ
0484 not some fifty species of British brambles are TRUE species will cease. Systematists will have onl
0084 then, that nature's productions should be far TRUER in character than man's productions; that the
0166 nge of form or character. When the oldest and TRUEST breeds of various colours are crossed, we see
0058 this statement is indeed scarcely more than a TRUISM, for if a variety were found to have a wider
0013 f strange and rare deviations of structure are TRULY inherited, less strange and commoner deviatio
0017 dog, which we all know propagate their kind so TRULY, were the offspring of any single species, th
0023 especially as in each of these breeds several TRULY inherited sub breeds, or species as he might
0025 some semi domestic breeds and some apparently TRULY wild breeds have, besides the two black bars,
0062 two canine animals in a time of dearth, may be TRULY said to struggle with each other which shall
0063 verage only one comes to maturity, may be more TRULY said to struggle with the plants of the same
0063 ng close together on the same branch, may more TRULY be said to struggle with each other. As the m
0080 ry tendency is. Under domestication, it may be TRULY said that the whole organisation becomes in s
0128 f all organic beings may be explained. It is a TRULY wonderful fact, the wonder of which we are ap
0148 this is the case with the male Ibla, and in a TRULY extraordinary manner with the Proteolepas: fo
0152 depart widely from the standard. There may be TRULY said to be a constant struggle going on betwe
0179 exist having every intermediate grade between TRULY aquatic and strictly terrestrial habits; and
0202 he death of some few members. If we admire the TRULY wonderful power of scent by which the males o
0209 layed a tune with no practice at all, he might TRULY be said to have done so instinctively. But it
0227 we must suppose the Melipona to make her cells TRULY spherical, and of equal sizes; and this would
0280 , which have formerly existed on the earth, be TRULY enormous. Why then is not every geological fo
0308 t present must remain inexplicable; and may be TRULY urged as a valid argument against the views h
0486 hey can be so made, genealogies; and will then TRULY give what may be called the plan of creation.
0021 , much elongated wing and tail feathers. The TRUMPETER and Laugher, as their names express, utter
0145 n pelargoniums, that the central flower of the TRUSS often loses the patches of darker colour in t
0002 horities for my several statements; and I must TRUST to the reader reposing some confidence in my
0051 ks between his doubtful forms, he will have to TRUST almost entirely to analogy, and his difficult
0096 is impossible here to enter on details, I must TRUST to some general considerations alone. In the
0164 leg stripes; and a man, whom I can implicitly TRUST, has examined for me a small dun Welch pony w
0176 n the forms which they connect. Now, if we may TRUST these facts and inferences, and therefore con
0312 s we know, on the face of the earth. If we may TRUST the observations of Philippi in Sicily, the s
0414 tance! We must not, therefore, in classifying, TRUST to resemblances in parts of the organisation,
0013 uch more important rule, which I think may be TRUSTED, that is if the genera of animals are as dis
0252 plants. If our systematic arrangements can be TRUSTED, I know not; but the forms experimentised on
0270 are greater. How far these experiments may be TRUSTED, it must be admitted that scarcely a single
0357 indeed the arguments used by Forbes are to be TRUSTED, we have direct evidence of glacial action i
0373 f ore account which has been published can be TRUSTED. I can here give o
0002 though I hope I have always been cautious in TRUSTING to good authorities alone. I can here give o
0220 n on this and other subjects. Although fully TRUSTING to the statements of Huber and Mr. Smith, I
0014 there would be great difficulty in proving its TRUTH: we may safely conclude that very many of the
0051 e expense of admitting much variation, and the TRUTH of this admission will often be disputed by o
0055 y species, than in one having few. To test the TRUTH of this anticipation I have arranged the plan
0062 . nothing is easier than to admit in words the TRUTH of the universal struggle for life, or more d
0105 e variations. If we turn to nature to test the TRUTH of these remarks, and look at any small isola
0114 from each other, take the rank of species. The TRUTH of the principle, that the greatest amount of
0129 tree. I believe this simile largely speaks the TRUTH. The green and budding twigs may represent ex
0131 t be here given would be necessary to show the TRUTH of the remark, that the reproductive system i
0146 ommon to whole groups of species, and which in TRUTH are simply due to inheritance; for an ancient
0147 ers, more especially botanists, believe in its TRUTH. I will not, however, here give any instances
0149 een stated by some authors, and I believe with TRUTH, are apt to be highly variable. We shall have
0150 hopeless to attempt to convince any one of the TRUTH of this proposition without giving the long a
0151 s would seriously have shaken my belief in its TRUTH, had not the great variability in plants made
0156 t of difference between their females; and the TRUTH of this proposition will be granted. The caus
0194 ch no transitional grade is known to lead. The TRUTH of this remark is indeed shown by that old ca
0220 s any one may well be excused for doubting the TRUTH of so extraordinary and odious an instinct as
0254 probable, and I am inclined to believe in its TRUTH, although it rests on no direct evidence. I b
0289 plement to Lyell's Manual, will bring home the TRUTH, how accidental and rare is their preservatio
0292 ed intermittent. I feel much confidence in the TRUTH of these views; for they are in strict accord
0301 inct species. If then, there be some degree of TRUTH in these remarks, we have no right to expect
0316 opposed to such views as I maintain) admit its TRUTH; and the rule strictly accords with my theory
0330 l ages; and it would be difficult to prove the TRUTH of the proposition, for every now and then ev
0331 same classes, we must admit that there is some TRUTH in the remark. Let us see how far these sever
0334 ive formations. It is no real objection to the TRUTH of the statement, that the fauna of each peri
0338 follow Pictet and Huxley in thinking that the TRUTH of this doctrine is very far from proved. Yet
0346 merica alone would almost suffice to prove its TRUTH: for if we exclude the northern parts where t
0389 l consider some other facts, which bear on the TRUTH of the two theories of independent creation a
0444 in the parent. These two principles, if their TRUTH be admitted, will, I believe, explain all the
0459 itable deviation of structure or instinct. The TRUTH of these propositions cannot, I think, be dis
0460 ystems of the intercrossed species. We see the TRUTH of this conclusion in the vast difference in
0481 erations. Although I am fully convinced of the TRUTH of the views given in this volume under the f
```

Page **(Key Word)**

truth

Page ***************************************(Key Word)***************************************

```
0034  obvious. At the present time, eminent breeders TRY by methodical selection, with a distinct objec
0039  some slight degree by nature. No man would ever TRY to make a fantail, till he saw a pigeon with a
0078  t of competitors or enemies. It is good thus to TRY in our imagination to give any form some advan
0102  a nearly common standard of perfection, and all TRY to get and breed from the best animals, much i
0251  periment during five years, and he continued to TPY it during several subsequent years, and always
0413  is in my judgment fully explained. Naturalists  TRY to arrange the species, genera, and families i
0034  unconscious, and which results from every one   TRYING to possess and breed from the best individual
0039  s attention. But to use such an expression as   TRYING to make a fantail, is, I have no doubt, in mo
0090  nscious selection which results from each man    TRYING to keep the best dogs without any thought to
0101  many cases of difficulty, some of which I am     TRYING to investigate. Finally then, we may conclude
0125  lie between the larger groups, which are all     TRYING to increase in number. One large group will s
0147  namely, that natural selection is continually    TRYING to economise in every part of the organisatio
0177  at the base; and that the inhabitants are all    TRYING with equal steadiness and skill to improve th
0204  d, bearing in mind that each organic being is    TRYING to live wherever it can live, how it has aris
0232  e same relative distance from each other, all    TRYING to sweep equal spheres, and then building up,
0412  from the varying descendants of each species     TRYING to occupy as many and as different places as
0472  s. but on the view of each species constantly    TRYING to increase in number, with natural selection
0095  lover to have a shorter or more deeply divided   TUBE to its corolla, so that the hive bee could vis
0144  the union of the petals of the corolla into a    TUBE. Hard parts seem to affect the form of adjoini
0033  wer garden; the diversity of leaves, pods, or    TUBERS, or whatever part is valued, in the kitchen g
0267  ardeners in their frequent exchanges of seed,    TUBERS, etc, from one soil or climate to another, an
0094  o a similar slight deviation of structure. The   TUBES of the corollas of the common red and incarna
0225  to hold honey, sometimes adding to them short    TUBES of wax, and likewise making separate and very
0263  plant having a pistil too long for the pollen    TUBES to reach the ovarium. It has also been observ
0263  a distantly allied species, though the pollen    TUBES protrude, they do not penetrate the stigmatic
0451  he office of the pistil is to allow the pollen   TUBES to reach the ovules protected in the ovarium
0137  ion. In South America, a burrowing rodent, the   TUCO tuco, or Ctenomys, is even more subterranean i
0137  in south America, a burrowing rodent, the tuco   TUCO, or Ctenomys, is even more subterranean in its
0473  estic duck; or when we look at the burrowing     TUCUTUCU, which is occasionally blind, and then at ce
0090  nalogous cases under nature, for instance, the   TUFT of hair on the breast of the turkey cock, whic
0090  ul or ornamental to this bird; indeed, had the   TUFT appeared under domestication, it would have be
0147  y in size and quality. In our poultry, a large   TUFT of feathers on the head is generally accompani
0214  ably could have taught, the tumbler pigeon to    TUMBLE, an action which, as I have witnessed, is per
0214  by young birds, that have never seen a pigeon    TUMBLE. We may believe that some one pigeon showed a
0021  mpare the English carrier and the short faced    TUMBLER, and see the wonderful difference in their b
0021  ls, and a wide gape of mouth. The short faced    TUMBLER has a beak in outline almost like that of a
0021  e almost like that of a finch; and the common    TUMBLER has the singular and strictly inherited habi
0023  ld place the English carrier, the short faced    TUMBLER, the runt, the barb, pouter, and fantail in
0024  e english carrier, or that of the short faced    TUMBLER, or barb; for reversed feathers like those o
0027  y, although an English carrier or short faced    TUMBLER differs immensely in certain characters from
0027  of the carrier, the shortness of that of the    TUMBLER, and the number of tail feathers in the fant
0029  se, a greyhound and bloodhound, a carrier and    TUMBLER pigeon. One of the most remarkable features
0087  been asserted, that of the best short beaked     TUMBLER pigeons more perish in the egg than are able
0112  hey both go on (as has actually occurred with    TUMBLER pigeons) choosing and breeding from birds wi
0152  even in the sub breeds, as in the short faced    TUMBLER, it is notoriously difficult to breed them n
0153  far as to breed a bird as coarse as a common    TUMBLER from a good short faced strain. But as long
0214  f teaching or probably could have taught, the    TUMBLER pigeon to tumble, an action which, as I have
0423  e allied in the greatest number of points. In    TUMBLER pigeons, though some sub varieties differ fr
0446  o this rule, for the young of the short faced    TUMBLER differed from the young of the wild rock pig
0446  early period. But the case of the short faced    TUMBLER, which when twelve hours old had acquired it
0447  arec. In either case (as with the short faced    TUMBLER) the young or embryo would closely resemble
0026  ecies, aboriginally as distinct as carriers,    TUMBLERS, pouters, and fantails now are, should yield
0036  iven in old pigeon treatises of carriers and    TUMBLERS with these breeds as now existing in Britain
0152  erence there is in the beak of the different    TUMBLERS, in the beak and wattle of the different car
0214  t individuals in successive generations made    TUMBLERS what they now are: and near Glasgow there ar
0214  ey now are: and near Glasgow there are house    TUMBLERS, as I hear from Mr. Brent, which cannot fly
0335  of beak originated earlier than short beaked    TUMBLERS, which are at the opposite end of the series
0424  reasoning or thinking on the subject, these    TUMBLERS are kept in the same group, because allied i
0445  ntails, runts, barbs, dragons, carriers, and   TUMBLERS. Now some of these birds, when mature, diffe
0021  ng at a great height in a compact flock, and   TUMBLING in the air head over heels. The runt is a bi
0424  ept together from having the common habit of   TUMBLING; but the short faced breed has nearly or qui
0209  with wonderfully little practice, had played a  TUNE with no practice at all, he might truly be sai
0028  taken with the court. The monarchs of Iran and  TURAN sent him some very rare birds; and, continues
0021  ll excite astonishment and even laughter. The   TURBIT has a very short and conical beak, with a lin
0039  n did not inflate its crop much more than the   TURBIT now does the upper part of its oesophagus, a
0067  re destroyed, chiefly by slugs and insects. If  TURF which has long been mown, and the case would b
0067  been mown, and the case would be the same with  TURF closely browsed by quadrupeds, be let to grow,
0068  of twenty species growing on a little plot of   TURF (three feet by four) nine species perished fro
0114  bitants. For instance, I found that a piece of  TURF, three feet by four in size, which had been ex
0090  stance, the tuft of hair on the breast of the   TURKEY cock, which can hardly be either useful or or
0197  he skin on the head of the clean feeding male   TURKEY is likewise naked. The sutures in the skulls
0361  njured through even the digestive organs of a   TURKEY. In the course of two months, I picked up in
0216  chuckle, they will run (more especially young   TURKEYS) from under her, and conceal themselves in t
0144  ation between the hair and teeth in the naked   TURKISH dog, though here probably homology comes int
0083  fference of structure or constitution may well  TURN the nicely balanced scale in the struggle for
0094  ion of the sexes would be affected. Let us now   TURN to the nectar feeding insects in our imaginary
0099  have found, of the seedlings thus raised will   TURN out mongrels: for instance, I raised 233 seedl
0105  the appearance of favourable variations. If we  TURN to nature to test the truth of these remarks,
0153  s and the enlarged crop of pouters. Now let us  TURN to nature. When a part has been developed in a
0162  that the rock pigeon was not feather footed as  TURN crowned; we could not have told, whether these
0165  he most distant parts of the world. Now let us  TURN to the effects of crossing the several species
0217  nd aberrant habit of their mother and in their  TURN would be apt to lay their eggs in other birds'
0235  nomical instinct to new swarms, which in their  TURN will have had the best chance of succeeding in
0249  interbreeding having been avoided. Now let us   TURN to the results arrived at by the third most ex
0268  ature will assuredly have to be granted. If we  TURN to varieties, produced, or supposed to have be
0282  species, now generally extinct, have in their   TURN been similarly connected with more ancient spe
```

Page ***************************************(Key Word)***************************************

Page ***(Key Word)***

```
0287  eded each other in the long roll of years! Now  TURN  to our richest geological museums, and what a
0317  slowly converted into species, which in their  TURN  produce by equally slow steps other species, a
0325  ey add, if struck by this strange sequence, we  TURN  our attention to North America, and there disc
0331  somewhat complex, I must request the reader to  TURN  to the diagram in the fourth chapter. We may s
0412  , and preceding forms. I request the reader to  TURN  to the diagram illustrating the action, as for
0443  born, what its merits or form will ultimately  TURN  out. We see this plainly in our own children;
0466  heory of descent with modification. Now Let us  TURN  to the other side of the argument. Under domes
0468  e to the surrounding physical conditions, will  TURN  the balance. With animals having separated sex
0469  pposed difficulties and objections: now let us  TURN  to the special facts and arguments in favour o
0015  rossing, that only a single variety should be  TURNED  loose in its new home. Nevertheless, as our v
0190  fish Cobites. In the Hydra, the animal may be  TURNED  inside out, and the exterior surface will the
0100  rees simply to call attention to the subject.  TURNING  for a very brief space to animals: on the la
0182  lide far through the air, slightly rising and  TURNING  by the aid of their fluttering fins, might h
0196  pper, an organ of prehension, or as an aid in  TURNING, as with the dog, though the aid must be sli
0227  in wood many insects can make, apparently by  TURNING  round on a fixed point. We must suppose the
0348  those characteristic of distinct continents.  TURNING  to the sea, we find the same law. No two mar
0432  ment, would be possible. We shall see this by  TURNING  to the diagram: the letters, A to L, may rep
0461  ween less modified forms, increase fertility.  TURNING  to geographical distribution, the difficulti
0159  , or roots as commonly called, of the Swedish  TURNIP  and Ruta baga, plants which several botanists
0159  hese a third may be added, namely, the common  TURNIP. According to the ordinary view of each speci
0427  the thickened stems of the common and swedish  TURNIP. The resemblance of the greyhound and racehor
0423  identical; no one puts the swedish and common  TURNIPS  together, though the esculent and thickened
0030  a seedling. So it has probably been with the  TURNSPIT  dog; and this is known to have been the case
0005  c beings throughout time; in the eleventh and  TWELFTH, their geographical distribution throughout
0021  hirty or even forty tail feathers, instead of  TWELVE  or fourteen, the normal number in all members
0055  is anticipation I have arranged the plants of  TWELVE  countries, and the coleopterous insects of tw
0071  of the heath plants were wholly changed, but  TWELVE  species of plants (not counting grasses and c
0160  a dozen or even a score of generations. After  TWELVE  generations, the proportion of blood, to use
0233  n experimentally found that no less than from  TWELVE  to fifteen pounds of dry sugar are consumed b
0362  eir prey whole, and after an interval of from  TWELVE  to twenty hours, disgorge pellets, which, as
0362  r, and beet germinated after having been from  TWELVE  to twenty one hours in the stomachs of differ
0385  urvived on the duck's feet, in damp air, from  TWELVE  to twenty hours; and in this length of time a
0445  pared young pigeons of various breeds, within  TWELVE  hours after being hatched; I carefully measur
0446  e case of the short faced tumbler, which when  TWELVE  hours old had acquired its proper proportions
0026  d by the rock pigeon: I say within a dozen or  TWENTY  generations, for we know of no fact countenan
0032  sent day are compared with drawings made only  TWENTY  or thirty years ago. When a race of plants is
0064  e pair. Even slow breeding man has doubled in  TWENTY  five years, and at this rate, in a few thousa
0064  gs next year produced two, and so on, then in  TWENTY  years there would be a million plants. The el
0068  rous, though fully grown, plants: thus out of  TWENTY  species growing on a little plot of turf (thr
0068  ot one head of game were shot during the next  TWENTY  years in England, and at the same time, if no
0071  of exactly the same nature had been enclosed  TWENTY  five years previously and planted with Scotch
0072  judging from the rings of growth, had during  TWENTY  six years tried to raise its head above the s
0093  yards from a male tree, I put the stigmas of  TWENTY  flowers, taken from different branches, under
0114  ars to exactly the same conditions, supported  TWENTY  species of plants, and these belonged to eigh
0135  n wings that they cannot fly; and that of the  TWENTY  nine endemic genera, no less than twenty thre
0135  the twenty nine endemic genera, no less than  TWENTY  three genera have all their species in this c
0218  in one day's hunting I picked up no less than  TWENTY  Lost and wasted eggs. Many bees are parasitic
0221  long the same road to a tall Scotch fir tree,  TWENTY  five yards distant, which they ascended toget
0221  ried their dead bodies as food to their nest,  TWENTY  nine yards distant, but they were prevented f
0231  always make the first rough wall from ten to  TWENTY  times thicker than the excessively thin finis
0247  difficulty in the manipulation) half of these  TWENTY  plants had their fertility in some degree imp
0286  coast at the rate on one yard in nearly every  TWENTY  two years. I doubt whether any rock, even as
0286  do not believe that any line of coast, ten or  TWENTY  miles in length, ever suffers degradation at
0362  hole, and after an interval of from twelve to  TWENTY  hours, disgorge pellets; which, as I know fro
0362  t germinated after having been from twelve to  TWENTY  one hours in the stomachs of different birds
0362  es adheres to them: in one instance I removed  TWENTY  two grains of dry argillaceus earth from one
0365  by similar means. I do not doubt that out of  TWENTY  seeds or animals transported to an island, ev
0376  ern forms of fish. Dr. Hooker informs me that  TWENTY  five species of Algae are common to New Zeala
0378  intense, when arctic forms had migrated some  TWENTY  five degrees of latitude from their native co
0385  the duck's feet, in damp air, from twelve to  TWENTY  hours; and in this length of time a duck or h
0397  ad again hybernated I put it in sea water for  TWENTY  days, and it perfectly recovered. As this spe
0398  le stamp of the American continent. There are  TWENTY  six land birds, and twenty five of these are
0398  ntinent. There are twenty six Land birds, and  TWENTY  five of these are ranked by Mr. Gould as dist
0231  the plates of the pyramidal basis being about  TWICE  as thick. By this singular manner of building
0247  n to believe to be varieties; and only once or  TWICE  succeeded in getting fertile seed; as he foun
0262  sorbus, when grafted on other species, yielded  TWICE  as much fruit as when on their own roots. We
0283  t cases reach the cliffs only for a short time  TWICE  a day, and the waves eat into them only when
0293  that the average duration of each formation is  TWICE  or thrice as long as the average duration of
0381  man's agency in a foreign land; why one ranges  TWICE  or thrice as far, and is twice or thrice as c
0381  why one ranges twice or thrice as far, and is  TWICE  or thrice as common, as another species withi
0385  ges from a pond covered with duck weed, I have  TWICE  seen these little plants adhering to its back
0129  argely speaks the truth. The green and budding  TWIGS  may represent existing species; and those pro
0129  cies. At each period of growth all the growing  TWIGS  have tried to branch out on all sides, and to
0129  sides, and to overtop and kill the surrounding  TWIGS  and branches, in the same manner as species a
0129  mselves once, when the tree was small, budding  TWIGS; and this connexion of the former and present
0129  s in groups subordinate to groups. Of the many  TWIGS  which flourished when the tree was a mere bus
0138  . next follow those that are constructed for  TWILIGHT: and, last of all, those destined for total
0012  in some small degree from that of the parental  TYPE. Any variation which is not inherited is unimp
0104  on destroying any which depart from the proper  TYPE; but if their conditions of life change and th
0124  s have gone on diverging in character from the  TYPE  of their parents, the new species (F14) will n
0171  cases absolutely perfect. The Law of Unity of  TYPE  and of the Conditions of Existence embraced by
0184  onsiderably modified from that of their proper  TYPE. And such instances do occur in nature. Can a
0186  as pleased the Creator to cause a being of one  TYPE  to take the place of one of another type: but
0186  f one type to take the place of one of another  TYPE; but this seems to me only restating the fact
0206  s have been formed on two great laws, Unity of  TYPE, and the Conditions of Existence. By unity of
0206  , and the Conditions of Existence. By unity of  TYPE  is meant that fundamental agreement in structu
```

Page ***(Key Word)***

type

0206 f their habits of life. On my theory, unity of TYPE is explained by unity of descent. The expressi
0206 itance of former adaptations, that of Unity of TYPE. Chapter VII. Instinct. Instincts comparable w
0349 rodents, but they plainly display an American TYPE of structure. We ascend the lofty peaks of the
0349 he coypu and capybara, rodents of the American TYPE. Innumerable other instances could be given. I
0417 important points of structure from the proper TYPE of the order, yet M. Richard sagaciously saw,
0434 blance is often expressed by the term unity of TYPE; or by saying that the several parts and organ
0485 alists of affinity, relationship, community of TYPE, paternity, morphology, adaptive characters, r
0124 intermediate between them, but rather between TYPES of the two groups; and every naturalist will
0182 ammals, flying insects of the most diversified TYPES, and formerly had flying reptiles, it is conc
0312 f ancient forms. On the succession of the same TYPES within the same areas. Summary of preceding a
0338 e is very small. On the Succession of the same TYPES within the same areas, during the later terti
0339 in such numbers, are related to South American TYPES. This relationship is even more clearly seen
0339 839 and 1845, on this law of the succession of TYPES, on this wonderful relationship in the same c
0339 s remarkable law of the succession of the same TYPES within the same areas mean? He would be a bol
0340 of conditions, for the uniformity of the same TYPES in each during the later tertiary periods. No
0340 australia; or that Edentata and other American TYPES should have been solely procuced in South Ame
0340 ing, but not immutable, succession of the same TYPES within the same areas, is at once explained;
0345 ll be intelligible. The succession of the same TYPES of structure within the same areas during the
0349 hown in the last chapter, and we find American TYPES then prevalent on the American continent and
0432 fine the several groups; but we could pick out TYPES, or forms, representing most of the character
0433 le pape'r, on the high importance of looking to TYPES, whether or not we can separate and define th
0433 n separate and define the groups to which such TYPES belong. Finally, we have seen that natural se
0338 ee less distinct from each other than are the TYPICAL members of the same groups at the present da
0183 tances could be given: I have often watched a TYRANT flycatcher (Saurophagus sulphuratus) in South
0221 y were eagerly seized, and carried off by the TYRANTS, who perhaps fancied that, after all, they h
0304 was a Chthamalus, a very commom, large, and UBIQUITOUS genus, of which not one specimen has as yet
0011 t. the great and inherited development of the UDDERS in cows and goats in countries where they are
0451 ur developed and two rudimentary teats in the UDDERS of the genus Bos, but in our domestic cows th
0439 nd in plants: thus the embryonic leaves of the ULEX or furze, and the first leaves of the phyllodi
0231 ntinually given to the comb, with the utmost UDDERS economy of wax. It seems at first to add to
0031 rked and classed, so that the very best may ULTIMATELY be selectec for breeding. What English bree
0061 ich I have called incipient species, become ULTIMATELY converted into good and distinct species, w
0126 continue to increase. But which groups will ULTIMATELY prevail, no man can predict; for we well kn
0202 community for any other end, and which are ULTIMATELY slaughtered by their industrious and steril
0231 thin finished wall of the cell, which will ULTIMATELY be left. We shall understand how they work,
0239 he profitable modification, all the neuters ULTIMATELY came to have the desired character. On this
0298 st civen rise, first to local varieties and ULTIMATELY to new species; and this again would greatl
0317 keeping for a space of equal thickness, and ULTIMATELY thins out in the upper beds, marking the de
0320 ed on the belief that each new variety, and ULTIMATELY each new species, is produced and maintaine
0368 he returning warmth, first at the bases and ULTIMATELY on the summits of the mountains, the case w
0392 and thus convert them first into bushes and ULTIMATELY into trees. With respect to the absence of
0405 s offspring, firstly into new varieties and ULTIMATELY into new species. In considering the wide d
0411 ieties, or incipient species, thus produced ULTIMATELY become converted, as I believe, into new an
0436 d as to be finally lost, by the atrophy and ULTIMATELY by the complete abortion of certain parts,
0442 arts in the same individual embryo, which ULTIMATELY become very unlike and serve for diverse pu
0443 has been born, what its merits or form will ULTIMATELY turn out. We see this plainly in our own ch
0454 seldom been forced to take flight, and have ULTIMATELY lost the power of flying. Again, an organ u
0489 the larger and dominant groups, which will ULTIMATELY prevail and procreate new and dominant spec
0145 the central and exterior flowers of a head or UMBEL, I do not feel at all sure that C. C. Sprenge
0145 ea; but, in the case of the corolla of the UMBELLIFERAE, it is by no means, as Dr. Hooker informs
0146 way advantageous to the plant: yet in the UMBELLIFERAE these differences are of such apparent imp
0144 and inner flowers in some Compositous and UMBELLIFEROUS plants. Every one knows the difference in
0168 of crowth, the nature of which we are utterly UNABLE to understand. Multiple parts are variable in
0198 alluded to them only to show that, if we are UNABLE to account for the characteristic differences
0281 other. Hence in all such cases, we should be UNABLE to recognise the parent form of any two or mo
0461 e one part to the others. We are often wholly UNABLE even to conjecture how this could have been e
0468 y variability under nature, it would be an UNACCOUNTABLE fact if natural selection had not come in
0121 may for a long period continue transmitting UNALTERED descendants; and this is shown in the diagr
0124 but to have retained the form of (F), either UNALTERED or altered only in a slight degree. In this
0187 ed from the earlier stages of descent, in an UNALTERED or little altered condition. Amongst existi
0281 one form had remained for a very long period UNALTERED, whilst its descendants had undergone a vas
0288 m of the sea not rarely lying for ages in an UNALTERED condition. The remains which co become embe
0314 as the marine shells and birds have remained UNALTERED. We can perhaps understand the apparently q
0338 process, whilst it leaves the embryo almost UNALTERED, continually adds, in the course of success
0381 ew groups of forms, and others have remained UNALTERED. We cannot hope to explain such facts, unti
0489 hat not one living species will transmit its UNALTERED likeness to a distant futurity. And of the
0310 , as Lyell, Murchison, Sedgwick, etc., have UNANIMOUSLY, often vehemently, maintained the immutabi
0481 ecies were immutable productions was almost UNAVOIDABLE as long as the history of the world was th
0450 ut through the gums, in the upper jaws of our UNBORN calves. It has even been stated on good autho
0316 stance, must have continuously existed by an UNBROKEN succession of generations, from the lowest S
0015 their acquired characters, whilst kept under UNCHANGED conditions, and whilst kept in a considerab
0035 erve for comparison. In some cases, however, UNCHANGED or but little changed individuals of the sa
0488 ing in a body might remain for a long period UNCHANGED, whilst within this same period, several of
0038 pe, nor any other region inhabited by quite UNCIVILISED man, has afforded us a single plant worth
0038 . in regard to the domestic animals kept by UNCIVILISED man, it should not be overlooked that they
0140 domestic animals were originally chosen by UNCIVILISED man because they were useful and bred read
0048 rieties of this doubtful nature are far from UNCOMMON cannot be disputed. Compare the several flor
0218 e same or of a distinct species, is not very UNCOMMON with the Gallinaceae; and this perhaps expla
0007 ently followed, its Effects. Methodical and UNCONSCIOUS Selection. Unknown Origin of our Domestic
0034 e, a kind of Selection, which may be called UNCONSCIOUS, and which results from every one trying t
0036 that in this case there woulc be a kind of UNCONSCIOUS selection going on. We see the value set o
0039 would become through long continued, partly UNCONSCIOUS and partly methodical selection. Perhaps I
0040 ged, the principle, as I have called it, of UNCONSCIOUS selection will always tend, perhaps more a
0083 oduced a great result by his methodical and UNCONSCIOUS means of selection, what may not nature ef
0090 areful and methodical selection, or by that UNCONSCIOUS selection which results from each man tryi
0102 fication surely but slowly follow from this UNCONSCIOUS process of selection, notwithstanding a la

Page **(Key Word)**

0214 eneration would soon complete the work; and UNCONSCIOUS selection is still at work, as each man tr
0269 under domestication by man's methodical and UNCONSCIOUS power of selection, for his own use and pl
0012 ary peculiarity, he will almost certainly UNCONSCIOUSLY modify other parts of the structure, owin
0035 lieve that King Charles's spaniel has been UNCONSCIOUSLY modified to a large extent since the time
0035 s us is, that the change has been effected UNCONSCIOUSLY and gradually, and yet so effectually, th
0036 e of selection, which may be considered as UNCONSCIOUSLY followed, in so far that the breeders cou
0037 ult is concerned, has been followed almost UNCONSCIOUSLY. It has consisted in always cultivating t
0037 in our cultivated plants, thus slowly and UNCONSCIOUSLY accumulated, explains, as I believe, the
0043 applied methodically and more quickly, or UNCONSCIOUSLY and more slowly, but more efficiently, is
0208 n is performed, but not of its origin. How UNCONSCIOUSLY many habitual actions are performed, inde
0216 f selection, pursued both methodically and UNCONSCIOUSLY; but in most cases, probably, habit and s
0420 he hidden bond which naturalists have been UNCONSCIOUSLY seeking, and not some unknown plan of cre
0425 not this same element of descent have been UNCONSCIOUSLY used in grouping species under genera, an
0425 me to complete? I believe it has thus been UNCONSCIOUSLY used; and only thus can I understand the
0467 may do this methodically, or he may do it UNCONSCIOUSLY by preserving the individuals most useful
0081 s of its inhabitants would almost immediately UNDERGO a change, and some species might become exti
0104 t if their conditions of life change and they UNDERGO modification,uniformity of character can be
0107 a large continental area, which will probably UNDERGO many oscillations of level, and which conseq
0296 gical period. So that if such species were to UNDERGO a considerable amount of modification during
0343 the same degree; yet in the long run that all UNDERGO modification to some extent. The extinction
0350 xposed to new conditions, and will frequently UNDERGO further modification and improvement; and th
0441 oper place on which to become attached and to UNDERGO their final metamorphosis. When this is comp
0454 for I doubt whether species under nature ever UNDERGO abrupt changes. I believe that disuse has be
0081 l selection by taking the case of a country UNDERGOING some physical change, for instance, of clim
0108 d by some of the inhabitants of the country UNDERGOING modification of some kind. The existence of
0110 ich stand in closest competition with those UNDERGOING modification and improvement, will naturall
0113 can succeed in increasing (the country not UNDERGOING any change in its conditions) only by its v
0152 those points, which at the present time are UNDERGOING rapid change by continued selection, are al
0153 ted to be much variability in the structure UNDERGOING modification. It further deserves notice th
0197 ubsequently taken advantage of by the plant UNDERGOING further modification and becoming a climber
0283 few miles any line of rocky cliff, which is UNDERGOING degradation, we find that it is only here a
0295 ave to be a very thick one; and the species UNDERGOING modification will have had to live on the s
0354 rom a common progenitor, can have migrated (UNDERGOING modification during some part of their migr
0405 the unmodified parent should range widely, UNDERGOING modification during its diffusion, and shou
0448 final cause of the young in these cases not UNDERGOING any metamorphosis, or closely resembling th
0448 her explanation, however, of the embryo not UNDERGOING any metamorphosis is perhaps requisite. If,
0463 ason to believe that only a few species are UNDERGOING change at any one period; and all changes a
0028 mense amount of variation which pigeons have UNDERGONE, will be obvious when we treat of Selection
0082 the conditions of life are supposed to have UNDERGONE a change, and this would manifestly be favo
0153 me genus, we may conclude that this part has UNDERGONE an extraordinary amount of modification, si
0281 period unaltered, whilst its descendants had UNDERGONE a vast amount of change; and the principle
0287 ations of level, which we know this area has UNDERGONE, the surface may have existed for millions
0291 cal facts tell us plainly that each area has UNDERGONE numerous slow oscillations of level, and ap
0309 s weight of superincumbent water, might have UNDERGONE far more metamorphic action than strata whi
0313 tions, all the species will be found to have UNDERGONE some change. When a species has once disapp
0330 ust be restricted to those groups which have UNDERGONE much change in the course of geological age
0333 , which have within known geological periods UNDERGONE much modification, should in the older form
0351 or. In the case of those species, which have UNDERGONE during whole geological periods but little
0420 rent degrees of modification which they have UNDERGONE; and this is expressed by the forms being r
0422 modification which the different groups have UNDERGONE, have to be expressed by ranking them under
0425 s been used in classing varieties which have UNDERGONE a certain, and sometimes a considerable amo
0430 progenitor, and that both groups have since UNDERGONE much modification in divergent directions.
0447 ature animal; the limbs in the latter having UNDERGONE much modification at a rather late period o
0466 h modification our domestic productions have UNDERGONE: but we may safely infer that the amount ha
0468 geology plainly proclaims that each land has UNDERGONE great physical changes, we might have expec
0474 to variation; but, on my view, this part has UNDERGONE, since the several species branched off fro
0481 ence of the mutation of species, if they had UNDERGONE mutation. But the chief cause of our natura
0286 e geologists suppose, a range of older rocks UNDERLIES the Weald, on the flanks of which the overl
0288 by another and later formation, without the UNDERLYING bed having suffered in the interval any wea
0297 species, and a third, A, to be found in an UNDERLYING bed; even if A were strictly intermediate b
0324 ose forms which are only found in the older UNDERLYING deposits, would be correctly ranked as simu
0334 n those of the overlying carboniferous, and UNDERLYING Silurian system. But each fauna is not nece
0283 away rock. At last the base of the cliff is UNDERMINED, huge fragments fall down, and these remain
0245 generally sterile, has, I think, been much UNDERRATED by some late writers. On the theory of natu
0025 l feathers, as any wild rock pigeon! We can UNDERSTAND these facts, on the well known principle of
0038 esent standard of usefulness to man; we can UNDERSTAND how it is that neither Australia, the Cape
0038 wants or fancies. We can, I think, further UNDERSTAND the frequently abnormal character of our do
0059 analogy with varieties. And we can clearly UNDERSTAND these analogies, if species have once exist
0081 e species called polymorphic. We shall best UNDERSTAND the probable course of natural selection by
0095 ive bee could visit its flowers. Thus I can UNDERSTAND how a flower and a bee might slowly become,
0097 t this is a law of nature, we can, I think, UNDERSTAND several large classes of facts, such as the
0100 rial animal which fertilises itself. We can UNDERSTAND this remarkable fact, which offers so stron
0106 nic world. We can, perhaps, on these views, UNDERSTAND some facts which will be again alluded to i
0124 aracter between existing groups; and we can UNDERSTAND this fact, for the extinct species lived at
0126 of the same species making a class, we can UNDERSTAND how it is that there exist but very few cla
0148 ng up an useless structure. I can thus only UNDERSTAND a fact with which I was much struck when ex
0168 he nature of which we are utterly unable to UNDERSTAND. Multiple parts are variable in number and
0169 tion since the genus arose; and thus we can UNDERSTAND why it should often still be variable in a
0172 as the eye, of which we hardly as yet fully UNDERSTAND the inimitable perfection? Thirdly, can ins
0176 s which they connect, then, I think, we can UNDERSTAND why intermediate varieties should not endur
0191 n's interesting description of these parts, UNDERSTAND the strange fact that every particle of foo
0194 theory of natural selection, we can clearly UNDERSTAND why she should not; for natural selection c
0202 ls subsequently intensified, we can perhaps UNDERSTAND how it is that the use of the sting should
0204 hose of its nearest congeners. Hence we can UNDERSTAND, bearing in mind that each organic being is
0206 theory of natural selection we can clearly UNDERSTAND the full meaning of that old canon in natur
0216 ave acted together. We shall, perhaps, best UNDERSTAND how instincts in a state of nature have bec

Page **(Key Word)**

Page ************************************(Key Word)************************************

```
0231  ll, which will ultimately be left. We shall UNDERSTAND how they work, by supposing masons first to
0243  ly the same instincts. For instance, we can UNDERSTAND on the principle of inheritance, how it is
0266  must, however, be confessed that we cannot UNDERSTAND, excepting on vague hypotheses, several fac
0292  one formation, it becomes more difficult to UNDERSTAND, why we do not therein find closely graduat
0314  rds have remained unaltered. We can perhaps UNDERSTAND the apparently quicker rate of change in te
0314  y have become modified and improved, we can UNDERSTAND, on the principle of competition, and on th
0315  in a slowly changing drama. We can clearly UNDERSTAND why a species when once lost should never r
0322  esumption in imagining for a moment that we UNDERSTAND the many complex contingencies, on which th
0331  r from its ancient progenitor. Hence we can UNDERSTAND the rule that the most ancient fossils diff
0343  tion through natural selection. We can thus UNDERSTAND how it is that new species come in slowly a
0343  ence of the production of new forms. We can UNDERSTAND why when a species has once disappeared it
0344  link of generation has been broken. We can UNDERSTAND how the spreading of the dominant forms of
0344  pear to have changed simultaneously. We can UNDERSTAND how it is that all the forms of life, ancie
0344  for all are connected by generation. We can UNDERSTAND, from the continued tendency to divergence
0350  le of inheritance with modification, we can UNDERSTAND how it is that sections of genera, whole ge
0355  ry will be strengthened; for we can clearly UNDERSTAND, on the principle of modification, why the
0367  ic regions of both hemispheres. Thus we can UNDERSTAND the identity of many plants at points so im
0367  ited States and of Europe. We can thus also UNDERSTAND the fact that the Alpine plants of each mou
0371  and the United States. On this view we can UNDERSTAND the relationship, with very little identity
0371  ation by the Atlantic Ocean. We can further UNDERSTAND the singular fact remarked on by several ob
0372  completely sundered. Thus, I think, we can UNDERSTAND the presence of many existing and tertiary
0385  water, as are the adults. I could not even UNDERSTAND how some naturalised species have rapidly s
0395  relation to their new position, and we can UNDERSTAND the presence of endemic bats on islands, wi
0396  slands separated by deeper channels, we can UNDERSTAND the frequent relation between the depth of
0408  he multiplication of new forms. We can thus UNDERSTAND the high importance of barriers, whether r
0408  ogical and botanical provinces. We can thus UNDERSTAND the localisation of sub genera, genera and
0409  the world. On these same principles, we can UNDERSTAND, as I have endeavoured to show, why oceanic
0420  , sections, or orders. The reader will best UNDERSTAND what is meant, if he will take the trouble
0425  een unconsciously used; and only thus can I UNDERSTAND the several rules and guides which have bee
0426  ve especial value in classification. We can UNDERSTAND why a species or a group of species may dep
0427  ity cescended from the same parents. We can UNDERSTAND, on these views, the very important distinc
0427  far as they reveal descent, we can clearly UNDERSTAND why analogical or adaptive character, altho
0427  their proper lines of descent. We can also UNDERSTAND the apparent paradox, that the very same ch
0428  nts of land, air, and water, we can perhaps UNDERSTAND how it is that a numerical parallelism has
0431  itance of some characters in common, we can UNDERSTAND the excessively complex and radiating affin
0431  ossible to do this without this aid, we can UNDERSTAND the extraordinary difficulty which naturali
0433  erms genera, families, orders, etc., we can UNDERSTAND the rules which we are compelled to follow
0433  led to follow in our classification. We can UNDERSTAND why we value certain resemblances far more
0438  another in the same individual. And we can UNDERSTAND this fact; for in molluscs, even in the low
0449  of the natural system. On this view we can UNDERSTAND how it is that, in the eyes of most natural
0455  ect or reduce it in the embryo. Thus we can UNDERSTAND the greater relative size of rudimentary or
0455  h has long existed, to be inherited, we can UNDERSTAND, on the genealogical view of classification
0456  e of similarity in organic beings, we shall UNDERSTAND what is meant by the natural system: it is
0457  inherited at a corresponding period, we can UNDERSTAND the great leading facts in Embryology; name
0469  by secondary laws. On this same view we can UNDERSTAND how it is that in each region where many sp
0473  e become in a high degree permanent, we can UNDERSTAND this fact; for they have already varied sin
0474  , but profitable modifications. We can thus UNDERSTAND why nature moves by graduated steps in endo
0474  and having inherited much in common, we can UNDERSTAND how it is that allied species, when placed
0476  and unknown means of dispersal, then we can UNDERSTAND, on the theory of descent with modification
0477  ned in most cases with modification, we can UNDERSTAND, by the aid of the Glacial period, the iden
0480  ng conditions of life; and we can clearly UNDERSTAND on this view the meaning of rudimentary org
0480  n, which it seems that we wilfully will not UNDERSTAND. I have now recapitulated the chief facts a
0005  st, the difficulties of transitions, or in UNDERSTANDING how a simple being or a simple organ can
0060  ation for the work, helps us but little in UNDERSTANDING how species arise in nature. How have all
0116  t. the accompanying diagram will aid us in UNDERSTANDING this rather perplexing subject. Let A to
0195  , i have sometimes felt much difficulty in UNDERSTANDING the origin of simple parts, of which the
0231  seems at first to add to the difficulty of UNDERSTANDING how the cells are made, that a multitude
0237  insect communities: the difficulty lies in UNDERSTANDING how such correlated modifications of stru
0307  fe at these periods. But the difficulty of UNDERSTANDING how the absence of vast piles of fossiliferou
0352  very many cases of extreme difficulty, in UNDERSTANDING how the same species could possibly have
0396  t there are many and grave difficulties in UNDERSTANDING how several of the inhabitants of the mor
0207  mmonly embraced by this term: but every one UNDERSTANDS what is meant, when it is said that instin
0143  a very important subject, most imperfectly UNDERSTOOD. The most obvious case is, that modificatio
0150  e made due allowance for them. It should be UNDERSTOOD that the rule by no means applies to any pa
0159  or. These propositions will be most readily UNDERSTOOD by looking to our domestic races. The most
0327  suppose that the inhabitants of each region UNDERWENT a considerable amount of mocification and e
0379  vidence that the whole body of arctic shells UNDERWENT scarcely any modification during their long
0387  ellet containing the seeds of the Nelumbium UNDIGESTED; or the seeds might be dropped by the bird
0026  an see to the contrary, may be transmitted UNDIMINISHED for an indefinite number of generations. T
0485  the vain search for the undiscovered and UNDISCOVERABLE essence of the term species. The other an
0485  east be freed from the vain search for the UNDISCOVERED and undiscoverable essence of the term spe
0048  ave been ranked by one eminent naturalist as UNDOUBTED species, and by another as varieties, or a
0049  es, whereas the greater number rank it as an UNDOUBTED species peculiar to Great Britain. A wide d
0268  n crossed; and he consequently ranks them as UNDOUBTED species. If we thus argue in a circle, the
0057  forms, that is, round their parent species? UNDOUBTEDLY there is one most important point of diffe
0080  , seeing that variations useful to man have UNDOUBTEDLY occurred, that other variations useful in
0162  e other members of the same group. And this UNDOUBTEDLY is the case in nature. A considerable part
0192  by successive transitional gradations, yet, UNDOUBTEDLY, grave cases of difficulty occur, some of
0203  rn, or to perish herself in the combat; for UNDOUBTEDLY this is for the good of the community; and
0310  ations beneath the Silurian strata, are all UNDOUBTEDLY of the gravest nature. We see this in the
0310  s of other kinds given in this volume, will UNDOUBTEDLY at once reject my theory. For my part, fol
0312  al. In some of the most recent beds, though UNDOUBTEDLY of high antiquity if measured by years, on
0334  wances, the fauna of each geological period UNDOUBTEDLY is intermediate in character, between the
0352  one or more points of the earth's surface. UNDOUBTEDLY there are very many cases of extreme diffi
0353  conditions permitted, is the most probable. UNDOUBTEDLY many cases occur, in which we cannot expla
0401  spread from one island to another, it would UNDOUBTEDLY be exposed to different conditions of life
```

Page ************************************(Key Word)************************************

774   undoubtedly

Page **(Key Word)***

```
0402  ory, when put into free intercommunication. UNDOUBTEDLY if one species has any advantage whatever
0416  igh physiological or vital importance; yet, UNDOUBTEDLY, organs in this condition are often of hig
0184  ouring, in the harsh tone of its voice, and UNDULATORY flight, told me plainly of its close blood
0032  differences absolutely inappreciable by an UNEDUCATED eye, differences which I for one have vainl
0339  imilar relationship is manifest, even to an UNEDUCATED eye, in the gigantic pieces of armour like
0467  o slight as to be quite inappreciable by an UNEDUCATED eye. This process of selection has been the
0072  ce 1 could examine hundreds of acres of the UNENCLOSED heath, and literally I could not see a sing
0116  pecies are supposed to resemble each other in UNEQUAL degrees, as is so generally the case in natu
0116  ted in the diagram by the letters standing at UNEQUAL distances. I have said a large genus, becaus
0117  . the little fan of diverging dotted lines of UNEQUAL lengths proceeding from (A), may represent i
0122  genus were supposed to resemble each other in UNEQUAL degrees, as is so generally the case in natu
0266  o the sterility of hybrids; for instance, the UNEQUAL fertility of hybrids produced from reciproca
0318  ies and whole groups of species last for very UNEQUAL periods; some groups, as we have seen, havin
0333  umerous, they will have endured for extremely UNEQUAL lengths of time, and will have been modified
0334  a is not necessarily exactly intermediate, as UNEQUAL intervals of time have elapsed between conse
0343  es increase in numbers slowly, and endure for UNEQUAL periods of time; for the process of modifica
0415  i apprehend, in every natural family, is very UNEQUAL, and in some cases seems to be entirely lost
0416  these two divisions of the same order are of UNEQUAL physiological importance. Any number of inst
0044  esemble varieties in being very closely, but UNEQUALLY, related to each other, and in having restr
0057  and what are varieties but groups of forms, UNEQUALLY related to each other, and clustered round
0059  enera the species are apt to be closely, but UNEQUALLY, allied together, forming little clusters r
0120  ill have come to differ largely, but perhaps UNEQUALLY, from each other and from their common pare
0128  , species of the same genus less closely and UNEQUALLY related together, forming sections and sub
0419  hers, and so onwards, can be recognised as UNEQUIVOCALLY belonging to this, and to no other class
0189  ly, and natural selection will pick out with UNERRING skill each improvement. Let this process go
0071  cases are on record showing how complex and UNEXPECTED are the checks and relations between organi
0383  er productions of ranging widely, though so UNEXPECTED, can, I think, in most cases be explained b
0004  physical conditions of life, untouched and UNEXPLAINED. It is, therefore, of the highest importan
0006  t to feel surprise at much remaining as yet UNEXPLAINED in regard to the origin of species and var
0482  position leads him to attach more weight to UNEXPLAINED difficulties than to the explanation of a
0308  osited, or again as the bed of an open and UNFATHOMABLE sea. Looking to the existing oceans, which
0080  same species. Circumstances favourable and UNFAVOURABLE to Natural Selection, namely, intercrossin
0097  e inexplicable. Every hybridizer knows how UNFAVOURABLE exposure to wet is to the fertilisation of
0107  to sum up the circumstances favourable and UNFAVOURABLE to natural selection, as far as the extrem
0194  , and by the destruction of those with any UNFAVOURABLE deviation of structure, I have sometimes f
0256  and of hybrids, is more easily affected by UNFAVOURABLE conditions, than is the fertility of pure
0259  ng eminently susceptible to favourable and UNFAVOURABLE conditions, is innately variable. That it
0276  is eminently susceptible of favourable and UNFAVOURABLE conditions. The degree of sterility does n
0319  ecies is rare, we answer that something is UNFAVOURABLE in its conditions of life; but what that s
0319  inent. But we could not have told what the UNFAVOURABLE conditions were which checked its increase
0004  species in a state of nature; but I shall, UNFORTUNATELY, be compelled to treat this subject far t
0013  riters alone. When a deviation appears not UNFREQUENTLY, and we see it in the father and child, we
0229  plates of the vermilion wax having been left UNGNAWED, were situated, as far as the eye could judg
0232  al spheres, and then building up, or leaving UNGNAWED, the planes of intersection between these sp
0236  imal in the ordinary state, I should have UNHESITATINGLY assumed that all its characters had been
0247  tinct species, quite fertile together, he UNHESITATINGLY ranks them as varieties. Gartner, also, m
0007  g been raised under conditions of life not so UNIFORM as, and somewhat different from, those to wh
0020  imes (as I have found with pigeons) extremely UNIFORM, and everything seems simple enough; but whe
0046  e larvae of certain insects are very far from UNIFORM. Authors sometimes argue in a circle when th
0073  ely balanced, that the face of nature remains UNIFORM for long periods of time, though assuredly t
0103  ame species, or of the same variety, true and UNIFORM in character. It will obviously thus act far
0104  s of life will generally be in a great degree UNIFORM; so that natural selection will tend to modi
0114  t is with the plants and insects on small and UNIFORM islets; and so in small ponds of fresh water
0152  r the whole breed will cease to have a nearly UNIFORM character. The breed will then be said to ha
0156  should not have been rendered as constant and UNIFORM as other parts of the organisation; for seco
0162  forded by parts or organs of an important and UNIFORM nature occasionally varying so as to acquire
0174  hin recent times in a far less continuous and UNIFORM condition than at present. But I will pass o
0301  d differ from their former state, in a nearly UNIFORM, though perhaps extremely slight degree, the
0351  modification in different species will be no UNIFORM quantity. If, for instance, a number of spec
0356  als of the species will have been kept nearly UNIFORM by intercrossing; so that many individuals w
0366  old had reached its maximum, we should have a UNIFORM arctic fauna and flora, covering the centra!
0367  verywhere travelled southward, are remarkably UNIFORM round the world. We may suppose that the Gla
0369  s commencement the arctic productions were as UNIFORM round the polar regions as they are at the p
0372  or even a somewhat earlier period, was nearly UNIFORM along the continuous shores of the Polar Cir
0389  sed so great a difference in number. Even the UNIFORM county of Cambridge has 847 plants, and the
0418  ular species. If they find a character nearly UNIFORM, and common to a great number of forms, and
0418  se for propagating the race, are found nearly UNIFORM, they are considered as highly serviceable i
0015  ch domestic race, as already remarked, less UNIFORMITY of character than in true species. Domestic
0104  xist organic beings which never intercross, UNIFORMITY of character can be retained amongst them,
0104  f life change and they undergo modification, UNIFORMITY of character can be given to their modified
0273  both of hybrids and mongrels long retaining UNIFORMITY of character could be given. The variabilit
0340  hand, by similarity of conditions, for the UNIFORMITY of the same types in each during the later
0369  d how I account for the necessary degree of UNIFORMITY of the sub arctic and northern temperate fo
0370  limate, I attribute the necessary amount of UNIFORMITY in the sub arctic and northern temperate pr
0025  ese characters; for instance, I crossed some UNIFORMLY white fantails with some uniformly black ba
0025  ssed some uniformly white fantails with some UNIFORMLY black barbs, and they produced mottled brow
0269  er their general habits of life. Nature acts UNIFORMLY and slowly during vast periods of time on t
0172  en varieties are crossed their fertility is UNIMPAIRED? The two first heads shall be here discusse
0265  mes resist great changes of conditions with UNIMPAIRED fertility; and certain species in a group w
0010  same conditions of life; and this shows how UNIMPORTANT the direct effects of the conditions of li
0012  pe. Any variation which is not inherited is UNIMPORTANT for us. But the number and diversity of in
0045  generally affect what naturalists consider UNIMPORTANT parts; but I could show by a long catalogu
0085  rance permits us to judge, seem to be quite UNIMPORTANT, we must not forget that climate, food, et
0197  ignorant of the causes producing slight and UNIMPORTANT variations; and we are immediately made co
0205  bled to assert that any part or organ is so UNIMPORTANT for the welfare of a species, that modific
0261  grafted or budded on another is so entirely UNIMPORTANT for its welfare in a state of nature, I pr
```

Page **(Key Word)***

```
0263  fic forms, in the case of grafting it is   UNIMPORTANT  for their welfare. Causes of the Sterility
0272  find very few and, as it seems to me, quite UNIMPORTANT  differences between the so called hybrid o
0274  ents made by Kolreuter. These alone are the UNIMPORTANT  differences, which Gartner is able to poin
0331  nd too few species being given, but this is UNIMPORTANT  for us. The horizontal lines may represent
0426  t number of characters, let them be ever so UNIMPORTANT, betrays the hidden bond of community of d
0443  d protected by its parent, it must be quite UNIMPORTANT  whether most of its characters are fully a
0281  s of life will tend to supplant the old and UNIMPROVED  forms. By the theory of natural selection a
0217  st laid would have to be left for some time  UNINCUBATED, or there would be eggs and young birds of
0212  ss of our large birds to this cause; for in  UNINHABITED islands large birds are not more fearful t
0361  of a bird; but hard seeds of fruit will pass UNINJURED  through even the digestive organs of a turk
0397  several species did in this state withstand  UNINJURED  an immersion in sea water during seven days
0093  highly attractive to insects, they would,    UNINTENTIONALLY on their part, regularly carry pollen fr
0223  might become developed; and the ants thus    UNINTENTIONALLY reared would then follow their proper in
0385  ne aquarium to another, that I have quite    UNINTENTIONALLY stocked the one with fresh water shells
0390  n created on oceanic islands; for man has    UNINTENTIONALLY stocked them from various sources far mo
0467  not actually produce variability; he only    UNINTENTIONALLY exposes organic beings to new conditions
0132  exual elements seem to be affected before that UNION  takes place which is to form a new being. In
0143  us plants; and nothing is more common than the UNION  of homologous parts in normal structures, as
0143  homologous parts in normal structures, as the UNION  of the petals of the corolla into a tube. Har
0260  le difference, in the facility of effecting an UNION. The hybrids, moreover, produced from recipro
0260  strictly related to the facility of the first UNION  between their parents, seems to be a strange
0262  ciprocal crosses, the facility of effecting an UNION  is often very far from equal, so it sometimes
0262  n the mere adhesion of grafted stocks, and the UNION  of the male and female elements in the act of
0263  tally different, for, as just remarked, in the UNION  of two pure species the male and female sexua
0263  he greater or lesser difficulty in effecting a UNION  apparently depends on several distinct causes
0275  much or little from each other, namely in the  UNION  of individuals of the same variety, or of dif
0101  ome of them hermaphrodites, and some of them   UNISEXUAL. But if, in fact, all hermaphrodites do occ
0101  s, the difference between hermaphrodites and   UNISEXUAL species, as far as function is concerned, b
0008  ven in the many cases when the male and female UNITE. How many animals there are which will not br
0047  ue species. Practically, when a naturalist can UNITE  two forms together by others having intermedi
0048  e of 139 doubtful forms! Amongst animals which UNITE  for each birth, and which are highly locomoti
0096  ourse obvious that two individuals must always UNITE  for each birth; but in the case of hermaphrod
0096  umber pair; that is, two individuals regularly UNITE  for reproduction, which is all that concerns
0103  rcrossing will most affect those animals which UNITE  for each birth, which wander much, and which
0103  ly occasionally, and likewise in animals which UNITE  for each birth, but which wander little and w
0103  en in the case of slow breeding animals, which UNITE  for each birth, we must not overrate the effe
0104  far more efficiently with those animals which  UNITE  for each birth; but I have already attempted
0178  rms, more especially amongst the classes which UNITE  for each birth and wander much, may have sepa
0218  at least in the case of the American species,  UNITE  and lay first a few eggs in one nest and then
0257  of very closely allied species which will not  UNITE, or only with extreme difficulty; and on the
0257  the other hand of very distinct species which  UNITE  with the utmost facility. In the same family
0276  the greater or less facility of one species to UNITE  with another, is incidental on unknown differ
0333  wide intervals in the natural system, and thus UNITE  distinct families or orders. All that we have
0356  s, namely, with all organisms which habitually UNITE  for each birth, or which often intercross, I
0432  that branch, though at the actual fork the two UNITE  and blend together. We could not, as I have s
0048  floras of Great Britain, of France or of the   UNITED  States, drawn up by different botanists, and
0050  fically distinct. On the other hand, they are  UNITED  by many intermediate links, and it is very do
0074  on the ancient Indian mounds, in the Southern  UNITED  States, display the same beautiful diversity
0076  his in the recent extension over parts of the  UNITED  States of one species of swallow having cause
0085  xcellent horticulturist, Downing, that in the  UNITED  States smooth skinned fruits suffer far more
0091  wolf inhabiting the Catskill Mountains in the  UNITED  States, one with a light greyhound like form,
0100  of new Zealand, and Dr. Asa Gray those of the  UNITED  States, and the result was as I anticipated.
0115  sa gray's Manual of the Flora of the Northern  UNITED  States, 260 naturalised plants are enumerated
0142  hown in works on fruit trees published in the  UNITED  States, in which certain varieties are habitu
0180  ave their limbs and even the base of the tail  UNITED  by a broad expanse of skin, which serves as a
0212  the same species in the northern and southern  UNITED  States. Fear of any particular enemy is certa
0226  ssarily the case, the three flat surfaces are  UNITED  into a pyramid; and this pyramid, as Huber ha
0227  ill result a double layer of hexagonal prisms  UNITED  together by pyrimidal bases formed of three r
0256  many cases, in which two pure species can be   UNITED  with unusual facility, and produce numerous h
0259  considered as good and distinct species, are  UNITED, their fertility graduated from zero to perfe
0299  ned with any accuracy, excepting those of the  UNITED  States of America. I fully agree with Mr. God
0302  he ancient archipelagoes of Europe and of the  UNITED  States. We do not make due allowance for the
0306  untries beyond the confines of Europe and the  UNITED  States; and from the revolution in our palaeo
0308  n the several formations of Europe and the     UNITED  States; and from the amount of sediment, mile
0308  successive formations; whether Europe and the  UNITED  States during these intervals existed as dry
0324  believe that the existing productions of the   UNITED  States are more closely related to those whic
0332  together that they probably would have to be   UNITED  into one great family, in nearly the same man
0332  y, a14, etc., and b14, etc.) would have to be  UNITED  into one family; and the two other families (
0346  ican continent, from the central parts of the  UNITED  States to its extreme southern point, we meet
0352  h cases as Great Britain having been formerly  UNITED  to Europe, and consequently possessing the sa
0357  thetically bridged over every ocean, and have  UNITED  almost every island to some mainland. If thin
0357  gle island exists which has not recently been  UNITED  to some continent. This view cuts the Gordian
0357  s in their position and extension, as to have  UNITED  them within the recent period to each other a
0357  ve been continuously, or almost continuously,  UNITED  with each other, and with the many existing o
0365  hat the plants on the White Mountains, in the  UNITED  States of America, are all the same with thos
0366  ine and maize. Throughout a large part of the  UNITED  States, erratic boulders, and rocks scored by
0367  into Spain. The now temperate regions of the   UNITED  States would likewise be covered by arctic pl
0367  o immensely remote as on the mountains of the  UNITED  States. We can thus also unders
0368  plants of northern Scandinavia; those of the   UNITED  States to Labrador; those of the mountains of
0368  er than at present (as some geologists in the  UNITED  States believe to have been the case, chiefly
0371  ition, in the central parts of Europe and the  UNITED  States. On this view we can understand the re
0371  new Worlds will have been almost continuously  UNITED  by land, serving as a bridge, since rendered
0396  els are more likely to have been continuously  UNITED  within a recent period to the mainland than I
0401  y have at any former period been continuously  UNITED. The currents of the sea are rapid and sweep
0403  nearly the same physical conditions, and are   UNITED  by continuous land, yet they are inhabited by
0413  e to, sub families, families, and orders, all  UNITED  into one class. Thus, the grand fact in natur
```

Page **(Key Word)**

Page ***(Key Word)**

```
0456 p, by which all living and extinct beings are UNITED by complex, radiating, and circuitous lines o
0227 eres so as to intersect largely; and then she UNITES the points of intersection by perfectly flat
0268 ated, for instance, that the German Spitz dog UNITES more easily than other dogs with foxes, or th
0257 and, by very closely allied species generally UNITING with facility. But the correspondence betwee
0262  a very different case from the difficulty of UNITING two pure species, which have their reproduc
0277 t surprising that the degree of difficulty in UNITING two species, and the degree of sterility of
0171 ot in all cases absolutely perfect. The Law of UNITY of Type and of the Conditions of Existence em
0206 nic beings have been formed on two great laws, UNITY of Type, and the Conditions of Existence. By
0206 y of Type, and the Conditions of Existence. By UNITY of type is meant that fundamental agreement i
0206 pendent of their habits of life. On my theory, UNITY of type is explained by unity of descent. The
0206 e. on my theory, unity of type is explained by UNITY of descent. The expression of conditions of e
0434 the inheritance of former adaptations, that of UNITY of Type. Chapter VII. Instinct. Instincts com
0482 his resemblance is often expressed by the term UNITY of type; or by saying that the several parts
0060 nder such expressions as the plan of creation, UNITY of design, etc., and to think that we give an
0060 ature of the checks to increase. Competition   UNIVERSAL. Effects of climate. Protection from the nu
0062 sier than to admit in words the truth of the   UNIVERSAL struggle for life, or more difficult, at le
0096 acts, showing, in accordance with the almost   UNIVERSAL belief of breeders, that with animals and p
0147  can hardly be maintained that the law is of   UNIVERSAL application: but many good observers, more
0208 hat none of these characters of instinct are   UNIVERSAL. A little dose, as Pierre Huber expresses i
0245 of hybrids. Sterility various in degree, not   UNIVERSAL, affected by close interbreeding, removed b
0245 n crossed and of their mongrel offspring not   UNIVERSAL. Hybrids and mongrels compared independentl
0246 egree of sterility. Kolreuter makes the rule   UNIVERSAL; but then he cuts the knot, for in ten case
0247 eties. Gartner, also, makes the rule equally   UNIVERSAL; and he disputes the entire fertility of Ko
0249 cannot doubt the correctness of this almost   UNIVERSAL belief amongst breeders. Hybrids are seldom
0250 e of sterility between distinct species is a   UNIVERSAL law of nature. He experimentised on some of
0254 must either give up the belief of the almost   UNIVERSAL sterility of distinct species of animals wh
0254 te of knowledge, be considered as absolutely   UNIVERSAL. Laws governing the Sterility of first Cros
0266 rent class of facts. It is an old and almost   UNIVERSAL belief, founded, I think, on a considerable
0272 ertility of varieties can be proved to be of   UNIVERSAL occurence, or to form a fundamental distinc
0354 transport, the belief that this has been the   UNIVERSAL law, seems to me incomparably the safest. I
0399 facts could be given: indeed it is an almost   UNIVERSAL rule that the endemic productions of island
0417 high physiological importance and of almost   UNIVERSAL prevalence, and yet leave us in no doubt wh
0433 nant parent species, explains that great and   UNIVERSAL feature in the affinities of all organic be
0442 embryology, namely the very general, but not   UNIVERSAL difference in structure between the embryo
0446 per proportions, proves that this is not the   UNIVERSAL rule: for here the characteristic differenc
0452 s wholly disappear. It is also, I believe, a   UNIVERSAL rule, that a rudimentary part or organ is o
0460 can be mastered. With respect to the almost   UNIVERSAL sterility of species when first crossed, wh
0460 rms so remarkable a contrast with the almost   UNIVERSAL fertility of varieties when crossed, I must
0460 ir mongrel offspring cannot be considered as   UNIVERSAL; nor is their very general fertility surpri
0044 ne, but here community of descent is almost   UNIVERSALLY implied, though it can rarely be proved. W
0050  his attention, varieties of it will almost   UNIVERSALLY be found recorded. These varieties, moreov
0058 n as doubtful species, but which are almost   UNIVERSALLY ranked by British botanists as good and tr
0209 ossibly have been thus acquired. It will be   UNIVERSALLY admitted that instincts are as important a
0215 . familiarity alone prevents our seeing how   UNIVERSALLY and largely the minds of our domestic anim
0276 their hybrids, are very generally, but not   UNIVERSALLY, sterile. The sterility is of all degrees,
0277 ffspring, are very generally, but not quite   UNIVERSALLY, fertile. Nor is this nearly general and p
0297 nd that the embedded fossils, though almost   UNIVERSALLY ranked as specifically different, yet are
0352 and calls in the agency of a miracle. It is   UNIVERSALLY admitted, that in most cases the area inha
0358 uity with continents. Nor does their almost   UNIVERSALLY volcanic composition favour the admission
0403 ds of the Galapagos Archipelago, not having   UNIVERSALLY spread from island to island. In many othe
0404  could be given. And it will, I believe, be   UNIVERSALLY found to be true, that wherever in two reg
0416 ing physiological importance, but which are   UNIVERSALLY admitted as highly serviceable in the defi
0417 failed; for no part of the organisation is   UNIVERSALLY constant. The importance of an aggregate o
0423 ree, a genealogical classification would be   UNIVERSALLY preferred; and it has been attempted by so
0425 ose resemblance or affinity. As descent has   UNIVERSALLY been used in classing together the individ
0442 s within the same class, generally, but not   UNIVERSALLY, resembling each other; of the structure o
0456 n mind that the element of descent has been   UNIVERSALLY used in ranking together the sexes, ages,
0059 . and thus, the forms of life throughout the   UNIVERSE become divided into groups subordinate to gr
0002 ery many naturalists, some of them personally   UNKNOWN to me. I cannot, however, let this opportuni
0003 on would, I presume, say that after a certain   UNKNOWN number of generations, some bird had given b
0007 ffects. Methodical and Unconscious Selection.   UNKNOWN Origin of our Domestic Productions. Circumst
0012 n of growth. The result of the various, quite   UNKNOWN, or dimly seen laws of variation is infinite
0013 aly. The laws governing inheritance are quite   UNKNOWN; no one can say why the same peculiarity in
0023 they were originally domesticated, and yet be   UNKNOWN to ornithologists: and this, considering the
0024 very species have since all become extinct or   UNKNOWN. So many strange contingencies seem to me im
0027 stiction: these supposed species being quite   UNKNOWN in a wild state, and their becoming nowhere
0043 of reversion. Variability is governed by many   UNKNOWN laws, more especially by that of correlation
0044 of a species. Generally the term includes the   UNKNOWN element of a distinct act of creation. The t
0050 ences the study of a group of organisms quite   UNKNOWN to him, he is at first much perplexed to det
0085 necessary to bear in mind that there are many   UNKNOWN laws of correlation of growth, which, when o
0124 s (a) and (I), now supposed to be extinct and   UNKNOWN, it will be in some degree intermediate in c
0125 from one species of a still more ancient and   UNKNOWN genus. We have seen that in each country it
0146 atists as of high value, may be wholly due to   UNKNOWN laws of correlated growth, and without being
0153 sometimes become attached, from causes quite   UNKNOWN to us, more to one sex than to the other, ge
0159 ndency to variation, when acted on by similar   UNKNOWN influences. In the vegetable kingdom we have
0161 e character in question, which at last, under   UNKNOWN favorable conditions, gains an ascendancy. F
0166 lost character, and that this tendency, from   UNKNOWN causes, sometimes prevails. And we have just
0167 ject a real for an unreal, or at least for an   UNKNOWN, cause. It makes the works of God a mere moc
0172 at each species as descended from some other   UNKNOWN form, both the parent and all the transition
0181 en genera of birds had become extinct or were   UNKNOWN, who would have ventured to have surmised th
0194 of living and known forms to the extinct and   UNKNOWN is very small, I have been astonished how ra
0196 has primarily arisen from the above or other   UNKNOWN causes, it may at first have been of no adva
0197 the hooks on the bamboo may have arisen from   UNKNOWN laws of growth, and have been subsequently t
0198 relative importance of the several known and   UNKNOWN laws of variation: and I have here alluded t
0199 sen from correlation of growth, or from other   UNKNOWN cause, may reappear from the law of reversio
0209 s; that is of variations produced by the same   UNKNOWN causes which produce slight deviations of bo
```

Page ***(Key Word)**

0212 untry inhabited, but often from causes wholly UNKNOWN to us: Audubon has given several remarkable
0238 eld on their heads, the use of which is quite UNKNOWN: in the Mexican Myrmecocystus, the workers o
0260 hybrids is simply incidental or dependent on UNKNOWN differences, chiefly in the reproductive sys
0263 can be grafted on each other as incidental on UNKNOWN differences in their vegetative systems, so
0263 facility of first crosses, are incidental on UNKNOWN differences, chiefly in their reproductive s
0267 m to be connected together by some common but UNKNOWN bond, which is essentially related to the pr
0276 o take on another, is incidental on generally UNKNOWN differences in their vegetative systems, so
0276 ecies to unite with another, is incidental on UNKNOWN differences in their reproductive systems. T
0280 mediate between each species and a common but UNKNOWN progenitor; and the progenitor will generall
0281 mediate between them, but between each and an UNKNOWN common parent. The common parent will have h
0307 nt day; and that during these vast, yet quite UNKNOWN periods of time, the world swarmed with liv
0320 to wonder and to suspect that he died by some UNKNOWN deed of violence. The theory of natural sele
0358 ss facilities may be said to be almost wholly UNKNOWN. Until I tried, with Mr. Berkeley's aid, a f
0387 of the smaller fresh water animals. Other and UNKNOWN agencies probably have also played a part. I
0397 ed by it. Yet there must be, on my view, some UNKNOWN, but highly efficient means for their transp
0414 he view that classification either gives some UNKNOWN plan of creation, or is simply a scheme for
0420 have been unconsciously seeking, and not some UNKNOWN plan of creation, or the enunciation of gene
0420 descended from a species which existed at an UNKNOWN anterior period. Species of three of these g
0432 ents; or those parents from their ancient and UNKNOWN progenitor. Yet the natural arrangement in t
0434 tinct object in view, and do not look to some UNKNOWN plan of creation, we may hope to make sure b
0437 ms; therefore we may readily believe that the UNKNOWN progenitor of the vertebrata possessed many
0437 the vertebrata possessed many vertebrae; the UNKNOWN progenitor of the articulata, many segments;
0437 tor of the articulata, many segments; the UNKNOWN progenitor of flowering plants, many spiral
0453 ese vestiges of nails have appeared, not from UNKNOWN laws of growth, but in order to excrete horn
0464 e been deposited at these ancient and utterly UNKNOWN epochs in the world's history. I can answer
0464 most of the links between any two species are UNKNOWN, if any one link or intermediate variety be
0476 phical changes and to the many occasional and UNKNOWN means of dispersal, then we can understand,
0023 d a pouter be produced by crossing two breeds UNLESS one of the parent stocks possessed the charac
0035 hanges of this kind could never be recognised UNLESS actual measurements or careful drawings of th
0041 he qualities or structure of each individual. UNLESS such attention be paid nothing can be effecte
0062 nstantly to bear this conclusion in mind. Yet UNLESS it be thoroughly engrained in the mind, I am
0082 hance of profitable variations occurring; and UNLESS profitable variations do occur, natural selec
0108 eing thus disturbed. Nothing can be effected, UNLESS favourable variations occur, and variation it
0109 ms are continually and slowly being produced, UNLESS we believe that the number of specific forms
0150 ies to any part, however unusually developed, UNLESS it be unusually developed in comparison with
0162 either varieties or species; and this shows, UNLESS all these forms be considered as independentl
0181 d decrease in numbers or become exterminated, UNLESS they also became modified and improved in str
0281 parent with that of its modified descendants, UNLESS at the same time we had a nearly perfect chai
0297 een any two forms, they rank both as species, UNLESS they are enabled to connect them together by
0297 ly be ranked as a third and distinct species, UNLESS at the same time it could be most closely con
0297 the lower and upper beds of a formation, and UNLESS we obtained numerous transitional gradations,
0305 ld be an insuperable difficulty on my theory, UNLESS it could likewise be shown that the species o
0328 , considering the proximity of the two areas, UNLESS, indeed, it be assumed that an isthmus separa
0766 ions would at the same time travel southward, UNLESS they were stopped by barriers, in which case
0413 that it reveals the plan of the Creator: but UNLESS it be specified whether order in time or spac
0464 s if we were to examine them ever so closely, UNLESS we likewise possessed many of the intermediat
0019 , the bull dog, or Blenheim spaniel, etc., so UNLIKE all wild Canidae, ever existed freely in a st
0033 are, and how extremely alike the flowers, how UNLIKE the flowers of the heartsease are, and how al
0151 the different species being sometimes wholly UNLIKE in shape; and the amount of variation in the
0204 ave diversified habits, with some habits very UNLIKE those of its nearest congeners. Hence we can
0243 s of our distinct Kitty wrens, a habit wholly UNLIKE that of any other known bird. Finally, it may
0413 like, and for separating those which are most UNLIKE; or as an artificial means for enunciating, a
0442 vividual embryo, which ultimately become very UNLIKE and serve for diverse purposes; being at this
0479 should be so closely alike, and should be so UNLIKE the adult forms. We may cease marvelling at t
0023 breeding on precipices, and good fliers, are UNLIKELY to be exterminated, and the common rock pige
0304 t me a drawing of a perfect specimen of an UNMISTAKEABLE sessile cirripede, which he had himself e
0323 organic remains in certain beds present an UNMISTAKEABLE degree of resemblance to those of the Cha
0398 ry product of the land and water bears the UNMISTAKEABLE stamp of the American continent. There ar
0440 the larva shows this to be the case in an UNMISTAKEABLE manner. So again the two main divisions o
0451 aning of rudimentary organs is often quite UNMISTAKEABLE: for instance there are beetles of the sa
0316 either new and modified or the same old and UNMODIFIED forms. Species of the genus Lingula, for in
0405 e very widely; for it is necessary that the UNMODIFIED parent should range widely, undergoing modi
0447 g mature age. Whereas the young will remain UNMODIFIED, or be modified in a lesser degree, by the
0008 geoffroy St. Hilaire's experiments show that UNNATURAL treatment of the embryo causes monstrositie
0009 anisms will breed most freely under the most UNNATURAL conditions (for instance, the rabbit and fe
0229 n the opposed basins, but the work, from the UNNATURAL state of things, had not been neatly perfor
0251 on, we must infer that the plants were in an UNNATURAL state. Nevertheless these facts show on wha
0264 ngs seem eminently sensitive to injurious or UNNATURAL conditions of life. In regard to the steril
0265 and plants are rendered impotent by the same UNNATURAL conditions; and whole groups of species ten
0265 when organic beings are placed under new and UNNATURAL conditions, and when hybrids are produced b
0265 itions, and when hybrids are produced by the UNNATURAL crossing of two species, the reproductive s
0266 s offered why an organism, when placed under UNNATURAL conditions, is rendered sterile. All that I
0077 seeds may be widely distributed and fall on UNOCCUPIED ground. In the water beetle, the structure
0082 n, is actually necessary to produce new and UNOCCUPIED places for natural selection to fill up by
0102 ferent degrees, so as better to fill up the UNOCCUPIED place. But if the area be large, its severa
0119 o the nature of the places which are either UNOCCUPIED or not perfectly occupied by other beings:
0388 for instance, on a rising islet, it will be UNOCCUPIED; and a single seed or egg will have a good
0472 e slowly varying descendants of each to any UNOCCUPIED or ill occupied place in nature, these fact
0318 country and has increased in numbers at an UNPARALLELED rate, I asked myself what could so recentl
0009 eproductive system so seriously affected by UNPERCEIVED causes as to fail in acting, we need not b
0319 living being is constantly being checked by UNPERCEIVED injurious agencies; and that these same un
0319 ved injurious agencies: and that these same UNPERCEIVED agencies are amply sufficient to cause rar
0064 d only two seeds, and there is no plant so UNPRODUCTIVE as this, and their seedlings next year pro
0167 s, as it seems to me, to reject a real for an UNREAL, or at least for an unknown, cause. It makes
0003 onclusion, even if well founded, would be UNSATISFACTORY, until it could be shown how the innumera
0412 , for all have descended from one ancient but UNSEEN parent, and, consequently, have inherited som

Page **(Key Word)**

```
0088  he females; the result is not death to the UNSUCCESSFUL competitor, but few or no offspring. Sexua
0264  may be exposed to conditions in some degree UNSUITABLE, and consequently be liable to perish at an
0026  cies closely related together, though it is UNSUPPORTED by a single experiment. But to extend the
0458  hesitation adopt this view, even if it were UNSUPPORTED by other facts or arguments. Chapter XIV.
0004  er and to their physical conditions of life, UNTOUCHED and unexplained. It is, therefore, of the h
0480  hereas in the calf, the teeth have been left UNTOUCHED by selection or disuse, and on the principl
0486  f natural history become! A grand and almost UNTRODDEN field of inquiry will be opened, on the cau
0039  a tail developed in some slight degree in an UNUSUAL manner, or a pouter.till he saw a pigeon wit
0039  till he saw a pigeon with a crop of somewhat UNUSUAL size; and the more abnormal or unusual any c
0039  mewhat unusual size; and the more abnormal or UNUSUAL any character was when it first appeared, th
0082  great physical change, as of climate, or any UNUSUAL degree of isolation to check immigration, is
0131  ed structures variable. Parts developed in an UNUSUAL manner are highly variable: specific charact
0150  dary sexual characters, when displayed in any UNUSUAL manner. The term, secondary sexual character
0151  e characters, whether or not displayed in any UNUSUAL manner, of which fact I think there can be l
0154  character, and its variation would be a more UNUSUAL circumstance. I have chosen this example bec
0157  the differences in these cases are of a very UNUSUAL nature, the relation can hardly be accidenta
0256  in which two pure species can be united with UNUSUAL facility, and produce numerous hybrid offspr
0429  etitors, with a few members preserved by some UNUSUAL coincidence of favourable circumstances. Mr.
0474  ry of creation why a part developed in a very UNUSUAL manner in any one species of a genus, and th
0474  ies branched off from a common progenitor, an UNUSUAL amount of variability and modification, and
0474  able. But a part may be developed in the most UNUSUAL manner, like the wing of a bat, and yet not
0487  n will be recognised as having depended on an UNUSUAL concurrence of circumstances, and the blank
0150  ule by no means applies to any part, however UNUSUALLY developed, unless it be unusually developed
0150  t, however unusually developed, unless it be UNUSUALLY developed in comparison with the same part
0153  raordinary amount of modification implies an UNUSUALLY large and long continued amount of variabil
0221  h of July, I came across a community with an UNUSUALLY large stock of slaves, and I observed a few
0265  and certain species in a group will produce UNUSUALLY fertile hybrids. No one can tell, till he t
0322  . moreover, when by sudden immigration or by UNUSUALLY rapid development, many species of a new gr
0264  rility in first crosses. I was at first very UNWILLING to believe in this view: as hybrids, when o
0185  its manner of swimming, and of flying when UNWILLINGLY it takes flight, would be mistaken by any
0481  tation. But the chief cause of our natural UNWILLINGNESS to admit that one species has given birth
0285  reat cracks along which the strata have been UPHEAVED on one side, or thrown down on the other, to
0309  tear; and would have been at least partially UPHEAVED by the oscillations of level, which we may f
0355  en stocked. A volcanic island, for instance, UPHEAVED and formed at the distance of a few hundreds
0365  geological time, whilst an island was being UPHEAVED and formed, and before it had become fully s
0404  obvious that a mountain, as it became slowly UPHEAVED, would naturally be colonised from the surro
0185  geese are formed for swimming? yet there are UPLAND geese with webbed feet which rarely or never
0185  g change of structure. The webbed feet of the UPLAND goose may be said to have become rudimentary
0199  an hardly believe that the webbed feet of the UPLAND goose or of the frigate bird are of special u
0200  to inheritance. But to the progenitor of the UPLAND goose and of the frigate bird, webbed feet no
0204  it can live, how it has arisen that there are UPLAND geese with webbed feet, ground woodpeckers, d
0471  reated to prey on insects on the ground; that UPLAND geese, which never or rarely swim should have
0031  says: It would seem as if they had chalked out UPON a wall a form perfect in itself, and then had
0228  basins, so that each hexagonal prism was built UPON the festooned edge of a smooth basin, instead
0256  er the same circumstances, but depends in part UPON the constitution of the individuals which happ
0282  redly, if this theory be true, such have lived UPON this earth. On the lapse of Time. Independentl
0456  alists in their classifications; the value set UPON characters, if constant and prevalent, whether
0293  en we find a species disappearing before the UPPERMOST layers have been deposited, it would be equ
0331  al formations, and all the forms beneath the UPPERMOST line may be considered as extinct. The thre
0332  e series, the three existing families on the UPPERMOST line would be rendered less distinct from e
0412  to groups. In the diagram each letter on the UPPERMOST line may represent a genus including severa
0420  ed by the fifteen genera (a14 to z14) on the UPPERMOST horizontal line. Now all these modified des
0287  n the gently inclined Wealden district, when UPRAISED, could hardly have been great, but it would
0288  if in sand or gravel, will when the beds are UPRAISED generally be dissolved by the percolation of
0290  f the South American coasts, which have been UPRAISED several hundred feet within the recent perio
0290  he incessant action of the waves, when first UPRAISED and during subsequent oscillations of level.
0291  by extremely few animals, and the mass when UPRAISED will give a most imperfect record of the for
0291  a fossiliferous formation thick enough, when UPRAISED, to resist any amount of degradation, may be
0292  ccumulated will have been destroyed by being UPRAISED and brought within the limits of the coast a
0294  e part of the glacial period shall have been UPRAISED, organic remains will probably first appear
0296  of the lower beds of a formation having been UPRAISED, denuded, submerged, and then re covered by
0296  nce in great fossilised trees, still standing UPRIGHT as they grew, of many long intervals of time
0231  shall thus have a thin wall steadily growing UPWARD! but always crowned by a gigantic coping. Fro
0036  d from the original stock of Mr. Bakewell for UPWARDS of fifty years. There is not a suspicion exi
0121  diagram by the dotted lines not prolonged far UPWARDS from want of space. But during the process o
0285  . the Craven fault, for instance, extends for UPWARDS of 30 miles, and along this line the vertica
0317  int, but abruptly; it then gradually thickens UPWARDS, sometimes keeping for a space of equal thic
0303  mencement of this great series. Cuvier used to URGE that no monkey occurred in any tertiary stratu
0343  one, as that of Europe, are considered; he may URGE the apparent, but often falsely apparent, sudd
0001  d as my health is far from strong, I have been URGED to publish this Abstract. I have more especia
0095  pen to the same objections which were at first URGED against Sir Charles Lyell's noble views on th
0173  ervals of time immensely remote. But it may be URGED that when several closely allied species inha
0203  f the difficulties and objections which may be URGED against my theory. Many of them are very grav
0267  sed, and of their Mongrel offspring. It may be URGED, as a most forcible argument, that there must
0279  ted the chief objections which might be justly URGED against the views maintained in this volume.
0280  ost obvious and gravest objection which can be URGED against my theory. The explanation lies, as I
0299  bvious of all the many objections which may be URGED against my views. Hence it will be worth whil
0302  uddenly appear in certain formations, has been URGED by several palaeontologists, for instance, by
0308  ent must remain inexplicable; and may be truly URGED as a valid argument against the views here en
0418  he economy of nature. Yet it has been strongly URGED by those great naturalists, Milne Edwards and
0463  d forcible of the many objections which may be URGED against my theory. Why, again, do whole group
0465  bjections and difficulties which may justly be URGED against my theory; and I have now briefly rec
0202  age instinctive hatred of the queen bee, which URGES her instantly to destroy the young queens her
0030  d, but to man's use or fancy. Some variations USEFUL to him have probably arisen suddenly, or by o
0030  chard, and flower garden races of plants most USEFUL to man at different seasons and for different
0030  eeds were suddenly produced as perfect and as USEFUL as we now see them; indeed, in several cases,
```

Page **(Key Word)**

0030 tions; man adds them up in certain directions USEFUL to him. In this sense he may be said to make
0030 this sense he may be said to make for himself USEFUL breeds. The great power of this principle of
0036 stic animals, yet any one animal particularly USEFUL to them, for any special purpose, would be ca
0038 e chance possess the aboriginal stocks of any USEFUL plants, but that the native plants have not b
0041 sired direction. But as variations manifestly USEFUL or pleasing to man appear only occasionally,
0041 that the animal or plant should be so highly USEFUL to man, or so much valued by him, that the cl
0044 cture in one part, either injurious to or not USEFUL to the species, and not generally propagated.
0050 nimal or plant in a state of nature be highly USEFUL to man, or from any cause closely attract his
0061 principle, by which each slight variation, if USEFUL, is preserved, by the term of Natural Selecti
0061 uses, through the accumulation of slight but USEFUL variations, given to him by the hand of Natur
0080 be thought improbable, seeing that variations USEFUL to man have undoubtedly occurred, that other
0080 e undoubtedly occurred, that other variations USEFUL in some way to each being in the great and co
0081 i call Natural Selection. Variations neither USEFUL nor injurious would not be affected by natura
0083 appearances, except in so far as they may be USEFUL to any being. She can act on every internal o
0083 ent enough to catch his eye, or to be plainly USEFUL to him. Under nature, the slightest differenc
0090 s, etc.) which we cannot believe to be either USEFUL to the males in battle, or attractive to the
0090 f the turkey cock, which can hardly be either USEFUL or ornamental to this bird; indeed, had the t
0098 ensure self fertilisation; and no doubt it is USEFUL for this end: but, the agency of insects is o
0127 dinary fact if no variation ever had occurred USEFUL to each being's own welfare, in the same way
0127 same way as so many variations have occurred USEFUL to man. But if variations useful to any organ
0127 ave occurred useful to man. But if variations USEFUL to any organic being do occur, assuredly indi
0127 g. sexual selection will also give characters USEFUL to the males alone, in their struggles with o
0140 y chosen by uncivilised man because they were USEFUL and bred readily under confinement, and not b
0148 changed conditions of life a structure before USEFUL becomes less useful, any diminution, however
0148 f life a structure before useful becomes less USEFUL, any diminution, however slight, in its devel
0181 ller flank membranes, each modification being USEFUL, each being propagated, until by the accumula
0181 ls; and that each grade of structure had been USEFUL to its possessor. Nor can I see any insuperab
0186 e very imperfect and simple, each grade being USEFUL to its possessor, can be shown to exist: if f
0186 ariation or modification in the organ be ever USEFUL to an animal under changing conditions of lif
0199 no doubt played a most important part, and a USEFUL modification of one part will often have enta
0199 use. So again characters which formerly were USEFUL, or which formerly had arisen from correlatio
0199 in beauty to charm the females, can be called USEFUL only in rather a forced sense. But by far the
0200 he frigate bird, webbed feet no doubt were as USEFUL as they now are to the most aquatic of existi
0202 for if on the whole the power of stinging be USEFUL to the community, it will fulfil all the requ
0205 produce parts, organs, and excretions highly USEFUL or even indispensable, or highly injurious to
0205 er species; but in all cases at the same time USEFUL to the owner. Natural selection in each well
0223 hat work they could. If their presence proved USEFUL to the species which had seized them, if it w
0240 that if the smaller workers had been the most USEFUL to the community, and those males and females
0241 d then the extreme forms, from being the most USEFUL to the community, having been produced in gre
0241 their parents, has originated. We can see how USEFUL their production may have been to a social co
0242 same principle that the division of labour is USEFUL to civilised man. As ants work,by inherited i
0243 difications of instinct to any extent, in any USEFUL direction. In some cases habit or use and dis
0383 heir having become fitted, in a manner highly USEFUL to them, for short and frequent migrations fr
0423 griculturist Marshall says the horns are very USEFUL for this purpose with cattle, because they ar
0454 ely lost the power of flying. Again, an organ USEFUL under certain conditions, might become injuri
0455 terials forming any part or structure, if not USEFUL to the possessor, will be saved as far as is
0455 systematists have found rudimentary parts as USEFUL as, or even sometimes more useful than, parts
0455 ry parts as useful as, or even sometimes more USEFUL than, parts of high physiological importance.
0467 onsciously by preserving the individuals most USEFUL to him at the time, without any thought of al
0467 cy in the production of the most distinct and USEFUL domestic breeds. That many of the breeds prod
0469 hy should we doubt that variations in any way USEFUL to beings, under their excessively complex re
0469 if man can by patience select variations most USEFUL to himself, should nature fail in selecting v
0469 f, should nature fail in selecting variations USEFUL, under changing conditions of life, to her li
0486 as the summing up of many contrivances, each USEFUL to the possessor, nearly in the same way as w
0427 phical distribution may sometimes be brought USEFULLY into play in classing large and widely distr
0038 our plants up to their present standard of USEFULNESS to man, we can understand how it is that ne
0099 lisation, should in so many cases be mutually USELESS to each other! How simply are these facts ex
0137 it is difficult to imagine that eyes, though USELESS, could be in any way injurious to animals li
0148 o have its nutriment wasted in building up an USELESS structure. I can thus only understand a fact
0148 g wasted in developing a structure now become USELESS. Thus, as I believe, natural selection will
0161 endless number of generations, than in quite USELESS or rudimentary organs being, as we all know
0168 in the scale. Rudimentary organs, from being USELESS, will be disregarded by natural selection, a
0202 ose of thousands of drones, which are utterly USELESS to the community for any other end, and whic
0216 instinct retained by our chickens has become USELESS under domestication, for the mother hen has
0392 ecs, would form an endemic species, having as USELESS an appendage as any rudimentary organ, for i
0433 ; why we are permitted to use rudimentary and USELESS organs, or others of trifling physiological
0448 f locomotion or of the senses, etc., would be USELESS; and in this case the final metamorphosis wo
0453 entary or atrophied organs, are imperfect and USELESS. In works on natural history rudimentary org
0454 rgan rendered, during changed habits of life, USELESS or injurious for one purpose, might easily b
0455 its former functions. An organ, when rendered USELESS, may well be variable, for its variations ca
0455 d, still retained in the spelling, but become USELESS in the pronunciation; but which serve as a c
0455 ce of organs in a rudimentary, imperfect, and USELESS condition, or quite aborted, far from presen
0479 n tend to reduce an organ, when it has become USELESS by changed habits or under changed condition
0149 heir variability seems to be owing to their USELESSNESS, and therefore to natural selection having
0061 sults, and can adapt organic beings to his own USES, through the accumulation of slight but useful
0185 diving, grasping the stones with its feet and USING its wings under water. He who believes that e
0040 eviation of structure, or takes more care than USUAL in matching his best animals and thus improve
0057 for they differ from each other by a less than USUAL amount of difference. Moreover, the species o
0221 to search for aphides. This difference in the USUAL habits of the masters and slaves in the two c
0259 certain hybrids which instead of having, as is USUAL, an intermediate character between their two
0446 either have appeared at an earlier period than USUAL, or, if not so, the differences must have bee
0076 lly sorted. As species of the same genus have USUALLY, though by no means invariably, some similar
0203 asons assigned, the intermediate variety will USUALLY exist in lesser numbers than the two forms w
0207 knowing for what purpose it is performed, is USUALLY said to be instinctive. But I could show tha
0223 r the community. In England the masters alone USUALLY leave the nest to collect building materials
0248 reatly decreased. I do not doubt that this is USUALLY the case, and that the fertility often sudde

Page ***************************************(Key Word)***************************************

0253 g. on the contrary, brothers and sisters have USUALLY been crossed in each successive generation,
0259 y sterile. So again amongst hybrids which are USUALLY intermediate in structure between their pare
0261 ecies, and varieties of the same species, can USUALLY, but not invariably, be grafted with ease. B
0295 palaeontologist, that very thick deposits are USUALLY barren of organic remains, except near their
0199 ately made by some naturalists, against the UTILITARIAN doctrine that every detail of structure ha
0144 ifying important structures, independently of UTILITY and, therefore, of natural selection, than t
0413 given of each kind of dog. The ingenuity and UTILITY of this system are indisputable. But many na
0419 closely allied forms. Temminck insists on the UTILITY or even necessity of this practice in certai
0435 y of pattern in members of the same class, by UTILITY or by the doctrine of final causes. The hope
0008 s; but how many cultivated plants display the UTMOST vigour, and yet rarely or never seed! In some
0027 igeons have been watched, and tended with the UTMOST care, and loved by many people. They have bee
0066 g around us may be said to be striving to the UTMOST to increase in numbers: that each lives by a
0113 and thus, as it may be said, is striving its UTMOST to increase its numbers. Consequently, I cann
0114 ature), and may be said to be striving to the UTMOST to live there; but, it is seen, that where th
0231 th is continually given to the comb, with the UTMOST ultimate economy of wax. It seems at first to
0257 of very distinct species which unite with the UTMOST facility. In the same family there may be a g
0427 ogical or adaptive character, although of the UTMOST importance to the welfare of the being, are a
0020 ke, and then the extreme difficulty, or rather UTTER hopelessness, of the task becomes apparent. C
0021 trumpeter and laugher, as their names express, UTTER a very different coo from the other breeds. T
0070 exist together, and thus save each other from UTTER destruction. I should add that the good effec
0078 in the arctic regions or on the borders of an UTTER desert, will competition cease. The land may
0109 he number of its enemies, run a good chance of UTTER extinction. But we may go further than this:
0175 rey, or in the seasons, be extremely liable to UTTER extermination: and thus its geographical rang
0255 revent their crossing and blending together in UTTER confusion. The following rules and conclusion
0321 still inhabit our fresh waters. Therefore the UTTER extinction of a group is generally, as we hav
0344 ed offspring on the face of the earth. But the UTTER extinction of a whole group of species may of
0009 fertile eggs. Many exotic plants have pollen UTTERLY worthless, in the same exact condition as in
0029 n his treatise on pears and apples, shows how UTTERLY he disbelieves that the several sorts, for i
0039 fantail, is, I have no doubt, in most cases, UTTERLY incorrect. The man who first selected a pige
0059 thus originated: whereas, these analogies are UTTERLY inexplicable if each species has been indepe
0097 o believe that it is a general law of nature (UTTERLY ignorant though we be of the meaning of the
0126 ps, a multitude of smaller groups will become UTTERLY extinct, and leave no modified descendants:
0168 lations of growth, the nature of which we are UTTERLY unable to understand. Multiple parts are var
0191 position. But it is conceivable that the now UTTERLY lost branchiae might have been gradually wor
0202 gle purpose of thousands of drones, which are UTTERLY useless to the community for any other end,
0203 he theory of independent acts of creation are UTTERLY obscure. We have seen that species at any on
0218 other observers, the European cuckoo has not UTTERLY lost all maternal love and care for her own
0219 ctually carry their masters in their jaws. So UTTERLY helpless are the masters, that when Huber sh
0242 nt of exercise, or habit, or volition, in the UTTERLY sterile members of a community could possibl
0258 longiflora with the pollen of M. jalappa, and UTTERLY failed. Several other equally striking cases
0259 parents; and these hybrids are almost always UTTERLY sterile, even when the other hybrids raised
0268 hough resembling each other most closely, are UTTERLY sterile when intercrossed. Several considera
0319 ons of life apparently so favourable. But how UTTERLY groundless was my astonishment! Professor Ow
0322 ed by us, the whole economy of nature will be UTTERLY obscured. Whenever we can precisely say why
0341 cies and the other whole genera having become UTTERLY extinct. In failing orders, with the genera
0347 ble to point out three faunas and floras more UTTERLY dissimilar. Or again we may compare the prod
0372 ysical conditions, but with their inhabitants UTTERLY dissimilar. But we must return to our more i
0406 o those of the nearest mainland, are I think, UTTERLY inexplicable on the ordinary view of the ind
0429 which on my theory have been exterminated and UTTERLY lost. And we have some evidence of aberrant
0431 ief that many ancient forms of life have been UTTERLY lost, through which the early progenitors of
0451 s do we see wings so reduced in size as to be UTTERLY incapable of flight, and not rarely lying up
0451 for two purposes, many become rudimentary or UTTERLY aborted for one, even the more important pur
0452 in certain groups. Rudimentary organs may be UTTERLY aborted: and this implies, that we find in a
0464 here have been deposited at these ancient and UTTERLY unknown epochs in the world's history. I can
0464 the lapse of time has been so great as to be UTTERLY inappreciable by the human intellect. The nu
0471 he grouping of all organic beings seems to me UTTERLY inexplicable on the theory of creation. As n
0478 of all other mammals, on oceanic islands, are UTTERLY inexplicable on the theory of independent ac
0480 rate organ having been specially created, how UTTERLY inexplicable it is that parts, like the teet
0489 ra, have left no descendants, but have become UTTERLY extinct. We can so far take a prophetic glan
0253 on to believe that the hybrids from Cervulus VAGINALIS and Reevesii, and from Phasianus colchicus
0018 egypt? The whole subject must, I think, remain VAGUE; nevertheless, I may, without here entering o
0048 rican mainland, I was much struck how entirely VAGUE and arbitrary is the distinction between spec
0266 fessed that we cannot understand, excepting on VAGUE hypotheses, several facts with respect to the
0345 scale of nature; and this may account for that VAGUE yet ill defined sentiment, felt by many palae
0476 orms are generally looked at as being, in some VAGUE sense, higher than ancient and extinct forms;
0044 d all naturalists, yet every naturalist knows VAGUELY what he means when he speaks of a species. G
0014 ces to species in a state of nature. I have in VAIN endeavoured to discover on what decisive facts
0024 other parts of their structure, we may look in VAIN throughout the whole great family of Columbida
0095 o that whole fields of the red clover offer in VAIN an abundant supply of precious nectar to the h
0285 my mind almost in the same manner as does the VAIN endeavour to grapple with the idea of eternity
0338 he same groups at the present day, it would be VAIN to look for animals having the common embryolo
0342 htly reject my whole theory. For he may ask in VAIN where are the numberless transitional links wh
0485 ; but we shall have at least be freed from the VAIN search for the undiscovered and undiscoverable
0032 ducated eye, differences which I for one have VAINLY attempted to appreciate. Not one man in a tho
0049 f these terms has been generally accepted, is VAINLY to beat the air. Many of the cases of strongl
0384 to live in fresh water; and, according to VALENCIENNES, there is hardly a single group of fishes
0100 limited sense. I believe this objection to be VALID, but that nature has largely provided against
0308 main inexplicable; and may be truly urged as a VALID argument against the views here entertained.
0330 en two living forms, the objection is probably VALID. But I apprehend that in a perfectly natural
0365 aturalised. Put this, as it seems to me, is no VALID argument against what would be effected by oc
0018 d to have manufactured pottery existed in the VALLEY of the Nile thirteen or fourteen thousand yea
0095 ished such views as the excavation of a great VALLEY by a single diluvial wave, so will natural se
0373 etritus, about 800 feet in height, crossing a VALLEY of the Andes; and this I now feel convinced w
0095 e, when applied to the excavation of gigantic VALLEYS or to the formation of the longest lines of
0366 boulders, of the icy streams with which their VALLEYS were lately filled. So greatly has the clima
0481 s of inland cliffs had been formed, and great VALLEYS excavated, by the slow action of the coast w

Page ***************************************(Key Word)***************************************

valleys

Page **(Key Word)**

```
0010  recent experiments on plants seem extremely VALUABLE. When all or nearly all the individuals expo
0040  ill get recognised as something distinct and VALUABLE, and will then probably first receive a prov
0041  successful than amateurs in getting new and  VALUABLE varieties. The keeping of a large number of
0042  a large stock not kept in geese, from being  VALUABLE only for two purposes, food and feathers, an
0053  h. c. Watson, to whom I am much indebted for VALUABLE advice and assistance on this subject, soon
0361  roots of drifted trees, these stones being a VALUABLE royal tax. I find on examination, that when
0419  ogy, we shall see why such characters are so VALUABLE, on the view of classification tacitly inclu
0479  nd why embryological characters are the most VALUABLE of all. The real affinities of all organic b
0428  sses (and all our experience shows that this VALUATION has hitherto been arbitrary), could easily
0016  t characters are of generic value; all such  VALUATIONS being at present empirical. Moreover, on th
0004  y venture to express my conviction of the high VALUE of such studies, although they have been very
0016  iffer from each other in characters of generic VALUE. I think it could be shown that this statemen
0016  in determining what characters are of generic VALUE: all such valuations being at present empiric
0017  hat these capacities have added largely to the VALUE of most of our domesticated productions; but
0018  large and varied stores of knowledge, I should VALUE more than that of almost any one, thinks that
0036  of unconscious selection going on. We see the VALUE set on animals even by the barbarians of Tier
0036  heir old women, in times of dearth, as of less VALUE than their dogs. In plants the same gradual p
0039  mall differences, and it is in human nature to VALUE any novelty, however slight, in one's own pos
0039  slight, in one's own possession. Nor must the VALUE which would formerly be set on any slight dif
0039  duals of the same species, be judged of by the VALUE which would now be set on them, after several
0040  ll be a slow process. As soon as the points of VALUE of the new sub breed are once fully acknowled
0050  find the greatest number of forms of doubtful VALUE. I have been struck with the fact, that if an
0058  y allied to other species as to be of doubtful VALUE: these 63 reputed species range on an average
0146  f structure, viewed by systematists as of high VALUE, may be wholly due to unknown laws of correla
0155  hat a character, which is generally of generic VALUE, when it sinks in value and becomes only of s
0155  s generally of generic value, when it sinks in VALUE and becomes only of specific value, often bec
0155  it sinks in value and becomes only of specific VALUE, often becomes variable, though its physiolog
0376  and in the southern hemisphere are of doubtful VALUE, being ranked by some naturalists as specific
0415  an organ does not determine its classificatory VALUE, is almost shown by the one fact, that in all
0415  to suppose, has nearly the same physiological VALUE, its classificatory value is widely different
0415  e same physiological value, its classificatory VALUE is widely different. No naturalist can have w
0416  , and the differences are of quite subordinate VALUE in classification; yet no one probably will s
0416  ly, organs in this condition are often of high VALUE in classification. No one will dispute that t
0417  her characters of more or less importance. The VALUE indeed of an aggregate of characters is very
0418  not trouble themselves about the physiological VALUE of the characters which they use in defining
0418  t common to others, they use it as one of high VALUE; if common to some lesser number, they use it
0418  lesser number, they use it as of subordinated VALUE. This principle has been broadly confessed by
0418  exion can be discovered between them, especial VALUE is set on them. As in most groups of animals,
0418  found to offer characters of quite subordinate VALUE. We can see why characters derived from the e
0419  ists. Finally, with respect to the comparative VALUE of the various groups of species, such as ord
0419  ers, have strongly insisted on their arbitrary VALUE. Instances could be given amongst plants and
0421  long to Silurian genera. So that the amount or VALUE of the differences between organic beings all
0426  very different habits of life, it assumes high VALUE; for we can account for its presence in so ma
0426  related or aggregated characters have especial VALUE in classification. We can understand why a sp
0426  enerally the most constant, we attach especial VALUE to them; but if these same organs, in another
0426  a group, are found to differ much, we at once VALUE them less in our classification. We shall her
0428  e class, by arbitrarily raising or sinking the VALUE of the groups in other classes (and all our e
0432  or small, and thus give a general idea of the VALUE of the differences between them. This is what
0433  n our classification. We can understand why we VALUE certain resemblances far more than others; wh
0446  that the characteristic differences which give VALUE to each breed, and which have been accumulate
0456  d by naturalists in their classifications; the VALUE set upon characters, if constant and prevalen
0456  gans, of no importance; the wide opposition in VALUE between analogical or adaptive characters, an
0479  to the being, are often of high classificatory VALUE; and why embryological characters are the mos
0485  we shall be led to weigh more carefully and to VALUE higher the actual amount of difference betwee
0486  of domestic productions will rise immensely in VALUE. A new variety raised by man will be a far mo
0028  up their pedigree and race. Pigeons were much VALUED by Akber Khan in India, about the year 1600;
0033  leaves, pods, or tubers, or whatever part is  VALUED, in the kitchen garden, in comparison with th
0034  stone shows how much good domestic breeds are VALUED by the negroes of the interior of Africa who
0040  a distinct name, and from being only slightly VALUED, their history will be disregarded. When furt
0041  should be so highly useful to man, or so much VALUED by him, that the closest attention should be
0042  its, cannot be matched, and, although so much VALUED by women and children, we hardly ever see a d
0427  ance to the welfare of the being, are almost  VALUELESS to the systematist. For animals, belonging
0304  from the ease with which even a fragment of a VALVE can be recognised; from all these circumstanc
0151  ule in its largest application. The opercular VALVES of sessile cirripedes (rock barnacles) are, i
0151  several species of one genus, Pyrgoma, these  VALVES present a marvellous amount of diversificatio
0151  ous amount of diversification: the homologous VALVES in the different species being sometimes whol
0151  ch other in the characters of these important VALVES than do other species of distinct genera. As
0415  e or absence of albumen, in the imbricate or  VALVULAR aestivation. Any one of these characters sin
0029  in breed was descended from a distinct species. VAN Mons, in his treatise on pears and apples, sho
0004  ight variations. I will pass on to the        VARIABILITY of species in a state of nature; but I sha
0007  on under Domestication Chapter I. Causes of   VARIABILITY. Effects of Habit. Correlation of Growth.
0007  we are driven to conclude that this greater   VARIABILITY is simply due to our domestic productions
0007  view propounded by Andrew Knight, that this   VARIABILITY may be partly connected with excess of foo
0008  sputed at what period of life the causes of   VARIABILITY, whatever they may be, generally act; whet
0008  to suspect that the most frequent cause of    VARIABILITY may be attributed to the male and female r
0009  ne of horticulture; but on this view we owe   VARIABILITY to the same cause which produces sterility
0009  he same cause which produces sterility; and   VARIABILITY is the source of all the choicest producti
0010  that, in fact, sports support my view, that   VARIABILITY may be largely attributed to the ovules or
0017  would endure other climates? Has the little   VARIABILITY of the ass or guinea fowl, or the small po
0030  we must, I think, look further than to mere   VARIABILITY. We cannot suppose that all the breeds wer
0040  man's power of selection. A high degree of    VARIABILITY is obviously favourable, as freely giving
0043  so far of the highest importance as causing   VARIABILITY. I do not believe that variability is an i
0043  causing variability. I do not believe that   VARIABILITY is an inherent and necessary contingency,
0043  s some authors have thought. The effects of   VARIABILITY are modified by various degrees of inherit
0043  us degrees of inheritance and of reversion.  VARIABILITY is governed by many unknown laws, more esp
0043  ultivator here quite disregards the extreme   VARIABILITY both of hybrids and mongrels, and the freq
```

Page **(Key Word)**

Page ***(Key Word)***

```
0044  power. Chapter II. Variation Under Nature. VARIABILITY. Individual differences. Doubtful species.
0045  be surprised at the number of the cases of VARIABILITY, even in important parts of structure, whi
0045  ystematists are far from pleased at finding VARIABILITY in important characters, and that there ar
0046  recently Mr. Lubbock has shown a degree of VARIABILITY in these main nerves in Coccus, which may
0046  ng, for they seem to show that this kind of VARIABILITY is independent of the conditions of life.
0060  a state of nature there is some individual VARIABILITY; indeed I am not aware that this has ever
0060  itted. But the mere existence of individual VARIABILITY; and of some few well marked varieties, tho
0082  he reproductive system, causes or increases VARIABILITY; and in the foregoing case the conditions
0082  t that, as I believe, any extreme amount of VARIABILITY is necessary; as man can certainly produce
0102  large amount of inheritable and diversified VARIABILITY is favourable, but I believe mere individu
0102  ons, will compensate for a lesser amount of VARIABILITY in each individual, and is, I believe, an
0118  their parents variable, and the tendency to VARIABILITY is in itself hereditary, consequently they
0131  hild like its parents. But the much greater VARIABILITY, as well as the greater frequency of monst
0134  e reproductive system, and in thus inducing VARIABILITY; and natural selection will then accumulat
0149  organs; and I will here only add that their VARIABILITY seems to be owing to their uselessness, an
0151  sexual characters, may be due to the great VARIABILITY of these characters, whether or not displa
0151  n my belief in its truth, had not the great VARIABILITY in plants made it particularly difficult t
0151  ficult to compare their relative degrees of VARIABILITY. When we see any part or organ developed i
0153  e, as well as an innate tendency to further VARIABILITY of all kinds and on the other hand, the po
0153  on, there may always be expected to be much VARIABILITY in the structure undergoing modification.
0153  nusually large and long continued amount of VARIABILITY, which has continually been accumulated by
0153  for the benefit of the species. But as the VARIABILITY of the extraordinarily developed part or o
0153  s a general rule, expect still to find more VARIABILITY in such parts than in other parts of the o
0153  one hand, and the tendency to reversion and VARIABILITY on the other hand, will in the course of t
0154  great that we ought to find the generative VARIABILITY, as it may be called, still present in a h
0154  sent in a high degree. For in this case the VARIABILITY will seldom as yet have been fixed by the
0156  will be granted. The cause of the original VARIABILITY of secondary sexual characters is not mani
0157  red males. Whatever the cause may be of the VARIABILITY of secondary sexual characters, as they ar
0158  finally, then, I conclude that the greater VARIABILITY of specific characters, or those which dis
0158  ossess in common; that the frequent extreme VARIABILITY of any part which is developed in a specie
0158  its congeners; and the not great degree of VARIABILITY in a part, however extraordinarily it may
0158  to a whole group of species; that the great VARIABILITY of secondary sexual characters, and the gr
0158  ed the tendency to reversion and to further VARIABILITY to sexual selection being less rigid than
0169  ffer much in the species of the same group. VARIABILITY in the same parts of the organisation has
0169  ad time to overcome the tendency to further VARIABILITY and to reversion to a less modified state.
0272  shows that the difference in the degree of VARIABILITY graduates away. When mongrels and the more
0272  or several generations an extreme amount of VARIABILITY in their offspring is notorious; but some
0273  uniformity of character could be given. The VARIABILITY, however, in the successive generations of
0273  haps, greater than in hybrids. This greater VARIABILITY of mongrels than of hybrids does not seem
0273  es in most cases that there has been recent VARIABILITY; and therefore we might expect that such v
0273  ty; and therefore we might expect that such VARIABILITY would often continue and be super added to
0273  mere act of crossing. The slight degree of VARIABILITY in hybrids from the first cross or in the
0273  generation, in contrast with their extreme VARIABILITY in the succeeding generations, is a curiou
0273  which I have taken on the cause of ordinary VARIABILITY; namely, that it is due to the reproductiv
0297  ecies and varieties; they grant some little VARIABILITY to each species, but when they meet with a
0314  of modification must be extremely slow. The VARIABILITY of each species is quite independent of th
0314  pendent of that of all others. Whether such VARIABILITY be taken advantage of by natural selection
0314  pends on many complex contingencies, on the VARIABILITY being of a beneficial nature, on the power
0351  in no law of necessary development. As the VARIABILITY of each species is an independent property
0465  latter periods there will probably be more VARIABILITY in the forms of life; during periods of su
0466  e argument. Under domestication we see much VARIABILITY. This seems to be mainly due to the reprod
0466  uce offspring exactly like the parent form. VARIABILITY is governed by many complex laws, by corre
0466  ns. On the other hand we have evidence that VARIABILITY, when it has once come into play, does not
0466  productions. Man does not actually produce VARIABILITY; he only unintentionally exposes organic b
0467  nature acts on the organisation, and causes VARIABILITY. But man can and does select the variation
0468  tions of domestication. And if there be any VARIABILITY under nature, it would be an unaccountable
0469  north America. If then we have under nature VARIABILITY and a powerful agent always ready to act a
0474  m a common progenitor, an unusual amount of VARIABILITY and modification, and therefore we might e
0489  ce which is almost implied by reproduction; VARIABILITY from the indirect and direct action of the
0008  many generations. No case is on record of a VARIABLE being ceasing to be variable under cultivati
0008  on record of a variable being ceasing to be VARIABLE under cultivation. Our oldest cultivated pla
0009  ffspring not perfectly like their parents or VARIABLE. Sterility has been said to be the bane of h
0022  e of the apertures in the sternum are highly VARIABLE; so is the degree of divergence and relative
0022  toes, are all points of structure which are VARIABLE. The period at which the perfect plumage is
0027  in the fantail, are in each breed eminently VARIABLE; and the explanation of this fact will be ob
0045  entral ganglion of an insect would have been VARIABLE in the same species; I should have expected
0117  the same conditions which made their parents VARIABLE, and the tendency to variability is in itsel
0118  varieties. If, then, these two varieties be VARIABLE, the most divergent of their variations will
0131  rudimentary, and lowly organised structures VARIABLE. Parts developed in an unusual manner are hi
0131  ts developed in an unusual manner are highly VARIABLE: specific characters more variable than gene
0131  re highly variable: specific characters more VARIABLE than generic: secondary sexual characters va
0131  le than generic: secondary sexual characters VARIABLE. Species of the same genus vary in an analog
0149  tamens in polyandrous flowers) the number is VARIABLE; whereas the number of the same part or orga
0149  t beings low in the scale of nature are more VARIABLE than those which are higher. I presume that
0149  ork, we can perhaps see why it should remain VARIABLE, that is, why natural selection should have
0149  d i believe with truth, are apt to be highly VARIABLE. We shall have to recur to the general subje
0150  e part in allied species, tends to be highly VARIABLE. Several years ago I was much struck with a
0153  ation. It further deserves notice that these VARIABLE characters, produced by man's selection, som
0154  same state; and thus it comes to be no more VARIABLE than any other structure. It is only in thos
0154  notorious that specific characters are more VARIABLE than generic. To explain by a simple example
0154  e, namely, that specific characters are more VARIABLE than generic, because they are taken from pa
0155  statement, that specific characters are more VARIABLE than generic; but I have repeatedly noticed
0155  osely allied species, that it has also, been VARIABLE in the individuals of some of the species. A
0155  ecomes only of specific value, often becomes VARIABLE, though its physiological importance may rem
0155  y created species of the same genus, be more VARIABLE than those parts which are closely alike in
0156  at they should still often be in some degree VARIABLE, at least more variable than those parts of
```

Page ***(Key Word)***

variable

Page ***(Key Word)***

```
0156 en be in some degree variable, at least more  VARIABLE than those parts of the organisation which h
0156 s, that secondary sexual characters are very   VARIABLE; I think it also will be admitted that speci
0157 ondary sexual characters, as they are highly   VARIABLE, sexual selection will have had a wide scope
0157 genitor, or of its early descendants, became   VARIABLE; variations of this part would, it is highly
0162 able part of the difficulty in recognising a   VARIABLE species in our systematic works, is due to i
0163 ouble. The shoulder stripe is certainly very   VARIABLE in length and outline. A white ass, but not
0168 rly unable to understand. Multiple parts are   VARIABLE in number and in structure, perhaps arising
0168 c beings low in the scale of nature are more   VARIABLE than those which have their whole organisati
0168 by natural selection, and hence probably are   VARIABLE. Specific characters, that is, the character
0168 branched off from a common parent, are more   VARIABLE than generic characters, or those which have
0168 erred to special parts or organs being still   VARIABLE, because they have recently varied and thus
0169 cies. Secondary sexual characters are highly   VARIABLE, and such characters differ much in the spec
0169  can understand why it should often still be   VARIABLE in a much higher degree than other parts; fo
0203 ecies at any one period are not indefinitely   VARIABLE, and are not linked together by a multitude
0207 incts graduated. Aphides and ants. Instincts   VARIABLE. Domestic instincts, their origin. Natural i
0256 the degree of fertility is likewise innately   VARIABLE; for it is not always the same when the same
0259 ble and unfavourable conditions, is innately   VARIABLE. That it is by no means always the same in d
0260 y should the degree of sterility be innately   VARIABLE in the individuals of the same species? Why
0266 that their sterility, though in some degree   VARIABLE, rarely diminishes. It must, however, be con
0272 at in the first generation mongrels are more   VARIABLE than hybrids; but Gartner admits that hybrid
0272 es which have long been cultivated are often   VARIABLE in the first generation; and I have myself s
0272 between very closely allied species are more   VARIABLE than those from very distinct species; and t
0273 ystems in any way affected, and they are not   VARIABLE; but hybrids themselves have their reproduct
0273 y affected, and their descendants are highly   VARIABLE. But to return to our comparison of mongrels
0276 orms by this test. The sterility is innately   VARIABLE in individuals of the same species, and is e
0423 s purpose with cattle, because they are less   VARIABLE than the shape or colour of the body, etc.;
0455 an organ, when rendered useless, may well be   VARIABLE, for its variations cannot be ohecked by nat
0459 d instincts are, in ever so slight a degree,  VARIABLE, and, lastly, that there is a struggle for e
0473 e same genus differ from each other, be more   VARIABLE than the generic characters in which they al
0474 characters would be more likely still to be   VARIABLE than the generic characters which have been
0474 might expect this part generally to be still   VARIABLE. But a part may be developed in the most unu
0474 like the wing of a bat, and yet not be more   VARIABLE than any other structure, if the part be com
0003 e, food, etc., as the only possible cause of   VARIATION. In one very limited sense, as we shall her
0004 at our knowledge, imperfect though it be, of   VARIATION under domestication, afforded the best and
0004 devote the first chapter of this Abstract to   VARIATION under Domestication. We shall thus see that
0004 ss what circumstances are most favourable to   VARIATION. In the next chapter the Struggle for Exist
0005 discuss the complex and little known laws of   VARIATION and of correlation of growth. In the four s
0007 ain but not exclusive means of modification.  VARIATION Under Domestication Chapter I. Causes of Va
0007 s of life to cause any appreciable amount of   VARIATION; and that when the organisation has once be
0010 of conception. These cases anyhow show that   VARIATION is not necessarily connected, as some autho
0010 anner. To judge how much, in the case of any   VARIATION, we should attribute to the direct action o
0011 ems probable. There are many laws regulating   VARIATION, some few of which can be dimly seen, and w
0012 arious, quite unknown, or dimly seen laws of   VARIATION is infinitely complex and diversified. It i
0012 l degree from that of the parental type. Any   VARIATION which is not inherited is unimportant for u
0019 here has been an immense amount of inherited   VARIATION. Who can believe that animals closely resem
0028 erations in explaining the immense amount of   VARIATION which pigeons have undergone, will be obvio
0032 t productions have been produced by a single   VARIATION from the aboriginal stock. We have proofs t
0044 is by far the predominant Power. Chapter II.  VARIATION Under Nature. Variability. Individual diffe
0044 cuss whether these latter are subject to any   VARIATION. To treat this subject at all properly, a l
0044 erally propagated. Some authors use the term   VARIATION in a technical sense, as implying a modific
0046  the species present an inordinate amount of   VARIATION; and hardly two naturalists can agree which
0049 , from geographical distribution, analogical  VARIATION, hybridism, etc., have been brought to bear
0050 r he knows nothing of the amount and kind of   VARIATION to which the group is subject; and this sho
0050 , at least, how very generally there is some   VARIATION. But if he confine his attention to one cla
0051 e has little general knowledge of analogical   VARIATION in other groups and in other countries, by
0051 eed in this at the expense of admitting much   VARIATION, and the truth of this admission will often
0055 species of a genus have been formed through   VARIATION, circumstances have been favourable for var
0055 tion, circumstances have been favourable for   VARIATION; and hence we might expect that the circums
0055 ances would generally be still favourable to   VARIATION. On the other hand, if we look at each spec
0061 r life. Owing to this struggle for life, any   VARIATION, however slight and from whatever cause pro
0061 called this principle, by which each slight   VARIATION, if useful, is preserved, by the term of Na
0062 ribution, rarity, abundance, extinction, and   VARIATION, will be dimly seen or quite misunderstood.
0080 riefly in the last chapter, act in regard to   VARIATION? Can the principle of selection, which we h
0081 on the other hand, we may feel sure that any   VARIATION in the least degree injurious would be rigi
0084 ly scrutinising, throughout the world, every   VARIATION, even the slightest; rejecting that which i
0085 part of the organisation is modified through   VARIATION, and the modifications are accumulated by n
0108 ted, unless favourable variations occur, and   VARIATION itself is apparently always a very slow pro
0117 mall numbered letter, a sufficient amount of   VARIATION is supposed to have been accumulated to hav
0119 lation of a considerable amount of divergent   VARIATION. As all the modified descendants from a com
0125 mbers, and thus lessen its chance of further   VARIATION and improvement. Within the same large grou
0127 it would be a most extraordinary fact if no   VARIATION ever had occurred useful to each being's ow
0131 beautiful ramifications. Chapter V. Laws of   VARIATION. Effects of external conditions. Use and di
0131 ur ignorance of the cause of each particular   VARIATION. Some authors believe it to be as much the
0133 t this other shell became bright coloured by   VARIATION when it ranged into warmer or shallower wat
0133 nged into warmer or shallower waters. When a   VARIATION is of the slightest use to a being, we cann
0149 that multiple parts are also very liable to   VARIATION in structure. Inasmuch as this vegetative r
0151 es wholly unlike in shape; and the amount of   VARIATION in the individuals of several of the specie
0152 the part in this case is eminently liable to   VARIATION. Why should this be so? On the view that ea
0152 nued selection, are also eminently liable to   VARIATION. Look at the breeds of the pigeon; see what
0154 ur would become a generic character, and its   VARIATION would be a more unusual circumstance. I hav
0159 athers in the pouter, may be considered as a   VARIATION representing the normal structure of anothe
0159 parent the same constitution and tendency to   VARIATION, when acted on by similar unknown influence
0159 egetable kingdom we have a case of analogous   VARIATION, in the enlarged stems, or roots as commonl
0159 t so, the case will then be one of analogous   VARIATION in two so called distinct species; and that
0160 of reversion, and not of a new yet analogous   VARIATION appearing in the several breeds. We may I t
0162 obable would all appear together from simple   VARIATION. More especially we might have inferred thi
```

Page ***(Key Word)***

```
0162  ers (either from reversion or from analogous  VARIATION) which already occur in some other members
0166  ecies of the horse genus becoming, by simple   VARIATION, striped on the legs like a zebra, or strip
0166  marks; and when any breed assumes by simple    VARIATION a bluish tint, these bars and other marks i
0167  shore. Summary. Our ignorance of the laws of   VARIATION is profound. Not in one case out of a hundr
0169  d, that is, where there has been much former   VARIATION and differentiation, or where the manufacto
0169  n a much higher degree than other parts; for   VARIATION is a long continued and slow process, and n
0169  s may not arise from reversion and analogous   VARIATION, such modifications will add to the beautif
0177  se new varieties are very slowly formed, for   VARIATION is a very slow process; and natural selecti
0179  numbers will, in the aggregate, present more   VARIATION, and thus be further improved through natur
0186  ted, which is certainly the case: and if any   VARIATION or modification in the organ be ever useful
0189  old ones to be destroyed. In living bodies,    VARIATION will cause the slight alterations, generati
0194  servation of individuals with any favourable   VARIATION, and by the destruction of those with any u
0198  he pelvis; and then by the law of homologous   VARIATION, the front limbs and even the head would pr
0198  nce of the several known and unknown laws of   VARIATION; and I have here alluded to them only to sh
0211  y may be here passed over. As some degree of   VARIATION in instincts under a state of nature, and t
0295  iver sufficient time for the slow process of   VARIATION; hence the deposit will generally have to b
0300  he periods of elevation, there would be much   VARIATION, but the geological record would then be le
0342  n during the periods of subsidence, and more  VARIATION during the periods of elevation, and during
0345  roved forms of life, produced by the laws of   VARIATION still acting round us, and preserved by Nat
0405  ecies thus has an immense range: but, if the   VARIATION had been a little greater, the two varietie
0410  in time and space; in both cases the laws of   VARIATION have been the same, and modifications have
0411  abits. In our second and fourth chapters, on   VARIATION and on Natural Selection, I have attempted
0443  question is not, at what period of life any    VARIATION has been caused, but at what period it is f
0443  d, even before the embryo is formed; and the   VARIATION may be due to the male and female sexual e
0444  render it probable, that at whatever age any   VARIATION first appears in the parent, it tends to re
0446  en, from the many slight successive steps of   VARIATION having supervened at a rather late age, and
0447  se. In certain cases the successive steps of   VARIATION might supervene, from causes of which we ar
0468  quite incapable of proof, that the amount of   VARIATION under nature is a strictly limited quantity
0472  the complex and little known laws governing    VARIATION are the same, as far as we can see, with th
0474  o the species, should be eminently liable to   VARIATION; but, on my view, this part has undergone,
0481  eings in a state of nature are subject to no   VARIATION; it cannot be proved that the amount of var
0481  tion; it cannot be proved that the amount of   VARIATION in the course of long ages is a limited qua
0482  they admit that these have been produced by   VARIATION, but they refuse to extend the same view to
0482  those produced by secondary laws. They admit  VARIATION as a vera causa in one case, they arbitrari
0486  ry will be opened, on the causes and laws of   VARIATION, on correlation of growth, on the effects o
0004  mulating by his Selection successive slight   VARIATIONS. I will then pass on to the variability of
0008  by any clear line of distinction from mere   VARIATIONS. But I am strongly inclined to suspect that
0015  er nature the conditions of life do change,   VARIATIONS and reversions of character probably do occ
0030  s own good, but to man's use or fancy. Some   VARIATIONS useful to him have probably arisen suddenly
0030  mulative selection: nature gives successive  VARIATIONS; man adds them up in certain directions use
0032  es are followed by horticulturists; but the  VARIATIONS are here often more abrupt. No one supposes
0033  oubt that the continued selection of slight  VARIATIONS, either in the leaves, the flowers, or the
0038  he can never act by selection, excepting on  VARIATIONS which are first given to him in some slight
0041  ion in almost any desired direction. But as  VARIATIONS manifestly useful or pleasing to man appear
0044  due to the physical conditions of life, and  VARIATIONS in this sense are supposed not to be inheri
0046  ect that we see in these polymorphic genera  VARIATIONS in points of structure which are of no serv
0061  rough the accumulation of slight but useful  VARIATIONS, given to him by the hand of Nature. But Na
0080  t, then, be thought improbable, seeing that  VARIATIONS useful to man have undoubtedly occurred, th
0080  o man have undoubtedly occurred, that other  VARIATIONS useful in some way to each being in the gre
0081  destroyed. This preservation of favourable   VARIATIONS and the rejection of injurious variations,
0081  e variations and the rejection of injurious  VARIATIONS, I call Natural Selection. Variations neith
0081  rious variations, I call Natural Selection.  VARIATIONS neither useful nor injurious would not be a
0082  on, by giving a better chance of profitable  VARIATIONS occurring; and unless profitable variations
0082  variations occurring; and unless profitable  VARIATIONS do occur, natural selection can do nothing.
0086  most expected nature. As we see that those   VARIATIONS which under domestication appear at any par
0086  any age, by the accumulation of profitable   VARIATIONS at that age; and by their inheritance at a
0102  rance within any given period of profitable  VARIATIONS, will compensate for a lesser amount of var
0104  al selection preserving the same favourable  VARIATIONS. Isolation, also, is an important element i
0105  the chance of the appearance of favourable   VARIATIONS. If we turn to nature to test the truth of
0105  will there be a better chance of favourable  VARIATIONS arising from the large number of individual
0108  nothing can be effected, unless favourable   VARIATIONS occur, and variation itself is apparently a
0109  ion acts solely through the preservation of  VARIATIONS in some way advantageous, which consequentl
0110  roducing within any given period favourable  VARIATIONS. We have evidence of this, in the facts giv
0117  ), may represent its varying offspring. The  VARIATIONS are supposed to be extremely slight, but of
0117  sed to endure for equal periods. Only those  VARIATIONS which are in some way profitable will be pr
0117  lly lead to the most different or divergent  VARIATIONS (represented by the outer dotted lines) bei
0118  es be variable, the most divergent of their  VARIATIONS will generally be preserved during the next
0127  o's own welfare, in the same way as so many  VARIATIONS have occurred useful to man. But if variati
0127  iations have occurred useful to man. But if  VARIATIONS useful to any organic being do occur, assur
0131  i have hitherto sometimes spoken as if the   VARIATIONS, so common and multiform in organic beings
0134  lection will then accumulate all profitable  VARIATIONS, however slight, until they become plainly
0143  ts growth and development, that when slight   VARIATIONS in any one part occur, and are accumulated
0157  of its early descendants, became variable:  VARIATIONS of this part would, it is highly probable,
0158  less rigid than ordinary selection, and to   VARIATIONS in the same parts having been accumulated b
0159  urposes. Distinct species present analogous  VARIATIONS; and a variety of one species often assumes
0159  ginal rock pigeon: these then are analogous  VARIATIONS in two or more distinct races. The frequent
0159  t no one will doubt that all such analogous  VARIATIONS are due to the several races of the pigeon
0162  ic breeds were reversions or only analogous  VARIATIONS; but we might have inferred that the bluene
0162  ng character and what are new but analogous  VARIATIONS, yet we ought, on my theory, sometimes to f
0169  es will naturally tend to present analogous  VARIATIONS, and these same species may occasionally re
0177  en period, of presenting further favourable  VARIATIONS for natural selection to seize on, than wil
0177  l selection can do nothing until favourable  VARIATIONS chance to occur, and until a place in the n
0186  the eye does vary ever so slightly, and the  VARIATIONS be inherited, which is certainly the case:
0194  ach being and taking advantage of analogous  VARIATIONS, has sometimes modified in very nearly the
0194  ly by taking advantage of slight successive  VARIATIONS; she can never take a leap, but must advanc
0197  the causes producing slight and unimportant  VARIATIONS; and we are immediately made conscious of t
```

Page ***(Key Word)***

```
0205 ts solely by the preservation of profitable VARIATIONS in the struggle for life. Natural selection
0209 ion preserving and continually accumulating VARIATIONS of instinct to any extent that may be profi
0209 selection of what may be called accidental VARIATIONS of instincts: that is of variations produce
0209 idental variations of instincts: that is of VARIATIONS produced by the same unknown causes which p
0210 lation of numerous, slight, yet profitable, VARIATIONS. Hence, as in the case of corporeal structu
0211 tate of nature, and the inheritance of such VARIATIONS, are indispensable for the action of natura
0212 sibility, or even probability, of inherited VARIATIONS of instinct in a state of nature will be st
0213 t and the selection of so called accidental VARIATIONS have played in modifying the mental qualiti
0239 afely conclude from the analogy of ordinary VARIATIONS, that each successive, slight, profitable m
0242 light, and as we must call them accidental, VARIATIONS, which are in any manner profitable, withou
0242 of our domestic animals vary, and that the VARIATIONS are inherited. Still more briefly I have at
0314 ge of by natural selection, and whether the VARIATIONS be accumulated to a greater or lesser amoun
0326 tter chance of the appearance of favourable VARIATIONS, and that severe competition with many alre
0338 the adult differs from its embryo, owing to VARIATIONS supervening at a not early age, and being i
0411 dividual. Embryology, laws of, explained by VARIATIONS not supervening at an early age, and being
0436 y the doubling or multiplication of others, VARIATIONS which we know to be within the limits of po
0443 embryo at a very early period, that slight VARIATIONS necessarily appear at an equally early peri
0444 corresponding age in the offspring. Certain VARIATIONS can only appear at corresponding ages; for
0444 t full grown cattle. But further than this, VARIATIONS which, for all that we can see, might have
0444 case; and I could give a good many cases of VARIATIONS (taking the word in the largest sense) whic
0449 been obliterated, either by the successive VARIATIONS in a long course of modification having sup
0450 c supervened at a very early age, or by the VARIATIONS having been inherited at an earlier period
0455 ered useless, may well be variable, for its VARIATIONS cannot be checked by natural selection. At
0457 lant. On the principle of successive slight VARIATIONS, not necessarily or generally supervening a
0459 t by the accumulation of innumerable slight VARIATIONS, each good for the individual possessor. Ne
0467 ariability. But man can and does select the VARIATIONS given to him by nature, and thus accumulate
0469 to act and select, why should we doubt that VARIATIONS in any way useful to beings, under their ex
0469 herited? Why, if man can by patience select VARIATIONS most useful to himself, should nature fail
0469 to himself, should nature fail in selecting VARIATIONS useful, under changing conditions of life,
0471 accumulating slight, successive, favourable VARIATIONS, it can produce no great or sudden modifica
0479 each class. On the principle of successive VARIATIONS not always supervening at an early age, and
0480 ccumulation of successive slight favourable VARIATIONS. Why, it may be asked, have all the most em
0481 nd perceive the full effects of many slight VARIATIONS, accumulated during an almost infinite numb
0007 ls which have been cultivated, and which have VARIED during all ages under the most different clim
0010 nditions been direct, if any of the young had VARIED, all would probably have varied in the same m
0010 the young had varied, all would probably have VARIED in the same manner. To judge how much, in the
0017 of our existing domesticated productions have VARIED. In the case of most of our anciently domesti
0018 mr. Blyth, whose opinion, from his large and VARIED stores of knowledge, I should value more than
0041 his plant. No doubt the strawberry had always VARIED since it was cultivated, but the slight varie
0118 ry in nearly the same manner as their parents VARIED. Moreover, these two varieties, being only sl
0121 reme species (I), as those which have largely VARIED, and have given rise to new varieties and spe
0155 in those parts of their structure which have VARIED within a moderately recent period, and which
0156 common progenitor, and subsequently have not VARIED or come to differ in any degree, or only in a
0156 acters; and as these specific characters have VARIED and come to differ within the period of the b
0158 mon, to parts which have recently and largely VARIED being more likely still to go on varying than
0158 s which have long been inherited and have not VARIED, to natural selection having more or less com
0168 ng still variable, because they have recently VARIED and thus come to differ; but we have also see
0189 refully selecting each alteration which under VARIED circumstances, may in any way, or in any degr
0237 being inherited by its offspring, which again VARIED and were again selected, and so onwards. But
0267 es and females of the same species which have VARIED and become slightly different, give vigour an
0285 e the vertical displacement of the strata has VARIED from 600 to 3000 feet. Prof. Ramsay has publi
0342 t widely ranging species are those which have VARIED most, and have oftenest given rise to new spe
0353 e to migrate, whereas some plants, from their VARIED means of dispersal, have migrated across the
0389 for insects. If we look to the large size and VARIED stations of New Zealand, extending over 780 m
0401 cks of somewhat different enemies. If then it VARIED, natural selection would probably favour diff
0466 ; it cannot be pretended that we know all the VARIED means of Distribution during the long lapse o
0468 have expected that organic beings would have VARIED under nature, in the same way as they general
0468 ature, in the same way as they generally have VARIED under the changed conditions of domesticatio
0473 n understand this fact: for they have already VARIED since they branched off from a common progeni
0022 dth and length of the ramus of the lower jaw, VARIES in a highly remarkable manner. The number of
0022 riod at which the perfect plumage is acquired VARIES, as does the state of the down with which the
0029 ces; and though they well know that each race VARIES slightly, for they win their prizes by select
0157 engidae, as Westwood has remarked, the number VARIES greatly: and the number likewise differs in t
0279 formations. On the absence of intermediate VARIETEIES in any one formation. On the sudden appeara
0003 dependently created, but had descended, like VARIETIES, from other species. Nevertheless, such a c
0005 infertility of species and the fertility of VARIETIES when intercrossed: and fourthly, the imperf
0006 ained in regard to the origin of species and VARIETIES, if he makes due allowance for our profound
0006 cies, in the same manner as the acknowledged VARIETIES of any one species are the descendants of t
0007 f growth. Inheritance. Character of Domestic VARIETIES. Difficulty of distinguishing between Varie
0007 ieties. Difficulty of distinguishing between VARIETIES and Species. Origin of Domestic Varieties f
0007 en varieties and Species. Origin of Domestic VARIETIES from one or more Species. Domestic Pigeons,
0008 plants, such as wheat, still often yield new VARIETIES: our oldest domesticated animals are still
0012 s in structure and constitution in which the VARIETIES and sub varieties differ slightly from each
0012 constitution in which the varieties and sub VARIETIES differ slightly from each other. The whole
0014 de by naturalists, namely, that our domestic VARIETIES, when run wild, gradually but certainly rev
0014 ry many of the most strongly marked domestic VARIETIES could not possibly live in a wild state. In
0015 loose in its new home. Nevertheless, as our VARIETIES certainly do occasionally revert in some of
0015 nged. If it could be shown that our domestic VARIETIES manifested a strong tendency to reversion,
0015 t that we could deduce nothing from domestic VARIETIES in regard to species. But there is not a sh
0015 be preserved. When we look to the hereditary VARIETIES or races of our domestic animals and plants
0016 s (and with that of the perfect fertility of VARIETIES when crossed, a subject hereafter to be dis
0016 been ranked by some competent judges as mere VARIETIES, and by other competent judges as the desce
0023 fly give them. If the several breeds are not VARIETIES, and have not proceeded from the rock pigeo
0033 ng the diversity of flowers in the different VARIETIES of the same species in the flower garden; i
0033 , in comparison with the flowers of the same VARIETIES; and the diversity of fruit of the same spe
0033 th the leaves and flowers of the same set of VARIETIES. See how different the leaves of the cabbag
```

Page ***(Key Word)***

```
0033  very slight differences. It is not that the  VARIETIES which differ largely in some one point do n
0036  have the appearance of being quite different  VARIETIES. If there exist savages so barbarous as nev
0037  ranked at their first appearance as distinct  VARIETIES, and whether or not two or more species or
0037  ased size and beauty which we now see in the  VARIETIES of the heartsease, rose, pelargonium, dahli
0037  d other plants, when compared with the older  VARIETIES or with their parent stocks. No one would e
0037  ving naturally chosen and preserved the best  VARIETIES they could anywhere find. A large amount of
0038  n remarked by some authors, namely, that the  VARIETIES kept by savages have more of the character
0038  ve more of the character of species than the  VARIETIES kept in civilised countries. On the view he
0039  ommon goose has not given rise to any marked  VARIETIES; hence the Thoulouse and the common breed,
0041  ul than amateurs in getting new and valuable  VARIETIES. The keeping of a large number of individua
0041  ried since it was cultivated; but the slight  VARIETIES had been neglected. As soon, however, as ga
0042  with distinct species) those many admirable  VARIETIES of the strawberry which have been raised du
0043  reeds; but the importance of the crossing of  VARIETIES has, I believe, been greatly exaggerated, b
0043  the crossing both of distinct species and of  VARIETIES is immense; for the cultivator here quite d
0044  of the species of the larger genera resemble  VARIETIES in being very closely, but unequally, relat
0044  called monstrosities; but they graduate into  VARIETIES. By a monstrosity I presume is meant some c
0046  which forms to rank as species and which as  VARIETIES. We may instance Rubus, Rosa, and Hieracium
0047  uralists, for few well marked and well known  VARIETIES can be named which have not been ranked as
0048  cies by at least some competent judges. That  VARIETIES of this doubtful nature are far from uncomm
0048  nist as good species, and by another as mere  VARIETIES. Mr. H. C. Watson, to whom I lie under deep
0048  sh plants, which are generally considered as  VARIETIES, but which have all been ranked by botanist
0048  aking this list he has omitted many trifling  VARIETIES, but which nevertheless have been ranked by
0048  list as undoubted species; and by another as  VARIETIES, or, as they are often called, as geographi
0048  trary is the distinction between species and  VARIETIES. On the islets of the little Madeira group
0048  are many insects which are characterized as  VARIETIES in Mr. Wollaston's admirable work, but whic
0049  has a few animals, now generally regarded as  VARIETIES, but which have been ranked as species by s
0049  ms, considered by highly competent judges as  VARIETIES, have so perfectly the character of species
0049  s whether they are rightly called species or  VARIETIES, before any definition of these terms has b
0049  he air. Many of the cases of strongly marked  VARIETIES or doubtful species well deserve considerat
0050  parents, and consequently must be ranked as  VARIETIES. Close investigation, in most cases, will b
0050  rom any cause closely attract his attention,  VARIETIES of it will almost universally be found reco
0050  almost universally be found recorded. These  VARIETIES, moreover, will be often ranked by some aut
0050  orms, which are very generally considered as  VARIETIES, and in this country the highest botanical
0050  are either good and distinct species or mere  VARIETIES. When a young naturalist commences the stud
0050  erences to consider as specific, and what as  VARIETIES; for he knows nothing of the amount and kin
0051  nabled to make up his own mind which to call  VARIETIES and which species; but he will succeed in t
0051  , again, between sub species and well marked  VARIETIES, or between lesser varieties and individual
0051  and well marked varieties, or between lesser  VARIETIES and individual differences. These differenc
0051  as being the first step towards such slight  VARIETIES as are barely thought worth recording in wo
0051  c in works or natural history. And I look at  VARIETIES which are in any degree more distinct and p
0052  o to more strongly marked and more permanent  VARIETIES; and at these latter, as leading to sub spe
0052  this work. It need not be supposed that all  VARIETIES or incipient species necessarily attain the
0052  state become extinct, or they may endure as  VARIETIES for very long periods, as has been shown to
0052  own to be the case by Mr. Wollaston with the  VARIETIES of certain fossil land shells in Madeira. I
0053  ecies which vary most, by tabulating all the  VARIETIES in several well worked floras. At first thi
0053  hich have very wide ranges generally present  VARIETIES; and this might have been expected, as they
0053  extent from commonness), often give rise to  VARIETIES sufficiently well marked to have been recor
0054  ividuals, which oftenest produce well marked  VARIETIES, or, as I consider them, incipient species.
0054  erhaps, might have been anticipated; for, as  VARIETIES, in order to become in any degree permanent
0055  ies as only strongly marked and well defined  VARIETIES, I was led to anticipate that the species o
0055  genera in each country would oftener present  VARIETIES, than the species of the smaller genera; fo
0055  es of the same genus) have been formed, many  VARIETIES or incipient species ought, as a general ru
0055  eation, there is no apparent reason why more  VARIETIES should occur in a group having many species
0055  ies on the side of the larger genera present  VARIETIES, than on the side of the smaller genera. Mo
0055  pecies of the large genera which present any  VARIETIES, invariably present a larger average number
0055  nvariably present a larger average number of  VARIETIES than do the species of the small genera. Bo
0056  ecies are only strongly marked and permanent  VARIETIES; for wherever many species of the same genu
0056  slow one. And this certainly is the case, if  VARIETIES be looked at as incipient species; for my t
0056  he species of that genus present a number of  VARIETIES, that is of incipient species, beyond the a
0056  e species of large genera and their recorded  VARIETIES which deserve notice. We have seen that the
0056  which to distinguish species and well marked  VARIETIES; and in those cases in which intermediate l
0057  her two forms should be ranked as species or  VARIETIES. Now Fries has remarked in regard to plants
0057  e, the species of the larger genera resemble  VARIETIES, more than do the species of the smaller ge
0057  t in the larger genera, in which a number of  VARIETIES or incipient species greater than the avera
0057  ufactured still to a certain extent resemble  VARIETIES, for they differ from each other by a less
0057  ted to each other, in the same manner as the  VARIETIES of any one species are related to each othe
0057  s around certain other species. And what as  VARIETIES but groups of forms, unequally related to e
0057  e most important point of difference between  VARIETIES and species; namely, that the amount of dif
0057  amely, that the amount of difference between  VARIETIES, when compared with each other or with thei
0058  ined, and how the lesser differences between  VARIETIES will tend to increase into the greater diff
0058  other point which seems to me worth notice.  VARIETIES generally have much restricted ranges: this
0058  ied to other species, and in so far resemble  VARIETIES, often have much restricted ranges. For ins
0058  now, in this same catalogue, 53 acknowledged  VARIETIES are recorded, and these range over 7.7 prov
0058  ovinces; whereas, the species to which these  VARIETIES belong range over 14.3 provinces. So that t
0058  ver 14.3 provinces. So that the acknowledged  VARIETIES have very nearly the same restricted averag
0058  sts as good and true species. Finally, then,  VARIETIES have the same general characters as species
0059  ffering very little, are generally ranked as  VARIETIES, notwithstanding that intermediate linking
0059  genera have more than the average number of  VARIETIES. In large genera the species are apt to be
0059  f large genera present a strong analogy with  VARIETIES. And we can clearly understand these analog
0059  e analogies, if species have once existed as  VARIETIES, and have thus originated: whereas, these a
0059  er genera which on an average vary most; and  VARIETIES, as we shall hereafter see, tend to become
0060  for life most severe between individuals and  VARIETIES of the same species; often severe between s
0060  ul forms be called species or sub species or  VARIETIES; what rank, for instance, the two or three
0060  to hold, if the existence of any well marked  VARIETIES be admitted. But the mere existence of indi
0060  dual variability and of some few well marked  VARIETIES, though necessary as the foundation for the
0061  crlc. Again, it may be asked, how is it that  VARIETIES, which I have called incipient species, bec
```

Page ***(Key Word)**

```
0061  differ from each other far more than do the  VARIETIES of the same species? How do those groups of
0075  exposed to the same dangers. In the case of   VARIETIES of the same species, the struggle will gene
0075  ntest soon decided: for instance, if several  VARIETIES of wheat be sown together, and the mixed se
0075  r, and the mixed seed be resown, some of the  VARIETIES which best suit the soil or climate, or are
0075  ntly in a few years quite supplant the other  VARIETIES. To keep up a mixed stock of even such extr
0075  p a mixed stock of even such extremely close  VARIETIES as the variously coloured sweet peas, they
0076  in numbers and disappear. So again with the   VARIETIES of sheep: it has been asserted that certain
0076  : it has been asserted that certain mountain   VARIETIES will starve out other mountain varieties, s
0076  ain varieties will starve out other mountain  VARIETIES, so that they cannot be kept together. The
0076  has followed from keeping together different  VARIETIES of the medicinal leech. It may even be doub
0076  al leech. It may even be doubted whether the   VARIETIES of any one of our domestic plants or animal
0085  great difference in cultivating the several   VARIETIES, assuredly, in a state of nature, where the
0086  riod; for instance, in the seeds of the many  VARIETIES of our culinary and agricultural plants: in
0086  in the caterpillar and cocoon stages of the   VARIETIES of the silkworm; in the eggs of poultry, an
0091  dividuals best fitted for the two sites, and  VARIETIES might slowly be formed. These varieties wou
0091  two varieties might slowly be formec. These   VARIETIES would cross and blend where they met: but t
0091  that, according to Mr. Pierce, there are two  VARIETIES of the wolf inhabiting the Catskill Mountai
0096  animals and plants a cross between different  VARIETIES, or between individuals of the same variety
0098  l known that if very closely allied forms or  VARIETIES are planted near each other, it is hardly p
0099  ng advantageous or indispensable! If several  VARIETIES of the cabbage, radish, onion, and some oth
0099  dling cabbages from some plants of different  VARIETIES growing near each other, and of these only
0103  imals of this nature, for instance in birds,  VARIETIES will generally be confined to separated cou
0103  y, as the chance of intercrossing with other  VARIETIES is thus lessened. Even in the case of slow
0103  of facts, showing that within the same area,  VARIETIES of the same animal can long remain distinct
0103  eding at slightly different seasons, or from  VARIETIES of the same kind preferring to pair togethe
0106  read most wicely, will give rise to most new  VARIETIES and species, and will thus play an importan
0108  nhabitants; and time will be allowed for the  VARIETIES in each to become well modified and perfect
0108  e competition: the most favoured or improved  VARIETIES will be enabled to spread: there will be mu
0110  which afford the greatest number of recorded  VARIETIES, or incipient species. Hence, rare species
0110  ce that it is the most closely allied forms,  VARIETIES of the same species, and species of the sam
0111  eds of cattle, sheep, and other animals, and  VARIETIES of flowers, take the place of older and inf
0111  several important facts. In the first place,  VARIETIES, even strongly marked ones, though having s
0111  species. Nevertheless, according to my view,  VARIETIES are species in the process of formation, or
0111  ow, then, does the lesser difference between  VARIETIES become augmented into the greater differenc
0111  presenting well marked differences; whereas  VARIETIES, the supposed prototypes and parents of fut
0111  arge an amount of difference as that between  VARIETIES of the same species and species of the same
0113  hen first one variety and then several mixed  VARIETIES of wheat have been sown on equal spaces of
0113  es of grass were to go on varying, and then  VARIETIES were continually selected which differed fr
0114  thousands of generations, the most distinct  VARIETIES of any one species of grass would always ha
0114  s, and thus cf supplanting the less cistinct  VARIETIFS; and varieties, when rendered very distinct
0114  supplanting the less distinct varieties; and  VARIETIES, when rendered very distinct from each othe
0117  the large genera present a greater number of  VARIETIES. We have, also, seen that the species, whic
0117  osec to have produced two fairly well marked  VARIETIES, namely al and ml. These two varieties will
0117  arked varieties, namely al and ml. These two  VARIETIES will generally continue to be exposed to th
0118  as their parents varied. Moreover, these two  VARIETIES, being only slightly modified forms, will t
0118  ow to be favourable to the production of new  VARIETIES. If, then, these two varieties be variable,
0118  uction of new varieties. If, then, these two  VARIETIES be variable, the most divergent of their va
0118  variety ml is supposed to have produced two  VARIETIES, namely m2 and s2, differing from each othe
0118  ar steps for any length of time; some of the  VARIETIES, after each thousand generations, producing
0118  ified condition, some producing two cr three  VARIETIES, and some failing to produce any. Thus the
0118  s, and some failing to produce any. Thus the  VARIETIES or modified descendants, proceeding from th
0119  am far from thinking that the most divergent  VARIETIES will invariably prevail and multiply: a med
0119  come sufficiently distinct to be recorded as  VARIETIES. But these breaks are imaginary, and might
0120  se three forms may still be crly well marked  VARIETIES; or they may have arrived at the doubtful c
0120  y which the small differences distinguishing  VARIETIES are increased into the larger differences d
0120  thousand generations, either two well marked  VARIETIES (w10 and z10) or two species, according to
0121  e largely varied, and have given rise to new  VARIETIES and species. The other nine species (marked
0122  odification, species (A) and all the earlier  VARIETIES will have become extinct, having been repla
0125  of the larger genera which oftenest present  VARIETIES or incipient species. This, indeed, might h
0128  e, thus the small differences distinguishing  VARIETIES of the same species, will steadily tend to
0128  e manner which we everywhere behold, namely,  VARIETIES of the same species most closely related to
0132  other such cases could be given. The fact of  VARIETIES of one species, when they range into the zo
0133  all kinds are only well marked and permanent  VARIETIES. Thus the species of shells which are confi
0133  ceived: and, on the other hand, of different  VARIETIES being produced from the same species under
0141  it, and how much to the natural selection of  VARIETIES having different innate constitutions, and
0142  on many kinds of cultivated plants, and some  VARIETIES are said to withstand certain climates bett
0142  ished in the United States, in which certain  VARIETIES are habitually recommended for the northern
0142  or the southern States: and as most of these  VARIETIES are of recent origin, they cannot owe their
0142  gated by seed, and of which consequently new  VARIETIES have not been produced, has even been advan
0147  to give milk and to fatten readily. The same  VARIETIES of the cabbage do not yield abundant and nu
0149  emarked by Is. Geoffroy St. Hilaire, both in  VARIETIES and in species, that when any part or organ
0151  that it is no exaggeration to state that the  VARIETIES differ more from each other in the characte
0155  species being only strongly marked and fixed  VARIETIES, we might surely expect to find them still
0159  countries most widely set apart, present sub  VARIETIES with reversed feathers on the head and feat
0159  baga, plants which several botanists rank as  VARIETIES produced by cultivation from a common paren
0162  ecies in our systematic works, is due to its  VARIETIES mocking, as it were, some of the other spec
0162  emselves must be doubtfully ranked as either  VARIETIES or species; and this shows, unless all thes
0167  in producing the lesser differences between  VARIETIES of the same species, and the greater differ
0169  work, there, on an average, we now find most  VARIETIES or incipient species. Secondary sexual char
0171  ansitions. Absence or rarity of transitional  VARIETIES. Transitions in habits of life. Diversified
0172  d producing sterile offspring, whereas, when  VARIETIES are crossed their fertility is unimpaired?
0172  rs. On the absence or rarity of transitional  VARIETIES. As natural selection acts solely by the pr
0172  rm, both the parent and all the transitional  VARIETIES will generally have been exterminated by th
0173  its original parent and all the transitional  VARIETIES between its past and present states. Hence
0174  sent time to meet with numerous transitional  VARIETIES in each region, though they must have exist
0174  o we not now find close linking intermediate  VARIETIES? This difficulty for a long time quite conf
```

Page ***(Key Word)**

0174	rmed without the possibility of intermediate	VARIETIES existing in the intermediate zones. By chan
0175	me rather suddenly rarer and rarer; then, as	VARIETIES do not essentially differ from species, the
0176	very large area, we shall have to adapt two	VARIETIES to two large areas, and a third variety to
0176	as i can make out, this rule holds good with	VARIETIES in a state of nature. I have met with strik
0176	triking instances of the rule in the case of	VARIETIES intermediate between well marked varieties
0176	f varieties intermediate between well marked	VARIETIES in the genus Balanus. And it would appear f
0176	gray, and Mr. Wollaston, that generally when	VARIETIES intermediate between two other forms occur,
0176	and inferences, and therefore conclude that	VARIETIES linking two other varieties together have g
0176	re conclude that varieties linking two other	VARIETIES together have generally existed in lesser n
0176	i think, we can understand why intermediate	VARIETIES should not endure for very long periods: wh
0176	rocess of further modification, by which two	VARIETIES are supposed on my theory to be converted a
0177	n an average a greater number of well marked	VARIETIES than do the rarer species. I may illustrate
0177	ay illustrate what I mean by supposing three	VARIETIES of sheep to be kept, one adapted to an exte
0177	and intermediate links: firstly, because new	VARIETIES are very slowly formed, for variation is a
0178	entative species. In this case, intermediate	VARIETIES between the several representative species
0178	in a living state. Thirdly, when two or more	VARIETIES have been formed in different portions of a
0178	of a strictly continuous area, intermediate	VARIETIES will, it is probable, at first have been fo
0178	had a short duration. For these intermediate	VARIETIES will, from reasons already assigned (namely
0178	tative species, and likewise of acknowledged	VARIETIES), exist in the intermediate zones in lesser
0178	ntermediate zones in lesser numbers than the	VARIETIES which they tend to connect. From this cause
0179	nect. From this cause alone the intermediate	VARIETIES will be liable to accidental extermination;
0179	f my theory be true, numberless intermediate	VARIETIES, linking most closely all the species of th
0203	uate away from one part to another. When two	VARIETIES are formed in two districts of a continuous
0245	onditions of life and crossing. Fertility of	VARIETIES when crossed and of their mongrel offspring
0246	ce of our reasoning powers. The fertility of	VARIETIES, that is of the forms known or believed to
0246	o make a broad and clear distinction between	VARIETIES and species. First, for the sterility of sp
0247	le together, he unhesitatingly ranks them as	VARIETIES. Gartner, also, makes the rule equally univ
0247	ch we have such good reason to believe to be	VARIETIES, and only once or twice succeeded in gettin
0247	coerulea), which the best botanists rank as	VARIETIES, absolutely sterile together; and as he cam
0248	oubtful forms should be ranked as species or	VARIETIES, with the evidence from fertility adduced b
0248	ds any clear distinction between species and	VARIETIES. It is also a remarkable fact, that hybrids
0258	labra) that many botanists rank them only as	VARIETIES. It is also a remarkable fact, that hybrids
0261	the other hand, closely allied species, and	VARIETIES of the same species, can usually, but not i
0261	s a member of the same genus. Even different	VARIETIES of the pear take with different degrees of
0262	s of facility in the quince; so do different	VARIETIES of the apricot and peach on certain varieti
0262	arieties of the apricot and peach on certain	VARIETIES of the plum. As Gartner found that there wa
0267	lated to the principle of life. Fertility of	VARIETIES when crossed, and of their Mongrel offsprin
0268	me essential distinction between species and	VARIETIES, and that there must be some error in all t
0268	or in all the foregoing remarks, inasmuch as	VARIETIES, however much they may differ from each oth
0268	lmost invariably the case. But if we look to	VARIETIES produced under nature, we are immediately i
0268	ss difficulties; for if two hitherto reputed	VARIETIES be found in any degree sterile together, th
0268	considered by many of our best botanists as	VARIETIES, are said by Gartner not to be quite fertil
0268	thus argue in a circle, the fertility of all	VARIETIES produced under nature will assuredly have t
0268	assuredly have to be granted. If we turn to	VARIETIES, produced, or supposed to have been produce
0268	ss the perfect fertility of so many domestic	VARIETIES, differing widely from each other in appear
0268	s, however, render the fertility of domestic	VARIETIES less remarkable than at first appears. It c
0269	i and we may apply the same rule to domestic	VARIETIES. In the second place, some eminent naturali
0269	reproductive system. He supplies his several	VARIETIES with the same food: treats them in nearly t
0269	n the result. I have as yet spoken as if the	VARIETIES of the same species were invariably fertile
0270	no one, I believe, has suspected that these	VARIETIES of maize are distinct species; and it is im
0270	gartner did not venture to consider the two	VARIETIES as specifically distinct. Girou de Buzarein
0270	istinct. Girou de Buzareingues crossed three	VARIETIES of gourd, which like the maize has separate
0270	lassification by the test of infertility, as	VARIETIES. The following case is far more remarkable,
0270	s, as Gartner: namely, that yellow and white	VARIETIES of the same species of Verbascum when inter
0270	d produce less seed, than do either coloured	VARIETIES when fertilised with pollen from their own
0271	eover, he asserts that when yellow and white	VARIETIES of one species are crossed with yellow and
0271	ne species are crossed with yellow and white	VARIETIES of a distinct species, more seed is produce
0271	se which are differently coloured. Yet these	VARIETIES of Verbascum present no other difference be
0271	om observations which I have made on certain	VARIETIES of hollyhock, I am inclined to suspect that
0271	widely distinct species, than are the other	VARIETIES. He experimentised on five forms, which are
0271	five forms, which are commonly reputed to be	VARIETIES, and which he tested by the severest trial,
0271	ing perfectly fertile. But one of these five	VARIETIES, when used either as father or mother, and
0271	hose which were produced from the four other	VARIETIES when crossed with N. glutinosa. Hence the r
0271	ifficulty of ascertaining the infertility of	VARIETIES in a state of nature, for a supposed variet
0271	the production of the most distinct domestic	VARIETIES, and from not wishing or being able to prod
0272	not think that the very general fertility of	VARIETIES can be proved to be of universal occurence,
0272	or to form a fundamental distinction between	VARIETIES and species. The general fertility of varie
0272	ieties and species. The general fertility of	VARIETIES does not seem to me sufficient to overthrow
0272	the offspring of species when crossed and of	VARIETIES when crossed may be compared in several oth
0272	rked line of distinction between species and	VARIETIES, could find very few and, as it seems to me
0272	cies, and the so called mongrel offspring of	VARIETIES. And, on the other hand, they agree most cl
0273	surprising. For the parents of mongrels are	VARIETIES, and mostly domestic varieties (very few ex
0273	mongrels are varieties, and mostly domestic	VARIETIES (very few experiments having been tried on
0273	few experiments having been tried on natural	VARIETIES), and this implies in most cases that there
0274	rom each other; whereas if two very distinct	VARIETIES of one species are crossed with another spe
0274	n the hybrid; and so I believe it to be with	VARIETIES of plants. With animals one variety certain
0275	ccur with mongrels, which are descended from	VARIETIES often suddenly produced and semi monstrous
0275	viduals of the same variety, or of different	VARIETIES, or of distinct species. Laying aside the q
0275	offspring of crossed species, and of crossed	VARIETIES. If we look at species as having been speci
0275	ies as having been specially created, and at	VARIETIES as having been produced by secondary laws,
0276	no essential distinction between species and	VARIETIES. Summary of Chapter. First crosses between
0277	ies. First crosses between forms known to be	VARIETIES, or sufficiently alike to be considered as var
0277	ties, or sufficiently alike to considered as	VARIETIES, and their mongrel offspring, are very gene
0277	we are to argue in a circle with respect to	VARIETIES in a state of nature; and when we remember
0277	when we remember that the greater number of	VARIETIES have been produced under domestication by t
0279	fundamental distinction between species and	VARIETIES. Chapter IX. On the Imperfection of the Geo

Page ∗∗∗(Key Word)∗∗

```
0279  gical Record. On the absence of intermediate VARIETIES at the present day. On the nature of extinc
0279  t day. Cn the nature of extinct intermediate VARIETIES; on their number. On the vast lapse of time
0279  endeavoured, also, to show that intermediate VARIETIES, from existing in lesser numbers than the f
0280  cess of natural selection, through which new  VARIETIES continually take the places of and extermin
0280  us scale, so must the number of intermediate  VARIETIES, which have formerly existed on the earth,
0280  pigeon; if we possessed all the intermediate  VARIETIES which have ever existed, we should have an
0280  h and the rock pigeon; but we should have no  VARIETIES directly intermediate between the fantail a
0281  ferences not greater than we see between the  VARIETIES of the same species at the present day; and
0292  eviously explained, for the formation of new  VARIETIES and species; but during such periods there
0292  uch there will be much extinction, fewer new  VARIETIES or species will be formed; and it is during
0293  why we do not therein find closely graduated  VARIETIES between the allied species which lived at i
0293  cord of the same species presenting distinct  VARIETIES in the upper and lower parts of the same fo
0297  den rule by which to distinguish species and  VARIETIES; they grant some little variability to each
0297  ith either one or both forms by intermediate  VARIETIES. Nor shoulc it be forgotten, as before expl
0297  ies of D'Orbigny and others into the rank of  VARIETIES; and on this view we do find the kind of ev
0298  uspect, as we have formerly seen, that their  VARIETIES are generally at first local; and that such
0298  enerally at first local; and that such local  VARIETIES do not spread widely and supplant their par
0298  have the widest range, that oftenest present  VARIETIES; so that with shells and other marine anima
0298  ich have oftenest given rise, first to local  VARIETIES and ultimately to new species; and this aga
0298  orms can seldom be connected by intermediate  VARIETIES and thus proved to be the same species, unt
0299  ntatives, and by other conchologists as only  VARIETIES, are really varieties or are, as it is call
0299  conchologists as only varieties, are really  VARIETIES or are, as it is called, specifically disti
0299  hem together by numerous, fine, intermediate  VARIETIES; and this not having been effected, is prob
0301  tions were not fully preserved, transitional  VARIETIES would merely appear as so many distinct spe
0301  ing species which would oftenest produce new  VARIETIES; and the varieties would at first generally
0301  ould oftenest produce new varieties; and the  VARIETIES would at first generally be local or confin
0301  d and supplant their parent forms. When such  VARIETIES returned to their ancient homes, as they wo
0317  ne species giving rise first to two or three  VARIETIES, these being slowly converted into species,
0320  sed, it at first supplants the less improved  VARIETIES in the same neighbourhood; when much improv
0325  ral selection. New species are formed by new  VARIETIES arising, which have some advantage over old
0325  y, would naturally oftenest give rise to new  VARIETIES or incipient species; for these latter must
0325  , having produced the greatest number of new  VARIETIES. It is also natural that the dominant, vary
0326  , and of giving rise in new countries to new  VARIETIES and species. The process of diffusion may o
0335  r the parent rock pigeon now lives; and many  VARIETIES between the rock pigeon and the carrier hav
0336  y one or two formations all the intermediate  VARIETIES between the species which appeared at the c
0342  oftenest given rise to new species; and that  VARIETIES have at first often been local. All these c
0342  tent explain why we do not find interminable  VARIETIES, connecting together all the extinct and ex
0350  sms cuite like, or, as we see in the case of  VARIETIES nearly like each other. The dissimilarity o
0355  have descended from a succession of improved  VARIETIES, which will never have blended with other i
0355  never have blended with other individuals or  VARIETIES, but will have supplanted each other; so th
0369  ecies are identically the same, some present  VARIETIES, some are ranked as doubtful forms, and som
0376  lists as specifically distinct, by others as  VARIETIES; but some are certainly icentical, and many
0379  now exist in their new homes as well marked  VARIETIES or as distinct species. It is a remarkable
0401  al selection would probably favour different  VARIETIES in the different islands. Some species, how
0404  alists rank as distinct species, and some as  VARIETIES; these doubtful forms showing us the steps
0405  rmine their average range. For instance, two  VARIETIES of the same species inhabit America and Eur
0405  variation had been a little greater, the two  VARIETIES would have been ranked as distinct species,
0405  onversion of its offspring, firstly into new  VARIETIES and ultimately into new species. In conside
0409  on, in the presence of identical species, of  VARIETIES, of doubtful species, and of cistinct but r
0411  descent with modification. Classification of  VARIETIES. Descent always used in classification. Ana
0411  g to the larger genera, which vary most. The  VARIETIES, or incipient species, thus produced ultima
0423  view, let us glance at the classification of  VARIETIES, which are believed or known to have descen
0423  s. these are grouped under species, with sub  VARIETIES under varieties; and with our domestic prod
0423  uped under species, with sub varieties under  VARIETIES; and with our domestic productions, several
0423  oups subordinate to groups, is the same with  VARIETIES as with species, namely, closeness of desce
0423  y the same rules are followed in classifying  VARIETIES, as with species. Authors have insisted on
0423  s have insisted on the necessity of classing  VARIETIES on a natural instead of an artificial syste
0423  re cautioned, for instance, not to class two  VARIETIES of the pine apple together, merely because
0423  und to be most constant, is used in classing  VARIETIES: thus the great agriculturist Marshall says
0423  viceable, because less constant. In classing  VARIETIES, I apprehend if we had a real pedigree, a g
0423  points. In tumbler pigeons, though some sub  VARIETIES differ from the others in the important cha
0424  ndividual. He includes monsters; he includes  VARIETIES, not solely because they closely resemble t
0425  fferent; and as it has been used in classing  VARIETIES which have undergone a certain, and sometim
0427  ething of the same kind even in our domestic  VARIETIES, as in the thickened stems of the common an
0432  as fine as those between the finest existing  VARIETIES, nevertheless a natural classification, or
0432  ks to be as fine as those between the finest  VARIETIES. In this case it would be quite impossible
0433  ies; we use descent in classing acknowledged  VARIETIES, however different they may be from their p
0444  us look at a few analogous cases in domestic  VARIETIES. Some authors who have written on Dogs, mai
0444  a, though appearing so different, are really  VARIETIES most closely allied, and have probably desc
0446  o the later embryonic stages of our domestic  VARIETIES. Fanciers select their horses, dogs, and pi
0456  g together the sexes, ages, and acknowledged  VARIETIES of the same species, however different they
0456  s of acquired difference marked by the terms  VARIETIES, species, genera, families, orders, and cla
0460  trast with the almost universal fertility of  VARIETIES when crossed, I must refer the reader to th
0460  her and then as the mother. The fertility of  VARIETIES when intercrossed and of their mongrel offs
0461  n profoundly modified. Moreover, most of the  VARIETIES which have been experimentised on have been
0461  the offspring of slightly modified forms or  VARIETIES acquire from being crossed increased vigour
0462  h group by gradations as fine as our present  VARIETIES, it may be asked, Why do we not see these l
0463  t right to expect often to find intermediate  VARIETIES in the intermediate zone. For we have reaso
0463  ted. I have also shown that the intermediate  VARIETIES which will at first probably exist in the i
0463  oved at a quicker rate than the intermediate  VARIETIES, which exist in lesser numbers; so that the
0463  in lesser numbers; so that the intermediate  VARIETIES will, in the long run, be supplanted and ex
0464  tful forms could be named which are probably  VARIETIES; but who will pretend that in future ages s
0464  iew, whether or not these doubtful forms are  VARIETIES? As long as most of the links between any t
0464  umber. Widely ranging species vary most, and  VARIETIES are often at first local, both causes rende
0464  ery of intermediate links less likely. Local  VARIETIES will not spread into other and distant regi
0466  me into play, does not wholly cease; for new  VARIETIES are still occasionally produced by our most
```

Page ∗∗∗(Key Word)∗∗

Page **(Key Word)***

```
0467  ricable doubts whether very many of them are VARIETIES or aboriginal species. There is no obvious
0468  it will he almost equally severe between the VARIETIES of the same species, and next in severity b
0469  l naturalists have admitted the existence of VARIETIES, which they think sufficiently distinct to
0469  on between individual differences and slight VARIETIES: or between more plainly marked varieties a
0469  ht varieties: or between more plainly marked VARIETIES and sub species, and species. Let it be obs
0469  ecies are only strongly marked and permanent VARIETIES, and that each species first existed as a v
0469  en produced by special acts of creation, and VARIETIES which are acknowledged to have been produce
0470  rish, these same species should present many VARIETIES: for where the manufactory of species has b
0470  it still in action; and this is the case if VARIETIES be incipient species. Moreover, the species
0470  r genera, which afford the greater number of VARIETIES or incipient species, retain to a certain d
0470  retain to a certain degree the character of VARIETIES; for they differ from each other by a less
0470  her species, in which respects they resemble VARIETIES. These are strange relations on the view of
0470  intelligible if all species first existed as VARIETIES. As each species tends by its geometrical r
0470  n, the slight differences, characteristic of VARIETIES of the same species, tend to be augmented i
0470  species of the same genus. New and improved VARIETIES will inevitably supplant and exterminate th
0470  te the older, less improved and intermediate VARIETIES; and thus species are rendered to a large e
0472  produced but little direct effect; yet when VARIETIES enter any zone, they occasionally assume so
0472  of the species proper to that zone. In both VARIETIES and species, use and disuse seem to have pr
0473  he dark caves of America and Europe. In both VARIETIES and species correlation of growth seems to
0473  ther parts are necessarily modified. In both VARIETIES and species reversions to long lost charact
0473  red flowers? If species are only well marked VARIETIES, of which the characters have become in a h
0475  if species be only well marked and permanent VARIETIES, we can at once see why their crossed offsp
0475  as do the crossed offspring of acknowledged VARIETIES. On the other hand, these would be strange
0475  species have been independently created, and VARIETIES have been produced by secondary laws. If we
0478  t species occur, many doubtful forms and and VARIETIES of the same species likewise occur. It is a
0481  an be, drawn between species and well marked VARIETIES. It cannot be maintained that species when
0481  hen intercrossed are invariably sterile, and VARIETIES invariably fertile; or that sterility is a
0485  distinction between species and well marked VARIETIES is, that the latter are known, or believed,
0485  orms now generally acknowledged to be merely VARIETIES may hereafter be thought worthy of specific
0005  strong principle of inheritance, any selected VARIETY will tend to propagate its new and modified
0007  . when we look to the individuals of the same VARIETY or sub variety of our older cultivated plant
0007  to the individuals of the same variety or sub VARIETY of our older cultivated plants and animals,
0007  than do the individuals of any one species or VARIETY in a state of nature. When we reflect on the
0015  effects of intercrossing, that only a single VARIETY should be turned loose in its new home. Neve
0030  lled by any mechanical contrivance, is only a VARIETY of the wild Dipsacus; and this amount of cha
0032  isted merely in separating some very distinct VARIETY, and breeding from it, the principle would b
0037  onsisted in always cultivating the best known VARIETY, sowing its seeds, and, when a slightly bett
0037  sowing its seeds, and, when a slightly better VARIETY has chanced to appear, selecting it, and so
0044  ement of a distinct act of creation. The term VARIETY is almost equally difficult to define, but h
0045  ase I presume that the form would be called a VARIETY. Again, we have many slight differences whic
0047  termediate characters; he treats the one as a VARIETY of the other, ranking the most common, but s
0047  scribed, as the species, and the other as the VARIETY. But cases of great difficulty, which I will
0047  deciding whether or not to rank one form as a VARIETY of another, even when they are closely conne
0047  many cases, however, one form is ranked as a VARIETY of another, not because the intermediate lin
0047  her a form should be ranked as a species or a VARIETY, the opinion of naturalists having sound jud
0048  ne zoologist as a species and by another as a VARIETY, can rarely be found within the same country
0052  n this view; and I attribute the passage of a VARIETY, from a state in which it differs very sligh
0052  ite directions. Hence I believe a well marked VARIETY may be justly called an incipient species; b
0052  f certain fossil land shells in Madeira. If a VARIETY were to flourish so as to exceed in numbers
0052  n rank as the species, and the species as the VARIETY; or it might come to supplant and exterminat
0052  it does not essentially differ from the term VARIETY, which is given to less distinct and more fl
0052  distinct and more fluctuating forms. The term VARIETY, again, in comparison with mere individual d
0058  indeed scarcely more than a truism, for if a VARIETY were found to have a wider range than that o
0085  ch differences would effectually settle which VARIETY, whether a smooth or downy, a yellow or purp
0091  and by the repetition of this process, a new VARIETY might be formed which would either supplant
0096  varieties, or between individuals of the same VARIETY but of another strain, gives vigour and fert
0099  t it must arise from the pollen of a distinct VARIETY having a prepotent effect over a flower's ow
0103  ease at a very rapid rate, a new and improved VARIETY might be quickly formed on any one spot, and
0103  iefly between the individuals of the same new VARIETY. A local variety when once thus formed might
0103  individuals of the same new variety. A local VARIETY when once thus formed might subsequently slo
0103  seed from a large body of plants of the same VARIETY, as the chance of intercrossing with other v
0103  dividuals of the same species, or of the same VARIETY, true and uniform in character. It will obvi
0105  ently competition, will give time for any new VARIETY to be slowly improved; and this may sometime
0110  ition with each other. Consequently, each new VARIETY or species, during the progress of its forma
0111  re chance, as we may call it, might cause one VARIETY to differ in some character from its parents
0111  r from its parents, and the offspring of this VARIETY again to differ from its parent in the very
0113  me has been found to hold good when first one VARIETY and then several mixed varieties of wheat ha
0113  . and we well know that each species and each VARIETY of grass is annually sowing almost countless
0117  cumulated to have formed a fairly well marked VARIETY, such as would be thought worthy of record i
0118  housand generations. And after this interval, VARIETY a1 is supposed in the diagram to have produc
0118  1 is supposed in the diagram to have produced VARIETY a2, which will, owing to the principle of di
0118  of divergence, differ more from (A) than did VARIETY a1. Variety m1 is supposed to have produced
0118  ce, differ more from (A) than did variety a1. VARIETY m1 is supposed to have produced two varietie
0118  thousand generations, producing only a single VARIETY, but in a more and more modified condition,
0133  drupeds. Instances could be given of the same VARIETY being produced under conditions of life as d
0159  t species present analogous variations; and a VARIETY of one species often assumes some of the cha
0161  onally vary in an analogous manner; so that a VARIETY of one species would resemble in some of its
0161  g on my view only a well marked and permanent VARIETY. But characters thus gained would probably b
0176  two varieties to two large areas, and a third VARIETY to a narrow intermediate zone. The intermedi
0176  a narrow intermediate zone. The intermediate VARIETY, consequently, will exist in lesser numbers
0176  have a great advantage over the intermediate VARIETY, which exists in smaller numbers in a narrow
0177  position of the supplanted, intermediate hill VARIETY. To sum up, I believe that species come to b
0194  has well expressed it, nature is prodigal in VARIETY, but niggard in innovation. Why, on the theo
0199  ed for beauty in the eyes of man, or for mere VARIETY. This doctrine, if true, would be absolutely
0203  stricts of a continuous area, an intermediate VARIETY will often be formed, fitted for an intermed
0203  ; but from reasons assigned, the intermediate VARIETY will usually exist in lesser numbers than th
```

Page **(Key Word)***

Page ***(Key Word)***

```
0204  advantage over the less numerous intermediate  VARIETY, and will thus generally succeed in supplant
0238  nd confidently expects to get nearly the same   VARIETY; breeders of cattle wish the flesh and fat t
0249  ccasional cross with a distinct individual or   VARIETY increases fertility, that I cannot doubt the
0252  cy the several individuals of the same hybrid   VARIETY are allowed to freely cross with each other,
0270  f kind of maize with yellow seeds, and a tall   VARIETY with red seeds, growing near each other in h
0271  esides the mere colour of the flower; and one   VARIETY can sometimes be raised from the seed of the
0271  ver, has proved the remarkable fact, that one   VARIETY of the common tobacco is more fertile, when
0271  sa. hence the reproductive system of this one   VARIETY must have been in some manner and in some de
0271  arieties in a state of nature, for a supposed   VARIETY if infertile in any degree would generally b
0274  be with varieties of plants. With animals one   VARIETY certainly often has this prepotent power ove
0274  y often has this prepotent power over another   VARIETY. Hybrid plants produced from a reciprocal cr
0274  species is crossed with another, and when one  VARIETY is crossed with another variety. For instanc
0274  and when one variety is crossed with another   VARIETY. For instance, I think those authors are rig
0275  amely in the union of individuals of the same   VARIETY, or of different varieties, or of distinct s
0276  rafting trees, the capacity of one species or   VARIETY to take on another, is incidental on general
0320  ction is grounded on the belief that each new   VARIETY, and ultimately each new species, is produce
0320  productions: when a new and slightly improved  VARIETY has been raised, it at first supplants the v
0355  oes not differ much (by substituting the word   VARIETY for species) from that lately advanced in an
0355  and improvement, all the individuals of each   VARIETY will have descended from a single parent. Bu
0464  are unknown, if any one link or intermediate   VARIETY be discovered, it will simply be classed as
0467  ividual shall live and which shall die, which   VARIETY or species shall increase in number, and whi
0469  ies, and that each species first existed as a   VARIETY, we can see why it is that no line of demarc
0471  we can plainly see why nature is prodigal in   VARIETY, though niggard in innovation. But why this
0486  oductions will rise immensely in value. A new   VARIETY raised by man will be a far more important a
0012  the correlation of growth. The result of the   VARIOUS, quite unknown, or dimly seen laws of variat
0015  long and short horned cattle, and poultry of   VARIOUS breeds, and esculent vegetables, for an almo
0027  pects to the rock pigeon: the blue colour and  VARIOUS marks occasionally appearing in all the bree
0028  me much; namely, that all the breeders of the  VARIOUS domestic animals and the cultivators of plan
0030  and race horse, the dromedary and camel, the   VARIOUS breeds of sheep fitted either for cultivated
0043  t. the effects of variability are modified by  VARIOUS degrees of inheritance and of reversion. Var
0044  my future work. Nor shall I here discuss the   VARIOUS definitions which have been given of the ter
0064  cases of the astonishingly rapid increase of   VARIOUS animals in a state of nature, when circumsta
0067  lings, also, are destroyed in vast numbers by  VARIOUS enemies; for instance, on a piece of ground
0090  et us take the case of a wolf, which preys on   VARIOUS animals, securing some by craft, some by str
0108  and the relative proportional numbers of the   VARIOUS inhabitants of the renewed continent will ag
0127  cted in nature, in modifying and adapting the   VARIOUS forms of life to their several conditions an
0129  d and dropped off; and these lost branches of  VARIOUS sizes may represent those whole orders, fami
0143  of the constitution, and of the structure of  VARIOUS organs; but that the effects of use and disu
0150  entary parts are left to the free play of the  VARIOUS laws of growth, to the effects' of long conti
0164  tripes in horses of very different breeds, in  VARIOUS countries from Britain to Eastern China; and
0166  aracter. When the oldest and truest breeds of  VARIOUS colours are crossed, we see a strong tendenc
0196  e had a most important influence in modifying  VARIOUS structures; and finally, that sexual selecti
0212  lowly acquired, as I have elsewhere shown, by  VARIOUS animals inhabiting desert islands; and we ma
0217  vertheless, I could give several instances of  VARIOUS birds which have been known occasionally to
0220  h informs me that he has watched the nests at  VARIOUS hours during May, June and August, both in S
0245  ty of first crosses and of hybrids. Sterility  VARIOUS in degree, not universal, affected by close
0248  rtain, on the one hand, that the sterility of  VARIOUS species when crossed is so different in degr
0249  lity of pure species is so easily affected by  VARIOUS circumstances, that for all practical purpos
0253  ole flocks of these crossed geese are kept in  VARIOUS parts of the country; and as they are kept f
0260  ity be so extremely different in degree, when  VARIOUS species are crossed, all of which we must su
0263  culty of either grafting or crossing together  VARIOUS species has been a special endowment; althou
0265  iance. In both cases, the sterility occurs in  VARIOUS degrees: in both, the male element is the mo
0276  that species have been specially endowed with  VARIOUS degrees of sterility to prevent them crossin
0276  k that trees have been specially endowed with  VARIOUS and somewhat analogous degrees of difficulty
0294  ean seas. In examining the latest deposits of  VARIOUS quarters of the world, it has everywhere bee
0304  the Arctic regions to the equator, inhabiting  VARIOUS zones of depths from the upper tidal limits
0325  arallelism of the palaeozoic forms of life in  VARIOUS parts of Europe, they add, if struck by this
0325  e relation between the physical conditions of  VARIOUS countries, and the nature of their inhabitan
0333  ngths of time, and will have been modified in  VARIOUS degrees. As we possess only the last volume
0346  y nor the dissimilarity of the inhabitants of  VARIOUS regions can be accounted for by their climat
0347  to the differences between the productions of  VARIOUS regions. We see this in the great difference
0354  former climatal and geographical changes and  VARIOUS occasional means of transport, the belief th
0382  l epoch, a few forms were widely dispersed to  VARIOUS points of the southern hemisphere by occasio
0382  ands to a great height under the equator. The  VARIOUS beings thus left stranded may be compared wi
0390  for man has unintentionally stocked them from  VARIOUS sources far more fully and perfectly than ha
0413  ganic beings, is the bond, hidden as it is by  VARIOUS degrees of modification, which is partially
0419  with respect to the comparative value of the   VARIOUS groups of species, such as orders, sub order
0422  n would afford the best classification of the  VARIOUS languages now spoken throughout the world; a
0422  rise to many new languages and dialects. The  VARIOUS degrees of difference in the languages from
0423  th species, namely, closeness of descent with  VARIOUS degrees of modification. Nearly the same rul
0431  ansmitted some of its characters; modified in  VARIOUS ways and degrees, to all; and the several sp
0431  each other by circuitous lines of affinity of  VARIOUS lengths (as may be seen in the diagram so of
0431  describing, without the aid of a diagram, the  VARIOUS affinities which they perceive between the m
0443  otorious that breeders of cattle, horses, and  VARIOUS fancy animals, cannot positively tell, until
0445  one wild species, I compared young pigeons of  VARIOUS breeds, within twelve hours after being hatc
0454  in the cauliflower. We often see rudiments of  VARIOUS parts in monsters. But I doubt whether any o
0454  ssive generations to the gradual reduction of  VARIOUS organs, until they become rudimentary, as in
0462  s yet very ignorant of the full extent of the  VARIOUS climatal and geographical changes which have
0477  from some third source or from each other, at  VARIOUS periods and in different proportions, the co
0483  in the descendants. Throughout whole classes  VARIOUS structures are formed on the same pattern, an
0487  e sides of a continent, and the nature of the  VARIOUS inhabitants of that continent in relation to
0489  kinds, with birds singing on the bushes, with  VARIOUS insects flitting about, and with worms crawl
0075  f even such extremely close varieties as the   VARIOUSLY coloured sweet peas, they must be each year
0005  or existence, it follows that any being, if it VARY however slightly in any manner profitable to i
0007  d that when the organisation has once begun to VARY, it generally continues to vary for many gener
0007  once begun to vary, it generally continues to  VARY for many generations. No case is on record of
```

Page **(Key Word)***

Page ***(Key Word)***

0009 ts withstand domestication or cultivation, and VARY very slightly, perhaps hardly more than in a s
0017 s having an extraordinary inherent tendency to VARY, and likewise to withstand diverse climates. I
0017 hen he first tamed an animal, whether it would VARY in succeeding generations, and whether it woul
0017 of generations under domestication, they would VARY on an average as largely as the parent species
0022 the number of the caudal and sacral vertebrae VARY; as does the number of the ribs, together with
0022 d when hatched. The shape and size of the eggs VARY. The manner of flight differs remarkably; as d
0041 as most fortunate that the strawberry began to VARY just when gardeners began to attend closely to
0042 ough I do not doubt that some domestic animals VARY less than others, yet the rarity or absence of
0044 ide ranging, much diffused, and common species VARY most. Species of the larger genera in any coun
0044 t. species of the larger genera in any country VARY more than the species of the smaller genera. M
0045 cal or classificatory point of view, sometimes VARY in the individuals of the same species. I am c
0046 le when they state that important organs never VARY; for these same authors practically rank that
0046 alists have honestly confessed) which does not VARY; and, under this point of view, no instance of
0053 the nature and relations of the species which VARY most, by tabulating all the varieties in sever
0059 ecies of the larger genera which on an average VARY most; and varieties, as we shall hereafter see
0080 , and, in a lesser degree, those under nature, VARY; and how strong the hereditary tendency is. Un
0087 ted, the thickness of the shell being known to VARY like every other structure. Sexual Selection.
0113 hout all time to all animals, that is, if they VARY, for otherwise natural selection can do nothin
0117 an average more of the species of large genera VARY than of small genera; and the varying species
0117 re the commonest and the most widely diffused, VARY more than rare species with restricted ranges.
0118 elf hereditary, consequently they will tend to VARY, and generally to vary in nearly the same mann
0118 ently they will tend to vary, and generally to VARY in nearly the same manner as their parents var
0120 t is probable that more than one species would VARY. In the diagram I have assumed that a second s
0127 der varying conditions of life, organic beings VARY at all in the several parts of their organisat
0128 species, belonging to the larger genera, which VARY most; and these will tend to transmit to their
0131 characters variable. Species of the same genus VARY in an analogous manner. Reversions to long los
0132 system is disturbed, this or that part should VARY more or less, we are profoundly ignorant. Neve
0143 ly embryonic period, are alike, seem liable to VARY in an allied manner: we see this in the right
0151 tinct genera. As birds within the same country VARY in a remarkably small degree, I have particula
0155 expect to find them still often continuing to VARY in those parts of their structure which have v
0156 ht degree, it is not probable that they should VARY at the present day. On the other hand, the poi
0159 unity of descent, and a consequent tendency to VARY in a like manner, but to three separate yet cl
0161 might be expected that they would occasionally VARY in an analogous manner; so that a variety of o
0167 ch species has been created with a tendency to VARY, both under nature and under domestication, in
0168 ent in their effects. Homologous parts tend to VARY in the same way, and homologous parts tend to
0186 increase in numbers; and that if any one being VARY ever so little, either in habits or structure,
0186 an be shown to exist; if further, the eye does VARY ever so slightly, and the variations be inheri
0209 cies; and if it can be shown that instincts do VARY ever so little, then I can see no difficulty i
0211 i can only assert, that instincts certainly do VARY, for instance, the migratory instinct, both in
0212 loss. So it is with the nests of birds, which VARY partly in dependence on the situations chosen
0242 t the mental qualities of our domestic animals VARY, and that the variations are inherited. Still
0243 riefly I have attempted to show that instincts VARY slightly in a state of nature. No one will dis
0244 ement of all organic beings, namely, multiply, VARY, let the strongest live and the weakest die. C
0265 natural to them, they are extremely liable to VARY, which is due, as I believe, to their reproduc
0265 successive generations are eminently liable to VARY, as every experimentalist has observed. Thus w
0344 t forms of life, which are those that oftenest VARY, will in the long run tend to people the world
0405 the facility with which widely ranging species VARY and give rise to new forms will largely determ
0411 species belonging to the larger genera, which VARY most. The varieties, or incipient species, thu
0437 ts many times repeated are eminently liable to VARY in number and structure; consequently it is qu
0452 viduals of the same species are very liable to VARY in degree of development and in other respects
0464 st in any great number. Widely ranging species VARY most, and varieties are often at first local,
0473 hould the colour of a flower be more likely to VARY in any one species of a genus, if the other sp
0005 le to itself, under the complex and sometimes VARYING conditions of life, will have a better chanc
0040 ny record having been preserved of such slow, VARYING, and insensible changes. I must now say a fe
0046 int of view, no instance of an important part VARYING will ever be found: but under any other poin
0053 themselves of the proportional numbers of the VARYING species. Dr. Hooker permits me to add, that
0056 rage. It is not that all large genera are now VARYING much, and are thus increasing in the number
0056 heir species, or that no small genera are now VARYING and increasing; for if this had been so, it
0073 tle within battle must ever be recurring with VARYING success; and yet in the long run the forces
0082 ill up by modifying and improving some of the VARYING inhabitants. For as all the inhabitants of e
0083 ll inferior animals, but protects during each VARYING season, as far as lies in his power, all his
0102 l always tend to preserve all the individuals VARYING in the right direction, though in different
0104 will tend to modify all the individuals of a VARYING species throughout the area in the same mann
0113 ing any change in its conditions) only by its VARYING descendants seizing on places at present occ
0113 ce, if any one species of grass were to go on VARYING, and those varieties were continually select
0117 rge genera vary than of small genera; and the VARYING species of the large genera present a greate
0117 es. Let (A) be a common, widely diffused, and VARYING species, belonging to a genus large in its o
0117 engths proceeding from (A), may represent its VARYING offspring. The variations are supposed to be
0126 . if during the long course of ages and under VARYING conditions of life, organic beings vary at a
0132 urbed in the parents, I chiefly attribute the VARYING or plastic condition of the offspring. The m
0134 ry naturalist of species keeping true, or not VARYING at all, although living under the most oppos
0143 this in the right and left sides of the body VARYING in the same manner; in the front and hind le
0143 nd hind legs, and even in the jaws and limbs, VARYING together, for the lower jaw is believed to b
0154 by the continued selection of the individuals VARYING in the required manner and degree, and by th
0154 would be surprised at one of the blue species VARYING into red, or conversely; but if all the spec
0158 rgely varied being more likely still to go on VARYING than parts which have long been inherited an
0162 we ought, on my theory, sometimes to find the VARYING offspring of a species assuming characters (
0162 ndependently created species; that the one in VARYING has assumed some of the characters of the ot
0162 an important and uniform nature occasionally VARYING so as to acquire, in some degree, the charac
0176 ply to both; and if we in imagination adapt a VARYING species to a very large area, we shall have
0177 y one period present an inextricable chaos of VARYING and intermediate links: firstly, because new
0195 ent to cause the preservation of successively VARYING individuals. I have sometimes felt as much d
0206 ral selection acts by either now adapting the VARYING parts of each being to its organic and inorg
0314 eater or lesser amount of modification in the VARYING species, depends on many complex contingenci
0314 ature of the other inhabitants with which the VARYING species comes into competition. Hence it is
0317 family, be represented by a vertical line of VARYING thickness, crossing the successive geologica

Page ***(Key Word)***

Page **(Key Word)**

```
0318 represented, as before, by a vertical line of VARYING thickness, the line is found to taper more g
0326 ieties. It is also natural that the dominant, VARYING, and far spreading species, which already ha
0327 rmed by dominant species spreading widely and VARYING; the new species thus produced being themsel
0327 or over other species; these again spreading, VARYING, and producing new species. The forms which
0408 d according as the immigrants were capable of VARYING more or less rapidly, there would ensue in d
0412 ze. I further attempted to show that from the VARYING descendants of each species trying to occupy
0416 any number of instances could be given of the VARYING importance for the classification of the sam
0472 al selection always ready to adapt the slowly VARYING descendants of each to any unoccupied or ill
0190 for its supply, and being divided by highly VASCULAR partitions. In these cases, one of the two o
0007 y in a state of nature. When we reflect on the VAST diversity of the plants and animals which have
0037 , as I believe, the well known fact, that in a VAST number of cases we cannot recognise, and there
0067 ot invariably the case. With plants there is a VAST destruction of seeds, but, from some observati
0067 ther plants. Seedlings, also, are destroyed in VAST numbers by various enemies; for instance, on a
0096 which certainly do not habitually pair, and a VAST majority of plants are hermaphrodites. What re
0099 he same plant. How, then, comes it that such a VAST number of the seedlings are mongrelized? I sus
0102 rtant element of success. Though nature grants VAST periods of time for the work of natural select
0173 eological history. The crust of the earth is a VAST museum; but the natural collections have been
0269 life. Nature acts uniformly and slowly during VAST periods of time on the whole organisation, in
0279 ntermediate varieties; on their number. On the VAST lapse of time, as inferred from the rate of de
0281 ltered, whilst its descendants had undergone a VAST amount of change; and the principle of competi
0282 ience, yet does not admit how imcomprehensibly VAST have been the past periods of time, may at onc
0285 the action cf the sea, that no trace of these VAST dislocations is externally visible. The Craven
0289 nd shell is known belonging to either of these VAST periods, with one exception discovered by Sir
0290 t has been derived, accords with the belief of VAST intervals of time having elapsed between each
0293 ugh each formation has indisputably required a VAST number of years for its deposition, I can see
0294 marine animals can flourish; for we know what VAST geographical changes occurred in other parts o
0296 ignorant of this fact would have suspected the VAST lapse of time represented by the thinner forma
0307 age to the present day; and that during these VAST, yet quite unknown, periods of time, the world
0307 e question why we do not find records of these VAST primordial periods, I can give no satisfactory
0307 the difficulty of understanding the absence of VAST piles of fossiliferous strata, which on my the
0319 t its rarity; for rarity is the attribute of a VAST number of species of all classes, in all count
0327 ods of subsidence; and that blank intervals of VAST curation occurred during the periods when the
0332 onditions of life, and yet retain throughout a VAST period the same general characteristics. This
0346 new and Old Worlds; yet if we travel over the VAST American continent, from the central parts of
0348 we come to the shores of Africa; and over this VAST space we meet with no well defined and distinc
0351 period, so certain species have migrated over VAST spaces, and have not become greatly modified.
0351 migrated from the same region; for during the VAST geographical and climatal changes which will h
0353 d means of dispersal, have migrated across the VAST and broken interspace. The great and striking
0357 ns of level in our continents; but not of such VAST changes in their position and extension, as to
0361 ntly birds of many kinds are blown by gales to VAST distances across the ocean. We may I think saf
0373 l chile I was astonished at the structure of a VAST mound of detritus, about 800 feet in height, c
0377 been very long; and when we remember over what VAST spaces some naturalised plants and animals hav
0387 t transport the seeds of fresh water plants to VAST distances, and if consequently the range of th
0403 y continuous land, yet they are inhabited by a VAST number of distinct mammals, birds, and plants.
0460 es. We see the truth of this conclusion in the VAST difference in the result, when the same two sp
0472 's own death; at drones being produced in such VAST numbers for one single act, and being then sla
0487 etween the successive stages as having been of VAST duration. But we shall be able to gauge with s
0005 of Malthus, applied to the whole animal and VEGETABLE kingdoms. As many more individuals of each
0012 ffected from coloured individuals by certain VEGETABLE poisons. Hairless dogs have imperfect teeth
0063 with manifold force to the whole animal and VEGETABLE kingdoms; for in this case there can be no
0101 rongly inclined to suspect that, both in the VEGETABLE and animal kingdoms, an occasional intercro
0116 ts that a stomach by being adapted to digest VEGETABLE matter alone, or flesh alone, draws most nu
0126 sses in each main division of the animal and VEGETABLE kingdoms. Although extremely few of the mos
0159 ted on by similar unknown influences. In the VEGETABLE kingdom we have a case of analogous variati
0237 gain the desired end. Thus, a well flavoured VEGETABLE is cooked, and the individual is destroyed;
0379 d direction. We see, however, a few southern VEGETABLE forms on the mountains of Borneo and Abyssi
0382 lightly tinted by the same peculiar forms of VEGETABLE life. Sir C. Lyell in a striking passage ha
0411 the water; one to feed on flesh, another on VEGETABLE matter, and so on; but the case is widely d
0429 ct belonging to a new order; and that in the VEGETABLE kingdom, as I learn from Dr. Hooker, it has
0438 in the other great classes of the animal and VEGETABLE kingdoms. Naturalists frequently speak of t
0016 and poultry of various breeds, and esculent VEGETABLES, for an almost infinite number of generatio
0251 gress to maturity, and bore good seed, which VEGETATED freely. In a letter to me, in 1839, Mr. Her
0071 d with Scotch fir. The change in the native VEGETATION of the planted part of the heath was most r
0073 ave observed in parts of South America) the VEGETATION; this again would largely affect the insect
0074 erican forest is cut down, a very different VEGETATION springs up; but it has been observed that t
0180 all natural conditions. Let the climate and VEGETATION change; let other competing rodents or new
0283 ring. The appearance of the surface and the VEGETATION show that elsewhere years have elapsed sinc
0373 e frozen mammals and nature of the mountain VEGETATION, that Siberia was similarly affected. Along
0378 othed with a mingled tropical and temperate VEGETATION, like that now growing with strange luxuria
0399 arctic islancs, when they were clothed with VEGETATION, before the commencement of the Glacial per
0414 ts, how remarkable it is that the organs of VEGETATION, on which their whole life depends, are of
0149 to variation in structure. Inasmuch as this VEGETATIVE repetition, to use Prof. Owen's expression,
0263 incidental on unknown differences in their VEGETATIVE systems; so I believe that the still more c
0276 l on generally unknown differences in their VEGETATIVE systems; so in crossing, the greater or les
0310 on, Sedgwick, etc., have unanimously, often VEHEMENTLY, maintained the immutability of species. Bu
0004 ion, afforded the best and safest clue. I may VENTURE to express my conviction of the high value o
0167 se with that ofthe horse genus! For myself, I VENTURE confidently to look back thousands on thousa
0200 d for walking or grasping; and we may further VENTURE to believe that the several bones in the lim
0270 rfectly fertile; so that even Gartner did not VENTURE to consider the two varieties as specificall
0181 come extinct or were unknown, who would have VENTURED to have surmised that birds might have exist
0159 larged stems of these three plants, not to the VERA causa of community of descent, and a consequen
0352 vates the mind. He who rejects it, rejects the VERA causa of ordinary generation with subsequent m
0482 d by secondary laws. They admit variation as a VERA causa in one case, they arbitrarily reject it
0251 ome other genera, as Lobelia, Passiflora and VERBASCUM. Although the plants in these experiments a
0270 ts made during many years on nine species of VERBASCUM, by so good an observer and so hostile a wi
0270 w and white varieties of the same species of VERBASCUM when intercrossed produce less seed, than d
```

Page **(Key Word)**

0271 differently coloured. Yet these varieties of VERBASCUM present no other difference besides the mer
0398 the islands, between the Galapagos and Cape de VERDE Archipelagos: but what an entire and absolute
0398 ir inhabitants! The inhabitants of the Cape de VERDE Islands are related to those of Africa, like
0399 continuous land, from America; and the Cape de VERDE islands from Africa; and that such colonists
0478 ng american mainland; and those of the Cape de VERDE archipelago and other African islands to the
0393 eat oceans are studded. I have taken pains to VERIFY this assertion, and I have found it strictly
0049 wn one of the primrose and cowslip, or Primula VERIS and elatior. These plants differ considerably
0439 t be that of a mammal, bird, or reptile. The VERMIFORM larvae, of moths, flies, beetles, etc., rese
0442 pment of this insect, we see no trace of the VERMIFORM stage. How, then, can we explain these seve
0228 and narrow, knife edged ridge, coloured with VERMILION. The bees instantly began on both sides to
0229 bottoms, formed by thin little plates of the VERMILION wax having been left ungnawed, were situate
0229 e rate on the opposite sides of the ridge of VERMILION wax, as they circularly gnawed away and dee
0230 tten it. From the experiment of the ridge of VERMILION wax, we can clearly see that if the bees we
0232 comb, with an extremely thin layer of melted VERMILION wax; and I invariably found that the colour
0068 estate depends chiefly on the destruction of VERMIN. If not one head of game were shot during the
0068 years in England, and at the same time, if no VERMIN were destroyed, there would, in all probabili
0325 tly struck those admirable observers, MM. de VERNEUIL and d'Archiac. After referring to the parall
0253 nus colchicus with P. torquatus and with P. VERSICOLOR are perfectly fertile. The hybrids from the
0022 manner. The number of the caudal and sacral VERTEBRAE vary; as does the number of the ribs, toget
0149 the structure of the same individual (as the VERTEBRAE in snakes, and the stamens in polyandrous f
0436 , the elemental parts of a certain number of VERTEBRAE. The anterior and posterior parts of the
0437 the vertebrata, we see a series of internal VERTEBRAE bearing certain processes and appendages; i
0437 progenitor of the vertebrata possessed many VERTEBRAE; the unknown progenitor of the articulata,
0438 peak of the skull as formed of metamorphosed VERTEBRAE: the jaws of crabs as metamorphosed legs; t
0438 ley has remarked, to speak of both skull and VERTEBRAE, both jaws and legs, etc., as having been m
0438 e of descent, primordial organs of any kind, VERTEBRAE in the one case and legs in the other, have
0479 se, and leg of the horse, the same number of VERTEBRAE forming the neck of the giraffe and of the
0187 little altered condition. Amongst existing VERTEBRATA, we find but a small amount of gradation in
0191 which the glottis is closed. In the higher VERTEBRATA the branchiae have wholly disappeared, the
0329 nto one general system. With respect to the VERTEBRATA, whole pages could be filled with striking
0338 g the common embryological character of the VERTEBRATA, until beds far beneath the lowest Silurian
0416 ermal covering, as hair or feathers, in the VERTEBRATA. If the Ornithorhynchus had been covered wi
0437 tisfactorily answer these questions. In the VERTEBRATA, we see a series of internal vertebrae bear
0437 believe that the unknown progenitor of the VERTEBRATA possessed many vertebrae; the unknown proge
0440 nstance, suppose that in the embryos of the VERTEBRATA the peculiar loop like course of the arteri
019 ation. I can, indeed, hardly doubt that all VERTEBRATE animals having true lungs have descended by
0096 rials prepared for an ample discussion. All VERTEBRATE animals, all insects, and some other large
0191 and structure with the lungs of the higher VERTEBRATE animals: hence there seems to me to be no g
0431 ther, for instance, of birds from all other VERTEBRATE animals, by the belief that many ancient fo
0431 ted with the early progenitors of the other VERTEBRATE classes. There has been less entire extinct
0436 r and posterior limbs in each member of the VERTEBRATE and articulate classes are plainly homologo
0439 ving forgotten to ticket the embryo of some VERTEBRATE animal, he cannot now tell whether it be th
0285 upwards of 30 miles; and along this line the VERTICAL displacement of the strata has varied from 6
0317 the genera of a family, be represented by a VERTICAL line of varying thickness, crossing the succ
0318 p of species be represented, as before, by a VERTICAL line of varying thickness, the line is found
0484 n their chemical composition, their germinal VESICLES, their cellular structure, and their laws of
0454 s the stump of a limb in tailless breeds, the VESTIGE of an ear in earless breeds, the reappearanc
0003 ition of the plant itself. The author of the VESTIGES of Creation would, I presume, say that after
0453 e stumps: I could as soon believe that these VESTIGES of nails have appeared, not from unknown law
0363 e was a pebble quite as large as the seed of a VETCH. Thus seeds might occasionally be transported
0361 ng birds long retain their vitality: peas and VETCHES, for instance, are killed by even a few days
0187 ive to light, and likewise to those coarser VIBRATIONS of the air which produce sound. In looking
0336 es and conditions. Consider the prodigious VICISSITUDES of climate during the pleistocene period,
0088 ring. Sexual selection by always allowing the VICTOR to breed might surely give incomitable courag
0076 d we precisely say why one species has been VICTORIOUS over another in the great battle of life. A
0106 ced on large areas, which already have been VICTORIOUS over many competitors, will be those that w
0221 haps fancied that, after all, they had been VICTORIOUS in their late combat. At the same time I la
0325 incipient species; for these latter must be VICTORIOUS in a still higher degree in order to be pre
0326 birthplace and some from the other might be VICTORIOUS. But in the course of time, the forms domin
0327 and which yield their places to the new and VICTORIOUS forms, will generally be allied in groups,
0337 ner the organisation of the more recent and VICTORIOUS forms of life, in comparison with the ancie
0350 nti and thus they will become still further VICTORIOUS, and will produce groups of modified descen
0405 ders, but the more important power of being VICTORIOUS in distant lands in the struggle for life w
0073 uredly the merest trifle would often give the VICTORY to one organic being over another. Neverthel
0088 , will leave most progeny. But in many cases, VICTORY will depend not on general vigour, but on ha
0088 almon; for the shield may be as important for VICTORY, as the sword or spear. Amongst birds, the c
0468 les; and the slightest advantage will lead to VICTORY. As geology plainly proclaims that each land
0006 onate judgment of which I am capable, that the VIEW which most naturalists entertain, and which I
0007 ere is, also, I think, some probability in the VIEW propounded by Andrew Knight, that this variabi
0009 id to be the bane of horticulture; but on this VIEW we owe variability to the same cause which pro
0010 formation; so that, in fact, sports support my VIEW, that variability may be largely attributed to
0010 ore in the case of plants. Under this point of VIEW, Mr. Buckman's recent experiments on plants se
0011 ha's not in some country drooping ears; and the VIEW suggested by some authors, that the drooping i
0015 is not a shadow of evidence in favour of this VIEW: to assert that we could not breed our cart an
0016 s being at present empirical. Moreover, on the VIEW of the origin of genera which I shall presentl
0017 ve, or even strong, evidence in favour of this VIEW. It has often been assumed that man has chosen
0027 ts geographical sub species. In favour of this VIEW, I may add, firstly, that C. livia, or the roc
0034 ethodical selection, with a distinct object in VIEW, to make a new strain or sub breed, superior t
0038 varieties kept in civilised countries. On the VIEW here given of the all important part which sel
0045 der a physiological or classificatory point of VIEW, sometimes vary in the individuals of the same
0046 which does not vary; and, under this point of VIEW, no instance of an important part varying will
0046 ll ever be found: but under any other point of VIEW many instances assuredly can be given. There i
0052 ent regions; but I have not much faith in this VIEW; and I attribute the passage of a variety, fro
0056 these facts are of plain signification on the VIEW that species are only strongly marked and perm
0057 imperfect results go, they always confirm the VIEW. I have also consulted some sagacious and most
0057 , and, after deliberation, they concur in this VIEW. In this respect, therefore, the species of th

Page **(Key Word)**

```
0069 to its direct action. But this is a very false  VIEW: we forget that each species, even where it mo
0070 have in this case lost every single seed. This   VIEW of the necessity of a large stock of the same
0072 their numbers that I went to several points of   VIEW, whence I could examine hundreds of acres of t
0074 kinds to what we call chance. But how false a    VIEW is this! Every one has heard that when an Amer
0084 ng lapse of ages, and then so imperfect is our   VIEW into long past geological ages, that we only s
0089 enter on the details necessary to support this   VIEW; but if man can in a short time give elegant c
0089 plumage of the young, can be explained on the    VIEW of plumage having been chiefly modified by sex
0096 oncur for the reproduction of their kind. This   VIEW, I may add, was first suggested by Andrew Knig
0097 cts, such as the following, which on any other   VIEW are inexplicable. Every hybridizer knows how u
0099 r! how simply are these facts explained on the   VIEW of an occasional cross with a distinct individ
0100 ong a contrast with terrestrial plants, on the   VIEW of an occasional cross being indispensable, by
0101 e of very great difficulty under this point of   VIEW; but I have been enabled, by a fortunate chanc
0101 ature. I am well aware that there are, on this   VIEW, many cases of difficulty, some of which I am
0111 istinct species. Nevertheless, according to my   VIEW, varieties are species in the process of forma
0126 on Classification, but I may add that on this   VIEW of extremely few of the more ancient species h
0126 ies having transmitted descendants, and on the   VIEW of all the descendants of the same species mak
0129 ts, and so on in almost endless cycles. On the   VIEW that each species has been independently creat
0133 e characters of such species, accords with our   VIEW that species of all kinds are only well marked
0138 nearly similar climate; so that on the common   VIEW of the blind animals having been separately cr
0138 inhabitants of North America and Europe. On my   VIEW we must suppose that American animals, having
0138 ons, the deepest recesses, disuse will on this   VIEW have more or less perfectly obliterated its ey
0139 abitants of the two continents on the ordinary   VIEW of their independent creation. That several of
0139 s have descended from a single parent, if this   VIEW be correct, acclimatisation must be readily ef
0141 tion, which is common to most animals. On this   VIEW, the capacity of enduring the most different c
0152 le to variation. Why should this be so? On the   VIEW that each species has been independently creat
0152 see them, I can see no explanation. But on the   VIEW that groups of species have descended from oth
0155 it is to individual anomalies. On the ordinary   VIEW of each species having been independently crea
0155 that any explanation can be given. But on the   VIEW of species being only strongly marked and fixe
0157 ecies. This relation has a clear meaning on my   VIEW of the subject: I look at all the species of t
0159 , the common turnip. According to the ordinary   VIEW of each species having been independently crea
0161 ion in the plumage to assume this colour. This   VIEW is hypothetical, but could be supported by som
0161 nother species; this other species being on my   VIEW only a well marked and permanent variety. But
0167 but other species of the genus. To admit this   VIEW is, as it seems to me, to reject a real for an
0169 rent of many modified descendants, which on my   VIEW must be a very slow process, requiring a long
0191 s of certain fish, or, for I do not know which   VIEW is now generally held, a part of the auditory
0191 istinct purpose: in the same manner as, on the   VIEW entertained by some naturalists that the branc
0225 presence of adjoining cells; and the following   VIEW may, perhaps, be considered only as a modifica
0239 ly came to have the desired character. On this   VIEW we ought occasionally to find neuter insects o
0245 independently of their fertility. Summary. The   VIEW generally entertained by naturalists is that s
0245 event the confusion of all organic forms. This   VIEW certainly seems at first probable, for species
0254 es, since commingled by intercrossing. On this   VIEW, the aboriginal species must either at first h
0254 ust be considered as distinct species. On this   VIEW of the origin of many of our domestic animals,
0264 was at first very unwilling to believe in this   VIEW; as hybrids, when once born, are generally hea
0272 oes not seem to me sufficient to overthrow the   VIEW which I have taken with respect to the very ge
0273 ttention. For it bears on and corroborates the   VIEW which I have taken on the cause of ordinary va
0276 ing fact. But it harmonises perfectly with the   VIEW that there is no essential distinction between
0277 l conditions of life have been disturbed. This   VIEW is supported by a parallelism of another kind:
0278 me opposed to, but even rather to support the   VIEW, that there is no fundamental distinction betw
0280 diate between them. But this is a wholly false   VIEW: we should always look for forms intermediate
0288 eve we are continually taking a most erroneous   VIEW, when we tacitly admit to ourselves that sedim
0288 any wear and tear, seem explicable only on the   VIEW of the bottom of the sea not rarely lying for
0297 others into the rank of varieties: and on this   VIEW we do find the kind of evidence of change whic
0298 in some considerable degree. According to this   VIEW, the chance of discovering in a formation in a
0308 ussia and in North America, do not support the   VIEW, that the older a formation is, the more it ha
0311 ive, but widely separated, formations. On this   VIEW, the difficulties above discussed are greatly
0312 organic beings, better accord with the common   VIEW of the immutability of species, or with that o
0315 rmations is not equal. Each formation, on this   VIEW, does not mark a new and complete act of creat
0323 o the Old and New Worlds be kept wholly out of   VIEW, the general parallelism in the successive for
0333 and they are wholly inexplicable on any other   VIEW. On this same theory, it is evident that the f
0337 native plants and animals. Under this point of   VIEW, the productions of Great Britain may be said
0338 d less modified condition of each animal. This   VIEW may be true, and yet it may never be capable o
0352 now found. Nevertheless the simplicity of the   VIEW that each species was first produced within a
0353 d on distribution, is intelligible only on the   VIEW that the great majority of species have been p
0353 as it has to many other naturalists, that the   VIEW of each species having been produced in one ar
0354 ace, can in many instances be explained on the   VIEW of each species having migrated from a single
0355 le on the theory of independent creation. This   VIEW of the relation of species in one region to th
0357 t recently been united to some continent. This   VIEW cuts the Gordian knot of the dispersal of the
0358 the recent period, as are necessitated on the   VIEW advanced by Forbes and admitted by his many fo
0370 el, I am strongly inclined to extend the above   VIEW, and to infer that during some earlier and sti
0371 parts of Europe and the United States. On this   VIEW we can understand the relationship, with very
0372 e glacial period. I am convinced that Forbes's   VIEW may be largely extended. In Europe we have the
0374 long certain broad belts of longitude. On this   VIEW of the whole world, or at least of broad longi
0380 osing that all difficulties are removed on the   VIEW here given in regard to the range and affiniti
0382 f his great cycles of change; and that on this   VIEW, combined with modification through natural se
0389 ting the greatest amount of difficulty, on the   VIEW that all the individuals both of the same and
0389 y stated that I cannot honestly admit Forbes's   VIEW on continental extensions, which, if legitimat
0389 nearly or quite joined to some continent. This   VIEW would remove many difficulties, but it would n
0392 rupeds. This case presents no difficulty on my   VIEW, for a hooked seed might be transported to an
0393 n to be immediately killed by sea water, on my   VIEW we can see that there would be great difficult
0394 multiplied. It cannot be said, on the ordinary   VIEW of creation, that there has not been time for
0394 and no other mammals on remote islands? On my   VIEW this question can easily be answered; for no t
0396 ing continent, an inexplicable relation on the   VIEW of independent acts of creation. All the foreg
0396 es, etc., seem to me to accord better with the   VIEW of occasional means of transport having been l
0396 ient in the long course of time, than with the   VIEW of all our oceanic islands having been formerl
0396 with the nearest continent; for on this latter   VIEW the migration would probably have been more co
0397 and are killed by it. Yet there must be, on my   VIEW, some unknown, but highly efficient means for
```

Page **(Key Word)**

Page **(Key Word)**

0398 receive no sort of explanation on the ordinary VIEW of independent creation; whereas on the view h
0398 y view of independent creation; whereas on the VIEW here maintained, it is obvious that the Galapa
0399 ker's account, to those of America: but on the VIEW that this island has been mainly stocked by se
0399 . but this difficulty almost disappears on the VIEW that both New Zealand, South America, and othe
0400 is is just what might have been expected on my VIEW, for the islands are situated so near each oth
0401 ference might indeed have been expected on the VIEW of the islands having been stocked by occasion
0402 hers. But we often take, I think, an erroneous VICW of the probability of closely allied species i
0404 ies, showing, in accordance with the foregoing VIFW, that at some former period there has been int
0405 e for life with foreign associates. But on the VIEW of all the species of a genus having descended
0406 i think, utterly inexplicable on the ordinary VIEW of the independent creation of each species, b
0406 ion of each species, but are explicable on the VIEW of colonisation from the nearest and readiest
0414 the difficulties which are encountered on the VIEW that classification either gives some unknown
0415 tal or physiological importance. No doubt this VIEW of the classificatory importance of organs whi
0418 but it is by no means obvious, on the ordinary VIEW, why the structure of the embryo should be mor
0419 ee why such characters are so valuable, on the VIEW of classification tacitly including the idea o
0420 ed, if I do not greatly deceive myself, on the VIEW that the natural system is founded on descent
0422 gst the beings of the same group. Thus, on the VIEW which I hold, the natural system is genealogic
0422 sses. It may be worth while to illustrate this VIEW of classification, by taking the case of langu
0423 origin of each tongue. In confirmation of this VIEW, Let us glance at the classification of variet
0425 cently exposed. Rudimentary structures on this VIEW are as good as, or even sometimes better than,
0427 e authors between very distinct animals. On my VIEW of characters being of real importance for cla
0430 odification in divergent directions. On either VIEW we may suppose that the bizcacha has retained,
0434 e class; but when we have a distinct object in VIEW, and do not look to some unknown plan of creat
0435 g work on the Nature of Limbs. On the ordinary VIEW of the independent creation of each being, we
0436 ir intimate structure, are intelligible on the VIEW that they consist of metamorphosed leaves, arr
0437 w inexplicable are these facts on the ordinary VIEW of creation! Why should the brain be enclosed
0438 anguage having this plain signification. On my VIEW these terms may be used literally; and the won
0443 ese facts can be explained, as follows, on the VIEW of descent with modification. It is commonly a
0444 idence from our domestic animals supports this VIEW. But in other cases it is quite possible that
0447 een in the case of pigeons. We may extend this VIEW to whole families or even classes. The fore li
0449 nt, would be genealogical. Cescent being on my VIEW the hidden bond of connexion which naturalists
0449 under the term of the natural system. On this VIEW we can understand how it is that, in the eyes
0454 he manatee were formed for this purpose. On my VIEW of descent with modification, the origin of ru
0455 erited, we can understand, on the genealogical VIEW of classification, how it is that systematists
0455 s a clue in seeking for its derivation. On the VIEW of descent with modification, we may conclude
0456 other such rules; all naturally follow on the VIEW of the common parentage of those forms which a
0456 d divergence of character. In considering this VIEW of classification, it should be borne in mind
0456 a, families, orders, and classes. On this same VIEW of descent with modification, all the great fa
0457 gans in classification is intelligible, on the VIEW that an arrangement is only so far natural as
0458 t, that I should without hesitation adopt this VIEW, even if it were unsupported by other facts or
0464 uralists will be able to decide, on the common VIFW, whether or not these doubtful forms are varie
0469 and arguments in favour of the theory. On the VIFW that species are only strongly marked and perm
0469 been produced by secondary laws. On this same VIEW we can understand how it is that in each regio
0470 varieties. These are strange relations on this VIEW of each species having been independently crea
0471 ! and so on in endless other cases. But on the VIEW of each species constantly trying to increase
0472 s of any one country, although on the ordinary VIEW supposed to have been specially created and ad
0473 e blue and barred rock pigeon! On the ordinary VIEW of each species having been independently crea
0474 d be eminently liable to variation; but, on my VIEW, this part has undergone, since the several sp
0475 it the effects of long continued habit. On the VIEW of all the species of the same genus having de
0477 wo areas will inevitably be different. On this VIEW of instincts having been slowly acquired throu
0479 s of a flower, is likewise intelligible on the VIEW of migration, with subsequent modification, we
0480 of life; and we can clearly understand on this VIEW of the gradual modification of parts or organs
0480 rom a remote period to the present day. On the VIEW the meaning of rudimentary organs. But disuse
0480 iving naturalists and geologists rejected this VIEW of each organic being and each separate organ
0481 during a long course of years, from a point of VIEW of the mutability of species? It cannot be ass
0482 ng and rising naturalists; who will be able to VIEW directly opposite to mine. It is so easy to hi
0482 variation; but they refuse to extend the same VIEW both sides of the question with impartiality.
0486 the blunders of numerous workmen! when we thus VIEW to other and very slightly different forms. Ne
0486 come simpler when we have a definite object in VIEW. We possess no pedigrees or armorial bearings;
0488 t eminence seem to be fully satisfied with the VIEW that each species has been independently creat
0488 the birth and death of the individual. When I VIEW all beings not as special creations, but as th
0490 s, directly follows. There is grandeur in this VIEW of life, with its several powers, having been
0045 parts which must be called important, whether VIEWED under a physiological or classificatory point
0146 hence we see that modifications of structure, VIEWED by systematists as of high value, may be whol
0200 direct action of physical conditions) may be VIEWED, either as having been of special use to some
0292 cannot be doubted that the geological record, VIEWED as a whole, is extremely imperfect; but if we
0481 nds are stocked with a multitude of facts all VIEWED, during a long course of years, from a point
0013 to be inheritable. Perhaps the correct way of VIEWING the whole subject, would be, to look at the
0284 struck with the evidence of denudation, when VIEWING volcanic islands, which have been worn by th
0040 s distinct at our poultry shows. I think these VIEWS further explain what has sometimes been notic
0052 by the general weight of the several facts and VIEWS given throughout this work. It need not be su
0095 first urged against Sir Charles Lyell's noble VIEWS on the modern changes of the earth, as illust
0095 and as modern geology has almost banished such VIEWS as the excavation of a great valley by a sing
0106 f the organic world. We can, perhaps, on these VIEWS, understand some facts which will be again al
0179 re. It has been asked by the opponents of such VIEWS as I hold, how, for instance, a land carnivor
0243 re, and is plainly explicable on the foregoing VIEWS, but is otherwise inexplicable, all tend to c
0279 ctions which might be justly urged against the VIEWS maintained in this volume. Most of them have
0291 formed during subsidence. Since publishing my VIEWS on this subject in 1845, I have watched the p
0292 . i feel much confidence in the truth of these VIEWS, for they are in strict accordance with the g
0299 many objections which may be urged against my VIEWS. Hence it will be worth while to sum up the f
0308 be truly urged as a valid argument against the VIEWS here entertained. To show that it may hereaft
0316 woodward (though all strongly opposed to such VIEWS as I maintain) admit its truth; and the rule
0316 act, which if true would have been fatal to my VIEWS. But such cases are certainly exceptional; th
0317 mont, Murchison, Barrande, etc., whose general VIEWS would naturally lead them to this conclusion.
0342 e finest graduated steps. He who rejects these VIEWS on the nature of the geological record, will

Page **(Key Word)**

views

Page **(Key Word)**

```
0351 and have not become greatly modified. On these VIEWS, it is obvious, that the several species of t
0368 a to the arctic regions of that country. These VIEWS, grounded as they are on the perfectly well a
0400 islands may be used as an argument against my VIEWS; for it may be asked, how has it happened in
0427 the same parents. We can understand, on these VIEWS, the very important distinction between real
0481 hough I am fully convinced of the truth of the VIEWS given in this volume under the form of an abs
0484 , into which life was first breathed. When the VIEWS entertained in this volume on the origin of s
0484 me on the origin of species, or when analogous VIEWS are generally admitted, we can dimly foresee
0067 in number is most obscure. Look at the most VIGOROUS species; by as much as it swarms in numbers,
0068 wsed by quadrupeds, be let to grow, the more VIGOROUS plants gradually kill the less vioorous, tho
0068 more vigorous plants gradually kill the less VIGOROUS, though fully grown, plants: thus out of twe
0069 me cold, acts directly, it will be the least VIGOROUS, or those which have got least food through
0079 that death is generally prompt, and that the VIGOROUS, the healthy, and the happy survive and mult
0083 the same climate. He does not allow the most VIGOROUS males to struggle for the females. He does n
0088 than natural selection. Generally, the most VIGOROUS males, those which are best fitted for their
0092 e more fully alluded to), would produce very VIGOROUS seedlings, which consequently would have the
0127 ordinary selection, by assuring to the most VIGOROUS and best adapted males the greatest number o
0130 rise by growth to fresh buds, and these, if VIGOROUS, branch out and overtop on all sides many a
0217 sional habit, or if the young were made more VIGOROUS by advantage having been taken of the mistak
0251 impregnated by the pollen of the hybrid made VIGOROUS growth and rapid progress to maturity, and b
0344 as these are formed, the species of the less VIGOROUS groups, from their inferiority inherited fro
0377 intruders, that a certain number of the more VIGOROUS and dominant temperate forms might have pene
0468 ales for possession of the females. The most VIGOROUS individuals, or those which have most succes
0072 as enclosed, it became thickly clothed with VIGOROUSLY growing young firs. Yet the heath was so ex
0077 whilst struggling with other plants crowing VIGOROUSLY all around. Look at a plant in the midst of
0221 in search of food; they approached and were VIGOROUSLY repulsed by an independent community of the
0008 how many cultivated plants display the utmost VIGOUR, and yet rarely or never seed! In some few su
0088 any cases, victory will depend not on general VIGOUR, but on having special weapons, confined to t
0096 the same variety but of another strain, gives VIGOUR and fertility to the offspring; and on the ot
0096 her hand, that close interbreeding diminishes VIGOUR and fertility; that these facts alone incline
0104 the young thus produced will gain so much in VIGOUR and fertility over the offspring from long co
0267 ers of different strains or sub breeds, gives VIGOUR and fertility to the offspring. I believe, in
0267 ve varied and become slightly different, give VIGOUR and fertility to the offspring. But we have s
0277 only slightly different is favourable to the VIGOUR and fertility of their offspring; and that sl
0277 ions of life are apparently favourable to the VIGOUR and fertility of all organic beings. It is no
0461 ly opposite, class of facts; namely, that the VIGOUR and fertility of all organic beings are incre
0461 arieties accuire from being crossed increased VIGOUR and fertility. So that, on the one hand, cons
0074 he number of cats: and Mr. Newman says, near VILLAGES and small towns I have found the nests of hu
0393 of whole orders on oceanic islands, Bory St. VINCENT long ago remarked that Batrachians (frogs, t
0366 , left by old glaciers, are now clothed by the VINE and maize. Throughout a large part of the Unit
0073 nsable to the fertilisation of the heartsease (VIOLA tricolor), for other bees do not visit this f
0320 suspect that he died by some unknown deed of VIOLENCE. The theory of natural selection is grounded
0360 oat for a less time than those protected from VIOLENT movement as in our experiments. Therefore it
0074 and proportion of kinds as in the surrounding VIRGIN forests. What a struggle between the several
0144 the position of several of the most important VISCERA. The nature of the bond of correlation is ve
0202 es the death of the insect by tearing out its VISCERA? If we look at the sting of the bee, as havi
0211 experience. But as the excretion is extremely VISCID, it is probably a convenience to the aphides
0387 nd were altogether 537 in number; and yet the VISCID mud was all contained in a breakfast cup! Con
0038 of structure excepting such as is externally VISIBLE; and indeed he rarely cares for what is inte
0083 ture effect? Man can act only on external and VISIBLE characters: nature cares nothing for appeara
0285 race of these vast dislocations is externally VISIBLE. The Craven fault, for instance, extends for
0442 r proportion, as soon as any structure became VISIBLE in the embryo. And in some whole groups of a
0131 th natural selection; organs of flight and of VISION. Acclimatisation. Correlation of growth. Comp
0137 some days in the light, some slight power of VISION. In the same manner as in Madeira the wings o
0138 t american animals, having ordinary powers of VISION, slowly migrated by successive cenerations fr
0240 ring not only in size, but in their organs of VISION, yet connected by some few members in an inte
0073 tsease (Viola tricolor), for other bees do not VISIT this flower. From experiments which I have tr
0073 lisation of cur clovers; but humble bees alone VISIT the common red clover (Trifolium pratense), a
0095 ube to its corolla, so that the hive bee could VISIT its flowers. Thus I can understand how a flow
0211 t one excreted. Afterwards I allowed an ant to VISIT them, and it immediately seemed, hy its eager
0391 n birds, during their great annual micrations, VISIT either periodically or occasionally this isla
0394 rican species either regularly or occasionally VISIT Bermuda, at the distance of 600 miles from th
0073 ia fulgens, in this part of England, is never VISITED by insects, and consequently, from its pecul
0092 which excreted most nectar, would be oftenest VISITED by insects, and would be oftenest crossed; a
0092 ze and habits of the particular insects which VISITED them, so as to favour in any degree the tran
0095 ut not out of the common red clover, which is VISITED by humble bees alone; so that whole fields o
0098 to receive them; and as this flower is never VISITED, at least in my garden, by insects, it never
0098 species of Lobelia growing close by, which is VISITED by bees, seeds freely. In very many other ca
0222 eart and carried off the pupae. One evening I VISITED another community of F. sanguinea, and found
0092 ted. We might have taken the case of insects VISITING flowers for the sake of collecting pollen in
0095 fertility of clover greatly depends on bees VISITING and moving parts of the corolla, so as to pu
0073 ur orchidaceous plants absolutely recuire the VISITS of moths to remove their pollen masses and th
0073 nts which I have tried, I have found that the VISITS of bees, if not indispensable, are at least h
0097 len from another flower. So necessary are the VISITS of bees to papilionaceous flowers, that I hav
0097 heir fertility is greatly diminished if these VISITS be prevented. Now, it is scarcely possible th
0249 brids, generally grow in the same garden, the VISITS of insects must be carefully prevented during
0179 ts to its place in nature. Look at the Mustela VISON of North America, which has webbed feet and w
0415 atest stress on resemblances in organs of high VITAL or physiological importance. No doubt this vi
0416 atrophied organs are of high physiological or VITAL importance; yet, undoubtedly, organs in this
0418 roups of animals all these, the most important VITAL organs, are found to offer characters of quit
0456 rs, if constant and prevalent, whether of high VITAL importance, or of the most trifling importanc
0479 ost permanent characters, however slight their VITAL importance may be. The framework of bones bei
0361 he crops of floating birds long retain their VITALITY: peas and vetches, for instance, are killed
0364 eat distances; for seeds do not retain their VITALITY when exposed for a great length of time to t
0384 rely dropped by whirlwinds in India, and the VITALITY of their ova when removed from the water. Ru
0394 nowhere else in the world: Norfolk Island, the VITI Archipelago, the Bonin Islands, the Caroline a
0008 ive country! This is generally attributed to VITIATED instincts; but how many cultivated plants di
```

Page **(Key Word)**

Page ***(Key Word)***

```
0188  ir lower ends there seems to be an imperfect VITREOUS substance. With these facts, here far too br
0366  same belief, had not Agassiz and others called VIVID attention to the Glacial Period, which, as we
0144  ich are most abnormal in their dermal covering, VIZ. Cetacea (whales) and Edentata (armadilloes, s
0018  ommunicated to me by Mr. Blyth, on the habits, VOICE, and constitution, etc., of the humped Indian
0022  differs remarkably: as does in some breeds the VOICE and disposition. Lastly, in certain breeds, t
0024  gh agreeing generally in constitution, habits, VOICE, colouring, and in most parts of their struct
0184  ven in its colouring, in the harsh tone of its VOICE, and undulatory flight, told me plainly of it
0398  acter, in their habits, gestures, and tones of VOICE, was manifest. So it is with the other animal
0284  ith the evidence of denudation, when viewing VOLCANIC islands, which have been worn by the waves a
0355  nother region, whence it has been stocked. A VOLCANIC island, for instance, upheaved and formed at
0358  ontinents. Nor does their almost universally VOLCANIC composition favour the admission that they a
0358  ocks, instead of consisting of mere piles of VOLCANIC matter. I must now say a few words on what a
0375  india, on the heights of Ceylon, and on the VOLCANIC cones of Java, many plants occur, either ide
0394  been time for the creation of mammals; many VOLCANIC islands are sufficiently ancient, as shown b
0398  uralist, looking at the inhabitants of these VOLCANIC islands in the Pacific, distant several hund
0398  a considerable degree of resemblance in the VOLCANIC nature of the soil, in climate, height, and
0003  external conditions, or of habit, or of the VOLITION of the plant itself. The author of the Vesti
0242  lay. For no amount of exercise, or habit, or VOLITION, in the utterly sterile members of a communi
0002  ean Society, and it is published in the third VOLUME of the Journal of that Society. Sir C. Lyell
0002  scarcely a single point is discussed in this VOLUME on which facts cannot be adduced, often appar
0279  ly urged against the views maintained in this VOLUME. Most of them have now been discussed. One, n
0282  past periods of time, may at once close this VOLUME. Not that it suffices to study the principles
0309  urian period. The coloured map appended to my VOLUME on Coral Reefs, led me to conclude that the g
0310  ts and arguments of other kinds given in this VOLUME, will undoubtedly at once reject my theory. F
0310  dialect: of this history we possess the last VOLUME alone, relating only to two or three countrie
0310  ating only to two or three countries. Of this VOLUME, only here and there a short chapter has been
0333  various degrees. As we possess only the last VOLUME of the geological record, and that in a very
0368  rdance with the principles inculcated in this VOLUME, they will not have been liable to much modif
0459  al history. Concluding remarks. As this whole VOLUME is one long argument, it may be convenient to
0481  inced of the truth of the views given in this VOLUME under the form of an abstract, I by no means
0482  ability of species, may be influenced by this VOLUME; but I look with confidence to the future, to
0484  breathed. When the views entertained in this VOLUME on the origin of species, or when analogous v
0012  s. dr. Prosper Lucas's treatise, in two large VOLUMES, is the fullest and the best on this subject
0210  h which I am acquainted, is that of aphides VOLUNTARILY yielding their sweet excretion to ants: th
0210  ir sweet excretion to ants: that they do so VOLUNTARILY, the following facts show. I removed all t
0385  at a somewhat more advanced age they would VOLUNTARILY drop off. These just hatched molluscs, tho
0393  ar case. I have carefully searched the oldest VOYAGES, but have not finished my search; as yet I h
0197  ng a climber. The naked skin on the head of a VULTURE is generally looked at as a direct adaptatio
0386  some quantity to the feet and beaks of birds. WADING birds, which frequent the muddy edges of pond
0391  seaweed or floating timber, or to the feet of WADING birds, might be transported far more easily t
0061  hrough the water; in the plumed seed which is WAFTED by the gentlest breeze; in short, we see beau
0147  seeds which were a little better fitted to be WAFTED further, might get an advantage over those pr
0203  f pollen, in order that a few granules may be WAFTED by a chance breeze on to the ovules? Summary
0366  plainly, than do the mountains of Scotland and WALES, with their scored flanks, polished surfaces,
0011  ed to the domestic duck flying much less, and WALKING more, than its wild parent. The great and in
0185  t the long toes of grallatores are formed for WALKING over swamps and floating plants, yet the wat
0200  flipper, but a foot with five toes fitted for WALKING or grasping; and we may further venture to b
0031  t would seem as if they had chalked out upon a WALL a form perfect in itself, and then had given i
0230  f the bees were to build for themselves a thin WALL of wax, they could make their cells of the pro
0230  growing comb, do make a rough, circumferential WALL or rim all round the comb; and they gnaw into
0230  ll is excavated out of a little parallel sided WALL of wax, is not, as far as I have seen, strictl
0230  suppose that the bees cannot build up a rough WALL of wax in the proper position, that is, along
0231  this. Even in the rude circumferential rim or WALL of wax round a growing comb, flexures may some
0231  ic basal plates of future cells. But the rough WALL of wax has in every case to be finished off, b
0231  d is curious: they always make the first rough WALL from ten to twenty times thicker than the exce
0231  mes thicker than the excessively thin finished WALL of the cell, which will ultimately be left. We
0231  ides near the ground, till a smooth, very thin WALL is left in the middle; the masons always pilin
0231  summit of the ridge. We shall thus have a thin WALL steadily growing upward; but always crowned by
0232  s case the bees can lay the foundations of one WALL of a new hexagon, in its strictly proper place
0232  triking imaginary spheres, they can build up a WALL intermediate between two adjoining spheres; bu
0232  aying down under certain circumstances a rough WALL in its proper place between two just commenced
0234  r together, so as to intersect a little; for a WALL in common even to two adjoining cells, would s
0001  re especially been induced to do this, as Mr. WALLACE, who is now studying the natural history of
0002  by thinking it advisable to publish, with Mr. WALLACE's excellent memoir, some brief extracts from
0355  lately advanced in an ingenious paper by Mr. WALLACE, in which he concludes, that every species h
0395  o by the admirable zeal and researches of Mr. WALLACE. I have not as yet had time to follow up thi
0197  nerally looked at as a direct adaptation for WALLOWING in putridity; and so it may be, or it may p
0226  er permitted, the bees building perfectly flat WALLS of wax between the spheres which thus tend to
0226  s wax by this manner of building; for the flat WALLS between the adjoining cells are not double, b
0228  ceased to excavate, and began to build up flat WALLS of wax on the lines of intersection between t
0230  ges of the rhombic plates, until the hexagonal WALLS are commenced. Some of these statements diffe
0231  e comb without injuring the delicate hexagonal WALLS, which are only about one four hundredth of a
0231  s fact, by covering the edges of the hexagonal WALLS of a single cell, or the extreme margin of th
0232  elative distances from each other and from the WALLS of the last completed cells, and then, by str
0031  to change it altogether. It is the magician's WAND, by means of which he may summon into life wha
0103  ose animals which unite for each birth, which WANDER much, and which do not breed at a very quick
0103  animals which unite for each birth, but which WANDER little, and which can increase at a very rapid
0178  st the classes which unite for each birth and WANDER much, may have separately been rendered suffi
0282  ents of which we see around us. It is good to WANDER along lines of sea coast, when formed of mode
0364  and; but seeds could be transported by these WANDERERS only by one means, namely, in dirt sticking
0379  will have profited them. Thus many of these WANDERERS, though still plainly related by inheritanc
0386  ds of this order I can show are the greatest WANDERERS, and are occasionally found on the most rem
0042  respect enclosure of the land plays a part. WANDERING savages or the inhabitants of open plains r
0174  es, more especially with freely crossing and WANDERING animals. In looking at species as they are
0394  but bats can fly across. Bats have been seen WANDERING by day far over the Atlantic Ocean; and two
0394  nds. Hence we have only to suppose that such WANDERING species have been modified through natural
```

Page ***(Key Word)***

Page **(Key Word)**

```
0002  nnot possibly be here done. I much regret that WANT  of space prevents my having the satisfaction o
0056  maxima, declined, and disappeared. All that we WANT  to show is, that where many species of a genus
0121  he dotted lines not prolonged far upwards from WANT  of space. But during the process of modificati
0211  s interval, I felt sure that the aphides would WANT  to excrete. I watched them for some time throu
0211  as possible ought to have been here given; but WANT  of space prevents me. I can only assert, that
0472  y of natural selection, that more cases of the WANT  of absolute perfection have not been observed.
0397  ed and crawled away: but more experiments are WANTED on this head. The most striking and important
0032  ucceed, and may make great improvements; if he WANTS  any of these qualities, he will assuredly fai
0038  in their structure or in their habits to man's WANTS  or fancies. We can, I think, further understa
0448  y generations, having to provide for their own WANTS  at a very early stage of development, and sec
0457  life when the being has to provide for its own WANTS, and bearing in mind how strong is the princi
0075  ally scattering its seeds by the thousand; what WAR  between insect and insect, between insects, sn
0079  onsole ourselves with the full belief, that the WAR  of nature is not incessant, that no fear is fe
0088  ellowing, and whirling round, like Indians in a WAR  dance, for the possession of the females; male
0088  nds from the huge mandibles of other males. The WAR  is, perhaps, severest between males of polygam
0490  tinction of less improved forms. Thus, from the WAR  of nature, from famine and death, the most exa
0133  nstance, was created with bright colours for a WARM  sea; but that this other shell became bright c
0230  pposed cells and push and bend the ductile and WARM  wax (which as I have tried is easily done) int
0368  s the low intervening tracts, since become too WARM  for their existence. If the climate, since the
0370  migrate southwards as the climate became less WARM, long before the commencement of the Glacial p
0133  ght coloured by variation when it ranged into WARMER or shallower waters. When a variation is of t
0140  the number of plants and animals brought from WARMER countries which here enjoy good health. We ha
0140  imes having largely extended their range from WARMER to cooler latitudes, and conversely; but we d
0368  e glacial period, has ever been in any degree WARMER than at present (as some geologists in the Un
0368  ce with respect to this intercalated slightly WARMER period, since the Glacial period. The arctic
0370  specifically the same as now, the climate was WARMER than at the present day. Hence we may suppose
0370  d to infer that during some earlier and still WARMER period, such as the older Pliocene period, a
0371  hey are at the present time: for during these WARMER periods the northern parts of the Old and New
0374  he cordillera under the equator and under the WARMER temperate zones, and on both sides of the sou
0376  y of the forms living on the mountains of the WARMER regions of the earth and in the southern hemi
0377  inroads of competitors, can withstand a much WARMER climate than their own. Hence, it seems to me
0381  s in the northern hemisphere, to a former and WARMER period, before the commencement of the Glacia
0078  the individuals of the same species, for the WARMEST or dampest spots. Hence, also, we can see th
0133  how much of this difference may be due to the WARMEST clad individuals having been favoured and pr
0377  erature, more especially by escaping into the WARMEST spots. But the great fact to bear in mind is
0017  inea fowl, or the small power of endurance of WARMTH by the rein deer, or of cold by the common ca
0367  ake no difference in the final result. As the WARMTH returned, the arctic forms would retreat nort
0367  d, always ascending higher and higher, as the WARMTH increased, whilst their brethren were pursuin
0367  suine their northern journey. Hence, when the WARMTH had fully returned, the same arctic species,
0367  ame on, and the re migration on the returning WARMTH, will generally have been due south and north
0368  eft isolated from the moment of the returning WARMTH, first at the bases and ultimately on the sum
0371  eir inhabitants. During the slowly decreasing WARMTH of the Pliocene period, as soon as the specie
0378  ns, and some even crossed the equator. As the WARMTH returned, these temperate forms would natural
0201  d with a rattle for its own injury, namely, to WARN  its prey to escape. I would almost as soon bel
0201  its tail when preparing to spring, in order to WARN  the doomed mouse. But I have not space here to
0085  o, that on parts of the Continent persons are WARNED not to keep white pigeons, as being the most
0212  ot more fearful than small; and the magpie, so WARY  in England, is tame in Norway, as is the hoode
0359  imberi and it occurred to me that floods might WASH  down plants or branches, and that these might
0192  d as organs for preventing the ova from being WASHED out of the sack? Although we must be extremel
0283  elsewhere years have elapsed since the waters WASHED their base. He who most closely studies the a
0359  ks, and then by a fresh rise in the stream be WASHED into the sea. Hence I was led to dry stems in
0361  em, so perfectly that not a particle could be WASHED away in the longest transport: out of one sma
0386  ace of the sea, so that the dirt would not be WASHED off their feet; when making land, they would
0202  less perfect. Can we consider the sting of the WASP  or of the bee as perfect, which, when used aga
0218  n young. Some species, likewise, of Sphegidae (WASP  like insects) are parasitic on other species;
0233  amely, that the cells on the extreme margin of WASP  combs are sometimes strictly hexagonal; but I
0233  in a single insect (as in the case of a queen WASP) making hexagonal cells; if she work alternate
0472  d by their sterile sisters; at the astonishing WASTE  of pollen by our fir trees; at the instinctiv
0148  ofit the individual not to have its nutriment WASTED in building up a useless structure. I can th
0148  of supporting itself, by less nutriment being WASTED in developing a structure now become useless.
0218  ting I picked up no less than twenty lost and WASTED eggs. Many bees are parasitic, and always lay
0235  n economy of wax; that individual swarm which WASTED least honey in the secretion of wax; having s
0282  imself great piles of superimposed strata; and WATCH  the sea at work grinding down old rocks and m
0027  eat of selection. Fourthly, pigeons have been WATCHED, and tended with the utmost care, and loved
0028  length; because when I first kept pigeons and WATCHED the several kinds, knowing well how true the
0183  erable instances could be given: I have often WATCHED a tyrant flycatcher (Saurophagus sulphuratus
0211  ure that the aphides would want to excrete. I WATCHED them for some time through a lens, but not o
0220  e and July, on three successive years, I have WATCHED for many hours several nests in Surrey and S
0220  umerous; but Mr. Smith informs me that he has WATCHED the nests at various hours during May, June
0291  hing my views on this subject in 1845, I have WATCHED the progress of Geology, and have been surpr
0189  uppose that there is a power always intently WATCHING each slight accidental alteration in the tra
0238  g horns, could be slowly formed by carefully WATCHING which individual bulls and cows, when matche
0008  rifling chances, such as a little more or less WATER  at some particular period of growth, will det
0054  allude to only two causes of obscurity. Fresh WATER  and salt loving plants have generally very wi
0061  tructure of the beetle which dives through the WATER; in the plumed seed which is wafted by the ge
0077  , and in the flattened and fringed legs of the WATER  beetle, the relation seems at first confined
0077  s at first confined to the elements of air and WATER. Yet the advantage of plumec seeds no doubt s
0077  tributed and fall on unoccupied ground. In the WATER  beetle, the structure of its legs, so well ad
0100  ising hermaphrodites; there are currents in the WATER  offer an obvious means for an occasional cros
0107  he extinct tertiary flora of Europe. All fresh WATER  basins, taken together, make a small area com
0107  d, consequently, the competition between fresh WATER  productions will have been less severe than e
0107  s more slowly exterminated. And it is in fresh WATER  that we find seven genera of Ganoid fishes, r
0107  nts of a once proponderant order: and in fresh WATER  we find some of the most anomalous forms now
0113  tinc new stations, climbing trees, frequenting WATER, and some perhaps becoming less carnivorous.
0114  uniform islets; and so in small ponds of fresh WATER. Farmers find that they can raise most food b
0132  eir southern limit, and when living in shallow WATER, are more brightly coloured than those of the
```

Page **(Key Word)**

Page **(Key Word)**

0134	america can only flap along the surface of the WATER, and has its wings in nearly the same conditi
0182	ed duck (Micropterus of Eyton); as fins in the WATER and front legs on the land, like the penguin;
0182	re possible. Seeing that a few members of such WATER breathing classes as the Crustacea and Mollus
0183	any kinds in many ways, on the land and in the WATER, until their organs of flight had come to a h
0183	her times standing stationary on the margin of WATER, and then dashing like a kingfisher at a fish
0184	h, thus catching, like a whale, insects in the WATER. Even in so extreme a case as this, if the su
0185	est observer by examining the dead body of the WATER ouzel would never have suspected its sub aqua
0185	stones with its feet and using its wings under WATER. He who believes that each being has been cre
0185	webbed feet which rarely or never go near the WATER: and no one except Audubon has seen the friga
0185	lking over swamps and floating plants, yet the WATER hen is nearly as aquatic as the coot: and the
0186	n the dry land or most rarely alighting on the WATER; that there should be long toed corncrakes li
0190	ranchiae that breathe the air dissolved in the WATER, at the same time that they breathe free air
0283	bles; for there is reason to believe that pure WATER can effect little of nothing in wearing away
0287	ree hundred million years. The action of fresh WATER on the gently inclined Wealden district, when
0287	years, over the whole world, the land and the WATER has been peopled by hosts of living forms. Wh
0288	tion of the ocean, the bright blue tint of the WATER bespeaks its purity. The many cases on record
0288	erally be dissolved by the percolation of rain WATER and has been found fossil in Sicily, whereas
0288	gle Mediterranean species, which inhabits deep WATER near the mouth of the Mississippi during some
0294	e. when such beds as were deposited in shallow WATER, might have undergone far more metamorphic ac
0309	sed on by an enormous weight of superincumbent WATER change at distant points in the same parallel
0323	ether the productions of the land and of fresh WATER shells of the Aralo Caspian Sea. Now what doe
0339	a; and between the extinct and living brackish WATER, and independent of their physical conditions
0350	pace and time, over the same areas of land and WATER productions; and thirdly, the occurrence of t
0354	owing chapter), the wide distribution of fresh WATER. To my surprise I found that out of 87 kinds,
0358	seeds could resist the injurious action of sea WATER. Afterwards I tried some larger fruits, capsu
0359	, whether or not they were injured by the salt WATER. The majority sank quickly, but some which wh
0359	ants with ripe fruit, and to place them on sea WATER. On the other hand he did not previously dry
0360	resistance to the injurious action of the salt WATER; but some taken out of the crop of a pigeon,
0361	are killed by even a few days immersion in sea WATER for 30 days, to my surprise nearly all germin
0361	a pigeon, which had floated on artificial salt WATER fish, I find, eat seeds of many land and wate
0362	etained for two days and fourteen hours. Fresh WATER plants: fish are frequently devoured by birds
0362	water fish, I find, eat seeds of many land and WATER; nor could they be long carried in the crops
0364	or a great length of time to the action of sea WATER, they could not endure our climate. Almost ev
0364	if not killed by so long an immersion in salt WATER productions. On the inhabitants of oceanic is
0383	distribution, continued. Distribution of fresh WATER productions would not have ranged widely with
0383	of land, it might have been thought that fresh WATER species, belonging to quite different classes
0383	exactly the reverse. Not only have many fresh WATER insects, shells, etc., and at the dissimilari
0383	g much surprise at the similarity of the fresh WATER productions of ranging widely, though so unex
0383	with those of Britain. But this power in fresh WATER. But I am inclined to attribute the dispersal
0384	he vitality of their ova when removed from the WATER fish mainly to slight changes within the rece
0384	m inclined to attribute the dispersal of fresh WATER fish. The wide difference of the fish on op
0384	surface was peopled by existing land and fresh WATER shells. The wide difference of the fish on op
0384	same conclusion. With respect to allied fresh WATER fish occurring at very distant points of the
0384	cannot at present be explained: but some fresh WATER fish belong to very ancient forms, and in suc
0384	for much migration. In the second place, salt WATER fish can with care be slowly accustomed to li
0384	ith care be slowly accustomed to live in fresh WATER; and, according to Valenciennes, there is har
0384	group of fishes confined exclusively to fresh WATER, so that we may imagine that a marine member
0384	we may imagine that a marine member of a fresh WATER group might travel far along the shores of th
0385	aters of a distant land. Some species of fresh WATER shells have a very wide range, and allied spe
0385	birds, and they are immediately killed by sea WATER, as are the adults. I could not even understa
0385	ite unintentionally stocked the one with fresh WATER shells from the other. But another agency is
0385	pond, in an aquarium, where many ova of fresh WATER shells were hatching; and I found that number
0385	g to them so firmly that when taken out of the WATER they could not be jarred off, though at a som
0386	ticus has been caught with an Ancylus (a fresh WATER shell like a limpet) firmly adhering to it; a
0386	ll like a limpet) firmly adhering to it; and a WATER beetle of the same family, a Colymbetes, once
0386	ong been known what enormous ranges many fresh WATER and even marsh species have, both over contin
0386	ey would be sure to fly to their natural fresh WATER haunts. I do not believe that botanists are a
0386	ns of mud from three different points, beneath WATER, on the edge of a little pond: this mud when
0387	nk it would be an inexplicable circumstance if WATER birds did not transport the seeds of fresh wa
0387	ter birds did not transport the seeds of fresh WATER plants to vast distances, and if consequently
0387	lay with the eggs of some of the smaller fresh WATER animals. Other and unknown agencies probably
0387	e also played a part. I have stated that fresh WATER fish eat some kinds of seeds, though they rej
0387	allow seeds of moderate size, as of the yellow WATER lily and Potamogeton. Herons and other birds,
0387	i saw the great size of the seeds of that fine WATER lily the Nelumbium, and remembered Alph. de C
0387	that he found the seeds of the great southern WATER lily (probably, according to Dr. Hooker, the
0388	also, remember that some, perhaps many, fresh WATER productions are low in the scale of nature, a
0388	aving formerly ranged as continuously as fresh WATER productions ever can range, over immense area
0388	te regions. But the wide distribution of fresh WATER plants and of the lower animals, whether reta
0388	and eggs by animals, more especially by fresh WATER birds, which have large powers of flight, and
0388	from one to another and often distant piece of WATER. Nature, like a careful gardener, thus takes
0393	pawn are known to be immediately killed by sea WATER, on my view we can see that there would be gr
0397	gs, at least such as I have tried, sink in sea WATER and are killed by it. Yet there must be, on m
0397	state withstand uninjured an immersion in sea WATER during seven days: one of these shells was th
0397	after it had again hybernated I put it in sea WATER for twenty days, and it perfectly recovered.
0397	us one, I immersed it for fourteen days in sea WATER, and it recovered and crawled away: but more
0398	ica. Here almost every product of the land and WATER bears the unmistakeable stamp of the American
0404	terflies and beetles. So it is with most fresh WATER productions, in which so many genera range ov
0407	little length the means of dispersal of fresh WATER productions. If the difficulties be not insup
0408	igh importance of barriers, whether of land or WATER, which separate our several zoological and bo
0411	ly fitted to inhabit the land, and another the WATER; one to feed on flesh; another on vegetable m
0428	tions in both classes for swimming through the WATER; but the shape of the body and fin like limbs
0428	instance the three elements of land, air, and WATER, we can perhaps understand how it is that a n
0440	ed in a nest, and in the spawn of a frog under WATER. We have no more reason to believe in such a
0479	fish which has to breathe the air dissolved in WATER, by the aid of well developed branchiae. Disu
0116	each other, and feebly representing, as Mr. WATERHOUSE and others have remarked, our carnivorous,
0150	early to the above effect, published by Mr. WATERHOUSE. I infer also from an observation made by P

Page **(Key Word)**

Page **************************************(Key Word)**************************************

```
0151  re add, that I particularly attended to Mr. WATERHOUSE's remark, whilst investigating this Order,
0225  was led to investigate this subject by Mr. WATERHOUSE, who has shown that the form of the cell st
0429  oincidence of favourable circumstances. Mr. WATERHOUSE has remarked that, when a member belonging
0430  ral and not special; thus, according to Mr. WATERHOUSE, of all Rodents, the bizcacha is most nearl
0430  n the other hand, of all Marsupials, as Mr. WATERHOUSE has remarked, the phascolomys resembles mos
0288  which live on the beach between high and low WATERMARK are preserved. For instance, the several sp
0044  e dwarfed condition of shells in the brackish WATERS of the Baltic, or dwarfed plants on Alpine su
0133  ation when it ranged into warmer or shallower WATERS. When a variation is of the slightest use to
0180  t during the long winter it leaves the frozen WATERS, and preys like other polecats on mice and La
0283  w that elsewhere years have elapsed since the WATERS washed their base. He who most closely studie
0312  ne after another, both on the land and in the WATERS. Lyell has shown that it is hardly possible t
0321  roup of Ganoid fishes still inhabit our fresh WATERS. Therefore the utter extinction of a group is
0326  ies on the land, and another for those in the WATERS of the sea. If two great regions had been for
0349  an alpine species of bizcacha; we look to the WATERS, and we do not find the beaver or musk rat, b
0372  cifically distinct. As on the land, so in the WATERS of the sea, a slow southern migration of a ma
0382  ed forms of life can be explained. The living WATERS may be said to have flowed during one short p
0382  re the tide rises highest, so have the living WATERS left their living drift on our mountain summi
0383  remember, when first collecting in the fresh WATERS of Brazil, feeling much surprise at the simil
0384  hat the same species never occur in the fresh WATERS of distant continents. But on the same contin
0385  ntly become modified and adapted to the fresh WATERS of a distant land. Some species of fresh wate
0387  g fish; they then take flight and go to other WATERS, or are blown across the sea; and we have see
0388  al species; consequently an intruder from the WATERS of a foreign country, would have a better cha
0048  , and by another as mere varieties. Mr. H. C. WATSON, to whom I lie under deep obligation for assi
0053  irst this seemed a simple task; but Mr. H. C. WATSON, to whom I am much indebted for valuable advi
0058  ch restricted ranges. For instance, Mr. H. C. WATSON has marked for me in the well sifted London C
0058  rage over 6.5 of the provinces into which Mr. WATSON has divided Great Britain. Now, in this same
0058  ry closely allied forms, marked for me by Mr. WATSON as doubtful species, but which are almost uni
0140  gous observations have been made by Mr. H. C. WATSON on European species of plants brought from th
0176  would appear from information given me by Mr. WATSON, Dr. Asa Gray, and Mr. Wollaston, that genera
0363  o the mainland, and (as remarked by Mr. H. C. WATSON) from the somewhat northern character of the
0367  xample, of Scotland, as remarked by Mr. H. C. WATSON, and those of the Pyrenees, as remarked by Ra
0376  to the northern temperate zones. As Mr. H. C. WATSON has recently remarked, In receding from polar
0027  y distinctive of each breed, for instance the WATTLE and length of beak of the carrier, the shortn
0090  the male sex in our domestic animals (as the WATTLE in male carriers, horn like protuberances in
0152  ak of the different tumblers, in the beak and WATTLE of the different carriers, in the carriage an
0153  other, generally to the male sex, as with the WATTLE of carriers and the enlarged crop of pouters.
0147  inished comb, and a large beard by diminished WATTLES. With species in a state of nature it can ha
0095  avation of a great valley by a single diluvial WAVE, so will natural selection, if it be a true pr
0095  om hear the action, for instance, of the coast WAVES, called a trifling and insignificant cause, w
0283  ffs only for a short time twice a day, and the WAVES eat into them only when they are charged with
0283  educed in size they can be rolled about by the WAVES, and then are more quickly ground into pebble
0285  volcanic islands, which have been worn by the WAVES and pared all round into perpendicular cliffs
0287  likewise, have escaped the action of the coast WAVES. So that in all probability a far longer peri
0290  e land within the grinding action of the coast WAVES. We may, I think, safely conclude that sedime
0290  order to withstand the incessant action of the WAVES, when first upraised and during subsequent os
0360  but I do not doubt that plants exposed to the WAVES would float for a less time than those protec
0481  eys excavated, by the slow action of the coast WAVES. The mind cannot possibly grasp the full mean
0224  with the least possible consumption of precious WAX in their construction. It has been remarked th
0224  , would find it very difficult to make cells of WAX of the true form, though this is perfectly eff
0225  honey, sometimes adding to them short tubes of WAX, and likewise making separate and very irregul
0225  ng separate and very irregular rounded cells of WAX. At the other end of the series we have the ce
0225  hatched; and, in addition, some large cells of WAX for holding honey. These latter cells are near
0226  tted, the bees building perfectly flat walls of WAX between the spheres which thus tend to interse
0226  ng cells. It is obvious that the Melipona saves WAX by this manner of building; for the flat walls
0227  me way as the rude humble bee adds cylinders of WAX to the circular mouths of her old cocoons. By
0228  put between them a long thick, square strip of WAX: the bees instantly began to excavate minute c
0228  o excavate, and began to build up flat walls of WAX on the lines of intersection between the basin
0228  o the hive, instead of a thick, square piece of WAX, a thin and narrow, knife edged ridge, coloure
0228  er, in the same way as before; but the ridge of WAX was so thin, that the bottoms of the basins, i
0229  , formed by their little plates of the vermilion WAX having been left ungnawed, were situated, as f
0229  he basins on the opposite sides of the ridge of WAX. In parts, only little bits, in other parts, l
0229  on the opposite sides of the ridge of vermilion WAX, as they circularly gnawed away and deepened t
0229  of intersection. Considering how flexible thin WAX is, I do not see that there is any difficulty
0229  , whilst at work on the two sides of a strip of WAX, perceiving when they have gnawed the wax away
0229  ip of wax, perceiving when they have gnawed the WAX away to the proper thinness, and then stopping
0230  ed cells and push and bend the ductile and warm WAX (which as I have tried is easily done) into it
0230  . from the experiment of the ridge of vermilion WAX, we can clearly see that if the bees were to b
0230  ees were to build for themselves a thin wall of WAX, they could make their cells of the proper sha
0230  xcavated out of a little parallel sided wall of WAX, is not, as far as I have seen, strictly corre
0230  ommencement having always been a little hood of WAX; but I will not here enter on these details. W
0230  e that the bees cannot build up a rough wall of WAX in the proper position, that is, along the pla
0231  even in the rude circumferential rim or wall of WAX around a growing comb, flexures may sometimes b
0231  l plates of future cells. But the rough wall of WAX has in every case to be finished off, by being
0231  leted, being thus crowned by a strong coping of WAX, the bees can cluster and crawl over the comb
0231  o the comb, with the utmost ultimate economy of WAX. It seems at first to add to the difficulty of
0232  ith an extremely thin layer of melted vermilion WAX; and I invariably found that the colour was mo
0232  e done with his brush, by atoms of the coloured WAX having been taken from the spot on which it ha
0233  hive of bees for the secretion of each pound of WAX: so that a prodigious quantity of fluid nectar
0234  by the bees in a hive for the secretion of the WAX necessary for the construction of their combs.
0234  er of bees being supported. Hence the saving of WAX by largely saving honey must be a most importa
0234  to two adjoining cells, would save some little WAX. Hence it would continually be more and more a
0234  cell would serve to bound other cells, and much WAX would be saved. Again, from the same cause, it
0235  e can see, is absolutely perfect in economising WAX. Thus, as I believe, the most wonderful of all
0235  double layer, and to build up and excavate the WAX along the planes of intersection. The bees, of
0235  ess of natural selection having been economy of WAX: that individual swarm which wasted least hone
0235  rm which wasted least honey in the secretion of WAX, having succeeded best, and having transmitted
```

Page **************************************(Key Word)**************************************

Page **(Key Word)**

```
0225  lated to the latter: it forms a nearly regular WAXEN comb of cylindrical cells, in which the young
0234  dification of her instinct led her to make her WAXEN cells near together, so as to intersect a lit
0003  st fifteen years has aided me in every possible WAY by his large stores of knowledge and his excel
0010  to certain conditions are affected in the same WAY, the change at first appears to be directly du
0013  admitted to be inheritable. Perhaps the correct WAY of viewing the whole subject, would be, to loo
0035  would improve and modify any breed, in the same WAY as Bakewell, Collins, etc., by this very same
0057  alter genera. Or the case may be put in another WAY, and it may be said, that in the larger genera
0066  cases is an early one. If an animal can in any WAY protect its own eggs or young, a small number
0078  ome, we should have to modify it in a different WAY to what we should have done in its native coun
0080  occurred, that other variations useful in some WAY to each being in the great and complex battle
0082  urse of ages chanced to arise, and which in any WAY favoured the individuals of any of the species
0109  through the preservation of variations in some WAY advantageous, which consequently endure. But a
0114  could live on it (supposing it not to be in any WAY peculiar in its nature), and may be said to be
0117  eriods. Only those variations which are in some WAY profitable will be preserved or naturally sele
0120  ved, excepting that from a1 to a10. In the same WAY, for instance, the English race horse and Engl
0127  useful to each being's own welfare, in the same WAY as so many variations have occurred useful to
0131  elieve that deviations of structure are in some WAY due to the nature of the conditions of life, t
0137  gine that eyes, though useless, could be in any WAY injurious to animals living in darkness, I att
0146  rs, it seems impossible that they can be in any WAY advantageous to the plant: yet in the Umbellif
0147  , here give any instances, for I see hardly any WAY of distinguishing between the effects, on the
0168  erve for one special purpose alone. In the same WAY that a knife which has to cut all sorts of thi
0174  ects. Homologous parts tend to vary in the same WAY, and homologous parts tend to cohere. Modifica
0189  tion than at present. But I will pass over this WAY of escaping from the difficulty: for I believe
0193  on which under varied circumstances, may in any WAY, or in any degree, tend to produce a distincte
0205  am inclined to believe that in nearly the same WAY as two men have sometimes independently hit on
0207  y due to the laws of growth, and at first in no WAY advantageous to a species, have been subsequen
0211  when performed by many individuals in the same WAY, without their knowing for what purpose it is
0214  t them, and it immediately seemed, by its eager WAY of running about, to be well aware what a rich
0215  howed a trace of its wild parentage only in one WAY, by not coming in a straight line to his maste
0227  was originally instinctive in them, in the same WAY as it is so plainly instinctive in young pheas
0228  uisite to hold the stock of honey; in the same WAY as the rude humble bee adds cylinders of wax t
0234  e little basins near to each other; in the same WAY as before: but the ridge of wax was so thin, t
0269  ells closer together, and more regular in every WAY than at present: for then, as we have seen, th
0273  riods of time on the whole organisation, in any WAY which may be for each creature's own good: and
0337  have not had their reproductive systems in any WAY affected, and they are not variable: but hybri
0380  the ancient and beaten forms: but I can see no WAY of testing this sort of progress. Crustaceans,
0387  the larger areas of the north, just in the same WAY as the productions of real islands have everyw
0404  the bird whilst feeding its young, in the same WAY as fish are known sometimes to be dropped. In
0435  ied to it, is shown in another and more general WAY. Mr. Gould remarked to me long ago, that in th
0443  ons, each modification being profitable in some WAY to the modified form, but often affecting by c
0468  ad, indeed the evidence rather points the other WAY: for it is notorious that breeders of cattle,
0469  nos would have varied under nature, in the same WAY as they generally have varied under the change
0486  ect, why should we doubt that variations in any WAY useful to beings, under their excessively comp
0030  ach useful to the possessor, nearly in the same WAY as when we look at any great mechanical invent
0183  s of dogs, each good for man in very different WAYS: when we compare the game cock, so pertinaciou
0232  e forms, for taking prey of many kinds in many WAYS, on the land and in the water, until their org
0255  ld entirely pull down and rebuild in different WAYS the same cell, sometimes recurring to a shape
0290  ertility. It is surprising in how many curious WAYS this gradation can be shown to exist: but only
0327  accumulations of sediment may be formed in two WAYS: either, in profound depths of the sea, in whi
0431  the world: and the succession of forms in both WAYS will everywhere tend to correspond. There is o
0009  ed some of its characters, modified in various WAYS and degrees, to all: and the several species w
0087  domesticated animals and plants, though often WEAK and sickly, yet breeding quite freely under co
0089  most powerful and hardest beaks, for all with WEAK beaks would inevitably perish: or, more delica
0160  ish to attribute any effect to such apparently WEAK means: I cannot here enter on the details nece
0168  tor possessed, the tendency, whether strong or WEAK, to reproduce the lost character might be, as
0076  ces, and use in strengthening, and disuse in WEAKENING and diminishing organs, seem to have been m
0211  d then mixed in due proportion, otherwise the WEAKER kinds will steadily decrease in numbers and d
0244  cts of others, as each takes advantage of the WEAKER bodily structure of others. So again, in some
0267  ultiply, vary, let the strongest live and the WEAKEST die. Chapter VIII. Hybridism. Distinction be
0285  the same conditions of life, always induces WEAKNESS and sterility in the progeny. Hence it seems
0285  e, the well known one of the denudation of the WEALD. Though it must be admitted that the denudati
0286  it must be admitted that the denudation of the WEALD has been a mere trifle, in comparison with th
0286  t dome of rocks which must have covered up the WEALD within so limited a period as since the latte
0286  suppose, a range of older rocks underlies the WEALD, on the flanks of which the overlying sedimen
0287  measure the time requisite to have denuded the WEALD. This, of course, cannot be done: but we may,
0287  rate, on the above data, the denudation of the WEALD must have required 306,662,400 years: or say
0088  action of fresh water on the gently inclined WEALDEN district, when upraised, could hardly have b
0088  not on general vigour, but on having special WEAPONS, confined to the male sex. A hornless stag o
0090  and these seem oftenest provided with special WEAPONS. The males of carnivorous animals are alread
0242  e slight advantage over other males, in their WEAPONS, means of defence, or charms: and have trans
0468  inherited instincts and by inherited tools or WEAPONS, and not by acquired knowledge and manufactu
0288  t success will often depend on having special WEAPONS or means of defence, or on the charms of the
0308  rlying bed having suffered in the interval any WEAR and tear, seem explicable only on the view of
0283  n accumulated from sediment derived from their WEAR and tear: and would have been at least partial
0286  at pure water can effect little of nothing in WEARING away rock. At last the base of the cliff is
0093  en, we knew the rate at which the sea commonly WEARS away a line of cliff of any given height, we
0097  pollen could not thus have been carried. The WEATHER had been cold and boisterous, and therefore
0073  heir anthers and stigmas fully exposed to the WEATHER! but if an occasional cross be indispensable
0434  l never, probably, disentangle the inextricable WEB of affinities between the members of any one c
0179  the Mustela vison of North America, which has WEBBED feet and which resembles an otter in its fur,
0185  agreement. What can be plainer than that the WEBBED feet of ducks and geese are formed for swimmi
0185  for swimming? yet there are upland geese with WEBBED feet which rarely or never go near the water:
0185  the frigate bird, which has all its four toes WEBBED, alight on the surface of the sea. On the oth
0185  hout a corresponding change of structure. The WEBBED feet of the upland goose may be said to have
0186  there should be geese and frigate birds with WEBBED feet, either living on the dry land or most r
```

Page **(Key Word)**

webbed

Page ***(Key Word)**

```
0199  species. Thus, we can hardly believe that the WEBBED feet of the upland goose or of the frigate bi
0200  of the upland goose and of the frigate bird, WEBBED feet no doubt were as useful as they now are
0204  t has arisen that there are upland geese with WEBBED feet, ground woodpeckers, diving thrushes, an
0435  thick membrane, so as to serve as a fin; or a WEBBED foot might have all its bones, or certain bon
0471  or rarely swim should have been created with WEBBED feet; that a thrush should have been created
0067  iven inwards by incessant blows, sometimes one WEDGE being struck, and then another with greater f
0067  o a yielding surface, with ten thousand sharp WEDGES packed close together and driven inwards by i
0385  suddenly emerges from a pond covered with duck WEED, I have twice seen these little plants adherin
0385  has happened to me, in removing a little duck WEED from one aquarium to another, that I have quit
0067  ants, I marked all the seedlings of our native WEEDS as they came up, and out of the 357 no less t
0258  et has observed the same fact with certain sea WEEDS or Fuci. Gartner, moreover, found that this d
0011  n the domestic duck that the bones of the wing WEIGH less and the bones of the leg more, in propor
0485  ions between any two forms, we shall be led to WEIGH more carefully and to value higher the actual
0386  the edge of a little pond; this mud when dry WEIGHED only 6 3/4 ounces; I kept it covered up in m
0017  gle species, then such facts would have great WEIGHT in making us doubt about the immutability of
0024  ent. An argument, as it seems to me, of great WEIGHT, and applicable in several other cases, is, t
0035  n how the cattle of England have increased in WEIGHT and in early maturity, compared with the stoc
0052  justifiable must be judged of by the general WEIGHT of the several facts and views given througho
0113  ses, a greater number of plants and a greater WEIGHT of dry herbage can thus be raised. The same h
0134  ations as these incline me to lay very little WEIGHT on the direct action of the conditions of lif
0135  reased in successive generations the size and WEIGHT of its body, its legs were used more, and its
0142  for a similar purpose, and with much greater WEIGHT: but until some one will sow, during a score
0180  t i think such difficulties have very little WEIGHT. Here, as on other occasions, I lie under a h
0201  cannot find even one which seems to me of any WEIGHT. It is admitted that the rattlesnake has a po
0293  y no means pretend to assign due proportional WEIGHT to the following considerations. Although eac
0309  and which had been pressed on by an enormous WEIGHT of superincumbent water, might have undergone
0310  ny degree perfect, and who do not attach much WEIGHT to the facts and arguments of other kinds giv
0466  too heavily during many years to doubt their WEIGHT. But it deserves especial notice that the mor
0482  ne whose disposition leads him to attach more WEIGHT to unexplained difficulties than to the expla
0483  in force. But some arguments of the greatest WEIGHT extend very far. All the members of whole cla
0035  s for the Goodwood Races, are favoured in the WEIGHTS they carry. Lord Spencer and others have sho
0164  licitly trust, has examined for me a small dun WELCH pony with three short parellel stripes on eac
0165  s so distinct as the heavy Belgian cart horse, WELCH ponies, cobs, the lanky Kattywar race, etc.,
0165  short shoulder stripes, like those on the dun WELCH pony, and even had some zebra like stripes on
0006  st importance, for they determine the present WELFARE, and, as I believe, the future success and m
0127  ever had occurred useful to each being's own WELFARE, in the same way as so many variations have
0205  t any part or organ is so unimportant for the WELFARE of a species, that modifications in its stru
0209  e as important as corporeal structure for the WELFARE of each species, under its present condition
0261  on another is so entirely unimportant for its WELFARE in a state of nature, I presume that no one
0263  case of grafting it is unimportant for their WELFARE. Causes of the Sterility of first Crosses an
0415  sation, however important they may be for the WELFARE of the being in relation to the outer world.
0427  ter, although of the utmost importance to the WELFARE of the being, are almost valueless to the sy
0443  shape of the horns of either parent. For the WELFARE of a very young animal, as long as it remain
0164  arellel stripes on each shoulder. In the north WEST part of India the Kattywar breed of horses is
0240  om the same nest of the driver ant (Anomma) of WEST Africa. The reader will perhaps best appreciat
0285  , remembering that at no great distance to the WEST the northern and southern escarpments meet and
0290  ven a short geological period. Along the whole WEST coast, which is inhabited by a peculiar marine
0291  hat the only ancient tertiary formation on the WEST coast of South America, which has been bulky a
0364  though they might and do bring seeds from the WEST Indies to our western shores, where, if not ki
0378  he tropical natives. The mountain ranges north WEST of the Himalaya, and the long line of the Cord
0395  arated by similar channels from Australia. The WEST Indian Islands stand on a deeply submerged ban
0403  itions of life. Thus, the south east and south WEST corners of Australia have nearly the same phys
0286  greatly affect the estimate as applied to the WESTERN extremity of the district. If, then, we knew
0290  ill explain why along the rising coast of the WESTERN side of South America, no extensive formatio
0323  l successive palaeozoic formations of Russia, WESTERN Europe and North America, a similar parallel
0347  racts of land in Australia, South Africa, and WESTERN South America, between latitudes 25 degrees
0348  crab in common, than those of the eastern and WESTERN shores of South and Central America; yet the
0348  named three approximate faunas of Eastern and WESTERN America and the eastern Pacific Islands, yet
0364  nd do bring seeds from the West Indies to our WESTERN shores, where, if not killed by so long an i
0364  ole Atlantic Ocean, from North America to the WESTERN shores of Ireland and England; but seeds cou
0370  r circle there is almost continuous land from WESTERN Europe, through Siberia, to eastern America.
0372  tiary representative forms on the eastern and WESTERN shores of temperate North America; and the s
0373  lainest evidence of the cold period, from the WESTERN whores of Britain to the Oural range, and so
0374  al action was simultaneous on the eastern and WESTERN sides of North America, in the Cordillera un
0393  er, icebergs formerly brought boulders to its WESTERN shores, and they may have formerly transport
0399  ooker is real, between the flora of the south WESTERN corner of Australia and of the Cape of Good
0348  e narrow, but impassable, isthmus of Panama. WESTWARD of the shores of America, a wide space of op
0348  on the other hand, proceeding still further WESTWARD from the eastern islands of the tropical par
0057  fries has remarked in regard to plants, and WESTWOOD in regard to insects, that in large genera t
0157  ge groups of beetles, but in the Engidae, as WESTWOOD has remarked, the number varies greatly; and
0415  ivision of the Hymenoptera, the antennae, as WESTWOOD has remarkec, are most constant in structure
0097  y hybridizer knows how unfavourable exposure to WET is to the fertilisation of a flower, yet what
0360  the actual sea, so that they were alternately WET and exposed to the air like really floating pl
0184  with widely open mouth, thus catching, like a WHALE, insects in the water. Even in so extreme a c
0184  till a creature was produced as monstrous as a WHALE. As we sometimes see individuals of a species
0303  he most striking case, however, is that of the WHALE family; as these animals have huge bones, are
0303  the world, the fact of not a single bone of a WHALE having been discovered in any secondary forma
0414  larity of a mouse to a shrew, of a dugong to a WHALE, of a whale to a fish, as of any importance.
0414  mouse to a shrew, of a dugong to a whale, of a WHALE to a fish, as of any importance. These resemb
0427  ong, which is a pachydermatous animal, and the WHALE, and between both these mammals and fishes, i
0428  ue affinity between the several members of the WHALE family; for these cetaceans agree in so many
0144  ormal in their dermal covering, viz. Cetacea (WHALES) and Edentata (armadilloes, scaly anteaters,
0304  d in 1858, clear evidence of the existence of WHALES in the upper greensand, some time before the
0428  y and fin like limbs are only analogical when WHALES are compared with fishes, being adaptations i
0450  for instance, the presence of teeth in foetal WHALES, which when crown up have not a tooth in thei
0452  ry organs, such as teeth in the upper jaws of WHALES and ruminants, can often be detected in the e
```

Page ***(Key Word)**

Page **(Key Word)**

```
0008 at period of life the causes of variability,  WHATEVER they may be, generally act: whether during t
0013 o look at the inheritance of every character  WHATEVER as the rule, and non inheritance as the anom
0013 e, which I think may be trusted, is that, at   WHATEVER period of life a peculiarity first appears,
0031 d, by means of which he may summon into life   WHATEVER form and mould he pleases. Lord Somerville,
0033 the diversity of leaves, pods, or tubers, or  WHATEVER part is valued, in the kitchen garden, in co
0040 to the characteristic features of the breed.  WHATEVER they may be. But the chance will be infinite
0041 any species are scanty, all the individuals,  WHATEVER their quality may be, will generally be allo
0061 life, any variation, however slight and from  WHATEVER cause proceeding, if it be in any degree pro
0071 f the introduction of a single tree, nothing  WHATEVER else having been done, with the exception th
0103 ght there maintain itself in a body, so that  WHATEVER intercrossing took place would be chiefly be
0157 fewer offspring to the less favoured males.   WHATEVER the cause may be of the variability of secon
0157 xes of any one of the species. Consequently,  WHATEVER part of the structure of the common progenit
0170 eautiful and harmonious diversity of nature. Grant WHATEVER the cause may be of each slight difference f
0224 crowd of bees working in a dark hive. Grant   WHATEVER instincts you please, and it seems at first
0261 a multitude of cases we can assign no reason  WHATEVER. Great diversity in the size of two plants,
0392 alph. de Cancolle has shown, generally have,  WHATEVER the cause may be, confined ranges. Hence the
0392 baceous plants when growing on an island, to  WHATEVER order they belonged, and thus convert them f
0402 undoubtedly if one species has any advantage  WHATEVER over another, it will in a very brief time w
0423 esculent and thickened stems are so similar.  WHATEVER part is found to be most constant, is used i
0435 tructed on the existing general pattern, for  WHATEVER purpose they served, we can at once perceive
0444 some evidence to render it probable, that at  WHATEVER age any variation first appears in the paren
0447 converted into hands, or paddles, or wings.   WHATEVER influence long continued exercise or use on
0455 s cannot be checked by natural selection. At  WHATEVER period of life disuse or selection reduces a
0457 ttern displayed in the homologous organs, to  WHATEVER purpose applied, of the different species of
0008 ivation. Our oldest cultivated plants, such as WHEAT, still often yield new varieties; our oldest
0070 s how troublesome it is to get seed from a few WHEAT or other such plants in a garden; I have in t
0075 decided: for instance, if several varieties of WHEAT be sown together, and the mixed seed be resow
0113 ne variety and then several mixed varieties of WHEAT have been sown on equal spaces of ground. Hen
0362 capable of germination. Some seeds of the oat, WHEAT, millet, canary, hemp, clover, and beet germi
0439 and stripes can be plainly distinguised in the WHELP of the Lion. We occasionally though rarely se
0440 e. no one will suppose that the stripes on the WHELP of a Lion, or the spots on the young blackbir
0019 omestic breeds have originated in Europe; for  WHENCE could they have been derived, as these severa
0072 umbers that I went to several points of view.  WHENCE I could examine hundreds of acres of the unen
0222 forty yards, to a very thick clump of heath,   WHENCE I saw the last individual of F. sanguinea eme
0290 es in the geography of the surrounding lands,  WHENCE the sediment has been derived, accords with t
0295 f subsidence will often tend to sink the area  WHENCE the sediment is derived, and thus diminish th
0308 irst to last large islands or tracts of land,  WHENCE the sediment was derived, occurred in the nei
0355 should be related to those of another region,  WHENCE it has been stocked. A volcanic island, for i
0403 nly related to the inhabitants of that region  WHENCE colonists could most readily have been derive
0409 hose of the nearest continent or other source WHENCE immigrants were probably derived. We can see
0478 ated to the inhabitants of the nearest source WHENCE immigrants might have been derived. We see th
0084 at is good; silently and insensibly working,  WHENEVER and wherever opportunity offers, at the impr
0166 ss. In the horse we see this tendency strong  WHENEVER a dun tint appears, a tint which approaches
0167 less, from the same part in the parents. But  WHENEVER we have the means of instituting a compariso
0249 ame plant, would be thus effected. Moreover,  WHENEVER complicated experiments are in progress, so
0322 economy of nature will be utterly obscured.   WHENEVER we can precisely say why this species is mor
0326 favourably circumstanced in an equal degree,  WHENEVER their inhabitants met, the battle would be p
0357 rings of coral or atolls standing over them.  WHENEVER it is fully admitted, as I believe it will s
0440 ty may come on earlier or later in life; but  WHENEVER it comes on, the adaptation of the larva to
0009 exception of the plantigrades or bear family; WHEREAS, carnivorous birds, with the rarest exceptio
0048 rphic forms, Mr. Babington gives 251 species; WHEREAS Mr. Bentham gives only 112, a difference of
0049 strongly marked race of a Norwegian species,  WHEREAS the greater number rank it as an undoubted s
0058 recorded, and these range over 7.7 provinces; WHEREAS, the species to which these varieties belong
0059 isted as varieties, and have thus originated; WHEREAS, these analogies are utterly inexplicable if
0095 ore from a certain disease than yellow plums;  WHEREAS another disease attacks yellow fleshed peach
0111 ut nature presenting well marked differences;  WHEREAS varieties, the supposed prototypes and paren
0136 if they had been able to swim still further,  WHEREAS it would have been better for the bad swimme
0138 t and to have increased the size of the eyes; WHEREAS with all the other inhabitants of the caves,
0141 s were capable of enduring a glacial climate, WHEREAS the living species are now all tropical or s
0149 polyandrous flowers) the number is variable;  WHEREAS the number of the same part or organ, when i
0172 eing sterile and producing sterile offspring, WHEREAS, when varieties are crossed their fertility
0222 y and instantly seized the pupae of F. fusca, WHEREAS they were much terrified when they came acro
0240 ch though small can be plainly distinguished, WHEREAS the smaller workers have their ocelli rudime
0251 grow, and after a few days perished entirely, WHEREAS the pod impregnated by the pollen of the hyb
0262 r instance, cannot be grafted on the currant, WHEREAS the currant will take, though with difficult
0263 male and female sexual elements are perfect,  WHEREAS in hybrids they are imperfect. Even in first
0274 hybrids are widely different from each other; WHEREAS if two very distinct varieties of one specie
0288 ep water and has been found fossil in Sicily, WHEREAS not one other species has hitherto been foun
0313 little from the living species of this genus; WHEREAS most of the other Silurian Molluscs and all
0314 ir nearest allies on the continent of Europe, WHEREAS the marine shells and birds have remained un
0353 , that mammals have not been able to migrate, WHEREAS some plants, from their varied means of disp
0380 have to a certain extent beaten the natives;  WHEREAS extremely few southern forms have become nat
0391 y a wonderful number of peculiar land shells, WHEREAS not one species of sea shell is confined to
0398 on the ordinary view of independent creation; WHEREAS on the view here maintained, it is obvious t
0400 ry as the most important for its inhabitants; WHEREAS it cannot, I think, be disputed that the nat
0414 ation, excepting in the first main divisions; WHEREAS the organs of reproduction, with their produ
0423 e then the shape or colour of the body, etc.; WHEREAS with sheep the horns are much less serviceab
0447 l be inherited at a corresponding mature age. WHEREAS the young will remain unmodified, or be modi
0461 rgans are more or less functionally impotent; WHEREAS in first crosses the organs on both sides ar
0480 atural selection to browse without their aid; WHEREAS in the calf, the teeth have been left untouc
0485 t the present day by intermediate gradations, WHEREAS species were formerly thus connected. Hence,
0055 than the species of the smaller genera: for  WHEREVER many closely related species (i.e. species o
0056 strongly marked and permanent varieties; for WHEREVER many species of the same genus have been for
0056 tables clearly show as a general rule that,  WHEREVER many species of a genus have been formed, th
0084 ilently and insensibly working, whenever and WHEREVER opportunity offers, at the improvement of ea
0204 nd that each organic being is trying to live WHEREVER it can live, how it has arisen that there ar
```

Page **(Key Word)**

Page **(Key Word)**

```
0228  t was most interesting to me to observe that  WHEREVER  several bees had begun to excavate these bas
0300  r sand, would not endure to a distant epoch.  WHEREVER  sediment did not accumulate on the bed of th
0327  e, the forms dominant in the highest degree,  WHEREVER  produced, would tend everywhere to prevail.
0404  lieve, be universally found to be true, that  WHEREVER  in two regions, let them be ever so distant,
0404  on or migration between the two regions. And  WHEREVER  many closely allied species occur, there wil
0407  hat all the individuals of the same species,  WHEREVER  located, have descended from the same parent
0478  th areas; and we almost invariably find that  WHEREVER  many closely allied species inhabit two area
0478  dentical species common to both still exist.  WHEREVER  many closely allied yet distinct species occ
0008  ability, whatever they may be, generally act;  WHETHER  during the early or late period of developme
0008  e particular period of growth, will determine  WHETHER  or not the plant sets a seed. I cannot here
0013  ee it in the father and child, we cannot tell  WHETHER  it may not be due to the same original cause
0014  e aboriginal stock was, and so could not tell  WHETHER  or not nearly perfect reversion had ensued.
0015  wholly, revert to the wild aboriginal stock.  WHETHER  or not the experiment would succeed, is not
0016  are soon involved in doubt, from not knowing  WHETHER  they have descended from one or several pare
0017  possibly know, when he first tamed an animal,  WHETHER  it would vary in succeeding generations, and
0017  it would vary in succeeding generations, and  WHETHER  it would endure other climates? Has the litt
0017  possible to come to any definite conclusion,  WHETHER  they have descended from one or several spec
0029  sked, a celebrated raiser of Hereford cattle,  WHETHER  his cattle might not have descended from lon
0036  asional preservation of the best individuals,  WHETHER  or not sufficiently distinct to be ranked at
0037  r first appearance as distinct varieties, and  WHETHER  or not two or more species or races have bec
0043  ed that the accumulative action of Selection,  WHETHER  applied methodically and more quickly, or un
0044  in a state of nature, we must briefly discuss  WHETHER  these latter are subject to any variation. T
0045  s, that parts which must be called important,  WHETHER  viewed under a physiological or classificato
0047  t here enumerate, sometimes occur in deciding  WHETHER  or not to rank one form as a variety of anot
0047  d conjecture is opened. Hence, in determining  WHETHER  a form should be ranked as a species or a va
0049  dges as good and true species. But to discuss  WHETHER  they are rightly called species or varieties
0050  y intermediate links, and it is very doubtful  WHETHER  these links are hybrids; and there is, as it
0052  ay be justly called an incipient species; but  WHETHER  this belief be justifiable must be judged of
0056  f difference between them, judging by analogy  WHETHER  or not the amount suffices to raise one or b
0056  e is one very important criterion in settling  WHETHER  two forms should be ranked as species or var
0060  s ever been disputed. It is immaterial for us  WHETHER  a multitude of doubtful forms be called spec
0068  most severe struggle between the individuals,  WHETHER  of the same or of distinct species, which su
0076  f the medicinal leech. It may even be doubted  WHETHER  the varieties of any one of our domestic pla
0085  ences would effectually settle which variety,  WHETHER  a smooth or downy, a yellow or purple fleshe
0105  s greatly deceive ourselves, for to ascertain  WHETHER  a small isolated area, or a large open area
0116  n structure. It may be doubted, for instance,  WHETHER  the Australian marsupials, which are divided
0127  s alone; in their struggles with other males.  WHETHER  natural selection has really thus acted in n
0136  large or to reduce the wings, would depend on  WHETHER  a greater number of individuals were saved b
0140  r this from our frequent inability to predict  WHETHER  or not an imported plant will endure our cli
0140  an, by adaptation to particular climates. But  WHETHER  or not the adaptation be generally very clos
0151  to the great variability of these characters,  WHETHER  or not displayed in any unusual manner, of w
0160  ich their progenitor possessed, the tendency,  WHETHER  strong or weak, to reproduce the lost charac
0162  oted or turn crowned, we could not have told,  WHETHER  these characters in our domestic breeds have
0166  om the ass and hemionus; to ask Colonel Poole  WHETHER  such face stripes ever occur in the eminentl
0167  ted, the common parent of our domestic horse,  WHETHER  or not it be descended from one or more wild
0183  is difficult to tell, and immaterial for us,  WHETHER  habits generally change first and structure
0183  lly change first and structure afterwards; or  WHETHER  slight modifications of structure lead to ch
0214  hout going head over heels. It may be doubted  WHETHER  any one would have thought of training a dog
0222  ising courage. Now I was curious to ascertain  WHETHER  F. sanguinea could distinguish the pupae of
0225  to the great principle of gradation, and see  WHETHER  Nature does not reveal to us her method of w
0247  to me that we may well be permitted to doubt  WHETHER  many other species are really so sterile, wh
0248  dvanced by our best botanists on the question  WHETHER  certain doubtful forms should be ranked as s
0252  themselves are; I think, more sterile. I doubt  WHETHER  any case of a perfectly fertile hybrid anima
0254  together; and analogy makes me greatly doubt,  WHETHER  the several aboriginal species would at firs
0255  d of hybrids. Our chief object will be to see  WHETHER  or not the rules indicate that species have
0265  tile hybrids. No one can tell, till he tries,  WHETHER  any particular animal will breed under confi
0265  nder culture; nor can he tell, till he tries,  WHETHER  any two species of a genus will produce more
0275  nce of the child to its parents are the same,  WHETHER  the two parents differ much or little from e
0281  their structure with that of the rock pigeon,  WHETHER  they had descended from this species or from
0286  ard in nearly every twenty two years. I doubt  WHETHER  any rock, even as soft as chalk, would yield
0294  is same glacial period. Yet it may be doubted  WHETHER  in any quarter of the world, sedimentary dep
0299  termediate, fossil links, by asking ourselves  WHETHER,  for instance, geologists at some future per
0299  or from several aboriginal stocks; or, again,  WHETHER  certain sea shells inhabiting the shores of
0300  ould then be least perfect. It may be doubted  WHETHER  the duration of any one great period of subs
0308  intervals between the successive formations;  WHETHER  Europe and the United States during these in
0312  receding and present chapters. Let us now see  WHETHER  the several facts and rules relating to the
0314  s is quite independent of that of all others.  WHETHER  such variability be taken advantage of by na
0314  taken advantage of by natural selection, and  WHETHER  the variations be accumulated to a greater o
0319  e conditions were which checked its increase.  WHETHER  some one or several contingencies; and at wh
0321  rom some inherited inferiority in common. But  WHETHER  it be species belonging to the same or to a
0323  e world: we have not sufficient data to judge  WHETHER  the productions of the land and of fresh wat
0323  nts in the same parallel manner. We may doubt  WHETHER  they have thus changed: if the Megatherium,
0324  kilful naturalist would hardly be able to say  WHETHER  the existing or the pleistocene inhabitants
0336  ancient Forms. There has been much discussion  WHETHER  recent forms are more highly developed than
0337  f new Zealand were set free in Great Britain,  WHETHER  any considerable number would be enabled to
0341  nt distribution. It may be asked in ridicule,  WHETHER  I suppose that the megatherium and other all
0352  een largely discussed by naturalists, namely,  WHETHER  species have been created at one or more poi
0353  y species. So that we are reduced to consider  WHETHER  the exceptions to continuity of range are so
0354  der a point equally important for us, namely,  WHETHER  the several distinct species of a genus, whi
0355  ectly bear on another allied question, namely  WHETHER  all the individuals of the same species have
0355  om a single pair, or single hermaphrodite, or  WHETHER,  as some authors suppose, from many individu
0359  not be floated across wide spaces of the sea,  WHETHER  or not they were injured by the salt water.
0363  t sir C. Lyell wrote to M. Hartung to inquire  WHETHER  he had observed erratic boulders on these is
0388  fresh water plants and of the lower animals,  WHETHER  retaining the same identical form or in some
0396  f the inhabitants of the more remote islands,  WHETHER  still retaining the same specific form or mo
0398  islands would be likely to receive colonists,  WHETHER  by occasional means of transport or by forme
```

Page **(Key Word)**

whether

Page **(Key Word)**

0408	s understand the high importance of barriers,	WHETHER	of land or water, which separate our several l
0410	roughout time and space are intelligible; for	WHETHER	we look to the forms of life which have chan
0413	an of the Creator; but unless it be specified	WHETHER	order in time or space, or what else is mean
0416	the definition of whole groups. For instance,	WHETHER	or not there is an open passage from the nos
0423	pted by some authors. For we might feel sure,	WHETHER	there had been more or less modification, th
0426	e manner in which an insect's wing is folded,	WHETHER	the skin be covered by hair or feathers, if
0432	senting most of the characters of each group,	WHETHER	large or small, and thus give a general idea
0433	, on the high importance of looking to types,	WHETHER	or not we can separate and define the groups
0439	of some vertebrate animal, he cannot now tell	WHETHER	it be that of a mammal, bird, or reptile. Th
0442	alled a metamorphosis. The larvae of insects,	WHETHER	adapted to the most diverse and active habit
0443	ly in our own children; we cannot always tell	WHETHER	the child will be tall or short, or what its
0443	d by its parent, it must be quite unimportant	WHETHER	most of its characters are fully acquired a
0444	obtained its food best by having a long beak,	WHETHER	or not it assumed a beak of this particular
0446	hey are nearly grown up: they are indifferent	WHETHER	the desired qualities and structures have be
0454	nts of various parts in monsters. But I doubt	WHETHER	any of these cases throw light on the origin
0454	g that rudiments can be produced: for I doubt	WHETHER	species under nature ever undergo abrupt cha
0456	t upon characters, if constant and prevalent,	WHETHER	of high vital importance, or of the most tri
0457	reat facts in Morphology become intelligible,	WHETHER	we look to the same pattern displayed in the
0464	s will be able to decide, on the common view,	WHETHER	or not these doubtful forms are varieties? A
0467	species, is shown by the inextricable doubts	WHETHER	very many of them are varieties or aborigina
0484	t be incessantly haunted by the shadowy doubt	WHETHER	this or that form be in essence a species. T
0484	ill be no slight relief. The endless disputes	WHETHER	or not some fifty species of British bramble
0484	e only to decide (not that this will be easy)	WHETHER	any form be sufficiently constant and distin
0484	o be capable of definition; and if definable,	WHETHER	the differences be sufficiently important to
0011	ead. Some instances of correlation are quite	WHIMSICAL	: thus cats with blue eyes are invariably de
0088	e been described as fighting, bellowing, and	WHIRLING	round, like Indians in a war dance, for the
0384	that of the live fish not rarely dropped by	WHIRLWINDS	in India, and the vitality of their ova whe
0012	facts collected by Heusinger, it appears that	WHITE	sheep and pigs are differently affected from
0025	the rock pigeon is of a slaty blue, and has a	WHITE	rump (the Indian sub species, C. intermedia o
0025	es of the outer feathers externally edged with	WHITE	; the wings have two black bars: some semi dom
0025	l bred birds, all the above marks, even to the	WHITE	edging of the outer tail feathers, sometimes
0025	acters; for instance, I crossed some uniformly	WHITE	fantails with some uniformly black barbs, and
0025	ossed together, and one grandchild of the pure	WHITE	fantail and pure black barb was of as beautif
0025	rb was of as beautiful a blue colour, with the	WHITE	rump, double black wing bar, and barred and w
0025	te rump, double black wing bar, and barred and	WHITE	edged tail feathers, as any wild rock pigeon!
0084	ark feeders mottled grey; the alpine ptarmigan	WHITE	in winter, the red grouse the colour of heath
0085	f the Continent persons are warned not to keep	WHITE	pigeons, as being the most liable to destruct
0085	uld remember how essential it is in a flock of	WHITE	sheep to destroy every lamb with the faintest
0159	blue birds with two black bars on the wings, a	WHITE	rump, a bar at the end of the tail, with the
0160	eathers externally edged near their bases with	WHITE	. As all these marks are characteristic of the
0163	rtainly very variable in length and outline. A	WHITE	ass, but not an albino, has been described wi
0213	now that he points to aid his master, than the	WHITE	butterfly knows why she lays her eggs on the
0270	a witness, as Gartner: namely, that yellow and	WHITE	varieties of the same species of Verbascum wh
0271	ers. Moreover, he asserts that when yellow and	WHITE	varieties of one species are crossed with yel
0271	ies of one species are crossed with yellow and	WHITE	varieties of a distinct species, more seed is
0365	is far more remarkable, that the plants on the	WHITE	Mountains, in the United States of America, a
0005	his is the doctrine of Malthus, applied to the	WHOLE	animal and vegetable kingdoms. As many more i
0006	ter I shall give a brief recapitulation of the	WHOLE	work, and a few concluding remarks. No one ou
0011	he bones of the leg more, in proportion to the	WHOLE	skeleton, than do the same bones in the wild
0012	varieties differ slightly from each other. The	WHOLE	organisation seems to have become plastic, an
0013	itable. Perhaps the correct way of viewing the	WHOLE	subject, would be, to look at the inheritance
0018	mestic dog, may not have existed in Egypt? The	WHOLE	subject must, I think, remain vague; neverthe
0019	. even in the case of the domestic dogs of the	WHOLE	world, which I fully admit have probably desc
0024	structure, we may look in vain throughout the	WHOLF	great family of Columbidae for a beak like th
0025	not occur together in any other species of the	WHOLE	family. Now, in every one of the domestic bre
0035	ess of selection, and by careful training, the	WHOLE	body of English racehorses have come to surpa
0053	ng statements are fairly well established. The	WHOLE	subject, however, treated as it necessarily h
0062	engrained in the mind, I am convinced that the	WHOLE	economy of nature, with every fact on distrib
0063	of Malthus applied with manifold force to the	WHOLE	animal and vegetable kingdoms; for in this ca
0064	ced plants which have become common throughout	WHOLE	islands in a period of less than ten years. S
0066	ears to people, under favourable conditions, a	WHOLF	district, let it be ever so large. The condor
0069	g conspicuous, we are tempted to attribute the	WHOLE	effect to its direct action. But this is a ve
0073	r. hence I have very little doubt, that if the	WHOLE	genus of humble bees became extinct or very r
0080	r domestication), it may be truly said that the	WHOLE	organisation becomes in some degree plastic.
0083	ery shade of constitutional difference, on the	WHOLE	machinery of life. Man selects only for his o
0084	mpared with those accumulated by nature during	WHOLE	geological periods. Can we wonder, then, that
0087	ion. A structure used only once in an animal's	WHOLE	life, if of high importance to it, might be m
0093	produce stamens alone in one flower or on one	WHOLF	plant, and pistils alone in another flower or
0095	which is visited by humble bees alone; so that	WHOLE	fields of the red clover offer in vain an abu
0101	reeing closely with each other in almost their	WHOLE	organisation, yet are not rarely, some of the
0105	tance in the production of new species, on the	WHOLF	I am inclined to believe that largeness of ar
0121	come extinct. So it probably will be with many	WHOLE	collateral lines of descent, which will be co
0129	branches of various sizes may represent those	WHOLE	orders, families, and genera which have now n
0142	e seedlings appeared to be than others. On the	WHOLE	, I think we may conclude that habit, use, and
0143	of Growth. I mean by this expression that the	WHOLF	organisation is so tied together during its g
0143	ecting the early embryo, seriously affects the	WHOLE	organisation of the adult. The several parts
0146	tion of growth, structures which are common to	WHOLE	groups of species, and which in truth are sim
0146	wo modifications, having been transmitted to a	WHOLE	group of descendants with diverse habits, wou
0146	me apparent correlations, occurring throughout	WHOLE	orders, are entirely due to the manner alone
0148	n the parasitic and protected Proteolepas, the	WHOLE	anterior part of the head is reduced to the m
0150	rule would not here apply, because there is a	WHOLE	group of bats having wings; it would apply on
0152	. in our domestic animals, if any part, or the	WHOLE	animal, be neglected and no selection be appl
0152	instance, the comb in the Dorking fowl) or the	WHOLE	breed will cease to have a nearly uniform cha
0158	rily it may be developed, if it be common to a	WHOLE	group of species; that the great variability
0168	are more variable than those which have their	WHOLE	organisation more specialised, and are higher
0168	chapter that the same principle applies to the	WHOLE	individual; for in a district where many spec

Page **(Key Word)**

whole

Page **(Key Word)**

```
0192  sack. These cirripedes have no branchiae, the   WHOLE  surface of the body and sack, including the s
0195  ace, we are much too ignorant in regard to the   WHOLE  economy of any one organic being, to say what
0201  caused by each part, each will be found on the   WHOLE  advantageous. After the lapse of time, under
0202  en cause the insect's own death: for if on the   WHOLE  the power of stinging be useful to the commun
0207  rs, as a difficulty sufficient to overthrow my   WHOLE  theory. I must premise, that I have nothing t
0214  s: and a cross with a greyhound has given to a   WHOLE  family of shepherd dogs a tendency to hunt ha
0215  ness; and I presume that we must attribute the   WHOLE  of the inherited change from extreme wildness
0230  as they deepen each cell. They do not make the   WHOLE  three sided pyramidal base of any one cell at
0236  ed to me insuperable, and actually fatal to my   WHOLE  theory. I allude to the neuters or sterile fe
0253  udges, namely Mr. Blyth and Capt. Hutton, that   WHOLE  flocks of these crossed geese are kept in var
0258  ty, or of any recognisable difference in their   WHOLE  organisation. On the other hand, these cases
0265  a certain extent with systematic affinity, for   WHOLE  groups of animals and plants are rendered imp
0265  impotent by the same unnatural conditions; and   WHOLE  groups of species tend to produce sterile hyb
0269  and slowly during vast periods of time on the   WHOLE  organisation, in any way which may be for eac
0275  species slowly and naturally produced. On the   WHOLE  I entirely agree with Dr. Prosper Lucas, who,
0277  fect, and which have had this system and their   WHOLE  organisation disturbed by being compounded of
0281  parent. The common parent will have had in its   WHOLE  organisation much general resemblance to the
0286  ff one yard in height to be eaten back along a   WHOLE  line of coast at the rate on one yard in near
0286  suffers degradation at the same time along its   WHOLE  indented length; and we must remember that al
0287  , a denudation of one inch per century for the   WHOLE  length would be an ample allowance. At this r
0287  of years. During each of these years, over the   WHOLE  world, the land and the water has been people
0288  at sediment is being deposited over nearly the   WHOLE  bed of the sea, at a rate sufficiently quick
0290  for even a short geological period. Along the   WHOLE  west coast, which is inhabited by a peculiar
0292  oubted that the geological record, viewed as a   WHOLE  , is extremely imperfect; and this fact will
0294  glacial period, which forms only a part of one   WHOLE  geological period; and likewise to reflect on
0294  n accumulating within the same area during the   WHOLE  of this period. It is not, for instance, prob
0294  robable that sediment was deposited during the   WHOLE  of the glacial period near the mouth of the M
0295  e had to live on the same area throughout this   WHOLE  time. But we have seen that a thick fossilife
0295  ld seem that each separate formation, like the   WHOLE  pile of formations in any country, has genera
0296  hey have not lived on the same spot during the   WHOLE  period of deposition; but have disappeared an
0300  chipelago is one of the richest regions of the   WHOLE  world in organic beings; yet if all the speci
0300  of any one great period of subsidence over the   WHOLE  or part of the archipelago, together with a c
0302  rdly on my theory. On the sudden appearance of   WHOLE  groups of Allied Species. The abrupt manner i
0302  of Allied Species. The abrupt manner in which   WHOLE  groups of species suddenly appear in certain
0303  w how liable we are to error in supposing that   WHOLE  groups of species have suddenly been produced
0305  gists of the apparently sudden appearance of a   WHOLE  group of species, is that of the teleostean f
0305  really teleostean. Assuming, however, that the   WHOLE  of them did appear, as Agassiz believes, at t
0307  , as long as, or probably far longer than, the   WHOLE  interval from the Silurian age to the present
0310  st or have existed; the sudden manner in which   WHOLE  groups of species appear in our European form
0318  inally from the world. Both single species and   WHOLE  groups of species last for very unequal perio
0318  . in some cases, however, the extermination of   WHOLE  groups of beings, as of ammonites towards the
0318  ndary period, has been wonderfully sudden. The   WHOLE  subject of the extinction of species has been
0318  ards into South America, has run wild over the   WHOLE  country and has increased in numbers at an un
0319  it would in a very few years have stocked the   WHOLE  continent. But we could not have told what th
0321  pect to the apparently sudden extermination of   WHOLE  families or orders, as of Trilobites at the c
0322  to me, the manner in which single species and   WHOLE  groups of species become extinct, accords wel
0322  ays in action, yet seldom perceived by us, the   WHOLE  economy of nature will be utterly obscured. W
0324  ote period as measured by years, including the   WHOLE  glacial epoch), were to be compared with thos
0325  y, but depend on general laws which govern the   WHOLE  animal kingdom. M. Barrande has made forcible
0327  orms would be allied in groups by inheritance,   WHOLE  groups would tend slowly to disappear: though
0329  eneral system. With respect to the Vertebrata,   WHOLE  pages could be filled with striking illustrat
0329  any fossil links, that he has had to alter the   WHOLE  classification of these two orders; and has p
0334  statement, that the fauna of each period as a   WHOLE  is nearly intermediate in character between t
0336  ing the pleistocene period, which includes the   WHOLE  glacial period, and note how little the speci
0341  ew genera; the other old species and the other   WHOLE  genera having become utterly extinct. In fail
0342  the geological record, will rightly reject my   WHOLE  theory. For he may ask in vain where are the
0343  ut often falsely apparent, sudden coming in of   WHOLE  groups of species. He may ask where are the r
0344  ce of the earth. But the utter extinction of a   WHOLE  group of species may often be a very slow pro
0345  any palaeontologists, that organisation on the   WHOLE  has progressed. If it should hereafter be pro
0350  understand how it is that sections of genera,   WHOLE  genera, and even families are confined to the
0351  of those species, which have undergone during   WHOLE  geological periods but little modification, t
0356  have gone on simultaneously changing, and the   WHOLE  amount of modification will not have been due
0362  r arrival. Some hawks and owls bolt their prey   WHOLE  , and after an interval of from twelve to twen
0364  ar, one or two land birds are blown across the   WHOLE  Atlantic Ocean, from North America to the wes
0374  to avoid believing that the temperature of the   WHOLE  world was at this period simultaneously coole
0374  broad belts of longitude. On this view of the   WHOLE  world, or at least of broad longitudinal belt
0376  a large body of geological evidence, that the   WHOLE  world, or a large part of it, was during the
0379  n to believe from geological evidence that the   WHOLE  body of arctic shells underwent scarcely any
0392  ely into trees. With respect to the absence of   WHOLE  orders on oceanic islands, Bory St. Vincent l
0396  classes or sections of classes, the absence of   WHOLE  groups, as of batrachians, and of terrestrial
0404  nsport has given the same general forms to the   WHOLE  world. We see this same principle in the blin
0406  e striking relation of the inhabitants of each   WHOLE  archipelago or island to those of the nearest
0407  am fully convinced simultaneously affected the   WHOLE  world, or at least great meridional belts. As
0409  to other quarters of the world. We can see why   WHOLE  groups of organisms, as batrachians and terre
0414  ances, though so intimately connected with the   WHOLE  life of the being, are ranked as merely adapt
0414  that the organs of vegetation, on which their   WHOLE  life depends, are of little signification, ex
0416  ted as highly serviceable in the definition of   WHOLE  groups. For instance, whether or not there is
0421  all the descendants of the genus F, along its   WHOLE  line of descent, are supposed to have been bu
0425  d have to be classed under the bear genus. The   WHOLE  case is preposterous: for where there has bee
0431  connected together. For the common parent of a   WHOLE  family of species, now broken up by extinctio
0431  may thus account even for the distinctness of   WHOLE  classes from each other, for instance, of bir
0434  erent species of the class are homologous. The   WHOLE  subject is included under the general name of
0435  ogous construction of the limbs throughout the   WHOLE  class. So with the mouths of insects, we have
0438  them to the most diverse purposes. And as the   WHOLE  amount of modification will have been effecte
0442  ture became visible in the embryo. And in some   WHOLE  groups of animals and in certain members of o
0447  he case of pigeons. We may extend this view to   WHOLE  families or even classes. The fore limbs, for
```

Page **(Key Word)**

Page **(Key Word)**

```
0448  hat this is the rule of development in certain  WHOLE  groups of animals, as with cuttle fish and sp
0454  youatt, in young animals, and the state of the  WHOLE  flower in the cauliflower. We often see rudim
0459  f natural history. Concluding remarks. As this  WHOLE  volume is one long argument, it may be conven
0460  s in saying that any organ or instinct, or any  WHOLE  being, could not have arrived at its present
0463  may be urged against my theory. Why, again, do  WHOLE  groups of allied species appear, though certa
0469  during long ages and rigidly scrutinising the  WHOLE  constitution, structure, and habits of each c
0475  erent groups. The extinction of species and of  WHOLE  groups of species, which has played so conspi
0477  thern temperate zones, though separated by the  WHOLE  intertropical ocean. Although two areas may p
0483  of species, on their own side they ignore the  WHOLE  subject of the first appearance of species in
0483  est weight extend very far. All the members of  WHOLE  classes can be connected together by chains o
0483  of modification in the descendants. Throughout  WHOLE  classes various structures are formed on the
0487  he former migrations of the inhabitants of the  WHOLE  world. Even at present, by comparing the diff
0488  e may have been slow in an extreme degree. The  WHOLE  history of the world, as at present known, al
0489  roken, and that no cataclysm has desolated the  WHOLE  world. Hence we may look with some confidence
0015  ), that they would to a large extent, or even  WHOLLY  revert to the wild aboriginal stock. Whether
0071  proportional numbers of the heath plants were  WHOLLY  changed, but twelve species of plants (not co
0074  ase and red clover would become very rare, or  WHOLLY  disappear. The number of humble bees in any d
0086  rva of an insect to a score of contingencies,  WHOLLY  different from those which concern the mature
0087  relations to the other sex, or in relation to  WHOLLY  different habits of life in the two sexes, as
0102  definite object, and free intercrossing will  WHOLLY  stop his work. But when many men, without int
0108  hat these several causes are amply sufficient  WHOLLY  to stop the action of natural selection. I do
0131  had been due to chance. This, of course, is a  WHOLLY  incorrect expression, but it serves to acknow
0135  o disuse modifications of structure which are  WHOLLY, or mainly, due to natural selection. Mr. Wol
0137  ls living in darkness, I attribute their loss  WHOLLY  to disuse. In one of the blind animals, namel
0146  ewed by systematists as of high value, may be  WHOLLY  due to unknown laws of correlated growth, and
0151  lves in the different species being sometimes  WHOLLY  unlike in shape; and the amount of variation
0171  ormed by the modification of some animal with  WHOLLY  different habits? Can we believe that natural
0185  ber of the strictly terrestrial thrush family  WHOLLY  subsists by diving, grasping the stones with
0190  of the same organ performing at the same time  WHOLLY  distinct functions; thus the alimentary canal
0190  o functions, for one function alone, and thus  WHOLLY  change its nature by insensible steps. Two di
0190  ly flotation, may be converted into one for a  WHOLLY  different purpose, namely respiration. The sw
0191  . in the higher Vertebrata the branchiae have  WHOLLY  disappeared, the slits on the sides of the ne
0205  confidently believe that many modifications,  WHOLLY  due to the laws of growth, and at first in no
0212  the country inhabited, but often from causes  WHOLLY  unknown to us: Audubon has given several rema
0215  on the other hand, young chickens have lost,  WHOLLY  by habit, that fear of the dog and cat which
0234  as we have seen, the spherical surfaces would  WHOLLY  disappear, and would all be replaced by plane
0242  fore, discussed this case, at some little but  WHOLLY  insufficient length, in order to show the pow
0243  he males of our distinct Kitty wrens, a habit  WHOLLY  unlike that of any other known bird. Finally,
0280  ctly intermediate between them. But this is a  WHOLLY  false view; we should always look for forms i
0288  made every year in Europe prove. No organism  WHOLLY  soft can be preserved. Shells and bones will
0293  e equally rash to suppose that it then became  WHOLLY  extinct. We forget how small the area of Euro
0307  ry great. If these most ancient beds had been  WHOLLY  worn away by denudation, or obliterated by me
0320  ich have been exterminated, either locally or  WHOLLY, through man's agency. I may repeat what I pu
0323  are common to the Old and New Worlds be kept  WHOLLY  out of view, the general parallelism in the s
0333  lained in a satisfactory manner. And they are  WHOLLY  inexplicable on any other view. On this same
0341  t be admitted. These huge animals have become  WHOLLY  extinct, and have left no progeny. But in the
0343  e that period, the world may have presented a  WHOLLY  different aspect; and that the older continen
0344  nd isolated situations. When a group has once  WHOLLY  disappeared, it does not reappear; for the li
0348  arriers, either of land or open sea, they are  WHOLLY  distinct. On the other hand, proceeding still
0358  r or less facilities may be said to be almost  WHOLLY  unknown. Until I tried, with Mr. Berkeley's a
0379  e migration northward, the case may have been  WHOLLY  different with those intruding forms which se
0402  er over another, it will in a very brief time  WHOLLY  or in part supplant it; but if both are equal
0445  difference between these two breeds has been  WHOLLY  caused by selection under domestication; but
0447  might supervene, from causes of which we are  WHOLLY  ignorant, at a very early period of life, or
0452  ten be detected in the embryo, but afterwards  WHOLLY  disappear. It is also, I believe, a universal
0455  ssible) the rudimentary part would tend to be  WHOLLY  lost, and we should have a case of complete a
0461  rom some one part to the others. We are often  WHOLLY  unable even to conjecture how this could have
0466  ty, when it has once come into play, does not  WHOLLY  cease; for new varieties are still occasional
0485  as a savage looks at a ship, as at something  WHOLLY  beyond his comprehension; when we regard ever
0373  evidence of the cold period, from the western  WHORES  of Britain to the Oural range, and southward
0437  plants, we see a series of successive spiral  WHORLS  of leaves. An indefinite repetition of the sa
0437  n progenitor of flowering plants, many spiral  WHORLS  of leaves. We have formerly seen that parts m
0006  he beings which live around us. Who can explain  WHY  one species ranges widely and is very numerous
0006  species ranges widely and is very numerous, and  WHY  another allied species has a narrow range and
0013  g inheritance are quite unknown; no one can say  WHY  the same peculiarity in different individuals
0013  s, is sometimes inherited and sometimes not so;  WHY  the child often reverts in certain characters
0013  grandmother or other much more remote ancestor;  WHY  a peculiarity is often transmitted from one se
0014  sion, and that when there is no apparent reason  WHY  a peculiarity should appear at any particular
0055  al act of creation, there is no apparent reason  WHY  more varieties should occur in a group having
0076  nother, and so in other cases. We can dimly see  WHY  the competition should be most severe between
0076  probably in no one case could we precisely say  WHY  one species has been victorious over another i
0077  und. Look at a plant in the midst of its range,  WHY  does it not double or quadruple its numbers? W
0109  shows us plainly; and indeed we can see reason  WHY  they should not have thus increased, for the n
0132  sentially from an ovule, is alone affected. But  WHY, because the reproductive system is disturbed,
0149  to perform diversified work, we can perhaps see  WHY  it should remain variable, that is, why natura
0149  aps see why it should remain variable, that is,  WHY  natural selection should have preserved or rej
0152  in this case is eminently liable to variation.  WHY  should this be so? On the view that each speci
0155  each species having been independently created,  WHY  should that part of the structure, which diffe
0156  xual characters is not manifest; but we can see  WHY  these characters should not have been rendered
0167  f a hundred can we pretend to assign any reason  WHY  this or that part differs, more or less, from
0169  nce the genus arose; and thus we can understand  WHY  it should often still be variable in a much hi
0171  be classed under the following heads: Firstly,  WHY  if species have descended from other species b
0171  everywhere see innumerable transitional forms?  WHY  is not all nature in confusion instead of the
0172  numerable transitional forms must have existed,  WHY  do we not find them embedded in countless numb
0174  region, having intermediate conditions of life,  WHY  do we not now find close linking intermediate
0176  they connect, then, I think, we can understand  WHY  intermediate varieties should not endure for v
```

Page **(Key Word)**

```
0176 ieties should not endure for very long periods;  WHY as a general rule they should be exterminated
0194 prodigal in variety, but niggard in innovation.  WHY, on the theory of Creation, should this be so?
0194 , on the theory of Creation, should this be so?  WHY should all the parts and organs of many indepe
0194 invariably linked together by graduated steps?  WHY should not Nature have taken a leap from struc
0194 of natural selection, we can clearly understand  WHY she should not; for natural selection can act
0213 aid his master, than the white butterfly knows  WHY she lays her eggs on the leaf of the cabbage,
0260 becoming confounded in nature? I think not. For  WHY should the sterility be so extremely different
0260 ually important to keep from blending together?  WHY should the degree of sterility be innately var
0260 ariable in the individuals of the same species?  WHY should some species cross with facility, and y
0260 iculty, and yet produce fairly fertile hybrids?  WHY should there often be so great a difference in
0260 reciprocal cross between the same two species?  WHY, it may even be asked, has the production of h
0261 the two plants. We can sometimes see the reason  WHY one tree will not take on another, from differ
0264 tion can be given of these facts, any more than  WHY certain trees cannot be grafted on others. Las
0266 e root of the matter: no explanation is offered  WHY an organism, when placed under unnatural condi
0279 s a very obvious difficulty. I assigned reasons  WHY such links do not commonly occur at the presen
0280 rmerly existed on the earth, be truly enormous.  WHY then is not every geological formation and eve
0290 etween each formation. But we can, I think, see  WHY the geological formations of each region are a
0290 a distant age. A little reflection will explain  WHY along the rising coast of the western side of
0293 ation, it becomes more difficult to understand,  WHY we do not therein find closely graduated varie
0293 s for its deposition, I can see several reasons  WHY each should not include a graduated series of
0307 swarmed with living creatures. To the question  WHY we do not find records of these vast primordia
0315 be liable to be exterminated. Hence we can see  WHY all the species in the same region do at last,
0315 lowly changing drama. We can clearly understand  WHY a species when once lost should never reappear
0319 classes, in all countries. If we ask ourselves  WHY this or that species is rare, we answer that s
0322 utterly obscured. Whenever we can precisely say  WHY this species is more abundant in individuals t
0322 cies is more abundant in individuals than that;  WHY this species and not another can be naturalise
0322 and not till then, we may justly feel surprise  WHY we cannot account for the extinction of this p
0342 y imperfect, and will to a large extent explain  WHY we do not find interminable varieties, connect
0343 the production of new forms. We can understand  WHY when a species has once disappeared it never r
0344 continued tendency to divergence of character,  WHY the more ancient a form is, the more it genera
0344 ore it generally differs from those now living.  WHY ancient and extinct forms often tend to fill u
0345 ct and very different forms. We can clearly see  WHY the organic remains of closely consecutive for
0345 nked together by generation! we can clearly see  WHY the remains of an intermediate formation are i
0352 species can be produced at two separate points,  WHY do we not find a single mammal common to Europ
0355 y understand, on the principle of modification,  WHY the inhabitants of a region should be related
0380 act lines and means of migration, or the reason  WHY certain speciesand not other have migrated; wh
0381 why certain speciesand not other have migrated;  WHY certain species have been modified and have gi
0381 ot hope to explain such facts, until we can say  WHY one species and not another becomes naturalise
0381 naturalised by man's agency in a foreign land!  WHY one ranges twice or thrice as far, and is twic
0393 their transportal across the sea, and therefore  WHY they do not exist on any oceanic island. But w
0393 hy they do not exist on any oceanic island. But  WHY, on the theory of creation, they should not ha
0394 and Mauritius, all possess their peculiar bats.  WHY, it may be asked, has the supposed creative fo
0398 yet feels that he is standing on American land.  WHY should this be so? Why should the species whic
0398 anding on American land. Why should this be so?  WHY should the species which are supposed to have
0402 arles Island, which has its own mocking thrush!  WHY should it succeed in establishing itself there
0408 anism are of the highest importance, we can see  WHY two areas having nearly the same physical cond
0409 can understand, as I have endeavoured to show,  WHY oceanic islands should have few inhabitants, b
0409 great number should be endemic or peculiar; and  WHY, in relation to the means of migration, one or
0409 mmon to other quarters of the world. We can see  WHY whole groups of organisms, as batrachians and
0409 r species of aerial mammals or bats. We can see  WHY there should be some relation between the pres
0409 an island and the mainland. We can clearly see  WHY all the inhabitants of an archipelago, though
0409 ce immigrants were probably derived. We can see  WHY in two areas, however distant from each other,
0418 aracters of quite subordinate value. We can see  WHY characters derived from the embryo should be o
0418 t is by no means obvious, on the ordinary view,  WHY the structure of the embryo should be more imp
0419 . in our discussion on embryology, we shall see  WHY such characters are so valuable, on the view o
0426 cial value in classification. We can understand  WHY a species or a group of species may depart, in
0426 ation. We shall hereafter, I think, clearly see  WHY embryological characters are of such high clas
0427 they reveal descent, we can clearly understand  WHY analogical or adaptive character, although of
0433 follow in our classification. We can understand  WHY we value certain resemblances far more than ot
0433 alue certain resemblances far more than others;  WHY we are permitted to use rudimentary and useles
0433 or others of trifling physiological importance;  WHY, in comparing one group with a distinct group,
0437 e these facts on the ordinary view of creation!  WHY should the brain be enclosed in a box composed
0437 n the same construction in the skulls of birds?  WHY should similar bones have been created in the
0437 s they are for such totally different purposes?  WHY should one crustacean, which has an extremely
0437 sely, those with many legs have simpler mouths?  WHY should the sepals, petals, stamens, and pistil
0442 anner on growth. But there is no obvious reason  WHY, for instance, the wing of a bat, or the fin o
0449 odified ancient progenitors, we can clearly see  WHY ancient and extinct forms of life should resem
0462 fine as our present varieties, it may be asked,  WHY do we not see these linking forms all around u
0462 o we not see these linking forms all around us?  WHY are not all organic beings blended together in
0463 od between the extinct and still older species,  WHY is not every geological formation charged with
0463 y geological formation charged with such links?  WHY does not every collection of fossil remains af
0463 bjections which may be urged against my theory.  WHY, again, do whole groups of allied species appe
0463 e in suddenly on the several geological stages?  WHY do we not find great piles of strata beneath t
0467 aboriginal species. There is no obvious reason  WHY the principles which have acted so efficiently
0469 powerful agent always ready to act and select,  WHY should we doubt that variations in any way use
0469 would be preserved, accumulated, and inherited?  WHY, if man can by patience select variations most
0469 species first existed as a variety, we can see  WHY it is that no line of demarcation can be drawn
0471 theory simply intelligible. We can plainly see  WHY nature is prodigal in variety, though niggard
0471 l in variety, though niggard in innovation. But  WHY this should be a law of nature if each species
0473 each species having been independently created,  WHY should the specific characters, or those by wh
0473 the generic characters in which they all agree?  WHY, for instance, should the colour of a flower b
0474 d. it is inexplicable on the theory of creation  WHY a part developed in a very unusual manner in a
0474 rofitable modifications. We can thus understand  WHY nature moves by graduated steps in endowing di
0475 e, we should follow nearly the same instincts;  WHY the thrush of South America, for instance, lin
0475 ked and permanent varieties, we can at once see  WHY their crossed offspring should follow the same
0476 with its later descendants; and thus we can see  WHY the more ancient a fossil is, the oftener it s
```

Page ***(Key Word)***

```
0476 great leading facts in Distribution. We can see  WHY there should be so striking a parallelism in t
0477 ation, with subsequent modification, we can see   WHY oceanic islands should be inhabited by few spe
0477 hat many should be peculiar. We can clearly see   WHY those animals which cannot cross wide spaces o
0477 ammals, should not inhabit oceanic islands; and   WHY, on the other hand, new and peculiar species o
0478 ach class are so complex and circuitous. We see   WHY certain characters are far more serviceable th
0478 ore serviceable than others for classification;   WHY adaptive characters, though of paramount impor
0479 are of hardly any importance in classification;   WHY characters derived from rudimentary parts, tho
0479 ng, are often of high classificatory value; and   WHY embryological characters are the most valuable
0479 ng not early period of life, we can clearly see   WHY the embryos of mammals, birds, reptiles, and f
0480 ion of successive slight favourable variations.   WHY, it may be asked, have all the most eminent li
0021 large external orifices to the nostrils, and a   WIDE gape of mouth. The short faced tumbler has a b
0044 ity. Individual differences. Doubtful species.   WIDE ranging, much diffused, and common species var
0047 xist, or may formerly have existed; and here a   WIDE door for the entry of doubt and conjecture is
0047 inion of naturalists having sound judgment and   WIDE experience seems the only guide to follow. We
0049 undoubted species peculiar to Great Britain. A   WIDE distance between the homes of two doubtful for
0053 others have shown that plants which have very   WIDE ranges generally present varieties; and this m
0053 ry (and this is a different consideration from   WIDE range, and to a certain extent from commonness
0054 ter and salt loving plants have generally very   WIDE ranges and are much diffused, but this seems t
0060 bears on natural selection. The term used in a   WIDE sense. Geometrical powers of increase. Rapid i
0065 veral of the plants now most numerous over the   WIDE plains of La Plata, clothing square leagues of
0065 xplains the extraordinarily rapid increase and   WIDE diffusion of naturalised productions in their
0067 , on a piece of ground three feet long and two   WIDE, dug and cleared, and where there could be no
0141 mate as a quality readily grafted on an innate   WIDE flexibility of constitution, which is common t
0157 hly variable, sexual selection will have had a   WIDE scope for action, and may thus readily have su
0174 at species.as they are now distributed over a   WIDE area, we generally find them tolerably numerou
0175 , are generally so distributed that each has a   WIDE range, with a comparatively narrow neutral ter
0177 paratively narrow, hilly tract; and a third to   WIDE plains at the base; and that the inhabitants a
0180 with the posterior part of their bodies rather   WIDE and with the skin on their flanks rather full,
0181 lsely ranked amongst bats. It has an extremely   WIDE flank membrane, stretching from the corners of
0284 be quite erroneous; yet, considering over what   WIDE spaces very fine sediment is transported by th
0289 formations being separated from each other by   WIDE intervals of time. When we see the formations
0289 sir R. Murchison's great work in Russia, what   WIDE gaps there are in that country between the sup
0291 nd apparently these oscillations have affected   WIDE spaces. Consequently formations rich in fossil
0291 sequent degradation, may have been formed over   WIDE spaces during periods of subsidence, but only
0296 eds of the same formation; facts, showing what   WIDE, yet easily overlooked, intervals have occurre
0298 d to some one spot. Most marine animals have a   WIDE range; and we have seen that with plants it is
0299 ade the intervals between some few groups less   WIDE than they otherwise would have been, yet has d
0299 , with its numerous large islands separated by   WIDE and shallow seas, probably represents the form
0315 s in the same region do at last, if we look to   WIDE enough intervals of time, become modified; for
0315 ns have been almost necessarily accumulated at   WIDE and irregularly intermittent intervals; conseq
0321 ber what has been already said on the probable   WIDE intervals of time between our consecutive form
0328 rmations have often been accumulated over very   WIDE spaces in the same quarter of the world; but w
0329 the extinct forms of life help to fill up the   WIDE intervals between existing genera, families, a
0330 he dissolves by fine gradations the apparently   WIDE difference between the pig and the camel. In r
0333 expect, except in very rare cases, to fill up   WIDE intervals in the natural system, and thus unit
0339 e law holds good with sea shells; but from the   WIDE distribution of most genera of molluscs, it is
0348 f panama. Westward of the shores of America, a   WIDE space of open ocean extends, with not an islan
0354 ; and secondly (in the following chapter), the   WIDE distribution of fresh water productions; and t
0358 or that plant is stated to be ill adapted for   WIDE dissemination; but for transport across the se
0359 n a few days, they could not be floated across   WIDE spaces of the sea, whether or not they were in
0374 re found by Gardner, which do not exist in the   WIDE intervening hot countries. So on the Silla of
0383 nd, or from stream to stream; and liability to   WIDE dispersal would follow from this capacity as a
0384 d by existing land and fresh water shells. The   WIDE difference of the fish on opposite sides of co
0385 some species of fresh water shells have a very   WIDE range, and allied species, which, on my theory
0386 ately to acquire, as if in consequence, a very   WIDE range. I think favourable means of dispersal e
0388 ecome extinct in intermediate regions. But the   WIDE distribution of fresh water plants and of the
0388 gree modified, I believe mainly depends on the   WIDE dispersal of their seeds and eggs by animals,
0394 terrestrial mammal can be transported across a   WIDE space of sea, but bats can fly across. Bats ha
0397 in chinks of drifted timber across moderately   WIDE arms of the sea. And I found that several spec
0404 over the world, many of the species have very   WIDE ranges. I can hardly doubt that this rule is g
0405 in world ranging genera all the species have a   WIDE range, or even that they have on an average a
0405 range, or even that they have on an average a   WIDE range; but only that some of the species range
0405 ltimately into new species. In considering the   WIDE distribution of certain genera, we should bear
0407 a species may have ranged continuously over a   WIDE area, and then have become extinct in the inte
0428 y), could easily extend the parallelism over a   WIDE range; and thus the septenary, quinary, quater
0456 s in rudimentary organs, of no importance; the   WIDE opposition in value between analogical or adap
0462 stress ought not to be laid on the occasional   WIDE diffusion of the same species; for during very
0462 of time there will always be a good chance for   WIDE migration by many means. A broken or interrupt
0462 period; and consequently the difficulty of the   WIDE diffusion of species of the same genus is in s
0463 nd some extinct and supplanted form. Even on a   WIDE area, which has during a long period remained
0477 early see why those animals which cannot cross   WIDE spaces of ocean, as frogs and terrestrial mamm
0483 fossil remains sometimes tend to fill up very   WIDE intervals between existing orders. Organs in a
0006 nd us. Who can explain why one species ranges   WIDELY and is very numerous, and why another allied
0016 s hardly correct; but naturalists differ most   WIDELY in determining what characters are of generic
0040 ow and gradual process, they will spread more   WIDELY, and will get recognised as something distinc
0051 sely allied forms. But if his observations are   WIDELY extended, he will in the end generally be ena
0053 n individuals, and the species which are most   WIDELY diffused within their own country (and this i
0054 lled, the dominant species, those which range   WIDELY over the world, are the most diffused in thei
0055 scale of organisation are generally much more   WIDELY diffused than plants higher in the scale; and
0055 . the cause of lowly organised plants ranging   WIDELY will be discussed in our chapter on geographi
0074 e species. In some cases it can be shown that   WIDELY different checks act on the same species in d
0077 hed by other plants; so that the seeds may be   WIDELY distributed and fall on unoccupied ground. In
0086 rofit a plant to have its seeds more and more   WIDELY disseminated by the wind, I can see no greate
0105 enduring for a long period, and of spreading   WIDELY. Throughout a great and open area, not only w
0106 petitors, will be those that will spread most   WIDELY, will give rise to most new varieties and spe
0107 ssils, connect to a certain extent orders now   WIDELY separated in the natural scale. These anomalo
```

Page ***(Key Word)***

```
0107  of life, likely to endure long and to spread  WIDELY. For the area will first have existed as a co
0112  l they be better enabled to seize on many and  WIDELY diversified places in the polity of nature, a
0116  in the general economy of any land, the more  WIDELY and perfectly the animals and plants are dive
0117  species, which are the commonest and the most  WIDELY diffused, vary more than rare species with re
0117  with restricted ranges. Let (A) be a common,  WIDELY diffused, and varying species, belonging to a
0119  ll the modified descendants from a common and  WIDELY diffused species, belonging to a large genus,
0121  will have the best chance of filling new and  WIDELY different places in the polity of nature: hen
0122  (i), were also supposed to be very common and  WIDELY diffused species, so that they must originall
0123  , moreover, will be allied to each other in a  WIDELY different manner. Of the eight descendants fr
0123  ement of the process of modification, will be  WIDELY different from the other five species, and ma
0125  eritance from a different parent, will differ  WIDELY from the three genera descended from (A), the
0128  nera. We have seen that it is the common, the  WIDELY diffused, and widely ranging species, belongi
0128  at it is the common, the widely diffused, and  WIDELY ranging species, belonging to the larger gene
0141  te of nature, could easily be brought to bear  WIDELY different climates. We must not, however, pus
0152  frequently individuals are born which depart  WIDELY from the standard. There may be truly said to
0156  ified several species, fitted to more or less  WIDELY different habits, in exactly the same manner:
0156  of the same group differ from each other more  WIDELY in their secondary sexual characters, than in
0159  distinct breeds of pigeons, in countries most  WIDELY set apart, present sub varieties with reverse
0171  bits in the same species. Species with habits  WIDELY different from those of their allies. Organs
0184  ar was seen by Hearne swimming for hours with  WIDELY open mouth, thus catching, like a whale, inse
0184  see individuals of a species following habits  WIDELY different from those both of their own specie
0193  ly about a dozen fishes, of which several are  WIDELY remote in their affinities. Generally when th
0200  han they now are to these animals having such  WIDELY diversified habits. Therefore we may infer th
0236  t communities: for these neuters often differ  WIDELY in instinct and in structure from both the ma
0236  difficulty lies in the working ants differing  WIDELY from both the males and the fertile females i
0241  duate insensibly into each other, as does the  WIDELY different structure of their jaws. I speak co
0241  jaw, or all of small size with jaws having a  WIDELY different structure; or lastly, and this is o
0241  erile workers existing in the same nest, both  WIDELY different from each other and from their pare
0251  ia integrifolia and plantaginea, species most  WIDELY dissimilar in general habit, reproduced itsel
0252  f plants, then we may infer that animals more  WIDELY separated in the scale of nature can be more
0257  es crossing. It can be shown that plants most  WIDELY different in habit and general appearance, an
0261  en and the other deciduous, and adaptation to  WIDELY different climates, does not always prevent t
0267  s between males and females which have become  WIDELY or specifically different, produce hybrids wh
0268  lity of so many domestic varieties, differing  WIDELY from each other in appearance, for instance o
0271  tobacco is more fertile, when crossed with a  WIDELY distinct species, than are the other varietie
0274  crossed with a third species, the hybrids are  WIDELY different from each other; whereas if two ver
0276  aws. It is generally different, and sometimes  WIDELY different, in reciprocal crosses between the
0277  ugh this latter capacity evidently depends on  WIDELY different circumstances; should all run, to a
0298  each other than are the species found in more  WIDELY separated formations; but to this subject I s
0298  ; and that such local varieties do not spread  WIDELY and supplant their parent forms until they ha
0303  ms, which would be able to spread rapidly and  WIDELY throughout the world. I will now give a few e
0305  developed in some one sea, might have spread  WIDELY. Nor have we any right to suppose that the se
0311  rms of life, entombed in our consecutive, but  WIDELY separated, formations. On this view, the diff
0323  uccessive forms of life, in the stages of the  WIDELY separated palaeozoic and tertiary periods, wo
0325  re commonest in their own homes, and are most  WIDELY diffused, having produced the greatest number
0327  ing been formed by dominant species spreading  WIDELY and varying; the new species thus produced be
0330  connect by some of its characters groups now  WIDELY separated from each other. This remark no dou
0332  by a long and circuitous course through many  WIDELY different forms. If many extinct forms were t
0342  new forms in any one area and formation; that  WIDELY ranging species are those which have varied m
0345  the common progenitor of groups, since become  WIDELY divergent. Extinct forms are seldom directly
0347  the conditions of the Old and New Worlds, how  WIDELY different are their living productions! In th
0350  ss of modification through natural selection.  WIDELY ranging species, abounding in individuals, wh
0350  triumphed over many competitors in their own  WIDELY extended homes will have the best chance of s
0364  llous fact if many plants had not thus become  WIDELY transported. These means of transport are som
0373  in new Zealand; and the same plants, found on  WIDELY separated mountains in this island, tell the
0382  inated by the Glacial epoch, a few forms were  WIDELY dispersed to various points of the southern h
0383  fresh water productions would not have ranged  WIDELY within the same country, and as the sea is ap
0383  s power in fresh water productions of ranging  WIDELY, though so unexpected, can, I think, in most
0384  on the same continent the species often range  WIDELY and almost capriciously: for two river system
0395  e of deep ocean; and this space separates two  WIDELY distinct mammalian faunas. On either side the
0401  s we see on continents some species spreading  WIDELY and remaining the same. The really surprising
0403  he same forms, chiefly of plants, have spread  WIDELY throughout the world during the recent Glacia
0405  but only that some of the species range very  WIDELY: for the facility with which widely ranging s
0405  ange very widely; for the facility with which  WIDELY ranging species vary and give rise to new for
0405  the capacity of crossing barriers and ranging  WIDELY, as in the case of certain powerfully winged
0405  werfully winged birds, will necessarily range  WIDELY; for we should never forget that to range wid
0405  ely; for we should never forget that to range  WIDELY implies not only the power of crossing barrie
0405  that some at least of the species range very  WIDELY; for it is necessary that the unmodified pare
0405  ssary that the unmodified parent should range  WIDELY, undergoing modification during its diffusion
0406  orms will have had a better chance of ranging  WIDELY and of still retaining the same specific char
0406  the lower any group of organisms is, the more  WIDELY it is apt to range. The relations just discus
0406  ow and slowly changing organisms ranging more  WIDELY than the high, some of the species of widely
0406  widely than the high, some of the species of  WIDELY ranging genera themselves ranging widely, suc
0406  s of widely ranging genera themselves ranging  WIDELY, such facts, as alpine, lacustrine, and marsh
0411  vegetable matter, and so on: but the case is  WIDELY different in nature; for it is notorious tha
0411  tion, I have attempted to show that it is the  WIDELY ranging, the much diffused and common, that i
0415  siological value, its classificatory value is  WIDELY different. No naturalist can have worked at a
0421  to the same millionth degree; yet they differ  WIDELY and in different degrees from each other. The
0421  r in the same degree in blood, has come to be  WIDELY different. Nevertheless their genealogical ar
0427  ught usefully into play in classing large and  WIDELY distributed genera, because all the species o
0428  ents dominant, they are almost sure to spread  WIDELY, and to seize on more and more places in the
0429  few of the higher groups are in number, and how  WIDELY spread they are throughout the world, the fac
0434  names can be given to the homologous bones in  WIDELY different animals. We see the same great law
0437  any individual flower, though fitted for such  WIDELY different purposes, be all constructed on the
0439  s in the individual, which when mature become  WIDELY different and serve for different purposes, a
0439  points of structure, in which the embryos of  WIDELY different animals of the same class resemble
```

Page **(Key Word)**

0440 s, the pedunculated and sessile, which differ WIDELY in external appearance, have larvae in all th
0442 likewise a close similarity in the embryos of WIDELY different animals within the same class, that
0442 ups, the embryo does not at any period differ WIDELY from the adult: thus Owen has remarked in reg
0457 ologous parts, which when matured will become WIDELY different from each other in structure and fu
0464 ssil condition, at least in any great number. WIDELY ranging species vary most, and varieties are
0470 ure, so as to be enabled to seize on many and WIDELY different places in the economy of nature, th
0475 of change, after equal intervals of time, is WIDELY different in different groups. The extinction
0477 d feel no surprise at their inhabitants being WIDELY different, if they have been for a long perio
0489 as to foretel that it will be the common and WIDELY spread species, belonging to the larger and d
0435 ts. The bones of a limb might be shortened or WIDENED to any extent, and become gradually envelope
0431 has played an important part in defiring and WIDENING the intervals between the several groups in
0014 ther facts make me believe that the rule has a WIDER extension, and that when there is no apparent
0058 truism, for if a variety were found to have a WIDER range than that of its supposed parent specie
0141 to many parts of the world, and now have a far WIDER range than any other rodent, living free unde
0228 hey deepened these little pits, they make them WIDER and wider until they were converted into shal
0228 ed these little pits, they make them wider and WIDER until they were converted into shallow basins
0297 ought to be found. Moreover, if we look to rather WIDER intervals, namely, to distinct but consecutiv
0370 t have been still more completely separated by WIDER spaces of ocean. I believe the above difficul
0401 parated by deep arms of the sea, in most cases WIDER than the British Channel, and there is no rea
0258 been reciprocally crossed. There is often the WIDEST possible difference in the facility of making
0260 nerally some difference, and occasionally the WIDEST possible difference, in the facility of effec
0298 n that with plants it is those which have the WIDEST range, that oftenest present varieties: so th
0298 mals, it is probably those which have had the WIDEST range, far exceeding the limits of the known
0349 erent points and stations. It is a law of the WIDEST generality, and every continent offers innume
0360 st islands, even on those in the midst of the WIDEST oceans: and the natives of the coral islands
0403 d better fitted to their new homes, is of the WIDEST application throughout nature. We see this on
0022 the two arms of the furcula. The proportional WIDTH of the gape of mouth, the proportional length
0228 time the basins had acquired the above stated WIDTH (i.e. about the width of an ordinary cell), a
0228 cquired the above stated width (i.e. about the WIDTH of an ordinary cell), and were in depth about
0360 be floated across a space of sea 900 miles in WIDTH, and would then germinate. The fact of the la
0445 (but will not here give details) of the beak, WIDTH of mouth, length of nostril and of eyelid, si
0445 oints of difference, for instance, that of the WIDTH of mouth, could hardly be detected in the you
0011 whole skeleton, than do the same bones in the WILD duck; and I presume that this change may be sa
0011 k flying much less, and walking more, than its WILD parent. The great and inheritec development of
0014 namely, that our domestic varieties, when run WILD, gradually but certainly revert in character t
0014 omestic varieties could not possibly live in a WILD state. In many cases we do not know what the a
0015 a large extent, or even wholly, revert to the WILD aboriginal stock. Whether or not the experimen
0017 that all our domestic dogs have descended from any one WILD species: but, in the case of some other domest
0018 our domestic dogs have descended from several WILD species. In regard to sheep and goats I can fo
0018 lieve that these latter have had more than one WILD parent. With respect to horses, from reasons w
0018 rs, that all the races have descended from one WILD stock. Mr. Blyth, whose opinion, from his larg
0018 eeds of poultry have proceeded from the common WILD Indian fowl (Gallus bankiva). In regard to duc
0019 t that they all have descended from the common WILD duck and rabbit. The doctrine of the origin of
0019 tive characters be ever so slight, has had its WILD prototype. At this rate there must have existe
0019 st have existed at least a score of species of WILD cattle, as many sheep, and several goats in Fu
0019 there formerly existed in Great Britain eleven WILD species of sheep peculiar to it! When we bear
0019 lly admit have probably descended from several WILD species, I cannot doubt that there has been an
0019 dog, or Blenheim spaniel, etc., so unlike all WILD Canidae, ever existed freely in a state of nat
0020 greyhound, bloodhound, bull dog, etc., in the WILD state. Moreover, the possibility of making dis
0022 ornithologist, and he were told that they were WILD birds, would certainly, I think, be ranked by
0023 bable; or they must have become extinct in the WILD state. But birds breeding on precipices, and g
0024 ir native country: but not one has ever become WILD or feral, though the dovecot pigeon, which is
0024 nce shows that it is most difficult to get any WILD animal to breed freely under domestication: ye
0024 and in most parts of their structure, with the WILD rock pigeon, yet are certainly highly abnormal
0025 semi domestic breeds and some apparently truly WILD breeds have, besides the two black bars, the w
0025 d barred and white edged tail feathers, as any WILD rock pigeon! We can understand these facts, on
0027 hese supposed species being quite unknown in a WILD state, and their becoming nowhere feral: these
0030 chanical contrivance, is only a variety of the WILD Dipsacus! and this amount of change may have s
0034 o. savages now sometimes cross their dogs with WILD canine animals, to improve the breed, and they
0037 t rate heartsease or dahlia from the seed of a WILD plant. No one would expect to raise a first ra
0037 a first rate melting pear from the seed of the WILD pear, though he might succeed from a poor seed
0037 he might succeed from a poor seedling growing WILD, if it had come from a garden stock. The pear,
0037 nnot recognise, and therefore do not know, the WILD parent stocks of the plants which have been lo
0064 domestic animals of many kinds which have run WILD in several parts of the world: if the statemen
0072 ither cattle nor horses nor dogs have ever run WILD, though they swarm southward and northward in
0141 n of some of our domestic animals from several WILD stocks: the blood, for instance, of a tropical
0141 for instance, of a tropical and arctic wolf or WILD dog may perhaps be mingled in our domestic bre
0167 hether or not it be descended from one or more WILD stocks, or the ass, the hemionus, quagga, and
0214 was a wolf, and this dog showed a trace of its WILD parentage only in one way, by not coming in a
0215 s more difficult to tame than the young of the WILD rabbit: scarcely any animal is tamer than the
0216 instinctive purpose of allowing, as we see in WILD ground birds, their mother to fly away. But th
0254 nce, that our dogs have descended from several WILD stocks: yet, with perhaps the exception of cer
0318 n by the Spaniards into South America, has run WILD over the whole country and has increased in nu
0337 itant of the southern hemisphere having become WILD in any part of Europe, we may doubt, if all th
0445 ied, and have probably descended from the same WILD stock: hence I was curious to see how far thei
0445 estic breeds of Pigeon have descended from one WILD species, I compared young pigeons of various b
0445 eyelid, size of feet and length of leg, in the WILD stock, in pouters, fantails, runts, barbs, dra
0446 t faced tumbler differed from the young of the WILD rock pigeon and of the other breeds, in all it
0484 the gall fly produces monstrous growths on the WILD rose or oak tree. Therefore I should infer fro
0212 ance of this even in England, in the greater WILDNESS of all our large birds than of our small bir
0212 by man. We may safely attribute the greater WILDNESS of our large birds to this cause: for in uni
0215 e whole of the inherited change from extreme WILDNESS to extreme tameness, simply to habit and lon
0480 heme of modification, which it seems that we WILFULLY will not understand. I have now recapitulate
0001 ion. My work is now nearly finished: but as it WILL take me two or three more years to complete it
0002 ome confidence in my accuracy. No doubt errors WILL have crept in, though I hope I have always bee
0002 illustration, but which, I hope, in most cases WILL suffice. No one can feel more sensible than I

Page **(Key Word)**

Page **(Key Word)**

0004 his Selection successive slight variations. I WILL then pass on to the variability of species in
0005 rom their high geometrical powers of increase, WILL be treated of. This is the doctrine of Malthus
0005 plex and sometimes varying conditions of life, WILL have a better chance of surviving, and thus be
0005 principle of inheritance, any selected variety WILL tend to propagate its new and modified form. T
0005 this fundamental subject of Natural Selection WILL be treated at some length in the fourth chapte
0005 pparent and gravest difficulties on the theory WILL be given: namely, first, the difficulties of t
0006 ts history. Although much remains obscure, and WILL long remain obscure, I can entertain no doubt,
0008 female unite. How many animals there are which WILL not breed, though living long under not very c
0008 ess water at some particular period of growth, WILL cetermine whether or not the plant sets a seed
0009 the garden. I may add, that as some organisms WILL breed most freely under the most unnatural con
0009 oductive system has not been thus affected; so WILL some animals and plants withstand domesticatio
0011 tion, some few of which can be dimly seen, and WILL be hereafter briefly mentioned. I will here on
0011 en, and will be hereafter briefly mentioned. I WILL here only allude to what may be called correla
0011 n of growth. Any change in the embryo or larva WILL almost certainly entail changes in the mature
0012 ting, and thus augmenting, any peculiarity, he WILL almost certainly unconsciously modify other pa
0015 r probably do occur; but natural selection, as WILL hereafter be explained, will determine how far
0015 ral selection, as will hereafter be explained, WILL determine how far the new characters thus aris
0018 irteen or fourteen thousand years ago, and who WILL pretend to say how long before these ancient p
0020 er for several generations, hardly two of them WILL be alike, and then the extreme difficulty, or
0023 re in some degree applicable in other cases, I WILL here briefly give them. If the several breeds
0026 rsion to any character derived from such cross WILL naturally become less and less, as in each suc
0026 d less, as in each succeeding generation there WILL be less of the foreign blood: but when there h
0027 tly variable: and the explanation of this fact WILL be obvious when we come to treat of selection.
0028 unt of variation which pigeons have undergone, WILL be obvious when we treat of Selection. We shal
0029 ght not have descended from long horns, and he WILL laugh you to scorn. I have never met a pigeon,
0032 fetire to it with indomitable perseverance, he WILL succeed, and may make great improvements: if h
0032 ements! if he wants any of these qualities, he WILL assuredly fail. Few would readily believe in t
0033 mportance of which should never be overlooked, WILL ensure some differences; but, as a general rul
0033 ther in the leaves, the flowers, or the fruit, WILL produce races differing from each other chiefl
0038 and thus by a process of natural selection, as WILL hereafter be more fully explained, two sub bre
0040 n the immediate neighbourhood. But as yet they WILL hardly have a distinct name, and from being on
0040 from being only slightly valued, their history WILL be disregarded. When further improved by the s
0040 ved by the same slow and gradual process, they WILL spread more widely, and will get recognised as
0040 ual process, they will spread more widely, and WILL get recognised as something distinct and valua
0040 gnised as something distinct and valuable, and WILL then probably first receive a provincial name.
0040 e spreading and knowledge of any new sub breed WILL be a slow process. As soon as the points of va
0040 as I have called it, of unconscious selection WILL always tend, perhaps more at one period than a
0040 he breed, whatever they may be. But the chance WILL be infinitely small of any record having been
0041 y occasionally, the chance of their appearance WILL be much increased by a large number of individ
0041 he individuals, whatever their quality may be, WILL generally be allowed to breec, and this will e
0041 , will generally be allowed to breed, and this WILL effectually prevent selection. But probably th
0043 haracters, and that there are not many men who WILL laboriously examine internal and important org
0046 view, no instance of an important part varying WILL ever be found: but under any other point of vi
0046 ed definite by natural selection, as hereafter WILL be explained. Those forms which possess in som
0047 ariety. But cases of great difficulty, which I WILL not here enumerate, sometimes occur in decidin
0047 e closely connected by intermediate links; nor WILL the commonly assumed hybrid nature of the inte
0049 esi but what distance, it has been well asked, WILL suffice? If that between America and Europe is
0049 ? if that between America and Europe is ample, WILL that between the Continent and the Azores, or
0049 bear on the attempt to determine their rank. I WILL here give only a single instance, the well kno
0050 varieties. Close investigation, in most cases, WILL bring naturalists to an agreement how to rank
0050 closely attract his attention, varieties of it WILL almost universally be found recorded. These va
0050 be found recorded. These varieties, moreover, WILL be often ranked by some authers as species. Lo
0050 attention to one class within one country, he WILL soon make up his mind how to rank most of the
0051 st of the doubtful forms. His general tendency WILL be to make many species, for he will become im
0051 tendency will be to make many species, for he WILL become impressed, just like the pigeon or poul
0051 s he extends the range of his observations, he WILL meet with more cases of difficulty; for he wil
0051 ill meet with more cases of difficulty; for he WILL encounter a greater number of closely allied f
0051 but if his observations be widely extended, he WILL in the end generally be enabled to make up his
0051 ch to call varieties and which species; but he WILL succeed in this at the expense of admitting mu
0051 uch variation, and the truth of this admission WILL often be disputed by other naturalists. When,
0051 ermediate links between his doubtful forms, hc WILL have to trust almost entirely to analogy, and
0051 most entirely to analogy, and his difficulties WILL rise to a climax. Certainly no clear line of d
0052 ction of natural selection in accumulating (as WILL hereafter be more fully explained) differences
0052 return to this subject. From these remarks it WILL be seen that I look at the term species, as on
0054 ountry, the species which are already dominant WILL be the most likely to yield offspring which, t
0054 which, though in some slight degree modified, WILL still inherit those advantages that enabled th
0054 y common and much diffused or dominant species WILL be found on the side of the larger genera. Thi
0054 l majority on the side of the larger genera. I WILL here allude to only two causes of obscurity. F
0055 cause of lowly organised plants ranging widely WILL be discussed in our chapter on geographical di
0058 d how the lesser differences between varieties WILL tend to increase into the greater differences
0061 o other organic beings and to external nature, WILL tend to the preservation of that individual, a
0061 nd to the preservation of that individual, and WILL generally be inherited by its offspring. The o
0061 herited by its offspring. The offspring, also, WILL thus have a better chance of surviving, for, o
0062 as the works of Nature are to those of Art. We WILL now discuss in a little more detail the strugg
0062 rarity, abundance, extinction, and variation, WILL be dimly seen or quite misunderstood. We behol
0063 y of these parasites grow on the same tree, it WILL languish and cie. But several seedling misslet
0064 probable minimum rate of natural increase: it WILL be under the mark to assume that it breeds whe
0066 strcyed, many must be produced, or the species WILL become extinct. It would suffice to keep up th
0067 ever so little, and the number of the species WILL almost instantaneously increase to any amount.
0067 by as much as it swarms in numbers, by so much WILL its tendency to increase be still further incr
0067 he checks are in even one single instance. Nor WILL this surprise any one who reflects how ignoran
0067 to the feral animals of South America. Here I WILL make only a few remarks, just to recall to the
0069 , for instance extreme cold, acts directly; it WILL be the least vigorous, or those which have got
0069 least food through the advancing winter, which WILL suffer most. When we travel from south to nort
0069 favoured by any slight change of climate, they WILL increase in numbers, and, as each area is alre
0069 ly stocked with inhabitants, the other species WILL decrease. When we travel southward and see a s

Page **(Key Word)**

Page **(Key Word)**

0071	f these cases; but on this intricate subject I	WILL	not here enlarge. Many cases are on record sho
0071	ve to struggle together in the same country. I	WILL	give only a single instance, which, though a s
0075	quadrupeds. But the struggle almost invariably	WILL	be most severe between the individuals of the
0075	of varieties of the same species, the struggle	WILL	generally be almost equally severe, and we som
0075	or climate, or are naturally the most fertile,	WILL	beat the others and so yield more seed, and wi
0075	ll beat the others and so yield more seed, and	WILL	consequently in a few years quite supplant the
0076	in due proportion, otherwise the weaker kinds	WILL	steadily decrease in numbers and disappear. So
0076	been asserted that certain mountain varieties	WILL	starve out other mountain varieties, so that t
0076	itution, and always in structure, the struggle	WILL	generally be more severe between species of th
0076	it its great congener. One species of charlock	WILL	supplant another, and so in other cases. We ca
0078	regions or on the borders of an utter desert,	WILL	competition cease. The land may be extremely c
0078	e land may be extremely cold or dry, yet there	WILL	be competition between some few species, or be
0078	ts former home, yet the conditions of its life	WILL	generally be changed in an essential manner. I
0078	hould we know what to do, so as to succeed. But	WILL	convince us of our ignorance on the mutual rel
0080	plains the grouping of all organic beings. How	WILL	the struggle for existence, discussed too brie
0084	how short his time! and consequently how poor	WILL	his products be, compared with those accumulat
0086	y natural selection for the good of the being,	WILL	cause other modifications, often of the most e
0086	it: so in a state of nature, natural selection	WILL	be enabled to act on and modify organic beings
0086	concern the mature insect. These modifications	WILL	no doubt affect, through the laws of correlati
0086	ae. So, conversely, modifications in the adult	WILL	probably often affect the structure of the lar
0086	the larva; but in all cases natural selection	WILL	ensure that modifications consequent on other
0086	e extinction of the species. Natural selection	WILL	modify the structure of the young in relation
0087	in relation to the young. In social animals it	WILL	adapt the structure of each individual for the
0087	natural history, I cannot find one case which	WILL	bear investigation. A structure used only once
0087	urs under nature, and if so, natural selection	WILL	be able to modify one sex in its functional re
0088	ch are best fitted for their places in nature,	WILL	leave most progeny. But in many cases, victory
0088	leave most progeny. But in many cases, victory	WILL	depend not on general vigour, but on having so
0092	f crossing, we have good reason to believe (as	WILL	hereafter be more fully alluded to), would pro
0093	ould easily show by many striking instances. I	WILL	give only one, not as a very striking case, bu
0095	f a great valley by a single diluvial wave, so	WILL	natural selection, if it be a true principle,
0097	the entrance of pollen from another individual	WILL	explain this state of exposure, more especiall
0097	e great good, as I believe, of the plant. Bees	WILL	act like a camel hair pencil, and it is quite
0098	en and pollen from another species, the former	WILL	have such a prepotent effect, that it will inv
0098	mer will have such a prepotent effect, that it	WILL	invariably and completely destroy, as has been
0099	as I have found, of the seedlings thus raised	WILL	turn out mongrels: for instance, I raised 233
0100	ularly carried from flower to flower; and this	WILL	give a better chance of pollen being occasiona
0102	hin any given period of profitable variations,	WILL	compensate for a lesser amount of variability
0102	corresponding degree with its competitors, it	WILL	soon be exterminated. In man's methodical sele
0102	r some definite object, and free intercrossing	WILL	wholly stop his work. But when many men, witho
0102	unt of crossing with inferior animals. Thus it	WILL	be in nature; for within a confined area, with
0102	fectly occupied as might be, natural selection	WILL	always tend to preserve all the individuals va
0102	ut if the area be large, its several districts	WILL	almost certainly present different conditions
0102	ving a species in the several districts, there	WILL	be intercrossing with the other individuals of
0103	each; for in a continuous area, the conditions	WILL	generally graduate away insensibly from one di
0103	rom one district to another. The intercrossing	WILL	most affect those animals which unite for each
0103	this nature, for instance in birds, varieties	WILL	generally be confined to separated countries;
0103	ame variety, true and uniform in character. It	WILL	obviously thus act far more efficiently with t
0104	s, I am convinced that the young thus produced	WILL	gain so much in vigour and fertility over the
0104	m long continued self fertilisation, that they	WILL	have a better chance of surviving and propagat
0104	luence of intercrosses even at rare intervals,	WILL	be great. If there exist organic beings which
0104	, the organic and inorganic conditions of life	WILL	generally be in a great degree uniform: so tha
0104	reat degree uniform: so that natural selection	WILL	tend to modify all the individuals of a varyin
0104	nding and differently circumstanced districts,	WILL	be prevented. But isolation probably acts more
0105	king immigration and consequently competition,	WILL	give time for any new variety to be slowly imp
0105	otal number of the individuals supported on it	WILL	necessarily be very small; and fewness of indi
0105	rily be very small; and fewness of individuals	WILL	greatly retard the production of new species t
0105	the total number of the species inhabiting it,	WILL	be found to be small, as we shall see in our c
0105	especially in the production of species, which	WILL	prove capable of enduring for a long period, a
0105	ly. Throughout a great and open area, not only	WILL	there be a better chance of favourable variati
0106	y species become modified and improved, others	WILL	have to be improved in a corresponding degree
0106	be improved in a corresponding degree or they	WILL	be exterminated. Each new form, also, as soon
0106	m, also, as soon as it has been much improved,	WILL	be able to spread over the open and continuous
0106	spread over the open and continuous area, and	WILL	thus come into competition with many others. H
0106	tition with many others. Hence more new places	WILL	be formed, and the competition to fill them wi
0106	ll be formed, and the competition to fill them	WILL	be more severe, on a large than on a small and
0106	ow continuous, owing to oscillations of level,	WILL	often have recently existed in a broken condit
0106	ndition, so that the good effects of isolation	WILL	generally, to a certain extent, have concurred
0106	w species, yet that the course of modification	WILL	generally have been more rapid on large areas;
0106	dy have been victorious over many competitors,	WILL	be those that will spread most widely, will gi
0106	ious over many competitors, will be those that	WILL	spread most widely, will give rise to most new
0106	s, will be those that will spread most widely,	WILL	give rise to most new varieties and species, a
0106	ve rise to most new varieties and species, and	WILL	thus play an important part in the changing hi
0106	s, on these views, understand some facts which	WILL	be again alluded to in our chapter on geograph
0106	islands. On a small island, the race for life	WILL	have been less severe, and there will have bee
0106	for life will have been less severe, and there	WILL	have been less modification and less extermina
0107	he competition between fresh water productions	WILL	have been less severe than elsewhere; new form
0107	ave been less severe than elsewhere; new forms	WILL	have been more slowly formed, and old forms mo
0107	al productions a large continental area, which	WILL	probably undergo many oscillations of level, a
0107	oscillations of level, and which consequently	WILL	exist for long periods in a broken condition,
0107	exist for long periods in a broken condition,	WILL	be the most favourable for the production of m
0107	endure long and to spread widely. For the area	WILL	first have existed as a continent, and the inh
0107	this period numerous in individuals and kinds,	WILL	have been subjected to very severe competition
0107	subsidence into large separate islands, there	WILL	still exist many individuals of the same speci
0107	o on the confines of the range of each species	WILL	thus be checked: after physical changes of any
0107	fter physical changes of any kind, immigration	WILL	be prevented, so that new places in the polity
0108	o that new places in the polity of each island	WILL	have to be filled up by modifications of the o

Page **(Key Word)**

0108 modifications of the old inhabitants; and time WILL be allowed for the varieties in each to become
0108 be re converted into a continental area, there WILL again be severe competition: the most favoured
0108 ition: the most favoured or improved varieties WILL be enabled to spread: there will be much extin
0108 ved varieties will be enabled to spread: there WILL be much extinction of the less improved forms,
0108 e various inhabitants of the renewed continent WILL again be changed: and again there will be a fa
0108 ntinent will again be changed; and again there WILL be a fair field for natural selection to impro
0108 us produce new species. That natural selection WILL always act with extreme slowness, I fully admi
0108 ion of some kind. The existence of such places WILL often depend on physical changes, which are ge
0108 n checked. But the action of natural selection WILL probably still oftener depend on some of the i
0108 rently always a very slow process. The process WILL often be greatly retarded by free intercrossin
0108 e greatly retarded by free intercrossing. Many WILL exclaim that these several causes are amply su
0108 ther hand, I do believe that natural selection WILL always act very slowly, often only at long int
0109 s power of selection. Extinction. This subject WILL be more fully discussed in our chapter on Geol
0109 cted and favoured form increases in number, so WILL the less favoured forms decrease and become ra
0109 e that any form represented by few individuals WILL, during fluctuations in the seasons or in the
0110 species which are most numerous in individuals WILL have the best chance of producing within any g
0110 ies, or incipient species. Hence, rare species WILL be less quickly modified or improved within an
0110 or improved within any given period, and they WILL consequently be beaten in the race for life by
0110 e are formed through natural selection, others WILL become rarer and rarer, and finally extinct. T
0110 those undergoing modification and improvement, WILL naturally suffer most. And we have seen in the
0110 species, during the progress of its formation, WILL generally press hardest on its nearest kindred
0112 cknowledged principle that fanciers do not and WILL not admire a medium standard, but like extreme
0112 ers, being neither very swift nor very strong, WILL have been neglected, and will have tended to d
0112 nor very strong, will have been neglected, and WILL have tended to disappear. Here, then, we see i
0112 tructure, constitution, and habits; by so much WILL they be better enabled to seize on many and wi
0113 enabled to occupy. What applies to one animal WILL apply throughout all time to all animals, that
0113 erwise natural selection can do nothing. So it WILL be with plants. It has been experimentally pro
0116 e diversified for different habits of life, so WILL a greater number of individuals be capable of
0116 at the modified descendants of any one species WILL succeed by so much the better as they become m
0116 ciples of natural selection and of extinction, WILL tend to act. The accompanying diagram will aid
0116 on, will tend to act. The accompanying diagram WILL aid us in understanding this rather perplexing
0117 se variations which are in some way profitable WILL be preserved or naturally selected. And here t
0117 rom divergence of character comes in; for this WILL generally lead to the most different or diverg
0117 rieties, namely a1 and m1. These two varieties WILL generally continue to be exposed to the same c
0118 ity is in itself hereditary, consequently they WILL tend to vary, and generally to vary in nearly
0118 varieties, being only slightly modified forms, WILL tend to inherit those advantages which made th
0118 he other inhabitants of the same country; they WILL likewise partake of those more general advanta
0118 riable, the most divergent of their variations WILL generally be preserved during the next thousan
0118 the diagram to have produced variety a2, which WILL, owing to the principle of divergence, differ
0118 ndants, proceeding from the common parent (A), WILL generally go on increasing in number and diver
0119 rom thinking that the most divergent varieties WILL invariably prevail and multiply: a medium form
0119 one modified descendant; for natural selection WILL always act according to the nature of the plac
0119 t perfectly occupied by other beings; and this WILL depend on infinitely complex relations. But as
0119 species can be rendered, the more places they WILL be enabled to seize on, and the more their mod
0119 seize on, and the more their modified progeny WILL be increased. In our diagram the line of succe
0119 diffused species, belonging to a large genus, WILL tend to partake of the same advantages which m
0119 ich made their parent successful in life, they WILL generally go on multiplying in number as well
0119 hly improved branches in the lines of descent, WILL, it is probable, often take the place of, and
0119 do not doubt that the process of modification WILL be confined to a single line of descent, and t
0119 of descent, and the number of the descendants WILL not be increased: although the amount of diver
0120 n character during the successive generations, WILL have come to differ largely, but perhaps unequ
0121 are already extremely different in character, WILL generally tend to produce the greatest number
0121 test number of modified descendants; for these WILL have the best chance of filling new and widely
0121 of our principles, namely that of extinction, WILL have played an important part. As in each full
0121 the struggle for life over other forms, there WILL be a constant tendency in the improved descend
0121 r it should be remembered that the competition WILL generally be most severe between those forms w
0121 as well as the original parent species itself, WILL generally tend to become extinct. So it probab
0121 nerally tend to become extinct. So it probably WILL be with many whole collateral lines of descent
0121 many whole collateral lines of descent, which WILL be conquered by later and improved lines of de
0122 ion, species (A) and all the earlier varieties WILL have become extinct, having been replaced by e
0122 ced by eight new species (a14 to m14); and (I) WILL have been replaced by six (n14 to z14) new spe
0122 number at the fourteen thousandth generation, WILL probably have inherited some of the same advan
0122 therefore, to me extremely probable that they WILL have taken the places of, and thus exterminate
0122 arents. Hence very few of the original species WILL have transmitted offspring to the fourteen tho
0123 am descended from the original eleven species, WILL now be fifteen in number. Owing to the diverge
0123 rence in character between species a14 and z14 WILL be much greater than that between the most dif
0123 nal eleven species. The new species, moreover, WILL be allied to each other in a widely different
0123 dants from (A) the three marked a14, c14, p14, WILL be nearly related from having recently branche
0123 having diverged at an earlier period from a5, WILL be in some degree distinct from the three firs
0123 named species; and lastly, o14, e14, and m14, WILL be nearly related one to the other, but from h
0123 t commencement of the process of modification, WILL be widely different from the other five specie
0123 a distinct genus. The six descendants from (I) WILL form two sub genera or even genera. But as the
0123 e original genus, the six descendants from (I) WILL, owing to inheritance, differ considerably fro
0123 ed from (I), and the eight descended from (A), WILL have to be ranked as very distinct genera, or
0124 s affinities to the other fourteen new species WILL be of a curious and circuitous nature. Having
0124 i), now supposed to be extinct and unknown, it WILL be in some degree intermediate in character be
0124 e type of their parents, the new species (F14) WILL not be directly intermediate between them, but
0124 types of the two groups; and every naturalist WILL be able to bring some such case before his min
0125 rked b14 and f14, and those marked o14 to m14, WILL form three very distinct genera. We shall also
0125 and from inheritance from a different parent, WILL differ widely from the three genera descended
0125 nded from (A), the two little groups of genera WILL form two distinct families, or even orders, ac
0125 diagram. And the two new families, or orders, WILL have descended from two species of the origina
0125 other forms in the struggle for existence, it WILL chiefly act on those which already have some a
0125 he production of new and modified descendants, WILL mainly lie between the larger groups, which ar
0125 trying to increase in number. One large group WILL slowly conquer another large group, reduce its
0126 ng on many new places in the polity of Nature, WILL constantly tend to supplant and destroy the ea

```
0126  groups. Small and broken groups and sub groups  WILL  finally tend to disappear. Looking to the futu
0126  , which as yet have suffered least extinction,   WILL  for a long period continue to increase. But wh
0126  period continue to increase. But which groups    WILL  ultimately prevail, no man can predict; for we
0126  e larger groups, a multitude of smaller groups   WILL  become utterly extinct, and leave no modified
0126  pecies living at any one period, extremely few   WILL  transmit descendants to a remote futurity. I s
0127  ccur, assuredly individuals thus characterised   WILL  have the best chance of being preserved in the
0127  from the strong principle of inheritance they    WILL  tend to produce offspring similarly characteri
0127  adult. Amongst many animals, sexual selection    WILL  give its aid to ordinary selection, by assurin
0127  greatest number of offspring. Sexual selection   WILL  also give characters useful to the males alone
0128  versified these descendants become, the better   WILL  be their chance of succeeding in the battle of
0128  distinguishing varieties of the same species,    WILL  steadily tend to increase till they come to eq
0128  the larger genera, which vary most; and these    WILL  tend to transmit to their modified offspring t
0133  who believes in the creation of each species,    WILL  have to say that this shell, for instance, was
0134  us inducing variability; and natural selection   WILL  then accumulate all profitable variations, how
0136  ss perfectly developed or from indolent habit,   WILL  have had the best chance of surviving from not
0136  hose beetles which most readily took to flight   WILL  oftenest have been blown to sea and thus have
0138  less generations, the deepest recesses, disuse   WILL  on this view have more or less perfectly oblit
0138  ly obliterated its eyes, and natural selection   WILL  often have effected other changes, such as an
0139  to which the inhabitants of these dark abodes    WILL  probably have been exposed. Acclimatisation. H
0140  ty to predict whether or not an imported plant   WILL  endure our climate, and from the number of pla
0142  see no reason to doubt that natural selection    WILL  continually tend to preserve those individuals
0142  d with much greater weight; but until some one   WILL  sow, during a score of generations, his kidney
0143  ted solely for the good of the young or larva,   WILL  it may safely be concluded, affect the struct
0147  especially botanists, believe in its truth. I    WILL  not, however, here give any instances; for I s
0148  iminution, however slight, in its development,   WILL  be seized on by natural selection, for it will
0148  will be seized on by natural selection, for it   WILL  profit the individual not to have its nutrimen
0148  useless. Thus, as I believe, natural selection   WILL  always succeed in the long run in reducing and
0149  bject of rudimentary and aborted organs; and I   WILL  here only add that their variability seems to
0151  k, give a list of the more remarkable cases; I   WILL  here only briefly give one, as it illustrates
0152  e comb in the Dorking fowl) or the whole breed   WILL  cease to have a nearly uniform character. The
0152  to have a nearly uniform character. The breed    WILL  then be said to have degenerated. In rudimenta
0153  he common progenitor of the genus. This period   WILL  seldom be remote in any extreme degree, as spe
0153  o reversion and variability on the other hand,   WILL  in the course of time cease; and that the most
0154  high degree. For in this case the variability   WILL  seldom as yet have been fixed by the continued
0156  an rarely have happened that natural selection   WILL  have modified several species, fitted to more
0156  tant. In connexion with the present subject, I   WILL  make only two other remarks. I think it will b
0156  i will make only two other remarks. I think it   WILL  be admitted, without my entering on details, t
0156  characters are very variable: I think it also    WILL  be admitted that species of the same group dif
0157  eir females; and the truth of this proposition   WILL  be granted. The cause of the original variabil
0157  as they are highly variable, sexual selection    WILL  have had a wide scope for action, and may thus
0157  e genus differ from each other. Of this fact I   WILL  give in illustration two instances, the first
0159  ers of an early progenitor. These propositions   WILL  be most readily understood by looking to our d
0159  other race, the fantail. I presume that no one   WILL  doubt that all such analogous variations are d
0159  m a common parent: if this be not so, the case   WILL  then be one of analogous variation in two so c
0160  the parent rock pigeon, I presume that no one   WILL  doubt that this is a case of reversion, and no
0161  , for the presence of all important characters   WILL  be governed by natural selection. in accordanc
0161  ce with the diverse habits of the species, and   WILL  not be left to the mutual action of the condit
0163  ly do occur, and seem to me very remarkable. I   WILL, however, give one curious and complex case, n
0167  each equine species was independently created,   WILL, I presume, assert that each species has been
0168  be saved without detriment to the individual,    WILL  be saved. Changes of structure at an early age
0168  be saved. Changes of structure at an early age   WILL  generally affect parts subsequently developed;
0168  scale. Rudimentary organs, from being useless,   WILL  be disregarded by natural selection, and hence
0169  tinued and slow process, and natural selection   WILL  in such cases not as yet have had time to over
0169  ommon parent and exposed to similar influences   WILL  naturally tend to present analogous variations
0169  on and analogous variation, such modifications   WILL  add to the beautiful and harmonious diversity
0171  this part of my work, a crowd of difficulties    WILL  have occurred to the reader. Some of them are
0172  ion of profitable modifications, each new form   WILL  tend in a fully stocked country to take the pl
0172  etition. Thus extinction and natural selection   WILL, as we have seen, go hand in hand. Hence, if w
0172  the parent and all the transitional varieties   WILL  generally have been exterminated by the very p
0172  ountless numbers in the crust of the earth? It   WILL  be much more convenient to discuss this questi
0172  e imperfection of the geological record; and I   WILL  here only state that I believe the answer main
0173  whilst it slowly subsides. These contingencies   WILL  concur only rarely, and after enormously long
0173  very little sediment is being deposited, there   WILL  be blanks in our geological history. The crust
0174  s and uniform condition than at present. But I   WILL  pass over this way of escaping from the diffic
0175  , depending as it does on the range of others,   WILL  tend to be sharply defined. Moreover, each spe
0175  ts range, where it exists in lessened numbers,   WILL, during fluctuations in the number of its enem
0175  extermination; and thus its geographical range   WILL  come to be still more sharply defined. If I am
0176  essentially differ from species, the same rule   WILL  probably apply to both; and if we in imaginati
0176  zone. The intermediate variety, consequently,   WILL  exist in lesser numbers from inhabiting a narr
0176  n larger numbers from inhabiting larger areas,   WILL  have a great advantage over the intermediate v
0177  ations for natural selection to seize on, than   WILL  always have a better chance, within any given
0177  , the more common forms, in the race for life,   WILL  the rarer forms which exist in lesser numbers.
0177  and supplant the less common forms, for these   WILL  tend to beat and supplant the less common form
0177  stocks by selection; the chances in this case   WILL  be more slowly modified and improved. It is th
0177  sequently the improved mountain or plain breed   WILL  be strongly in favour of the great holders on
0177  , which originally existed in greater numbers,   WILL  soon take the place of the less improved hill
0178  r more of its inhabitants. And such new places   WILL  come into close contact with each other, witho
0178  ch broken portion of the land, but these links  WILL  depend on slow changes of climate, or on the o
0178  the process of natural selection, so that they  WILL  have been supplanted and exterminated during t
0178  rictly continuous area, intermediate varieties  WILL  no longer exist in a living state. Thirdly, wh
0178  een formed in the intermediate zones, but they  WILL, it is probable, at first have been formed in
0178  ort duration. For these intermediate varieties  WILL  generally have had a short duration. For these
0179  om this cause alone the intermediate varieties  WILL, from reasons already assigned (namely from wh
0179  r modification through natural selection, they  WILL  be liable to accidental extermination; and dur
0179  ct; for these from existing in greater numbers  WILL  almost certainly be beaten and supplanted by t
                                                     WILL, in the aggregate, present more variation, and
```

will

Page ***************************************(Key Word)***

0182	ing early transitional grades of the structure	WILL	seldom continue to exist to the present day, f
0182	continue to exist to the present day, for they	WILL	have been supplanted by the very process of pe
0183	tures fitted for very different habits of life	WILL	rarely have been developed at an early period
0183	onal grades of structure in a fossil condition	WILL	always be less, from their having existed in t
0183	of species with fully developed structures. I	WILL	now give two or three instances of diversified
0183	simultaneously. Of cases of changed habits it	WILL	suffice merely to allude to that of the many B
0185	s in separate and innumerable acts of creation	WILL	say, that in these cases it has pleased the Cr
0186	nce and in the principle of natural selection,	WILL	acknowledge that every organic being is consta
0186	over some other inhabitant of the country, it	WILL	seize on the place of that inhabitant, however
0186	fferent it may be from its own place. Hence it	WILL	cause him no surprise that there should be gee
0188	y member of the great Articulate class. He who	WILL	go thus far, if he find on finishing this trea
0189	s to be destroyed. In living bodies, variation	WILL	cause the slight alterations, generation will
0189	will cause the slight alterations, generation	WILL	multiply them almost infinitely, and natural s
0189	them almost infinitely, and natural selection	WILL	pick out with unerring skill each improvement.
0190	be turned inside out, and the exterior surface	WILL	then digest and the stomach respire. In such c
0191	onversion from one function to another, that I	WILL	give one more instance. Pedunculated cirripede
0192	ave large folded branchiae. Now I think no one	WILL	dispute that the ovigerous frena in the one fa
0192	grave cases of difficulty occur, some of which	WILL	be discussed in my future work. One of the gra
0192	er the males or fertile females: but this case	WILL	be treated of in the next chapter. The electri
0196	tually injurious deviations in their structure	WILL	always have been checked by natural selection.
0196	e law of reversion; that correlation of growth	WILL	have had a most important influence in modifyi
0196	structures; and finally, that sexual selection	WILL	often have largely modified the external chara
0196	ed the external characters of animals having a	WILL	, to give one male an advantage in fighting wit
0199	nt part, and a useful modification of one part	WILL	often have entailed on other parts diversified
0201	n this and other such cases. Natural selection	WILL	never produce in a being anything injurious to
0201	s solely by and for the good of each. No organ	WILL	be formed, as Paley has remarked, for the purp
0201	en the good and evil caused by each part, each	WILL	be found on the whole advantageous. After the
0201	of life, if any part comes to be injurious, it	WILL	be modified; or if it be not so, the being wil
0201	ill be modified; or if it be not so, the being	WILL	become extinct, as myriads have become extinct
0202	mals introduced from Europe. Natural selection	WILL	not produce absolute perfection, nor do we alw
0202	wer of stinging be useful to the community, it	WILL	fulfil all the requirements of natural selecti
0203	artly because the process of natural selection	WILL	always be very slow, and will act, at any one
0203	atural selection will always be very slow, and	WILL	act, at any one time, only on a very few forms
0203	of a continuous area, an intermediate variety	WILL	often be formed, fitted for an intermediate zo
0203	rom reasons assigned, the intermediate variety	WILL	usually exist in lesser numbers than the two f
0204	odification, from existing in greater numbers,	WILL	have a great advantage over the less numerous
0204	er the less numerous intermediate variety, and	WILL	thus generally succeed in supplanting and exte
0205	ns in the struggle for life. Natural selection	WILL	produce nothing in one species for the exclusi
0205	inhabitants one with another, and consequently	WILL	produce perfection, or strength in the battle
0206	nts of one country, generally the smaller one,	WILL	often yield, as we see they do yield, to the i
0206	arger country. For in the larger country there	WILL	have existed more individuals, and more divers
0206	nd more diversified forms, and the competition	WILL	have been severer, and thus the standard of pe
0206	n severer, and thus the standard of perfection	WILL	have been rendered higher. Natural selection w
0206	l have been rendered higher. Natural selection	WILL	not necessarily produce absolute perfection; n
0207	tinct as that of the hive bee making its cells	WILL	probably have occurred to many readers, as a d
0207	qualities of animals within the same class. I	WILL	not attempt any definition of instinct. It wou
0208	t rarely in direct opposition to our conscious	WILL!	yet they may be modified by the will or reaso
0208	onscious will! yet they may be modified by the	WILL	or reason. Habits easily become associated wit
0209	could not possibly have been thus acquired. It	WILL	be universally admitted that instincts are as
0212	ed variations of instinct in a state of nature	WILL	be strengthened by briefly considering a few c
0213	nters (I have myself seen a striking instance)	WILL	sometimes point and even back other dogs the v
0215	out intending to improve the breed, dogs which	WILL	stand and hunt best. On the other hand, habit
0216	for if the hen gives the danger chuckle, they	WILL	run (more especially young turkeys) from under
0216	ed by selection, by considering a few cases. I	WILL	select only three, out of the several which I
0220	an instinct as that of making slaves. Hence I	WILL	give the observations which I have myself made
0223	teps the instinct of F. sanguinea originated I	WILL	not pretend to conjecture. But as ants, which
0223	ture. But as ants, which are not slave makers,	WILL	as I have seen, carry off pupae of other speci
0224	scens. Cell making instinct of the Hive Bee. I	WILL	not here enter on minute details on this subje
0224	enter on minute details on this subject, but I	WILL	merely give an outline of the conclusions at w
0227	everal spheres in both layers be formed, there	WILL	result a double layer of hexagonal prisms unit
0227	e rhombs and the sides of the hexagonal prisms	WILL	have every angle identically the same with the
0230	having always been a little hood of wax; but I	WILL	not here enter on these details. We see how im
0231	essively thin finished wall of the cell, which	WILL	ultimately be left. We shall understand how th
0235	al instinct to new swarms, which in their turn	WILL	have had the best chance of succeeding in the
0236	ed by independent acts of natural selection. I	WILL	not here enter on these several cases, but wil
0236	ill not here enter on these several cases, but	WILL	confine myself to one special difficulty, whic
0236	eserves to be discussed at great length, but I	WILL	here take only a single case, that of working
0239	which our European ants guard or imprison. It	WILL	indeed be thought that I have an overweening c
0240	driver ant (Anomma) of West Africa. The reader	WILL	perhaps best appreciate the amount of differen
0243	cts vary slightly in a state of nature. No one	WILL	dispute that instincts are of the highest impo
0249	ted during the flowering season: hence hybrids	WILL	generally be fertilised during each generation
0250	is command. Of his many important statements I	WILL	here give only a single one as an example, nam
0252	rhocodendrons, which produce no pollen, for he	WILL	find on their stigmas plenty of pollen brought
0254	sterility of first Crosses and of Hybrids. We	WILL	now consider a little more in cetail the circu
0255	first crosses and of hybrids. Our chief object	WILL	be to see whether or not the rules indicate th
0255	lity, beyond that which the plant's own pollen	WILL	produce. So in hybrids themselves, there are s
0257	be given of very closely allied species which	WILL	not unite, or only with extreme difficulty; an
0261	two species the male sexual element of the one	WILL	often freely act on the female sexual element
0261	the other, but not in a reversed direction. It	WILL	be advisable to explain a little more fully by
0261	re in a state of nature, I presume that no one	WILL	suppose that this capacity is a specially endo
0261	s capacity is a specially endowed quality, but	WILL	admit that it is incidental on differences in
0261	. we can sometimes see the reason why one tree	WILL	not take on another, from differences in their
0261	ther, in other cases species of the same genus	WILL	not take on each other. The pear can be grafte
0262	be grafted on the currant, whereas the currant	WILL	take, though with difficulty, on the gooseberr
0265	ids. On the other hand, one species in a group	WILL	sometimes resist great changes of conditions w
0265	ired fertility; and certain species in a group	WILL	produce unusually fertile hybrids. No one can

Page **(Key Word)**

```
0265  , till he tries, whether any particular animal  WILL  breed under confinement or any plant seed free
0265  l he tries, whether any two species of a genus  WILL  produce more or less sterile hybrids. Lastly,
0268  rtility of all varieties produced under nature  WILL  assuredly have to be granted. If we turn to va
0268  ross with European dogs, the explanation which  WILL  occur to everyone, and probably the true one,
0269  sterility in the few following cases, which I  WILL  briefly abstract. The evidence is at least as
0279  ser numbers than the forms which they connect,  WILL  generally be beaten out and exterminated durin
0280  mon but unknown progenitor; and the progenitor  WILL  generally have differed in some respects from
0281  nd an unknown common parent. The common parent  WILL  have had in its whole organisation much genera
0281  ir; and in this case direct intermediate links  WILL  have existed between them. But such a case wou
0281  ganism and organism; between child and parent,  WILL  render this a very rare event; for in all case
0281  n all cases the new and improved forms of life  WILL  tend to supplant the old and unimproved forms.
0282  onnecting links, it may be objected, that time  WILL  not have sufficed for so great an amount of or
0282  nciples of Geology, which the future historian  WILL  recognise as having produced a revolution in n
0283  y studies the action of the sea on our shores,  WILL,  I believe, be most deeply impressed with the
0286  ght at the rate of one inch in a century. This  WILL  at first appear much too small an allowance; b
0288  wholly soft can be preserved. Shells and bones  WILL  decay and disappear when left on the bottom of
0288  hich do become embedded, if in sand or gravel,  WILL  when the beds are upraised generally be dissol
0289  published in the Supplement to Lyell's Manual,  WILL  bring home the truth, how accidental and rare
0290  several successive and peculiar marine faunas  WILL  probably be preserved to a distant age. A litt
0290  reserved to a distant age. A little reflection  WILL  explain why along the rising coast of the west
0291  of E. Forbes, we may conclude that the bottom  WILL  be inhabited by extremely few animals, and the
0291  remely few animals, and the mass when upraised  WILL  give a most imperfect record of the forms of l
0291  of sediment nearly balance each other, the sea  WILL  remain shallow and favourable for life, and th
0291  gradation as it has as yet suffered, but which  WILL  hardly last to a distant geological age, was c
0292  curately, the beds which were then accumulated  WILL  have been destroyed by being upraised and brou
0292  f the coast action. Thus the geological record  WILL  almost necessarily be rendered intermittent. I
0292  nd and of the adjoining shoal parts of the sea  WILL  be increased, and new stations will often be f
0292  of the sea will be increased, and new stations  WILL  often be formed; all circumstances most favour
0292  ies and species; but during such periods there  WILL  generally be a blank in the geological record.
0292  , the inhabited area and number of inhabitants  WILL  decrease (excepting the productions on the sho
0292  d consequently during subsidence, though there  WILL  be much extinction, fewer new varieties or spe
0292  uch extinction, fewer new varieties or species  WILL  be formed; and it is during these very periods
0294  riod shall have been upraised, organic remains  WILL  probably first appear and disappear at differe
0295  e slow process of variation; hence the deposit  WILL  generally have to be a very thick one; and the
0295  k one; and the species undergoing modification  WILL  have had to live on the same area throughout t
0295  bsidence. But this same movement of subsidence  WILL  often tend to sink the area whence the sedimen
0295  and a supply of sediment of a different nature  WILL  generally have been due to geographical change
0296  geographical changes requiring much time. Nor  WILL  the closest inspection of a formation give any
0299  for instance, geologists at some future period  WILL  be able to prove, that our different breeds of
0299  which may be urged against my views. Hence it  WILL  be worth while to sum up the foregoing remarks
0303  ccumulation of each formation. These intervals  WILL  have given time for the multiplication of spec
0303  ; and in the succeeding formation such species  WILL  appear as if suddenly created. I may here reca
0303  ead rapidly and widely throughout the world. I  WILL  now give a few examples to illustrate these re
0305  d by running through Pictet's Palaeontology it  WILL  be seen that very few species are known from s
0308  t it may hereafter receive some explanation, I  WILL  give the following hypothesis. From the nature
0310  arguments of other kinds given in this volume,  WILL  undoubtedly at once reject my theory. For my p
0313  st closely related formations, all the species  WILL  be found to have undergone some change. When a
0315  t become in some degree modified and improved,  WILL  be liable to be exterminated. Hence we can see
0315  become modified; for those which do not change  WILL  become extinct. In members of the same class t
0315  e every reason to believe that the parent form  WILL  generally be supplanted and exterminated by it
0317  tions in which the species are found, the line  WILL  sometimes falsely appear to begin at its lower
0320  the same number of old forms. The competition  WILL  generally be most severe, as formerly explaine
0321  improved and modified descendants of a species  WILL  generally cause the extermination of the paren
0321  t species, i.e. the species of the same genus,  WILL  be the most liable to extermination. Thus, as
0321  that a new species belonging to some one group  WILL  have seized on the place occupied by a species
0321  e developed from the successful intruder, many  WILL  have to yield their places; and it will genera
0321  , many will have to yield their places; and it  WILL  generally be allied forms, which will suffer f
0321  ; and it will generally be allied forms, which  WILL  suffer from some inherited inferiority in comm
0322  roup have taken possession of a new area, they  WILL  have exterminated in a correspondingly rapid m
0322  s; and the forms which thus yield their places  WILL  commonly be allied, for they will partake of s
0322  their places will commonly be allied, for they  WILL  partake of some inferiority in common. Thus, a
0322  m perceived by us, the whole economy of nature  WILL  be utterly obscured. Whenever we can precisely
0325  e discover a series of analogous phenomena, it  WILL  appear certain that all these modifications of
0326  idents, but in the long run the dominant forms  WILL  generally succeed in spreading. The diffusion
0327  their places to the new and victorious forms,  WILL  generally be allied in groups, from inheriting
0327  groups spread throughout the world, old groups  WILL  disappear from the world; and the succession o
0327  orld; and the succession of forms in both ways  WILL  everywhere tend to correspond. There is one ot
0328  pecies would not exactly correspond; for there  WILL  have been a little more time in the one region
0331  nct. The three existing genera, a14, c14, p14,  WILL  form a small family; b14 and f14 a closely all
0331  s of descent diverging from the parent form A,  WILL  form an order; for all will have inherited som
0331  the parent form A, will form an order; for all  WILL  have inherited something in common from their
0331  gram, the more recent any form is, the more it  WILL  generally differ from its ancient progenitor.
0333  their collateral relations. In nature the case  WILL  be far more complicated than is represented in
0333  is represented in the diagram; for the groups  WILL  have been more numerous, they will have endure
0333  the groups will have been more numerous, they  WILL  have endured for extremely unequal lengths of
0333  red for extremely unequal lengths of time, and  WILL  have been modified in various degrees. As we p
0333  una of any great period in the earth's history  WILL  be intermediate in general character between t
0335  stribution of existing species over the globe. I  WILL  not attempt to account for the close resemblan
0336  orms are more highly developed than ancient. I  WILL  not here enter on this subject, for naturalist
0340  r the inhabitants of each quarter of the world  WILL  obviously tend to leave in that quarter, durin
0340  ed greatly from those of another continent, so  WILL  their modified descendants still differ in nea
0340  , permitting much inter migration, the feebler  WILL  yield to the more dominant forms, and there wi
0340  ll yield to the more dominant forms, and there  WILL  be nothing immutable in the laws of past and p
0341  of two or three alone of the six older genera  WILL  have been the parents of the six new genera; t
0341  south America, still fewer genera and species  WILL  have left modified blood descendants. Summary
0342  of elevation, and during the latter the record  WILL  have been least perfectly kept; that each sing
```

Page **(Key Word)**

0342 the geological record extremely imperfect, and WILL to a large extent explain why we do not find i
0342 views on the nature of the geological record, WILL rightly reject my whole theory. For he may ask
0344 s of life, which are those that oftenest vary, WILL in the long run tend to people the world with
0344 h allied, but modified, descendants; and these WILL generally succeed in taking the places of thos
0344 ntervals of time, the productions of the world WILL appear to have changed simultaneously. We can
0344 the more ancient a form is, the more nearly it WILL be related to, and consequently resemble, the
0345 ore recent animals of the same class, the fact WILL be intelligible. The succession of the same ty
0347 bit a considerably different climate, and they WILL be found incomparably more closely related to
0350 ysical conditions. The degree of dissimilarity WILL depend on the migration of the more dominant f
0350 competitors in their own widely extended homes WILL have the best chance of seizing on new places,
0350 ad into new countries. In their new homes they WILL be exposed to new conditions, and will frequen
0350 es they will be exposed to new conditions, and WILL frequently undergo further modification and im
0350 er modification and improvement; and thus they WILL become still further victorious, and will prod
0350 they will become still further victorious, and WILL produce groups of modified descendants. On thi
0351 f each species is an independent property, and WILL be taken advantage of by natural selection, on
0351 he degree of modification in different species WILL be no uniform quantity. If, for instance, a nu
0351 to a new and afterwards isolated country, they WILL be little liable to modification: for neither
0351 e vast geographical and climatal changes which WILL have supervened since ancient times, almost an
0352 ting distant points of the world. No geologist WILL feel any difficulty in such cases as Great Bri
0354 h cases. But after some preliminary remarks, I WILL discuss a few of the most striking classes of
0355 d immigrants from this other region, my theory WILL be strengthened; for we can clearly understand
0355 from a succession of improved varieties, which WILL never have blended with other individuals or v
0355 ended with other individuals or varieties, but WILL have supplanted each other; so that, at each s
0355 provement, all the individuals of each variety WILL have descended from a single parent. But in th
0356 of modification the individuals of the species WILL have been kept nearly uniform by intercrossing
0356 orm by intercrossing; so that many individuals WILL have gone on simultaneously changing, and the
0356 changing, and the whole amount of modification WILL not have been due, at each stage, to descent f
0356 rmerly have been submerged, and the two faunas WILL now blend or may formerly have blended: where
0357 pass from one to the other. No other geologist WILL dispute that great mutations of level, have oc
0357 whenever it is fully admitted, as I believe it WILL some day be, that each species has proceeded f
0357 sion of the land. But I do not believe that it WILL ever be proved that within the recent period c
0361 intestines of a bird; but hard seeds of fruit WILL pass uninjured through even the digestive orga
0367 r or later in North America than in Europe, so WILL the southern migration there have been a littl
0367 have been a little earlier or later; but this WILL make no difference in the final result. As the
0367 and the re migration on the returning warmth, WILL generally have been due south and north. The A
0368 on); then the arctic and temperate productions WILL at a very late period have marched a little fu
0368 southern migration and re migration northward, WILL have been exposed to nearly the same climate,
0368 ate; and, as is especially to be noticed, they WILL have kept in a body together: consequently the
0368 together; consequently their mutual relations WILL not have been much disturbed, and, in accordan
0368 the principles inculcated in this volume, they WILL not have been liable to much modification. But
0368 tely on the summits of the mountains, the case WILL have been somewhat different: for it is not li
0369 is not likely that all the same arctic species WILL have been left on mountain ranges distant from
0369 ther, and have survived there ever since; they WILL, also, in all probability have become mingled
0369 ial epoch, and which during its coldest period WILL have been temporarily driven down to the plain
0369 en temporarily driven down to the plains; they WILL, also, have been exposed to somewhat different
0369 nt climatal influences. Their mutual relations WILL thus have been in some degree disturbed; conse
0369 en in some degree disturbed; consequently they WILL have been liable to modification; and this we
0371 s the northern parts of the Old and New Worlds WILL have been almost continuously united by land,
0371 he plants and animals migrated southward, they WILL have become mingled in the one great region wi
0372 ong the continuous shores of the Polar Circle, WILL account, on the theory of modification, for ma
0377 ave spread within a few centuries, this period WILL have been ample for any amount of migration. A
0377 all the tropical plants and other productions WILL have retreated from both sides towards the equ
0377 bear in mind is, that all tropical productions WILL have suffered to a certain extent. On the othe
0377 r migrating nearer to the equator, though they WILL have been placed under somewhat new conditions
0378 ave been placed under somewhat new conditions, WILL have suffered less. And it is certain that man
0379 emisphere. These being surrounded by strangers WILL have had to compete with many new forms of lif
0379 in their structure, habits, and constitutions WILL have profited them. Thus many of these wandere
0381 tic regions. These cannot be here discussed. I WILL only say that as far as regards the occurrence
0384 and almost capriciously; for two river systems WILL have some fish in common and some different. A
0384 to very ancient forms, and in such cases there WILL have been ample time for great geographical ch
0386 : i have tried several little experiments, but WILL here give only the most striking case: I took
0388 st formed, for instance, on a rising islet, it WILL be unoccupied; and a single seed or egg will h
0388 t will be unoccupied; and a single seed or egg WILL have a good chance of succeeding. Although there
0388 ve a good chance of succeeding. Although there WILL always be a struggle for life between the indi
0388 mpared with those on the land, the competition WILL probably be less severe between aquatic than b
0388 modified less quickly than the high; and this WILL give longer time than the average for the migr
0390 rine of the creation of each separate species, WILL have to admit that a sufficient number of the
0390 ct, and having to compete with new associates, WILL be eminently liable to modification, and will
0390 will be eminently liable to modification, and WILL often produce groups of modified descendants.
0391 and when settled in their new homes, each kind WILL have been kept by the others to their proper p
0391 others to their proper places and habits, and WILL consequently have been little liable to modifi
0394 ions. Yet it cannot be said that small islands WILL not support small mammals, for they occur in m
0397 a wreck now remains, must not be overlooked. I WILL here give a single instance of one of the case
0397 erous instances could be given of this fact. I WILL give only one, that of the Galapagos Archipela
0399 t this affinity is confined to the plants, and WILL, I do not doubt, be some day explained. The la
0402 es has any advantage whatever over another, it WILL in a very brief time wholly or in part supplan
0402 for their own places in nature, both probably WILL hold their own places and keep separate for al
0404 . other analogous facts could be given. And it WILL, I believe, be universally found to be true, t
0404 allied or representative species occur, there WILL likewise be found some identical species, show
0404 rever many closely allied species occur, there WILL be found many forms which some naturalists ran
0405 ancing species vary and give rise to new forms WILL largely determine their average range. For ins
0405 n the case of certain powerfully winged birds, WILL necessarily range widely: for we should never
0405 at a remote epoch: so that in such cases there WILL have been ample time for great climatal and ge
0406 higher forms: and consequently the lower forms WILL have had a better chance of ranging widely and
0410 wo forms are related in blood, the nearer they WILL generally stand to each other in time and spac

Page **************************************(Key Word)**************************************

```
0412 explained, of these several principles; and he WILL see that the inevitable result is that the mod
0415 ged in the writings of almost every author. It WILL suffice to quote the highest authority, Robert
0416 e value in classification; yet no one probably WILL say that the antennae in these two divisions o
0416 within the same group of beings. Again, no one WILL say that rudimentary or atrophied organs are o
0416 often of high value in classification. No one WILL dispute that the rudimentary teeth in the uppe
0420 era, families, sections, or orders. The reader WILL best understand what is meant, if he will take
0420 ader will best understand what is meant, if he WILL take the trouble of referring to the diagram i
0420 rring to the diagram in the fourth chapter. We WILL suppose the letters A to L to represent allied
0421 sed to have been but slightly modified; and it WILL then rank with the parent genus F; just as som
0421 f descent. All the modified descendants from A WILL have inherited something in common from their
0421 mething in common from their common parent, as WILL all the descendants from I; so will it be with
0421 parent, as will all the descendants from I; so WILL it be with each subordinate branch of descenda
0421 case, their places in a natural classification WILL have been more or less completely lost, as som
0421 e genus. But this genus, though much isolated, WILL still occupy its proper intermediate position;
0421 several genera descended from these two genera WILL have inherited to a certain extent their chara
0425 there has been close descent in common, there WILL certainly be close resemblance or affinity. As
0427 se external resemblance; but such resemblances WILL not reveal, will rather tend to conceal their
0427 blance; but such resemblances will not reveal, WILL rather tend to conceal their blood relationshi
0429 e present day descendants but little modified, WILL give to us our so called osculant or aberrant
0430 ed off from some very ancient Marsupial, which WILL have had a character in some degree intermedia
0430 itor than have other Rodents; and therefore it WILL not be specially related to any one existing M
0431 xtinction into distinct groups and sub groups, WILL have transmitted some of its characters, modif
0431 s and degrees, to all; and the several species WILL consequently be related to each other by circu
0435 organisation. In changes of this nature, there WILL be little or no tendency to modify the origina
0435 in all this great amount of modification there WILL be no tendency to alter the framework of bones
0436 ery simple in form; and then natural selection WILL account for the infinite diversity in structur
0437 e pieces in the act of parturition of mammals, WILL by no means explain the same construction in t
0438 poses. And as the whole amount of modification WILL have been effected by slight successive steps,
0440 related to similar conditions of life. No one WILL suppose that the stripes on the whelp of a lio
0441 ion being higher or lower. But no one probably WILL dispute that the butterfly is higher than the
0443 animal has been born, what its merits or form WILL ultimately turn out. We see this plainly in ou
0443 ldren; we cannot always tell whether the child WILL be tall or short, or what its precise features
0443 be tall or short, or what its precise features WILL be. The question is not, at what period of lif
0444 se two principles, if their truth be admitted, WILL, I believe, explain all the above specified le
0445 hed; I carefully measured the proportions (but WILL not here give details) of the beak, width of m
0447 young of the new species of our supposed genus WILL manifestly tend to resemble each other much mo
0447 the several descendants of the parent species WILL still resemble each other closely; for they wi
0447 ll still resemble each other closely, for they WILL not have been modified. But in each individual
0447 dividual new species, the embryonic fore limbs WILL differ greatly from the fore limbs in the matu
0447 may have in modifying an organ, such influence WILL mainly affect the mature animal, which has com
0447 its own living; and the effects thus produced WILL be inherited at a corresponding mature age. Wh
0449 nic structure reveals community of descent. It WILL remain unmodified, or be modified in a lesser
0451 her most closely in all respects, one of which WILL reveal this community of descent, however much
0455 disuse or selection reduces an organ, and this WILL have full sized wings, and another mere rudime
0455 principle of inheritance at corresponding ages WILL generally be when the being has come to maturi
0455 educed state at the same age, and consequently WILL reproduce the organ in its reduced state at th
0455 or structure, if not useful to the possessor, WILL seldom affect or reduce it in the embryo. Thus
0455 ossessor, will be saved as far as is possible, WILL be saved as far as is possible, will probably
0455 , will probably often come into play; and this WILL probably often come into play; and this will t
0457 yo of the homologous parts, which when matured WILL tend to cause the entire obliteration of a rud
0457 d in size, either from disuse or selection, it WILL become widely different from each other in str
0462 es; for during very long periods of time there WILL generally be at that period of life when the b
0462 earth during modern periods; and such changes WILL always be a good chance for wide migration by
0462 essarily been slow, all the means of migration WILL obviously have greatly facilitated migration.
0463 so shown that the intermediate varieties which WILL have been possible during a very long period;
0463 irst probably exist in the intermediate zones, WILL at first probably exist in the intermediate zo
0463 the latter, from existing in greater numbers, WILL be liable to be supplanted by the allied forms
0463 er numbers; so that the intermediate varieties WILL generally be modified and improved at a quicke
0464 be named which are probably varieties; but who WILL, in the long run, be supplanted and exterminat
0464 etend that in future ages so many fossil links WILL pretend that in future ages so many fossil lin
0464 sil links will be discovered, that naturalists WILL be discovered, that naturalists will be able t
0464 link or intermediate variety be discovered, it WILL be able to decide, on the common view, whether
0464 ntermediate links less likely. Local varieties WILL simply be classed as another and distinct spec
0465 if discovered in a geological formation, they WILL not spread into other and distant regions unti
0465 will appear as if suddenly created there, and WILL appear as if suddenly created there, and will
0465 f elevation and of stationary level the record WILL be simply classed as new species. Most formati
0465 ll be blank. During these latter periods there WILL be blank. During these latter periods there wi
0465 r. that the geological record is imperfect all WILL probably be more variability in the forms of l
0465 s imperfect to the degree which I require, few WILL admit; but that it is imperfect to the degree
0467 n can possibly survive. A grain in the balance WILL be inclined to admit. If we look to long enoug
0468 sest competition with each other, the struggle WILL determine which individual shall live and whic
0468 will generally be most severe between them; it WILL generally be most severe between them; it will
0468 he species of the same genus. But the struggle WILL be almost equally severe between the varieties
0468 degree to the surrounding physical conditions, WILL often be very severe between beings most remot
0468 nce. With animals having separated sexes there WILL turn the balance. With animals having separate
0468 fully struggled with their conditions of life, WILL in most cases be a struggle between males for
0468 will generally leave most progeny. But success WILL generally leave most progeny. But success will
0468 arms of the males; and the slightest advantage WILL often depend on having special weapons or mean
0470 nd as the modified descendants of each species WILL lead to victory. As geology plainly proclaims
0470 fferent places in the economy of nature, there WILL be enabled to increase by so much the more as
0470 of the same genus. New and improved varieties WILL be a constant tendency in natural selection to
0474 ed for a very long period; for in this case it WILL inevitably supplant and exterminate the older,
0475 principle of natural selection; for old forms WILL have been rendered constant by long continued
0476 ter, the progenitor with its early descendants WILL be supplanted by new and improved forms. Neith
0476 a confined country, the recent and the extinct WILL often be intermediate in character in comparis
                                                 WILL naturally be allied by descent. Looking to geo
```

Page **************************************(Key Word)**************************************

```
0477  each great class are plainly related; for they  WILL  generally be descendants of the same progenito
0477  portant of all relations, and as the two areas  WILL  have received colonists from some third source
0477  s, the course of modification in the two areas  WILL  inevitably be different. On this view of migra
0479  disuse, aided sometimes by natural selection,  WILL  often tend to reduce an organ, when it has bec
0480  f rudimentary organs. But disuse and selection  WILL  generally act on each creature, when it has co
0480  s full part in the struggle for existence, and  WILL  thus have little power of acting on an organ d
0480  on an organ during early life; hence the organ  WILL  not be much reduced or rendered rudimentary at
0480  modification, which it seems that we wilfully  WILL  not understand. I have now recapitulated the c
0482  o the explanation of a certain number of facts  WILL  certainly reject my theory. A few naturalists,
0482  e future, to young and rising naturalists, who  WILL  be able to view both sides of the question wit
0482  ver is led to believe that species are mutable  WILL  do good service by conscientiously expressing
0482  ning any distinction in the two cases. The day  WILL  come when this will be given as a curious illu
0482  in the two cases. The day will come when this  WILL  be given as a curious illustration of the blin
0484  ally admitted, we can dimly foresee that there  WILL  be a considerable revolution in natural histor
0484  le revolution in natural history. Systematists  WILL  be able to pursue their labours as at present;
0484  o pursue their labours as at present; but they  WILL  not be incessantly haunted by the shadowy doub
0484  his I feel sure, and I speak after experience,  WILL  be no slight relief. The endless disputes whet
0484  y species of British brambles are true species  WILL  cease. Systematists will have only to decide (
0484  bles are true species will cease. Systematists  WILL  have only to decide (not that this will be eas
0484  atists will have only to decide (not that this  WILL  be easy) whether any form be sufficiently cons
0484  to deserve a specific name. This latter point  WILL  become a far more essential consideration than
0485  nd in this case scientific and common language  WILL  come into accordance. In short, we shall have
0485  nd more general departments of natural history  WILL  rise greatly in interest. The terms used by na
0485  racters, rudimentary and aborted organs, etc.,  WILL  cease to be metaphorical, anc will have a plai
0485  gans, etc., will cease to be metaphorical, and  WILL  have a plain signification. When we no longer
0486  far more interesting, I speak from experience,  WILL  the study of natural history become! A grand a
0486  a grand and almost untrodden field of inquiry  WILL  be opened, on the causes and laws of variation
0486  nd so forth. The study of domestic productions  WILL  rise immensely in value. A new variety raised
0486  mmensely in value. A new variety raised by man  WILL  be a far more important and interesting subjec
0486  already recorded species. Our classifications  WILL  come to be, as far as they can be so made, gen
0486  s far as they can be so made, genealogies; and  WILL  then truly give what may be called the plan of
0486  he plan of creation. The rules for classifying  WILL  no doubt become simpler when we have a definit
0486  h have long been inherited. Rudimentary organs  WILL  speak infallibly with respect to the nature of
0486  which may fancifully be called living fossils,  WILL  aid us in forming a picture of the ancient for
0486  cture of the ancient forms of life. Embryology  WILL  reveal to us the structure, in some degree obs
0487  en, by the light which geology now throws, and  WILL  continue to throw, on former changes of climat
0487  mutation of each great fossiliferous formation  WILL  be recognised as having depended on an unusual
0488  ough of a length quite incomprehensible by us,  WILL  hereafter be recognised as a mere fragment of
0488  for far more important researches. Psychology  WILL  be based on a new foundation, that of the nece
0488  mental power and capacity by gradation. Light  WILL  be thrown on the origin of man and his history
0489  e may safely infer that not one living species  WILL  transmit its unaltered likeness to a distant f
0489  turity. And of the species now living very few  WILL  transmit progeny of any kind to a far distant
0489  tic glance into futurity as to foretel that it  WILL  be the common and widely spread species, belon
0489  nging to the larger and dominant groups, which  WILL  ultimately prevail and procreate new and domin
0489  ach being, all corporeal and mental endowments  WILL  tend to progress towards perfection. It is int
0023  been rock pigeons, that is, not breeding or  WILLINGLY  perching on trees. But besides C. Livia, wi
0029  l know that each race varies slightly, for they  WIN  their prizes by selecting such slight differen
0086  seeds more and more widely disseminated by the  WIND,  I can see no greater difficulty in this being
0093  ins, and on some a profusion of pollen. As the  WIND  had set for several days from the female to th
0100  analogous to the action of insects and of the  WIND  in the case of plants, by which, an occasional
0136  y mr. Wollaston, lie much concealed, until the  WIND  lulls and the sun shines: that the proportion
0364  al, nor is the direction of prevalent gales of  WIND.  It should be observed that scarcely any means
0401  and sweep across the archipelago, and gales of  WIND  are extraordinarily rare: so that the islands
0136  s were saved by successfully battling with the  WINDS,  or by giving up the attempt and rarely or ne
0395  ies in a more or less modified condition. Mr.  WINDSOR  Earl has made some striking observations on
0011  ind in the domestic duck that the bones of the  WING  weigh less and the bones of the leg more, in p
0021  as, proportionally to its size, much elongated  WING  and tail feathers. The trumpeter and laugher,
0022  on of the oil gland; the number of the primary  WING  and caudal feathers; the relative length of wi
0022  ng and caudal feathers; the relative length of  WING  and tail to each other and to the body; the re
0025  blue colour, with the white rump, double black  WING  bar, and barred and white edged tail feathers,
0088  urage, length to the spur, and strength to the  WING  to strike in the spurred leg, as well as the b
0150  art in closely allied species. Thus, the bat's  WING  is a most abnormal structure in the class mamm
0154  ny modified descendants, as in the case of the  WING  of the bat, it must have existed, according to
0181  convert it into a bat. In bats which have the  WING  membrane extended from the top of the shoulder
0182  d from these remarks that any of the grades of  WING  structure here alluded to, which perhaps may a
0184  ith elongated wings which chase insects on the  WING;  and on the plains of La Plata, where not a tr
0200  e monkey, in the fore leg of the horse, in the  WING  of the bat, and in the flipper of the seal, ar
0426  le of the jaw, the manner in which an insect's  WING  is folded, whether the skin be covered by hair
0434  the horse, the paddle of the porpoise, and the  WING  of the bat, should all be constructed on the s
0435  m increased to any extent, so as to serve as a  WING:  yet in all this great amount of modification
0437  ones have been created in the formation of the  WING  and leg of a bat, used as they are for such to
0440  ieve that the same bones in the hand of a man,  WING  of a bat, and fin of a porpoise, are related t
0442  re is no obvious reason why, for instance, the  WING  of a bat, or the fin of a porpoise, should not
0450  e males of mammals: I presume that the bastard  WING  in birds may be safely considered as a digit i
0451  ncapable of flight, and not rarely lying under  WING  cases, firmly soldered together! The meaning o
0474  developed in the most unusual manner, like the  WING  of a bat, and yet not be more variable than an
0479  of bones being the same in the hand of a man,  WING  of bat, fin of the porpoise, and leg of the ho
0480  r like the shrivelled wings under the soldered  WING  covers of some beetles, should thus so frequen
0191  annelids are homologous with the wings and  WINGCOVERS  of insects, it is probable that organs whic
0091  cat, according to Mr. St. John, bringing home  WINGED  game, another hares or rabbits, and another h
0146  instance, Alph. de Candolle has remarked that  WINGED  seeds are never found in fruits which do not
0146  he fact that seeds could not gradually become  WINGED  through natural selection, except in fruits w
0182  fins, might have been modified into perfectly  WINGED  animals. If this had been effected, who would
0405  widely, as in the case of certain powerfully  WINGED  birds, will necessarily range widely; for we
0134  to escape danger, I believe that the nearly  WINGLESS  condition of several birds, which now inhabi
```

Page **************************************(Key Word)**************************************

```
0136  s and the sun shines! that the proportion of WINGLESS beetles is larger on the exposed Dezertas th
0136  considerations have made me believe that the WINGLESS condition of so many Madeira beetles is main
0391  slands reptiles, and in New Zealand gigantic WINGLESS birds, take the place of mammals. In the pla
0021  f runts have very long necks, others very long WINGS and tails, others singularly short tails. The
0021  oad one. The pouter has a much elongated body, and legs! and its enormously developed crop,
0025  uter feathers externally edged with white! the WINGS have two black bars! some semi domestic breed
0025  d breeds have, besides the two black bars, the WINGS chequered with black. These several marks do
0134  ap along the surface of the water, and has its WINGS in nearly the same condition as the domestic
0135  of its body, its legs were used more, and its WINGS less, until they became incapable of flight.
0135  es inhabiting Madeira, are so far deficient in WINGS that they cannot fly! and that of the twenty
0136  idual beetle which flew least, either from its WINGS having been ever so little less perfectly dev
0136  era and Lepidoptera, must habitually use their WINGS to gain their subsistence, have, as Mr. Wolla
0136  stence, have, as Mr. Wollaston suspects, their WINGS not at all reduced, but even enlarged. This i
0136  natural selection to enlarge or to reduce the WINGS, would depend on whether a greater number of
0137  f vision. In the same manner as in Madeira the WINGS of some of the insects have been enlarged, an
0137  ome of the insects have been enlarged, and the WINGS of others have been reduced by natural select
0150  because there is a whole group of bats having WINGS! it would apply only if some one species of b
0150  apply only if some one species of bat had its WINGS developed in some remarkable manner in compar
0157  al hymenoptera, the manner of neuration of the WINGS is a character of the highest importance, bec
0159  of slaty blue birds with two black bars on the WINGS, a white rump, a bar at the end of the tail,
0182  that birds might have existed which used their WINGS solely as flappers, like the logger headed du
0182  hly perfected for any particular habit, as the WINGS of a bird for flight, we should bear in mind
0184  ed largely on fruit, and others with elongated WINGS which chase insects on the wing! and on the p
0185  rasping the stones with its feet and using its WINGS under water. He who believes that each being
0191  sal scales of Annelids are homologous with the WINGS and wingcovers of insects, it is probable tha
0236  shape of the thorax and in being destitute of WINGS and sometimes of eyes, and in instinct. As fa
0392  mentary organ, for instance, as the shrivelled WINGS under the soldered elytra of many insular bee
0416  aws of the Marsupials, the manner in which the WINGS of insects are folded, mere colour in certain
0447  as hands, in another as paddles, in another as WINGS! and on the above two principles, namely of e
0447  thus been converted into hands, or paddles, or WINGS. Whatever influence long continued exercise o
0451  ryonic birds. Nothing can be plainer than that WINGS are formed for flight, yet in how many insect
0451  for flight, yet in how many insects do we see WINGS so reduced in size as to be utterly incapable
0451  ll respects, one of which will have full sized WINGS, and another mere rudiments of membrane! and
0451  ossible to doubt, that the rudiments represent WINGS. Rudimentary organs sometimes retain their po
0452  r fact is well exemplified in the state of the WINGS of the female moths in certain groups. Rudime
0454  of animals inhabiting dark caverns, and of the WINGS of birds inhabiting oceanic islands, which ha
0454  ght become injurious under others, as with the WINGS of beetles living on small and exposed island
0473  instance, at the logger headed duck, which has WINGS incapable of flight, in nearly the same condi
0480  h in the embryonic calf or like the shrivelled WINGS under the soldered wing covers of some beetle
0068  effective of all checks. I estimated that the WINTER of 1854 55 destroyed four fifths of the birds
0069  ich have got least food through the advancing WINTER, which will suffer most. When we travel from
0070  of seed, as their numbers are checked during WINTER: but any one who has tried, knows how trouble
0084  s mottled grey; the alpine ptarmigan white in WINTER, the red grouse the colour of heather, and th
0179  es for and preys on fish, but during the long WINTER it leaves the frozen waters, and preys like o
0234  e to support a large stock of bees during the WINTER: and the security of the hive is known mainly
0234  ppose that the community lived throughout the WINTER, and consequently required a store of honey:
0034  s breeds from his own best dogs, but he has no WISH or expectation of permanently altering the bre
0050  sed only with much difficulty. We could hardly WISH for better evidence of the two forms being spe
0090  ages to their male offspring. Yet, I would not WISH to attribute all such sexual differences to th
0215  rarely or never become broody, that is, never WISH to sit on their eggs. Familiarity alone preven
0238  et nearly the same variety: breeders of cattle WISH the flesh and fat to be well marbled together:
0269  s them in nearly the same manner, and does not WISH to alter their general habits of life. Nature
0272  several other respects. Gartner, whose strong WISH was to draw a marked line of distinction betwe
0036  eeders could never have expected or even have WISHED to have produced the result which ensued, nam
0078  s. In this case we can clearly see that if we WISHED in imagination to give the plant the power of
0078  ally be changed in an essential manner. If we WISHED to increase its average numbers in its new ho
0084  fe, and so be preserved. How fleeting are the WISHES and efforts of man! how short his time! and c
0269  ion, for his own use and pleasure! he neither WISHES to select, nor could select, slight differenc
0271  ost distinct domestic varieties, and from not WISHING or being able to produce recondite and funct
0147  disuse, and, on the other hand, the actual WITHDRAWAL of nutriment from one part owing to the exc
0202  ed against many attacking animals, cannot be WITHDRAWN, owing to the backward serratures, and so i
0255  t species causing the flower of the hybrid to WITHER earlier than it otherwise would have done! an
0255  it otherwise would have done! and the early WITHERING of the flower is well known to be a sign of
0009  us affected; so will some animals and plants WITHSTAND domestication or cultivation, and vary very
0017  y inherent tendency to vary, and likewise to WITHSTAND diverse climates. I do not dispute that the
0077  numbers? We know that it can perfectly well WITHSTAND a little more heat or cold, dampness or dry
0142  ivated plants, certain varieties are said to WITHSTAND certain climates better than others: this i
0173  sediment sufficiently thick and extensive to WITHSTAND an enormous amount of future degradation; a
0290  ick, solid, or extensive masses, in order to WITHSTAND the incessant action of the waves, when fir
0377  ow that many tropical plants and animals can WITHSTAND a considerable amount of cold, many might h
0377  otected from the inroads of competitors, can WITHSTAND a much warmer climate than their own. Hence
0397  found that several species did in this state WITHSTAND uninjured an immersion in sea water during
0140  pacity in our domestic animals of not only WITHSTANDING the most different climates but of being p
0221  in England. One day I fortunately chanced to WITNESS a migration from one nest to another, and it
0270  scum, by so good an observer and so hostile a WITNESS, as Gartner: namely, that yellow and white v
0214  pigeon to tumble, an action which, as I have WITNESSED, is performed by young birds, that have nev
0270  the evidence is, also, derived from hostile WITNESSES, who in all other cases consider fertility
0090  inary illustrations. Let us take the case of a WOLF, which preys on various animals, securing some
0090  mbers, during that season of the year when the WOLF is hardest pressed for food. I can under such
0091  oportional numbers of the animals on which our WOLF preyed, a cub might be born with an innate ten
0091  habit or of structure benefited an individual WOLF, it would have the best chance of surviving an
0091  er supplant or coexist with the parent form of WOLF. Or, again, the wolves inhabiting a mountainou
0091  to Mr. Pierce, there are two varieties of the WOLF inhabiting the Catskill Mountains in the Unite
0141  blood, for instance, of a tropical and arctic WOLF or wild dog may perhaps be mingled in our dome
0213  true instincts. If we were to see one kind of WOLF, when young and without any training, as soon
0213  ward with a peculiar gait; and another kind of WOLF rushing round, instead of at, a herd of deer,
```

Page **************************************(Key Word)**************************************

0214 describes a dog, whose great grandfather was a WOLF, and this dog showed a trace of its wild paren
0393 the Falkland Islands, which are inhabited by a WOLF like fox, come nearest to an exception; but th
0048 which are characterized as varieties in Mr. WOLLASTON's admirable work, but which it cannot be do
0052 ods, as has been shown to be the case by Mr. WOLLASTON with the varieties of certain fossil land s
0132 islands or near the coast. So with insects, WOLLASTON is convinced that residence near the sea af
0135 ly, or mainly, due to natural selection. Mr. WOLLASTON has discovered the remarkable fact that 200
0135 t the beetles in Madeira, as observed by Mr. WOLLASTON, lie much concealed, until the wind lulls a
0136 rdinary fact, so strongly insisted on by Mr. WOLLASTON, of the almost entire absence of certain la
0136 ings to gain their subsistence, have, as Mr. WOLLASTON suspects, their wings not at all reduced, b
0176 iven me by Mr. Watson, Dr. Asa Gray, and Mr. WOLLASTON, that generally when varieties intermediate
0389 lph. de Candolle admits this for plants, and WOLLASTON for insects. If we look to the large size a
0402 liar to Chatham Island. Sir C. Lyell and Mr. WOLLASTON have communicated to me a remarkable fact b
0090 eason to doubt that the swiftest and slimmest WOLVES would have the best chance of surviving, and
0091 with the parent form of wolf. Or, again, the WOLVES inhabiting a mountainous district, and those
0215 of men has become instinctive in the dog. All WOLVES, foxes, jackals, and species of the cat genus
0198 sure the shape of the head of the young in the WOMB. The laborious breathing necessary in high reg
0264 as long as it is nourished within its mother's WOMB or within the egg or seed produced by the moth
0425 perfect kangaroo were seen to come out of the WOMB of a bear? According to all analogy, it would
0440 in the young mammal which is nourished in the WOMB of its mother, in the egg of the bird which is
0443 animal, as long as it remains in its mother's WOMB, or in the egg, or as long as it is nourished
0483 e false marks of nourishment from the mother's WOMB? Although naturalists very properly demand a f
0036 uego, by their killing and devouring their old WOMEN, in times of dearth, as of less value than th
0042 ot be matched, and, although so much valued by WOMEN and children, we hardly ever see a distinct b
0072 ve the stems of the heath, and had failed. No WONDER that, as soon as the land was enclosed, it be
0084 ature during whole geological periods. Can we WONDER, then, that nature's productions should be fa
0128 explained. It is a truly wonderful fact, the WONDER of which we are apt to overlook from familiar
0320 e at sickness, but when the sick man dies, to WONDER and to suspect that he died by some unknown d
0438 ected by slight successive steps, we need not WONDER at discovering in such parts or organs, a cer
0472 of caterpillars; and at other such cases. The WONDER indeed is, on the theory of natural selection
0021 ier and the short faced tumbler, and see the WONDERFUL difference in their beaks, entailing corres
0021 y the male bird, is also remarkable from the WONDERFUL development of the carunculated skin about
0037 rise expressed in horticultural works at the WONDERFUL skill of gardeners, in having produced such
0128 ganic beings may be explained. It is a truly WONDERFUL fact, the wonder of which we are apt to ove
0172 apper, and on the other hand, organs of such WONDERFUL structure, as the eye, of which we hardly a
0202 of some few members. If we admire the truly WONDERFUL power of scent by which the males of many i
0204 gans and their intermediate states show that WONDERFUL metamorphoses in function are at least poss
0207 eat the subject separately; especially as so WONDERFUL an instinct as that of the hive bee making
0209 tions. It can be clearly shown that the most WONDERFUL instincts with which we are acquainted, nam
0209 as I believe, that all the most complex and WONDERFUL instincts have originated. As modifications
0216 stly, been ranked by naturalists as the most WONDERFUL of all known instincts. It is now commonly
0219 have been hopeless to have speculated how so WONDERFUL an instinct could have been perfected. Form
0223 ot need confirmation by me, in regard to the WONDERFUL instinct of making slaves. Let it be observ
0227 by the Melipona, and in themselves not very WONDERFUL, this bee would make a structure as wonderf
0227 ications of instincts in themselves not very WONDERFUL, hardly more wonderful than those which gui
0227 n themselves not very wonderful, hardly more WONDERFUL than those which guide a bird to make its n
0235 conomising wax. Thus, as I believe, the most WONDERFUL sort of shield on their heads, the use of w
0238 erus, the workers of one caste alone carry a WONDERFUL of all known instincts, that of the hive be
0239 ral selection, when I do not admit that such WONDERFUL and well established facts at once annihila
0241 cture were produced. Thus, as I believe, the WONDERFUL fact of two distinctly defined castes of st
0339 elationship is even more clearly seen in the WONDERFUL collection of fossil bones made by MM. Lund
0339 this law of the succession of types, on this WONDERFUL relationship in the same continent between
0376 uthority, Prof. Dana, that it is certainly a WONDERFUL fact that New Zealand should have a closer
0391 ification. Madeira, again, is inhabited by a WONDERFUL number of peculiar land shells, whereas not
0439 w these terms may be used literally; and the WONDERFUL fact of the jaws, for instance, of a crab r
0477 een the same. We see the full meaning of the WONDERFUL fact, which must have struck every travelle
0490 inning endless forms most beautiful and most WONDERFUL have been, and are being, evolved.
0209 ying the pianoforte at three years old with WONDERFULLY little practice, had played a tune with no
0227 nderful, this bee would make a structure as WONDERFULLY perfect as that of the hive bee. We must s
0241 working ants of the several sizes differed WONDERFULLY in shape, and in the form and number of th
0318 the close of the secondary period, has been WONDERFULLY sudden. The whole subject of the extinctio
0417 years, only degraded flowers, departing so WONDERFULLY in a number of the most important points o
0431 in that of the Crustacea, for here the most WONDERFULLY diverse forms are still tied together by a
0436 ogous. We see the same law in comparing the WONDERFULLY complex jaws and legs in crustaceans. It i
0192 s impossible to conceive by what steps these WONDROUS organs have been produced but; as Owen and
0227 d seeing what perfectly cylindrical burrows in WOOD many insects can make, apparently by turning r
0232 itions for working, for instance, on a slip of WOOD, placed directly under the middle of a comb or
0261 their rate of growth, in the hardness of their WOOD, in the period of the flow or nature of their
0361 l portion of earth thus completely enclosed by WOOD in an oak about 50 years old; three dicotyledo
0091 on marshy ground and almost nightly catching WOODCOCKS or snipes. The tendency to catch rats rathe
0003 itions, the structure, for instance, of the WOODPECKER, with its feet, tail, beak, and tongue, so
0004 generations, some bird had given birth to a WOODPECKER, and some plant to the misseltoe, and that
0060 eautiful co adaptations most plainly in the WOODPECKER and misseltoe; and only a little less plain
0184 tance of adaptation be given than that of a WOODPECKER for climbing trees and for seizing insects
0184 a plata, where not a tree grows, there is a WOODPECKER, which in every essential part of its organ
0184 tionship to our common species; yet it is a WOODPECKER which never climbs a tree! Petrels are the
0471 trance it is that a bird, under the form of a WOODPECKER, should have been created to prey on insect
0184 of the bark? Yet in North America there are WOODPECKERS which feed largely on fruit, and others wi
0186 instead of in swamps; that there should be WOODPECKERS where not a tree grows; that there should
0197 o illustrate these latter remarks. If green WOODPECKERS alone had existed, and we did not know tha
0204 e are upland geese with webbed feet, ground WOODPECKERS, diving thrushes, and petrels with the hab
0293 e worthy of much deference, namely Bronn and WOODWARD, have concluded that the average duration of
0316 gly few, so few, that E. Forbes, Pictet, and WOODWARD (though all strongly opposed to such views a
0339 lso in the birds of the caves of Brazil. Mr. WOODWARD has shown that the same law holds good with
0261 diversity in the size of two plants, one being WOODY and the other herbaceous, one being evergreen
0030 cultivated land or mountain pasture, with the WOOL of one breed good for one purpose, and that of
0083 r manner; he exposes sheep with long and short WOOL to the same climate. He does not allow the mos

```
0380  turalised in any part of Europe, though hides,  WOOL, and other objects likely to carry seeds have
0392  ptation of hooked seeds for transportal by the  WOOL and fur of quadrupeds. This case presents no d
0151  es (rock barnacles) are, in every sense of the  WORD, very important structures, and they differ ex
0311  ch page, only here and there a few lines. Each  WORD of the slowly changing language, in which the
0355  her, does not differ much (by substituting the  WORD variety for species) from that lately advanced
0444  ve a good many cases of variations (taking the  WORD in the largest sense) which have supervened at
0455  y organs may be compared with the letters in a  WORD, still retained in the spelling, but become us
0040  , and insensible changes. I must now say a few  WORDS on the circumstances, favourable, or the reve
0062  knowledge. Nothing is easier than to admit in  WORDS the truth of the universal struggle for life,
0088  e with insects. And this leads me to say a few  WORDS on what I call Sexual Selection. This depends
0111  ere swept away by the short horns (I quote the  WORDS of an agricultural writer) as if by some murd
0139  of sleep, etc., and this leads me to say a few  WORDS on acclimatisation. As it is extremely common
0199  us. The foregoing remarks lead me to say a few  WORDS on the protest lately made by some naturalist
0356  f single centres of creation, I must say a few  WORDS on the means of dispersal. Means of Dispersal
0358  piles of volcanic matter. I must now say a few  WORDS on what are called accidental means, but whic
0001  ibly have any bearing on it. After five years'  WORK I allowed myself to speculate on the subject,
0001  ave not been hasty in coming to a decision. My  WORK is now nearly finished; but as it will take me
0002  r c. Lyell and Dr. Hooker, who both knew of my  WORK, the latter having read my sketch of 1844, hon
0002  ons have been grounded; and I hope in a future  WORK to do this. For I am well aware that scarcely
0006  shall give a brief recapitulation of the whole  WORK, and a few concluding remarks. No one ought to
0011  given in Isidore Geoffroy St. Hilaire's great  WORK on this subject. Breeders believe that long li
0040  s freely giving the materials for selection to  WORK on: not that mere individual differences are n
0044  given, but these I shall reserve for my future  WORK. Nor shall I here discuss the various definiti
0048  ized as varieties in Mr. Wollaston's admirable  WORK, but which it cannot be doubted would be ranke
0052  several facts and views given throughout this  WORK. It need not be supposed that all varieties or
0053  stronger terms. I shall reserve for my future  WORK the discussion of these difficulties, and the
0060  es, though necessary as the foundation for the  WORK, helps us but little in understanding how spec
0062  etail the struggle for existence. In my future  WORK this subject shall be treated, as it well dese
0067  by several authors; and I shall, in my future  WORK, discuss some of the checks at considerable le
0082  l selection would thus have free scope for the  WORK of improvement. We have reason to believe, as
0102  ve mere individual differences suffice for the  WORK. A large number of individuals, by giving a be
0102  ugh nature grants vast periods of time for the  WORK of natural selection, she does not grant an in
0102  t, and free intercrossing will wholly stop his  WORK. But when many men, without intending to alter
0115  e has well remarked in his great and admirable  WORK, that floras gain by naturalisation, proportio
0117  ld be thought worthy of record in a systematic  WORK. The intervals between the horizontal lines in
0138  caves, disuse by itself seems to have done its  WORK. It is difficult to imagine conditions of life
0149  ng as the same part has to perform diversified  WORK, we can perhaps see why it should remain varia
0151  ds good with cirripedes. I shall, in my future  WORK, give a list of the more remarkable cases; I w
0169  ory of new specific forms has been actively at  WORK, there, on an average, we now find most variet
0171  long before having arrived at this part of my  WORK, a crowd of difficulties will have occurred to
0190  odified and perfected so as to perform all the  WORK by itself, being aided during the process of m
0192  , some of which will be discussed in my future  WORK. One of the gravest is that of neuter insects,
0208  hed up to the sixth stage, so that much of its  WORK was already done for it, far from feeling the
0208  nd thus tried to complete the already finished  WORK. If we suppose any habitual action to become i
0214  successive generation would soon complete the  WORK; and unconscious selection is still at work, a
0215  he work; and unconscious selection is still at  WORK, as each man tries to procure, without intendi
0216  ral which I shall have to discuss in my future  WORK, namely, the instinct which leads the cuckoo t
0219  ngle year. The males and fertile females do no  WORK. The workers or sterile females, though most e
0219  nd courageous in capturing slaves, do no other  WORK. They are incapable of making their own nests,
0219  th their larvae and pupae to stimulate them to  WORK, they did nothing; they could not even feed th
0219  gle slave (F. fusca), and she instantly set to  WORK, fed and saved the survivors; made some cells
0220  d the larvae and pupae are exposed, the slaves  WORK energetically with their masters in carrying t
0221  ervation, in Switzerland the slaves habitually  WORK with their masters in making the nest, and the
0223  ditions. In Switzerland the slaves and masters  WORK together, making and bringing materials for th
0223  hen follow their proper instincts, and do what  WORK they could. If their presence proved useful to
0224  eat as it at first appears: all this beautiful  WORK can be shown, I think, to follow from a few ve
0225  her Nature does not reveal to us her method of  WORK. At one end of a short series we have humble b
0228  ese basins near together, they had begun their  WORK at such a distance from each other, that by th
0229  been left between the opposed basins, but the  WORK, from the unnatural state of things, had not b
0229  ng flat plates between the basins, by stopping  WORK along the intermediate planes or planes of int
0229  there is any difficulty in the bees, whilst at  WORK on the two sides of a strip of wax, perceiving
0229  o the proper thinness, and then stopping their  WORK. In ordinary combs it has appeared to me that
0231  timately be left. We shall understand how they  WORK, by supposing masons first to pile up a broad
0231  e cells are made, that a multitude of bees all  WORK together; one bee after working a short time a
0231  t, as Huber has stated, a score of individuals  WORK even at the commencement of the first cell. I
0232  the growing edges of the cells all round. The  WORK of construction seems to be a sort of balance
0233  f a queen wasp) making hexagonal cells, if she  WORK alternately on the inside and outside of two o
0236  mber should have been annually born capable of  WORK, but incapable of procreation, I can see no ve
0242  of labour is useful to civilised man. As ants  WORK by inherited instincts and by inherited tools
0255  are chiefly drawn up from Gartner's admirable  WORK on the hybridisation of plants. I have taken m
0282  ime. He who can read Sir Charles Lyell's grand  WORK on the Principles of Geology, which the future
0282  s of superimposed strata, and watch the sea at  WORK grinding down old rocks and making fresh sedim
0289  w, for instance, from Sir R. Murchison's great  WORK in Russia, what wide gaps there are in that co
0304  appearance of a great group of species. But my  WORK had hardly been published, when a skilful pala
0372  crustaceans (as described in Dana's admirable  WORK), of some fish and other marine animals, in th
0415  es seems to be entirely lost. Again in another  WORK he says, the genera of the Connaraceae differ
0417  y founded. Practically when naturalists are at  WORK, they do not trouble themselves about the phys
0435  essly admitted by Owen in his most interesting  WORK on the Nature of Limbs. On the ordinary view o
0053  tabulating all the varieties in several well  WORKED floras. At first this seemed a simple task; b
0190  respiration. The swimbladder has, also, been  WORKED in as an accessory to the auditory organs of
0191  ld, a part of the auditory apparatus has been  WORKED in as a complement to the swimbladder. All ph
0191  erly lost branchiae might have been gradually  WORKED in by natural selection for some quite distin
0196  atic animal, it might subsequently come to be  WORKED in for all sorts of purposes; as a fly flappe
0207  mary. The subject of instinct might have been  WORKED into the previous chapters; but I have though
0229  not been neatly performed. The bees must have  WORKED at very nearly the same rate on the opposite
0229  onvex on the opposed side, where the bees had  WORKED less quickly. In one well marked instance, I
```

Page **(Key Word)**

0232 rom the spot on which it had been placed, and WORKED into the growing edges of the cells all round
0415 e is widely different. No naturalist can have WORKED at any group without being struck with this f
0219 the males and fertile females do no work. The WORKERS or sterile females, though most energetic an
0224 more advantageous to this species to capture WORKERS than to procreate them, the habit of collect
0236 ase, that of working or sterile ants. How the WORKERS have been rendered sterile is a difficulty:
0236 igious difference in this respect between the WORKERS and the perfect females, would have been far
0238 xtraordinarily different: in Cryptocerus, the WORKERS of one caste alone carry a wonderful sort of
0239 te unknown: in the Mexican Myrmecocystus, the WORKERS of one caste never leave the nest: they are
0239 ste never leave the nest: they are fed by the WORKERS of another caste, and they have an enormousl
0239 happens that the larger or the smaller sized WORKERS are the most numerous: or that both large an
0240 numbers, Formica flava has larger and smaller WORKERS, with some of intermediate size: and, in thi
0240 ies, as Mr. F. Smith has observed, the larger WORKERS have simple eyes (ocelli), which though smal
0240 be plainly distinguished, whereas the smaller WORKERS have their ocelli rudimentary. Having carefu
0240 arefully dissected several specimens of these WORKERS, I can affirm that the eyes are far more rud
0240 eyes are far more rudimentary in the smaller WORKERS than can be accounted for merely by their pr
0240 ugh I dare not assert so positively, that the WORKERS of intermediate size have their ocelli in an
0240 n. so that we here have two bodies of sterile WORKERS in the same nest, differing not only in size
0240 i may digress by adding, that if the smaller WORKERS had been the most useful to the community, a
0240 , which produced more and more of the smaller WORKERS, until all the workers had come to be in thi
0240 nd more of the smaller workers, until all the WORKERS had come to be in this condition: we should
0240 same condition with those of Myrmica. For the WORKERS of Myrmica have not even rudiments of ocelli
0240 appreciate the amount of difference in these WORKERS, by my giving not the actual measurements, b
0241 the important fact for us is, that though the WORKERS can be grouped into castes of different size
0241 da of the jaws which I had dissected from the WORKERS of the several sizes. With these facts befor
0241 this is our climax of difficulty, one set of WORKERS of one size and structure, and simultaneousl
0241 structure, and simultaneously another set of WORKERS of a different size and structure: a graduat
0241 t of two distinctly defined castes of sterile WORKERS existing in the same nest, both widely diffe
0242 abour could be effected with them only by the WORKERS being sterile: for had they been fertile, th
0460 e existence of two or three defined castes of WORKERS or sterile females in the same community of
0084 up all that is good: silently and insensibly WORKING, whenever and wherever opportunity offers, a
0194 he very same invention, so natural selection, WORKING for the good of each being and taking advant
0224 this is perfectly effected by a crowd of bees WORKING in a dark hive. Grant whatever instincts you
0229 to me that the bees do not always succeed in WORKING at exactly the same rate from the opposite s
0229 into the hive, and allowed the bees to go on WORKING for a short time, and again examined the cel
0230 naw into this from the opposite sides, always WORKING circularly as they deepen each cell. They do
0231 tude of bees all work together: one bee after WORKING a short time at one cell going to another, s
0232 they can stand in their proper positions for WORKING, for instance, on a slip of wood, placed dir
0236 i will here take only a single case, that of WORKING or sterile ants. How the workers have been r
0236 difficulty. The great difficulty lies in the WORKING ants differing widely from both the males an
0236 far better exemplified by the hive bee. If a WORKING ant or other neuter insect had been an anima
0237 again selected, and so onwards. But with the WORKING ant we have an insect differing greatly from
0238 of the same family. Thus in Eciton, there are WORKING and soldier neuters, with jaws and instincts
0241 five times as big. The jaws, moreover, of the WORKING ants of the several sizes ciffered wonderful
0224 truction. It has been remarked that a skilful WORKMAN, with fitting tools and measures, would find
0084 should plainly bear the stamp of far higher WORKMANSHIP? It may be said that natural selection is
0240 ce was the same as if we were to see a set of WORKMEN building a house of whom many were five feet
0241 eet high: but we must suppose that the larger WORKMEN had heads four instead of three times as big
0486 the reason, and even the blunders of numerous WORKMEN: when we thus view each organic being, how f
0031 t, who was probably better acquainted with the WORKS of agriculturists than almost any other indiv
0033 ledgment of the importance of the principle in WORKS of high antiquity. In rude and barbarous peri
0037 seen great surprise expressed in horticultural WORKS at the wonderful skill of gardeners, in havin
0051 eties as are barely thought worth recording in WORKS on natural history. And I look at varieties w
0053 well marked to have been recorded in botanical WORKS. Hence it is the most flourishing, or, as the
0061 rably superior to man's feeble efforts, as the WORKS of Nature are to those of Art. We will now di
0087 ough statements to this effect may be found in WORKS of natural history, I cannot find one case wh
0141 rom the incessant advice given in agricultural WORKS, even in the ancient Encyclopaedias of China,
0142 than others: this is very strikingly shown in WORKS on fruit trees published in the United States
0155 than generic: but I have repeatedly noticed in WORKS on natural history, that when an author has r
0162 cognising a variable species in our systematic WORKS, is due to its varieties mocking, as it were,
0167 r at least for an unknown, cause. It makes the WORKS of God a mere mockery and deception; I would
0188 ? have we any right to assume that the Creator WORKS by intellectual powers like those of man? If
0189 be formed as superior to one of glass, as the WORKS of the Creator are to those of man? If it cou
0201 tion. Although many statements may be found in WORKS on natural history to this effect, I cannot f
0246 is impossible to study the several memoirs and WORKS of those two conscientious and admirable obse
0289 hen we see the formations tabulated in written WORKS, or when we follow them in nature, it is diff
0358 ll here confine myself to plants. In botanical WORKS, this or that plant is stated to be ill adapt
0381 rable clearness by Dr. Hooker in his botanical WORKS on the antarctic regions. These cannot be her
0453 trophied organs, are imperfect and useless. In WORKS on natural history rudimentary organs are gen
0469 distinct to be worthy of record in systematic WORKS. No one can draw any clear distinction betwee
0489 inappreciable length. And as natural selection WORKS solely by and for the good of each being, all
0380 rated in the larger areas and more efficient WORKSHOPS of the north. In many islands the native pr
0003 wn how the innumerable species inhabiting this WORLD have been modified, so as to acquire that per
0004 ence amongst all organic beings throughout the WORLD, which inevitably follows from their high geo
0006 s and modification of every inhabitant of this WORLD. Still less do we know of the mutual relation
0006 elations of the innumerable inhabitants of the WORLD during the many past geological epochs in its
0017 ny foxes, inhabiting different quarters of the WORLD. I do not believe, as we shall presently see,
0019 in the case of the domestic dogs of the whole WORLD, which I fully admit have probably descended
0020 voured with skins from several quarters of the WORLD, more especially by the Hon. W. Elliot from I
0024 eeds have been transported to all parts of the WORLD; and, therefore, some of them must have been
0027 thousands of years in several quarters of the WORLD; the earliest known record of pigeons is in t
0031 w been exported to almost every quarter of the WORLD. The improvement is by no means generally due
0054 ant species, those which range widely over the WORLD, are the most diffused in their own country,
0061 ns everywhere and in every part of the organic WORLD. Again, it may be asked, how is it that varie
0064 rapidly, in numbers, all cannot do so; for the WORLD would not hold them. There is no exception to
0064 ds which have run wild in several parts of the WORLD: if the statements of the rate of increase of
0066 s believed to be the most numerous bird in the WORLD. One fly deposits hundreds of eggs, and anoth

Page **(Key Word)**

Page **(Key Word)**

```
0072  of the Scotch fir; but in several parts of the  WORLD insects determine the existence of cattle. Pe
0073  he cause, we invoke cataclysms to desolate the  WORLD, or invent laws on the duration of the forms
0084  daily and hourly scrutinising, throughout the   WORLD, every variation, even the slightest; rejecti
0106  nt part in the changing history of the organic  WORLD. We can, perhaps, on these views, understand
0107  e of the most anomalous forms now known in the  WORLD, as the Ornithorhynchus and Lepidosiren, whic
0109  te and manner at which the inhabitants of this  WORLD have changed. Slow though the process of sele
0110  wded together than in any other quarter of the  WORLD, some foreign plants have become naturalised,
0127  n; and how largely extinction has acted in the  WORLD's history, geology plainly declares. Natural
0135  cts, namely, that beetles in many parts of the  WORLD are very frequently blown to sea and perish;
0138  rated by successive generations from the outer  WORLD into the deeper and deeper recesses of the Ke
0141  e been transported by man to many parts of the  WORLD, and now have a far wider range than any othe
0164  archipelago in the south. In all parts of the   WORLD these stripes occur far oftenest in duns and
0165  etc., inhabiting the most distant parts of the  WORLD. Now let us turn to the effects of crossing t
0167  ith species inhabiting distant quarters of the  WORLD, to produce hybrids resembling in their strip
0206  we look only to the present inhabitants of the  WORLD, is not strictly correct, but if we include a
0211  though I do not believe that any animal in the  WORLD performs an action for the exclusive good of
0243  species, when inhabiting distant parts of the   WORLD and living under considerably different condi
0287  rs. During each of these years, over the whole  WORLD, the land and the water has been peopled by h
0288  essile cirripedes) coat the rocks all over the  WORLD in infinite numbers: they are all strictly li
0289  north America, and in many other parts of the   WORLD. The most skilful geologist, if his attention
0293  rea of Europe is compared with the rest of the  WORLD; nor have the several stages of the same form
0294  the latest deposits of various quarters of the  WORLD, it has everywhere been noted, that some few
0294  t may be doubted whether in any quarter of the  WORLD, sedimentary deposits, including fossil remai
0300  ago is one of the richest regions of the whole  WORLD in organic beings: yet if all the species wer
0300  ould they represent the natural history of the  WORLD! But we have every reason to believe that the
0302  hat stage. We continually forget how large the  WORLD is, compared with the area over which our geo
0303  le to spread rapidly and widely throughout the  WORLD. I will now give a few examples to illustrate
0303  ave huge bones, are marine, and range over the  WORLD, the fact of not a single bone of a whale hav
0304  f the individuals of many species all over the  WORLD, from the Arctic regions to the equator, inha
0305  red suddenly and simultaneously throughout the  WORLD at this same period. It is almost superfluous
0305  e we any right to suppose that the seas of the  WORLD have always been so freely open from south to
0306  he succession of organic beings throughout the  WORLD, as it would be for a naturalist to Land for
0307  vast, yet quite unknown, periods of time, the   WORLD swarmed with living creatures. To the questio
0307  ld not forget that only a small portion of the  WORLD is known with accuracy. M. Barrande has latel
0309  urface. The immense areas in some parts of the  WORLD, for instance in South America, of bare metam
0310  natural geological record, as a history of the  WORLD imperfectly kept, and written in a changing d
0312  us changes in the forms of life throughout the  WORLD. On the affinities of extinct species to each
0318  e spot, the from another, and finally from the  WORLD. Both single species and whole groups of spec
0322  changing almost simultaneously throughout the   WORLD. Scarcely any palaeontological discovery is m
0322  fe change almost simultaneously throughout the  WORLD. Thus our European Chalk formation can be rec
0322  can be recognised in many distant parts of the  WORLD, under the most different climates, where not
0323  imilarly absent at these distant points of the  WORLD. In the several successive palaeozoic formati
0323  the marine inhabitants of distant parts of the  WORLD: we have not sufficient data to judge whether
0324  s having changed simultaneously throughout the  WORLD, it must not be supposed that this expression
0324  the above large sense, at distant parts of the  WORLD, has greatly struck those admirable observers
0325  mutations in the forms of life throughout the   WORLD, under the most different climates. We must,
0325  succession of the forms of life throughout the  WORLD, is explicable on the theory of natural selec
0326  rable, as before explained. One quarter of the  WORLD may have been most favourable for the product
0327  ssion of the same forms of life throughout the  WORLD, accords well with the principle of new speci
0327  new and improved groups spread throughout the   WORLD, old groups will disappear from the world; an
0327  the world, old groups will disappear from the   WORLD; and the succession of forms in both ways wil
0328  ere was much migration from other parts of the  WORLD. As we have reason to believe that large area
0328  er very wide spaces in the same quarter of the  WORLD; but we are far from having any right to conc
0336  e changed almost simultaneously throughout the  WORLD, and therefore under the most different clima
0337  , the eocene inhabitants of one quarter of the  WORLD were put into competition with the existing i
0339  same generalisation to the mammals of the Old   WORLD. We see the same law in this author's restora
0340  edi for the inhabitants of each quarter of the  WORLD will obviously tend to leave in that quarter,
0343  n epoch; but that long before that period, the  WORLD may have presented a wholly different aspect;
0344  vary, will in the long run tend to people the   WORLD with allied, but modified, descendants; and t
0344  long intervals of time, the productions of the  WORLD will appear to have changed simultaneously. W
0345  e irhabitants of each successive period in the  WORLD's history have beaten their predecessors in t
0346  uring the Glacial period co extensive with the  WORLD. In considering the distribution of organic b
0346  re is hardly a climate or condition in the Old  WORLD which cannot be paralleled in the New, at lea
0347  t degree; for instance, small areas in the Old  WORLD could be pointed out hotter than any in the N
0347  ould be pointed out hotter than any in the New  WORLD, yet these are not inhabitec by a peculiar fa
0351  gh inhabiting the most distant quarters of the  WORLD, must originally have proceeded from the same
0352  e same mammal inhabiting distant points of the  WORLD. No geologist will feel any difficulty in suc
0367  ed southward, are remarkably uniform round the  WORLD. We may suppose that the Glacial period came
0369  arctic and northern temperate forms round the   WORLD, at the commencement of the Glacial period. A
0370  whilst the majority of the inhabitants of the   WORLD were specifically the same as now, the climat
0371  the other great region, with those of the Old   WORLD. Consequently we have here everything favoura
0372  merica with the southern continents of the Old  WORLD, we use countries closely corresponding in al
0373  al far distant points on opposite sides of the  WORLD. But we have good evidence in almost every ca
0374  e period, actually simultaneous throughout the  WORLD. Without some distinct evidence to the contra
0374  id believing that the temperature of the whole  WORLD was at this period simultaneously cooler. But
0374  belts of longitude. On this view of the whole   WORLD, or at least of broad longitudinal belts, hav
0375  large island. Hence we see that throughout the  WORLD, the plants growing on the more lofty mountai
0376  n, its antipode, than to any other part of the  WORLD. Sir J. Richardson, also, speaks of the re ap
0376  ge body of geological evidence, that the whole  WORLD, or a large part of it, was during the Glacia
0382  geographical distribution. I believe that the   WORLD has recently felt one of his great cycles of
0383  prevail in a remarkable manner throughout the   WORLD. I well remember, when first collecting in th
0384  r fish occurring at very distant points of the  WORLD, no doubt there are many cases which cannot a
0385  d from a single source, prevail throughout the  WORLD. Their distribution at first perplexed me muc
0390  species (i.e. those found nowhere else in the   WORLD) is often extremely large. If we compare, for
0394  l mammals, for they occur in many parts of the  WORLD on very small islands, if close to a continen
0394  d possesses two bats found nowhere else in the  WORLD: Norfolk Island, the Viti Archipelago, the Bo
```

Page **(Key Word)**

```
0395  w up this subject in all other quarters of the  WORLD: but as far as I have gone, the relation gene
0400  an to the inhabitants of any other part of the  WORLD. And this is just what might have been expect
0400  hipelago which are found in other parts of the  WORLD (laying on one side for the moment the endemi
0401  pecies, both those found in other parts of the  WORLD and those confined to the archipelago, are co
0403  y of plants, have spread widely throughout the  WORLD during the recent Glacial epoch, are related
0404  has given the same general forms to the whole  WORLD. We see this same principle in the blind anim
0404  ons, and the existence at remote points of the  WORLD of other species allied to it, is shown in an
0404  in those genera of birds which range over the  WORLD, many of the species have very wide ranges. I
0404  ctions, in which so many genera range over the  WORLD, and many individual species have enormous ra
0404  have enormous ranges. It is not meant that in  WORLD ranging genera all the species have a wide ra
0405  w distributed to the most remote points of the  WORLD, we ought to find, and I believe as a general
0406  f some of the species into all quarters of the  WORLD, where they may have become slightly modified
0407  ly convinced simultaneously affected the whole  WORLD, or at least great meridional belts. As showi
0409  different great geographical provinces of the  WORLD. On these same principles, we can understand,
0409  ll its species common to other quarters of the  WORLD. We can see why whole groups of organisms, as
0410  ow looking to distant provinces throughout the  WORLD, we find that some organisms differ little, w
0410  successive ages within the same quarter of the  WORLD, or to those which have changed after having
0415  welfare of the being in relation to the outer  WORLD. Perhaps from this cause it has partly arisen
0422  he various languages now spoken throughout the  WORLD: and if all extinct languages, and all interm
0429  and how widely spread they are throughout the  WORLD, the fact is striking, that the discovery of
0457  nd families of organic beings, with which this  WORLD is peopled, have all descended, each within i
0461  , in however distant and isolated parts of the  WORLD they are now found, they must in the course o
0462  e and of representative species throughout the  WORLD. We are as yet profoundly ignorant of the man
0463  ween the living and extinct inhabitants of the  WORLD, and at each successive period between the ex
0464  hese ancient and utterly unknown epochs in the  WORLD's history. I can answer these questions and c
0464  distinct species. Only a small portion of the  WORLD has been geologically explored. Only organic
0471  ot thus succeed in increasing in size, for the  WORLD would not hold them, the more dominant groups
0475  nspicuous a part in the history of the organic  WORLD, almost inevitably follows on the principle o
0475  they had changed simultaneously throughout the  WORLD. The fact of the fossil remains of each forma
0476  se of ages much migration from one part of the  WORLD to another, owing to former climatal and geog
0481  most unavoidable as long as the history of the  WORLD was thought to be of short duration; and now
0487  mer migrations of the inhabitants of the whole  WORLD. Even at present, by comparing the difference
0488  in an extreme degree. The whole history of the  WORLD, as at present known, although of a length qu
0488  ion of the past and present inhabitants of the  WORLD should have been due to secondary causes, lik
0489  and that no cataclysm has desolated the whole  WORLD. Hence we may look with some confidence to a
0139  e inhabitants of the caves of the Old and New  WORLDS should be closely related, we might expect fr
0323  l species which are common to the Old and New  WORLDS be kept wholly out of view, the general paral
0346  distribution is that between the New and Old  WORLDS; yet if we travel over the vast American cont
0347  allelism in the conditions of the Old and New  WORLDS, how widely different are their living produc
0347  he terrestrial productions of the New and Old  WORLDS, excepting in the northern parts, where the l
0367  y together on the lowlands of the Old and New  WORLDS, would be left isolated on distant mountain s
0369  hern temperate productions of the Old and New  WORLDS are separated from each other by the Atlantic
0370  riod, when the inhabitants of the Old and New  WORLDS lived further southwards than at present, the
0370  hern temperate productions of the Old and New  WORLDS, at a period anterior to the Glacial epoch. B
0370  e plants and animals, both in the Old and New  WORLDS, began slowly to migrate southwards as the cl
0371  periods the northern parts of the Old and New  WORLDS will have been almost continuously united by
0371  es in common, which inhabited the New and Old  WORLDS, migrated south of the Polar Circle, they mus
0371  ain ranges and on the arctic lands of the two  WORLDS. Hence it has come, that when we compare the
0371  s of the temperate regions of the New and Old  WORLDS, we find very few identical species (though A
0442  triment, yet nearly all pass through a similar  WORM like stage of development; but in some few cas
0070  called epidemics appear to be due to parasitic  WORMS, which have from some cause, possibly in part
0100  ome hermaphrodites, as land mollusca and earth  WORMS: but these all pair. As yet I have not found
0489  with various insects flitting about, and with  WORMS crawling through the damp earth, and to refle
0283  ll down, and these remaining fixed, have to be  WORN away, atom by atom, until reduced in size they
0283  with the slowness with which rocky coasts are  WORN away. The observations on this head by Hugh Mi
0283  many other deposits, yet, from being formed of  WORN and rounded pebbles, each of which bears the s
0285  when viewing volcanic islands, which have been  WORN by the waves and pared all round into perpendi
0290  oral and sub littoral deposits are continually  WORN away, as soon as they are brought up by the sl
0307  at. If these most ancient beds had been wholly  WORN away by denudation, or obliterated by metamorp
0033  hardly any one is so careless as to allow his  WORST animals to breed. In regard to plants, there
0012  infinitely complex and diversified. It is well  WORTH while carefully to study the several treatise
0032  principle would be so obvious as hardly to be  WORTH notice; but its importance consists in the gr
0038  ncivilised man, has afforded us a single plant  WORTH culture. It is not that these countries, so r
0051  ds such slight varieties as are barely thought  WORTH recording in works on natural history. And I
0058  es. There is one other point which seems to me  WORTH notice. Varieties generally have much restric
0124  f our several new sub genera and genera. It is  WORTH while to reflect for a moment on the characte
0292  ed at a similar conclusion. One remark is here  WORTH a passing notice. During periods of elevation
0298  following chapter. One other consideration is  WORTH notice: with animals and plants that can prop
0299  ay be urged against my views. Hence it will be  WORTH while to sum up the foregoing remarks, under
0327  s one other remark connected with this subject  WORTH making. I have given my reasons for believing
0422  lies, sections, orders, and classes. It may be  WORTH while to illustrate this view of classificati
0009  eggs. Many exotic plants have pollen utterly  WORTHLESS, in the same exact condition as in the most
0117  well marked variety, such as would be thought  WORTHY of record in a systematic work. The intervals
0293  that two palaeontologists, whose opinions are  WORTHY of much deference, namely Bronn and Woodward,
0442  leted; and again in spiders, there is nothing  WORTHY to be called a metamorphosis. The larvae of i
0469  which they think sufficiently distinct to be  WORTHY of record in systematic works. No one can dra
0485  be merely varieties may hereafter be thought  WORTHY of specific names, as with the primrose and c
0088  ng all day long; male stag beetles often bear  WOUNDS from the huge mandibles of other males. The w
0136  been able to swim at all and had stuck to the  WRECK. The eyes of moles and of some burrowing rode
0396  ving existed as halting places, of which not a  WRECK now remains, must not be overlooked. I will h
0139  iles of Europe, I am only surprised that more  WRECKS of ancient life have not been preserved, owin
0358  sition favour the admission that they are the  WRECKS of sunken continents; if they had originally
0243  es cur British thrush: how it is that the male  WRENS (Troglodytes) of North America, build cock ne
0243  roost in, like the males of our distinct Kitty  WRENS, a habit wholly unlike that of any other know
0111  t horns (I quote the words of an agricultural  WRITER) as if by some murderous pestilence. Divergen
0012  been thrown on this principle by theoretical  WRITERS alone. When a deviation appears not unfreque
```

Page **(Key Word)**

```
0034   are laid down by some of the Roman classical WRITERS. From passages in Genesis, it is clear that
0245   s, I think, been much underrated by some late WRITERS. On the theory of natural selection the case
0330   in such distinct groups as they now are. Some WRITERS have objected to any extinct species or grou
0098   ecial contrivances, as I could show from the WRITINGS of C. C. Sprengel and from my own observatio
0194   t saltum. We meet with this admission in the WRITINGS of almost every experienced naturalist; or,
0415   d it has been most fully acknowledged in the WRITINGS of almost every author. It will suffice to c
0164   am aware that Colonel Hamilton Smith, who has WRITTEN on this subject, believes that the several b
0289   time. When we see the formations tabulated in WRITTEN works, or when we follow them in nature, it
0310   a history of the world imperfectly kept, and WRITTEN in a changing dialect; of this history we po
0311   guage, in which the history is supposed to be WRITTEN, being more or less different in the interru
0422   ed, and only the names of the groups had been WRITTEN in a linear series, it would have been still
0425   followed by our best systematists. We have no WRITTEN pedigrees; we have to make out community of
0444   in domestic varieties. Some authors who have WRITTEN on Dogs, maintain that the greyhound and bul
0226   ect as the comb of the hive bee. Accordingly I WROTE to Professor Miller, of Cambridge, and this g
0363   the glacial epoch. At my request Sir C. Lyell WROTE to M. Hartung to inquire whether he had obser
0072   ally browsed down by the cattle. In one square YARD, at a point some hundred yards distant from on
0286   s the same as if we were to assume a cliff one YARD in height to be eaten back along a whole line
0286   along a whole line of coast at the rate on one YARD in nearly every twenty two years. I doubt whet
0072   e. In one square yard, at a point some hundred YARDS distant from one of the old clumps, I counted
0093   cted. Having found a female tree exactly sixty YARDS from a male tree, I put the stigmas of twenty
0221   me road to a tall Scotch fir tree, twenty five YARDS distant, which they ascended together probabl
0221   dead bodies as food to their nest, twenty nine YARDS distant; but they were prevented from getting
0222   ing file burthened with booty, for about forty YARDS, to a very thick clump of heath, whence I saw
0002   ons that I have on the origin of species. Last YEAR he sent to me a memoir on this subject, with a
0028   much valued by Akber Khan in India, about the YEAR 1600; never less than 20,000 pigeons were take
0062   it is not so at all seasons of each recurring YEAR. I should premise that I use the term Struggle
0063   its life, and during some season or occasional YEAR, otherwise, on the principle of geometrical in
0064   unproductive as this, and their seedlings next YEAR produced two, and so on, then in twenty years
0075   riously coloured sweet peas, they must be each YEAR harvested separately, and the seed then mixed
0079   period of its life, during some season of the YEAR, during each generation or at intervals, has t
0090   ecreased in numbers, during that season of the YEAR when the wolf is hardest pressed for food. I c
0090   ir prey at this or at some other period of the YEAR, when they might be compelled to prey on other
0127   rease of each species, at some age, season, or YEAR, a severe struggle for life, and this certainl
0129   species; and those produced during each former YEAR may represent the long succession of extinct s
0189   millions on millions of years; and during each YEAR on millions of individuals of many kinds; and
0210   eriods of life, or at different seasons of the YEAR, or when placed under different circumstances,
0219   ies would certainly become extinct in a single YEAR. The males and fertile females do no work. The
0220   est, and food of all kinds. During the present YEAR, however, in the month of July, I came across
0288   care, as the important discoveries made every YEAR in Europe prove. No organism wholly soft can b
0312   case of the several tertiary stages; and every YEAR tends to fill up the blanks between them, and
0324   s to the same thousandth or hundred thousandth YEAR, or even that it has a very strict geological
0363   t remain to be discovered; have been in action YEAR after year, for centuries and tens of thousand
0363   be discovered, have been in action year after YEAR, for centuries and tens of thousands of years,
0364   hey could not endure our climate. Almost every YEAR, one or two land birds are blown across the wh
0391   ny european and African birds are almost every YEAR blown there, as I am informed by Mr. E. V. Har
0001   ld possibly have any bearing on it. After five YEARS' work I allowed myself to speculate on the su
0001   shed; but as it will take me two or three more YEARS to complete it, and as my health is far from
0003   ations to Dr. Hooker, who for the last fifteen YEARS has aided me in every possible way by his lar
0018   breeds originated there, four or five thousand YEARS ago? But Mr. Horner's researches have rendere
0018   lley of the Nile thirteen or fourteen thousand YEARS ago, and who will pretend to say how long bef
0027   . they have been domesticated for thousands of YEARS in several quarters of the world: the earlies
0031   at he would produce any given feather in three YEARS, but it would take him six years to obtain he
0031   ther in three years, but it would take him six YEARS to obtain head and beak. In Saxony the import
0032   hese qualities, and he studies his subject for YEARS, and devotes his lifetime to it with indomita
0032   ld readily believe in the natural capacity and YEARS of practice requisite to become even a skilfu
0032   pared with drawings made only twenty or thirty YEARS ago. When a race of plants is once pretty wel
0033   it has certainly been more attended to of late YEARS, and many treatises have been published on th
0036   nal stock of Mr. Bakewell for upwards of fifty YEARS. There is not a suspicion existing in the min
0037   ens. If it has taken centuries or thousands of YEARS to improve or modify most of our plants up to
0042   ve been raised during the last thirty or forty YEARS. In the case of animals with separate sexes,
0045   ority, as I have collected, during a course of YEARS. It should be remembered that systematists ar
0048   are often called, as geographical races! Many YEARS ago, when comparing, and seeing others compar
0049   very numerous experiments made during several YEARS by that most careful observer Gartner, they c
0064   n slow breeding man has doubled in twenty five YEARS, and at this rate, in a few thousand years, t
0064   ive years, and at this rate, in a few thousand YEARS, there would literally not be standing room f
0064   t year produced two, and so on, then in twenty YEARS there would be a million plants. The elephant
0064   the mark to assume that it breeds when thirty YEARS old, and goes on breeding till ninety years o
0064   ty years old, and goes on breeding till ninety YEARS old, bringing forth three pair of young in th
0064   out whole islands in a period of less than ten YEARS. Several of the plants now most numerous over
0065   hat the slow breeders would require a few more YEARS to people, under favourable conditions, a who
0066   tree, which lived on an average for a thousand YEARS, if a single seed were produced once in a tho
0066   a single seed were produced once in a thousand YEARS, supposing that this seed were never destroye
0068   head of game were shot during the next twenty YEARS in England, and at the same time, if no vermi
0071   the same nature had been enclosed twenty five YEARS previously and planted with Scotch fir. The c
0071   s on the distant hilltops: within the last ten YEARS large spaces have been enclosed, and self sow
0072   rom the rings of growth, had during twenty six YEARS tried to raise its head above the stems of th
0074   riods of life, and during different seasons or YEARS, probably come into play; some one check or s
0075   ield more seed, and will consequently in a few YEARS quite supplant the other varieties. To keep u
0114   four in size, which had been exposed for many YEARS to exactly the same conditions, supported twe
0150   species; tends to be highly variable. Several YEARS ago I was much struck with a remark, nearly t
0189   this process go on for millions on millions of YEARS; and during each year on millions of individu
0209   rt, instead of playing the pianoforte at three YEARS old with wonderfully little practice, had pla
0220   e months of June and July, on three successive YEARS, I have watched for many hours several nests
0247   impaired. Moreover, as Gartner during several YEARS repeatedly crossed the primrose and cowslip,
0248   author, from experiments made during different YEARS. It can thus be shown that neither sterility
0251   t he had then tried the experiment during five YEARS, and he continued to try it during several su
```

Page **(Key Word)**

Page ***(Key Word)***

```
0251  continued to try it during several subsequent  YEARS, and always with the same result. This result
0258  than two hundred times, during eight following  YEARS, to fertilise reciprocally M. longiflora with
0270  cific distinction. Gartner kept during several  YEARS a dwarf kind of maize with yellow seeds, and
0270  nishing number of experiments made during many  YEARS on nine species of Verbascum, by so good an o
0282  formation or even each stratum. A man must for  YEARS examine for himself great piles of superimpos
0283  surface and the vegetation show that elsewhere  YEARS have elapsed since the waters washed their ba
0284  he rate of only 600 feet in a hundred thousand  YEARS. This estimate may be quite erroneous; yet, c
0286  he rate on one yard in nearly every twenty two  YEARS. I doubt whether any rock, even as soft as ch
0287  on of the Weald must have required 306,662,400  YEARS; or say three hundred million years. The acti
0287  06,662,400 years; or say three hundred million  YEARS. The action of fresh water on the gently incl
0287  , the surface may have existed for millions of  YEARS as land, and thus have escaped the action of
0287  obability a far longer period than 300 million  YEARS has elapsed since the latter part of the Seco
0287  ome notion, however imperfect, of the lapse of  YEARS. During each of these years, over the whole w
0287  t, of the lapse of years. During each of these  YEARS, over the whole world, the land and the water
0287  have succeeded each other in the long roll of  YEARS! Now turn to our richest geological museums,
0293  ion has indisputably required a vast number of  YEARS for its deposition, I can see several reasons
0293  h each formation may mark a very long lapse of  YEARS, each perhaps is short compared with the peri
0303  at in geological treatises, published not many  YEARS ago, the great class of mammals was always sp
0306  , which the discoveries of even the last dozen  YEARS have effected, it seems to me to be about as
0312  h undoubtedly of high antiquity if measured by  YEARS, only one or two species are lost forms, and
0319  e favourable conditions it would in a very few  YEARS have stocked the whole continent. But we coul
0324  od (an enormously remote period as measured by  YEARS, including the whole glacial epoch), were to
0336  find after intervals, very long as measured by  YEARS, but only moderately long as measured geologi
0339  ing the later tertiary periods. Mr. Clift many  YEARS ago showed that the fossil mammals from the A
0361  completely enclosed by wood in an oak about 50  YEARS old, three dicotyledonous plants germinated:
0364  r year, for centuries and tens of thousands of  YEARS, it would I think be a marvellous fact if man
0374  t endured for an enormous time, as measured by  YEARS, at each point. The cold may have come on, or
0377  at present. The Glacial period, as measured by  YEARS, must have been very long; and when we rememb
0380  La Plata, and during the last thirty or forty  YEARS from Australia. Something of the same kind mu
0417  n aspicarpa produced in France, during several  YEARS, only degraded flowers, departing so wonderfu
0462  y long periods, enormously long as measured by  YEARS, too much stress ought not to be laid on the
0465  these difficulties far too heavily during many  YEARS to doubt their weight. But it deserves especi
0466  means of Distribution during the long lapse of  YEARS, or that we know how imperfect the Geological
0481  full meaning of the term of a hundred million  YEARS; it cannot add up and perceive the full effec
0481  e of facts all viewed, during a long course of  YEARS, from a point of view directly opposite to mi
0085  s suffer far more from a certain disease than  YELLOW plums; whereas another disease attacks yellow
0085  yellow plums; whereas another disease attacks  YELLOW fleshed peaches far more than those with othe
0085  e which variety, whether a smooth or downy, a  YELLOW or purple fleshed fruit, should succeed. In l
0222  species, F. flava, with a few of these little  YELLOW ants still clinging to the fragments of the n
0222  rter of an hour, shortly after all the little  YELLOW ants had crawled away, they took heart and ca
0270  ring several years a dwarf kind of maize with  YELLOW seeds, and a tall variety with red seeds, gro
0270  o hostile a witness, as Gartner: namely, that  YELLOW and white varieties of the same species of Ve
0271   oured flowers. Moreover, he asserts that when  YELLOW and white varieties of one species are crosse
0271  ite varieties of one species are crossed with  YELLOW and white varieties of a distinct species, mo
0387  ish swallow seeds of moderate size, as of the  YELLOW water Lily and Potamogeton. Herons and other
0006  one ought to feel surprise at much remaining as  YET  unexplained in regard to the origin of species
0006  allied species has a narrow range and is rare?  YET  these relations are of the highest importance,
0008  ultivated plants display the utmost vigour, and  YET  rarely or never seed! In some few such cases i
0009  imals and plants, though often weak and sickly,  YET  breeding quite freely under confinement; and w
0009  thy (of which I could give numerous instances),  YET  having their reproductive system so seriously
0014  ecularity should appear at any particular age,  YET  that it does tend to appear in the offspring a
0023  es where they were originally domesticated, and  YET  be unknown to ornithologists; and this, consid
0024  ild animal to breed freely under domestication;  YET  on the hypothesis of the multiple origin of ou
0024  of their structure, with the wild rock pigeon,  YET  are certainly highly abnormal in other parts o
0027  ely in certain characters from the rock pigeon,  YET  by comparing the several sub breeds of these b
0028  he probable origin of domestic pigeons at some,  YET  quite insufficient, length; because when I fir
0029  ir prizes by selecting such slight differences,  YET  they ignore all general arguments, and refuse
0029  ntermediate links in the long lines of descent,  YET  admit that many of our domestic races have des
0033  ffer in size, colour, shape, and hairiness, and  YET  the flowers present very slight differences, I
0035  been effected unconsciously and gradually, and  YET  so effectually, that, though the old Spanish p
0036  rom the pure blood of Mr. Bakewell's flock, and  YET  the difference between the sheep possessed by
0036  ter of the offspring of their domestic animals,  YET  any one animal particularly useful to them, fo
0040  y spread in the immediate neighbourhood. But as  YET  they will hardly have a distinct name, and fro
0042  at some domestic animals vary less than others,  YET  the rarity or absence of distinct breeds of th
0044  n of the term species. No one definition has as  YET  satisfied all naturalists, yet every naturalis
0044  efinition has as yet satisfied all naturalists,  YET  every naturalist knows vaguely what he means w
0046  could have been effected only by slow degrees:  YET  quite recently Mr. Lubbock has shown a degree
0050  sts to an agreement how to rank doubtful forms.  YET  it must be confessed, that it is in the best k
0050  he common oak, how closely it has been studied;  YET  a German author makes more than a dozen specie
0051  . certainly no clear line of demarcation has as  YET  been drawn between species and sub species, th
0062  han constantly to bear this conclusion in mind.  YET  unless it be thoroughly engrained in the mind,
0066  s a couple of eggs and the ostrich a score, and  YET  in the same country the condor may be the more
0066  of the two: the Fulmar petrel lays but one egg,  YET  it is believed to be the most numerous bird in
0066  s or young, a small number may be produced, and  YET  the average stock be fully kept up; but if man
0072  kly clothed with vigorously growing young firs.  YET  the heath was so extremely barren and so exten
0073  ust ever be recurring with varying success: and  YET  in the long run the forces are so nicely balan
0077  organic being is related, in the most essential  YET  often hidden manner, to that of all other orga
0077  irst confined to the elements of air and water.  YET  the advantage of plumed seeds no doubt stands
0078  n cease. The land may be extremely cold or dry,  YET  there will be competition between some few spe
0078  may be exactly the same as in its former home,  YET  the conditions of its life will generally be c
0084  ct only through and for the good of each being,  YET  characters and structures, which we are apt to
0085  as characters of the most trifling importance:  YET  we hear from an excellent horticulturist, Down
0090  itted these advantages to their male offspring.  YET, I would not wish to attribute all such sexual
0092  destruction appears a simple loss to the plant;  YET  if a little pollen were carried, at first occa
0094  t on a hasty glance appear to differ in length;  YET  the hive bee can easily suck the nectar out of
0097  ure to wet is to the fertilisation of a flower,  YET  what a multitude of flowers have their anthers
```

Page ***(Key Word)***

Page ***(Key Word)***

```
0098  he stigma of a flower receiving its own pollen,   YET, as C. C. Sprengel has shown, and as I can con
0099  and some even of these were not perfectly true.   YET the pistil of each cabbage flower is surrounde
0100  ollusca and earth worms; but these all pair. As   YET I have not found a single case of a terrestria
0100  ross. And, as in the case of flowers, I have as   YET failed, after consultation with one of the hig
0101  each other in almost their whole organisation,   YET are not rarely, some of them hermaphrodites, a
0105  ee in our chapter on geographical distribution;   YET of these species a very large proportion are e
0106  y favourable for the production of new species,   YET that the course of modification will generally
0110  any means of knowing that any one region has as   YET got its maximum of species. Probably no region
0110  ts maximum of species. Probably no region is as   YET fully stocked, for at the Cape of Good Hope, w
0111  hopeless doubts in many cases how to rank them,   YET certainly differ from each other far less than
0124  or families, or genera, with those now living,   YET are often, in some degree, intermediate in cha
0126  nd which are least broken up, that is, which as   YET have suffered least extinction, will for a lon
0126  s may now have living and modified descendants,   YET at the most remote geological period, the eart
0129  ly two or three, now grown into great branches,   YET survive and bear all the other branches; so wi
0134  anomaly in nature than a bird that cannot fly;   YET there are several in this state. The loggerhea
0146  ey can be in any way advantageous to the plant:   YET in the Umbelliferae these differences are of s
0154  for in this case the variability will seldom as   YET have been fixed by the continued selection of
0154  a genera. I believe this explanation is partly,   YET only indirectly, true; I shall, however, have
0159  to vary in a like manner, but to three separate   YET closely related acts of creation. With pigeons
0160  t this is a case of reversion, and not of a new   YET analogous variation appearing in the several b
0160  on, of any one ancestor, is only 1 in 2048; and   YET, we see, it is generally believed that a tende
0162  cter and what are new but analogous variations,   YET we ought, on my theory, sometimes to find the
0169  and natural selection will in such cases not as   YET have had time to overcome the tendency to furt
0172  ul structure, as the eye, of which we hardly as   YET fully understand the inimitable perfection? Th
0180  re difficult, and I could have given no answer.   YET I think such difficulties have very little we
0181  ect the Galeopithecus with the other Lemuridae,   YET I can see no difficulty in supposing that such
0182  d functionally for no purpose like the Apteryx.   YET the structure of each of these birds is good f
0184  for seizing insects in the chinks of the bark?   YET in North America there are woodpeckers which f
0184  close blood relationship to our common species;   YET it is a woodpecker which never climbs a tree!
0184  trels are the most aerial and oceanic of birds,   YET in the quiet Sounds of Tierra del Fuego, the P
0185  ld never have suspected its sub aquatic habits;   YET this anomalous member of the strictly terrestr
0185  eet of ducks and geese are formed for swimming?   YET there are upland geese with webbed feet which
0185  ed for walking over swamps and floating plants,   YET the water hen is nearly as acutic as the coot
0186  confess, absurd in the highest possible degree.   YET reason tells me, that if numerous gradations f
0192  produced by successive transitional gradations,   YET, undoubtedly, grave cases of difficulty occur,
0193  losely analogous to the electric apparatus, and   YET do not, as Matteuchi asserts, discharge any el
0193  ance and function of the organ may be the same,   YET some fundamental difference can generally be d
0194  organ could have arrived at its present state;   YET, considering that the proportion of living and
0195  or so trifling an object as driving away flies;   YET we should pause before being too positive even
0199  f true, would be absolutely fatal to my theory.   YET I fully admit that many structures are of no d
0204  ection, is more than enough to stagger any one!   YET in the case of any organ, if we know of a long
0208  ely in direct opposition to our conscious will!   YET they may be modified by the will or reason. Ha
0210  w and gradual accumulation of numerous, slight,   YET profitable, variations. Hence, as in the case
0211  xclusive good of another of a distinct species,   YET each species tries to take advantage of the in
0218  ct, however, of the American ostrich has not as   YET been perfected; for a surprising number of egg
0218  h paralysed prey for its own larvae to feed on,   YET that when this insect finds a burrow already m
0226  thickness as the outer spherical portions, and   YET each flat portion forms a part of two cells. R
0236  re from both the males and fertile females, and   YET, from being sterile, they cannot propagate the
0237  e an insect differing greatly from its parents;   YET absolutely sterile: so that it could never hav
0238  ched, produced oxen with the longest horns; and   YET no one ox could ever have propagated its kind.
0238  see in many social insects. But we have not as   YET touched on the climax of the difficulty; namel
0240  ot only in size, but in their organs of vision,   YET connected by some few members in an intermedia
0241  can be grouped into castes of different sizes,   YET they graduate insensibly into each other, as d
0243  nder considerably different conditions of life,   YET often retaining nearly the same instincts. For
0246  organs of reproduction in a perfect condition,   YET when intercrossed they produce either few or n
0248  or seven, and in one case for ten generations,   YET he asserts positively that their fertility nev
0251  e perfectly good with respect to other species,   YET as they were functionally imperfect in their m
0251  petunia, Rhododendron, etc., have been crossed,   YET many of these hybrids seed freely. For instanc
0254  r dogs have descended from several wild stocks;   YET, with perhaps the exception of certain indigen
0256  acility, and produce numerous hybrid offspring,   YET these hybrids are remarkably sterile. On the o
0260  hy should some species cross with facility, and   YET produce very sterile hybrids; and other specie
0260  ther species cross with extreme difficulty, and   YET produce fairly fertile hybrids? Why should the
0262  , which have their reproductive organs perfect;   YET these two distinct cases run to a certain exte
0262  and female elements in the act of reproduction;   YET that there is a rude degree of parallelism in
0266  that a similar parallelism extends to an allied   YET very different class of facts. It is an old an
0269  sed at some difference in the result. I have as   YET spoken as if the varieties of the same species
0271  n between those which are differently coloured.   YET these varieties of Verbascum present no other
0275  wn that this does sometimes occur with hybrids;   YET I grant much less frequently with hybrids than
0282  aving produced a revolution in natural science,   YET does not admit how imcomprehensibly vast have
0283  med at a quicker rate than many other deposits.   YET, from being formed of worn and rounded pebbles
0284  me which has elapsed during their accumulation;   YET what time this must have consumed! Good observ
0284  nd years. This estimate may be quite erroneous;   YET, considering over what wide spaces very fine s
0285  there is one in Merionethshire of 12,000 feet;   YET in these cases there is nothing on the surface
0285  prof. Ramsay's masterly memoir on this subject.   YET it is an admirable lesson to stand on the Nort
0288  hitherto been found in any tertiary formation:   YET it is now known that the genus Chthamalus exis
0291  enough to resist such degradation as it has as   YET suffered, but which will hardly last to a dist
0294  , all included within this same glacial period.   YET it may be doubted whether in any quarter of th
0296  ired an enormous period for their accumulation;   YET no one ignorant of this fact would have suspec
0296  f the same formation; facts, showing what wide,   YET easily overlooked, intervals have occurred in
0297  might be the actual progenitor of B and C, and   YET might not at all necessarily be strictly inter
0298  t universally ranked as specifically different,   YET are far more closely allied to each other than
0299  less wide than they otherwise would have been,   YET has done scarcely anything in breaking down th
0300  t regions of the whole world in organic beings;   YET if all the species were to be collected which
0304  quitous genus, of which not one specimen has as   YET been found even in any tertiary stratum. Hence
0305  ch older fishes, of which the affinities are as   YET imperfectly known, are really teleostean. Assu
0307  to the present day; and that during these vast,   YET quite unknown, periods of time, the world swar
```

Page ***(Key Word)***

Page ***(Key Word)***

0308	many islands; but not one oceanic island is as	YET	known to afford even a remnant of any palaeozo
0313	have seldom changed ir exactly the same degree.	YET	if we compare any but the most closely related
0315	in the economy of nature, and thus supplant it;	YET	the two forms, the old and the new, would not
0317	group becomes large. On Extinction. We have as	YET	spoken only incidentally of the cisappearance
0319	e assuredly should not have perceivec the fact,	YET	the fossil horse would certainly have become r
0320	eel no surprise at the rarity of a species, and	YET	to marvel greatly when it ceases to exist, is
0322	ately, and that some check is always in action,	YET	seldom perceived by us, the whole economy of n
0328	rs of the species belonging to the same genera,	YET	the species themselves differ in a manner very
0330	finities directed towards very distinct groups.	YET	if we compare the older Reptiles and Batrachia
0332	to its slightly altered conditions of life, and	YET	retain throughout a vast period the same gener
0332	as has occurred with ruminants and pachyderms.	YET	he who objected to call the extinct genera, wh
0332	ow including five genera, and o14 to m14) would	YET	remain distinct. These two families, however,
0336	er on this subject; for naturalists have not as	YET	defined to each other's satisfaction what is m
0337	be said to be higher than those of New Zealand.	YET	the most skilful naturalist from an examinatio
0338	truth of this doctrine is very far from proved.	YET	I fully expect to see it hereafter confirmed,
0338	tion of each animal. This view may be true, and	YET	it may never be capable of full proof. Seeing,
0343	er, or at the same rate, or in the same degree;	YET	in the long run that all undergo modification
0345	of nature; and this may account for that vague	YET	ill defined sentiment, felt by many palaeontol
0346	ibution is that between the New and Cld Worlds;	YET	if we travel over the vast American continent,
0347	e pointed out hotter than any in the New World,	YET	these are not inhabited by a peculiar fauna or
0347	arts extremely similar in all their conditions,	YET	it would not be possible to point out three fa
0348	nd western shores of South and Central America;	YET	these great faunas are separated only by the n
0348	estern America and the eastern Pacific Islands,	YET	many fish range from the Pacific into the Indi
0349	essive groups of beings, specifically distinct,	YET	clearly related, replace each other. He hears
0349	place each other. He hears from closely allied,	YET	distinct kinds of birds, notes nearly similar,
0369	ed as doubtful forms, and some few are distinct	YET	closely allied or representative 'species. In i
0387	y kinds, and were altogether 537 in number; and	YET	the viscid mud was all contained in a breakfas
0387	ron's stomach; although I do not know the fact,	YET	analogy makes me believe that a heron flying t
0388	ecies, however few, already occupying any pond;	YET	as the number of kinds is small, compared with
0390	possessed under half a dozen flowering plants;	YET	many have become naturalised on it, as they ha
0391	gh we do not know how sea shells are dispersed;	YET	we can see that their eggs or larvae, perhaps
0392	e endemic plants have beautifully hooked seeds;	YET	few relations are more striking than the adapt
0393	st voyages, but have not finished my search; as	YET	I have not found a single instance, free from
0394	o frequently now happens in the arctic regions.	YET	it cannot be said that small islands will not
0395	al and researches of Mr. Wallace. I have not as	YET	had time to follow up this subject in all othe
0397	tried, sink in sea water and are killed by it.	YET	there must be, on my view, some unknown, but h
0398	ct species, supposed to have been created here;	YET	the close affinity of most of these birds to A
0398	stant several hundred miles from the continent,	YET	feels that he is standing on American land. Wh
0400	t, we sometimes see displayed on a small scale;	YET	in a most interesting manner, within the limit
0401	slands. Some species, however, might spread and	YET	retain the same character throughout the group
0403	nually transported from Porto Santo to Madeira,	YET	this latter island has not become colonised by
0403	conditions, and are united by continuous land,	YET	they are inhabited by a vast number of distinc
0403	the inhabitants, when not identically the same,	YET	are plainly related to the inhabitants of that
0410	y may fairly be attributed to our not having as	YET	discovered in an intermediate deposit the form
0416	e of quite subordinate value in classification;	YET	no one probably will say that the antennae in
0416	are of high physiological or vital importance;	YET,	undoubtedly, organs in this condition are oft
0417	ortance and of almost universal prevalence, and	YET	leave us in no doubt where it should be ranked
0417	of structure from the proper type of the order,	YET	M. Richard sagaciously saw, as Jussieu observe
0418	e plays its full part in the economy of nature.	YET	it has been strongly urged by those great natu
0419	eries, which have hardly a character in common;	YET	the species at both ends, from being plainly a
0421	be called cousins to the same millionth degree;	YET	they differ widely and in different degrees fr
0421	osed to have been but little modified, and they	YET	form a single genus. But this genus, though mu
0422	ement would, I think, be the only possible one.	YET	it might be that some very ancient language ha
0424	he important character of having a longer beak,	YET	all are kept together from having the common h
0424	hrodites of certain cirripedes, when adult, and	YET	no one dreams of separating them. The naturali
0426	important characteristics, from its allies, and	YET	be safely classed with them. This may be safel
0426	wo forms have not a single character in common,	YET	if these extreme forms are connected together
0432	ents from their ancient and unknown progenitor.	YET	the natural arrangement in the diagram would s
0433	y reject analogical or adaptive characters, and	YET	use these same characters within the limits of
0434	ange to almost any extent in form and size, and	YET	they always remain connected together in the s
0434	f a bee or bug, and the great jaws of a beetle?	YET	all these organs, serving for such different p
0435	reased to any extent, so as to serve as a wing;	YET	in all this great amount of modification there
0438	ave actually been modified into skulls or jaws.	YET	so strong is the appearance of a modification
0442	nts or placed in the midst of proper nutriment,	YET	nearly all pass through a similar worm like st
0445	of them could be distinguished from each other,	YET	their proportional differences in the above sp
0450	ryonic stages of recent forms, may be true, but	YET,	owing to the geological record not extending
0451	plainer than that wings are formed for flight,	YET	in how many insects do we see wings so reduced
0462	o conjecture how this could have been effected.	YET,	as we have reason to believe that some specie
0462	ate regions. It cannot be denied that we are as	YET	very ignorant of the full extent of the variou
0462	ntative species throughout the world. We are as	YET	profoundly ignorant of the many occasional mea
0472	seem to have produced but little direct effect;	YET	when varieties enter any zone, they occasional
0474	ost unusual manner, like the wing of a bat, and	YET	not be more variable than any other structure,
0475	nder considerably different conditions of life,	YET	should follow nearly the same instincts; why t
0478	both still exist. Wherever many closely allied	YET	distinct species occur, many doubtful forms an
0184	es seen and heard it hammering the seeds of the	YEW	on a branch; and thus breaking them like a nut
0008	cultivated plants, such as wheat, still often	YIELD	new varieties: our oldest domesticated animal
0026	umblers, pouters, and fantails now are, should	YIELD	offspring perfectly fertile, inter se, seems
0054	re already dominant will be the most likely to	YIELD	offspring which, though in some slight degree
0075	the most fertile, will beat the others and so	YIELD	more seed, and will consequently in a few yea
0147	dily. The same varieties of the cabbage do not	YIELD	abundant and nutritious foliage and a copious
0205	country, generally the smaller one, will often	YIELD,	as we see they do yield, to the inhabitants
0205	aller one, will often yield, as we see they do	YIELD,	to the inhabitants of another and generally
0250	wn pollen. For these plants have been found to	YIELD	seed to the pollen of a distinct species, tho
0268	l appearance, cross with perfect facility, and	YIELD	perfectly fertile offspring. I fully admit th
0286	whether any rock, even as soft as chalk, would	YIELD	at this rate excepting on the most exposed co
0321	rom the successful intruder, many will have to	YIELD	their places; and it will generally be allied

Page ***(Key Word)***

```
0321  ning to the same or to a distinct class, which  YIELD  their places to other species which have been
0322  the old inhabitants; and the forms which thus  YIELD  their places will commonly be allied, for the
0327  species. The forms which are beaten and which  YIELD  their places to the new and victorious forms,
0340  mitting much inter migration, the feebler will  YIELD  to the more dominant forms, and there will be
0106  smaller continent of Australia have formerly  YIELDED, and apparently are now yielding, before tho
0262  ies of Sorbus, when grafted on other species,  YIELDED  twice as much fruit as when on their own roo
0271  crossed with the Nicotiana glutinosa, always  YIELDED  hybrids not so sterile as those which were p
0380  rms; but these have almost everywhere largely  YIELDED  to the more dominant forms, generated in the
0380  the productions of these islands on the land  YIELDED  to those produced within the larger areas of
0380  ctions of real islands have everywhere lately  YIELDED  to continental forms, naturalised by man's a
0067  unt. The face of Nature may be compared to a  YIELDING  surface, with ten thousand sharp wedges pack
0106  ave formerly yielded, and apparently are now  YIELDING, before those of the larger Europaeo Asiatic
0201  pared with another; but they are now rapidly  YIELDING  before the advancing legions of plants and a
0210  m accuainted, is that of aphides voluntarily  YIELDING  their sweet excretion to ants: that they do
0238  do not doubt that a breed of cattle, always  YIELDING  oxen with extraordinarily long horns, could
0437  n has remarked, the benefit derived from the  YIELDING  of the separate pieces in the act of parturi
0255  us applied to the stigma of some one species,  YIELDS  a perfect gradation in the number of seeds pr
0041  arked, with respect to the sheep of parts of  YORKSHIRE, that as they generally belong to poor peop
0111  ke the place of older and inferior kinds. In  YORKSHIRE, it is historically known that the ancient
0031  his effect from highly competent authorities.  YOUATT, who was probably better acquainted with the
0036  me to differ so greatly from the rock pigeon.  YOUATT  gives an excellent illustration of the effect
0036  p kept by Mr. Buckley and Mr. Burgess, as Mr.  YOUATT  remarks, have been purely bred from the origi
0454  eeds of cattle, more especially, according to  YOUATT, in young animals, and the state of the whole
0009  e other hand, we see individuals, though taken  YOUNG  from a state of nature, perfectly tamed, long
0010  ration. Seedlings from the same fruit, and the  YOUNG  of the same litter, sometimes differ consider
0010  considerably from each other, though both the  YOUNG  and the parents, as Muller has remarked, have
0010  n of the conditions been direct, if any of the  YOUNG  had varied, all would probably have varied in
0050  and distinct species or mere varieties. When a  YOUNG  naturalist commences the study of a group of
0064  ninety years old, bringing forth three pair of  YOUNG  in this interval; if this be so, at the end o
0065  sequently been less destruction of the old and  YOUNG, and that nearly all the young have been enab
0065  of the old and young, and that nearly all the  YOUNG  have been enabled to breed. In such cases the
0066  animal can in any way protect its own eggs or  YOUNG, a small number may be produced, and yet the
0066  ge stock be fully kept up; but if many eggs or  YOUNG  are destroyed, many must be produced, or the
0066  avy destruction inevitably falls either on the  YOUNG  or old, during each generation or at recurren
0067  's mind some of the chief points. Eggs or very  YOUNG  animals seem generally to suffer most, but th
0068  e tiger in India most rarely dares to attack a  YOUNG  elephant protected by its dam. Climate plays
0072  all cannot live. When I ascertained that these  YOUNG  trees had not been sown or planted, I was so
0072  became thickly clothed with vigorously growing  YOUNG  firs. Yet the heath was so extremely barren a
0076  eings in a state of nature, and if the seed or  YOUNG  were not annually sorted. As species of the s
0077  to other plants. But from the strong growth of  YOUNG  plants produced from such seeds (as peas and
0077  ent in the seed is to favour the growth of the  YOUNG  seedling, whilst struggling with other plants
0087  ral selection will modify the structure of the  YOUNG  in relation to the parent, and of the parent
0087  e parent, and of the parent in relation to the  YOUNG. In social animals it will adapt the structur
0087  ultaneously the most rigorous selection of the  YOUNG  birds within the egg, which had the most powe
0099  e birds, in comparison with the plumage of the  YOUNG, can be explained on the view of plumage havi
0104  nly at long intervals, I am convinced that the  YOUNG  would probably inherit the same habits or str
0127  rresponding ages, can modify the egg, seed, or  YOUNG, as easily as the adult. Amongst many animals
0143  cations accumulated solely for the good of the  YOUNG  or larva, will, it may safely be concluded, a
0144  , and the presence of more or less down on the  YOUNG  birds when first hatched, with the future col
0166  character, is, that there is a tendency in the  YOUNG  of each successive generation to produce the
0166  either plainer or appear more commonly in the  YOUNG  than in the old. Call the breeds of pigeons,
0197  s likewise naked. The sutures in the skulls of  YOUNG  mammals have been advanced as a beautiful ada
0197  his act; but as sutures occur in the skulls of  YOUNG  birds and reptiles, which have only to escape
0198  ffect by pressure the shape of the head of the  YOUNG  in the womb. The laborious breathing necessar
0203  bee, which urges her instantly to destroy the  YOUNG  queens her daughters as soon as born, or to p
0207  formed by an animal, more especially by a very  YOUNG  one, without any experience, and when perform
0211  as eagerly devoured by the ant. Even the quite  YOUNG  aphides behaved in this manner, showing that
0213  eral breeds of dogs: it cannot be doubted that  YOUNG  pointers (I have myself seen a striking insta
0213  e actions, performed without experience by the  YOUNG, and in nearly the same manner by each indivi
0213  reed, and without the end being known, for the  YOUNG  pointer can no more know that he points to ai
0213  ncts. If we were to see one kind of wolf, when  YOUNG  and without any training, as soon as it scent
0214  on which, as I have witnessed, is performed by  YOUNG  birds, that have never seen a pigeon tumble.
0215  ; no animal is more difficult to tame than the  YOUNG  of the wild rabbit; scarcely any animal is ta
0215  rabbit; scarcely any animal is tamer than the  YOUNG  of the tame rabbit; but I do not suppose that
0215  r hand, do our civilised dogs, even when quite  YOUNG, require to be taught not to attack poultry,
0215  ng by inheritance our dogs. On the other hand,  YOUNG  chickens have lost, wholly by habit, that fea
0216  he same way as it is so plainly instinctive in  YOUNG  pheasants, though reared under a hen. It is n
0216  danger chuckle, they will run (more especially  YOUNG  turkeys) from under her, and conceal themselv
0217  e time unincubated, or there would be eggs and  YOUNG  birds of different ages in the same nest. If
0217  at a very early period; and the first hatched  YOUNG  would probably have to be fed by the male alo
0217  ti for she makes her own nest and has eggs and  YOUNG  successively hatched, all at the same time. I
0217  d profited by this occasional habit, or if the  YOUNG  were made more vigorous by advantage having b
0217  s she can hardly fail to be by having eggs and  YOUNG  of different ages at the same time; then the
0217  same time; then the old birds or the fostered  YOUNG  would gain an advantage. And analogy would le
0217  and analogy would lead me to believe, that the  YOUNG  thus reared would be apt to follow by inherit
0217  nests, and thus be successful in rearing their  YOUNG. By a continued process of this nature, I bel
0218  essary if they had to store food for their own  YOUNG. Some species, likewise, of Sphegidae (wasp l
0223  tions, does not collect food for itself or its  YOUNG, and cannot even feed itself: it is absolutel
0225  waxen comb of cylindrical cells, in which the  YOUNG  are hatched, and, in addition, some large cel
0243  satisfactory to look at such instincts as the  YOUNG  cuckoo ejecting its foster brothers, ants mak
0264  t an early period; more especially as all very  YOUNG  beings seem eminently sensitive to injurious
0387  ight be dropped by the bird whilst feeding its  YOUNG, in the same way as fish are known sometimes
0397  for their transportal. Would the just hatched  YOUNG  occasionally crawl on and adhere to the feet
0416  hat the rudimentary teeth in the upper jaws of  YOUNG  ruminants, and certain rudimentary bones of t
0440  lits are related to similar conditions, in the  YOUNG  mammal which is nourished in the womb of its
```

0440 es on the whelp of a lion, or the spots on the YOUNG blackbird, are of any use to these animals, o
0443 ns of either parent. For the welfare of a very YOUNG animal, as long as it remains in its mother's
0445 ve descended from one wild species, I compared YOUNG pigeons of various breeds, within twelve hour
0446 idth of mouth, could hardly be detected in the YOUNG. But there was one remarkable exception to th
0446 one remarkable exception to this rule, for the YOUNG of the short faced tumbler differed from the
0446 g of the short faced tumbler differed from the YOUNG of the wild rock pigeon and of the other bree
0447 ing been inherited at a corresponding age, the YOUNG of the new species of our supposed genus will
0447 ted at a corresponding mature age. Whereas the YOUNG will remain unmodified, or be modified in a l
0447 her case (as with the short faced tumbler) the YOUNG or embryo would closely resemble the mature p
0448 aphis. With respect to the final cause of the YOUNG in these cases not undergoing any metamorphos
0448 two following contingencies; firstly, from the YOUNG, during a course of modification carried on f
0448 uisite. If, on the other hand, it profited the YOUNG to follow habits of life in any degree differ
0448 inheritance at corresponding ages, the active YOUNG or larvae might easily be rendered by natural
0454 ttle, more especially, according to Youatt, in YOUNG animals, and the state of the whole flower in
0482 i but I look with confidence to the future, to YOUNG and rising naturalists, who will be able to v
0395 l history of this archipelago by the admirable ZEAL and researches of Mr. Wallace. I have not as y
0100 request Dr. Hooker tabulated the trees of New ZEALAND, and Dr. Asa Gray those of the United States
0201 under nature. The endemic productions of New ZEALAND, for instance, are perfect one compared with
0337 ean productions have recently spread over New ZEALAND, and have seized on places which must have b
0337 plants of Great Britain were set free in New ZEALAND, that in the course of time a multitude of B
0337 r hand, from what we see now occurring in New ZEALAND, and from hardly a single inhabitant of the
0337 , we may doubt, if all the productions of New ZEALAND were set free in Great Britain, whether any
0337 in may be said to be higher than those of New ZEALAND. Yet the most skilful naturalist from an exa
0339 rect evidence of former glacial action in New ZEALAND. We see it also in the birds of the caves of
0373 he admirable Introduction to the Flora of New ZEALAND; and the same plants, found on widely separa
0375 hat it is certainly a wonderful fact that New ZEALAND, by Dr. Hooker, analogous and striking facts
0376 aks of the re appearance on the shores of New ZEALAND should have a closer resemblance in its crus
0376 wenty five species of Algae are common to New ZEALAND, Tasmania, etc., of northern forms of fish.
0381 s so enormously remote as Kerguelen Land, New ZEALAND and to Europe, but have not been found in th
0382 he southern shores of America, Australia, New ZEALAND, and Fuegia, I believe that towards the clos
0389 to the large size and varied stations of New ZEALAND have become slightly tinted by the same pecu
0390 become naturalised on it, as they have on New ZEALAND, extending over 780 miles of latitude, and c
0391 in the Galapagos Islands reptiles, and in New ZEALAND and on every other oceanic island which can
0393 s on the mountains of the great island of New ZEALAND gigantic wingless birds, take the place of m
0394 mammals do occur on almost every island. New ZEALAND; but I suspect that this exception (if the i
0399 ailing currents, this anomaly disappears. New ZEALAND possesses two bats found nowhere else in the
0399 y almost disappears on the view that both New ZEALAND in its endemic plants is much more closely r
0163 ars on its legs, like those on the legs of the ZEBRA: it has been asserted that these are plainest
0163 r. the quagga, though so plainly barred like a ZEBRA over the body, is without bars on the legs: b
0163 ay has figured one specimen with very distinct ZEBRA like bars on the hocks. With respect to the h
0165 hought that it must have been the product of a ZEBRA; and Mr. W. C. Martin, in his excellent treat
0165 ch i have seen, of hybrids between the ass and ZEBRA, the legs were much more plainly barred than
0165 those on the dun Welch pony, and even had some ZEBRA like stripes on the sides of its face. With r
0166 y simple variation, striped on the legs like a ZEBRA, or striped on the shoulders like an ass. In
0167 enerations, and I see an animal striped like a ZEBRA, but perhaps otherwise very differently const
0167 stocks, or the ass, the hemionus, cuagga, and ZEBRA. He who believes that each equine species was
0255 f first crosses and of hybrids, graduates from ZERO to perfect fertility. It is surprising in how
0255 han so much inorganic dust. From this absolute ZERO of fertility, the pollen of different species
0259 es, are united, their fertility graduates from ZERO to perfect fertility, or even to fertility und
0133 eties of one species, when they range into the ZONE of habitation of other species, often acquirin
0176 , and a third variety to a narrow intermediate ZONE. The intermediate variety, consequently, will
0177 n smaller numbers in a narrow and intermediate ZONE. For forms existing in larger numbers will alw
0203 ll often be formed, fitted for an intermediate ZONE: but from reasons assigned, the intermediate v
0307 d beds beneath Barrande's so called primordial ZONE. The presence of phosphatic nodules and bitumi
0366 as the cold came on, and as each more southern ZONE became fitted for arctic beings and ill fitted
0463 ind intermediate varieties in the intermediate ZONE. For we have reason to believe that only a few
0472 le direct effect; yet when varieties enter any ZONE, they occasionally assume some of the characte
0472 f the characters of the species proper to that ZONE. In both varieties and species, use and disuse
0141 n the south, and on many islands in the torrid ZONES. Hence I am inclined to look at adaptation to
0174 mediate varieties existing in the intermediate ZONES. By changes in the form of the land and of cl
0178 at first have been formed in the intermediate ZONES, but they will generally have had a short dur
0178 owledged varieties), exist in the intermediate ZONES in lesser numbers than the varieties which th
0304 tic regions to the equator, inhabiting various ZONES of depths from the upper tidal limits to 50 f
0374 der the equator and under the warmer temperate ZONES, and on both sides of the southern extremity
0376 t arctic, but belong to the northern temperate ZONES. As Mr. H. C. Watson has recently remarked, I
0378 eriod from the northern and southern temperate ZONES into the intertropical regions, and some even
0380 ch live in the northern and southern temperate ZONES and on the mountains of the intertropical reg
0463 ll at first probably exist in the intermediate ZONES, will be liable to be supplanted by the allie
0477 the sea in the northern and southern temperate ZONES, though separated by the whole intertropical
0362 ich, as I know from experiments made in the ZOOLOGICAL Gardens, include seeds capable of germinati
0408 f land or water, which separate our several ZOOLOGICAL and botanical provinces. We can thus unders
0048 ly locomotive, doubtful forms, ranked by one ZOOLOGIST as a species and by another as a variety, c
0049 t which have been ranked as species by some ZOOLOGISTS. Several most experienced ornithologists co

Library of Congress Cataloging in Publication Data

Barrett, Paul H.
  A concordance to Darwin's Origin of species, first
edition.

  1. Darwin, Charles, 1809-1882. On the origin of
species--Concordances. I. Weinshank, Donald J. (Donald
Jerome), 1937-   . II. Gottleber, Timothy T.
III. Darwin, Charles, 1809-1882. On the origin of
species. IV. Title.
QH365.08B37   575.01'62   80-66893
ISBN 0-8014-1319-2   AACR2